D1358807

Show Your Students How an Integrated Approach Proves that . . .

MathMatters.

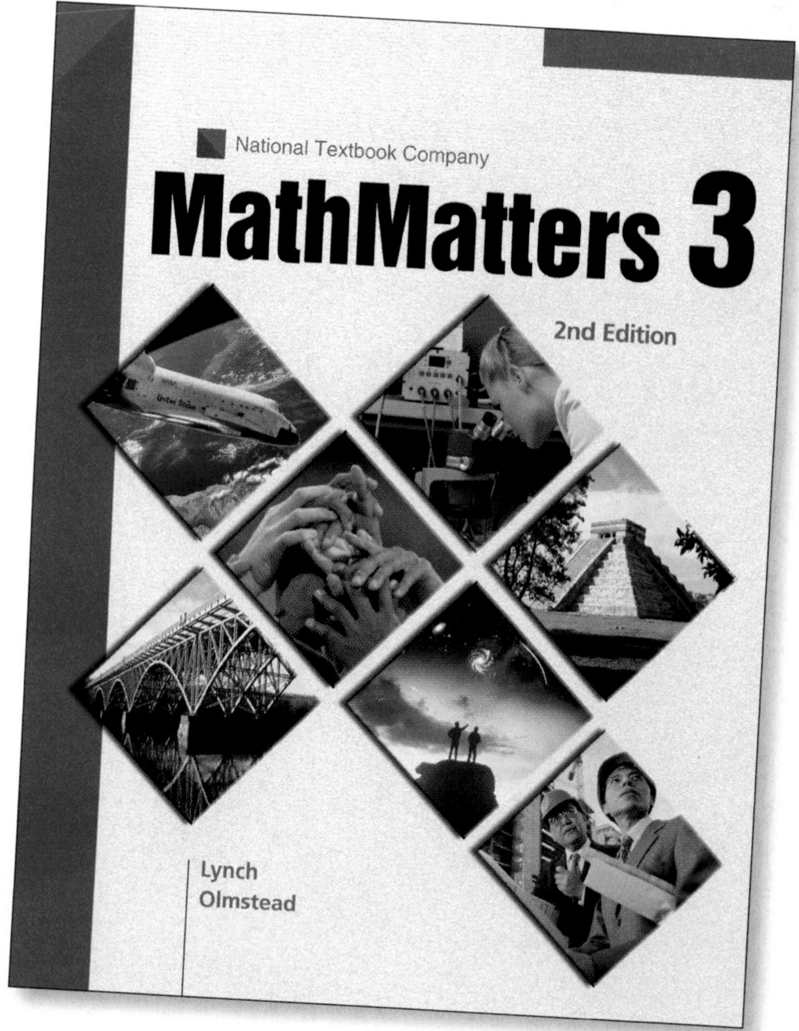

Connects to the World

Integrates mathematics with everyday life through chapter themes and application exercises to motivate students to learn.

Develops Problem Solving and Critical Thinking Skills

Develops students' problem solving and critical thinking skills though step-by-step explanations and expanded student exercises.

Meets NCTM Standards

Provides your students with the math curriculum they need to succeed in school–and beyond.

Provides Comprehensive Support Package

Gives you all the supplemental materials you need.

Lesson Walk-Throughs

Integrate Mathematics with the Real World

Integrating mathematics with everyday life through chapter themes and application exercises motivates students to learn.

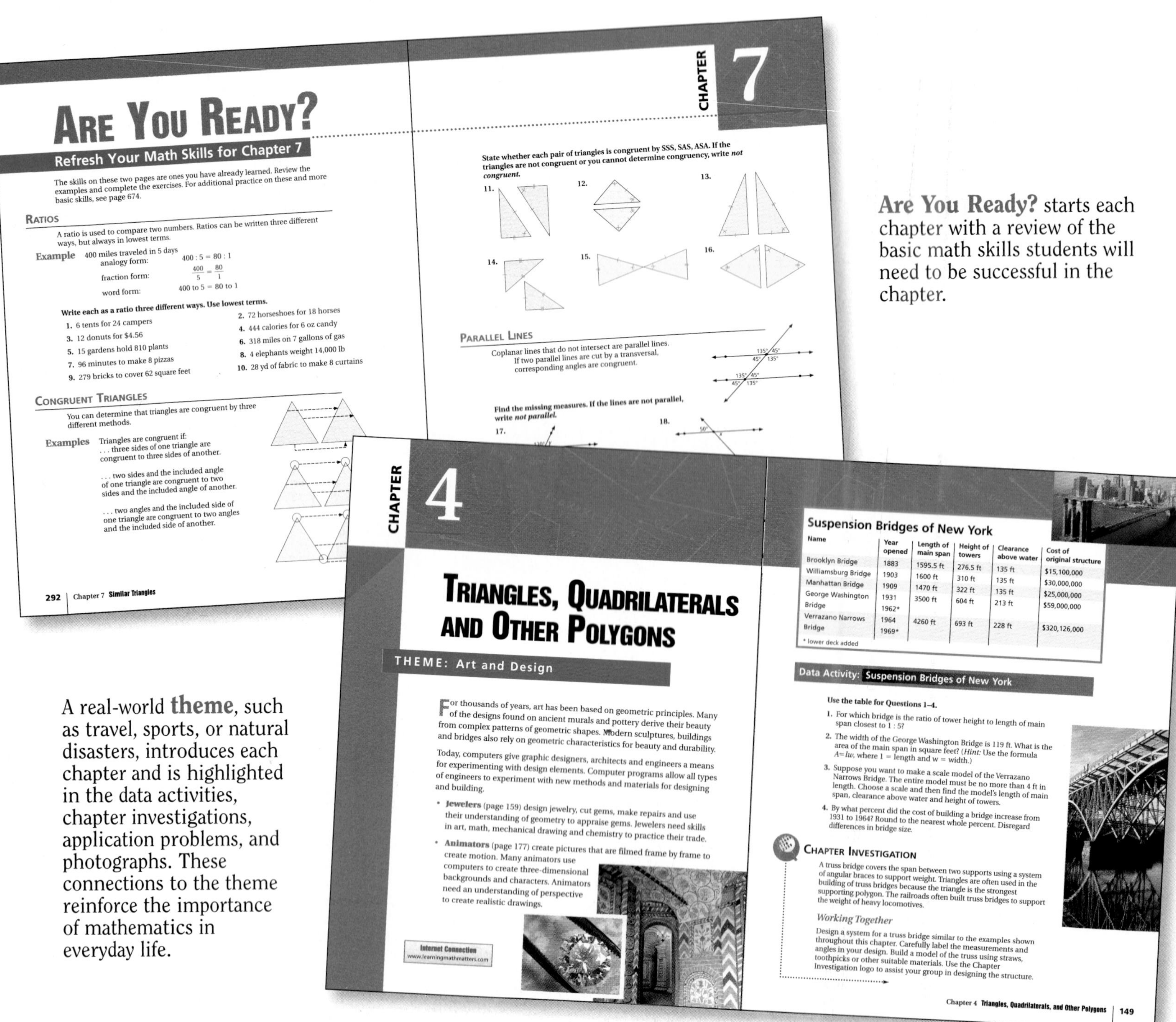

CHAPTER 7

ARE YOU READY?

Refresh Your Math Skills for Chapter 7

The skills on these two pages are ones you have already learned. Review the examples and complete the exercises. For additional practice on these and more basic skills, see page 674.

RATIOS

A ratio is used to compare two numbers. Ratios can be written three different ways, but always in lowest terms.

Example 400 miles traveled in 5 days

analogy form: 400 : 5 = 80 : 1

fraction form: $\frac{400}{5} = \frac{80}{1}$

word form: 400 to 5 = 80 to 1

Write each as a ratio three different ways. Use lowest terms.

1. 6 tents for 24 campers
2. 72 horseshoes for 18 horses
3. 12 donuts for $4.56
4. 444 calories for 6 oz candy
5. 15 gardens hold 810 plants
6. 318 miles on 7 gallons of gas
7. 96 minutes to make 8 pizzas
8. 4 elephants weight 14,000 lb
9. 279 bricks to cover 62 square feet
10. 28 yd of fabric to make 8 curtains

CONGRUENT TRIANGLES

You can determine that triangles are congruent by three different methods.

Examples Triangles are congruent if:

. . . three sides of one triangle are congruent to three sides of another.

. . . two sides and the included angle of one triangle are congruent to two sides and the included angle of another.

. . . two angles and the included side of one triangle are congruent to two angles and the included side of another.

292 | Chapter 7 **Similar Triangles**

State whether each pair of triangles is congruent by SSS, SAS, ASA. If the triangles are not congruent or you cannot determine congruency, write *not congruent*.

11. 12. 13. 14. 15. 16.

PARALLEL LINES

Coplanar lines that do not intersect are parallel lines. If two parallel lines are cut by a transversal, corresponding angles are congruent.

Find the missing measures. If the lines are not parallel, write *not parallel*.

17. 18.

CHAPTER 4

TRIANGLES, QUADRILATERALS AND OTHER POLYGONS

THEME: Art and Design

For thousands of years, art has been based on geometric principles. Many of the designs found on ancient murals and pottery derive their beauty from complex patterns of geometric shapes. Modern sculptures, buildings and bridges also rely on geometric characteristics for beauty and durability.

Today, computers give graphic designers, architects and engineers a means for experimenting with design elements. Computer programs allow all types of engineers to experiment with new methods and materials for designing and building.

- **Jewelers** (page 159) design jewelry, cut gems, make repairs and use their understanding of geometry to appraise gems. Jewelers need skills in art, math, mechanical drawing and chemistry to practice their trade.
- **Animators** (page 177) create pictures that are filmed frame by frame to create motion. Many animators use computers to create three-dimensional backgrounds and characters. Animators need an understanding of perspective to create realistic drawings.

Internet Connection
www.learningmathmatters.com

Suspension Bridges of New York

Name	Year opened	Length of main span	Height of towers	Clearance above water	Cost of original structure
Brooklyn Bridge	1883	1595.5 ft	276.5 ft	135 ft	$15,100,000
Williamsburg Bridge	1903	1600 ft	310 ft	135 ft	$30,000,000
Manhattan Bridge	1909	1470 ft	322 ft	135 ft	$25,000,000
George Washington Bridge	1931 1962*	3500 ft	604 ft	213 ft	$59,000,000
Verrazano Narrows Bridge	1964 1969*	4260 ft	693 ft	228 ft	$320,126,000

* lower deck added

Data Activity: Suspension Bridges of New York

Use the table for Questions 1–4.

1. For which bridge is the ratio of tower height to length of main span closest to 1 : 5?
2. The width of the George Washington Bridge is 119 ft. What is the area of the main span in square feet? (*Hint:* Use the formula $A = lw$, where l = length and w = width.)
3. Suppose you want to make a scale model of the Verrazano Narrows Bridge. The entire model must be no more than 4 ft in length. Choose a scale and then find the model's length of main span, clearance above water and height of towers.
4. By what percent did the cost of building a bridge increase from 1931 to 1964? Round to the nearest whole percent. Disregard differences in bridge size.

CHAPTER INVESTIGATION

A truss bridge covers the span between two supports using a system of angular braces to support weight. Triangles are often used in the building of truss bridges because the triangle is the strongest supporting polygon. The railroads often built truss bridges to support the weight of heavy locomotives.

Working Together

Design a system for a truss bridge similar to the examples shown throughout this chapter. Carefully label the measurements and angles in your design. Build a model of the truss using straws, toothpicks or other suitable materials. Use the Chapter Investigation logo to assist your group in designing the structure.

Chapter 4 **Triangles, Quadrilaterals, and Other Polygons** | 149

Are You Ready? starts each chapter with a review of the basic math skills students will need to be successful in the chapter.

A real-world **theme**, such as travel, sports, or natural disasters, introduces each chapter and is highlighted in the data activities, chapter investigations, application problems, and photographs. These connections to the theme reinforce the importance of mathematics in everyday life.

All-American Girls Professional Baseball League Batting Champions, 1943-1954

Year	Player, Team	At-bats	Average
1943	Gladys Davis, Rockford	349	.332
1944	Betsy Jochum, South Bend	433	.296
1945	Helen Callahan, Fort Wayne	408	.299
1946	Dorothy Kamenshek, Rockford	408	.316
1947	Dorothy Kamenshek, Rockford	366	.306
1948*	Audrey Wagner, Kenosha	417	.312
1949	Doris Sams, Muskegon	408	.279
1950	Betty Weaver Foss, Fort Wayne	361	.346
1951	Betty Weaver Foss, Fort Wayne	342	.368
1952	Joanne Weaver, Fort Wayne	314	.344
1953	Joanne Weaver, Fort Wayne	410	.346
1954	Joanne Weaver, Fort Wayne	333	.429

*First year overhand pitching was allowed.
Source: *A Whole New Ballgame*, Sue Macy, Henry Holt and Company, New York, 1993.

South Bend, Indiana Blue Sox

Olympic Record Times for 400-m Freestyle Swimming, in Minutes

Year	1924	1928	1932	1936	1948	1952	1956	1960	1964
Male	5:04.2	5:01.6	4:48.4	4:44.5	4:41.0	4:30.7	4:27.3	4:18.3	4:12.2
Female	6:02.2	5:42.8	5:28.5	5:26.4	5:17.8	5:12.1	4:54.6	4:50.6	4:43.3

Year	1968	1972	1976	1980	1984	1988	1992	1996
Male	4:09.0	4:00.27	3:51.93	3:51.31	3:51.23	3:46.25	3:45.00	3:47.97
Female	4:31.8	4:19.44	4:09.89	4:08.76	4:07.10	4:03.85	4:07.18	4:07.25

648 | Data Files

Data File

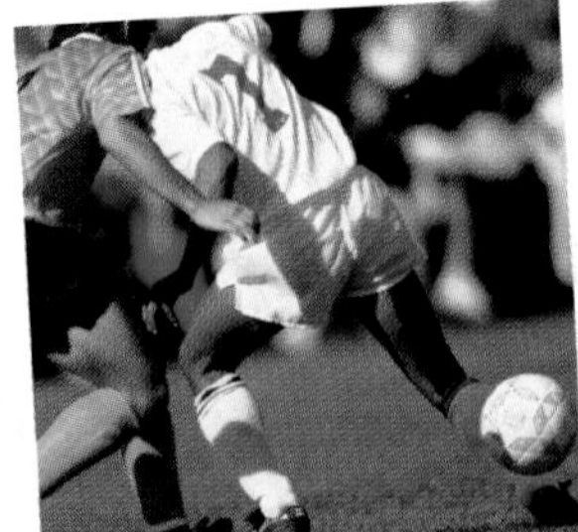

Sizes and Weights of Balls Used in Various Sports

Type	Diameter (centimeters)	Average weight (grams)
Baseball	7.6	145
Basketball	24.0	596
Croquet ball	8.6	340
Field hockey ball	7.6	160
Golf ball	4.3	46
Handball	4.8	65
Soccer ball	22.0	425
Softball, large	13.0	279
Softball, small	9.8	187
Table tennis ball	3.7	2
Tennis ball	6.5	57
Volleyball	21.9	256

Soccer Playboard

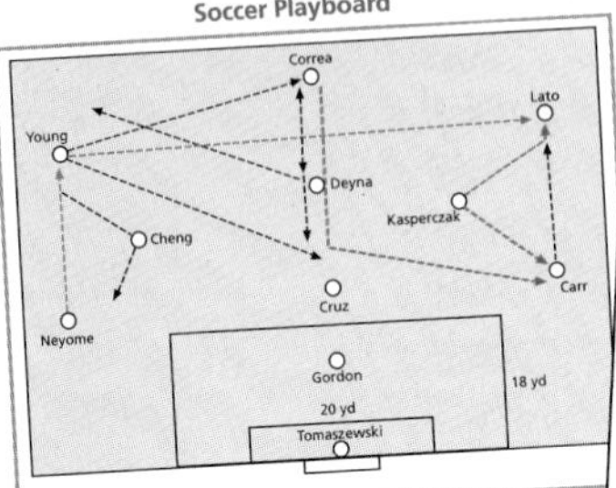

Data Files provide students with real-world statistics. Data File Exercises teach students to solve problems by reading data from a variety of visual displays.

8. Ralph is a tour guide. In how many ways can he choose 3 museums to visit from the 8 museums in a city?
9. In how many ways can a disk jockey play 5 of the 20 top hits?
10. **ERROR ALERT** On a test, students must choose 3 out of 5 essay questions to answer. Dale calculates that there are 60 ways to do this. What has Dale done wrong?
11. **SPORTS** Bill, Phil, and Jill are among 12 players competing for 3 spots on a table-tennis team. Every player has an equal chance of making the team. Find the probability that all three will make the team.
12. What is n, if ${}_nP_3 = 120$?
13. Suppose that license plates contain three letters. What is the probability that Meg's plates will spell her name?
14. **SPORTS** How many ways can a batting order be made for 9 players if you know that one player has already been designated to bat first and another to bat fourth?
15. **WRITING MATH** Find the number of permutations of the letters in the word *shutout*. Explain how you did it.
16. Two students out of 8 will be chosen to speak at a school assembly. How many different outcomes are possible?

Technology Note

Some calculators have a factorial key, marked [x!]. To find 7!, enter 7, then [x!]. On graphing calculators, you can choose the factorial function from a displayed mathematical menu. This menu may also include operations for finding permutations and combinations.

USING DATA For Exercise 17–18, use the Data Index on pages 632–633 to locate information about the All-American Girls Professional Baseball League.

17. Suppose you could interview all eight women who were batting champions of the All-American Girls Professional Baseball League to discover what playing in this league was like. How many orders for these interviews would be possible?
18. If you were to interview only four of the eight women, what is the probability that you would first interview a player from either Fort Wayne or Rockford?

Extended Practice Exercises

19. Compare the values of ${}_8C_5$ and ${}_8C_3$. What do you notice?
20. Find the values of ${}_7C_4$ and ${}_7C_3$. What do you notice?
21. What can you say about the sum of the number of items taken at one time for each combination shown in Exercises 18 and 19?
22. **CRITICAL THINKING** Use what you have discovered to quickly find ${}_{67}C_{64}$.

Mixed Review Exercises

Add. (Lesson 8-5)

23. $\begin{bmatrix} 2 & 4 \\ 3 & 0 \end{bmatrix} + \begin{bmatrix} 1 & 3 \\ -2 & 4 \end{bmatrix}$
24. $\begin{bmatrix} 6 & 0 & 3 \\ -5 & 2 & 1 \end{bmatrix} + \begin{bmatrix} 3 & -4 & -2 \\ 7 & 1 & 3 \end{bmatrix}$

Find the measures of the angles. (Lesson 3-2)

25. $\angle BZC$
26. $\angle CZD$
27. $\angle AZC$

B, C, $(4x - 9)°$, $(3x - 6)°$, A, Z, D

Lesson 9-5 **Permutations and Combinations** | 405

Practice ■ Lesson 2-1–Lesson 2-2

Write each as a set of ordered pairs. Given the domain and range. (Lesson 2-2)

36.

x	2	4	6	8
y	11	8	5	2

37.

x	0	−1	1	−2	2
y	3	4	4	7	7

Determine the next three terms or figures in each pattern. (Lesson 2-1)

38. 243, 260, 277, 294, ___, ___, ___
39. −748, −595, −442, −289, ___, ___, ___
40. 32, 48, 72, 108, ___, ___, ___
41. Z, ZY, ZYX, ZYXW, ___, ___, ___
42. *, **, ***, ****, ___, ___, ___
43. $10, -10^2, 10^3, -10^4$, ___, ___, ___

Given $f(x) = 3x^2$ and $g(x) = -3x + 1\frac{1}{2}$, find each value. (Lesson 2-2)

44. $f(2)$
45. $f(-2)$
46. $g\left(\frac{1}{2}\right)$
47. $f(1) + g(1)$

Name the quadrant in which each point is located. (Lesson 2-2)

48. $W(9, -1.5)$
49. $X(-8, 17)$
50. $Z(-6.5, -6.5)$

MathWorks — Career – Environmental Journalism

Workplace Knowhow

Environmental journalists inform the public about what is being done to harm and to save the planet. They work with a great deal of mathematical information such as, how much pollution a smokestack emits or how much garbage is added to a landfill each year.

Environmental journalists study data gathered by scientists and the government. They must be able to analyze raw data, understand and create graphs, spot trends in numbers, and draw conclusions about what is happening to the environment.

1. The United States has 17 commercial incinerators that burn hazardous waste and release pollutants into the air. The hazardous waste size has increased from 1.3 billion to 2.4 billion to 4.6 billion pounds of waste. Project the growth for the next five years. (Hint: Plot the ordered pairs as (0, 0) for starting point, (1. difference of 2.4 billion and 1.3 billion) for increase at end of year 1, etc.)
2. Create a graph to go with your story. Place years on the x-axis and billions of pounds of waste (in 0.5 billion increments) on the y-axis.
3. A television reporter claims that fewer people in your town are recycling. But is the report true? Last year, 30% of the people in your town recycled regularly. This year recycling is down to 27%. Last year, your town had a population of 2,500, but this year the population is 3,100. Is the reporter's claim true? Explain your thinking.

Chapter 2 **Review and Practice Your Skills** | 61

MathWorks features mathematics at work in various professions, from meteorologist to truck driver, from fitness trainer to financial advisor. Students will learn how a knowledge of mathematics is instrumental in a broad spectrum of careers.

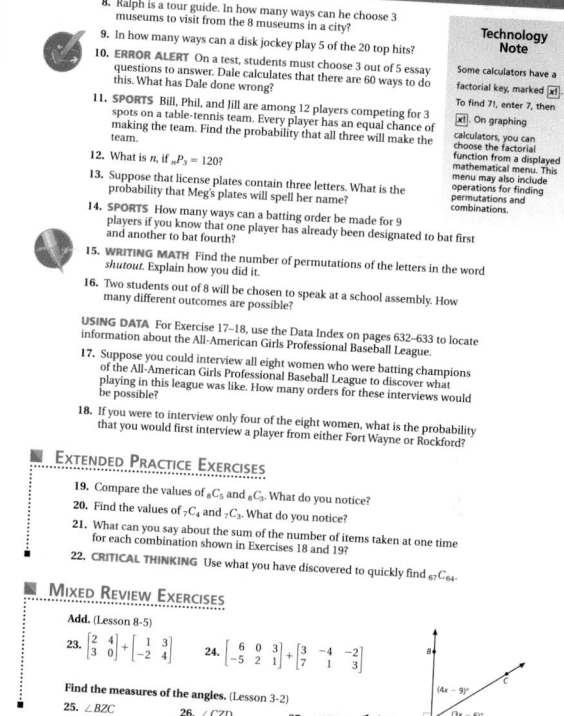

Lesson Walk-Throughs

Lessons

Clear explanations and a wealth of examples build strong understanding of concepts.

Goals and **Applications** are clearly stated at the beginning of each lesson, so students know exactly what is covered in that section.

Build Understanding sections provide clear and concise explanations of concepts.

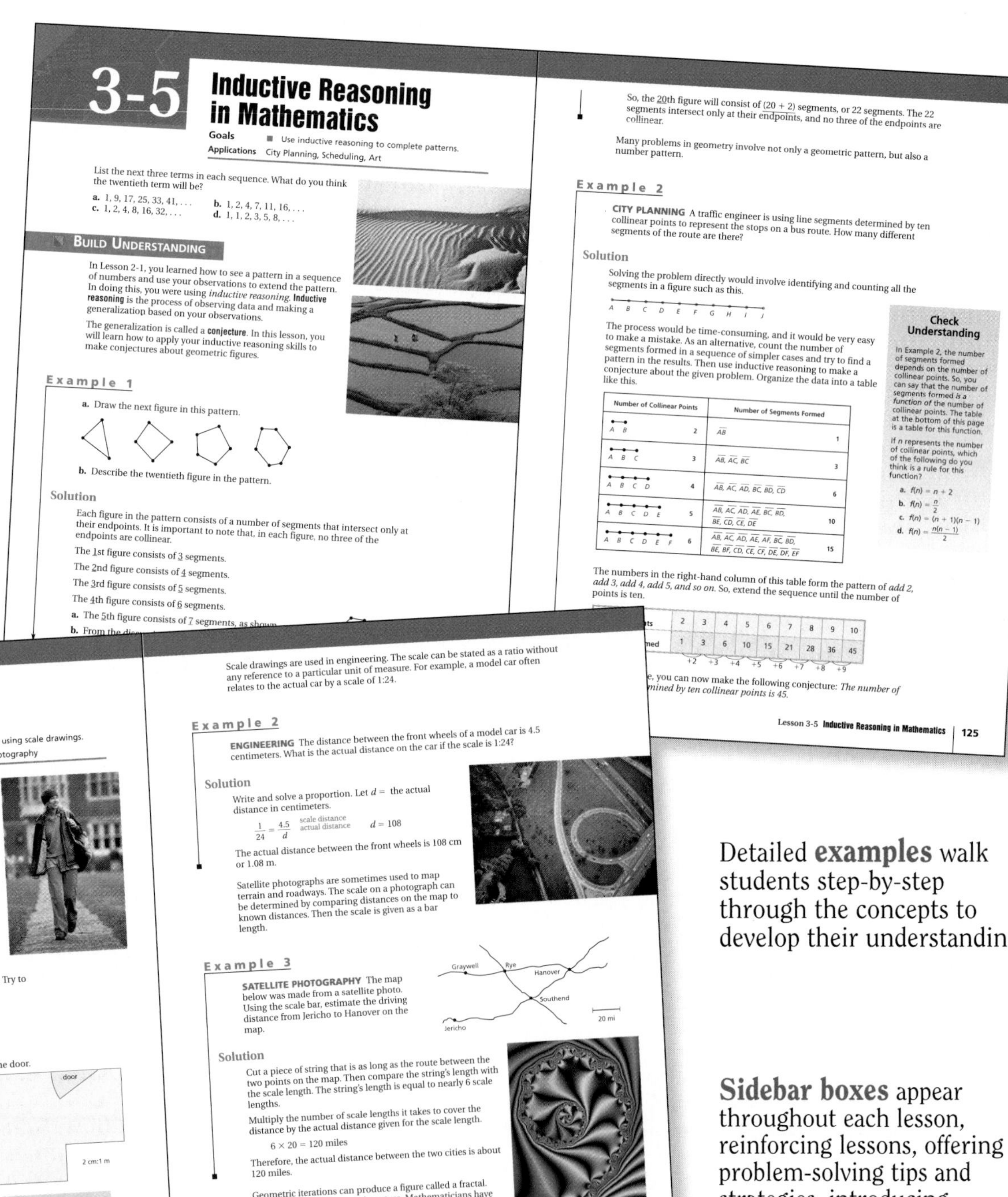

3-5 Inductive Reasoning in Mathematics

Goals ■ Use inductive reasoning to complete patterns.
Applications City Planning, Scheduling, Art

List the next three terms in each sequence. What do you think the twentieth term will be?

a. 1, 9, 17, 25, 33, 41, . . .
b. 1, 2, 4, 7, 11, 16, . . .
c. 1, 2, 4, 8, 16, 32, . . .
d. 1, 1, 2, 3, 5, 8, . . .

BUILD UNDERSTANDING

In Lesson 2-1, you learned how to see a pattern in a sequence of numbers and use your observations to extend the pattern. In doing this, you were using *inductive reasoning*. **Inductive reasoning** is the process of observing data and making a generalization based on your observations.

The generalization is called a **conjecture**. In this lesson, you will learn how to apply your inductive reasoning skills to make conjectures about geometric figures.

Example 1

a. Draw the next figure in this pattern.

b. Describe the twentieth figure in the pattern.

Solution

Each figure in the pattern consists of a number of segments that intersect only at their endpoints. It is important to note that, in each figure, no three of the endpoints are collinear.

The 1st figure consists of 3 segments.
The 2nd figure consists of 4 segments.
The 3rd figure consists of 5 segments.
The 4th figure consists of 6 segments.
a. The 5th figure consists of 7 segments, as shown.

So, the 20th figure will consist of (20 + 2) segments, or 22 segments. The 22 segments intersect only at their endpoints, and no three of the endpoints are collinear.

Many problems in geometry involve not only a geometric pattern, but also a number pattern.

Example 2

CITY PLANNING A traffic engineer is using line segments determined by ten collinear points to represent the stops on a bus route. How many different segments of the route are there?

Solution

Solving the problem directly would involve identifying and counting all the segments in a figure such as this.

The process would be time-consuming, and it would be very easy to make a mistake. As an alternative, count the number of segments formed in a sequence of simpler cases and try to find a pattern in the results. Then use inductive reasoning to make a conjecture about the given problem. Organize the data into a table like this.

Number of Collinear Points		Number of Segments Formed	
A B	2	AB	1
A B C	3	AB, AC, BC	3
A B C D	4	AB, AC, AD, BC, BD, CD	6
A B C D E	5	AB, AC, AD, AE, BC, BD, BE, CD, CE, DE	10
A B C D E F	6	AB, AC, AD, AE, AF, BC, BD, BE, BF, CD, CE, CF, DE, DF, EF	15

The numbers in the right-hand column of this table form the pattern of *add 2, add 3, add 4, add 5, and so on.* So, extend the sequence until the number of points is ten.

Check Understanding

In Example 2, the number of segments formed depends on the number of collinear points. So, you can say that the number of segments formed *is a function of* the number of collinear points. The table at the bottom of this page is a table for this function.

If *n* represents the number of collinear points, which of the following do you think is a rule for this function?

a. $f(n) = n + 2$
b. $f(n) = \frac{n}{2}$
c. $f(n) = (n + 1)(n - 1)$
d. $f(n) = \frac{n(n-1)}{2}$

Lesson 3-5 Inductive Reasoning in Mathematics 125

7-3 Scale Drawings

Goals ■ Find actual or scale length using scale drawings.
Applications Architecture, Engineering, Photography

Suppose a new student has joined your class. Sketch a map of the school building. Try to represent distances accurately. For example, if the principal's office is further from the gym than the library, the distance from the gym on the map should be greater also.

BUILD UNDERSTANDING

A **scale drawing** is a representation of a real object. All lengths on the drawing are proportional to actual lengths of the object. The **scale** of the drawing is the ratio of the size of the drawing to the actual size.

Work with a group of 2 or 3 students.

1. Think of two well-known locations within your community. For instance, you might choose your school and the public library.
2. Sketch a map showing how to get from one location to the other. Try to represent distances accurately on the map.

Example 1

ARCHITECTURE This is a scale drawing of a room in a house.

Find the actual distance along the wall between the window and the door.

Solution

The ratio of the scale drawing to the actual size of the room is 2 cm:1 m. The first step is to measure the drawing. Because the scale is given in terms of centimeters, measure the distance to the nearest centimeter.

The scale distance between the window and the door is 5 cm. Write and solve a proportion.

Let *x* = the actual distance in meters.

$\frac{2 \text{ cm}}{1 \text{ m}} = \frac{5 \text{ cm}}{x}$ scale distance / actual distance $2x = 5 \rightarrow x = 2.5$

The actual distance between the window and the door is 2.5 meters.

Check Understanding

How would the scale drawing of this room change if the scale were changed to 1 cm:1 m?

306 Chapter 7 Similar Triangles

Scale drawings are used in engineering. The scale can be stated as a ratio without any reference to a particular unit of measure. For example, a model car often relates to the actual car by a scale of 1:24.

Example 2

ENGINEERING The distance between the front wheels of a model car is 4.5 centimeters. What is the actual distance on the car if the scale is 1:24?

Solution

Write and solve a proportion. Let *d* = the actual distance in centimeters.

$\frac{1}{24} = \frac{4.5}{d}$ scale distance / actual distance $d = 108$

The actual distance between the front wheels is 108 cm or 1.08 m.

Satellite photographs are sometimes used to map terrain and roadways. The scale on a photograph can be determined by comparing distances on the map to known distances. Then the scale is given as a bar length.

Example 3

SATELLITE PHOTOGRAPHY The map below was made from a satellite photo. Using the scale bar, estimate the driving distance from Jericho to Hanover on the map.

Solution

Cut a piece of string that is as long as the route between the two points on the map. Then compare the string's length with the scale length. The string's length is equal to nearly 6 scale lengths.

Multiply the number of scale lengths it takes to cover the distance by the actual distance given for the scale length.

$6 \times 20 = 120$ miles

Therefore, the actual distance between the two cities is about 120 miles.

Geometric iterations can produce a figure called a fractal. Fractal shapes also appear in nature. Mathematicians have discovered that coastlines are better described as fractals than as smooth curves.

Computer generated fractal

Lesson 7-3 Scale Drawings 307

Detailed **examples** walk students step-by-step through the concepts to develop their understanding.

Sidebar boxes appear throughout each lesson, reinforcing lessons, offering problem-solving tips and strategies, introducing technology, and giving self-assessment activities.

Exercises

Exercise sets for each section contain a variety of exercises, from drill through critical thinking and review. These exercises emphasize problem-solving and critical-thinking skills.

Try These Exercises offer students an introduction to the types of exercises that appear throughout each set. These can also be used as guided in-class practice.

Practice Exercises give students lots of practice in the specific skills and concepts taught in the section.

Icons point out technology, manipulative exercises, writing math, error analysis, and chapter investigation problems throughout each exercise set.

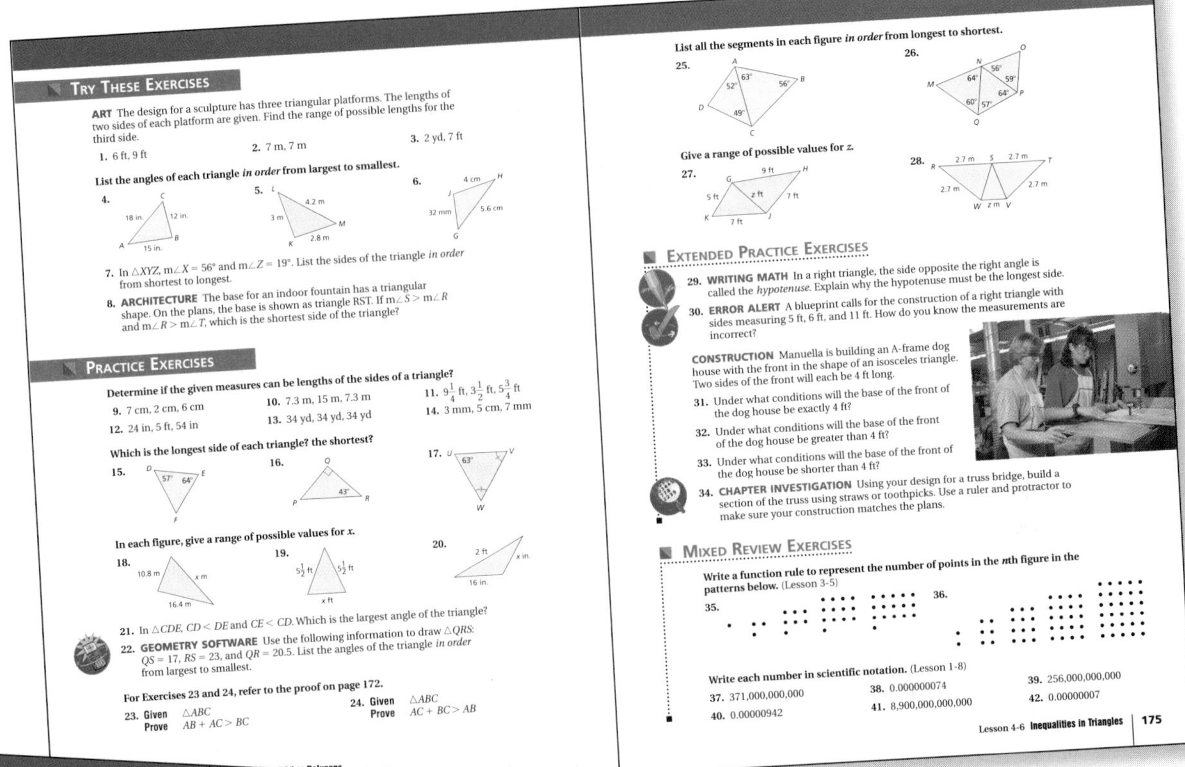

TRY THESE EXERCISES

ART The design for a sculpture has three triangular platforms. The lengths of two sides of each platform are given. Find the range of possible lengths for the third side.

1. 6 ft, 9 ft
2. 7 m, 7 m
3. 2 yd, 7 ft

List the angles of each triangle *in order* from largest to smallest.

4.
5.
6.

7. In $\triangle XYZ$, $m\angle X = 56°$ and $m\angle Z = 19°$. List the sides of the triangle *in order* from shortest to longest.
8. **ARCHITECTURE** The base for an indoor fountain has a triangular shape. On the plans, the base is shown as triangle RST. If $m\angle S > m\angle R$ and $m\angle R > m\angle T$, which is the shortest side of the triangle?

PRACTICE EXERCISES

Determine if the given measures can be lengths of the sides of a triangle?

9. 7 cm, 2 cm, 6 cm
10. 7.3 m, 15 m, 7.3 m
11. $9\frac{1}{4}$ ft, $3\frac{1}{2}$ ft, $5\frac{3}{4}$ ft
12. 24 in, 5 ft, 54 in
13. 34 yd, 34 yd, 34 yd
14. 3 mm, 5 cm, 7 mm

Which is the longest side of each triangle? the shortest?

15.
16.
17.

In each figure, give a range of possible values for *x*.

18.
19.
20.

21. In $\triangle CDE$, $CD < DE$ and $CE < CD$. Which is the largest angle of the triangle?
22. **GEOMETRY SOFTWARE** Use the following information to draw $\triangle QRS$: $QS = 17$, $RS = 23$, and $QR = 20.5$. List the angles of the triangle *in order* from largest to smallest.

For Exercises 23 and 24, refer to the proof on page 172.

23. **Given** $\triangle ABC$ **Prove** $AB + AC > BC$
24. **Given** $\triangle ABC$ **Prove** $AC + BC > AB$

List all the segments in each figure *in order* from longest to shortest.

25.
26.

Give a range of possible values for *z*.

27.
28.

EXTENDED PRACTICE EXERCISES

29. **WRITING MATH** In a right triangle, the side opposite the right angle is called the *hypotenuse*. Explain why the hypotenuse must be the longest side.
30. **ERROR ALERT** A blueprint calls for the construction of a right triangle with sides measuring 5 ft, 6 ft, and 11 ft. How do you know the measurements are incorrect?

CONSTRUCTION Manuella is building an A-frame dog house with the front in the shape of an isosceles triangle. Two sides of the front will each be 4 ft long.

31. Under what conditions will the base of the front of the dog house be exactly 4 ft?
32. Under what conditions will the base of the front of the dog house be greater than 4 ft?
33. Under what conditions will the base of the front of the dog house be shorter than 4 ft?
34. **CHAPTER INVESTIGATION** Using your design for a truss bridge, build a section of the truss using straws or toothpicks. Use a ruler and protractor to make sure your construction matches the plans.

MIXED REVIEW EXERCISES

Write a function rule to represent the number of points in the *n*th figure in the patterns below. (Lesson 3-5)

35.
36.

Write each number in scientific notation. (Lesson 1-8)

37. 371,000,000,000
38. 0.000000074
39. 256,000,000,000
40. 0.00000942
41. 8,900,000,000,000
42. 0.00000007

Lesson 4-6 **Inequalities in Triangles** 175

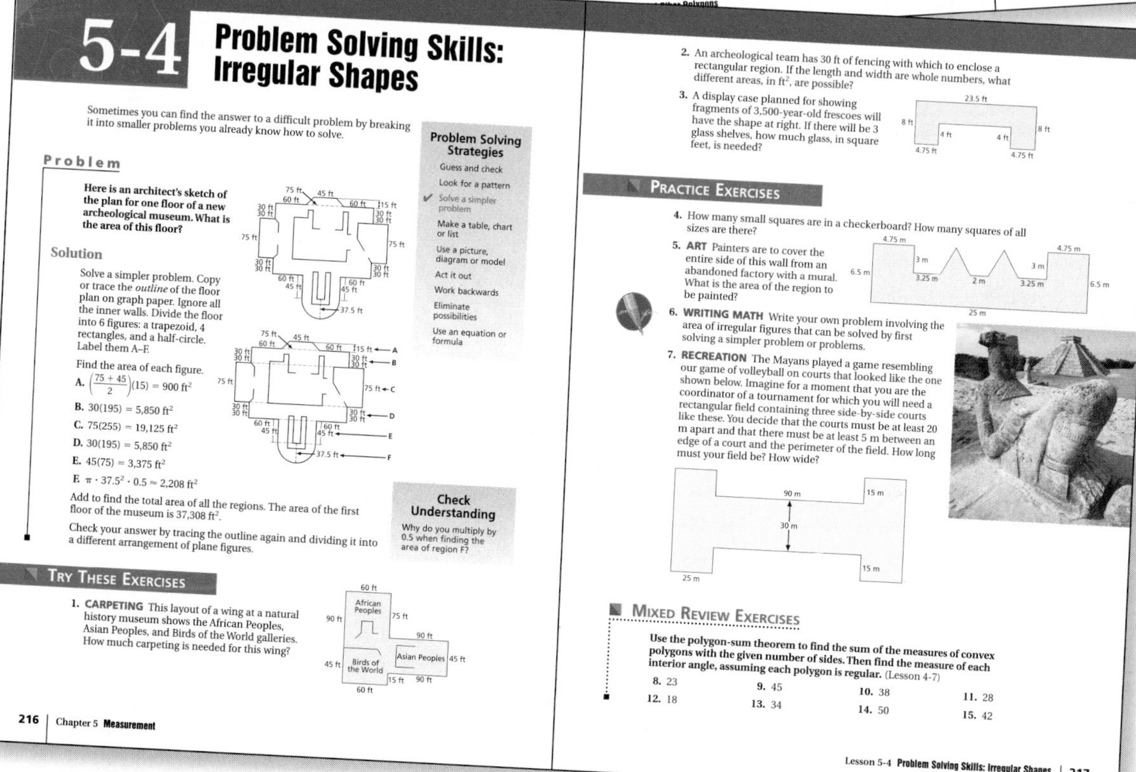

5-4 Problem Solving Skills: Irregular Shapes

Sometimes you can find the answer to a difficult problem by breaking it into smaller problems you already know how to solve.

Problem Solving Strategies

Guess and check
Look for a pattern
✓ Solve a simpler problem
Make a table, chart or list
Use a picture, diagram or model
Act it out
Work backwards
Eliminate possibilities
Use an equation or formula

Problem

Here is an architect's sketch of the plan for one floor of a new archeological museum. What is the area of this floor?

Solution

Solve a simpler problem. Copy or trace the *outline* of the floor plan on graph paper. Ignore all the inner walls. Divide the floor into 6 figures: a trapezoid, 4 rectangles, and a half-circle. Label them A–F.

Find the area of each figure.

A. $\left(\frac{75 + 45}{2}\right)(15) = 900 \text{ ft}^2$

B. $30(195) = 5{,}850 \text{ ft}^2$

C. $75(255) = 19{,}125 \text{ ft}^2$

D. $30(195) = 5{,}850 \text{ ft}^2$

E. $45(75) = 3{,}375 \text{ ft}^2$

F. $\pi \cdot 37.5^2 \cdot 0.5 \approx 2{,}208 \text{ ft}^2$

Add to find the total area of all the regions. The area of the first floor of the museum is 37,308 ft².

Check your answer by tracing the outline again and dividing it into a different arrangement of plane figures.

Check Understanding

Why do you multiply by 0.5 when finding the area of region F?

TRY THESE EXERCISES

1. **CARPETING** This layout of a wing at a natural history museum shows the African Peoples, Asian Peoples, and Birds of the World galleries. How much carpeting is needed for this wing?

216 | Chapter 5 **Measurement**

2. An archeological team has 30 ft of fencing with which to enclose a rectangular region. If the length and width are whole numbers, what different areas, in ft², are possible?
3. A display case planned for showing fragments of 3,500-year-old frescoes will have the shape at right. If there will be 3 glass shelves, how much glass, in square feet, is needed?

PRACTICE EXERCISES

4. How many small squares are in a checkerboard? How many squares of all sizes are there?
5. **ART** Painters are to cover the entire side of this wall from an abandoned factory with a mural. What is the area of the region to be painted?
6. **WRITING MATH** Write your own problem involving the area of irregular figures that can be solved by first solving a simpler problem or problems.
7. **RECREATION** The Mayans played a game resembling our game of volleyball on courts that looked like the one shown below. Imagine for a moment that you are the coordinator of a tournament for which you will need a rectangular field containing three side-by-side courts like these. You decide that the courts must be at least 20 m apart and that there must be at least 5 m between an edge of a court and the perimeter of the field. How long must your field be? How wide?

MIXED REVIEW EXERCISES

Use the polygon-sum theorem to find the sum of the measures of convex polygons with the given number of sides. Then find the measure of each interior angle, assuming each polygon is regular. (Lesson 4-7)

8. 23
9. 45
10. 38
11. 28
12. 18
13. 34
14. 50
15. 42

Lesson 5-4 **Problem Solving Skills: Irregular Shapes** | 217

Extended Practice Exercises take students one step further in developing their critical thinking skills and applying the concepts to real-life problems.

Mixed Review Exercises are brief reviews of the skills and concepts taught in previous sections.

Additional **Problem Solving** lessons appear once each chapter to enhance students' problem solving skills.

Lesson Walk-Throughs

Review

Multiple review sections give students opportunities to reinforce the concepts they have learned.

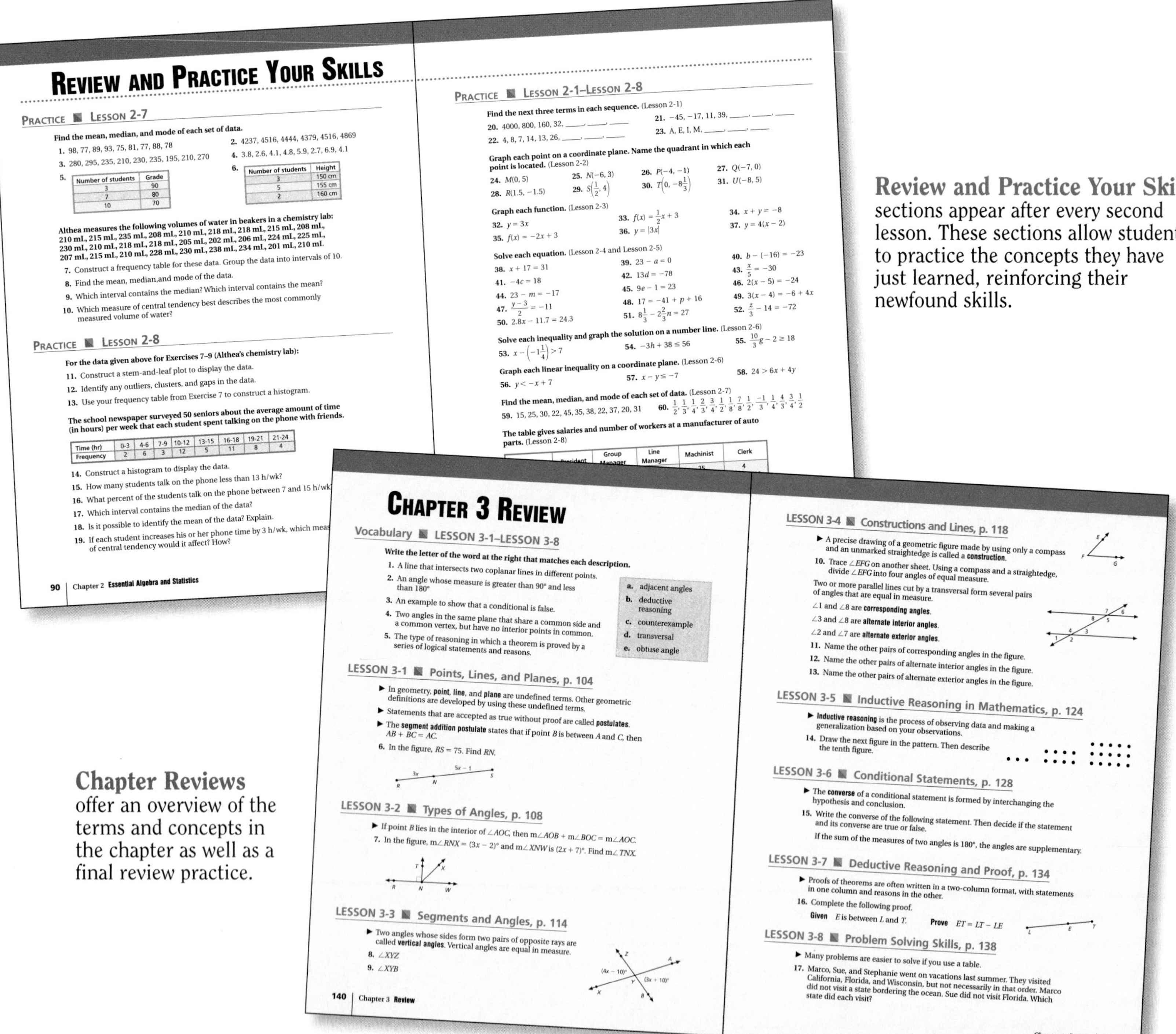

Review and Practice Your Skills sections appear after every second lesson. These sections allow students to practice the concepts they have just learned, reinforcing their newfound skills.

Chapter Reviews offer an overview of the terms and concepts in the chapter as well as a final review practice.

Assessment

Standard and alternative assessment options give both teachers and students the opportunity to evaluate students' progress in learning mathematical concepts.

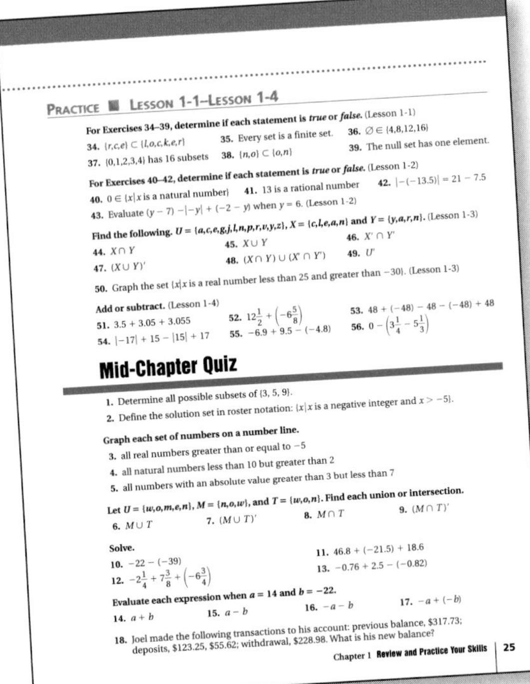

PRACTICE ■ LESSON 1-1–LESSON 1-4

For Exercises 34–39, determine if each statement is *true* or *false*. (Lesson 1-1)

34. {r,c,e} ⊂ {l,o,c,k,e,r} 35. Every set is a finite set. 36. ∅ ∈ {4,8,12,16}
37. {0,1,2,3,4} has 16 subsets 38. {n,o} ⊂ {o,n} 39. The null set has one element.

For Exercises 40–42, determine if each statement is *true* or *false*. (Lesson 1-2)

40. 0 ∈ {x|x is a natural number} 41. 13 is a rational number 42. |−(−13.5)| = 21 − 7.5
43. Evaluate (y − 7) −|−y| + (−2 − y) when y = 6. (Lesson 1-2)

Find the following. U = {a,c,e,g,j,l,n,p,r,v,y,z}, X = {c,l,e,a,n} and Y = {y,a,r,n}. (Lesson 1-3)

44. X ∩ Y 45. X ∪ Y 46. X′ ∩ Y′
47. (X ∪ Y)′ 48. (X ∩ Y) ∪ (X′ ∩ Y′) 49. U′

50. Graph the set {x|x is a real number less than 25 and greater than −30}. (Lesson 1-3)

Add or subtract. (Lesson 1-4)

51. 3.5 + 3.05 + 3.055 52. $12\frac{1}{2} + \left(-6\frac{5}{8}\right)$ 53. 48 + (−48) − 48 − (−48) + 48
54. |−17| + 15 − |15| + 17 55. −6.9 + 9.5 − (−4.8) 56. $0 - \left(3\frac{1}{4} - 5\frac{1}{3}\right)$

Mid-Chapter Quiz

1. Determine all possible subsets of {3, 5, 9}.
2. Define the solution set in roster notation: {x|x is a negative integer and x > −5}.

Graph each set of numbers on a number line.

3. all real numbers greater than or equal to −5
4. all natural numbers less than 10 but greater than 2
5. all numbers with an absolute value greater than 3 but less than 7

Let U = {w,o,m,e,n}, M = {n,o,w}, and T = {w,o,n}. Find each union or intersection.

6. M ∪ T 7. (M ∪ T)′ 8. M ∩ T 9. (M ∩ T)′

Solve.

10. −22 − (−39) 11. 46.8 + (−21.5) + 18.6
12. $-2\frac{1}{4} + 7\frac{3}{8} + \left(-6\frac{3}{4}\right)$ 13. −0.76 + 2.5 − (−0.82)

Evaluate each expression when a = 14 and b = −22.

14. a + b 15. a − b 16. −a − b 17. −a + (−b)

18. Joel made the following transactions to his account: previous balance, $317.73; deposits, $123.25, $55.62; withdrawal, $228.98. What is his new balance?

Chapter 1 Review and Practice Your Skills 25

Mid-Chapter Quizzes give students the opportunity to assess their progress in learning the chapter material at the halfway point.

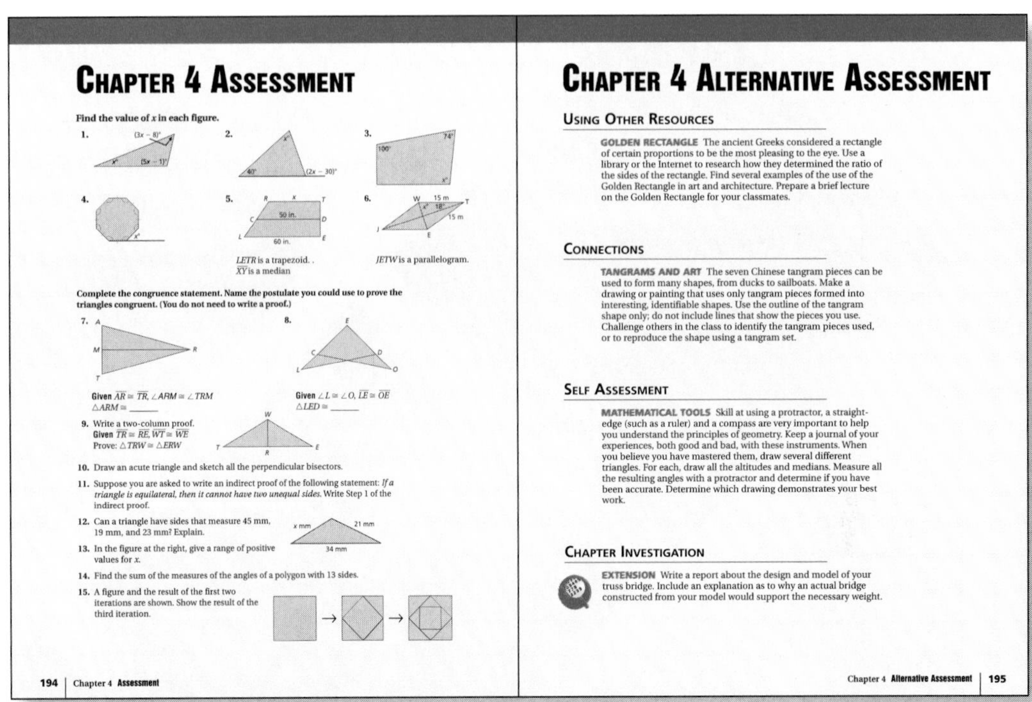

CHAPTER 4 ASSESSMENT

Find the value of x in each figure.

1. 2. 3. 4. 5. 6.

LETR is a trapezoid. XY is a median. JETW is a parallelogram.

Complete the congruence statement. Name the postulate you could use to prove the triangles congruent. (You do not need to write a proof.)

7. Given AR ≅ TR, ∠ARM ≅ ∠TRM; △ARM ≅ ____
8. Given ∠L ≅ ∠O, LE ≅ OE; △LED ≅ ____
9. Write a two-column proof. Given TR ≅ RE, WT ≅ WE. Prove: △TRW ≅ △ERW
10. Draw an acute triangle and sketch all the perpendicular bisectors.
11. Suppose you are asked to write an indirect proof of the following statement: *If a triangle is equilateral, then it cannot have two unequal sides.* Write Step 1 of the indirect proof.
12. Can a triangle have sides that measure 45 mm, 19 mm, and 23 mm? Explain.
13. In the figure at the right, give a range of positive values for x.
14. Find the sum of the measures of the angles of a polygon with 13 sides.
15. A figure and the result of the first two iterations are shown. Show the result of the third iteration.

194 Chapter 4 Assessment

CHAPTER 4 ALTERNATIVE ASSESSMENT

USING OTHER RESOURCES

GOLDEN RECTANGLE The ancient Greeks considered a rectangle of certain proportions to be the most pleasing to the eye. Use a library or the Internet to research how they determined the ratio of the sides of the rectangle. Find several examples of the use of the Golden Rectangle in art and architecture. Prepare a brief lecture on the Golden Rectangle for your classmates.

CONNECTIONS

TANGRAMS AND ART The seven Chinese tangram pieces can be used to form many shapes, from ducks to sailboats. Make a drawing or painting that uses only tangram pieces formed into interesting, identifiable shapes. Use the outline of the tangram shape only; do not include lines that show the tangram pieces you use. Challenge others in the class to identify the tangram pieces used, or to reproduce the shape using a tangram set.

SELF ASSESSMENT

MATHEMATICAL TOOLS Skill at using a protractor, a straight-edge (such as a ruler) and a compass are very important to help you understand the principles of geometry. Keep a journal of your experiences, both good and bad, with these instruments. When you believe you have mastered them, draw several different triangles. For each, draw all the altitudes and medians. Measure all the resulting angles with a protractor and determine if you have been accurate. Determine which drawing demonstrates your best work.

CHAPTER INVESTIGATION

EXTENSION Write a report about the design and model of your truss bridge. Include an explanation as to why an actual bridge constructed from your model would support the necessary weight.

Chapter 4 Alternative Assessment 195

Assessments and **Alternative Assessments** offer standard test questions as well as alternative means of evaluating students' progress in learning the concepts and skills presented.

Cumulative Reviews at the end of each chapter give students open-ended problems to solve that draw on cumulative material from the previous chapters.

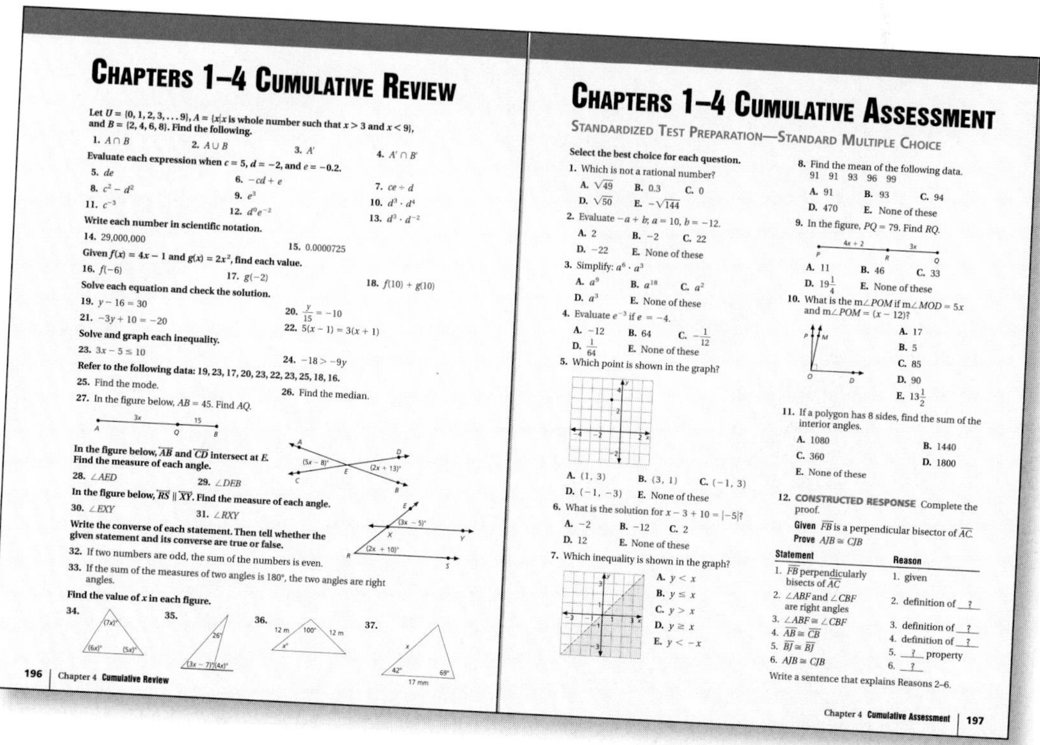

CHAPTERS 1–4 CUMULATIVE REVIEW

Let U = {0, 1, 2, 3, . . . 9}, A = {x|x is whole number such that x > 3 and x < 9}, and B = {2, 4, 6, 8}. Find the following.

1. A ∩ B 2. A ∪ B 3. A′ 4. A′ ∩ B′

Evaluate each expression when c = 5, d = −2, and e = −0.2.

5. de 6. −cd + e 7. ce ÷ d
8. $c^2 - d^2$ 9. e^3 10. $d^3 \cdot d^4$
11. c^{-3} 12. d^0e^{-2} 13. $d^3 \cdot d^{-2}$

Write each number in scientific notation.

14. 29,000,000 15. 0.0000725

Given f(x) = 4x − 1 and g(x) = 2x², find each value.

16. f(−6) 17. g(−2) 18. f(10) + g(10)

Solve each equation and check the solution.

19. y − 16 = 30 20. $\frac{y}{15} = -10$
21. −3y + 10 = −20 22. 5(x − 1) = 3(x + 1)

Solve and graph each inequality.

23. 3x − 5 ≤ 10 24. −18 > −9y

Refer to the following data: 19, 23, 17, 20, 23, 22, 23, 25, 18, 16.

25. Find the mode. 26. Find the median.
27. In the figure below, AB = 45. Find AQ.

In the figure below, AB and CD intersect at E. Find the measure of each angle.

28. ∠AED 29. ∠DEB

In the figure below, RS ∥ XY. Find the measure of each angle.

30. ∠EXY 31. ∠RXY

Write the converse of each statement. Then tell whether the given statement and its converse are true or false.

32. If two numbers are odd, the sum of the numbers is even.
33. If the sum of the measures of two angles is 180°, the two angles are right angles.

Find the value of x in each figure.

34. 35. 36. 37.

196 Chapter 4 Cumulative Review

CHAPTERS 1–4 CUMULATIVE ASSESSMENT

STANDARDIZED TEST PREPARATION—STANDARD MULTIPLE CHOICE

Select the best choice for each question.

1. Which is not a rational number?
A. √49 B. 0.3 C. 0 D. √50 E. −√144
2. Evaluate −a + b; a = 10, b = −12.
A. 2 B. −2 C. 22 D. −22 E. None of these
3. Simplify: $a^6 \cdot a^3$
A. a^9 B. a^{18} C. a^2 D. a^3 E. None of these
4. Evaluate e^{-3} if e = −4.
A. −12 B. 64 C. $-\frac{1}{12}$ D. $\frac{1}{64}$ E. None of these
5. Which point is shown in the graph?
A. (1, 3) B. (3, 1) C. (−1, 3) D. (−1, −3) E. None of these
6. What is the solution for x − 3 + 10 = |−5|?
A. −2 B. −12 C. 2 D. 12 E. None of these
7. Which inequality is shown in the graph?
A. y < x B. y ≤ x C. y > x D. y ≥ x E. y < −x
8. Find the mean of the following data. 91 91 93 96 99
A. 91 B. 93 C. 94 D. 470 E. None of these
9. In the figure, PQ = 79. Find RQ.
A. 11 B. 46 C. 33 D. $19\frac{1}{4}$ E. None of these
10. What is the m∠POM if m∠MOD = 5x and m∠POM = (x − 12)?
A. 17 B. 5 C. 85 D. 90 E. $13\frac{1}{2}$
11. If a polygon has 8 sides, find the sum of the interior angles.
A. 1080 B. 1440 C. 360 D. 1800 E. None of these
12. CONSTRUCTED RESPONSE Complete the proof.
Given FB is a perpendicular bisector of AC
Prove AJB ≅ CJB

Statement	Reason
1. FB perpendicularly bisects of AC	1. given
2. ∠ABF and ∠CBF are right angles	2. definition of ?
3. ∠ABF ≅ ∠CBF	3. definition of ?
4. AB ≅ CB	4. definition of ?
5. BJ ≅ BJ	5. ? property
6. AJB ≅ CJB	6. ?

Write a sentence that explains Reasons 2–6.

Chapter 4 Cumulative Assessment 197

Cumulative Assessments at the end of each chapter covers quantitative comparison, grid response, and constructed response questions to give teachers the opportunity to assess students' progress in a variety of ways.

MathMatters 3, 2nd Edition
An Integrated Approach

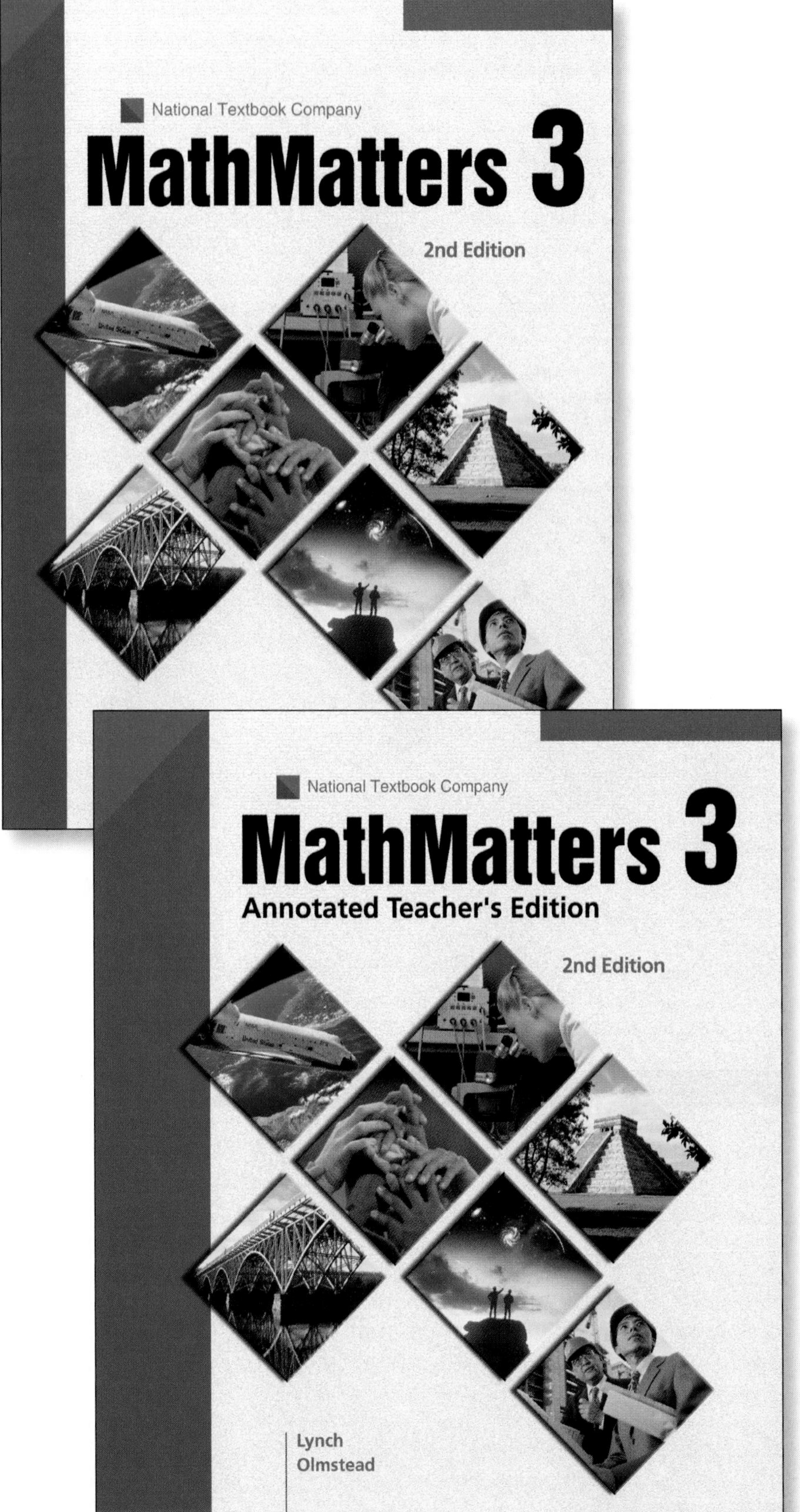

Table of Contents

MathMatters 3, 2nd Edition An Integrated Approach

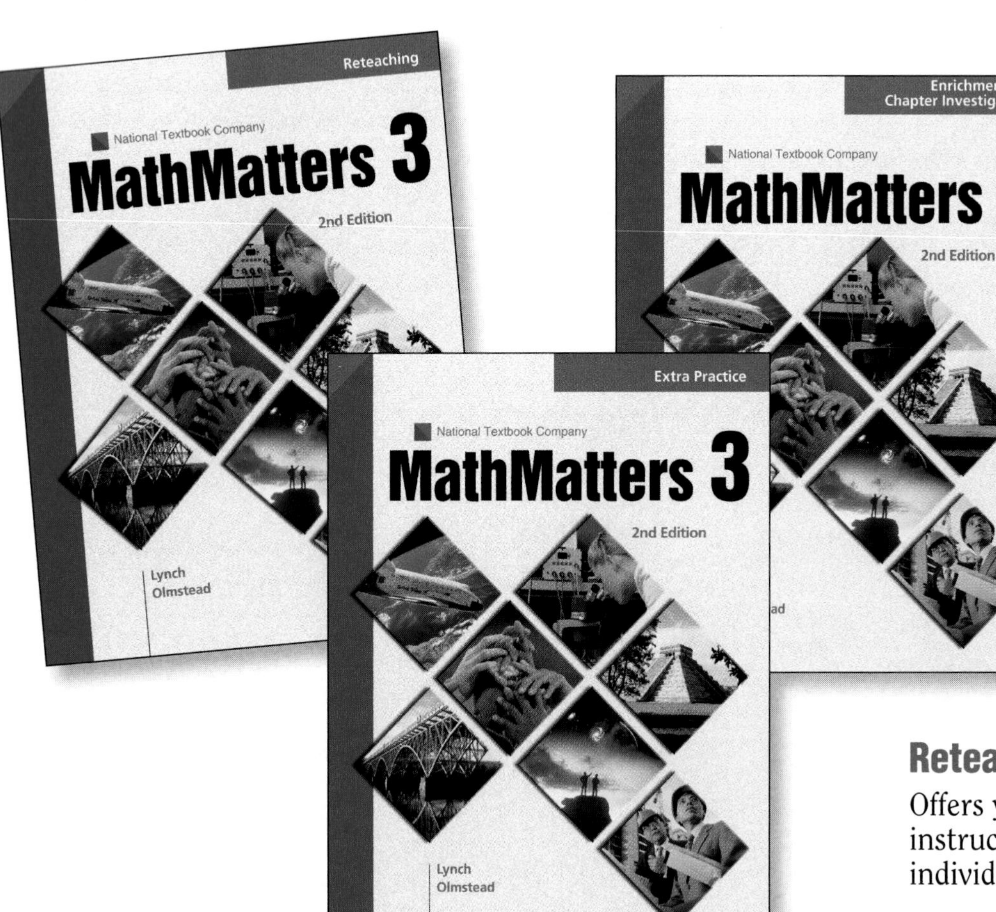

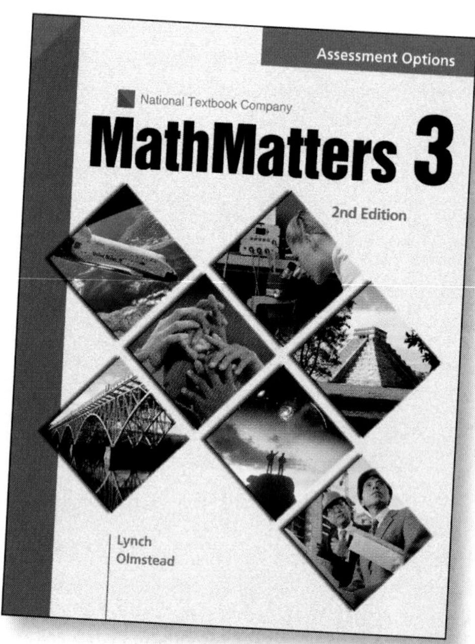

TEACHER'S RESOURCE BOX

Includes:

Annotated Teacher's Edition
Reteaching
Extra Practice
Enrichment and Chapter Investigations
Assessment Options
Technology Activities
Lesson Warm-Ups and Color Transparencies
Study Skills Activities
Multicultural Connections

Reteaching

Offers you alternate teaching strategies–including instructions, examples, and exercises–for classroom or individual instruction.

Extra Practice

Many more exercises to give your students additional practice with mathematical concepts.

Enrichment and Chapter Investigations

Go beyond the scope of the student text to give your students interesting and challenging hands-on activities to further their knowledge of mathematics.

Assessment Options

Multiple options for assessing students' progress, including two additional chapter tests, a cumulative test, scoring rubrics, and journal prompts for each chapter in the book.

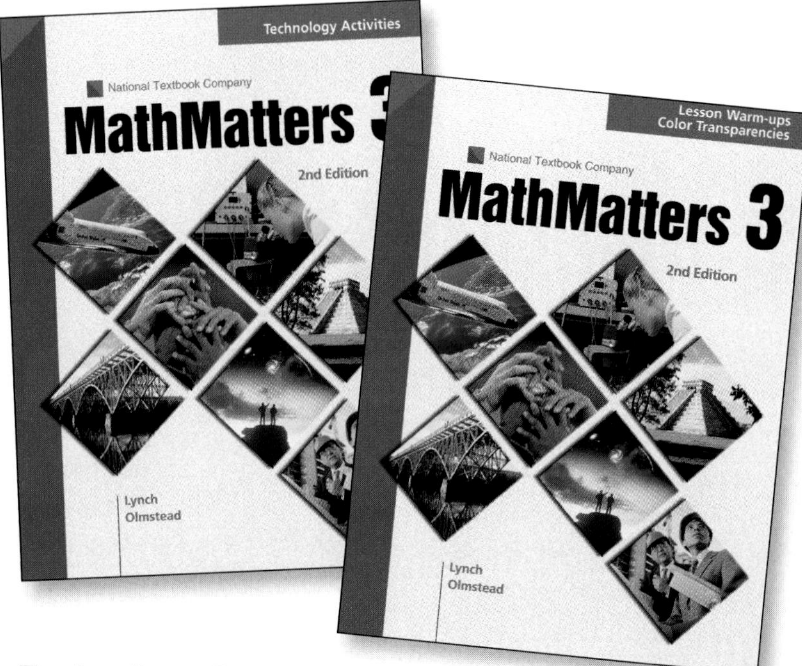

Technology Activities

Additional instructions and exercises specifically for students to use with their graphing utilities or in a computer lab.

Lesson Warm-Ups and Color Transparencies

Lesson Warm-Ups are five-minute activities that you can use to "warm up" your class. The color transparencies are lesson-specific that you can use for reference. In addition to the Transparency Toolkit that is included, you will find the standard transparencies, such as number lines and graphs, extremely helpful.

Study Skills Activities

Includes:

- Listening Skills
- Test Taking Skills
- Reading Skills
- Cooperative Learning Skills

Multicultural Connections

Includes contributions made to algebra by many cultural groups, different processes used by various cultural groups, and cultural legends. Also provides opportunities to showcase explorations of mathematical history, the development of mathematical symbols, and women in mathematics.

Algeblocks

Algeblocks provide your students with colorful, three-dimensional activities that connect concrete and abstract mathematical concepts.

Algeblocks Activity Cards

Your students will demonstrate mathematical concepts and promote their understanding visually through a set of 16 self-directing activity cards. Also includes teachers' notes, activity mats, special challenges, extension problems, and answers.

Computer Interactive Algeblocks

Computer Interactive Algeblocks will give your students computer practice and further their comprehension in mathematics with this drag-and-drop technique. Available in both Macintosh and Windows format.

Computerized Testing

This new software assessment bank allows you to customize assessment examinations by choosing which questions best suit your classroom and teaching style. Available for both Window and Macintosh.

Discover
How A Real World Approach Matters!

Title	ISBN
MathMatters 3, 2nd Edition	
Student Edition	68663-4
Teacher's Resource Box includes:	01233-9
Annotated Teacher's Edition	68664-2
Reteaching	69471-8
Extra Practice	69473-4
Enrichment and Chapter Investigations	69470-X
Assessment Options	69474-2
Technology Activities	69472-6
Lesson Warm-Ups and Color Transparencies	69475-0
Study Skills Activities	69355-X
Multicultural Connections	69354-1
Computerized Testing	01230-4
Algeblocks	
Expressions and Operations Set	66837-7
Study Group Set	65879-7
Overhead Manipulative Set	62495-7
Starter Set	65890-8

Title	ISBN
Algeblocks Activity Cards	
Expressions and Operations Set	65884-3
Equations and Inequalities Set	65885-1
Integers and Operations Set	65886-X
Making Connections Set	65887-8
Card Collection *(all card set)*	65883-5
Computer Interactive Algeblocks, Vol.1	
Windows Single-User Version	64691-8
Macintosh Single-User	64692-6
Windows Network Version	65587-9
Macintosh Network Version	65586-0
Computer Interactive Algeblocks, Vol.2	
Windows Single-User Version	65384-1
Macintosh Single-User	65385-X
Windows Network Version	65589-5
Macintosh Network Version	65588-7
Algeblocks Staff Development	
Video 1	65881-9
Video 2	65882-7

Need More Information?
Would You Like a Sample?
Who's Your Local Sales Rep?

Contact Us At:

1-800-323-4900 Phone or 1-800-998-3103 Fax or www.ntc-school.com

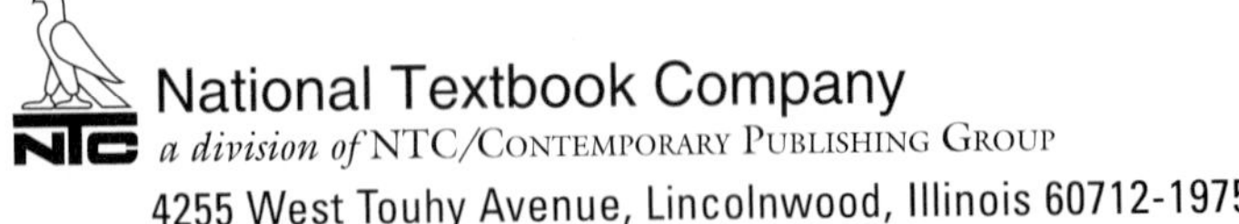

NTC
National Textbook Company
a division of NTC/Contemporary Publishing Group
4255 West Touhy Avenue, Lincolnwood, Illinois 60712-1975

National Textbook Company

MathMatters 3

Annotated Teacher's Edition

2nd Edition

Chicha Lynch
Capuchino High School
San Bruno, California

Eugene Olmstead
Elmira Free Academy
Elmira, New York

National Textbook Company
a division of NTC/Contemporary Publishing Group
Lincolnwood, Illinois USA

Senior Editor: Lisa A. De Mol
Development and Project Management: MATHQueue, Inc.—Tamara S. Jones, Suzanne M. Weisker, Edna Stroble, James Bowers, and Lisa Carmona
Internal Design: Grannan Graphic Design
Design Coordinator and Cover Design: Pagliaro Design
Composition: Better Graphics, Inc.

Acknowledgments begin on page 793, which is to be considered an extension of this copyright page.

ISBN: 0-538-68664-2

Published by National Textbook Company,
A division of NTC/Contemporary Publishing Group, Inc.,
4255 West Touhy Avenue,
Lincolnwood (Chicago), Illinois 60712-1975 U.S.A.

Manufactured in the United States of America.

04 05 06 07 08 09 QB 0 9 8 7 6 5 4 3 2

Authors

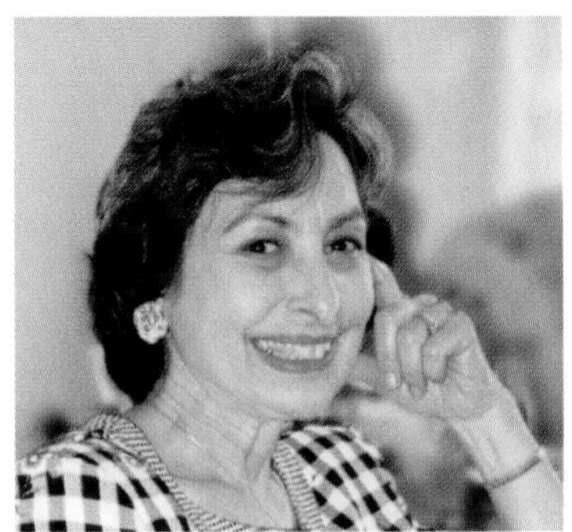

Chicha Lynch has taught mathematics at every level in grades 8 through 12. She currently teaches at Capuchino High School in San Bruno, California. Ms. Lynch is a graduate of the University of Florida and received the LaBoskey Award in 1988 from Stanford University for her contributions to its teacher education program. She was a state finalist in 1988 for the Presidential Award for Excellence in mathematics teaching. Also in 1988, Ms. Lynch began the first of three two-year terms as a California State Mentor teacher specifically in the area of Math A and Math B. Over the years, Ms. Lynch has devoted much of her professional expertise on how disadvantaged students learn math best. At National Council of Teachers of Mathematics conferences, Ms. Lynch presents workshops for local, regional, and national meetings. Her expertise has also been shared by way of video. Twice, in 1995 and again in 1998, she was a featured teacher in an in-service video produced by Annenburg Institute of School Reform.

Eugene Olmstead is a mathematics teacher at Elmira Free Academy in Elmira, New York. He has taught courses from algebra to calculus in public schools since 1972. Mr. Olmstead earned his B.S. in Mathematics at State University College at Geneseo in New York. In addition to teaching high school, he is an instructor for T^3, Teachers Teaching with Technology, and has participated in writing several of the T^3 Institutes. Mr. Olmstead frequently presents sessions and workshops on using technology in the classroom. He is a published author of articles on this topic as well. In 1991 and 1992, Mr. Olmstead was selected as a state finalist for the Presidential Award for Excellence in mathematics teaching.

Reviewers and Consultants

Educators that served as reviewers for the MathMatters series.

Tamara L. Amundsen
Teacher
Windsor Forest High School
Savannah, Georgia

Kyle A. Anderson
Mathematics Teacher
Waiakea High School
Milo, Hawaii

Murney Bell
Mathematics and Science Teacher
Anchor Bay High School
New Baltimore, Michigan

Fay Bonacorsi
High School Math Teacher
Lafayette High School
Brooklyn, New York

Boon C. Boonyapat
Mathematics Department Chairman
Henry W. Grady High School
Atlanta, Georgia

Peggy A. Bosworth
Retired Math Teacher
Plymouth-Canton High School
Canton, Michigan

Sandra C. Burke
Mathematics Teacher
Page High School
Page, Arizona

Jill Conrad
Math Teacher
Crete Public Schools
Crete, Nebraska

Nancy S. Cross
Math Educator
Merritt High School
Merritt Island, Florida

Mary G. Evangelista
Chairperson, Mathematics Department
Grove High School
Garden City, Georgia

Timothy J. Farrell
Teacher of Mathematics and Physical Science
Perth Amboy Adult School
Perth Amboy, New Jersey

Greg A. Faulhaber
Mathematics and Computer Science Teacher
Winton Woods High School
Cincinnati, Ohio

Leisa Findley
Math Teacher
Carson High School
Carson City, Nevada

Linda K. Fiscus
Mathematics Teacher
New Oxford High School
New Oxford, Pennsylvania

Louise M. Foster
Teacher and Mathematics Department Chairperson
Frederick Douglass High School
Altanta, Georgia

Darleen L. Gearhart
Mathematics Curriculum Specialist
Newark Public Schools
Newark, New Jersey

Faye Gunn
Teacher
Douglass High School
Atlanta, Georgia

Dave Harris
Math Department Head
Cedar Falls High School
Cedar Falls, Iowa

Barbara Heinrich
Teacher
Wauconda High School
Wauconda, Illinois

Margie Hill
District Coordinating Teacher Mathematics, K-12
Blue Valley School District USD229
Overland Park, Kansas

Suzanne E. Hills
Mathematics Teacher
Halifax Area High School
Halifax, Pennsylvania

Diane S. Holder
Mathematics Teacher
Larry A. Ryle High School
Union, Kentucky

Robert J. Holman
Mathematics Department
St. John's Jesuit High School
Toledo, Ohio

Eric Howe
Applied Math Graduate Student
Air Force Institute of Technology
Dayton, Ohio

Daniel R. Hudson
Mathematics Teacher
Northwest Local School District
Cincinnati, Ohio

Susan Hunt
Math Teacher
Del Norte High School
Albuquerque, New Mexico

Todd J. Jorgenson
Secondary Mathematics Instructor
Brookings High School
Brookings, South Dakota

Susan H. Kohnowich
Math Teacher
Hartford High School
White River Junction, Vermont

Mercedes Kriese
Chairperson, Mathematics Department
Neenah High School
Neenah, Wisconsin

Kathrine Lauer
Mathematics Teacher
Decatur High School
Federal Way, Washington

Laurene Lee
Mathematics Instructor
Hood River Valley High School
Hood River, Oregon

Randall P. Lieberman
Math Teacher
Lafayette High School
Brooklyn, New York

Scott Louis
Mathematics Teacher
Elder High School
Cincinnati, Ohio

Gary W. Lundquist
Teacher
Macomb Community College
Warren, Michigan

Dan Lufkin
Mathematics Instructor
Foothill High School
Pleasanton, California

Evelyn A. McDaniel
Mathematics Teacher
Natrona County High School
Casper, Wyoming

Lin McMullin
Educational Consultant
Ballston Spa, New York

Margaret H. Morris
Mathematics Instructor
Saratoga Springs Senior High School
Saratoga Springs, New York

Tom Muchlinski
Mathematics Resource Teacher
Wayzata Public Schools
Plymouth, Minnesota

Andy Murr
Mathematics
Wasilla High School
Wasilla, Alaska

Janice R. Oliva
Mathematics Teacher
Maury High School
Norfolk, VA

Fernando Rendon
Mathematics Teacher
Tucson High Magnet School/Tucson Unified School District #1
Tucson, Arizona

Candace Resmini
Mathematics Teacher
Belfast Area High School
Belfast, Maine

Kathleen A. Rooney
Chairperson, Mathematics Department
Yorktown High School
Arlington, Virginia

Mark D. Rubio
Mathematics Teacher
Hoover High School—GUSD
Glendale, California

Tony Santilli
Chairperson, Mathematics Department
Godwin Heights High School
Wyoming, Michigan

Michael Schlomer
Mathematics Department Chair
Elder High School
Cincinnati, Ohio

Jane E. Swanson
Math Teacher
Warren Township High School
Gurnee, Illinois

Martha Taylor
Teacher
Jesuit College Preparatory School
Dallas, Texas

Cheryl A. Turner
Chairperson, Mathematics Department
LaQuinta High School
LaQuinta, California

Linda Wadman
Instructor, Mathematics
Cut Bank High School
Cut Bank, Montana

George K. Wells
Coordinator of Mathematics
Mt. Mansfield Union High School
Jericho, Vermont

TEACHER PREFACE

Learn How To Use This Book

Welcome to *MathMatters.* You may find this textbook different from other math books you have used because *MathMatters* combines mathematic topics in an integrated program. Number sense, algebra, geometry, statistics, and logic are presented as tools for investigating phenomena and exploring new math concepts. Each chapter has a theme. This theme is the focus of many application problems, the chapter career features, the chapter data activity and the chapter investigation, as well as many of the photographs in the chapter. The theme and its relationship with mathematics makes for an interesting classroom discussion and an excellent way to get started with a chapter.

As you work through *MathMatters,* help your students find new ways to use estimation and approximation skills so that you can use computers and calculators more effectively. Students will discover that algebraic concepts can enhance their critical thinking skills. They will also see how geometry relates to reasoning and problem solving. Your students can also become good at evaluating the meaning of statistics that are presented on TV and in newspapers. In short, help your students learn how to use mathematics to their advantage in school, at home, and at work.

MathMatters has a consistent organization and many recurring features. Knowing what to expect will allow you to focus efficiently on what your students need to learn. Take time now to become familiar with these features of your text, and you will save valuable time later on.

Are You Ready? At the beginning of each chapter, there are two pages that contain basic math skills that have been learned in previous math courses or skills that have been learned earlier in this course. The topics presented on these two pages are skills that are needed to be successful in the chapter. This is not a pretest, but rather an assessment of readiness for the new skills that will be presented in the chapter. For your students who need remediation, there are limited practice pages near the back of the book or reference the *Basic Mathematics Review* worktext. *Basic Mathematics Review,* Second Edition is available from NTC/Contemporary also and is referenced for additional practice.

Chapter Opener The Chapter Opener sets the stage for the topics that are to be studied. It tells you about the subject matter and theme of the chapter. It also presents some fascinating data and visual information. Each opener requires that students use the data presented to answer questions. Also presented on these pages is the beginning of an overarching question posed as a chapter investigation. This question is then revisited several times throughout the lessons. By the end of the chapter, your students will have solved the problem posed at the beginning of the chapter. By starting each chapter with the opener, students can find a reason to study the mathematics of the chapter. Incorporating the data activity into your lesson provides your student with year-long review of data analysis and statistical topics that are often found in proficiency-type tests.

Introductory Activities Each four-page lesson begins with an activity that provides an introduction to the mathematical concepts or skills to be studied. These activities often involve hands-on investigations that lead to meaningful mathematical insights. Many of the Introductory Activities require students to work cooperatively with other students. Review each introduction before your class begins.

Build Understanding This section explains the concepts and skills to be mastered. Here you will find helpful discussions, along with essential definitions. The Build Understanding section presents the key points of the lesson through examples and worked-out solutions for students to study and apply later on. It is helpful for students to keep a list of each unfamiliar word, definition, or property. Write the correct definition as given in the text or glossary. Also include an example if that will help. Review the vocabulary list periodically to maintain a clear idea of what each term means. The vocabulary words are printed in a bold type that looks different than the other words that surround it. Formulas and properties are usually presented in a light purple box in the Build Understanding section of the lesson. It is helpful to keep a list of these as well.

Try These Exercises These exercises are problems that you will likely discuss in your classroom. The items in Try These Exercises are similar to the ones presented in the Building Understanding section, so they are an excellent way for you to determine if students have absorbed the key points of each lesson. If a student has trouble with any Try These Exercises items, it is worthwhile to take a second look at the relevant examples.

Practice Exercises The Practice Exercises provide numerous opportunities to practice and apply the mathematical concepts and skills studied in each lesson. These problems also make connections with other areas of mathematics, other subject areas, and the real world.

Extended Practice Exercises Each four-page lesson has an exercise set with a few questions that stretch your students' skills. Critical Thinking questions may appear in this section, as do the chapter investigations questions. A more advanced connection or a look ahead to the new lesson will also appear in these exercises.

Mixed Review Exercises Students are given the opportunity to constantly review basic math skills and skills learned in previous lessons. Doing these exercises means that students are preparing for the year-end exam throughout the year.

Problem Solving Skills and Strategies Once in each chapter, a two-page Problem Solving Skills lesson focuses on one important problem-solving strategy. Each Problem Solving Skills lesson has a brief introduction about the topic and the strategy followed by a model problem and solution. Try These Exercises are brief and are followed by Practice Exercises and Mixed Review Exercises.

Data Files This section, located near the back of the book, provides data organized around major themes, such as Health and Fitness, Economics, Ecology, and so on. The charts are needed to answer exercises that appear under the heading Using Data.

MathWorks MathWorks features a career related to the theme of the chapter. Questions are given to illustrate the mathematics used by people who perform that type of work.

Chapter Investigation Icon Identifies the chapter investigation and the questions throughout the chapter that guide students to a solution.

Writing Math Icon Identifies where students are asked to explain, describe, and summarize mathematical procedures and problems in writing.

Manipulative Icon Identifies questions that require the use of manipulatives including Algeblocks (or other algebra manipulatives) and other items that can be used as concrete models.

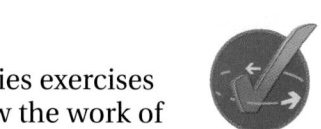

Technology Icon Identifies exercises that incorporate the use of scientific and graphing calculators, plus spreadsheets and geometry software.

Error Analysis Icon Identifies exercises where students are to either review the work of others or their own work to check for possible errors. These questions invite students to take on the role of an evaluator.

CONTENTS

Chapter 1 Essential Mathematics

THEME: Language and Communication

Real World Applications: finance, retail, communication, temperature, advertising, language, logic, recreation, construction, food service, fitness, number sense, sales

Interdisciplinary Connections: geography, science, chemistry, astronomy, business, health

Tools: paper, pencil, index cards, calculator, counters, number lines, spreadsheet software, grid paper

Contents

Chapter 2 Essential Algebra and Statistics

THEME: News Media

Real World Applications: media, finance, engineering, television, journalism, temperature, retail sales, photography, news media, advertising, recreation, sports, education, marketing, entertainment

Interdisciplinary Connections: biology, business, health

Tools: paper, pencil, blocks, grid paper, Algeblocks, straightedge, number lines, calculator

CONTENTS

Chapter 3 Geometry and Reasoning

THEME: Geography

Real World Applications: carpentry, tiling, architecture, engineering, map making, surveying, navigation, city planning, scheduling, drafting, sports

Interdisciplinary Connections: art, geography

Tools: paper, pencil, protractor, ruler, grid paper, compass, straightedge, geometric drawing software, counters, straws

CONTENTS

Chapter 4 Triangles, Quadrilaterals and Other Polygons

THEME: Art and Design

Real World Applications: technical art, construction, engineering, animation, design, architecture, service

Interdisciplinary Connections: physics, art

Tools: grid paper, scissors, straightedges, protractors, metric ruler, compass, geometric drawing software, paper, pencil, heavy paper

CONTENTS

Chapter 5 Measurement

THEME: Lost Cities of Ancient Worlds

Real World Applications: engineering, number sense, archeology, stage design, sports, recreation, games, weather, architecture, packaging, manufacturing, astronomy

Interdisciplinary Connections: history, art, science

Tools: calculator, compass, centimeter ruler, protractor, paper, pencil, grid paper, scissors, straightedge, construction paper, tape

CONTENTS

Chapter 6 Linear Systems of Equations

THEME: Manufacturing Industry

Real World Applications: manufacturing, finance, recreation, electronics, map making, product design, fitness, real estate, income tax, shipping, packaging, entertainment, community service, budgeting

Interdisciplinary Connections: business, health

Tools: geoboards, rubber bands, grid paper, straightedge, paper, pencil, graphing calculator, Algeblocks

CONTENTS

Chapter 7 Similar Triangles

THEME: Photography

Real World Applications: recreation, real estate, retail, photography, architecture, framing, engineering, surveying, scale models, model building

Interdisciplinary Connections: business, art

Tools: calculator, centimeter ruler, protractor, string, straightedge, compass, geometric drawing software, mirror

CONTENTS

Chapter 8 Transformations

THEME: Amusement Parks

Real World Applications: recreation, amusement park design, ride management, computer graphics, art, graphic design, engineering, souvenir sales, cryptography, manufacturing, ticket sales, encryption, inventory, ride design

Interdisciplinary Connections: art, business

Tools: grid paper, ruler, scissors, geometric drawing software, graphing calculator

CONTENTS

Chapter 9 Probability and Statistics

THEME: Sports

Real World Applications: sports, games, test taking, surveys, scheduling, travel, office work, manufacturing, sales, education

Interdisciplinary Connections: business

Tools: calculators, blank spinners, coins, computers, paper bags, dice, grid paper, straightedges

Contents

Chapter 10 Right Angles and Circles

THEME: Architecture

Real World Applications: architecture, construction, home repairs, road planning, plumbing, landscape, navigation, surveying, design

Interdisciplinary Connections: art

Tools: calculator, grid paper, straightedges, scissors, geometric drawing software, computer, compasses, protractors

CONTENTS

Chapter 11 Polynomials

THEME: Consumerism

Real World Applications: packaging, shipping, advertising, landscaping, payroll, manufacturing, sculpture, small business, product development, sales, construction

Interdisciplinary Connections: chemistry, art

Tools: paper, pencil, Algeblocks, computers

CONTENTS

Chapter 12 Quadratic Functions

THEME: Gravity

Real World Applications: small business, astronomy, archery, sports, science aeronautics, skydiving, space exploration, archeology

Interdisciplinary Connections: geology, physics

Tools: graphing calculators, grid paper, Algeblocks, paper, pencil

CONTENTS

Chapter 13 Quadratic Relations

THEME: Astronomy

Real World Applications astronomy, sports, architecture, satellite communications, energy, oceanography, communications, food prices, space exploration, travel, computer design

Interdisciplinary Connections physics, science

Tools graphing calculators, grid paper, pencil, paper, cardboard, scissors, string, thumbtacks, calculators

CONTENTS

Chapter 14 Trigonometry

THEME: Navigation

Real World Applications: navigation, construction, city planning, surveying, safety, communications, population, flight, medicine

Interdisciplinary Connections: health, music

Tools: calculators, rulers, graphing calculators

Chapter 1 Pacing

Lesson, Pages	Lesson Title	45 min class	Assignments Basic, Enriched	90 min class	Assignments Basic, Enriched	___ min class	Assignments Basic, Enriched
1–1 6–9	AYR, Opener, The Language of Mathematics	Day 1	B: 1–34, 41–52 E: 1–52	Day 1	B: 1–34, 41–52 E: 1–52		B: 1–34, 41–52 E: 1–52
1–2 10–13	Real Numbers, R&PYS	Day 2–3	B: 1–32, 36–46 E: 1–46	Day 1–2	B: 1–32, 36–46 E: 1–46		B: 1–32, 36–46 E: 1–46
1–3 16–19	Union and Intersection of Sets	Day 4	B: 1–34, 43–56 E: 1–56	Day 2	B: 1–34, 43–56 E: 1–56		B: 1–34, 43–56 E: 1–56
1–4 20–23	Addition, Subtraction and Estimation, R&PYS	Day 5–6	B: 1–28, 31–44 E: 1–44	Day 3	B: 1–28, 31–44 E: 1–44		B: 1–28, 31–44 E: 1–44
1–5 26–29	Multiplication and Division	Day 7	B: 1–40, 46–66 E: 1–66	Day 4	B: 1–40, 46–66 E: 1–66		B: 1–40, 46–66 E: 1–66
1–6 30–31	Problem Solving Skills: Use Technology, R&PYS	Day 8–9	B: 1–38 E: 1–38	Day 4–5	B: 1–38 E: 1–38		B: 1–38 E: 1–38
1–7 34–37	Distributive Properties and Properties of Exponents	Day 10	B: 1–44, 49–52 E: 1–52	Day 5	B: 1–44, 49–52 E: 1–52		B: 1–44, 49–52 E: 1–52
1–8 38–41	Exponents and Scientific Notation	Day 11	B: 1–40, 46–69 E: 1–69	Day 6	B: 1–40, 46–69 E: 1–69		B: 1–40, 46–69 E: 1–69
Review/ Assess		Day 12–14		Day 6–7			

	Chapter 1 Days	Chapter Cumulative Days
Yearly Pacing (45 min class)	14 days	14 days
Yearly Pacing (90 min class)	7 days	7 days

Chapter 2 Pacing

Lesson, Pages	Lesson Title	45 min class	Assignments Basic, Enriched	90 min class	Assignments Basic, Enriched	___ min class	Assignments Basic, Enriched
2–1 52–55	AYR, Opener, Patterns and Iterations	Day 15	B: 1–20, 23–24 E: 1–34	Day 8	B: 1–20, 23–24 E: 1–34		B: 1–20, 23–24 E: 1–34
2–2 56–59	The Coordinate Plane, Relations and Functions, R&PYS	Day 16–17	B: 1–29, 36–44 E: 1–44	Day 8–9	B: 1–29, 36–44 E: 1–44		B: 1–29, 36–44 E: 1–44
2–3 62–65	Linear Functions	Day 18	B: 1–20, 25–38 E: 1–38	Day 9	B: 1–20, 25–38 E: 1–38		B: 1–20, 25–38 E: 1–38
2–4 66–69	Solve One-Step Equations, R&PYS	Day 19–20	B: 1–38, 46–56 E: 1–56	Day 10	B: 1–38, 46–56 E: 1–56		B: 1–38, 46–56 E: 1–56
2–5 72–75	Solve Multi-Step Equations	Day 21	B: 1–34, 38–55 E: 1–55	Day 11	B: 1–34, 38–55 E: 1–55		B: 1–34, 38–55 E: 1–55
2–6 76–79	Solve Linear Inequalities, R&PYS	Day 22–23	B: 1–33, 37–42 E: 1–42	Day 11–12	B: 1–33, 37–42 E: 1–42		B: 1–33, 37–42 E: 1–42
2–7 82–85	Data and Measures of Central Tendency	Day 24	B: 1–14, 17–26 E: 1–26	Day 12	B: 1–14, 17–26 E: 1–26		B: 1–14, 17–26 E: 1–26
2–8 86–89	Display Data, R&PYS	Day 25–26	B: 1–18, 20–35 E: 1–35	Day 13	B: 1–18, 20–35 E: 1–35		B: 1–18, 20–35 E: 1–35
2–9 92–93	Problem Solving Skills: Misleading Graphs	Day 27	B: 1–4 E: 1–6	Day 14	B: 1–4 E: 1–6		B: 1–4 E: 1–6
Review/ Assess		Day 28–30		Day 14–15			

	Chapter 1 Days	Chapter Cumulative Days
Yearly Pacing (45 min class)	16 days	30 days
Yearly Pacing (90 min class)	9 days	15 days

Chapter 3 Pacing

Lesson, Pages	Lesson Title	45 min class	Assignments Basic, Enriched	90 min class	Assignments Basic, Enriched	___ min class	Assignments Basic, Enriched
3–1 104–107	AYR, Opener, Points, Lines and Planes	Day 31	B: 1–18, 21–34 E: 1–34	Day 16	B: 1–18, 21–34 E: 1–34		B: 1–18, 21–34 E: 1–34
3–2 108–111	Types of Angles, R&PYS	Day 32–33	B: 1–25, 29–42 E: 1–42	Day 16–17	B: 1–25, 29–42 E: 1–42		B: 1–25, 29–42 E: 1–42
3–3 114–117	Segments and Angles	Day 34	B: 1–29, 36–45 E: 1–45	Day 17	B: 1–29, 36–45 E: 1–45		B: 1–29, 36–45 E: 1–45
3–4 118–121	Constructions, Perpendicular and Parallel Lines, R&PYS	Day 35–36	B: 1–14, 19, 21–22 E: 1–22	Day 18	B: 1–14, 19, 21–22 E: 1–22		B: 1–14, 19, 21–22 E: 1–22
3–5 124–127	Inductive Reasoning in Mathematics	Day 37	B: 1–11, 15–24 E: 1–24	Day 19	B: 1–11, 15–24 E: 1–24		B: 1–11, 15–24 E: 1–24
3–6 128–131	Conditional Statements, R&PYS	Day 38–39	B: 1–26, 31, 32 E: 1–32	Day 19–20	B: 1–26, 31, 32 E: 1–32		B: 1–26, 31, 32 E: 1–32
3–7 134–137	Deductive Reasoning and Proof	Day 40	B: 1–6, 10–30 E: 1–30	Day 20	B: 1–6, 10–30 E: 1–30		B: 1–6, 10–30 E: 1–30
3–8 138–139	Problem Solving Skills: Use Logical Reasoning, R&PYS	Day 41	B: 1–14 E: 1–14	Day 21	B: 1–14 E: 1–14		B: 1–14 E: 1–14
Review/ Assess		Day 42–44		Day 21–22			

	Chapter 1 Days	Chapter Cumulative Days
Yearly Pacing (45 min class)	14 days	44 days
Yearly Pacing (90 min class)	7 days	22 days

Chapter 4 and Pacing

Lesson, Pages	Lesson Title	45 min class	Assignments Basic, Enriched	90 min class	Assignments Basic, Enriched	___ min class	Assignments Basic, Enriched
4–1 150–153	AYR, Opener, Triangles and Triangle Theorems	Day 45	B: 1–18, 23–26 E: 1–26	Day 23	B: 1–18, 23–26 E: 1–26		B: 1–18, 23–26 E: 1–26
4–2 154–157	Congruent Triangles, R&PYS	Day 46–47	B: 1–10, 13–16 E: 1–16	Day 23–24	B: 1–10, 13–16 E: 1–16		B: 1–10, 13–16 E: 1–16
4–3 160–163	Congruent Triangles and Proofs	Day 48	B: 1–16, 19–22 E: 1–22	Day 24	B: 1–16, 19–22 E: 1–22		B: 1–16, 19–22 E: 1–22
4–4 164–167	Altitudes, Medians and Perpendicular Bisectors, R&PYS	Day 49–50	B: 1–22, 24–36 E: 1–36	Day 25	B: 1–22, 24–36 E: 1–36		B: 1–22, 24–36 E: 1–36
4–5 170–171	Problem Solving Skills: Write an Indirect Proof	Day 51	B: 1–6 E: 1–6	Day 26	B: 1–6 E: 1–6		B: 1–6 E: 1–6
4–6 172–175	Inequalities in Triangles, R&PYS	Day 52–53	B: 1–28, 35–45 E: 1–45	Day 26–27	B: 1–28, 35–45 E: 1–45		B: 1–28, 35–45 E: 1–45
4–7 178–181	Polygons and Angles	Day 54	B: 1–23, 29–33 E: 1–33	Day 27	B: 1–23, 29–33 E: 1–33		B: 1–23, 29–33 E: 1–33
4–8 182–185	Special Quadrilaterals: Parallelograms, R&PYS	Day 55–56	B: 1–24, 27–33 E: 1–33	Day 28	B: 1–24, 27–33 E: 1–33		B: 1–24, 27–33 E: 1–33
4–9 188–191	Special Quadrilaterals: Trapezoids	Day 57	B: 1–28, 31–36 E: 1–36	Day 29	B: 1–28, 31–36 E: 1–36		B: 1–28, 31–36 E: 1–36
Review/ Assess		Day 58–60		Day 29–30			

	Chapter 1 Days	Chapter Cumulative Days
Yearly Pacing (45 min class)	15 days	60 days
Yearly Pacing (90 min class)	8 days	30 days

Chapter 5 Pacing

Lesson, Pages	Lesson Title	45 min class	Assignments Basic, Enriched	90 min class	Assignments Basic, Enriched	___ min class	Assignments Basic, Enriched
5–1 202–205	AYR, Opener, Ratios and Units of Measure	Day 61	B: 1–34, 37–38 E: 1–38	Day 31	B: 1–34, 37–38 E: 1–38		B: 1–34, 37–38 E: 1–38
5–2 206–209	Perimeter, Circumference and Area, R&PYS	Day 62–63	B: 1–22, 26–40 E: 1–40	Day 31–32	B: 1–22, 26–40 E: 1–40		B: 1–22, 26–40 E: 1–40
5–3 212–215	Probability and Area	Day 64	B: 1–26, 29–40 E: 1–40	Day 32	B: 1–26, 29–40 E: 1–40		B: 1–26, 29–40 E: 1–40
5–4 216–217	Problem Solving Skills: Irregular Shapes, R&PYS	Day 65–66	B: 1–7, 8–15 E: 1–15	Day 33	B: 1–7, 8–15 E: 1–15		B: 1–7, 8–15 E: 1–15
5–5 220–223	Three-dimensional Figures and Loci	Day 67	B: 1–18, 21–29 E: 1–29	Day 34	B: 1–18, 21–29 E: 1–29		B: 1–18, 21–29 E: 1–29
5–6 224–227	Surface Area of Three-dimensional Figures, R&PYS	Day 68–69	B: 1–18, 23–24 E: 1–24	Day 34–35	B: 1–18, 23–24 E: 1–24		B: 1–18, 23–24 E: 1–24
5–7 230–233	Volume of Three-dimensional Figures	Day 70	B: 1–22, 26–29 E: 1–29	Day 35	B: 1–22, 26–29 E: 1–29		B: 1–22, 26–29 E: 1–29
Review/ Assess		Day 71–73		Day 36–37			

	Chapter 1 Days	Chapter Cumulative Days
Yearly Pacing (45 min class)	13 days	73 days
Yearly Pacing (90 min class)	7 days	37 days

Chapter 6 Pacing

Lesson, Pages	Lesson Title	45 min class	Assignments Basic, Enriched	90 min class	Assignments Basic, Enriched	___ min class	Assignments Basic, Enriched
6–1 244–247	AYR, Opener, Slope of a Line and Slope-intercept Form	Day 74	B: 1–41, 48–64 E: 1–64	Day 38	B: 1–41, 48–64 E: 1–64		B: 1–41, 48–64 E: 1–64
6–2 248–251	Parallel and Perpendicular Lines, R&PYS	Day 75–76	B: 1–25, 31–33 E: 1–33	Day 38–39	B: 1–25, 31–33 E: 1–33		B: 1–25, 31–33 E: 1–33
6–3 254–257	Write Equations for Lines	Day 77	B: 1–22, 26–34 E: 1–34	Day 39	B: 1–22, 26–34 E: 1–34		B: 1–22, 26–34 E: 1–34
6–4 258–261	Systems of Equations, R&PYS	Day 78–79	B: 1–20, 23–29 E: 1–29	Day 40	B: 1–20, 23–29 E: 1–29		B: 1–20, 23–29 E: 1–29
6–5 264–267	Solve Systems by Substitution	Day 80	B: 1–20, 26–33 E: 1–33	Day 41	B: 1–20, 26–33 E: 1–33		B: 1–20, 26–33 E: 1–33
6–6 268–271	Solve Systems by Adding and Multiplying, R&PYS	Day 81–82	B: 1–25, 31–44 E: 1–44	Day 41–42	B: 1–25, 31–44 E: 1–44		B: 1–25, 31–44 E: 1–44
6–7 274–275	Problem Solving Skills: Determinants and Matrices	Day 83	B: 1–12 E: 1–12	Day 42	B: 1–12 E: 1–12		B: 1–12 E: 1–12
6–8 276–279	Systems of Inequalities, R&PYS	Day 84–84	B: 1–24, 27–30 E: 1–30	Day 43	B: 1–24, 27–30 E: 1–30		B: 1–24, 27–30 E: 1–30
6–9 282–285	Linear Programming	Day 85	B: 1–24, 27–32 E: 1–32	Day 44	B: 1–24, 27–32 E: 1–32		B: 1–24, 27–32 E: 1–32
Review/ Assess		Day 86–88		Day 44–45			

	Chapter 1 Days	Chapter Cumulative Days
Yearly Pacing (45 min class)	16 days	88 days
Yearly Pacing (90 min class)	8 days	45 days

Chapter 7 Pacing

Lesson, Pages	Lesson Title	45 min class	Assignments Basic, Enriched	90 min class	Assignments Basic, Enriched	___ min class	Assignments Basic, Enriched
7–1 296–299	AYR, Opener, Ratios and Proportions	Day 89	B: 1–34, 37–56 E: 1–56	Day 46	B: 1–34, 37–56 E: 1–56		B: 1–34, 37–56 E: 1–56
7–2 300–303	Similar Polygons, R&PYS	Day 90–91	B: 1–15, 19–27 E: 1–27	Day 46–47	B: 1–15, 19–27 E: 1–27		B: 1–15, 19–27 E: 1–27
7–3 306–309	Scale Drawings	Day 92	B: 1–19, 22–24 E: 1–24	Day 47	B: 1–19, 22–24 E: 1–24		B: 1–19, 22–24 E: 1–24
7–4 310–313	Postulates for Similar Triangles, R&PYS	Day 93–94	B: 1–10, 15–26 E: 1–26	Day 48	B: 1–10, 15–26 E: 1–26		B: 1–10, 15–26 E: 1–26
7–5 316–319	Triangles and Proportional Segments	Day 95	B: 1–10, 14–20 E: 1–20	Day 49	B: 1–10, 14–20 E: 1–20		B: 1–10, 14–20 E: 1–20
7–6 320–323	Parallel Lines and Proportional Segments, R&PYS	Day 96–97	B: 1–14, 19–26 E: 1–26	Day 49–50	B: 1–14, 19–26 E: 1–26		B: 1–14, 19–26 E: 1–26
7–7 326–327	Problem Solving Skills: Indirect Measurement	Day 98	B: 1–5, 6–14 E: 1–14	Day 50	B: 1–5, 6–14 E: 1–14		B: 1–5, 6–14 E: 1–14
Review/ Assess		Day 99–101		Day 51–52			

	Chapter 1 Days	Chapter Cumulative Days
Yearly Pacing (45 min class)	12 days	101 days
Yearly Pacing (90 min class)	7 days	52 days

Chapter 8 Pacing

Lesson, Pages	Lesson Title	45 min class	Assignments Basic, Enriched	90 min class	Assignments Basic, Enriched	___ min class	Assignments Basic, Enriched
8–1 338–341	AYR, Opener, Translations and Reflections	Day 102	B: 1–15, 19–34 E: 1–34	Day 53	B: 1–15, 19–34 E: 1–34		B: 1–15, 19–34 E: 1–34
8–2 342–345	Rotations in the Coordinate Plane, R&PYS	Day 103–104	B: 1–14, 17–30 E: 1–30	Day 53–54	B: 1–14, 17–30 E: 1–30		B: 1–14, 17–30 E: 1–30
8–3 348–351	Dilations in the Coordinate Plane	Day 105	B: 1–16, 19–31 E: 1–31	Day 54	B: 1–16, 19–31 E: 1–31		B: 1–16, 19–31 E: 1–31
8–4 352–355	Multiple Transformations, R&PYS	Day 106–107	B: 1–17, 21–24 E: 1–24	Day 55	B: 1–17, 21–24 E: 1–24		B: 1–17, 21–24 E: 1–24
8–5 358–361	Addition and Multiplication with Matrices	Day 108	B: 1–27, 32–40 E: 1–40	Day 56	B: 1–27, 32–40 E: 1–40		B: 1–27, 32–40 E: 1–40
8–6 362–365	More Operations on Matrices, R&PYS	Day 109–110	B: 1–33, 40–45 E: 1–45	Day 56–57	B: 1–33, 40–45 E: 1–45		B: 1–33, 40–45 E: 1–45
8–7 368–371	Transformations and Matrices	Day 111	B: 1–24, 28–34 E: 1–34	Day 57	B: 1–24, 28–34 E: 1–34		B: 1–24, 28–34 E: 1–34
8–8 372–373	Problem Solving Skills: Use a Matrix, R&PYS	Day 112	B: 1–12 E: 1–12	Day 58	B: 1–12 E: 1–12		B: 1–12 E: 1–12
Review/ Assess		Day 113–115		Day 58–59			

	Chapter 1 Days	Chapter Cumulative Days
Yearly Pacing (45 min class)	14 days	115 days
Yearly Pacing (90 min class)	7 days	59 days

Chapter 9 Pacing

Lesson, Pages	Lesson Title	45 min class	Assignments Basic, Enriched	90 min class	Assignments Basic, Enriched	___ min class	Assignments Basic, Enriched
9–1 384–387	AYR, Opener, Review Percents and Probability	Day 115	B: 1–20, 25–28 E: 1–28	Day 59	B: 1–20, 25–28 E: 1–28		B: 1–20, 25–28 E: 1–28
9–2 388–389	Problem Solving Skills: Simulations, R&PYS	Day 116–117	B: 1–25 E: 1–25	Day 59–60	B: 1–25 E: 1–25		B: 1–25 E: 1–25
9–3 392–395	Compound Events	Day 118	B: 1–26, 31–42 E: 1–42	Day 60	B: 1–26, 31–42 E: 1–42		B: 1–26, 31–42 E: 1–42
9–4 396–399	Independent and Dependent Events, R&PYS	Day 119–120	B: 1–25, 32–34 E: 1–34	Day 61	B: 1–25, 32–34 E: 1–34		B: 1–25, 32–34 E: 1–34
9–5 402–405	Permutations and Combinations	Day 121	B: 1–18, 23–27 E: 1–27	Day 62	B: 1–18, 23–27 E: 1–27		B: 1–18, 23–27 E: 1–27
9–6 406–409	Scatter Plots and Boxplots, R&PYS	Day 122–123	B: 1–12, 16–18 E: 1–18	Day 63	B: 1–12, 16–18 E: 1–18		B: 1–12, 16–18 E: 1–18
9–7 412–415	Standard Deviation	Day 124	B: 1–20, 25–30 E: 1–30	Day 63–64	B: 1–20, 25–30 E: 1–30		B: 1–20, 25–30 E: 1–30
Review/ Assess		Day 125–126		Day 64–66			

	Chapter 1 Days	Chapter Cumulative Days
Yearly Pacing (45 min class)	12 days	126 days
Yearly Pacing (90 min class)	8 days	66 days

Chapter 10 Pacing

Lesson, Pages	Lesson Title	45 min class	Assignments Basic, Enriched	90 min class	Assignments Basic, Enriched	___ min class	Assignments Basic, Enriched
10–1 426–429	AYR, Opener, Irrational Numbers	Day 127	B: 1–40, 45–50 E: 1–50	Day 67	B: 1–40, 45–50 E: 1–50		B: 1–40, 45–50 E: 1–50
10–2 430–433	The Pythagorean Theorem, R&PYS	Day 128–129	B: 1–22, 26–30 E: 1–30	Day 67–68	B: 1–22, 26–30 E: 1–30		B: 1–22, 26–30 E: 1–30
10–3 436–439	Special Right Triangles	Day 130	B: 1–15, 19–28 E: 1–28	Day 68	B: 1–15, 19–28 E: 1–28		B: 1–15, 19–28 E: 1–28
10–4 440–443	Circles, Angles, and Arcs, R&PYS	Day 131–132	B: 1–10, 13–25 E: 1–25	Day 69	B: 1–10, 13–25 E: 1–25		B: 1–10, 13–25 E: 1–25
10–5 446–447	Problem Solving Skills: Circle Graphs	Day 133	B: 1–10 E: 1–10	Day 70	B: 1–10 E: 1–10		B: 1–10 E: 1–10
10–6 448–451	Circles and Segments, R&PYS	Day 134–135	B: 1–20, 23–25 E: 1–25	Day 70–71	B: 1–20, 23–25 E: 1–25		B: 1–20, 23–25 E: 1–25
10–7 454–457	Construction with Circles	Day 136	B: 1–19, 21–26 E: 1–26	Day 71	B: 1–19, 21–26 E: 1–26		B: 1–19, 21–26 E: 1–26
Review/ Assess		Day 137–138		Day 72–73			

	Chapter 1 Days	Chapter Cumulative Days
Yearly Pacing (45 min class)	12 days	137 days
Yearly Pacing (90 min class)	7 days	73 days

Chapter 11 Pacing

Lesson, Pages	Lesson Title	45 min class	Assignments Basic, Enriched	90 min class	Assignments Basic, Enriched	___ min class	Assignments Basic, Enriched
11–1 468–471	AYR, Opener, Add and Subtract Polynomials	Day 139	B: 1–35, 40–54 E: 1–54	Day 74	B: 1–35, 40–54 E: 1–54		B: 1–35, 40–54 E: 1–54
11–2 472–475	Multiply by a Monomial	Day 139–140	B: 1–52, 55 E: 1–55	Day 74–75	B: 1–52, 55 E: 1–55		B: 1–52, 55 E: 1–55
11–3 478–481	Divide and Find Factors, R&PYS	Day 141	B: 1–37, 42–44 E: 1–44	Day 75	B: 1–37, 42–44 E: 1–44		B: 1–37, 42–44 E: 1–44
11–4 482–485	Multiply Two Binomials, R&PYS	Day 142–143	B: 1–45, 49–60 E: 1–60	Day 76	B: 1–45, 49–60 E: 1–60		B: 1–45, 49–60 E: 1–60
11–5 488–491	Find Binomial Factors in a Polynomial	Day 144	B: 1–38, 43–56 E: 1–56	Day 77	B: 1–38, 43–56 E: 1–56		B: 1–38, 43–56 E: 1–56
11–6 492–495	Special Factoring Patterns, R&PYS	Day 145–146	B: 1–40, 43–50 E: 1–50	Day 77–78	B: 1–40, 43–50 E: 1–50		B: 1–40, 43–50 E: 1–50
11–7 498–501	Factor Trinomials	Day 147	B: 1–53, 55–74 E: 1–74	Day 78	B: 1–53, 55–74 E: 1–74		B: 1–53, 55–74 E: 1–74
11–8 502–503	Problem Solving Skills: The General Case, R&PYS	Day 148	B: 1–16 E: 1–16	Day 79	B: 1–16 E: 1–16		B: 1–16 E: 1–16
11–9 506–509	More on Factoring Trinomials	Day 149	B: 1–42, 45–68 E: 1–68	Day 80	B: 1–42, 45–68 E: 1–68		B: 1–42, 45–68 E: 1–68
Review/ Assess		Day 150–151		Day 80–82			

	Chapter 1 Days	Chapter Cumulative Days
Yearly Pacing (45 min class)	13 days	151 days
Yearly Pacing (90 min class)	9 days	82 days

Chapter 12 Pacing

Lesson, Pages	Lesson Title	45 min class	Assignments Basic, Enriched	90 min class	Assignments Basic, Enriched	___ min class	Assignments Basic, Enriched
12–1 520–523	AYR, Opener, Graph Parabolas	Day 152	B: 1–18, 26–34 E: 1–34	Day 83	B: 1–18, 26–34 E: 1–34		B: 1–18, 26–34 E: 1–34
12–2 524–527	The General Quadratic Function, R&PYS	Day 153–154	B: 1–20, 24–35 E: 1–35	Day 83–84	B: 1–20, 24–35 E: 1–35		B: 1–20, 24–35 E: 1–35
12–3 530–533	Factor and Graph	Day 155	B: 1–31, 37–50 E: 1–50	Day 84	B: 1–31, 37–50 E: 1–50		B: 1–31, 37–50 E: 1–50
12–4 534–537	Complete the Square, R&PYS	Day 156–157	B: 1–40, 47–67 E: 1–67	Day 85	B: 1–40, 47–67 E: 1–67		B: 1–40, 47–67 E: 1–67
12–5 540–543	The Quadratic Formula	Day 158	B: 1–30, 35–48 E: 1–48	Day 86	B: 1–30, 35–48 E: 1–48		B: 1–30, 35–48 E: 1–48
12–6 544–547	Use the Pythagorean Theorem, R&PYS	Day 159–160	B: 1–25, 29–40 E: 1–40	Day 86–87	B: 1–25, 29–40 E: 1–40		B: 1–25, 29–40 E: 1–40
12–7 550–551	Problem Solving Skills: Use Graphs to Write Equations	Day 161	B: 1–14 E: 1–14	Day 87	B: 1–14 E: 1–14		B: 1–14 E: 1–14
Review/ Assess		Day 162–163		Day 88–89			

	Chapter 1 Days	Chapter Cumulative Days
Yearly Pacing (45 min class)	12 days	163 days
Yearly Pacing (90 min class)	7 days	89 days

Chapter 13 Pacing

Lesson, Pages	Lesson Title	45 min class	Assignments Basic, Enriched	90 min class	Assignments Basic, Enriched	___ min class	Assignments Basic, Enriched
13–1 562–565	AYR, Opener, The Standard Equation of a Circle	Day 164	B: 1–26, 31–45 E: 1–45	Day 90	B: 1–26, 31–45 E: 1–45		B: 1–26, 31–45 E: 1–45
13–2 566–569	More on Parabolas, R&PYS	Day 165–166	B: 1–22, 27–49 E: 1–49	Day 90–91	B: 1–22, 27–49 E: 1–49		B: 1–22, 27–49 E: 1–49
13–3 572–573	Problem Solving Skills: Visual Thinking	Day 167	B: 1–16 E: 1–16	Day 91	B: 1–16 E: 1–16		B: 1–16 E: 1–16
13–4 574–577	Ellipses and Hyperbolas, R&PYS	Day 168–169	B: 1–14, 17–22 E: 1–22	Day 92	B: 1–14, 17–22 E: 1–22		B: 1–14, 17–22 E: 1–22
13–5 580–583	Direct Variation	Day 170	B: 1–24, 31–44 E: 1–44	Day 93	B: 1–24, 31–44 E: 1–44		B: 1–24, 31–44 E: 1–44
13–6 584–587	Inverse Variation, R&PYS	Day 171–172	B: 1–16, 23–31 E: 1–31	Day 93–94	B: 1–16, 23–31 E: 1–31		B: 1–16, 23–31 E: 1–31
13–7 590–593	Quadratic Inequalities	Day 173	B: 1–22, 27–45 E: 1–45	Day 94	B: 1–22, 27–45 E: 1–45		B: 1–22, 27–45 E: 1–45
Review/ Assess		Day 174–175		Day 95–96			

	Chapter 1 Days	Chapter Cumulative Days
Yearly Pacing (45 min class)	12 days	175 days
Yearly Pacing (90 min class)	7 days	96 days

Chapter 14 Pacing

Lesson, Pages	Lesson Title	45 min class	Assignments Basic, Enriched	90 min class	Assignments Basic, Enriched	___ min class	Assignments Basic, Enriched
14–1 604–607	Basic Trigonometric Ratios	Day 176	B: 1–26, 31–43 E: 1–43	Day 97	B: 1–26, 31–43 E: 1–43		B: 1–26, 31–43 E: 1–43
14–2 608–611	Solve Right Triangles	Day 177–178	B: 1–25, 31–36 E: 1–36	Day 97–98	B: 1–25, 31–36 E: 1–36		B: 1–25, 31–36 E: 1–36
14–3 614–617	Graph the Sine Function	Day 179	B: 1–42, 49–54 E: 1–54	Day 98	B: 1–42, 49–54 E: 1–54		B: 1–42, 49–54 E: 1–54
14–4 618–621	Experiment with the Sine Function	Day 180–181	B: 1–23, 31–42 E: 1–42	Day 99	B: 1–23, 31–42 E: 1–42		B: 1–23, 31–42 E: 1–42
14–5 624–625	Problem Solving Skills: Choose a Strategy	Day 182	B: 1–27 E: 1–27	Day 100	B: 1–27 E: 1–27		B: 1–27 E: 1–27
Review/ Assess		Day 183–184		Day 100–102			

	Chapter 1 Days	Chapter Cumulative Days
Yearly Pacing (45 min class)	9 days	184 days
Yearly Pacing (90 min class)	6 days	102 days

Chapter 1—Essential Mathematics

Theme: Language and Communication
Chapter Investigation: Number Line, pages 5, 13, 23, 31, 45
Careers: Cryptographers, page 15; Cashiers, page 33
Data Activity: Cryptology

Content and Connections

Lesson, Pages	Lesson Objectives	NCTM Standards	State/Local Objectives	Interdisciplinary Connections	Real World Applications
1–1 6–9	• Use mathematical symbols to describe sets • Describe relationships among sets and elements of sets	Number Operation Patterns, Functions & Algebra Problem Solving Connections			Finance, Retail, Communication
1–2 10–13	• Identify and graph real numbers	Number Operation Patterns, Functions & Algebra Connections			Temperature, Geography, Earth Science
1–3 16–19	• Identify unions and intersections of sets • Use Venn diagrams to solve problems	Number Operation Patterns, Functions & Algebra Problem Solving Reasoning & Proof Connections			Advertising, Language, Logic
1–4 20–23	• Add and subtract rational numbers • Estimate answers to addition and subtraction problems	Number Operation Patterns, Functions & Algebra Connections			Recreation, Finance, Construction
1–5 26–29	• Multiply and divide rational numbers	Number Operation Patterns, Functions & Algebra Problem Solving Connections		Science	Science, Food service, Health, Fitness
1–6 30–31	• Solve a problem using technology • Use an equation or formula	Number Operation Patterns, Functions & Algebra Problem Solving Reasoning & Proof Connections			
1–7 34–37	• Use the distributive property to evaluate and simplify expressions • Use properties of exponents to evaluate and simplify expressions	Number Operation Patterns, Functions & Algebra Reasoning & Proof Connections		Science	Science, Number Sense Sales
1–8 38–41	• Evaluate variable expressions with negative exponents • Write numbers in scientific notation	Number Operation Patterns, Functions & Algebra Reasoning & Proof Connections		Chemistry	Chemistry, Astronomy, Business

Planning and Resources

Lesson, Pages	Tools/Materials Needed	Trans-parency	Learning/Teaching Styles Options	Assignments: Basic Enriched	Additional Practice in SE	Reteaching, Extra Practice, Enrichment	Other Resources
1–1 6–9	paper/pencil	Warm up 1	ESL/LEP	B: 1–34, 41–52 E: 1–52	Page 14–15, 25, 33, 42, 684	R: page 1 EP: page 1 E: page 5	SS: Teacher's Choice
1–2 10–13	paper/pencil index cards number lines calculator counters	Warm up 1 Trans TK-6	Visual Learner Real World	B: 1–32, 36–46 E: 1–46	Page 14–15, 25, 33, 42, 684	R: page 3 EP: page 3 E: page 7	SS: Teacher's Choice
1–3 16–19	paper/pencil	Warm up 1	Real World	B: 1–34, 43–56 E: 1–56	Page 24–25, 33, 42, 685	R: page 5 EP: page 5 E: page 9	SS: Teacher's Choice

1–4 20–23	paper/pencil number lines	Warm up 2 Trans RF-1, TK-6	Visual Learners Real World Prior Knowledge	B: 1–28, 31–44 E: 1–44	Page 24–25, 33, 43, 685	R: page 7 EP: page 7 E: page 11	SS: Teacher's Choice
1–5 26–29	paper/pencil calculator	Warm up 2 Trans RF-2	Real World	B: 1–40, 46–66 E: 1–66	Page 32–33, 43, 686	R: page 9 EP: page 9 E: page 13	SS: Teacher's Choice
1–6 30–31	paper/pencil calculator	Warm up 2 Trans RF-3, TK-1	Visual Learners Real World	B: 1–38 E: 1–38	Page 32–33, 43	R: page 11 E: page 15	
1–7 34–37	paper/pencil calculator	Warm up 3 Trans RF-4, 5	ESL/LEP Prior Knowledge	B: 1–44, 49–52 E: 1–52	Page 43, 686	R: page 13 EP: page 11 E: page 15	
1–8 38–41	calculator graph paper	Warm up 3 Trans RF-6		B: 1–40, 46–69 E: 1–69	Page 43, 687	R: page 15 EP: page 13 E: page 19	MC: 3

Planning and Pacing

Lesson, Pages	Lesson Title	45 min class	Assignments Basic, Enriched	90 min class	Assignments Basic, Enriched	___ min class	Assignments Basic, Enriched
1–1 6–9	AYR, Opener, The Language of Mathematics	Day 1	B: 1–34, 41–52 E: 1–52	Day 1	B: 1–34, 41–52 E: 1–52		B: 1–34, 41–52 E: 1–52
1–2 10–13	Real Numbers, R&PYS	Day 2–3	B: 1–32, 36–46 E: 1–46	Day 1–2	B: 1–32, 36–46 E: 1–46		B: 1–32, 36–46 E: 1–46
1–3 16–19	Union and Intersection of Sets	Day 4	B: 1–34, 43–56 E: 1–56	Day 2	B: 1–34, 43–56 E: 1–56		B: 1–34, 43–56 E: 1–56
1–4 20–23	Addition, Subtraction and Estimation, R&PYS	Day 5–6	B: 1–28, 31–44 E: 1–44	Day 3	B: 1–28, 31–44 E: 1–44		B: 1–28, 31–44 E: 1–44
1–5 26–29	Multiplication and Division	Day 7	B: 1–40, 46–66 E: 1–66	Day 4	B: 1–40, 46–66 E: 1–66		B: 1–40, 46–66 E: 1–66
1–6 30–31	Problem Solving Skills: Use Technology, R&PYS	Day 8–9	B: 1–38 E: 1–38	Day 4–5	B: 1–38 E: 1–38		B: 1–38 E: 1–38
1–7 34–37	Distributive Properties and Properties of Exponents	Day 10	B: 1–44, 49–52 E: 1–52	Day 5	B: 1–44, 49–52 E: 1–52		B: 1–44, 49–52 E: 1–52
1–8 38–41	Exponents and Scientific Notation	Day 11	B: 1–40, 46–69 E: 1–69	Day 6	B: 1–40, 46–69 E: 1–69		B: 1–40, 46–69 E: 1–69
Review/ Assess		Day 12–14		Day 6–7			

	Chapter 1 Days	Chapter Cumulative Days
Yearly Pacing (45 min class)	14 days	14 days
Yearly Pacing (90 min class)	7 days	7 days

Assessment Options

Assessment in Student Edition	Assessment in Teacher's Edition	Pages in Assessment Book	Software Generated Assessment
Are You Ready?, pages 2–3; Writing Math, pages 8, 13, 19, 22, 28, 31, 37, 41; Mixed Review, pages 9, 13, 19, 23, 29, 31, 37, 41; Check Understanding pages 6, 10, 16, 20, 26, 34, 38; Mid-Chapter Quiz, page 25, Chapter Review, page 42; Chapter Assessment, page 44; Alternative Assessment, page 45; Cumulative Review, page 46; Cumulative Assessment, page 47	5-minute Warm ups, pages 6, 10, 16, 20, 26, 30, 34, 38; Quick Assessment, pages 8, 12, 18, 22, 28, 31, 36, 40; Scoring Rubrics, page 45	Mid-Chapter Quiz, page 7; Test Form A, pages 9, 11; Test Form B, pages 13, 15; Cumulative Test 17, 19; Math Journal prompt, page 21, 22;	Chapter 1

Refresher Skills

The skills on these two pages are skills that have been taught in previous math courses. Continuous review of basic math skills will make stronger math students. These skills are identified as necessary to be successful in Chapter 1.

Skills Correlation Chart

Skill	Lesson Number
Addition and Subtraction	All lessons
Multiplication and Division	All lessons
Word Problems	All lessons

Vocabulary

lowest terms
round
remainder
whole numbers
numerical

ARE YOU READY?

Refresh Your Math Skills for Chapter 1

The skills on these two pages are ones you have already learned. Stretch your memory and complete the exercises. For additional practice on these and more basic skills, see page 674.

ADDITION AND SUBTRACTION

Add or subtract.

1. 8563 + 9476 — 18,039
2. 6905 − 4381 — 2524
3. 2765 + 7949 — 10,714
4. 8614 − 2030 — 6584
5. 7158 + 6235 — 13,393
6. 8190 − 5766 — 2424
7. 5374 + 7928 — 13,302
8. 6000 − 4173 — 1827
9. 24.86 + 13.92 — 38.78
10. 58.43 − 27.86 — 30.57
11. 30.247 + 64.892 — 95.139
12. 71.056 − 38.173 — 32.883
13. 4.638 + 8.469 — 13.0728
14. 4.76 − 3.0892 — 1.6708
15. 5.958 + 8.0537 — 14.0117
16. 6 − 1.8428 — 4.1572
17. $1\frac{7}{8} + 3\frac{3}{4}$ — $5\frac{5}{8}$
18. $5\frac{1}{3} - 2\frac{5}{6}$ — $2\frac{1}{2}$
19. $3\frac{5}{12} + 7\frac{3}{16}$ — $10\frac{29}{48}$
20. $8\frac{1}{15} - 4\frac{7}{8}$ — $3\frac{23}{120}$

MULTIPLICATION AND DIVISION

Multiply or divide. Give remainders in whole-number division. In decimal exercises, round to the nearest hundredth. Reduce fractions to lowest terms.

21. 6438 × 45 — 289,710
22. $32\overline{)1874}$ — 58R18
23. 5038 × 73 — 367,774
24. $24\overline{)5964}$ — 248R12
25. 397 × 482 — 191,354
26. $17\overline{)1394}$ — 82
27. 604 × 295 — 178,180
28. $56\overline{)7000}$ — 125
29. 3.48 × 2.5 — 8.7
30. $2.8\overline{)7.56}$ — 2.7
31. 6.143 × 0.25 — 1.54
32. $1.8\overline{)53.92}$ — 29.96
33. 7.641 × 0.03 — 0.23
34. $0.05\overline{)9.765}$ — 195.3
35. 5.05 × 0.0076 — 0.04
36. $8.08\overline{)62}$ — 7.67
37. $3\frac{2}{3} \times 5\frac{1}{8}$ — $18\frac{17}{24}$
38. $9\frac{7}{12} \div 2\frac{1}{4}$ — $4\frac{7}{27}$
39. $4\frac{3}{4} \times 6\frac{13}{16}$ — $14\frac{35}{64}$
40. $10\frac{1}{6} \div 2\frac{5}{8}$ — $3\frac{55}{63}$

CHAPTER 1

Solving Word Problems

In real life, problems are not always presented in numerical form. More often you have to decide what kind of answer you are looking for and how to go about finding it.

Remember the 5-Step Plan: Read - Plan - Solve - Answer - Check.

41. Keshawn bought a CD for $12.95 and a package of batteries for $3.79. How much change should he receive from a $20 bill? $3.26

42. Amie belongs to a book club. This month she ordered 3 books at $9.95 each. Shipping and handling for the 3 books came to $5.95. What is Amie's book bill this month? $35.80

43. Mr. Sanders pays $575 rent for his apartment. This month he also paid $57.82 for electricity, $12.86 for gas, $48.86 for telephone service, and $30 for garbage pickup. What were his total expenses this month? $724.54

44. Emily has a cat and a dog. A bag of dog food costs $8.99 and will feed the dog for 15 days. A bag of cat food costs $10.89 and will feed the cat for 17 days. Which animal is more costly to feed? the cat, since $0.64 > $0.60

45. The local movie theater charges $8 for adults and $5.50 for children. How much is the cost for 82 adults and 153 children to watch a movie? $1497.50

46. Darius drove 496 miles on one tank of gas. His car gets 32 mi/gal. About how many gallons of gas does the car's tank hold? 15.5

47. One supermarket sold 72 cases of *Zing* soap, but only one-third as many cases of *Essence* soap. If each case holds 156 bars of soap, how many bars were sold in all? 14,976

48. The Allen family is planning a 1,200-mile trip. Assuming they travel at exactly 55 mi/h for 6 h/day, how many days will it take them to reach their destination? 4 days

49. A display in a store is between 5 feet and 6 feet tall. There is a stack of boxes. Each box is 17 inches tall. There are also stacks of cans. Each can is 4 inches tall. The stacks of boxes and cans are the same height. How tall are the stacks? 68 in.

50. Jan has 3-inch by 5-inch cards. She placed them next to each other to form a square (same length and width). What is the least number of cards Jan could have had? 15 cards.

Chalkboard Examples

Working with Whole Numbers

Add or subtract.

1. $7523 + 9468$ = 16,991
2. $2952 - 1876$ = 1076

Rounding Numbers

Round each number to the nearest tenth.

1. 3.82 3.8
2. 6.99 7.0
3. 8.25 8.3
4. 9.0821 9.1

Working with Decimals

Multiply of divide.

1. 2.3×7.6 17.48
2. $7.03\overline{)18.276}$ 2.6

Problem Solving Five-Step Plan

Solve.

1. Mitchell bought 6 packages of pencils for $0.89 each. How much change should he receive from a $10-bill? $4.66

Refresher Wrap-up

Quick Assessment

Ask the following questions to determine if students have mastered the basic skills reviewed on these pages.

Answer *true* or *false*.

1. The remainder in a division problem should be less than the divisor. True
2. 23/8 = 3 1/8 False
3. Dividing by 2 is the same as multiplying by 1/2. False

Additional Practice

Refer to the Extra Practice Index on page 674 of this text.

Chapter Opener

NCTM Standards/Strands

- Number & Operation
- Patterns, functions, & Algebra
- Problem Solving
- Reasoning and Proof
- Communication
- Connections
- Representation

Vocabulary

cryptology, cipher, encode, decode

Theme Connections

Tables and graphs are often used to organize and display statistics about language. In the data activity, a table is used to show the frequency in which letters of the alphabet occur in written English. Discuss with students how this data may have been compiled.

Career Opportunities

Many careers related to language require the use of data and graphs. Two are highlighted in the MathWorks features. Others include interpreter, translator, teacher, writer, instructional aide, salesperson, computer programmer, and language development specialist.

- Cryptographer, page 15
- Cashier, page 33

Internet Connection

Theme Activities

Learningmathmatters.com provides web addresses to search that help students gather information about the use of math in the real world, particularly data and graphs. To search for additional addresses, begin a search using the keyword *language*. Then within that search, use such key words as *codes, ciphers*, and *translator*. Students can brainstorm in small groups other key words.

Chapter Investigation

Go to Learningmathmatters.com to locate additional information about language.

CHAPTER 1

Essential Mathematics

THEME: Language and Communication

Have you ever had a great idea for a book, a television show, or a movie? Your idea could be worth a great deal of money if you can put it into words.

Translating ideas into language that others can understand has always been a challenge. You may be surprised to learn that mathematics is a language. People living 4,000 years ago could solve difficult algebra and geometry problems, but they could not communicate their ideas to others because many of the symbols for mathematics had not been invented. Today, people in many careers use mathematical language to communicate their ideas.

- **Cryptographers** (page 15) encrypt important information so that messages sent electronically can be kept private. They must be able to apply special step-by-step mathematical processes called algorithms. Cryptographers use computers to speed up their work.
- **Cashiers** (page 33) use communication and math skills to help customers find the cost of purchases, apply discounts and find the best value for their money. Cashiers must be able to use estimation and mental math skills to guard against errors in their work.

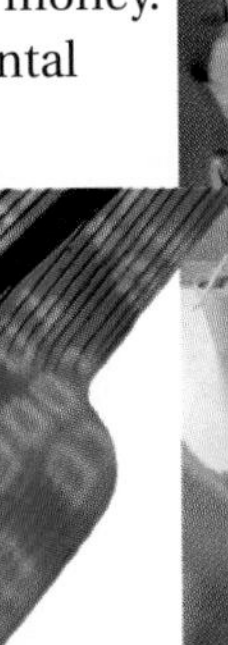

Internet Connection
www.learningmathmatters.com

The Science of Communication

Percent of Letters in Written English

E	13%	D	4%	G	1.5%
T	9%	L	3.5%	W	1.5%
A	8%	C	3%	V	1%
O	8%	M	3%	J	0.5%
N	7%	U	3%	K	0.5%
R	6.5%	F	2.5%	X	0.5%
I	6.5%	P	2%	Q	0.3%
S	6%	Y	2%	Z	0.2%
H	5.5%	B	1.5%		

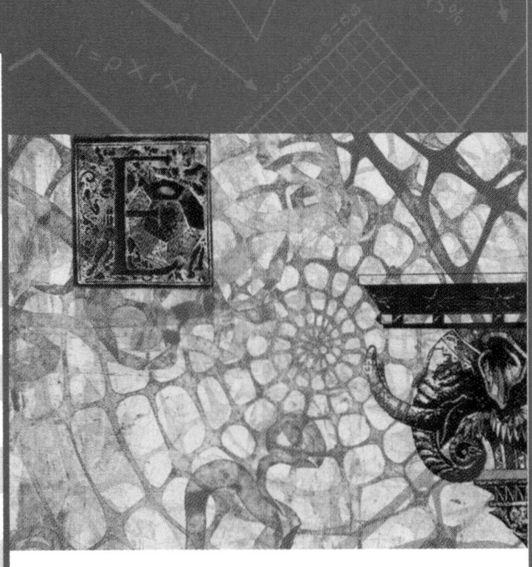

Data Activity: Cryptology–The Science of Secret Communication

Use the table for Questions 1–3.

1. A secret message was encoded by replacing the letters of the alphabet with symbols. The message contains 2,200 symbols. How many times would you expect the symbol representing the letter E to occur? 286
2. In any large passage, what percent of the letters would you expect to be vowels, excluding Y? 38.5%
3. Which letter appears 20 times more often than the letter Q? S

Chapter Investigation

A cipher is a secret method of writing. Many ciphers substitute numbers or symbols for the letters in a message. The person who receives the message must have a key in order to read it.

Working Together

Create a cipher using a number line. Then encode a short message, five or six sentences in length. Use the table above and logic to decode the message. Can you discover the key? Use the Chapter Investigation logo to guide your group.

Project Planning Calendar

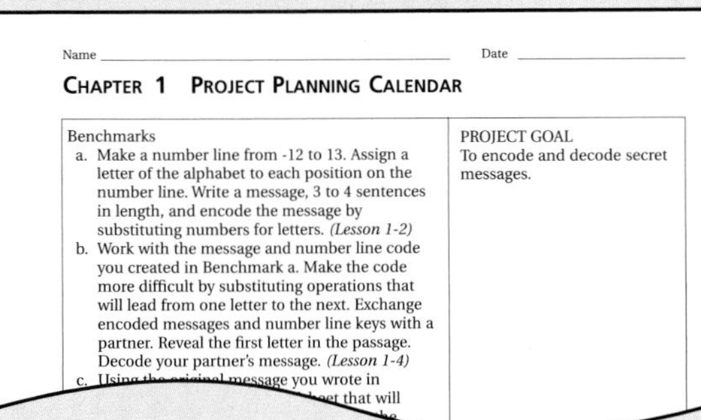

Name ______ Date ______

CHAPTER 1 PROJECT PLANNING CALENDAR

Benchmarks
a. Make a number line from -12 to 13. Assign a letter of the alphabet to each position on the number line. Write a message, 3 to 4 sentences in length, and encode the message by substituting numbers for letters. *(Lesson 1-2)*
b. Work with the message and number line code you created in Benchmark a. Make the code more difficult by substituting operations that will lead from one letter to the next. Exchange encoded messages and number line keys with a partner. Reveal the first letter in the passage. Decode your partner's message. *(Lesson 1-4)*

PROJECT GOAL
To encode and decode secret messages.

Group Project Planner

Name ______ Date ______

CHAPTER 1 GROUP PROJECT PLANNER

Assignment ______ Objective ______

Group Members
1) ______
2) ______
3) ______
4) ______
5) ______

Assigned Roles

Data Activity

Discuss with students the use of codes and encryption. Ask students if they know of any uses of codes in real life. Ask whether they think the type of information presented in the table would ever be helpful to them in their lives.

Assign students Questions 1–3.

Extend the Data Activity

REAL WORLD CONNECTION Students can select passages from a newspaper or magazine article and create a frequency table of letters. They can then compare their percents with those shown in this table. Discuss: How would you account for differences between your table and the one shown here?

Chapter Investigation

As an Overarching Problem

Bring several books about secret codes to class. Have students work in small groups to skim the books and record facts about codes. Students should also make a list of general knowledge that they have about codes and their uses. Remind students to consider computer applications as they work. Ask students to share what they have recorded with the class. Students will continue to work on the investigation as they complete the exercises identified by the Investigation Logo that are found throughout the chapter. These exercises will guide students through the task described in *Working Together*. Encourage students to keep all of their work on the Investigation together. Have students use the suggestions in the Chapter Investigation Extension to summarize their work.

As a Chapter Project

The goal of this project to encode and decode secret messages. Students can use the Team Project Planner on page 3 and the Project Planning Calendar on page 4 in the Enrichment text to complete the project. Benchmarks **a**, **b**, and **c** should be completed after the lesson listed in parentheses has been studied. Benchmark **d** should be completed at the end of the chapter.

Lesson Planning

NCTM Standards/Strands

- Number & Operation
- Patterns, functions, & Algebra
- Problem Solving
- Reasoning and Proof
- Communication
- Connections
- Representation

Vocabulary

set	infinite set
finite set	subset
empty set	null set
variable	equation
open sentence	replacement set
solution	solution set

Materials Needed

paper/pencil
word processor (optional)

Lesson Resources

Warm-up Transparency 1
Reteaching 1-1
Extra Practice 1-1
Enrichment 1-1

Assignment Guide

Basic: 1–34, 41–44
Enriched: 1–44

Getting Started

5-Minute Warm-Up

Determine if each statement is true (T) or false (F).

1. $4 \times 12 < 58$ T
2. $23 < 51/3$ F
3. $19 + 34 = 78 - 24$ F

Introduction to Lesson 1-1

Have each student write his or her name on a slip of paper. Collect the slips and place in a bag. Then have each student draw one slip. Students should make a list to describe the classmate they selected. Have students exchange papers to identify each other's "mystery person."

1-1 The Language of Mathematics

Goals
- Use mathematical symbols to describe sets.
- Describe relationships among sets and elements of sets.

Applications Finance, Retail, Communication

Work in groups of two students each.

1. Make a list of the different ways you can use language to identify who you are. You might say: "I am the son of Dennis Williams. My name is Jacob Ryan Williams. My nickname is Jake. I am a member of the family at 36 W. Main. I am Jill's brother."
2. Compare lists with your partner.

Build Understanding

You can find many different kinds of **sets** in your life, such as chess sets and salt-and-pepper sets. You will even find some salt-and-pepper chess sets! In mathematics, you will study many different kinds of sets.

A *well-defined* set makes it possible to determine whether an item is a member of that set. A set whose elements cannot be counted or listed is called an **infinite set**. If all the elements of a set can be counted or listed, it is called a **finite set**.

You can identify or describe yourself in different ways. You can also use different mathematical symbols to describe sets as well as relationships between sets and/or elements of sets.

You can use *description notation* such as "the set of all natural numbers," *roster notation* such as $S = \{1, 2, 3, \ldots\}$, or *set-builder notation* such as $\{x | x \text{ is a natural number}\}$. The symbol $\in$ can be used to show that an element is a member of a set. If every element of set a is also an element of set B, then A is called a **subset** *of* B, written as $A \subset B$.

If $a = \{e, f, g, h, i, j\}$ and $B = \{e, i\}$, you can write $f \in A$ and $f \notin B$, which means that f is a member of set a and f is not a member of set B. You can also write $B \subset A$, which means that set B is a subset of set A. Since every set is a subset of itself, you can also write $A \subset A$.

Consider the set D, which is days containing 26 hours. Since no day contains 26 hours, D has no elements. A set having no elements is called an **empty set** or the **null set**. To indicate that D is an empty set, write either $D = \{\ \}$ or $D = \varnothing$. The null set is a subset of every set.

Teaching Strategies

ESL/LEP When introducing the opening activity, encourage ESL students to use their native languages to identify themselves. You may want to write a series of sentences on the board with blanks. Invite ESL students to fill in the blank and write each sentence in their native languages.

Examples: I attend __________ school.
My favorite sport is __________.
One thing I do after school is ___________.

Example 1

Determine all the possible subsets of the set {1, 5}.

Solution

Each single element of {1, 5} is a subset: {1}, {5}. The set itself is a subset: {1, 5}. The null set is a subset: ∅. So, {1, 5} has four subsets: {1}, {5}, {1, 5}, ∅.

Sets are useful in the study of mathematical sentences. It is often necessary to use a **variable** to represent an unknown number or quantity. Usually, a letter such as *a, n, x,* or *y* is used for the variable.

You can use **sentences** or **equations** of different types to describe the truth of a statement. An **equation** is a statement that two numbers or expressions are equal. It can be true, false, or neither true nor false.

Any sentence that contains one or more variables is called an **open sentence**. The set of all possible values for the variable in an open sentence is called the **replacement set**. An open sentence can be true or false, depending on what values are substituted for the variables.

A value of the variable that makes an equation true is called the **solution** of the equation. The **solution set** of an open sentence is the set of all elements in the replacement set that makes the sentence true.

Check Understanding

How many elements are in the word *language*? Are *a* and *g* listed more than once?

6 elements, no

Example 2

Which of the values −2 and 4 is a solution of the equation $4x + 3 = 19$?

Solution

Substitute −2 for *x* in the equation.

$$\begin{aligned} 4x + 3 &= 4(-2) + 3 \\ &= -8 + 3 \\ &= -5 \end{aligned}$$

Remember the order of operations: Multiply first, then add.

So −2 is not a solution.

Substitute 4 for *x* in the equation.

$$\begin{aligned} 4x + 3 &= 4(4) + 3 \\ &= 16 + 3 \\ &= 19 \end{aligned}$$

So 4 is a solution.

Example 3

Use mental math to solve the equation $x - 8 = 11$.

Solution

Think: Eight subtracted from what number equals 11? You know that $19 - 8 = 11$, so $x = 19$.

Chalkboard Examples

Supplementary Example 1
Determine all the possible subsets of the set {b, e}. {b}, {e}, {b, e}, ∅

Supplementary Example 2
Which of the values 6 and −11 is a solution of the equation $17 - 2y = 5$? 6

Supplementary Example 3
Use mental math to solve the equation $x - 13 = 10$. 23

Supplementary Example 4
Using the equation $153 = 3x - 15$ and {48, 51, 56} for *x*, find the solution to the equation. 56

Teaching Strategies

SIMULATIONS Students may need to create a visual model such as a diagram before using mental math to solve an equation. Invite students to make such a visual to represent any equation presented in the lesson. Students may also use Algeblocks or objects to represent the numbers in the equation.

Reteaching Worksheet 1-1

Name ______________________ Date ____________

RETEACHING 1-1

THE LANGUAGE OF MATHEMATICS

A set is a well-defined collection of objects called elements. The symbol ∈ can be used to show that an element is a member of a set. If every element of a set *A* is also an element of another set *B*, then set *A* is a subset of set *B*. Sets are usually described in one of these three ways:

Description notation	Example: S = set of even whole numbers
Roster notation	Example: {0, 2, 4, 6, 8 . . .}
Set-builder notation	Example: $S = \{x \mid x$ is an even whole number$\}$

Example 1

Use these sets: $A = \{1, 3, 5\}$, $B = \{1, 2\}$, and $C = \{-2, -1, 0, 1, 2\}$. Read each statement and tell if it is true or false.

a. $1 \in A$
b. $A \subset C$
c. $B = \varnothing$

Solution

a. 1 is an element of *A*. True
b. *A* is a subset of *C*. False
c. *B* is an empty or null set. False

Example 2

Which value, 1 or −1, is a solution of the equation $2x + 3 = 1$?

Solution

Substitute 1 for *x* in the equation.
$2(1) + 3 = 1$
$2 + 3 \neq 1$, so 1 is not a solution.
Substitute −1 for *x* in the equation.
$2(-1) + 3 = 1$
$-2 + 3 = 1$, so −1 is a solution.

EXERCISES

Use the following sets $A = \{2, 4\}$, $B = \varnothing$, $C = \{2, 4, 6, 8, 10\}$, and $D = \{-2, -1, 0, 1, 2\}$ for Exercises 1–8. Tell if each statement is true or false.

1. $1 \in D$ T **2.** $1 \in A$ F **3.** $-2 \in D$ T **4.** $-3 \in D$ F
5. $A \subset C$ T **6.** $B \subset D$ T **7.** $C \subset A$ F **8.** $B \subset A$ T
9. Write all the possible subsets of the set {*x*, *y*}. {x}, {y}, {x, y}, ∅

Define each set using roster notation.

10. odd numbers greater than 5 {7, 9, 11, 13, . . .}
11. even negative numbers with a value less than −3 {. . ., −10, −8, −6, −4}

Which of the given values is a solution of each equation?

12. $6 - m = 4$; −2, 2 2
13. $4n + 6 = 30$; 6, 9 6
14. $b + 7 = -8$; −1, −15 −15
15. $j \div 12 = 2$; 6, 24 24

2 RETEACHING *LESSON 1-1*

Lesson Wrap-up

Quick Assessment

Ask the following questions to determine if students understand the content presented in this lesson.

1. Use mental math to solve the equation $22 = t - 7$. 11
2. What are three different ways to describe a set? description notation, roster notation, set-builder notation
3. Identify a set that has only one subset. the null set

Assignment Guide

Basic: 1–34, 41–44
Enriched: 1–44
Additional Practice: See Extra Practice Index on page 674.

Example 4

FINANCE Josh's earnings equaled the sum of Aimee's and twice Nora's earnings. Josh earned \$104 and Aimee earned \$32. Using the equation $104 = 32 + 2x$ and {32, 36, 40} for x, find the amount Nora earned.

Solution

$104 = 32 + 2x$

Substitute:

32 for x in the equation.
$104 = 32 + 2(32)$
$104 = 32 + 64$
$104 \neq 96$ ≠ means "is not equal to."

36 for x in the equation.
$104 = 32 + 2(36)$
$104 = 32 + 72$
$104 = 104$ 36 is a solution.

40 for x in the equation.
$104 = 32 + 2(40)$
$104 = 32 + 80$
$104 \neq 112$

So Nora earned \$36.

Try These Exercises

1. Use set notation to write the following: 6 is not an element of {1, 3, 5, 7, 9}. $6 \notin \{1, 3, 5, 7, 9\}$
2. Determine all the possible subsets of {*M*, *A*, *N*}. {M, A, N}, {M, A}, {M, N}, {A, N}, {M}, {A}, {N}, ∅
3. Which of the given values is a solution of the equation $2x + 1 = -1$? −2, −1, 0, 1, 2 −1

Use mental math to solve the following equations.

4. $x + 15 = 13$ −2
5. $z - 2 = 9$ 11

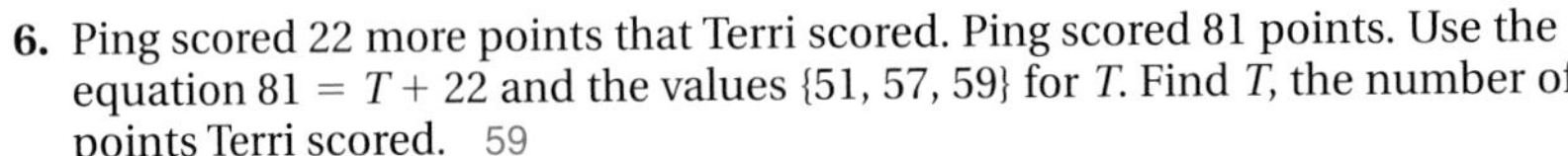

6. Ping scored 22 more points that Terri scored. Ping scored 81 points. Use the equation $81 = T + 22$ and the values {51, 57, 59} for T. Find T, the number of points Terri scored. 59

7. **YOU MAKE THE CALL** Ray says that any set containing three members has exactly three subsets. Is he correct? Explain. No. There are 8. The subsets include ∅, the members alone and any possible combinations of members.

Practice Exercises

Define each set using roster notation.

8. even natural numbers greater than 9 {10, 12, 14, 16, . . .}
9. days having 27 hours ∅

Determine if each statement is *true* or *false*.

10. $0 \in \{x \mid x \text{ is a natural number}\}$ False
11. $9 \in \{-3, 0, 3, 6, \ldots\}$ True
12. $\{a, b, c\} \subset \{a, e, i, o, u\}$ False

Write all the subsets of each set.

13. {*a*} {a} ∅
14. {*b*, *c*} {b, c}, {b}, {c}, ∅
15. {*t*, *e*, *n*} {t, e}, {t, n}, {e,n}, {t, e, n}, {t}, {e}, {n}
16. **WRITING MATH** For Exercise 15, you could write $\{a\} \not\subset \{t, e, n\}$. Write a definition of the symbol $\not\subset$. The symbol $\not\subset$ means "is not a subset of."

Extend the Lesson

ONGOING ASSESSMENT Give each student an index card. Tell students to think of a particular set of numbers such as "numbers that are divisible by two." Have students describe this set on the index card using roster notation. Students should then swap cards and analyze the elements given to identify their classmate's set using description notation.

8 Chapter 1 Essential Mathematics

Which of the given values is a solution of the equation?

17. $a - 9 = -5; 4, 5$ 4

18. $c + 3 = -1; -4, 4$ −4

19. $5n + 3 = 8; -1, 0, 1$ 1

20. $\frac{3r}{2} = -3; -2, -4, 2$ −2

Use mental math to solve each equation.

21. $y + 8 = 2$ −6

22. $m - 5 = 2$ 7

23. $6p = -18$ −3

Determine whether the statement is *true* or *false*.

24. $\varnothing$ is not a subset of itself. True

25. $\{2, 3\}$ is a subset of $\varnothing$. False

Rewrite each statement so that it is correct.

26. $\{r\} \in \{o, r\}$ $r \in \{o, r\}$ or $\{r\} \subset \{o, r\}$

27. $\{\varnothing\} \subset \{1, 2\}$ $\{\ \} \subset \{1, 2\}$ or $\varnothing \subset \{1, 2\}$

Define each set in roster notation.

28. $\{x | x$ is negative integer and $x > -3\}$ $x = \{-2, -1\}$

29. $\{x | x$ is a whole number and $4x + 6 = 2\}$ $x = \{\ \}$

Which of the given values is a solution of the equation.

30. $3b = -2\left(1 - \frac{1}{2}\right); \frac{1}{2}, -\frac{1}{3}, -1$ $-\frac{1}{3}$

31. $-\frac{1}{3} + m = -\frac{1}{4}; \frac{1}{12}, -\frac{1}{12}, \frac{1}{4}$ $\frac{1}{12}$

32. A stack of quarters is worth \$2.25. Use the equation $0.25q = 2.25$ and the values $\{7, 8, 9\}$ for the replacement set. Find q, the number of quarters in the stack. 9

33. **RETAIL** The sum of John and Sarah's sales equals $\frac{1}{2}$ the sum of Cora and Dan's sales. Sarah made 15 sales, Cora made 20, and Dan made 22. Use the equation $J + 15 = \frac{1}{2}(20 + 22)$ and the replacement set $\{17, 11, 6\}$ to find J, the number of John's sales. 6

34. **COMMUNICATION** A newspaper charges \$3 per word for its classified advertisements. Don paid \$54 to put an ad in the newspaper. Use the equation $\$3w = \$54$ and the values $\{16, 17, \text{and } 18\}$ for the replacement set to find w, the number of words in the ad. 18

Extended Practice Exercises

Determine the number of subsets for each set.

35. $\{a, b, c\}$ 8

36. $\{a, b, c, d\}$ 16

37. $\{a, b, c, d, e\}$ 32

38. How many subsets do you think a set of 6 elements has? A set of 7 elements? 64, 128

Use mental math to find two solutions of these equations.

39. $x^2 = 36$ 6, −6

40. $r^2 - 4 = 12$ 4, −4

Mixed Review Exercises

Use set notation to write each of the following. (Basic Math Skills)

41. The set P has no members. $P = \{\ \}$ or $\varnothing$

42. The set G is a subset of the set K. $G \subset K$

43. The number 8 is not a member of the set R. $8 \in R$

44. The letter m is a member of the set W. $m \in W$

Extend the Lesson

MATH JOURNAL Give students this example: Maria and Joe have jointly saved a total of \$40.50. Maria contributed twice as much as Joe. Using the equation $2x + x = 40.50$ and the replacement set {\$9.50, \$11.75, \$13.50}, find how much Joe saved. Have students write the steps they would use to find the solution set. Discuss: How many subsets does the solution set have? What are they? two; the solution set itself and the null set

Extra Practice Worksheet 1-1

Name ____________ Date ____________

EXTRA PRACTICE 1-1

THE LANGUAGE OF MATHEMATICS

EXERCISES

Define each set using roster notation.

1. odd natural numbers greater than 10 $\{11, 13, 15, 17, \ldots\}$
2. weeks having 8 days $\varnothing$
3. integers less than −3 $\{\ldots, -7, -6, -5, -4\}$
4. $\{x | x$ is a positive integer and $x > 4\}$ $\{5, 6, 7, 8, \ldots\}$

Determine if each statement is *true* or *false*.

5. $-2 \in \{x | x$ is a whole number$\}$ false
6. $\{4, 7, 9\} \subset \{1, 2, 3, \ldots\}$ true
7. $5 \in \{x | x$ is a natural number and $3 \le x < 5\}$ false

Write all the subsets of each set.

8. $\{3\}$ $\{3\}, \varnothing$
9. $\{-1, 0, 5\}$ $\{-1\}, \{0\}, \{5\}, \{-1, 0\}, \{0, 5\}, \{-1, 5\}, \{-1, 0, 5\}, \varnothing$
10. $\{a, v\}$ $\{a\}, \{v\}, \{a, v\}, \varnothing$

Which of the given values is a solution of the equation?

11. $r - 4 = -3; -1, 1$ 1
12. $9 + x = 4; -5, 13$ −5
13. $a + 2 = -8; -10, -6$ −10
14. $-2d = 14; -28, -7$ −7
15. $\frac{m}{6} = 4; 2, 24$ 24
16. $2w + 2 = 20; 9, 11$ 9
17. $5 - 3z = -10; 5, 30, 45$ 5
18. $\frac{2p}{5} = -4, -18, -10, 2$ −10

Use mental math to solve each equation.

19. $h + 3 = 10$ 7
20. $m - 2 = 6$ 8
21. $7w = -21$ −3
22. $\frac{h}{-5} = -2$ 10
23. $4 + j = -2$ −6
24. $8 - p = -7$ 15

2 EXTRA PRACTICE *LESSON 1-1*

Enrichment Worksheet 1-1

Name ____________ Date ____________

ENRICHMENT 1-1

PROBLEM SOLVING WITH DOMINOES

Many games and puzzles involve domino tiles, rectangular pieces of wood, plastic, ivory, or some other material. Each domino is a rectangle divided into two squares, called half-tiles. And, each half-tile is marked with some number of pips. The double-6 set of dominoes includes all the domino tiles from 0-0 to 6-6.

6-6

0-3 or 3-0

2-5 or 5-2

EXERCISES

Find the number of dominoes in each set. If you do not have a set of dominoes, you can use small pieces of paper instead.

1. The double-3 10
2. The double-6 28
3. The double-9 55
4. The double-12 91

Let n-n stand for the largest domino in a set. Find an algebraic expression for each of the following.

5. The total number of dominoes in the set $\frac{1}{2}(n + 1)(n + 2)$
6. The number of times each number from 0 to n occurs $n + 2$
7. The sum of all the half-tiles in the set $(n + 2) \cdot \frac{1}{2}n(n + 1)$

To form a *chain* of dominoes, the tiles are laid end-to-end so that the numbers on the ends of adjacent dominoes match.

8. Use all the dominoes in the double-2 set to form a chain. Repeat with the double-4 set. What do you notice?
The numbers at the ends of each chain are the same.
9. Remove the 3-4 domino from the double-4 set and form a chain with the rest of the dominoes. What happens at the ends of the chain? Will this always be true?
The numbers at the ends will be 3 and 4. Yes, it is always true.
10. Experiment and make a conjecture. For what values of n is it possible to form a chain from a double-n set of dominoes?
n must be equal to 1 or even.

6 ENRICHMENT *LESSON 1-1*

Lesson 1-1 **The Language of Mathematics** 9

Lesson Planning

NCTM Standards/Strands

- Number & Operation
- Patterns, functions, & Algebra
- Problem Solving
- Reasoning and Proof
- Communication
- Connections

Representation Vocabulary

natural numbers
whole numbers
real numbers
coordinate
rational numbers
irrational numbers
integers
inequality
absolute value

Materials Needed

paper/pencil
index cards
counters or markers
number lines
calculator

Lesson Resources

Warm-up Transparency 1
Reteaching 1-2
Extra Practice 1-2
Enrichment 1-2
Transparency TK-6

Assignment Guide

Basic: 1–32, 36–43
Enriched: 1–43

Getting Started

5-Minute Warm-up

Replace each ? with <, >, or =.

1. 3 1/7 ? 42/14 >
2. 2.6 ? 53/5 <
3. 98/3 ? 0.32 >
4. 0.125 ? 1/8 =

Introduction to Lesson 1-2

Have students use a number line to solve this problem: Ming entered an elevator on the fourteenth floor. The elevator then proceeded to go up four floors, down eight floors, up seven floors, and down two floors. At this point, where is Ming in relation to the spot at which she boarded the elevator? one floor above her starting point

1-2 Real Numbers

Goals ■ Identify and graph real numbers.

Applications Temperature, Geography, Earth Science

Work in groups of four.

1. Make a set of cards and a number line as shown.
2. To play, each member draws a card, uses a calculator to find the answer, and places a marker on the number line to show the location of the answer.
3. The player whose marker is farthest to the right scores one point. Remove the markers and play again. The winner is the first person to score 5 points.

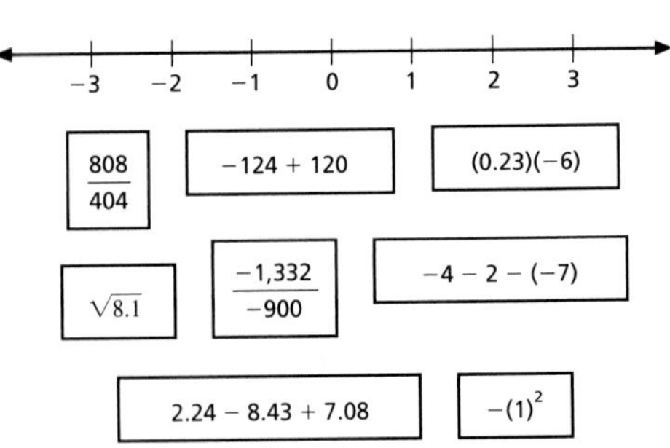

Build Understanding

The set of **natural** or counting numbers consists of {1, 2, 3, . . .}. The set of **whole numbers** consists of the natural numbers plus zero {0, 1, 2, 3, . . .}. The set of **integers** consists of the whole numbers and their **opposites** {. . . , −3, −2, −1, 0, 1, 2, 3, . . .}.

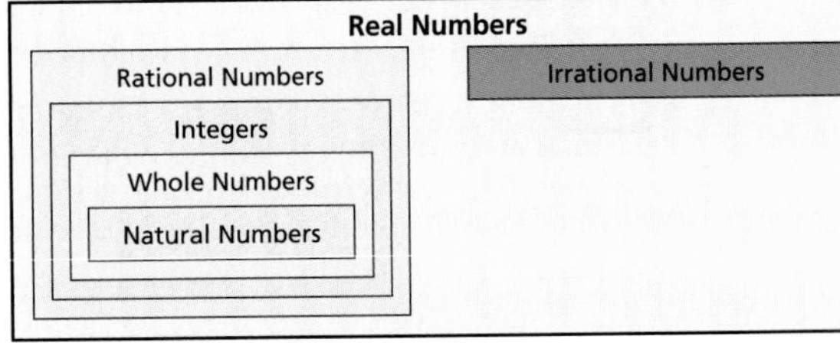

Numbers other than integers can also be shown on a number line. Numbers such as $\frac{1}{4}$, $-\frac{1}{2}$, 1.75, and −6.3 can be represented as locations that lie between points represented by integers. This type of number along with the integers represents another set of numbers, the **rational** numbers.

A **rational number** is one that can be expressed as a ratio of two integers a and b, where b is not equal to zero. This is usually written $\frac{a}{b}$, $b \neq 0$. The symbol ≠ means "is not equal to."

Any rational number can be written as a fraction, and any fraction can be written as either a *terminating* decimal or a *repeating* decimal.

Some numbers, such as π (pi), the square root of 2, or 8.112111211112 . . . , are nonterminating and nonrepeating decimals. Such numbers are called **irrational** numbers.

An **irrational number** is one that *cannot* be written as $\frac{a}{b}$, $b \neq 0$, but can still be designated by a point on the number line.

Together, the set of rational numbers and the set of irrational numbers make up a set of **real numbers**.

Reading Math

Positive integers are greater than zero. **Negative integers** are less than zero. The integer zero is *neither* positive nor negative.

Extend the Lesson

MATH JOURNAL Have students describe the relationship that exists between rational numbers, irrational numbers, and real numbers. The sets of rational and irrational numbers make up the set of real numbers.

Technical Math Test

Name______________________________

Write in scientific notation.

1) 4,870,000,000 4.87×10^{9}

2) 120.04 1.2004×10^{2}

3) 0.0000000543 5.43×10^{-8}

Evaluate each expression. Let a = 3, b = 6, and c = 8

4) 7 + a – b

7 + 3 – 6

4

5) 4c – a x 6

4(8) – 3·6

32 – 18

14

6) c ÷ 2 x b + a

8 ÷ 2 x 6 + 3

4 x 6 + 3

24 + 3

27

Write each answer in exponential form.

7) $6^9 \times 6 \times 6^2 \times 6^3$ 6^{15}

8) $y^{32} \div y^{16}$ y^{16}

9) $(a^5)^{11}$ a^{55}

10) $m^9 \div m^3 \bullet m^3$

$m^6 \cdot m^3$

m^9

11) $(x^4)^3(x)(x^3)$

$x^{12}(x)(x^3)$

x^{16}

Write in standard form.

12) 3.42×10^{-9} .00000000342

13) 8.1063×10^6 8,106,300

Complete the pattern. Write the rule for each.

14) 1, ½ , ¼ , 1/8 , 1/16 , 1/32 , 1/64   Rule: x ½

15) 3, 6, 5, 8, 7, 10 , 9 , 12   +3, –1

Simpilfy.

16) $6(y + 5)$
$6y+30$

17) $5(a + b)$
$5a+5b$

18) $6(f - g + 2)$
$6f-6g+12$

Solve.

19) Amber is trying to get a job. Amber makes a deal about how much money she is to be paid. She will receive \$0.01 for the first day, \$0.02, for the second, \$0.04, for the third, \$0.08 for the fourth, and so on. Using this pattern, how much money will she receive on the 15th day? On the ~~30th~~ 20th day?

15th – \$163.84

20th – \$5,242.88

Simplify. Show all work to get credit.

20) $5 - 8 \div 2 + 3$

$5-4+3$

$1+3$

(4)

21) $4 \times (18 - 12 \div 4) - 5 + 3$

$4 \times (18-3) -5+3$

$4 \times 15 -5+3$

$60-5+3$ (58)

22) $8 + 2 + 18 \div 6 \div 3 \times 3$

$8+2+3 \div 3 \times 3$

$8+2+1 \times 3$

$8+2+3$

(13)

23) $(3 + 3)^2 - 4 \times 2 + (4 \div 2)^2$

$6^2 - 4 \times 2 + 2^2$

$36 - 8 + 4$

$28 + 4 = 32$

State which property is used for each equation.

24) $5 \times (2 \times 12) = (5 \times 2) \times 12$ assoc.

25) $6 + 0 = 6$ identity

26) $4(x - 4)$ Distributive

27) $0 \times 756 = 0$ prop of zero

All real numbers can be graphed on a number line. The number that corresponds to a point on a number line is called the **coordinate of the point**. Each real number corresponds to exactly one point on a number line. The point that corresponds to a number is called the **graph of the number**, and is indicated by a solid dot.

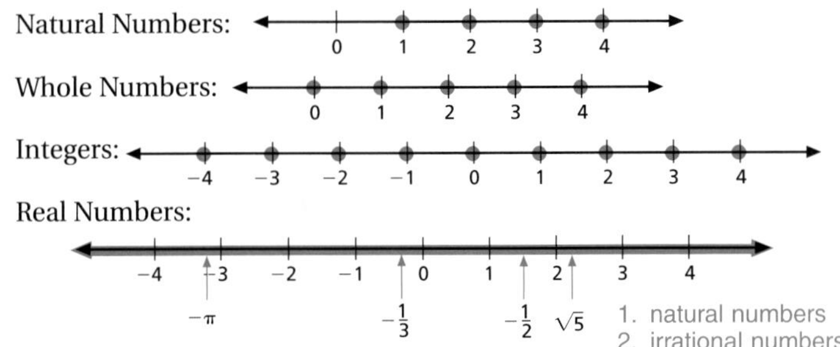

Check Understanding

Name the set or sets of numbers for each of the following:

1. $\sqrt{36}$
2. $-\sqrt{3}$
3. 8.63
4. 2.15115 . . .

1. natural numbers
2. irrational numbers
3. rational numbers
4. irrational numbers

Example 1

Graph this set of numbers on a number line.

$\left(-1.75,\ 3,\ 2\frac{1}{4},\ -2\frac{2}{3},\ \sqrt{3},\ \frac{\pi}{4}\right)$

Solution

Draw a number line. Use a solid dot to graph each number.

$-2\frac{2}{3}$ −1.75 $\frac{\pi}{4}$ $\sqrt{3}$ $2\frac{1}{4}$ 3

−3 −2 −1 0 1 2 3

Numbers on a number line *increase* as you move from left to right. A number to the left of another number on a number line is *less.* Likewise, a number to the right of another number is *greater.*

The mathematical symbols $<$, $>$, $\geq$, and $\leq$ are used to express these relationships. A mathematical sentence that contains one of the symbols $<$, $>$, $\geq$, or $\leq$ is called an **inequality**. The inequality symbols, $<$ or $>$, and the equal sign are used to compare numbers.

$<$ means *is less than*	$\leq$ means *is less than or equal to*
$>$ means *is greater than*	$\geq$ means *is greater than or equal to*
$=$ means *is equal to*	

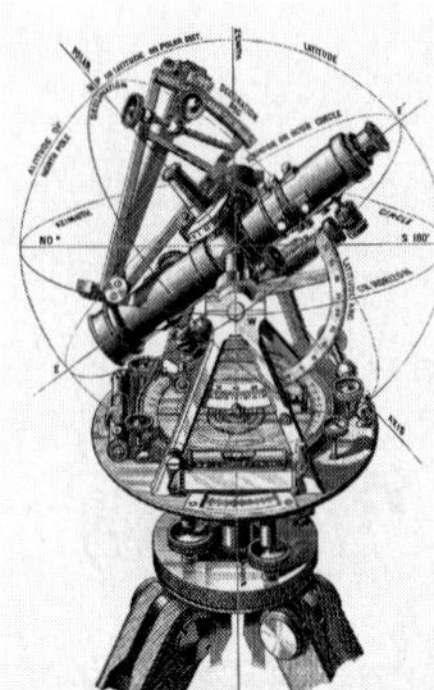

GRAPHIC ILLUSTRATION OF THE SOLAR APPARATUS

Math: Who Where, When

A sixteenth century English mathematician, Thomas Harriot, was the first to use the signs $>$ and $<$. He was considered one of the founders of algebra as we know it today. He was sent to survey and map what was called Virginia. That region is now North Carolina.

Example 2

Graph each set of numbers on a number line.

a. the set of integers from −2 to 3 inclusive

b. the set of real numbers from −2 to 3 inclusive

c. {all real numbers less than or equal to 2}

d. {all real numbers greater than −2}

Solution

a. The set consists of −2, −1, 0, 1, 2, 3.

b. The set consists of −2 and 3 and all the real numbers between.

Learning Styles

VISUAL LEARNER Reinforce students' understanding of number lines by having them create a time line of their daily tasks. Develop the idea that each type of visual consists of a line divided into equal segments. Encourage students to carry a time line for an entire day and mark the line to show how they spend their time. Stress the idea that moving to the right on a timeline indicates an increase in value or time passing. While moving to the left indicates a decrease in value.

Chalkboard Examples

Supplementary Example 1

Order this set of numbers from least to greatest:

$\left(-2.5,\ 4,\ 1\frac{3}{4},\ -1\frac{1}{3},\ \sqrt{5}\right).$

$-2.5,\ -1\frac{1}{3},\ 1\frac{3}{4},\ \sqrt{5},\ 4$

Supplementary Example 2

Graph each set of numbers on a number line.

a. the set of integers from 2 to 5

b. the set of real numbers greater than or equal to 3

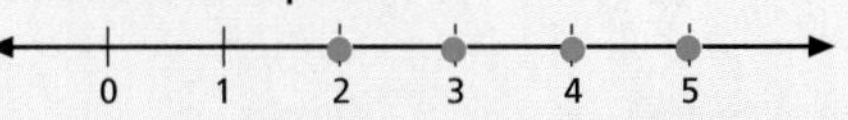

c. {all numbers with an absolute value greater than 2 but less than 4}

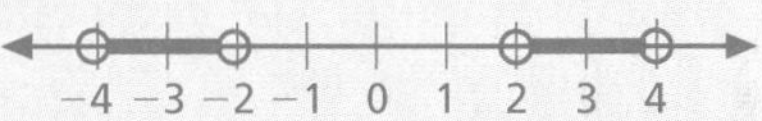

Supplementary Example 3

Evaluate each expression when $x = -8$.

a. $-x$ 8 **b.** $|x|$ 8

c. $|-x|$ 8 **d.** $-|-x|$ −8

Reteaching Worksheet 1-2

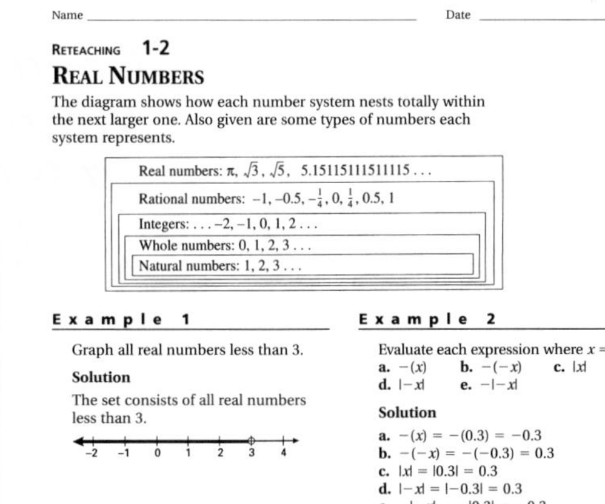

Name __________ Date __________

RETEACHING 1-2

REAL NUMBERS

The diagram shows how each number system nests totally within the next larger one. Also given are some types of numbers each system represents.

Real numbers: π, $\sqrt{3}$, $\sqrt{5}$, 5.15115111511115 . . .
Rational numbers: −1, −0.5, $-\frac{1}{4}$, 0, $\frac{1}{4}$, 0.5, 1
Integers: . . . −2, −1, 0, 1, 2 . . .
Whole numbers: 0, 1, 2, 3 . . .
Natural numbers: 1, 2, 3 . . .

Example 1

Graph all real numbers less than 3.

Solution

The set consists of all real numbers less than 3.

−2 −1 0 1 2 3 4

Example 2

Evaluate each expression where $x = 0.3$.

a. $-(x)$ **b.** $-(-x)$ **c.** $|x|$
d. $|-x|$ **e.** $-|-x|$

Solution

a. $-(x) = -(0.3) = -0.3$
b. $-(-x) = -(-0.3) = 0.3$
c. $|x| = |0.3| = 0.3$
d. $|-x| = |-0.3| = 0.3$
e. $-|-x| = -|0.3| = -0.3$

EXERCISES

Tell whether each statement is true or false.

1. $\sqrt{2}$ is a rational number. F
2. −42 is an integer. T
3. 0 is a natural number. F
4. $-\frac{3}{5}$ is an integer. F
5. 213 is a whole number. T
6. 0.31131113 is an irrational number. F

Graph each set of numbers on a number line.

7. $\left\{\frac{1}{5}, -1\frac{3}{8}, \sqrt{2}, 3.9\right\}$

−2 −1 0 1 2 3 4

8. real numbers less than or equal to −1

−4 −3 −2 −1 0 1 2

Evaluate each expression where $b = -0.8$.

9. $-b$ 0.8
10. $-(-b)$ −0.8
11. $|b|$ 0.8
12. $-|b|$ −0.8
13. $|-b|$ 0.8
14. $-|-b|$ −0.8

4 RETEACHING *LESSON 1-2*

Lesson Wrap-up

Quick Assessment

Ask the following questions to determine if students understand the content presented in this lesson.

1. Evaluate each expression when $p = 10$.
   a. $-p$ -10 b. $|p|$ 10
   c. $|-p|$ 10 d. $-|-10|$ -10
2. Determine whether each statement is true (T) or false (F).
   a. Zero is a natural number. false
   b. -5 is an integer. true
   c. The square root of 2 is a rational number. false

Assignment Guide

Basic: 1–32, 36–43
Enriched: 1–43
Additional Practice: See Extra Practice Index on page 674.

Additional Answers

5.

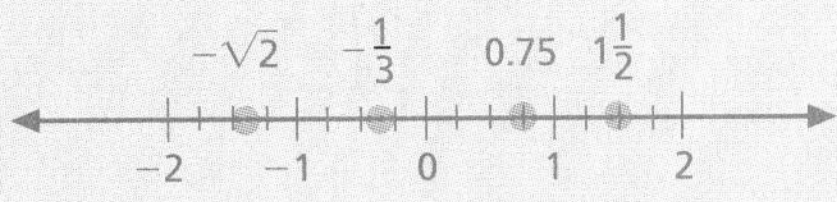

6.

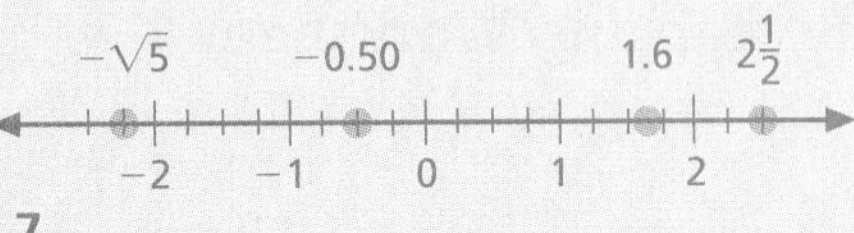

7.

8.

15.

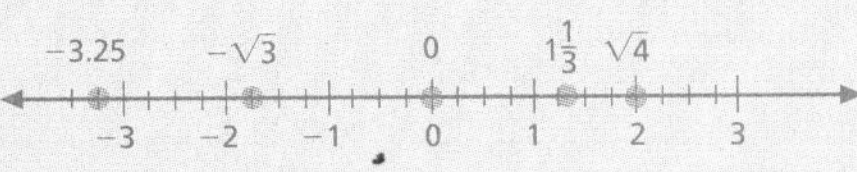

16.

17.

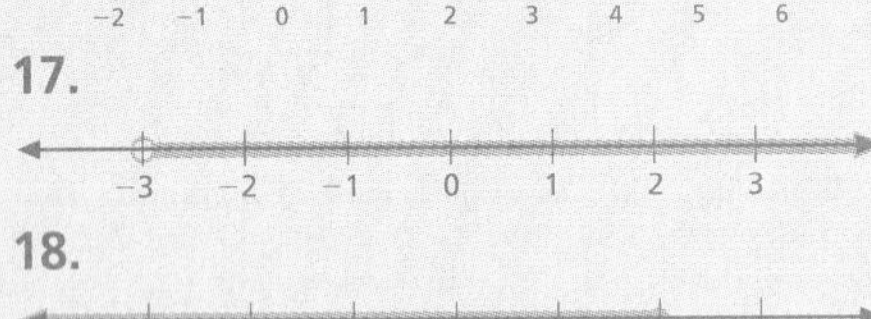

18.

33.

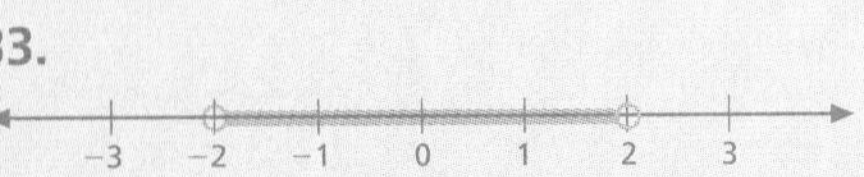

c. The set consists of 2 and all real numbers less than 2.

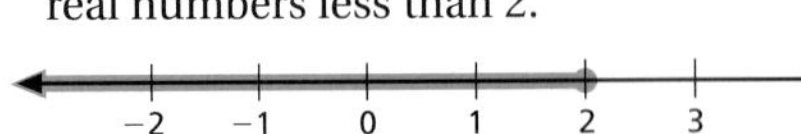

d. The set consists of all real numbers greater than -2.

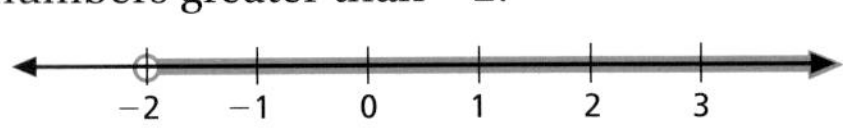

The **absolute value** of a number is the distance that number is from zero on the number line. Since opposite numbers are the same distance from zero, opposite numbers have the same absolute value. The absolute value of a number a is written as $|a|$.

> **Problem Solving Tip**
>
> When working with real numbers, the **opposite of the opposite** property can be useful. For every real number n, $-(-n) = n$.

Example 3

Evaluate each expression when $m = -5$.

a. $-m$ **b.** $-(-m)$ **c.** $|m|$ **d.** $|-m|$ **e.** $-|-m|$

Solution

a. Since $m = -5$, $-m = -(-5) = 5$.
b. Since $-m = 5$, then $-(-m) = -5$.
c. Since $m = -5$, then $|m| = |-5| = 5$.
d. Since $-m = 5$, then $|-m| = |5| = 5$.
e. Since $|-m| = 5$, then $-|-m| = -5$.

Try These Exercises

Determine if each statement is *true* or *false*.

1. $\sqrt{6}$ is a rational number. False
2. $-\sqrt{25}$ is an integer. True
3. $\frac{13}{16}$ is an irrational number. False
4. 00.10 is a whole number. False

Graph the given sets of numbers on a number line. See additional answers.

5. $\left\{-\sqrt{2}, -\frac{1}{3}, 0.75, 1\frac{1}{2}\right\}$
6. $\left\{1.\overline{6}, 2\frac{1}{2}, -\sqrt{5}, -0.50\right\}$

Graph each set of numbers on a number line. See additional answers.

7. the integers from -2 to 5 inclusive
8. all real numbers less than -1

Evaluate each expression.

9. $|t|$, when $t = -15$ 15
10. $-|-a|$, when $a = -5$ -5

Practice Exercises

Determine if each statement is *true* or *false*.

11. $-1.010010001\ldots$ is an irrational number. True
12. 2.17 is a rational number. True
13. 0 is a natural number. False
14. $\sqrt{3\frac{1}{8}}$ is not a real number. True

Graph each set of numbers on a number line. See additional answers.

15. $\left\{-3.25, -\sqrt{3}, 0, 1\frac{1}{3}, \sqrt{4}\right\}$
16. whole numbers less than 6
17. real numbers greater than -3
18. real numbers less than or equal to 2

Extend the Lesson

REAL WORLD CONNECTION Display different types of thermometers to the class. Have students consider how a thermometer is similar to a number line. Ask students to identify directions that indicate an increase or decrease in temperature.

Evaluate each expression.

19. $|r|$, when $r = -9$ 9

20. $-|-(-n)|$, when $n = 12$ −12

Replace each ● with <, >, or =.

21. $-(-3)$ ● $|-3|$ =

22. $|-\sqrt{16}|$ ● $|-\sqrt{25}|$ <

23. $|-(-1.5)|$ ● -1.5 >

List or describe the numbers that are graphed on each number line.

24.–26.
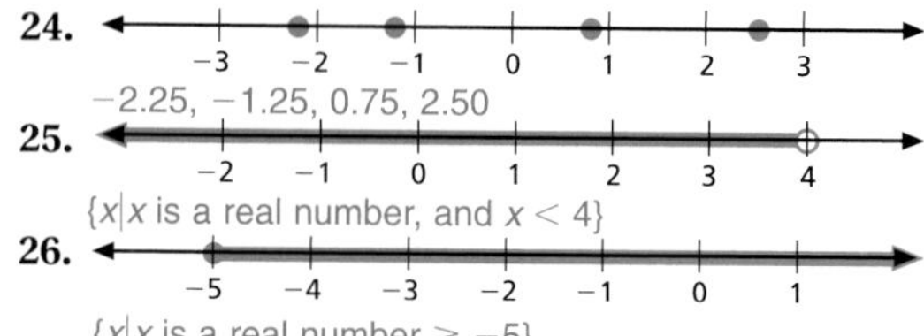

24. −2.25, −1.25, 0.75, 2.50

25. {x|x is a real number, and $x < 4$}

26. {x|x is a real number ≥ -5}

Use <, >, or ≠ to write an inequality for each statement.

27. **TEMPERATURE** Water freezes at 32°F and boils at 212°F. $32 < 212$

28. **GEOGRAPHY** The depth of the Red Sea is −1,764 ft and the depth of the Black Sea is −3,906 ft. $-1{,}764 > -3{,}906$

29. **LANGUAGE** It is estimated that there are 284 million people who speak Russian and 126 million who speak Japanese. 284 million > 126 million

30. **WRITING MATH** When comparing two real numbers, will the number with the greater absolute value always be the greater number? Explain your thinking. No, absolute value does not indicate the real value of the number. It simply gives the distance the number is from zero on the number line.

31. **DATA FILE** Use the Data Index on pages 632–633 to find the lengths of the Volga and Mackenzie rivers. Compare to find which river is longer. Volga < Mackenzie or 2,914 < 2,635.

32. **CHAPTER INVESTIGATION** Make a number line from −12 to 13. Assign a letter of the alphabet to each position on the number line. Write a message, 3 to 4 sentences in length, and encode the message by substituting numbers for letters. The message will look like a series of math problems.

Extended Practice Exercises

33. Graph the values that make $|x| < 2$ true. See additional answers.

34. **CRITICAL THINKING** *True* or *false*. For any real number n, where $n < 0$, $|-n| = n$. True

35. If there are real numbers, are there non-real numbers? Research imaginary numbers just to find out if they exist. Answers will vary.

Mixed Review Exercises

Which of the given values is a solution of the equation? (Lesson 1-1)

36. $m - 3 = 8$; 5, 11 11

37. $g + 6 = -4$; −10, 10 −10

38. $p - 6 = -2$; −8, −4, 4 −8

39. $a + 3 = 1$; −4, −2, 4 −2

40. $3b + 5 = -1$; $-2, \frac{1}{5}, 2$ −2

41. $\frac{(2x)}{3} = 2$; −3, 1, 3 3

42. $4w - 2 = -10$; −3, −2, 2 −2

43. $5f + 6 = 6$; $-\frac{1}{5}, 0, \frac{1}{5}$ 0

Extend the Lesson

ONGOING ASSESSMENT Have students work in pairs to show the following on a number line.

1. all the integers with an absolute value of 11 −11, 11
2. all the integers with an absolute value greater than 4, but less than 8 −5, 5, −6, 6, −7, 7

Extra Practice Worksheet 1-2

Name ______________________ Date ____________

EXTRA PRACTICE 1-2

REAL NUMBERS

EXERCISES

Determine if each statement is *true* or *false*.

1. 8 is an irrational number. false
2. 5 is a real number. true
3. $\sqrt{4}$ is not a natural number. true
4. −3.1235278... is an integer. false
5. $\frac{6}{7}$ is a rational number. true
6. π is not a real number. false
7. All irrational numbers are real numbers. true
9. A natural number is a rational number. true
8. Some rational numbers are integers. true

Graph each set of numbers on a number line on your own paper. Check students' graphs.

10. $\{-2\frac{1}{2}, -1, 0, \sqrt{5}, 4.75, 8\frac{2}{3}\}$
11. natural numbers less than 5
12. real numbers less than or equal to −2
13. real numbers greater than 8
14. whole numbers less than 9
15. integers greater than or equal to −4
16. integers from −5 to 2 inclusive
17. real numbers from −7 to −1 inclusive

Evaluate each expression.

18. $|w|$, when $w = -3$ 3
19. $-|f|$, when $f = -5$ −5
20. $|-(-m)|$, when $m = 7$ 7
21. $-|-b|$, when $b = 6$ −6
22. $-(-|t|)$, when $t = -2$ 2
23. $-|-(-y)|$, when $y = -4$ −4

Replace each — with <, >, or =.

24. $-(-4)$ = $|-4|$
25. $|-6|$ > $-|6|$
26. $-|-3|$ = $-|3|$
27. $|-5|$ = $|-(-5)|$
28. $|-7|$ > $|2|$
29. $-|-8|$ < $|-(-8)|$

Use <, >, or = to write an inequality for each statement.

30. The low temperatures on Monday and Tuesday were −4°F and 10°F. $-4 < 10$
31. Thad rushed for −3 yards and 6 yards on the last two plays. $-3 < 6$

4 EXTRA PRACTICE *LESSON 1-2*

Enrichment Worksheet 1-2

Name ______________________ Date ____________

ENRICHMENT 1-2

THE GOLDEN RATIO

The irrational number ϕ (phi) is equal to half the sum of 1 plus the square root of 5. Phi is the ratio of the sides in the Golden Rectangle, a shape thought to be particularly attractive by early mathematicians.

The Golden Rectangle

$\phi = \frac{1 + \sqrt{5}}{2}$

Here are two different series that get closer and closer to ϕ as you evaluate more and more terms in the series.

Fibonacci Fractions

$\frac{1}{1}, \frac{2}{1}, \frac{3}{2}, \frac{5}{3}, \frac{8}{5}, \frac{13}{8} \ldots$

Continued Radicals

$\phi = \sqrt{1 + \sqrt{1 + \sqrt{1 + \sqrt{1 + \ldots}}}}$

Each numerator equals the sum of the two before it. The same is true of the denominators. The decimal equivalents approach ϕ.

With a calculator: √ (1 + √ (1 + √ ...

EXERCISES

1. Describe what happens to the decimal equivalents in the Fibonacci fractions. They get closer to ϕ, alternating between being a little greater and a little less.
2. Which is the first Fibonacci fraction to equal ϕ to three decimal places? $\frac{89}{55}$
3. Describe what happens to the decimal equivalents in the continued radicals. The result gets closer and closer to ϕ, starting with 1.414 . . .
4. With the continued radicals, how many times do you need to use the square root key to get ϕ to three decimal places? 9 times

Make an enlarged copy on your own paper of each design, using 10 cm for the lengths marked ①. Lengths and angles that look equal are equal. Check students' drawings.

5. Whirling Golden Rectangles
6. Five-Pointed Star
7. Whirling Golden Triangles

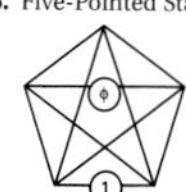

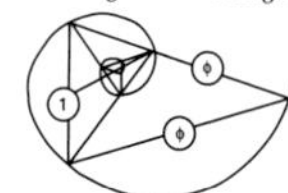

8 ENRICHMENT *LESSON 1-2*

Skills Practice

Vocabulary Review

Lesson 1–1

set	infinite set
finite set	subset
empty set	null set
variable	equation
open sentence	replacement set
solution	solution set

Lesson 1–2

natural numbers
whole numbers
integers
rational numbers
irrational numbers

real numbers	coordinate
inequality	absolute value

Assignment Guide

All students: 1–54
Additional Practice: Refer to the Extra Practice Index on page 674 of this text.

Additional Answers

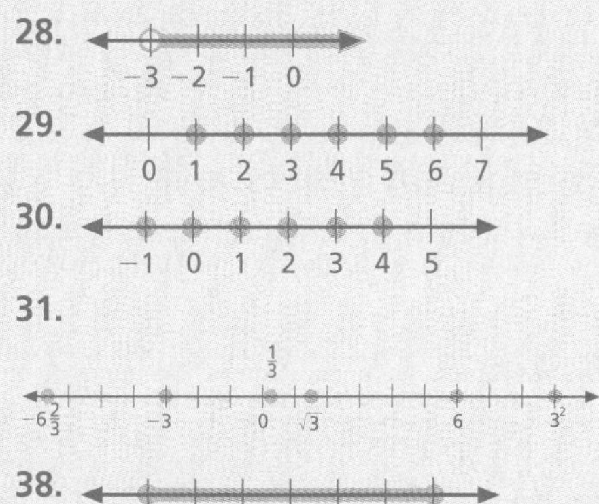

Review and Practice Your Skills

Practice ■ Lesson 1-1

1. List all subsets of the set {10, 15}. ∅, {10}, {15}, {10, 15}
2. For "the set of positive even numbers less than 8":
   a. Define this set using roster notation $S = \{2, 4, 6\}$
   b. Define this set using set-builder notation $\{x|x$ is a positive even number less than 8$\}$
   c. List all the subsets of this set ∅, {2}, {4}, {6}, {2, 4}, {2, 6}, {4, 6}, {2, 4, 6}

Use mental math to solve each equation. Name all solutions.

3. $x + 9 = 4$ −5
4. $21 - x = 7$ 14
5. $x^2 = 64$ 8, −8
6. $\frac{1}{2}x = 5$ 10
7. $x^2 = 6^2$ 6, −6
8. $2x + 2 = 10$ 4
9. A stack of nickels is worth \$1.65. Use the equation $0.05n = 1.65$ and the values {31, 33, 35} for the replacement set. Find n, the number of nickels in the stack. 33 nickels
10. Define **null set**. Explain why the null set is a subset of all sets. A null set is a set having no elements. Answers will vary.

Which of the given values is a solution of the equation?

11. $n - 17 = -9$; 26, 8, −8 8
12. $\frac{3}{2}d = -12$; −9, −18, −8 −8
13. $-2p = 11$; −5.5, 22 −5.5

Use set notation to write the following.

14. 5 is not an element of {2, 4, 6, 8} $5 \notin \{2, 4, 6, 8\}$
15. The null set is a subset of {b, u, g} $\varnothing \subset \{b, u, g\}$

Practice ■ Lesson 1-2

Determine if each statement is *true* or *false*.

16. $3\frac{1}{3}$ is a rational number. true
17. $-\sqrt{49}$ is an integer. true
18. −3 is a natural number false
19. π is a rational number. false
20. π is a real number. true
21. $\sqrt{12}$ is an irrational number true
22. 3.51 is a rational number true
23. 0 is a natural number false
24. $\sqrt{16}$ is an irrational number false

Evaluate each expression when $b = -8$.

25. $|b|$ 8
26. $-|-b|$ −8
27. $-(-(-b))$ 8

Graph each set of numbers on a number line. See additional answers.

28. real numbers greater than −3
29. natural numbers less than or equal to 6
30. the integers from −1 to 4 inclusive
31. $\{\sqrt{3}, \frac{1}{3}, -3, -6\frac{2}{3}, 3^2, 6\}$

Solve, using <, >, or =.

32. $-\sqrt{9}$ __?__ $-\pi$ >
33. $|5| - |-5|$ __?__ 10 <
34. $-\sqrt{9}$ __?__ $-\sqrt{16}$ >
35. $-|3.75 - |$ __?__ $|-3.75|$ <
36. 6 __?__ $-(-(-(-6)))$ =
37. $\frac{1}{3}$ __?__ 0.3 >
38. Graph the values that make $|x| \leq 3$ true. See additional answers.

Learning Styles

VISUAL LEARNER Compare the characteristics of a replacement set with a solution set. Develop the idea that both types of sets contain possible values for the variables in an open sentence. However, all the elements in a solution set are values for variables that make the open sentence true. A solution set is actually a subset of a replacement set. Diagram this relationship by drawing a Venn diagram in which a larger circle (the replacement set) contains a smaller circle (the solution set).
You may want to use the following example:
Open sentence: $x - 3 = 10$ Replacement set: {8, 10, 13, 15}
Solution set: {13}

PRACTICE ■ LESSON 1-1–LESSON 1-2

39. How many elements are in the word *maximize*? Which elements are used more than once? (Lesson 1-1) 6, *m* and *i*

For Exercises 2–7, use $A = \{m, n, p, q, r\}$, $B = \{n, q, r\}$, $C = \{x, y, z\}$. Classify each statement as *true* or *false*. (Lesson 1-1)

40. $m \in A$ true
41. $m \in B$ false
42. $\varnothing \subset B$ true
43. $\{x, r\} \subset C$ false
44. B has 6 subsets false
45. $B \subset A$ true
46. C contains 3 elements true
47. A contains no vowels true
48. A has 32 subsets true

List all sets of numbers to which each number belongs. (Lesson 1-2)

49. $\sqrt{10}$ irrational, real
50. -17 integer, rational, real
51. 0 integer, rational, real
52. 4.3232323232 . . . . rational, real
53. 8.71765392 . . . . irrational, real
54. $-\frac{3}{5}$ rational, real

MathWorks Workplace Knowhow — Career – Cryptographer

Cryptography is a branch of mathematics that combines communication with math. Cryptographers create and try to break secret codes using step-by-step mathematical processes called algorithms.

In the past, encryption was used mainly by the military to send secret messages. Today, businesses and governments employ cryptographers to create codes to protect the privacy of their electronic messages and financial data.

Cryptographers encrypt text using a key. The key works like a recipe with steps for encoding and decoding messages. The key is supposed to be the only way to read a coded message. Cryptographers frequently write computer programs to encode and decode messages.

Use this encoding key for Exercises 1–3.

Key: Each letter of the alphabet is assigned an odd positive integer beginning with 1 for Z and proceeding backwards to 51 for A. Then each odd positive integer is added to the even positive integer one number higher. Finally, subtract the sum from 105. Assign the result of this algorithm to each letter. Example: T corresponds to 13. Add: $13 + 14 = 27$. Subtract: $105 - 27 = 78$. T is represented by the number 78.

1. What number represents the letter L? 46

2. Encrypt the following message: Your bank account is overdrawn.
7 58 82 70 6 2 54 42 2 10 10 58 82 54 78 34 74 58 86 18 70 14 70 2 90 54

3. Use the key to decipher this message:
Codes are fun. You broke my best code.
10 58 14 18 74 2 70 18 22 82 54
98 58 82 6 70 58 42 18 50 98 6 18 74 78 10 58 14 18

Teaching Strategies

ESL/LEP Review the meaning of the word "inclusive." Point out its similarity to the word "included." Stress the fact that a closed dot must be drawn on a point to show that the point is included in the set being graphed while an open dot means that the point is not included in the set.

Chalkboard Examples

Example from Lesson 1-1

Write the solution set in roster notation. $\{x \mid x$ is a positive whole number and $2 < x < 9\}$
$\{3, 4, 5, 6, 7, 8\}$

Example from Lesson 1-2

On a number line, graph all real numbers less than or equal to -2.

Career Opportunity

Describe the kind of work cryptologists do and the types of places they work. Have students brainstorm about times in their lives, (past, present, and future) when a cryptology may have been or be of value to them.

Students should answer Questions 1–3 to better understand how cryptologists use mathematics in performing their job.

Students who are interested in learning more about this profession can go to Learningmathmatters.com. Guidance Counselors should have information about school and training requirements.

Chapter 1 **Review and Practice Your Skills** 15

Lesson Planning

NCTM Standards/Strands

- Number & Operation
- Patterns, functions, & Algebra
- Problem Solving
- Reasoning and Proof
- Communication
- Connections

Representation Vocabulary

union
intersection
universe
complement
compound inequality
Venn diagram
disjoint set
universal set

Materials Needed

paper/pencil

Lesson Resources

Warm-Up Transparency 1
Reteaching 1-3
Extra Practice 1-3
Enrichment 1-3

Assignment Guide

Basic: 1–34, 43–46
Enriched: 1–46

Getting Started

5-Minute Warm-Up

Have students determine all the possible subsets of {T, E, N} and of {W, E. N, T}. Identify the subsets that belong to both sets. {T}, {E}, {N}, {TE}, {EN}, {TN}, {TEN}, { }

Introduction to Lesson 1-3

Poll the class to determine the numbers of students who were members of a soccer team and/or a baseball or softball team during the past year. Ask a volunteer to display the results in a table on the board. Then have the class create a Venn diagram to illustrate the data.

1-3 Union and Intersection of Sets

Goals
- Identify unions and intersections of sets.
- Use Venn diagrams to solve problems.

Applications Advertising, Language, Logic

Suppose that in your homeroom 8 students are on the basketball team and 11 students are on the soccer team. Five of the students are on both teams. Another 12 students are not on either team. How many students in total are on both teams? Show your answer by copying and completing the Venn diagram at the right. 14

U = ?
Basketball
Soccer

Build Understanding

Two or more sets can be used to form new sets. Two of these new sets are the union and intersection of the sets. The **union** of any two sets A and B is symbolized as $A \cup B$ (read as "A union B").

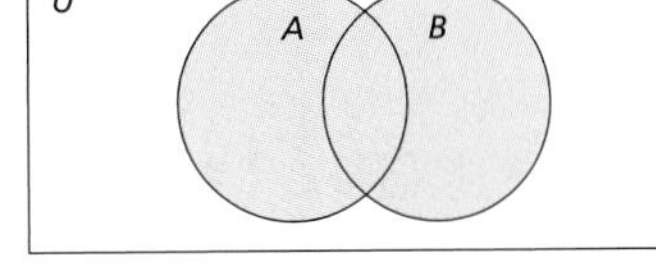

The set $A \cup B$ contains all the elements that are in A, in B, or in both.

$A \cup B = \{x | x \in A \text{ or } x \in B\}$

At the right is a Venn diagram of $A \cup B$. A **Venn diagram** uses circles inside a rectangle to represent the union and intersection of sets. The rectangle, labeled U, represents the **universe** or the **universal set**.

The **intersection** of two sets A and B is symbolized by $A \cap B$, read as "A intersect B." The set $A \cap B$ contains the elements that are common to both A and B.

$A \cap B = \{x | x \in A \text{ and } x \in B\}$

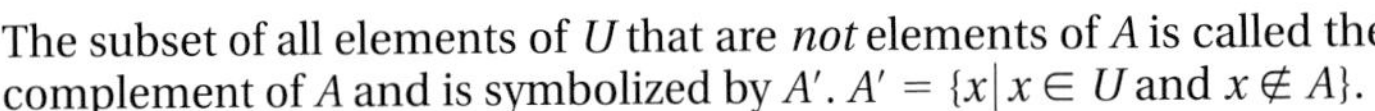

The subset of all elements of U that are *not* elements of A is called the complement of A and is symbolized by A'. $A' = \{x | x \in U \text{ and } x \notin A\}$.

If two sets have no elements in common, their intersection will be the empty set, $\varnothing$. Two sets whose intersection is the empty set are called **disjoint sets**.

Example 1

Refer to the diagram to find each.
Let $A = \{1, 3, 5\}$, $B = \{3, 6\}$, and $C = \{2, 4\}$.

a. C'
b. $A \cup B$
c. $A \cap B$
d. $A \cap C$

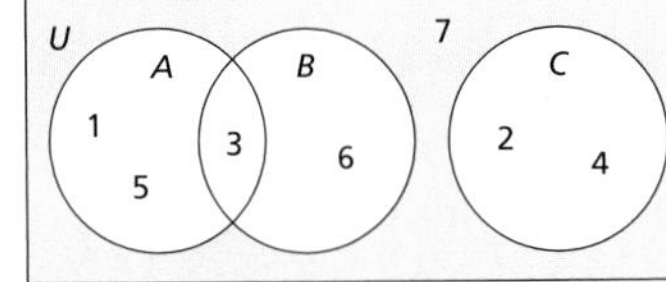

Teaching Strategies

TEAM LEARNING Have students work in pairs or small groups to answer the following questions:

1. Can the union of two sets ever be the empty set? Explain and give an example, if possible. Yes, if both sets are empty sets. Examples will vary.
2. When are Venn diagrams useful for solving problems? Answers will vary. Students' answers may include the following ideas: Venn diagrams are a good way to view the relationship among sets. They are useful whenever a member of one set is also a member of another.

Solution

a. C' is the set of those elements in U and *not* in C. From the diagram, $U = \{1, 2, 3, 4, 5, 6, 7\}$. So, $C' = \{1, 3, 5, 6, 7\}$.

b. $A \cup B$ is the set of those elements that are in A, in B, or in both.
$\{1, 3, 5\} \cup \{3, 6\} = \{1, 3, 5, 6\}$
$A \cup B = \{1, 3, 5, 6\}$

If an element is in both, it is still listed *only once* in a union.

c. $A \cap B$ is the set of elements common in A and B. The only element common to both A and B is 3.
$\{1, 3, 5\} \cap \{3, 6\} = \{3\}$
So, $A \cap B = \{3\}$

d. $A \cap C$ have *no* elements in common. So, $A \cap C = \varnothing$.

Math: Who Where, When

Although the English mathematician John Venn is credited with creating the rectangular regions to represent the universal set U, and using the circular regions within the rectangles to represent subsets of U, Leonhard Euler (1707–1783) also used circles, referred to as Euler circles, to represent his work. He became one of the greatest mathematicians and astronomers of his generation.

An inequality that combines two inequalities is called a **compound inequality**. The solution set for a compound inequality containing the word *and* can be found from the *intersection* of the graphs of the individual inequality. The solution set for a compound inequality containing the word *or* can be found from the *union* of the graphs of the individual inequalities.

Example 2

Using the replacement set of the real numbers, find the solution set for $x \geq -1$ and $x < 4$.

Let $A = \{x \mid x \text{ is a real number and } x \geq -1\}$
$B = \{x \mid x \text{ is a real number and } x < 4\}$
Graph $A \cap B$ on a number line.

Solution

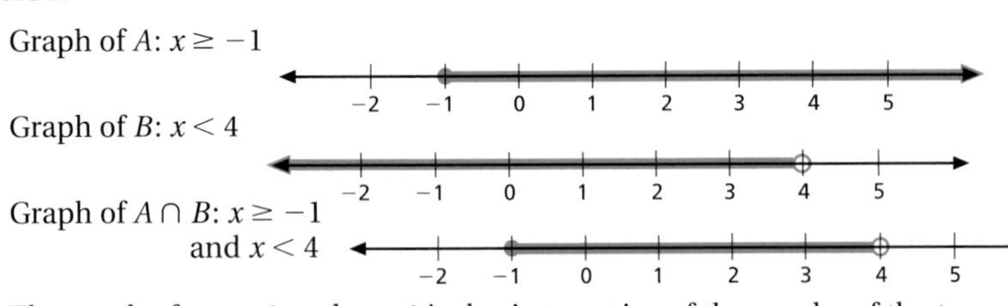

The graph of $x \geq -1$ and $x < 4$ is the *intersection* of the graphs of the two inequalities, $A \cap B$.

This can be written as $-1 \leq x < 4$.

The solution set is $\{x \mid x \text{ is a real number and } -1 \leq x < 4\}$.

Example 3

Using the replacement set of the real numbers, find the solution set for $x \geq 4$ or $x < -1$.

Let $A = \{x \mid x \text{ is a real number and } x \geq 4\}$
$B = \{x \mid x \text{ is a real number and } x < -1\}$

Graph $A \cup B$ on a number line.

Chalkboard Examples

Supplementary Example 1

Refer to the diagram in Example 1. Find:

a. B' $\{1, 2, 4, 5\}$ b. A' $\{2, 4, 6\}$
c. $B \cap C$ $\varnothing$

Supplementary Example 2

Let $A = \{x \mid x \text{ is a real number and } \geq -3\}$ and $B = \{x \mid x \text{ is a real number and } x < 2\}$. Describe the graph of $A \cap B$.

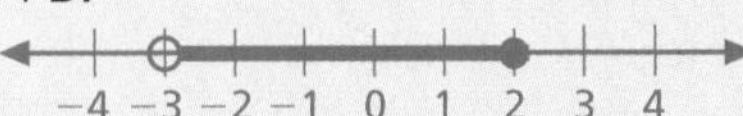

Supplementary Example 3

Let $A = \{x \mid x \text{ is a real number and } x \geq -3\}$ and $B = \{x \mid x \text{ is a real number and } x > 2\}$. Describe the graph $A \cup B$.

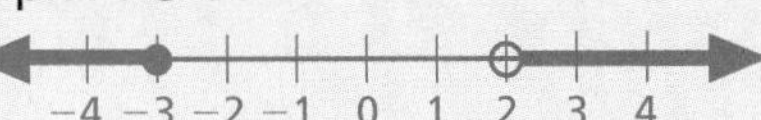

Extend the Lesson

REAL WORLD CONNECTIONS Generally a particular product will be advertised in a number of different magazines. Have students examine the advertisements contained in two different magazines. Tell them to select a minimum of ten ads from each publication. Have students make a Venn diagram to illustrate how many of the products were advertised in both publications.

Reteaching Worksheet 1-3

Name ______________ Date ______________

RETEACHING 1-3

UNION AND INTERSECTION OF SETS

This Venn diagram represents the union and intersection of two sets and their complements.

$A \cup B = \{2, 4, 6, 8, 10\}$ $A' = \{8, 10\}$
$A \cap B = \{6\}$ $B' = \{2, 4\}$

Example 1

Graph the solution set for $x < 1$ and $x \geq -2$.

Solution

Graph of A if $A = x < 1$
Graph of B if $B = x \geq -2$
Graph of $A \cap B$: $x < 1$ and $x \geq -2$

SOLUTION SET
Roster notation: $\{-2$, and all real numbers between -2 and $1\}$
Set-builder notation: $\{x \mid x$ is a real number and $-2 \leq x < 1\}$

Example 2

Graph the solution set for $x \leq -2$ or $x > 1$.

Solution

Graph of A if $A = x \leq -2$
Graph of B if $B = x > 1$
Graph of $A \cup B$: $x \leq -2$ or $x > 1$

SOLUTION SET
Roster notation: $\{-2$, and all real numbers less than 2 or greater than $1\}$
Set-builder notation: $\{x \mid x \leq -2 \text{ or } x > 1\}$

EXERCISES

Refer to the diagram. Find the sets named by listing the members.

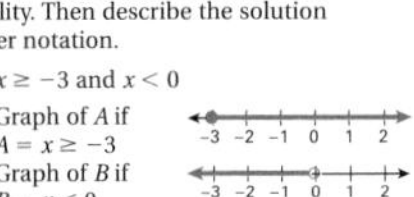

1. A' $\{7, 10\}$ 2. $A \cap B$ $\{2\}$ 3. $A \cup B$ $\{2,3,5,7,10\}$

Graph the solution sets for each compound inequality. Then describe the solution set in two ways using roster notation and set-builder notation.

4. $x > 2$ or $x \leq -1$
Graph of A if $A = x > 2$
Graph of B if $B = x \leq -1$
Graph of $A \cup B$: $x > 2$ or $x \leq -1$
Roster notation: $\{-1$, and all real numbers less than -1 or greater than $2\}$
Set-builder notation: $\{x \mid x$ is a real number and $x > 0$ or $x \leq 1\}$

5. $x \geq -3$ and $x < 0$
Graph of A if $A = x \geq -3$
Graph of B if $B = x < 0$
Graph of $A \cap B$: $x \geq -3$ and $x < 0$
Roster notation: $\{-3$, and all real numbers between -3 and $0\}$
Set-builder notation: $\{x \mid x$ is a real number and $-3 \leq x < 0\}$

6 RETEACHING *LESSON 1-3*

Lesson Wrap-up

Quick Assessment

Ask the following questions to determine if students understand the content presented in this lesson. Suppose that $U = \{1, 3, 5, 7, 9\}$, $A = \{3, 7\}$ and $B = \{1, 9\}$.

1. Find $A \cup B$. $\{1, 3, 7, 9\}$
2. Find $(A \cup B)'$. $\{5\}$
3. Find $A \cap B$. the null set
4. Find A'. $\{1, 5, 9\}$
5. Find B'. $\{3, 5, 7\}$

Assignment Guide

Basic: 1–34, 43–46
Advanced: 1–46
Additional Practice: See Extra Practice Index on page 674.

Additional Answers

5. 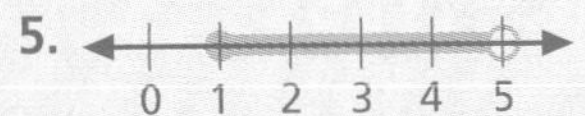

6. −2 −1 0 1 2 3 4 5 6

41. 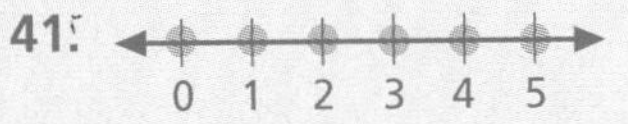

42. 0 1 2 3 4 5

43. 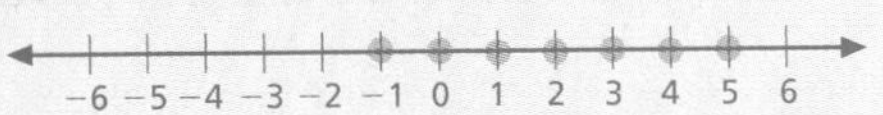

44. −5 −4 −3 −2 −1 0 1 2 3 4 5 6

45. 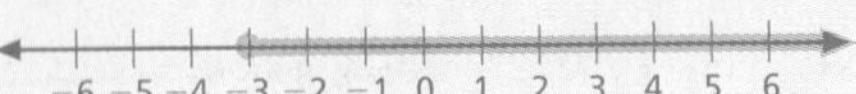

46. −6 −5 −4 −3 −2 −1 0 1 2 3 4 5 6

Solution

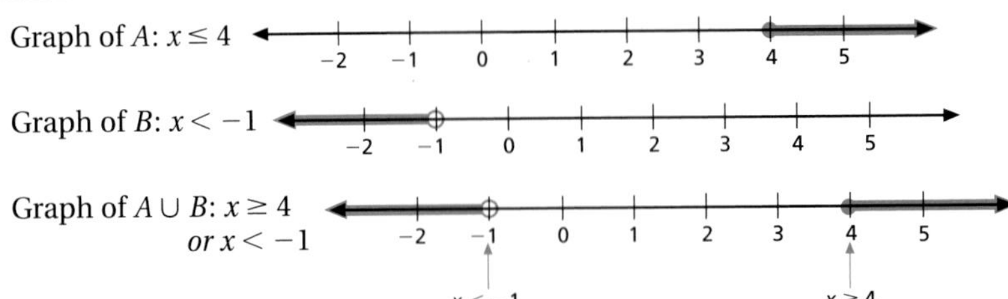

The graph of $x \geq 4$ *or* $x < -1$ is the *union* of the graphs of the two inequalities, $A \cup B$. The solution set is $\{x | x < -1 \text{ or } x \geq 4\}$.

If you graph the solution set for $x \geq 4$ and $x < -1$ in Example 3, you would need to find the *intersection* of the two graphs, $A \cap B$. The graphs of the sets have *no* values in common. They *do not* intersect. The solution set would be { } or $\varnothing$.

Try These Exercises

Refer to the diagram. Find the sets named by listing the members.

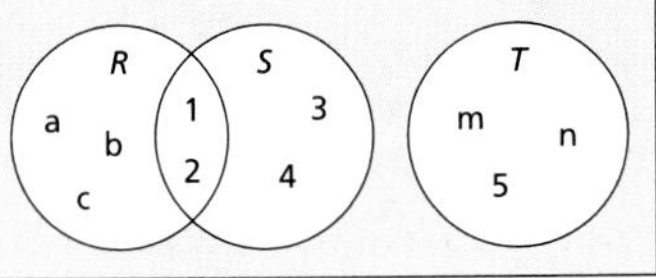

1. T' $\{1, 2, 3, 4, a, b, c\}$
2. $R \cap S$  $\{1, 2\}$
3. $R \cup S$  $\{1, 2\}$
4. $R \cap T$ $\varnothing$

Using the replacement set of the real numbers, find the solution set for each compound inequality. See additional answers.

5. $x \geq 1$ and $x < 5$
6. $x > 5$ or $x \leq -1$

Practice Exercises

Refer to the diagram. Find the set named by listing the members.

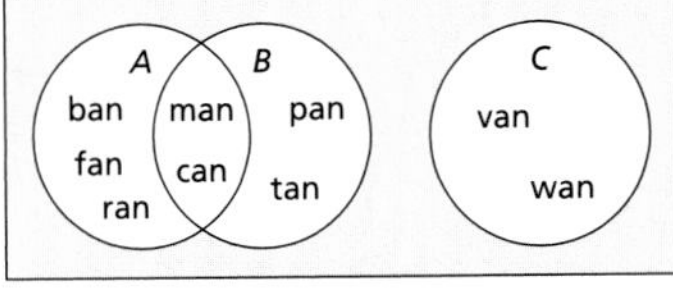

7. $A \cup B$ {ban, can, fan, man, pan, ran, tan}
8. $A \cap B$ {can, man}
9. $A \cup C$ {ban, can, man, ran, van, wan}
10. $A \cap C$ $\varnothing$ or { }

Let U = {c, h, a, r, t}, M = {h, a, r, t}, and N = {a, r, t}. Find each union or intersection.

11. M' {c}
12. $M' \cup N'$ {c, h}
13. $(M \cup N)'$ {c}
14. $M \cap N$ {a, r, t}
15. $(M \cap N)'$ {c, h}
16. Let A = {f, o, r, g, e} and B = {m, a, j, o, r}. Find $A \cap B$. {o, r}
17. Let $P = \{3, 6, 9\}$ and $Q = \{4, 8\}$. Find $P \cap Q$. { } or $\varnothing$

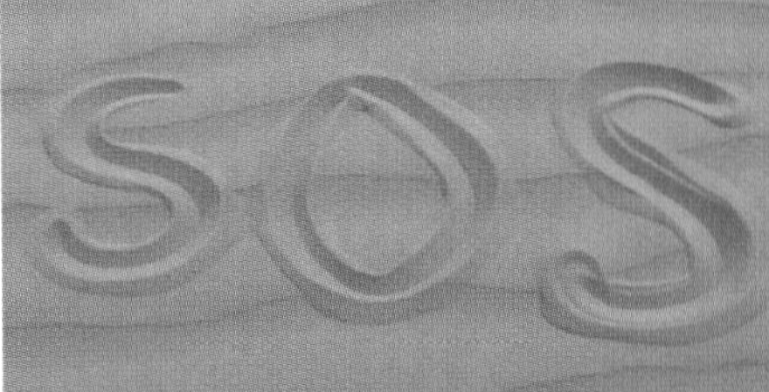

Extend the Lesson

COOPERATIVE LEARNING Have students work in pairs to create their own word problems similar to Exercise 33. After recording the problem on a sheet of paper, student pairs should trade papers and solve each others' problems. Discuss: How do you know when you can use a Venn diagram to solve a problem?

Use the replacement set of real numbers for the solution set for each compound inequality.

18. $x \le 3$ and $x \ge 3$
$\{x|x$ is a real number and $3 \le x \le 3\}$

19. $x \le 1$ or $x \ge 2$
$\{x|x \le 1$ or $x \ge 2\}$

For Exercises 20–31 let $U = \{0, 1, 2, 3, \ldots, 9\}$, $A = \{x|x$ is a whole number such that $x > 2$ *and* $x < 8\}$, $B = \{1, 3, 5, 7, 9\}$, $C = \{0, 1, 2, 3, 4, 5\}$, and $D = \{0, 3, 6, 9\}$.

20. A' $\{0, 1, 9\}$

21. D' $\{1, 2, 4, 5, 7, 8\}$

22. $A \cap B$ $\{3, 5, 7\}$

23. $C \cup D$ $\{0, 1, 2, 3, 4, 5, 6, 9\}$

24. $B' \cup C$ $\{0, 1, 2, 3, 4, 5, 6, 8\}$

25. $A' \cap B$ $\{1, 9\}$

26. $B \cap (C \cap D)$ $\{3\}$

27. $B \cup (C \cap D)$ $\{0, 1, 3, 5, 7, 9\}$

28. $A \cap (B \cup C)$ $\{0, 1, 2, 3, 4, 5, 7, 9\}$

29. $(A \cap B) \cap C$ $\{3, 5\}$

30. $(A \cap B) \cap (C \cap D)$ $\{3\}$

31. $(A \cap B)' \cap (B \cap C)$ $\{1\}$

32. WRITING MATH Let $A = \{x|x$ is a real number and $x > -3\}$
$B = \{x|x$ is a real number and $x \le 1\}$
Find $A \cap B$ and write a brief description of the graph of $A \cap B$.
$A \cap B = \{x|x$ is a real number and $-3 < x \le 1\}$

Draw Venn diagrams to solve these problems.

33. LANGUAGE Twenty-four students are on a tour of the United Nations building. Twelve of these students speak Russian. Six speak German, and 15 speak Spanish. Only one student speaks all three languages. Two speak Russian and German, and two different students speak German and Spanish. How many students speak Russian and Spanish, but not German? 3

34. ADVERTISING A company runs commercials on radio and television for the 18 products it makes. Sixteen of the products are advertised on television and 14 are advertised on the radio. Four of the products are advertised on television only. How many products are advertised only on the radio? 2

Extend Practice Exercises

Determine whether each statement is *true* or *false*. If the statement is false, provide the correct answer.

$U = \{0, 1, 2, 3, 4, 5, 6, \ldots\}$

$W = \{$whole numbers$\}$ $N = \{$natural numbers$\}$

$P = \{$positive integers$\}$ $M = \{1, 3, 5, 7\}$

35. $P \cup N' = W$ True

36. $P \cap N = \varnothing$ False

37. $N \cup M = M$ False

38. $N \cap W = N$[True

39. $P \cup N = P$ True

40. $P \cap M = M$ True

41. Graph $A = \{x|x \in W$ or $x \in N\}$
See additional answers.

42. Graph $B = \{x|x \in P$ and $x \in W\}$
See additional answers.

Mixed Review Exercises

Graph each set of numbers on a number line. (Lesson 1-2)
For 43–46, see additional answers.

43. the set of integers from -1 to 5, inclusive

44. the set of real numbers from -4 to 3, inclusive.

45. the set of all real numbers greater than or equal to -3

46. the set of all real numbers less than -2

Lesson 1-3 **Union and Intersections of Sets** 19

Extend the Lesson

MATH JOURNAL Have students use the following example. Let $A = \{x|x$ is a real number and $x > -2\}$ and $B = \{x|x$ is a real number and $x < 5\}$. Have students explain how to find $A \cup B$ and $A \cap B$. Have them show the relationships using number lines and Venn diagrams.

Extra Practice Worksheet 1-3

Name ________________ Date ________

EXTRA PRACTICE 1-3

UNION AND INTERSECTION OF SETS

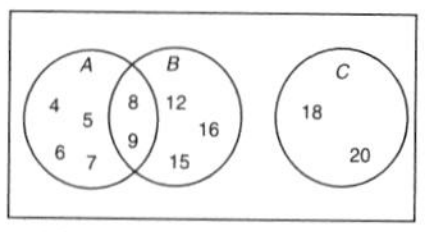

EXERCISES

Refer to the diagram. Find the set named by listing the members.

1. $A \cup B$ $\{4, 5, 6, 7, 8, 9, 12, 15, 16\}$

2. $A \cap B$ $\{8, 9\}$

3. $A \cup C$ $\{4, 5, 6, 7, 18, 20\}$

4. $A \cap C$ $\varnothing$

5. $B \cup C$ $\{8, 9, 12, 15, 16, 18, 20\}$

6. $B \cap C$ $\varnothing$

7. $(A \cup B) \cup C$ $\{4, 5, 6, 7, 8, 9, 15, 16, 18, 20\}$

8. $(A \cap C) \cup B$ $\{8, 9, 12, 15, 16\}$

Let $U = \{0, 2, 4, 6, 8, 10, 12\}$, $X = \{2, 4, 6, 8, 10\}$, $Y = \{4, 6, 8\}$ and $Z = \{2, 10\}$. Find each union or intersection.

9. X' $\{0, 12\}$

10. Y' $\{0, 2, 10, 12\}$

11. Z' $\{0, 4, 6, 8, 12\}$

12. $X \cup Y$ $\{2, 4, 6, 8, 10\}$

13. $Y \cap Z$ $\varnothing$

14. $(X \cup Y)'$ $\{0, 12\}$

15. $X' \cup Y'$ $\{0, 2, 10, 12\}$

16. $(X \cap Y)'$ $\{0, 2, 10, 12\}$

17. $Y' \cap Z'$ $\{0, 12\}$

18. $(X \cup Z)'$ $\{0, 12\}$

19. $(X \cup Y) \cap Z$ $\{2, 10\}$

20. $(X \cup Z) \cap Y$ $\{4, 6, 8\}$

Use the replacement set of real numbers to find the solution set for each compound inequality. Graph the solution set on a number line on your own paper. Check students' graphs.

21. $x < 2$ and $x > 5$

22. $x \le -2$ or $x \ge 2$

23. $x \le -4$ or $x \ge 1$

24. $x \le -3$ and $x > 1$

6 EXTRA PRACTICE *LESSON 1-3*

Enrichment Worksheet 1-3

Name ________________ Date ________

ENRICHMENT 1-3

USING PRIME FACTORS FOR GCF AND LCM

One application of intersection and union is in finding the greatest common factor (GCF) and the least common multiple (LCM).

Example 1

Find the greatest common factor of 380, 475, and 855.

Solution

Write the prime factorization of each number. The common prime factors are the intersection of the factors. The GCF is the product of the numbers in the intersection.

$380 = 2 \cdot 2 \cdot 5 \cdot 19$ Set A = {2, 2, 5, 19}
$475 = 5 \cdot 5 \cdot 19$ Set B = {5, 5, 19}
$855 = 3 \cdot 3 \cdot 5 \cdot 19$ Set C = {3, 3, 5, 19}

$A \cap B \cap C = \{5, 19\}$

The GCF of 380, 475, and 855 is 95.

Example 2

Find the least common multiple of 380, 475, and 855.

Solution

Now, find the union of the prime factors, using a factor the maximum number of times it appears in any one factorization. The LCM is the product of the numbers in the union.

$380 = 2 \cdot 2 \cdot 5 \cdot 19$ Set A = {2, 2, 5, 19}
$475 = 5 \cdot 5 \cdot 19$ Set B = {5, 5, 19}
$855 = 3 \cdot 3 \cdot 5 \cdot 19$ Set C = {3, 3, 5, 19}

$A \cup B \cup C = \{2, 2, 3, 3, 5, 5, 19\}$

The LCM of 380, 475, and 855 is 17,100.

EXERCISES

Use the methods shown above to find the GCF and LCM.

1. 92; 483; 138
Set A = {2, 2, 23}
Set B = {3, 7, 23}
Set C = (2, 3, 23}
GCF = 23
LCM = 1932

2. 175; 390; 485
Set A = {5, 5, 7}
Set B = {2, 3, 5, 13}
Set C = {5, 97}
GCF = 5
LCM = 1,324,050

3. 99; 330; 462
Set A = {3, 3, 11}
Set B = {2, 3, 5, 11}
Set C = {2, 3, 7, 11}
GCF = 33
LCM = 6930

4. 490; 1365; 4235
Set A = {2, 5, 7, 7}
Set B = {3, 5, 7, 13}
Set C = {5, 7, 11, 11}
GCF = 35
LCM = 2,312,310

5. 1870; 408; 1564
Set A = {2, 5, 11, 17}
Set B = {2, 2, 2, 3, 17}
Set C = {2, 2, 17, 23}
GCF = 34
LCM = 516,120

6. 13,156; 29,315; 51,623
Set A = {2, 2, 11, 13, 23}
Set B = {5, 11, 13, 41}
Set C = {11, 13, 19, 19}
GCF = 143
LCM = 973,609,780

10 ENRICHMENT *LESSON 1-3*

Lesson Planning

NCTM Standards/Strands

- Number & Operation
- Patterns, functions, & Algebra
- Problem Solving
- Reasoning and Proof
- Communication
- Connections

Representation Vocabulary

addition property of opposites
additive inverse
associative property
closure property
commutative property
identity property
inverse property

Materials Needed

paper/pencil   number lines

Lesson Resources

Warm-Up Transparency 2
Transparency TK-6, RF-1
Reteaching 1-4
Extra Practice 1-4
Enrichment 1-4

Assignment Guide

Basic: 1–28, 31–58
Enriched: 1–58

Getting Started

5-Minute Warm-up

Write an integer for each description. Then write the opposite integer.

1. a decrease of 12°   −12, 12
2. a deposit of $15.00 $15, −$15

Introduction to Lesson 1-4

Have students draw a vertical number line to represent the floors of a building and label it from 2 through 12. Have each student use a clip or counter to represent an elevator. Describe an up-and-down ride that starts at 0 and consists of no more than three segments. Have students determine the final location of the elevator. Write corresponding addition sentences.

1-4 Addition, Subtraction, and Estimation

Goals
- Add and subtract rational numbers.
- Estimate answers to addition and subtraction problems.

Applications Recreation, Finance, Construction

You can use a number line to add rational numbers. To add a positive number, move right. To add a negative number, move left.

For example, to show $1\frac{1}{2} + \left(-2\frac{3}{4}\right)$ on a number line, start at 0. Move right $1\frac{1}{2}$ units, then move left $2\frac{3}{4}$ unit.

$-2\frac{3}{4}$, $+1\frac{1}{2}$; −2, −1, 0, 1, 2; $-1\frac{1}{4}$; $1\frac{1}{2} + \left(-2\frac{3}{4}\right) = -1\frac{1}{4}$

Add the following pairs of rational numbers. Locate each answer and its corresponding letter on the number line below. Then complete these generalizations using your answers.

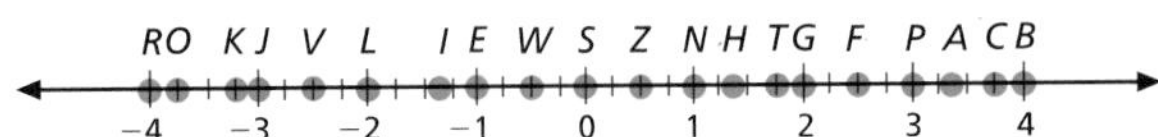

1. −3.1, 3.1
2. 1.5, 0.5
3. $-2, -\frac{1}{2}$
4. $-1\frac{1}{2}, 2\frac{1}{2}$
5. −4.75, 6.50
6. $-1.25, -2\frac{3}{4}$
7. $-\frac{7}{8}, 4\frac{1}{4}$
8. $3\frac{1}{8}, -4\frac{1}{2}$
9. −0.4, −0.6
10. $4\frac{3}{4}, -8.5$
11. $1\frac{1}{3}, 1\frac{2}{3}$
12. The sum of two positive numbers is _ _ _ _ _ _ _ _ .
    11 10 1 8 5 8 3 9
13. The sum of two negative numbers is _ _ _ _ _ _ _ _ .
    4 9 2 7 5 8 3 9
14. The sign of the sum of a positive and a negative number is the same sign as the number with the same sign as the number with the _ _ _ _ _ _ _ absolute value.
    2 6 9 7 5 9 6

Build Understanding

Using a number line to add and subtract numbers can become quite cumbersome. Rules that apply to all numbers make it easier.

If the numbers have the **same** sign, add the absolute values and use the sign of the addends.

If the numbers have **different** signs, subtract the absolute values and use the sign of the addend with the greater absolute value.

All the properties of addition of integers also apply to rational numbers. For every real number a, $a + (-a) = 0$. In the table, a, b, and c represent rational numbers.

Learning Styles

VISUAL LEARNERS These students may need to create a visual model to understand the concept of positive and negative. Have students use masking tape to make a walk-on number line along the classroom floor. Then have each student act as a human marker to complete simple computation problems with positive and negative numbers.

Additive inverses are also used when you subtract integers.

$-22 + 9 = -13$
$-22 - (-9) = -13$ (same result)

Subtracting an integer is the same as adding its inverse.

Addition Properties

Property	In Symbols	Examples
Closure	$a + b$ is unique	$2 - 4 = -2$ $\frac{1}{2} + \frac{7}{8} = 1\frac{3}{8}$
Commutative	$a + b = b + a$	$14 + 16 = 16 + 14$
Associative	$(a + b) + c = a + (b + c)$	$(\frac{1}{8} + \frac{1}{4}) + \frac{1}{2} = \frac{1}{8} + (\frac{1}{4} + \frac{1}{2})$
Identity	$a + 0 = 0 + a = a$	$-7.3 + 0 = 0 + (-7.3) = -7.3$
Inverse	$a + (-a) = 0$	$13.6 + (-13.6) = 0$ $\frac{1}{2} + (-\frac{1}{2}) = 0$

Example 1

Add or subtract.

a. $-3.6 - 4.9$ **b.** $3.5 + (-0.875)$

Solution

a. $-3.6 - 4.9 = -3.6 + (-4.9)$ Add the additive inverse of 4.9.
$= -8.5$ Add the numbers.

b. $3.5 + (-0.875) = 3.5 - 0.875$ Signs are different, so subtract absolute values.
$= 2.625$ Use the sign of the addend with the larger absolute value.

Evaluating expressions may involve adding or subtracting integers.

Example 2

Evaluate each expression when $a = -32$ and $b = 19$.

a. $a + b$ **b.** $b - a$ **c.** $a - b$

Solution

a. $a + b$
$-32 + 19$
-13

b. $b - a$
$19 - (-32)$
$19 + 32$
51

c. $a - b$
$-32 - 19$
$-32 + (-19)$
-51

Reading Math

Suppose you are to find the sum of $-5 + 5$. The addition involves a number and its opposite. The pair of numbers is called **additive inverses.** Additive inverses are used to state the **addition property of opposites.**

Estimating the answer *before* you start your calculations can give you a "ballpark" figure which you can use to check the reasonableness of your answer.

Example 3

FINANCE Jeremy had \$295.48 in his bank account. He deposited a check for \$196.68 and withdrew \$65.00. How much is in his account now?

Solution

Before solving the problem, estimate the answer. Begin by rounding the amounts.

\$295.48 → \$300.00 \$196.68 → \$200.00 \$65.00 → \$70.00

Then use the numbers to get an estimate of the total. \$300 + \$200 − \$70 = \$430

After completing the transactions, Jeremy will have about \$430.00 in his account.

Chalkboard Examples

Supplementary Example 1

a. $-\frac{3}{4} - \frac{1}{2}$ $-\frac{5}{4}$ or $-1\frac{1}{4}$

b. $2.6 + (-3.25)$ -0.65

Supplementary Example 2

Evaluate each expression when $a = -28$ and $b = 25$.

a. $a + b$ -3 **b.** $b - a$ 53

c. $a - b$ -53

Supplementary Example 3

Linda has \$126.74 in her checking account. She deposits \$84.66 and withdraws \$120.00. How much is in her account after these transactions? \$91.40

Extend the Lesson

REAL WORLD CONNECTIONS Working in small groups, have students select a stock from the business section of a newspaper. The group should chart the opening and closing prices of their stock each day for a week and write an expression for the difference between the opening and closing prices. At the end of the week, have them write the difference between the opening price on the first day and the closing price on the final day. Groups should compare their findings to determine the stock that realized the greatest gain as well as the stock that suffered the greatest loss.

Reteaching Worksheet 1-4

Name ______________ Date ______________

RETEACHING 1-4

ADDITION, SUBTRACTION AND ESTIMATION

When you add or subtract rational numbers, you use the same rules that apply to adding or subtracting integers.

ADDITION

Rule 1: When the numbers have the same sign, add the absolute values and give the sum the same sign as the addends.

Rule 2: When the numbers have different signs, subtract the absolute values and give the sum the sign of the number with the greater absolute value.

SUBTRACTION

Rule 3: When you subtract a number, add its opposite, or its additive inverse. Be sure to use either Rule 1 or Rule 2 when adding to find the sum.

Example 1

Find each answer.

a. $2.51 + (-0.36)$ **b.** $-\frac{7}{9} - \left(-\frac{5}{9}\right)$

Solution

a. $2.51 + (-0.36) =$ Use Rule 2.
$2.51 - 0.36 = 2.15$

b. $-\frac{7}{9} - \left(-\frac{5}{9}\right) =$ Use Rule 3.
$-\frac{7}{9} + \frac{5}{9} = -\frac{2}{9}$

Example 2

Evaluate each expression when $x = -1.6$ and $y = 2.1$.

a. $x - y$ **b.** $y + x$

Solution

a. $x - y$
$-1.6 - 2.1 = -3.7$

b. $y + x$
$2.1 + (-1.6) = 2.1 - 1.6 = 0.5$

EXERCISES

Add or subtract.

1. $0.23 - (-0.16)$ 0.39

2. $-3.58 + (-1.32)$ −4.9

3. $-1.58 + 2.21 - (-3.67)$ 4.3

4. $1\frac{3}{7} - 1\frac{5}{7}$ $-\frac{2}{7}$

5. $-3\frac{1}{5} - \left(-1\frac{3}{5}\right)$ $-1\frac{3}{5}$

6. $-1\frac{1}{10} - 2\frac{1}{5} - \left(-1\frac{1}{2}\right)$ $-1\frac{4}{5}$

Evaluate each expression when $c = -1.3$ and $d = 2.4$.

7. $c - d$ −3.7

8. $d + c$ 1.1

9. $-d - c$ −1.1

10. $-c + d$ 3.7

8 RETEACHING *LESSON 1-4*

Lesson Wrap-up

Quick Assessment

Ask the following question to determine if students understand the content presented in this lesson.

How can you use absolute value to find the sign of the sum of two rational numbers? The sign of the addend with the greater absolute value determines the sign of the sum.

Assignment Guide

Basic: 1–28, 31–58
Advanced: 1–58
Additional Practice: See Extra Practice Index on page 674.

Try These Exercises

Add or subtract.

1. $-12 + (-15)$ −27
2. $-3.7 - (-2.4)$ −1.3
3. $1\frac{1}{2} - \left(-9\frac{3}{4}\right)$ $11\frac{1}{4}$

Evaluate each expression when $m = -2.1$ and $n = 1.5$.

4. $m + n$ −0.6
5. $m - n$ −3.6
6. $-m - n$ 0.6
7. $n - m$ 3.6

8. **FINANCE** The balance in Dawn's checking account is $251.92. She wrote a check for $129.63. What's her new balance? $122.29

9. **ERROR ALERT** A submarine drove to a depth of 195 feet below the surface of the ocean. It then dove another 186 feet before climbing 229 feet. How many feet below the surface is the submarine? Clay solved the problem this way: $195 + (-186) + 229 = 238$ Is he correct? 235

10. **RECREATION** Carla is playing a game in which she needs 250 points to win. She started the game with no points and lost 180 points on the first round. She gained 165 points on the second round. How many points does she need to gain on the third round to win the game? No, the submarine begins by diving. The expression should read $-195 + (-186) + 229$ which equals −152.

11. **WRITING MATH** The cost of sending four packages is $4.86, $7.65, $2.12, and $8.61. You have $25. Do you have enough money to cover the total cost of sending the packages. How do you know?
Yes, round the amounts and add:
$5 + $8 + $2 + $9 = $24

Practice Exercises

Add or subtract.

12. $-9 + (-45)$ −54
13. $-23.6 + 19.7$ −3.9
14. $-1\frac{3}{4} - 3\frac{1}{3}$ −5.0833
15. $-28.2 - (-3.8)$ −24.4
16. $-10\frac{2}{3} + \left(-5\frac{1}{6}\right) + \left(-2\frac{1}{3}\right)$ $-18\frac{1}{6}$
17. $-7.6 + (-12.5) + 2.4 + 9.7$ −8
18. $-12 + 18 + 29 + (-16)$ 19

Evaluate each expression when $a = 27$ and $b = -16$.

19. $a - b$ 43
20. $a + b$ 11
21. $-a - b$ −11

22. **FINANCE** Kendra makes the following transactions to her savings account. Previous balance, $816.85; Deposit, $135; Withdrawal, $185; Deposit, $239.58; Withdrawal, $395.50. What is her new balance? $610.93

Replace each ■ with <, >, or =.

23. $-4\frac{3}{4} + \left(-6\frac{1}{2}\right)$ ■ $-18.0 - 6.75$ >
24. $-23.6 + (-44.7)$ ■ $14.8 + 53.5$ <
25. $-89 + (-77)$ ■ $-143 - (-35)$ <
26. $131.0 - (-112.6)$ ■ $272.0 - 33.7$ >

Extend the Lesson

CONNECTING TO PRIOR KNOWLEDGE Have students work in small groups. Have groups discuss, why, in some situations, it is best to round up, rather than to the nearest dollar or whole number. Have students brainstorm a list of at least five situations in which it is best to round up. Also discuss: Is there a situation in which it would be best to round down?

27. **CONSTRUCTION** Tom wants to construct a garden next to his house with 75 ft of fencing, using the house as one side of the fence. The lengths of the sides of the garden using the fencing are 25.6 ft, 18.9 ft, and 24.8 ft. How much fencing will he have left? 5.7 ft

28. **CHAPTER INVESTIGATION** Code makers often use layers of encryption to make the code harder to break. Work with the message and number line code you created in Lesson 1-2. Make the code more difficult by substituting operations that will lead from one letter to the next.

Example: If the word CAT was first encoded as $+5\ -3\ +12$, it becomes $5 - 8 + 15$. The operation $-\ 8$ shows the change from $+5$ to -3. The operation $+\ 15$ shows the change from -3 to $+12$.

Exchange encoded messages and number line keys with a partner. Reveal the first letter in the passage. Decode your partner's message.

Extended Practice Exercises

29. **CRITICAL THINKING** Evaluate $|-a - (a - b(a - b(a - b)))|$ when $a = -3$ and $b = -2$. 39

30. **TEMPERATURE** The temperature in Cheyenne, Wyoming was 3°F at 4:00 AM. During the next five hours, the temperature changed as follows:

5 AM, + 8°F; 6 AM, + 6°F; 7 AM, + 13°F;
8 AM, −10°F; 9 AM, −2°F.

What was the temperature at the end of the five hours? 18°F

Mixed Review Exercises

Refer to the diagram at the right. Let A = {0, 4, 5, 7}, B = {0, 3, 4, 6}, C = {1, 8} and U = {0, 1, 2, 3, ...9} (Lesson 1-3)

U A B C 5 4 0 7 3 6 2 1 8 9

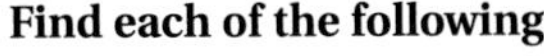

Find each of the following.

31. B' {1, 2, 5, 7, 8, 9}
32. C' {0, 2, 3, 4, 5, 6, 7, 9}
33. $A \cup B$ {0, 3, 4, 5, 6, 7}
34. $A \cap B$ {0, 4}
35. $A' \cap C$ {1, 8}
36. $B \cap C$ { } or ∅

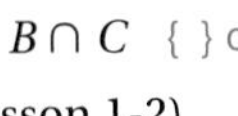

Evaluate each expression when $x = -3$. (Lesson 1-2)

37. $|x|$ 3
38. $|-x|$ 3
39. $-|x|$ −9
40. $-(x)$ 3
41. $-|-x|$ −3
42. $-(-x)$ −3
43. $-|-(-x)|$ −3
44. $-(-|x|)$ 3

Simplify each numerical expression. (Basic Math Skills)

45. $4 \times 3 + 8$ 20
46. $16 \div 2 - 2 \times 3$ 2
47. $5 \times 6 + 12 - 4$ 38
48. $25 \div 5 + (7 \times 3)$ 26
49. $(4^2 + 3) \times 2$ 38
50. $6^2 - 24 \div (5 + 3)$ 33
51. $5 + 9 \div 2 - 3$ $\frac{13}{2}$ or 6.5
52. $4^2 \div 2^2 + (2 \times 8)$ 20

Evaluate each expression when $x = -12$. (Lesson 1-2)

53. $0 - x$ 12
54. $-|x|$ −12
55. $x + |x|$ 0
56. $\frac{x}{|-x|}$ −1
57. $12 + x - |2x|$ −24
58. $x^2 + |x|^2 + x^2$ 2016

Extend the Lesson

REAL WORLD CONNECTION Have small groups of students research the changes in population for at least ten major cities in different parts of the country. Have them use the data from the last two census counts. Students should express the change in population for each city as both an addition and a subtraction expression. Have students draw conclusions about population trends.

Extra Practice Worksheet 1-4

Name ______ Date ______

EXTRA PRACTICE 1-4

ADDITION, SUBTRACTION AND ESTIMATION

EXERCISES

Add or subtract.

1. $-3 + 14$ 11
2. $-7 + (-6.7)$ −13.7
3. $3.245 - 8.18$ −4.935
4. $-2\frac{1}{3} + 1\frac{1}{2}$ $-\frac{1}{6}$
5. $-10.6 - 7.2$ −17.8
6. $-23.1 - (-25.3)$ 2.2
7. $4\frac{3}{5} - \left(-6\frac{7}{10}\right)$ $11\frac{3}{10}$
8. $-7\frac{4}{9} - 2\frac{1}{6}$ $-9\frac{11}{18}$
9. $-14 + 18 - (-13) + 12$ 29
10. $-16 + 9 - 7 - 11$ −25
11. $1.5 + 5.6 - 3.2 - (-2.1)$ 6
12. $-5\frac{5}{6} + 2\frac{4}{9} - 3\frac{1}{2}$ $-6\frac{8}{9}$

Evaluate each expression when $m = 15$ and $n = -12$.

13. $m - n$ 27
14. $-m - n$ −3
15. $n - m$ −27
16. $-m - (-n)$ 27
17. $-m + n$ −27
18. $-n - m$ −3
19. $m + n$ 3
20. $m - (-n)$ 27
21. $-n + m$ 27

Replace each ____ with <, >, or =.

22. $-15.3 + 3.5$ $<$ $12.6 - 9.6$
23. $-90 - (-76)$ $>$ $23 - 50$
24. $-2\frac{3}{4} - 6\frac{1}{2}$ $=$ $1\frac{1}{4} - 10\frac{1}{2}$
25. $-6\frac{2}{3} - 2\frac{1}{6}$ $<$ $3\frac{1}{3} - 7\frac{5}{6}$
26. Anna makes the following transactions to her checking account: previous balance, $654.89; deposit, $235.00; withdrawal, $146.50; check 1401, $56.89; deposit, $325.80. What is her new balance? $1,012.30
27. Ron has landscape timbers that are 4.5 ft,. 5.6 ft, 8.3 ft and 9.4 ft long. He wants to place the timbers end-to-end along the front of a flower bed that is 30 ft long. Will the timbers be long enough? If not, how much too short are they? If so, how much will he have left over? No; 2.2 ft
28. The temperature at midnight on Saturday was 15°F. Between midnight and noon, the temperature then fell 4 °, rose 1°, fell 8° and rose 3°. What was the temperature at noon on Saturday? 7°F

8 EXTRA PRACTICE *LESSON 1-4*

Enrichment Worksheet 1-4

Name ______ Date ______

ENRICHMENT 1-4

THE TOPPLING TOWER?

Consider five boards, each 1-ft long. A tower is built with the boards on the top of a table as shown in the illustration below. The fractions in the illustration give the amount of overlap in fractions of a foot. Do you think the stack of boards will topple over?

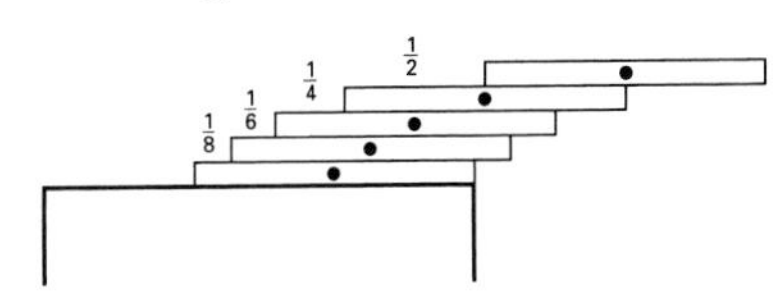

EXERCISES

Starting with the bottom board, how far does the right end of each board extend out over the right edge of the tabletop? Give all answers in fractions of feet.

1. Board 1 0 ft
2. Board 2 $\frac{1}{8}$ ft
3. Board 3 $\frac{7}{24}$ ft
4. Board 4 $\frac{13}{24}$ ft
5. Board 5 $1\frac{1}{24}$ ft

Now compute the distance from the right edge of the table to the center of each board. If the distance extends to the right, write it as a negative number.

6. Board 1 $\frac{1}{2}$ ft
7. Board 2 $\frac{3}{8}$ ft
8. Board 3 $\frac{5}{24}$ ft
9. Board 4 $-\frac{1}{24}$ ft
10. Board 5 $-\frac{13}{24}$ ft

11. Find the center of gravity of the stack of boards by averaging the answers to Exercises 6 through 10. What does this number tell you about whether or not the stack of boards will topple over?
$\frac{1}{10}$ ft; the center of gravity is $\frac{1}{10}$ of a foot to the left of the edge of the table. So, the stack is stable and will not topple. Students might enjoy building the stack to check.

12 ENRICHMENT *LESSON 1-4*

Skills Practice

Vocabulary Review

Lesson 1–3

union
intersection
universe
complement
compound inequality
Venn diagram
disjoint set
universal set

Lesson 1–4

addition property of opposites
additive inverse
associative property
closure property
commutative property
identity property
inverse property

Assignment Guide

All students: 1–56

Additional Practice: Refer to the Extra Practice Index on page 674 of this text.

Additional Answers

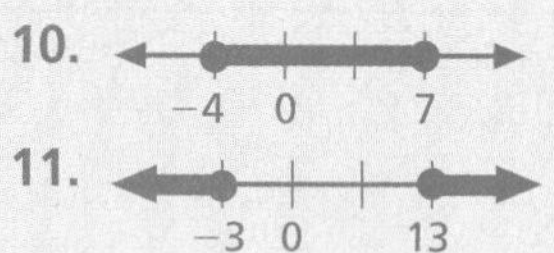

12. −11 −4

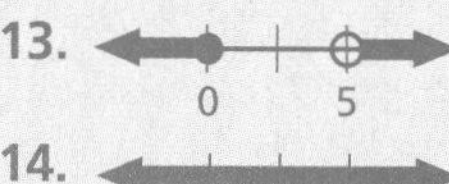

14. 6

15. −1

17. 10

Review and Practice Your Skills

Practice ■ Lesson 1-3

Let $U = \{c, o, m, p, u, t, e, r\}$, $A = \{c, o, m\}$, $B = \{o, m, p, u, t\}$, and $C = \{c, e, r\}$. Find each set.

1. A' $\{p, u, t, e, r\}$

2. $A \cup B'$ $\{c, o, m, e, r\}$

3. $A' \cap C$ $\{e, r\}$

4. $A \cap A'$ $\varnothing$

5. $B \cap C$ $\varnothing$

6. $(A' \cap B') \cup A$ $\{c, o, m, e, r\}$

7. $A' \cap (B \cup C)$ $\{p, u, t, e, r\}$

8. $(A \cup C) \cup (B \cap C)$ $\{c, o, m, e, r\}$

9. $(B' \cap C') \cup (B' \cup C')$ $\{c, o, m, p, u, t, e, r\}$

Use the replacement set of real numbers to graph the solution of each inequality.
See additional answers.

10. $x \le 7$ and $x \ge -4$

11. $x \le -3$ or $x \ge 13$

12. $-11 < x \le -4$

13. $x > 5$ or $x \le 0$

14. $x \ge 6$ or $x \le 6$

15. $x \le -1$ and $x \ge -1$

16. $x \le 3$ and $x > 9$ $\varnothing$

17. $x > 10$ or $x < 10$

18. $x < -3$ and $x > -3$ $\varnothing$

Draw a Venn diagram to solve this problem.

19. Eighty-nine students are registered for beginning art electives. Forty-five of these students are enrolled in Drawing I. Thirty-four are enrolled in Painting I, and forty are enrolled in Photography. Two students are enrolled in all three courses. Five students are enrolled in both Drawing I and Painting I. Six students are enrolled in both Painting I and Photography. How many students are enrolled in Drawing I and Photography, but not Painting I? 15

Practice ■ Lesson 1-4

Add or subtract.

20. $-3.6 + 4.2 + (-8.7) - 2.5 + 11.3$ 0.7

21. $133 - (-77) + (-110) + 100$ 200

22. $3\frac{3}{4} + 6\frac{7}{8} - \left(-2\frac{1}{4}\right) + 8\frac{1}{8} - 10$ 11

23. $10 - 9 + 8 - 7 + 6 - 5 + 4 - 3 + 2 - 1$ 5

Evaluate each expression when $a = -48$ and $b = 19$.

24. $-a - b$ 29

25. $-b + a$ −67

26. $|a| - |-b|$ 29

27. $a - b(b + a) + a$ 455

28. $a(a - b) + a(b - a)$ 0

29. $a + b - b - (-a) + (-b)$ −115

30. Sam makes the following transactions to his savings account. Previous balance, $1345.67; Deposits, $228.48, $74.36; Withdrawals, $435, $110. What is his new balance? $1103.51

For Exercises 31–33, estimate to find the answer.

31. Arlene wants to install a fence around a play area for her dog. She will use her house as one side of the fence. She has 110 ft of fencing available. The lengths of the sides of the play area using the fencing are 33.1 ft, 27.8 ft, and 27.8 ft. How much fencing will she have left? 21.3 ft

32. You have $20 to spend at the store. Do you have enough money to purchase five items, costing $4.47, $6.12, $2.76, $3.97, and $3.35? No

33. The cost of sending four packages is $7.82, $2.92, $5.29, and $11.13. Is $25 enough to cover the total cost of sending the packages? No

Learning Styles

TACTILE/KINESTHETIC LEARNER These students may benefit from working with two-color counters to add or subtract integers. If two-color counters are unavailable, cut 1-inch squares from cardstock and mark a red X on one side and a yellow X on the other side of each square. Use the yellow side for positive and the red side for negative.

Model simple operations using the counters. Stress that since yellow (+1) plus red (−1) = 0, students can remove an equal number of red and yellow counters to simplify their work.

Example: 5 + (−3)

Model: Y Y Y Y Y R R R After removing 3 red-yellow pairs: YY

Answer: 5 + (−3) = 2

24 Chapter 1 **Essential Mathematics**

Practice ■ Lesson 1-1–Lesson 1-4

For Exercises 34–39, determine if each statement is *true* or *false*. (Lesson 1-1)

34. $\{r,c,e\} \subset \{l,o,c,k,e,r\}$ true **35.** Every set is a finite set. false **36.** $\varnothing \in \{4,8,12,16\}$ true

37. $\{0,1,2,3,4\}$ has 16 subsets false **38.** $\{n,o\} \subset \{o,n\}$ true **39.** The null set has one element. false

For Exercises 40–42, determine if each statement is *true* or *false*. (Lesson 1-2)

40. $0 \in \{x|x$ is a natural number$\}$ false **41.** 13 is a rational number true **42.** $|-(-13.5)| = 21 - 7.5$ true

43. Evaluate $(y - 7) - |-y| + (-2 - y)$ when $y = 6$. (Lesson 1-2) -15

Find the following. $U = \{a,c,e,g,j,l,n,p,r,v,y,z\}$, $X = \{c,l,e,a,n\}$ **and** $Y = \{y,a,r,n\}$. (Lesson 1-3)

44. $X \cap Y$ $\{a, n\}$ **45.** $X \cup Y$ $\{c, l, e, a, n, y, r\}$ **46.** $X' \cap Y'$ $\{g, j, p, v, z\}$

47. $(X \cup Y)'$ $\{g, j, p, v, z\}$ **48.** $(X \cap Y) \cup (X' \cap Y')$ $\{a, n, g, j, p, v, z\}$ **49.** U' $\varnothing$

50. Graph the set $\{x|x$ is a real number less than 25 and greater than $-30\}$. (Lesson 1-3)

Add or subtract. (Lesson 1-4)

51. $3.5 + 3.05 + 3.055$ 9.605 **52.** $12\frac{1}{2} + \left(-6\frac{5}{8}\right)$ $5\frac{7}{8}$ **53.** $48 + (-48) - 48 - (-48) + 48$ 48

54. $|-17| + 15 - |15| + 17$ 34 **55.** $-6.9 + 9.5 - (-4.8)$ 7.4 **56.** $0 - \left(3\frac{1}{4} - 5\frac{1}{3}\right)$ $2\frac{1}{12}$

Mid-Chapter Quiz

1. Determine all possible subsets of $\{3, 5, 9\}$. The null set, the set itself, $\{3\}$, $\{5\}$, $\{9\}$, $\{3.5\}$, $\{5, 9\}$, $\{3, 9\}$
2. Define the solution set in roster notation: $\{x|x$ is a negative integer and $x > -5\}$. $\{-4, -3, -2, -1\}$

Graph each set of numbers on a number line. See additional answers.

3. all real numbers greater than or equal to -5
4. all natural numbers less than 10 but greater than 2
5. all numbers with an absolute value greater than 3 but less than 7

Let $U = \{w,o,m,e,n\}$, $M = \{n,o,w\}$, **and** $T = \{w,o,n\}$. **Find each union or intersection.**

6. $M \cup T$ $\{n, o, w\}$ **7.** $(M \cup T)'$ $\{m, e\}$ **8.** $M \cap T$ $\{n, o, w\}$ **9.** $(M \cap T)'$ $\{m, e\}$

Solve.

10. $-22 - (-39)$ 17 **11.** $46.8 + (-21.5) + 18.6$ 43.9

12. $-2\frac{1}{4} + 7\frac{3}{8} + \left(-6\frac{3}{4}\right)$ $-1\frac{5}{8}$ **13.** $-0.76 + 2.5 - (-0.82)$ 4.08

Evaluate each expression when $a = 14$ and $b = -22$.

14. $a + b$ -8 **15.** $a - b$ 36 **16.** $-a - b$ 8 **17.** $-a + (-b)$ 8

18. Joel made the following transactions to his account: previous balance, \$317.73; deposits, \$123.25, \$55.62; withdrawal, \$228.98. What is his new balance? \$267.62

Chalkboard Examples

Example from Lesson 1-3

Let $U = \{c, i, p, h, e, r\}$, $A = \{i, e, r\}$ and $B = \{c, h, r\}$. Find each set.

1. A' $\{c, p, h\}$
2. $A \cup B$ $\{i, e, r, c, h\}$
3. $A \cap B$ $\{r\}$

Example from Lesson 1-4

Evaluate each expression when $x = -9$ and $y = 5$.

1. $x + y - (-y)$ 1
2. $(x + y) - (x - y)$ 10

Mid-Chapter Quiz

Correlation Chart

Question Number	Lesson Number
1–2	1-1
3–5	1-2
6–9	1-3
10–18	1-4

3.

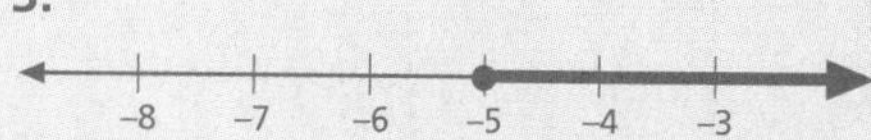

4.

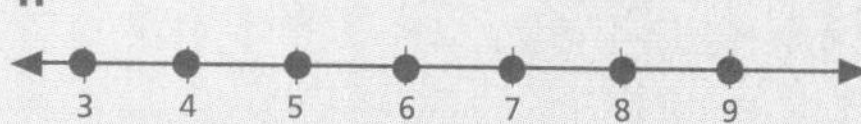

5.

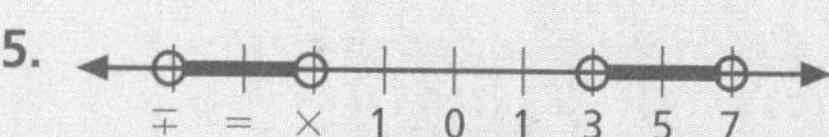

Teaching Strategies

ESL/LEP Have students brainstorm a list of situations in which numerical changes could be represented by positive and negative integers. You may want to have students write a possible problem for each situation. Students can exchange problems for problem-solving practice. Some possible answers: change in score during a card game, temperature changes, depositing and withdrawing money from an account, changes in stock prices.

Lesson Planning

NCTM Standards/Strands

- Number & Operation
- Patterns, functions, & Algebra
- Problem Solving
- Reasoning and Proof
- Communication
- Connections
- Representation

Vocabulary

reciprocals
multiplicative inverse

Materials Needed

paper/pencil calculator

Lesson Resources

Warm-Up Transparency 2
Transparency RF-2
Reteaching 1-5
Extra Practice 1-5
Enrichment 1-5

Assignment Guide

Basic: 1–40, 46–65
Enriched: 1–65

Getting Started

5-Minute Warm-up

Use mental math to solve.

1. $8 \times 3 \times 10$ 240
2. $360 \div 30$ 12
3. $550 \div 25$ 22
4. $9 \times 6 \times 5$ 270

Introduction to Lesson 1-5

After students complete the opening activity, have them write a multiplication and a division sentence that models each square of the diagram. Remind them that the numbers they use do not have to be integers.

1-5 Multiplication and Division

Goals ■ Multiply and divide rational numbers.

Applications Science, Food service, Health, Fitness

1. Use the diagram to represent the product or quotient of two numbers. Begin at zero. Let the horizontal values represent the first number and the vertical values represent the second number.

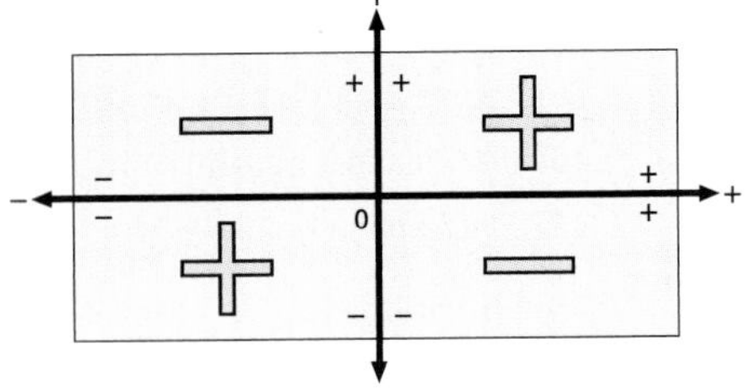

Square	First Number	Second Number	Product or Quotient
1	+	+	□
2	−	+	□
3	−	−	□
4	+	−	□

2. What pattern did you use to complete the sentences?
   a. If two positive numbers are multiplied together, what is the sign of the product? +
   b. If two negative numbers are multiplied together, what is the sign of the product? +
   c. If a positive number and a negative number are multiplied together, what is the sign of the product? −

Build Understanding

When you multiply and divide with rational numbers, you follow the same rules you use when you multiply and divide with integers.

The product or quotient of two numbers with the *same* signs is positive.

The product or quotient of two numbers with *different* signs is negative.

All the properties of multiplication of integers also apply to rational numbers. Multiplication and division are inverse or opposite operations.

The associative and commutative properties of multiplication may be used to arrange factors in the order easiest for multiplying.

Teaching Strategies

COOPERATIVE LEARNING To reinforce the fact that the sum of two negative numbers is negative, while the product of two negative numbers is positive, write a set of negative numbers on index cards. Also create a "+" card and a "×" card. Have a volunteer select two cards and first create an addition sentence, then create a multiplication sentence. Have the class, as a group, give answers.

In the table, a, b, and c represent rational numbers.

Two numbers whose product is 1 are **reciprocals** of each other. Another name for reciprocal is **multiplicative inverse.** Zero does *not* have a reciprocal.

Multiplication Properties

Property	In Symbols	Examples
Closure	ab is unique	$(-5)\cdot(4)=-20$ is unique
Commutative	$ab=ba$	$25\cdot 5=5\cdot 25$
Associative	$(ab)c=a(bc)$	$\left(\frac{3}{8}\cdot\frac{2}{5}\right)\cdot\frac{10}{9}$ $=\frac{3}{8}\cdot\left(\frac{2}{5}\cdot\frac{10}{9}\right)$
Identity	$a\cdot 1=1\cdot a=a$	$26\cdot 1=1\cdot 26=26$
Property of Zero	$a\cdot 0=0\cdot a=a$	$83.2\cdot 0=0\cdot 83.2=0$
Inverse	$a\cdot\frac{1}{a}=\frac{1}{a}\cdot a=1,$ $a\neq 0$	$5\cdot\frac{1}{5}=\frac{1}{5}\cdot 5=1$

Example 1

Find each product or quotient.

a. $\left(\frac{1}{7}\right)\left(-\frac{5}{6}\right)\left(-\frac{7}{25}\right)$ **b.** $(-5)(-0.9)(0.4)(-20)$

c. $-8.4\div 7$ **d.** $-\frac{3}{29}\div\left(-\frac{8}{9}\right)$

Solution

a. $\left(\frac{1}{7}\right)\left(-\frac{5}{6}\right)\left(-\frac{7}{25}\right)=\left[\left(\frac{1}{7}\right)\left(-\frac{7}{25}\right)\left(-\frac{5}{6}\right)\right]$

$=\left(-\frac{1}{25}\right)\left(-\frac{5}{6}\right)$

$=\left(\frac{1}{30}\right)$

b. $(-5)(-0.9)(0.4)(-20)$

$=[(-0.9)(-0.4)][(-5)(20)]$

$=(0.36)(-100)$

$=-36$

c. Since the signs are different, the quotient is negative. $-8.4\div 7=-1.2$

d. Since the signs are the same, the quotient is positive. $-\frac{32}{9}\div\left(-\frac{8}{9}\right)=4$

Remember to use the order of operations when simplifying expressions.

Example 2

Evaluate each expression when $q=4$, $r=-0.5$, and $s=-8$.

a. rs **b.** $-qs+r$ **c.** $r\left(\frac{s}{q}\right)$

Solution

a. rs

$(-0.5)(-8)$

$=4$

b. $-qs+r$

$(-4)(-8)+(-0.5)$

$=32+(-0.5)$

$=31.5$

c. $r\left(\frac{s}{q}\right)$

$=-0.5\left(\frac{-8}{4}\right)$

$=-0.5(-2)$

$=1$

Check Understanding

Name the property.

1. $2+(5+7)=$ $2+(7+5)=$
2. $2(5\cdot 13)=(2\cdot 5)\,13$
3. $10\cdot 1=10$
4. $92\cdot 0=0$
5. $25\cdot\left(\frac{1}{25}\right)=1$

1. commutative
2. associative
3. identity
4. property of zero
5. inverse

Example 3

Simplify.

a. $-0.8+3.9\div 3$ **b.** $(-4.5+7)\div 5$ **c.** $5+25\div 5\div 5\cdot 2-16\div 8$

Chalkboard Examples

Supplementary Example 1

Find each product or quotient.

a. $\left(\frac{1}{2}\right)\left(-\frac{3}{8}\right)\left(-\frac{1}{4}\right)$ $\frac{3}{64}$

b. $(-3)(-0.2)(5)(-0.1)$ -0.3

c. $-9.6\div 8$ -1.2

d. $-\frac{48}{7}\div\left(-\frac{6}{7}\right)$ 8

Supplementary Example 2

Simplify.

a. $8.5\div 5+(-1.5)$ 0.2

b. $(-1.3+0.8)\div 0.2$ -2.5

Supplementary Example 3

Evaluate each expression when $a=6$, $b=-0.25$ and $c=-4$.

a. bc 1 **b.** $-ab-c$ 2.5

c. $b\left(\frac{a}{c}\right)$ 0.375

Reteaching Worksheet 1-5

Name ______________________ Date ____________

RETEACHING 1-5

MULTIPLICATION AND DIVISION

When you multiply or divide two rational numbers, you use the same rules that apply to multiplying and dividing with integers.

Rule 1: When the numbers have the same sign, find the product or quotient of their absolute values. The product or quotient is always positive.

Rule 2: When the numbers have different signs, find the product or quotient of their absolute values. The product or quotient is always negative.

Example 1

Find each answer.

a. $(-6.1)(-3.2)(10)$ **b.** $-\frac{1}{12}\div\left(-\frac{3}{8}\right)$

Solution

a. $(-6.1)(-3.2)(10)$

$=(19.52)(10)$ Use Rule 1.

$=195.2$

b. $-\frac{1}{12}\div\left(-\frac{3}{8}\right)=$ Use Rule 2.

$-\frac{1}{3}\cdot-\frac{2}{3}=\frac{2}{9}$

Example 2

Evaluate each expression when $x=-0.3$, $y=-1.8$, and $z=\frac{1}{3}$.

a. xz **b.** $z\left(\frac{y}{x}\right)$

Solution

a. xz

$(-0.3)\left(\frac{1}{3}\right)=$

-0.1

b. $z\left(\frac{y}{x}\right)$

$\left(\frac{1}{3}\right)\left(\frac{-1.8}{-0.3}\right)=$

$\left(\frac{1}{3}\right)(6)=2$

EXERCISES

Find each answer.

1. $(-7.2)(1.8)$ -12.96
2. $-13.5\div 0.5$ -27
3. $-1.25\div -2.5$ 0.5
4. $(-1.8)(-3.5)$ 6.3
5. $-2\frac{1}{2}\div -1\frac{3}{4}$ $1\frac{3}{7}$
6. $\left(1\frac{1}{2}\right)\left(-\frac{1}{3}\right)$ $-\frac{1}{2}$
7. $-4\frac{1}{8}\div 2\frac{1}{4}$ $-1\frac{5}{6}$
8. $\left(-4\frac{1}{2}\right)\left(2\frac{1}{12}\right)$ $-9\frac{3}{8}$
9. $(1.84)(-0.1)(-3)$ 0.552
10. $(-2.5)(-10)(-0.2)$ -5
11. $(-0.5)(-0.5)\div 10$ 0.025

Evaluate each expression when $x=-0.1$, $y=10$, and $z=-\frac{1}{5}$.

12. xyz 0.2
13. xz 0.02
14. $xy\div z$ 5
15. $yz\div -x$ -20

10 RETEACHING *LESSON 1-5*

Teaching Strategies

TEAM LEARNING Working in pairs or small groups, challenge students to use guess-and-check and work-backward strategies to identify at least ten different sets of three factors whose product is -100. Some possible answers include: $(-10, 10, 1)$; $(-10, -10, -1)$; $(2, 5, 10)$; $(-2, -5, -10)$; $(-5, 4, 5)$; $(-5, -4, -5)$; $(-25, -2, -2)$; $(-2, 2, 25)$; $(-50, -2, -1)$; $(-1, -5, -20)$)

Lesson Wrap-up

Quick Assessment

Ask the following questions to determine if students understand the content presented in this lesson.

1. What is the product of a negative rational number and its multiplicative inverse? Give an example. 1, $-\frac{3}{4}\left(-\frac{4}{3}\right) = \frac{12}{12} = 1$
2. What is the sign of the product of four negative factors? of five negative numbers? positive, negative

Assignment Guide

Basic: 1–40, 46–66
Advanced: 1–66
Additional Practice: See Extra Practice Index on page 674.

Solution

a. $-0.8 + 3.9 \div 3$
$= -0.8 + 1.3$
$= 0.5$

b. $(-4.5 + 7) \div 5$
$= 2.5 \div 5$
$= 0.5$

c. $5 + 25 \div 5 \div 5 \cdot 2 - 16 \div 8$
$= 5 + 5 \div 5 \cdot 2 - 16 \div 8$
$= 5 + 1 \cdot 2 - 2$
$= 5 + 2 - 2$
$= 5$

You will need to use the properties of multiplication along with the rules for multiplying and dividing integers to evaluate expressions.

Try These Exercises

Find each product or each quotient.

1. $(-1.7)(2.9)$ -4.93
2. $-4\frac{1}{4} \div \frac{3}{4}$ $-5\frac{2}{3}$
3. $3\frac{1}{8}\left(-\frac{8}{25}\right)$ -1
4. $-98 \div -14$ 7
5. $(3.2)(-5)(2)$ -32
6. $225 \div -15$ -15

Simplify.

7. $6 \div (-3) + 7.1$ 5.1
8. $-4(-9) - (-30.5)$ 66.5
9. $(7.6 - 1.6)(-0.2)$ -1.2
10. $2[-7.7 \div (-99.5 + 98.5)]$ 15.4

Evaluate each expression when $a = 8$, $b = 0.5$, and $c = \frac{1}{4}$.

11. ac 2
12. $(a + b)c$ $2\frac{1}{8}$
13. $a + bc$ $8\frac{1}{8}$
14. $ab \div c$ 16

15. TEMPERATURE From 1 AM to 5 AM, the temperature dropped 16°F. What was the average change in temperature per hour? -4°F

16. WRITING MATH When is the product of three rational numbers negative? When either one or three of the factors is negative

Practice Exercises

Perform the indicated operations.

17. $8.4(-2.6)$ -21.84
18. $2.7(5)(-2)$ -27
19. $-48 \div (-1.2)$ 40
20. $2\frac{1}{4} \div \left(-\frac{3}{8}\right)$ $\frac{-4}{3}$
21. $(-125) \div (0.25)$ -500
22. $2\frac{1}{5} + (4)(-6)$ 21.6
23. $5.4 \div 1.8 - 2.6$ 0.4
24. $-25 \div (-10) + \left(-3\frac{1}{2}\right)$ 1
25. $8\frac{1}{3} \div (-5) + 6$ 4.334

Evaluate each expression when $m = -5$, $n = -1.6$, and $p = \frac{5}{8}$.

26. mn 8
27. $m + np$ -6
28. $m(n + p)$ 4.875
29. $(m + n) \div p$ -10.56

Teaching Strategies

TEAM LEARNING Have students work in pairs. On twelve different index cards, the students should write six positive and six negative rational numbers. Have them shuffle the cards and place them face down on a flat surface. One student then turns over two cards and calculates the product of the numbers displayed. The product becomes this player's score. The other student then turns over two other cards, calculating their product. Play continues in this manner until all cards have been used. The player with the greatest total score wins.

30. **SCIENCE** At the beginning of an experiment, the temperature of a solution was 68°C. During a 12-hour period, the temperature dropped 1.5 degrees each hour. Write an expression to show the temperature change during the entire time period. $12(-1,5) = -18$

CALCULATOR Use a calculator to perform the indicated operation. Express your answer as a decimal rounded to the nearest tenth.

31. $-1\frac{3}{16} \cdot 5.16$ −6.1
32. $\frac{4}{9} \cdot \frac{3}{16}$ 0.1
33. $\frac{3}{8}\left(-\frac{5}{8}\right)$ −0.2
34. $-10\frac{7}{8} \div (-1.3)$ 8.4
35. $4\frac{9}{32} - (-3.8502)$ 8.1
36. $\left(5\frac{8}{9}\right)^2$ 34.7
37. **FOOD SERVICE** Sharon is planning a party for 25 people. If she caters the food, the cost will be \$8.75 per person (tax and tip are included). Sharon has \$200 to spend. Does she have enough money for 25 guests? If not, how many guests can she afford to invite? no; 22
38. **COMMUNICATION** Martin is writing a magazine article. He will be paid \$0.12 per word or \$100, whichever is greater. How many words long will the article need to be in order for Martin to earn more than \$100? 834 words

DATA FILE For Exercises 39–40, use the Data Index on pages 632–633 to find information about calories used by people of different weights.

39. How many calories does a 154-lb person use playing racquetball for 15 min? 193.75 calories
40. Denise consumed 470 calories at lunch. If she weighs 110 lb, how long must she walk at 3 mi/hr to use up those calories? 2 hours

Extended Practice Exercises

Replace each ■ with +, −, ×, or ÷ to make a true sentence.

41. (−9 ■ 5) ■ (−4) = 1   + ÷
42. −6 ■ 12 ■ 8 = −4   ÷ ×
43. 21 ■ (−7) ■ 3 + 5 ■ −5 = −1   ÷ ÷ ÷
44. (64 ■ (−8)) ■ 9 ■ (−8) = 0   − ÷ +
45. Write 0.016 as a quotient of two integers. $\frac{1}{1000} \times 16 = \frac{16}{1000} = \frac{4}{250} = \frac{2}{125}$

Mixed Review Exercises

Evaluate each expression when $a = 12$ and $b = -9$. (Lesson 1-4)

46. $a + b$ 3
47. $b - a$ −21
48. $2a - 3b$ 51
49. $3a \div b$ −4
50. $(b + 6) \div a$ $-\frac{1}{4}$
51. $4a + 8b$ −24
52. $2b - 6a$ −90
53. $4b \div 2a$ $-\frac{3}{2}$
54. $2(a + 4) - 3b$ 59
55. $3(b - 1) + a$ −18
56. $2(b + 1) - 4a$ −24
57. $9a + 12b$ 0

Let $U = \{0, 1, 2, 3, ... 9\}$. Let $A = \{x | x$ is a whole number such that $x \geq 2$ and $x \leq 5\}$. Let $B = \{0, 2, 4, 6, 8\}$, $C = \{1, 3, 5, 7, 9\}$, and $D = \{0, 3, 5, 8\}$. Find each of the following. (Lesson 1-3)

58. D' {1, 2, 4, 6, 7, 9}
59. $C' \cup A$ {0, 2, 3, 5, 6, 8}
60. $A \cap B$ {2, 4}
61. $D \cup A$ {1, 2, 3, 4, 5, 6, 7, 9}
62. $(A \cap B) \cup D$ ∅
63. $(B' \cap A) \cup C$ {1, 3, 5, 7, 9}
64. $(C \cap D)' \cup A$ U
65. $(B \cap D) \cup (A' \cap C)$ {0, 1, 7, 9}

Teaching Strategies

REAL WORLD CONNECTION Science experiments often involve changes in temperature. Have students work in pairs to solve the following problem. At the beginning of an experiment, the temperature of a solution was 68°C. During a 14-hour period, the temperature dropped 1.5 degrees each hour.

1. Write an expression to show the temperature change during the entire time period. $12(-1.5) = -21$
2. What was the temperature of the solution at the end of the experiment?. 47°C

Extra Practice Worksheet 1-5

Name ____________ Date ____________

EXTRA PRACTICE 1-5

Multiplication and Division

Exercises

Perform the indicated operations.

1. $6(-3)$ −18
2. $-14 \div 7$ −2
3. $-5(-8)$ 40
4. $-63 \div (-9)$ 7
5. $3(4.5)(-2.8)$ −37.8
6. $-100 \div (-2.5)$ 40
7. $-1\frac{1}{2} \div 2\frac{1}{4}$ $-\frac{2}{3}$
8. $\frac{1}{3}\left(-\frac{5}{6}\right)\left(3\frac{3}{10}\right)$ $-\frac{11}{12}$
9. $8.5 + (-9)(-6.3)$ 65.2
10. $5\frac{9}{10} \div (-10) + (-2)$ $-2\frac{59}{100}$
11. $6.3 \div 0.2 - 5.6$ 25.9
12. $3[-2.2 + 3.2 \div (-0.8)]$ −18.6
13. $(3.5 - 8.9)(-1.6)$ 8.64
14. $(-3.7 - 4.1) \div (-0.2)$ 39

Evaluate each expression when $a = -4$, $b = -2.2$, and $c = \frac{4}{5}$.

15. ab 8.8
16. bc −1.76
17. $b + ac$ −5.4
18. $ab + bc$ 7.04
19. $a(b + c)$ 5.6
20. $ab - c$ 8
21. $(a + b) \div c$ −7.75
22. $c(a - b)$ −1.44
23. $a \div b + c$ 2.61818...
24. $b \div c + a$ −6.75
25. $\frac{a}{c} + b$ −7.2
26. $c - \frac{b}{a}$ 0.25
27. On Sunday, the temperature dropped 24° in 6 hours. What was the average change in temperature per hour? −4°
28. Raul has \$400 to spend on stereo equipment. He buys two speakers for \$78.99 each and a receiver for \$198.59. How much money does he have left? \$43.43
29. A plane descended an average of 200 feet per second for 15 seconds. How far had the plane descended at the end of the 15 seconds? 3000 ft
30. At noon, the temperature was 59°. The temperature rose an average of 2° per hour for the next 6 hours. What was the temperature at 6:00 p.m.? 71°

10   EXTRA PRACTICE *LESSON 1-5*

Enrichment Worksheet 1-5

Name ____________ Date ____________

ENRICHMENT 1-5

The Non-Zero Digits and Fractions

Here are two problems involving the nine non-zero digits from 1 through 9. A fractions calculator may not help you much here, as all the numerators and denominators have 3, 4, or 5 digits!

Exercises

As a warm-up, express each of the fractions from $\frac{1}{2}$ through $\frac{1}{9}$ using the nine digits. For each fraction, you must use all nine digits once, and only once.

1. $\frac{1}{2} = \frac{6729}{13,458}$
2. $\frac{1}{3} = \frac{5823}{17,469}$
3. $\frac{1}{4} = \frac{3942}{15,768}$
4. $\frac{1}{5} = \frac{2697}{13,485}$
5. $\frac{1}{6} = \frac{2943}{17,658}$
6. $\frac{1}{7} = \frac{2394}{16,758}$
7. $\frac{1}{8} = \frac{3187}{25,496}$
8. $\frac{1}{9} = \frac{6381}{57,429}$

Now, express the number 100 as a mixed number. (The fraction part will be improper.) There are 11 different solutions. Find them all. Here are a few hints: in ten of the solutions, the whole number part is between 80 and 97. In the eleventh solution, the whole number part equals 3. Two sets of three solutions each have the same whole number part, 96 or 91.

9. $100 = 96\frac{2148}{537}$
10. $100 = 96\frac{1752}{438}$
11. $100 = 96\frac{1428}{357}$
12. $100 = 94\frac{1578}{263}$
13. $100 = 91\frac{7524}{836}$
14. $100 = 91\frac{5823}{647}$
15. $100 = 91\frac{5742}{638}$
16. $100 = 82\frac{3546}{197}$
17. $100 = 81\frac{7524}{396}$
18. $100 = 81\frac{5643}{297}$
19. $100 = 3\frac{69,258}{714}$

14   ENRICHMENT *LESSON 1-5*

Lesson Planning

NCTM Standards/Strands

- Number & Operation
- Patterns, functions, & Algebra
- Problem Solving
- Reasoning and Proof
- Communication
- Connections
- Representation

Vocabulary

cell spreadsheet
formula

Materials Needed

paper/pencil calculator
spreadsheet software

Lesson Resources

Warm-Up Transparency 2
Transparency TK-1, RF-3
Reteaching 1-6
Enrichment 1-6

Assignment Guide

Basic: 1–28
Enriched: 1–28

Getting Started

5-Minute Warm-up

Find the value of $P = 2l + 2w$

1. $l = 12$ in., $w = 3$ in 30 in
2. $l = 40$ cm, $w = 25$ cm 130 cm
3. $l = 8$ yd, $w = 6$ yd 28 yd

Use the Five-step Plan and the Strategy

THE FIVE-STEP PLAN **Read**—ask questions of the students to help them understand the problem. **Plan**—guide students to related problems and previously mastered skills and strategies. **Solve**—students solve problem on their own. **Answer**—write solution in format that answers the question. **Check**—review work, check for reasonableness and review strategy.
THE STRATEGY *Use technology*—this strategy helps students perform time-consuming or repetitive calculations quickly.

1-6 Problem Solving Skills: Use Technology

The use of calculators and computers is becoming routine in solving real-world problems. Graphing calculators can be used for a wide range of operations. They can also be used to store data, formulas and equations.

A computer spreadsheet is an application that stores data in cells formed by vertical columns and horizontal rows. The cells are identified by a row and a column. The columns are named by letters and the rows by numbers. A spreadsheet will do calculations automatically by using a "hidden" formula.

Both calculators and computers follow the order of operations. Always think through the order of operations carefully when writing formulas.

Problem Solving Strategies

- Guess and check
- Look for a pattern
- Solve a simpler problem
- Make a table, chart or list
- Use a picture, diagram or model
- Act it out
- Work backwards
- Eliminate possibilities
- ✔ Use an equation or formula

PROBLEM

A printer is printing the bylaws of a civic organization. The organization wants 3 sizes, each of which has a width that is half the length. If the lengths they have chosen are 1 ft, 2 ft, and 3 ft, find each width, perimeter, and area.

Solve the Problem

To find the values of the perimeter and area for a rectangle with length equal to whole numbers 1, 2, and 3 and widths equal to half the length, a spreadsheet can be used. The rule or formula for generating each value is shown in each cell. Note that Row 1 and Column A are used for headings. The other cells are used to store data and formulas.

	A	B	C	D	E
1		Length	Width	Perimeter	Area
2	R1	1	B2 * 0.5	2 * B2 + 2 * C2	B2 * C2
3	R2	B2 + 1	B3 * 0.5	2 * B3 + 2 * C3	B3 * C3
4	R3	B3 + 1	B4 * 0.5	2 * B4 + 2 * C4	B4 * C4
5	R4	B4 + 1	B5 * 0.5	2 * B5 + 2 * C5	B5 * C5
6	R5	B5 + 1	B6 * 0.5	2 * B6 + 2 * C6	B6 * C6

After entering the formulas and data in the spreadsheet, your results should look like this.

	A	B	C	D	E
1		Length	Width	Perimeter	Area
2	R1	1	0.5	3	10.5
3	R2	2	1	6	2
4	R3	3	1.5	9	4.5
5	R4	4	2	12	8
6	R5	5	2.5	15	12.5

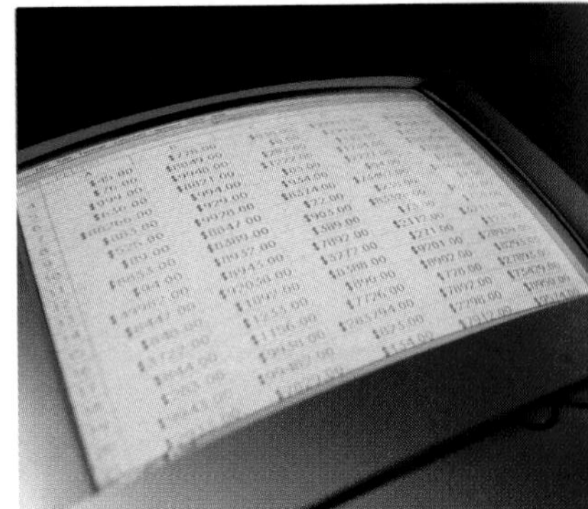

Learning Styles

VISUAL LEARNERS Students may benefit from first drawing a table of the spreadsheet and using a calculator to fill in cells. Then the table can be transferred to the spreadsheet.
When solving area and perimeter problems, visual learners may find it helpful to make a sketch of the object and label its sides before beginning calculations.

Try These Exercises

Find the value in each cell named.

	A	B	C	D
1	1	2	5	10
2	2 * A1	2 * A2	B2 + 3	2 * C2
3	2 * A2	2 * A3	B3 + 3	2 * C3
4	2 * A3	2 * A4	B4 + 3	2 * C4

1. A2, B2, C2, D2 2 4 7 14
2. A3, B3, C3, D3 4 8 11 22
3. A4, B4, C4, D4 8 16 19 38
4. Write a formula to find the difference between twice a number (x) and the sum of 5 and that number. What is the result if 8 is the number? If 15 is the number? $2x - (5 + x)$; 3; 10

5. **WRITING MATH** The civic organization decides to print their bylaws in a larger size with a 4 ft length. Explain how you would use a calculator to find the perimeter and area of the bylaws. Store the value 4 for L. Enter these formulas using the variable L for length. Perimeter 2L + 2L * 0.5, Area: L * L * 0.5. You may have written the formulas differently. The solutions should be perimeter 12 ft and area 8 sq ft.

Five-step Plan

1 Read
2 Plan
3 Solve
4 Answer
5 Check

Practice Exercises

SEWING You are making school pennants in the shape of isosceles triangles. The sizes will vary so that the height of each triangle is three times the length of the base. You also know that an additional 10% of the material will be used for hems or waste.

6. Write a formula for a calculator that can be used to find the material needed for any base (B). $\frac{1}{2}B * 3B + 0.1 * \frac{1}{2}B * 3B$ or $\frac{3}{2}B^2$ or $0.1\left(\frac{3}{2}B^2\right)$

Make a spreadsheet to find the material needed (including 10% extra) for each of the following bases.

7. 9 in. 133.65 in.^{2}
8. 1 ft 1.65 ft^2
9. 1 yd 1.65 yd^2
10. 5 ft 41.25 ft^2

DATA FILE For Exercises 11–13, use the Data Index on pages 632–633 to find information on customary measurements.

11. For the pennant with a 9-in. base, would more than or less than 1 square foot of material be used? less than
12. For the pennant with a 5-ft base, about how many square yards of material would be needed? $2.\overline{77}$ sq yds
13. For the pennant with a 1-yd base, about how many square feet of material would be needed? 9 sq ft

14. **CHAPTER INVESTIGATION** Using the original message you wrote in Lesson 1-2, make a spreadsheet that will automatically calculate the percent for the letters E, T, A, O, N, R, I, S and H. You will need to know the total number of letters in the passage and the number of times each letter appears. Compare your percents to the table on page 5.

Mixed Review Exercises

Perform the indicated operations. Round your answers to the nearest hundredth if necessary. (Lesson 1-5)

15. $4.3(-2.8)$ −9.89
16. $9.60 \div 3.2$ 3
17. -7.65×4.04 −30.91
18. $-8.54 \div 2.91$ −2.93
19. $-3.68(-6.14)$ 22.60
20. $-8.5 \div 2.47$ 3.44
21. 0.52×-5.73 −2.98
22. $38 \div 1.7$ −22.35

Chalkboard Examples

Supplementary Problem

On a landscaping plan, a flower garden is 8 feet wide. It can be either 12, 15, or 18 feet in length depending on the owner's preference. Find the perimeter and area of the garden for each length.
Perimeter: 40 ft, 46 ft, 52 ft;
Area: 96 ft^2, 120 ft^2, and 144 ft^2.

Lesson Wrap-up

Quick Assessment

Ask the following questions to determine if students understand the content presented in this lesson.

1. How are cells named in a spreadsheet? by letter of column followed by number of row
2. How do you make a spreadsheet program perform a calculation in a particular cell? by inserting a formula that combines cell numbers and arithmetic operations
3. What formula would be entered in cell C3 if you want C3 to be the sum of A3 and twice B3? A3 + (2 × B3)

Reteaching Worksheet 1-6

Name ______________________ Date ____________

Reteaching 1-6

Problem Solving Skills: Use Technology

A spreadsheet is composed of columns and rows, whose intersections are called cells. You may enter data into each cell or you can enter a formula into the cell that will perform calculations on data in other cells.

Example

A company budget estimates gross profit (sales less cost of sales) for one of its products. Each unit sells for $15.00. Cost for one unit is $9.50. Find gross profit for sales from 1000 units to 20,000 units using 1000 unit intervals.

a. Make a spreadsheet. Show the first five lines (including headings).
b. Find the value in each cell in your spreadsheet.

Solution

a.

	A	B	C	D
1	Units	Sales	Cost	Profit
2	1000	A2*15	A2*9.5	B2 – C2
3	A2 + 1000	A3*15	A3*9.5	B3 – C3
4	A3 + 1000	A4*15	A4*9.5	B4 – C4
5	A4 + 1000	A5*15	A5*9.5	B5 – C5

b.

	A	B	C	D
1	Units	Sales	Cost	Profit
2	1000	15,000	9500	5500
3	2000	30,000	19,000	11,000
4	3000	45,000	28,500	16,500
5	4000	60,000	38,000	22,000

Note: * represents the multiplication sign.

Exercises

1. Find the value in each cell.

	A	B	C	D
1	20	A1*5	B1*0.3	B1 – C1
2	A1 + 20	A2*5	B2*0.3	B2 – C2
3	A2 + 20	A3*5	B3*0.3	B3 – C3

	A	B	C	D
1	20	100	30	70
2	40	200	60	140
3	60	300	90	210

2. Complete the first three rows of this spreadsheet and find the value for each cell. Individual gifts are 60% of corporate gifts and government grants are 20% of corporate gifts. Use $5000 intervals to find total donations if corporate gifts range from $5000 to $100,000.

	A	B	C	D
1	Corp	Ind	Govt	Total
2	5000	A2*0.6	A2*0.2	A2+B2+C2
3	A2 + 5000	A3*0.6	A3*0.2	A3+B3+C3

	A	B	C	D
1	Corp	Ind	Govt	Total
2	5000	3000	1000	9000
3	10,000	6000	2000	18,000

12 Reteaching *Lesson 1-6*

Extend the Lesson

REAL WORLD CONNECTION Have students imagine that they sell video games on commission. They earn an 8% commission on their total sales each day. Have students determine what records they would need to keep to track their total sales and commission for each day, and what calculations they would need to do each day. Discuss: How could spreadsheet software be used to make record-keeping easier?

Enrichment Worksheet 1-6 not shown.

Skills Practice

Vocabulary Review

Lesson 1–5
reciprocals
multiplicative inverse

Lesson 1–6
cell
formula
spreadsheet

Assignment Guide

All students: 1–46
Additional Practice: Refer to the Extra Practice Index on page 674 of this text.

Additional Answers

32.

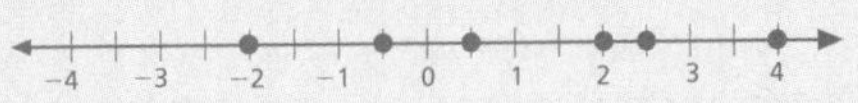

Review and Practice Your Skills

Practice ■ Lesson 1-5

For Exercises 1–9, perform the indicated operations.

1. $(-5.6)(3.9)$ -21.84
2. $\left(4\frac{2}{3}\right)\left(3\frac{4}{5}\right)$ $17\frac{11}{15}$
3. $\frac{4}{15} \div \frac{8}{3}$ $\frac{1}{10}$
4. $8\frac{2}{7} + (3)\left(-7\frac{1}{2}\right)$ $-14\frac{3}{14}$
5. $18 \div (-4) + 3\frac{1}{2}$ -1
6. $\frac{1}{2} \div \frac{3}{4} \cdot \left(-\frac{5}{9}\right)$ $\frac{-10}{27}$
7. $(-9)(0.4)(-0.8)(-3)$ -8.64
8. $5[16 - (-5)]$ 105
9. $\left(-\frac{3}{8}\right)^2$ $\frac{9}{64}$

Evaluate each expression when $a = -6\frac{1}{2}$, $b = -\frac{1}{3}$, and $c = 9$.

10. $c(a - b)$ $-37\frac{1}{2}$
11. $-bc + ac$ $-37\frac{1}{2}$
12. $a \div b$ $2\frac{11}{14}$
13. $b^2 - c$ $-3\frac{5}{9}$
14. $bc + |(-bc)|$ 0
15. $abc \div ac$ $-2\frac{1}{3}$

16. Steve earns \$8.50/h in his 40-h work week. For each hour over 40 h, he earns $1\frac{1}{2}$ times his hourly pay. How much does he earn by working 47 h in one week? \$429.25
17. Write 0.625 as a quotient of two integers. $\frac{5}{8}$
18. You are organizing a banquet. Expenses include: Food, \$355; Decorations, \$35; Trophies, \$85; Sound system rental, \$42.50. You expect 90 people to attend. What should you charge each person to cover expenses? \$5.75

Practice ■ Lesson 1-6

19. Ben earns \$8.82 for the first 40 hours he works each week and 1.5 times his hourly pay for any hours over 40. Write a formula that can calculate the total weekly pay for Ben for any number of hours.
Letting H = the number of hours worked, Ben's pay is equal to 8.82H + (H − 40)(1.5)(8.82).

Use your calculator to find the total weekly pay for Ben if he works the following number of hour.

20. 42 \$379.26
21. 36 \$317.52
22. 45 \$418.95
23. 31 \$273.42

Find the value in each cell.

		A	B	C	
		A	**B**	**C**	
24.	1	0.5	A1 + 5	A1 − 2*B1	5.5, −10,5
25.	2	2*A1	A2 + 5	A2 − 2*B2	1, 6, −11
26.	3	3*A1	A3 + 5	A3 − 2*B3	1.5, 6.5, −11.5
27.	4	4*A1	A4 + 5	A4 − 2*B4	2, 7, −12

28. The volume of a box is found using the formula $V = l \cdot w \cdot h$. The formula to find the surface area of a box is $S = 2lw + 2wh + 2lh$. Make a spreadsheet which uses these formulas to calculate the volume and surface area of boxes with the following dimensions: Answers will vary. Check students' spreadsheets for formulas.

a. $l = 1, w = 2, h = 3$ V = 6, SA = 22
b. $l = 2, w = 3, h = 4$ V = 24, SA = 52
c. $l = 3, w = 4, h = 5$ V = 60, SA = 94
d. $l = 4, w = 5, h = 6$ V = 120; SA = 148

Teaching Strategies

ESL/LEP Review the uses of parentheses in mathematics. Point out that parentheses can be used both to enclose operations and to enclose factors. Have students find two examples of each usage in Lesson 1-5.

PRACTICE ■ LESSON 1-1–LESSON 1-6

Write all subsets of the following sets. (Lesson 1-1) 30. ∅, {c}, {a}, {n}, {e}, {c, a}, {c, n}, {c, e}, {a, n}, {a, e}, {n, e}, {c, a, n}, {c, n, e}, {a, n, e}, {c, a, e}, {c, a, n, e}.

29. $\left\{\frac{1}{2}, \frac{3}{5}\right\}$ ∅, $\left\{\frac{1}{2}\right\}$, $\left\{\frac{3}{5}\right\}$, $\left\{\frac{1}{2}, \frac{3}{5}\right\}$ **30.** {c, a, n, e} **31.** {0} ∅, {0}

32. Graph the set $\left\{-2, -\frac{1}{2}, -2, |-2|, 2\frac{1}{2}, 2^2\right\}$ on a number line. (Lesson 1-2) See additional answers.

Let $U = \{x | x \text{ is an integer}\}$, $A = \{x | x \text{ is a multiple of } 4\}$, $B = \{x | x \text{ is a natural number}\}$, $C = \{0\}$. For Exercises 32-37, find each set. (Lesson 1-3)

33. $A \cap B$ {4, 8, 12, 16, 20, . . .} **34.** B' {0, −1, −2, −3, −4, . . .} **35.** $A' \cup B$ **36.** $A \cap B'$ {0, −4, −8, −12, −16, . . .} **37.** $A \cap C$ {0} **38.** $B' \cup C$ {0, −1, −2, −3, −4, . . .}

39. Jean goes to the grocery store to purchase five items that cost $3.95, $5.12, $0.78, $7.05, and $14.88. She has $35 in her pocket. Use estimation to determine whether or not she has enough money to buy all five items. (Lesson 1-4) Yes

35. {x|x is a natural number or a negative non-multiple of 4}

Simplify. (Lessons 1-4, 1-5)

40. $-35.7 - (-26.3)$ −9.4 **41.** $(-35.7)(-26.3)$ 938.91 **42.** $(2)(-3.5) + \left(-8\frac{1}{2}\right)(6) - (-72 \div 12)$ 50

43. $\frac{10}{3} - \frac{1}{3}[12 + (-3)]$ 65 **44.** $\frac{3}{4} + 2\left(5 - 1\frac{1}{2}\right)$ $7\frac{3}{4}$ **45.** $\frac{10}{3} - \frac{1}{3}[12 + (-3)]$ $\frac{1}{3}$

MathWorks Workplace Knowhow — Career – Cashier

Cashiers are employed by stores to add up your purchases, take your money or credit card, and give you change when you pay with cash. Communication and math skills are two large components of a cashier's job no matter the kind of business.

In addition to working with money, cashiers work with weights, length, area, and volume. Estimation and mental math skills enable cashiers to spot errors and check the reasonableness of cash register calculations.

1. Suppose you are a cashier at a pet store. A young boy has $20. He wants to buy as many cans of cat food as he can. Brand A is $0.49 per can; however, for every four cans you buy, the fifth one is free. Brand B is $0.39 per can. Which brand should the boy purchase? Brand B will give him 51 cans; Brand A gives only 50 cans.
2. In a clothing store, a customer asks you to estimate whether she has enough money for two blouses. She has $30. The first blouse is $17.00 with a 40% discount. The second is $20.00 with a 25% discount. Find the total cost of the blouses. $25.20
3. You are a cashier in a state in which the sales tax is 8.25%. The subtotal of a customer's purchases is $112.78. The register then adds a sales tax of $0.93 for a total of $113.71. Does the result seem reasonable? Explain your thinking. No. The tax is not nearly enough. Ten percent of the subtotal is $11.28, so 8.25% must be somewhat less. The actual tax should be $9.30.

Chalkboard Examples

Example from Lesson 1-5

Evaluate each expression when $x = 3$, $y = -4$, and $z = -2$.

1. $x(y - z)$ −6 2. $-xy + xz$ 2. 6

Example from Lesson 1-6

On a spreadsheet, cell A1 contains the value 3.5 and cell B1 contains the value −5. C1 contains the formula (A1 − B1) * (A1 + B1). What is the value of C1? −12.75

Career Opportunity

Describe the kind of work cashiers do and the types of places they work. Discuss the importance of language, communication, and an understanding of mathematics in working as a cashier. Students should answer Questions 1–3 to better understand how cashiers use mathematics in performing their job.

Students who are interested in learning more about this profession can go to Learningmathmatters.com. Guidance Counselors should have information about school and training requirements.

Extend the Lesson

REAL WORLD CONNECTION A cash register receipt is similar to a spreadsheet. Have students brainstorm what might be written in the cells for a receipt. Remind them to include cells for subtotals, tax, amount paid, and change.

Lesson Planning

NCTM Standards/Strands

- Number & Operation
- Patterns, functions, & Algebra
- Problem Solving
- Reasoning and Proof
- Communication
- Connections
- Representation

Vocabulary

exponent distributive property
base exponential form

Materials Needed

paper/pencil calculator

Lesson Resources

Warm-Up Transparency 3
Transparency RF-4, 5
Reteaching 1-7
Extra Practice 1-7
Enrichment 1-7

ASSIGNMENT GUIDE

Basic: 1–44, 49–52
Enriched: 1–52

Getting Started

5-MINUTE WARM-UP

Find each product.

1. $5 \times 5 \times 5$ 125
2. $6 \times 6 \times 3 \times 3$ 324
3. $(7 \times 7 \times 7) \times 4$ 1,372
4. $8 \times 8 \times (2 \times 2)$ 256

Introduction to Lesson 1-7

Have students present the methods they used to determine the length of the term paper. Discuss which method seems easier and why. If students have difficulty with the problem, you may want to guide them through the 5-step plan shown in the Teacher's Guide on page 30.

1-7 Distributive Properties and Properties of Exponents

Goals
- Use the distributive property to evaluate and simplify expressions.
- Use properties of exponents to evaluate and simplify expressions.

Applications Science, Number Sense, Sales

Suppose you are typing a term paper, and your typing rate is 4 pages per hour. On Monday night you type for 3.5 hours, and on Tuesday night you complete the paper by typing 2.5 hours. How long is the term paper? Show two ways to determine the total number of pages. 24 pages; $4(3.5) + 4(2.5)$; $4(3.5 + 2.5)$

BUILD UNDERSTANDING

As you can see from this question, you may have more than one way to solve a problem. Examine the expressions:

$$3(6 + 5) = 3 \cdot 6 + 3 \cdot 5 = 18 + 15 = 33 \qquad 3(6 + 5) = 3(11) = 33$$

Both ways have the same answer. This is an example of the distributive property.

The **distributive property** relates addition and multiplication. It states that a factor outside parentheses can be used to multiply each term within the parentheses.

Distributive Property	For any real numbers, a, b, and c, $a(b + c) = ab + ac$

Other mathematical properties, including the distributive, are summarized below. Each applies to all real numbers a, b, and c.

These mathematical properties are useful in simplifying mathematical expressions. Some expressions will also involve exponents. There are special properties that apply to exponents. Exponents are used as a short way to indicate repeated multiplication of a number, or factor.

A number written in **exponential form** has a base and an exponent.

The **base** tells what factor is being multiplied. The **exponent** tells how many equal factors there are. The expression a^4 is read as "a to the fourth power."

Other Mathematical Properties

Property	In Symbols	Examples
Distributive	$a(b + c) = ab + ac$ or $a(b - c) = ab - ac$	$3(5 + 4) = 3 \cdot 5 + 3 \cdot 4$; $3(5 - 4) = 3 \cdot 5 - 3 \cdot 4$
Transitive	If $a = b$ and $b = c$, then $a = c$	If $3 + 5 = 4 + 4$ and $4 + 4 = 2 + 6$, then $3 + 5 = 2 + 6$
Reflexive	$a = a$ any number is equal to itself	$8 = 8$
Substitution	If $a = b$, then we can substitute b for a or substitute a for b in any statement.	If $x = 4$, then $x + 2 = 4 + 2$
Symmetric	If $a = b$, then $b = a$	If $4 + 6 = 5 + 5$, then $5 + 5 = 4 + 6$

Teaching Strategies

ESL/LEP Help these students focus on the basic concepts of the lesson by having them make flash cards that illustrate each type of exponent problem. Have students then work in pairs with one student displaying a card while the other describes the process it shows.

When the base is a negative number, enclose it in parentheses. The expressions $(-3)^2$ and -3^2 represent different numbers.

$(-3)^2 = (-3)(-3) = 9$ and $-3^2 = -(3)(3) = -9$

The exponents zero and one are special. Any number raised to the first power is the number itself.

$n^1 = n$, so $13^1 = 13$

Any number, except 0, raised to the zero power is equal to one. For every nonzero real number a, $a^0 = 1$.

$2^0 = 1$, $4^0 = 1$

Mental Math Tip

Use the Distributive property to multiply numbers mentally
$15 \cdot 8 + 15 \cdot 12$

Think: $15 \cdot (8 + 12)$

$15(20)$

300

Example 1

Evaluate each expression. Let $a = 3$ and $b = -2$.

a. a^2 **b.** b^3 **c.** a^2b

Solution

a. $a^2 = 3^2 = 9$

b. $b^3 = (-2)^3 = (-2)(-2)(-2) = -8$

c. $a^2b = (3)^2(-2) = (9)(-2) = -18$

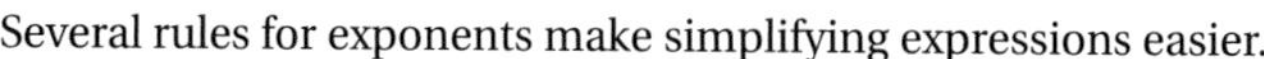

Several rules for exponents make simplifying expressions easier.

Properties of Exponents for Multiplication	For all real numbers a and b, if m and n are integers, then $a^m \cdot a^n = a^{m+n}$ $(a^m)^n = a^{mn}$ $(ab)^m = a^mb^m$

Example 2

Simplify.

a. $x^2 \cdot x^7$ **b.** $(a^3)^5$ **c.** $(5^2 \cdot n)^2$

Solution

a. $x^2 \cdot x^7 = x^{2+7} = x^9$

b. $(a^3)^5 = a^{3 \cdot 5} = a^{15}$

c. $(5^2 \cdot n)^2 = (5^{2 \cdot 2})(n)^2 = 5^4n^2$

As with multiplication, some rules make simplifying division expressions easier.

Properties of Exponents for Division	For all real numbers a and b, if m and n are integers, then $\frac{a^m}{a^n} = a^{m-n}$, if $a \neq 0$ $\left(\frac{a}{b}\right)^m = \frac{a^m}{b^m}$, if $b \neq 0$

Chalkboard Examples

Supplementary Example 1

Evaluate. Let $x = -4$ and $y = 3$.

a. x^2 16 **b.** y^3 27 **c.** xy^2 −36

Supplementary Example 2

Simplify.

a. $x^5 \cdot x^3$ x^8 **b.** $(x^4)^3$ x^{12} **c.** $(2^3 \cdot y)^3$ 2^9y^3 or $512y^3$

Supplementary Example 3

Simplify.

a. $\frac{x^5}{x^4}$ x **b.** $\left(\frac{y}{4}\right)^3$ $\frac{y^3}{4^3}$ or $\frac{y^3}{64}$

Teaching Strategies

COOPERATIVE LEARNING Have students work in pairs or small groups. Each student should work backward to create a problem with one or more missing exponents. Then students exchange problems and solve. Example: $(a^5b^4)^{\square} = a^{10}b^8$ The missing exponent is 2.

Reteaching Worksheet 1-7

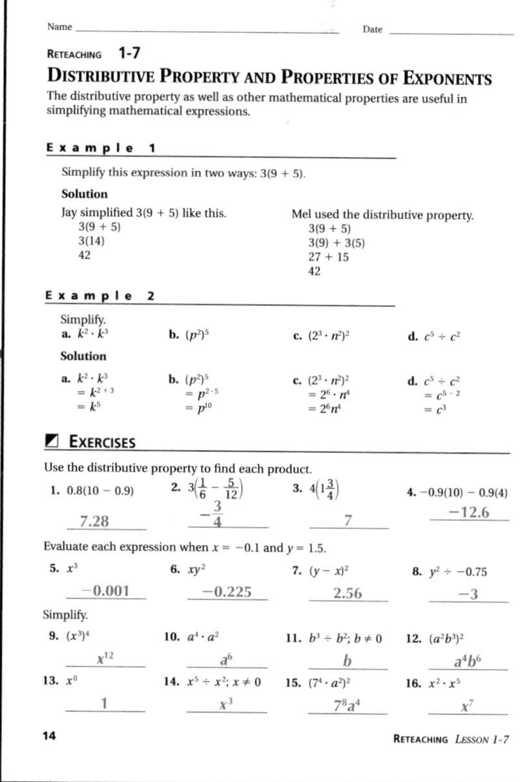

Name ______________ Date ______________

RETEACHING 1-7

DISTRIBUTIVE PROPERTY AND PROPERTIES OF EXPONENTS

The distributive property as well as other mathematical properties are useful in simplifying mathematical expressions.

Example 1

Simplify this expression in two ways: $3(9 + 5)$.

Solution

Jay simplified $3(9 + 5)$ like this.
$3(9 + 5)$
$3(14)$
42

Mel used the distributive property.
$3(9 + 5)$
$3(9) + 3(5)$
$27 + 15$
42

Example 2

Simplify.

a. $k^2 \cdot k^3$ **b.** $(p^2)^5$ **c.** $(2^3 \cdot n^2)^2$ **d.** $c^5 \div c^2$

Solution

a. $k^2 \cdot k^3 = k^{2+3} = k^5$

b. $(p^2)^5 = p^{2 \cdot 5} = p^{10}$

c. $(2^3 \cdot n^2)^2 = 2^6 \cdot n^4 = 2^6n^4$

d. $c^5 \div c^2 = c^{5-2} = c^3$

EXERCISES

Use the distributive property to find each product.

1. $0.8(10 - 0.9)$ 7.28
2. $3\left(\frac{1}{6} - \frac{5}{12}\right)$ $-\frac{3}{4}$
3. $4\left(1\frac{3}{4}\right)$ 7
4. $-0.9(10) - 0.9(4)$ −12.6

Evaluate each expression when $x = -0.1$ and $y = 1.5$.

5. x^3 −0.001
6. xy^2 −0.225
7. $(y - x)^2$ 2.56
8. $y^2 \div -0.75$ −3

Simplify.

9. $(x^3)^4$ x^{12}
10. $a^4 \cdot a^2$ a^6
11. $b^3 \div b^2; b \neq 0$ b
12. $(a^2b^3)^2$ a^4b^6
13. x^0 1
14. $x^5 \div x^2; x \neq 0$ x^3
15. $(7^4 \cdot a^2)^2$ 7^8a^4
16. $x^2 \cdot x^5$ x^7

14 RETEACHING *LESSON 1-7*

Lesson Wrap-up

Quick Assessment

Ask the following questions to determine if students understand the content presented in this lesson.

1. Simplify: $\frac{15r^8}{3r^3}$ $5r^5$
2. Fill in the blank:
   To find the power of a quotient, find the power of each number and ________. divide
3. Simplify: $(w^3)(w^4)(w^2)$ w^9

Assignment Guide

Basic: 1–44, 49–52
Advanced: 1–52
Additional Practice: See Extra Practice Index on page 674.

To divide numbers with the same base, subtract the exponents.

$$a^m \div a^n = a^{m-n}$$

$$2^5 \div 2^3 = \frac{2^5}{2^3} = \frac{2^1 \cdot 2^1 \cdot 2^1 \cdot 2 \cdot 2}{2_1 \cdot 2_1 \cdot 2_1} = 2^{5-3} = 2^2$$

To find the power of a quotient, find the power of each number and divide.

$$\left(\frac{a}{b}\right)^m = \frac{a^m}{b^m} \quad \left(\frac{3}{4}\right)^4 = \frac{3}{4} \cdot \frac{3}{4} \cdot \frac{3}{4} \cdot \frac{3}{4} = \frac{3^4}{4^4}$$

Example 3

Simplify.

a. $\frac{a^7}{a^3}$ **b.** $\left(\frac{t}{3}\right)^4$

Solution

a. $\frac{a^7}{a^3} = a^{7-3}$
$= a^4$

b. $\left(\frac{t}{3}\right)^4 = \frac{t^4}{3^4}$
$= \frac{t^4}{3^4}$

Try These Exercises

Use the distributive property to find each product.

1. $16 \cdot 15 - 16 \cdot 5$ 160
2. $18 \cdot 92$ 1,656
3. $28\left(10\frac{1}{7}\right)$ 284

Evaluate each expression when $x = -4$ and $y = 3$.

4. y^2 9
5. x^3 −64
6. xy^2 −36
7. x^3y^2 −576

Simplify.

8. $(-2)^3(-2)^2$ $(-2)^5$ or -32
9. $\frac{c^9}{c^5}, c \neq 0$ c^4
10. $(4^2 \cdot 3)^5$ $4^{10} \cdot 3^5$
11. $c^{10} \cdot c^5$ c^{15}
12. $(d^6)^3$ d^{18}
13. $(a^2b^3)^4$ a^8b^{12}
14. 3^0 1
15. $x^9 \cdot x^2$ x^{11}

Practice Exercises

Use the distributive property to find each product.

16. $8.5 \cdot 92 + 8.5 \cdot 8$ 850
17. $82 \cdot 53$ 4,346
18. $45\left(2\frac{4}{9}\right)$ 110

Evaluate each expression when $a = -3$ and $b = 4$.

19. a^2 9
20. $a^2 - b^2$ −7
21. b^3 64
22. ab^2 −48
23. $-2\,ab^2$ 96
24. $(-2ab)^2$ 576
25. $(2 - b)^3$ −8
26. $(a^2 - 7)^3$ 8

Simplify.

27. $2^{10} \cdot 2^{15}$ 2^{25}
28. $\frac{a^9}{a^6}, a \neq 0$ a^3
29. $r^5 \cdot r^4$ r^9
30. $(a^9)^6$ a^{54}
31. $\left(\frac{1}{d}\right)^9$ $\frac{1}{d^9}$
32. $m^{15} \cdot m^3$ m^{18}
33. $(a^5)^5$ a^{25}
34. $(n^4)(n^5)(n^6)$ n^{15}

Extend the Lesson

CONNECTING TO PRIOR KNOWLEDGE Have students substitute values for *x* and *y* in the expressions below. Students should test cases in which both numbers are positive, both are negative, and one is positive and one is negative.

1. $x^2 - y^2$
2. $(-x)^2 + (-y)^2$

36 Chapter 1 **Essential Mathematics**

Evaluate mentally each sum or product when $a = 3.5$, $b = 2$, and $c = 0$. Use the properties of mathematics as needed.

35. $5ab$ 35

36. $165a^2bc$ 0

37. $(6.5 + a)(1.4b) + c$ 28

Simplify. You may need to use more than one of the properties of exponents.

38. $m^4(m^5m^6)^2$ m^{26}

39. $\dfrac{(c^6 \cdot c^4)^2}{c^2}$ c^{18}

40. $\left(\dfrac{18g^6}{6g^3}\right)^2$ $9g^6$

41. NUMBER SENSE Without multiplying, tell which of these numbers is the greatest. 100^5 $1{,}000^4$ $10{,}000^3$ $100{,}000^2$
Explain your reasoning. (*Hint:* $100 = 10^2$; $1{,}000 = 10^3$; $10{,}000 = 10^4$; $100{,}000 = 10^5$.) $1{,}000^4$ and $10{,}000^3$ both equal 10^{12}

42. SALES John and Brian bought 125 pieces of candy for $43.75. They sold each piece for $0.50. How much profit did they make on their total sale of candy? $18.75

43. The volume of a cube is the product of its length, width, and height. What is the volume of a cube with side length 3 in? 5 cm? e cm? g in? 27 in.^{3}, 125 cm^3, e^3 cm^3, g^3 in.^{3}

44. WRITING MATH Does $3^2 = 2^3$? Explain your answer.
No, $3^2 = 3 \times 3 = 9$ and $2^3 = 2 \times 2 \times 2 = 8$

Extended Practice Exercises

CRITICAL THINKING Write *true* or *false*. Give examples to support your answer.

45. If a and b are negative integers and $a < b$, then $a^2 > b^2$. True

46. If a and b are negative integers and $a < b$, then $a^3 > b^3$. False

47. SCIENCE A certain type of bacteria doubles in number every hour. At 10 AM, there were 250 bacteria. How many bacteria will be present at 1 PM? 2,000

48. WRITING MATH Decide whether the distributive property works with the intersection and union of sets. For example, does $A \cap (B \cup C) = (A \cap B) \cup (A \cap C)$? Give examples to support your answer. Yes. Examples will vary.

Mixed Review Exercises

Solve. (Lesson 1-6)

49. Adult passes to the amusement park cost $27.95 each, while a child's pass costs $19.95. What would the total cost be for a family of four adults and six children to buy passes for the park? $259.45

50. Kordell drove his car 396 miles on 15 gallons of gas. Gas costs $1.32 per gallon. How much did Kordell pay for gas per mile driven? $0.05

51. By buying a washing machine on sale, Natasha saved 30% off the regular price. The washer regularly sold for $418. How much did Natasha pay for the washer? $292.60

52. The paint crew has 2,000 ft^2 of wall to paint in the office building. They painted 384 ft^2 on Monday, and 377 ft^2 on Tuesday. How much will they have to paint each day to finish the job on Friday? 413 square feet per day

Extend the Lesson

ONGOING ASSESSMENT Ask students to write the greatest number that can be expressed using three 5s. $(5^5)^5$

Extra Practice Worksheet 1-7

Name ______________________ Date ____________

EXTRA PRACTICE **1-7**

DISTRIBUTIVE PROPERTY AND PROPERTIES OF EXPONENTS

EXERCISES

Use the distributive property to find each product.

1. $12 \cdot 8 + 12 \cdot 2$ 120

2. $19 \cdot 24$ 456

3. $4.5 \cdot 12 - 4.5 \cdot 2$ 45

4. $26 \cdot 78$ 2028

5. $18\left(2\frac{1}{2}\right)$ 45

6. $54\left(10\frac{5}{9}\right)$ 570

7. $1.6 \cdot 12$ 19.2

8. $\frac{4}{7} \cdot 3 + \frac{4}{7} \cdot 4$ 4

Evaluate each expression when $x = -4$ and $y = 5$.

9. x^2 16

10. y^2 25

11. $x^2 - y^2$ −9

12. x^3 −64

13. y^3 125

14. xy^2 −100

15. $-x^2y$ −80

16. $(-xy)^2$ 400

17. $(3 - y)^2$ 4

18. $(x^2 + 3)^2$ 361

Simplify.

19. $3^2 \cdot 3^5$ 2187

20. $a^6 \cdot a^3$ a^9

21. $\frac{m^7}{m^4}$, $m \neq 0$ m^3

22. $(t^3)^5$ t^{15}

23. $q^4 \cdot q^9$ q^{13}

24. $\left(\frac{2}{d}\right)^4$, $d \neq 0$ $\frac{16}{d^4}$

25. $(w^4)^3$ w^{12}

26. $(k^3)(k^4)(k^6)$ k^{13}

27. $(g^4g^6)^3$ g^{30}

28. $(-3p^6p^2)^2$ $9p^{16}$

Evaluate mentally each sum or product when $r = 2.5$, $s = 0$ and $t = 3$. Use the properties of mathematics as needed.

29. rst 0

30. $-2r^2s^2t^2$ 0

31. $(3.5 + r)(2t) + s$ 36

31. $\frac{s^2}{rt^2}$ 0

33. $s(12r + 2t)$ 0

34. $r(s + t)$ 7.5

Enrichment Worksheet 1-7

Name ______________________ Date ____________

ENRICHMENT **1-7**

EXPONENTS AND TESSELLATIONS

In a regular tessellation, the plane is completely covered with congruent regular polygons. In a semi-regular tessellation, there is more than one type of polygon at each vertex. A form of exponential notation is sometimes used to name regular and semi-regular tessellations.

Example 1

Use exponents to name this tessellation.

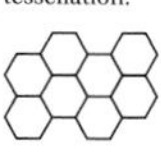

Solution

At each vertex there are three regular 6-sided figures (hexagons). So, the tessellation is named 6^3.

Example 2

Use exponents to name this tessellation.

Solution

Here, there are two 3-sided polygons and two 6-sided polygons at each vertex. The name of the tessellation is $3^2 . 6^2$. A period is used between the two kinds of polygons.

EXERCISES

Use exponents to name each tessellation.

1.

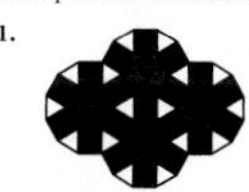

$3 . 4^2 . 6$

2.

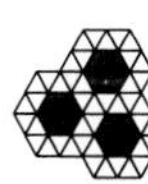

$3^4 . 6$

Describe the polygons at each vertex of these tessellations. Then, on another sheet of paper, make a drawing of each tessellation. Check students' drawings.

3. $4 . 8^2$ 1 square, 2 octagons

4. $3^3 . 4^2$ 3 equilateral triangles, 2 squares

5. If m is the number of polygons at a vertex, and n is the number of sides in each polygon, the equation $(m - 2)(n - 2) = 4$ can be used to find all the regular tessellations. Find them. Draw the ones not shown on this page on your own paper. Check students' drawings. They are 6^3, 4^4, and 3^6. The second is made of squares; the third of equilateral triangles.

1-8 Exponents and Scientific Notation

Goals
- Evaluate variable expressions with negative exponents.
- Write numbers in scientific notation.

Applications Chemistry, Astronomy, Business

Lesson Planning

NCTM Standards/Strands
- Number & Operation
- Patterns, functions, & Algebra
- Reasoning and Proof
- Communication
- Connections
- Representation

Vocabulary
negative exponents
scientific notation

Materials Needed
calculator
graph paper

Lesson Resources
Warm-Up Transparency 3
Transparency RF-6
Reteaching 1-8
Extra Practice 1-8
Enrichment 1-8
Multicultural Connection 3

Assignment Guide
Basic: 1–40, 46–69
Enriched: 1–69

Getting Started

5-Minute Warm-up
Write each in standard form.
1. 7^3 343
2. $8^3 \div 2^4$ 32
3. $5^2 \times 3^3$ 675
4. $16^2 \div 4^3$ 4

Introduction to Lesson 1-8
On graph paper, have students draw a square with an area of 81 square units. Have them write the length of a side and the area as exponential numbers with a base of 3. $3^2, 3^4$

Then have them draw a square with an area of 9 square units and write the new side and area using exponents with a base of 3. $3^1, 3^2$

Explore the meaning of exponents using a number line.

1. Draw a number line from 0 to 16. Check students' number line.
2. Graph 2^4, 2^3, 2^2, 2^1 and 2^0 on the number line.
3. Describe the pattern shown by the graphed points. 2^4 is twice as large as 2^3. 2^3 is twice as large as 2^2. 2^2 is twice as large as 2^1. As the exponent decreases by 1, the value is halved.

Build Understanding

Exponents can be negative numbers. The quotient property of exponents can help you understand the meaning of negative exponents.

$$\frac{2^3}{2^4} = \frac{\cancel{2}^1 \cdot \cancel{2}^1 \cdot \cancel{2}^1}{\cancel{2}_1 \cdot \cancel{2}_1 \cdot \cancel{2}_1 \cdot 2} = \frac{1}{2}$$

If you use the quotient property, you obtain $\frac{2^3}{2^4} = 2^{3-4} = 2^{-1}$.

So, $2^{-1} = \frac{1}{2}$.

For every nonzero real number a, if n is an integer, than $a^{-n} = \frac{1}{a^n}$.

So, for any nonzero real number a, a^{-n} is the reciprocal, or multiplicative inverse, of a^n.

Check Understanding
Write each expression as a fraction.

1. 4^{-3}
2. $(-3)^{-3}$
3. $(-3)^{-x}$
4. 2^{-5}

1. $\frac{1}{64}$ 2. $\frac{1}{27}$ 3. $\frac{1}{81}$ 4. $-\frac{1}{32}$

Example 1

Simplify each expression, using the properties of exponents.

a. $a^9 \div a^5$ **b.** $x^4 \cdot x^{-3}$ **c.** $(c^2)^{-5}$

Solution

a. $a^9 \div a^5 = a^{9-(-5)} = a^{14}$

b. $x^4 \cdot x^{-3} = x^{4+(-53)} = x$

c. $(c^2)^{-5} = c^{2x(-5)} = c^{-10}$

You can use negative exponents to evaluate variable expressions.

Example 2

Evaluate a^{-5} when $a = 2$.

Solution

$a^{-5} = 2^{-5} = \frac{1}{2^5} = \frac{1}{32}$

Extend the Lesson

ONGOING ASSESSMENT Have students write 7.8×10^4 in scientific notation. The correct answer is 78,000, but some students may write 780,000. These students have affixed the number of zeros in the exponent, rather than moving the decimal point the correct number of places.

Exponential expressions having a base of 10 serve a very useful purpose in science.

Scientists often deal with very large or very small numbers. To save time and space and be able to write and compute such numbers more easily, the system of **scientific notation** was developed. The method of scientific notation uses powers of 10 to write in a more concise manner.

A number written in scientific notation has two factors. The first factor is greater than or equal to 1 and less than 10. The second is a power of 10.

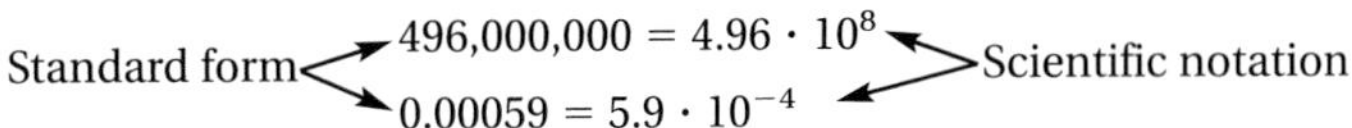

Mental Math Tip

To multiply 10^n when n is a positive integer, move the decimal point n places to the right.

$8.302 \times 10^5 = 8.30200$

To multiply by 10^n when n is a negative integer, move the decimal point n places to the left.

$6.03 \times 10^{-3} = 0.006.03$

Example 3

CHEMISTRY The mass of an oxygen atom is $2.66 \cdot 10^{-23}$ gram (g). What is the approximate mass of 1 billion oxygen atoms?

Solution

The mass of an oxygen atom $= 2.66 \cdot 10^{-23}$. Recall,
1 billion $= 1 \cdot 10^9$
The mass of 1 billion oxygen atoms
$= 1$ billion $\cdot$ the mass of 1 oxygen atom
$= (10^9)(2.66 \cdot 10^{-23})$ g
$= (2.66)(10^{9-23})$ g
$= 2.66 \cdot 10^{-14}$ g

If your calculator has an EE or an EXP key, you can use it as a quick way to enter a number in scientific notation. For example, to enter 2.6×10^{-23}, use the key sequence:

2.6 [EE] [(−)] 23

Now try the multiplication from Example 3:

1 [EE] 9 · 2.6 [EE] [(−)] 23 [=] 2.66 E −14,

which equals 2.66×10^{-14}.

The ability to measure quantities such as length, area, volume, mass and elapsed time is of vital importance. The measurement system used most often in scientific work is the metric system (also called the SI system).

The metric system is based on the decimal system. It uses prefixes to indicate multiples of 10. You can use the prefixes shown on the table to change from one unit to another.

Prefix	Multiple	Scientific notation
mega	1,000,000	10^6
kilo	1,000	10^3
hecto	100	10^2
deka	10	10^1
base	1	10^0
deci	0.1	10^{-1}
centi	0.01	10^{-2}
milli	0.001	10^{-3}
micro	0.0000001	10^{-6}
nano	0.000000001	10^{-9}
pico		10^{-12}
femto		10^{-15}

Interdisciplinary Connections

SCIENCE Have pairs of students research various features of a particular planet. Possible features include average distance from the sun, circumference of the body, distance to the closest planet, and density. Students should write their data using scientific notation. Encourage students to share their data with the class on a poster. Display the posters about the room and have the class compare the data.

Chalkboard Examples

Supplementary Example 1

Simplify each expression.
a. $x^7 \div x^2$ x^5 b. $y^2 \cdot y^{-4}$ y^{-2}
c. $(z^3)^{-2}$ z^{-6}

Supplementary Example 2

Evaluate x^{-3} when $x = 4$. $\frac{1}{64}$

Supplementary Example 3

A molecule of carbon monoxide (CO) weighs 12.01 amu (atomic mass units). Find the weight of one million molecules of CO. 12.01×10^7

Supplementary Example 4

How many centimeters are in five kilometers? 500,000 cm or 5×10^5 cm

Reteaching Worksheet 1-8

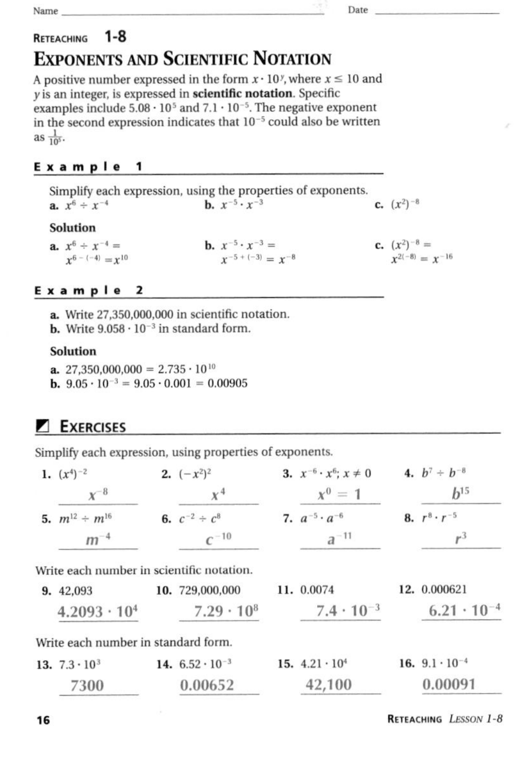

Name ______________ Date ______________

RETEACHING 1-8

EXPONENTS AND SCIENTIFIC NOTATION

A positive number expressed in the form $x \cdot 10^y$, where $x \leq 10$ and y is an integer, is expressed in **scientific notation**. Specific examples include $5.08 \cdot 10^5$ and $7.1 \cdot 10^{-5}$. The negative exponent in the second expression indicates that 10^{-5} could also be written as $\frac{1}{10^5}$.

Example 1

Simplify each expression, using the properties of exponents.
a. $x^6 \div x^{-4}$ b. $x^{-5} \cdot x^{-3}$ c. $(x^2)^{-8}$

Solution

a. $x^6 \div x^{-4} = x^{6-(-4)} = x^{10}$
b. $x^{-5} \cdot x^{-3} = x^{-5+(-3)} = x^{-8}$
c. $(x^2)^{-8} = x^{2(-8)} = x^{-16}$

Example 2

a. Write 27,350,000,000 in scientific notation.
b. Write $9.058 \cdot 10^{-3}$ in standard form.

Solution

a. $27{,}350{,}000{,}000 = 2.735 \cdot 10^{10}$
b. $9.05 \cdot 10^{-3} = 9.05 \cdot 0.001 = 0.00905$

EXERCISES

Simplify each expression, using properties of exponents.

1. $(x^4)^{-2}$ x^{-8}
2. $(-x^2)^2$ x^4
3. $x^{-6} \cdot x^6$; $x \neq 0$ $x^0 = 1$
4. $b^7 \div b^{-8}$ b^{15}
5. $m^{12} \div m^{16}$ m^{-4}
6. $c^{-2} \div c^8$ c^{-10}
7. $a^{-5} \cdot a^{-6}$ a^{-11}
8. $r^8 \cdot r^{-5}$ r^3

Write each number in scientific notation.

9. 42,093 $4.2093 \cdot 10^4$
10. 729,000,000 $7.29 \cdot 10^8$
11. 0.0074 $7.4 \cdot 10^{-3}$
12. 0.000621 $6.21 \cdot 10^{-4}$

Write each number in standard form.

13. $7.3 \cdot 10^3$ 7300
14. $6.52 \cdot 10^{-3}$ 0.00652
15. $4.21 \cdot 10^4$ 42,100
16. $9.1 \cdot 10^{-4}$ 0.00091

16 RETEACHING *LESSON 1-8*

Lesson Wrap-up

Quick Assessment

Ask the following questions to determine if students understand the content presented in this lesson.

1. Write 8,732,100,000 in scientific notation. 8.7321 × 109
2. Write 2.905×10^{-4} in standard form. 0.0002905
3. Write in order from least to greatest: 5^{-1}, 72^0, $-(3^2)$ $-(3^2)$, 5^{-1}, 72^0

Assignment Guide

Basic: 1–40, 46–69
Advanced: 1–69
Additional Practice: See Extra Practice Index on page 674.

Ratios of equal quantities can be used to change from one unit to another.

Example 4

MEASUREMENT Find the number of milligrams in 1 kilogram.

Solution

$$1\text{ kg} \cdot \frac{10^3\text{ g}}{1\text{ kg}} \cdot \frac{1\text{ mg}}{10^{-3}\text{ g}} = 10^6\text{ mg}$$

Try These Exercises

Simplify.

1. $(-2)^{-4}$ $\frac{1}{16}$
2. $(-1)^{-5}$ -1
3. $r^8 \div r^{12}$ r^{-4} or $\frac{1}{r^4}$
4. $x^{-9} \cdot x^9, x \neq 0$ 1
5. $(m^7)^{-3}$ m^{-21} or $\frac{1}{m^{21}}$
6. $w^5 \cdot w^{-13}$ w^{-8} or $\frac{1}{w^8}$
7. $d^6 \div d^{-11}$ d^{17}
8. $(x^2y)^{-5}$ $\frac{1}{x^{10}y^5}$

Evaluate each expression when $x = -3$ and $y = 4$.

8. x^{-2} $\frac{1}{9}$
9. y^{-4} $\frac{1}{256}$
10. $x^0y^{-2}y^0$ $\frac{1}{16}$
11. $x^3x^{-6}y$ $-\frac{4}{27}$

Write each number in scientific notation.

12. 59,300,000,000 $5.9 \cdot 10^{10}$
13. 0.000059 5.9×10^{-5}
14. 0.00006052 $6.052 \cdot 10^{-5}$

Write each number in standard form.

15. $3.6 \cdot 10^5$ 360,000
16. $4.3 \cdot 10^{-4}$ 0.00043
17. $2.09 \cdot 10^{-7}$ 0.000000209

18. **ASTRONOMY** The distance from the planet Mercury to the Sun is about 36,000,000 miles. Write the distance in scientific notation. $3.6 \cdot 10^7$

Practice Exercises

Simplify.

19. $(-1)^{-3} + (-1)^{-4}$ 0
20. $d^{-15} \div d^{-3}$ $\frac{1}{d^{12}}$
21. $y^{-5} \cdot y^{-4}$ $\frac{1}{y^9}$

Evaluate each expression when $p = -2$ and $q = 2$.

22. p^{-6} $\frac{1}{64}$
23. q^{-3} $\frac{1}{8}$
24. $(pq)^{-3}$ $-\frac{1}{64}$
25. $q^3q^{-2}p^{-3}$ $\frac{1}{4}$

Write each number in scientific notation.

26. 9,300 $9.3 \cdot 10^3$
27. 0.00356 $3.56 \cdot 10^{-3}$
28. 0.00000215 $2.15 \cdot 10^{-6}$

CALCULATOR Your calculator displays these answers. Write the answers in scientific notation and standard form.

29. 2.7 E6 — $2.7 \cdot 10^6$, 2,700,000
30. 3.9 E−5 — $3.9 \cdot 10^{-5}$, 0.000039

31. **LANGUAGE** It is estimated that there are 750,000 words in the English language. Write the number in scientific notation. $7.5 \cdot 10^5$

Extend the Lesson

MATH JOURNAL Have students describe how they know when to use positive integers and when to use negative integers as exponents when writing a number in scientific notation. Possible answer: If the number in standard form is between −1 and 1, use negative exponents. If the number in standard form is greater than or equal to 10 or less than or equal to −10, use positive exponents.

Solve. Write your answer in scientific notation.

32. The speed of light is $3.00 \cdot 10^{10}$ meters per second. How far does light travel in 1 day? $2.592 \cdot 10^{15} m$

33. **WRITING MATH** Which is greater $4.12 \cdot 10^3$ or $1.5 \cdot 10^4$? Write a rule for comparing numbers written in scientific notation. $1.5 \cdot 10^4$. If the powers are not equal, the number with the greatest power is equal. If the powers are equal, compare the mixed decimal portions of the numbers.

Simplify. You may need to use more than one of the properties of exponents.

34. $\left(\frac{y^4}{y^3}\right)^{-2}, y \neq 0$ y^{-2} or $\frac{1}{y^2}$

35. $\left(\frac{m^{-9}}{m}\right)^{-4}, m \neq 0$ m^{40}

36. $\frac{r^{-9}(r^2)^3}{r^{-6}r^{-5}}, r \neq 0$ $\frac{1}{r^2}$

Solve. Write your answer in scientific notation.

37. Simplify $0.000136 \times 18.05 \div 0.001$. 2.45×10^0

38. **BUSINESS** Greater Automotive employs $3.2 \cdot 10^2$ people. Each hour, the company pays $\$2.88 \cdot 10^3$ in wages. What is the average wage per hour for each employee? $9.00

39. **TIME** Find the number of nanoseconds in 5 milliseconds. 5,000,000

40. **MEASUREMENT** Find the number of centimeters (cm) in 1 foot (ft). Use the equivalent: 1 meter equals 3.2808 ft. 30.48 cm

Extended Practice Exercises

41. **CRITICAL THINKING** You know that $3^2 = 9$ and $3^{-2} = \frac{1}{9}$. What do you think is the value of $\frac{1}{3^{-2}}$? If a is a nonzero real number and n is an integer, what does $\frac{1}{a^{-n}}$ represent? 9, a^n

Use the results of Exercise 20. Rewrite each with positive exponents.

42. $\frac{5^{-2}}{r^{-5}}$ $\frac{r^5}{25}$

43. $\frac{a^{-3}b}{b^{-2}a^{-1}}$ $\frac{b^3}{a^2}$

44. $\left(\frac{m^5n^{-3}n}{n^{-2}m^{-2}}\right)^{-1}$ $\frac{n}{m^7}$

45. $\left[\frac{a^{-6}(b^2)^{-4}}{b^3a^0}\right]^{-2}, a, b \neq 0$ $a^{12}b^{22}$

Mixed Review Exercises

Evaluate each expression when $a = 5$ and $b = -3$. (Lesson 1-7)

46. a^2 25
47. $a^2 - b^2$ 16
48. $b^2 - a^2$ −16
49. ab^2 −75
50. $(ab)^2$ 225
51. $3a^2b$ 135
52. $-3ab^2$ 225
53. $(b^2 - 4)^3$ 9261
54. $a^2 + b^2$ 34
55. $-a^3$ 27
56. $-(b^3)$ −125
57. $-(a^2 - b^2)$ 16

Simplify. (Lesson 1-7)

58. $m^3 \cdot m^7$ m^{10}
59. $(m^4)^3$ m^{12}
60. $\frac{m^8}{m^5}, m^0$ m^3
61. $\left(\frac{1}{m}\right)^7$ $\frac{1}{m^7}$
62. $m^2(m^4 \cdot m^3)^5$ m^{37}
63. $\frac{m^{10}}{m^2}, m^0$ m^8
64. $\left(\frac{m^3}{m^2}\right)^8, m^0$ m^8
65. $m^5 \cdot m^8 \cdot m^{10}$ m^{23}
66. $m^4(m^3)^4$ m^{16}
67. $m^4(((m^2)^3)^4)^2$ m^{52}
68. $m^5 \times m^6 \times m^3$ m^{14}
69. $m^2(m^6)^5$ m^{32}

Extend the Lesson

ONGOING ASSESSMENT Ask students to explain why $\frac{10^6}{5^2} \neq 2^4$.

To apply the quotient of powers rule, the terms must have like bases. $2^4 = 16$, $\frac{10^6}{5^2} = \frac{1{,}000{,}000}{25} = 40{,}000$.

Extra Practice Worksheet 1-8

Name ______ Date ______

EXTRA PRACTICE 1-8

EXPONENTS AND SCIENTIFIC NOTATION

EXERCISES

Simplify.

1. $(-3)^{-2}$ $\frac{1}{9}$
2. $(-1)^{-10}$ 1
3. 2^{-4} $\frac{1}{16}$
4. d^{-3} $\frac{1}{d^3}$
5. $w^4 \div w^8$ $\frac{1}{w^4}$
6. $m^{-10} \div m^{-3}$ $\frac{1}{m^7}$
7. $y^{-3} \div y^{-6}$ y^3
8. $r^{-5} \cdot r^{-4}$ $\frac{1}{r^9}$
9. $(k^4)^{-2}$ $\frac{1}{k^8}$
10. $(t^{-3})^6$ $\frac{1}{t^{18}}$
11. $(h^{-6})^{-4}$ h^{24}
12. $c^4 \div c^{-5}$ c^9

Evaluate each expression when $n = 3$ and $m = -2$.

13. n^{-3} $\frac{1}{27}$
14. m^{-2} $-\frac{1}{4}$
15. $(mn)^{-2}$ $-\frac{1}{36}$
16. m^2n^{-2} $\frac{4}{9}$
17. $m^{-4}n^2$ $\frac{9}{16}$
18. $m^0n^{-3}n^0$ $\frac{1}{27}$
19. $m^4m^{-5}n^{-2}$ $-\frac{1}{18}$
20. $m^4n^{-5}n^2$ $\frac{16}{27}$
21. $(m^{-2}n^{-2})^{-3}$ 46,656
22. $(m^3n^{-4})^0$ 1

Write each number in scientific notation.

23. 8500 $8.5 \cdot 10^3$
24. 0.00098 $9.8 \cdot 10^{-4}$
25. 455,000 $4.55 \cdot 10^5$
26. 67,920,000 $6.792 \cdot 10^7$
27. 0.00000764 $7.64 \cdot 10^{-6}$
28. 0.00000703 $7.03 \cdot 10^{-6}$
29. 2,000,000 $2 \cdot 10^6$
30. 0.0000008 $8 \cdot 10^{-7}$

Write each number in standard form.

31. $4.9 \cdot 10^{-3}$ 0.0049
32. $6.2 \cdot 10^7$ 62,000,000
33. $8.94 \cdot 10^{-5}$ 0.0000894
34. $9.18 \cdot 10^{-6}$ 0.00000918
35. $6.7321 \cdot 10^6$ 6,732,100
36. $1.954 \cdot 10^3$ 1954
37. $9 \cdot 10^{-9}$ 0.000000009
38. $2 \cdot 10^8$ 200,000,000

14 EXTRA PRACTICE *LESSON 1-8*

Enrichment Worksheet 1-8

Name ______ Date ______

ENRICHMENT 1-8

FERMAT NUMBERS

Numbers of the form $2^{2^n} + 1$ are called Fermat numbers after the French mathematician Pierre de Fermat (1601–1665). Many people have spent a great deal of time trying to find out which Fermat numbers are prime.

You should, of course, use a calculator for the exercises on this page. (Fermat and Euler didn't have calculators!) However, you will need some cleverness in some of the exercises to overcome the limited number of digits your calculator displays.

EXERCISES

Complete this chart for the first seven Fermat numbers. Use the values 0 through 6 for n.

	Fermat Number Exponential Form	Fermat Number Scientific Notation	Fermat Number Standard Form
1.	$2^{2^0} + 1$	$3.0 \cdot 10^0$	3
2.	$2^{2^1} + 1$	$5.0 \cdot 10^0$	5
3.	$2^{2^2} + 1$	$1.7 \cdot 10^1$	17
4.	$2^{2^3} + 1$	$2.57 \cdot 10^2$	257
5.	$2^{2^4} + 1$	$6.55 \cdot 10^4$	65,537
6.	$2^{2^5} + 1$	$4.29 \cdot 10^9$	4,294,967,297
7.	$2^{2^6} + 1$	$1.84 \cdot 10^{19}$	18,446,744,073,709,551,617

Fermat made a conjecture that all of his numbers are prime. But he was wrong!

8. In 1732, Euler factored the Fermat number with $n = 5$. One prime factor is 641. Find the other prime factor. 6,700,417

9. One prime factor of the Fermat number with $n = 6$ is 274,177. Find the other prime factor. 67,280,421,310,721

20 ENRICHMENT *LESSON 1-8*

Lesson 1-8 **Exponents and Scientific Notation** 41

Assessment Planning

Vocabulary Review

Lesson 1-1
set, infinite set, finite set, subset, empty set, null set, variable, equation, open sentence, replacement set, solution, solution set

Lesson 1-2
natural numbers, whole numbers, integers, coordinate, rational numbers, inequality, irrational numbers, real numbers, absolute value

Lesson 1-3
union, Venn diagram, intersection, disjoint set, universe, universal set, complement, compound inequality

Lesson 1-4
addition property of opposites, additive inverse, associative property, closure property, commutative property, identity property, inverse property

Lesson 1-5
reciprocals, multiplicative inverse

Lesson 1-6
cell, spreadsheet, formula

Lesson 1-7
exponent, distributive property, base, exponential form

Lesson 1-8
negative exponents, scientific notation

Assessment Options

Chapter Assessment
Alternative Assessment
Chapter Tests Forms A and B
Cumulative Assessment
Achievement Test
Writing Prompts

Assignment Guide

All students: 1–24
Additional Practice: Refer to the Extra Practice Index on page 674.

Chapter 1 Review

Vocabulary ■ LESSON 1-1–LESSON 1-8

Choose the word from the list that completes each statement.

a. distributive
b. absolute value
c. rational
d. scientific notation

1. A __?__ number can be expressed as either a terminating or a repeating decimal. c
2. __?__ uses powers of 10 to write a number in a more concise manner. d
3. The __?__ property relates addition and multiplication. a
4. The __?__ of a number is the distance that number is from zero. b

LESSON 1-1 ■ The Language of Mathematics, p. 6

▶ A **set** is a well-defined collection of items whose elements can be finite or infinite.

5. Which of the given values is a solution of the equation $3x + 2 = 5$? $-1, 0, 1, 2$
1

LESSON 1-2 ■ Real Numbers, p. 10

▶ A **rational** number can be represented as a ratio of two integers, $\frac{a}{b}$, $b \neq 0$. An **irrational** number cannot. See additional answers.

6. Graph the set of all whole numbers less than 5 on a number line.
7. Graph the set of all real numbers greater than -4 on a number line.

LESSON 1-3 ■ Union and Intersection of Sets, p. 16

▶ The **union** of two sets A and B, $A \cup B$, contains the elements in A, in B, or in both. The **intersection** of two sets A and B, $A \cap B$, contains elements common to both A and B.

8. Let $A = \{1, 3, 5\}$ and $B = \{2, 3, 4\}$. Find $A \cup B$. $\{1, 2, 3, 4, 5\}$
9. Given that $A = \{x|x \text{ is a real number and } x \geq 0\}$
$B = \{x|x \text{ is a real number and } x < 5\}$, graph $A \cup B$.
See additional answers.

LESSON 1-4 ■ Addition, Subtraction and Estimation, p. 20

▶ To add and subtract rational numbers, use the same rules as for integers.

Closure	Commutative	Associative	Identity	Inverse
$a \rightarrow b$ is unique	$a + b = b + a$	$(a + b) + c = a + (b + c)$	$a + 0 = 0 + a = a$	$a + (-a) = 0$

Add or subtract.

10. $-5.4 + (-9.8)$ -15.2
11. $-2\frac{1}{2} + 3\frac{1}{8}$ $\frac{5}{8}$
12. $-39.6 - (-23.9)$ -15.7

Additional Answers

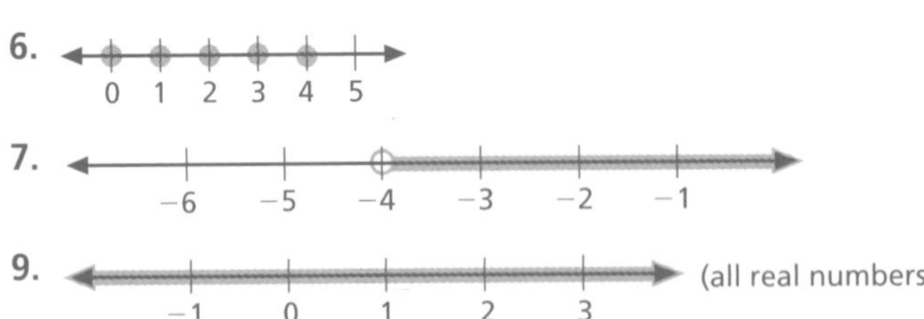

LESSON 1-5 ■ Multiplication and Division, p. 26

- ▶ The product or quotient of two rational numbers with the same sign is positive.
- ▶ The product or quotient of two rational numbers with different signs is negative.

Closure	Commutative	Associative	Identity	Inverse	Property of Zero
ab is unique	$ab = ba$	$(ab)c = a(bc)$	$a \cdot 1 = 1 \cdot a = a$	$a \cdot \left(\frac{1}{a}\right) = \left(\frac{1}{a}\right) \cdot a = 1$	$a \cdot 0 = 0 \cdot a = 0$

Multiply or divide.

13. $(-5.1)(3.2)(10)$ -163.2

14. $(20)(-6.4)(-2.5 \div -0.5)$ -640

LESSONS 1-6 ■ Problem Solving Skills: Use Technology, p. 30

- ▶ Decide which computational method is best for solving a given problem—estimation, mental math, pencil and paper, or calculator. Then use your method to solve the problem.

15. The jazz band is selling water bottles and license plate holders to pay for a trip. The band earns a profit of \$2.00 for each water bottle sold and \$1.50 for each license plate holder sold. Each band member must earn \$150. If Maria sells 50 license plate holders, and the remainder of the money comes from selling water bottles, how many water bottles does she sell? 38

LESSON 1-7 ■ Distributive Properties and Properties of Exponents, p. 34

Distributive property: $a(b + c) = ab + ac$

Properties of one and zero: $a^1 = a$, $a^0 = 1$, $a \neq 0$

Multiplication: $a^m \cdot a^n = a^{m+n}$, $(a^m)^n = a^{mn}$, $(ab)^m = a^m b^m$

Division: $\frac{a^m}{a^n} = a^{m-n}$, $a \neq 0$, $\left(\frac{a}{b}\right)^m = \frac{a^m}{b^m}$, $b \neq 0$

Simplify.

16. $x^3 \cdot x^8$ x^{11}

17. $\frac{z^6}{z^9}$, $z \neq 0$ z^{-3} or $\frac{1}{z^3}$

18. $\left(\frac{a^2b^3}{c}\right)^2$, $c \neq 0$ $\frac{a^4b^6}{c^2}$

LESSON 1-8 ■ Exponents and Scientific Notation, p. 38

- ▶ For every nonzero real number a, $a^{-n} = \frac{1}{a^n}$.
- ▶ A number written in scientific notation has two factors. The first factor is greater than or equal to 1 and less than 10. The second factor is a power of 10.

Simplify.

19. $x^5 \cdot x^{-6}$ x^{-1} or $\frac{1}{x}$

20. $a^{-2} \div a^{-5}$ a^3

21. $(-2)^{-3}$ $-\frac{1}{8}$

22. Write 0.000392 in scientific notation. 3.92×10^{-4}

23. Write 4.6×10^4 in standard notation. 46,000

USING DATA Solve using the data on page 3.

24. Decode this message. GUVF VF GUR RAQ BS PUNCGRE BAR.
This is the end of chapter one.

Teaching Strategies

COOPERATIVE LEARNING Have students work in small groups to create a magic square using expressions with a single variable. In a magic square, the sum for each row, column and diagonal is the same. Put the following example on the board. Have students substitute the value 3 for x. The sum of each column, row and diagonal is 60.

Example:

$6x + 6$	$7x + 7$	$2x + 2$
$x + 1$	$5x + 5$	$9x + 9$
$8x + 8$	$3x + 3$	$4x + 4$

Chapter 1 Test Form A, page 1

CHAPTER 1
ESSENTIAL MATHEMATICS
ASSESSMENT FORM A, PAGE 1

Name ______
Date ______

Scoring Record	
Possible: 52	Earned:

1. Use roster notation to define this set: odd numbers greater than 4. $\{5, 7, 9, 11, \ldots\}$
2. Write all the subsets of the set $\{a, b\}$. $\{a\}, \{b\}, \{a, b\}, \varnothing$

Which of the given values is a solution for each equation?

3. $x - 2 = -10$; -8, -12 -8
4. $3a \div 4 = -6$; 8, -8 -8

Graph each given set of numbers on a number line.

5. $\left\{\sqrt{3}, -\frac{1}{2}, -1.6, 2\frac{2}{5}\right\}$
   -2 -1 0 1 2 3 4
6. All real numbers greater than or equal to -3
   -4 -3 -2 -1 0 1 2

Evaluate each expression where $a = -3$.

7. $-(a)$ -3
8. $|-a|$ 3
9. $-|-a|$ -3

Refer to the diagram. Find the sets named by listing the members.

10. $A \cup B$ $\{1, 2, 3, 5, 9\}$
11. $A \cap C$ $\varnothing$

U A 1 5 3 9 2 B 6 4 C

Using the replacement set of the real numbers, find the solution set for each compound inequality. Describe the solution sets using set-builder notation.

12. $x > 2$ *and* $x < 4$

Let $A = \{x | x$ is a real number and $x > 2\}$
Let $B = \{x | x$ is a real number and $x < 4\}$

Graph $A \cap B$ on the number line.
Graph of A: -1 0 1 2 3 4
Graph of B: -1 0 1 2 3 4
Graph of $A \cap B$: -1 0 1 2 3 4

Solution Set: $\{x | x$ is a real number and $2 < x < 4\}$

13. $x > 2$ *or* $x \leq -4$

Let $A = \{x | x$ is a real number and $x > 1\}$
Let $B = \{x | x$ is a real number and $x \leq -2\}$

Graph $A \cup B$ on the number line.
Graph of A: -3 -2 -1 0 1 2
Graph of B: -3 -2 -1 0 1 2
Graph of $A \cup B$: -3 -2 -1 0 1 2

Solution Set: $\{x | x$ is a real number and $x > 1$ or $x \leq -2\}$

2 ASSESSMENT *CHAPTER 1 FORM A*

Chapter 1 Test Form A, page 2

Name ______ Date ______

Add or subtract.

14. $0.35 - (-2.07)$ 2.42
15. $-2.67 + 3.21 - (-1.54)$ 2.08
16. $2\frac{5}{12} - \left(-1\frac{1}{2}\right)$ $3\frac{11}{12}$
17. $-2\frac{1}{2} - 1\frac{1}{3} - \left(-1\frac{1}{6}\right)$ $-2\frac{2}{3}$

Find each product or quotient.

18. $(2.6)(-1.3)$ -3.38
19. $-7.2 \div 12$ -0.6
20. $-1\frac{1}{2} \div 1\frac{1}{3}$ $-1\frac{1}{8}$
21. $\left(-1\frac{1}{6}\right)\left(-1\frac{1}{2}\right)$ $-1\frac{3}{4}$

Evaluate each expression when $x = -0.2$, $y = 0.1$ and $z = -6$.

22. $x + y$ -0.1
23. $y - x$ 0.3
24. $x - (-y)$ -0.1
25. $-y + x$ -0.3
26. xyz 0.12
27. xz 1.2
28. $yz \div x$ 3
29. $xz \div y$ 12

Evaluate each expression when $a = 2$ and $b = -3$.

30. a^2 4
31. b^3 -27
32. ab^2 18
33. a^2b 12

Simplify.

34. $2.4 \div (-8) + (-0.6)$ -0.9
35. $2(-6) - (-14)$ 2
36. $x^3 \cdot x^5$ x^8
37. $(y^3)^2$ y^6
38. $(ab)^4$ a^4b^4
39. $a^5 \div a^2$; $a \neq 0$ a^3
40. $(m^2)^{-3}$ m^{-6}
41. $a^2 \cdot a^{-10}$ a^{-8}
42. $k^{10} \div k^{-3}$; $k \neq 0$ k^{13}
43. $n^{-8} \div n^6$; $n \neq 0$ n^{-14}

Evaluate each expressin when $x = -4$ and $y = 2$.

44. x^{-2} $\frac{1}{16}$
45. y^{-3} $\frac{1}{8}$
46. x^0y^{-3} $\frac{1}{8}$
47. $y^2 - x^{-2}$ $3\frac{15}{16}$

Write in scientific notation.

48. 129,300,000 $1.293 \cdot 10^8$
49. 0.0065 $6.5 \cdot 10^{-3}$

Write in standard form.

50. $4.2 \cdot 10^4$ 42,000
51. $2.13 \cdot 10^{-2}$ 0.0213

52. Find the value in each cell.

	A	B	C	D
1	30	A1*4	A1*0.5	B1 – C1
2	A1 + 30	A2*4	A2*0.5	B2 – C2
3	A2 + 30	A3*4	A3*0.5	B3 – C3
4	A3 + 30	A4*4	A4*0.5	B4 – C4

	A	B	C	D
1	30	120	15	105
2	60	240	30	210
3	90	360	45	315
4	120	480	60	420

4 ASSESSMENT *CHAPTER 1 FORM A*

Chapter 1 Test Form B, page 1

CHAPTER 1

ESSENTIAL MATHEMATICS

ASSESSMENT FORM B, PAGE 1

Name ______________

Date ______________

Scoring Record	
Possible: 52	Earned:

1. Use roster notation to define this set: even integers less than 5. {... −2, 0, 2, 4}
2. Write all the subsets of the set $\{m, n\}$. {m}, {n}, {m ,n}, ≠

Which of the given values is a solution for each equation?

3. $3m = -12$; −36, −4 −4
4. $-2d \div 8 = -4$; −16, 16 16

Graph each given set of numbers on a number line.

5. $\left\{\sqrt{2}, -1\frac{1}{2}, 3\frac{1}{12}, 4.5\right\}$
6. All real numbers less than or equal to −1

Evaluate each expression where $a = -4$.

7. $-(a)$ 4
8. $|-a|$ 4
9. $-|-a|$ −4

Refer to the diagram. Find the sets named by listing the members.

10. $A \cap B$ {3}
11. C' {1, 2, 3, 6, 8}

Using the replacement set of the real numbers, find the solution set for each compound inequality. Describe the solution sets using set-builder notation.

12. $x > -1$ *and* $x \leq 2$

Let $A = \{x|x$ is a real number and $x > -1\}$

Let $B = \{x|x$ is a real number and $x \leq 2\}$

Graph $A \cap B$ on the number line.

Graph of A:

Graph of B:

Graph of $A \cap B$:

Solution Set: $\{x \mid x$ is a real number and $-1 < x \leq 2\}$

13. $x \geq 2$ *or* $x < -1$

Let $A = \{x|x$ is a real number and $x \geq 2\}$

Let $B = \{x|x$ is a real number and $x < -1\}$

Graph $A \cup B$ on the number line.

Graph of A:

Graph of B:

Graph of $A \cup B$:

Solution Set: $\{x \mid x$ is a real number and $x \geq 2$ or $x < -1\}$

6 ASSESSMENT *CHAPTER 1 FORM B*

Chapter 1 Test Form B, page 2

Name ______________ Date ______________

Add or subtract.

14. $0.71 + (-3.09)$ −2.38
15. $2.76 - (-2.35) + (-1.62)$ 3.49
16. $-4\frac{5}{24} - \left(-1\frac{2}{3}\right)$ $-2\frac{13}{24}$
17. $-1\frac{1}{2} + \left(-1\frac{1}{3}\right) - \left(-3\frac{1}{4}\right)$ $\frac{5}{12}$

Find each product or quotient.

18. $(-1.8)(-2.7)$ 4.86
19. $-14.4 \div 12$ −1.2
20. $-1\frac{1}{4} \div 1\frac{1}{2}$ $-\frac{5}{6}$
21. $\left(1\frac{1}{5}\right)\left(-1\frac{2}{3}\right)$ −2

Evaluate each expression when $x = -1.6$, $y = 0.5$ and $z = -10$.

22. $x + y$ −1.1
23. $y - x$ 2.1
24. $x - (-y)$ −1.1
25. $-y + x$ −2.1
26. xyz 8
27. xz 16
28. $xz \div y$ 32
29. $xy \div z$ 0.08

Evaluate each expression when $a = -2$ and $b = 3$.

30. a^2 4
31. b^3 27
32. ab^2 −18
33. a^2b 12

Simplify.

34. $-7.2 \div (-9) + (-0.3)$ 0.5
35. $4(-8) - (-10)$ −22
36. $x^8 \cdot x^{-3}$ x^5
37. $(y^4)^2$ y^8
38. $(ab)^6$ a^6b^6
39. $a^8 \div a^{-2}; a \neq 0$ a^{10}
40. $(m^{-2})^8$ m^{-16}
41. $a^{-4} \cdot a^{-3}$ a^{-7}
42. $k^9 \div k^{-5}; k \neq 0$ k^{14}
43. $n^{-4} \div n^{-7}$ n^3

Evaluate each expressin when $x = 3$ and $y = -4$.

44. x^{-2} $\frac{1}{9}$
45. y^{-3} $-\frac{1}{64}$
46. x^0y^{-2} $\frac{1}{16}$
47. $y^2 - x^{-2}$ $15\frac{8}{9}$

Write in scientific notation.

48. 0.00073 $7.3 \cdot 10^{-4}$
49. 1,745,320,000 $1.74532 \cdot 10^9$

Write in standard form.

50. $8.21 \cdot 10^{-3}$ 0.00821
51. $2.4163 \cdot 10^4$ 24,160
52. Find the value in each cell.

	A	B	C	D
1	150	A1/5	A1 + B1	C1*0.2
2	A1 − 25	A2/5	A2 + B2	C2*0.2
3	A2 − 25	A3/5	A3 + B3	C3*0.2
4	A3 − 25	A4/5	A4 + B4	C4*0.2

	A	B	C	D
1	150	30	180	36
2	125	25	150	30
3	100	20	120	24
4	75	15	90	18

8 ASSESSMENT *CHAPTER 1 FORM B*

CHAPTER 1 ASSESSMENT

1. Define the set of positive integers greater than 2. {3, 4, 5, . . .}
2. Determine all the possible subsets of [*s*, *e*, *t*}. {s, e, t}, {se},{s, t}, {e, t}, {s}, {e}, {t}, { }
3. Which of the given values is a solution of the equation? −4

   $4x + 5 = -11$;   −4, 0, 2, 4
4. Use mental math to solve $a - 9 = -2$. 7
5. *True* or *false.* 0 is a whole number. true
6. Graph the set of all natural numbers greater than −1. See additional answers.
7. Graph the set of all real numbers less than or equal to 2. See additional answers.
8. Evaluate $-|r|$, when $r = -2.6$ −2.6

Let $U = \{d, r, a, k, e\}$, $A = \{r, a, k, e\}$, $B = \{r, e, d\}$, and $C = \{e, d\}$. Find each union or intersection.

9. $A \cap B$ {r, e}
10. $B \cup C$ {r, e, d}
11. $A \cap C$ {e}
12. $A' \cap B'$ ∅

Given:

$A = \{x|x$ is a negative integer and $x > -4\}$

$B = \{x|x$ is a real number and $x \leq 0\}$

Graph the following. For 13–14, see additional answers.

13. $A \cap B$
14. $A \cup B$

Add or subtract.

15. $-18.3 + (-17.8)$ −36.1
16. $-85.6 - (-79.7)$ −5.9

Multiply or divide.

17. $(-9.6) \div (-1.6)$ 6
18. $\left(-8\frac{1}{3}\right)\left(\frac{12}{5}\right)(-5)(2)$ 200

Simplify.

19. $z^{-9} \cdot z^{12}$ z^3
20. $\frac{x^5}{x^6}$ x^{-1} or $\frac{1}{x}$
21. $\frac{(a^2b^{-1})^2}{a^0b^{-4}}$ a^4b^2
22. Evaluate $-m - n$ when $m = -8\frac{1}{2}$ and $n = -5.7$. 14.2
23. Evaluate $m^{-2}n^5$ when $m = -3$ and $n = -2$. $-\frac{32}{9}$ or $-3\frac{5}{9}$
24. Write 0.0000035 in scientific notation. 3.5×10^{-6}
25. Write 2.79×10^7 in standard notation. 27,900,000
26. Sharie, Cal, and Nick each wrote a report for literature class. Sharie's report was twice as long as Nick's. Cal's report was 4 pages shorter than Nick's. If Cal's report was 8 pages, how long was Sharie's? 24 pages

44 | Chapter 1 **Assessment**

ADDITIONAL ANSWERS

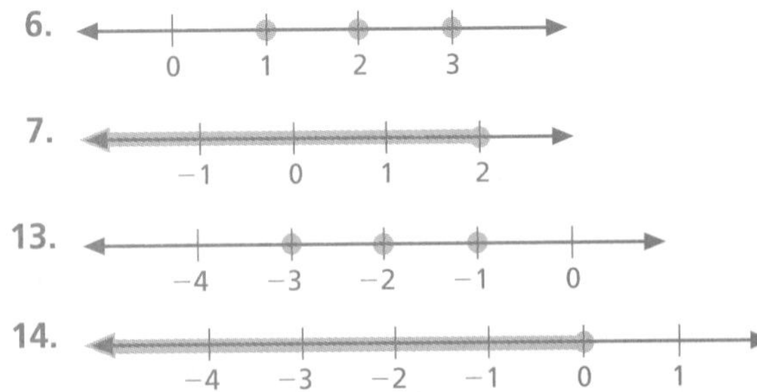

SCORING RUBRIC 5—Create more than 6 graphs. Provide clear report comparing all aspects of graphs. **4** Create 5 or 6 graphs. Provide clear report comparing all aspects of graphs. **3**—Create 5 or 6 graphs. Provide report comparing some aspects of graphs. **2**—Create 3 or 4 graphs. Provide report comparing some aspects of graphs with some errors. **1**—Create less than 3 graphs. Provide report comparing some aspects of graphs with many errors. **0**—Students make no graphs.

Chapter 1 Alternative Assessment

Cooperative Learning

BOOK OF NUMBERS Work with a group. Search out all the kinds of numbers you can find in newspapers, magazines, advertisements, etc. Make a notebook of these numbers, including what kind of numbers they are (decimal, money, fraction, rational). Find as many names as possible for each number and as many examples as possible for each name.

Modeling/Using Manipulatives

GAME OF NUMBERS Design and play a game of numbers. The game should involve four to six players reaching a goal number by adding, subtracting, multiplying and/or dividing signed numbers. Players may draw cards, roll a number cube of your design or determine numbers in other ways. List all the rules of the game and be sure all players know these before play begins. Write a brief explanation of how signed numbers are used in your game.

Technology

COMPUTER LANGUAGE From the very first to the most recent models of laptops and desktops, computers function using a binary system of numbers. Research the binary system to discover how computers use this system and whether it is in use in any other field today. Prepare a class presentation on how the binary number system works, giving examples of numbers from 0 to 20. Demonstrate your local area code and zip code as they would be written in binary numbers, digit by digit.

Chapter Investigation

EXTENSION Write a report about how you encoded your message and how you decoded your partner's message. Describe how your code differed from your partner's code.

Chapter Investigation

The benchmarks and expectations for this extension are as follows.

- Students assign the letters of the alphabet to a position on a number line. They write a message, and encode the message.
- Students make their code more difficult by substituting operations that will lead from one letter to the next. They exchange encoded messages and number line keys with a partner. They reveal the first letter of their passage to their partner. Then they decode their partner's message.
- Students use the original message they wrote and make a spreadsheet that will automatically calculate the percent for the letters E, T, A, O, N, R, I, S, and H. They compare their percents to the table on page 5.
- Students write a report about how they encoded their message.

Cooperative Learning

Assign students to groups of 4 or 5. Suggest that each student make 1 or 2 graph. The graphs can be compared and discussed by the group.
SCORING RUBRIC **5**—All make graphs. Provide commentary about all graph's main feature and usefulness. **4**—All make graphs. Provide commentary about most of graph's main features and usefulness. **3**—All make graphs. Provide summary of main features and usefulness of few graphs. **2**—All make graphs. One student provides summary of main features and usefulness of few graphs. **1**—Few make graphs. One student provides summary of main features and usefulness of few graphs. **0**—Students make no graphs.

Critical Thinking

You may want students to work in groups of 2 or 3 for this project. Encourage students to consider all the options before they decide which option will be the most persuasive. You may want students to present their findings to the class and receive feedback concerning how convincing their report is before sending it to the principal or student council. If possible, give students time to report any responses they receive.
SCORING RUBRIC **5**—Conducts survey with appropriate questions. Provides very convincing report using measures of central tendency and one or more graphs. **4**—Conducts survey with appropriate questions. Provides fairly convincing report using measures of central tendency and one graph. **3**—Conducts survey with some appropriate questions. Makes minor errors in fairly convincing report using measures of central tendency and one graph. **2**—Conducts survey. Makes several errors in report using measures of central tendency and one graph. **1**—Conducts survey. Provides report that contains many errors. **0**—Students do not conduct survey or write report.

Technology

If software is not available, or graphing options are not in spreadsheet package, student can use graphing software or graphing calculator. Most packages include a few graphs with which students may not be familiar. Encourage them to investigate these graphs and determine their unique purposes.
See page 44 for Scoring Rubric.

Cumulative Review

Cumulative review is the best way to maintain previously taught skills and concepts. This will keep students prepared for new lessons that build on previously covered skills.

Cumulative Review covers:

Lesson 1-1
Lesson 1-2
Lesson 1-3
Lesson 1-4
Lesson 1-5
Lesson 1-7

Additional Answers

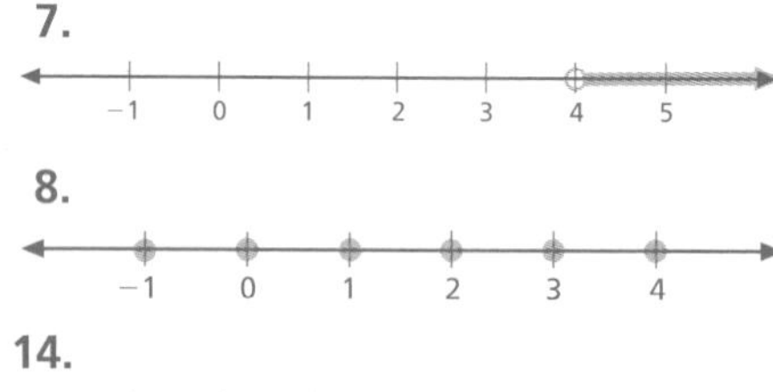

Chapter 1 Cumulative Review

1. List all the subsets of $\{A, Z\}$. $\{A\}, \{Z\}, \{A, Z\}, \varnothing$

2. Define the following set in roster notation. $\{x \mid x \text{ is a whole number and } x < 6\}$ $\{0, 1, 2, 3, 4, 5\}$

Determine if each statement is *true* or *false*.

3. 4 is a solution of the equation $3x - 2 = 12$. false

4. $\{1, 2\} \subset \{1, 2\}$ true

5. 3.14 is a rational number. true

6. $\sqrt{5}$ is an integer. false

Graph each set of numbers on a number line.

7. real numbers greater than 4 See additional answers.

8. integers less than or equal to 4 See additional answers.

9. Evaluate $-|n|$ if $n = -6$. -6

Let $U = \{6, 7, 8, 9, 10, 11, 12\}$, $A = \{6, 7\}$, $B = \{8, 9, 10\}$, and $C = \{8, 9, 10, 11\}$. Find each intersection or union.

10. $A \cap B$ $\varnothing$

11. $B \cup C$ $\{8, 9, 10, 11\}$

12. $A' \cap C'$ $\{12\}$

13. $A \cup C$ $\{6, 7, 8, 9, 10, 11\}$

14. Graph $R \cap S$ on a number line, given that
$R = \{x \mid x \text{ is a whole number and } x \le 0\}$
$S = \{x \mid x \text{ is a real number and } x > -2\}$. See additional answers.

Add or subtract.

15. $-8 + (-20)$ -28

16. $-17.9 + 13.4$ -4.5

17. $-1\frac{3}{4} - \left(-2\frac{2}{3}\right)$ $\frac{11}{12}$

18. $74.3 - (-45.9)$ 120.2

Multiply or divide.

19. $(-2.7)(1.8)$ -4.86

20. $\left(-2\frac{1}{3}\right) \div \frac{1}{2}$ $-4\frac{2}{3}$

21. $-3\frac{1}{5}\left(-1\frac{7}{8}\right)$ 6

22. $-70 \div (-0.2)$ 350

Simplify.

23. $x^2 \cdot x^{12}$ x^{14}

24. $\frac{m^{15}}{m^{10}}, m \neq 0$ m^5

25. $(y^4)^3$ y^{12}

26. $\frac{(m^{-1}n)^2}{m^0 n^{-3}}$ $\frac{n^5}{m^2}$

Evalute when $x = -4$ and $y = 3$.

27. x^{-3} $-\frac{1}{64}$

28. y^{-4} $\frac{1}{81}$

29. $(xy)^{-2}$ $\frac{1}{144}$

30. x^2y^{-2} $\frac{16}{9}$

Write in scientific notation.

31. 768,000,000 7.68×10^8

32. 0.0000189 1.89×10^{-5}

Write in standard notation.

33. 6.38×10^7 63,800,000

34. 3.2×10^{-4} 0.00032

Teaching Tip

If students have difficulty remembering the rules of exponents in Exercises 11–13, you may want to suggest that they write out the factors before multiplying or dividing.
Examples:
$x^2 \cdot x^3 = (x \cdot x)(x \cdot x \cdot x) = x^5$
$(x^2)^3 = (x \cdot x)(x \cdot x)(x \cdot x) = x^6$

CHAPTER 1 CUMULATIVE ASSESSMENT

STANDARDIZED TEST PREPARATION—STANDARD MULTIPLE CHOICE

Choose the best solution for each problem.

1. Which is not a subset of {D, O, G}?
D
A. ∅ **B.** {D, O, G}
C. {O, G} **D.** {D, E}
E. {O}

2. Which is a solution of $3x - 5 = 16$?
A
A. 7 **B.** −7
C. $3\frac{2}{3}$ **D.** $10\frac{1}{3}$
E. none of these

3. Which of the following is correct roster notation for $\{x | x$ is a negative integer and $x < -3\}$?
C
A. {−3, −2, −1} **B.** {−2, −1, 0}
C. {−2, −1} **D.** {−4, −5, −6, . . .}
E. none of these

4. Which is not a rational number?
C
A. sqrt(25) **B.** −3.14
C. sqrt(28) **D.** $-\frac{3}{8}$
E. 4/sqrt(4)

5. Evaluate $-|m|$ when $m = -5$.
B
A. 5 **B.** −5
C. 0 **D.** −151
E. none of these

6. Which graph represents the set of real numbers less than 5?
B
A. −3 −2 −1 0 1 2 3 4 5 6
B. −3 −2 −1 0 1 2 3 4 5 6
C. −3 −2 −1 0 1 2 3 4 5 6
D. −3 −2 −1 0 1 2 3 4 5 6
E. none of these

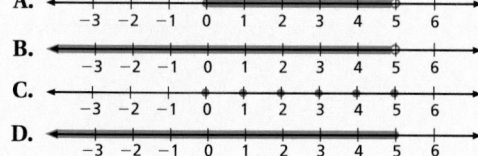

7. Divide: −150 ÷ (0.5)
C
A. −75 **B.** 300
C. −300 **D.** 75
E. −30

8. Given that $R = \{2, 4, 6, 8\}$ and $S = \{1, 2, 3\}$, find $R \cup S$.
B
A. {2} **B.** {1, 2, 3, 4, 6, 8}
C. {1, 3, 4, 6, 8} **D.** {2, 4, 6, 8}
E. none of these

9. Subtract: $\left(-1\frac{3}{5}\right) - \frac{3}{10}$
D
A. $1\frac{9}{10}$ **B.** $-1\frac{3}{10}$
C. $1\frac{3}{10}$ **D.** $-1\frac{9}{10}$
E. none of these

10. Multiply: $\frac{5}{16}x - \left(\frac{4}{15}\right)$
B
A. $\frac{1}{12}$ **B.** $-\frac{1}{12}$
C. $\frac{1}{7}$ **D.** $-\frac{1}{7}$
E. $-1\frac{1}{4}$

11. Simplify: $x^4 \cdot x^8$
B
A. x^{32} **B.** x^{12}
C. x^4 **D.** x^2
E. none of these

12. Simplify: $\frac{x^{12}}{x^2}$
A
A. x^{10} **B.** x^6
C. x^{14} **D.** x^{24}
E. none of these

13. Simplify: $(m^2)^6$
A
A. m^{12} **B.** m^8
C. m^{26} **D.** m^3
E. none of these

14. CONSTRUCTED RESPONSE Describe how to use a Venn diagram to solve this problem. Then give the solution. Shari has a flower garden that contains 24 flowers. Half of the flowers are grown from bulbs. Eight flowers are over 3 ft tall. One-third of the flowers are her favorite color—pink. Only 1 pink flower comes from a bulb and is over 3 ft tall. Five flowers from bulbs are over 3 ft. tall. How many flowers over 3 ft tall are not from bulbs?
See additional answers.

STANDARDIZED TEST PREPARATION

STANDARD MULTIPLE CHOICE— One strategy for completing multiple-choice questions is to use mental math and eliminate unreasonable answers as choices. Often, a couple of the answer choices are values that you may get while working out the problem. For instance, if the solution requires two or three steps, one answer choice may be the value or expression that you have at the end of the first step and another choice may be the value or expression that you have at the end of the second step.

Sometimes all but the correct answer can be eliminated using only mental math and reasoning. If this is the case, verify your answer choice by working out the problem.

TESTING STRATEGIES

Remind students to double check their work. In Exercise 10, students may think "Since −15 divided by 5 is 3, the answer must be −30 since I'm really dividing −150 not −15." Actually, the answer is −300. Division problems can be quickly checked using multiplication.

ADDITIONAL ANSWERS

14. Possible answer: Draw three circles labeled "Bulbs," "Over 3 ft," and "Pink". Use the clues given in the problem to write the number of flowers in each section. Revise the numbers as more hints are given always remembering to keep the total to 24. Shari has 3 flowers that are not from bulbs that are over 3 ft tall.

Teaching Strategies

Some students assume that a Venn diagram always consists of three overlapping circles. Have students try this exercise:
A city recreation center has three sports programs: golf, tennis, and basketball. Every person who signed up for golf also signed up for tennis. Half of the people who signed up for basketball also signed up for tennis. No one who signed up for basketball signed up for golf. Twenty people signed up for golf and 30 people signed up for basketball. Half of the people who signed up for tennis signed up for golf. How many people signed up for only tennis? Five; the Venn diagram should be two interlocking circles, one labeled *Tennis* the other labeled *Basketball*. Inside the circle labeled *Tennis* is another circle labeled *Golf*.

Chapter 2—Essential Algebra and Statistics

Theme: News Media
Chapter Investigation: Media Plan, pages 51, 65, 75, 89, 97
Careers: Environmental Journalists, page 61; Transcriptionists, page 81
Data Activity: Where Can You Find the News?

Content and Connections

Lesson, Pages	Lesson Objectives	NCTM Standards	State/Local Objectives	Interdisciplinary Connections	Real World Applications
2–1 52–55	• Identify the next terms in a sequence and the rule in an iterative process	Patterns, Functions & Algebra Reasoning & Proof Connections; Representation			Media, Finance, Business
2–2 56–59	• Identify relations and their domains and ranges • Identify and evaluate functions	Patterns, Functions & Algebra Geometry & Spatial Sense Representation; Num Oper; Prob Solv; Connections		Biology	Engineering, Biology, Television
2–3 62–65	• Graph linear functions • Evaluate absolute value functions	Geometry & Spatial Sense Num Oper; Patterns, Functions & Algebra; Reasoning & Proof Connections			Retail sales, Photography
2–4 66–69	• Use the Addition or Multiplication Properties of Equality to solve one-step equations	Number Operation Patterns, Functions & Algebra			Business, Finance, News media
2–5 72–75	• Solve equations with more than one step	Number Operation Patterns, Functions & Algebra Prob Solv; Connections			Advertising, Finance, Recreation
2–6 76–79	• Solve an inequality in one or two variables • Graph the solution to an inequality in one or two variables	Number Operation Patterns, Functions & Algebra Prob Solv; Reasoning & Proof Representation			News media, Sales
2–7 82–85	• Construct frequency tables for data • Determine mean, median and mode for a set of data	Data Analysis, Statistics & Prob Prob Solv; Communication Connections; Representation			Sports, Education, Marketing, News media
2–8 86–89	• Construct stem-and-leaf plots • Construct histograms for a data set	Patterns, Functions & Algebra Data Analysis, Stat & Prob Num Oper; Problem Solving Reasoning & Proof Connections; Representation			Health, News, Entertainment
2–9 92–93	• Solve a problem with misleading graphs • Make a table, chart or list • Use an equation or formula				Sports, Advertising, Business

Planning and Resources

Lesson, Pages	Tools/Materials Needed	Trans-parency	Learning/Teaching Styles Options	Assignments: Basic Enriched	Additional Practice in SE	Reteaching, Extra Practice, Enrichment	Other Resources
2–1 52–55	paper/pencil blocks or tiles	Warm up 4 Trans RF-7	Prior Knowledge	B: 1–20, 23–24 E: 1–34	Page 60–61, 71, 81, 91, 94, 687	R: page 17 EP: page 15 E: page 23	SS: Teacher's Choice
2–2 56–59	graph paper	Warm up 4 Trans TK-7	Visual Learner Real World	B: 1–29, 36–44 E: 1–44	Page 60–61, 71, 81, 91, 94, 688	R: page 19 EP: page 17 E: page 25	SS: Teacher's Choice
2–3 62–65	paper/pencil Algeblocks straightedge graph paper	Warm up 4 Trans TK-2–3, 5, 7–10		B: 1–20, 25–38 E: 1–38	Page 70–71, 81, 91, 94, 688	R: page 21 EP: page 19 E: page 27	MC: 7
2–4 66–69	paper/pencil Algeblocks	Warm up 5 Trans TK-2–3, 5	ESL/LEP Visual Learners	B: 1–38, 46–56 E: 1–56	Page 70–71, 81, 91, 95, 689	R: page 23 EP: page 21 E: page 29	SS: Teacher's Choice

2–5 72–75	paper/pencil Algeblocks	Warm up 5 Trans TK-2-3, 5	Prior Knowledge	B: 1–34, 38–55 E: 1–55	Page 80–81, 91, 95, 689	R: page 25 EP: page 23 E: page 31	
2–6 76–79	paper/pencil number lines grid paper	Warm up 5 Trans TK-6-13		B: 1–33, 37–42 E: 1–42	Page 80–81, 91, 95, 690	R: page 27 EP: page 25 E: page 33	MC: 9
2–7 82–85	paper/pencil calculator	Warm up 6 Trans RF-8	Real World	B: 1–14, 17–26 E: 1–26	Page 90–91, 95, 690	R: page 29 EP: page 27 E: page 35	
2–8 86–89	paper/pencil grid paper	Warm up 6 Trans TK-1	Real World Visual Learners	B: 1–18, 20–35 E: 1–35	Page 90–91, 95, 691	R: page 31 EP: page 29 E: page 37	
2–9 92–93	paper/pencil graph paper straightedges	Warm up 6 Trans TK-1, RF-36	Real World	B: 1–4 E: 1–6	Page 94–95	R: page 33 E: page 39	MC: 1

Planning and Pacing

Lesson, Pages	Lesson Title	45 min class	Assignments Basic, Enriched	90 min class	Assignments Basic, Enriched	___ min class	Assignments Basic, Enriched
2–1 52–55	AYR, Opener, Patterns and Iterations	Day 15	B: 1–20, 23–24 E: 1–34	Day 8	B: 1–20, 23–24 E: 1–34		B: 1–20, 23–24 E: 1–34
2–2 56–59	The Coordinate Plane, Relations and Functions, R&PYS	Day 16–17	B: 1–29, 36–44 E: 1–44	Day 8–9	B: 1–29, 36–44 E: 1–44		B: 1–29, 36–44 E: 1–44
2–3 62–65	Linear Functions	Day 18	B: 1–20, 25–38 E: 1–38	Day 9	B: 1–20, 25–38 E: 1–38		B: 1–20, 25–38 E: 1–38
2–4 66–69	Solve One-Step Equations, R&PYS	Day 19–20	B: 1–38, 46–56 E: 1–56	Day 10	B: 1–38, 46–56 E: 1–56		B: 1–38, 46–56 E: 1–56
2–5 72–75	Solve Multi-Step Equations	Day 21	B: 1–34, 38–55 E: 1–55	Day 11	B: 1–34, 38–55 E: 1–55		B: 1–34, 38–55 E: 1–55
2–6 76–79	Solve Linear Inequalities, R&PYS	Day 22–23	B: 1–33, 37–42 E: 1–42	Day 11–12	B: 1–33, 37–42 E: 1–42		B: 1–33, 37–42 E: 1–42
2–7 82–85	Data and Measures of Central Tendency	Day 24	B: 1–14, 17–26 E: 1–26	Day 12	B: 1–14, 17–26 E: 1–26		B: 1–14, 17–26 E: 1–26
2–8 86–89	Display Data, R&PYS	Day 25–26	B: 1–18, 20–35 E: 1–35	Day 13	B: 1–18, 20–35 E: 1–35		B: 1–18, 20–35 E: 1–35
2–9 92–93	Problem Solving Skills: Misleading Graphs	Day 27	B: 1–4 E: 1–6	Day 14	B: 1–4 E: 1–6		B: 1–4 E: 1–6
Review/ Assess		Day 28–30		Day 14–15			

	Chapter 1 Days	Chapter Cumulative Days
Yearly Pacing (45 min class)	16 days	30 days
Yearly Pacing (90 min class)	9 days	15 days

Assessment Options

Assessment in Student Edition	Assessment in Teacher's Edition	Pages in Assessment Book	Software Generated Assessment
Are You Ready?, pages 48–49; Writing Math, pages 55, 59, 64, 68, 74, 75, 79, 85, 88, 89, 93; Mixed Review, pages 55, 59, 65, 69, 75, 79, 85, 89, 93; Check Understanding pages 52, 56, 62, 66, 72, 76, 82, 86; Mid-Chapter Quiz, page 71, Chapter Review, page 94; Chapter Assessment, page 96; Alternative Assessment, page 97; Cumulative Review, page 98; Cumulative Assessment, page 99	5-minute Warm ups, pages 52, 56, 62, 66, 72, 76, 82, 86, 92; Quick Assessment, pages 54, 58, 64, 68, 74, 78, 84, 88, 91; Scoring Rubrics, page 97	Mid-Chapter Quiz, page 23; Test Form A, pages 25, 27; Test Form B, pages 29, 31; Cumulative Test 33, 35; Math Journal prompt, page 37, 38;	Chapter 2

Refresher Skills

The skills on these two pages are skills that have been taught in previous math courses. Continuous review of basic math skills will make stronger math students. These skills are identified as necessary to be successful in Chapter 2.

Skills Correlation Chart

Skill	Lesson Number
Graphing Inequalities	2-2, 2-3, 2-6
Points on a Coordinate Plane	2-2, 2-3, 2-6
Measures of Central Tendency	2-7, 2-8, 2-9
Language of Mathematics	All lessons

Vocabulary

inequality
number line
mean
median
mode
range
coordinate plane
linear equations
measures of central tendency

Additional Answers

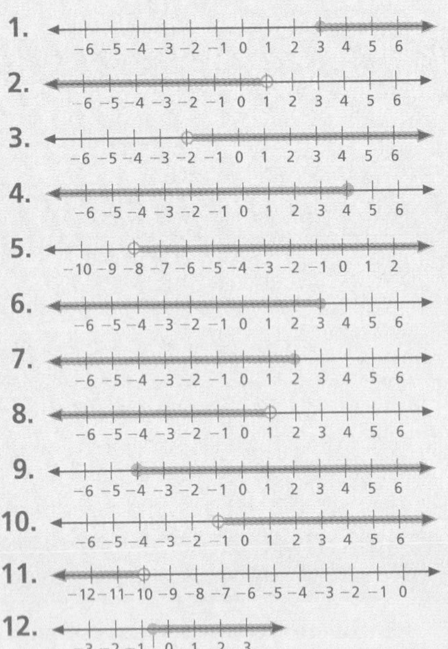

Are You Ready?

Refresh Your Math Skills for Chapter 2

The skills on these two pages are ones you have already learned. Use the examples to refresh your memory and complete the exercises. For additional practice on these and more basic skills, see page 674.

Graphing Inequalities

Graphing an inequality on a number line can give you an easily understandable visual record of all the solutions for the inequality.

Example Graph the inequality $3x \geq -9$ on a number line.

Solve the related equation.

$3x \geq -9 \rightarrow 3x = -9$

$$\frac{3x}{3} = -\frac{9}{3}$$

$x = -3$

Graph the solution.

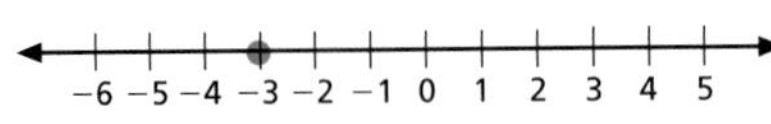

Use a closed dot if that number is included in the solution set. Use an open dot if it is not included.

Graph each inequality on a number line. See additional answers.

1. $2a \geq 6$ **2.** $2(m + 3) < 8$ **3.** $\frac{k}{2} > -1$

4. $4b - 3 \leq 13$ **5.** $\left(\frac{1}{2}\right)x > -4$ **6.** $3z + 5 \leq 14$

7. $3(b - 3) \leq -3$ **8.** $8c - 3 < 5c$ **9.** $\frac{m}{4} - 7 \geq -8$

10. $5(w - 2) > -15$ **11.** $2y + 7 < -13$ **12.** $9d + 2 \geq -4$

Points on a Coordinate Plane

You will need to know how to plot points on the coordinate plane in order to graph linear equations and inequalities.

Plot each point on a coordinate plane. Label each point with its letter. See additional answers.

13. $A(-3, 2)$ **14.** $B(1, 8)$ **15.** $C(8, -1)$

16. $D(2, -6)$ **17.** $E(3, 4)$ **18.** $F(-7, 5)$

19. $G(-5, -5)$ **20.** $H(6, 2)$ **21.** $I(0, 8)$

22. $J(-7, 0)$ **23.** $K(7, -4)$ **24.** $L(-4, -3)$

13–24.

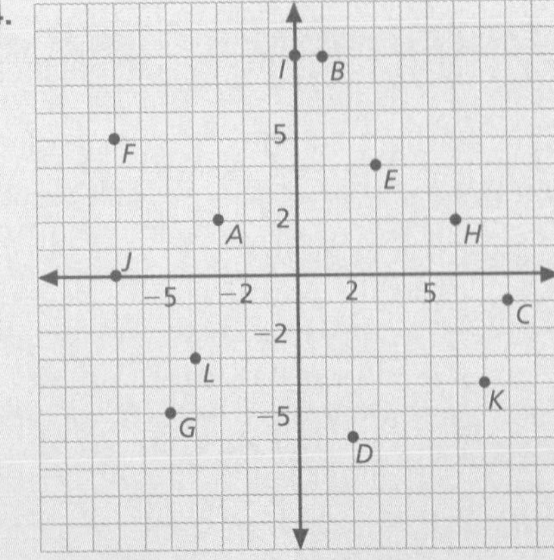

Learning Styles

AUDITORY LEARNER When working with coordinate grids, have students think of a mnemonic device to help them remember that the *x*-coordinate is first. One possibility: A baby crawls on the floor (the *x*-axis) before it stands (the *y*-axis).

CHAPTER 2

Measures of Central Tendency

Example Find the mean, median, mode and range of this data:

8 10 9 9 8 7 5 12 8 6 6

The *mean* is the sum of the data divided by the number of data.

$(8 + 10 + 9 + 9 + 8 + 7 + 5 + 12 + 8 + 6 + 6) \div 11 = 8$

The *median* is the middle value when the data is arranged in numerical order.

5 6 6 7 **8** 8 8 9 9 10 12

If the number of data items is even, the median is the average of the two middle numbers.

The *mode* is the number that occurs most often in the set of data.

5 6 6 7 **8 8 8** 9 9 10 12

The *range* is the difference between the greatest and least values in a set of data.

5 6 6 7 8 8 8 9 9 10 **12**

$12 - 5 = 7$

Find the mean, median, mode and range of each set of data.

25. 2 3 8 7 8 10
8 12 1 2 5 6; 7; 8; 11

26. 5 8 11 13 9 2 4
6 5 7; 5; 6; 11

27. 22 31 16 19 15 24
30 27 27 14 31 32
24; 25.5; 27, 31; 18

28. 42 40 38 46 51 28 37
44 30 29 45 36 27 43
34 36; 38; none; 24

Language of Mathematics

Write each phrase as a numerical expression. Use *n* for a number.

29. seven less than five times a number $5n - 7$

30. the product of two and the quantity a number plus eight $2(n + 8)$

31. the quotient of three times a number and five $\frac{3n}{5}$

32. a number less than the sum of thirteen and eight. $(13 + 8) - n$

33. fourteen decreased by five times a number $14 - 5n$

34. five times the difference of a number and ten $5(n - 10)$

35. three increased by the quotient of a number and two $3 + \frac{n}{2}$

Teaching Strategies

ESL/LEP Have students brainstorm a list of words that could be synonyms for *average*. Encourage students to include words from their native languages. Some English words are *common, typical, usual,* and *normal*. Have students discuss how they would go about finding the average height of a student in the class. Encourage both formal and informal measures of central tendency.

Chalkboard Examples

Graphing Inequalities

Graph the inequality $3y < 15$.

Points on a Coordinate Plane

Plot $A(-3, 2)$, $B(-1, -3)$, $C(3, 3)$, *and* $D(3, -2)$ on a coordinate plane.

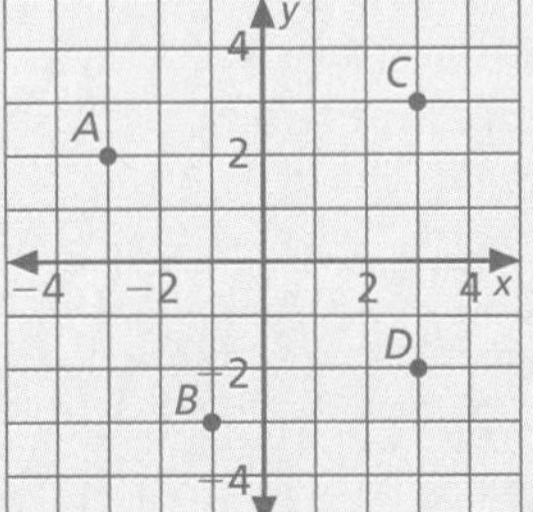

Measures of Central Tendency

Find the mean (to the nearest tenth), median, mode, and range of this data. \$15, \$12, \$18, \$15, \$22, \$13, \$21 **mean: \$16.6; median: \$15; mode: \$15; range: \$10**

Language of Mathematics

Write the phrase "the product of three and the difference of a number and five" as a numerical expression. Use *n* for "a number."
$3(n - 5)$

Refresher Wrap-up

Quick Assessment

Ask the following questions to determine if students have mastered the basic skills reviewed on these pages.

1. Write an inequality equivalent to "*x* is not less than 3." $x \geq 3$
2. Give a situation in which the mode could be the most useful measure of central tendency. **Answers will vary. Example: A shoe store owner deciding how many to order of each shoe size.**

Additional Practice

Refer to the Extra Practice Index on page 674 of this text.

Chapter Opener

NCTM Standards/Strands

- Number & Operation
- Patterns, Functions, & Algebra
- Geometry & Spatial Sense
- Data Analysis, Statistics, & Probability
- Problem Solving
- Reasoning and Proof
- Communication
- Connections
- Representation

Vocabulary

media	trend
news copy	graphic

Theme Connections

Newspapers and television news shows are the primary source for news information for people today. Both use tables, charts, graphs, and other types of graphics to engage their audience and communicate information effectively. Survey students to find out what sources they rely on for news. Have students use the data to create a graphic of the type found in newspapers or magazines.

Career Opportunities

Many careers related to the news media require the use of data and graphs. Two are highlighted in the MathWorks features. Others include reporter, statistician, graphic artist, editor, computer operator, and subject-matter expert.

- Environmental journalist, page 61
- Transcriptionist, page 81

Internet Connection

Theme Activities

Learningmathmatters.com provides web addresses to search that help students gather information about the use of math in the real world, particularly data and graphs. To search for additional addresses, begin a search using the keyword *news*. Then within that search, use such key words as *newspaper, journalism, magazines*, and *television*. Students can brainstorm in small groups other key words.

CHAPTER 2

Essential Algebra and Statistics

THEME: News Media

you? Newspapers and print magazines communicate information in words, symbols, and pictures. Many feature colorful graphs and charts to make data easier to understand.

Television is a primary source of news information for many people. Reporters research the important facts of a story and explain the facts to the viewing audience. Pictures, lists, and colorful graphics help the viewers make sense of the story.

- **Environmental journalists** (page 61) research what is being done to harm and save our planet. They must be able to interpret data, understand graphs, and notice trends in numbers to present the facts without inserting opinions.
- **Transcriptionists or prompter operators** (page 81) prepare the copy that television newscasters read as they report the news. They must be able to work quickly and accurately.

Internet Connection
www.learningmathmatters.com

Chapter Investigation

Go to Learningmathmatters.com to locate additional information about news media.

Where Can You Find the News?
News Coverage by Story Type, 1997

Story type	Newspapers	Print news magazines	TV news	TV news magazines
Consumer News	7.8%	14.2%	12.3%	12.2%
Health, Medicine, and Lifestyle	9.3%	9.7%	14.8%	24.3%
Celebrity, Entertainment and Personal Profiles	3.6%	20.0%	5.6%	18.6%
Government and Economics	32.8%	21.3%	16.9%	0.6%
Foreign Affairs and Military	17.8%	7.7%	16.1%	2.5%
Crime and Justice	13.3%	9.1%	15.1%	23.1%
Accidents and Disasters	1.8%	1.3%	4.9%	4.5%
Science	2.7%	9.7%	9.4%	5.2%
Other	10.9%	7.0%	4.9%	9.0%

Data Activity: Where Can You Find the News?

Use the table for Questions 1–3.

1. Which type of news media gives the most coverage to the daily happenings in the U.S. Congress? newspapers
2. A TV news magazine show is about how many more times as likely as a newspaper to give coverage to problems in a rock star's marriage? About 5 times
3. A print news magazine devotes an average of 4 pages per magazine to its coverage of foreign affairs and the military. About how many pages would you expect to find in the section entitled *Consumer News*? Between 7 and 8 pages

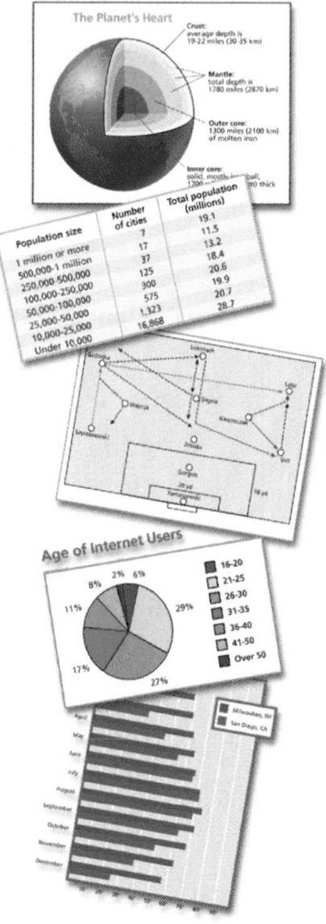

Chapter Investigation

The news media use many different ways to present data to the public. Newspapers and magazines use charts, tables, and graphs. Television uses animated diagrams and graphs to make data appealing. Some types of presentations are more effective than others. Some graphs and charts may even be biased or misleading.

Working Together

Think of something that would improve your school or community. Discuss what kind of data would be needed to convince voters and community leaders to make the change. Develop a media plan that would include using television, newspapers, and magazines to encourage support for your plan. Use the Chapter Investigation logo to guide your group.

Data Activity

Discuss with students the different categories of news stories covered in newspapers and on television. For each story type shown in the table, have them think of one or two articles that might be found in that section. Discuss whether they consider the news sources in the table to be equally reliable.

Extend the Data Activity

REAL WORLD CONNECTION Have students choose a widely covered current event and find samples of coverage in each of the news sources from the table. Discuss how the sources handled the story differently.

Chapter Investigation

As an Overarching Problem

Display several newspaper advertisements that promote an idea or proposal. Ask students to study the advertisements and think about why the ad would sway someone to support the idea presented. Have students share their thoughts with the class. Students will continue to work on the investigation as they complete the exercises identified by the Investigation Logo that are found throughout the chapter. These exercises will guide students through the task described in *Working Together*. Encourage students to keep all of their work on the Investigation together. Have students use the suggestions in the Chapter Investigation Extension to summarize their work.

As a Chapter Project

The goal of this project is to collect and use data to write an advertisement to promote a proposed change to your school or community. Students can use the Team Project Planner on page 21 and the Project Planning Calendar on page 22 in the Enrichment text to complete the project. Benchmarks **a**, **b**, and **c** should be completed after the lesson listed in parentheses has been studied. Benchmark **d** should be completed at the end of the chapter.

Project Planning Calendar

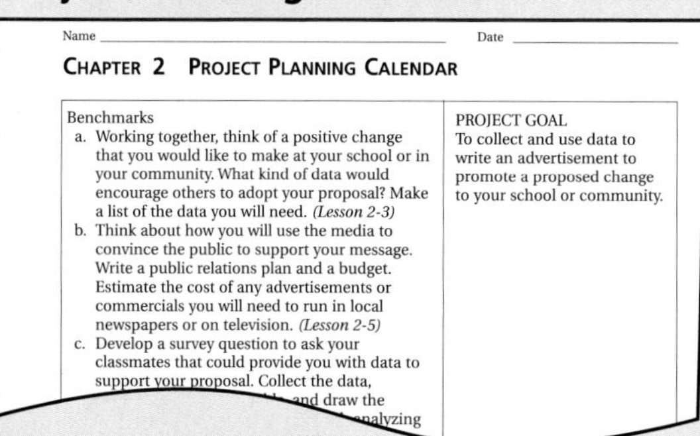

Name ______ Date ______

CHAPTER 2 PROJECT PLANNING CALENDAR

Benchmarks
a. Working together, think of a positive change that you would like to make at your school or in your community. What kind of data would encourage others to adopt your proposal? Make a list of the data you will need. *(Lesson 2-3)*
b. Think about how you will use the media to convince the public to support your message. Write a public relations plan and a budget. Estimate the cost of any advertisements or commercials you will need to run in local newspapers or on television. *(Lesson 2-5)*
c. Develop a survey question to ask your classmates that could provide you with data to support your proposal. Collect the data, ... and draw the ... analyzing

PROJECT GOAL
To collect and use data to write an advertisement to promote a proposed change to your school or community.

Group Project Planner

Name ______ Date ______

CHAPTER 2 GROUP PROJECT PLANNER

Assignment ______ Objective ______

Group Members: 1) ______ 2) ______ 3) ______ 4) ______ 5) ______

Assigned Roles

Deadlines ______ Done ☐

Lesson Planning

NCTM Standards/Strands

- Number & Operation
- Patterns, Functions, & Algebra
- Problem Solving
- Reasoning and Proof
- Communication
- Connections
- Representation

Vocabulary

pattern
term
sequence
iteration

Materials Needed

paper/pencil
blocks or tiles
word processor (optional)

Lesson Resources

Warm-Up Transparency 4
Reteaching 2-1
Extra Practice 2-1
Enrichment 2-1
Transparency RF-7

Assignment Guide

Basic: 1–20, 23–34
Enriched: 1–34

Getting Started

5-Minute Warm-up

Evaluate $x + 4$ for each x.

1. 9 13 **2.** −3 1 **3.** 5 9

Evaluate $5x + 4$ for each x.

4. 2 14 **5.** −1 −1 **6.** −3 −11

Introduction to Lesson 2-1

Have students state the rule they discover in their own words and in as many different ways as they can. Note: Some students may need to build the 6th and 7th figures in order to find the answer.

2-1 Patterns and Iterations

Goals ■ Identify the next terms in a sequence and the rule in an iterative process.

Applications Media, Finance, Business

1. Use blocks or tiles to make the next two figures in this pattern.
2. Make a chart like this to show the number of pieces used in each figure.
3. How many pieces will it take to make the 7th figure? How do you know your answer is correct?
   28; Answers will vary.

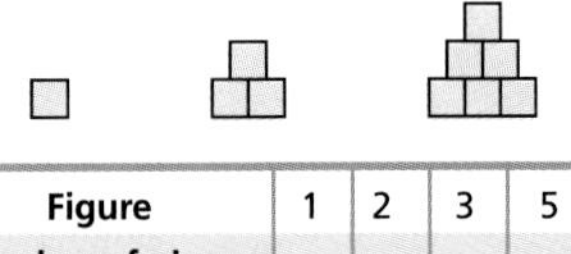

Figure	1	2	3	5
Number of pieces				

Build Understanding

Mathematicians have always been interested in number patterns. Many such patterns exist naturally, both in mathematics and in nature. A **pattern**, or **sequence**, is an arrangement of numbers in a particular order. The numbers are called **terms**, and the pattern is formed by applying a rule. If a pattern exists, a prediction can be made about the terms in the pattern. For example, the numbers 2, 7, 12, 17, 22, . . . are arranged in a pattern. Each number is 5 more than the preceding number. The rule is "add 5."

Example 1

Identify the pattern 1, 2, 4, 7, _____, _____, _____, and find the next three terms.

Solution

In this pattern, the number being added each time is 1 more than the number that was added previously.

1, 2, 4, 7, 11, 16, 22
+1 +2 +3 +4 +5 +6

The next three terms are 11, 16, and 22.

In some patterns, a term can be calculated by applying a rule to the term's position number.

Teaching Strategies

TEAM LEARNING Have groups of students make up their own sequences of at least 4 iterations and record them on index cards. On the back of each card, they should write the rule used to make the sequence. Have groups exchange cards and attempt to find the rule and the next three items in the sequence for each card.

Example 2

In the sequence 1, 4, 9, 16, . . . , identify the rule relating each term to its position number. Then find the 10th term and the 15th term.

Solution

Each term in the sequence 1, 4, 9, 16, . . . , is found by taking the square of its position number. The 10th term is 10^2, or 100. Likewise, the 15th term is $15^2 = 225$.

Position number	1	2	3	4
Term	$1^2 = 1$	$2^2 = 4$	$3^2 = 9$	$4^2 = 16$

In mathematics, the term **iteration** is used to describe a process that is repeated over and over. You have already seen how iterations can be used to create number patterns. For example, the sequence 1, 2, 4, 8, . . . , is generated by using the iterative process of multiplying by 2. An iteration diagram can also be used to model the sequence.

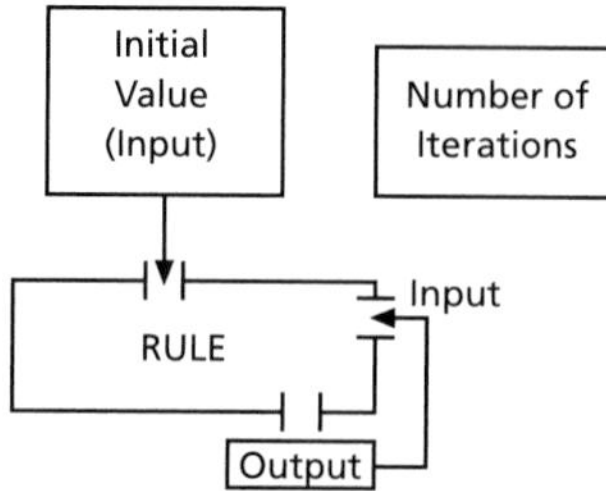

Example 3

The sequence 1, 3, 9, 27, . . . , can be modeled using an iteration diagram. Draw the diagram and calculate the output for 7 iterations.

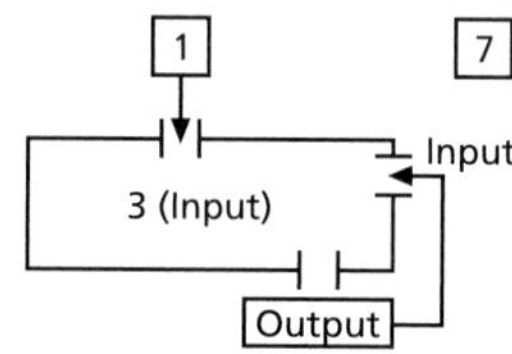

Solution

Initial value (input): 1 Number of iterations: 7

Rule: Multiply input by 3 Output: 3, 9, 27, 81, 243, 729, 2,187

Many occupations require the use of iterative processes. Most assets a business owns, such as a car or a piece of business equipment, become less valuable over the time they are used. This is called **depreciation**. For example, if a new car was purchased for $13,000 and sold three years later for $5,000, the value of the car has depreciated $8,000 in value.

There are different methods to calculate depreciation. Many such methods are iterative, such as the one called the declining-balance method.

Example 4

NEWSPAPER A printing machine has an expected life of five years, a beginning book value (cost when bought) of $50,000, and a depreciation rate of 30% per year. Find the ending book value after five years.

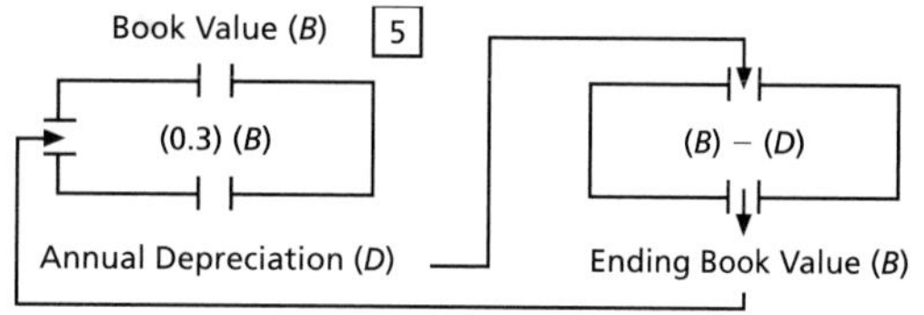

Extend the Lesson

MAKING CONNECTIONS The sequence, 1, 1, 2, 3, 5, 8, . . . is called the Fibonacci sequence. It occurs many times in nature and has a connection to the Golden Ratio. Have students investigate either the occurrences in nature of the sequence or its relationship to the Golden Ratio and use their findings to prepare a classroom display.

Chalkboard Examples

Supplementary Example 1

Identify the pattern 2, 8, 14, 20, __, __, __, and find the next three terms. add 6; 26, 32, 38

Supplementary Example 2

In the sequence, −2, −4, −6, −8, . . . , identify the rule relating each term to its position number. Then find the 10th term and the 15th term. multiply the position number by −2; −20, −30

Supplementary Example 3

The sequence 1, −2, 4, −8, . . . , can be modeled using an iteration diagram. Draw the diagram and calculate the output for 7 iterations. Rule: Multiply by −2; Output: 1, −2, 4, −8, 16, −32, 64

Supplementary Example 4

A large machine was purchased at a cost of $100,000. It has an expected life of 15 years. Using a depreciation rate of 10% per year, find the book value of the machine at the end of five years. $59,049

Reteaching Worksheet 2-1

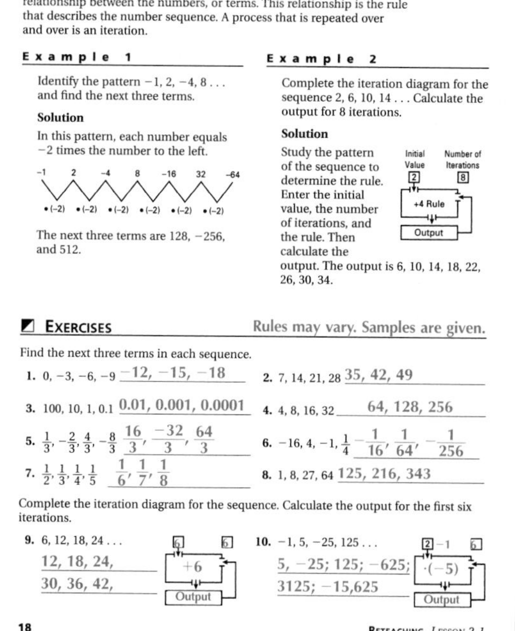

Name __________ Date __________

RETEACHING 2-1

PATTERNS AND ITERATIONS

You can determine the number pattern in a sequence by finding the relationship between the numbers, or terms. This relationship is the rule that describes the number sequence. A process that is repeated over and over is an iteration.

Example 1

Identify the pattern −1, 2, −4, 8 . . . and find the next three terms.

Solution

In this pattern, each number equals −2 times the number to the left.

−1 2 −4 8 −16 32 −64

•(−2) •(−2) •(−2) •(−2) •(−2) •(−2)

The next three terms are 128, −256, and 512.

Example 2

Complete the iteration diagram for the sequence 2, 6, 10, 14 . . . Calculate the output for 8 iterations.

Solution

Study the pattern of the sequence to determine the rule. Enter the initial value, the number of iterations, and the rule. Then calculate the output. The output is 6, 10, 14, 18, 22, 26, 30, 34.

Initial Value 2 Number of Iterations 8 +4 Rule Output

EXERCISES Rules may vary. Samples are given.

Find the next three terms in each sequence.

1. 0, −3, −6, −9 −12, −15, −18
2. 7, 14, 21, 28 35, 42, 49
3. 100, 10, 1, 0.1 0.01, 0.001, 0.0001
4. 4, 8, 16, 32 64, 128, 256
5. $\frac{1}{3}, -\frac{2}{3}, \frac{4}{3}, -\frac{8}{3}$ $\frac{16}{3}, \frac{-32}{3}, \frac{64}{3}$
6. $-16, 4, -1, \frac{1}{4}$ $-\frac{1}{16}, \frac{1}{64}, -\frac{1}{256}$
7. $\frac{1}{2}, \frac{1}{3}, \frac{1}{4}, \frac{1}{5}$ $\frac{1}{6}, \frac{1}{7}, \frac{1}{8}$
8. 1, 8, 27, 64 125, 216, 343

Complete the iteration diagram for the sequence. Calculate the output for the first six iterations.

9. 6, 12, 18, 24 . . . 12, 18, 24, 30, 36, 42, [6] [6] +6 Output
10. −1, 5, −25, 125 . . . 5, −25; 125; −625; 3125; −15,625 [2] −1 [6] ·(−5) Output

18 RETEACHING *LESSON 2-1*

Lesson 2-1 **Patterns and Iterations** 53

Lesson Wrap-up

Quick Assessment

Ask the following questions to determine if students understand the content presented in this lesson.

1. Determine the rule and the next three terms in each sequence.
   a. 4, 12, 20, 28 . . . Add 8; 26, 44, 52
   b. 2, 6, 18, 54 . . . Multiply by 3; 162, 486, 1,458
   c. 100, −50, 25, −12.5 . . . Divide by −2; 6.26, −3.125, 1.5625
2. Start with 3. Use the rule "add 5" six times. 3, 8, 13, 18, 23, 28, 33

Assignment Guide

Basic: 1–20, 23–34
Advanced: 1–34
Additional Practice: See Extra Practice Index on page 674.

Additional Answers

16.

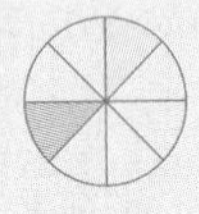

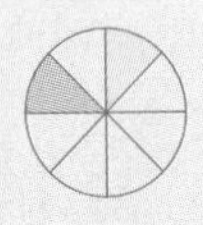

17.

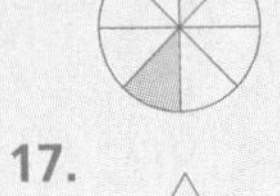

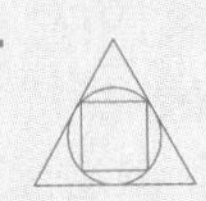

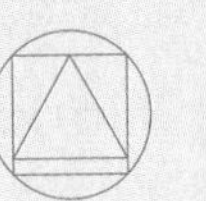

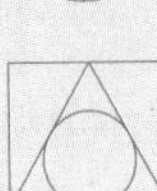

23.

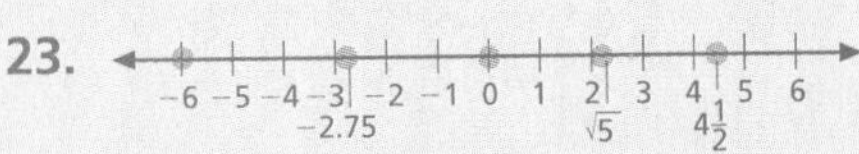

24.

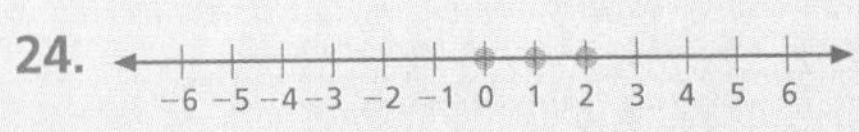

25.

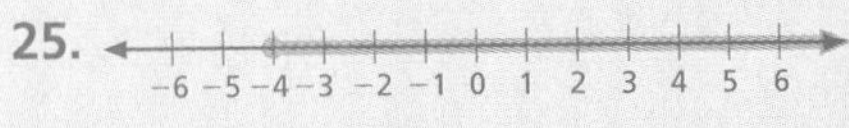

26.

27.

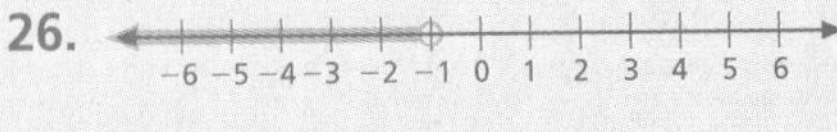

28.

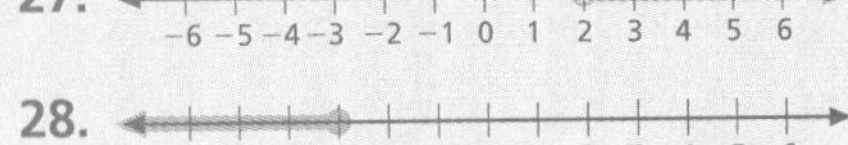

29.

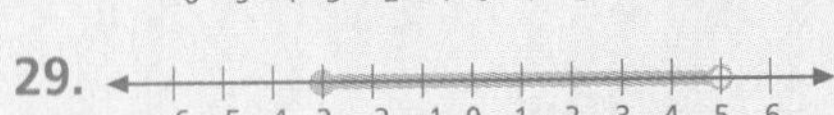

30.

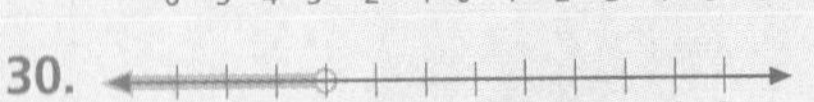

31.

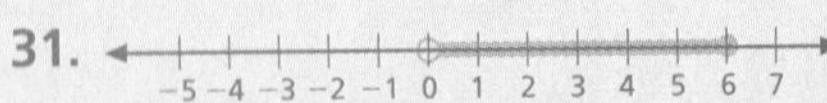

32.

33.

34.

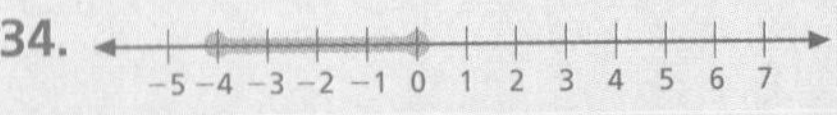

Solution

Calculate the output for 5 iterations.

Year 1	Beginning book value	Depreciation rate	Annual depreciation	Ending book value
1	$50,000.00	0.3	$15,000.00	$35,000.00
2	35,000.00	0.3	10,500.00	24,500.00
3	24,500.00	0.3	7,350.00	17,150.00
4	17,150.00	0.3	5,145.00	12,005.00
5	12,005.00	0.3	3,601.50	8,403.50

The ending book value is $8,403.50.

Try These Exercises

Identify the rule relating each term. Find the next three terms in each sequence.

1. −15, −11, −7, −3, ____, ____, ____ Add 4; 1, 5, 9

2. 1, 7, 13, 19, ____, ____, ____ Add 6; 25, 31, 37

3. 5, 2, −1, −4, ____, ____, ____ Subtract 3; −7, −10, −13

4. −1, 3, −9, 27, ____, ____, ____ Multiply by −3; −81, 243, −729

5. 2, 6, 18, 54, ____, ____, ____ Multiply by 3; 162, 486, 1,458

6. Draw the iteration diagram for this sequence:

16, 4, 8, 2, ____, ____, ____.

Identify each part of the diagram. Then calculate the output for the first 7 iterations.
Answers may vary. Sample: 16, 4, 8, 2, 4, 1, 2

Practice Exercises

Determine the next three terms in each sequence.

7. 2, 3, 5, 8, ____, ____, ____ 17, 17, 23

8. 1, 2, 5, 10, ____, ____, ____ 17, 26, 37

9. 1, 3, 7, 13, ____, ____, ____ 21, 31, 43

10. 20, 8, −4, −16, ____, ____, ____ −28, −40, −52

11. 1, 8, 27, 64, ____, ____, ____ 125, 216, 343

12. $-1, \frac{1}{2}, -\frac{1}{4}, \frac{1}{8}$, ____, ____, ____ $-\frac{1}{16}, \frac{1}{32}, -\frac{1}{64}$

13. **ERROR ALERT** Ryan determines that the next term in the sequence 1.7, 6.9, 12.1, 17.3, . . . is 23.5. Explain what Ryan did wrong.
Add 5.2 to continue the pattern.

14. Draw the iteration diagram for the sequence 1, 2, 4, 8, 16, ... Calculate the output for the first 7 iterations.
32, 64, 128

15. Determine the output for the given iteration. (Round each answer to the nearest cent.)
300; 450; 1,012.50; 1,518.75

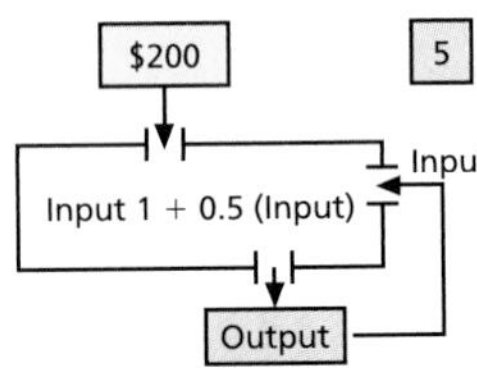

Extend the Lesson

CONNECTING TO PRIOR KNOWLEDGE To find the book value after a number of years, students can save time by using the complement of the of the depreciation rate. Instead of multiplying by 30% ands subtracting the depreciation, they can multiply by 100%—30%, or 70%, to find the ending book value for that year. By using a calculator and repeated multiplication, students can quickly find the ending book value after a certain number of years.

Draw the next three figures in the pattern.
See additional answers.

16. 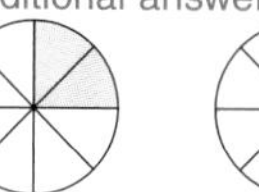 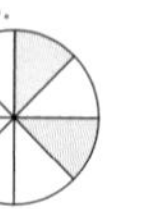 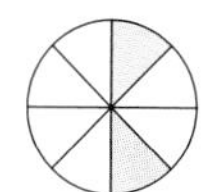

17. 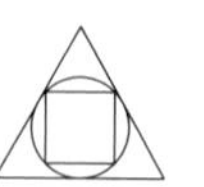 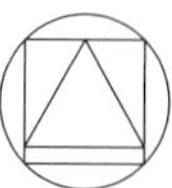 

TELEVISION Use the declining-balance method to find the book value at the end of the expected life for each of these assets.

18. sound system for \$18,000, expected life of five years, depreciation rate of 30% per year \$3,025.26

19. video equipment for \$12,000, expected life of three years, depreciation rate of 50% per year \$1500

20. NEWSPAPER A local newspaper hopes to increase its circulation at the rate of 15% per year for the next five years. If its current circulation is 15,640, what will its circulation be at the end of five years. Round your answer to the nearest whole number. 31,458

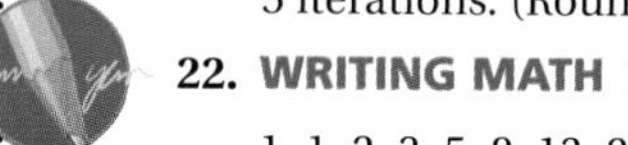

Extended Practice Exercises

21. FINANCE John deposits \$500 in his savings account which earns interest at a 7.5% rate. At the end of each year, he also adds \$100 to his account. Determine the iteration diagram and the output for 5 iterations. (Round each answer to the nearest cent.) \$637.50, \$785.31, \$944.21, \$1,115.03, \$1,298.66

22. WRITING MATH Examine the following sequence.

1, 1, 2, 3, 5, 8, 13, 21, . . .

Identify the rule being applied to the pattern, and find the next three terms.
n = sum of previous two numbers; 34, 55, 89

Mixed Review

Graph each set of numbers on a number line. (Lesson 1-1)

23. $\left\{-6, -2.75, 0, \sqrt{5}, 4\frac{1}{2}\right\}$ See additional answers.

24. whole numbers less than 3

25. real numbers greater than or equal to -4

26. real numbers less than -1

27. real numbers greater than 2

28. real numbers less than or equal to -3

Graph the intersection of each pair of sets. (Lesson 1-3) See additional answers.

29. $A = \{x|x \leq -3\}$
$B = \{x|x < 5\}$

30. $A = \{x|x > -1$
$B = \{x|x \geq 4\}$

31. $A = \{x|x > 0\}$
$B = \{x|x \leq 5\}$

32. $A = \{x|x < 2\}$
$B = \{x|x > 1\}$

33. $A = \{x|x > -5\}$
$B = \{x|x \leq 2\}$

34. $A = \{x|x \geq 0\}$
$B = \{x|x \leq -4\}$

Learning Styles

TACTILE/KINESTHETIC LEARNER Students who are having difficulty seeing how the rule determines the next number in a sequence may benefit from drawing an iteration diagram. You may have students write input and output numbers on small slips of paper and move them through the diagram to illustrate the process.

Extra Practice Worksheet 2-1

Name ______ Date ______

EXTRA PRACTICE 2-1
PATTERNS AND ITERATION

EXERCISES

Identify the rule relating each term. Find the next three terms in each sequence.

1. 4, 6, 8, 10, 12, 14, 16 — rule: add 2
2. 1, 3, 7, 13, 21, 31, 43, 57 — rule: add 2, 4, 6, . . .
3. 200, 190, 180, 170, 160, 150, 140 — rule: subtract 10
4. −16, −13, −10, −7, −4, −1, 2 — rule: add 3
5. 4, 8, 16, 32, 64, 128, 256 — rule: multiply by 2
6. 9, 6, 3, 0, −3, −6, −9 — rule: subtract 3
7. 8, −24, 72, −216, 648, −1944, 5832 — rule: multiply by −3
8. 15, 14, 12, 9, 5, 0, −6 — rule: subtract 1, 2, 3, . . .
9. 4, 2, 1, $\frac{1}{2}$, $\frac{1}{4}$, $\frac{1}{8}$, $\frac{1}{16}$ — rule: divide by 2
10. 2, 1.5, 1, 0.5, 0, −0.5, −1 — rule: subtract 0.5
11. 6, 0, −6, −12, −18, −24, −30 — rule: subtract 6
12. −4, 16, −64, 256, −1024, 4096, −16,384 — rule: multiply by −4
13. 4, −1, $\frac{1}{4}$, $-\frac{1}{16}$, $\frac{1}{64}$, $-\frac{1}{256}$, $\frac{1}{1024}$ — rule: divide by −4
14. 6, 12, 18, 24, 30, 36, 42 — rule: add 6
15. 8, −4, 2, −1, $\frac{1}{2}$, $-\frac{1}{4}$, $\frac{1}{8}$ — rule: divide by −2
16. Start with 4. Use the rule "subtract 2" six times. 4, 2, 0, −2, −6, −8
17. Start with 100. Use the rule "add 25" six times. 100, 125, 150, 175, 200
18. Start with 30. Use the rule "multiply by −2" six times. 30, −60, 120, −240, 480, −960
19. Start with 500. Use the rule "divide by 5" six times. 500, 100, 20, 4, $\frac{4}{5}$, $\frac{4}{25}$

16 EXTRA PRACTICE *LESSON 2-1*

Enrichment Worksheet 2-1

Name ______ Date ______

ENRICHMENT 2-1
TOOTHPICK PUZZLES

EXERCISES

The puzzles on this page involve only a box of toothpicks.

1. Move 4 toothpicks to make 3 equilateral triangles.
2. Remove 3 toothpicks to make 3 equilateral triangles.
3. Move 4 toothpicks to make 3 identical squares.
4. Remove 5 toothpicks to make 5 congruent triangles.
5. Move 6 toothpicks to make a six-pointed star.
6. Move 2 toothpicks to make 6 equilateral triangles.

Use toothpicks to make a 3-by-3 square divided into 9 smaller squares. Then solve each of these puzzles. Sketch your answers below.

7. Remove 4 toothpicks to make 5 identical squares.
8. Remove 8 toothpicks to make 3 squares.
9. Remove 6 toothpicks to make 3 squares.

In each puzzle, make a true equation by moving just one toothpick.

10. X − I = I I X I = I
11. I − III = II I = III − II

24 ENRICHMENT *LESSON 2-1*

Lesson Planning

NCTM Standards/Strands

- Number & Operation
- Patterns, Functions, & Algebra
- Geometry & Spatial Sense
- Problem Solving
- Reasoning and Proof
- Communication
- Connections
- Representation

Vocabulary

coordinate plane, origin, coordinate, domain, mapping, dependent variable, independent variable, vertical line test, x-axis and y-axis, quadrant, ordered pair, relation, range, function

Materials Needed

graph paper or coordinate grids

Lesson Resources

Warm-Up Transparency 4
Transparency TK-7
Reteaching 2-2
Extra Practice 2-2
Enrichment 2-2

Assignment Guide

Basic: 1–29, 36–44
Enriched: 1–44

Getting Started

5-Minute Warm-up

Evaluate each expression.

1. $3x + 7$ for $x = -2$ 1
2. $5a^2$ for $a = 3$ 45
3. $-4d - 11$ for $d = -4$ 5
4. $-2x^2 + 10$ for $x = 5$ −40

Introduction to Lesson 2-2

Discuss how copiers work to enlarge or reduce an original. Many copiers allow reduction of the copy by a given percent. Ask students what percent reduction they would set the copier at to reduce a figure to 2/3 of its original size. 67%

2-2 The Coordinate Plane, Relations and Functions

Goals
- Identify relations and their domains and ranges.
- Identify and evaluate functions.

Applications Engineering, Biology, Television

Determine the final text size by completing the table. 6 in., $3\frac{1}{2}$ in., 5.6 cm

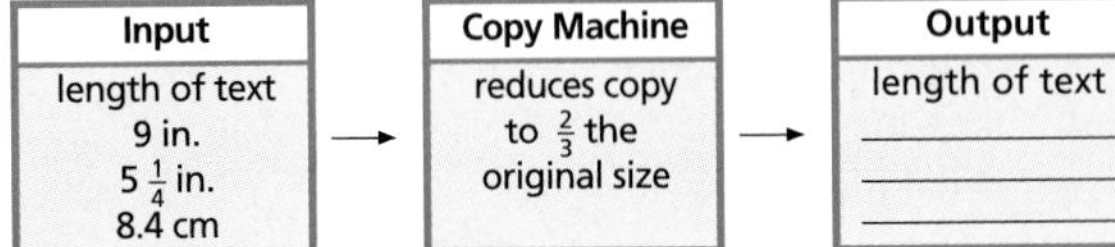

Build Understanding

You know that real numbers can be graphed on a number line. You can graph pairs of numbers on a grid system called a **coordinate plane**. A coordinate plane consists of two perpendicular number lines, dividing the plane into four regions called **quadrants**. The horizontal number line is the *x-axis,* and the vertical number line is the *y-axis.* The point where the axes cross is the **origin**. Points on the axes are not part of the quadrants.

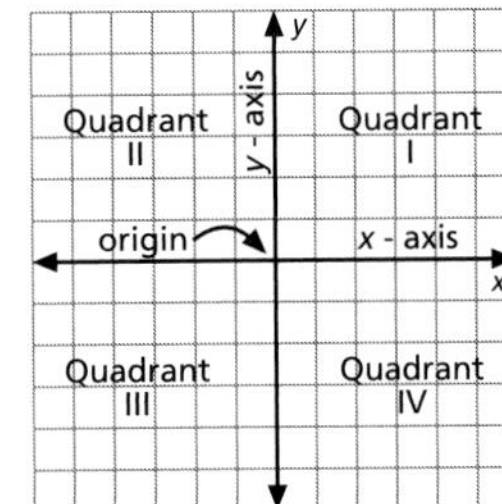

There are infinitely many points in the plane. Each point is unique and is assigned an **ordered pair** of real numbers, consisting of one **x-coordinate** and one **y-coordinate**. For example, the point $(2, -5)$ has an x-coordinate of 2, and a y-coordinate of -5. The x-coordinate, or *abscissa,* determines the horizontal location of the point, while the y-coordinate, or *ordinate,* determines the vertical location. The order of the numbers is important—$(2, -5)$ and $(-5, 2)$ locate two different points.

> **Reading Math**
>
> The x-coordinate is *always* the first coordinate in ordered pairs. The y-coordinate is the second coordinate.
>
> (x, y)

A set of ordered pairs is defined as a **relation**. You can represent a relation by a table of values, a graph, or a set of ordered pairs.

The **domain** of a relation is the set of all possible x-coordinates. The **range** of a relation is the set of all possible y-coordinates. A **mapping** is the relationship between the elements of the domain and the range.

For the set of ordered pairs $\{(0, 1), (2, -1), (3, 2)\}$, a mapping shows the relationship between the x-coordinates (domain) and the y-coordinates (range).

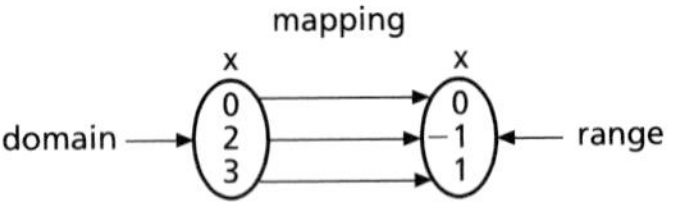

A special kind of relation that is important in mathematics is called a *function.* A **function** is a set of ordered pairs in which each element of the domain is paired with *exactly one* element in the range.

Example 1

Determine whether each relation is a function. State the domain and range of each.

a. 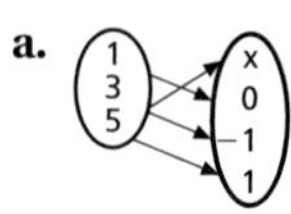

b.

x	-4	-2	0	2
y	2	0	2	4

c. $\{(0, 3), (1, 4), (-3, 0)\}$

Solution

a. No; the element 3 in the domain is paired with two elements in the range, -3 and 3. Domain: $\{1, 3, 5\}$, Range: $\{-3, 0, 3, 4\}$

b. Yes; each element of the domain is paired with exactly one element of the range. However, you will note that one element of the *range* can be paired with more than one element of the domain.
Domain: $\{-4, -2, 0, 2\}$, Range $\{0, 2, 4\}$

c. Yes. Domain: $\{-3, 0, 1\}$, Range: $\{0, 3, 4\}$

Below is another method to determine if a relation is a function.

Vertical Line Test: When a vertical line is drawn through the graph of a relation, the graph is *not a function* if the vertical line intersects the graph in more than one point.

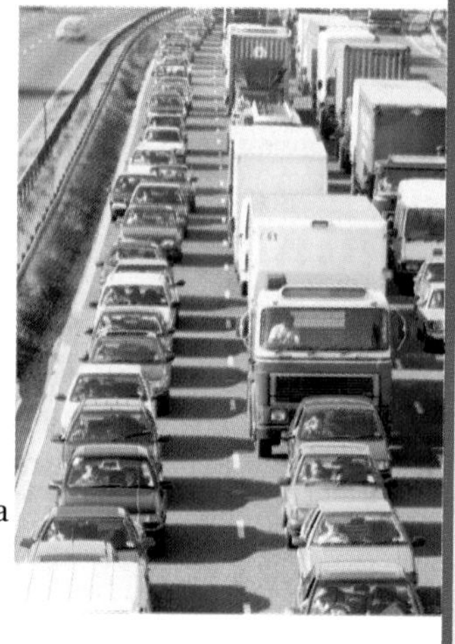

Example 2

Determine whether $\{(2, 2), (0, 0), (2, -2), (4, 4), (4, -4)\}$ is a function by using the vertical line test.

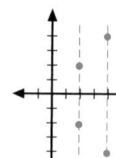

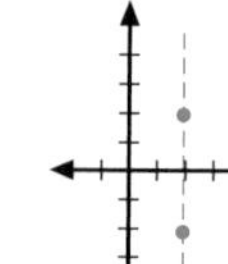

Solution

The relation is *not* a function. A vertical line passes through more than one point.

Examples of functions in everyday life are the relationship between the numbers of hamburgers sold and the price of each, or the weight of a package and the cost of postage. In mathematics, functions are usually given as *rules* that show the relationship of elements of the domain (input values) to elements of the range (output values). *Input* values are *independent* variables, and *output* values are *dependent* variables.

Function notation can represent the rule that associates the input value with the output value, or dependent variable. The most commonly used function notation is called the "*f* of *x*" notation. If f is the function that assigns to each real number x the value $x + 1$, then $f{:}x \rightarrow x + 1$ or $f(x) = x + 1$.

Rule represented by	example	is read
equation in two variables	$y = x + 1$	"*y* is a function of *x* equal to *x* + 1" or "*y* equals *x* + 1"
function notation	$g{:}x \rightarrow x + 1$	"the function *g* that assigns x with *x* + 1"
f of *x* notation	$f(x) = x + 1$	"*f* of *x* equals *x* + 1"

Check Understanding

Write the set of ordered pairs as a table and a mapping; then graph each point.

Learning Styles

VISUAL LEARNERS Have students use colored pencils when drawing coordinate axes to distinguish between *x*-coordinates and *y*-coordinates. For example, they may draw the *x*-axis in blue and the *y*-axis in red. Then, when plotting points, have them underline or circle the *x*-coordinate in blue and the *y*-coordinate in red.

Chalkboard Examples

Supplementary Example 1

Determine whether each relation is a function. State the domain and range of each.

a.

x	1	3	5	8
y	2	4	6	4

Yes. Domain: $\{1, 3, 5, 8\}$, Range: $\{2, 4, 6\}$

b. $\{(1, 2), (-2, 1), (1, 4), (3, 3)\}$

No. Domain: $\{-2, 1, 3\}$, Range: $\{1, 2, 3, 4\}$

Supplementary Example 2

Determine whether $\{(-3, 1), (-2, 3), (0, 5), (2, 3), (3, 1)\}$ is a function by using the vertical line test.

Yes. Verticals intersect the graph in only one point.

Supplementary Example 3

Evaluate $f(x) = 2x^2 + 2x$; $x(3)$ 24

Supplementary Example 4

A school sells magazines to raise money. The school is paid 1/2 the selling price of each magazine. The formula for this function is $f(x) = \frac{x}{2}$. How much will the school make on sales of $12,496? $7,248

Reteaching Worksheet 2-2

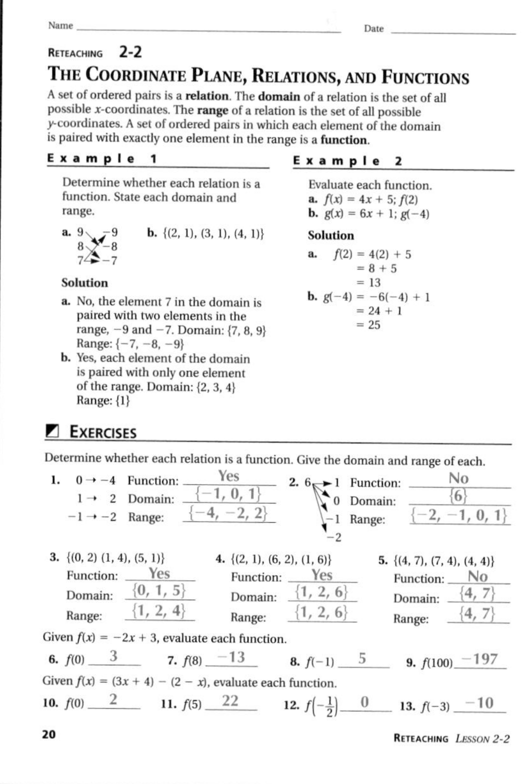

Name ______________________ Date ____________

RETEACHING 2-2

THE COORDINATE PLANE, RELATIONS, AND FUNCTIONS

A set of ordered pairs is a **relation**. The **domain** of a relation is the set of all possible *x*-coordinates. The **range** of a relation is the set of all possible *y*-coordinates. A set of ordered pairs in which each element of the domain is paired with exactly one element in the range is a **function**.

Example 1

Determine whether each relation is a function. State each domain and range.

a. 9 → −9, 8 → −8, 7 → −7 (mapping)

b. $\{(2, 1), (3, 1), (4, 1)\}$

Solution

a. No, the element 7 in the domain is paired with two elements in the range, -9 and -7. Domain: $\{7, 8, 9\}$ Range: $\{-7, -8, -9\}$

b. Yes, each element of the domain is paired with only one element of the range. Domain: $\{2, 3, 4\}$ Range: $\{1\}$

Example 2

Evaluate each function.

a. $f(x) = 4x + 5$; $f(2)$

b. $g(x) = 6x + 1$; $g(-4)$

Solution

a. $f(2) = 4(2) + 5 = 8 + 5 = 13$

b. $g(-4) = -6(-4) + 1 = 24 + 1 = 25$

EXERCISES

Determine whether each relation is a function. Give the domain and range of each.

1. $0 \rightarrow -4$, $1 \rightarrow 2$, $-1 \rightarrow -2$ Function: Yes Domain: $\{-1, 0, 1\}$ Range: $\{-4, -2, 2\}$
2. 6 → 1, 0, −1, −2 Function: No Domain: $\{6\}$ Range: $\{-2, -1, 0, 1\}$
3. $\{(0, 2) (1, 4), (5, 1)\}$ Function: Yes Domain: $\{0, 1, 5\}$ Range: $\{1, 2, 4\}$
4. $\{(2, 1), (6, 2), (1, 6)\}$ Function: Yes Domain: $\{1, 2, 6\}$ Range: $\{1, 2, 6\}$
5. $\{(4, 7), (7, 4), (4, 4)\}$ Function: No Domain: $\{4, 7\}$ Range: $\{4, 7\}$

Given $f(x) = -2x + 3$, evaluate each function.

6. $f(0)$ 3
7. $f(8)$ −13
8. $f(-1)$ 5
9. $f(100)$ −197

Given $f(x) = (3x + 4) - (2 - x)$, evaluate each function.

10. $f(0)$ 2
11. $f(5)$ 22
12. $f\left(-\frac{1}{2}\right)$ 0
13. $f(-3)$ −10

20 RETEACHING *LESSON 2-2*

Lesson Wrap-up

Quick Assessment

Ask the following questions to determine if students understand the content presented in this lesson.

1. Is a straight line the graph of a function? Explain. All straight lines, except vertical lines, are graphs of functions.
2. Is the relation {(0, −2), (3, −2), (6, -2)} a function? Yes
3. Given $f(x) = 4x - 7$. Find $f(-2)$. −15

Assignment Guide

Basic: 1–29, 39–55
Advanced: 1–55
Additional Practice: See Extra Practice Index on page 674.

Additional Answers

1–4.

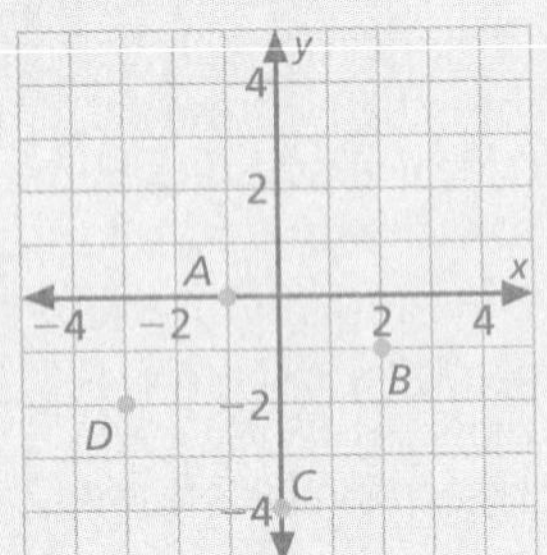

11–14.

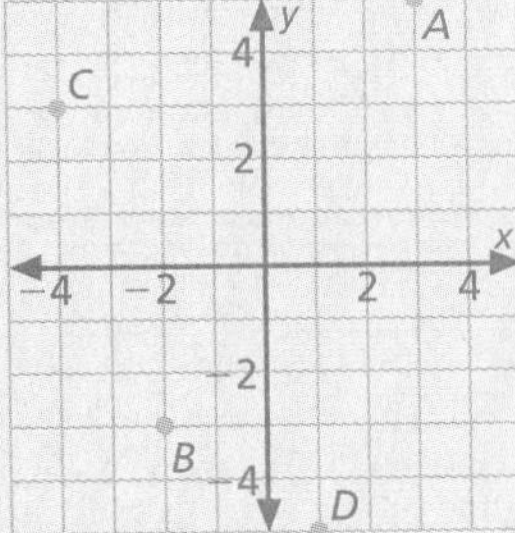

Example 3

Evaluate each function.

a. $f(x) = 3x + 2; f(6)$ **b.** $g(x) = x^2 - 1; g(-1)$

Solution

a. $f(6) = 3(6) + 2$
$= 18 + 2$
$= 20$

b. $g(-1) = (-1)^2 - 1$
$= 1 - 1$
$= 0$

Because functions define the mathematical relationship between two variables, they are often used to model real-world problems.

Example 4

ENGINEERING The air conditioner in a car should produce air that is 26 degrees below the temperature outside the car. The formula for this function is $T(x) = x - 26$, where x is the outside air temperature. What is the temperature inside a car when the temperature outside is 92°F?

Solution

$T(92) = 92 - 26 = 66$ The temperature is 66° inside the car.

Try These Exercises

Graph each point on a coordinate plane. See additional answers.

1. $A(-1, 0)$ **2.** $B(2, -1)$ **3.** $C(0, -4)$ **4.** $D(-3, -2)$

Given $f(x) = x - 5$, evaluate each function.

5. $f(3)$ −2 **6.** $f(0)$ −5 **7.** $f(-2)$ −7 **8.** $f(11)$ 6

Determine if each relation is a function. Give the domain and range.

9.

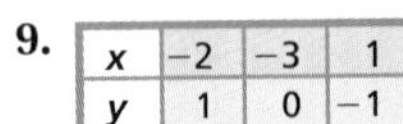

x	−2	−3	1	0	1
y	1	0	−1	1	−3

no; domain: {−3, −2, 0, 1}; range: {−3, −1, 0, 1}

10.

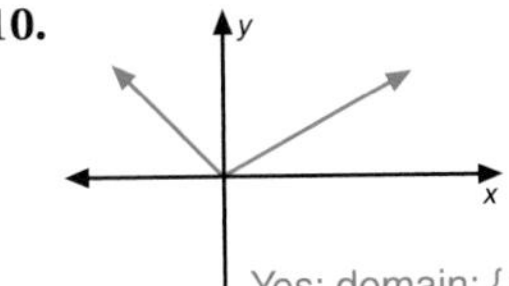

Yes; domain: {x|x is a real number}; range: {y|y is a real number and y ≥ 0}

Practice Exercises

Graph each point on a coordinate plane. Name the quadrant in which each point is located. See additional answers.

11. $A(3, 5)$ I **12.** $B(-2, -3)$ III **13.** $C(-4, 3)$ II **14.** $D(1, -5)$ IV

Given $f(x) = 4x - 1$, evaluate each function.

15. $f(-4)$ −17 **16.** $f(0)$ −1 **17.** $f(2)$ 7

Extend the Lesson

REAL WORLD CONNECTION It may be helpful to give other out-of-class examples of mappings that might be functions. For example, ask students if a list that matches every student to his or her homeroom is an example of a function. Yes

Then ask about a mapping of a list of common first names with a list of last names of students in the school. Is that list a function? Probably not

Encourage students to think of other examples.

Write each as a set of ordered pairs. Give the domain and range.

18. 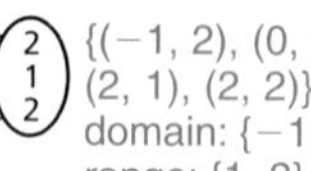
{(−1, 2), (0, 1), (2, 1), (2, 2)}
domain: {−1, 0, 2}
range: {1, 2}

19. 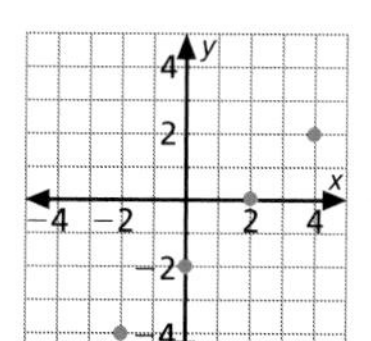
{(−2, −4), (0, −2), (2, 0), (4, 2)};
domain: {−2, 0, 2, 4};
range: {−4, −2, 0, 2}

20.

x	0	1	2
y	−1	1	3

{(0, −1), (1, 1), (2, 3)};
domain: {0, 1, 2}
range: {−1, 1, 3}

BIOLOGY Biologists have determined that the number of chirps made by a cricket in one minute is a function of the temperature (t) measured in degrees Fahrenheit. This relationship is modeled by the function $c(t) = \left(\frac{1}{4}\right)t + 37$. Calculate the number of chirps per minute for the given temperatures.

21. 48°F 49 **22.** 92°F 60

Given $f(x) = 5x + 2$, $g(x) = -2x + 1$, and $h(x) = 3x^2$, find each value.

23. $f(3)$ 17 **24.** $g(5)$ −9 **25.** $h(4)$ 48 **26.** $f(1) + g(1)$ 6

Extended Practice Exercises

Given $f(x) = ax + b$, $g(x) = cx^2$. Find each value.

27. $f\left(\frac{1}{a}\right)$ $1 + b$ **28.** $f\left(\frac{-b}{a}\right)$ $\varnothing$ **29.** $g(c)$ c^3 **30.** $g\left(\frac{-1}{c}\right)$ $\frac{1}{c}$

Use the Vertical Line Test to determine if each graph is a function.

31. 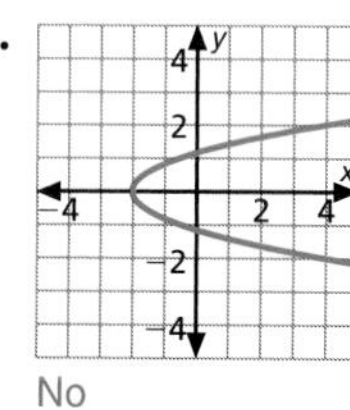
No

32. 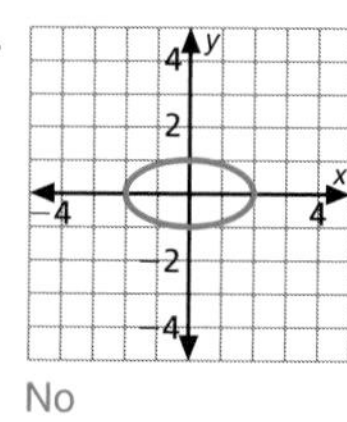
No

33. 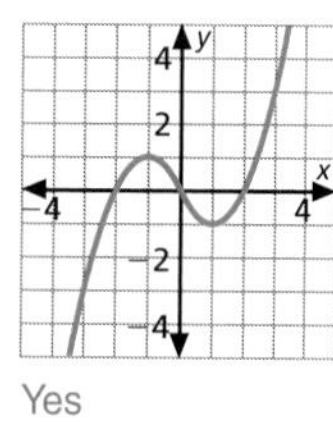
Yes

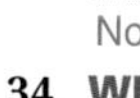

34. WRITING MATH Explain *why* the Vertical Line Test works.
A vertical line shows when an x-value is matched with two y-values.

35. TELEVISION A video technician charges $80 for the first hour. Each additional half-hour or part of a half-hour costs $30. What is the total charge for a $3\frac{1}{4}$-hour session? $230

Mixed Review Exercises

Let $U = \{0, 3, 5, 6, 8, 10, 13, 14, 18, 20\}$. Let $A = \{0, 3, 6, 10, 14, 20\}$. Let $B = \{3, 5, 10, 13, 18\}$ and $C = \{0, 3, 6, 8, 14, 20\}$. Find the following: (Lesson 1-3)

36. $A' \cup B'$ {0, 5, 6, 8, 13, 14, 18, 20}
37. $B' \cap C$ {0, 8, 14, 20}
38. $(A \cup C) \cap B'$ {0, 6, 8, 14, 20}
39. $(B \cap C') \cup A$ {0, 3, 5, 6, 14, 18, 20}
40. $(B' \cap C') \cup A$ {0, 3, 6, 10, 14, 20}
41. $A \cup B' \cap C$ {0, 3, 6, 14, 20}
42. $A \cup B \cap C'$ {5, 10, 18}
43. $(B' \cup C) \cap (A \cup B')$ {0, 3, 6, 8, 10, 20}
44. $(A' \cap C) \cup (C' \cap B)$ {5, 8, 10, 13, 18}

Extend the Lesson

MATH JOURNAL Have students explain the relationship between functions and relations in their own words. Answers will vary. One possible answer: All functions are relations, but not all relations are functions. A function is a relation in which each element in the domain is paired with one and only one element in the range.

Extra Practice Worksheet 2-2

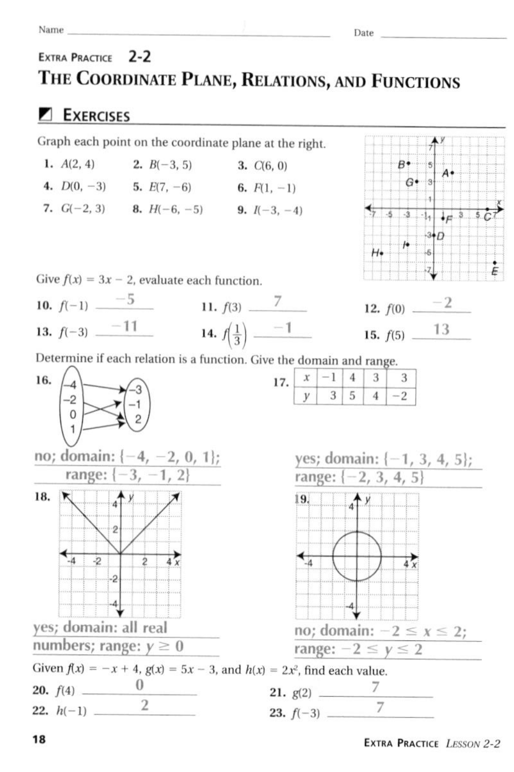
Name ______ Date ______

EXTRA PRACTICE 2-2
THE COORDINATE PLANE, RELATIONS, AND FUNCTIONS

EXERCISES

Graph each point on the coordinate plane at the right.

1. $A(2, 4)$ 2. $B(-3, 5)$ 3. $C(6, 0)$
4. $D(0, -3)$ 5. $E(7, -6)$ 6. $F(1, -1)$
7. $G(-2, 3)$ 8. $H(-6, -5)$ 9. $I(-3, -4)$

Give $f(x) = 3x - 2$, evaluate each function.

10. $f(-1)$ −5 11. $f(3)$ 7 12. $f(0)$ −2
13. $f(-3)$ −11 14. $f\left(\frac{1}{3}\right)$ −1 15. $f(5)$ 13

Determine if each relation is a function. Give the domain and range.

16. no; domain: {−4, −2, 0, 1}; range: {−3, −1, 2}

17.

x	−1	4	3	3
y	3	5	4	−2

yes; domain: {−1, 3, 4, 5}; range: {−2, 3, 4, 5}

18. yes; domain: all real numbers; range: $y \geq 0$

19. no; domain: $-2 \leq x \leq 2$; range: $-2 \leq y \leq 2$

Given $f(x) = -x + 4$, $g(x) = 5x - 3$, and $h(x) = 2x^2$, find each value.

20. $f(4)$ 0 21. $g(2)$ 7
22. $h(-1)$ 2 23. $f(-3)$ 7

18 EXTRA PRACTICE *LESSON 2-2*

Enrichment Worksheet 2-2

Name ______ Date ______

ENRICHMENT 2-2
SECRET CODE

A coordinate graph can be used to make a secret code. Letter labels on the points show the order of the letters in the message. For example, if $(1, 0) = H$, $(2, -2) = E$, $(1, -1) = L$, and $(0, 0) = P$, the grid at the right would show the word "HELP."

EXERCISES

Choose a different pair of coordinates for each letter of the alphabet below. Use values from −2 through 2 for both x and y. Then graph points to show each secret message.

A	B	C	D	E	F	G

Answers may vary. Check students' work.

H	I	J	K	L	M	N	O	P

R	S	T	U	V	W	X	Y	Z

1. SUCCESS 2. MEET AT TWO 3. NEED A CAR
4. SEND MONEY 5. THANK YOU 6. BRING FOOD

26 ENRICHMENT *LESSON 2-2*

Skills Practice

Vocabulary Review

Lesson 2–1
pattern, sequence, term, iteration

Lesson 2–2
coordinate plane, quadrant, origin, ordered pair, coordinate, relation, domain, range, mapping, function, dependent variable, independent variable, vertical line test, *x*-axis and *y*-axis

Assignment Guide

All students: 1–50
Additional Practice: Refer to the Extra Practice Index on page 674 of this text.

Additional Answers

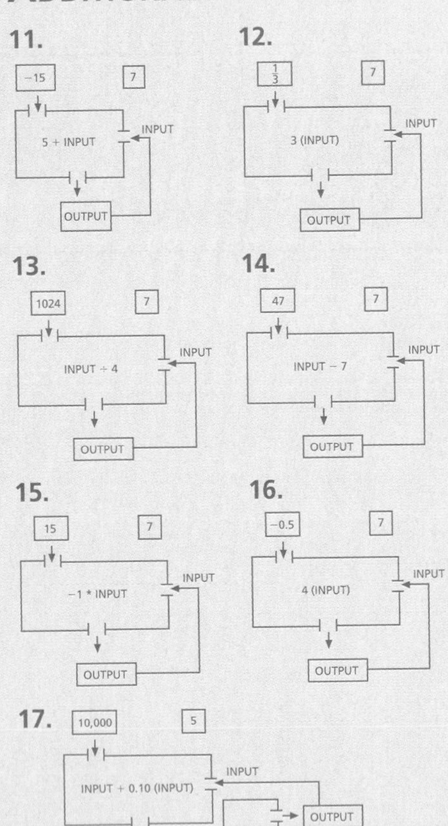

Review and Practice Your Skills

Practice ■ Lesson 2-1

Find the next three terms in each sequence.

1. 5, −10, 20, −40, ____, ____, ____ 80, −160, 320
2. 2187, 729, 243, 81, ____, ____, ____ 27, 9, 3
3. 3, 5, 9, 15, ____, ____, ____ 23, 33, 45
4. 1, 8, 5, 12, ____, ____, ____ 9, 16, 13
5. 5, 2, −1, −4, ____, ____, ____ −7, −10, −13
6. $\frac{1}{2}, \frac{1}{3}, \frac{1}{4}, \frac{1}{5}$, ____, ____, ____ $\frac{1}{6}, \frac{1}{7}, \frac{1}{8}$
7. $\frac{1}{7}, \frac{2}{7}, \frac{4}{7}, \frac{8}{7}$, ____, ____, ____ $\frac{16}{7}, \frac{32}{7}, \frac{64}{7}$
8. 1, 4, 9, 16, ____, ____, ____ 25, 36, 49
9. 2, 1, 0.5, 0.25, ____, ____, ____ 0.125, 0.0625, 0.03125
10. 38, 65, 92, 119, ____, ____, ____ 146, 173, 200

Draw the iteration diagram for each sequence. Calculate the output for the first 7 iterations. See additional answers.

11. −15, −10, −5, 0, ____, ____, ____ 5, 10, 15
12. $\frac{1}{3}$, 1, 3, 9, ____, ____, ____ 27, 81, 243
13. 1024, 256, 64, 16, ____, ____, ____ 4, 1, $\frac{1}{4}$
14. 47, 40, 33, 26, ____, ____, ____ 19, 12, 5
15. 15, −15, 15, −15, ____, ____, ____ 15, −15, 15
16. −0.5, 2, −8, 32, ____, ____, ____ −128, 512, −2048
17. Glenda's money market account has a starting balance of $10,000. Annual interest rate is 10%. At the end of each year, Glenda also deposits $500 into her account. Determine the iteration diagram and the output for 5 iterations. (Round to the nearest cent.) See additional answers. $11,500; $13,150; $14,965; $16,961.50; $19,157.65

Practice ■ Lesson 2-2

Graph each point on a coordinate plane. See additional answers.

18. $J(3, -5)$
19. $K(5, -3)$
20. $L(0, 6)$
21. $M(-6, 0)$
22. $A(-1, -2)$
23. $B(-2, 7)$
24. $C(-5, 3)$
25. $D(0, -8)$

Given $f(x) = 11 - 7x$, evaluate each function.

26. $f(2)$ −3
27. $f(-2)$ 25
28. $f(0)$ 11
29. $f(5)$ −24

Given $f(x) = x^2 + 5x + 6$, evaluate each function.

30. $f(-2)$ 0
31. $f(-3)$ 0
32. $f(0)$ 6
33. $f(10)$ 156

Determine if each relation is a function. Give the domain and range.

34.

x	y
1	5
−1	6
−3	5
−5	6

D:{−5, −3, −1, 1}
R: {5, 6}
Yes, it is a function.

35.

x	y
0	0
3	6
3	12
4	18

D: {0, 3, 4}
R: {0, 6, 12, 18}
Not a function

Learning Styles

VISUAL LEARNER Compare the characteristics of a replacement set with a solution set. Develop the idea that both types of sets contain possible values for the variables in an open sentence. However, all the elements in a solution set are values for variables that make the open sentence true. A solution set is actually a subset of a replacement set. Diagram this relationship by drawing a Venn diagram in which a larger circle (the replacement set) contains a smaller circle (the solution set).
You may want to use the following example:
Open sentence: $x - 3 = 10$ Replacement set: {8, 10, 13, 15}
Solution set: {13}

PRACTICE ■ LESSON 2-1–LESSON 2-2

Write each as a set of ordered pairs. Given the domain and range. (Lesson 2-2)

36.

x	2	4	6	8
y	11	8	5	2

(2, 11), (4, 8), (6, 5), (8, 2); D: {2, 4, 6, 8}; R: {2, 5, 8, 11}

37.

x	0	−1	1	−2	2
y	3	4	4	7	7

(0, 3), (−1, 4), (1, 4), (−2, 7), (2, 7); D: {−2, −1, 0, 1, 2}, R: {3, 4, 7}

Determine the next three terms or figures in each pattern. (Lesson 2-1)

38. 243, 260, 277, 294, 311, 328, 345

39. −748, −595, −442, −289, −136, 17, 170

40. 32, 48, 72, 108, 162, 243, 364.5

41. *Z, ZY, ZYX, ZYXW,* ZYXWV, ZYXWVU, ZYXWVUT

42. *, **, ***, ****, *****, ******, *******

43. 10, -10^2, 10^3, -10^4, 10^5, -10^6, 10^7

Given $f(x) = 3x^2$ and $g(x) = -3x + 1\frac{1}{2}$, find each value. (Lesson 2-2)

44. $f(2)$ 12

45. $f(-2)$ 12

46. $g\left(\frac{1}{2}\right)$ 0

47. $f(1) + g(1)$ $1\frac{1}{2}$

Name the quadrant in which each point is located. (Lesson 2-2)

48. $W(9, -1.5)$ IV

49. $X(-8, 17)$ II

50. $Z(-6.5, -6.5)$ III

MathWorks
Workplace Knowhow

Career – Environmental Journalism

Environmental journalists inform the public about what is being done to harm and to save the planet. They work with a great deal of mathematical information such as, how much pollution a smokestack emits or how much garbage is added to a landfill each year.

Environmental journalists study data gathered by scientists and the government. They must be able to analyze raw data, understand and create graphs, spot trends in numbers, and draw conclusions about what is happening to the environment.

1. The United States has 17 commercial incinerators that burn hazardous waste and release pollutants into the air. The hazardous waste size has increased from 1.3 billion to 2.4 billion to 4.6 billion pounds of waste. Project the growth for the next five years. (Hint: Plot the ordered pairs as (0, 0) for starting point, (1. difference of 2.4 billion and 1.3 billion) for increase at end of year 1, etc.) See additional answers.

year	growth
3	3.3
4	4.4
5	5.5
6	6.6
7	7.7

2. Create a graph to go with your story. Place years on the x-axis and billions of pounds of waste (in 0.5 billion increments) on the y-axis. Answers will vary.
3. A television reporter claims that fewer people in your town are recycling. But is the report true? Last year, 30% of the people in your town recycled regularly. This year recycling is down to 27%. Last year, your town had a population of 2,500, but this year the population is 3,100. Is the reporter's claim true? Explain your thinking. See additional answers.

Teaching Strategies

ESL/LEP Review the meaning of the word "inclusive." Point out its similarity to the word "included." Stress the fact that a closed dot must be drawn on a point to show that the point is included in the set being graphed while an open dot means that the point is not included in the set.

Chalkboard Examples

Example from Lesson 2-1

Determine the next three items in the sequence: 48, −24, 12, −6, . . . 3, $-\frac{3}{2}$, $\frac{3}{4}$

Example from Lesson 2-2

Answer questions about the relation: {(0, 3), (2, 5), (5, 8)}.

1. What is the domain? {0, 2, 5}
2. What is the range? {3, 5, 8}
3. Is this a function? yes

Career Opportunity

Describe the kind of work environmental journalists do and the types of places they work. Have students brainstorm a list of issues that an environmental journalist might cover. Discuss the importance of a remaining objective when reporting a news story.

Students should answer Questions 1–3 to better understand how environmental journalists use mathematics in performing their job.

Students who are interested in learning more about this profession can go to Learningmathmatters.com. Guidance Counselors should have information about school and training requirements.

18–25.

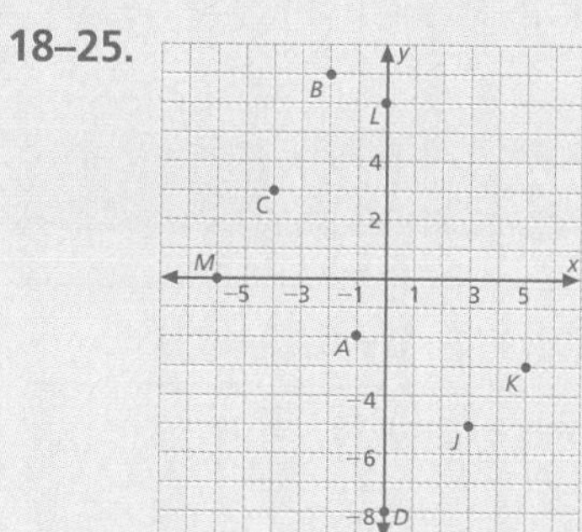

Math Works

1. (0, 0), $y = mx + b$, $m = \frac{2.2 - 1.1}{2 - 1} =$ 1.1, (1, 1.1), $y = 1.1x + b$ $1.1 = (1.1)(1) + b$, $b = 0$, $y = 1.1x$, (2, 2.2)
3. (2,500)(0.30) = 750 people recycled, (3,100)(0.27) = 837 people recycled. Reporter's claim is false. Percentage is lower but since population is higher, the actual number of people recycling increased.

Lesson Planning

NCTM Standards/Strands

- Number & Operation
- Patterns, Functions, & Algebra
- Geometry & Spatial Sense
- Problem Solving
- Reasoning and Proof
- Communication
- Connections
- Representation

Vocabulary

linear equation
linear function
constant
absolute value function

Materials Needed

paper/pencil
Algeblocks
straightedges
graph paper or coordinate grids

Lesson Resources

Warm-Up Transparency 4
Reteaching 2-3
Extra Practice 2-3
Enrichment 2-3
Multicultural Connection 7
Transparency TK-2, 3, 5, 7-10

Assignment Guide

Basic: 1–20, 25–38
Enriched: 1–38

Getting Started

5-Minute Warm-Up

Find each absolute value.

1. $|4|$ 4
2. $|3.4|$ 3.4
3. $|-2|$ 2
4. $|-11.3|$ 11.3
5. $|0|$ 0
6. $|5-3|$ 3
7. $|4+3|$ 7
8. $|3(-2)|$ 6

Introduction to Lesson 2-3

After students have completed the activity, have them make up a few more exercises to model the use of the mats and unit pieces. This helps prepare them to use Algeblocks as a problem-solving tool.

2-3 Linear Functions

Goals
- Graph linear functions
- Evaluate absolute value functions.

Applications Retail sales, Photography

Use Algeblocks to represent and simplify expressions.

The green blocks in a set of Algeblocks can be used to represent integers. Blocks on the top half of the mat represent positive integers, and blocks on the bottom half represent negative integers. An equal number of positive and negative blocks add to zero and may be removed from the mat. These pairs are called **zero pairs**. For example, $+3 + (-7)$ is simplified to -4.

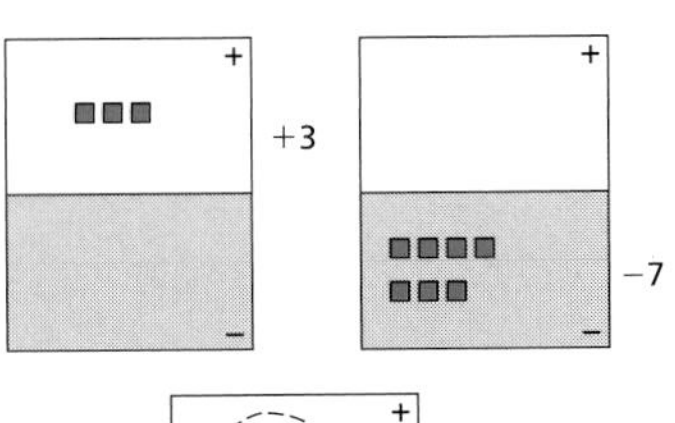

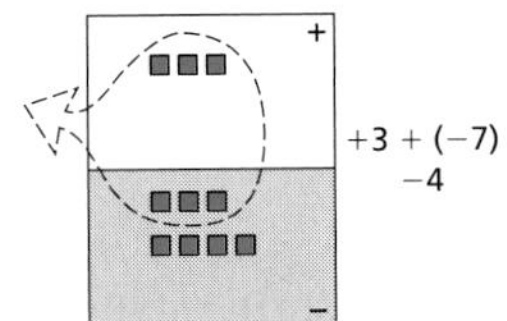

Use the units pieces on a mat to show how these expressions can be simplified.

a. $4 + -1$

b. $-6 + 2$

c. $7 - 4$ (rewrite as addition)

d. $6 - 8$

Build Understanding

In Section 2-2 you learned about number pairs produced by functions. Such pairs can be plotted on a coordinate plane and used to construct a graphical representation of the function. Equations in two variables that can be written in the form $y = ax + b$, where a and b are constants, are called **linear equations**. Graphs of such equations are *straight* lines. Such functions are called **linear functions**.

Example 1

Graph $y = 2x + 3$.

Solution

x	$2x + 3$	y
−2	2(−2) + 3	−1
0	2(0) + 3	3
1	2(1) + 3	5

Choose at least three values for x, calculate the corresponding y-values, and make a table to show the ordered pairs. Then plot the points and draw the line containing them. The domain is the set of all real numbers. The range is the set of all real numbers.

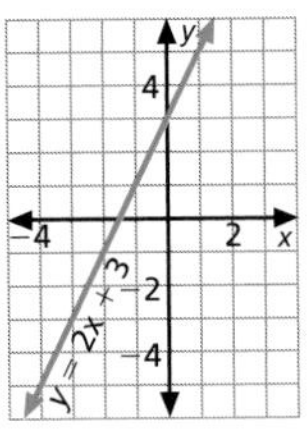

A **constant** function is a special linear function where the domain consists of all real numbers and where the range consists of only one value. The graph of a constant function is a horizontal line.

62 Chapter 2 **Essential Algebra and Statistics**

Learning Styles

TACTILE/KINESTHETIC LEARNER Algeblocks provide a tool for students to model some basic concepts of algebra. Within the exercises in this lesson, students learn to represent equations that include x, y, and numbers, positive and negative. Students can see how like terms are combined to simplify such expressions. After students have completed these problems, have them make up a few of their own expressions and model them.

Example 2

Graph each relation. Determine if the relation is a function. Then determine the domain and range.

a. $x = 2$ **b.** $f(x) = -1$

Solution

a. Any value of y results in an x value equal to 2.
$x = 2$ is not a function.
Domain: $x = 2$ Range: set of all real numbers.

b. Any value of x results in an $f(x)$ value equal to (-1).
$f(x) = -1$ is a linear constant function.
Domain: set of all real numbers Range: $y = -1$

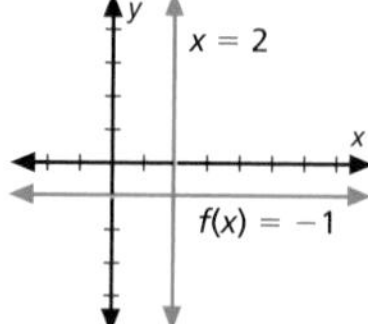

Example 3

TEMPERATURE The relationship between the scales used to measure temperature in degrees Fahrenheit (F) and degrees Celsius (C) can be represented by the linear equation $F = \frac{9}{5}C + 32$. Graph this function and determine the Fahrenheit temperature that is equivalent to 35°C.

Solution

Select three values for C. Calculate the corresponding F-values. Then plot the points and draw the line.

Find the point that has an x-coordinate of 35. The second coordinate of that point, 95, is the equivalent temperature measured in degrees Fahrenheit.

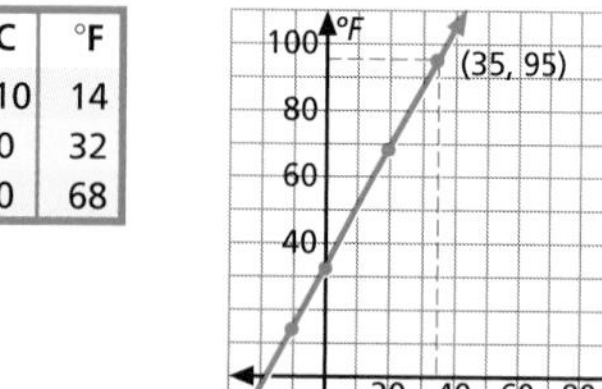

°C	°F
−10	14
0	32
20	68

An **absolute value function** is defined as:

$$g(x) = |x| = \begin{cases} x \text{ if } x \geq 0 \\ -x \text{ if } x < 0 \end{cases}$$

For example, $g(-3) = |-3| = -(-3)$, because $-3 < 0$.

Example 4

Given $h(x) = |x + 2|$, evaluate the given function.

a. $h(-3)$ **b.** $h(0)$

Solution

a. $h(-3) = |-3 + 2|$
$= |-1|$
$= 1$, because $-1 < 0$

b. $h(0) = |0 + 2|$
$= |2|$
$= 2$, because $2 \geq 0$

Extend the Lesson

COOPERATIVE LEARNING Have students work in small groups to create two sets of index cards. For the first set, they should write one linear function on each card. The second set of cards should show the corresponding graphs of the functions, one per card. Students can challenge their classmates to select a card from the function set and match it with the correct graph card.

Chalkboard Examples

Supplementary Example 1

Graph $y = 5x$.

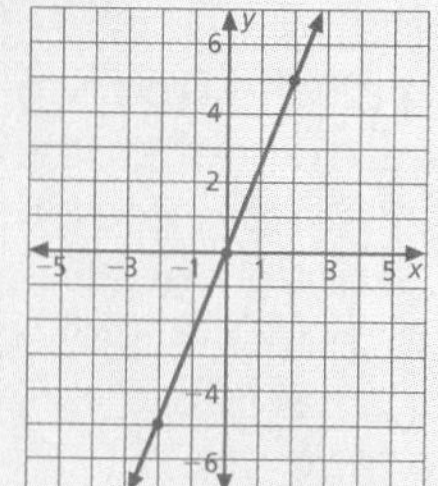

Supplementary Example 2

Graph $x = -1$. Is the relation a function? Determine the domain and range. **Yes. Domain: $x = -1$. Range: set of all real numbers.**

Supplementary Example 3

Use the graph for Example 3. Determine the Fahrenheit temperature that is equivalent to 50°C. **122°F**

Supplementary Example 4

Given $g(x) = |x + 1|$, evaluate the given function.

1. $g(0)$ **1** **2.** $g(-4)$ **3**

Supplementary Example 5

Use Algeblocks to represent and simplify the equation $x + 3 - 4 = 2y - y + 3$. $x - 1 = y + 3$

Reteaching Worksheet 2-3

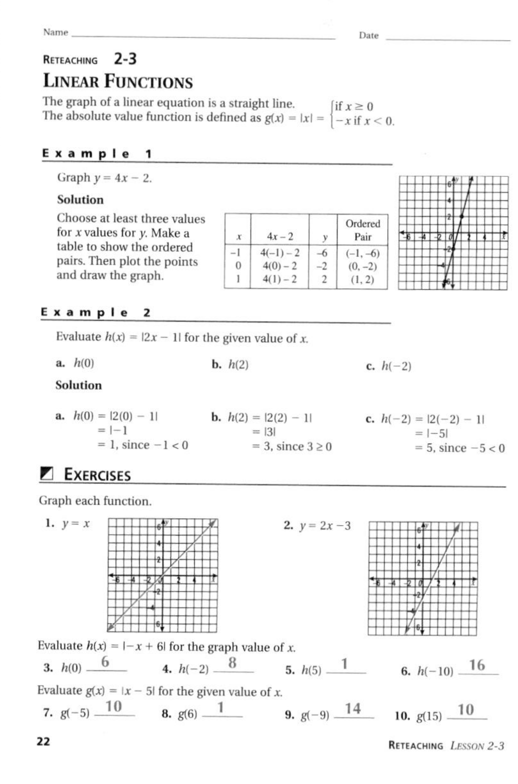

Name ______________________ Date ____________

RETEACHING 2-3

LINEAR FUNCTIONS

The graph of a linear equation is a straight line.
The absolute value function is defined as $g(x) = |x| = \begin{cases} \text{if } x \geq 0 \\ -x \text{ if } x < 0. \end{cases}$

Example 1

Graph $y = 4x - 2$.

Solution

Choose at least three values for x values for y. Make a table to show the ordered pairs. Then plot the points and draw the graph.

x	$4x - 2$	y	Ordered Pair
−1	4(−1) − 2	−6	(−1, −6)
0	4(0) − 2	−2	(0, −2)
1	4(1) − 2	2	(1, 2)

Example 2

Evaluate $h(x) = |2x - 1|$ for the given value of x.

a. $h(0)$ **b.** $h(2)$ **c.** $h(-2)$

Solution

a. $h(0) = |2(0) - 1|$
$= |-1$
$= 1$, since $-1 < 0$

b. $h(2) = |2(2) - 1|$
$= |3|$
$= 3$, since $3 \geq 0$

c. $h(-2) = |2(-2) - 1|$
$= |-5|$
$= 5$, since $-5 < 0$

EXERCISES

Graph each function.

1. $y = x$ **2.** $y = 2x - 3$

Evaluate $h(x) = |-x + 6|$ for the graph value of x.

3. $h(0)$ 6 **4.** $h(-2)$ 8 **5.** $h(5)$ 1 **6.** $h(-10)$ 16

Evaluate $g(x) = |x - 5|$ for the given value of x.

7. $g(-5)$ 10 **8.** $g(6)$ 1 **9.** $g(-9)$ 14 **10.** $g(15)$ 10

22 RETEACHING *LESSON 2-3*

Lesson Wrap-up

Quick Assessment

Ask the following questions to determine if students understand the content presented in this lesson.

Answer these questions about the function $y = 3x$.

1. What is the value of this function for $x = -1$? -3
2. Write three ordered pairs that can be graphed for this function. Answers will vary.
3. Is this a linear function? Explain. Yes. The Graph is a straight line.

Assignment Guide

Basic: 1–20, 25–41
Advanced: 1–41
Additional Practice: See Extra Practice Index on page 674.

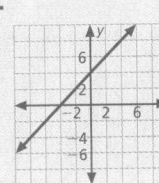

A graphing calculator is useful for solving real-world problems. Functions and equations can be graphed more quickly than on paper. Follow these rules when using a graphing calculator.

Step 1 Use the viewing window to select the minimum and maximum values for the x- and y-axes. These values will be determined by the problem.

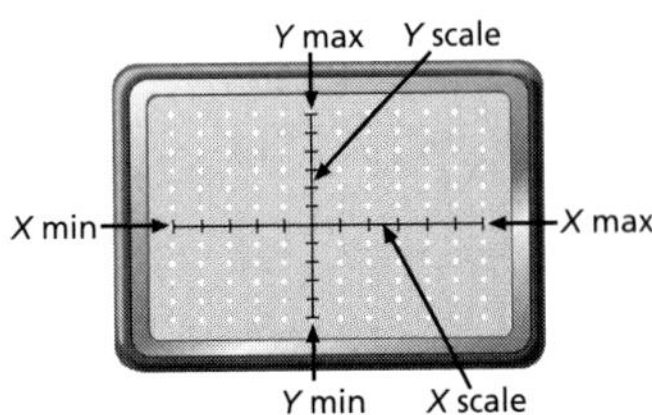

The minimum values (min) for x and y refer to the extreme negative values on the x- and y-axes. Likewise, the maximum values (max) refer to the extreme positive values on the x- and y-axes. The scale refers to what each tick-mark on the axes represents.

Step 2 Enter the equation into the calculator and graph.

Example 5

GRAPHING Use a graphing calculator to graph $y = x + 3$.

Solution

Calculator input:

Settings for Viewing Window

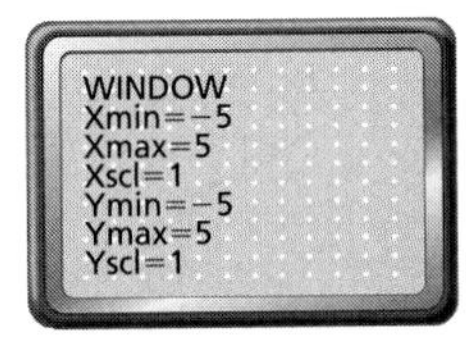

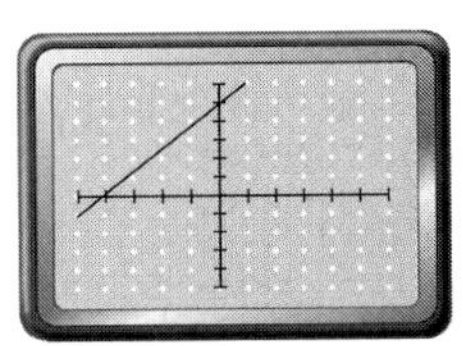

Try These Exercises

MODELING Use Algeblocks to represent each equation. Simplify where possible. Sketch your answer.

1. $3 - 2x + x = y - y + 3$ $(3 - x = 3$
2. $3x + x + 1 = -4 - 2$ $4x + 1 = -6$

Evaluate $g(x) = |2x - 1|$ for the given values of x.

3. $g(1)$ 1
4. $g(-2)$ 5
5. **RETAIL** Rich's weekly salary is based on the number of pairs of shoes he sells. He is paid a base salary of \$25, plus \$5 for every pair of shoes he sells. The relationship between his pay (p) and pairs of shoes (s) sold can be represented by the linear equation $p = 25 + 5s$. Graph this function, and determine Rich's pay for a week in which he sold 7 pairs of shoes. \$60, see additional answers.

6. **WRITING MATH** Why should you use three values for x when graphing a linear equation or function? While only 2 points are needed to determine a line, plotting a third point assures no mistake was made.

Additional Answers

5.

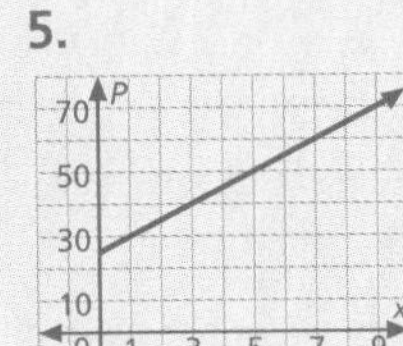

7.

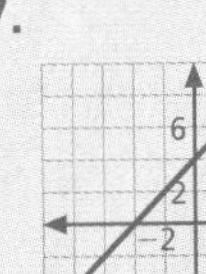

8.

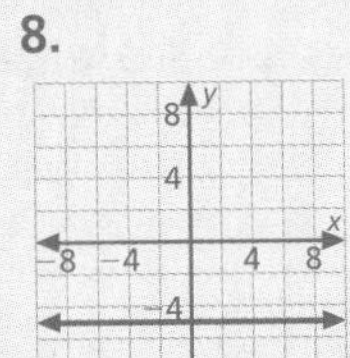

9.

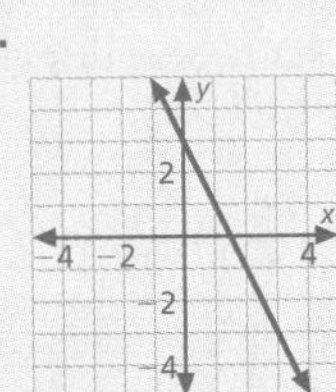

12.

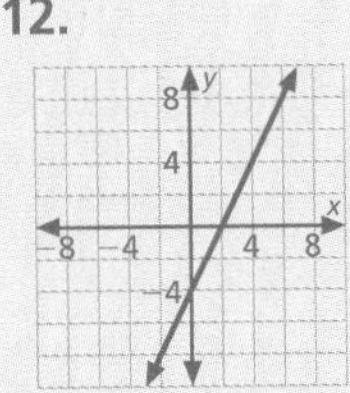

13.

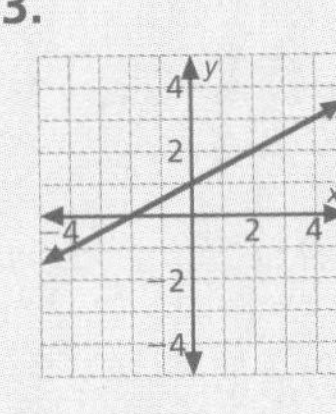

14.

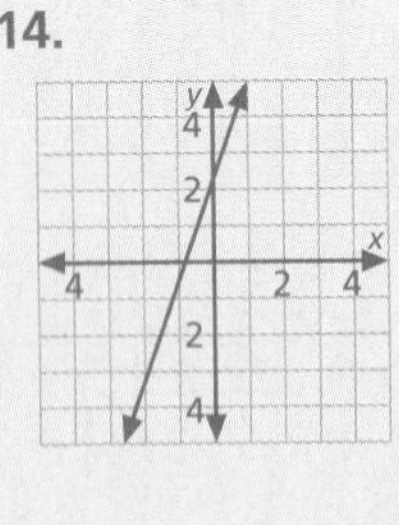

17.

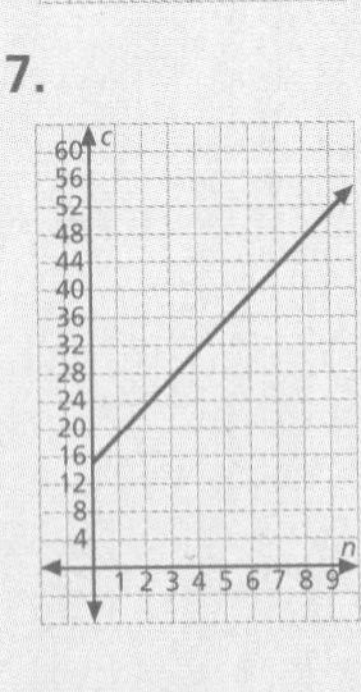

21.

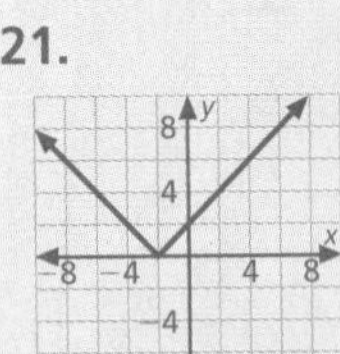

22.

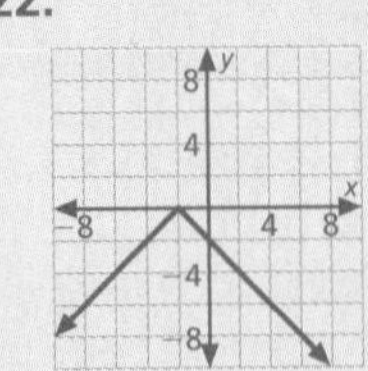

23.

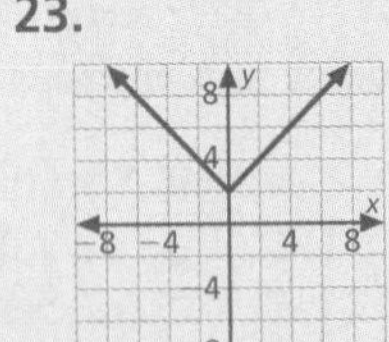

24.

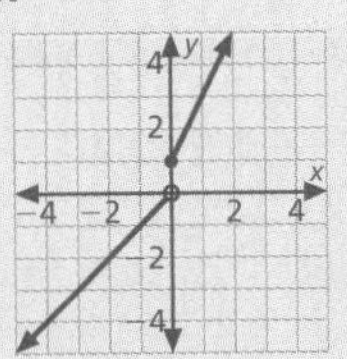

Practice Exercises

Graph each function. See additional answers.

7. $y = x + 4$
8. $f(x) = -5$
9. $y = -2x + 3$

Evaluate $h(x) = |-2x + 3|$ for the given value of x.

10. $h(-6)$ 15
11. $h(2)$ 1

GRAPHING Use a graphing calculator to graph each function. See additional answers.

12. $y + 1 = 2x - 3$
13. $2x + 4 = 4y$
14. $y - 2 = 3x$

Evaluate $F(x) = 2|2x| - 3|x + 1|$ for the given value of x.

15. $F(-4)$ 7
16. $F(0)$ -3

PHOTOGRAPHY A photographer charges a sitting fee of \$15, and charges \$4 for each 5″ × 7″ photograph the customer orders. The linear function $c = 15 + 4n$ can be used to calculate the customer's cost (c) based on the number of photographs (n) purchased.

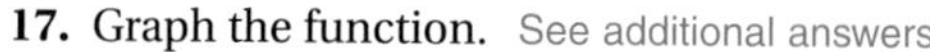

17. Graph the function. See additional answers.
18. Use your graph to determine the total cost of 6 photographs. \$39
19. How many photographs can be purchased if you cannot spend more than \$50.00? 8

20. **CHAPTER INVESTIGATION** Working together, think of a positive change that you would like to make at your school or in your community. What kind of data would encourage others to adopt your proposal? Make a list.

Extended Practice Exercises

Graph each function. See additional answers.

21. $y = |x + 2|$
22. $y = -|x + 2|$
23. $y = |x| + 2$
24. Graph $y = \begin{cases} x & \text{for } x < 0 \\ 2x + 1 & \text{for } x \geq 0 \end{cases}$

Mixed Review Exercises

Add or subtract. (Lesson 1-4)

25. $-6 + (-3) - 12 - (-5)$ -16
26. $(-3) + (-2) + (-6)$ -11
27. $-(-4) - (-8)$ 12
28. $5 + (-4) - 16 + (-2)$ -17
29. $(-8) + (-(-3)) + 2$ -3
30. $6 - 12 + (-7) - 4$ -17

Estimate each sum or difference. (Lesson 1-4)

31. 5382 + 7649 → 13,000
32. 9764 − 3478 → 7,000
33. 5894 + 9763 → 16,000
34. 8043 − 5612 → 2,000
35. \$78.64 + 85.06 → \$170.00
36. \$83.98 − 36.52 → \$40.00
37. \$94.76 + 75.15 → \$170.00
38. \$52.25 − 18.96 → \$30.00

Extend the Lesson

MATH JOURNAL In their math journals, have students write a description of each type of function—linear, constant, and absolute value—and given an example of each.

Extra Practice Worksheet 2-3

Name ______________________ Date __________

EXTRA PRACTICE 2-3

Linear Functions

Exercises

Graph each function.

1. $y = x + 2$
2. $y = -3x$
3. $y - 2 = -x + 1$
4. $2y = -4x + 2$
5. $y = |x - 1|$
6. $y = \begin{cases} 2 & \text{for } x < -2 \\ -x - 1 & \text{for } x \geq -2 \end{cases}$

Evaluate $F(x) = |-2x - 1|$ for the give value of x.

7. $F(-2)$ 3
8. $F(0)$ 1
9. $F(1)$ 3
10. $F(-4)$ 7

20 EXTRA PRACTICE *LESSON 2-3*

Enrichment Worksheet 2-3

Name ______________________ Date __________

ENRICHMENT 2-3

The Step Function

The symbol $[x]$ is used to stand for the greatest integer that is less than or equal to a given real number x. For example, $[4.3] = 4$ because 4 is the greatest integer less than or equal to 4.3. The greatest integer function can be written as $G(x) = [x]$.

Exercises

Find $[x]$ for each value of x.

1. $[6.2]$ 6
2. $[6]$ 6
3. $[-6.2]$ -7
4. $[-6]$ -6
5. $[0.2]$ 0
6. $[-0.2]$ -1
7. $[0]$ 0
8. $[\pi]$ 3

Graph each function.

9. $G(x) = [x]$
10. $G(x) = [x] + 1$
11. $G(x) = [x + 1]$

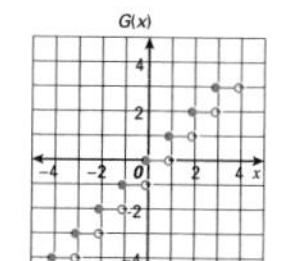

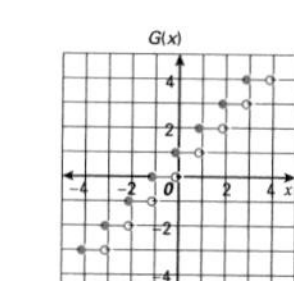

12. The cost of a telephone dial-in weather service is \$2.50 for the first minute (or less), and \$1.00 for each additional minute (or part of a minute). Draw a graph of this function on the grid at the right. Use t for the time in minutes and d for the cost in dollars.

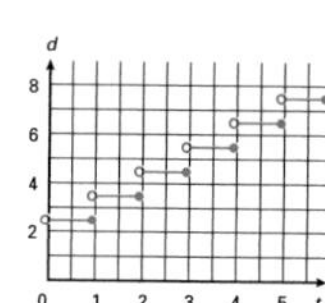

13. Why is the greatest integer function sometimes called the "step function"?
Its graph looks like a series of steps.

28 ENRICHMENT *LESSON 2-3*

Lesson Planning

NCTM Standards/Strands

- Number & Operation
- Patterns, Functions, & Algebra
- Problem Solving
- Reasoning and Proof
- Communication
- Connections
- Representation

Vocabulary

inverse operation
equivalent equation
addition property of equality
multiplication property of equality
reciprocal

Materials Needed

paper/pencil   Algeblocks

Lesson Resources

Warm-Up Transparency 5
Transparency TK-2, 3, 5
Reteaching 2-4
Extra Practice 2-4
Enrichment 2-4

Assignment Guide

Basic: 1–38, 46–56
Enriched: 1–56

Getting Started

5-Minute Warm-up

Simplify.

1. $4.1 - 4.1$ 0   **2.** $3 - 4$ -1

3. $8 \times \frac{1}{8}$ 1   **4.** $6 \times \frac{1}{8}$ $\frac{3}{4}$

Introduction to Lesson 2-4

Have students work in pairs and make up two more equations like the ones in the activity. Have the students model and solve the equations using Algeblocks.

2-4 Solve One-Step Equations

Goals ■ Use the Addition or Multiplication Properties of Equality to solve one-step equations.

Applications Business, Finance, News media

In Lesson 2–3, you used Algeblocks to model equations. Model $x + 5 = 3$. Then add -5 to both sides and simplify each side. What is the result on both sides? Sketch your answer and complete the equations.

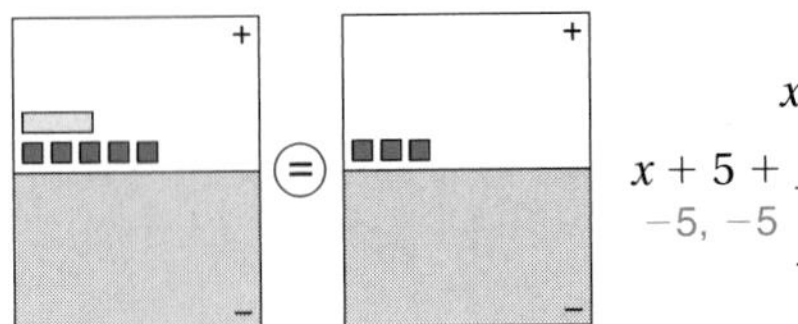

$x + 5 = 3$

$x + 5 +$ ____ $= 3 +$ ____ (-5, -5)

____ $=$ ____

$x = -2$

Build Understanding

In the activity above, you used the opposite of a number to simplify and solve an equation. In the same way, you can use opposite, or **inverse operations** to get a variable alone on one side of an equation.

Reading Math

Mathematical notation can be used to show the steps in solving each equation in Example 1.

$x + 3 = 10$

$x + 3 - 3 = 10 - 3$

$x = 7$

Add the opposite. x is alone.

Example 1

Use Algeblocks to solve $x + 3 = 10$.

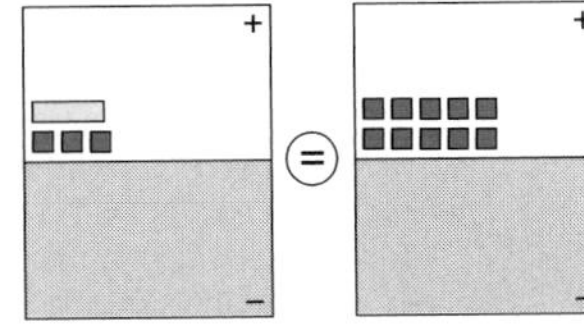

Solution

Represent the equation. Adding -3 to each mat will result in zeros and leave the x-piece alone on one mat.

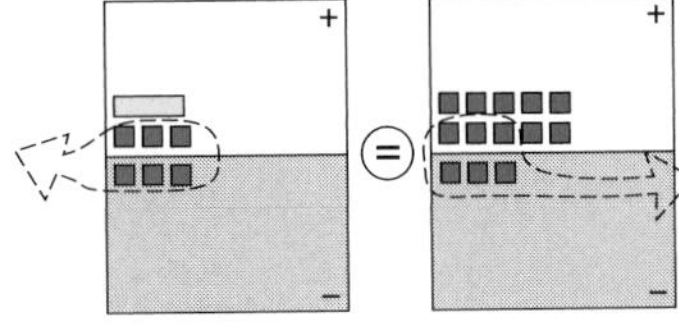

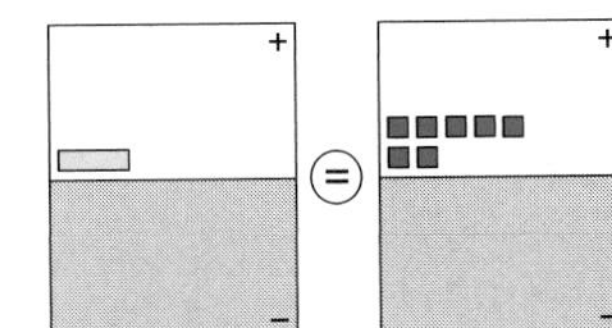

Read the answer, $x = 7$.

When equations involve the inverse operations of addition and subtraction, you can use opposites and the addition property of equality to solve them. This property states that adding the same number to both sides of an equation results in an equivalent equation.

Addition Property of Equality	For all real numbers a, b, and c, if $a = b$, then $a + c = b + c$ and $c + a = c + b$.

Teaching Strategies

ESL/LEP Have students make a list of mathematical operations. Then discuss sentence structure and vocabulary cues that may signal the use of each operation. For example, the word "is" often translates to an equals sign. The word "of," when it follows a fraction, indicates multiplication.

Example 2

Use mathematical notation to show the steps in solving the equation.

$x - 3.7 = -0.1$

Solution

$$x - 3.7 = -0.1$$
$$x - 3.7 + 3.7 = -0.1 + 3.7$$
$$x = 3.6$$

opposites: $-2, 5, -\frac{2}{3}, -0.5$

reciprocals: $\frac{1}{2}, -\frac{1}{5}, \frac{3}{2}, 2$

Check Understanding

Find the opposite and reciprocal of each number:

$2, -5, \frac{2}{3}, 0.5$

In a similar manner, reciprocals and the multiplication property of equality are used in solving equations involving multiplication. The multiplication property of equality states that multiplying both sides of an equation by the same number still maintains the equality.

Multiplication Property of Equality	For all real numbers *a*, *b*, and *c*, if $a = b$, then $ac = bc$ and $ca = cb$.

Example 3

Solve each equation.

a. $\left(\frac{2}{5}\right)y = 12$ **b.** $-3q = 45$

Solution

a.
$$\left(\frac{2}{5}\right)y = 12$$
$$\left(\frac{5}{2}\right)\left(\frac{2}{5}\right)y = \left(\frac{5}{2}\right)12$$
$$1y = 30$$
$$y = 30$$

b.
$$-3q = 45$$
$$\left(-\frac{1}{3}\right)(-3q) = \left(-\frac{1}{3}\right)45$$
$$1q = -15$$
$$q = -15$$

You may need to simplify one or both sides of an equation before applying the properties of equality.

Reading Math

In Example 3:

a. Multiply both sides of the equation by the reciprocal of $\frac{2}{5}$.

b. Multiply both sides of the equation by the reciprocal of -3.

In part b, dividing by -3 is the same as multiplying by the reciprocal of -3.

Example 4

Solve: $w - 9 + 23 = |-3 - 5|$

Solution

$$w - 9 + 23 = |-3 - 5|$$
$$w + 14 = 8$$
$$w + 14 + (-14) = 8 + (-14)$$
$$w = -6$$

The solution is -6.

Teaching Strategies

SIMULATIONS Get a working balance scale, and use weights or paper clips to balance a small box. Demonstrate how adding the same number of weights to each side keeps the scale in balance. Have students explore the effects of subtractions the same number of weights. To show the multiplication property of equality, use two or more boxes that have the same weight.

Chalkboard Examples

Supplementary Example 1

Use Algeblocks to solve $x - 2 = 4$. $x = 6$

Supplementary Example 2

Use mathematical notation to show the steps in solving the equation: $x + 2.5 = 1.2$.

$$x + 2.5 = 1.2$$
$$x + 2.5 - 2.5 = 1.2 - 2.5$$
$$x = -1.3$$

Supplementary Example 3

Solve each equation.

a. $\frac{3}{4}x = 15$ $x = 20$

b. $-4y = 16$ $y = -4$

Supplementary Example 4

Solve: $x + 4 - 15 = |5 - 6|$ $x = 12$

Supplementary Example 5

Translate into an equation and solve. The product of a number and 12 is 15 more than 45. $12n = 15 + 45; n = 5$

Reteaching Worksheet 2-4

Name ______________________ Date ____________

RETEACHING 2-4

SOLVE ONE-STEP EQUATIONS

For all real numbers, *a*, *b*, and *c*, if $a = b$ then:

$a + c = b + c$ and $c + a = c + b$ ← Addition Property of Equality

$ac = bc$ and $ca = cb$ ← Multiplication Property of Equality

Example 1

Solve each equation. Use Algeblocks to model each step if you wish.

a. $p + 7 - 15 = 9 + 15$ **b.** $\frac{1}{3}d = 5(60)$

Solution

a.

$p + 7 - 15 = 9 + 15$	Represent equation.
$p - 8 = 24$	Simplify.
$p - 8 + 8 = 24 + 8$	Add 8 to each side.
$p = 32$	

b.

$\frac{1}{3}d = 5(60)$	Represent equation.
$3\left(\frac{1}{3}\right)d = 300(3)$	Multiply each side by 3.
$d = 900$	

Example 2

Translate each sentence into an equation using *n* to represent the unknown number. Then solve the equation of *n*.

a. One tenth of 40 is the same as the sum of -3 and some number.

b. Increasing a number by 43 yields the same result as multiplying 6 by 12.

Solution

a. The equation is $\frac{1}{10}(4) = -3 + n$.
Solve. $4 = -3 + n$
$7 = n$

b. The equation is $n + 43 = 6(12)$.
Solve. $n + 43 = 72$
$n = 29$

EXERCISES

Solve each equation. Use Algeblocks to model each step if you wish.

1. $-12 + a = 32$ <u>44</u>
2. $12 - b = 16$ <u>-4</u>
3. $-5 = c + 9$ <u>-14</u>
4. $2d = 16$ <u>8</u>
5. $\frac{m}{9} = -5$ <u>-45</u>
6. $-\frac{1}{3}f = 24$ <u>-72</u>
7. $g + \frac{1}{2} = 2$ <u>$1\frac{1}{2}$</u>
8. $\frac{h}{7} = 9 + 5$ <u>98</u>
9. $-(-8)(5) = 2j$ <u>20</u>

Translate each sentence into an equation using *n* to represent the unknown number. Then solve the equation for *n*.

10. A number subtracted from 46 is -21. <u>$46 - n = -21; n = 67$</u>
11. The quotient of a number divided by 11 is 0.5. <u>$\frac{n}{11} = 0.5; n = 5.5$</u>

24 RETEACHING *LESSON 2-4*

Solving problems often involves translating a verbal problem into an algebraic equation. These equations can then be solved using the properties of equality.

Lesson Wrap-up

Quick Assessment

Ask the following questions to determine if students understand the content presented in this lesson.

1. When you are given an equation to solve, how do you decide which property of equality to use? Look at what operation is being performed on the variable.
2. How do inverse operations help you to solve equations? Using inverse operations and the properties of equality, you can form equivalent equations in which the solution can be read easily.

Assignment Guide

Basic: 1–38, 46–56
Advanced: 1–56
Additional Practice: See Extra Practice Index on page 674.

Example 5

Translate the sentence into an equation using n to represent the unknown number. Then solve the equation for n.

When a number is decreased by 31, the result is the square of 3.

Solution

When a number is decreased by 31, the result is the square of 3.
The equation is: $n - 31 = 3^2$

$$n - 31 = 9$$

$$n - 31 + (31) = 9 + (31) \quad \text{The number is 40.}$$

$$n = 40$$

Try These Exercises

Solve each equation.

1. $q + 18 = 32$ 14
2. $r - 5 = -2$ 3
3. $-4z = 36$ −9
4. $-16 = 21 + h$ −37
5. $-7 = \frac{v}{8}$ −56
6. $\left(\frac{5}{3}\right)k = 30$ 18
7. $e + \frac{3}{4} = 1$ $\frac{1}{4}$
8. $\frac{2}{5} = m - \frac{1}{2}$ $\frac{9}{10}$
9. $w - 1.7 = -4.2$ −2.5

WRITING MATH Translate each sentence into an equation using n to represent the unknown number. Do not solve.

10. The product of −8 and a number is the same as the square of −4. $-8n = (-4)^2$
11. Increasing a number by 15 yields the same result as taking half of 72. $n + 15 = \left(\frac{1}{2}\right)(72)$
12. The quotient of a number and 5 is 0.2. $\frac{n}{5} = 0.2$
13. The difference between a number and 26 is −9. $n - 26 = -9$
14. **DATA FILE** Use the Data Index on pages 632–633 to locate information about average daily temperatures. On November 16, the temperature in San Diego climbed 9° higher than the average daily temperature in that city for November and then dropped 12°. What was the temperature on November 16? 59°

Practice Exercises

Solve each equation.

15. $f + 19 = 41$ 22
16. $7m = -35$ −5
17. $21 + a = -4$ −25
18. $-10 = p - 1$ −9
19. $25n = -10$ −0.4
20. $0.9u = 0.63$ 0.7
21. $12 = \left(\frac{-4}{3}\right)y$ −9
22. $\left(\frac{3}{8}\right)x = 6$ 16
23. $5.74 = j - 3.6$ 9.34

Extend the Lesson

MATH JOURNAL Write the following equations on the board. Have students translate the equations into words.

1. $8n = 12 - 4$ Eight times a number is equal to the difference of 12 and 4.
2. $n + 3 = 2(-5)$ Three more than a number is twice −5.
3. $4 + n = \frac{12}{2}$ Four more than a number is the quotient of 12 and 2.

24. **YOU MAKE THE CALL** Anthony says that $(-4)^2$ and -4^2 are equal. What do you think. $(-4)^2 = 16$ and $-4^2 = -16$

Translate each sentence into an equation using n to represent the unknown number. Then solve the equation for n.

25. **FINANCE** When an account balance is increased by \$25, the result is −\$15. −40
26. The difference between a number and 26 is the square of −3. 35
27. The quotient of a number and 8 is 0.7. 5.6
28. One-third of −81 is the same as the product of 3 and some number. −9

Solve each equation.

29. $(-2)(-3)(-4) = 12c$ −2
30. $|13 - 19| = \left(\frac{-4}{5}\right)y$ −7.5
31. $(2.5)(5) = m - 17 + 4$ 25.5
32. $w + 3^4 = 4^3$ −17
33. $a - 7 + 25 = 2^3$ −10
34. $0.01k = (1 + 2 + 3 + 4)^2$ 10,000

Find all solutions in each equation.

35. $|x| + 5 = 11$ 6, −6
36. $-48 = -4|z|$ 12, −12
37. $|w| - 3 = -3$ 0

38. **NEWS MEDIA** A television news magazine has 48 minutes of airtime to fill. The producer decides to run an 8-minute health segment and a 9-minute science segment. At the last minute, a 12-minute celebrity feature is canceled. The producer decides to add a 20-minute segment. What length segment is needed to complete the broadcast. Write an equation to model the situation and solve. $s + 8 + 9 - 12 + 20 = 48$; a 23-min segment

Extended Practice Exercises

Replace each __?__ so that the equation will have the given solution.

39. $x +$ __?__ $= -4$; The solution is −15. 11
40. __?__ $x = \frac{1}{2}$; The solution is $\frac{1}{12}$. 6
41. $-24 = x -$ __?__; The solution is 16. 40
42. $-0.27 =$ __?__ x; The solution is 0.9. −0.3
43. Write an equation that has no solution. Answers will vary; $|x| = -3$
44. Write an equation that has infinitely many solutions. Answers wil vary; $y = x$
45. **BUSINESS** The cost of making a camera is 60% of its selling price (p). The rest is profit. If the camera cost \$101.25 to make, how much is its selling price? Write an equation and solve. $0.6p = \$101.25$; \$168.75

Mixed Review Exercises

Find each product or each quotient. (Lesson 1-5)

46. $(16)(-3.9)$ −62.4
47. $-5\frac{1}{3} \div \frac{2}{3}$ −8
48. $(-7)(-0.5)(-2)$ −7
49. $345 \div -15$ −23
50. $\left(-\frac{3}{4}\right)\left(\frac{1}{2}\right)\left(-\frac{5}{8}\right)$ $\frac{15}{64}$
51. $-63 \div (-0.7)$ 90
52. $(-3)\left(-\frac{7}{8}\right)(-9)$ $-23\frac{5}{8}$
53. $54.6 \div (-4.2)$ −13

Evaluate each expression when $a = 6$ and $= -4$. (Lesson 1-4)

54. $a + (-b)$ 10
55. $4a + 3b$ 12
56. $6b - (-2a)$ −12

Learning Styles

VISUAL LEARNER Model the solving of a one-step equation using Algeblocks. Then point out that the key to solving an equation with Algeblocks is to get the variable alone on one side of the mat. In the same way, students must use inverse operations to isolate the variable on one side of the equation. Have students model several equations using Algeblocks and then solve the same equations by writing out the steps used to apply inverse operations.

Extra Practice Worksheet 2-4

Name ____________ Date ________

EXTRA PRACTICE 2-4

SOLVE ONE-STEP EQUATIONS

EXERCISES

Solve each equation.

1. $n + 13 = 24$ 11
2. $35 - b = 19$ 16
3. $5r = -45$ −9
4. $-12 = q - 3$ −9
5. $15j = 30$ 2
6. $0.4h = 1.6$ 0.64
7. $14 = \left(-\frac{7}{8}\right)x$ −16
8. $\left(\frac{5}{9}\right)d = 20$ 36
9. $6.32 = t - 4.16$ 10.48
10. $\frac{3}{4} = f + \frac{1}{2}$ $\frac{1}{4}$

Translate each sentence into an equation using n to represent the unknown number. Then solve the equation for n.

11. When a number is decreased by 13, the result is −2. $n - 13 = 2$; 11
12. Twelve more than a number is the product of −3 and 6. $12 + n = (-3)(6)$; −30
13. One-fourth of a number is the same as the square of −3. $\frac{1}{4}n = (-3)^2$; 36
14. Sixteen is the same as the quotient of a number and 12. $16 = \frac{n}{12}$; 192
15. Increasing a number by 14 yields the same result as taking one-half of 40. $n + 14 = \frac{1}{2}(40)$; 6
16. The quotient of a number of −2 is the same as the sum of −4 and 10. $\frac{n}{-2} = -4 + 10$; −12

Solve each equation.

17. $(-1)(-4)(5) = 20d$ $\frac{1}{5}$
18. $|14 - 22| = 4y$ 2
19. $2^2 + v = 3^2$ 5
20. $(1.5)(10) = f - 13 + 7$ 21
21. $15 + 4 - s = (-3)(-2)$ 13
22. $0.02g = (0.5)(4.2)$ 105

Find all solutions in each equation.

23. $|x| + 4 = 12$ −8, 8
24. $-16 = -2|q|$ −8, 8
25. $|p| - 5 = 10$ −15, 15
26. $8 - |r| = -2$ −10, 10

22 EXTRA PRACTICE *LESSON 2-4*

Enrichment Worksheet 2-4

Name ____________ Date ________

ENRICHMENT 2-4

HEXA-MAZE

This figure is called a *hexa-maze* because each cell has the shape of a hexagon.

To solve the maze, start with the number in the top left corner. This number must be the solution of the equation in the next cell. The number in the new cell will then be the solution to the equation in the next cell. At each move, you may only move to an adjacent cell. Each cell is used only once.

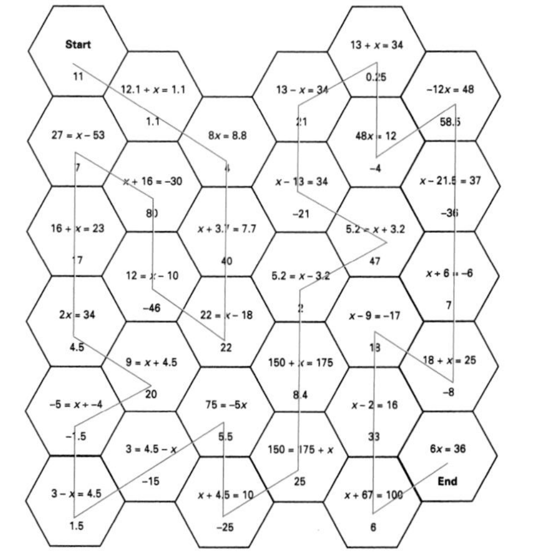

30 ENRICHMENT *LESSON 2-4*

Lesson 2-4 **Solve One-Step Equations** 69

Skills Practice

Vocabulary Review

Lesson 2–3
linear equation constant
linear function
absolute value function

Lesson 2–4
inverse operation
equivalent equation
addition property of equality
multiplication property of equality
reciprocal

Assignment Guide

All students: 1–65
Additional Practice: Refer to the Extra Practice Index on page 674 of this text.

Additional Answers

1.
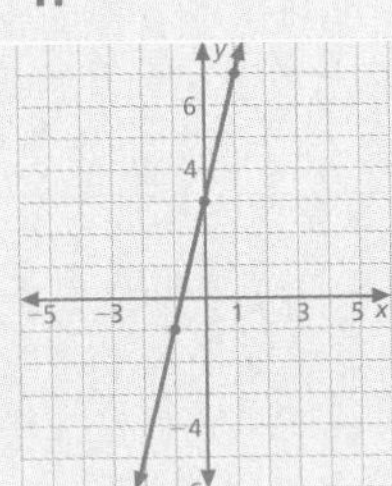

2.
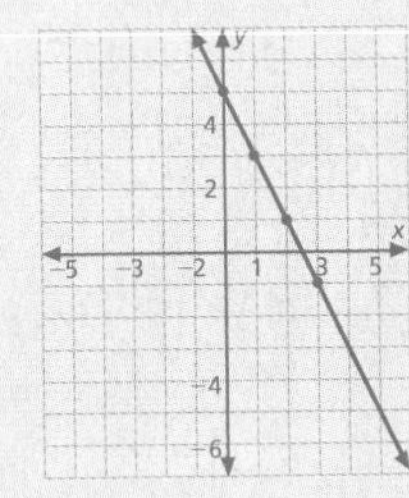

3.
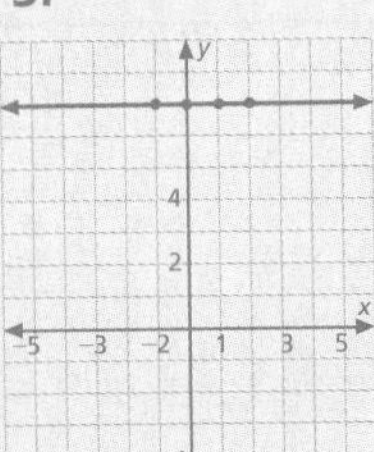

4.
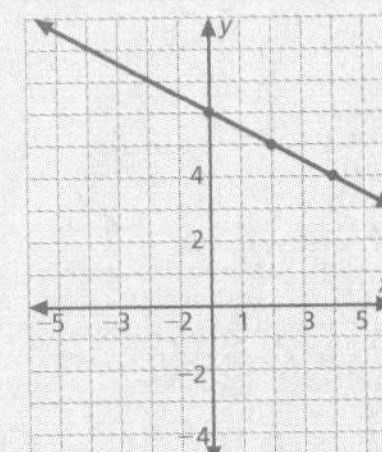

5.
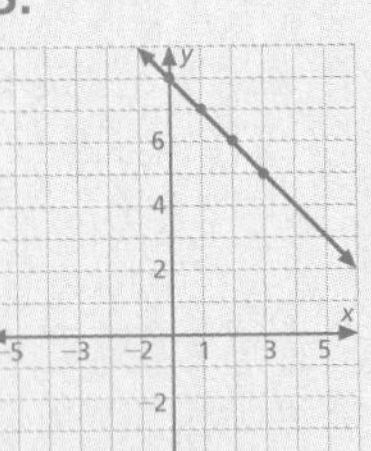

6.
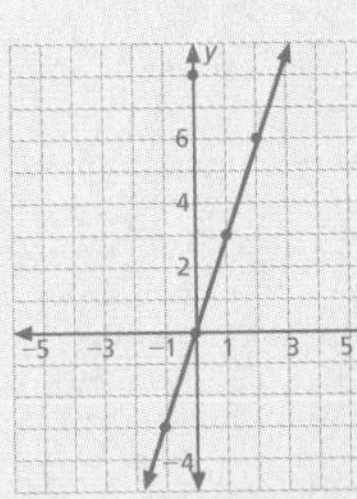

7.
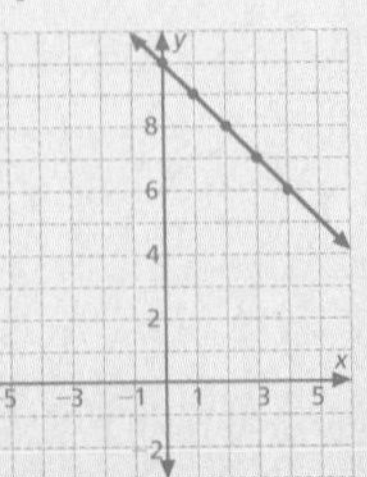

8.
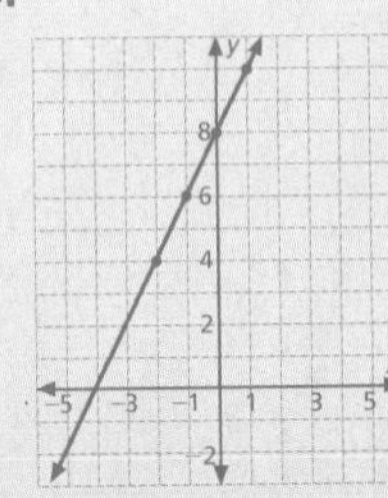

Review and Practice Your Skills

Practice ■ Lesson 2-3

Graph each function. See additional answers.

1. $y = 4x + 3$
2. $f(x) = -2x + 5$
3. $y = 7$
4. $f(x) = -\frac{1}{2}x + 6$
5. $y = 8 - x$
6. $y = 3x$
7. $x + y = 10$
8. $y - 2 = 2x + 6$
9. $y = x$

Evaluate $g(x) = |-3x - 2|$ for the given value of x.

10. $g(0)$ 2
11. $g(-5)$ 13
12. $g(3)$ 11

Evaluate $F(x) = 3|x| - 2|2x - 5|$ for the given value of x.

13. $F(0)$ −10
14. $F(-3)$ −13
15. $F(3)$ 7

A car rental agency charges a flat fee of $30 to rent a car, and $21 for each day the car is rented. The linear function $c = 30 + 21d$ can be used to calculate the customer's cost (c) based on the number of days (d) the car is rented.

16. Graph the function. See additional answers.
17. Determine the cost for a 7-day rental. $177
18. What is the maximum number of days Lakesha can rent a car if she has only $140 to spend? 5

Practice ■ Lesson 2-4

Solve each equation.

19. $4x = -12$ 3
20. $x - 5 = -4$ 1
21. $l + 8 = 11$ 3
22. $y + 5 = 7.2$ 2.2
23. $\frac{1}{3}p = -2$ −6
24. $-5b = 65$ −13
25. $\frac{2}{3} = -\frac{7}{12} + m$ $1\frac{1}{4}$
26. $0.8t = 9.6$ 12
27. $2u = \frac{1}{2}$ $\frac{1}{4}$
28. $-11 = n - 4$ −7
29. $45 + m = 71$ 26
30. $-1\frac{1}{8}x = 1$ $-\frac{8}{9}$
31. $62.4 + k = -39.9$ −102.3
32. $x - 4 = -4$ 0
33. $\frac{y}{-3} = 27$ −81
34. $38 = -43 + x$ 81
35. $-8.4 = 0.12x$ −70
36. $d - (-13) = 25$ 12
37. $\frac{6}{13}a = 52$ $112\frac{2}{3}$
38. $2.18 = r + 3.59$ −1.41
39. $b - 5 = -41$ −36
40. $\frac{e}{4} = -7$ −28
41. $p - \frac{4}{5} = 1\frac{1}{5}$ 2
42. $-6 = y - 6$ 0
43. $-12n = 3$ $-\frac{1}{4}$
44. $-12 + n = 3$ 15
45. $\frac{n}{-12} = 3$ −36
46. $3n = -12$ −4
47. $\frac{m}{-3} = \frac{4}{9}$ $-\frac{4}{3}$
48. $\frac{7}{3}x = 21$ 9
49. $-4.7 = n - 2.5$ −2.2
50. $0 = d + 11$ −11
51. $-x = 16$ −16

Teaching Strategies

CHALLENGE Have students look for a pattern in the graphs of these absolute value functions: $y = |x|$, $y = |x + 2|$, $y = |x - 2|$.
The graph of $y = |x + 2|$ is the same as the graph of $y = |x|$ moved up two units. The graph of $y = |x - 2|$ is the same as the graph of $y = |x|$ moved down two units.

PRACTICE ■ LESSON 2-1–LESSON 2-4

Determine the next three terms in each sequence. (Lesson 2-1)

52. 1600, 400, 100, 25, $\frac{25}{4}$, $\frac{25}{16}$, $\frac{25}{64}$

53. 47, 36, 25, 14, ____, ____, ____

54. a, c, e, g, ____, ____, ____ i, k, m

55. −174, −148, −122, −96, ____, ____, ____ 70, 44, 18

3, −8, −19

Graph each function. (Lesson 2-3) See additional answers.

56. $f(x) = -x + 1$

57. $y = |x|$

58. $y = 9$

Translate each sentence into an equality using *n* to represent the unknown number. Then solve the equation for *n*. (Lesson 2-4)

59. When n is increased by 13, the result is −29. $n + 13 = -29$, $n = -42$

60. The product of a number and 8 is the same as the square of −7. $8n = (-7)^2$, $n = 6\frac{1}{8}$

61. The quotient of a number and −4 is 11. $\frac{n}{-4} = 11$, $n = -44$

62. The difference between n and 17 is 25. $n - 17 = 25$, $n = 42$

Solve each equation. (Lesson 2-4)

63. $t + 1\frac{3}{8} = 3\frac{3}{4}$ $2\frac{3}{8}$

64. $|x| + 4 = 10$ −6, 6

65. $|21 - 27| = -\frac{3}{2}y$ −4

Mid-Chapter Quiz

Determine the next three terms in each sequence.

1. 4, 13, 22, 31, . . . 40, 49, 58

2. 2, −10, 50, −250, . . . 1,250, −6,250, 31,250

3. $-1, \frac{1}{3}, -\frac{1}{9}, \frac{1}{27}, \ldots$ $-\frac{1}{81}, \frac{1}{243}, -\frac{1}{729}$

4. Determine the output for the first five iterations: The initial input is 6; the rule is "add −4." 6, 2, −2, −6, −10, −14

Use the relation, {(−1, 2), (0, 3), (2, 5)} for Exercises 5–7.

5. What is the domain? {−1, 0, 2}

6. What is the range? {2, 3, 5}

7. Is it a function? yes

Solve.

8. $f(2)$ if $f(x) = -3x + 1$. −5

9. $f(-4)$ if $f(x) = 2$. 2

10. $f(-3)$ if $f(x) = |-x|$. 3

Solve each equation.

11. $9 = -3 + j$ 12

12. $-4m = 1$ $-\frac{1}{4}$

13. $\frac{2n}{3} = 7$ $\frac{21}{2}$

Translate the sentence into an equation using *n* to represent the unknown number. Then solve the equation for *n*.

14. The difference between a number and 6 is the product of 3 and 8. $n - 6 = (3)(8)$; 30

15. The product of −5 and −4 is the product of 8 and a number. $(-5)(-4) = 8n$; 2.5

Chalkboard Examples

Example from Lesson 2-3

Evaluate $f(x) = |x - 2|$ for the given values.

1. $f(5)$ 3 **2.** $f(-4)$ 6

Example from Lesson 2-4

Solve each equation.

1. $a - 2 = -18$ −16

2. $d/7 = -3$ −21

Mid-Chapter Quiz

Correlation Chart

Question Number	Lesson Number
1–4	2-1
5–8	2-2
9–10	2-3
11–15	2-4

9.

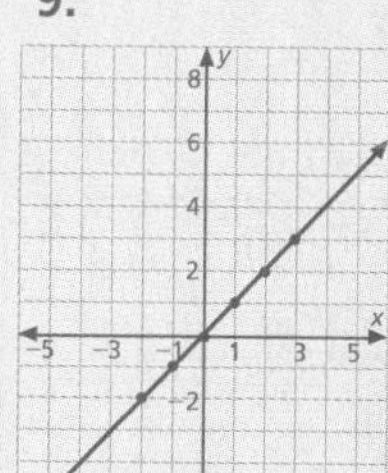

16.

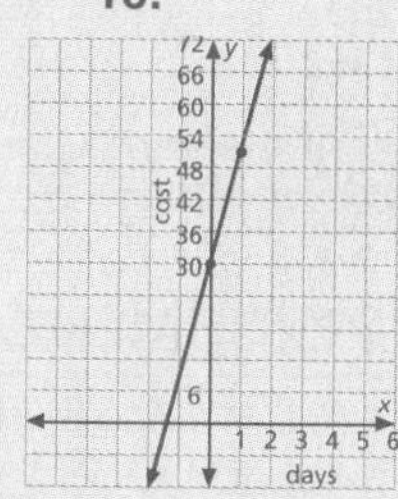

56.

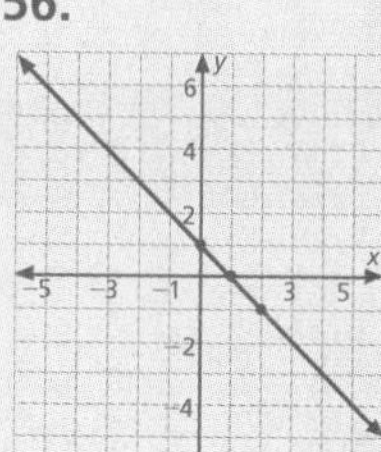

57.

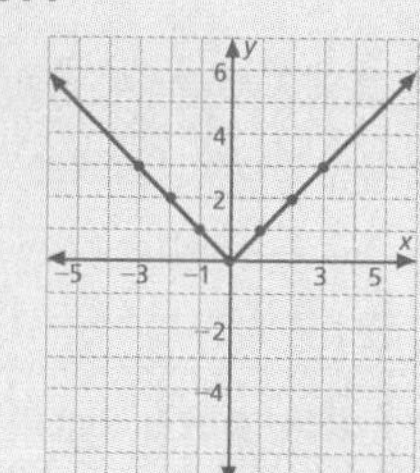

58.

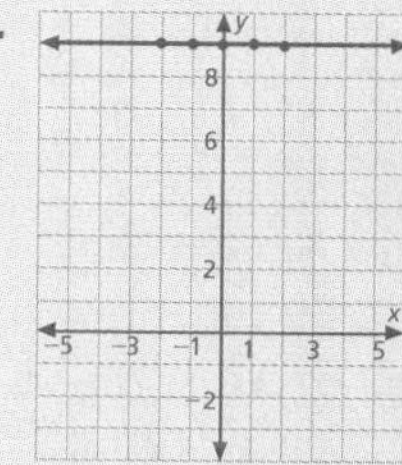

Teaching Strategies

COOPERATIVE LEARNING Have students write an operation on an index card such as "add to 12.," or "divide by −3." Post the cards where they can be seen by all students. Ask for a volunteer to think of a secret number. This student should perform each operation on the secret number and announce the results aloud. Have students raise their hands as soon as they know the secret number.

Lesson Planning

NCTM Standards/Strands
- Number & Operation
- Patterns, Functions, & Algebra
- Geometry & Spatial Sense
- Data Analysis, Statistics, & Probability
- Problem Solving
- Reasoning and Proof
- Communication
- Connections
- Representation

Vocabulary
like or similar terms

Materials Needed
paper/pencil   Algeblocks

Lesson Resources
Warm-Up Transparency 5
Transparency TK-2, 3, 5
Reteaching 2-5
Extra Practice 2-5
Enrichment 2-5

Assignment Guide
Basic: 1–34, 38–55
Enriched: 1–55

Getting Started

5-Minute Warm-Up
Solve each equation.
1. $x + 11 = -3$ -14
2. $6 - x = 9$ -3
3. $5x = -75$ -15
4. $x/2 = -4$ -8

Introduction to Lesson 2-5
Have students work with the Algeblocks in small groups. Make sure they represent each step of the problem-solving process algebraically.

2-5 Solve Multi-Step Equations

Goals ■ Solve equations with more than one step.

Applications Advertising, Finance, Recreation

Algeblocks can be used to solve two-step equations. Complete the equation to show algebraically the steps taken to solve $3x - 2 = 4$.

Algeblocks	*Algebraically*	
a. Represent the equation.	**a.** Write the equation.	**a.** $3x - 2 = 4$
	b. Add the opposite of -2 to both sides and simplify.	**b.** $3x - 2 + \square = 4 + \square$ 2, 2 $\square = \square$ 3x, 6
	c. Divide each side into three groups.	**c.** $\frac{3x}{\square} = \frac{\square}{\square}$ 3, $\frac{6}{3}$
	d. Read the solution.	**d.** $x = \square$ 2

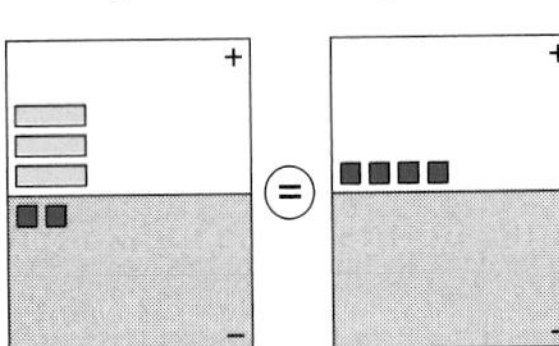

Build Understanding

To solve some equations, it may take two or more steps to get the variable alone on one side of the equation. When solving these equations, use the addition property of equality first. Then use the multiplication property of equality.

Example 1

MODELING Solve $2x - 7 = -1$. Along with using Algeblocks, explain and represent each step algebraically.

Solution

Use Algeblocks to represent the equation.

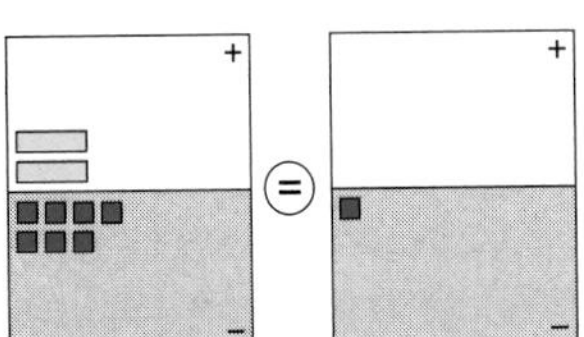

Add + 7 to each side of the equation. Simplify.

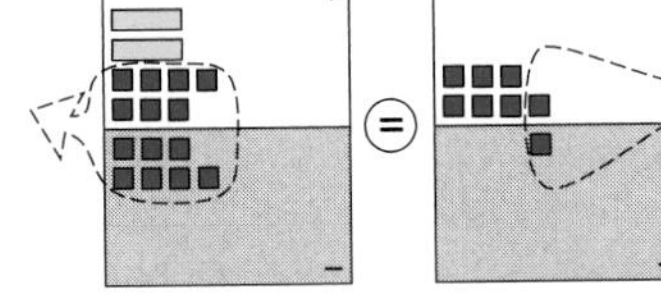

Separate into 2 groups.

The solution is $x = 3$.

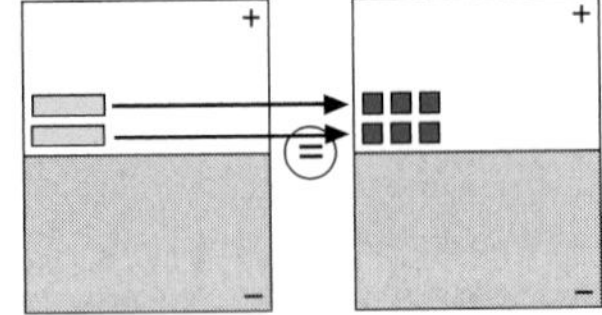

Teaching Strategies

COOPERATIVE LEARNING Have students work together to explain how this number puzzle works. "Think of a number, add two to it, multiply the result by four, subtract eight, subtract twice the number you started with, then divide the result by two. You should get the original number!"
Simplify the algebraic expression for the steps. When x is the number, $\frac{4(x + 2) - 8 - 2x}{2} = x$

Some equations contain variables on both sides. For these equations, simplify the equation by using the addition property of equality to move like terms to the same side of the equation. Terms that have exactly the same variables are called *like* or *similar* terms.

Example 2

Solve $x + 5 = 2x - 3$. Check the solution.

Solution

$x + 5 = 2x - 3$	
$x + 5 + -2x = 2x + -2x - 3$	Add $-2x$ to each side.
$-x + 5 = -3$	
$-x + 5 - 5 = -3 + -5$	Add -5 to each side.
$-x = -8$	
$-x(-1) = -8(-1)$	Multiply each side by -1.
$x = 8$	

The solution is 8.

Check

$x + 5 = 2x - 3$

$8 + 5 = 2(8) - 3$

$13 = 16 - 3$

$13 = 13$ ✓

Sometimes you will need to simplify each side of an equation before applying the properties of equality.

Example 3

Solve $6(2x - 1) = -36 + 6$. Check the solution.

Solution

$6(2x - 1) = -36 + 6$	Apply the distributive property.
$12x - 6 = -36 + 6$	
$12x - 6 = -30$	
$12x - 6 + 6 = -30 + 6$	Add 6 to each side.
$12x = -24$	
$\left(\frac{1}{12}\right)(12x) = \left(\frac{1}{12}\right)(-24)$	Multiply each side by $\frac{1}{12}$.
$x = -2$	

The solution is -2.

Check Be sure to follow the order of operations.

$6(2x - 1) = -36 + 6$

$6(2(-2) - 1) = -30$

$6(-4 - 1) = -30$

$6(-5) = -30$

$-30 = -30$ ✓

Problem Solving Tip

Example 3 can also be solved by dividing both sides by 6 first.

$6(2x - 1) = -30$

$2x - 1 = -5$

Then the equation is easier to solve.

$2x - 1 + 1 = -5 + 1$

$2x = -4$

$x = -2$

Extend the Lesson

ONGOING ASSESSMENT Write the following solution on the board. Challenge the students to find the error.

$-2(x - 3) = 8$

$-2x - 6 = 8$

$-2x = 14$

$x = -7$

The student did not multiply the second term inside the parentheses by -2. The answer should be $x = -1$.

Chalkboard Examples

Supplementary Example 1

Use Algeblocks to solve the equation: $3x + 8 = 17$. Represent each step algebraically.

Write the equation.	$3x + 8 = 17$
Add -8 to each side.	$3x + 8 - 8 = 17 - 8$
Simplify.	$3x = 9$
Separate each side into 3 groups.	$\frac{3x}{3} = \frac{9}{3}$
The solution is:	$x = 3$

Supplementary Example 2

Solve for x.

$x - 4 = 2x + 8$ $x = -12$

Supplementary Example 3

Solve $2(6 - s) = -18 + 20$. Check the solution. $x = 5$

Supplementary Example 4

A small business spent $360 one week on newspaper advertisements. An ad in the Sunday paper costs 5 times as much as an ad that is run on a weekday. The business ran one Sunday ad and four weekday ads. What is the cost of a weekday ad? $40

Reteaching Worksheet 2-5

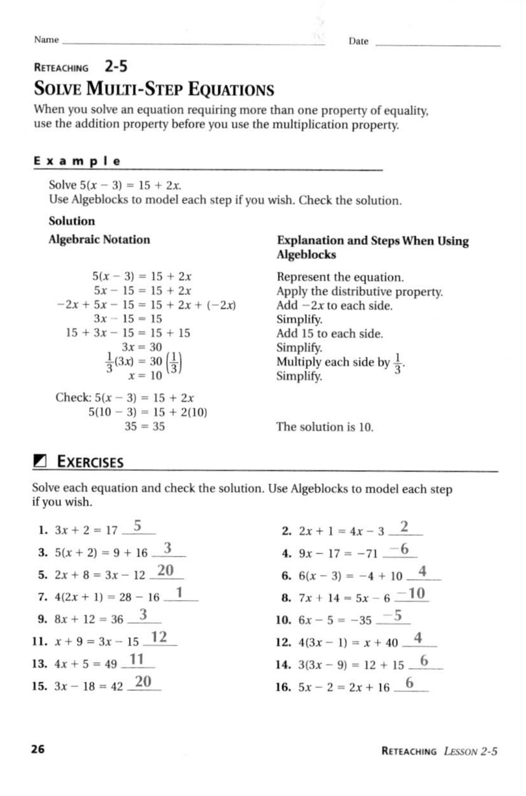

Name ______________________ Date ______________

RETEACHING 2-5

SOLVE MULTI-STEP EQUATIONS

When you solve an equation requiring more than one property of equality, use the addition property before you use the multiplication property.

Example

Solve $5(x - 3) = 15 + 2x$.
Use Algeblocks to model each step if you wish. Check the solution.

Solution

Algebraic Notation	Explanation and Steps When Using Algeblocks
$5(x - 3) = 15 + 2x$	Represent the equation.
$5x - 15 = 15 + 2x$	Apply the distributive property.
$-2x + 5x - 15 = 15 + 2x + (-2x)$	Add $-2x$ to each side.
$3x - 15 = 15$	Simplify.
$15 + 3x - 15 = 15 + 15$	Add 15 to each side.
$3x = 30$	Simplify.
$\frac{1}{3}(3x) = 30\left(\frac{1}{3}\right)$	Multiply each side by $\frac{1}{3}$.
$x = 10$	Simplify.

Check: $5(x - 3) = 15 + 2x$

$5(10 - 3) = 15 + 2(10)$

$35 = 35$

The solution is 10.

EXERCISES

Solve each equation and check the solution. Use Algeblocks to model each step if you wish.

1. $3x + 2 = 17$ 5
2. $2x + 1 = 4x - 3$ 2
3. $5(x + 2) = 9 + 16$ 3
4. $9x - 17 = -71$ −6
5. $2x + 8 = 3x - 12$ 20
6. $6(x - 3) = -4 + 10$ 4
7. $4(2x + 1) = 28 - 16$ 1
8. $7x + 14 = 5x - 6$ −10
9. $8x + 12 = 36$ 3
10. $6x - 5 = -35$ −5
11. $x + 9 = 3x - 15$ 12
12. $4(3x - 1) = x + 40$ 4
13. $4x + 5 = 49$ 11
14. $3(3x - 9) = 12 + 15$ 6
15. $3x - 18 = 42$ 20
16. $5x - 2 = 2x + 16$ 6

26 RETEACHING *LESSON 2-5*

Lesson Wrap-up

Quick Assessment

Ask the following questions to determine if students understand the content presented in this lesson.

1. Tell the first step in solving the equation $4(x - 2) = 3$. Simplify.
2. Write an equation and solve. When the sum of a number and 6 is multiplied by 5, the result is 25. What is the number? $5(n + 6) = 25, x = -1$
3. How do the properties of equality help us solve problems? The properties allow us to form equivalent equations to which the solution is easy to read.

Assignment Guide

Basic: 1–34, 38–55
Advanced: 1–55
Additional Practice: See Extra Practice Index on page 674.

Example 4

ADVERTISING A local newspaper sells all classified ads for the same price. Larger boxed ads cost \$24.50. Eun Ah bought three classified ads and one boxed ad. If the total cost for the ads was \$79.25, what was the price of each classified ad?

Solution

Let a represent the price of each classified ad.

$$3a + 24.50 = 79.25$$
$$3a + 24.50 + (-24.50) = 79.25 + (-24.50)$$
$$3a = 54.75$$
$$\left(\frac{1}{3}\right)(3a) = \left(\frac{1}{3}\right)(54.75)$$
$$a = 18.25$$

Check 3 classified ads = 3(\$18.25): \$54.75
1 larger ad: \$24.50
Total: \$79.25 ✓

Each classified ad costs \$18.25.

Try These Exercises

1. **MODELING** Use Algeblocks to solve $2x - 5 = 7$. Show each step algebraically. 6

Solve each equation and check the solution.

2. $3a - 5 = 7$ 4
3. $-4x + 1 = 25$ −6
4. $52 = 4(2j + 5)$ 4
5. $4u - 5 = 2u - 13$ −46
6. $-6b + 9 = 4b - 41$ 5
7. Twice n plus 14 is −8. −11
8. **YOU MAKE THE CALL** Maggie says that multiplying by the reciprocal of a number is the same as dividing by the number. Is Maggie correct? \$1.75
9. **RECREATION** A carnival pass costs \$15, and buys unlimited access to 10 rides. This pass costs \$2.50 less than paying the individual price for each of the 10 rides. What is the individual price of each ride? Yes, for example, $\frac{1}{3}x$ is the same as $\frac{x}{3}$.

10. **WRITING MATH** Write a multi-step equation in which x equals −4. Answers may vary.

Practice Exercises

Solve each equation and check the solution.

11. $4n + 3 = 15$ 3
12. $-2d - 16 = 4$ −10
13. $-28 = 3r - 7$ −7
14. $-14 = 18 - 8e$ 4
15. $2(5z - 3) = 34$ 4
16. $-3(2h - 1) = 3$ 0
17. $-5p - 1 = 3p + 15$ −2
18. $4 - 7a = -1 - 2a$ 1
19. $6v + 3 - 2v = 1 + 5v$ 2
20. $9 - 4c + 15 = 0$ 6
21. $\left(\frac{1}{2}\right)(12f + 30) = 9$ −1
22. $8(1.25 - q) = 6$ $\frac{1}{2}$

Teaching Strategies

TEAM LEARNING Have small groups of students start with a value for a variable and make a multi-step equation for it. After each group has made two or three, have them exchange with other groups and solve.

Translate each sentence into an equation. Then solve.

23. Four more than 3 times a number is 31. Find the number. 9

24. When 12 is decreased by twice a number, the result is -14. Find the number. 13

Solve each equation and check the solution.

25. $-3(d-5) = 2(4d-9)$ 3

26. $\left(\frac{1}{3}\right)(15z-21) = \left(\frac{2}{5}\right)(10z-35)$ -7

27. $4(5-3m) - 9 = 3m - 4$ -1

28. $-2(4k+1) + k = 8 - 5k$ -5

29. $9(a+4) - 2a = 19 - 3(a+6)$ -3.5

30. $15x - 4(4+3x) = -5(2x-5) + 11$ 4

Translate each sentence into an equation. Then solve.

31. Fifteen more than twice a number is the same as 7 less than four times the number. Find the number. 11

32. When the sum of twice a number and 3 is multiplied by 5, the result is the same as decreasing the product of 6 and the number by 1. Find the number. -4

33. FINANCE The stereo system Doug wants to buy can be purchased by paying a \$50.00 down payment, and paying the rest in equal monthly installments over the next 6 months. If the total cost of the stereo system is \$228.50, what will be the amount of each monthly payment? \$29.75

34. CHAPTER INVESTIGATION Think about how you will use the media to convince the public to support your message. Write a public relations plan and a budget. Estimate the cost of any advertisements or commercials you will need to run in local newspapers or on television.

Extended Practice Exercises

35. Solve $2(3x+2) + x = 3x + 4 + 4x$. Explain your solution.
Answers may vary.

36. Solve $5 + 4(2x-1) = 3(x+1) + 5x$. Explain your solution.
The solution is {all real numbers}. Any real number will satisfy the equation.

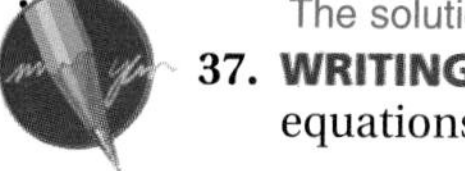

37. WRITING MATH Summarize, in writing, the steps used to solve equations. Since $1 \neq 3$, no real numbers satisfy this equation.

Mixed Review Exercises

Evaluate each expression when $a = 2$ and $b = -3$. (Lesson 1-7)

38. ab^2 18

39. a^3b^2 72

40. $a^3 - b^3$ 35

41. $a^2 + b^2$ 13

42. $(a^2 - b)^2$ 49

43. $-4ab^3$ 216

44. $(a^2)(b^2)$ 36

45. $(a^3 - 5)^3$ -27

46. $-(b^2)(a^3)$ -72

47. $-4a^3b$ 96

48. $(b+8)(a^2)$ 20

49. $(a^3+2)^2$ 100

Write each number in scientific notation. (Lesson 1-8)

50. 8,640,000,000,000 8.64×10^{12}

51. 0.000000045 4.5×10^{-8}

52. 0.0000017 1.7×10^{-6}

53. 0.000000000039 3.9×10^{-11}

54. 128,000,000,000,000 1.28×10^{14}

55. 0.0000000026 2.6×10^{-9}

Extend the Lesson

CONNECTING TO PRIOR KNOWLEDGE Review the order of operations.
1. Perform operations inside grouping symbols.
2. Do multiplication and division from left to right.
3. Do addition and subtraction from left to right.

Discuss the importance of using the order of operations when checking the solution to an equation. Students may notice that in solving an equation, they must generally use the addition property of equality before they use the multiplication property in order to isolate the variable.

Extra Practice Worksheet 2-5

Name ____________ Date ____________

EXTRA PRACTICE 2-5

SOLVE MULTI-STEP EQUATIONS

EXERCISES

Solve each equation and check the solution.

1. $2c + 3 = 15$ 6
2. $-3s + 4 = -2$ 2
3. $-14 = 4d + 6$ -5
4. $19 = 25 - 3w$ 2
5. $2(b+3) = 2$ -2
6. $5y + 3 = 2y + 12$ 3
7. $5 - 2x = x - 19$ 8
8. $7t - 5 + 3t = 15$ 2
9. $2 - 3(m+4) = 2$ -4
10. $1 - 6r = -4 - 3r$ $1\frac{2}{3}$
11. $\frac{1}{3}(6p - 12) = 5$ $4\frac{1}{2}$
12. $4(0.5 - w) = -18$ 5

Translate each sentence into an equation. Then solve.

13. Six more than twice a number is 16. Find the number. $6 + 2n = 16; 5$
14. Four times a number decreased by 12 is 8. Find the number. $4n - 12 = 8; 5$
15. When 15 is decreased by three times a number, the result is 21. Find the number. $15 - 3n = 21; 2$
16. Eight more than five times a number is the same as one less than eight times the number. Find the number. $8 + 5n = 8n - 1; 3$
17. When the sum of twice a number and 2 is multiplied by 3, the result is the same as 4 times the sum of the number and 4. Find the number. $3(2n+2) = 4(n+4); 5$

Solve each equation and check the solutions.

18. $-3(r+5) = 3(r-1)$ -2
19. $\frac{1}{2}(4m+8) = \frac{1}{3}(3m-3)$ -5
20. $6(2-3x) + 8 = 2 - 9x$ 2
21. $3 - 10k = -3(5k+2) - 4k$ -1
22. Juan bought 4 T-shirts and a leather jacket. The T-shirts were all the same price, and the price of the leather jacket was 6 times the cost of one T-shirt. If the total cost of the T-shirts and the leather jacket was \$209.00, what was the price of each T-shirt? \$20.90

24 EXTRA PRACTICE *LESSON 2-5*

Enrichment Worksheet 2-5

Name ____________ Date ____________

ENRICHMENT 2-5

QUANTITATIVE COMPARISONS

A quantitative comparison is a type of problem found on some multiple-choice tests. In each question, you are given two quantities, one in Column A and one in Column B. You are to compare the two quantities and shade one of the four circles on an answer sheet.

Shade circle A if the quantity in Column A is greater.
Shade circle B if the quantity in Column B is greater.
Shade circle C if the two quantities are equal.
Shade circle D if the relationship cannot be determined from the information given.

EXERCISES

Shade the correct circle to the left of each problem number.

		Column A	Column B
A. B. C. D.	1.	Solution to $-17 = 9x - 8$	Solution to $4x - 7 = -11$
A. B. C. D.	2.	Solution to $3x + 9 = 17$	Solution to $8x = 22 - 19$
A. B. C. D.	3.	Solution to $18 - 4x = 9$	Solution to $3x + 9 = 17$
A. B. C. D.	4.	$-x$ if x is greater than 0	Solution to $15x - 7 = -37$
A. B. C. D.	5.	The solution to $7 = 4x + 19$	$-x$ if x is less than 0
A. B. C. D.	6.	$\lvert x\rvert$ if x is greater than 0	Solution to $10x - 9 = -9$
A. B. C. D.	7.	$\sqrt{x}$ if x is between 0 and 10	Solution to $8 = 10x - 32$
A. B. C. D.	8.	Solution to $10x - 9 = -9$	$\lvert x\rvert - \lvert y\rvert$ if $y = x$
A. B. C. D.	9.	$\lvert x\rvert - \lvert y\rvert$ if y is greater than x	Solution to $-36 = 10x + 14$
A. B. C. D.	10.	Solution to $13x - 4x = -14 + 2x$	$\lvert x\rvert - \lvert y\rvert$ if x is greater than y

32 ENRICHMENT *LESSON 2-5*

Lesson Planning

NCTM Standards/Strands

- Number & Operation
- Patterns, Functions, & Algebra
- Problem Solving
- Reasoning and Proof
- Communication
- Connections
- Representation

Vocabulary

linear inequality half-plane
boundary
transitive property
open half-plane
closed half-plane

Materials Needed

paper/pencil
number lines
grid paper

Lesson Resources

Warm-Up Transparency 5
Transparency TK-6-13
Reteaching 2-6
Extra Practice 2-6
Enrichment 2-6
Multicultural Connection 9

Assignment Guide

Basic: 1–33, 37–42
Enriched: 1–42

Getting Started

5-Minute Warm-up

Solve each equation.

1. $2x - 3 = 9$ 6
2. $7x + 6 = 6x - 11$ -17
3. $4(x + 3) = 28$ 4
4. $7 - 2(3x + 1) = -x$ 1

Introduction to Lesson 2-6

Be certain that students know what is meant by the phrase "below the line." Have them share their results from this activity with the class.

2-6 Solve Linear Inequalities

Goals
- Solve an inequality in one or two variables.
- Graph the solution to an inequality in one or two variables.

Applications News media, Sales

Consider the graph of the equation $x + y = 5$. The points that lie on this line have coordinates whose sum is 5. For example, (0, 5), (1, 4), and (−2, 7) are points that lie on the line $x + y = 5$.

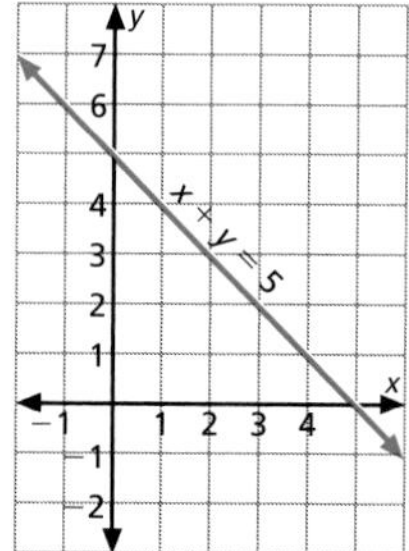

a. Are there any points not on the line that have coordinates whose sum is 5? no

b. Select any three points below the line and find the sum of the coordinates. How do these sums compare with 5? sum < 5

c. Select any three points above the line and find the sum of the coordinates. How do these sums compare with 5? sum > 5

Build Understanding

A mathematical sentence that contains one of the symbols <, >, ≤, or ≥ is an inequality. Inequalities are used to indicate the *order* of a comparison between two quantities.

A linear inequality in *one variable* is an inequality only in x or y. The techniques used to solve an inequality are similar to those used to solve equations. The addition property of inequality states that adding the same number to both sides of an inequality maintains the order of the inequality.

Addition Property of Inequality	For all real numbers a, b, and c: If $a < b$, then $a + c < b + c$. If $a > b$, then $a + c > b + c$.

The multiplication property of inequality states that multiplying both sides of an inequality by the same positive number still maintains the order of the inequality. However, if the number you are multiplying by is a negative number, you must reverse the order of the inequality.

Multiplication Property of Inequality	For all real numbers a, b, and c: If $a < b$ and $c > 0$, then $ac < bc$. If $a < b$ and $c < 0$, then $ac > bc$. If $a > b$ and $c > 0$, then $ac > bc$. If $a > b$ and $c < 0$, then $ac < bc$.

Check Understanding

Multiply the inequality $3 > -1$ by -2. Discuss the result.

Multiply the inequality by 2. Discuss the result.

76 Chapter 2 **Essential Algebra and Statistics**

The transitive property of inequality shows how two inequalities can be combined to produce a third.

Transitive Property of Inequality	For all real numbers a, b, and c: If $a < b$ and $b < c$, then $a < c$. If $a > b$ and $b > c$, then $a > c$.

For example, if $x + 3 < y$ and $y < 7$, then $x + 3 < 7$. This inequality can then be solved for x.

Example 1

Solve each inequality and graph its solution on a number line.

a. $3x + 10 < 4$ **b.** $23 \geq 8 - 5y$

Solution

a.

$$3x + 10 < 4$$
$$3x + 10 + (-10) < 4 + (-10)$$
$$3x < -6$$
$$\left(\frac{1}{3}\right)3x < \left(\frac{1}{3}\right)(-6)$$
$$x < -2$$

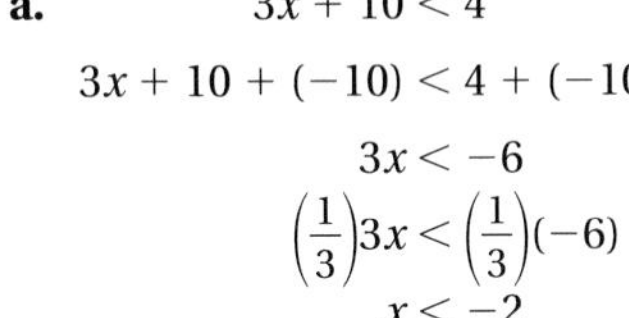

The open circle indicates that −2 is not part of the solution.

b.

$$23 \geq 8 - 5y$$
$$23 + (-8) \geq 8 + (-8) - 5y$$
$$15 \geq -5y$$
$$\left(-\frac{1}{5}\right)15 \leq \left(-\frac{1}{5}\right)(-5y)$$
$$-3 \leq y$$
$$y \geq -3$$

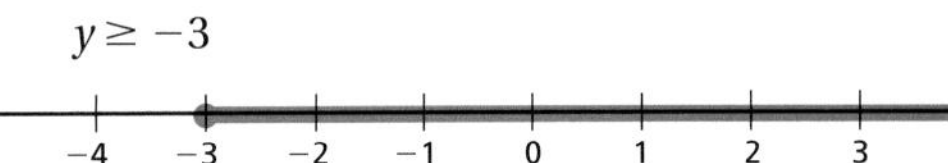

The closed circle indicates that −3 is part of the solution.

A solution of a linear inequality in two variables, such as $2x + y < 4$, is an ordered pair that makes the inequality true. The graph of all such solutions is a region called a **half-plane**.

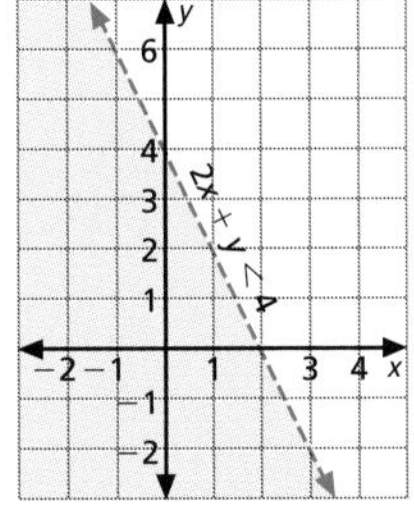

The edge of the half-plane is called the **boundary**. If the inequality is a strict inequality (< or >), then the region is an **open half-plane**, and the boundary is not part of the solution. If the inequality is inclusive (≤ or ≥), then the region is a **closed half-plane**, and the boundary is part of the solution.

To graph an inequality in two variables, first graph the related equation. This line will serve as the boundary. If the solution will be a closed half-plane, draw the boundary as a solid line. Otherwise, draw it with a broken line. Then shade the half-plane that contains the solutions of the inequality.

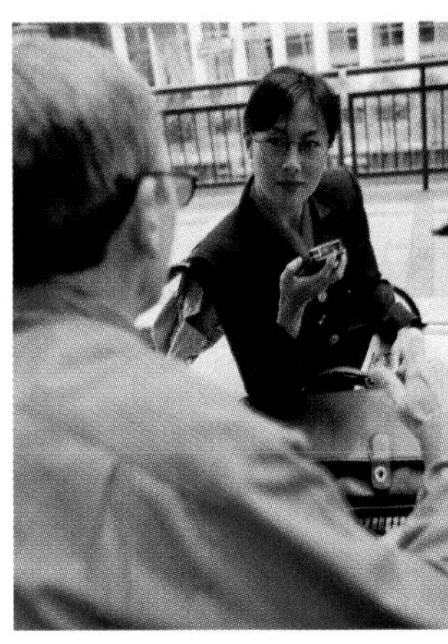

Extend the Lesson

TEAM LEARNING Have students do a scavenger hunt of everyday examples that can be written as inequalities. For example, for a car, a student may write $p \leq 6$, where p is the number of passengers.

Chalkboard Examples

Supplementary Example 1

Solve each inequality and graph its solution on a number line.

a. $2x + 7 \leq 5$ $x \leq -1$

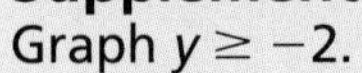

b. $17 < 23 - 3x$ $2 > x$

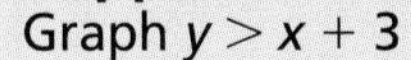

Supplementary Example 2

Graph $y \geq -2$.

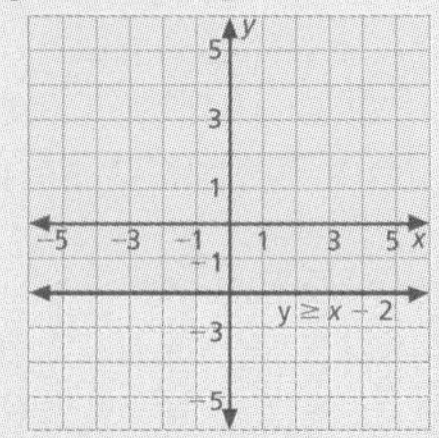

Supplementary Example 3

Graph $y > x + 3$

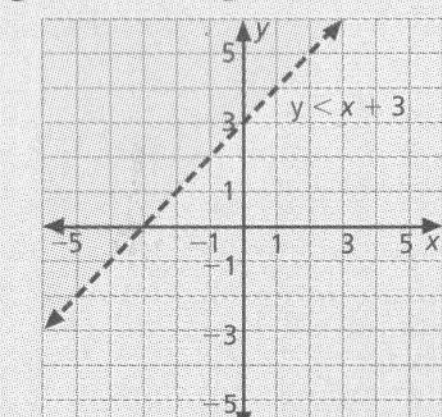

Reteaching Worksheet 2-6

Name ______ Date ______

RETEACHING 2-6

SOLVE LINEAR INEQUALITIES

The steps used to solve an inequality are similar to those used to solve an equation. When an inequality is multiplied by a negative number, the order of the inequality is reversed.

Example 1

Solve $2x < 10$ and graph its solution on a number line.

Solution

Solve the inequality. Graph the inequality.

$2x < 10$

$\frac{1}{2}(2x) < 10\left(\frac{1}{2}\right)$

$x < 5$

The solution is $x < 5$.

Example 2

Graph $y > 4x - 2$ on the coordinate plane.

Solution

Write the related equation.

$y = 4x - 2$

Make a table of values (ordered pairs) to graph the boundary.

x	y
−1	−6
0	−2
1	2

The boundary is a broken line because the boundary is not part of the solution set.

Determine the shading.

Test point: (0, 0) $y > 4x - 2 \rightarrow 0 > 4(0) - 2 \rightarrow 0 > -2$.

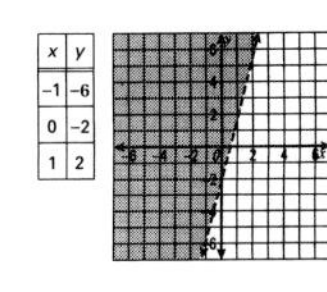

EXERCISES

Solve each inequality and graph its solution on the number line.

1. $3x - 2 \leq 7$ $x \leq 3$
2. $9 < 5x - 1$ $x > 2$

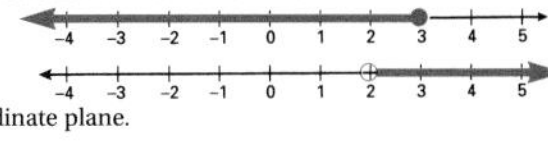

Graph each inequality on the coordinate plane.

3. $2x - 2y \geq -2$
4. $y < 2x - 3$

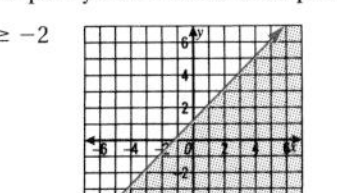

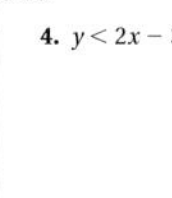

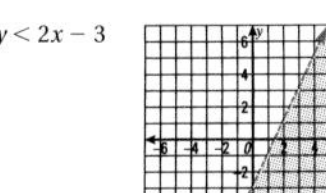

28 RETEACHING LESSON 2-6

Lesson Wrap-up

Quick Assessment

Ask the following questions to determine if students understand the content presented in this lesson.

1. How is the graph of an inequality in one variable different from the graph of an inequality in two variables? For one variable, the graph is a number line; for two variables the graph is part of the coordinate plane.
2. In solving an inequality, when is the inequality sign reversed? when each side is multiplied by a negative number

Assignment Guide

Basic: 1–33, 37–42
Advanced: 1–42
Additional Practice: See Extra Practice Index on page 674.

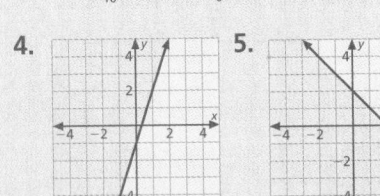

Example 2

Graph $y \le 4x$.

x	y
−1	−4
0	0
1	4

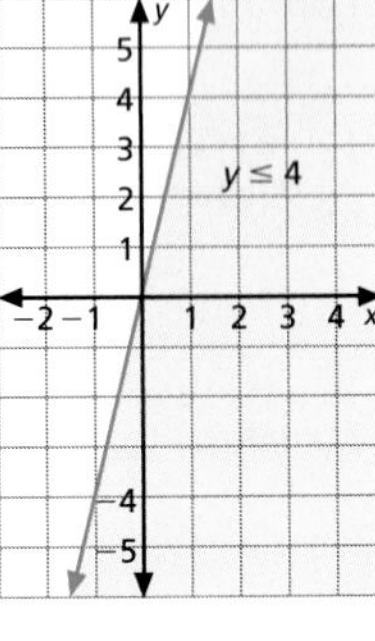

Solution

The related equation is $y = 4x$. Make a table of values that can be used to graph the boundary.

Note that the boundary is part of the solution set, and is drawn as a solid line. To decide which half-plane to shade, use a test-point not on the boundary. If it *is* a solution, then all points on that half-plane will also be solutions; so, shade that side. If the point is not a solution, shade the half-plane that does not contain the test point.

Test Point: $(-1, 1)$ $\quad y \le 4x$

$1 \le 4(-1)$

$1 \le -4 \quad$ × (false)

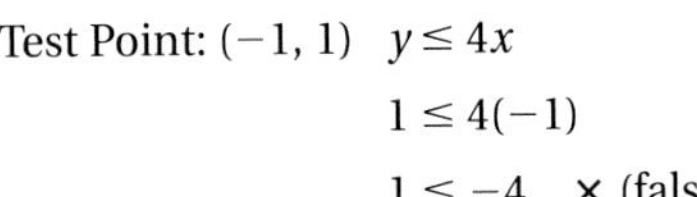

Because 1 *is not* less than or equal to −4, shade the half-plane that does not contain $(-1, 1)$.

Example 3

Graph $y > \frac{3}{2}x - 4$.

Solution

The related equation is $y = \frac{3}{2}x - 4$.

Make a table of values that can be used to graph the boundary. Note that the boundary is not included in the solution set, and is drawn as a broken line.

$y = \frac{3}{2}x - 4$

Test Point: $(0, 0)$ $\quad y > \frac{3}{2}x - 4$

$(0) > \frac{3}{2}(0) - 4$

$0 > -4$

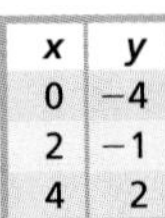

x	y
0	−4
2	−1
4	2

Because 0 *is* less than 8, shade the half-plane containing (0,0).

Try These Exercises

Solve each inequality and graph the solution on a number line. See additional answers.

1. $4m + 5 > 25$ $\quad m > 5$
2. $-2k + 9 \ge 1$ $\quad k \le 4$
3. $\left(\frac{3}{4}\right)c - 4 > -16$ $\quad c > -16$

Graph each inequality on the coordinate plane. See additional answers.

4. $y \le 3x - 1$
5. $y \ge -x + 2$
6. $2x - y < 3$
7. $x + 2y < 10$
8. $4x + 3y \ge -6$
9. $2x - 5y \ge 15$

Additional Answers

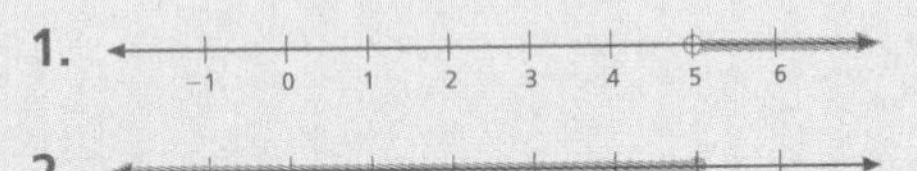

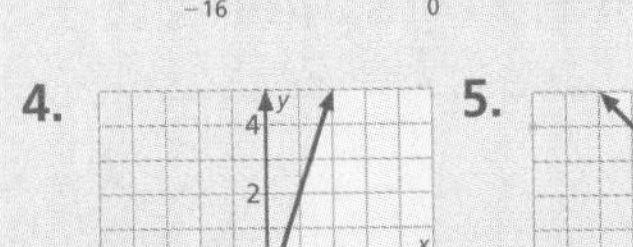

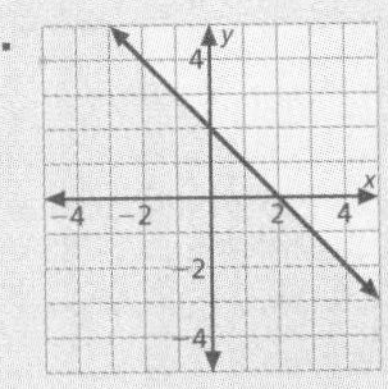

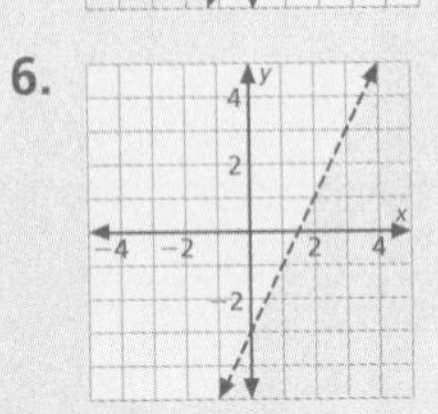

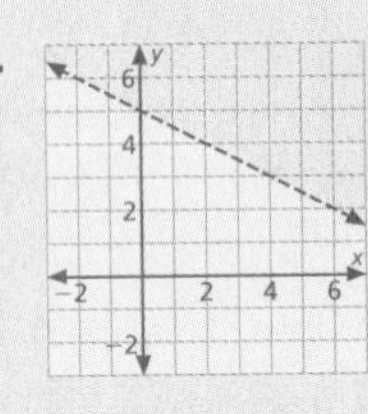

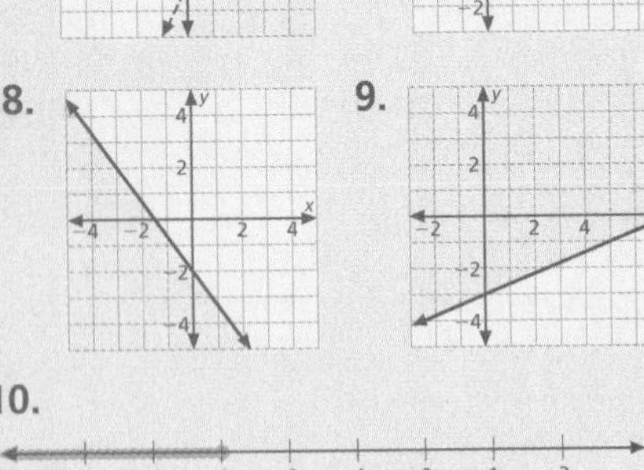

10.

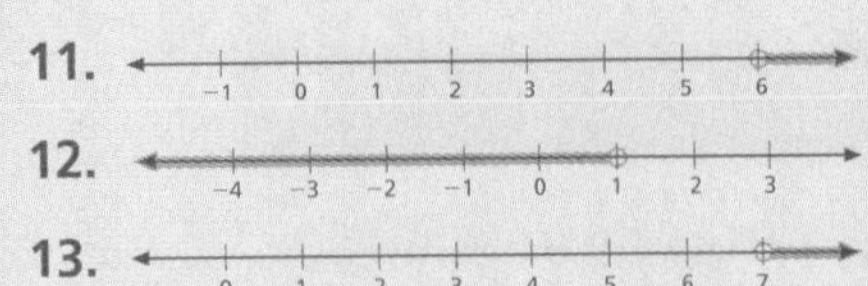

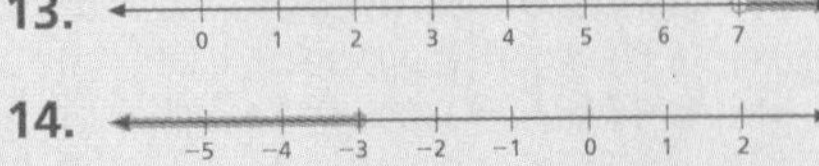

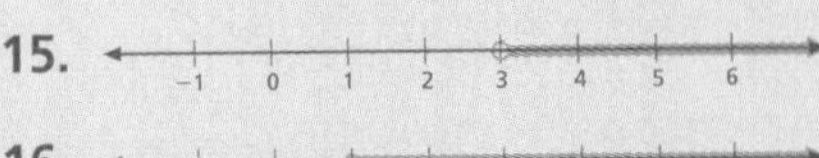

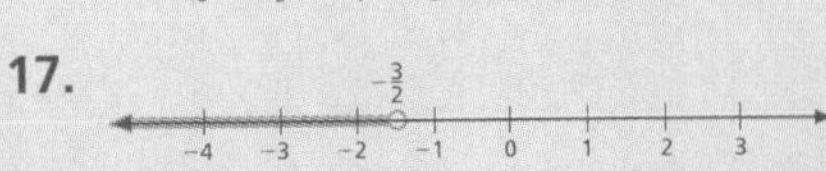

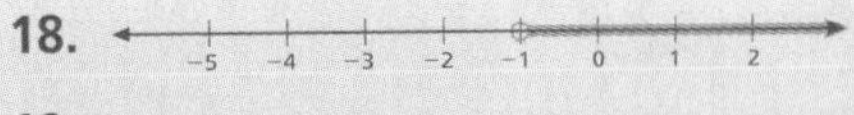

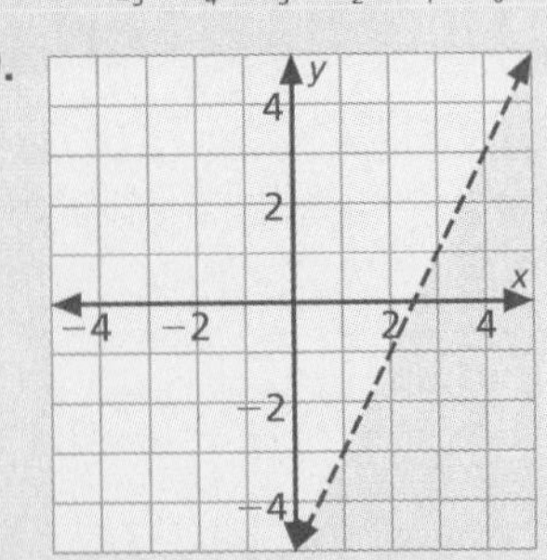

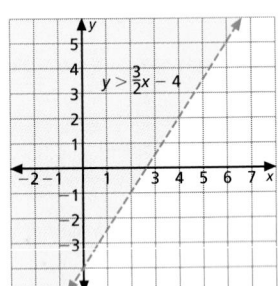

Practice Exercises

Solve each inequality and graph the solution on a number line. See additional answers.

10. $5b + 4 \leq -11$ $b \leq -3$ **11.** $\left(\frac{1}{2}\right)p - 10 > -7$ $p > 6$ **12.** $9 - 4r > 5$ $r < 1$

13. $13 < 3a - 8$ $a > 7$ **14.** $26 \leq -9n - 1$ $n \leq -3$ **15.** $-31 > 14 - 15z$ $z > 3$

16. $\left(\frac{3}{2}\right)h + 12 \geq 6$ $h \geq -4$ **17.** $3(4e + 3) < -9$ $e < -\frac{3}{2}$ **18.** $8k - 7 > 6k - 9$ $k > -1$

Graph each inequality on the coordinate plane. See additional answers.

19. $y > 2x - 5$ **20.** $y \leq -3x + 4$ **21.** $x + y \geq -3$

22. $x - y > 2$ **23.** $2x - 3y < 6$ **24.** $6 > 2x - \left(\frac{2}{3}\right)y$

25. NEWS MEDIA A reporter estimates that $\frac{2}{3}$ of the hours (h) spent on a story increased by 15 h is less than 27 h. What values are possible for h? $h < 18$ h

26. SALES A jacket sells for $55. Decreasing the price of the jacket by a discount amount (d) yields a result greater than one-half the sum of the discount and $10. What are the possible values for the discount? $d < \$20$

Solve each inequality and graph the solution on a number line. See additional answers.

27. $0.5c - 7.4 \geq 0.35 + 1.75c$ **28.** $6 - (5m + 7) < 3(2m + 1) - 10m$

29. $4(3d + 1) - 5d \leq 8 - 2(5d + 2)$ **30.** $-10 \leq \left(\frac{2}{5}\right)(10 - 5q)$ $q \leq 7$

Write the inequality represented by each graph.

31.

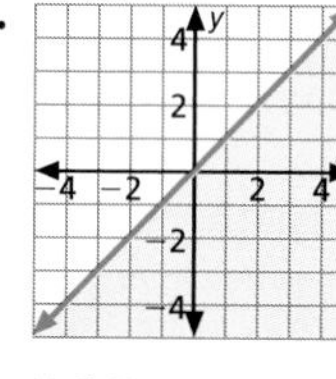

$y \leq x$

32.

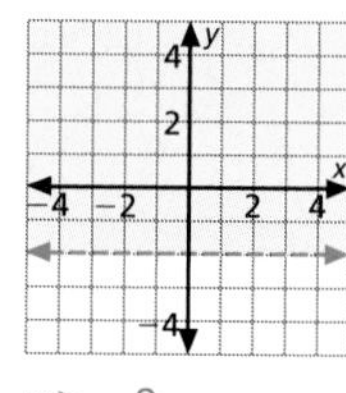

$y > -2$

33.

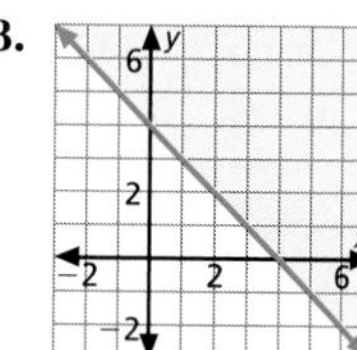

$x + y \geq 4$

Extended Practice Exercises

34. Graph the solution to the inequality $|x| > 2$ on a number line. See additional answers.

35. WRITING MATH Write a paragraph in which you discuss two different ways to interpret and graph the inequality $y < 5$. How are these two interpretations and their graphs related? See additional answers.

36. Name three points that are solutions of both the inequality $x + y < 1$ and the equation $-3x + 2y = 10$. Answers will vary.

Mixed Review Exercises

Identify the rule relating each term. Determine the next three terms in each sequence. (Lesson 2-1)

37. 1, 4, 16, 64, 256, ...
rule: × 4; 1024; 4096; 16,384

38. 100, 50, 25, 12.5, 6.25, ...
rule: ÷ 2; 3.125; 1.5625; 0.78125

39. 1, 4, 7, 10, 13, ... rule: + 3; 16, 19, 22

40. 200, 193, 186, 179, 172, ...
rule: − 7; 165, 158, 151

41. 1, −3, 9, −27, 81, ...
rule: × −3; −243, 729, −2187

42. 50,000, −10,000, 2,000, −400, 80, ...
rule: ÷ −5; −16, 3.2, −0.64

20.

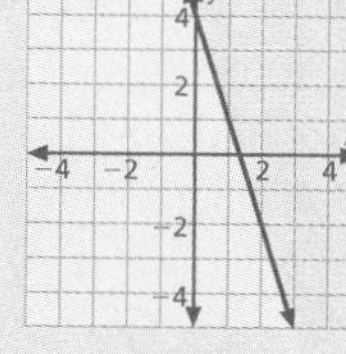

21.

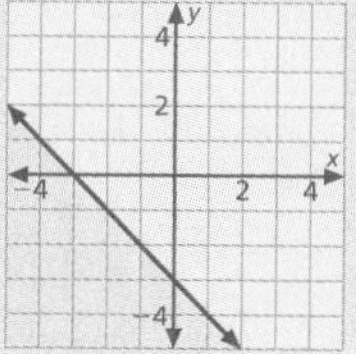

22.

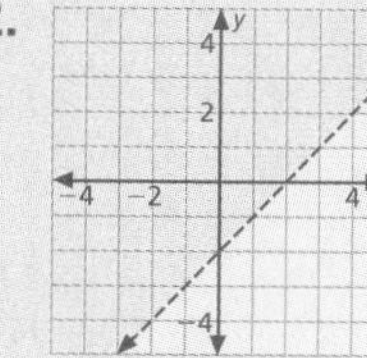

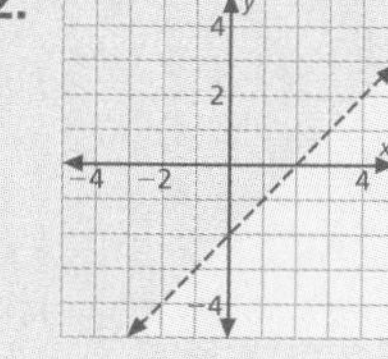

23.

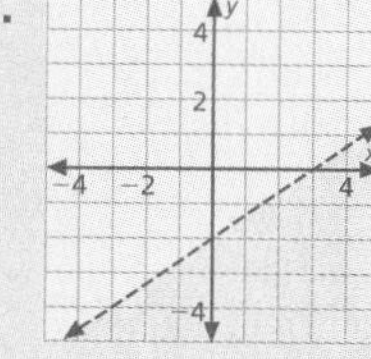

24.

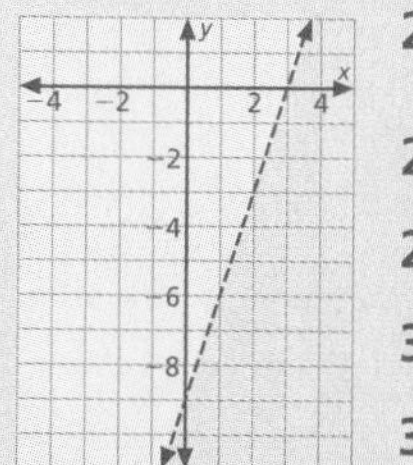

27. **28.** **29.** **30.** **34.**

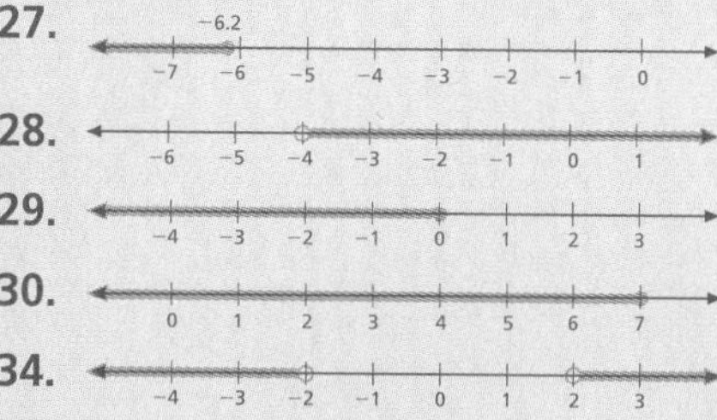

35. Students should discuss and compare the solutions as they appear on both a number line and a coordinate plane.

Extra Practice Worksheet 2-6

Name ______________________ Date __________

EXTRA PRACTICE 2-6

Solve Linear Inequalities

Exercises

Solve each inequality and graph the solution on a number line. Use your own paper.

1. $2d + 1 \geq 13$ $d \geq 6$
2. $8 - 3r < -4$ $r > 4$
3. $14 \leq 5a + 4$ $a \geq 2$
4. $-5k - 3 > 12$ $k < -3$
5. $-10 - 6z \leq 20$ $z \geq -5$
6. $6r - 4 < -10$ $r < -1$
7. $\frac{1}{2}q + 4 > 1$ $q > -6$
8. $5 - 2k \leq -19$ $k \geq 12$
9. $5n - 6 \leq 12 - n$ $n \leq 3$
10. $8 - z > 2z - 10$ $z < 6$

Graph each inequality on the coordinate plane. Use your own paper.
Check students' graphs.

11. $y < 2x$
12. $y \leq 5$
13. $x \leq -1$
14. $y \geq x + 3$
15. $y \leq -x - 2$
16. $x - y < 4$
17. $x \leq 2y - 6$
18. $9 < 6x - 3y$
19. $4x + 2y \leq 10$
20. $y < \frac{3}{4}x + 1$
21. $2x + \left(\frac{2}{3}\right)y \leq 4$
22. $8 < 5x - 3y$
23. Five times some number n decreased by 3 is greater than 7. What values are possible for n? $n > 2$
24. Twelve less than four times some number n is at least three more than the number. What values are possible for n? $n \geq 5$
25. Six minus twice a number n is less than or equal to the opposite of one-half the number. What values are possible for n? $n \geq 4$

Solve each inequality and graph the solution on a number line. Use your own paper.
Check students' graphs.

26. $0.6m - 8.3 \geq 0.9 - 91.4m$ $m \geq 0.1$
27. $4 - (r + 3) \leq 3(2r - 13) - 2$ $r \geq 6$
28. $3(2a + 5) - 6a < 2(3 - 3a)$ $a < 1.5$
29. $4(2b - 1) > 4b + 3(b + 2) - 7$ $b > 3$

26 EXTRA PRACTICE *Lesson 2-6*

Enrichment Worksheet 2-6

Name ______________________ Date __________

ENRICHMENT 2-6

Consecutive Integers

Consecutive integers are integers that follow in order. For example, −3, −1, 1, 3 is a set of consecutive odd integers. You can use algebraic expressions to solve problems about consecutive integers. In solving these problems, you will need to translate certain statements into expressions. Here are some examples.

Statement "sum of five consecutive integers"
Expression $n + (n + 1) + (n + 2) + (n + 3) + (n + 4)$

Statement "sum of five consecutive even integers"
Expression $2n + (2n + 2) + (2n + 4) + (2n + 6) + (2n + 8)$

Statement "sum of five consecutive odd integers"
Expression $(2n + 1) + (2n + 3) + (2n + 5) + (2n + 7) + (2n + 9)$

Exercises

Solve each problem.

1. Find three consecutive integers that have a sum of −12. −5, −4, −3
2. Find two consecutive integers that have a sum of 17. 8, 9
3. Find three consecutive integers that have a sum of 72. 23, 24, 25
4. Find two consecutive integers that have a sum of 95. 47, 48
5. Find three consecutive odd integers with a sum of 615. 203, 205, 207
6. Find four consecutive odd integers with a sum of −80. −23, −21, −19, −17
7. Find two consecutive even integers with a sum of 70. 34, 36
8. Find three consecutive even integers with a sum of −18. −8, −6, −4
9. The larger of two consecutive even integers is 6 less than 3 times the smaller integer. Find the integers. 4, 6
10. Find four consecutive even integers such that the largest is twice as large as the smallest. 6, 8, 10, 12

34 ENRICHMENT *Lesson 2-6*

Skills Practice

Vocabulary Review

Lesson 2–5
like or similar terms

Lesson 2–6
linear inequality
half-plane boundary
transitive property
open half-plane
closed half-plane

Assignment Guide

All students: 1–85
Additional Practice: Refer to the Extra Practice Index on page 674 of this text.

Additional Answers

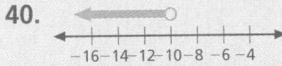

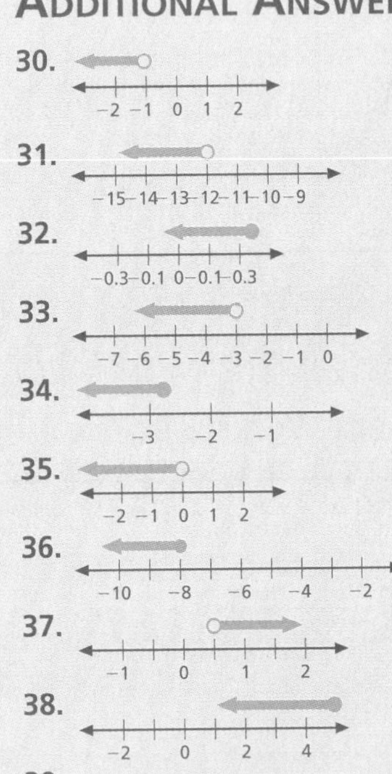

40.

41.

42.

43.

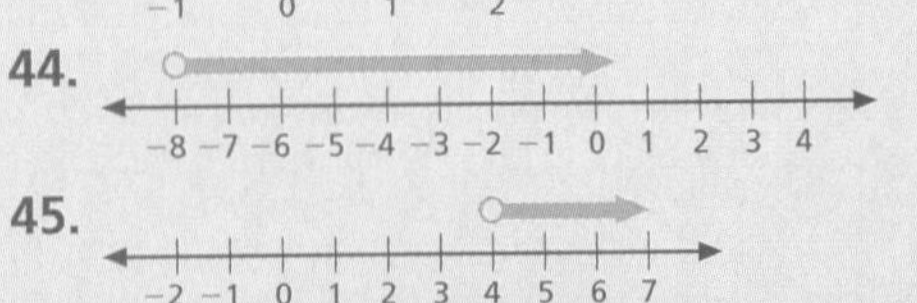

Review and Practice Your Skills

Practice ■ Lesson 2-5

Solve each multi-step equation.

1. $4r - 1 = 35$ 9
2. $5g + 1 = -29$ −6
3. $-4q - 5 = 7$ −3
4. $2(x - 3) = 14$ 10
5. $6x - 13 = -13$ 0
6. $\frac{1}{2}x + 5 = 16$ 22
7. $0.4x - 3.8 = 4.2$ 20
8. $-7(m - 3) = 2(4m + 3)$ 1
9. $0.2(1.8 + z) = 0.3z$ 3.6
10. $10b - 6b - 3 = 9$ 3
11. $\frac{z}{3} + 70 = 98$ 84
12. $12 - 3m = -15$ 9
13. $7 - x = 23$ −16
14. $-4x + 23 = 75$ −13
15. $15(2 + x) - 3x = 114$ 7
16. $14 = 8r - 58$ 9
17. $\frac{2}{5}x - 7 = 11$ 45
18. $7(x - 2) = -14$ 0
19. $12 - 3(4 - 7x) = 9(3x + 2) + x$ $-2\frac{4}{7}$
20. $\frac{1}{3}(18y - 6) = -\frac{5}{6}(12 - 6y)$ −8
21. $-m + 3(m + 1) = 11$ 4
22. $-8 - 2w = 11$ −9.5
23. $112 = 12 + 8y$ 12.5
24. $\frac{4}{7}x = -(21 + 14)$ $-61\frac{1}{4}$

Translate each sentence into an equation. Then solve.

25. Six more than 5 times a number is −29. Find the number. $6 + 5n = -29, n = -7$
26. When 47 is decreased by twice a number, the result is −75. Find the number. $47 - 2n = -75, n = 61$
27. The sum of one-third of a number and $\frac{1}{2}$ is $3\frac{1}{2}$. Find the number. $\frac{1}{3}n + \frac{1}{2} = 3\frac{1}{2}, n = 9$
28. 7 less than twice a number is 14. Find the number. $2n - 7 = 14, n = 10.5$
29. Three times the sum of a number and 2 is −27. Find the number. $3(n + 2) = -27, n = -11$

Practice ■ Lesson 2-6

Solve each inequality and graph the solution on a number line. See additional answers for graphs.

30. $3 + x < 2$ $x < -1$
31. $-g - 9 > 3$ $g < -12$
32. $2x - 0.3 \le 0.5$ $x \le 0.4$
33. $n + 5 < 2$ $n < -3$
34. $-4x + 6 \ge 17$ $x \le -2.75$
35. $5d - 8 < -8$ $d < 0$
36. $\frac{r}{4} \le -2$ $r \le -8$
37. $c + \frac{2}{3} > 1\frac{1}{3}$ $c > \frac{2}{3}$
38. $2 - (3 - s) \le 4$ $s \le 5$
39. $5 > \frac{1}{3}k + 14$ $-27 > k$
40. $-6(k + 2) > 48$ $k < -10$
41. $72 \le -3h + 4 - 5h$ $-8.5 \ge h$
42. $5(7 + r) \ge 12r$ $5 \ge r$
43. $2x - 5(x + 3) \le -20$ $x \ge \frac{5}{3}$
44. $\frac{5}{8}e - 3 - \frac{3}{8}e > -5$ $e > -8$

Solve each inequality and graph the solution on a number line. See additional answers for graphs.

45. $m - 19 > -15$ $m > 4$
46. $\frac{1}{3}x - 2 \ge 11$ $x \ge 39$
47. $-3x - 2 \ge 4.5$ $x \le -2\frac{1}{6}$

Graph each inequality on the coordinate plane. See additional answers.

48. $y \le x - 2$
49. $y > \frac{4}{5}x - 4$
50. $y < -2x$
51. $y \ge 5 - 3x$
52. $5x + 10y < -30$
53. $x - y > 4$
54. $y \ge -3$
55. $x < 8$
56. $y \ge -3x + 2$
57. $y < x$
58. $-2y \le x$
59. $4x + 3y \le 12$
60. $-5x + 4y > 20$
61. $0.2x - 0.8y \le 3.2$
62. $-\frac{1}{4}x - \frac{1}{2}y > 2\frac{1}{2}$

47. 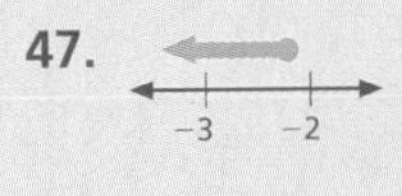

48. 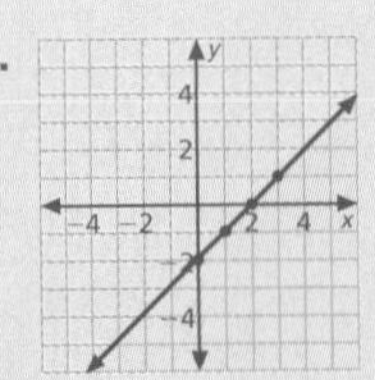

49. 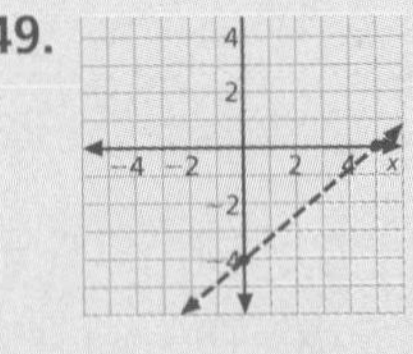

50. 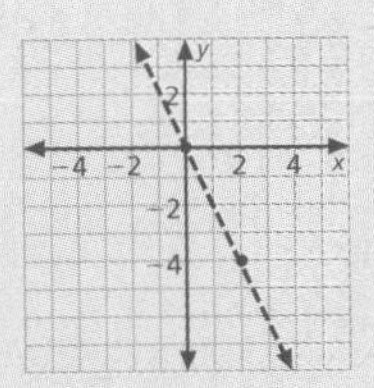

51. 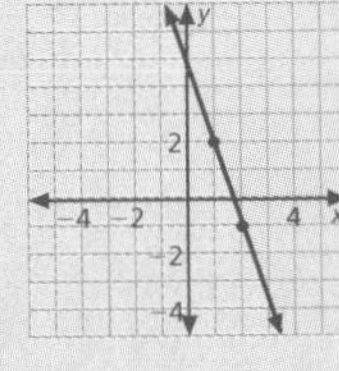

52. 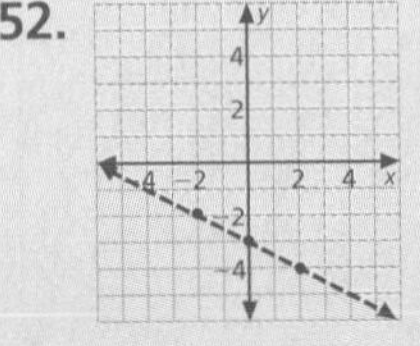

53. 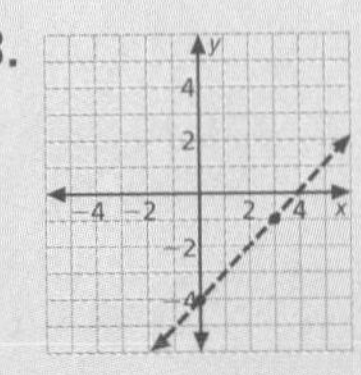

54. 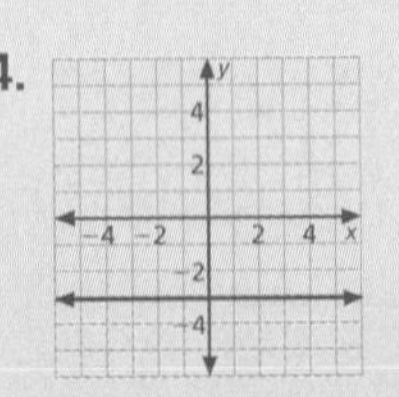

55. 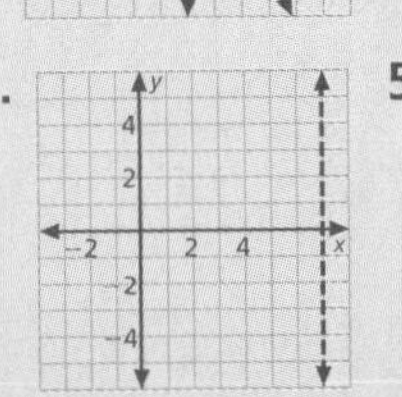

56. 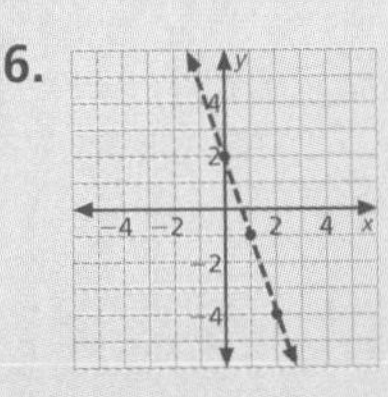

80 Chapter 2 **Essential Algebra and Statistics**

PRACTICE ■ LESSON 2-1–LESSON 2-6

Find the next three terms in each sequence. (Lesson 2-1)

63. 1.2, 1.5, 1.8, 2.1, 2.4, 2.7, 3.0

64. 1.2, 2.4, 4.8, 9.6, 19.2, 38.4, 76.8

65. 1.2, −1.2, −3.6, −6.0, −8.4, −10.8, −13.2

66. 1.2, −6, 30, −150, −750, 3750, −18750

67. 1.2, 1.21, 1.212, 1.2121, ____, ____, ____ 1.21212, 1.1212121, 1.2121212

68. 1.2, 1.44, 1.728, 2.0736, ____, ____, ____ 2.48832, 2.985984, 3.5831808

Write each as a set of ordered pairs. Graph on a coordinate plane. List the domain and range. Determine if each is a function. (Lessons 2-2 and 2-3) See additional answers for graphs.

69.

x	−3	−4	−5	−6
y	1	6	1	6

(−3, 1), (−4, 6), (−5, 1), (−6, 6); D: {−6, −5, −4, −3}; R: {1, 6}; Yes, it is a function.

70.

x	10	13	10	7
y	−5	−6	−7	−8

(10, −5), (13, −6), (10, −7), (7, −8), D: {7, 10, 13}; R: {−8, −7, −6, −5}; Not a function.

Solve each equation. (Lesson 2-4)

71. $x - 3 = 0$ 3

72. $9 = (-4) + f$ 13

73. $c + \frac{3}{5} = -1\frac{4}{5}$ $-2\frac{2}{5}$

74. $-3.2d = 48$ −15

75. $\frac{w}{7} = -14$ −98

76. $\frac{2}{3}x = 24$ −36

Solve each equation. (Lesson 2-5)

77. $-8 - x = 14$ −22

78. $-4x + 5 = 33$ −7

79. $\frac{t}{-3} - 9 = -1$ −24

MathWorks Workplace Knowhow — Career – Transcriptionist

TV newscasters can't possibly remember everything they need to say in a newscast. Instead, they read from a device known as a teleprompter. A teleprompter is a television screen that scrolls slowly through the script, for the show. Transcriptionists make sure the copy is typed accurately and on time for the program.

1. You hire a transcription assistant at the rate of \$4/page of typed copy. You also pay her a base salary of \$25 per day. Her total earnings is represented by $e = 4p + 25$ when e is the total earnings and p is the number of pages. If you can afford to pay her up to \$150 for one day, how many pages of copy can you ask her to type? Up to 31 pages
2. Samatha has six hours to get four tasks done. She spends 45 min talking to the producer, 1 h 45 min talking to a repair technician, and 2 h 15 min proofreading the copy of a speech. She still has to type 1,800 words of copy for the evening newscast. How many words per minute must she type? 24 words per min
3. Victor knows that he can type 30 words/min. He has 60 pages of copy to type, and there are about 100 words/page. How long will it take him to finish the job? 3 h 20 min

Chalkboard Examples

Example from Lesson 2-5

Solve each equation.

1. $4x - 3 = -11$ −2
2. $3(y - 1) = y + 13$ 8

Example from Lesson 2-6

Solve each inequality.

1. $7x + 2 \leq -47$ $x \leq -7$
2. $2 - (x - 4) < 2(3x + 1) - 6$ $x > \frac{10}{7}$

Career Opportunity

Describe the kind of work transcriptionists or prompter operators do and the types of places they work. Discuss the importance of computer skills and mathematics in working as a transcriptionist. Students should answer Questions 1–3 to better understand how transcriptionists use mathematics in performing their job.

69.

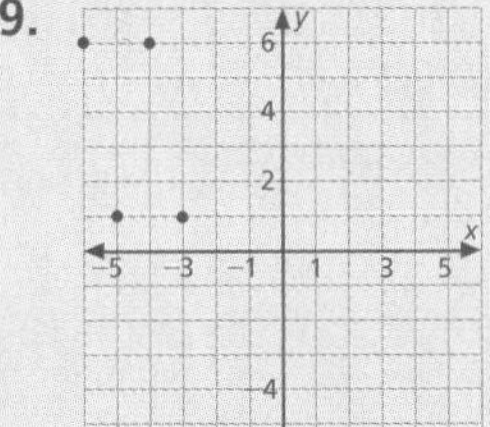

70.

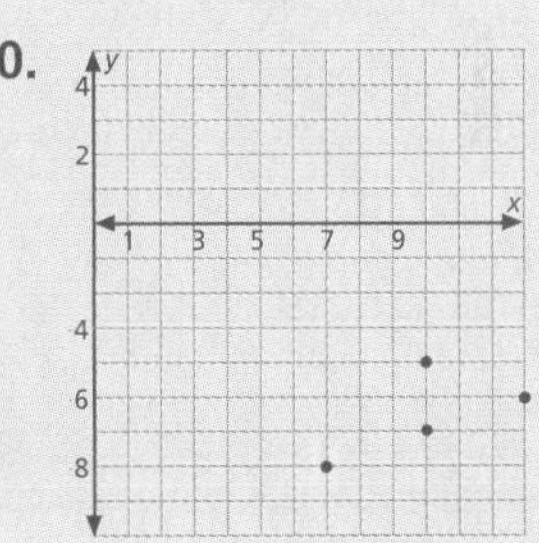

57.

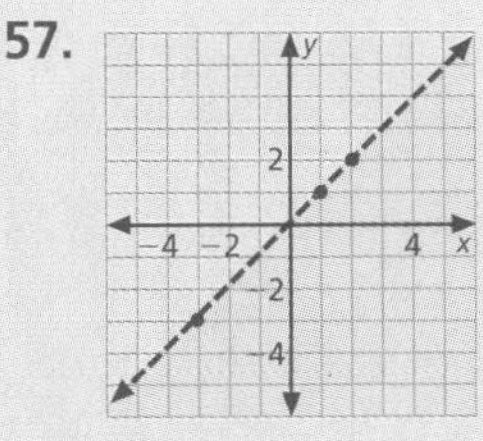

58.

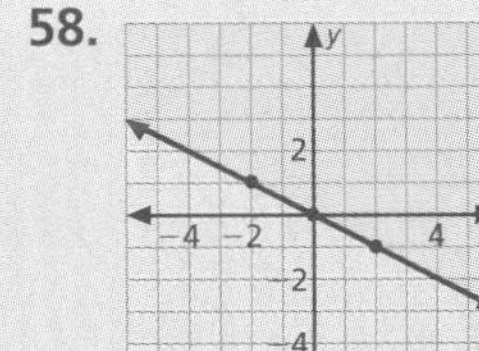

59.

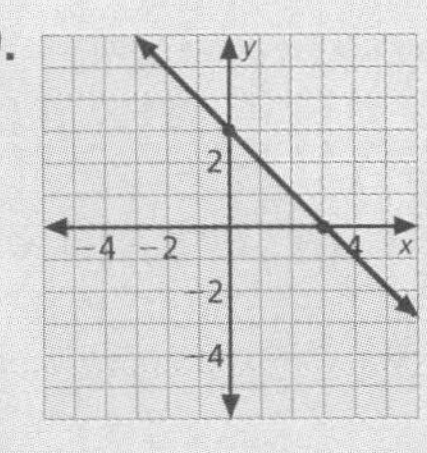

60.

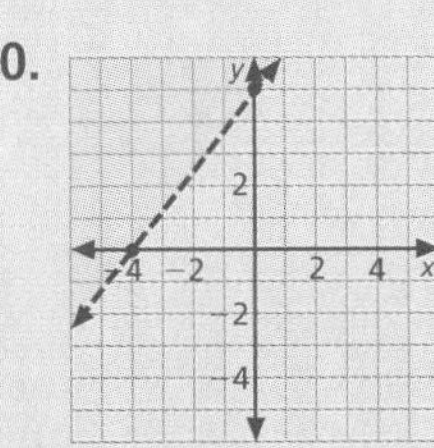

61.

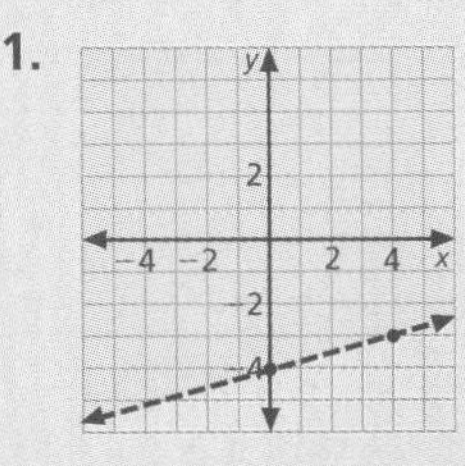

62.

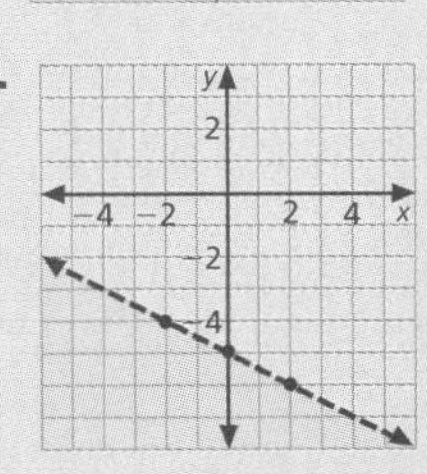

Lesson Planning

NCTM Standards/Strands

- Number & Operation
- Patterns, Functions, & Algebra
- Geometry & Spatial Sense
- Data Analysis, Statistics, & Probability
- Problem Solving
- Reasoning and Proof
- Communication
- Connections
- Representation

Vocabulary

statistics, data, population, sample, frequency table, measures of central tendency, mean, median, mode

Materials Needed

paper/pencil, calculator

Lesson Resources

Warm-Up Transparency 6
Transparency RF-8
Reteaching 2-7
Extra Practice 2-7
Enrichment 2-7

Assignment Guide

Basic: 1–14, 17–26
Enriched: 1–26

Getting Started

5-Minute Warm-up

Arrange these numbers in ascending order.

1. 34, 23, 42, 35, 41, 19, 23
   19, 23, 23, 34, 35, 41, 42
2. 7.3, 4.02, 5   4.02, 5, 7.3

Introduction to Lesson 2-7

After students read and answer these questions, they may want to discuss current advertisements that use statistics and consider whether or not the statistics are misleading.

2-7 Data and Measures of Central Tendency

Goals
- Construct frequency tables for data.
- Determine mean, median, and mode for a set of data.

Applications Sports, Education, Marketing, News media

Consumers are constantly bombarded with facts and figures from advertisers who use statistics to entice people to buy their products. Newspaper and television advertisements are full of these statistics. For example, a television commercial makes the following claim:

"In a national taste-test, 7 out of 10 teenagers preferred our brand of cola to our competitor's." For a and b, answers will vary.

a. How is this advertiser trying to influence consumers?
b. Is it possible that this statistic is not truly representative of the nation's preference? Explain.

Think Back

There are four common methods of sampling.

Random sampling: each member of the population has an equal chance of being selected.

Cluster sampling: members of the population are randomly selected from particular parts of the population and surveyed in groups.

Convenience sampling: members of a population are selected because they are readily available.

Systematic sampling: members of a population that has been ordered in some way are selected according to a pattern.

Build Understanding

Statistics is a branch of mathematics that involves the study of **data**. Statisticians study methods of collecting, organizing, and interpreting data. The purpose of a statistical study is to reach a conclusion or make a decision about an entire group called a **population**. Often, it is not possible to survey or poll an entire population. In these cases, a **sample**, or representative part, of the population is used.

Once the sample is selected and the data are collected, the data must be organized so it can be analyzed. One common way to organize data is a **frequency table** or tally system. When data sets contain a wide range of items, it is sometimes useful to group the data into intervals.

Example 1

SPORTS In preparing a sports report for the newspaper, Juan recorded the batting averages of 3 baseball players systematically sampled from each of the ten teams in the league. Construct a frequency table for this data.

.243	.281	.255	.296	.278	.248	.267
.303	.254	.292	.304	.269	.253	.241
.249	.281	.277	.295	.244	.294	.266
.251	.270	.268	.261	.302	.276	.265

Solution

The lowest batting average is .241, and the highest is .304. Group the data into intervals. Then mark a tally for each data piece in the appropriate interval, and total the frequencies.

Batting Average	Tally	Frequency
.240–.249	𝍸	5
.250–.259	||||	4
.260–.269	𝍸 |	6
.270–.279	||||	4
.280–.289	||	2
.290–.299	||||	4
.300–.309	|||	3

Learning Styles

TACTILE/KINESTHETIC LEARNER Have students write each item of data on an index card. This will make it easier to arrange the numbers in numerical order. It will also facilitate making the frequency table, as well as finding the median.

Once data has been organized, it can then be analyzed statistically. Three **measures of central tendency** that can be calculated are the mean, median, and mode.

The **mean**, or arithmetic average, is the sum of the data divided by the number of data. The mean is the most accurate measure of central tendency for data sets that do not contain extreme values.

The **median** is the middle value of the data when arranged in numerical order. If the number of data items is even, the median is the average of the two middle numbers. The median is the most accurate measure of central tendency for data sets that contain extreme values.

The **mode** is the number that occurs most in the set of data. A set of data may contain one mode, more than one mode, or no mode. The mode is used to describe the most characteristic value of a set of data.

Example 2

TEST TAKING The S.A.T. mathematics scores for 8 high school students are listed below.

539 541 576 505 548 576 565 558

a. Find the mean of the data.

b. Find the median of the data.

c. Find the mode of the data.

d. Which measure of central tendency is the best indicator of the typical S.A.T. mathematics score for these students?

Solution

a. To find the mean, add the data and divide by the number of data.

$$\frac{539 + 541 + 576 + 505 + 548 + 576 + 565 + 558}{8} = \frac{4{,}408}{8} = 551$$

The mean is 551.

b. To find the median, rewrite the data in numerical order.

505 539 541 548 558 565 576 576

Because there is an even number of data, the median is the average of the two middle numbers.

$$548 + \frac{558}{2} = \frac{1106}{2} = 553$$

The median is 553.

c. The mode is the number that occurs the most. So the mode is 576.

d. The best indicator of the typical S.A.T. mathematics score for the students is the median, 553, which is not affected by the extreme value (505).

Chalkboard Examples

Supplementary Example 1

Juan also recorded the runs scored by two players systematically sampled from each of the ten teams in the league. Construct a frequency table

27	65	95	14	30
57	31	26	23	9
108	29	11	43	89
60	100	69	72	34

Runs Scored	Frequency
0–10	1
11–20	2
21–30	5
31–40	2
41–50	1
51–60	2
61–70	2
71–80	1
81–90	1
91–100	2
101–110	1

Supplementary Example 2

Daily attendance at a school for five days was 1,258; 1,336; 1,269; 1,294; and 1,271. Using a calculator, find the mean and median of the data. **mean: 1285.6; median: 1,271**

Reteaching Worksheet 2-7

Name __________ Date __________

RETEACHING 2-7

DATA AND MEASURES OF CENTRAL TENDENCY

Data can be organized using a tally system in a frequency table. Mean, median, and mode are measures of central tendency and can be used to analyze data.

Example

The following are a baseball team's scores for ten games: 4, 3, 0, 3, 2, 0, 1, 1, 3, 3.

a. Construct a frequency table for the data.

b. Find the mean, median and mode for the data.

Solution

a. List each different data item. Make a tally to record the number of times each data item occurs. Total the tally marks to compute the frequency of each response.

Runs	Tally	Freq
0	II	2
1	II	2
2	I	1
3	IIII	4
4	I	1

b. **Mean**: divide the sum of the data by the number of data items.

$$\frac{0 + 0 + 1 + 1 + 2 + 3 + 3 + 3 + 3 + 4}{10} = \frac{20}{10} = 2$$

Median: the middle value of the data.
0, 0, 1, 1, 2, 3, 3, 3, 3, 4 Since the number of data items is even, the median is the average of the two middle numbers:

$$\frac{2 + 3}{2} = \frac{5}{2} = 2.5.$$

Mode: the number that occurs most frequently. The mode is 3.

EXERCISES

Construct a frequency table for each set of data. Then find the mean, median, and mode.

1. 100, 103, 103, 100, 103, 103, 100, 104
Mean: 102
Median: 103
Mode: 103

No.	Tally	Freq
100	III	3
101		
102		
103	IIII	4
104	I	I

2. 215, 214, 212, 211, 212, 215, 215, 210
Mean: 213
Median: 213
Mode: 215

No.	Tally	Freq
210	I	1
211	I	1
212	II	2
213		
214	I	1
215	III	3

30 RETEACHING *LESSON 2-7*

Extend the Lesson

REAL WORLD CONNECTION Some applications use weighted averages. A weighted average is one in which some scores count more than others. For example, in Steve's class, test scores are worth twice as much as quiz scores. Have students find Steve's mean score (to the nearest tenth) if he has quiz scores of 90, 95, and 92 and test scores of 83 and 85. **87.6**

Lesson Wrap-up

Quick Assessment

Ask the following questions to determine if students understand the content presented in this lesson.

Here are the number of goals scored in all of Betty's teams' games this season.

1 3 3 2 4 5 2 1 4 3 4 3

1. Construct a frequency table for the data.

Goals	1	2	3	4	5
Frequency	2	2	4	3	1

2. Find the mean (to the nearest tenth), median, and mode for these data. mean: 2.9; median: 3; mode: 3

Assignment Guide

Basic: 1–14, 17–26
Advanced: 1–26
Additional Practice: See Extra Practice Index on page 674.

Example 3

CALCULATOR A photographer sold photos to a magazine for the following: \$150, \$225, \$175, \$350, \$635, \$120, and \$550. Find the mean and median of the amounts.

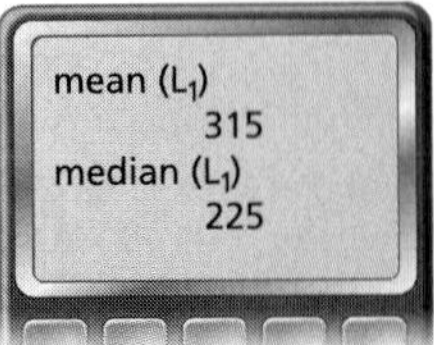

Solution

Use the list feature to create and store a new list (L1). After entering the data, choose MATH from the LIST menu to find the mean and median of the new list.

The mean of the list is \$315 and the median is \$225.

Try These Exercises

EDUCATION A random sample of 20 student records was sampled to determine the average number of absences per student during the school year. The number of absences on each record is listed below.

3	2	1	4	3	1	2	1	0	2
2	3	5	1	4	8	9	0	4	1

1. Construct a frequency table for these data. See additional answers.
2. Find the mean, median, and mode of the data. 2.8, 2, 1
3. Which measure of central tendency is the best indicator of the average number of absences per student for this school year? median

NEWS MEDIA The manager of the school newspaper researched and recorded the number of issues of each edition of the newspaper that were sold.

362	398	409	377	421	351	399	358	406	388
379	412	423	361	414	420	409	387	361	425
366	401	392	387	390	371	405	417	399	358

4. Construct a frequency table for these data. Group the data into intervals of 10. See additional answers.
5. Determine the interval that contains the median of the data. 390–399

Practice Exercises

MARKETING Thirty families were randomly sampled and surveyed as to the number of magazines to which they subscribe. The results are listed below.

3	1	0	0	2	3
1	4	5	1	0	2
2	0	1	1	1	4
3	2	1	3	4	4
1	0	2	3	2	1

6. Construct a frequency table for these data. See additional answers.
7. Find the mean, median, and mode of the data. 1.9, 2, 1

Additional Answers

1. NUMBER OF ABSENCES PER STUDENT

Absences	Tally	Frequency
0	𝍷𝍷	2
1	𝍸	5
2	𝍷𝍷𝍷𝍷	4
3	𝍷𝍷𝍷	3
4	𝍷𝍷𝍷	3
5	𝍷	1
6		0
7		0
8	𝍷	1
9	𝍷	1

4. SCHOOL NEWSPAPER SALES

Issues Sold	Tally	Frequency
350–359	𝍷𝍷𝍷	3
360–369	𝍷𝍷𝍷𝍷	4
370–379	𝍷𝍷𝍷	3
380–389	𝍷𝍷𝍷	3
390–399	𝍸	5
400–409	𝍸	5
410–419	𝍷𝍷𝍷	3
420–429	𝍷𝍷𝍷𝍷	4

6. MAGAZINES SUBSCRIPTIONS

Number of Subscriptions	Tally	Frequency
0	𝍸	5
1	𝍸𝍷𝍷𝍷𝍷	9
2	𝍸𝍷	6
3	𝍸	5
4	𝍷𝍷𝍷𝍷	4
5	𝍷	1

8. MONEY SPENT ON LUNCH

\$ Spent	Tally	Frequency
\$2.60–\$2.69	𝍷𝍷	2
\$2.70–\$2.79	𝍷𝍷	2
\$2.80–\$2.89	𝍷𝍷𝍷𝍷	4
\$2.90–\$2.99	𝍸𝍷	6
\$3.00–\$3.09	𝍷𝍷𝍷	3
\$3.10–\$3.19	𝍷𝍷𝍷𝍷	4
\$3.20–\$3.29	𝍷𝍷	2
\$3.30–\$3.39	𝍷	1

JOURNALISM For the survey she was taking for the school newspaper, Amanda systematically sampled every tenth student on the cafeteria lunch line to record the amount of money spent on lunch that day. The results of her survey are listed below.

$2.95 $3.10 $2.85 $2.95 $3.35 $3.15 $3.15 $2.80

$2.60 $2.85 $3.15 $2.70 $3.25 $3.00 $2.95 $3.20

$2.85 $2.95 $2.90 $3.00 $2.95 $2.65 $3.05 $2.75

8. Construct a frequency table for these data. Group the data into intervals. See additional answers.

9. Which interval contains the median of the data? $2.90–$2.99

SPORTS The heights, in inches, of the members of the Hills High School Boys' Basketball Team are listed below.

75 74 66 76 71 74 78 77 67 76 77 74

10. Use a calculator to find the mean and median of the data. 73.75, 74.5

11. Find the mode of the data. 74

12. Which measure of central tendency is the best indicator of the typical height of a member of the basketball team? mode

WEATHER For a television documentary on desert environments, a meteorologist recorded the highest temperature for each day of June in Death Valley, California. The data are displayed in the frequency table.

Temperature	Tally	Frequency
80–89	III	3
90–99	𝍸 III	8
100–109	𝍸 IIII	9
110–119	𝍸 II	7
120–129	III	3

13. Find the interval that contains the median. 100–109

14. To the nearest percent, on what percent of the days was the recorded temperature at least 100°F? 63%

Death Valley, California

Extended Practice Exercises

15. Do you think the mean of the daily high temperatures for Death Valley is greater than or less than 100°F? Explain. greater than

16. **WRITING MATH** Suppose you were interested in determining the *average* temperature during June (as opposed to the average daily high temperature). What sampling method would you use, and how would you collect the data? Answers will vary.

Mixed Review Exercises

Evaluate each function.

17. $f(x) = 2x - 1; f(3)$ 5

18. $f(x) = \frac{1}{2}x + 3; f(4)$ 5

19. $f(x) = 3x + 5; f(7)$ 26

20. $f(x) = \frac{x}{3} + 4; f(-9)$ 1

21. $f(x) = 2x + 6; f(-3)$ 0

22. $f(x) = x + 4; f(2)$ 6

23. $f(x) = 3x - 2; f(6)$ 16

24. $f(x) = -3x + 2; f(-4)$ 14

25. $f(x) = 5x + 8; f(-2)$ −2

26. Clarks Plumbing and Heating purchased a new computer for $6,000. The depreciation rate for this computer is 30%. Use a declining-balance method to find the ending book value after the fourth year. (Lesson 2-2) $1440.60

Extra Practice Worksheet 2-7

Name ________________ Date ________

EXTRA PRACTICE 2-7

DATA AND MEASURES OF CENTRAL TENDENCY

EXERCISES

Twenty students were randomly sampled and surveyed as to the number of hours per week they study. The results are shown below.

5 6 5 2 4 6 3 2 1 5

4 4 6 3 3 5 2 1 6 4

1. Construct a frequency table for these data.

# of hours	Tally	Frequency
1	II	2
2	III	3
3	III	3
4	IIII	4
5	IIII	4
6	IIII	4

2. Find the mean, median, and mode of the data mean: 3.85; median: 4; modes: 4, 5, 6

The number of hours worked in one week by employees at The Print Shop are listed below.

25 36 18 43 40 38 39 40 16

20 24 29 30 45 42 19 20 28

3. Construct a frequency table for these data. Group the data into intervals.

# of hours	Tally	Frequency
10–19	III	3
20–29	𝍸 I	6
30–39	IIII	4
40–49	𝍸	5

4. Which interval contains the median of the data? 30–39

Enrichment Worksheet 2-7

Name ________________ Date ________

ENRICHMENT 2-7

THE AVERAGE AREA

One way to estimate the area of a curved shape is to find the arithmetic average of two areas—the inner area and the outer area.

Example

Estimate the area of the circle drawn on the grid.

Solution

Inner Area 37 sq units

Outer Area 69 sq units

Average
37 + 69 = 106
106 ÷ 2 = 53
53 sq units

EXERCISES

Estimate the area of each figure. Answers may vary. Possible answers are given.

1. 69 sq units

2. 66 sq units

3. 53 sq units

4. 63 sq units

5. 61 sq units

6. 68 sq units

For each figure, estimate the area by averaging. Then compute the area using the formula $A = lw$. How do your results compare?

7. 9, 5, 5, 9 — 45 sq units

8. 6, 6, 6, 6 — 36 sq units

9. 4, 4, 2, 6, 10, 6 — 52 sq units

Lesson Planning

NCTM Standards/Strands

- Number & Operation
- Patterns, Functions, & Algebra
- Geometry & Spatial Sense
- Data Analysis, Statistics, & Probability
- Problem Solving
- Reasoning and Proof
- Communication
- Connections
- Representation

Vocabulary

stem-and-leaf plot
outlier
gap
cluster
histogram

Materials Needed

paper/pencil
lined or grid paper

Lesson Resources

Warm-Up Transparency 6
Transparency TK-1
Reteaching 2-8
Extra Practice 2-8
Enrichment 2-8

Assignment Guide

Basic: 1–18, 20–35
Enriched: 1–35

Getting Started

5-Minute Warm-up

Find the median of each set of data.

1. 7, 3, 2, 8, 7, 9, 10, 1, 2 7
2. 32, 31, 65, 72, 34, 40 37
3. 98, 67, 89, 91, 100 91

Introduction to Lesson 2-8

When the students have completed the activity, have them work together to create a classroom display from the examples.

2-8 Display Data

Goals
- Construct stem-and-leaf plots.
- Construct histograms for a data set.

Applications Health, News, Entertainment

Work in small groups.

From a newspaper or magazine, find a table of information. Study the data presented in the table. Discuss whether a different type of display might have been more effective. If so, sketch your idea.

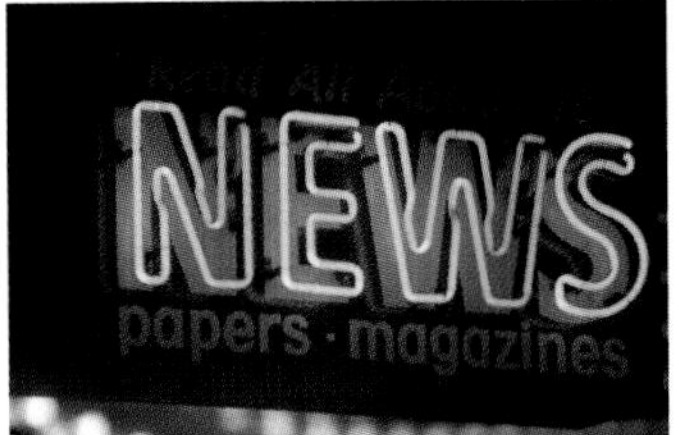

Build Understanding

Graphs and plots are often used to present a picture of the data. These types of displays provide visual representations of the distribution of the data. They also display characteristics about the data that are sometimes difficult to identify from charts and tables.

One type of visual data display is the **stem-and-leaf plot**. To construct a stem-and-leaf plot, first divide each piece of data into two parts: a stem and a leaf. The last digit of each number is referred to as its **leaf**; the remaining digits comprise the **stem**. The data is then organized by grouping together data items that have common stems.

Example 1

HEALTH For an article she was preparing for a women's health magazine, Sharon recorded the cholesterol levels of the twenty women on the magazine staff.

185	234	208	197	259
177	192	188	208	200
215	199	209	234	208
146	216	201	232	186

Construct a stem-and-leaf plot to display the data. Interpret the data using your plot.

Solution

For these data, the digits in the hundreds and tens columns form the stem, and the units digit is the leaf. Sort the data according to stems, and arrange the leaves in numerical order.

Be sure to provide a title and a key for your plot.

Cholesterol Levels of Female Staff Members

Stems	Leaves
14	6
15	
16	
17	7
18	5 6 8
19	2 7 9
20	0 1 8 8 8 9
21	5 6
22	
23	2 4 4
24	
25	9

Key: 14/6 represents a cholesterol level of 146 mg/dL

Extend the Lesson

REAL WORLD CONNECTION Have students examine newspapers over a period of one or two weeks. Have them sort and tally the types of graphs used in that particular newspaper (for example, line, bar, pictograph, histogram, stem-and-leaf, circle). Have them make either a stem-and-leaf plot or a histogram of their findings.

Because 146 is much less than the other data, and 259 is much greater, these data pieces are considered to be **outliers**. There is a large data **cluster** for cholesterol levels between 177 and 216, and a smaller cluster for levels in the lower 230s. **Gaps** exist between these two clusters, and between the outliers and the rest of the data. The mode is 208. The median cholesterol level is 204.5 mg/dL (the average of the tenth and eleventh data pieces in the plot).

A **histogram** is a type of bar graph used to display data. The height of the bars of the graph are used to measure frequency. Histograms are frequently used to display data that have been grouped into intervals.

Example 2

The Town Gazette surveyed 40 families, and asked them to record the number of hours per week their television was in use. The results are shown in this frequency table. Construct a histogram to display these data.

Television Use

Number of hours	Frequency
0–9	1
10–19	6
20–29	15
30–39	12
40–49	2
50–59	4

Solution

Let the horizontal axis represent the number of hours, and the vertical axis represent the frequency. Draw each bar so that its height corresponds to the frequency of the interval it represents.

A spreadsheet program can quickly create a variety of charts and graphs from a set of data.

Television Use

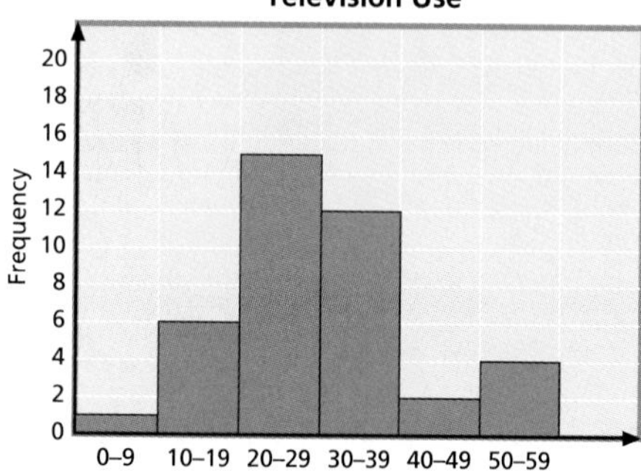

Example 3

SPREADSHEET A class earned the following scores on a science quiz: 89, 88, 72, 66, 89, 90, 94, 78, 95, 82, 84. Make a frequency table and a histogram of the data.

	A	B
1	Scores	Frequency
2	61-70	1
3	71-80	2
4	81-90	5
5	91-100	3

Solution

Create the frequency table on a spreadsheet. Use the intervals 61–70, 71–80, 81–90, and 91–100.

Highlight the cells and select **CHART** from the **INSERT** menu. From the list of charts and graphs, choose **bar graph**. Add titles and your histogram is complete.

Science Scores

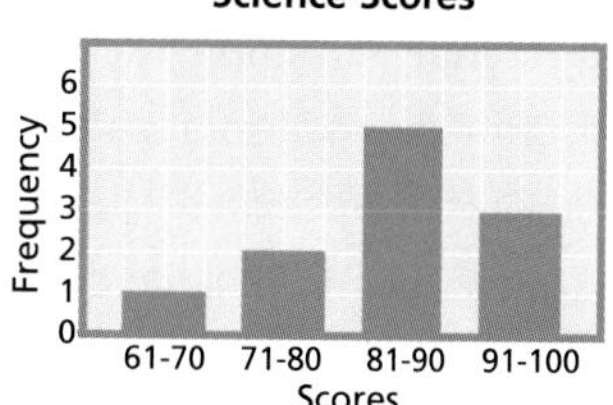

Chalkboard Examples

Supplementary Example 1

Here are the scores on a science quiz:

89, 88, 72, 66, 89, 90, 94, 78, 95, 82.

Construct a stem-and-leaf plot for these data.

6	6
7	2 8
8	2 4 8 9 9
9	0 4 5

Supplementary Example 2 and 3

Make a histogram for this frequency table.

Weight (lbs)	Students
90–99	11
100–109	15
110–119	9
120–129	7

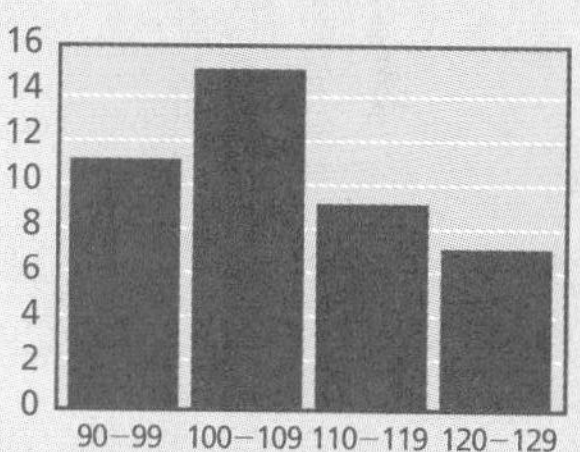

Reteaching Worksheet 2-8

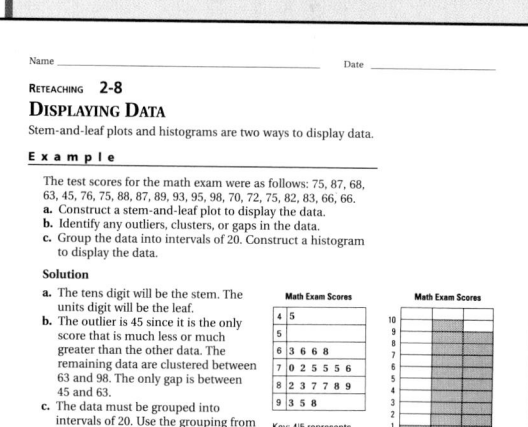

Name ________________ Date ________________

RETEACHING 2-8

DISPLAYING DATA

Stem-and-leaf plots and histograms are two ways to display data.

Example

The test scores for the math exam were as follows: 75, 87, 68, 63, 45, 76, 75, 88, 87, 89, 93, 95, 98, 70, 72, 75, 82, 83, 66, 66.

a. Construct a stem-and-leaf plot to display the data.
b. Identify any outliers, clusters, or gaps in the data.
c. Group the data into intervals of 20. Construct a histogram to display the data.

Solution

a. The tens digit will be the stem. The units digit will be the leaf.
b. The outlier is 45 since it is the only score that is much less or much greater than the other data. The remaining data are clustered between 63 and 98. The only gap is between 45 and 63.
c. The data must be grouped into intervals of 20. Use the grouping from the stem-and-leaf plot, combining **stems** 4 and 5, 6 and 7, and 8 and 9.

Math Exam Scores

4	5
5	
6	3 6 6 8
7	0 2 5 5 5 6
8	2 3 7 7 8 9
9	3 5 8

Key: 4|5 represents a score of 45.

Math Exam Scores

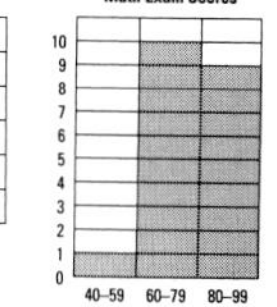

EXERCISES

For each data group below, **a.** construct a stem-and-leaf plot to display the data, **b.** identify any outliers, clusters, or gaps in the data, and **c.** use the stem-and-leaf plot to help group the data into intervals of 20. Then construct a histogram on your own paper.

1. Ages at Recital
17, 28, 36, 32, 65,
45, 21, 16, 29, 17,
23, 34, 42, 44, 19

2. Cost of in-line skates (in dollars)
142, 163, 125, 126, 128,
146, 101, 131, 130, 129,
131, 125, 143, 132, 141

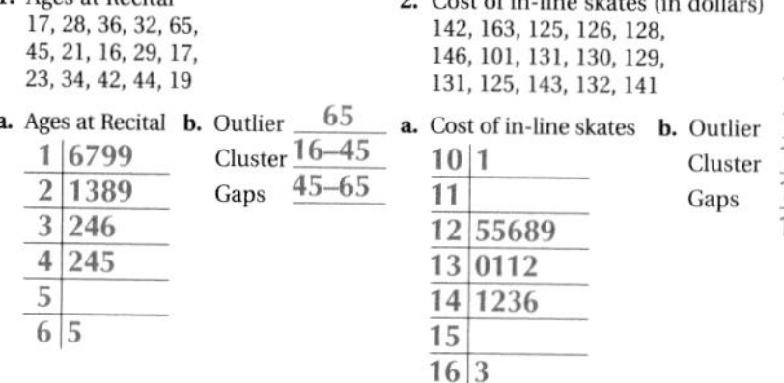

a. Ages at Recital

1	6799
2	1389
3	246
4	245
5	
6	5

b. Outlier 65
Cluster 16–45
Gaps 45–65

a. Cost of in-line skates

10	1
11	
12	55689
13	0112
14	1236
15	
16	3

b. Outlier 101 and 163
Cluster 125–146
Gaps 101–125, 146–163

32 RETEACHING *LESSON 2-8*

Lesson Wrap-up

Quick Assessment

1. How are data arranged in a stem-and-leaf-plot? All but the last digits are arranged numerically in a column on the left; the last digits are written in a row along the corresponding stem and are also arranged numerically.
2. What can a glance at the stem-and-leaf plot tell you? The location of clusters, gaps, and outliers; usually the mode can be decided by inspection.

Assignment Guide

Basic: 1–18, 20–35
Advanced: 1–35
Additional Practice: See Extra Practice Index on page 674.

Additional Answers

1. Aptitude Test Scores

```
1 | 6
2 | 7 8 8
3 | 2 3 4 8 8 8
4 | 0
5 | 0 2 5 6 6 7 9
6 | 0 0 1 2 5 6 9
7 | 1 1 4 5
8 | 6
```

Key
1|6 represents a score of 16 on a scale of 0–100

2. oultiers: 16 and 86; clusters: 27–40 and 50–75; gaps: between 16 and 27, 40 and 50, 75 and 86

6.

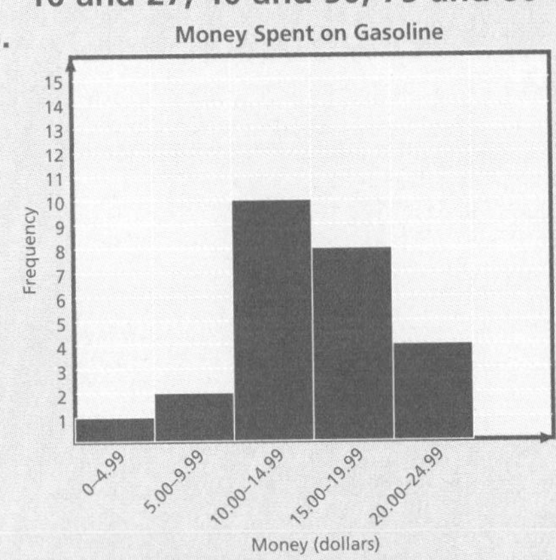

7. Time Spent on Homework

```
1 | 7
2 |
3 | 0 0 1 5 8
4 | 1 3 3 3 5 5 5 6 6 7
5 | 0 2 3 5 5 8 8
6 |
7 | 2 5 5 7 8
8 | 1
9 | 8
```

Key
1|7 represents 17 minutes

Try These Exercises

On an aptitude test measuring reasoning ability on a scale of 0 to 100, a class of 30 students received the following scores.

38	75	28	34	56	32	61	28	71	27
62	50	66	40	38	71	60	52	33	59
74	69	86	57	65	16	60	56	55	38

1. Construct a stem-and-leaf plot to display the data. See additional answers.
2. Identify any outliers, clusters, and gaps in the data. See additional answers.
3. Find the mode of the data. 38
4. Find the median of the data. 56
5. Find the mean of the data. 51.9

6. **NEWS MEDIA** A newspaper report on the price of gasoline contained this frequency table showing the amount of money spent weekly at the gas pump by 25 people surveyed. Construct a histogram to represent these data. See additional answers.

Money Spent on Gasoline

Amount of money	Frequency
0–4.99	1
5.00–9.99	2
10.00–14.99	10
15.00–19.99	8
20.00–24.99	4

Practice Exercises

REPORTING For an article he was writing for the school newspaper, Norman surveyed 30 students about the average amount of time (in minutes) each student spent on homework on a weeknight. He recorded the following table.

30	43	58	50	41	98	75	30	72	45
38	75	81	45	17	43	55	52	78	47
31	45	46	55	77	53	58	46	43	35

7. Construct a stem-and-leaf plot to display the data. See additional answers.
8. Identify any outliers, clusters, and gaps in the data. See additional answers.
9. Find the mode of the data. 43 and 45
10. Find the median of the data. 46.5
11. Find the mean of the data to the nearest tenth. 52.1

12. **WRITING MATH** What conclusions could Norman draw from the data? Write the lead paragraph of his newspaper article. Answers will vary.

8. oultiers: 17 and 98; clusters: 30–58 and 72–81; gaps: between 17 and 30, between 58 and 72, between 81 and 98
14. Check students' stem-and-leaf plots. Sample paragraph: There is a large data cluster between 21 and 26. A gap between 26 and 28. A smaller cluster between 28 and 31. 58, 46, and 40 are outliers.

13.

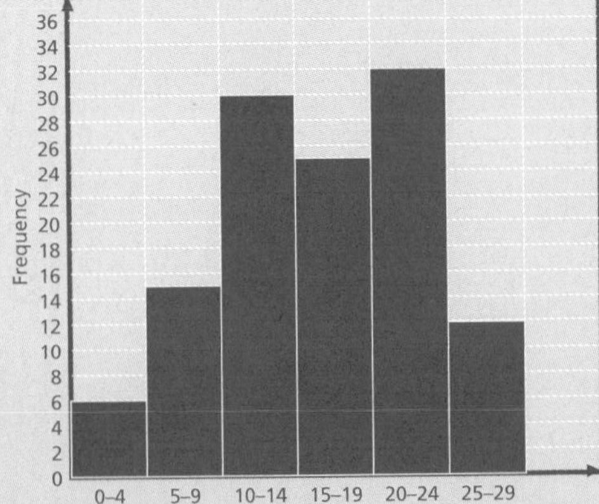

13. **ENTERTAINMENT** An entertainment magazine surveyed a sample of its readers about the average number of movies they see in a year. The data are recorded in this frequency table. Construct a histogram to display the data. See additional answers.

Number of Movies Seen in One Year

Number of movies	Frequency
0–4	6
5–9	15
10–14	30
15–19	25
20–24	32
25–29	12

14. **DATA FILE** Use the Data Index on pages 632–633 to find information on box office grosses for the 20 top-grossing feature films. Make a stem-and-leaf plot of the data. Then write a paragraph describing the data. See additional answers.

Refer to the histogram for Exercises 15–18.

15. How many students earn less than \$60.00 per week? 21

16. What percent of the students earn between \$60.00 and \$99.99 per week? 25

17. Which interval contains the median amount of earnings? 60.00–79.99

18. Is it possible to identify the mode of the data? Explain. no

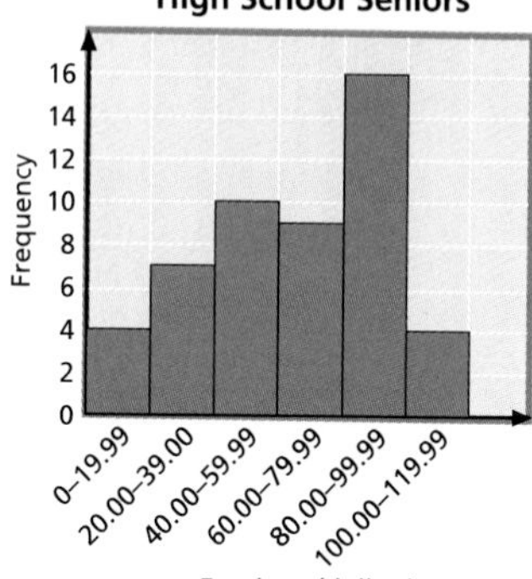

Extended Practice Exercises

19. **WRITING MATH** Write a paragraph comparing stem-and-leaf plots and histograms. Include in your comparison a discussion of the kinds of data for which each type of display is best suited, and describe the statistical conclusions that can be deduced from analyzing each type of display. Answers will vary.

20. **CHAPTER INVESTIGATION** Develop a survey question to ask your classmates that could provide you with data to support your proposal. Collect the data, construct a frequency table, and draw the related histogram. Write a paragraph analyzing your results. Answers will vary.

Mixed Review Exercises

Graph each function. (Lesson 2-3) See additional answers.

21. $y = 3x - 2$

22. $y = \frac{1}{2}x + 3$

23. $y = 2x + 1$

24. $y = x - 4$

25. $y = -2x + 1$

26. $y = x - 3$

Solve each equation. (Lesson 2-4)

27. $m - 13 = 28$ 41

28. $6n = -42$ −7

29. $8 + g = -2$ −10

30. $25 = p + 47$ −22

31. $0.9x = 7.2$ 8

32. $-7h = 28$ −4

33. $\frac{a}{4} = 1.2$ 4.8

34. $\left(\frac{2}{3}\right)c = 12$ 18

35. $0.4w = -6$ −15

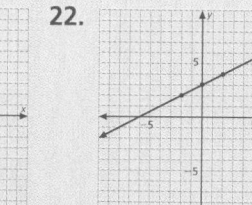

21. 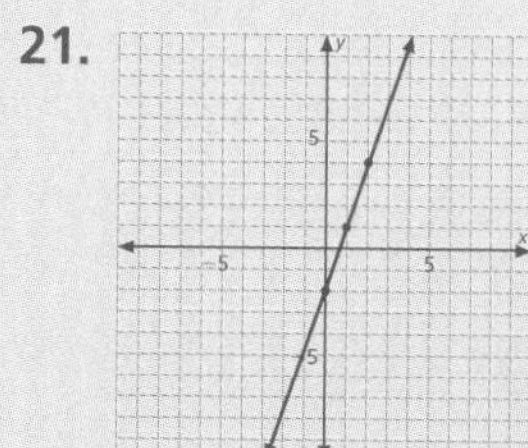
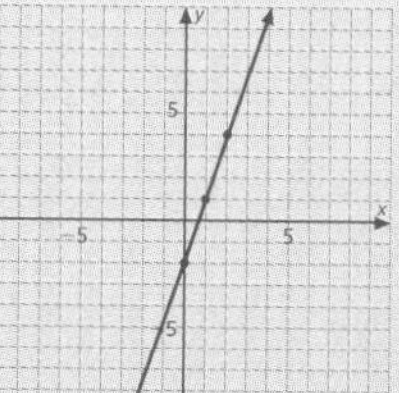
22. 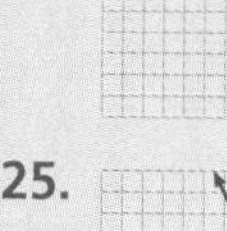
23. 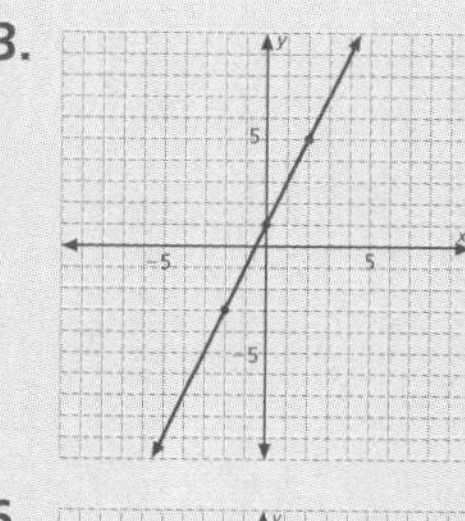
24. 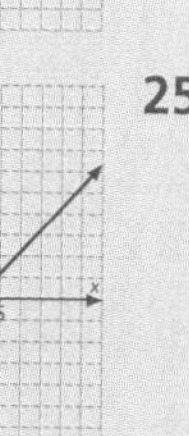
25. 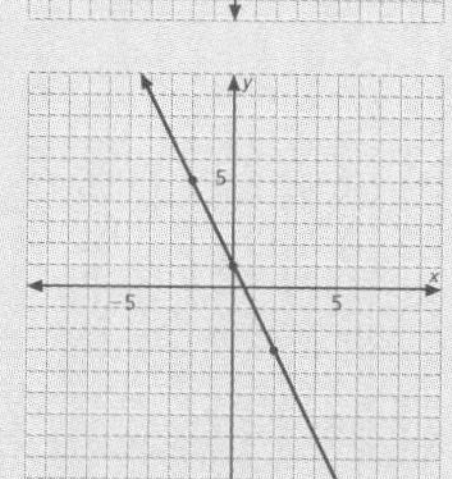
26. 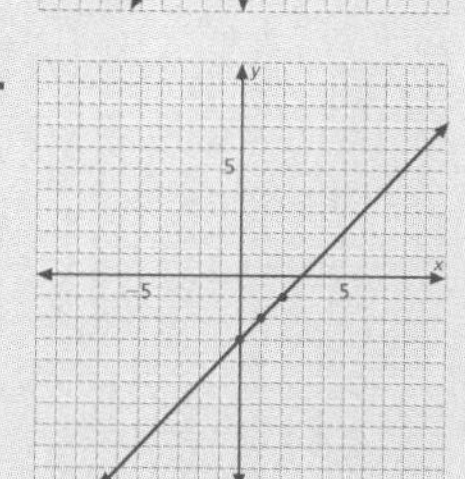

Extra Practice Worksheet 2-8

Name ______ Date ______

EXTRA PRACTICE 2-8

DISPLAYING DATA

EXERCISES

The ages of instructors at a health club are listed below.

24 28 29 35 37 22 48 56 42 47
20 66 43 40 19 18 25 29 30 32

1. Construct a stem-and-leaf plot to display the data.

stem	leaves
1	8 9
2	0 2 4 5 8 9 9
3	0 2 5 7
4	0 2 3 7 8
5	6
6	6

2. Identify any outliers, clusters, and gaps in the data. no outliers; Data cluster from 20 to 48.; no gaps

3. Find the mode of the data. 29

4. Find the median of the data. 31

5. Find the mean of the data. 33.5

Refer to the histogram for Exercises 6–9.

6. How many players scored 15 points or fewer? 11

7. What percent of the players scored between 11 and 25 points? about 57%

8. Which interval contains the median number of points scored? 16–20

9. Is it possible to identify the mean of the data? Explain. No; each piece of data is not known.

Points Scored During a Tournament by Members of the Monroe High School Boy's and Girl's Basketball Team

Frequency; Number of Points: 1–5, 6–10, 11–15, 16–20, 21–25, 26–30, 31–35

30 EXTRA PRACTICE *LESSON 2-8*

Enrichment Worksheet 2-8

Name ______ Date ______

ENRICHMENT 2-8

THE TOWER OF HANOI

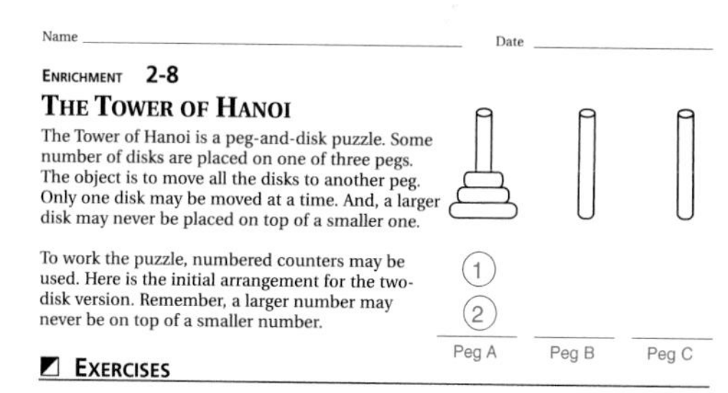

The Tower of Hanoi is a peg-and-disk puzzle. Some number of disks are placed on one of three pegs. The object is to move all the disks to another peg. Only one disk may be moved at a time. And, a larger disk may never be placed on top of a smaller one.

To work the puzzle, numbered counters may be used. Here is the initial arrangement for the two-disk version. Remember, a larger number may never be on top of a smaller number.

Peg A (1, 2) | Peg B | Peg C

EXERCISES

1. Solve the puzzle for 2 disks. Find two different solutions.

Peg A	Peg B	Peg C
1 2		
2	1	
	1	2
		1 2

Peg A	Peg B	Peg C
1 2		
2		1
	2	1
	1 2	

2. Solve the puzzle for 3 disks. Possible solution is given.

Peg A	Peg B	Peg C
1 2 3		
2 3		1
3	2	1
3	1 2	
	1 2	3
1	2	3
1		2 3
		1 2 3

38 ENRICHMENT *LESSON 2-8*

Skills Practice

Vocabulary Review

Lesson 2-7
statistics data
population sample
frequency table
measures of central tendency
mean median
mode

Lesson 2-8
stem-and-leaf plot outlier
cluster gap
histogram

Assignment Guide

All students: 1–62
Additional Practice: Refer to the Extra Practice Index on page 674 of this text.

Review and Practice Your Skills

Practice ■ Lesson 2-7

Find the mean, median, and mode of each set of data.

1. 98, 77, 89, 93, 75, 81, 77, 88, 78 84, 81, none
2. 4237, 4516, 4444, 4379, 4516, 4869 44, 93.5, 4480, 4516
3. 280, 295, 235, 210, 230, 235, 195, 210, 270 240, 235, 235 and 210
4. 3.8, 2.6, 4.1, 4.8, 5.9, 2.7, 6.9, 4.1 4.4, 4.1, 4.1
5.

Number of students	Grade
3	90
7	80
10	70

76.5, 75, 70

6.

Number of students	Height
3	150 cm
5	155 cm
2	160 cm

154.5, 155, 155

Althea measures the following volumes of water in beakers in a chemistry lab: 210 mL, 215 mL, 235 mL, 208 mL, 210 mL, 218 mL, 218 mL, 215 mL, 208 mL, 230 mL, 210 mL, 218 mL, 218 mL, 205 mL, 202 mL, 206 mL, 224 mL, 225 mL, 207 mL, 215 mL, 210 mL, 228 mL, 230 mL, 238 mL, 234 mL, 201 mL, 210 mL

7. Construct a frequency table for these data. Group the data into intervals of 10. 201–210 : 12, 211–220 : 7, 221–230 : 5, 231–240 : 3
8. Find the mean, median,and mode of the data. 216.6, 215, 210
9. Which interval contains the median? Which interval contains the mean? 211–220; 211–220
10. Which measure of central tendency best describes the most commonly measured volume of water? Median, the mean is affected by the few large values.

Practice ■ Lesson 2-8

For the data given above for Exercises 7–9 (Althea's chemistry lab):

11. Construct a stem-and-leaf plot to display the data. See additional answers.
12. Identify any outliers, clusters, and gaps in the data. clusters: 210, 215, 218, 205–209 gaps: 210–215, 218–224
13. Use your frequency table from Exercise 7 to construct a histogram. See additional answers.

The school newspaper surveyed 50 seniors about the average amount of time (in hours) per week that each student spent talking on the phone with friends.

Time (hr)	0-3	4-6	7-9	10-12	13-15	16-18	19-21	21-24
Frequency	2	6	3	12	5	11	8	4

14. Construct a histogram to display the data. See additional answers.
15. How many students talk on the phone less than 13 h/wk? 23
16. What percent of the students talk on the phone between 7 and 15 h/wk? 40%
17. Which interval contains the median of the data? 13–15
18. Is it possible to identify the mean of the data? Explain. No, you do not know the exact value for each senior.
19. If each student increases his or her phone time by 3 h/wk, which measures of central tendency would it affect? How? mean: increased by 3 hours; median: increased by 3 hours; mode: increased by 3 hours.

Additional Answers

11.

Stem	Leaves
20	1 2 5 6 7 8 8
21	0 0 0 0 0 5 5 5 8 8
22	4 5 8
23	0 0 4 5 8

13.

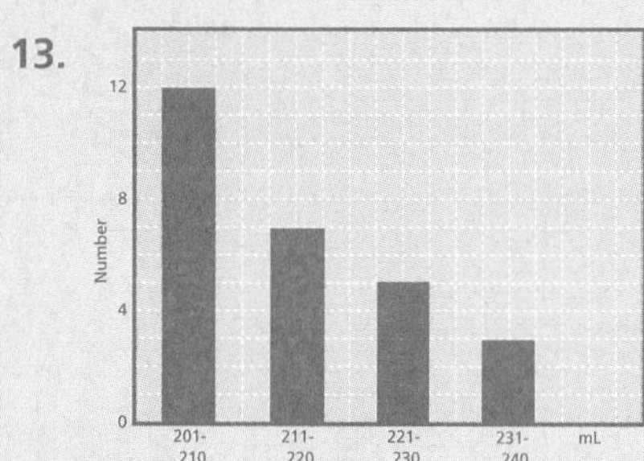

14.

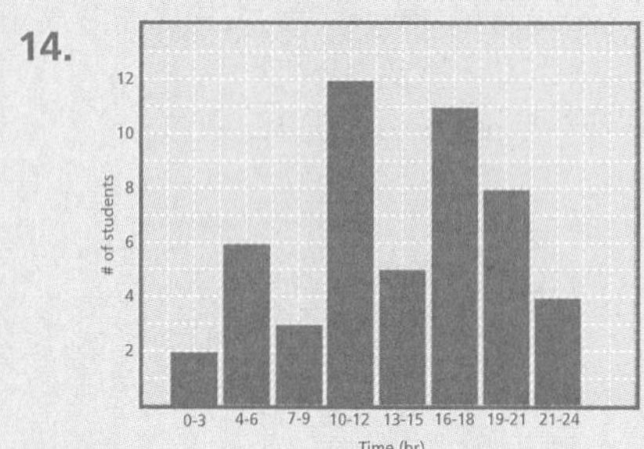

24-31.

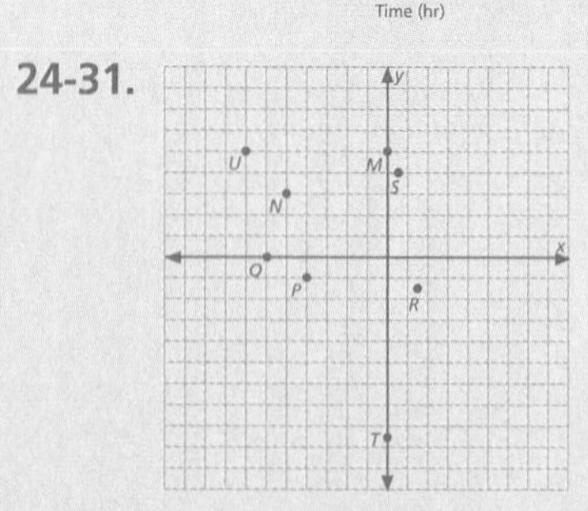

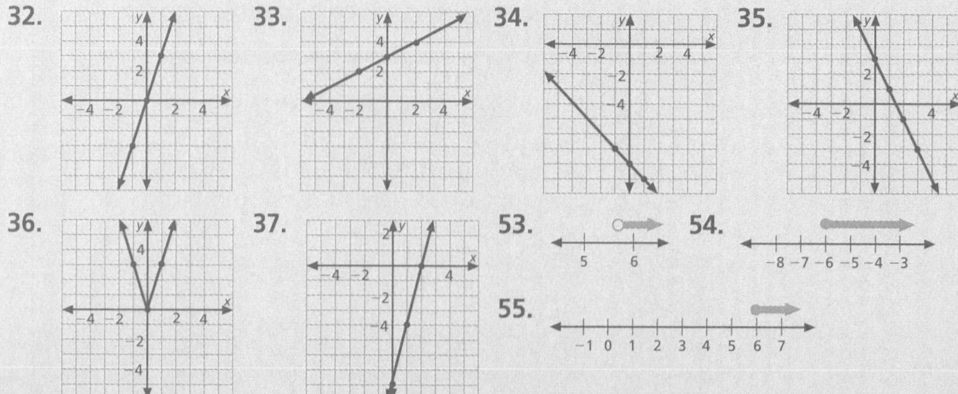

PRACTICE ◣ LESSON 2-1–LESSON 2-8

Find the next three terms in each sequence. (Lesson 2-1)

20. 4000, 800, 160, 32, $\frac{32}{5}$, $\frac{32}{25}$, $\frac{32}{125}$ **21.** −45, −17, 11, 39, 67, 95, 123

22. 4, 8, 7, 14, 13, 26, 25, 50, 49 **23.** A, E, I, M, Q, U, Y

Graph each point on a coordinate plane. Name the quadrant in which each point is located. (Lesson 2-2) See additional answers for graph.

24. $M(0, 5)$ *y*-axis **25.** $N(-6, 3)$ II **26.** $P(-4, -1)$ III **27.** $Q(-7, 0)$ *x*-axis

28. $R(1.5, -1.5)$ IV **29.** $S\left(\frac{1}{2}, 4\right)$ I **30.** $T\left(0, -8\frac{1}{3}\right)$ *y*-axis **31.** $U(-8, 5)$ II

Graph each function. (Lesson 2-3) See additional answers.

32. $y = 3x$ **33.** $f(x) = \frac{1}{2}x + 3$ **34.** $x + y = -8$

35. $f(x) = -2x + 3$ **36.** $y = |3x|$ **37.** $y = 4(x - 2)$

Solve each equation. (Lesson 2-4 and Lesson 2-5)

38. $x + 17 = 31$ 14 **39.** $23 - a = 0$ 23 **40.** $b - (-16) = -23$ −39

41. $-4c = 18$ $-4\frac{1}{2}$ **42.** $13d = -78$ −6 **43.** $\frac{x}{5} = -30$ −150

44. $23 - m = -17$ 40 **45.** $9e - 1 = 23$ $2\frac{2}{3}$ **46.** $2(x - 5) = -24$ −7

47. $\frac{y - 3}{2} = -11$ −19 **48.** $17 = -41 + p + 16$ 42 **49.** $3(x - 4) = -6 + 4x$ −6

50. $2.8x - 11.7 = 24.3$ 12.9 **51.** $8\frac{1}{3} - 2\frac{2}{3}n = 27$ −7 **52.** $\frac{z}{3} - 14 = -72$ −174

Solve each inequality and graph the solution on a number line. (Lesson 2-6) See additional answers for graphs.

53. $x - \left(-1\frac{1}{4}\right) > 7$ $x > 5\frac{3}{4}$ **54.** $-3h + 38 \le 56$ $h \ge -6$ **55.** $\frac{10}{3}g - 2 \ge 18$ $g \ge 6$

Graph each linear inequality on a coordinate plane. (Lesson 2-6) See additional answers.

56. $y < -x + 7$ **57.** $x - y \le -7$ **58.** $24 > 6x + 4y$

Find the mean, median, and mode of each set of data. (Lesson 2-7)

59. 15, 25, 30, 22, 45, 35, 38, 22, 37, 20, 31
29.1, 30, 22

60. $\frac{1}{2}, \frac{1}{3}, \frac{1}{4}, \frac{2}{3}, \frac{3}{4}, \frac{1}{2}, \frac{1}{8}, \frac{7}{8}, \frac{1}{2}, \frac{-1}{3}, \frac{1}{4}, \frac{4}{3}, \frac{3}{4}, \frac{1}{2}$ $\frac{1}{2}, \frac{1}{2}, \frac{1}{2}$

The table gives salaries and number of workers at a manufacturer of auto parts. (Lesson 2-8)

Job	President	Group Manager	Line Manager	Machinist	Clerk
Number	1	2	5	25	4
Salary Range	$81-95,000	$66-80,000	$51-65,000	$35-50,000	$20-34,000

61. In a labor dispute, which measures of central tendency might the president use to show that the worker's wages were already high enough? Why? mean; it is influenced by the salaries of president and group manager and will be higher than median salary

62. What percent of the workers earn $51,000 per year or more?
21.6%

Chalkboard Examples

Example from Lesson 2-7

At a small-town gallery the number of paintings sold each day for the last ten days is:

2 2 1 0 3 2 2 0 0 1

1. Construct a frequency table.

Number	0	1	2	3
Frequency	3	2	4	1

2. Find the mean, median, and mode for these data. mean: 1.3; median: 1.5; mode:2

Example from Lesson 2-8

Survey results on minutes of exercise per day:

15 20 25 15 25 30 40
45 60 20 60 40 35

1. Construct a stem-and-leaf plot for these data.

1	5 5
2	0 0 5 5
3	0 5
4	0 0 5
5	
6	0 0

2. Identify clusters, outliers, and gaps. clusters: 15–45; outlier: 60; gaps: 45–60

56.

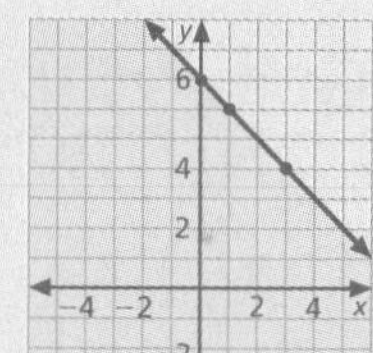

57.

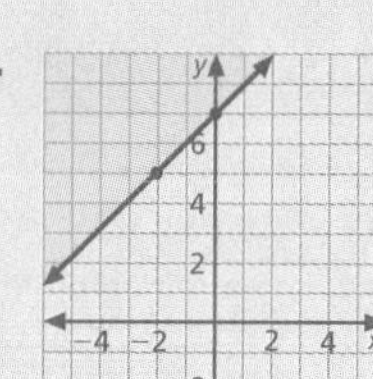

58.

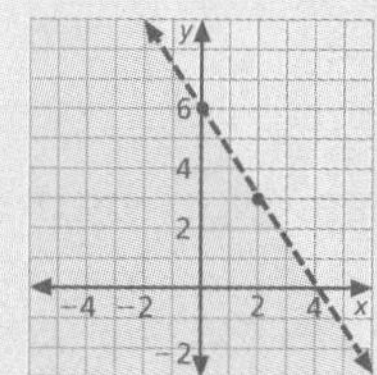

Learning Styles

VISUAL LEARNER Students may benefit from color-coding data before they create stem-and-leaf plots. With colored pencils, have students circle the last digit in each number and make a color code for the other digits of the numbers. Example: for 75, 72, 69, and 66, have students circle the 5, 2, 9, and 6. Then have them use red to underline the 7s and green to underline the 6s. When working with many items of data, this method will make it easier to arrange the data in the stem-and-leaf plot.

Lesson Planning

NCTM Standards/ Strands
- Number & Operation
- Patterns, Functions, & Algebra
- Geometry & Spatial Sense
- Data Analysis, Statistics, & Probability
- Problem Solving
- Reasoning and Proof
- Communication
- Connections
- Representation

Vocabulary
measures of central tendency

Materials Needed
paper/pencil
straightedges
graph paper

Lesson Resources
Warm-Up Transparency 6
Transparency TK-1, RF-36
Reteaching 2-9
Multicultural Connection 1

Assignment Guide
Basic: 1–4, 7–23
Enriched: 1–23

Getting Started

5-Minute Warm-Up
Draw a bar graph to display student grades; seven students got As, 16 got Bs, 13 got Cs, and three got Ds.

2-9 Problem Solving Skills: Misleading Graphs

When examining or reading statistics, think critically. Although informative and useful, statistics can be misleading. Graphs can mislead when scales or dimensions are changed. Measures of central tendency can be misleading if they do not accurately represent the data. Advertising claims can be misleading if they are vague, omit information, or hint at something that may not be true.

Problem Solving Strategies
- Guess and check
- Look for a pattern
- Solve a simpler problem
- ✔ Make a table, chart or list
- Use a picture, diagram or model
- Act it out
- Work backwards
- Eliminate possibilities
- Use an equation or formula

PROBLEM

A sales representative for a new soft drink is trying to convince a chain of food stores to order a much larger quantity. The salesperson uses the graph at the right in the sales pitch.

1. What can you say about sales of the drink?
2. What is deceptive about the graph?
3. How would you change the graph so that it is not misleading?

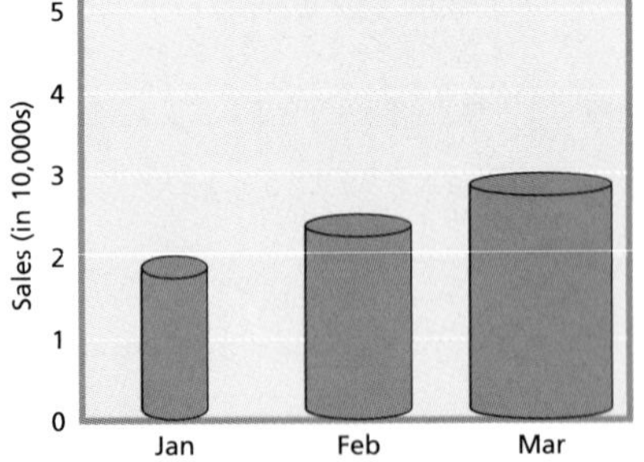

Solve the Problem

1. Read the vertical scale. Sales are increasing. Sales in March are 1.5 times as great as the sales in January.
2. Dimensions other than height have been changed. Sales for February and March appear to be greater than they actually are.
3. Diameters should be the same, so only the heights are compared.

Try These Exercises

1. **SPORTS** "Just look at the graph," the player's agent said. "Phil's hits have doubled since last year. We are looking for a large raise and a long contract."

   Look at the graph. Is the agent's claim misleading? Why?
   See additional answers.

Phil's Hits
This year
Last year
100 120 140 160 180 200 220
Hits

2. **ADVERTISING** Tell why you think each of the following advertisements is misleading.
   See additional answers.
   a. Eighty percent of all dentists surveyed agree: Zilch toothpaste tastes best.
   b. In the past 5 years, 25,000 cola drinkers have switched to Koala Kola.
   c. Thousands of teenagers wake in the morning to a glass of Zest. It has 20% real fruit juice and 100% bounce.

Use the Five-step Plan and the Strategy
THE FIVE-STEP PLAN **Read**—ask questions of the students to help them understand the problem. **Plan**—guide students to related problems and previously mastered skills and strategies. **Solve**—students solve problem on their own. **Answer**—write solution in format that answers the question. **Check**—review work, check for reasonableness and review strategy used. Students will benefit from the experience of verbalizing their methods.

THE STRATEGY **Use a diagram**—This strategy helps students use statistics to spot misleading graphs.

Additional Answers
1. Yes; the horizontal scale does not have uniform increments starting with zero—although one line is twice the length of the other, it does not show twice the number of hits.
2. Answers will vary. Sample answers are given. a. *How many* were surveyed? *Why* does it matter that dentists like the taste? *What* does it taste better than? What *else* did they say about the toothpaste? b. 25,000 may not be a large figure for the soft drink business. How many have switched *away* from Koala Kola? c. How many thousands? What is in the other 80%? Do they drink Zest every moring, as implied? What does *bounce* mean?

PRACTICE EXERCISE

Five-step Plan
1 Read
2 Plan
3 Solve
4 Answer
5 Check

These data show sales of the *Earn A Million-A-Day At Home* video. Read the table, and then examine the graphs below.

Week	1	2	3	4	5
Sales	252	246	265	276	280

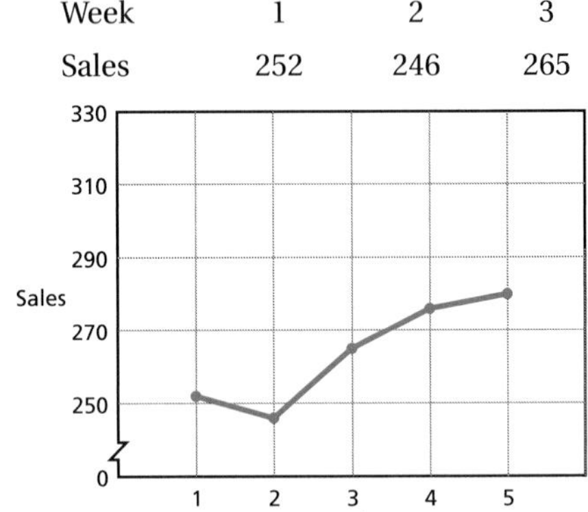

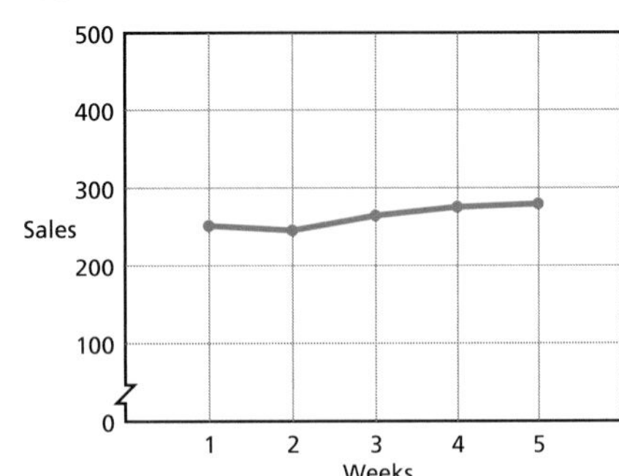

3. **WRITING MATH** The graphs show the same data. Why do they have different appearances? Different size intervals on the vertical scale; different horizontal distances.
4. Suppose you were a sales representative for the video and wanted to convince stores to stock more copies. Which graph would you use? Why? The one on the left; it indicates a more marked increase in sales.
5. **BUSINESS** Employee annual wages at a plant rose steadily, but very gradually, from one year to the next during one 5-year period. Make two graphs to show these changes, one from the perspective of the factory owner who wants to show that workers' wages are rising rapidly, and one from the perspective of an employee representative who wants to show that wages are rising minimally. Graphs will vary.

6. **WRITING MATH** Create your own advertisement that contains misleading statistics, or find one in a newspaper or magazine. See if a classmate can tell what is misleading about the ad. Check students' work.

MIXED REVIEW EXERCISES

Solve each equation. (Lesson 2-5)

7. $4b - 3 = 17$ 5
8. $2(m + 5) = -9$ $-9\frac{1}{2}$
9. $-4d + 6 = 2d - 8$ $2\frac{1}{3}$
10. $\left(\frac{1}{2}\right)(8x - 6) = -7$ -1
11. $-12 + 3a + 14 = 4$ $\frac{2}{3}$
12. $(0.75)(8w - 12) = 9$ 3
13. $-5k - 3 = -38$ 7
14. $3(2m - 4) = 2m - 22$ $-\frac{5}{2}$
15. $3(3x + 6.5) = -3$ -2.5

Evaluate $g(x) = |3x - 4|$ for the given values of x. (Lesson 2-3)

16. $g(2)$ 2
17. $g(-4)$ 16
18. $g(3)$ 5
19. $g(-2)$ 10
20. $g(-1)$ 7
21. $g(7)$ 17
22. $g(0.5)$ 2.5
23. $g(1.3)$ 0.1

Chalkboard Examples

Supplementary Problem

A circle graph with radius of 1 unit was used to show company sales of one million dollars in April. In May, sales doubled to two million dollars. What size circle would be appropriate to show May sales compared to April? **Students may suggest a circle with radius $\sqrt{2}$ units, so area is double the April circle.**

Lesson Wrap-up

QUICK ASSESSMENT

Look at the graph in Exercise 1. Redraw it so it appears that Phil's hits have quadrupled since last year. **Students may use the same scale and make the top bar twice as thick, or they may alter the scale so the top bar appears four times as long as the bottom bar.**

Enrichment Worksheet 2-9 not shown.

Reteaching Worksheet 2-9

Name ______________________ Date ______________

RETEACHING 2-9

PROBLEM SOLVING SKILLS: MISLEADING GRAPHS

Statistics can be useful when used properly. However, they can be misleading when the scales of graphs are manipulated, measures of central tendency are used that don't accurately relate the data, or advertisements omit data.

Example

A high school principal used the graph at the right to illustrate the gains in mean test scores over a four-year period.

a. What is deceptive about the graph?
b. How would you change the graph so that it is not misleading?

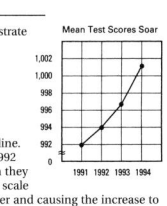

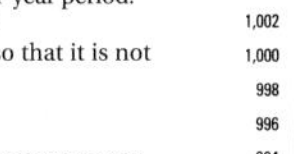

Solution

a. The small gains in the mean scores appear more significant in the graph because of the slope of the line. This happens because: a) the break between 0 and 992 makes the first 992 points visually appear as though they are equal to 2 points, b) the dates on the horizontal scale are squeezed closely together making the line steeper and causing the increase to appear more significant.
b. You could use a vertical scale that has consistent intervals from 0–1002. The dates on the horizontal scale could be written somewhat farther apart. You might also eliminate the word 'soar' from the title of the graph.

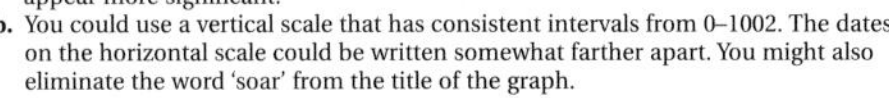

EXERCISES

Solve. **Answers may vary. Sample answers are given.**

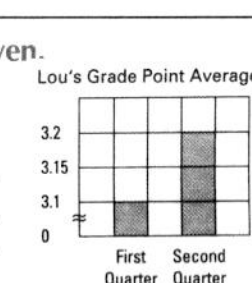

1. Describe how the graph at the right is deceptive. How would you change it so that it is not misleading? **It appears Lou's grade point average for the second quarter is 3 times his first quarter average. Change the intervals on the vertical scale so they are consistent.**
2. On another sheet of paper, make two graphs that show the increase in a high-school population for the next five years if it is estimated that there will be a 25-student increase per year. Use one graph to support claims of overcrowding and the other to support claims that the growth is controlled. Explain why each graph supports each position. **Check students' graphs. The graph to support the position that the school will be overcrowded will have small increments on the scale representing number of students. This graph would have a steeper line than the other, making the increase to appear dramatic.**

34 RETEACHING *LESSON 2-9*

Extend the Lesson

REAL WORLD CONNECTION Have students collect misleading advertisements, slogans, graphs, and other visual displays from newspapers, magazines, pamphlets, and fliers. Have other classmates identify what is misleading about each. Have students work in small groups to create collages entitled "Misleading Media."

Assessment Planning

Vocabulary Review

Lesson 2-1
pattern, sequence, term, iteration

Lesson 2-2
coordinate plane, quadrant, origin, ordered pair, coordinate, relation, domain, range, mapping, function, dependent variable, independent variable, vertical line test, *x*-axis and *y*-axis

Lesson 2-3
linear equation, linear function, constant, absolute value function

Lesson 2-4
inverse operation, reciprocal, equivalent equation, addition property of equality, multiplication property of equality

Lesson 2-5
like or similar terms

Lesson 2-6
linear inequality, half-plane, transitive property, boundary, open half-plane, closed half-plane

Lesson 2-7
statistics, data, population, sample, frequency table, mean, median, mode, measures of central tendency

Lesson 2-8
stem-and-leaf plot, outlier, cluster, gap, histogram

Lesson 2-9
measures of central tendency

Assessment Options

Chapter Assessment
Alternative Assessment
Chapter Tests Forms A and B
Cumulative Assessment
Achievement Test
Writing Prompts

Assignment Guide

All students: 1–29
Additional Practice: Refer to the Extra Practice Index on page 674 of this text.

Chapter 2 Review

Vocabulary ■ LESSON 2-1–LESSON 2-8

Choose the word from the list at the right that completes each statement.

a. function
b. median
c. histogram
d. mean

1. A type of bar graph used to display data is called a __?__. c
2. A __?__ is a relationship in which each element of one set is paired with *exactly one* element of another set. a
3. The middle value of data when the data are arranged in numerical order is called the __?__. b
4. To calculate the __?__, divide the sum of the data by the number of data pieces. d

LESSON 2-1 ■ Patterns and Iterations, p. 52

▶ An arrangement of numbers in a pattern is a **sequence**.

▶ An **iteration** is a proven that is repeated over and over again.

Find the next three terms in each pattern. Identify the rule.

5. 64, 16, 4, . . . $1, \frac{1}{4}, \frac{1}{16}$, divide by 4
6. $1, \frac{3}{2}, \frac{9}{4}, \ldots$ $\frac{27}{8}, \frac{81}{16}, \frac{243}{32}$, multiply by $\frac{3}{2}$
7. 1, −2, 4, −8, . . . 16, −32, 64, multiply by −2

LESSON 2-2 ■ The Coordinate Plane, Relations, and Functions, p. 56

▶ A set of ordered pairs is defined as a **relation**. The **domain** of a relation is the set of all possible *x*-coordinates. The **range** of a relation is the set of all possible *y*-coordinates.

▶ A **function** is a set of ordered pairs in which each element of the domain is paired with exactly one element in the range.

Determine whether each relation is a function. Give the domain and range of each.

8. {(2, 3), (−1, 4), (0, 2), (1, 2)} function; domain: {−1, 0, 1, 2}, range: {2, 3, 4}
9. {(2, −1), (1, 0), (0, 1)} function; domain: {0, 1, 2}; range; {−1, 0, 1}

Given $f(x) = 5x - 9$, evaluate each function.

10. $f(1)$ −4
11. $f(0)$ −9
12. $f(-1)$ −14
13. $f(-2)$ −19

LESSON 2-3 ■ Linear Functions, p. 62

▶ Equations or two variables that can be written in the form $y = ax + b$ where a and b are constants are called **linear equations**. A linear equation represents a **linear function**. Graphs of such equations are *straight* lines.

An **absolute value** function is $g(x) = |x| = \begin{cases} x \text{ if } x \geq 0 \\ -x \text{ if } x < 0 \end{cases}$

Graph each function. See additional answers.

14. $y = 2x - 3$
15. $y = -x + 2$
16. $y = 2x - 7$

Additional Answers

14. 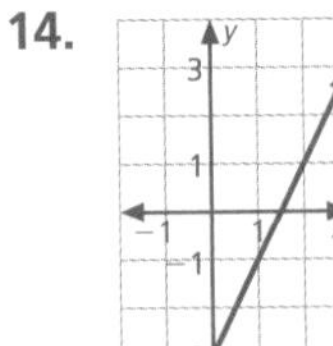

15. 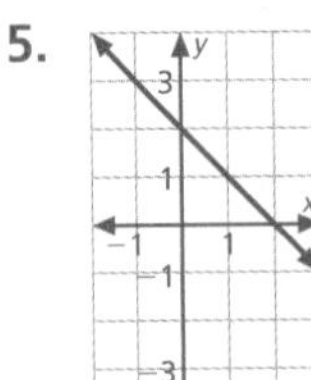

16. 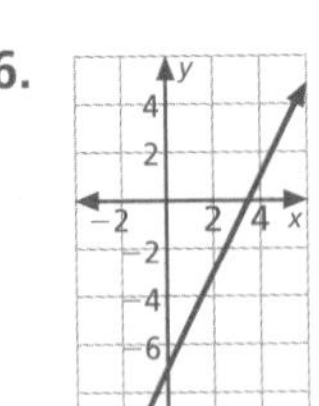

19. 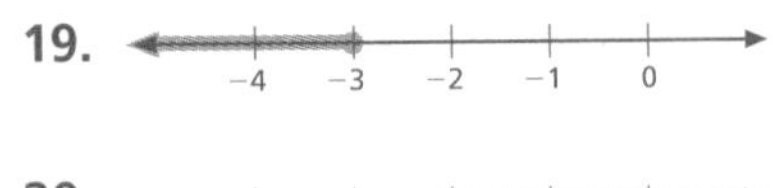

20. 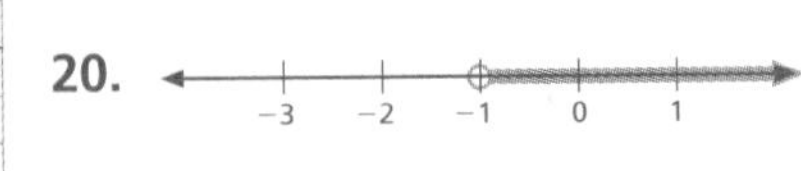

21. 

LESSON 2-4 and 2-5 ■ Solve One-step, Multi-step Equations, p. 66

▶ You can add (subtract) the same number to (from) each side of an equation and/or multiply (divide) each side by the same number. Remember to perform the same operations on each side until the variable is alone on one side of the equation.

Solve each equation.

17. $3a + 9 = -15$ $a = -8$ **18.** $\frac{2}{5}x - 1 = 3$ $x = 10$

LESSON 2-6 ■ Solve Linear Inequalities, p. 76

▶ The **graph of an inequality** is the set of all ordered pairs that make the inequality true.

Solve each inequality, and graph the solution on a number line. See additional answers.

19. $-3x + 4 \geq 13$ **20.** $5x + 1 > -4$ **21.** $-6x + 2 < 12 - x$

Graph each inequality on the coordinate plane. See additional answers.

22. $y > 2x + 1$ **23.** $y \leq -x + 3$ **24.** $y - x > 2$

LESSON 2-7 ■ Data and Measures of Central Tendency, p. 82

▶ The **mean** of a set of data is the sum of the items divided by the number of items.

▶ The **median** is the middle value (or the mean of the two middle values) of a set of data arranged in numerical order.

▶ The **mode** is the number that occurs the most often in a set of data.

The results of the last math test are:

81 78 90 85 62 59 86 94 93 92 85 82 90 80 86 85

25. Construct a frequency table. See additional answers.

26. Find the mean, median, and mode. 83; 85; 85

LESSON 2-8 ■ Display Data, p. 86

▶ Individual items can be displayed in **stem-and-leaf plots**. In these, the digit farthest to the right in a number is the leaf. The other digits make up the stem.

▶ The frequency of data can be displayed in a type of bar graph called a **histogram**.

27. Construct a stem-and-leaf plot to display the data in Exercises 26–27. See additional answers.

28. Identify any outliers, clusters, and gaps in the data. Outliers: 59 and 62; clusters: 78–94; gaps: between 62 and 78

LESSON 2-9 ■ Problem Solving Skills: Misleading Graphs, p. 92

▶ Statistics can be helpful when trying to make a decision. However, they can be misleading.

29. A television advertisement says, "Over 100 dentists can't be wrong. XYZ toothpaste is the one you should use for a healthier smile." Tell why this advertisement might be misleading. Answers will vary.

22.
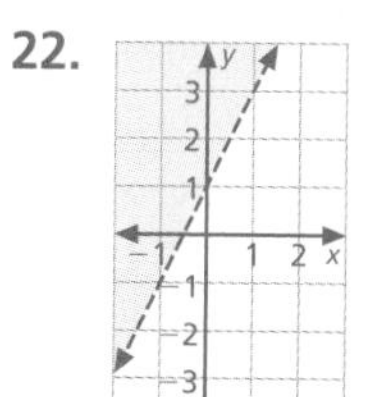

23.
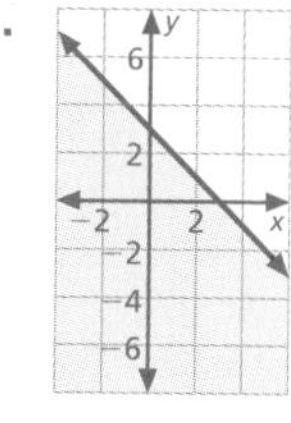
24.
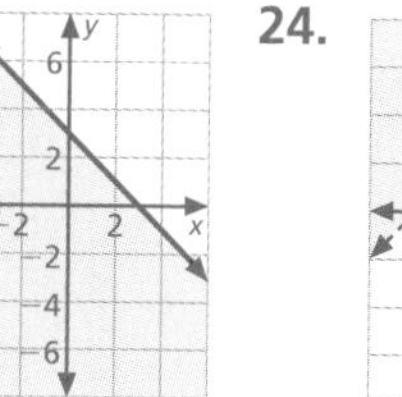
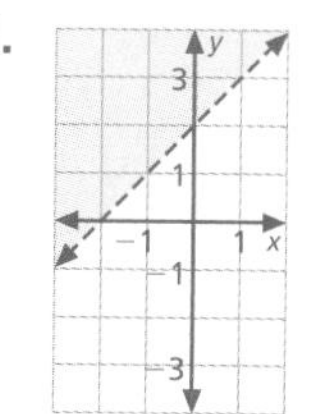
25.
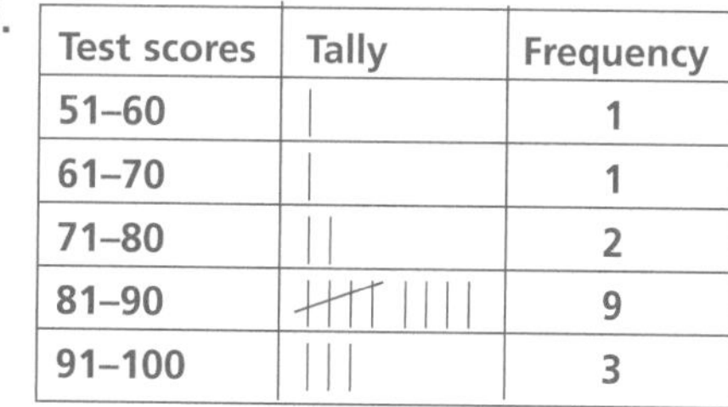

Test scores	Tally	Frequency
51–60	|	1
61–70	|	1
71–80	||	2
81–90	卌 ||||	9
91–100	|||	3

27.

5	9
6	2
7	8
8	0 1 2 5 5 5 6 6
9	0 0 2 3 4

Chapter 2 Test Form A, page 1

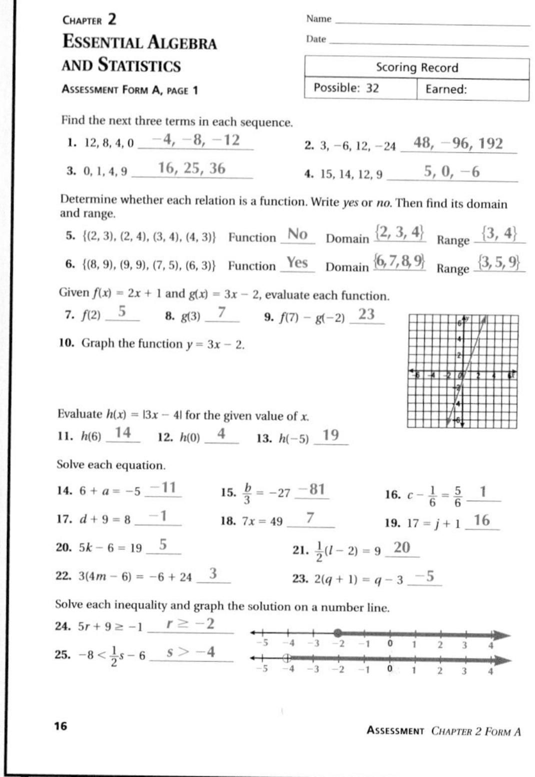

CHAPTER 2
ESSENTIAL ALGEBRA AND STATISTICS
ASSESSMENT FORM A, PAGE 1

Name ______ Date ______

Scoring Record	
Possible: 32	Earned:

Find the next three terms in each sequence.

1. 12, 8, 4, 0 −4, −8, −12
2. 3, −6, 12, −24 48, −96, 192
3. 0, 1, 4, 9 16, 25, 36
4. 15, 14, 12, 9 5, 0, −6

Determine whether each relation is a function. Write *yes* or *no*. Then find its domain and range.

5. {(2, 3), (2, 4), (3, 4), (4, 3)} Function No Domain {2, 3, 4} Range {3, 4}
6. {(8, 9), (9, 9), (7, 5), (6, 3)} Function Yes Domain {6, 7, 8, 9} Range {3, 5, 9}

Given $f(x) = 2x + 1$ and $g(x) = 3x - 2$, evaluate each function.

7. $f(2)$ 5 8. $g(3)$ 7 9. $f(7) - g(-2)$ 23

10. Graph the function $y = 3x - 2$.

Evaluate $h(x) = |3x - 4|$ for the given value of x.

11. $h(6)$ 14 12. $h(0)$ 4 13. $h(-5)$ 19

Solve each equation.

14. $6 + a = -5$ −11 15. $\frac{b}{3} = -27$ −81 16. $c - \frac{1}{6} = \frac{5}{6}$ 1
17. $d + 9 = 8$ −1 18. $7x = 49$ 7 19. $17 = j + 1$ 16
20. $5k - 6 = 19$ 5 21. $\frac{1}{2}(l - 2) = 9$ 20
22. $3(4m - 6) = -6 + 24$ 3 23. $2(q + 1) = q - 3$ −5

Solve each inequality and graph the solution on a number line.

24. $5r + 9 \geq -1$ $r \geq -2$
25. $-8 < \frac{1}{2}s - 6$ $s > -4$

16 ASSESSMENT *CHAPTER 2 FORM A*

Chapter 2 Test Form A, page 2

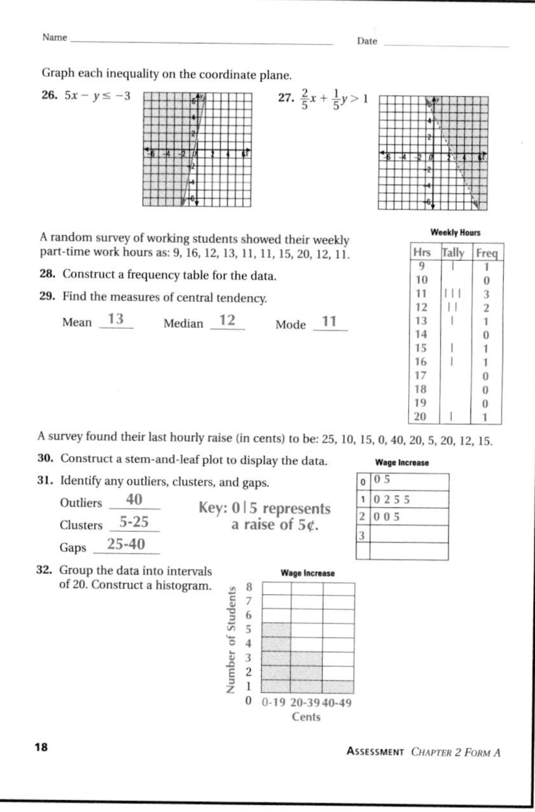

Name ______ Date ______

Graph each inequality on the coordinate plane.

26. $5x - y \leq -3$ 27. $\frac{2}{5}x + \frac{1}{5}y > 1$

A random survey of working students showed their weekly part-time work hours as: 9, 16, 12, 13, 11, 11, 15, 20, 12, 11.

28. Construct a frequency table for the data.
29. Find the measures of central tendency.

Mean 13 Median 12 Mode 11

Weekly Hours

Hrs	Tally	Freq
9	|	1
10		0
11	|||	3
12	||	2
13	|	1
14		0
15	|	1
16	|	1
17		0
18		0
19		0
20	|	1

A survey found their last hourly raise (in cents) to be: 25, 10, 15, 0, 40, 20, 5, 20, 12, 15.

30. Construct a stem-and-leaf plot to display the data.
31. Identify any outliers, clusters, and gaps.

Outliers 40
Clusters 5-25
Gaps 25-40

Wage Increase

0	0 5
1	0 2 5 5
2	0 0 5
3	

Key: 0 | 5 represents a raise of 5¢.

32. Group the data into intervals of 20. Construct a histogram.

18 ASSESSMENT *CHAPTER 2 FORM A*

Chapter 2 Test Form B, page 1

CHAPTER 2
ESSENTIAL ALGEBRA AND STATISTICS
ASSESSMENT FORM B, PAGE 1

Name ____________
Date ____________

Scoring Record	
Possible: 32	Earned:

Find the next three terms of each sequence.

1. 1, 4, 7, 10 __13, 16, 19__
2. 1, −3, 9, −27 __81, −243, 729__
3. 0, 1, 8, 27 __64, 125, 216__
4. 1, 3, 6, 10 __15, 21, 28__

Determine whether each relation is a function. Write *yes* or *no*. Then find its domain and range.

5. {(4,1), (5, 1), (1, 5), (2, 2)} Function __Yes__ Domain __{1,2,4,5}__ Range __{1, 2, 5}__
6. {(1, 5), (1, 6), (2, 7), (3, 8)} Function __No__ Domain __{1, 2, 3}__ Range __{5,6,7,8}__

Given $f(x) = 4x - 2$ and $g(x) = x + 3$, evaluate each function.

7. $f(-1)$ __−6__
8. $g(9)$ __12__
9. $f(3) - g(4)$ __3__
10. Graph the function $y = 4x + 1$.

Evaluate $h(x) = |x + 1|$ for the given value of x.

11. $h(4)$ __5__
12. $h(0)$ __1__
13. $h(-2)$ __1__

Solve each equation.

14. $9 - a = -3$ __12__
15. $\frac{b}{4} = 16$ __64__
16. $c - \frac{2}{5} = \frac{3}{5}$ __1__
17. $d - 15 = 23$ __38__
18. $9x = 108$ __12__
19. $53 = j + 17$ __36__
20. $4k + 7 = 39$ __8__
21. $\frac{1}{3}(l - 6) = 19$ __63__
22. $2(2m + 5) = 75 + 15$ __20__
23. $7(q + 2) = q + 14$ __0__

Solve each inequality and graph the solution on a number line.

24. $4 \le 2r - 6$ __$r \ge 5$__
25. $-\frac{1}{4}s + 4 > 3$ __$s < 4$__

20 ASSESSMENT *CHAPTER 2 FORM B*

Chapter 2 Test Form B, page 2

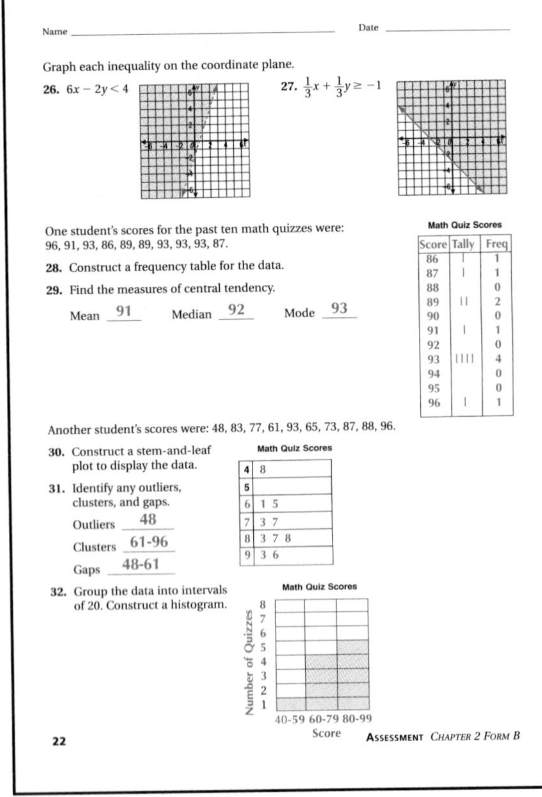
Name ____________ Date ____________

Graph each inequality on the coordinate plane.

26. $6x - 2y < 4$
27. $\frac{1}{3}x + \frac{1}{3}y \ge -1$

One student's scores for the past ten math quizzes were: 96, 91, 93, 86, 89, 89, 93, 93, 93, 87.

28. Construct a frequency table for the data.
29. Find the measures of central tendency.

Mean __91__ Median __92__ Mode __93__

Math Quiz Scores

Score	Tally	Freq
86	|	1
87	|	1
88		0
89	||	2
90		0
91	|	1
92		0
93	||||	4
94		0
95		0
96	|	1

Another student's scores were: 48, 83, 77, 61, 93, 65, 73, 87, 88, 96.

30. Construct a stem-and-leaf plot to display the data.
31. Identify any outliers, clusters, and gaps.

Outliers __48__
Clusters __61-96__
Gaps __48-61__

Math Quiz Scores

4	8
5	
6	1 5
7	3 7
8	3 7 8
9	3 6

32. Group the data into intervals of 20. Construct a histogram.

22 ASSESSMENT *CHAPTER 2 FORM B*

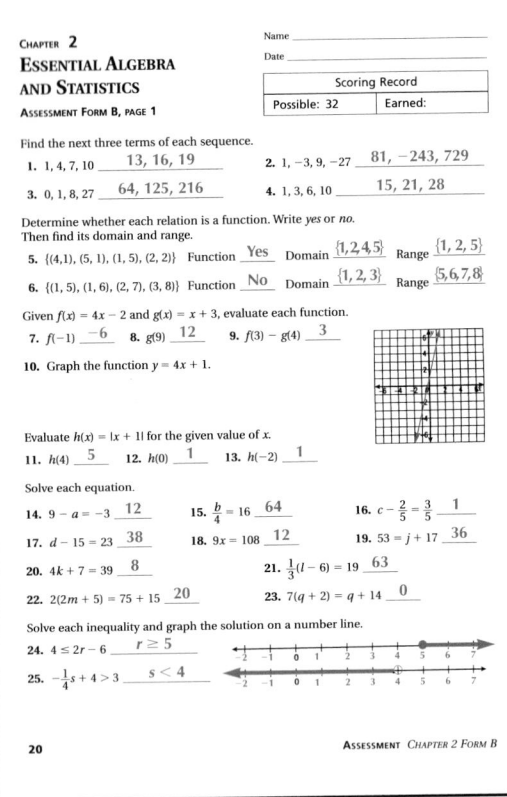

CHAPTER 2 ASSESSMENT

1. Find the next three terms in the pattern 1, 2, 5, 10, ... 17, 26, 37
2. Determine whether {(−1, −2), (1, 2), (2, 2)} is a function. Give the domain and range. function; domain: {−1, 1, 2}; range: {−2, 2}
3. Given $f(x) = -3x + 5$, find $f(-1)$. 8
4. Graph $y = 2x - 3$. See additional answers.
5. Evaluate $g(x) = -|2x - 1| + |-x|$ for $g(-1)$. −2

Solve each equation.

6. $7x - 5 = -4$ $\frac{1}{7}$
7. $\frac{2}{5}x + 1 = 3$ 5

Translate each sentence into an equation. Then solve.

8. When three times a number is increased by 2, the result is 17. Find the number. $3x + 2 = 17,\ x = 5$
9. When three-sevenths of a number is decreased by 1, the result is 5. Find the number. $\frac{3}{7}x - 1 = 5,\ x = 14$

Solve and graph each equation. For 10–13, see additional answers for graphs.

10. $8x + 5 > 12$ $x > \frac{7}{8}$
11. $-3x + 8 \ge 11$ $x \le -1$
12. $y \le 3x - 4$
13. $y > -x + 3$
14. Solve $ax = -a + ba + c$ for a $a = \frac{c}{x + 1 - b}$

Teenagers polled about the number of evening meals they ate at home in one week reported the following number of meals.

5 7 1 3 5 4 5 4 6 2 6 4

15. Construct a frequency table for the data. See additional answers.
16. Find the mean of the data. ≈ 4.3
17. Find the median of the data. 4.5
18. Find the mode of the data. 4 and 5

Another group of teenagers polled about the number of evening meals they ate at home in one four-week period reported the following number of meals.

20 35 31 17 3 28 18 28 30 30 26 28

19. Construct a stem-and-leaf plot to display the data.
20. Identify the outliers, clusters, and gaps in the data. outliers: 3, clusters: 17–35, gaps: between 3 and 17

19.

0	3
1	7 8
2	0 6 8 8 8
3	0 0 1 5

ADDITIONAL ANSWERS

4.

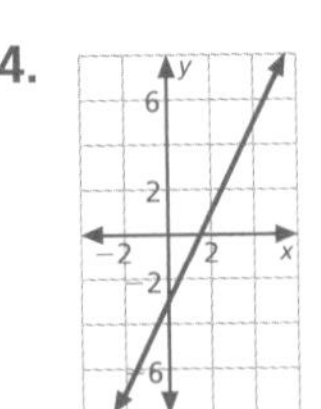

10.

11.

12.

13.

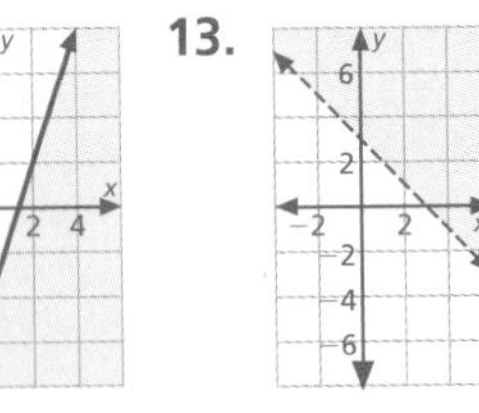

15.

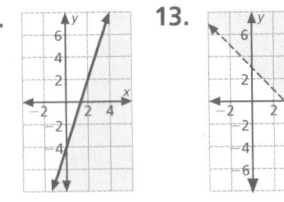

Meals	Tally	Frequency
1–2	||	2
3–4	||||	4
5–6	~~||||~~	5
7–8	|	1

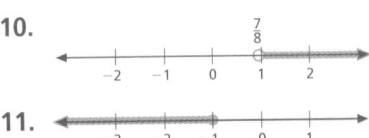

Chapter 2 Alternative Assessment

Modeling/Using Manipulatives

TEACHING WITH ALGEBLOCKS Prepare a lesson for your class on solving one-step equations using Algeblocks. Prepare at least 10 original equations for use during your teaching period. Write a lesson plan that includes the equations, what questions you might anticipate from your students and how you expect to answer those questions.

Cooperative Learning

COMPUTER SURVEY Work with a group. Poll 50 students in your school about the number of hours they have spent using a computer in the past week. Make a frequency table of the results of your survey. Display the data in both a stem-and-leaf plot and a histogram. Find the mean, median, mode and range of the data, as well as any outliers, clusters and gaps. Discuss the best measure of central tendency of this data, and the most appropriate method of display.

Technology

COMPUTER LANGUAGE Use a computer graphing program to graph at least 10 original linear equations and 5 linear inequalities. Print out each graph to make a notebook that also shows the calculated answers. If no computer or graphing program is available, the graphs may be hand-drawn.

Chapter Investigation

EXTENSION Create an advertisement for your school or community newspaper to promote your proposal. Include your survey question, the histogram of the data you collected, and a paragraph that will convince people to support your proposal.

Chapter Investigation

The benchmarks and expectations for this extension are as follows.

- Students work together to think of a positive change that they would like to make at their school or in their community. They make a list of data that would encourage others to adopt their proposal.
- Students think about how they will use the media to convince the public to support their message. They write a public relations plan and a budget.
- Students develop a survey question to ask their classmates. They collect the data, construct a frequency table, and draw the related histogram. They write a paragraph analyzing their results.
- Students create an advertisement to promote their proposal.

Modeling/Using Manipulatives

Have students make presentations to an audience unfamiliar with Algeblocks, to more accurately gauge their success.

SCORING RUBRIC **5**—Prepares a thorough, well-thought out lesson plan and successfully teaches other students the use of Algeblocks in solving equations. **4**—Prepares a good lesson plan and successfully teaches the material. **3**—Prepares a good lesson plan and makes a good effort in teaching others to use Algeblocks. **2**—Prepares a fair lesson plan and makes some effort to explain the use of Algeblocks. **1**—Student prepares a partial lesson plan but makes little effort to teach other students the use of Algeblocks. **0**—Student does not prepare lesson plan.

Cooperative Learning

If several groups choose this assignment, compare the different results. As a class activity, students can find the mean, median, mode, and range of the groups' data.

SCORING RUBRIC **5**—Conducts a complete survey, correctly computes the measures of central tendency, prepares excellent displays of the data. **4**—Conducts a complete survey, correctly computes the measures of central tendency, prepares accurate display of data. **3**—Conducts a partial survey, correctly computes most of the measures of central tendency, prepares adequate display of data. **2**—Conducts a partial survey, correctly computes some of the measures of central tendency, prepares inadequate display of data. **1**—Attempts to conduct survey, computes one of the measures of central tendency, prepares incomplete display of data. **0**—Makes no attempt to survey or analyze data.

Technology

If needed, schedule computer lab time.

SCORING RUBRIC **5**—Writes linear equations and inequalities, graphs them correctly, displays all work in a well-organized notebook. **4**—Writes linear equations and inequalities, graphs them correctly, displays most work in a notebook. **3**—Writes some linear equations and inequalities, graphs most correctly, displays most graphs and calculations in a notebook. **2**—Writes a few linear equations and inequalities, graphs most correctly, displays work in a notebook. **1**—Write a few linear equations and inequalities, graphs some correctly. **0**—Makes no attempt to write and graph equations and inequalities, does not prepare display of work.

Cumulative Review

Cumulative review is the best way to maintain previously taught skills and concepts. This will keep students prepared for new lessons that build on previously covered skills.

Cumulative Review covers:

Lesson 1-1
Lesson 1-2
Lesson 1-3
Lesson 1-4
Lesson 1-5
Lesson 1-7
Lesson 1-8
Lesson 2-1
Lesson 2-2
Lesson 2-4
Lesson 2-5
Lesson 2-6
Lesson 2-7
Lesson 2-8

Additional Answers

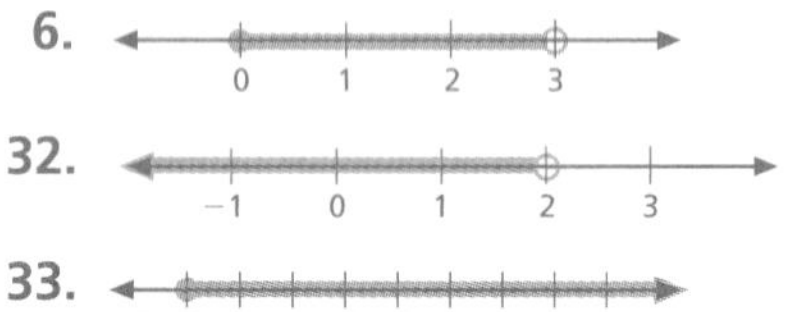

35.

Stem	Leaf
16	8
17	8 7 7 7 5
18	8 6 6 5 4 4
19	5 3 2 2 1 0

Key: 19 | 2 means 192

Cumulative Assessment

14.

Stem	Leaf
8	7 6 8 5
9	0 4 7 1 2
10	5 0 1
11	1 3
12	1 2
13	1
14	
15	0

Possible interpretation: Most of the data clusters between 85 and 105. There is a smaller cluster between 111 and 131. There is a gap between 131 and 150. 150 is considered an outlier.

Chapters 1–2 Cumulative Review

1. Write all the subsets of $\{M, E, T\}$. $\{\ \}, \{M\}, \{E\}, \{T\}, \{M, E\}, \{M, T\}, \{E, T\}, \{M, E, T\}$

Determine if each statement is *true* or *false*.

2. $\{2, 4\} \subset \{2, 4\}$ true

3. $15 \in \{\ldots -10, -5, 0, 5, 10, \ldots\}$ true

4. $\sqrt{36}$ is a rational number. true

5. $-\sqrt{7}$ is an integer. false

6. Graph $A \cap B$ on a number line, given that $A = \{x|x$ is a real number and $x < 3\}$ and $B = \{x|x$ is a real number and $x \geq 0\}$. See additional answers.

Perform the indicated operation.

7. $-19 - (-6)$ -13

8. $-12 + (-10)$ -22

9. $4\frac{1}{5} - \left(-\frac{3}{10}\right)$ $4\frac{1}{2}$

10. $-5\frac{7}{16} + 2\frac{1}{4}$ $-3\frac{3}{16}$

11. $(-7)(-9)$ 63

12. $72 \div (-9)$ -8

13. $(-4)\left(-\frac{1}{8}\right)$ $\frac{1}{2}$

14. $3\frac{1}{5} \div (-4)$ $-\frac{4}{5}$

Simplify.

15. $x^5 \cdot x^{15}$ x^{20}

16. $\frac{y^{16}}{y^4}$ y^{12}

17. $(a^4)^5$ a^{20}

18. $(3x^2)^3$ $27x^6$

19. $\frac{m^5}{m^{-3}}$ m^8

20. $x^{-3} \cdot x^{-2}$ $\frac{1}{x^5}$

21. Evaluate a^{-4} when $a = 3$. $\frac{1}{81}$

22. Write 0.00000076 in scientific notation. 7.6×10^{-7}

23. Write 6.25×10^7 in standard form. 62,500,000

Find the next three terms in each sequence.

24. $21, 7, \frac{7}{3}, \frac{7}{9}, \ldots$ $\frac{7}{27}, \frac{7}{81}, \frac{7}{243}$

25. 1, 6, 11, 16, . . . 21, 26, 31

Given $f(x) = 2x^3$ and $g(x) = 4x - 3$, find each value.

26. $f(-2)$ -16

27. $f(2) + g(-2)$ 5

Solve each equation and check the solution.

28. $x + 21 = 48$ 27

29. $0.7x = 4.2$ 6

30. $6x + 5 = 4x - 1$ -3

31. $3(2x - 1) = 5 - x$ $\frac{8}{7}$

Solve each inequality and graph on a number line. For 32–33. See additional answers.

32. $3x - 5 < 1$

33. $17 \geq 5 - 2y$

34. Find the mean, median, and mode of the following test scores: mean, 85; median, 85; mode, 76 and 90

89 76 76 90 93

97 80 78 90 82 85

35. Use the following data to construct a stem-and-leaf plot. See additional answers.

192 168 178 175 186 184

195 191 177 185 184 186

190 192 177 188 193 177

Teaching Strategies

Remind students that the rules for solving equations and inequalities are identical, except that the sign of the inequality must be reverse when multiplying or dividing both sides of an inequality by a negative number.

CHAPTERS 1–2 CUMULATIVE ASSESSMENT

STANDARDIZED TEST PREPARATION—QUANTITATIVE COMPARISON

In each question, compare the quantity in Column 1 with the quantity in Column 2. Select the letter of the correct answer from these choices:

A. The quantity in Column 1 is greater.

B. The quantity in Column 2 is greater.

C. The two quantities are equal.

D. The relationship cannot be determined by the information given.

Note: In some questions, information which refers to one or both columns is centered over both columns. A symbol used in both columns has the same meaning in each column. All variables represent real numbers. Most figures are not drawn to scale.

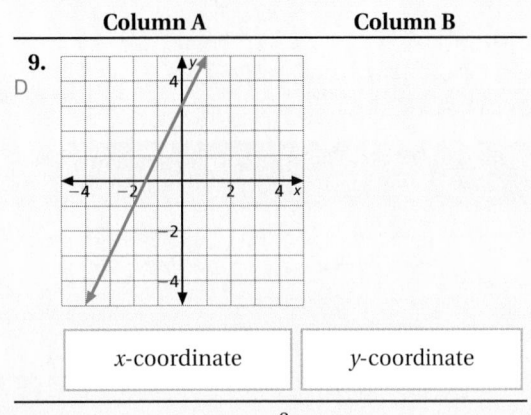

	Column A	Column B
1. $\{x \mid x \text{ is a negative integer and } x > -5\}$ B	x	5
2. B	$\sqrt{(10)}$	3.5
3. A	$-\frac{1}{3}(15)$	$-\frac{1}{15}(3)$
4. $x < 0$ A	$x^6 \cdot x^8$	$\frac{x^8}{x^6}$
5. $y > 0$ A	$(y^4)^2$	y^6
6. A	x^{-4} when $x = -2$	$-\frac{1}{16}$
7. B	$f(-3)$ when $f(x) = 3x - 5$	$g(-3)$ when $g(x) = 3x^2$
8. $3x - 5 > 7$ A	x	0
9. D (graph)	x-coordinate	y-coordinate
10. $\frac{3}{5}x = 15$ A	x	15
11. $5x - 3 = 2x + 9$ C	x^2	4^2
12. A (graph)	x-coordinate	y-coordinate
13. 22, 17, 23, 24, 24, 18, 19 A	median	mean

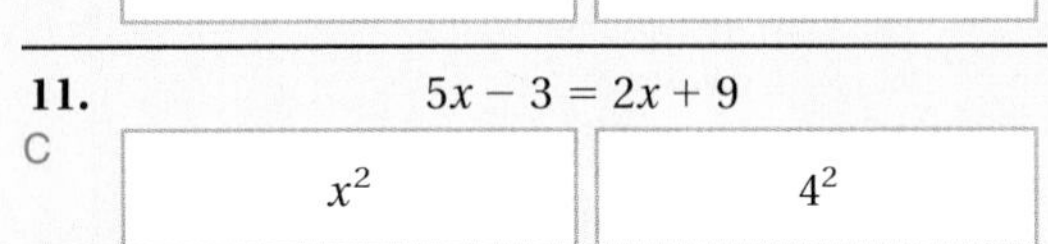

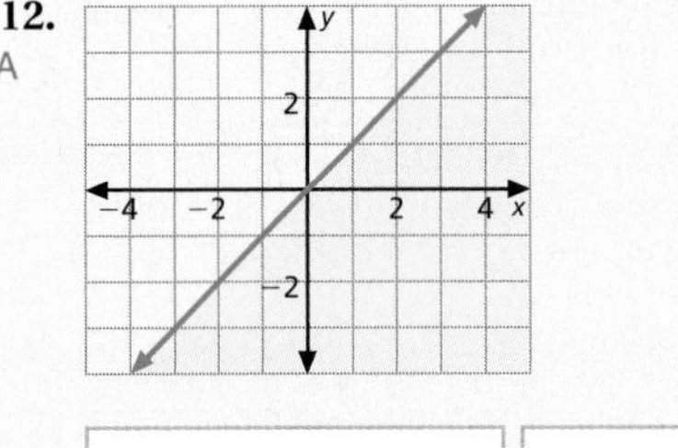

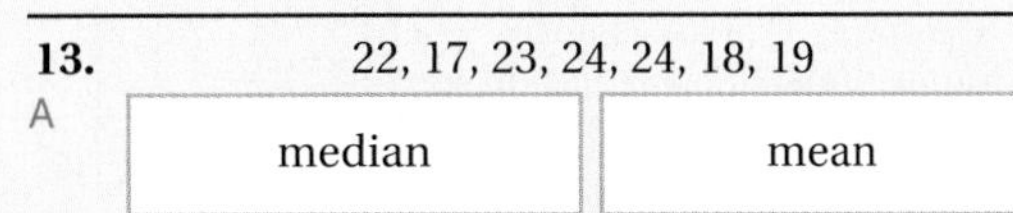

14. CONSTRUCTED RESPONSE Construct a stem-and-leaf plot to display the following data. Then interpret the data. 90, 87, 111, 92, 105, 121, 86, 94, 113, 122, 85, 88, 100, 97, 150, 131, 101, 91 See additional answers.

STANDARDIZED TEST PREPARATION

Quantitative Comparison Questions

Quantitative comparison questions are generally written with fewer words and require less time to compute. There are four answer choices that describe the relationships of equality and inequality. You are required to compare the quantity from Column A with the quantity from Column B and then determine their relationship. Students should carefully read the four answer choices often during the testing period.

Horizontal rules are used as separators between questions and to clarify when figures and additional information accompanies one or more problems.

Understanding the directions for quantitative comparison questions is critical for success since this type of question is not as commonplace as other types of questions.

Teaching Strategies

As a quick review, have students use Algeblocks or draw diagrams to model each equation below and its solution.

1. $5x + 4 = 29$ $x = 5$
2. $7x - 5 = 3x + 7$ $x = 3$
3. $x + 8 = 2x - 9$ $x = 17$

Chapter 3—Geometry and Reasoning

Theme: Geography
Chapter Investigation: Plot Course, pages 103, 121, 137
Careers: Cross-country bus drivers, page 113; Cartographers, page 133
Data Activity: Latitude and Longitude of World Cities

Content and Connections

Lesson, Pages	Lesson Objectives	NCTM Standards	State/Local Objectives	Interdisciplinary Connections	Real World Applications
3–1 104–107	• Define basic terms • Apply postulates about points, lines and planes	Geometry & Spatial Sense Measurement Number Operation Patterns, Functions & Algebra Reasoning & Proof Representation			Geography, Carpentry
3–2 108–111	• Classify and measure angles	Geometry & Spatial Sense Patterns, Functions & Algebra Measurement Problem Solving Reasoning & Proof Connections; Representation			Tiling, Architecture
3–3 114–117	• Identify bisectors of angles and segments • Apply theorems about midpoints, angle bisectors and vertical angles	Geometry & Spatial Sense Patterns, Functions & Algebra Measurement Reasoning & Proof Representation			Engineering, Art, Map Making
3–4 118–121	• Use constructions of segment bisectors and angle bisectors to solve problems • Identify parallel, perpendicular and skew lines • Identify congruent angles formed by parallel lines and a transversal	Patterns, Functions & Algebra Geometry & Spatial Sense Measurement Connections Representation			Architecture, Surveying, Navigation
3–5 124–127	• Use inductive reasoning to complete patterns	Problem Solving Reasoning & Proof Patterns, Functions & Algebra Geometry & Spatial Sense Connections; Representation		Art	City Planning, Scheduling, Art
3–6 128–131	• Identify and evaluate conditional statements • Identify converses and biconditionals	Reasoning & Proof Geometry & Spatial Sense Problem Solving Communication Connections; Representation			Drafting, Sports, Geography
3–7 134–137	• Write geometric proofs in two-column format	Reasoning & Proof Number Operation Geometry & Spatial Sense Problem Solving Connections; Representation			Surveying, Carpentry
3–8 138–139	• Solve a problem using logical reasoning • Eliminate possibilities	Problem Solving Communication Connections; Representation Reasoning & Proof			Travel

Planning and Resources

Lesson, Pages	Tools/Materials Needed	Trans-parency	Learning/Teaching Styles Options	Assignments: Basic Enriched	Additional Practice in SE	Reteaching, Extra Practice, Enrichment	Other Resources
3–1 104–107	paper/pencil small flat objects	Warm up 7 Trans RF-9, 10	Visual Learner Challenge	B: 1–18, 21–34 E: 1–34	Page 112–113, 123, 133, 140, 691	R: page 35 EP: page 31 E: page 43	SS: Teacher's Choice
3–2 108–111	paper/pencil protractor	Warm up 7 Trans RF-11, 12	Visual Learner Challenge	B: 1–25, 29–42 E: 1–42	Page 112–113, 123, 133, 140, 692	R: page 37 EP: page 33 E: page 45	SS: Teacher's Choice

3–3 114–117	paper/pencil protractor ruler	Warm up 7	Visual Learner Challenge	B: 1–29, 36–45 E: 1–45	Page 122–123, 133, 140, 692	R: page 39 EP: page 35 E: page 47	SS: Teacher's Choice
3–4 118–121	paper/pencil graph paper compass protractor straightedge	Warm up 8 Trans RF-13	ESL/LEP Challenge	B: 1–14, 19, 21–22 E: 1–22	Page 122–123, 133, 140, 693	R: page 41 EP: page 37 E: page 47	MC: 6
3–5 124–127	paper/pencil straightedges	Warm up 8	Visual Learner Challenge	B: 1–11, 15–24 E: 1–24	Page 132–133, 141, 693	R: page 43 EP: page 39 E: page 51	
3–6 128–131	paper/pencil	Warm up 8		B: 1–26, 31, 32 E: 1–32	Page 132–133, 141, 694	R: page 45 EP: page 41 E: page 53	
3–7 134–137	paper/pencil compass straightedge	Warm up 9	ESL/LEP	B: 1–6, 10–30 E: 1–30	Page 141, 694	R: page 47 EP: page 43 E: page 55	
3–8 138–139	paper/pencil	Warm up 9	Challenge	B: 1–14 E: 1–14	Page 141	R: page 49 E: page 57	

Planning and Pacing

Lesson, Pages	Lesson Title	45 min class	Assignments Basic, Enriched	90 min class	Assignments Basic, Enriched	___ min class	Assignments Basic, Enriched
3–1 104–107	AYR, Opener, Points, Lines and Planes	Day 31	B: 1–18, 21–34 E: 1–34	Day 16	B: 1–18, 21–34 E: 1–34		B: 1–18, 21–34 E: 1–34
3–2 108–111	Types of Angles, R&PYS	Day 32–33	B: 1–25, 29–42 E: 1–42	Day 16–17	B: 1–25, 29–42 E: 1–42		B: 1–25, 29–42 E: 1–42
3–3 114–117	Segments and Angles	Day 34	B: 1–29, 36–45 E: 1–45	Day 17	B: 1–29, 36–45 E: 1–45		B: 1–29, 36–45 E: 1–45
3–4 118–121	Constructions, Perpendicular and Parallel Lines, R&PYS	Day 35–36	B: 1–14, 19, 21–22 E: 1–22	Day 18	B: 1–14, 19, 21–22 E: 1–22		B: 1–14, 19, 21–22 E: 1–22
3–5 124–127	Inductive Reasoning in Mathematics	Day 37	B: 1–11, 15–24 E: 1–24	Day 19	B: 1–11, 15–24 E: 1–24		B: 1–11, 15–24 E: 1–24
3–6 128–131	Conditional Statements, R&PYS	Day 38–39	B: 1–26, 31, 32 E: 1–32	Day 19–20	B: 1–26, 31, 32 E: 1–32		B: 1–26, 31, 32 E: 1–32
3–7 134–137	Deductive Reasoning and Proof	Day 40	B: 1–6, 10–30 E: 1–30	Day 20	B: 1–6, 10–30 E: 1–30		B: 1–6, 10–30 E: 1–30
3–8 138–139	Problem Solving Skills: Use Logical Reasoning, R&PYS	Day 41	B: 1–14 E: 1–14	Day 21	B: 1–14 E: 1–14		B: 1–14 E: 1–14
Review/ Assess		Day 42–44		Day 21–22			

	Chapter 1 Days	Chapter Cumulative Days
Yearly Pacing (45 min class)	14 days	44 days
Yearly Pacing (90 min class)	7 days	22 days

Assessment Options

Assessment in Student Edition	Assessment in Teacher's Edition	Pages in Assessment Book	Software Generated Assessment
Are You Ready?, pages 100–101; Writing Math, pages 106, 110, 111, 117, 121, 127, 131, 136; Mixed Review, pages 107, 111, 117, 121, 127, 131, 137; Check Understanding pages 104, 108, 114, 118, 124, 128, 134; Mid-Chapter Quiz, page 123, Chapter Review, page 140; Chapter Assessment, page 142; Alternative Assessment, page 143; Cumulative Review, page 144; Cumulative Assessment, page 145	5-minute Warm ups, pages 104, 108, 114, 118, 124, 128, 134; Quick Assessment, pages 106, 110, 116, 120, 126, 130, 137, 139; Scoring Rubrics, page 143	Mid-Chapter Quiz, page 39; Test Form A, pages 41, 43; Test Form B, pages 45, 47; Cumulative Test 49, 51; Math Journal prompt, page 53, 54;	Chapter 3

Refresher Skills

The skills on these two pages are skills that have been taught in previous math courses. Continuous review of basic math skills will make stronger math students. These skills are identified as necessary to be successful in Chapter 3.

Skills Correlation Chart

Skill	Lesson Number
Identifying Angles	3-2, 3-3, 3-4, 3-6, 3-7
Measuring and Drawing Angles	3-2, 3-3, 3-4
Logical Reasoning	All lessons

Vocabulary

perpendicular lines
adjacent angles
vertical angles
acute angle
protractor
deductive reasoning
conclusion
right angle
obtuse angle
premise

Additional Answers

16.
17.
18.
19.

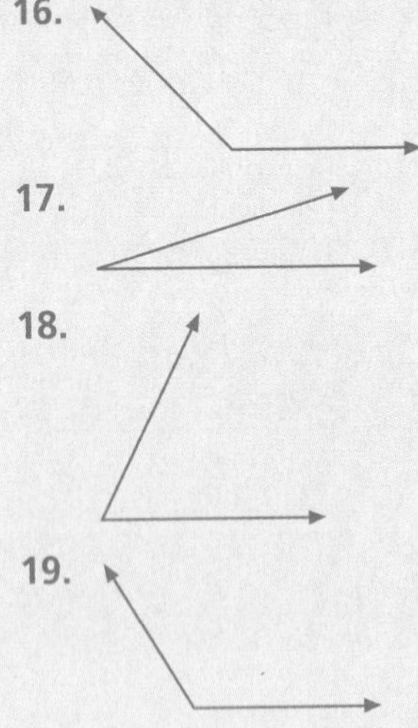

Are You Ready?

Refresh Your Math Skills for Chapter 3

The skills on these two pages are ones you have already learned. Use the examples to jog your memory and complete the exercises. For additional practice on these and more basic skills, see page 674.

As you look forward to learning more about geometry, it might be helpful to refresh your memory about lines and angles.

Identifying Angles

Examples Perpendicular lines are two lines that intersect to form adjacent right angles.

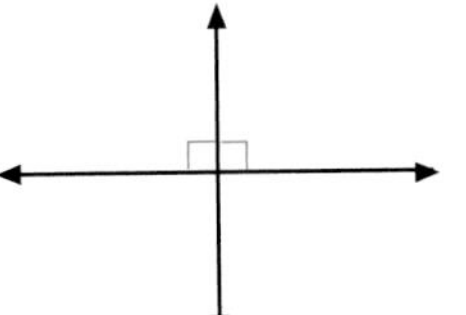

Vertical angles are angles that are not adjacent to each other when two lines intersect. Vertical angles are congruent.

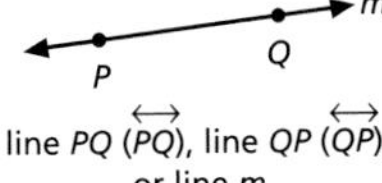

Use the figure at the right to name the following:

1. all pairs of perpendicular lines $\overleftrightarrow{LP}$ and $\overleftrightarrow{RN}$
2. all right angles $\angle ROL$, $\angle ROP$, $\angle LON$, $\angle NOP$
3. all pairs of vertical angles that are not right angles $\angle LMO$ and $\angle QOP$, $\angle MON$ and $\angle ROQ$

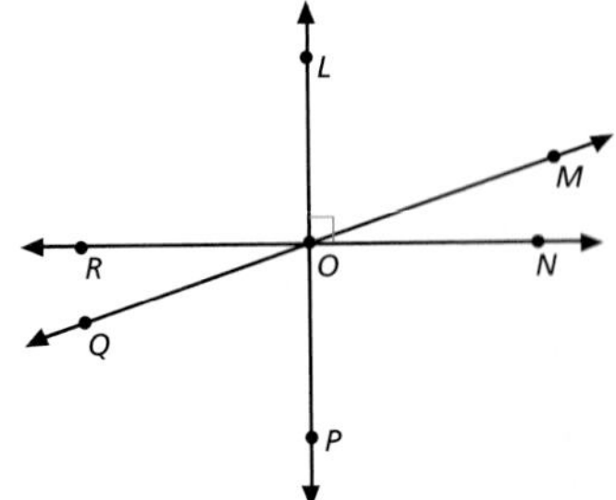

Identify each angle as right, acute or obtuse.

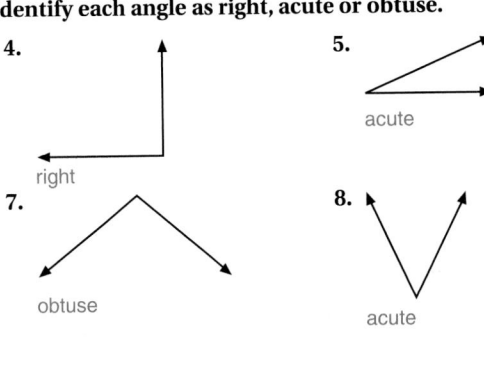

4. right
5. acute
6. obtuse

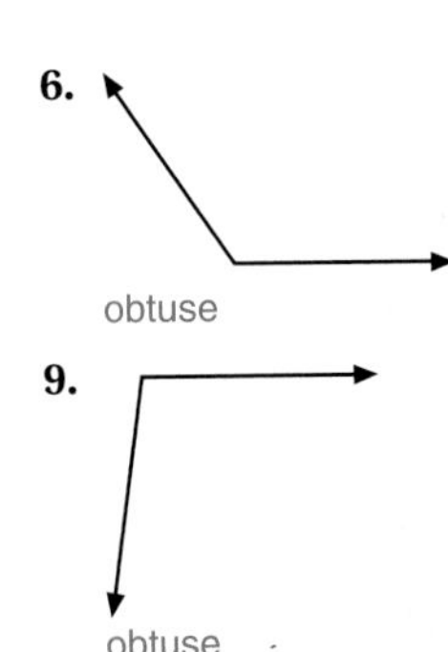

7. obtuse
8. acute
9. obtuse

Teaching Strategies

COOPERATIVE LEARNING Call out an angle measurement and have students using a straightedge only attempt to draw the angle. Then have them use a protractor to check their guess. You may want to make this activity into a game. To score the game, students receive a number of points equal to the difference between their guess and the given angle measurement. Low score wins after five turns.

CHAPTER 3

Measuring and Drawing Angles

Use a protractor to measure each angle.

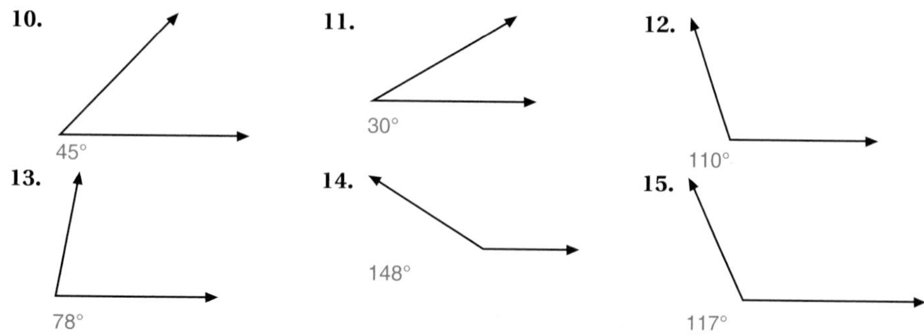

Use a protractor to draw an angle with the given measure.
For 16–19, check students' drawings.

16. 135° **17.** 15° **18.** 63° **19.** 122°

Logical Reasoning

For each set of premises, write the conclusion that follows by deductive reasoning from the premises. If no conclusion is possible, write *none*.

20. If an animal is a horse, it has a tail.
This animal is a horse. This animal has a tail.

21. A person who wears white gloves to tea has impeccable manners.
Miss Jones wears white gloves to tea. Miss Jones has impeccable manners.

22. All parakeets sing with a sweet voice.
This bird has a lousy voice. This bird is not a parakeet.

23. Flowers that are red attract hummingbirds.
Daffodils are yellow. Daffodils do not attract hummingbirds.

24. A red sports car is faster than a green sports car.
My car is blue. none

25. Michael is taller than Scotty.
Kareem is shorter than Scotty. Michael is taller than Kareem.

26. Every day at 5 PM, my dogs expect their evening meal.
It is 3:47 PM right now. My dogs do not expect to be fed right now.

Chalkboard Examples

Identifying Angles

Have students find two examples each of right angles, acute angles, and obtuse angles in the classroom. Answers will vary.

Measuring and Drawing Angles

Have students draw a large uppercase K on a sheet of paper and measure the three angles created with a protractor. Ask: What other upper case letters contain obtuse angles? X and Y

Logical Reasoning

Pat, Nita, and Cathy had a race. Cathy did not finish third. Nita finished either first or third. Pat did not finish first or second. Who finished first? Nita

Refresher Wrap-up

Quick Assessment

Ask the following questions to determine if students have mastered the basic skills reviewed on these pages.

1. Draw two adjacent right angles. Check students' work.
2. Describe how to use a protractor to measure an obtuse angle. Answers will vary. Students should include how to position the protractor on the vertex and how to read the scale.

Additional Practice

All topics: Refer to the Extra Practice Index on page 674 of this text.

Extend the Lesson

MATH JOURNAL Have students describe one real-life situation in which they used logical reasoning to solve a problem. For example, you have lost your sunglasses, and you can think of three places you have been since you last remember having them. If there are only three places they can be and you have checked two of them, the sunglasses must be at the third place.

Chapter Opener

NCTM Standards/Strands

- Number & Operation
- Patterns, Functions, & Algebra
- Geometry & Spatial Sense
- Measurement
- Problem Solving
- Reasoning and Proof
- Communication
- Connections
- Representation

Vocabulary

geography
latitude
longitude

Theme Connections

Tables and graphs are often used to organize and display geographical data. In this activity, a table is used to show the latitude and longitude of 12 world cities. Explain that mapmakers use not only degrees but also fractions of degrees, called minutes, and fractions of minutes, called seconds. Discuss the symbols used to show degrees (°), minutes (′) and seconds (″).

Career Opportunities

Many careers related to geography require the use of data and graphs. Two are highlighted in the MathWorks features. Others include pilot, navigator, geologist, surveyor, and astronaut.

- Cross-country bus driver, page 113
- Cartographer, page 133

Internet Connection

Theme Activities

Learningmathmatters.com provides web addresses to search that help students gather information about the use of math in the real world, particularly data and graphs. To search for additional addresses, begin a search using the keyword *geography*. Then within that search, use such key words as *navigation*, *mapmaking*, and *atlas*. Students can brainstorm in small groups other key words.

CHAPTER 3

Geometry and Reasoning

THEME: Geography

Have you ever wondered how the early explorers found their way on their journeys? Navigators used their understanding of planetary objects and angle measurements to calculate how many degrees north or south their location was from the equator. Today, aircraft pilots and astronauts use these same ideas to identify positions in air and in space.

To locate a point on the Earth's surface, navigators lay an imaginary grid on the Earth's surface. The latitude of a point is given as the number of degrees (°) north or south of the equator. The longitude of a point is given as the number of degrees east or west of the prime meridian. For more accurate measurements, degrees are divided into minutes (′) which are divided into seconds (″).

- **Cross-country bus drivers** (page 113) read maps, plan routes and make calculations to conserve fuel and stay on schedule.
- **Cartographers** (page 133) make maps. They must be able to interpret data from satellites and computers and translate actual distance to map distances. They use a knowledge of geometry to draw three-dimensional surfaces on a flat plane.

Internet Connection
www.learningmathmatters.com

Chapter Investigation

Go to Learningmathmatters.com to locate additional information about newsmedia.

Latitude and Longitude of World Cities

City	Latitude	Longitude
Washington, D.C.	38° 54′ 18″ N	77° 00′ 58″ W
Sydney, Australia	33° 55′ 00″ S	151° 17′ 00″ E
Rio de Janeiro, Brazil	22° 27′ 00″ S	42° 43′ 01″ W
Greenwich, England	51° 28′ 00″ N	0° 00′ 00″
Athens, Greece	38° 01′ 36″ N	23° 44′ 00″ E
Honolulu, Hawaii	21° 19′ 02″ N	157° 48′ 15″ W
Johannesburg, South Africa	26° 08′ 00″ S	27° 54′ 00″ E
Beijing, China	39° 55′ 00″ N	116° 23′ 00″ E
Salt Lake City, Utah	40° 46′ 38″ N	111° 55′ 48″ W
Moscow, Russia	55° 45′ 00″ N	37° 37′ 00″ E
Tokyo, Japan	35° 41′ 00″ N	139° 44′ 00″ E
Panama City, Panama	9° 04′ 01″ N	79° 22′ 59″ W

Data Activity: Latitude and Longitude of World Cities

Use the table for Questions 1–6.

1. The prime meridian passes through the poles and Greenwich, England. Name one city that lies east of Greenwich. Sydney, Athens, Johannesburg, Beijing, Moscow or Tokyo
2. Which city from the table lies south of the equator and west of the prime meridian? Rio de Janeiro
3. What is the latitude of a point that lies on the equator? 0° 00′ 00″
4. Which is farther north: Washington, D.C. or Athens, Greece? How much farther? (*Hint:* There are 60 seconds in a minute and 60 minutes in a degree.) Washington, D.C.; 0° 52′ 42″
5. *True or False*: All the "lines" of longitude are circles of the same size, and they intersect in just two points. True; explanations will vary.
6. *True or False*: All the "lines" of latitude are circles of the same size parallel to the equator. False, explanations will vary.

Chapter Investigation

Navigators use grid lines and angles to plot a course from one location to another. By drawing a ray due north from a destination, navigators can determine the bearing, or angle, they need to travel to arrive at the new destination.

Working Together

Draw a map of your neighborhood showing at least ten specific points of interest. Draw and label latitude and longitude lines on the map. then plot a course from one location to another. Create a navigator's log in which you specify the direction and bearing for each leg of the journey.

Project Planning Calendar

Name ________ Date ________

CHAPTER 3 PROJECT PLANNING CALENDAR

Benchmarks	PROJECT GOAL
a. Draw a map of your community showing at least ten specific points of interest. Make distances as accurate as possible. Overlay your map with a grid of latitude (east-west) and longitude (north-south) lines. *(Lesson 3-2)* b. Using the map you made, sketch a course to travel between two points. Write directions in which you specify the bearing of each leg of the journey. *(Lesson 3-4)* c. Exercises 7–9 on page 137 state three important theorems related to perpendicular and parallel lines. Find an example of the situation in the theorems on your map. Then write a proof of ... of your map.	To create a map of your neighborhood and a navigator's log for a course shown on the map.

Group Project Planner

Name ________ Date ________

CHAPTER 3 GROUP PROJECT PLANNER

Assignment ________ Objective ________

Group Members / Assigned Roles
1) ________
2) ________
3) ________
4) ________
5) ________

Deadlines Done

Data Activity

Discuss with students how longitude and latitude are used to specify an exact location. Discuss how minutes and seconds are used to refine the measurements. Ask students whether or not lines of longitude and latitude intersect at right angles. Students should realize that although lines of latitude are parallel to one another, lines of longitude are not. Assign students Questions 1–6.

Chapter Investigation

As an Overarching Problem

Introduce this chapter investigation by discussing a map you display at the front of your classroom. Point out the lines of latitude and longitude on the maps. Ask students to choose two points on one of the maps and describe a course they could use to travel between the two points. Students will continue to work on the investigation as they complete the exercises identified by the Investigation Logo that are found throughout the chapter. These exercises will guide students through the task described in *Working Together*. Encourage students to keep all of their work on the Investigation together. Have students use the suggestions in the Chapter Investigation Extension to summarize their work.

As a Chapter Project

The goal of this project is for students to create a map of their neighborhood and a navigator's log for a course shown on the map. Students can use the Team Project Planner on page 41 and the Project Planning Calendar on page 42 in the Enrichment text to complete the project. Benchmarks **a, b,** and **c** should be completed after the lesson listed in parentheses has been studied. Benchmark **d** should be completed at the end of the chapter.

Lesson Planning

NCTM Standards/Strands

- Number & Operation
- Patterns, Functions, & Algebra
- Geometry & Spatial Sense
- Measurement
- Problem Solving
- Reasoning and Proof
- Communication
- Connections
- Representation

Vocabulary

point	line
plane	space
collinear points	coplanar points
intersection	postulate
ray	endpoint
line segment	

Materials Needed

paper/pencil
small flat objects (counters or algebra tiles)

Lesson Resources

Warm-Up Transparency 7
Transparency RF-9, 10
Reteaching 3-1
Extra Practice 3-1
Enrichment 3-1

Assignment Guide

Basic: 1–18, 21–34
Enriched: 1–34

Getting Started

5-Minute Warm-Up

Find the absolute value.

1. $|-12|$ 12
2. $|3|$ 3
3. $|4-9|$ 5
4. $|5-(-3)|$ 8
5. $|-3-4|$ 7

Introduction to Lesson 3-1

This activity gives students a concrete experience they can use to see that two points in a plane determine a line but that three points may or may not lie on the same line. The activity can be done on the floor or on top of a deck. Students may use coins as well as the other objects suggested.

3-1 Points, Lines, and Planes

Goals
- Define basic terms.
- Apply postulates about points, lines, and planes.

Applications Geography and Carpentry

Work in groups of 2 or 3 students.

You will need three small, flat objects that will not roll when dropped.

1. Drop two of the objects onto a flat surface. Determine whether you can place a yardstick on the surface so that it touches both objects. Repeat the experiment ten times and record the result. It will always be possible to place a yardstick so that it touches both objects.
2. Perform the experiment using three objects. Repeat ten times.
3. Compare the results from the two experiments. What generalization can you make from your results? Answers will vary.

2. Sometimes it will be possible to place a yardstick so that it touches all three objects, but most often this will not be possible.

Build Understanding

In geometry, *point, line,* and *plane* are intuitive ideas. It is impossible to give a precise definition of these words, so they are *undefined terms.*

A **point** indicates a specific location. Although you use a dot to picture a point, it actually has no dimensions.

point R

A **line** is a set of points that extends without end in two opposite directions. When you picture a line, it appears to have some thickness, but actually it has none.

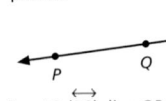

line PQ ($\overleftrightarrow{PQ}$), line QP ($\overleftrightarrow{QP}$), or line m

A **plane** is a set of points that extends without end in all directions along a flat surface. Like a line, it has no thickness. Although a plane has no edges, you usually picture a plane as a four-sided shape.

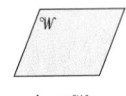

plane $\mathcal{W}$

Using these undefined terms as a base, it is now possible to develop *definitions* of other geometric terms. For instance, a figure is defined as any set of points. The set of all possible points is called **space.**

Collinear points are points that lie on the same line. Points that do not lie on the same line are called *noncollinear* points.

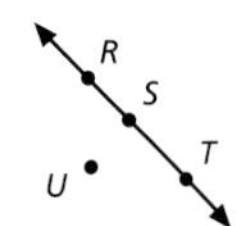

R, S, and T are collinear.
R, S, and U are noncollinear.

Coplanar points are points that lie in the same plane. Points that do not lie in the same plane are called *noncoplanar* points.

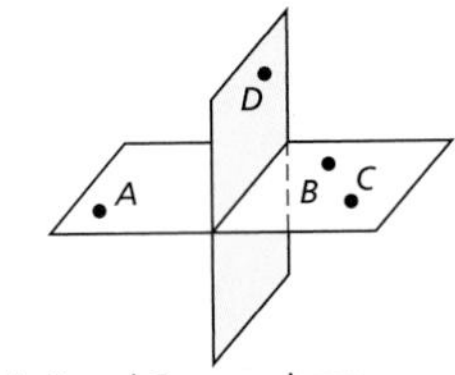

A, B, and C are coplanar.
A, B, C, and D are noncoplanar

Learning Styles

VISUAL LEARNER If students have difficulty reading the diagram in Example 1, have them cut a slit in one sheet of paper and place another sheet in the slit so that the construction looks like the intersecting planes in the drawing.

The **intersection** of two geometric figures is the set of all points common to both figures.

Just as the meanings of some words are accepted without definition, **postulates** are accepted as true without proof.

Postulate 1	**The Unique Line Postulate** Through any two points, there is exactly one line. (This is sometimes stated: *Two points determine a line.*)
Postulate 2	**The Unique Plane Postulate** Through any three noncollinear points, there is exactly one plane. (*Three noncollinear points determine a plane.*)
Postulate 3	If two points lie in a plane, then the line joining them lies in that plane.
Postulate 4	If two planes intersect, then their intersection is a line.

Example 1

Which postulate justifies the answer to each question?

a. Name three points that determine plane *J*.

b. Name the intersection of planes *J* and *K*.

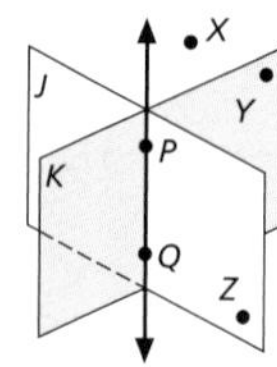

Solution

a. According to Postulate 2, three noncollinear points determine a plane. Three points that determine plane *J* are points *P*, *Q*, and *Z*.

b. According to Postulate 4, the intersection of two planes is a line. Planes *J* and *K* intersect in line *PQ* ($\overleftrightarrow{PQ}$).

A **ray** is part of a line that begins at one point, called the **endpoint**, and extends without end in one direction.

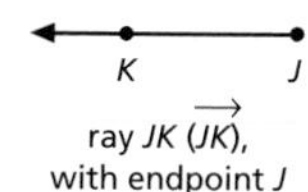

ray *JK* ($\overrightarrow{JK}$), with endpoint *J*

A **line segment,** more simply called a **segment,** is part of a line that begins at one endpoint and ends at another.

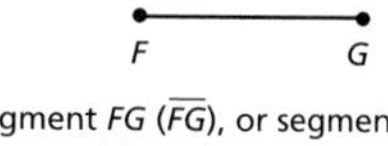

segment *FG* ($\overline{FG}$), or segment *GF* ($\overline{GF}$), with endpoints *F* and *G*

Reading Math

Whenever you use a ruler to measure a length, you are using the ruler postulate. The points paired with 0 and 1 on your ruler depend on the unit of measure you are using—inches, centimeters, and the like.

The *ruler postulate* is a basic assumption about segments.

Postulate 5	**The Ruler Postulate** The points on any line can be paired with the real numbers in such a way that any point can be paired with 0 and any other point can be paired with 1. The real number paired with each point is the **coordinate** of that point. The **distance** between any two points on the line is equal to the absolute value of the difference of their coordinates.

Extend the Lesson

MATH JOURNAL Have students make a list of all the terms presented in this lesson in their journals. They should write a brief definition or draw a diagram for each.

Chalkboard Examples

Supplementary Example 1

Draw the figure below on the board.

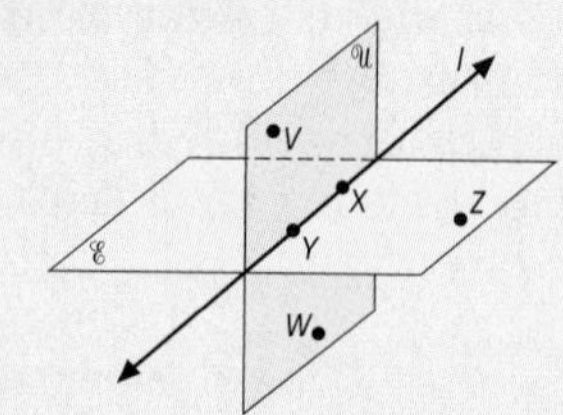

1. Name three points that determine plane *U*. points *V, X, W*
2. Name the intersection of planes *A* and *E*. line *XY*

Supplementary Example 2

Using the number line for Example 2, find the length of $\overline{PR}$. 5

Supplementary Example 3

In the figure below, *DF* = 32. Find *EF*. 18

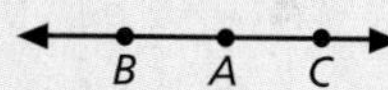

Reteaching Worksheet 3-1

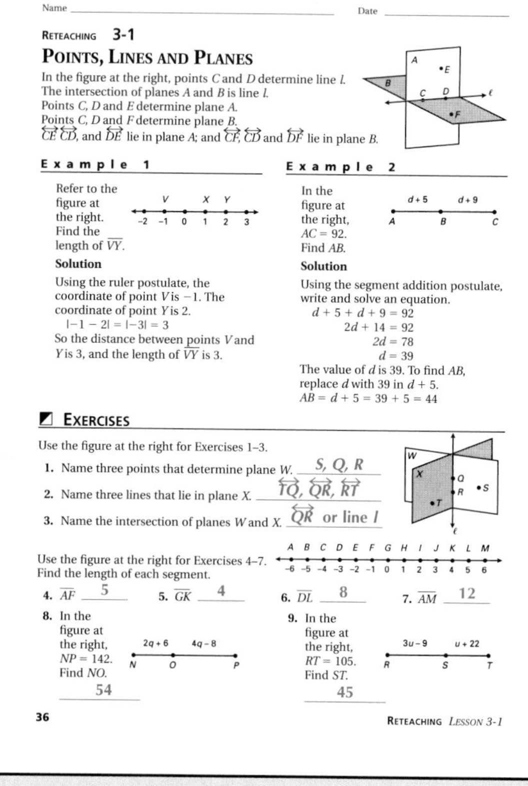

Name ______________________ Date ____________

RETEACHING 3-1

POINTS, LINES AND PLANES

In the figure at the right, points *C* and *D* determine line *l*.
The intersection of planes *A* and *B* is line *l*.
Points *C*, *D* and *E* determine plane *A*.
Points *C*, *D* and *F* determine plane *B*.
$\overleftrightarrow{CE}$ $\overleftrightarrow{CD}$, and $\overleftrightarrow{DE}$ lie in plane *A*; and $\overleftrightarrow{CF}$, $\overleftrightarrow{CD}$ and $\overleftrightarrow{DF}$ lie in plane *B*.

Example 1

Refer to the figure at the right. Find the length of $\overline{VY}$.

Solution

Using the ruler postulate, the coordinate of point *V* is −1. The coordinate of point *Y* is 2.

$|-1 - 2| = |-3| = 3$

So the distance between points *V* and *Y* is 3, and the length of $\overline{VY}$ is 3.

Example 2

In the figure at the right, *AC* = 92. Find *AB*.

Solution

Using the segment addition postulate, write and solve an equation.

$d + 5 + d + 9 = 92$
$2d + 14 = 92$
$2d = 78$
$d = 39$

The value of *d* is 39. To find *AB*, replace *d* with 39 in *d* + 5.
$AB = d + 5 = 39 + 5 = 44$

EXERCISES

Use the figure at the right for Exercises 1–3.

1. Name three points that determine plane *W*. S, Q, R
2. Name three lines that lie in plane *X*. $\overleftrightarrow{TQ}$, $\overleftrightarrow{QR}$, $\overleftrightarrow{RT}$
3. Name the intersection of planes *W* and *X*. $\overleftrightarrow{QR}$ or line *l*

Use the figure at the right for Exercises 4–7. Find the length of each segment.

4. $\overline{AF}$ 5
5. $\overline{GK}$ 4
6. $\overline{DL}$ 8
7. $\overline{AM}$ 12

8. In the figure at the right, *NP* = 142. Find *NO*. 54
9. In the figure at the right, *RT* = 105. Find *ST*. 45

36 RETEACHING *LESSON 3-1*

Lesson Wrap-up

Quick Assessment

Ask the following questions to determine if students understand the content presented in this lesson.

Draw each of the following and label it using the correct symbols. Check students' drawings.

1. line *AB* $\overleftrightarrow{AB}$
2. ray *CD* $\overrightarrow{CD}$
3. segment *EF* $\overline{EF}$

Assignment Guide

Basic: 1–18, 21–34
Advanced: 1–34
Additional Practice: See Extra Practice Index on page 674.

Additional Answers

1. There are three possible answers: $\overline{XY}$ (or $\overline{YX}$), $\overline{XZ}$ (or $\overline{ZX}$), $\overline{RV}$ (or $\overline{VR}$), and $\overline{YZ}$ (or $\overline{ZY}$). Each answer is justified by Postulate 3.
4. The notation $\overline{JL}$ represents "line segment JL." The symbol with no overbar, *JL*, means the length of JL.
5. points *R* and *S*; Postulate 1.
6. There are four possible answers: points *U*, *R*, and *S*; points *U*, *R*, and *T*; points *U*, *S*, and *T*; and points *R*, *S*, and *T*. each answer is justified by Postulate 2.
7. $\overline{RS}$ (or $\overline{SR}$); Postulate 4.
8. $\overline{RS}$ (or $\overline{SR}$), $\overline{RV}$ (or $\overline{VR}$), and $\overline{SV}$ (or $\overline{VS}$); Postulate 3.
20. Think of the end of each leg as a point. When a table has four legs, the ends of the legs represent four points. If the legs are of different lengths, the four points are noncoplanar, and the table wobbles. When a table has three legs, the ends of the legs represent three points. Since Postulate 2 guarantees that any three points are coplanar, the length of the legs does not matter, and so the table does not wobble.

Example 2

Using the number line, find the length of QS.

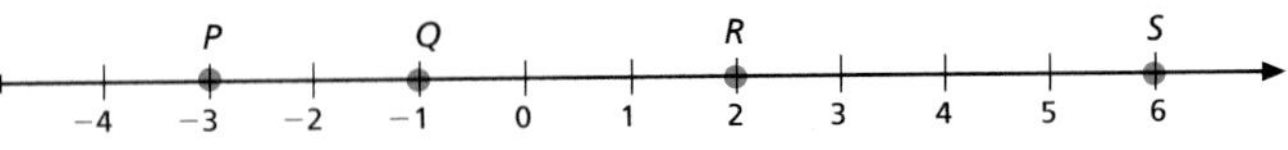

Solution

The coordinate of point Q is -1. The coordinate of point S is 6.

$|-1-6| = |-7| = 7$ or $|6-(-1)| = |7| = 7$

So, the distance between points Q and S is 7. This means that the length of $\overline{QS}$ is 7 or $QS = 7$.

A point B is **between** points A and C if and only if the coordinate of B is between the coordinates of A and C. This leads to the *segment addition postulate.*

Postulate 6 — **The Segment Addition Postulate** If point B is between points A and C, then $AB + BC = AC$.

Example 3

In the figure at the right, $AC = 47$. Find AB.

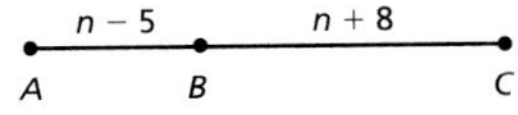

Solution

From the figure, $AB = n - 5$ and $BC = n + 8$. You are given $AC = 47$. Use the segment addition postulate to write and solve an equation.

$$AB + BC = AC$$
$$n - 5 + n + 8 = 47 \quad \text{Combine like terms.}$$
$$2n + 3 = 47 \quad \text{Add } -3 \text{ to each side.}$$
$$2n = 44 \quad \text{Multiply each side by } \tfrac{1}{2}.$$
$$n = 22$$

The value of n is 22. To find AB, replace n with 22 in $n - 5$.

$$AB = n - 5 = 22 - 5 = 17$$

Technology Note

Geometric software can help you draw geometric figures quickly and easily. Most software gives you a quick reading of the length of a segment. To learn how to use this feature on your software, try this activity.

1. Draw a segment. Label its endpoints *A* and *B*.
2. Find *AB*, the length of the segment. (The length should appear on the screen.)
3. Locate a point *C* between *A* and *B*.
4. Find *AC* and *CB*.
5. Calculate *AC* + *CB*.
6. Change the location of point *C* and repeat Steps 4 and 5.

Try These Exercises

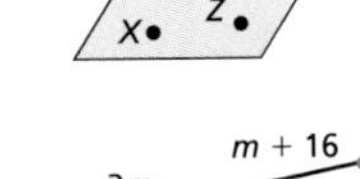

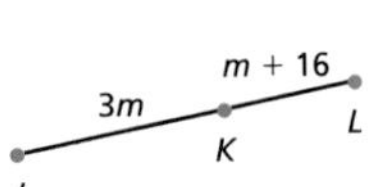

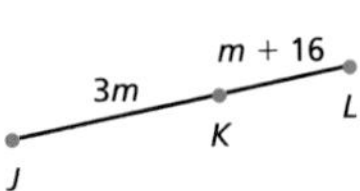

1. Refer to the figure at the right. Name a line that lies in plane G. Which postulate justifies your answer? See additional answers.
2. Refer to the number line in Example 2 above. Find PR. 5
3. In the figure at the right, $JL = 88$. Find KL. 34

4. **WRITING MATH** The length of $\overline{JL}$ is 88, or $JL = 88$. What is the difference in meaning between the notations $\overline{JL}$ and JL? See additional answers.

21. 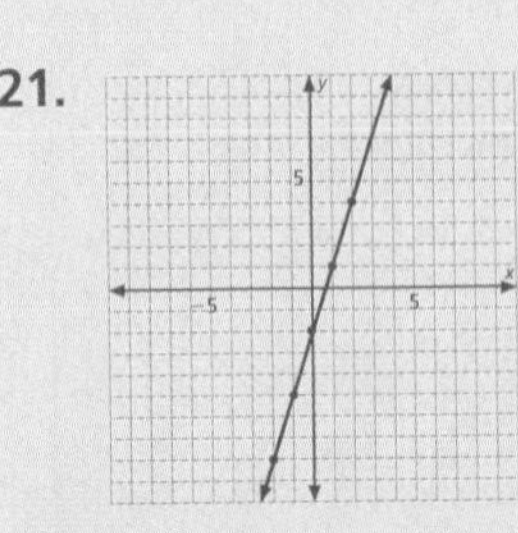

22. 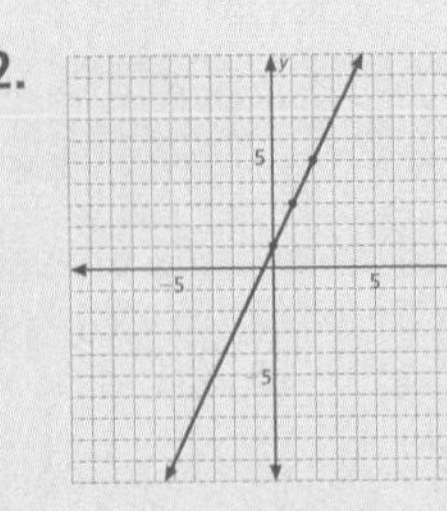

23. 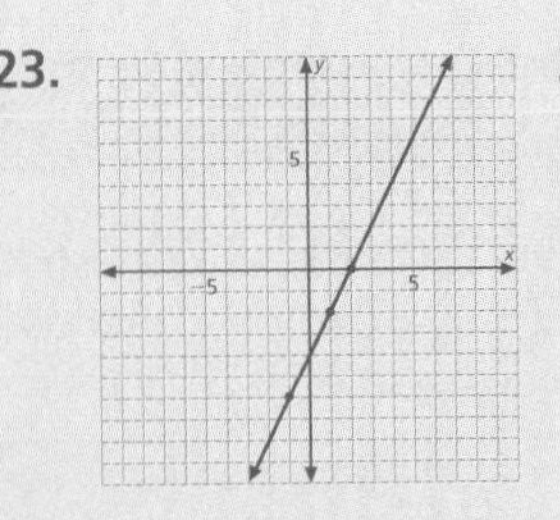

24. 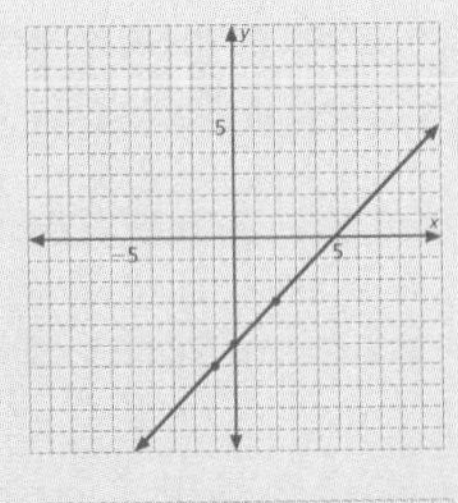

25. 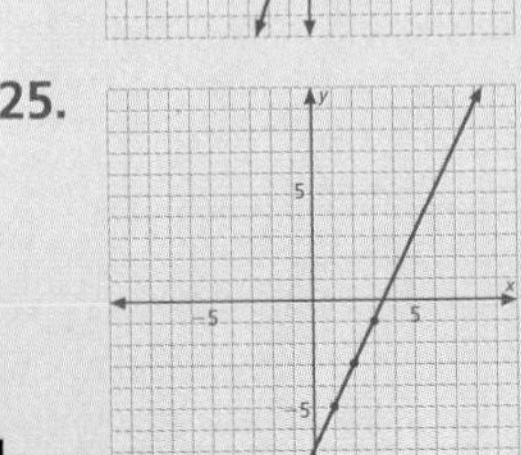

26. 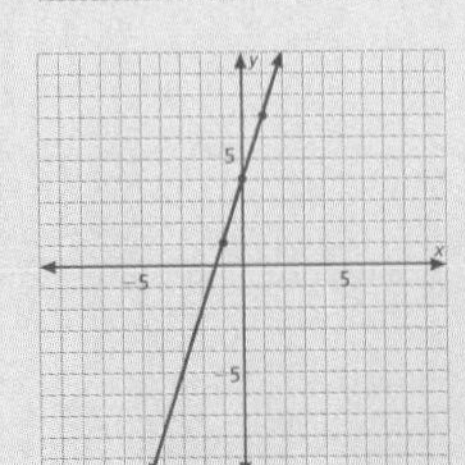

27. 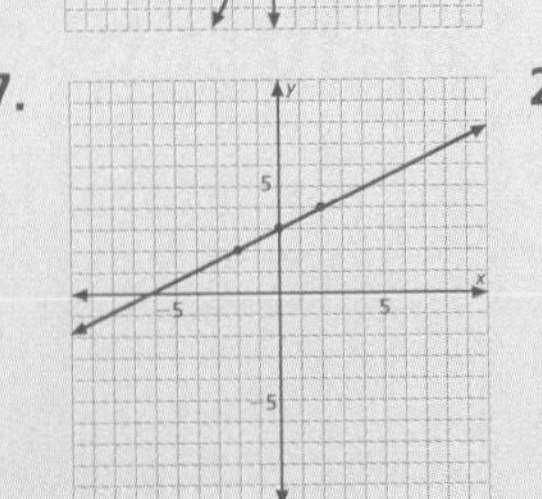

28. 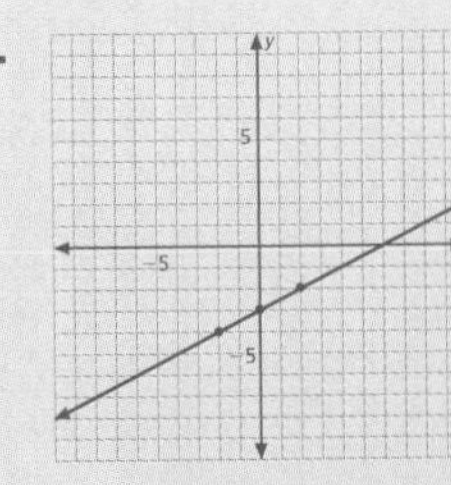

106 Chapter 3 **Geometry and Reasoning**

Practice Exercises

Use the figure at the right for Exercises 5–8. Which postulate justifies your answer?
For 5–8, see additional answers.

5. Name two points that determine line ℓ.

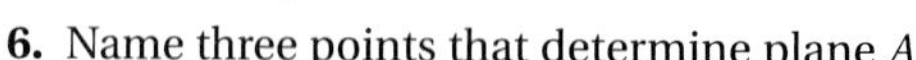

6. Name three points that determine plane A.
7. Name the intersection of planes A and B.
8. Name three lines that lie in plane B.

Use the number line at the right for Exercises 9–12. Find each length.

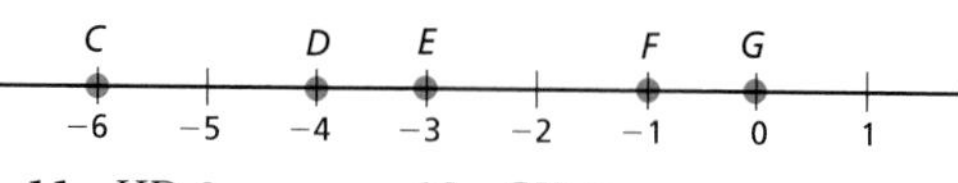

9. CF 5
10. GE 3
11. HD 6
12. GH 2

13. In the figure below, $MP = 104$. Find NP. 56

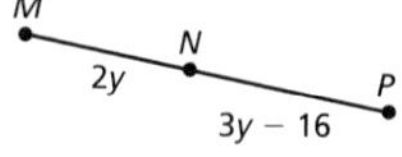

14. In the figure below, $XZ = 61$. Find YX. 33.5

15. On a number line, the coordinate of point S is -8. The length of ST is 17. Give two possible coordinates for point T. -25 and 9
16. On a number line, the distance between points V and W is 39. The coordinate of point W is -3.25. Give two possible coordinates for point V. -42.25 and 35.75
17. **GEOGRAPHY** On a map, three cities, represented by points N, P, and Q, lie on a straight line, and N lies between P and Q. The distance from N to P is twice the distance from N to Q. The actual distance between P and Q is 51 miles. Find PN and NQ. $PN = 34$ and $NQ = 17$
18. Point Z is between points J and K. The length of $\overline{KZ}$ is two less than three times the length $\overline{JZ}$, and $JK = 18$. Find JZ and KZ. $JZ = 5$; $KZ = 13$

Extended Practice Exercises

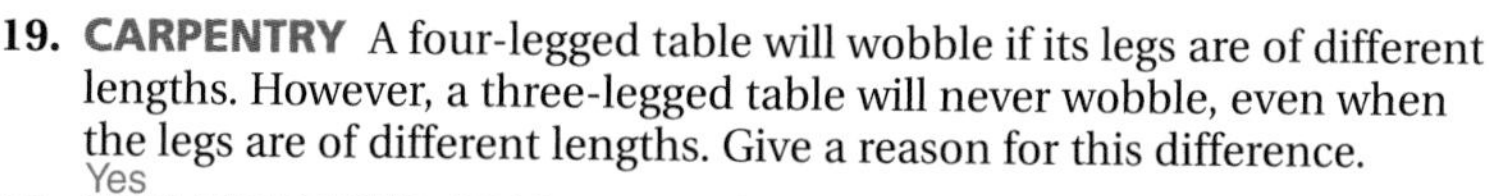

19. **CARPENTRY** A four-legged table will wobble if its legs are of different lengths. However, a three-legged table will never wobble, even when the legs are of different lengths. Give a reason for this difference.

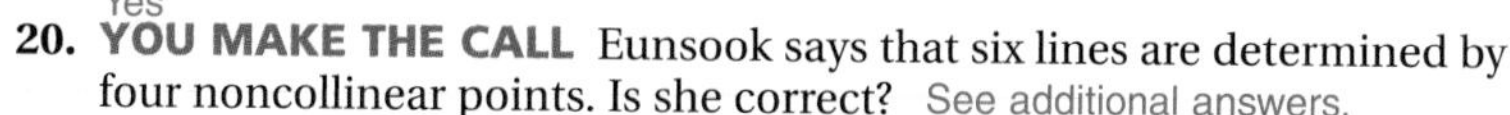

20. **YOU MAKE THE CALL** Eunsook says that six lines are determined by four noncollinear points. Is she correct? Yes See additional answers.

Mixed Review Exercises

Graph each function on the coordinate plane. (Lesson 2-3)
For 21–28, see additional answers.

21. $y = 3x - 2$
22. $y = 2x + 1$
23. $y = 2(x - 2)$
24. $y = x - 5$
25. $y = 2x - 7$
26. $y = 3x + 4$
27. $y = \left(\frac{1}{2}\right)x + 3$
28. $y = \left(\frac{1}{2}\right)x - 3$

Solve each equation. (Lesson 2-5)

29. $4m - 6 = -14$ -2
30. $3(d + 1) = 2d - 2$ -5
31. $-3a + 4 = 16$ -4
32. $-2(4p + 2) = 8$ $-\frac{3}{2}$
33. $-16 = 5x + 4$ -4
34. $3 - 2n = 5n + 4$ $\frac{-1}{7}$

Extend the Lesson

CHALLENGE How many lines are determined by three noncollinear points? 3
How many lines are determined by four noncollinear points? 6
After students have answered these questions, have them look for a pattern or formula that relates the maximum number of lines that can be formed from a given number of noncollinear points. The pattern is the same as that for triangular numbers, 1, 3, 6, 10, 15 . . . A formula that can be used is $l = \frac{p \cdot (p - 1)}{2}$, where l is the number of lines, and p is the number of points.

Extra Practice Worksheet 3-1

Name ______ Date ______

EXTRA PRACTICE 3-1
POINTS, LINES AND PLANES

EXERCISES

Use the figure at the right for Exercises 1–5. Which postulate justifies your answer?

1. Name two points that determine line k. points S and T; Postulate 1
2. Name three points that determine plane X. Possible answer: points M, N, V; Postulate 2
3. Name three points that determine plane Y. Possible answer: points P, Q, R; Postulate 2
4. Name the intersection of planes X and Y. line k; Postulate 4
5. Name three lines that line in plane Y. Possible answer: lines k, PS, TR; Postulate 3

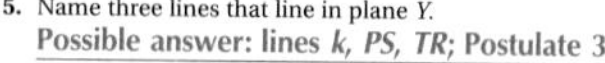

Use the number line below for Exercises 6–13. Find each length.

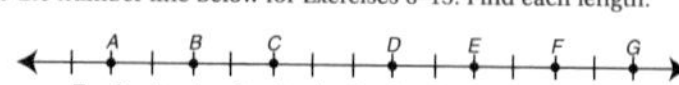

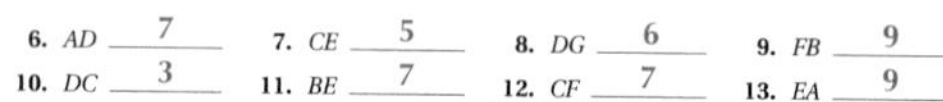

6. AD 7
7. CE 5
8. DG 6
9. FB 9
10. DC 3
11. BE 7
12. CF 7
13. EA 9

14. In the figure at the right, $RT = 52$. Find RS. 31

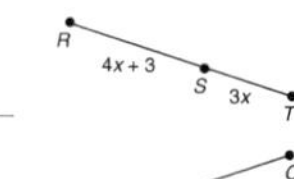

15. In the figure at the right, $AC = 86$. Find BC. 46

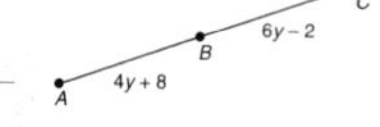

16. On a number line, the coordinate of point Q is -6. The length of PQ is 15. Give two possible coordinates for point P. -21; 9

32 EXTRA PRACTICE LESSON 3-1

Enrichment Worksheet 3-1

Name ______ Date ______

ENRICHMENT 3-1
TETRABOLOES

The polygons at the right are called tetraboloes. Each one is made of four congruent isosceles right triangles. There are 14 different tetraboloes. Reflections and rotations do not count as different figures.

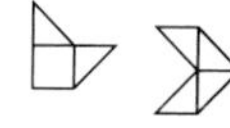

EXERCISES

1. Find the other 12 tetraboloes. Sketch all 14 figures below.

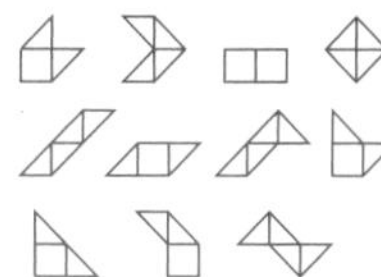

2. Prove that it is impossible to use all 14 tetraboloes to make a square. Possible answer: If the squares are 1 in. on a side, the total area of the 14 pieces is 28 in.². Since 28 is not a perfect square, it is impossible.
3. Make yourself a set of tetraboloes. A good size is 1 in. for the sides of the squares. Check students' figures.
4. Use a subset of the 14 pieces to make a square like the one shown. Sketch your answer at the right.
5. Use the four tetraboloes at the right to make larger copies of as many of the 14 tetraboloes as you can. Sketch your answers below. There are 8 possible solutions.

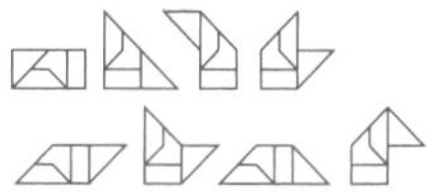

44 ENRICHMENT LESSON 3-1

Lesson Planning

NCTM Standards/Strands

- Number & Operation
- Patterns, Functions, & Algebra
- Geometry & Spatial Sense
- Measurement
- Problem Solving
- Reasoning and Proof
- Communication
- Connections
- Representation

Vocabulary

angle	vertex
protractor	degree measure
acute angle	right angle
obtuse angle	straight angle
complementary	supplementary
adjacent	exterior sides

Materials Needed

paper/pencil
protractors

Lesson Resources

Warm-Up Transparency 7
Transparency RF-11, 12
Reteaching 3-2
Extra Practice 3-2
Enrichment 3-2
Study Skills Activities

Assignment Guide

Basic: 1–25, 29–42
Enriched: 1–42

Getting Started

5-Minute Warm-Up

Solve for *x*.

1. $x + 6 + 3x = 90$ 21
2. $4x + x = 90$ 18
3. $2x - 30 + x = 180$ 70
4. $3x + 11 + 2x - 6 = 180$ 35

Introduction to Lesson 3-2

Students who do not understand the symbols used in this activity should review the previous lesson. Have students share and compare their results for Exercise 3.

3-2 Types of Angles

Goals ■ Classify and measure angles.
Applications Tiling and Architecture

Draw and label a representation of each figure. See additional answers.

1. $\overline{JK}$
2. $\overleftrightarrow{JK}$
3. $\overrightarrow{JK}$
4. $\overrightarrow{KJ}$
5. $\overrightarrow{VX}$ and $\overrightarrow{VY}$, so that V is between X and Y
6. $\overrightarrow{VX}$ and $\overrightarrow{VY}$, so that V, X, and Y are noncollinear

Build Understanding

An **angle** is the union of two rays with a common endpoint. The endpoint is the **vertex** of the angle, and each ray is a **side** of the angle.

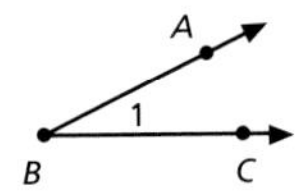

angle *ABC* (∠*ABC*), angle *CBA* (∠*CBA*), angle *B* (∠*B*), or angle 1 (∠1)

The *protractor postulate* is a basic assumption about angles.

Postulate 7

The Protractor Postulate Let O be a point on $\overleftrightarrow{AB}$ such that O is between A and B.

Consider $\overrightarrow{OA}$, $\overrightarrow{OB}$, and all the rays that can be drawn from O on one side of $\overleftrightarrow{AB}$. These rays can be paired with the real numbers from 0 to 180 in such a way that:

1. $\overrightarrow{OA}$ is paired with 0 and $\overrightarrow{OB}$ is paired with 180.
2. If $\overrightarrow{OP}$ is paired with x and $\overrightarrow{OQ}$ is paired with y, then the number paired with m∠POQ is $|x - y|$. This number is called the **measure**, or the **degree measure**, of ∠POQ.

Example 1

Use the figure, find the measure of ∠*POQ* (m∠*POQ*).

Solution

Notice that the protractor has two scales.

Using the inner scale, $\overrightarrow{OP}$ is paired with 140 and $\overrightarrow{OQ}$ is paired with 50.

$|140 - 5\,| = |90| = 90$

Using the outer scale, $\overrightarrow{OP}$ is paired with 40 and $\overrightarrow{OQ}$ is paired with 130.

$|40 - 130| = |-90| = 90$

In either case, the measure of ∠POQ is 90° or m∠POQ = 90°.

Learning Styles

TACTILE/KINESTHETIC LEARNER Take two thin strips of cardstock and join them together with a brad so that they can move freely. Have students use this angle model with a protractor to show different angle measures and to find complements and supplements of angles.

GEOMETRY SOFTWARE You can draw an angle and find its measure using geometry software. Using your software, draw an angle. Then select and measure the angle. The measure of the angle will appear on the screen.

From your work in previous courses, recall that angles are classified according to their measures. Using the figure in Example 1, the chart below summarizes the types of angles and gives an example of each.

Type of angle	Measure	Example
acute	greater than 0° and less than 90°	∠QOA
right	equal to 90°	∠POQ
obtuse	greater than 90° and less than 180°	∠POA
straight	equal to 180°	∠BOA

An angle with a measure less than 180° divides a plane into three sets of points: the angle itself, the points in the *interior* of the angle, and the points in the *exterior* of the angle.

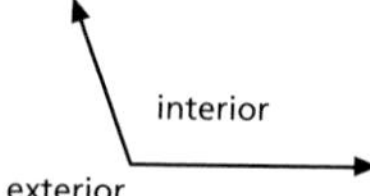

Sometimes, pairs of angles are classified by the *sum* of their measures.

Two angles are **complementary** if the sum of their measures is 90°. Each angle is called the *complement* of the other.

Two angles are **supplementary** if the sum of their measures is 180°. Each angle is called the *supplement* of the other.

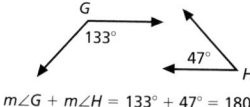

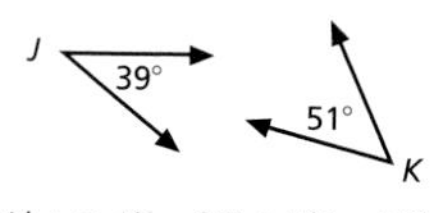

$m\angle J + m\angle K = 39° + 51° = 90°$
$\angle J$ and $\angle K$ are complementary.

G
133°
47°
H

$m\angle G + m\angle H = 133° + 47° = 180°$
$\angle G$ and $\angle H$ are supplementary.

Another basic assumption about angles is the *angle addition postulate.*

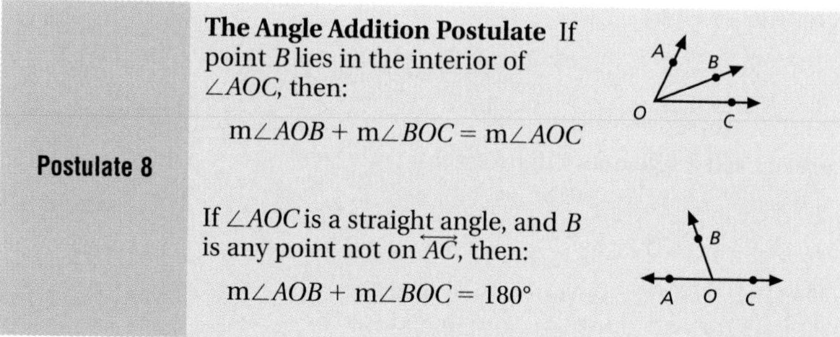

Postulate 8

The Angle Addition Postulate If point B lies in the interior of $\angle AOC$, then:

$$m\angle AOB + m\angle BOC = m\angle AOC$$

If $\angle AOC$ is a straight angle, and B is any point not on $\overleftrightarrow{AC}$, then:

$$m\angle AOB + m\angle BOC = 180°$$

In each of the two figures that illustrate the angle addition postulate, $\angle AOB$ and $\angle BOC$ form a pair of *adjacent angles.* **Adjacent angles** are two angles in the same plane that share a common side and a common vertex, but have no interior points in common. The sides of the two adjacent angles that are not common to the adjacent angles are called **exterior sides**.

Teaching Strategies

COOPERATIVE LEARNING Have students work in pairs. Each pair should draw examples of each type of angle listed in the table on page 109. Which angles have both a complement and a supplement? Which angles have only one of these? Have students share their conclusions. All acute angles have both complements and supplements; right and obtuse angles have only supplements. Straight angles have neither.

Chalkboard Examples

Supplementary Example 1

Use the figure in Example 1. Find the measure of $\angle POA$. 140°

Supplementary Example 2

The figure below shows another portion of the tile border. The $m\angle QRS$ is $x°$ and $m\angle SRT$ is $(x + 34)°$. Find $m\angle SRT$. 62°

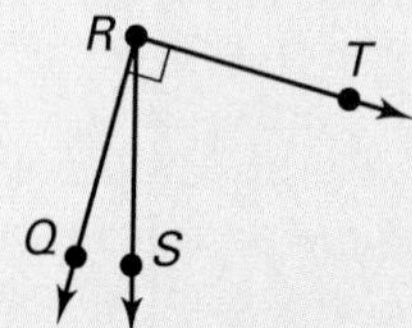

Additional Answers

1., 2., 3., 4.

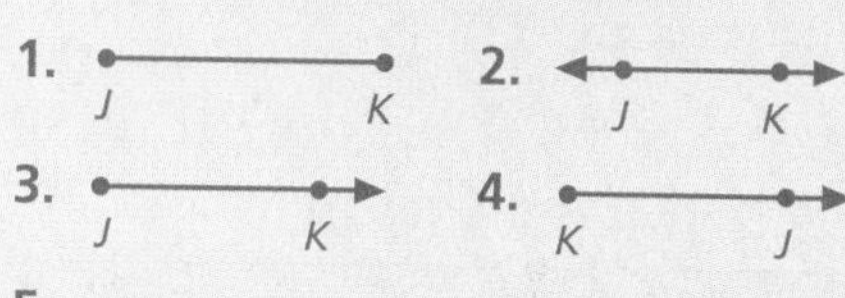

5.

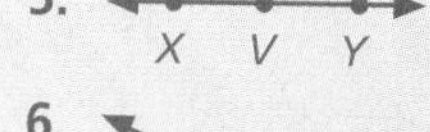

6.

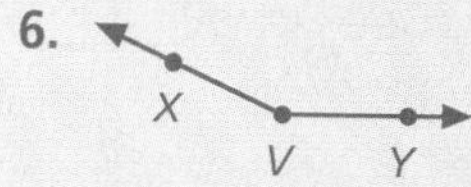

Reteaching Worksheet 3-2

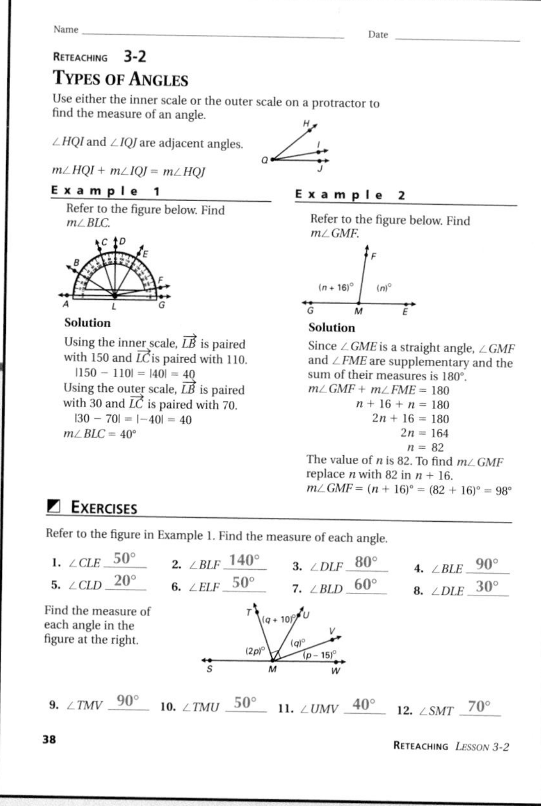

Name ______________________ Date ____________

RETEACHING 3-2

TYPES OF ANGLES

Use either the inner scale or the outer scale on a protractor to find the measure of an angle.

$\angle HQI$ and $\angle IQJ$ are adjacent angles.

$m\angle HQI + m\angle IQJ = m\angle HQJ$

Example 1

Refer to the figure below. Find $m\angle BLC$.

Solution

Using the inner scale, $\overrightarrow{LB}$ is paired with 150 and $\overrightarrow{LC}$ is paired with 110.

$|150 - 110| = |40| = 40$

Using the outer scale, $\overrightarrow{LB}$ is paired with 30 and $\overrightarrow{LC}$ is paired with 70.

$|30 - 70| = |-40| = 40$

$m\angle BLC = 40°$

Example 2

Refer to the figure below. Find $m\angle GMF$.

Solution

Since $\angle GME$ is a straight angle, $\angle GMF$ and $\angle FME$ are supplementary and the sum of their measures is 180°.

$$m\angle GMF + m\angle FME = 180$$
$$n + 16 + n = 180$$
$$2n + 16 = 180$$
$$2n = 164$$
$$n = 82$$

The value of n is 82. To find $m\angle GMF$ replace n with 82 in $n + 16$.

$m\angle GMF = (n + 16)° = (82 + 16)° = 98°$

EXERCISES

Refer to the figure in Example 1. Find the measure of each angle.

1. ∠CLE 50°
2. ∠BLF 140°
3. ∠DLF 80°
4. ∠BLE 90°
5. ∠CLD 20°
6. ∠ELF 50°
7. ∠BLD 60°
8. ∠DLE 30°

Find the measure of each angle in the figure at the right.

9. ∠TMV 90°
10. ∠TMU 50°
11. ∠UMV 40°
12. ∠SMT 70°

38 RETEACHING LESSON 3-2

Lesson Wrap-up

Quick Assessment

Ask the following questions to determine if students understand the content presented in this lesson.

1. How can you use a protractor to find the measure of an angle? Answers may vary. Explanations should include mention of finding the two points on a protractor that intersect the rays of the angle and subtracting the degree measures for those points.
2. When are two angles complementary? when the sum of their measures is 90°
3. When are two angles supplementary? when the sum of their measures is 180°

Assignment Guide

Basic: 1–25, 29–42
Advanced: 1–42
Additional Practice: See Extra Practice Index on page 674.

Additional Answers

9. Answers will vary. Both postulates refer to the pairing of real numbers with geometric figures in a systematic way; both involve taking the absolute value of the difference of two real numbers.

When complementary angles are adjacent, the exterior sides form a right angle.

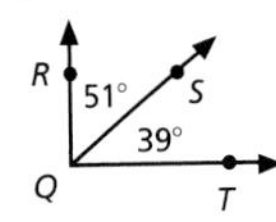

$\angle RQT$ is a right angle.

When supplementary angles are adjacent, the exterior sides form a straight angle.

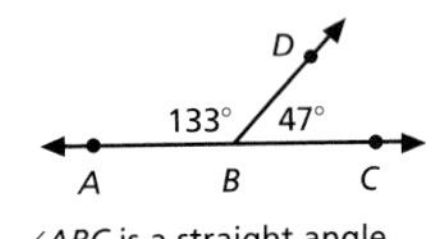

$\angle ABC$ is a straight angle.

Reading Math

Many of the words used in this lesson have related meanings in daily life. Match each word in this list with an "everyday meaning" from the choices below.

1. acute
2. adjacent
3. angle
4. obtuse
5. supplement

a. adjoining
b. attachment
c. point of view
d. sharp
e. blunt

1. d
2. a
3. c
4. e
5. b

Example 2

TILING To make a decorative tile border, two tiles must be joined in the angle shown in the figure. The square bracket (⌐) indicates that $\angle UYW$ is a right angle. Find $m\angle UYV$.

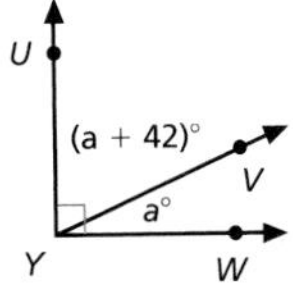

Solution

Since $\angle UYW$ is a right angle, the adjacent angles $\angle UYV$ and $\angle VYW$ are complementary. This means that the sum of their measures is 90°. Use this fact to write and solve an equation.

$$m\angle UYV + m\angle VYW = 90°$$

$$a + 42 + a = 90 \quad \text{Combine like terms.}$$

$$2a + 42 = 90 \quad \text{Add } -42 \text{ to each side.}$$

$$2a = 48 \quad \text{Multiply each side by } \tfrac{1}{2}.$$

$$a = 24$$

So, the value of a is 24. From the figure, $m\angle UYV = (a + 42)°$. Substituting 24 for a, $m\angle UYV = (24 + 42)° = 66°$.

Try These Exercises

Find the measure of each angle. Then classify each angle as acute, obtuse, or right.

1. $\angle HXL$ 120°; obtuse
2. $\angle JXF$ 90°; right
3. $\angle KXG$ 120°; obtuse
4. $\angle GXJ$ 65°; acute
5. $\angle FXL$ 180°; straight
6. $\angle HXK$ 85°; acute

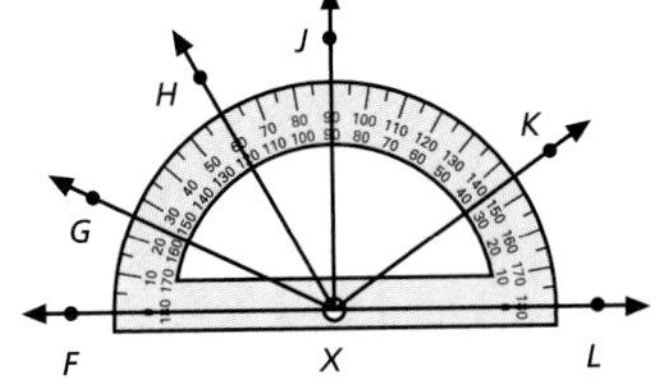

Find the measure of each angle in the figure at the right.

7. $\angle MNP$ 79°
8. $\angle PNO$ 101°

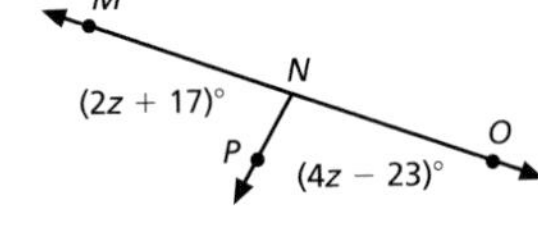

9. **WRITING MATH** In Lesson 3-1, you studied the ruler postulate. How are the ruler postulate and the protractor postulate alike? See additional answers.
10. **CHAPTER INVESTIGATION** Draw a map of your community showing at least ten specific points of interest. Make distances as accurate as possible. Overlay your map with a grid of latitude (east-west) and longitude (north-south) lines. Check students' work.

Learning Styles

VISUAL LEARNER Using an overhead projector, draw a right angle with heavy black lines on a transparency. Draw two to four acute angles on another transparency using thinner lines. Invite students to measure one of the acute angles with a protractor, then place it over the right-angle transparency and measure its complement.

PRACTICE EXERCISES

Exercises 11–14 refer to the protractor at the right.

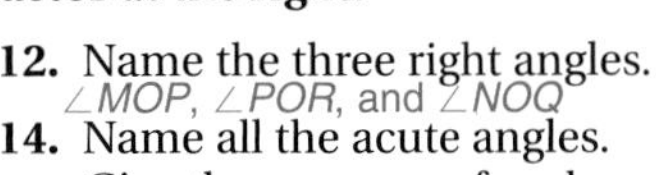

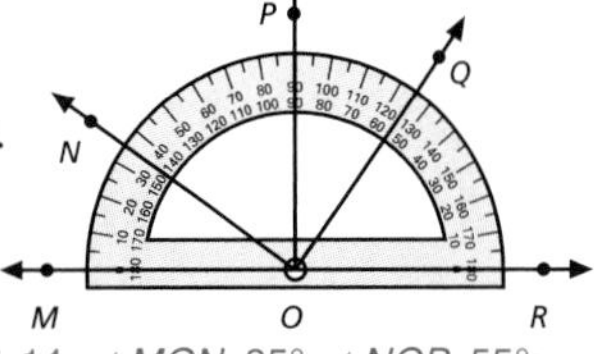

11. Name the straight angle. $\angle MOR$

12. Name the three right angles. $\angle MOP$, $\angle POR$, and $\angle NOQ$

13. Name all the obtuse angles. Give the measure of each. $\angle MOQ$, 125° and $\angle NOR$, 145°

14. Name all the acute angles. Give the measure of each. 14. $\angle MON$, 35°; $\angle NOP$, 55°; $\angle POQ$, 35°; and $\angle QOR$, 55°

15. In the figure below, m$\angle FJG$ is y° and m$\angle GJH$ is $4y$°. Find m$\angle GJH$. 72°

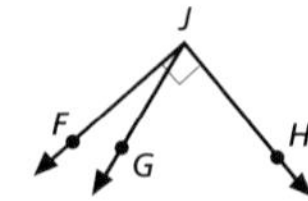

16. In the figure below, m$\angle AZC$ is $(5x + 8)$° and m$\angle CZD$ is $(2x - 17)$°. Find m$\angle BZC$. 53°

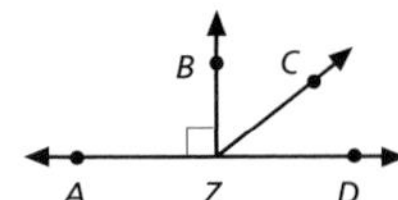

SPATIAL SENSE Without measuring, draw an angle that you think has the given measure. Then use a protractor to determine the actual measure of your angle. For 17–21, check students' drawings.

17. 90° **18.** 45° **19.** 60° **20.** 135° **21.** 120°

22. The measure of $\angle JKL$ is 34° more than the measure of its complement. Find m$\angle JKL$. 62°

23. The measure of $\angle XYZ$ is 15° less than twice the measure of its supplement. Find m$\angle XYZ$. 115°

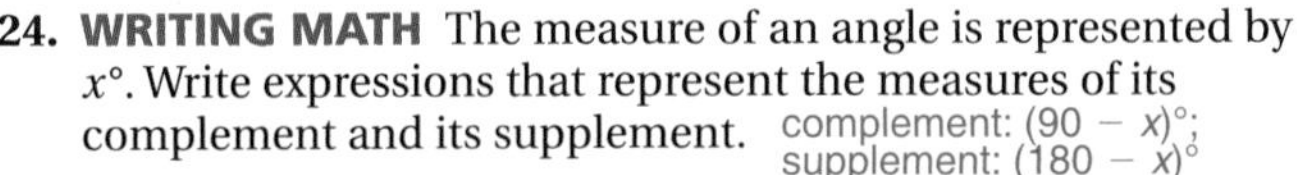

24. **WRITING MATH** The measure of an angle is represented by x°. Write expressions that represent the measures of its complement and its supplement. complement: $(90 - x)$°; supplement: $(180 - x)$°

25. **ARCHITECTURE** On the plans for a building, the sum of the measure of the complement of an angle and the measure of its supplement is 136°. Find the measure of the angle. 67°

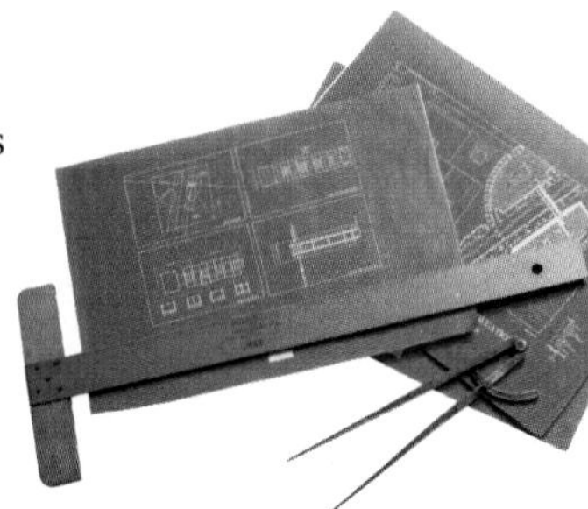

EXTENDED PRACTICE EXERCISES

Determine whether each statement is *always, sometimes,* or *never* true.

26. The supplement of an obtuse angle is an acute angle. always

27. The complement of an acute angle is an obtuse angle. never

28. Complementary angles are also adjacent angles. sometimes

MIXED REVIEW EXERCISES

Evaluate each expression when $a = -5$ and $b = 4$. (Lesson 1-7)

29. $a^2 + b^2$ 41 **30.** $a^3 - b$ −129 **31.** $(a + b)^2$ 1 **32.** $\frac{a^5}{a^2} - b^3, a \neq 0$ −189

33. $-2(a^2b^3)$ −3200 **34.** $3(ab)^2$ 1200 **35.** $(a^3 - 2) \cdot b$ −508 **36.** $\frac{b^5}{b} + a^2, b \neq 0$ 281

37. $-4(a^2b)^2$ −40,000 **38.** $a - b^3$ −69 **39.** $(a^2 - b^3)^2$ 1521 **40.** $\frac{a^4}{a^2} \cdot b^2, a \neq 0$ 400

Write each number in scientific notation. (Lesson 1-8)

41. 24,000,000,000,000 2.4×10^{13} **42.** 0.0000000000301 3.01×10^{-11}

Teaching Strategies

CHALLENGE How many times each day do the hands of a clock form a right angle? a straight angle? 44; 22 Discuss how the minutes and seconds on a clock relate to the degrees used in measuring angles.

Extra Practice Worksheet 3-2

Name ______ Date ______

EXTRA PRACTICE 3-2
TYPES OF ANGLES

EXERCISES

Refer to the figure at the right for Exercises 1–8. Use a protractor to find the measure of each angle. Then classify each angle as acute, right, or obtuse.

1. $\angle BFA$ 57°; acute
2. $\angle CFA$ 90°; right
3. $\angle EFD$ 27°; acute
4. $\angle DFB$ 96°; obtuse
5. $\angle EFB$ 123°; obtuse
7. $\angle DFC$ 63°; acute
6. $\angle EFC$ 90°; right
8. $\angle CFB$ 33°; acute

In the figure at the right, $m\angle PQN = (3x)$° and $m\angle NQM = (4x + 5)$°. Find the measure of each angle.

9. $\angle PQN$ 75°
10. $\angle NQM$ 105°

In the figure at the right, $m\angle ADB = (7x + 2)$° and $m\angle CDB = (3x - 2)$°. Find the measure of each angle.

11. $\angle ADB$ 65°
12. $\angle CDB$ 25°

13. The measure of $\angle XYZ$ is 46° more than the measure of its supplement. Find $m\angle XYZ$. 113°

14. The measure of $\angle RST$ is 6° less than three times the measure of its complement. Find $m\angle RST$. 66°

34 EXTRA PRACTICE *LESSON 3-2*

Enrichment Worksheet 3-2

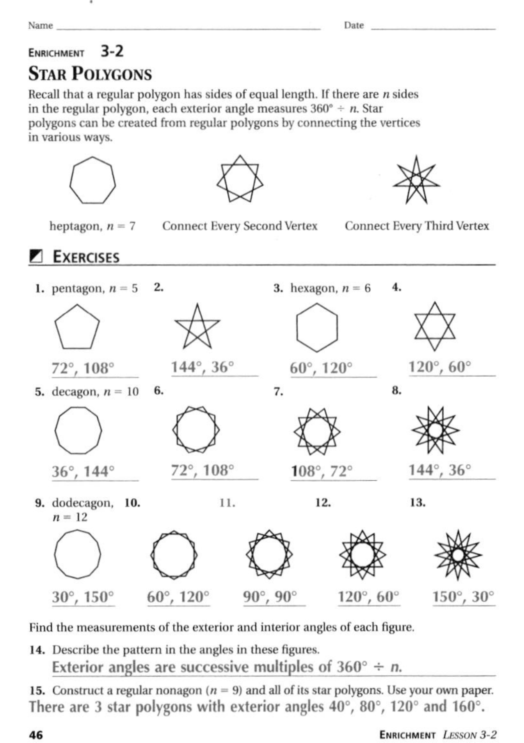

Name ______ Date ______

ENRICHMENT 3-2
STAR POLYGONS

Recall that a regular polygon has sides of equal length. If there are n sides in the regular polygon, each exterior angle measures $360° \div n$. Star polygons can be created from regular polygons by connecting the vertices in various ways.

heptagon, $n = 7$ Connect Every Second Vertex Connect Every Third Vertex

EXERCISES

1. pentagon, $n = 5$ 72°, 108°
2. 144°, 36°
3. hexagon, $n = 6$ 60°, 120°
4. 120°, 60°
5. decagon, $n = 10$ 36°, 144°
6. 72°, 108°
7. 108°, 72°
8. 144°, 36°
9. dodecagon, $n = 12$ 30°, 150°
10. 60°, 120°
11. 90°, 90°
12. 120°, 60°
13. 150°, 30°

Find the measurements of the exterior and interior angles of each figure.

14. Describe the pattern in the angles in these figures. Exterior angles are successive multiples of $360° \div n$.

15. Construct a regular nonagon ($n = 9$) and all of its star polygons. Use your own paper. There are 3 star polygons with exterior angles 40°, 80°, 120° and 160°.

46 ENRICHMENT *LESSON 3-2*

Skills Practice

Vocabulary Review

Lesson 3-1

point	line
plane	space
collinear points	coplanar points
intersection	postulate
ray	endpoint
line segment	

Lesson 3-2

angle	vertex
protractor	degree measure
acute angle	right angle
obtuse angle	straight angle
complementary	supplementary
adjacent	exterior sides

Assignment Guide

All students: 1–42

Additional Practice: Refer to the Extra Practice Index on page 674 of this text.

Review and Practice Your Skills

Practice ■ Lesson 3-1

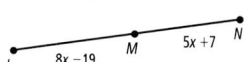

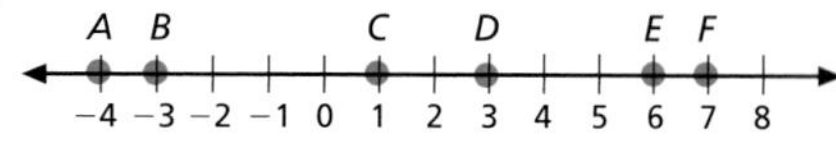

Use the number line at the right for Exercises 1–4. Find each length.

1. *AC* 5 **2.** *BE* 9 **3.** *FC* 6 **4.** *DA* 7

5. In the figure below, *QS* = 78. Find *QR*. 60

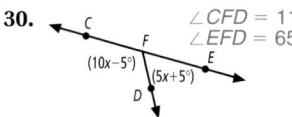
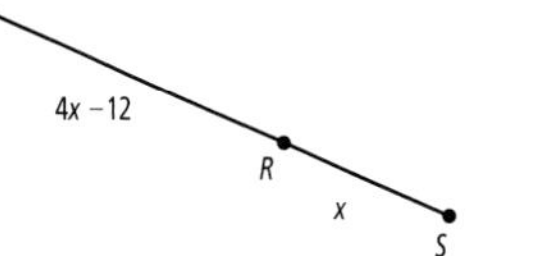

6. In the figure below, *LN* = 131. Find *MN*. 62

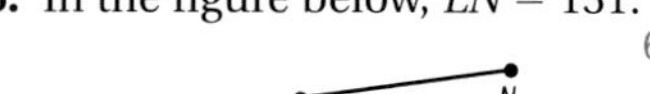

7. On a number line, the coordinate of point *G* is −5. The length of $\overline{GH}$ is 13. Give two possible coordinates for point *H*. −18, 8

8. On a number line, the distance from *R* to *S* is twice the distance from *S* to *T*. The coordinate of point *T* is 15 and the coordinate of point *S* is −3. Give two possible coordinates for point *R*. −39, 33

9. *True* or *false*: A plane has four edges. False

10. *True* or *false*: Two points may be collinear and also noncoplanar. True

11. *True* or *false*: Two planes intersect in a line. True

12. *True* or *false*: The distance between two points on a number line is negative when moving from right to left. False

Practice ■ Lesson 3-2

Find the complement of each angle.

13. 78° 12° **14.** 27° 63° **15.** 6° 84° **16.** 51° 39°

17. 89° 1° **18.** 35° 55° **19.** 42° 48° **20.** 104° none

Find the supplement of each angle.

21. 83° 97° **22.** 55° 125° **23.** 126° 54° **24.** 3° 177°

25. 30.5° 149.5° **26.** 77° 103° **27.** 18° 162° **28.** 180° 0°

Find the measure of each angle in the given figure.

29.

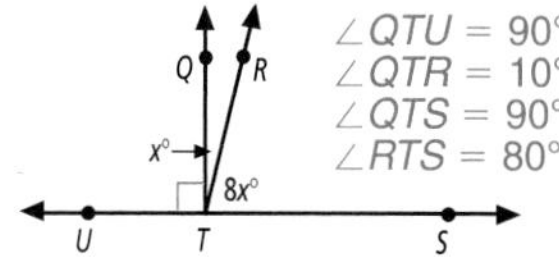

∠*QTU* = 90°
∠*QTR* = 10°
∠*QTS* = 90°
∠*RTS* = 80°

30. C, F, E, D; (10x − 5°), (5x + 5°)

∠*CFD* = 115°
∠*EFD* = 65°

31.

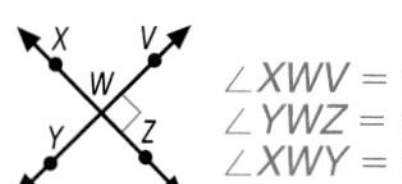

∠*XWV* = 90°
∠*YWZ* = 90°
∠*XWY* = 90°

32. The measure of ∠*ABC* is 28° more than the measure of its complement. Find m∠*ABC*. 59°

33. The measure of ∠*DEF* is 48° less than twice its supplement. Find m∠*DEF*. 104°

Learning Styles

VISUAL LEARNER For items such as Exercises 7 and 8, students may benefit from making a quick sketch of the situation. Even when a diagram is given, as in Exercises 5 and 6, students may gain a greater understanding of the relationships shown by redrawing and labeling the diagram on scratch paper.

PRACTICE ■ LESSON 3-1–LESSON 3-2

Use the figure at the right for Exercises 1–4. Which postulate justifies your answer? (Lesson 3-1)

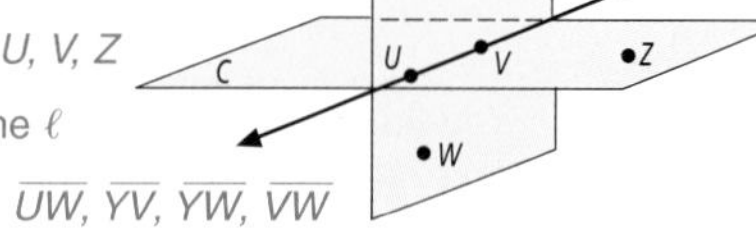

34. Name two points that determine line ℓ. U, V

35. Name three points that determine plane *C*. U, V, Z

36. Name the intersection of planes *C* and *D*. line ℓ

37. Name three lines that lie in plane *D*. $\overline{UY}$, $\overline{UV}$, $\overline{UW}$, $\overline{YV}$, $\overline{YW}$, $\overline{VW}$

38. In the figure below, $PR = 36$. Find PQ. (Lesson 3-1) 20

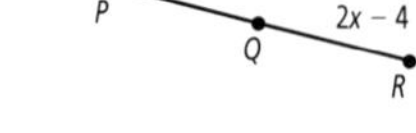

39. In the figure below, $GJ = 354$. Find GH. (Lesson 3-1) 176

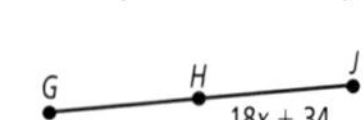

40. The measure of the supplement $\angle JKL$ is 2.5 times its complement. Find $m\angle JKL$. (Lesson 3-2) 30°

41. The measure of $\angle GHI$ is 5 times the measure of its complement. Find $m\angle GHI$. (Lesson 3-2) 75°

42. *True* or *false*: An angle is always smaller than its supplement. (Lesson 3-2) false

MathWorks Workplace Knowhow — Career – Cross-Country Bus Driver

Bus drivers do not only drive school buses and cross-town buses. Some drive great distances to towns and cities all over the country. These bus drivers are responsible for the safe and timely arrival of their passengers to the correct location. Bus drivers need to be able to read road maps, follow directions, and use a compass. They use math to calculate the quickest route to their destination and to know how far they can drive before they need to refuel.

1. A bus driver's drives from New York to Chicago on a route that goes through Pittsburgh and Cleveland. The entire route is 885 mi. The line segment below represents the driver's route. Using the Segment Addition Postulate, what is the driving distance between Pittsburgh and Cleveland? 140 mi

390 mi | 355 mi
New York City — Pittsburgh — Cleveland — Chicago

2. A driver needs to drive from point *B* to point *A* (see the diagram). There are two alternate routes: route 1 passes goes directly from *B* to *A*, and route 2 passes through point *P*. Calculate which route will take less time. Use the formula: rate × time = distance.
Route *BPA* is faster.

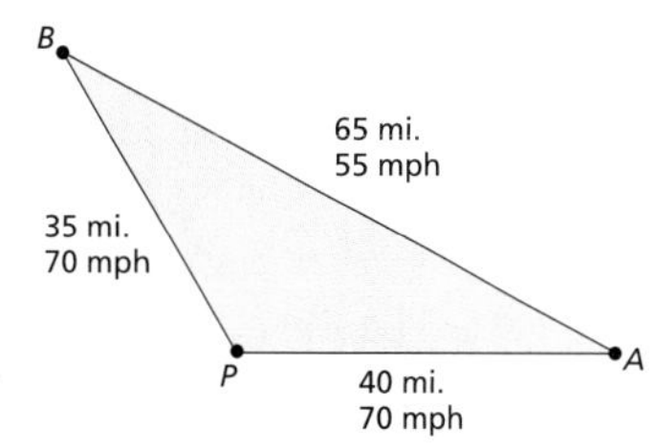

Chalkboard Examples

Example from Lesson 3-1

Point *R* is between points *K* and *L*. The length $\overline{KR}$ of is twice the length of *RL*, and $KL = 45$. Find *KR*. 30

Example from Lesson 3-2

The measure of the supplement of an angle is 12° more than twice the measure of the angle. Find the measure of the angle. 56°

Career Opportunity

Describe the kind of long distance drivers do and the types of companies that hire these workers. Have students brainstorm a list of the challenges that a driver might face in his or her work. Explain that drivers must be in good physical health, have good record-keeping skills, and be able to adapt to changing conditions quickly.

Students should answer Questions 1–2 to better understand how drivers use mathematics in performing their job.

Students who are interested in learning more about this profession can go to Learningmathmatters.com. Guidance Counselors should have information about school and training requirements.

Teaching Strategies

AUDITORY LEARNERS To determine the validity of statements, these students may benefit from reading the statement aloud. Have them practice reading quietly. Encourage them to emphasize key words. For practice, have them determine whether each statement below is *always*, *sometimes*, or *never true*.

1. The measure of the supplement of an angle is greater than the measure of the angle. sometimes
2. The complement of an angle is an obtuse angle. never

Chapter 3 **Review and Practice Your Skills** 113

Lesson Planning

NCTM Standards/Strands

- Number & Operation
- Patterns, Functions, & Algebra
- Geometry & Spatial Sense
- Measurement
- Problem Solving
- Reasoning and Proof
- Communication
- Connections
- Representation

Vocabulary

midpoint, bisector, theorem, opposite rays, vertical angles

Materials Needed

paper/pencil, protractor, ruler

Lesson Resources

Warm-Up Transparency 7
Reteaching 3-3
Extra Practice 3-3
Enrichment 3-3

Assignment Guide

Basic: 1–29, 36–45
Enriched: 1–45

Getting Started

5-Minute Warm-Up

Solve for x.

1. $6x - 25 = 2x + 27$ 13
2. $4x + 35 = 3x + 80$ 45
3. $7x - 11 = 5x + 35$ 23
4. $3x - 71 = x - 13$ 29

Introduction to Lesson 3-3

For the first part, suggest that students use a small unit of measurement, such as millimeters or eighths of an inch. For the second part of the activity, some students may need help using a protractor correctly. Be sure they align the vertex of the angle with the center point of the protractor.

3-3 Segments and Angles

Goals
- Identify bisectors of angles and segments.
- Apply theorems about midpoints, angle bisectors, and vertical angles.

Applications Engineering, Art, Map Making

Work with a partner.

1. On paper, draw any two points R and S. Then draw $\overline{RS}$. Fold the paper so point R falls on top of point S. Unfold. Label the point where the fold intersects $\overline{RS}$ as point Y. With a ruler, find RS, RY, and SY. What do you notice? $RV = SV = \frac{1}{2}RS$
2. On a sheet of paper, draw $\angle JKL$. Fold the paper so $\overrightarrow{KJ}$ falls on top of $\overrightarrow{KL}$. Unfold the paper. Choose a point along the fold that falls in the interior of $\angle JKL$, and label it Z. Draw $\overrightarrow{KZ}$. With a protractor, find $m\angle JKL$, $m\angle JKZ$, and $m\angle LKZ$. What do you notice? $m\angle JKZ = m\angle LKZ = \frac{1}{2}m\angle JKL$

Build Understanding

The **midpoint** of a segment is the point that divides it into two segments of equal length. A **bisector of a segment** is any line, segment, ray, or plane that intersects the segment at its midpoint.

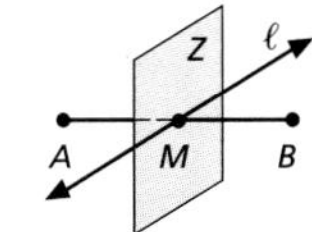

M is the midpoint of $\overline{AB}$.
$AM = MB$
Line ℓ and plane Z bisect $\overline{AB}$.

In Lessons 3-1 and 3-2, you studied several fundamental postulates of geometry. In this lesson, you will look at other statements that are called *theorems*. Whereas a postulate is an assumption that is accepted as true without proof, a **theorem** is a statement that can be proved true.

The Midpoint Theorem If point M is the midpoint of $\overline{AB}$, then $AM = \frac{1}{2}AB$ and $MB = \frac{1}{2}AB$

Example 1

Use the figure below. Find the midpoint of $\overline{SY}$.

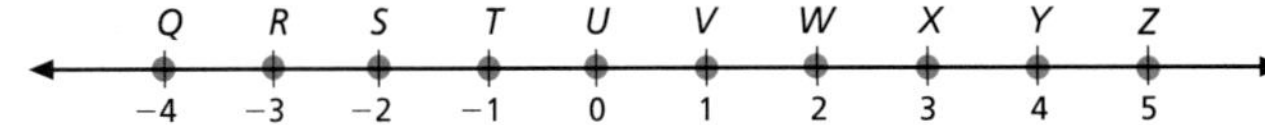

Solution

The coordinate of point S is -2. The coordinate of point Y is 4. By the ruler postulate, $SY = |-2 - 4| = |-6| = 6$.

Since $\frac{1}{2}(6) = 3$, the midpoint of $\overline{SY}$ is 3 units to the right of point S.

The coordinate of this point would be $-2 + 3 = 1$.
So, the midpoint of $\overline{SY}$ is point V.

Check Understanding

Explain the difference between the *midpoint* of a segment and a *bisector* of a segment.

See additional answers.

Learning Styles

VISUAL LEARNER These students may find it helpful to make their own sketch of the figures presented in the text. They can then add angle measurements as they are calculated. In the process of sketching the figure, students more easily focus on the relationships among the geometric elements.

The **bisector of an angle** is the ray that divides the angle into two adjacent angles that are equal in measure. This definition leads to the following theorem.

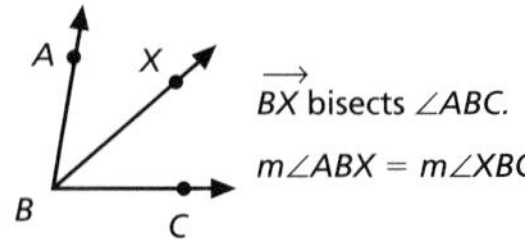

$\overrightarrow{BX}$ bisects $\angle ABC$.

$m\angle ABX = m\angle XBC$

The Angle Bisector Theorem	If $\overrightarrow{BX}$ is the bisector of $\angle ABC$, then: $m\angle ABX = \frac{1}{2} m\angle ABC$ and $m\angle XBC = \frac{1}{2} m\angle ABC$

Example 2

ENGINEERING The figure shows a portion of the plans for adding angular supports to a bridge. In the figure, $\angle JKL$ is a straight angle, and $\overrightarrow{KH}$ bisects $\angle GKL$. Find $m\angle HKL$.

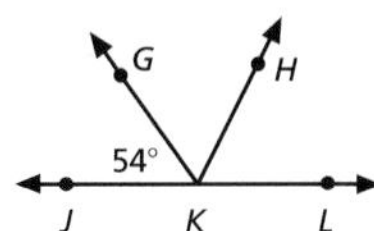

Solution

Since $\angle JKL$ is a straight angle, $m\angle JKL = 180°$.
By the angle addition postulate, $m\angle JKG + m\angle GKL = 180°$.

From the figure, $m\angle JKG = 54°$.
So, $54° + m\angle GKL = 180°$, and $m\angle GKL = 180° - 54° = 126°$.

Since $\overrightarrow{KH}$ bisects $\angle GKL$, $m\angle HKL = \frac{1}{2} m\angle GKL = \frac{1}{2}(126°) = 63°$.

Two rays $\overrightarrow{BA}$ and $\overrightarrow{BC}$ are called **opposite rays** if point B is on $\overleftrightarrow{AC}$ and is between points A and C.

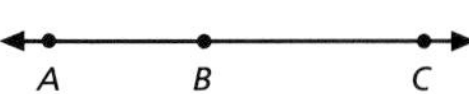

opposite rays $\overrightarrow{BA}$ and $\overrightarrow{BC}$

Two angles whose sides form two pairs of opposite rays are called **vertical angles**. The following theorem is an important statement about vertical angles.

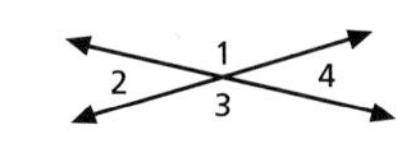

$\angle 1$ and $\angle 3$ are vertical angles.

$\angle 2$ and $\angle 4$ are vertical angles.

For 1–2, answers will vary.
3. Measures will vary, but $m\angle AEC$ is always equal to $m\angle BED$, and $m\angle CEB$ is always equal to $m\angle DEA$.

The Vertical Angles Theorem	If two angles are vertical angles, then they are equal in measure.

Technology Note

Use geometric software to explore vertical angles. Try this activity.

1. Draw two intersecting lines. Label points *A*, *B*, *C*, *D*, and *E* as shown.

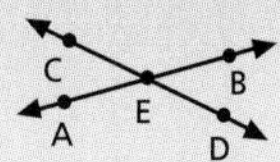

2. Find $m\angle AEC$, $m\angle CEB$, $m\angle BED$, and $m\angle DEA$.
3. Change the position of each line and repeat Step 2.

Learning Styles

TACTILE/KINESTHETIC LEARNER To demonstrate the theorem about vertical angles, have students draw a pair of intersecting lines on a sheet of paper. Have them label the angles 1, 2, 3, and 4 as shown at the bottom of page 115. Then have them cut out all four angles and compare their measures. They should be able to verify that $m\angle 1 = m\angle 3$ and $m\angle 2 = m\angle 4$.

Chalkboard Examples

Supplementary Example 1

Use the figure in Example 1. Find the midpoint of $\overline{QU}$ and $\overline{TZ}$. **S; W**

Supplementary Example 2 and 3

In another part of the plans (diagram below), $\angle EBA$ is a straight angle, and $\overrightarrow{BC}$ bisects $\angle ABD$. Find $m\angle EBC$.
135°

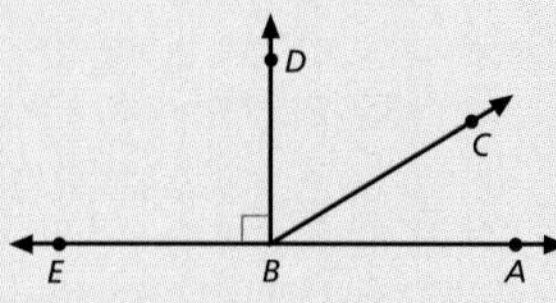

Additional Answers

Check Understanding
The midpoint is that pint of the segment that is exactly halfway between the endpoints. A bisector is a figure that passes through this point.

Reteaching Worksheet 3-3

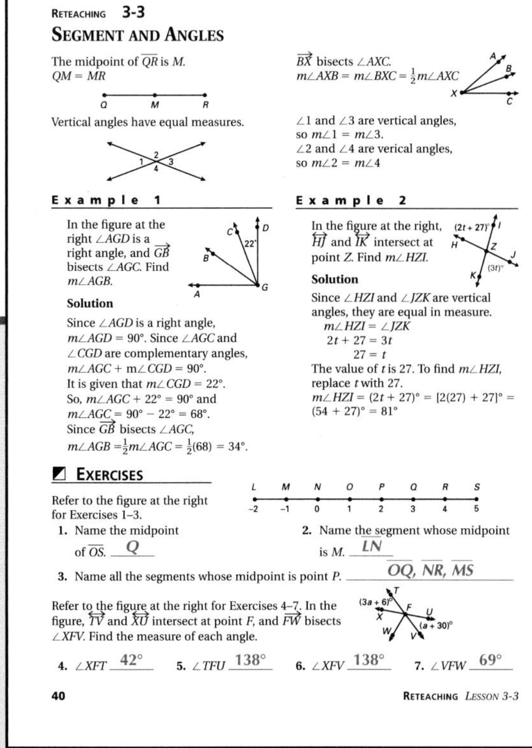

Name ______________________ Date ______________

RETEACHING 3-3

SEGMENT AND ANGLES

The midpoint of $\overline{QR}$ is *M*.
$QM = MR$

$\overrightarrow{BX}$ bisects $\angle AXC$.
$m\angle AXB = m\angle BXC = \frac{1}{2} m\angle AXC$

Vertical angles have equal measures.

$\angle 1$ and $\angle 3$ are vertical angles, so $m\angle 1 = m\angle 3$.
$\angle 2$ and $\angle 4$ are verical angles, so $m\angle 2 = m\angle 4$

Example 1

In the figure at the right $\angle AGD$ is a right angle, and $\overrightarrow{GB}$ bisects $\angle AGC$. Find $m\angle AGB$.

Solution

Since $\angle AGD$ is a right angle, $m\angle AGD = 90°$. Since $\angle AGC$ and $\angle CGD$ are complementary angles, $m\angle AGC + m\angle CGD = 90°$.
It is given that $m\angle CGD = 22°$.
So, $m\angle AGC + 22° = 90°$ and $m\angle AGC = 90° - 22° = 68°$.
Since $\overrightarrow{GB}$ bisects $\angle AGC$,
$m\angle AGB = \frac{1}{2} m\angle AGC = \frac{1}{2}(68) = 34°$.

Example 2

In the figure at the right, $\overleftrightarrow{HJ}$ and $\overleftrightarrow{IK}$ intersect at point *Z*. Find $m\angle HZI$.

Solution

Since $\angle HZI$ and $\angle JZK$ are vertical angles, they are equal in measure.
$m\angle HZI = \angle JZK$
$2t + 27 = 3t$
$27 = t$
The value of *t* is 27. To find $m\angle HZI$, replace *t* with 27.
$m\angle HZI = (2t + 27)° = [2(27) + 27]° = (54 + 27)° = 81°$

EXERCISES

Refer to the figure at the right for Exercises 1–3.

1. Name the midpoint of $\overline{OS}$. Q
2. Name the segment whose midpoint is *M*. $\overline{LN}$
3. Name all the segments whose midpoint is point *P*. $\overline{OQ}$, $\overline{NR}$, $\overline{MS}$

Refer to the figure at the right for Exercises 4–7. In the figure, $\overleftrightarrow{TV}$ and $\overleftrightarrow{XU}$ intersect at point *F*, and $\overrightarrow{FW}$ bisects $\angle XFV$. Find the measure of each angle.

4. $\angle XFT$ 42°
5. $\angle TFU$ 138°
6. $\angle XFV$ 138°
7. $\angle VFW$ 69°

40 RETEACHING *LESSON 3-3*

Lesson 3-3 **Segments and Angles** 115

Lesson Wrap-up

Quick Assessment

Ask the following questions to determine if students understand the content presented in this lesson.

1. $\angle ABC$ and $\angle DBE$ are vertical angles. $m\angle ABC = (5x + 15)°$ and $m\angle DBE = (4x + 20)°$. Find their measures. 40°
2. Are $\angle AXB$ and $\angle EXD$ vertical angles? Explain. Yes; their sides form two pairs of opposite rays.

Assignment Guide

Basic: 1–29, 36–45
Advanced: 1–45
Additional Practice: See Extra Practice Index on page 674.

Additional Answers

Check Understanding
Since $\angle DXA$ and $\angle BXC$ are a pair of vertical angles, they must be equal in measure. So, when $n = 16$, the value of $2n + 20$, must be equal to 52. Check: $2n + 20 = 2(16) + 20 = 32 + 20 = 52$.

30. Since $m\angle ZXY + m\angle YXW + m\angle WXV = 180°$, $\angle ZXV$ is a straight angle. Therefore, $\overrightarrow{XZ}$ and $\overrightarrow{XV}$ are opposite rays.
31. Since $m\angle YXW + m\angle WXV + m\angle VXU = 176°$, $\angle AXB$ is an obtuse angle. Therefore, $\overrightarrow{XY}$ and $\overrightarrow{XU}$ are not opposite rays.
32. The labels on the figure show that each of $\angle UXV$ and $\angle VXW$ has a measure of 43°. Since the angles are equal in measure, $\overrightarrow{XV}$ bisects $\angle UXW$.
33. Since $\angle TXV$ is a right angle, $\angle TXU$ and $\angle UXV$ are complementary, and $\angle TXU = 90° - 43° = 47°$. Since $\angle TXU$ and $\angle UXV$ are not equal in measure, $\overrightarrow{XU}$ does not bisect $\angle TXV$.
34. The sides of the angles do not form two pairs of opposite rays.
35. No information is given to indicate that $XZ = XV$, so it is not possible to identify X as the midpoint of $\overline{ZV}$.

Example 3

In the figure at the right, $\overleftrightarrow{AC}$ and $\overleftrightarrow{DB}$ intersect at point X. Find $m\angle DXA$.

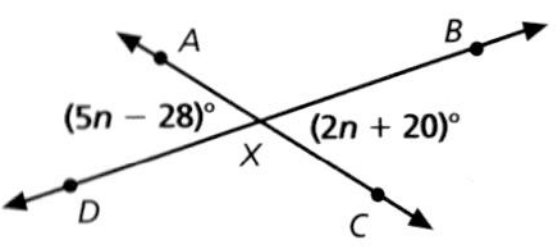

Solution

Since $\angle DXA$ and $\angle BXC$ are a pair of vertical angles, they are equal in measure. Use this fact to write and solve an equation.

$$m\angle DXA = m\angle BXC$$
$$5n - 28 = 2n + 20 \quad \text{Add } -2n \text{ to each side.}$$
$$3n - 28 = 20 \quad \text{Add 28 to each side.}$$
$$3n = 48 \quad \text{Multiply each side by } \tfrac{1}{3}.$$
$$n = 16$$

So, the value of n is 16. From the figure, $m\angle DXA = (5n - 28)°$. Substituting 16 for n, $m\angle DXA = (5 \cdot 16 - 28)° = (80 - 28)° = 52°$.

Check Understanding

In Example 3, how can you use the expression $2n + 20$ to check that the answer is correct?

See additional answers.

Try These Exercises

Refer to the figure below. Find the midpoint of each segment.

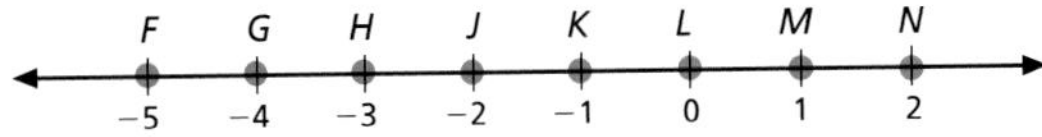

1. $\overline{FM}$ point J
2. $\overline{JN}$ point L
3. $\overline{HM}$ point K
4. In the figure below, $\overrightarrow{RT}$ bisects $\angle SRU$. Find $m\angle URV$. 16°
5. In the figure below, find $m\angle QZW$ and $m\angle QZY$. 119°; 61°

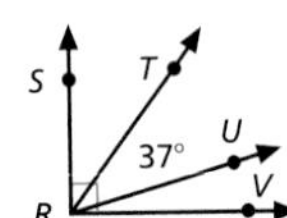

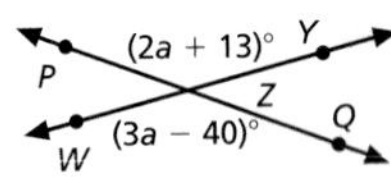

Practice Exercises

Exercises 6–9 refer to the figure at the right.

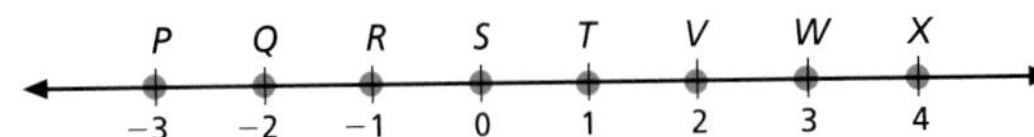

6. Name the midpoint of $\overline{PT}$. point R
7. Name the segment whose midpoint is point W. $\overline{VX}$
8. Name all the segments whose midpoint is point T. $\overline{SV}$, $\overline{RW}$, and $\overline{QX}$
9. Assume that point Y is the midpoint of $\overline{PV}$. What is its coordinate? −0.5

ART A company logo has five line segments that seem to radiate from a single point. An artist is recreating the logo on a computer using the figure on the right. In the figure, $\overleftrightarrow{AD}$ and $\overleftrightarrow{BE}$ intersect at point F, and $\overrightarrow{FC}$ bisects $\angle BFD$. Find the measure of each angle.

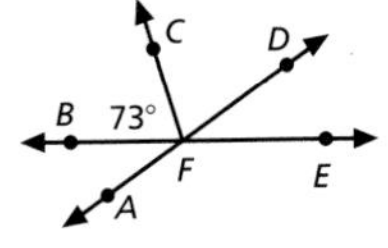

10. $\angle BFD$ 146°
11. $\angle AFE$ 146°
12. $\angle BFA$ 34°

In the figure at the right, $\overleftrightarrow{MN}$, $\overleftrightarrow{PQ}$, and $\overleftrightarrow{RS}$ intersect at point T, and $\overrightarrow{TP}$ bisects $\angle MTR$. Find the measure of each angle.

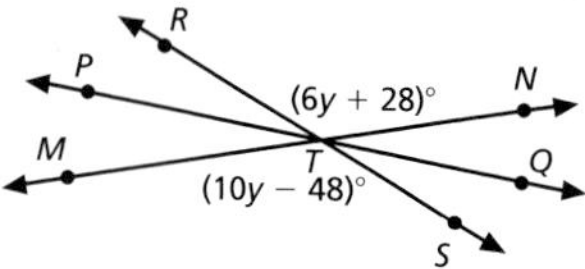

13. $\angle MTS$ 142° **14.** $\angle MTR$ 38° **15.** $\angle NTQ$ 19°

MAP MAKING The figure at the right represents the distances between four cities along a bus route. City O is located at the midpoint between cities P and Q. Find the distance between these cities.

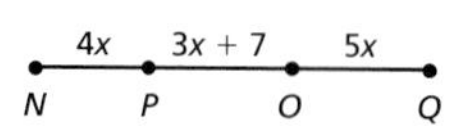

16. O and Q 17.5 **17.** N and P 14 **18.** N and O 31.5 **19.** N and Q 49

In the figure at the right, $\overleftrightarrow{GJ}$ and $\overleftrightarrow{HL}$ intersect at point M, and $\overrightarrow{MK}$ bisects $\angle JML$. Find the measure of each angle.

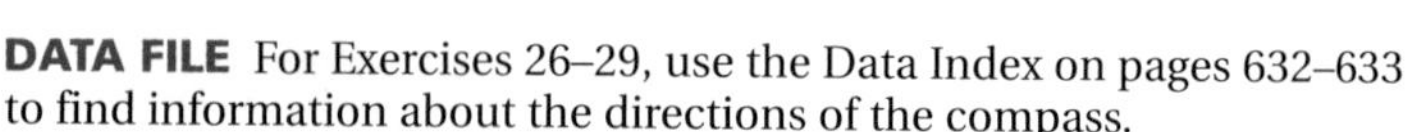

20. $\angle KML$ 59° **21.** $\angle JML$ 118° **22.** $\angle HMJ$ 62°

23. $\angle GML$ 62° **24.** $\angle HMK$ 121° **25.** $\angle KMG$ 121°

DATA FILE For Exercises 26–29, use the Data Index on pages 632–633 to find information about the directions of the compass.

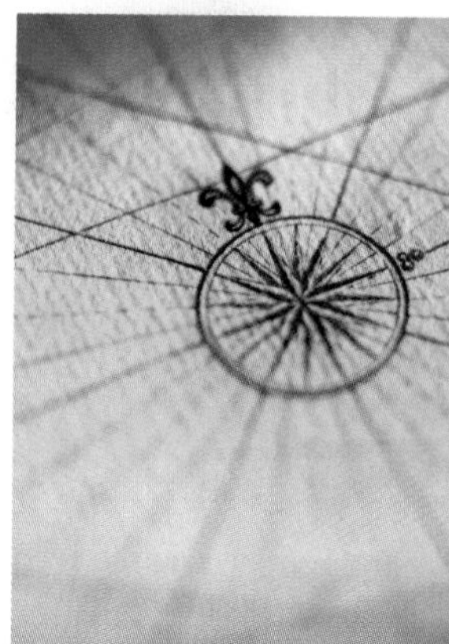

26. Which direction is directly opposite NE? SW

27. What is the measure of the angle formed by NW and W? 45°

28. What is the measure of the angle formed by SE and SW? 90°

29. Which two directions form a 135° angle with NW? E and S

Extended Practice Exercise

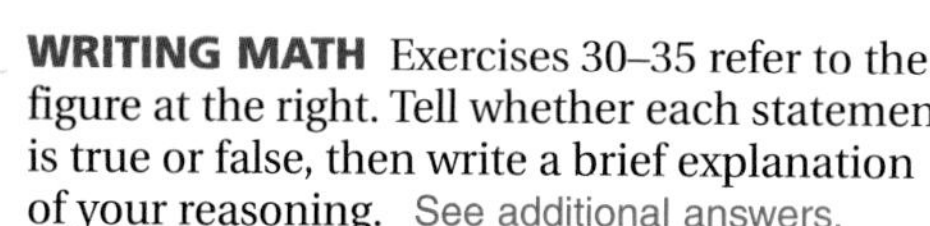

WRITING MATH Exercises 30–35 refer to the figure at the right. Tell whether each statement is true or false, then write a brief explanation of your reasoning. See additional answers.

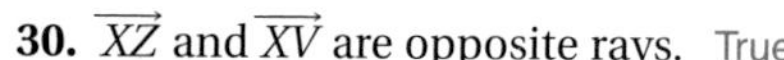

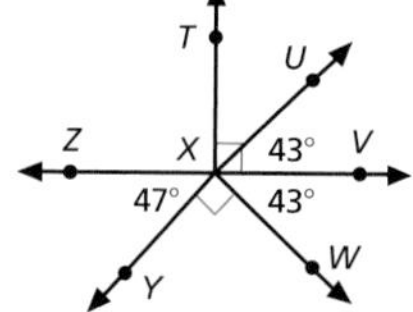

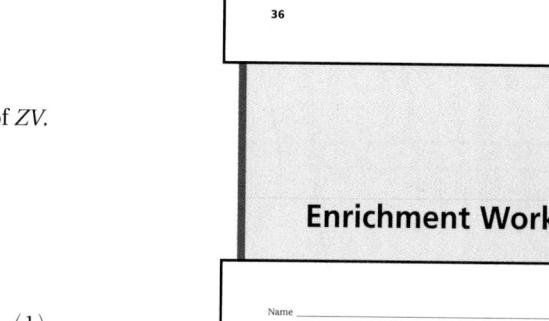

30. $\overrightarrow{XZ}$ and $\overrightarrow{XV}$ are opposite rays. True

31. $\overrightarrow{XY}$ and $\overrightarrow{XU}$ are opposite rays. False

32. $\overrightarrow{XV}$ bisects $\angle UXW$. True

33. $\overrightarrow{XU}$ bisects $\angle TXV$. False

34. $\angle TXU$ and $\angle ZXY$ are a pair of vertical angles. False

35. Point X is the midpoint of ZV. Cannot tell.

Mixed Review Exercises

Use a graphing calculator. Solve each equation by graphing each side of the equation and finding the point of intersection. (Lesson 1-6)

36. $x - 6 = 2$ 8 **37.** $2x + 3 = 8$ 2.5 **38.** $-2x - 3 = 7$ −5 **39.** $\left(\frac{1}{2}\right)x + 4 = -3$ −14

Identify the rule relating each term. Find the next three terms in each sequence. (Lesson 2-1)

40. 1 2 4 8 16 32
rule: × 2; 64, 128, 256

41. 1 −3 9 −27 81
rule: × −3; −243, 729, −2187

42. 1 13 25 37 49
rule: + 12; 61, 73, 85

43. 100 92.5 85 77.5 70
rule: − 7.5; 62.5, 55, 47.5

44. 1000 −500 250 −125 62.5
rule: ÷ −2; −31.25, 15.625, −7.8125

45. 50 54 27 31 15.5
rule: + 4, ÷ 2; 19.5, 9.75, 13.75

Teaching Strategies

CHALLENGE Since the statement about vertical angles is a theorem, it must be able to be proven. Have students draw an example of vertical angles, and write a paragraph to prove that one of the pairs of vertical angles is equal in measure. (Hint: Look for supplementary angles.) The proof lies in the fact that two vertical angles are each supplementary to the same angle.

Extra Practice Worksheet 3-3

Name ______________ Date ______________

EXTRA PRACTICE **3-3**

SEGMENTS AND ANGLES

EXERCISES

For Exercises 1–7 refer to the figure below.

A B C D E F G H I J K L M N P
−7 −6 −5 −4 −3 −2 −1 0 1 2 3 4 5 6 7

1. Name the midpoint of $\overline{FN}$. point J
2. Find the midpoint of $\overline{DL}$. point H
3. Find the midpoint of $\overline{GM}$. point J
4. Name all the segments whose midpoint is point M. $\overline{LN}$, $\overline{KP}$
5. Name all the segments whose midpoint is point D. $\overline{CE}$, $\overline{BF}$, $\overline{AG}$
6. Assume that point X is the midpoint of $\overline{GL}$. What is its coordinate? 1.5
7. Assume that point Y is the midpoint of $\overline{CH}$. What is its coordinate? −2.5

In the figure at the right, $\overleftrightarrow{ML}$, $\overleftrightarrow{NK}$, and $\overleftrightarrow{PJ}$ intersect at point H, $\overleftrightarrow{HK}$ bisects $\angle LHJ$, $\overleftrightarrow{HN}$ bisects $\angle PHM$, and $m\angle PHL = 136°$. Find the measure of each angle.

8. $\angle LHJ$ 44°
9. $\angle MHJ$ 136°
10. $\angle PHM$ 44°
11. $\angle LHK$ 22°
12. $\angle PHN$ 22°
13. $\angle NHM$ 22°

In the figure at the right, point N is the midpoint of $\overline{PH}$. Find the length of each segment.

P 8x + 3 N 9x − 6 H 5x + 5 J

14. $\overline{PN}$ 75
15. $\overline{NH}$ 75
16. $\overline{HJ}$ 50
17. $\overline{PJ}$ 200

36 EXTRA PRACTICE *LESSON 3-3*

Enrichment Worksheet 3-3

Name ______________ Date ______________

ENRICHMENT **3-3**

IMAGES IN PLANE MIRRORS

In this experiment, you are seeking a relationship between the angle formed by two mirrors and the number of images you see in the mirrors.

Materials

two plane mirrors • masking tape • coin or other small object • protractor

Experimental Set Up

Hinge the mirrors together with the masking tape so that you can vary the angle between them. Start by placing the mirrors on top of the protractor at a 90° angle, with the coin on the 45° bisector line. This arrangement is shown in the diagram. You should see four coins in all, the real coin and three images.

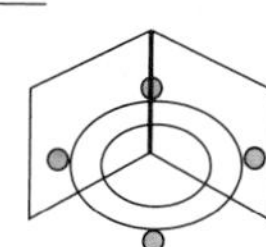

EXERCISES

1. Vary the angle between the mirrors and count the number of coins you see. Always place the coin on the bisector of the angle. Record your data in this chart.

Data will vary, depending on the angles students choose.

Angle Between Mirrors (θ)	Total Number of Coins (n = 1)	Number of Images (n)

2. Write an equation relating θ and n. $n = \frac{360°}{\theta} - 1$

48 ENRICHMENT *LESSON 3-3*

Lesson 3-3 **Segments and Angles** 117

Lesson Planning

NCTM Standards/Strands

- Number & Operation
- Patterns, Functions, & Algebra
- Geometry & Spatial Sense
- Measurement
- Problem Solving
- Reasoning and Proof
- Communication
- Connections
- Representation

Vocabulary

compass
straightedge
perpendicular lines
parallel lines
skew lines
alternate interior angles
transversal
alternate exterior angles
bearings
corresponding angles

Materials Needed

paper/pencil
grid or graph paper
compass, protractor, straightedge
geometric drawing software

Lesson Resources

Warm-Up Transparency 8
Transparency RF-13
Reteaching 3-4
Extra Practice 3-4
Enrichment 3-4
Multicultural Connection 6

Assignment Guide

Basic: 1–14, 19, 21–22
Enriched: 1–22

3-4 Constructions and Lines

Goals
- Use constructions of segment bisectors and angle bisectors to solve problems.
- Identify parallel, perpendicular and skew lines.
- Identify congruent angles formed by parallel lines and a transversal.

Applications Architecture, Surveying, Navigation

Work with a partner.

Draw a line segment on a piece of paper, and find its midpoint. Can you find a solution that does not require measuring the segment?
Check students' work. Possible answer: Use paper folding to locate the midpoint.

Build Understanding

A geometric construction is a precise drawing of a geometric figure made with the aid of only two tools: a compass and a straightedge. One of the most basic geometric constructions is the segment bisector construction described below.

Example 1

Divide line segment into two segments of equal length.

Solution

Step 1: Start with a line segment, $\overline{AB}$.

Step 2: Open the compass to a radius that is more than half AB. With the compass tip at A, draw one arc above and one arc below $\overline{AB}$.

Step 3: Place the compass tip at B. Using the same radius as in Step 2, draw arcs above and below $\overline{AB}$.

Step 4: Label the points where the two pairs of arcs intersect X and Y. Using a straightedge, draw $\overleftrightarrow{XY}$. Label the point where $\overleftrightarrow{XY}$ intersects $\overline{AB}$ as point M. This is the midpoint of $\overline{AB}$.

Reading Math

A **compass** is used to draw circles and arcs. The **straightedge** is used to draw lines, rays, and segments. You may use a ruler as a straightedge but you must ignore the markings.

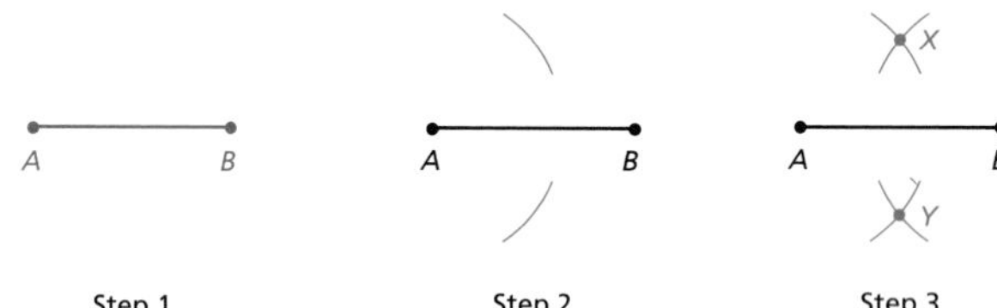

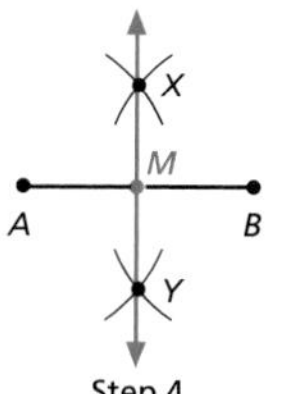

Another important geometric construction is the angle bisector construction.

Getting Started

5-Minute Warm-up

Find the supplement of each.
1. 35° 145° **2.** 110° 70°
Solve for *x*.
3. $4x - 18 = 2x + 60$ 39
4. $3x + 20 = 2x + 75$ 55

Introduction to Lesson 3-4

Students may use a variety of methods to bisect the segment including paper folding and making successive approximations using various objects. Have students share their methods with the class.

Teaching Strategies

ESL/LEP These students may benefit from making a vocabulary list of the terms presented in this chapter, including an illustration of each term. They should keep the list handy to refer to when reading the lessons and working on exercises; they can add to it as necessary.

Example 2

ARCHITECTURE On the plans for a building, the angle of the roof is bisected by a length of decorative molding. Draw the bisector using a compass and a straightedge.

Step 1: Start with an angle, $\angle ABC$.

Step 2: With the compass point at B, draw an arc that intersects $\overrightarrow{BA}$ and $\overrightarrow{BC}$. Label the points of intersection P and Q.

Step 3: Use a radius that is more than half PQ. Place the compass tip at P. Draw an arc in the interior of $\angle ABC$. With the compass set to the same radius, place the compass tip at Q and draw an arc that intersects the first arc. Label the point of intersection X.

Step 4: Using a straightedge, draw $\overrightarrow{BX}$. This is the bisector of $\angle ABC$.

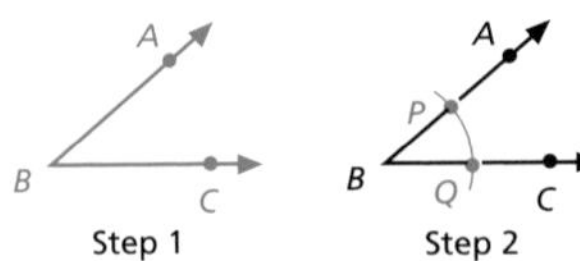

Step 1   Step 2

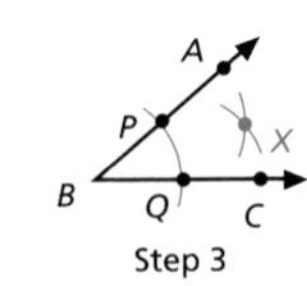

Step 3

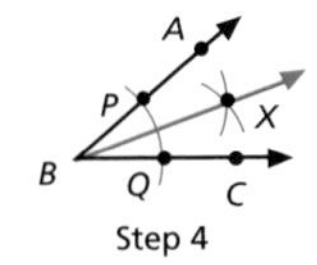

Step 4

Technology Note

Another way to bisect segments and angles is using geometric-drawing software. Try this activity.

1. Draw $\angle JKL$ with a measure of 60°.
2. Select $\angle JKL$ and use the construction menu to draw the angle bisector. Label point M on the bisector.
3. Measure the new angles: $\angle JKM$ and $\angle KML$. Each should measure 30°.

Two lines that intersect to form right angles are called **perpendicular lines**. Because a ray and a segment are each part of a line, the word *perpendicular* (shown by $\perp$) also can refer to rays and segments.

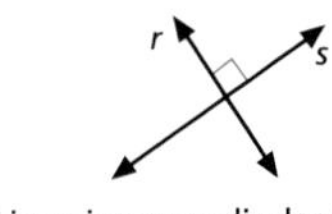

Line r is perpendicular to line s.

$r \perp s$

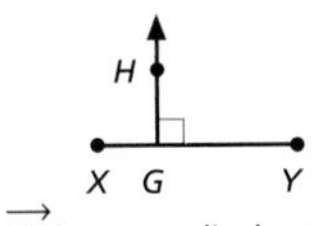

$\overrightarrow{GH}$ is perpendicular to $\overline{XY}$.

$\overrightarrow{GH} \perp \overline{XY}$

Example 3

In the figure, $\overleftrightarrow{AD} \perp \overleftrightarrow{BE}$. Find m$\angle AZF$.

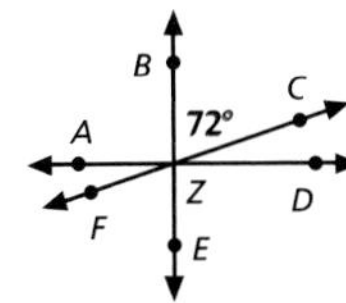

Solution

Since $\overleftrightarrow{AD} \perp \overleftrightarrow{BE}$, $\angle BZD$ is a right angle. The exterior sides of $\angle BZC$ and $\angle CZD$ form a right angle, so the angles are complementary. So, $72° + m\angle CZD = 90°$, and $m\angle CZD = 90° - 72° = 18°$. Because $\angle AZF$ and $\angle CZD$ are vertical angles, $m\angle AZF = m\angle CZD$. So, $m\angle AZF = 18°$.

Not all lines intersect.

Parallel lines ($\|$) are *coplanar* lines that do not intersect.

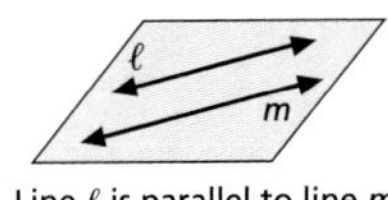

Line ℓ is parallel to line m.

$\ell \| m$

Noncoplanar lines are called **skew lines**.

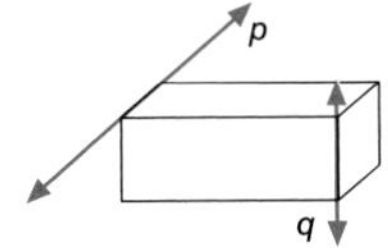

Lines p and q are skew lines.

Interdisciplinary Connections

HISTORY From ancient times, mathematicians have looked without success for a way to trisect an angle using only a compass and straightedge. Some ancient scholars were able to trisect the angle using other means. For example, Hippias (425 B.C.) used a quadratrix, and Archimedes (250 B.C.) used a spiral. Some students may be interested in researching this topic.

Chalkboard Examples

Supplementary Example 1

Have students draw a large triangle and bisect one of the sides of the triangle using the method explained in Example 1. Check students' drawings.

Supplementary Example 2

Have students draw an obtuse triangle and bisect the obtuse angle using the method described in Example 2. Check students' drawings.

Supplementary Example 3

In the figure below, $\overline{AD} \perp \overline{FC}$. Find $m\angle DGE$. 60°

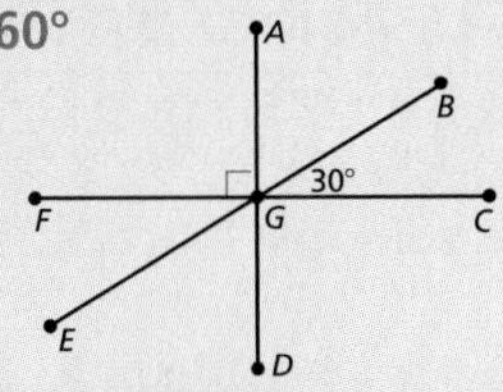

Supplementary Example 4

In the figure below, $\overleftrightarrow{AB} \| \overleftrightarrow{CD}$. Find $m\angle AGH$. 54°

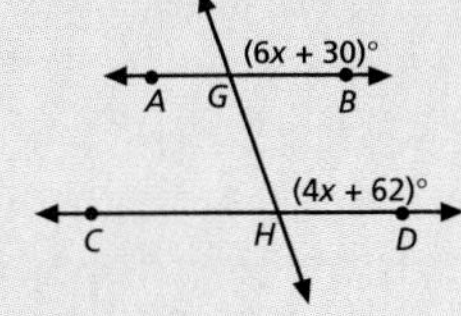

Reteaching Worksheet 3-4

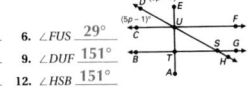

Name ______________________ Date ______________

RETEACHING 3-4

CONSTRUCTION AND LINES

If two parallel lines are cut by a transversal, then corresponding angles are equal in measure.
$\angle 1$ and $\angle 5$ are corresponding angles, so $m\angle 1 = m\angle 5$.
$\angle 2$ and $\angle 6$ are corresponding angles, so $m\angle 2 = m\angle 6$.
$\angle 3$ and $\angle 7$ are corresponding angles, so $m\angle 3 = m\angle 7$.
$\angle 4$ and $\angle 8$ are corresponding angles, so $m\angle 4 = m\angle 8$.
$6h - 5 = 4h + 35$

Example 1

In the figure at the right, $\overline{MO} \perp \overrightarrow{AP}$. Find $m\angle OAN$.

Solution

Since $\overline{MO} \perp \overrightarrow{AP}$,
$\angle PAM$ is a right angle and $m\angle PAM = 90°$.
Since $m\angle PAQ$ and $m\angle QAM$ are complementary,
$m\angle PAQ + m\angle QAM = 90°$.
It is given that $m\angle PAQ = 26°$.
So, $m\angle QAM + 26° = 90°$ and
$m\angle QAM = 90° - 26° = 64°$.
Since $\angle QAM$ and $\angle OAN$ are vertical angles, they are equal in measure.
$m\angle OAN = m\angle QAM = 64°$

Example 2

In the figure at the right, $\overline{AE} \| \overline{BD}$. Find $m\angle CTD$ and $m\angle CSE$.

Solution

Since $\angle BTF$ and $\angle CTD$ are vertical angles, they are equal in measure.
$6h - 5 = 4h + 35$
$2h = 40$
$h = 20$
$m\angle CTD = (4h + 35)° =$
$[4(20) + 35]° = (80 + 35)° = 115°$
$\angle CSE$ and $\angle CTD$ are corresponding angles, so they are equal in measure.
$m\angle CSE = m\angle CTD = 115°$

EXERCISES

In the figure at the right, $\overline{AE} \perp \overline{FC}$. Find the measure of each angle.

1. $\angle AYB$ 71°
2. $\angle DYE$ 52°
3. $\angle GYF$ 38°

In the figure at the right, $\overleftrightarrow{BG} \| \overleftrightarrow{CF}$ and $\overline{AE} \perp \overleftrightarrow{BG}$. Find the measure of each angle.

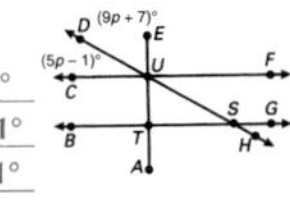

4. $\angle CUE$ 90°
5. $\angle BTS$ 180°
6. $\angle FUS$ 29°
7. $\angle CUD$ 29°
8. $\angle GSH$ 29°
9. $\angle DUF$ 151°
10. $\angle EUS$ 119°
11. $\angle EUD$ 61°
12. $\angle HSB$ 151°

42   RETEACHING *LESSON 3-4*

Lesson Wrap-up

Quick Assessment

Ask the following questions to determine if students understand the content presented in this lesson.

1. If two lines are perpendicular, what do you know about the angles formed by the lines? They are four right angles.
2. If two parallel lines are intersected by a transversal, which pairs of angles have equal measure? corresponding, alternate interior, and alternate exterior angles

Assignment Guide

Basic: 1–14, 19, 21–22
Advanced: 1–22
Additional Practice: See Extra Practice Index on page 674.

Additional Answers

7. Check students' drawings. Begin with the segment bisector construction shown above. Because point M is the midpoint of $\overline{AB}$, $AN = MB$. Now use the segment bisector construction two more times to find the midpoint L of $\overline{AM}$ and midpoint N of $\overline{MB}$. Points L, M, and N divide $\overline{AB}$ into four segments of equal length: $AL = LM = MN = NB$.

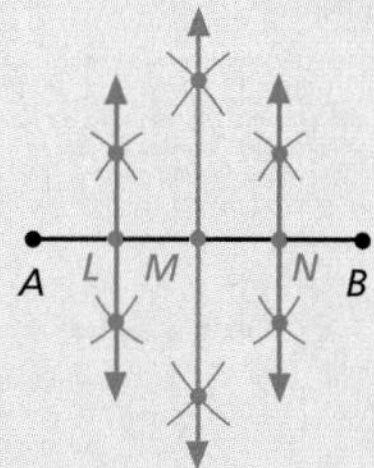

8. No. Three segments of equal length are not possible. Using the segment bisector construction, it is only possible to divide a segment into an even number of segments of equal length. Using the segment bisector construction, the number of segments of equal length that you can obtain is 2, 4, 8, 16, 32, . . . , or $2^1, 2^2, 2^3, 2^4, 2^5, \ldots$. So, it is possible to use the segment bisector construction to divide a segment into 2^n segments of equal length, where n is any positive integer.

As with the word perpendicular, *parallel* also is used in reference to rays and segments.

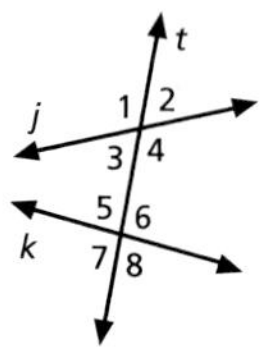

The figure at the right shows lines j and k intersected by a third line, t, called a *transversal.* A **transversal** is a line that intersects two or more coplanar lines in different points. The four angles between the lines—$\angle 3$, $\angle 4$, $\angle 5$, and $\angle 6$—are called **interior angles.** The other four angles are called **exterior angles.**

Alternate interior angles are nonadjacent interior angles on opposite sides of the transversal. There are two pairs of alternate interior angles shown above.

$\angle 3$ and $\angle 6$ $\quad$ $\angle 4$ and $\angle 5$

Alternate exterior angles are nonadjacent exterior angles on opposite sides of the transversal. There are two pairs of alternate exterior angles.

$\angle 1$ and $\angle 8$ $\quad$ $\angle 2$ and $\angle 7$

Corresponding angles are in the same position relative to the transversal and the lines. The figure above has four pairs of corresponding angles.

$\angle 1$ and $\angle 5$ $\quad$ $\angle 2$ and $\angle 6$ $\quad$ $\angle 3$ and $\angle 7$ $\quad$ $\angle 4$ and $\angle 8$

Postulate 9	**The Parallel Lines Postulate** If two parallel lines are cut by a transversal, then corresponding angles are equal in measure.

Example 4

In the figure, $\overleftrightarrow{GH} \parallel \overleftrightarrow{JK}$. Find $m\angle HQR$.

Solution

Since $\angle PQG$ and $\angle QRJ$ are corresponding angles, they are equal in measure.

$$m\angle PQG = m\angle QRJ$$
$$3c + 40 = 7c - 72 \quad \text{Add } -3c \text{ to each side.}$$
$$40 = 4c - 72 \quad \text{Add 72 to each side.}$$
$$112 = 4c \quad \text{Multiply each side by } \tfrac{1}{4}.$$
$$28 = c$$

Substituting 28 for c, $m\angle PQG = (3 \cdot 28 + 40)° = (84 + 40)° = 124°$.

Since $\angle PQG$ and $\angle HQR$ are vertical angles, $m\angle HQR = 124°$.

Check Understanding

In the figure below, which line is a transversal?

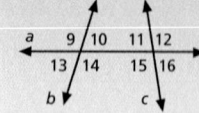

Which pairs of angles are alternate interior angles? alternate exterior angles? corresponding angles?

See additional answers.

Try These Exercises

In the figure at the right, $m\angle TZV = m\angle UZS$, and $\overleftrightarrow{RS} \perp \overleftrightarrow{VW}$. Find the measure of each angle.

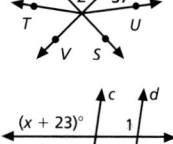

1. $\angle SZV$ 90°
2. $\angle RZT$ 37°
3. $\angle TZV$ 53°

In the figure at the right, $c \parallel d$. Find the measure of each angle.

4. $\angle 1$ 97°
5. $\angle 2$ 97°
6. $\angle 3$ 83°

15–18

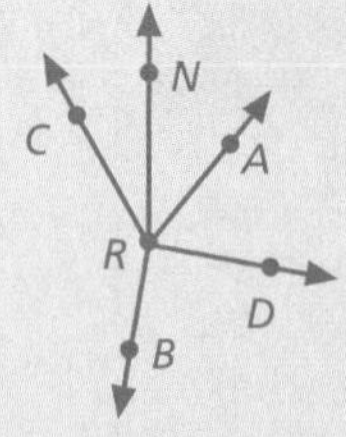

Check Understanding
Line a is a transversal. Alternate interior angles are 10, 11, 14, and 15. Alternate exterior angles are 9, 12, 13, and 16. Corresponding angles are 9 and 11, 10 and 12, 13 and 15, and 14 and 16.

PRACTICE EXERCISE

7. **GEOMETRY SOFTWARE** First, on paper using only a compass and a straightedge, divide a line segment, $\overline{AB}$, into four segments of equal length. Try the same construction using geometric-drawing software.
Check students' drawings.

8. **WRITING MATH** Can the segment bisector construction be used to divide a segment into three segments of equal length? Explain your thinking.
See additional answers.

SURVEYING City streets form the angles shown at the right. In the figure, $\overleftrightarrow{CF} \parallel \overleftrightarrow{BG}$ and $\overleftrightarrow{AE} \perp \overleftrightarrow{DH}$. Find the measure of each angle.

9. $\angle ELF$ 51°
10. $\angle LKJ$ 51°
11. $\angle DLC$ 39°
12. $\angle LJK$ 39°
13. $\angle KJH$ 141°
14. $\angle CLJ$ 141°

NAVIGATION The **bearing** of a ship is the measure of the angle formed by a ray pointing due north from the harbor, $\overrightarrow{HN}$, and the ray that represents the ship's course out of the harbor, $\overrightarrow{HS}$, measured in a clockwise direction. A bearing may be any measure from 0° up to 360°, but may not include 360°. In the figure, the ship's bearing is $180° + 60° = 240°$

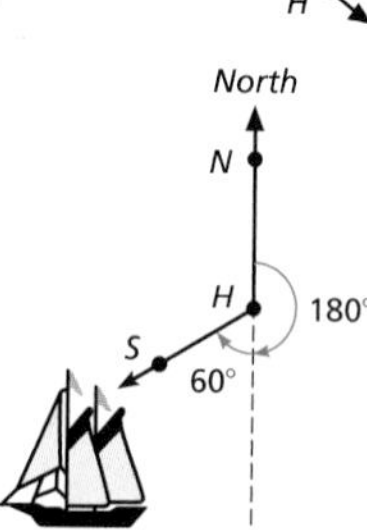

For Problems 15–18, see additional answers.

Draw a point R near the middle of a sheet of paper. Draw $\overrightarrow{RN}$ to show due north from R. Using a protractor, draw a ray to show each bearing.

15. $\overrightarrow{RA}$, 40°
16. $\overrightarrow{RB}$, 190°
17. $\overrightarrow{RC}$, 330°
18. $\overrightarrow{RD}$, 100°

EXTENDED PRACTICE EXERCISES

19. **CHAPTER INVESTIGATION** Using the map you made, sketch a course to travel between two points. Write directions in which you specify the bearing of each leg of the journey. Answers will vary.

20. **CRITICAL THINKING** Janeira says that three lines that intersect in one point are always coplanar. Do you agree? No. Three lines intersecting in one point are sometimes coplanar.

MIXED REVIEW EXERCISES

Use the Vertical Line Test to determine if each graph is a function. (Lesson 2-2)

21.

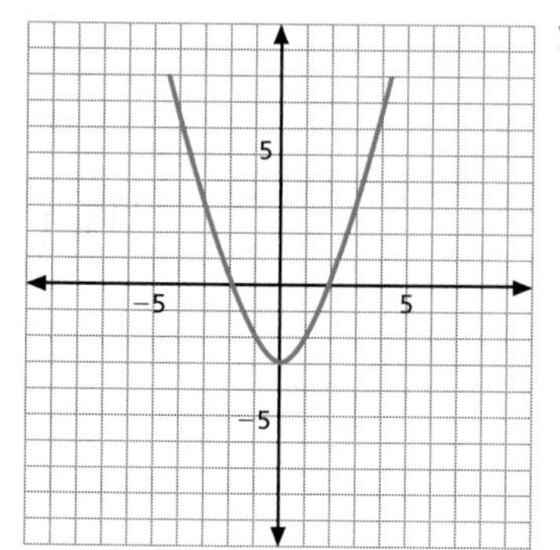

yes

22.

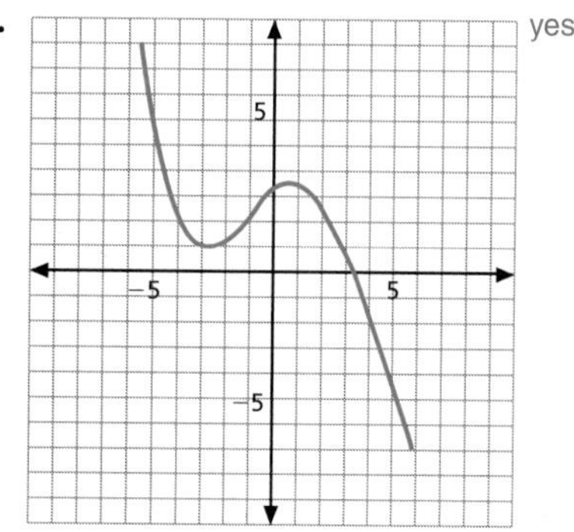

yes

Teaching Strategies

ESL/LEP To make sure students understand how bearings are measured, have them stand and face north. Then turn to show a bearing of 90°, 180°, and 270°. Have them identify the direction each bearing represents. east; south; west

Extra Practice Worksheet 3-4

Name ______ Date ______

EXTRA PRACTICE 3-4
CONSTRUCTION AND LINES

EXERCISES

1. Trace $\overline{PM}$ onto a sheet of paper. Using a compass and straightedge, divide it into four segments of equal length. Use a ruler to check the accuracy of your construction. Check students' work.

2. Trace $\angle JKL$ onto a sheet of paper. Using a compass and a straightedge, divide $\angle JKL$ into four angles of equal measure. Use a protractor to check the accuracy of your construction. Check students' work.

In the figure at the right, $\overleftrightarrow{BE} \perp \overleftrightarrow{GD}$ and $m\angle CAD = 60°$.

3. $\angle BAC$ 30°
4. $\angle GAF$ 60°
5. $\angle EAF$ 30°
6. $\angle DAE$ 90°
7. $\angle GAB$ 90°
8. $\angle FAD$ 120°

Find the measure of each angle.

9. $\angle 6$ 68°
10. $\angle 8$ 68°
11. $\angle 4$ 68°
12. $\angle 1$ 112°
13. $\angle 5$ 112°
14. $\angle 3$ 112°
15. $\angle 7$ 112°

38 EXTRA PRACTICE *LESSON 3-4*

Enrichment Worksheet 3-4

Name ______ Date ______

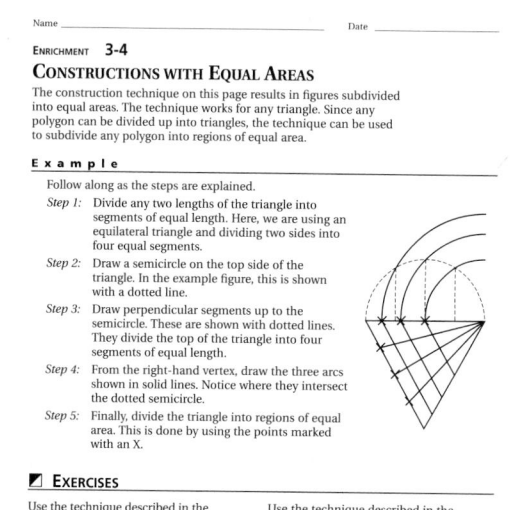

ENRICHMENT 3-4
CONSTRUCTIONS WITH EQUAL AREAS

The construction technique on this page results in figures subdivided into equal areas. The technique works for any triangle. Since any polygon can be divided up into triangles, the technique can be used to subdivide any polygon into regions of equal area.

Example

Follow along as the steps are explained.

Step 1: Divide any two lengths of the triangle into segments of equal length. Here, we are using an equilateral triangle and dividing two sides into four equal segments.

Step 2: Draw a semicircle on the top side of the triangle. In the example figure, this is shown with a dotted line.

Step 3: Draw perpendicular segments up to the semicircle. These are shown with dotted lines. They divide the top of the triangle into four segments of equal length.

Step 4: From the right-hand vertex, draw the three arcs shown in solid lines. Notice where they intersect the dotted semicircle.

Step 5: Finally, divide the triangle into regions of equal area. This is done by using the points marked with an X.

EXERCISES

Use the technique described in the example to create a regular hexagon divided into equal areas on your own paper. Each side of the hexagon will be divided into four segments of equal length. Before you begin, answer these questions.

1. How many equilateral triangles will you need? 6
2. How many regions of equal area will you get? 96

Use the technique described in the example to create a square divided into equal areas on your own paper. Each side of the square will be divided into eight segments of equal length. Before you begin, answer these questions.

3. How many isosceles right triangles will you need? 8
4. How many regions of equal area will you get? 128

Note: The designs are very attractive when colored in checkerboard-type patterns.

50 ENRICHMENT *LESSON 3-4*

Skills Practice

Vocabulary Review

Lesson 3-3
midpoint, bisector, theorem, opposite rays, vertical angles

Lesson 3-4
compass, straightedge, perpendicular lines, parallel lines, skew lines, alternate interior angles, alternate exterior angles, corresponding angles, bearings, transversal

Assignment Guide

All students: 1–35
Additional Practice: Refer to the Extra Practice Index on page 674 of this text.

Review and Practice Your Skills

Practice ■ Lesson 3-3

Exercises 1–4 refer to the figure at the right.

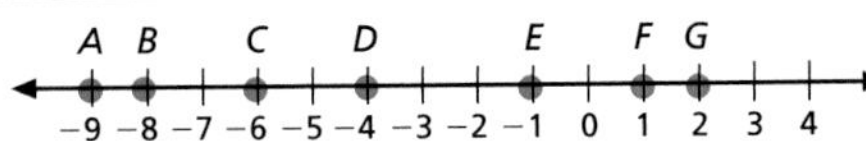

1. Name the midpoint of $\overline{BF}$. D
2. Name the segment whose midpoint is point E. $\overline{DG}$
3. Name all the segments whose midpoint is point C. $\overline{BD}$
4. Assume the point Z is the midpoint of $\overline{AG}$. Find its coordinate. −3.5

In the figure at the right, $\overleftrightarrow{GH}$ bisects $\angle TUV$. Find the measure of each angle.

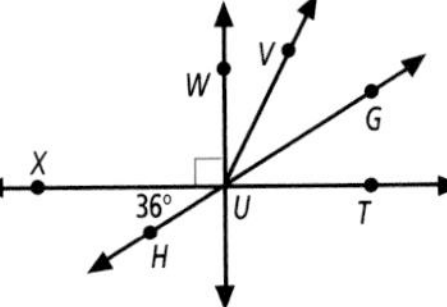

5. $\angle VUX$ 108°
6. $\angle GUW$ 54°
7. $\angle TUH$ 144°
8. $\angle WUV$ 18°

In the figure at the right, point *B* is the midpoint of $\overline{AC}$. Find the length of each segment.

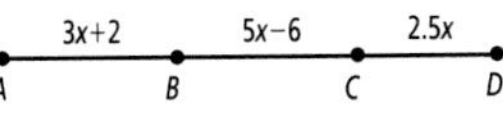

9. $\overline{BC}$ 14
10. $\overline{CD}$ 10
11. $\overline{AC}$ 28
12. $\overline{AD}$ 38

13. *True* or *false*: Only a line segment can be a bisector of a segment. false
14. *True* or *false*: Vertical angles are never supplements of each other. false

Practice ■ Lesson 3-4

15. Draw line segment $\overline{AB}$ 2 inches long. Using a compass and straightedge, construct a perpendicular bisector to this segment. Check students' work.
16. Trace $\overline{MN}$ onto a sheet of paper. Using a compass and straightedge, divide it into eight segments of equal length. Check students' work.

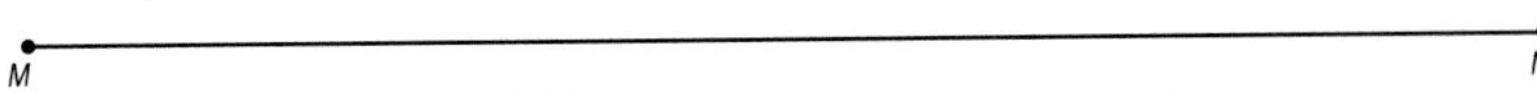

17. Using a compass and straightedge, construct a 22.5° angle. Check students' work.

In the figure at the right, $\overleftrightarrow{DE} \parallel \overleftrightarrow{RS}$ and $\overleftrightarrow{DE} \perp \overleftrightarrow{GH}$. Find the measure of each angle.

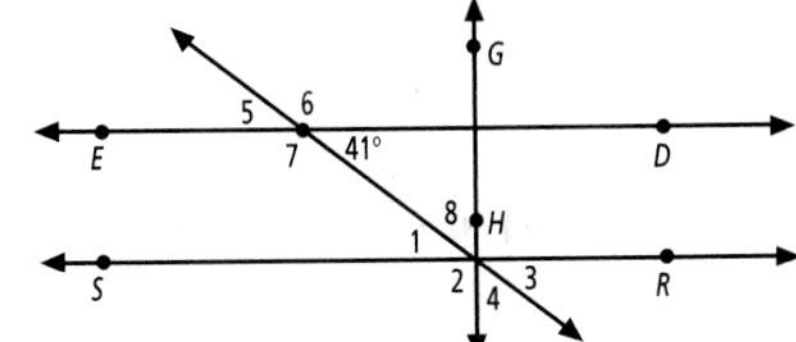

18. $\angle 1$ 41°
19. $\angle 2$ 90°
20. $\angle 3$ 41°
21. $\angle 4$ 49°
22. $\angle 5$ 41°
23. $\angle 6$ 139°
24. $\angle 7$ 139°
25. $\angle 8$ 49°

26. *True* or *false*: When two parallel lines are cut by a transversal, the alternate interior angles are supplements of each other. false
27. *True* or *false*: When two parallel lines are cut by a transversal, the alternate exterior angles are supplements of each other. true
28. *True* or *false*: Parallel lines have the same slope. true

Teaching Strategies

For Exercise 17 of Practice Your Skills, remind students to use what they know to construct the angle. Since they have learned to bisect angles and line segments, they must use these skills to construct the small angle. Many students may feel the item is impossible to answer without using a protractor. Encourage them to discover what series of bisections would be needed to arrive at an angle of 22.5°.

PRACTICE ■ LESSON 3-1–LESSON 3-4

29. In the figure below, $PR = 200$. Find PQ. (Lesson 3-1) 128

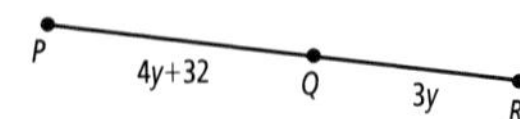

30. In the figure below, $GJ = 35$. Find GH. (Lesson 3-1) 16

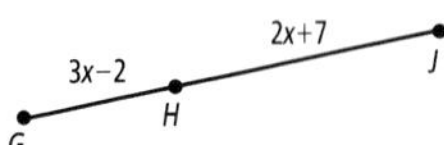

31. In the figure below, find $m\angle CDE$. (Lesson 3-2) 54°

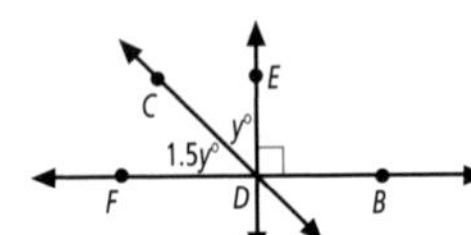

32. In the figure below, find $m\angle TPN$. (Lesson 3-2) 22°

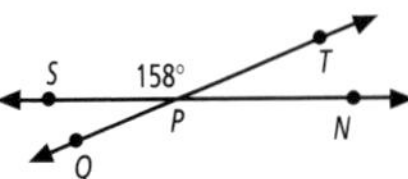

For Exercises 33–35, refer to the rectangular prism on the right. (Lesson 3-4)

33. Name three segments parallel to $\overline{AB}$. $\overline{CD}$, $\overline{EJ}$, $\overline{FH}$

34. Name four segments perpendicular to $\overline{EF}$. $\overline{EJ}$, $\overline{FH}$, $\overline{DF}$, $\overline{AE}$

35. Name four segments skew to $\overline{EJ}$. $\overline{AD}$, $\overline{BC}$, $\overline{DF}$, $\overline{CH}$

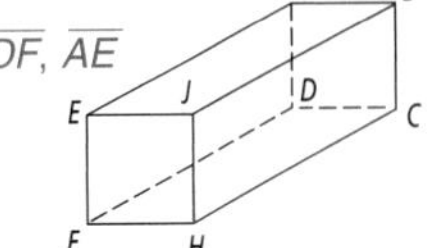

Mid-Chapter Quiz

Identify each of the following.

1. part of a line with one endpoint that extends infinitely in the other direction ray
2. a statement that is assumed to be true postulate
3. points on the same plane coplanar

Find the complement of each angle.

4. 56° 34°
5. $x°$ $(90 - x)°$
6. $3x°$ $(90 - 3x)°$

Find the supplement of each angle.

7. 72° 108°
8. $b°$ $(180 - b)°$
9. $(2x + 1)°$ $(179 - 2x)°$

Find the midpoint of each segment.

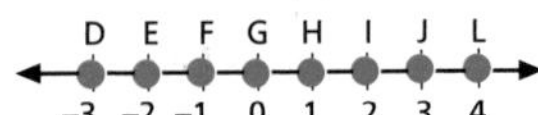

10. $\overline{HL}$ J

11. $\overline{DJ}$ G

***QV* is the bisector of ∠*UQW*. Find the measure of each angle.**

12. $m\angle TQX$ 150°
13. $m\angle UQV$ 75°
14. $m\angle VQX$ 105°

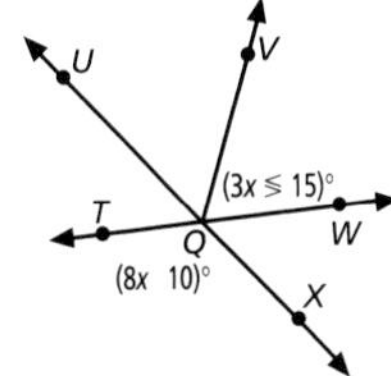

15. Draw two parallel lines and a transversal intersecting both.
See additional answers.

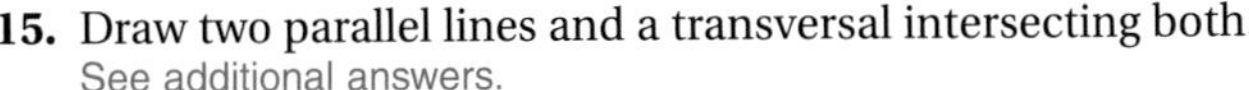

Chalkboard Examples

Example from Lesson 3-3

∠*XYZ* and ∠*QRS* are vertical angles. Find their measures if $m\angle XYZ = 4x - 5$ and $m\angle QRS = x + 25$. $m\angle XYZ = m\angle QRS = 35$

Example from Lesson 3-4

In the figure below, $\overleftrightarrow{AB} \parallel \overleftrightarrow{CD}$, and is perpendicular to $\overleftrightarrow{GH}$. Find the measure of each angle.

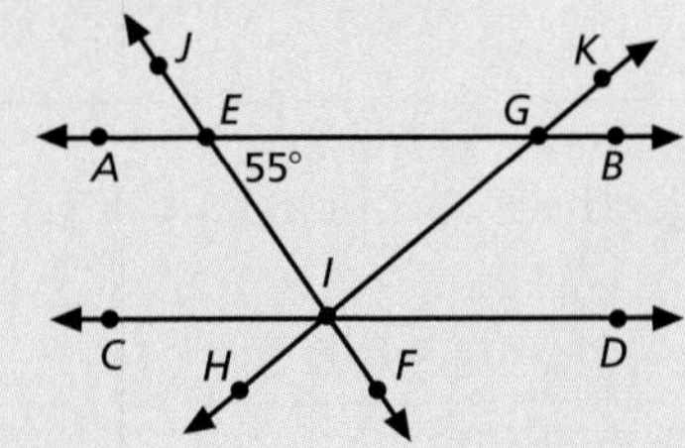

1. ∠*CIH* 35°
2. ∠*DIG* 35°
3. ∠*CIE* 55°
4. ∠*EIG* 90°
5. ∠*GID* 35°
6. ∠*AEJ* 55°

Mid-Chapter Quiz

Correlation Chart

Question Number	Lesson Number
1–3	3-1
4–9	3-2
10–14	3-3
15	3-4

ADDITIONAL ANSWERS

15. Check students drawings. Parallel lines should have indicator on them that the lines are indeed parallel.

Lesson Planning

NCTM Standards/Strands

- Number & Operation
- Patterns, Functions, & Algebra
- Geometry & Spatial Sense
- Problem Solving
- Reasoning and Proof
- Communication
- Connections
- Representation

Vocabulary

inductive reasoning
conjecture sequence

Materials Needed

paper/pencil
straightedges
counters, straws, blocks (optional)

Lesson Resources

Warm-Up Transparency 8
Reteaching 3-5
Extra Practice 3-5
Enrichment 3-5

Assignment Guide

Basic: 1–11, 15–24
Enriched: 1–24

Getting Started

5-Minute Warm-up

Determine the next three terms in each sequence.

1. 3, 9, 15, 21, . . . 27, 33, 39
2. 128, 64, 32, 16, . . . 8, 4, 2
3. 4, 7, 11, 16, . . . 22, 29, 37

Introduction to Lesson 3-5

The rule for the last pattern may be very difficult for students to determine. The sequence is called the Fibonacci sequence. Each additional term in the sequence is the sum of the preceding two terms. You may invite students to invent sequences of their own to challenge the class.

3-5 Inductive Reasoning in Mathematics

Goals ■ Use inductive reasoning to complete patterns.
Applications City Planning, Scheduling, Art

List the next three terms in each sequence. What do you think the twentieth term will be?

a. 1, 9, 17, 25, 33, 41, . . . 49, 57, 65; 153
b. 1, 2, 4, 7, 11, 16, . . . 22, 29, 37; 191
c. 1, 2, 4, 8, 16, 32, . . . 64, 128, 256; 524,288
d. 1, 1, 2, 3, 5, 8, . . . 13, 21, 34; 6,765

Build Understanding

In Lesson 2-1, you learned how to see a pattern in a sequence of numbers and use your observations to extend the pattern. In doing this, you were using *inductive reasoning*. **Inductive reasoning** is the process of observing data and making a generalization based on your observations.

The generalization is called a **conjecture**. In this lesson, you will learn how to apply your inductive reasoning skills to make conjectures about geometric figures.

Example 1

a. Draw the next figure in this pattern.

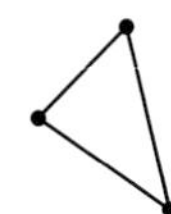 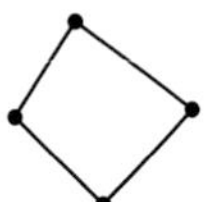  

b. Describe the twentieth figure in the pattern.

Solution

Each figure in the pattern consists of a number of segments that intersect only at their endpoints. It is important to note that, in each figure, no three of the endpoints are collinear.

The 1st figure consists of 3 segments.

The 2nd figure consists of 4 segments.

The 3rd figure consists of 5 segments.

The 4th figure consists of 6 segments.

a. The 5th figure consists of 7 segments, as shown.

b. From the discussion above, you see that the nth figure consists of $(n + 2)$ segments.

Teaching Strategies

COOPERATIVE LEARNING Have students work in pairs to solve the following problem.

A restaurant has square tables that fit one person on each side. When two tables are placed together, six people can sit. How many people can be seated if ten tables are placed together into one long table?

Ask students to find a rule for the number of people who can sit if n tables are joined to form one long table. 22; $2n + 2$

So, the 20th figure will consist of (20 + 2) segments, or 22 segments. The 22 segments intersect only at their endpoints, and no three of the endpoints are collinear.

Many problems in geometry involve not only a geometric pattern, but also a number pattern.

Example 2

CITY PLANNING A traffic engineer is using line segments determined by ten collinear points to represent the stops on a bus route. How many different segments of the route are there?

Solution

Solving the problem directly would involve identifying and counting all the segments in a figure such as this.

The process would be time-consuming, and it would be very easy to make a mistake. As an alternative, count the number of segments formed in a sequence of simpler cases and try to find a pattern in the results. Then use inductive reasoning to make a conjecture about the given problem. Organize the data into a table like this.

Number of Collinear Points		Number of Segments Formed	
A B	2	$\overline{AB}$	1
A B C	3	$\overline{AB}, \overline{AC}, \overline{BC}$	3
A B C D	4	$\overline{AB}, \overline{AC}, \overline{AD}, \overline{BC}, \overline{BD}, \overline{CD}$	6
A B C D E	5	$\overline{AB}, \overline{AC}, \overline{AD}, \overline{AE}, \overline{BC}, \overline{BD}, \overline{BE}, \overline{CD}, \overline{CE}, \overline{DE}$	10
A B C D E F	6	$\overline{AB}, \overline{AC}, \overline{AD}, \overline{AE}, \overline{AF}, \overline{BC}, \overline{BD}, \overline{BE}, \overline{BF}, \overline{CD}, \overline{CE}, \overline{CF}, \overline{DE}, \overline{DF}, \overline{EF}$	15

Check Understanding

In Example 2, the number of segments formed depends on the number of collinear points. So, you can say that the number of segments formed *is a function of* the number of collinear points. The table at the bottom of this page is a table for this function.

If n represents the number of collinear points, which of the following do you think is a rule for this function?

a. $f(n) = n + 2$

b. $f(n) = \frac{n}{2}$

c. $f(n) = (n + 1)(n - 1)$

d. $f(n) = \frac{n(n - 1)}{2}$

d

The numbers in the right-hand column of this table form the pattern of *add 2, add 3, add 4, add 5, and so on.* So, extend the sequence until the number of points is ten.

Number of collinear points	2	3	4	5	6	7	8	9	10
Number of segments formed	1	3	6	10	15	21	28	36	45
	+2	+3	+4	+5	+6	+7	+8	+9	

Using this table, you can now make the following conjecture: *The number of segments determined by ten collinear points is 45.*

Chalkboard Examples

Supplementary Example 1

Draw the next figure in the pattern. The next figure would contain 20 dots.

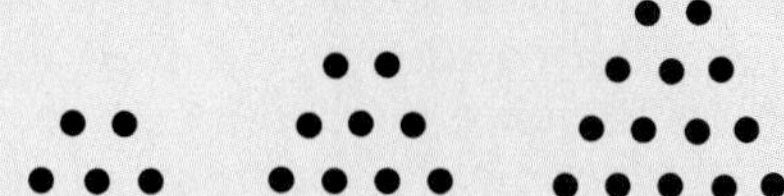

Describe the tenth figure in the pattern. The tenth figure would contain 77 dots.

Supplementary Example 2

How many squares of any size can be found on a checkerboard with 64 squares? 204; Through investigation, students should realize that as you add 1 unit to the side of the square, the number of additional squares that can be made from the arrangement is the square of the measure of the side.

Reteaching Worksheet 3-5

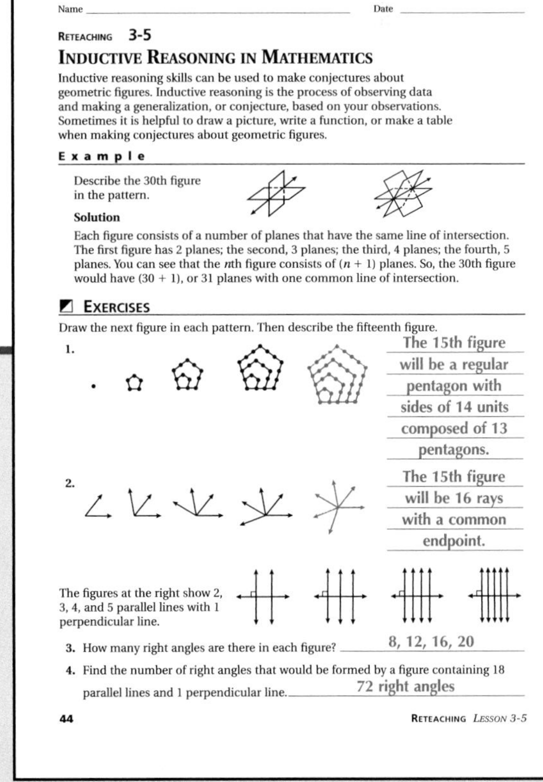

Name ______________________ Date ____________

RETEACHING 3-5

INDUCTIVE REASONING IN MATHEMATICS

Inductive reasoning skills can be used to make conjectures about geometric figures. Inductive reasoning is the process of observing data and making a generalization, or conjecture, based on your observations. Sometimes it is helpful to draw a picture, write a function, or make a table when making conjectures about geometric figures.

Example

Describe the 30th figure in the pattern.

Solution

Each figure consists of a number of planes that have the same line of intersection. The first figure has 2 planes; the second, 3 planes; the third, 4 planes; the fourth, 5 planes. You can see that the nth figure consists of $(n + 1)$ planes. So, the 30th figure would have $(30 + 1)$, or 31 planes with one common line of intersection.

EXERCISES

Draw the next figure in each pattern. Then describe the fifteenth figure.

1. The 15th figure will be a regular pentagon with sides of 14 units composed of 13 pentagons.

2. The 15th figure will be 16 rays with a common endpoint.

The figures at the right show 2, 3, 4, and 5 parallel lines with 1 perpendicular line.

3. How many right angles are there in each figure? 8, 12, 16, 20

4. Find the number of right angles that would be formed by a figure containing 18 parallel lines and 1 perpendicular line. 72 right angles

44 RETEACHING *LESSON 3-5*

Teaching Strategies

COOPERATIVE LEARNING Encourage students, working in groups of 2 or 3 students, to experiment with dots or lines to form patterns. The first pattern can relate to rectangles, the second can relate to squares, while the third relates to cubes. Have students determine the rules for their patterns. Allow students to share one pattern with the class. You may want to have the class try to determine the tenth figure in the pattern. Then have groups present the rule in order to check their classmates' work.

Lesson 3-5 **Inductive Reasoning in Mathematics** 125

Lesson Wrap-up

Quick Assessment

Ask the following questions to determine if students understand the content presented in this lesson.

1. Begin with a square. The next figure is a square divided into two equal sections. Next is that square with each of the two parts divided into two equal sections. Draw the next three figures in this sequence. square with 8 equal sections, 16 equal sections, 32 equal sections
2. For the sequence above, how would you determine the number of equal sections in the tenth figure? Write the numbers in a table to see a pattern. The number of sections is $2n - 1$, or 2^9, or 512.

Assignment Guide

Basic: 1–11, 15–24
Advanced: 1–24
Additional Practice: See Extra Practice Index on page 674.

Additional Answers

1. 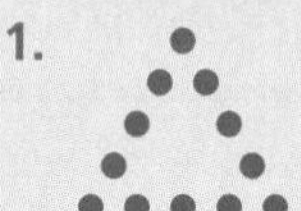

4. Next figure: 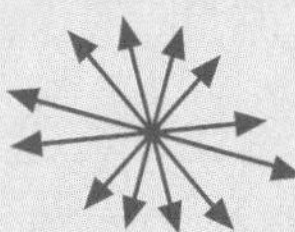

The fourteenth figure will consist of fifteen lines that intersect at one point.

5. Next figure: 

The fourteenth figure will be a rectangular arrangement of 210 points, with 14 points along one side of the rectangle and 15 points along the other.

15. $-\frac{10}{3}$ 0

16. 0 2

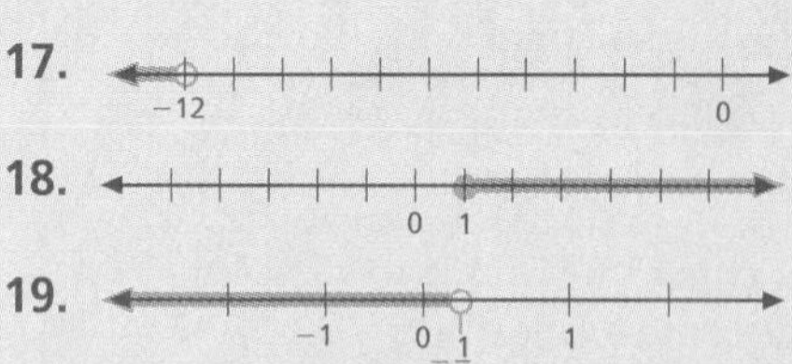

21.

Interval	Tally	Frequency
90-99	|||	3
80-89	||||| |	6
70-79	||||| |||	8
60-69	|||	3
50-59	|	1
40-49	|	1

22.

Interval	Tally	Frequency
125-129	||	2
120-124	||	2
115-119	|||	3
110-114	||||	4
105-109	||||| |	6
100-104	||||| ||||	9

23.

Interval	Tally	Frequency
85-87	||||| ||||	9
82-84	||||| |	6
79-81	|||||	5
76-78	||	2
73-75	|	1
70-72	||	2

24.

Interval	Tally	Frequency
90-97	||	2
82-89	||||| |	6
74-81	||||| |||||	10
66-73	||||	4
58-65	||||	4
50-57	|	1

Try These Exercises

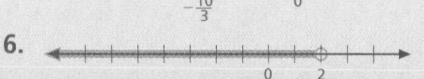

1. Draw the next figure in this pattern of points. Then describe the sixteenth figure in the pattern. See additional answers for next figure.

 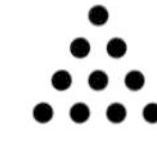

sixteenth figure: triangle with 16 dots on each side

SCHEDULING Employees work in pairs to accomplish certain tasks. To find the number of possible pairings for any number of workers, the supervisor draws segments determined by noncollinear points. Each segment represents a pairing of two employees. The figures below show the segments determined by three, four, five and six noncollinear points.

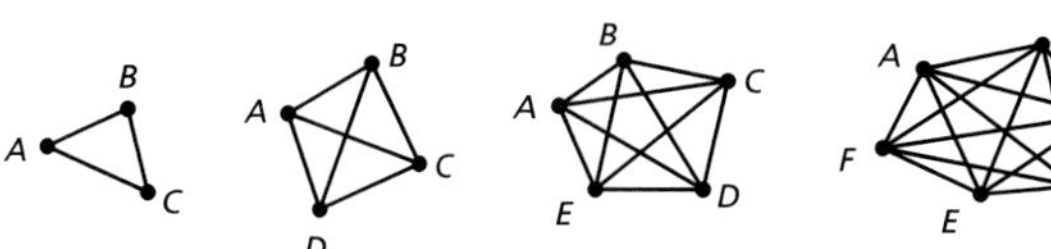

2. How many segments are in each of the figures shown? 3; 6; 10; 15
3. How many pairings are possible for 12 employees? *Hint:* Find the number of segments determined by 12 noncollinear points. 66

Practice Exercises

Draw the next figure in each pattern. Then describe the fourteenth figure in the pattern. See additional answers.

4. 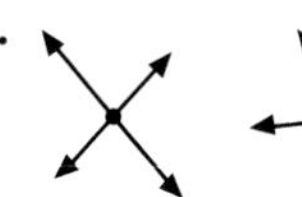 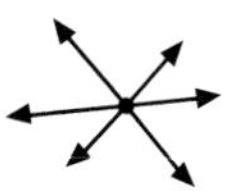 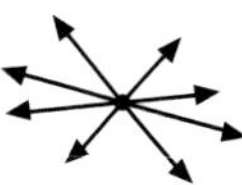 

5. 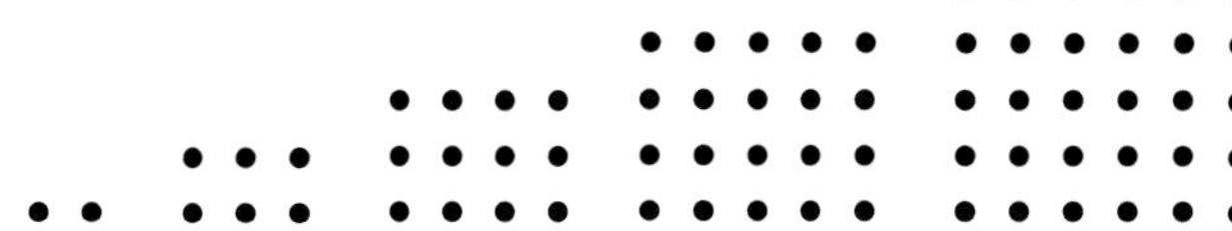

The figures below show one, two, three, and four lines passing through the center of a circle.

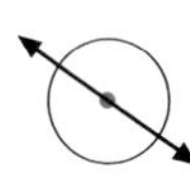 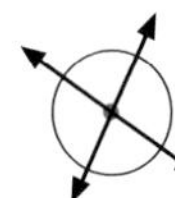 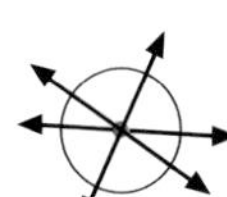 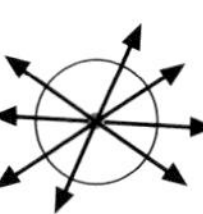

6. In each figure, the lines divide the interior of the circle into regions. How many regions are formed in each of the figures shown? 2; 4; 6; 8
7. Find the number of regions that would be formed when eleven lines pass through the center of a circle. 22

126 Chapter 3 **Geometry and Reasoning**

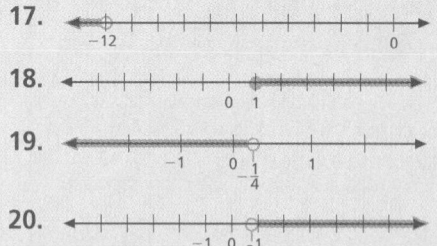

The figures show two, three, four, and five rays sharing a common endpoint.

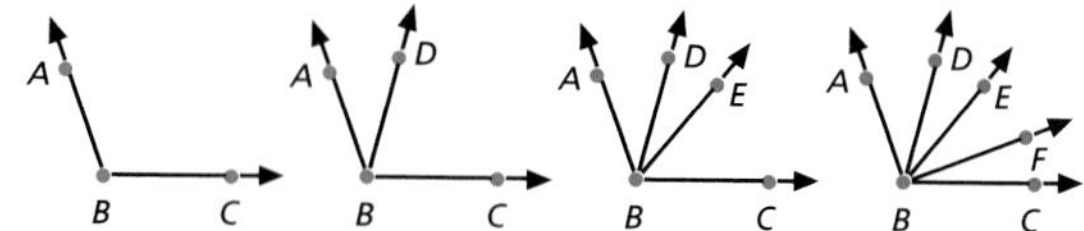

8. How many angles are formed in each of the figures shown? 1; 3; 6; 10

9. Find the number of angles that would be formed when eighteen rays share a common endpoint. 153

10. ART A block structure similar to those shown below will be at the center of a fountain. How many blocks will the artist need to build the tenth figure in the pattern? 385

  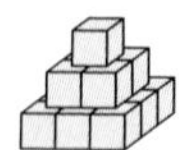 

11. WRITING MATH Describe the pattern below. Then write a function rule to represent the number of points in the nth figure in the pattern.
$f(n) = n^2 + 1$

Extended Practice Exercises

Create a geometric pattern for each number pattern. Answers will vary.

12. 1, 3, 9, 27, . . . **13.** 1, 4, 9, 16, . . . **14.** 1, 8, 27, 64, . . .

Mixed Review Exercises

Graph each inequality on the number line. (Lesson 2-6) See additional answers.

15. $3a + 2 \le -8$ $a \le -\frac{10}{3}$ **16.** $7 - 2b > 3$ $b < 2$ **17.** $\left(\frac{1}{2}\right)c + 4 < -2$ $c < -12$

18. $4d - 6 \ge -2$ $d \ge 1$ **19.** $4x + 3 < 4$ $x < \frac{1}{4}$ **20.** $2m - \frac{3}{2} > -3$ $m > -\frac{3}{4}$

Construct a frequency table for each set of data, using the given interval. (Lesson 2-7) See additional answers.

21. interval = 10
62 71 59 94 86 67 82
75 74 81 79 40 93 72
79 80 78 99 85 65 75
83

22. interval = 5
124 117 114 122 103 102 100
107 123 112 129 106 103 108
107 100 104 117 116 102 108
112 101 113 108 103

23. interval = 3
85 72 87 77 70 79 82
85 83 86 73 85 87 84
77 80 85 82 79 81 83
86 80 83 87

24. interval = 8
80 59 50 92 77 79 82 97 84
84 75 68 70 59 63 68 75 85
77 89 79 72 76 84 79 78 63

Extend the Lesson

CHALLENGE Ask students to find the next three terms in the sequence: O, T, T, F, F, S, ... If students have difficulty, point out that while every pattern follows a rule, the rule may not be mathematical. S, E, N; The pattern is the first letter of the word name for the counting numbers.

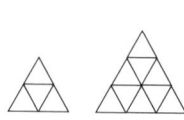

Extra Practice Worksheet 3-5

Name ______________ Date ______

EXTRA PRACTICE 3-5
INDUCTIVE REASONING IN MATHEMATICS

EXERCISES

Draw the next figure in each pattern on your own paper. Then describe the tenth figure in each pattern. Check students' drawings.

1.

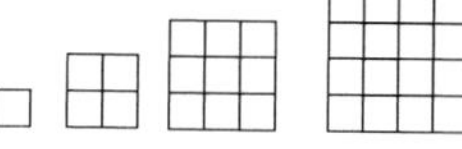

tenth figure: 100 small squares forming a larger 10 by 10 square

2.

 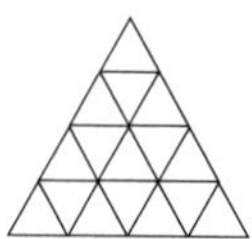

tenth figure: 100 small triangles forming a larger triangle with 10 small triangles along the base

3.

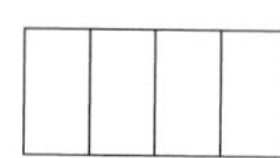 

tenth figure: rectangle divided into equal parts with 1024 segments

Create a geometric pattern for each number pattern. Use your own paper. Check students' work.

4. 2, 4, 6, 8, ... 5. 1, 5, 9, 13, ... 6. 2, 5, 9, 14, ...

40 EXTRA PRACTICE *LESSON 3-5*

Enrichment Worksheet 3-5

Name ______________ Date ______

ENRICHMENT 3-5
PATTERNS WITH DISSECTIONS

There are five ways to dissect a convex five-sided polygon into three triangles.

    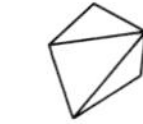

EXERCISES

1. Show all the different ways to dissect a convex four-sided polygon into 2 triangles. How many are there? 2 ways

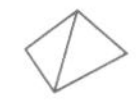 

2. Show all the different ways to dissect a convex six-sided polygon into 4 triangles. How many are there? 14 ways

Study the patterns in this chart. Then fill in the missing values.

	Number of Sides	Number of Dissectios	Formula for D_n
	3	$D_3 = 1$	
	4	$D_4 = 2$	$D_3 + D_3$
	5	$D_5 = 5$	$D_4 + D_3D_3 + D_4$
3.	6	$D_6 = 14$	$D_5 + D_3D_4 + D_4D_3 + D_5$
4.	7	$D_7 = 42$	$D_6 + D_3D_5 + D_4D_4 + D_5D_3 + D_6$
5.	8	$D_8 = 132$	$D_7 + D_3D_6 + D_4D_5 + D_5D_4 + D_6D_3 + D_7$
6.	9	$D_9 = 429$	$D_8 + D_3D_7 + D_4D_6 + D_5D_5 + D_6D_4 + D_7D_3 + D_8$
7.	10	$D_{10} = 1430$	$D_9 + D_3D_8 + D_4D_7 + D_5D_6 + D_6D_5 + D_7D_4 + D_8D_3 + D_9$

52 ENRICHMENT *LESSON 3-5*

Lesson Planning

NCTM Standards/Strands
- Number & Operation
- Patterns, Functions, & Algebra
- Geometry & Spatial Sense
- Problem Solving
- Reasoning and Proof
- Communication
- Connections
- Representation

Vocabulary
conditional
conclusion
biconditional
hypothesis
counterexample

Materials Needed
paper/pencil

Lesson Resources
Warm-Up Transparency 8
Reteaching 3-6
Extra Practice 3-6
Enrichment 3-6

Assignment Guide
Basic: 1–26, 31, 32
Enriched: 1–32

Getting Started

5-Minute Warm-up
Is each statement true or false?
1. If $x = 3$, then $2x = 6$. true
2. If $x = -2$, then $x + 11 = 9$. true
3. If $x = 5$, then $4x - 1 = 3$. false

Introduction to Lesson 3-6
Have students work in small groups to answer the questions. Point out that while the first two statements are either true or false, the third cannot be determined without knowing more about lines *m* and *n*.

3-6 Conditional Statements

Goals
- Identify and evaluate conditional statements.
- Identify converses and biconditionals.

Applications Drafting, Sports, Geography

Do you think each statement is true or false? Explain your reasoning.

a. Denver is the capital of Colorado. True; this is a fact
b. If $x = 2$, then $2x = 5$. False; if $x = 2$ then $2x = 4$.
c. Lines *m* and *n* are parallel.
Cannot tell; there is no information given to identify line *m* and *n*.

Denver, CO

Build Understanding

Many of the statements in this chapter are written in *if–then* form. Statements like these are called **conditional statements**, or simply **conditionals**. The clause following "if" is called the **hypothesis** of the conditional. The clause following "then" is called the **conclusion**. For example, the *parallel lines postulate* was presented as a conditional.

If [two parallel lines are cut by a transversal], — hypothesis

then [corresponding angles are equal in measure]. — conclusion

A conditional is either true or false. When a conditional is true, you can justify it in a variety of ways. For instance, you may be able to show that the conditional is true because it follows directly from a definition. When a conditional is a postulate, such as the parallel lines postulate, it is *assumed* to be true. Still other conditionals are theorems, and these must be *proved* true.

To demonstrate that a conditional is false, you need to find only one example for which the hypothesis is true but the conclusion is false. An example like this is called a **counterexample**.

Example 1

Tell whether each conditional is true or false.

a. If two lines are parallel, then they are coplanar.
b. If two lines do not intersect, then they are parallel.

Solution

a. Parallel lines are defined as *coplanar* lines that do not intersect. So, the conditional is true.

b. Consider skew lines *k* and *ℓ* shown. By the definition of skew lines, *k* and *ℓ* do not intersect, and so the hypothesis is true. However, also by the definition of skew lines, *k* and *ℓ* are noncoplanar. Lines *k* and *ℓ* cannot be parallel, and so the conclusion is false. Therefore, lines *k* and *ℓ* are a counterexample, and the conditional is false.

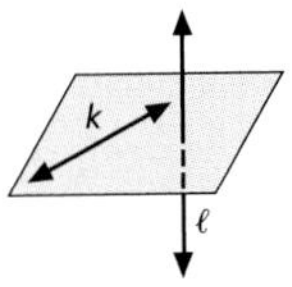

Additional Answers

7.
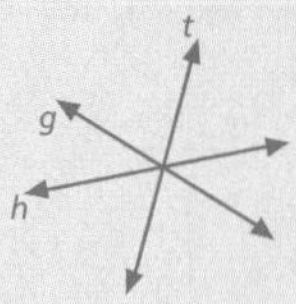

8.
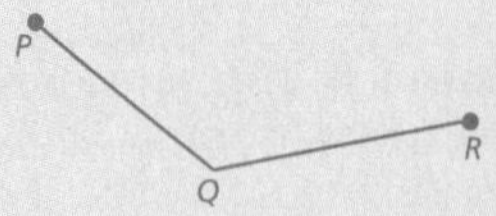

9.
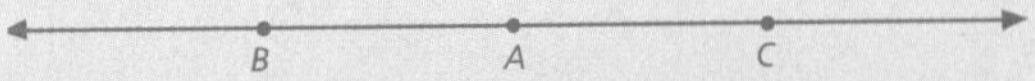

10. There are two possible counterexamples. In the figures below, $\angle AXB$ and $\angle AXC$ share a common side and a common vertex, but they also have interior points in common. Therefore, they are not adjacent angles.

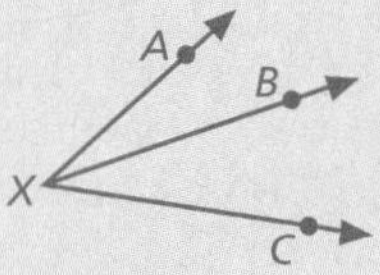

In the figure below, $\angle MON$ and $\angle NOP$ share a common side and a common vertex, but they do not lie in the same plane. Therefore, they are not adjacent.

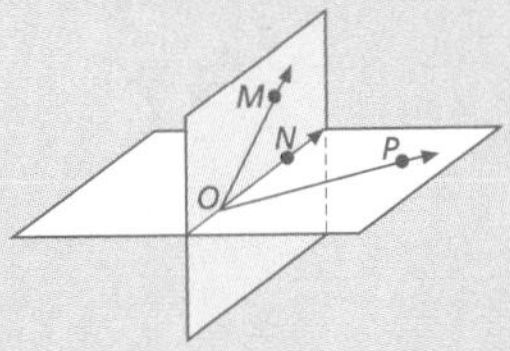

The **converse** of a conditional is formed by interchanging the hypothesis and the conclusion. The fact that a conditional is true is no guarantee that its converse is true.

Example 2

DRAFTING People who draw plans must apply this true statement: If two lines are parallel, then they do not intersect. Write the converse of the statement. Is it also true?

Solution

Interchange the hypothesis and the conclusion of the given statement.

Statement: If [two lines are parallel] (hypothesis), then [they do not intersect] (conclusion).

Converse: If [two lines do not intersect] (hypothesis), then [they are parallel] (conclusion).

By definition, parallel lines do not intersect, and so the given statement is *true.* Part b of Example 1 demonstrated that lines that do not intersect are not necessarily parallel, and so the converse is *false.*

The converse of the parallel lines postulate also is assumed to be true. It is stated as the *corresponding angles postulate* in the following manner.

Postulate 10 **The Corresponding Angles Postulate** If two lines are cut by a transversal so that corresponding angles are equal in measure, then the lines are parallel.

When a statement and its converse are both true, they can be combined into an "if and only if" statement. This type of statement is called a **biconditional statement**, or simply a **biconditional**. Every definition can be written as a biconditional.

Example 3

Write this definition as two conditionals and as a single biconditional.

A right angle is an angle whose measure is exactly 90°.

Solution

The definition leads to two true conditionals.

If an angle is a right angle, then its measure is exactly 90°.

If the measure of an angle is exactly 90°, then it is a right angle.

These can be combined into a single biconditional as follows.

An angle is a right angle if and only if its measure is exactly 90°.

Reading Math

Many statements about everyday situations can be expressed as conditionals. For instance, the following is a true conditional.

If it is raining, then it is cloudy.

Its converse is false.

If it is cloudy, then it is raining.

Problem Solving Tip

When writing the converse of a conditional, you may need to change the wording of the hypotheses and the conclusion to make the converse read clearly.

Chalkboard Examples

Supplementary Example 1

Tell whether this conditional is true or false: If two lines are coplanar, then they are not skew lines. true

Supplementary Example 2

Write the converse of this statement: If a number is divisible by six, then it is divisible by three. Then decide whether the statement and its converse are true or false. If a number is divisible by three, then it is divisible by six. The statement is true; the converse is false.

Supplementary Example 3

Write this definition as two conditionals and a single biconditional: A triangle is a polygon with three sides. If a polygon is a triangle, then it has three sides. If a polygon has three sides, then it is a triangle. A polygon is a triangle if and only if it has three sides.

Reteaching Worksheet 3-6

Name ______ Date ______

RETEACHING 3-6

CONDITIONAL STATEMENTS

Conditional statements are written in an *if-then* (hypothesis/conclusion) form.
The converse of a conditional statement is formed by interchanging the hypothesis and the conclusion.
A counterexample proves a conditional or converse is false.

Example

Write the converse of this statement: If $\overline{RS} \parallel \overline{TU}$, then $\overline{RS}$, $\overline{TU}$ and $\overline{RT}$ are coplanar. Then decide whether the statement and its converse are true or false. If false, give a counterexample.

Solution

Converse: If $\overline{RS}$, $\overline{TU}$ and $\overline{RT}$ are coplanar, then $\overline{RS} \parallel \overline{TU}$.
Original statement is true, since parallel lines are coplanar by definition and for any two points in a plane, the line joining them lies in the plane (Postulate 3).
Converse is false. Counterexample: Each pair of coplanar lines could intersect to form a triangle.

EXERCISES

Write the converse of each statement. Then tell whether the given statement and its converse are true or false. If false, give a counterexample.

1. If points A, B, C and D lie on both plane L and plane M, then points A, B, C and D are collinear.
   If points A, B, C and D are collinear, then points A, B, C and D lie in both plane L and plane M.; true; false; Counterexample: $\overleftrightarrow{AD}$ lies on plane L and plane L is parallel to plane M.
2. If $m\angle IQJ + m\angle HQJ > 180°$, then $\angle IQJ$ and $\angle HQJ$ are obtuse angles.
   If $\angle IQJ$ and $\angle HQJ$ are obtuse angles, then $m\angle IQJ + m\angle HQJ > 180°$.; false; Counterexample: If $m\angle IQJ = 30°$ and $m\angle HQJ = 170°$, then $m\angle IQJ + m\angle HQJ = 200°$ 180°.; true
3. If three lines have one point in common, they are coplanar.
   If three lines are coplanar, then they have one point in common. false; Counterexample: 2 lines can be coplanar and a third line can intersect that plane at only one point.; false; 3 parallel lines
4. If two lines are skew, then they are not coplanar.
   If two lines are not coplanar, then they are skew.; true; true

46 RETEACHING *LESSON 3-6*

11. Converse; If points J, K, and L are collinear, then they are coplanar. The given statement is false. Its converse is true.
12. Converse: If $XY + YZ = XZ$, then point Y is the midpoint of XZ. The given statement is true. Its converse is false.
13. Converse: If two angles are complementary, then the sum of their measures is 90°, Both the given statement and its converse are true.
14. Converse: If the sum of the measures of two angles is greater than 90°, then the angles are supplementary. The given statement is true. Its converse is false.
15. Converse: If two lines do not intersect, then they are perpendicular. Both the given statement and its converse are false.
16. Converse: If RS bisects $\angle QST$, then $m\angle QRS = m\angle SRT$. The given statement is false. Its converse is true.

Try These Exercises

1. **TALK ABOUT IT** Decide whether this conditional is true or false.
   False; it is possible that the two lines are noncoplanar.
   If two lines are each perpendicular to a third line, then they are parallel to each other.

   Discuss your reasoning with a classmate.

2. Write the converse of this statement.
   If two angles are equal in measure, then they are vertical angles.
   If two angles are vertical angles, then they are equal in measure.

   Are the given statement and its converse true or false?
   Given statement is true. Converse is false.

3. Write this definition as two conditionals and as a single biconditional. See additional answers.

   The bisector of an angle is the ray that divides the angle into two adjacent angles that are equal in measure.

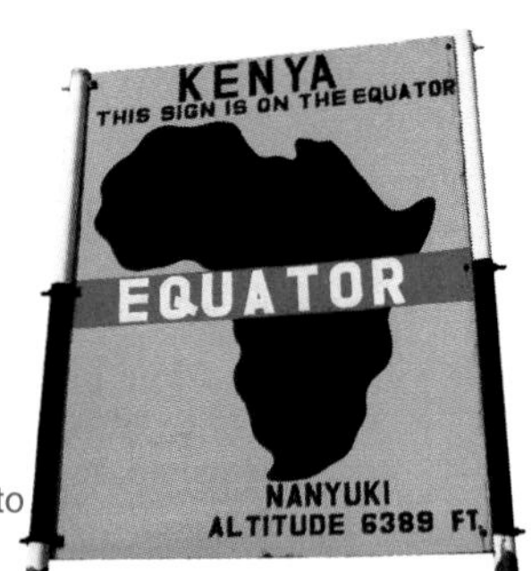

4. **NUMBER SENSE** Tell whether this conditional is true or false.
   False. Negative integers are less than 1.
   If a number is less than one, the number is a proper fraction.

   Write the converse of the statement. Is the converse true or false?
   Converse: If the number is a proper fraction, then it is less than 1. True.

5. **SPORTS** If a shortstop makes a bad throw to first base, the error is charged to the shortstop. This statement is true. Write the converse of the statement. Is it true or false? If an error is charged to the shortstop, then the shortstop made a bad throw to first base. False.

6. **GEOGRAPHY** If a point is located north of the equator, it has a northern latitude. This statement is true. Write the converse of the statement. Is it true or false?
   If a point has a northern latitude, then the point is located north of the equator. True.

Practice Exercises

Sketch a counterexample that shows why each conditional is false.
See additional answers.

7. If line t intersects lines g and h, then line t is a transversal.
8. If $PQ = QR$, then point Q is the midpoint of $\overline{PR}$.
9. If points A, B, and C are collinear, then $\overrightarrow{BA}$ and $\overrightarrow{BC}$ are opposite rays.
10. If two angles share a common side and a common vertex, then they are adjacent angles.

Write the converse of each statement. Then tell whether the given statement and its converse are true or false. See additional answers.

11. If points J, K, and L are coplanar, then they are collinear.
12. If point Y is the midpoint of $\overline{XZ}$, then $XY + YZ = XZ$.
13. If the sum of the measures of two angles is 90°, then the angles are complementary.
14. If two angles are supplementary, then the sum of their measures is greater than 90°.
15. If two lines are perpendicular, then they do not intersect.
16. If $m\angle QRS = m\angle SRT$, then $\overrightarrow{RS}$ bisects $\angle QRT$.

Lesson Wrap-up

Quick Assessment

Ask the following questions to determine if students understand the content presented in this lesson.

1. When is a conditional statement true? when it is always true; when it can be shown that if the hypothesis is true, then the conclusion must also be true
2. Give an example of a mathematical statement that is true although its converse is false. Answers may vary.
3. Explain how a definition and a biconditional statement are related. Any definition can be written as two true conditional statements (the statement and its converse), which can be written as a biconditional.

Assignment Guide

Basic: 1–26, 31, 32
Advanced: 1–32
Additional Practice: See Extra Practice Index on page 674.

Additional Answers

17. Conditionals: If a point is the midpoint of a segment, then it divides the segment into two segments of equal length; if a point divides a segment into two segments of equal length, then it is the midpoint of the segment. Biconditional: A point is the midpoint of a segment if and only if it divides the segment into two segments of equal length.
18. Conditionals: If two lines are perpendicular, then they intersect to form right angles; if two lines intersect to form right angles, then they are perpendicular. Biconditional: Two lines are perpendicular if and only if they intersect to form right angles.
19. Conditionals: If a line is a transversal, then it intersects two or more coplanar lines in different points; if a line intersects two or more coplanar lines in different points, then it is a transversal. Biconditional: A line is a transversal if and only if it intersects two or more coplanar lines in different points.
20. Conditionals: If two angles are vertical angles, then their sides form two pairs of opposite rays; then they are vertical angles. Biconditional: Two angles are vertical angles if and only if their sides form two pairs of opposite rays.
27. Write given definitions as two conditionals: *If two angles are vertical angles, then their sides form opposite rays is true.* However, *If the sides of two angles form opposite rays, then the angles are vertical angles is false.* Here is a counterexample in which the sides of ∠1 and ∠2 form a pair of opposite rays, but the angles are not vertical angles. For this reason, it is necessary to define vertical angles as two angles whose sides form two pairs of opposite rays.
28. Write the given definition as two conditionals: *If a figure is a line segment, then it is part of a line is true.* How-

Write each definition as two conditionals and as a single biconditional.
See additional answers.

17. The midpoint of a segment is the point that divides it into two segments of equal length.
18. Perpendicular lines are two lines that intersect to form right angles.
19. A transversal is a line that intersects two or more coplanar lines in different points.
20. Vertical angles are two angles whose sides form two pairs of opposite rays.

GEOMETRIC CONSTRUCTION The corresponding angles postulate provides a method for constructing parallel lines.

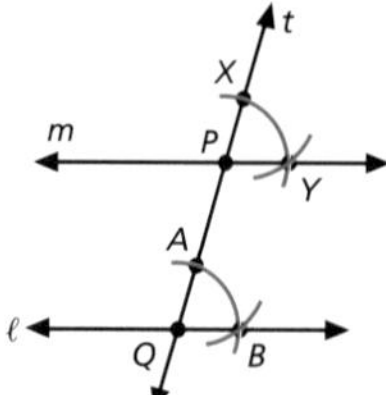

In the figure at the right, you see the finished construction of a line parallel to line ℓ through point *P*. Trace line ℓ and point *P* onto a sheet of paper and repeat the construction. Then complete the statements below that outline the steps of the construction.

21. *Step 1:* Using a straightedge, draw any line __?__ through point *P* intersecting line ℓ. Label the intersection point __?__. *t, Q*
22. *Step 2:* With the compass point at __?__, draw an arc intersecting lines *t* and ℓ. Label the intersection points __?__ and __?__. *Q, A, B*
23. *Step 3:* Using the same radius as in Step 2, place the compass point at point __?__ and draw an arc intersecting line *t*. Label the intersection point __?__. *P, X*
24. *Step 4:* Place the compass point at point __?__ and the pencil at point __?__. Using this radius, draw an arc that intersects line ℓ. *A, B*
25. *Step 5:* Using the same radius as in Step 4, place the compass point at point __?__ and draw an arc that intersects the arc you drew in Step 3. Label the intersection point __?__. *X, Y*
26. *Step 6:* Draw line __?__ through points *P* and *Y*. m∠__?__ = m∠__?__, and so __?__ ∥ __?__. *m, XPY, AQB, m, ℓ*

Extended Practice Exercises

WRITING MATH Explain why each of the following is *not* a good definition.
See additional answers.

27. Vertical angles are two angles whose sides form opposite rays.
28. A line segment is part of a line.
29. Complementary angles are adjacent angles whose exterior sides form a right angle.
30. Skew lines are noncoplanar lines that do not intersect.

Mixed Review Exercises

Find each length. (Lesson 3-1)

31. In the figure below, $AC = 130$. Find BC. 72

A —— $3x + 4$ —— B —— $4x$ —— C

32. In the figure below, $LM = 94$. Find LN. 70

L —— $7p + 14$ —— N —— $3p$ —— M

Extra Practice Worksheet 3-6

Name ______________ Date ______________

EXTRA PRACTICE 3-6
CONDITIONAL STATEMENTS

EXERCISES

Sketch a counterexample that shows why each conditional is false. Use your own paper. Check students' drawings.

1. If point *A* is the midpoint of $\overline{CD}$, then $\overline{CA} \perp \overline{DA}$.
2. If $\overrightarrow{XY}$ and $\overrightarrow{XZ}$ are opposite rays, then point *X* is the midpoint of $\overline{YZ}$.
3. If ∠*RST* and ∠*TSV*, then $\overrightarrow{ST}$ bisects ∠*RSV*.

Write the converse of each statement. Then tell whether the given statement and its converse are true or false.

4. If points *A*, *B*, and *C* are collinear, then *B* is the midpoint of $\overline{AC}$. If *B* is the midpoint of *AC*, then points *A*, *B*, and *C* are collinear; false; true
5. If point *X* is the vertex of ∠1 and ∠2, then ∠1 and ∠2 are adjacent angles. If ∠1 and ∠2 are adjacent angles, then they have a common vertex; false; true
6. If two angles are complementary, then both of the angles are acute. If two angles are acute, then the angles are complementary.; true; false
7. If two lines intersect, then they are parallel. If two lines are parallel, then the line intersect.; false; false
8. If an angle is obtuse, then its supplement is acute. If the supplement of an angle is acute, then the angle is obtuse.; true; true

Write each definition as two conditionals and as a single biconditional.

9. Coplanar points are points that lie in the same plane. If points are coplanar, then they lie in the same plane. If points lie in the same plane, then the points are coplanar. Points are coplanar if and only if they lie in the same plane
10. The intersection of two geometric figures is the set of all points common to both figures. If a figure is a segment, then it is a part of a line that begins at one endpoint and ends at another. If a figure is a part of a line that begins at one endpoint and ends at another, then the figure is a segment. A figure is a segment if and only if it is part of a line that begins at one endpoint and ends at another.

42 EXTRA PRACTICE *LESSON 3-6*

Enrichment Worksheet 3-6

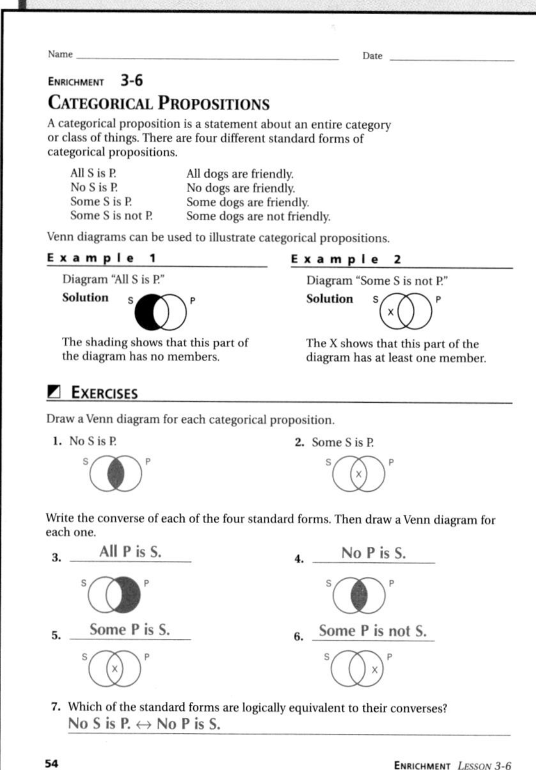
Name ______________ Date ______________

ENRICHMENT 3-6
CATEGORICAL PROPOSITIONS

A categorical proposition is a statement about an entire category or class of things. There are four different standard forms of categorical propositions.

All S is P.	All dogs are friendly.
No S is P.	No dogs are friendly.
Some S is P.	Some dogs are friendly.
Some S is not P.	Some dogs are not friendly.

Venn diagrams can be used to illustrate categorical propositions.

Example 1
Diagram "All S is P."
Solution S P
The shading shows that this part of the diagram has no members.

Example 2
Diagram "Some S is not P."
Solution S X P
The X shows that this part of the diagram has at least one member.

EXERCISES

Draw a Venn diagram for each categorical proposition.

1. No S is P.
2. Some S is P.

Write the converse of each of the four standard forms. Then draw a Venn diagram for each one.

3. All P is S.
4. No P is S.
5. Some P is S.
6. Some P is not S.
7. Which of the standard forms are logically equivalent to their converses? No S is P. ↔ No P is S.

54 ENRICHMENT *LESSON 3-6*

ever, *If a figure is part of a line, then it is a line segment is false.* A ray also is part of a line. It is necessary to specify that a line segment is part of a line that begins at one endpoint and ends at another.

29. Write the given definitions as two conditionals: *If two angles are adjacent angles whose exterior sides form a right angle, then they are complementary* is true. However, *If two angles are complementary, then they are adjacent angles whose exterior sides form a right angle* is false. Complementary angles are not necessarily adjacent.

30. The given definition is a true statement, but it contains too much information to be a good definition. It is not necessary to include the phrase *that do not intersect* because the term noncoplanar lines already indicates that the tow lines do not intersect.

Skills Practice

Vocabulary Review

Lesson 3-5
inductive reasoning conjecture
sequence

Lesson 3-6
conditional hypothesis
conclusion counterexample
biconditional

Assignment Guide

All students: 1–24
Additional Practice: Refer to the Extra Practice Index on page 674 of this text.

Additional Answers

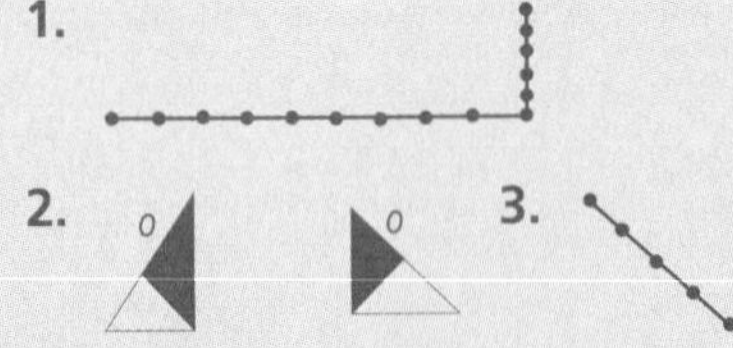

1\.

2\. 3.

4\.

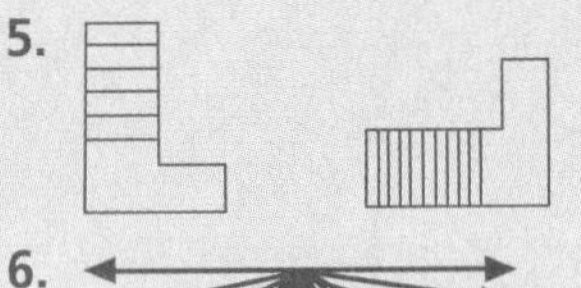

5\.

6\.

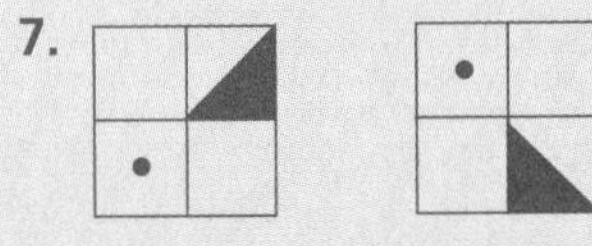

7\.

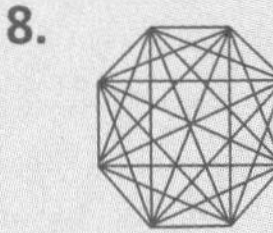

8\.

Review and Practice Your Skills

Practice ■ Lesson 3-5

Draw the next figure in each pattern. Then describe the tenth figure in the pattern. Check students' figures.

1.

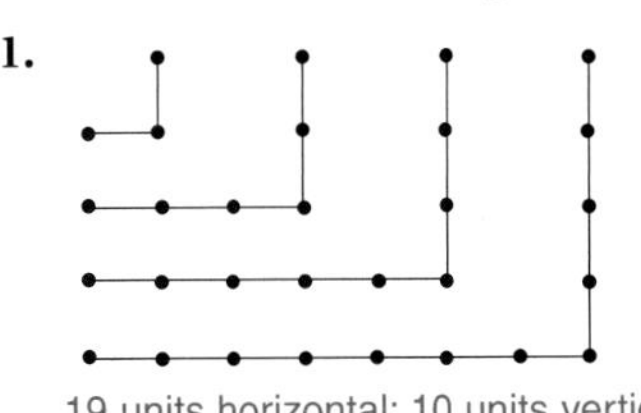

19 units horizontal; 10 units vertical

2.

Same as second figure.

3.

10 units long

4.

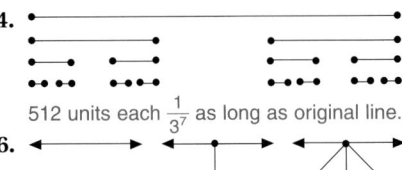

512 units each $\frac{1}{3^7}$ as long as original line.

5.

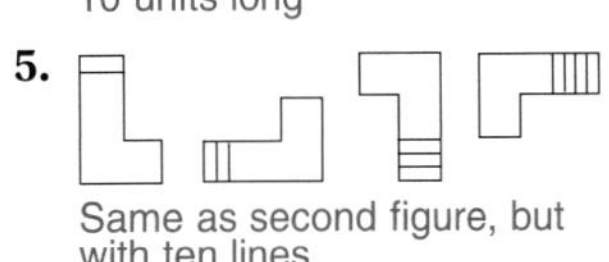

Same as second figure, but with ten lines.

6.

512 angles, each $\frac{180}{512}°$

7.

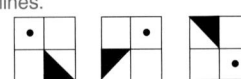

Same as second figure.

8.

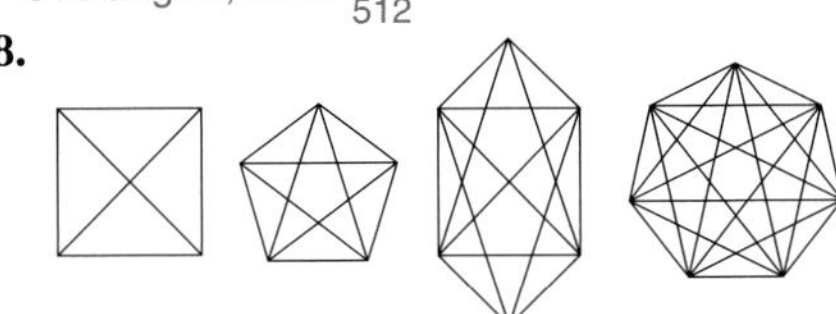

13 sided polygon with all diagonals

Practice ■ Lesson 3-6

Sketch a counterexample to show why each conditional is false.
Answers will vary.

9. If $\angle ABC$ and $\angle DEF$ are supplements, then $m\angle ABC > m\angle DEF$.

10. If two point are coplanar, then they are collinear.

11. If two lines are skew, then they intersect.

Write the converse of each statement. Then tell whether the given statement and its converse are true or false.

12. If two lines intersect, then they are perpendicular. If two lines are perpendicular, then they intersect. false, true

13. If C is the midpoint of $\overline{AB}$, then $AB = 2(AC)$. If $AB = 2(AC)$, then C is the midpoint of $\overline{AB}$. true, false

14. If two angles are vertical angles, then their supplements are equal.
If the supplements of two angles are equal, then they are vertical angles. true, false

Write each definition as two conditionals and as a single biconditional.

15. Perpendicular lines are lines that intersect to form right angles. If two lines are perpendicular, then they intersect to form right angles. If two lines intersect to form right angles, then they are perpendicular. Two lines are perpendicular if and only if they intersect to form right angles.

16. Skew lines are noncoplanar lines.
If lines are skew, they are noncoplanar.
If lines are noncoplanar, then they are skew.
Two lines are skew if and only if they are noncoplanar.

PRACTICE ■ LESSON 3-1–LESSON 3-6

17. On a number line, the coordinate of point G is -6. The length of $\overline{GH}$ is 13. Give two possible coordinates for point H. (Lesson 3-1) $-19, 7$

18. Point Z is between points X and Y. The length of $\overline{XZ}$ is three times the length of $\overline{ZY}$, and $XY = 48$. Find XZ and ZY. (Lesson 3-1) 36, 12

19. In the figure below, find $m\angle PYV$ and $m\angle PYX$. (Lesson 3-3) 148°, 32°

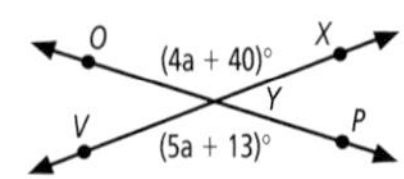

20. In the figure below, $\overline{QS}$ bisects $\angle RQT$. Find $m\angle TQU$. (Lesson 3-3) 34°

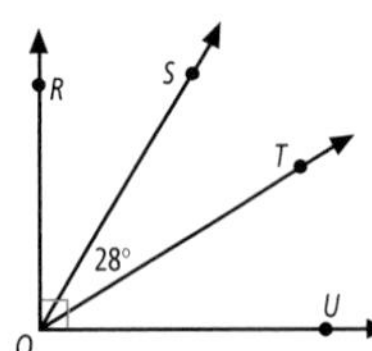

In the figure at the right, $\overrightarrow{NP} \parallel \overrightarrow{MQ}$. Find the measure of each angle. (Lesson 3-4)

21. $\angle LMQ$ 164°

22. $\angle NMQ$ 16°

23. $\angle MNP$ 164°

24. $\angle ONP$ 16°

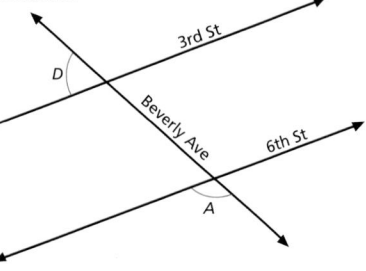

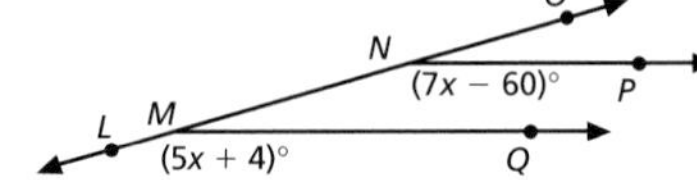

MathWorks Workplace Knowhow

Career – Cartographer

A cartographer is a mapmaker. Before there were ways to see the Earth from above, cartographers made maps by visiting the places they were mapping. Lewis and Clark mapped their route all the way to the Pacific Ocean by drawing the features of the landscape as they traveled.

Today, a cartographer has many tools to make maps more accurate. Satellite photos give cartographers a birds-eye view of the Earth and its features. Mathematics is an essential tool for cartographers. Mapmakers must be able to convert distances in miles to distances in inches or centimeters. They must understand the principles of geometry to represent three-dimensional objects on a flat plane and check the accuracy of angle measurements.

1. A cartographer is making a road map. On the map, Beverly Avenue is transversed by two smaller roads, 3rd Street and 6th Street. On the map, if angle A measures 104 degrees, what is the measure of angle D? 76°

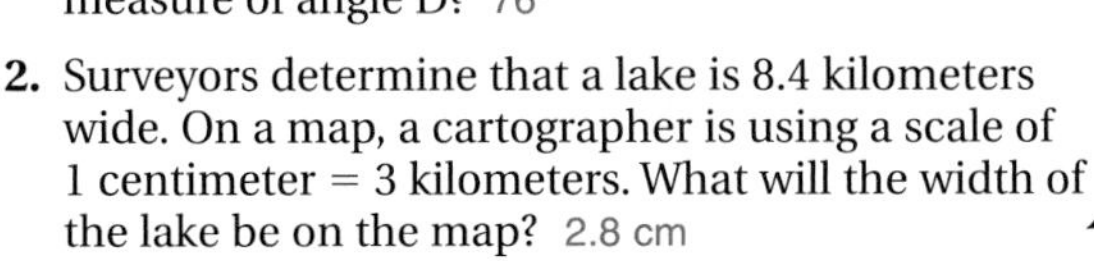

2. Surveyors determine that a lake is 8.4 kilometers wide. On a map, a cartographer is using a scale of 1 centimeter = 3 kilometers. What will the width of the lake be on the map? 2.8 cm

3. On a map of Oregon, Cottonwood Mountain is about $2\frac{1}{2}$ inches north of Cedar Mountain. The map scale shows that $\frac{7}{8}$ inch = 20 miles. To the nearest mile, how many miles apart are the mountains? 57 miles

Chalkboard Examples

Example from Lesson 3-5 and 3-6

1. Write the converse of this statement: If two lines are parallel, then they intersect. If two lines intersect, then they are parallel.
2. Explain why this is not a good definition: Two lines intersect if they have at least one point in common. Two lines may have all points in common, in which case, they are the same line and not intersecting lines.

Career Opportunity

Describe the kind of work cartographers do and the types of places they work. Discuss the importance of mathematics, art, and computer skills in working as a cartographer. Students should answer Questions 1–3 to better understand how cartographers use mathematics in performing their job.

Students who are interested in learning more about this profession can go to Learningmathmatters.com. Guidance Counselors should have information about school and training requirements.

Lesson Planning

NCTM Standards/Strands

- Number & Operation
- Patterns, Functions, & Algebra
- Geometry & Spatial Sense
- Problem Solving
- Reasoning and Proof
- Communication
- Connections
- Representation

Vocabulary

proof, theorem, deductive reasoning, two-column proof

Materials Needed

paper/pencil, compass, straightedge

Lesson Resources

Warm-Up Transparency 9
Reteaching 3-7
Extra Practice 3-7
Enrichment 3-7

Assignment Guide

Basic: 1–6, 10–30
Enriched: 1–30

Getting Started

5-Minute Warm-up

If $x = y$, and $y = -3$, what is the value of x? Explain. -3; transitive property of equality: if $a = b$ and $b = c$, then $a = c$

Introduction to Lesson 3-7

Encourage students to share their work so that they can see that it does not matter how the six points were placed on the circle. Have students compare their conjectures and discuss their reasoning.

3-7 Deductive Reasoning and Proof

Goals ■ Write geometric proofs in two-column format.

Applications Surveying and Carpentry

Study this geometric pattern.

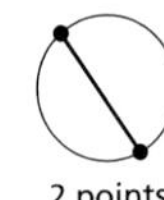
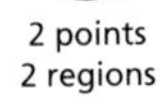
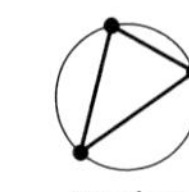

2 points, 2 regions | 3 points, 4 regions | 4 points, 8 regions | 5 points, 16 regions

Use inductive reasoning to make a conjecture about the number of regions formed by six points on a circle. 32

Build Understanding

A theorem is a statement that can be proved true. To do this, you start with the hypothesis of the theorem and make a series of logical statements that demonstrates why the conclusion is true. Each statement must be accompanied by a reason—a definition, postulate, theorem, or algebraic property that justifies the statement.

The set of statements and reasons is called a **proof of the theorem**. When you prove a theorem by this process, you are using **deductive reasoning**.

Example 1 shows a two-column proof of the vertical angles theorem.

Theorem If two angles are vertical angles, then they are equal in measure.

Example 1

Given $\angle 1$ and $\angle 3$ are vertical angles.

Prove $m\angle 1 = m\angle 3$

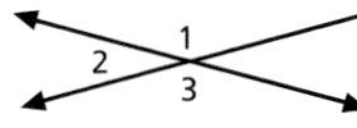

Solution

Statements	Reasons
1. $\angle 1$ and $\angle 3$ are vertical angles.	1. given
2. $m\angle 1 + m\angle 2 = 180°$ $m\angle 3 + m\angle 2 = 180°$	2. angle addition postulate
3. $m\angle 1 + m\angle 2 = m\angle 3 + m\angle 2$	3. transitive property of equality
4. $m\angle 1 = m\angle 3$	4. addition property of equality

Reading Math

When a proof is written in the two-column format, the hypothesis is usually labeled *Given*.

Teaching Strategies

ESL/LEP Have students work in small groups. For each example in the lesson, have students explain the logic of the proof informally to one another. Have students make index cards with the formal language of the proof on one side and an informal description in their own words on the back of the card. Allow students to use these cards has they work on the proofs in the exercises.

Once a theorem has been proved true, you can use it to justify statements in proofs that follow. For instance, Example 2 shows how the vertical angles theorem is used to prove the next theorem about parallel lines.

Theorem	If two parallel lines are cut by a transversal, then alternate interior angles are equal in measure.

Example 2

SURVEYING Two parallel property lines are cut by a third line. How can the surveyor know that $m\angle 1 = m\angle 3$?

Given $p \parallel q$

Prove $m\angle 1 = m\angle 3$

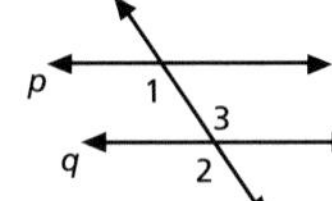

Solution

Statements	Reasons
1. $p \parallel q$	1. given
2. $m\angle 1 = m\angle 2$	2. parallel lines postulate
3. $m\angle 3 = m\angle 2$	3. vertical angles theorem
4. $m\angle 1 = m\angle 3$	4. transitive property of equality

Example 3 shows how the vertical angles theorem also can be used to prove the converse of the theorem above.

Theorem	If two lines are cut by a transversal so that alternate interior angles are equal in measure, then the lines are parallel.

Example 3

CARPENTRY A cross-brace is nailed to two wooden beams so that the alternate interior angles formed are equal in measure. How can the carpenter know that the beams are parallel?

Given $m\angle 1 = m\angle 3$

Prove $p \parallel q$

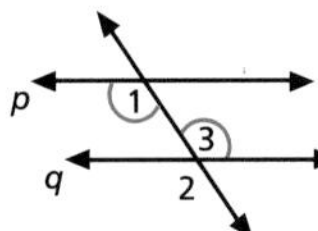

Solution

Statements	Reasons
1. $m\angle 1 = m\angle 3$	1. given
2. $m\angle 2 = m\angle 3$	2. vertical angles theorem
3. $m\angle 1 = m\angle 2$	3. transitive property of equality
4. $p \parallel q$	4. corresponding angles postulate

Reading Math

How does deductive reasoning differ from inductive reasoning?

Inductive reasoning often is described as "particular to general": you observe several particular cases and arrive at a general conclusion. The conclusion is *probably* true, but, as illustrated in the Explore activity, it is not *necessarily* true. In geometry, inductive reasoning is used primarily to explore ideas and to make conjectures.

In contrast, deductive reasoning is described as "general to particular": you use several general statements to arrive at a particular conclusion. If all the statements in a deductive argument are true, the conclusion *must* be true. For this reason, most geometric proofs are based in deductive reasoning.

Chalkboard Examples

Supplementary Example 1

Given: $\angle 2$ and $\angle 3$ are vertical angles.

Prove: $m\angle 2 = m\angle 3$

Statements	Reasons
1. $\angle 2$ and $\angle 3$ are vertical angles.	1. given
2. $m\angle 2 + m\angle 1 = 180°$ $m\angle 3 + m\angle 1 = 180°$	2. angle addition postulate
3. $m\angle 2 + m\angle 1 = m\angle 3 + m\angle 1$	3. transitive property of equality
4. $m\angle 2 = m\angle 3$	4. add. prop. of equality

Supplementary Example 2 and 3

Two parallel roads are cut by a third road. How can a mapmaker know that $\angle 1$ and $\angle 3$ have the same measure?

Given: $a \parallel b$; Prove: $m\angle 1 = m\angle 3$.

Statements	Reasons
1. $a \parallel b$	1. given
2. $m\angle 1 = m\angle 2$	2. parallel lines post.
3. $m\angle 3 = m\angle 2$	3. vert. angles theorem
4. $m\angle 1 = m\angle 3$	4. trans. prop. of equality

Reteaching Worksheet 3-7

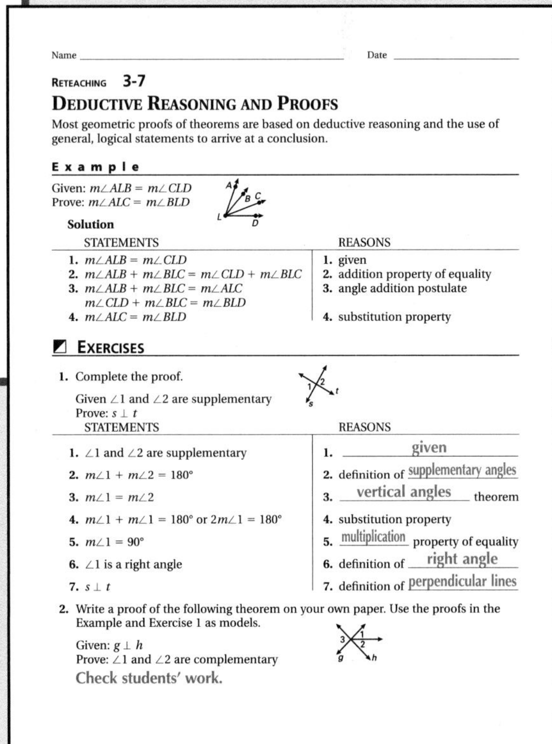

Name ____________ Date ____________

RETEACHING 3-7

DEDUCTIVE REASONING AND PROOFS

Most geometric proofs of theorems are based on deductive reasoning and the use of general, logical statements to arrive at a conclusion.

Example

Given: $m\angle ALB = m\angle CLD$

Prove: $m\angle ALC = m\angle BLD$

Solution

STATEMENTS	REASONS
1. $m\angle ALB = m\angle CLD$	1. given
2. $m\angle ALB + m\angle BLC = m\angle CLD + m\angle BLC$	2. addition property of equality
3. $m\angle ALB + m\angle BLC = m\angle ALC$ $m\angle CLD + m\angle BLC = m\angle BLD$	3. angle addition postulate
4. $m\angle ALC = m\angle BLD$	4. substitution property

EXERCISES

1. Complete the proof.

Given $\angle 1$ and $\angle 2$ are supplementary

Prove: $s \perp t$

STATEMENTS	REASONS
1. $\angle 1$ and $\angle 2$ are supplementary	1. given
2. $m\angle 1 + m\angle 2 = 180°$	2. definition of supplementary angles
3. $m\angle 1 = m\angle 2$	3. vertical angles theorem
4. $m\angle 1 + m\angle 1 = 180°$ or $2m\angle 1 = 180°$	4. substitution property
5. $m\angle 1 = 90°$	5. multiplication property of equality
6. $\angle 1$ is a right angle	6. definition of right angle
7. $s \perp t$	7. definition of perpendicular lines

2. Write a proof of the following theorem on your own paper. Use the proofs in the Example and Exercise 1 as models.

Given: $g \perp h$

Prove: $\angle 1$ and $\angle 2$ are complementary

Check students' work.

48 RETEACHING *LESSON 3-7*

Extend the Lesson

MATH JOURNAL Have students write a paragraph to tell how they could prove the following. Given: $\angle 1$ and $\angle 6$ are supplementary. Prove $p \parallel q$.

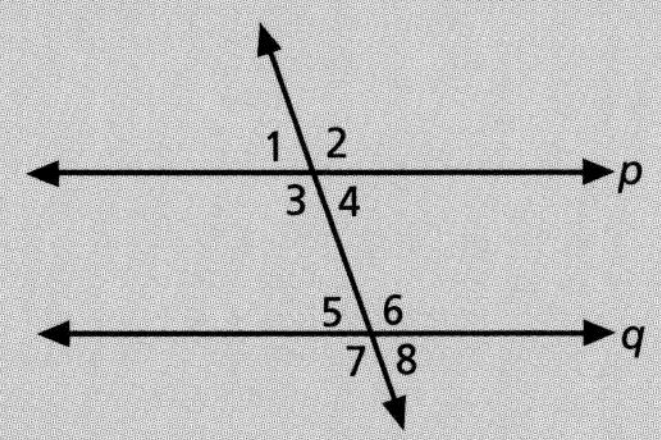

Show that $\angle 5$ is supplementary to $\angle 6$. Then show $m\angle 1 = m\angle 5$. Lines are parallel by corresponding angles postulate.

You will be asked to prove two related theorems in *Try These Exercises.*

Theorem	If two lines are cut by a transversal so that alternate exterior angles are equal in measure, then the lines are parallel.

Try These Exercises

Theorem	If two parallel lines are cut by a transversal, then alternate exterior angles are equal in measure.

Copy and complete each proof.

1. Given $p \parallel q$

Prove $m\angle 1 = m\angle 2$

Statements	Reasons
1. $p \parallel q$	1. ___?___ given
2. $m\angle 1 = m\angle 3$	2. ___?___ parallel lines postulate
3. ___?___ $m\angle 2 = m\angle 3$	3. vertical angles theorem
4. $m\angle 1 = m\angle 2$	4. ___?___ transitive property of equality

2. Given $m\angle 1 = m\angle 2$

Prove $p \parallel q$

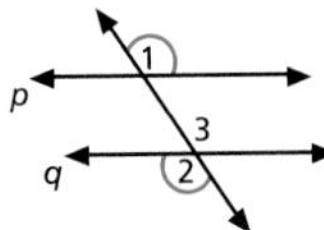

Statements	Reasons
1. ___?___ $m\angle 1 = m\angle 2$	1. ___?___ given
2. ___?___ $m\angle 3 = m\angle 2$	2. vertical angles theorem
3. $m\angle 1 = m\angle 3$	3. ___?___ transitive property of equality
4. ___?___ $p \parallel q$	4. ___?___ corresponding angles postulate

Practice Exercises

3. **WRITING MATH** Describe a real-life example of an occasion when you used inductive reasoning to solve a problem. Then describe a situation when you used deductive reasoning. Answers will vary.

4. **YOU MAKE THE CALL** Given: $\angle DEF$ and $\angle FEG$ are adjacent angles. Based on the given information, Dana makes the statement that $m\angle DEF + m\angle FEG = m\angle DEG$. She justifies the statement using the definition of complementary angles. Is her reasoning correct? Explain. No. The correct justification is the angle addition postulate.

Lesson Wrap-up

Quick Assessment

Ask the following questions to determine if students understand the content presented in this lesson.

1. Write the two theorems about alternate interior angles as a biconditional statement. Two lines are parallel if and only if alternate interior angles are equal in measure.
2. In Example 1, explain how the addition property of equality was used. In Statement 3, $m\angle 2$ was added to each side of the equation.
3. In Example 2, what are angles 1 and 2 called in this situation? corresponding angles

Assignment Guide

Basic: 1–6, 10–30
Advanced: 1–30
Additional Practice: See Extra Practice Index on page 674.

Additional Answers

5. Statement: 1: $\angle 1$ is complementary to $\angle 2$. $\angle 3$ is complementary to $\angle 2$. Statement 2: $m\angle 3 + m\angle 2 = 90°$. Statement 3: $m\angle 3 + m\angle 2$. Statement 4: $m\angle 1 = m\angle 3$.
6. Given: $\angle 1$ is supplementary to $\angle 2$. $\angle 3$ is supplementary to $\angle 2$.
Prove: $m\angle 1 = m\angle 3$

Statements	Reasons
1. $\angle 1$ is suppl. to $\angle 2$. $\angle 3$ is suppl. to $\angle 2$.	1. given
2. $m\angle 1 + m\angle 2 = 180°$ $m\angle 3 + m\angle 2 = 180°$	2. def. of suppl. angles
3. $m\angle 1 + m\angle 2 = m\angle 3 + m\angle 2$	3. transitive property of equality
4. $m\angle 1 = m\angle 3$	4. add. prop. of equality

7. Given: $k \parallel m$, $l \parallel m$
Prove: $k \parallel l$

Statements	Reasons
1. $k \parallel m$, $l \parallel m$	1. given
2. $m\angle 1 = m\angle 2$; $m\angle 3 = m\angle 2$	2. parallel lines post.
3. $m\angle 1 = m\angle 3 =$	3. trans. property of eq.
4. $k \parallel l$	4. corres. Angles post.

8. Given: $k \perp m$, $l \perp m$
Prove: $k \parallel l$

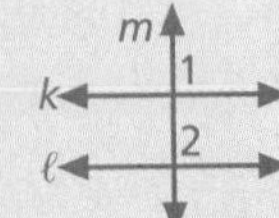

Statements	Reasons
1. $k \perp m$; $l \perp m$	1. given
2. $\angle 1$ is a right angle; $\angle 2$ is a right angle	2. definition of perpendicular lines
3. $m\angle 1 = 90°$; $m\angle 2 = 90°$	3. def. of right angle
4. $m\angle 1 = m\angle 2$	4. trans. prop. of equality
5. $k \parallel l$	5. corres. angles post.

5. Copy and complete the following proof of this theorem.

If two angles are complementary to the same angle, then they are equal in measure to each other.

Given $\angle 1$ is complementary to $\angle 2$.
$\angle 3$ is complementary to $\angle 2$.

Prove $\underline{\ ?\ }$ $m\angle 1 = m\angle 3$

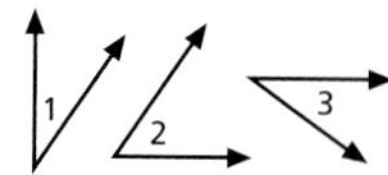

Statements	Reasons
1. ?	1. ? given
2. $m\angle 1 + m\angle 2 = 90°$; $m\angle 3 + m\angle 2 = 90°$	2. ? definition of complementary angles
3. $m\angle 1 + m\angle 2 = 90°$	3. ? transitive property of equality
4. ? See additional answers.	4. addition property of equality

Problem Solving Tip

For Exercise 5, you probably will find it helpful to refer to the proof of the vertical angles theorem in Example 1 on page 134.

6. Write a proof of the following theorem. See additional answers.

Theorem If two angles are supplementary to the same angle, then they are equal in measure to each other.

(*Hint:* Use the proof in Exercise 5 as a model.)

Extended Practice Exercises

Exercises 7–9 state three important theorems related to perpendicular and parallel lines. Write a proof of each theorem. See additional answers.

7. If two coplanar lines are each parallel to a third line, then the lines are parallel to each other.

8. If two coplanar lines are each perpendicular to a third line, then the lines are parallel to each other.

9. If a transversal is perpendicular to one of two parallel lines, then it is perpendicular to the other.

10. CHAPTER INVESTIGATION List at least two geometric theorems or postulates that are evident in your neighborhood map.

Mixed Review Exercises

Write each number in standard notation. (Lesson 1-8)

11. 1.46×10^{-8} 0.0000000146
12. 3.7×10^{5} 370,000
13. 7.02×10^{7} 70,200,000
14. 6.82×10^{-11} 0.0000000000682
15. 5.9×10^{-9} 0.0000000059
16. 8.13×10^{8} 813,000,000
17. 2.1×10^{4} 21,000
18. 4×10^{-7} 0.0000004
19. 3.97×10^{-4} 0.000397
20. 6.48×10^{7} 64,800,000
21. 5.12×10^{9} 5,120,000,000
22. 7.3×10^{-10} 0.00000000073

Find the complement and supplement of each angle. (Lesson 3-2)

23. $m\angle AB = 57°$ 33°, 123°
24. $m\angle CD = 27°$ 63°, 153°
25. $m\angle EF = 75°$ 15°, 105°
26. $m\angle GH = 40°$ 50°, 140°
27. $m\angle JK = 83°$ 7°, 97°
28. $m\angle LM = 34°$ 56°, 146°
29. $m\angle NO = 61°$ 29°, 119°
30. $m\angle PR = 18°$ 72°, 162°

9. Given: $l \parallel m$, $k \perp l$
Prove: $k \perp m$

Statements	Reasons
1. $l \parallel m$, $k \perp l$	1. given
2. $\angle 1$ is a right angle	2. def. of perpendicular lines
3. $m\angle 1 = 90°$	3. def. of right angle
4. $m\angle 1 = m\angle 2$	4. parallel lines postulate
5. $m\angle 2 = 90°$	5. substitution property
6. $\angle 2$ is a right angle	6. def. of right angle
7. $k \perp m$	7. def. of perpendicular lines

Extra Practice Worksheet 3-7

Name ____________ Date ____________

EXTRA PRACTICE 3-7

DEDUCTIVE REASONING AND PROOFS

EXERCISES

Complete each proof.

1. Given: $s \parallel t$
Prove: $m\angle 1 = m\angle 7$

Statements	Reasons
1. $s \parallel t$	1. given
2. $m\angle 1 = m\angle 5$	2. parallel lines postulate
3. $m\angle 5 = m\angle 7$	3. vertical angles theorem
4. $m\angle 1 = m\angle 7$	4. transitive property of equality

2. Given: $m\angle 1 = m\angle 2$; $m\angle 2 = 45°$
Prove: $\angle 1$ is complementary to $\angle 2$.

Statements	Reasons
1. $m\angle 1 = m\angle 2$; $m\angle 2 = 45°$	1. given
2. $m\angle 1 = 45°$	2. transitive property of equality
3. $m\angle 1 + m\angle 2 = 45° + 45°$	3. addition property of equality
4. $m\angle 1 + m\angle 2 = 90°$	4. addition
5. $\angle 1$ is complementary to $\angle 2$.	5. definition of complementary angles

44 EXTRA PRACTICE *LESSON 3-7*

Enrichment Worksheet 3-7

Name ____________ Date ____________

ENRICHMENT 3-7

CATEGORICAL SYLLOGISMS

A categorical syllogism is a deductive argument made from three categorical propositions. Some forms of this type of argument are valid and some are not.

Example

Show that this form of argument is not valid.

All T is P.
No S is T.
Therefore, no S is P.

Solution

Experiment with specific examples until you find one that is obviously false.

All triangles are polygons.
No squares are triangles.
Therefore, no squares are polygons.

EXERCISES

Each of these forms of argument is valid. Write an example for each. Answers may vary.

1. All H is M.
All S is H.
Therefore, all S is M.

2. All H is M.
Some S is H.
Therefore, some S is M.

3. No P is M.
Some M is S.
Therefore, some S is not P.

4. No M is D.
All B is D.
Therefore, no B is M.

Write an example to show that each of these forms is not valid.

5. All S is M.
All S is H.
Therefore, some H is M.

6. Some S is M.
Some S is H.
Therefore, some H is M.

56 ENRICHMENT *LESSON 3-7*

Lesson Planning

NCTM Standards/Strands

- Problem Solving
- Reasoning and Proof
- Communication
- Connections
- Representation

Vocabulary

process of elimination

Materials Needed

paper/pencil

Lesson Resources

Warm-Up Transparency 9
Reteaching 3-8
Enrichment 3-8

Assignment Guide

Basic: 1–14
Enriched: 1–14

Getting Started

5-Minute Warm-Up

Find a two-digit number that is odd, that is divisible by 3 and has a ones digit that is one more than its tens digit. 45

Use the Five-step Plan and the Strategy

THE FIVE-STEP PLAN Read—ask questions of the students to help them understand the problem. **Plan**—guide students to related problems and previously mastered skills and strategies. **Solve**—students solve problem on their own. **Answer**—write solution in format that answers the question. **Check**—review work, check for reasonableness and review strategy used. Students will benefit from the experience of verbalizing their methods. **THE STRATEGY Eliminate possibilities**—Making a table and using the process of elimination allows students away of organizing the conclusions they draw from the given clues.

3-8 Problem Solving Skills: Use Logical Reasoning

To solve some problems, you can use logical reasoning without using any numbers. Making a table and using the *process of elimination* is one way to solve some problems.

Problem Solving Strategies

- Guess and check
- Look for a pattern
- Solve a simpler problem
- Make a table, chart or list
- Use a picture, diagram or model
- Act it out
- Work backwards
- 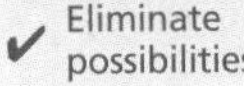 Eliminate possibilities
- Use an equation or formula

Problem

TRAVEL Four friends talked about their trips in the United States. Andrea and Don discussed the differences between the states they visited, even though each state has the word *New* in its name.

Natalie told Edward that she liked South Dakota a lot. Edward visited New Mexico or California; Don visited the other. One person talked about sightseeing in New York. Where had each person gone on vacation?

Solve the Problem

Make a table listing each person and each state.

Andrea and Don went to states with *New* in their names. Put an **X** in the boxes to show what states they could *not* have visited.

	Andrea	Don	Natalie	Edward
California	x	x		
New Mexico				
New York				
South Dakota	x	x		

Natalie visited South Dakota. Use an **O** to show that information. Notice that the rest of each row and column is filled with **X**s.

	Andrea	Don	Natalie	Edward
California	x	x	x	
New Mexico			x	
New York			x	
South Dakota	x	x	o	x

Notice, then, that Ed is the only one who could have visited California.

Finally, Don visited either New Mexico or California. Because *New* was part of the name, it must have been New Mexico. So, by process of elimination, Andrea must have visited New York.

	Andrea	Don	Natalie	Edward
California	x	x	x	o
New Mexico			x	x
New York			x	x
South Dakota	x	x	o	x

Additional Answers

1.

	Cory	Srey	Molly	Mao
Des Moines	x	x	o	x
Pittsburg	x	o	x	x
Santa Clara	o	x	x	x
Seattle	x	x	x	o

2.

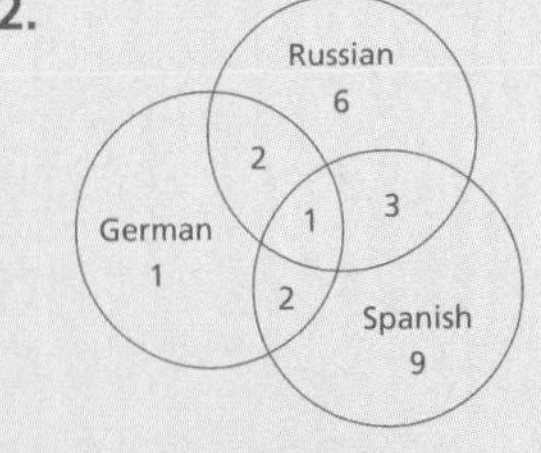

Try These Exercises

1. Make a table to help solve this problem. Cory, Srey, Molly, and Mao each visited a different city: Santa Clara, Seattle, Des Moines, and Pittsburgh. Cory and Molly visited cities with two words in their names. Mao and Cory visited cities whose names start with a **S**. Who visited each city? See additional answers.
2. Draw a Venn diagram to solve this problem. Twenty-four students are touring the United Nations building. Twelve of these students speak Russian, six speak German, and 15 speak Spanish; only one student speaks *all three* languages. Two students speak only Russian and German, and two others speak only German and Spanish. How many students speak Russian and Spanish, but not German? 3; See additional answers.

Five-step Plan
1 Read
2 Plan
3 Solve
4 Answer
5 Check

Practice Exercise

3. Ned—Miami; Carina—Dallas; Pedro—San Francisco; and Eva—San Diego

3. **DATA FILE** Use the Data Index on pages 632–633 to find information about major zoos in the United States. Then use the data to solve this problem. You may want to copy and complete the table below.

Pedro, Carina, Ned, and Eva each visited different zoos during their vacations. The zoos were the San Diego Zoo, the San Francisco Zoo, the Miami Metrozoo, and the Dallas Zoo.

Name	San Diego	San Francisco	Miami	Dallas
Pedro				
Carina				
Ned				
Eva				

- ▶ Eva, Carina, and Pedro saw outstanding gorilla exhibits.
- ▶ Both Carina and Pedro went to zoos with fewer than 400 species.
- ▶ The zoo that Pedro visited has a greater budget than the zoo that Carina visited.

Which zoo did each person visit?

4. Use logical reasoning to solve this problem. Xenia corresponded on a regular basis with three pen pals. She wrote to all three on one Friday in 1993. She knew she would next write to Daksha in India on Thursday two weeks after that, then on Wednesday in the fourth week. She would write to Leonel in the Dominican Republic every Friday. Finally, she would write to Catherine every fourth day, since they had been good friends before Catherine moved to England. On what date in 1993 did Xenia write for the second time to all three pen pals on the same day? December 31, 1993; See additional answers.

Mixed Review Exercises

Graph each point on the coordinate plane. Label each point. See additional answers.

5. $A(3, 6)$
6. $B(7, 3)$
7. $C(2, -2)$
8. $D(-6, 6)$
9. $E(-2, 3)$
10. $F(-7, -3)$
11. $G(8, -5)$
12. $H(-2, -7)$

Solve each equation.

13. $3(4x - 3) + 8 = 15x - 2(x + 1)$ 1
14. $7(x + 4) - 8 = -3x + 8\left(x + \frac{1}{2}\right)$ −2

Chalkboard Examples

Supplementary Problem

Lisa, Kendra, Abby, and Allison had a race. Abby finished behind Kendra but not last. The winner's name did not begin with an A. Lisa was faster than Kendra. In which order did they finish the race?
Lisa, Kendra, Abby, Allison

Lesson Wrap-up

Quick Assessment

Ask the following question to determine if students understand the content presented in this lesson.

How would you go about setting up a table in order to use the process of elimination? **Answers may vary. Students should include the idea that they would categorize the types of information presented in the problem and then create a table with one category as row headings and the other category as column headings.**

Enrichment Worksheet 3-8 not shown.

Reteaching Worksheet 3-8

Name ______________________ Date __________

RETEACHING 3-8

PROBLEM SOLVING SKILLS: USING LOGICAL REASONING

It can be useful to make a table to solve some logic problems.

Example

Four students were talking about their Friday night dates. Amy went skating. Juan and Toi each saw a movie or play, but not necessarily in that order. Toi and Jamal saw live performances. One student went to a football game.

Solution

Make a table listing each student and each activity. Put a ✔ in a box to indicate a fact is true. Put an ✘ in a box for any fact that is not true.

Since Amy went skating, place a ✔ in the box where the row *Amy* meets the column *skating* and an ✘ in the other boxes in the column and row. Juan and Toi saw a movie or play, so place an ✘ in the boxes where the rows *Juan* and *Toi* meet the columns *football* and *skating.* Thus Jamal must have seen the football game. Place a ✔ in that box. Put an ✘ in the other boxes in the row.

	movie	play	skating	football game
Juan			✘	✘
Toi			✘	✘
Jamal	✘	✘	✘	✓
Amy	✘	✘	✓	✘

Toi did see a live performance, so she saw a play. Place a ✔ in the box where the row *Toi* meets the column *play* and an ✘ in the other boxes in that column and row. This leaves Juan seeing the movie. Mark the box with a ✔.

	movie	play	skating	football game
Juan		✘	✘	✘
Toi	✘	✓	✘	✘
Jamal	✘	✘	✘	✓
Amy	✘	✘	✓	✘

EXERCISES

Make a table to help solve the problem.

1. Maria, Keisha, Jo, and Leah wear jerseys numbered 5, 8, 13 and 36, but not necessarily in that order. The girl wearing 8 has a better jumpshot than Maria or Leah. Jo's jersey number is divisible by 6. The number of letters in one name is the same as her number. What number does each wear?
Maria: 5; Keisha: 8; Leah: 13; Jo: 36
2. A grocery, a cafe, a boutique, and a dance studio are all on one street. Their addresses are 500 Elm, 524 Elm, 538 Elm, and 544 Elm, but not necessarily in that order. The boutique is between the cafe and the studio. The studio has the highest address. What is the address of each business?
Grocery: 500 Elm; Boutique: 538 Elm; Cafe: 524 Elm; Studio: 544 Elm

50 RETEACHING *LESSON 3-8*

4. **Xenia will write to Daksha every 13 days, to Leonel every 7 days, and to Catherine every 4 days. The next time she will write to all three on the same day is (13 × 7 × 4) days, or 364 days. Since 1993 is not a leap year, this occurs on December 31, 1993.**

5–12.

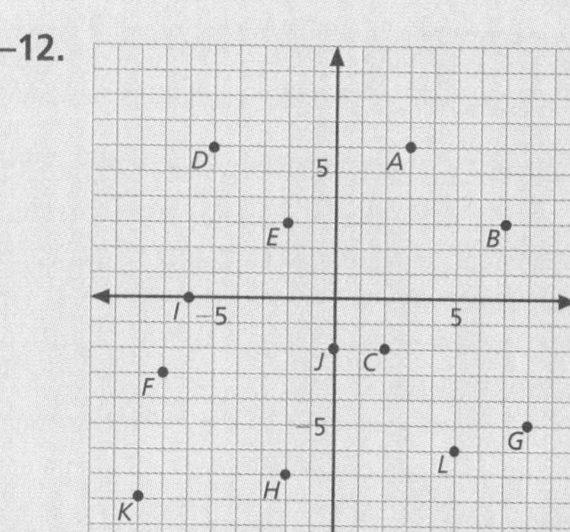

Assessment Planning

Vocabulary Review

Lesson 3-1
point, line, plane, collinear points, space, coplanar points, intersection, postulate, ray, endpoint, line segment

Lesson 3-2
angle, vertex, protractor, degree measure, acute angle, right angle, obtuse angle, straight angle, complementary, supplementary, adjacent, exterior sides

Lesson 3-3
midpoint, bisector, theorem, opposite rays, vertical angles

Lesson 3-4
compass, straightedge, parallel lines, skew lines, transversal, bearings, perpendicular lines, alternate interior angles, alternate exterior angles, corresponding angles

Lesson 3-5
sequence, conjecture, inductive reasoning

Lesson 3-6
conditional, hypothesis, conclusion, counterexample, biconditional

Lesson 3-7
proof, theorem, deductive reasoning, two-column proof

Lesson 3-8
process of elimination

Assessment Options

Chapter Assessment
Alternative Assessment
Chapter Tests, Forms A and B
Cumulative Assessment
Achievement Test
Writing Prompts

Assignment Guide

All students: 1–17
Additional Practice: Refer to the Extra Practice Index on page 674 of this text.

Chapter 3 Review

Vocabulary ■ LESSON 3-1–LESSON 3-8

Write the letter of the word at the right that matches each description.

1. A line that intersects two coplanar lines in different points. d
2. An angle whose measure is greater than 90° and less than 180° e
3. An example to show that a conditional is false. c
4. Two angles in the same plane that share a common side and a common vertex, but have no interior points in common. a
5. The type of reasoning in which a theorem is proved by a series of logical statements and reasons. b

a. adjacent angles
b. deductive reasoning
c. counterexample
d. transversal
e. obtuse angle

LESSON 3-1 ■ Points, Lines, and Planes, p. 104

- In geometry, **point**, **line**, and **plane** are undefined terms. Other geometric definitions are developed by using these undefined terms.
- Statements that are accepted as true without proof are called **postulates**.
- The **segment addition postulate** states that if point B is between A and C, then $AB + BC = AC$.

6. In the figure, $RS = 75$. Find RN. 28.5

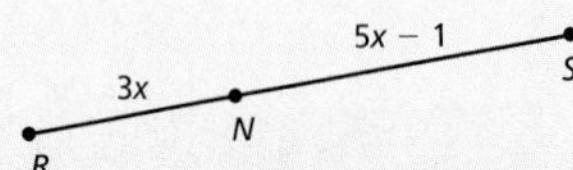

LESSON 3-2 ■ Types of Angles, p. 108

- If point B lies in the interior of $\angle AOC$, then $m\angle AOB + m\angle BOC = m\angle AOC$.

7. In the figure, $m\angle RNX = (3x - 2)°$ and $m\angle XNW$ is $(2x + 7)°$. Find $m\angle TNX$. 13°

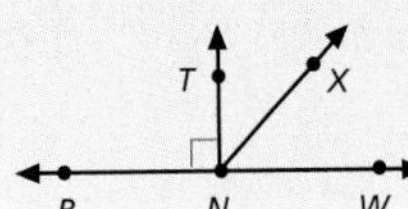

LESSON 3-3 ■ Segments and Angles, p. 114

- Two angles whose sides form two pairs of opposite rays are called **vertical angles**. Vertical angles are equal in measure.

8. $\angle XYZ$ 70°
9. $\angle XYB$ 110°

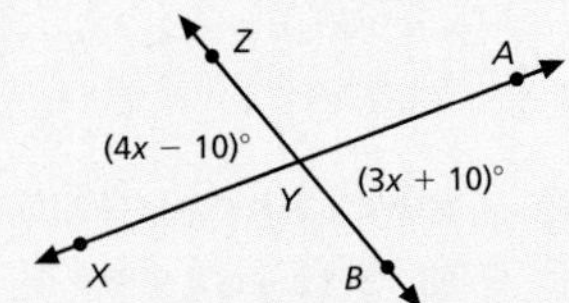

Teaching Strategies

VISUAL LEARNERS Sets of flashcards are especially helpful for reviewing geometry relationships and postulates. Have students make one card for each postulate in the chapter. Include the definition of the postulate and a diagram to show how the postulate is used. As these learners select a diagram to draw, they are reinforcing the connection between the words and the drawing.

LESSON 3-4 ■ Constructions and Lines, p. 118

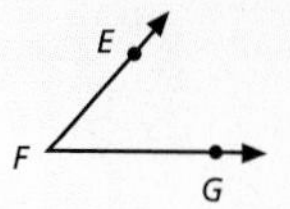

► A precise drawing of a geometric figure made by using only a compass and an unmarked straightedge is called a **construction**.

10. Trace $\angle EFG$ on another sheet. Using a compass and a straightedge, divide $\angle EFG$ into four angles of equal measure. Check students' work.

Two or more parallel lines cut by a transversal form several pairs of angles that are equal in measure.

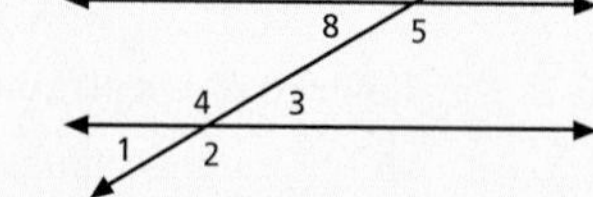

$\angle 1$ and $\angle 8$ are **corresponding angles**.

$\angle 3$ and $\angle 8$ are **alternate interior angles**.

$\angle 2$ and $\angle 7$ are **alternate exterior angles**.

11. Name the other pairs of corresponding angles in the figure. $\angle 2$ and $\angle 5$, $\angle 3$ and $\angle 6$, $\angle 4$ and $\angle 7$

12. Name the other pairs of alternate interior angles in the figure. $\angle 4$ and $\angle 5$

13. Name the other pairs of alternate exterior angles in the figure. $\angle 1$ and $\angle 6$

LESSON 3-5 ■ Inductive Reasoning in Mathematics, p. 124

► **Inductive reasoning** is the process of observing data and making a generalization based on your observations.

See additional answers for next figure.

14. Draw the next figure in the pattern. Then describe the tenth figure. The tenth figure will be a rectangular arrangement of 120 dots, with 10 rows of 12 dots each.

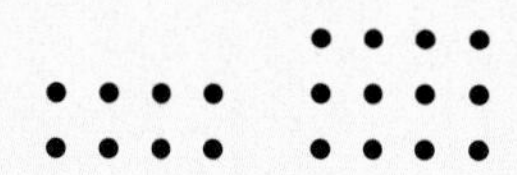

LESSON 3-6 ■ Conditional Statements, p. 128

► The **converse** of a conditional statement is formed by interchanging the hypothesis and conclusion.

15. Write the converse of the following statement. Then decide if the statement and its converse are true or false.

If the sum of the measures of two angles is 180°, the angles are supplementary.

If two angles are supplementary, the sum of the measures is 180°. Both statements are true.

LESSON 3-7 ■ Deductive Reasoning and Proof, p. 134

► Proofs of theorems are often written in a two-column format, with statements in one column and reasons in the other.

16. Complete the following proof.

Given E is between L and T. **Prove** $ET = LT - LE$

See additional answers.

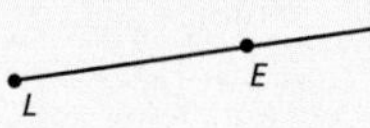

LESSON 3-8 ■ Problem Solving Skills, p. 138

► Many problems are easier to solve if you use a table.

17. Marco, Sue, and Stephanie went on vacations last summer. They visited California, Florida, and Wisconsin, but not necessarily in that order. Marco did not visit a state bordering the ocean. Sue did not visit Florida. Which state did each visit? Maro: Wisconsin; Sue: California; Stephanie: Florida

ADDITIONAL ANSWERS

14. • • • • • •
• • • • • •
• • • • • •
• • • • • •

16.

Statements	Reasons
1. E is between L and T	1. given
2. $LE + ET = LT$	2. segment add. post.
3. $ET = LT - LE$	3. add. prop. of equality

Chapter 3 Test Form A, page 1

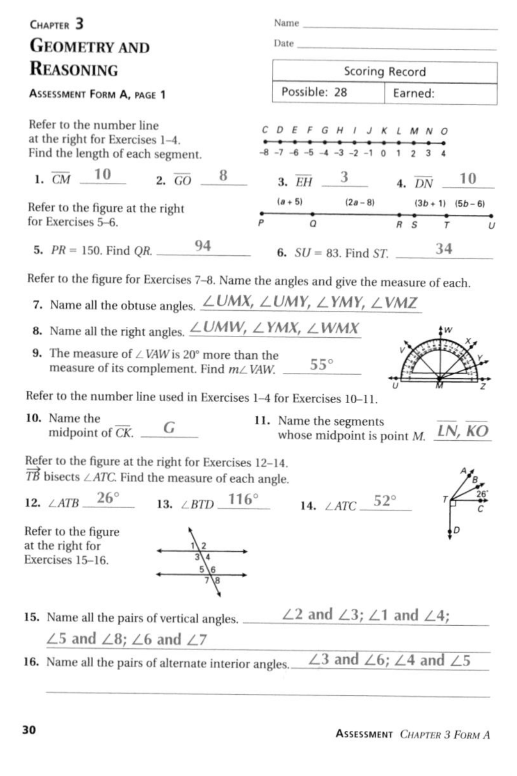

CHAPTER 3
GEOMETRY AND REASONING
ASSESSMENT FORM A, PAGE 1

Name ______
Date ______

Scoring Record	
Possible: 28	Earned:

Refer to the number line at the right for Exercises 1–4. Find the length of each segment.

C D E F G H I J K L M N O
−8 −7 −6 −5 −4 −3 −2 −1 0 1 2 3 4

1. $\overline{CM}$ 10 2. $\overline{GO}$ 8 3. $\overline{EH}$ 3 4. $\overline{DN}$ 10

Refer to the figure at the right for Exercises 5–6.

$(a + 5)$ $(2a - 8)$ $(3b + 1)$ $(5b - 6)$
P Q R S T U

5. $PR = 150$. Find QR. 94 6. $SU = 83$. Find ST. 34

Refer to the figure for Exercises 7–8. Name the angles and give the measure of each.

7. Name all the obtuse angles. $\angle UMX$, $\angle UMY$, $\angle YMY$, $\angle VMZ$

8. Name all the right angles. $\angle UMW$, $\angle YMX$, $\angle WMX$

9. The measure of $\angle VAW$ is 20° more than the measure of its complement. Find $m\angle VAW$. 55°

W V X Y U M Z

Refer to the number line used in Exercises 1–4 for Exercises 10–11.

10. Name the midpoint of $\overline{CK}$. G

11. Name the segments whose midpoint is point M. $\overline{LN}$, $\overline{KO}$

Refer to the figure at the right for Exercises 12–14. $\overrightarrow{TB}$ bisects $\angle ATC$. Find the measure of each angle.

12. $\angle ATB$ 26° 13. $\angle BTD$ 116° 14. $\angle ATC$ 52°

A B 26° T C D

Refer to the figure at the right for Exercises 15–16.

1 2 3 4 5 6 7 8

15. Name all the pairs of vertical angles. $\angle 2$ and $\angle 3$; $\angle 1$ and $\angle 4$; $\angle 5$ and $\angle 8$; $\angle 6$ and $\angle 7$

16. Name all the pairs of alternate interior angles. $\angle 3$ and $\angle 6$; $\angle 4$ and $\angle 5$

30 ASSESSMENT *CHAPTER 3 FORM A*

Chapter 3 Test Form A, page 2

Name ______ Date ______

Refer to the figure at the right for Exercises 17–22. $\overline{EH} \parallel \overline{LI}$ and $\overline{FJ} \perp \overline{GK}$. Find the measure of each angle.

F G E P Q H S L I 50° K J

17. $\angle KSF$ 90° 18. $\angle GQH$ 40° 19. $\angle JSG$ 90°

20. $\angle GQE$ 140° 21. $\angle FPE$ 50° 22. $\angle LSJ$ 130°

23. Draw the next figure in the pattern. Then describe the 30th figure.

The 30th figure will be 30 rows of dots, or 30^2 dots.

24. Write the converse of this statement: If two parallel lines are cut by a transversal, then all vertical angles are equal. Then decide whether the statement and its converse are true or false. If false, give a counterexample.

If all the vertical angles are equal, then two parallel lines are cut by a transversal; true; false; 2 intersecting lines

25. Use a compass and a straightedge to draw an obtuse angle on the back of this page. Then bisect it. Check students' drawings.

26. Use a compass and a straightedge to draw a line segment of about 6 in. on the back of this page. Then bisect it. Check students' drawings.

27. Complete this proof.
Given: $\overline{XA} \perp \overline{XB}$
$\angle 1$ and $\angle 2$ are adjacent
Prove: $m\angle 1 + m\angle 2 = 90°$

A 1 2 X B

STATEMENTS	REASONS
1. $\overline{XA} \perp \overline{XB}$, $\angle 1$ and $\angle 2$ are adjacent	1. given
2. $\angle AXB$ is a right angle	2. definition of perpendicular
3. $m\angle AXB$ = 90°	3. definition of right angle
4. $m\angle$ AXB = $m\angle 1$ + $m\angle 2$	4. angle addition postulate
5. $m\angle 1 + m\angle 2$ = 90°	5. substitution property

28. A health care presentation had four speakers. Andy and Luisa were a psychiatrist and a surgeon. Neither the nurse nor Carl spoke about AIDS awareness. The counselor and his wife, a psychiatrist, have a practice together. Ann spoke first. What was each person's profession? Andy: surgeon; Luisa: psychiatrist; Carl: counselor; Ann: nurse

32 ASSESSMENT *CHAPTER 3 FORM A*

Chapter 3 **Review** 141

Chapter 3 Test Form B, page 1

CHAPTER 3

GEOMETRY AND REASONING

ASSESSMENT FORM B, PAGE 1

Name ____________

Date ____________

Scoring Record	
Possible: 28	Earned:

Refer to the number line at the right for Exercises 1–4. Find the length of each segment.

A B C D E F G H I J
–3 –2 –1 0 1 2 3 4 5 6

1. $\overline{AF}$ 5 2. $\overline{DG}$ 3 3. $\overline{BI}$ 7 4. $\overline{HJ}$ 2

Refer to the figure at the right for Exercises 5–6.

(4x + 3) (2x – 2) (8c – 16) (4c + 13)
J K L M N

5. $JL = 133$. Find JK. 91
6. $LN = 57$. Find MN. 33

Refer to the figure for Exercises 7–8. Name the angles and give the measure of each.

7. Name four acute angles. Possible answers: ∠OAP, ∠OAQ, ∠PAQ, ∠PAR, ∠QAS, ∠RAS, ∠RAT, ∠SAT, ∠QAR
8. Name all the straight angles. ∠OAT
9. The measure of ∠TDU is 48° less than the measure of its supplement. Find m∠TDU. 66°

Refer to the number line used in Exercises 1–4 for Exercises 10–11.

10. Name the midpoint of $\overline{EI}$. G
11. Name the segments whose midpoint is point D. $\overline{AG}$, $\overline{BF}$, $\overline{CE}$

Refer to the figure at the right for Exercises 12–14. $\overline{AU}$ bisects ∠TAV. Find the measure of each angle.

12. ∠TAU 45° 13. ∠TAV 90° 14. ∠TAW 180°

Refer to the figure at the right for Exercises 15–16.

15. Name all the pairs of corresponding angles. ∠1 and ∠5; ∠2 and ∠6; ∠3 and ∠7; ∠4 and ∠8
16. Name all the pairs of alternate exterior angles. ∠1 and ∠8; ∠2 and ∠7

34 ASSESSMENT CHAPTER 3 FORM B

Chapter 3 Test Form B, page 2

Name ____________ Date ____________

Refer to the figure at the right for Exercises 17–22. $\overline{AF} \parallel \overline{BE}$ and $\overline{HD} \perp \overline{GC}$. Find the measure of each angle.

17. ∠END 60° 18. ∠GPA 150° 19. ∠BNE 180°
20. ∠GNB 150° 21. ∠HOA 60° 22. ∠CPA 30°

23. Study the pattern. Write how many squares are in each figure. Then find the number of squares in the fifteenth figure. 120 squares
24. Write the converse of this statement: If two lines intersect, then the vertical angles are equal. Then decide whether the statement and its converse are true or false. If false, give a counterexample. If vertical angles are equal, then two lines intersect; true; true
25. Use a compass and a straightedge to draw an acute angle on the back of this page. Then bisect it. Check students' drawings.
26. Use a compass and a straightedge to draw a line segment of about 4 in. on the back of this page. Then bisect it. Check students' drawings.
27. Complete this proof.
Given: $\overrightarrow{XM}$ intersects $\overleftrightarrow{LN}$; ∠1 and ∠2 are adjacent
Prove: $m\angle 1 + m\angle 2 = 180°$

STATEMENTS	REASONS
1. $\overrightarrow{XM}$ intersects $\overleftrightarrow{LN}$; ∠1 and ∠2 are adjacent	1. given
2. ∠LXN is a straight angle	2. definition of straight angle
3. m∠LXN = 180°	3. definition of straight angle
4. m∠LXN = m∠1 + m∠2	4. adjacent angles postulate
5. m∠1 + m∠2 = 180°	5. substitution property

28. Four students were discussing what they were doing after school. One had cheerleading tryouts. Kirk and Alicia had baseball and tennis practice, but not necessarily in that order. Marsha did not have drama rehearsal. Chad thought Alicia would make it to the state tennis finals. What did each person do? Kirk: baseball; Alicia: tennis; Marsha: cheerleading; Chad: drama

36 ASSESSMENT CHAPTER 3 FORM B

CHAPTER 3 ASSESSMENT

Use the number line at the right to find each length.

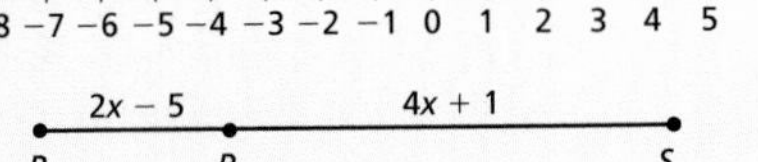

1. AB 4 2. AC 12 3. BD 5
4. In the figure at the right, $RS = 110$. Find RP. 33

2x – 5 4x + 1
R P S

Refer to the protractor at the right.

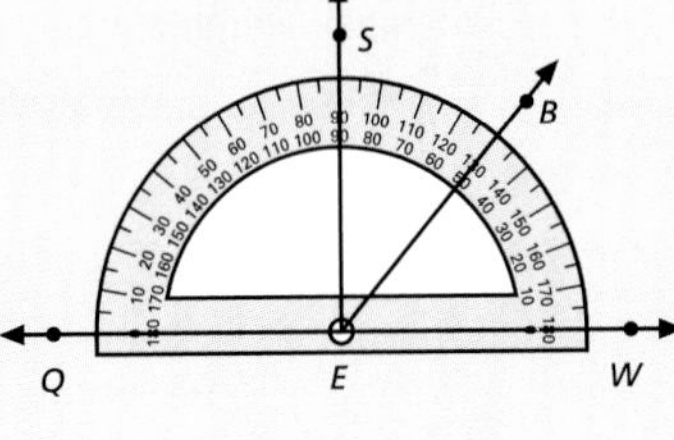

5. Name the right angles. ∠QES, ∠SEW
6. Name the obtuse angles and give the measure of each. ∠QEB, 130°
7. In the figure below, m∠RSY is $(4x + 6)°$ and m∠YSX is $(2x - 12)°$. Find m∠TSY. 40°

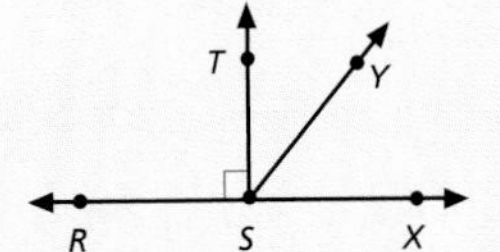

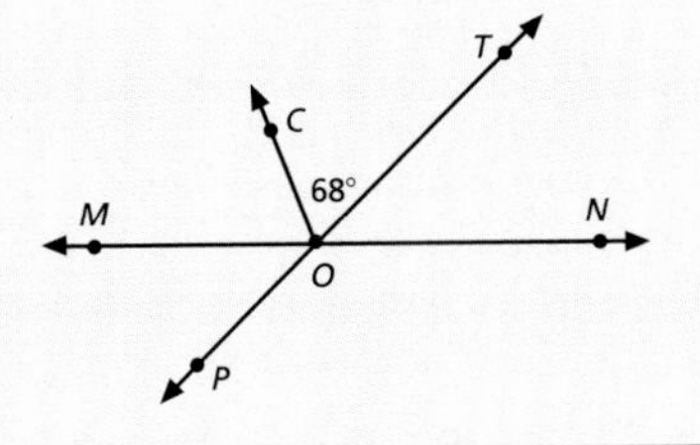

8. In the figure at the right, $\overleftrightarrow{MN}$ and $\overleftrightarrow{PT}$ intersect at point O. $\overrightarrow{OC}$ bisects ∠MOT. Find the measure of ∠MOP. 44°
9. Trace $\overline{FP}$ onto a sheet of paper. Using a compass and a straightedge, divide it into four segments of equal length. Check students' work.

F P

In the figure at the right $\overleftrightarrow{AB} \parallel \overleftrightarrow{CD}$ and $\overleftrightarrow{XY} \perp \overleftrightarrow{EF}$ at G. Find the measure of each angle.

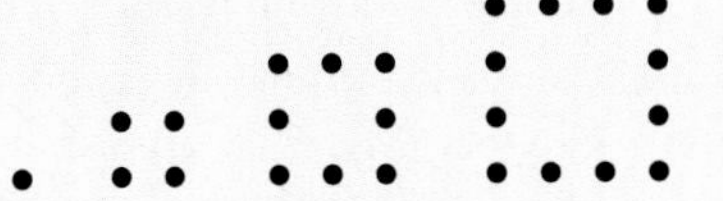

10. ∠AGE 42° 11. ∠XKH 132° 12. ∠DHF 42°
13. Write the converse of the following statement. Then tell if the given statement and the converse are true or false. If B is the midpoint of AC, then AB + BC = AC; original statement, false; converse, true.
If $AB + BC = AC$, then B is the midpoint of $\overline{AC}$.
14. In a two-column proof, the hypothesis is labeled __?__ and the conclusion is labeled __?__. given, prove
15. Draw the next figure in the pattern of dota. See additional answers.
16. Three students named Alf, Beth, and Chan play three different sports. One plays baseball, one plays tennis, and one plays hockey. Chan plays a sport that does not require a ball. Beth does not play baseball. What sport does each play? Alf, baseball; Beth, tennis; Chan, hockey

ADDITIONAL ANSWERS

15.

Chapter 3 Alternative Assessment

Portfolio

GEOMETRY ENCYCLOPEDIA Create your own Geometry Encyclopedia. Choose at least 5 theorems or postulates from the lessons in this chapter. Record each, then illustrate them in an original example for each. List vocabulary words and their meanings, then illustrate these also. Include an explanation of each principle or word in you own words.

Connections

LITERATURE AND LOGIC Read a detective/mystery book. Write a synopsis of the plot, along with a brief description of each of the main characters. List at least 10 conditional statements the detective may have used to solve the case. The statements should be listed in order, so that it would be easy to understand how the crime was solved.

Cooperative Learning

WRITING AND LOGIC Work in a group. Make up 4 logic puzzles, similar to those in your text. Puzzles should have a brief introduction, setting the scene, then several statements about the arrangement depicted. Trade puzzles with other groups to solve.

Critical Thinking

PATTERNS Draw at least 3 new geometric patterns, such as those found in your text. Draw or illustrate the first 5 elements in the pattern. Write a function rule for each and represent each numerically as well.

Chapter Investigation

EXTENSION Make a presentation to the class of your map. Show that the navigator's log you created accurately specifies the direction and bearing for each leg of the course you plotted.

Chapter Investigation

The benchmarks and expectations for this extension are as follows.

- Students draw a map of their community showing at least ten specific points of interest. They make distances as accurate as possible. They overlay their map with a grid of latitude and longitude lines.
- Students use the maps they made and sketch a course to travel between two points. They write directions in which they specify the bearing of each leg of the journey.
- Students find an example on their map of the situations described in the theorems stated in Exercises 7–9. They write a proof of each.
- Students make a presentation to the class of their map. They show that the navigator's log they created accurately specifies the direction and bearing for each leg of the course they plotted.

Portfolio

Check students' illustrations of the postulates and theorems. If explanations are unclear, ask questions that will clarify their level of comprehension.

SCORING RUBRIC **5**—Creates a geometric encyclopedia that is both accurate and comprehensive. **4**—Creates a geometric encyclopedia that is accurate and comprehensive, but some explanations could be improved. **3**—Creates a geometric encyclopedia that is accurate, but not comprehensive. **2**—Creates a geometric encyclopedia with some errors in understanding. **1**—Creates a geometric encyclopedia, but exhibits serious errors in understanding. **0**—Make no attempt to create a geometric encyclopedia.

Connections

Assist students in selecting a detective/mystery book to read. Novels, such as Agatha Christie's *The Murder of Roger Ackroyd*, lend themselves well to this activity.

SCORING RUBRIC **5**—Reads assigned book, insightfully analyzes plot and characters, then forms ten logical conditional statements. **4**—Read assigned book, analyzes plot and characters satisfactorily, then forms ten logical conditional statements. **3**—Read assigned book, analyzes satisfactorily, then forms less than ten conditional statements; logic may be fuzzy. **2**—Reads assigned book, reviews plot and characters, then forms a few conditional statements; logic is not clear. **1**—Reads assigned book, reviews the plot and some of the characters, does not form conditional statements. **0**—Makes no attempt to complete task.

Cooperative Learning

Caution students that making their logic problems too obvious or easy to solve defeats the purpose of the activity.

SCORING RUBRIC **5**—Writes four original, clever logic puzzles and solves all from other groups. **4**—Writes four original logic puzzles and solves all from other groups. **3**—Write four original logic puzzles, solves most from other groups. **2**—Writes less than four logic puzzles, solves a few puzzles from other groups. **1**—Attempts to write logic puzzles, cannot solve those from other groups. **0**—Makes no attempt to write and solve logic puzzles.

See page 144 for information and Scoring Rubric for Critical Thinking.

Cumulative Review

Cumulative review is the best way to maintain previously taught skills and concepts. This will keep students prepared for new lessons that build on previously covered skills.

Cumulative Review covers:

Lesson 1-3
Lesson 1-5
Lesson 1-7
Lesson 1-8
Lesson 2-2
Lesson 2-4
Lesson 2-5
Lesson 2-6
Lesson 2-7
Lesson 2-8
Lesson 3-3
Lesson 3-4
Lesson 3-6

Additional Answers

27. 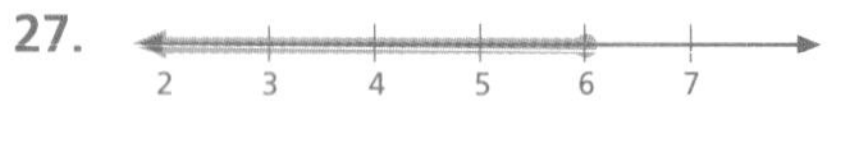

28. −9 −8 −7 −6 −5 −4

36. If $OT + TB = OB$, then T is the midpoint of OB; original statement, true; converse, false.
37. If the measure of an angle is less than 90°, the angle is acute; both statements true.

Critical Thinking

To more correctly gauge students' level of understanding, you may wish to have students explain each of their patterns to you in an individual interview.

SCORING RUBRIC 5—Student devises 3 original geometric patterns, correctly writes a function rule for each. 4—Student devises 3 geometric patterns, correctly writes a function rule for each. 3—Student devises 3 geometric patterns, writes a function rule for each with a few minor errors. 2—Student devises less than 3 geometric patterns, writes a function rule for each with a few minor errors. 1—Student devises less than 3 geometric patterns, attempts to write function rules for each. May contain serious errors. 0—Student makes no attempt to devise original geometric patterns.

Chapters 1–3 Cumulative Review

Let $U = \{r, s, t, u, v, w, x, y, z\}$, $M = \{x, y, z\}$, and $N = \{u, v, w, x\}$. Find each union or intersection.

1. $M \cup N$ $\{u, v, w, x, y, z\}$ **2.** $M \cap N$ $\{x\}$ **3.** $M' \cup N'$ $\{r, s, t, u, v, w, y, z\}$ **4.** $(M \cap N)'$ $\{r, s, t, u, v, w, y, z\}$

Peform each indicated operation.

5. $(-1.8)(2.5)$ -4.5 **6.** $-60 \div (-1.2)$ 50 **7.** $6\frac{2}{3} \div (-3) + 5$ $2\frac{7}{9}$ **8.** $3\frac{1}{8} + (-6)(-2)$ $15\frac{1}{8}$

Simplify.

9. $r^6 \cdot r^3$ r^9 **10.** $(m^3)^5$ m^{15} **11.** $(x^3)(x^3)(x^3)$ x^9 **12.** $\left(\frac{1}{y}\right)^4$ $\frac{1}{y^4}$

Evaluate when $x = -4$ and $y = 3$.

13. x^{-2} $\frac{1}{16}$ **14.** y^{-3} $\frac{1}{27}$ **15.** x^0y^{-2} $\frac{1}{9}$ **16.** $x^2x^{-3}y$ $-\frac{3}{4}$

Write each number in scientific notation.

17. 0.000007124 7.124×10^{-6} **18.** 86,000,000,000 8.6×10^{10}

Given $f(x) = 3x - 2$ and $g(x) = 2x^2$, find each value.

19. $f(-2)$ -8 **20.** $g(3)$ 18 **21.** $f(1) + g(1)$ 3 **22.** $g(10) - f(10)$ 172

Solve each equation and check the solution.

23. $x + 15 = -27$ -42 **24.** $\frac{m}{6} = 14$ 84 **25.** $6x - 1 = 3x + 4$ $\frac{5}{3}$ **26.** $3(x - 2) = 2(x + 1)$ 8

Solve each inequality and graph the solution on a number line.
See additional answers.

27. $3x - 5 \le 13$ **28.** $-25 > 2x - 11$

Refer to the following data.

26 33 43 30 28
36 42 29 46 43
39 35 47 46 43

29. Make a stem-and-leaf plot.

2	6 8 9
3	0 3 5 6 9
4	2 3 3 3 6 6 7

30. Find the mode. 43

31. Find the median. 39

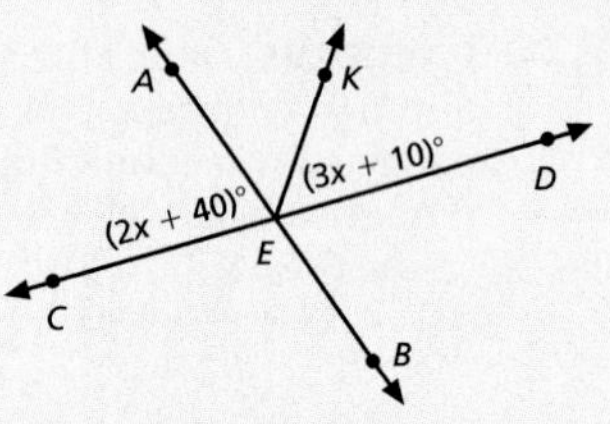

In the figure, $\overleftrightarrow{AB}$ and $\overleftrightarrow{CD}$ intersect at point E, and $\overrightarrow{EK}$ bisects $\angle AED$. Find the measure of each angle.

32. $\angle AEC$ 70° **33.** $\angle AEK$ 55° **34.** $\angle CEB$ 110° **35.** $\angle KED$ 55°

Write the converse of each statement, then tell if the given statement and its converse are true or false.
36.–37. See additional answers.

36. If T is the midpoint of OB, then $OT + TB = OB$.

37. If an angle is acute, then the measure of the angle is less than 90°.

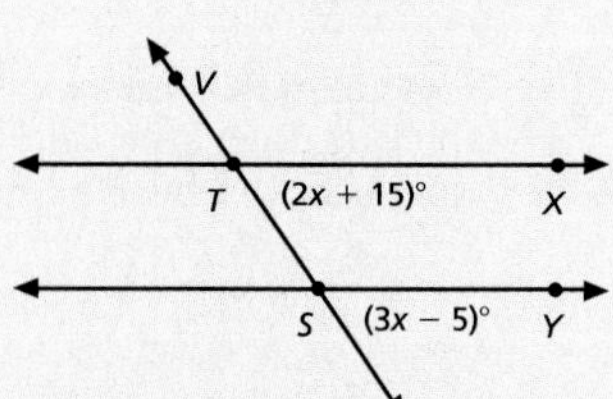

In the figure, $\overleftrightarrow{TX} \parallel \overleftrightarrow{SY}$. Find the measure of each angle.

38. $\angle VTX$ 125° **39.** $\angle STX$ 55°

Chapters 1–3 Cumulative Assessment

Standardized Test Preparation—Gridded Response

Solve each question. Write your answer at the top of the answer grid and fill in the ovals.

Notes: Mixed numbers such as $1\frac{1}{2}$ must be gridded as 1.5 or $\frac{3}{2}$. Grid only one answer per question. If your answer is a decimal, enter the most accurate value the grid will accommodate.

1. Grid the absolute value of the least element of this set.
   $\{x \mid x \text{ is a negative integer and } x > -3\}$ 2
2. Let $A = \{2, 4, 6, 8\}$ and let $B = \{1, 2, 3\}$. Grid the only element of $A \cap B$. 2
3. Evaluate $|-m| - |m + 1|$ if $m = -5$. 1
4. Evaluate $a - b$ of $a = 16$ and $b = -9$. 25
5. Evaluate $m(n + p)$ if $m = -3$, $n = -2.4$, and $p = \frac{3}{4}$. 4.96
6. Evaluate x^3y^2 when $x = 3$ and $y = -1$. 27
7. Evaluate x^{-2} if $x = 5$. $\frac{1}{25}$
8. What is the exponent on 10 in the scientific notation for 3,607,000,000,000? 12
9. What is the next term in the following sequence? 28

   3, 6, 10, 15, 21

10. What is the solution for $m - 1.2 = 3.4$? 4.6
11. Grid the least y-value that will satisfy this inequality when $x = 8$. 10

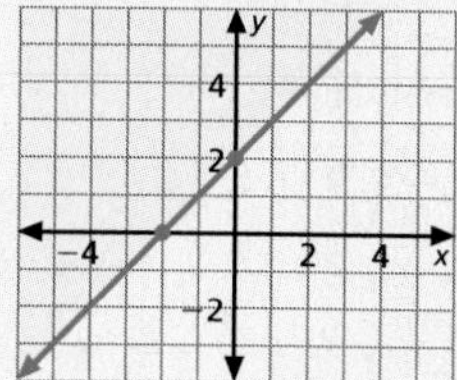

12. Grid the median for this set of data. 20

21	19	16	13	15
21	18	22	24	21

13. In the figure, $AC = 33$. Find AB. 18

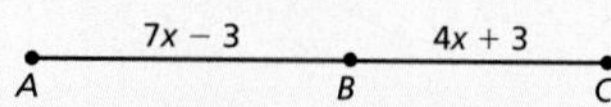

14. Grid the number of dots in the next figure in the pattern. 28

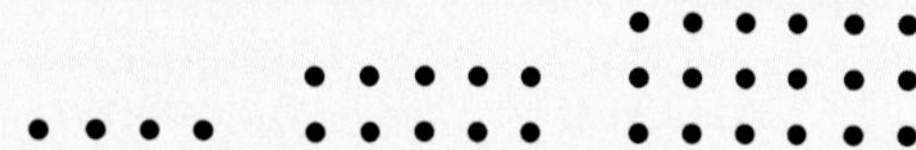

15. What is the least integer in the solution set for this inequality? 11

$$-25 + 3x > 7$$

16. Find a two-digit number with two unique even digits. The sum of the digits is an even two-digit number. The sum of the ones digit and 6 equals the tens digit. 82
17. Grid 1 for true, 2 for false. The complement of an acute angle is always acute. 1

Find the supplement of each angle.

18. 53° 127°
19. 117° 63°
20. 90° 90°
21. 142° 38°
22. **CONSTRUCTED RESPONSE** Describe each segment in the figure as either parallel, perpendicular, or skew to segment AB. If a segment could be extended to intersect AB, describe the segment as intersecting.

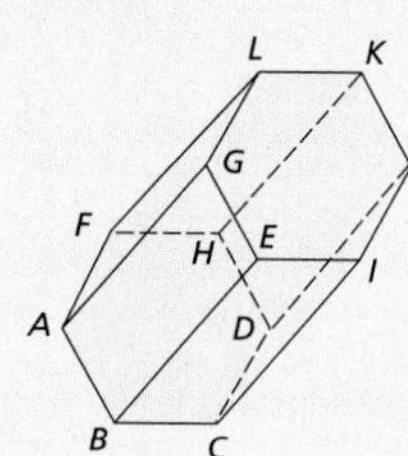

Segments *ED, GH*, and *JK* are parallel to segment *AB*; segment *GA* and *GB* are perpendicular to segment *AB*; segment *CI, DJ, EK, FL, HI, IJ, LK, gl* are skew to segment *AB*; and lines *AF, BC, CD*, and *EF* intersect line *AB*.

Standardized Test Preparation

GRIDDED RESPONSE Questions that require you fill in grids on an answer sheet, but do not have answer choices provided are called gridded response or students-produced response. You must solve the problem and then fill in the answer on a special grid. Your answer consists of two parts-the numbers that you write in the answer boxes and the ovals that you fill in. Ovals that correspond to the digits 0–9, negative sign, fraction bars and decimal points are listed in columns under the answer boxes.

You should grid the form of the answer that you obtain from solving the problem. Enter the most accurate value the grid will accommodate. The grid will only hold numbers that range form 0–9999. The grid will not accommodate a mixed number. Change mixed numbers to either decimals or improper fractions.

Columns not needed should be left blank. It does not matter whether the extra columns are on the left or the right of the answer.

Testing Strategies

For questions 5 and 7 have students discuss how their answers should be entered in the grid.

Chapter 3 **Cumulative Assessment** 145

Chapter 4—Triangles, Quadrilaterals and Other Polygons

Theme: Art and Design
Chapter Investigation: Design a Bridge, pages 149, 157, 167, 175
Careers: Jewelers, page 159; Animators, page 177
Data Activity: Suspension Bridges of New York

Content and Connections

Lesson, Pages	Lesson Objectives	NCTM Standards	State/Local Objectives	Interdisciplinary Connections	Real World Applications
4–1 150–153	• Solve algabraic equations to find measure of angles in triangles • Classify triangles according to their sides or angles	Number Operation Patterns, Functions & Algebra Reasoning & Proof Connections Representation			Technical Art, Construction, Engineering
4–2 154–157	• Prove triangles are congruent	Geometry & Spatial Sense Measurement Prob Solv; Reasoning & Proof Representation			Animation, Construction, Engineering
4–3 160–163	• Establish congruence between two triangles to show that corresponding parts are congruent • Find missing angle and side measures of isosceles and equilateral triangles	Geometry & Spatial Sense Patterns, Functions & Algebra Measurement Prob Solv; Reasoning & Proof		Connections; Representation	Design, Architecture, Construction, Engineering
4–4 164–167	• Identify and sketch altitudes and medians of a triangle and perpendicular bisectors of sides of a triangle	Geometry & Spatial Sense Measurement Prob Solv; Reasoning & Proof Connections		Physics	Architecture, Physics, Service
4–5 170–171	• Write an Indirect Proof • Guess and check	Geometry & Spatial Sense Prob Solv; Reasoning & Proof Connections; Representation		Art	Art, Architecture
4–6 172–175	• Understand relationships among sides and angles of a triangle	Patterns, Functions & Algebra Geometry & Spatial Sense Prob Solv; Reasoning & Proof			Construction, Art, Architecture
4–7 178–181	• Find the measures of interior and exterior angles of polygons	Geometry & Spatial Sense Measurement; Prob Solv			Surveying, Sign Making, Games, Sports
4–8 182–185	• Apply properties of parallelograms to find missing lengths and angle measures	Geometry & Spatial Sense Prob Solv; Measurement Reasoning & Proof Connections; Representation			Art, Construction, Engineering, Architecture
4–9 188–191	• Apply properties of trapezoids to find missing lengths and angle measures			Art	Stage Design, Construction, Art

Planning and Resources

Lesson, Pages	Tools/Materials Needed	Trans-parency	Learning/Teaching Styles Options	Assignments: Basic Enriched	Additional Practice in SE	Reteaching, Extra Practice, Enrichment	Other Resources
4–1 150–153	graph paper straightedge scissors	Warm up 10 Trans RF-14	ESL/LEP Visual Learner	B: 1–18, 23–26 E: 1–26	Page 158–159, 169, 177, 187, 192, 695	R: page 51 EP: page 45 E: page 61	SS: Teacher's Choice
4–2 154–157	graph paper protractor metric ruler	Warm up 10 Trans RF-15	Real World	B: 1–10, 13–16 E: 1–16	Page 158–159, 169, 177, 187, 192, 695	R: page 53 EP: page 47 E: page 63	MC: 8
4–3 160–163	scissors	Warm up 10	Challenge Real World	B: 1–16, 19–22 E: 1–22	Page 168–169, 177, 187, 192, 696	R: page 55 EP: page 49 E: page 65	SS: Teacher's Choice

4–4 164–167	compass scissors rulers straightedges	Warm up 11	ESL/LEP	B: 1–22, 24–36 E: 1–36	Page 168–169, 177, 187, 193, 696	R: page 57 EP: page 51 E: page 67	MC: 2
4–5 170–171	paper/pencil	Warm up 11	Real World Challenge	B: 1–6 E: 1–6	Page 176–177, 187, 193,	R: page 59 E: page 69	
4–6 172–175	paper/pencil	Warm up 11	Challenge	B: 1–28, 35–45 E: 1–45	Page 176–177, 187, 193, 697	R: page 61 EP: page 53 E: page 71	
4–7 178–181	paper/pencil scissors	Warm up 12 Trans RF-16	ESL/LEP Challenge	B: 1–23, 29–33 E: 1–33	Page 186–187, 193, 697	R: page 63 EP: page 55 E: page 73	
4–8 182–185	compass straightedge	Warm up 12 Trans RF-17	Prior Knowledge Challenge	B: 1–24, 27–33 E: 1–33	Page 186, 187, 193, 698	R: page 65 EP: page 57 E: page 73	
4–9 188–191	scissors heavy paper	Warm up 12 Trans RF-17	Challenge	B: 1–28, 31–36 E: 1–36	Page 193, 698	R: page 67 EP: page 59 E: page 77	

Planning and Pacing

Lesson, Pages	Lesson Title	45 min class	Assignments Basic, Enriched	90 min class	Assignments Basic, Enriched	___ min class	Assignments Basic, Enriched
4–1 150–153	AYR, Opener, Triangles and Triangle Theorems	Day 45	B: 1–18, 23–26 E: 1–26	Day 23	B: 1–18, 23–26 E: 1–26		B: 1–18, 23–26 E: 1–26
4–2 154–157	Congruent Triangles, R&PYS	Day 46–47	B: 1–10, 13–16 E: 1–16	Day 23–24	B: 1–10, 13–16 E: 1–16		B: 1–10, 13–16 E: 1–16
4–3 160–163	Congruent Triangles and Proofs	Day 48	B: 1–16, 19–22 E: 1–22	Day 24	B: 1–16, 19–22 E: 1–22		B: 1–16, 19–22 E: 1–22
4–4 164–167	Altitudes, Medians and Perpendicular Bisectors, R&PYS	Day 49–50	B: 1–22, 24–36 E: 1–36	Day 25	B: 1–22, 24–36 E: 1–36		B: 1–22, 24–36 E: 1–36
4–5 170–171	Problem Solving Skills: Write an Indirect Proof	Day 51	B: 1–6 E: 1–6	Day 26	B: 1–6 E: 1–6		B: 1–6 E: 1–6
4–6 172–175	Inequalities in Triangles, R&PYS	Day 52–53	B: 1–28, 35–45 E: 1–45	Day 26–27	B: 1–28, 35–45 E: 1–45		B: 1–28, 35–45 E: 1–45
4–7 178–181	Polygons and Angles	Day 54	B: 1–23, 29–33 E: 1–33	Day 27	B: 1–23, 29–33 E: 1–33		B: 1–23, 29–33 E: 1–33
4–8 182–185	Special Quadrilaterals: Parallelograms, R&PYS	Day 55–56	B: 1–24, 27–33 E: 1–33	Day 28	B: 1–24, 27–33 E: 1–33		B: 1–24, 27–33 E: 1–33
4–9 188–191	Special Quadrilaterals: Trapezoids	Day 57	B: 1–28, 31–36 E: 1–36	Day 29	B: 1–28, 31–36 E: 1–36		B: 1–28, 31–36 E: 1–36
Review/ Assess		Day 58–60		Day 29–30			

	Chapter 1 Days	Chapter Cumulative Days
Yearly Pacing (45 min class)	15 days	60 days
Yearly Pacing (90 min class)	8 days	30 days

Assessment Options

Assessment in Student Edition	Assessment in Teacher's Edition	Pages in Assessment Book	Software Generated Assessment
Are You Ready?, pages 146–147; Writing Math, pages 152, 157, 163, 167, 171, 175, 181, 185, 190; Mixed Review, pages 153, 157, 163, 167, 175, 181, 185, 191; Check Understanding pages 150, 154, 160, 164, 172, 178, 182, 188; Mid-Chapter Quiz, page 169, Chapter Review, page 192; Chapter Assessment, page 194; Alternative Assessment, page 195; Cumulative Review, page 196; Cumulative Assessment, page 197	5-minute Warm ups, pages 150, 154, 160, 164, 170, 172, 178, 182, 188; Quick Assessment, pages 152, 156, 162, 166, 171, 174, 180, 184, 190; Scoring Rubrics, page 195	Mid-Chapter Quiz, page 55; Test Form A, pages 57, 59; Test Form B, pages 61, 63; Cumulative Test 65, 67; Math Journal prompt, page 69, 70;	Chapter 4

Refresher Skills

The skills on these two pages are skills that have been taught in previous math courses. Continuous review of basic math skills will make stronger math students. These skills are identified as necessary to be successful in Chapter 3.

Skills Correlation Chart

Skill	Lesson Number
Polygons	4-1, 4-2, 4-7, 4-8, 4-9
Congruent Triangles	4-1, 4-2, 4-3, 4-6
Angles of Triangles	4-1, 4-2, 4-3, 4-4, 4-5, 4-6

Vocabulary

polygon
congruent
triangle
interior angle

Are You Ready?

Refresh Your Math Skills for Chapter 4

The skills on these two pages are ones you have already learned. Stretch your memory and complete the exercises. For additional practice on these and more basic skills, see page 674.

You will be learning more about geometric shapes and their properties. Now is a good time to review what you already have learned about polygons and triangles.

Polygons

A **polygon** is a closed plane figure formed by joining 3 or more line segments at their endpoints. Polygons are named for the number of their sides.

Tell whether each figure is a polygon. If not, give a reason.

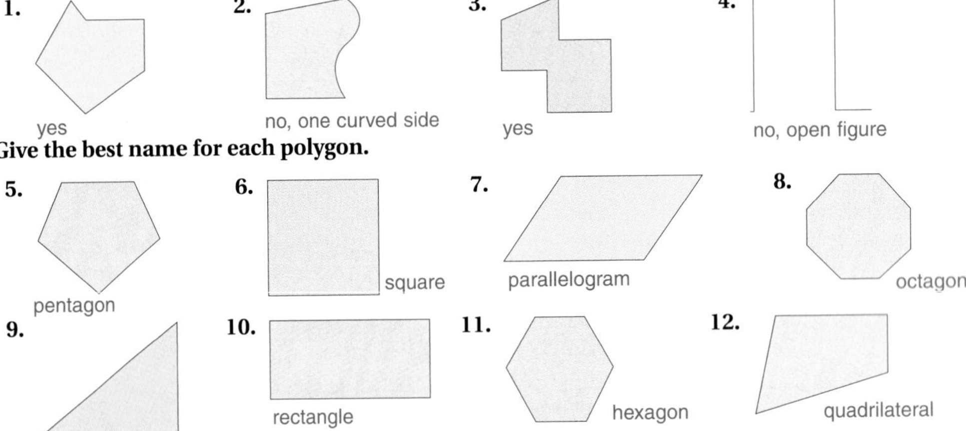

1. yes
2. no, one curved side
3. yes
4. no, open figure

Give the best name for each polygon.

5. pentagon
6. square
7. parallelogram
8. octagon
9. triangle
10. rectangle
11. hexagon
12. quadrilateral

Congruent Triangles

Triangles can be determined to be congruent, or having the same size and shape, by three tests:

Triangles with the same measure of two angles and the included side are congruent.

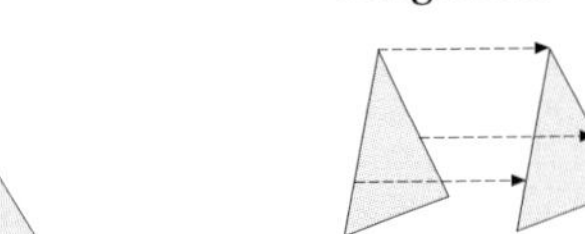

angle-side-angle
ASA

Triangles with the same measure of two sides and the included angle are congruent.

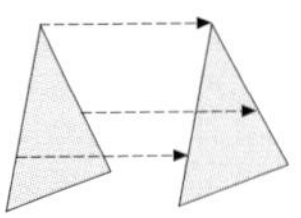

side-angle-side
SAS

Triangles with the same measures of three sides are congruent.

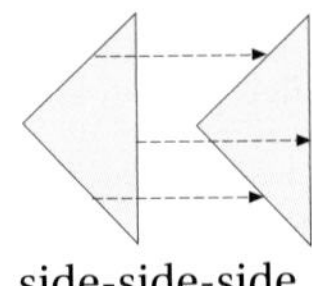

side-side-side
SSS

Extend the Lesson

MATH JOURNAL If two triangles share a common side, what is the least information you need to determine whether they triangles are congruent? the measures of the other sides; the measures of the angles that include the common die; the measures of one pair of corresponding sides and the angle included between the common side and each corresponding side

CHAPTER 4

Tell whether each pair of triangles is congruent. If they are congruent, identify how you determined congruency.

13. 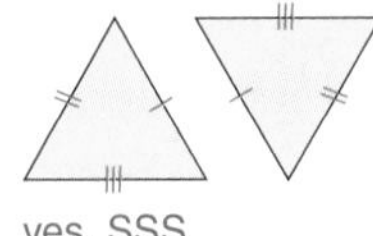
yes, SSS

14. 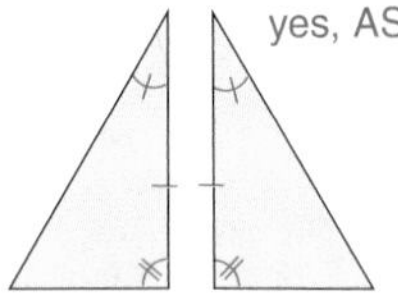
yes, ASA

15. 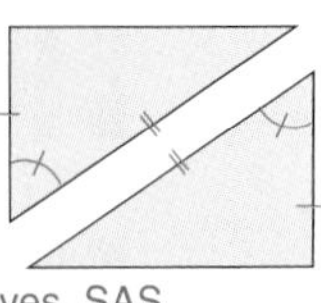
yes, SAS

16. 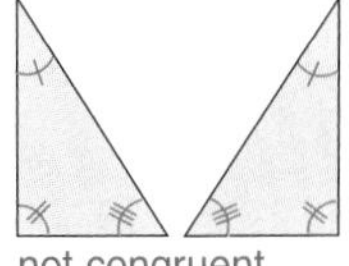
not congruent

17. 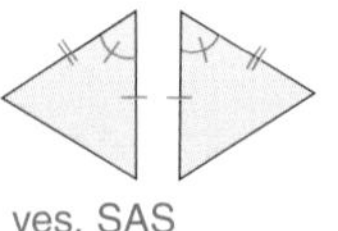
yes, SAS

18. 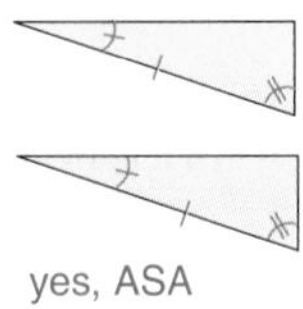
yes, ASA

Angles of Triangles

Example The sum of the interior angles of any triangle is 180°. Find the unknown measure.

$$53° + 73 + x = 180°$$
$$126° + x = 180°$$
$$x = 54°$$

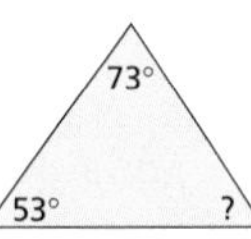

Find the unknown measures in each triangle.

19. 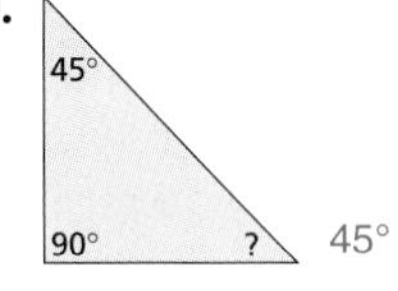

45°

20. 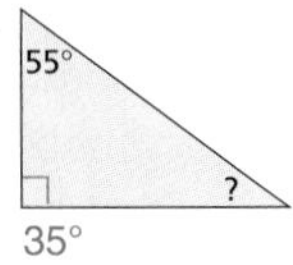

35°

21. 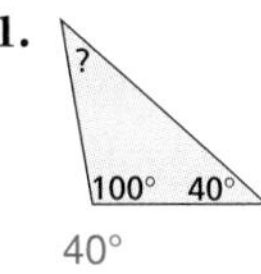

40°

22. 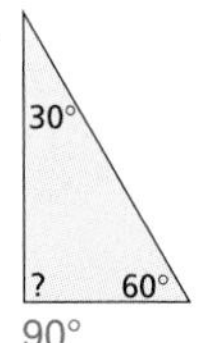

90°

23. 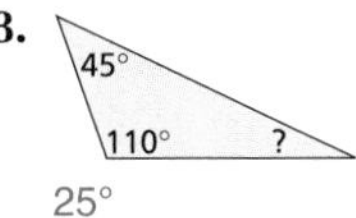

25°

24. 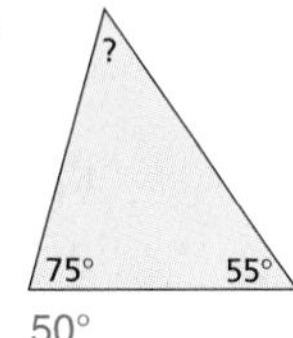

50°

Chalkboard Examples

Polygons

Have students draw:

1. a figure which is not a polygon
2. a triangle
3. a pentagon

Check students' drawings.

Congruent Triangles

Draw an illustration of congruent triangles for each of three ways that triangles can be proven congruent. **Students' drawings will vary but should show an example of ASA, SAS and SSS.**

Angles of Triangles

Given a triangle with angles measuring 73° and 21°, find the measure of the third angle. **180° − 73° − 21° = 86° The measure of the third angle is 86°.**

Refresher Wrap-up

Quick Assessment

Ask the following questions to determine if students have mastered the basic skills reviewed on these pages.

1. Two interior angles of a triangle measure 80° and 24°. What is the measure of the third interior angle? **76°**
2. Draw a pair of congruent triangles. Show a minimum of measurements needed to prove the triangles are needed. **Check students' work.**

Additional Practice

All topics: Refer to the Extra Practice Index on page 674 of this text.

Teaching Strategies

COOPERATIVE LEARNING Have students work in pairs. Using geometric-drawing software, students should construct a triangle and measure each interior angle. Use the calculation features of the software to find the sum of each interior angle. Then have students change the angle measures by moving one point of the triangle. How are the angle measured changed? **Answers will vary.** What happens to the sum of the interior angles? **It remains 180°**

Chapter Opener

NCTM Standards/Strands

- Number & Operation
- Patterns, Functions, & Algebra
- Geometry & Spatial Sense
- Measurement
- Data Analysis, Statistics, & Probability
- Problem Solving
- Reasoning and Proof
- Communication
- Connections
- Representation

Vocabulary

pattern	span
truss bridge	suspension bridge

Theme Connections

Tables and graphs are used to organize data about art and design. Tables provide a way to compare statistics in a wide range of situations. One example is in engineering or architecture, where the specifications of various structures can be compared. Discuss how data and graphs are related to art and design.

Career Opportunities

Many careers related to art and design the use of data and graphs. Two are highlighted in the Math-Works features. Others include graphics artists, advertisers, artists, sculptures, architects and engineers.

- Jeweler, page 159
- Animator, page 177

Internet Connection

Theme Activities

Learningmathmatters.com provides web addresses to search that help students gather information about the use of math in the real world, particularly data and graphs. To search for additional addresses, begin a search using the keyword art. Then within that search, use such key words as *design, architecture, bridge* and *sculpture*. Students can brainstorm in small groups other key words.

Chapter Investigation

Go to Learningmathmatters.com to locate additional information about art and design.

CHAPTER 4

Triangles, Quadrilaterals and Other Polygons

THEME: Art and Design

For thousands of years, art has been based on geometric principles. Many of the designs found on ancient murals and pottery derive their beauty from complex patterns of geometric shapes. Modern sculptures, buildings and bridges also rely on geometric characteristics for beauty and durability.

Today, computers give graphic designers, architects and engineers a means for experimenting with design elements. Computer programs allow all types of engineers to experiment with new methods and materials for designing and building.

- **Jewelers** (page 159) design jewelry, cut gems, make repairs and use their understanding of geometry to appraise gems. Jewelers need skills in art, math, mechanical drawing and chemistry to practice their trade.
- **Animators** (page 177) create pictures that are filmed frame by frame to create motion. Many animators use computers to create three-dimensional backgrounds and characters. Animators need an understanding of perspective to create realistic drawings.

Suspension Bridges of New York

Name	Year opened	Length of main span	Height of towers	Clearance above water	Cost of original structure
Brooklyn Bridge	1883	1595.5 ft	276.5 ft	135 ft	$15,100,000
Williamsburg Bridge	1903	1600 ft	310 ft	135 ft	$30,000,000
Manhattan Bridge	1909	1470 ft	322 ft	135 ft	$25,000,000
George Washington Bridge	1931 1962*	3500 ft	604 ft	213 ft	$59,000,000
Verrazano Narrows Bridge	1964 1969*	4260 ft	693 ft	228 ft	$320,126,000

* lower deck added

Data Activity: Suspension Bridges of New York

Use the table for Questions 1–4.

1. For which bridge is the ratio of tower height to length of main span closest to 1 : 5? Williamsburg Bridge

2. The width of the George Washington Bridge is 119 ft. What is the area of the main span in square feet? (*Hint:* Use the formula $A = lw$, where 1 = length and w = width.) 416,500 ft^2

3. Suppose you want to make a scale model of the Verrazano Narrows Bridge. The entire model must be no more than 4 ft in length. Choose a scale and then find the model's length of main span, clearance above water and height of towers.

Answers may vary. If the main span is 48 in. in length, the scale is 1 in. = 88.75 ft. The clearance above water is about 2.6 in. and the height of each tower is about 7.8 in.

4. By what percent did the cost of building a bridge increase from 1931 to 1964? Round to the nearest whole percent. Disregard differences in bridge size. 443% increase

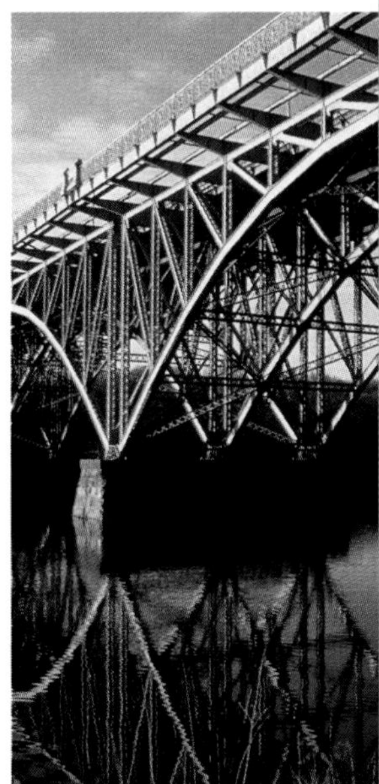

Chapter Investigation

A truss bridge covers the span between two supports using a system of angular braces to support weight. Triangles are often used in the building of truss bridges because the triangle is the strongest supporting polygon. The railroads often built truss bridges to support the weight of heavy locomotives.

Working Together

Design a system for a truss bridge similar to the examples shown throughout this chapter. Carefully label the measurements and angles in your design. Build a model of the truss using straws, toothpicks or other suitable materials. Use the Chapter Investigation logo to assist your group in designing the structure.

Project Planning Calendar

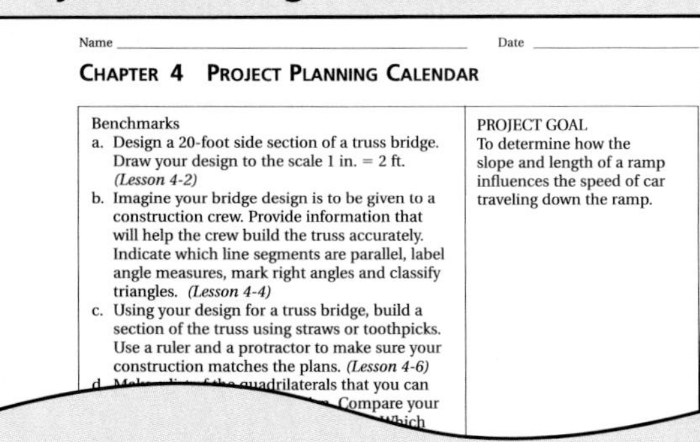

Name ______ Date ______

CHAPTER 4 PROJECT PLANNING CALENDAR

Benchmarks
a. Design a 20-foot side section of a truss bridge. Draw your design to the scale 1 in. = 2 ft. (*Lesson 4-2*)
b. Imagine your bridge design is to be given to a construction crew. Provide information that will help the crew build the truss accurately. Indicate which line segments are parallel, label angle measures, mark right angles and classify triangles. (*Lesson 4-4*)
c. Using your design for a truss bridge, build a section of the truss using straws or toothpicks. Use a ruler and a protractor to make sure your construction matches the plans. (*Lesson 4-6*)
d. ...quadrilaterals that you can ... Compare your ... Which

PROJECT GOAL
To determine how the slope and length of a ramp influences the speed of car traveling down the ramp.

Group Project Planner

Name ______ Date ______

CHAPTER 4 GROUP PROJECT PLANNER

Assignment ______ Objective ______

Group Members
1) ______
2) ______
3) ______
4) ______
5) ______

Assigned Roles

Deadlines Done

Data Activity

Discuss with students how suspension bridges are built. Explain that the roadway is suspended from two or more towers using strong metal cables. These structures have proven to be strong and relatively easy to maintain. Have students look at a map of New York if they are unfamiliar with the geography of the state. Have them locate each bridge listed in the table. Students should realize that the natural geography of the region determines the main span of the bridge. Assign students Questions 1–4.

Chapter Investigation

As an Overarching Problem

Bring several photos and/or drawings of truss bridges to class. Ask students to identify parallel line segments, special angles, triangles, and quadrilaterals used in the design of each bridge. Discuss possible reasons for the design of each bridge. Encourage students to record ideas shared during this opening activity so they can use them in their work throughout the chapter. Students will continue to work on the investigation as they complete the exercises identified by the Investigation Logo that are found throughout the chapter. These exercises will guide students through the task described in *Working Together*. Encourage students to keep all of their work on the Investigation together. Have students use the suggestions in the Chapter Investigation Extension to summarize their work.

As a Chapter Project

The goal of this project is to build a model of a truss bridge. Students can use the Team Project Planner on page 59 and the Project Planning Calendar on page 60 in the Enrichment text to complete the project. Benchmarks **a**, **b**, **c**, and **d** should be completed after the lesson listed in parentheses has been studied. Benchmark **e** should be completed at the end of the chapter.

Lesson Planning

NCTM Standards/Strands

- Number & Operation
- Patterns, Functions, & Algebra
- Geometry & Spatial Sense
- Measurement
- Reasoning and Proof
- Communication
- Connections
- Representation

Vocabulary

triangle
side
vertex
interior angle
equilateral triangle
isosceles triangle
scalene triangle
acute triangle
obtuse triangle
right triangle
equiangular triangle
auxiliary line
exterior angle

Materials Needed

graph paper
scissors
straightedges

Lesson Resources

Warm-Up Transparency 10
Reteaching 4-1
Extra Practice 4-1
Enrichment 4-1
Transparency RF-14

Assignment Guide

Basic: 1–18, 23–26
Enriched: 1–26

Getting Started

5-Minute Warm-up

Find the value of x.

1. $x + 2x + 3x = 180$ 30
2. $x + x + 40 = 180$ 70
3. $2x + (x + 1) + 35 = 180$ 48

Introduction to Lesson 4-1

Have students repeat this activity using a right triangle and an obtuse triangle. Discuss whether their conjectures changed.

4-1 Triangles and Triangle Theorems

Goals
- Solve equations to find measure of angles.
- Classify triangles according to their sides or angles.

Applications Technical Art, Construction, Engineering

Work with a partner.

1. Using a pencil and a straightedge, draw and label a triangle as shown below. Carefully cut on the straight lines. Then tear off the four labeled angles.

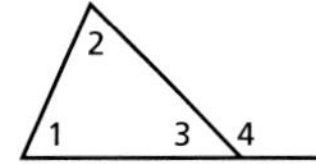

⇒

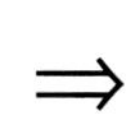

For a–b, see additional answers.

a. What relationships can you discover among the four angles?

b. Using these relationships, make at least two conjectures that you think apply to all triangles.

Build Understanding

A **triangle** is the figure formed by the segments that join three noncollinear points. Each segment is called a **side** of the triangle. Each point is called a **vertex** (plural: *vertices*). The angles determined by the sides are called the **interior angles** of the triangle.

triangle ABC ($\triangle ABC$)

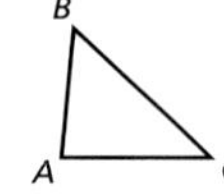

sides: $\overline{AB}$, $\overline{BC}$ and $\overline{AC}$
vertices: points A, B and C
interior angles: $\angle A$, $\angle B$ and $\angle C$

Often a triangle is classified by relationships among its sides.

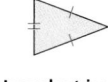

Equilateral triangle three sides of equal length

Isosceles triangle at least two sides of equal length

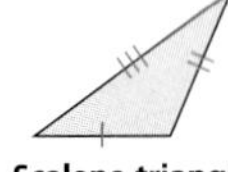

Scalene triangle no sides of equal length

A triangle also can be classified by its angles.

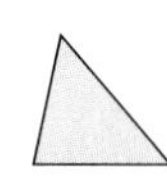

Acute triangle three acute angles

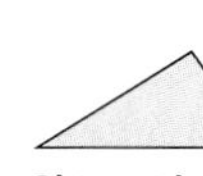

Obtuse triangle one obtuse angle

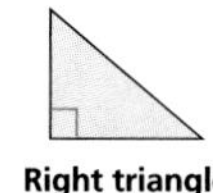

Right triangle one right angle

Equiangular triangle three angles equal in measure

You probably remember a special property of the measures of the angles of a triangle. Since the fact is a theorem, it can be proved true.

The Triangle-Sum Theorem	The sum of the measures of the angles of a triangle is 180°.

Teaching Strategies

ESL/LEP To help students understand the many new terms in this lesson, relate them to everyday items. Examples include the *interior* and *exterior* of a house, an *acute* pain, and a hospital *auxiliary*.

As you read the proof of the theorem, notice that it makes use of a line that intersects one of the vertices of the triangle and is parallel to the opposite side. This line, which has been added to the diagram to help in the proof, is called an **auxiliary line**.

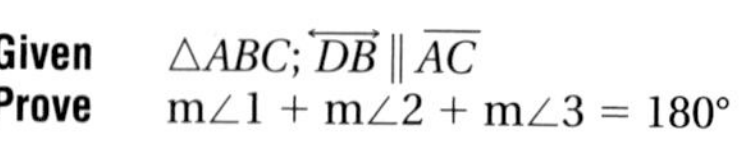

Given $\triangle ABC$; $\overleftrightarrow{DB} \parallel \overline{AC}$
Prove $m\angle 1 + m\angle 2 + m\angle 3 = 180°$

Statements	**Reasons**
1. $\triangle ABC$; $\overleftrightarrow{DB} \parallel \overline{AC}$	1. given
2. $m\angle 4 + m\angle 2 = m\angle DBC$ $m\angle DBC + m\angle 5 = 180°$	2. angle addition postulate
3. $m\angle 4 + m\angle 2 + m\angle 5 = 180°$	3. substitution property
4. $m\angle 4 = m\angle 1$ $m\angle 5 = m\angle 3$	4. If two parallel lines are cut by a transversal, then alternate interior angles are equal in measure.
5. $m\angle 1 + m\angle 2 + m\angle 3 = 180°$	5. substitution property

The triangle-sum theorem is useful in art and design.

Check Understanding

Explain how the substitution property was used in both Step 3 and Step 5 of the proof.

In Step 3, the exprssion $m\angle 4 + m\angle 2$ was substituted for $m\angle DBC$ in the equation $m\angle DBC + m\angle 5 = 180°$. In Step 5, $m\angle 1$ was substituted for $m\angle 4$ and $m\angle 3$ was substituted for $m\angle 5$ in the equation $m\angle 4 + m\angle 2 + m\angle 5 = 180°$.

Example 1

TECHNICAL ART An artist is using the figure at the right to create a diagram for a publication. Using the triangle-sum theorem, find $m\angle Q$.

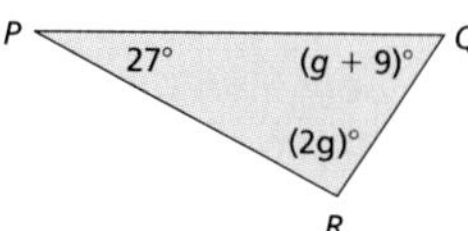

Solution

From the triangle-sum theorem, you know that the sum of the measures of the angles of a triangle is 180°. Use this fact to write and solve an equation.

$$m\angle P + m\angle Q + m\angle R = 180$$
$$27 + (g + 9) + 2g = 180 \quad \text{Combine like terms.}$$
$$3g + 36 = 180 \quad \text{Add } -36 \text{ to each side.}$$
$$3g = 144 \quad \text{Multiply each side by } \tfrac{1}{3}.$$
$$g = 48$$

So, the value of g is 48. From the figure $m\angle Q = (g + 9)°$. Substituting 48 for g, $m\angle Q = (48 + 9)° = 57°$.

An **exterior angle** of a triangle is an angle that is both adjacent to and supplementary to an interior angle, as shown at the right. The following is an important theorem concerning exterior angles.

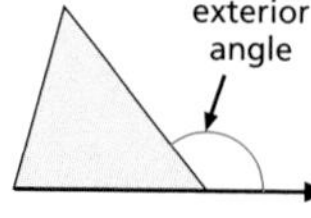

The Exterior Angle Theorem	The measure of an exterior angle of a triangle is equal to the sum of the measures of the two nonadjacent (remote) interior angles.

Teaching Strategies

COOPERATIVE LEARNING Working in small groups, have students suggest different small sets of information they could be given that would enable them to find the measure of each interior and exterior angle of a triangle. Be sure they consider working with actual angle measures, descriptions of the relation between the measure of one angle and the measure of another, and triangle classifications.

Chalkboard Examples

Supplementary Example 1

A second triangle has the following interior angle measures: 48°, $(2x + 17)°$, and $(3x)°$. Find the measure of the three angles. **48°, 63°, and 69°**

Supplementary Example 2

Use the drawing that accompanies the proof on page 151. Let $m\angle 4 = 48°$, $m\angle 5 = 88°$, $m\angle 1 = 4x°$, and $m\angle 3 = (4x + 8)°$. Find $m\angle 3$. **72°**

Additional Answers

a. Answers will vary. Some possible answers: $m\angle 1 + m\angle 2 + m\angle 3 = 180°$; $m\angle 3 + m\angle 4 = 180°$; $m\angle 1 + m\angle 2 = m\angle 4$. b. Answers will vary. Some possible answers: The sum of the measures of the angles of a triangle is always 180°; the sum of the measures of $\angle 3$ and $\angle 4$ will always be 180°; the sum of the measures of $\angle 1$ and $\angle 2$ will always equal the measure of $\angle 3$.

Reteaching Worksheet 4-1

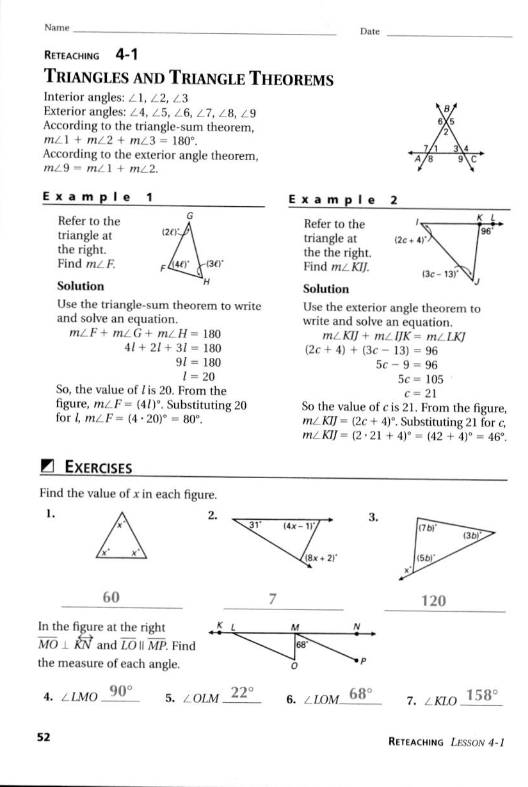

Name ______________________ Date ____________

RETEACHING 4-1

TRIANGLES AND TRIANGLE THEOREMS

Interior angles: $\angle 1, \angle 2, \angle 3$
Exterior angles: $\angle 4, \angle 5, \angle 6, \angle 7, \angle 8, \angle 9$
According to the triangle-sum theorem, $m\angle 1 + m\angle 2 + m\angle 3 = 180°$.
According to the exterior angle theorem, $m\angle 9 = m\angle 1 + m\angle 2$.

Example 1

Refer to the triangle at the right. Find $m\angle F$.

Solution

Use the triangle-sum theorem to write and solve an equation.

$$m\angle F + m\angle G + m\angle H = 180$$
$$4l + 2l + 3l = 180$$
$$9l = 180$$
$$l = 20$$

So, the value of l is 20. From the figure, $m\angle F = (4l)°$. Substituting 20 for l, $m\angle F = (4 \cdot 20)° = 80°$.

Example 2

Refer to the triangle at the the right. Find $m\angle KIJ$.

Solution

Use the exterior angle theorem to write and solve an equation.

$$m\angle KIJ + m\angle IJK = m\angle LKJ$$
$$(2c + 4) + (3c - 13) = 96$$
$$5c - 9 = 96$$
$$5c = 105$$
$$c = 21$$

So the value of c is 21. From the figure, $m\angle KIJ = (2c + 4)°$. Substituting 21 for c, $m\angle KIJ = (2 \cdot 21 + 4)° = (42 + 4)° = 46°$.

EXERCISES

Find the value of x in each figure.

1. 60

2. 7

3. 120

In the figure at the right $\overline{MO} \perp \overleftrightarrow{KN}$ and $\overline{LO} \parallel \overline{MP}$. Find the measure of each angle.

4. $\angle LMO$ 90°
5. $\angle OLM$ 22°
6. $\angle LOM$ 68°
7. $\angle KLO$ 158°

52 RETEACHING *LESSON 4-1*

Lesson Wrap-up

Quick Assessment

Ask the following questions to determine if students understand the content presented in this lesson.

1. Suppose you are asked to find the measures of the three interior angles and one exterior angle of a triangle. Which two angle measures would you need to be given? the measures of the exterior and one nonadjacent interior angle or the measures of both nonadjacent interior angles
2. In an acute isosceles triangle whose angle measures are all whole numbers, what is the smallest measure possible for one of the congruent angles? 46°

Assignment Guide

Basic: 1–18, 23–26
Advanced: 1–26
Additional Practice: See Extra Practice Index on page 674.

Additional Answers

10. There are six exterior angles. Labeling will vary. In the figure below, the exterior angles are labeled ∠1, ∠2, ∠3, ∠4, ∠5, and ∠6.

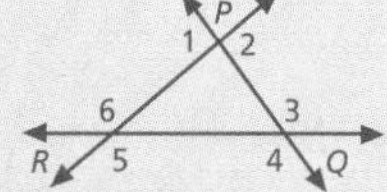

14.

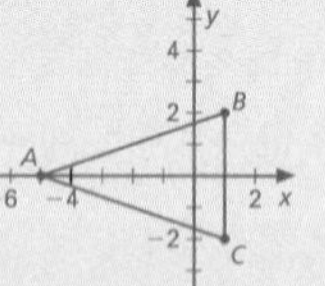

15.

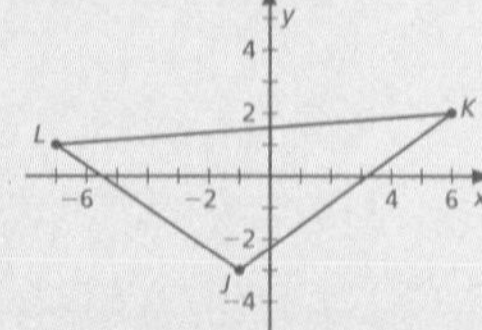

16. 17.

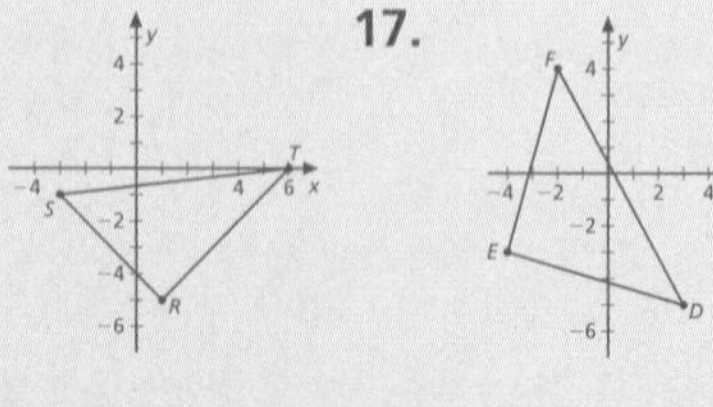

You will have an opportunity to prove the exterior angle theorem in Exercise 12 on the following page. Example 2 shows one way the theorem can be applied.

Example 2

In the figure at the right, find m∠EFG.

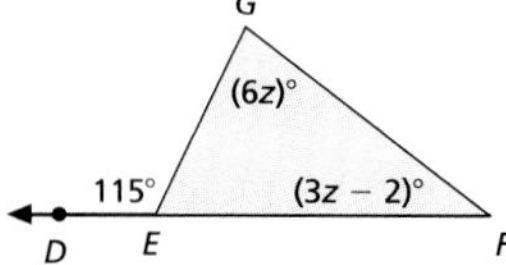

Solution

Notice that ∠*DEG* is an exterior angle, while ∠*EGF* and ∠*EFG* are nonadjacent interior angles. Use the exterior angle theorem to write and solve an equation.

m∠*DEG* = m∠*EGF* + m∠*EFG*

$115 = 6z + (3z - 2)$ Combine like terms.

$115 = 9z - 2$ Add 2 to each side.

$117 = 9z$ Multiply each side by $\frac{1}{9}$.

$13 = z$

So, the value of z is 13. From the figure, m∠*EFG* = $(3z - 2)°$. Substituting 13 for z, m∠*EFG* = $(3 \cdot 13 - 2)° = (39 - 2)° = 37°$.

Try These Exercises

Refer to △*RST* below. Find the measure of each angle.

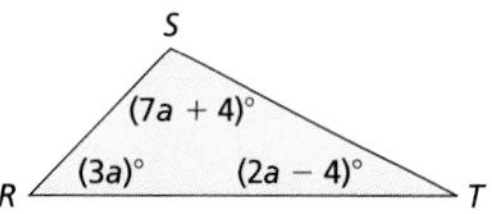

1. ∠*R* 45°
2. ∠*S* 109°
3. ∠*T* 26°

Refer to △*XYZ* below. Find the measure of each angle.

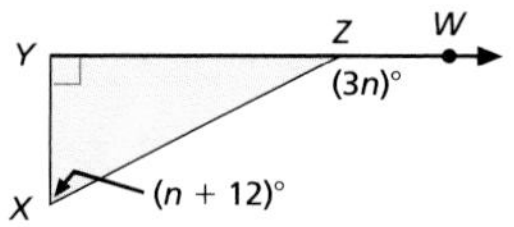

4. ∠*YXZ* 63°
5. ∠*XZW* 153°
6. ∠*XZY* 27°

Practice Exercises

Find the value of *x* in each figure.

7.

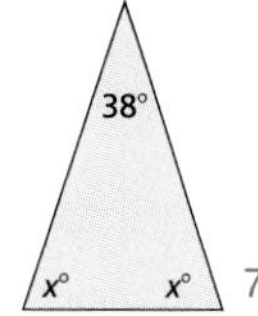

71

8.

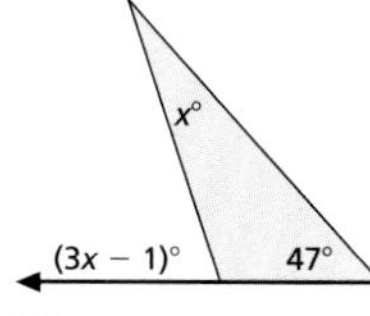

24

9.

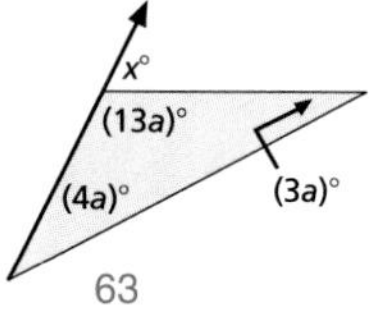

63

10. **WRITING MATH** How many exterior angles does a triangle have? Draw a triangle and label all its exterior angles. See additional answers.
11. The measure of the largest angle of a triangle is twice the measure of the smallest angle. The measure of the third angle is 10° less than the measure of the largest angle. Find all three measures. 38°, 76°, 66°

Learning Styles

VISUAL LEARNERS Encourage students to draw pictures for Exercises 19–22. If the students cannot find a way to draw the figure, they can be reasonably sure that the answer is never. If the students draw the figure but can also draw a counterexample, they can be certain the answer is *sometimes*.

12. In the figure below, $\overleftrightarrow{PQ} \parallel \overleftrightarrow{RS}$. Find m∠*RPS*.

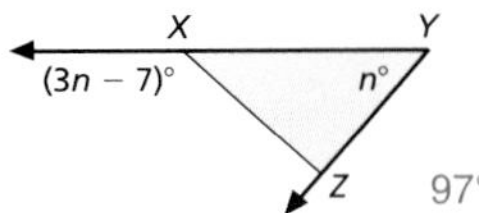

13. BRIDGE BUILDING The plans for a bridge call for the addition of triangular bracing to increase the amount of weight the bridge can hold. On the plans, △*FGH* is drawn so that the m∠*F* is 14° less than three times the m∠*G*, and ∠*H* is a right angle. Find the measure of each angle. m∠F = 64°; m∠G = 26°, m∠H = 90°

GEOMETRY SOFTWARE On a coordinate plane, draw the triangle with the given vertices. Measure all sides and angles. Then classify the triangle, first by its sides, then by its angles. See additional answers.

14. $A(-5, 0)$; $B(1, 2)$; $C(1, -2)$ isosceles, acute

15. $J(-1, -3)$; $K(6, 2)$; $L(-7, 1)$ scalene, obtuse

16. $R(1, -5)$; $S(-3, -1)$; $T(6, 0)$ scalene, right

17. $D(3, -5)$; $E(-4, -3)$; $F(-2, 4)$ isosceles, right

18. Copy and complete this proof of the exterior angle theorem.

Given △*ABC*, with ∠4 an exterior angle

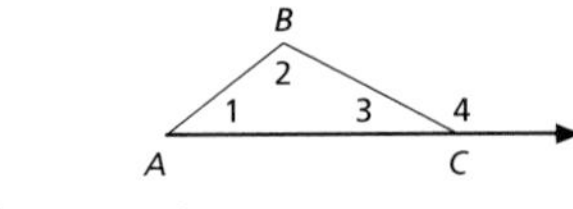

Prove ? m∠1 + m∠2 = m∠4

Statements	Reasons
1. ? △*ABC* with ∠4 an exterior angle	1. ?
2. m∠1 + m∠2 + m∠3 = 180°	2. ?
3. m∠4 + m∠3 = 180°	3. ?
4. m∠1 + m∠2 + m∠3 = m∠4 + m∠3	4. ?
5. ? m∠1 + m∠2 = m∠4	5. ?

1. given
2. triangle—sum theorem
3. angle add postulate
4. trans.prop. of equal.
5. add. prop. of equal.

Extended Practice Exercises

Determine whether each statement is *always, sometimes,* or *never* true.

19. Two interior angles of a triangle are obtuse angles. never

20. Two interior angles of a triangle are acute angles. always

21. An exterior angle of a triangle is an obtuse angle. sometimes

22. The measure of an exterior angle of a triangle is greater than the measure of either nonadjacent interior angle. always

Mixed Review Exercises

Find each length. (Lesson 3-1)

23. In the figure below, *AC* = 75. Find *BC*. 52

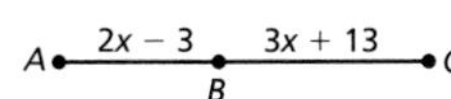

24. In the figure below, *PR* = 138. Find *PQ*. 105

P — 2x + 5 — Q — x − 17 — R

Find the mean, median and mode of each set of data. (Lesson 2-7)

25. 4 8 7 10 8 8 3 7 9 14 3 5 mean: $7\frac{1}{6}$; median: $7\frac{1}{2}$; mode: 8

26. 8 7 3 9 9 5 7 9 1 3 2 6 9 1 4 mean: $5\frac{8}{15}$; median: 6; mode: 9

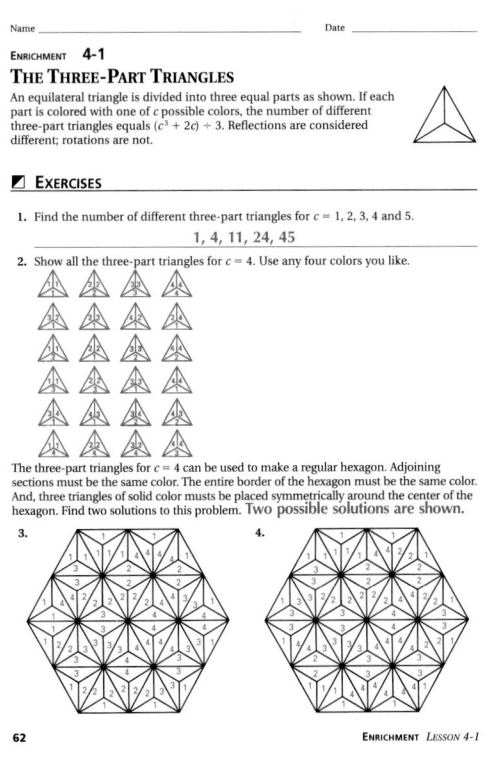

Extra Practice Worksheet 4-1

Name ______ Date ______

EXTRA PRACTICE 4-1

TRIANGLES AND TRIANGLE THEOREM

EXERCISES

Find the value of *x* in each figure.

1. 46°, x°, x° — 67°
2. 26°, x°, (4x − 4)° — 10
3. (6x + 15)°, (x + 12)°, (2x)° — 17

4. In the figure at the right, $\overleftrightarrow{AB} \parallel \overleftrightarrow{CD}$. Find *m*∠*ACD*. 40°

5. In the figure at the right, $\overleftrightarrow{NM} \perp \overleftrightarrow{PM}$. Find *m*∠*MNP*. 54°

6. In △*RST*, *m*∠*R* is 20° more than two times *m*∠*S*, and *m*∠*S* = *m*∠*T*. Find the measure of each angle. *m*∠*S* = *m*∠*T* = 40°, *m*∠*R* = 100°

7. The measure of the smallest angle of a triangle is 34° less than the measure of the largest angle of the triangle. The measure of the third angle is 14° more than the measure of the smallest angle. Find all three measures. 78°, 44°, 58°

On a coordinate plane on your own paper, sketch the triangle with the given vertices. Then classify the triangle, first by its sides, then by its angles.

8. $A(-2, -2)$, $B(0, 4)$, $C(2, -2)$ isosceles, acute

9. $R(-5, 1)$, $S(-2, 3)$, $T(6, 0)$ scalene, obtuse

46 EXTRA PRACTICE *LESSON 4-1*

Enrichment Worksheet 4-1

Name ______ Date ______

ENRICHMENT 4-1

THE THREE-PART TRIANGLES

An equilateral triangle is divided into three equal parts as shown. If each part is colored with one of *c* possible colors, the number of different three-part triangles equals $(c^3 + 2c) \div 3$. Reflections are considered different; rotations are not.

EXERCISES

1. Find the number of different three-part triangles for *c* = 1, 2, 3, 4 and 5. 1, 4, 11, 24, 45

2. Show all the three-part triangles for *c* = 4. Use any four colors you like.

The three-part triangles for *c* = 4 can be used to make a regular hexagon. Adjoining sections must be the same color. The entire border of the hexagon must be the same color. And, three triangles of solid color musts be placed symmetrically around the center of the hexagon. Find two solutions to this problem. Two possible solutions are shown.

3.

4.

62 ENRICHMENT *LESSON 4-1*

Learning Styles

TACTILE/KINESTHETIC LEARNERS Have students work in small groups. Have each student draw any triangle and extend all sides to create two exterior angles at each vertex. Then have them carefully cut out each exterior angle and look for relationships among them. Lead them to see that the sum of their measures is 360°.

Lesson Planning

NCTM Standards/Strands

- Number & Operation
- Patterns, Functions, & Algebra
- Geometry & Spatial Sense
- Measurement
- Problem Solving
- Reasoning and Proof
- Communication
- Connections
- Representation

Vocabulary

congruent	SSS postulate
SAS postulate	ASA postulate
included side	included angle

Materials Needed

graph paper	metric rulers
protractors	

Lesson Resources

Warm-Up Transparency 10
Reteaching 4-2
Extra Practice 4-2
Enrichment 4-2
Multicultural Connection 8
Transparency RF-15

Assignment Guide

Basic: 1–10, 13–16
Enriched: 1–16

Getting Started

5-Minute Warm-Up

Classify each triangle.

1. a 42° and a 57° angle   acute
2. two sides 5 cm long   isosceles
3. two 60° angles   equiangular

Introduction to Lesson 4-2

Have students work together to construct the triangles described. To construct the triangle in Step (c), some students may find it helpful to use pipe cleaners bent to form the given angles. To construct the triangle in Step (d), some students may find it helpful to use straws cut to the given lengths. Once students have had the opportunity to compare their triangles, discuss why some triangles are the same while others are different.

4-2 Congruent Triangles

Goals ■ Prove triangles are congruent.

Applications Animation, Construction, Engineering

For the following activity, use a protractor and a metric ruler, or use geometric drawing software. Give lengths to the nearest tenth of a centimeter, and give angle measures to the nearest degree.

a. Draw $\triangle ABC$, with $m\angle A = 40°$, $AB = 7$ cm, and $m\angle B = 60°$. What is the measure of $\angle C$? What is the length of $\overline{AC}$? of $\overline{BC}$?
$m\angle C$ 80°; AC 6.2 cm; BC 4.5 cm

b. Draw $\triangle DEF$, with $DF = 5$ cm, $m\angle D = 120°$, and $DE = 6$ cm. What is the measure of $\angle E$? of $\angle F$? What is the length of $\overline{EF}$?
$m\angle E$ 27°; $m\angle F$ 33°

c. Draw $\triangle GHJ$, with $m\angle G = 35°$, $m\angle H = 45°$, and $m\angle J = 100°$. What is the length of $\overline{GH}$? of $\overline{GJ}$? of $\overline{HJ}$?
Answers will vary.

d. Draw $\triangle KLM$, with $KM = 3$ cm, $KL = 6$ cm, and $LM = 4$ cm. What is the measure of $\angle K$? of $\angle L$? of $\angle M$?
$m\angle K$ 36°, $m\angle L$ 26°, $m\angle M$ 117°

Check Understanding

Name all the pairs of congruent parts in the triangles below. Then state the congruence between the triangles.

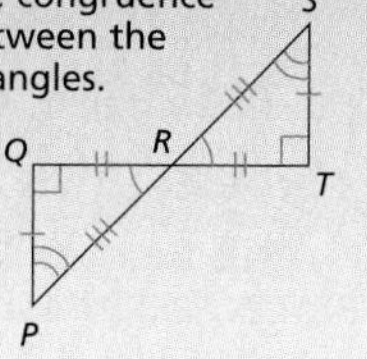

Is there a different way to state the congruence?

Check Understanding
$\angle QRP \cong \angle TRS$, $\angle P \cong \angle S$, $\angle Q \cong \angle T$, $PQ \cong ST$, $QR \cong TR$, $PR = SR$. We have six possible ways to state the congruence: $\triangle PQR \cong \triangle STR$, $\triangle PRQ \cong \triangle SRT$, $\triangle QRP \cong \triangle TRS$, $\triangle QPR \cong \triangle TSR$, $\triangle RQP \cong \triangle RTS$, and $\triangle RPQ \cong \triangle RST$.

Build Understanding

When two geometric figures have the same size and shape, they are said to be **congruent**. The symbol for *congruence* is $\cong$.

It is fairly easy to recognize when segments and angles are congruent. **Congruent segments** are segments with the same length. **Congruent angles** are angles with the same measure.

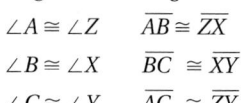

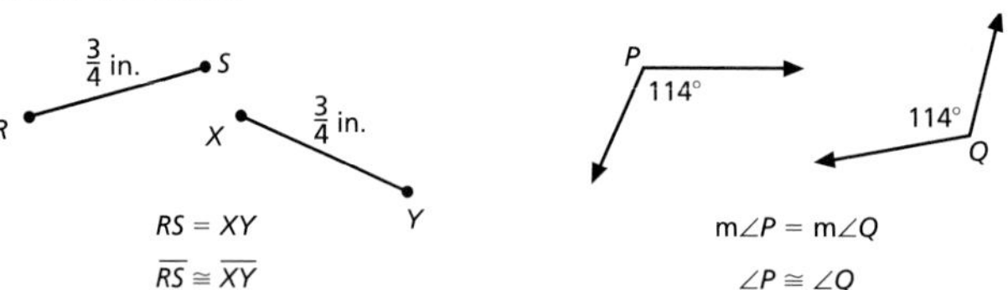

Congruent triangles are two triangles whose vertices can be paired in such a way that each angle and side of one triangle is congruent to a *corresponding angle* or *corresponding side* of the other. For instance, the markings in the triangles at the right indicate these six congruences.

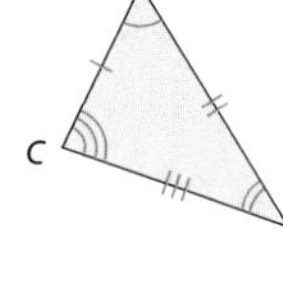

$\angle A \cong \angle Z$ $\overline{AB} \cong \overline{ZX}$

$\angle B \cong \angle X$ $\overline{BC} \cong \overline{XY}$

$\angle C \cong \angle Y$ $\overline{AC} \cong \overline{ZY}$

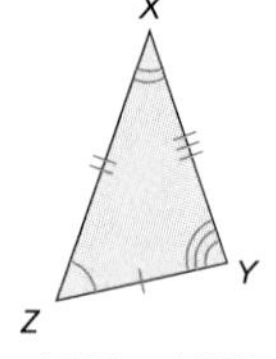

$\triangle ABC \cong \triangle ZXY$

So, the triangles are congruent, and you can pair the vertices in the following correspondence.

$A \leftrightarrow Z$ $B \leftrightarrow X$ $C \leftrightarrow Y$

To state the congruence between the triangles, you list the vertices of each triangle in the **same order** as this correspondence.

Teaching Strategies

Some students may have difficulty understanding that there is no SSA, AAS, or AAA postulate for congruence of triangles. These students should be given additional practice building triangles with given side and/or angle measures using geoboards, dot paper, or other manipulatives.

You can use given information to prove that two triangles are congruent. One way to do this is to show the triangles are congruent *by definition.* That is, you prove that all six parts of one triangle are congruent to six corresponding parts of the other. However, this can be quite cumbersome.

Fortunately, triangles have special properties that allow you to prove triangles congruent by identifying only *three* sets of corresponding parts. The first way to do this is summarized in the *SSS postulate.*

Postulate 11	**The SSS Postulate** If three sides of one triangle are congruent to three sides of another triangle, then the triangles are congruent.

Reading Math

The statements $\overline{AB} \cong \overline{CD}$ and $AB = CD$ are equivalent, and can be used interchangeably. The same is true of the statements $\angle A \cong \angle B$ and $m\angle A = m\angle B$.

Example 1

ANIMATION The figure shown is part of a perspective drawing for a background scene of a city. How can the artist be sure that the two triangles are congruent?

Given $JK \cong JM$; $KL \cong ML$

Prove $\triangle JKL \cong \triangle JML$

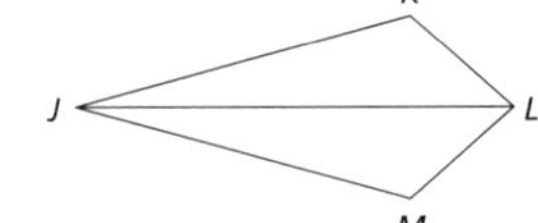

Solution

Statements	Reasons
1. $\overline{JK} \cong \overline{JM}$; $\overline{KL} \cong \overline{ML}$	1. given
2. $\overline{JL} \cong \overline{JL}$	2. reflexive property
3. $\triangle JKL \cong \triangle JML$	3. SSS postulate

GEOMETRY SOFTWARE Use geometric drawing software to explore the postulate. Draw two triangles so that the three sides of one triangle are congruent to the three sides of the other triangle. Measure the interior angles of both triangles. They are also congruent.

B
A C

$\angle A$ is included between $\overline{AB}$ and $\overline{AC}$.
$\overline{AB}$ is included between $\angle A$ and $\angle B$.

Sometimes it is helpful to describe the parts of a triangle in terms of their relative positions.

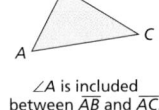

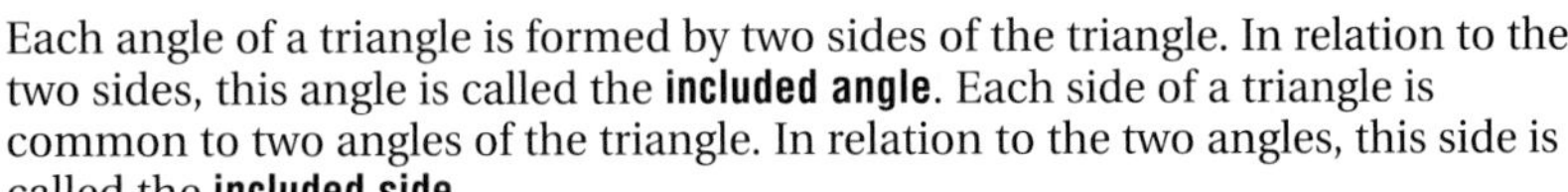

Each angle of a triangle is formed by two sides of the triangle. In relation to the two sides, this angle is called the **included angle**. Each side of a triangle is common to two angles of the triangle. In relation to the two angles, this side is called the **included side**.

Using these terms, it is now possible to describe two additional ways of showing that two triangles are congruent.

Postulate 12	**The SAS Postulate** If two sides and the included angle of one triangle are congruent to two sides and the included angle of another triangle, then the triangles are congruent.
Postulate 13	**The ASA Postulate** If two angles and the included side of one triangle are congruent to two angles and the included side of another triangle, then the triangles are congruent.

Learning Styles

TACTILE/KINESTHETIC LEARNERS Some students may find it helpful to copy the figures and mark each pair of corresponding congruent parts with a different colored pencil.

Chalkboard Examples

Supplementary Example 1

In another part of the background, the artist finds two triangles that are part of a parallelogram. Draw the figure.

Given: $\overline{AB} \cong \overline{CD}$; $\overline{AD} \cong \overline{BC}$

Prove: $\triangle ABD \cong \triangle BCD$

Statements	Reasons
1. $\overline{AB} \cong \overline{CD}$; $\overline{AD} \cong \overline{BC}$	1. given
2. $\overline{BD} \cong \overline{BD}$	2. reflexive property
3. $\triangle ABD \cong \triangle BCD$	3. SSS postulate

Supplementary Example 2

Draw the figure. These two triangles share a common vertex of *F*, the midpoint of segment *EG*.

Given: $\overline{CE} \cong \overline{DG}$; $\angle C \cong \angle D$; $\overline{CE} \perp \overline{EG}$; $\overline{DG} \perp \overline{EG}$ Prove: $\triangle CEF \cong \triangle DFG$

Statements	Reasons
1. $\overline{CE} \cong \overline{DG}$; $\angle C \cong \angle D$; $\overline{CE} \perp \overline{EG}$; $\overline{DG} \perp \overline{EG}$	1. given
2. $\angle E$ and $\angle G$ are right angles.	2. definition of of perpendicular lines
3. $m\angle E = 90°$; $m\angle G = 90°$	3. definition of right angle
4. $m\angle E = m\angle G$, or $\angle E \cong \angle G$	4. transitive property of equality
5. $\triangle CEF \cong \triangle DFG$	5. ASA postulate

Reteaching Worksheet 4-2

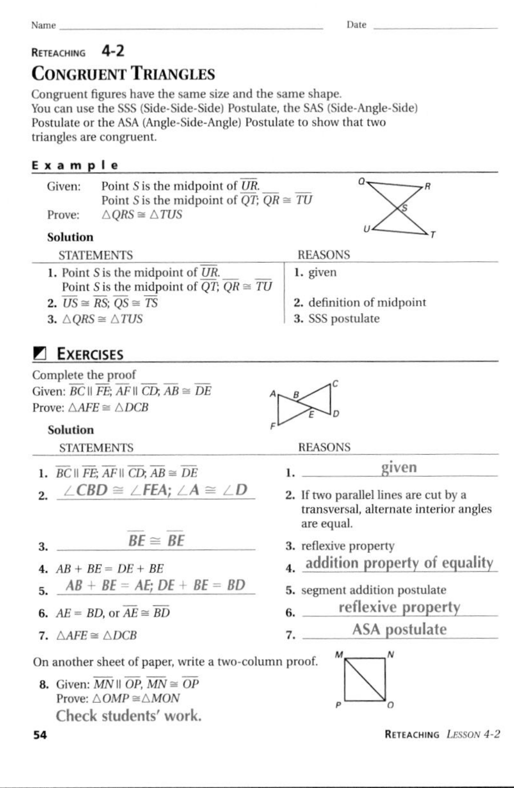

Name ______________________ Date ____________

RETEACHING 4-2

CONGRUENT TRIANGLES

Congruent figures have the same size and the same shape.
You can use the SSS (Side-Side-Side) Postulate, the SAS (Side-Angle-Side) Postulate or the ASA (Angle-Side-Angle) Postulate to show that two triangles are congruent.

Example

Given: Point *S* is the midpoint of $\overline{UR}$.
Point *S* is the midpoint of $\overline{QT}$; $\overline{QR} \cong \overline{TU}$

Prove: $\triangle QRS \cong \triangle TUS$

Solution

STATEMENTS	REASONS
1. Point *S* is the midpoint of $\overline{UR}$. Point *S* is the midpoint of $\overline{QT}$; $\overline{QR} \cong \overline{TU}$	1. given
2. $\overline{US} \cong \overline{RS}$; $\overline{QS} \cong \overline{TS}$	2. definition of midpoint
3. $\triangle QRS \cong \triangle TUS$	3. SSS postulate

EXERCISES

Complete the proof

Given: $\overline{BC} \parallel \overline{FE}$; $\overline{AF} \parallel \overline{CD}$; $\overline{AB} \cong \overline{DE}$

Prove: $\triangle AFE \cong \triangle DCB$

Solution

STATEMENTS	REASONS
1. $\overline{BC} \parallel \overline{FE}$; $\overline{AF} \parallel \overline{CD}$; $\overline{AB} \cong \overline{DE}$	1. given
2. $\angle CBD \cong \angle FEA$; $\angle A \cong \angle D$	2. If two parallel lines are cut by a transversal, alternate interior angles are equal.
3. $\overline{BE} \cong \overline{BE}$	3. reflexive property
4. $AB + BE = DE + BE$	4. addition property of equality
5. $AB + BE = AE$; $DE + BE = BD$	5. segment addition postulate
6. $AE = BD$, or $\overline{AE} \cong \overline{BD}$	6. reflexive property
7. $\triangle AFE \cong \triangle DCB$	7. ASA postulate

On another sheet of paper, write a two-column proof.

8. Given: $\overline{MN} \parallel \overline{OP}$, $\overline{MN} \cong \overline{OP}$
Prove: $\triangle OMP \cong \triangle MON$
Check students' work.

54 RETEACHING *LESSON 4-2*

Lesson Wrap-up

Quick Assessment

Ask the following questions to determine if students understand the content presented in this lesson.

1. Examine the triangles in Example 2. If triangle *ZYX* had been rotated so that side *XY* was at the top, would the two triangles still be congruent? Yes; position does not matter; two corresponding angles and the included side of each triangle are still congruent.
2. Suppose you needed to make a copy of a triangle. What is the least information you would need? the measures of all three sides, the measures of two sides and the included angle, or the measures of two angles and the included side

Assignment Guide

Basic: 1–10, 13–16
Advanced: 1–16
Additional Practice: See Extra Practice Index on page 674.

Additional Answers

2. Proofs may vary. A sample proof is given.

Statements	Reasons
1. $\overline{GL} \cong \overline{JK}$; $\overline{HL} \cong \overline{HK}$ Point H is the midpoint of $\overline{GJ}$.	1. given
2. $\overline{GH} \cong \overline{JK}$	2. definition of midpoint
3. $\triangle GHL \cong \triangle JHK$	3. SSS postulate

3. Proofs may vary. A sample proof is given.

Statements	Reasons
1. $\overline{AB} \cong \overline{CB}$; $\overline{EB} \cong \overline{DB}$ $\overline{AD}$ and $\overline{CE}$ intersect at point *B*.	1. given
2. $\angle ABE$ and $\angle CBD$ are vertical angles.	2. definition of vertical angles
3. $\angle ABE \cong \angle CBD$	3. vertical angle theorem
4. $\triangle ABE \cong \triangle CBD$	4. SAS postulate

Example 2

Given $\overline{VW} \cong \overline{ZY}$; $\angle V \cong \angle Z$
$\overline{VW} \perp \overline{WY}$; $\overline{ZY} \perp \overline{WY}$

Prove $\triangle VWX \cong \triangle ZYK$

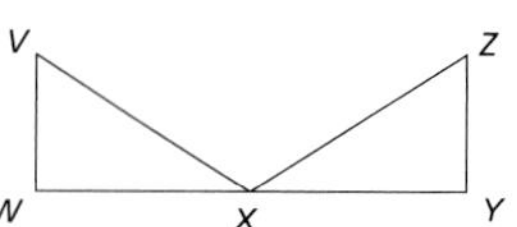

Solution

Statements	Reasons
1. $\overline{VW} \cong \overline{ZY}$; $\angle V \cong \angle Z$ $\overline{VW} \perp \overline{WY}$; $\overline{ZY} \perp \overline{WY}$	1. given
2. $\angle W$ and $\angle Y$ are right angles.	2. definition of perpendicular lines
3. $m\angle W = 90°$; $m\angle Y = 90°$	3. definition of right angle
4. $m\angle W = m\angle Y$, or $\angle W \cong \angle Y$	4. transitive property of equality
5. $\triangle VWX \cong \triangle ZYK$	5. ASA postulate

Problem Solving Tip

Before you start to write a proof, it is a good idea to develop a *plan* for the proof. When the proof involves congruent triangles, many students find it helpful to first copy the figure and mark as many congruences as they can. For instance, after reviewing the given information, the figure for Example 2 would be copied and marked as follows.

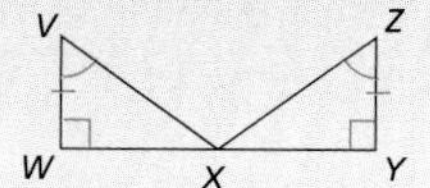

Once the figure is marked, it becomes fairly clear that the plan for proof will involve the ASA Postulate.

Try These Exercises

1. Copy and complete this proof.

Given $\overline{RQ} \cong \overline{RS}$; $\overrightarrow{RT}$ bisects $\angle QRS$.

Prove $\triangle QRT \cong \triangle SRT$

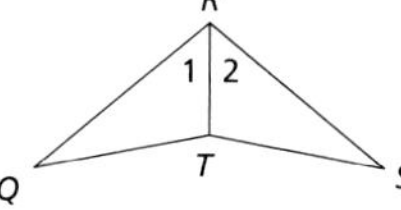

Statements	Reasons
1. ___?___ $\overline{RQ} \cong \overline{RS}$	1. ___?___ $\overrightarrow{RT}$ bisects $\angle QRS$ given.
2. $m\angle 1 = m\angle 2$, or $\angle 1 \cong \angle 2$	2. definition of ___?___ angle bisector
3. ___?___ $\overline{RT} \cong \overline{RT}$	3. ___?___ property reflexive
4. $\triangle QRT \cong \triangle SRT$	4. ___?___ SAS postulate

2. **CONSTRUCTION** Plans call for triangular bracing to be added to a horizontal beam. Prove the triangles are congruent by writing a two-column proof.

Given $\overline{GL} \cong \overline{JK}$; $\overline{HL} \cong \overline{HK}$
Point *H* is the midpoint of $\overline{GJ}$.

Prove $\triangle GHL \cong \triangle JHK$
See additional answers.

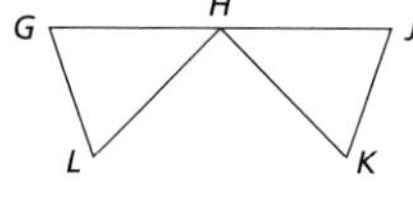

Practice Exercises

3. Write a two-column proof.

Given $\overline{AB} \cong \overline{CB}$; $\overline{EB} \cong \overline{DB}$
$\overline{AD}$ and $\overline{CE}$ intersect at point *B*.

Prove $\triangle ABE \cong \triangle CBD$
See additional answers.

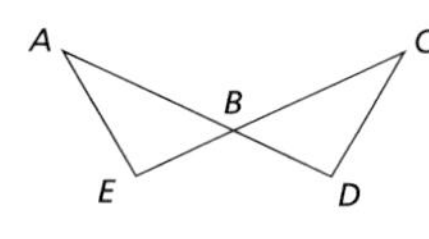

11. Proofs may vary. A sample proof is given.
Given: $\overline{AB} \cong \overline{XY}$; $\overline{BC} \cong \overline{YZ}$
$\angle B$ and $\angle Y$ are right angles.
Proof: $\triangle ABC \cong \triangle XYZ$

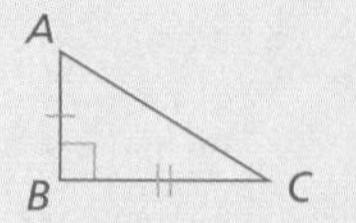

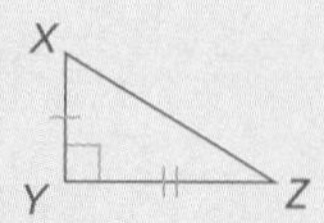

12. Answers will vary. Check students' work. The argument should be based on the fact that, when two sides and a not included angle of one triangle are congruent to two sides and a not included angle of another triangle, the triangles are not necessarily congruent. This is illustrated on the next page.

ENGINEERING The figures in Exercises 4–7 are taken from truss bridge designs. Each figure contains a pair of congruent overlapping triangles.

Use the given information to complete the congruence statement. Then name the postulate that would be used to prove the congruence. (You do *not* need to write the proof.)

4. 

$\triangle SRP$; SSS postulate

Given $\overline{PQ} \cong \overline{SR}$; $\overline{QS} \cong \overline{RP}$ $\triangle PQS \cong$ ___?___

5. 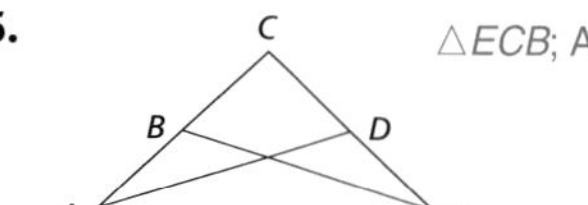

$\triangle ECB$; ASA postulate

Given $\overline{AC} \cong \overline{EC}$; $\angle A \cong \angle E$ $\triangle ACD \cong$ ___?___

6. 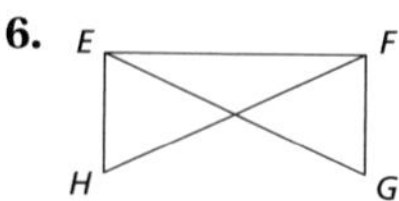

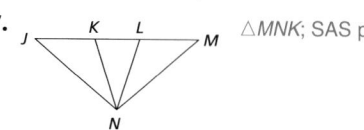

$\triangle GFE$; SAS postulate

Given $\overline{EH} \cong \overline{FG}$; $\overline{EH} \perp \overline{EF}$; $\overline{FG} \perp \overline{EF}$ $\triangle HEF \cong$ ___?___

7. 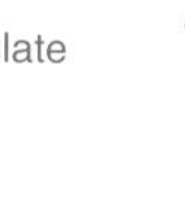 

$\triangle MNK$; SAS postulate

Given $\overline{JL} \cong \overline{MK}$; $\angle J \cong \angle M$; $\triangle JNM$ is isosceles $\triangle JNL \cong$ ___?___

GEOMETRY SOFTWARE Use geometric drawing software or paper and pencil to draw the figures in Exercises 8–9 on a coordinate plane.

8. Draw $\triangle MNP$ with vertices $M(-5, 5)$, $N(3, 5)$, and $P(3, -6)$. Then draw $\triangle QRS$ with vertices $Q(-4, 2)$, $R(7, -6)$ and $S(-4, -6)$. Explain how you know that the triangles are congruent. Then state the congruence.
Students will most likely cite the SAS postulate $\triangle MNP \cong \triangle QSR$

9. Draw $\triangle ABC$ with vertices $A(-3, 5)$, $B(6, 5)$, and $C(6, -8)$. Then graph points $X(2, 2)$ and $Y(-7, 2)$. Find two possible coordinates of a point Z so that $\triangle ABC \cong \triangle XYZ$. $(-7, -11)$ or $(-7, 15)$

10. CHAPTER INVESTIGATION Design a 20-foot side section of a truss bridge. Draw your design to the scale 1 in. = 2 ft.

EXTENDED PRACTICE EXERCISES

11. Write a proof of the following statement.
See additional answers.
If two legs of one right triangle are congruent to two legs of another right triangle, then the triangles are congruent.

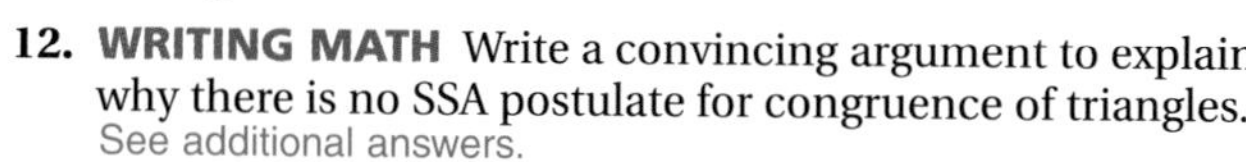

12. WRITING MATH Write a convincing argument to explain why there is no SSA postulate for congruence of triangles.
See additional answers.

MIXED REVIEW EXERCISES

Find the measure of each angle. (Lesson 3-2)

13. $\angle ABD$
104°

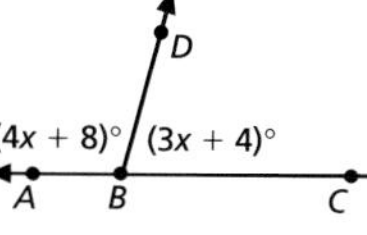

14. $\angle EFH$
71°

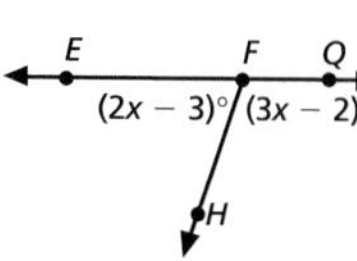

15. $\angle CBD$ 76°

16. $\angle GFH$ 109°

Extra Practice Worksheet 4-2

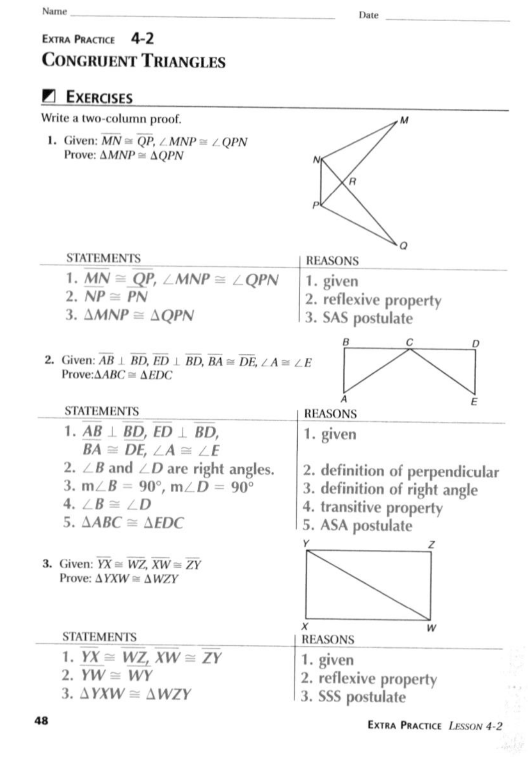

Name ______ Date ______

EXTRA PRACTICE 4-2
CONGRUENT TRIANGLES

EXERCISES

Write a two-column proof.

1. Given: $\overline{MN} \cong \overline{QP}$, $\angle MNP \cong \angle QPN$
Prove: $\triangle MNP \cong \triangle QPN$

STATEMENTS	REASONS
1. $\overline{MN} \cong \overline{QP}$, $\angle MNP \cong \angle QPN$	1. given
2. $\overline{NP} \cong \overline{PN}$	2. reflexive property
3. $\triangle MNP \cong \triangle QPN$	3. SAS postulate

2. Given: $\overline{AB} \perp \overline{BD}$, $\overline{ED} \perp \overline{BD}$, $\overline{BA} \cong \overline{DE}$, $\angle A \cong \angle E$
Prove: $\triangle ABC \cong \triangle EDC$

STATEMENTS	REASONS
1. $\overline{AB} \perp \overline{BD}$, $\overline{ED} \perp \overline{BD}$, $\overline{BA} \cong \overline{DE}$, $\angle A \cong \angle E$	1. given
2. $\angle B$ and $\angle D$ are right angles.	2. definition of perpendicular
3. $m\angle B = 90°$, $m\angle D = 90°$	3. definition of right angle
4. $\angle B \cong \angle D$	4. transitive property
5. $\triangle ABC \cong \triangle EDC$	5. ASA postulate

3. Given: $\overline{YX} \cong \overline{WZ}$, $\overline{XW} \cong \overline{ZY}$
Prove: $\triangle YXW \cong \triangle WZY$

STATEMENTS	REASONS
1. $\overline{YX} \cong \overline{WZ}$, $\overline{XW} \cong \overline{ZY}$	1. given
2. $\overline{YW} \cong \overline{WY}$	2. reflexive property
3. $\triangle YXW \cong \triangle WZY$	3. SSS postulate

48 EXTRA PRACTICE *LESSON 4-2*

Enrichment Worksheet 4-2

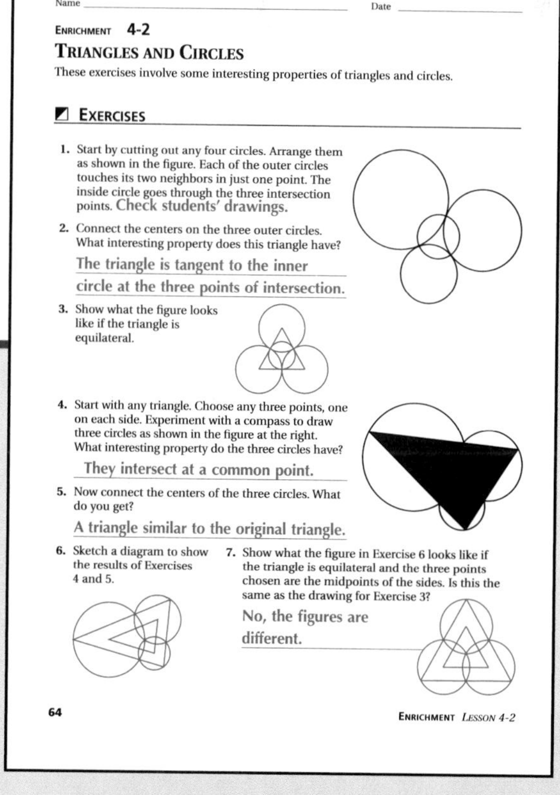

Name ______ Date ______

ENRICHMENT 4-2
TRIANGLES AND CIRCLES

These exercises involve some interesting properties of triangles and circles.

EXERCISES

1. Start by cutting out any four circles. Arrange them as shown in the figure. Each of the outer circles touches its two neighbors in just one point. The inside circle goes through the three intersection points. Check students' drawings.
2. Connect the centers on the three outer circles. What interesting property does this triangle have?
The triangle is tangent to the inner circle at the three points of intersection.
3. Show what the figure looks like if the triangle is equilateral.
4. Start with any triangle. Choose any three points, one on each side. Experiment with a compass to draw three circles as shown in the figure at the right. What interesting property do the three circles have?
They intersect at a common point.
5. Now connect the centers of the three circles. What do you get?
A triangle similar to the original triangle.
6. Sketch a diagram to show the results of Exercises 4 and 5.
7. Show what the figure in Exercise 6 looks like if the triangle is equilateral and the three points chosen are the midpoints of the sides. Is this the same as the drawing for Exercise 3?
No, the figures are different.

64 ENRICHMENT *LESSON 4-2*

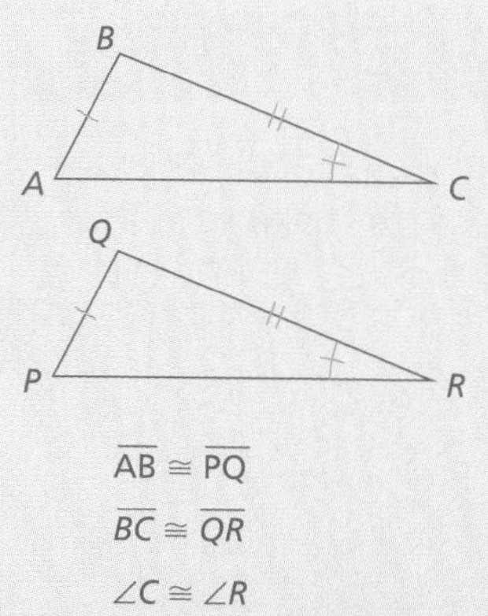

$\overline{AB} \cong \overline{PQ}$
$\overline{BC} \cong \overline{QR}$
$\angle C \cong \angle R$
$\triangle ABC \cong \triangle PQR$

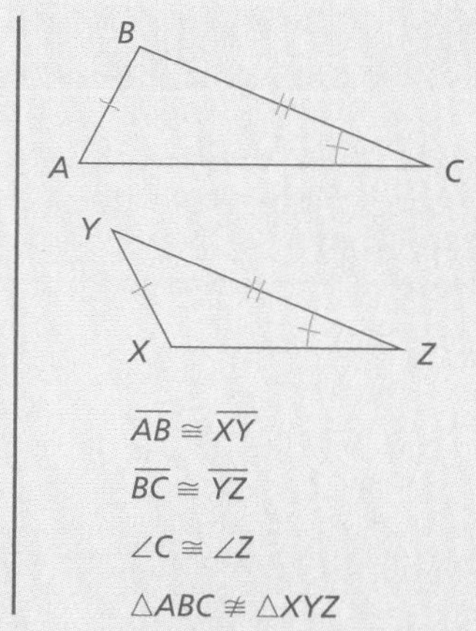

$\overline{AB} \cong \overline{XY}$
$\overline{BC} \cong \overline{YZ}$
$\angle C \cong \angle Z$
$\triangle ABC \ncong \triangle XYZ$

Skills Practice

Vocabulary Review

Lesson 4-1

triangle	vertex
interior angle	side
equilateral triangle	
isosceles triangle	
scalene triangle	acute triangle
obtuse triangle	right triangle
auxiliary line	exterior angle
equiangular triangle	

Lesson 4-2

congruent	SSS postulate
SAS postulate	ASA postulate
included side	included angle

Assignment Guide

All students: 1–22
Additional Practice: Refer to the Extra Practice Index on page 674 of this text.

Additional Answers

11.
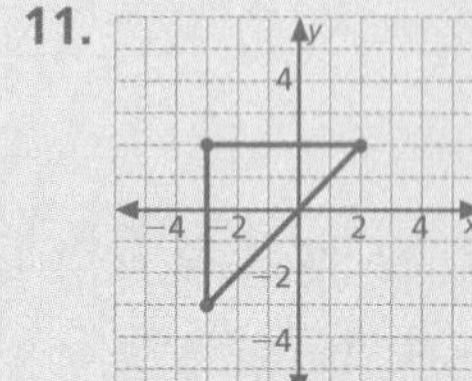

12.
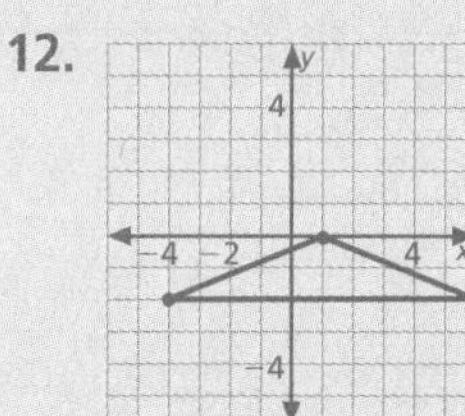

13.
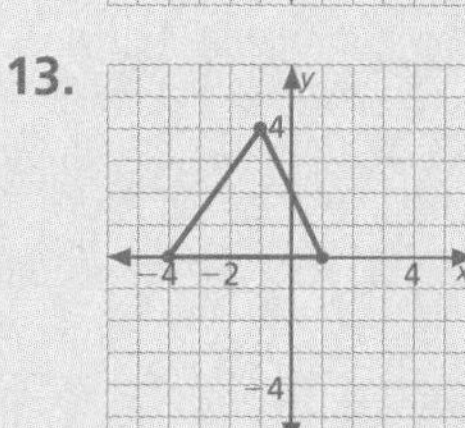

15. $\overline{RS} \cong \overline{VU}$, $\overline{RT} \cong \overline{VT}$, $\overline{ST} \cong \overline{UT}$, $\angle SRT \cong \angle UVT$, $\angle RTS \cong \angle VTU$, $\angle TSR \cong \angle TUV$, $\triangle RST \cong \triangle VUT$

Review and Practice Your Skills

Practice ■ Lesson 4-1

Find the value of x in each figure.

1.
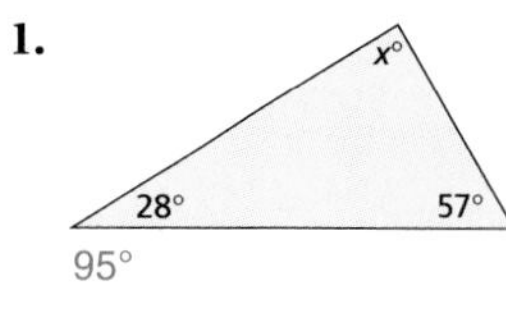

2.
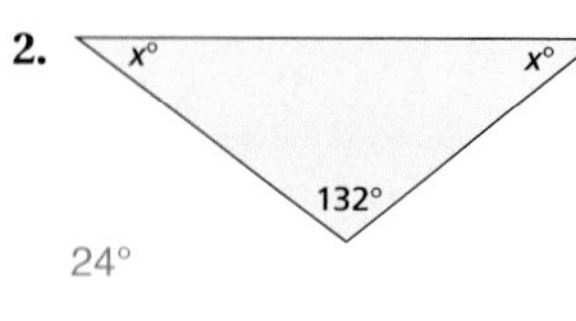

3.
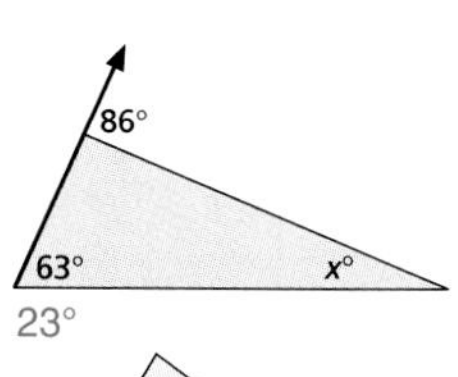

4.
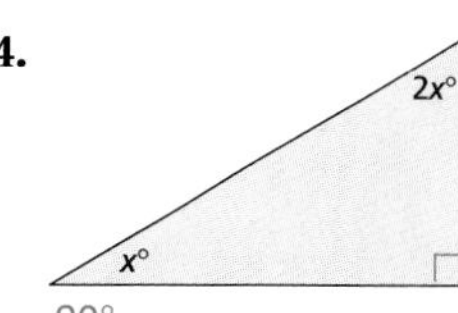

5.
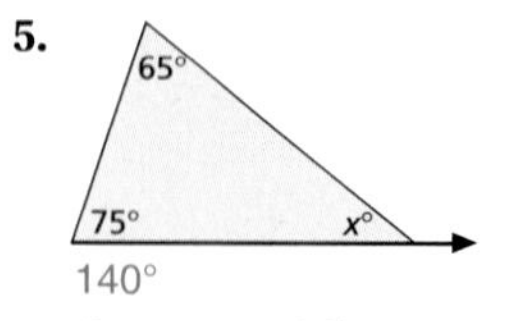

6.
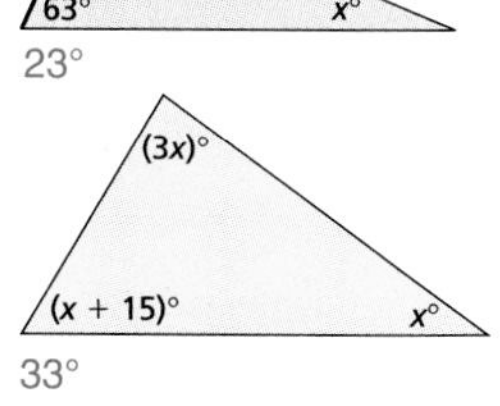

Determine whether each statement is *true* or *false*.

7. If two angles in a triangle are acute, then the third angle is always obtuse. false
8. If one angle in a triangle is obtuse, then the other two angles are always acute. true
9. If one exterior angle of a triangle is obtuse, then all three interior angles are acute. false
10. If two angles in a triangle are equal, then the triangle is equiangular. false

On a coordinate plane, sketch the triangle with the given vertices. Then classify the triangle, first by its sides, then by its angles. See additional answers for sketches.

11. $A(2, 2)$; $B(-3, -3)$; $C(-3, 2)$ right isosceles
12. $X(6, -2)$; $Y(-4, -2)$; $Z(1, 0)$ isosceles obtuse
13. $M(-1, 4)$; $N(1, 0)$; $P(-4, 0)$ scalene acute

Practice ■ Lesson 4-2

14. Copy and complete this proof.

Given $\overline{AB} \parallel \overline{DE}$; C is the midpoint of $\overline{BD}$
Prove $\triangle ABC \cong \triangle EDC$

Statements	Reasons
1. $\overline{AB} \parallel \overline{DE}$	1. __?__ Given
2. $\angle B \cong \angle D$	2. __?__ Alt. Interior Ang. Theorem
3. $\angle BCA \cong \angle ECD$	3. __?__ vertical angles Theorem
4. __?__ C is midpoint of $\overline{BD}$	4. Given
5. $\overline{BC} \cong \overline{CD}$	5. __?__ midpoint theorem
6. __?__ $\triangle ABC \cong \triangle EDC$	6. ASA Postulate

15. Name all the pairs of congruent parts in these triangles. Then state the congruence between the triangles. See additional answers.

16. *True* or *false*: If three angles of one triangle are congruent to three angles of another triangle, then the triangles are congruent. false

Teaching Strategies

For Exercises 1–6, have students check their work by using the value of x to find the measure of each interior angle. Have them check to make sure that the sum of the interior angles is 180°.

PRACTICE ■ LESSON 4-1–LESSON 4-2

Determine whether each statement is *always, sometimes,* or *never* true. (Lesson 4-1)

17. There are two exterior angles at each vertex of a triangle. Always
18. An exterior angle of a triangle is an acute angle. Sometimes
19. Two of the three angles in a triangle are complementary angles. Sometimes
20. The sum of the measures of the angles in a triangle is 90°. Never

Use the given information to complete each congruence statement. Then name the postulate that would be used to prove the congruence. (You do *not* need to write the proof.) (Lesson 4-2)

21. **Given** $\overline{PQ} \cong \overline{NO}$; $\overline{QR} \cong \overline{MO}$; $\angle Q \cong \angle O$
Prove $\triangle PQR \cong$? $\triangle NOM$

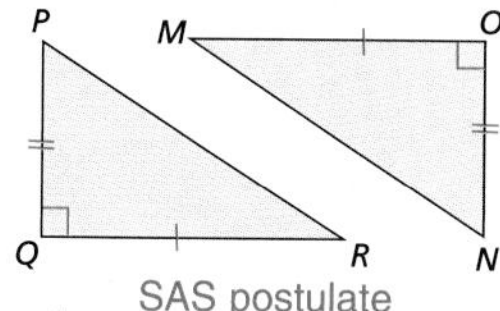

SAS postulate

22. **Given** $\overline{EY} \cong \overline{EF}$; $\overline{DY} \cong \overline{LF}$; $\overline{DE} \cong \overline{LE}$
Prove $\triangle DYE \cong$? $\triangle LFE$

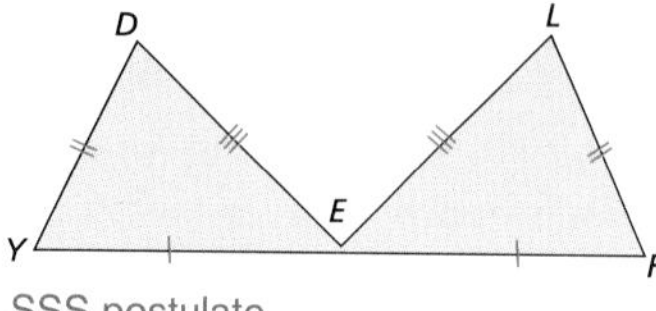

SSS postulate

MathWorks Workplace Knowhow — Career – Jeweler

A jeweler designs and repairs jewelry, cuts gems and appraises the value of gemstones and jewelry. Most jewelers go through an apprenticeship program where they work under an experience jeweler to hone their skills and learn new techniques. A background in art, math, mechanical drawing and chemistry are all useful when working with gems and precious metals. Math skills help a jeweler in the areas of design and gem cutting. Jewelers use computer-aided design (CAD) programs to design jewelry to meet a customer's expectations. A symmetrically cut gem is a valuable gem. A poorly cut gem becomes a wasted investment for the jeweler.

In the gem cut shown to the right, all triangles shown can be classified as isosceles triangles.

1. What additional classifications can be given to triangle *ABC*? equilateral and equiangular
2. What is the measure of $\angle BCE$? 104°
3. Sides *CE* and *DE* are congruent and $\angle BCE$ and $\angle EDF$ are congruent. Angle *DEF* measures 38°. Are triangles *BCE* and *DEF* congruent? If so, what postulate could be used to prove the congruence?
Yes. ASA Postulate

Chalkboard Examples

Example from Lesson 4-1

Classify this triangle by its angles. A triangle has an exterior angle of $(x + 20)°$ and an adjacent interior angle of $(x - 70)°$. Is the triangle *acute, equilateral, right* or *obtuse*? **obtuse**

Example from Lesson 4-2

Based on the information in the diagram, what postulate can you use to show the triangles are congruent? **ASA postulate**

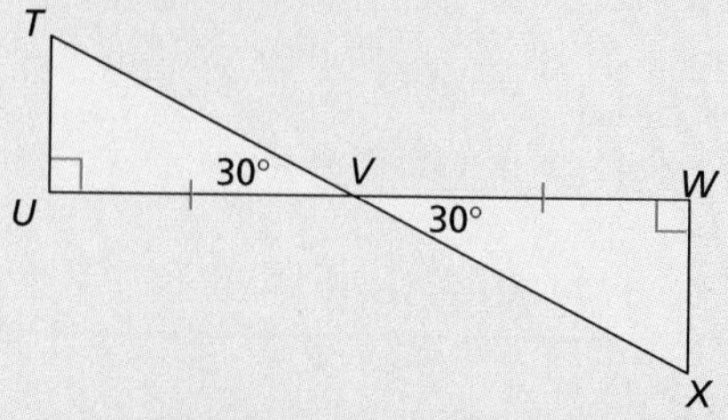

Career Opportunity

Describe the kind of work jewelers do and the types of companies that hire these workers. Have students discuss the importance of geometry and measurement in doing this type of job. Explain that jewelers must understand the principles of art and design to create settings that will increase the value of their work. Students should answer Questions 1–3 to better understand how jewelers use mathematics in performing their job.

Students who are interested in learning more about this profession can go to learningmathmatters.com. Guidance Counselors should have information about school and training requirements.

Teaching Strategies

VISUAL LEARNERS To answer true/false questions as in Exercise 7–10 and 16, have students assume the conclusion is true and then try to draw a counterexample. If a counterexample can be found, the conclusion must be false.

Lesson Planning

NCTM Standards/Strands

- Number & Operation
- Patterns, Functions, & Algebra
- Geometry & Spatial Sense
- Measurement
- Problem Solving
- Reasoning and Proof
- Communication
- Connections
- Representation

Vocabulary

base	base angles
legs	vertex angle
corollary	
isosceles triangle theorem	

Materials Needed

scissors

Lesson Resources

Warm-Up Transparency 10
Reteaching 4-3
Extra Practice 4-3
Enrichment 4-3

Assignment Guide

Basic: 1–16, 19–22
Enriched: 1–22

Getting Started

5-Minute Warm-Up

In relation to triangles, what do the following abbreviations stand for?

1. SAS two sides and the included angle
2. ASA two angles and the included side
3. SSS three sides

Introduction to Lesson 4-3

After students have completed this activity, discuss when else they may have used this method to cut out objects and why. Most students have probably used a paper with one fold to cut out a heart and a paper with more than one fold to cut out snowflakes to insure that all parts are the same.

4-3 Congruent Triangles and Proofs

Goals
- Establish congruence between two triangles to show that corresponding parts are congruent.
- Find angle and side measures of triangles.

Applications Design, Architecture, Construction, Engineering

Fold a piece of paper and draw a segment on it as shown. Now cut both thicknesses of paper along the segment. Unfold and label the triangle.

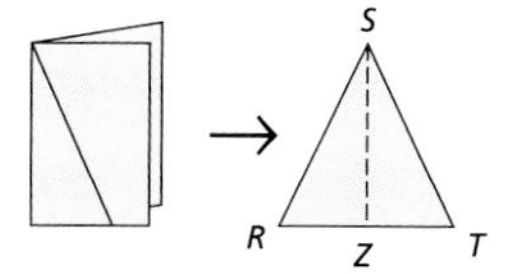

1. Are there any perpendicular segments on the triangle? Yes; $\overline{SZ} \perp \overline{RT}$
2. Does any segment lie on an angle bisector of the triangle? Yes; $\overline{SZ}$ lies on the bisector of $\angle RST$
3. List as many congruences as you can among the segments, angles, and triangles that you see on the folded triangle.
Answers will vary. Possible responses: $\overline{SR} \cong \overline{ST}$; $\overline{RZ} \cong \overline{TZ}$; $\angle R \cong \angle T$; $\angle RSZ \cong \angle TSZ$; $\angle SZR \cong \angle SZT$; $\triangle RSZ \cong \triangle TSZ$

Build Understanding

The SSS, SAS, and ASA postulates help you determine a congruence between two triangles by identifying just three pairs of corresponding parts. Once you establish a congruence, you may conclude that *all* pairs of corresponding parts are congruent. Example 1 shows how this fact can be used to show that two angles are congruent.

Example 1

Given $\overline{AB} \cong \overline{CB}$; $\overline{AD} \cong \overline{CD}$

Prove $\angle A \cong \angle C$

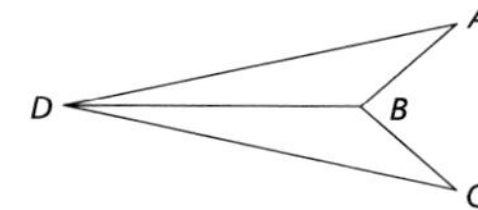

Solution

Statements	Reasons
1. $\overline{AB} \cong \overline{CB}$; $\overline{AD} \cong \overline{CD}$	1. given
2. $\overline{BD} \cong \overline{BD}$	2. reflexive property
3. $\triangle ABD \cong \triangle CBD$	3. SSS postulate
4. $\angle A \cong \angle C$	4. Corresponding parts of congruent triangles are congruent.

Reading Math

The final reason of the proof in Example 1 is *Corresponding parts of congruent triangles are congruent.* This fact is used so often that it is commonly abbreviated *CPCTC.*

Corresponding parts of congruent triangles are used in the proofs of many theorems. For example, an isosceles triangle is a triangle with two **legs** of equal length. The third side is the **base**. The angles at the base are called the **base angles**, and the third angle is the **vertex angle**. *CPCTC* can be used to prove the following theorem about base angles.

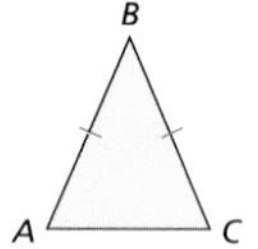

legs: $\overline{AB}$, $\overline{CB}$
base: $\overline{AC}$
base angles: $\angle A$, $\angle C$
vertex angle: $\angle B$

Teaching Strategies

CHALLENGE ABCDE is a regular pentagon. Prove that $\triangle ABG$ is an isosceles triangle.

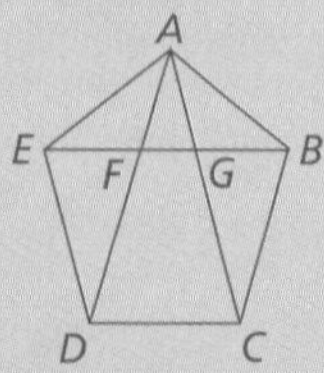

$\triangle ABC \cong \triangle ABE$ by SAS, so, $\angle CAB \cong \triangle$

The Isosceles Triangle Theorem	If two sides of a triangle are congruent, then the angles opposite those sides are congruent. This is sometimes stated: *Base angles of an isosceles triangle are congruent.*

Given $\triangle ABC$ is isosceles, with base $\overline{AC}$. $\overline{BX}$ bisects $\angle ABC$.

Prove $\angle A \cong \angle C$

Statements	**Reasons**
1. $\triangle ABC$ is isosceles, with base $\overline{AC}$. $\overline{BX}$ bisects $\angle ABC$.	1. given
2. $AB = CB$, or $\overline{AB} \cong \overline{CB}$	2. definition of isosceles $\triangle$
3. $m\angle 1 = m\angle 2$, or $\angle 1 \cong \angle 2$	3. definition of $\angle$ bisector
4. $\overline{BX} \cong \overline{BX}$	4. reflexive property
5. $\triangle AXB \cong \triangle CXB$	5. SAS postulate
6. $\angle A \cong \angle C$	6. CPCTC

Example 2

DESIGN An artist is positioning the design elements for a new company logo. At the center of the logo is the triangle shown in the figure. Find $m\angle P$.

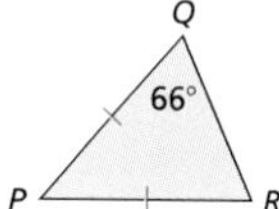

Solution

Since $PQ = PR$, $\triangle PQR$ is isosceles with base $\overline{QR}$. By the isosceles triangle theorem, $m\angle R = m\angle Q = 66°$. By the triangle-sum theorem, $m\angle P + 66° + 66° = 180°$, or $m\angle P = 48°$.

A statement that follows directly from a theorem is called a **corollary**. The following are corollaries to the isosceles triangle theorem.

Corollary 1	If a triangle is equilateral, then it is equiangular.
Corollary 2	The measure of each angle of an equilateral triangle is 60°.

Check Understanding

How would the solution of Example 2 be different if the measure of $\angle Q$ were 54°?

The measure of $\angle R$ would be 54°, and so the measure of $\angle P$ would be $180° - (54° + 54°) = 72°$.

The converse of the isosceles triangle theorem is the *base angles theorem.*

The Base Angles Theorem	If two angles of a triangle are congruent, then the sides opposite those angles are congruent.
Corollary	If a triangle is equiangular, then it is equilateral.

Extend the Lesson

REAL WORLD CONNECTION Invite an architect or builder to visit your class and discuss the use of triangular shapes in construction. Then have students find examples of triangular shapes used in the construction of your school or other structures in your community.

Chalkboard Examples

Supplementary Example 1

Given: $EF \cong FG$; $FH \cong GH$

Prove: $\angle E \cong \angle G$

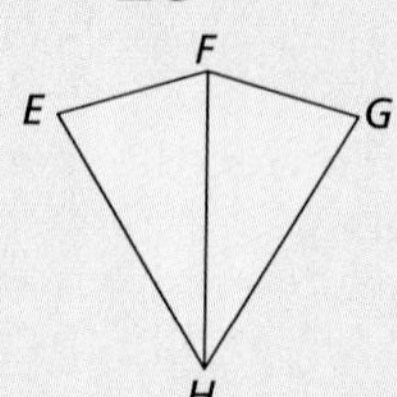

Statements	Reasons
1. $EF \cong FG$; $FH \cong GH$	1. given
2. $FH \cong FH$	2. reflexive property
3. $\triangle EFH \cong \triangle FGH$	3. SSS postulate
4. $\angle E \cong \angle G$	4. CPCTC

Supplementary Example 2

For another logo, the designer draws an isosceles triangle ABC where $AC = BC$ and one of the base angles is 73°. Find the $m\angle C$. **34°**

Reteaching Worksheet 4-3

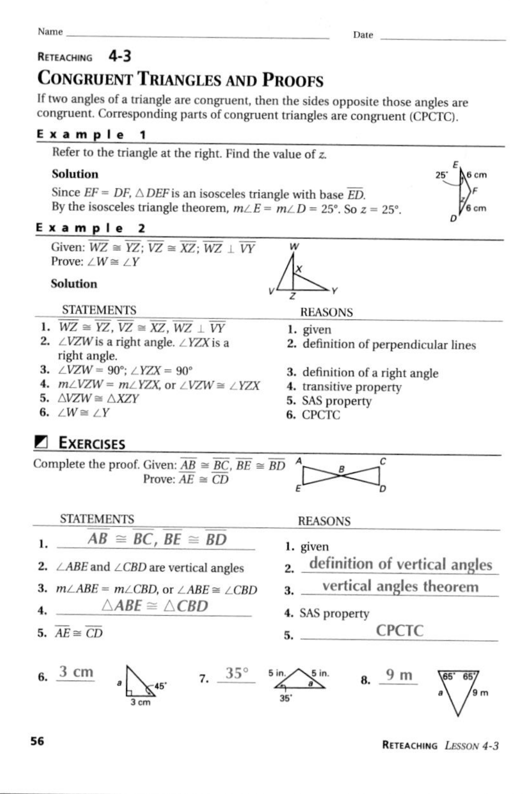

Name ________ Date ________

RETEACHING 4-3

CONGRUENT TRIANGLES AND PROOFS

If two angles of a triangle are congruent, then the sides opposite those angles are congruent. Corresponding parts of congruent triangles are congruent (CPCTC).

Example 1

Refer to the triangle at the right. Find the value of z.

Solution

Since $EF = DF$, $\triangle DEF$ is an isosceles triangle with base $\overline{ED}$. By the isosceles triangle theorem, $m\angle E = m\angle D = 25°$. So $z = 25°$.

Example 2

Given: $\overline{WZ} \cong \overline{YZ}$; $\overline{VZ} \cong \overline{XZ}$; $\overline{WZ} \perp \overline{VY}$

Prove: $\angle W \cong \angle Y$

Solution

STATEMENTS	REASONS
1. $\overline{WZ} \cong \overline{YZ}$, $\overline{VZ} \cong \overline{XZ}$, $\overline{WZ} \perp \overline{VY}$	1. given
2. $\angle VZW$ is a right angle. $\angle YZX$ is a right angle.	2. definition of perpendicular lines
3. $\angle VZW = 90°$; $\angle YZX = 90°$	3. definition of a right angle
4. $m\angle VZW = m\angle YZX$, or $\angle VZW \cong \angle YZX$	4. transitive property
5. $\triangle VZW \cong \triangle XZY$	5. SAS property
6. $\angle W \cong \angle Y$	6. CPCTC

EXERCISES

Complete the proof. Given: $\overline{AB} \cong \overline{BC}$, $\overline{BE} \cong \overline{BD}$
Prove: $\overline{AE} \cong \overline{CD}$

STATEMENTS	REASONS
1. $\overline{AB} \cong \overline{BC}$, $\overline{BE} \cong \overline{BD}$	1. given
2. $\angle ABE$ and $\angle CBD$ are vertical angles	2. definition of vertical angles
3. $m\angle ABE = m\angle CBD$, or $\angle ABE \cong \angle CBD$	3. vertical angles theorem
4. $\triangle ABE \cong \triangle CBD$	4. SAS property
5. $\overline{AE} \cong \overline{CD}$	5. CPCTC

6. 3 cm
7. 35°
8. 9 m

56 RETEACHING *LESSON 4-3*

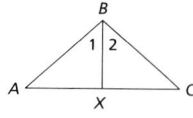

Lesson Wrap-up

Quick Assessment

Ask the following questions to determine if students understand the content presented in this lesson.

1. If you were given the measure of only one angle of an isosceles triangle, which would it have to be for you to know that the triangle is also equilateral? Any angle measure; if it is 60°, the other angles would also be 60°.
2. When can you use the reason CPCTC to prove that two parts of a triangle are congruent? after you have been given or have proved that the triangles are congruent

Assignment Guide

Basic: 1–16, 19–22
Advanced: 1–22
Additional Practice: See Extra Practice Index on page 674.

Additional Answers

12. Statement 1: Point H is the midpoint of $\overline{GK}$. Point H is the midpoint of $\overline{LJ}$. Reason 7: Two lines are cut by a transversal so that alternate interior angles are equal in measure, the lines are parallel.

13.

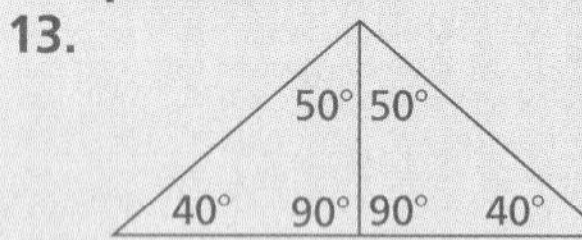

14.

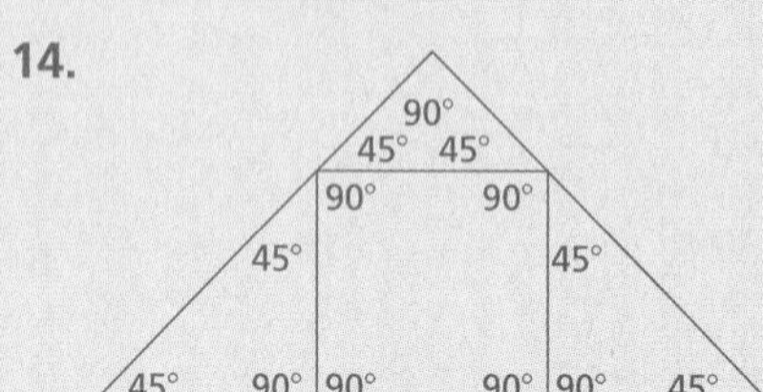

Try These Exercises

1. Copy and complete this proof.

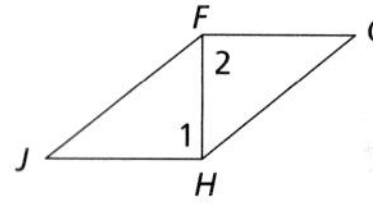

Given $\overline{FG} \cong \overline{HJ}$; $\overline{FG} \perp \overline{FH}$; $\overline{JH} \perp \overline{FH}$

Prove $\angle J \cong \angle G$

Statements	Reasons
1. ? $\overline{FG} \cong \overline{HJ}$; $\overline{FG} \perp \overline{FH}$; $\overline{JH} \perp \overline{FH}$	1. ? given
2. $\angle 1$ and $\angle 2$ are right angles.	2. definition of ? $\perp$ lines
3. $m\angle 1 = 90°$; $m\angle 2 = 90°$	3. definition of ? right $\angle$
4. ? $m\angle 1 = m\angle 2$, or $\angle 1 \cong \angle 2$	4. transitive property of equality
5. ? $\overline{FH} \cong \overline{HF}$	5. reflexive property
6. $\triangle JFH \cong \triangle GHF$	6. ? SAS postulate
7. ? $\angle J \cong \angle G$	7. ? CPCTC

Find the value of *n* in each figure.

2.

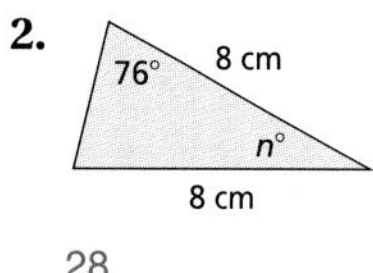

28

3.

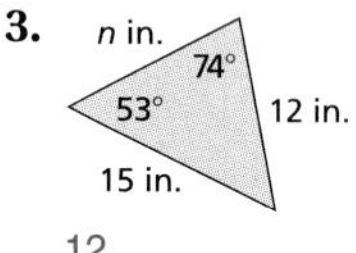

12

4.

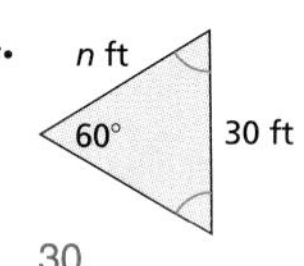

30

Practice Exercises

Find the value of *x* in each figure.

5.

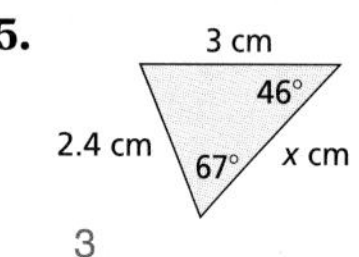

3

6.

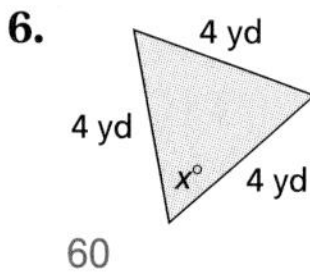

60

7.

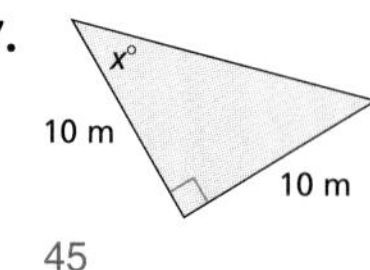

45

8. **ARCHITECTURE** An architect sees the figure at the right on a set of building plans. The architect wants to be certain that $\angle T \cong \angle R$. Copy and complete this proof.

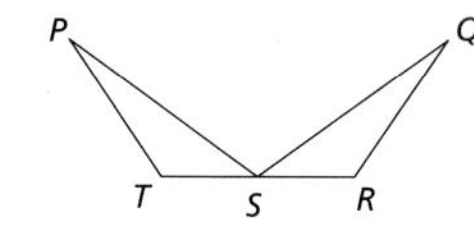

Given $\overline{PS} \cong \overline{QS}$; $\overline{PT} \cong \overline{QR}$

Point S is the midpoint of $\overline{TR}$.

Prove $\angle T \cong \angle R$

Statements	Reasons
1. ? $\overline{PS} \cong \overline{QS}$; $\overline{PT} \cong \overline{QR}$ Point S is the midpoint of $\overline{TR}$.	1. ? given
2. ? midpoint of $\overline{TR}$.	2. definition of ? given
3. ? $\triangle PTS \cong \triangle QRS$	3. SSS postulate
4. $\angle T \cong \angle R$	4. ? CPCTC

9. **YOU MAKE THE CALL** A base angle of an isosceles triangle measures 70°. Cina says the two remaining angles must each measure 55°. What mistake has Cina made? There are two base angles. If each measures 70°, the vertex angle must measure 40°.

Additional Answers

15.

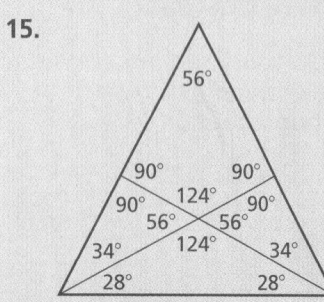

16.

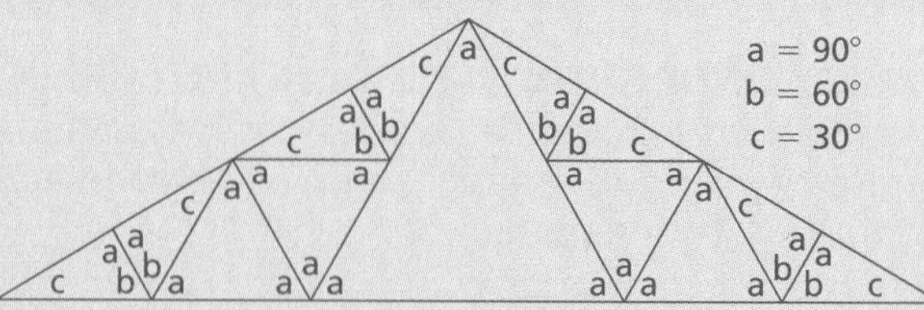

Name all the pairs of congruent angles in each figure.

10.

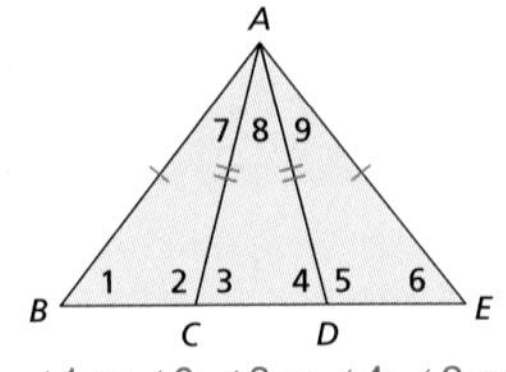

∠1 ≅ ∠6; ∠3 ≅ ∠4; ∠2 ≅ ∠5; ∠7 ≅ 9; ∠BAD ≅ ∠EAC

11.

∠1 ≅ ∠2; ∠4 ≅ ∠5; ∠3 ≅ ∠6; ∠8 ≅ ∠7; ∠XZY ≅ ∠XYZ

12. BRIDGE BUILDING On a truss bridge, steel cables cross as shown in the figure at the right. The inspector needs to be certain that $\overline{GL}$ and $\overline{JK}$ are parallel. Copy and complete the proof.

Given Point H is the midpoint of $\overline{GK}$.
Point H is the midpoint of $\overline{LJ}$.

Prove $\overline{GL} \parallel \overline{JK}$

Statements	Reasons
1. __?__ See additional answers.	1. __?__ given
2. $GH \cong KH$; __?__ $LH \cong JH$	2. __?__ def. of midpoint
3. ∠1 and ∠2 are vertical angles.	3. definition of __?__ vert ∠s
4. __?__ m∠1 = m∠2, or ∠1 ≅ ∠2	4. __?__ theorem vert. ∠s
5. __?__ △GHL ≅ △KHJ	5. SAS postulate
6. ∠G ≅ ∠K, or m∠G = m∠K	6. __?__ CPCTC
7. $\overline{GL} \parallel \overline{JK}$	7. If __?__, then __?__. See additional answers.

DATA FILE For Exercises 13–16, use the Data Index on pages 632–633 to locate information about the types of structural supports used in architecture. For each type of support, find the measure of each angle in the diagram. See additional answers.

13. king-post **14.** queen-post **15.** scissors **16.** Fink

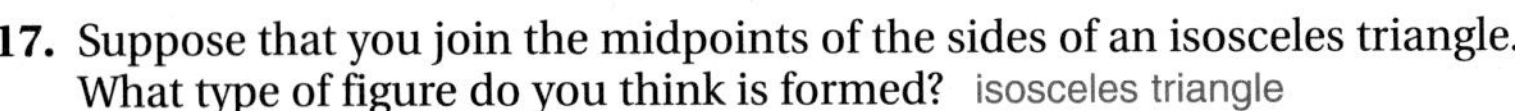

Extended Practice Exercises

17. Suppose that you join the midpoints of the sides of an isosceles triangle. What type of figure do you think is formed? isosceles triangle

18. WRITING MATH Write a proof of the second corollary to the isosceles triangle theorem: *The measure of each angle of an equilateral triangle is 60°.* See additional answers.

Mixed Review Exercises

Exercises 19–22 refer to the protractor at the right. (Lesson 3-2)

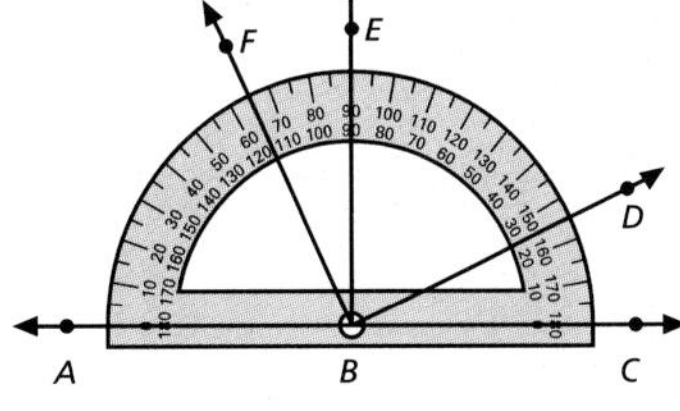

19. Name the straight angle. ∠ABC

20. Name the three right angles. ∠ABE, ∠EBC, ∠FBD

21. Name all the obtuse angles and give the measure of each. ∠FBC = 115°, ∠ABD = 155°

22. Name all the acute angles and give the measure of each. ∠ABF = 65°, ∠FBE = 25°, ∠EBD = 65°, ∠DBC = 25°

18. Proofs may vary. A sample proof is given.
Given: △ABC is equilateral.
Prove: m∠A = m∠B = m∠C = 60°

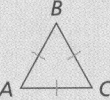

Statements	Reasons
1. △ABC is equilateral.	1. given
2. AB = BC = AC	2. def. of equilateral △
3. m∠A = m∠B = m∠C	3. isosceles △ theorem
4. m∠A + m∠B + m∠C = 180°	4. triangle-sum theorem
5. m∠A + m∠A + m∠A = 180° or 3m∠A = 180°	5. substitution property
6. 1/3(3m∠A) = 1/3(180°), or 3m∠A = 60°	6. mult. prop. of equality
7. m∠B = 60°; m∠C = 60°	7. substitution property

Extra Practice Worksheet 4-3

Name ________________ Date ________

EXTRA PRACTICE 4-3

CONGRUENT TRIANGLES AND PROOFS

EXERCISES

Find the value of x in each figure.

1. 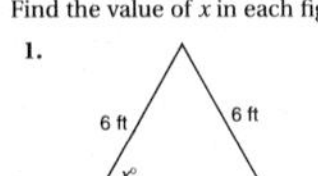 60

2. 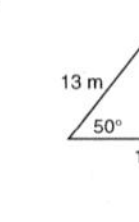 13

3. 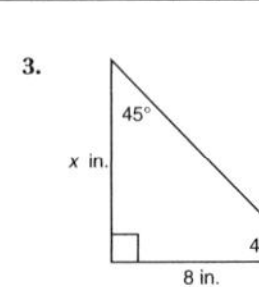 8

Complete each proof.

4. Given: $\overline{RS} \parallel \overline{UT}$, $\overline{RU} \parallel \overline{ST}$
Prove: $\overline{RS} \cong \overline{TU}$

STATEMENTS	REASONS
1. $\overline{RS} \parallel \overline{UT}$, $\overline{RU} \parallel \overline{ST}$	1. given
2. ∠SRT ≅ ∠UTR; ∠STR ≅ ∠URT	2. alt. int. ∠s are = in measure
3. $\overline{RT} \cong \overline{TR}$	3. reflexive property
4. ΔSRT ≅ ΔUTR	4. ASA postulate
5. $\overline{RS} \cong \overline{TU}$	5. CPCTC

5. Given: $\overline{XZ} \perp \overline{WY}$, Point Z is the midpoint of $\overline{WY}$
Prove: ∠W ≅ ∠Y

STATEMENTS	REASONS
1. $\overline{XZ} \perp \overline{WY}$, Point Z is the midpoint of $\overline{WY}$	1. given
2. ∠WZX and ∠YZX are right angles.	2. definition of perpendicular
3. $\overline{WZ} \cong \overline{YZ}$	3. definition of midpoint
4. $\overline{XZ} \cong \overline{XZ}$	4. reflexive property
5. ΔXZW ≅ ΔXZY	5. SAS postulate
6. ∠W ≅ ∠Y	6. CPCTC

50 EXTRA PRACTICE *LESSON 4-3*

Enrichment Worksheet 4-3

Name ________________ Date ________

ENRICHMENT 4-3

GEOMETRIC FALLACIES: TRIANGLES

A fallacy is a demonstration or proof which leads to an impossible result. In a geometric fallacy, the "trick" or error is often found in an incorrectly drawn figure. Sometimes, the "proof" contains an incorrect assumption.

EXERCISES

1. Follow along with this "proof," marking the figure as directed and completing the statements.

Prove: All triangles are isosceles.

1. Given any triangle ABC, let D be a point between B and C, such that $\overline{BD} \cong \overline{DC}$ and $\overline{AD} \perp \overline{BC}$.
2. Then, ∠ADB ≅ ∠ADC because both are right angles.
3. Therefore, △ADB ≅ △ADC because of the SAS Postulate.
4. Therefore, ∠B ≅ ∠C.

2. Use the figure at the right to explain what is wrong with the proof in Exercise 1.
The midpoint of $\overline{BC}$ is not always the same point as the perpendicular from A to $\overline{BC}$.

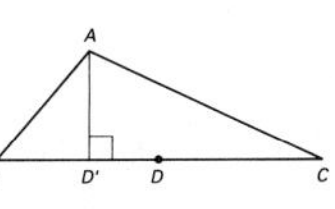

3. Make a drawing for this "proof." Then, find the error.

Prove: Any point in the interior of an angle is on the bisector of that angle.

1. On the sides of ∠A, choose points B and C so that $\overline{AB} \cong \overline{AC}$.
2. Choose any point D in the interior of the angle. Draw the ray which bisects ∠A and contains point D.
3. Draw $\overline{DC}$ and $\overline{DB}$.
4. △ADC ≅ △ADB by the SAS Postulate.
5. Therefore, $\overline{DB} \cong \overline{DC}$ and D is equidistant from B and C.
6. Therefore, D is on the bisector of ∠A.
The ray that bisects ∠A does not necessarily contain the point D.

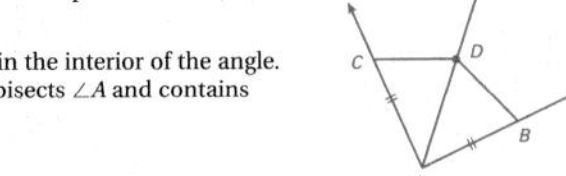

66 ENRICHMENT *LESSON 4-3*

Lesson Planning

NCTM Standards/Strands

- Patterns, Functions, & Algebra
- Geometry & Spatial Sense
- Measurement
- Problem Solving
- Reasoning and Proof
- Communication
- Connections

Vocabulary

altitude
median
concurrent lines
center of gravity
perpendicular bisector

Materials Needed

compass
rulers
scissors
straightedges
geometric drawing software, if available

Lesson Resources

Warm-Up Transparency 11
Reteaching 4-4
Extra Practice 4-4
Enrichment 4-4
Multicultural Connection 2

Assignment Guide

Basic: 1–22, 24–36
Enriched: 1–36

Getting Started

5-Minute Warm-up

What does each statement tell about the relationship of line segments XY and AB?

1. $\overline{XY}$ is perpendicular to $\overline{AB}$. They intersect and form right angles.
2. $\overline{XY}$ bisects $\overline{AB}$. $\overline{XY}$ divides $\overline{AB}$ into two equal parts.

Introduction to Lesson 4-4

Watch for students who say that one or more of the lines bisect a side of the triangle. Some may look as if they do, but unless the triangle is isosceles or equilateral, none will. Have students check by measuring the parts of the side they think is bisected.

4-4 Altitudes, Medians, and Perpendicular Bisectors

Goals ■ Identify and sketch altitudes and medians of a triangle and perpendicular bisectors of sides of a triangle.

Applications Architecture, Physics, Service

Working with a partner, draw a large acute triangle *ABC*, as shown at the right.
1–6 Check students' drawings.

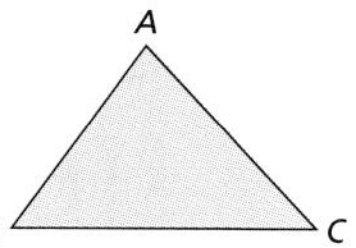

1. With compass tip at point *A*, draw two arcs of equal radii that intersect $\overline{BC}$. Label the points of intersection *X* and *Y*.
2. Choose a suitable radius of the compass. With compass tip first at point *X*, then at point *Y*, draw two arcs that intersect at *Z*.
3. Using a straightedge, draw $\overleftrightarrow{AZ}$.
4. Label point D where $\overleftrightarrow{AZ}$ intersects $\overline{BC}$.
5. Repeat steps **a** through **d**, but this time place the compass tip at point *B* and construct a line that intersects $\overline{AC}$ at point *E*.
6. Repeat steps **a** through **d** again, but now place the compass tip at point *C* and construct a line that intersects $\overline{AB}$ at point *F*.
7. What observations do you make about the lines you constructed?
Answers will vary. Possible responses; Each line is perpendicular to a side of the triangle; the three lines intersect in one point.

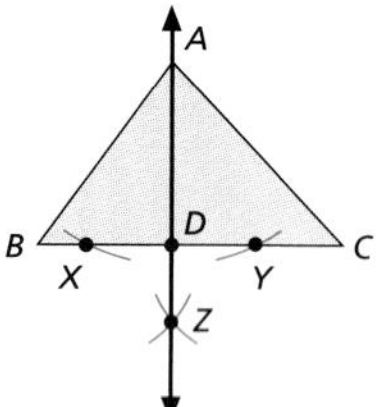

Build Understanding

There are several special segments related to triangles. An **altitude** is a perpendicular segment from a vertex to the line containing the opposite side. A **median** is a segment with endpoints that are a vertex of the triangle and the midpoint of the opposite side.

Example 1

Sketch all the altitudes and medians of $\triangle ABC$:

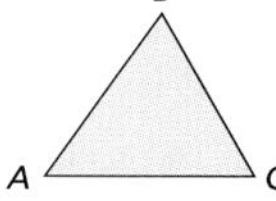

Solution

There are three altitudes, shown below in red.

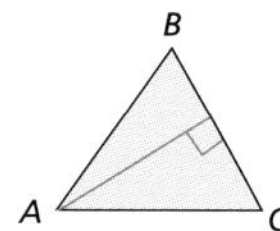

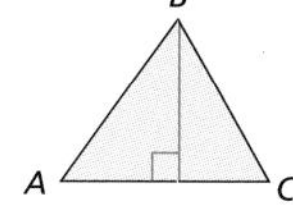

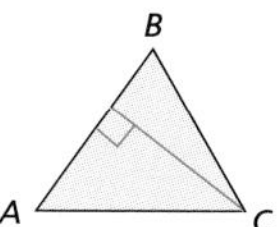

Similarly, there are three medians, shown below in blue.

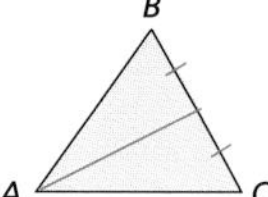

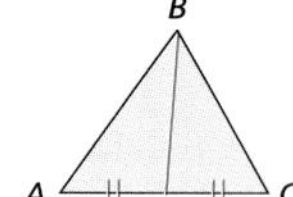

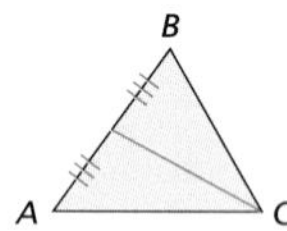

Teaching Strategies

ESL/LEP To help students understand altitude and median, discuss the *altitude* of a person climbing a mountain (distance is measured perpendicular to sea level) and the *median* of a highway (usually an equal number of lanes on either side).

Any line, ray, or segment that is perpendicular to a segment at its midpoint is called a **perpendicular bisector** of the segment. In a given plane, however, there is exactly one *line* perpendicular to a segment at its midpoint. That line usually is called *the* perpendicular bisector of the segment. The following is an important theorem concerning perpendicular bisectors.

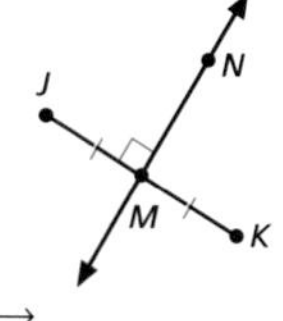

$\overleftrightarrow{MN}$ is the perpendicular bisector of $\overline{JK}$.

The Perpendicular Bisector Theorem	If a point lies on the perpendicular bisector of a segment, then the point is equidistant from the endpoints of the segment.

You will have a chance to prove this theorem in Exercise 22 on page 167.

Example 2

ARCHITECTURE A triangular construction is shown on a set of plans. The architect has determined that $\overline{DF}$ is a perpendicular bisector of $\overline{GE}$. She needs to know whether these statements are true or false.

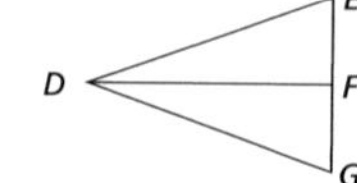

a. $\overline{EF} \cong \overline{GF}$ **b.** $\overline{DE} \cong \overline{DG}$

Solution

a. By the definition of perpendicular bisector, F is the midpoint of $\overline{GE}$. By the definition of midpoint, this means that $EF = GF$, or $\overline{EF} \cong \overline{GF}$. The given statement is *true.*

b. Point D is a point on the perpendicular bisector of $\overline{GE}$. By the perpendicular bisector theorem, this means that point D is equidistant from points G and E. That is, $DE = DG$, or $\overline{DE} \cong \overline{DG}$. The given statement is *true.*

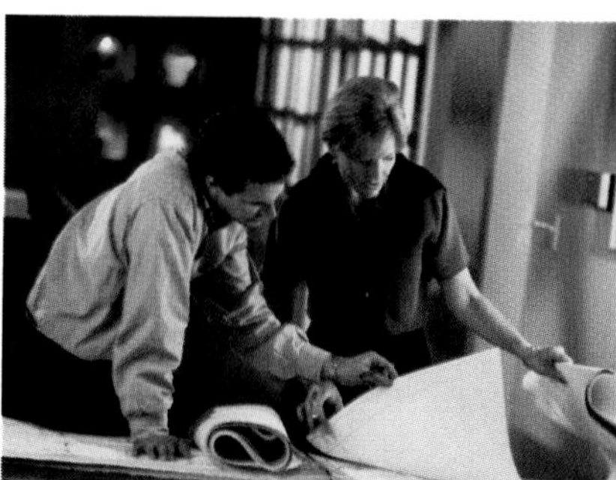

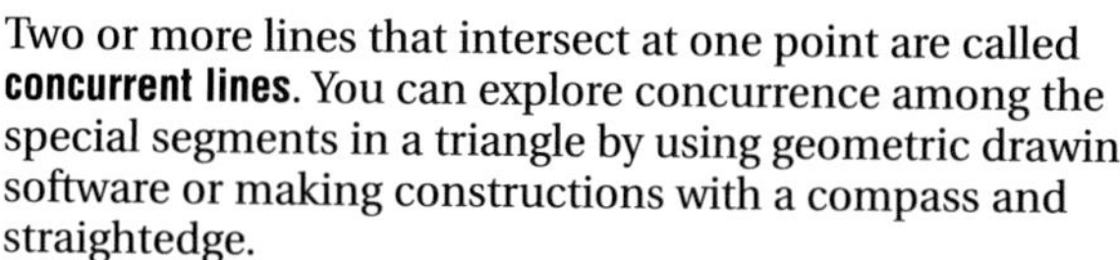

Two or more lines that intersect at one point are called **concurrent lines**. You can explore concurrence among the special segments in a triangle by using geometric drawing software or making constructions with a compass and straightedge.

Example 3

GEOMETRY SOFTWARE Draw a scalene, acute triangle ABC. Locate the midpoints of $\overline{AB}$, $\overline{BC}$ and $\overline{AC}$. Draw the three medians of the triangle. What do you notice?

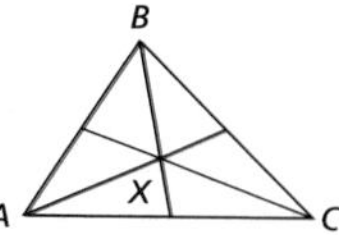

Solution

The medians are concurrent. Label the point of concurrence X.

Chalkboard Examples

Supplementary Example 1

Have students draw a triangle with an obtuse angle. Have them sketch all the altitudes and medians of the triangle. Check students' drawings.

Supplementary Example 2

An architect plans to add a brace to form a perpendicular bisector to a triangular construction. Once the brace is added, will the following statements be true or false. $\overline{BC} \cong \overline{CD}$ true $\overline{AB} \cong \overline{AD}$ true

Supplementary Example 3

Using geometry software, copy the triangle drawn in Example 3. Construct the three altitudes of the figure. Are the altitudes concurrent? yes

Teaching Strategies

COOPERATIVE LEARNING Have students work in pairs. Ask each student to draw a triangle and one or more of its altitudes or medians, label all points, and write whether the triangle is scalene, isosceles, equilateral, and/or right and whether altitudes and medians are shown. Then have students exchange triangles. Based on the information given, ask each student to record any right angles and any congruent parts in the figure.

Reteaching Worksheet 4-4

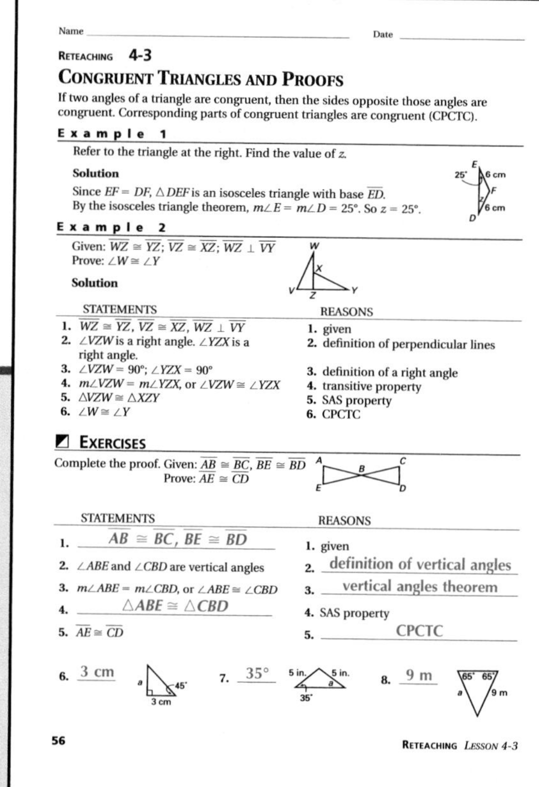

Name ____________ Date ____________

RETEACHING 4-3

CONGRUENT TRIANGLES AND PROOFS

If two angles of a triangle are congruent, then the sides opposite those angles are congruent. Corresponding parts of congruent triangles are congruent (CPCTC).

Example 1

Refer to the triangle at the right. Find the value of z.

Solution

Since $EF = DF$, $\triangle DEF$ is an isosceles triangle with base $\overline{ED}$.
By the isosceles triangle theorem, $m\angle E = m\angle D = 25°$. So $z = 25°$.

Example 2

Given: $\overline{WZ} \cong \overline{YZ}$; $\overline{VZ} \cong \overline{XZ}$; $\overline{WZ} \perp \overline{VY}$
Prove: $\angle W \cong \angle Y$

Solution

STATEMENTS	REASONS
1. $\overline{WZ} \cong \overline{YZ}$, $\overline{VZ} \cong \overline{XZ}$, $\overline{WZ} \perp \overline{VY}$	1. given
2. $\angle VZW$ is a right angle. $\angle YZX$ is a right angle.	2. definition of perpendicular lines
3. $\angle VZW = 90°$; $\angle YZX = 90°$	3. definition of a right angle
4. $m\angle VZW = m\angle YZX$, or $\angle VZW \cong \angle YZX$	4. transitive property
5. $\triangle VZW \cong \triangle XZY$	5. SAS property
6. $\angle W \cong \angle Y$	6. CPCTC

EXERCISES

Complete the proof. Given: $\overline{AB} \cong \overline{BC}$, $\overline{BE} \cong \overline{BD}$
Prove: $\overline{AE} \cong \overline{CD}$

STATEMENTS	REASONS
1. $\overline{AB} \cong \overline{BC}$, $\overline{BE} \cong \overline{BD}$	1. given
2. $\angle ABE$ and $\angle CBD$ are vertical angles	2. definition of vertical angles
3. $m\angle ABE = m\angle CBD$, or $\angle ABE \cong \angle CBD$	3. vertical angles theorem
4. $\triangle ABE \cong \triangle CBD$	4. SAS property
5. $\overline{AE} \cong \overline{CD}$	5. CPCTC

6. 3 cm 7. 35° 8. 9 m

56 RETEACHING *LESSON 4-3*

Lesson 4-4 **Altitudes, Medians, and Perpendicular Bisectors** 165

Lesson Wrap-up

Quick Assessment

Ask the following questions to determine if students understand the content presented in this lesson.

1. How could you use a compass and straightedge to construct the medians of a triangle? For each median, use the compass and straightedge to construct the bisector of one side of the triangle. Use the straightedge to connect the point where the bisector intersects the side to the vertex of the angle opposite that side.
2. To find the area of any triangle, which measure do you need—the measure of the perpendicular bisector of the base, the median extending from the base, or the altitude extending from the base? the altitude

Assignment Guide

Basic: 1–22, 25–36
Advanced: 1–36
Additional Practice: See Extra Practice Index on page 674.

Additional Answers

1.

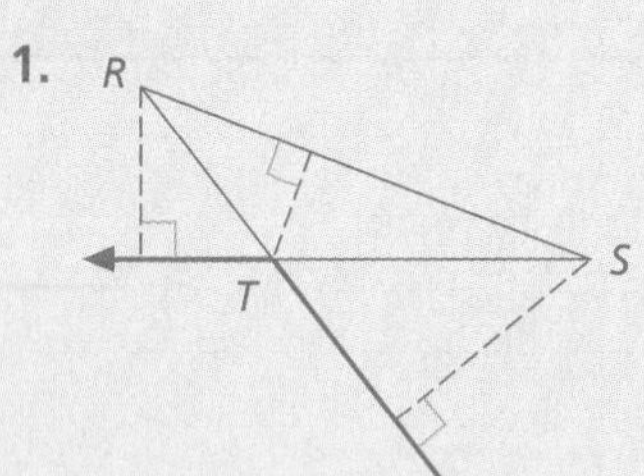

2.

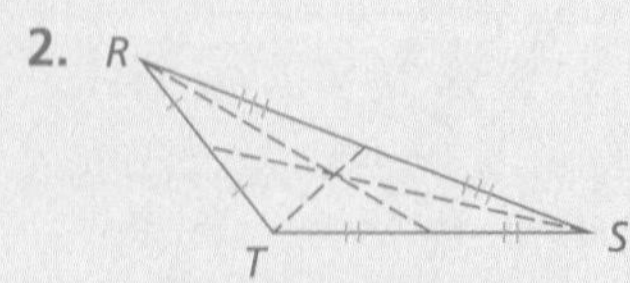

8. In an isosceles triangle, the altitude from the vertex angle to the base is the same segment as the median from the vertex angle to the base. In an equilateral triangle, the altitude from any vertex to the opposite side is the same segment as the median from the same vertex to the opposite side.

Try These Exercises

Trace $\triangle RST$ onto a sheet of paper. See additional answers.

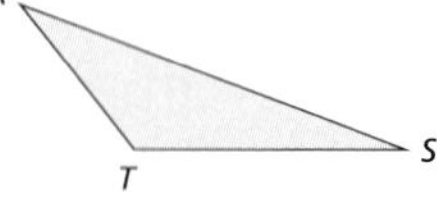

1. Sketch all the altitudes.
2. Sketch all the medians.

In $\triangle XYZ$, $\overline{YW}$ is an altitude. Tell whether each statement is true or false.

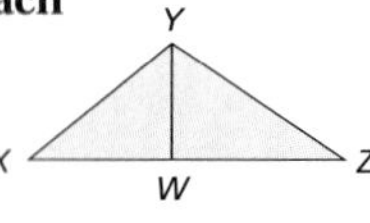

3. $\overline{YW} \perp \overline{XZ}$ true
4. $\overline{XW} \cong \overline{ZW}$ false
5. $\overline{XY} \cong \overline{ZY}$ false
6. $\angle XWY \cong \angle ZWY$ true
7. **GEOMETRY SOFTWARE** Draw a scalene, acute triangle QPR. Construct its three altitudes. What do you observe? The altitudes are concurrent lines.
8. **TALK ABOUT IT** Ezra says that an altitude and a median of a triangle could possibly be the same segment. Do you think Ezra's thinking is correct? Discuss the idea with a partner. See additional answers.

Problem Solving Tip

Notice that an altitude of a triangle is defined as a segment from a vertex *to the line containing the opposite side*. So, for $\triangle RST$, you sketch the altitude from vertex R by first extending side TS as shown below.

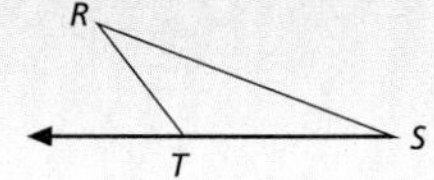

Practice Exercises

Trace $\triangle JKL$, at the right, onto a sheet of paper. See additional answers.

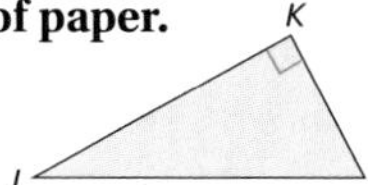

9. Sketch all the altitudes.
10. Sketch all the medians.

Exercises 11–17 refer to $\triangle EFG$, at the right. Tell whether each statement is true or false.

11. $\overline{EG} \cong \overline{EF}$ false
12. $\angle EHG \cong \angle EHF$ false
13. $\overline{GH} \cong \overline{FH}$ true
14. $\angle GEH \cong \angle FEH$ false
15. $\overline{EH}$ is median of $\triangle EFG$. false
16. $\triangle EGH \cong \triangle EFH$ true
17. $\overline{EH}$ is an altitude of $\triangle EFG$. false

GEOMETRIC CONSTRUCTIONS **Draw two copies of a scalene, acute triangle.**

18. Label vertices A, B and C. Bisect $\angle A$, $\angle B$ and $\angle C$. Label the point of concurrence Z. Now measure the perpendicular distance from point Z to each side of the triangle. What do you observe? The three distances are equal.
19. Draw the perpendicular bisectors of $\overline{AB}$, $\overline{BC}$ and $\overline{AC}$. Label the point of concurrence W. Measure the distance from point W to each vertex of the triangle. What do you observe? The three distances are equal.
20. **PHYSICS** The **center of gravity** of an object is the point at which the weight of the object is in perfect balance. Which point of concurrence do you think is the center of gravity of a triangle? Cut a large triangle out of cardboard. Using compass and straightedge, draw medians, altitudes, angle bisectors, and perpendicular bisectors. Place the eraser tip of a pencil at each point of concurrence. When the triangle balances, you have located the center of gravity. See additional answers.

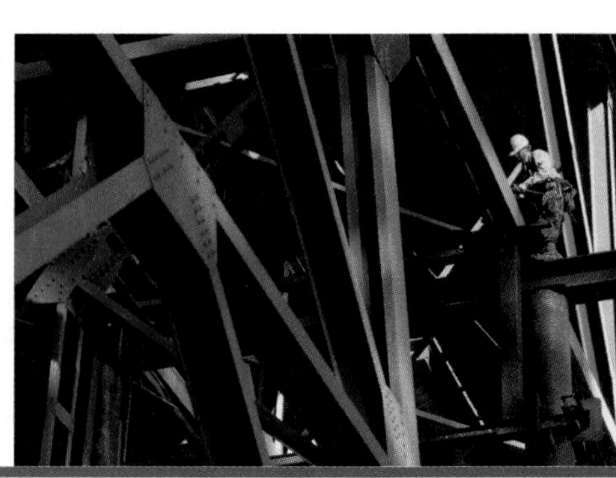

9. and 10.

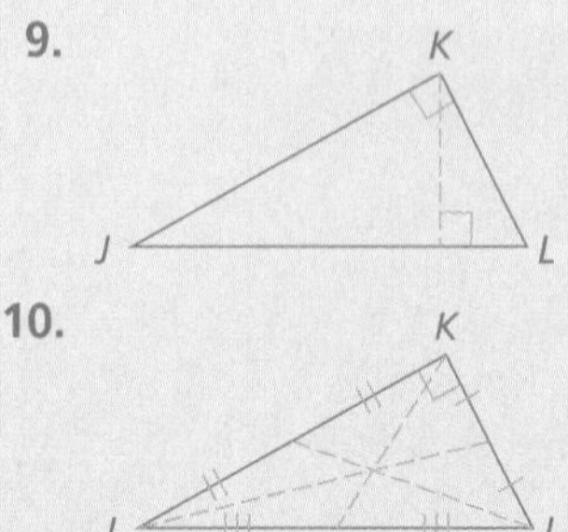

20. The point where the medians intersect is the center of gravity.

21. In a right triangle, the two sides that are perpendicular (the legs) are two altitudes of the triangle. It is never true that a side of a triangle is also a median.

23. Answers will vary. Possible responses: $\triangle PQR$ is isosceles; $QR \cong PR$; $QS \cong PS$; $\angle RQP \cong \angle RPQ$; $\angle QRS \cong \angle PRS$; $\triangle QRS \cong \angle PRS$; $\overline{QT}$ is an altitude of $\triangle PQR$; RS is an altitude of $\triangle PQR$; $\overline{RS}$ is a median of $\triangle PQR$; $\overline{RS}$ is a perpendicular bisector of $\overline{QP}$; $\overline{RS}$ lies on the bisector of $\angle QRP$; $\angle RSQ$, $\angle RSP$, $\angle QTP$, and $\angle QTR$ are right angles.

21. **WRITING MATH** Can the side of a triangle also be an altitude or a median of the triangle? Explain your reasoning. See additional answers.

22. Copy and complete this proof of the perpendicular bisector theorem.

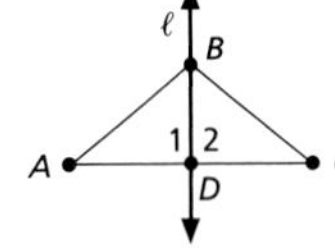

Given Line ℓ is the perpendicular bisector of $\overline{AC}$. Point B lies on ℓ.

Prove $AB = BC$

Statements	Reasons
1. ___?___	1. ___?___ given
2. Point D is midpoint of $\overline{AC}$.	2. definition of ___?___ ⊥ bisector
3. $AD = CD$, or $\overline{AD} \cong \overline{CD}$	3. definition of ___?___ midpoint
4. $\ell \perp \overline{AC}$	4. definition of ___?___ ⊥ bisector
5. $\angle 1$ and $\angle 2$ are right angles.	5. definition of ___?___ ⊥ lines
6. $m\angle 1 = 90°$; $m\angle 2 = 90°$	6. definition of ___?___ right angle
7. $m\angle 1 = m\angle 2$, or $\angle 1 \cong \angle 2$	7. ___?___ transitive property
8. $BD = BD$, or $BD \cong BD$	8. ___?___ reflective property
9. ___?___ $\triangle ABD \cong \triangle CBD$	9. SAS postulate
10. $\overline{AB} \cong \overline{BC}$, or $AB = BC$	10. ___?___ CPCTC

Statement 1: line ℓ is the perpendicular bisector of $\overline{AC}$; Point B lies on ℓ.

Extended Practice Exercises

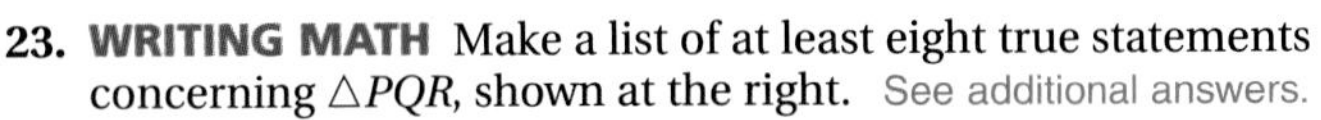

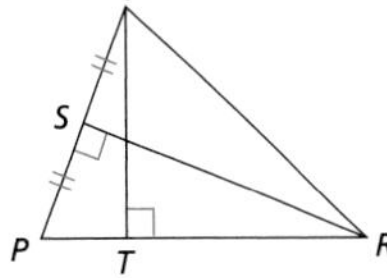

23. **WRITING MATH** Make a list of at least eight true statements concerning $\triangle PQR$, shown at the right. See additional answers.

24. **CHAPTER INVESTIGATION** Imagine your bridge design is to be given to a construction crew. Provide information that will help the crew build the truss accurately. Indicate which line segments are parallel, label angle measures, mark right angles, and classify triangles.

Mixed Review Exercises

Exercises 25–28 refer to the figure at the right. (Lesson 4-4)

25. Name the midpoint of $\overline{AG}$. D

26. Name the segment whose midpoint is J. $\overline{IK}$

27. Name all the segments whose midpoint is E. $\overline{DF}$, $\overline{CG}$, $\overline{BH}$, $\overline{AI}$

28. Assume that L is the midpoint of $\overline{BI}$. What is its coordinate? −0.5

Given $f(x) = 3(x - 2)$, evaluate each function. (Lesson 3-3)

29. $f(-3)$ −15
30. $f(2)$ 0
31. $f(-5)$ −21
32. $f(8)$ 18

Given $f(x) = -2(x - 3)$, evaluate each function. (Lesson 2-2)

33. $f(3)$ 0
34. $f(-4)$ 14
35. $f(-2)$ 10
36. $f(9)$ −12

Extra Practice Worksheet 4-4

Name ______ Date ______

EXTRA PRACTICE 4-4

Altitudes, Medians, and Perpendicular Bisectors

Exercises

Trace ΔEFG, at the right, onto a sheet of paper. Check students' drawings.

1. Sketch all the altitudes.
2. Sketch all the medians.

Exercises 3–9 refer to ΔACD, at the right If you know that $\overline{DB}$ is the perpendicular bisector of $\overline{AC}$, tell whether each statement is true or false.

3. $\overline{AD} \cong \overline{CD}$ true
4. $\angle A \cong \angle C$ true
5. $\angle ABD \cong \angle CBD$ true
6. $\overline{AB} \cong \overline{CB}$ true
7. $\overline{DB}$ is an altitude of ΔACD. true
8. $\overline{AC}$ is a median of ΔACD. true
9. $\Delta ABD \cong \Delta CBD$ true

Exercises 10–18 refer to ΔLNP, at the right. If you know that M is the midpoint of $\overline{LN}$ tell whether each statement is true or false.

10. $\overline{LM} \cong \overline{NM}$ true
11. $\angle L \cong \angle N$ false
12. $\angle LMP \cong \angle NMP$ false
13. $\overline{LP} \cong \overline{NP}$ false
14. $\overline{PM} \perp \overline{LN}$ false
15. $\angle MLP \cong \angle MNP$ false
16. $\overline{MP}$ is an altitude of ΔLNP. false
17. $\overline{MP}$ is a median of ΔLNP. true
18. $\Delta LMP \cong \Delta NMP$ false

52 EXTRA PRACTICE *LESSON 4-4*

Enrichment Worksheet 4-4

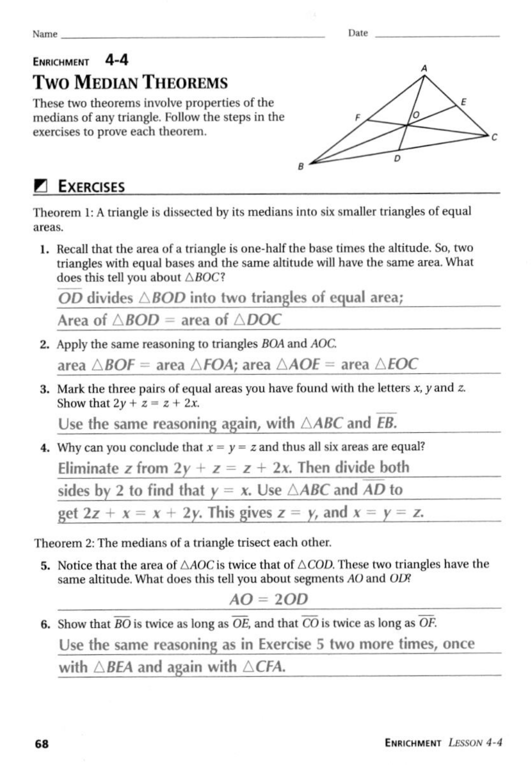

Name ______ Date ______

ENRICHMENT 4-4

Two Median Theorems

These two theorems involve properties of the medians of any triangle. Follow the steps in the exercises to prove each theorem.

Exercises

Theorem 1: A triangle is dissected by its medians into six smaller triangles of equal areas.

1. Recall that the area of a triangle is one-half the base times the altitude. So, two triangles with equal bases and the same altitude will have the same area. What does this tell you about $\triangle BOC$?
   $\overline{OD}$ divides $\triangle BOD$ into two triangles of equal area; Area of $\triangle BOD$ = area of $\triangle DOC$
2. Apply the same reasoning to triangles BOA and AOC.
   area $\triangle BOF$ = area $\triangle FOA$; area $\triangle AOE$ = area $\triangle EOC$
3. Mark the three pairs of equal areas you have found with the letters x, y and z. Show that $2y + z = z + 2x$.
   Use the same reasoning again, with $\triangle ABC$ and $\overline{EB}$.
4. Why can you conclude that $x = y = z$ and thus all six areas are equal?
   Eliminate z from $2y + z = z + 2x$. Then divide both sides by 2 to find that $y = x$. Use $\triangle ABC$ and $\overline{AD}$ to get $2z + x = x + 2y$. This gives $z = y$, and $x = y = z$.

Theorem 2: The medians of a triangle trisect each other.

5. Notice that the area of $\triangle AOC$ is twice that of $\triangle COD$. These two triangles have the same altitude. What does this tell you about segments AO and OD?
   $AO = 2OD$
6. Show that $\overline{BO}$ is twice as long as $\overline{OE}$, and that $\overline{CO}$ is twice as long as $\overline{OF}$.
   Use the same reasoning as in Exercise 5 two more times, once with $\triangle BEA$ and again with $\triangle CFA$.

68 ENRICHMENT *LESSON 4-4*

Skills Practice

Vocabulary Review

Lesson 4-3
base
base angles
legs
vertex angle
corollary
isosceles triangle theorem

Lesson 4-4
altitude
median
perpendicular bisector
concurrent lines
center of gravity

Assignment Guide

All students: 1–26
Additional Practice: Refer to the Extra Practice Index on page 674 of this text.

Review and Practice Your Skills

Practice ■ Lesson 4-3

Find the value of *n* in each figure.

1. 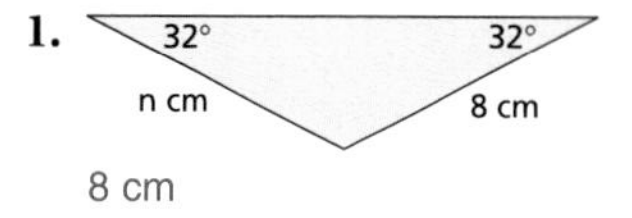

8 cm

2. 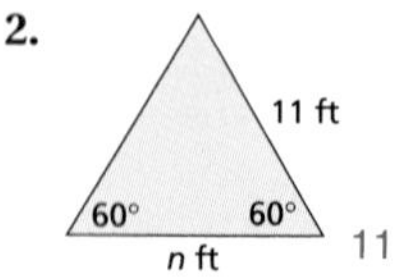

11 ft

3. 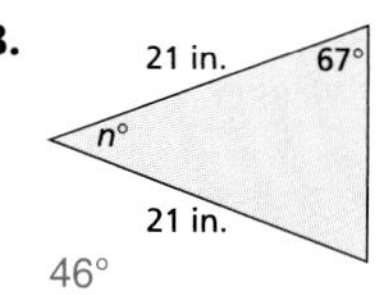

46°

Name all pairs of congruent angles in each figure.

4. 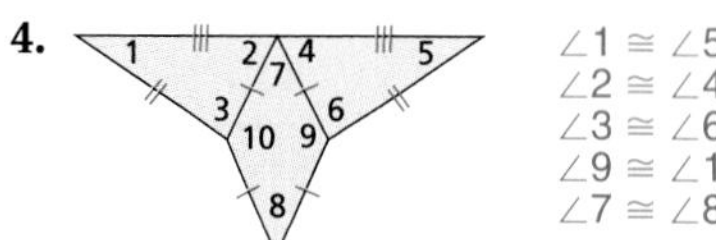

$\angle 1 \cong \angle 5$
$\angle 2 \cong \angle 4$
$\angle 3 \cong \angle 6$
$\angle 9 \cong \angle 10$
$\angle 7 \cong \angle 8$

5. 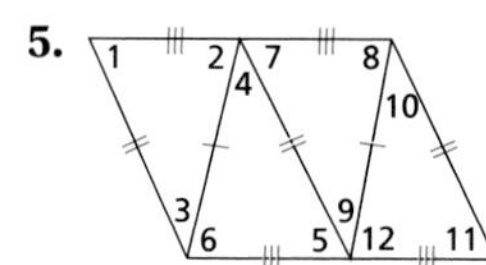

$\angle 3 \cong \angle 4$ $\angle 2 \cong \angle 6$ $\angle 1 \cong \angle 5$
$\angle 3 \cong \angle 9$ $\angle 2 \cong \angle 8$ $\angle 1 \cong \angle 7$
$\angle 3 \cong \angle 10$ $\angle 2 \cong \angle 12$ $\angle 1 \cong \angle 1$
$\angle 4 \cong \angle 9$ $\angle 6 \cong \angle 8$ $\angle 5 \cong \angle 7$
$\angle 4 \cong \angle 10$ $\angle 6 \cong \angle 12$ $\angle 5 \cong \angle 1$
$\angle 4 \cong \angle 10$ $\angle 8 \cong \angle 12$ $\angle 7 \cong \angle 1$

Determine whether each statement is *true* or *false*.

6. All equiangular triangles are equilateral. true

7. If two sides and one angle in a triangle are congruent to two sides and one angle in another triangle, then the triangles are congruent. false (must be "included angle")

8. The Symmetric Property applies to both congruent sides and congruent angles. true

9. If two triangles share a common side, then they are congruent. false

10. Given $\triangle ABC \cong \triangle DEF$, it can be shown that $\angle B \cong \angle F$ and $\overline{AB} \cong \overline{DE}$. false ($\angle R \cong \angle E$)

Practice ■ Lesson 4-4

Trace each triangle onto a sheet of paper. Sketch all the altitudes and all the medians.

11. 

12. 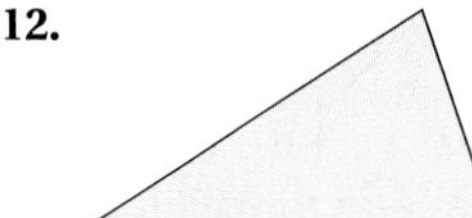

13. 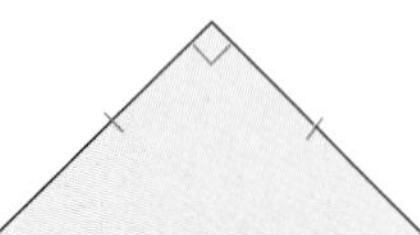

For Exercises 14–19, refer to $\triangle DEG$ at the right with altitude $\overline{DF}$. Tell whether each statement is *true* or *false*.

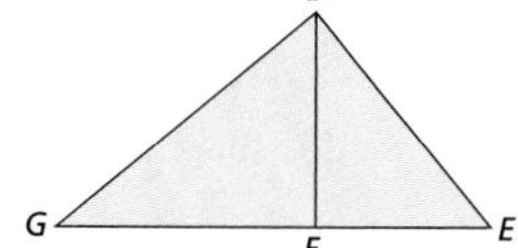

14. $\angle E \cong \angle G$ false **15.** $\angle DFE \cong \angle GFD$ true **16.** $m\angle EFD = 90°$ true

17. $\overline{GF} \cong \overline{EF}$ false **18.** $\angle FDG \cong \angle FDE$ false **19.** $\overline{DF} \perp \overline{GE}$ true

20. Copy and complete this proof.

Given $\overline{JK} \cong \overline{ML}$; $\overline{KM} \cong \overline{LJ}$

Prove __?__ $\cong \triangle LMJ$ $\triangle KJM$

Statements	Reasons
1. __?__ $\overline{JK} \cong \overline{ML}$; $\overline{KM} \cong \overline{LJ}$	1. Given
2. $\overline{JM} \cong \overline{JM}$	2. __?__ Reflexive Property
3. __?__ $\cong \triangle LMJ$	3. SSS Postulate

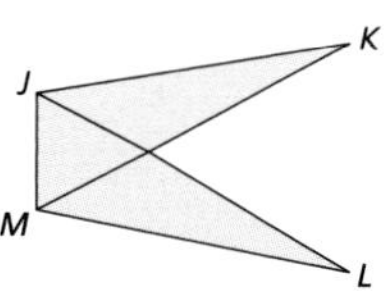

Extend the Lesson

MATH JOURNAL Have students name ways in which altitudes and medians are the same and ways in this they are different. Same: Both connect the line containing a side of a triangle to the opposite angle. Different: Altitudes are always perpendicular to the line containing the particular side, while medians are sometimes perpendicular to it; all medians of any triangle intersect at a point inside the triangle, while all altitudes of a triangle may or may not intersect inside the triangle.

PRACTICE ■ LESSON 4-1–LESSON 4-4

Find the value of x in each figure. (Lesson 4-1)

21. 38° $x°$ 71°

22.

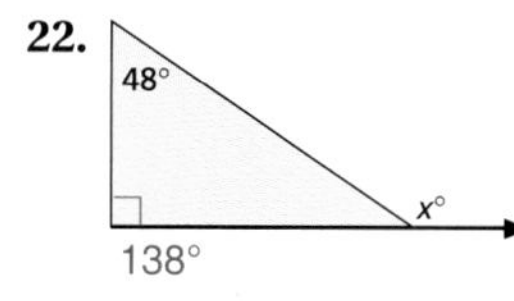

138°

23.

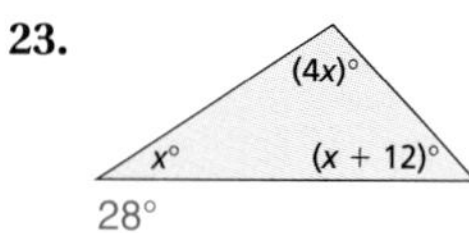

28°

Find m∠XYZ in each figure. (Lesson 4-2)

24.

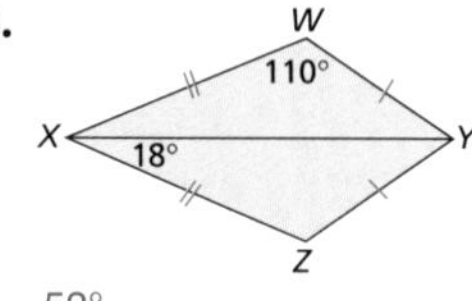

52°

25.

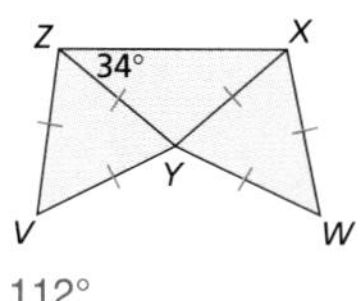

112°

26.

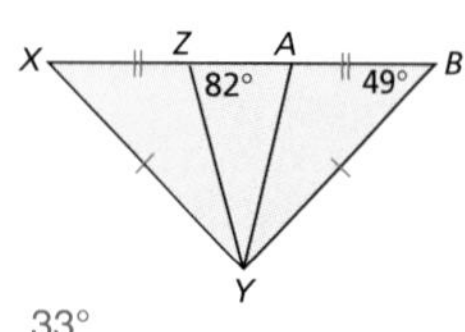

33°

Mid-Chapter Quiz

Find the unknown interior angles of each triangle.

1. an isosceles triangle with an interior angle of 100° 40°, 40°
2. a triangle with interior angles of $x°$, $(2x + 10)°$, and $(2x - 5)°$ 35°, 80°, 65°

Sketch each pair of triangles and state either that they are congruent or that no conclusion is possible. If they are congruent, name the postulate that could be used to prove the congruence.

3. Triangles *ABD* and *CBD* share side *BD*. Side *BD* is the perpendicular bisector of side *AC*. congruent; SAS
4. Triangles *EFH* and *GFH* share side *FH*. Sides *EF* and *GF* are congruent. Angles *E* and *G* are congruent. no conclusion possible
5. Triangles *RSU* and *TSU* share side *SU*. Side *SU* bisects angles *RST* and *RUT*. congruent; ASA

Write True or False

6. Two isosceles triangles with congruent vertex angles always have congruent base angles. true
7. If two sides of a triangle are congruent, then the base angles and the vertex angle must be congruent. false

Decide whether each statement is *always, sometimes,* or *never* true.

8. An altitude divides the side of a triangle into two congruent parts. sometimes
9. A median is perpendicular to the side of a triangle. sometimes
10. An altitude is a line inside a triangle. sometimes

Learning Styles

TACTILE/KINESTHETIC LEARNERS Provide students with a variety of triangles, each showing a median, altitude or a line inside the triangle that is neither a median nor an altitude. Have students fold each triangle along the interior line and/or use a ruler and protractor to help determine whether that line is a median, altitude or neither.

Chalkboard Examples

Example from Lesson 4-3

Find the measures of the other angles of these isosceles triangles.

1. vertex angle: 74° 53°, 53°
2. base angle: 65° 65°, 50°
3. right angle: 90° 45°, 45°

Example from Lesson 4-4

Sketch and label each figure.

1. an acute triangle ABC with median *AD*. Figure should include a line from angle A that bisects *BC*.
2. an acute scalene triangle PQR with altitude *RS*. Figure should include a line from angle R perpendicular to *PQ*.

Mid-Chapter Quiz

Correlation Chart

Question Number	Lesson Number
1, 2, 3	4-1
7–9	4-2
4–6	4-3
10–15	4-4

Chapter 4 **Review and Practice Your Skills** 169

Lesson Planning

NCTM Standards/Strands

- Patterns, Functions, & Algebra
- Geometry & Spatial Sense
- Problem Solving
- Reasoning and Proof
- Connections
- Representation

Vocabulary

indirect proof

Materials Needed

paper/pencil

Lesson Resources

Warm-Up Transparency 11
Reteaching 4-5
Enrichment 4-5

Assignment Guide

Basic: 1–12
Enriched: 1–12

Getting Started

5-Minute Warm-up

Write the contradiction of the following statement: An equilateral triangle cannot have a right angle. **An equilateral triangle can have a right triangle.**

Use the Five-step Plan and the Strategy

THE FIVE-STEP PLAN **Read**—ask questions of the students to help them understand the problem. **Plan**—guide students to related problems and previously mastered skills and strategies. **Solve**—students solve problem on their own. **Answer**—write solution in format that answers the question. **Check**—review work, check for reasonableness and review strategy used.

THE STRATEGY **Guess and check**—Making an assumption and testing it is a useful way of solving a problem. Students often have trouble knowing where to start in solving a problem. When a problem is difficult, guess and check is a good way to get started and a valuable source of information.

4-5 Problem Solving Skills: Write an Indirect Proof

The proofs that you have studied so far in this book have been *direct proofs*. That is, the proofs proceeded logically from a hypothesis and known facts to show that a desired conclusion is true. In this lesson, you will study *indirect proof.* In an **indirect proof**, you begin with the desired conclusion and assume that it is *not* true. You then reason logically until you reach a contradiction of the hypothesis or of a known fact.

Problem Solving Strategies

- ✔ Guess and check
- Look for a pattern
- Solve a simpler problem
- Make a table, chart or list
- Use a picture, diagram or model
- Act it out
- Work backwards
- Eliminate possibilities
- Use an equation or formula

Problem

Prove the following statement.

If a figure is a triangle, then it cannot have two right angles.

Solve the Problem

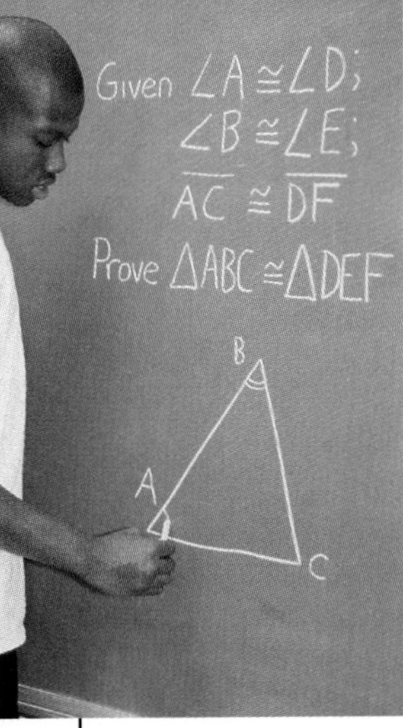

Begin by drawing a representative triangle, such as $\triangle ABC$ at the right.

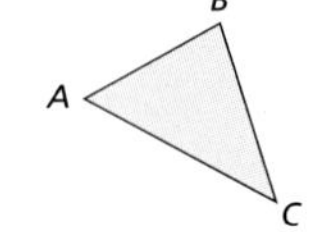

Step 1: Assume that the conclusion is false. That is, assume that a triangle *can* have two right angles. In particular, in $\triangle ABC$, assume that $\angle A$ and $\angle B$ are right angles.

Step 2: Reason logically from the assumption, as follows. By the definition of a right angle, $m\angle A = 90°$ and $m\angle B = 90°$. By the addition property of equality, $m\angle A + m\angle B = 90° + 90° = 180°$. By the protractor postulate, $m\angle C = n°$, where n is a positive number less than or equal to 180. By the addition property of equality, $m\angle A + m\angle B + m\angle C = 180° + n°$. By the triangle-sum theorem, $m\angle A + m\angle B + m\angle C = 180°$.

Step 3: Note that the last two statements in Step 2 are contradictory. Therefore, the assumption that a triangle can have two right angles is false. The given statement must be true.

The solution of the problem above illustrates the following general method for writing an indirect proof.

Writing an Indirect Proof

Step 1 Assume temporarily that the conclusion is false.

Step 2 Reason logically until you arrive at a contradiction of the hypothesis or a contradiction of a known fact (a definition, a postulate, or a previously proved theorem).

Step 3 State that the temporary assumption must be false, and that the given statement must be true.

Additional Answers

5. *Step 1:* Assume that there are two lines through point *P* perpendicular to the given line. In particular, in the figure to the right, assume that $m \perp l$ and $n \perp l$.
Step 2: By definition of right angle, $m\angle 1 = 90°$ and $m\angle 2 = 90°$ By the transitive property of equality, $m\angle 1 = m\angle 2$. By the corresponding angles postulate, since $m\angle 1 = m\angle 2$, it follows that m_n.
By definition of intersecting lines, since *m* and *n* each pass through point *P*, *m* and *n* are intersecting lines.
Step 3: The last two statements in *Step 2* are contradictory.
Therefore, the assumption that there can be two lines perpendicular to a given line through a point outside the line is false. The given statement must be true.

TRY THESE EXERCISES

Five-step Plan
1 Read
2 Plan
3 Solve
4 Answer
5 Check

Suppose you are asked to write an indirect proof of each statement. Write Step 1 of the proof.

1. **ART** If the triangle in a sculpture is a right triangle, then it cannot be an obtuse triangle. Assume that the triangle *can* be an obtuse triangle
2. **ARCHITECTURE** If the triangle in a building design is equilateral, then it is an isosceles triangle. Assume that the triangle is *not* isosceles

PRACTICE EXERCISE

Copy and complete the indirect proof of each theorem.

3. Theorem: *If two lines intersect, then they intersect in one point.*

   Step 1: Assume that two lines can intersect in __?__ points. In particular, in the figure at the right, assume that there are lines __?__ and __?__ that intersect at points __?__ and __?__. two; *r*; *s*; *x*; *y*

   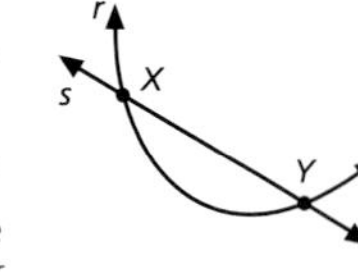

   Step 2: By the unique line postulate (postulate 1), there is exactly __?__ line through points *X* and *Y*. one

   Step 3: The statements in Step 1 and Step 2 are __?__. Therefore, the assumption that two lines can intersect in two points is __?__. The given statement must be __?__. contradictory; false; true

4. Theorem: *Through a point outside a line, there is exactly one line parallel to the given line.*

   Step 1: Assume that there are __?__ lines parallel to the given line. In particular, in the figure at the right, assume that, through point *P*, __?__ ∥ ℓ and __?__ ∥ ℓ. two; *m*; *n*

   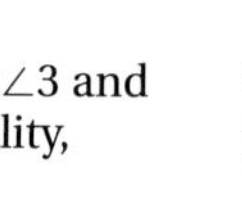

   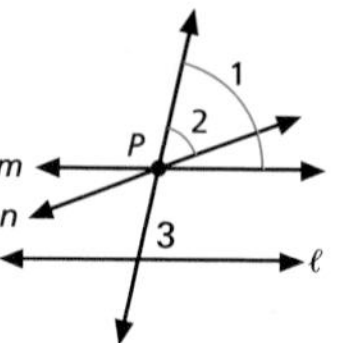

   Step 2: By the parallel lines postulate, m∠ __?__ = m∠3 and m∠ __?__ = m∠3. By the transitive property of equality, m∠ __?__ = m∠ __?__. 1; 2

   However, because *m* and *n* are different lines, m∠1 ≠ m∠2.

   Step 3: The last two statements in Step 2 are __?__. Therefore, the assumption that there can be two lines parallel to a given line through a point outside the line is __?__. The given statement must be __?__. contradictory; false; true

5. Write an indirect proof of the following theorem. *Through a point outside a line, there is exactly one line perpendicular to the given line.* (*Hint:* Use the proof in Exercise 4 above as a model.) See additional answers.

6. **WRITING MATH** Write what you would do to prove indirectly that a triangle cannot have two obtuse angles. See additional answers.

MIXED REVIEW EXERCISES

Simplify each expression. (Lesson 1-4)

7. $-3 + 4 - (-6) + (-2)$ 5
8. $-9 + (-4) - (-3) + 8$ −2
9. $4 - (-6) + 2 + (-3)$ 9
10. $(4) + 9 - 3 - (-2)$ 12
11. $8 - (-3) + 2 - 3 + 9$ 19
12. $2 - (-8) - (-(-12))$ −2

Chalkboard Examples

Supplementary Problem

Prove:
If two angles are supplementary to the same angle, then they are congruent. Assume: If two angles are supplementary to the same angle, then they are not congruent. Reason: ∠1 is supplementary to ∠2. ∠3 is supplementary to ∠2. By the definition of supplementary, $m\angle 1 + m\angle 2 = 180°$, and $m\angle 3 + m\angle 2 = 180°$. Using the transitive property of equality, $m\angle 1 = m\angle 3$. By the definition of congruence, $\angle 1 \cong \angle 3$, which contradicts the assumption. Therefore, the given statement is true.

Lesson Wrap-up

QUICK ASSESSMENT

Have students write what they would do in each of three steps to prove indirectly that if a triangle is isosceles, then it cannot be scalene. Students should go through the three steps modeled on page 170.

Enrichment Worksheet 4-5

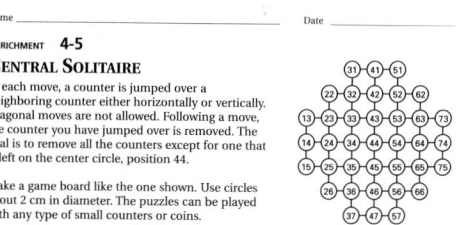

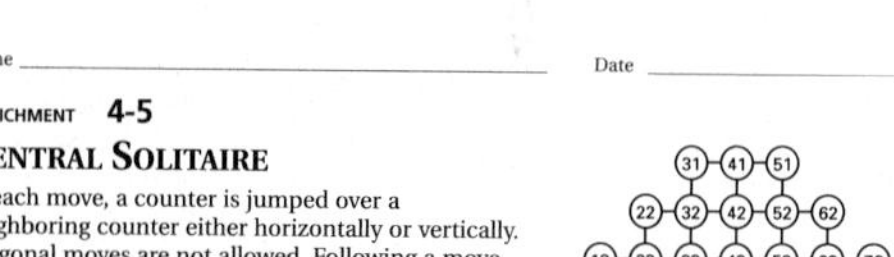

Name ________________ Date ________

ENRICHMENT 4-5

CENTRAL SOLITAIRE

In each move, a counter is jumped over a neighboring counter either horizontally or vertically. Diagonal moves are not allowed. Following a move, the counter you have jumped over is removed. The goal is to remove all the counters except for one that is left on the center circle, position 44.

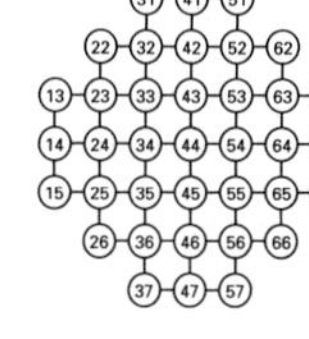

Make a game board like the one shown. Use circles about 2 cm in diameter. The puzzles can be played with any type of small counters or coins.

Example

Start with counters in positions 33, 53, 42, 43, 44 and 45.

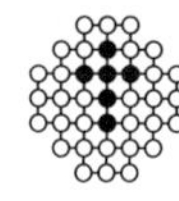

Solution

Each move is described by using the numbers of the circles occupied before and after a move. A solution to this puzzle is 43–23, 45–43, 53–33, 23–43, 42–44.

EXERCISES

Here are three more puzzles for you to try. In each case, the last counter must be left on circle 44. Try to find solutions with the fewest number of moves.

Puzzle 1

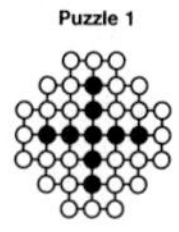

Puzzle 2

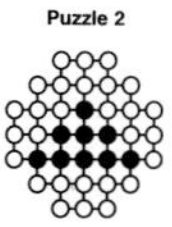

Puzzle 3

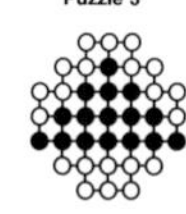

Solutions

1. 45–47, 43–45, 64–44–46, 24–44, 47–45–43, 42–44
2. 55–53–33, 35–55, 65–45, 44–46, 33–35, 25–45, 46–44
3. 34–36, 15–35, 54–56, 75–55, 36–34–54–52, 42–44, 56–54–34, 24–44, 45–43, 33–53, 52–54, 64–44 or 34–14, 43–23, 44–46, 54–56–36–34, 75–55, 15–13–33–35, 25–45–65–63–43, 42–44

70 ENRICHMENT *LESSON 4-5*

Enrichment Worksheet 4-5 not shown.

6. Begin by drawing a representative triangle such as △*ABC* at the right.
Step 1: Assume that a triangle can have two obtuse angles. In particular, in △*ABC*, assume that ∠*A* and ∠*B* are obtuse angles.
Step 2: Reason logically from the assumption, as follows.
By the definition of an obtuse angle, $m\angle A > 90°$ and $m\angle B > 90°$.
A property of inequality states that, if $a > b$ and $c > d$, then $a + c > b + d$. So, $m\angle A + m\angle B > 90° + 90°$, or $m\angle A + m\angle B > 180°$.
By the protractor postulate, $m\angle C = n°$, where *n* is a positive number less than or equal to 180.
By the addition property of inequality, $m\angle A + m\angle B + m\angle C > 180° + n$.
By the triangle sum theorem, $m\angle A + m\angle B + m\angle C = 180°$.
Step 3: The last two statements in *Step 2* are contradictory. Therefore, the assumption that a triangle can have two obtuse angles is false. The given statement must be true.

Lesson Planning

NCTM Standards/Strands

- Number & Operation
- Patterns, Functions, & Algebra
- Geometry & Spatial Sense
- Measurement
- Problem Solving
- Reasoning and Proof
- Communication
- Connections
- Representation

Vocabulary

shortest path postulate
triangle inequality theorem
unequal sides theorem
unequal angles theorem

Materials Needed

paper/pencil

Lesson Resources

Warm-Up Transparency 11
Reteaching 4-6
Extra Practice 4-6
Enrichment 4-6

Assignment Guide

Basic: 1–28, 35–42
Enriched: 1–42

Getting Started

5-Minute Warm-Up

In triangle *DEF*, name the angle opposite each side.

1. $\overline{DE}$ $\angle F$
2. $\overline{EF}$ $\angle D$
3. $\overline{DF}$ $\angle E$

Introduction to Lesson 4-6

When students have completed the activity, have them make conjectures about the shortest path between two points and the longest path between two points. Possible answer: The shortest path is a straight line; no longest path exists, because no matter how long a path is drawn, a longer path could still be drawn.

4-6 Inequalities in Triangles

Goals ■ Understand relationships among sides and angles of a triangle.

Applications Construction, Art, Architecture

Work in groups of two or three students.

The figure at the right shows four paths that ants took from point A to point B.

For 1–3, see additional answers.

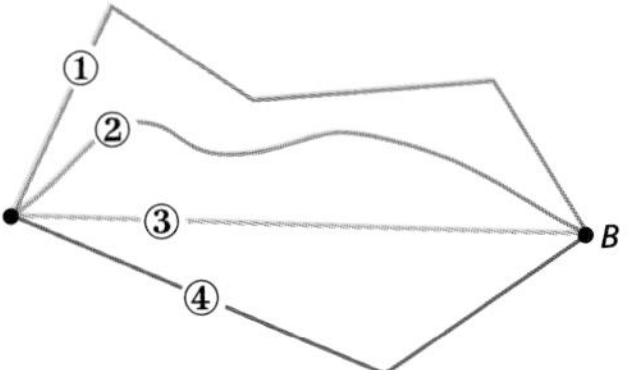

1. Using a centimeter ruler, find the length of each path. (You will need to use some ingenuity to measure path ②!)
2. Trace points A and B onto a sheet of paper. Can you draw a path from point A to point B that is *longer* than any of the given paths? Use the ruler to find the length of your path.
3. Can you draw a path from point A to point B that is *shorter* than any of the given paths? Use the ruler to find the length of your path.

Build Understanding

In the activity above, you had an opportunity to investigate yet another fundamental postulate of geometry.

Postulate 14	**The Shortest Path Postulate** The length of the segment connecting two points is shorter than the length of any other path connecting the points.

The shortest path postulate leads to some important conclusions about triangles. As an example, consider the following proof.

Reading Math

Just as the symbol $\neq$ means *is not equal to*, the symbol $\ngtr$ means *is not greater than*. What do you think the symbol $\nless$ means? is not less than

Given $\triangle ABC$

Prove $AB + BC > AC$

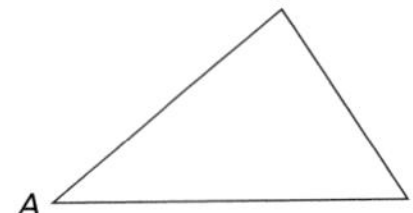

Proof

Assume that $AB + BC \ngtr AC$. Then by the property of comparison, one of these two statements must be true:
$AB + BC = AC$ or $AB + BC < AC$.

If $AB + BC = AC$, then there is a path connecting points A and C that is equal in length to $\overline{AC}$; this contradicts the shortest path postulate.

Similarly, if $AB + BC < AC$, then there is a path connecting points A and C that is shorter than $\overline{AC}$; this also contradicts the shortest path postulate.

Therefore, the assumption $AB + BC \ngtr AC$ must be false. It follows that the desired conclusion, $AB + BC > AC$, is true.

Teaching Strategies

COOPERATIVE LEARNING Have students work in pairs. Tell them that two sides of a triangle are 5 cm and 8 cm. Have them discuss whether these two sides could be the shorter sides, the longer sides, and /or the shortest and longest sides. For each case, have them predict the appearance of the triangle. Finally, have them construct models using straws to check their predictions.

In Exercises 15 and 16 on page 149, you will prove that $AB + AC > BC$ and $AC + BC > AB$ are true statements also. So, you will have completed the proof of the following theorem.

The Triangle Inequality Theorem	The sum of the lengths of any two sides of a triangle is greater than the length of the third side.

Example 1

CONSTRUCTION A frame must be built to pour a triangular cement slab to complete a walkway. The lengths of two sides of the triangle are 5 ft and 9 ft. Find the range of possible lengths for the third side.

Solution

Use the variable n to represent the length in feet of the third side. By the triangle inequality theorem, these three inequalities must be true.

I. $5 + 9 > n$ **II.** $5 + n > 9$ **III.** $9 + n > 5$

$14 > n$ $n > 4$ $n > -4$

Inequality **III** is not useful, since a length must be a positive number.

From inequalities **I** and **II**, you obtain the combined inequality $14 > n > 4$.

So, the length of the third side must be less than 14 ft and greater than 4 ft.

The following two theorems also involve inequalities in triangles. In this book, we will accept these theorems as true without proof.

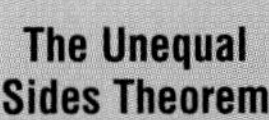

The Unequal Sides Theorem	If two sides of a triangle are unequal in length, then the angles opposite those sides are unequal in measure, in the same order.
The Unequal Angles Theorem	If two angles of a triangle are unequal in measure, then the sides opposite those angles are unequal in length, in the same order.

Example 2

In $\triangle KLM$, $KL = 8$ in, $LM = 10$ in, and $KM = 7$ in. List the angles of the triangle *in order* from largest to smallest.

Solution

Draw and label $\triangle KLM$, as shown at the right.

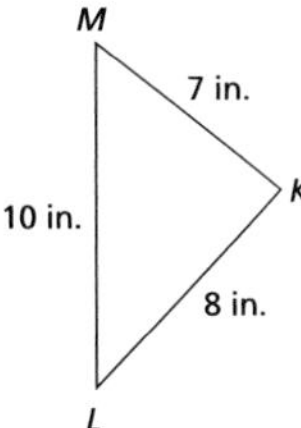

The angle opposite $\overline{LM}$ is $\angle K$.
The angle opposite $\overline{KL}$ is $\angle M$.

Since $10 > 8$, $LM > KL$.

So, by the unequal sides theorem, $m\angle K > m\angle M$.
By similar logic, $m\angle M > m\angle L$.
So, from largest to smallest, the angles are $\angle K$, $\angle M$, and $\angle L$.

Additional Answers

1. path ①: 7 cm; path ②: about 5.7 cm; path ③: 5 cm; path ④: 5.5 cm
2. Answers will vary.
3. There is no path that is shorter than 5 cm.

Chalkboard Examples

Supplementary Example 1

The lengths of two sides of a triangular flower garden are 7 cm and 15 cm. Find the range of possible lengths for the third side.
$22 > n > 8$

Supplementary Example 2

In $\triangle ABC$, $AB = 9$ cm, $BC = 16$ cm and $AC = 24$ cm. List the angles of the triangle in order from largest to smallest. $\angle B$, $\angle A$, $\angle C$

Reteaching Worksheet 4-6

Name ______________________ Date ____________

RETEACHING 4-6

INEQUALITIES IN TRIANGLES

If two sides of a triangle are unequal in length, then the angle opposite the longer side has the greater measure. If two angles of a triangle are unequal in measure, then the side opposite the angle with the greater measure has the longer length.

Example 1

Two sides of a triangle measure $6\frac{1}{2}$ in. and $3\frac{1}{2}$ in. Find the range of lengths for the third side.

Solution

Use the variable l to represent the length in inches of the third side. By the triangle inequality theorem, these three inequalities must be true.

a. $6\frac{1}{2} + 3\frac{1}{2} > l$, $10 > l$ **b.** $6\frac{1}{2} + l > 3\frac{1}{2}$, $l > -3$ **c.** $3\frac{1}{2} + l > 6\frac{1}{2}$, $l > 3$

Inequality b is not useful since a length must be a positive number. Combining inequalities a and c results in the inequality $10 > l > 3$, so the third side must be less than 10 in. and greater than 3 in.

Example 2

Refer to $\triangle DEF$. List the sides from longest to shortest.

E, 85°, D, 65°, 30°, F

Solution

List the angles from the largest to smallest:
$m\angle E > m\angle D > m\angle F$. By the unequal angles theorem, opposite sides of unequal angles will be unequal in length in the same order.
So, $DF > EF > DE$.

EXERCISES

The lengths of two sides of a triangle are given. Find the range of lengths for the third side.

1. 3 mm, 5 mm $2 \text{ mm} < x < 8\text{mm}$
2. 1.2 in., 4.6 in. $3.4 \text{ in.} < x < 5.8 \text{ in.}$
3. 2.5 cm, 8 cm $5.5 \text{ cm} < x < 10.5 \text{ cm}$
4. In $\triangle GHI$, $GH = 3$ cm, $HI = 4$ cm, and $GI = 3.5$ cm. List the angles of the triangle in order from smallest to largest. $\angle I, \angle H, \angle G$
5. In $\triangle MNO$, $m\angle M > m\angle O$ and $m\angle N < m\angle O$. List the sides of the triangle in order from shortest to longest. $\overline{MO}, \overline{MN}, \overline{NO}$

Which side is the longest side of each triangle? Which side is the shortest?

6. (Q, 78°, 29°, P, R) Longest $\overline{PR}$ Shortest $\overline{QR}$
7. (S, 40°, U, T) Longest $\overline{UT}$ Shortest $\overline{ST}$

62 RETEACHING *LESSON 4-6*

Lesson Wrap-up

Quick Assessment

Ask the following questions to determine if students understand the content presented in this lesson.

1. In any triangle, how are the lengths of its sides related to each other? The sum of the lengths of any two sides is greater than the length of the third side.
2. In any triangle with different length sides, how are the lengths of its sides, related to its angles? If the sides have different measures, the angle opposite the longest side is the largest angle, and the angle opposite the shortest side is the smallest.

Assignment Guide

Basic: 1–28, 35–42
Advanced: 1–42
Additional Practice: See Extra Practice Index on page 674.

Additional Answers

23. Assume that $AB + AC \ngtr BC$. Then, by the property of comparison, one of these two statements must be true:
$AB + AC = BC$ or $AB + AC < BC$
If $AB + AC = BC$, then there is a path connecting points B and C that is equal in length to $\overline{BC}$; this contradicts the shortest path postulate.
Similarly, if $AB + AC < BC$, then there must be a path connecting points B and C that is shorter than $\overline{BC}$; this also contradicts the shortest path postulate. Therefore, the assumption $AB + AC \ngtr BC$ must be false. It follows that the desired conclusion, $AB + AC < BC$, is true.

Try These Exercises

ART The design for a sculpture has three triangular platforms. The lengths of two sides of each platform are given. Find the range of possible lengths for the third side.

1. 6 ft, 9 ft — between 3 ft and 15 ft.
2. 7 m, 7 m — between 0 ft and 14 ft.
3. 2 yd, 7 ft — between 1 ft and 13 ft.

List the angles of each triangle *in order* from largest to smallest.

4. 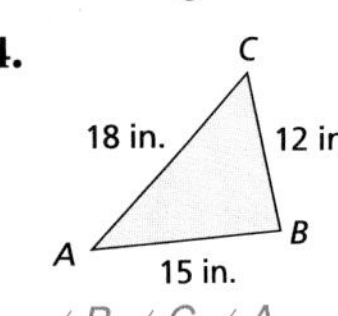

$\angle B, \angle C, \angle A$

5. 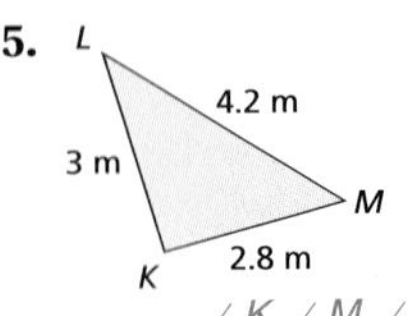

$\angle K, \angle M, \angle L$

6. 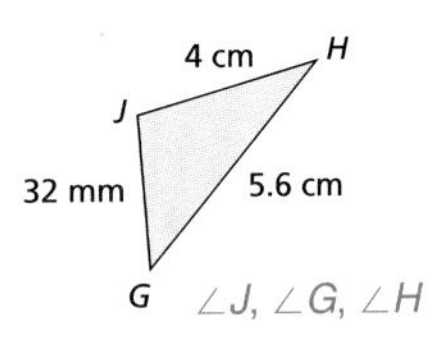

$\angle J, \angle G, \angle H$

7. In $\triangle XYZ$, $m\angle X = 56°$ and $m\angle Z = 19°$. List the sides of the triangle *in order* from shortest to longest. $\overline{XY}, \overline{YZ}, \overline{XZ}$

8. **ARCHITECTURE** The base for an indoor fountain has a triangular shape. On the plans, the base is shown as triangle RST. If $m\angle S > m\angle R$ and $m\angle R > m\angle T$, which is the shortest side of the triangle? $\overline{RS}$

Practice Exercises

Determine if the given measures can be lengths of the sides of a triangle?

9. 7 cm, 2 cm, 6 cm yes
10. 7.3 m, 15 m, 7.3 m no
11. $9\frac{1}{4}$ ft, $3\frac{1}{2}$ ft, $5\frac{3}{4}$ ft no
12. 24 in, 5 ft, 54 in yes
13. 34 yd, 34 yd, 34 yd yes
14. 3 mm, 5 cm, 7 mm no

Which is the longest side of each triangle? the shortest?

15. 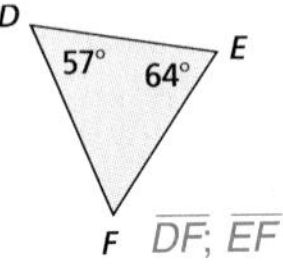

$\overline{DF}$; $\overline{EF}$

16. 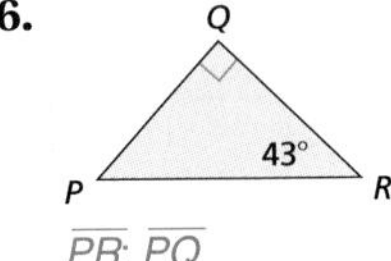

$\overline{PR}$; $\overline{PQ}$

17. 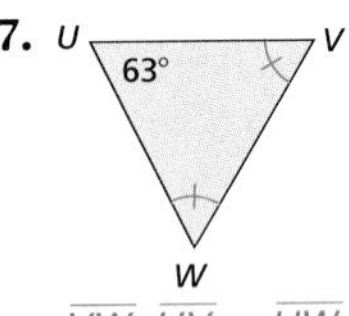

$\overline{VW}$, $\overline{UV}$ or $\overline{UW}$

In each figure, give a range of possible values for *x*.

18. 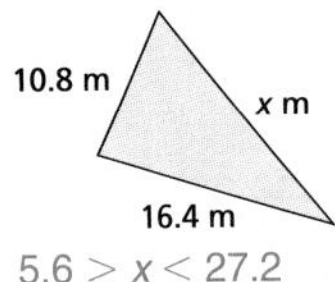

$5.6 > x < 27.2$

19. $5\frac{1}{2}$ ft, $5\frac{1}{2}$ ft, x ft — $0 < x < 11$

20. 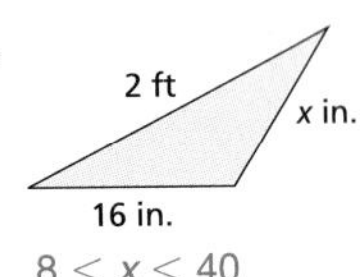

$8 < x < 40$

21. In $\triangle CDE$, $CD < DE$ and $CE < CD$. Which is the largest angle of the triangle? $\angle C$

22. **GEOMETRY SOFTWARE** Use the following information to draw $\triangle QRS$: $QS = 17$, $RS = 23$, and $QR = 20.5$. List the angles of the triangle *in order* from largest to smallest. $\angle Q, \angle S, \angle R$

For Exercises 23 and 24, refer to the proof on page 172. See additional answers.

23. **Given** $\triangle ABC$
Prove $AB + AC > BC$

24. **Given** $\triangle ABC$
Prove $AC + BC > AB$

24. Assume that $AC + BC \ngtr AB$. Then, by the property of comparison, one of these two statements must be true:
$AC + BC = AB$ or $AC + BC < AB$
If $AC + BC = AB$, then there is a path connecting points A and B that is equal in length to $\overline{AB}$; this contradicts the shortest path postulate.
Similarly, if $AC + BC < AB$, then there must be a path connecting points A and B that is shorter than $\overline{AB}$ this also contradicts the shortest path postulate. Therefore, the assumption $AC + BC \ngtr AB$ must be false. It follows that the desired conclusion, $AC + BC < AB$, is true.

29. A right triangle can have only one right angle, and it cannot have an obtuse angle. Therefore, the one right angle is the largest angle. By the unequal angles theorem, the side opposite that angle, the hypotenuse, is the longest side.

List all the segments in each figure *in order* from longest to shortest.

25.

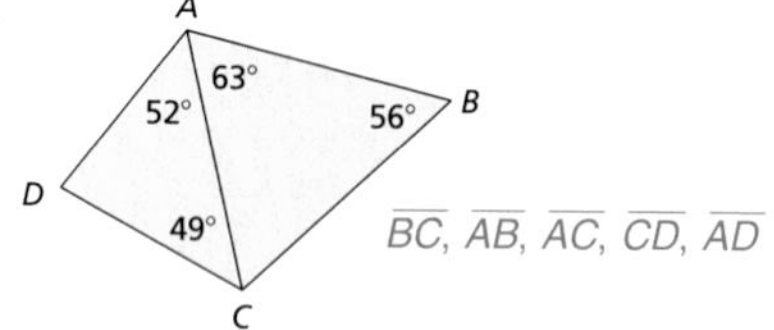

$\overline{BC}, \overline{AB}, \overline{AC}, \overline{CD}, \overline{AD}$

26.

M N O P Q; 64°, 56°, 59°, 64°, 60°, 57°

$\overline{MQ}, \overline{MN}, \overline{NQ}, \overline{QP}, \overline{NP}, \overline{ND}, \overline{OP}$

Give a range of possible values for *z*.

27.

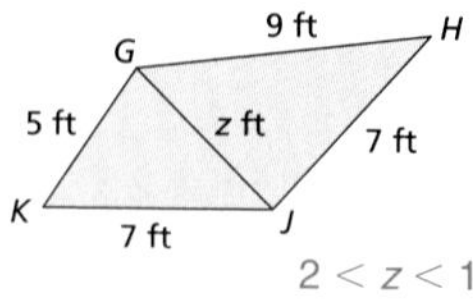

$2 < z < 12$

28.

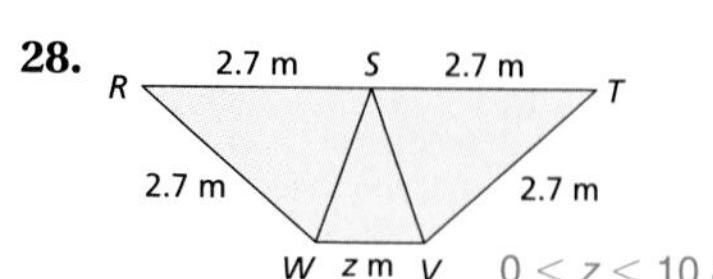

$0 < z < 10.8$

Extended Practice Exercises

29. WRITING MATH In a right triangle, the side opposite the right angle is called the *hypotenuse*. Explain why the hypotenuse must be the longest side.
See additional answers.

30. ERROR ALERT A blueprint calls for the construction of a right triangle with sides measuring 5 ft, 6 ft, and 11 ft. How do you know the measurements are incorrect? See additional answers.

CONSTRUCTION Manuella is building an A-frame dog house with the front in the shape of an isosceles triangle. Two sides of the front will each be 4 ft long.

31. Under what conditions will the base of the front of the dog house be exactly 4 ft?
When the top angle = 60°

32. Under what conditions will the base of the front of the dog house be greater than 4 ft?
When the top angle > 60°

33. Under what conditions will the base of the front of the dog house be shorter than 4 ft?
When the top angle < 60°

34. CHAPTER INVESTIGATION Using your design for a truss bridge, build a section of the truss using straws or toothpicks. Use a ruler and protractor to make sure your construction matches the plans.

Mixed Review Exercises

Write a function rule to represent the number of points in the *n*th figure in the patterns below. (Lesson 3-5)

35.

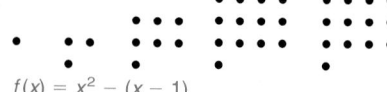

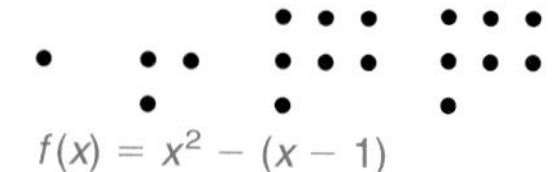

$f(x) = x^2 - (x - 1)$

36.

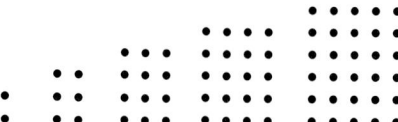

$f(x) = x^2 + x$

Write each number in scientific notation. (Lesson 1-8)

37. 371,000,000,000 3.71×10^{11}

38. 0.000000074 7.4×10^{-8}

39. 256,000,000,000 2.56×10^{11}

40. 0.00000942 9.42×10^{-6}

41. 8,900,000,000,000 8.9×10^{12}

42. 0.00000007 7×10^{-8}

30. Using the triangle inequality theorem, the sum of any two sides of a triangle must be greater than the length of the third side. Since 5 + 6 is not greater than 11, the measurements cannot form a triangle.

Extra Practice Worksheet 4-6

Name ______ Date ______

EXTRA PRACTICE **4-6**

INEQUALITIES IN TRIANGLES

EXERCISES

Can the given measures be the lengths of the sides of a triangle?

1. 4 cm, 5 cm, 6 cm yes
2. 9.1 m, 5.6 m, 7.5 m yes
3. 15 in., 24 in., 19 in. yes
4. 8 cm, 12 mm, 4 cm no
5. $5\frac{3}{4}$ in., 9 in., $12\frac{1}{8}$ in. no
6. 3.5 yd, 3.5 yd, 3.5 yd yes
7. 4 ft, 3 yd, 6 ft no
8. 5 m, 0.5 km, 3 m no

Which is the longest side of each triangle? the shortest?

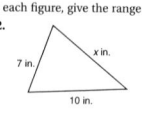

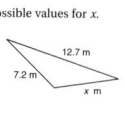

9. N, M, P; 74°, 45° — $\overline{NP}$; $\overline{MP}$
10. X, Z, Y; 36° — $\overline{XY}$; $\overline{ZY}$
11. D, C, E; 46°, 46° — $\overline{CE}$; $\overline{DC}$ and $\overline{DE}$

In each figure, give the range of possible values for *x*.

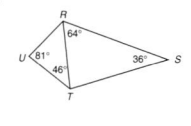

12. 7 in., *x* in., 10 in. — $3 < x < 17$
13. 12.7 m, 7.2 m, *x* m — $5.5 < x < 19.9$
14. 2.5 ft, 2.5 ft, *x* ft — $0 < x < 5$

List all the segments in each figure *in order* from longest to shortest.

15. R, U, S, T; 64°, 81°, 46°, 36° — $\overline{RS}, \overline{ST}, \overline{RT}, \overline{UT}, \overline{RU}$
16. A, B, C, D, E; 34°, 24°, 44°, 28°, 115°, 114° — $\overline{AB}, \overline{DB}, \overline{AD}, \overline{BC}, \overline{AE}, \overline{CD}, \overline{ED}$

54 EXTRA PRACTICE *LESSON 4-6*

Enrichment Worksheet 4-6

Name ______ Date ______

ENRICHMENT **4-6**

TRIANGLE INEQUALITY NUMBERS

If the lengths corresponding to three numbers will form a triangle, the numbers are called **triangle inequality numbers**.

Example

Find all the possible triangles that can be made with the lengths in the set {1, 2, 3, 4}.

Solution

1 scalene (2, 3, 4)
4 equilateral (1, 1, 1), (2, 2, 2), (3, 3, 3), (4, 4, 4)
8 isosceles (2, 2, 1), (2, 2, 3), (3, 3, 1), (3, 3, 2), (3, 3, 4), (4, 4, 1), (4, 4, 2), (4, 4, 3)
Total number 1 + 4 + 8 = 13

EXERCISES

Complete the chart to show the number of possible triangles for each set

	Set	Scalene	Equilateral	Isosceles	Total
1.	{1, 2, 3}	0	3	4	7
	{1, 2, 3, 4}	1	4	8	13
2.	{1, 2, 3, 4, 5}	3	5	14	22
3.	{1, 2, 3, 4, 5, 6}	7	6	21	34
4.	{1, 2, 3, 4, 5, 6, 7}	13	7	30	50
5.	{1, 2, 3, 4, 5, 6, 7, 8}	22	8	40	70
6.	{1, 2, 3, 4, 5, 6, 7, 8, 9}	34	9	52	95
7.	{1, 2, 3, 4, 5, 6, 7, 8, 9, 10}	50	10	65	125

Match each quantity with its formula. The letter *n* stands for the last number in each set.

8. d number of equilateral — a. $[m(4m - 5)(m - 1)] \div 6$
9. f number of isosceles, *n* even, $n = 2m$ — b. $3m^2 + m$
10. b number of isosceles, *n* odd, $n = 2m + 1$ — c. $[(m + 2)(4m + 3)(m + 1)] \div 6$
11. a number of scalene, *n* even, $n = 2m$ — d. n
12. g number of scalene, *n* odd, $n = 2m + 1$ — e. $[m(4m + 5)(m + 1)] \div 6$
13. e total, *n* even, $n = 2m$ — f. $3m^2 - 2m$
14. c total, *n* odd, $n = 2m + 1$ — g. $[m(4m + 1)(m - 1)] \div 6$

72 ENRICHMENT *LESSON 4-6*

Skills Practice

Vocabulary Review

Lesson 4-5
indirect proof

Lesson 4-6
shortest path postulate
triangle inequality theorem
unequal sides theorem
unequal angles theorem

Assignment Guide

All students: 1–32
Additional Practice: Refer to the Extra Practice Index on page 674 of this text.

Review and Practice Your Skills

Practice ■ Lesson 4-5

Write Step 1 of an indirect proof of each statement.

1. If a triangle is not isosceles, then it is not equilateral. If a triangle is not isoscels, then it *is* equilateral.
2. If a point lies on the perpendicular bisector of a segment, then the point is equidistant from the endpoints of the segment. Assume the point is *not* equidistant from the end points of the segment.
3. If two angles are vertical angles, then they are equal in measure. Assume they are *not* equal in measure.
4. If two parallel lines are cut by a transversal, then alternate interior angles are equal in measure. Assume the alternate interior angles are *not* equal in measure.
5. If two lines are perpendicular, then they do not intersect. Assume they *do* intersect.
6. The sum of the measures of the angles of a triangle is 180°. Assume the sum of the measures is *not* 180°.

Write an indirect proof of each statement. For 7–10, answers will vary.

7. If a triangle is a right triangle, then it cannot be an obtuse triangle.
8. If a triangle is equilateral, then it is isosceles.
9. If two angles are vertical angles, then they are equal in measure.
10. If two sides of a triangle are not congruent, then the angles opposite those sides are not congruent.

Practice ■ Lesson 4-6

Can the given measures be the lengths of the sides of a triangle?

11. 5.5 ft, 8.2 ft, 12.9 ft yes
12. 14 cm, 35 cm, 21 cm no
13. 21 m, 13.2 m, 7 m no

In each figure, give a range of possible values for *x*.

14. 12.5 m, *X* m, 16.5 m

$4 < x < 29$

15.

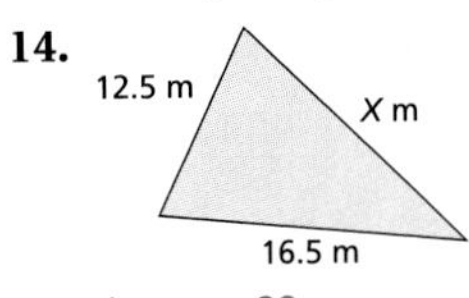

$0 < x < 14$

16.

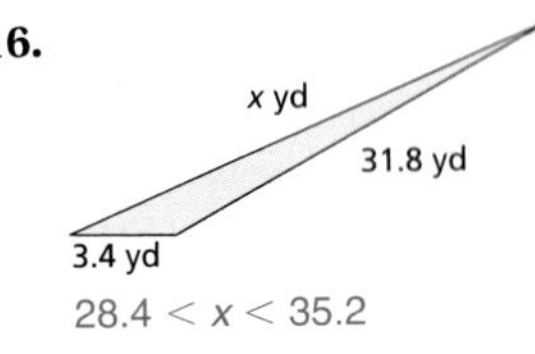

$28.4 < x < 35.2$

List all the segments in each figure in order from shortest to longest.

17. *M*, *P*, *N*, *O*; 86°, 61°, 57°, 48°

$\overline{PM}$, $\overline{MO}$, $\overline{PO}$, $\overline{MN}$, $\overline{NO}$

18. *A*, *B*, *C*, *D*; 74°, 71°, 72°, 39°

$\overline{BC}$, $\overline{AD}$, $\overline{CD}$, $\overline{AB}$, $\overline{AC}$

19.

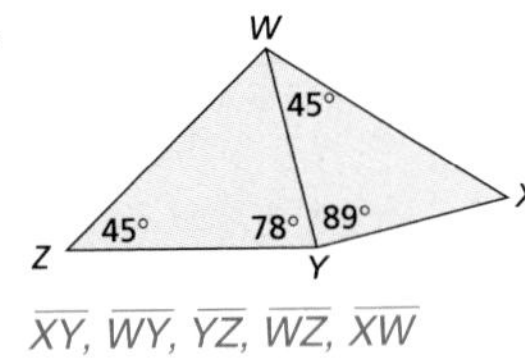

$\overline{XY}$, $\overline{WY}$, $\overline{YZ}$, $\overline{WZ}$, $\overline{XW}$

Determine whether each statement is *true* or *false*.

20. In a scalene triangle, no two angles are equal in measure. (Lesson 4-6) true
21. A triangle can have sides of length 178 cm, 259 cm, and 440 cm. (Lesson 4-6) false

Extend the Lesson

ONGOING ASSESSMENT In triangle XYZ, $XY = 7$ cm and $YZ = 9$ cm. Tell whether each statement is *always, sometimes,* or *never true.*

1. The length of XZ is between 2 cm and 16 cm. always
2. Angle Z is the largest angle. never
3. Angle Y is the smallest angle. sometimes
4. Triangle XYZ is isosceles with legs of 7 cm. sometimes
5. $\overline{XY}$ is the hypotenuse of a right triangle. never

PRACTICE ■ LESSON 4-1–LESSON 4-6

Determine whether each statement is *true* or *false*.

22. All equilateral triangles are also isosceles triangles. (Lesson 4-1) true
23. If $\triangle ABC \cong \triangle DEF$ it can also be stated that $\triangle BAC \cong \triangle EDF$. (Lesson 4-2) true
24. If two angles of a triangle are congruent, then the triangle is isosceles. true
25. All altitudes are drawn within the boundaries of a triangle. (Lesson 4-4) false
26. In an indirect proof, one starts by assuming that the conclusion is false. (Lesson 4-5) true

Find the value of x in each figure. (Find the range of possible values for x in Exercise 32.)

27. (Lesson 4-1)

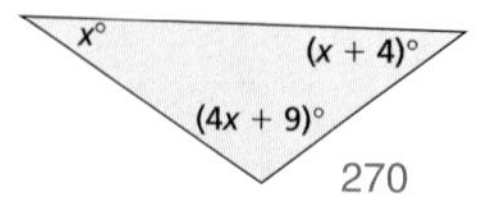

270

28. (Lesson 4-1)

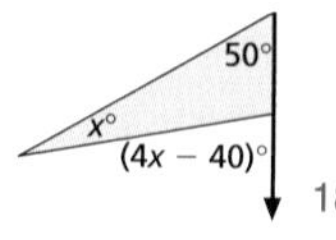

180

29. (Lesson 4-2)

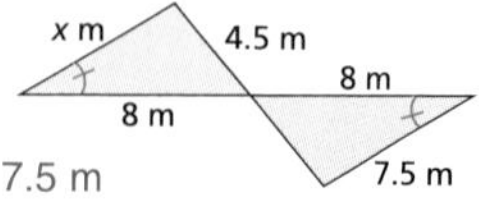

7.5 m

30. (Lesson 4-3)

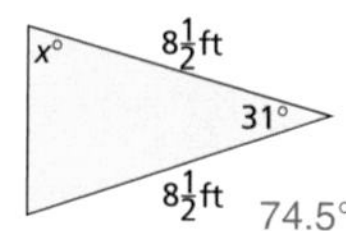

74.5°

31. (Lesson 4-4)

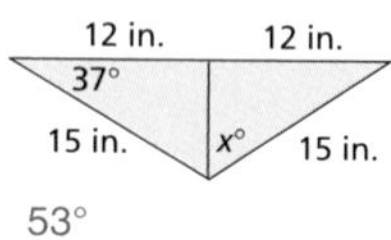

53°

32. (Lesson 4-6)

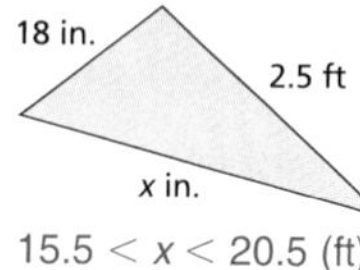

$15.5 < x < 20.5$ (ft)

MathWorks Workplace Knowhow — Career – Animator

Traditional animation involves making many hand-drawn pictures with slight differences and filming them frame by frame to create the illusion of motion. The newest form of animation is computer-assisted animation. Knowledge of coordinates, area of curved surfaces, conics and polygons are all important pieces of an animator's tool kit for drawing great pictures.

To give objects depth, animators use perspective drawing. For instance, to make a house look three-dimensional, it must be drawn so that the house's front walls look larger than those in the rear of the house.

1. The front wall of the house in the drawing has a perimeter of $6\frac{1}{4}$ in. Find the measure of x. $1\frac{1}{8}$ in.
2. The roof panels and side wall shown are parallelograms. Find the measures of a, b, c, and d. $m\angle a = 18°$, $m\angle b = 108°$; $m\angle c = 135°$, $m\angle d = 45°$
3. The altitude of the triangle is 0.5 in. Find the length of the sides of the triangle to the nearest hundredth inch. 1.12 in.

Chalkboard Examples

Example from Lesson 3-5

Write an indirect proof of the following statement: If any angle of an isosceles triangle measures 60°, the triangle is equiangular. **Answers will vary. Assume the triangle is not equiangular. If the vertex is 60°, the remaining angles, which must be equal, have a sum of 120°. Since 120/2 = 60°, the assumption would be false. If a base angle is 60°, the other base angle must be 60° and the remaining angle must equal 60°. Again, the assumption must be false. Therefore, the original statement is true.**

Example from Lesson 3-6

The lengths of two sides of a triangle are 9 inches and 5 inches. Find the range of possible lengths for the third side of the triangle if the angle opposite the 5-inch side is the smallest angle. **< 14 in. and > 5 in.**

Career Opportunity

Describe the kind of work animators do and the types of places they work. Discuss the importance of mathematics, art, and computer skills in working as an animator. Students should answer Questions 1–3 to better understand how animators use mathematics in performing their job.

Students who are interested in learning more about this profession can go to learningmathmatters.com. Guidance Counselors should have information about school and training requirements.

Learning Styles

VISUAL LEARNERS Have students use geometric drawing software to create triangles to illustrate each of the theorems on page 173. Students should draw a triangle, measure the lengths of each side and angle and then move vertices to change the lengths of sides and measures of angles.

Lesson Planning

NCTM Standards/Strands

- Number & Operation
- Patterns, Functions, & Algebra
- Geometry & Spatial Sense
- Measurement
- Problem Solving
- Reasoning and Proof
- Communication
- Connections
- Representation

Vocabulary

polygon, convex, concave, consecutive sides, consecutive vertices, diagonal, interior angle, exterior angle, equiangular, equilateral polygon, polygon, regular polygon, polyhedron

Materials Needed

paper/pencil, scissors, geometric drawing software

Lesson Resources

Warm-Up Transparency 12
Reteaching 4-7
Extra Practice 4-7
Enrichment 4-7
Transparency RF-16

Assignment Guide

Basic: 1–23, 29–32
Enriched: 1–32

Getting Started

5-Minute Warm-Up

In triangle ABC, angle A measure 48° and angle B measures 36°. Find each measure.

1. angle A + angle B + angle C 180°
2. angle C 96°
3. exterior angle ABD 144°

Introduction to Lesson 4-7

Have students compare the relationships found by using different size or shape pentagons and hexagons. Ask them to speculate about the relationship among the exterior angles of any polygon.

4-7 Polygons and Angles

Goals
- Find the measures of interior angles of polygons.
- Find the measures of exterior angles of polygons.

Applications Surveying, Sign making, Games, Sports

Work with a partner.

Draw and label a pentagon as shown at the left below. Then cut out the five exterior angles and arrange them as shown at the right.

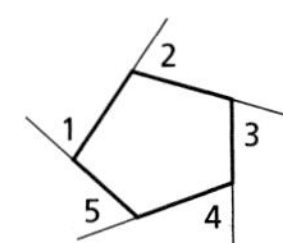

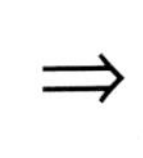

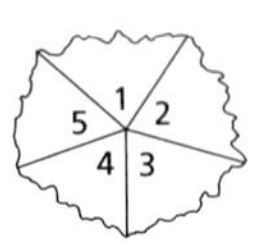

1. What is the relationship among the five exterior angles?
   The sum of the measures of the exterior angles is 360°.
2. Repeat the experiment, this time drawing a hexagon and labeling six exterior angles. What is relationship among the exterior angles?
   The relationship is the same: The sum of the measures of the exterior angles is 360°.

Reading Math

From your previous work, you should recall that a polygon can be classified by its number of sides.

Number of Sides	Name of Polygon
3	triangle
4	quadrilateral
5	pentagon
6	hexagon
7	heptagon
8	octagon
9	nonagon
10	decagon
n	*n*-gon

You can remember names of the polygons by associating them with everyday words that have the same *prefix*. An octopus has eight tentacles, and an octagon has eight sides.

Can you think of everyday words to associate with the other names of polygons?

See additional answers.

Build Understanding

A **polygon** is a closed plane figure that is formed by joining three or more coplanar segments at their endpoints. Each segment is called a **side** of the polygon. Each side intersects exactly two other sides, one at each endpoint. The point at which two sides meet is called a **vertex** of the polygon. The angles determined by the sides are called the **angles**, or the **interior angles**, of the polygon.

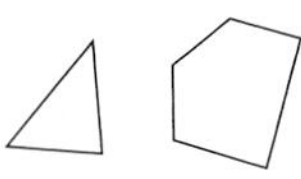

polygons

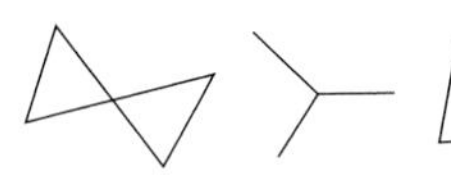

not polygons

A polygon is **convex** if each line containing a side contains no points in the interior of the polygon. A polygon that is not convex is called **concave**. In this book, when the word polygon is used, assume the polygon is *convex*.

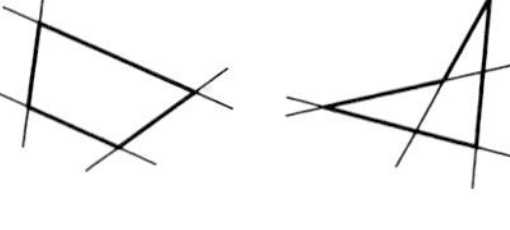

Two sides of a polygon that intersect are called **consecutive sides.** The endpoints of any side of a polygon are **consecutive vertices.** When naming a polygon, you list consecutive vertices in order. For example, two names for the pentagon at the right are "pentagon *ABCDE*" and "pentagon *BCDEA*." It is *not* correct to call the figure "pentagon *ABCED*."

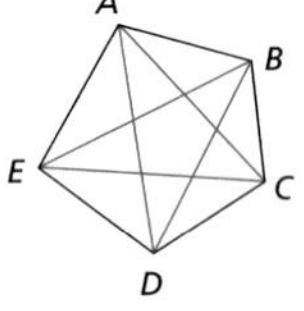

A **diagonal** of a polygon is a segment that joints two *nonconsecutive* vertices. In pentagon *ABCDE*, the diagonals are shown in red.

Additional Answers

Reading Math
Answers will vary depending upon students' experiences. If students are not familiar with related words for a given prefix, you may wish to suggest that they research related words in a dictionary. Sample answers are given.
triangle: triple, triplet, tricycle, trilogy
quadrilateral: quadruple, quadruplet, quadrangle
pentagon: pentathlon, Pentagon(government office building.)
hexagon: hexapod, hexose
heptagon: heptathlon, heptameter, heptarchy
octagon: octave, octet, octameter, octogenarian
nonagon: nonagenarian
decagon: decade, decathlon, decimal

If you draw all the diagonals from just one vertex of a polygon, you divide the interior of the polygon into nonoverlapping triangular regions. The sum of the measures of the angles of the polygon is the product of the number of triangular regions formed and 180°.

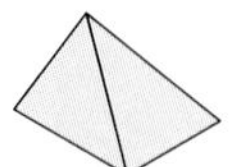

4 sides
2 triangular regions
$2 \times 180° = 360°$

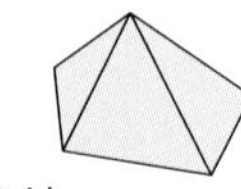

5 sides
3 triangular regions
$3 \times 180° = 540°$

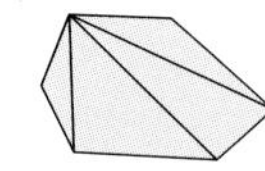

6 sides
4 triangular regions
$4 \times 180° = 720°$

In each case, the number of triangular regions formed is two fewer than the number of sides of the polygon. This leads to the following theorem.

The Polygon-Sum Theorem The sum of the measures of the angles of a convex polygon with n sides is $(n - 2)180°$.

Check Understanding

Refer to pentagon *ABCDE*, on page 178. Name the following.

- all the sides
- all the angles
- all the vertices
- all the diagonals

Give at least two names for the pentagon other than those names given in the text.

See additional answers.

Example 1

SURVEYING A playground has the shape shown in the figure to the right. A surveyor measures six of the angles of the playground. Find the unknown angle.

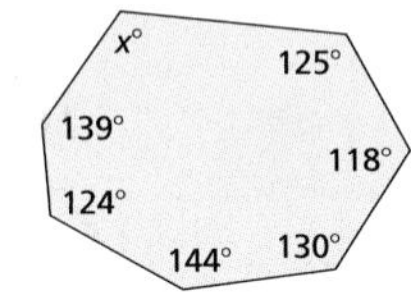

Solution

The polygon has 7 sides. Use the polygon-sum theorem to find the sum of the angle measures.

$$(n - 2)180° = (7 - 2)180° = (5)180° = 900°$$

Add the *known* angle measures.

$$139° + 124° + 144° + 130° + 118° + 125° = 780°$$

Subtract this sum from 900°: $900° - 780° = 120°$

The unknown angle measure is 120°.

An **exterior angle** of any polygon is an angle both adjacent to and supplementary to an interior angle. Since the sum of the *interior* angles of a polygon depends on the number of sides of the polygon, you might expect that the same would be true for the exterior angles. So, the following theorem about exterior angles may come as a surprise to you.

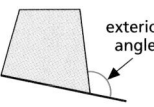

exterior angle

The Polygon Exterior Angle Theorem The sum of the measures of the exterior angles of a convex polygon, one angle at each vertex, is 360°.

A polygon with all sides of equal length is called an **equilateral polygon.** A polygon with all angles of equal measure is an **equiangular polygon.** A **regular polygon** is a polygon that is *both* equilateral and equiangular.

Technology Note

Explore the theorem using geometric software.

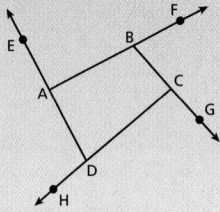

1. Draw four rays to form a polygon. Mark and label a point on each ray outside the polygon.
2. Use the software to measure each exterior angle.
3. Calculate the total of the angles.
4. Change the positions of the rays to change the measures of the angles. What happens to the sum?

Teaching Strategies

COOPERATIVE LEARNING Have students work in pairs to write an expression for the total number of diagonals that can be drawn from all the vertices of a polygon with n sides. Encourage them to make models or sketches of a triangle, quadrilateral, pentagon, and so on, draw all the different diagonals and then look for a pattern. $\frac{n(n-3)}{2}$

Chalkboard Examples

Supplementary Example 1

A parking lot is in the shape of a polygon with 6 sides. A surveyor measures 5 angles. They measure 56°, 84°, 138°, 165° and 150°. What is the measure of the remaining angle? 127°

Supplementary Example 2

Find the measure of each interior angle of a regular nonagon (9-sided polygon). 140°

Find the measure of each exterior angle of a regular nonagon. 40°

Check Understanding
Sides: *AB, BC, CD, DE, EA*
Angles: $\angle A, \angle B, \angle C, \angle D, \angle E$
Vertices: *A, B, C, D, E*
Diagonals: *AC, BE, CE, DA, BD*
CDEAB, DEABC

Reteaching Worksheet 4-7

Name ______________________ Date ______________

RETEACHING 4-7

POLYGONS AND ANGLES

Polygons are closed plane figures formed by joining three or more coplanar segments at their endpoints. A polygon is complex when each line containing a side contains no points in the interior of the polygon. The sum of the measures of the angles of a convex polygon with n sides can be found by solving the equation $(n - 2)(180)°$. The sum of the measures of the exterior angles of a convex polygon, one angle at each vertex, is 360°.

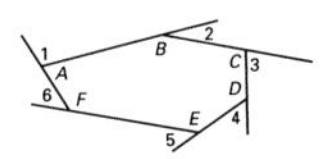

$m\angle A + m\angle B + m\angle C + m\angle D + m\angle E + m\angle F = (n - 2)180°$ or $(6 - 2)180° = (4)180° = 720°$

Exterior angles: $\angle 1, \angle 2, \angle 3, \angle 4, \angle 5, \angle 6$

$m\angle 1 + m\angle 2 + m\angle 3 + m\angle 4 + m\angle 5 + m\angle 6 = 360°$

Example

a. Find the measure of each interior angle of a regular pentagon.
b. Find the measure of each exterior angle of a regular pentagon.

Solution

a. A pentagon has five sides. Use the polygon-sum theorem to find the sum of the measures of the interior angles.
$(n - 2)180° = (5 - 2)180° = (3)180° = 540°$
Because the pentagon is regular, each interior angle is equal in measure. So, the measure of one interior angle is $540° \div 5 = 108°$.
b. By the polygon exterior angle theorem, the sum of the measures of the exterior angles is 360°. Because the pentagon is regular, each exterior angle is equal in measure. So, the measure of one exterior angle is $360° \div 5 = 72°$.

EXERCISES

Find the unknown angle measure or measures in each figure.

1.

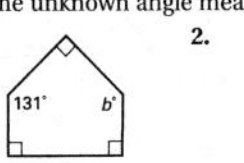

$b = 139°$

2.

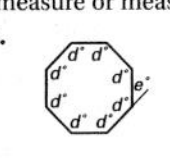

$d = 135$
$e = 45$

3.

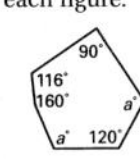

$a = 117$

4.

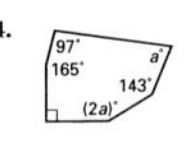

$a = 75$
$2a = 150$

5. Find the measure of each interior angle of a regular decagon. 144°
6. Find the measure of each exterior angle of a regular hexagon. 60°
7. Find the sum of the measures of the interior angles of a polygon with 22 sides. 3600°

64 RETEACHING *LESSON 4-7*

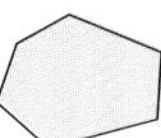
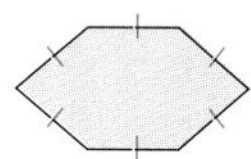
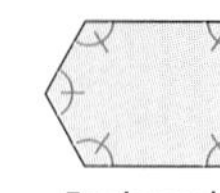
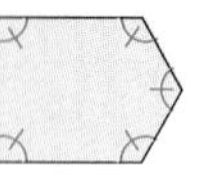
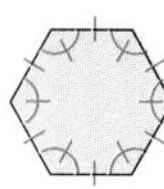

Hexagon | Equilateral hexagon | Equiangular hexagon | Regular hexagon

Example 2

a. Find the measure of each interior angle of a regular octagon.

b. Find the measure of each exterior angle of a regular octagon.

Solution

a. Using the polygon-sum theorem, the sum of the measures of the interior angles is $(n - 2)180° = (8 - 2)180° = (6)180° = 1080°$.

Because the octagon is regular, each interior angle is equal in measure.

So, the measure of one interior angle is $1080° \div 8 = 135°$.

b. By the polygon exterior angle theorem, the sum of the measures of the exterior angles is 360°.

So, the measure of one exterior angle is $360° \div 8 = 45°$.

Try These Exercises

Find the unknown angle measure or measures in each figure.

1.

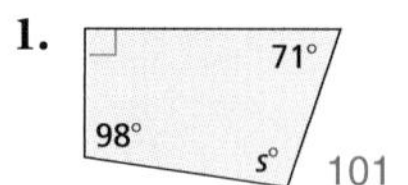

101

2.

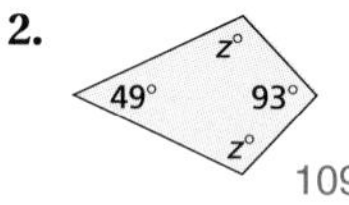

109

3.

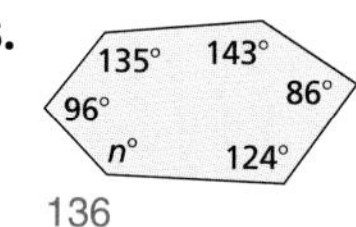

136

4. Find the measure of each interior angle of a regular polygon with 15 sides. 156°

5. Find the measure of each exterior angle of a regular decagon. 36°

Practice Exercises

Find the unknown angle measure or measures in each figure.

6.

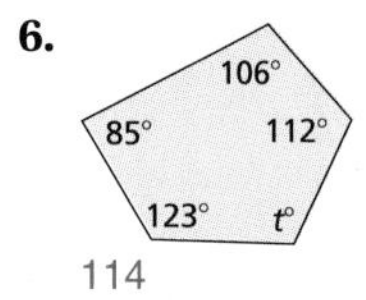

114

7.

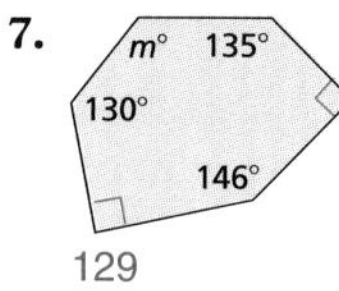

129

8.

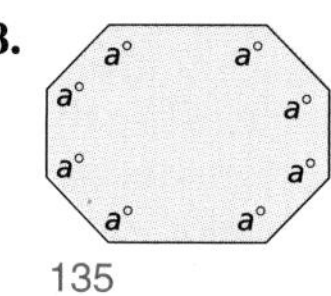

135

9.

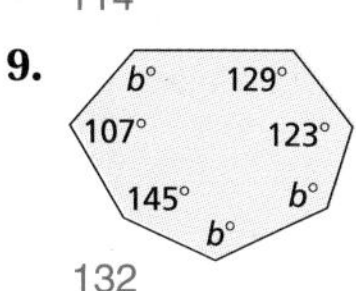

132

10.

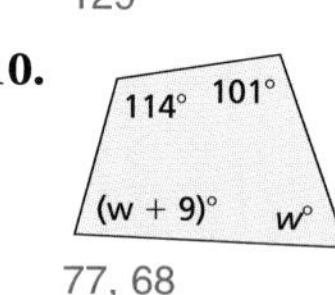

77, 68

11.

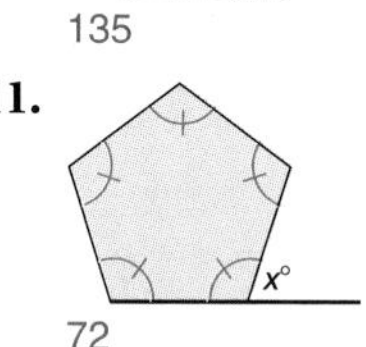

72

12. A road sign is in the shape of a regular hexagon. Find the measure of each interior angle. 120°

Lesson Wrap-up

Quick Assessment

Ask the following questions to determine if students understand the content presented in this lesson.

1. How are the measures of the interior angles of a polygon related? The sum of their measures always equals two fewer than the number of sides of the polygon times 180°.
2. How are the measures of the exterior angles of a polygon related? The sum of their measures always equals 360°.

Assignment Guide

Basic: 1–23, 29–32
Advanced: 1–32
Additional Practice: See Extra Practice Index on page 674.

Teaching Strategies

For Exercises 19–21, you may want to point out that students could use the strategy "Work Backward" or write and solve equations based on the polygon sum theorem and the polygon exterior angle theorem.

13. **RECREATION** A game board is in the shape of a regular polygon with 18 sides. Find the sum of the measures of the interior angles. 2,880°

14. Find the sum of the measures of the exterior angles of a regular nonagon. 360°

15. Find the measure of each exterior angle of a regular polygon with 24 sides. 15°

Each figure is a regular polygon. Find the values of *x*, *y*, and *z*.

16. 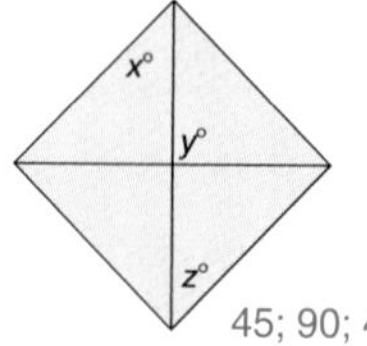

45; 90; 45

17. 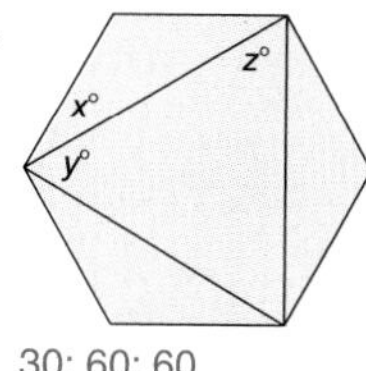

30; 60; 60

18. 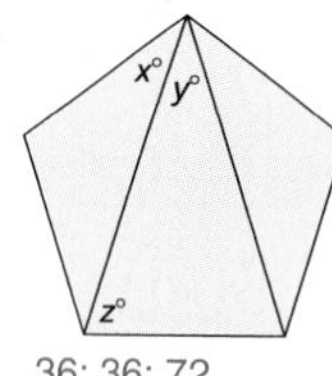

36; 36; 72

Find the number of sides of each regular polygon.

19. The measure of each exterior angle is 9°. 40

20. The sum of the measures of the interior angles is 1,980°. 13

21. The measure of each interior angle is 162°. 20

DATA FILE For Exercises 22–23, use the Data Index on pages 632–633 to locate information about convex regular polyhedrons.

22. From your previous work, recall that a **polyhedron** is a closed three-dimensional figure in which each surface is a polygon. Why do you think these are called *regular* polyhedrons?
Answers will vary. Possible response: All the surfaces are identical regular polygons.

23. **SPORTS** At the right is a soccer ball. It is shaped like a polyhedron with faces that are all regular polygons. However, this shape is not pictured with the convex regular polyhedrons. Explain.
Although the faces are all regular polygons, there are two different types of faces, pentagons and hexagons.

Extended Practice Exercises

For Exercises 24 and 25, consider a regular polygon with *n* sides. Write an expression to represent each quantity.

24. the measure in degrees of one exterior angle $\frac{360}{n}$

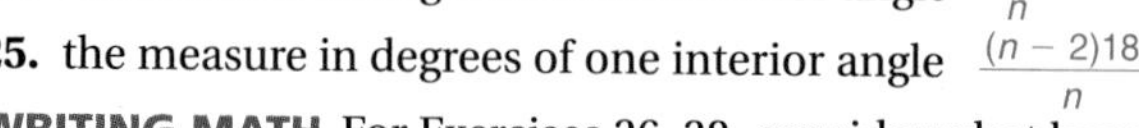

25. the measure in degrees of one interior angle $\frac{(n-2)180}{n}$

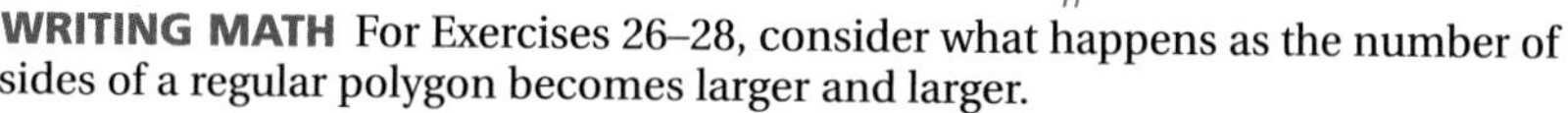

WRITING MATH For Exercises 26–28, consider what happens as the number of sides of a regular polygon becomes larger and larger.

26. What happens to the measure of each exterior angle? Approaches 0°

27. What happens to the measure of each interior angle? Approaches 180°

28. What happens to the overall appearance of the polygon? begins to resemble a circle.

Mixed Review Exercises

Classify each triangle by its sides and angles.

29. 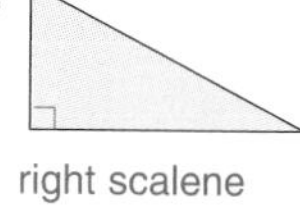

right scalene

30. 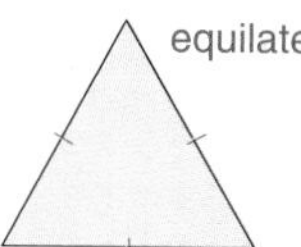

equilateral

31. 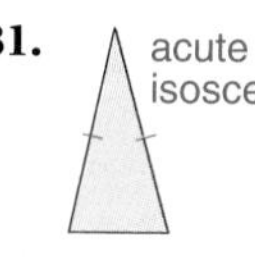

acute isosceles

32. 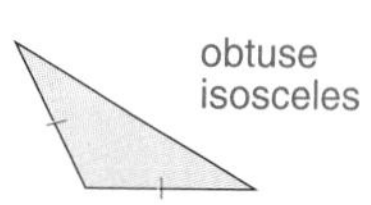

obtuse isosceles

Teaching Strategies

CHALLENGE Would it be possible to have a regular pentagon with an acute angle? Explain your thinking. No; the sum of the measures of the interior angles is 3 × 180°, or 540°. Therefore, each angle would have to measure 540° ÷ 5, or 108°.

Extra Practice Worksheet 4-7

Name ______ Date ______

EXTRA PRACTICE 4-7

POLYGONS AND ANGLES

EXERCISES

Find the unknown angle measure in each figure.

1. 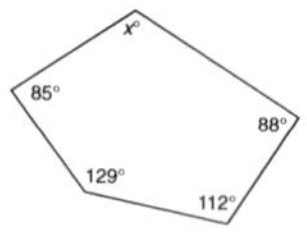 (pentagon: x°, 85°, 88°, 129°, 112°) 126°

2. 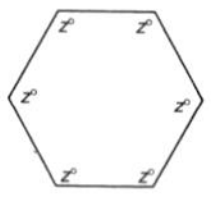 (hexagon: z° each angle) 120°

3. 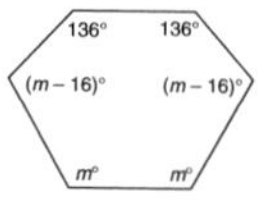 (hexagon: 136°, 136°, (m − 16)°, (m − 16)°, m°, m°) 120°, 120°, 104°, 104°

4. 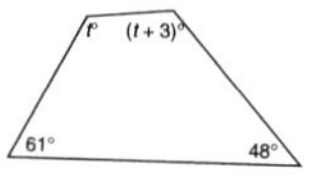 (quadrilateral: t°, (t + 3)°, 61°, 48°) 124°, 127°

5. Find the measure of each interior angle of a regular decagon. 144°

6. Find the sum of the measures of the interior angles of a regular polygon with 20 sides. 3240°

7. Find the sum of the measures of the exterior angles of a regular heptagon. 360°

8. Find the measure of each exterior angle of a regular polygon with 30 sides. 12°

9. Find the measure of each interior angle of a regular polygon with 24 sides. 165°

10. Find the sum of the measures of the interior angles of a regular polygon with 30 sides. 5040°

Find the number of sides of each regular polygon.

11. The measure of each exterior angle is 4°. 90

12. The sum of the measures of the interior angles is 6120°. 36

13. The measure of each interior angle is 160°. 18

14. The sum of the measures of the interior angles is 2520°. 16

15. The measure of each interior angle is 170°. 36

56 EXTRA PRACTICE *LESSON 4-7*

Enrichment Worksheet 4-7

Name ______ Date ______

ENRICHMENT 4-7

TESSELLATING REGULAR POLYGONS WITH RHOMBUSES

Any regular polygon with an even number of sides can be tessellated with some number of rhombuses. The case for $n = 12$ is shown.

EXERCISES

Use the figure for $n = 12$ for these exercises.

1. The tessellation is made up of bands of rhombuses that surround the center of the polygon. The innermost band forms a 12-pointed star. How many bands of rhombuses are there? 5 bands

2. How many rhombuses are in each band? 12 rhombuses

3. Describe the angles in the rhombuses in each band.
Numbering outward from the center, the 1st and 5th bands contain 30°–150° rhombuses; the 2nd and 4th bands contain 60°–120° rhombuses; and the 3rd band contains 90°–90° rhombuses (squares).

Draw each of these cases.

4. $n = 8$

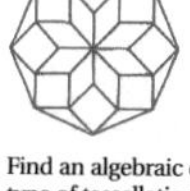

5. $n = 6$

6. $n = 4$

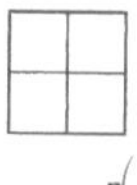

7. Find an algebraic expression for the total number of rhombuses in this type of tessellation. Use n for the number of sides of the regular polygon. $n\left(\frac{n}{2} - 1\right)$

8. Does your algebraic expression work when n is odd? Draw one example and give reasons for your answer.
No, when n is odd, the total is $n\frac{(n-1)}{2}$ and the first band of rhombuses overlaps in the center. Examples may vary. Check students' drawings.

74 ENRICHMENT *LESSON 4-7*

Lesson Planning

NCTM Standards/Strands

- Patterns, Functions, & Algebra
- Geometry & Spatial Sense
- Measurement
- Problem Solving
- Reasoning and Proof
- Communication
- Connections
- Representation

Vocabulary

opposite sides	opposite angles
parallelogram	rectangle
rhombus	square
diagonal	

Materials Needed

compasses straightedges

Lesson Resources

Warm-Up Transparency 12
Reteaching 4-8
Extra Practice 4-8
Enrichment 4-8
Transparency RF-17

Assignment Guide

Basic: 1–24, 27–32
Enriched: 1–32

Getting Started

5-Minute Warm-Up

Find a value for the variable.

1. $125 + x = 180$ 55
2. $360 - y = 250$ 110
3. $95 = 180 - a$ 85
4. $\frac{5\frac{1}{2}}{2} = b$ $2\frac{3}{4}$

Introduction to Lesson 4-8

Have students cut two pairs of congruent segments from four straws. Have them connect the segments to form a rectangle, using pipe cleaners or clay for the connections. Then direct them to move the sides of the rectangle until they have formed a parallelogram. Ask them to speculate about the relationship between a rectangle and a parallelogram.

4-8 Special Quadrilaterals: Parallelograms

Goals ■ Apply properties of parallelograms to find missing lengths and angle measures.

Applications Art, Construction, Engineering, Architecture

On a sheet of paper, draw line ℓ. Identify point A on line ℓ as shown at the right. For 1–3, check students' work.

1. With compass tip at point A, draw two arcs of equal radii that intersect ℓ. Label the points of intersection X and Y.
2. With compass tip first at point X, then at point Y, draw two arcs that intersect at Z.
3. Using a straightedge, draw $\overleftrightarrow{AZ}$. What do you observe about the line you constructed? It is perpendicular to line ℓ.
4. Use this method to construct a rectangle. Using a straightedge, draw the diagonals of your rectangle. What observations do you make about the diagonals? Answers will vary. Possible responses: they are congruent; they divide the interior into pairs of congruent triangles. See additional answers.

Build Understanding

Opposite sides of a quadrilateral are two sides that do not share a common endpoint. **Opposite angles** are two angles that do not share a common side.

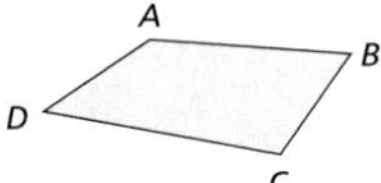

opposite sides	opposite angles
$\overline{AB}$ and $\overline{CD}$	$\angle A$ and $\angle C$
$\overline{BC}$ and $\overline{DA}$	$\angle B$ and $\angle D$

A **parallelogram** is a quadrilateral with both pairs of opposite sides parallel.

parallelogram $PQRS$
$\square PQRS$

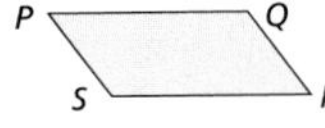

$\overline{PQ} \parallel \overline{SR}$
$\overline{PS} \parallel \overline{QR}$

The following theorems identify some properties of all parallelograms.

The Parallelogram-Side Theorem	If a quadrilateral is a parallelogram, then its opposite sides are equal in length.
The Parallelogram-Angle Theorem	If a quadrilateral is a parallelogram, then its opposite angles are equal in measure.
The Parallelogram-Diagonal Theorem	If a quadrilateral is a parallelogram, then its diagonals bisect each other.

Teaching Strategies

Some students have difficulty understanding that a figure can represent several special quadrilaterals simultaneously. Use an analogy to the residence of students to help clarify this idea. Each student is a resident of a particular town. Students in your town and in neighboring towns are residents of different towns but are residents of the same state. Encourage students to think of other analogies that show how a person, place or object can be members of more than one set.

The proofs of these theorems are based on properties of parallel lines and congruent triangles. You will have a chance to prove them in Exercises 24–27 on page 185.

Example 1

Find m$\angle J$ in $\square$ *JKLM*, at the right.

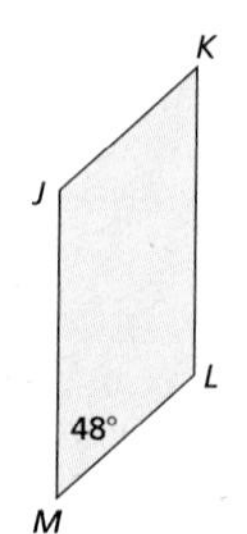

Solution

Since $\angle K$ and $\angle M$ are opposite angles, by the parallelogram-angle theorem, m$\angle K$ = m$\angle M$ = 48°.

Use the polygon-sum theorem to find the sum of the measures of the interior angles.

$(n - 2)180° = (4 - 2)180° = 2(180°) = 360°$

Notice that m$\angle M$ + m$\angle K$ = 48° + 48° = 96°. It follows that m$\angle J$ + m$\angle L$ = 360° − 96° = 264°.

Since $\angle J$ and $\angle L$ are opposite angles, by the parallelogram-angle theorem, m$\angle J$ = 264° ÷ 2 = 132°.

Other special quadrilaterals are *rectangles, rhombuses,* and *squares.*

A **rectangle** is a quadrilateral with four right angles.

A **rhombus** is a quadrilateral with four sides of equal length.

A **square** is a quadrilateral with four right angles *and* four sides of equal length.

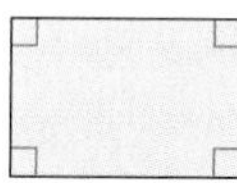
Rectangle

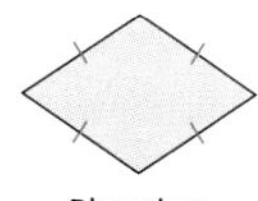
Rhombus

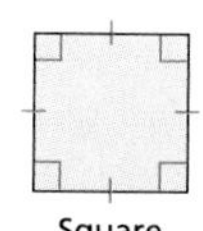
Square

Rectangles, rhombuses, and squares are all parallelograms, and so they have all the properties of parallelograms. In addition, however, they have the special properties summarized in the following theorems. In this book, these theorems will be accepted as true without proof.

The Rectangle-Diagonal Theorem	If a quadrilateral is a rectangle, then its diagonals are equal in length.
The Rhombus-Diagonal Theorem	If a quadrilateral is a rhombus, then its diagonals are perpendicular and bisect the vertex angles.

Example 2

ART A rectangular mural is reinforced from the back using wire diagonals. The diagram at the right shows how the wires are attached. If ZO = 8 ft, find WY.

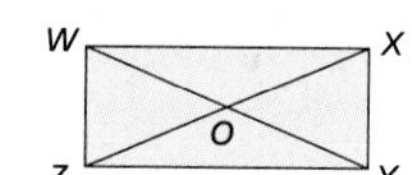

Mental Math Tip

Most students find it helpful to memorize the fact that the sum of the interior angles of a quadrilateral is 360°. If you remember this, problems such as Example 1 involve far less work.

Math: Who, Where, When

In 1981, when she was a 21-year-old senior at Yale University, Maya Ying Lin won a nationwide competition to design the proposed Vietnam Veterans Memorial in Washington, D.C. At the time, her design was criticized as being too simple—two large walls of polished black granite, joined at a 130° angle and engraved with the names of all those killed or missing in the conflict. However, since its official dedication on November 11, 1982, the memorial has become the most visited monument in the nation's capital. Lin has since graduated from Yale and become a highly respected architect and sculptor.

Chalkboard Examples

Supplementary Example 1

Find m$\angle B$ in parallelogram ABCD where m$\angle A$ = 36°. 144°

Supplementary Example 2

The frame of a shelving unit is reinforced by adding diagonal supports, DF and EG, to its rectangular shape. Label the rectangle DEFG and the point where the diagonals intersection O. If GO = 4 ft, what is the measure of DF? 8 ft

Answer for Introduction on page 182.

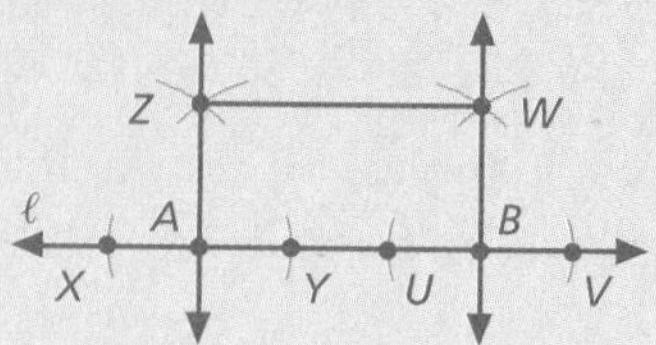

Extend the Lesson

CONNECTING TO PRIOR KNOWLEDGE Have students draw Venn diagrams or flowcharts to show the relationships of different types of parallelograms.

Reteaching Worksheet 4-8

Name ____________ Date ________

RETEACHING 4-8

SPECIAL QUADRILATERALS: PARALLELOGRAMS

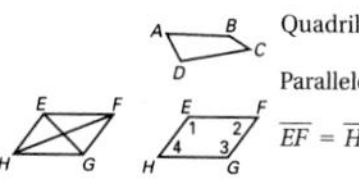

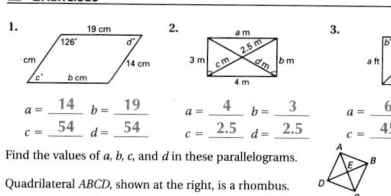

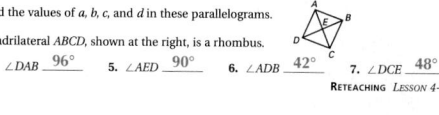

Quadrilateral — Opposite sides: $\overline{AB}$ and $\overline{CD}$, $\overline{BC}$ and $\overline{DA}$; Opposite angles: $\angle A$ and $\angle C$, $\angle B$ and $\angle D$

Parallelogram — $\overline{EF} \parallel \overline{HG}$ m$\angle 1$ = m$\angle 3$; $\overline{EH} \parallel \overline{FG}$ m$\angle 2$ = m$\angle 4$

$\overline{EF} = \overline{HG}$ — $\overline{HF}$ bisects $\overline{EG}$, and $\overline{EG}$ bisects $\overline{HF}$.

Example 1

Refer to $\square$*IJKL* at the right. Find m$\angle I$ in $\square$*IJKL*.

Solution

Since $\angle L$ and $\angle J$ are opposite angles, by the parallelogram-angle theorem, m$\angle L$ = m$\angle J$ = 55°.
Use the polygon-sum theorem to find the sum of the measures of the interior angles.
$(n - 2)180° = (4 - 2)180° = (2)180° = 360°$
Notice that $\angle L + \angle J$ = 55° + 55° = 110°.
It follows that
m$\angle I$ + m$\angle K$ = 360° − 110° = 250°.
Since $\angle I$ and $\angle K$ are opposite angles, by the parallelogram-angle theorem,
m$\angle I$ = m$\angle K$ = 250° ÷ 2 = 125°.

Example 2

In the figure at the right, *MNOP* is a square and PN = 9 cm. Find OZ.

Solution

A square is also a rectangle. By the rectangle-diagonal theorem, the diagonals are equal in length. So, $PN = OM$ = 9 cm.
By the parallelogram-diagonal theorem, the diagonals bisect each other.
So, $QM = OZ + MZ = 2(OZ)$.
Substitute 9 for OM and solve.
$2OZ = 9$
$OZ = 4.5$
The length of $\overline{OZ}$ is 4.5 cm.

EXERCISES

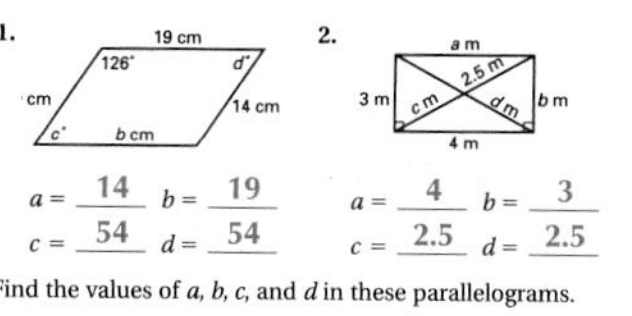

1. a = 14 b = 19 c = 54 d = 54

2. a = 4 b = 3 c = 2.5 d = 2.5

3. a = 6 b = 90 c = 45 d = 6

Find the values of a, b, c, and d in these parallelograms.

Quadrilateral *ABCD*, shown at the right, is a rhombus.

4. $\angle DAB$ 96° 5. $\angle AED$ 90° 6. $\angle ADB$ 42° 7. $\angle DCE$ 48°

66 RETEACHING *LESSON 4-8*

Lesson Wrap-up

Quick Assessment

Ask the following questions to determine if students understand the content presented in this lesson.

Match the property named with all the parallelograms that have that property. Choose from *rectangle, rhombus*, and *square*.

1. Opposite sides are congruent. all
2. All sides are congruent. rhombus, square
3. Diagonals are congruent. rectangle, square
4. Diagonals are perpendicular bisectors of each other. rhombus, square
5. Opposite angles are congruent, but consecutive angles are not. rhombus

Assignment Guide

Basic: 1–24, 27–32
Advanced: 1–32
Additional Practice: See Extra Practice Index on page 674.

Additional Answers

23. Given: *ABCD* is a parallelogram.
Prove: $m\angle B = m\angle D$

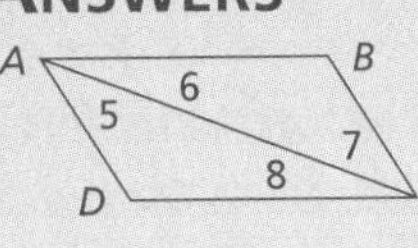

Statements	Reasons
1. *ABCD* is a parallelogram	1. given
2. $\overline{AB} \parallel \overline{DC}$; $\overline{AD} \parallel \overline{BC}$	2. definition of $\parallel$-ogram
3. $m\angle 5 = m\angle 7$, or $\angle 5 \cong \angle 7$; $m\angle 6 = m\angle 8$, or $\angle 6 \cong \angle 8$	3. If two $\parallel$ lines are cut by a tran., then then alt. int. $\angle s$ are = in measure.
4. $\overline{AC} \cong \overline{CA}$	4. reflexive property
5. $\triangle ABC \cong \triangle CDA$	5. ASA postulate
6. $\angle B \cong \angle D$, or $m\angle B \cong m\angle D$	6. CPCTC

Solution

A rectangle is a parallelogram. By the parallelogram-diagonal theorem, the diagonals bisect each other. So, $XZ = 2(ZO) = 2(8 \text{ ft}) = 16$ ft.

Then, by the rectangle-diagonal theorem, you know that the diagonals are equal in length. So, $WY = XZ = 16$ ft.

Try These Exercises

In Exercises 1–2, each figure is a parallelogram. Find the values of *x* and *z*.

1. 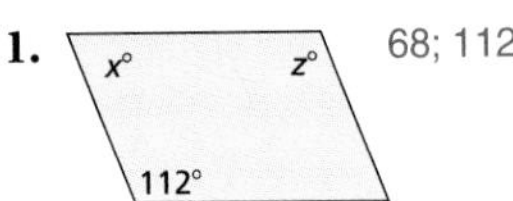

68; 112

2. 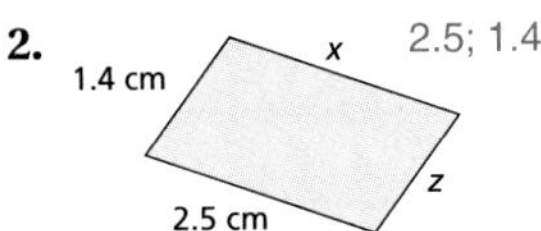

2.5; 1.4

BRIDGE BUILDING A portion of a truss bridge forms quadrilateral *XYZW*, shown at the right. Given that *XYZW* is a rhombus and $m\angle YXZ = 32°$, find the measure of each angle.

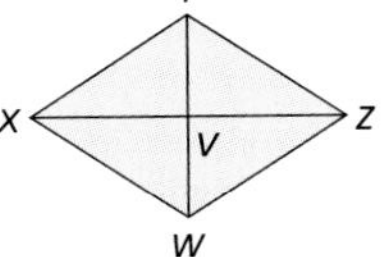

3. $\angle YXW$ 64° **4.** $\angle XYW$ 58° **5.** $\angle XVY$ 90°
6. $\angle YZW$ 64° **7.** $\angle YVZ$ 90° **8.** $\angle XWZ$ 116°

Practice Exercises

ARCHITECTURE The parallelograms in Exercises 9–12 are from building plans. Find the values of *a, b, c*, and *d*.

9. 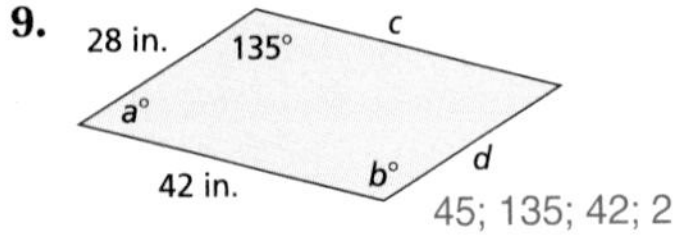

45; 135; 42; 28

10. 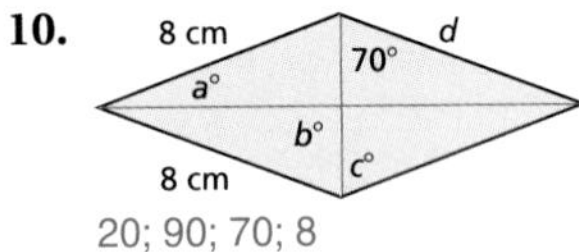

20; 90; 70; 8

11. $MJ = 1\frac{1}{2}$ yd $MK = 3\frac{1}{4}$ yd

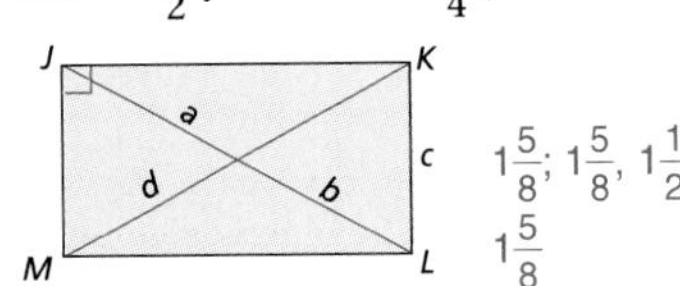

$1\frac{5}{8}$; $1\frac{5}{8}$; $1\frac{1}{2}$; $1\frac{5}{8}$

12. $RS = 4.9$ mm $RQ = 5.6$ mm $RP = 9.7$ mm

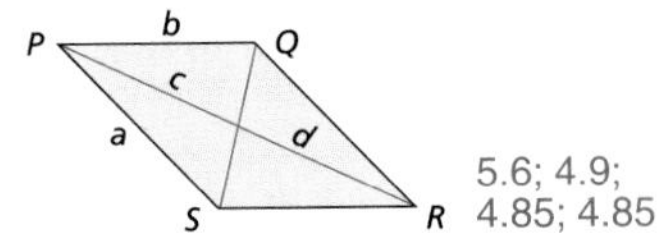

5.6; 4.9; 4.85; 4.85

ERROR ALERT Dillon made the following statements about quadrilaterals. Decide whether each statement is *true* or *false*.

13. A rectangle is a parallelogram. true
14. No rhombus is a square. false
15. Every quadrilateral is a parallelogram. false
16. Some rectangles are rhombuses. true
17. The diagonals of a square are not equal in length. false
18. Consecutive angles of a parallelogram are supplementary. true

24. Given: ABCD is a parallelogram.
Prove: $AB = CD$; $AD = CB$

Statements	Reasons
1. *ABCD* is a parallelogram	1. given
2. $\overline{AB} \parallel \overline{DC}$; $\overline{AD} \parallel \overline{BC}$	2. definition of $\parallel$-ogram
3. $m\angle 1 = m\angle 3$, or $\angle 1 \cong \angle 3$; $m\angle 2 = m\angle 4$, or $\angle 2 \cong \angle 4$	3. If two $\parallel$ lines are cut by a tran., then alt. int. $\angle s$ are = in measure.
4. $\overline{DB} \cong \overline{BD}$	4. reflexive property
5. $\triangle ABD \cong \triangle CDB$	5. ASA postulate
6. $\overline{AB} \cong \overline{CD}$, or $AB = CD$; $\overline{AD} \cong \overline{CB}$, or $AD = CB$	6. CPCTC

25. Sample response: Draw the parallelogram and both diagonals. The goal is to show that *BD* bisects *AC*, and that *AC* bisects *BD*. By the definition of parallelogram, $AB \parallel DC$ and $AD \parallel BC$. When parallel lines are cut by a transversal, alternate interior angles are equal in measure, so there are four pairs of angles that are equal in measure: $\angle 1$ and $\angle 3$; $\angle 2$ and $\angle 4$; $\angle 5$ and $\angle 7$; and $\angle 6$ and $\angle 8$. Because the parallelogram-side theorem

WRITING MATH Do you think that the given figure is a parallelogram? Write *yes* or *no*. Then explain your reasoning.

19.

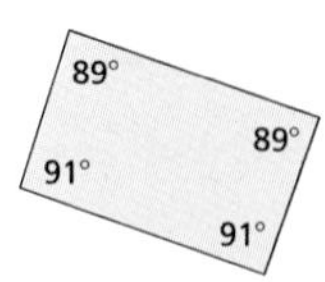

No; the angles that are equal in measure are not opposite angles.

20. No; the diagonals do not bisect each other.

21. Yes, the figure is a square, and a square is, by definition, a parallelogram.

22. Copy and complete this proof.
Given *ABCD* is a parallelogram.
Prove $m\angle A = m\angle C$

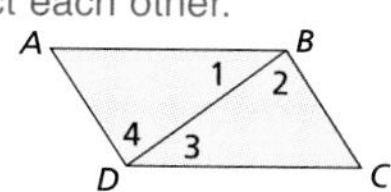

Statements	**Reasons**
1. __?__ *ABCD* is a ▱.	1. __?__ given
2. $\overline{AB} \parallel \overline{DC}$; $\overline{AD} \parallel \overline{BC}$	2. definition of __?__ parallelogram
3. $m\angle 1 = m\angle 3$, or $\angle 1 \cong \angle 3$; $m\angle 2 = m\angle 4$, or $\angle 2 \cong \angle 4$	3. If __?__, then __?__ 2 ∥ lines are cut by a trans.; alt int. ∠s are = in measure.
4. __?__ $\overline{DB} \cong \overline{BD}$	4. reflexive property
5. __?__ $\triangle ABD \cong \triangle CDB$	5. ASA postulate
6. $m\angle A = m\angle C$	6. __?__ CPCTC

23. The proof in Exercise 14 is the beginning of a proof of the parallelogram-angle theorem. Using this proof as a model, write the second part of the proof. That is, prove $m\angle B = m\angle D$. For 23–24, see additional answers.

24. Write a proof of the parallelogram-side theorem.

Extended Practice Exercises

25. WRITING MATH Suppose that you are asked to prove the parallelogram-diagonal theorem. Write a paragraph that explains how you would proceed. (Do not write the two-column proof.)
For 25–26, see additional answers.

26. DESIGN Suppose you need to describe the figure at the right to a graphics designer. State as many facts as you can about the figure.

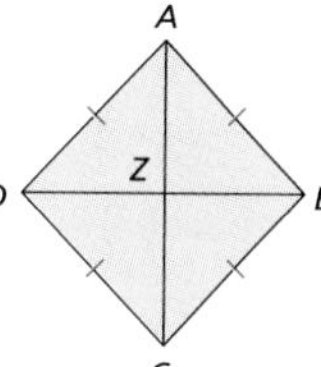

Mixed Review Exercises

Refer to the figure at the right for Exercises 1–3.

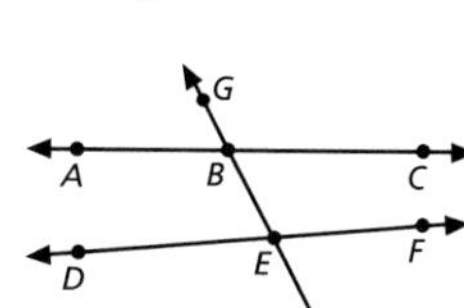

27. Name all the alternate exterior angles. $\angle ABG$ and $\angle FEH$, $\angle GBC$ and $\angle DEH$

28. Name all the corresponding angles. $\angle ABG$ and $\angle DEG$, $\angle GBC$ and $\angle GEF$, $\angle DEH$ and $\angle ABH$, $\angle CBH$ and $\angle FEH$

29. Name all the alternate interior angles. $\angle ABH$ and $\angle GEF$, $\angle CBH$ and $\angle GED$

Determine if each relation is a function. Give the domain and range.

30.

a	0	1	2	2	3
b	−1	−3	−4	−5	−6

no; domain: {−1, −3, −4, −5, −6}; range: {0, 1, 2, 3}

31.

x	2	3	4	5	6
y	4.5	6.5	8.5	10.5	12.5

yes; domain: {2, 3, 4, 5, 6}; range: {4.5, 6.5, 8.5, 10.5, 12.5}

32.

m	−1	0	1	0	−1
n	−4	−1	0	2	5

no; domain: {−4, −1, 0, 2, 5}; range: {−1, 0, 1}

Extra Practice Worksheet 4-8

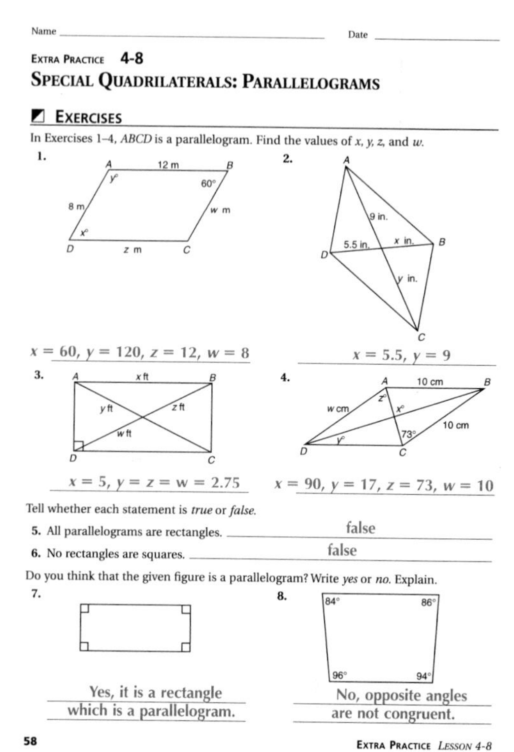

Name ____________ Date ________

EXTRA PRACTICE 4-8
SPECIAL QUADRILATERALS: PARALLELOGRAMS

EXERCISES

In Exercises 1–4, *ABCD* is a parallelogram. Find the values of *x, y, z,* and *w.*

1. $x = 60$, $y = 120$, $z = 12$, $w = 8$
2. $x = 5.5$, $y = 9$
3. $x = 5$, $y = z = w = 2.75$
4. $x = 90$, $y = 17$, $z = 73$, $w = 10$

Tell whether each statement is *true* or *false.*

5. All parallelograms are rectangles. false
6. No rectangles are squares. false

Do you think that the given figure is a parallelogram? Write *yes* or *no.* Explain.

7. Yes, it is a rectangle which is a parallelogram.
8. No, opposite angles are not congruent.

58 EXTRA PRACTICE *LESSON 4-8*

Enrichment Worksheet 4-8

Name ____________ Date ________

ENRICHMENT 4-8
A GEOMETRIC FALLACY: QUADRILATERALS

In this geometric fallacy, the theorem is true for some cases and not for others.

EXERCISES

1. Follow along with this "proof," marking the figure as directed and completing the statements.
Prove: If a quadrilateral has an opposite pair of congruent angles and an opposite pair of congruent sides, then the quadrilateral is a parallelogram.
   1. Given quadrilateral *ABCD* with $\angle A \cong \angle C$, and $\overline{AB} \cong \overline{CD}$.
   2. Draw $\overline{BX}$ perpendicular to $\overline{AD}$, and $\overline{DY}$ perpendicular to $\overline{BC}$. Join *B* and *D*.
   3. The right triangles *ABX* and *CDY* are congruent, so $\overline{BX} \cong$ __DY__ and $\overline{AX} \cong$ __CY__
   4. Therefore, $\triangle BXD \cong \triangle$ __DYB__.
   5. Therefore, $\overline{XD} \cong$ __YB__.
   6. $AX + XD = CY + YB$, so $AD =$ __CB__.
   7. $AB = CD$ and $AD = CB$, so the quadrilateral is a parallelogram.
2. In the figure above, both *X* and *Y* are on the quadrilateral. For which of these other cases is the proof valid? Draw the figures for the third and fourth cases.
X is on the quadrilateral, but *Y* is not. — Neither *X* nor *Y* are on the quadrilateral. — *Y* is on the quadrilateral, but *X* is not.
The proof is valid for the middle case, but not for the other two.
3. Explain why the proof is a fallacy.
The proof does not hold when *X* is on the quadrilateral and *Y* is not, or vice versa. It is true if both *X* and *Y* are either on or off the quadrilateral.

76 ENRICHMENT *LESSON 4-8*

has been proved, it is known that $AB = CD$ and $AD = CB$. Therefore, by the ASA postulate, there are two pairs of congruent triangles: $\triangle BEA$ and $\triangle DEC$, and $\triangle BEC$ and $\triangle DEA$. Because corresponding parts of congruent triangles are congruent, it follows that $AE \cong CE$, or $AD = DE$, and that $DE \cong BE$, or $DE = BE$. Therefore, point *E* is the midpoint of both *AC* and *BD*. It follows that *BD* bisects *AC* and *AC* bisects *BD*.

26. Answers will vary. Possible responses: *ABCD* is a rhombus; *ABCD* is a parallelogram; *ABCD* is a quadrilateral; $AB \cong BC \cong CD \cong DA$; $AZ \cong CZ$; $DZ \cong BZ$; $\angle DAB \cong \angle BCD$; $\angle ADC \cong \angle CBA$; $\angle DAZ \cong \angle BAZ \cong \angle BCZ \cong \angle DCZ$; $\angle ADB \cong \angle CDB \cong \angle CBD \cong \angle ABD$; $\triangle AZD \cong \triangle AZB \cong \triangle CZB \cong \triangle CZD$; $\triangle DAB \cong \triangle BCD$; $\triangle ADC \cong \triangle CBA$; $\angle AZD$, $\angle AZB$, $\angle CZB$, and $\angle CZD$ are right angles; $AC \perp DB$; *AC* is a perpendicular bisector of *DB*; *BD* is a perpendicular bisector of *AC*. Watch that students do not make the false assumption that *ABCD* is a square.

Skills Practice

Vocabulary Review

Lesson 4-7

polygon, convex, concave, consecutive sides, consecutive vertices, diagonal, interior angle, exterior angle, equilateral polygon, equiangular polygon, regular polygon, polyhedron

Lesson 4-8

opposite sides, opposite angles, parallelogram, rectangle, rhombus, square, diagonal

Assignment Guide

All students: 1–35

Additional Practice: Refer to the Extra Practice Index on page 674 of this text.

Review and Practice Your Skills

Practice ■ Lesson 4-7

Find the unknown angle measure or measures in each figure.

1. 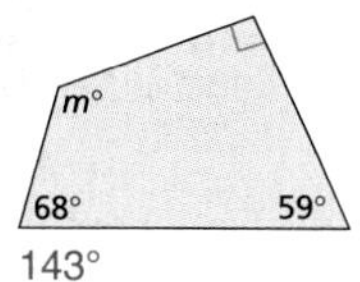

143°

2. 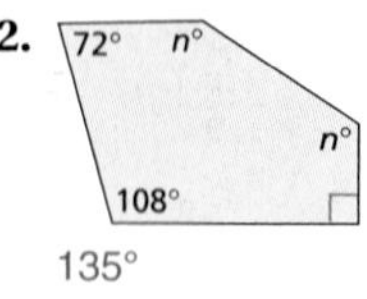

135°

3. 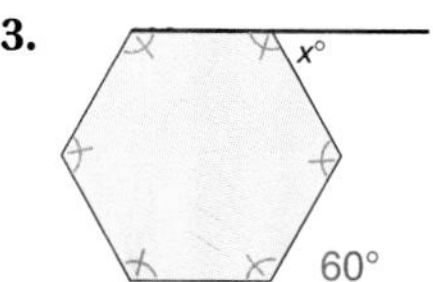

60°

4. Find the measure of each interior angle of a regular polygon with 13 sides. 152.3°
5. Find the measure of each exterior angle of a regular polygon with 20 sides. 18°
6. Find the sum of the measures of the interior angles of a regular heptagon. 900°
7. Find the sum of the measures of the exterior angles of a regular heptagon. 360°
8. Using diagonals from one vertex, into how many nonoverlapping triangular regions can you divide a nonagon? a polygon with 21 sides? 7, 19

Find the number of sides of each regular polygon.

9. The measure of each exterior angle is 40°. 9
10. The sum of the measures of interior angles is 2160°. 14
11. The measure of each interior angle is 165°. 24

Practice ■ Lesson 4-8

Determine whether each statement is *true* or *false*.

12. The diagonals of a rhombus are equal in length. false
13. Every square is a rhombus. true
14. Quadrilaterals include squares, parallelograms, pentagons, and rectangles. false
15. A square is a regular polygon. true
16. In all quadrilaterals, the opposite sides are equal in length. false

For the following parallelograms, find the values of *a*, *b*, *c*, and *d*.

17. 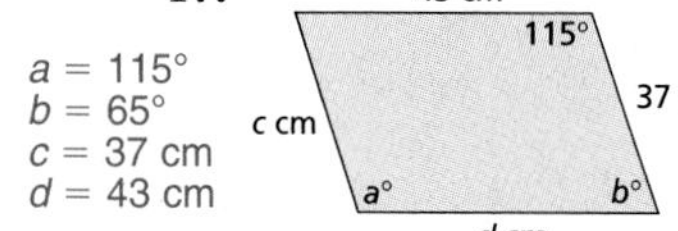

$a = 115°$
$b = 65°$
$c = 37$ cm
$d = 43$ cm

18. 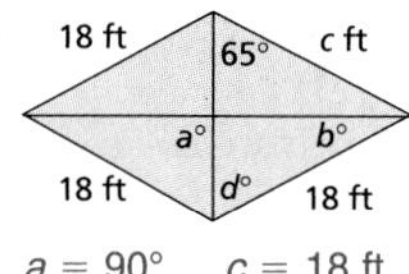

$a = 90°$, $c = 18$ ft
$b = 25°$, $d = 65°$

19. 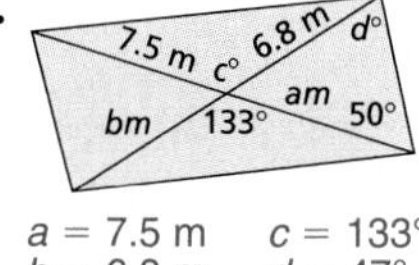

$a = 7.5$ m, $c = 133°$
$b = 6.8$ m, $d = 47°$

Is the given figure a parallelogram? Write *yes* or *no*. Then explain your reasoning.

20. 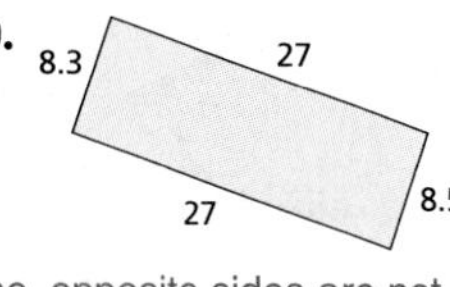

no, opposite sides are not equal

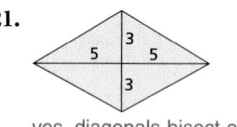

21. 3, 5, 5, 3
yes, diagonals bisect each other.

22. 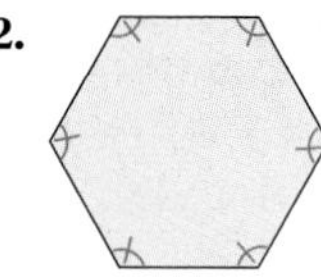
no, has six sides.

PRACTICE ■ LESSON 4-1–LESSON 4-8

Find the value of x in each figure. (Lesson 4-1)

23.

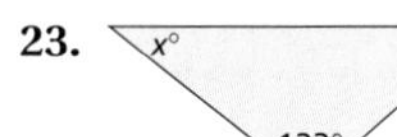

24.

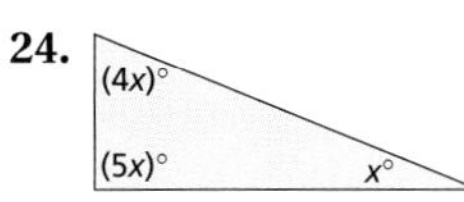

25.

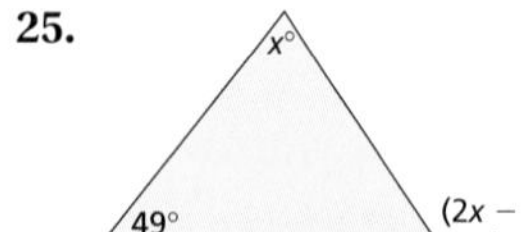

26. Copy and complete this proof. (Lesson 4-2)

Given $\overline{AE} \cong \overline{EC}$; $\overline{DE} \cong \overline{EB}$
Prove $\triangle DAE \cong$ ___?___

Statements	**Reasons**
1. ___?___ $\overline{AE} \cong \overline{EC}$; $\overline{DE} \cong \overline{EB}$	1. Given
2. ___?___ $\angle AED \cong \angle BEC$	2. Vertical Angles Theorem
3. $\triangle DAE \cong$ ___?___ $ABCE$	3. ___?___ SAS Postulate

Find the value of n in each figure. (Lesson 4-3)

27.

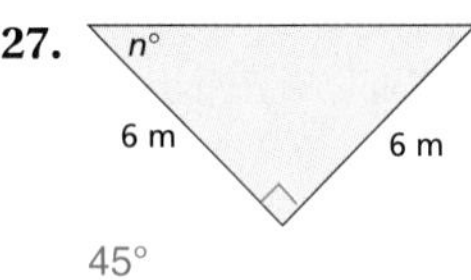

45°

28.

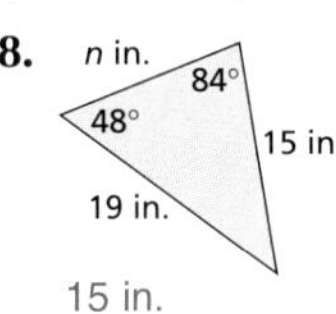

15 in.

29.

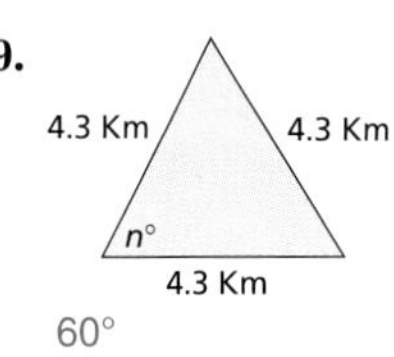

60°

Give the range of possible values for x in each figure. (Lesson 4-6)

30.

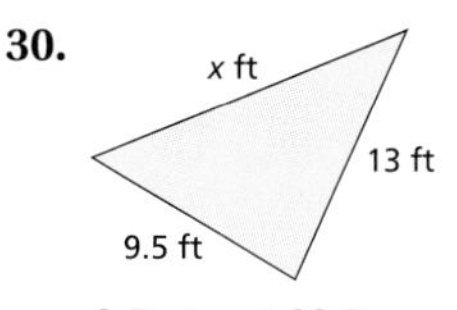

$3.5 < x < 22.5$

31.

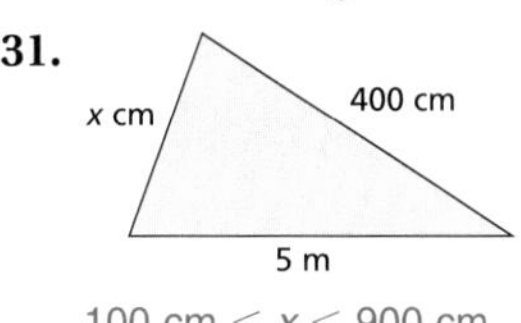

$100 \text{ cm} < x < 900 \text{ cm}$

32.

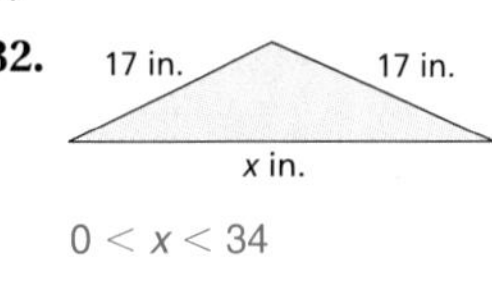

$0 < x < 34$

Find the unknown angle measure or measures in each figure. (Lesson 4-7)

33.

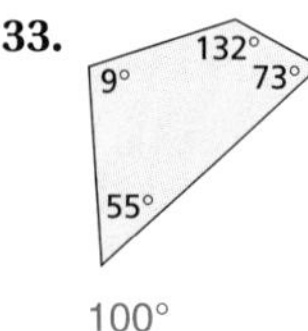

100°

34.

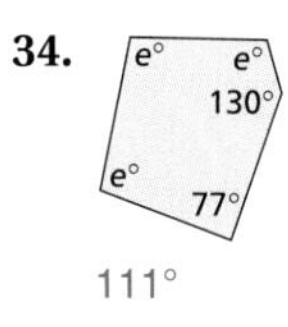

111°

35.

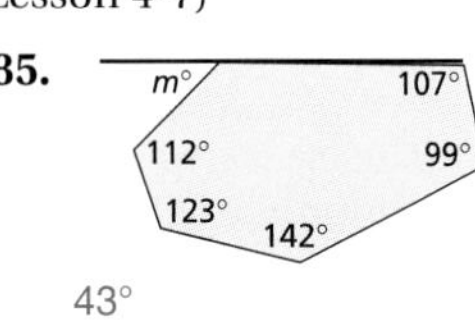

43°

Chalkboard Examples

Example from Lesson 4-7

Find the unknown angle in a hexagon with angles measuring 60°, 45°, 150°, 160°, and 160°. 145°

Example from Lesson 4-8

One angle of a parallelogram measures 72°. What are the measures of its other three angles? 72°, 108°, 108°

Lesson Planning

NCTM Standards/Strands

- Number & Operation
- Patterns, Functions, & Algebra
- Geometry & Spatial Sense
- Measurement
- Data Analy6sis, Statistics, & Probability
- Problem Solving
- Reasoning and Proof
- Communication
- Connections
- Representation

Vocabulary

trapezoid, base angles, median, kite, isosceles trapezoid

Materials Needed

scissors and heavy paper or sets of tangram pieces

Lesson Resources

Warm-Up Transparency 12
Reteaching 4-9
Extra Practice 4-9
Enrichment 4-9
Transparency RF-17

Assignment Guide

Basic: 1–28, 31–36
Enriched: 1–36

Getting Started

5-Minute Warm-Up

Find a value for the variable.

1. $1/2\,(a + 16) = 20$ 24
2. $1/2\,(b + 2b) = 15$ 10
3. $c + 28 = 2c - 12$ 40
4. $3d - 19 = 2d + 6$ 25

Introduction to Lesson 4-9

If sets of tangram pieces are not available, have students trace and cut the figure very precisely. Larger pieces can be made by cutting out a large square and carefully folding and cutting. Try to have each student in a group use a different colored set to avoid mixing up pieces from different sets.

4-9 Special Quadrilaterals: Trapezoids

Goals ■ Apply properties of trapezoids to find missing lengths and angle measures.

Applications Stage Design, Construction, Art

Work with a partner.

The *tangram* is an ancient Chinese puzzle consisting of the seven pieces shown at the right. Use a manufactured set of tangram pieces or trace the figure onto a sheet of paper and then cut out the pieces along the lines.

For 1–6, see additional answers.

1. Arrange pieces *E, F,* and *G* to form a rectangle.
2. Arrange *E, F,* and *G* to form a parallelogram that is not a rectangle.
3. Arrange *E, F,* and *G* to form a quadrilateral that is not a parallelogram.
4. Arrange pieces *A, C, E,* and *G* to form a square.
5. Arrange all seven tangram pieces to form a quadrilateral that is not a parallelogram.
6. Form as many different rectangles that are not squares as possible. (For each rectangle, use as many tangram pieces as needed.)

Build Understanding

A **trapezoid** is a quadrilateral with exactly one pair of parallel sides. The parallel sides are called the **bases** of the trapezoid. Two consecutive angles that share a base form a pair of **base angles**; every trapezoid has two pairs of base angles. The nonparallel sides are called the **legs**.

$\overline{AB} \parallel \overline{DC}$

ABCD is a trapezoid.

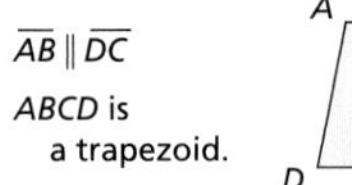

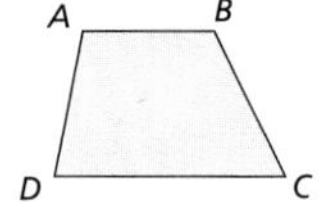

bases: $\overline{AB}$ and $\overline{DC}$
legs: $\overline{AD}$ and $\overline{BC}$
base angles: $\angle A$ and $\angle B$; $\angle D$ and $\angle C$

The **median** of a trapezoid is the segment that joins the midpoints of the legs. Two important properties of the median are stated in the following theorem, which will be accepted as true without proof.

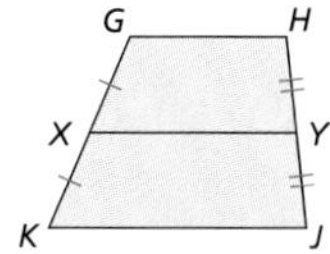

$\overline{GH} \parallel \overline{KJ}$
$\overline{XY}$ is the median of trapezoid *GHJK*.

The Trapezoid-Median Theorem

If a segment is the median of a trapezoid, then it is:

1. parallel to the bases; and
2. equal in length to one half the sum of the lengths of the bases.

Teaching Strategies

Have students trace the trapezoids 188–189 and draw their diagonals. Ask: Why do you think there are no theorems included in this lesson regarding the diagonals of a trapezoid? Unless a trapezoid is isosceles, its diagonals are different lengths, and they do not bisect each other.

Example 1

STAGE DESIGN The plans for two panels of a stage setting are shown in the figure at the right. In the figure, $\overline{QT} \parallel \overline{RS}$. Find AB.

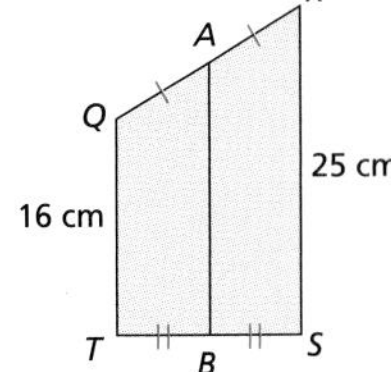

Solution

Quadrilateral $QRST$ is a trapezoid.
$\overline{QT}$ and $\overline{RS}$ are the bases, and $\overline{AB}$ is the median.
To find AB, apply the trapezoid-median theorem.

$$AB = \frac{1}{2}(QT + RS)$$
$$AB = \frac{1}{2}(16 + 25)$$
$$AB = \frac{1}{2}(41)$$
$$AB = 20.5$$

So, the length of $\overline{AB}$ is 20.5 cm.

A trapezoid with legs of equal length is called an **isosceles trapezoid**.

$\overline{PQ} \parallel \overline{SR}$
$PS = QR$
$PQRS$ is an isosceles trapezoid.

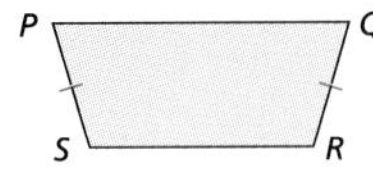

The following theorem states an important fact about isosceles trapezoids. This theorem also will be accepted as true without proof.

The Isosceles Trapezoid Theorem	If a quadrilateral is an isosceles trapezoid, then its base angles are equal in measure.

Technology Note

A **kite** is a quadrilateral that has two distinct pairs of consecutive sides of the same length.

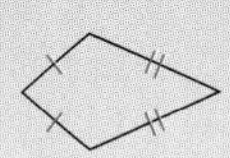

Draw a kite using geometric drawing software. Use the figure to explore the following questions.

1. What relationship exists among the angles of a kite?
2. What relationship exist between the diagonals of a kite?
3. Connect the midpoints of the sides of the kite. What type of figure do you obtain?

1. One pair of opposite angles is equal in measure.
2. They are perpendicular; one diagonal bisects the other.
3. a rectangle.

Example 2

In the figure at the right, $\overline{ST} \parallel \overline{WV}$. Find $m\angle V$.

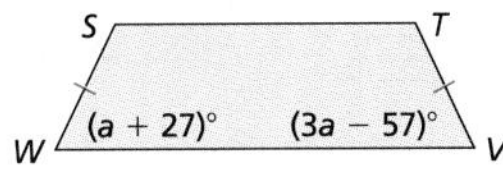

Solution

Quadrilateral $STVW$ is an isosceles trapezoid, with bases $\overline{ST}$ and $\overline{WV}$. So, $\angle W$ and $\angle V$ are a pair of base angles, and they are equal in measure. Use this fact to write and solve an equation.

$a + 27 = 3a - 57$	Add $-a$ to each side.
$a + 27 + (-a) = 3a - 57 + (-a)$	Combine like terms.
$27 = 2a - 57$	Add 57 to each side.
$27 + 57 = 2a - 57 + 57$	
$84 = 2a$	Multiply each side by $\frac{1}{2}$.
$42 = a$	

So, the value of a is 42. From the figure, $m\angle V = (3a - 57)°$.
Substituting 42 for a, $m\angle V = (3 \cdot 42 - 57)° = (126 - 57)° = 69°$.

Check Understanding

In Example 2, what is the measure of $\angle S$? $\angle T$?

111°; 111°

Chalkboard Examples

Supplementary Example 1

The plan for the stage setting also shows the panels below. Use the figure at the bottom of page 188. $\overline{GN} \parallel \overline{KJ}$. $GH = 3.4$ cm. $KJ = 6.6$ cm. Find XY. 5 cm

Supplementary Example 2

Use the figure from Example 2. Change $m\angle W$ to $(8x + 2x)°$ and $m\angle V$ to $(10x - 6)°$. Find $m\angle D$. 44°

Additional Answers

1. 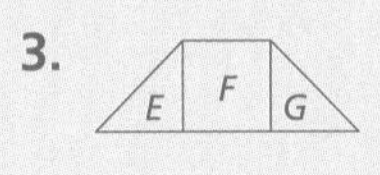

2. 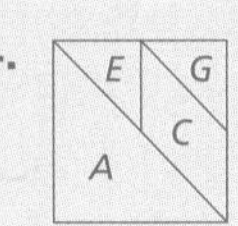

3. 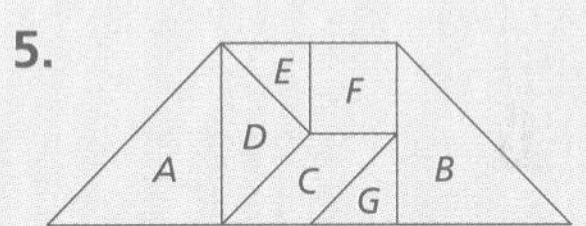

4.

5.

6.

Reteaching Worksheet 4-9

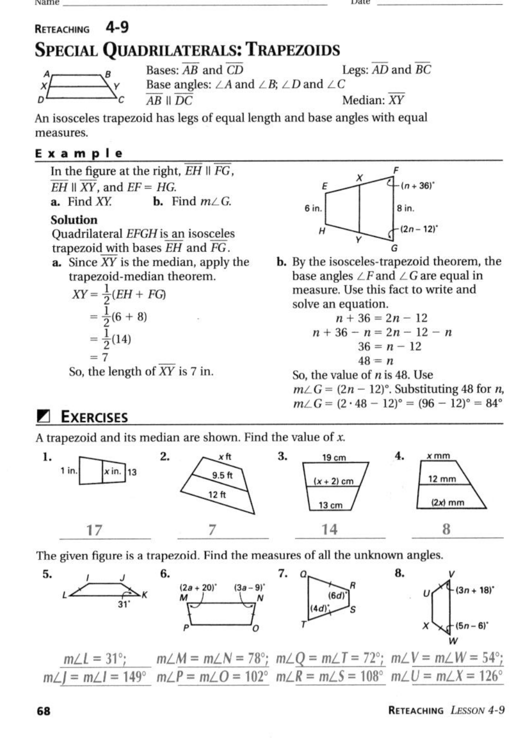

Name ______________ Date ______________

RETEACHING 4-9

SPECIAL QUADRILATERALS: TRAPEZOIDS

Bases: $\overline{AB}$ and $\overline{CD}$ Legs: $\overline{AD}$ and $\overline{BC}$
Base angles: $\angle A$ and $\angle B$; $\angle D$ and $\angle C$
$\overline{AB} \parallel \overline{DC}$ Median: $\overline{XY}$

An isosceles trapezoid has legs of equal length and base angles with equal measures.

Example

In the figure at the right, $\overline{EH} \parallel \overline{FG}$, $\overline{EH} \parallel \overline{XY}$, and $EF = HG$.

a. Find XY. b. Find $m\angle G$.

Solution

Quadrilateral $EFGH$ is an isosceles trapezoid with bases $\overline{EH}$ and $\overline{FG}$.

a. Since $\overline{XY}$ is the median, apply the trapezoid-median theorem.

$$XY = \frac{1}{2}(EH + FG) = \frac{1}{2}(6 + 8) = \frac{1}{2}(14) = 7$$

So, the length of $\overline{XY}$ is 7 in.

b. By the isosceles-trapezoid theorem, the base angles $\angle F$ and $\angle G$ are equal in measure. Use this fact to write and solve an equation.

$$n + 36 = 2n - 12$$
$$n + 36 - n = 2n - 12 - n$$
$$36 = n - 12$$
$$48 = n$$

So, the value of n is 48. Use $m\angle G = (2n - 12)°$. Substituting 48 for n, $m\angle G = (2 \cdot 48 - 12)° = (96 - 12)° = 84°$

EXERCISES

A trapezoid and its median are shown. Find the value of x.

1. 17
2. 7
3. 14
4. 8

The given figure is a trapezoid. Find the measures of all the unknown angles.

5. $m\angle L = 31°$; $m\angle J = m\angle I = 149°$
6. $m\angle M = m\angle N = 78°$; $m\angle P = m\angle O = 102°$
7. $m\angle Q = m\angle T = 72°$; $m\angle R = m\angle S = 108°$
8. $m\angle V = m\angle W = 54°$; $m\angle U = m\angle X = 126°$

68 RETEACHING *LESSON* 4-9

Lesson Wrap-up

Quick Assessment

Ask the following questions to determine if students understand the content presented in this lesson.

Match the property named with all the parallelograms that have that property. Choose from *parallelogram, rectangle, rhombus, square, trapezoid* and *isosceles trapezoid*.

1. at least one pair of parallel sides all
2. all opposite angles equal parallelogram, rectangle, rhombus, square
3. two pairs of equal consecutive angles rectangle, square, isosceles trapezoid
4. two right angles possible parallelogram, rectangle, square, trapezoid
5. diagonals equal rectangle, square, isosceles trapezoid
6. exactly one pair of congruent sides isosceles trapezoid

Assignment Guide

Basic: 1–28, 31–36
Advanced: 1–36
Additional Practice: See Extra Practice Index on page 674.

Additional Answers

16. rectangle, parallelogram, quadrilateral
17. quadrilateral
18. rhombus, parallelogram, quadrilateral
19. trapezoid, quadrilateral
20. parallelogram, quadrilateral
21. isosceles triangle, trapezoid, quadrilateral

Try These Exercises

A trapezoid and its median are shown. Find the value of x.

1. 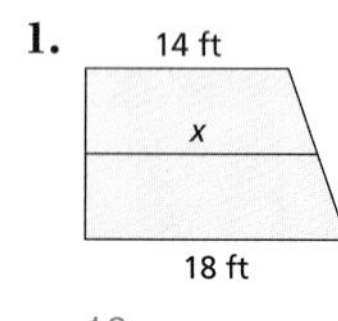

16

2. 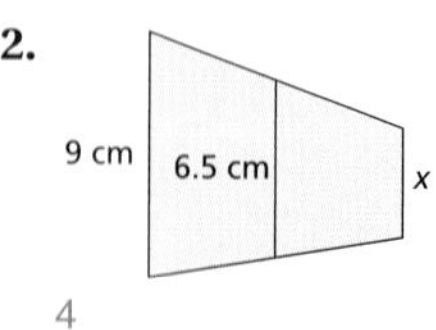

4

3. 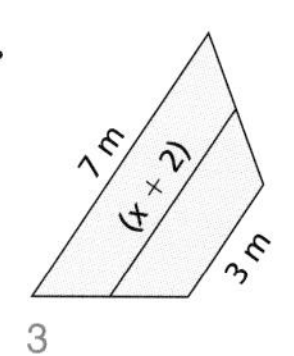

3

CONSTRUCTION The given figures are part of a design for a wrought-iron railing. Find all unknown angle measures.

4. 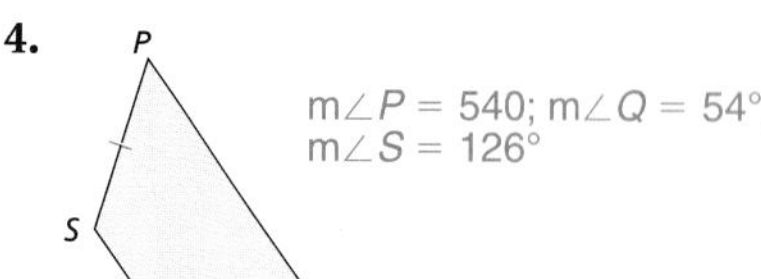

$m\angle P = 540$; $m\angle Q = 54°$; $m\angle S = 126°$

5. 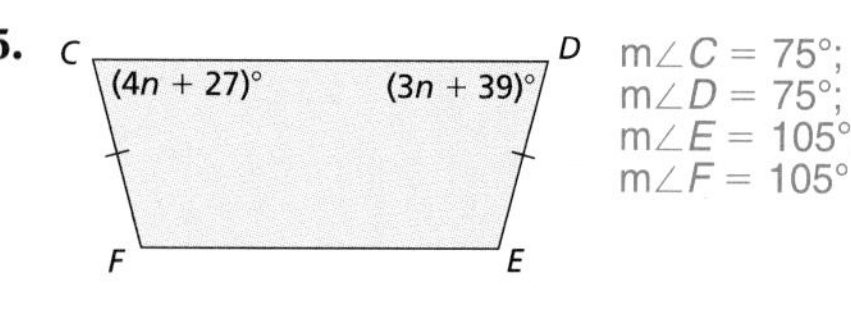

$m\angle C = 75°$; $m\angle D = 75°$; $m\angle E = 105°$; $m\angle F = 105°$

Practice Exercises

A trapezoid and its median are shown. Find the value of z.

6. 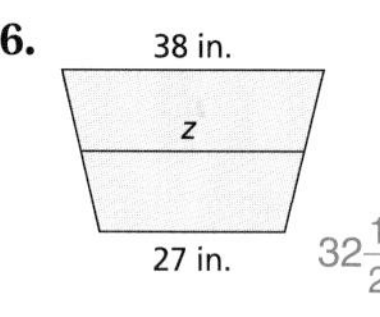

$32\frac{1}{2}$

7. 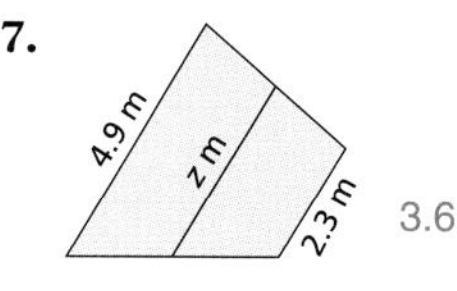

3.6

8. 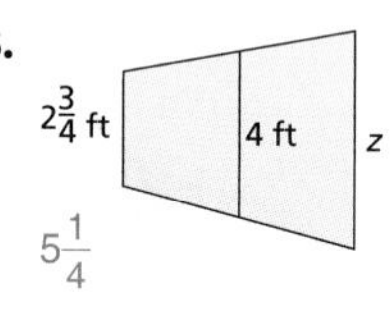

$5\frac{1}{4}$

9. 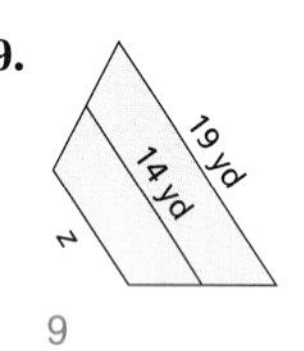

9

10. 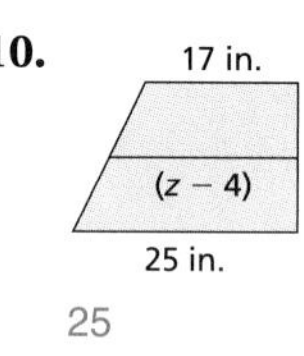

25

11. 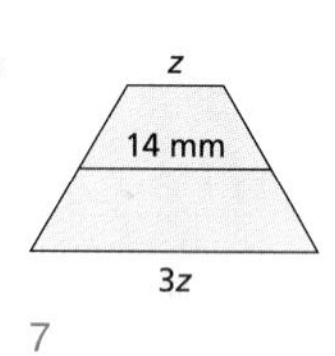

7

The given figure is a trapezoid. Find all the unknown angle measures.

12. 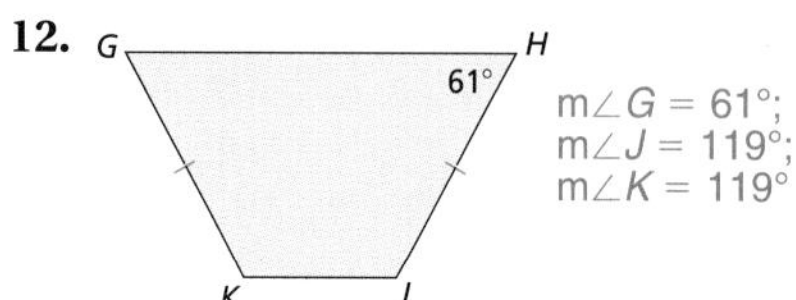

$m\angle G = 61°$; $m\angle J = 119°$; $m\angle K = 119°$

13. 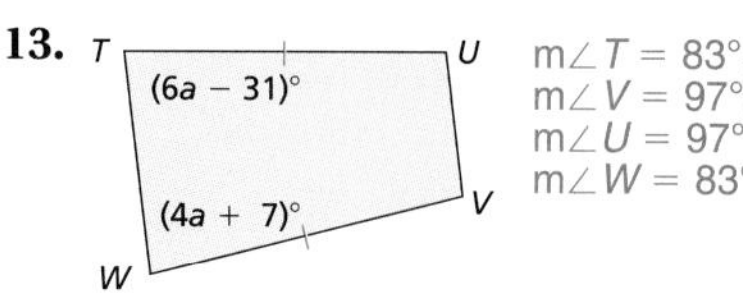

$m\angle T = 83°$; $m\angle V = 97°$; $m\angle U = 97°$; $m\angle W = 83°$

14. CHAPTER INVESTIGATION Make a list of the quadrilaterals that you can see in your truss bridge design. Compare your design with those of your classmates. Which design do you think will support the most weight? Why?

15. WRITING MATH Compare the median of a trapezoid to the median of a triangle. How are they alike? How are they different?
Answers may vary. *Possible likeness*: Each has an endpoint at the midpoint of a side of the figure. *Possible differences:* A median of a triangle has one endpoint that is also a vertex of the figure, whereas the median of a trapezoid does not; a triangle has three medians, whereas a trapezoid has only one.

29. Given: *ABCD* is a trapezoid.
Prove: *ADC* and *DAB* are supplementary.

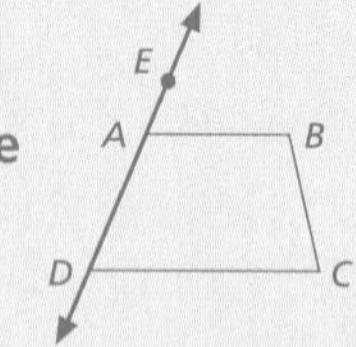

Statements	Reasons
1. *ABCD* is a trapezoid	1. given
2. $\overline{AB} \parallel \overline{DC}$	2. definition of trapezoid
3. $m\angle EAB = m\angle ADC$	3. corr. $\angle s$ postulate
4. $m\angle EAB + m\angle DAB = 180°$	4. angle addition postulate
5. $m\angle ADC + m\angle DAB = 180°$	5. substitution property
6. $\angle ADC$ and $\angle DAC$ are supplementary	6. Definition of supplementary angles

In Exercises 16–21, give as many names as are appropriate for the given figure. Choose from *quadrilateral, parallelogram, rhombus, rectangle, square, trapezoid,* and *isosceles trapezoid.* Then underline the *best* name for the figure.
For 16–21, see additional answers.

16.
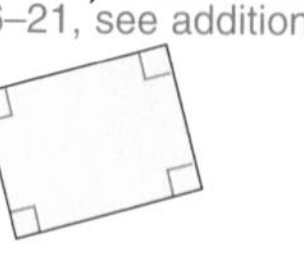

17.
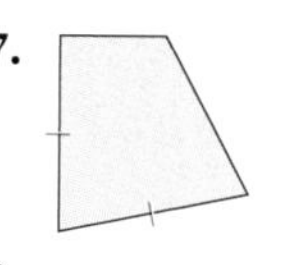

18.
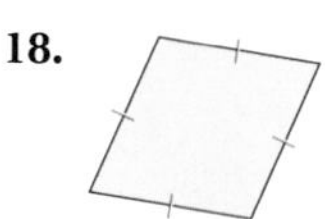

19.
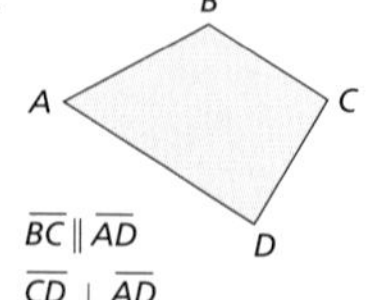

20.
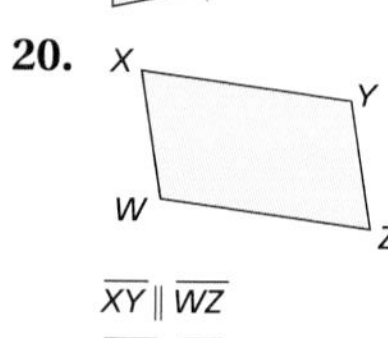

21.
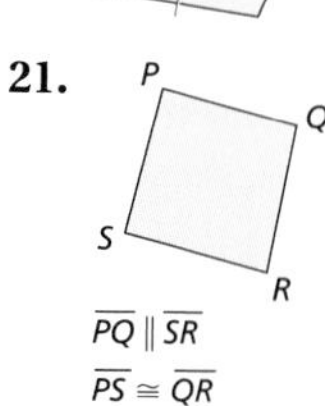

Copy and complete the following table that summarizes what you have learned about quadrilaterals. For each entry, write *yes* or *no.*

	Property	Quadrilateral	Parallelogram	Rectangle	Rhombus	Square	Trapezoid	Isos. Trap.
22.	sum of interior angles 360°	■ yes	■ yes	■ yes	■ yes	■ yes	■ yes	■ yes
23.	all opposite sides equal in length	■ no	■ yes	■ yes	■ yes	■ yes	■ no	■ no
24.	all opposite angles equal in measure	■ no	■ yes	■ yes	■ yes	■ yes	■ no	■ no
25.	diagonals bisect each other	■ no	■ yes	■ yes	■ yes	■ yes	■ no	■ no
26.	diagonals are perpendicular	■ no	■ no	■ no	■ yes	■ yes	■ no	■ no
27.	diagonals equal in length	■ no	■ no	■ yes	■ no	■ yes	■ no	■ yes
28.	diagonals bisect vertex angles	■ no	■ no	■ no	■ yes	■ yes	■ no	■ no

Extended Practice Exercises

29. ART The side view of the marble base of a statue is a trapezoid, shown at the right. Prove that $\angle A$ and $\angle D$ are supplementary. (*Hint:* Extend $\overline{AD}$ to show $\overleftrightarrow{AD}$.) See additional answers.

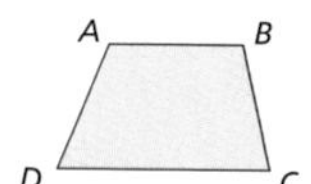

30. What type of figure do you obtain if you join the midpoints of the sides of an isosceles trapezoid? rhombus

Mixed Review Exercises

Use the number line at the right for Exercises 31–36. Find each length. (Lesson 3-1)

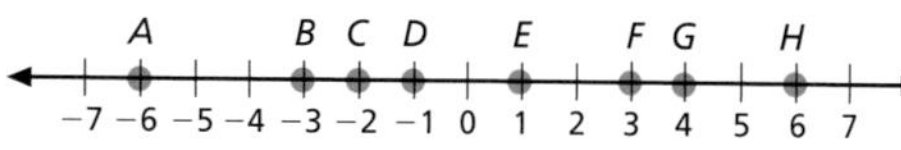

31. $\overline{AF}$ 9 **32.** $\overline{BE}$ 4 **33.** $\overline{DG}$ 5

34. $\overline{AH}$ 12 **35.** $\overline{CH}$ 8 **36.** $\overline{DF}$ 4

Teaching Strategies

CHALLENGE Have students work in pairs with sets of tangram pieces. Ask them to find which of these shapes can be made using all 7 pieces in the set: square, triangle, rectangle, parallelogram, trapezoid, pentagon, and hexagon. Check students' work. All can be made although the pentagon and hexagon will be irregular.

Extra Practice Worksheet 4-9

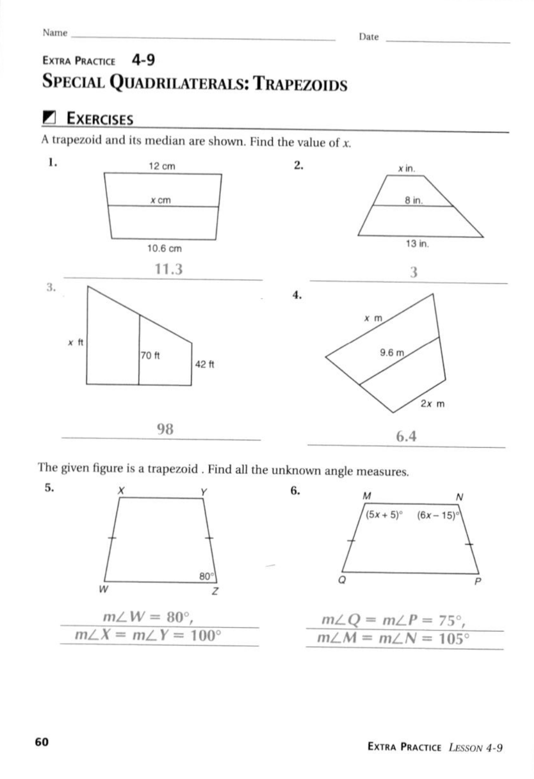
Name ______ Date ______

EXTRA PRACTICE 4-9
SPECIAL QUADRILATERALS: TRAPEZOIDS

EXERCISES

A trapezoid and its median are shown. Find the value of x.

1. 12 cm; x cm; 10.6 cm — 11.3
2. x in.; 8 in.; 13 in. — 3
3. x ft; 70 ft; 42 ft — 98
4. x m; 9.6 m; $2x$ m — 6.4

The given figure is a trapezoid. Find all the unknown angle measures.

5. X, Y, W, Z; 80° — $m\angle W = 80°$, $m\angle X = m\angle Y = 100°$
6. M, N, Q, P; $(5x + 5)°$, $(6x - 15)°$ — $m\angle Q = m\angle P = 75°$, $m\angle M = m\angle N = 105°$

60 EXTRA PRACTICE *LESSON 4-9*

Enrichment Worksheet 4-9

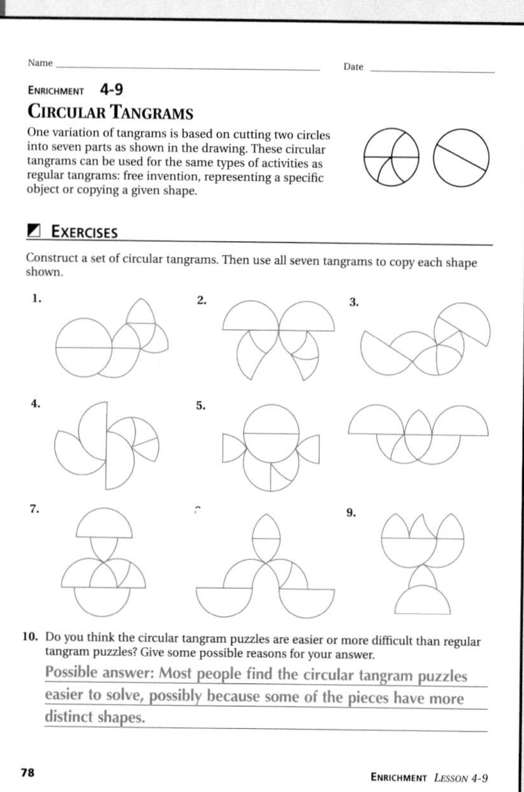
Name ______ Date ______

ENRICHMENT 4-9
CIRCULAR TANGRAMS

One variation of tangrams is based on cutting two circles into seven parts as shown in the drawing. These circular tangrams can be used for the same types of activities as regular tangrams: free invention, representing a specific object or copying a given shape.

EXERCISES

Construct a set of circular tangrams. Then use all seven tangrams to copy each shape shown.

1. 2. 3. 4. 5. 7. 9.

10. Do you think the circular tangram puzzles are easier or more difficult than regular tangram puzzles? Give some possible reasons for your answer.
Possible answer: Most people find the circular tangram puzzles easier to solve, possibly because some of the pieces have more distinct shapes.

78 ENRICHMENT *LESSON 4-9*

Assessment Planning

Vocabulary Review

Lesson 4-1
triangle, side, vertex, interior angle, obtuse triangle, right triangle, scalene triangle, acute triangle, equilateral triangle, isosceles triangle, equiangular triangle, auxiliary line, exterior angle

Lesson 4-2
congruent, SSS postulate, SAS postulate, ASA postulate, included side, included angle

Lesson 4-3
base, base angles, legs, vertex angle, corollary, isosceles triangle theorem

Lesson 4-4
altitude, median, perpendicular bisector, concurrent lines, center of gravity

Lesson 4-5
indirect proof

Lesson 4-6
shortest path postulate, triangle inequality theorem, unequal sides theorem, unequal angles theorem

Lesson 4-7
polygon, convex, concave, consecutive sides, consecutive vertices, diagonal, interior angle, exterior angle, equilateral polygon, equiangular polygon, regular polygon, polyhedron

Lesson 4-8
opposite sides, opposite angles, parallelogram, rectangle, rhombus, square, diagonal

Lesson 4-9
trapezoid, base angles, median, kite, isosceles trapezoid

Assessment Options

Chapter Assessment
Alternative Assessment
Chapter Tests, Forms A and B

Chapter 4 Review

Vocabulary ■ LESSON 4-1–LESSON 4-9

Choose the word from the list that best completes each statement.

a. midpoint
b. trapezoid
c. parallelogram
d. congruent
e. diagonal

1. When two geometric figures have the same size and shape, they are said to be __?__. d
2. If a point lies on the __?__ of a segment, then the point is equidistant from the endpoints of the segment. a
3. A __?__ is a quadrilateral with both pairs of opposite sides parallel. c
4. A __?__ of a polygon is a segment that joins two nonconsecutive vertices. e
5. A __?__ is a quadrilateral with exactly one pair of parallel sides. b

LESSON 4-1 ■ Triangles and Triangle Theorems, p. 150

▶ The sum of the measures of the angles of a triangle is 180°.

▶ The measure of the exterior angle of a triangle is equal to the sum of the measures of the two nonadjacent (remote) interior angles.

Find the value of x in each figure.

6. 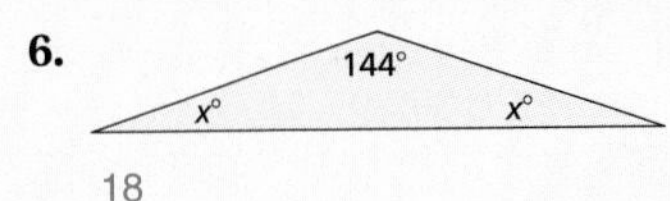

18

7. 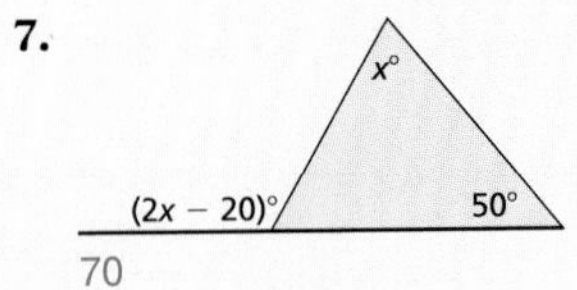

70

8. 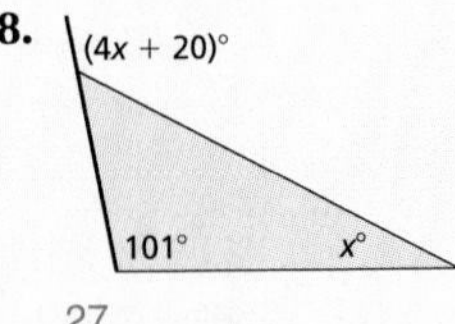

27

LESSON 4-2 ■ Congruent Triangles, p. 154

▶ Three postulates for proving that two triangles are congruent are the SSS (Side-Side-Side) Postulate, the SAS (Side-Angle-Side) Postulate, and the ASA (Angle-Side-Angle) Postulate.

In each case, name a pair of congruent triangles. Then name the postulate you could use to prove the triangles congruent. You do not need to write a proof.

9. 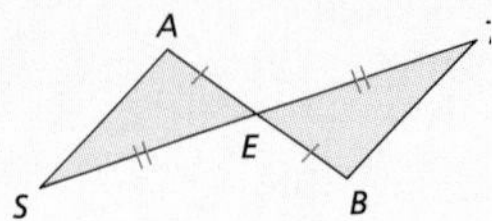

$\triangle AES \cong \triangle BET$, SAS

10. M N 1 2 3 4 T S
$\overline{MN} \parallel \overline{TS}$, $\angle 1 \cong \angle 4$, $\angle 2 \cong \angle 3$ $\triangle MNS \cong \triangle STM$, ASA

11. 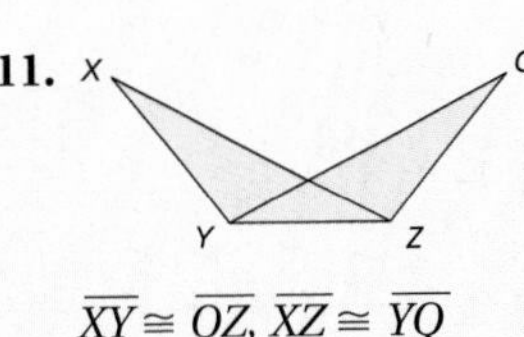

$\overline{XY} \cong \overline{QZ}$, $\overline{XZ} \cong \overline{YQ}$
$\triangle XYZ \cong \triangle QZY$, SSS

Cumulative Assessment
Achievement Test
Writing Prompts

Assignment Guide

All students: 1–20
Additional Practice: Refer to the Extra Practice Index on page 674 of this text.

Additional Answers

15. Sample answer;

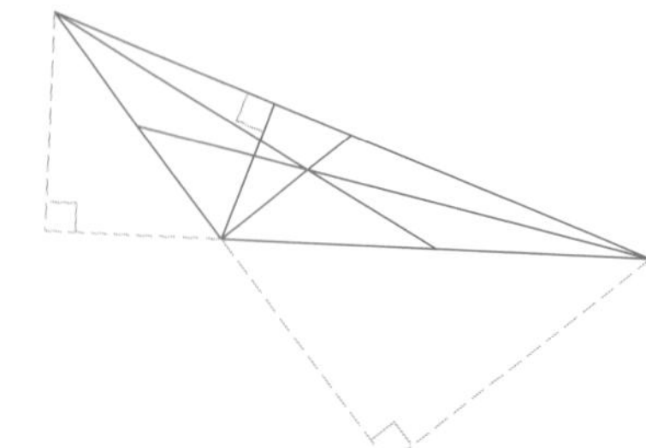

LESSON 4-3 ■ Congruent Triangles and Proofs, p. 160

▶ Base angles of an isosceles triangle are congruent.

Find the value of x in each figure.

12.

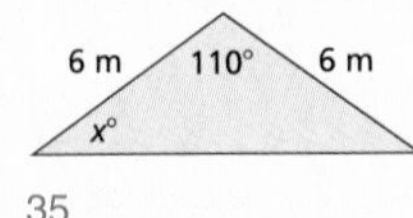

35

13.

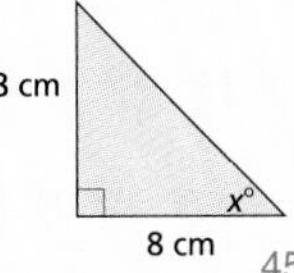

45

14.

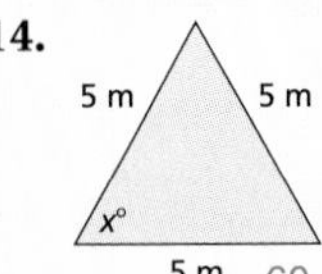

60

LESSON 4-4 ■ Altitudes, Medians, and Perpendicular Bisectors, p. 164

▶ An altitude of a triangle is the perpendicular segment from a vertex to the line containing the opposite side. A median of a triangle is a segment whose endpoints are on a vertex of the triangle and the midpoint of the opposite side.

15. Draw an obtuse triangle. Sketch all the altitudes and the medians.
See additional answers.

LESSON 4-5 ■ Problem Solving Skills: Write an Indirect Proof, p. 170

▶ To write an indirect proof, the first step is to assume temporarily that the conclusion is false.

16. Suppose you are asked to write an indirect proof of this statement: *If a triangle is obtuse, then it cannot have a right angle.* Write Step 1 of the proof.
Assume a triangle has a right angle

LESSON 4-6 ■ Inequalities in Triangles, p. 172

▶ The sum of the lengths of two sides in a triangle is greater than the length of the third side.

17. Can 19 cm, 10 cm, and 8 cm be the lengths of the sides of a triangle? Explain.
No; $10 + 8 < 19$

LESSON 4-7 ■ Polygons and Angles, p. 178

▶ The sum of the measures of the angles of a convex polygon with n sides is $(n - 2)180°$.

18. Find the sum of the measures of the angles of a polygon with 11 sides. 1,620°

LESSONS 4-8 and 4-9 ■ Parallelograms and Trapezoids, p. 182, p. 188

▶ If a figure is a parallelogram, the opposite sides are equal in length, the opposite angles are equal in measure, and the diagonals bisect each other.

▶ The length of the median of a trapezoid equals half the sum of the lengths of the bases.

19. The median of a trapezoid is 25 cm in length and one of the bases is 20 cm in length. Find the length of the other base. 30 cm

20. In parallelogram $WXYZ$, $WX = 8$ cm and $WZ = 8$ cm. Diagonals $\overline{XY}$ and $\overline{WY}$ intersect at point O, and $m\angle WXY = 70°$. Find $m\angle WXO$, $m\angle XWZ$, and $m\angle XYO$. What type of parallelogram is $WXYZ$? 35°, 110°, 55°, rhombus

Extend the Lesson

MATH JOURNAL Have students write a short explanation of how they could use information about sides, angles, and diagonals to determine whether a parallelogram is a rectangle, rhombus, square, or none of these.

Chapter 4 Test Form A, page 1

CHAPTER 4

TRIANGLES, QUADRILATERALS, AND OTHER POLYGONS

ASSESSMENT FORM A, PAGE 1

Name ____________

Date ____________

Scoring Record	
Possible: 21	Earned:

Find the value of x in each figure.

1. 120

2. 68

3. Complete the two-column proof.

Given: $\overline{QN} \cong \overline{ON}$
Point N is the midpoint of $\overline{MP}$
Prove: $\triangle MNQ \cong \triangle PNO$

STATEMENTS	REASONS
1. $\overline{QN} \cong \overline{ON}$ Point N is midpoint of $\overline{MP}$	1. given
2. $\overline{MN} \cong \overline{PN}$, or $MN = PN$	2. definition of midpoint
3. $\angle MNQ$ and $\angle PNO$ are vertical angles	3. definition of vertical angles
4. $\angle MNQ \cong \angle PNO$	4. vertical $\angle$s theorem
5. $\triangle MNQ \cong \triangle PNO$	5. SAS postulate

Find the value of y in each figure.

4. 9

5. 70

Refer to $\triangle ABC$ at the right. Tell whether each statement is true or false.

6. $\overline{AD} \cong \overline{DC}$ true

7. $\overline{AB} \cong \overline{BC}$ true

8. $\overline{DB} \cong \overline{BC}$ false

9. $\triangle ABD \cong \triangle CBD$ true

10. $\overline{DB}$ is a perpendicular bisector. true

44 ASSESSMENT *CHAPTER 4 FORM A*

Chapter 4 Test Form A, page 2

Name ____________ Date ____________

The lengths of two sides of a triangle are given. Find the range of lengths for the third side.

11. 6 cm, 9 cm $3 \text{ cm} < x < 15 \text{ cm}$

12. 1.2 m, 5 m $3.8 \text{ m} < x < 6.2 \text{ m}$

13. $1\frac{1}{3}$ yd, $2\frac{1}{3}$ yd $1 \text{ yd} < x < 3\frac{2}{3} \text{ yd}$

14. In $\triangle CDF$, $m\angle C = 63°$ and $m\angle F = 88°$. List the sides of the triangle in order from longest to shortest. $\overline{CD}, \overline{DF}, \overline{CF}$

15. Find the measures of each interior angle of a regular polygon with 40 sides. 171°

16. Find the measures of each exterior angle of a regular polygon with 16 sides. 22.5°

Find the values of a, b, c and d in each parallelogram.

17. $a = 3, b = 123, c = 57, d = 5$

18. $a = 45, b = 90, c = 3.5, d = 3.5$

19. Refer to the trapezoid and median below. Find the value of r.
9

20. Refer to the trapezoid below. Find the value of all the unknown angle measures.
$m\angle G = m\angle H = 68°$; $m\angle J = m\angle I = 112°$

21. Suppose you are asked to write an indirect proof of this statement.
Write Step 1 of the proof.
If $\overline{AB}$ is congruent to $\overline{DB}$, then $\overline{AB}$ is not congruent to $\overline{BC}$.
If $\overline{AB} \cong \overline{DB}$, then $\overline{AB} \cong \overline{BC}$.

46 ASSESSMENT *CHAPTER 4 FORM A*

Chapter 4 Test Form B, page 1

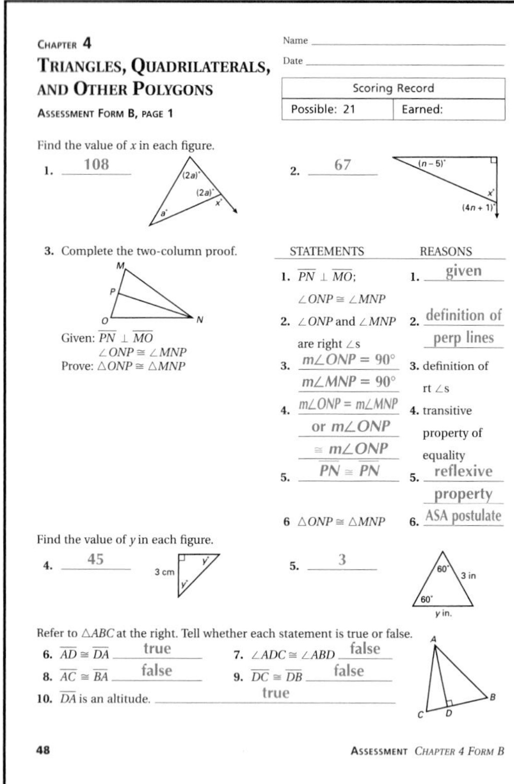

CHAPTER 4
TRIANGLES, QUADRILATERALS, AND OTHER POLYGONS
ASSESSMENT FORM B, PAGE 1

Name ____________
Date ____________

Scoring Record	
Possible: 21	Earned:

Find the value of x in each figure.

1. 108
2. 67

3. Complete the two-column proof.

Given: $\overline{PN} \perp \overline{MO}$, $\angle ONP \cong \angle MNP$
Prove: $\triangle ONP \cong \triangle MNP$

STATEMENTS	REASONS
1. $\overline{PN} \perp \overline{MO}$; $\angle ONP \cong \angle MNP$	1. given
2. $\angle ONP$ and $\angle MNP$ are right $\angle$s	2. definition of perp lines
3. $m\angle ONP = 90°$, $m\angle MNP = 90°$	3. definition of rt $\angle$s
4. $m\angle ONP = m\angle MNP$ or $m\angle ONP = m\angle ONP$	4. transitive property of equality
5. $\overline{PN} \cong \overline{PN}$	5. reflexive property
6. $\triangle ONP \cong \triangle MNP$	6. ASA postulate

Find the value of y in each figure.

4. 45
5. 3

Refer to $\triangle ABC$ at the right. Tell whether each statement is true or false.

6. $\overline{AD} \cong \overline{DA}$ true
7. $\angle ADC \cong \angle ABD$ false
8. $\overline{AC} \cong \overline{BA}$ false
9. $\overline{DC} \cong \overline{DB}$ false
10. $\overline{DA}$ is an altitude. true

48 ASSESSMENT CHAPTER 4 FORM B

Chapter 4 Test Form B, page 2

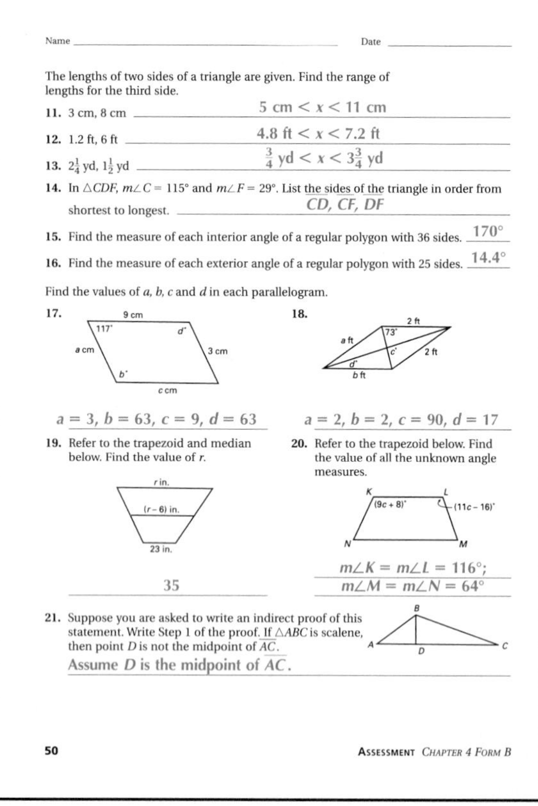

Name ____________ Date ____________

The lengths of two sides of a triangle are given. Find the range of lengths for the third side.

11. 3 cm, 8 cm $5 \text{ cm} < x < 11 \text{ cm}$
12. 1.2 ft, 6 ft $4.8 \text{ ft} < x < 7.2 \text{ ft}$
13. $2\frac{1}{4}$ yd, $1\frac{1}{2}$ yd $\frac{3}{4} \text{ yd} < x < 3\frac{3}{4} \text{ yd}$
14. In $\triangle CDF$, $m\angle C = 115°$ and $m\angle F = 29°$. List the sides of the triangle in order from shortest to longest. CD, CF, DF
15. Find the measure of each interior angle of a regular polygon with 36 sides. 170°
16. Find the measure of each exterior angle of a regular polygon with 25 sides. 14.4°

Find the values of a, b, c and d in each parallelogram.

17. $a = 3$, $b = 63$, $c = 9$, $d = 63$
18. $a = 2$, $b = 2$, $c = 90$, $d = 17$

19. Refer to the trapezoid and median below. Find the value of r. 35
20. Refer to the trapezoid below. Find the value of all the unknown angle measures. $m\angle K = m\angle L = 116°$; $m\angle M = m\angle N = 64°$
21. Suppose you are asked to write an indirect proof of this statement. Write Step 1 of the proof. If $\triangle ABC$ is scalene, then point D is not the midpoint of $\overline{AC}$. Assume D is the midpoint of $\overline{AC}$.

50 ASSESSMENT CHAPTER 4 FORM B

CHAPTER 4 ASSESSMENT

Find the value of x in each figure.

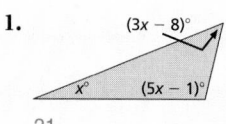

1. 21

2. 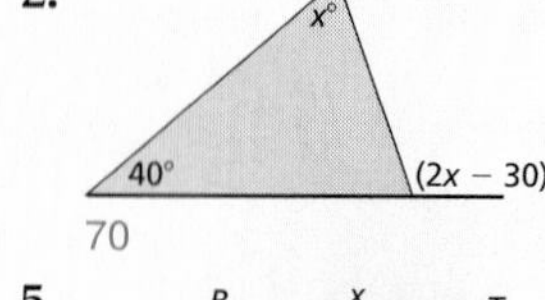

70

3. 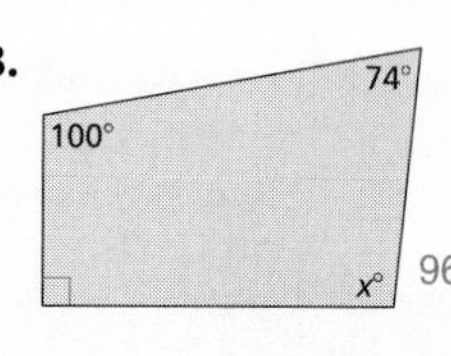

96

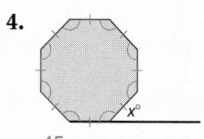

4. 45

5. 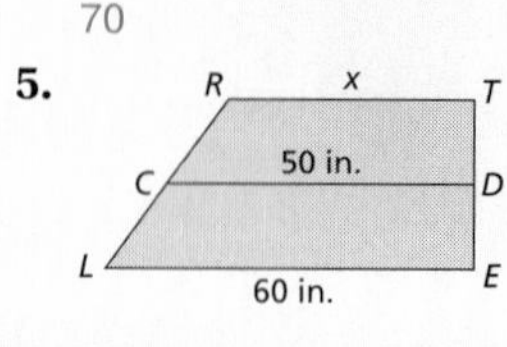

$LETR$ is a trapezoid. .
$\overline{XY}$ is a median 40

6. 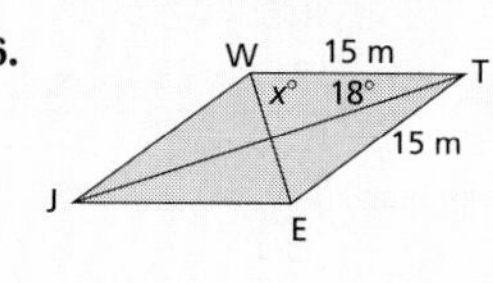

$JETW$ is a parallelogram.
72

Complete the congruence statement. Name the postulate you could use to prove the triangles congruent. (You do not need to write a proof.)

7. 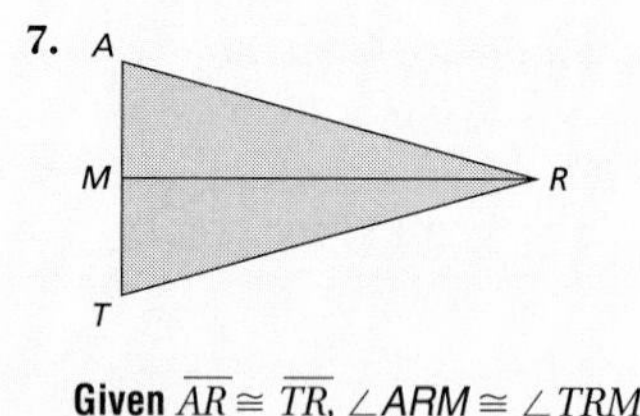

Given $\overline{AR} \cong \overline{TR}$, $\angle ARM \cong \angle TRM$
$\triangle ARM \cong$ ______ $\triangle TRM$, SAS

8. 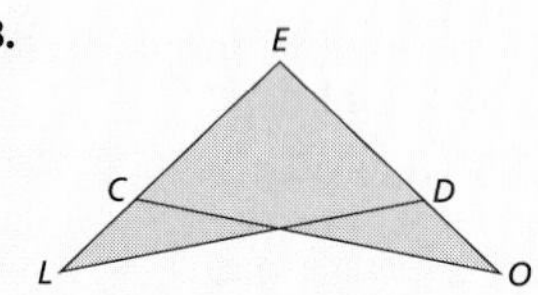

Given $\angle L \cong \angle O$, $\overline{LE} \cong \overline{OE}$
$\triangle LED \cong$ ______ $\triangle OEC$, ASA

9. Write a two-column proof.
Given $\overline{TR} \cong \overline{RE}$, $\overline{WT} \cong \overline{WE}$
Prove: $\triangle TRW \cong \triangle ERW$
See additional answers.

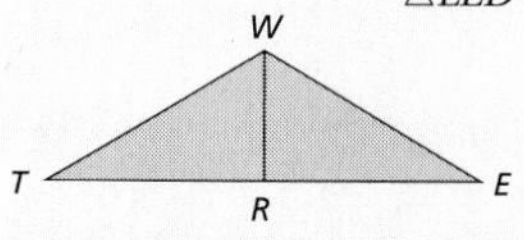

10. Draw an acute triangle and sketch all the perpendicular bisectors.
See additional answers.

11. Suppose you are asked to write an indirect proof of the following statement: *If a triangle is equilateral, then it cannot have two unequal sides.* Write Step 1 of the indirect proof. Assume a triangle has two unequal sides

12. Can a triangle have sides that measure 45 mm, 19 mm, and 23 mm? Explain.
No; $19 + 23 < 45$

13. In the figure at the right, give a range of positive values for x. $13 < x < 55$

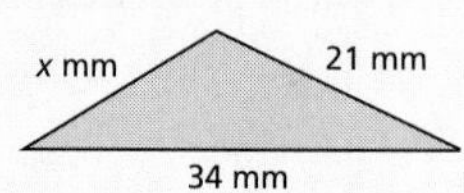

14. Find the sum of the measures of the angles of a polygon with 13 sides. 1,980°

15. A figure and the result of the first two iterations are shown. Show the result of the third iteration. See additional answers.

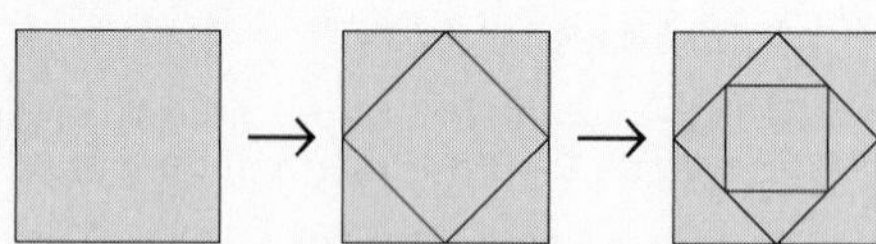

ADDITIONAL ANSWERS

9. Statements	Reasons
1. $TR \cong RE$, $WT \cong WE$	1. given
2. $WR \cong WR$	2. reflexive property
3. $\triangle TRW \cong \triangle ERW$	3. SSS postulate

10. Sample answer:

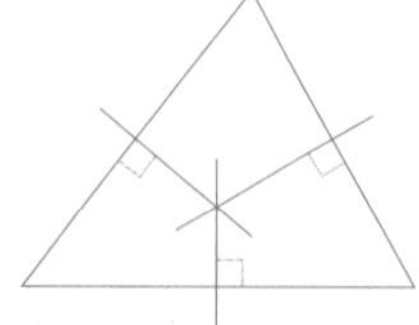

15.

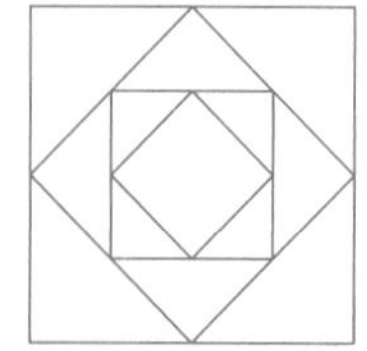

Chapter 4 Alternative Assessment

Using Other Resources

GOLDEN RECTANGLE The ancient Greeks considered a rectangle of certain proportions to be the most pleasing to the eye. Use a library or the Internet to research how they determined the ratio of the sides of the rectangle. Find several examples of the use of the Golden Rectangle in art and architecture. Prepare a brief lecture on the Golden Rectangle for your classmates.

Connections

TANGRAMS AND ART The seven Chinese tangram pieces can be used to form many shapes, from ducks to sailboats. Make a drawing or painting that uses only tangram pieces formed into interesting, identifiable shapes. Use the outline of the tangram shape only; do not include lines that show the pieces you use. Challenge others in the class to identify the tangram pieces used, or to reproduce the shape using a tangram set.

Self Assessment

MATHEMATICAL TOOLS Skill at using a protractor, a straightedge (such as a ruler) and a compass are very important to help you understand the principles of geometry. Keep a journal of your experiences, both good and bad, with these instruments. When you believe you have mastered them, draw several different triangles. For each, draw all the altitudes and medians. Measure all the resulting angles with a protractor and determine if you have been accurate. Determine which drawing demonstrates your best work.

Chapter Investigation

EXTENSION Write a report about the design and model of your truss bridge. Include an explanation as to why an actual bridge constructed from your model would support the necessary weight.

Using Other Resources

Examples of the Golden Rectangle and its use can be found in many Algebra and Geometry texts. You may want to have several available for use in the classroom.

SCORING RUBRIC 5—Student researches the Golden Rectangle thoroughly, and is able to explain its principle with details, giving many examples of its use in art and architecture. **4**—Student researches adequately, explains principle, giving many examples. **3**—Student researches adequately, explains principle, giving some examples. **2**—Student performs some research, attempts to explain the principle, giving a few examples. **1**—Student performs some research, but cannot explain its principle or give examples. **0**—Student makes no attempt to research the Golden Rectangle.

Connections

Commercial tangram sets can usually be borrowed from an elementary school. If these are not available, provide reasonably heavy cardboard for students to make their own sets.

SCORING RUBRIC 5—Student produces a painting or drawing that incorporates several original and imaginative tangram shapes. **4**—Student produces a painting or drawing that incorporates a few original and imaginative tangram shapes. **3**—Student produces a painting or drawing that incorporates one or two original tangram shapes. **2**—Student produces a painting or drawing that incorporates at least one original tangram shape. **1**—Student attempts to produce a painting or drawing , but cannot form new tangram shapes. **0**—Student makes no attempt to produce a drawing.

See page 196 for Self Assessment information and Scoring Rubric.

Chapter Investigation

The benchmarks and expectations for this extension are as follows.

- Students design a 20-foot side section of a truss bridge. They draw their design to the scale 1 in. = 2 ft.
- Students provide construction information. They indicate which line segments are parallel, label angle measures, mark right angles, and classify triangles.
- Students use their design for a truss bridge to build a section of the truss using straws or toothpicks. They use a ruler and a protractor to make sure their construction matches the plans.
- Students make a list of the quadrilaterals that they can see in their truss bridge design. They compare their design with those of their classmates. They decide which design will support the most weight and explain why.
- Students write a report about the design and model of their truss bridge. They include an explanation as to why an actual bridge constructed from their model would support the necessary weight.

Cumulative Review

Cumulative review is the best way to maintain previously taught skills and concepts. This will keep students prepared for new lessons that build on previously covered skills.

Cumulative Review covers:

Lesson 1-3
Lesson 1-4
Lesson 1-5
Lesson 1-7
Lesson 1-8
Lesson 2-2
Lesson 2-4
Lesson 2-5
Lesson 2-7
Lesson 2-9
Lesson 3-1
Lesson 3-2
Lesson 3-7
Lesson 4-1

Additional Answers

23. $x < 5$

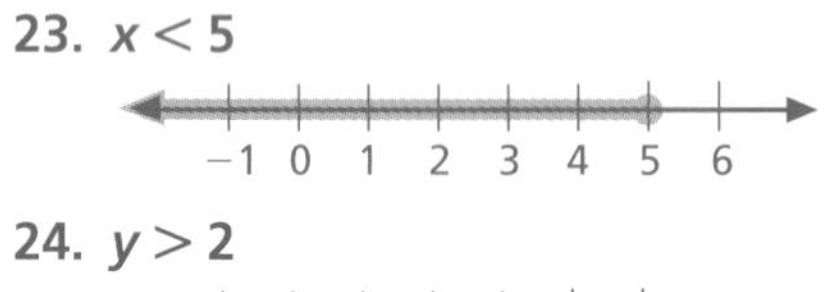

24. $y > 2$

32. If the sum of two numbers is even, then the numbers are odd. Original: true; converse: false
33. If two angles are right angles, the sum of the measures of the angles is 180°. Original: false; converse: true.

Self Assessment

Stress to students that neatness and accuracy go hand in hand. **SCORING RUBRIC 5**—Student keeps a complete journal, demonstrating progress and mastery of the mathematical tools, and is able to produce accurate drawings. **4**—Student keeps a journal, demonstrating progress in the use of mathematical tools, and is able to produce accurate drawings. **3**—Student keeps a journal, demonstrating progress in the use of mathematical tools, and is able to produce reasonably accurate drawings. **2**—Student keeps a journal, demonstrating some progress in the use of mathematical tools, and is able to produce a few generally accurate drawings. **1**—Student keeps a haphazard journal, demonstrating minimal progress in the use of mathematical tools, and is able to produce a few drawings. **0**—Student makes no attempt to keep a journal or produce drawings.

Chapters 1–4 Cumulative Review

Let $U = \{0, 1, 2, 3, \ldots 9\}$, $A = \{x | x$ is whole number such that $x > 3$ and $x < 9\}$, and $B = \{2, 4, 6, 8\}$. Find the following.

1. $A \cap B$ {4, 6, 8}
2. $A \cup B$ {2, 4, 5, 6, 7, 8}
3. A' {0, 1, 2, 3, 9}
4. $A' \cap B'$ {0, 1, 3, 9}

Evaluate each expression when $c = 5$, $d = -2$, and $e = -0.2$.

5. de 0.4
6. $-cd + e$ 9.8
7. $ce \div d$ 0.5
8. $c^2 - d^2$ 21
9. e^3 −0.008
10. $d^3 \cdot d^4$ −128
11. c^{-3} $\frac{1}{125}$
12. d^0e^{-2} 25
13. $d^3 \cdot d^{-2}$ −2

Write each number in scientific notation.

14. 29,000,000 2.9×10^7
15. 0.0000725 7.25×10^{-5}

Given $f(x) = 4x - 1$ and $g(x) = 2x^2$, find each value.

16. $f(-6)$ −25
17. $g(-2)$ 8
18. $f(10) + g(10)$ 239

Solve each equation and check the solution.

19. $y - 16 = 30$ 46
20. $\frac{y}{15} = -10$ −150
21. $-3y + 10 = -20$ 10
22. $5(x - 1) = 3(x + 1)$ 4

Solve and graph each inequality. For 23–24, see additional answers.

23. $3x - 5 \le 10$
24. $-18 > -9y$

Refer to the following data: 19, 23, 17, 20, 23, 22, 23, 25, 18, 16.

25. Find the mode. 23
26. Find the median. 21
27. In the figure below, $AB = 45$. Find AQ. 30

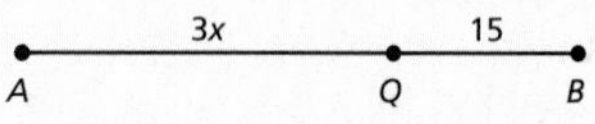

In the figure below, $\overleftrightarrow{AB}$ and $\overleftrightarrow{CD}$ intersect at E. Find the measure of each angle.

28. $\angle AED$ 153°
29. $\angle DEB$ 27°

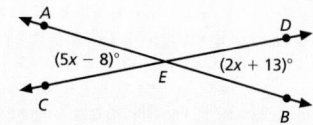

In the figure below, $\overleftrightarrow{RS} \parallel \overleftrightarrow{XY}$. Find the measure of each angle.

30. $\angle EXY$ 40°
31. $\angle RXY$ 140°

Write the converse of each statement. Then tell whether the given statement and its converse are true or false. For 32–33, see additional answers.

32. If two numbers are odd, the sum of the numbers is even.
33. If the sum of the measures of two angles is 180°, the two angles are right angles.

Find the value of x in each figure.

34. 10

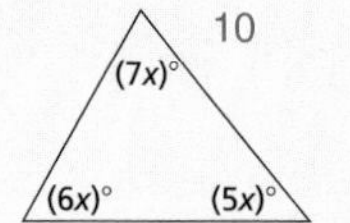

35. 19

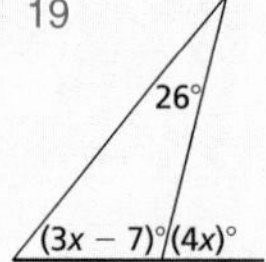

36. 40

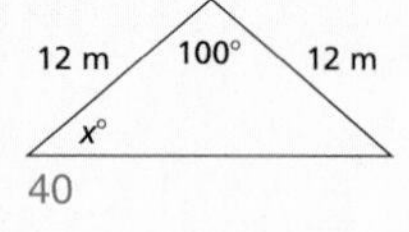

37. 17

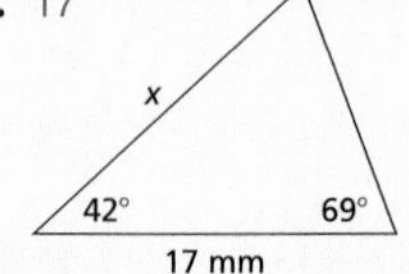

Chapters 1–4 Cumulative Assessment

Standardized Test Preparation—Standard Multiple Choice

Select the best choice for each question.

1. Which is not a rational number?
D
A. $\sqrt{49}$ **B.** 0.3 **C.** 0
D. $\sqrt{50}$ **E.** $-\sqrt{144}$

2. Evaluate $-a + b$; $a = 10$, $b = -12$.
D
A. 2 **B.** -2 **C.** 22
D. -22 **E.** None of these

3. Simplify: $a^6 \cdot a^3$
A
A. a^9 **B.** a^{18} **C.** a^2
D. a^3 **E.** None of these

4. Evaluate e^{-3} if $e = -4$.
E
A. -12 **B.** 64 **C.** $-\frac{1}{12}$
D. $\frac{1}{64}$ **E.** None of these

5. Which point is shown in the graph?
C

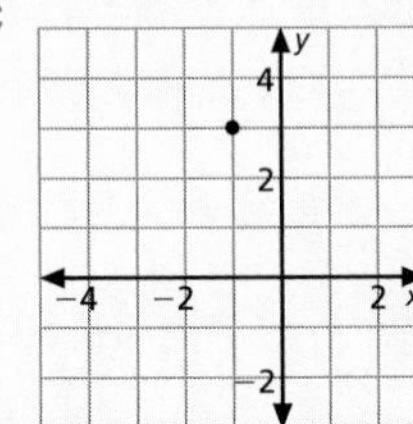

A. (1, 3) **B.** (3, 1) **C.** $(-1, 3)$
D. $(-1, -3)$ **E.** None of these

6. What is the solution for $x - 3 + 10 = |-5|$?
A
A. -2 **B.** -12 **C.** 2
D. 12 **E.** None of these

7. Which inequality is shown in the graph?
A

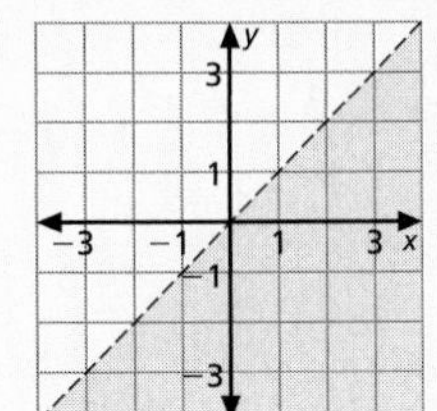

A. $y < x$
B. $y \leq x$
C. $y > x$
D. $y \geq x$
E. $y < -x$

8. Find the mean of the following data.
C 91 91 93 96 99
A. 91 **B.** 93 **C.** 94
D. 470 **E.** None of these

9. In the figure, $PQ = 79$. Find RQ.
C

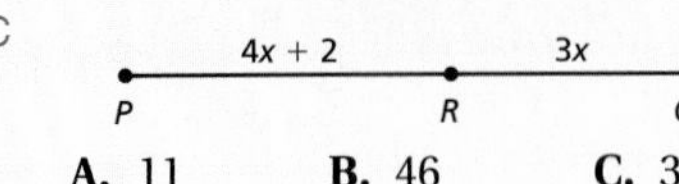

A. 11 **B.** 46 **C.** 33
D. $19\frac{1}{4}$ **E.** None of these

10. What is the m$\angle POM$ if m$\angle MOD = 5x$ and m$\angle POM = (x - 12)$?
B

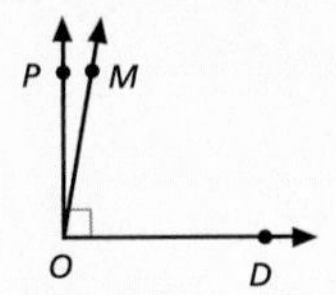

A. 17
B. 5
C. 85
D. 90
E. $13\frac{1}{2}$

11. If a polygon has 8 sides, find the sum of the interior angles.
A
A. 1080 **B.** 1440
C. 360 **D.** 1800
E. None of these

12. **CONSTRUCTED RESPONSE** Complete the proof.

Given $\overline{FB}$ is a perpendicular bisector of $\overline{AC}$.
Prove $AJB \cong CJB$

Statement	Reason
1. $\overline{FB}$ perpendicularly bisects of $\overline{AC}$	1. given
2. $\angle ABF$ and $\angle CBF$ are right angles	2. definition of ? perpenticular bisector
3. $\angle ABF \cong \angle CBF$	3. definition of ? right angles
4. $\overline{AB} \cong \overline{CB}$	4. definition of ? bisector
5. $\overline{BJ} \cong \overline{BJ}$	5. ? property reflexive
6. $AJB \cong CJB$	6. ? SAS triangle property

Write a sentence that explains Reasons 2–6.

Standardized Test Preparation

STANDARD MULTIPLE CHOICE— One strategy for completing multiple-choice questions is to use mental math and eliminate unreasonable answers as choices. Often, a couple of the answer choices are values that you may get while working out the problem. For instance, if the solution requires two or three steps, one answer choice may be the value or expression that you have at the end of the first step and another choice may be the value or expression that you have at the end of the second step.

Sometimes all but the correct answer can be eliminated using only mental math and reasoning. If this is the case, verify your answer choice by working out the problem.

Chapter 4 **Cumulative Assessment** 197

CHAPTER 5—MEASUREMENT

Theme: Lost Cities of Ancient Worlds
Chapter Investigation: Structures, pages 201, 209, 223, 227, 233, 237
Careers: Heavy Equipment Operators, page 211; Archeologists, page 229
Data Activity: Seven Wonders of the World

Content and Connections

Lesson, Pages	Lesson Objectives	NCTM Standards	State/Local Objectives	Interdisciplinary Connections	Real World Applications
5–1 202–205	• Use ratios and rates to solve problems	Measurement Number Operation Problem Solving Connections; Representation		History	Engineering, Number Sense, Archeology, History
5–2 206–209	• Apply perimeter, circumference and area formulas	Number Operation Patterns, Functions & Algebra Geometry & Spatial Sense Measurement Problem Solving Connections; Representation			Archeology, Stage Design, Sports, Recreation
5–3 212–215	• Determine probabilities using areas	Data Analysis, Statistics & Probability Number Operation Geometry & Spatial Sense Measurement Problem Solving Reasoning & Proof Connections; Representation			Games, Archeology, Weather
5–4 216–217	• Solve a problem with irregular shapes • Solve a simpler problem	Problem Solving Reasoning & Proof Connections Number Operation Geometry & Spatial Sense Measurement		Art	Carpeting, Art, Recreation
5–5 220–223	• Analyze space figures	Geometry & Spatial Sense Measurement Problem Solving Connections; Representation		Art	Archeology, Architecture, Art
5–6 224–227	• Find surface areas of space figures	Geometry & Spatial Sense Problem Solving Number Operation Patterns, Functions & Algebra Measurement Reasoning & Proof Communication Connections; Representation		Art	Packaging, Manufacturing, Sports, Astronomy
5–7 230–233	• Find the volume of space figures	Geometry & Spatial Sense Number Operation Patterns, Functions & Algebra Measurement Problem Solving Reasoning & Proof Communication Connections; Representation			Manufacturing, Astronomy, Archeology

Planning and Resources

Lesson, Pages	Tools/Materials Needed	Trans-parency	Learning/Teaching Styles Options	Assignments: Basic Enriched	Additional Practice in SE	Reteaching, Extra Practice, Enrichment	Other Resources
5–1 202–205	calculator	Warm up 13	Real World Challenge	B: 1–34, 37–38 E: 1–38	Page 210–211, 219, 229, 234, 699	R: page 69 EP: page 62 E: page 81	SS: Teacher's Choice

5–2 206–209	calculator compass centimeter ruler protractor	Warm up 13 Trans RF-18	Real World Challenge	B: 1–22, 26–40 E: 1–40	Page 210–211, 219, 229, 234, 699	R: page 71 EP: page 63 E: page 83	SS: Teacher's Choice
5–3 212–215	calculator compass	Warm up 13 Trans RF-18		B: 1–26, 29–40 E: 1–40	Page 218–219, 229, 234, 700	R: page 73 EP: page 65 E: page 85	SS: Teacher's Choice
5–4 216–217	paper/pencil	Warm up 14 Trans RF-18	Real World	B: 1–7, 8–15 E: 1–15	Page 218–219, 229, 235	R: page 75 E: page 87	MC: 10
5–5 220–223	graph paper scissors	Warm up 14 Trans RF-19, 20	ESL/LEP Challenge	B: 1–18, 21–29 E: 1–29	Page 228–229, 235, 700	R: page 77 EP: page 67 E: page 89	
5–6 224–227	calculator compass straightedge construction paper scissors tape	Warm up 15 Trans RF-18		B: 1–18, 23–24 E: 1–24	Page 228–229, 235, 701	R: page 79 EP: page 69 E: page 91	
5–7 230–233	calculator	Warm up 15 Trans RF-18	Visual Learner Challenge	B: 1–22, 26–29 E: 1–29	Page 235, 701	R: page 81 EP: page 71 E: page 93	

Planning and Pacing

Lesson, Pages	Lesson Title	45 min class	Assignments Basic, Enriched	90 min class	Assignments Basic, Enriched	___ min class	Assignments Basic, Enriched
5–1 202–205	AYR, Opener, Ratios and Units of Measure	Day 61	B: 1–34, 37–38 E: 1–38	Day 31	B: 1–34, 37–38 E: 1–38		B: 1–34, 37–38 E: 1–38
5–2 206–209	Perimeter, Circumference and Area, R&PYS	Day 62–63	B: 1–22, 26–40 E: 1–40	Day 31–32	B: 1–22, 26–40 E: 1–40		B: 1–22, 26–40 E: 1–40
5–3 212–215	Probability and Area	Day 64	B: 1–26, 29–40 E: 1–40	Day 32	B: 1–26, 29–40 E: 1–40		B: 1–26, 29–40 E: 1–40
5–4 216–217	Problem Solving Skills: Irregular Shapes, R&PYS	Day 65–66	B: 1–7, 8–15 E: 1–15	Day 33	B: 1–7, 8–15 E: 1–15		B: 1–7, 8–15 E: 1–15
5–5 220–223	Three-dimensional Figures and Loci	Day 67	B: 1–18, 21–29 E: 1–29	Day 34	B: 1–18, 21–29 E: 1–29		B: 1–18, 21–29 E: 1–29
5–6 224–227	Surface Area of Three-dimensional Figures, R&PYS	Day 68–69	B: 1–18, 23–24 E: 1–24	Day 34–35	B: 1–18, 23–24 E: 1–24		B: 1–18, 23–24 E: 1–24
5–7 230–233	Volume of Three-dimensional Figures	Day 70	B: 1–22, 26–29 E: 1–29	Day 35	B: 1–22, 26–29 E: 1–29		B: 1–22, 26–29 E: 1–29
Review/ Assess		Day 71–73		Day 36–37			

	Chapter 1 Days	Chapter Cumulative Days
Yearly Pacing (45 min class)	13 days	73 days
Yearly Pacing (90 min class)	7 days	37 days

Assessment Options

Assessment in Student Edition	Assessment in Teacher's Edition	Pages in Assessment Book	Software Generated Assessment
Are You Ready?, pages 198–199; Writing Math, pages 204, 208, 215, 223, 227, 232; Mixed Review, pages 205, 209, 215, 217, 223, 227, 233; Check Understanding pages 202, 206, 212, 220, 224, 230; Mid-Chapter Quiz, page 219, Chapter Review, page 234; Chapter Assessment, page 236; Alternative Assessment, page 237; Cumulative Review, page 238; Cumulative Assessment, page 239	5-minute Warm ups, pages 202, 206, 212, 216, 220, 224, 230; Quick Assessment, pages 204, 208, 214, 217, 222, 226, 232; Scoring Rubrics, page 237	Mid-Chapter Quiz, page 71; Test Form A, pages 73, 75; Test Form B, pages 77, 79; Cumulative Test 81, 83; Math Journal prompt, page 85, 86;	Chapter 5

Refresher Skills

The skills on these two pages are skills that have been taught in previous math courses. Continuous review of basic math skills will make stronger math students. These skills are identified as necessary to be successful in Chapter 5.

Skills Correlation Chart

Skill	Lesson Number
Perimeter, Circumference and Area	5-1, 5-2, 5-4, 5-6
Probability	5-3
Volume	5-7

Vocabulary

perimeter
area
probability
circumference
volume
outcome

Assignment Guide

All students: 1–23
Additional Practice: Refer to the Extra Practice Index on page 674 of this text.

Are You Ready?

Refresh Your Math Skills for Chapter 5

The skills on these two pages are ones you have already learned. Stretch your memory and complete the exercises. For additional practice on these and more basic skills, see page 674

In this chapter you will solve problems involving perimeter, circumference and area. It may be helpful to review a few of the basic formulas.

Perimeter, Circumference, and Area

You can use these formulas to find the perimeter and area of a rectangle, the circumference and area of a circle, and the perimeter and area of a triangle.

Examples

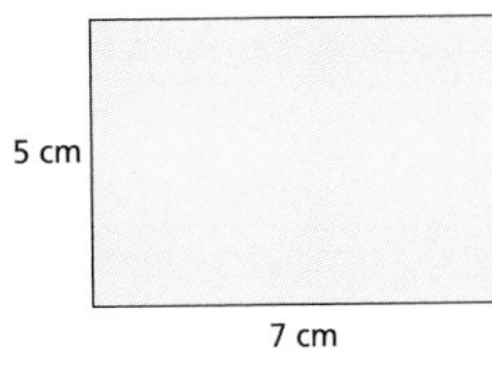

$P = 2b + 2h$

$P = 2(7) + 2(5)$

$P = 14 + 10$

$P = 24$ cm

$A = b \times h$

$A = 7 \times 5$

$A = 35\text{ cm}^2$

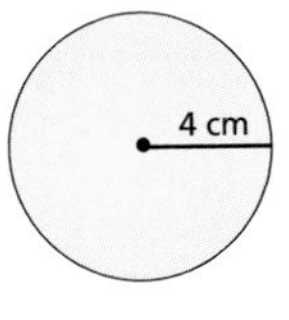

$C = 2\pi r$

$C = 2 \times 3.14 \times 4$

$C = 25.13$ cm

$A = \pi r^2$

$A = 3.14 \times 42$

$A = 3.14 \times 14$

$A = 5.27\text{ cm}^2$

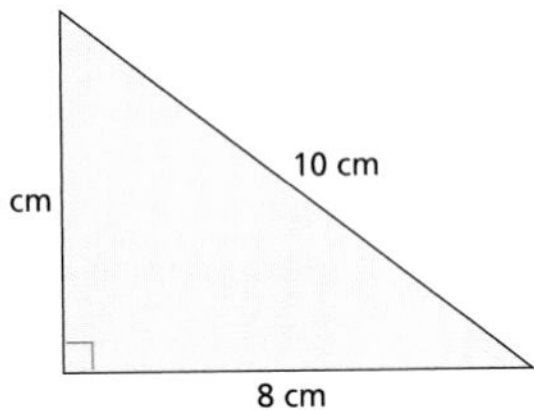

$P = a + b + c$

$P = 6 + 8 + 10$

$P = 24$ cm

$A = \left(\frac{1}{2}\right)(b \times h)$

$A = \left(\frac{1}{2}\right)(8 \times 6)$

$A = \left(\frac{1}{2}\right)(48)$

$A = 24\text{ cm}^2$

Find the perimeter or circumference and the area of each figure. Round answers to the nearest hundredth if necessary.

1.

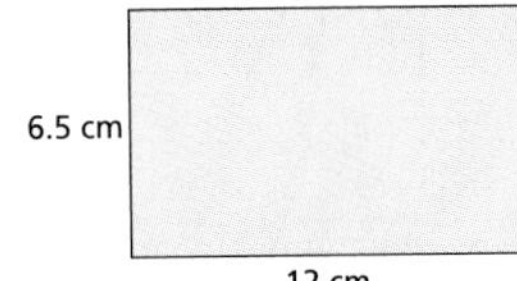

$P = 37$ cm; $A = 78\text{ cm}^2$

2.

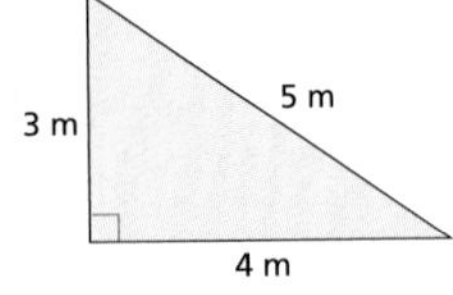

$P = 12$ m; $A = 6\text{ m}^2$

3.

5.66 ft
5.66 ft
4 ft
4 ft
4 ft

$P = 19.32$ ft; $A = 16\text{ ft}^2$

Extend the Lesson

STUDENT PORTFOLIO Have students draw a circle with a diameter of 12 units and two perpendicular diameters. Have them connect two consecutive endpoints on the circle to form a triangle. Have students find the area of the triangle and explain their methods in a paragraph. 18 units

4.

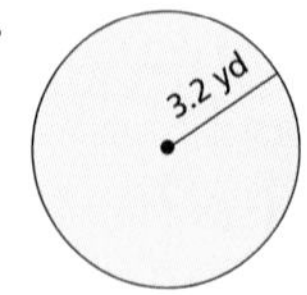

$C = 20.11$ yd; $A = 32.17$ yd^2

5.

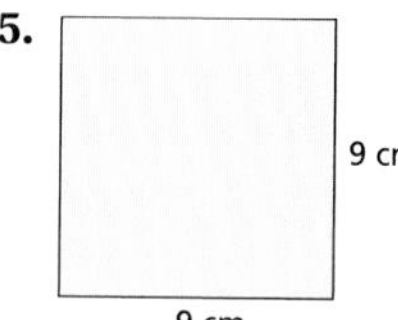

$P = 36$ cm; $A = 81$ cm^2

6.

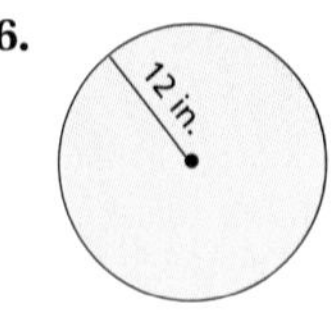

$C = 75.4$ in.; $A = 452.4$ in.^{2}

7.

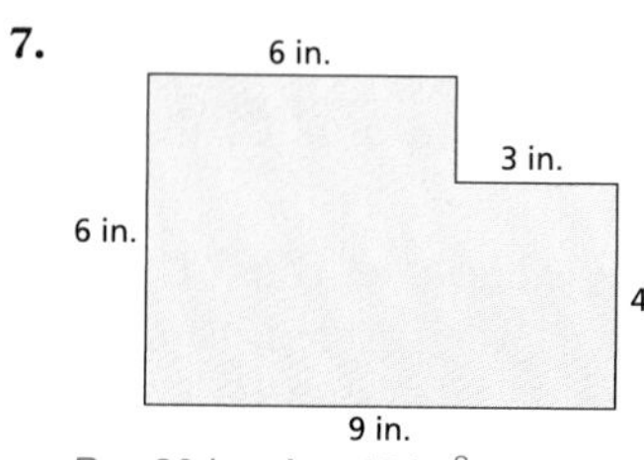

$P = 30$ in.; $A = 48$ in.^{2}

8.

8.4 cm

$C = 26.39$ cm; $A = 55.42$ cm^2

9.

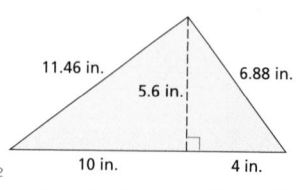

11.46 in.
6.88 in.
5.6 in.
10 in.
4 in.

$P = 32.34$ in.; $A = 39.2$ in.^{2}

PROBABILITY

The probability of an event can be expressed as a ratio:

$$P\text{ (any event)} = \frac{\text{number of ways an event can occur}}{\text{total number of possible outcomes}}$$

Example You toss a number cube. What is the probability that it will show an even number?

Number of ways to show an even number: {2, 4, 6} = 3

Total number of possible numbers: {1, 2, 3, 4, 5, 6} = 6

$$\text{P (even number)} = \frac{3}{6} = \frac{1}{2}$$

Find the probability of each event.

10. tossing a "heads" on a coin $\frac{1}{2}$
11. tossing a 3 on a number cube $\frac{1}{6}$
12. tossing a number less than 5 on a number cube $\frac{4}{6} = \frac{2}{3}$
13. picking a "diamond" from a standard deck of cards $\frac{13}{52} = \frac{1}{4}$
14. picking a "5" from a standard deck of cards $\frac{4}{52} = \frac{1}{13}$
15. picking a "jack of clubs" from a standard deck of cards $\frac{1}{52}$
16. picking a red marble from a bag of 8 blue marbles and 7 red marbles $\frac{7}{15}$
17. picking a brown sock from a drawer of 12 black socks and 3 brown socks $\frac{3}{15} = \frac{1}{5}$

Chalkboard Examples

Perimeter, Circumference and Area

Draw a circle with a radius of 4 units. Shade one-half of the circle. Find:

1. the circumference of the circle about 25.12 units
2. the area of the shaded part about 50.24 units

Probability

Find the probability of picking a blue marble from a bag containing 10 red and 20 blue marbles. $\frac{2}{3}$

Volume

A box has a length of 5 in., a width of 3 in. and a height of 4 in. Find its volume. 60 in.^{3}

Refresher Wrap-up

QUICK ASSESSMENT

Ask the following questions to determine if students have mastered the basic skills reviewed on these pages.

1. Four marbles are in a bag. Two are red and two are black. What is the probability of selecting a red marble? $\frac{1}{2}$ Suppose a red marble is drawn but not replaced. What is the probability of drawing red? $\frac{1}{3}$
2. How can you find the volume of any rectangular prism? multiply length by width by height

ADDITIONAL PRACTICE

All topics: Refer to the Extra Practice Index on page 674 of this text.

Teaching Strategies

COOPERATIVE LEARNING Have students work together to create three probability problems involving common objects (dice, spinners, counters). Then have students exchange problems and solve. Students may actually use objects or draw illustrations to model the problem.

Chapter Opener

NCTM Standards/Strands

- Number & Operation
- Patterns, Functions, & Algebra
- Geometry & Spatial Sense
- Measurement
- Data Analysis, Statistics, & Probability
- Problem Solving
- Reasoning and Proof
- Communication
- Connections
- Representation

Vocabulary

excavate	archelogy
ancient	artifacts

Theme Connections

Tables and graphs are used to organize data for the purposes of contrast and comparison. Archeologists use data to draw conclusions about the past. For example, data found in writings about ancient structures can give clues to a society's religion, values and scientific knowledge. Discuss how tables and graphs are used to compare and contrast scientific data.

Career Opportunities

Many careers related to archeology make use of data and graphs. Two are highlighted in the MathWorks features. Others include linguists, paleontologists and anthropologists.

- Equipment Operator, page 211
- Archeologist, page 229

Internet Connection

Theme Activities

Learningmathmatters.com provides web addresses to search that help students gather information about the use of math in the real world, particularly data and graphs. To search for additional addresses, begin a search using the keyword *archeology.* Then within that search, use such key words as *ancient structures, sculpture, artifacts and wonders of the world.* Students can brainstorm in small groups other key words.

CHAPTER 5

MEASUREMENT

THEME: Lost Cities of Ancient Worlds

What happens to a city once the people are gone? Often, it lies buried in the earth waiting for discovery. Once it is excavated, its roads, buildings, sewers and drains provide clues to the habits and ingenuity of the people who lived there. Fragments of pottery, sculptures, paintings and toys offer glimpses into the society's values, beliefs and daily lives.

Archeologists are scientists who work like detectives to discover and decipher these clues from the past. Through painstaking digging and sifting through the remains of ancient cities, science has given us a remarkable portrait of the world as it once was and the wonders of the past.

- **Heavy Equipment Operators** (page 211) clear the land, dig, and move dirt, debris, rock and water to uncover archeological ruins. These equipment operators must be able to operate many types of heavy machinery, supervise others, and follow plans precisely.
- **Archeologists** (page 229) trace the histories of ancient civilizations by studying maps, artifacts and the surviving writings of ancient people. Archeologists use math to measure their finds and create models. They must be able to catalog items and draw conclusions from the clues to the past that they find.

Internet Connection
www.learningmathmatters.com

Chapter Investigation

Go to Learningmathmatters.com to locate additional information about archeology.

Data Activity

Sample Answer: Using the formula for volume, determine the length of one side of its square base—about 1,400 ft; multiply by 4 to find the perimeter of the base—about 1 mi; it takes about 15–20 min to walk 1 mi.

The Seven Wonders of the Ancient World

Name	Location	Date built	Size
The Great Pyramid of Khufu	Giza	c. 2700–2500 B.C.	height: 480 ft width of square base: 756 ft
The Hanging Gardens of Babylon	Baghdad	c. 600 B.C.	height: 75 ft width of square base: 400 ft
The Statue of Zeus at Olympia	Greece	c. 457 B.C.	height: 40 ft
The Colossus of Rhodes	Greece	c. 290 B.C.	height: 120 ft
The Temple of Artemis	Turkey	c. 550 B.C.	length: 370 ft; width: 170 ft
The Mausoleum at Halicarnassus	Turkey	c. 353 B.C.	length: 126 ft; width: 105 ft; height: 140 ft
Lighthouse of Alexandria	Egypt	c. 270 B.C.	height: 400 ft

(c. stands for *circa*, meaning *about* or *approximately*)

Data Activity: The Seven Wonders of the Ancient World

Use the table for Questions 1–3.

1. Find the volume in cubic feet of the Great Pyramid of Khufu. Use the formula for volume of a pyramid: $V = \frac{1}{3}Bh$, where B = base and h = height. 120,960 ft^3
2. The Lighthouse of Alexandria was toppled by an earthquake in the fourteenth century A.D. Approximately how long did it stand? About 1670 yr
3. The largest pyramid ever built is not the Pyramid of Khufu. It is the Quetzalcoatl, located in the ancient city of Cholula in modern-day Central America. This monument, about 177 ft tall, has a volume estimated at about 116.5 million ft^3. About how long would it take to walk around it? Explain how you figured it out. See additional answers.

Chapter Investigation

Archeologists often build three-dimensional models as an aid to understanding how an artifact, monument, or building may have looked at the time it was built. Scientists use ancient writings and their knowledge of the customs of the people to make educated guesses about the features and functions of structures that no longer exist.

Working Together

Choose an ancient structure for further research. Gather data about the measurements and known features of the structure. Then make a scale model or drawing of the structure. Using your data, estimate the exterior surface area and volume of the structure. Use the Chapter Investigation logo to guide your group.

Data Activity

Explain that the Seven Wonders of the Ancient World were ancient structures that were artistic and architectural triumphs of the people living at that time. A few are still standing. We have found the ruins of most. Others are known from the descriptions in ancient writings. Assign students Questions 1–3.

Extend the Data Activity

REAL WORLD CONNECTION Have students research the Modern Wonders of the World. Have them create a table comparing the significant data about these modern structures.

Chapter Investigation

As an Overarching Problem

Display several pictures of manmade structures. Have students describe the features of the structures shown in each picture. Ask students why knowing the measurements of different parts of the structure would be important. Wrap up the discussion by asking students to estimate the surface area and volume of one of the structures. Students will continue to work on the investigation as they complete the exercises identified by the Investigation Logo that are found throughout the chapter. These exercises will guide students through the task described in *Working Together*. Encourage students to keep all of their work on the Investigation together. Have students use the suggestions in the Chapter Investigation Extension to summarize their work.

As a Chapter Project

The goal of this project is to make a three-dimensional model of a manmade structure, and estimate the surface area and volume of the structure. Students can use the Team Project Planner on page 79 and the Project Planning Calendar on page 80 in the Enrichment text to complete the project. Benchmarks **a, b, c,** and **d** should be completed after the lesson listed in parentheses. Benchmark **e** should be completed at the end of the chapter.

Project Planning Calendar

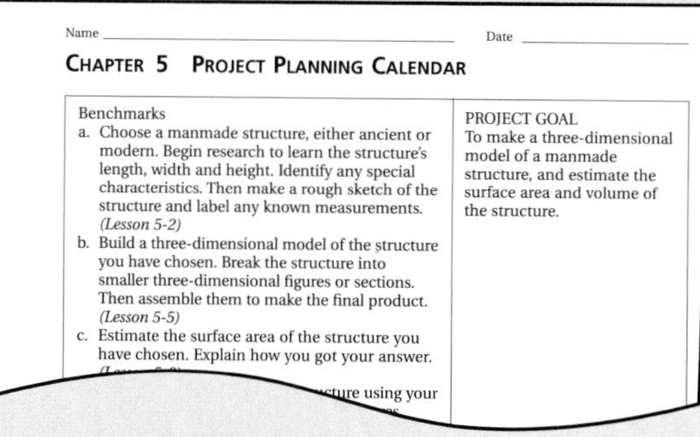

Name Date

CHAPTER 5 PROJECT PLANNING CALENDAR

Benchmarks
a. Choose a manmade structure, either ancient or modern. Begin research to learn the structure's length, width and height. Identify any special characteristics. Then make a rough sketch of the structure and label any known measurements. (Lesson 5-2)
b. Build a three-dimensional model of the structure you have chosen. Break the structure into smaller three-dimensional figures or sections. Then assemble them to make the final product. (Lesson 5-5)
c. Estimate the surface area of the structure you have chosen. Explain how you got your answer.

PROJECT GOAL
To make a three-dimensional model of a manmade structure, and estimate the surface area and volume of the structure.

Group Project Planner

Name Date

CHAPTER 5 GROUP PROJECT PLANNER

Assignment Objective

Group Members Assigned Roles
1)
2)
3)
4)
5)

Deadlines Done

Lesson Planning

NCTM Standards/Strands

- Number & Operation
- Patterns, Functions, & Algebra
- Measurement
- Problem Solving
- Reasoning and Proof
- Connections
- Representation

Vocabulary

ratio	customary units
metric units	like units
rate	unit rate
unit price	

Materials Needed

calculators

Lesson Resources

Warm-Up Transparency 13
Reteaching 5-1
Extra Practice 5-1
Enrichment 5-1

Assignment Guide

Basic: 1–34, 37–38
Enriched: 1–38

Getting Started

5-Minute Warm-up

Write a ratio for each.

1. days in a week 7 to 1
2. weeks in a year 52 to 1
3. hours in a day 24 to 1
4. seconds in a minute 60 to 1

Introduction to Lesson 5-1

After students have sketched their rooms and contents on graph paper, have them examine their sketches for reasonableness. Have them consider whether the size relationships in the sketch make sense. For example, is the bed in the sketch larger than a table? Is the ratio of the size of the desk to a chair in the sketch close to the ratio of the real objects?

5-1 Ratios and Units of Measure

Goals ■ Use ratios and rates to solve problems.
Applications Engineering, Number Sense, Archeology, History

Imagine you are an archeologist of the 25th century, and you have discovered your room, or another room in your home, looking exactly as it does today! List what you would find there. Describe how you would measure the contents. Then sketch the room as it would look when seen from above. Use a ruler and graph paper. Show all furniture and any rugs or other features you would see. Make your drawing as accurate as you can.

Build Understanding

Measurement is a process we use to find size, quantities, or amounts. When you make measurements, you can use either **customary** or **metric** units. We can measure to varying degrees of accuracy. Different instruments are used to make different measurements.

The **compass** is used for drawing curved lines and circles and for measuring distances. The **protractor** is an instrument for measuring and drawing angles. **Steel scales**, or rules, measure length.

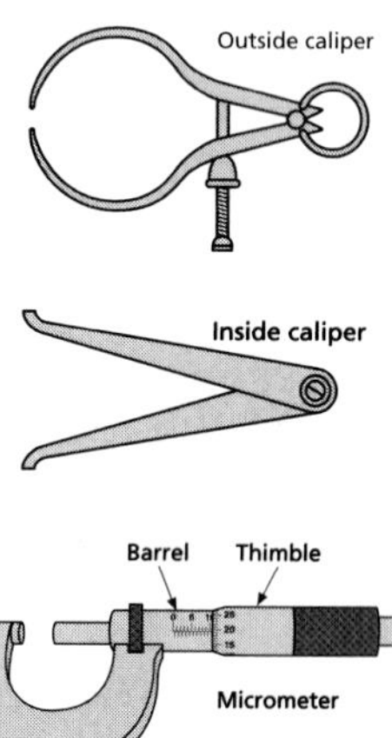

Tool-and-die makers use **calipers** and **micrometers** to make precise measurements. Outside calipers are used to transfer the measurement of an object to a scale or drawing. Inside calipers are often used to measure diameters of objects. Micrometers are used to measure length and/or thickness.

The **precision** of a measurement is related to the unit of measure used. The smaller the unit of measure, the more precise the measurement. The **greatest possible error** (GPE) of any measurement is $\frac{1}{2}$ the smallest unit used to make the measurement.

Example 1

ENGINEERING An engineer is using a steel scale. the smallest markings on the scale are $\frac{1}{64}$″. What is the GPE of any measurement the engineer makes with the scale?

Solution

Find half of $\frac{1}{64}$: $\frac{1}{64} \times \frac{1}{2} = \frac{1}{128}$. The GPE is $\frac{1}{128}$″.

Measurements are often made in order to compare quantities. To compare quantities in the same unit, such as the length of a table in a drawing with the length of the actual table, you are using a ratio. A **ratio** is a quotient of two numbers that compares one number with the other.

Extend the Lesson

REAL WORLD CONNECTION Comparisons of many real-life rates can be made by scanning daily newspapers. Have students select an item such as a car rental, car lease, supermarket sale, or airline trip and look for several companies offering what appear to be similar rates. Then have them carefully read the fine print to determine which, if any, is actually a better rate.

For both customary and metric measures, you divide to change from a smaller unit of measure to an equivalent larger unit.

You have three ways to write a ratio. The order in which the terms appear is important. Each form below reads "six to eleven."

analogy form	*fraction form*	*word form*
6:11	$\frac{6}{11}$	6 to 11

Example 2

Change 17 ft to yards.

Solution

3 ft = 1 yd

Divide 17 by 3 to find how many yards are in 17 ft.

$$17 \div 3 = 5\frac{2}{3}$$

So, 17 ft = $5\frac{2}{3}$ yd.

Use multiplication to change from a larger unit of measure to an equivalent measure in a smaller unit.

Example 3

Change 3.426 kg to grams.

Solution

1 kg = 1,000 g

Multiply 3.426 by 1,000 to find how many grams are in 3.426 kg.

$$(3.426)(1{,}000) = 3{,}426$$

So, 3.426 kg = 3,426 g.

When you write ratios involving measurements, it is sometimes necessary to rename measurements using **like units**.

Mental Math Tip

To multiply by a multiple of 10, move the decimal point to the right one place for each zero in the multiplier. To divide by a multiple of 10, move the decimal point to the left one place for each zero in the divisor.

Example 4

Write the ratio of measurements 3 in. to 20 ft in lowest terms.

Solution

$\frac{3\text{ in.}}{20\text{ ft}}$ Write the ratio as a fraction.

$\frac{3\text{ in.}}{20\text{ ft}} = \frac{3\text{ in.}}{240\text{ in.}}$ Rename the measurements using the same units.

$\frac{3 \div 3}{240 \div 3} = \frac{1}{80}$ Divide to write the fraction in lowest terms.

The ratio of measurements is 1 to 80.

Chalkboard Examples

Supplementary Example 1

On a scale, the smallest markings are $\frac{1}{32}$ inch. What is the GPE of any measurement made with the scale? Half of $\frac{1}{32}$ which equals $\frac{1}{64}$

Supplementary Example 2

Change 45 days to weeks. $6\frac{3}{7}$ weeks

Supplementary Example 3

Change $3\frac{1}{2}$ miles to feet. (Remind students that 5,280 feet = 1 mile.) 18,480 feet

Supplementary Example 4

Write the ratio of measurements 6 cups to 3 gallons in lowest terms. (16 cups = 1 gallon) 1 to 8

Supplementary Example 5

An 8-oz can of tomato sauce is \$0.54. A 12-oz can is \$0.72. Which is the better buy? The 12-oz can at \$0.06 per ounce.

Extend the Lesson

MATH JOURNAL Have students make a list of situations in which the determination of rates could play an important role. You may want to assign them to think of one situation for each of the following areas: sports, cooking, commuting and recreation. Examples include finding the yards per carry for a football player, finding the servings per pound for a recipe, finding miles per gallon of gas and finding the average score per game for a recreational activity.

Reteaching Worksheet 5-1

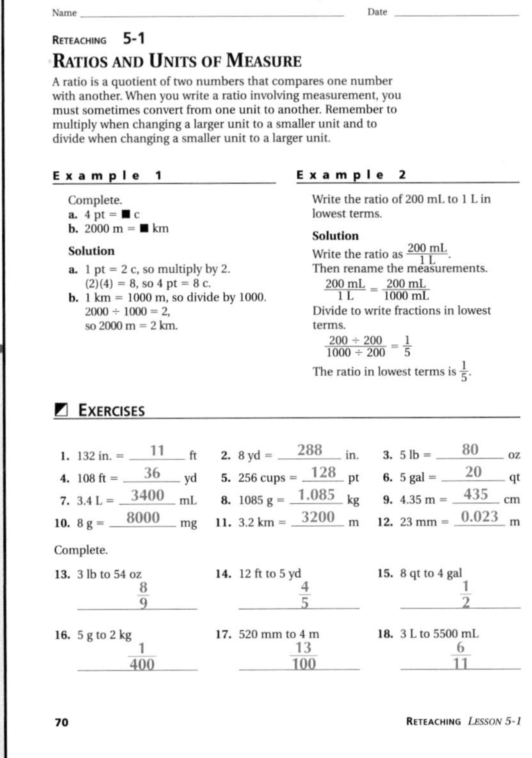

Name ______________________ Date ____________

RETEACHING 5-1

RATIOS AND UNITS OF MEASURE

A ratio is a quotient of two numbers that compares one number with another. When you write a ratio involving measurement, you must sometimes convert from one unit to another. Remember to multiply when changing a larger unit to a smaller unit and to divide when changing a smaller unit to a larger unit.

Example 1

Complete.

a. 4 pt = ■ c

b. 2000 m = ■ km

Solution

a. 1 pt = 2 c, so multiply by 2. (2)(4) = 8, so 4 pt = 8 c.

b. 1 km = 1000 m, so divide by 1000. 2000 ÷ 1000 = 2, so 2000 m = 2 km.

Example 2

Write the ratio of 200 mL to 1 L in lowest terms.

Solution

Write the ratio as $\frac{200\text{ mL}}{1\text{ L}}$.

Then rename the measurements.

$\frac{200\text{ mL}}{1\text{ L}} = \frac{200\text{ mL}}{1000\text{ mL}}$

Divide to write fractions in lowest terms.

$\frac{200 \div 200}{1000 \div 200} = \frac{1}{5}$

The ratio in lowest terms is $\frac{1}{5}$.

EXERCISES

1. 132 in. = 11 ft
2. 8 yd = 288 in.
3. 5 lb = 80 oz
4. 108 ft = 36 yd
5. 256 cups = 128 pt
6. 5 gal = 20 qt
7. 3.4 L = 3400 mL
8. 1085 g = 1.085 kg
9. 4.35 m = 435 cm
10. 8 g = 8000 mg
11. 3.2 km = 3200 m
12. 23 mm = 0.023 m

Complete.

13. 3 lb to 54 oz $\frac{8}{9}$
14. 12 ft to 5 yd $\frac{4}{5}$
15. 8 qt to 4 gal $\frac{1}{2}$
16. 5 g to 2 kg $\frac{1}{400}$
17. 520 mm to 4 m $\frac{13}{100}$
18. 3 L to 5500 mL $\frac{6}{11}$

70 RETEACHING *LESSON 5-1*

A ratio that compares two *different* quantities is called a **rate**. When you compare a quantity to *one unit* of that quantity, you are finding the **unit rate**. The **unit price** of an item is its cost per unit. Consumers can determine which of two items is the *better buy* by comparing the unit prices.

Example 5

COST ANALYSIS A 10-oz box of Cat Cravings costs \$1.80. A 16-oz box of Kitty Yummies sells for \$2.56. Which box is the better buy?

Solution

Write a ratio of price per weight for each product to find each unit price. Then compare prices.

$$\frac{1.80}{10} = \frac{0.18}{1} = 0.18 \quad \text{Cat Cravings}$$

$$\frac{2.56}{16} = \frac{0.16}{1} = 0.16 \quad \text{Kitty Yummies}$$

↑ unit rate ↑ unit price

Kitty Yummies costs \$0.16-oz. Cat Cravings costs \$0.18-oz. Kitty Yummies is the better buy.

Lock Ness, Scotland

Try These Exercises

Change each unit of measure as indicated.

1. 2 gal to cups 32 c
2. 162 in. to yards $4\frac{1}{2}$ yd
3. 3.5 L to milliliters 3,500 mL
4. 6.25 km to meters 6,250 m

Write each ratio in lowest terms.

5. 12 m to 30 m $\frac{2}{5}$
6. 9 yd:4 ft $\frac{27}{4}$
7. $\frac{135 \text{ g}}{15 \text{ g}}$ 9:1
8. 6 m : 25 cm 24:1
9. **ARCHEOLOGY** It took a team of archeological volunteers 24 days of steady work to excavate a 30-ft wall. At that rate, how much did the volunteers excavate each day? $1\frac{1}{4}$ ft per day

10. **WRITING MATH** Which measurement is more precise, 3 in. or $3\frac{1}{4}$ in.? Explain. $3\frac{1}{4}$ in.; it is measured to a smaller unit.

Practice Exercises

Complete.

11. 8 qt = __?__ c 32
12. 444 in. = __?__ yd __?__ ft 12; 1
13. 2 gal = __?__ fl oz 256
14. 3.2 T = __?__ oz 102,400
15. 0.4 cm = __?__ m 0.004
16. 300 mg = __?__ g 0.3
17. 0.006 kg = __?__ g 6
18. 8.7 mL = __?__ L 0.0087

Lesson Wrap-up

Quick Assessment

Ask the following questions to determine if students understand the content presented in this lesson.

1. What does it mean to say that a rectangle has a ratio of 2 : 1? One edge is twice the length of the other.
2. How does finding the unit rate help you determine the better buy? The unit rate compares the price to the quantity of the product. Using unit rate, you can compare products with differing quantities.

Assignment Guide

Basic: 1–34, 37–38
Advanced: 1-38
Additional Practice: See Extra Practice Index on page 674.

Teaching Strategies

CHALLENGE Often rates in real-life situations are expressed in different units. For example, the swift, a very fast bird, can fly at 155 feet per second. A fastball thrown by a major league pitcher can travel 95 miles per hour. Which is faster? The swift is faster: 155 ft/s is about 106 mi/h; 1 h = 3,600 s; 155 ft/s = 558,000 ft/h; 558,000 ft/yh/5,280 is about 106 mile per hour.

204 Chapter 5 **Measurement**

Name the best customary unit for expressing the measure of each.

19. weight of a TV lb

20. height of a room ft or yd

Name the best metric unit for expressing the measure of each.

21. capacity of a reservoir KL

22. mass of a shovel Kg

Write each ratio in lowest terms.

23. 14 kg : 35 kg 2:5

24. 80 m to 400 cm 20:1

25. $\frac{16 \text{ h}}{2 \text{ days}}$ 1:3

Find each unit rate.

26. 135 mi in 3 h 45 mi per h

27. $15 for 250 copies $0.06 per copy

28. **HISTORY** An Egyptian merchant ship from 1500 B.C. was about 90 ft long and a Roman galley was about 235 ft long. Write a ratio in lowest terms to express the relationship between lengths of the two ships. 18:47

29. Which holds more liquid, a 4-L vase or a 3,500-mL vase? 4-L

30. Use a centimeter ruler. Measure the width of your desktop to the nearest centimeter. What is the measurement? What is the GPE of the measurement? ±0.5 cm

31. **DATA FILE** Use the Data Index on pages 632–633 to locate a listing of rectangular structures. What is the GPE for the measurements of the length and width of the Wat Kukut Temple? of the Parthenon? ±0.5 m; ± 0.05 m

32. **NUMBER SENSE** The capacity of a cup is either 0.25 L, 2.5 L, or 25 L. Which measurement makes the most sense? 0.25 L

33. Loch Ness has a capacity of about 2,000 billion gal of water. If you were to drain the lake to look for the "monster," how many quarts of water would you have to remove? 8,000 billion qt

34. The ratio of boys to girls at O'Neal High is 5:6. If there are 308 students, how many are boys and how many are girls? 140 boys and 168 girls

Extended Practice Exercise

35. The 1.5-mi walking trail through the main ruins area at Bandelier National Monument in New Mexico passes by caves, cliff ruins, petroglyphs, and rock carvings from this ancient village. Most people walk the trail in 45 min. What is their walking speed in miles per hour? 2 mi/h

36. According to the early Greeks, if the ratio of the length to the width of a rectangle is 1.6:1, it is a *Golden Rectangle.* Why do you think rectangles with this shape are "golden"? What objects in the classroom or in daily life have nearly this same shape? Sample answers: The shape is pleasing to the eye; flag, index cards, notebooks,and so on.

Mixed Review Exercise

Find the measure of each angle. (Lesson 4-1)

37. $\angle ABC$ 48°

$\angle BCA$ 70°

$\angle CAB$ 62°

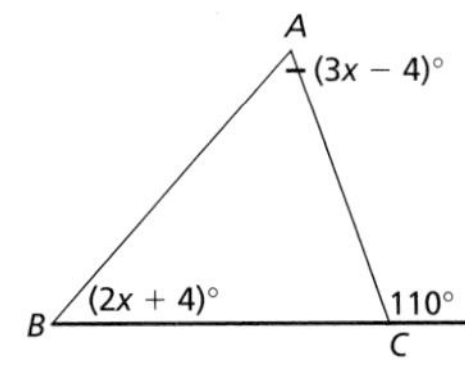

38. $\angle DEF$ 40°

$\angle EFD$ 35°

$\angle FDE$ 105°

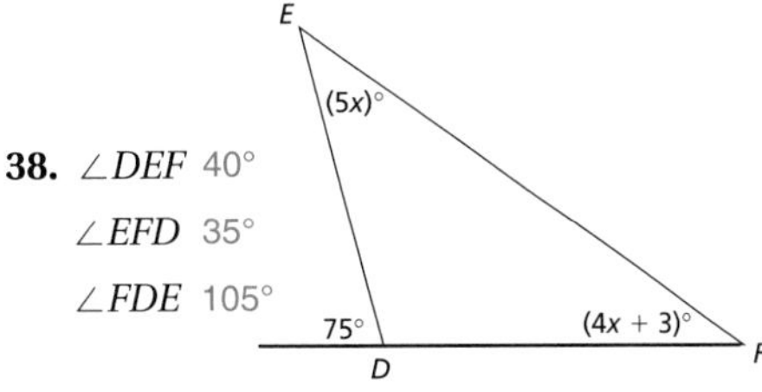

Extend the Lesson

REAL-WORLD CONNECTION Have students discuss why store-bought items are often sold in packages that contain amounts such as 15 oz or 23 oz. If unit pricing laws exist in your state, have students discuss the value of these laws to consumers. On their next shopping trip, have students look for examples of items that are packaged in a way that makes determination of unit price particularly difficult.

Extra Practice Worksheet 5-1

Name ______ Date ______

EXTRA PRACTICE 5-1

RATIO AND UNITS OF MEASURE

EXERCISES

Complete.

1. 5 qt = 20 c
2. 38 in. = 3 ft 2 in.
3. 5 kg = 5000 g
4. 2.5 T = 5000 lb
5. 40 m = 40,000 mm
6. 8 gal = 1024 oz
7. 0.005 mL = 5000 kL
8. 15 yd = 5 ft
9. 70 oz = 8 c 6 oz
10. 1300 cm = 0.0013 km

Name the best customary unit for expressing the measure of each.

11. height of a flag pole ft or yd
12. capacity of a large bucket qt

Name the best metric unit for expressing the measure of each.

13. mass of a pen g
14. length of a driveway m

Write each ratio in lowest terms.

15. 16 mg : 48 mg 1 : 3
16. 16 in. to 3 ft 4 : 9
17. $\frac{4 \text{ h}}{45 \text{ min}}$ 16 : 3
18. 8 km to 500 m 16 : 1

Find each unit rate.

19. $52 for 8 hours of work $6.50 per h
20. 32 minutes to run 4 miles 8 min per mi
21. $10.71 for 9 gallons of gas $1.19 per gal

Solve.

22. Which is a better buy, 5 apples for $2.10, or 7 apples for $3.01? 5 for $2.10
23. Which is a better buy, 8 oz of bottled water for $0.72, or 12 oz of bottled water for $0.96? 12 oz for $0.96

Choose the best estimate for each.

24. weight of a dog a. 15 oz b. 15 lb c. 15 T b
25. height of a door a. 8 in. b. 8 ft c. 8 mi b

62 EXTRA PRACTICE *LESSON 5-1*

Enrichment Worksheet 5-1

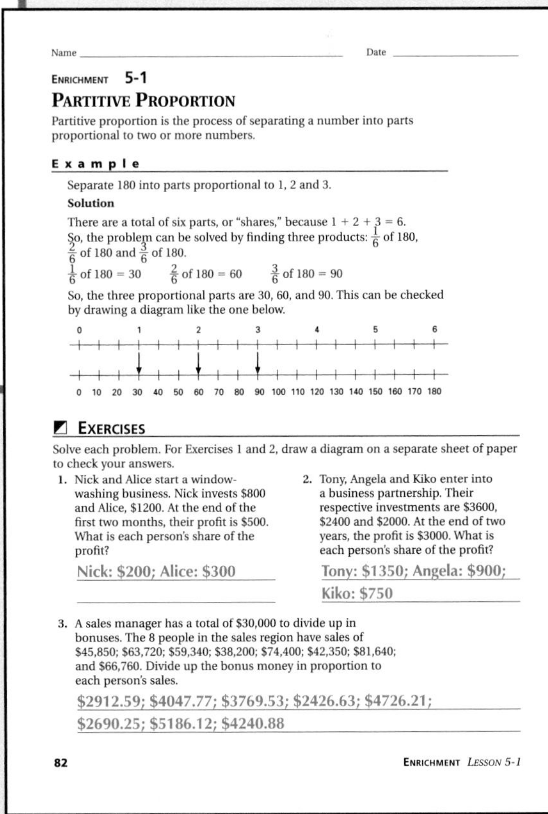
Name ______ Date ______

ENRICHMENT 5-1

PARTITIVE PROPORTION

Partitive proportion is the process of separating a number into parts proportional to two or more numbers.

Example

Separate 180 into parts proportional to 1, 2 and 3.

Solution

There are a total of six parts, or "shares," because 1 + 2 + 3 = 6. So, the problem can be solved by finding three products: $\frac{1}{6}$ of 180, $\frac{2}{6}$ of 180 and $\frac{3}{6}$ of 180.

$\frac{1}{6}$ of 180 = 30 $\frac{2}{6}$ of 180 = 60 $\frac{3}{6}$ of 180 = 90

So, the three proportional parts are 30, 60, and 90. This can be checked by drawing a diagram like the one below.

EXERCISES

Solve each problem. For Exercises 1 and 2, draw a diagram on a separate sheet of paper to check your answers.

1. Nick and Alice start a window-washing business. Nick invests $800 and Alice, $1200. At the end of the first two months, their profit is $500. What is each person's share of the profit? Nick: $200; Alice: $300

2. Tony, Angela and Kiko enter into a business partnership. Their respective investments are $3600, $2400 and $2000. At the end of two years, the profit is $3000. What is each person's share of the profit? Tony: $1350; Angela: $900; Kiko: $750

3. A sales manager has a total of $30,000 to divide up in bonuses. The 8 people in the sales region have sales of $45,850; $63,720; $59,340; $38,200; $74,400; $42,350; $81,640; and $66,760. Divide up the bonus money in proportion to each person's sales. $2912.59; $4047.77; $3769.53; $2426.63; $4726.21; $2690.25; $5186.12; $4240.88

82 ENRICHMENT *LESSON 5-1*

Lesson Planning

NCTM Standards/Strands

- Number & Operation
- Patterns, Functions, & Algebra
- Geometry & Spatial Sense
- Measurement
- Problem Solving
- Reasoning and Proof
- Connections
- Representation

Vocabulary

area
circumference
perimeter

Materials Needed

calculators
centimeter rulers
compasses
protractors

Lesson Resources

Warm-Up Transparency 13
Reteaching 5-2
Extra Practice 5-2
Enrichment 5-2
Transparency RF-18

Assignment Guide

Basic: 1–22, 26–40
Enriched: 1–40

Getting Started

5-Minute Warm-up

Solve for *x*.

1. $3x + 24 = 72$ $x = 16$
2. $2(3.14)x = 86.92$ $x = 14$
3. $3.14x^2 = 78.5$ $x = 5$
4. $4x + 5(21) = 141$ $x = 9$

Introduction to Lesson 5-2

You may wish to have students research the area of these sports fields and determine whether each would have fit on the floor of the Roman Colosseum: tennis court (2,808 ft^2), baseball diamond (8,100 ft^2), ice hockey rink (19,998 ft^2), field hockey field (54,000 ft^2), soccer field (79,200 ft^2).

5-2 Perimeter, Circumference and Area

Goals ■ Apply perimeter, circumference and area formulas.
Applications Archeology, Stage Design, Sports and Recreation

Work with a partner.

What if the ancient Romans had invented basketball or football? Imagine spectators cheering slam dunks and touchdowns instead of gladiator fights!

The arena within the Roman Colosseum was oval-shaped and had an area of about 40,000 ft^2. Could a basketball court fit within the arena? Was the arena large enough for a football field? Explain your thinking.

Yes; a basketball court is a rectangle with approximate dimensions of 90 ft by 50 ft.
No, a football field has an area of about 50,000 ft^2

Roman Colosseum, Italy

Build Understanding

When solving a problem involving measuring a plane figure, you may need to decide whether the problem requires finding the distance around the figure, or the amount of surface the figure covers, or both. When you know which measurement you want, apply the correct formula.

Recall that the distance around a polygon is its **perimeter**, the distance around a circle is its **circumference**, and the amount of surface a figure covers is its **area**.

Example 1

ARCHEOLOGY What is the width of the fence around an archeological dig if the region enclosed is a rectangle with a perimeter of 68 m and a length of 24.4 m?

Solution

The situation involves perimeter. Use the formula $P = 2l + 2w$.

$P = 2l + 2w$

$68 = 2(24.4) + 2w$ Substitute.

$68 = 48.8 + 2w$ Subtract 48.8 from each side.

$19.2 = 2w$ Multiply each side by $\frac{1}{2}$.

$9.6 = w$

The fence is 9.6 m wide.

> **Reading Math**
> The perimeter of a figure means "the measure all around it." The term comes from two Greek words—*peri*, meaning "all around," and *metron*, "measure." What other words can you think of that are derived from *peri* and/or *metron*? Check your choices with a dictionary.

Extend the Lesson

MATH JOURNAL Have students write the minimum amount of information they need to find the area of each of the following figures: parallelogram, triangle, trapezoid and circle.

Example 2

The largest pizza ever made measured 122 ft 8 in. in diameter. If your classmates were to share this pizza equally, about how many square inches of pizza would each get?

Solution

First, find the area of the pizza. Use the formula $A = \pi r^2$.

$A = \pi r^2$

$A \approx 3.14 \times 736^2$ Use a calculator.

$A \approx 1{,}700{,}925.44$ Round your answer.

The pizza has an area of about 1,700,925 in.^{2}. Divide by the number of students in your class to find out how much pizza each gets.

Example 3

Find the area of this figure.

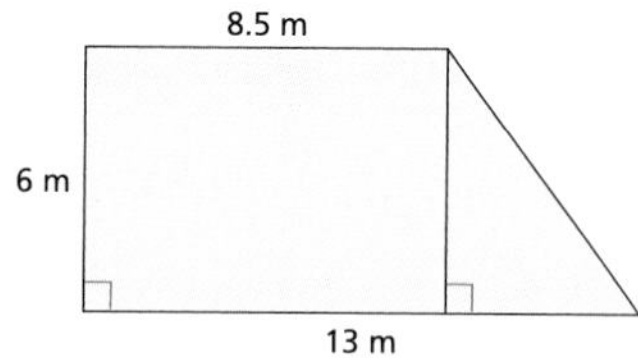

Solution

The figure can be divided into a rectangle and a triangle.

rectangle	**triangle**
$A = lw$	$A = \frac{1}{2}bh$
$A = (8.5)(6)$	$A = (0.5)(4.5)(6)$
$A = 51$	$A = 13.5$

The area of the figure is the sum of the areas of the rectangle and triangle. The area is 51 m^2 + 13.5 m^2, or 64.5 m^2.

Example 4

What is the area of the shaded region of this figure?

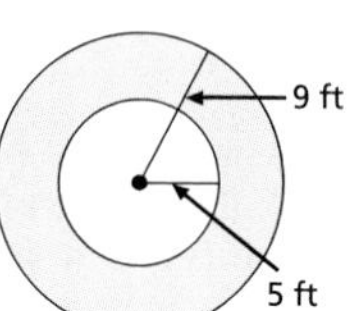

Solution

The shaded area is the difference between the areas of the circles.

$A = \pi r^2$	$A = \pi r^2$
$A \approx 3.14(9^2)$	$A \approx 3.14(5^2)$
$A \approx 254$ ft^2	$A \approx 79$ ft^2

Subtract. $254 - 79 = 175$

The area of the shaded region is about 175 ft^2.

Math: Who, Where, When

In traditional African societies, as well as in traditional societies elsewhere, the circular house is a common shape. Among the many reasons for this is a geometric one. You can consider the example of the *Kikuyu* house to see why. The diameter of the base of one of these Kenyan houses is typically about 14 ft and its circumference is about 44 ft. First, find what the area would be for a square with a perimeter of 44 ft, and then for a different rectangle with that same perimeter. What do you notice.

Extend the Lesson

REAL WORLD CONNECTION Have students choose three items in their home environment, take appropriate measurements and find area. Have students make sketches to show their work.

Chalkboard Examples

Supplementary Example 1

What is the length around an archeological dig if the region enclosed is a rectangle with a perimeter of 28.6 m and a width of 6.8 m? 70.8 m

Supplementary Example 2

A circular flower garden has a diameter of 14 ft 6 in. What is the area of the garden to the nearest square foot? 165 sq ft

Supplementary Example 3

Find the area of a figure that is a rectangle 16.8 cm by 11.2 cm with a triangle on top of the rectangle that has a base of 16.8 cm and a height of 8.4 cm. 258.72 cm^2

Reteaching Worksheet 5-2

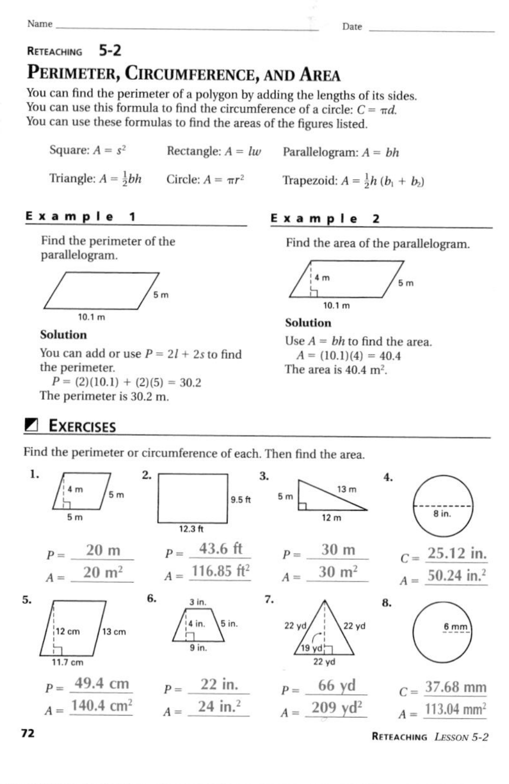

Name ____________ Date ________

RETEACHING 5-2

PERIMETER, CIRCUMFERENCE, AND AREA

You can find the perimeter of a polygon by adding the lengths of its sides.
You can use this formula to find the circumference of a circle: $C = \pi d$.
You can use these formulas to find the areas of the figures listed.

Square: $A = s^2$	Rectangle: $A = lw$	Parallelogram: $A = bh$
Triangle: $A = \frac{1}{2}bh$	Circle: $A = \pi r^2$	Trapezoid: $A = \frac{1}{2}h(b_1 + b_2)$

Example 1

Find the perimeter of the parallelogram.

5 m
10.1 m

Solution

You can add or use $P = 2l + 2s$ to find the perimeter.
$P = (2)(10.1) + (2)(5) = 30.2$
The perimeter is 30.2 m.

Example 2

Find the area of the parallelogram.

4 m
5 m
10.1 m

Solution

Use $A = bh$ to find the area.
$A = (10.1)(4) = 40.4$
The area is 40.4 m^2.

EXERCISES

Find the perimeter or circumference of each. Then find the area.

1. 4 m, 5 m, 5 m — $P =$ 20 m; $A =$ 20 m^2
2. 9.5 ft, 12.3 ft — $P =$ 43.6 ft; $A =$ 116.85 ft^2
3. 5 m, 13 m, 12 m — $P =$ 30 m; $A =$ 30 m^2
4. 8 in. — $C =$ 25.12 in.; $A =$ 50.24 in.^{2}
5. 12 cm, 13 cm, 11.7 cm — $P =$ 49.4 cm; $A =$ 140.4 cm^2
6. 3 in., 4 in., 5 in., 9 in. — $P =$ 22 in.; $A =$ 24 in.^{2}
7. 22 yd, 22 yd, 19 yd, 22 yd — $P =$ 66 yd; $A =$ 209 yd^2
8. 6 mm — $C =$ 37.68 mm; $A =$ 113.04 mm^2

72 RETEACHING *LESSON 5-2*

Lesson Wrap-up

Quick Assessment

Ask the following questions to determine if students understand the content presented in this lesson.

1. As the area of a circle gets smaller, what happens to the ratio of the length of the circumference to the length of the diameter? The ratio is always constant.
2. Draw a circle and two perpendicular diameters. Connect two consecutive endpoints on the circle to form a triangle. If the diameter of the circle is 12 units, what is the area of the triangle? 18 units^2

Assignment Guide

Basic: 1–22, 26–40
Advanced: 1–40
Additional Practice: See Extra Practice Index on page 674.

Additional Answers

23. Sample answers: Area of wall space, not including windows, doors, etc.; number and price of cans of paint needed based on the area one can of paint covers; painting speed and number of painters involved.
24. The perimeter is $2x + 2y$ and is the same as the total rectangle, no matter what the size of the shaded region.

Try These Exercises

Find the perimeter or circumference of each. Then find the area of each. If necessary, round answers to the nearest whole number.

1.
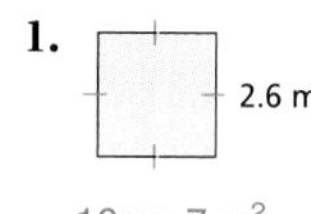

10 m; 7 m^2

2.
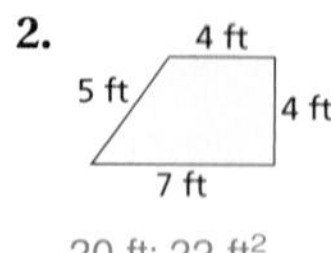

20 ft; 22 ft^2

3.
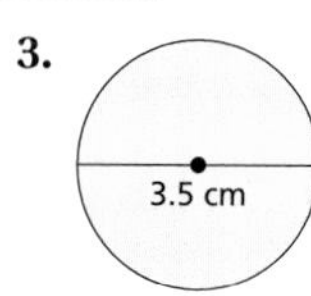

11 cm; 10 cm^2

4.
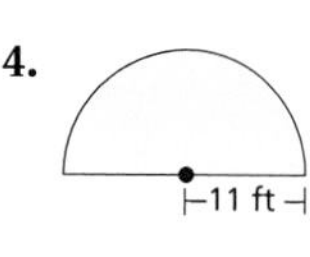

56 ft; 190 ft^2

Practice Exercises

5. What is the perimeter of a regular octagon with 5-cm sides? 40 cm
6. What is the circumference of a circle with a radius of 6.6 m? about 41.4 m
7. Find the height of a triangle if area = 24 cm^2 and base = 10 cm. 4.8 cm

Find the area of the shaded region of each figure.

8.
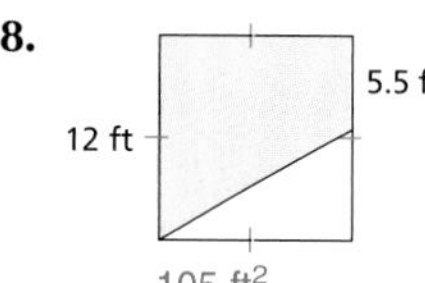

105 ft^2

9.
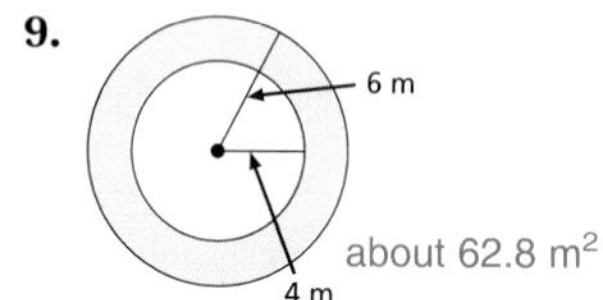

about 62.8 m^2

10.
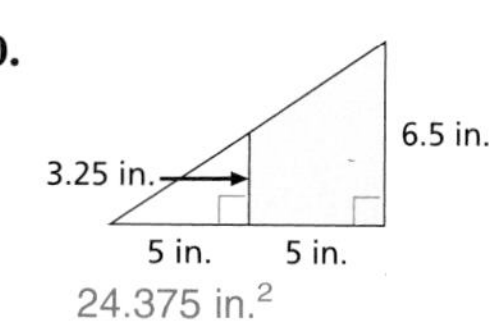

24.375 in.^{2}

11. If you triple the length of the radius of a circle, how does the circumference change? It triples too.
12. **DATA FILE** use the Data Index on pages 632–633 to find the location of the information needed to answer this question. Which has the greater area, the base of the Ziggurat of Ur or the Parthenon in Athens? Use mental math. Ziggurat of Ur
13. **ARCHEOLOGY** A 1,000-year-old Anasazi *kiva* is in the shape of a circle. If the area of the *kiva* is about 1,661 ft^2, what is the distance around it? about 144 ft
14. **STAGE DESIGN** A stage from an ancient amphitheatre is shaped like a trapezoid. The front of the stage is 30 ft across, the back is 40 ft across, and the distance from front to back is 25 ft. If a circular region of the stage, 6 ft across, is designated as a pond, what is the area of the space left for actors to walk in? about 846.74 ft^2
15. **SPORTS** The distance from one base to the next in a standard baseball diamond is 90 ft. If the ratio of that length to the length of a basepath in a Little League diamond is 3:2, what is the area of a Little League diamond? 3,600 ft^2

16. **WRITING MATH** Since π is an irrational number, most solutions to calculations involving π are found using the approximations 3.14 or $\frac{22}{7}$. When might it be easier to use $\frac{22}{7}$ rather than 3.14 to estimate area or circumference? When data involves numbers that are multiples of 7.

17. **TALK ABOUT IT** Irene and Luis both used calculators to find the area of a circle with a radius measuring 2.5 cm. Irene got 19.634954 cm^2 and Luis got 19.625 cm^2. How can you account for the difference in their answers? Luis entered 3.14 for π. Irene used the π key on her calculator.
18. The side of a square is equal to the diameter of a circle. Which figure will have the greater area, the square or the circle? Square

Teaching Strategies

CHALLENGE Have students draw a 4 × 9 unit rectangle on either graph or dot paper. Then have them try to change the figure: (1) Keep the perimeter the same, but make the area smaller; (2) keep the perimeter the same, but make the area larger; (3) keep the area the same, but make the perimeter smaller; and (4) keep the area the same, but make the perimeter larger. Check students' work.

RECREATION A community has set aside a rectangular area 250 ft by 200 ft to use for a swim center. The board of directors wants the park to have four pools:

- Pool A—a circular wading pool for small children
- Pool B—a large rectangular lap pool
- Pool C—a smaller L-shaped pool
- Pool D—a pool with an appealing irregular shape that has semi-circular as well as rectangular regions

The park should also have a small circular fountain and a shower area. It may also have picnic tables, chairs, and a food concession.

19. Imagine that you are on the planning board. Design a park layout. Submit a detailed sketch showing the size and location of each pool and feature. (You may want to use graph paper.) Check students' work.

20. Find the area of each pool. Answers will vary.

21. Suppose you decide to make the pools safer by placing a border of 1-ft^2 tiles around each. If each tile costs $5, what will be the total cost of the number of tiles you need? Answers will vary.

22. **CHAPTER INVESTIGATION** Choose a manmade structure, either ancient or modern. Begin research to learn the structure's length, width, and height. Identify any special characteristics. Then make a rough sketch of the structure and label any known measurements. Check students' work.

Extended Practice Exercises

23. Suppose you were going to paint all the walls of your classroom. What must you know to find how much it will cost and how long it will take?

23. Sample answers: Area of wall space, not including windows, doors, etc.; number and price of cans of paint needed based on the area one can of paint covers; painting speed and number of painters involved.

24. If all sides of the figure at the right are either parallel or perpendicular, then what is the perimeter of the unshaded portion? Does it matter what size the shaded region is?

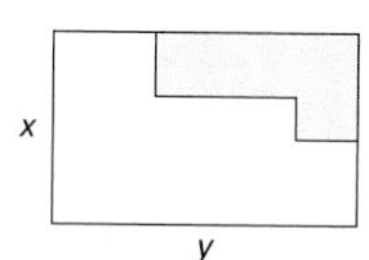

24. The perimeter is $2x + 2y$ and is the same as the total rectangle, no matter what the size of the shaded region.

25. A farmer has 100 ft of fence. Can the farmer enclose more pasture for grazing with a square or a circular enclosure? Is this true for any length of fence? The circle ($r = 15.9$ ft and $A = 793.8$ ft^2) always has more area than the square (625 ft^2).

Mixed Review Exercises

Can the given measures be the lengths of the sides of a triangle? (Lesson 4-6)

26. 5 cm, 8 cm, 10 cm yes **27.** 6 in., 9 in., 5 in. yes

28. 8 ft, 9 ft, 17 ft no **29.** 3 m, 9 m, 11 m yes

30. 14 yd, 18 yd, 36 yd no **31.** 7 km, 3 km, 5 km yes

32. 9 dm, 9 dm, 16 dm yes **33.** 4 mi, 10 mi, 10 mi yes **34.** 8.6 m, 5.8 m, 15.3 m no

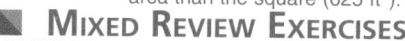

Use the number line at the right for Exercises 35–40. Find each length. (Lesson 3-1)

35. $\overline{MP}$ 5 **36.** $\overline{QS}$ 4 **37.** $\overline{PS}$ 6

38. $\overline{MR}$ 10 **39.** $\overline{NR}$ 6 **40.** $\overline{MS}$ 11

M N P Q R S
−6 −5 −4 −3 −2 −1 0 1 2 3 4 5 6

Extra Practice Worksheet 5-2

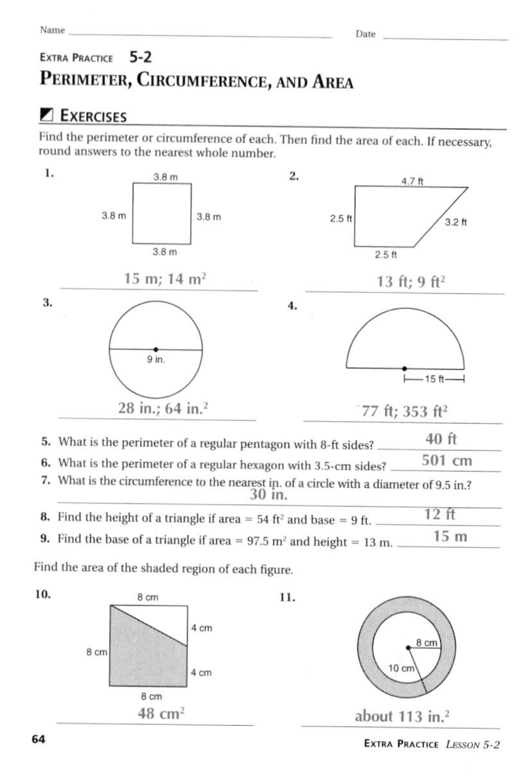

Name ______ Date ______

EXTRA PRACTICE 5-2

PERIMETER, CIRCUMFERENCE, AND AREA

EXERCISES

Find the perimeter or circumference of each. Then find the area of each. If necessary, round answers to the nearest whole number.

1. (square, 3.8 m sides) 15 m; 14 m^2
2. (4.7 ft, 2.5 ft, 3.2 ft, 2.5 ft) 13 ft; 9 ft^2
3. (circle, 9 in.) 28 in.; 64 in.^{2}
4. (semicircle, 15 ft) 77 ft; 353 ft^2

5. What is the perimeter of a regular pentagon with 8-ft sides? 40 ft
6. What is the perimeter of a regular hexagon with 3.5-cm sides? 501 cm
7. What is the circumference to the nearest in. of a circle with a diameter of 9.5 in.? 30 in.
8. Find the height of a triangle if area = 54 ft^2 and base = 9 ft. 12 ft
9. Find the base of a triangle if area = 97.5 m^2 and height = 13 m. 15 m

Find the area of the shaded region of each figure.

10. (8 cm, 4 cm, 4 cm, 8 cm, 8 cm) 48 cm^2
11. (8 cm, 10 cm) about 113 in.^{2}

64 EXTRA PRACTICE *LESSON 5-2*

Enrichment Worksheet 5-2

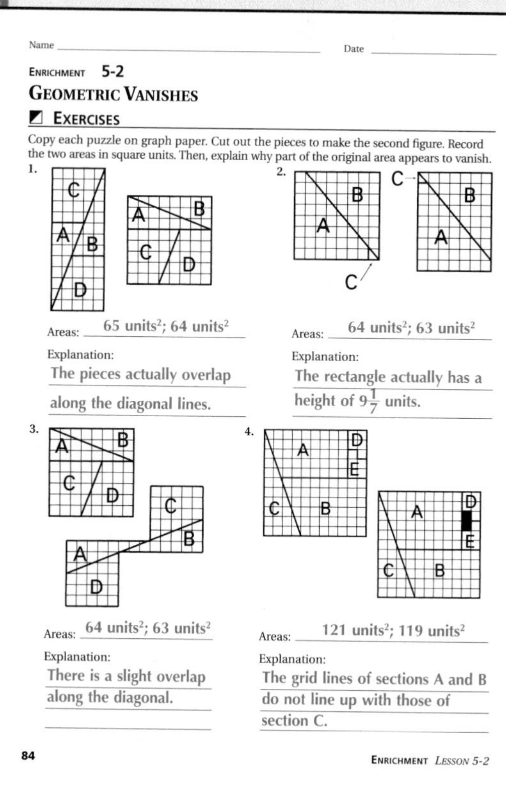

Name ______ Date ______

ENRICHMENT 5-2

GEOMETRIC VANISHES

EXERCISES

Copy each puzzle on graph paper. Cut out the pieces to make the second figure. Record the two areas in square units. Then, explain why part of the original area appears to vanish.

1. Areas: 65 units^2; 64 units^2
Explanation: The pieces actually overlap along the diagonal lines.

2. Areas: 64 units^2; 63 units^2
Explanation: The rectangle actually has a height of $9\frac{1}{7}$ units.

3. Areas: 64 units^2; 63 units^2
Explanation: There is a slight overlap along the diagonal.

4. Areas: 121 units^2; 119 units^2
Explanation: The grid lines of sections A and B do not line up with those of section C.

84 ENRICHMENT *LESSON 5-2*

Teaching Strategies

COOPERATIVE LEARNING Have students work in pairs to solve these problems. What happens to the area of a rectangle if you double the length of one side? both sides? area doubles, area is multiplied by 4 What happens to the area of a circle if you double the radius? triple the radius? area is multiplied by 4, area is multiplied by 9 You may wish to have students write about the patterns they see.

Skills Practice

Vocabulary Review

Lesson 5-1
ratio
customary units
metric units
like units
rate
unit rate
unit price

Lesson 5-2
area
circumference
perimeter

Assignment Guide

All students: 1–28
Additional Practice: Refer to the Extra Practice Index on page 674 of this text.

Review and Practice Your Skills

Practice ■ Lesson 5-1

Complete.

1. 48 c = 12 qt **2.** 512 fl oz = 4 gal **3.** 11,600 oz = 0.3625 T
4. 7000 mm = 700 cm **5.** 7.03 L = 7030 mL **6.** 48 g = 0.048 kg

Write each ratio in lowest terms.

7. 120 m : 2700 cm 40:9 **8.** 3 yd : 48 in 9:4 **9.** 7 days to 120 hours 7:5
10. 5 kg to 5,000,000 mg 1:1 **11.** 1200 min : 1 day 5:6 **12.** 1 yd^2 : 3 ft^2 3:1

Choose the best estimate for each.

13. length of basketball court b	**a.** 90 in.	**b.** 90 ft	**c.** 90 yd
14. weight of an infant a	**a.** 9 kg	**b.** 9 g	**c.** 9 mg
15. length of a straw c	**a.** 20 m	**b.** 20 mm	**c.** 20 cm

Solve.

16. What is the unit rate for a train which travels 805 miles in 7 hours? 115 mph
17. Which is the better buy, 1 gallon of milk for $1.79, or 1 pint of milk for $0.25? 1 gal for $1.79
18. The ratio of girls to boys at South High School is 3:4. If there are 980 students, how many are boys? 560

Practice ■ Lesson 5-2

Find the perimeter or circumference of each.

19.
9 m
15 m
12 m
P = 36 in.

20.
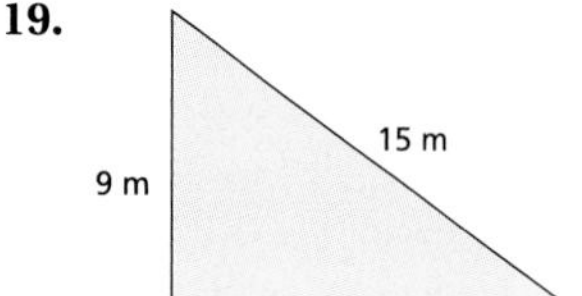

P = 60 ft

21.

C = 94.2 in.

22–24. Find the area of each figure in Exercises 19–21. 22. $A = 54\ m^2$ 23. $A = 204\ ft^2$ 24. $A = 706.5\ in.^2$

Find the area of the shaded region in each figure.

25.
4 in.
4 in.
6 in.
8 in.
96 $in.^2$

26.
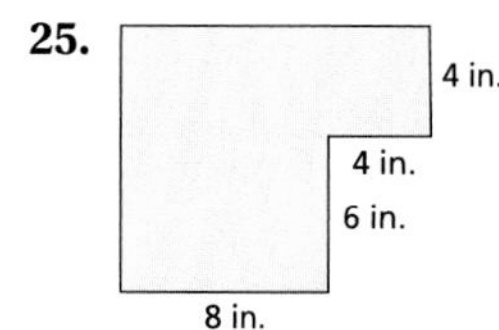

279 m^2

27.
7 mi
7 mi
27.93 mi^2

Teaching Strategies

For Exercises 1–15, review the measurement abbreviations used. Refer students to the measurement conversion tables in the Data File. You may want to allow students to use these tables to find unit rates to complete these exercises.

PRACTICE ■ LESSON 5-1–LESSON 5-2

28. If you triple the length of the radius of a circle, how does the area change? (Lesson 5-2) multiplied by 9

Complete. (Lesson 5-1)

29. 6.5 gal = 832 fl oz
30. $6\frac{5}{8}$ kg = 6625 g
31. 816 in. = 22 yd 2 ft
32. 2,500,000 mL = 2.5 kL
33. 0.04 cm = 0.0004 m
34. 56 c = 3 gal 2 qt

Write each ratio in lowest terms. (Lesson 5-1)

35. 1000 m : 10 km 1:10
36. 8.5 T to 34,000 lb 1:2
37. 75 cL : 25 mL 30:1
38. 10 yd to 540 in. 2:3
39. 9.2 mL : 9.2 L 1:1000
40. 1 gal : 1 fl oz 128:1

41. Which holds more liquid, a 200 L barrel or a 22,500 mL barrel? (Lesson 5-1) 200 L barrel

Find the perimeter or circumference of each figure. Then find the area of each figure. (Lesson 5-2)

42.

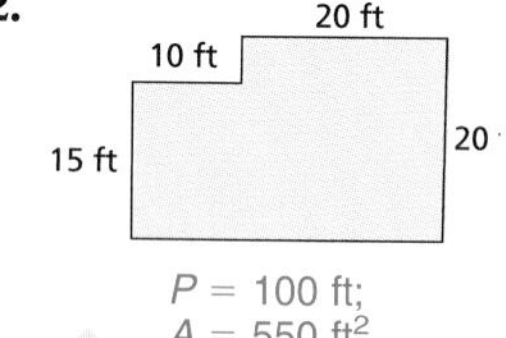

$P = 100$ ft; $A = 550$ ft^2

43.

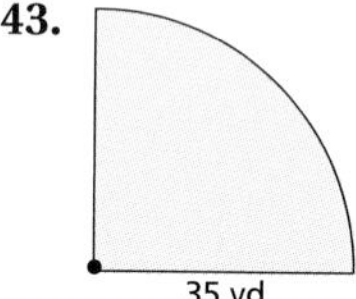

35 yd

$P = 124.95$ yd; $A = 961.625$ yd^2

44.

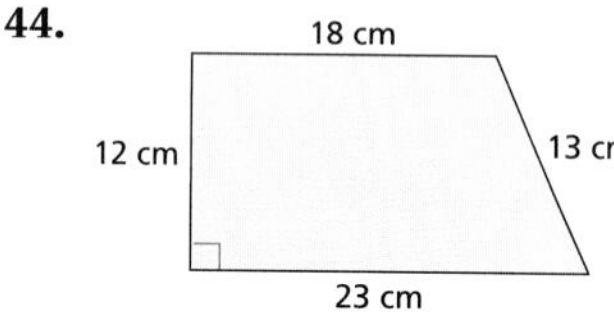

$P = 66$ cm; $A = 246$ cm^2

MathWorks Workplace Knowhow

Career – Equipment Operators

Ancient city ruins often lay under many tons of dirt, debris, rock or even water. Some sites are overgrown with vegetation. To get to this ruins, archeologists must employ workers who can operate earth-moving machines such as bulldozers, conveyors, trench excavators, hoists, winches, backhoes, and cranes. Earth-moving contractors help unlock the past through their efforts.

To remove trees from the ground area above ancient ruins, the project site has been divided into three sections—a rectangle, a triangle and a half circle.

1. Find the area in square feet of each section. rectangle = 5000 ft^2, triangle = 3000 ft^2, half circle ≈ 1962.5 ft^2

To clear 10 ft^2 of vegetation requires 3 workers paid \$12.50-h and 2 machine operators paid \$16 h. Working together, these workers can clear 10 ft^2 in 2 h.

2. How long will it take the workers to clear each section? rectangle, 1000 hrs; triangle, 600 hrs; half circle, 392.5 hrs
3. What is the cost to clear the triangular section of vegetation? \$41,700

Chalkboard Examples

Example from Lesson 5-1

Motion picture film is usually shot at 24 frames per second. At that rate, how long is a film with 43,200 frames? 1,800 sec or 30 min

Example from Lesson 5-2

A circle has a radius of 8 cm. A smaller circle inside the first circle has a radius of 5 cm. What is the area of the area outside the smaller circle but inside the larger one? about 122.46 cm^2

Career Opportunity

Describe the kind of work equipment operators do and the types of companies that hire these workers. Have students discuss the importance of measurement and ratio and proportion in doing this type of job. Explain that equipment operators must be able to read and interpret plans, understand the scale of the plans and how to transfer angles and measurements to the actual landscape. Students should answer Questions 1-3 to better understand how equipment operators use mathematics in performing their jobs.

Students who are interested in learning more about this profession can go to learningmathmatters.com. Guidance Counselors should have information about school and training requirements.

Teaching Strategies

Point out that there are two ways to find area when only a portion of a figure is shaded. Students can either use information in the problem to find the area of the shaded portion directly or they can find the area of the entire figure and then subtract the unshaded portion of the figure. Since either method gives the same answer, students should be encouraged to use the method that seems easier in each situation.

Lesson Planning

NCTM Standards/Strands

- Number & Operation
- Patterns, Functions, & Algebra
- Geometry & Spatial Sense
- Measurement
- Data Analysis, Statistics, & Probability
- Problem Solving
- Reasoning and Proof
- Communication
- Connections
- Representation

Representation Vocabulary

probability random
favorable outcome

Materials Needed

calculators compasses

Lesson Resources

Warm-Up Transparency 13
Reteaching 5-3
Extra Practice 5-3
Enrichment 5-3
Transparency RF-18

Assignment Guide

Basic: 1–26, 29–40
Enriched: 1–40

Getting Started

5-Minute Warm-up

Write each probability as a fraction and a decimal.

1. Roll a prime number on a number cube. $\frac{1}{2}$ or 0.5
2. Spin a 6 on a spinner marked 1-8. $\frac{1}{8}$ or 0.125

Introduction to Lesson 5-3

After students have played several rounds of the game, ask them to estimate how large a square they would have to place in the box so that the cube would have a $\frac{1}{4}$ or $\frac{1}{2}$ chance of landing in the square. Use the game to review basic probability concepts by asking students to discuss how large the square would have to be for them to be certain that it would land within it and how small so that it would be impossible.

5-3 Probability and Area

Goals ■ Determine probabilities using areas.
Applications Games, Archeology, Weather

Play this game in groups of 3–4 students. You will need the bottom of a large 16-in. pizza box, a 4-in. square paper, and a counter or coin.

1. One person plays against the other members of the group. This person puts the box on a table and places the paper square anywhere inside the box.
2. The remaining members of the group sit on the floor a few feet away (so that the placement of the square cannot be seen). These players take five turns each tossing the counter into the box.
3. A player on the floor wins if the counter comes to rest completely within the paper square once. The person placing the square wins if a counter never lands on the paper square.
4. Is there a way to determine the chance of landing within the paper square? Explain. How can you find the chance of landing elsewhere within the box? Do you think the game is fair? Explain your thinking. Answers will vary.

Build Understanding

You can use what you know about **probability** to solve problems like the one above, in which you need to determine the likelihood that an event will occur. Recall that the probability of an event can be expressed as a ratio:

$$P(\text{any event}) = \frac{\text{number of favorable outcomes}}{\text{number of possible outcomes}}$$

Check Understanding

The probability of an event is a number between 0 and 1. What is the probability of an event that will always occur? What is the probability of an impossible event? Give an example of each.

1; 0; Examples will vary.

Example 1

Let A = where the number cube lands in M, the pizza box, and let N = the paper square. Find $P(A$ lands within region $N)$.

Solution

Find the probability.

$$P = \frac{\text{area of } N}{\text{area of } M} \quad \text{area of } N = \text{favorable outcome}, \ \text{area of } M = \text{possible outcome}$$

$$= \frac{16}{256}$$

$$= \frac{1}{16} \text{ or } 0.0625$$

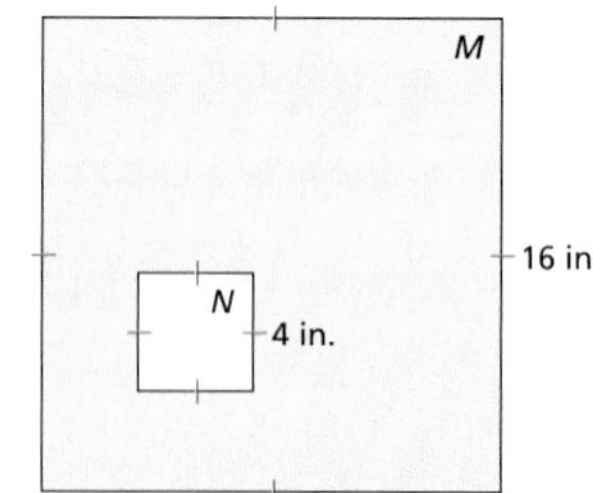

So the probability that the number cube will land on the paper square is $\frac{1}{16}$, or 0.0625.

Teaching Strategies

Have students use coordinate grid paper when drawing polygons. The area relationships will be clearer. Determining the actual areas will also be easier if students can test their calculations by counting square units.

Example 2

A coin is dropped into a can whose bottom is shown at the right. What is the probability that the coin lands in the blue region? In the green region?

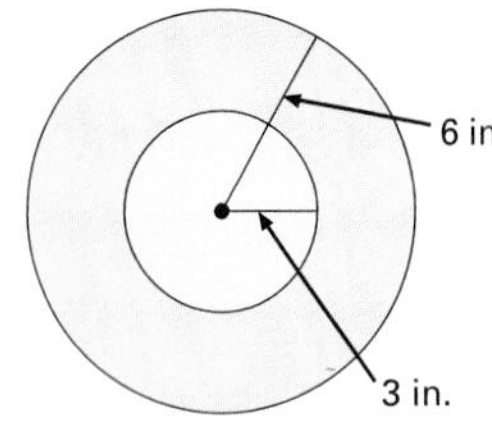

Solution

$$P(\text{blue}) = \frac{\text{area of blue circle}}{\text{area of larger circle}}$$

$$= \frac{(\pi)(3^2)}{(\pi)(6^2)}$$ Since π is a common factor, you do not have to calculate each area.

$$= \frac{9}{36} = \frac{1}{4}$$

Since $P(\text{green or blue}) = 1$, then $P(\text{green}) = 1 - P(\text{blue})$. $P(\text{green}) = \frac{3}{4}$.

Example 3

A treasure chest was buried long ago beneath what is now school property. No one knows where the chest lies. If the school property is a rectangle measuring 600 ft by 540 ft, what is the probability that the chest could be found by excavating the baseball diamond, a square with sides of 90 ft each?

Solution

$$P(\text{chest in diamond}) = \frac{\text{area of diamond}}{\text{area of property}}$$

area of diamond	**area of property**
$A = s^2$	$A = lw$
$A = 90^2 = 8{,}100$	$A = (600)(540) = 324{,}000$

$$P(\text{chest in diamond}) = \frac{8{,}100}{324{,}000} = \frac{1}{40} = 0.025$$

The probability is $\frac{1}{40}$ or 0.025.

Example 4

GAMES Twenty-five darts are thrown at random at a circular dartboard. Four hit the bull's-eye. If the diameter of the bull's-eye is 24 cm, what is the probable area of the target?

Solution

Since 4 of 25 darts landed in the bull's-eye, the probability of a single dart hitting the bull's-eye is $\frac{4}{25}$.

$$P(\text{dart landing in bull's-eye}) = \frac{\text{area of bull's-eye}}{\text{area of target}} = \frac{4}{25}$$

Area of bull's-eye $= \pi r^2 \approx 3.14 \cdot 12^2 \approx 452.16 \text{ cm}^2$

Let x = area of target. $\frac{4}{25} \approx \frac{452.16}{x}$; $x \approx 2{,}826$

The probable area of the target is about $2{,}826 \text{ cm}^2$.

Problem Solving Tip

In Example 4, you know that the ratio of the area of the bull's-eye to the area of the whole target is 4:25. How many times greater is the area of the target? How can you use this information to solve the problem?

6.25 times greater; find the area of the bull's-eye and multiply it by 6.25.

Extend the Lesson

MATH JOURNAL Have students draw two figures one inside the other and explain the relationship between the areas of the two figures and the probability that an object dropped at random will land inside the smaller figure.

Chalkboard Examples

Supplementary Example 1

Find $P(A$ lands within region $N)$ if the side of the larger square is 12 and the side of the smaller square is 6. $\frac{1}{4}$ or 0.25

Supplementary Example 2

Using the diagram for Example 2, if the radius of the blue region is 2 in., what is the probability that the coin lands in the blue region? $\frac{1}{9}$ In the yellow region? $\frac{8}{9}$

Supplementary Example 3

A school yard is in the shape of a rectangle measuring 800 ft by 660 ft. A broken irrigation line lies somewhere under the yard. What is the probability that the break in the line could be found by excavating a soccer field measuring 240 ft by 330 ft. $\frac{3}{20}$ or 0.15

Supplementary Example 4

Twenty darts are thrown at random at a circular dartboard. Five hit the bull's-eye. If the diameter of the bull's eye is 10 cm, what is the probable area of the target? probable area is about 314 cm^2

Reteaching Worksheet 5-3

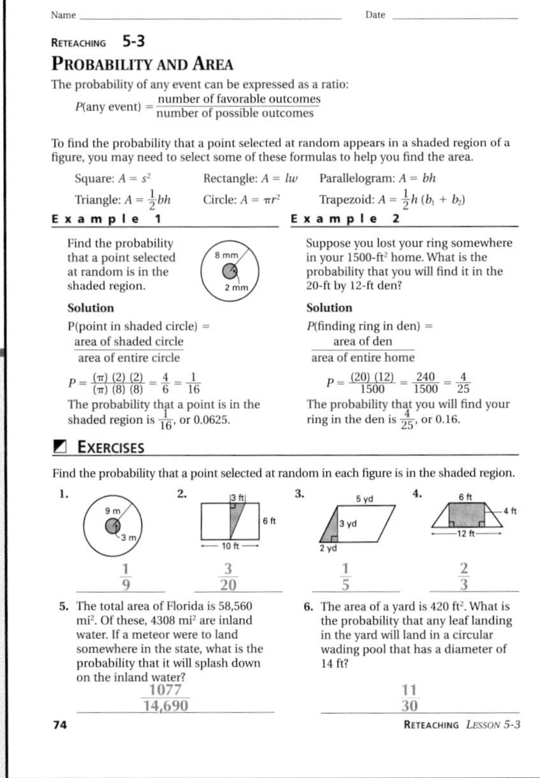

Name ______________________ Date ____________

Reteaching 5-3

Probability and Area

The probability of any event can be expressed as a ratio:

$$P(\text{any event}) = \frac{\text{number of favorable outcomes}}{\text{number of possible outcomes}}$$

To find the probability that a point selected at random appears in a shaded region of a figure, you may need to select some of these formulas to help you find the area.

Square: $A = s^2$	Rectangle: $A = lw$	Parallelogram: $A = bh$
Triangle: $A = \frac{1}{2}bh$	Circle: $A = \pi r^2$	Trapezoid: $A = \frac{1}{2}h(b_1 + b_2)$

Example 1

Find the probability that a point selected at random is in the shaded region.

Solution

P(point in shaded circle) = $\frac{\text{area of shaded circle}}{\text{area of entire circle}}$

$$P = \frac{(\pi)(2)(2)}{(\pi)(8)(8)} = \frac{4}{6} = \frac{1}{16}$$

The probability that a point is in the shaded region is $\frac{1}{16}$, or 0.0625.

Example 2

Suppose you lost your ring somewhere in your 1500-ft² home. What is the probability that you will find it in the 20-ft by 12-ft den?

Solution

$P(\text{finding ring in den}) = \frac{\text{area of den}}{\text{area of entire home}}$

$$P = \frac{(20)(12)}{1500} = \frac{240}{1500} = \frac{4}{25}$$

The probability that you will find your ring in the den is $\frac{4}{25}$, or 0.16.

Exercises

Find the probability that a point selected at random in each figure is in the shaded region.

1. $\frac{1}{9}$

2. $\frac{3}{20}$

3. $\frac{1}{5}$

4. $\frac{2}{3}$

5. The total area of Florida is 58,560 mi². Of these, 4308 mi² are inland water. If a meteor were to land somewhere in the state, what is the probability that it will splash down on the inland water? $\frac{1077}{14{,}690}$

6. The area of a yard is 420 ft². What is the probability that any leaf landing in the yard will land in a circular wading pool that has a diameter of 14 ft? $\frac{11}{30}$

74 Reteaching *Lesson 5-3*

Lesson 5-3 **Probability and Area** 213

Lesson Wrap-up

Quick Assessment

Ask the following questions to determine if students understand the content presented in this lesson.

1. You are trying to find the experimental probability of a dart's missing the bull's-eye on a circular target. During the experiment, what event would you consider to be a favorable outcome? missing the bull's eye
2. In a game, eight out of 25 cubes tossed into a square box land in a smaller square drawn on the bottom. If the box is a 10 in. × 14 in. rectangle, what is the probable area of the small square? 44.8 in.^{2}

Assignment Guide

Basic: 1–26, 29–40
Advanced: 1–40
Additional Practice: See Extra Practice Index on page 674.

Additional Answers

29.
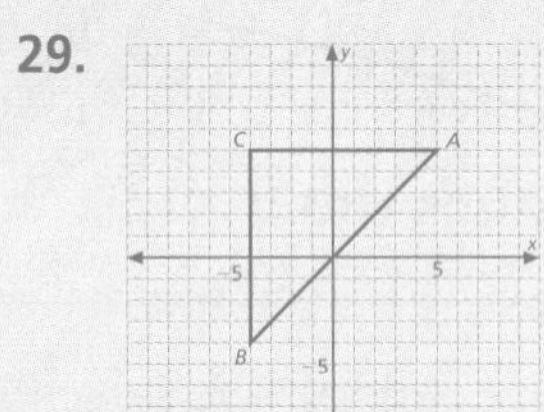

30.
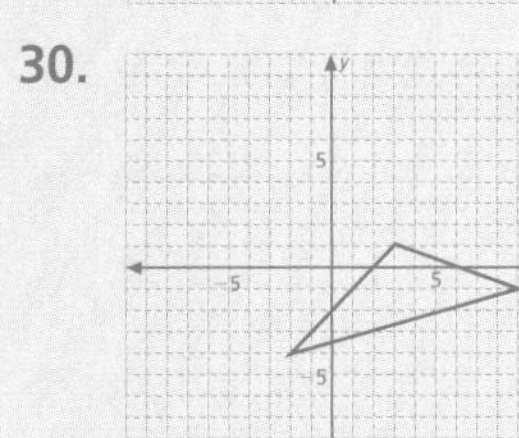

31.
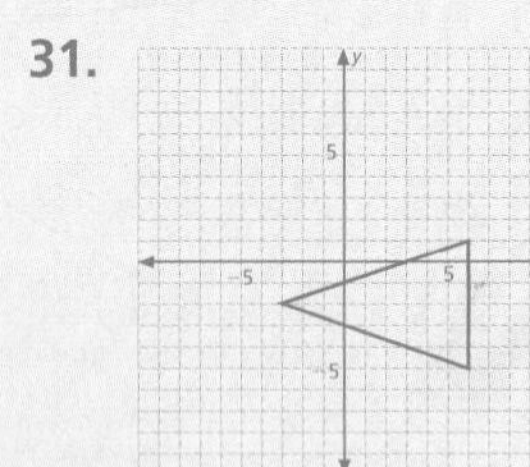

32.
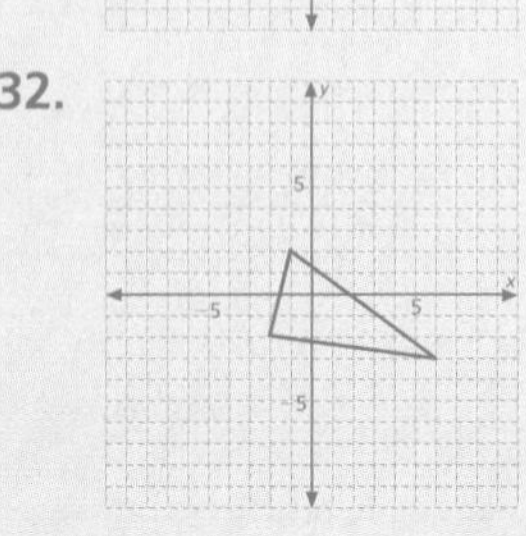

Try These Exercises

Find the probability that a point selected at random is in the shaded region.

1.
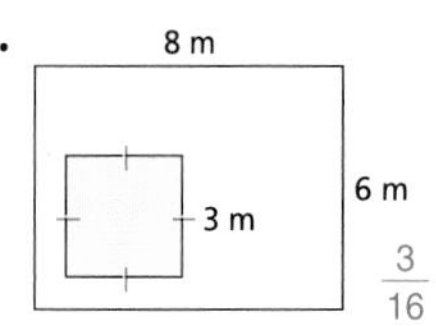

$\frac{3}{16}$

2.
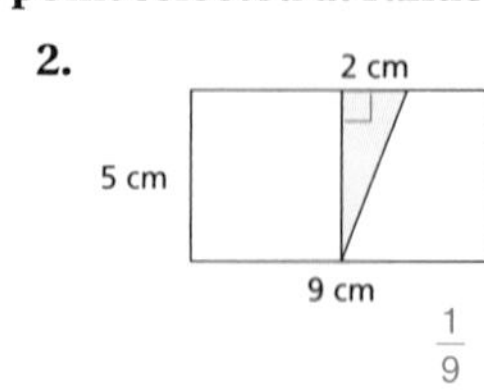

$\frac{1}{9}$

3.
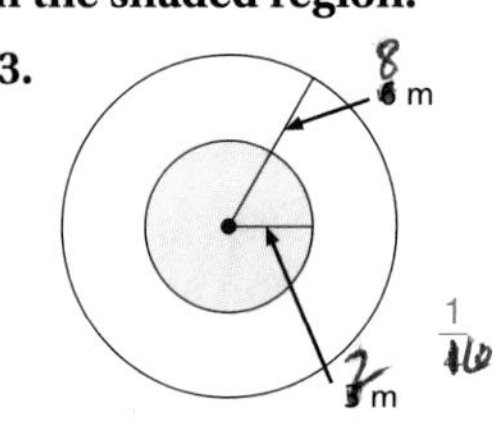

$\frac{1}{16}$

4. **GAMES** Think back to the game in the opening situation. What is the probability of landing the number cube within the small square if the pizza box is a 12-in. box? $\frac{1}{9}$

Practice Exercises

GAMES A standard deck of playing cards has 52 cards. A card is drawn at random from a shuffled deck. Find each probability.

5. P(queen) $\frac{1}{13}$ **6.** P(red card) $\frac{1}{2}$ **7.** P(black face card) $\frac{3}{26}$

Find the probability that a point selected at random is in the shaded region.

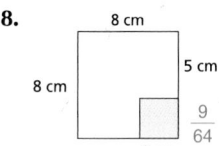

8. 8 cm; 8 cm; 5 cm; 3 cm $\frac{9}{64}$

9.
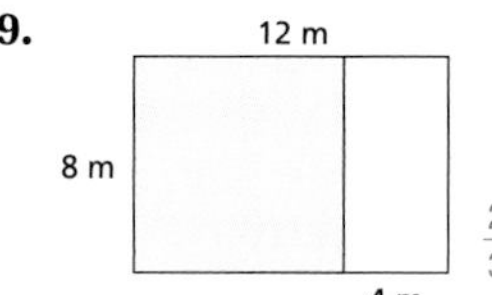

$\frac{2}{3}$

10.
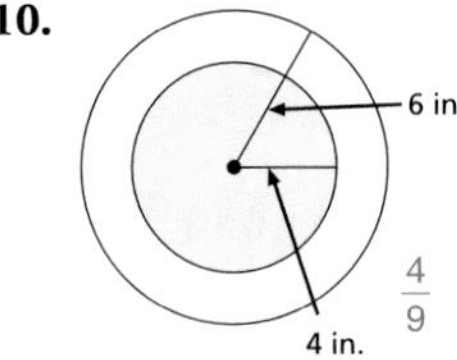

$\frac{4}{9}$

11.
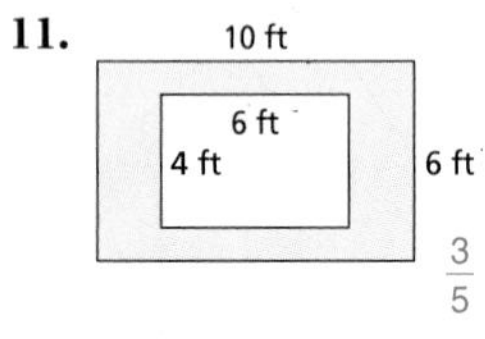

$\frac{3}{5}$

12.
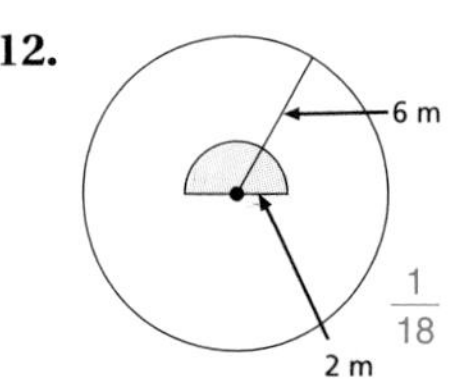

$\frac{1}{18}$

13.
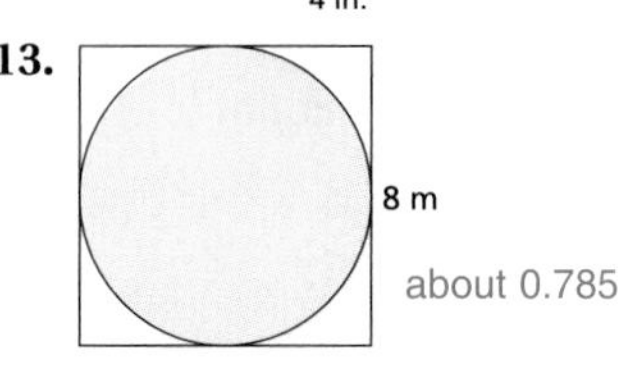

about 0.785

14. Suppose a cordless phone has been left somewhere within a 2,000 ft^2 house. What is the probability it is in the 20-ft by 15-ft living room? $\frac{3}{20}$

15. The total area of the state of Oklahoma is 69,919 mi^2. The area of its capital, Oklahoma City, is 604 mi^2. If a meteor were to land somewhere in the state, estimate the probability that it would land within the city limits of the capital. Possible answer: about 0.009 or $\frac{9}{1,000}$

16. **ARCHEOLOGY** The rectangular foundation for an ancient building measures 150 ft by 80 ft. The foundation is made from stone cubes, with sides measuring 2 ft each. A scroll is hidden within one of the blocks. What is the probability of finding the scroll if a block is chosen randomly? $\frac{1}{326}$, or about 0.3%

Oklahoma City

Teaching Strategies

COOPERATIVE LEARNING Have students work in small groups to solve this problem mathematically. You may wish to have them construct the game and test their solutions. A carnival game consists of a rectangular board measuring 5 ft × 8 ft. On the board are 15 circles, each with a radius of 6 in. You can toss any coin onto the board. If the coin lands in a circle, you double your money. Is the game fair? Who has the advantage? The probability of winning is slightly greater than $\frac{1}{4}$. The game is unfair. The carnival has the advantage.

17. **YOU MAKE THE CALL** A square (side = 2 in.) is placed inside a larger square (side = 6 in.). Evan says that the probability of selecting a point at random within the smaller square is $\frac{1}{3}$, since the ratio of the smaller square's side to the larger square's side is 2:6, or 1:3. Do you agree with Evan's thinking? Explain your reasoning. No. That actual probability is $\frac{1}{9}$, which is $\left(\frac{1}{3}\right)^2$

Describe the probability of each event by writing *certain, likely, unlikely,* or *impossible.*

18. **WEATHER** It will snow in July where you live. unlikely
19. You will roll a sum of 5 or greater using two number cubes. likely
20. You left your pencil in one of your 5 classrooms, but you don't know which one. You find it in the first room you search. unlikely
21. There is life on other planets. likely

Find the probability that a point selected at random in each figure is in the shaded region.

22. 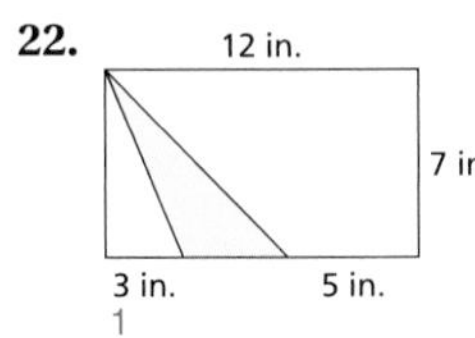

$\frac{1}{6}$

23. 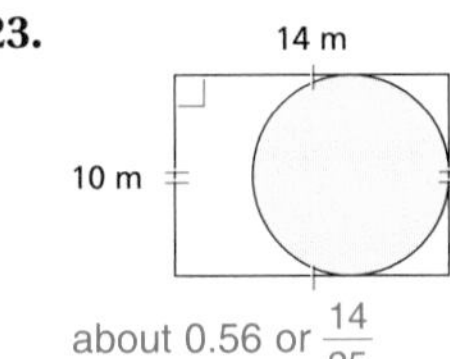

about 0.56 or $\frac{14}{25}$

24. 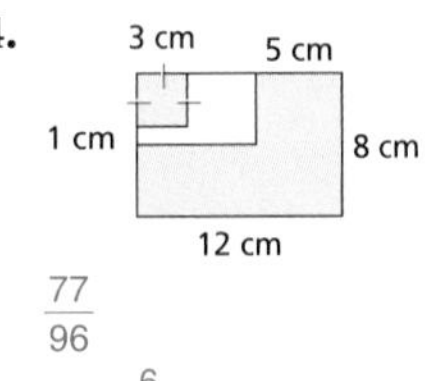

$\frac{77}{96}$

25. Suppose a leak occurs from above the room shown at the right. What is the probability the leak will be over the carpet? $\frac{6}{31}$

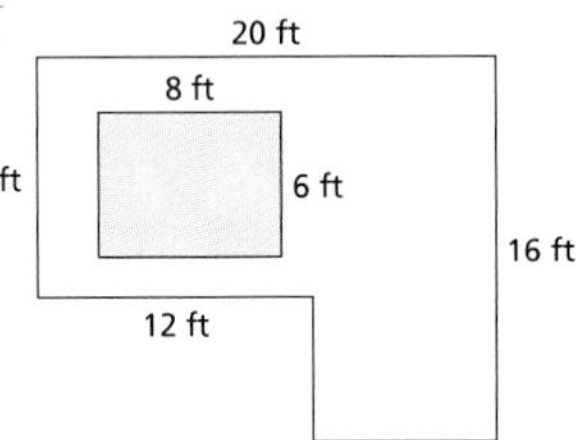

26. Draw a figure containing a shaded region, so that the probability is 1 out of 6 that a point selected at random will be in the shaded region. Figures will vary, but for each the area of the shaded region should be $\frac{1}{6}$ the area of the entire region.

Extended Practice Exercises

27. **WRITING MATH** In a scale drawing of the ruins of Pompeii, 1 in. = 400 ft. A student is erasing a pencil mark accidentally made somewhere on the drawing. What additional information do you need to know to find the probability that the mark was made on the Palaestra? The map or actual dimensions of the Palaestra and the total map or actual area of the ruins.
28. Suppose a square target looks like this. What is the probability of hitting the shaded area? 0.05357; The target area is 2 ft by 2 ft or 4 ft^2 and the shaded area is 0.215 ft^2.

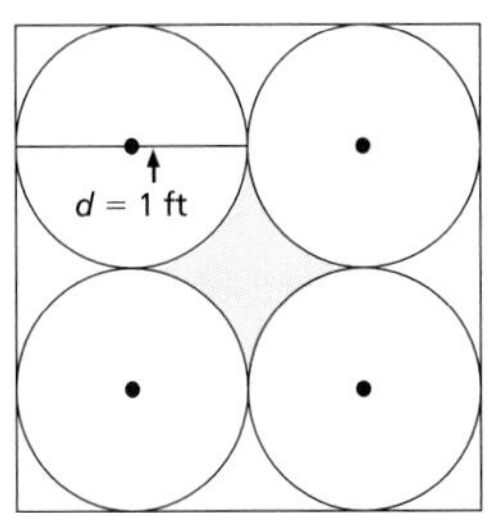

Mixed Review Exercises

On a coordinate plane, sketch the triangle with the given coordinates. Then classify the triangle by both its angles and its sides. (Lesson 4-1) For 29–32, see additional answers.

29. $A(5, 5), B(-4, -4), C(-4, 5)$ right isosceles
30. $L(3, 1), M(9, -1), N(-2, -4)$ obtuse scalene
31. $R(6, 1), S(-3, -2), T(6, -5)$ acute isosceles
32. $X(-2, -2), Y(-1, 2), Z(6, -3)$ acute scalene

Given $f(x) = 5x - 8$ and $g(x) = 1.3x - 4.7$, find each value. (Lesson 2-2)

33. $f(-5)$ −33
34. $f(13)$ 57
35. $f(-8)$ −48
36. $f(9)$ 37
37. $g(12)$ 10.9
38. $g(-17)$ −26.8
39. $g(34)$ 39.5
40. $g(-27)$ −39.8

Extend the Lesson

ONGOING ASSESSMENT Have students work the following problem: A circle has a radius of 8 cm. A smaller circle within the larger circle has a radius of 5 cm. Find the probability that a random point in the figure lies within the smaller circle. The solution is $\frac{25}{64}$.
Watch for the common mistake of using the radii rather than the areas to write the probability. Students who make this error will get the answer $\frac{5}{8}$.

Extra Practice Worksheet 5-3

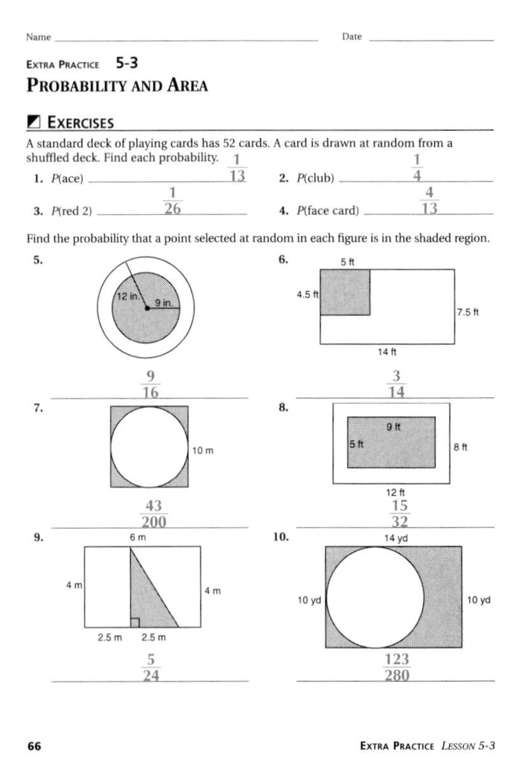

Name ____________ Date ____________

EXTRA PRACTICE 5-3

PROBABILITY AND AREA

EXERCISES

A standard deck of playing cards has 52 cards. A card is drawn at random from a shuffled deck. Find each probability.

1. P(ace) $\frac{1}{13}$
2. P(club) $\frac{1}{4}$
3. P(red 2) $\frac{1}{26}$
4. P(face card) $\frac{4}{13}$

Find the probability that a point selected at random in each figure is in the shaded region.

5. (12 in., 9 in.) $\frac{9}{16}$
6. (5 ft, 4.5 ft, 7.5 ft, 14 ft) $\frac{3}{14}$
7. (10 m) $\frac{43}{200}$
8. (9 ft, 5 ft, 8 ft, 12 ft) $\frac{15}{32}$
9. (6 m, 4 m, 4 m, 2.5 m, 2.5 m) $\frac{5}{24}$
10. (14 yd, 10 yd, 10 yd) $\frac{123}{280}$

66 EXTRA PRACTICE *LESSON 5-3*

Enrichment Worksheet 5-3

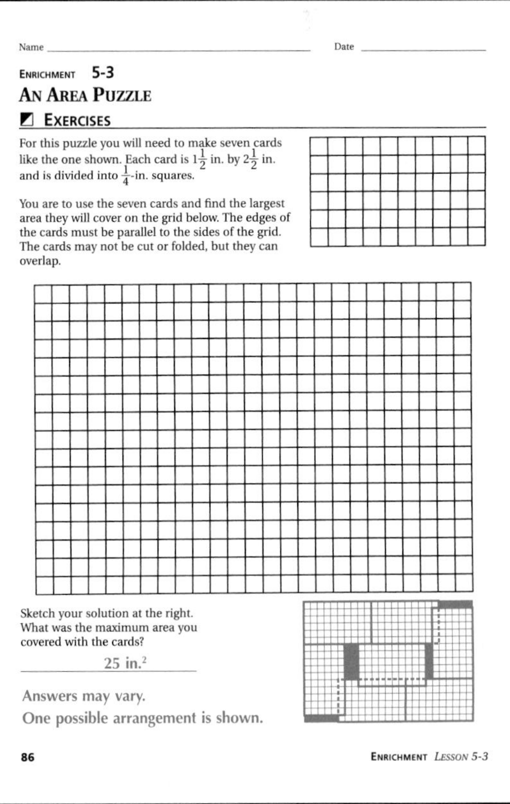

Name ____________ Date ____________

ENRICHMENT 5-3

AN AREA PUZZLE

EXERCISES

For this puzzle you will need to make seven cards like the one shown. Each card is $1\frac{1}{2}$ in. by $2\frac{1}{2}$ in. and is divided into $\frac{1}{4}$-in. squares.

You are to use the seven cards and find the largest area they will cover on the grid below. The edges of the cards must be parallel to the sides of the grid. The cards may not be cut or folded, but they can overlap.

Sketch your solution at the right. What was the maximum area you covered with the cards? 25 in.^{2}

Answers may vary. One possible arrangement is shown.

86 ENRICHMENT *LESSON 5-3*

Lesson Planning

NCTM Standards/Strands

- Number & Operation
- Patterns, Functions, & Algebra
- Geometry & Spatial Sense
- Measurement
- Problem Solving
- Reasoning and Proof
- Connections
- Representation

Vocabulary

indirect proof

Materials Needed

paper/pencil

Lesson Resources

Warm-Up Transparency 14
Enrichment 5-4
Reteaching 5-4
Multicultural Connection 10
Transparency RF-18

Assignment Guide

Basic: 1–7, 8–15
Enriched: 1–15

Getting Started

5-Minute Warm-up

Find the area of each figure.

1. square with side of 65 ft
   $4{,}225\ ft^2$
2. trapezoid with bases of 8 m and 14 m and height of 6 m
   $66\ m^2$
3. circle with diameter of 26 ft
   $830.66\ ft^2$

5-4 Problem Solving Skills: Irregular Shapes

Sometimes you can find the answer to a difficult problem by breaking it into smaller problems you already know how to solve.

Problem Solving Strategies

- Guess and check
- Look for a pattern
- ✓ Solve a simpler problem
- Make a table, chart or list
- Use a picture, diagram or model
- Act it out
- Work backwards
- Eliminate possibilities
- Use an equation or formula

Problem

Here is an architect's sketch of the plan for one floor of a new archeological museum. What is the area of this floor?

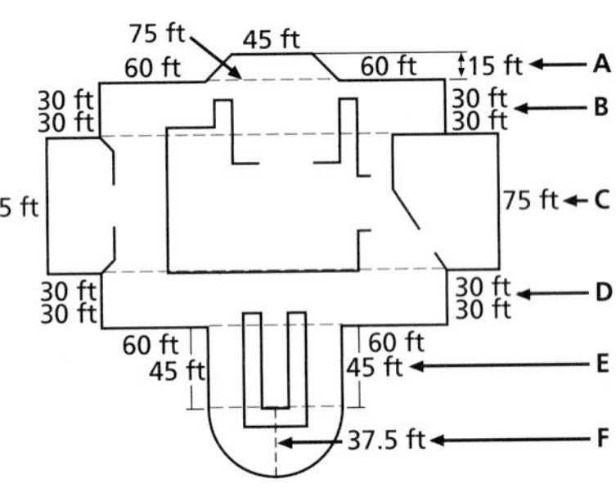

Solution

Solve a simpler problem. Copy or trace the *outline* of the floor plan on graph paper. Ignore all the inner walls. Divide the floor into 6 figures: a trapezoid, 4 rectangles, and a half-circle. Label them A–F.

75 ft, 45 ft, 60 ft, 60 ft, 15 ft ← A; 30 ft, 30 ft ← B; 75 ft, 75 ft ← C; 30 ft, 30 ft ← D; 60 ft, 45 ft, 60 ft, 45 ft ← E; 37.5 ft ← F

Find the area of each figure.

A. $\left(\frac{75 + 45}{2}\right)(15) = 900\ ft^2$

B. $30(195) = 5{,}850\ ft^2$

C. $75(255) = 19{,}125\ ft^2$

D. $30(195) = 5{,}850\ ft^2$

E. $45(75) = 3{,}375\ ft^2$

F. $\pi \cdot 37.5^2 \cdot 0.5 \approx 2{,}208\ ft^2$

Add to find the total area of all the regions. The area of the first floor of the museum is $37{,}308\ ft^2$.

Check your answer by tracing the outline again and dividing it into a different arrangement of plane figures.

Check Understanding

Why do you multiply by 0.5 when finding the area of region F?

The figure is a half-circle.

Try These Exercises

1. **CARPETING** This layout of a wing at a natural history museum shows the African Peoples, Asian Peoples, and Birds of the World galleries. How much carpeting is needed for this wing?
   $12{,}150\ ft^2$

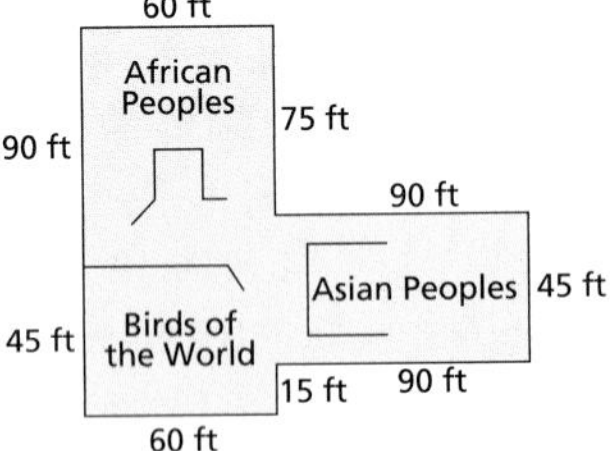

Use the Five-step Plan and the Strategy

THE FIVE-STEP PLAN **Read**—ask questions of the students to help them understand the problem. **Plan**—guide students to related problems and previously mastered skills and strategies. **Solve**—students solve problem on their own. **Answer**—write solution in format that answers the question. **Check**—review work, check for reasonableness and review strategy used. Students will benefit from the experience of verbalizing their methods.

THE STRATEGY Solve a simpler problem—Many complex problems can be divided into smaller pieces that are easier to solve. Then the solutions of the simpler problems are combined to form the solution of the more difficult problem.

Teaching Strategies

For Exercises 4, 6 and 7, students may find it helpful to keep a table of data and draw a diagram for each simpler problem they solve.

2. An archeological team has 30 ft of fencing with which to enclose a rectangular region. If the length and width are whole numbers, what different areas, in ft^2, are possible? 14, 26, 36, 44, 50, 54, 56

3. A display case planned for showing fragments of 3,500-year-old frescoes will have the shape at right. If there will be 3 glass shelves, how much glass, in square feet, is needed? 396 ft^2

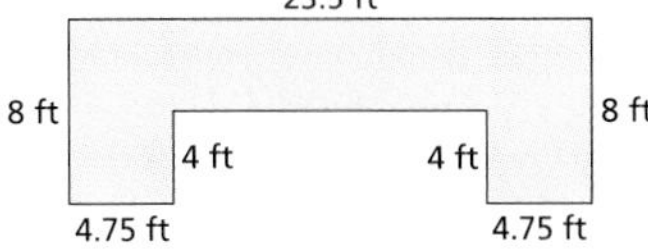

Practice Exercises

4. How many small squares are in a checkerboard? How many squares of all sizes are there? 64; 204

5. **ART** Painters are to cover the entire side of this wall from an abandoned factory with a mural. What is the area of the region to be painted? 126.5 m^2

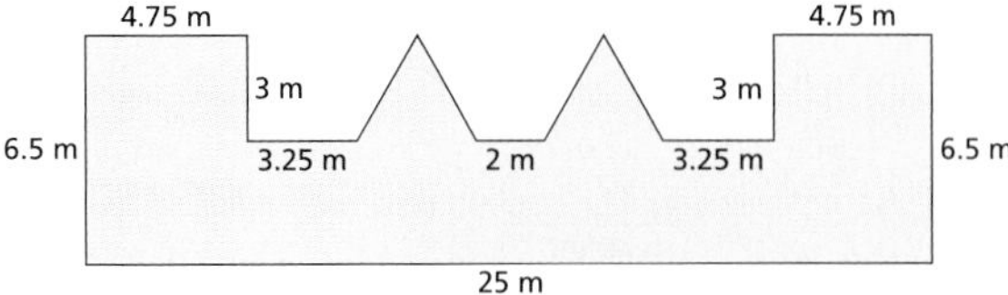

6. **WRITING MATH** Write your own problem involving the area of irregular figures that can be solved by first solving a simpler problem or problems. Check students' work.

7. **RECREATION** The Mayans played a game resembling our game of volleyball on courts that looked like the one shown below. Imagine for a moment that you are the coordinator of a tournament for which you will need a rectangular field containing three side-by-side courts like these. You decide that the courts must be at least 20 m apart and that there must be at least 5 m between an edge of a court and the perimeter of the field. How long must your field be? How wide? at least 230 ft long, at least 150 ft wide.

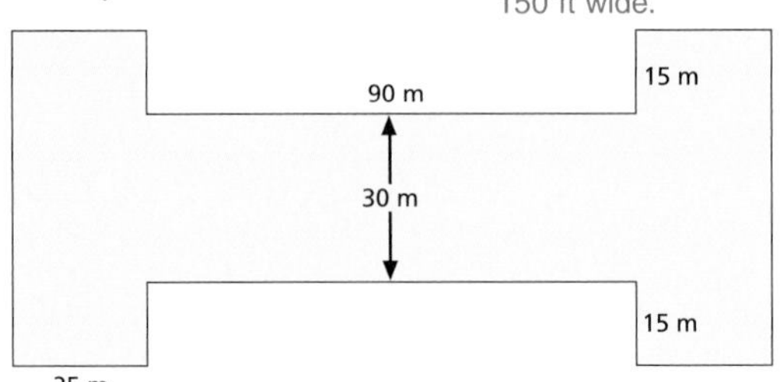

Mixed Review Exercises

Use the polygon-sum theorem to find the sum of the measures of convex polygons with the given number of sides. Then find the measure of each interior angle, assuming each polygon is regular. (Lesson 4-7)

8. 23 3780°, 164.35°
9. 45 7740°, 172°
10. 38 6480°, 170.53°
11. 28 4680°, 167.14°
12. 18 2880°, 160°
13. 34 5760°, 169.41°
14. 50 8640°, 172.8°
15. 42 7200°, 171.43°

Chalkboard Examples

Supplementary Problem

A day care center needs a new fence. Use the art in Exercise 3 to represent the yard. How many feet of fencing is needed? 71 ft

Lesson Wrap-up

Quick Assessment

Ask the following question to determine if students understand the content presented in this lesson.

How do you know whether solving a simpler problem is a good strategy to follow? Answers will vary. Generally, this is a good strategy whenever a problem breaks naturally into smaller pieces. A problem can also be made less difficult by working with simpler numbers. For example, using 3.14 for pi.

Enrichment Worksheet 5-4 not shown.

Reteaching Worksheet 5-4

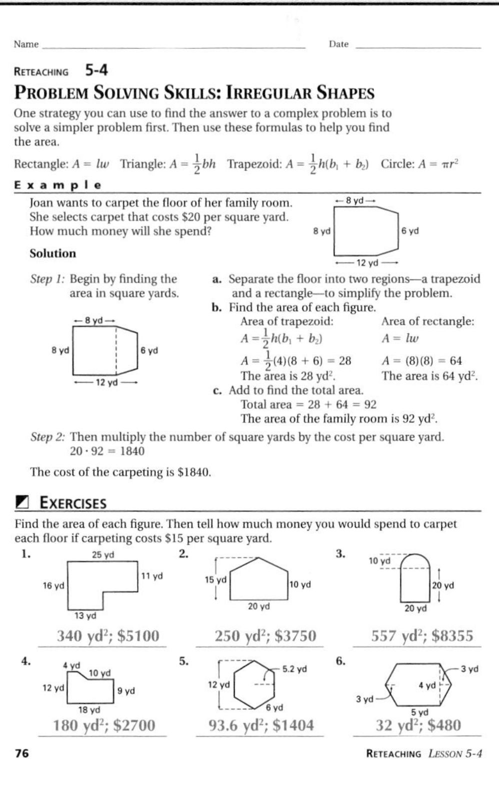

Name ______________________ Date ____________

Reteaching 5-4

Problem Solving Skills: Irregular Shapes

One strategy you can use to find the answer to a complex problem is to solve a simpler problem first. Then use these formulas to help you find the area.

Rectangle: $A = lw$ Triangle: $A = \frac{1}{2}bh$ Trapezoid: $A = \frac{1}{2}h(b_1 + b_2)$ Circle: $A = \pi r^2$

Example

Joan wants to carpet the floor of her family room. She selects carpet that costs \$20 per square yard. How much money will she spend?

Solution

Step 1: Begin by finding the area in square yards.

a. Separate the floor into two regions—a trapezoid and a rectangle—to simplify the problem.

b. Find the area of each figure.

Area of trapezoid: $A = \frac{1}{2}h(b_1 + b_2)$; $A = \frac{1}{2}(4)(8 + 6) = 28$. The area is 28 yd^2.

Area of rectangle: $A = lw$; $A = (8)(8) = 64$. The area is 64 yd^2.

c. Add to find the total area.
Total area = 28 + 64 = 92
The area of the family room is 92 yd^2.

Step 2: Then multiply the number of square yards by the cost per square yard.
$20 \cdot 92 = 1840$

The cost of the carpeting is \$1840.

Exercises

Find the area of each figure. Then tell how much money you would spend to carpet each floor if carpeting costs \$15 per square yard.

1. 340 yd^2; \$5100
2. 250 yd^2; \$3750
3. 557 yd^2; \$8355
4. 180 yd^2; \$2700
5. 93.6 yd^2; \$1404
6. 32 yd^2; \$480

76 Reteaching *Lesson 5-4*

Teaching Strategies

REAL WORLD CONNECTIONS Have students find the area of several rooms in a hallway or wing of the school. You may want to provide students with the necessary measurements or arrange for students to do the measuring in small groups. As an added challenge, explain that 1 yd^2 = 9 ft^2. Ask them to find out how many square yards of carpeting would be needed to carpet the rooms.

Skills Practice

Vocabulary Review

Lesson 5-3
probability random
favorable outcome

Lesson 5-4
indirect proof

Assignment Guide

All students: 1–24
Additional Practice: Refer to the Extra Practice Index on page 674 of this text.

Review and Practice Your Skills

Practice ■ Lesson 5-3

Find the probability that a point selected at random is in the shaded region.

1. 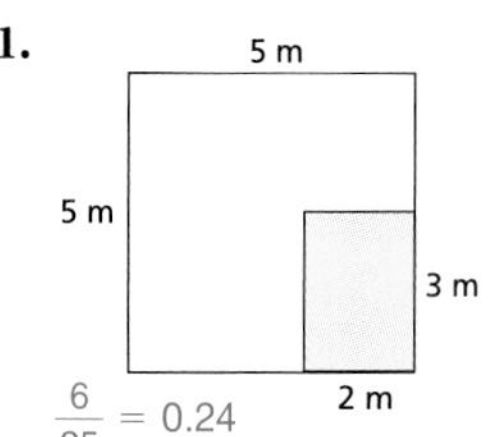

$\frac{6}{25} = 0.24$

2. 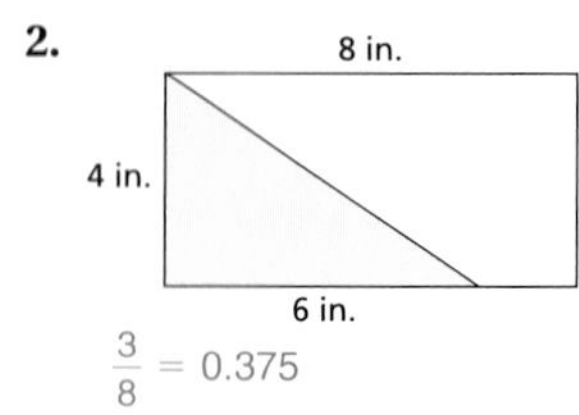

$\frac{3}{8} = 0.375$

3. 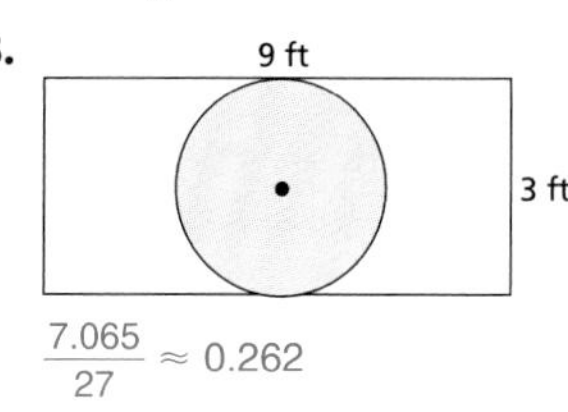

$\frac{7.065}{27} \approx 0.262$

4. 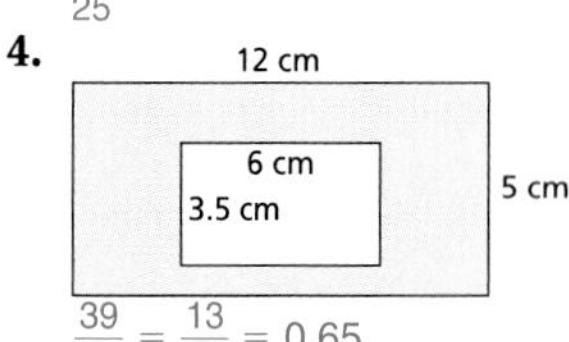

$\frac{39}{60} = \frac{13}{20} = 0.65$

5. 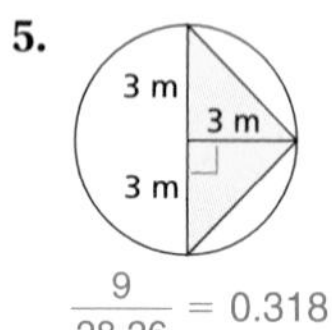

$\frac{9}{28.26} = 0.318$

6. 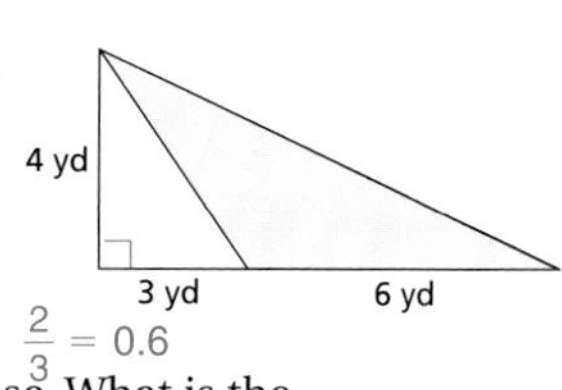

$\frac{2}{3} = 0.6$

7. Suppose a book has been left somewhere within a 2700 ft^2 house. What is the probability that the book is either in the 15-ft by 10-ft den or a 12-ft by 10-ft bedroom? $\frac{1}{10} = 0.1$

Describe the possibility of each event by writing *certain, likely, unlikely,* or *impossible.*

8. The temperature will hit 90 degrees Farenheit in December. unlikely
9. You will draw a red face card from a standard, shuffled deck of cards. unlikely
10. The area of a rectangular region is calculated by multiplying its length by its width. certain
11. You will leave school grounds before 6 P.M. this evening. Answers will vary. Most students will answer "likely."

Practice ■ Lesson 5-4

Find the area of each figure.

12. 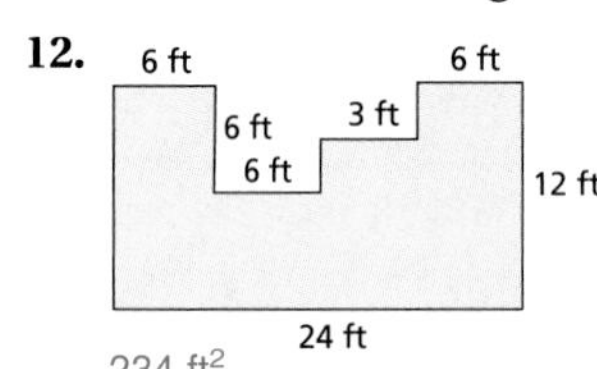

234 ft^2

13. 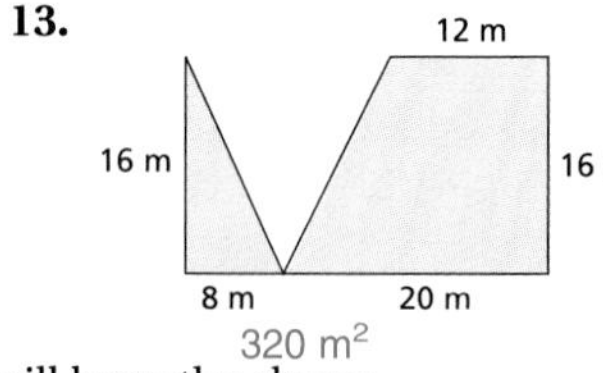

320 m^2

14. 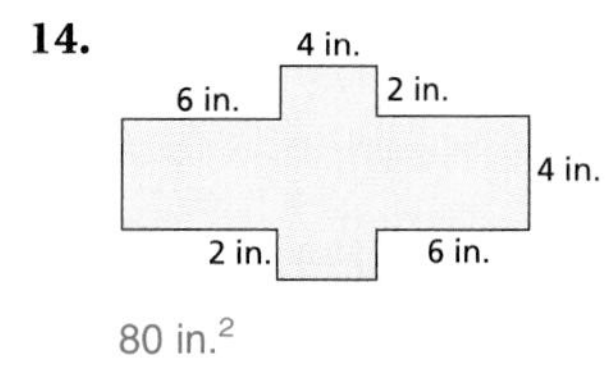

80 $in.^2$

15. A counter top for a kitchen will have the shape shown. How much laminate material, in square inches, is needed to make this counter top? 46.25 ft^2

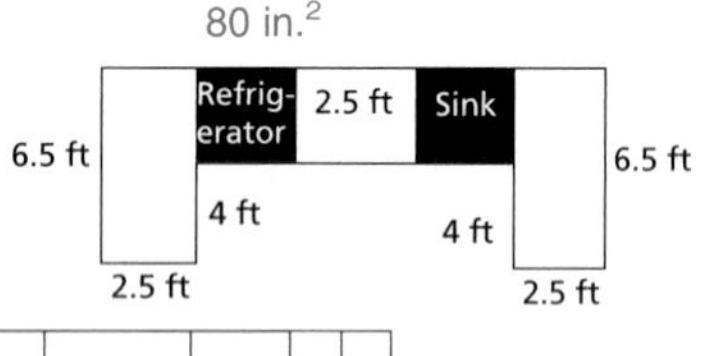

16. How many rectangles of all sizes are in the figure shown? 63

Teaching Strategies

Discuss the difference between experimental and theoretical probability. In Exercises 1–6, the students are expected to find the theoretical probability based on the ratio of a smaller target to a larger figure. Ask: How might the actual results of trying to hit the target differ from the theoretical probability? Students may raise the idea that if one is trying to hit the target, the event is not longer completely random. In the same way, knowledge allows a student to eliminate wrong answer choices on a multiple-choice test and increase the probability of them guessing the correct answer.

PRACTICE ■ LESSON 5-1–LESSON 5-4

Write each ratio in lowest terms. (Lesson 5-1)

17. 98 m:49 km 1:500 **18.** 13 c to 4 fl oz 26:1 **19.** 1760 yd:10560 ft 1:2

Find the perimeter or circumference of each figure. Then find the area of each figure. (Lesson 5-2)

20.

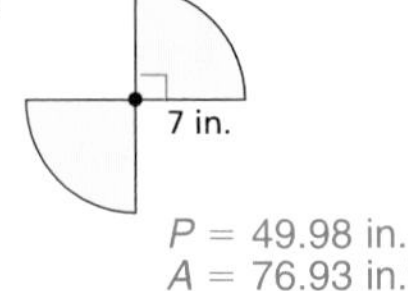

$P = 49.98$ in.
$A = 76.93$ in.^{2}

21.

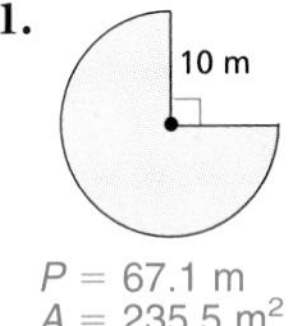

$P = 67.1$ m
$A = 235.5$ m^2

22.

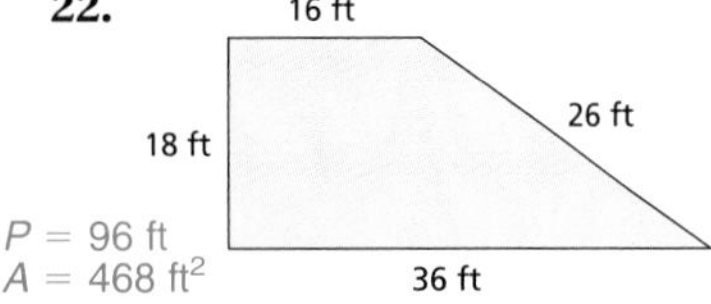

$P = 96$ ft
$A = 468$ ft^2

23. Suppose a parachutist will be landing in the region at the right. What is the probability that she will land in the shaded part of the region? (Lesson 5-3) $\frac{2}{5} = 0.40$

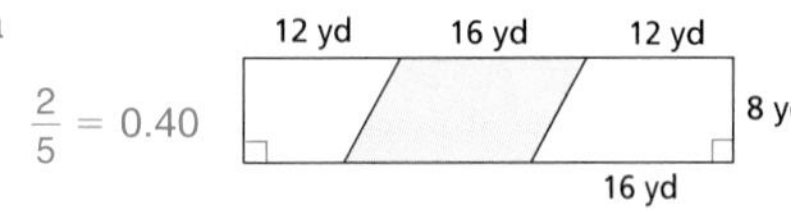

24. If everyone in this classroom shakes hands with everybody else exactly once, how many handshakes will occur? (Lesson 5-4)
Answers will vary.

Mid-Chapter Quiz

Write each ratio in lowest terms.

1. 25 in. to 5 ft 5:12
2. 48:320 oz 6:5
3. Which is the better buy, 19 oz of cereal for \$.266 or 12 oz of the same cereal for \$1.92? 19 oz
4. The diagonal of a rectangle is 25 cm. If the long side of the rectangle is 24 cm, what is the ratio of the short side to the diagonal? 7:25

Find the perimeter and area of each figure. Round answers to the nearest tenth.

5.

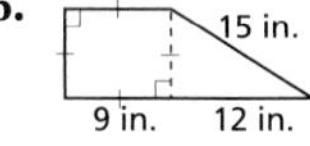

54 in; 135 in.^{2}

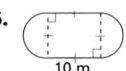

6. 10 m

51.4 m; 178.5 m^2

Draw the described figures. Find the probability that a random point lies inside the larger figure but outside the shaded area.

7. Draw a circle with a radius of 6 cm. Draw a smaller circle inside the first with a radius of 3 cm. Shade the smaller circle. about $\frac{1}{4}$ or 0.25
8. Draw a square with sides measuring 10 cm each. Find the midpoint of one side. Connect the point to either opposite vertex. Shade the triangle. $\frac{3}{4}$ or 0.75

Chalkboard Examples

Example from Lesson 5-3

A square with side 9 m is drawn inside a square with side 12 m. What is the probability that a randomly picked point lies inside the large square but outside the smaller one? $\frac{7}{16}$

Example from Lesson 5-4

A hotel suite has the following layout. Find the area of the suite. 652 square units

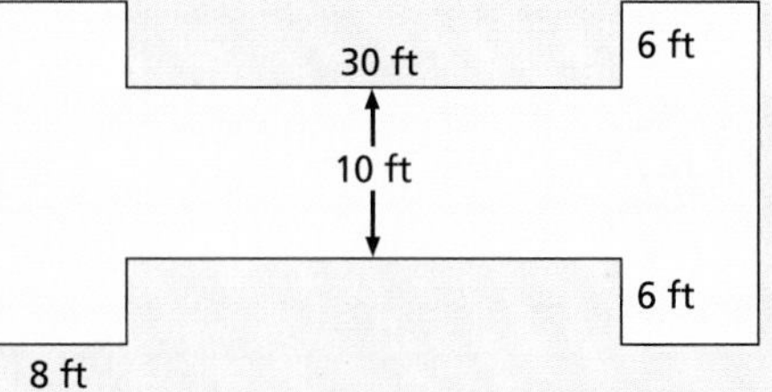

Mid-Chapter Quiz

Correlation Chart

Question Number	Lesson Number
1–5	5-1
6–7	5-2
8–10	5-3 and 5-4

Extend the Lesson

MATH JOURNAL Have students solve the following problem and then write a paragraph explaining the procedures they used.
The perimeter of a rectangle is 200 m. The length of a short side is 35 m. What is the ratio of the length of the short side to the length of the long side? 7:13

Lesson Planning

NCTM Standards/Strands

- Patterns, Functions, & Algebra
- Geometry & Spatial Sense
- Measurement
- Problem Solving
- Reasoning and Proof
- Connections
- Representation

Vocabulary

polyhedron	vertex
prism	pyramid
lateral faces	lateral edges
cylinder	axis
cone	sphere
locus	

Materials Needed

graph paper scissors

Lesson Resources

Warm-Up Transparency 14
Reteaching 5-5
Extra Practice 5-5
Enrichment 5-5
Transparency RF-19, 20

Assignment Guide

Basic: 1–18, 21–25
Enriched: 1–25

Getting Started

5-Minute Warm-up

Name several examples of each.

1. Space figures in which all faces are polygons. all prisms, pyramids
2. Space figures in which some faces are not polygons. cylinder, cone, sphere

Introduction to Lesson 5-5

After students have had an opportunity to assemble the figures, ask them to describe the differences between those that form polyhedra and those that do not. Students should realize that to form polyhedra, the polygons must meet at vertices without any overlaps.

5-5 Three-dimensional Figures and Loci

Goals ■ Analyze space figures.
Applications Archeology, Architecture, Art

Trace and cut out each pattern below. Try to fold each to form a three-dimensional figure. What do you notice? Answers will vary.

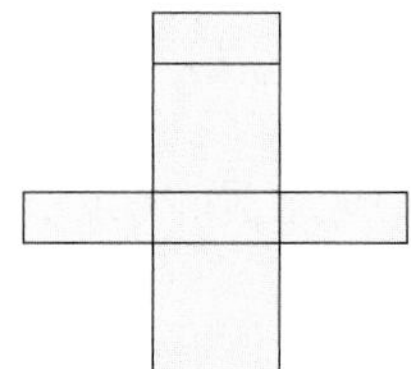
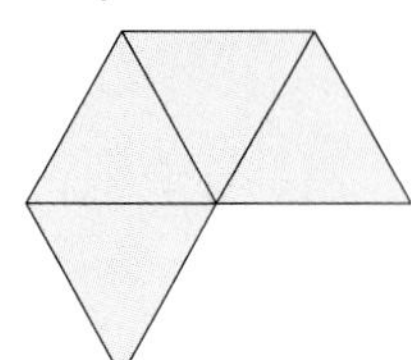
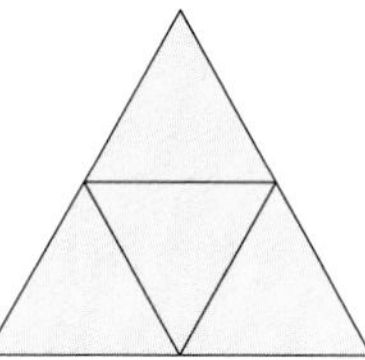
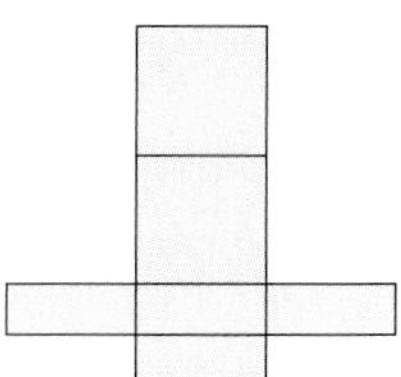

Build Understanding

A **polyhedron** (plural: *polyhedra*) is a three-dimensional figure in which each surface is a polygon. The surfaces are called **faces**. Two faces intersect at an **edge.** A **vertex** is a point where three or more edges intersect.

A polyhedron with two identical parallel faces is called a **prism**. Each of these faces is called a **base**. Every other face is a parallelogram. A **pyramid** is a polyhedron with only one base. The other faces are triangles that meet at a vertex. A prism is named by the shape of its bases and a pyramid by the shape of its base. The **lateral faces** are those that are not bases. The edges of these faces are called **lateral edges** and can be parallel, intersecting, or skew.

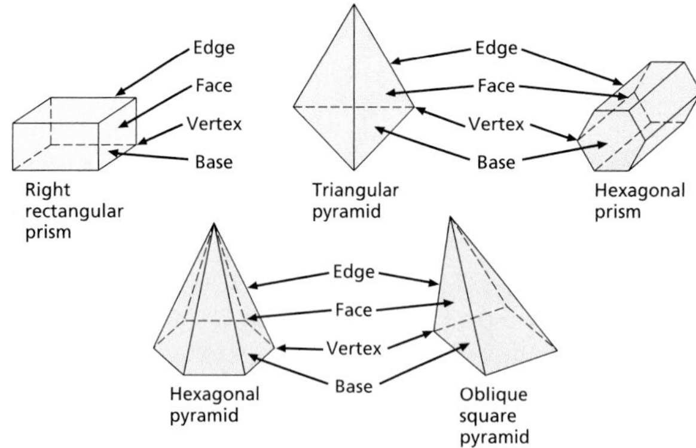

Some three-dimensional figures have flat *and* curved surfaces. A **cylinder** has a curved region and two parallel congruent circular bases. Its **axis** joins the centers of the two bases.

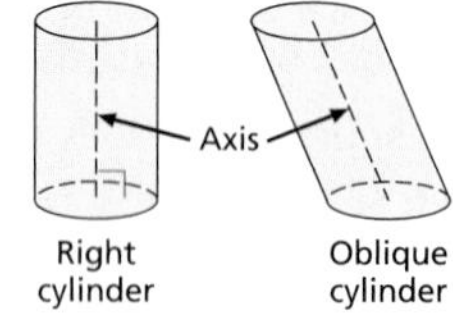

Teaching Strategies

ESL/LEP The prefixes *poly-*, *penta-*, and *octa-* may be comparable to their forms in students' primary languages. Have students keep a chart comparing the meanings of the prefixes in both their primary language and English. Have them list other objects in both languages that use the prefixes.

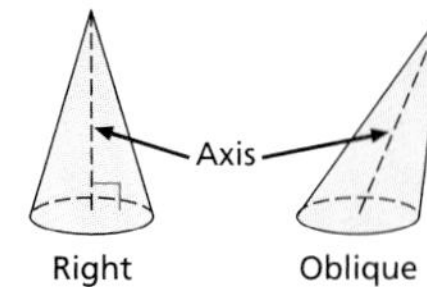

A **cone** is a three-dimensional figure with a curved surface and one circular base. Its axis is a segment from the vertex to the center of the base.

A **sphere** is the set of points in space that are the same distance from a given point called the **center** of the sphere.

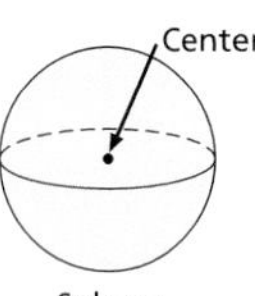

Example 1

Identify the figure.

a.

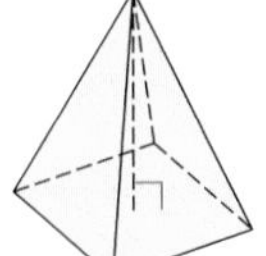

b.

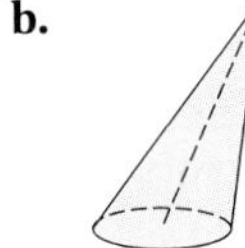

Solution

a. Square pyramid—it has one square base and triangular faces.

b. Cone—it has a curved surface and one circular base.

Example 2

For the pentagonal prism at the right, identify the bases, a pair of intersecting faces and the edge at which they intersect, and a pair of skew edges.

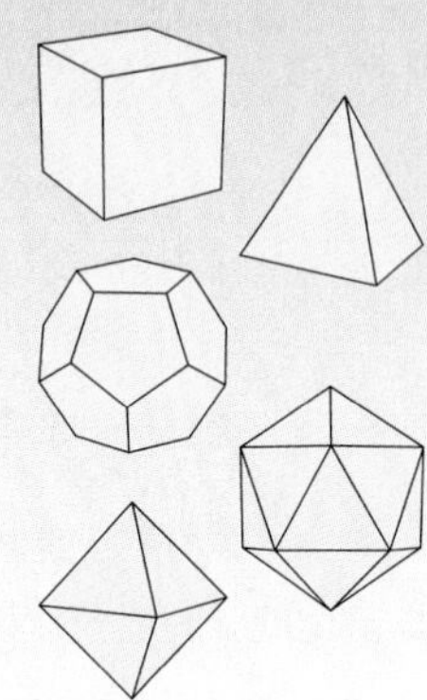

Solution

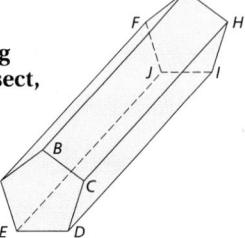

Some answers may vary.

Bases: *ABCDE* and *FGHIJ*

Pair of intersecting faces: *AFGB* and *BGHC*

Edge where these two faces intersect: $\overline{BG}$

Pair of skew edges: $\overline{CH}$ and $\overline{ED}$

Example 3

ARCHEOLOGY An archeologist says that a Greek artifact is in the shape of a right hexagonal prism. Draw the prism.

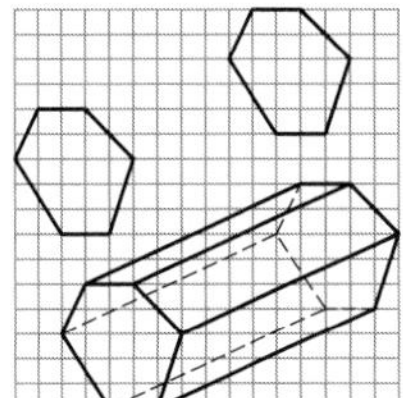

Solution

Step 1: Draw two congruent hexagons on graph paper.

Step 2: Use a straightedge to connect the corresponding vertices. Use dotted lines to show the unseen lateral edges.

Reading Math

A polyhedron is a **regular polyhedron** if all its faces are congruent regular polygons. The Greek scholar, Plato, studied these figures, also known as the five **Platonic solids.**

Extend the Lesson

MATH JOURNAL Have students select several of the space figures discussed. In their journals, have them write about where in the real world they see these figures and why, in their opinion, the particular figure was chosen for the purpose.

Chalkboard Examples

Supplementary Example 1 and 2

Identify the bases, a pair of intersecting faces and the edge at which they intersect, and a pair of skew edges. Bases: *ABC* and *DEF*; pair of intersecting faces: *ACDF* and *BDEF*; edge at which they intersect: ; pair of skew edges: $\overline{AC}$ and $\overline{BE}$

Supplementary Example 3

A pedestal for a statue is in the shape of a right octagonal prism. Draw the prism. Check students' drawings.

Supplementary Example 3

Line segment *AB* measures 1 inch. Draw a picture to show the locus of points $\frac{1}{2}$ inch from the segment. Check students' drawings.

Reteaching Worksheet 5-5

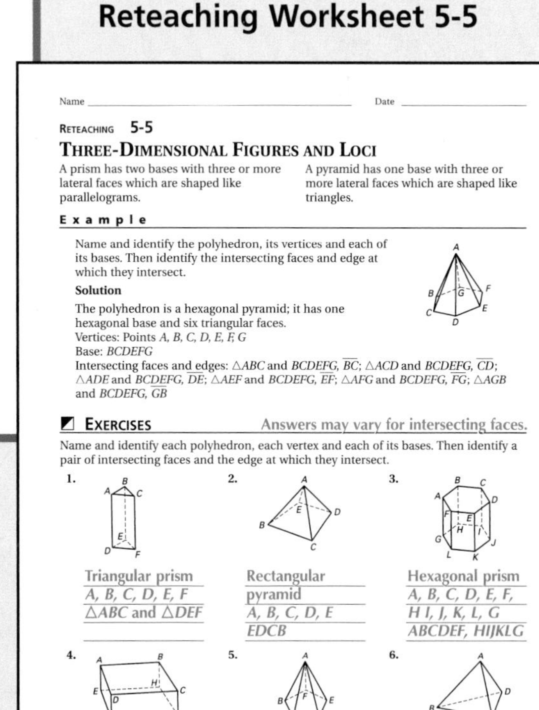

Name ______________________ Date ______________

RETEACHING 5-5

THREE-DIMENSIONAL FIGURES AND LOCI

A prism has two bases with three or more lateral faces which are shaped like parallelograms.

A pyramid has one base with three or more lateral faces which are shaped like triangles.

Example

Name and identify the polyhedron, its vertices and each of its bases. Then identify the intersecting faces and edge at which they intersect.

Solution

The polyhedron is a hexagonal pyramid; it has one hexagonal base and six triangular faces.
Vertices: Points *A, B, C, D, E, F, G*
Base: *BCDEFG*
Intersecting faces and edges: $\triangle ABC$ and *BCDEFG*, $\overline{BC}$; $\triangle ACD$ and *BCDEFG*, $\overline{CD}$; $\triangle ADE$ and *BCDEFG*, $\overline{DE}$; $\triangle AEF$ and *BCDEFG*, $\overline{EF}$; $\triangle AFG$ and *BCDEFG*, $\overline{FG}$; $\triangle AGB$ and *BCDEFG*, $\overline{GB}$

EXERCISES Answers may vary for intersecting faces.

Name and identify each polyhedron, each vertex and each of its bases. Then identify a pair of intersecting faces and the edge at which they intersect.

1. Triangular prism; *A, B, C, D, E, F*; $\triangle ABC$ and $\triangle DEF$
2. Rectangular pyramid; *A, B, C, D, E*; *EDCB*
3. Hexagonal prism; *A, B, C, D, E, F, H I, J, K, L, G*; *ABCDEF, HIJKLG*
4. Rectangular prism; *A, B, C, D, E, H, G, F*; *ABCD, EHGH*
5. Pentagonal pyramid; *A, B, F, E, D, C*; *BFEDC*
6. Triangular pyramid; *A, B, D, C*; $\triangle BDC$

78 RETEACHING *LESSON 5-5*

Lesson Wrap-up

Quick Assessment

Ask the following questions to determine if students understand the content presented in this lesson.

1. Describe the differences between the lateral faces of a pyramid and those of a prism. Lateral faces of a pyramid are triangles that meet at a vertex. Lateral faces of a prism are rectangles.
2. Describe the similarities and differences between a pyramid and a cone. Both have one base; a pyramid is a polyhedron, with all faces polygons, while a cone is not a polyhedron.

Assignment Guide

Basic: 1–18, 21–25
Advanced: 1–25
Additional Practice: See Extra Practice Index on page 674.

Additional Answers

1. Triangular prism; bases *ABC* and *EFD*; Remaining answers will vary. Sample answers include parallel edges $\overline{AB}$ and $\overline{EF}$; intersectin faces *BCDF* and *ABC*; intersection edges $\overline{EF}$ and $\overline{FD}$.
2. Rectangular pyramid; base *GKIJ*; Remaining answers will vary. Sample answers include parallel edges $\overline{GK}$ and $\overline{JI}$; intersection faces *GHK* and *GHJ*; intersecting edges $\overline{HI}$ and $\overline{JI}$.
3. Triangular pyramid; base *LNO*; no parallel edges; Remaining answers will vary. Sample answers include intersecting faces *LMO* and *LNO*; intersecting edges $\overline{OM}$ and $\overline{OL}$.

12. 13.

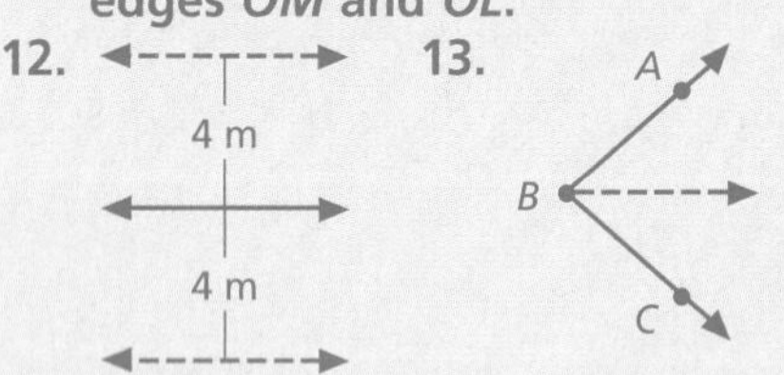

Sometimes you will be asked to describe or identify a set of points that meets particular requirements. The mathematical term for specifying points is **locus,** the set of all points that satisfy a given set of conditions. The word *locus* comes from Latin and means *place*; its plural is *loci.*

Example 4

Describe the locus of points 6 cm from a given point, *P*. All points lie within the same plane.

Solution

Draw point P on a sheet of paper. Locate and mark several points 6 cm from it. If you continue to add points to the drawing, what figure is formed? A circle with a radius of 6 cm.

Try These Exercises

Identify each figure. Then identify the base(s), a pair of parallel edges, intersecting faces, and intersecting edges. See additional answers.

1.

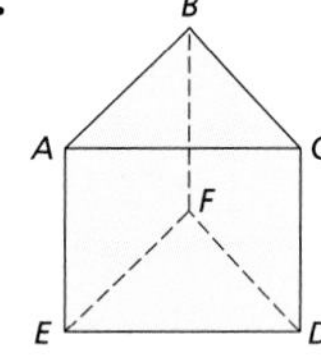

2.

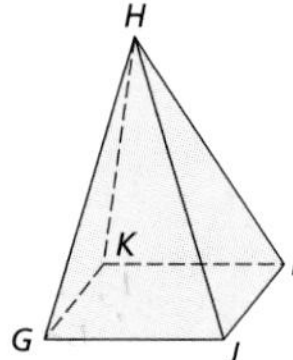

3.

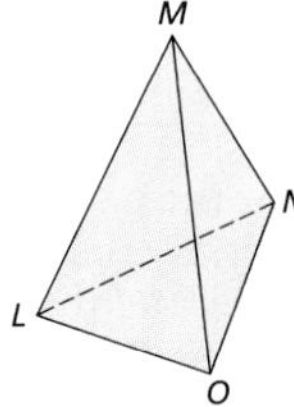

4. Draw a cone. Check students' drawings.
5. Describe the locus of points equidistant from two parallel lines. a coplanar line halfway between them.

Practice Exercises

Name the polyhedra shown below. Then state the number of faces, vertices, and edges each has.

6.

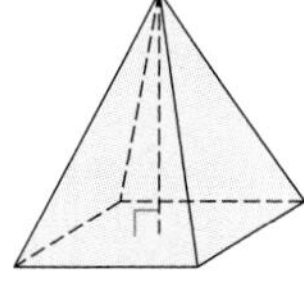

rectangular pyramid; 5; 5; 8

7.

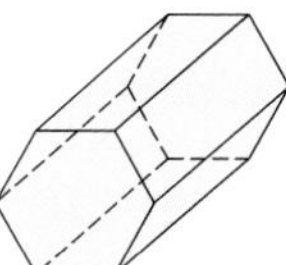

hexagonal prism; 8; 12; 18

8.

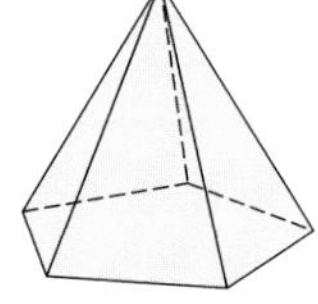

pentagonal pyramid; 6; 6; 10

Draw the figure. Check students' drawings.

9. triangular prism
10. cylinder

222 Chapter 5 Measurement

Teaching Strategies

COOPERATIVE LEARNING From sheets of paper, have pairs of students construct several cylinders. Then, using scissors, have them cut through each of the cylinders, creating as many different cross sections as possible. After a short while, have students compare their results. Cross sections should include circle, rectangle, and oval.

11. The sum of he faces and ertices is 2 more han the number of edges. Possible ule: $e = f + v - 2$.

11. **WRITING MATH** Examine your answers for Exercises 6–8. For each polyhedron, what can you say about how the sum of its faces and vertices compares with the number of edges? Write a rule to describe the relationship among the faces, vertices and edges of a polyhedron.

12. Describe and draw the locus of point 4 m from a given line. The locus is 2 lines, each 4 m from the given line. See additional answers.

13. Draw a picture to show the locus of points equidistant from the two sides of $\angle ABC$. See additional answers.

14. Describe the locus of points *in space* 3 ft from point O. (*Hint:* A locus of points in space may form a three-dimensional object.) Sphere with center O and radius = 3 ft

15. Describe the locus of points in space that are the same distance from a given line. a cylindrical surface

16. **ARCHITECTURE** A 6-story building is 72 ft high. Each story is the same height. Describe the locus of points 24 ft from the floor of the fourth floor of the building. floor of 6th and 2nd floors

Blarney Castle, Ireland

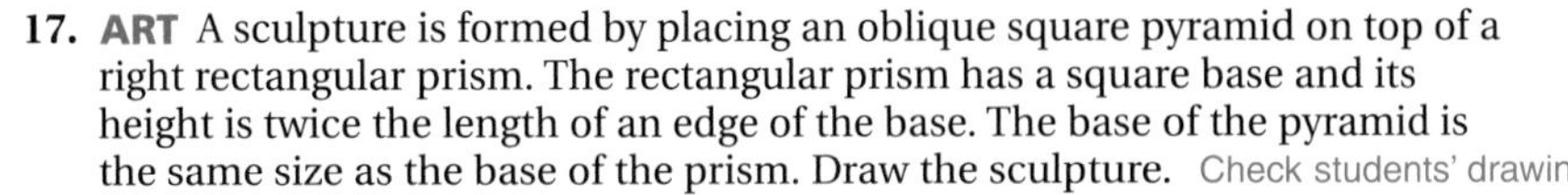

17. **ART** A sculpture is formed by placing an oblique square pyramid on top of a right rectangular prism. The rectangular prism has a square base and its height is twice the length of an edge of the base. The base of the pyramid is the same size as the base of the prism. Draw the sculpture. Check students' drawings.

18. **CHAPTER INVESTIGATION** Build a three-dimensional model of the structure you have chosen. Break the structure into smaller three-dimensional figures or sections. Then assemble them to make the final product. Check students' work.

Extended Practice Exercises

19. A **cross section** is the two-dimensional figure formed when you cut a three-dimensional shape with a plane. If you cut a cross section of a square pyramid parallel to the base, what polygon will be formed? a square

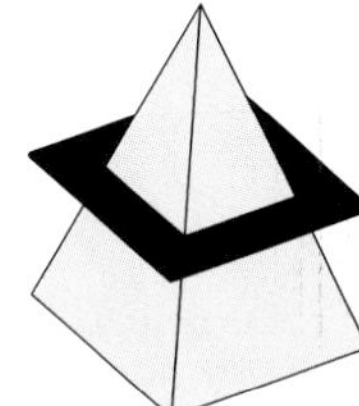

20. What would a triangular prism look like if seen from the side? What would it look like from above? side-rectangle; above-triangle

Mixed Review Exercises

Find the value of x in each figure. (Lesson 4-3)

21.

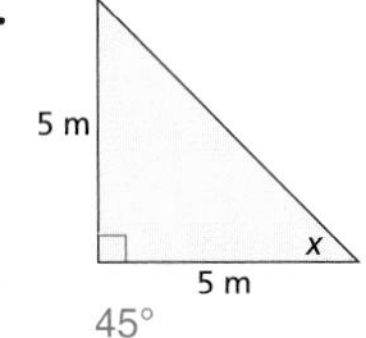

45°

22.

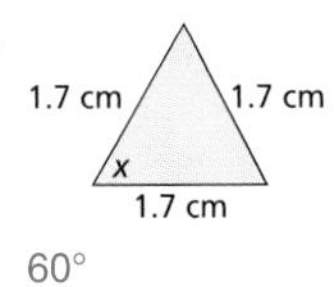

60°

23.

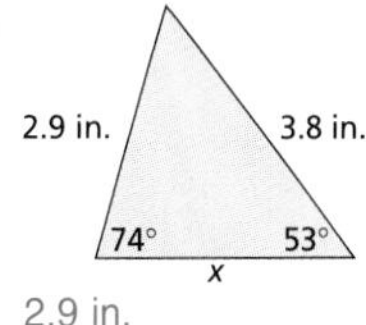

2.9 in.

Solve each equation. (Lesson 2-5)

24. $3(2x + 1) - 8 = 5x + 2(x + 1)$ -7

25. $-3x + 4(x - 1) = 6 - 4(x + 2)$ $\frac{2}{5}$

26. $-2(x - 3) + 5 = -5x + 9$ $-\frac{2}{3}$

27. $3(x - 3) + 2x = 7x - 3(x + 1)$ 6

28. $5 - 4(x - 8) = -5(4x - 1)$ -2

29. $-4(x - 2) + 3 = x + 2(x - 5)$ 3

Extra Practice Worksheet 5-5

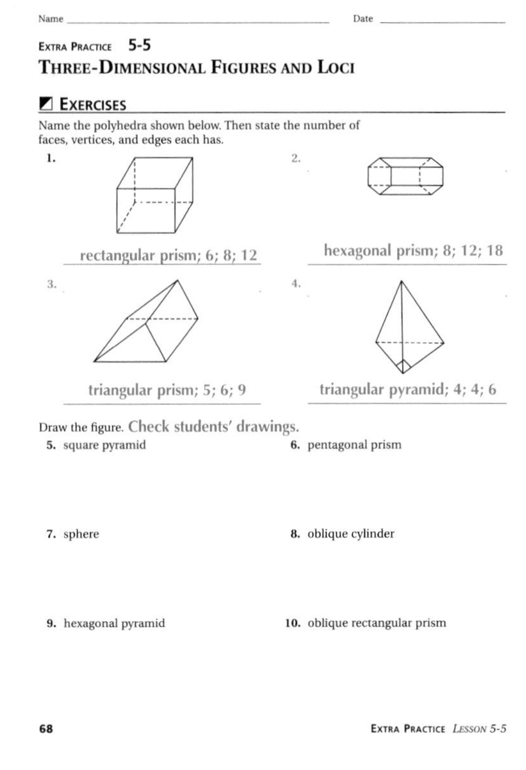

Name ______ Date ______

EXTRA PRACTICE 5-5

THREE-DIMENSIONAL FIGURES AND LOCI

EXERCISES

Name the polyhedra shown below. Then state the number of faces, vertices, and edges each has.

1. rectangular prism; 6; 8; 12
2. hexagonal prism; 8; 12; 18
3. triangular prism; 5; 6; 9
4. triangular pyramid; 4; 4; 6

Draw the figure. Check students' drawings.

5. square pyramid
6. pentagonal prism
7. sphere
8. oblique cylinder
9. hexagonal pyramid
10. oblique rectangular prism

68 EXTRA PRACTICE *LESSON 5-5*

Enrichment Worksheet 5-5

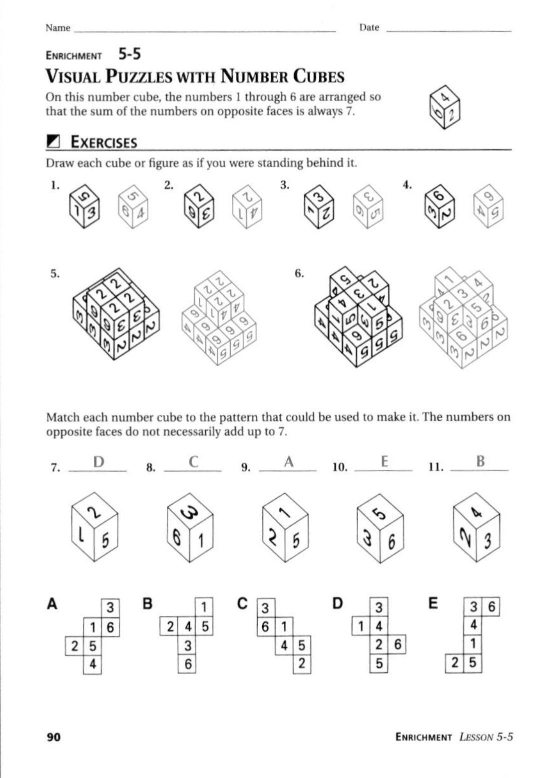

Name ______ Date ______

ENRICHMENT 5-5

VISUAL PUZZLES WITH NUMBER CUBES

On this number cube, the numbers 1 through 6 are arranged so that the sum of the numbers on opposite faces is always 7.

EXERCISES

Draw each cube or figure as if you were standing behind it.

1. 2. 3. 4.

5. 6.

Match each number cube to the pattern that could be used to make it. The numbers on opposite faces do not necessarily add up to 7.

7. D 8. C 9. A 10. E 11. B

90 ENRICHMENT *LESSON 5-5*

Teaching Strategies

CHALLENGE When you look at a standard die, you can see 1, 2 or 3 faces. It is possible to see any total number of spots from 1 through 15, with one exception. Have students examine a die to determine which total cannot be seen and discuss why. Thirteen cannot be seen since only 6, 5, 2 and 5, 4, 3 add to 13; 5 and 2 or 4 and 3 cannot both be seen, since opposite faces of a die total 7.

Lesson Planning

NCTM Standards/Strands

- Number & Operation
- Patterns, Functions, & Algebra
- Geometry & Spatial Sense
- Measurement
- Problem Solving
- Reasoning and Proof
- Communication
- Connections
- Representation

Vocabulary

surface area

Materials Needed

calculators	construction paper
compasses	scissors
straightedges	tape

Lesson Resources

Warm-Up Transparency 15
Reteaching 5-6
Extra Practice 5-6
Enrichment 5-6
Transparency RF-18

Assignment Guide

Basic: 1–18, 23–24
Enriched: 1–24

Getting Started

5-Minute Warm-up

Write the formula for the area of each polygon.

1. rectangle length × width, lw
2. square side × side, s^2
3. triangle base × height × $\frac{1}{2}$, $\frac{1}{2}bh$

Introduction to Lesson 5-6

Students should not use a marked ruler for this construction. You may wish to review using a compass both to mark a specified length and to construct the legs of the equilateral triangle on each side of the square.

5-6 Surface Area of Three-dimensional Figures

Goals ■ Find surface areas of space figures.

Applications Packaging, Manufacturing, Sports, Astronomy

Work with a partner.

Construct a square pyramid out of construction paper, using only the following: straightedge, compass, scissors, and tape. Write a description of how you did it. Possible description: Use the compass and straightedge to construct a square; with compass opening equal to length of side of square, draw arcs with the compass point on each corner; each point of intersection of the arcs is the third vertex of each equilateral triangle.

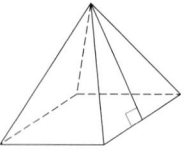

Pyramids in Egypt

Build Understanding

When you are asked to find the **surface area** of a three-dimensional figure, think about whether the figure is a prism, pyramid, cylinder, cone, or sphere, or whether it is a combination of figures. To help you identify the shape of each surface, think about what the figure would look like if it were cut apart. Notice whether any surfaces are congruent.

Example 1

PACKAGING A box of Teen Chow cereal is 11.5 in. high, 7.5 in. wide, and 2.5 in. deep. What is the surface area of the box?

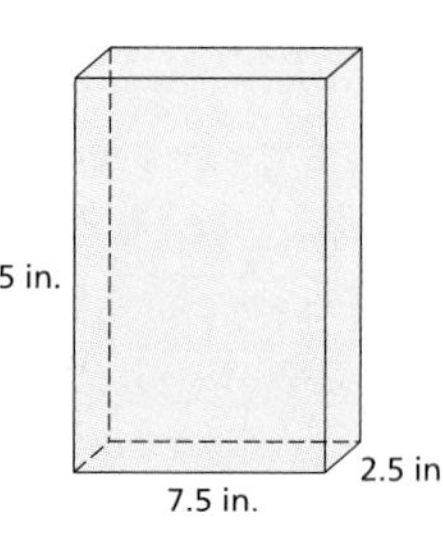

Solution

The cereal box is a rectangular prism, so it has 3 pairs of congruent rectangular faces. To find its surface area, find the area of each face. Use the formula $A = lw$.

SA means "surface area."

$SA = 2(\text{area of front}) + 2(\text{area of side}) + 2(\text{area of top})$

$= 2(11.5 \times 7.5) + 2(11.5 \times 2.5) + 2(7.5 \times 2.5)$

$= 172.5 + 57.5 + 37.5$

$= 267.5$

The surface area is 267.5 in.^{2}

Example 2

MANUFACTURING At Farrow's Ceramic Factory, the salt and pepper shakers are all ceramic replicas of the Great Pyramid at Giza. Each shaker has a square base 10 cm in length, and triangular faces each with a height of 12 cm. Farrow's plans to paint the shakers. What is the surface area of each?

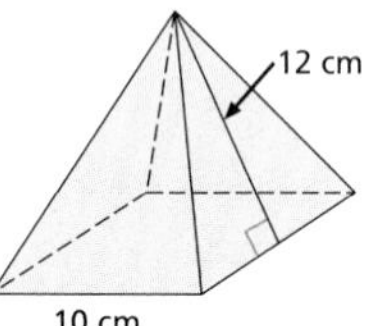

Teaching Strategies

TEAM LEARNING Topology is the study of figures that do not change when twisted and stretched out of shape. A Möbius strip, named after German mathematician and astronomer August Ferdinand Möbius, is shown below. To make a Möbius strip, cut out a long thin, strip of paper, give it a half-twist and then tape the ends together. Work with a partner to experiment with Möbius strips. What happens if you draw a line down the length of the middle? The line covers both sides of the paper. What happens if you try to cut your strip in half lengthwise? It forms one strip, double in length to the original.

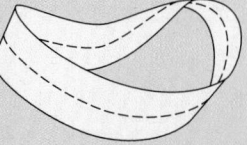

Solution

The model is a square pyramid. Each of the four triangular faces has the same area. To find its surface area, find the area of each face and of the base.

$SA = 4(\text{area of triangular face}) + \text{area of square base}$

area of triangular face	**area of square base**
$A = \frac{1}{2}bh$	$A = s^2$
$A = \left(\frac{1}{2}\right)(10)(12)$	$A = 10^2$
$A = 60$	$A = 100$

$SA = 4(60) + 100 = 340$

The surface area is 340 cm^2.

Example 3

A can of bread crumbs is 14 cm high and 8 cm across. What is the surface area of the can?

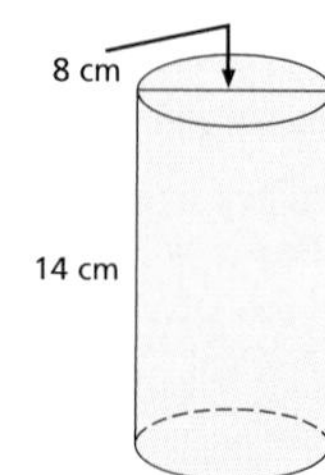

Solution

The can is a cylinder. To find its surface area, add the area of the curved surface to the area of the two bases.

area of the curved surface	**area of each circular base**
$A = 2\pi rh$	$A = \pi r^2$
$A = (2)(\pi)(4)(14)$	$A = (\pi)(16)$
$A \approx 351.68$	$A \approx 50.24$

The can has two congruent circular bases. $(2)(50.24) = 100.48$

$SA \approx 351.68 + 100.48 \approx 452.16$

The surface area of the can is approximately equal to 452.16 cm^2.

Check Understanding

If the curved surface of the cylinder is laid flat, it forms a rectangle. What dimension of the cylinder equals the length of this rectangle? To what dimension is the height equal?

length = circumference of cylinder; height = height of the cylinder.

Example 4

A tent in the shape of a teepee is 4 m across with a slant height of 2.6 m. What is the surface area of the canvas, including the floor?

Solution

The tent is a cone. To find its surface area, add the area of the curved surface to the area of the base.

$SA = \pi rs + \pi r^2$ s = slant height

$SA = \pi(2)(2.6) + \pi(2)^2$

$SA \approx 16.328 + 12.56$

$SA \approx 29$

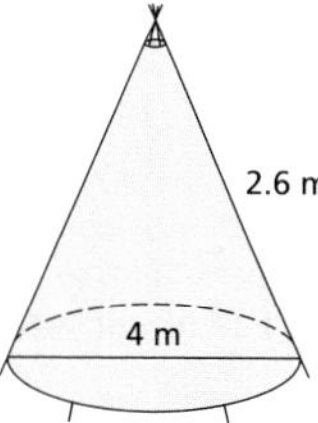

The surface area of the tent is approximately equal to 29 m^2.

Chalkboard Examples

Supplementary Example 1

A small box is $3\frac{1}{2}$ in. high, 4 in. wide, and $2\frac{3}{4}$ in. deep. What is the surface area of the box? 69.25 in.^{2}

Supplementary Example 2

A hair products company sells its shampoo and conditioner in pyramid-shaped containers. Each container has a square base 7 cm in length, and triangular faces, each with a height of 14 cm. Find the surface area of the container. 245 cm^2

Supplementary Example 3

A can of lemonade mix is 15 cm across and 18 cm high. What is the surface area of the can? about 1201.05 cm^2

Supplementary Example 4

Another tent, also in the shape of a teepee, is 6 m across with a slant height of 4.8 m. What is the surface area of the canvas, including the floor? about 73.465 m^2

Supplementary Example 5

What is the surface area of a globe with a diameter of about 12 inches? 452.16 in.^{2}

Reteaching Worksheet 5-6

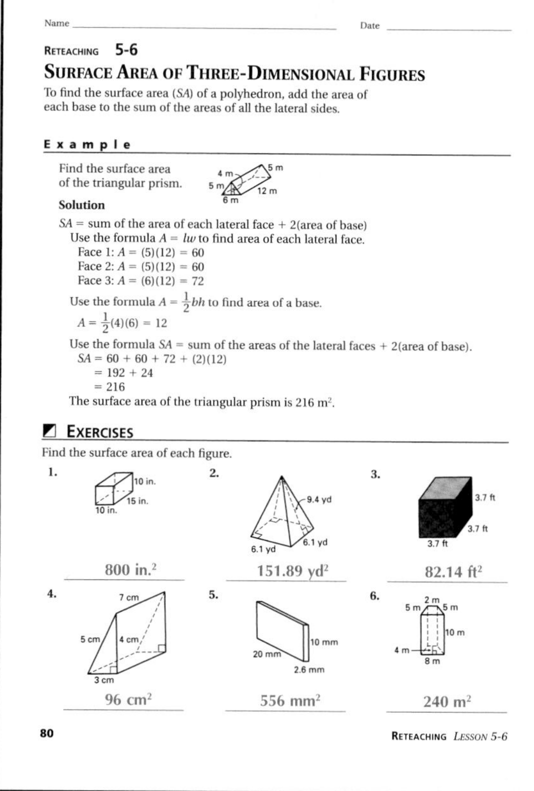

Name ______________________ Date ______________

RETEACHING 5-6

SURFACE AREA OF THREE-DIMENSIONAL FIGURES

To find the surface area (SA) of a polyhedron, add the area of each base to the sum of the areas of all the lateral sides.

Example

Find the surface area of the triangular prism.

(4 m, 5 m, 5 m, 6 m, 12 m)

Solution

SA = sum of the area of each lateral face + 2(area of base)

Use the formula $A = lw$ to find area of each lateral face.

Face 1: $A = (5)(12) = 60$

Face 2: $A = (5)(12) = 60$

Face 3: $A = (6)(12) = 72$

Use the formula $A = \frac{1}{2}bh$ to find area of a base.

$A = \frac{1}{2}(4)(6) = 12$

Use the formula SA = sum of the areas of the lateral faces + 2(area of base).

$SA = 60 + 60 + 72 + (2)(12)$

$= 192 + 24$

$= 216$

The surface area of the triangular prism is 216 m^2.

EXERCISES

Find the surface area of each figure.

1. (10 in., 15 in., 10 in.) 800 in.^{2}
2. (9.4 yd, 6.1 yd, 6.1 yd) 151.89 yd^2
3. (3.7 ft, 3.7 ft, 3.7 ft) 82.14 ft^2
4. (7 cm, 5 cm, 4 cm, 3 cm) 96 cm^2
5. (10 mm, 20 mm, 2.6 mm) 556 mm^2
6. (2 m, 5 m, 5 m, 10 m, 4 m, 8 m) 240 m^2

80 RETEACHING *LESSON 5-6*

Extend the Lesson

ONGOING ASSESSMENT Students may assume that the faces on an object are congruent. Give students the problem below. Look for students who assume the three lateral faces are congruent. These students will get the incorrect answer 93 cm^2.

Find the surface area of the triangular prism. 120 cm^2

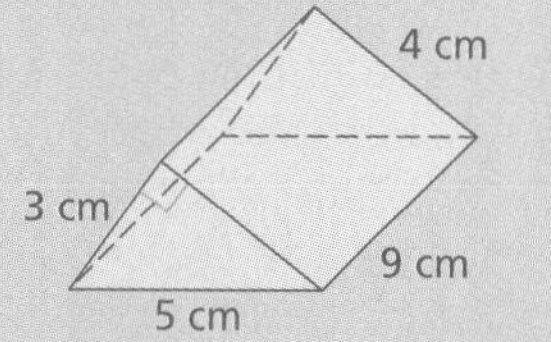

Lesson Wrap-up

Quick Assessment

Ask the following questions to determine if students understand the content presented in this lesson.

1. If you know the vertical height and radius of a cone, how can you find the slant height? Use the Pythagorean theorem to find the third side of a right triangle.
2. If you know the surface area of a sphere, how can you find its radius? Substitute the value of the surface area in the formula $SA = 4\pi r^2$, and solve for r.

Assignment Guide

Basic: 1–18, 23–24
Advanced: 1–24
Additional Practice: See Extra Practice Index on page 674.

Example 5

SPORTS What is the surface area of a soccer ball with a diameter of about 9 in.?

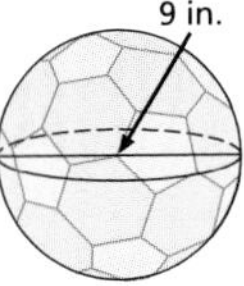

Solution

The soccer ball is a sphere. To find its surface area, use the formula $SA = 4\pi r^2$.

$SA \approx (4)(3.14)(4.5)^2$

$SA \approx 254$

The soccer ball has a surface area of about 254 in.^{2}

Technology Note

You can store the value 3.14 in your calculator's memory. To compute a value such as 15π, enter the following key sequence.

Does your calculator have a special key for π? How do you use it?

Try These Exercises

Find the surface area of each figure. Assume that all pyramids are regular pyramids. Use $\pi = 3.14$.

1. 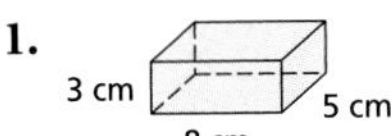

158 cm^2

2. 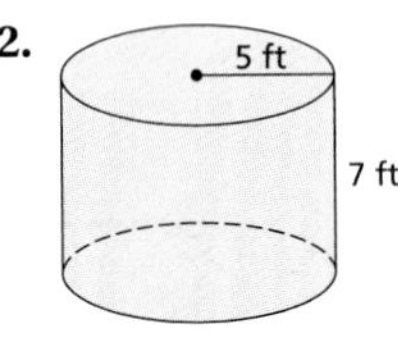

about 376.8 ft^2

3. 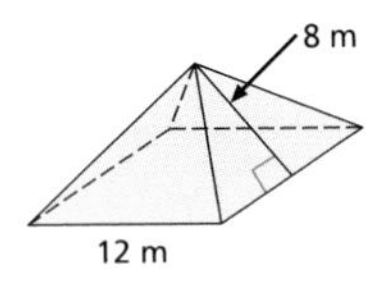

336 m^2

4. **DATA FILE** Use the Data Index on pages 632–633 to locate information about the sizes and weights of various balls used in sports. Calculate the surface area of a volleyball. About 1506 in.^{3}

Practice Exercises

Find the surface area of each figure. Assume that all pyramids are regular pyramids. Use $\pi = 3.14$. Round answers to the nearest whole number.

5. 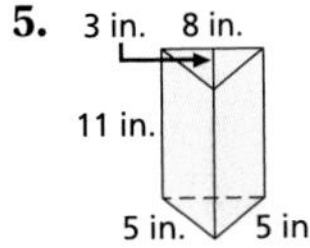

222 in.^{2}

6. 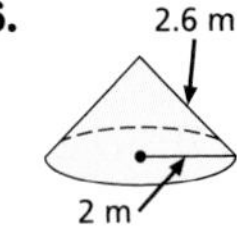

about 28.9 m^2

7. 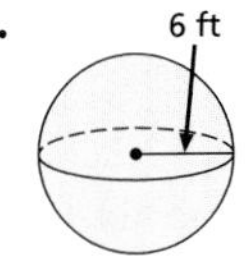

452 ft^2

8. 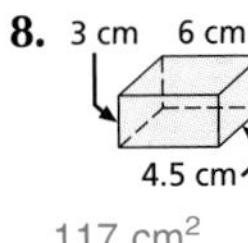

117 cm^2

9. What is the surface area of a square pyramid whose base length is 8 m and whose faces have heights of 6.4 m? 166.4 m^2

10. **ARCHEOLOGY** The ancient Mayan observatory at Chichen Itza was a cylindrical tower about 41 ft high with a diameter of 37 ft. What was the surface area of the exposed part of the observatory? about 5,838 ft^2

Chichen Itza ruins, Mexico

Teaching Strategies

COOPERATIVE LEARNING Have small groups of students research and report on the difference in surface area between the following commonly used sports balls. The diameters are given for your convenience. Have the students determine the diameter through measurement and research.

baseball (approximate diameter 7.6 cm)
basketball (24 cm)
golf ball (4.3 cm)
soccer ball (22 cm)
table tennis ball (6.5 cm)
volley ball (21.9 cm)

Find the surface area of each. Use $\pi = 3.14$. Round answers to the nearest whole number.

11. 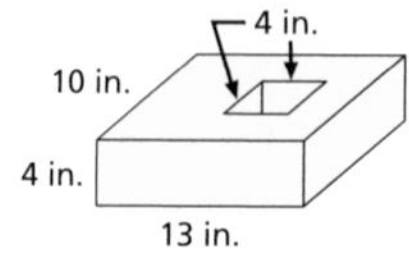

476 in.^{2}

12. 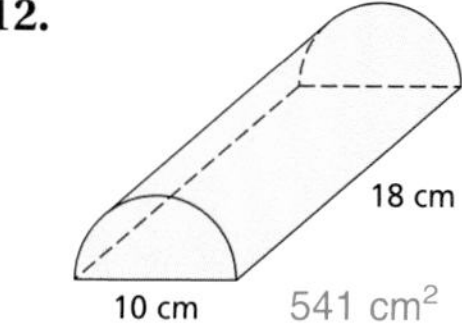

541 cm^2

13. 4 m, 20 m, 14.5 m

735 m^2

14. Use mental math. Which has the greater surface area, the can or the box?

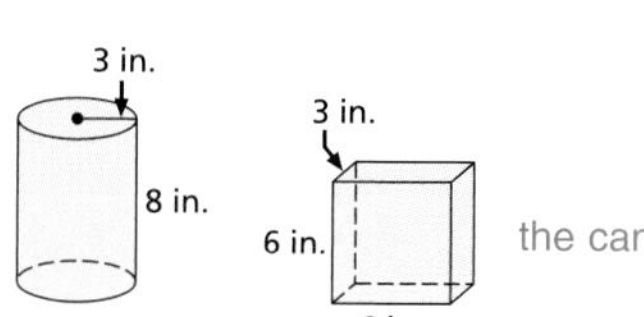
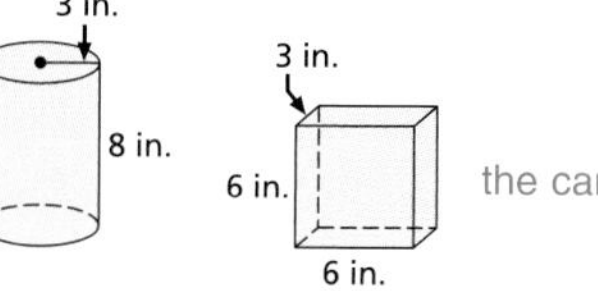

the can

15. ART The base of a sculpture is a regular pentagonal prism with sides 10 cm high and 6 cm wide. What additional information do you need to find the surface area of the base? the area of the base

16. WRITING MATH The surface area of a rectangular prism is 178 in.^{2} What is the height of the figure if its length is 3 in. and its width is 4 in.? Explain how you got your answer. 11 in.; possible explanation: make a sketch, then guess and check.

17. SPORTS The "shots" that shot-putters toss are heavy spheres that range in diameter from 95 mm to 130 mm. What is the difference in surface area between the largest and smallest shot? About 24,727.5 mm^2

18. ASTRONOMY Jupiter, the largest planet, has an equator with a diameter of about 88,000 miles. To the nearest million miles, what is the surface area of Jupiter? Assume that it is a sphere. About 24,320 million mi^2

Extended Practice Exercises

19. What happens to the surface area of a cube if you (a) double the length of a side, or (b) divide the length of a side by 3? a) SA is 4 times as great; b) SA is $\frac{1}{9}$ times as great.

20. One way to express the formula for finding the surface area of a cylinder is $SA = 2\pi rh + 2\pi r^2$. How else can this be expressed? one way: $2\pi r(h + r)$

21. To paint the sides of a cube, 1 quart of paint is used. Suppose two such cubes are glued together to form a rectangular solid. How much paint will it take to paint the new rectangular solid? $1\frac{2}{3}$ qt

22. CHAPTER INVESTIGATION Estimate the surface area of the ancient structure you have chosen. Explain how you got your answer. Check students' work.

Mixed Review Exercises

Find each length. (Lesson 3-1)

23. In the figure below, $\overline{AC} = 106$. Find $\overline{AB}$. 45

A —— $2x + 7$ —— B —— $3x + 4$ —— C

24. In the figure below, $\overline{RT} = 170$. Find $\overline{ST}$. 50

R —— $4(x + 3)$ —— S —— $2(x - 2)$ —— T

Extend the Lesson

MATH JOURNAL For each of the following figures, have students write the minimum amount of information they would need to find the surface area: square pyramid, rectangular prism, cube, cylinder and cone.

Extra Practice Worksheet 5-6

Name ______ Date ______

EXTRA PRACTICE 5-6

SURFACE AREA OF THREE-DIMENSIONAL FIGURES

EXERCISES

Find the surface area of each figure. Assume that all pyramids are regular pyramids. Use $\pi = 3.14$. Round answers to the nearest whole number.

1. 8 in., 4 in., 6 in. — 152 in.^{2}
2. 19 m, 6 m — 471 m^2
3. 14 ft, 18 ft — 1099 ft^2
4. 5 m, 3.5 m — 37 m^2
5. 5 in., 5 in. — 150 in.^{2}
6. 8 cm, 16 cm, 6 cm — 544 cm^2
7. What is the surface area of a ball with a diameter of about 15 cm? 706.5 cm^2
8. What is the surface area of a square pyramid whose base length is 8 ft and whose faces have heights of 12 ft? 256 ft^2
9. What is the surface area of a box with a height of 5 in., a length of 8 in. and a width of 4 in.? 184 in.^{2}
10. What is the surface area of a cube if each edge has length 3.5 ft? 73.5 ft^2
11. Can you find the surface area of a cylinder if all you know is its radius? Explain. No, you must also know the height of the cylinder.

70 EXTRA PRACTICE *LESSON 5-6*

Enrichment Worksheet 5-6

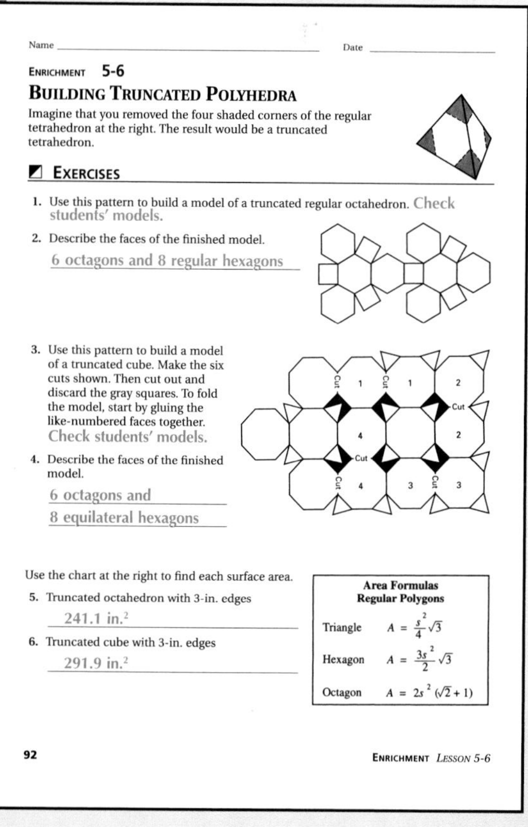

Name ______ Date ______

ENRICHMENT 5-6

BUILDING TRUNCATED POLYHEDRA

Imagine that you removed the four shaded corners of the regular tetrahedron at the right. The result would be a truncated tetrahedron.

EXERCISES

1. Use this pattern to build a model of a truncated regular octahedron. Check students' models.
2. Describe the faces of the finished model. 6 octagons and 8 regular hexagons
3. Use this pattern to build a model of a truncated cube. Make the six cuts shown. Then cut out and discard the gray squares. To fold the model, start by gluing the like-numbered faces together. Check students' models.
4. Describe the faces of the finished model. 6 octagons and 8 equilateral hexagons

Use the chart at the right to find each surface area.

5. Truncated octahedron with 3-in. edges 241.1 in.^{2}
6. Truncated cube with 3-in. edges 291.9 in.^{2}

Area Formulas Regular Polygons	
Triangle	$A = \frac{s^2}{4}\sqrt{3}$
Hexagon	$A = \frac{3s^2}{2}\sqrt{3}$
Octagon	$A = 2s^2(\sqrt{2} + 1)$

92 ENRICHMENT *LESSON 5-6*

Skills Practice

Vocabulary Review

Lesson 5-5
polyhedron
prism
lateral faces
cylinder
cone
locus
vertex
pyramid
lateral edges
axis
sphere

Lesson 5-6
surface area

Assignment Guide

All students: 1–20
Additional Practice: 1–28

Additional Answers

7. 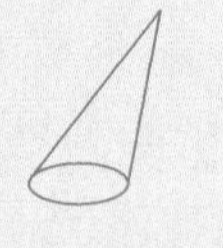

8. 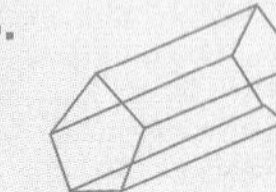

9. 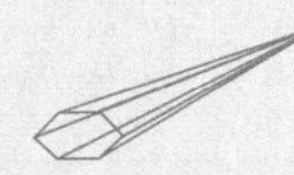

10. 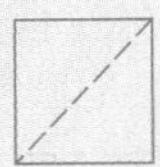

11.

Review and Practice Your Skills

Practice ■ Lesson 5-5

Name the polyhedra shown below. Then state the number of faces, vertices, and edges for each.

1. 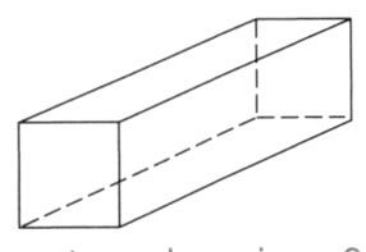
rectangular prism; 6 faces, 8 vertices, 12 edges

2. 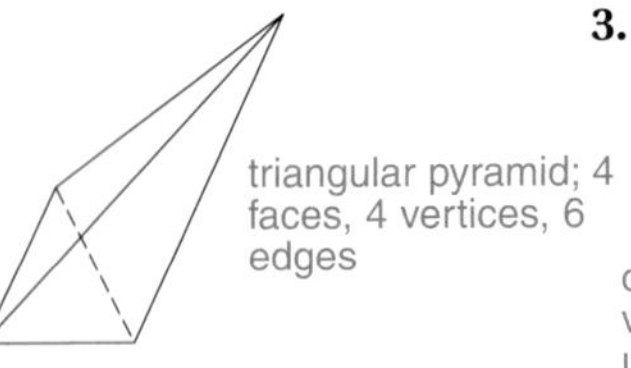
triangular pyramid; 4 faces, 4 vertices, 6 edges

3. oblique cone; faces, vertices, and edges are undefined for cones.

4–6. For the figures in Exercises 1–3, identify the following: Answers will vary.
  a. base(s)
  b. a pair of parallel edges
  c. a pair of intersecting faces
  d. a pair of intersecting edges

Draw each figure. For 7–11, see additional answers.

7. oblique cone
8. pentagonal prism
9. oblique hexagonal pyramid
10. Describe and draw the locus of points equidistant from two adjacent sides of a square.
11. Describe and draw the locus of points equidistant from a line and a point that is not on the line.

Practice ■ Lesson 5-6

Find the surface area of each figure. Assume all pyramids are regular pyramids. Round answers to the nearest tenth.

12. 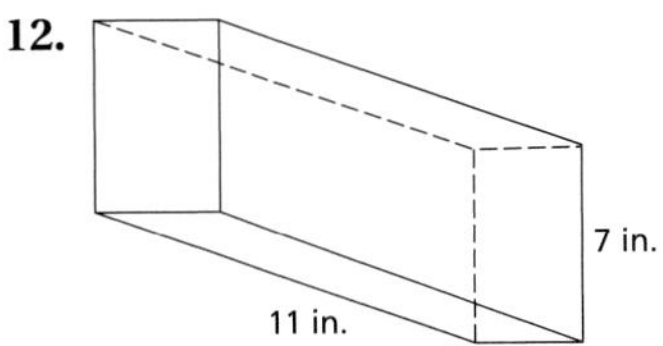

316 in.^{2}

13. 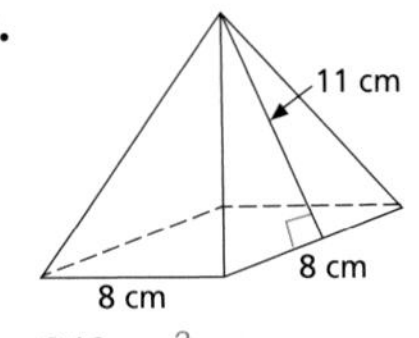

240 cm^2

14. 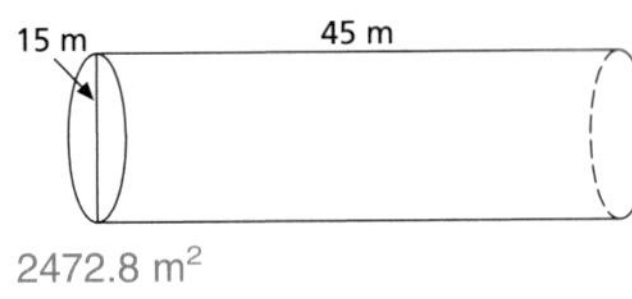

2472.8 m^2

Find the surface area of each figure. Round answers to the nearest tenth.

15. 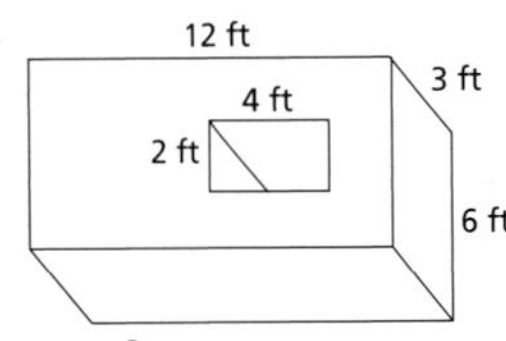

272 ft^2

16. 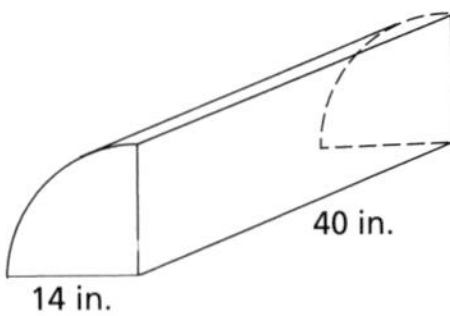

1746.9 in.^{2}

17. 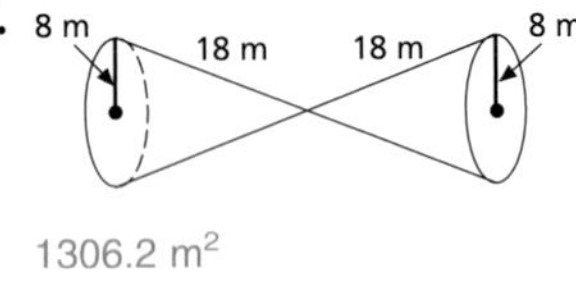

1306.2 m^2

18. Find the surface area of a square pyramid with base length = 10 m and faces heights of 8.2 m. 264 m^2

Extend the Lesson

MATH JOURNAL Have students draw and label a pentagonal prism. Then by referring to the drawing, have them write a paragraph explaining the difference between parallel edges, intersecting edges, and skew edges.

228 Chapter 5 **Measurement**

PRACTICE ■ LESSON 5-1–LESSON 5-6

19. What happens to the surface area of a sphere if you triple the radius? (Lesson 5-6) 9 times as large
20. What happens to the surface area of a cone if you double the radius? (Lesson 5-6) surface area of base is multiplied by 4; surface area of curved surface is doubled.

Find each unit rate. (Lesson 5-1)

21. 260 mi in 4 h $65\ \frac{mi}{h}$
22. $42 for 140 stamps $0.30 per stamp
23. 51 gal in 1.5 min $34\ \frac{gal}{min}$

Find the probability that a point selected at random in each figure is in the shaded region. (Lesson 5-3)

24. $\frac{5}{12} = 0.417$

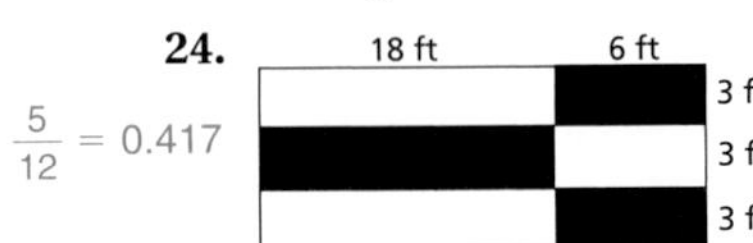

25.

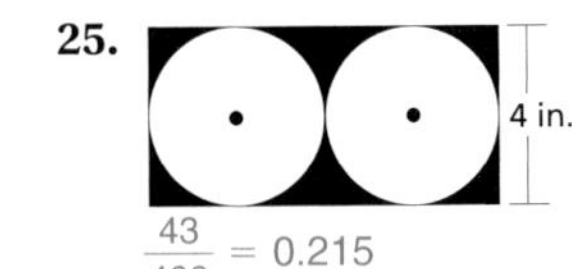

$\frac{43}{400} = 0.215$

26.

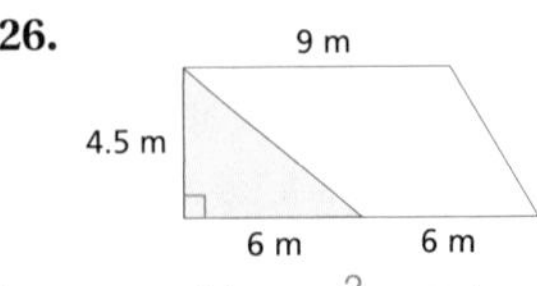

$\frac{2}{5} = 0.4$

Find the surface area of each figure. Assume all pyramids are regular pyramids. Round answers to the nearest tenth. (Lesson 5-6)

27.

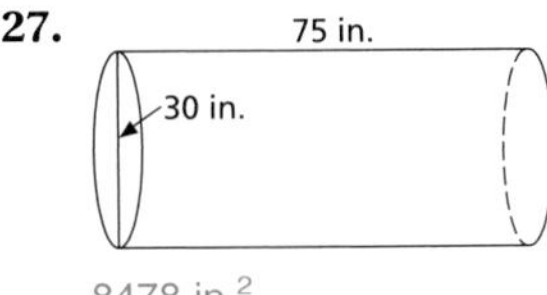

8478 in.^{2}

28.

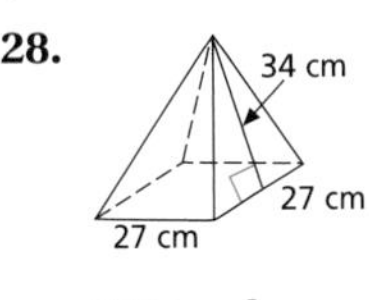

2565 cm^2

MathWorks
Workplace Knowhow

Career – Archeologist

Archeologists trace the histories of ancient civilizations by studying ancient records and artifacts. Many hours are spent in the field searching for buried cities and artifacts.

Archeologists make detailed maps of each site they discover. They analyze the layout of buildings and rooms. Each item found is labeled and catalogued. Archeologists note the exact location artifacts are found in order to determine their purpose.

You are excavating a building buried under several feet of ancient volcanic ash and silt. By reading inscriptions on stones, you expect to find an altar, 3 ft in diameter, somewhere in the interior of the room.

1. If the building is circular with a diameter of 25 feet, what is the probability of finding the altar if a dig site is chosen randomly? $\frac{9}{100}$ or 0.09
2. The site can be divided into two rectangular sections: the first measuring 15 ft by 13 ft, and the second, 20 ft by 20 ft. Find the total area of the site. 595 ft^2

Chalkboard Examples

Example from Lesson 5-5
Draw a triangular prism. Then write the number of faces, vertices and edges. faces = 5, vertices = 6, edges = 9

Example from Lesson 5-6
What is the difference in surface area between two cones, both with a base radius of 7 cm, with one having a slant height of 6 cm and the other 8 cm? approximately 44 cm^2

Career Opportunity

Describe the kind of work archeologist do and the types of places they work. Discuss the importance of geometry, measurement, logical reasoning and observation in working as an archeologist. Students should answer Questions 1–3 to better understand how archeologists use mathematics in performing their job.

Students who are interested in learning more about this profession can go to learningmathmatters.com. Guidance Counselors should have information about school and training requirements.

Extend the Lesson

REAL WORLD CONNECTION Have students bring two containers from home. Then using the principles they have learned in this chapter, have them calculate the outer surface area of each container.

Lesson Planning

NCTM Standards/Strands

- Number & Operation
- Patterns, Functions, & Algebra
- Geometry & Spatial Sense
- Measurement
- Problem Solving
- Reasoning and Proof
- Communication
- Connections
- Representation

Vocabulary

volume

Materials Needed

calculators

Lesson Resources

Warm-Up Transparency 15
Reteaching 5-7
Extra Practice 5-7
Enrichment 5-7
Transparency RF-18

Assignment Guide

Basic: 1-22, 26–29
Enriched: 1–29

Getting Started

5-Minute Warm-up

Write the area formula needed for the base or bases.

1. cone πr^2
2. triangular prism $\frac{1}{2}bh$
3. rectangular prism lw

Introduction to Lesson 5-7

Have students share the results of their investigation. Discuss why a company may have chosen a particular packaging shape. Encourage students to share their ideas about why companies are using too much packaging. Some reasons may include to protect the public from product tampering, for consumer convenience and to meet government regulations.

5-7 Volume of Three-dimensional Figures

Goals ■ Find the volume of space figures.

Applications Manufacturing, Astronomy, Archeology

Work in groups of three or four students.

Many environmental groups criticize manufacturers for over-packaging their products. On the other hand, over-packaging is one way to make customers think they are getting more for their money. Check students' work.

1. Choose a product that you think uses too much packaging.
2. Develop a new way to package the product that uses less packaging material. Remember, the packaging must keep the product from breaking, fit neatly in shipping cartons, and look appealing to the consumer.
3. Make a packaging sample for display.

Build Understanding

Recall that volume is a measure of the number of cubic units needed to fill a region of space. To find the volume of a three-dimensional figure, first you must determine whether the figure is a prism, pyramid, cylinder, cone, sphere, or combination of any of these shapes. Then you apply the appropriate formula or formulas for volume.

Example 1

Find the volume of the figure at the right.

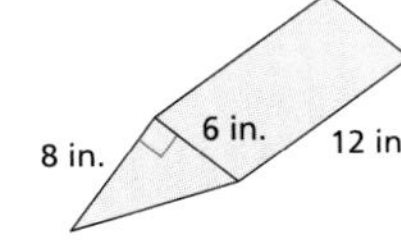

Solution

The figure is a prism. To find the volume (V) of any prism, multiply the area of the base (B) by the height (h) of the prism. First find the area of the base, which is a right triangle.

$$B = \frac{1}{2}bh$$

$$B = \left(\frac{1}{2}\right)(8)(6)$$

$$B = 24$$

The area of the base is 24 in.^{2} Then use the volume formula.

$$V = Bh$$

$$V = (24)(12)$$

$$V = 288$$

The volume is 288 in.^{3}

Learning Styles

VISUAL LEARNERS Reinforce the volume relationships by having students use cardboard to construct a prism and a pyramid with the same base and height. Have them also construct a cylinder and a cone with the same base and height. By using sand, salt or sugar, students can see that they must fill the pyramid or cone three times to equal the volume of the prism or cylinder respectively.

A three-dimensional figure may be a combination of shapes. Mentally break the figure into smaller pieces. Then find the volume of each piece. Finally, use the information to solve the problem.

Mental Math Tip

If either the area of the base or the height of a pyramid is evenly divisible by 3, you can use mental math to do that division, then multiply the remaining factors to find the volume. For example, what is V, if $B = 18 \text{ m}^2$ and $h = 11$ m?

66 m³

Example 2

Find the volume of the shaded part of the figure shown.

Solution

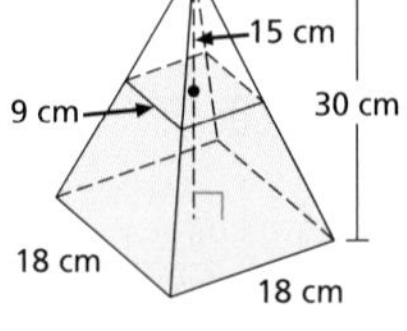

To find the volume of the shaded part, find the difference between the volume of the small pyramid and the volume of the large pyramid.

To find the volume of any pyramid, multiply $\frac{1}{3}$ of the area of its base (B) by its height (h).

The base of each of these pyramids is a square.

Find the volume of the small pyramid.

$$V = \frac{1}{3}Bh$$
$$V = \left(\frac{1}{3}\right)(9^2)(15)$$
$$V = 405$$

Find the volume of the large pyramid.

$$V = \frac{1}{3}Bh$$
$$V = \left(\frac{1}{3}\right)(18^2)(30)$$
$$V = 3{,}240$$

The volume of the large pyramid is 3,240 cm³. The volume of the small pyramid is 405 cm³.

$$3{,}240 - 405 = 2{,}835$$

The volume of the shaded region is 2,835 cm³.

Example 3

MANUFACTURING A candy company decides to sell its new Blast Off candy bars in a package shaped like a rocket. The body of the rocket is shown at the right. Find the volume of the figure.

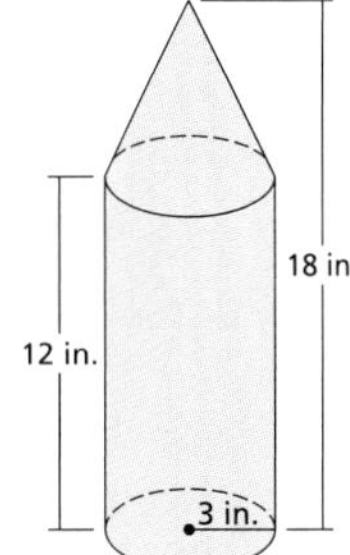

Solution

A cylinder and cone combine to form the figure shown. Add the volume of the cone to the volume of the cylinder.

Find the volume of the cylinder.

$$V = \pi r^2 h$$
$$V = (\pi)(3^2)(12)$$
$$V \approx 339$$

The volume of the cylinder is about 339 in³.

Find the volume of the cone.

$$V = \frac{1}{3}\pi r^2 h$$
$$V = \left(\frac{1}{3}\right)(\pi)(3^2)(6)$$
$$V \approx 57$$

The volume of the cone is about 57 in³.

$$339 + 57 = 396$$

The volume of the figure is about 396 in³.

Chalkboard Examples

Supplementary Example 1 and 2

Find the volume of the cylindrical part of the figure below. about 125.6 in.³

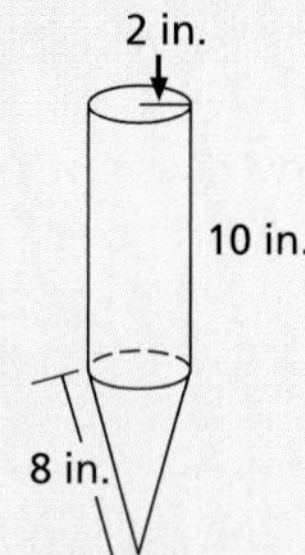

Supplementary Example 3

The company decides to make a smaller version of the rocket. The radius of the circular base is $1\frac{1}{2}$ in. The height of the cylinder is 6 in. and the height of the entire container is 8 in. Find the volume of the rocket. about 47.1 in.³

Supplementary Example 4

Find the volume of an asteroid with a diameter of 158 miles. about 26,128 mi³

Teaching Strategies

COOPERATIVE LEARNING Have students work in small groups to answer this question: Earth's diameter is about 3.7 times that of the moon. Earth's volume, however, is about 50 times that of the moon. What accounts for this dramatic difference? The volume of a sphere is a function of the cube of the radius, rather than the square, as in the case of cylinders and cones.

Reteaching Worksheet 5-7

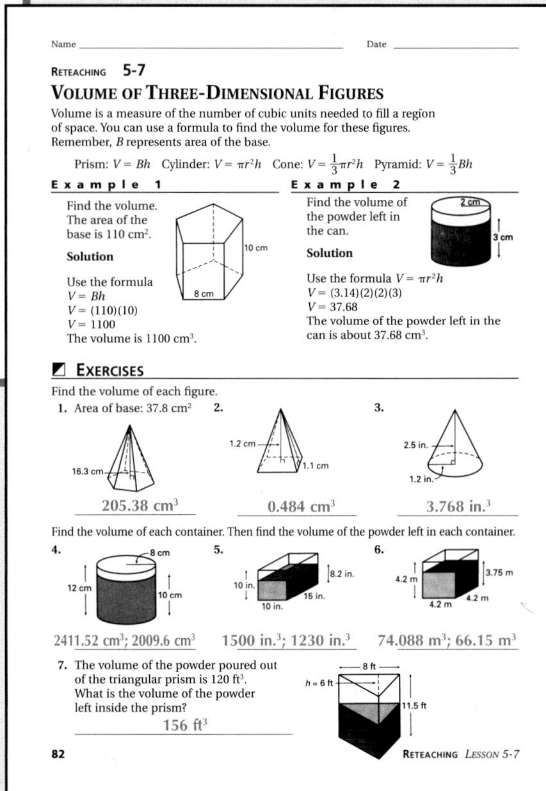

Name ______________________ Date ____________

RETEACHING 5-7

VOLUME OF THREE-DIMENSIONAL FIGURES

Volume is a measure of the number of cubic units needed to fill a region of space. You can use a formula to find the volume for these figures. Remember, B represents area of the base.

Prism: $V = Bh$ Cylinder: $V = \pi r^2 h$ Cone: $V = \frac{1}{3}\pi r^2 h$ Pyramid: $V = \frac{1}{3}Bh$

Example 1

Find the volume. The area of the base is 110 cm². (10 cm, 8 cm)

Solution

Use the formula
$V = Bh$
$V = (110)(10)$
$V = 1100$
The volume is 1100 cm³.

Example 2

Find the volume of the powder left in the can. (2 cm, 3 cm)

Solution

Use the formula $V = \pi r^2 h$
$V = (3.14)(2)(2)(3)$
$V = 37.68$
The volume of the powder left in the can is about 37.68 cm³.

EXERCISES

Find the volume of each figure.

1. Area of base: 37.8 cm² (16.3 cm) 205.38 cm³
2. (1.2 cm, 1.1 cm) 0.484 cm³
3. (2.5 in., 1.2 in.) 3.768 in.³

Find the volume of each container. Then find the volume of the powder left in each container.

4. (8 cm, 12 cm, 10 cm) 2411.52 cm³; 2009.6 cm³
5. (10 in., 10 in., 15 in., 8.2 in.) 1500 in.³; 1230 in.³
6. (4.2 m, 4.2 m, 4.2 m, 3.75 m) 74.088 m³; 66.15 m³
7. The volume of the powder poured out of the triangular prism is 120 ft³. What is the volume of the powder left inside the prism? (8 ft, h = 6 ft, 11.5 ft) 156 ft³

82 RETEACHING *LESSON 5-7*

Lesson Wrap-up

Quick Assessment

Ask the following questions to determine if students understand the content presented in this lesson.

1. How is volume measured? in cubic units since it is a measure of how many unit cubes fill a given space
2. How can you find the volume of any prism? multiply the area of the base by the height of the prism
3. A cylindrical cake pan has a diameter of 10 in. and is 5 in. high. A tube in the center of the pan has a diameter of 3 in. What is the greatest volume of cake batter that can be poured into the pan? about 357 in.^{3}

Assignment Guide

Basic: 1–22, 26–29
Advanced: 1–29
Additional Practice: See Extra Practice Index on page 674.

Additional Answers

25. Cutting 4-cm squares gives the greatest volume—16 cm × 16 cm × 4 = 1,024 cm^3. List all cuts and the resulting dimensions and volumes in a table, until the volumes begin to decrease.

Example 4

ASTRONOMY The only asteroid visible to the naked eye is 4 Vesta, discovered in 1807. Its diameter is 323 mi. What is its volume?

Solution

Assume that 4 Vesta is a sphere. To find the volume of a sphere, use the formula $V = \frac{4}{3}\pi r^3$.

$$V = \frac{4}{3}\pi r^3$$

$$V \approx \left(\frac{4}{3}\right)(3.14)(161.5^3) \approx 17{,}635{,}426$$

The volume of 4 Vesta is approximately 17,635,426 mi^3.

Try These Exercises

Find the volume of each. Use 3.14 for π. Round answers to the nearest whole number.

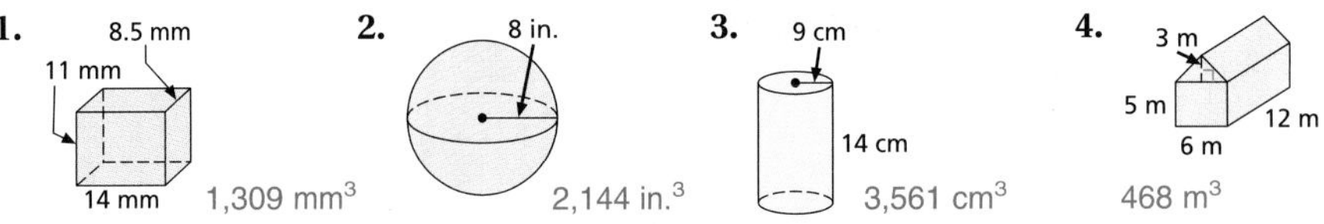

1,309 mm^3 2,144 in.^{3} 3,561 cm^3 468 m^3

5. A prism has a hexagonal base with an area of 24 cm^2. If the volume of the figure is 144 cm^3, what is its height? 6 cm

Practice Exercises

Find the volume to the nearest whole number. Use 3.14 for π.

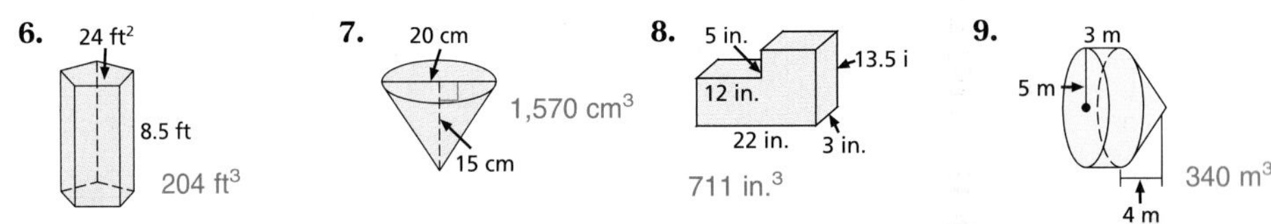

204 ft^3 1,570 cm^3 711 in.^{3} 340 m^3

10. **ARCHEOLOGY** One room in a 12th-century cliff dwelling is in the shape of a rectangular prism. The floor measures 10 ft by 12.5 ft, and the ceiling is 7 ft high. What is the volume of the room? 875 ft^3
11. How many cubic meters of water can a water tank hold, if the tank is a cylinder 9 m high and 6 m in diameter? about 354 m^3
12. A pyramid has a base with an area of 64 cm^2 and a volume of 384 cm^3. What is the height of the figure? 18 cm

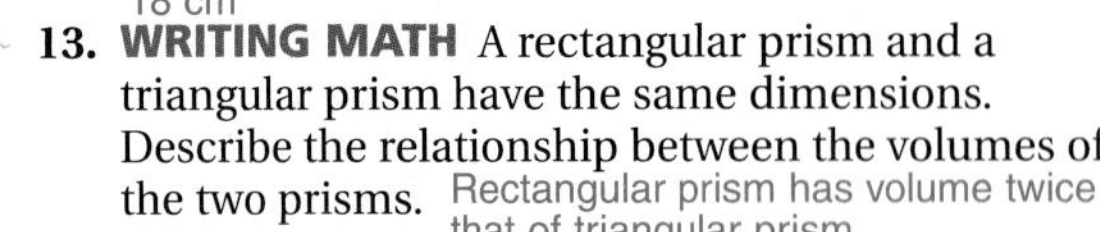

13. **WRITING MATH** A rectangular prism and a triangular prism have the same dimensions. Describe the relationship between the volumes of the two prisms. Rectangular prism has volume twice that of triangular prism

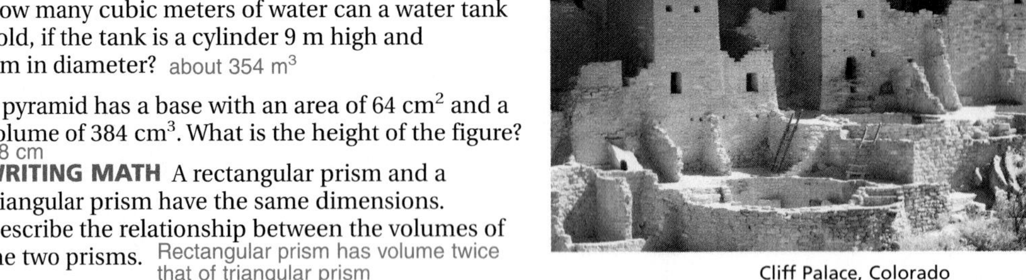

Cliff Palace, Colorado

Extend the Lesson

MATH JOURNAL Have students explain how to find a missing dimension if they know the volume and another dimension. For example, if they know the volume and base area of a prism, they can find the height. $V = Bh$; then $h = V/B$.

Find the volume of each. Use 3.14 for π. Round answers to the nearest whole number.

14.

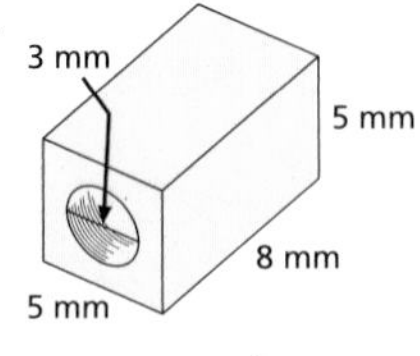

V = 143 mm^3

15.

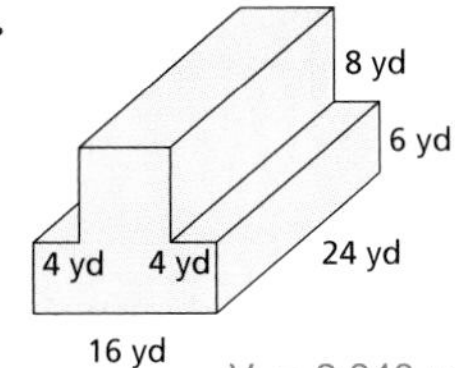

V = 3,840 yd^3

16.

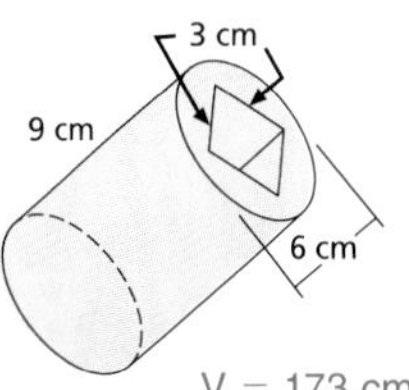

V = 173 cm^3

17. ARCHEOLOGY The height of the pyramid at Cheops is higher than the height of the Quetzalcoatl pyramid at Cholula, but its volume is less. How can that be? One answer: Area of base of Cheops is smaller than the area of the base of Cholula.

18. The base of a rectangular pyramid has sides of 3.14 m and 4 m. A cone has the same height and volume. What is its radius? *Hint*: Write and solve an equation; use 3.14 for π. r = 2 m

19. What happens to the volume of a cone if its height is tripled? It is tripled.

20. If the radius of a cylinder is tripled, what happens to its volume? It is 9 times greater.

21. DATA FILE Use the Data Index on pages 632–633 to locate information on crystal systems. Describe how you could find the volume of a crystal belonging to the tetragonal system.

22. CHAPTER INVESTIGATION Estimate the volume of your structure using your understanding of three-dimensional figures. Using an index card, make a fact list about your structure. Include the date the structure was built, the name of the architect, materials used, dimensions, surface area and volume. Display your model and fact list. Check students' work.

21. Tetragonal system is combination of rectangular prism and two square pyramids; find volume of separate shapes and add.

Extended Practice Exercises

23. COST ANALYSIS A can of Iguana Goodies that is 14 cm high with a radius of 6 cm sells for $1.70. Another can is the same height, but with a radius of 3 cm. It sells for $0.40. Which can of Iguana Goodies is the better buy? the $0.40 can

24. A rectangular prism is 12 ft long and 3 ft wide. If its volume is 288 ft^3, what is its surface area? 312 ft^2

25. PACKAGING Suppose you have a square sheet of cardboard 24 cm on a side. You want to cut equal squares from the corners of the large square to fold the cardboard into an open box. If you cut corners of 1 cm, 2 cm, 3 cm and so on, which size cuts will give you the largest box? See additional answers.

Mixed Review Exercises

Find the perimeter or circumference and the area of each. Round answers to the nearest hundredth if necessary. (Lesson 5-2)

26.

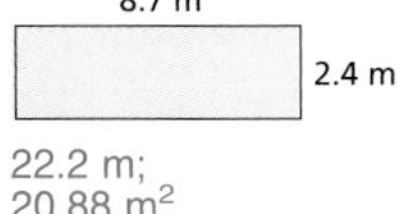

22.2 m; 20.88 m^2

27.

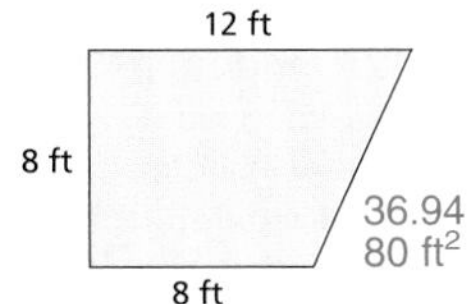

36.94 ft; 80 ft^2

28.

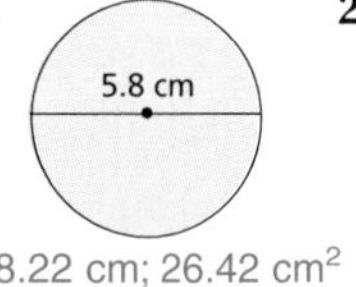

18.22 cm; 26.42 cm^2

29.

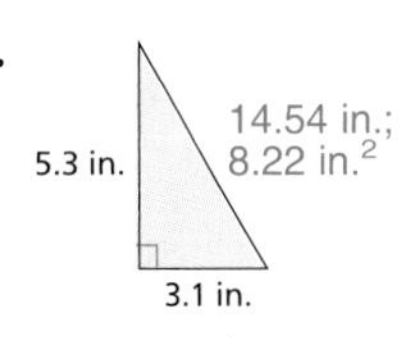

14.54 in.; 8.22 in.^{2}

Teaching Strategies

CHALLENGE A piece of wood in the shape of a rectangular prism has a cylindrical hole drilled halfway through its length. The wood is 18 in. long and has a square base 2 in. on a side. If the hole has a radius of 0.5 in., what is the volume of the piece of wood? about 65 in.^{3}

Extra Practice Worksheet 5-7

Name ______ Date ______

EXTRA PRACTICE 5-7

VOLUME OF THREE-DIMENSIONAL FIGURES

EXERCISES

Find the volume to the nearest whole number. Use 3.14 for π.

1. (5 ft, 8 ft, 5 ft) 200 ft^3
2. (24 in., 12 in.) 10,852 in.^{3}
3. (15 cm, 10 cm, 12.5 cm) 313 cm^3
4. (21 m, 14 m, 16 m, 16 m, 14 m) 1568 m^3
5. How many cubic meters of water can a water tank hold, if the tank is a cylinder 10 m high and 12 m in diameter? 1130.4 m^3
6. A rectangular prism has a volume of 382.5 in.^{3} and a base area of 45 in.^{2}. What is the height of the prism? 8.5 in.
7. A sphere has a radius of 4 m. What is the volume of the sphere? 200.96 m^3
8. A cylinder with a diameter of 8 in. and a height of 5 in. fits completely inside a rectangular box with a height of 5 in., a length of 10 in., and a width of 8 in. What is the volume of the box outside of the cylinder? 148.8 in.^{3}
9. What happens to the volume of a rectangular prism if its height is doubled? The volume is doubled.
10. A container is made by placing a triangular pyramid with a base area of 12 cm^2 and a height of 8 cm on top of a rectangular prism with a height of 6 cm, a length of 8 cm, and a width of 4 cm. What is the volume of the container? 224 cm^3

72 EXTRA PRACTICE *LESSON 5-7*

Enrichment Worksheet 5-7

Name ______ Date ______

ENRICHMENT 5-7

SOLID PENTOMINOES

EXERCISES

Each of the 12 possible pentominoes is made from five squares joined together. Solid versions of the pentominoes can be made by glueing cubes together.

Build a set of solid pentominoes for these activities.

1. Use all 12 solid pentominoes to make a 2 by 5 by 6 rectangular prism. Sketch the two layers of the prism below. Answers may vary. Possible answer is given.
2. Now use all 12 pieces to make a 3 by 4 by 5 prism. The drawing at the right shows one way to do this. Find at least three different solutions. Answers may vary.
3. Use all pieces to create a solid figure of your own. Sketch your figure in the space below. Answers may vary.

94 ENRICHMENT *LESSON 5-7*

Assessment Planning

Vocabulary Review

Lesson 5-1
ratio
metric units
rate
unit price
customary units
like units
unit rate

Lesson 5-2
area
circumference
perimeter

Lesson 5-3
probability
favorable outcome
random

Lesson 5-4
indirect proof

Lesson 5-5
polyhedron
prism
lateral faces
cylinder
cone
locus
vertex
pyramid
lateral edges
axis
sphere

Lesson 5-6
surface area

Lesson 5-7
volume

Assessment Options

Chapter Assessment
Alternative Assessment
Chapter Test Forms A and B
Cumulative Assessment
Achievement Test
Writing Prompts

Assignment Guide

All students: 1–22
Additional Practice: Refer to the Extra Practice Index on page 674 of this text.

Chapter 5 Review

Vocabulary ■ LESSON 5-1–LESSON 5-7

Choose the word from the list that best completes each statement.

a. pyramid
b. edge
c. cone
d. tolerance
e. rate

1. Two faces of a polyhedron intersect at an __?__. b
2. The amount a manufactured item may vary from a specified size is called the __?__. d
3. A ratio that compares two different quantities is a __?__. e
4. A polyhedron with one base is called a __?__. a
5. A three-dimensional figure with a curved surface and one circular base is called a __?__. c

LESSON 5-1 ■ Ratios and Units of Measure, p. 202

▶ A ratio compares two numbers by division and can be written as 4 : 5, $\frac{4}{5}$, or 4 to 5.

▶ For both customary and metric measures, multiply to change from a larger unit to an equivalent smaller unit. Divide to convert a smaller unit to an equivalent larger one.

6. Write 8:20 in lowest terms in two other ways. $\frac{2}{5}$; 2 to 5
7. How many grams are in 3 kilograms? 3,000
8. The greatest possible error (GPE) of any measurement is $\frac{1}{2}$ the smallest unit used to make the measurement. What is the GPE of a measurement of 2.46 m? 0.005 m

LESSON 5-2 ■ Perimeter, Circumference and Area, p. 206

▶ To solve a problem involving the distance around a plane figure or the surface covered by it, you must choose the correct formula.

Find the perimeter or circumference and area of each.

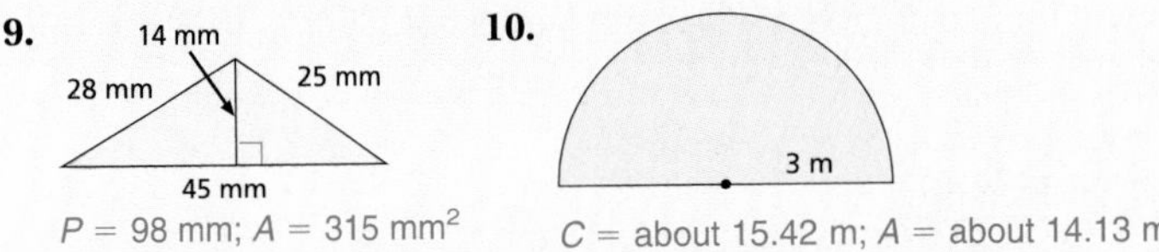

9. P = 98 mm; A = 315 mm^2
10. C = about 15.42 m; A = about 14.13 m^2

LESSON 5-3 ■ Probability and Area, p. 212

▶ To determine the probability that a random point in a figure will be within a given region, use the ratio: $\frac{\text{area of given region}}{\text{total area of figure}}$

11. An ice skater skates in a rectangular rink 40 m by 125.6 m. If she were to fall, what is the probability she will fall within a circle with a diameter of 8 m? 1:100

Extend the Lesson

ONGOING ASSESSMENT Which word should go in the blank, *sometimes, always* or *never*?

1. The sum of the areas of the flat surfaces of a solid figure ____ equals the figure's area. sometimes
2. Two figures with the same volume _____ have the same surface area. sometimes
3. A cylinder with the same height and base as a cone ______ has a greater volume than the cone. always
4. Doubling the radius of a sphere ________ doubles the surface area. never

LESSON 5-4 ■ Irregular Shapes, p. 216

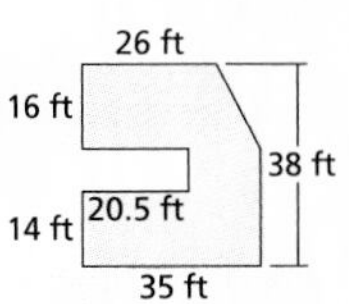

▶ To find the total area of an irregular figure, first break it down into smaller, recognizable shapes and then add the areas.

12. A wing of a new house has the shape shown at the right. What is the area of the wing? 1,094 ft^2

LESSON 5-5 ■ Three-dimensional Figures and Loci, p. 220

▶ A polyhedron is a three-dimensional figure in which each face is a polygon.

Identify each figure. Then state the number of faces, vertices, and edges each has.

13.

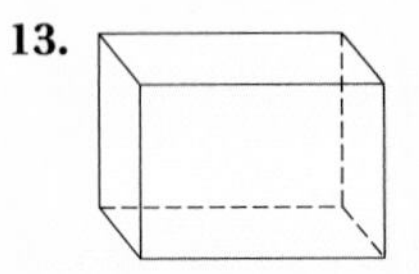

triangular prism; 5 faces; 6 vertices; 9 edges

14.

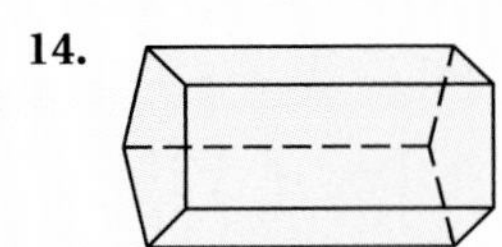

rectangular pyramid; 5 faces; 5 vertices; 8 edges.

LESSON 5-6 ■ Surface Area of Three-dimensional Figures, p. 224

▶ To find the surface area of a three-dimensional figure, choose the correct formula for the area of each surface of the figure.

Find the surface area of each figure to the nearest tenth.

15.

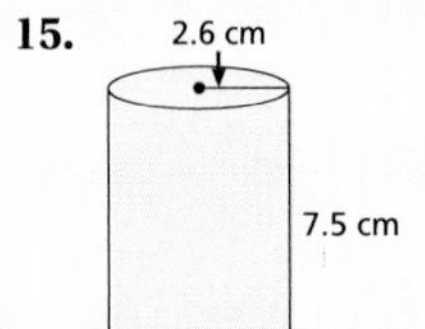

164.9 cm^2

16.

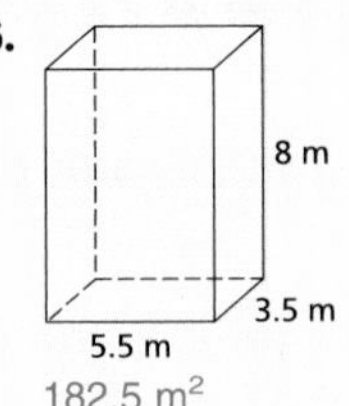

182.5 m^2

17.

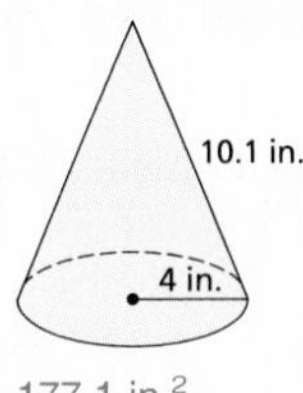

177.1 in.^{2}

18. A cube has a surface area of 294 cm^2. What is the length of a side? 7 cm

LESSON 5-7 ■ Volume of Three-dimensional Figures, p. 230

▶ To find the volume of a three-dimensional figure, use the correct formula or formulas.

Find the volume of each figure to the nearest tenth.

19.

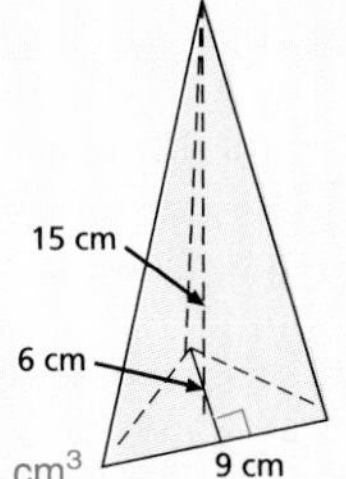

135 cm^3

20.

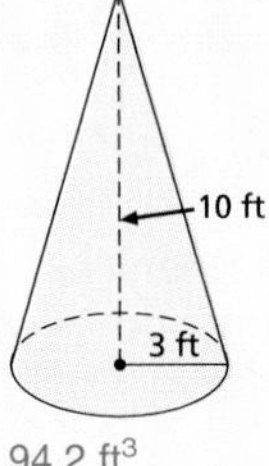

94.2 ft^3

21.

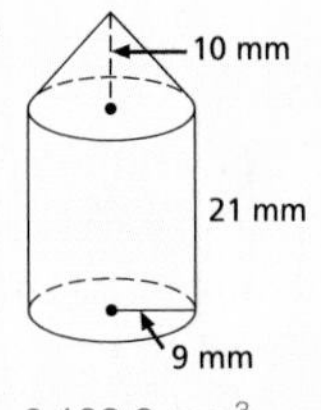

6,188.9 mm^3

Teaching Strategies

To help students remember the formulas for volume of cones and pyramids, explain that volume can be thought of as making a stack of bases. So, B (area of base) should be found first. For cones and pyramids, the area of the bases decrease as the height increases, so a fractional multiplier is required in the volume formula. It can be shown that the fraction in each case is $\frac{1}{3}$.

Chapter 5 Test Form A, page 1

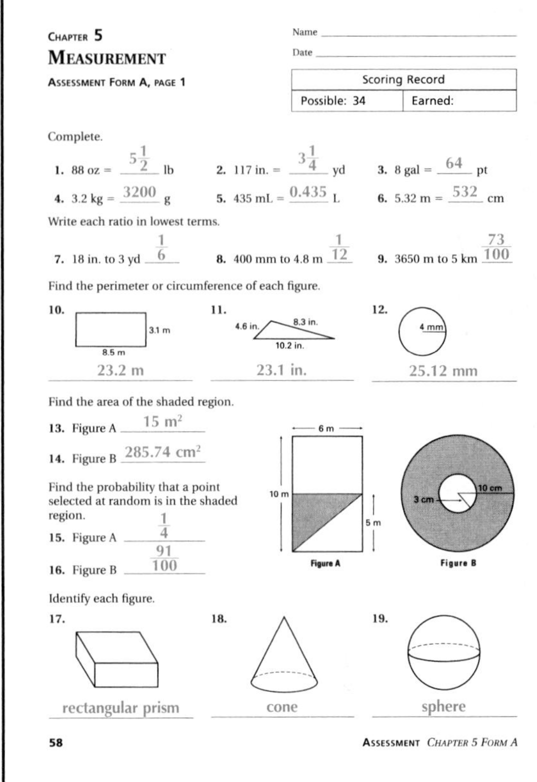

CHAPTER 5
MEASUREMENT
ASSESSMENT FORM A, PAGE 1

Name ____________
Date ____________

Scoring Record	
Possible: 34	Earned:

Complete.

1. 88 oz = $5\frac{1}{2}$ lb
2. 117 in. = $3\frac{1}{4}$ yd
3. 8 gal = 64 pt
4. 3.2 kg = 3200 g
5. 435 mL = 0.435 L
6. 5.32 m = 532 cm

Write each ratio in lowest terms.

7. 18 in. to 3 yd $\frac{1}{6}$
8. 400 mm to 4.8 m $\frac{1}{12}$
9. 3650 m to 5 km $\frac{73}{100}$

Find the perimeter or circumference of each figure.

10. (3.1 m, 8.5 m) 23.2 m
11. (4.6 in., 8.3 in., 10.2 in.) 23.1 in.
12. (4 mm) 25.12 mm

Find the area of the shaded region.

13. Figure A 15 m^2
14. Figure B 285.74 cm^2

Find the probability that a point selected at random is in the shaded region.

15. Figure A $\frac{1}{4}$
16. Figure B $\frac{91}{100}$

Identify each figure.

17. rectangular prism
18. cone
19. sphere

58 ASSESSMENT *CHAPTER 5 FORM A*

Chapter 5 Test Form A, page 2

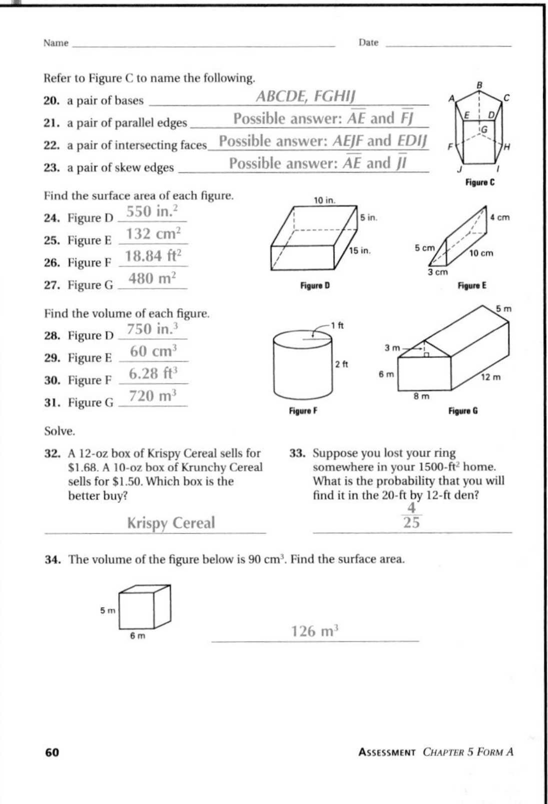

Name ____________ Date ____________

Refer to Figure C to name the following.

20. a pair of bases *ABCDE, FGHIJ*
21. a pair of parallel edges Possible answer: $\overline{AE}$ and $\overline{FJ}$
22. a pair of intersecting faces Possible answer: *AEJF* and *EDIJ*
23. a pair of skew edges Possible answer: $\overline{AE}$ and $\overline{JI}$

Find the surface area of each figure.

24. Figure D 550 in.^{2}
25. Figure E 132 cm^2
26. Figure F 18.84 ft^2
27. Figure G 480 m^2

Find the volume of each figure.

28. Figure D 750 in.^{3}
29. Figure E 60 cm^3
30. Figure F 6.28 ft^3
31. Figure G 720 m^3

Solve.

32. A 12-oz box of Krispy Cereal sells for \$1.68. A 10-oz box of Krunchy Cereal sells for \$1.50. Which box is the better buy? Krispy Cereal

33. Suppose you lost your ring somewhere in your 1500-ft^2 home. What is the probability that you will find it in the 20-ft by 12-ft den? $\frac{4}{25}$

34. The volume of the figure below is 90 cm^3. Find the surface area. (5 m, 6 m) 126 m^3

60 ASSESSMENT *CHAPTER 5 FORM A*

Chapter 5 Test Form B, page 1

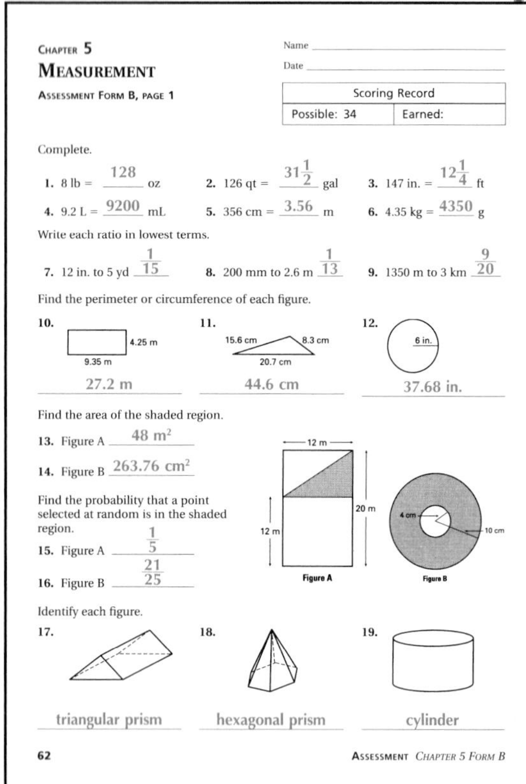

CHAPTER 5
MEASUREMENT
ASSESSMENT FORM B, PAGE 1

Name
Date

Scoring Record	
Possible: 34	Earned:

Complete.

1. 8 lb = 128 oz
2. 126 qt = $31\frac{1}{2}$ gal
3. 147 in. = $12\frac{1}{4}$ ft
4. 9.2 L = 9200 mL
5. 356 cm = 3.56 m
6. 4.35 kg = 4350 g

Write each ratio in lowest terms.

7. 12 in. to 5 yd $\frac{1}{15}$
8. 200 mm to 2.6 m $\frac{1}{13}$
9. 1350 m to 3 km $\frac{9}{20}$

Find the perimeter or circumference of each figure.

10. 4.25 m; 9.35 m — 27.2 m
11. 15.6 cm; 8.3 cm; 20.7 cm — 44.6 cm
12. 6 in. — 37.68 in.

Find the area of the shaded region.

13. Figure A 48 m²
14. Figure B 263.76 cm²

Find the probability that a point selected at random is in the shaded region.

15. Figure A $\frac{1}{5}$
16. Figure B $\frac{21}{25}$

12 m; 20 m; 12 m; Figure A; 4 cm; 10 cm; Figure B

Identify each figure.

17. triangular prism
18. hexagonal prism
19. cylinder

62 ASSESSMENT *CHAPTER 5 FORM B*

Chapter 5 Test Form B, page 2

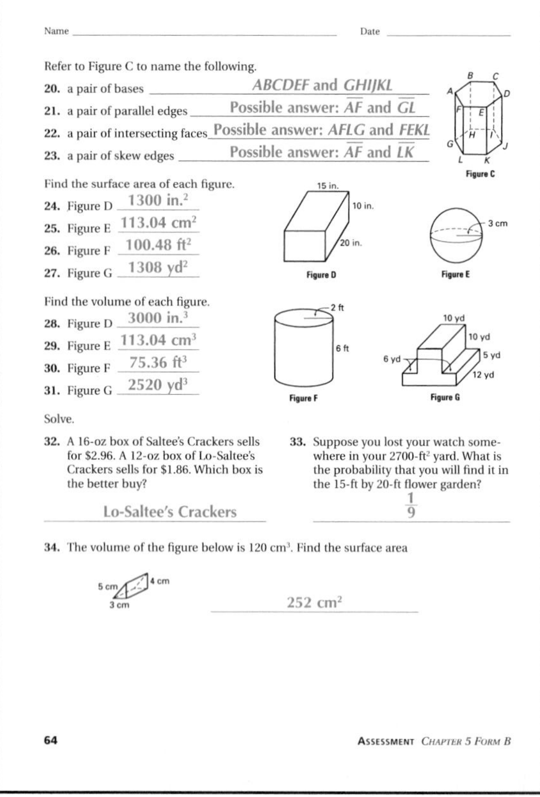

Name
Date

Refer to Figure C to name the following.

20. a pair of bases *ABCDEF* and *GHIJKL*
21. a pair of parallel edges Possible answer: $\overline{AF}$ and $\overline{GL}$
22. a pair of intersecting faces Possible answer: *AFLG* and *FEKL*
23. a pair of skew edges Possible answer: $\overline{AF}$ and $\overline{LK}$

Figure C

Find the surface area of each figure.

24. Figure D 1300 in.²
25. Figure E 113.04 cm²
26. Figure F 100.48 ft²
27. Figure G 1308 yd²

15 in.; 10 in.; 20 in.; Figure D; 3 cm; Figure E

Find the volume of each figure.

28. Figure D 3000 in.³
29. Figure E 113.04 cm³
30. Figure F 75.36 ft³
31. Figure G 2520 yd³

2 ft; 6 ft; Figure F; 10 yd; 10 yd; 5 yd; 6 yd; 12 yd; Figure G

Solve.

32. A 16-oz box of Saltee's Crackers sells for \$2.96. A 12-oz box of Lo-Saltee's Crackers sells for \$1.86. Which box is the better buy? Lo-Saltee's Crackers

33. Suppose you lost your watch somewhere in your 2700-ft² yard. What is the probability that you will find it in the 15-ft by 20-ft flower garden? $\frac{1}{9}$

34. The volume of the figure below is 120 cm³. Find the surface area 252 cm²

5 cm; 4 cm; 3 cm

64 ASSESSMENT *CHAPTER 5 FORM B*

CHAPTER 5 ASSESSMENT

Write each ratio in lowest terms.

1. 4 yd to 16 ft $\frac{3}{4}$
2. 400 m to 2 km $\frac{1}{5}$
3. 6 qt to 9 gal $\frac{1}{6}$
4. What is the height of a triangle with an area of 40 cm² and a base of 12 cm? $6\frac{2}{3}$
5. How many faces, vertices, and edges does a hexagonal prism have? faces: 8; vertices: 12; edges: 18

For Exercises 6–9, use $\pi \approx 3.14$. Round your answers to the nearest tenth.

Find the area of the shaded region of each.

6. 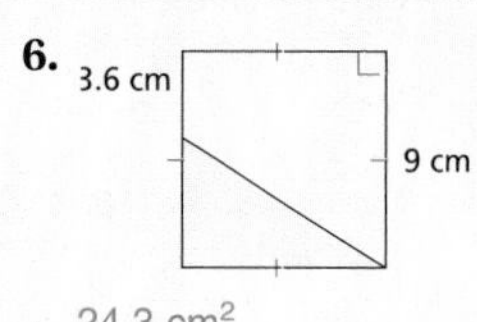

24.3 cm²

7. 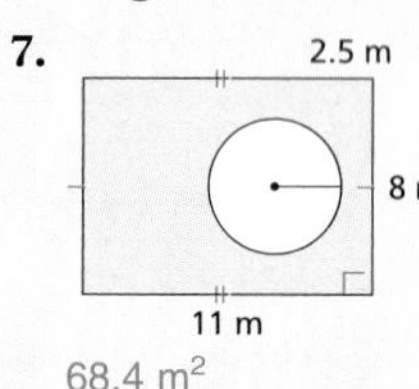

68.4 m²

Find the surface area of each.

8. 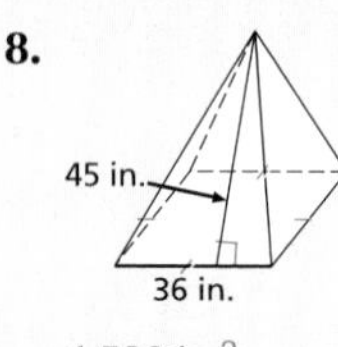

4,536 in.²

9. 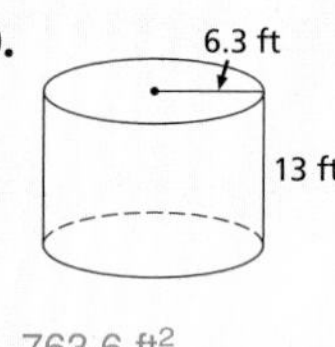

763.6 ft²

Find the probability that a point selected at random in each figure is in the shaded region.

10. 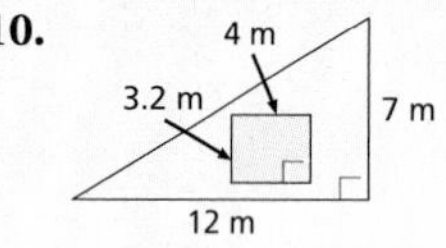

$P = \frac{32}{105}$

11. 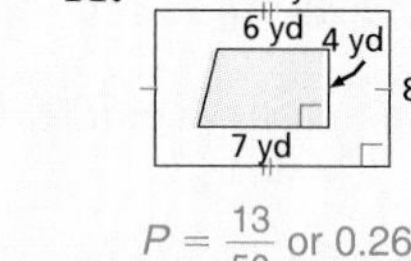

$P = \frac{13}{50}$ or 0.26

12. 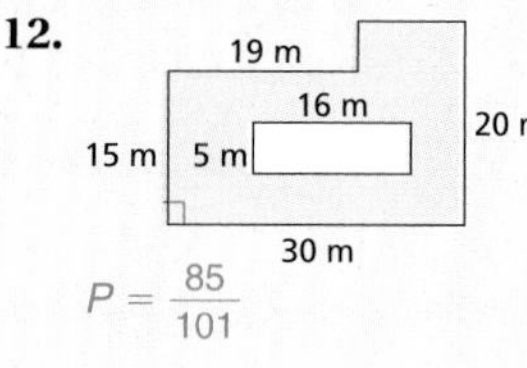

$P = \frac{85}{101}$

Find the volume of each figure to the nearest tenth. Use $\pi \approx 3.14$.

13. 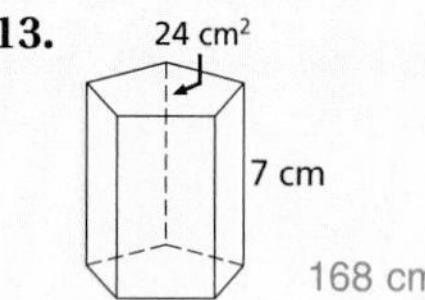

168 cm³

14. 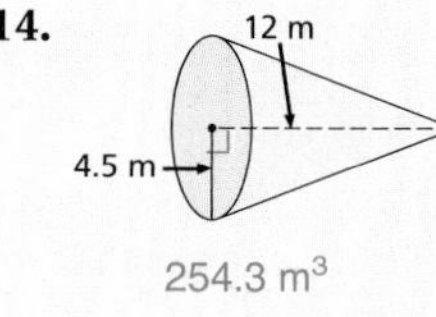

254.3 m³

15. 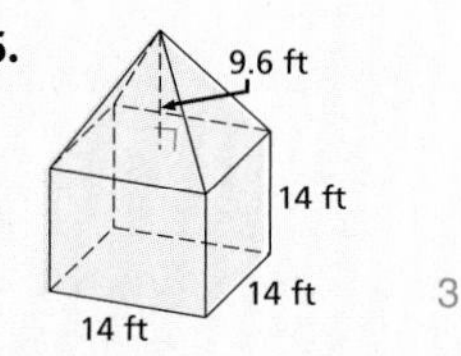

3,371.2 ft³

16. You want to paint the figure shown below. Which formula will you use to know how much paint you will need? $SA = \pi rs + \pi r^2$

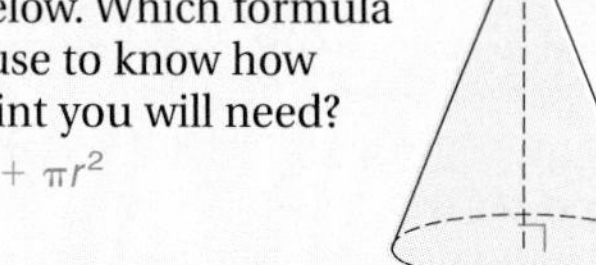

17. How many 1-ft² floor tiles are needed to cover the floor shown below? 1,270 tiles

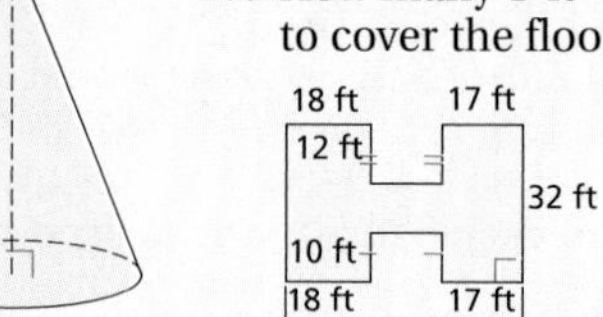

18. Which measurement is more precise, 4.003 L or 4.99 L? 4.003L
19. What is the GPE of the measurement 100.9 cm? 0.05 cm
20. A micrometer setting shows the measurement 24.46 mm. What are the upper and lower limits for the measurement if the tolerance is ±0.03 mm? upper: 24.49 mm; lower: 24.43 mm

Critical Thinking

Students may be interested in comparing their results to find different shapes, and to find if their answers are in agreement.

SCORING RUBRIC 5—Student submits a wide variety of innovative shapes for the fencing allowed, completes all calculations correctly, and draws a valid conclusion. **4**—Student submits a wide variety of shapes for the fencing allowed, completes all calculations correctly, and draws a valid conclusion. **3**—Student submits a reasonable variety of shapes for the fencing allowed, completes all calculations with only minor errors, and draws a valid conclusion. **2**—Student submits a few shapes for the fencing allowed, completes all calculations with several errors, and draws a partially valid conclusion. **1**—Student submits a few shapes for the fencing allowed, completes calculations with several significant errors, and draws no conclusion. **0**—Student makes no attempt to make drawings or perform the calculations.

Chapter 5 Alternative Assessment

Portfolio

PYRAMIDS Look up the dimensions of five pyramids other than Quetzalcoatl, which is mentioned in your text book. Draw a scale diagram of each, then find the perimeter, the base area, the surface area and the volume of each pyramid. Be sure to show complete calculations of each measure. Compare the pyramids using each measure. Are the rankings the same throughout? If not, give an explanation.

Modeling/Using Manipulatives

DEMONSTRATION Prepare a class demonstration. Using small blocks, demonstrate how the perimeter, area surface area and volume of a rectangular prism are related. Demonstrate also why area is measured in square units and volume is measured in cubic units. If no blocks are available, use three-dimensional drawings, prepared in advance, for your presentation.

Critical Thinking

FENCING Given 1000 feet of fencing, what is the greatest area that could be enclosed? Experiment with different shapes, each with a perimeter of exactly 1000 feet. Keep a record of all the various shapes you tried, and sketch the basic shape of each. Write a brief paragraph explaining and supporting your conclusions.

Chapter Investigation

EXTENSION Write a report about the structure you chose. Be sure to include the measurements and known features of the structure. Explain why you chose the scale you did for your model. State your estimate of the exterior surface area and volume of your structure, and explain what these estimates tell you about the structure.

Portfolio

Enthusiastic students may be encouraged to make scale models of the pyramids used in this activity.

SCORING RUBRIC **5**—Student completes the calculations for five pyramids, demonstrating a thorough knowledge of the formulas and methods of calculation. **4**—Student completes the calculations for five pyramids, demonstrating a working knowledge of formulas and methods. **3**—Student completes the calculations for five pyramids, demonstrating a general knowledge of the formulas and methods, with a few minor errors. **2**—Student completes the calculations for fewer than five pyramids, demonstrating some knowledge of the formulas and methods, with minor errors. **1**—Student completes the calculations for one or two pyramids, demonstrating some knowledge of the formulas and methods, but with significant errors. **0**—Student makes no attempt to complete the calculations.

Modeling/Using Manipulatives

Small blocks may be borrowed from an elementary school. Other suitable materials include commercially made snap cubes, pattern blocks, color tiles and centimeter cubes.

SCORING RUBRIC **5**—Student prepares and gives a class demonstration, exhibiting a thorough knowledge and complete understanding of the relationships between the measures. **4**—Student prepares and gives demonstration, exhibiting a general knowledge and understanding. **3**—Student prepares and gives a class demonstration, exhibiting some knowledge and understanding. **2**—Student prepares and gives a partial or incomplete class demonstration, exhibiting minor knowledge and understanding. **1**—Student prepares and gives an incomplete class demonstration, exhibiting little knowledge and understanding. **0**—Student makes no attempt to prepare or give a class demonstration.

See page 236 for information and Scoring Rubric for Critical Thinking.

Chapter Investigation

The benchmarks and expectations for this extension are as follows.

- Students choose a manmade structure. They research the structure's length, width, height, and special characteristics. Students then make a rough sketch of the structure and label any known measurements.
- Students build a three-dimensional model of the structure they have chosen. They break the structure into smaller three-dimensional figures or sections. They then assemble the smaller sections to make the final product.
- Students estimate the surface area of the structure they have chosen. They explain how they got their answer.
- Students estimate the volume of their structure using your understanding of three-dimensional figures. They make a fact list, including the date the structure was built, the name of the architect, materials used, dimensions, surface area, and volume. They display their model and fact list.
- Students write a report about the structure, including the measurements and known features of the structure. They explain why they chose the scale they did for their model. They state their estimate of the exterior surface area and volume of their structure, and they explain what these estimates tell them about the structure.

Cumulative Review

Cumulative review is the best way to maintain previously taught skills and concepts. This will keep students prepared for new lessons that build on previously covered skills.

Cumulative Review covers:

Lesson 1-1
Lesson 1-2
Lesson 1-5
Lesson 1-7
Lesson 1-8
Lesson 2-2
Lesson 2-4
Lesson 2-5
Lesson 2-6
Lesson 2-7
Lesson 3-1
Lesson 3-4
Lesson 3-5
Lesson 4-1
Lesson 4-2
Lesson 5-6
Lesson 5-7

Additional Answers

25. 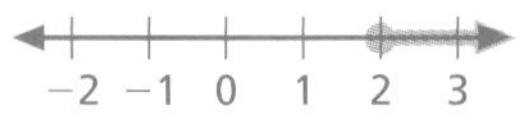 (number line: −2 −1 0 1 2 3)

26. 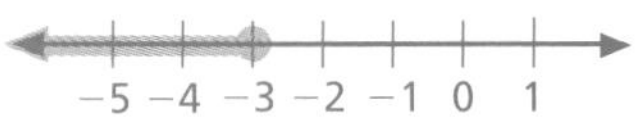 (number line: −5 −4 −3 −2 −1 0 1)

27.

The tenth figure will be a rectangular array with 11 × 12 or 132 dots.

Chapters 1–5 Cumulative Review

Evaluate each expression.

1. $|s|$ if $s = -10$ 10

2. $-|-n|$ if $n = 12$ −12

Perform each operation.

3. $(-4.6)(-1.9)$ 8.74

4. $-75 \div (-5)$ 15

5. $10\frac{1}{2} \div (-3) + 2$ $-1\frac{1}{2}$

6. $2\frac{1}{5} + (-2)(3)$ $-3\frac{4}{5}$

Evaluate when $x = -3$ and $y = 4$.

7. $6x^3$ −162

8. $x^{-4}y^2$ $\frac{16}{81}$

Given $f(x) = x^3$ and $g(x) = 2x^2$. Find each value.

9. $f(-2)$ −8

10. $g(4)$ 32

Solve each equation and check.

11. $x - (-17) = -50$ −67

12. $1\frac{2}{3}m = 20$ 12

13. $5w - 35 = 40$ 15

14. $3(x - 5) = 2(x + 2)$ 19

Refer to the following data.

16 23 21 17 17 20 22 17 14 23

15. Find the mean. 19

16. Find the mode. 17

17. In the figure, $MN = 120$. Find ME. 102

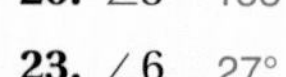

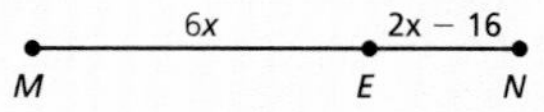

Find the measure of each angle. $\overleftrightarrow{RS} \parallel \overleftrightarrow{GH}$

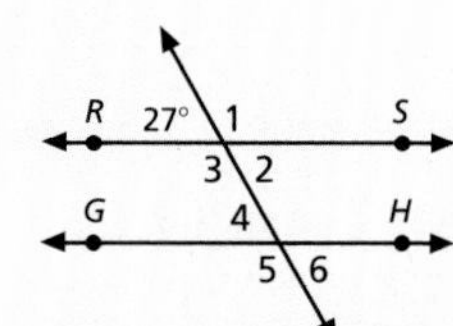

18. ∠1 153°

19. ∠2 27°

20. ∠3 153°

21. ∠4 27°

22. ∠5 153°

23. ∠6 27°

Solve each inequality and graph. For 24–25, see additional answers.

24. $7x - 12 \geq 2$

25. $5m < -15$

26. Draw the next figure in the pattern. Then describe the tenth figure in the pattern. See additional answers.

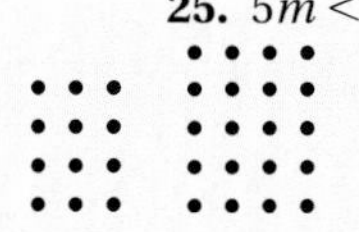

27. Given: $\overline{KR} \cong \overline{RN}$, $\overline{ZR} \cong \overline{RW}$. $\overleftrightarrow{ZW}$ and $\overleftrightarrow{KN}$ are intersection lines. Name the postulate you could use to prove $\triangle KRZ$ is congruent to $\triangle NRW$. SAS

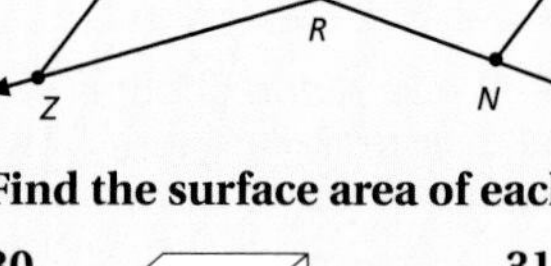

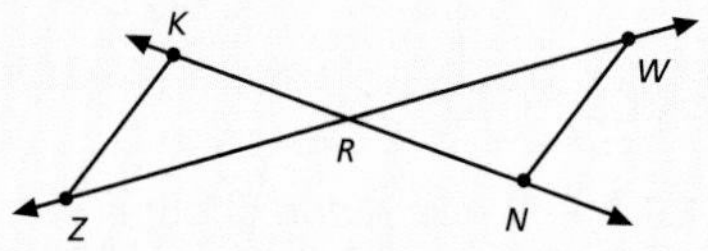

Find the value of x.

28.

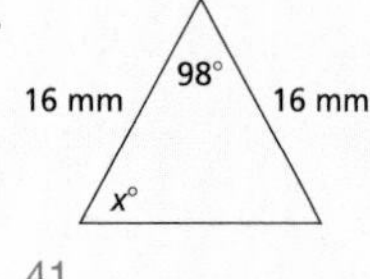

41

29.

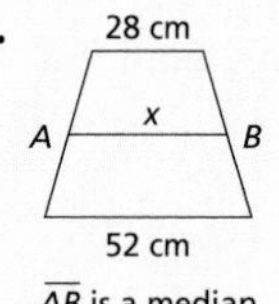

$\overline{AB}$ is a median

40

Find the surface area of each figure.

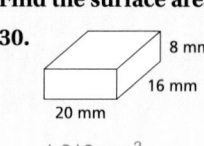

30. (rectangular prism: 8 mm, 16 mm, 20 mm)

1,216 mm²

31.

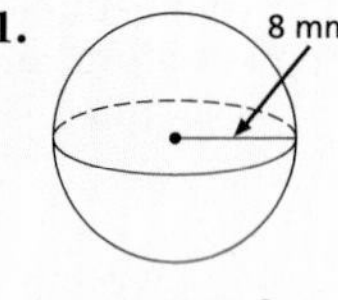

803.84 mm²

Teaching Strategies

Remind students to consider all the information given in the problem. For example, in Exercise 29 of the Cumulative Review, students may focus only on the interior angle measures; however, the fact that the two sides are equal is necessary in order to recognize that the triangle is isosceles and the two base angles are equal.

Chapters 1–5 Cumulative Assessment

Standardized Test Preparation—Quantitative Comparison

In each question, compare the quantity in Column 1 with the quantity in Column 2. Select the letter of the correct answer from these choices:

A. The quantity in Column 1 is greater.

B. The quantity in Column 2 is greater.

C. The two quantities are equal.

D. The relationship cannot be determined by the information given.

Notes: In some questions, information that refers to one or both columns is centered over both columns. A symbol used in both columns has the same meaning in each column. All variables represent real numbers. Most figures are not drawn to scale.

	Column 1	Column 2
1. A	32 16 31 / 28 31 19	38 30 27 / 25 32 26
	mean	median
2. A	$a = -5, b = -0.4$	
	$-a + ab$	$a - ab$
3. A	$x = -6$	
	x^{-3}	x^3
4. B	$5x + 8 = 2x - 13$	
	x	$-x$
5. A	$4x + 1 < 9$	
	9	x
6. C		
	117.5	m∠TOZ

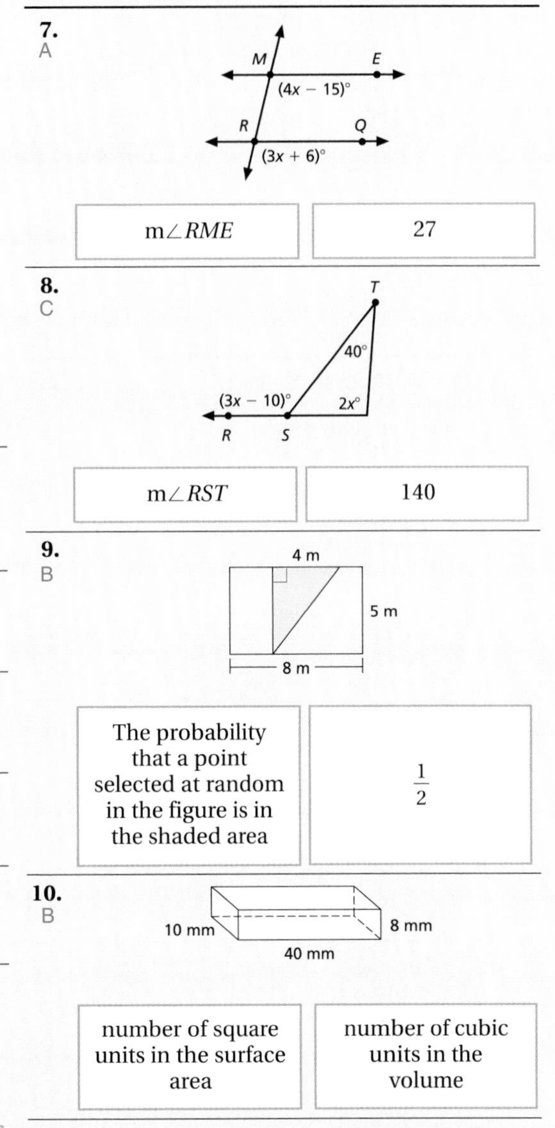

11. CONSTRUCTED RESPONSE Describe a rectangular prism that has the same number of units in its surface area as its volume.

Possible answer: a cube with sides that measure 6 units each.

Standardized Test Preparation

QUANTITATIVE COMPARISON QUESTIONS Quantitative comparison questions are generally written with fewer words and require less time to compute. There are four answer choices that describe the relationships of equality and inequality. You are required to compare the quantity from Column A with the quantity from Column B and then determine their relationship. Students should carefully read the four answer choices often during the testing period.

Horizontal rules are used as separators between questions and to clarify when figures and additional information accompanies one or more problems.

Understanding the directions for quantitative comparison questions is critical for success since this type of question is not as commonplace as other types of questions.

Testing Strategies

To save time, students should compare the two quantities without lengthy solving procedures. For example, in Exercise 5, you only need to know whether x is positive or negative. This can be determined quickly by combining like terms. From $3x = -21$, you know that x must be negative, so $-x$ must be greater than x. The correct answer choice is B. After students have completed the exercise, discuss the minimum information needed for other items to compare the two quantities.

Teaching Strategies

Exercise 10 of the Cumulative Assessment requires students to use area to determine probability. Students often make the common error of writing a ratio comparing the shaded area to the unshaded area. Remind these students that the correct ratio compares the shaded area to the area of the entire figure (shaded + unshaded).

Chapter 6—Linear Systems of Equations

Theme: Manufacturing Industry
Chapter Investigation: Manufacturing, pages 243, 251, 261, 285
Careers: Precision Assemblers, page 253; Engineering Technicians, page 273
Data Activity: U.S. Goods–Imports and Exports

Content and Connections

Lesson, Pages	Lesson Objectives	NCTM Standards	State/Local Objectives	Interdisciplinary Connections	Real World Applications
6–1 244–247	• Find the slope of a line • Write the slope–intercept form of an equation and graph the equation	Patterns, Functions & Algebra Number Operation Prob Solv; Reasoning & Proof Connections			Manufacturing, Finance, Recreation
6–2 248–251	• Use slope to determine whether two lines are parallel or perpendicular	Patterns, Functions & Algebra Geometry & Spatial Sense Prob Solv; Reasoning & Proof Connections; Representation			Electronics, Mapmaking, Manufacturing
6–3 254–257	• Write equations for lines in slope-intercept and point-slope forms	Patterns, Functions & Algebra Prob Solv; Reasoning & Proof Communication Connections; Representation			Product design, Fitness, Real Estate
6–4 258–261	• Solve a system of equations by graphing	Patterns, Functions & Algebra Reasoning & Proof Representation Number Operation; Prob Solv		Art	Income Tax, Manufacturing, Finance
6–5 264–267	• Solve systems of equations using substitution	Patterns, Functions & Algebra Prob Solv; Reasoning & Proof Connections; Representation Number Operation			Shipping, Packaging, Recreation
6–6 268–271	• Solve systems of equations by adding, subtracting and multiplying	Patterns, Functions & Algebra Reasoning & Proof Number Operation Problem Solving Connections; Representation			Entertainment, Community Service, Manufacturing
6–7 274–275	• Solve a problem with detreminants and matrices • Use a formula	Patterns, Functions & Algebra Reasoning & Proof Problem Solving Connections; Representation			Finance, Business
6–8 276–279	• Use graphing to solve systems of linear inequalities	Patterns, Functions & Algebra Number Operation Prob Solv; Reasoning & Proof Connections; Representation		Manufacturing, Health, Budgeting	
6–9 282–285	• Write, minimize and maximize an objective function Patterns, Functions & Algebra	Prob Solv; Reasoning & Proof Geometry & Spatial Sense Connections; Representation		Business, Manufacturing,	Farming

Planning and Resources

Lesson, Pages	Tools/Materials Needed	Trans-parency	Learning/Teaching Styles Options	Assignments: Basic Enriched	Additional Practice in SE	Reteaching, Extra Practice, Enrichment	Other Resources
6–1 244–247	geoboards tubber bands graph paper straightedges	Warm up 16 Trans TK-7–10, RF-21–23	Real World	B: 1–41, 48–64 E: 1–64	Page 252–253, 263, 273, 281, 286, 701	R: page 83 EP: page 73 E: page 97	SS: Teacher's Choice
6–2 248–251	graph paper straightedges	Warm up 16 Trans TK 7–10	ESL/LEP Challenge	B: 1–25, 31–33 E: 1–33	Page 252–253, 263, 273, 281, 286, 702	R: page 85 EP: page 75 E: page 99	SS: Teacher's Choice
6–3 254–257	paper/pencil	Warm up 16 Trans TK-7–10, RF-24		B: 1–22, 26–34 E: 1–34	Page 262–263, 273, 281, 286, 703	R: page 87 EP: page 77 E: page 101	SS: Teacher's Choice

6–4 258–261	graphing calculator graph paper straightedge	Warm up 17 Trans TK-7–10	ESL/LEP Challenge	B: 1–20, 23–29 E: 1–29	Page 262–263, 273, 281, 287, 703	R: page 89 EP: page 79 E: page 103	
6–5 264–267	paper/pencil	Warm up 17 Trans TK-7–10, RF-25		B: 1–20, 26–33 E: 1–33	Page 272–273, 281, 287, 704	R: page 91 EP: page 81 E: page 105	
6–6 268–271	Algeblocks	Warm up 17 Trans TK-7–10	Prior Knowledge Challenge	B: 1–25, 31–44 E: 1–44	Page 272–273, 281, 287, 704	R: page 93 EP: page 83 E: page 107	
6–7 274–275	pencil/paper	Warm up 18 Trans TK-7–10, RF-26		B: 1–12 E: 1–12	Page 280–281, 287	R: page 95 E: page 109	
6–8 276–279	graph paper straightedge	Warm up 18 Trans TK-7–13	Prior Knowledge	B: 1–24, 27–30 E: 1–30	Page 280–281, 287. 704	R: page 97 EP: page 85 E: page 111	
6–9 282–285	graphing calculator	Warm up 18 Trans TK-7–10	Real World	B: 1–24, 27–32 E: 1–32	Page 287	R: page 99 EP: page 87 E: page 113	

Planning and Pacing

Lesson, Pages	Lesson Title	45 min class	Assignments Basic, Enriched	90 min class	Assignments Basic, Enriched	___ min class	Assignments Basic, Enriched
6–1 244–247	AYR, Opener, Slope of a Line and Slope-intercept Form	Day 74	B: 1–41, 48–64 E: 1–64	Day 38	B: 1–41, 48–64 E: 1–64		B: 1–41, 48–64 E: 1–64
6–2 248–251	Parallel and Perpendicular Lines, R&PYS	Day 75–76	B: 1–25, 31–33 E: 1–33	Day 38–39	B: 1–25, 31–33 E: 1–33		B: 1–25, 31–33 E: 1–33
6–3 254–257	Write Equations for Lines	Day 77	B: 1–22, 26–34 E: 1–34	Day 39	B: 1–22, 26–34 E: 1–34		B: 1–22, 26–34 E: 1–34
6–4 258–261	Systems of Equations, R&PYS	Day 78–79	B: 1–20, 23–29 E: 1–29	Day 40	B: 1–20, 23–29 E: 1–29		B: 1–20, 23–29 E: 1–29
6–5 264–267	Solve Systems by Substitution	Day 80	B: 1–20, 26–33 E: 1–33	Day 41	B: 1–20, 26–33 E: 1–33		B: 1–20, 26–33 E: 1–33
6–6 268–271	Solve Systems by Adding and Multiplying, R&PYS	Day 81–82	B: 1–25, 31–44 E: 1–44	Day 41–42	B: 1–25, 31–44 E: 1–44		B: 1–25, 31–44 E: 1–44
6–7 274–275	Problem Solving Skills: Determinants and Matrices	Day 83	B: 1–12 E: 1–12	Day 42	B: 1–12 E: 1–12		B: 1–12 E: 1–12
6–8 276–279	Systems of Inequalities, R&PYS	Day 84–84	B: 1–24, 27–30 E: 1–30	Day 43	B: 1–24, 27–30 E: 1–30		B: 1–24, 27–30 E: 1–30
6–9 282–285	Linear Programming	Day 85	B: 1–24, 27–32 E: 1–32	Day 44	B: 1–24, 27–32 E: 1–32		B: 1–24, 27–32 E: 1–32
Review/ Assess		Day 86–88		Day 44–45			

	Chapter 1 Days	Chapter Cumulative Days
Yearly Pacing (45 min class)	16 days	88 days
Yearly Pacing (90 min class)	8 days	45 days

Assessment Options

Assessment in Student Edition	Assessment in Teacher's Edition	Pages in Assessment Book	Software Generated Assessment
Are You Ready?, pages 240–241; Writing Math, pages 247, 251, 257, 261, 266, 270, 278, 284; Mixed Review, pages 247, 251, 257, 261, 267, 271, 275, 279, 285; Check Understanding pages 244, 248, 254, 258, 264, 268, 276, 282; Mid-Chapter Quiz, page 263, Chapter Review, page 286; Chapter Assessment, page 288; Alternative Assessment, page 289; Cumulative Review, page 290; Cumulative Assessment, page 291	5-minute Warm ups, pages 244, 248, 254, 258, 264, 268, 274, 276, 282; Quick Assessment, pages 246, 250, 256, 260, 266, 270, 275, 278, 284; Scoring Rubrics, page 289	Mid-Chapter Quiz, page 87; Test Form A, pages 89, 91; Test Form B, pages 93, 95; Cumulative Test 97, 99; Math Journal prompt, page 101, 102;	Chapter 6

Refresher Skills

The skills on these two pages are skills that have been taught in previous math courses. Continuous review of basic math skills will make stronger math students. These skills are identified as necessary to be successful in Chapter 6.

Skills Correlation Chart

Skill	Lesson Number
Graphing Linear Equations	6-1, 6-2, 6-3, 6-4
Solving Equations	6-4, 6-5, 6-6, 6-8
Solving Compound Inequatlies	6-8

Vocabulary

linear equation   coordinate plane
compound inequality

Additional Answers

1. 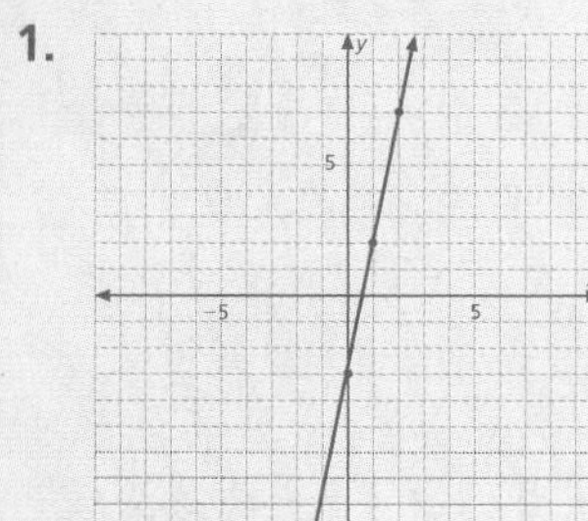

2. 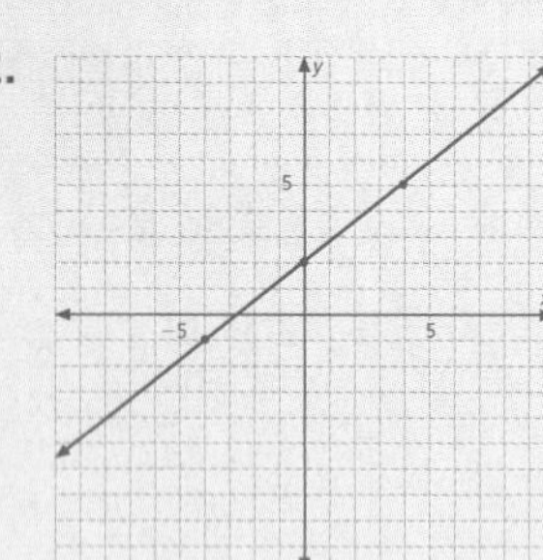

3. 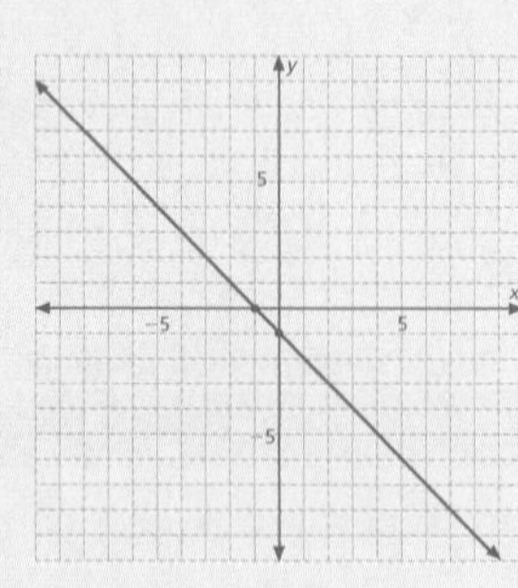

Are You Ready?

Refresh Your Math Skills for Chapter 6

The skills on these two pages are ones you have already learned. Read the examples and complete the exercises. For additional practice on these and more basic skills, see page 674.

Graphing Linear Equations

In this chapter you will learn more about graphing equations on the coordinate plane. It will be helpful to review some of these basic skills.

Example Graph the equation $3x - y = -5$.

Change the equation to the form $y = mx + b$:

$$3x - y = -5$$
$$-y = -3x - 5$$
$$y = 3x + 5$$

Make a table. Substitute each number for x to find the value of y.

x	y
1	8
0	5
−1	2

Plot the points and draw a line.

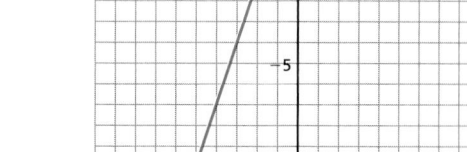

Graph each equation on the coordinate plane.

See additional answers.

1. $y = 5x - 3$
2. $3x + 8 = 4y$
3. $-4x - 3y = 4$
4. $3y - x = 7$
5. $-3x + 2y = -7$
6. $y = -4x - 3$
7. $3y = -5x + 7$
8. $\frac{1}{2}(x - 3) = 2y$
9. $-2x = y + 3$

Solving Equations

In this chapter you will be working with various forms of equations. Remember to always apply the Order of Operations.

Solve each equation.

10. $5a - 3 = 12$ 3
11. $-2a + 4 = 4a - 2$ 1
12. $3(a - 2) = -3a + 12$ 3
13. $-4(a - 2) = 16$ −2
14. $12 - 3a = 2a + 2$ 2
15. $-2(a + 4) = 16$ −12
16. $\frac{1}{2}(a + 3) - 4 = -2$ 1
17. $8 - 2a + 7 = 5$ 5
18. $4a - 6 - a = 2$ $\frac{8}{3}$
19. $3(a + 3) - 2 = 4(a - 6) + a$ $\frac{31}{2}$
20. $2(a - 4) + 6 = 3(a + 2) - 4$ 4
21. $a + 2(3a - 6) = 30 - 2(a + 3)$ 4
22. $-4(a - 2) + 8 = -2a + 3(a - 2)$ $\frac{22}{5}$

4. 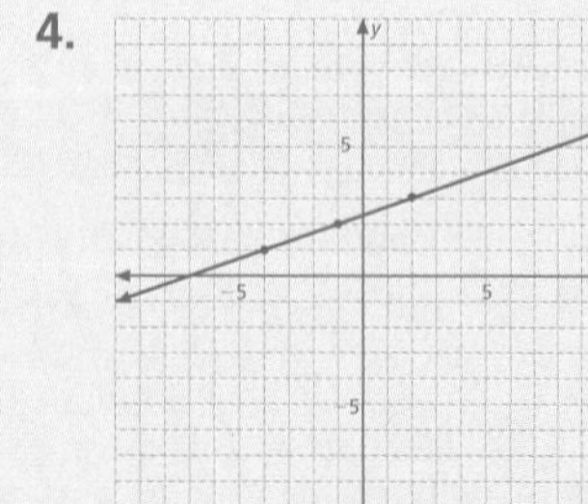

5. 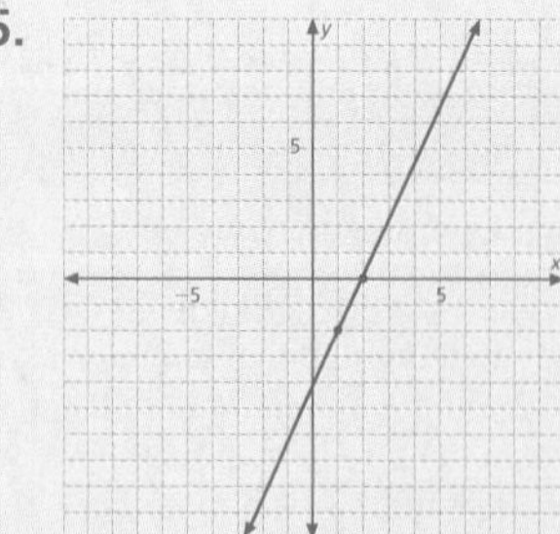

6. 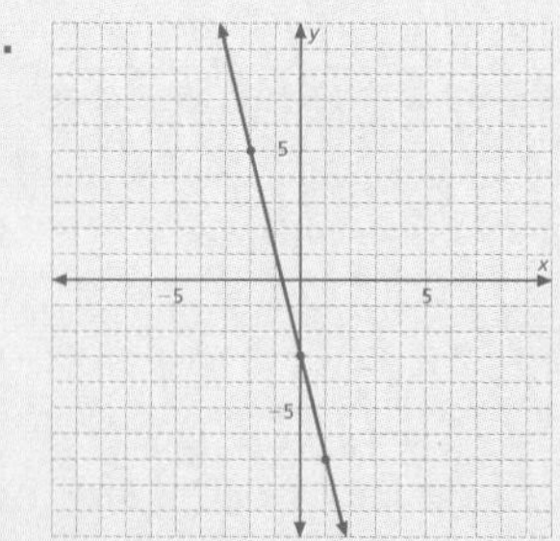

240 Chapter 6 **Linear Systems of Equations**

CHAPTER 6

Solving Compound Inequalities

Example Using the replacement set of the real numbers, graph the solution set for x 3 and $x > -5$.

Graph each inequality: x 3

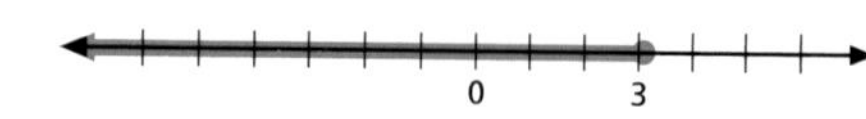

$x > -5$

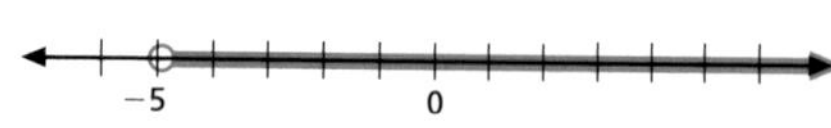

Graph the intersection:

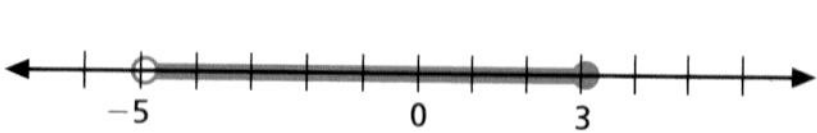

Graph the solution set for each compound inequality on a number line. Use the replacement set of the real numbers. See additional answers.

23. $x > 3$ and $x < 9$

24. $x \leq -1$ and $x > -8$

25. $x > -2$ and $x \geq 1$

26. $x \leq 5$ and $x \geq -3$

27. $x < 4$ and $x \leq 2$

28. $x > 5$ and $x < -2$ $\varnothing$

29. $x \geq -4$ and $x < -3$

30. $x \geq -2$ and $x < 5$

31. $x < 4$ and $x > 1$

Ordering Rational Numbers

To compare rational numbers that are in mixed formats, you need to convert all values to the same format. Decimal form is often the preferred way to compare and order numbers.

Convert the rational numbers to decimal form. Then write the numbers in order from least to greatest value.

32. $1.156, 1\frac{1}{8}, \frac{7}{6}, \sqrt{2}, -0.2$ $-0.2, 1\frac{1}{8}, 1.156, \frac{7}{6}, \sqrt{2}$

33. $0.7856, \frac{15}{16}, \frac{5}{7}, \sqrt{0.49}, -1, 1$ $-1, \sqrt{0.49}, \frac{5}{7}, 0.7856, \frac{15}{6}, 1$

34. $-6.42, -6\frac{5}{8}, -0.6, \frac{32}{5}, -6.042$ $-6\frac{5}{8}, -6.42, -6.042, -0.6, \frac{32}{5}$

35. $0.0025, 0.14500, 0.19, \frac{1}{7}, \sqrt{\frac{4}{5}}$ $0.0025, \frac{1}{7}, 0.14500, 0.19, \sqrt{\frac{4}{5}}$

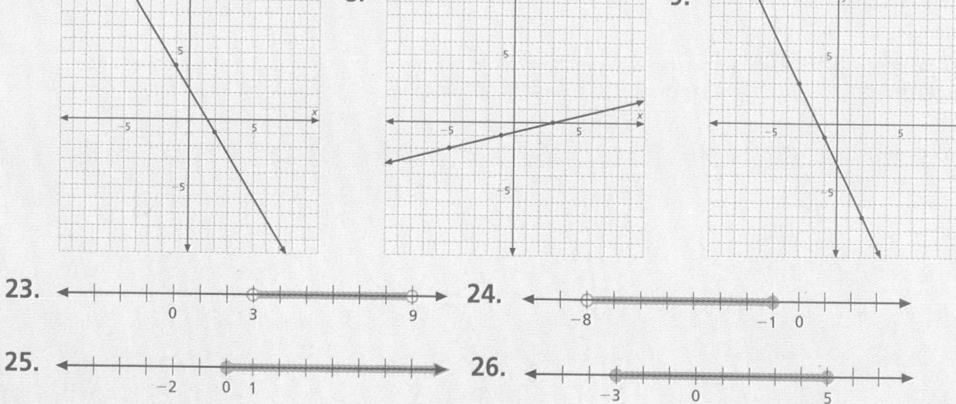

Chalkboard Examples

Graphing Linear Equations

Graph the equation $2x - 3y = -8$ on the coordinate plane. Check students' graphs. It should contain points $(-1, 2)$ and $(2, 4)$.

Solving Equations

Solve for x: $7x + 5 = 3x - 15$
$x = -5$

Solving Compound Inequalities

Describe the solution set for the compound inequality on a number line. Use the replacement set of real numbers.
$x > -2$ and $x \leq 5$ It should contain an open circle at -2 and closed circle at 5 and all the points between them.

Refresher Wrap-up

Quick Assessment

Ask the following questions to determine if students have mastered the basic skills reviewed on these pages.

1. When should you use a open dot when graphing an inequality on a number line? when the solution does not include the number represented by the dot or point
2. Describe the steps necessary for graphing a linear equation on a coordinate grid. Answers may vary depending on the students' experience with graphing.

Additional Practice

All topics: Refer to the Extra Practice Index on page 674 of this text.

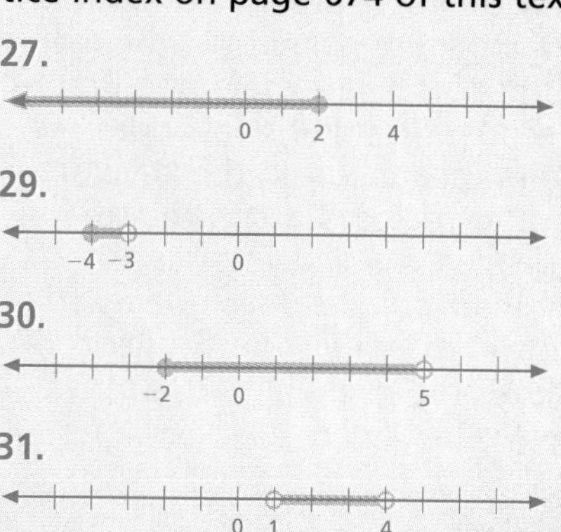

Chapter Opener

NCTM Standards/Strands

- Number & Operation
- Patterns, Functions, & Algebra
- Geometry & Spatial Sense
- Problem Solving
- Reasoning and Proof
- Communication
- Representation

Vocabulary

manufacturing	specification
assembly line	schematics
technician	

Theme Connections

Tables and graphs are used to organize data about manufacturing and product development. Tables provide a way to organize numerical information. Many types of data are kept in regards to manufacturing. Companies are concerned with quality, performance, sales, shipping and production quotas. Graphs are often used to show trends in sales and performance. Discuss how data and graphs are used in manufacturing.

Career Opportunities

Many careers related to manufacturing make use of data and graphs. Two are highlighted in the Math-Works features. Others include product developers, packagers, designers, and sales representatives.

- Precision Assembler, page 253
- Engineering Technicians, page 273

Internet Connection

Theme Activities

Learningmathmatters.com provides web addresses to search that help students gather information about the use of math in the real world, particularly data and graphs. To search for additional addresses, begin a search using the keyword *manufacturing*. Then within that search, use such key words as *design, product safety, quality control* and *assembly line*. Students can brainstorm in small groups other key words.

CHAPTER 6

Linear Systems of Equations

THEME: Manufacturing Industry

Think back through your day and mentally make a list of all the products that you have used. It's an eye-opening experience to realize that every one of those products was designed, tested and built by a manufacturer.

Once a new product is designed and thoroughly tested, engineers break the assembly of the product into steps. Each step will be performed by workers on an assembly line. The engineers, technicians, and line workers use math to ensure that each product is manufactured according to the design specifications.

- **Precision Assemblers** (page 253) work in factories to produce complex goods and must know how to use specialized measuring instruments. They read engineering diagrams called *schematics*.
- **Engineering Technicians** (page 273) design and develop new products to meet all required safety standards. Engineering technicians test designs for product quality and look for ways to keep costs down to make new products affordable to the consumer.

Internet Connection
www.learningmathmatters.com

Chapter Investigation

Go to learningmathmatters.com to locate additional information about manufacturing.

U.S. Goods–Imports and Exports, 1997
(in millions of dollars)

Category	Exports	Imports	Category	Exports	Imports
Airplanes	25,552	4,557	TVs, VCRs	24,093	36,771
Clothing	8,396	48,408	Fabric	8,975	11,951
Footwear	800	14,026	Vehicles	55,669	112,926
Paper	10,283	11,697	Toys and Games	3,827	17,374
Pottery	101	1,683	Scientific Instruments	24,039	13,969
Magnetic media	6,815	4,137	Printed Materials	4,605	2,871

Source: U.S. Department of Commerce

Data Activity: U.S. Goods–Imports and Exports

Use the three for Questions 1–4.

1. What is the dollar value of the clothing exports? $8,396,000
2. Much of the pottery sold in the United States is manufactured in other countries. How much greater is the value of pottery imports than pottery exports? $1,582,000
3. Of total exports and imports of scientific instruments, what percent is exported? Round to the nearest percent. 63%
4. For which category are exports and imports most nearly balanced? Paper

Chapter Investigation

Manufactured items go through an extensive design and development process. Many companies use focus groups made up of panels of consumers to find new ideas for product improvement. Before a new product is sold in stores, it is tested by selected consumers.

Working Together

Choose a simple product that you use frequently. Gather a focus group of four to five students and brainstorm ideas for product improvement. Finally, draw a model or prototype of the new product and determine how the improvements will add to the cost of the product. Use the Chapter Investigation logo to guide your group.

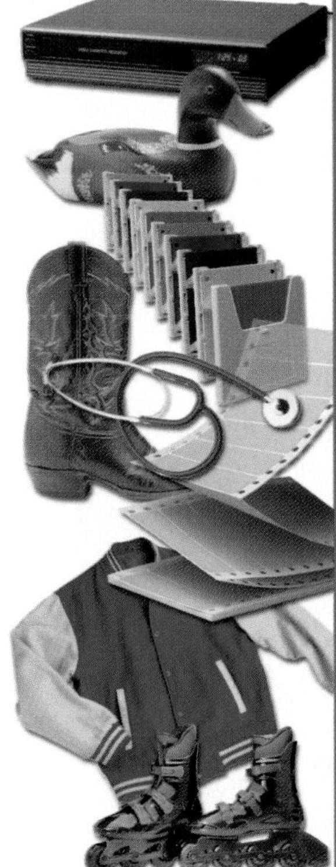

Project Planning Calendar

Name ______ Date ______

Chapter 6 Project Planning Calendar

Benchmarks
a. Brainstorm a list of products that the members of your group have used during the past week. Select a product that all agree is in need of improvement. Prepare a list of questions that could be used to find out whether consumers share your concerns. *(Lesson 6-2)*
b. Make a detailed drawing or build a prototype of your new product. Choose at least five selling points to emphasize in your marketing materials. *(Lesson 6-4)*
c. Compare your improved product to the original product. Write a report about how your improvements will make the product better. ... how the improve- ... increased

PROJECT GOAL
To create a drawing or prototype of an improved consumer product.

Group Project Planner

Name ______ Date ______

Chapter 6 Group Project Planner

Assignment ______ Objective ______

Group Members
1) ______
2) ______
3) ______
4) ______
5) ______

Assigned Roles

Deadlines Done

Data Activity

Discuss with students the importance of imports and exports to the U.S. economy. Point out that a country's natural resources often determine the types of goods that country must export and import. Discuss the importance of reading any titles or legends on a table or graph. For example, the amounts shown in this table represent millions of dollars. Have students rewrite several of the amounts in millions of dollars and determine the actual amount. Assign students Questions 1–4.

Extend the Data Activity

REAL WORLD CONNECTION Have students research what types of goods are imported and exported from their state and create a table similar to the one shown here.

Chapter Investigation

As an Overarching Problem

Display a few consumer products that students might use often. Ask students to choose one product and think about the features of the product they like and the features they would change. Have students share some of their ideas with the class. Students will continue to work on the investigation as they complete the exercises identified by the Investigation Logo that are found throughout the chapter. These exercises will guide students through the task described in *Working Together*. Encourage students to keep all of their work on the Investigation together. Have students use the suggestions in the Chapter Investigation Extension to summarize their work.

As a Chapter Project

The goal of this project is to create a drawing or prototype of an improved consumer product. Students can use the Team Project Planner on page 95 and the Project Planning Calendar on page 96 in the Enrichment text to complete the project. Benchmarks **a** and **b** should be completed after the lesson listed in parentheses has been studied. Benchmark **c** should be completed at the end of the chapter.

Lesson Planning

NCTM Standards/Strands

- Number & Operation
- Patterns, Functions, & Algebra
- Geometry & Spatial Sense
- Problem Solving
- Reasoning and Proof
- Connections
- Representation

Vocabulary

slope	rise
run	slope-intercept form

Materials Needed

geoboards	graph paper
rubber bands	straightedges

Lesson Resources

Warm-Up Transparency 16
Transparency TK-7–10, RF-21–23
Reteaching 6-1
Extra Practice 6-1
Enrichment 6-1

Assignment Guide

Basic: 1–41, 48–64
Enriched: 1–64

Getting Started

5-Minute Warm-up

Simplify.

1. $\frac{-3+(-2)}{9-3}$ $-\frac{5}{6}$
2. $\frac{4-(-7)}{-7-5}$ $-\frac{11}{12}$

Introduction to Lesson 6-1

It will be easier for students to find the lengths of the legs of their triangles if they make their original line segments on the diagonal. After students have completed this activity, have them share their findings. Discuss possible reasons why the ratios are the same.

6-1 Slope of a Line and Slope-intercept Form

Goals
- Find the slope of a line.
- Write the slope-intercept form of an equation and graph the equation.

Applications Manufacturing, Finance, Recreation

Work with a partner.

Place identical line segments on geoboards. Lengthen or shorten one of the line segments. Make a right triangle on each geoboard using the line segment as the hypotenuse. Write the ratio of the length of the vertical side to the length of the horizontal side for each triangle. How do the ratios compare? The ratios are the same.

Build Understanding

Many examples of slope exist in everyday life. We measure the **slope**, or steepness, of roads, stairs and ramps. In mathematics, an important aspect of a line is its slope. We measure slope as a ratio of the vertical change (rise or fall) to the horizontal change (run).

$$\text{slope} = \frac{\text{rise}}{\text{run}} = \frac{\text{vertical change (change in } y\text{-coordinates)}}{\text{horizontal change (change in } x\text{-coordinates)}}$$

You can find the slope of a line on a coordinate plane by counting the units of change between the coordinates of any two points on the line. The vertical change, or change in y, is found by determining the difference of the y-coordinates. Likewise, the horizontal change, or change in x, is found by determining the difference in corresponding x-coordinates of the same two points.

The slope of a line segment containing the point $A(x_1, y_1)$ and $B(x_2, y_2)$ is

$$m = \frac{y_2 - y_1}{x_2 - x_1} \quad \text{or} \quad \frac{y_1 - y_2}{x_1 - x_2}$$

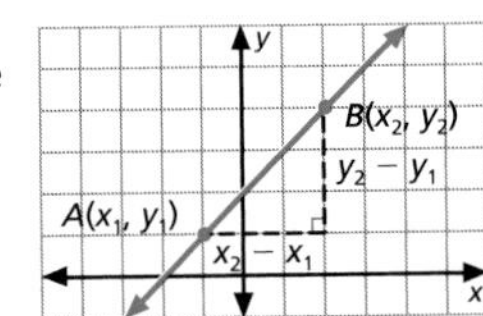

Example 1

Find the slope of $\overleftrightarrow{AB}$ containing points $A(-1, 2)$ and $B(3, -4)$.

Solution

$$m = \frac{y_2 - y_1}{x_2 - x_1} = \frac{-4-2}{3-(-1)} = \frac{-6}{4} = \frac{-3}{2}$$

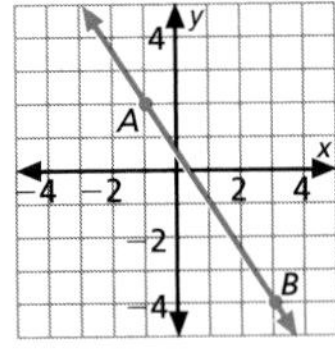

Example 1 shows that a line with a *negative* slope falls from left to right. As the value of x increases, the value of y decreases. A line with a *positive* slope rises from left to right. As the value of x increases, the value of y increases.

Extend the Lesson

MATH JOURNAL Have students write about the different types of information they could be given that would enable them to find the slope of a line. For each different type, they should write a short explanation or give an example of how they would use it.

If you know one point on a line and the slope of the line, you can also graph the line.

Example 2

Graph the line that passes through $G(1, 1)$ and has a slope of $\frac{-3}{4}$.

Solution

First, plot the point $G(1, 1)$. The slope is $\frac{-3}{4}$, so from G go *down* 3 units (because the rise is -3) and *right* 4 units (because the run is 4). The point is $(1 + 4, 1 - 3)$ or $(5, -2)$.

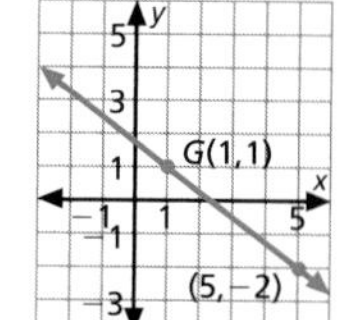

Draw a line through $(1, 1)$ and $(5, -2)$.

Check Understanding

A horizontal line has a slope of 0 because all *y*-values are the same. The slope of a vertical line is undefined because all *x*-values are the same.

Find the slope of a line containing these points:

1. $(-2, 3)$ and $(4, 3)$
2. $(1, 3)$ and $(1, 2)$

0, undefined

In a linear equation, you can find the **x-intercept**, or point at which the line crosses the x-axis, by substituting 0 for y and solving for x. You can find the **y-intercept**, or point at which the line crosses the y-axis, by substituting 0 for x and solving for y. You can then draw a line through these two points to graph the equation. Writing a linear equation in **slope-intercept** form is another way to easily find the slope and the y-intercept of the line. In the slope-intercept form, $y = mx + b$, m represents the slope of the line and b represents the y-intercept.

Example 3

Find the slope and y-intercept for the line with the equation $5x + 6y = 6$.

Solution

$5x + 6y = 6$

$6y = 5x + 6$ First, write the equation in slope-intercept form.

$y = \frac{-5}{6x} + 1$

$m = \frac{-5}{6}$ and $b = 1$

The slope is $\frac{-5}{6}$, and the line crosses the y-axis at coordinates $(0, 1)$.

Technology Note

Write a program for your graphing calculator that can find the slope of a line when the coordinates of two points are known.

1. Prompt for the values of x_1, x_2, y_1 and y_2. You may want to call these values *X, Y, A* and *B*.
2. Enter an expression that compares the ratio of the rise to the run:

   (B − Y) / (A − X)
3. Run the program to find the slope of a line passing through points (1, 4) and (−2, 3). Did your calculator correctly display a slope of .333 or $\frac{1}{3}$?

Example 4

MANUFACTURING Production figures for an assembly plant are represented by a line with a slope of $\frac{1}{2}$ and a y-intercept of -1. Find the equation of the line. Then draw the graph of the line.

Solution

$m = \frac{1}{2}$ and $b = -1$

$y = mx + b$

$y = \frac{1}{2}x - 1$ slope-intercept form

To draw the graph, start at point $(0, -1)$. Then using a slope of $\frac{1}{2}$, locate a point 1 unit up and 2 units right at $(2, 0)$. Draw a line through the points.

Chalkboard Examples

Supplementary Example 1

Find the slope CD of containing points $C(-2, 2)$ and $D(2, 4)$. $\frac{1}{2}$

Supplementary Example 2

Graph the line that passes through $M(3, 2)$ and has a slope of $-\frac{1}{3}$. Check students' graphs.

Supplementary Example 3

Find the slope and y-intercept for the line with the equation $3x + 4y = 12$. slope: $-\frac{3}{4}$; y-intercept: $(0, 3)$

Supplementary Example 4

Sales figures for a new product are represented by a line with a slope of 2 and a y-intercept of -3. Find the equation of the line. $y = 2x - 3$

Reteaching Worksheet 6-1

Name ____________ Date ____________

RETEACHING 6-1

SLOPE OF A LINE AND SLOPE-INTERCEPT FORM

The slope of a line is measured as a ratio.

$$\text{Slope} = \frac{\text{rise}}{\text{run}} = \frac{\text{vertical change (change in } y\text{-coordinates)}}{\text{horizontal change (change in } x\text{-coordinates)}}$$

The slope of a line containing point $A(x_1, y_1)$ and point $B(x_2, y_2)$ is $\frac{y_2 - y_1}{x_2 - x_1}$.

Example 1

Find the slope of $\overline{AB}$ containing points $A(-4, 7)$ and $B(4, -7)$.

Solution

Substituting the values into the formula, $m = \frac{y_2 - y_1}{x_2 - x_1}$, you find $\frac{-7 - 7}{4 - (-4)} = \frac{-14}{8} = -\frac{7}{4}$.

Example 2

Find the slope and the y-intercept for the line $-2x + y = 4$.

Solution

Write the equation in slope-intercept form: $y = 2x + 4$, where $m = 2$ and $b = 4$. The slope (m) is 2 and the y-intercept (b) is 4.

EXERCISES

Find the slope of the line containing the given points.

1. $E(2, 6)$, $F(4, 4)$ **−1**
2. $G(3, 1)$, $H(5, 1)$ 0
3. $I(-1, -2)$, $J(1, 2)$ 2
4. $K(7, -1)$, $L(6, 2)$ −3
5. $M(8, 4)$, $N(4, 2)$ $\frac{1}{2}$
6. $O(3, 4)$, $P(-1, -2)$ $\frac{3}{2}$
7. $Q(5, 2)$, $R(5, 8)$ undefined
8. $S(9, 6)$, $T(3, -2)$ $\frac{4}{3}$
9. $U(-1, -2)$, $V(-3, -4)$ 1

Find the slope and y-intercept for each line.

10. $y = -2x$ $m =$ −2 $b =$ 0
11. $3x + y = 10$ $m =$ −3 $b =$ 10
12. $x = \frac{1}{3}$ $m =$ undefined $b =$ none
13. $y = x + 6$ $m =$ 1 $b =$ 6
14. $4x + 2y = 12$ $m =$ −2 $b =$ 6
15. $y = 3x + 1$ $m =$ 3 $b =$ 1
16. $\frac{1}{3}x + \frac{1}{3}y = 1$ $m =$ −1 $b =$ 3
17. $y = \frac{1}{4}x + \frac{1}{2}$ $m =$ $\frac{1}{4}$ $b =$ $\frac{1}{2}$
18. $y = 8$ $m =$ 0 $b =$ 8
19. $4x + 2y = 5$ $m =$ −2 $b =$ $\frac{5}{2}$

84 RETEACHING *LESSON 6-1*

Extend the Lesson

REAL-WORLD CONNECTION On graph paper, have students draw a cross section of part of a staircase showing the rise and the run of each step to scale. Have them connect a point at the edge of each step to form a line. Then have them find the slope of the staircase by finding the slope of that line. Have students who used different staircases compare the slopes.

Lesson Wrap-up

Quick Assessment

Ask the following questions to determine if students understand the content presented in this lesson.

1. How will the slope of a line when it passes through the first quadrant compare to its slope when it passes through the third quadrant? It will be the same, because the slope of a line remains constant.
2. How could you write an equation for a line if you are given only its graph? By counting the units up and across from one point to another on the graph, you could find its slope. You can read the y-intercept from the graph. You could then substitute the slope and the y-intercept in the slope-intercept form $y = mx + b$.

Assignment Guide

Basic: 1–42, 48–64
Advanced: 1–64
Additional Practice: See Extra Practice Index on page 674.

Try These Exercises

Find the slope of the line containing the given points.

1. $A(3, 5)$ and $B(0, 6)$ $-\frac{1}{3}$
2. $P(-2, -6)$ and $Q(2, 1)$ $\frac{7}{4}$
3. $G(-2, 1)$ and $H(-2, 10)$ undefined

Graph the line that passes through the given point P and has the given slope. See additional answers.

4. $P(-1, 2)$, slope $= -1$
5. $P(4, -1)$, slope $= \frac{2}{3}$

Find the slope and y-intercepts for each line.

6. $-5x + 13y = 15$ $m = \frac{5}{13}, b = \frac{15}{13}$
7. $y = -6x + 4$ $m = -6, b = 4$
8. $y = 5x - 1$ $m = 5, b = -1$
9. $y = -\frac{1}{10}x + \frac{1}{5}$ $m = -\frac{1}{10}, b = \frac{1}{5}$

Find an equation of the line with the given slope and y-intercept.

10. slope $= -1$, y-intercept $= 3$ $y = -x + 3$
11. slope $= 4$, y-intercept $= -2$ $y = 4x - 2$

Graph each equation. See additional answers.

12. $-2x + 3y = 3$
13. $2x - y = 4$
14. $y = 1$

Practice Exercises

Find the slope of each line.

15. 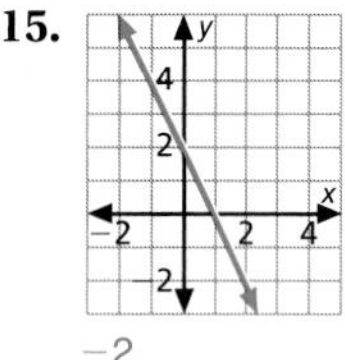

-2

16. 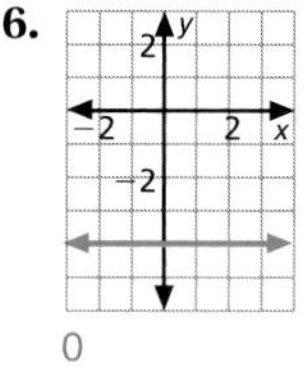

0

17. 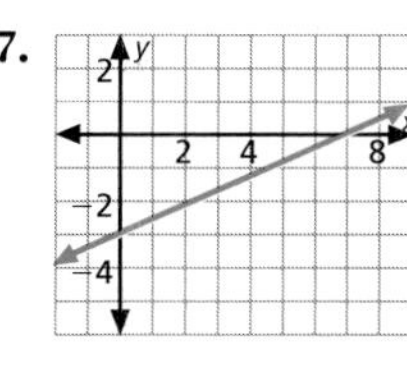

$\frac{3}{7}$

18. Find the slope of the line containing the points $A\left(\frac{1}{2}, \frac{-3}{4}\right)$ and $B\left(\frac{2}{3}, \frac{5}{8}\right)$. $\frac{33}{4}$

GRAPHING Plot the point $P(1, 1)$. Then graph the lines on the same axes that pass through P with the given slope. See additional answers.

19. $m = 0$
20. $m = 1$
21. $m = -1$

DATA FILE For Exercises 23–25, use the Data Index on pages 632–633 to find the location of a table showing the desired weight range based on corresponding height for men and women.

22. Use y as the *lower male* weight (in pounds) and x as the height (in inches). Plot the corresponding weights for $x = 62, 63, 64$. Connect the points. Label them $P_1(62, w_1)$, $P_2(63, w_2)$, and $P_3(64, w_3)$. See additional answers.
23. Determine the slope for the line segment between P_1 and P_2 and between P_2 and P_3. Are the points collinear? 2; yes
24. Repeat Exercise 23 using y as the *lower female* weight (in pounds) and x as the height (in inches). Plot $(62, w_1)$, $(63, w_2)$, $(64, w_3)$. Connect the points. See additional answers.

Find the slope and y-intercept for each line.

25. $y = 4x$ $m = 4, b = 0$
26. $-6x - y = 12$ $m = -6, b = -12$
27. $x = -9$ $m =$ undefined, $b =$ none
28. $-\frac{x}{2} + 2y = 0$ $m = \frac{1}{4}, b = 0$
29. $x = 2$ $m =$ undefined, $b =$ none
30. $3y = -x + \frac{1}{2}$ $m = \frac{-1}{3}, b = \frac{1}{6}$

246 Chapter 6 **Linear Systems of Equations**

Additional Answers

4. 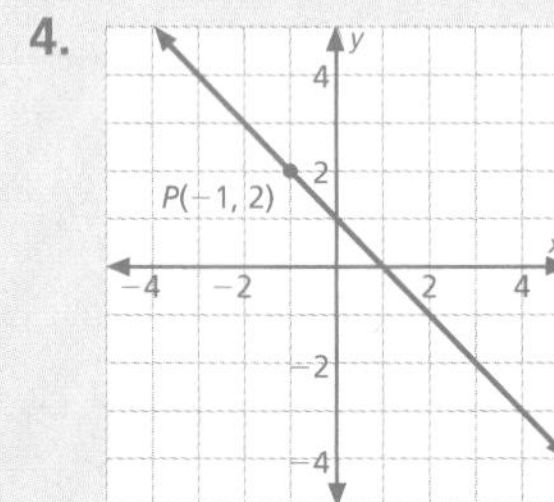

5. 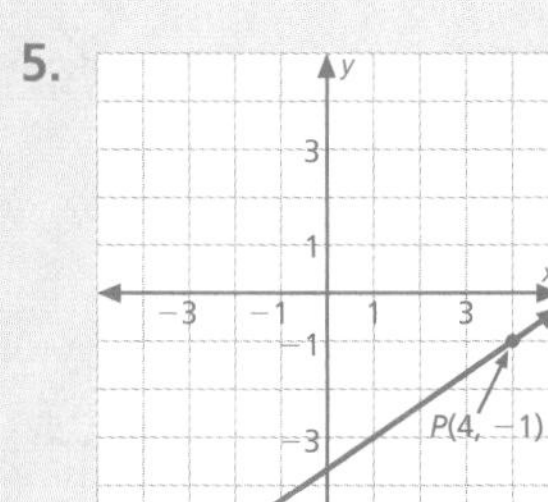

12. 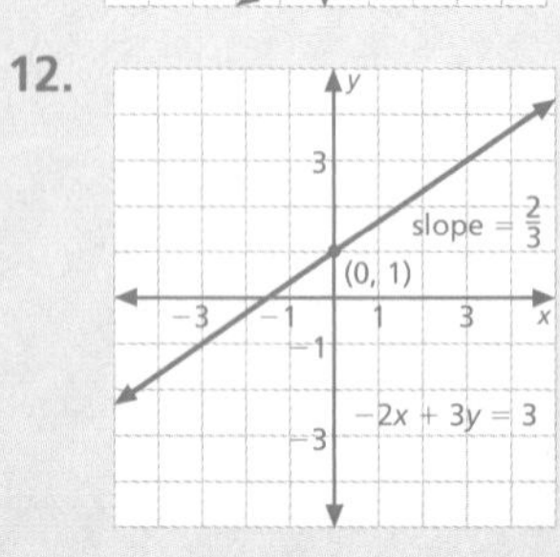

13. 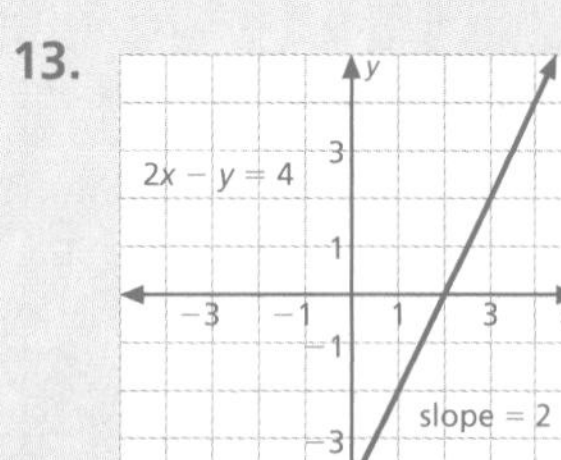

14. 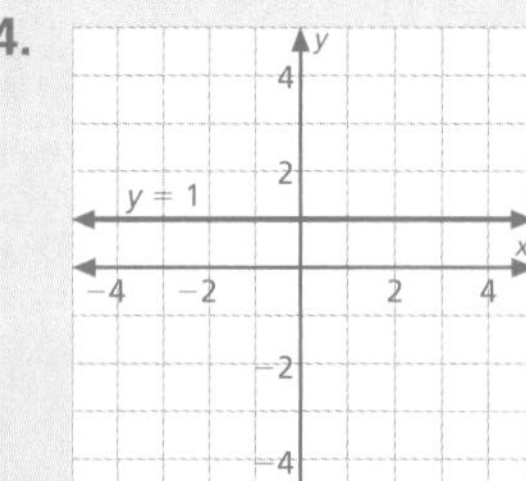

19–21. 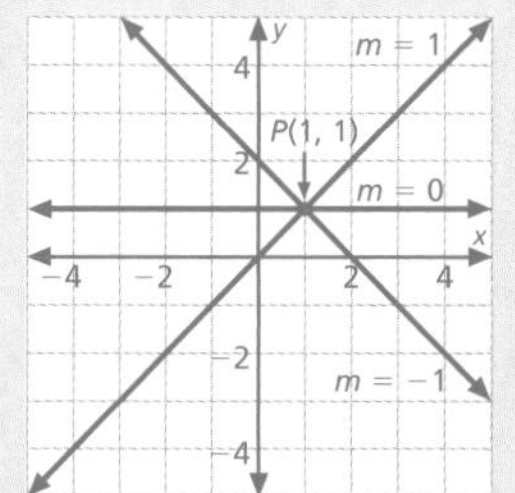

Write an equation of the line with the given slope and y-intercept.

31. $m = -5, b = 4$ $y = -5x + 4$

32. $m = 2, b = -\frac{3}{4}$ $y = 2x - \frac{3}{4}$

33. $m = 6, b = 0$ $y = 6x$

34. $m = 0, b = -1$ $y = -1$

CALCULATOR Graph each equation. See additional answers.

35. $2x + 3y = -6$

36. $-x + y = 3$

37. $y = -3x + 2$

FINANCE Dave receives a salary of $200 a week plus a commission of 10% of his weekly sales. An equation $y = mx + b$ represents Dave's weekly earnings. The y-intercept is Dave's base salary. The slope of the line is his commission.

38. Write an equation representing Dave's weekly earnings. $y = 0.1x + 200$

39. Graph the equation. See additional answers.

40. If Dave sells $1500 of goods for one week, what is his salary for the week? $350

41. What value of goods does Dave need to sell in one week to receive a weekly earning of $500? $3,000

Extended Practice Exercises

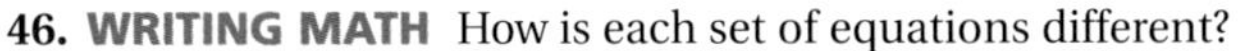

42. RECREATION A ski resort is being built a new ski slope. The desired slope of the new hill is 0.8. For that hill, the horizontal distance from the crest of the hill to the bottom is 400 ft. What should the height of the hill be? 320 ft

Find the slope and y-intercept for each line.

43. $y = ax + r - m$
slope $= a$, y-intercept $= r - m$

44. $\frac{b}{a}x + \frac{c}{b}y = 1$
slope $= -\frac{b^2}{ac}$, y-intercept $= \frac{b}{c}$

45. The standard form of an equation is $Ax + By + C = 0$. Write the slope-intercept form of the equation. $y = -\frac{A}{B}x - \frac{C}{B}$

46. WRITING MATH How is each set of equations different?

a. $y = 2x$ and $y = -2x$
different slopes

b. $y = x - 3$ and $y = x + 3$
different y-intercepts

47. YOU MAKE THE CALL A line passes through $P(-5, 2)$ and $Q(-5, 5)$. Geoff says the slope of the line must be 0 because there is no change in x. Does Geoff's reasoning make sense?
No, since there is no change in x, the line must be vertical and the slope is undefined.

Mixed Review Exercises

Find each unit rate or unit price. Round answers to the nearest tenth if necessary. (Lesson 5-1)

48. $3.87 for 18 oz of soup
21.54 per oz

49. 496 mi in 8 h
62 mi/h

50. 487.5 ft in 325 steps
1.5 ft/step

51. $5.95 for 5 lb nuts
$1.19 per lb

52. 426.4 mi on 13 gal gas
32.8 mi/gal

53. 9 mi in 2 h 15 min
4 mi/h

54. 128 cookies in 8 sacks
16 cookies per sack

55. $96.67 for 58 jars jam
$1.67 per jar

56. 528 fish in 6 tanks
88 fish per tank

Find all the solutions for each equation. (Lesson 2-4)

57. $|x| - 6 = 8$
14, −14

58. $-24 = -3 \times |x|$
8, −8

59. $|x| + 4 = 8$
4, −4

60. $3 \times |x| = 21$
7, −7

61. $\frac{1}{2}|x| = 13$
26, −26

62. $36 = 4 \times |x|$
9, −9

63. $|x| - 7 = 8$
15, −15

64. $-48 = -8 \times |x|$
6, −6

Extra Practice Worksheet 6-1

Name ______ Date ______

EXTRA PRACTICE 6-1
SLOPE OF A LINE AND SLOPE-INTERCEPT FORM

EXERCISES

Find the slope of each line.

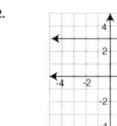

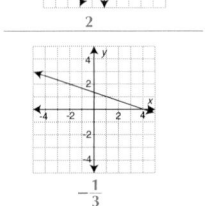

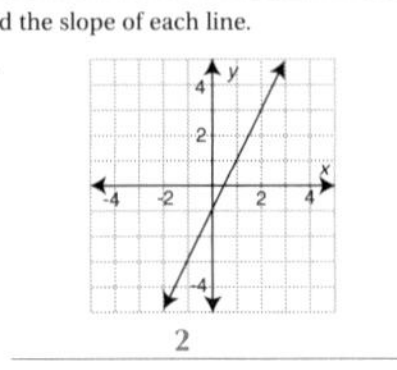

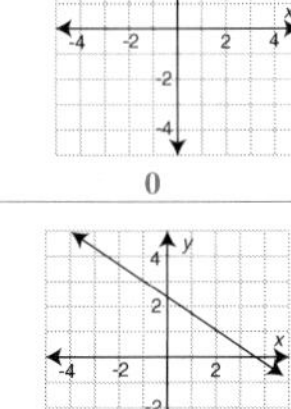

1. 2
2. 0
3. $-\frac{1}{3}$
4. $-\frac{2}{3}$

Find the slope of the line containing the given points.

5. $A(-2, 2)$ and $B(0, 4)$ 1
6. $R(5, -1)$ and $S(2, 3)$ $-\frac{4}{3}$
7. $M(4, -8)$ and $N(6, -1)$ $\frac{7}{2}$
8. $F(-5, 5)$ and $G(-1, 6)$ $\frac{1}{4}$

Find the slope and y-intercept for each line.

9. $y = 2x + 1$ $m = 2, b = 1$
10. $-4x + y = 2$ $m = 4, b = 2$
11. $y = -1$ $m = 0$, no y-intercept
12. $3x - 3y = 6$ $m = 1, b = -2$

Write an equation of the line with the given slope and y-intercept.

13. $m = -1, b = 3$ $y = -x + 3$
14. $m = 4, b = -2$ $y = 4x - 2$

Graph each equation on your own paper. Check students' graphs.

15. $-x + y = 1$
16. $y = -2x - 5$
17. $4x - 2y = -4$

74 EXTRA PRACTICE *LESSON 6-1*

Enrichment Worksheet 6-1 not shown.

22. and 24.

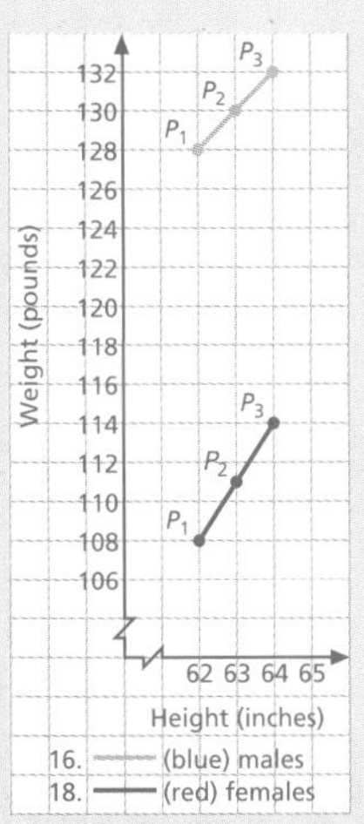

35.

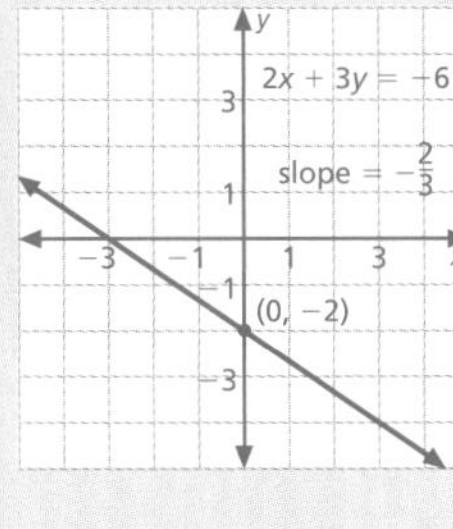

36.

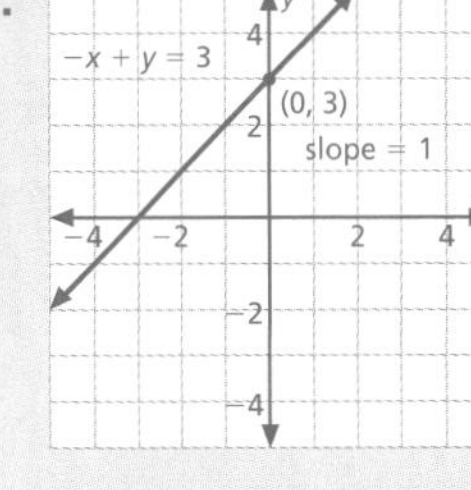

37.

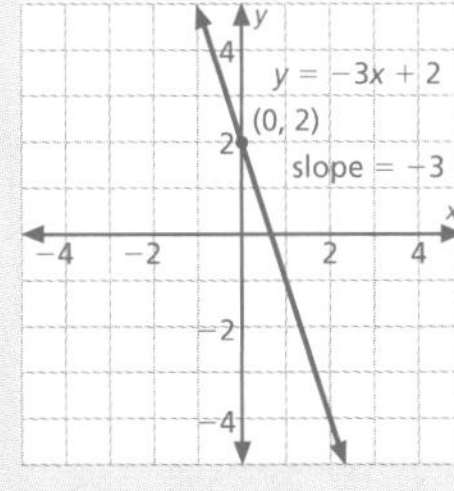

39.

Lesson Planning

NCTM Standards/Strands

- Number & Operation
- Patterns, Functions, & Algebra
- Geometry & Spatial Sense
- Problem Solving
- Reasoning and Proof
- Connections
- Representation

Vocabulary

parallel perpendicular

Materials Needed

graph paper straightedges

Lesson Resources

Warm-Up Transparency 16
Transparency TK-7–10
Reteaching 6-2
Extra Practice 6-2
Enrichment 6-2

Assignment Guide

Basic: 1–25, 31–33
Enriched: 1–33

Getting Started

5-Minute Warm-Up

Find the negative reciprocal of each number.

1. $\frac{1}{2}$ -2
2. $-\frac{3}{4}$ $\frac{4}{3}$
3. -5 $\frac{1}{5}$
4. 3 $-\frac{1}{3}$

Introduction to Lesson 6-2

Have students graph each equation on the same coordinate grid. After they have answered the questions independently, have them share their observations with the class.

6-2 Parallel and Perpendicular Lines

Goals ■ Use slope to determine whether two lines are parallel or perpendicular.

Applications Electronics, Mapmaking, Manufacturing

GRAPHING Use the zoom feature of your graphing utility to change the viewing window to a square. Then graph the following equations on the same screen.

$y = 2x - 1$	$y = 2x$	$y = 2x + 1$
$y = -\frac{1}{2}x + 1$	$y = -\frac{1}{2}x + 2$	$y = -\frac{1}{2}x - 2$

Which lines seem to be parallel? Which seem to be perpendicular? What do you notice about the slopes and y-intercepts of parallel and perpendicular lines? Each of the two sets is parallel. Their slopes are the same. Each line of one set is perpendicular to all the lines of the other set. The product of their slopes is -1.

Build Understanding

From the activity above, you can see that two lines with the same slope and different y-intercepts are parallel. Conversely, two lines are parallel if they have the same slope and different y-intercepts. Vertical lines are parallel and have slopes that are undefined.

Two lines are perpendicular if the product of their slopes is -1. Conversely, if the slopes of two lines are negative reciprocals of each other, then the lines are perpendicular.

Example 1

Find the slope of a line parallel (m_1) to the given line and a line perpendicular (m_2) to the given line.

a. The line containing points $A(-2, 5)$ and $B(0, -1)$.

b. The line containing points $C(4, -1)$ and $D(-5, -1)$.

c. $x = 2$

Think Back

Two rational numbers that have a product of -1 are called negative reciprocals.

Solution

	m	m_1	m_2
a.	$\frac{-1-5}{0-(-2)} = \frac{-6}{2} = -3$	-3	$\frac{-1}{-3} = \frac{1}{3}$
b.	$\frac{-1-(-1)}{-5-4} = \frac{0}{-9} = 0$	0	undefined
c.	undefined	undefined	0

Teaching Strategies

ESL/LEP Have students find five examples of parallel and perpendicular lines within the classroom. Encourage them to look for examples where more than one line is parallel or perpendicular to a given line.

Example 2

Determine whether each pair of lines is *parallel, perpendicular,* or *neither.*

a. $7x + 2y = 14$
$7y = 2x - 5$

b. $-5x + 3y = 2$
$3x - 5y = 15$

c. $2x - 3y = 6$
$\frac{8}{3}x - 4y = 4$

Solution

Rewrite each equation in slope-intercept form and find the slope of each line.

a. $7x + 2y = 14 \rightarrow y = -\frac{7}{2}x + 7 \qquad m_1 = -\frac{7}{2}$

$7y = 2x - 5 \rightarrow y = \frac{2}{7}x - \frac{5}{7} \qquad m_2 = \frac{2}{7}$

$m_1 \neq m_2$

$m_1 \cdot m_2 = -\frac{7}{2} \cdot \frac{2}{7} = -1$

Because $m_1 \cdot m_2 = -1$, the lines are perpendicular.

b. $-5x + 3y = 2 \rightarrow y = \frac{5}{3}x + \frac{2}{3} \qquad m_1 = \frac{5}{3}$

$3x - 5y = 15 \rightarrow y = \frac{3}{5}x - 3 \qquad m_2 = \frac{3}{5}$

$m_1 \neq m_2$

$m_1 \cdot m_2 = \frac{5}{3} \cdot \frac{3}{5} = 1 \neq -1$

Because $m_1 \neq m_2$ and $m_1 \cdot m_2 \neq -1$, the lines are neither parallel nor perpendicular.

c. $2x - 3y = 6 \rightarrow y = \frac{2}{3}x - 2 \qquad m_1 = \frac{2}{3}$

$\frac{8}{3}x - 4y = 4 \rightarrow y = \frac{2}{3}x + 1 \qquad m_2 = \frac{2}{3}$

$m_1 = m_2$

Because $m_1 = m_2$, the lines are parallel.

Example 3

ELECTRONICS A manufacturer of circuit boards uses a grid system to insert connecting pins. The design for a board requires pins at points $M(0, 1)$, $N(3, 0)$, $P(5, 6)$ and $Q(2, 7)$. When the pins are connected, will $MNPQ$ be a parallelogram?

Solution

Plot the points and draw the quadrilateral. Find the slope of $\overline{NP}$ and $\overline{MQ}$, and the slope of $\overline{MN}$ and $\overline{QP}$.

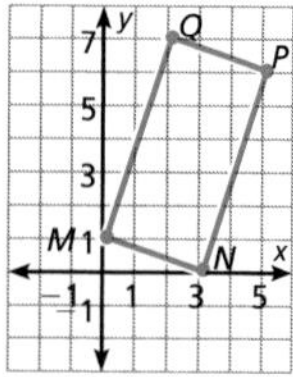

$m_{NP} = \frac{6 - 0}{5 - 3} = \frac{6}{2} = 3; \qquad m_{MN} = \frac{0 - 1}{3 - 0} = \frac{-1}{3}$

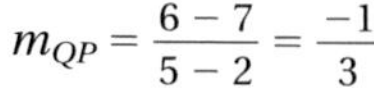

$m_{MQ} = \frac{7 - 1}{2 - 0} = \frac{6}{2} = 3; \qquad m_{QP} = \frac{6 - 7}{5 - 2} = \frac{-1}{3}$

Learning Styles

TACTILE/KINESTHETIC LEARNERS Have students make two parallel line segments on a geoboard and find the slope of each by counting the number of units in the rise and run from one end of the segment to the other. Then have them compare slopes. Repeat this activity for other parallel line segments and then for perpendicular line segments.

Chalkboard Examples

Supplementary Example 1

Find the slope of a line parallel to a line containing points $E(-1, 3)$ and $F(0, -2)$ and a line perpendicular to it. -5; $\frac{1}{5}$

Supplementary Example 2

Determine whether each pair of lines is *parallel, perpendicular,* or *neither.*

a. $3x + 2y = 5$ parallel
$2y = -3x + 6$

b. $2x - y = 8$ neither
$3y = -x + 2$

c. $6y = -4x + 7$ perpendicular
$6x - 4y = 8$

Supplementary Example 3

Connecting pins are also inserted at points $A(2, 2)$, $B(6, 2)$, $C(2, -2)$ and $D(6, -2)$. When the pins are connected, what figure will be formed? **square**

Reteaching Worksheet 6-2

Name ______________________ Date ____________

RETEACHING 6-2

PARALLEL AND PERPENDICULAR LINES

Two lines are *parallel* when they have the same slope and different y-intercepts; $m_1 = m_2$.
Two lines are *perpendicular* if the product of their slopes is -1: $m_1 \cdot m_2 = -1$.

Slopes of perpendicular lines are negative reciprocals of each other: $m_1 = -\frac{1}{m_2}$.

Example 1

Line AB contains points $A(4, 0)$ and $B(-3, 2)$. Find the slope of a line

a. parallel to $\overleftrightarrow{AB}$.
b. perpendicular to $\overleftrightarrow{AB}$.

Solution

The slope of $\overleftrightarrow{AB}$ is $\frac{0 - 2}{4 - (-3)} = \frac{-2}{7}$.

a. Parallel lines have the same slope, so $m = -\frac{2}{7}$.

b. Since slopes of perpendicular lines are negative reciprocals, the slope is $\frac{7}{2}$.

Example 2

Determine whether the pair of lines is *parallel, perpendicular* or *neither.*

$4x + 2y = -8$
$6y = 12x + 36$

Solution

Write each in slope-intercept form.
$4x - 2y = -8 \rightarrow y = 2x + 4 \qquad m_1 = 2$
$6y = 12x + 36 \rightarrow y = 2x + 6 \qquad m_2 = 2$
Since $m_1 = m_2$, the lines are parallel.

EXERCISES

Find the slope of a line parallel to the given line and of a line perpendicular to the given line.

1. The line containing points $G(2, 0)$ and $H(-2, -4)$. Parallel 1 Perpendicular -1
2. The line containining points $I(0, 5)$ and $J(8, -4)$ Parallel $-\frac{9}{8}$ Perpendicular $\frac{8}{9}$
3. $3x + 4y = 5$ Parallel $-\frac{3}{4}$ Perpendicular $\frac{4}{3}$
4. $7y = x + 7$ Parallel $\frac{1}{7}$ Perpendicular -7

Determine whether each pair of lines is *parallel, perpendicular* or *neither.*

5. The line containing points $K(-1, 0)$ and $L(2, 6)$.
The line containing points $M(4, 2)$ and $N(3, 0)$. parallel
6. $-5x + 7 = 3y$
$y = 3$ neither
7. $3x - 6 = 2y$
$\frac{3}{2}x + y = 5$ parallel

86 RETEACHING *LESSON* 6-2

Lesson Wrap-up

Quick Assessment

Ask the following questions to determine if students understand the content presented in this lesson.

1. How many different equations are possible for lines that are parallel or perpendicular to a given line? Explain. An infinite number; the slope would remain the same in each, but there are an infinite number of possible y-intercepts.
2. Without using a ruler or protractor, how could you determine whether two lines graphed on the same coordinate grid are parallel, perpendicular or neither? Use the coordinates of two points on each line to find their slopes, then compare slopes. If they are equal, the lines are parallel. If they are negative reciprocals, the lines are perpendicular. Otherwise, the lines are neither.

Assignment Guide

Basic: 1–25, 31–33
Advanced: 1–33
Additional Practice: See Extra Practice Index on page 674.

Additional Answers

27.

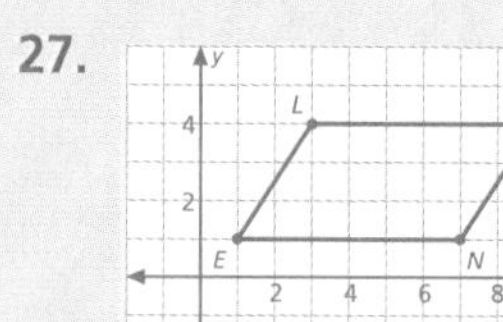

28. slope of $\overline{LI}$ = slope of $\overline{NE} = 0$; slope of $\overline{IN}$ = slope of $\overline{EL} = \frac{3}{2}$; by definition a quadrilateral with opposite sides parallel is a parallelogram.
29. slope of $\overline{LN} = -\frac{3}{4}$ and slope of $\overline{EI} = \frac{3}{8}$; $\left(-\frac{3}{4}\right)\left(\frac{3}{8}\right) \neq -1$, so $\overline{LN}$ is not perpendicular to $\overline{EI}$.

Because the slopes of $\overline{NP}$ and $\overline{MQ}$ are the same, $\overline{NP} \parallel \overline{MQ}$. Because the slopes of $\overline{MN}$ and $\overline{QP}$ are the same, $\overline{MN} \parallel \overline{QP}$. Thus, opposite sides of the quadrilateral are parallel. Therefore, $MNPQ$ is a parallelogram.

Try These Exercises

Find the slope of a line parallel to the given line and a line perpendicular to the given line.

1. The line containing points $M(8, 3)$ and $N(-1, 5)$ $-\frac{2}{9}, \frac{9}{2}$
2. The line containing points $A(0, 5)$ and $B(-6, 5)$ 0, undefined
3. The line containing points $(2, -1)$ and $(4, 7)$ $4, -\frac{1}{4}$
4. The line containing points $P(3, -5)$ and $Q(3, 2)$ undefined, 0

Determine whether each pair of lines is *parallel*, *perpendicular*, or *neither*.

5. The line containing points $A(9, 5)$ and $B(-1, 6)$
   The line containing points $C(-8, 2)$ and $D(12, 4)$ neither
6. $-6x - 14y = 5$
   $7x = 3y + 21$ perpendicular
7. $-5x + 2y = 10$
   $3x - 5y = 15$ neither
8. **CARTOGRAPHY** A road atlas is laid out on a grid. A mapmaker notes intersections at points $A(1, 1)$, $B(2, 0)$, $C(3, 2)$ and $D(2, 3)$. If the points are connected, will $ABCD$ form a parallelogram? yes

Practice Exercises

Find the slope of a line parallel to the given line and a line perpendicular to the given line.

9. The line containing $(6, -3)$ and $(4, 5)$ $-4, \frac{1}{4}$
10. The line containing $(-3, 2)$ and $(-7, 8)$ $-\frac{3}{2}, \frac{2}{3}$
11. $10x + 12y = -6$ $-\frac{5}{6}, \frac{6}{5}$
12. $-3y + 2x = 7$ $\frac{2}{3}, -\frac{3}{2}$
13. $y = -7x + 6$ $-7, \frac{1}{7}$

Determine whether each pair of lines is *parallel*, *perpendicular*, or *neither*.

14. The line containing points $P(9, -4)$ and $Q(-2, 7)$
    The line containing points $M(14, 8)$ and $N(19, 13)$ perpendicular
15. $-4x + 6y = 6$ neither
    $5x - 2y = 10$
16. $8y = 12x - 20$ parallel
    $9x = 6y + 5$
17. **MANUFACTURING** A machine is used to apply a heat-sensitive transfer to a product. The transfer is placed by locating three points on a grid. The points are $A(0, 0)$, $B(5, 3)$ and $C(7, -3)$. Do the connected points form a right triangle? no

Math: Who, Where, When

Mathematicians as early as the fifth century have formulated ideas and postulates regarding parallel and perpendicular lines. Euclid defined parallel lines in his *parallel postulate* as straight coplanar lines that will not meet each other regardless of how far they extend. Mathematicians since then have tried to replace Euclid's definition. In the early 1800s, a Scottish mathematician, John Playfair, derived a parallel postulate used in most present-day geometry texts. It states that through a given point not on a given line, only one line can be drawn parallel to the given line.

Extend the Lesson

ONGOING ASSESSMENT Have students write a set of equations for four lines, the intersection of which would create a rectangle. Then have students graph the lines to check their work. Check students' work.

30. Drawings may vary, but one quadrilateral should be a rhombus and the other a square. Slopes of sides and diagonals will vary, but parallel sides should have the same slope and perpendicular lines should have slopes that are negative reciprocals of each other.

Determine whether each pair of lines is *parallel, perpendicular,* or *neither.*

18. $1.5x - 4.5y = x + 15$ perpendicular
$2 - 3x = 13 + \frac{1}{3}y$

19. $\frac{2}{3}y - \frac{1}{4}x = 1$ parallel
$15 + 3x = 15 + 8y$

Determine the value of *x* so that the line containing the given points is parallel to another line whose slope is also given.

20. $A(x, 9)$ and $B(-5, 6)$ $\frac{-13}{2}$
slope $= -2$

21. $M(-1, 7)$ and $N(x, -1)$ 23
slope $= -\frac{1}{3}$

Determine the value of *y* so that the line containing the given points is perpendicular to another line whose slope is also given.

22. $A\left(\frac{-5}{6}, y\right)$ and $B\left(\frac{2}{3}, 1\right)$ $-\frac{7}{2}$
slope $= \frac{-1}{3}$

23. $C\left(\frac{1}{3}, -\frac{2}{5}\right)$ and $D(1, y)$ $-\frac{2}{5}$
slope is undefined

24. WRITING MATH Given the coordinates of four points which are the vertices of a polygon, how can you determine if the polygon is a rectangle?

25. CHAPTER INVESTIGATION Brainstorm a list of products that the members of your group have used during the past week. Select a product that all agree is in need of improvement. Prepare a list of questions that could be used to find out whether consumers share your concerns.

24. Compare slopes to determine whether opposite sides are parallel. Compare slopes at one vertex to determine whether connecting sides are perpendicular.

Extended Practice Exercises

26. The vertices of a triangle are $A(a, 0)$, $B(a + b, c)$ and $C(c + a + b, c - b)$. Determine if $\triangle ABC$ is a right triangle. yes

Plot the given points. Then use your graphs for Exercises 28–30. See additional answers.

27. $L(3, 4)$; $I(9, 4)$; $N(7, 1)$; $E(1, 1)$

28. Use what you know about the slopes of lines to show that figure *LINE* is a parallelogram.

29. Draw diagonals *LN* and *EI*. Use the slopes of the diagonals to show whether they are perpendicular to each other.

30. On graph paper, draw two different shaped quadrilaterals in which the diagonals are perpendicular to each other. Name each type of quadrilateral. List the slope of each side and each diagonal for each quadrilateral.

Mixed Review Exercises

Find the area of the shaded region of each figure. (Lesson 5-2)

31.

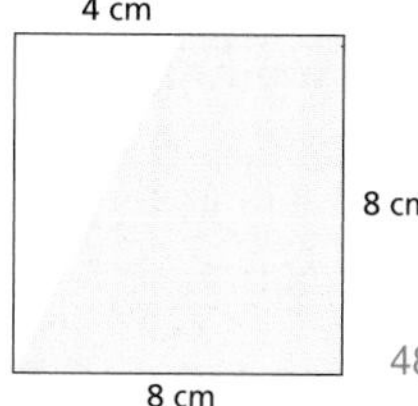

48 cm²

32.

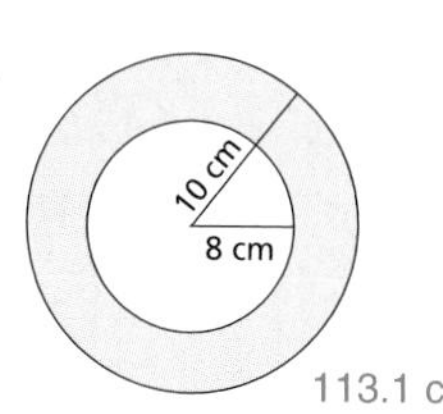

113.1 cm²

33.

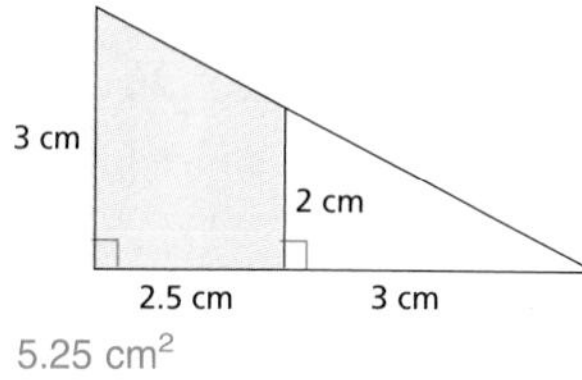

5.25 cm²

Extra Practice Worksheet 6-2

Name ____________ Date ____________

EXTRA PRACTICE 6-2
PARALLEL AND PERPENDICULAR LINES

EXERCISES

Find the slope of a line parallel to the given line and a line perpendicular to the given line.

1. The line containing (2, −4) and (5, 1). $\frac{5}{3}, -\frac{3}{5}$
2. The line containing (0, 3) and (−2, −6). $\frac{9}{2}, -\frac{2}{9}$
3. $5x - 5y = 10$ 1, −1
4. $-2x + y = 2$ 2, $-\frac{1}{2}$
5. $y = \frac{3}{4}x - 2$ $\frac{3}{4}, -\frac{4}{3}$

Determine whether each pair of lines is *parallel, perpendicular* or *neither*.

6. The line containing points X(8, −3) and Y(−4, −6). The line containing points R(1, 4) and S(−3, −3). neither
7. The line containing points A(2, −1) and B(3, −4). The line containing points C(1, 6) and D(3, 0). parallel
8. $y = 2x + 1$, $2y = -x - 1$ perpendicular
9. $4x - y = 2$, $8x - 2y = -6$ parallel
10. Plot and connect the points X(−2, 1), Y(−1, −1) and Z(3, 1) on your own paper. Determine if XYZ is a right triangle. yes
11. Plot and connect the points A(−4, 3), B(1, 4), C(2, 0) and D(−3, −1) on your own paper. Determine if ABCD is a rectangle. no

Determine the value of *x* so that the line containing the given points is parallel to another line whose slope is also given.

12. A(x, 5) and B(−4, 3) slope = −1 −6
13. R(3, −5) and S(1, x) slope = $-\frac{1}{2}$ −4

Determine the value of *y* so that the line containing the given points is perpendicular to another line whose slope is also given.

14. A(4, y) and B(−2, 6) slope = 2 3
15. R(2, −4) and S(y, 8) slope = 1 −10

76 EXTRA PRACTICE *LESSON 6-2*

Enrichment Worksheet 6-2

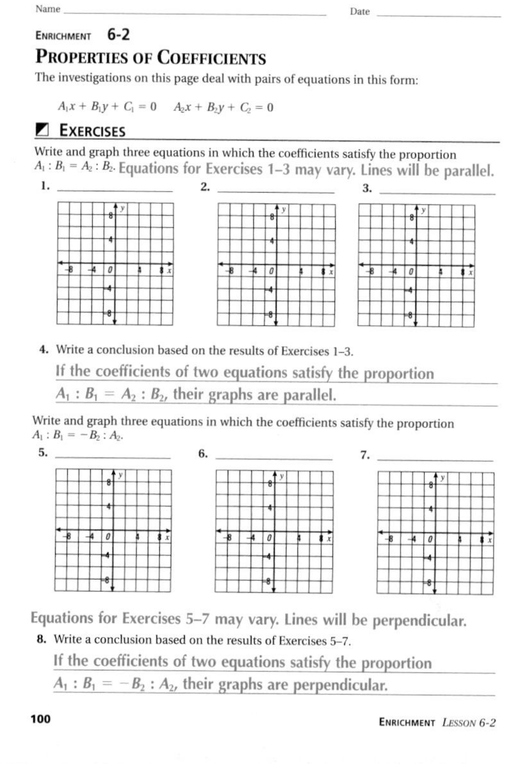

Name ____________ Date ____________

ENRICHMENT 6-2
PROPERTIES OF COEFFICIENTS

The investigations on this page deal with pairs of equations in this form:

$A_1x + B_1y + C_1 = 0 \quad A_2x + B_2y + C_2 = 0$

EXERCISES

Write and graph three equations in which the coefficients satisfy the proportion $A_1 : B_1 = A_2 : B_2$. Equations for Exercises 1–3 may vary. Lines will be parallel.

1. ______ 2. ______ 3. ______

4. Write a conclusion based on the results of Exercises 1–3.
If the coefficients of two equations satisfy the proportion $A_1 : B_1 = A_2 : B_2$, their graphs are parallel.

Write and graph three equations in which the coefficients satisfy the proportion $A_1 : B_1 = -B_2 : A_2$.

5. ______ 6. ______ 7. ______

Equations for Exercises 5–7 may vary. Lines will be perpendicular.

8. Write a conclusion based on the results of Exercises 5–7.
If the coefficients of two equations satisfy the proportion $A_1 : B_1 = -B_2 : A_2$, their graphs are perpendicular.

100 ENRICHMENT *LESSON 6-2*

Teaching Strategies

CHALLENGE Find the equation of the line that lies midway between the graphs of $2x + 3y = 5$ and $2x + 3y = 8$. $2x + 3y = 6.5$

Skills Practice

Vocabulary Review

Lesson 6-1

slope — rise
run — slope-intercept form

Lesson 6-2

parallel — perpendicular

Assignment Guide

All students: 1–67
Additional Practice: Refer to the Extra Practice Index on page 674 of this text.

Additional Answers

10.

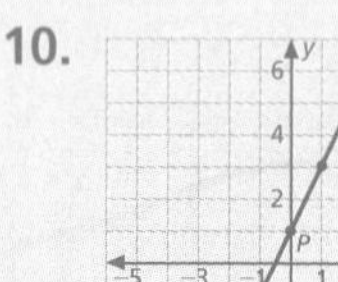
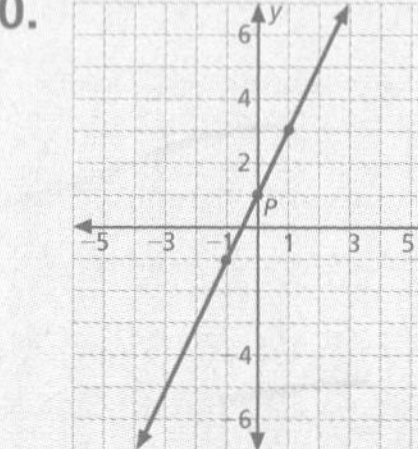

11.

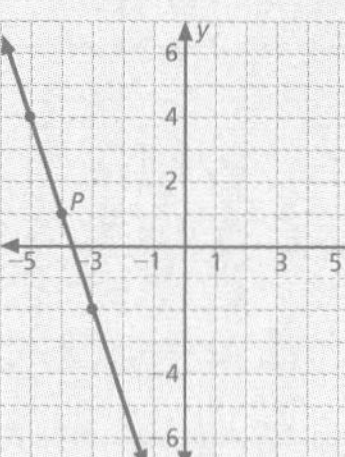

12.

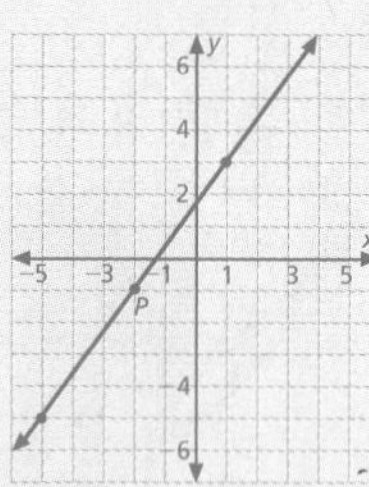

13.

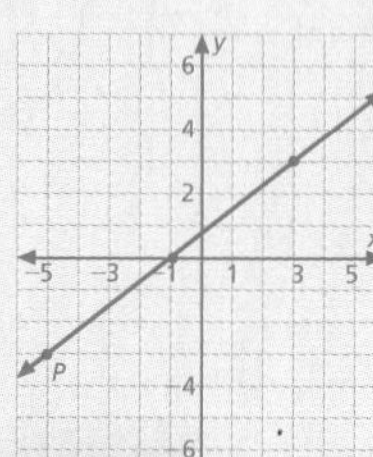

14.

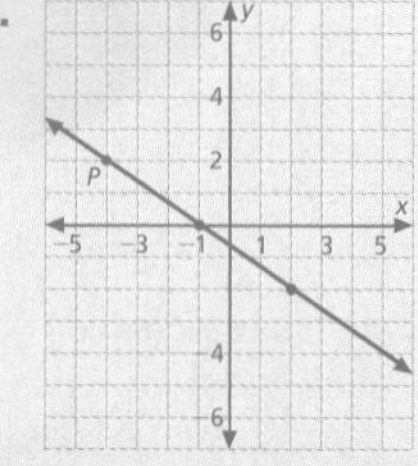

15.

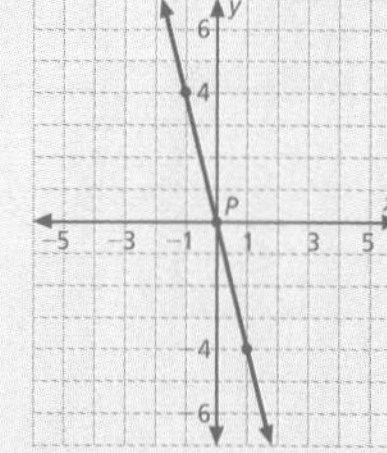

30.

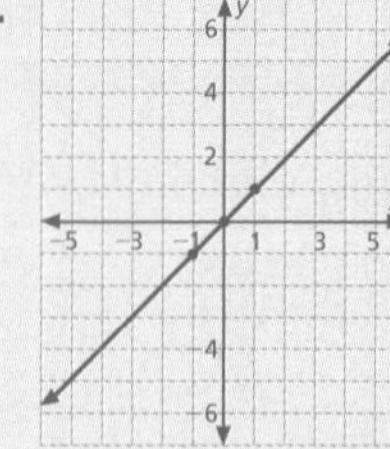

31.

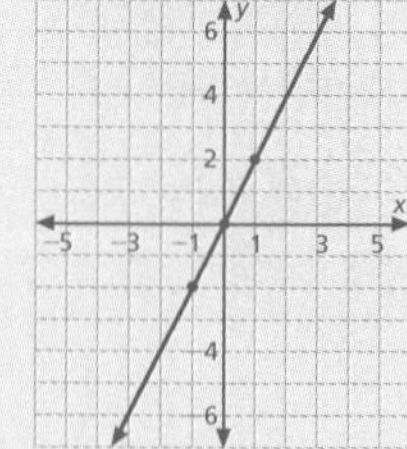

32.

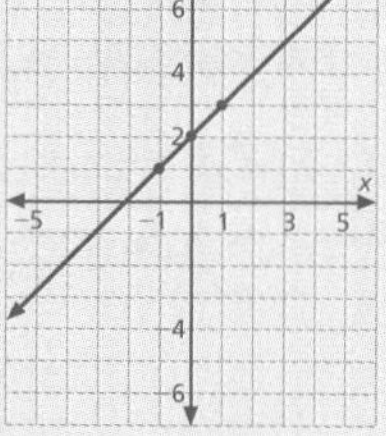

Review and Practice Your Skills

Practice ■ Lesson 6-1

Find the slope of the line containing the given points.

1. $A(2, 4)$ and $B(1, 3)$ 1
2. $C(3, 2)$ and $D(5, 6)$ 2
3. $E(0, 5)$ and $F(4, 0)$ $-\frac{5}{4}$
4. $G(-3, 5)$ and $H(-5, 6)$ $-\frac{1}{2}$
5. $Z(6, -2)$ and $Y(-3, 2)$ $-\frac{4}{9}$
6. $X(-2, -2)$ and $W(-12, -8)$ $\frac{3}{5}$
7. $V(5, 7)$ and $U(-4, 7)$ 0
8. $T(0, 0)$ and $S(-6, 5)$ $-\frac{5}{6}$
9. $R(-2.4, 6)$ and $Q(-1, -2.4)$ -6

Graph the line that passes through the given point P and has the given slope. See additional answers.

10. $P(0, 1)$, $m = 2$
11. $P(-4, 1)$, $m = -3$
12. $P(-2, -1)$, $m = \frac{4}{3}$
13. $P(-5, -3)$, $m = 0$
14. $P(-4, 2)$, $m = -\frac{2}{3}$
15. $P(0, 0)$, $m = -4$

Find the slope and y-intercept for each line.

16. $y = 3x + 2$ $m = 3, b = 2$
17. $y = -5x + 9$ $m = -5, b = 9$
18. $y = x$ $m = 1, b = 0$
19. $y = \frac{2}{3}x - 1$ $m = \frac{2}{3}, b = -1$
20. $x = -5$ slope undefined, no y-intercept
21. $2x + 3y = 12$ $m = -\frac{2}{3}, b = 4$
22. $5x - 2y = 20$ $m = \frac{5}{2}, b = -10$
23. $\frac{4}{3}x + 4y = 1$ $m = -\frac{1}{3}, b = \frac{1}{4}$

Write an equation of the line with the given slope and y-intercept.

24. $m = 4, b = -4$ $y = 4x - 4$
25. $m = 0, b = 15$ $y = 15$
26. $m = \frac{1}{3}, b = -2$ $y = \frac{1}{3}x - 2$
27. $m = -\frac{7}{2}, b = -3$ $y = -\frac{7}{2}x - 3$
28. $m = 2, b = \frac{1}{2}$ $y = 2x + \frac{1}{2}$
29. $m = -20, (0, 5)$ $y = -20x + 5$

Graph each equation. See additional answers.

30. $y = x$
31. $y = 2x$
32. $y = x + 2$
33. $5x - y = 10$
34. $x + 5y = 10$
35. $5x - y = 0$

36. Determine the equation of the line that contains the points $C(-5, -3)$ and $D(5, 9)$. $y = \frac{6}{5}x + 3$

Practice ■ Lesson 6-2

Find the slope of a line parallel to the given line and a line perpendicular to the given line.

37. The line containing $(4, -9)$ and $(-3, 5)$ $-2, \frac{1}{2}$
38. The line containing $(8, 13)$ and $(15, -6)$ $-\frac{19}{7}, \frac{7}{19}$
39. The line containing $(-3, -6)$ and $(5, 2)$ $1, -1$
40. The line containing $(-5, 10)$ and $(-5, -17)$ undefined, 0
41. The line containing $(2, -9)$ and $(15, -14)$ $-\frac{5}{13}, \frac{13}{5}$
42. The line containing $(6, 6)$ and $(-7, 7)$ $-\frac{1}{13}, 13$
43. $y = -3x + 10$ $-3, \frac{1}{3}$
44. $y = \frac{-3}{5}x - 7$ $-\frac{3}{5}, \frac{5}{3}$
45. $-12x - 5y = -1$ $-\frac{12}{5}, \frac{5}{12}$

Determine whether each pair of lines is *parallel, perpendicular,* or *neither.*

46. The line containing $(-3, -1)$ and $(4, -3)$
The line containing $(13, 9)$ and $(27, 5)$ parallel
47. The line containing $(7, -2)$ and $(-8, -2)$
The line containing $(0, 14)$ and $(0, -31)$ perpendicular
48. $y = \frac{1}{4}x - 6$ and $x + 4y = -8$ neither
49. $y = \frac{7}{2}x + 17$ and $7x - 2y = 28$ parallel
50. $x + y = 10$ and $y = 4 - x$ parallel
51. $\frac{1}{5}x + \frac{2}{5}y = -2$ and $2y = x - 14$ neither

Practice ■ Lessons 6-1–6-2

Find the slope of the line containing the given points. (Lesson 6-1)

52. $A(0, 6)$ and $B(-9, 0)$ $\frac{2}{3}$

53. $C(-3, 14)$ and $D(-3, 8)$ undefined

54. $E(-4, -8)$ and $F(-8, -4)$ -1

55. $G(-725, 630)$ and $H(435, -240)$ $\frac{3}{4}$

Find the slope and y-intercept for each line. (Lesson 6-1)

56. $y = x - 4.5$ $m = 1, b = -4.5$

57. $y = -\frac{1}{4}x + 7$ $m = -\frac{1}{4}, b = 7$

58. $y = 8 - 2x$ $m = -2, b = 8$

59. $9x - 3y = 12$ $m = 3, b = -4$

60. $\frac{1}{2}x + 2y = 0$ $m = -\frac{1}{4}, b = 0$

61. $-\frac{5}{6}x + \frac{10}{3}y = 30$ $m = \frac{1}{4}, b = 9$

Find the slope of a line parallel to the given line and a line perpendicular to the given line. (Lesson 6-2)

62. The line containing $(9, -4)$ and $(-5, 3)$ $-\frac{1}{2}, 2$

63. The line containing $(7, 12)$ and $(16, -7)$ $-\frac{19}{9}, \frac{9}{19}$

64. The line containing $(-5, -6)$ and $(1, 0)$ $1, -1$

65. The line containing $(-8, 13)$ and $(-8, -22)$ undefined, 0

66. The line containing $(4, -18)$ and $(30, -28)$ $\frac{-5}{13}, \frac{13}{5}$

67. The line containing $(6.8, 6.5)$ and $(2.8, 7)$ $-\frac{1}{8}, 8$

MathWorks
Workplace Knowhow

Career – Precision Assemblers

Precision assemblers work in factories to produce manufactured goods. These workers must be able to follow complex instructions and complete many steps with very little error.

Precision assemblers work on the assembly of aircraft, automobiles, computers and many other electrical and electronic devices. They work closely with engineers to build prototypes.

Precision assemblers must know how to read engineering schematics and how to use specialized and precise measuring instruments. An incorrect measurement may cause a product to work improperly or to endanger its users.

1. You must be certain that the slope of an airplane wing is exactly 0.13. If the wing is 5 feet long, how many inches should it rise from the fuselage to the wing tip? 7.8 in.
2. You are mounting a Global Positioning System (GPS) onto the dashboard of a car. The slope of the GPS must be identical to the slope of the dashboard. Otherwise, the GPS will not work properly. The dashboard has a depth of 18 inches wide and rises 6 inches. Find the slope of the dashboard. $\frac{1}{3}$
3. The GPS unit has a depth of $4\frac{3}{8}$ inches. How much should it rise in order to match the slope of the dashboard? $1\frac{11}{24}$ in.

Chalkboard Examples

Example from Lesson 6-1

Find the slope of the line containing the points $C(-1, 2)$ and $D(-3, -6)$. 4

Example from Lesson 6-2

Within the set of lines, name the pair of lines that are parallel and the pair that are perpendicular.

a. the line containing points $A(4-1)$ and $B(6,0)$

b. the line containing points $C(-2, -1)$ and $D(-1, 1)$

c. the line containing points $E(3, 5)$ and $F(4, 3)$

d. the line containing points $G(0, -4)$ and $H(2, 0)$ parallel: b and d; perpendicular: a and c

Career Opportunity

Describe the kind of work precision assemblers do and the types of companies that hire these workers. Have students discuss the importance of measurement, geometry and algebraic thinking in doing this type of job. Explain that precision assemblers must be able to follow complex directions and interpret mathematical symbols as they read schematics. Students should answer Questions 1–3 to better understand how precision assemblers use mathematics in performing their job.

Students who are interested in learning more about this profession can go to learningmathmatters.com. Guidance Counselors should have information about school and training requirements.

Teaching Strategies

COOPERATIVE LEARNING Have each student draw four lines on a coordinate grid to form a quadrilateral. Then, on another sheet of paper or index card, have them describe the lines using the following form: Line A contains the points $(-2, 1)$ and $(4, 2)$. Have students exchange line descriptions and determine the type of figure drawn without graphing the lines.

33.

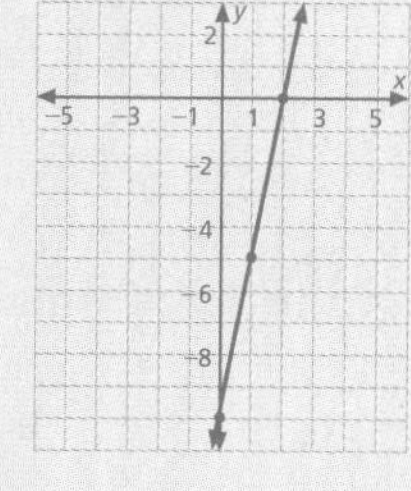

34.

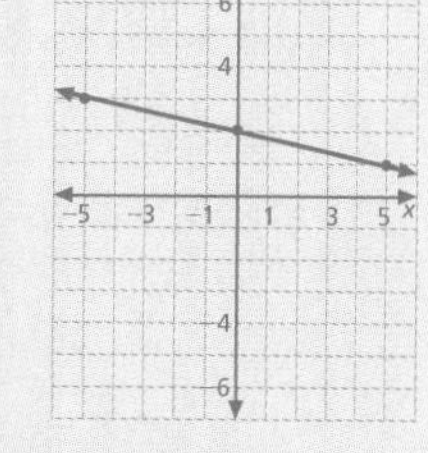

35.

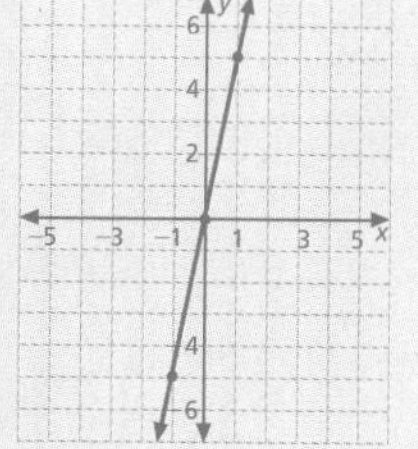

Chapter 6 **Review and Practice Your Skills** 253

NCTM Standards/Strands
- Number & Operation
- Patterns, Functions, & Algebra
- Geometry & Spatial Sense
- Problem Solving
- Reasoning and Proof
- Communication
- Connections
- Representation

Vocabulary
point-slope form

Materials Needed
paper/pencil

Lesson Resources
Warm-Up Transparency 16
Reteaching 6-3
Extra Practice 6-3
Enrichment 6-3
Transparency TK-7–10, RF-24

ASSIGNMENT GUIDE
Basic: 1–22, 26–34
Enriched: 1–34

Getting Started

5-MINUTE WARM-UP
Find the slope and y-intercept of each line.

1. $y = 4x - 2$ 4; −2
2. $y = -\frac{1}{3}x + 5$ $-\frac{1}{3}$; 5
3. $y = 3x$ 3; 0

Introduction to Lesson 6-3
After students have competed the table, have volunteers share their strategies with the class. Then ask which line of the table was easiest/hardest to complete and why.

6-3 Write Equations for Lines

Goals ■ Write equations for lines in slope-intercept and point-slope forms.

Applications Product design, Fitness, Real Estate

Complete the table by matching the correct slope and x- and y-intercept with the linear equation.

	Equation	Slope	x-intercept	y-intercept
a.				
b.	$y = \frac{1}{2}x + 2$			
		−2		−1
	$y = 2x - 1$			
			0	0

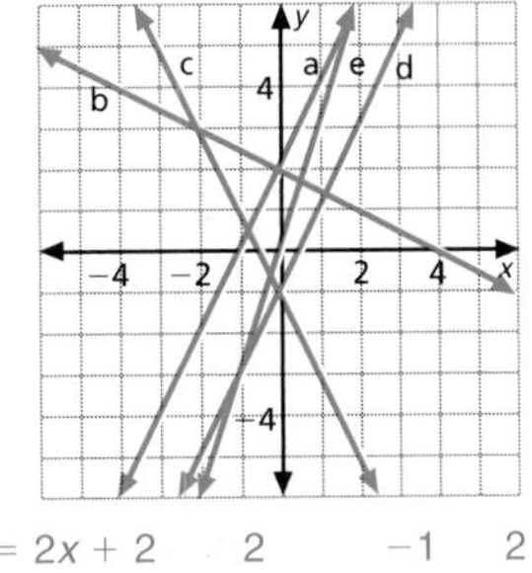

$y = 2x + 2$ 2 −1 2
$-\frac{1}{2}$ 4 2
c. $y = -2x - 1$ $-\frac{1}{2}$
d. 2 $\frac{1}{2}$ −1
e. $y = 3x$ 3

BUILD UNDERSTANDING

You can write an equation of a line if you know the information summarized in the following table.

You wrote equations using the slope-intercept form in Section 6–1. You can write an equation in **point-slope form** if you know the slope of the line and the coordinates of any point of the line.

If you know	You can write an equation in
1. The slope m and y-intercept b	1. Slope-intercept form: $y = mx + b$
2. A point (x_1, y_1) and the slope m	2. Point slope form: $y - y_1 = m(x - x_1)$
3. Two points (x_1, y_1) and (x_2, y_2)	3. Point slope form: $y - y_1 = m(x - x_1)$ or $y - y_2 = m(x - x_2)$ where $m = \frac{y_2 - y_1}{x_2 - x_1}$
4. The graph with points $A(x_1, y_1)$ and $B(x_2, y_2)$	4. Same as 3.

Example 1

Write an equation of the line with a slope of −2 and containing point $P(-1, 3)$.

Solution

$y - y_1 = m(x - x_1)$ Point-slope form.
$y - 3 = -2(x - (-1))$ Substitute for 2 for m, −1 for x_1, and 3 for y_1.
$y - 3 = -2(x + 1)$ Solve for y.
$y - 3 = -2x - 2$
$y = -2x + 1$ the equation in slope-intercept form

Teaching Strategies

To help students decide which form of an equation to write, have them organize given information in the first two columns of the following tables, or use the information to complete the third column.

point(s)	slope	point-slope form

y-intercept	slope	slope-intercept form

You can also write an equation in point-slope form, if you know the coordinates of two points on the line.

Example 2

Write an equation of the line that contains the points $A(1, -3)$ and $B(3, 2)$.

Solution

Given: $x_1 = 1, y_1 = -3, x_2 = 3, y_2 = 2$

Find the slope: $m = \frac{y_2 - y_1}{x_2 - x_1} = \frac{2 - (-3)}{3 - 1} = \frac{5}{2}$

Find the equation using the point-slope form.

$$y - y_1 = m(x - x_1)$$

$$y - (-3) = \frac{5}{2}(x - 1) \quad \text{Solve for } y.$$

$$y + 3 = \frac{5}{2}x - \frac{5}{2}$$

$$y = \frac{5}{2}x - \frac{11}{2} \quad \text{Write equation in slope-intercept form.}$$

An equation of the line is $y = \frac{5}{2}x - \frac{11}{2}$.

Check Understanding

Find the slope of the line.

1. the line containing $P(0, 0)$ and $Q(-5, 2)$
2. $9y = 2x + 3$
3. $\frac{6}{5}x + \frac{2}{3}y = 1$

1. $-\frac{2}{5}$ 2. $\frac{2}{9}$ 3. $-\frac{4}{5}$

You can write an equation for a line if you are given the graph of the line and are able to read information from the graph.

Example 3

PRODUCT DESIGN A technician is using a coordinate grid to design a schematic for a circuit board. A connection aligns with the line shown at the right. Write an equation of the line.

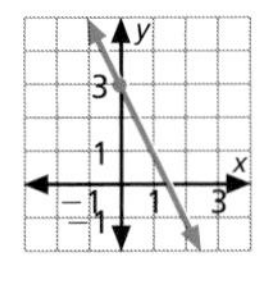

Solution

y-intercepts: The line intersects the y-axis at the point $(0, 3)$. The y-intercept is 3.

slope: Use two points on the line whose coordinates are easily determined. Use $(x_1, y_1) = (0, 3)$ and $(x_2, y_2) = (2, -1)$.

$$m = \frac{-1 - 3}{2 - 0} = \frac{-4}{2} = -2$$

An equation of the line: $y = mx + b$

$$y = -2x + 3 \quad \text{Slope-intercept form.}$$

GRAPHING When writing an equation from a graph, check your work using a graphing utility. Using the graphing keys, enter the slope-intercept form of the equation and display the graph. You may need to adjust the size or shape of the viewing window.

Extend the Lesson

MATH JOURNAL Have students write a description of the slope-intercept form and the point-slope form of an equation. Then have them tell under what circumstances they would use each form.

Chalkboard Examples

Supplementary Example 1

Write an equation of the line with a slope of $-\frac{1}{2}$ and containing point $B(-2, 1)$. $y = -\frac{1}{2}x$

Supplementary Example 2

Write an equation of the line that contains the points $P(-2, -3)$ and $Q(11, 2)$. $y = \frac{7}{2}x + 4$

Supplementary Example 3

A product logo is in the shape of a square intersected by a line. An artist draws the logo on a coordinate grid with vertices of the rectangle at $(3, 4)$, $(-3, 4)$, $(-3, -2)$ and $(3, -2)$. The line passes through $(1, 5)$ and $(-1, -1)$. Write an equation of the line. $y = 3x + 2$

Supplementary Example 4

Write an equation of a line parallel to $y = \frac{1}{2}x + 1$ containing the point $S(4, 2)$. $y = \frac{1}{2}x$

Reteaching Worksheet 6-3

Name ______________________ Date ____________

RETEACHING 6-3

WRITING EQUATIONS FOR LINES

An equation can be written in slope-intercept form, $y = mx + b$, when you know the slope, m, and the y-intercept, b.

An equation can be written in the point-slope form, $y_2 - y_1 = m(x_2 - x_1)$ when you know:

1. a point (x_1, y_1) and the slope, m,
2. two points (x_1, y_1) and (x_2, y_2), or
3. the graph with points $A(x_1, y_1)$ and $B(x_2, y_2)$.

Example 1

Write an equation of the line whose slope is 9 and contains the point $C(-5, 2)$.

Solution

Given a point and the slope, write an equation in point-slope form, substituting 9 for m, -5 for x_1, and 2 for y_1.

$y_2 - y_1 = m(x_2 - x_1)$ **Point-slope form**

$y_2 - 2 = 9(x_2 - (-5))$ **Substitute for m, x_1, and y_1.**

$y_2 - 2 = 9x_2 + 45$ **Solve for y_2.**

$y_2 = 9x_2 + 47$ **Write in slope-intercept form.**

An equation of the line is $y = 9x + 47$.

Example 2

Write an equation of the line that contains points $D(3, 7)$ and $E(4, 1)$.

Solution

Given two points, use the ratio to determine slope.

$m = \frac{y_2 - y_1}{x_2 - x_1} = \frac{1 - 7}{4 - 3} = \frac{-6}{1} = -6$

Then write an equation in point-slope form, substituting -6 for m, 7 for y_1, and 3 for x_1.

$y_2 - y_1 = m(x_2 - x_1)$ **Point-slope form**

$y_2 - 7 = -6(x_2 - 3)$ **Substitute for m, x_1, and y_1.**

$y_2 - 7 = -6x_2 + 18$ **Solve for y_2.**

$y_2 = -6x_2 + 25$ **Write in slope-intercept form.**

An equation of the line is $y = -6x + 25$.

EXERCISES

Write an equation of the line that contains the given point and slope.

1. $m = \frac{2}{3}$, $V(1, 1)$ $y = \frac{2}{3}x + \frac{1}{3}$
2. $m = -2$, $W(-2, 0)$ $y = -2x - 4$
3. $m = 0$, $X(0, 5)$ $y = 5$
4. $m = -3$, $A(-1, -4)$ $y = -3x - 7$
5. $m = \frac{1}{2}$, $B(3, 2)$ $y = \frac{1}{2}x + \frac{1}{2}$
6. $m = \frac{1}{3}$, $C(9, -1)$ $y = \frac{1}{3}x - 4$

Write an equation of the line that contains the given points.

7. $F(6, -6)$, $G(-7, 7)$ $y = -x$
8. $H(-1, -3)$, $I(2, 0)$ $y = x - 2$

88 RETEACHING *LESSON 6-3*

Lesson Wrap-up

QUICK ASSESSMENT

Ask the following questions to determine if students understand the content presented in this lesson.

1. In Example 2, what difference, if any, would it have made to substitute the coordinates of point *B* in the point-slope form of the equation instead of the coordinates of point *A*? no difference
2. Point *C*(3, 4) lies on a line. What other information do you need to know to write the equation of the line? either the slope or one other point on the line

ASSIGNMENT GUIDE

Basic: 1–22, 26–34
Advanced: 1–34
Additional Practice: See Extra Practice Index on page 674.

Example 4

Write an equation of a line parallel to $y = -\frac{1}{3}x + 1$ containing point $R(1, 1)$.

Solution

$y = -\frac{1}{3}x + 1 \quad m = -\frac{1}{3}$ Slope of line

Because parallel lines have equal slopes, $m = -\frac{1}{3}$.

$y - y_1 = m(x - x_1)$ Point-slope form.

$y - 1 = -\frac{1}{3}(x - 1)$ $x_1 = 1, y_1 = 1$

$y - 1 = -\frac{1}{3}x + \frac{1}{3}$

$y = -\frac{1}{3}x + \frac{4}{3}$

An equation of the line is $y = -\frac{1}{3}x + \frac{4}{3}$.

Check Understanding

Write an equation of the line whose slope and point on the line are given.

1. $m = \frac{3}{4}$, $P(-1, 6)$
2. $m = 0$, $R(2, -2)$

1. $y = -\frac{3}{4}x + \frac{21}{4}$; 2. $y = -2$

TRY THESE EXERCISES

Write an equation of the line with the given slope and *y*-intercept.

1. $m = -3$, $b = -2$ $y = -3x - 2$
2. $m = \frac{-3}{5}$, $b = 0$ $y = \frac{-3}{5}x$

Write an equation of the line with the given slope and point on the line.

3. $m = 7$, $Q(-1, -5)$ $y = 7x + 2$
4. $m = \frac{3}{7}$, $S(5, 3)$ $y = \frac{3}{7}x + \frac{6}{7}$

Write an equation of the line that contains the given points.

5. $M(0, -1)$ and $N(-1, 4)$ $y = -5x - 1$
6. $T(-2, -3)$ and $V(2, 2)$ $y = \frac{5}{4}x - \frac{1}{2}$
7. $R(-6, 9)$ and $S(9, 9)$ $y = 9$
8. $A(-5, 6)$ and $B(-5, 10)$ $x = -5$

Write an equation of the lines for the graphs shown below.

9.

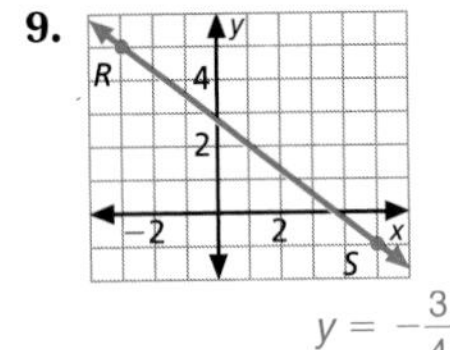

$y = -\frac{3}{4}x + \frac{11}{4}$

10.

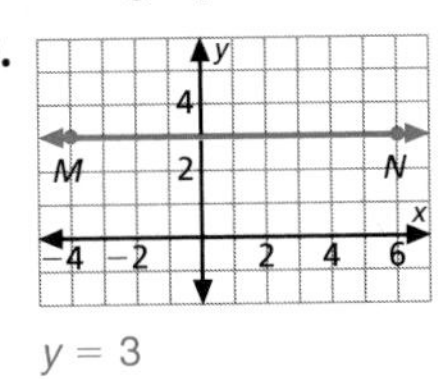

$y = 3$

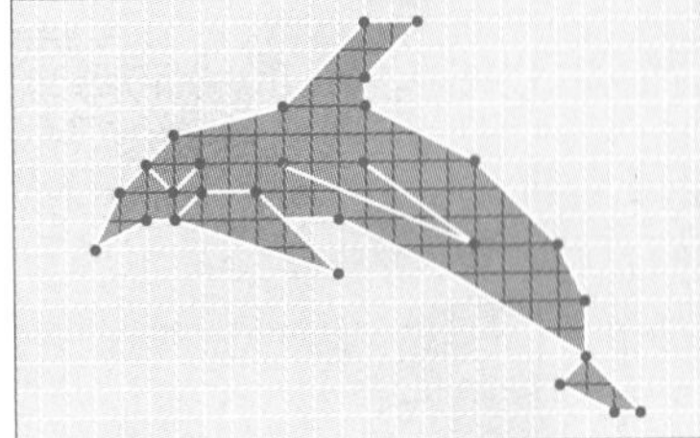

PRACTICE EXERCISES

Write an equation of the line with the given slope and *y*-intercept.

11. $m = \frac{2}{3}$, $b = -3$ $y = \frac{2}{3}x - 3$
12. $m = 7$, $b = 2$ $y = 7x + 2$

13. Write an equation of the line that is parallel to $3y + x = 6$ containing the point $P(1, -1)$. $y = \frac{-1}{3}x - \frac{2}{3}$

14. **TEMPERATURE** The temperature of water at the freezing point is 0°C or 32°F. The temperature of water at the boiling point is 100°C or 212°F. Use these two data points to find an equation to convert the temperature from Celsius to Fahrenheit. $F = \frac{9}{5}C + 32$

Teaching Strategies

TEAM LEARNING Have each student draw a simple object consisting of straight lines on a coordinate grid and write clues that would help another person draw each line of the object. Clues may include slope, *y*-intercept, points on the line, and equations for parallel or perpendicular lines. Then have students trade clues and try to draw the objects.

Write an equation of the line with the given information.

15. $m = -\frac{1}{2}$, $B\left(0, -\frac{2}{5}\right)$ $y = \frac{-1}{2}x - \frac{2}{5}$

16. m is undefined, $D\left(\frac{1}{3}, \frac{1}{2}\right)$ $x = \frac{1}{3}$

17. $P(4, 0)$ and $Q(-9, 3)$ $y = \frac{-3}{13}x + \frac{12}{13}$

18. $M(2, -5)$ and $N(1, 3)$ $y = -8x + 11$

19.

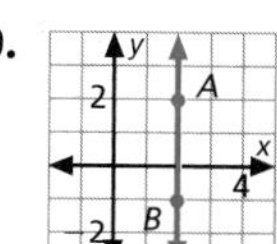

$x = 2$

20.

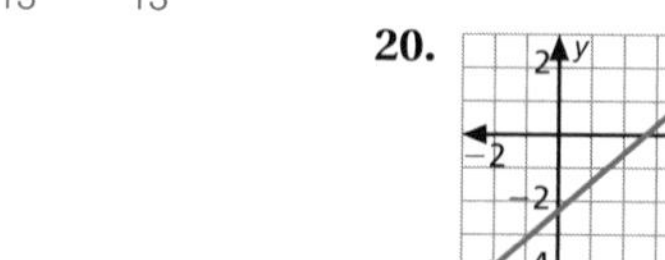

$y = \frac{6}{7}x - \frac{16}{7}$

21. FITNESS Kim's average walking speed is 6 feet every 3 seconds. Complete the table. Then determine an equation for the distance Kim walks in a given time. $d = 2t$

Distance Kim walks (d) feet	0 2 6 12
Time (t) sec	

22. TALK ABOUT IT Marta is asked to write an equation of a line with a slope of 3 that contains point $D(2, 5)$. Marta writes $y - 2 = 3(x - 5)$, so $y = 3x - 13$. Where did Marta make her mistake?

Martha switched the x and y values in the point-slope form of the equation. The equation should read $y - 5 = 3(x - 2)$, which equals $y = 2x - 1$.

Extended Practice Exercises

The coordinates of the vertices of a parallelogram are (2, 6), (7, 2), (3, 0), and (−2, 4).

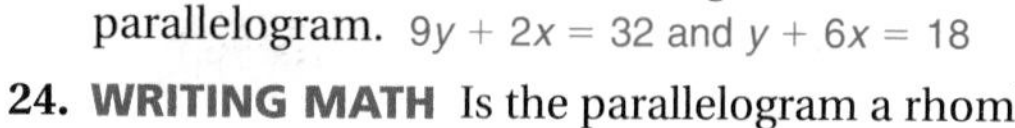

23. Write the equations for the diagonals of the parallelogram. $9y + 2x = 32$ and $y + 6x = 18$

24. WRITING MATH Is the parallelogram a rhombus? Justify your answer. No, the diagonals are not perpendicular.

25. REAL ESTATE A realtor is paid a fixed amount per week plus a commission on the total sales. If the fixed amount is F and the commission rate is p%, find an equation to represent the amount of pay he receives each week.

$y = 0.01px + F$ where x = sales y = pay

Mixed Review Exercises

Find the surface area of each figure. Assume that all pyramids are regular pyramids. Use $\pi = 3.14$. (Lesson 5-6)

26.

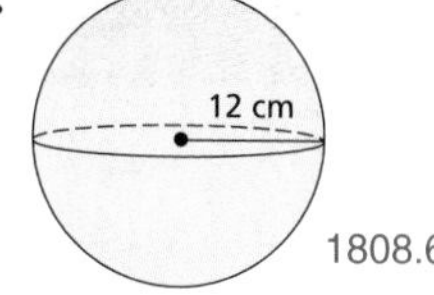

1808.64 cm²

27.

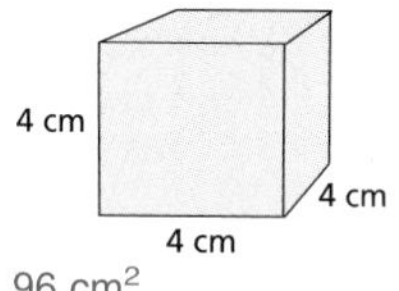

96 cm²

28.

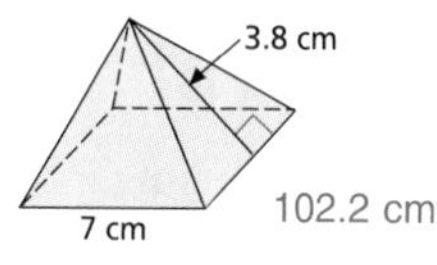

102.2 cm²

Refer to the diagram. Find the set named by listing the numbers. (Lesson 1-3)

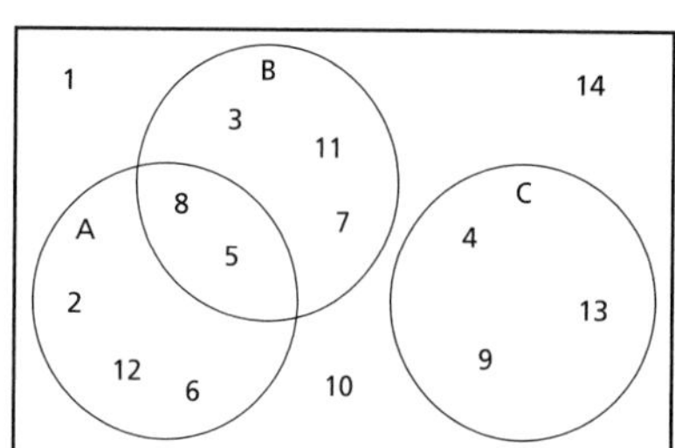

29. $A \cup B$ {2, 3, 5, 6, 7, 8, 11, 12}

30. $B \cup C$ {3, 4, 5, 7, 8, 9, 11, 13}

31. $A' \cap C$ {4, 9, 13}

32. $B' \cap C'$ {1, 2, 6, 10, 12, 14}

33. $A' \cap B'$ {1, 4, 9, 10, 13, 14}

34. $A' \cup C'$ {1, 2, 3, 4, 5, 6, 7, 8, 9, 10, 11, 12, 13, 14}

Teaching Strategies

Instead of having students become focused on memorizing formulas for the slope-intercept and point-slope forms of an equation, guide them in how to derive both from the formula for slope if needed.

Extra Practice Worksheet 6-3

Name ______ Date ______

EXTRA PRACTICE 6-3

Write Equations for Lines

Exercises

Write an equation of the line with the given slope and y-intercept.

1. $m = -2, b = 4$ $y = -2x + 4$

2. $m = -\frac{2}{5}, b = 1$ $y = -\frac{2}{5}x + 1$

3. $m = -5, b = -2$ $y = -5x - 2$

4. $m = 1, b = \frac{3}{4}$ $y = x + \frac{3}{4}$

Write an equation of the line with the given information.

5. $m = 0$, $C(-1, 4)$ $y = 4$

6. $m = \frac{1}{3}$, $W\left(\frac{1}{2}, 2\right)$ $y = \frac{1}{3}x + \frac{11}{6}$

7. m is undefined, $T(5, -6)$ $x = 5$

8. $m = -4$, $S\left(\frac{3}{5}, -\frac{1}{5}\right)$ $y = -4x + \frac{11}{5}$

9. $A(3, -1)$ and $B(2, 4)$ $y = -5x + 14$

10. $M(-6, 4)$ and $N(0, -5)$ $y = -\frac{3}{2}x - 5$

11. $R(6, -1)$ and $S(-3, 0)$ $y = -\frac{1}{9}x - \frac{1}{3}$

12. $F(1, -8)$ and $G(3, 2)$ $y = 5x - 13$

13. $x = -3$

14. $y = x + 1$

15. Parallel to $x + y = 4$ and passes through $M(3, 2)$. $y = -x + 5$

16. Parallel to $5x - 2y = -3$ and passes through $Q(-4, 0)$. $y = \frac{5}{2}x + 10$

17. Perpendicular to $3x - y = 1$ and passes through $R(-1, 1)$. $y = -\frac{1}{3}x + \frac{2}{3}$

18. Perpendicular to $4y = 2x + 3$ and has y-intercept -4. $y = -2x - 4$

19. Parallel to the line through points $J(4, -2)$ and $K(0, 1)$ and has y-intercept 2. $y = -\frac{3}{4}x + 2$

20. Perpendicular to the line $2y = 6 - 2x$ and has x-intercept -1. $y = x + 1$

78 EXTRA PRACTICE *LESSON 6-3*

Enrichment Worksheet 6-3

Name ______ Date ______

ENRICHMENT 6-3

Writing Equations for Planes

In two-dimensional space, the intercept form of an equation is $\frac{x}{a} + \frac{y}{b} = 1$

In three-dimensional space, a plane can also be described with an intercept form equation.

Exercises

Write an equation for each plane.

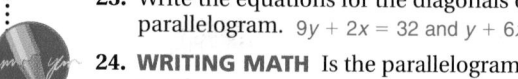

1.

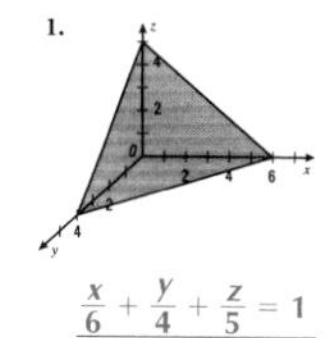

$\frac{x}{6} + \frac{y}{4} + \frac{z}{5} = 1$

2.

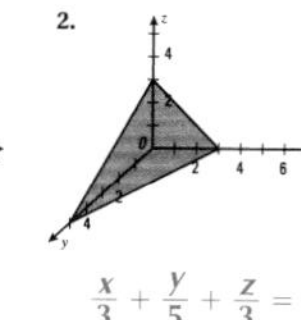

$\frac{x}{3} + \frac{y}{5} + \frac{z}{3} = 1$

3.

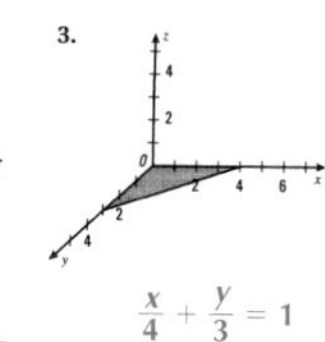

$\frac{x}{4} + \frac{y}{3} = 1$

4.

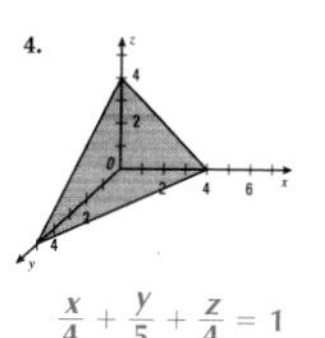

$\frac{x}{4} + \frac{y}{5} + \frac{z}{4} = 1$

5.

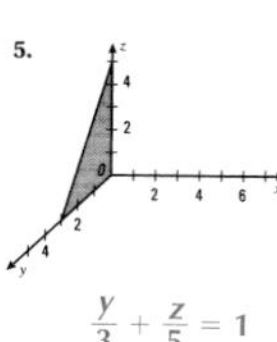

$\frac{y}{3} + \frac{z}{5} = 1$

6.

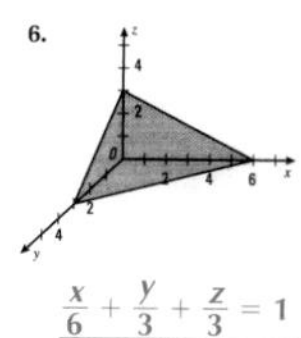

$\frac{x}{6} + \frac{y}{3} + \frac{z}{3} = 1$

7. Write the equation of a plane that passes through the point (3, 0, 0) and is parallel to the yz-plane. $x = 3$

8. Write the equation of a plane that passes through the point (0, 0, −1) and is parallel to the xy-plane. $z = -1$

102 ENRICHMENT *LESSON 6-3*

Lesson Planning

NCTM Standards/Strands

- Number & Operation
- Patterns, Functions, & Algebra
- Geometry & Spatial Sense
- Problem Solving
- Reasoning and Proof
- Connections
- Representation

Vocabulary

system of equations
solution
independent system
inconsistent system
dependent system

Materials Needed

graphing calculators
graph paper
straightedges

Lesson Resources

Warm-Up Transparency 17
Transparency TK-7–10
Reteaching 6-4
Extra Practice 6-4
Enrichment 6-4

Assignment Guide

Basic: 1–20, 23–29
Enriched: 1–29

Getting Started

5-Minute Warm-Up

Is (2, −1) a solution of the equation?

1. $2y = x$ no
2. $y = x - 3$ yes
3. $4y = -2x$ yes

Introduction to Lesson 6-4

Have students speculate what the ordered pair for each point of intersection represents and why there is no point of intersection for the equations in Pair 5.

6-4 Systems of Equations

Goals ■ Solve a system of equations by graphing.
Applications Income Tax, Manufacturing, Finance

Graph the following pairs of equations on the same coordinates system. Then complete the table.

Equations	Point of Intersection (x, y)	
1. $y = x$ and $y = 1$	(,)	(1, 1)
2. $2y = x + 2$ and $y = x$	(,)	(2, 2)
3. $y = 2x + 1$ and $y = x$	(,)	(−1, −1)
4. $y = x + 1$ and $y = x$	(,)	no ordered pair

Build Understanding

Two linear equations with the same two variables form a **system of equations**. The *solution* of a system of equations is the ordered pair that makes both equations true. One way to solve a system of two equations is to graph both equations on the same coordinate system. The *point of intersection* of the two lines is the **solution** of the system. If the graph of each equation intersects in *one* point, the system is known as an **independent system**.

Example 1

Solve the system of equations by graphing. $y = -2x + 1$
$y = -3x + 4$

Solution

Graph each equation. Then read the solution.

$y = -2x + 1$ $b = 1, m = -2$; use $(1, -1)$
$y = -3x + 4$ $b = 4, m = -3$; use $(1, 1)$

The lines intersect at the point $(3, -5)$.
The solution of system equations is $(3, -5)$.

Check: Substitute $(3, -5)$ in each equation.

$y = -2x + 1$	$y = -3x + 4$
$-5 = -2(3) + 1$	$-5 = -3(3) + 4$
$-5 = -6 + 1$	$-5 = -9 + 4$
$-5 = -5$ ✓	$-5 = -5$ ✓

If the graphs of each equation *do not intersect*, the system is known as an **inconsistent system**. The lines are parallel and have no common points. There is *no solution* to this system of equations.

Technology Note

Solve a system of equations using a graphing calculator.

1. Use the graphing features to enter all equations in slope-intercept form. Then press graph.
2. Use the trace feature to find the point of intersection.
3. Use the zoom feature to read the x and y values of the intersection point.
4. Continue to zoom and trace until you reach the desirded degree of accuracy.

Teaching Strategies

ESL/LEP Discuss everyday usages of the word "independent," "inconsistent," and "dependent," to help students understand why these terms were applied to particular types of systems of equations.

Example 2

Solve the system of equations by graphing. $y = \frac{1}{2}x + 3$
$2y = x - 2$

Solution

Graph each equation, then read the solution from the graph.

$y = \frac{1}{2}x + 3$ $\quad b = 3,\ m = \frac{1}{2}$; use $(-2, 2)$

$y = \frac{1}{2}x - 1$ $\quad$ Rewritten in slope-intercept form.

$b = -1,\ m = \frac{1}{2}$; use $(2, 0)$

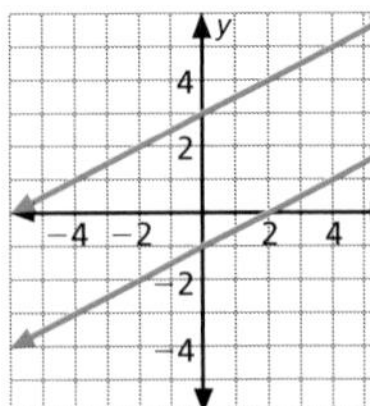

The lines are parallel and do not intersect. Therefore, there is no solution.

If the graph of each equation is the same line—i.e., the lines coincide—the system is a **dependent system**. Any point on the line is a solution. There is an *infinite number of solutions.*

Problem Solving Tip

If all the numerical values in an equation have a common factor, dividing by that common factor results in an equation that is easier to work with.

Example 3

Solve the system of equations by graphing. $4x + 2y = 8$
$3y = -6x + 12$

Solution

Graph each equation. Then read the solution.

$4x + 2y = 8$

$y = -2x + 4$ $\quad$ Divide both sides by 2.

$3y = -6x + 12$

$y = -2x + 4$ $\quad$ Divide both sides by 3.

The equations are equivalent. The graphs are on the same line. The lines coincide. The system has an infinite number of solutions.

To summarize:

If the graph of two lines:	then the system is:	and has:
intersect	independant	one solution
do not intersect	inconsistent	no solution
coincide	dependant	an infinite number of solutions

Equations can give you a better understanding of everyday life. For example, the rate of taxation is a function of the amount of taxable income earned. The table below shows the tax rate schedule for single individuals for 1999.

If you are single . . .

Line	Taxabe Income		Federal Income Tax	
	At least	But less than	You pay	of the amount over
1.	$0	$25,750	15%	$0
2.	$25,750	$62,450	$3,862.50 + 28%	$25,750
3.	$62,450	$130,250	$14,138.50 + 31%	$62,450

Chalkboard Examples

Supplementary Example 1

Solve the system of equations by graphing.
$a + b = 8$
$a - b = 4$ $\quad$ (6, 2)

Supplementary Example 2

Solve the system of equations by graphing.
$x + 2y = 5$
$2x + 4y = 2$ $\quad$ no solution

Supplementary Example 3

Solve the system of equations by graphing.
$4x - 2y = -6$ $\quad$ infinite number
$-y = -3 - 2x$ $\quad$ of solutions

Supplementary Example 4

Janette earned $32,500 in 1999. Find the amount of income tax she owed. $6,311.50

Teaching Strategies

CHALLENGE The solution of the following system of equations is $(1, -2)$:
$Px + y = 5$
$Px + Qy = 3$
What are the values of P and Q? 7; 2

Reteaching Worksheet 6-4

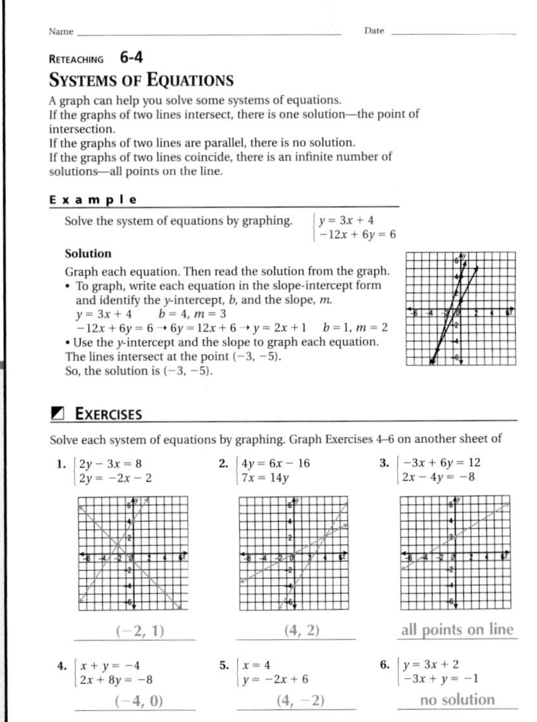

Name ______ Date ______

RETEACHING 6-4

SYSTEMS OF EQUATIONS

A graph can help you solve some systems of equations.
If the graphs of two lines intersect, there is one solution—the point of intersection.
If the graphs of two lines are parallel, there is no solution.
If the graphs of two lines coincide, there is an infinite number of solutions—all points on the line.

Example

Solve the system of equations by graphing. $\begin{cases} y = 3x + 4 \\ -12x + 6y = 6 \end{cases}$

Solution

Graph each equation. Then read the solution from the graph.
- To graph, write each equation in the slope-intercept form and identify the y-intercept, b, and the slope, m.
  $y = 3x + 4$ $\quad b = 4,\ m = 3$
  $-12x + 6y = 6 \rightarrow 6y = 12x + 6 \rightarrow y = 2x + 1$ $\quad b = 1,\ m = 2$
- Use the y-intercept and the slope to graph each equation.

The lines intersect at the point $(-3, -5)$.
So, the solution is $(-3, -5)$.

EXERCISES

Solve each system of equations by graphing. Graph Exercises 4–6 on another sheet of

1. $\begin{cases} 2y - 3x = 8 \\ 2y = -2x - 2 \end{cases}$ $\quad$ (−2, 1)
2. $\begin{cases} 4y = 6x - 16 \\ 7x = 14y \end{cases}$ $\quad$ (4, 2)
3. $\begin{cases} -3x + 6y = 12 \\ 2x - 4y = -8 \end{cases}$ $\quad$ all points on line
4. $\begin{cases} x + y = -4 \\ 2x + 8y = -8 \end{cases}$ $\quad$ (−4, 0)
5. $\begin{cases} x = 4 \\ y = -2x + 6 \end{cases}$ $\quad$ (4, −2)
6. $\begin{cases} y = 3x + 2 \\ -3x + y = -1 \end{cases}$ $\quad$ no solution

90 $\quad$ RETEACHING *LESSON 6-4*

Lesson Wrap-up

Quick Assessment

Ask the following questions to determine if students understand the content presented in this lesson.

1. What information about the solution can you get from looking at the graphs of both equations on the same coordinate grid? If the lines intersect, the ordered pair for the point of intersection is the solution of the system. If the lines are parallel, you know that there is no solution. if the lines are the same, you know that there is an infinite number of solutions and that the ordered pair for any point on the lines is a solution.
2. Is it possible to have a system of two equations of lines with exactly two solutions? Explain. No; two straight lines either intersect at one point or are the same line with all points in common.

Assignment Guide

Basic: 1–20, 23–29
Advanced: 1–29
Additional Practice: See Extra Practice Index on page 674.

Additional Answers

3.

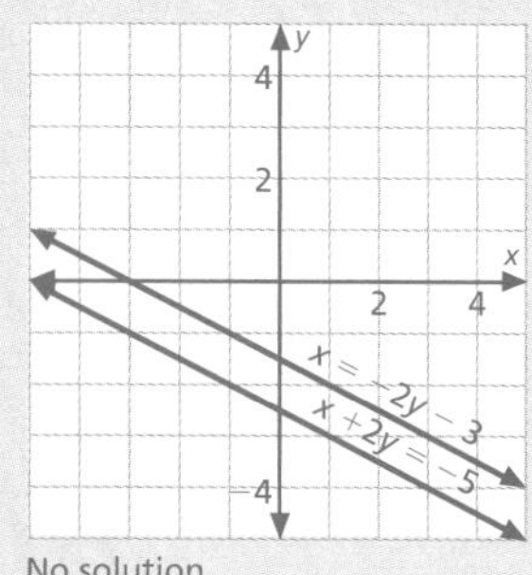

No solution

You can graph the equations defined in the table to gain an understanding of how the tax law works. Let T represent the amount of income tax owed and x represent the amount of taxable income. Using these variables, construct three linear equations that reflect the information contained in the table. Notice that equation has a different slope.

Line 1: If $0 < x \le 25{,}750$, then $T = 0.15x$

Line 2: If $25{,}750 < x \le 62{,}450$, then $T = 3862.50 + 0.28(x - 25{,}750)$, or $T = 0.28x - 3347.50$

Line 3: If $62{,}450 < x \le 130{,}250$, then $T = 14{,}138.50 + 0.31(x - 62{,}450)$, or $T = 0.31x - 5221$

Federal Income Tax ($)
14,138.50
3,862.50
25,750
62,450
Taxable Income

Example 4

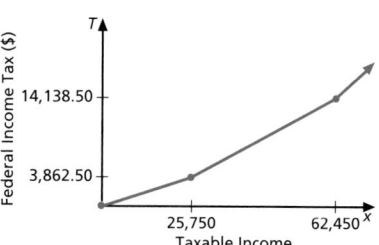

INCOME TAX Rick's taxable income in 1999 was $18,000. Find the amount of income tax he will owe.

Solution

$T = 0.15x$ His income is less than $25,750.

$T = 0.15(18{,}000)$

$T = \$2700$

Try These Exercises

Determine whether the given ordered pair is a solution.

1. (2, 1) $5y = 3x - 1$; $2x - 3y = 1$ yes
2. (−1, 3) $7x + 2y = -1$; $-4x + y = 15$ no

Solve each system of equations by graphing. See additional answers.

3. $x = -2y - 3$; $x + 2y = -5$
4. $x = \frac{4}{3}y - 1$; $4y = 3x + 3$
5. $-x + \frac{3y}{2} = 1$; $4x = 2y + 4$

Practice Exercises

Determine the solution of each system of equations.

6. 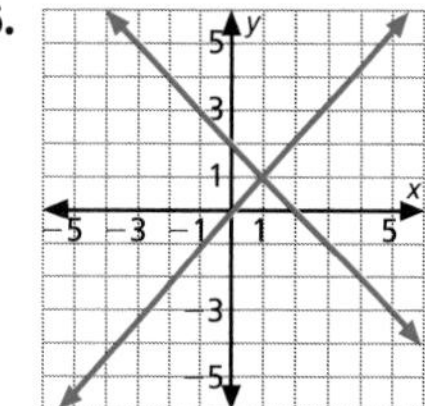 (1, 1)

7. 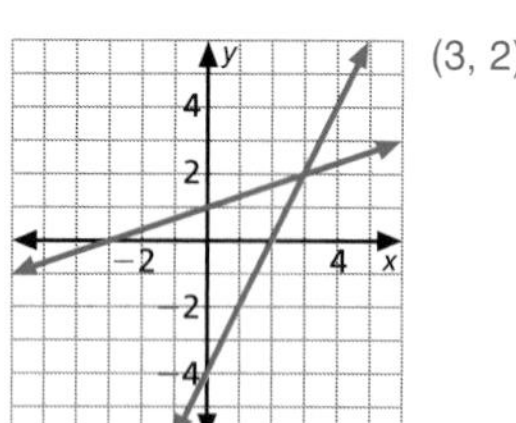 (3, 2)

4.

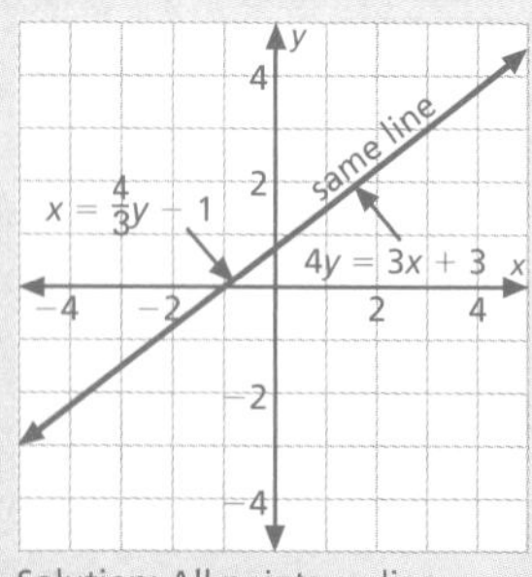

Solution: All points on line.

5.

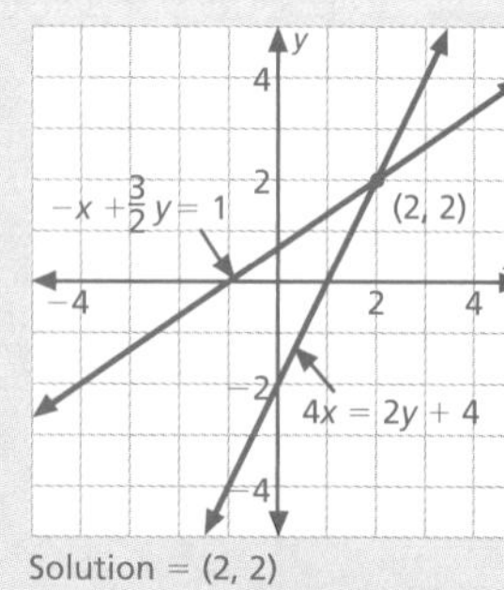

Solution = (2, 2)

8.

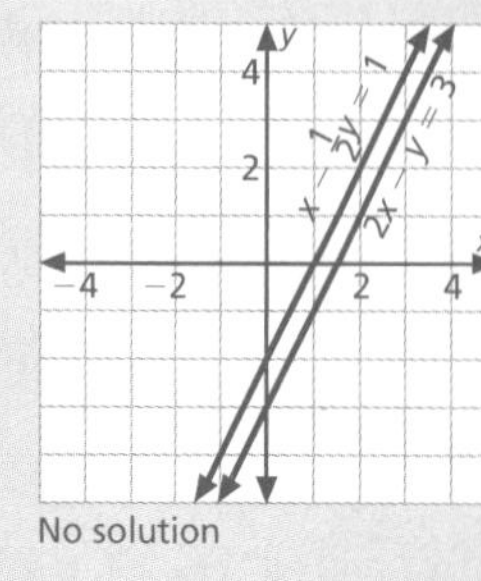

No solution

9.

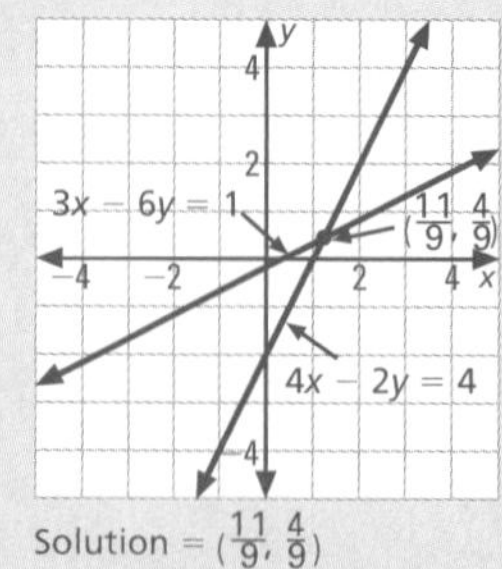

Solution = $(\frac{11}{9}, \frac{4}{9})$

10.

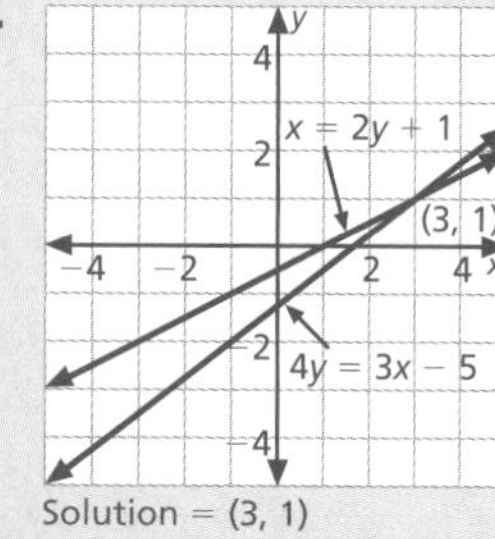

Solution = (3, 1)

GRAPHING **Solve each system of equations by graphing.** See additional answers.

8. $2x - y = 3$
$x - \frac{1}{2}y = 1$

9. $4x - 2y = 4$
$3x - 6y = 1$

10. $4y = 3x - 5$
$x = 2y + 1$

11. $\frac{x}{3} + \frac{y}{5} = \frac{7}{15}$
$\frac{x}{4} + \frac{3y}{8} = \frac{1}{8}$

12. $\frac{c}{2} = \frac{d}{3} + \frac{1}{6}$
$\frac{2}{3}d = c + \frac{1}{2}$

13. MANUFACTURING The Food Division and Personal Care Division introduced a total of 10 new products this year. The Food Division introduced 4 more products than the Personal Care Division. How many products did each group introduce this year? food, 7; personal care, 3

14. FINANCE Candice's monthly savings is twice the amount that she spends on transportation each month. The total of her monthly savings and transportation bill is $135. Find both amounts. Savings, $90; transportation, $45

15. INCOME TAX Phil's taxable income in 1999 was $48,000. Use the table on page 259. Find the amount of income tax he had to pay. $10,092.50

16. WRITING MATH Is it possible for a system of equations to have exactly two solutions? Explain your thinking. No. Two straight lines either intersect at one point or are the same line with all points in common.

17. CHAPTER INVESTIGATION Make a detailed drawing or build a prototype of your new product. Choose at least five selling points to emphasize in your marketing materials.

Determine the number of solutions for each system. Do not graph.

18. $4x + 5y = 3$ no solution
$3x - 2y = 8$

19. $\frac{a}{2} = \frac{b}{3} + \frac{1}{6}$ no solution
$\frac{2}{3}b = a + \frac{1}{2}$

20. $y - 2 = \frac{1}{2}x$ Infinite solution
$\frac{x}{4} - \frac{1}{2}y = -1$

Extended Practice Exercises

21. The equations $2x - y = 1$, $x + 2y = 8$, and $x = 4$ form a triangle. Graphically determine the location of each vertex. See additional answers.

22. Determine whether the ordered pair $\left(-a, \frac{b}{2}\right)$ is a solution to the system of equations.

$4ay = -3bx - ab$ yes
$5bx - 2ay = -6ab$

Mixed Review Exercises

Find the probability that a point selected at random in each figure is in the shaded region.

23. 15.48 in.²

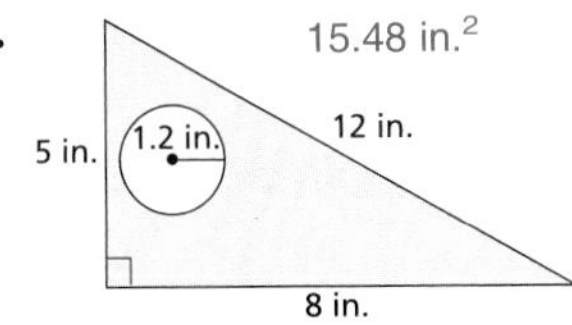

24.

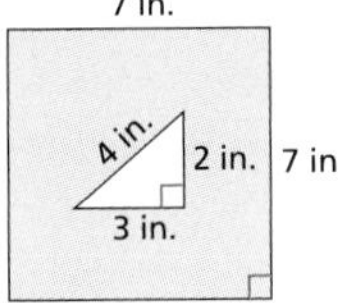

46 in.²

25.

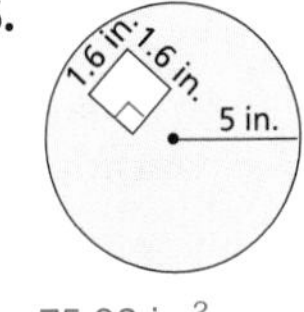

75.98 in.²

Write each number in standard form.

26. 3.84×10^{-6} 0.00000384

27. 1.9×10^{8} 190,000,000

28. 7×10^{-9} 0.000000001

29. 6.52×10^{12} 6,520,000,000,000

11.

Solution = (2, −1)

12.

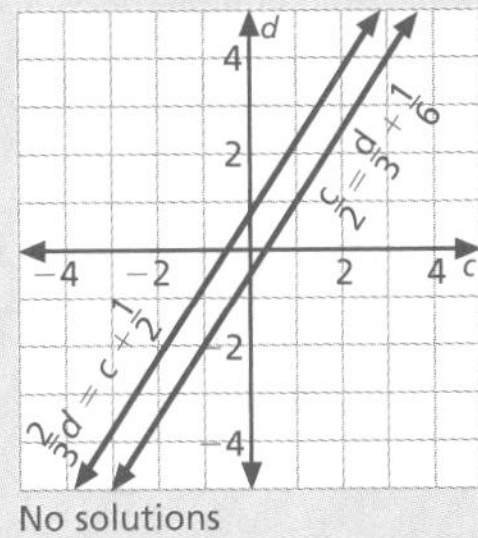

No solutions

21.

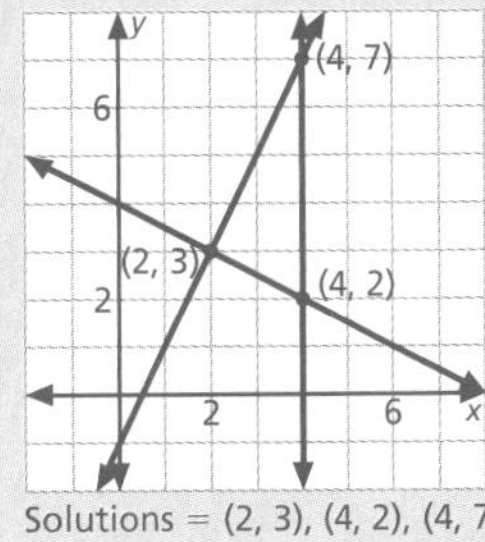

Solutions = (2, 3), (4, 2), (4, 7)

Extra Practice Worksheet 6-4

Name ______ Date ______

EXTRA PRACTICE 6-4

SYSTEMS OF EQUATIONS

EXERCISES

Determine the solution of each system of equations.

1.

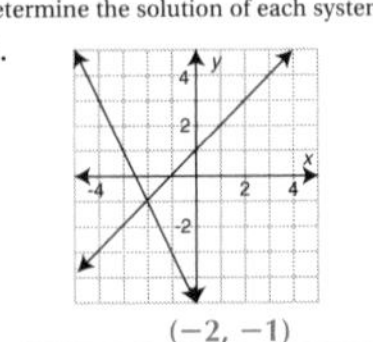

(−2, −1)

2.

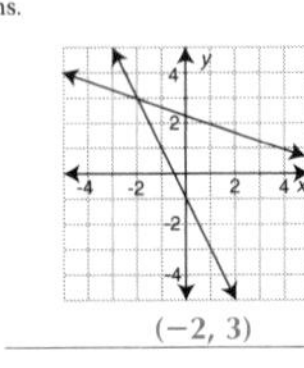

(−2, 3)

Solve each system of equations by graphing. Use your own paper.

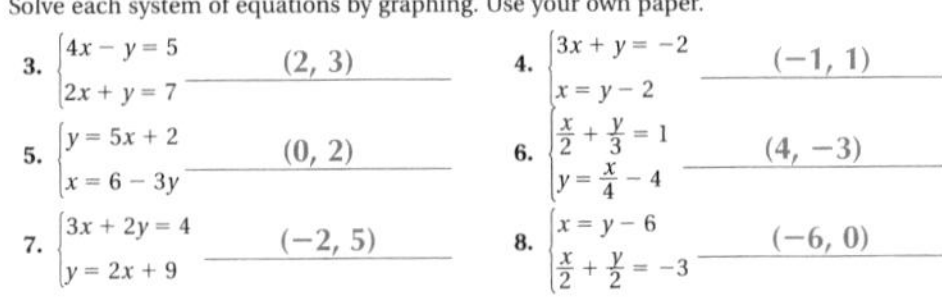

3. $4x - y = 5$, $2x + y = 7$ (2, 3)

4. $3x + y = -2$, $x = y - 2$ (−1, 1)

5. $y = 5x + 2$, $x = 6 - 3y$ (0, 2)

6. $\frac{x}{2} + \frac{y}{3} = 1$, $y = \frac{x}{4} - 4$ (4, −3)

7. $3x + 2y = 4$, $y = 2x + 9$ (−2, 5)

8. $x = y - 6$, $\frac{x}{2} + \frac{y}{2} = -3$ (−6, 0)

9. The sum of two numbers is 2. Their difference is 10. Find the numbers. 6, 4

10. One number is three times another number. The difference of the two numbers is 2. Find the numbers. 1, 3

Determine the number of solutions for each system. Do not graph.

11. $2x - y = -4$, $3x - y = 1$ one solution

12. $3x + 6y = 12$, $x + 2y = 10$ no solution

13. $8x + 6y = 10$, $3y = 5 - 4x$ infinite solutions

14. $12x + 4y = 1$, $3x + y = 0$ no solution

80 EXTRA PRACTICE *Lesson 6-4*

Enrichment Worksheet 6-4

Name ______ Date ______

ENRICHMENT 6-4

SYSTEMS OF LINES

The slope-intercept form of a line has two constants, *m* for the slope and *b* for the *y*-intercept. If one of these constants is fixed and the other is left arbitrary, you get a *system of lines*. The arbitrary constant is called the *parameter* of the system.

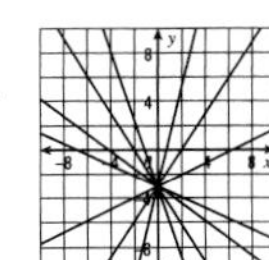

For example, the graph shows part of the system for the equation $y = kx - 3$. Every line in this system passes through (0, −3). But, each line can have a different slope. The parameter of the system is the slope *k*.

A system of lines is also called a *pencil* of lines.

EXERCISES

Write an equation to describe each system of lines. Use *k* for the parameter.

1.

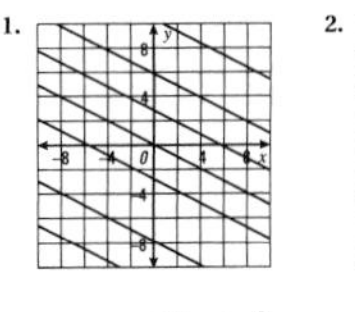

$y = -0.5x + k$

2.

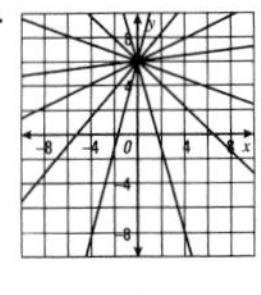

$y = kx + 6$

3.

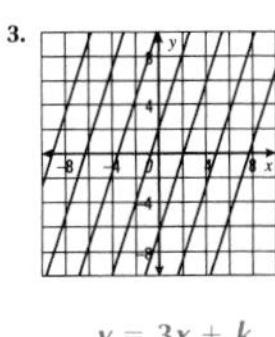

$y = 3x + k$

4.

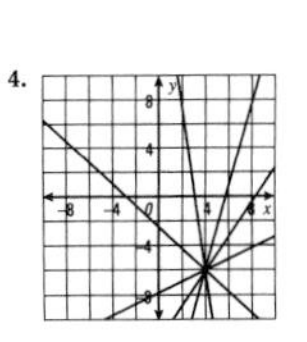

$y + 6 = k(x - 4)$

5.

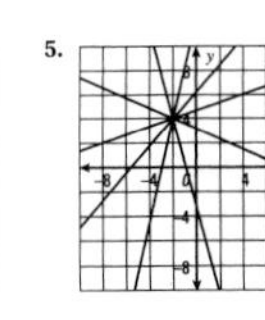

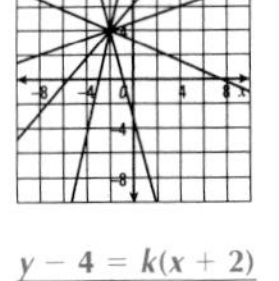

$y - 4 = k(x + 2)$

6.

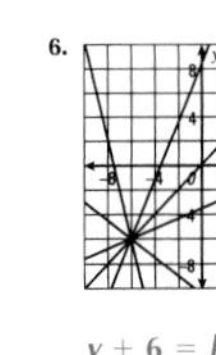

$y + 6 = k(x + 6)$

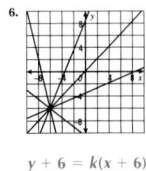

104 ENRICHMENT *Lesson 6-4*

Skills Practice

Vocabulary Review

Lesson 6-3
point-slope form

Lesson 6-4
system of equations
solution
independent system
inconsistent system
dependent system

Assignment Guide

All students: 1–62
Additional Practice: Refere to the Extra Practice Index on page 674 of this text.

Review and Practice Your Skills

Practice ■ Lesson 6-3

Write an equation of the line with the given slope and *y*-intercept.

1. $m = 4.5, b = -7.5$ $y = 4.5x - 7.5$
2. $m = -3, (0, 16)$ $y = -3x + 16$
3. $m = 0, b = -4$ $y = -4$
4. $m = \frac{1}{6}, b = 6$ $y = \frac{1}{6}x + 6$
5. $m = \frac{5}{4}, b = 13$ $y = -\frac{5}{4}x + 13$
6. $m = \frac{11}{8}, (0, -44)$ $y = \frac{11}{8}x - 44$
7. $m = 1, b = 0.05$ $y = x + 0.5$
8. $m = -9.25, b = 0$ $y = -9.25x$
9. $m = 300, b = 530$ $y = 300x + 530$

Write an equation of the line with the given slope and point on the line.

10. $m = \frac{-2}{3}, G(0, 0)$ $y = \frac{-2}{3}x$
11. $m = 2, H(-4, -1)$ $y = 2x + 7$
12. $m = \frac{3}{5}, J(5, 0)$ $y = \frac{3}{5}x - 3$
13. $m = \frac{7}{2}, K(2, 9)$ $y = \frac{7}{2}x + 2$
14. $m = \frac{1}{2}, L(-8, 1)$ $y = \frac{-1}{2}x - 3$
15. $m = -4, N(3, -6)$ $y = -4x + 6$
16. $m = 0. P(-8, 9)$ $y = 9$
17. slope undefined, $Q(13, -12)$ $x = 13$
18. $m = 1.25, R(12, -15)$ $y = \frac{-5}{4}x$

Write an equation of the line that contains the given points.

19. $A(-8, 3)$ and $B(4, 9)$ $y = \frac{1}{2}x + 7$
20. $C(2, -9)$ and $D(-3, 11)$ $y = -4x - 1$
21. $E(3, -5)$ and $F(3, 7)$ $x = 3$
22. $G(-5, 3)$ and $H(10, -6)$ $y = \frac{-3}{5}x$
23. $J(2, 8)$ and $K(-3, -7)$ $y = 3x + 2$
24. $L(4, -2)$ and $M(-12, -2)$ $y = -2$
25. $P(-7, -8)$ and $Q(6, 5)$ $y = 9$
26. $R(12, -7)$ and $S(-6, -4)$ $y = \frac{-1}{6}x - 5$
27. $T(-4, -2)$ and $U(8, 7)$ $y = \frac{3}{4}x + 1$

Write an equation for the line with the given information.

28. Parallel to $-2x + y = -14$ and passes through $Z(5, 3)$. $y = 2x - 7$
29. Perpendicular to $3y = -2x + 18$ and has *x*-intercept 8. $y = \frac{3}{2}x - 12$
30. Parallel to line through points $G(-4, 8)$ and $H(2, 5)$ and has same *y*-intercept as $y = 8x - 11$. $y = -\frac{1}{2}x - 11$

Practice ■ Lesson 6-4

Solve each system of equations by graphing.

31. $y = \frac{2}{3}x - 2$, $x = -3$ $(-3, -4)$
32. $y = -\frac{1}{2}x + 4$, $y = 3x - 3$ $(2, 3)$
33. $y = 8$, $x - y = 0$ $(8, 8)$
34. $y = 2x - 4$, $y = \frac{1}{7}x + 9$ $(7, 10)$
35. $4x - 3y = 12$, $y = \frac{3}{4}x - 5$ parallel lines (no solution)
36. $y = x + 2$, $y = -3x - 18$ $(-5, -3)$
37. $x - 3y = -3$, $3y = 2x - 6$ $(9, 4)$
38. $y = 5 - x$, $y = 4x - 20$ $(5, 0)$
39. $3x + 12y = -12$, $5x + 2y = 16$ $(4, -2)$

Determine the number of solutions for each system. Do not graph.

40. $y = 3x - 8$, $3x - 4 = -5$ 0
41. $y = -x + 4$, $-3x + 2y = 8$ 1
42. $y = \frac{1}{4}x - 1$, $2x - 8y = 8$ infinite
43. The sum of two numbers is -3. Their difference is 13. Find the numbers. 5, −8

Extend the Lesson

MATH JOURNAL Have students write an explanation of the steps needed to solve the following problem. Have them include in their answer whether there is more than one way to solve the problem.

Write an equation of the line whose slope is $-\frac{1}{4}$ and contains the point (8, 2). Answers will vary. The equation is $y = -\frac{1}{4}x + 4$.

262 Chapter 6 **Linear Systems of Equations**

PRACTICE ■ LESSON 6-1–LESSON 6-4

Find the slope and y-intercept for each line. (Lesson 6-1)

44. $y = -\frac{1}{5}x + 4$ $m = \frac{-1}{5}, b = 4$ **45.** $y = x - 3\frac{1}{2}$ $m = 1, b = -3\frac{1}{2}$ **46.** $y = 15 - \frac{2}{3}x$ $m = \frac{-2}{3}, b = 15$

47. $6x - 4y = 28$ $m = \frac{3}{2}, b = -7$ **48.** $3x + 7y = -28$ $m = \frac{-3}{7}, b = -4$ **49.** $y = -12$ $m = 0, b = -12$

Determine whether each pair of lines is *parallel, perpendicular,* or *neither.* (Lesson 6-2)

50. $y = 4x - 6$ and $4x + y = 8$ neither **51.** $y = 7 - 2x$ and $4x + 2y = 16$ parallel

52. $y = -\frac{3}{8}x + 11$ and $42x - 9y = 1$ perpendicular **53.** $y = 16$ and $x = -3$ perpendicular

Write an equation for the line with the given information. (Lesson 6-3)

54. $m = -\frac{1}{8}, B(0, -7)$ $y = \frac{-1}{8}x - 7$ **55.** $M(-1, -7), N(2, -10)$ $y = -x - 8$ **56.** m undefined, $Z(-8, 6.5)$ $x = -8$

57. $m = 0, Q(-2.7, 36)$ $y = 36$ **58.** $m = 4, b = -11.4$ $y = 4x - 11.4$ **59.** $m = 3$, x-intercept -5 $y = 3x + 15$

Solve each system of equations by graphing. (Lesson 6-4)

60. $y = -3x$ (2, −6)
$y = \frac{1}{2}x - 7$

61. $y = x - 9$ (45, 36)
$y = \frac{4}{5}x$

62. $x - 2y = 6$ (4, −1)
$y = -\frac{3}{2}x + 5$

Mid-Chapter Quiz

Find the slope of the line containing the given points.

1. $K(7, 1)$ and $L(4, 7)$ −2 **2.** $I(2, 9)$ and $J(-2, 6)$ $\frac{3}{4}$

Write an equation of the line with the given slope and y-intercept.

3. $m = -5, b = 3$ $y = -5x + 3$ **4.** $m = \frac{3}{4}, b = -2$ $y = \frac{3}{4}x - 2$

Graph each equation. See additional answers.

5. $x = -3$ **6.** $x + 2y = 4$

Determine whether each pair of lines is parallel, perpendicular or neither.

7. $4x + y = 3$ and $x - 4y = -8$ perpendicular **8.** $x + 2y = 4$ and $y = -2x + 2$ neither

9. Plot and connect the points $J(-3, 6)$, $K(-5, 0)$, $L(1, -2)$ and $M(3, 4)$. Determine if $JKLM$ is a rectangle. yes

Write an equation of the line:

10. that contains the points $Y(-3, -2)$ and $Z(-1, 4)$. $y = 3x + 7$

11. that is perpendicular to $x - 3y = 2$ and contains the point (2, 4). $y = -3x + 10$

12. that is parallel to $2x + 5y = 4$ and contains the point (2, 1). $y = -\frac{2}{5}x + \frac{9}{5}$

Solve each system of equations by graphing.

13. $3m + n = -8$ (−3, 1)
$m + 6n = -3$

14. $x + 2y = 5$ (4, 11)
$-6y = 3x - 15$

15. $p - 2q = 4$ no solution
$2q = p + 2$

Chalkboard Examples

Example from Lesson 6-3

Write an equation of the line that is perpendicular to $3x + 8y = 4$ and contains the point (0, 4).

$y = \frac{8}{3}x + 4$

Example from Lesson 6-4

Solve this system of equations by graphing.

$d = -3c$

$4c + d = 2$ (2, −6)

Mid-Chapter Quiz

Correlation Chart

Question Number	Lesson Number
1–6	5-1
7–9	5-2
10–12	5-3
13–15	5-4

ADDITIONAL ANSWERS

5.

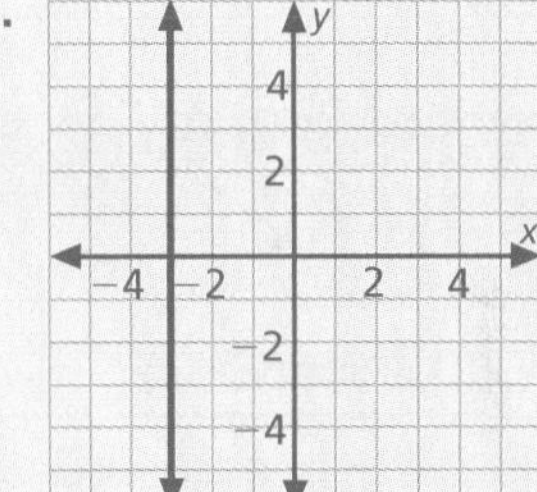

6.

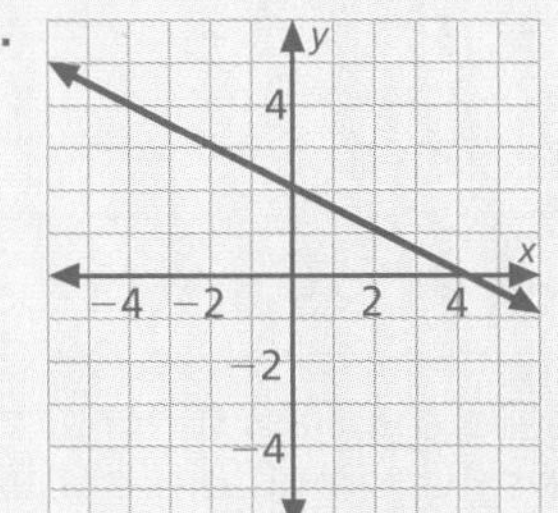

Teaching Strategies

Encourage students to use one or both of these methods for checking an equation: sketch its graph to check for reasonableness and substitute the coordinates of a second point, when available, in the equation.

6-5 Solve Systems by Substitution

Goals ■ Solve systems of equations using substitution.

Applications Shipping, Packaging, Recreation

Work with a partner to practice using the distributive property.

Write one of the equations below on a piece of paper. Pass the paper back and forth, adding the next line to solve the equation.

1. $x + 2(3x - 6) = 2$ 2 **2.** $-(4x - 2) = 2(x + 7)$ −2

3. $5(2x + 4) - 10 = 70$ 6 **4.** $3(2x + 9) = 81$ 9

Build Understanding

You can use algebraic methods to solve a system of equations. One of these methods is **substitution**. This method is useful when one equation has already been solved for one of the variables. In any algebraic method, you need to eliminate one variable so you will have the equation in one variable to solve.

Example 1

Find the solution to the system of equations: $3x - y = 6$, $x + 2y = 2$.

Solution

$$\begin{aligned} 3x - y &= 6 \\ y &= 3x - 6 && \text{Solve the first equation for } y \text{ in terms of } x. \\ x + 2y &= 2 && \text{Write the second equation.} \\ x + 2(3x - 6) &= 2 && \text{Substitute } (3x - 6) \text{ for } y. \\ x + 6x - 12 &= 2 && \text{Solve for } x. \\ 7x &= 14 \\ x &= 2 \end{aligned}$$

Choose one of the original equations.

$$\begin{aligned} 3x - y &= 6 \\ 3(2) - y &= 6 && \text{Substitute 2 for } x. \\ 6 - y &= 6 && \text{Solve for } y. \\ y &= 0 \end{aligned}$$

Check $x = 2$, $y = 0$ in each original equation.

$$\begin{aligned} 3x - y &= 6 \\ 3(2) - 0 &= 6 \\ 6 - 0 &= 6 \\ 6 &= 6 \ \checkmark \end{aligned} \qquad \begin{aligned} x + 2y &= 2 \\ 2 + 2(0) &= 2 \\ 2 + 0 &= 2 \\ 2 &= 2 \ \checkmark \end{aligned}$$

The solution is (2, 0).

The substitution method is most useful when one of the coefficients is 1 or −1.

Check Understanding

If you solve the second equation for x in Example 1, does the solution change?

no

Lesson Planning

NCTM Standards/Strands

- Number & Operation
- Patterns, Functions, & Algebra
- Problem Solving
- Reasoning and Proof
- Connections
- Representation

Vocabulary

substitution method

Materials Needed

paper/pencil

Lesson Resources

Warm-Up Transparency 17
Reteaching 6-5
Extra Practice 6-5
Enrichment 6-5
Transparency TK-7–10, RF-25

Assignment Guide

Basic: 1–20, 26–33
Enriched: 1-33

Getting Started

5-Minute Warm-Up

Find the value of y in each equation if $x = 2$.

1. $x + y = 5$ 3
2. $2x - y = -2$ 6
3. $6x - 4y = 12$ 0

Introduction to Lesson 6-5

After students have solved the equations, have the class summarize the steps for solving equations using the distributive property.

Teaching Strategies

You may wish to suggest that, when students are deciding which equation to use to solve for one variable, they look for the equation, that most easily lends itself to being put into the slope-intercept form, $y = mx + b$. Students are familiar with this form, and this form already shows the solution for y in terms of x.

Example 2

Find the solution to the system of equations: $2x + 3y = 6$, $4x + 6y = 6$.

Solution

$$2x + 3y = 6$$
$$3y = -2x + 6 \quad \text{Solve for } y.$$
$$y = -\frac{2}{3}x + 2$$

$$4x + 6y = 6$$
$$4x + 6\left(-\frac{2}{3}x + 2\right) = 6 \quad \text{Substitute for } y.$$
$$4x - 4x + 12 = 6$$
$$12 = 6$$

There is no solution. The lines are parallel.

Example 3

Solve the system of equations: $4x - 2y = 10$, $-2x + y = -5$.

Solution

$$-2x + y = -5$$
$$y = 2x - 5 \quad \text{Solve for } y.$$
$$4x - 2y = 10$$
$$4x - 2(2x - 5) = 10 \quad \text{Substitute for } y.$$
$$4x - 4x + 10 = 10$$
$$10 = 10$$

The lines are the same. There is an infinite number of solutions that satisfy the equation $y = 2x - 5$.

Problems in everyday life can lead to a system of equations that can be solved using the substitution method.

Example 4

SHIPPING An appliance store delivers large appliances using vans and trucks. When loaded, each van holds 4 appliances and each truck holds 6. If 42 appliances are delivered and 8 vehicles are full, how many vans and trucks are used?

Solution

Define each of the variables.

Write and solve a system of equations relating to the variables.

Let t = number of trucks used
v = number of vans used

Problem Solving Tip

It is not necessary to always use x and y as the variables. Use letters that will help you remember what the variables represent. Writing them in cursive or caps helps separate them from math symbols.

Extend the Lesson

COOPERATIVE LEARNING Have each student write a system of equations that has one solution. Then have students trade systems, solve and compare solutions found with the solutions the writers intended.

Chalkboard Examples

Supplementary Example 1

Solve by substitution: $x + 3y = 7$, $-2x + 4y = 6$

(1, 2)

Supplementary Example 2

Solve by substitution: $3x - 2y = 2$, $9x - 6y = 2$

no solution; the lines are parallel

Supplementary Example 3

Solve by substitution: $x + 4y = 10$, $\frac{1}{2}x + 2y = 5$

infinite number of solutions

Supplementary Example 4

Dante bought two tapes and one compact disc for \$25. Marc bought one tape and three compact discs for \$40. Find the cost of each tape and each compact disc. \$7; \$11

Reteaching Worksheet 6-5

Name ____________________ Date __________

RETEACHING 6-5

SOLVING SYSTEMS BY SUBSTITUTION

You can use algebraic methods to find the solution to a system of equations.

Example

Solve. $\begin{cases} 3x + 6y = 12 \\ 3x - 6y = 12 \end{cases}$

Solution

Solve the first equation for y in terms of x.

$$3x + 6y = 12$$
$$6y = -3x + 12$$
$$y = -\frac{1}{2}x + 2$$

Write the second equation.
Substitute $\left(-\frac{1}{2}x + 2\right)$ for y.
Solve for x.

$$3x - 6y = 12$$
$$3x - 6\left(-\frac{1}{2}x + 2\right) = 12$$
$$3x + 3x - 12 = 12$$
$$6x = 24$$
$$x = 4$$

Choose one of the original equations; then substitute 4 for x.

$$3x + 6y = 12$$
$$3(4) + 6y = 12$$
$$12 + 6y = 12$$
$$6y = 0$$
$$y = 0$$

The solution is (4, 0).
You can check the solution by substituting 4 for x and 0 for y in each equation.

EXERCISES

Solve each system of equations by the substitution method. Check the solution.

1. $\begin{cases} x + y = 4 \\ 2x - y = 5 \end{cases}$ (3, 1)
2. $\begin{cases} 5x + y = 0 \\ x - 2y = 11 \end{cases}$ (1, −5)
3. $\begin{cases} -4x + 2y = 8 \\ 2x + 2y = 6 \end{cases}$ $\left(-\frac{1}{3}, \frac{10}{3}\right)$
4. $\begin{cases} 2x + \frac{1}{2}y = 25 \\ -x - y = 10 \end{cases}$ (20, −30)
5. $\begin{cases} x = 4y + 16 \\ 2x - y = 53 \end{cases}$ (28, 3)
6. $\begin{cases} y = 7x + 21 \\ 3x - y = 11 \end{cases}$ (−8, −35)
7. $\begin{cases} 2x + 3y = 3 \\ x - 4y = 7 \end{cases}$ (3, −1)
8. $\begin{cases} -6x + \frac{1}{2}y = 30 \\ -x - \frac{1}{2}y = -2 \end{cases}$ (−4, 12)

92 RETEACHING *LESSON 6-5*

Lesson Wrap-up

Quick Assessment

Ask the following questions to determine if students understand the content presented in this lesson.

1. In Example 2, why does $12 \neq 6$ mean there is no solution the system? Regardless of the value substituted for x in the second equation, the equation will not be true.
2. In Example 3, why does $10 = 10$ mean there is an infinite number of solutions? Regardless of the value substituted for x in the second equation, the equation will always be true.
3. What methods could you use to check the solution to a system of equations? Substitute the values in each equation, or graph the equations and compare the point of intersection to the solution.

Assignment Guide

Basic: 1-20, 26–33
Advanced: 1-33
Additional Practice: See Extra Practice Index on page 674.

There are 8 vehicles.

$$v + t = 8$$
$$t = 8 - v \quad \text{Solve for } t.$$

42 appliances are delivered; 6 in each truck, and 4 in each van.

$$4v + 6t = 42$$
$$4v + 6(8 - v) = 42 \quad \text{Substitute } 8 - v \text{ for } t.$$
$$4v + 48 - 6v = 42$$
$$-2v = -6 \quad \text{Solve for } v.$$
$$v = 3$$

$$v + t = 8$$
$$3 + t = 8 \quad \text{Substitute for } v.$$
$$t = 5 \quad \text{Solve for } t.$$

There are 3 vans and 5 trucks delivering appliances.

Problem Solving Tip

Steps to follow when using the substitution method.

1. Solve one of the equations for one variable in terms of the other.
2. Substitute that expression in the other equation and solve.
3. Substitute that value in one of the original equations and solve.
4. Check the solution in both of the original equations.

Try These Exercises

Solve and check each system of equations by the substitution method.

1. $2x + y = 0$, $x - 5y = -11$ (−1, 2)
2. $x + 3y = -9$, $-5x - 2y = -7$ (3, −4)
3. $x = \frac{1}{2}y$, $-x + 6y = -11$ (−1, −2)
4. $x - 5y = 6$, $y = -\frac{1}{2}x - \frac{1}{2}$ (1, −1)
5. **PACKAGING** The perimeter of a rectangular package is 78 cm. If the width is $\frac{5}{8}$ of the length, find the dimensions of the package. length 24 cm; width, 15 cm

Practice Exercises

Solve and check each system of equations by the substitution method.

6. $3x = y + 9$, $2x - 4y = 16$ (2, −3)
7. $-4x + 3y = -16$, $-x + 2y = -4$ (4, 0)
8. $\frac{x}{2} - y = \frac{5}{4}$, $8x + 3y = 1$ $\left(\frac{1}{2}, -1\right)$
9. $5x + 2y = 1$, $3x + 4y = -5$ (1, −2)
10. $6x - 3y = -9$, $13x - 5y = -15$ (0, 3)
11. $10x - 5y = 65$, $10y - 5x = -55$ (5, −3)
12. **RECREATION** Jake is going on a 20-day vacation to the beach and to the mountains. He wants the time spent at the beach to be $\frac{2}{3}$ the time spent in the mountains. How many days will he spend at the beach, and how many in the mountains? 8 days, 12 days

13. **WRITING MATH** A friend asks you how to know which variable to solve for first when solving a system of equations by substitution. What advice would you give? Since it doesn't matter which variable you solve for first, choose the variable that will be easy to isolate in one of the equations.

Extend the Lesson

MATH JOURNAL Have students select a system of equations that they solved in Lesson 6-4 by graphing. Have them write an explanation of how they solved the system using graphing, how they could solve it using substitution, which method they prefer and why.

14. **FINANCE** Shari has 17 coins consisting of dimes and quarters worth $3.35. How many quarters and how many dimes does she have? 11 quarters, 6 dimes

15. At the Hearty Hut, Chad bought 4 hamburgers and 5 fries and paid $8.71. Alisa bought 1 hamburger and 3 fries and paid $3.56. Find the cost of each hamburger and each order of fries. $1.19, $0.79

Solve each system of equations by the substitution method. Check the solutions.

16. $4y + 5x - 5 = 2x + 5y$ (2, 1)
$-9x + 2y = -4x - 8y$

17. $\frac{-3a + 5b}{2} = 1, \frac{6a - 5b}{2} = 1$ $\left(\frac{4}{3}, \frac{6}{5}\right)$

DATA FILE For Exercises 18–20, use the Data Index on pages 632–633 to find the location of a table showing the **shrinking value of the dollar**. Use a system of equations to find the year in which these prices existed.

18. 3 quarts of milk and 5 pounds of round steak cost $5.33; 5 quarts of milk and 1 pound of round steak cost $1.99 1950

19. 20 pounds of potatoes and 15 pounds of flour cost $4.92; 10 pounds of potatoes and 25 pounds of flour cost $5.89 1975

20. 5 pounds of flour and 4 quarts of milk cost $0.79; 15 pounds of flour and 5 quarts of milk cost $1.39 1930

Extended Practice Exercises

Solve each system of equations by the substitution method. Check the solutions.

21. $\frac{1}{x} + \frac{1}{y} = -1$, $\frac{4}{x} - \frac{5}{y} = 23$ $\left(\frac{1}{2}, -\frac{1}{3}\right)$

22. $\frac{2}{p} - \frac{1}{q} = -11$, $\frac{-3}{p} - \frac{2}{q} = 13$ $\left(\frac{-1}{5}, 1\right)$

23. $-x + y - z = 4$, $y = -x$, $x - 4z = 7$ (−1, 1, −2)

24. $-2a - 3b + c = 6$, $c = a + 2b + 5$, $a + 2b + 3c = -1$ (2, −3, 1)

25. **FINANCE** Coins consisting of nickels, dimes, and quarters total $2.40. The number of dimes is equal to one less than $\frac{2}{3}$ the number of nickels. Three times the number of quarters plus the number of dimes is 18. How many of each coin are there? 15 nickels, 9 dimes, 3 quarters

Mixed Review Exercises

26. Find the volume of the figure to the nearest thousandth. (Lesson 5-7) 729 cm^3

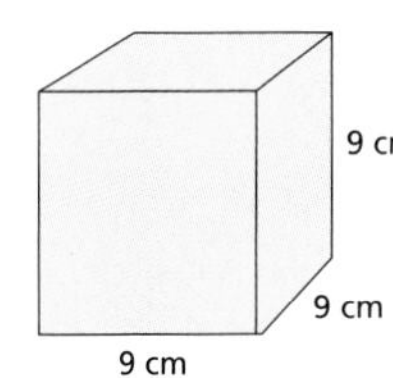

Evaluate each expression when $a = -4$ and $b = 2$. (Lesson 1-8)

27. $a^2 - b^2$ 12
28. a^2b^2 64
29. $(a - b)^2$ 36
30. $(a^2 + b^2)^2$ 400
31. $a^3 - b^3$ −72
32. $4(ab)^3$ −2048
33. $-3(a^3 - ab)^2$ −9408
34. $2(ab^2 + a^2b)$ 32

Teaching Strategies

Students may find it easier to work with an equation that contains fractions if they first eliminate the fractions by multiplying both sides of the equation by the denominator or the LCM of the denominators (if they are different).

Extra Practice Worksheet 6-5

Name ______ Date ______

EXTRA PRACTICE 6-5
SOLVE SYSTEMS BY SUBSTITUTION

EXERCISES

Solve and check each system of equations by the substitution method.

1. $4x + y = -2$, $2x = -y$ (−1, 2)
2. $2x - y = 3$, $2y = x - 3$ (1, −1)
3. $4x + y = 0$, $x - y = 10$ (2, −8)
4. $4x = 2y$, $x + 2 = y$ (2, 4)
5. $2x - 3y = -2$, $4x + y = 3$ $\left(\frac{1}{2}, 1\right)$
6. $y + 4 = 2x$, $\frac{x}{2} + \frac{y}{2} = -2$ (0, −4)
7. $x - 2y = -2$, $3x - 5y = -7$ (−4, −1)
8. $3x + 8y = 3$, $x + 4y = 2$ $\left(-1, \frac{3}{4}\right)$
9. $3x - 2y = -2$, $5x + y = -12$ (−2, −2)
10. $-2y - 6x = 6$, $x + y = 3$ (−3, 6)
11. $x + y = -1$, $3x + 2y = 2$ (4, −5)
12. $2x + y = 7$, $x - 2y = -14$ (0, 7)
13. $x = y - 2$, $3x - 4y = -13$ (5, 7)
14. $2x + 5y = -3$, $x + 8y = 4$ (−4, 1)

15. Ryan has 10 coins consisting of dimes and nickels worth $0.70. How many dimes and how many nickels does he have? 4 dimes, 6 nickels

16. Brooke spent $94.92 at the music store. She bought some cassette tapes for $9.99 each and some CDs for $12.99 each. How many cassette tapes and how many CDs did she buy? 3 cassette tapes, 5 CDs

17. Hannah jogs and walks for 48 minutes everyday. She spends 3 times as many minutes jogging as she does walking. How many minutes does she jog each day, and how many minutes does she walk? 36 min jogging, 12 min walking

18. Trevor spent $48.50 on 13 pounds of steaks and ground beef for his cookout. The steaks cost $6.50 a pound and the ground beef cost $2.50 a pound. How much steak and how much ground beef did Trevor buy? 4 lb, 9 lb

82 EXTRA PRACTICE *LESSON 6-5*

Enrichment Worksheet 6-5

Name ______ Date ______

ENRICHMENT 6-5
POINT OF DIVISION FORMULAS

Here are two equations that can be used to divide any line segment into two proportional parts. Let the ratio between the parts be $r_1 : r_2$. If the two endpoints of the line segment are (x_1, y_1) and (x_2, y_2), then the point of division is given by the pair (x_0, y_0) where

$$x_0 = \frac{r_2x_1 + r_1x_2}{r_1 + r_2} \quad y_0 = \frac{r_2y_1 + r_1y_2}{r_1 + r_2}$$

EXERCISES

Graph the line segment that connects each pair of points. Then use the point of division formulas to find the coordinates of the point three-fourths of the distance from the first to the second point. Label the "dividing" point *P*.

1. (−6, 6) and (2, 4) P(0, 4.5)

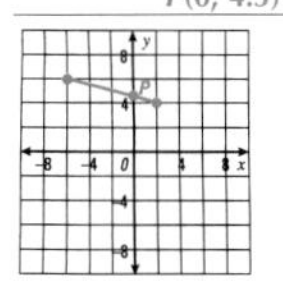

2. (−4, 8) and (6, 0) P(3.5, −2)

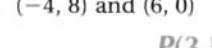

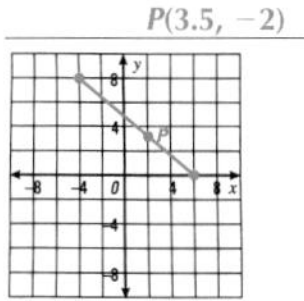

3. (−1, 5) and (3, −4) P(2, −1.75)

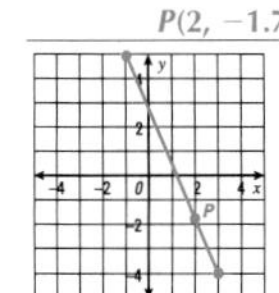

4. (−6, 8) and (3, −2) P(0.75, 0.5)

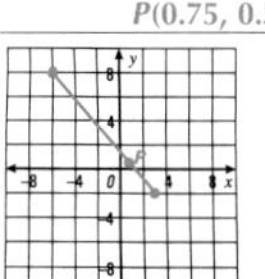

106 ENRICHMENT *LESSON 6-5*

Lesson Planning

NCTM Standards/Strands

- Number & Operation
- Patterns, Functions, & Algebra
- Problem Solving
- Reasoning and Proof
- Connections
- Representation

Vocabulary

addition/subtraction method
multiplication and addition method

Materials Needed

Algeblocks

Lesson Resources

Warm-Up Transparency 17
Transparency TK-7–10
Reteaching 6-6
Extra Practice 6-6
Enrichment 6-6

Assignment Guide

Basic: 1–25, 31–44
Enriched: 1–44

Getting Started

5-Minute Warm-Up

Find the LCM of each pair of numbers.

1. 2 and 5 10
2. 3 and 9 9
3. 4 and 6 12
4. 6 and 8 24

Introduction to Lesson 6-6

After students have simplified the expressions, ask them to describe the results of the simplifying. One variable disappeared. Remind students that, when adding and subtracting monomials, only coefficients are added or subtracted.

6-6 Solve Systems by Adding and Multiplying

Goals ■ Solve systems of equations by adding, subtracting and multiplying.

Applications Entertainment, Community Service, Manufacturing

Work with a partner and use Algeblocks to simplify these expressions.

a. $(3x - y - 6) - (3x - 2y - 6)$ y

b. $(4x + y + 4) + (2x - y - 10)$ $6x - 6$

Build Understanding

Another algebraic method for solving a system of equations is the **addition/subtraction method**. To eliminate one of the variables to get one equation in one variable, you can add or subtract the two equations. If the *coefficients* of one of the variables are *opposites* or the *same*, simply adding or subtracting the equations eliminates one of the variables.

Steps to follow when using the addition/subtraction method:

1. If the coefficients of one of the variables are opposites, add the equations to eliminate one of the variables.
   If the coefficients of one of the variables are the same, subtract the equations to eliminate one of the variables.
2. Solve the resulting equation for the remaining variable.
3. Substitute the value for the variable in one of the original equations and solve for the unknown variable.
4. Check the solution in both of the original equations.

Problem Solving Tip

The addition or subtraction is best done with equations written in standard form:
$Ax + By = C$

Example 1

Solve: $2x + 7y = -5$
$-5x + 7y = -12$

Solution

$2x + 7y = -5$
$-5x + 7y = -12$ The y-coefficients are the *same*, so *subtract* the equations.

$2x + 7y = -5$ $\rightarrow$ $2x + 7y = -5$
$-(-5x + 7y = -12)$ $\quad 5x - 7y = 12$
$7x = 7$
$x = 1$

Remember: to subtract, you add the opposite of each term.

Choose one of the original equations.

$2x + 7y = -5$
$2(1) + 7y = -5$ Substitute for x.
$7y = -7$
$y = -1$ The check is left to you. The solution is $(1, -1)$.

Extend the Lesson

MATH JOURNAL Have students choose a system of equations they solved in a previous lesson and describe how it could be solved by adding, subtracting and/or multiplying.

Unless the coefficients of one variable are the same or are opposites, you will still have an equation with two variables when you add or subtract a system of equations. In that case, you need to multiply one or both of the equations by a number to obtain an equivalent system of equations where the coefficients of one of the variables of the equivalent system are the same or opposite.

This method combines the multiplication property of equations with the addition/subtraction method, and is known as the **multiplication and addition method**.

Both the addition method and the multiplication and addition method are best used when the equations are written in standard form.

Example 2

Solve: $3x - 4y = 10$
$3y = 2x - 7$

Solution

$3x - 4y = 10$
$-2x + 3y = -7$ Rewrite the second equation in standard form.

$3x - 4y = 10 \rightarrow 2(3x - 4y = 10) \rightarrow 6x - 8y = 20$
$-2x + 3y = -7 \rightarrow 3(-2x + 3y = -7) \rightarrow -6x + 9y = -21$

Add. $\rightarrow y = -1$

$3x - 4y = 10$ Choose one of the original equations.

$3x - 4(-1) = 10$ Substitute for y.

$3x + 4 = 10$

$3x = 6$

$x = 2$

The check is left to you. The solution is $(2, -1)$.

Problem Solving Tip

The least common multiple of 3 and 2 is 6. We can eliminate the x terms if we multiply the first equation by 2,and the second equation by 3. The coefficients will be 6 and -6.

Example 3

ENTERTAINMENT Tim sold 25 movie tickets for a total of $132. If each adult ticket sold for $6 and each children's ticket sold for $4, how many of each kind did he sell?

Solution

Make a chart for the number and values of the tickets.

	Adult	Child	Total
Number	A	C	$A + C$
Value	$6A$	$4C$	$6A + 4C$

Write and solve a system of equations to represent the problem.

Chalkboard Examples

Supplementary Example 1

Solve: $x - y = -5$
$x + y = 15$ (5, 10)

Supplementary Example 2

Solve: $x - 5y = 0$
$3y = 2x - 7$ (5, 1)

Supplementary Example 3

Carol sold 40 tickets to a school play for a total of $245. If each adult ticket sold for $8 and each children's ticket sold for $3, how many of each kind did she sell?
25 adult and 15 children's tickets

Teaching Strategies

COOPERATIVE LEARNING Have students write a word problem based on the value of coins, similar to Exercise 18. Then have students trade problems and solve.
Writing their own problems may help students understand how these types of problems are constructed and help them recognize them in their work.

Reteaching Worksheet 6-6

Name ______________________ Date ____________

RETEACHING 6-6

SOLVING SYSTEMS BY ADDING, SUBTRACTING, AND MULTIPLYING

Algebraic methods of solving a system of equations include the addition/subtraction method and the multiplication and addition method.

Example

Solve. $4x - 3y = 11$
$2x = 4y + 3$

Solution

Write each equation in standard form.
$4x - 3y = 11$
$2x = 4y + 3 \rightarrow 2x - 4y = 3$

Multiply to obtain an equivalent system of equations in which the coefficients of one of the variables are either the same or the opposite. Then add or subtract.
$4x - 3y = 11 \rightarrow 4x - 3y = 11 \rightarrow 4x - 3y = 11$
$2x - 4y = 3 \rightarrow -2(2x - 4y) = 3(-2) \rightarrow -4x + 8y = -6$
$5y = 5$, so $y = 1$

Choose one of the original equations, for example $4x - 3y = 11$. Substitute 1 for y.
$4x - 3(1) = 11$
$4x - 3 = 11$
$4x = 14$
$x = 3\frac{1}{2}$

The solution is $\left(3\frac{1}{2}, 1\right)$.

You can check the solution by substituting $3\frac{1}{2}$ for x and 1 for y in each equation.

EXERCISES

Solve each system of equations. Check the solution.

1. $2x - y = 4$; $x + 2y = 7$ (3, 2)
2. $3x - y = 4$; $4y = -2x + 12$ (2, 2)
3. $x = 3y - 6$; $6y = x + 3$ (−9, −1)
4. $-x + 3y = 8$; $y = 2x - 4$ (4, 4)
5. $6x + 6y = 6$; $x + y = 1$ infinite solutions
6. $5x - y = 4$; $3x = -2y + 18$ (2, 6)
7. $3y = 2x - 9$; $6x + 13 = y$ (−3, −5)
8. $10x + 5y = 20$; $x = y + 2$ (2, 0)
9. The perimeter of a rectangle is 24 in. The length is twice the width. Find the dimensions. 4 in. by 8 in.

94 RETEACHING *LESSON* 6-6

Lesson Wrap-up

Quick Assessment

Ask the following questions to determine if students understand the content presented in this lesson.

1. How do you know when to use the multiplication and addition method? when the addition/subtraction method cannot be used to eliminate a variable
2. In Example 2, what could you do differently in order to eliminate the *y* variable instead of the *x*? How will this affect the solution? Multiply the first equation by 3 and the second by 4 to eliminate the y terms. The final solution is the same.

Assignment Guide

Basic: 1–25, 31–44
Advanced: 1–44
Additional Practice: See Extra Practice Index on page 674.

The number of tickets sold is 25. $A + C = 25$
The value of the tickets is \$132. $6A + 4C = 132$

$$A + C = 25 \rightarrow -4A - 4C = -100 \quad \text{Multiply the first equation by } -4.$$
$$6A + 4C = 132 \rightarrow 6A + 4C = 132$$
$$2A = 32$$
$$A = 16$$

$$A + C = 25 \quad \text{Choose one of the original equations.}$$
$$16 + C = 25 \quad \text{Substitute for } A.$$
$$C = 9$$

Tim sold 16 adult tickets and 9 children's tickets.

Try These Exercises

Solve each system of equations. Check the solutions.

1. $x - y = -1$, $x + y = 9$ (4, 5)
2. $3x + y = -7$, $5x - y = -9$ (−2, −1)
3. $-3x - 5y = 4$, $2x + y = -5$ (−3, 1)
4. $4x = 2y - 12$, $3y = x + 13$ (−1, 4)
5. **COMMUNITY SERVICE** Adult tickets for a benefit breakfast cost \$2.50. Children's tickets cost \$1.50. If 56 tickets were sold for total sales of \$97, how many of each kind were sold? 13 adults, 43 children

6. **FINANCE** Afton invested \$5,400 in two products, a new mouthwash and a new line of frozen dinners. She invested $\frac{1}{2}$ as much money in mouthwash as she did in frozen dinners. How much did she invest in each product? mouthwash, \$1,800; dinners, \$3,600.
7. **MANUFACTURING** It takes 18 months from the time a shoe design is approved for the shoes to arrive in stores for sale. The actual assembly of the shoe takes 3 months less than the time spent on development. How many months does the assembly take? $7\frac{1}{2}$ months

8. **WRITING MATH** Using the multiplication and addition method, how can you know when a system has an infinite number of solutions? The final step is 0 = 0

Practice Exercises

Solve each system of equations. Check the solutions.

9. $-x + 2y = 3$, $3x + 2y = -1$ (−1, 1)
10. $x = 5y + 7$, $5y = 9x + 15y - 8$ (2, −1)
11. $-2x + 6y = 10$, $2x - 9y = -19$ (4, 3)
12. $-10x + 6y = 25$, $9y = -2x + 12$ $\left(\frac{-3}{2}, \frac{5}{3}\right)$
13. $4x - 9y = -1$, $9y = 8x + 1$ $\left(0, \frac{1}{9}\right)$
14. $7x + 4y = 6$, $5y + 7x = 15$ $\left(\frac{-30}{7}, 9\right)$
15. $8x + 3y = -4$, $6x + 5y = 8$ (−2, 4)
16. $12x - 3y = 18$, $8x = 2y + 12$ infinite number of solutions
17. **FARMING** The Deckerts grow wheat and barley on their 1,200-acre farm. The amount of wheat they plant is 200 acres more than 3 times the number of acres of barley. How many acres of wheat and how many acres of barley do they plant? wheat, 950 acres; barley, 250 acres
18. Jason has 15 coins in his pocket, consisting of nickels and dimes. The total value of the coins is \$1.15. How many of each coin does Jason have? 7 nickels, 8 dimes

270 Chapter 6 **Linear Systems of Equations**

Additional Answers

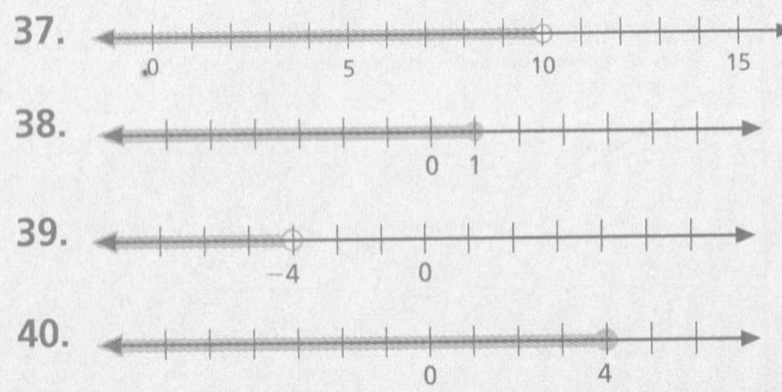

41. 0 6
42. −4 0
43. 0 5 10 15 20

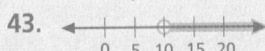

Solve each system of equations. Check the solutions.

19. $\frac{x}{2} - \frac{4y}{3} = -3$ (2, 3)
$-3x + 4y = 6$

20. $2x - 5 = 3y - 7$ (2, 2)
$6y + 4 = 5x + 6$

21. $\frac{-x}{3} + \frac{2}{9}y = 1$ (3, 9)
$\frac{2}{3}x + \frac{y}{18} = \frac{5}{2}$

22. $\frac{1}{3}a - \frac{1}{2}b = -9$ (−12, 10)
$\frac{1}{5}b = \frac{1}{4}a + 5$

23. $\frac{5x + 3y}{10} = \frac{-1}{2}$ (−4, 5)
$\frac{9x}{16} + \frac{7y}{25} = \frac{-17}{20}$

24. $\frac{s}{2} - \frac{r + 3}{4} = \frac{-1}{2}$ (1, 1)
$\frac{2r - 1}{3} - \frac{3s}{4} = \frac{-5}{12}$

25. A number divided by 3 plus another number divided by 9 have a sum of 7. If the first number is multiplied by 4 and divided by 5 and then subtracted *from* the second number divided by 2, the result is −3. What are the numbers? 15 and 18

Extended Practice Exercises

Solve each system of equations. Check the solutions.

26. $\frac{9}{x} + \frac{4}{y} = -19$ $\left(-\frac{1}{3}, \frac{1}{2}\right)$
$\frac{-6}{x} - \frac{2}{y} = 14$

27. $\frac{7}{s} = \frac{2}{r} + \frac{2}{15}$ (10, 21)
$\frac{5}{r} = \frac{3}{s} + \frac{5}{14}$

28. $3x - 2y + z = 9$ (2, −1, 1)
$2x + y - z = 2$
$4x + 5y + 2z = 5$

29. $-a + 3b - 2c = 9$ (−1, 2, −1)
$9a + 4b = 2c + 1$
$12a + 4b + 8c = -12$

30. $y = ax^2 + bx + c$ is an equation for a quadratic function. (0, 0), (1, −2) and (2, 3) are solutions to the equation. Find *a*, *b*, and *c*. $a = \frac{7}{2}$, $b = -\frac{11}{2}$, $c = 0$

Mixed Review Exercises

Complete the table. Use π = 3.14. Assume each planet is a sphere. Round answers to the nearest million. (Lesson 5-6)

Planet	Diameter at equator	Radius	Surface area
31. Mercury	3031 mi	1515.5 mi	
32. Mars	4200 mi		
33. Saturn	71,000 mi		
34. Uranus	32,000 mi		
35. Neptune	30,600 mi		
36. Pluto	715 mi		

Radius	Surface area
	29,000,000
2100	55,000,000
35,500	15,837,000,000
16,000	3,216,000,000
15,300	2,942,000,000
715	6,000,000

Solve each inequality. Graph the solution on the number line. (Lesson 2-6) See additional answers.

37. $7 - x > -3$

38. $6 + 5x \leq 11$

39. $x - 4 < -8$

40. $8 - x \geq 4$

41. $3x - 5 < 13$

42. $4(x + 3) \geq -4$

43. $\frac{1}{2}(x - 4) > 3$

44. $2(3 - x) \leq -1$

Teaching Strategies

CHALLENGE In what type(s) of systems of equations could both variables be eliminated by addition? Give examples to support your answer. inconsistent system (no solution) and dependent system (infinite solutions)

Extra Practice Worksheet 6-6

Name ______________ Date ______________

EXTRA PRACTICE 6-6

Solve Systems by Adding, Subtracting, and Multiplying

Exercises

Solve each system of equations. Check the solutions.

1. $-x + 5y = 8$, $x - y = 4$ (2, −2)

2. $x + 2y = -2$, $3x - 2y = -14$ (−4, 1)

3. $2x - 3y = -1$, $-2x + 4y = 4$ (4, 3)

4. $3x + 2y = -4$, $x - 2y = 4$ (0, −2)

5. $x - 3y = -3$, $y = 3x - 1$ $\left(\frac{3}{4}, \frac{5}{4}\right)$

6. $3y + 2 = 2x$, $x + y = 1$ (1, 0)

7. $4x + y = -1$, $x - 2y = 3$ $\left(\frac{1}{9}, -\frac{13}{9}\right)$

8. $5x + 2y = 1$, $x - 4y = 2$ $\left(\frac{4}{11}, -\frac{9}{22}\right)$

9. $4x - y = -10$, $x + 2y = 20$ (0, 10)

10. $3x - 5y = 2$, $2y = x - 3$ (−11, −7)

11. $x + y = 7$, $2x + 3y = 22$ (−1, 8)

12. $2y = 1 - x$, $2x - y = 4$ $\left(\frac{9}{5}, -\frac{2}{5}\right)$

13. $3x = 3y + 1$, $2x - 6y = 4$ $\left(-\frac{1}{2}, -\frac{5}{6}\right)$

14. $2x + 5y = 2$, $4x + y = 4$ (1, 0)

15. $\frac{x}{2} - \frac{y}{2} = -3$, $-x + 2y = 8$ (−4, 2)

16. $4x + 2 = 3y + 5$, $2y - 1 = x + 2$ (3, 3)

17. $-\frac{x}{3} - \frac{3}{4}y = 4$, $\frac{2}{3}x + \frac{y}{8} = 3$ (6, −8)

18. $\frac{2}{5}x + \frac{5}{6}y = 6$, $\frac{y}{3} = \frac{x}{2} + 9$ (10, 12)

19. Mika paid \$68 for 14 movie tickets. Some of the tickets were adult tickets, and some were student tickets. Each adult ticket cost \$6 and each student ticket cost \$4. How many adult tickets and how many student tickets did she buy?
6 adult tickets, 8 student tickets

Enrichment Worksheet 6-6

Name ______________ Date ______________

ENRICHMENT 6-6

Systems with a System!

Exercises

Connect each set of points in order to form a systematic design on the grid at the right.

Set A

1. Where $y = 3x + 12$ meets $y = -2x + 27$ (3, 21)
2. Where $y = -x + 24$ meets $y = 2x - 12$ (12, 12)
3. Where $y = ax$ meets $y = -ax$ (0, 0)
4. Where $y = -x$ meets $x - 2y = 36$ (12, −12)
5. Where $-15 = 2x + y$ meets $-6 = 5x + y$ (3, −21)
6. Where $60 = -x + 3y$ meets $75 = 4x + 3y$ (3, 21)

Set B

7. Where $x = 3y + 72$ meets $-5x + 9y = -234$ (9, −21)
8. Where $y = 3x - 12$ meets $2y + 24 = x$ (0, −12)
9. Where $4y = -x + 12$ meets $y = x - 12$ (12, 0)
10. Where $12 = 2x + y$ meets $5y - 60 = x$ (0, 12)
11. Where $y = 2x + 3$ meets $x + y = 30$ (9, 21)
12. Where $y - x + 30 = 0$ meets $x + y + 12 = 0$ (9, −21)

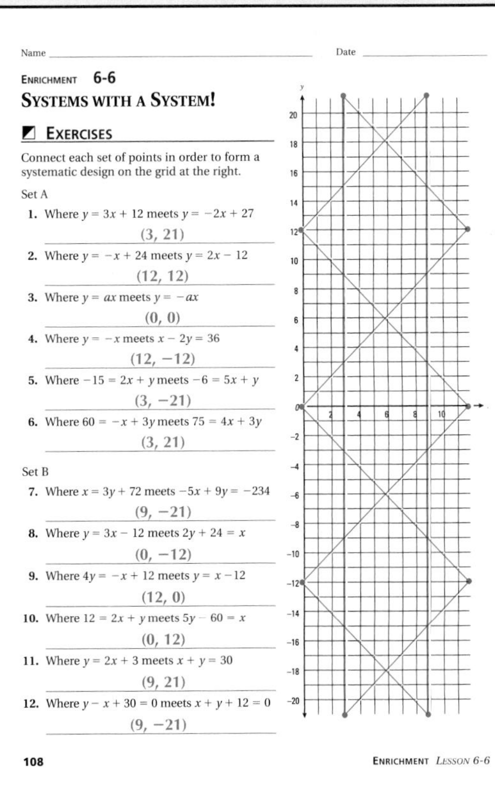

Skills Practice

Vocabulary Review

Lesson 6-5
substitution method

Lesson 6-6
addition/subtraction method
multiplication and addition method

Assignment Guide

All students: 1–55
Additional Practice: Refer to the Extra Practice Index on page 674 of this text.

Additional Answers

36.

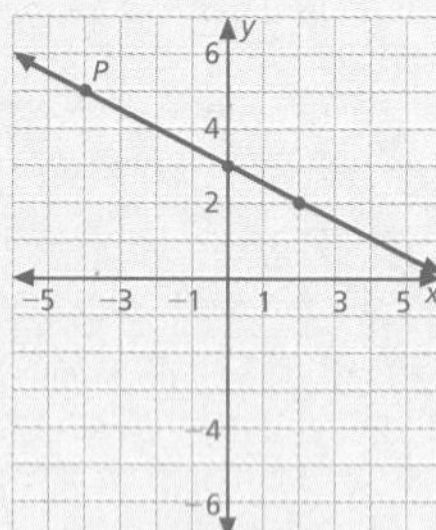

37.

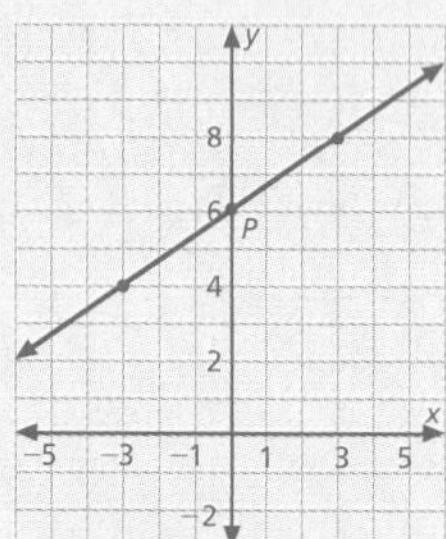

38.

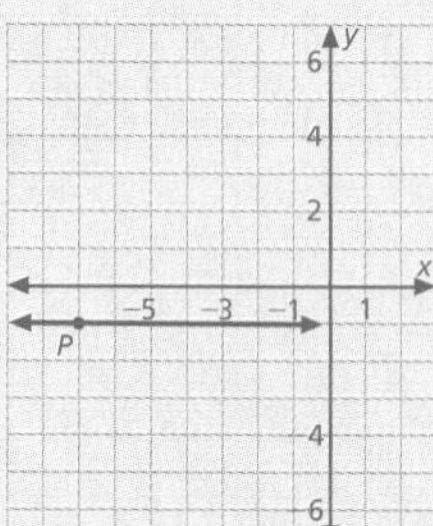

Review and Practice Your Skills

Practice ■ Lesson 6-5

Solve and check each system of equations by the substitution method.

1. $x = 5y$, $x - 3y = 6$ (15, 3)
2. $m = -2m - 2$, $3m - 2m = 10$ (2, −2)
3. $x - y = 1$, $y = -x + 5$ (3, 2)
4. $y + 6 = 3x$, $9x - 2y = 3$ (−3, −15)
5. $3x - 2y = -3$, $3x + y = 3$ $\left(\frac{1}{3}, 2\right)$
6. $3p - 4q = 8$, $4p + q = 17$ (4, 1)
7. $x - 2y = 16$, $4x + y = 1$ (2, −7)
8. $s + 2t = 6$, $4s + 3t = 4$ (−2, 4)
9. $y = -4x + 5$, $2x - 3y = 13$ (2, −3)
10. $5n - v = -23$, $3n - v = -15$ (−4, 3)
11. $-2x + y = 2$, $2x + 3y = 6$ (0, 2)
12. $x + 5y = 11$, $4x - y = 2$ (1, 2)
13. $2x + 3y = 11$, $3x + 3y = 18$ (7, −1)
14. $5a + 3b = 4$, $4a - 2b = 1$ (0.5, 0.5)
15. $5x + 7y = -3$, $2x + 14y = 2$ $\left(-1, \frac{2}{7}\right)$
16. Leonard has 23 coins consisting of quarters and nickels worth \$4.15. How many quarters and how many nickels does he have? 15 quarters, 8 nickels
17. Lisa bought 7 bagels and 4 peaches and paid \$5.25. Emily bought 1 bagel and 6 peaches and paid \$2.65. Find the cost of each bagel and each peach. bagel, \$0.55; peach, \$0.35

Practice ■ Lesson 6-6

Solve each system of equations. Check the solutions.

18. $x - y = -5$, $x + y = 1$ (−2, 3)
19. $2r + s = -1$, $-2r + s = 3$ (−1, 1)
20. $x + y = 5$, $x + 2y = 8$ (2, 3)
21. $3m + n = -6$, $m - n = -2$ (−2, 0)
22. $3x + 4y = 8$, $5x - 4y = 24$ (4, −1)
23. $x + 2y = 0$, $x - y = -3$ (−2, 1)
24. $8x - 3y = 17$, $-7x + 6y = 2$ (4, 5)
25. $7p - 10q = -1$, $3p + 2q = -13$ (−3, −2)
26. $4x - 3y = 15$, $8x + 2y = -10$ (0, −5)
27. $2a + 8b = -1$, $-10a + 4b = 16$ (−1.5, 0.25)
28. $5x + 3y + 9 = 0$, $3x - 4y + 17 = 0$ (−3, 2)
29. $6g - 5 = 2g - 7h$, $2g = 5h - 6$ $\left(-\frac{1}{2}, 1\right)$
30. $6x + 7y = -11$, $4x - 7y = 21$ $\left(1, \frac{-17}{7}\right)$
31. $3c + 4d = -15$, $3c + 9d = 10$ $\left(-11\frac{2}{3}, 5\right)$
32. $3x - 5y = -1$, $6x + 2y = 10$ $\left(\frac{4}{3}, 1\right)$
33. $-6x + 2y = -10$, $-5x + 7y = -35$ (0, −5)
34. $3x + 3y + 4 = 0$, $9x - 5y = -20 - 6x$ $\left(-\frac{4}{3}, 0\right)$
35. $3x - y = 15$, $y = 3(x - 7)$ no solution

Practice ■ Lessons 6-1–Lesson 6-6

Graph the line that passes through the given point *P* and has the given slope. (Lesson 6-1) See additional answers.

36. $P(-4, 5)$, $m = -\frac{1}{2}$
37. $P(0, 6)$, $m = \frac{2}{3}$
38. $P(-7, -1)$, $m = 0$
39. $P(4, -8)$, $m = 3$
40. $P(9, 7)$, $m = -5$
41. $P(1.5. 1.5)$, $m = 1.5$

39.

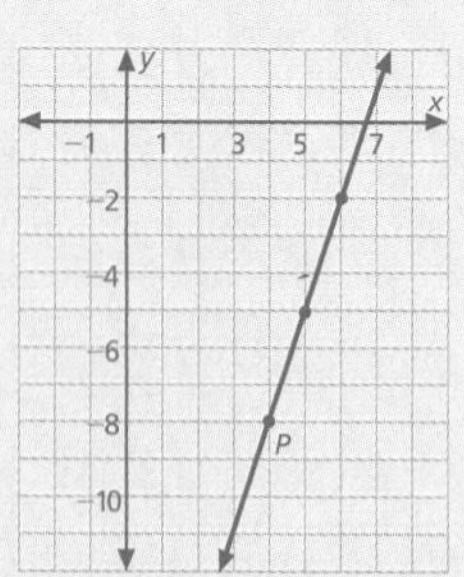

40.

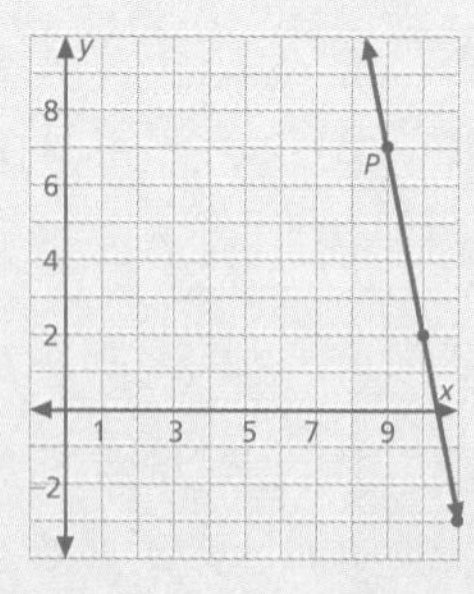

41.

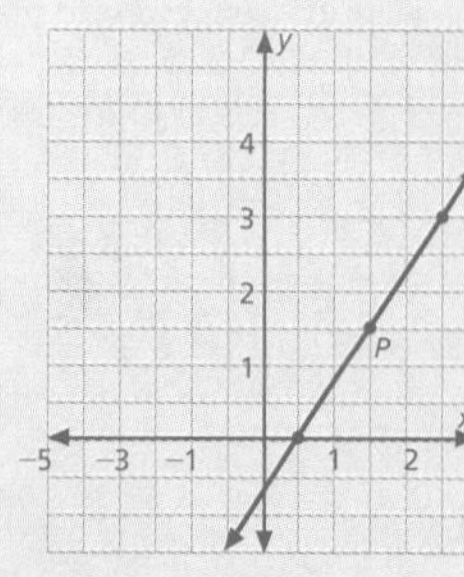

Teaching Strategies

Help students see that when both original equations have been multiplied, there are many places where errors may occur. Discuss the importance of substituting the ordered pair in both original equations as a check.

Determine whether each pair of lines is *parallel, perpendicular,* or *neither.* (Lesson 6-2)

42. The line containing $A(0,5)$ and $B(-3, 7)$
The line containing $Y(10, 2)$ and $Z(4, 6)$ parallel

43. The line containing $C(-4, -9)$ and $D(5, 9)$
The line containing $M(-7, -1)$ and $N(1, -5)$ perpendicular

44. $y = 4.5x - 6.3$ parallel
$18x - 4y = 36$

45. $y = -13$ neither
$y = -13x$

46. $3x + y = 0$ perpendicular
$3y = x$

Write an equation for the line with the given information. (Lesson 6-3)

47. $m = 3$, $Q(8, 5)$ $y = -3x + 29$

48. $m = \frac{3}{4}$, $K(8, -1)$ $y = \frac{3}{4}x - 7$

49. $m = 1.5$, x-intercept -4 $y = 1.5x + 6$

50. $G(10, 3)$ and $H(4, -3)$ $y = x - 7$

51. $P(-5, 2)$ and $Q(10, -1)$ $y = \frac{-1}{5}x + 1$

52. $T(-4, 7.5)$ and $U(-4, -12)$ $x = -4$

Solve each system of equations by graphing. (Lesson 6-4)

53. $x + y = 2$ (3, −1)
$y = x - 4$

54. $x + y = 1$ (−2, 3)
$y = x + 5$

55. $3x + 2y = -8$ (−2, −1)
$2x - 3y = -1$

MathWorks
Workplace Knowhow

Career – Engineering Technician

Engineering technicians help design and build new products. They employ mathematics, engineering and science to solve technical problems. They may be asked to create specifications for materials, establish quality testing procedures and improve manufacturing efficiency.

Engineering technicians work in laboratories, offices, industrial plants and construction sites. They use math to make precise measurements and create schematics. They must be able to apply mathematical reasoning to find ways to cut costs and save time.

1. You are testing two products for your company. Your boss gives you a budget of $800 to spend on the testing. You ended up spending $\frac{1}{4}$ as much on product A as on product B. If you spent the whole amount, how much did you spend on each product? product A, $160; product B, $640

You study efficiency in your plant by using the formula:

$$\text{efficiency rate} = \frac{\text{\# of products}}{\text{\# of hours}}$$

2. You discover that the efficiency rating for the second shift at the plant is 61.125 for $7\frac{1}{2}$ hours of work. How many products are they producing each day? 458 products

3. After research, you discover that the efficiency rating of the second shift increases by 1.75 points when the workers are given three 15-minutes breaks instead of only two. Recalculate the number of products the shift produces based on the loss of 15 minutes from the shift but a gain of 1.75 in efficiency rating. Should the plant increase the number of breaks? 455 products; no

Chalkboard Examples

Example from Lesson 6-5

Solve this system of equations using the substitution method.

$2x + y = 25$

$y = \frac{1}{3}x - 3$ (12, 1)

Example from Lesson 6-6

Solve these systems of equations by adding, subtracting and multiplying.

1. $5x + 4y = 12$
$4y = 3x - 4$ $\left(2, \frac{1}{2}\right)$

2. $y = 3x + 8$
$4x - 2y = -14$ (−1, 5)

Career Opportunity

Describe the kind of work engineering technicians do and the types of places they work. Discuss the importance of mechanical ability, geometry, measurement and algebraic thinking in working as an engineering technician. Students should answer Questions 1–3 to better understand how engineering technicians use mathematics in performing their job.

Students who are interested in learning more about this profession can go to learningmathmatters.com. Guidance Counselors should have information about school and training requirements.

Extend the Lesson

MATH JOURNAL Have students write an original word problem of the type modeled in Exercises 16 and 17. As a class, discuss how they can recognize word problems that can be solved by writing a system of equations.

Lesson Planning

NCTM Standards/Strands

- Number & Operation
- Patterns, Functions, & Algebra
- Problem Solving
- Reasoning and Proof
- Connections
- Representation

Vocabulary

determinant of a system of equations
matrix

Materials Needed

pencil/paper

Lesson Resources

Warm-Up Transparency 18
Reteaching 6-7
Enrichment 6-7
Transparency TK-7–10, RF-26

Assignment Guide

Basic: 1–12
Enriched: 1–12

Getting Started

5-Minute Warm-Up

Simplify.

1. $\dfrac{(3)(5) - (-3)(-3)}{(-3)(4) - (-2)(7)}$ 3
2. $\dfrac{(-2)(3) - (-1)(1)}{(1)(4) - (2)(6)}$ $\frac{5}{8}$

Use the Five-step Plan and the Strategy

THE FIVE-STEP PLAN **Read**—ask questions of the students to help them understand the problem. **Plan**—guide students to related problems and previously mastered skills and strategies. **Solve**—students solve problem on their own. **Answer**—write solution in format that answers the question. **Check**—review work, check for reasonableness and review strategy used. Students will benefit from the experience of verbalizing their methods.

6-7 Problem Solving Skills: Determinants & Matrices

Another method of solving a system of equations is the method of determinants. A 2 × 2 **determinant** is a square array in the form

$\begin{vmatrix} a & b \\ c & c \end{vmatrix}$ consisting of two rows and 2 columns.

The **determinant of a system of equations**, det A, is formed using the coefficient of the variables when the equations are written in standard form.

System of equations $\begin{cases} ax + by = e \\ cx + dy = f \end{cases}$ ← $\det A = \begin{vmatrix} a & b \\ c & d \end{vmatrix}$ Determinant of coefficients

The value of the determinant is given by $\det A = ad - bc$, which is the difference of the product of the diagonals.

You can find the solution to a system of equations using this determinant and another determinant formed by replacing the x or y column of the determinant with the constant column.

$A_x = \begin{vmatrix} e & b \\ f & d \end{vmatrix}$ Replace the x-column with the constant column.

$A_y = \begin{vmatrix} a & e \\ c & f \end{vmatrix}$ Replace the y-column with the constant column.

To find x, divide A_x by determinant A.

$$x = \frac{A_x}{A} = \frac{\begin{vmatrix} e & b \\ f & d \end{vmatrix}}{\begin{vmatrix} a & b \\ c & d \end{vmatrix}} = \frac{ed - bf}{ad - bc}$$

To find y, divide A_y by determinant A.

$$y = \frac{A_y}{A} = \frac{\begin{vmatrix} a & e \\ c & f \end{vmatrix}}{\begin{vmatrix} a & b \\ c & d \end{vmatrix}} = \frac{af - ec}{ad - bc}$$

Problem Solving Strategies

- Guess and check
- Look for a pattern
- Solve a simpler problem
- Make a table, chart or list
- Use a picture, diagram or model
- Act it out
- Work backwards
- Eliminate possibilities
- ✔ Use an equation or formula

Problem

Solve: $x + 3y = 4$ by the method of determinants.
$-2x + y = -1$

Solution

$$x = \frac{A_x}{A} = \frac{\begin{vmatrix} 4 & 3 \\ -1 & 1 \end{vmatrix}}{\begin{vmatrix} 1 & 3 \\ -2 & 1 \end{vmatrix}} = \frac{4(1) - (3)(-1)}{1(1) - 3(-2)} = \frac{4 + 3}{1 + 6} = \frac{7}{7} = 1$$

$$y = \frac{A_y}{A} = \frac{\begin{vmatrix} 1 & 4 \\ -2 & -1 \end{vmatrix}}{\begin{vmatrix} 1 & 3 \\ -2 & 1 \end{vmatrix}} = \frac{1(-1) - (4)(-2)}{1(1) - 3(-2)} = \frac{-1 + 8}{1 + 6} = \frac{7}{7} = 1$$

The solution is (1, 1). The check is left to you.

Reading Math

The e and f in these equations represent a numerical value. They are not part of variable expressions. They are called **constants**. The constant column is
e
f

THE STRATEGY Use a formula—Formulas are a useful way to help students remember complex procedures.

Additional Answers

1. $\begin{bmatrix} 5 & 1 \\ -3 & 4 \end{bmatrix} \begin{bmatrix} x \\ y \end{bmatrix} = \begin{bmatrix} 6 \\ 2 \end{bmatrix}$; det = 23
   Solution $\left(\frac{22}{23}, \frac{28}{23}\right)$
2. $\begin{bmatrix} 3 & 1 \\ 1 & 2 \end{bmatrix} \begin{bmatrix} x \\ y \end{bmatrix} = \begin{bmatrix} 2 \\ -4 \end{bmatrix}$; det = 7
   Solution (0, −2)
3. $\begin{bmatrix} 4 & -7 \\ -2 & 1 \end{bmatrix} \begin{bmatrix} x \\ y \end{bmatrix} = \begin{bmatrix} 2 \\ -4 \end{bmatrix}$; det = −10
   Solution $\left(\frac{13}{5}, \frac{6}{5}\right)$

A **matrix** is also an array of numbers. Each number in the array is called an **element**. A *column matrix* is an array of only one column. A *row matrix* is an array of only one row. A *square matrix* is an array with the same number of rows and columns. Examples are:

Column matrix: $\begin{bmatrix} 2 \\ 1 \end{bmatrix}$, Row matrix: $[1 \;\; 2]$, Square matrix: $\begin{bmatrix} 2 & 1 \\ -1 & 5 \end{bmatrix}$

A system of equations $2x - y = 4$, $-3x + 2y = 5$ can be written in matrix form by using a square matrix and 2 column matrices.

The matrix equation is $AX = B$. $\begin{bmatrix} 2 & -1 \\ -3 & 2 \end{bmatrix} \begin{bmatrix} x \\ y \end{bmatrix} = \begin{bmatrix} 4 \\ 5 \end{bmatrix}$

Reading Math

The matrix comes from the coefficient when the equations are in standard form.

$A = \begin{bmatrix} 2 & -1 \\ -3 & 2 \end{bmatrix}$ is the square matrix of coefficients.

$X = \begin{bmatrix} x \\ y \end{bmatrix}$ is a column matrix of variables.

$B = \begin{bmatrix} 4 \\ 5 \end{bmatrix}$ is a column matrix of constants.

Try These Exercises

Write the matrix equations for 1–5. Then for each of the systems of equations, determine the determinant of the coefficients, and find the solution of the system using determinants. For Problems 6 and 7, also define the system of equations.

For 1–5, see additional answers.

1. $5x + y = 6$, $-3x + 4y = 2$
2. $3x - y = 2$, $x + 2y = -4$
3. $4x - 7y = 2$, $-2x + y = -4$
4. **FINANCE** Deanna has \$2.15 in dimes and quarters. If the dimes were nickels and the quarters were dimes, she would have \$1.25 less. How many of each coin does Deanna have?
5. **BUSINESS** Car Rental Company A charges \$25 per day plus \$0.35 per mile. Company B charges \$35 per day, plus \$0.25 per mile. Wayne Know-it-all determines the cost of a trip he will take will be \$230 for Company A and \$250 for Company B. How many miles and for how many days will Mr. Know-it-all's trip be?

6. **WRITING MATH** If det $A = 0$, what is the solution of a system of equations? Why? No solution because division by 0 is undefined.

Mixed Review Exercises

Find the perimeter of each figure. (Lesson 5-2)

7. 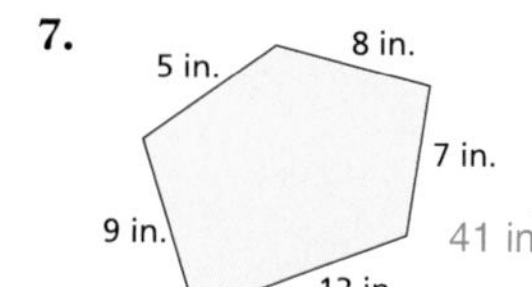

41 in.

8. 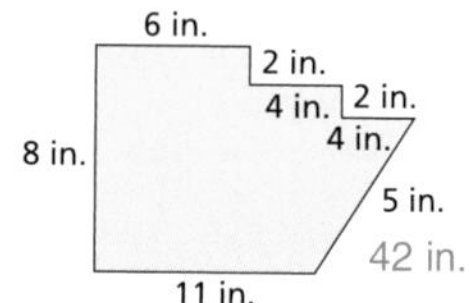

42 in.

9. 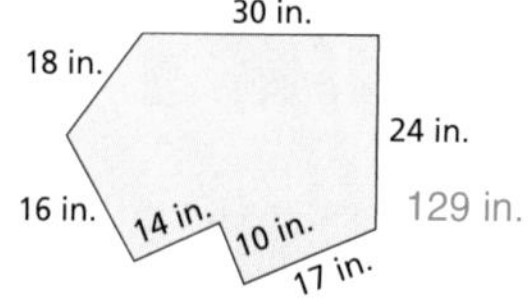

129 in.

10. 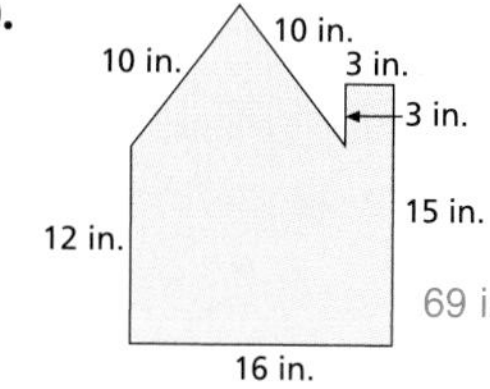

69 in.

11. 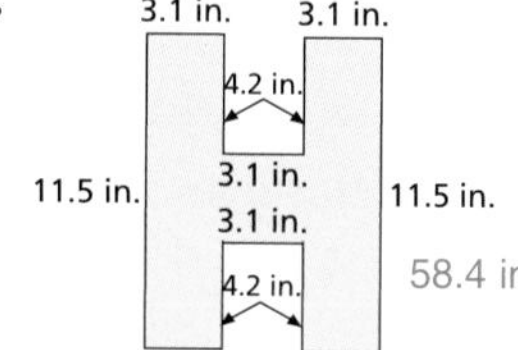

58.4 in.

12. 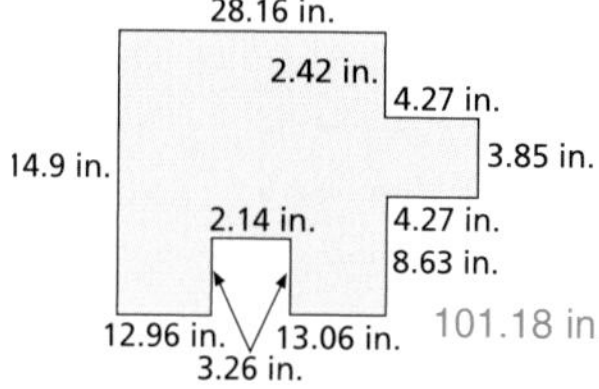

101.18 in.

4. $\begin{matrix} 10d + 25Q = 215 \\ 5d + 10Q = 90 \end{matrix}$ $\begin{bmatrix} 10 & 25 \\ 5 & 10 \end{bmatrix} \begin{bmatrix} d \\ Q \end{bmatrix} = \begin{bmatrix} 215 \\ 90 \end{bmatrix}$

det = −25; 4 dimes and 7 quarters

5. $\begin{matrix} 25D + 0.35M = 230 \\ 35D + 0.25M = 250 \end{matrix}$ $\begin{bmatrix} 25 & 0.35 \\ 35 & 0.25 \end{bmatrix} \begin{bmatrix} D \\ M \end{bmatrix} = \begin{bmatrix} 230 \\ 250 \end{bmatrix}$

det = −6; $D = 5$ days; $M = 300$ miles

Chalkboard Examples

Supplementary Problem

Solve: $\begin{matrix} x + 2y = -7 \\ 3x - y = 14 \end{matrix}$ by the method of determinants (3, 5)

$$x = \frac{A_x}{A} = \frac{\begin{vmatrix} -7 & 2 \\ 14 & -1 \end{vmatrix}}{\begin{vmatrix} 1 & 2 \\ 3 & -1 \end{vmatrix}} = \frac{-7(-1) - 2(14)}{1(-1) - 2(3)}$$

$$= \frac{-21}{-7} = 3$$

$$y = \frac{A_y}{A} = \frac{\begin{vmatrix} 1 & -7 \\ 3 & 14 \end{vmatrix}}{\begin{vmatrix} 1 & 2 \\ 3 & -1 \end{vmatrix}} = \frac{1(14) - (-7)(3)}{1(-1) - 2(3)} = \frac{35}{-7}$$

$$= -5$$

Lesson Wrap-up

Quick Assessment

How does writing a matrix equation make it easier to find the determinants and solution? Writing a matrix equation helps to organize the coefficients, variables and constants.

Enrichment Worksheet 6-7 not shown.

Reteaching Worksheet 6-7

Name ______ Date ______

RETEACHING 6-7

PROBLEM SOLVING SKILLS: DETERMINANTS AND MATRICES

A system of equations can be written using the **matrix equation, $AX = B$**, where A is a matrix of the coefficients, X is a matrix of variables, and B is a matrix of constants. The method of determinants can be used to solve some systems of equations.

Given: $ax + by = e$, $cx + dy = f$

Matrix Equation: $\begin{bmatrix} a & b \\ c & d \end{bmatrix} \begin{bmatrix} x \\ y \end{bmatrix} = \begin{bmatrix} e \\ f \end{bmatrix}$ det $A = \begin{vmatrix} a & b \\ c & d \end{vmatrix} = ad - bc$

Example

For the system of equations, $\begin{cases} 2x + 4y = 6 \\ 3x - 2y = 4 \end{cases}$

a. write the matrix equation.
b. solve using the method of determinants.

Solution

a. $\begin{bmatrix} 2 & 4 \\ 3 & -2 \end{bmatrix} \begin{bmatrix} x \\ y \end{bmatrix} = \begin{bmatrix} 6 \\ 4 \end{bmatrix}$

b. $x = \frac{A_x}{A} = \frac{\begin{vmatrix} 6 & 4 \\ 4 & -2 \end{vmatrix}}{\begin{vmatrix} 2 & 4 \\ 3 & -2 \end{vmatrix}} = \frac{-12 - 16}{-4 - 12} = \frac{-28}{-16} = 1\frac{3}{4}$

$y = \frac{A_y}{A} = \frac{\begin{vmatrix} 2 & 6 \\ 3 & 4 \end{vmatrix}}{\begin{vmatrix} 2 & 4 \\ 3 & -2 \end{vmatrix}} = \frac{8 - 18}{-4 - 12} = \frac{-10}{-16} = \frac{5}{8}$

The solution is $\left(1\frac{3}{4}, \frac{5}{8}\right)$. Check by substituting $1\frac{3}{4}$ for x and $\frac{5}{8}$ for y in the original equations.

EXERCISES

For each system of equations, **a.** write the matrix equation and **b.** solve using the method of determinants.

1. $\begin{cases} -3x + 4y = 12 \\ x - 2y = 6 \end{cases}$ a. $\begin{bmatrix} -3 & 4 \\ 1 & -2 \end{bmatrix} \begin{bmatrix} x \\ y \end{bmatrix} = \begin{bmatrix} 12 \\ 6 \end{bmatrix}$ b. (−24, −15)
2. $\begin{cases} 5x + y = 10 \\ x - y = 5 \end{cases}$ a. $\begin{bmatrix} 5 & 1 \\ 1 & -1 \end{bmatrix} \begin{bmatrix} x \\ y \end{bmatrix} = \begin{bmatrix} 10 \\ 5 \end{bmatrix}$ b. $\left(2\frac{1}{2}, -2\frac{1}{2}\right)$
3. $\begin{cases} x - y = 16 \\ x + y = 10 \end{cases}$ a. $\begin{bmatrix} 1 & -1 \\ 1 & 1 \end{bmatrix} \begin{bmatrix} x \\ y \end{bmatrix} = \begin{bmatrix} 16 \\ 10 \end{bmatrix}$ b. (13, −3)
4. $\begin{cases} 2x - 2y = 8 \\ -x + 3y = 12 \end{cases}$ a. $\begin{bmatrix} 2 & -2 \\ -1 & 3 \end{bmatrix} \begin{bmatrix} x \\ y \end{bmatrix} = \begin{bmatrix} 8 \\ 12 \end{bmatrix}$ b. (12,8)

96 RETEACHING *LESSON 6-7*

Lesson Planning

NCTM Standards/Strands

- Number & Operation
- Patterns, Functions, & Algebra
- Problem Solving
- Reasoning and Proof
- Connections
- Representation

Vocabulary

system of linear inequalities
boundary half-plane
solution set

Materials Needed

graph paper straightedges

Lesson Resources

Warm-Up Transparency 18
Reteaching 6-8
Extra Practice 6-8
Enrichment 6-8
Transparency TK-7–13

Assignment Guide

Basic: 1–24, 27–30
Enriched: 1–30

Getting Started

5-Minute Warm-up

Tell whether each is true (T) or false (F).

1. $-6 < 3$ T **2.** $7 > 11$ F
3. $-4 \geq -5$ T **4.** $6 \leq 6$ T

Introduction to Lesson 6-8

After students have completed the activity at the beginning of the lesson, have them share how they determined what regions to shade and what conclusions they drew.

Additional Answers

1–5.

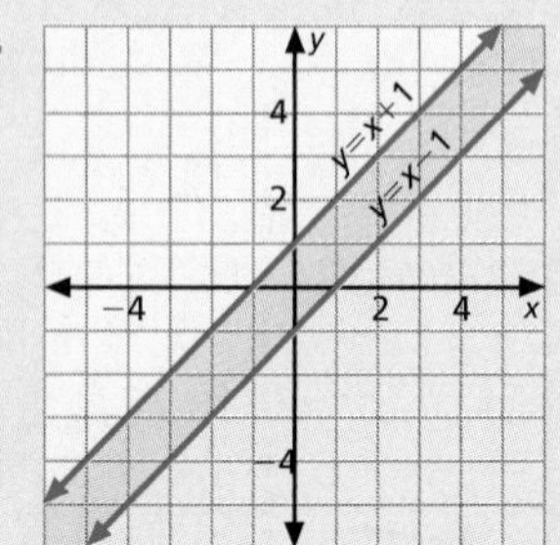

6-8 Systems of Inequalities

Goals ■ Use graphing to solve systems of linear inequalities.
Applications Manufacturing, Health, Budgeting

Work with a partner.

1. On a coordinate plane, graph $y = x + 1$ and $y = x - 1$.
   See additional answers for graph.
2. Complete the table. Replace □ with <, =, or >.
3. On the graph, shade the region for Column A where $y < x + 1$.
4. For Column B, shade the region where $y > x - 1$.
5. Find the section on the graph where the shading overlaps. What conclusion can you draw about this area?
   Both inequalities are true in the region between the lines.

Point	Column A	Column B
	y □ $x + 1$	y □ $x - 1$
(0, 0)	<	>
(1, 1)	<	>
(1, −1)	<	<
(−1, 1)	>	>

Build Understanding

The activity above shows a system of linear inequalities. A **system of linear inequalities** can be solved by graphing each equation and determining the region where the inequality is true. The intersection of the graphs of the inequalities is the solution set.

> **Reading Math**
>
> The graph of a linear equation separates the coordinate plane into two regions—one above the line, one below the line—and points on the line. The line is called the *boundary* of the region, and the regions are called *half-planes*.

Example 1

Determine whether the given ordered pair is a solution to the given system of inequalities.

a. (3, 1); $x + 2y < 5$
$2x - 3y \leq 1$

b. (2, −5); $4x - y \geq 5$
$8x + 5y \leq 3$

c. (1, 2); $x + y \geq 3$
$3x - y < 1$

Solution

Substitute for x and y in each system of inequalities.

a. $x = 3, y = 1$

$x + 2y > 5$
$3 + 2 > 5$ False

$2x - 3y \leq 1$
$6 - 3 \leq 1$ False

The point is not a solution for either inequality. Therefore, (3, 1) is not a solution for this system.

b. $x = 2, y = -5$

$4x - y \geq 5$
$8 + 5 \geq 5$
$13 \geq 5$ True

$8x + 5y \leq 3$
$16 - 25 \leq 3$
$-9 \leq 3$ True

The point is a solution for both inequalities. Therefore, (2, −5) is a solution for this system.

Teaching Strategies

To help students recognize the shaded region that is common to both inequalities in a system, have them shade the entire solution set for each inequality with a different-colored pencil. The solution set of the system then appears as the region in a third color.

c. $x = 1, y = 2$ $\quad x + y \geq 3 \quad\quad 3x - y < 1$

$1 + 2 \geq 3 \quad\quad 3 - 2 < 1$

$3 \geq 3$ True $\quad\quad 1 < 1$ False

The point is a solution for only one of the inequalities. Therefore, (1, 2) is not a solution for this system.

Example 2

Write a system of linear inequalities for the graph at the right.

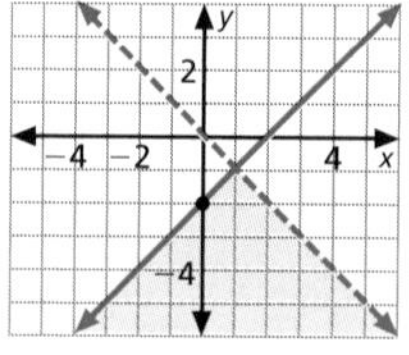

Solution

1. Determine the equation of each line	2. Determine shading	3. Determine Inequality symbol	4. Inequality
line *l*: $b = -2, m = 1$; $y = x - 2$	below and including line	$\leq$	$y \leq x - 2$
line *m*: $b = 0, m = -1$; $y = -x$	below line	$<$	$y < -x$

The system of linear inequalities for the graph is
$y \leq x - 2$
$y < -x$

Problem Solving Tip

A solid line is used for $\leq$, and a dashed line for $<$. The dashed line shows the boundary without including it in the solution.

Example 3

MANUFACTURING A company writes a system of inequalities, shown below, to analyze how changes in plastic and paper affect a product's cost. Graph the solution set of the system.

$$2x - 3y \leq 6 \quad x + 2y < 2$$

Solution

Graph each inequality by graphing the equation of each line. Write each inequality in slope-intercept form. Then make a chart to use for graphing.

$2x - 3y \leq 6 \quad\quad x + 2y < 2$

$-3y \leq -2x + 6 \quad\quad 2y < -x + 2$

$y \geq \frac{2}{3}x - 2 \quad\quad y < -\frac{1}{2}x + 1$

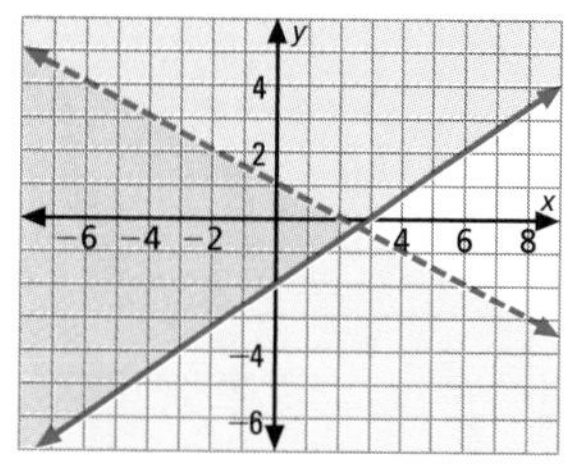

The solution set consists of all the points in the region that has been doubly shaded. The solution includes points on the solid boundary line

$y = \frac{2}{3}x - 2$, but not on the dashed boundary line $y = -\frac{1}{2}x + 1$.

Chalkboard Examples

Supplementary Example 1

Determine whether the given ordered pair is a solution to the given system of inequalities.

a. (4, −1) Yes
$x + 3y < 2$
$3x - 2y \leq 14$

b. (−1, 3) No
$3x - y? -8$
$2x - 5y \leq -18$

Supplementary Example 2 and 3

This system of inequalities is used by a company to determine how changes in the cost of materials affects the cost of a product. Where will the shaded area be in relationship to the corresponding lines.
$y > 2x + 3$
$y \leq -2x - 4$
The shaded area of the first inequality is above and to the left of the dotted line. The shaded area for the second inequality is below and to the left of the solid line.

Extend the Lesson

CONNECTING TO PRIOR KNOWLEDGE Have students write systems of inequalities whose solution sets would form regions shaped like these geometric figures: an isosceles triangle, a square, a rectangle, a parallelogram, a trapezoid and a pentagon. Students may check their work using a graphing calculator. Check students' work.

Reteaching Worksheet 6-8

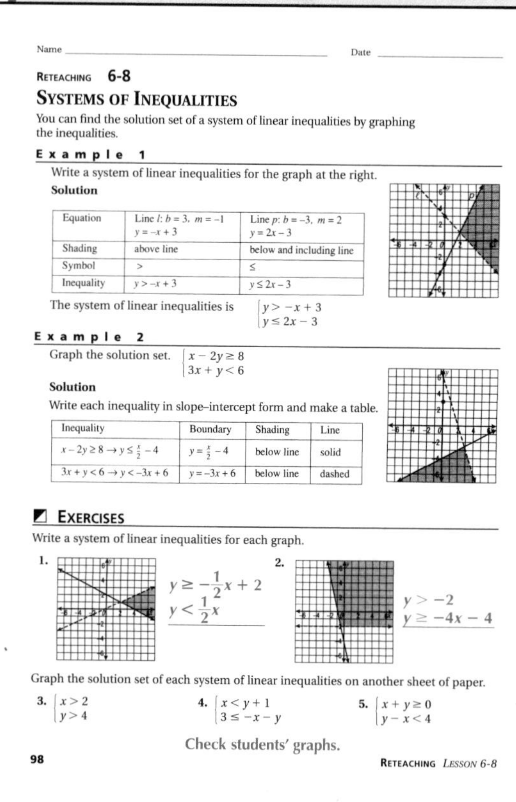

Name ______________ Date ______________

RETEACHING 6-8

SYSTEMS OF INEQUALITIES

You can find the solution set of a system of linear inequalities by graphing the inequalities.

Example 1

Write a system of linear inequalities for the graph at the right.

Solution

Equation	Line *l*: $b = 3, m = -1$; $y = -x + 3$	Line *p*: $b = -3, m = 2$; $y = 2x - 3$
Shading	above line	below and including line
Symbol	$>$	$\leq$
Inequality	$y > -x + 3$	$y \leq 2x - 3$

The system of linear inequalities is $\begin{cases} y > -x + 3 \\ y \leq 2x - 3 \end{cases}$

Example 2

Graph the solution set. $\begin{cases} x - 2y \geq 8 \\ 3x + y < 6 \end{cases}$

Solution

Write each inequality in slope–intercept form and make a table.

Inequality	Boundary	Shading	Line
$x - 2y \geq 8 \rightarrow y \leq \frac{x}{2} - 4$	$y = \frac{x}{2} - 4$	below line	solid
$3x + y < 6 \rightarrow y < -3x + 6$	$y = -3x + 6$	below line	dashed

EXERCISES

Write a system of linear inequalities for each graph.

1. $y \geq -\frac{1}{2}x + 2$, $y < \frac{1}{2}x$

2. $y > -2$, $y \geq -4x - 4$

Graph the solution set of each system of linear inequalities on another sheet of paper.

3. $\begin{cases} x > 2 \\ y > 4 \end{cases}$ 4. $\begin{cases} x < y + 1 \\ 3 \leq -x - y \end{cases}$ 5. $\begin{cases} x + y \geq 0 \\ y - x < 4 \end{cases}$

Check students' graphs.

98 RETEACHING *LESSON 6-8*

Lesson Wrap-up

Quick Assessment

Ask the following questions to determine if students understand the content presented in this lesson.

1. How can you check that you have shaded the correct region? Select a point in the shaded region, and substitute its ordered pair in each inequality in the system.
2. When is a solid line used and when is a dashed line used in graphing inequalities? Points on a solid line are part of the solution set, and points on a dashed line are not.

Assignment Guide

Basic: 1–24, 27–30
Advanced: 1–30
Additional Practice: See Extra Practice Index on page 674.

Additional Answers

5.

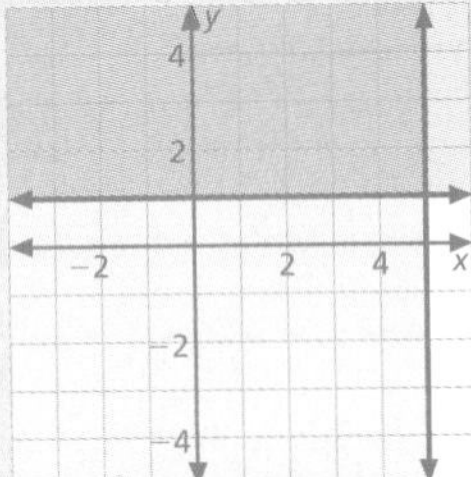

6.

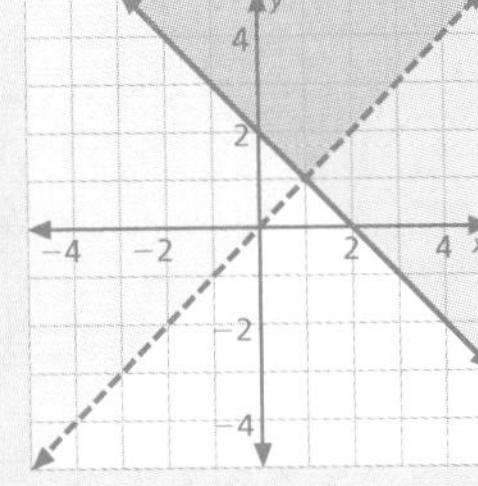

7.

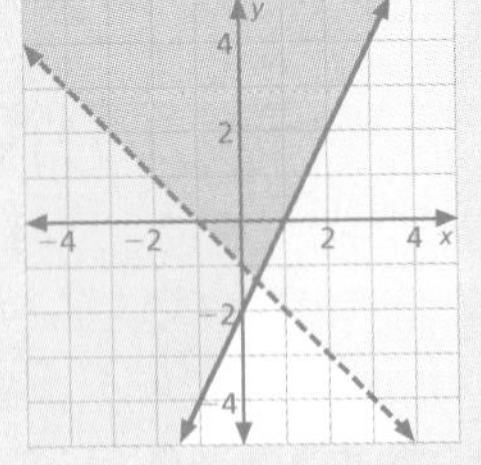

8.

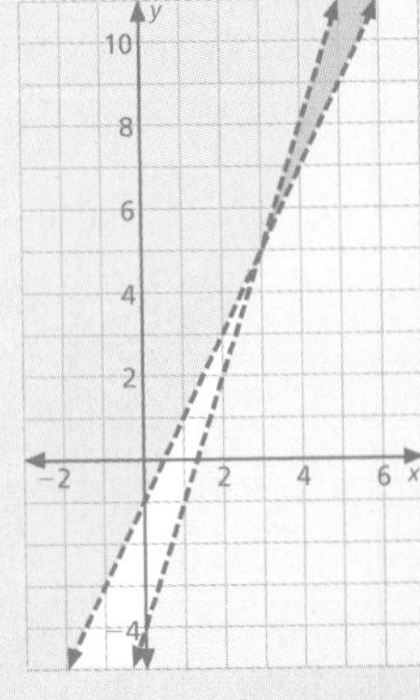

13.

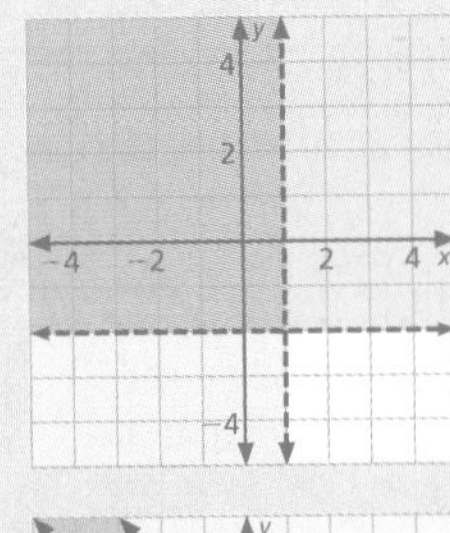

14.

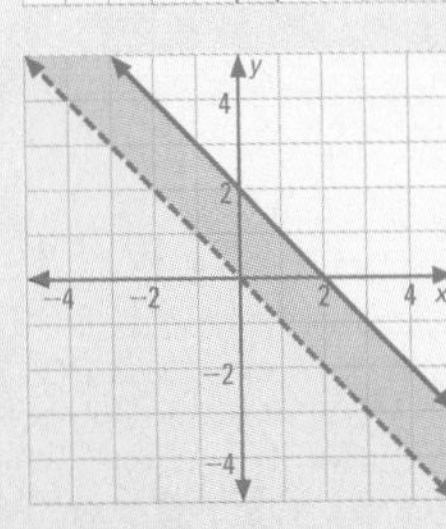

15.

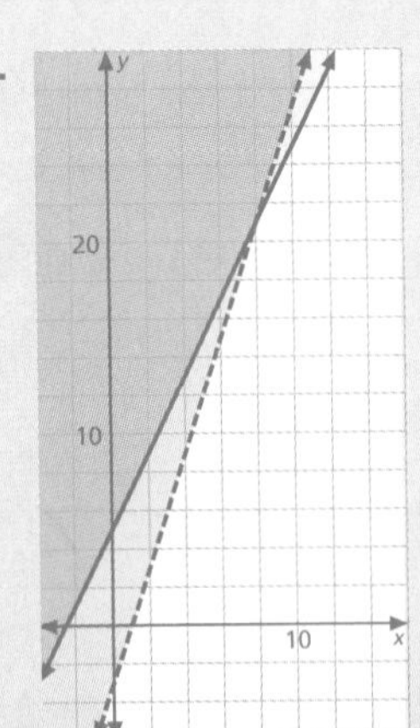

16.

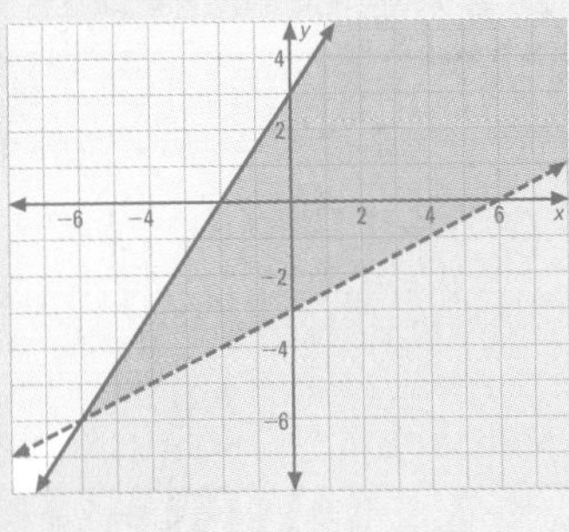

Try These Exercises

Determine whether the given ordered pair is a solution to the given system of inequalities.

1. $(1, -3);\ 3x + 4y \leq 12$, $-5x + y \leq 5$ yes
2. $(-2, 1);\ 2x + y < 4$, $2x - 2y \geq 3$ no

Write a system of linear inequalities for the given graph.

3. 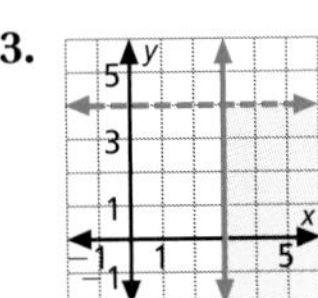

$y < 4,\ x \geq 3$

4. 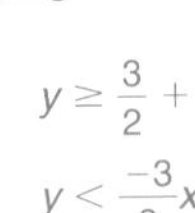

$y \geq \frac{3}{2}x + 3$

$y < \frac{-3}{2}x + 3$

Graph the solution set of the system of linear inequalities. See additional answers.

5. $x \leq 5$, $y \geq 1$
6. $x < y$, $x + y \geq 2$
7. $2x - y \leq 2$, $x + y > -1$
8. $4 < 3x - y$, $y > 2x - 1$

Technology Note

Most graphing calculators allow you to shade portions of a graph. Perform these steps on the Y= screen.

1. Write the inequality in slope-intercept form and enter as an equation.
2. If the inequality contains < or > symbols, change the display to show a dotted line.
3. Determine which side of the line must be shaded and choose a shading option.

Practice Exercises

Determine whether the given ordered pair is a solution to the given system of inequalities.

9. $(3, 5);\ x - y \leq -4$, $x - 2y \leq 1$ no
10. $(-2, -1);\ x + 3y \leq 6$, $4x - 2y \geq 4$ no

Write a system of linear inequalities for the given graph.

11. 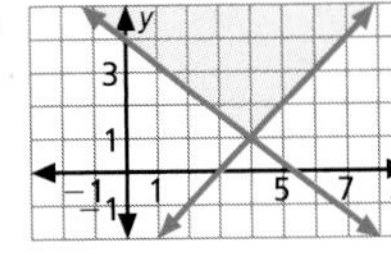

$y \geq x - 3$

$y \geq \frac{-3}{4}x + 4$

12. 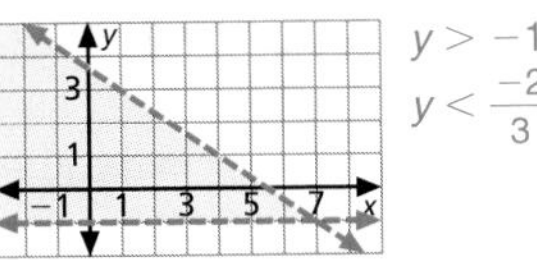

$y > -1$

$y < \frac{-2}{3}x + \frac{11}{3}$

Graph the solution set of the system of linear inequalities. See additional answers.

13. $y > -2$, $x < 1$
14. $y > -x$, $x + y \leq 2$
15. $y \leq 2x + 5$, $x - \frac{1}{3}y < 1$
16. $x - 2y < 6$, $3x \geq 2y - 6$

17. **WRITING MATH** How is the solution set of $y < 2x + 3$ different from the solution set of $y \leq 2x + 3$? Solutions of $y < 2x + 3$ do not include solutions where $y = 2x + 3$.

Write a system of linear inequalities for the given graph.

18. 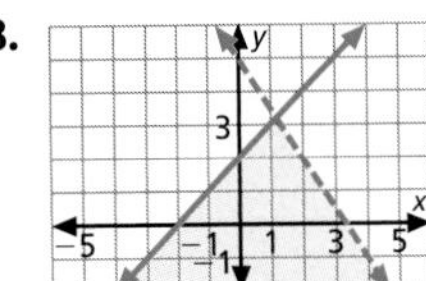

$y < \frac{-3}{2}x + 5$

$y \leq x + 2$

19. 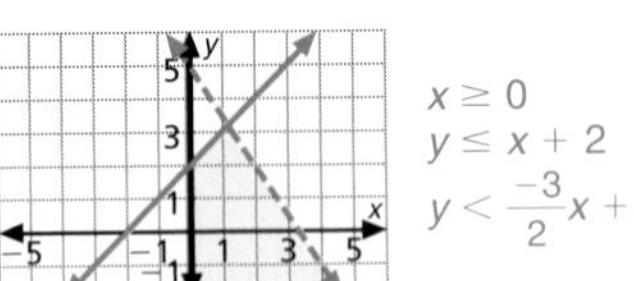

$x \geq 0$

$y \leq x + 2$

$y < \frac{-3}{2}x + 5$

278 Chapter 6 Linear Systems of Equations

DATA FILE Use the Data Index on pages 632–633 to locate height and weight tables for men and women. For Exercises 20–21, use y to represent weight (in pounds) and x to represent height (in inches).

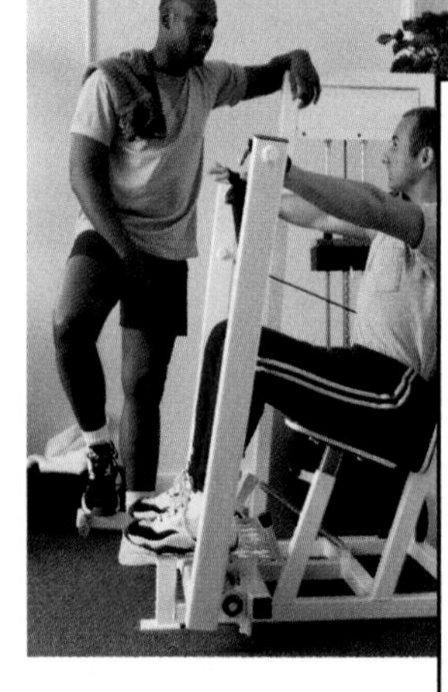

20. HEALTH Determine an equation for the lower male weights for heights up to 64 inches. Determine an equation for the upper male weights for heights up to 64 inches. Using the equations, write and graph a system of inequalities by shading the corresponding range of weights. $y \geq 2x + 4$; $y \leq 3x - 36$

21. HEALTH Determine equations for the lower and upper female weight for heights up to 60 in. Write a system of inequalities. $y \geq x + 44$; $y \geq 2x - 16$

Graph the solution set of the system of linear inequalities. See additional answers.

22. $3 < x + y < 6$
$x \geq 0$
$y \geq 0$

23. $2 \leq 2x + y \leq 6$
$x > 1$
$y \geq 0$

24. BUDGETING Jasmine needs to earn at least $100 this week. She earns $6 per hour doing gardening and $8 per hour as part-time receptionist. She has only 18 hours available to work during the week. Write and graph a system of linear inequalities that models the weekly number of hours Jasmine can work at each job and how much money she needs to earn.
Let x represent gardening hours and y represent receptionist hours. $6x + 8y \geq 100$; $x + y \leq 18$

Extended Practice Exercises

25. NUMBER THEORY Find all numbers such that the ordered pairs (x, y) have the following conditions: See additional answers.
1. x is greater than 1;
2. y is greater than 0; 3. the sum of the two numbers is less than 9, and the sum of $x + 3y$ is at least 6.
(*Hint:* Graph the solution set of the system of inequalities.)

26. Graph the system of inequalities and identify the figure.

$y \geq -2x + 4$
$y < -2x + 8$
$-3 < x - y \leq 3$
$1 < x \leq 9$

Mixed Review Exercises

Find the measure of the indicated angle in each isosceles trapezoid. (Lesson 4-9)

Solve. (Lesson 5-1)

27. On one farm, the ratio of brown eggs to white eggs produced by the chickens is 1:3. If 312 eggs are produced, how many are brown? 78

28. At the amusement park, the ratio of children to adults is 5:2. If 63,000 people visit the park, how many are children? 45,000

29. At the mall, the ratio of people buying to people just looking is 8:7. If 9000 people come to the mall, how many will buy something? 4800

30. In Seattle, WA, the ratio of rainy days to non-rainy days is approximately 2:3. In 365 days, how many days will it rain in Seattle? 146

Extra Practice Worksheet 6-8

Name ______________ Date ______________

EXTRA PRACTICE 6-8
SYSTEMS OF INEQUALITIES

EXERCISES

Determine whether the given ordered pair is a solution to the given system of inequalities.

1. (1, 2); $\begin{cases} x + y \leq 1 \\ x - y > 2 \end{cases}$ no
2. (0, 3); $\begin{cases} x + 2y > 4 \\ 2x - y \leq 2 \end{cases}$ yes
3. (−3, −2); $\begin{cases} x + 4y \geq 4 \\ 2x + 2y \leq 2 \end{cases}$ no
4. (−1, 1); $\begin{cases} 2x + y > -2 \\ x - 4y \leq 3 \end{cases}$ yes

Write a system of linear inequalities for the given graph.

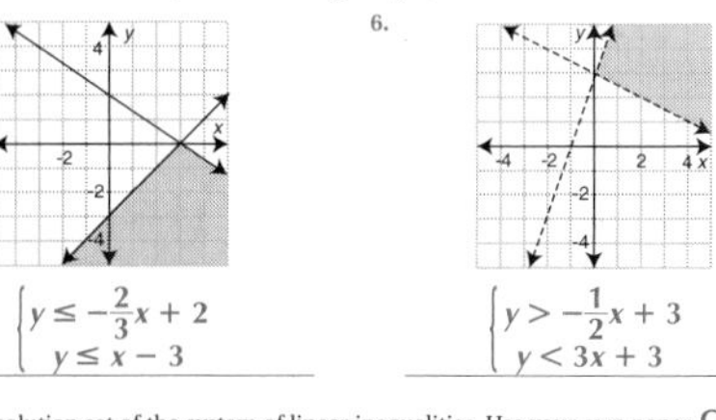

5. $\begin{cases} y \leq -\frac{2}{3}x + 2 \\ y \leq x - 3 \end{cases}$
6. $\begin{cases} y > -\frac{1}{2}x + 3 \\ y < 3x + 3 \end{cases}$

Graph the solution set of the system of linear inequalities. Use your own paper. Check students' graphs.

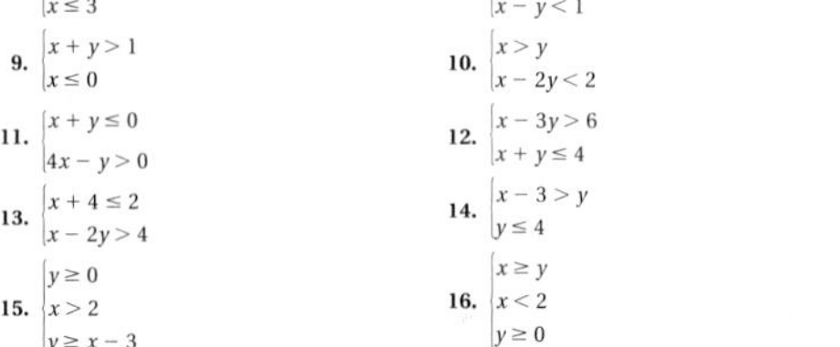

7. $\begin{cases} y > 1 \\ x \leq 3 \end{cases}$
8. $\begin{cases} x + y > 2 \\ x - y < 1 \end{cases}$
9. $\begin{cases} x + y > 1 \\ x \leq 0 \end{cases}$
10. $\begin{cases} x > y \\ x - 2y < 2 \end{cases}$
11. $\begin{cases} x + y \leq 0 \\ 4x - y > 0 \end{cases}$
12. $\begin{cases} x - 3y > 6 \\ x + y \leq 4 \end{cases}$
13. $\begin{cases} x + 4 \leq 2 \\ x - 2y > 4 \end{cases}$
14. $\begin{cases} x - 3 > y \\ y \leq 4 \end{cases}$
15. $\begin{cases} y \geq 0 \\ x > 2 \\ y \geq x - 3 \end{cases}$
16. $\begin{cases} x \geq y \\ x < 2 \\ y \geq 0 \end{cases}$

86 EXTRA PRACTICE *LESSON 6-8*

Enrichment Worksheet 6-8 not shown.

22.
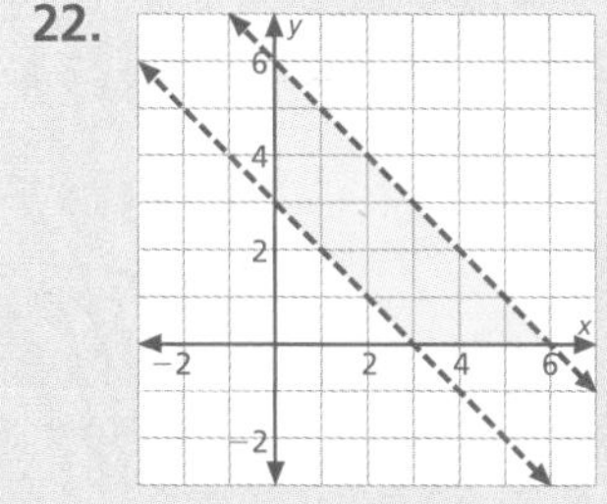

23.
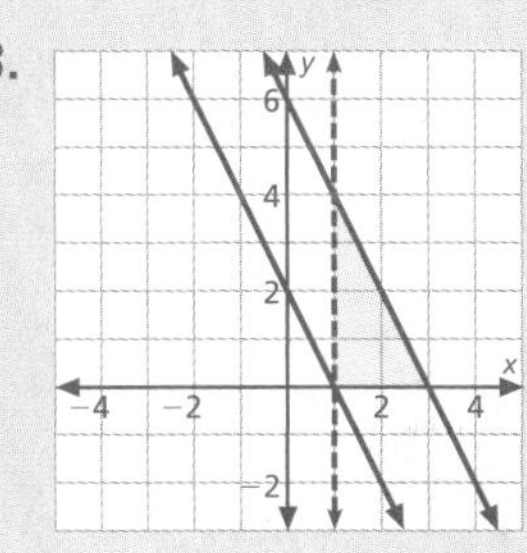

25.
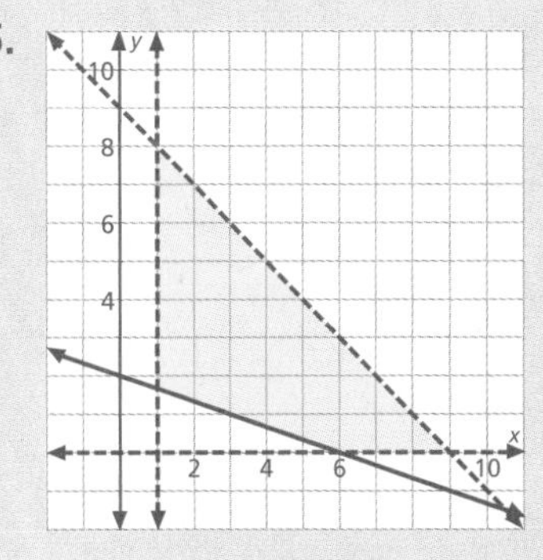

26.
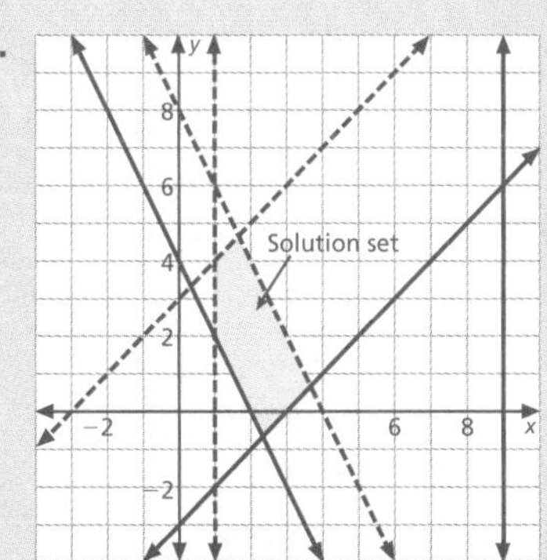

Skills Practice

Vocabulary Review

Lesson 6-7
determinant of a system of equations
matrix

Lesson 6-8
system of linear inequalities
boundary
half-plane
solution set

Assignment Guide

All students: 1–76
Additional Practice: Refer to the Extra Practice Index on page 674 of this text.

Additional Answers

20.
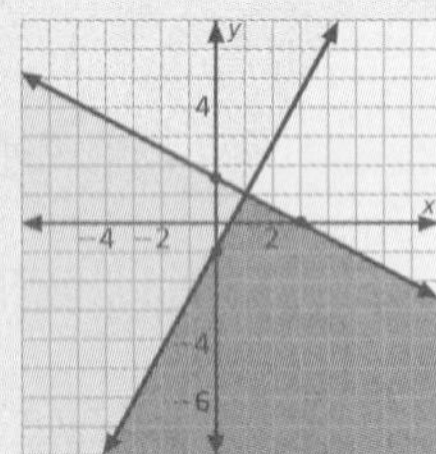

21.
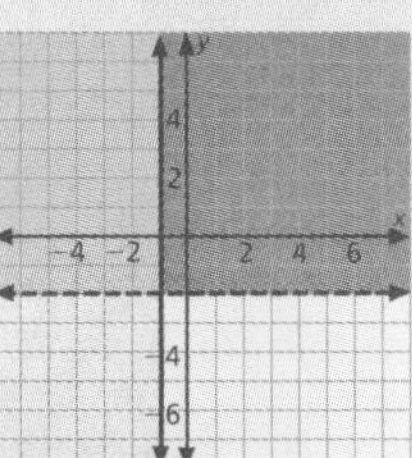

22.
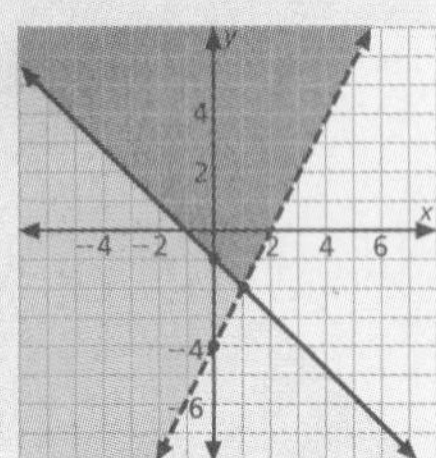

Review and Practice Your Skills

Practice ■ Lesson 6-7

Use determinants to solve each system of equations. Check your answers.

1. $4x - y = 6$ (1, −5)
$x - 3y = 16$

2. $3x - 5y = -23$ (−1, 4)
$5x + 4y = 11$

3. $x + 4y = 13$ (1, 3)
$5x - 7y = -16$

4. $-2x + 3y = 10$ (−92, −58)
$3x - 5y = 14$

5. $-x - y = -15$ (7, 8)
$2x - y = 6$

6. $2x - 7y = -18$ (−2, 2)
$x - 2y = -6$

7. $3x + y = -1$ (−1, 2)
$2x - 3y = -8$

8. $2y + 3 = 9x$ (−3, −15)
$3x - y = 6$

9. $x - 7 = 3y$ $\left(0, -\frac{7}{3}\right)$
$2(3y + 7) = 5x$

10. $3x - 7y = 2$ $\left(\frac{2}{3}, 0\right)$
$6x - 4 = 13v$

11. $6x = 6y$ (6, 9)
$3x - 4y = -18$

12. $x - 6y = 3$ $\left(\frac{9}{2}, \frac{1}{4}\right)$
$x + 2y = 5$

13. $4x + 6y = 16$ (2.8, 0.8)
$x = 2y + 1.2$

14. $-2x + 3y = 15$ no solution
$2x - 3y = 6$

15. $4y = 10 - 5x$ (−2, 5)
$6x + 22 = 2v$

16. Nadine has $2.35 in dimes and quarters. She has 6 fewer quarters than dimes. How many of each type of coin does she have? 11 dimes, 5 quarters

Practice ■ Lesson 6-8

Determine whether the given ordered pair is a solution to the given system of inequalities.

17. (−2, 3); ($2x + y < 4$ yes
$-2x + y \geq 2$

18. (−2, 3); $x + 3y \leq 3$ no
$x + 3y > 3$

19. (−2, 3); $x < 2$ no
$y \geq 3$

Graph the solution set of the system of linear inequalities. See additional answers.

20. $x + 2y \leq 3$
$2x - y \geq 1$

21. $x \geq -1$
$y > -2$

22. $y > 2x - 4$
$y \geq -x - 1$

23. $x + y > 2$
$x - y < -2$

24. $2x - 3y \leq 9$
$x + 2y < 6$

25. $y > 2x$
$y - x \leq 5$

26. $x + y < -1$
$3x - y > 4$

27. $5x + 2y \geq 12$
$2x + 3y \leq 10$

28. $y > -3$
$y < 2x + 4$

29. $y < -3x + 2$
$3y \geq x + 15$

30. $y > x$
$x \leq -3$
$-x - 3y > 3$

31. $2y - x \geq 0$
$x + 5y < 15$
$y \geq x + 1$

32. $0.1x + 0.4y \geq -0.8$
$4x - 8y \geq -24$

33. $y < -2x + 2$
$6x - 3y \leq -6$

34. $0.5x - y < 1$
$x - 2y \geq -6$

23.

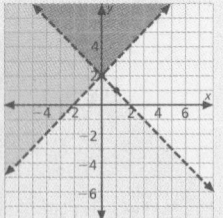

24.
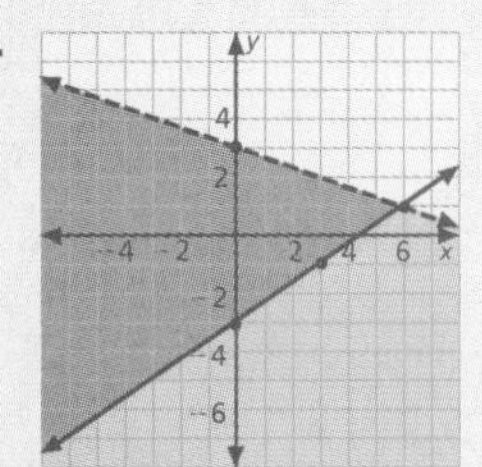

25.
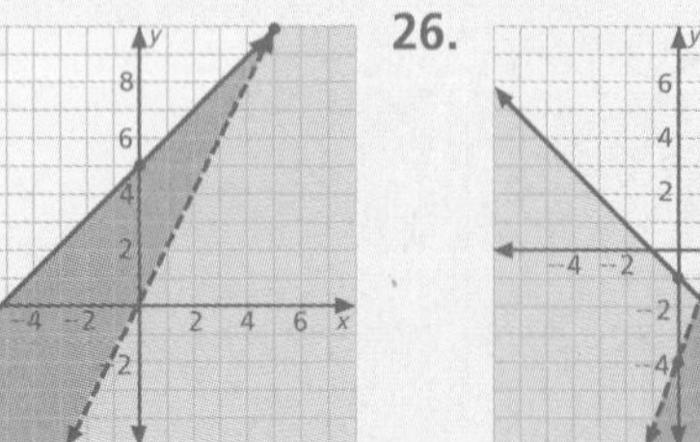

26.
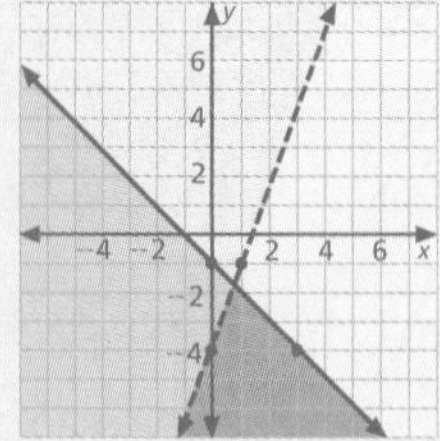

27.
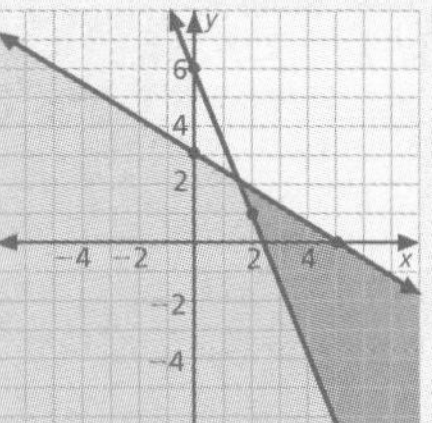

280 Chapter 6 Linear Systems of Equations

PRACTICE ◣ LESSON 6-1–LESSON 6-8

Find the slope and *y*-intercept for each line. (Lesson 6-1)

35. $y = \frac{5}{7}x - 11$ $m = \frac{5}{7}, b = -11$ **36.** $5x - 7y = 77$ $m = \frac{5}{7}, b = -11$ **37.** $154 + 10x = 14y$ $m = \frac{5}{7}, b = 11$

38. $y = -3x$ $m = -3, b = 0$ **39.** $6x + 2y = 26$ $m = -3, b = 13$ **40.** $0.012x + 0.0004y = 0.096$ $m = -30, b = 240$

Determine whether each pair of lines is *parallel, perpendicular,* or *neither*. (Lesson 6-2)

41. $y = 3.2x - 64$ parallel
$16x - 5y = 30$

42. $y = -4x$ neither
$y = 4x$

43. $-7x + 2y = 0$
$3.5y = -x$ perpendicular

Write an equation for the line with the given information. (Lesson 6-3)

44. $m = -\frac{1}{8}$, $Q(-8, 3)$ $y = \frac{-1}{8}x + 2$

45. $m = \frac{2}{3}$, $K(12, -5)$ $y = \frac{2}{3}x - 13$

46. $m = 7.5$, $b = -2$ $y = 7.5x - 2$

47. $G(10, 3)$, $b = 7$ $y = \frac{-2}{5}x + 7$

48. $P(-6, 2)$ and $Q(6, -2)$ $y = \frac{-1}{3}x$

49. $T(-3, 3.5)$ and $U(2, 16)$ $y = 2.5x + 11$

Solve each system of equations by graphing. (Lesson 6-4)

50. $x + y = -1$ (−3, 2)
$y = x + 5$

51. $x + y = 5$ (5, 0)
$y = 2x - 10$

52. $3x + 2y = 13$ (7, −4)
$2x - 3y = 26$

53. $x + 4y = -2$ infinite number of solutions
$y = -0.25x - 0.5$

54. $4x - 2y = 18$ (1.5, −6)
$y = 4x - 12$

55. $3x - 6y = 6$ no solution
$-x = -2y + 15$

Solve and check each system of equations by the substitution method. (Lesson 6-5)

56. $y = 2x$ (−3, −6)
$x + y = -9$

57. $n = 2m - 6$ (4, 2)
$2m + n = 10$

58. $y = 2x + 5$ (−6, −7)
$x - 2y = 8$

59. $r + 3s = -5$ (−5, 0)
$3r + 2s = -15$

60. $4x = y + 3$ infinite number of solutions
$9 = 12x - 3y$

61. $3a + 4b = -12$ (−8, 3)
$2b - a = 14$

Solve each system of equations. Check the solutions. (Lesson 6-6)

62. $2x + y = 8$ (0, 8)
$-2x - 3y = -24$

63. $4x + 3y = 8$ (5, −4)
$4x - 3y = 32$

64. $2x + y = 4$ (−3, 10)
$2x + 3y = 24$

65. $2x + y = -8$ (−2, −4)
$0.1x + 0.2y = 1.0$

66. $2x - y = 7$ (5, 3)
$0.03x + 0.20y = 0.21$

67. $6x + 3y = 0$ (2, −4)
$-4y = 2x + 12$

Use determinants to solve each system of equations. Check your answers. (Lesson 6-7)

68. $x + 3y = 11$ (8, 1)
$2x - 3y = 13$

69. $5x - y = 16$ (3, −1)
$5x + 2y = 13$

70. $x = 4y$ (−8, −2)
$4x + 2y = -36$

71. $y = 7x - 1$ (−6, −43)
$42x - 7y = 49$

72. $2y - 3x = -10$ (2, −2)
$2x - y = 6$

73. $0.2x + 0.2y = 0.6$ (1, 2)
$-9x + 3y = 0$

Graph the solution set of the system of linear inequalities. (Lesson 6-8)
See additional answers.

74. $2x + y < 7$
$y \geq 2(1 - x)$

75. $x > -2$
$4x - 4y < 12$

76. $y > \frac{2}{3}x + 7$
$2x + 3y \leq 9$

Chalkboard Examples

Supplementary Example 6-7

Give the system of equations:
$3x - y = 15$
$x + y = 1$
Name determinant A

$$A = \begin{vmatrix} 3 & -1 \\ 1 & 1 \end{vmatrix}$$

Supplementary Example 6-8

Given: $2x - y > 5$
$x \leq -2$
Is the ordered pair (−4, 5) a solution? no

ADDITIONAL ANSWERS

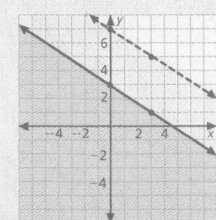

28.

29.
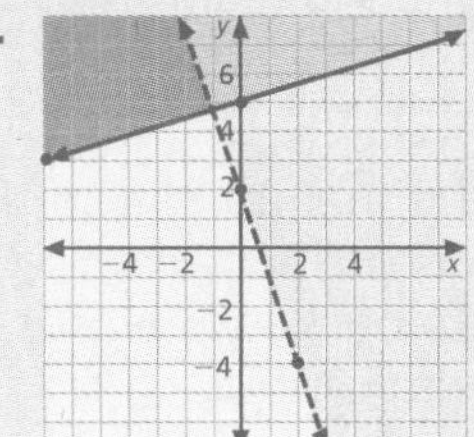

30.
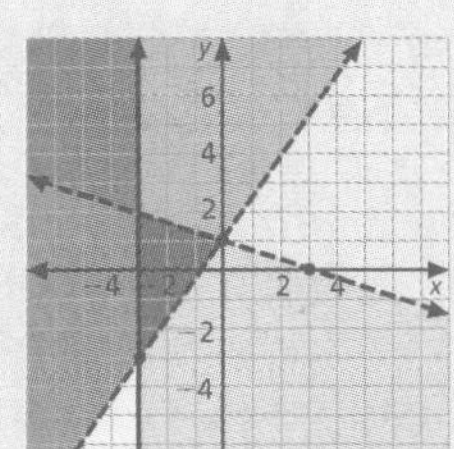

31.
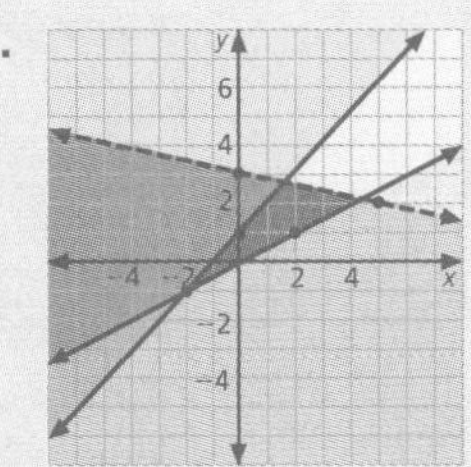

32.
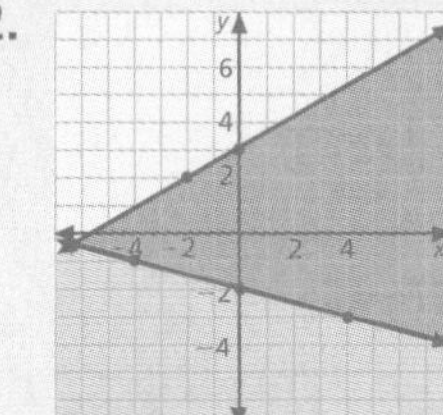

33.
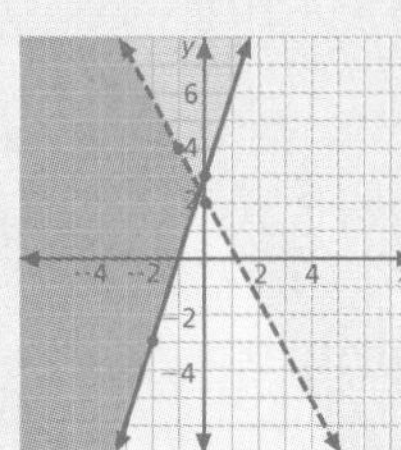

34.
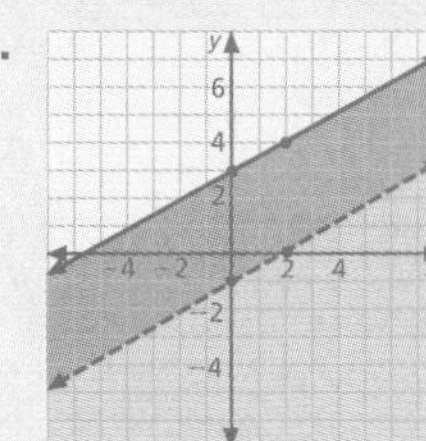

74.
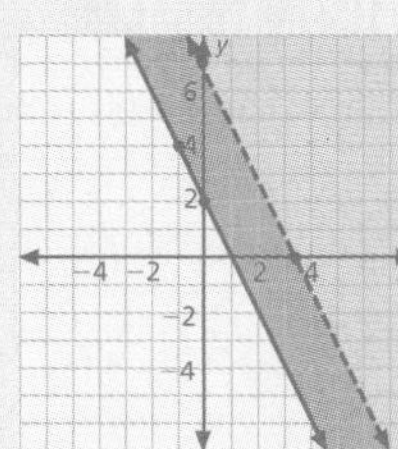

75.
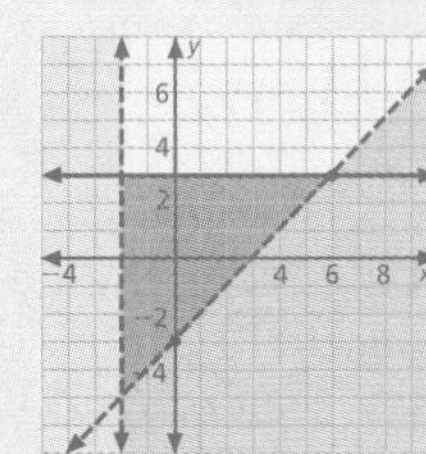

76.

6-9 Linear Programming

Goals ■ Write, minimize and maximize an objective function.

Applications Business, Manufacturing, Farming

Work with a partner. For 1 and 2, check students' graphs.

1. On graph paper, find the region defined by the following inequalities.

   $x \geq 0$ $y \geq 0$ $y \leq 5$ $y + x \leq 10$

2. On the same coordinate axes, draw the line $y = -\frac{1}{2}x$.
3. Place a pencil on its side over the line $y = -\frac{1}{2}x$. Slowly slide the pencil over the polygonal region keeping it parallel to line $y = -\frac{1}{2}x$. Name the coordinates of the last point in the region that the pencil passes over? (5, 5)
4. Place the pencil on its side anywhere outside of the polygonal region but not parallel to any of its sides. Slowly slide the pencil over the region. What are the coordinates of the last point in the polygonal region that the pencil passes over? Any of these points: (0, 0), (0, 5), (10, 0)
5. What conclusion can you draw about the last point in the polygonal region that the pencil passes over? The elements of its ordered pair are solutions to each of the inequalities in Question 1.

Build Understanding

Linear programming is a method used by business and government to help manage resources and time. Limits to available resources are called **constraints.** In linear programming, such constraints are represented by inequalities. The intersection of the graphs of a system of constraints is known as a **feasible region.** The feasible region includes all the possible solutions to the system.

In the activity above, you determined that the last point in the polygonal region that the pencil passed over was located at a vertex. The line represented by the pencil is known as the **objective function.** The equation of this line can represent quantities such as revenue, profit or cost. In business, the objective function is used to determine how to make the maximum profit with minimum cost.

Example 1

MANUFACTURING High Tops Corporation makes two types of athletic shoes: running shoes and basketball shoes. The shoes are assembled by machine and then finished by hand. It takes 0.25 hour for the machine assembly and 0.1 hour by hand to make a running shoe. It takes 0.15 hour on the machine and 0.2 hour by hand to make the basketball shoe. At their manufacturing plant, the company can allocate no more than 900 machine hours and 500 hand hours per day. The profit is $10 on each type of running shoe and $15 on each basketball shoe. How many of each type of shoe should be made to maximize the profit?

Lesson Planning

NCTM Standards/Strands
- Number & Operation
- Patterns, Functions, & Algebra
- Geometry & Spatial Sense
- Problem Solving
- Reasoning and Proof
- Connections
- Representation

Vocabulary
linear programming
vertices
feasible region
objective function

Materials Needed
graphing calculators

Lesson Resources
Warm-Up Transparency 18
Transparency TK-7-10
Reteaching 6-9
Extra Practice 6-9
Enrichment 6-9

Assignment Guide
Basic: 1–24, 27–32
Enriched: 1–32

Getting Started

5-Minute Warm-Up
Describe the graph of each inequality.
1. $x + y > 3$ shaded above broken line for $y = -x + 3$
2. $y \leq 2x - 4$ shaded below solid line for $y = 2x - 4$

Introduction to Lesson 6-9
The pencil represents an objective function, that is, a function of x and y that is held constant. As the pencil moves out in a parallel fashion, the constant increases. Explain that objective functions are used in business and manufacturing. A common purpose is to maximize profits while minimizing costs.

Teaching Strategies

Some students may write constraint inequalities in the wrong direction, confusing the inequality symbols. Suggest that after writing each inequality, they test it with numbers, translate back into words, and make sure that the inequality expresses the correct relationship.

Solution

If x represents the number of running shoes made and y represents the number of basketball shoes made, then the profit objective function (P) is $P = 10x + 15y$.

We can write inequalities to represent each constraint.

Machine hours: $0.25x + 0.15y \leq 900$

Hand hours: $0.1x + 0.2y \leq 500$

To make sure the feasible region is completely within the first quadrant of the coordinate plane, we include the constraints $x \geq 0$ and $y \geq 0$. Graph the system of inequalities.

The vertices of the feasible region are at (0, 0), (3600, 0), (0, 2500) and (3000, 1000). Evaluate the objective function at each of the vertices of the feasible region.

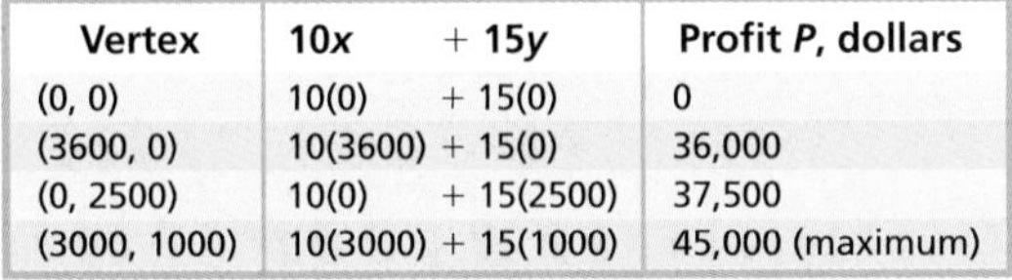

Vertex	10x + 15y	Profit P, dollars
(0, 0)	10(0) + 15(0)	0
(3600, 0)	10(3600) + 15(0)	36,000
(0, 2500)	10(0) + 15(2500)	37,500
(3000, 1000)	10(3000) + 15(1000)	45,000 (maximum)

Under the given daily constraints, the maximum daily profit the shoe company should expect to make is \$45,000. To do this, they would have to produce and sell 3000 running shoes and 1000 basketball shoes per day.

Graphing utilities can make calculations easier. Using a graphing utility, you should write the constraints as equations.

Example 2

GRAPHING Using a graphing utility, graph the solution set of the system of inequalities below to determine the maxmimum value of $P = 6x + 2y$.

$x + y \leq 3$ $\quad x \geq 0$

$y \geq -x + 1$ $\quad y \geq 0$

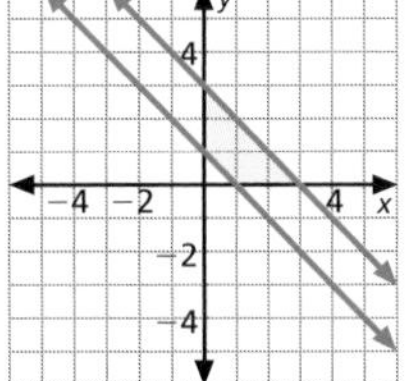

Solution

Graph each inequality by graphing the equation of each line. Then locate the vertices of the feasible region using the trace, zoom and intersect features to determine the coordinates. Make a table of the vertices and the value of P for each vertex.

inequality	$x + y \leq 3$	$y \geq -x + 1$	$x \geq 0$	$y \geq 0$
boundary equation	$x + y = 3$	$y = -x + 1$	$x = 0$	$y = 0$
shading	below	above	right of y-axis	above x-axis
line	solid	solid	solid	solid

Vertex	6x + 2y	P
(0,1)	6(0) + 2(1)	2
(0,3)	6(0) + 2(3)	6
(3,0)	6(3) + 2(0)	18
(1,0)	6(1) + 2(0)	6

The maximum value of P.

The maximum value of P is 18 when $x = 3$ and $y = 0$.

Technology Note

You cannot graph equations such as $x = 0$ on most graphing calculators. You may need to transfer a drawing from the screen to paper to complete it.

Teaching Strategies

COOPERATIVE LEARNING Have students work in pairs to think of other situations in which linear programming might be a helpful problem-solving strategy. Then ask pairs to share their ideas with the class.

Chalkboard Examples

Supplementary Example 1

A research company is planning a week long survey of people's television viewing habits. The company knows they will need statisticians and interviewers for the survey. Each statistician receives \$900 per week, and each interviewer receives \$550 per week. The data collection phase of the survey will require at least 210 hr and the data analysis phase will require at least 150 hr. A statistician spends 10 hr per week collecting data and 30 hr per week analyzing data. An interviewer spends 30 hr per week collecting data and 10 hr per week analyzing data. How many of each type of worker should be assigned to minimize the cost of the survey? **3 statisticians, 6 interviewers**

Supplementary Example 2

Using a graphing calculator, graph the solution set of the system of inequalities below to determine the maximum value of $P = 5x + 6y$.

$y + 3x \leq 6$

$y + 2x \leq 5$

$x \geq 0$

$y \geq 0$ **maximum 30 at (0, 5)**

Reteaching Worksheet 6-9

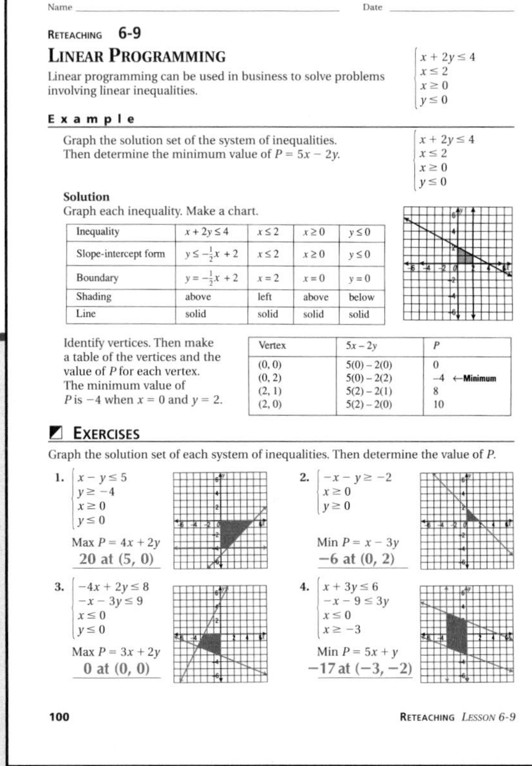

Name ________ Date ________

RETEACHING 6-9

LINEAR PROGRAMMING

Linear programming can be used in business to solve problems involving linear inequalities.

$x + 2y \leq 4$, $x \leq 2$, $x \geq 0$, $y \leq 0$

Example

Graph the solution set of the system of inequalities. Then determine the minimum value of $P = 5x - 2y$.

$x + 2y \leq 4$, $x \leq 2$, $x \geq 0$, $y \leq 0$

Solution

Graph each inequality. Make a chart.

Inequality	$x + 2y \leq 4$	$x \leq 2$	$x \geq 0$	$y \leq 0$
Slope-intercept form	$y \leq -\frac{1}{2}x + 2$	$x \leq 2$	$x \geq 0$	$y \leq 0$
Boundary	$y = -\frac{1}{2}x + 2$	$x = 2$	$x = 0$	$y = 0$
Shading	above	left	above	below
Line	solid	solid	solid	solid

Identify vertices. Then make a table of the vertices and the value of P for each vertex. The minimum value of P is −4 when $x = 0$ and $y = 2$.

Vertex	5x − 2y	P
(0, 0)	5(0) − 2(0)	0
(0, 2)	5(0) − 2(2)	−4 ←Minimum
(2, 1)	5(2) − 2(1)	8
(2, 0)	5(2) − 2(0)	10

EXERCISES

Graph the solution set of each system of inequalities. Then determine the value of P.

1. $x - y \leq 5$, $y \geq -4$, $x \geq 0$, $y \leq 0$ — Max $P = 4x + 2y$: **20 at (5, 0)**
2. $-x - y \geq -2$, $x \geq 0$, $y \geq 0$ — Min $P = x - 3y$: **−6 at (0, 2)**
3. $-4x + 2y \leq 8$, $-x - 3y \leq 9$, $x \leq 0$, $y \leq 0$ — Max $P = 3x + 2y$: **0 at (0, 0)**
4. $x + 3y \leq 6$, $-x - 9 \leq 3y$, $x \leq 0$, $x \geq -3$ — Min $P = 5x + y$: **−17 at (−3, −2)**

100 RETEACHING LESSON 6-9

Lesson Wrap-up

Quick Assessment

Ask the following questions to determine if students understand the content presented in this lesson.

1. How is the use of the word *vertex* in linear programming similar to its use in polygons? In both cases, the word refers to the point at which two segments meet.
2. Give an example of how linear programming is used in business. Answers will vary. Possible answer: to find the maximum profit given a set of manufacturing constraints.

Assignment Guide

Basic: 1-24, 27–32
Advanced: 1–32
Additional Practice: See Extra Practice Index on page 674.

Additional Answers

7.

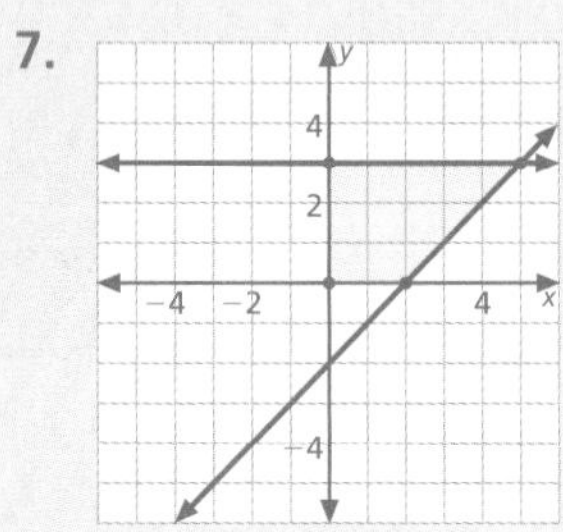

Max P = 28 at (5, 3)

8.

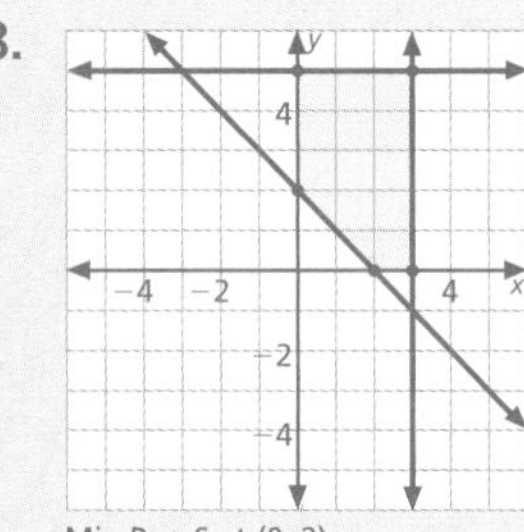

Min P = 6 at (0, 2)

19.

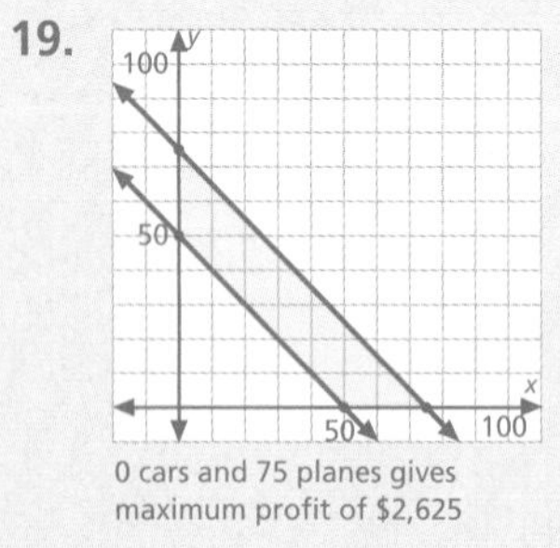

0 cars and 75 planes gives maximum profit of $2,625

Try These Exercises

Determine if each point is within the feasible region $x \geq 0$, $y \geq 0$ and $5x + 2y \leq 30$.

1. (10, 7) no **2.** (1, 6) yes **3.** (2.75, 6.1) yes **4.** (5, 2.5) yes

Determine the maximum value of $P = 15x + 12y$ for each feasible region.

5.

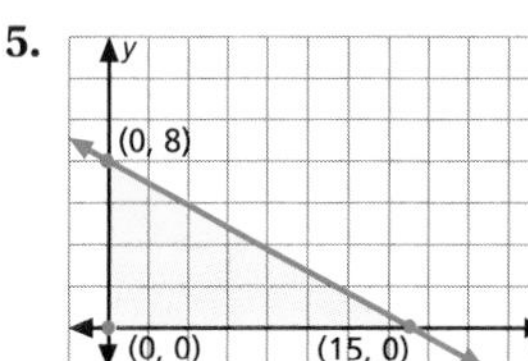

maximum 225 at (15, 0)

6.

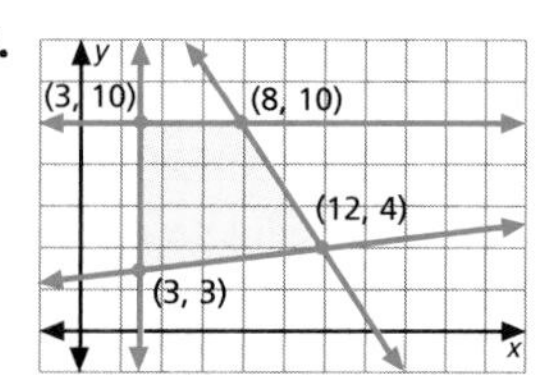

maximum 240 at (8, 10)

Find the feasible region for each system of constraints. Determine the maximum or minimum value of P as directed. See additional answers.

7. $-x + y \geq -2$ $x \geq 0$
$y \leq 3$ $y \geq 0$
maximum $P = 5x + y$

8. $x + y \geq 2$ $x \geq 0$
$y \leq 5$ $y \geq 0$
$x \leq 3$
minimum $P = 4x + 3y$

Practice Exercises

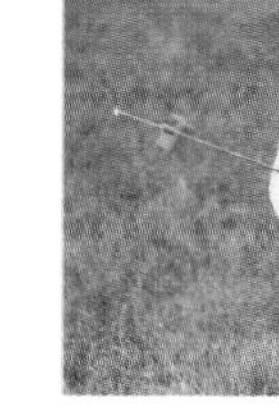

For the feasible region determine the minimum and maximum value of the objective function and identify the coordinates at which they occur.

9. $P = 10x + 6y$ (0, 10) (5, 15) (8, 8) (12, 0)
maximum 140 at (5, 15); minimum 60 at (0, 10)

10. $P = 4x + 5y$ (2, 5) (2, 9) (6, 11) (8, 5)
maximum 79 at (6, 11); minimum 33 at (2, 5)

11. $P = 1.25x + 0.75y$ (0, 4) (9, 15) (20, 2)
maximum 26.5 at (20, 2); minimum 3 at (0, 4)

12. $P = 120x + 180y$ (6, 6) (6, 10) (10, 12) (13, 11) (13, 6)
maximum 3540 at (13, 11); minimum 1800 at (6, 6)

13. WRITING MATH Why do you think each system of constraints in this lesson contains the inequalities $x \geq 0$ and $y \geq 0$? What do these constraints accomplish? See additional answers.

Identify the vertices of the feasible region defined by the constraints.

14. $x \geq 0$; $y \geq 0$; $7x + 9y \leq 63$ (0, 0), (0, 7), (9, 0)

15. $x \geq 0$; $y \geq 0$; $y + 2x \geq 8$ (0, 8), (4, 0), (8, 0)

16. $x \geq 0$; $y \geq 0$; $y + x \leq 10$ (0, 6), $\left(2\frac{2}{3}, 7\frac{1}{3}\right)$, (10, 0), (0, 0)

Determine the minimum value of the objective function $C = 3x + 2y$ for the graph of each feasible region.

17.

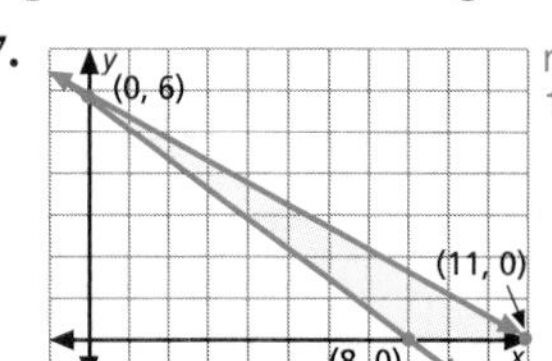

minimum 12 at (0, 6)

18.

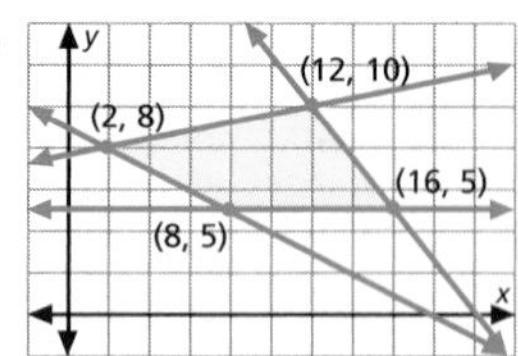

minimum 22 at (2, 8)

Extend the Lesson

REAL WORLD CONNECTION Invite the chief financial officer of a business in your area to visit your class and discuss how his or her company uses graphs to make business decisions and how those graphs are generated.

13. These constraints insure that the feasible region will be confined to the first quadrant. This eliminates the possibility of a negative coordinate that would skew the results of certain computations and applications.

19. **MANUFACTURING** Glimmering Hobbies manufactures remote control cars and airplanes. Let x represent a car and y represent an airplane. The plant manufactures at least 50 items but not more than 75 items each week. If the profit is determined by $P = 25x + 35y$, where x is the number of cars and y is the number of planes manufactured, determine the optimum number of cars and airplanes that can be manufactured to maximize the profit.
See additional answers.

SMALL BUSINESS A group of students are making and selling custom-printed T-shirts and sweatshirts. Their costs are \$3.00/T-shirts and \$5.00/sweatshirts. A local store owner has agreed to sell their shirts but will only take up to a total of 50 T-shirts and sweatshirts combined. In addition, the store owner said they must sell at least 15 T-shirts and 10 sweatshirts to continue selling at the store. Let x equal the number of T-shirts and y equal the number of seatshirts.

20. The students need to minimize cost.Write the objective function for cost (C). $C = 3x + 5y$

21. Write the inequalities that express the constraints. $x \geq 15$; $y \geq 10$; $x + y \leq 50$

22. Graph the inequalities and determine coordinates of the vertices of the feasible region. See additional answers.

23. How many of each type must they sell to minimize cost?
They should sell 15 T-shirts and 10 sweatshirts.

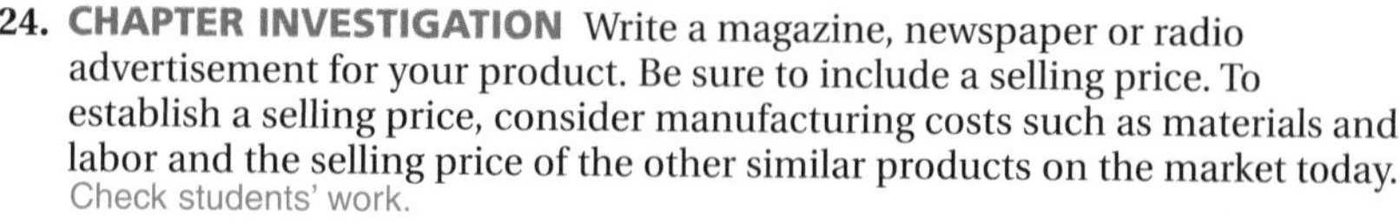

24. **CHAPTER INVESTIGATION** Write a magazine, newspaper or radio advertisement for your product. Be sure to include a selling price. To establish a selling price, consider manufacturing costs such as materials and labor and the selling price of the other similar products on the market today.
Check students' work.

Extended Practice Exercises

For 25–26, see additional answers.

25. **AGRICULTURE** Valley Farms owns a 3600-acre field. The farmers want to plant Iceberg lettuce which yields \$200 per acre and Romaine lettuce which yields \$250 per acre. To prevent loss due to disease, the farmers should plant no more than 3000 acres of Iceberg and no more than 2500 acres of Romaine. How many acres of each crop should Valley Farms plant in order to maximize profits? What is the maximum profit?

26. **SMALL BUSINESS** Sasha owns and operates the Stand-In-Line Skate Shop. She makes a profit of \$40 on each pair of adult skates and \$20 on each pair of child skates sold. Sasha can stock at most 80 pairs of skates on her shelves. Sasha orders skates once every 6 weeks and can order up to 50 pairs of each type of skate. How many of each type of skate must Sasha stock and sell in a 6-week period in order to maximize her profits? What is the maximum profit?

Mixed Review Exercises

Find each product or quotient. (Lesson 1-5)

27. $(-3.9)(-4.8)(-7.6)$ -142.272

28. $-387 \div (4 \times -3)$ 32.25

29. $125 \div -10 + 38$ 25.5

30. $(8.36)(9.74)(-3.85)$ -313.49164

31. $(-4 \times -6) \div -2$ -12

32. $(-60) \div (-3 \times -5)$ -4

22. (15, 10), (15, 35), (40, 10)

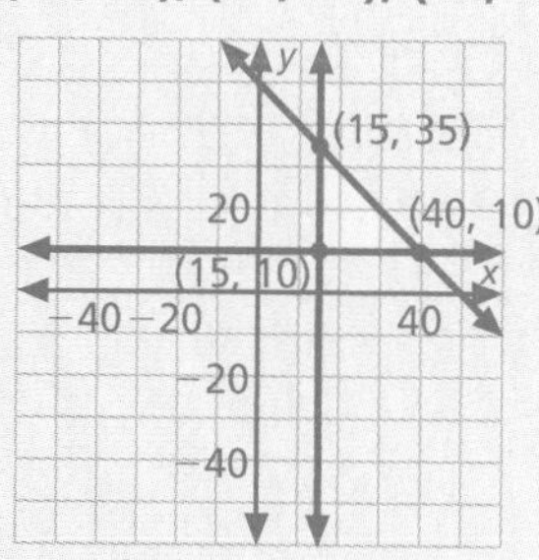

25. Plant 1100 acres of Iceberg and 2500 acres of Romaine to yield a maximum profit of \$845,000.

26. She must stock and sell 50 adult and 30 child skates for a maximum profit of \$2600.

Extra Practice Worksheet 6-9

Name ______ Date ______

EXTRA PRACTICE 6-9

LINEAR PROGRAMMING

EXERCISES

Graph the solution set of the system of inequalities. Use your own paper. Determine the maximum or minimum value of P. Check students' graphs.

1. $\begin{cases} x + y \leq 3 \\ y \leq 3 \\ x \geq 0 \\ y \geq 0 \end{cases}$
maximum $P = 3x + y$ (3, 0)

2. $\begin{cases} 2x + y \leq 4 \\ x + 2y \geq 2 \\ x \geq 0 \\ y \geq 0 \end{cases}$
maximum $P = 4x - y$ (2, 0)

3. $\begin{cases} x + 2y \geq 8 \\ y \leq 1 \\ x \geq 0 \\ y \geq 0 \end{cases}$
minimum $P = x + 2y$ (0, 0)

4. $\begin{cases} x + 3y \geq 6 \\ 4x + 3y \leq 12 \\ x \geq 0 \\ y \geq 0 \end{cases}$
minimum $P = x + 3y$ (0, 2)

5. $\begin{cases} 2x + 2y \leq 6 \\ x + y \geq 2 \\ x \geq 0 \\ y \geq 0 \end{cases}$
maximum $P = 4x + y$ (2, 0)

6. $\begin{cases} x - y \leq 4 \\ x + y \leq 4 \\ y \geq 0 \end{cases}$
maximum $P = 4x + y$ (4, 0)

7. $\begin{cases} x + y \leq 6 \\ y \leq 4 \\ x \leq 3 \\ x \geq 0 \\ y \geq 0 \end{cases}$
minimum $P = 2x + 2y$ (0, 0)

8. $-x - y \geq -10$
$\begin{cases} y \leq 8 \\ x \leq 6 \\ x \geq 0 \\ y \geq 0 \end{cases}$
maximum $P = 6x + y$ (6, 4)

9. $\begin{cases} 2x + y \leq 4 \\ x + y \geq 2 \\ y \geq 0 \\ x \geq 0 \end{cases}$
maximum $P = 2x - 2y$ (2, 0)

10. $\begin{cases} 4x - 2y \geq 8 \\ x - y \leq 4 \\ x \leq 5 \\ y \geq 0 \end{cases}$
minimum $P = 3x + 2y$ (0, 2)

88 EXTRA PRACTICE *LESSON* 6-9

Enrichment Worksheet 6-9

Name ______ Date ______

ENRICHMENT 6-9

A RECTANGLE DISSECTION PUZZLE

In a dissection puzzle, one figure is cut into pieces so that the pieces can be rearranged to form another figure.

In this dissection, you are to cut apart the rectangle shown and use the pieces to make a six-pointed star.

EXERCISES

1. Analyze the rectangle by describing the polygons into which it is dissected.
3 different right triangles, 2 large congruent equilateral triangles, 2 smaller congruent equilateral triangles, 2 congruent trapezoids

2. Make a careful copy of the rectangle and cut apart the pieces. Check students' work.

3. To get started, show how the three right triangles can be put together to form an equilateral triangle. Show your answer on the drawing below.

4. Now use the results from Exercise 3 to finish solving the puzzle. Show your answer on the drawing below.

114 ENRICHMENT *LESSON* 6-9

Assessment Planning

Vocabulary Review

Lesson 6-1
slope
rise
run
slope-intercept form

Lesson 6-2
parallel
perpendicular

Lesson 6-3
point-slope form

Lesson 6-4
system of equations solution
independent system
inconsistent system
dependent system

Lesson 6-5
substitution method

Lesson 6-6
addition/subtraction method
multiplication and addition method

Lesson 6-7
determinant of a system of equations
matrix

Lesson 6-8
system of linear inequalities
boundary
half-plane
solution set

Lesson 6-9
linear programming vertices
feasible region
objective function

Assessment Options

Chapter Assessment
Alternative Assessment
Chapter Test Forms A and B
Cumulative Assessment
Achievement Test
Writing Prompts

Assignment Guide

All students: 1–28
Additional Practice: Refer to the Extra Practice Index on page 674 of this text.

Chapter 6 Review

Vocabulary ■ LESSON 6-1–LESSON 6-9

Choose the word from the list at the right that completes each statement.

1. The __?__ is the ratio of the vertical change to the horizontal change. c

2. The graphs of an __?__ system do not intersect. d

3. The graphs of an __?__ system of equations intersect in one point. a

4. Two lines are __?__ if the product of the slope is -1. b

a. independent system
b. perpendicular
c. slope
d. inconsistent system

LESSON 6-1 ■ Slope of a Line and Slope-Intercept Form, p. 244

▶ slope $= \frac{\text{rise}}{\text{run}} = \frac{\text{vertical change (change in } y\text{-coordinates)}}{\text{horizontal change (change in } x\text{-coordinates)}}$

▶ A horizontal line has a 0 slope. The slope of a vertical line is undefined.

▶ The slope-intercept form of an equation of a line is written as $y = mx + b$, where m represents the slope of the line and b represents the y-intercept.

5. Find the slope of the line containing the points $A(3, -2)$ and $B(2, 9)$. -11

6. Graph the line that passes through the point $P(2, -1)$ with a slope of $\frac{1}{2}$. See additional answers.

7. Graph the line that passes through $P(8, -6)$ with a slope of $-\frac{7}{8}$. (0, 1)

8. Find the slope and y-intercept for $4y = 2x - 8$. Then graph the equation. $m = \frac{1}{2}$; $b = -2$

9. Find an equation of the line with slope $= -2$ and y-intercept $= 1$. $y = -2x + 1$

LESSON 6-2 ■ Parallel and Perpendicular Lines, p. 248

▶ Two lines with the same slope and different y-intercepts are parallel.

▶ Two lines are perpendicular if the product of their slopes is -1.

10. $\overleftrightarrow{MN}$ contains points $M(4, 6)$ and $N(-1, 3)$. Find the slope of a line parallel to $\overleftrightarrow{MN}$ and the slope of a line perpendicular to $\overleftrightarrow{MN}$. $\parallel: \frac{3}{5}$; $\perp: \frac{-5}{3}$

11. Are $7x - 3y = 21$ and $7y = 3x + 4$ parallel, perpendicular, or neither? neither

12. Determine whether the line containing the points $T(0, 3)$ and $U(3, 0)$ is perpendicular or parallel to the line containing the points $V(7, 1)$ and $W(1, 7)$. parallel

LESSON 6-3 ■ Write Equations for Lines, p. 254

▶ An equation of a line can be written in point-slope form if you know the slope of the line and the coordinates of any point on the line or the coordinates of two points on the line.

Additional Answers

6.

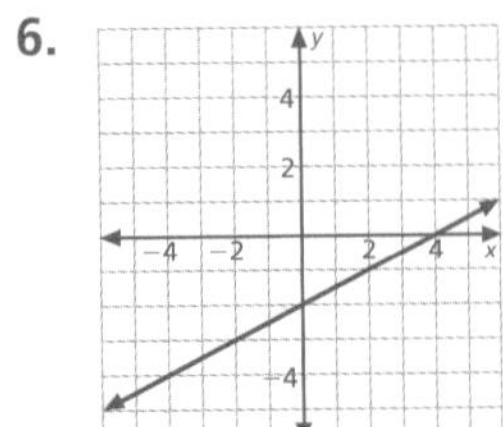

17.

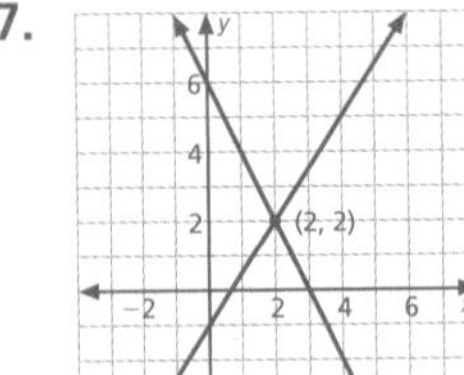

18.

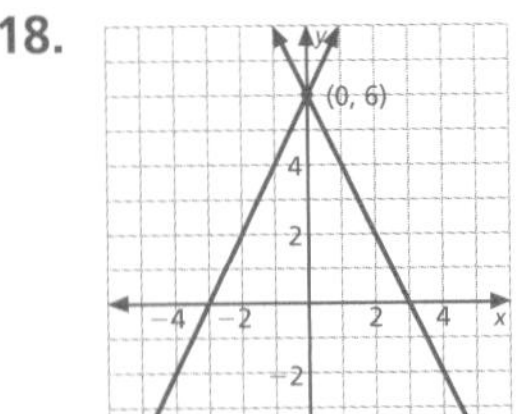

25.

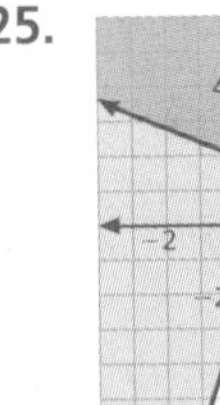

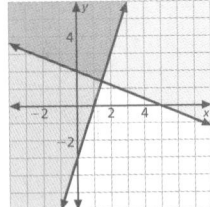

Write an equation of the line with the given information.

13. $m = -1$; $A(2, 3)$ $y = -x + 5$

14. $C(3, 6)$ and $D(1, -2)$ $y = 4x - 6$

15. A line that is perpendicular to $4y - 3x = 12$ containing the point $Z(2, -1)$. $y = -\frac{4}{3}x - \frac{5}{3}$

16. Write an equation of the line whose slope is $-\frac{1}{4}$ and contains the point (8, 2). $y = \frac{-1}{4}x + 4$

LESSON 6-4 ■ Systems of Equations, p. 258

- ▶ Two linear equations with the same two variables form a system of equations. The solution for the system is the ordered pair that makes both equations true. Graphing both equations can be used to solve the system; the point of intersection of the two lines is the solution.
- ▶ In an independent system, the graphs intersect in one point. In an inconsistent system, the graphs do not intersect. In a dependent system, the lines coincide.

Solve each system of equations by graphing. See additional answers.

17. $\begin{cases} 2x + y = 6 \\ 2y + 2 = 3x \end{cases}$

18. $\begin{cases} -6x + 3y = 18 \\ y = 6 - 2x \end{cases}$

LESSONS 6-5 and 6-6 ■ Solving Systems of Equations, p. 264

- ▶ A system of equations can be solved algebraically. You can choose one of three methods: substitution, addition and subtraction, or multiplication and addition.

Solve.

19. $\begin{cases} 3x + 2y = -3 \\ -4x + 2y = 4 \end{cases}$ (−1, 0)

20. $\begin{cases} 2x + 7y = -1 \\ 3x = -2y + 7 \end{cases}$ (3, −1)

21. $\begin{cases} 3x - 5y = -1 \\ -5x + 7y = -1 \end{cases}$ (3, 2)

LESSON 6-7 ■ Problem Solving, p. 274

- ▶ Determinants can be used to solve a system of equations.

$$\begin{cases} ax + by = e \\ cx + dx = f \end{cases} \rightarrow \det A = ad - bc; \quad x = \frac{ed - bf}{ad - bc}, \quad y = \frac{af - ec}{ad - bc}$$

Solve the system of equations using determinants.

22. $\begin{cases} 5y + 3x = 4 \\ 2x - 3y = 5 \end{cases}$ $\frac{37}{19}, -\frac{7}{19}$

23. $\begin{cases} 2y = 4x - 10 \\ 3y + x = -1 \end{cases}$ (2, −1)

24. Danielle is 5 years less than twice Mario's age. In 15 years, Mario will be the same age as Danielle is now. Find the ages of Danielle and Mario in 5 years.
Mario, 25; Danielle, 40

LESSONS 6-8 and 6-9 ■ Systems of Inequalities, p. 276

- ▶ A system of linear inequalities can be solved graphically. The intersection of the graphs of the inequalities is the solution set of the system.
- ▶ Linear programming can be used to solve business-related linear inequalities.

Graph the solution set of the system of linear inequalities. See additional answers.

25. $\begin{cases} 2x + 5y \geq 10 \\ 3x - y \leq 3 \end{cases}$

26. $\begin{cases} 2x - 3y \leq 6 \\ x + y \geq 1 \end{cases}$

27. $\begin{cases} 3x + y \geq 2 \\ y \geq 2x - 1 \end{cases}$

28. $\begin{cases} -y < x \\ x + \frac{1}{3}y \geq 1 \end{cases}$

26.

27.

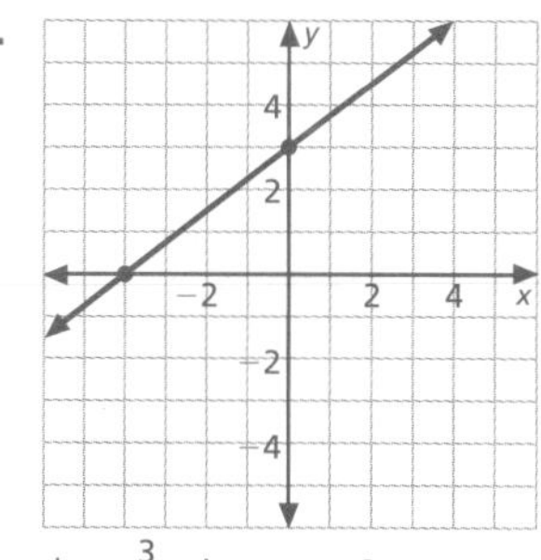

slope: $\frac{3}{4}$; y-intercept: 3

28.

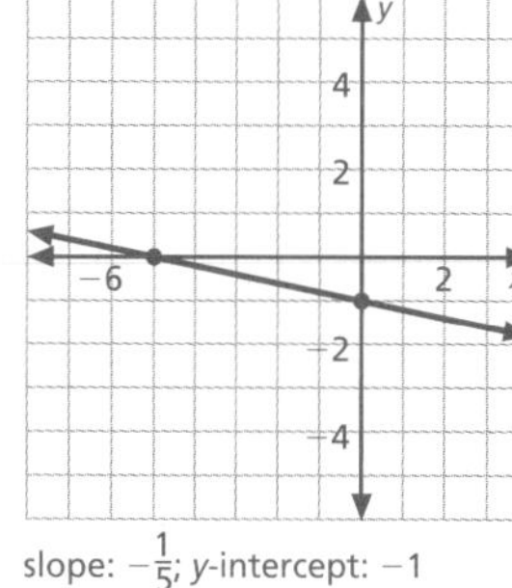

slope: $-\frac{1}{5}$; y-intercept: −1

Chapter 6 Test Form A, page 1

CHAPTER 6

SLOPES AND SYSTEMS OF EQUATIONS

ASSESSMENT FORM A, PAGE 1

Name ________

Date ________

Scoring Record	
Possible: 21	Earned:

1. Find the slope of a line containing points A (3, 6) and B (−2, 0). $\frac{6}{5}$
2. Find the slope of the line $-12x - 6y = 3$. -2
3. Find the slope and y-intercept for the line $7x - y = 2$. $m = 7, b = -2$
4. Write an equation of the line with slope $= \frac{1}{2}$ and y-intercept $= -\frac{1}{6}$. $y = \frac{1}{2}x - \frac{1}{6}$

Find the slope of each line. Determine whether each pair is *parallel, perpendicular* or *neither.*

5. The line containing F (6, 2) and G (3, 9). The line containing H (6, −2) and I (13, 1). $\overleftrightarrow{FG} : m = -\frac{7}{3}; \overleftrightarrow{HI} : m = \frac{3}{7}$; perpendicular
6. Line l $2x - 3y = 7$ Line n $-3x + y = -2$ $l : m = \frac{2}{3}; n : m = 3$; neither
7. Line p $7x = 6 - 2y$ Line q $4y = -14x + 1$ $p : m = -\frac{7}{2}; q : m = -\frac{7}{2}$; parallel

Write an equation for each line with the given information.

8. $m = -1$ and J (4, −1) $y = -x + 3$
9. K (−5, 1) and L (5, −3) $y = \frac{2}{5}x - 1$
10. Line parallel to $y = \frac{1}{4}x - 3$ and containing point M (4, 0) $y = \frac{1}{4}x - 1$
11. Write an equation of the line whose graph is shown. $y = \frac{1}{4}x - 3$
12. Solve the system of equations by graphing. $\begin{cases} y = -4x + 2 \\ -y = -3x + 5 \end{cases}$ (1, −2)

Solve by substitution.

13. $\begin{cases} 8x + 2y = 12 \\ \frac{1}{2}y = -2x + 3 \end{cases}$ infinite solutions
14. $\begin{cases} x = 10 - y \\ 2x - y = 8 \end{cases}$ (6, 4)
15. $\begin{cases} -\frac{1}{2}x = y \\ -6y = 3x + 9 \end{cases}$ no solution

72 ASSESSMENT *CHAPTER 6 FORM A*

Chapter 6 Test Form A, page 2

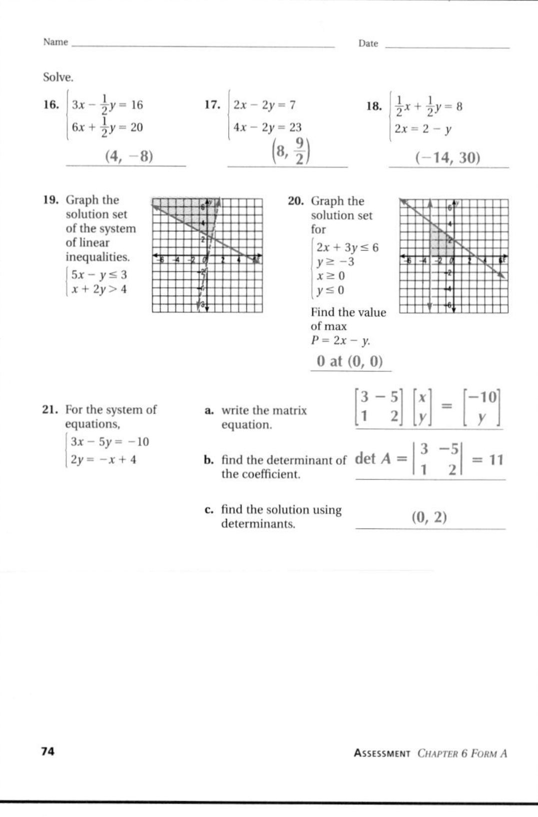

Name ________ Date ________

Solve.

16. $\begin{cases} 3x - \frac{1}{2}y = 16 \\ 6x + \frac{1}{2}y = 20 \end{cases}$ (4, −8)
17. $\begin{cases} 2x - 2y = 7 \\ 4x - 2y = 23 \end{cases}$ $\left(8, \frac{9}{2}\right)$
18. $\begin{cases} \frac{1}{2}x + \frac{1}{2}y = 8 \\ 2x = 2 - y \end{cases}$ (−14, 30)
19. Graph the solution set of the system of linear inequalities. $\begin{cases} 5x - y \leq 3 \\ x + 2y > 4 \end{cases}$
20. Graph the solution set for $\begin{cases} 2x + 3y \leq 6 \\ y \geq -3 \\ x \geq 0 \\ y \leq 0 \end{cases}$ Find the value of max $P = 2x - y$. 0 at (0, 0)
21. For the system of equations, $\begin{cases} 3x - 5y = -10 \\ 2y = -x + 4 \end{cases}$
    a. write the matrix equation. $\begin{bmatrix} 3 & -5 \\ 1 & 2 \end{bmatrix} \begin{bmatrix} x \\ y \end{bmatrix} = \begin{bmatrix} -10 \\ y \end{bmatrix}$
    b. find the determinant of the coefficient. $\det A = \begin{vmatrix} 3 & -5 \\ 1 & 2 \end{vmatrix} = 11$
    c. find the solution using determinants. (0, 2)

74 ASSESSMENT *CHAPTER 6 FORM A*

Chapter 6 **Review** 287

Chapter 6 Test Form B, page 1

CHAPTER 6
SLOPES AND SYSTEMS OF EQUATIONS
ASSESSMENT FORM B, PAGE 1

Name ______
Date ______

Scoring Record	
Possible: 21	Earned:

1. Find the slope of a line containing points $A(-1, -8)$ and $B(5, 4)$. 2
2. Find the slope of the line $2x = 6y + 3$. $\frac{1}{3}$
3. Find the slope and y-intercept for the line $2x + 7y = -7$. $m = \frac{2}{7}, b = -1$
4. Write an equation of the line with slope $= \frac{3}{4}$ and y-intercept $= \frac{5}{8}$. $y = \frac{3}{4}x + \frac{5}{8}$

Find the slope of each line. Determine whether each pair is *parallel, perpendicular* or *neither.*

5. The line containing $F(-2, 4)$ and $G(1, 3)$. The line containing $H(2, 9)$ and $I(4, 3)$. neither $\overleftrightarrow{FG}: m = -\frac{1}{3}; \overleftrightarrow{HI}: m = -3;$
6. Line l $x + y = 8$; Line n $-3y = 3x + 7$ $l: m = -1; n: m = -1$; parallel
7. Line p $2x = -9 + 4y$; Line q $5y = 15x - 9$ $p: m = \frac{1}{2}; q: m = 3$; neither

Write an equation for each line with the given information.

8. $m = 7$ and $J(3, 0)$ $y = 7x - 21$
9. $K(2, -3)$ and $L(-3, 2)$ $y = -x - 1$
10. Line parallel to $y = \frac{1}{3}x + 4$ and containing point $M(2, 1)$ $y = \frac{1}{3}x + \frac{1}{3}$
11. Write an equation of the line whose graph is shown. $y = \frac{2}{3}x - 1$
12. Solve the system of equations by graphing. $\begin{cases} y = -x - 2 \\ -y = x + 6 \end{cases}$ no solution

Solve by substitution.

13. $\begin{cases} x = 12 - y \\ 6x - 3y = 9 \end{cases}$ (5, 7)
14. $\begin{cases} 2x + 3y = 12 \\ -x = \frac{3}{2}y - 6 \end{cases}$ infinite solutions
15. $\begin{cases} \frac{2}{3}x = y \\ 9x - 6y = 15 \end{cases}$ (3, 2)

76 ASSESSMENT *CHAPTER 6 FORM B*

Chapter 6 Test Form B, page 2

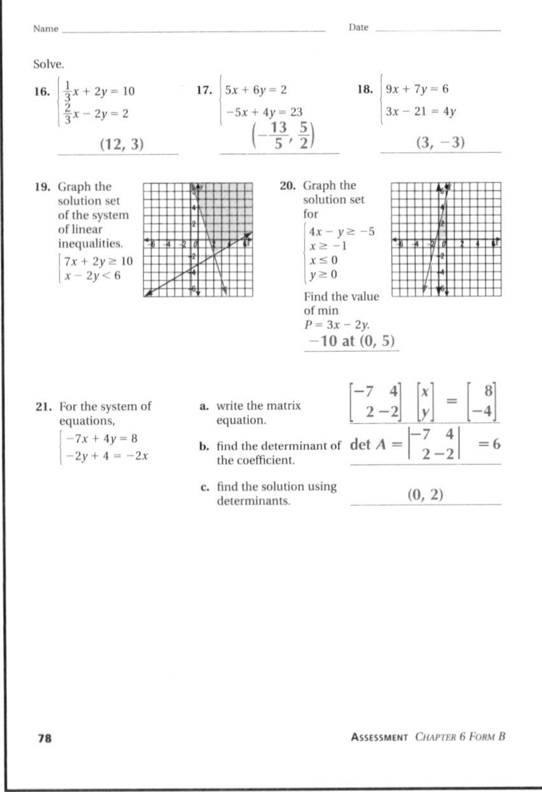

Name ______ Date ______

Solve.

16. $\begin{cases} \frac{1}{3}x + 2y = 10 \\ \frac{2}{3}x - 2y = 2 \end{cases}$ (12, 3)
17. $\begin{cases} 5x + 6y = 2 \\ -5x + 4y = 23 \end{cases}$ $\left(-\frac{13}{5}, \frac{5}{2}\right)$
18. $\begin{cases} 9x + 7y = 6 \\ 3x - 21 = 4y \end{cases}$ (3, −3)
19. Graph the solution set of the system of linear inequalities. $\begin{cases} 7x + 2y \geq 10 \\ x - 2y < 6 \end{cases}$
20. Graph the solution set for $\begin{cases} 4x - y \geq -5 \\ x \geq -1 \\ x \leq 0 \\ y \geq 0 \end{cases}$ Find the value of min $P = 3x - 2y$. −10 at (0, 5)
21. For the system of equations, $\begin{cases} -7x + 4y = 8 \\ -2y + 4 = -2x \end{cases}$
    a. write the matrix equation. $\begin{bmatrix} -7 & 4 \\ 2 & -2 \end{bmatrix}\begin{bmatrix} x \\ y \end{bmatrix} = \begin{bmatrix} 8 \\ -4 \end{bmatrix}$
    b. find the determinant of the coefficient. $\det A = \begin{vmatrix} -7 & 4 \\ 2 & -2 \end{vmatrix} = 6$
    c. find the solution using determinants. (0, 2)

78 ASSESSMENT *CHAPTER 6 FORM B*

CHAPTER 6 ASSESSMENT

Find the slope and y-intercept for each line. Then graph the equation. See additional answers.

1. $-3x + 4y = 12$
2. $x + 5y = -5$

Find the slope for each line. then identify the slope of a line parallel to the given line and a line perpendicular to the given line.

3. The line containing points $R(1, -1)$ and $S(-4, 1)$. $m = \frac{-2}{5}; \parallel: \frac{-2}{5}; \perp: \frac{5}{2}$
4. $4x - y = 8$ $m = 4; \parallel: 4; \perp: \frac{-1}{4}$
5. Graph the line that passes through $P(2, -1)$ with a slope of $-\frac{1}{2}$. See additional answers.

Write an equation of the line with the given information.

6. 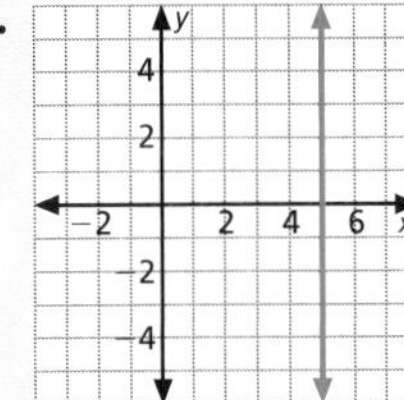 $x = 5$
7. $m = -\frac{4}{5}, b = 2$ $y = \frac{-4}{5}x + 2$
8. $m = \frac{1}{2}, A(-1, 2)$ $y = \frac{1}{2}x + 2\frac{1}{2}$
9. $C(-4, 1)$ and $D(1, 4)$ $y = \frac{3}{5}x + 3\frac{2}{5}$

Use a graph to solve each system. For 10–13, see additional answers.

10. $\begin{cases} 8x - 2y = 6 \\ 3x = 4y - 1 \end{cases}$
11. $\begin{cases} 2y = 3x \\ 3y - 2x = 10 \end{cases}$
12. $\begin{cases} x - 2y < 4 \\ 3 \geq x + y \end{cases}$
13. $\begin{cases} y - 3 < 0 \\ x + 2y > 2 \\ x \leq 2 \end{cases}$

Solve.

14. $\begin{cases} -4x - 5y = -2 \\ 6y = -x + 10 \end{cases}$ (−2, 2)
15. $\begin{cases} x + 3y = -9 \\ 7 - 2y = 5x \end{cases}$ (3, −4)
16. $\begin{cases} 5x - 2y = 9 \\ x - 4y = 9 \end{cases}$ (1, −2)
17. $\begin{cases} -9x - 3y = 9 \\ x = \frac{y}{3} + 1 \end{cases}$ (0, −3)
18. The sum of the digits of a two-digit number is 5. The units digit is one more than 3 times the tens digit. Find the original number. 14
19. Kari is 6 years older than Adam. In 9 years, $\frac{1}{2}$ of Adam's age will equal $\frac{1}{3}$ Kari's age. Find the ages of Kari and Adam in 2 years. Kari:11; Adam: 5
20. One solution contains a 40% acid solution. Another contains a 60% acid solution. Determine the number of liters of each solution needed to make 25 liters of a 56% solution. 40% Acid; 5 liters; 60% Acid: 20 liters

ADDITIONAL ANSWERS

1.

slope: $\frac{3}{4}$; y-intercept: 3

2.

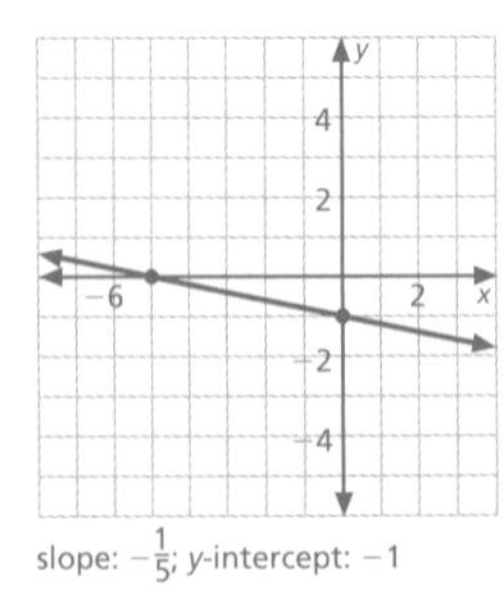

slope: $-\frac{1}{5}$; y-intercept: −1

5.

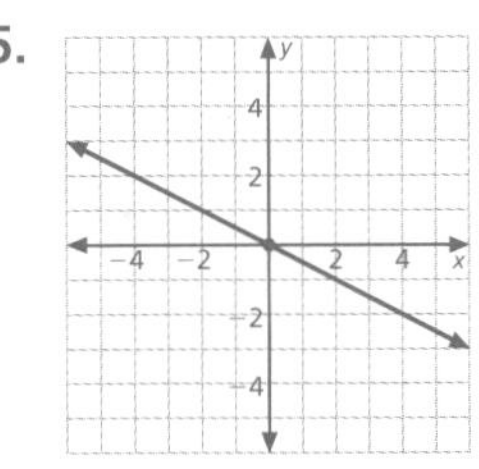

10.

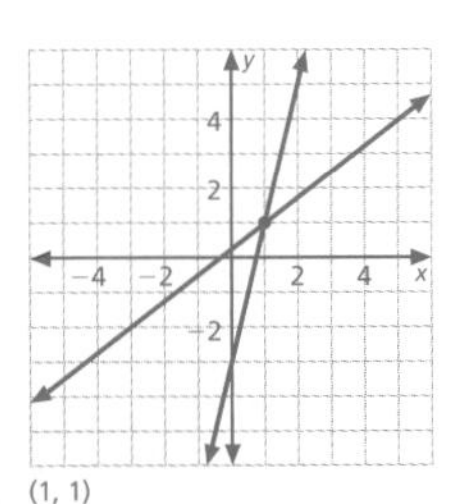

(1, 1)

Additional answers continued on page 290.

Chapter 6 Alternative Assessment

Self Assessment

METHODS OF SOLUTION Select a system of equations from your textbook, or make your own. Solve the system by graphing; substitution; and addition, subtraction and multiplication. For each method, explain the steps you use to solve the system. Select what you feel is the easiest method of solution and tell why you chose that method.

Using Other Resources

HILLS AND VALLEYS The slope of steep hills on our highways is usually given as a percent, such as "5% grade." Contact your Department of Transportation, or research in the library or on the Internet to find how this percent is determined. How do they measure the slope? Is there a maximum slope that is considered safe? Highways are usually "banked," or slightly tilted to the left or right, around curves. Find out if the slope of the bank is determined by the sharpness of the curve, the speed permitted on the road, or both.

Technology

IN THE WORKPLACE Today there are calculators and computers that will perform many of the calculations presented in this chapter. The understanding of the mathematics is still a requirement. Interview five people who work in different occupations to find what math they use in their jobs. Do they use a computer or calculator for this math? How do they use graphs in their jobs? How are these graphs produced? If possible, demonstrate on a computer or calculator for the class how some of this math is done.

Chapter Investigation

EXTENSION Compare your improved product to the original product. Write a report about how your improvements will make the product better. Include an explanation as to how the improvements to the product will justify the increased cost of the product.

Chapter Investigation

The benchmarks and expectations for this extension are as follows.

- Students brainstorm a list of products that the members of their group have used during the past week. They select a product that they all agree is in need of improvement. They prepare a list of questions that could be used to find out whether consumers share their concern.
- Students make a detailed drawing or build a prototype of their new product. They choose at least five selling points to emphasize in their marketing materials.
- Students compare their improved product to the original product. They write a report about how their improvements will make the product better. They include an explanation as to how the improvements to the product will justify the increased cost of the product.

Self Assessment

Instead of allowing students to select a system of equations, you may wish to provide a system.
SCORING RUBRIC 5—Student correctly solves a system of equations by three methods, and writes detailed and complete steps for solution. **4**—Student correctly solves a system of equations by three methods, and writes complete steps for solution. **3**—Student correctly solves a system of equations by three methods, and writes most of the steps for solution. **2**—Student solves a system of equations by three methods with minor errors, and writes most of the steps for solution. **1**—Student attempts to solve a system of equations by one or two methods with significant errors, and most of the steps for solution. **0**—Student makes no attempt solve a system of equations.

Using Other Resources

Students may also find useful information by contacting a local college engineering department or contractors who specialize in road construction and repair.
SCORING RUBRIC 5—Student contacts more than two informational agencies, determines and interprets the standards of highway gradations, and presents a detailed report. **4**—Student contacts two informational agencies, determines and interprets the standards of highway gradations, and presents a report. **3**—Student contacts two informational agencies, determines the standards of highway gradations, and presents a report. **2**—Student contacts at least one informational agency, determines the standards of highway gradations, and presents a partial or incomplete report. **1**—Student contacts at least one informational agency, attempts to determine the standards of highway gradations, but presents no report. **0**—Student makes no attempt to contact agencies or present a report.

Critical Thinking

You may want to direct students to more math-oriented occupations, such as engineers, architects, and surveyors.
SCORING RUBRIC 5—Student interviews five people from different occupations and competently demonstrates the technology for the class. **4**—Student interviews five people from different occupations and demonstrates the technology for the class. **3**—Student interviews less than five people from different occupations and demonstrates the

Scoring Rubric continued on page 291.

Cumulative Review

Cumulative review is the best way to maintain previously taught skills and concepts. This will keep students prepared for new lessons that build on previously covered skills.

Cumulative Review covers:

Additional answers continued from page 288.

11.

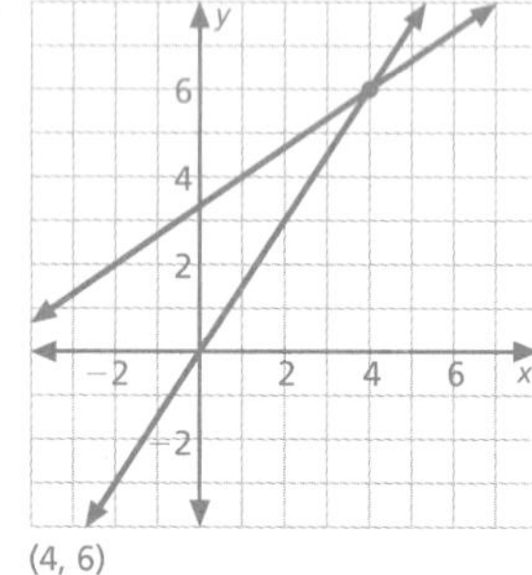

12.

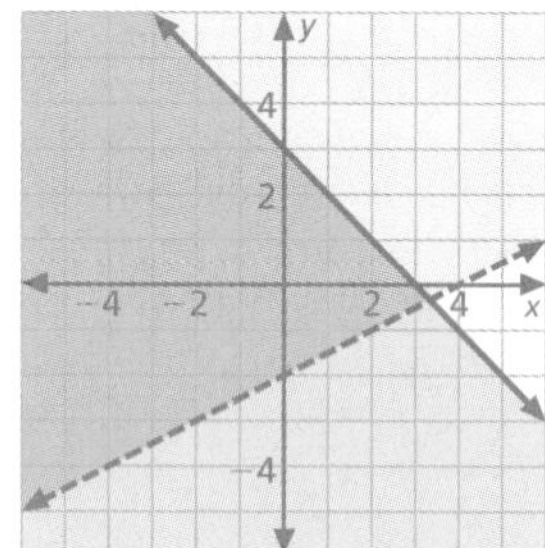

13.

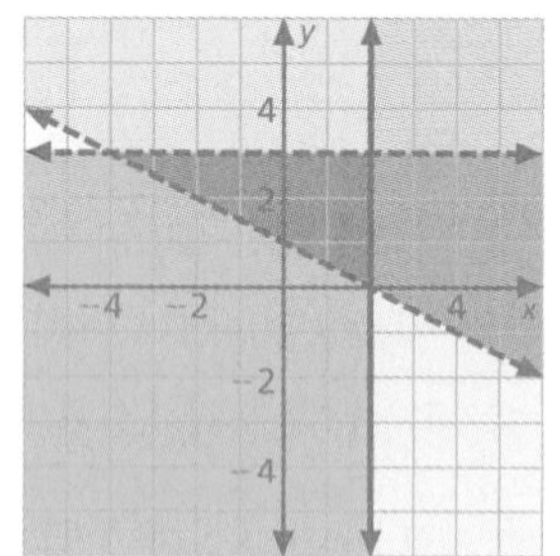

Chapters 1–6 Cumulative Review

1. Write 0.0003206 in scientific notation. 3.206×10^{-4}
2. Express the product in scientific notation: $(3 \times 10^9)(6.3 \times 10^8)$. 1.89×10^{18}
3. Simplify. $(2x^{-3}y^2)(x^5y^3z^{-9})^2$ $\frac{2x^7y^8}{z^{18}}$
4. Find the next three terms in this pattern: $1, -\frac{1}{2}, \frac{1}{4}, -\frac{1}{8}, \ldots$ $\frac{1}{16}, -\frac{1}{32}, \frac{1}{64}$

Solve.

5. $3x - 7 = -7x - 15$ $x = -\frac{4}{5}$
6. $-\frac{3}{4}x + 2 = 17$ $x = -20$
7. B is the midpoint of $\overline{AC}$. Find x if $AC = 26$ and $AB = 3x - 5$. $x = 6$
8. $\overline{AB} \perp \overline{CD}$ at point E. If $m\angle AEG = 81°$, find $m\angle DEF$. 171°

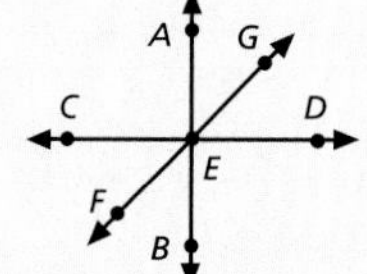

In the proof below, supply the reasons for the indicated statements.

Given: $n \parallel m$.

Prove: $\angle 1 \cong \angle 3$.

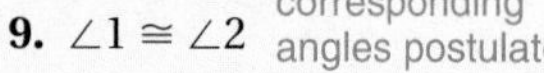

9. $\angle 1 \cong \angle 2$ corresponding angles postulate
10. $\angle 2 \cong \angle 3$ vertical angles are congruent
11. $\angle 1 \cong \angle 3$ transitive property

Given the information below, what congruency postulates will be used to prove the indicated triangles congruent?

12. E is a midpoint of $\overline{AC}$; $\overline{AC} \perp \overline{BE}$. SAS

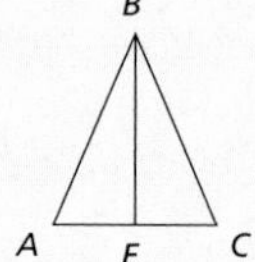

13. $BA \cong BC$, $\angle A \cong \angle C$ ASA

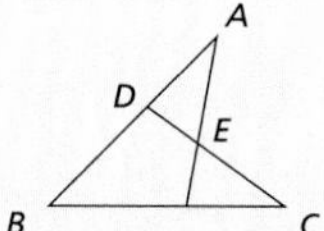

14. Two sides of a triangle measure 3 feet and 16 inches. What is the range of possible values of the third side? 20 in. $< x <$ 52 in.
15. Find n. 160

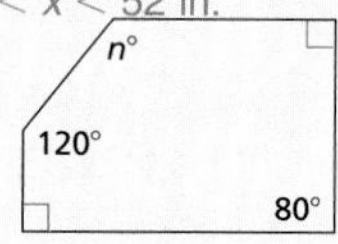

Find the shaded area in the figures below.

16. 12 cm³

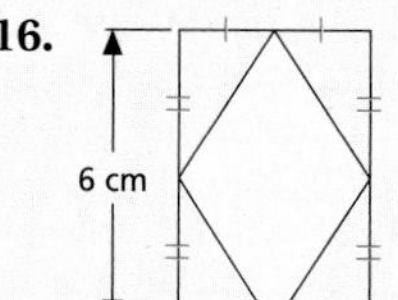

17. 15.7 cm²

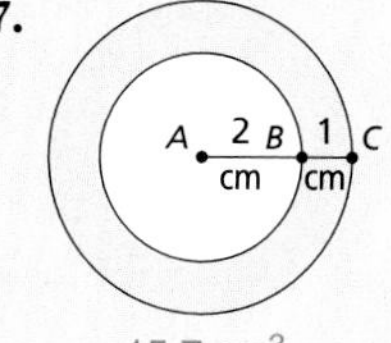

18. If the area of $\triangle ABC$ is 15 cm², find x. 6 cm

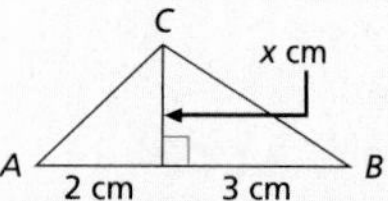

19. Find the volume and surface area of the figure. V = 282.6 cm³; SA = 244.92 cm²

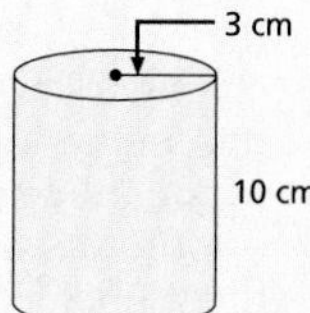

20. Write the equation of the line with slope $= -\frac{1}{2}$ through point (4, 4). $y = -\frac{1}{2}x + 6$
21. Write the equation of the line that goes through the points (−2, −3) and (1, 3). $y = 2x + 1$
22. Solve the system of equations. (0, 4)

$$\begin{cases} 2x + 3y = 12 \\ x - 2y = -8 \end{cases}$$

Teaching Strategies

In Exercise 19, students may have trouble remembering the formula for finding the surface area of a cylinder. Encourage students to think of the surface area as the sum of the area of the two bases and a rectangle wrapped around the bases. The length of the rectangle is equal to the circumference of a base and the width is given in the diagram as 10 cm. Compare this process with the actual formula to help students see the relationship.

Chapters 1–6 Cumulative Assessment

Standardized Test Preparation—Gridded Response

Solve each question. Write your answer at the top of the answer grid and fill in the ovals.

Notes: Mixed numbers such as $1\frac{1}{2}$ must be gridded as 1.5 or $\frac{3}{2}$. Grid only one answer per question. If your answer is a decimal, enter the most accurate value the grid will accommodate.

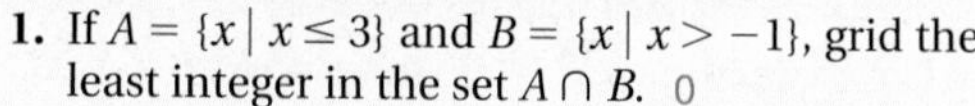

1. If $A = \{x \mid x \le 3\}$ and $B = \{x \mid x > -1\}$, grid the least integer in the set $A \cap B$. 0
2. Grid the solution of $-(3x^4y^3)(x^2y^3z^3)^2$ if $x = -\frac{2}{3}$, $y = -1$, and $z = 2$. $\frac{3}{64}$
3. If $g(x) = -3x^2 - |x| + 6$, grid $g(2)$. 16
4. If $-\frac{5}{6}x + 6 = -9$, grid the value of x. 18
5. If $BF^* \parallel EC^*$, $m\angle ABE = 70°$, and $m\angle FBE = 40°$, grid $m\angle BCD$. 150

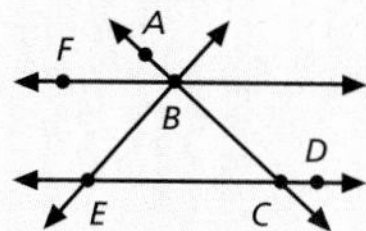

6. Given $AC^* \perp DB^*$, $m\angle EBC = 4x + 4$, and $m\angle DBE = 3x + 9$, grid $m\angle DBE$. 42

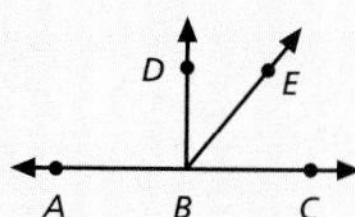

7. Give $AB^* \parallel DC^*$ and $AB^* \cong DC^*$, grid $m\angle DBC$. 60

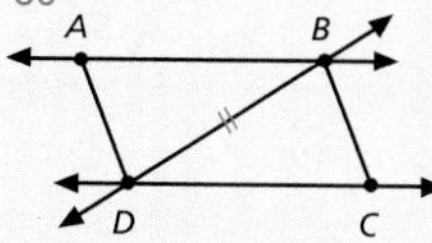

8. Nine-sided irregular polygon, with angles marked as follows: 132°, 177°, 125°, 143°, 138°, 156°,103°, 127°, x. Find x. 159°
9. Twelve-sided irregular polygon, with angles marked as follows: 156°, 155°, 154°, 153°, 152°, 151°,150°, 149°, 148°, 147°, 146°, x. Find x. 139°
10. Find the sum of the angles of a ten-sided polygon. 1440
11. Find the measure of angle C in this figure. 74

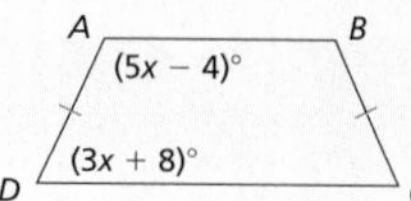

12. Find the surface area of this figure. 132 cm²

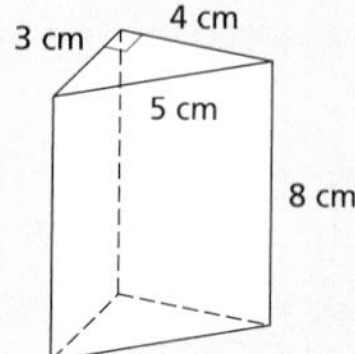

13. Find the height, x, if the volume of the rectangular box is 220 cm³. 16 cm

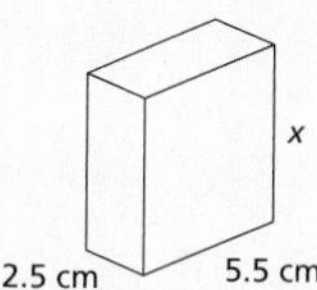

14. The line, $y = 4x + 11$, passes through the points $(-2, 3)$ and $(-1, c)$. Grid c. 7
15. What is the x-coordinate in the solution for this system of equations? $4x + y = 9$ and $-5x - 2y = 3$ 7
16. Which region(s) should be shaded to represent the solution set? Grid 1 for e, 2 for f, 3 for g, 4 for h, 5 for j, 6 for k, 7 for n. If more than one region should be shaded, grid the sum of the region numbers.
$x \ge 0$; $x \le 7$;
$y \ge 0$; $2y \ge x + 8$ 3

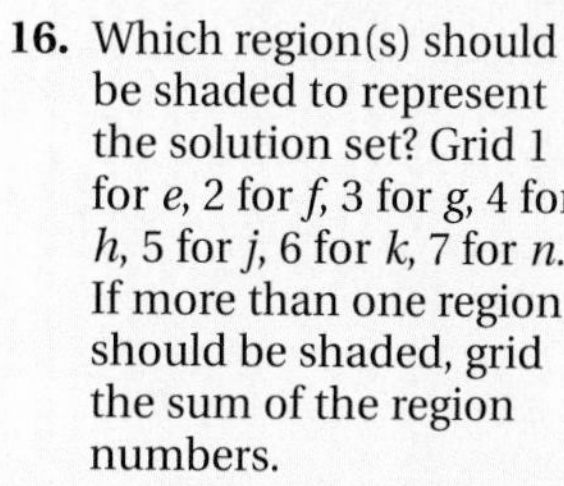

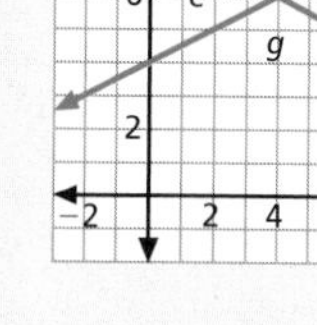

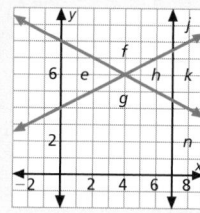

17. **CONSTRUCTED RESPONSE** Write the equation of the line perpendicular to the line that passes through (0, 6) and (−4, 2) and crosses the x-axis at 10.
$y = -x + 10$

Teaching Strategies

Have students share their methods for solving Exercise 12 of the Cumulative Assessment. One method is to graph the equation and visually inspect the graph to find the value of c. Students may also use the slope formula to find c. Discuss the importance of understanding how the slope formula relates to equations in both the slope-intercept and point-slope forms.

Standardized Test Preparation

GRIDDED RESPONSE Questions that require you fill in grids on an answer sheet, but do not have answer choices provided are called gridded response or students-produced response. You must solve the problem and then fill in the answer on a special grid. Your answer consists of two parts-the numbers that you write in the answer boxes and the ovals that you fill in. Ovals that correspond to the digits 0–9, negative sign, fraction bars and decimal points are listed in columns under the answer boxes.

You should grid the form of the answer that you obtain from solving the problem. Enter the most accurate value the grid will accommodate. The grid will only hold numbers that range from 0–9999. The grid will not accommodate a mixed number. Change mixed numbers to either decimals or improper fractions.

Columns not needed should be left blank. It does not matter whether the extra columns are on the left or the right of the answer.

Testing Strategies

Caution students to read the directions carefully when solving grid-response items. For example, in Exercise 14, the sections are numbered. Instead of gridding the value of either x or y, students are asked to grid the number that corresponds to the correct region. If more than one region should be shaded, the student should grid the sum of the corresponding numbers.

Critical Thinking Scoring Rubric continued from page 290.

technology for the class. **2**—Student interviews less than five people from different occupations and attempts to demonstrate the technology for the class. **1**—Student interviews one or two people from different occupations but gives no demonstration to the class. **0**—Student makes no attempt to interview people or demonstrate technology.

Chapter 7—Similar Triangles

Theme: Photography
Chapter Investigation: Sketches, pages 295, 299, 309, 323
Careers: Police Photographers, page 305; Photographic Process Workers, page 325
Data Activity: Camera Settings and Image Sizes

Content and Connections

Lesson, Pages	Lesson Objectives	NCTM Standards	State/Local Objectives	Interdisciplinary Connections	Real World Applications
7–1 296–299	• Find equivalent ratios • Use rations and proportions to solve problems	Number Operation Patterns, Functions & Algebra Problem Solving Reasoning & proof Connections; Representation		Art	Recreation, Real Estate, Retail, Business, Art
7–2 300–303	• Identify similar polygons • Find missing measures of similar polygons	Number Operation Patterns, Functions & Algebra Geometry & Spatial Sense Measurement Problem Solving Reasoning & Proof Connections; Representation			Photography, Architecture, Framing
7–3 306–309	• Find actual or scale length using scale drawings	Measurement Number Operation Patterns, Functions & Algebra Problem Solving Reasoning & Proof Representation			Architecture, Engineering, Photography
7–4 310–313	• Use the AA, SSS and SAS similarity postulates to determine if two triangles are similar	Problem Solving Number Operation Geometry & Spatial Sense Reasoning & Proof Representation			Art, Surveying, Photography
7–5 316–319	• Prove theorems involving similar triangles • Use theorems to find unknown lengths of sides of triangles	Problem Solving Reasoning & Proof Number Operation Patterns, Functions & Algebra Geometry & Spatial Sense Measurement Representation			Scale models, Photography
7–6 320–323	• Use theorems involving parallel lines and proportional segments to find unknown lengths • Divide a line segment into congruent parts	Geometry & Spatial Sense Problem Solving Reasoning & Proof Number Operation Patterns, Functions & Algebra Measurement Connections Representation			Model Building, Architecture, Real Estate
7–7 326–327	• Solve a problem using indirect measurement • Use a mode or a picture	Problem Solving Reasoning & Proof Connections Number Operation Patterns, Functions & Algebra Geometry & Spatial Sense Representation			Photography

Planning and Resources

Lesson, Pages	Tools/Materials Needed	Trans-parency	Learning/Teaching Styles Options	Assignments: Basic Enriched	Additional Practice in SE	Reteaching, Extra Practice, Enrichment	Other Resources
7–1 296–299	calculator	Warm up 19 Trans RF-27	Challenge	B: 1–34, 37–56 E: 1–56	Page 304–305, 315, 325, 328, 705	R: page 101 EP: page 89 E: page 117	

7–2 300–303	centimeter ruler protractor compass	Warm up 19	Visual Learners ESL/LEP Challenge	B: 1–15, 19–27 E: 1–27	Page 304–305, 315, 325, 328, 705	R: page 103 EP: page 91 E: page 119	SS: Teacher's Choice
7–3 306–309	string centimeter ruler	Warm up 19	Real World Challenge	B: 1–19, 22–24 E: 1–24	Page 314–315, 325, 328, 706	R: page 105 EP: page 93 E: page 121	SS: Teacher's Choice
7–4 310–313	calculator straightedge compass	Warm up 20 Trans RF-28		B: 1–10, 15–26 E: 1–26	Page 314–315, 325, 329, 706	R: page 107 EP: page 95 E: page 123	
7–5 316–319	calculator straightedge compass	Warm up 20	ESL/LEP Challenge	B: 1–10, 14–20 E: 1–20	Page 324–325, 329, 707	R: page 109 EP: page 97 E: page 125	
7–6 320–323	compass straightedge	Warm up 21		B: 1–14, 19–26 E: 1–26	Page 324–325, 329	R: page 111 EP: page 99 E: page 127	MC: 5
7–7 326–327	measuring tape mirror	Warm up 21	Real World	B: 1–5, 6–14 E: 1–14	Page 329, 707	R: page 113 E: page 129	

Planning and Pacing

Lesson, Pages	Lesson Title	45 min class	Assignments Basic, Enriched	90 min class	Assignments Basic, Enriched	___ min class	Assignments Basic, Enriched
7–1 296–299	AYR, Opener, Ratios and Proportions	Day 89	B: 1–34, 37–56 E: 1–56	Day 46	B: 1–34, 37–56 E: 1–56		B: 1–34, 37–56 E: 1–56
7–2 300–303	Similar Polygons, R&PYS	Day 90–91	B: 1–15, 19–27 E: 1–27	Day 46–47	B: 1–15, 19–27 E: 1–27		B: 1–15, 19–27 E: 1–27
7–3 306–309	Scale Drawings	Day 92	B: 1–19, 22–24 E: 1–24	Day 47	B: 1–19, 22–24 E: 1–24		B: 1–19, 22–24 E: 1–24
7–4 310–313	Postulates for Similar Triangles, R&PYS	Day 93–94	B: 1–10, 15–26 E: 1–26	Day 48	B: 1–10, 15–26 E: 1–26		B: 1–10, 15–26 E: 1–26
7–5 316–319	Triangles and Proportional Segments	Day 95	B: 1–10, 14–20 E: 1–20	Day 49	B: 1–10, 14–20 E: 1–20		B: 1–10, 14–20 E: 1–20
7–6 320–323	Parallel Lines and Proportional Segments, R&PYS	Day 96–97	B: 1–14, 19–26 E: 1–26	Day 49–50	B: 1–14, 19–26 E: 1–26		B: 1–14, 19–26 E: 1–26
7–7 326–327	Problem Solving Skills: Indirect Measurement	Day 98	B: 1–5, 6–14 E: 1–14	Day 50	B: 1–5, 6–14 E: 1–14		B: 1–5, 6–14 E: 1–14
Review/ Assess		Day 99–101		Day 51–52			

	Chapter 1 Days	Chapter Cumulative Days
Yearly Pacing (45 min class)	12 days	101 days
Yearly Pacing (90 min class)	7 days	52 days

Assessment Options

Assessment in Student Edition	Assessment in Teacher's Edition	Pages in Assessment Book	Software Generated Assessment
Are You Ready?, pages 292–293; Writing Math, pages 299, 303, 309, 313, 319, 323; Mixed Review, pages 299, 303, 309, 313, 319, 323, 327; Check Understanding pages 296, 300, 306, 310, 316, 320; Mid-Chapter Quiz, page 315, Chapter Review, page 328; Chapter Assessment, page 330; Alternative Assessment, page 331; Cumulative Review, page 332; Cumulative Assessment, page 333	5-minute Warm ups, pages 296, 300, 306, 310, 316, 320, 326; Quick Assessment, pages 298, 302, 308, 312, 318, 321, 327; Scoring Rubrics, page 331	Mid-Chapter Quiz, page 103; Test Form A, pages 105, 107; Test Form B, pages 109, 111; Cumulative Test 113, 115; Math Journal prompt, page 117, 118;	Chapter 7

Refresher Skills

The skills on these two pages are skills that have been taught in previous math courses. Continuous review of basic math skills will make stronger math students. These skills are identified as necessary to be successful in Chapter 7.

Skills Correlation Chart

Skill	Lesson Number
Ratios	7-1, 7-2, 7-3, 7-4, 7-5, 7-6, 7-7
Congruent Triangles	7-2, 7-4, 7-5, 7-6
Parallel Lines	7-4, 7-5, 7-6

Vocabulary

ratio
congruent
transversal

ARE YOU READY?

Refresh Your Math Skills for Chapter 7

The skills on these two pages are ones you have already learned. Review the examples and complete the exercises. For additional practice on these and more basic skills, see page 674.

RATIOS

A ratio is used to compare two numbers. Ratios can be written three different ways, but always in lowest terms.

Example 400 miles traveled in 5 days

analogy form: 400 : 5 = 80 : 1

fraction form: $\frac{400}{5} = \frac{80}{1}$

word form: 400 to 5 = 80 to 1

Write each as a ratio three different ways. Use lowest terms.

1. 6 tents for 24 campers 1 : 4, $\frac{1}{4}$, 1 to 4
2. 72 horseshoes for 18 horses 4 : 1, $\frac{4}{1}$, 4 to 1
3. 12 donuts for $4.56 1 : 38¢, $\frac{1}{38¢}$, 1 to 38¢
4. 444 calories for 6 oz candy 74 : 1, $\frac{74}{1}$, 74 to 1
5. 15 gardens hold 810 plants 1 : 54, $\frac{1}{54}$, 1 to 54
6. 318 miles on 7 gallons of gas 318 : 7, $\frac{318}{7}$, 318 to 7
7. 96 minutes to make 8 pizzas 12 : 1, $\frac{12}{1}$, 12 to 1
8. 4 elephants weight 14,000 lb 1 : 3500, $\frac{1}{3500}$, 1 to 3500
9. 279 bricks to cover 62 square feet 9 : 2, $\frac{9}{2}$, 9 to 2
10. 28 yd of fabric to make 8 curtains 7 : 2, $\frac{7}{2}$, 7 to 2

CONGRUENT TRIANGLES

You can determine that triangles are congruent by three different methods.

Examples Triangles are congruent if:

. . . three sides of one triangle are congruent to three sides of another.

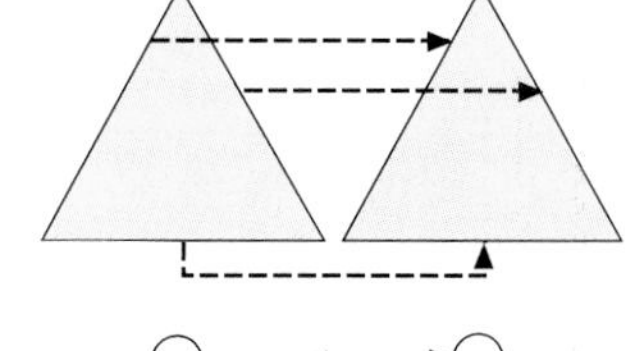

. . . two sides and the included angle of one triangle are congruent to two sides and the included angle of another.

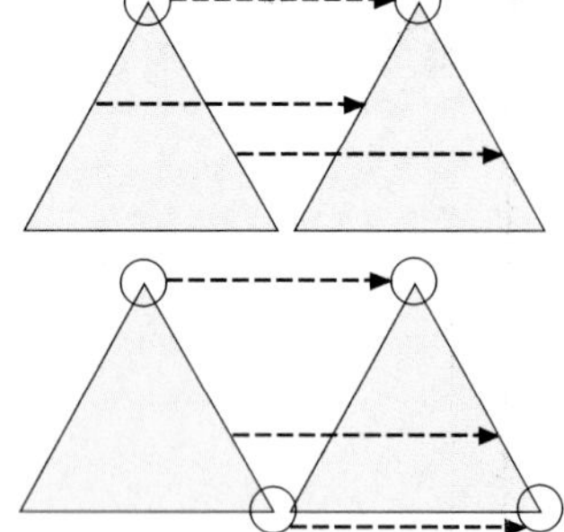

. . . two angles and the included side of one triangle are congruent to two angles and the included side of another.

Extend the Lesson

COOPERATIVE LEARNING Have students work in pairs to search the newspaper to find at least five ratios. Discuss how ratios are used in sales, sports, business and other areas.

CHAPTER 7

State whether each pair of triangles is congruent by SSS, SAS, ASA. If the triangles are not congruent or you cannot determine congruency, write *not congruent.*

11.
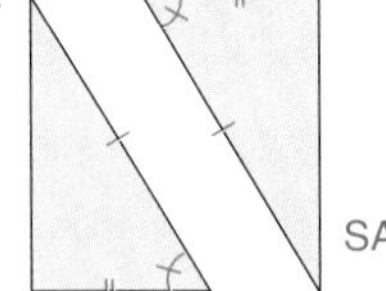
SAS

12.
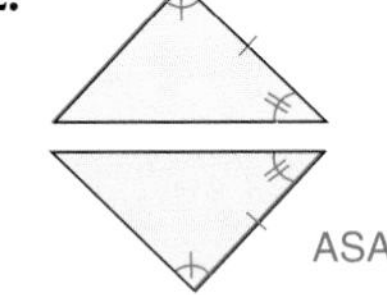
ASA

13.
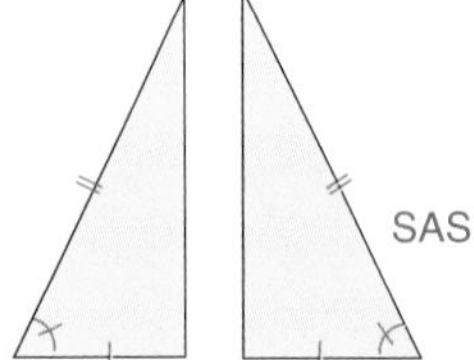
SAS

14.
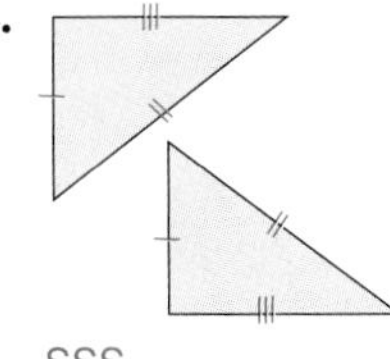
SSS

15.
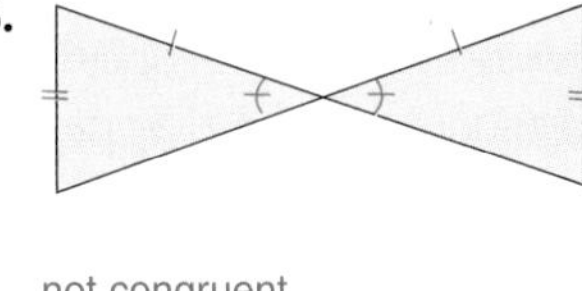
not congruent

16.
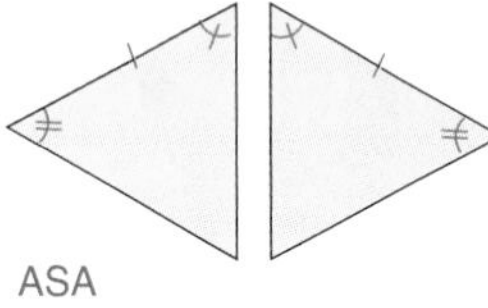
ASA

Parallel Lines

Coplanar lines that do not intersect are parallel lines.
If two parallel lines are cut by a transversal, corresponding angles are congruent.

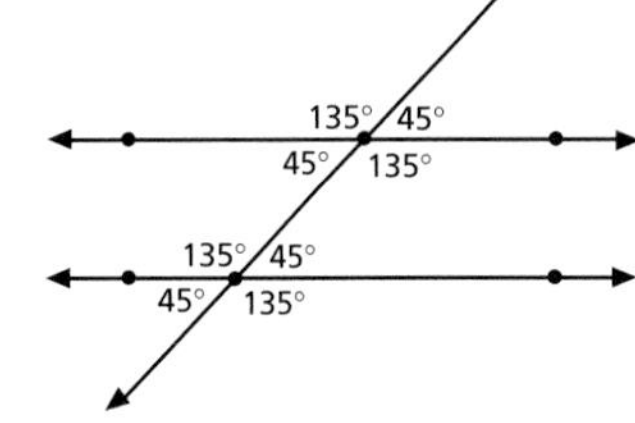

Find the missing measures. If the lines are not parallel, write *not parallel.*

17.
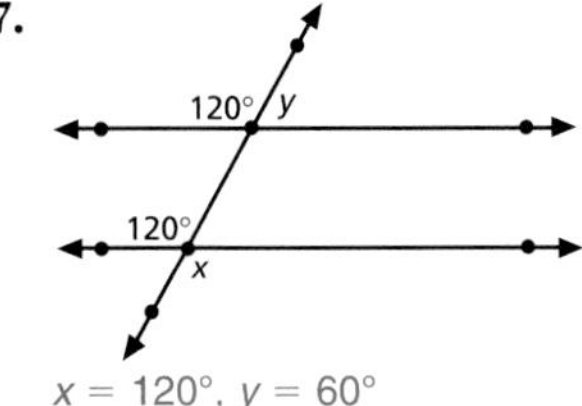

$x = 120°$, $y = 60°$

18.
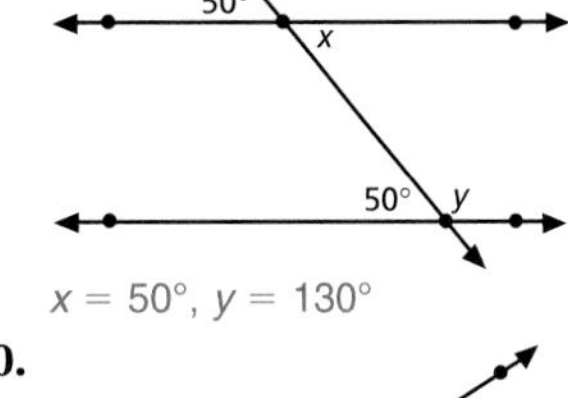

$x = 50°$, $y = 130°$

19.
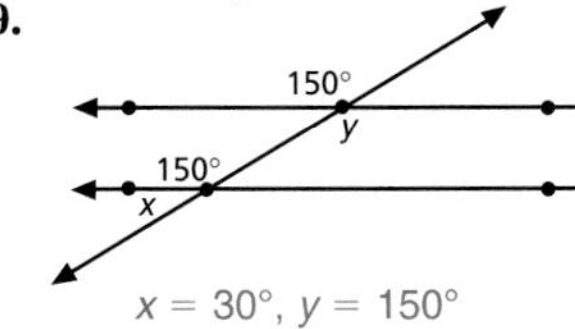

$x = 30°$, $y = 150°$

20.
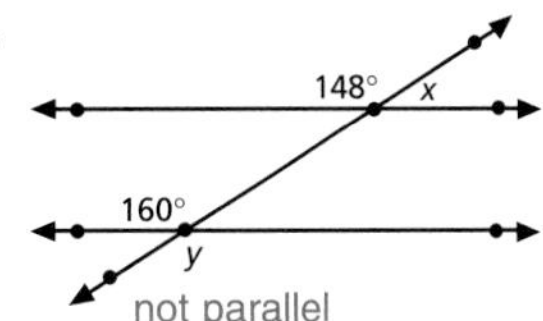

not parallel

Extend the Lesson

STUDENT PORTFOLIO Have each student draw a triangle on a sheet of paper. Then using a ruler and protractor, students should make three copies of the triangle. For each drawing, have them use a different postulate. For each congruent triangle they have drawn, have them record the measurements that were made and the postulate they used.

Chalkboard Examples

Ratios

Write the following ratio three different ways. Use lowest terms.

17 bottles hold 70 ounces

1 : 4, $\frac{1}{4}$, 1 to 4

Congruent Triangles

State the postulate used to determine that the triangles below are congruent.

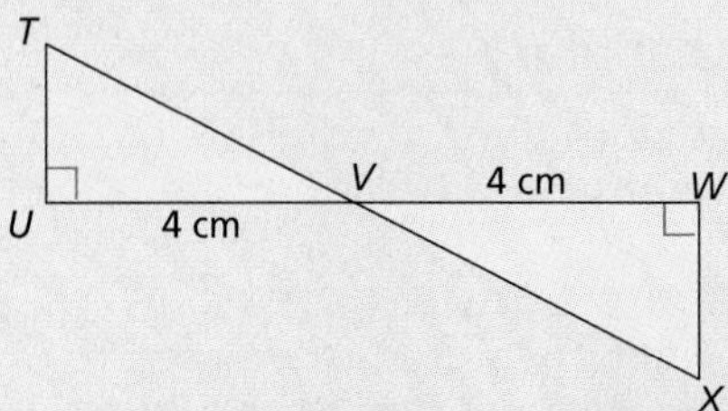

∠B and ∠E both measure 90°. ∠ACB and ∠DCE are equal in measure since they are vertical angles. For both, the included sides measure 4 cm. By the ASA postulate, the triangles are congruent.

Parallel Lines

Two parallel lines are cut by a transversal so that an interior angle is 104°. What is the measure of the other interior on the same side of the transversal as the 104° angle. **76°**

Refresher Wrap-up

Quick Assessment

Ask the following question to determine if students have mastered the basic skills reviewed on these pages.

If you need to copy a triangle, what is the least information you will need? **the measures of all sides, the measures of two sides and the included angle, or the measures of two angles and the included side.**

Additional Practice

All topics: Refer to the Extra Practice Index on page 674 of this text.

Chapter Opener

NCTM Standards/Strands

- Number & Operation
- Patterns, Functions, & Algebra
- Geometry & Spatial Sense
- Measurement
- Problem Solving
- Reasoning and Proof
- Connections
- Representation

Vocabulary

composition	digital image
forensic	aperture

Theme Connections

Tables and graphs are used to organize data about the photographic process. Photographers and other workers use the tables to find the settings and developing instructions they need. Tables and graphs are also used to compare film types, lenses, cameras and other photographic products. Discuss how data and graphs are used in photography.

Career Opportunities

Many careers related to photography make use of data and graphs. Two are highlighted in the MathWorks features. Others include salespeople, art directors, cinematographers, lighting directors and designers.

- Police Photography, page 305
- Photographic Processors, page 325

Internet Connection

Theme Activities

Learningmathmatters.com provides web addresses to search that help students gather information about the use of math in the real world, particularly data and graphs. To search for additional addresses, begin a search using the keyword *photography*. Then within that search, use such key words as *lighting conditions*, *lenses* and *cameras*. Students can brainstorm in small groups other key words.

CHAPTER 7

SIMILAR TRIANGLES

THEME: Photography

Photography is a blend of science and art. A camera produces an image on film by allowing light from an object to pass through a lens in a dark box. The amount of light is controlled by the size of the opening and the amount of time that the shutter, which covers the hole, is open. A photographer uses light and time, as well as color and composition, to create an artistic work.

Today, photographers use highly sophisticated cameras and computers to manipulate images. Digital cameras store images electronically. Digital images can be easily edited for special effect. They can be instantly transmitted using the Internet.

- **Photographic Process Workers** (page 325) develop film, make prints or slide and enlarge or retouch photographs. Photographic process workers use computers to enhance or alter photographs. They use their knowledge of ratio and proportion to make sure images look right.
- **Police Photographers** (page 305) work with forensic scientists to record details at a crime scene. Police photographers must take pictures from all angles to record all possible clues. They must make adjustments for lighting conditions to make objects recognizable.

Internet Connection
www.learningmathmatters.com

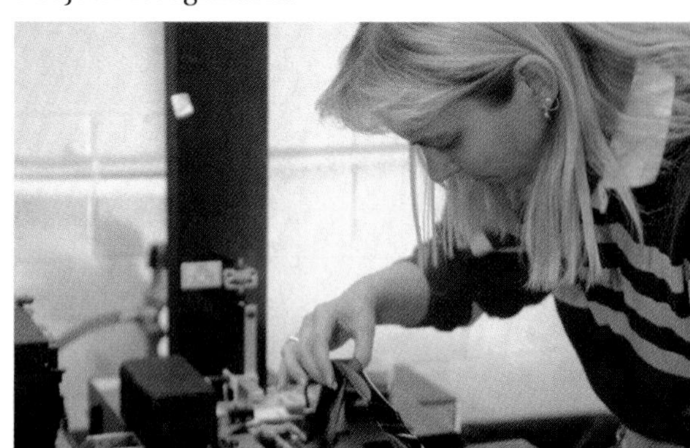

Chapter Investigation

Go to learningmathmatters.com to locate additional information about photography.

Extend the Lesson

ART Have students study composition in photographs by selecting one photograph published in a book or magazine. The photograph should show a variety of objects. Then have them draw a similar rectangle and make a quick sketch showing the location of objects within the photograph. Have students share their work and discuss the importance of composition in a work of art.

Camera Settings for Outdoor Lighting Conditions

Lighting	Very bright	Bright	Partly cloudy	Overcast
Shutter Speed	1/125	1/60	1/60	1/60
Aperture	f/16	f/16	f/8	f/5.6
Shutter Speed	1/250	1/125	1/125	1/125
Aperture	f/11	f/11	f/5.6	f/4

Image Sizes

Film format	Image size ratio
Disc	4 : 4
110	13 : 17
126	1 : 1
135	2 : 3
Panoramic	1 : 3

Data Activity: Camera Settings and Image Sizes

Use the tables for Questions 1–4.

1. Shutter speed is measured in fractions of a second. On a partly cloudy day, which combination of settings should a photographer use if a faster shutter speed is desired? shutter speed = $\frac{1}{125}$ s, aperture f/5.6

2. Suppose the aperture in a camera is stuck at f/5.6. What shutter speed should be used if the day is overcast? $\frac{1}{60}$ s

3. An image from a roll of 135 film is enlarged so that the width is 5 in. If the image is not cropped, what is the length of the print? 7.5 in.

4. A print from a roll of panoramic film must be reduced to fit in a magazine layout. The layout space is 4.5 in. in length. Find the width of the reduced image. 1.5 in.

Chapter Investigation

Artists often work from photographs to paint realistic portraits and murals. The image shown in a photograph can be enlarged using a ratio or scale. If the artwork is to appear realistic, the larger work must be proportional to the photograph.

Working Together

Choose a photograph (either an actual photograph or a photograph published in a magazine). Make a proportional sketch of the subject of the photograph. Enlarge the photograph by a factor of 5. Use the Chapter Investigation logo to guide your group.

Data Activity

Discuss with students the technical knowledge needed to be a good photographer. Point out that photographers need both artistic and technical knowledge to be successful. Discuss the importance of reading any titles or legends on a table or graph. Assign students Questions 1–4.

Extend the Data Activity

REAL WORLD CONNECTION Have students research the features available on different types of cameras. Have them create a table comparing these features for three models.

Chapter Investigation

As an Overarching Problem

Display a photograph and its enlargement and have students study them. Ask students to record any observations they make. Have students share their observations with the class. Students will continue to work on the investigation as they complete the exercises identified by the Investigation Logo that are found throughout the chapter. These exercises will guide students through the task described in *Working Together*. Encourage students to keep all of their work on the Investigation together. Have students use the suggestions in the Chapter Investigation Extension to summarize their work.

As a Chapter Project

The goal of this project is to make and describe a proportional sketch of the subject of a photograph. Students can use the Group Project Planner on page 115 and the Project Planning Calendar on page 116 in the Enrichment text to complete the project. Benchmarks **a**, **b**, and **c**. should be completed after the lesson listed in parentheses has been studied. Benchmark d should be completed at the end of the chapter.

Project Planning Calendar

Name ______ Date ______

CHAPTER 7 PROJECT PLANNING CALENDAR

Benchmarks	PROJECT GOAL
a. Prepare to make an enlargement of a photograph. Choose an actual photograph or a photograph published in a magazine. Measure the outer dimensions of the photograph and increase the dimensions by a factor of 5. Cut a piece of poster paper the size of the enlargement. *(Lesson 7-1)* b. Select at least five prominent features in the photograph and plot their locations on your enlargement. You may want to draw a coordinate grid over the surface of the photograph and draw a corresponding grid on your enlargement. Once the main features of ... correctly on the ... details	To make and describe a proportional sketch of the subject of a photograph.

Group Project Planner

Name ______ Date ______

CHAPTER 7 GROUP PROJECT PLANNER

Assignment ______ Objective ______

Group Members
1) ______
2) ______
3) ______
4) ______
5) ______

Assigned Roles

Deadlines Done

Lesson Planning

NCTM Standards/Strands
- Number & Operation
- Patterns, Functions, & Algebra
- Problem Solving
- Reasoning and Proof
- Connections
- Representation

Vocabulary
equivalent ratios	proportion
extremes	means
cross products	terms

Materials Needed
calculators

Lesson Resources
Warm-Up Transparency 19
Transparency RF-27
Reteaching 7-1
Extra Practice 7-1
Enrichment 7-1

Assignment Guide
Basic: 1–34, 37–56
Enriched: 1–56

Getting Started

5-Minute Warm-up
Solve for *x*.
1. $7x = 56(20)$ 160
2. $24x = 42(18)$ 31.5
3. $2x + 14x = 48$ 3
4. $800x + 250x = 2{,}100$ 2

Introduction to Lesson 7-1
You may wish to ask students to repeat the exercise if Glenisle has 315 technicians and 28 supervisors and Skatetown has 462 technicians and 42 supervisors. Have them discuss any problems they encounter and suggest ways to solve them. Since 45:4 and 11:1 are not easily compared, use a ratio equivalent to 11:1 such as 44:4 for comparison.

7-1 Ratios and Proportions

Goals
- Find equivalent ratios.
- Use ratios and proportions to solve problems.

Applications Recreation, Real Estate, Retail, Business, Art

Work with a partner.

The photo lab at Glenisle has 315 technicians and 30 supervisors. The lab at Skatetown has 450 technicians and 36 supervisors.

1. Write the ratio of technicians to supervisors in one lab while your partner does the same for the other lab. Write each ratio in lowest terms. Glenisle: 21 : 2; Skatetown: 25 : 2
2. Determine which lab has more technicians per supervisor. Skatetown

Build Understanding

Two ratios that can both be named by the same fraction are called **equivalent ratios.** 4:8 and 7:14 are equivalent ratios because they can each be written as the fraction $\frac{1}{2}$.

A **proportion** is an equation that states that two ratios are equivalent.

$$a{:}b = c{:}d \qquad \frac{a}{b} = \frac{c}{d}$$

The four numbers *a*, *b*, *c*, and *d* that are related in the proportion are called its **terms.** The first and last terms are called the **extremes.** The second and third are called the **means.**

extremes (a and d), means (b and c): $a{:}b = c{:}d$

extreme (a), mean (c), mean (b), extreme (d): $\frac{a}{b} = \frac{c}{d}$

In a proportion, the product of the extremes equals the product of the means. This is the same as saying that the **cross products** are equal.

$$\frac{a}{b} = \frac{c}{d} \qquad ad = bc \qquad \frac{3}{4} = \frac{12}{16} \qquad \begin{aligned} 3(16) &= 4(12) \\ 48 &= 48 \end{aligned}$$

Use cross products to find the missing term in a proportion.

Example 1

Find *x*: $\frac{x}{18} = \frac{12}{27}$

Solution

Use cross products to write another equation. Solve that equation for *x*.

$$\frac{x}{18} = \frac{12}{27} \qquad 27x = 18(12) \qquad 27x = 216$$
$$x = 8$$

Extend the Lesson

MATH JOURNAL Have students write an example of two equivalent ratios and a rule explaining when two ratios are equivalent. possible answer: $\frac{14}{21}, \frac{16}{24}$; when both ratios can be named by the same fraction or when the cross products are equal

Check your answer by substituting it in the original proportion.

$$\frac{8}{18} \stackrel{?}{=} \frac{12}{27}$$

$$\frac{8}{18} = \frac{8 \div 2}{18 \div 2} = \frac{4}{9} \qquad \frac{12}{27} = \frac{12 \div 3}{27 \div 3} = \frac{4}{9}$$

Because the ratios are equivalent when $x = 8$, the proportion is solved.

You can use proportions to solve a wide variety of problems.

Problem Solving Tip

When writing a proportion to solve a problem, it may help to write a word ratio first. In Example 2, the word ratio, enlargements:cost, makes it easier to write the terms in the correct place.

Example 2

PHOTO PROCESSING Fine Photo charges \$3 for 2 enlargements. How much does the company charge for 5 enlargements?

Solution

Write a proportion. Let $x =$ the cost of 5 enlargements.

$$\frac{2}{3} = \frac{5}{x} \quad \begin{matrix}\text{enlargements}\\ \text{cost}\end{matrix}$$

$$2x = 15$$

$$x = 7.5$$

So, the company charges \$7.50 for 5 enlargements.

Sometimes the information you are given in a problem is a ratio of two quantities.

Example 3

RECREATION The ratio of counselors to campers is 2:15. There are 102 people at a camp. How many are counselors?

Solution

Let $2x$ represent the number of counselors. Let $15x$ represent the number of campers.

The ratio of counselors to campers is $2x{:}15x$, which is the same as 2:15. Write an equation for the total number of people at camp.

$$2x + 15x = 102$$

$$17x = 102$$

$$x = 6$$

Because $2x$ represents the number of counselors, the answer is 12.

Chalkboard Examples

Supplementary Example 1

Find x: $\frac{x}{15} = \frac{60}{75}$. $x = 12$

Supplementary Example 2

Sam paid \$10.50 for 5 blank video tapes. At the same rate, how much would he pay for 12 blank tapes? **\$25.20**

Supplementary Example 3

At another camp, the ratio of counselors to campers is 2:7. If there are 108 people at the camp, how many are counselors? **24**

Teaching Strategies

Have students write statements expressing equivalent ratios such as the following: four batteries in one pack means 12 batteries in three packs; one movie takes two hours means three movies take six hours. Have students erase one number in each statement and substitute x. Then have them write proportions and solve for x. (e.g., $\frac{4}{1} = \frac{x}{12}$, $x = 48$).

Reteaching Worksheet 7-1

Name ______________________ Date ____________

RETEACHING 7-1

RATIOS AND PROPORTIONS

Equivalent ratios can be named by the same fraction. A proportion is an equation stating that two ratios are equivalent. Cross-products of a proportion are equal. Use them to find the missing term in a proportion.

$$a : b = c : d \rightarrow \frac{a}{b} = \frac{c}{d} \rightarrow ad = bc$$

Example 1

Solve the proportion by finding x.

$$\frac{x}{21} = \frac{5}{35}$$

Solution

Use cross-products to write another equation. Then solve for x.

$35x = 5(21)$ Find cross-products.

$35x = 105$ Simplify.

$\frac{1}{35}(35x) = 105\left(\frac{1}{35}\right)$ Multiply each side by $\frac{1}{35}$.

$x = 3$ Simplify.

Substitute 3 for x in the original proportion. Then determine if the ratios are equivalent.

$$\frac{3}{21} = \frac{3 \div 3}{21 \div 3} = \frac{1}{7} \qquad \frac{5}{35} = \frac{5 \div 5}{35 \div 5} = \frac{1}{7}$$

Since the ratios are equivalent, the proportion is solved for $x = 3$.

Example 2

A pattern calls for black tiles and white tiles in the ratio of 4 : 5. How many white tiles are needed if 200 black tiles are used?

Solution

Write a proportion.
Let $w =$ number of white tiles.

$\frac{4}{5} = \frac{200}{w}$ ←black tiles / ←white tiles

Use cross-products to write another equation and solve.

$4w = 5(200)$ Find cross-products.

$4w = 1000$ Simplify.

$\frac{1}{4}(4w) = 1000\left(\frac{1}{4}\right)$ Multiply by $\frac{1}{4}$.

$w = 250$ Simplify.

So 250 white tiles are needed.

EXERCISES

Solve each proportion.

1. $\frac{a}{8} = \frac{9}{12}$ 6
2. $\frac{15}{b} = \frac{6}{14}$ 35
3. $\frac{6}{27} = \frac{14}{c}$ 63
4. $\frac{12}{42} = \frac{d}{28}$ 8
5. $\frac{16}{24} = \frac{18}{y}$ 27
6. $\frac{f}{44} = \frac{16}{64}$ 11

Solve.

7. A pattern calls for green tiles and blue tiles in a ratio of 3 : 4. How many green tiles are needed if 100 blue tiles are used? 75 tiles
8. Two entrepreneurs imported crafts for resale. The ratio of their investment was 3 : 5. What was each person's share of the \$12,000 sales income? \$4500 and \$7500

102 RETEACHING *LESSON 7-1*

Lesson Wrap-up

Quick Assessment

Ask the following questions to determine if students understand the content presented in this lesson.

1. Use cross products to express the proportion $4 : x = y : 8$. $4(8) = xy$ or $32 = xy$
2. Write two equivalent ratios for 6.7 : 13. possible answer: 13.4 : 26, 20.1 : 39
3. State a rule for finding a ratio equivalent to $a : b$. To find an equivalent ratio, multiply or divide a and b each by the same number.

Assignment Guide

Basic: 1–34, 37–56
Advanced: 1–56
Additional Practice: See Extra Practice Index on page 674.

Additional Answers

32. If $a{:}b = c{:}d$, then $ab = bc$. By using the division property of equality, the statement $ad = bc$ can be rewritten as the proportion $a{:}c = b{:}d$.
$\frac{ad}{c} = \frac{bc}{c}$ $\frac{ad}{c} = b$
$\frac{ad}{cd} = \frac{b}{d}$ $\frac{a}{c} = \frac{b}{d}$

Try These Exercises

Is each pair of ratios equivalent? Write *yes* or *no*.

1. 28:49, 16:28 yes
2. 39 to 13, 36 to 9 no
3. $\frac{4}{5} = \frac{6}{7.5}$ yes

Solve each proportion.

4. $\frac{x}{21} = \frac{18}{27}$ 14
5. $\frac{9}{y} = \frac{36}{8}$ 2
6. $0.04{:}0.06 = m{:}0.24$ 0.16
7. **REAL ESTATE** Two families decide to split the cost of renting a vacation house in a ratio of 3:4. The total cost is $2,100. What will be each family's share of the cost? $900, $1200
8. **RETAIL** A used book store is selling 5 paperback books for $2. How much will 12 paperback books cost? $4.80

Practice Exercises

Is each pair of ratios equivalent? Write *yes* or *no*.

9. 3:6, 15:18 no
10. $\frac{3.5}{4.2}, \frac{10}{12}$ yes
11. $\frac{4}{5} : \frac{2}{5}$, 4:8 no

Solve each proportion.

12. $\frac{3}{18} = \frac{10}{s}$ 60
13. $\frac{n}{0.9} = \frac{0.7}{0.3}$ 2.1
14. $\frac{27}{81} = \frac{k}{45}$ 15

CALCULATOR **Use a calculator to solve these proportions.**

15. $\frac{119}{476} = \frac{247}{r}$ 988
16. $\frac{245}{372} = \frac{t}{1{,}488}$ 980
17. $\frac{426}{z} = \frac{1{,}491}{2{,}205}$ 630
18. A recipe for fruit punch calls for 3 parts pineapple juice to 5 parts orange juice. How much pineapple juice should be added to 16 liters of orange juice? 9.6 liters
19. **BUSINESS** The manager of Music World stocks audio cassettes and CDs in the ratio 2:7. This month, she is ordering 400 audio cassettes. How many CDs will she order? 1,400 CDs
20. **INVESTING** Two business partners purchased stock. The ratio of the money invested by one partner to the money invested by the other was 4:5. The stock earned $31,500. What is each partner's share? $14,000; $17,500
21. Ricardo mixes dried fruit and nuts in a 3:5 ratio. He wants to make 12 pounds of the mixture. How many pounds of nuts does he need? 7.5 pounds

Arrange the given terms to form a proportion. Supply the missing term.

22. 1.5, 5, 6 5: 1.5 = 6 : 1.8
23. 100, 3, 30 100 : 10 = 30 : 3
24. 35, 36, 14 14 : 35 = 36 : 90

Solve each proportion.

25. $\frac{2x}{3} = \frac{x+5}{4}$ 3
26. $\frac{4}{x+1} = \frac{5}{2x-1}$ 3
27. $\frac{3x}{25} = \frac{8x+2}{70}$ 5

Teaching Strategies

COOPERATIVE LEARNING Have students work in pairs to find a way to use proportions to solve the following problem: The ratio of boys to girls at the school concert was 6 : 7. There were 221 students at the concert. How many were girls? (You may wish to give the following hint: Use word ratios, and consider the ratio of girls to boys + girls.) $\frac{7}{(6+7)} = \frac{x}{221}$, $x = 119$

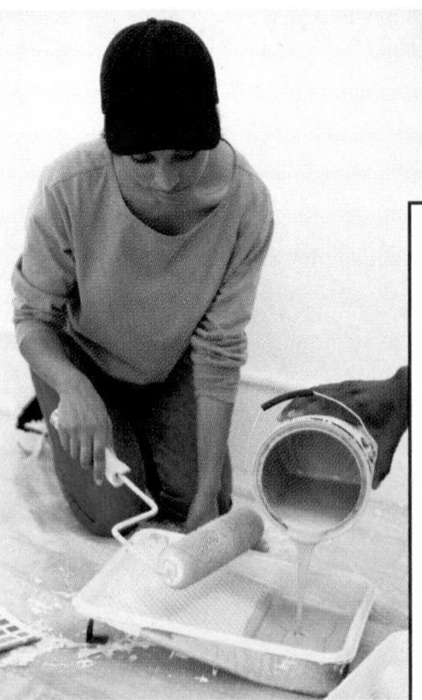

28. **ART** Louisa wants to mix 1 part yellow paint to 3 parts blue to make a certain shade of green. How many pints of blue paint will she need if she wants 1 gallon of green paint? 6 pints

29. Eddie must read a biography by the end of the month. The book has 317 pages. He found that he could read 10 pages in 15 minutes. Estimate how many hours it will take him to read the whole book. 8 hours

30. Otis and Steve bought an old car and fixed it up. Otis spent \$2,000 and Steve spent \$1,500. They were able to sell the car for \$4,900. How much should each receive from the profit made on the car? Otis, \$800; Steve, \$600

31. An angle and its complement are in the proportion 2:3. What are the measures of the two angles? 36°, 54°

32. **WRITING MATH** If $\frac{a}{b} = \frac{c}{d}$, is it always true that $\frac{a}{c} = \frac{b}{d}$? Explain. See additional answers.

33. **ERROR ALERT** A survey showed that 3 out of 10 students have a regular physical fitness program. Wanitta knows that 48 students said that they exercised regularly. To find out how many students were surveyed, she wrote the roportion $\frac{3}{10} = \frac{x}{48}$ and solved for x. Wanitta suspects and her solution of 14.4 cannot be correct. What went wrong? The terms in the two ratios are not written in the same order. The correct proportion is either 3\10 = 48\x or 10\3 = x\48.

34. **CHAPTER INVESTIGATION** Prepare to make an enlargement of a photograph. Choose an actual photograph or a photograph published in a magazine. Measure the outer dimensions of the photograph and increase the dimensions by a factor of 5. Cut a piece of poster paper the size of the enlargement.

Extended Practice Exercises

In the proportion $a{:}b = b{:}c$, b is called the **mean proportional** or the **geometric mean** between a and c.

35. Find b if $a = 3$ and $c = 12$. 6

36. If b is a positive integer, what must be true about the product, ac? The product must be a perfect square.

Mixed Review Exercises

Find the slope of the line containing the given points. (Lesson 6-1)

37. (−3, 2), (4, 8) $\frac{6}{7}$
38. (1, 6), (0, −2) 8
39. (4, −3), (1, 2) $-\frac{5}{3}$
40. (3, −4), (−2, −1) $\frac{-3}{5}$
41. (4, 6), (−2, −3) $\frac{3}{2}$
42. (−4, 1), (2, −3) $-\frac{2}{3}$
43. (3, −2), (3, 4) undefined
44. (5, 5), (4, −2) 7
45. (7, 4), (3, −3) $\frac{7}{4}$
46. (2, −1), (−5, −1) 0
47. (5, −2), (−1, 3) $-\frac{5}{6}$
48. (−3, 2), (3, −2) $-\frac{2}{3}$

The lengths of two sides of a triangle are given. Find the range of possible lengths for the third side. (Lesson 4-6)

49. 7 cm, 8 cm $0 < x < 15$ cm
50. 5 in., 12 in. $0 < x < 17$ in.
51. 14 cm, 13 cm $0 < x < 27$ cm
52. 7 m, 15 m $0 < x < 22$ m
53. 3 dm, 9 dm $0 < x < 12$ dm
54. 23 ft, 18 ft $0 < x < 41$ ft
55. 3 ft, 15 in. $0 < x < 51$ in.
56. 32 in., 2 ft $0 < x < 56$ in.

Teaching Strategies

CHALLENGE Some people were fishing and caught 78 fish in one day. Each person caught the same number of fish. If there had been three more people, they could have caught 96 fish. How many people were fishing?

$\frac{x}{78} = \frac{(x+3)}{96}$, $96x = 78(x + 3)$, $18x = 234$, $x = 13$; 13 people were fishing

Extra Practice Worksheet 7-1

Name ____________ Date ____________

EXTRA PRACTICE 7-1

Ratios and Proportions

Exercises

Is each pair of ratios equivalent? Write *yes* or *no*.

1. 2 : 8, 5 : 10 no
2. 14 to 28, 16 to 30 no
3. $\frac{8}{12}, \frac{12}{18}$ yes
4. $\frac{1}{2} : \frac{3}{4}$, 24 : 36 yes

Solve each proportion.

5. $\frac{9}{15} = \frac{x}{10}$ 6
6. $\frac{c}{14} = \frac{6}{21}$ 4
7. $\frac{16}{r} = \frac{10}{20}$ 32
8. $\frac{0.2}{1.5} = \frac{2}{d}$ 15
9. $\frac{24}{32} = \frac{v}{3}$ 2.25
10. $\frac{14}{m} = \frac{28}{48}$ 24
11. $\frac{2x}{6} = \frac{4x+8}{15}$ 8
12. $\frac{x+1}{10} = \frac{x-3}{2}$ 4
13. $\frac{4x}{18} = \frac{x+6}{36}$ 2

Use a calculator to solve these proportions.

14. $\frac{180}{450} = \frac{x}{440}$ 176
15. $\frac{204}{272} = \frac{225}{x}$ 300
16. $\frac{159}{x} = \frac{183}{427}$ 371

17. A recipe for muffins calls for 2 cups of flour for each 1.5 cups of sugar. How much sugar should be used if 8 cups of flour are used? 6 c

18. Maria is holding a garage sale. She is selling her old cassette tapes for \$2 for 8 tapes. How much will 20 tapes cost? \$5

19. Franco uses 2 teaspoons of fertilizer for each gallon of water. How many gallons can he make with 15 teaspoons of fertilizer? 7.5 gal

20. The areas of two rectangles are in the proportion 3 : 4. If the area of the larger rectangle is 64 in.^{2}, what is the area of the smaller rectangle? 48 in.^{2}

Arrange the given terms to form a proportion. Supply the missing term.

21. 3, 5, 8 3 : 5 = 4.8 : 8
22. 81, 3, 9 9 : 81 = 3 : 27
23. 6, 15, 4 4 : 6 = 10 : 15
24. 42, 24, 36 24 : 36 = 28 : 42

For Exercises 21–24, answers may vary. Samples are given.

90 EXTRA PRACTICE *LESSON 7-1*

Enrichment Worksheet 7-1

Name ____________ Date ____________

ENRICHMENT 7-1

Compound Proportion

A compound proportion involves more than two ratios. Here is one example. This notation can be read, "5 times 10 times 12 is to 8 times 9 times 25 as 1300 is to 3900."

$$\left\{\begin{matrix} 5 : 8 \\ 10 : 9 \\ 12 : 25 \end{matrix}\right\} = 1300 : 3900$$

Example

If 5 people make \$1300 working 10 hours a day for 12 days, how much can 8 people make working 9 hours a day for 25 days?

Solution

This situation can be represented by the compound proportion above. The key to the solution is the way the notation is read.

$$\left\{\begin{matrix} \text{5 people} : \text{8 people} \\ \text{10 hours} : \text{9 hours} \\ \text{12 days} : \text{25 days} \end{matrix}\right\} = \$1300 : n \qquad \frac{5 \cdot 10 \cdot 12}{8 \cdot 9 \cdot 25} = \frac{1300}{n} \qquad n = \$3900$$

Exercises

Write a compound proportion to solve each problem.

1. I f 7 horses eat 35 bushels of oats in 25 days, how many bushels will 15 horses eat in 21 days? 63 bushels

2. If it costs \$300 to build a walk 10 ft wide and 90 ft long, how much will it cost to make a walk 8 ft wide and 225 ft long? \$600

3. It takes 650 reams of paper to print 4000 copies of a 420-page book. How many reams are needed to print 7000 copies of a book containing 528 pages of the same size? 1430 reams

4. If 4 machines can produce 84 appliances in 7 days by working 10 hours a day, how many appliances can be made by 6 of the machines working 9 hours a day for 10 days? 162 appliances

5. If 1500 bricks, each 8 in. long and 4 in. wide, will make a walk, how many slabs of stone, each 2 ft long and 1 ft 4 in. wide, will be needed for the same walk? 125 slabs

118 ENRICHMENT *LESSON 7-1*

Lesson Planning

NCTM Standards/Strands

- Number & Operation
- Patterns, Functions, & Algebra
- Geometry & Spatial Sense
- Measurement
- Problem Solving
- Reasoning and Proof
- Connections
- Representation

Vocabulary

similar angles
congruence
corresponding angles
corresponding sides

Materials Needed

centimeter rulers
protractors
compasses (optional)

Lesson Resources

Warm-Up Transparency 19
Reteaching 7-2
Extra Practice 7-2
Enrichment 7-2

Assignment Guide

Basic: 1-15, 19-27
Enriched: 1-27

Getting Started

5-Minute Warm-Up

Is each pair of ratios equivalent?

1. $\frac{28}{32}, \frac{42}{56}$ no
2. $\frac{180}{125}, \frac{36}{25}$ yes
3. $\frac{6}{18}, \frac{1.5}{4}$ no
4. $\frac{52}{8}, \frac{2.6}{0.4}$ yes

Introduction to Lesson 7-2

Ask students to draw a third version of the logo with the upper and lower left-corner angles measures the same as those in both logos and the sides half the corresponding sides in the larger drawing. Have them compare all three logos.

7-2 Similar Polygons

Goals
- Identify similar polygons.
- Find missing measures of similar polygons.

Applications Photography, Architecture, Framing

Work with a partner. You will need a protractor and a centimeter ruler.

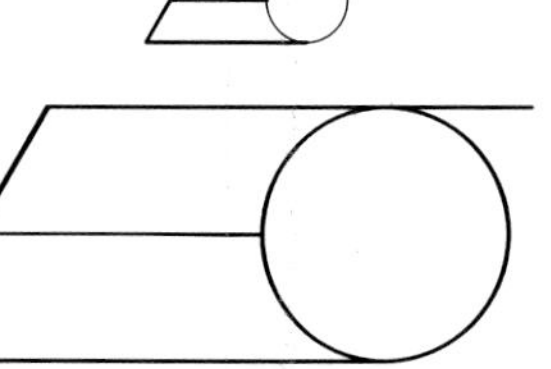

A copy machine is used to enlarge a company logo.

1. Measure the angles in the lower left and upper left corners of both drawings.
2. Measure the sides of both drawings.
3. Compare corresponding measurements. What do you notice? Angles are congruent; sides in larger drawing are 3 times the corresponding sides in the smaller drawing.

Build Understanding

Two figures are **similar** if they have the same shape. The figures may not be the same size. Two polygons are similar if all corresponding angles are congruent and the measures of all corresponding sides form the same ratio (are in proportion).

The symbol for similarity is ~. Polygon *ABCDE* ~ polygon *KLMNO.*

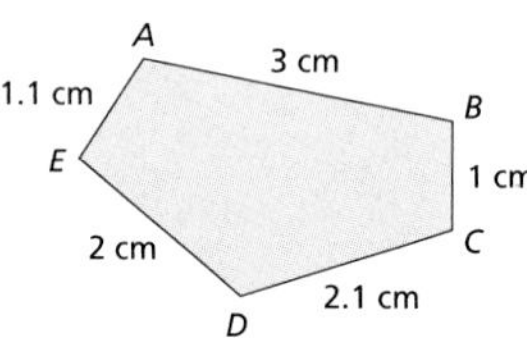

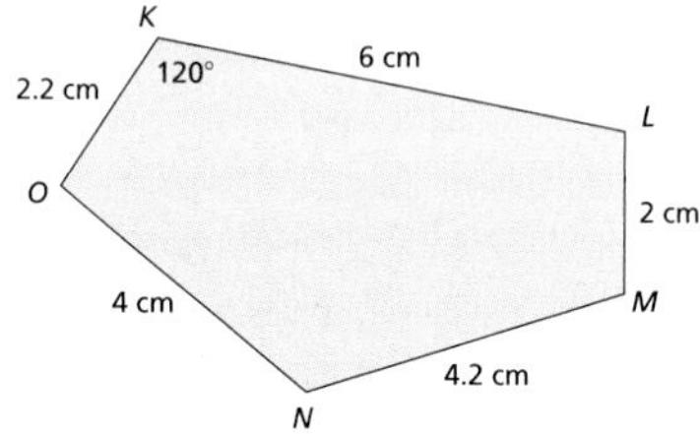

Corresponding angles are congruent.

$\angle A \cong \angle K, \angle B \cong \angle L, \angle C \cong \angle M, \angle D \cong \angle N, \angle E \cong \angle O,$

Corresponding sides are in proportion.

$$\frac{AB}{KL} = \frac{BC}{LM} = \frac{CD}{MN} = \frac{DE}{NO} = \frac{EA}{OK} = \frac{1}{2}$$

Example 1

Is *WXYZ* ~ *EFGH*?

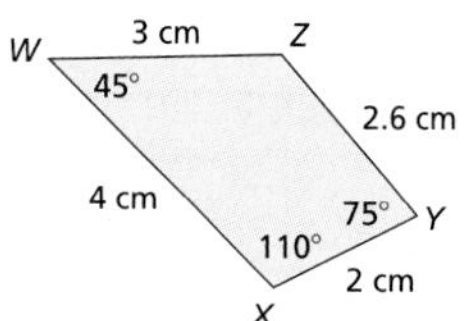

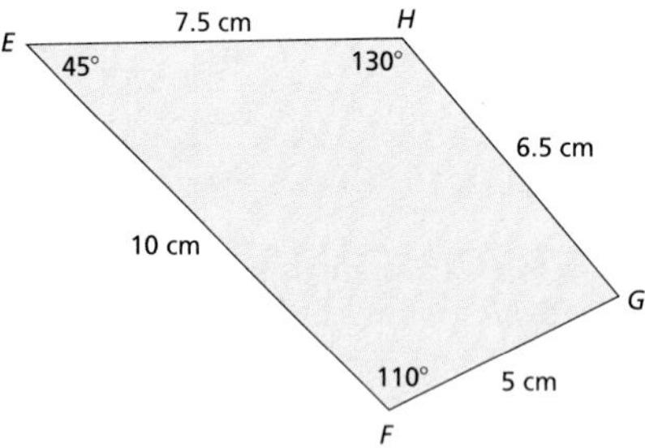

Learning Styles

VISUAL LEARNERS You may wish to use geoboards to help students visualize the concept of expanding, shrinking, and duplicating shapes to create similar figures. Have students construct any polygon on the geoboard. Encourage them to make irregular shapes. Have them construct similar (proportionately expanded or shrunken and duplicated) versions of the shape.

Solution

Find the missing angle measures.

$m\angle Z = 360° - (45° + 110° + 75°) = 130°$. So, $\angle Z \cong \angle H$.

$m\angle G = 360° - (45° + 110° + 130°) = 75°$. So, $\angle G \cong \angle Y$.

All four pairs of corresponding angles are congruent.

Find the ratios of all pairs of corresponding sides.

$$\frac{WX}{EF} = \frac{4}{10} = \frac{2}{5} \qquad \frac{XY}{FG} = \frac{2}{5}$$

$$\frac{YZ}{GH} = \frac{2.6}{6.5} = \frac{2}{5} \qquad \frac{WZ}{EH} = \frac{3}{7.5} = \frac{2}{5}$$

Each pair of corresponding sides has the same ratio. So, corresponding sides are in proportion. The two polygons are similar.

Example 2

PHOTOGRAPHY Two mats, shown at the right, are cut to display an array of photographs. $PS = 40$ cm, $TW = 60$ cm and $QR = 50$ cm. If the mats are similar figures, what is the measure of UV?

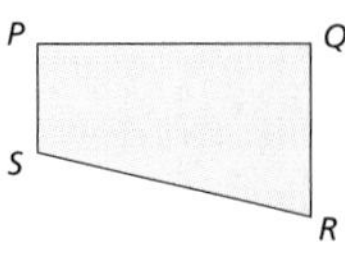

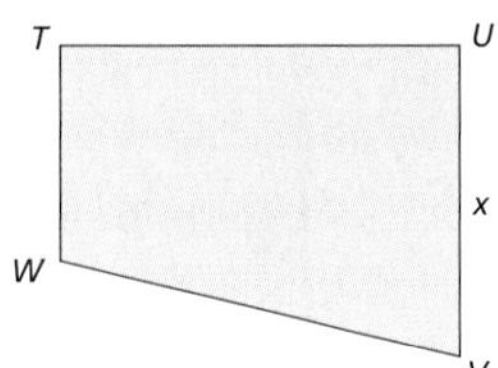

Solution

Because the figures are similar, their corresponding sides are in proportion. Write and solve a proportion to find x.

$$\frac{PS}{TW} = \frac{QR}{UV} \qquad \text{Ratios of corresponding sides are equivalent.}$$

$$\frac{40}{60} = \frac{50}{x}$$

$$40x = 3000$$

$$x = 75$$

So, $UV = 75$ cm.

Check Understanding

If $m\angle S = 115$, which angle of TUVW has that measure?

$\angle W$

Example 3

Name a pair of similar triangles.

Solution

$\angle L \cong \angle P$, $\angle M \cong \angle N$, $\angle MOL \cong \angle NOP$.

There are 3 pairs of congruent angles.

$$\frac{LM}{PN} = \frac{3}{6} = \frac{1}{2} \qquad \frac{MO}{NO} = \frac{2}{4} = \frac{1}{2} \qquad \frac{OL}{OP} = \frac{2.5}{5} = \frac{1}{2}$$

Corresponding sides are proportional. To name the similar triangles, name corresponding vertices in the same order.

$\triangle LOM \sim \triangle PON$

Chalkboard Examples

Supplementary Example 1

Is $ABCD \sim EFGH$? yes

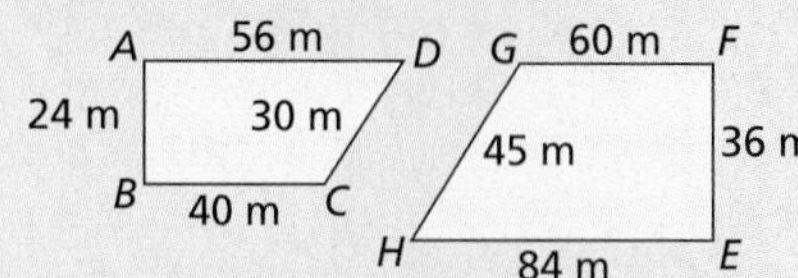

Supplementary Example 2 and 3

Two similar quadrilaterals $ABCD$ and $EFGH$ are cut from mat board to create a hall display. If $BC = 3.2$ m, $AD = 3.4$ m, $EF = 4.5$ m, $FG = 4.8$ m and $EH = 5.1$, what is the measure of AB? 3 m

Reteaching Worksheet 7-2

Name ______________ Date ______________

RETEACHING 7-2

SIMILAR POLYGONS

Similar figures have the same shape. In similar polygons, corresponding angles are congruent and the measures of all corresponding sides are in proportion.

Example 1

Is $ABCD \sim EFGH$?

Solution

Find the missing angle measures.

$m\angle D = 360° - (90° + 112° + 90°)$
$= 360° - 292° = 68°$, so $\angle D \cong \angle H$.
$m\angle F = 360° - (90° + 68° + 90°)$
$= 360° - 248° = 112°$, so $\angle F \cong \angle B$.

All corresponding angles are congruent.

Find the ratios of corresponding sides.

$$\frac{AB}{EF} = \frac{4}{6} = \frac{2}{3} \qquad \frac{BC}{FG} = \frac{4}{6} = \frac{2}{3}$$

$$\frac{CD}{GH} = \frac{6}{9} = \frac{2}{3} \qquad \frac{DA}{HE} = \frac{6}{9} = \frac{2}{3}$$

Each pair of corresponding sides has the same ratio, so they are in proportion.

The figures are similar.

Example 2

Given $JKLMP \sim NOPQ$, find the length of $\overline{NO}$.

Solution

In similar figures, corresponding sides ares in proportion. Determine the proportion between any two sides.

$$\frac{KL}{OP} = \frac{5}{10} = \frac{1}{2}$$

Write a ratio of corresponding sides, substituting known values. Solve by finding the cross-products.

$$\frac{JK}{NO} = \frac{KL}{OP} \qquad \frac{3}{x} = \frac{1}{2}$$

$$6 = x$$

So, $NO = 6$ in.

Explanations may vary. Sample answers are given.

EXERCISES

Use these figures to answer Exercises 1–4. Write yes or no and explain why.

1. Is $ABCD \sim EFGH$? No, corresponding ∠s not ≅.
2. Is $ABCD \sim IJKL$? Yes, corresponding ∠s ≅, corresponding sides proportional.
3. Is $EFGH \sim MNOP$? Yes, corresponding ∠ s ≅, corresponding sides proportional.
4. Is $IJKL \sim MNOP$? No, corresponding ∠s not ≅.

Find x in each pair of similar polygons.

5. 140
6. 15

104 RETEACHING *LESSON 7-2*

Teaching Strategies

ESL/LEP These students may have difficulty understanding the terms similar, congruent, and corresponding. Explain that similar means "like," congruent means "the same as" or "identical," and corresponding means "matching." Use diagrams to express each term, pointing out corresponding angles and sides. Then have students draw examples of each.

Lesson 7-2 **Similar Polygons** 301

Lesson Wrap-up

Quick Assessment

Ask the following questions to determine if students understand the content presented in this lesson.

1. How can you define similar polygons? **They are polygons that have the same shape.**
2. What properties do similar polygons have? **Pairs of corresponding angles are congruent and pairs of corresponding sides are in proportion.**
3. How can you determine which angles are corresponding angles? **They will be congruent and be between corresponding sides.**

Assignment Guide

Basic: 1–15, 19–27
Advanced: 1–27
Additional Practice: See Extra Practice Index on page 674.

Since congruent angles are important, you should know how to copy an angle by geometric construction. Follow these steps to copy $\angle ABC$.

Step 1: Draw a ray *DE*.

Step 2: Draw an arc with the center on point *B* so that it intersects both rays of the angle. Label the points *P* and *Q*.

Step 3: Place the compass point at *D* and use the same opening to draw an arc that intersects $\overline{DE}$ at *F*.

Step 4: Place the compass point on *F* and draw another arc that is the same measure as *PQ*. Label the point where the two arcs intersect point *G*. Draw ray *DG*. Now $\angle ABC \cong GDF$.

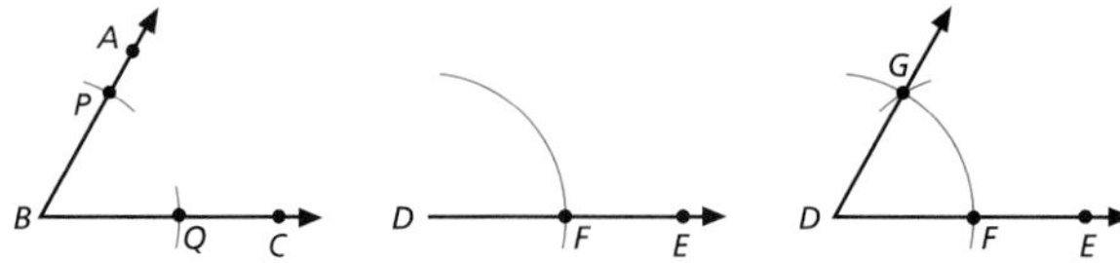

Try These Exercises

Determine if the polygons are similar. Write *yes* or *no*.

1. 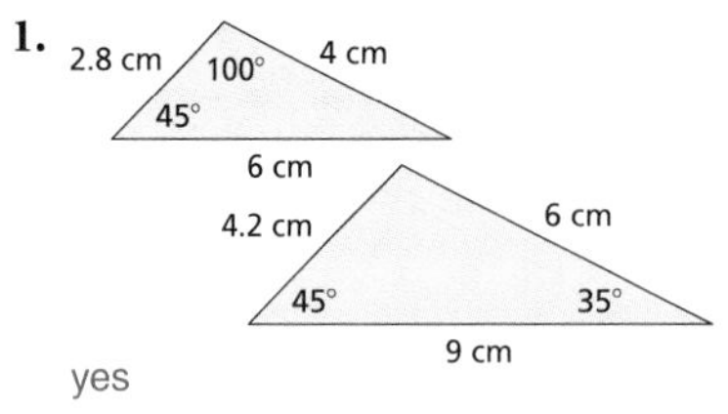

yes

2. 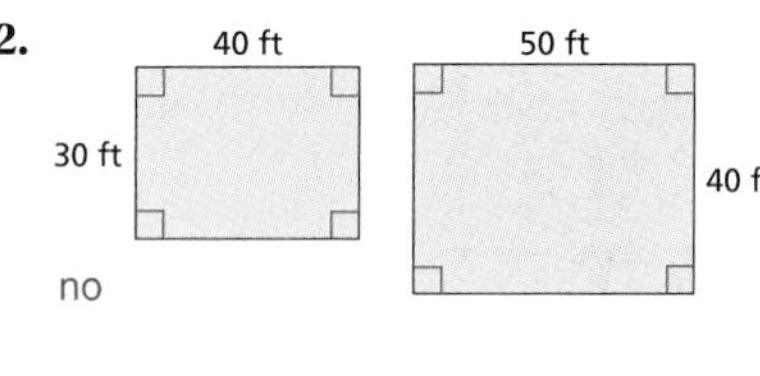

no

Find the measure of *x* in each pair of similar polygons.

3. 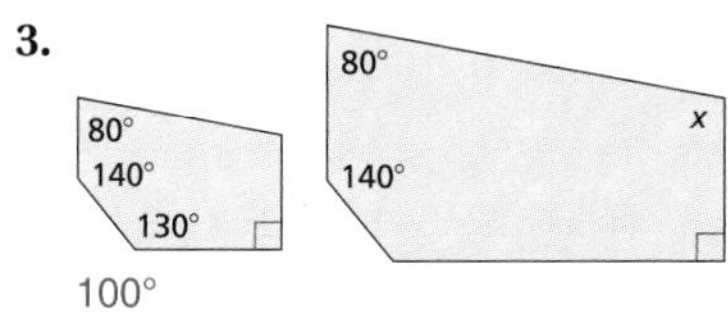

100°

4. 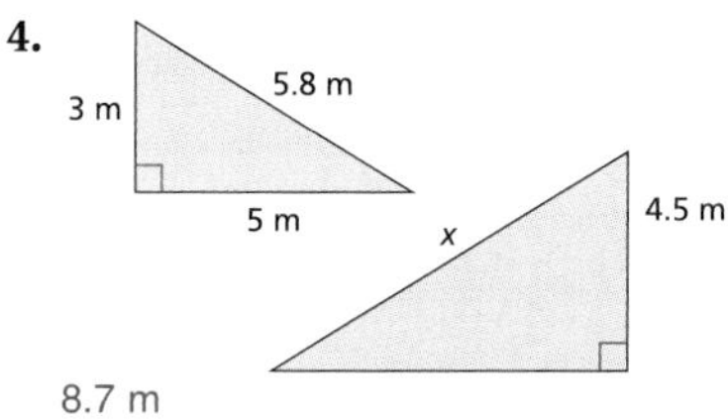

8.7 m

Practice Exercises

Determine if the polygons are similar. Write *yes* or *no*.

5. 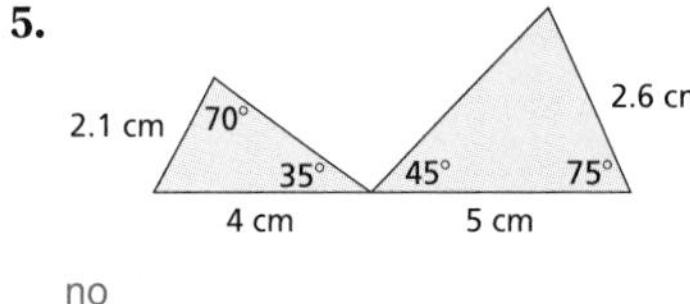

no

6. 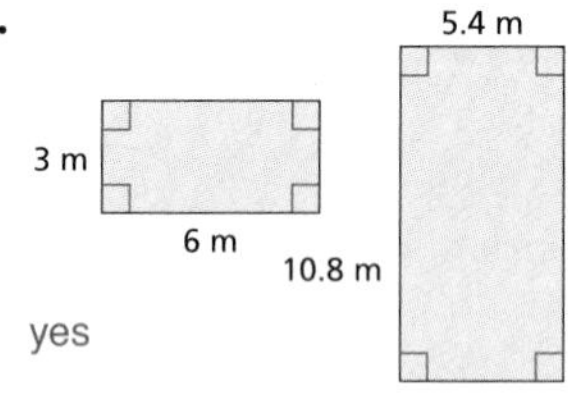

yes

Teaching Strategies

COOPERATIVE LEARNING Have students work in pairs to solve the following problem: Draw a square. Draw two line segments that intersect in the center and bisect opposite sides of the square. Draw one diagonal. Label the vertices, and name as many similar triangles and quadrilaterals as you can. **Answers may vary, but there will be five similar squares, four similar rectangles, and six similar triangles.**

Find the measure of x in each pair of similar figures.

7. 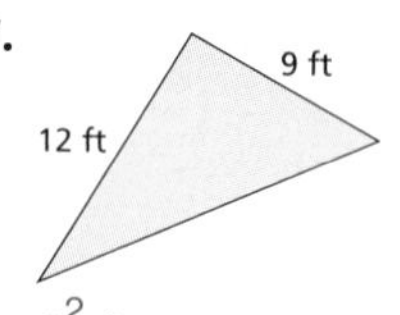

x 5 ft

8. 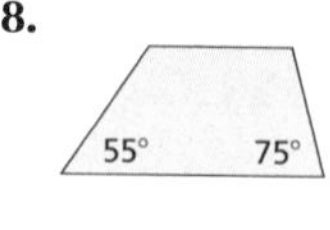

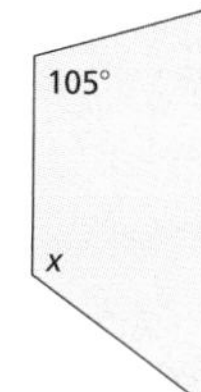

9. **WRITING MATH** When given two similar figures, how can you tell which angles are correspondng angles? Corresponding angles are between corresponding sides.
10. Draw any obtuse angle. Copy the angle using a straightedge and compass. Check students' work.
11. **PHOTOGRAPHY** A photograph that measures 3 in. by 4 in. is enlarged so that the 4-in. side measures 6 in. How long is the 3-in. side in the enlargement? 4.5 in.

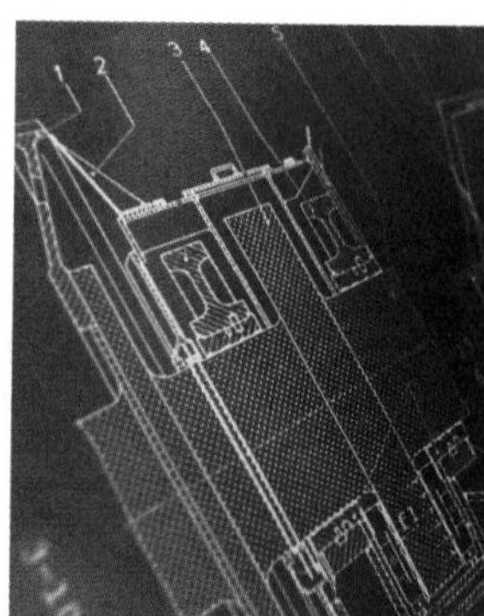

12. $ABCD$ and $STUV$ are similar rectangles. If $AB = 3$ cm, $BC = 8$ cm, and $ST = 6.6$ cm, what is the perimeter of $STUV$? 48.4 cm
13. **ARCHITECTURE** On a blueprint, a diagonal brace forms similar triangles RST and WXY. If $ST = 9$ ft and $XY = 12$ ft, what is the ratio of the perimeter of RST to the perimeter of WXY? 3 : 4
14. PQR and STU are similar triangles. $\angle Q$ and $\angle T$ are right angles. If $PQ = 5$ cm, $QR = 7$ cm, and $ST = 17.5$ cm, what is the area of $\triangle STU$? 214.375 cm^2
15. **FRAMING** Odetta has 34 in. of beautiful oak molding she would like to use for a picture frame. The photo she wants to frame measures 8 in. by 10 in. Find the dimensions of a reduced photo that would have the same shape as the original and would have a perimeter of exactly 34 in. $7\frac{5}{9}$ by $9\frac{4}{9}$

Extended Practice Exercises

16. Draw two hexagons that have the same angle measures but are not similar. Answers will vary.
17. Draw two quadrilaterals that have proportional corresponding sides but are not similar. Answers will vary. (One possible answer involves two rhombi with different angles)
18. Point D is said to divide $\overline{AB}$ *externally.* Two segments, $\overline{AD}$ and $\overline{BD}$, are formed. If $AD = 6$ and $\frac{AD}{AB} = \frac{3}{2}$, find AB. 4

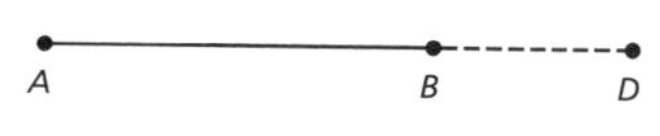

Mixed Review Exercises

Find the y-intercept for the line described by the equation. (Lesson 6-1)

19. $y = 4x - 3$ -3
20. $y = \frac{2}{3}x + 2$ 2
21. $2x - 8y = 12$ $-\frac{3}{2}$
22. $4y - 2x = 8$ 2
23. $4x - 3y = -2$ $\frac{2}{3}$
24. $7 - 2y = 5x$ $\frac{7}{2}$
25. $12x = 24y + 48$ -2
26. $7x - 4 = 3y$ $\frac{-4}{3}$
27. $15 - 3y = -2x$ 5

Teaching Strategies

CHALLENGE Parallelograms *ABCD* and *EFGH* are similar. *ABCD* has a perimeter of 34.8 ft. If *CD* = 2.4 ft and *GH* = 12 ft, what is the perimeter of *EFGH*? 174 ft

Extra Practice Worksheet 7-2

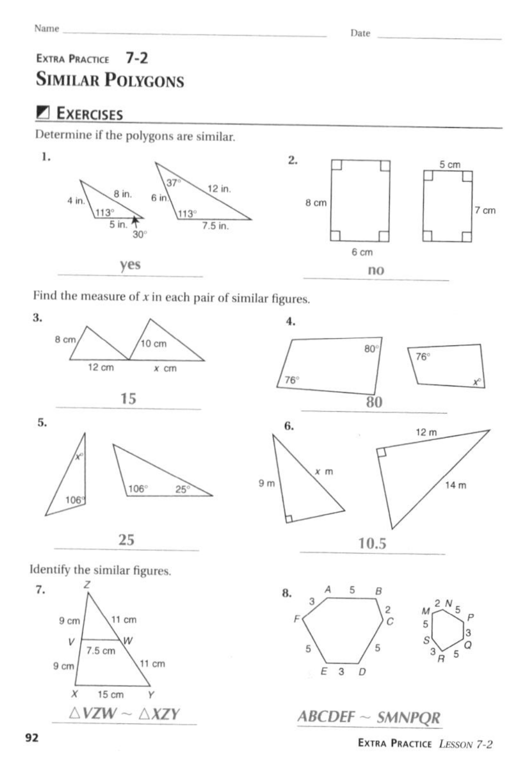

Name ______ Date ______

EXTRA PRACTICE 7-2
SIMILAR POLYGONS

EXERCISES

Determine if the polygons are similar.

1. yes
2. no

Find the measure of x in each pair of similar figures.

3. 15
4. 80
5. 25
6. 10.5

Identify the similar figures.

7. $\triangle VZW \sim \triangle XZY$
8. $ABCDEF \sim SMNPQR$

92 EXTRA PRACTICE *LESSON 7-2*

Enrichment Worksheet 7-2

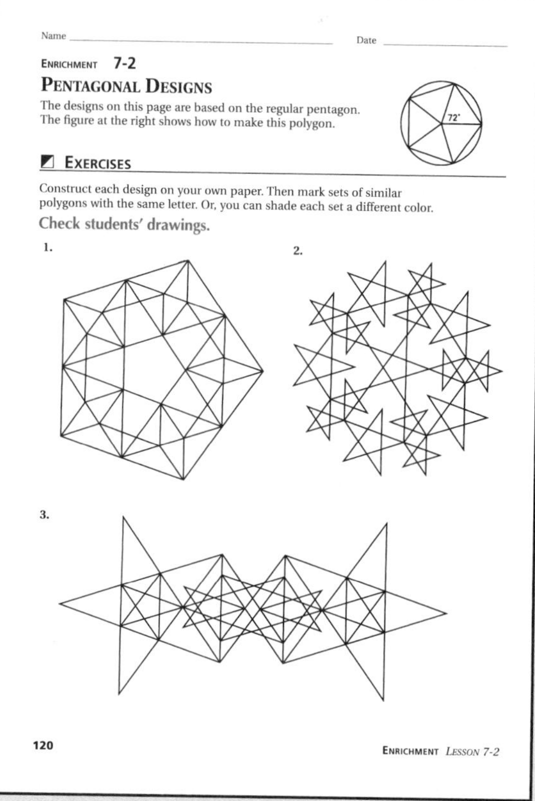

Name ______ Date ______

ENRICHMENT 7-2
PENTAGONAL DESIGNS

The designs on this page are based on the regular pentagon. The figure at the right shows how to make this polygon.

EXERCISES

Construct each design on your own paper. Then mark sets of similar polygons with the same letter. Or, you can shade each set a different color.

Check students' drawings.

1.
2.
3.

120 ENRICHMENT *LESSON 7-2*

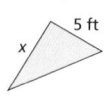

Skills Practice

Vocabulary Review

Lesson 7-1
equivalent ratios
proportion
extremes
means
cross products
terms

Lesson 7-2
similar angles
corresponding angles
corresponding sides
congruence

Assignment Guide

All students: 1–30
Additional Practice: Refer to the Extra Practice Index on page 674 of this text.

Review and Practice Your Skills

Practice ■ Lesson 7-1

Is each pair of ratios equivalent? Write *yes* or *no*.

1. 4 : 7, 12 : 14 no
2. 3.5 : 12, 14 : 48 yes
3. −3 : −2, 15 : 10 yes
4. $\frac{4.5}{13.5}, \frac{3}{1}$ no
5. $\frac{8}{20}, \frac{12}{30}$ yes
6. $\frac{3}{7} : \frac{4}{7}$, 1.8 : 2.4 yes

Solve each proportion.

7. $\frac{4}{7} = \frac{x}{42}$ 24
8. $\frac{3}{-2} = \frac{10}{d}$ $-6\frac{2}{3}$
9. $\frac{8}{m} = \frac{20}{12.5}$ 5
10. $\frac{p}{0.2} = \frac{0.5}{0.1}$ 1
11. $\frac{-9}{400} = \frac{-2.25}{f}$ 100
12. $\frac{1024}{12} = \frac{16}{3}$ 192
13. $\frac{117}{36} = \frac{585}{g}$ 180
14. $\frac{e}{47} = \frac{616}{5264}$ 5.5
15. $\frac{2}{300} = \frac{n}{200}$ $\frac{4}{3}$
16. $\frac{-13}{-39} = \frac{x}{-663}$ −221
17. $\frac{0.072}{z} = \frac{1.2}{0.6}$ 0.036
18. $\frac{p}{136{,}974} = \frac{10}{-1110}$ −1234
19. The ratio of flour to sugar in a recipe is 5:2. How much flour must be added to $\frac{1}{2}$ cup sugar? $1\frac{1}{4}$ cups
20. The ratio of blue to red in a paint mixture is 8 : 15. How many pints of red paint must be added to 72 pints of blue paint? 135 pints
21. An angle and its complement have the ratio 9 : 11. What are the measures of the two angles? 40.5^6 and 49.5°
22. The ratio of boys to girls in a class is 4 : 5. There are 27 students in the class. How many are boys? 12 boys

Practice ■ Lesson 7-2

Determine if the polygons are similar. Write *yes* or *no*.

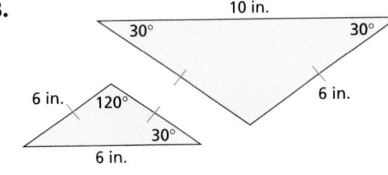

23. 10 in., 30°, 30°, 6 in., 6 in., 120°, 30°, 6 in.

yes

24. 2m, 4m

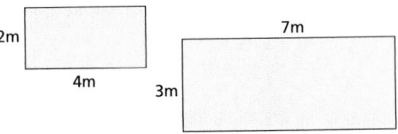

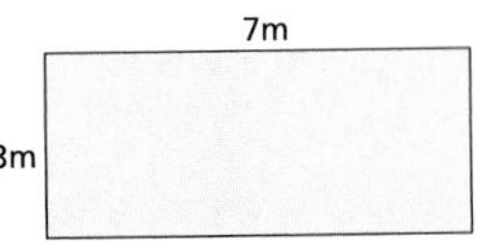

no

25.

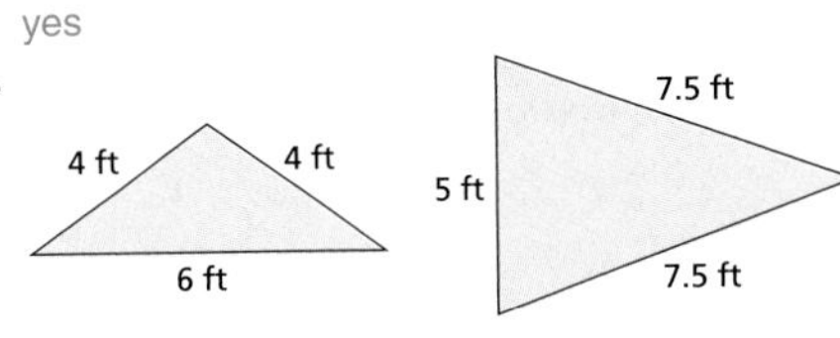

no

26.

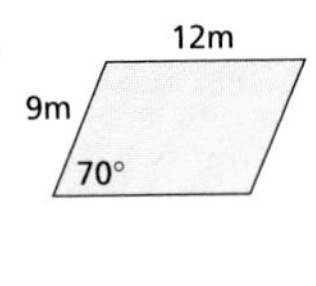

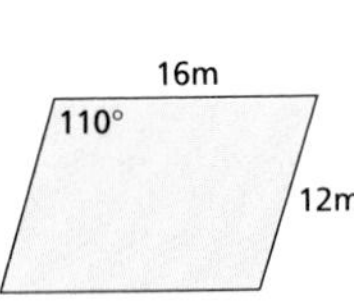

yes

Teaching Strategies

For Exercises 19–21, remind students to write the ratios in the proportion in the same order. Students may find it helpful to first write a ratio using words and then use the words to check the order of the numerical terms.

Find the measure of x in each pair of similar polygons.

27.

29.

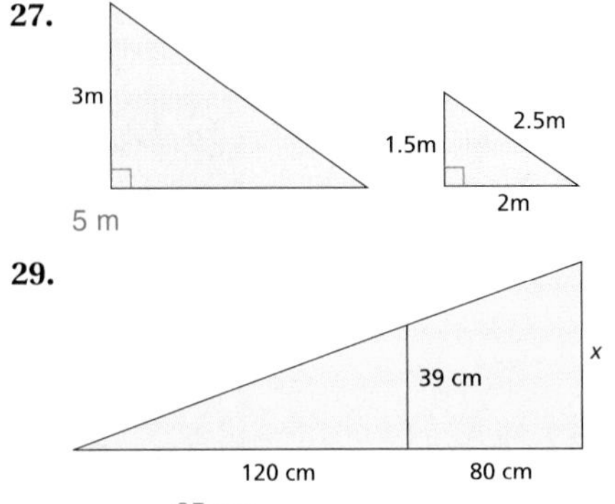

5 m

65 cm

28.

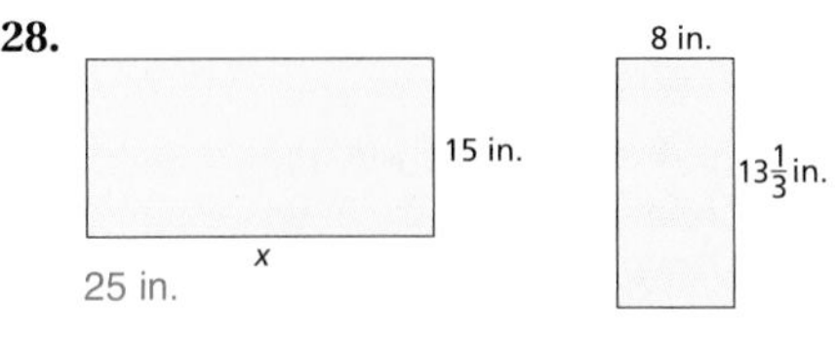

25 in.

30.

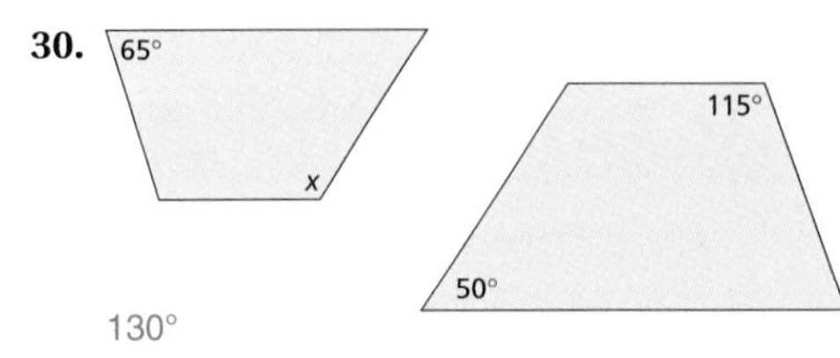

130°

MathWorks
Workplace Knowhow

Career – Police Photography

Police photographers record the details of a crime scene by taking photographs. Later, the photos are used by forensic scientists to search for clues. The solving of a crime may depend on a piece of information found in a crime photograph.

Police photographers must be prepared to work indoors or outdoors and in all types of weather and lighting conditions.

They must choose the best camera and film for the situation. Sometimes, these workers use special filters to enhance details in their subject.

Police photographers must be meticulous to make sure that every angle of a crime scene has been covered. Photos taken from odd angles will distort the perspective and slow down the investigation.

You have use of a surveillance video tape to produce several still photographs of a robbery suspect. The picture is taken at eye level and the suspect is standing in front of a counter with a known height.

1. The counter, which measures 3.5 ft in real life, measures 2.5 in. in the photograph. What is the scale of the photograph? 1 : 16.8
2. If the suspect's height is 4.2 in. in the photograph, how tall is the suspect in real life? 70 · 56 in. or 5.88 ft, about 5 ft $10\frac{1}{2}$ in.

In another picture, there are 4 objects on the floor between a desk and a wall safe that was robbed. The value is exactly $6\frac{1}{2}$ feet from the desk. The distance in the photograph is $3\frac{1}{4}$ inches.

3. What is the scale of the photograph? 1 : 24
4. A glove in the photo is 1.8 in. from the wall. What is its actual distance from the wall? 43.2 in. or 3.6 ft, about 3 ft 7 in.

Chalkboard Examples

Example from Lesson 7-1

A pharmacist mixes a medicine according to the ratio of three parts medicine to 50 parts saline solution. How much medicine should she add to 12 mL of saline solution? **0.72 mL**

Example from Lesson 7-2

For triangles ABC and DEF, $AB = 15$, $BC = 18$, $AC = 12$, $DE = 40$, $EF = 48$, $DF = 27$, $m\angle A = 84°$, $m\angle B = 42°$, $m\angle E = 42°$ and $m\angle F = 55°$. Are the triangles similar? **no**

Career Opportunity

Describe the kind of work police photographer do and the types of situations where their services are required. Have students discuss the importance of measurement and algebraic thinking in doing this type of job. Explain that this job requires excellent skills in using ratios and scales to assist in unveiling details of a crime. Students should answer Questions 1–4 to better understand how police photographers use mathematics in performing their job.

Students who are interested in learning more about this profession can go to learningmathmatters.com. Guidance Counselors should have information about school and training requirements.

Extend the Lesson

ONGOING ASSESSMENT Write always, sometimes or never for each statement.

1. Two different squares are similar. always
2. Similar polygons are congruent. always
3. Two different rectangles are similar. sometimes
4. Figures with different numbers of sides can be similar. never

Chapter 7 **Review and Practice Your Skills** 305

Lesson Planning

NCTM Standards/Strands

- Number & Operation
- Patterns, Functions, & Algebra
- Geometry & Spatial Sense
- Measurement
- Problem Solving
- Reasoning and Proof
- Connections
- Representation

Vocabulary

scale drawing
scale distance
actual distance
model

Materials Needed

string
centimeter ruler

Lesson Resources

Warm-Up Transparency 19
Reteaching 7-3
Extra Practice 7-3
Enrichment 7-3

Assignment Guide

Basic: 1–19, 22–24
Enriched: 1–24

Getting Started

5-Minute Warm-Up

Solve for *x*.

1. $\frac{1}{4} = \frac{15}{x}$ 60
2. $\frac{\left(\frac{2}{7}\right)}{3} = \frac{6}{x}$ 63
3. $\frac{1}{500} = \frac{x}{2{,}050}$ 4.1
4. $\frac{1.5}{75} = \frac{x}{512.5}$ 10.25

Introduction to Lesson 7-3

Ask students to share how they represented distances. Did they approximate, or did any of them have a system? You may wish to have students trade maps to see if another person can read and understand the map.

7-3 Scale Drawings

Goals ■ Find actual or scale length using scale drawings.

Applications Architecture, Engineering, Photography

Suppose a new student has joined your class. Sketch a map of the school building. Try to represent distances accurately. For example, if the principal's office is further from the gym than the library, the distance from the gym on the map should be greater also.

Build Understanding

A **scale drawing** is a representation of a real object. All lengths on the drawing are proportional to actual lengths of the object. The **scale** of the drawing is the ratio of the size of the drawing to the actual size.

Work with a group of 2 or 3 students.

1. Think of two well-known locations within your community. For instance,
you might choose your school and the public library.
2. Sketch a map showing how to get from one location to the other. Try to represent distances accurately on the map.

Example 1

ARCHITECTURE This is a scale drawing of a room in a house.

Find the actual distance along the wall between the window and the door.

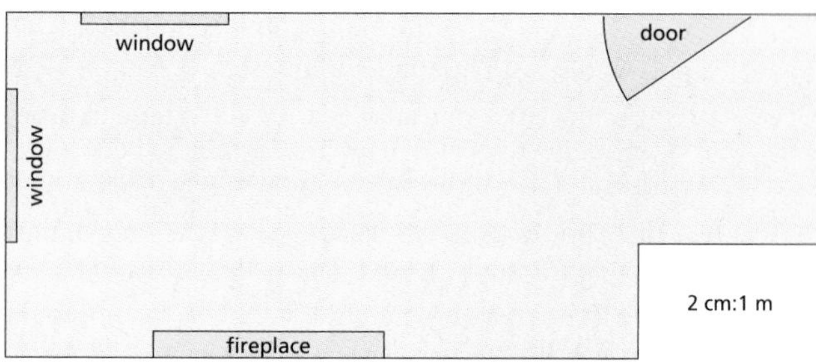

Solution

The ratio of the scale drawing to the actual size of the room is 2 cm:1 m. The first step is to measure the drawing. Because the scale is given in terms of centimeters, measure the distance to the nearest centimeter.

The scale distance between the window and the door is 5 cm. Write and solve a proportion.

Let x = the actual distance in meters.

$$\frac{2 \text{ cm}}{1 \text{ m}} = \frac{5 \text{ cm}}{x} \quad \begin{matrix}\text{scale distance}\\ \text{actual distance}\end{matrix} \qquad 2x = 5 \rightarrow x = 2.5$$

The actual distance between the window and the door is 2.5 meters.

Check Understanding

How would the scale drawing of this room change if the scale were changed to 1 cm:1 m?

The drawing will be smaller; all lines will be half as long as they are now.

Extend the Lesson

REAL WORLD CONNECTION A pantograph is an instrument used to copy plans, maps, or diagrams on any given scale. It consists of four bars joined by adjustable pins. Engineers, cartographers, and artists use pantographs by tracing the point of one bar over an outline to be duplicated, which causes another point to move in duplicate fashion, creating a similar outline. Have students research how a pantograph is built. You may want to assign them to build a small one using cardboard and brads.

Scale drawings are used in engineering. The scale can be stated as a ratio without any reference to a particular unit of measure. For example, a model car often relates to the actual car by a scale of 1:24.

Example 2

ENGINEERING The distance between the front wheels of a model car is 4.5 centimeters. What is the actual distance on the car if the scale is 1:24?

Solution

Write and solve a proportion. Let $d =$ the actual distance in centimeters.

$$\frac{1}{24} = \frac{4.5}{d} \quad \begin{matrix}\text{scale distance}\\ \text{actual distance}\end{matrix} \qquad d = 108$$

The actual distance between the front wheels is 108 cm or 1.08 m.

Satellite photographs are sometimes used to map terrain and roadways. The scale on a photograph can be determined by comparing distances on the map to known distances. Then the scale is given as a bar length.

Example 3

SATELLITE PHOTOGRAPHY The map below was made from a satellite photo. Using the scale bar, estimate the driving distance from Jericho to Hanover on the map.

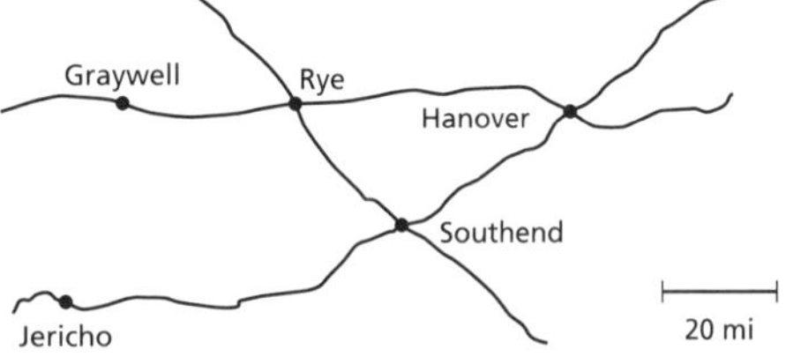

Solution

Cut a piece of string that is as long as the route between the two points on the map. Then compare the string's length with the scale length. The string's length is equal to nearly 6 scale lengths.

Multiply the number of scale lengths it takes to cover the distance by the actual distance given for the scale length.

$$6 \times 20 = 120 \text{ miles}$$

Therefore, the actual distance between the two cities is about 120 miles.

Geometric iterations can produce a figure called a fractal. Fractal shapes also appear in nature. Mathematicians have discovered that coastlines are better described as fractals than as smooth curves.

Computer generated fractal

Extend the Lesson

ONGOING ASSESSMENT Have students solve the following problem: The scale on a floor plan of a rectangular room is 3 in.: 1 ft. The length and width of the room on the floor plan measures $27\frac{1}{2}$ in. and $25\frac{1}{2}$. How many tiles will you need to cover the actual floor if each tile is $\frac{1}{2}$ ft $\times$ $\frac{1}{2}$ ft? **425 tiles**

Chalkboard Examples

Supplementary Example 1

Have students use the same scale drawing but change the scale to $\frac{3}{16}$ in. = 1 ft. Find the actual length in feet of the fireplace wall. **$13\frac{2}{3}$ ft or 13 ft 8 in.**

Supplementary Example 2

Gail is building a model of a theater stage using a scale of 1:36. On the model, the width of the stage is 15 in. What is the actual width of the stage? **540 inches or 45 feet**

Supplementary Example 3

Using the same drawing, estimate the distance from Graywell to Hanover. **80 miles**

Supplementary Example 4

Using the same drawing, what conclusion can you draw about the length of the nature trail? **The nature trail's length is greater than the bike path's length, but is less than the length of the actual coastline.**

Reteaching Worksheet 7-3

Name ____________ Date ____________

RETEACHING 7-3

SCALE DRAWINGS

A scale drawing is in direct proportion to the actual object. The scale gives you the ratio of the length in the drawing to the actual size of the object.

Example

In this scale drawing, the ratio of the scale to the actual boat is 5 cm : 40 m.

a. Find the actual overall length of the boat.

b. The actual length of the boat at the waterline is 26.5 m. Find the scale length of the boat at the waterline.

5.15 cm

Solution

The overall length of the boat in the scale drawing is 5.15 cm.
The actual length of the boat at the water line is 26.5 m.
The scale is 5 cm : 40 m.
Use these facts to write and solve a proportion.

a. Let $x =$ the actual length of the boat.

$\frac{5(cm)}{40(m)} = \frac{5.15(cm)}{x}$ ← Scale length / ← Actual length

$5x = 40(5.15)$ **Find cross-products.**

$5x = 206$ **Simplify.**

$\left(\frac{1}{5}\right) 5x = 206 \left(\frac{1}{5}\right)$ **Multiply by $\frac{1}{5}$.**

$x = 41.2$ **Simplify.**

The actual length of the boat is 41.2 m.

b. Let $y =$ scale length at waterline.

$\frac{5(cm)}{40(m)} = \frac{y}{26.5}$ ← Scale length / ← Actual length

$40y = 5(26.5)$ **Find cross-products.**

$40y = 132.5$ **Simplify.**

$\left(\frac{1}{40}\right) 40x = 132.5 \left(\frac{1}{40}\right)$ **Multiply by $\frac{1}{40}$.**

$y = 3.3125$ **Simplify.**

The scale length of the boat at the waterline is 3.3125 cm.

EXERCISES

Find the actual measure of each scale item.

1. scale length is 5 cm
scale is 1 cm : 3 m **15 m**
2. scale length is $1\frac{1}{2}$ in.
scale is 1 in. : 8 in. **12 in.**
3. scale distance is 2.5 cm
scale is 10 cm : 1 km **0.25 km**
4. scale distance is 15 in.
scale is 5 in. : 2 mi **6 mi**

Find the scale measure for each item.

5. actual length is 6 ft
scale is 1 in. : 15 ft **$\frac{2}{5}$ in.**
6. actual length is 45 m
scale is 3 cm : 10 m **13.5 cm**
7. actual distance is 75 mi
scale is 1 in. : 100 mi **$\frac{3}{4}$ in.**
8. actual distance is 165 km
scale is 15 cm : 75 km **33 cm**

106 RETEACHING *LESSON 7-3*

Lesson Wrap-up

Quick Assessment

Ask the following questions to determine if students understand the content presented in this lesson.

1. Why are scale drawings helpful? Very small objects can be represented showing detail that would be impossible to show in actual size; very large objects can be drawn in a workable size.
2. A scale drawing of a rectangle has a scale of 1 cm : 3 m. How does the length of a side of the scale drawing compare with the length of a side of the actual rectangle? It is $\frac{1}{300}$ the length of a side of the actual rectangle.

Assignment Guide

Basic: 1–19, 22–24
Advanced: 1–24
Additional Practice: See Extra Practice Index on page 674.

Additional Answers

20. Accept any reasonable answer. A convenient scale to use is 1 in. = 60 ft. The longest length will be represented by 10 in., and the width will be $5\frac{7}{12}$ in.

Example 4

Kent and Baywater are two villages along the coast. From the map below, we can see that the distance along the coast highway between the towns is about 2 miles. Is the bikepath between Kent and Baywater also 2 miles long?

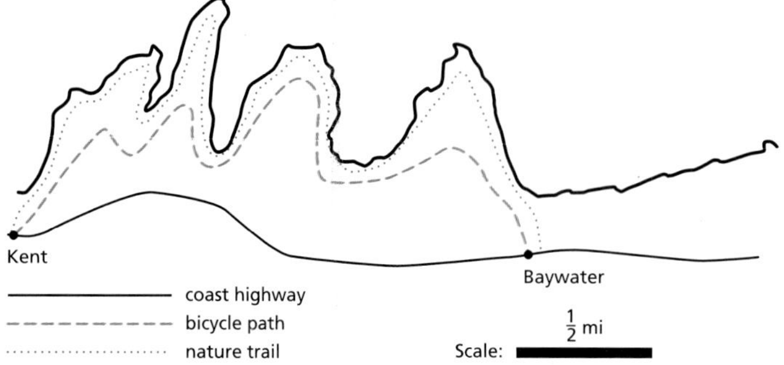

Solution

No. The bike path is closer to the actual coast than the highway. Using a piece of string, we can estimate that the distance along the bike path is approximately 5 miles.

Math: Who Where, When

The term fractal was coined by Benoit B. Mandelbrot in 1975. Mandelbrot was born in Warsaw, Poland, in 1924, educated in Paris, and came to the U.S. in 1958. He is a member of the research staff of IBM. He originated the theory of fractals, which is concerned with shapes and phenomena that are equally irregular or broken up at any scale.

Try These Exercises

Find the actual length of each of the following.

1. scale length is 2 in., scale is $\frac{1}{2}$ in.:3 ft 12 ft
2. scale length is 4 cm, scale is 1:200 800 cm

Find the scale length for each of the following.

3. actual length is 5 m 1.25 cm
   scale is 1 cm:4 m
4. actual distance is 175 mi 1.75 in.
   scale is $\frac{1}{4}$ in.:25 mi
5. Use the map in Example 3. Estimate the actual distance between Rye and Hanover. 40 mi
6. **PHOTOGRAPHY** A photo that measures 3.5 in. by 6 in. will be enlarged so that its width will be 8 in. Will the length of the enlargement be less than or greater than 15 in.? Less than
7. **MODEL BUILDING** A model plane has the scale of 1:500. The wingspan on the model is 6.4 cm. How many meters is the wingspan of the plane itself? 32 m
8. **CONSTRUCTION** The blueprint for a garage indicates that a wooden beam measures 4.1 cm. The scale of the plan is 1:300. What is the actual length of the beam in meters? 12.3 m

Extend the Lesson

STUDENT PORTFOLIO Have students use a road atlas and a city in the United States. Have them choose another city within the same state and plan which roadways they would use to make the trip. Have them use string and the map scale to find the distance between the cities. Then have them compare their calculations with the mileage shown on the road map.

Practice Exercises

Find the actual length of each of the following.

9. scale length is 3 cm, scale is 2 cm:5 m
   7.5 m
10. scale distance is 2.1 cm, scale is 1:300
    630 cm

Find the scale length for each of the following.

11. actual length is 10 m, scale is 1:20
    0.5 m
12. actual distance is 200 km, scale is 1.5 cm:25 km
    12 cm

Find the actual distances using the map at the right.

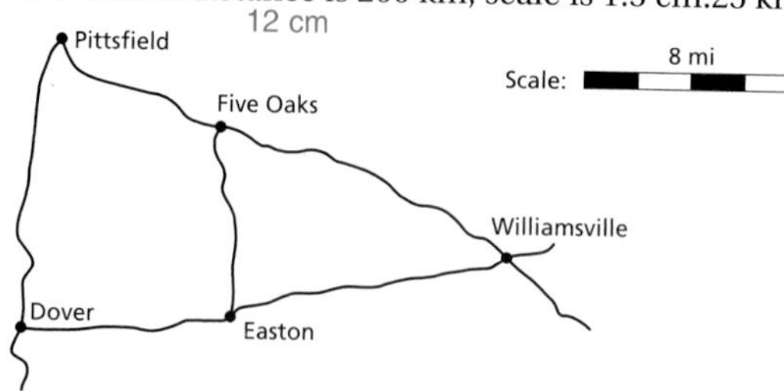

13. Easton to Williamsville 10 mi
14. Pittsfield to Five Oaks 7 mi
15. Dover to Williamsville 18 mi
16. Five Oaks to Williamsville 12 mi

17. **DATA FILE** Use the Data Index on pages 632–633 to locate information about the lengths of the St. Lawrence and Columbia rivers. On a map using a scale of 200 mi = $\frac{1}{4}$ in., what would be the lengths of the rivers on the map?
    St. Lawrence, 1 in.; Columbia, about $1\frac{1}{2}$ in.

18. **WRITING MATH** Use the map in Example 4. Using a piece of string, estimate the distance between Kent and Baywater using the nature trail. What do you think would happen to the length of the coast between Kent and Baywater if you got even closer to the water and used an inch as the measuring unit? The closer you get to the water's edge and the smaller the unit of measurement you use, the longer the length will be.

19. **CHAPTER INVESTIGATION** Select at least five prominent features in the photograph and plot their locations on your enlargement. You may want to draw a coordinate grid over the surface of the photograph and draw a corresponding grid on your enlargement. Once the main features of the photograph are placed correctly on the enlargement, sketch in the remaining details from the photograph.

Extended Practice Exercises

20. Lisa wants to make a map of the school that will fit on a sheet of paper that measures $8\frac{1}{2}$ in. by 11 in. The longest length of the school is 600 ft and its longest width is 350 ft. What would be a good scale to use so that the map is as large as possible, but will fit on the paper? See additional answers.
21. A **hectare** (abbreviated ha) is a metric unit of land area equal to 10,000 square meters. On a map, a rectangular plot of land measures 5 cm by 12 cm. The scale of the map is 1:5000. How many hectares does the plot include?
    15 ha

Mixed Review Exercises

Find the slope of a line parallel to the given line and the slope of a line perpendicular to the given line. (Lesson 6-2)

22. the line containing points $A(2, 3)$ and $B(-1, -4)$ $\parallel = \frac{7}{3}, \perp = \frac{-3}{7}$
23. the line containing points $C(1, 3)$ and $D(-3, 8)$ $\parallel = -\frac{5}{4}, \perp = \frac{4}{5}$
24. the line containing points $G(1, 0)$ and $H(-2, 5)$ $\parallel = \frac{-5}{3}, \perp = \frac{3}{5}$

Teaching Strategies

CHALLENGE Have students make a floor plan of an actual room. Have them take actual measurements and decide on an appropriate scale. Encourage them to draw the floor plan showing as much detail from the room as possible. For example, they can include bookcases, windows, doors, rugs, fireplaces, furniture, or any other interesting features of the room.

Extra Practice Worksheet 7-3

Name ______________ Date ______________

EXTRA PRACTICE 7-3

Scale Drawings

Exercises

Find the actual length of each of the following.

1. scale length is 5 cm
   scale is 2.5 cm : 4 m 8 m
2. scale distance is 3.5 cm
   scale is 1 cm : 100 km 350 km
3. scale length is 10 in.
   scale is 1: 50 500 in. or $41\frac{2}{3}$ ft
4. scale distance is 4.5 in.
   scale is 2 in. : 25 ft 56.25 ft
5. scale length is 4.6 cm
   scale is 1 : 600 2760 cm or 2.76 m
6. scale distance is 6.8 cm
   scale is 1 cm : 50 km 340 km

Find the scale length of each of the following.

7. actual length is 20 km
   scale is 2 cm : 10 km 4 cm
8. actual distance is 60 m
   scale is 2 cm : 15 m 8 cm
9. actual length is 36 ft
   scale is 3 in. : 24 ft 4.5 in.
10. actual distance is 64 mi
    scale is 1 in. : 10 mi 6.4 in.
11. actual length is 25 in.
    scale is 1 : 50 0.5 in.
12. actual distance is 150 cm
    scale is 1 : 100 1.5 cm
13. Laren has a map with a scale of 2 in. : 25 mi. There are 16 inches on the map between Laren's home and her aunt's home. If she can drive an average of 50 miles per hour, how long will it take Laren to get from her home to her aunt's home? 4 h
14. The floorplan of a house is drawn to a scale of 1 in. : 5 ft. The actual dimensions of the family room are 20 feet by 24 feet. What are the dimensions of the family room on the floorplan? 4 in. by 4.8 in.
15. The distance from the front to the back of a model bicycle is 3 centimeters. What is the actual distance on the bicycle if the scale is 1 : 100? 300 cm
16. The distance between two cities on a map is 5 cm. What is the actual distance between the cities if the scale is 1 cm : 50 km? 250 km

94 EXTRA PRACTICE *LESSON 7-3*

Enrichment Worksheet 7-3

Name ______________ Date ______________

ENRICHMENT 7-3

Rep-tiles

A *rep-tile* is a polygon that can be used to reproduce or replicate itself. For example, four congruent copies of the small triangle can be arranged to make a larger, similar triangle. Since four copies are used, this triangle is said to be *rep-4.*

Exercises

Here are three types of rep-4 hexagons. Dissect each figure to prove that it is rep-4. Then, describe how each hexagon is made.

1. 2 × 2 square minus 1 unit square
2. 2 × 3 rectangle minus 1 unit square
3. 2 × 3 rectangle minus 2 unit squares

Here are three types of rep-4 trapezoids. Dissect each figure to prove that it is rep-4. Then, describe the angles in each trapezoid.

4. 60°, 120°, 60°, 120°
5. 90°, 90°, 60°, 120°
6. 90°, 90°, 45°, 135°

7. Show that the figure in Exercise 6 is also rep-9.
8. Six equilateral triangles can be put together to make a rep-4 pentagon. Find this pentagon and sketch it below.

122 ENRICHMENT *LESSON 7-3*

Lesson Planning

NCTM Standards/Strands

- Number & Operation
- Patterns, Functions, & Algebra
- Geometry & Spatial Sense
- Measurement
- Problem Solving
- Reasoning and Proof
- Connections
- Representation

Vocabulary

Similarity Postulates: AA, SAS, SSS

Materials Needed

calculators compasses
straightedges

Lesson Resources

Warm-Up Transparency 20
Transparency RF-28
Reteaching 7-4
Extra Practice 7-4
Enrichment 7-4

Assignment Guide

Basic: 1–10, 15–26
Enriched: 1–26

Getting Started

5-Minute Warm-Up

Is each pair of ratios equivalent?

1. $\frac{5}{15}, \frac{12.4}{37.2}$ yes
2. $\frac{19.5}{13}, \frac{18.6}{6.2}$ no
3. $\frac{\left(\frac{1}{2} \times 7\right)}{7}, \frac{6}{12}$ yes

Introduction to Lesson 7-4

Ask students how they concluded that the triangles are similar. They are the same shape. How could you be sure they are the same shape? Measure all three angles and sides. They are similar if all corresponding angles are congruent and the measures of all corresponding sides form the same ratio.

7-4 Postulates for Similar Triangles

Goals ■ Use the AA, SSS and SAS similarity postulates to determine if two triangles are similar.

Applications Art, Surveying, Photography

Work with a partner. You will need a compass and a straightedge.

1. Using the straightedge, draw any triangle and label it *ABC*.
2. Draw a line segment on another sheet of paper that is longer than *AC*. Your partner should draw a line segment longer than *BC*.
3. Copy $\angle A$ at one end of your line segment and $\angle C$ at the other end. Have your partner copy $\angle B$ and $\angle C$ at the ends of his or her line segment.
4. Both you and your partner should now extend the outer rays of the angles you have drawn to form triangles.
5. Compare both triangles with the original triangle. What seems to be true?
The triangles are similar.

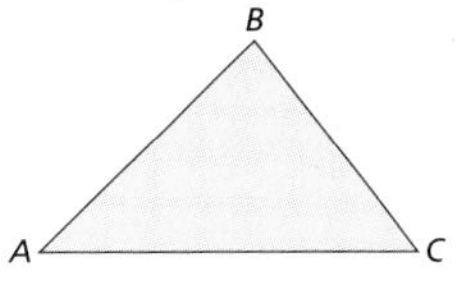

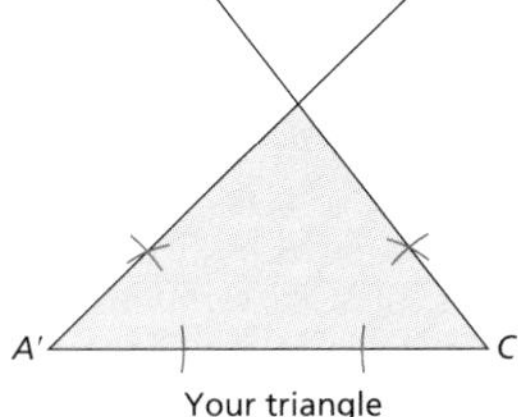

Your triangle

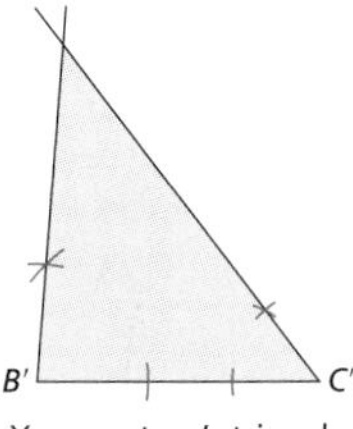

Your partner's triangle

Build Understanding

In Chapter 3, you learned that some statements in geometry are considered to be true without any proof. Usually, these statements are based on direct observation of principles that always work.

Postulate 15 (The AA Similarity Postulate)	If two angles of a triangle are congruent to two angles of another triangle, the two triangles are similar.

Example 1

Is $\triangle ABC \sim \triangle DEF$?

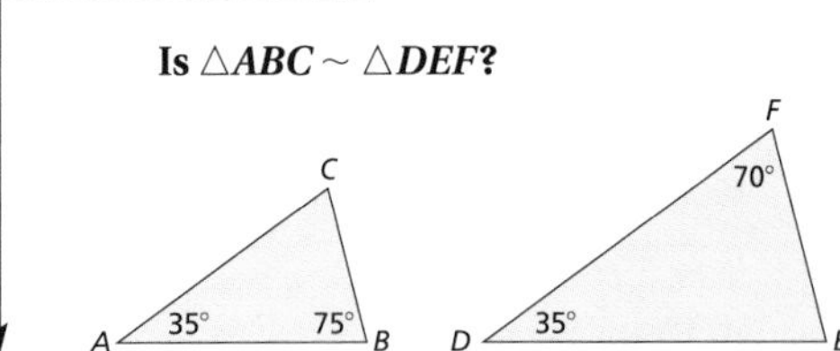

Teaching Strategies

COOPERATIVE LEARNING Draw the figure below on the chalkboard and have pairs of students work together to solve the following problem: Find two similar triangles and prove they are similar.

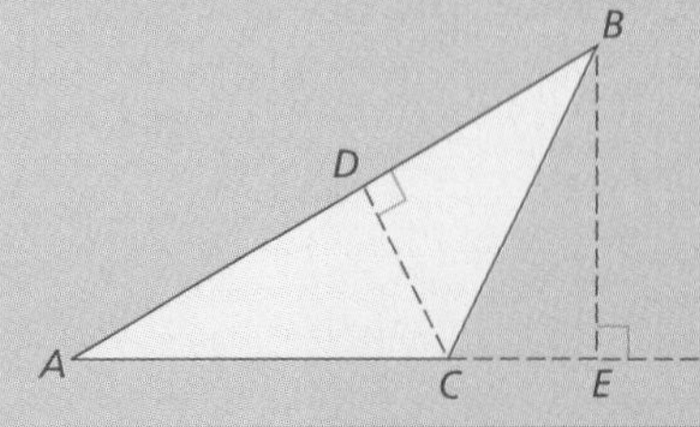

$\triangle ADC \sim \triangle AEB$ by AA;
$m\angle ADC = 90° = m\angle AEB$
and $\angle A \cong \angle A$

310 Chapter 7 Similar Triangles

Solution

Find one of the missing angle measures in either triangle. To find m∠*C*, subtract the sum of m∠*A* and m∠*B* from 180°.

$$180° - (35° + 75°) = 70°$$

Because ∠*A* ≅ ∠*D* and ∠*C* ≅ ∠*F*, the two triangles are similar by the AA Similarity Postulate. △*ABC* ~ △*DEF*

There are other ways to determine whether or not two triangles are similar.

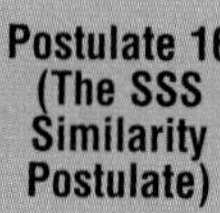

Postulate 16 (The SSS Similarity Postulate)	If the corresponding sides of two triangles are proportional, then the two triangles are similar.

Example 2

ART A wire sculpture is formed from triangles of the two sizes shown below. Is △*PQR* ~ △*STU*?

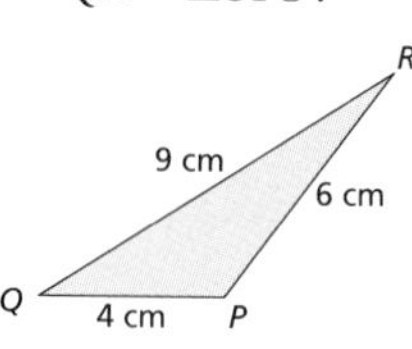

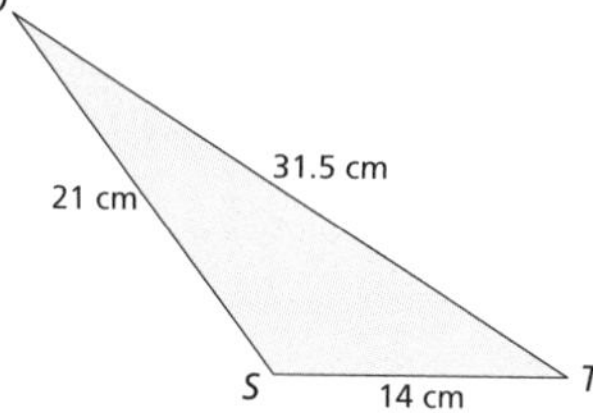

Solution

Find the ratio of each pair of corresponding sides.

$$\frac{PQ}{ST} = \frac{4}{14} = \frac{2}{7} \qquad \frac{QR}{TU} = \frac{9}{31.5} = \frac{2}{7} \qquad \frac{PR}{SU} = \frac{6}{21} = \frac{2}{7}$$

Because all three pairs of corresponding sides are proportional, the triangles are similar.

Another way of proving that two triangles are similar involves two pairs of corresponding sides and the angle between those sides.

Postulate 17 (The SAS Similarity Postulate)	If an angle of one triangle is congruent to an angle in another triangle, and the two sides that include that angle are proportional to the corresponding sides in the other triangle, then the two triangles are similar.

Example 3

If *PS* = 3*ST* and *XS* = 3*SY*, is △*PSX* ~ △*TSY*?

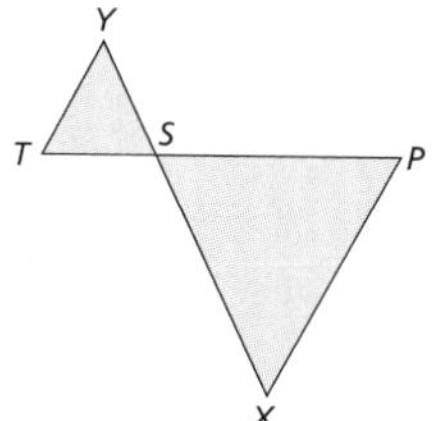

Technology Note

In Example 2, you could use a calculator to find each ratio is equal to 0.285714.

Check Understanding

If m∠*P* = 60, which angle in △*TSY* has a measure of 60°?

∠*T*

Chalkboard Examples

Supplementary Example 1

Is △*JKL* ~ △*MNO*? yes, by AA postulate

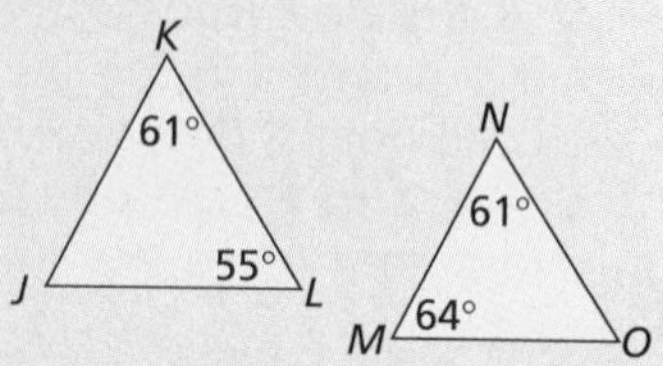

Supplementary Example 2

A tower is strengthened by adding triangles made from steel rods. On the plans for the tower, △*ABC* is smaller than △*DEF*. Given the measurements below, is △*ABC* ~ △*DEF*?

AB = 3, *BC* = 5 and *AC* = 4
DE = 7.5, *EF* = 12.5 and *DF* = 10
yes, by SSS postulate

Supplementary Example 3

If 2*CE* = *AC* and 2*CD* = *BC*, is △*ABC* ~ △*ECD*? yes, by SAS postulate

Teaching Strategies

In the diagram below, △*XYZ* ~ △*WXZ*. Have students name corresponding sides and angles. Suggest they refer to △*XYZ* ~ △*WXZ*, using the order of the letters to help them establish correct corresponding parts. For example, $\frac{\text{(first and third)}}{\text{(first and third)}} = \frac{XZ}{WZ}$, and second ≅ second means ∠*Y* ≅ ∠*WXZ*.

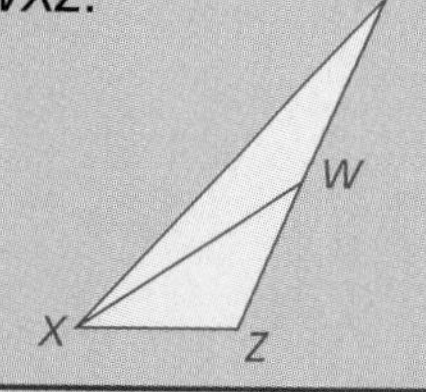

Reteaching Worksheet 7-4

Name ______________________ Date ____________

RETEACHING 7-4

POSTULATES FOR SIMILAR TRIANGLES

The AA (angle-angle) Similarity Postulate, the SSS (side-side-side) Similarity Postulate, or the SAS (side-angle-side) Similarity Postulate can be used to prove that two triangles are similar.

Example

Prove △*ABC* ~ △*DEF* using the
a. AA Similarity Postulate.
b. SSS Similarity Postulate.
c. SAS Similarity Postulate.

Solution

a. Find the measure of the missing angle in either triangle.
180° − (82° + 60°) = 180° − 142° = 38°
Since ∠*A* ≅ ∠*D* and ∠*B* ≅ ∠*E*, the triangles are similar by the AA Postulate.

b. Find the ratio of each pair of corresponding sides.

$$\frac{AC}{DF} = \frac{5}{7.5} = \frac{2}{3} \qquad \frac{AB}{DE} = \frac{7}{10.5} = \frac{2}{3} \qquad \frac{BC}{EF} = \frac{8}{12} = \frac{2}{3}$$

Since the ratios of all three pairs of corresponding sides are equal, the triangles are similar by the SSS Postulate.

c. Determine that one angle in one triangle is congruent to one angle in another triangle.
∠*A* ≅ ∠*D* since their measures are equal.
Then determine if the adjacent sides of ∠*A* are in proportion to those of ∠*D*.

$$\frac{AC}{DF} = \frac{AB}{DE} \text{ because } \frac{5}{7.5} = \frac{7}{10.5} = \frac{2}{3}$$

Since two pairs of corresponding sides are in proportion and the angles between those sides are congruent, the triangles are similar by the SAS Postulate.

EXERCISES

Reasons may vary. Samples are given.

Determine whether each pair of triangles is similar. If so, give a reason. Write AA, SSS or SAS.

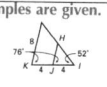

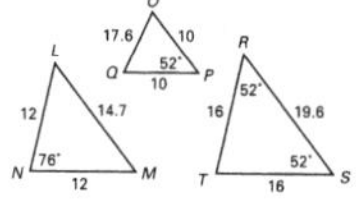

1. Is △*GIK* ~ △*HIJ*? yes; AA
2. Is △*GIK* ~ △*LMN*? yes; SAS
3. Is △*GIK* ~ △*OPQ*? no
4. Is △*LMN* ~ △*RST*? yes; AA
5. Is △*HIJ* ~ △*RST*? yes; SSS
6. Is △*OPQ* ~ △*RST*? no

108 RETEACHING *LESSON 7-4*

Lesson Wrap-up

Quick Assessment

Ask the following questions to determine if students understand the content presented in this lesson.

On the chalkboard, draw $\triangle ABC$, an altitude AE, and an altitude CD. Let F be the point where the altitudes intersect.

1. Is $\triangle BEA \sim \triangle BDC$? Explain. Yes by AA postulate; $m\angle BEA = 90° = m\angle BDC$ and $\angle B \cong \angle B$.
2. Is $\triangle ADF \sim \triangle CEF$? Explain. Yes by AA; $m\angle ADF = 90° = m\angle CEF$, and $\angle AFD \cong \angle CFE$ because they are vertical angles.

Assignment Guide

Basic: 1–10, 15–26
Advanced: 1–26
Additional Practice: See Extra Practice Index on page 674.

Additional Answers

7. The rays of the sun form the same angle with the ground for both the pole and tree. Therefore, $\angle A \cong \angle D$. Because $\angle B$ and $\angle E$ are both right angles, $\angle B \cong \angle E$. Therefore, the triangles are similar by AA Similarity.
9. Yes; because $\angle 1 \cong \angle 2$, $\angle D \cong \angle D$, there are 2 pairs of congruent angles. The AA Similarity Postulate applies.
10. Yes: $2x + 10 + x + 6x - 10 = 180$; $9x = 180$; $x = 20$. Then, $6x - 10 = 110$ and $2x = 50$. Because the smaller triangle includes angles of 110° and 50°, the triangles are similar by AA.

Solution

Use the given information to show that two pairs of corresponding sides are proportional.

$$\frac{PS}{ST} = \frac{3ST}{ST} = \frac{3}{1} \qquad \frac{XS}{SY} = \frac{3SY}{SY} = \frac{3}{1}$$

$\angle PSX \cong \angle TSY$ because they are vertical angles. Therefore, $\triangle PSX \sim \triangle TSY$ by the SAS Similarity Postulate.

Try These Exercises

Determine whether each pair of triangles is similar. If the triangles are similar, give a reason: write AA, SSS, or SAS.

1. no

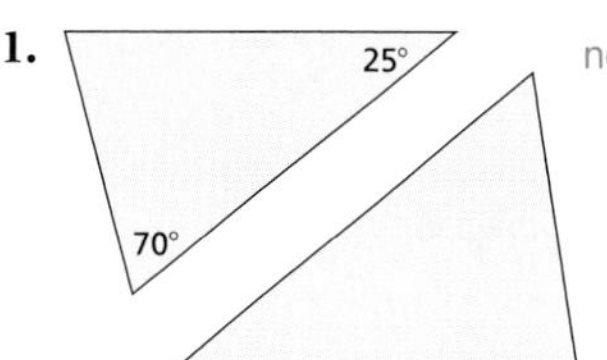

2. yes; SAS

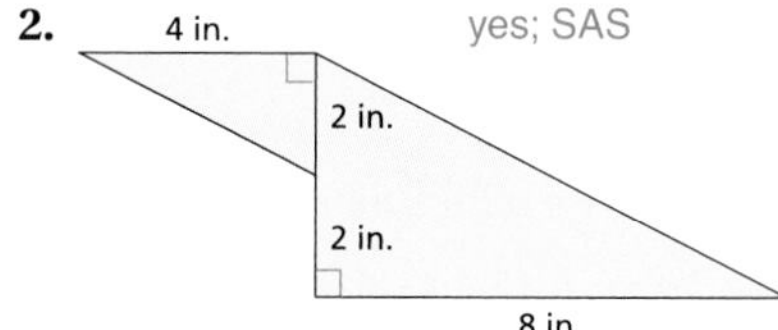

3. yes; AA

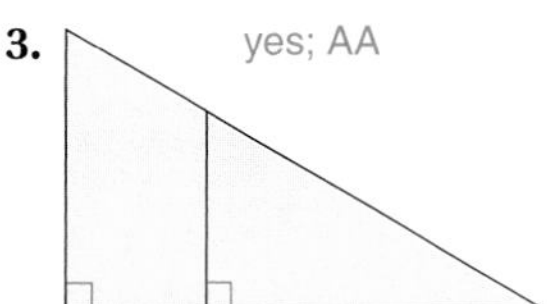

4. yes; SSS

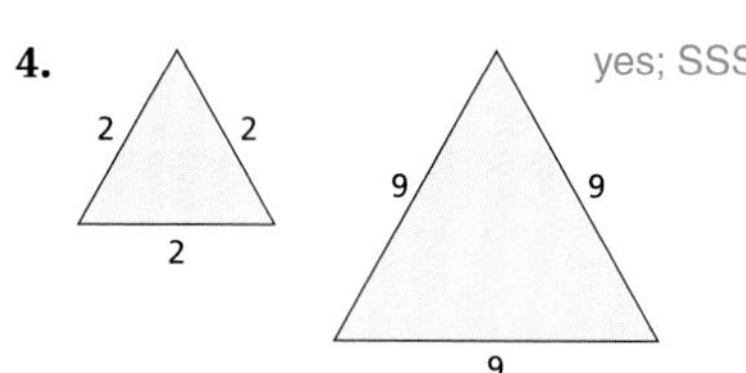

Practice Exercises

Determine whether each pair of triangles is similar. If the triangles are similar, give a reason: write AA, SSS, or SAS.

5. yes; AA

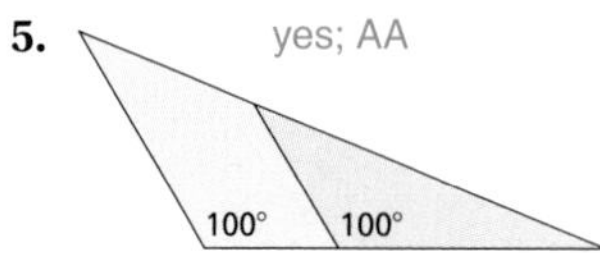

6. yes; SSS

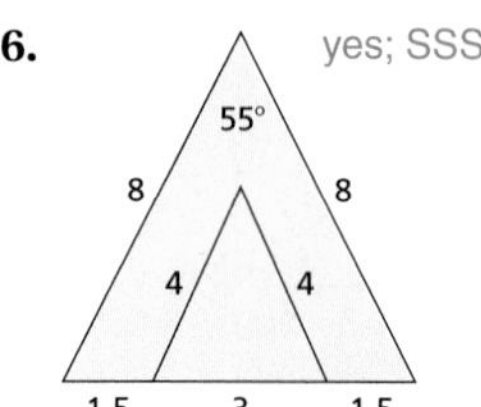

7. **SURVEYING** A surveyor measures the shadows cast by a tree and a pole at 4 P.M. and makes the drawing at the left. Explain why $\triangle ABC \sim \triangle DEF$.

The rays of the sun form the same angle with the ground for both the pole and tree. Therefore, $\angle A \cong \angle D$. Because $\angle B$ and $\angle E$ are both right angles, $\angle B \cong \angle E$. Therefore, the triangles are similar by AA Similarity.

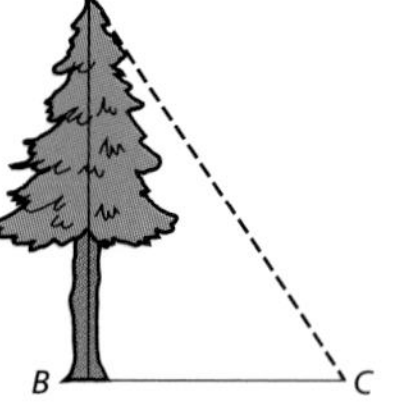

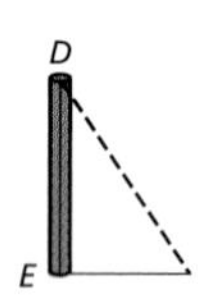

312 Chapter 7 Similar Triangles

Teaching Strategies

For Exercises 3, 5, 8 and 9, students may find it helpful to redraw the large and small triangles separately. This is also a useful strategy when working with similar triangles that form two vertical angles as in Example 3 of this lesson.

8. **PHOTOGRAPHY** For an exhibit, Bruce crops a photo in the shape of $\triangle PQR$. He wants to create a montage of smaller photos in the shape of similar triangles. To find a similar triangle, he marks S, the midpoint of PQ, and T, the midpoint of PR. Then he connects the midpoints. Is $\triangle PQR \sim \triangle PST$? Explain. yes; SAS

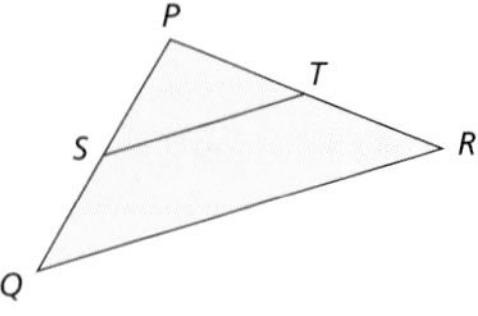

9. Suppose $\angle 1 \cong \angle 2$. Can you prove that $\triangle ABD \sim \triangle CED$? Explain. See additional answers.

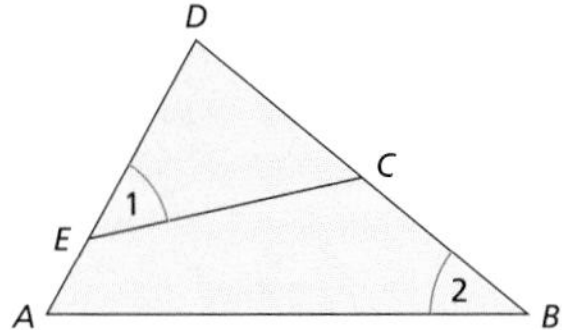

10. **WRITING MATH** Write a paragraph to prove that $\triangle DEF \sim \triangle FHG$? See additional answers.

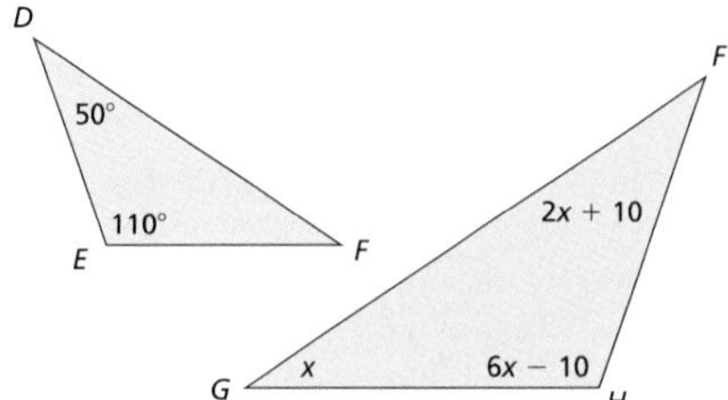

Extended Practice Exercises

Tell whether each statement is always true, sometimes true, or never true. Write *always, sometimes,* or *never.*

11. Two equilateral triangles are similar. always
12. Two isosceles triangles are similar. sometimes
13. Two isosceles triangles that each have a 45° angle are similar. sometimes
14. An acute triangle and a right triangle are similar. never

Mixed Review Exercises

Write the equation of the line with the given information. (Lesson 6-3)

15. $m = \frac{1}{2}, b = -1$ $y = \frac{1}{2}x - 1$
16. $P(-3, 1), Q(4, -2)$ $y = \frac{-3}{7}x - \frac{2}{7}$
17. $m = 1, b = \frac{1}{2}$ $y = x + \frac{1}{2}$
18. $m = -3, b = -2$ $y = 3x - 2$
19. $A(4, -5), B(-3, 4)$ $y = \frac{-9}{7}x + \frac{1}{7}$
20. $m = \frac{3}{4}, b = -3$ $y = \frac{3}{4}x - 3$
21. $m = 2, b = 4$ $y = 2x + 4$
22. $R(1,4), S(-2, -3)$ $y = \frac{7}{3}x + \frac{5}{3}$
23. $m = -2, b = -2$ $y = -2x - 2$

Trapezoids and their medians are shown. Find the value of x. (Lesson 4-9)

24.

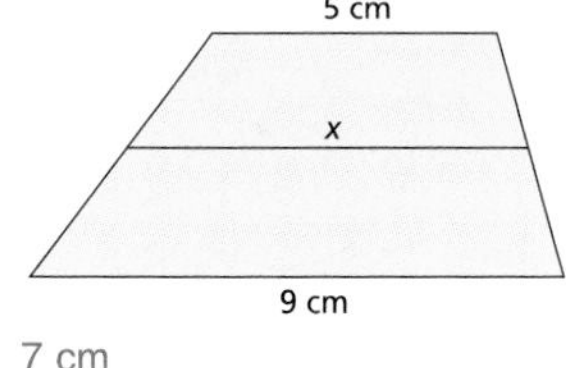

7 cm

25.

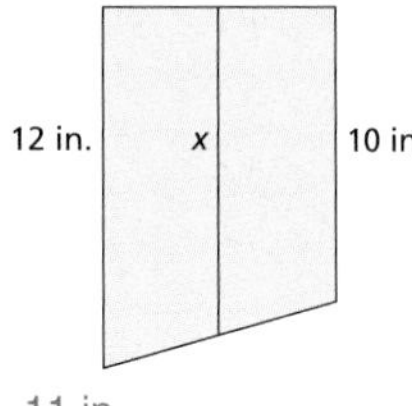

11 in.

26.

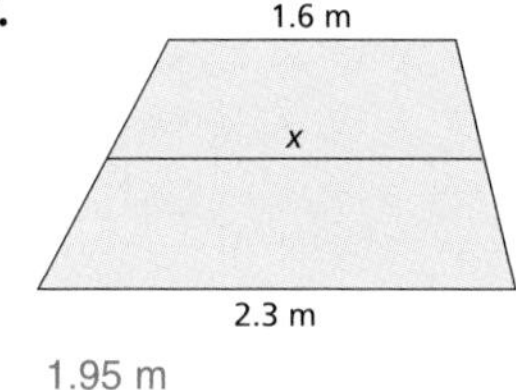

1.95 m

Extra Practice Worksheet 7-4

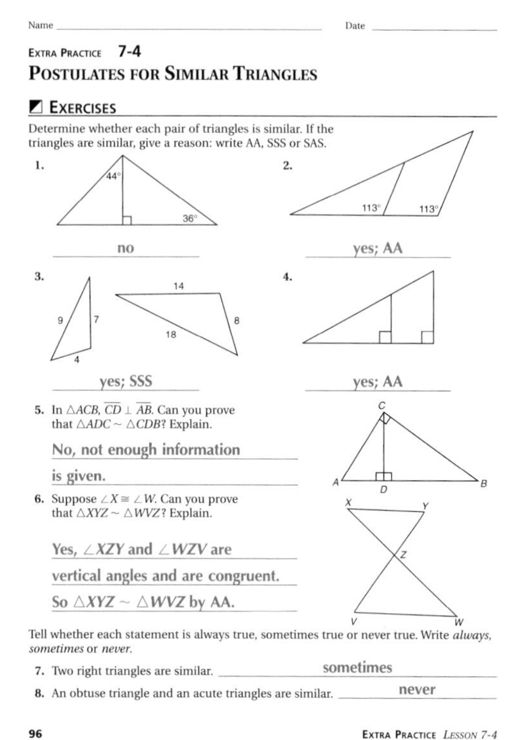

Name ____________ Date ____________

EXTRA PRACTICE 7-4

POSTULATES FOR SIMILAR TRIANGLES

EXERCISES

Determine whether each pair of triangles is similar. If the triangles are similar, give a reason: write AA, SSS or SAS.

1. (44°, 36°) no
2. (113°, 113°) yes; AA
3. (9, 7, 4; 14, 8, 18) yes; SSS
4. yes; AA
5. In $\triangle ACB$, $\overline{CD} \perp \overline{AB}$. Can you prove that $\triangle ADC \sim \triangle CDB$? Explain. No, not enough information is given.
6. Suppose $\angle X \cong \angle W$. Can you prove that $\triangle XYZ \sim \triangle WVZ$? Explain. Yes, $\angle XZY$ and $\angle WZV$ are vertical angles and are congruent. So $\triangle XYZ \sim \triangle WVZ$ by AA.

Tell whether each statement is always true, sometimes true or never true. Write *always, sometimes* or *never.*

7. Two right triangles are similar. sometimes
8. An obtuse triangle and an acute triangles are similar. never

96 EXTRA PRACTICE *LESSON 7-4*

Extend the Lesson

MATH JOURNAL Have students write the AA, SSS, and SAS Similarity Postulates in their journals and draw a figure for each.

Enrichment Worksheet 7-4

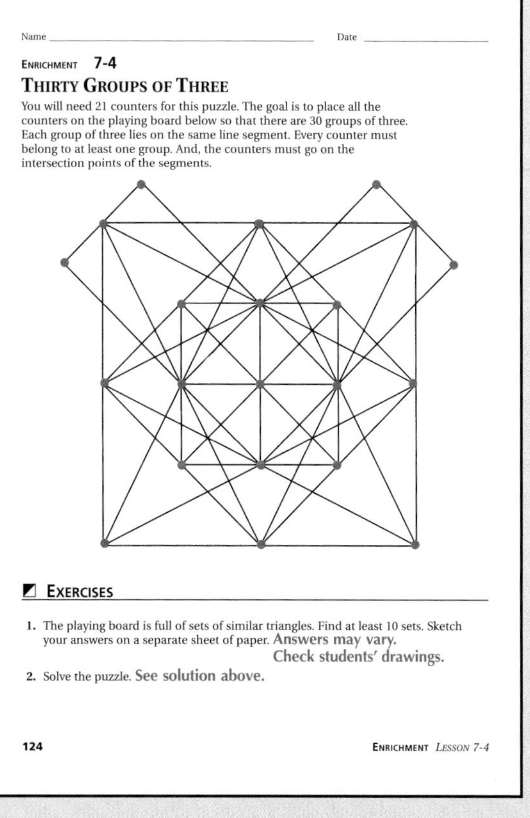

Name ____________ Date ____________

ENRICHMENT 7-4

THIRTY GROUPS OF THREE

You will need 21 counters for this puzzle. The goal is to place all the counters on the playing board below so that there are 30 groups of three. Each group of three lies on the same line segment. Every counter must belong to at least one group. And, the counters must go on the intersection points of the segments.

EXERCISES

1. The playing board is full of sets of similar triangles. Find at least 10 sets. Sketch your answers on a separate sheet of paper. Answers may vary. Check students' drawings.
2. Solve the puzzle. See solution above.

124 ENRICHMENT *LESSON 7-4*

Skills Practice

Vocabulary Review

Lesson 7-3
scale drawing
scale distance
actual distance
model

Lesson 7-4
Similarity Postulates:
AA, SAS, SSS

Assignment Guide

All students: 1–37
Additional Practice: Refer to the Extra Practice Index on page 674 of this text.

Review and Practice Your Skills

Practice ■ Lesson 7-3

Find the actual length of each of the following.

1. scale length is 40 cm 240 km
   scale is 2.5 cm: 15 km
2. scale length is $4\frac{3}{4}$ in.
   scale is $\frac{1}{4}$ in.:3 mi 57 mi
3. scale length is 14 ft
   scale is 700:1
   0.02 ft
4. scale length is 12.5 cm
   scale is 1:30 375 cm
5. scale length is $\frac{7}{16}$ in.
   scale is $\frac{1}{8}$ in.:15 mi 172.5 mi
6. scale length is 23 cm
   scale is 5 cm:28 m
   128.8 m
7. scale length is 37 yd 5.55 yd
   scale is 10 yd: $\frac{3}{2}$ yd
8. scale length is 1440 mm
   scale is 1 mm:0.001 m
   1.44 m
9. scale length is 6.5 in.
   scale is 1.5 in.:237 mi
   1027 mi

Find the scale length for each of the following.

10. actual length is 52 mi
    scale is 0.5 in.:4 mi 6.5 in.
11. actual length is 75 yd
    scale is 3 in.:18 yd 12.5 in.
12. actual length is 2450 mi
    scale is $\frac{3}{4}$ in.:5 mi 367.5 in.
13. actual length is 256 km
    scale is 5 cm:32 km 40 cm
14. actual length is 17,500 m
    scale is 2 cm:875 m 40 cm
15. actual length is 0.003 mm
    scale is 5000:1 15 mm
16. actual length is 817 mi
    scale is 4 in:19 mi 172 in.
17. actual length is 26 ft
    scale is 1:6.5 4 ft
18. actual length is 7500 mi
    scale is 10 in.:1.5 mi
    50,000 in.
19. The blueprints for a new house have a scale of $\frac{1}{2}$ in.:1.5 ft. The dimensions of one bedroom on the drawing are 6 in. by 4 in. What are the actual dimensions of the bedroom? 18 ft by 12 ft

Practice ■ Lesson 7-4

Determine whether each pair of triangles is similar. If the triangles are similar, give a reason: write AA, SSS, or SAS.

20. yes, SAS

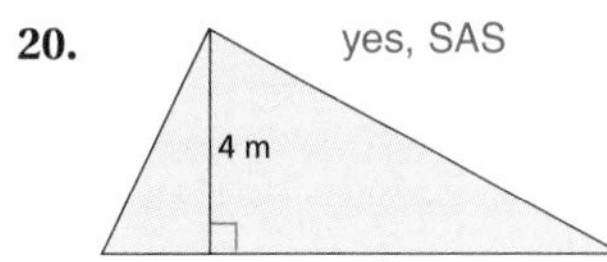

21. yes, AA

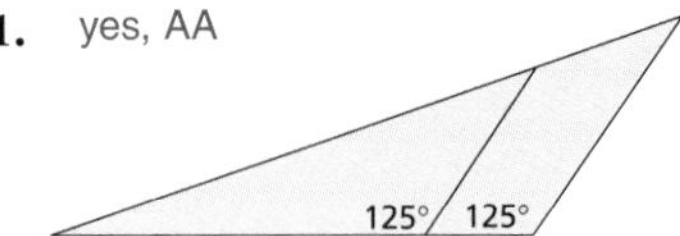

22.

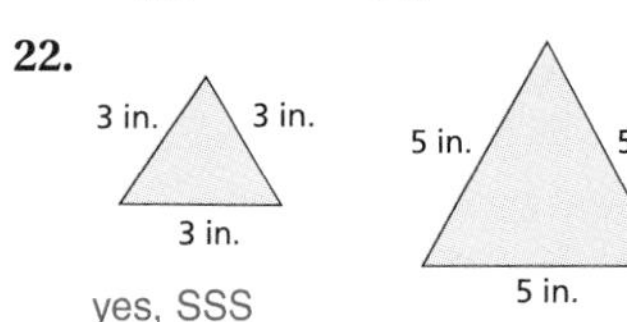

yes, SSS

23. no

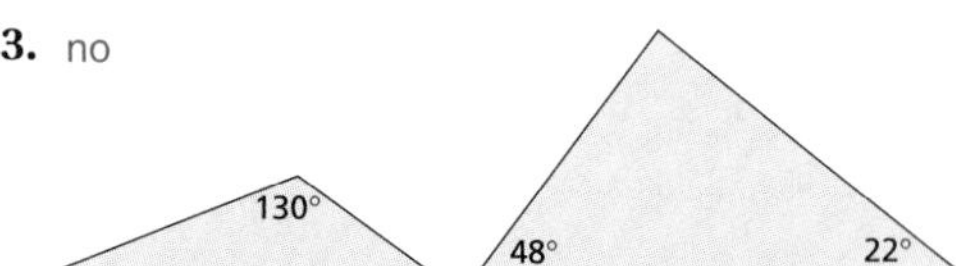

24.

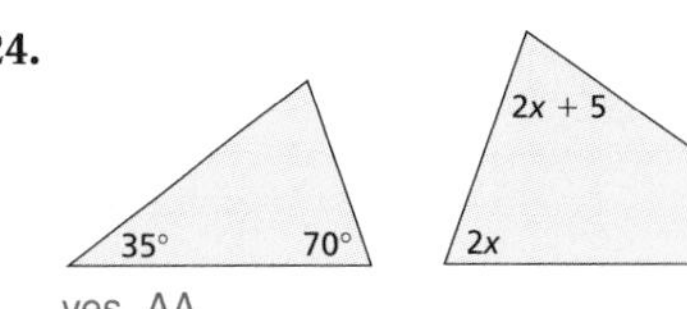

yes, AA

25.

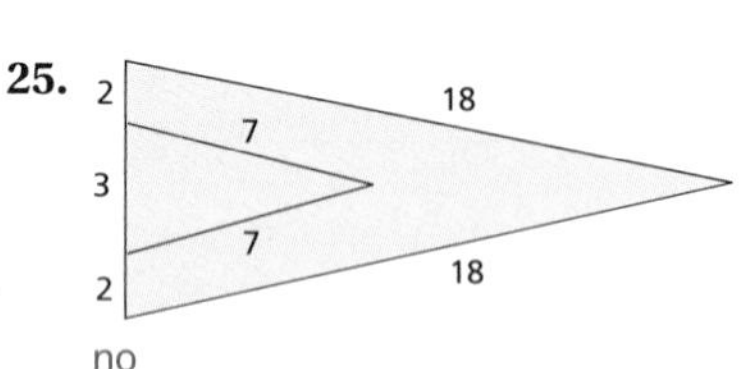

no

Teaching Strategies

CHALLENGE $\triangle ABC \sim \triangle RST$. Write and solve a proportion to find *BC, ST,* and *RT*.

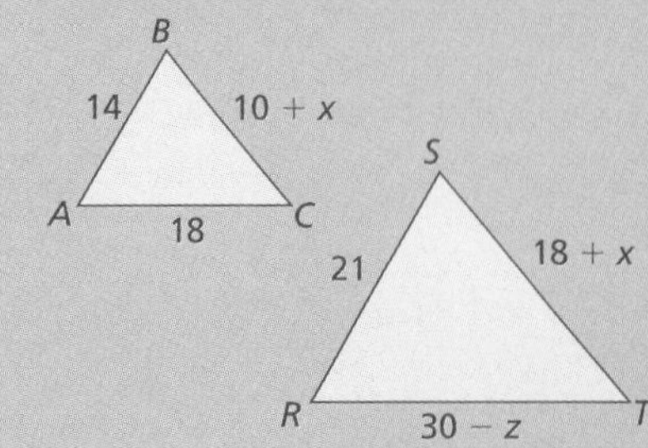

$\frac{14}{21} = \frac{(10 + x)}{(18 + x)}$, $x = 6$, $BC = 16$, $ST = 24$;

$\frac{14}{21} = \frac{18}{(30 - z)}$, $z = 3$, $RT = 27$

PRACTICE ▶ LESSON 7-1–LESSON 7-4

Solve each proportion. (Lesson 7-1)

26. $\frac{x}{4} = \frac{55}{22}$ 10 **27.** $\frac{156}{a} = \frac{-8}{3}$ −58.5 **28.** $-\frac{31}{112} = \frac{y}{-7}$ 1.9375

29. $\frac{0.6}{5} = \frac{0.72}{m}$ 6 **30.** $\frac{x+1}{36} = \frac{1}{3}$ 11 **31.** $\frac{x-1}{8} = \frac{65}{104}$ 6

32. A photograph that measures 4 in. by 6 in. is enlarged so that the 4 in. side measures 15 in. How long is the 6 in. side in the enlargement? (Lesson 7-2) 22.5 in.

33. *ABCD* and *JKLM* are similar rectangles. If *BC* = 22 cm, *CD* = 42 cm, and *KL* = 165 cm, what is the perimeter of *JKLM*? (Lesson 7-2) 960 cm

Find the actual distance using the map. (Lesson 7-3)

34. Clarktown to Pinckney ~36 miles

35. Pinckney to Grove City ~40 miles

36. Clarktown to Dwyer ~60 miles

37. Dwyer to Gurville ~56 miles

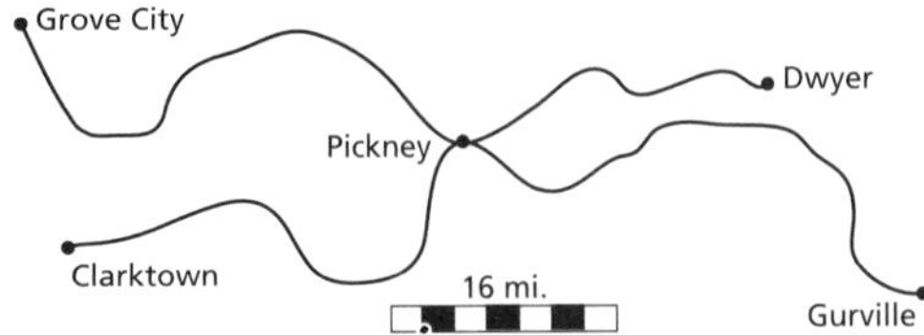

Mid-Chapter Quiz

Solve.

1. $\frac{0.45}{x} = \frac{0.9}{0.2}$ 0.1 **2.** $\frac{42}{112} = \frac{x}{24}$ 9

3. Tatiana and Simon bought art supplies. Tatiana spent $3.00 for every $2.00 Simon spent. If they spent $76.25 in all, how much did each one spend? $45.75, $30.50

4. Triangles *ABC* and *DEF* are similar. If $m\angle B = 55°$ and $m\angle C = 98°$, what is $m\angle D$? 27°

5. Rectangles *PQRS* and *ABCD* are similar. Find *QR* if *AB* = 72 cm, *BC* = 30 cm, and *RS* = 6 cm. 2.5 cm

6. A scale map shows a distance of 7.3 cm. The scale is 1:400. What is the actual distance in meters? 29.2 m

7. Find the scale length when the actual length is 3.75 km and the scale is 2 cm:15 km. 0.5 cm

If you can determine from the given information that the triangles are similar, write *yes*, and give a reason. Otherwise, write *no*.

8. $\triangle ABC$ has an altitude *BD* such that point *D* is between points *A* and *C*. If *AC* = 3, *BD* = 4, and $DC = 5\frac{1}{3}$, is $\triangle ADB \sim \triangle BDC$? yes; SAS

9. For $\triangle LMN$, $m\angle M = 105°$, *LM* = 15, *MN* = 8, and *LN* = 12. For $\triangle PQR$, $m\angle Q = 105°$, *PQ* = 45, *QR* = 24, and *PR* = 36. Are the triangles similar? yes, SSS or SAS

Chalkboard Examples

Example from Lesson 6-3

A map has a scale of 2 cm : 5 km. What is the actual distance if the scale distance is 4.1 cm? **10.25 km**

Example from Lesson 6-4

Determine whether this pair of triangles is similar. If the triangles are similar, give a reason: write AA, SSS, or SAS.

$\triangle XYZ$: $m\angle X = 47°$, $m\angle Z = 56°$

$\triangle PQR$: $m\angle Q = 77°$, $m\angle R = 56°$

yes; AA

Mid-Chapter Quiz

Correlation Chart

Question Number	Lesson Number
1–3	7-1
4–5	7-2
6–8	7-3
9–10	7-4

Extend the Lesson

ONGOING ASSESSMENT Have students solve the following problem and write a brief explanation of each step.

A map has a scale of $\frac{3}{4}$ in. : 8 mi. The closest town to the highway on the map is $2\frac{1}{4}$ in. from the highway. If Pasha travels at a rate of 40 mph, how long will it take her to reach the town? **$\frac{3}{5}$ h or 36 min**

Lesson Planning

NCTM Standards/Strands

- Number & Operation
- Patterns, Functions, & Algebra
- Geometry & Spatial Sense
- Measurement
- Problem Solving
- Reasoning and Proof
- Connections
- Representation

Vocabulary

theorem, midpoint, altitude, median

Materials Needed

calculators, compasses, straightedges, geometric drawing software

Lesson Resources

Warm-Up Transparency 20
Reteaching 7-5
Extra Practice 7-5
Enrichment 7-5

Assignment Guide

Basic: 1–10, 14–20
Enriched: 1–20

Getting Started

5-Minute Warm-Up

$\triangle RST \sim \triangle XYZ$. List corresponding congruent angles, and write equivalent ratios of corresponding sides. $\angle R \cong \angle X$, $\angle S \cong \angle Y$, $\angle T \cong \angle Z$, $\frac{RS}{XY} = \frac{ST}{YZ} = \frac{RT}{XZ}$

Introduction to Lesson 7-5

Students should conclude that $\triangle XYZ \sim \triangle XPQ$. Ask them to prove that the triangles are similar. Have them explain how they compare YZ to PQ. $\frac{XP}{XY} = \frac{XP}{(2XP)} = \frac{1}{2}$; $\frac{XQ}{XZ} = \frac{XQ}{(2XQ)} = \frac{1}{2}$, and $\angle X \cong \angle X$; $\triangle XYZ \sim \triangle XPQ$ by SAS; because the sides of similar triangles are proportional, $\frac{PQ}{YZ} = \frac{1}{2}$.

7-5 Triangles and Proportional Segments

Goals
- Prove theorems involving similar triangles.
- Find unknown lengths of sides of triangles.

Applications Scale models, Photography

You will need a compass and straightedge.

Draw any triangle XYZ. Construct the midpoint of $\overline{XY}$ and label it P. Construct the midpoint of $\overline{XZ}$ and label it Q. Then, use the straightedge to draw PQ.

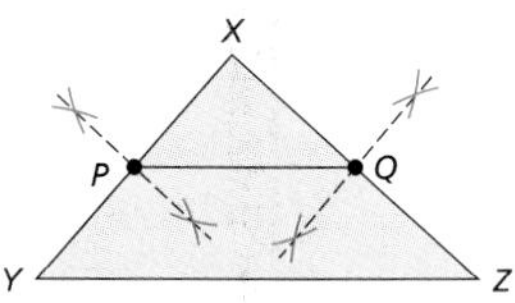

How does $\triangle XYZ$ compare to $\triangle XPQ$? How does the measure of $\overline{YZ}$ compare with the measure of $\overline{PQ}$?

Build Understanding

Recall that a theorem is a statement that can be proven true. In Example 1, a proof is given for the following theorem.

Theorem If a segment connects the midpoints of two sides of a triangle, then the length of the segment is equal to one-half the length of the third side.

Technology Note

Explore this theorem using geometric software. Try this activity.

1. Draw any triangle and label the vertices *A*, *B* and *C*.
2. Construct midpoints of segments *AB* and *BC* and label them *D* and *E*. Connect the midpoints.
3. Measure *AC* and *DE*.
4. Change the measure of the segments and angles by selecting and moving a vertex of $\triangle ABC$. What do you notice about the measures of *AC* and *DE*?

Example 1

Given P is the midpoint of $\overline{XY}$

Q is the midpoint of $\overline{XZ}$

Prove $PQ = \frac{1}{2}YZ$

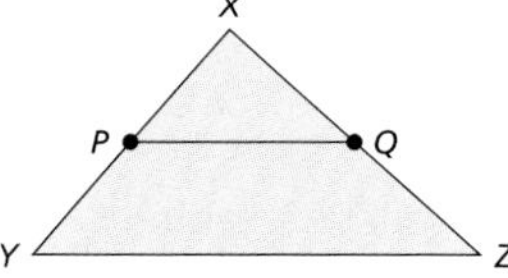

Solution

Statements	Reasons
1. P is the midpoint of XY; Q is the midpoint of XZ	1. given
2. $XP = \frac{1}{2}XY$; $XQ = \frac{1}{2}XZ$	2. definition of midpoint
3. $\frac{XP}{XY} = \frac{1}{2}$; $\frac{XQ}{XZ} = \frac{1}{2}$	3. division property of equality
4. $m\angle X = m\angle X$	4. reflexive property of equality
5. $\angle X \cong \angle X$	5. definition of congruent angles
6. $\triangle XYZ \cong \triangle XPQ$	6. SAS similarity postulate
7. $\frac{PQ}{YZ} = \frac{1}{2}$	7. corresponding parts of similar triangles are proportional
8. $PQ = \frac{1}{2}YZ$	8. multiplication property of equality

Teaching Strategies

ESL/LEP Some students may find the many terms in this lesson confusing. List the terms median and midpoint (middle), theorem (rule), altitude (height), perpendicular (90°), base (bottom), right triangles (90°), and hypotenuse. Draw a diagram to represent each. Have students take turns using each term in a statement (e.g., an altitude is perpendicular to the base).

There is also a theorem about altitudes of similar triangles.

Theorem	If two triangles are similar, their altitudes are in the same proportion as the sides of the triangles.

Example 2

SCALE MODELS Jan is building a scale model of a tower. Steel bracing forms large and small similar triangles throughout the structure. Jan believes that she will not need to measure the altitudes of all triangles in her model since the altitudes should be in the same proportion as the sides of the triangles. To be certain, she draws $\triangle ABC$ and $\triangle DEF$ and proves the theorem.

Given $\triangle ABC \sim \triangle DEF$
$\overline{AX} \perp \overline{BC}$, $\overline{DY} \perp \overline{EF}$

Prove $\frac{AX}{DY} = \frac{AB}{DE}$

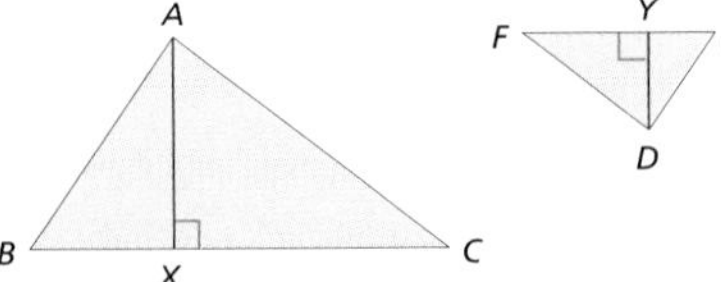

Solution

Statements	Reasons
1. $\triangle ABC \sim \triangle DEF$, $\overline{AX} \perp \overline{BC}$, $\overline{DY} \perp \overline{EF}$	1. given
2. $\angle AXB$ and $\angle DYE$ are right angles	2. definition of perpendicular lines
3. $m\angle AXB = 90$, $m\angle DYE = 90$	3. definition of right angles
4. $m\angle AXB = m\angle DYE$	4. substitution
5. $\angle AXB \cong \angle DYE$	5. definition of congruent angles
6. $\angle ABC \cong \angle DEF$	6. corresponding angles of similar triangles are congruent
7. $\triangle ABX \sim \triangle DEY$	7. AA similarity postulate
8. $\frac{AX}{DY} = \frac{AB}{DE}$	8. corresponding parts of similar triangles are in proportion

There is a similar theorem about the medians of similar triangles.

Theorem	If two triangles are similar, their medians are in the same proportion as the sides of the triangles.

In $\triangle ABC$ and $\triangle PQR$, median $\overline{AD}$ and median $\overline{PS}$ have the same ratio as any corresponding sides of the triangles.

$$\frac{AD}{PS} = \frac{AC}{PR}$$

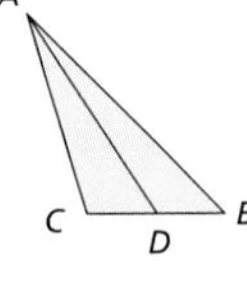

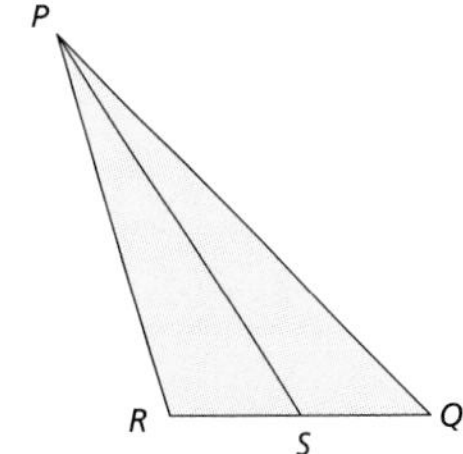

An altitude drawn to the hypotenuse of any right triangle always forms two similar triangles. The following theorem is used in Example 3 below to find the length of a missing segment.

Extend the Lesson

MATH JOURNAL Have students write the four theorems in their journals and give an example of each.

Chalkboard Examples

Supplementary Example 1

Draw the following figure on the chalkboard.

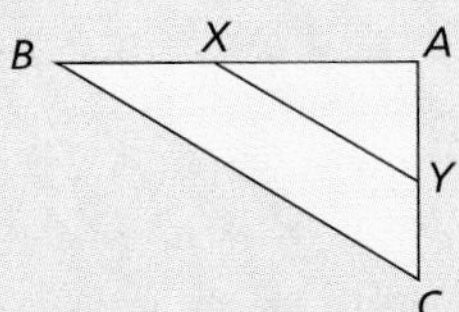

Given: X is the midpoint of AB
Y is the midpoint of AC

Prove: $XY = \frac{1}{2}BC$

Answers may vary. Students' steps should be similar to those shown in Example 1.

Supplementary Example 2 and 3

$\triangle DEF \sim \triangle GHI$

$DE = 2.4$ cm, $DF = 3.5$ cm and $GI = 12.25$ cm. What is the measure of GH? **8.4 cm**

Reteaching Worksheet 7-5

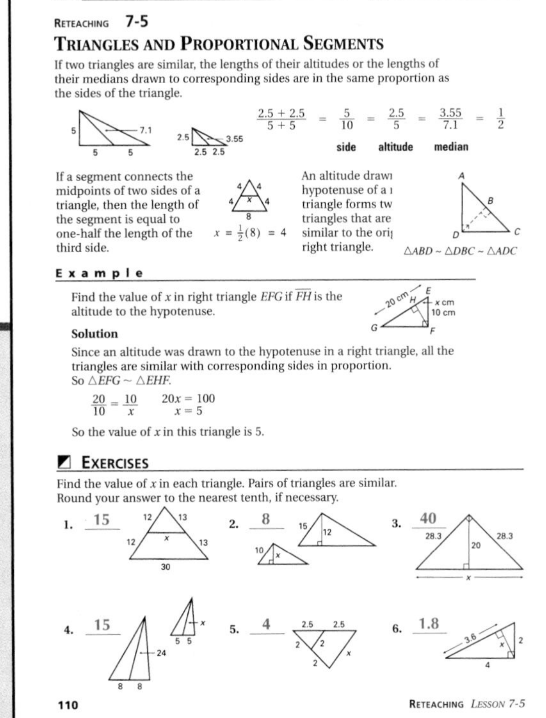

Name ______________________ Date ____________

RETEACHING 7-5

TRIANGLES AND PROPORTIONAL SEGMENTS

If two triangles are similar, the lengths of their altitudes or the lengths of their medians drawn to corresponding sides are in the same proportion as the sides of the triangle.

$$\frac{2.5 + 2.5}{5 + 5} = \frac{5}{10} = \frac{2.5}{5} = \frac{3.55}{7.1} = \frac{1}{2}$$

side altitude median

If a segment connects the midpoints of two sides of a triangle, then the length of the segment is equal to one-half the length of the third side. $x = \frac{1}{2}(8) = 4$

An altitude drawn to the hypotenuse of a right triangle forms two triangles that are similar to the original right triangle. $\triangle ABD \sim \triangle DBC \sim \triangle ADC$

Example

Find the value of x in right triangle EFG if $\overline{FH}$ is the altitude to the hypotenuse.

Solution

Since an altitude was drawn to the hypotenuse in a right triangle, all the triangles are similar with corresponding sides in proportion.
So $\triangle EFG \sim \triangle EHF$.

$\frac{20}{10} = \frac{10}{x}$ $20x = 100$, $x = 5$

So the value of x in this triangle is 5.

EXERCISES

Find the value of x in each triangle. Pairs of triangles are similar. Round your answer to the nearest tenth, if necessary.

1. 15
2. 8
3. 40
4. 15
5. 4
6. 1.8

110 RETEACHING *LESSON 7-5*

Lesson Wrap-up

Quick Assessment

Ask the following questions to determine if students understand the content presented in this lesson.

1. Two segments, AB and CD, trisect sides XY and YZ of $\triangle XYZ$. Find the length of AB and CD.
   $AY = \frac{1}{3}XY,\ BY = \frac{1}{3}ZY;$
   $\frac{AY}{XY} = \frac{\left(\frac{1}{3}\right)XY}{XY} = \frac{1}{3},\ \frac{BY}{ZY} = \frac{\left(\frac{1}{3}\right)ZY}{ZY} = \frac{1}{3};$
   $\angle Y \cong \angle Y;\ \triangle AYB \sim \triangle XYZ$ by SAS; so $\frac{AB}{XZ} = \frac{1}{3},\ AB = \frac{1}{3}XZ$, similarly $CD = \frac{2}{3}XZ$.
2. In Example 3, list the three pairs of similar triangles that are formed by the altitude SW in right triangle RST. $\triangle RWS \sim \triangle RST,\ \triangle SWT \sim \triangle RST,\ \triangle RWS \sim \triangle SWT$

Assignment Guide

Basic: 1–10, 14–20
Advanced: 1–20
Additional Practice: See Extra Practice Index on page 674.

Additional Answers

9. The given triangles are similar, so $RW{:}AK = RY{:}AL$. Because RX is half the length of RY, and AB is half the length of AL, then RX and AB have the same ratio, $\angle R \cong \angle A$, because they are corresponding parts of similar triangles. You now have $\triangle WRX \sim \triangle KAB$ by SAS Similarity Postulate. Then, $WX{:}KB = WR{:}KA$, because they are corresponding parts of similar triangles.
10. A diagonal in a rectangle divides the rectangle into two triangles. The corresponding sides of the rectangle are proportional and all its angles are congruent. The diagonals can be shown to be proportional to the sides because they are corresponding parts of two triangles that are similar by SAS.

Theorem	If the altitude to the hypotenuse of a right triangle is drawn, the altitude separates the original triangle into two triangles that are similar to the original triangle and to each other.

In right triangle ABC, $\overline{AD}$ is the altitude to the hypotenuse. Each of these pairs of triangles is similar.

$\triangle DAC \sim \triangle ABC$
$\triangle DBA \sim \triangle ABC$
$\triangle DAC \sim \triangle DBA$

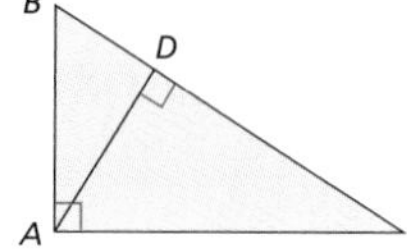

Example 3

Find x in right triangle RST if $\overline{SW}$ is the altitude to the hypotenuse.

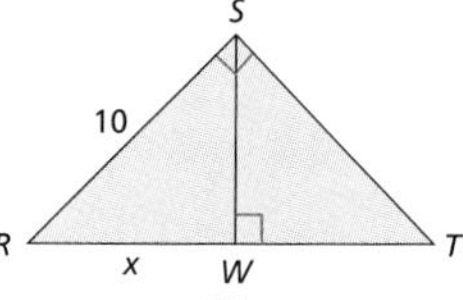

Solution

Identify which two of the three triangles include x. You may find it helpful to redraw the two triangles separately.

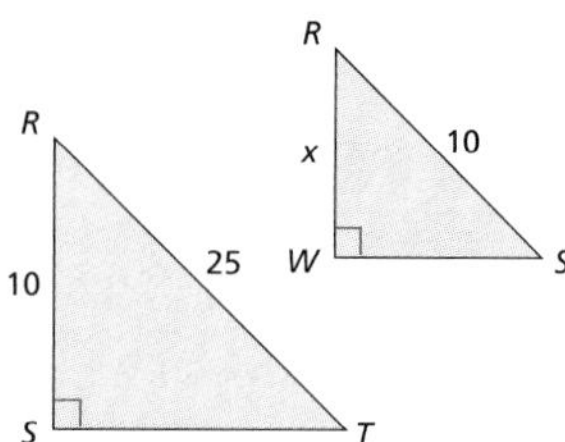

Because $\triangle RWS \sim \triangle RST$,

$$\frac{RW}{RS} = \frac{RS}{RT}$$
$$\frac{x}{10} = \frac{10}{25}$$
$$x = 4$$

So, the value of x in this triangle is 4.

Try These Exercises

Find x in each pair of similar triangles to the nearest tenth.

1.
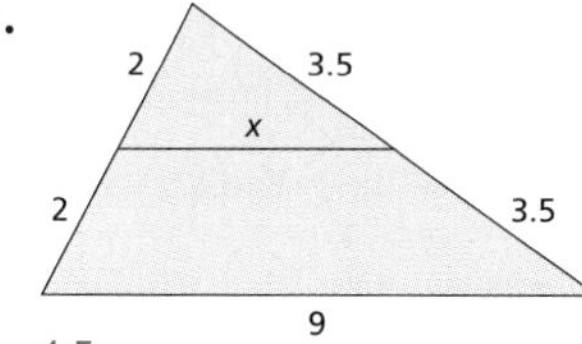

4.5

2.
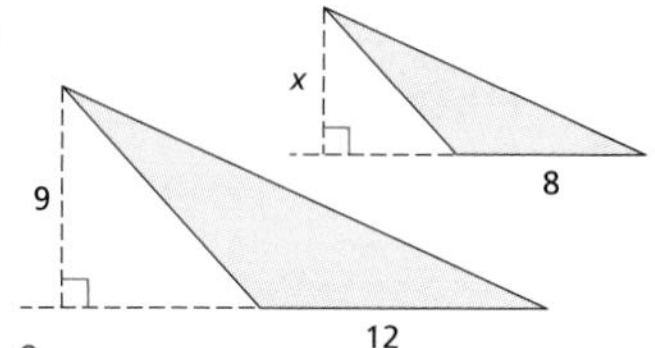

6

3.
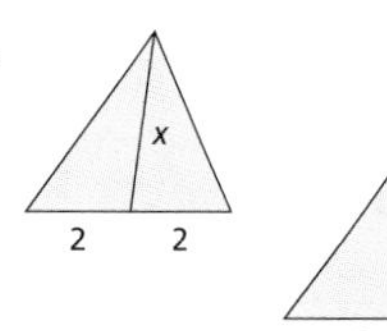

3.3

4.
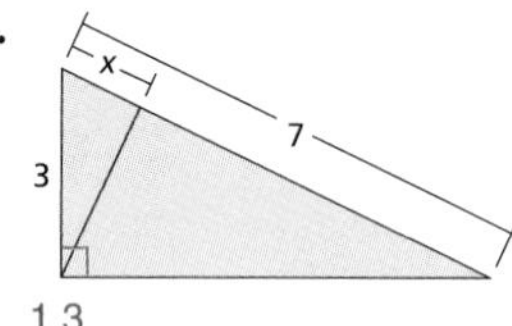

1.3

12. Answers may vary. Each face of the prism should be similar to the corresponding face in the similar prism. Corresponding edges should be proportional. Because both figures are rectangular prisms, all angles will be congruent.

Practice Exercises

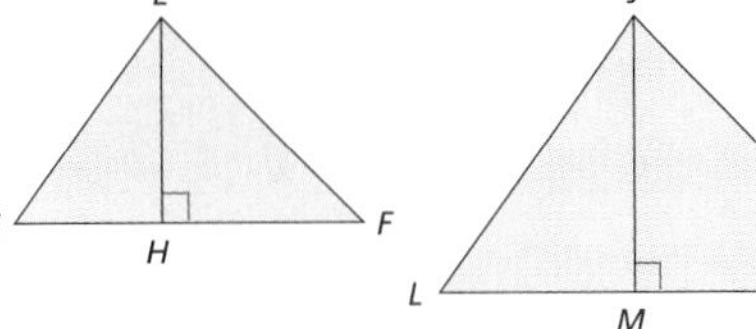

5. Copy and complete this proof.

Given $\triangle EFG \sim \triangle JKL$, $\overline{EH} \perp \overline{GF}$, $\overline{JM} \perp \overline{LK}$

Prove $\frac{EH}{JM} = \frac{FG}{KL}$

Statements	**Reasons**
1. $\triangle EFG \sim \triangle JKL$; $\overline{EH} \perp \overline{GF}$; $\overline{JM} \perp \overline{LK}$	1. __?__ given
2. $\overline{EH}$ is an altitude of $\triangle EFG$. $\overline{JM}$ is an altitude of $\triangle JKL$	2. __?__ definition of Altitude
3. $\frac{EH}{JM} = \frac{FG}{KL}$	3. __?__ altitudes of similar triangles in the same proportion as corresponding sides

Find x in each pair of similar triangles to the nearest tenth.

6.  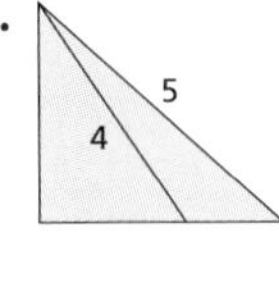 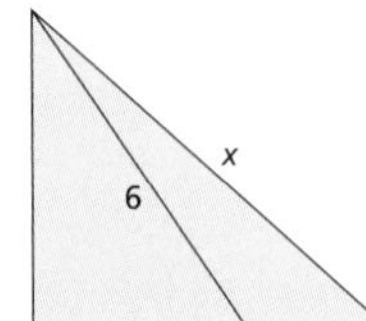

7.5

7. 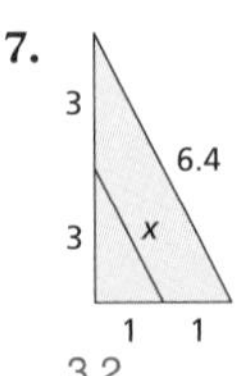

3.2

8. 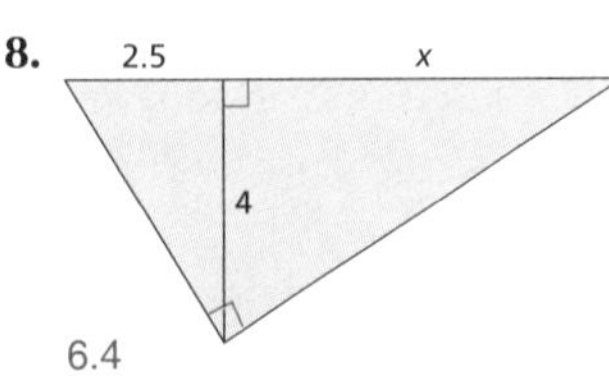

6.4

9. Prove the theorem stated after Example 2, about medians in similar triangles.
Given $\triangle WRY \sim \triangle KAL$, $\overline{WX}$ and $\overline{KB}$ are medians. **Prove** $\frac{WX}{KB} = \frac{WY}{KA}$ See additional answers.

10. **WRITING MATH** If two rectangles are similar, do you think their diagonals are proportional to corresponding sides? Explain your thinking.
See additional answers.

Extended Practice Exercises

11. $\triangle ABC$ has a base of x and a height of y. $\triangle DEF$ is similar. $AB{:}DE = 2{:}7$. What is the area of $\triangle DEF$ in terms of x and y?
$6.125\,xy$

12. **PHOTOGRAPHY** A photographer wants to create a series of similar rectangular prisms to display her work in an exhibit? How could the artist determine that two rectangular prisms are similar? See additional answers.

13. Two square pyramids are similar. One has a base of 3 cm and a height of 10 cm. The other's base is 10 cm. Find its height?
33.3 cm

Mixed Review Exercises

Solve each system of equations by graphing. (Lesson 6-4)

14. $y = 3x - 2$, $y = x + 5$ (3.5, 8.5)

15. $y = 2x + 2$, $y = -x - 3$ $\left(\frac{-5}{3}, \frac{-4}{3}\right)$

16. $2x - y = 4$, $3x + y = -1$ $\left(\frac{3}{5}, \frac{-14}{5}\right)$

Solve each equation. (Lesson 2-5)

17. $4(x - 3) + 8 = 3x + 2$ 6

18. $2(x - 5) - 4 = 3(x - 1)$ −11

19. $x - 3(x + 2) = 5(x - 2) + 1$ $\frac{3}{7}$

20. $-2(x - 4) = 5(x - 2) + 8$ $\frac{10}{7}$

Teaching Strategies

CHALLENGE Give students this theorem: The medians of a triangle intersect in one point called the centroid, which is $\frac{2}{3}$ of the distance from each vertex to the midpoint of the opposite side. Have them solve the following problem: $\triangle ABC \sim \triangle DEF$. The centroid of $\triangle ABC$ is P, and the centroid of $\triangle DEF$ is O. If $BC = 5$, $EF = 10$, and $EO = 5$, what is the length of the median from vertex E to side DF? What is the length of the median from vertex B to side AC?

Let x = median from E to DF: $\left(\frac{2}{3}\right)x = 5$, $x = 7.5$; $\frac{\frac{7}{5}}{\text{(median from B to AC)}} = \frac{10}{5}$, median = 3.75 or $\frac{5}{BP} = \frac{10}{5}$, $BP = 2.5$, $2.5 = \left(\frac{2}{3}\right)$median, 3.75 = median.

Extra Practice Worksheet 7-5

Name ______________________ Date __________

Extra Practice 7-5

Triangles and Proportional Segments

Exercises

Find x in each pair of similar triangles to the nearest tenth.

1. 4
2. 4.9
3. 24.5
4. $5\frac{1}{3}$

In Exercises 5–7, $\triangle ABC$ is a right triangle with right angle C and altitude $\overline{CD}$.

5. If $AD = 4$ and $DB = 9$, find CD. 6
6. If $CA = 12$, $BC = 15$, and $AD = 8$, find CD. 10
7. If $CD = 16$ and $BD = 32$, find AD. 8
8. If two parallelograms are similar, do you think their diagonals are proportional to corresponding sides? Explain your thinking. Yes, because each diagonal forms two pairs of similar triangles.
9. If two squares are similar, do you think their diagonals are proportional to corresponding sides? Explain your thinking. Yes, because each diagonal forms two pairs of similar triangles.

98 Extra Practice *Lesson* 7-5

Enrichment Worksheet 7-5

Name ______________________ Date __________

Enrichment 7-5

Concurrence of Angle Bisectors

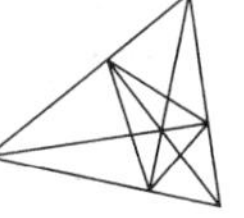

The angle bisectors of any triangle meet in a point, or are *concurrent*. The point is equidistant from each of the three sides.

This figure is constructed by first bisecting the angles of the inner triangle. Lines drawn perpendicular to these three bisectors form the sides of the outer triangle.

Exercises

1. Start with any scalene triangle. Construct the figure at the top of this page on your own paper. Check students' work.

2. How many pairs of similar triangles are there in your figure?
none
3. On your own paper, repeat the construction using the given isosceles triangle. Mark pairs of similar triangles.
Check students' work.

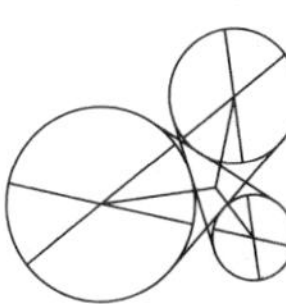

4. The design at the right is based on the figure at the top of this page. Construct the design on your own paper.
Check students' work.
5. Create a design of your own based on the figure at the top of this page. Use your own paper.
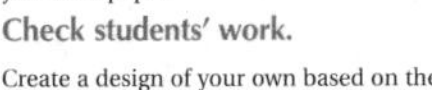
Check students' work.
6. Write a paragraph proof of this theorem on a separate sheet of paper.
Theorem: The angle bisectors of a triangle are concurrent in a point which is equidistant from the three sides.
Given: In any triangle ABC, let P be the intersection of the bisectors of angles A and B. Then P is in the interior of $\angle A$ and in the interior of $\angle B$ and is therefore in the interior of $\angle C$.
Any point on an angle bisector is equidistant from the two sides of the angle. So, P is equidistant from $\overline{AC}$ and $\overline{AB}$, and from $\overline{AB}$ and $\overline{BC}$. By the transitive property, P is equidistant from $\overline{AC}$ and $\overline{BC}$. P is on the bisector of $\angle C$ and the 3 bisectors are concurrent at P.

126 Enrichment *Lesson* 7-5

Lesson Planning

NCTM Standards/Strands

- Number & Operation
- Patterns, Functions, & Algebra
- Geometry & Spatial Sense
- Measurement
- Problem Solving
- Reasoning and Proof
- Connections
- Representation

Vocabulary

parallel	vertex
median	corresponding angles
transversal	trapezoid

Materials Needed

compasses straightedges

Lesson Resources

Warm-Up Transparency 21
Reteaching 7-6
Extra Practice 7-6
Enrichment 7-6
Multicultural Connection 5

Assignment Guide

Basic: 1-14, 19-26
Enriched: 1-26

Getting Started

5-Minute Warm-Up

Find the value of x to the nearest tenth.

1. $12 = \frac{(4 + x)}{2}$ 20
2. $\frac{(13 + 4)}{2} = x$ 8.5
3. $\frac{x}{18} = \frac{5}{7}$ 12.9

Introduction to Lesson 7-6

You may need to review with students how to construct a line through P that is parallel to YZ (e.g., by constructing an angle, $\angle XPQ$, congruent to $\angle XYZ$). Students should realize that $\triangle XYZ \sim \triangle XPQ$ (by AA), and therefore the given ratios are equivalent.

7-6 Parallel Lines and Proportional Segments

Goals
- Use theorems involving parallel lines and proportional segments to find unknown lengths.
- Divide a line segment into congruent parts.

Applications Model Building, Architecture, Real Estate

GEOMETRY SOFTWARE Use geometry software to construct similar triangles.

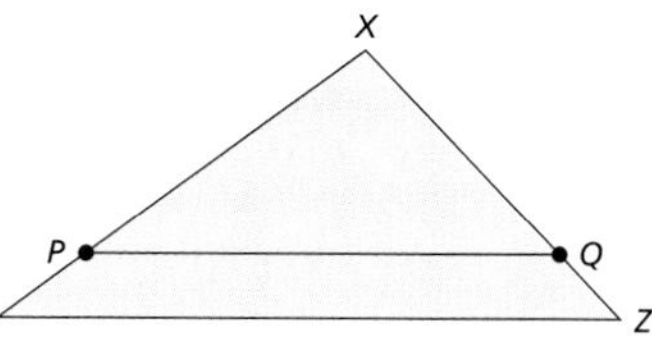

1. Draw any triangle XYZ. Mark any point P on $\overline{XY}$.
2. Construct a line through P that is parallel to $\overline{YZ}$. Label the point where the parallel line meets $\overline{XZ}$ as Q.
3. How can you use the software features to prove $\triangle XYZ \sim \triangle XPQ$?

Build Understanding

When a line segment intersects two sides of a triangle and is parallel to the third side, two similar triangles are formed.

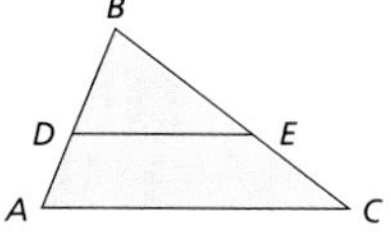

In $\triangle ABC$, $\overline{DE} \parallel \overline{AC}$. Notice that $\angle ABC \cong \angle DBE$ by the reflexive property, and $\angle BDE \cong \angle BAC$ because they are both corresponding angles of parallel lines. Therefore, $\triangle ABC \sim \triangle DBE$. Because the triangles are similar, $BD : BA = BE : BC$.

The same reasoning applies to a segment parallel to any side of the triangle, so you can state the following theorem:

Theorem	If a line is parallel to one side of a triangle and intersects the other sides at any points except the vertex, then the line divides the sides proportionally.

Example 1

Given $\triangle STR$, $\overline{WP} \parallel \overline{ST}$. Find the measure of x.

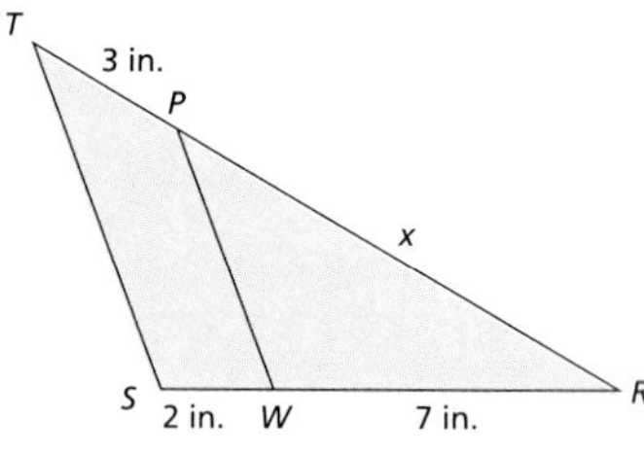

Solution

According to the theorem, because $\overline{WP} \parallel \overline{ST}$, P divides $\overline{TR}$ proportionally and W divides $\overline{SR}$ proportionally.

$$\frac{TP}{PR} = \frac{SW}{WR}$$

$$\frac{3}{x} = \frac{2}{7}$$

$$10.5 = x$$

So, the measure of x is 10.5 in.

Extend the Lesson

COOPERATIVE LEARNING Have students work in pairs to solve the following problem: Given right $\triangle ABC$ with $m\angle B = 90°$ and line segments DE and FG parallel to AB such that F and D are between A and C, and G and E are between B and C. $DE = 2.5$ cm, $CE = 2$ cm, and $CD = EG = GB$. Find the area of trapezoid $ABED$ and the area of $\triangle ABC$. (Hint: area of a trapezoid = [length of the median] × [length of the altitude]) 10 cm², 11.25 cm²

Recall that when a segment joins the midpoints of two sides of a triangle, that segment measures one-half the length of the third side of the triangle. Example 2 is a proof that such a segment is also parallel to the third side.

Example 2

MODEL BUILDING Len is building a miniature electrical tower for the filming of a movie. Working from a photograph, Len draws a diagram of the triangular tower. A catwalk, represented by $\overline{PQ}$ on the diagram shown to the right, seems to be parallel to the base of the tower.

Len measures to determine that the P is the midpoint of $\overline{XY}$ and Q is the midpoint of $\overline{XZ}$. How can he prove that $\overline{PQ}$ is parallel to $\overline{YZ}$?

Solution

As in the proof in Example 1 of Lesson 7–5, $\triangle XYZ \cong \triangle XPQ$ by the SAS postulate. Therefore, $\angle XPQ \cong \angle XYZ$, because they are corresponding angles of similar triangles. This fact also leads to the conclusion that $\overline{PQ} \parallel \overline{YZ}$, because corresponding angles formed by a transversal (XY) are congruent.

A similar theorem is true about medians in trapezoids. The **median** of a trapezoid is a segment that joins the midpoints of the legs.

Theorem	The median of a trapezoid is parallel to its bases, and its length is half the sum of the lengths of the bases.

Example 3

Given Trapezoid $WXYZ$, median $\overline{AB}$. Find AB.

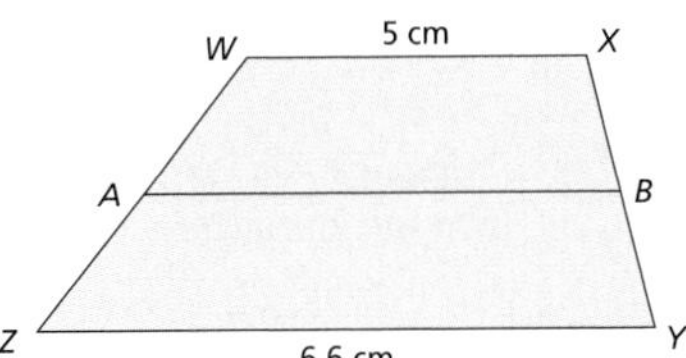

Solution

The length of the median is half the length of the sum of the bases.

$$AB = \frac{WX + YZ}{2}$$

$$AB = \frac{5 + 6.6}{2}$$

$$AB = \frac{11.6}{2} = 5.8$$

The length of the median is 5.8 cm.

Parallel segments can be used to divide a given segment into congruent parts using only a compass and straightedge.

Teaching Strategies

You may wish to suggest to students that a good way to check their answers when trying to find x is to set up a different proportion from the one they used. For example, given $\triangle CDE$ with $\overline{CD} \parallel \overline{AB}$, $AE = 4$, $BE = 3$, $BD = 9$, and $AC = x$, students may use any of the following ratios to find x: $\frac{x}{4} = \frac{9}{3}$, $\frac{4}{x} = \frac{3}{9}$, $\frac{4}{3} = \frac{x}{9}$, $\frac{3}{4} = \frac{9}{x}$.

Chalkboard Examples

Supplementary Example 1

Given: $\triangle ABC$, $\overline{AB} \parallel \overline{CD}$

Find the measure of x. 7.5

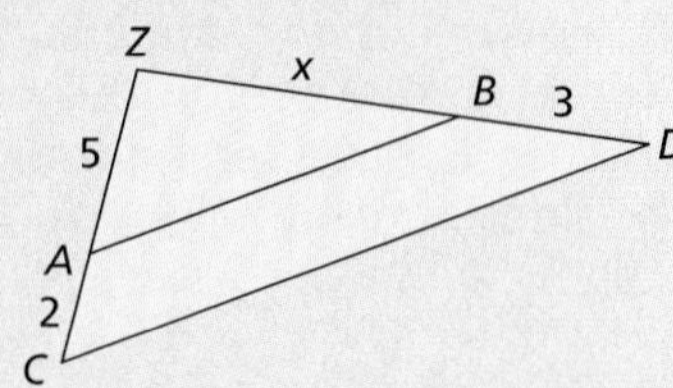

Supplementary Example 2

The figure below is from the plans for a steel sculpture. Given: M is the midpoint $\overline{DF}$. N is the midpoint of $\overline{EF}$. Proof: $\overline{MN} \parallel \overline{DE}$. The reasoning is identical to the proof stated in the solution to Example 2.

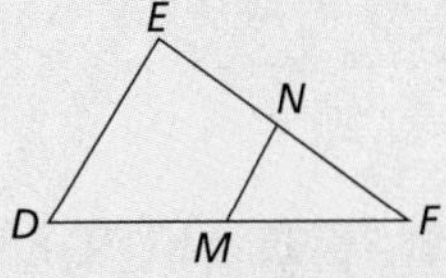

Supplementary Example 3

Refer to the figure in Example 3. Suppose the $WX = 8$ cm and $YZ = 15$ cm. Find AB. 11.5 cm

Reteaching Worksheet 7-6

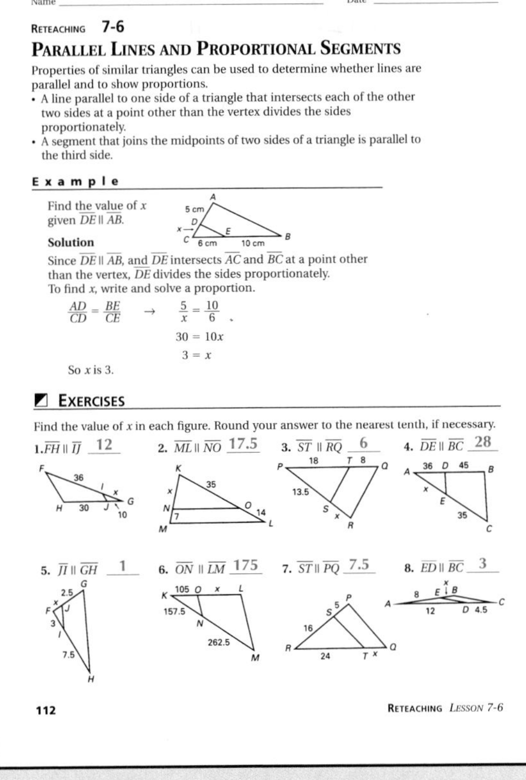

Name ____________ Date ____________

RETEACHING 7-6

PARALLEL LINES AND PROPORTIONAL SEGMENTS

Properties of similar triangles can be used to determine whether lines are parallel and to show proportions.

- A line parallel to one side of a triangle that intersects each of the other two sides at a point other than the vertex divides the sides proportionately.
- A segment that joins the midpoints of two sides of a triangle is parallel to the third side.

Example

Find the value of x given $\overline{DE} \parallel \overline{AB}$.

Solution

Since $\overline{DE} \parallel \overline{AB}$, and $\overline{DE}$ intersects $\overline{AC}$ and $\overline{BC}$ at a point other than the vertex, $\overline{DE}$ divides the sides proportionately.

To find x, write and solve a proportion.

$$\frac{AD}{CD} = \frac{BE}{CE} \rightarrow \frac{5}{x} = \frac{10}{6}$$

$$30 = 10x$$

$$3 = x$$

So x is 3.

EXERCISES

Find the value of x in each figure. Round your answer to the nearest tenth, if necessary.

1. $\overline{FH} \parallel \overline{IJ}$ 12
2. $\overline{ML} \parallel \overline{NO}$ 17.5
3. $\overline{ST} \parallel \overline{RQ}$ 6
4. $\overline{DE} \parallel \overline{BC}$ 28
5. $\overline{JI} \parallel \overline{GH}$ 1
6. $\overline{ON} \parallel \overline{LM}$ 175
7. $\overline{ST} \parallel \overline{PQ}$ 7.5
8. $\overline{ED} \parallel \overline{BC}$ 3

112 RETEACHING *LESSON 7-6*

Lesson Wrap-up

Quick Assessment

Ask the following questions to determine if students understand the content presented in this lesson.

1. When will a line segment divide the sides of a triangle proportionately? **when it is a parallel to one side of the triangle and intersects the other sides at any points except the vertex**
2. When will a line segment be parallel to one side of a triangle? **When it connects midpoints of two sides of the triangle, it will be parallel to the third side.**
3. What is the relationship of the median of a trapezoid to its bases? **Its length is half the sum of the lengths of the bases.**

Assignment Guide

Basic: 1–14, 19–36
Advanced: 1–36
Additional Practice: See Extra Practice Index on page 674.

Additional Answers

10.

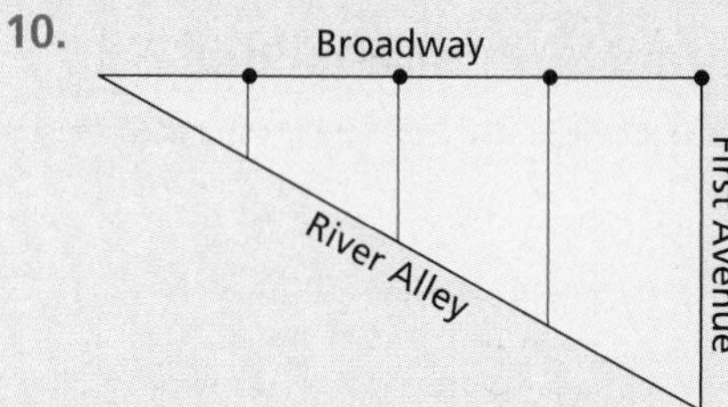

Example 4

Divide $\overline{AB}$ into three congruent parts.

Solution

Step 1: Draw a ray with A as an endpoint.

Step 2: Use a compass to mark off on the ray any shorter length AC. At C, mark off $\overline{CD} \cong \overline{AC}$. At D, mark off $\overline{DE} \cong \overline{AC}$.

Step 3: Draw $\overline{BE}$.

Step 4: At D, construct $\overline{DF} \parallel \overline{BE}$. At C, construct $\overline{CG} \parallel \overline{BE}$.

Because $\triangle ACG \sim \triangle ADF \sim \triangle AEB$ and $AC = CD = DE$, then $AG = GF = FB$.

This construction can be used to divide a segment into any given number of congruent parts.

Try These Exercises

In each figure, $AB \parallel CD$. Find the value of x to the nearest tenth.

1. 2.5

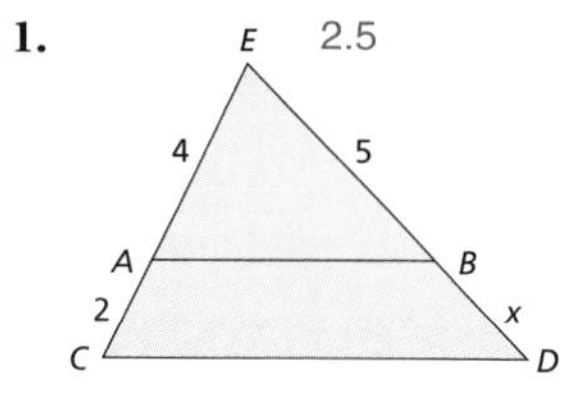

2. 10.7

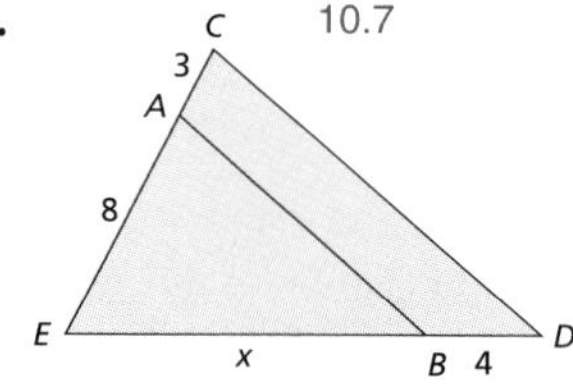

3. 6

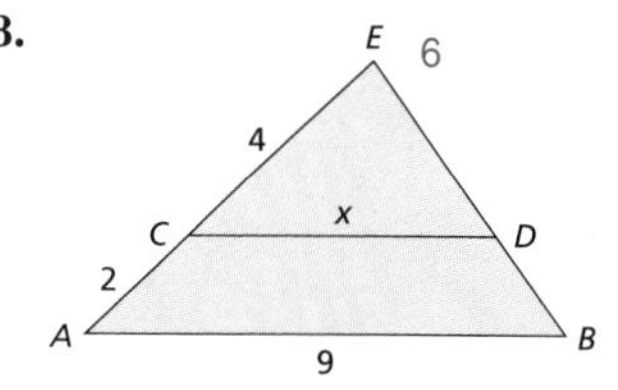

4. 12

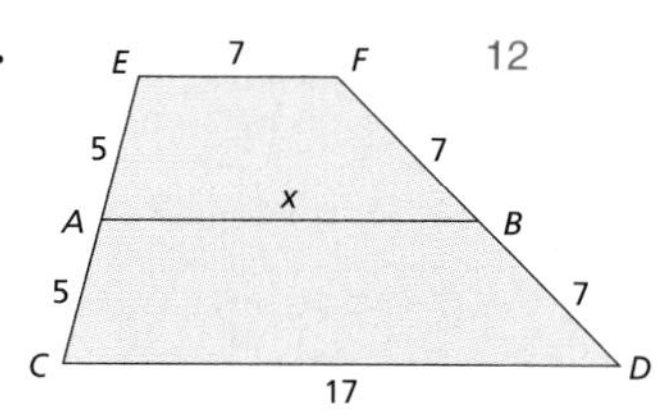

5. Draw any segment XY on a sheet of paper. Divide this segment into five congruent parts. Check students' work.

Practice Exercises

In each figure, $\overline{AB} \parallel \overline{CD}$. Find the value of x to the nearest tenth.

6. 4.5

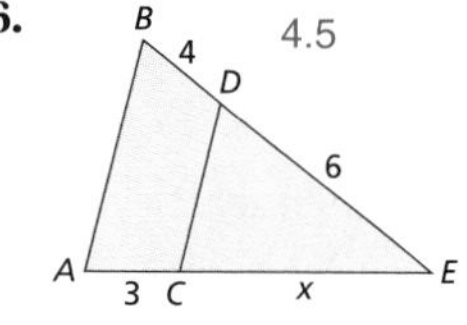

7. 3

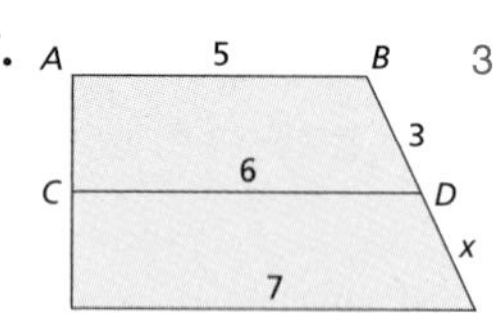

Extend the Lesson

ONGOING ASSESSMENT Have students solve the following problem: $\triangle ABE$ with $CD \parallel AB$, $AB = 15$, $BD = 3$ and $DE = 2$. Find CD.
Some students may get the incorrect answer $CD = 10$. These students have not use corresponding parts in the writing of their ratios. Remind students to be careful when writing proportions to use corresponding parts correctly.
Correct solution: $\frac{CD}{AB} = \frac{DE}{BE}, \frac{x}{15} = \frac{2}{5}, 5x = 2(15), 5x = 30, x = 6$

8. **WRITING MATH** Look at the trapezoid in Exercises 4 and 7. What is the relationship of the median of each trapezoid to its bases?
The median of a trapezoid is half the sum of the lengths of the bases.

9. **CHAPTER INVESTIGATION** Complete your drawing by adding details and shading. Does the sketch resemble the original photograph? Are there any areas which seem out of proportion? Check measurements and make corrections until the sketch is an accurate enlargement.

10. **REAL ESTATE** The map shows a triangular lot bought by a real-estate developer. Copy the map and construct proportional subdivisions along River Alley.
See additional answers.

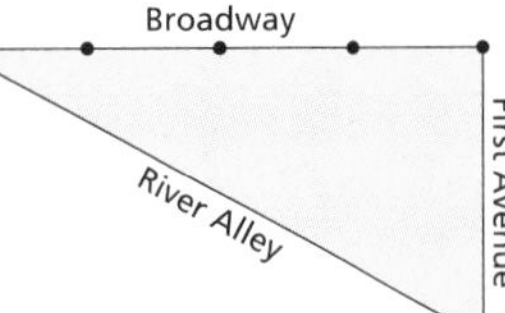

In each figure, $\overline{AB} \parallel \overline{CD}$. Find the value of x to the nearest tenth.

11. 8.6

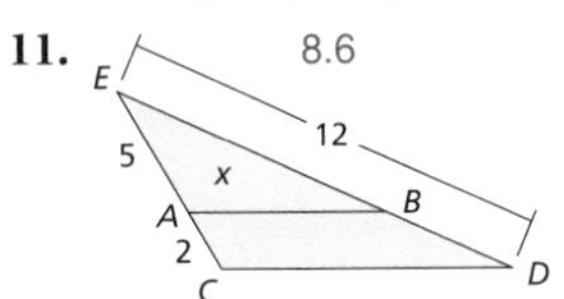

12. 7

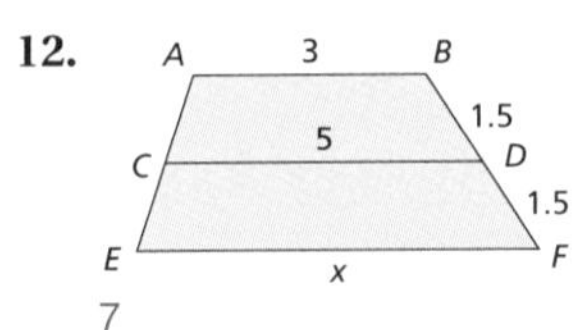

13. The angle bisector theorem states that any angle bisector in a triangle divides the opposite side into segments that are proportional to the other two sides of the triangle. Write a proportion based on this theorem for $\triangle XYZ$ if $\overline{XW}$ is an angle bisector. $yw : xy = zw : xz$

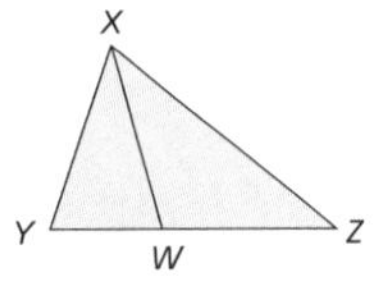

14. **ARCHITECTURE** A blueprint calls for an angle bisector to be added to a triangular structure as shown in the figure to the right. What is the measurement of YW, if $XY = 10$ ft, $WZ = 9$ ft, and $XZ = 12$ ft?
7.5 ft

Extended Practice Exercises

To prove the angle bisector theorem, a parallel line is drawn. Given that $\overline{BD}$ is the bisector of $\angle B$ and $\overline{CE} \parallel \overline{BD}$, use the drawing to answer Exercises 15–18.

15. For $\triangle ACE$, complete this proportion: $\frac{AD}{DC} = \frac{AB}{?}$ BE

16. Why is $\angle 3 \cong \angle 4$?

17. Why is $\overline{BC} \cong \overline{BE}$?

18. Why is $\frac{AD}{DC} = \frac{AB}{BC}$? Substitution of BC for BE

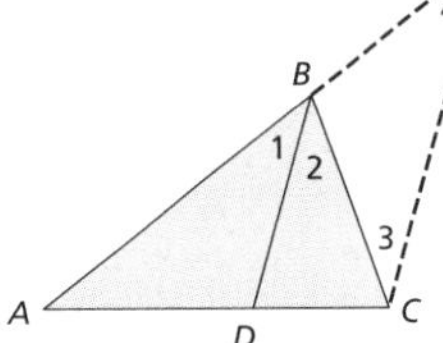

16. $\angle 1 \cong \angle 4$ and $\angle 2 \cong \angle 3$ because of parallel lines; $\angle 1 \cong \angle 2$ because they are each half the angle; so $\angle 3 \cong \angle 4$ by the transitive property.

17. congruent base angles mean triangle is isosceles

Mixed Review Exercises

Solve each system of equations by substitution. (Lesson 6-5)

19. $y = 3x - 2$, $2x + y = 5$ $\left(\frac{7}{5}, \frac{11}{5}\right)$

20. $-3y + 2x = 8$, $x - 2y = -7$ $(37, 22)$

21. $9 = 2y - 4x$, $3x + y = 4$ $\left(\frac{-1}{10}, \frac{43}{10}\right)$

22. $x - 5y = -7$, $3y + 4x = 2$ $\left(\frac{-11}{23}, \frac{30}{23}\right)$

23. $-2x - 3y = -9$, $4y + x = 3$ $\left(\frac{27}{5}, \frac{3}{5}\right)$

24. $y - 3x = 1$, $2x - 5y = -18$ $(1, 4)$

Find the mean, median and mode of each set of data. (Lesson 2-7)

25. 9 2 13 10 3 19 5 15 8 2 20 11 4 3 10 9 22 23 10 2
mean: 10; median: 9.5; mode: 10

26. 26 23 28 22 25 24 20 25 23 29 29 21 20 29 28
mean: 24.8; median: 25; mode: 29

Extra Practice Worksheet 7-6

Name ____________ Date ____________

EXTRA PRACTICE 7-6

PARALLEL LINES AND PROPORTIONAL SEGMENTS

EXERCISES

In each figure, $\overline{BC} \parallel \overline{DE}$. Find the value of x to the nearest tenth.

1. 4.5
2. 6
3. 12.5
4. 13
5. 10.8
6. 14.3
7. 12
8. 8

100 EXTRA PRACTICE *LESSON 7-6*

Enrichment Worksheet 7-6

Name ____________ Date ____________

ENRICHMENT 7-6

SIMILARITY ACROSTIC

To solve this acrostic, work back and forth between the clues and the puzzle box. For example, the first letter of the first missing word is T. So, write a T in box number 4. The solution to this acrostic puzzle is the last two lines of the limerick at the right.

There was an old man who said, "Do Tell me how I'm to add two and two? I'm not very sure . . .

1. To solve a problem with similar triangles, your first 4-3-34-15 might be to find the 20-14-32-24-38 of a pair of corresponding sides. TASK, RATIO
2. Triangles *ABC, DEF* and *XYZ* are similar. $AB = 1$, $DE = 5$ and $XY = 10$. So, *CA* is one-44-33-25-40-30 of *FD* and one-41-16-11-29-2 of *ZX*. FIFTH, TENTH
3. All isosceles right triangles are similar. True or false? 12-28-19-45 TRUE
4. If two isosceles triangles have a pair of equal angles, they are similar. True of false? 17-31-36-10-26 FALSE
5. The segment joining the midpoint of two sides of a triangle is one-half the length of the third 39-5-7-9. SIDE
6. How many pairs of corresponding angles are needed to guarantee similarity in two triangles? 1-46-8 TWO
7. Segments *UR* and *TA* are parallel. Points *U* and *T* are at the top of the figure. Segments *UA* and *RT* are drawn. They intersect in point *O*. What triangle is similar to *OUR*? 18-27-23 OAT
8. You are stuck on a ratio and proportion problem involving a household budget. Who might be able to help you with the answer? 37-43-13 MOM
9. The ratio of the sides of a Golden Rectangle is one-half the sum of the square root of 5 and 1, or 35-21-42-22-6 1.62. ABOUT

1 T	2 H	3 A	4 T	■	5 I	6 T	■	7 D	8 O	9 E	10 S	11 N	12 T
■	13 M	14 A	15 K	16 E	■	17 F	18 O	19 U	20 R	■	21 B	22 U	23 T
24 I	■	25 F	26 E	27 A	28 R	■	29 T	30 H	31 A	32 T	■	33 I	34 S
35 A	36 L	37 M	38 O	39 S	40 T	■	41 T	42 O	43 O	■	44 F	45 E	46 W

128 ENRICHMENT *LESSON 7-6*

Extend the Lesson

MATH JOURNAL Have students solve the following problem and write their explanations in their journals.
$\overline{ST}$ intersects two sides of $\triangle PQR$ such that S is the midpoint of $\overline{PQ}$ and T is the midpoint of PR. $\overline{QT}$ and $\overline{RS}$ intersect at point U. Is $\triangle STU \sim \triangle RQU$? Explain. **Yes; AA, because S and T are midpoints, $\overline{ST}$ is parallel to $\overline{QR}$ and by corresponding angles $\angle RQU \cong \angle STU$ and $\angle TSU \cong \angle QRU$. (Also, $\angle SUT \cong \angle RUT$ by vertical angles.)**

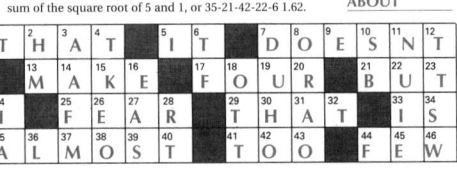

Skills Practice

Vocabulary Review

Lesson 7-5
calculators
compasses
straightedges
geometric drawing software

Lesson 7-6
parallel
vertex
median
corresponding angles
transversal
trapezoid

Assignment Guide

All students: 1–30
Additional Practice: Refer to the Extra Practice Index on page 674 of this text.

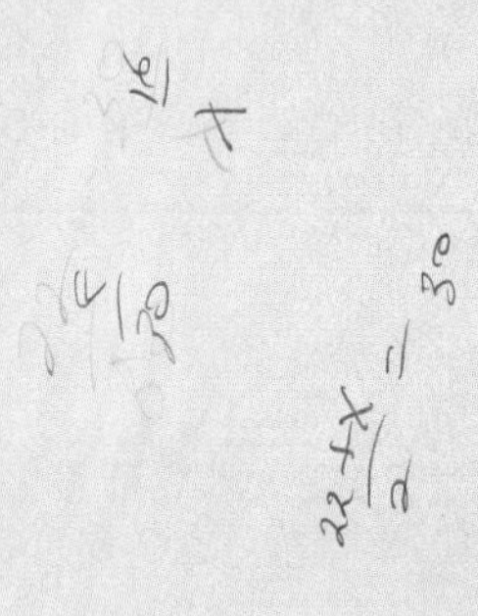

Review and Practice Your Skills

Practice ■ Lesson 7-5

Find x in each pair of similar triangles to the nearest tenth.

1.
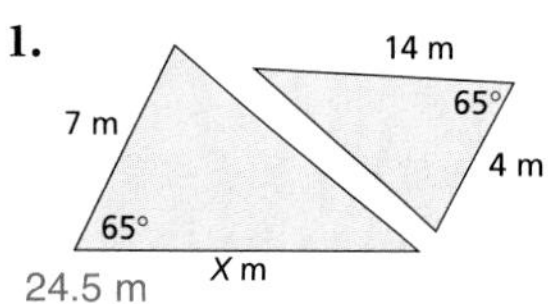

24.5 m

2.
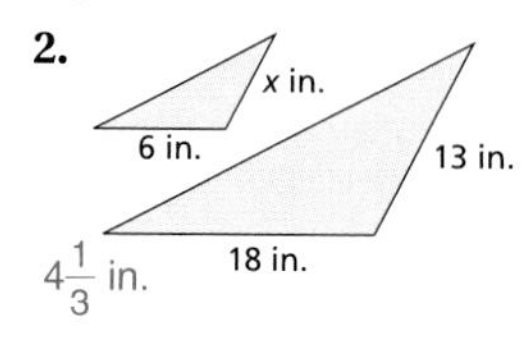

$4\frac{1}{3}$ in.

3.
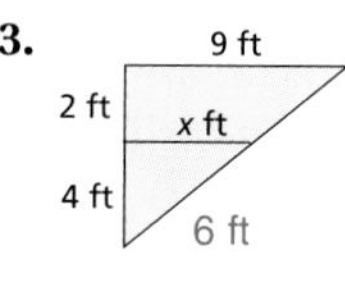

6 ft

4.
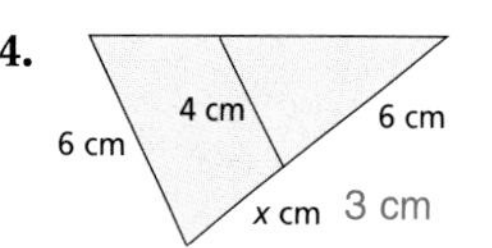

3 cm

5.
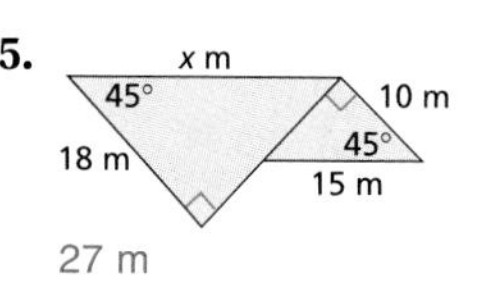

27 m

6.
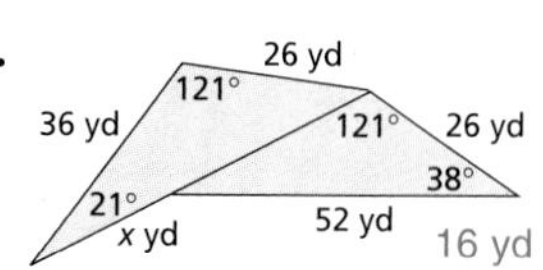

16 yd

7.
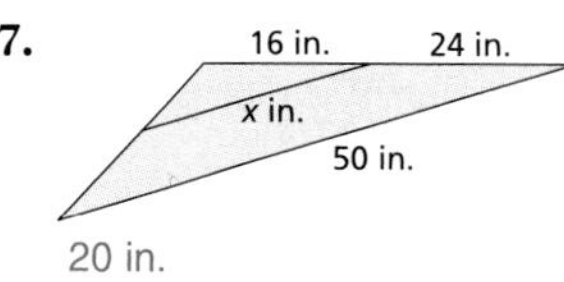

20 in.

8.
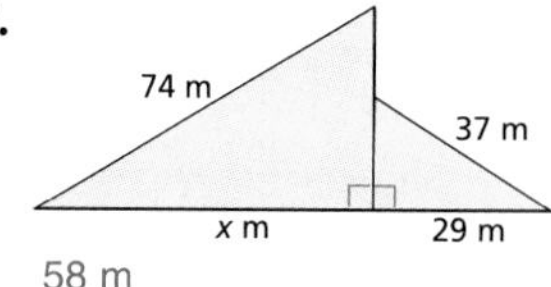

58 m

9.
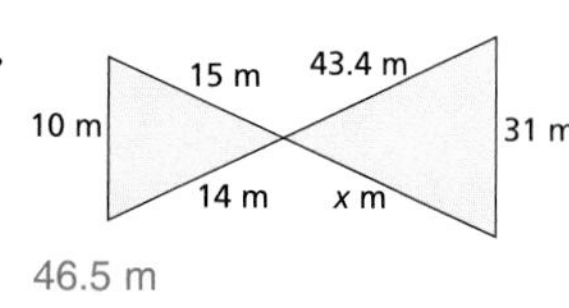

46.5 m

10. Two similar triangles have a 3:8 ratio of corresponding sides. What is the ratio of their areas? 9 : 64

Practice ■ Lesson 7-6

In each figure, $\overline{AB} \parallel \overline{CD}$. Find the value of x to the nearest tenth.

11.
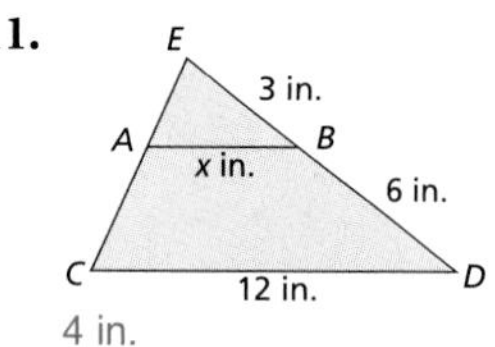

4 in.

12.
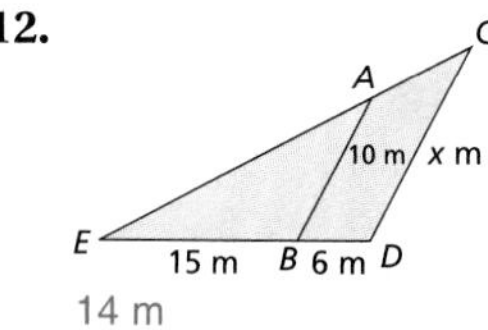

14 m

13.
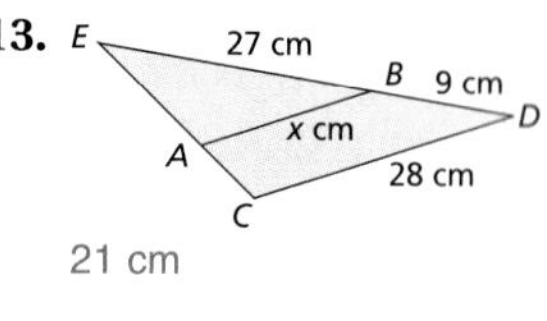

21 cm

14.
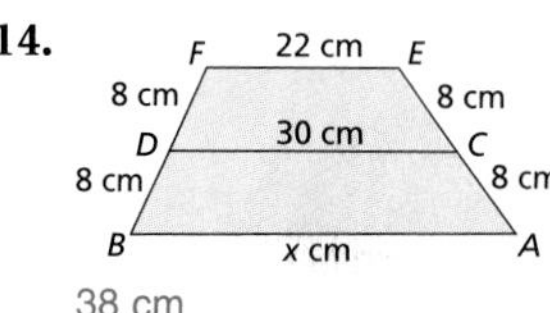

38 cm

15.
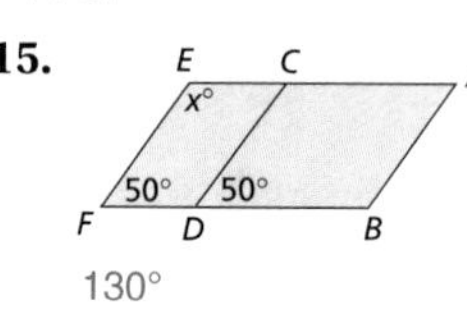

130°

16.
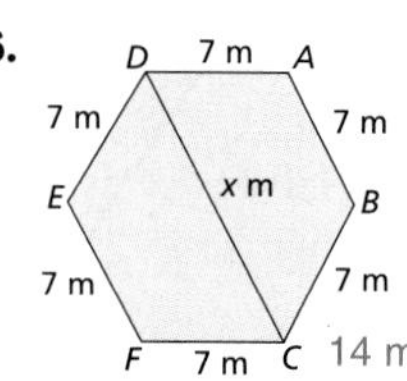

14 m

17. *True* or *false*: The line that joins the midpoints of two sides of a triangle is parallel to the third side. true

18. *True* or *false*: The line that joins the midpoints of two sides of a triangle creates a similar triangle whose perimeter is $\frac{1}{4}$ of the original triangle. false

324 Chapter 7 **Similar Triangles**

Practice ◾ Lesson 7-1–Lesson 7-6

Solve each proportion. (Lesson 7-1)

22. $\frac{b}{-9} = \frac{-4}{27}$ $\frac{4}{3}$

23. $\frac{13}{g} = \frac{78}{30}$ 5

24. $\frac{1.6}{4.5} = \frac{k}{14.4}$ 5.12

25. $\frac{-7}{26} = \frac{1.75}{m}$ −6.5

26. $\frac{x-3}{20} = \frac{3}{2}$ 33

27. $\frac{2x+1}{18} = \frac{26}{39}$ 5.5

Find the measure of x in each pair of similar polygons. (Lesson 7-2)

28.

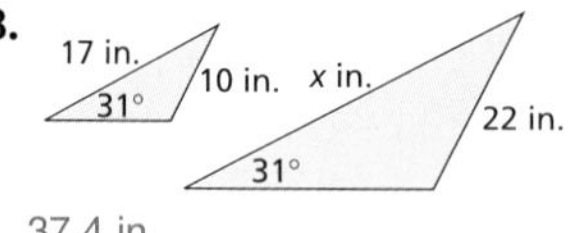

37.4 in.

29.

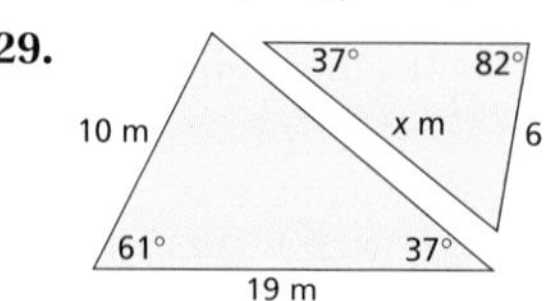

11.4 m

30.

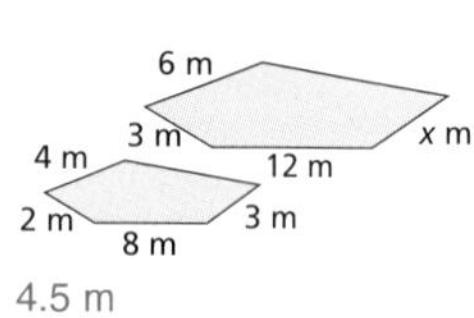

4.5 m

MathWorks Workplace Knowhow — Career – Photographic Processors

Photographic process workers develop film, make prints or slides and enlarge or retouch photographs. They operate many special types of machines. Specialized workers handle delicate tasks, such as retouching negatives and prints. They restore damaged and faded photographs and may color or shade drawings to enhance images using an airbrush.

Some photographic process workers use computers to enhance or alter images digitally. These workers may work for magazines to touch up portraits of models. They can also eliminate images from photographs or combine images from different photographs. To be successful in this field, workers must use ratio and proportion to make sure images look right.

1. A customer wants to crop and enlarge a portion of a 3 in. by 5 in. picture. The portion is 0.75 in. by 0.94 in. Find the ratio of the length to the width of the cropped portion. 15 : 19

Another customer brings in a group of old photographs in non-standard sizes for enlargement. You will convert the length to a standard photographic size. For each photo below, find the width of the enlarged photo when the length is convereted as indicated.

	Current photo dimensions	*New length*	
2.	$2\frac{1}{2}$ in. by $3\frac{1}{2}$ in.	5 in.	7 in.
3.	4 in. by 7 in.	7 in.	$12\frac{1}{4}$ in.
4.	7 in. by 9 in.	10 in.	10 in.
5.	6 in. by 9 in.	10 in.	15 in.

Chalkboard Examples

Example from Lesson 7-5

Two similar triangles have a 2 : 5 ratio of corresponding sides. What is the ratio of there medians? **2 : 5**

Example from Lesson 7-6

Given a trapezoid with bases of 12 m and 19 m, find the length of its median. **15.5 m**

Career Opportunity

Describe the kind of work photographic process workers do and the types of companies that hire these workers. Have students discuss the importance of measurement and algebraic thinking in doing this type of job. Explain that photographic process workers must use ratios and scales to process photographs according to a customer's instructions. Students should answer Questions 1–5 to better understand how photographic process workers use mathematics in performing their job.

Students who are interested in learning more about this profession can go to learningmathmatters.com. Guidance Counselors should have information about school and training requirements.

Lesson Planning

NCTM Standards/Strands

- Number & Operation
- Patterns, Functions, & Algebra
- Geometry & Spatial Sense
- Measurement
- Problem Solving
- Reasoning and Proof
- Connections
- Representation

Vocabulary

indirect measurement

Materials Needed

measuring tapes or centimeter rulers
mirrors

Lesson Resources

Warm-Up Transparency 21
Reteaching 7-7
Enrichment 7-7

Assignment Guide

Basic: 1–5, 6–14
Enriched: 1–14

Getting Started

5-Minute Warm-Up

Solve each proposition.

1. $\frac{x}{12} = \frac{14}{3}$ 56
2. $\frac{3}{x} = \frac{2.7}{4.5}$ 5
3. $\frac{0.25}{4} = \frac{5}{x}$ 80
4. $\frac{17}{153} = \frac{x}{81}$ 9

7-7 Problem Solving Skills: Indirect Measurement

Properties you have learned about similar triangles can be used to measure heights and distances indirectly. For example, you can find the height of a tree by measuring its shadow on a sunny day.

Indirect measurement can be used even when the sun is not shining.

Work with a partner to measure the height of your classroom in meters. You will need a mirror and a centimeter ruler.

1. Place the mirror on the floor so that you can see the place where the ceiling meets a classroom wall.
2. Working together, measure the distance in centimeters along the floor from the place you are standing to the wall (a). Measure the distance from the mirror to the wall (b). Measure the distance from the floor to your eye level (c). Draw a diagram. Record your measurements on the diagram.

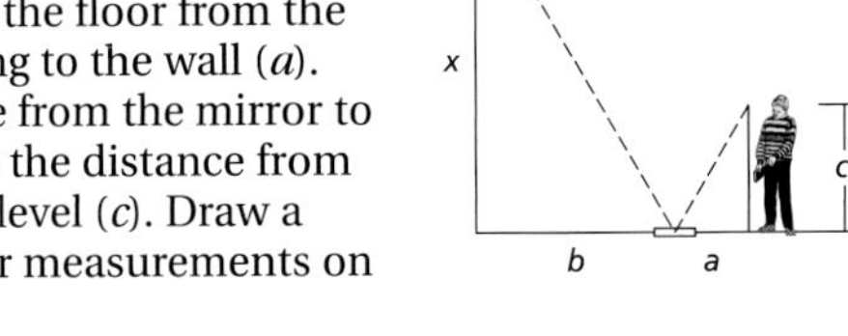

3. The triangles formed are similar because of a property of light reflection. Solve the proportion $a{:}b = c{:}x$ for x, the height of the classroom in centimeters. Change the measure to meters.

Problem Solving Strategies

Guess and check
Find a pattern
Solve a simpler problem
Make a table, chart or list
Use a picture, diagram or model
Act it out
Work backwards

✔ Use a model or a picture
Eliminate possibilities
Use an equation or formula

Problem

A tree casts a shadow 3.3 meters long. A meter stick placed perpendicular to the ground at the same time of day casts a shadow that is 0.75 meters long. How tall is the tree?

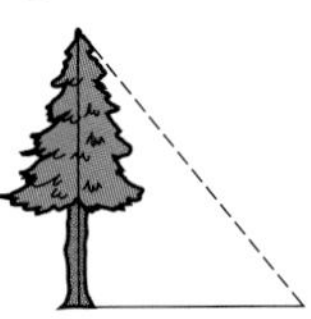

Solve the Problem

A sketch of the problem shows that the tree, the sun's rays, and the shadow form a right triangle similar to the triangle formed by the meter stick and its shadow.

Let h represent the height of the tree. Because the triangles are similar, $1{:}h = 0.75{:}3.3$. By cross multiplying, you get $0.75h = 3.3$ and $h = 4.4$. Therefore, the tree is 4.4 meters high. By using indirect measurement, you avoided having to climb the tree with a measuring tape.

Try These Exercises

1. Use the shadow method described above to find the height of a tree, flagpole, or streetlight near your home or school. Answers may vary.
2. Use the mirror method described above to find the height of your school, home, or other structure. Answers may vary.

Use the Five-step Plan and the Strategy

THE FIVE-STEP PLAN **Read**—ask questions of the students to help them understand the problem. **Plan**—guide students to related problems and previously mastered skills and strategies. **Solve**—students solve problem on their own. **Answer**—write solution in format that answers the question. **Check**—review work, check for reasonableness and review strategy used. Students will benefit from the experience of verbalizing their methods.

THE STRATEGY Use a model or picture—Drawing a picture to represent a situation or building a model is an effective way to make the details of a problem real to students.

Extend the Lesson

REAL WORLD CONNECTION Surveyors, navigators, astronomers, architects and artists are just a few of the professionals who use similar triangles and indirect measurement in their work. Have students choose a profession and investigate how similar triangles and indirect measurement are used.

Practice Exercises

3. **SURVEYING** The diagram below shows some measurements that a surveyor was able to take. Describe how she can find the width of the pond on her property using similar triangles. 8.4 m

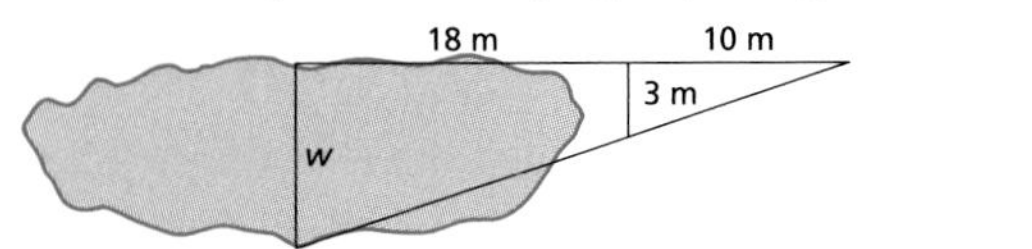

4. **PHOTOGRAPHY** A person is 150 cm from the camera lens. The film is 2.5 cm from the lens. If the person is 180 cm tall, how tall is his image on the film? 3 m

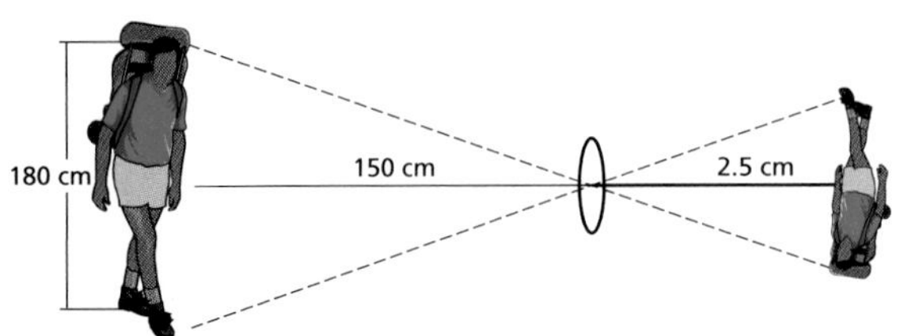

5. Ming took an index card and folded it exactly in half, to form a 45° angle. She walked back from a tree until she could sight the tree at the very edge of the card she was holding at eye level. Ming stated that her distance from the tree (*b*) plus the distance from the ground to the card (*a*) is equal to the height of the tree. Was she correct? Explain.
Answers will vary.

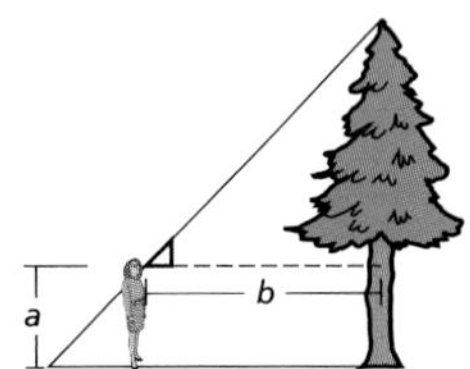

Five-step Plan

1 Read
2 Plan
3 Solve
4 Answer
5 Check

Mixed Review Exercises

Solve each system of equations by adding, subtracting and multiplying. (Lesson 6-6)

6. $y = 4 - 3x$
$2y - x = -3$ $\left(\frac{11}{7}, -\frac{5}{7}\right)$

7. $2y - 4x = -5$
$3x - y = 8$ (5.5, 8.5)

8. $-4x + 3y = -8$
$3x - 5y = 2$ $\left(\frac{34}{11}, \frac{16}{11}\right)$

9. $3y = 4x + 2$
$7x = y + 5$ (1, 2)

10. $-2y - 4x = 7$
$-3x = 7 - 2y$ $\left(-2, -\frac{1}{2}\right)$

11. $5y = 7 - 3x$
$2x - 5 = 2y$ $\left(\frac{39}{16}, \frac{-1}{16}\right)$

Find the probability that a point selected at random in each figure is in the shaded region. (Lesson 5-3)

12.

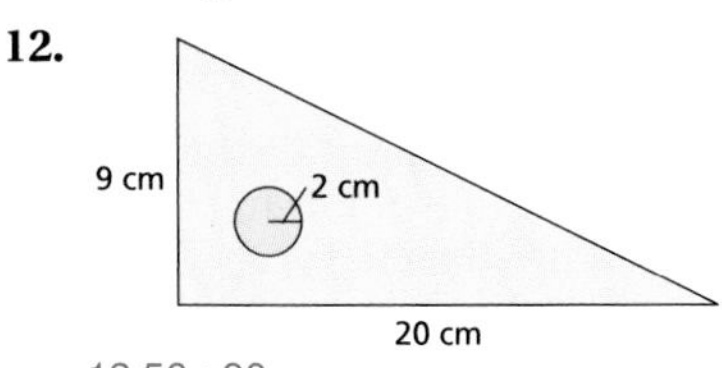

12.56 : 90

13.

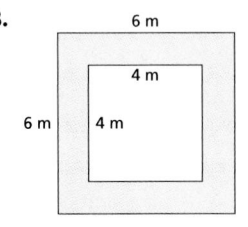

6 m
4 m
6 m
4 m

20 : 36

14.

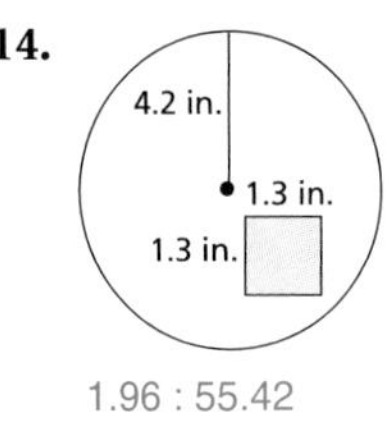

1.96 : 55.42

Chalkboard Examples

Supplementary Problem

A flagpole casts a shadow 18 m long. At the same time, Hassan who is 1.2 m tall casts a shadow that is 3 m long. How tall is the tree? 7.2 m

Lesson Wrap-up

Quick Assessment

Refer students to the sample problem in the lesson. Ask: How do you know that the right triangles in the problem are similar? The tree and the meter stick both form 90° angles with the ground and the sun's rays from the same angle with the ground; therefore, the two pairs of base angles are congruent. The triangles are similar by the AA Similarity Postulate.

Enrichment Worksheet 7-7 not shown.

Extend the Lesson

STUDENT PORTFOLIO Have students use indirect measurement to measure the height of a real object. Have them draw a diagram of the situation and write the steps they used to obtain the height of the object.

Reteaching Worksheet 7-7

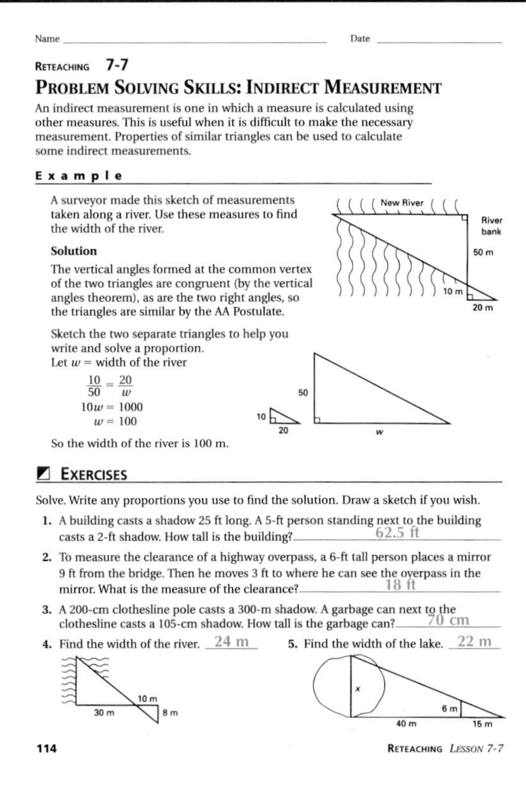

Name ______________ Date ______________

RETEACHING 7-7

PROBLEM SOLVING SKILLS: INDIRECT MEASUREMENT

An indirect measurement is one in which a measure is calculated using other measures. This is useful when it is difficult to make the necessary measurement. Properties of similar triangles can be used to calculate some indirect measurements.

Example

A surveyor made this sketch of measurements taken along a river. Use these measures to find the width of the river.

Solution

The vertical angles formed at the common vertex of the two triangles are congruent (by the vertical angles theorem), as are the two right angles, so the triangles are similar by the AA Postulate.

Sketch the two separate triangles to help you write and solve a proportion.
Let w = width of the river

$$\frac{10}{50} = \frac{20}{w}$$
$$10w = 1000$$
$$w = 100$$

So the width of the river is 100 m.

EXERCISES

Solve. Write any proportions you use to find the solution. Draw a sketch if you wish.

1. A building casts a shadow 25 ft long. A 5-ft person standing next to the building casts a 2-ft shadow. How tall is the building? 62.5 ft
2. To measure the clearance of a highway overpass, a 6-ft tall person places a mirror 9 ft from the bridge. Then he moves 3 ft to where he can see the overpass in the mirror. What is the measure of the clearance? 18 ft
3. A 200-cm clothesline pole casts a 300-m shadow. A garbage can next to the clothesline casts a 105-cm shadow. How tall is the garbage can? 70 cm
4. Find the width of the river. 24 m
5. Find the width of the lake. 22 m

114 RETEACHING *LESSON 7-7*

Assessment Planning

Vocabulary Review

Lesson 7-1
equivalent ratios, proportions, extreme, means, cross products, terms

Lesson 7-2
similar angles, congruence, corresponding angles, corresponding sides

Lesson 7-3
scale drawing, scale distance, actual distance, model

Lesson 7-4
Similarity Postulates: AA, SAS, SSS

Lesson 7-5
theorem, midpoint, altitude, median

Lesson 7-6
parallel, vertex, median, corresponding angles, transversal, trapezoid

Lesson 7-7
indirect measurement

Assessment Options

Chapter Assessment
Alternative Assessment
Chapter Test Forms A and B
Cumulative Assessment
Achievement Test
Writing Prompts

Assignment Guide

All students: 1–18
Additional Practice: Refer to the Extra Practice Index on page 674 of this text.

Chapter 7 Review

Vocabulary ■ LESSON 7-1–LESSON 7-7

Match the letter of the word in the right column with the description at the left.

1. terms in a proportion c
2. relationship between map distance and actual distance b
3. corresponding sides in similar polygons d
4. corresponding angles in similar polygons a

a. congruent
b. scale
c. means, extremes
d. proportional

LESSON 7-1 ■ Ratios and Proportions, p. 296

▶ **Equivalent ratios** can be named by the same fraction.

▶ A **proportion** is an equation that states two ratios are equivalent. In a proportion, the product of the extremes equals the product of the means.

If $\frac{a}{b} = \frac{c}{d}$, then $ad = bc$.

5. Solve for x. $x{:}7 = 5{:}8$. 4.375
6. A recipe calls for 2 cups of sugar to 5 cups of flour. How much flour would be added to 5 cups of sugar? 12.5 cups

LESSON 7-2 ■ Similar Polygons, p. 300

▶ Two figures are similar if they have the same shape. All corresponding angles are congruent. All corresponding sides are proportional.

Find x for each pair of similar figures.

7. 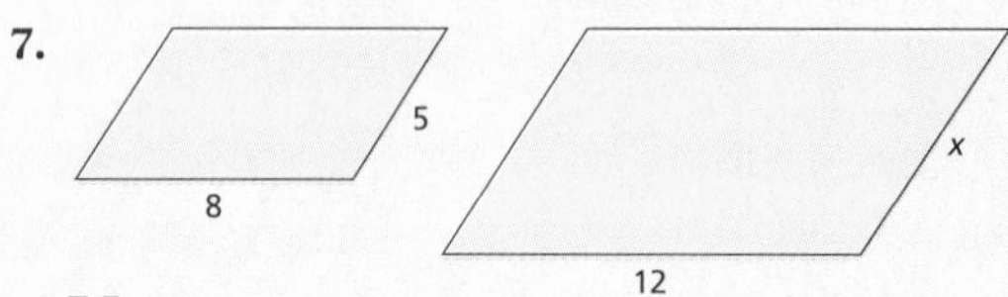

7.5

8. 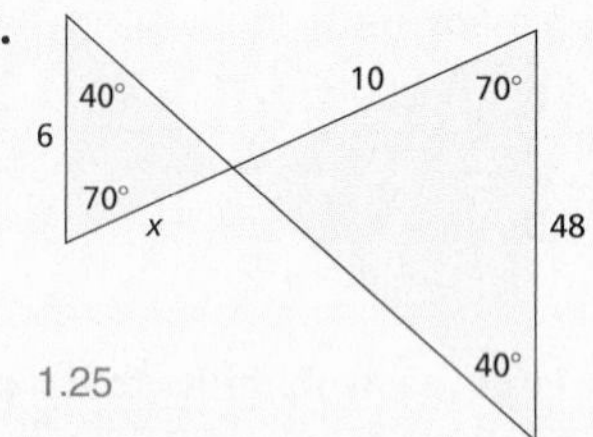

1.25

LESSON 7-3 ■ Scale Drawings, p. 306

▶ A **scale drawing** is a representation of a real object. All lengths on the drawing are proportional to actual lengths in the objects. The **scale** of the drawing is the ratio of the size of the drawing to the actual size of the object.

9. Find the actual length:
scale length: 4 cm
scale: 1 cm = 2.5 m 10 m

10. Find the scale length:
actual length: 2 ft
scale: $\frac{1}{2}$ in. = 4 ft $\frac{1}{4}$ in.

11. Find the actual length:
scale length: $\frac{1}{4}$ in.
scale: 2 in. = 420 mi $52\frac{1}{2}$ mi

LESSON 7-4 ◪ Postulates for Similar Triangles, p. 310

▶ Two triangles are similar if any of these conditions are true:

1. Two pairs of corresponding angles are congruent (AA Similarity Postulate).
2. All pairs of corresponding sides are proportional (SSS Similarity Postulate).
3. Two pairs of corresponding sides are proportional and the angle between those sides is congruent (SAS Similarity Postulate).

Find x to the nearest tenth for each pair of similar triangles.

12.

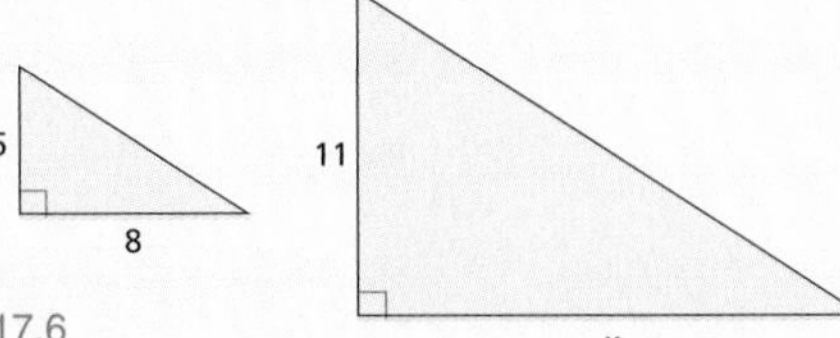

17.6

13.

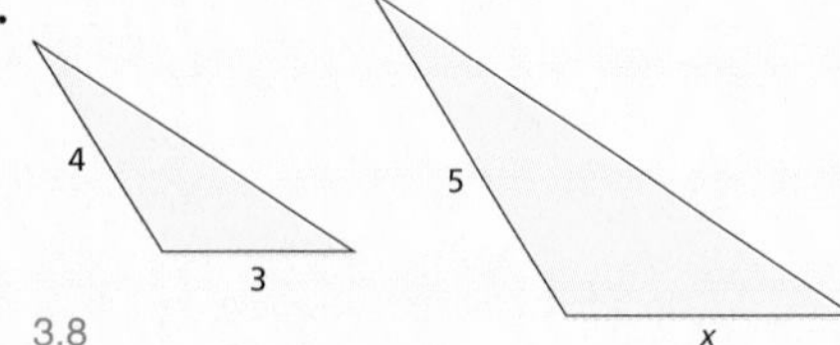

3.8

LESSONS 7-5 and 7-6 ◪ Proportional Segments, p. 316

▶ If a segment connects the midpoints of two sides of a triangle, then the length of the segment is equal to one-half the length of the third side.

▶ If two triangles are similar, their altitudes are in the same proportion, and their medians are in the same proportion as the sides of the triangle.

▶ The median of a trapezoid is parallel to its base, and its length is half the sum of the lengths of the bases.

Find x to the nearest tenth.

14.

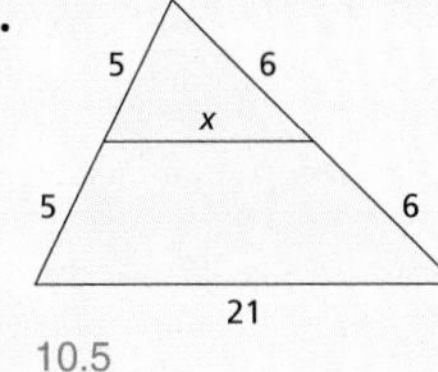

10.5

15.

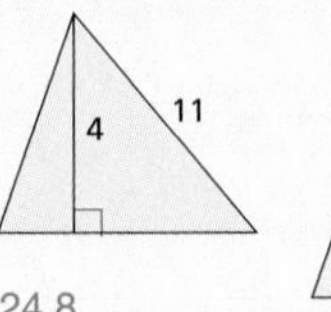

9 x

24.8

16.

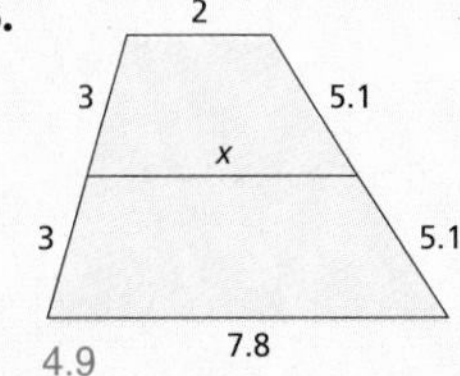

4.9

LESSON 7-7 ◪ Problem Solving: Indirect Measurement, p. 326

▶ Properties of similar figures can be used to measure lengths and distances indirectly.

Use a diagram to solve.

17. A flagpole casts a shadow 16 ft long. At the same time, a yardstick casts a shadow 4 ft long. How tall is the flagpole? 12 ft

18. To find the height of a tree, a forest ranger places a mirror on the ground 21 ft from the base of the tree. The ranger stands an additional 3 ft from the mirror so that she can see the top of the tree reflected in the mirror. If the ranger's eye level is 5 ft from the ground, what is the height of the tree? 30 ft

Chapter 7 Test Form A, page 1

CHAPTER 7
SIMILAR TRIANGLES
ASSESSMENT FORM A, PAGE 1

Name ______ Date ______

Scoring Record	
Possible: 27	Earned:

Solve each proportion.

1. $\frac{x}{36} = \frac{3}{27}$ 4
2. $\frac{6}{54} = \frac{9}{x}$ 81
3. $\frac{16}{x} = \frac{40}{115}$ 46
4. Is $ABCD \sim EFGH$? Write yes or no and explain why. Yes, corresponding ∠s are ≅ and corresponding sides are proportional.

Find x in each pair of similar polygons.

5. 10
6. 10.8

Find the actual measure of each scale item.

7. scale length is 5 cm, scale is 2 cm : 6 m — 15 m
8. scale distance is 3 in., scale is $\frac{1}{4}$ in. : 25 mi — 300 mi

Find the scale measure for each item.

9. actual length is 24 ft, scale is 3 in. : 9 ft — 8 in.
10. actual distance is 450 km, scale is 0.5 cm : 50 km — 4.5 cm

If the triangles are similar, write the reason: AA, SSS or SAS. If they are not similar, write no.

11. SSS
12. AA
13. no
14. SAS

86 ASSESSMENT CHAPTER 7 FORM A

Chapter 7 Test Form A, page 2

Name ______ Date ______

Find x in each pair of triangles to the nearest tenth.

15. 5
16. 8
17. 9.6
18. 6.8
19. 8
20. 18
21. $\overline{AB} \parallel \overline{ZY}$ 3.3
22. $\overline{FG} \parallel \overline{EC}$ 14.4
23. $\overline{KL} \parallel \overline{HJ}$ 4

Solve.

24. Peanuts and raisins are mixed in a 3 : 2 ratio. How many pounds of peanuts are needed to mix with 5 lb of raisins? 7.5 lb
25. A shade of pink paint is made by mixing white paint and red paint in a 1 : 3 ratio. How many fluid ounces of red paint would you need to make 12 fl oz of pink paint? 9 fl oz
26. A tree casts a shadow 8-ft long. A $5\frac{1}{2}$-ft person standing next to the tree casts a 2-ft shadow. How tall is the tree? 22 ft
27. Find the width of the lake 30 m

88 ASSESSMENT CHAPTER 7 FORM A

Chapter 7 Test Form B, page 1

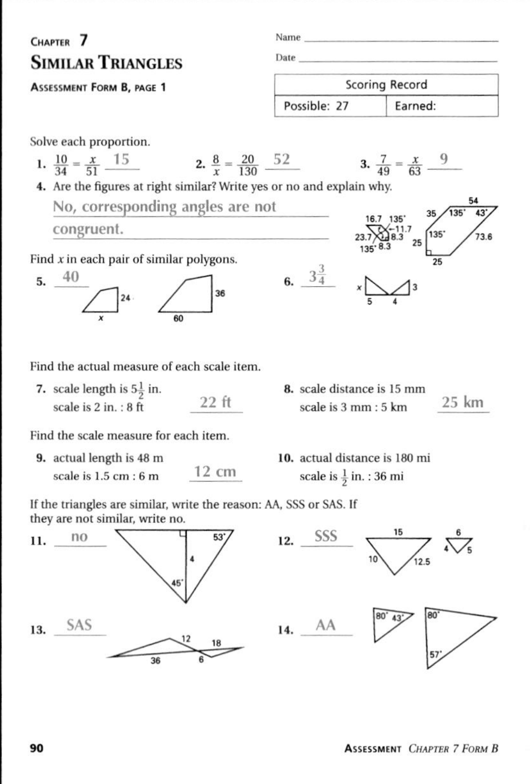

CHAPTER 7
SIMILAR TRIANGLES
ASSESSMENT FORM B, PAGE 1

Name ____________ Date ____________

Scoring Record	
Possible: 27	Earned:

Solve each proportion.

1. $\frac{10}{34} = \frac{x}{51}$ 15
2. $\frac{8}{x} = \frac{20}{130}$ 52
3. $\frac{7}{49} = \frac{x}{63}$ 9
4. Are the figures at right similar? Write yes or no and explain why. No, corresponding angles are not congruent.

Find x in each pair of similar polygons.

5. 40
6. $3\frac{3}{4}$

Find the actual measure of each scale item.

7. scale length is $5\frac{1}{2}$ in. scale is 2 in. : 8 ft 22 ft
8. scale distance is 15 mm scale is 3 mm : 5 km 25 km

Find the scale measure for each item.

9. actual length is 48 m scale is 1.5 cm : 6 m 12 cm
10. actual distance is 180 mi scale is $\frac{1}{2}$ in. : 36 mi

If the triangles are similar, write the reason: AA, SSS or SAS. If they are not similar, write no.

11. no
12. SSS
13. SAS
14. AA

90 ASSESSMENT CHAPTER 7 FORM B

Chapter 7 Test Form B, page 2

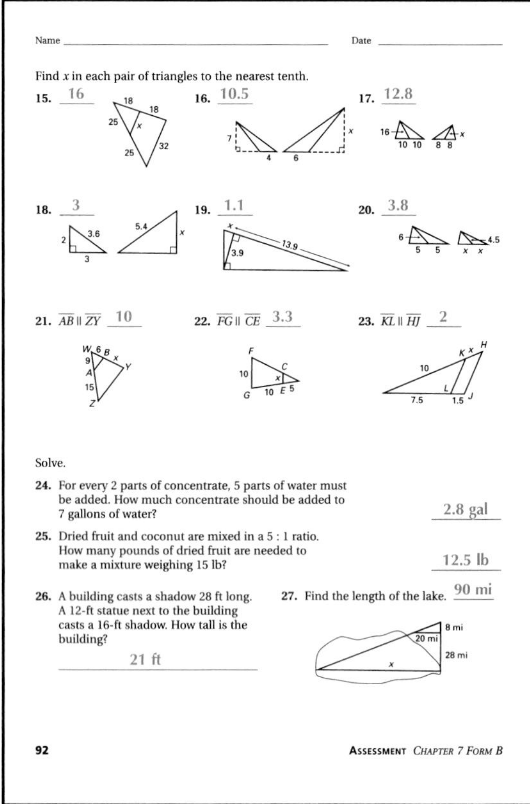

Name ____________ Date ____________

Find x in each pair of triangles to the nearest tenth.

15. 16
16. 10.5
17. 12.8
18. 3
19. 1.1
20. 3.8
21. $\overline{AB} \parallel \overline{ZY}$ 10
22. $\overline{FG} \parallel \overline{CE}$ 3.3
23. $\overline{KL} \parallel \overline{HJ}$ 2

Solve.

24. For every 2 parts of concentrate, 5 parts of water must be added. How much concentrate should be added to 7 gallons of water? 2.8 gal
25. Dried fruit and coconut are mixed in a 5 : 1 ratio. How many pounds of dried fruit are needed to make a mixture weighing 15 lb? 12.5 lb
26. A building casts a shadow 28 ft long. A 12-ft statue next to the building casts a 16-ft shadow. How tall is the building? 21 ft
27. Find the length of the lake. 90 mi

92 ASSESSMENT CHAPTER 7 FORM B

CHAPTER 7 ASSESSMENT

Solve each proportion.

1. $\frac{x}{7} = \frac{3}{21}$ 1
2. $\frac{4}{11} = \frac{6}{x}$ 16.5
3. $\frac{9}{16} = \frac{x}{2}$ 1.125

Determine if the polygons are similar. Write *yes* or *no*.

4.

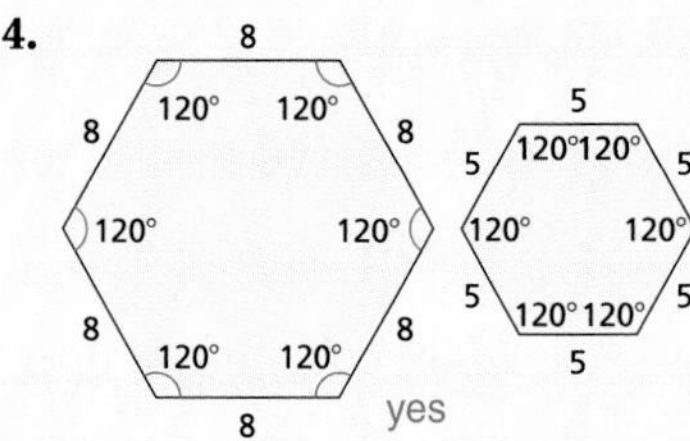

yes

5.

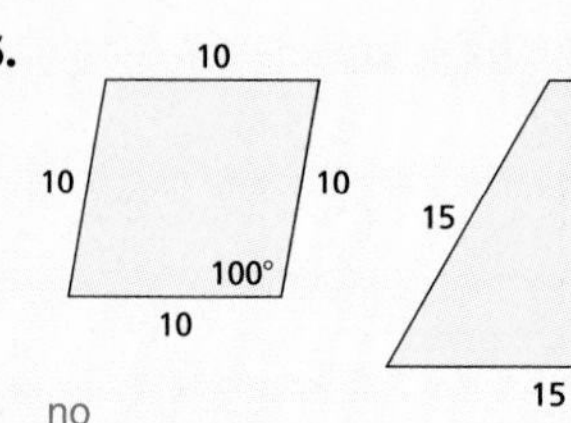

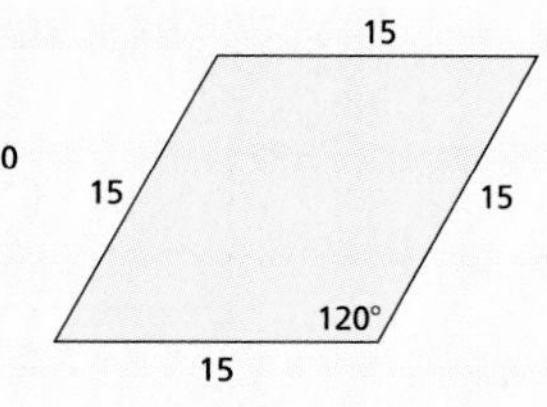

no

Find *x* in each pair of similar figures.

6.

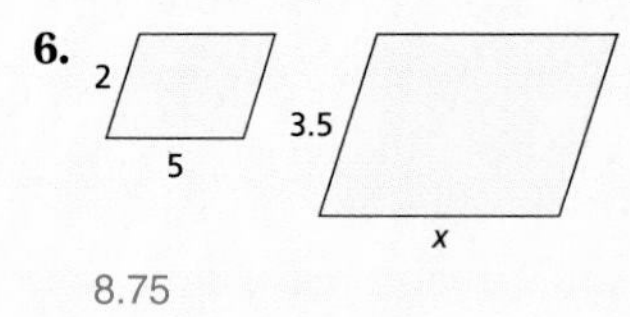

8.75

7.

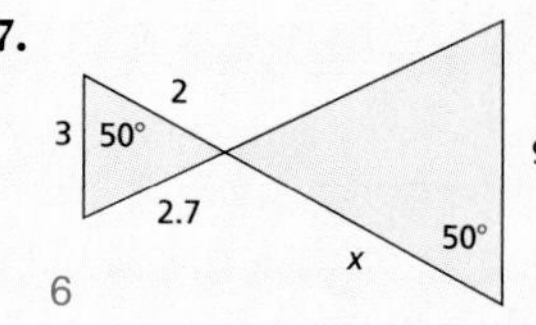

6

8.

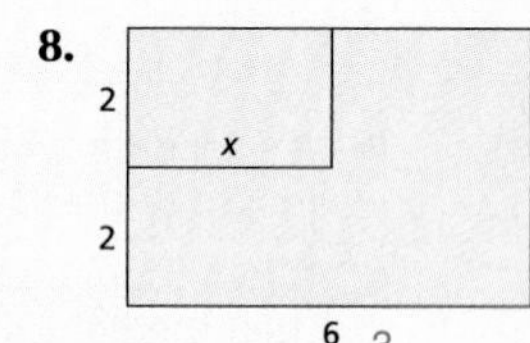

6 3

9. Find the actual length: scale length: 5 cm, scale: 1 cm = 3.5 m 17.5 m
10. Find the scale length: actual length: 2 mi, scale: $\frac{1}{4}$ in = 1 mi $\frac{1}{2}$ in.

Are the triangles similar?. If so, give a reason: write AA, SSS, or SAS.

11. no

12.

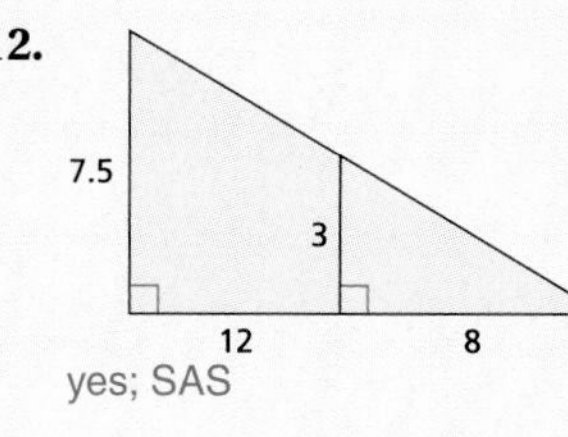

yes; SAS

13.

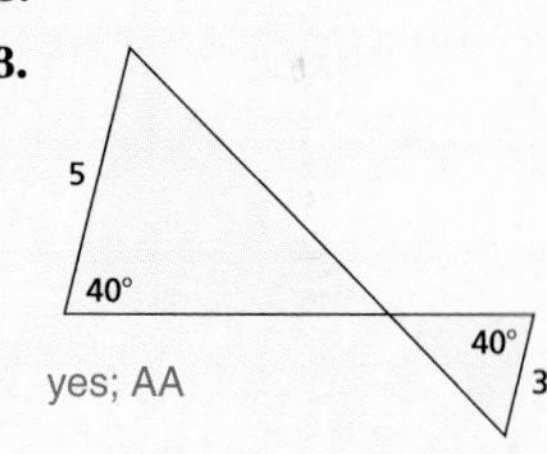

yes; AA

Find *x*.

14.

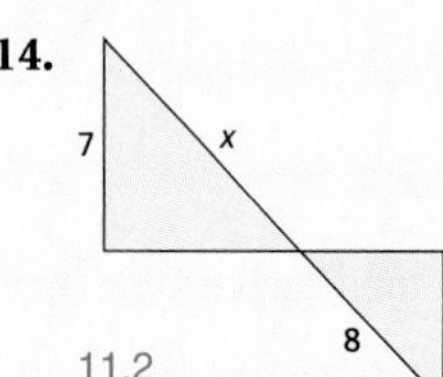

11.2

15.

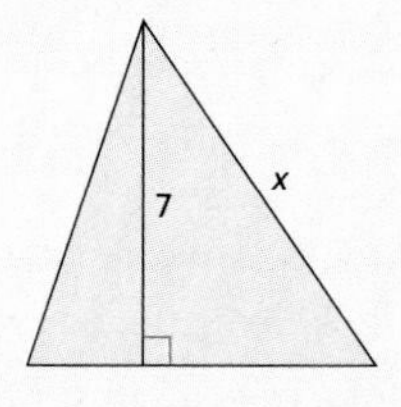

12.6

16. 9

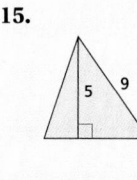

17. Luz placed a mirror on the ground and stood so that she could see the top of the tree. What is the height of the tree? 560 cm

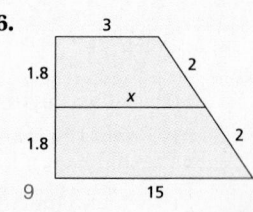

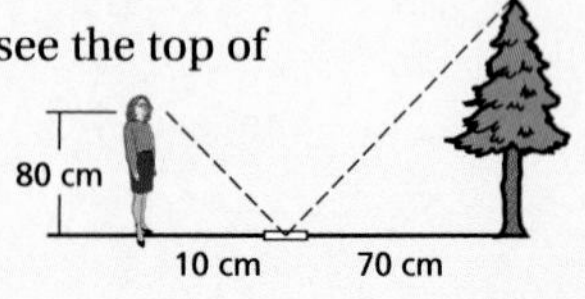

Portfolio

Students may compare their final drawings and discuss which methods were easier than others for completing the assignment.

SCORING RUBRIC 5—Student correctly and carefully enlarges the tangram pieces in two different proportions, and writes a complete and detailed step-by-step description of the process. **4**—Student enlarges the tangram pieces in two different proportions, and writes a complete step-by-step description of the process. **3**—Student enlarges the tangram pieces in two different proportions, and writes a reasonable step-by-step description of the process. **2**—Student enlarges some tangram pieces in two different proportions with minor errors, and writes a reasonable step-by-step description of the process. **1**—Student enlarges some of the tangram pieces with significant errors, and writes a partial step-by-step description of the process. **0**—Student makes no attempt to enlarge the tangram pieces.

Chapter 7 Alternative Assessment

Cooperative Learning

GIANT TANGRAMS Work in a group. Use the tangram set shown on page 188 in Chapter 4, Lesson 9 as a pattern. Measure the pieces, then enlarge them by a factor of 5, 6 or 7. Show all the calculations you used to determine the correct proportions and measurements. Cut the pieces out of heavy poster board or cardboard. Make a few tangram arrangements on the bulletin board using your new set of tangram pieces.

Connections

CAREERS Research to find five professions that involve the use of proportion in some way. You may find clues to some of these professions in this chapter. Interview people in those professions to find how they use proportions, how often they use them, and whether they generally enlarge or shrink the actual measures. Make a notebook of at least one example from each profession.

Portfolio

EXAMPLES Use the tangram pieces from page 188, Chapter 4, Lesson 9. Copy each in your notebook. Then enlarge each piece to fill a whole page. What proportions did you use? How did you determine the correct proportion? Write the steps you used to draw each shape. Then draw the pieces again, this time in a proportion to fill half a page. Describe the steps you took and name the proportions.

Chapter Investigation

EXTENSION Make a list of the similarities and differences between your photograph and proportional sketch. Write a short paragraph explaining how you determined the lengths to use in your sketch. Display your photograph, sketch, list, and paragraph on a piece of posterboard.

Cooperative Learning

After completing the assignment, you may want to save tangram sets for later assignments or use them to create art.

SCORING RUBRIC 5—Student correctly enlarges all the tangram pieces, shows detailed calculations, and makes several unique tangram arrangements with the pieces. **4**—Student correctly enlarges all the tangram pieces, shows all calculations, and makes several tangram arrangements with the pieces. **3**—Student correctly enlarges most of the tangram pieces, shows some calculations, and makes one tangram arrangement with the pieces that were completed. **2**—Student enlarges most of the tangram pieces with a few minor errors, shows some calculations, and makes one tangram arrangement with the pieces that were completed. **1**—Student enlarges one or two tangram pieces, shows a few calculations, but makes no tangram arrangements. **0**—Student makes no attempt to enlarge the tangram pieces.

Connections

Professions may include photographer, architect, caterer, doll maker, engineer, graphic artists, and so on.

SCORING RUBRIC 5—Student contacts people in five different professions, describes and gives several detailed examples of how proportions are use in those professions. **4**—Student contacts people in five different professions, describes and gives an example of how proportions are use in those professions. **3**—Student contacts people in five different professions and gives a few examples of how proportions are use in those professions. **2**—Student contacts people in less than five different professions, and gives poor or incomplete examples of how proportions are use in those professions. **1**—Student contacts a person in one profession, but cannot give an example of how proportion is used. **0**—Student makes no attempt to contact a person in any profession.

See page 330 for information and Scoring Rubric for Portfolio.

Chapter Investigation

The benchmarks and expectations for this extension are as follows.

- Students choose an actual photograph or a photograph published in a magazine. They measure the outer dimensions of the photograph and increase the dimensions by a factor of 5.
- Students select at least five prominent features in the photograph and plot the locations of the features on their enlargement. Once the main features of the photograph are placed correctly on the enlargement, they sketch the remaining details from the photograph.
- Students complete drawings by adding details and shading. They check measurements and make corrections until the sketch is accurate.
- Students make a list of the similarities and differences between their photograph and proportional sketch. They write a short paragraph explaining how they determined the lengths to use in their sketch. They display their photograph, sketch, list and paragraph on a piece of posterboard.

Cumulative Review

Cumulative review is the best way to maintain previously taught skills and concepts. This will keep students prepared for new lessons that build on previously covered skills.

Cumulative Review covers:

Lesson 1-4
Lesson 1-5
Lesson 1-7
Lesson 1-8
Lesson 2-1
Lesson 2-4
Lesson 2-5
Lesson 2-6
Lesson 2-7
Lesson 3-1
Lesson 3-2
Lesson 3-5
Lesson 4-1
Lesson 4-7
Lesson 4-9
Lesson 5-3
Lesson 5-7
Lesson 6-1
Lesson 6-6
Lesson 7-3
Lesson 7-5

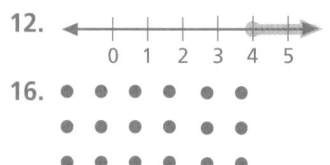

Additional Answers

12.

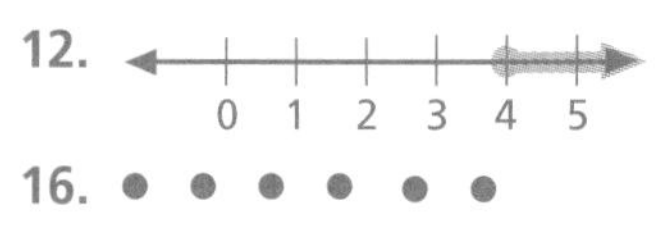

16.

Chapters 1–7 Cumulative Review

Evaluate each expression when $a = 12$, $b = -10$, and $c = -5$.

1. $a - (-b)$ 2
2. $-a + b$ −22
3. $(a + b)c$ −10
4. $ab \div c$ 24
5. b^3 −1,000
6. $(a - 2)^3$ 1,000
7. a^{-3} $\frac{1}{1728}$
8. $a^3a^{-2}b^{-3}$ −0.012

9. Find the next three terms in this sequence. 5, 6, 8, 11, . . . 15, 20, 26

Solve and check each equation.

10. $x - 2.4 = -6.3$ −3.9
11. $5(x - 1) = 5 + 2x$ $3\frac{1}{3}$

12. Graph $2 + y \geq 6$ on a number line. See additional answers.

13. Find the median and mode of the following set of data: 46, 48, 44, 40, 46, 48, 46, 49, 47, 50. 46.5, 46

14. In the figure below, LT = 220. Find ET. 152.5

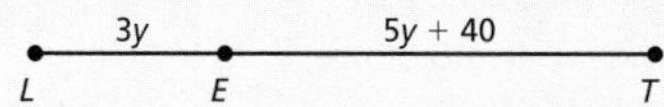

15. In the figure below, m∠*PQM* is $x°$ and m∠*MQN* is $5x°$. Find m∠*MQN*. 75°

16. Draw the next figure in the pattern. See additional answers.

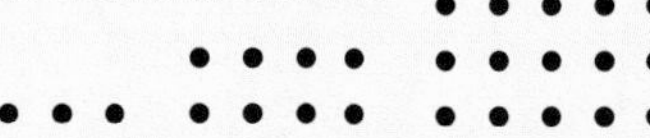

17. Find the value of *y*. 14

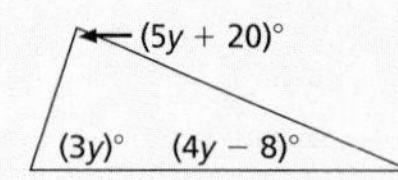

18. Find the unknown angle measure. 129°

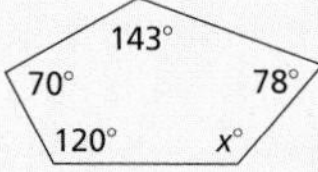

19. In the given figure, $\overline{AB} \parallel \overline{CD}$. find m∠*C*. 69°

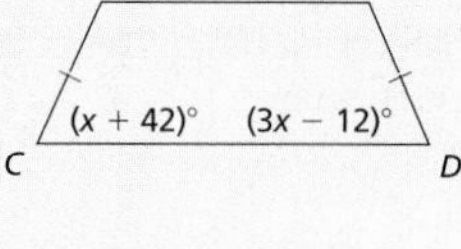

20. A house has an area of 1500 ft². If the family dog has access to the entire house, what is the probability it will be found in the 10-ft by 12-ft dining room. $\frac{2}{25}$

21. Find the volume of the triangular prism. 4,032 cm³

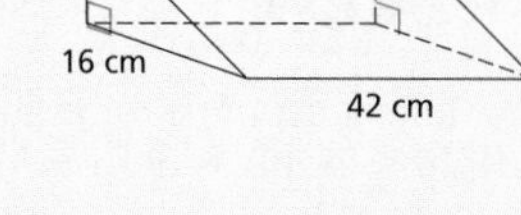

22. If the equation of a line is $6x + 2y = 18$, find its slope and *y*-intercept. −3, 9

23. Solve the system of equations: $x = 2$, $y = -4$

$$\begin{cases} 3x - 2y = 14 \\ 4x + y = 4 \end{cases}$$

24. Find the actual length: 125 km
map length: 2.5 cm, scale length: 1 cm = 50 km

25. Find *x* to the nearest tenth if $\overline{RM}$ is the altitude to the hypotenuse of right triangle *RST*. 8

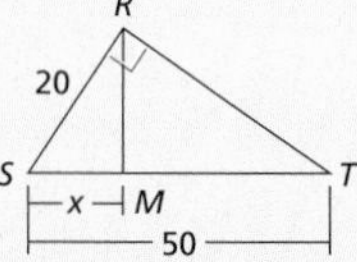

Teaching Strategies

If students have difficulty with Exercise 9 of the Cumulative Review, remind them that it is always a good idea to find the difference between the terms in a sequence; then study the differences to find a pattern. If the sequence cannot be found using subtraction, look for a multiplicative relationship or a combination of operations.

Chapters 1–7 Cumulative Assessment

Standardized Test Preparation—Standard Multiple Choice

Select the best choice for each question.

1. What is the correct scientific notation for 0.00000618?
B

A. 6.18×10^{-5} **B.** 6.18×10^{-6}
C. 6.18×10^{5} **D.** 6.18×10^{6}
E. none of these

2. Given $f(x) = -3x + 1$, find $f(-3)$.
A

A. 10 **B.** −3
C. −8 **D.** −32
E. none of these

3. Which equation is shown by the graph?
D

A. $y < x$ **B.** $y > x$
C. $y \leq x$ **D.** $y \geq x$
E. $y \leq -x$

4. Find the length of CH.
B

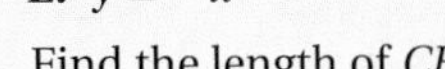

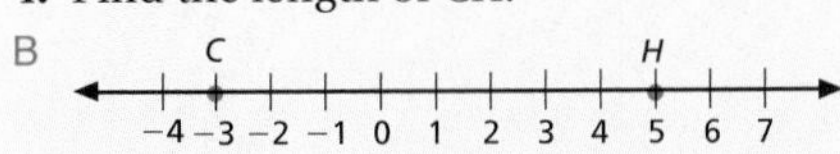

A. −8 **B.** 8
C. 2 **D.** 5
E. none of these

5. In the figure, $r \parallel s$. Find $m\angle 2$.
C

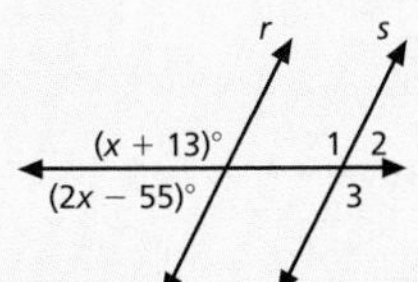

A. 74° **B.** 87°
C. 93° **D.** 68°
E. none of these

6. Which measures cannot be the lengths of the sides of a triangle?
C

A. 16 m, 12 m, 20 m **B.** 7 m, 7 m, 8 m
C. 6 ft, 3 ft, 8 ft **D.** 10 m, 10 m, 10 m
E. 12 cm, 8 cm, 21 cm

7. Given that $\overline{RL} \cong \overline{PA}$, $\overline{RL} \perp \overline{LA}$, and $\overline{PA} \perp \overline{LA}$, which postulate can be used to prove that $\triangle RLA$ is congruent to $\triangle PAL$?
B

A. ASA **B.** SAS
C. SSS **D.** AA
E. none of these

8. Which name best describes this figure?
A

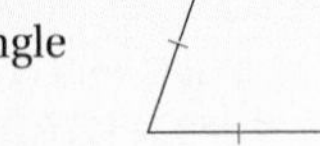

A. rhombus **B.** trapezoid
C. square **D.** rectangle
E. quadrilateral

9. Find the unit rate: \$6.40 for 80 copies.
B

A. \$0.80 per copy
B. \$0.08 per copy
C. 8 copies per dollar
D. 80 copies per dollar
E. none of these

10. How many edges are in this prism?
A

A. 15 **B.** 10
C. 2 **D.** 7
E. none of these

11. Find the surface area of the rectangular prism.
C

A. 1200 mm^2 **B.** 380 mm^2
C. 760 mm^2 **D.** 640 mm^2
E. none of these

12. CONSTRUCTED RESPONSE Given that $\overline{AB} \parallel \overline{CD}$, describe how you know that $\triangle ABE$ is similar to $\triangle CDE$. Then find the value of x.

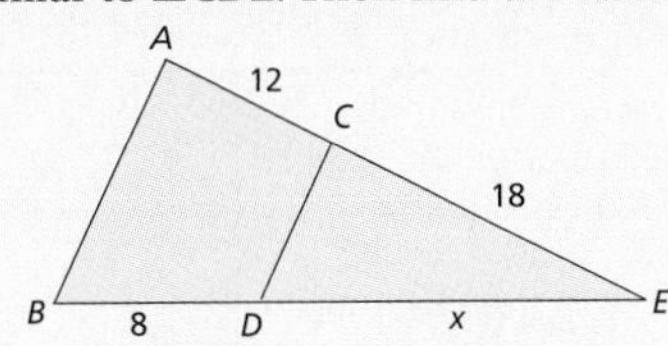

Possible answer: Since $\overline{AB} \parallel \overline{CD}$, $\triangle CC \sim \triangle DCE$ and $\triangle ABD \sim \triangle CDE$. Because of the AA postulate, the two triangles are similar: $x = 12$.

Standardized Test Preparation

STANDARD MULTIPLE CHOICE— One strategy for completing multiple-choice questions is to use mental math and eliminate unreasonable answers as choices. Often, a couple of the answer choices are values that you may get while working out the problem. For instance, if the solution requires two or three steps, one answer choice may be the value or expression that you have at the end of the first step and another choice may be the value or expression that you have at the end of the second step.

Sometimes all but the correct answer can be eliminated using only mental math and reasoning. If this is the case, verify your answer choice by working out the problem.

Testing Strategies

Remind students that if they have time left after completing a test, it is a good idea to go back and check their work. In a multiple-choice test, one way of checking answers is to look at all the incorrect answers and verify that they are incorrect. For example, in Exercise 9, students can determine the cost of 80 copies using each of the rates given. Only the correct rate will show that \$6.40 is the price of 80 copies.

Teaching Strategies

In Exercise 10 of the Cumulative Assessment, students may find it easier to count edges if they redraw the figure on scratch paper and then use a color to mark each edge as they count it. The answers to geometric problems often become apparent during the course of redrawing the figure.

Chapter 8—Transformations

Theme: Amusement Parks
Chapter Investigation: Design a Ride, pages 337, 345, 371, 377
Careers: Construction Supervisors, page 347; Aerospace Engineers, page 367
Data Activity: Classic Wooden Roller Coasters

Content and Connections

Lesson, Pages	Lesson Objectives	NCTM Standards	State/Local Objectives	Interdisciplinary Connections	Real World Applications
8–1 338–341	• Graph translation and reflections images on a coordinate plane	Patterns, Functions & Algebra Geometry & Spatial Sense Problem Solving Reasoning & Proof Representation		Art	Recreation, Art, Amusement Park Design
8–2 342–345	• Graph rotation images and identify center, angles and directions of rotations	Patterns, Functions & Algebra Geometry & Spatial Sense Reasoning & Proof Representation		Art	Ride Management, Computer Graphics, Art
8–3 348–351	• Draw dilation images on a coordinate plane	Patterns, Functions & Algebra Geometry & Spatial Sense Problem Solving Reasoning & Proof Connections; Representation		Art	Graphic Design, Business, Art
8–4 352–355	• Identify and find composites of transformations	Patterns, Functions & Algebra Geometry & Spatial Sense Problem Solving Reasoning & Proof Representation		Art	Engineering, Art, Graphics Design
8–5 358–361	• Identify matrices and elements within each matrix by rows and columns • Perform addition and scalar multiplication on matrices	Patterns, Functions & Algebra Reasoning & Proof Connections Representation			Souvenir Sales, Cryptography, Manufacturing
8–6 362–365	• Determine dimensions of product matrices in matrix multiplication • Perform row-by-column multiplication of matrices	Number Operation Patterns, Functions & Algebra Reasoning & Proof Representation			Ticket Sales, Encryption, Inventory
8–7 368–371	• Represent geometric figures on the coordinate plane by matrices • Identify and perform transformations with matrices	Patterns, Functions & Algebra Geometry & Spatial Sense Reasoning & Proof Representation			Graphic Art, Ride Design
8–8 372–373	• Solve problems using a matrix • Make a table, chart or list	Number Operation Patterns, Functions & Algebra Problem Solving Reasoning & Proof Connections; Representation			Food Distribution, Food Concessions, Souvenir Sales

Planning and Resources

Lesson, Pages	Tools/Materials Needed	Trans-parency	Learning/Teaching Styles Options	Assignments: Basic Enriched	Additional Practice in SE	Reteaching, Extra Practice, Enrichment	Other Resources
8–1 338–341	graph paper scissors ruler	Warm up 22 Trans TK-7–10, RF-29		B: 1–15, 19–34 E: 1–34	Page 346–347, 357, 367, 374, 708	R: page 115 EP: page 101 E: page 133	
8–2 342–345	graph paper ruler	Warm up 22 Trans TK-7–10, RF-30	Real World ESL/LEP	B: 1–14, 17–30 E: 1–30	Page 346–347, 357, 367, 374, 708	R: page 117 EP: page 103 E: page 135	SS: Teacher's Choice
8–3 348–351	Graph paper ruler	Warm up 22 Trans TK-7–10	Challenge	B: 1–16, 19–31 E: 1–31	Page 356–357, 367, 374, 709	R: page 119 EP: page 105 E: page 137	SS: Teacher's Choice

8–4 352–355	calculator ruler graph paper	Warm up 23 Trans TK-7–10		B: 1–17, 21–24 E: 1–24	Page 356–357, 367, 375, 709	R: page 121 EP: page 107 E: page 139	
8–5 358–361	graphing calculator	Warm up 23 Trans TK-7–10	ESL/LEP	B: 1–27, 32–40 E: 1–40	Page 366–367, 375, 710	R: page 123 EP: page 109 E: page 141	
8–6 362–365	graphing calculator	Warm up 23 Trans TK-7–10, RF-31	Real World	B: 1–33, 40–45 E: 1–45	Page 366–367, 375, 710	R: page 125 EP: page 111 E: page 143	SS: Teacher's Choice
8–7 368–371	graph paper ruler	Warm up 24 Trans Tk-7–10, RF-32	Challenge	B: 1–24, 28–34 E: 1–34	Page 375, 713	R: page 127 EP: page 113 E: page 145	
8–8 372–373	graphing calculator	Warm up 24 Trans TK-7–10	Real World	B: 1–12 E: 1–12	Page 375	R: page 129 E: page 147	

Planning and Pacing

Lesson, Pages	Lesson Title	45 min class	Assignments Basic, Enriched	90 min class	Assignments Basic, Enriched	___ min class	Assignments Basic, Enriched
8–1 338–341	AYR, Opener, Translations and Reflections	Day 102	B: 1–15, 19–34 E: 1–34	Day 53	B: 1–15, 19–34 E: 1–34		B: 1–15, 19–34 E: 1–34
8–2 342–345	Rotations in the Coordinate Plane, R&PYS	Day 103–104	B: 1–14, 17–30 E: 1–30	Day 53–54	B: 1–14, 17–30 E: 1–30		B: 1–14, 17–30 E: 1–30
8–3 348–351	Dilations in the Coordinate Plane	Day 105	B: 1–16, 19–31 E: 1–31	Day 54	B: 1–16, 19–31 E: 1–31		B: 1–16, 19–31 E: 1–31
8–4 352–355	Multiple Transformations, R&PYS	Day 106–107	B: 1–17, 21–24 E: 1–24	Day 55	B: 1–17, 21–24 E: 1–24		B: 1–17, 21–24 E: 1–24
8–5 358–361	Addition and Multiplication with Matrices	Day 108	B: 1–27, 32–40 E: 1–40	Day 56	B: 1–27, 32–40 E: 1–40		B: 1–27, 32–40 E: 1–40
8–6 362–365	More Operations on Matrices, R&PYS	Day 109–110	B: 1–33, 40–45 E: 1–45	Day 56–57	B: 1–33, 40–45 E: 1–45		B: 1–33, 40–45 E: 1–45
8–7 368–371	Transformations and Matrices	Day 111	B: 1–24, 28–34 E: 1–34	Day 57	B: 1–24, 28–34 E: 1–34		B: 1–24, 28–34 E: 1–34
8–8 372–373	Problem Solving Skills: Use a Matrix, R&PYS	Day 112	B: 1–12 E: 1–12	Day 58	B: 1–12 E: 1–12		B: 1–12 E: 1–12
Review/ Assess		Day 113–115		Day 58–59			

	Chapter 1 Days	Chapter Cumulative Days
Yearly Pacing (45 min class)	14 days	115 days
Yearly Pacing (90 min class)	7 days	59 days

Assessment Options

Assessment in Student Edition	Assessment in Teacher's Edition	Pages in Assessment Book	Software Generated Assessment
Are You Ready?, pages 334–335; Writing Math, pages, 340, 341, 344, 345, 353, 360, 365, 371; Mixed Review, pages 341, 345, 351, 355, 361, 365, 371, 373; Check Understanding pages 338, 342, 348, 352, 358, 362, 368; Mid-Chapter Quiz, page 357, Chapter Review, page 374; Chapter Assessment, page 376; Alternative Assessment, page 377; Cumulative Review, page 378; Cumulative Assessment, page 379	5-minute Warm ups, pages 338, 342, 348, 352, 358, 362, 368, 372; Quick Assessment, pages 341, 344, 350, 354, 360, 364, 370, 373; Scoring Rubrics, page 377	Mid-Chapter Quiz, page 119; Test Form A, pages 121, 123; Test Form B, pages 125, 127; Cumulative Test 129, 131; Math Journal prompt, page 133, 134;	Chapter 8

Refresher Skills

The skills on these two pages are skills that have been taught in previous math courses. Continuous review of basic math skills will make stronger math students. These skills are identified as necessary to be successful in Chapter 8.

Skills Correlation Chart

Skill	Lesson Number
Determinant of a Matrix	8-5, 8-6, 8-7, 8-8
Symmetry	8-1, 8-2, 8-3, 8-4, 8-7
Midpoint Formula	8-1, 8-2, 8-3, 8-4, 8-7

Vocabulary

determinant
symmetry
matrix
midpoint

Additional Answers

28.

29.

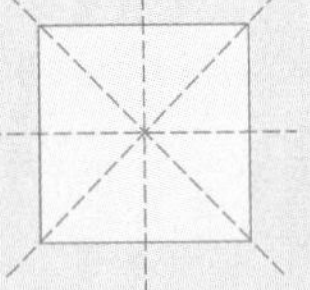

30.

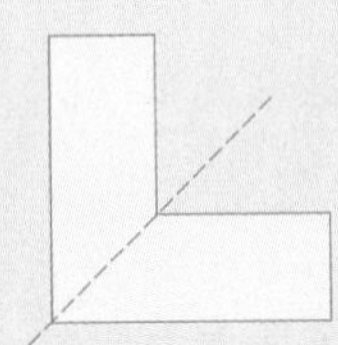

31. 33.

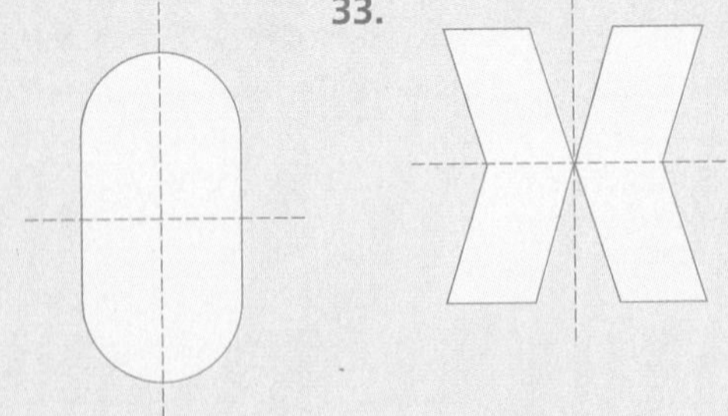

ARE YOU READY?

Refresh Your Math Skills for Chapter 8

The skills on these two pages are ones you have already learned. Stretch your memory and complete the exercises. For additional practice on these and more basic skills, see page 674.

In this chapter, you will be using matrices to solve equations. It is helpful to know how to find the determinant of a matrix.

Basic Operations with Integers

To perform basic operations with matrices, you must be able to do basic operations with integers. Recall that when adding integers whose signs are different, you actually subtract and use the sign of the larger number for the answer. When multiplying and dividing integers whose signs are different, your answer is negative.

Perform the indicated opeation.

1. $-6 + 10$ 4 **2.** $-8(-3)$ 24 **3.** $48 \div 2$ 24 **4.** $-13 - 8$ −21 **5.** -2×11 −22

6. $-19(0)$ 0 **7.** $4 - 21$ −17 **8.** $-100 - 1$ −101 **9.** $45 \div (-9)$ −6 **10.** $37 + 99$ 136

11. $-13(-1)$ 13 **12.** $-1 - 2$ −3 **13.** $-56 \div -8$ 7 **14.** $-3 + (-14)$ −17 **15.** $12 \times (-12)$ −144

Determinant of a Matrix

Example Find the determinant of this matrix: $\begin{bmatrix} 4 & -2 \\ 3 & -1 \end{bmatrix}$

To find the determinant, $\begin{vmatrix} a & c \\ b & d \end{vmatrix}$ use the formula $ad - bc$.

$$\begin{vmatrix} 4 & -2 \\ 3 & -1 \end{vmatrix} = 4(-1) - 3(-2) = -4 + 6 = 2$$

Find the determinant of each matrix.

16. $\begin{bmatrix} 5 & 3 \\ 2 & 4 \end{bmatrix}$ 14 **17.** $\begin{bmatrix} 8 & -7 \\ 3 & 2 \end{bmatrix}$ 37 **18.** $\begin{bmatrix} -5 & -1 \\ 1 & 4 \end{bmatrix}$ −19

19. $\begin{bmatrix} 9 & 4 \\ -3 & -2 \end{bmatrix}$ −6 **20.** $\begin{bmatrix} -1 & 6 \\ 8 & -7 \end{bmatrix}$ −41 **21.** $\begin{bmatrix} 4 & -5 \\ 2 & 6 \end{bmatrix}$ 34

22. $\begin{bmatrix} 8 & 2 \\ -5 & 6 \end{bmatrix}$ 58 **23.** $\begin{bmatrix} -7 & -3 \\ -2 & -5 \end{bmatrix}$ 29 **24.** $\begin{bmatrix} 3 & 0 \\ 4 & 2 \end{bmatrix}$ 6

25. $\begin{bmatrix} 0 & 5 \\ -6 & 2 \end{bmatrix}$ 30 **26.** $\begin{bmatrix} 1 & 1 \\ 1 & 1 \end{bmatrix}$ 0 **27.** $\begin{bmatrix} 0 & 3 \\ 2 & 0 \end{bmatrix}$ −6

Extend the Lesson

CHALLENGE The midpoint of segment *CD* is (2, −5). If *C* = (6, 0), what is the location of point *D*? *D*(−2, −10)

CHAPTER 8

Symmetry

A line of symmetry is a line on which a figure can be folded so that when one part is reflected over that line it matches the other part exactly.

Find all the possible lines of symmetry for each figure.
For 28–31, see additional answers.

28. 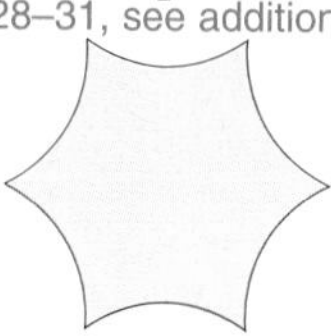

29. 

30. 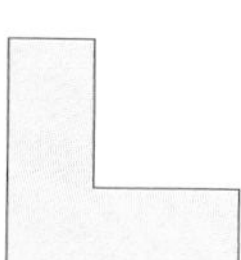

31. 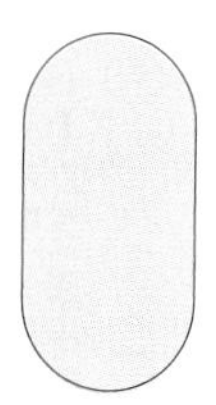

32. 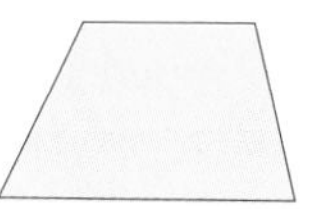
no lines of symmetry

33. 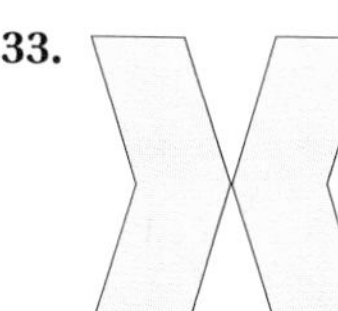
See additional answers.

Midpoint Formula

When you work on the coordinate graph, you may find the midpoint formula will come in handy.

Example Find the midpoint of the line *AB* when $A = (1, 6)$ and $B = (9, 2)$.

Remember to solve for both coordinates of the midpoint:

$x(\text{midpoint}) = x_1 + \frac{1}{2}(x_2 - x_1)$ $\qquad$ $y(\text{midpoint}) = y_1 + \frac{1}{2}(y_2 - y_1)$

$x_m = 1 + \frac{1}{2}(9 - 1)$ $\qquad$ $y_m = 6 + \frac{1}{2}(-2 - 6)$

$x_m = 1 + \frac{1}{2}(8)$ $\qquad$ $y_m = 6 + \frac{1}{2}(-8)$

$x_m = 1 + 4 = 5$ $\qquad$ $y_m = 6 + (-4) = 2$

The coordinates of the midpoint of *AB* are (5, 2).

Find the midpoint of each line.

34. *CD* when $C = (0, 3)$ $D = (-4, 9)$ (−2, 6)

35. *EF* when $E = (2, 5)$ $F = (8, -1)$ (5, 2)

36. *GH* when $G = (3, 3)$ $H = (7, 7)$ (5, 5)

37. *IJ* when $I = (1, 3)$ $J = (-5, 9)$ (−2, 6)

38. *KL* when $K = (4, 3)$ $L = (-6, -1)$ (−1, 1)

39. *MN* when $M = (8, 0)$ $N = (-4, 8)$ (2, 4)

40. *OP* when $O = (4, -3)$ $P = (4, -9)$ (4, −6)

41. *QR* when $Q = (6, -2)$ $R = (-2, 6)$ (2, 2)

Chalkboard Examples

Determinant of a Matrix

Find the determinant of the matrix below.

$\begin{bmatrix} 3 & -4 \\ 2 & -1 \end{bmatrix}$ 3

Symmetry

Draw a figure that has a horizontal line of symmetry. Answers will vary. Check students' drawings.

Midpoint Formula

Find the midpoint of segment *AB* for $A(7, -5)$ and $B(-3, 5)$. (2, 0)

Refresher Wrap-up

Quick Assessment

Ask the following questions to determine if students have mastered the basic skills reviewed on these pages.

1. Have students draw a figure with two lines of symmetry. Check students' work.
2. How could you find the midpoint of a horizontal segment. Find $\frac{1}{2}$ the sum of the *y*-coordinates. The *x*-coordinate remains the same.

Additional Practice

All topics: Refer to the Extra Practice Index on page 674 of this text.

Learning Styles

VISUAL LEARNERS To help students remember the order in which they should multiply the terms of the matrix, have them draw the symbol below over the top of each matrix with a colored pencil, starting in the upper left corner and finishing in the upper right corner.

Chapter 8 **Are You Ready?** 335

Chapter Opener

NCTM Standards/Strands

- Number & Operation
- Patterns, Functions, & Algebra
- Geometry & Spatial Sense
- Problem Solving
- Reasoning and Proof
- Connections
- Representation

Vocabulary

stress tolerance
gravitational forces
angle of descent
out-and-back coaster

Theme Connections

Tables and graphs are used to organize data about technical specifications. Technical specifications are used by engineers, designers, advertisers and the general public. For example, the type of data shown in the table can be used to compare roller coasters long after they have been torn down or rebuilt. Discuss how data and graphs are used to promote the public's interest in amusement park rides.

Career Opportunities

Many careers related to the amusement part industry make use of data and graphs. Two are highlighted in the MathWorks features. Others include artists, designers, safety specialists and marketing.

- Construction Supervisors, page 347
- Aerospace Engineers, page 367

Internet Connection

Theme Activities

Learningmathmatters.com provides web addresses to search that help students gather information about the use of math in the real world, particularly data and graphs. To search for additional addresses, begin a search using the keyword *amusement park*. Then within that search, use such key words as *roller coaster, kiddie rides, safety* and *specifications*. Students can brainstorm in small groups other key words.

CHAPTER 8

TRANSFORMATIONS

THEME: Amusement Parks

Suppose your family has the time and money for a week's vacation. What would you choose to do with your time? If you are like millions of Americans, your plans would probably include a day or two at an amusement park.

Coney Island, New York

The first amusement park in the United States was built in 1895 at Coney Island in New York City. Today, there are hundreds of amusement and theme parks throughout the country. More than 160 million people visit amusement parks across America each year. Many new companies specialize in the design and construction of new rides and adventures.

- **Construction Supervisors** (page 347) oversee the construction of new attractions. These workers must pay particular attention to details to assure public safety. Construction supervisors must follow complicated plans, oversee large budgets, and supervise carpenters, electricians, artists, and many other workers.
- **Aerospace Engineers** (page 367) design roller coasters. They use their knowledge of aerodynamics, propulsion, stress tolerances and gravitational forces to design roller coasters that are fast, exciting and safe to enjoy.

Internet Connection
www.learningmathmatters.com

Chapter Investigation

Go to learningmathmatters.com to locate additional information about the amusement park industry.

Answers to Data Activity

1. Mean Streak: 32.89 ft/sec; Beast 33.63 ft/sec. Beast is faster.
2. The reasoning is faulty since Rattler, the roller coaster with the greatest angle of descent, has a slower top speed than several coasters with less steep angles of descent.
3. Approximately 0.4 miles longer
4. About 51.6 seconds

Extend the Lesson

ART Have students study composition in photographs by selecting one photograph published in a book or magazine. The photograph should show a variety of objects. Then have them draw a similar rectangle and make a quick sketch showing the location of objects within the photograph. Have students share their work and discuss the importance of composition in a work of art.

Classic Wooden Roller Coasters

Coaster names	Top speed	Height	Track length	Ride duration	Angle of descent of first hill	Vertical drop
The Rattler	55 mi/h	179.6 ft	5080 ft	2:15	61.4°	124 ft
Shivering Timbers	65 mi/h	125 ft	5384 ft	2:30	53.25°	120 ft
Texas Giant	65 mi/h	143 ft	4920 ft	2:30	53°	137 ft
Mean Streak	65 mi/h	161 ft	5427 ft	2:45	52°	155 ft
Georgia Cyclone	50 mi/h	95 ft	2970 ft	1:48	53°	78.5 ft
The Beast	65 mi/h	135 ft	7400 ft	3:40	45°	141 ft

Data Activity: Classic Wooden Roller Coasters

Use the table for Questions 1–4.

1. Find the average speed in feet per second of Mean Streak and The Beast. Which coaster has the fastest average speed? (*Hint:* Use the formula $d = rt$, where d = distance, r = rate, and t = time.) Mean Streak: 32.89 ft/sec; Beast: 33.63 ft/sec; Beast is faster.
2. Some say that the characteristic that most influences the top speed of a coaster is the angle of descent of the first hill. The method for measuring the angle of descent is shown in the diagram at the right. Do you agree with this thinking? Explain your reasoning. See additional answers.
3. How many miles longer is The Beast than The Rattler? Approximately 0.4 miles longer
4. If Texas Giant could maintain its top speed for the entire length of the track, what would be the duration of the ride? About 51.6 seconds

78°

Chapter Investigation

Amusement parks are constantly building new rides to attract new and returning customers. Designing new rides requires an understanding of geometry, physics and construction techniques. Engineers are always looking for safe ways to provide greater thrills.

Working Together

Design an "out-and-back" roller coaster with eight hills. Make a scale drawing of the coaster indicating the height of each hill and the angle of descent. Estimate the track length and top speed of your coaster. Use the Chapter Investigation logo to guide your group's drawing.

Project Planning Calendar

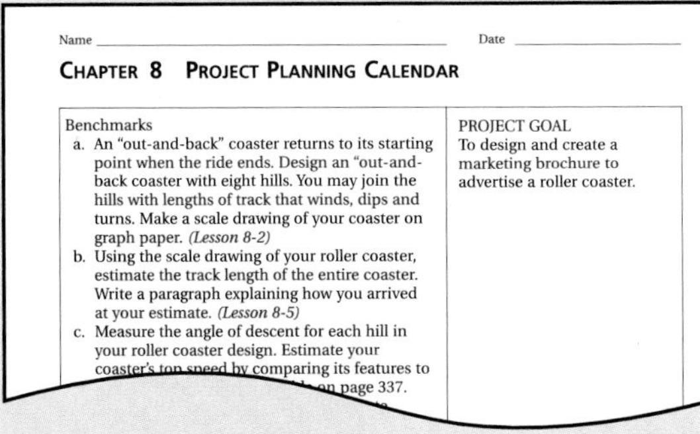

Name ____________ Date ____________

CHAPTER 8 PROJECT PLANNING CALENDAR

Benchmarks	PROJECT GOAL
a. An "out-and-back" coaster returns to its starting point when the ride ends. Design an "out-and-back coaster with eight hills. You may join the hills with lengths of track that winds, dips and turns. Make a scale drawing of your coaster on graph paper. *(Lesson 8-2)* b. Using the scale drawing of your roller coaster, estimate the track length of the entire coaster. Write a paragraph explaining how you arrived at your estimate. *(Lesson 8-5)* c. Measure the angle of descent for each hill in your roller coaster design. Estimate your coaster's top speed by comparing its features to ... on page 337.	To design and create a marketing brochure to advertise a roller coaster.

Group Project Planner

Name ____________ Date ____________

CHAPTER 8 GROUP PROJECT PLANNER

Assignment ____________ Objective ____________

Group Members ____________ Assigned Roles ____________

1)
2)
3)
4)
5)

Deadlines Done

Data Activity

Discuss with students the importance of measurements and other specifications in evaluating the safety and performance of amusement park rides. As parks compete for the public dollar, they attempt to build bigger and faster rides. Ask students to name some of their favorite thrill rides. Discuss what makes one ride more exciting than another. Assign students Questions 1–4.

Extend the Data Activity

REAL WORLD CONNECTION Have students find data about rides at various amusement parks they have visited or would like to visit. Using the data alone, which rides seem most exciting.

Chapter Investigation

As an Overarching Problem

Display several photographs of "out-and-back" coasters. Ask students to study the photographs and to describe what they see. Then have students discuss how they might design a coaster like those shown in the photographs. Students will continue to work on the investigation as they complete the exercises identified by the Investigation Logo that are found throughout the chapter. These exercises will guide students through the task described in *Working Together*. Encourage students to keep all of their work on the Investigation together. Have students use the suggestions in the Chapter Investigation Extension to summarize their work.

As a Chapter Project

The goal of this project is to design an out-and-back roller coaster and to create a marketing brochure to advertise the new ride. Students can use the Group Project Planner on page 131 and the Project Planning Calendar on page 132 in the Enrichment text to complete the project. Benchmarks **a**, **b**, and **c**. should be completed after the lesson listed in parentheses has been studied. Benchmark **d** should be completed at the end of the chapter.

Lesson Planning

NCTM Standards/Strands

- Number & Operation
- Patterns, Functions, & Algebra
- Geometry & Spatial Sense
- Problem Solving
- Reasoning and Proof
- Connections
- Representation

Vocabulary

translations, image, reflection, preimage, transformation, line of reflection

Materials Needed

graph paper
rulers
scissors

Lesson Resources

Warm-Up Transparency 22
Transparency TK-7–10, RF-29
Reteaching 8-1
Extra Practice 8-1
Enrichment 8-1

Assignment Guide

Basic: 1–15, 19–34
Enriched: 1–34

Getting Started

5-Minute Warm-Up

Let $x = 6$ and $y = -2$. Find the coordinates of each point.

1. $A(-x, y)$ $(-6, -2)$
2. $B(x, -y)$ $(6, 2)$
3. $C(x + 2, y - 3)$ $(8, -5)$

Introduction to Lesson 8-1

You may wish to review briefly the coordinate plane with students. Have them identify each quadrant and name a point in each one. Ask students how the three triangles are the same and how they are different. They are each the same size and shape (congruent); they are each in different positions on the coordinate plane.

8-1 Translations and Reflections

Goals ■ Graph translation and reflection images on a coordinate plane.

Applications Recreation, Art, Amusement Park Design

Work with a partner. You will need a piece of graph paper, a ruler, and scissors.

1. Using the ruler, draw an isosceles triangle near one bottom corner of your graph paper. Make the base $\overline{AB}$ 6 units long and the height 4 units. Cut out the triangle and label the vertices *A*, *B*, and *C*.
2. Draw a coordinate plane and label each axis from -10 to 10.
3. Place the triangle in the second quadrant so that $\overline{AB}$ is parallel to, but not on, the *x*-axis, and each vertex is at the intersection of a horizontal and a vertical line. Trace the triangle and label each vertex to match the original triangle. Label this figure "Triangle 1."
4. Slide your triangle 9 units straight to the right. Trace the triangle in this new position and label each vertex. Label this figure "Triangle 2."
5. Turn your triangle so vertex *C* is below $\overline{AB}$. Now place the triangle in the fourth quadrant with $\overline{AB}$ parallel to the *x*-axis. Trace this triangle.
6. Record the slopes of each side of the three triangles in a table like the one shown.
7. Compare the slopes of the sides of Triangles 1 and 2, Triangles 2 and 3, and Triangles 1 and 3. What do you notice? See additional answers.

	Triangle 1	Triangle 2	Triangle 3
$\overline{AB}$			
$\overline{AC}$			
$\overline{BC}$			

Build Understanding

A **translation** is a *slide* of a figure. It produces a new figure exactly like the original. The new figure is the **image** of the original figure, and the original figure is the **preimage** of the new figure. A translation is also known as a **transformation** of a figure.

Another kind of transformation that yields a congruent figure is a **reflection**, or *flip*. Under a reflection, a figure is *reflected*, or *flipped*, over a **line of reflection**.

If a line can be drawn through a geometric figure such that the figure on one side of the line is the reflection of the figure on the opposite side, the figure is said to exhibit line symmetry and the line is a **line of symmetry**, or *axis of symmetry*. A reflection and its preimage combined demonstrate line symmetry.

Check Understanding

Write the formula for finding the slope when given two points and their coordinates.

$\frac{y_2 - y_1}{x_2 - x_1}$ for (x_1, y_1) and (x_2, y_2)

Example 1

Graph the image of parallelogram *MNOP* with vertices *M*(2, 1), *N*(4, 7), *O*(7, 7), and *P*(5, 1) under each transformation from the original position.

a. 9 units down **b.** reflected across the *y*-axis.

Additional Answers

7. The slopes of corresponding sides of Triangles 1 and 2 are the same. The slopes of corresponding sides of Triangles 2 and 3 and of Triangles 1 and 3 are opposites.

Solution

a. To move the image 9 units down, subtract 9 from the y-coordinate of each vertex.

$M(2, 1) \to M'(2, 1 - 9) \to M'(2, -8)$

$N(4, 7) \to N'(4, 7 - 9) \to N'(4, -2)$

$O(7, 7) \to O'(7, 7 - 9) \to O'(7, -2)$

$P(5, 1) \to P'(5, 1 - 9) \to P'(5, -8)$

b. The reflection of the point (x, y) across the y-axis is the point $(-x, y)$.

$M(2, 1) \to M''(-2, 1)$ $\quad N(4, 7) \to N''(-4, 7)$

$O(7, 7) \to O''(-7, 7)$ $\quad P(5, 1) \to P''(-5, 1)$

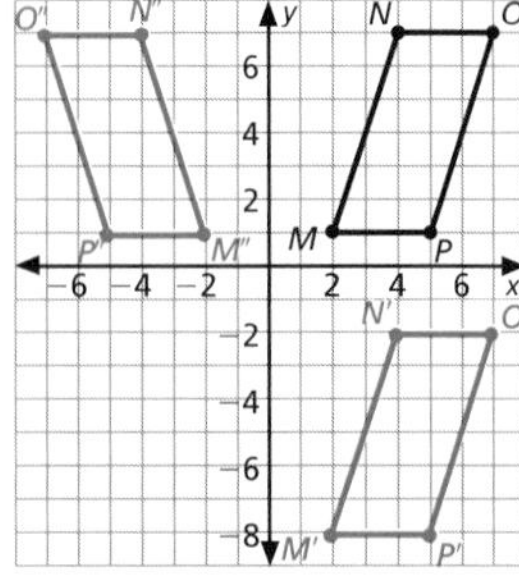

Example 2

Compare the slopes of corresponding non-horizontal sides for the preimage and each transformation in Example 1.

Solution

For the first transformation, compare the slopes of $\overline{MN}$ and $\overline{M'N'}$, as well as the slopes of $\overline{OP}$ and $\overline{O'P'}$. For the second transformation, compare the slopes of $\overline{MN}$ and $\overline{M''N''}$, as well as the slopes of $\overline{OP}$ and $\overline{O''P''}$.

Side	$\overline{MN}$	$\overline{M'N'}$	$\overline{OP}$	$\overline{O'P'}$	$\overline{M''N''}$	$\overline{O''P''}$
Slope	3	3	3	3	−3	−3

For the translation in part **a**, corresponding sides have equal slopes. For the reflection in part **b**, corresponding sides have opposite slopes.

Example 3

RECREATION At a miniature golf course, a hole is designed so that the ball must travel along a line of reflection between two congruent triangular blocks.

Graph the image of $\triangle ABC$ with vertices $A(3, 5)$, $B(5, 8)$, and $C(1, 7)$ under a reflection across the line whose equation is $y = x$. Compare the slopes of the corresponding sides of $\triangle ABC$ and $\triangle A'B'C'$.

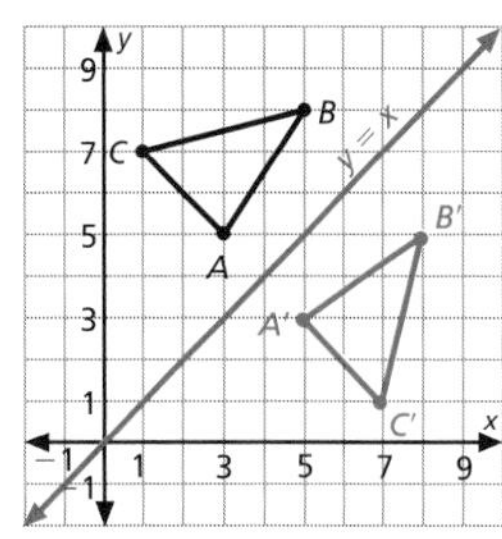

Solution

Graph the image and reflection as directed. Graph the line for $y = x$. Use the rule $(x, y) \to (y, x)$. Make a table to compare the slopes of the corresponding sides. The slopes of corresponding sides are reciprocals of each other.

Side	$\overline{AB}$	$\overline{BC}$	$\overline{CA}$	$\overline{A'B'}$	$\overline{B'C'}$	$\overline{C'A'}$
Slope	$\frac{3}{2}$	$\frac{1}{4}$	−1	$\frac{2}{3}$	4	−1

Chalkboard Examples

Supplementary Example 1

Graph the image of parallelogram $ABCD$ with vertices $A(1, 3)$, $B(3, 3)$, $C(3, 1)$ and $D(5, 1)$ under each transformation from the original position.

a. 4 units down Vertices are $A'(1, -1)$, $B'(3, -1)$, $C'(3, -3)$ and $D'(5, -3)$.

b. reflected across the y-axis Vertices are $A'(-1, 3)$, $B'(-3, 3)$, $C'(-3, 1)$ and $D'(-5, 1)$.

Supplementary Example 2

Compare the slopes of corresponding non-horizontal sides for each preimage and image above. For part a, corresponding sides have equal slopes (−1). For part b, corresponding sides have opposite slopes (−1 and 1).

Supplementary Example 3

The figure shown is half of a symmetrical design. Complete the figure. Check students' work.

Teaching Strategies

Some students may have difficulty identifying corresponding vertices of reflected or rotated (Lesson 8-2) images and those of the preimage. You may wish to suggest that students color-code the vertices to help them. For example, when reflecting $\triangle ABC$, students could use blue for A and A', red for B and B', and yellow for C and C'.

Reteaching Worksheet 8-1

Name ______ Date ______

RETEACHING 8-1

TRANSLATIONS AND REFLECTIONS

Two transformations that yield congruent figures include a **translation** or *slide* of a figure and a **reflection** or *flip* of a figure. When a line is reflected across the x-axis, $(x, y) \to (x, -y)$. When a line is reflected across the y-axis, $(x, y) \to (-x, y)$.

Example 1

Graph the image of $\triangle ABC$ under a translation of 8 units up.

Solution

Step 1: To find the coordinates of the vertices of the image, add 8 to each y-coordinate of the pre-image.

$A(-6, -4) \to A'(-6, 4)$ $\quad B(-2, -2) \to B'(-2, 6)$

$C(-3, -6) \to C'(-3, 2)$

Step 2: Graph these coordinates and draw the image.

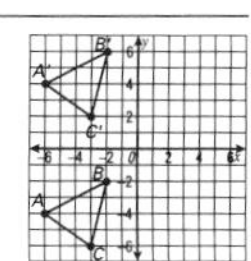

Example 2

Graph the image of $\triangle DEF$ under a reflection across the y-axis.

Solution

Step 1: To find the coordinates of the vertices of the image, use the rule $(x, y) \to (-x, y)$.

$D(-5, 4) \to D'(5, 4)$ $\quad E(-2, 5) \to E(-2, 5) \to E'(2, 5)$

$F(-3, 1) \to F'(3, 1)$

Step 2: Graph these coordinates and draw the image.

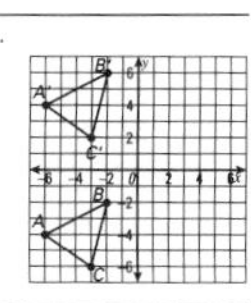

EXERCISES

Find the coordinates of the vertices of the image of $\triangle DEF$ in Example 2 under a translation of:

1. 5 units right. $D'(0, 4)$; $E'(3, 5)$; $F'(2, 1)$
2. 2 units left. $D'(-7, 4)$; $E'(-4, 5)$; $F'(-5, 1)$

Find the coordinates of the vertices of the image of $\triangle ABC$ in Example 1 under a reflection across:

3. the y-axis. $A'(6, -4)$; $B'(2, -2)$; $F'(3, -6)$
4. the x-axis. $A'(-6, 4)$; $B'(-2, 2)$; $C'(-3, 6)$

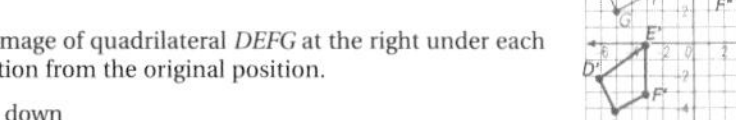

Graph the image of quadrilateral $DEFG$ at the right under each transformation from the original position.

5. 6 units down
6. reflected across the x-axis

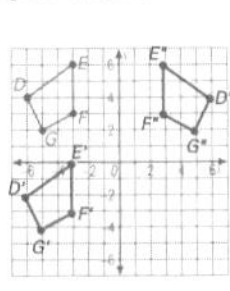

116

Lesson 8-1 **Translations and Reflections** 339

Lesson Wrap-up

Quick Assessment

Ask the following questions to determine if students understand the content presented in this lesson.

1. How do the slopes of corresponding sides in a translated image compare with those in the preimage? The slopes are the same. How do the slopes of corresponding sides in a reflected image compare with those in the preimage? The slopes are opposite or reciprocals.
2. A figure can have more than one line of symmetry. How many lines of symmetry does a square have? 4

Assignment Guide

Basic: 1–15, 19–34
Advanced: 1–34
Additional Practice: See Extra Practice Index on page 674.

Additional Answers

1–2.

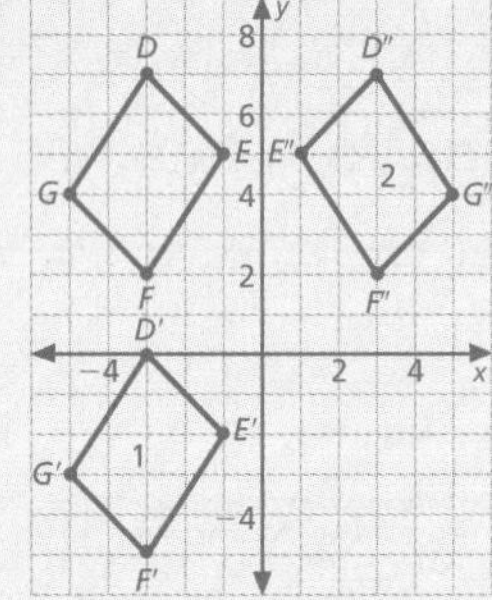

4.

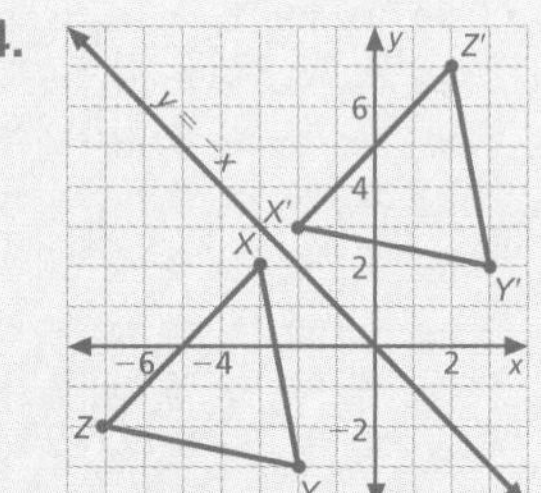

6–7.

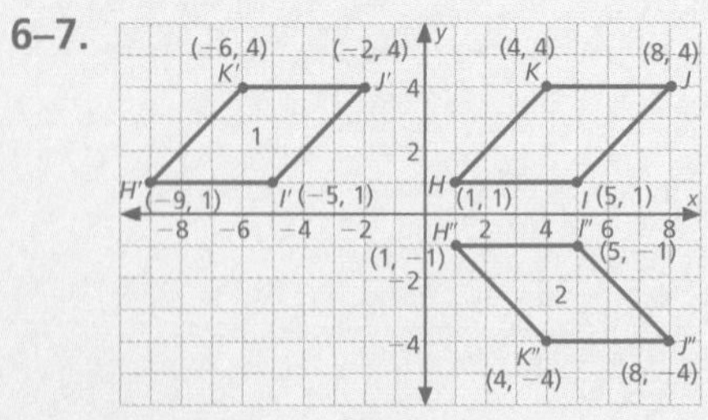

If an object is symmetrical and you only have half of it, you can draw the other half.

Example 4

ART The figure shown at the right is half of a symmetrical design. Complete the figure.

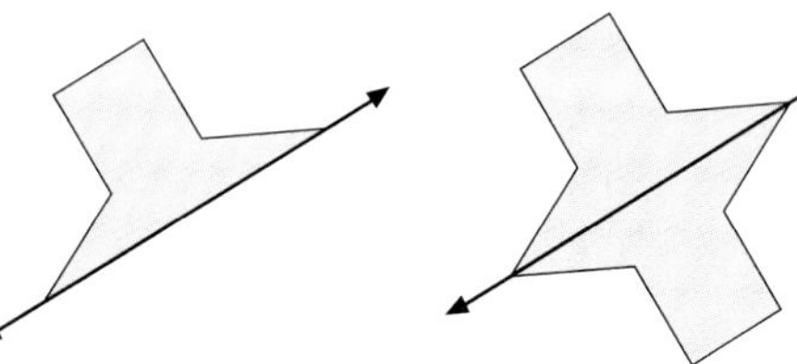

Solution

Draw a reflection across the line of symmetry for the half of the figure shown.

Try These Exercises

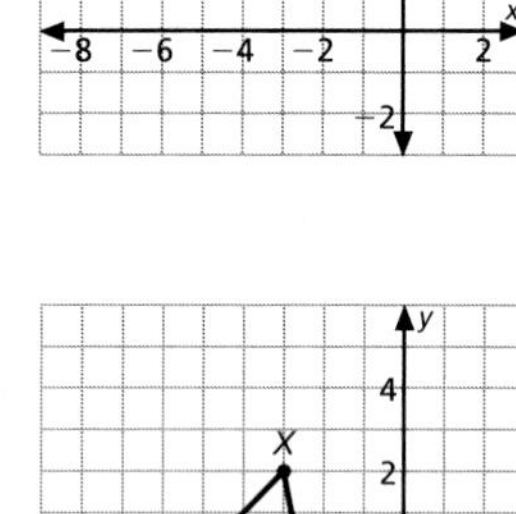

Copy parallelogram *DEFG* at the right on a coordinate plane. Then graph its image under each transformation from the original position.
For 1–3, see additional answers.

1. 7 units down ($D'E'F'G'$)
2. reflected across the y-axis ($D''E''F''G''$)
3. Copy and complete the chart below.

Side	$\overline{DE}$	$\overline{D'E'}$	$\overline{D''E''}$	$\overline{EF}$	$\overline{E'F'}$	$\overline{E''F''}$
Slope						

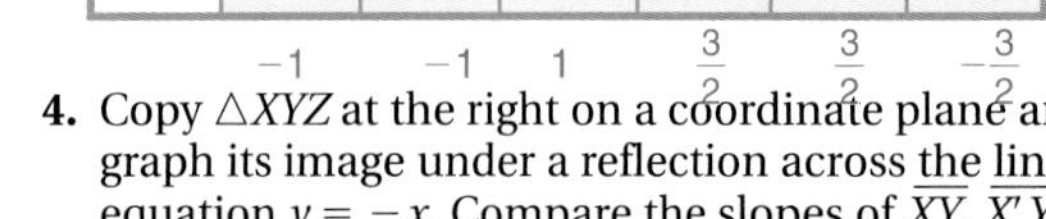

-1 -1 1 $\frac{3}{2}$ $\frac{3}{2}$ $-\frac{3}{2}$

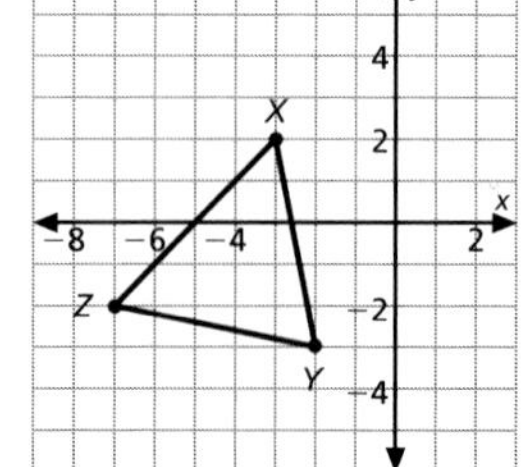

4. Copy $\triangle XYZ$ at the right on a coordinate plane and graph its image under a reflection across the line with equation $y = -x$. Compare the slopes of $\overline{XY}$, $\overline{X'Y'}$, $\overline{YZ}$, $\overline{Y'Z'}$, $\overline{XZ}$, and $\overline{X'Z'}$. See additional answers.

5. **WRITING MATH** How can you recognize a line of symmetry?
If the figure on one side of a line is the reflection of the figure on the opposite side, the line is a line of symmetry.

Practice Exercises

On a coordinate plane, graph parallelogram *HIJK* with vertices $H(1, 1)$, $I(5, 1)$, $J(8, 4)$, and $K(4, 4)$. Then graph its image under each transformation from the original position. For 6–9, see additional answers.

6. 10 units left
7. reflected across the x-axis
8. Compare the slopes of the non-horizontal sides of parallelogram *HIJK* in all three positions above.
9. Graph the image of $\triangle ABC$ with vertices $A(1, 4)$, $B(5, 6)$, and $C(2, 7)$ under a reflection across the line with equation $y = x$. Compare the slopes of the sides of $\triangle ABC$ and $\triangle A'B'C'$.

10. **YOU MAKE THE CALL** Jenna says that the rule $(x, y) \rightarrow (4 - x, y)$ can be used to translate an image 4 units to the left. Do you agree with Jenna's thinking?
Instead of subtracting x from 4, Jenna needs to subtract 4 from x. The correct rule is $(x, y) \rightarrow (x - 4, y)$.

8.

side	JI	KH	$J'I'$	$K'H'$	$J''I''$	$K''H''$
slope	1	1	1	1	−1	−1

9.

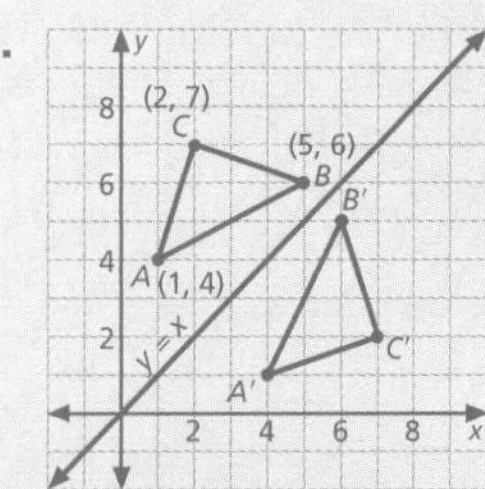

side	AB	BC	CA	$A'B'$	$B'C'$	$C'A'$
slope	$\frac{1}{2}$	$-\frac{1}{3}$	3	2	−3	$\frac{1}{3}$

11.

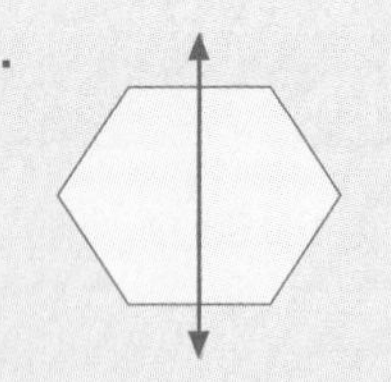

12.

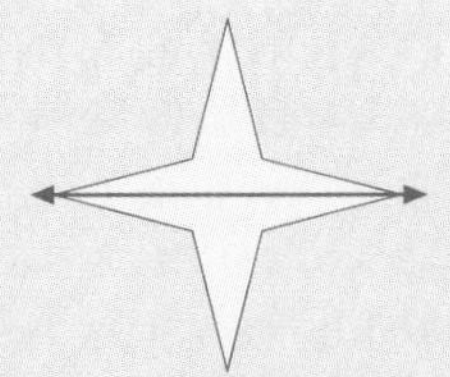

13.

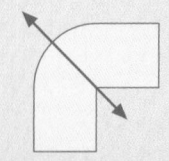

14.

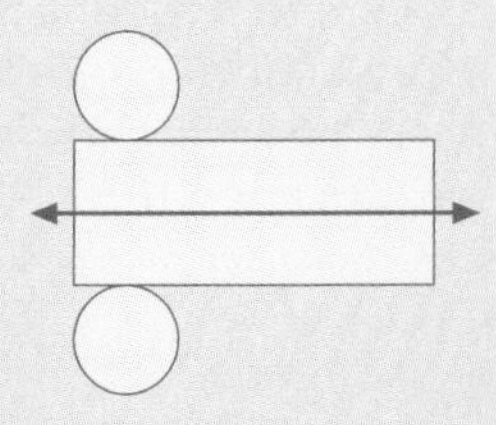

Copy the half shown for each figure below on graph paper along with its line of symmetry. Then complete the figure.
For 11–14, see additional answers.

11.

12.

13.

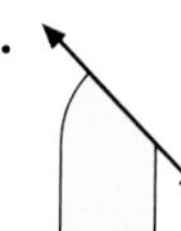

14.

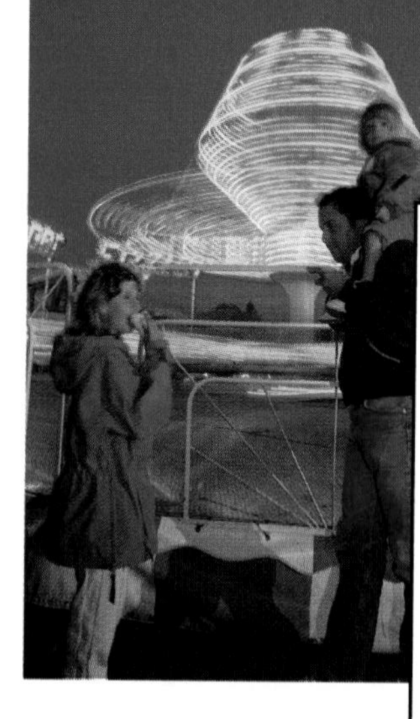

15. AMUSEMENT PARK DESIGN A new roller coaster has two entrances to the boarding platform. The eastern entrance is a reflection of the western entrance shown to the right. Copy the entrance on graph paper. Then draw its reflection.
See additional answers.

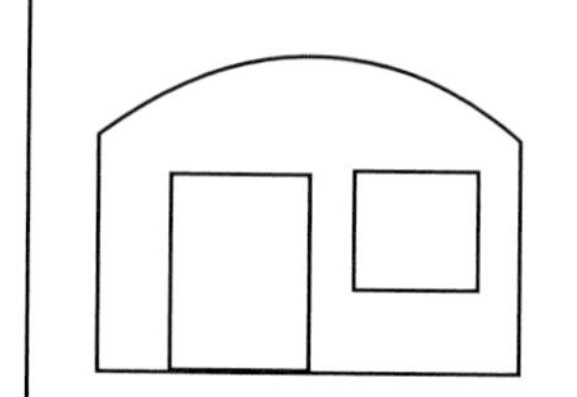

Extended Practice Exercises

16. Triangle $D'E'F'$ is the image of a figure that was translated under the rule $(x, y) \rightarrow (x + 3, y - 2)$. What are the vertices of the preimage of $\triangle D'E'F'$? What are the slopes of the sides $\overline{D'E'}$, $\overline{E'F'}$, and $\overline{D'F'}$? Are the slopes of the preimage the same? See additional answers.

17. How do the slopes of translated figures compare?
The slopes of translated figures are equal.

18. WRITING MATH How do the slopes of figures reflected across each line compare? Support your answer with an example for each. See additional answers.
a. x-axis   b. y-axis   c. $y = x$

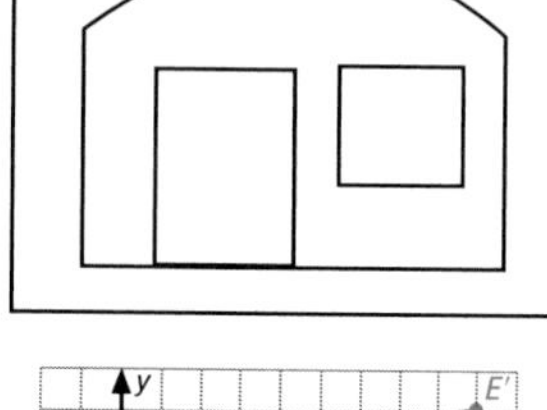

Mixed Review Exercises

Solve each proportion and find the value of x. Round to the nearest hundredth if necessary. (Lesson 7-1)

19. $\frac{2}{5} = \frac{6}{x}$ 15
20. $\frac{3}{8} = \frac{x}{24}$ 9
21. $\frac{75}{90} = \frac{15}{x}$ 18
22. $\frac{23}{48} = \frac{x}{60}$ 28, 75
23. $\frac{2x}{10} = \frac{6}{15}$ 2
24. $\frac{x-4}{12} = \frac{x}{8}$ −8
25. $\frac{3}{x+2} = \frac{18}{36}$ 4
26. $\frac{1}{3} = \frac{5x}{45}$ 3
27. $\frac{72}{48} = \frac{x+8}{4}$ −2
28. $\frac{125}{3x} = \frac{5}{3}$ 25
29. $\frac{169}{286} = \frac{13}{4x+2}$ 5
30. $\frac{x-2}{54} = \frac{8}{48}$ 11

Exercises 31–34 refer to the figure at the right. (Lesson 3-3)

31. Name the midpoint of $\overline{HJ}$. I

32. Name the segment whose midpoint is B. $\overline{AC}$

33. Name all the segments whose midpoint is E. $\overline{DF}$, $\overline{CG}$, $\overline{BH}$, $\overline{AI}$

34. Assume Z is the midpoint of $\overline{BE}$. What is its coordinate? −3.5

15.

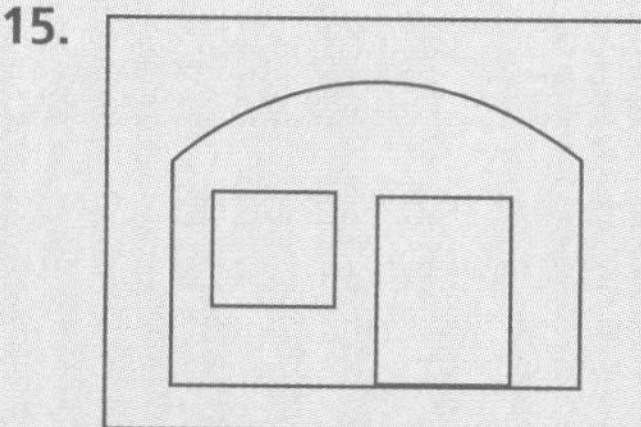

16. $D(0, 6)$, $E(6, 7)$, $F(5, 3)$

side	$D'E'$	$E'F'$	$D'F'$	DE	EF	DF
slope	$\frac{1}{6}$	4	$-\frac{3}{5}$	$\frac{1}{6}$	4	$-\frac{3}{5}$

Yes, slopes are same.

18. a. Slopes of figures reflected across the x-axis are negatives of each other.
b. Slopes of figures reflected across the y-axis are negatives of each other.
c. Slopes of figures reflected across the line $y = x$ are reciprocals of each other.

Extra Practice Worksheet 8-1

Name ________________ Date ________

EXTRA PRACTICE 8-1

TRANSLATIONS AND REFLECTIONS

EXERCISES

On a coordinate grid, graph parallelogram $ABCD$ with vertices $A(2, 4)$, $B(6,4)$, $C(5, 2)$ and $D(1, 2)$. Then graph its image under each transformation from the original position. Use your own paper. Check students' graphs.

1. 5 units left
2. 3 units right
3. reflected across the x-axis
4. reflected across the y-axis

5. Compare the slopes of the non-horizontal sides of parallelogram $ABCD$ in all five positions above. For the original parallelogram $ABCD$ and 1 and 2, the slope of $\overline{BC}$, $\overline{AD}$, $\overline{B'C'}$, and $\overline{A'D'}$ is 2. For 3 and 4, the slope of $\overline{B'C'}$ and $\overline{A'D'}$ is −2.

6. Graph the image of $\triangle MNP$ with vertices $M(3, 1)$, $N(4, 4)$ and $P(6, 2)$ under a reflection across the line with equation $y = x$. Use your own paper. Compare the slopes of the sides of $\triangle MNP$ and $\triangle M'N'P'$. slope of $\overline{MN} = 3$, slope of $\overline{NP} = -1$, slope of $\overline{PM} = -\frac{1}{3}$, slope of $\overline{M'N'} = \frac{1}{3}$, slope of $\overline{N'P'} = -1$, slope of $\overline{P'M'} = 3$

7. Triangle $A'B'C'$ is the image of a figure that was translated under the rule $(x, y) \rightarrow (x - 4, y + 1)$. What are the vertices of the preimage of $\triangle A'B'C'$? What are the slopes of the sides $\overline{A'B'}$, $\overline{B'C'}$ and $\overline{A'C'}$? Are the slopes of the preimage the same?

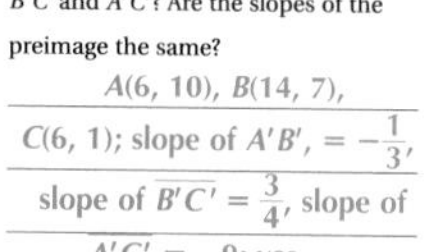

$A(6, 10)$, $B(14, 7)$, $C(6, 1)$; slope of $A'B'$, $= -\frac{1}{3}$, slope of $\overline{B'C'} = \frac{3}{4}$, slope of $\overline{A'C'} = -9$; yes

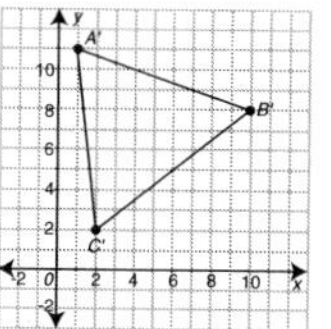

102   EXTRA PRACTICE LESSON 8-1

Enrichment Worksheet 8-1

Name ________________ Date ________

ENRICHMENT 8-1

MODIFIED TESSELLATIONS

The only regular polygons that will tessellate a plant surface are the equilateral triangle, the square and the regular hexagon. However, the shapes of these polygons can be modified in various ways to create different designs.

Example 1

Create a modified tessellation based on an equilateral triangle by replacing two sides of the basic unit with the same curved line.

Solution

Example 2

Create a modified tessellation based on a square by replacing each side of the basic unit with a line segment that has rotational symmetry about its midpoint.

Solution

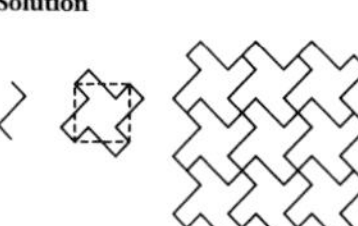

EXERCISES   Check student's tessellations.

1. Follow the example above. Use Curve A to make a modified triangular tessellation.
2. Follow the example above. Use Curve B to make a modified square tessellation.
3. Use Curve B. Make a modified triangular tessellation by replacing all three sides of the basic unit with Curve B.
4. Use Curve C to make a modified triangular tessellation. Replace all three sides of the basic unit. (There is more than one way to do this.)
5. Create an original modified triangular tessellation.
6. Create an original modified square tessellation.

Curve A

Curve B

Curve C

134   ENRICHMENT LESSON 8-1

Lesson Planning

NCTM Standards/Strands

- Number & Operation
- Patterns, Functions, & Algebra
- Geometry & Spatial Sense
- Problem Solving
- Reasoning and Proof
- Connections
- Representation

Vocabulary

clockwise
counterclockwise
rotation
origin
center of rotation
angle of rotation

Materials Needed

graph paper
rulers

Lesson Resources

Warm-Up Transparency 22
Transparency TK-7–10, RF-30
Reteaching 8-2
Extra Practice 8-2
Enrichment 8-2

ASSIGNMENT GUIDE

Basic: 1–14, 17–30
Enriched: 1–30

Getting Started

5-MINUTE WARM-UP

Name the portion of a circle for the number of degrees.

1. 360° whole circle
2. 180° semicircle
3. 90° quarter circle

Introduction to Lesson 8-2

Ask students how the number of degrees in a turn or fraction of a turn of the gears relates to the number of degrees in a circle or fraction of a circle. Ask them how many degrees the gear travels in Part 2 and what fractional part of a turn the gear travels in Part 3.

They are the same; 90°, $\frac{1}{2}$.

8-2 Rotations in the Coordinate Plane

Goals ■ Graph rotation images and identify centers, angles and directions of rotations.

Applications Ride Management, Computer Graphics, Art

Think about how gears mesh and turn one another. A gear is a mechanical device that transfers rotating motion and power from one part of a machine to another.

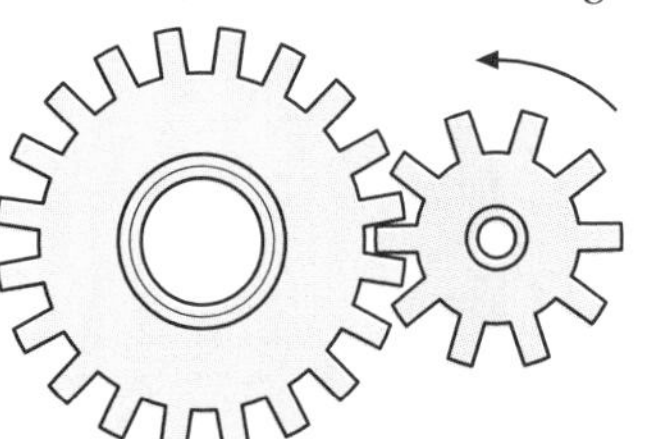

1. From the side at which you see the gears, would you say the larger gear turns clockwise or counterclockwise? clockwise
2. What fractional part of a turn does it take for a gear tooth to get from the top to a horizontal position? $\frac{1}{4}$
3. How many degrees does a gear tooth travel from the top to the bottom of the gear? 180°

BUILD UNDERSTANDING

Another transformation that produces a figure congruent to the original is a **rotation**, or *turn*. A figure is rotated, or turned, about a point.

Rotation is described by three pieces of information:

- the point about which the figure is rotated, or the **center of rotation**.
- the amount of turn expressed as a fractional part of a whole turn, or as an **angle of rotation** in degrees.
- the rotation direction—*clockwise* or *counterclockwise*.

When you rotate a point 180° clockwise about the origin, both the x-coordinate and the y-coordinate are transformed into their opposites.

Check Understanding

How many degrees are in a one-quarter turn? How many degrees are in a one-half turn? How many degrees are in a three-quarter turn? How many degrees are in a full turn?

90°, 180°, 270°, 360°

Example 1

Graph $\triangle QRS$ and its image after a 180° clockwise rotation about the origin. Then compare the slopes of $\overline{QR}$, $\overline{Q'R'}$, $\overline{QS}$, and $\overline{Q'S'}$.

Solution

Use the rule $(x, y) \rightarrow (-x, -y)$.

$Q(3, 4) \rightarrow Q'(-3, -4)$
$R(1, 1) \rightarrow R'(-1, -1)$
$S(5, 1) \rightarrow S'(-5, -1)$

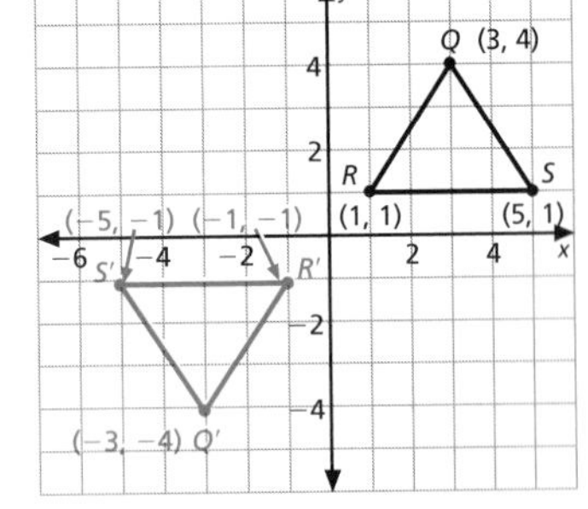

Side	$\overline{QR}$	$\overline{QS}$	$\overline{Q'R'}$	$\overline{Q'S'}$
Slope	$\frac{3}{2}$	$-\frac{3}{2}$	$\frac{3}{2}$	$-\frac{3}{2}$

The slopes are the same.

Teaching Strategies

COOPERATIVE LEARNING Have students work in pairs to solve the following problem: Draw a figure on a coordinate plane, and label all vertices. Develop a method to rotate the figure 45° and 60° clockwise and counterclockwise about the origin.

When you rotate a figure 90° counterclockwise, the y-coordinate is multiplied by -1, and then the x-coordinate and y-coordinate are transposed. That is, $(x, y) \rightarrow (-y, x)$.

Example 2

RIDE MANAGEMENT Computers are used to signal ride operators when it is safe to begin a new ride cycle. The ride can start when the screen shows a raised flag. A lowered flag tells the operator to wait. On the computer screen,the raised flag contains the points $A(0, 0)$, $B(-2, 2)$, $C(-5, 5)$, and $D(-5, 2)$. Graph the flag and its image after a 90° counterclockwise rotation about the origin. Label the points of the image A', B', C', and D'. Then compare the slope of $\overline{BC}$ with the slope of $\overline{B'C'}$.

Solution

Use the rule $(x, y) \rightarrow (-y, x)$

$A(0, 0) \quad \rightarrow A'(0, 0)$

$B(-2, 2) \rightarrow B'(-2, -2)$

$C(-5, 5) \rightarrow C'(-5, -5)$

$D(-5, 2) \rightarrow D'(-2, -5)$

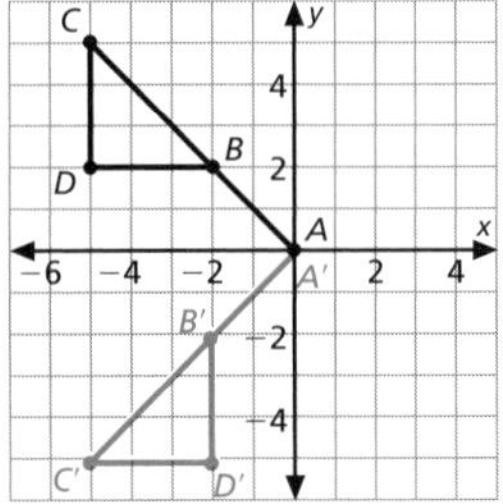

$$\text{slope of } \overline{BC} = \frac{5-2}{-5-(-2)} = \frac{3}{-3} = -1 \qquad \text{slope of } \overline{B'C'} = \frac{-5-(-2)}{-5-(-2)} = \frac{-3}{-3} = 1$$

The product of the slopes is -1. The lines are perpendicular.

Example 3

Triangle XYZ is rotated twice about the origin. Compare the slopes and determine the angle of rotation first for rotation 1 and then for rotation 2.

Original position		After rotation 1		After rotation 2	
Side	Slope	Side	Slope	Side	Slope
$\overline{XZ}$	2	$\overline{X'Z'}$	2	$\overline{X''Z''}$	$-\frac{1}{2}$
$\overline{YZ}$	$-\frac{3}{4}$	$\overline{Y'Z'}$	$-\frac{3}{4}$	$\overline{Y''Z''}$	$\frac{4}{3}$
$\overline{XY}$	$\frac{1}{6}$	$\overline{X'Y'}$	$\frac{1}{6}$	$\overline{X''Y''}$	$\frac{1}{6}$

Solution

The first rotation is 180° or 360°, because the slopes are equal to the slopes in the original position.

The second rotation is 90°, because the slopes are the negative reciprocals of the original position slopes. That is, the product of the slopes is -1.

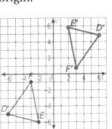

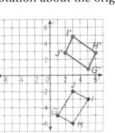

Teaching Strategies

Have students draw the four quadrants of a coordinate plane and write the slope, *a*, in quadrant I. Have them label quadrant II "90° counterclockwise." Ask them what slope they should write in that quadrant. Have them write the rotation and appropriate slope for each quadrant and then repeat for clockwise. have them compare charts.

Chalkboard Examples

Supplementary Example 1

Triangle MNO has vertices $M(-4, 3)$, $N(-1, 3)$ and $O(-2, 1)$. Graph the triangle and its image after a 180° clockwise rotation about the origin. Then compare the slopes of $\overline{MO}$, $\overline{M'O'}$, $\overline{NO}$ and $\overline{N'O'}$. **The corresponding slopes are the same.**

Supplementary Example 2

Rotate the new image containing points A', B', C' and D' 90° counterclockwise. Label the points of the image A'', B'', C'' and D''. Then compare the slope of $\overline{B'C'}$ with the slope of $\overline{B''C''}$. **Check students' work. The slopes are 1 and −1. The lines are perpendicular.**

Supplementary Example 3

$\triangle DEF$ is rotated twice about the origin. The slope of side DE is $\frac{2}{3}$, the slope of $D'E'$ after rotation 1 is $-\frac{3}{2}$, and the slope of $D''E''$ after rotation 2 is $\frac{2}{3}$. How much of a rotation was completed each time? **90°, 90°**

Reteaching Worksheet 8-2

Name ______________________ Date ____________

RETEACHING 8-2

ROTATIONS IN THE COORDINATE PLANE

Another type of transformation that yields congruent figures is a **rotation** or *turn.* A figure is rotated or turned about a point. To draw the image of a figure after rotation, you need to know three things: the center of rotation, the amount of turn expressed in degrees or as a fractional part of the whole turn and the direction of the rotation, *clockwise* or *counterclockwise.*

Use the following rules to find the coordinates of an image.
- When a figure is rotated 180°, $(x, y) \rightarrow (-x, -y)$.
- When a figure is rotated 90° clockwise, $(x, y) \rightarrow (y, -x)$.
- When a figure is rotated 90° counterclockwise, $(x, y) \rightarrow (-y, x)$.

Example

Graph the image of $\triangle ABC$ after a rotation of 90° clockwise about the origin.

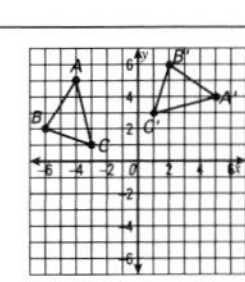

Solution

Find and graph the coordinates of A', B' and C' using the rule $(x, y) \rightarrow (y, -x)$.

$A(-4, 5) \rightarrow A'(5, 4)$

$B(-6, 2) \rightarrow B'(2, 6)$

$C(-3, 1) \rightarrow C'(1, 3)$

EXERCISES

Find the coordinates of A', B' and C' after the given rotation about the origin.

1. $A(-4, 3)$, $B(3, 4)$, $C(1, 2)$ rotated 90° clockwise $\underline{A'(3, 4); B'(4, 3); C'(2, -1)}$
2. $A(2, 1)$, $B(4, 2)$, $C(1, 3)$ rotated 180° $\underline{A'(-2,-1); B'(-4, -2); C'(-1, -3)}$
3. $A(1, -2)$, $B(3, -2)$, $C(4, 1)$ rotated 90° counterclockwise $\underline{A'(2, 1); B'(2, 3); C'(-1, 4)}$
4. Graph the image of $\triangle DEF$ after a 180° clockwise rotation about the origin.
5. Graph the image of quadrilateral $GHIJ$ after a 90° counterclockwise rotation about the origin.

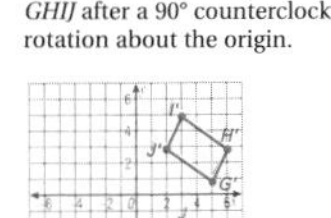

118 RETEACHING *LESSON 8-2*

Lesson Wrap-up

Quick Assessment

Ask the following questions to determine if students understand the content presented in this lesson.

1. What information is used to describe a rotation? center, angle, and direction
2. What do you notice about the rules for 90° rotations about the origin? The x-coordinate and y-coordinate always switch places; positive and negative values depend on quadrants.

Assignment Guide

Basic: 1–14, 17–30
Advanced: 1–30
Additional Practice: See Extra Practice Index on page 674.

Additional Answers

1.

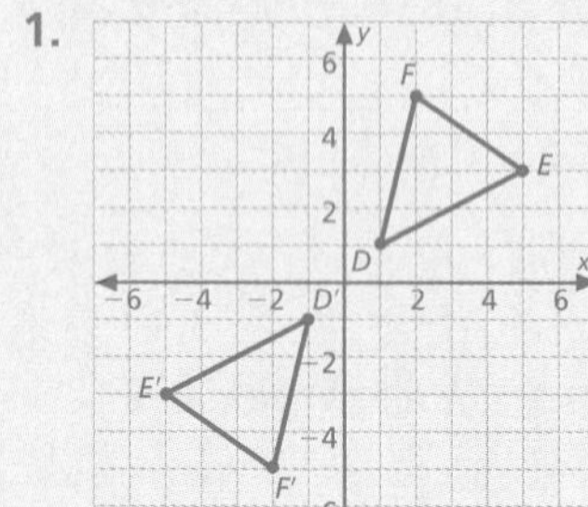

2.

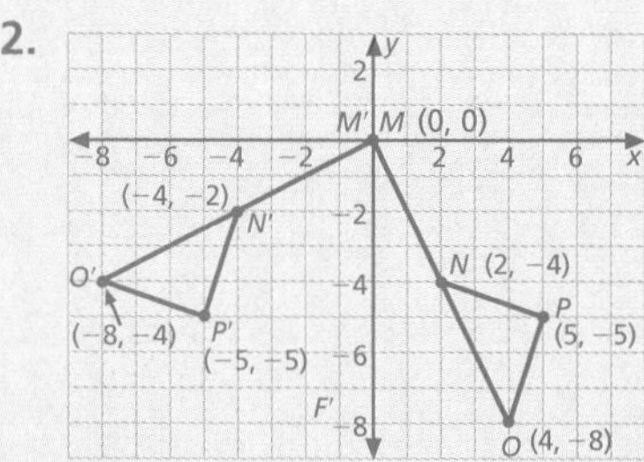

5.

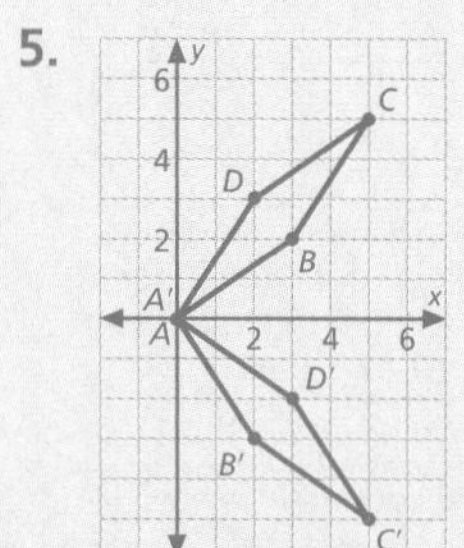

Try These Exercises

For 1–2, see additional answers.

1. Triangle DEF has vertices $D(1, 1)$, $E(5, 3)$, and $F(2, 5)$. Graph $\triangle DEF$ and its image after a 180° counterclockwise rotation about the origin. Then compare the slopes of the sides of $\triangle DEF$ before and after the rotation. The slopes are equal.

2. **COMPUTER GRAPHICS** A figure contains the points $M(0, 0)$, $N(2, -4)$, $O(4, -8)$, and $P(5, -5)$. Graph the figure and its image after a 90° clockwise rotation about the origin. Use the rule $(x, y) \rightarrow (y, -x)$. Then compare the slopes of $\overline{NO}$ and $\overline{N'O'}$. slope of $\overline{NO} = -2$; slope of $\overline{N'O'} = \frac{1}{2}$

3. **ANIMATION** For a television commercial, a triangular logo is animated so that it rotates twice about the origin, as shown in the table below. Compare the slopes and determine how much of a rotation was done each time.

Original position		After rotation 1		After rotation 2	
Side	Slope	Side	Slope	Side	Slope
$\overline{QR}$	−1	$\overline{Q'R'}$	1	$\overline{Q''R''}$	1
$\overline{RS}$	$\frac{3}{5}$	$\overline{R'S'}$	$-\frac{5}{3}$	$\overline{R''S''}$	$-\frac{5}{3}$
$\overline{QS}$	3	$\overline{Q'S'}$	$-\frac{1}{3}$	$\overline{Q''S''}$	$-\frac{1}{3}$

The first was a 90° rotation; the second was a 180° rotation.

4. **WRITING MATH** Describe how translations, reflections and rotations could be used to create a pattern. Sketch an example. Answers will vary.

Practice Exercises

For each figure, draw the image after the given rotation about the origin. Then calculate the slope of each side before and after the rotation.

For 5–7, see additional answers.

5. Use the rule $(x, y) \rightarrow (y, -x)$ for a 90° clockwise rotation.

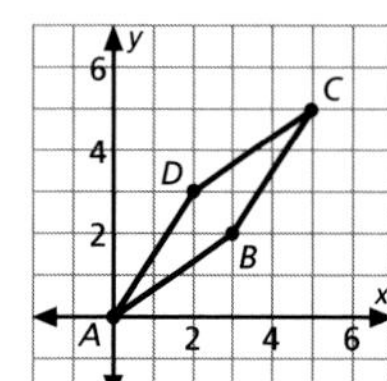

6. Use the rule $(x, y) \rightarrow (-x, -y)$ for a 180° clockwise rotation.

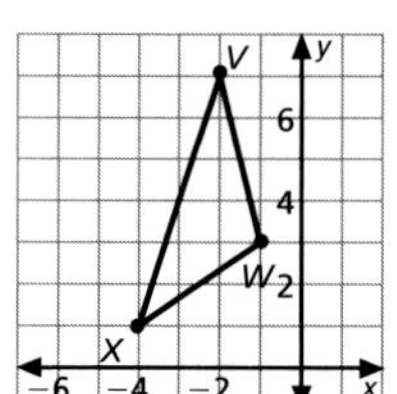

7. Use the rule $(x, y) \rightarrow (-y, x)$ for a 90° counter clockwise rotation.

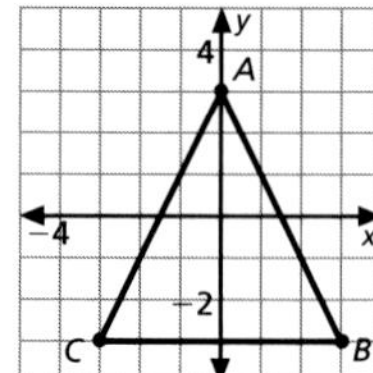

8. **ART** To create a pattern, $\triangle DEF$ is rotated twice about the origin, as shown. Compare the slopes to determine how much of a rotation was completed each time. first, 180°; second, 90°

Original position		After rotation 1		After rotation 2	
Side	Slope	Side	Slope	Side	Slope
$\overline{DE}$	−1	$\overline{D'E'}$	−1	$\overline{D''E''}$	1
$\overline{EF}$	$\frac{1}{7}$	$\overline{E'F'}$	$\frac{1}{7}$	$\overline{E''F''}$	−7
$\overline{DF}$	1	$\overline{D'F'}$	1	$\overline{D''F''}$	−1

6.

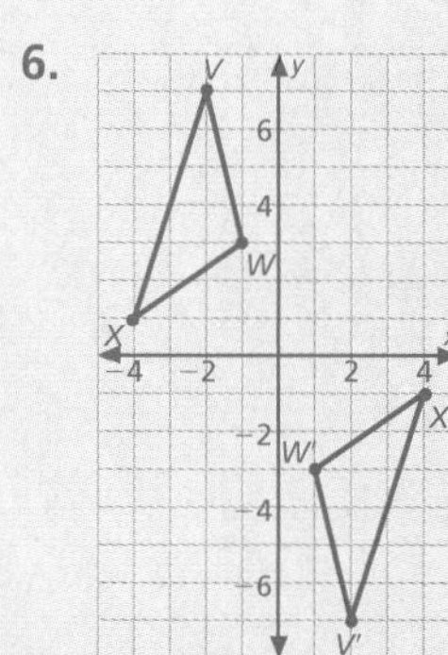

7.

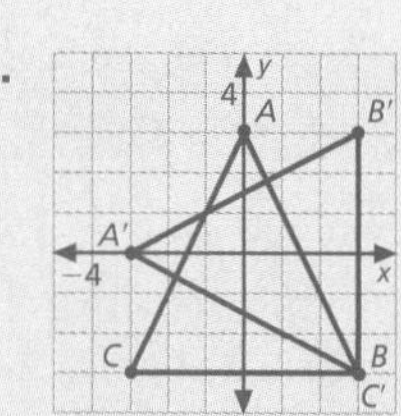

Use the figure at the right for Exercises 9–13.

9. Which triangle is the rotation image of Triangle 1 about the point $(-3, 0)$? Triangle 6

10. Which is the translation image of Triangle 1? Triangle 3

11. Which triangle is the reflection image of Triangle 1 across the x-axis? Triangle 5

12. Which triangle is the reflection image of Triangle 1 across the y-axis? Triangle 4

13. Which triangle is the rotation image of Triangle 1 180° clockwise about the origin? Triangle 8

14. **CHAPTER INVESTIGATION** An "out-and-back" coaster returns to its starting point when the ride ends. Design an "out-and-back" coaster with eight hills. You may join the hills with lengths of track that winds, dips and turns. Make a scale drawing of your coaster on graph paper.

Extended Practice Exercises

15. **WRITING MATH** Make a generalization about corresponding slopes in each situation.
    a. A triangle is rotated 180° clockwise about the origin. The slopes of corresponding sides are equal.
    b. A triangle is rotated 90° counterclockwise about the origin. The product of corresponding slopes is -1.
    c. A triangle is rotated 90° clockwise about the origin. The product of corresponding slopes is -1.

16. If you rotate a figure about its center and it fits back on top of itself in less than a 360° rotation, the figure is said to have **point symmetry**. For example, a square has point symmetry because, when it is rotated 90° about its center, each image vertex falls on top of one of the original vertices. Through how many degrees would you have to rotate a regular hexagon for this to happen? A regular pentagon? 60°, 72°

Mixed Review Exercises

Find the measure of x in each pair of similar polygons. (Lesson 7-2)

17. 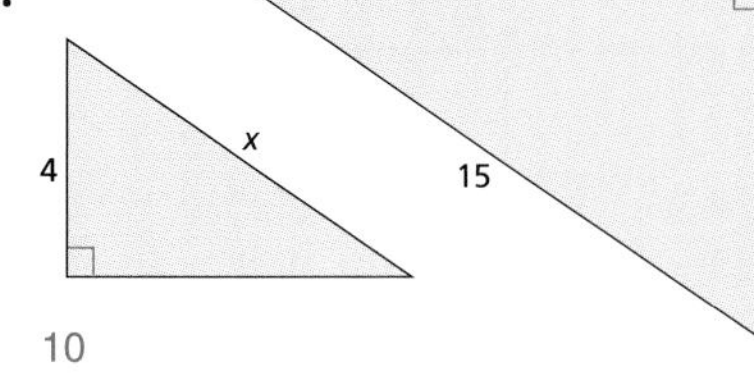

10

18. 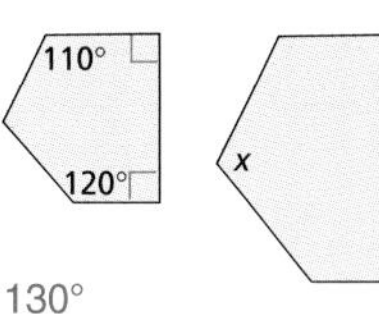

130°

Determine the slope of the line shown by each pair of points. (Lesson 6-1)

19. $A(-2, -3)$ $B(-9, -4)$ $\frac{1}{7}$
20. $C(3, 6)$ $D(-2, 8)$ $-\frac{2}{5}$
21. $E(-4, 5)$ $F(3, -1)$ $-\frac{6}{7}$
22. $G(4, 8)$ $H(-2, -5)$ $\frac{13}{6}$
23. $I(5, -2)$ $J(0, -8)$ $\frac{6}{5}$
24. $K(-4, 2)$ $L(3, 2)$ 0
25. $M(-1, 6)$ $N(5, 2)$ $-\frac{2}{3}$
26. $O(4, -2)$ $P(-6, 3)$ $-\frac{1}{2}$
27. $Q(-3, 1)$ $R(-3, -5)$ undefined slope
28. $S(-6, -2)$ $T(5, -3)$ $-\frac{1}{11}$
29. $U(6, -7)$ $V(2, -4)$ $-\frac{3}{4}$
30. $W(4, 3)$ $X(7, 12)$ 3

Extend the Lesson

REAL WORLD CONNECTION Translations, reflections and rotations are used extensively by artists to create patterns. Examples are all around you—even on the walls. Have students find examples of each, and copy some of the patterns into their journals. Have them label the kinds of transformation that each pattern uses.

Extra Practice Worksheet 8-2

Name ______________ Date ______________

EXTRA PRACTICE 8-2

ROTATIONS IN THE COORDINATE PLANE

EXERCISES

For each figure, draw the image after the given rotation about the origin. Use your own paper. Check students' graphs.

1. Use the rule $(x, -y) \rightarrow (y, x)$ for a 90° clockwise rotation.

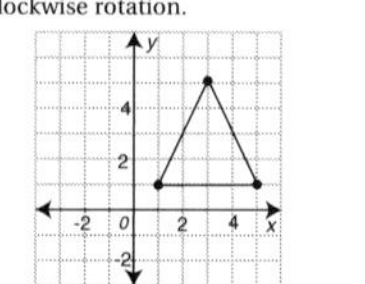

2. Use the rule $(-x, -y) \rightarrow (x, y)$ for a 180° clockwise rotation.

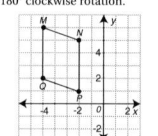

3. Use the rule $(x, y) \rightarrow (-y, x)$ for a 90° counter clockwise rotation.

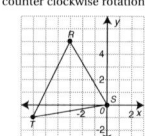

4. Use the rule $(x, y) \rightarrow (-x, -y)$ for a 180° clockwise rotation.

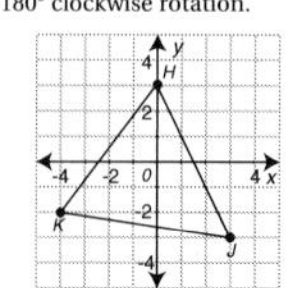

5. Triangle XYZ is rotated twice about the origin, as show in the table below. Compare the slope to determine how much of a rotation was completed each time. first; 90°; second 180°

ORIGINAL POSITION		AFTER ROTATION 1		AFTER ROTATION 2	
side	slope	side	slope	side	slope
$\overline{XY}$	3	$\overline{X'Y'}$	-3	$\overline{X''Y''}$	-3
$\overline{YZ}$	$\frac{1}{2}$	$\overline{Y'Z'}$	-2	$\overline{Y''Z''}$	-2
$\overline{XZ}$	-3	$\overline{X'Z'}$	2	$\overline{Y''Z''}$	2

104 EXTRA PRACTICE *LESSON 8-2*

Enrichment Worksheet 8-2

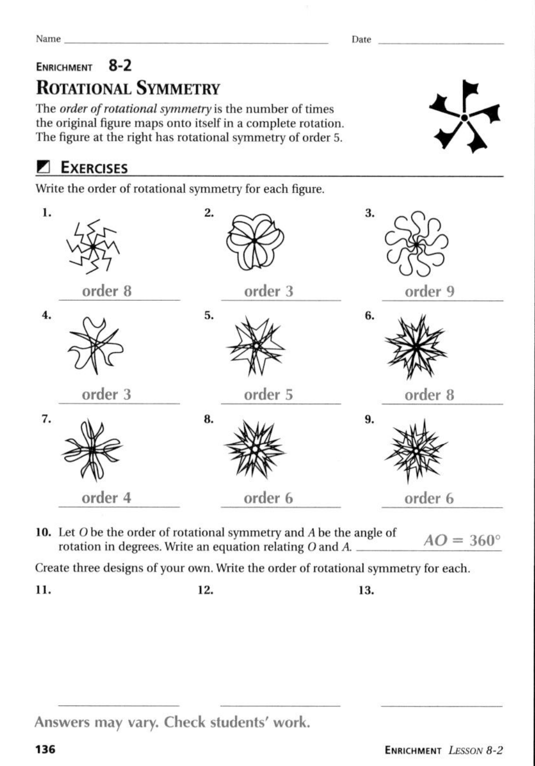
Name ______________ Date ______________

ENRICHMENT 8-2

ROTATIONAL SYMMETRY

The *order of rotational symmetry* is the number of times the original figure maps onto itself in a complete rotation. The figure at the right has rotational symmetry of order 5.

EXERCISES

Write the order of rotational symmetry for each figure.

1. order 8
2. order 3
3. order 9
4. order 3
5. order 5
6. order 8
7. order 4
8. order 6
9. order 6

10. Let O be the order of rotational symmetry and A be the angle of rotation in degrees. Write an equation relating O and A. $AO = 360°$

Create three designs of your own. Write the order of rotational symmetry for each.

11. 12. 13.

Answers may vary. Check students' work.

136 ENRICHMENT *LESSON 8-2*

Skills Practice

Vocabulary Review

Lesson 8-1

translation	preimage
image	transformation
reflection	line of reflection

Lesson 8-2

clockwise	counterclockwise
rotation	origin
center of rotation	
angle of rotation	

Assignment Guide

All students: 1–30
Additional Practice: Refer to the Extra Practice Index on page 674 of this text.

Additional Answers

1–3.

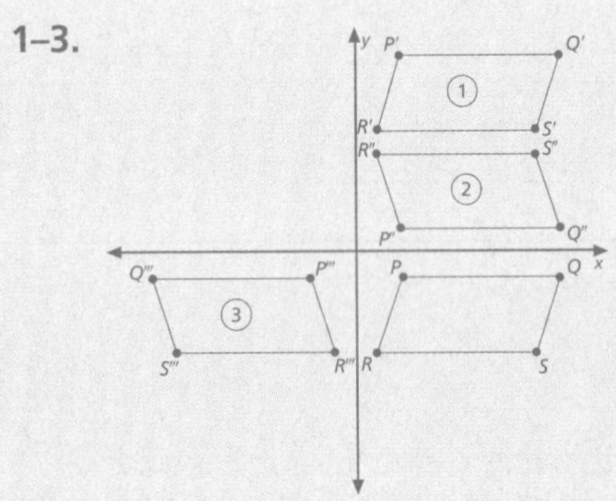

4. Slopes:

$\overline{PR}$	$\overline{P'R'}$	$\overline{P''R''}$	$\overline{P'''R'''}$	$\overline{QS}$	$\overline{Q'S'}$	$\overline{Q''S''}$	$\overline{Q'''S'''}$
3	3	−3	−3	3	3	−3	−3

For answers to 5–7, 9–11, 13–15, 17, 19–21, 23–25, and 27–29 see Selected Answers beginning on page 732.

Review and Practice Your Skills

Practice ■ Lesson 8-1

Graph the image of parallelogram *PQRS* with vertices *P*(2, −1), *Q*(9, −1), *R*(1, −4), and *S*(8, −4). Then graph its image under each transformation from the original position. For 1–4, see additional answers.

1. 9 units up **2.** reflected across *x*-axis **3.** reflected across *y*-axis

4. Compare the slopes of the non-horizontal sides of parallelogram *PQRS* in all four positions above. Translation maintains same slopes. Reflection across either axis creates opposite slopes.

Graph the image of triangle *DEF* with vertices *D*(1, 7), *E*(3, 4), and *F*(5, 11). Then graph its image under each transformation from the original position. For 5–8, see additional answers.

5. 15 units down **6.** reflected across *y*-axis **7.** reflected across $y = x$

8. Compare the slopes of triangle *DEF* in all four positions above.

Graph the image of trapezoid *MKLN* with vertices *K*(−4, 7), *L*(−3, 11), *M*(−2, 3), and *N*(0, 11). Then graph its image under each transformation from the original position. For 9–12, see additional answers.

9. reflected across *y*-axis **10.** reflected across $y = x$ **11.** reflected across $y = -x$

12. Compare the slopes of trapezoid *MKLN* in all four positions above.

Practice ■ Lesson 8-2

Triangle *ABC* has vertices *A*(−3, 1), *B*(−5, 2), and *C*(−1, 4). Graph the triangle and its image after each of the following rotations about the origin. For 13–18, see additional answers.

13. 90° clockwise **14.** 180° clockwise **15.** 270° clockwise

16. Reflect the original triangle *ABC* across the *y*-axis. Then reflect this new image across the *x*-axis. To which of the rotations in Exercises 13–15 does this double-reflection correspond? Exercise 14

17. Reflect the original triangle *ABC* across the line $y = x$. Then reflect this new image across the line $y = -x$. How does this double-reflection compare to the rotations in Exercises 16–18? Exercise 14

18. Triangle *FGH* is rotated twice about the origin, as shown in the table below. Compare the slopes and determine how much of a rotation was done each time. Rotation 1: 90°; Rotation 2: 90°

Original position		After rotation 1		After rotation 2	
side	slope	side	slope	side	slope
$\overline{FG}$	$-\frac{2}{3}$	$\overline{F'G'}$	$\frac{3}{2}$	$\overline{F''G''}$	$-\frac{2}{3}$
$\overline{GH}$	$\frac{1}{3}$	$\overline{G'H'}$	−3	$\overline{G''H''}$	$\frac{1}{3}$
$\overline{FH}$	$\frac{5}{6}$	$\overline{F'H'}$	$-\frac{6}{5}$	$\overline{F''H''}$	$\frac{5}{6}$

8. Slopes:

Side	Slope	Side	Slope	Side	Slope	Side	Slope
DE	$-\frac{3}{2}$	D'E'	$-\frac{3}{2}$	D''E''	$\frac{3}{2}$	D'''E'''	$-\frac{2}{3}$
EF	$\frac{7}{2}$	E'F'	$\frac{7}{2}$	E''F''	$-\frac{7}{2}$	E'''F'''	$\frac{2}{7}$
DF	1	D'F'	1	D''F''	−1	D'''F'''	1
Compared to original slopes:		Same		Opposite		Reciprocal	

12. Slopes:

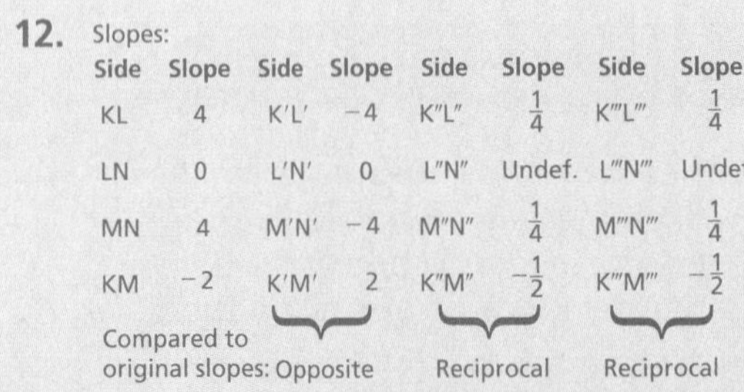

Side	Slope	Side	Slope	Side	Slope	Side	Slope
KL	4	K'L'	−4	K''L''	$\frac{1}{4}$	K'''L'''	$\frac{1}{4}$
LN	0	L'N'	0	L''N''	Undef.	L'''N'''	Undef.
MN	4	M'N'	−4	M''N''	$\frac{1}{4}$	M'''N'''	$\frac{1}{4}$
KM	−2	K'M'	2	K''M''	$-\frac{1}{2}$	K'''M'''	$-\frac{1}{2}$
Compared to original slopes:		Opposite		Reciprocal		Reciprocal	

16.

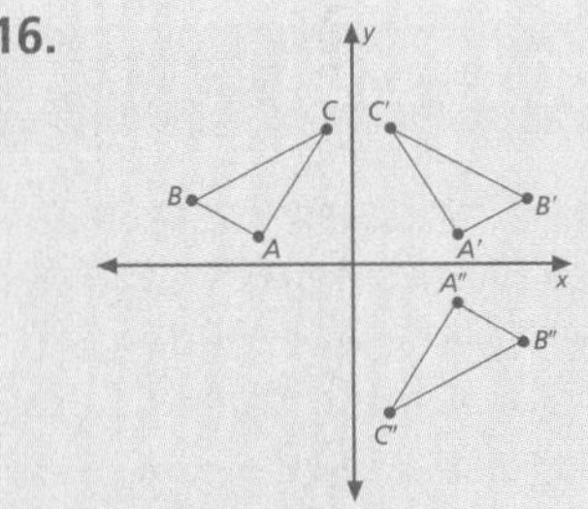

Practice ■ Lesson 8-1–Lesson 8-2

Graph the image of parallelogram *PQRS* with vertices $P(-1, 2)$, $Q(2, 3)$, $R(4, 7)$, and $S(1, 6)$. Then graph its image under each transformation from the original position. (Lesson 8-1) For 19–22, see additional answers.

19. 12 units to the left **20.** reflected across $y = x$ **21.** reflected across y-axis

22. Compare the slopes of the sides of parallelogram *PQRS* in all four positions above.

Graph the image of quadrilateral *ABCD* with vertices $A(5, 4)$, $B(8, -1)$, $C(0, 0)$, and $D(-2, 1)$. Then graph its image under each transformation from the original position. (Lesson 8-1) For 23–30, see additional answers.

23. 5 units to the left, 4 units down **24.** reflected across x-axis **25.** reflected across y-axis

26. Compare the slopes of the sides of quadrilateral *ABCD* in all four positions above.

Triangle *ABC* has vertices $A(2, 0)$, $B(8, -2)$, and $C(7, 3)$. Graph the triangle and its image after each of the following rotations about the origin. (Lesson 8-2)

27. 90° clockwise **28.** 180° clockwise **29.** 270° counterclockwise

30. Compare the slopes of the sides of triangle *ABC* in all four positions above.

MathWorks Workplace Knowhow

Career – Construction Supervisor

Amusement parks hire construction workers and supervisors to build new attractions. Whether the attraction is a roller coaster or a carousel, the construction supervisor must adhere to building and safety codes. The supervisor oversees costs, materials, labor, and transportation related to the project.

An engineer has designed a coaster with three hills. The final three hills are a reflection of the first three hills. In the diagram below, the first inclines have been straightened.

1. Identify the coordinates of apex of each of the three hills shown. $(-10, 7)$, $(-5, 6)$ and $(-2, 3)$
2. Draw the coaster's reflection over the y-axis. Write the coordinates of the reflections of the points you identified in Exercise 1.
Check students' drawings. (2, 3), (5, 6) and (10, 7)

Construction supervisors must be able to follow blueprints and design specifications. For high-speed rides, the supervisor may work closely with an engineer or specialist. When problems arise, the supervisor and the engineer work together to find solutions.

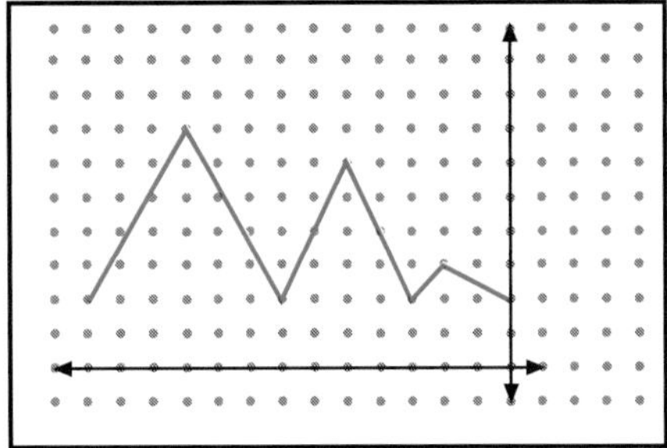

Chalkboard Examples

Example from Lesson 8-1

Graph $\triangle PQR$ with vertices $P(-7, 4)$, $Q(-5, 6)$ and $R(-3, 2)$, and its image reflected across $x = y$.
(4, −7), (6, −5), (2, −3)]

Example from Lesson 8-2

Graph $\triangle ABC$ with vertices $A(4, -2)$, $B(6, -4)$ and $C(1, -3)$ and the image after a 270° counterclockwise rotation. Calculate the slope of AB and $A'B'$. $A'(-2, -4)$, $B'(-4, -6)$, $C'(-3, -1)$; -1, 1

Career Opportunity

Describe the kind of work done by construction supervisors and the types of companies that hire these workers. Point out that these workers have many employment opportunities and that the demand for this profession is growing. Have students discuss the importance of measurement, geometry and algebraic thinking in doing this type of job. Explain that construction supervisors must have good management skills as well as construction knowledge. Students should answer Questions 1–2 to better understand how construction supervisors use mathematics in performing their jobs.

Students who are interested in learning more about this profession can go to learningmathmatters.com. Guidance Counselors should have information about school and training requirements.

22.

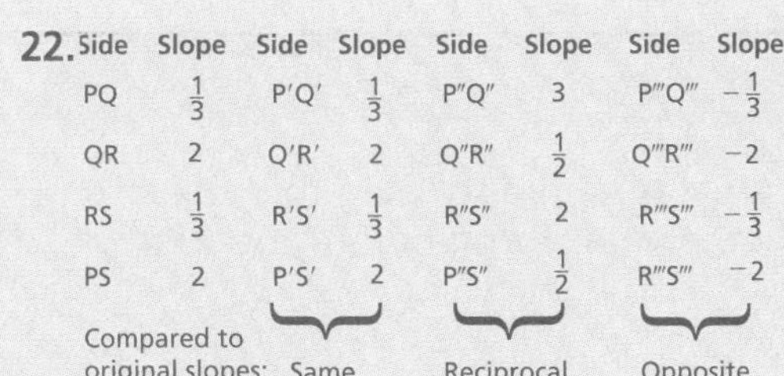

Side	Slope	Side	Slope	Side	Slope	Side	Slope
PQ	$\frac{1}{3}$	P′Q′	$\frac{1}{3}$	P″Q″	3	P‴Q‴	$-\frac{1}{3}$
QR	2	Q′R′	2	Q″R″	$\frac{1}{2}$	Q‴R‴	-2
RS	$\frac{1}{3}$	R′S′	$\frac{1}{3}$	R″S″	2	R‴S‴	$-\frac{1}{3}$
PS	2	P′S′	2	P″S″	$\frac{1}{2}$	R‴S‴	-2
Compared to original slopes:		Same		Reciprocal		Opposite	

26.

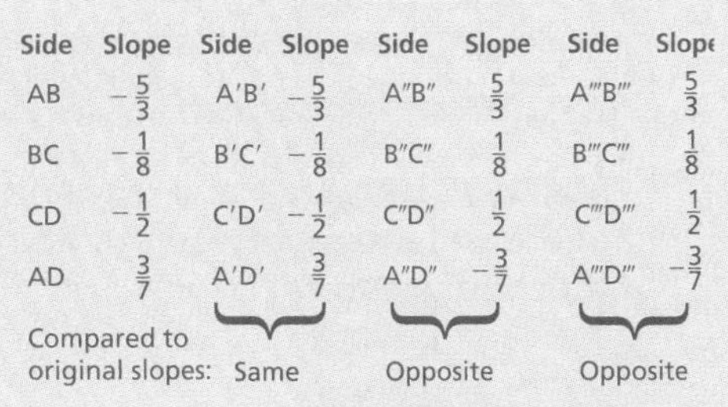

Side	Slope	Side	Slope	Side	Slope	Side	Slope
AB	$-\frac{5}{3}$	A′B′	$-\frac{5}{3}$	A″B″	$\frac{5}{3}$	A‴B‴	$\frac{5}{3}$
BC	$-\frac{1}{8}$	B′C′	$-\frac{1}{8}$	B″C″	$\frac{1}{8}$	B‴C‴	$\frac{1}{8}$
CD	$-\frac{1}{2}$	C′D′	$-\frac{1}{2}$	C″D″	$\frac{1}{2}$	C‴D‴	$\frac{1}{2}$
AD	$\frac{3}{7}$	A′D′	$\frac{3}{7}$	A″D″	$-\frac{3}{7}$	A‴D‴	$-\frac{3}{7}$
Compared to original slopes:		Same		Opposite		Opposite	

30.

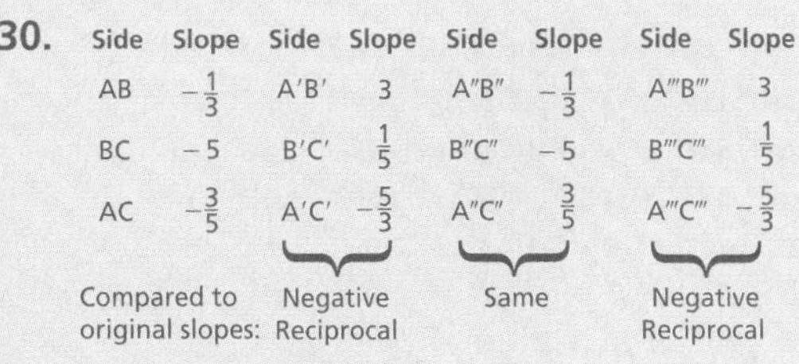

Side	Slope	Side	Slope	Side	Slope	Side	Slope
AB	$-\frac{1}{3}$	A′B′	3	A″B″	$-\frac{1}{3}$	A‴B‴	3
BC	-5	B′C′	$\frac{1}{5}$	B″C″	-5	B‴C‴	$\frac{1}{5}$
AC	$-\frac{3}{5}$	A′C′	$-\frac{5}{3}$	A″C″	$\frac{3}{5}$	A‴C‴	$-\frac{5}{3}$
Compared to original slopes:		Negative Reciprocal		Same		Negative Reciprocal	

8-3 Dilations in the Coordinate Plane

Goals ■ Draw dilation images on a coordinate plane.

Applications Graphic Design, Business, Art

Work with a partner to study the quilt pattern shown.

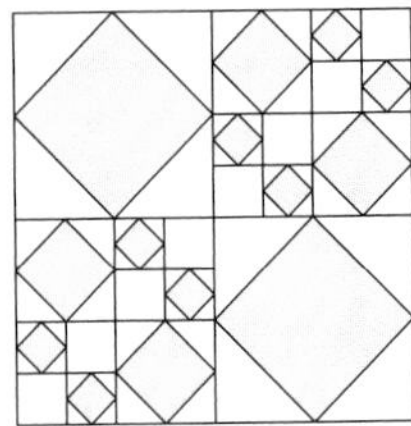

1. The basic pattern of a rotated square within a larger square is used repeatedly in the design. How many different sizes of the basic pattern appear? three
2. How does the length of a side in the largest pattern compare to the length of a side in the smallest? 4 times as long.

Build Understanding

A **dilation** is a transformation that produces an image in the same shape as the original figure, but a different size. An image larger than the original figure is called an **enlargement**. An image smaller than the original figure is called a **reduction**. A figure and its dilation image are similar.

> **Check Understanding**
>
> Explain the meaning of: "A figure and its dilation are similar."
>
> See additional answers.

A dilation image is obtained by multiplying the length of each side of the original figure by a number called the **scale factor**. If the scale factor is greater than 1, you create an enlargement. If the scale factor is smaller than 1, you create a reduction. The description of a dilation includes the scale factor and the **center of dilation**. The distance from the center of dilation to each point on the image is equal to the distance from the center of dilation to each corresponding point of the original figure times the scale factor.

When the center of dilation is at the origin, you use the following rule to locate points on the dilation image. Let (x, y) represent a point on the original figure, and let k represent the scale factor.

$(x, y) \rightarrow (kx, ky)$

Example 1

Draw the dilation image of quadrilateral *MNPQ* below with vertices at M(3, 1), *N*(6, 1), *P*(6, 3), and *Q*(3, 3). The center of dilation is the origin, and the scale factor is 2.

Solution

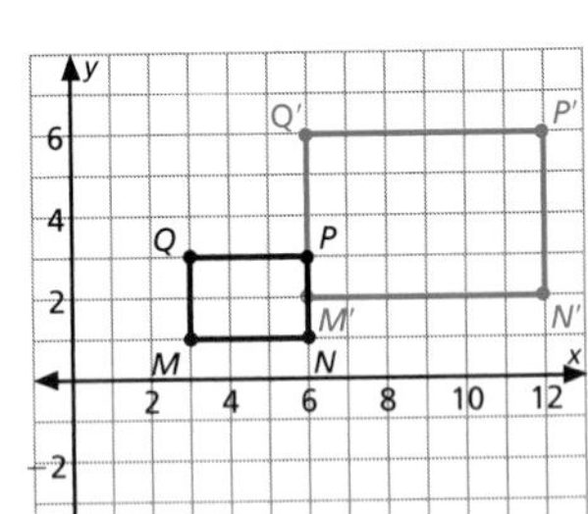

Because the center of dilation is at the origin, use the rule $(x, y) \rightarrow (2x, 2y)$.

$M(3, 1) \rightarrow M'(6, 2)$

$N(6, 1) \rightarrow N'(12, 2)$

$P(6, 3) \rightarrow P'(12, 6)$

$Q(3, 3) \rightarrow Q'(6, 6)$

Lesson Planning

NCTM Standards/Strands

- Number & Operation
- Patterns, Functions, & Algebra
- Geometry & Spatial Sense
- Problem Solving
- Reasoning and Proof
- Connections
- Representation

Vocabulary

dilation	enlargement
reduction	scale factor
center of dilation	

Materials Needed

graph paper   rulers
geometric drawing software (optional)

Lesson Resources

Warm-Up Transparency 22
Reteaching 8-3
Extra Practice 8-3
Enrichment 8-3
Transparency TK-7–10

Assignment Guide

Basic: 1–16, 19–31
Enriched: 1–31

Getting Started

5-Minute Warm-Up

Simplify.

1. $3 \times (4 - 1)$ 9
2. $\frac{1}{4} \times (10 - 4)$ $1\frac{1}{12}$
3. $5 \times (12 - (-7))$ 95
4. $\frac{2}{3} \times 3(-2 - (-8))$ 4

Introduction to Lesson 8-3

Ask students to describe how the three patterns are alike and how they are different. Ask them what conclusion they can make. They are the same shape but different sizes. They are similar figures.

Extend the Lesson

MATH JOURNAL Have students write a paragraph discussing the similarities and differences between dilations and other transformation (translations, reflections and rotations). Similarities: The shapes of preimages and images remain the same; translation images and dilations always face the same direction as preimages. Differences: Dilations produce images of different sizes, while the other transformations produce the same-size images.

Sometimes the center of dilation is not at the origin. For example, the center of dilation might be a vertex of the original figure. In this case, the center of dilation and the corresponding vertex of the dilation image are the same.

Example 2

GRAPHIC DESIGN A triangular flag is designed to promote a new attraction at an amusement park. The flags, represented by $\triangle ABC$ on the grid shown to the right, will also be manufactured in a smaller size. Draw a dilation image of $\triangle ABC$ with center of dilation at A and a scale factor of $\frac{2}{3}$.

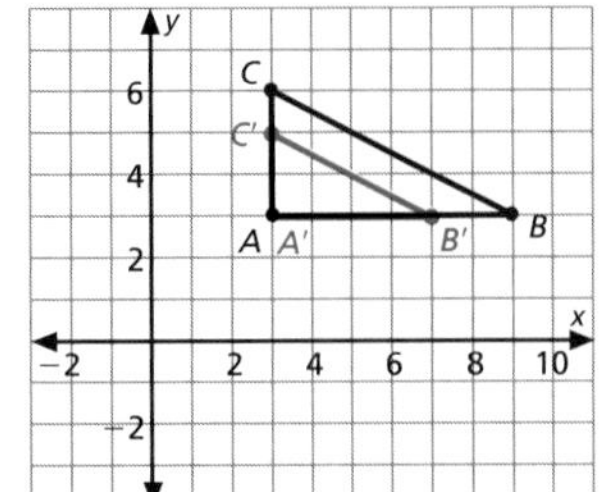

Solution

The distance from the center of dilation, A, to B is 6 units. So the distance from A to B' is $\frac{2}{3} \times 6$, or 4. Count over 4 units from A to locate B'. The distance from A to C' is $\frac{2}{3} \times 3$, or 2. Count up 2 units from A to locate C'. Points A and A' coincide because A is the center of dilation.

Technology Note

You can use geometric-drawing software to transform figures. To draw a dilation of a figure, follow these steps:

1. Plot the vertices of the figure and draw segments.
2. Select the center of dilation. Many programs allow you to select the center of dilation by double-clicking on the desired point.
3. Enter the scale factor as a ratio of new distance to old distance.
4. Draw the dilation.

TRY THESE EXERCISES

Refer to the figure at the right for Exercises 1–4.

1. What is the image of square $ABCD$? square $A'B'C'D'$
2. What is the center of dilation? origin
3. How do the lengths of the sides in the image compare to lengths of the sides in square $ABCD$? 4 times as long.
4. What is the scale factor? 4

Copy each graph on graph paper. Then draw each dilation image. For 5–6, see additional answers.

5. The center of dilation is the origin and the scale factor is 4. Use the rule $(x, y) \to (4x, 4y)$.

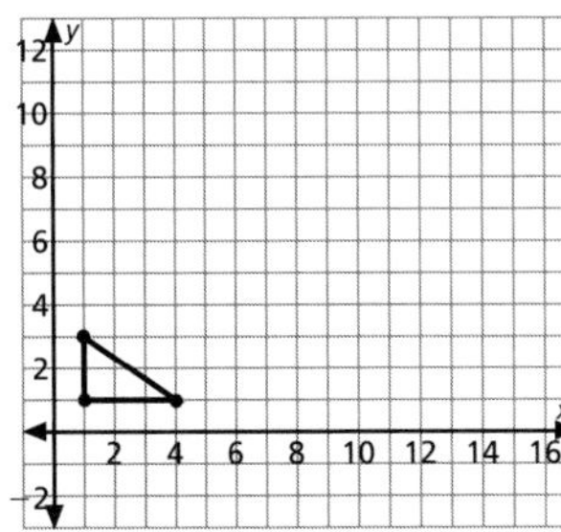

6. The center of dilation is the origin and the scale factor is $\frac{1}{2}$.

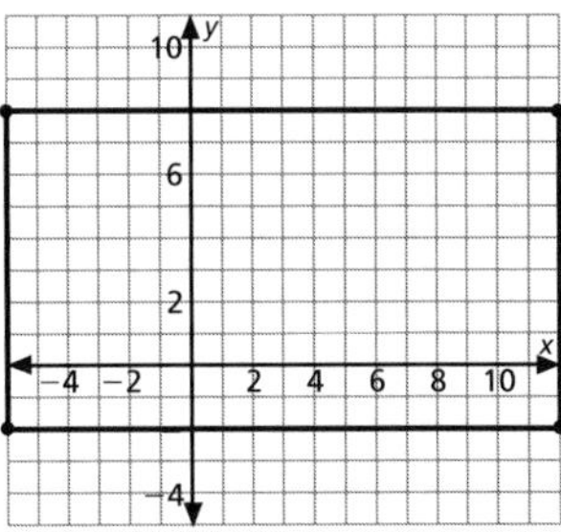

Chalkboard Examples

Supplementary Example 1

Graph $\triangle DEF$, with vertices $D(-3, 0)$, $E(4, 3)$ and $F(1, -2)$. The center of dilation is the origin and the scale factor is 2. $(-6, 0)$, $(8, 6)$, $(2, -4)$

Supplementary Example 2

A sign in the shape of a triangle will also be manufactured in two sizes. $\triangle XYZ$ has vertices $X(0, 0)$, $Y(4, 8)$ and $Z(12, 4)$. Draw the dilation image of $\triangle XYZ$ with center of dilation at Z and a scale factor of $\frac{1}{4}$. $X'(9, 3)$, $Y'(10, 5)$, $Z'(12, 4)$.

ADDITIONAL ANSWERS

5.

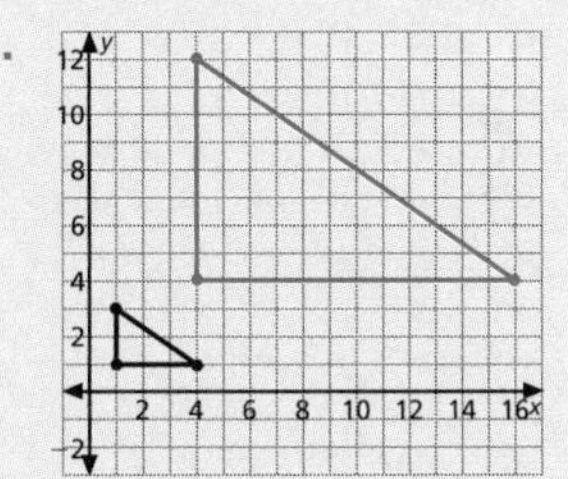

Reteaching Worksheet 8-3

Name ____________ Date ____________

RETEACHING **8-3**

DILATIONS IN THE COORDINATE PLANE

A **dilation** is a transformation under which a figure is enlarged or reduced to form an image that is similar to the original figure. You need to know the **center of dilation** and the **scale factor** to graph the image of a figure under dilation.

For a center of dilation at the origin and a scale factor of k, use the rule $(x, y) \to (kx, ky)$ to find the coordinates of the image.

Example

Graph the dilation image of $\triangle XYZ$ using a scale factor of 2 and the center of dilation at vertex X.

Solution

The distance from X to Y is 4 units, so the distance from X' to Y' is $4 \cdot 2$, or 8 units.
The distance from X to Z is 3 units, so the distance from X' to Z' is $3 \cdot 2$, or 6 units.
X and X' are at the same point since X is the center of the dilation.

EXERCISES

Find the coordinates of each point under the given dilation.

1. $(-3, 2)$; scale factor of 2, center of dilation at the origin $(-6, 4)$
2. $(8, 4)$; scale factor of $\frac{1}{4}$, center of dilation at the origin $(2, 1)$
3. Graph the dilation image of triangle ABC, using a scale factor of $\frac{1}{2}$ and the center of dilation at the origin.
4. Graph the dilation image of triangle $ABCD$, using a scale factor of 3 and the center of dilation at vertex D.
5. Graph the dilation image of figure $GHIJ$, using a scale factor of $\frac{1}{2}$ and the center of dilation at vertex J.

120 RETEACHING *LESSON 8-3*

ADDITIONAL ANSWERS

Check Understanding

They are similar because their angles are congruent and their sides are in proportion with each other.

6.

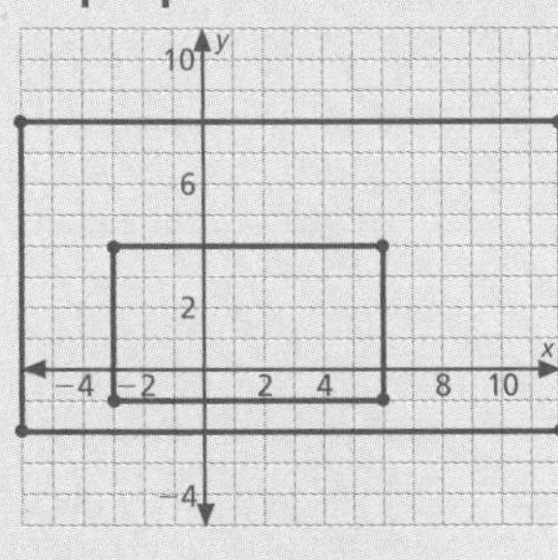

Teaching Strategies

CHALLENGE What is the effect of using a negative scale factor to create a dilation image? You may wish to have students explore this question using geometric drawing software. The image is the same as if the factor had been positive and the image rotated 180° about the center of the dilation.

Lesson Wrap-up

Quick Assessment

Ask the following questions to determine if students understand the content presented in this lesson.

1. What is the effect of a scale factor of 2 on the lengths of each side of a figure? The length of each side is doubled.
2. What is the effect of a fractional scale factor on a figure? The length of each side is the given fraction of the original length of the side.

Assignment Guide

Basic: 1–16, 19–31
Advanced: 1–31
Additional Practice: See Extra Practice Index on page 674.

Additional Answers

7. 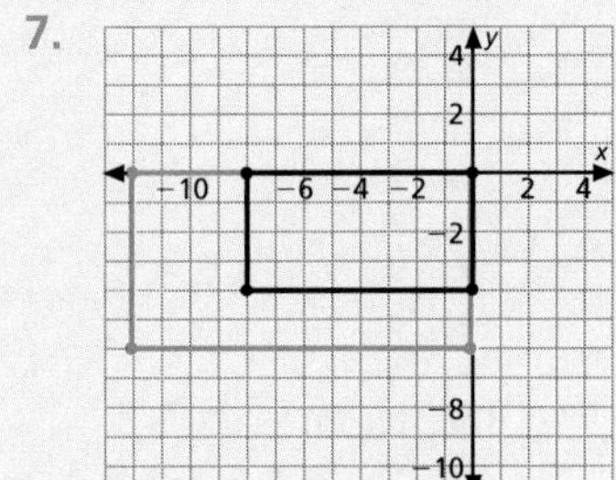

8. 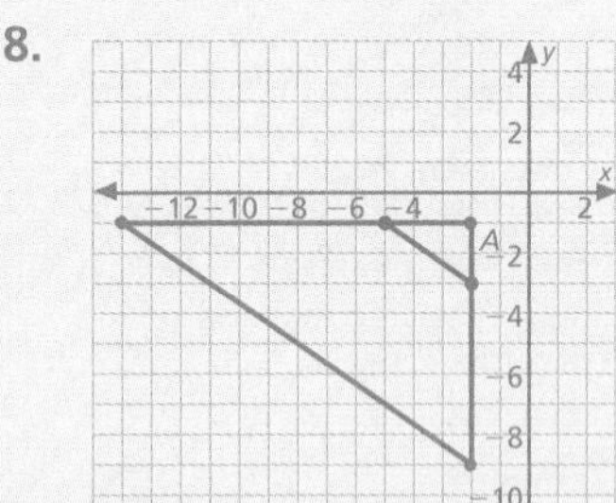

9. 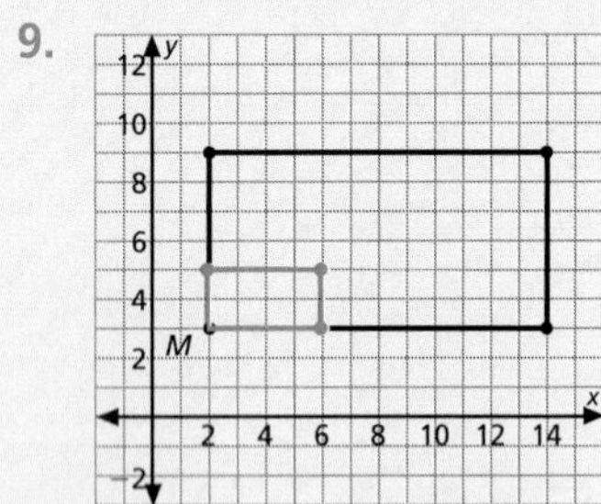

10. 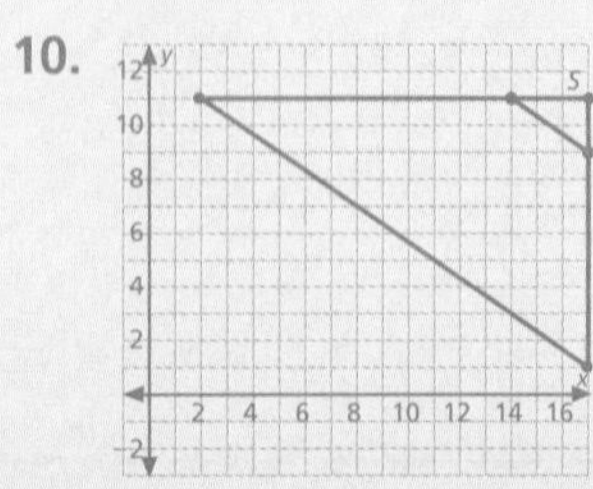

Practice Exercises

For 7–10, see additional answers.

7. The center of dilation is the origin and the scale factor is 1.5.

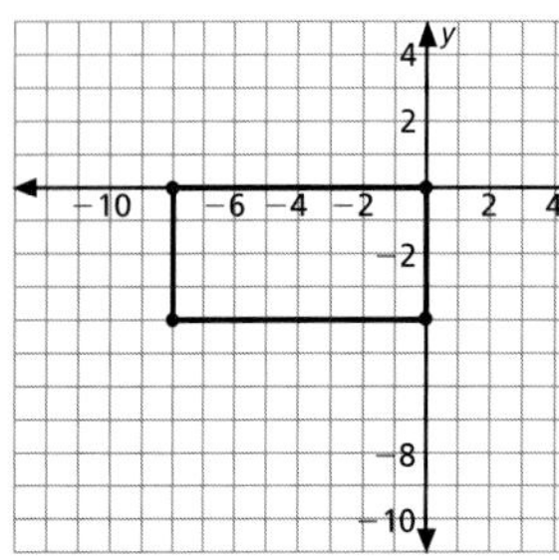

8. The center of dilation is point A and the scale factor is $\frac{1}{4}$.

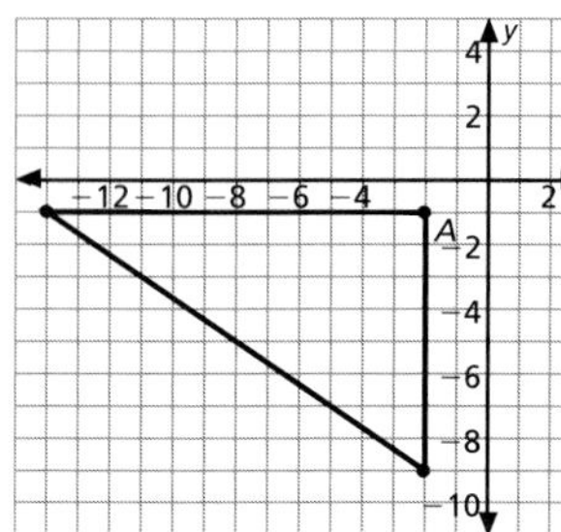

9. The center of dilation is point M and the scale factor is $\frac{1}{3}$.

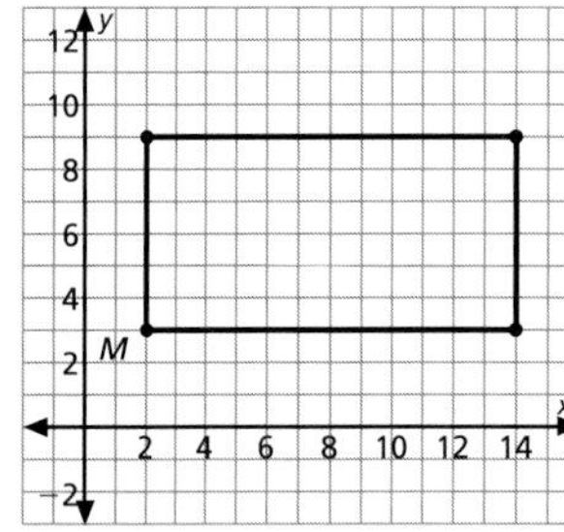

10. The center of dilation is point S and the scale factor is $\frac{1}{5}$.

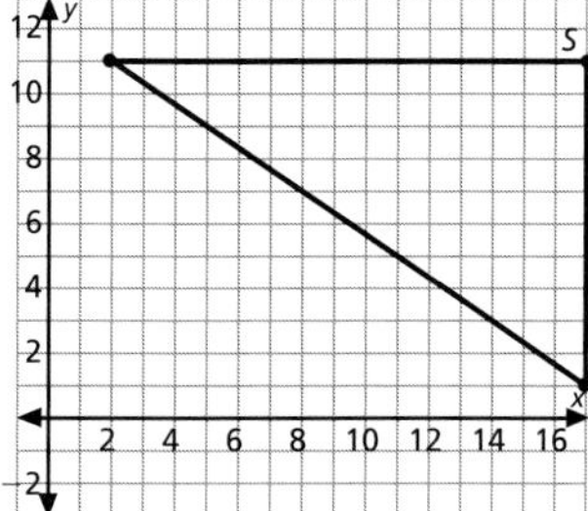

The following sets of points are the vertices of figures and their dilation images. For each two sets of points, give the scale factor.

11. $C(0, 0)$, $D(0, 4)$, $E(8, 0)$
$C'(0, 0)$, $D'(0, 5)$, $E'(10, 0)$ 1.25
12. $S(0, 0)$, $T(0, -6)$, $U(-3, -9)$
$S'(0, 0)$, $T'(0, -4)$, $U'(-2, -6)$ $\frac{2}{3}$
13. $A(0, 2)$, $B(2, 2)$, $C(2, -1)$, $D(0, -1)$
$A'(0, 6)$, $B'(6, 6)$, $C'(6, -3)$, $D'(0, -3)$ 3

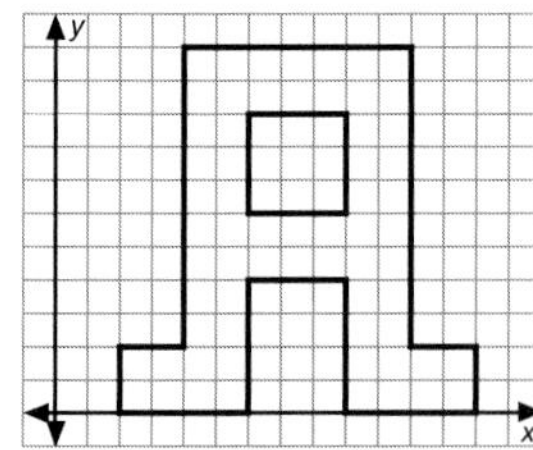

14. **BUSINESS** Alice designed a logo for her business. She drew it on a grid, as shown in the diagram at the right. She wants to create an enlargement that is 5 times as big. Draw the logo on a separate sheet of graph paper. Then draw the enlargement by multiplying all the coordinates by 5. Check students' drawings.

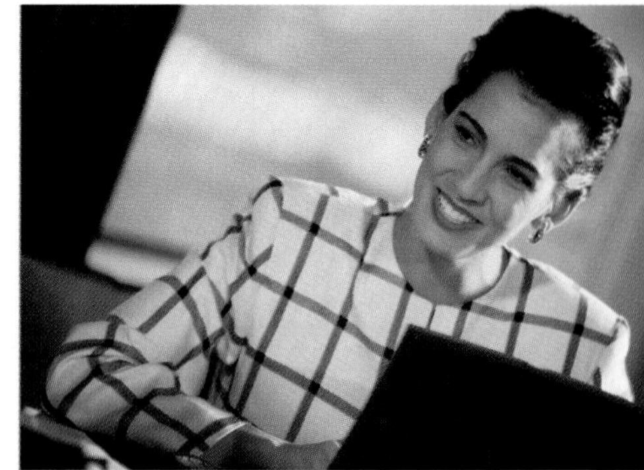

Teaching Strategies

TEAM LEARNING Have students work in pairs to find a way to make dilation images of a rectangle such that the distance from each corresponding side of the original rectangle is proportional to the distance of the corresponding sides of each dilation image. Draw examples, and give the equivalent ratios. Place the rectangle so that the origin is contained within the figure, and use the origin as the center of dilation.

15. **ART** An artist needs to enlarge the drawing shown at the right so that the resulting figure is twice as large. She uses another method for making enlargements and reductions which does not require a grid. To use this method, pick any convenient point outside the figure. Label the point O. Draw line segments from O to various points on the figure. For example, draw a segment from O to point A. Then extend the segment from A to point A' so that $OA = AA'$. Continue until you have enough points to complete the drawing. Check students' drawings.

16. Trace pentagon $ABCDE$ on another sheet. Use the method described in Exercise 15 to make an enlargement of the pentagon 3 times as big. *Hint:* The distance OA will be $\frac{1}{2}$ of the distance from A to A'. Check students' drawings.

Extended Practice Exercises

17. **GEOMETRY SOFTWARE** Draw rectangle $TUVW$ with vertices $T(2, 2)$, $U(2, 6)$, $V(8, 6)$, and $W(8, 2)$ on a coordinate plane using geometric-drawing software. Draw the dilation images of rectangle $TUVW$ with the center of dilation at T and scale factors of 2 and $\frac{1}{2}$.
    - **a.** Find the area of rectangle $TUVW$. 24 square units
    - **b.** Find the area of its enlargement. 96 square units
    - **c.** Find the area of its reduction. 6 square units
    - **d.** How does the area of the enlargement compare to the area of the original figure? 4 times as large
    - **e.** How does the area of the reduction compare to the area of the original figure? $\frac{1}{4}$ as large

18. **CRITICAL THINKING** Describe the effect a scale factor of $-\frac{1}{2}$ would have on a dilation image of rectangle $TUVW$ from Exercise 17. The image is the same as if the scale factor had been $\frac{1}{2}$ and the rectangle had rotated 180° about point T. The vertices of the new image are $T'(2, 2)$, $U'(-1, 2)$, $V'(-1, 0)$ and $W'(2, 0)$.

Mixed Review Exercises

Find the actual length of each of the following. (Lesson 7-3)

19. scale length = 6 in. scale is $\frac{1}{2}$ in. : 4 ft 48 ft
20. scale length = 4.5 cm scale is 1 cm : 8 mi 36 mi
21. scale length = 5 cm scale is $\frac{1}{2}$ cm : 6 km 60 km
22. scale length = 4 in. scale is 1 in. : 12 mi 48 mi
23. scale length = 9 cm scale is $\frac{1}{2}$ cm : 3 m 54 m
24. scale length = 3.75 in scale is $\frac{1}{4}$ in. : 4 ft 60 ft
25. scale length = 6.75 cm. scale is 1 cm. : 12 m 81 m
26. scale length = $3\frac{3}{4}$ in scale is $\frac{1}{2}$ in. : 3.8 ft 28.5 ft
27. scale length = 4.25 cm scale is 1 cm : 24 km 102 km

Determine whether each relation is a function. Give the domain and range of each. (Lesson 2-2)

28. $\{(1, 0)\ (0, -1)\ (1, -1)\ (2, 3)\}$ no; domain: $\{0, 1, 2\}$; range: $\{-1, 0, 3\}$
29. $\{(-1, 0)\ (-2, 1)\ (1, -1)\ (0, -2)\}$ yes; domain: $\{-2, -1, 0, 1\}$; range: $\{-2, -1, 0, 1\}$
30. $\{(-4, 2)\ (-5, 3)\ (-6, 4)\ (-7, 5)\}$ yes; domain: $\{-7, -6, -5, -4\}$; range: $\{2, 3, 4, 5\}$
31. $\{(2, 1)\ (3, 0)\ (-3, -1)\ (-2, -2)\}$ yes; domain: $\{-3, -2, 2, 3\}$; range: $\{-2, -1, 0, 1\}$

Extra Practice Worksheet 8-3

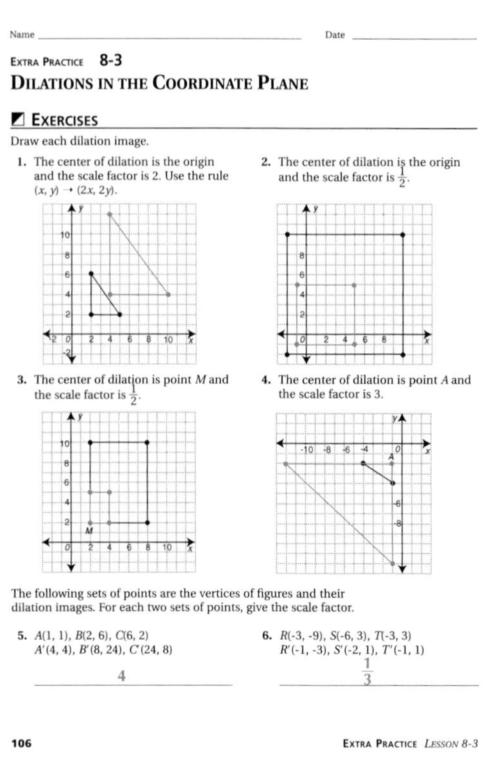

Name ______ Date ______

EXTRA PRACTICE 8-3

Dilations in the Coordinate Plane

EXERCISES

Draw each dilation image.

1. The center of dilation is the origin and the scale factor is 2. Use the rule $(x, y) \rightarrow (2x, 2y)$.
2. The center of dilation is the origin and the scale factor is $\frac{1}{2}$.
3. The center of dilation is point M and the scale factor is $\frac{1}{2}$.
4. The center of dilation is point A and the scale factor is 3.

The following sets of points are the vertices of figures and their dilation images. For each two sets of points, give the scale factor.

5. $A(1, 1)$, $B(2, 6)$, $C(6, 2)$; $A'(4, 4)$, $B'(8, 24)$, $C'(24, 8)$ 4
6. $R(-3, -9)$, $S(-6, 3)$, $T(-3, 3)$; $R'(-1, -3)$, $S'(-2, 1)$, $T'(-1, 1)$ $\frac{1}{3}$

106 EXTRA PRACTICE *Lesson 8-3*

Enrichment Worksheet 8-3

Name ______ Date ______

ENRICHMENT 8-3

Duplication of the Cube

The early mathematicians spent a great deal of time and effort on the problem of the "duplication of the cube." Given a cube with edge a, the problem is to find the length of the edge of a cube that is double the original in volume.

The original cube duplication problem was more difficult than you might at first suppose. An additional condition was imposed: the problem was to be solved by making a geometric construction that involved only circles and straight lines!

It turns out that there is no way to devise such a construction, so the problem in its original form cannot be solved.

EXERCISES

1. Carefully trace this pattern and fold it to make a cube.
2. What is the volume of your cube? $8\ cm^3$
3. What is double the volume of your cube? $16\ cm^3$
4. Construct a pattern that can be folded to make the cube that is double in volume. What is the length of each edge of this cube? ≈ 2.52 cm
5. Use a for the edge of the original cube. Write an equation giving a value for x, the edge of the doubled cube. $x = \sqrt[3]{2a^3}$ or $x = a\sqrt[3]{2}$

138 ENRICHMENT *Lesson 8-3*

Teaching Strategies

Encourage students to think about the size and position of each dilation before they actually solve the problem. Explain that this will help them avoid errors by having an approximate idea of what the image should look like.

Lesson Planning

NCTM Standards/Strands

- Number & Operation
- Patterns, Functions, & Algebra
- Geometry & Spatial Sense
- Problem Solving
- Reasoning and Proof
- Connections
- Representation

Vocabulary

composite of transformations
glide reflection

Materials Needed

graphing calculators rulers
geometric drawing software

Lesson Resources

Warm-Up Transparency 23
Transparency TK-7–10
Reteaching 8-4
Extra Practice 8-4
Enrichment 8-4

Assignment Guide

Basic: 1–17, 21–24
Enriched: 1–24

Getting Started

5-Minute Warm-up

△*ABC* has vertices *A*(2, 2), *B*(2, 5) and C(4, 2). Have students give examples of a reflection, a translations, a rotation, and a dilation. Answers will vary.

Introduction to Lesson 8-4

Ask students how the large pattern in the bottom right-hand corner could have been obtained from the pattern outlined in red. Translate down one unit, and then reflect across the vertical line, or vice versa.

8-4 Multiple Transformations

Goals ■ Identify and find composites of transformations.
Applications Engineering, Art, Graphics Design

Work with a partner.

Locate the basic pattern outlined in blue in the upper left corner.

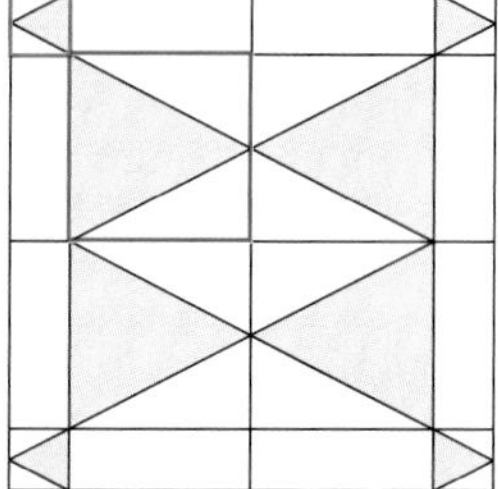

1. Could all the other images in the pattern be obtained by using only translations, rotations, or reflections? no
2. Describe how you could obtain the pattern outlined in red by applying two successive transformations to the pattern outlined in blue. 180° rotation and a dilation with scale factor of 3.

Build Understanding

Two or more successive transformations can be applied to a given figure. This is called a **composite of transformations**.

Example 1

Begin with square *PQRS* with vertices $P(-5, 2)$, $Q(-1, 2)$, $R(-1, 6)$, and $S(-5, 6)$. First, perform a reflection over the *x*-axis ($P'Q'R'S'$) followed by a dilation with center at the origin and a scale factor of 2 ($P''Q''R''S''$).

Solution

Start with square *PQRS*. Use the rule $(x, y) \rightarrow (x, -y)$.

$P(-5, 2) \rightarrow P'(-5, -2)$

$Q(-1, 2) \rightarrow Q'(-1, -2)$

$R(-1, 6) \rightarrow R'(-1, -6)$

$S(-5, 6) \rightarrow S'(-5, -6)$

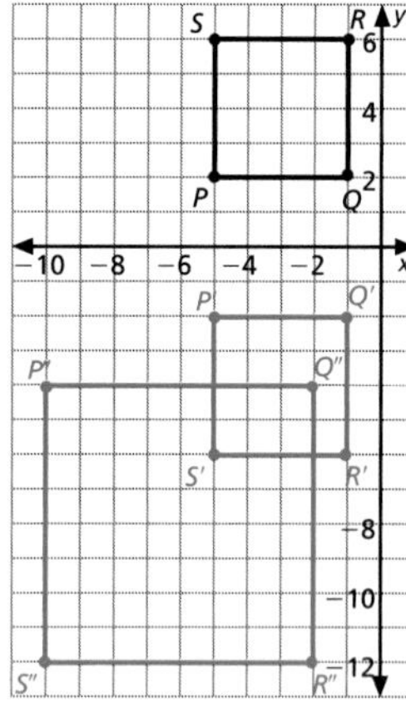

Then apply a dilation with center at the origin and a scale factor of 2 to square $P'Q'R'S'$. Use the rule $(x, y) \rightarrow (2x, 2y)$.

$P'(-5, -2) \rightarrow P''(-10, -4)$

$Q'(-1, -2) \rightarrow Q''(-2, -4)$

$R'(-1, -6) \rightarrow R''(-2, -12)$

$S'(-5, -6) \rightarrow S''(-10, -12)$

Additional Answers

1.

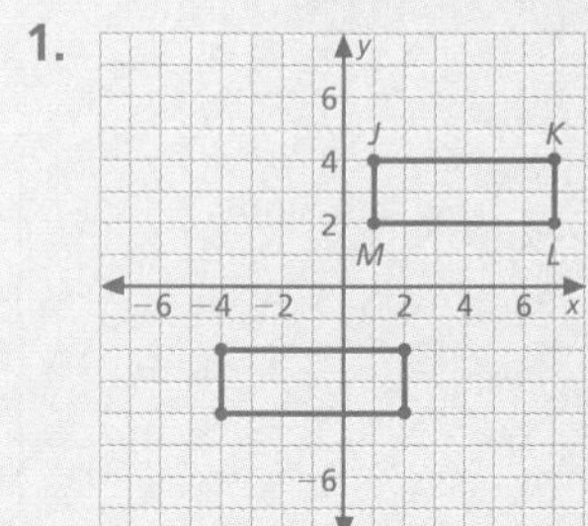

Example 2

ENGINEERING A ride designer is using a computer to map the movement of the car for a new amusement park ride. The two triangles at the right represent the car at the beginning and end of a short section of track after two transformations. Describe these transformations.

Solution

Think of the triangle in Quadrant III as the preimage and the triangle in Quadrant I as the image. Use what you know about transformations to identify the two transformations performed on the preimage.

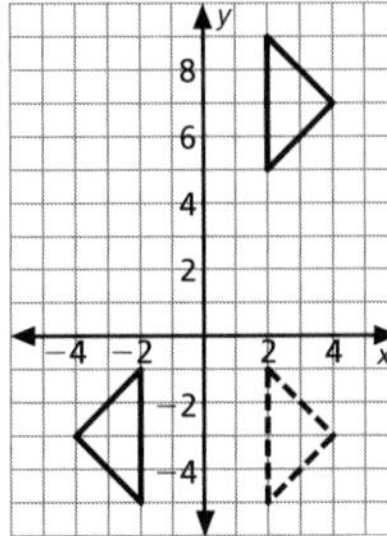

The preimage has been reflected, or flipped, across the *y*-axis. The triangle was then moved up, or slid, in a vertical direction for 10 units.

A composite of a reflection followed by a translation (or vice versa) is called a **glide reflection**.

Technology Note

To use geometric-drawing software to perform a reflection, first select a mirror line by double clicking. Then perform the transformation.

Try These Exercises

1. Perform the following two transformations on rectangle *JKLM*: a translation five units to the left followed by a reflection over the *x*-axis. See additional answers.

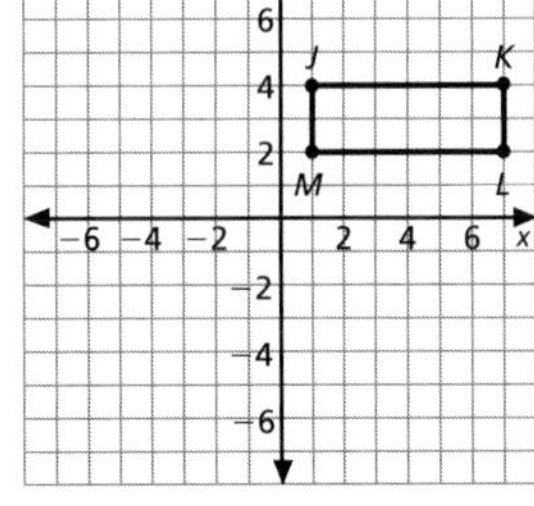

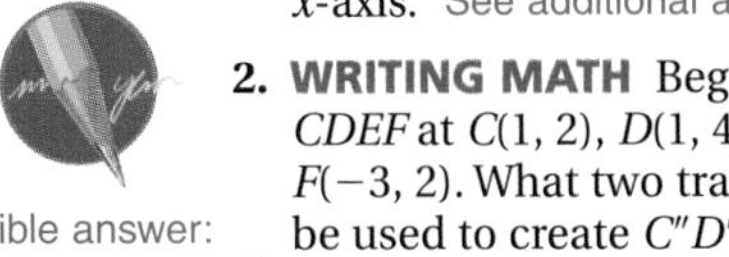

2. **WRITING MATH** Begin with rectangle *CDEF* at $C(1, 2)$, $D(1, 4)$, $E(-3, 4)$ and $F(-3, 2)$. What two transformations could be used to create $C''D''E''F''$ at $C''(9, -2)$, $D''(9, -4)$, $E''(5, -4)$ and $F''(5, -2)$?

Possible answer: reflection over *y*-axis and a translation 8 units down.

3. Does the order in which you perform the transformations for Exercise 3 affect the image produced?

reflection over *x*-axis and translation 8 units right

4. Describe two transformations that could be used to create the image in blue. no

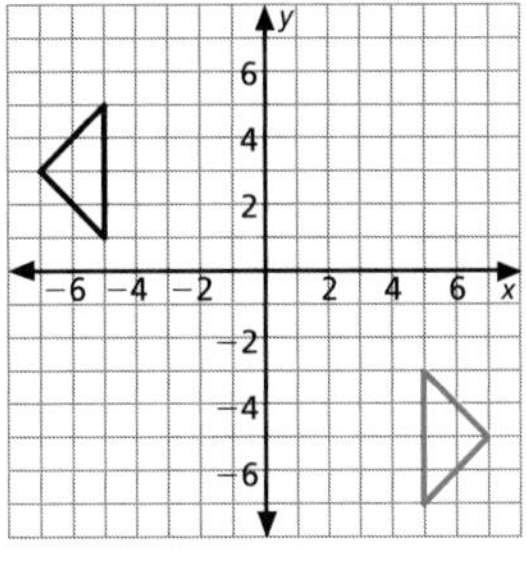

Teaching Strategies

After students have completed Exercises 1, ask them to name another composite of transformations that would produce the same image in Exercise 1. a reflection across the *x*-axis followed by a translation five units to the left

Chalkboard Examples

Supplementary Example 1

A rectangle *CDEF* has vertices $C(-4, 3)$, $D(-4, 2)$, $E(-1, 2)$ and $F(-1, 3)$. First, perform a reflection for the x-axis ($C'D'E'F'$) followed by a dilation with center at the origin and a scale factor of 2 ($C'D'E'F'$). $C'(4, 3)$, $D'(4, 2)$, $E'(1, 2)$, $F'(1, 3)$ $C''(8, 6)$, $D''(8, 4)$, $E''(2, 4)$, $F''(2, 6)$]

Supplementary Example 2

The two trapezoids represent a ride car at the beginning and end of a short section of a ride. Describe the two transformations needed to move from a figure in Quadrant I to a figure to Quadrant III. The preimage was moved down and reflected across the *y*-axis.

Reteaching Worksheet 8-4

Name ______________________ Date ____________

RETEACHING 8-4

MULTIPLE TRANSFORMATIONS

Sometimes two or more successive transformations can be applied to a given figure. This is called a **composite of transformations**.

Example

Use $\triangle ABC$. Perform a translation 3 units right followed by a reflection over the *y*-axis.

Solution

Find the coordinates for the translation.
$A(-6, 1) \rightarrow A'(-3, 1)$
$B(-4, 2) \rightarrow B'(-1, 2)$
$C(-6, -2) \rightarrow C'(-3, -2)$

Find the coordinates for the reflection
$A'(-3, 1) \rightarrow A''(3, 1)$
$B'(-1, 2) \rightarrow B''(1, 2)$
$C'(-3, -2) \rightarrow C''(3, -2)$

EXERCISES

For each exercise, perform and label the composite of transformations.

1. a rotation of 180° clockwise around the origin followed by a reflection over the *x*-axis
2. a dilation with center at the origin and a scale factor of 2 followed by a reflection over the *y*-axis
3. a translation 7 units left followed by a translation 5 units up
4. a reflection over the *y*-axis followed by a clockwise rotation of 90° around the origin
5. a translation 6 units to the left followed by a dilation with center at vertex A' and a scale factor of $\frac{1}{2}$
6. a counterclockwise rotation of 90° around the origin followed by a reflection over the *x*-axis

122 RETEACHING *LESSON 8-4*

PRACTICE EXERCISES

ART To produce a repeating pattern, an artist is asked to perform the following transformations. For Exercises 5–8, draw the result of the first transformation as a dashed figure and the result of the second transformation in red.

For 5–8, see additional answers.

5. a translation 8 units to the left, followed by a translation 3 units down.

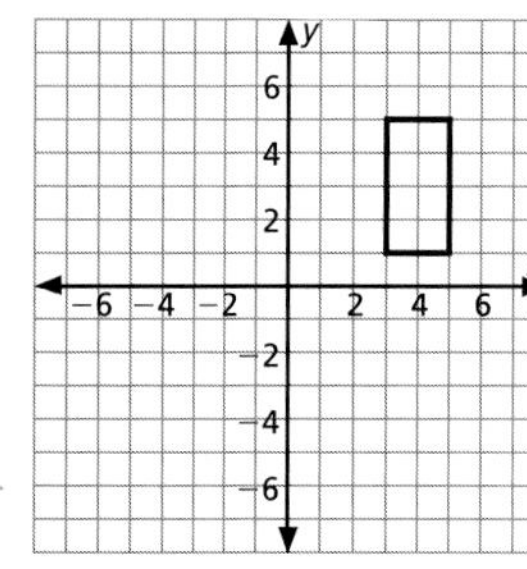

6. a translation 6 units up, followed by a reflection over the *y*-axis.

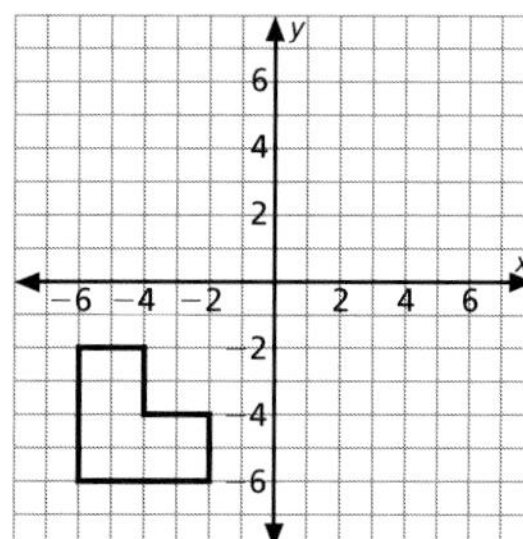

7. a reflection over the *x*-axis, followed by a dilation with center at the origin and a scale factor of 2.

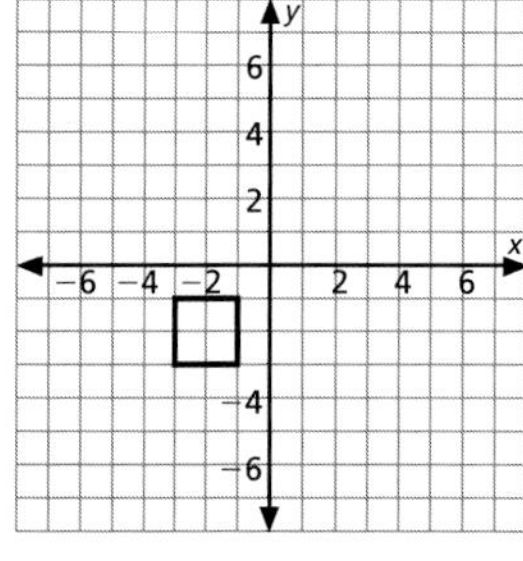

8. a clockwise rotation of 180° around the origin, followed by a clockwise rotation of 90° around the origin.

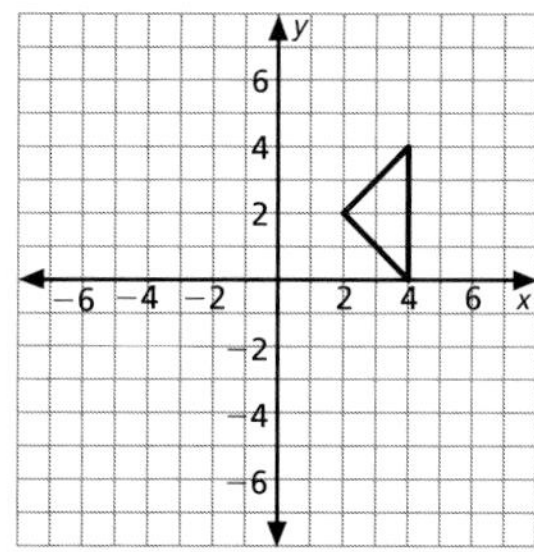

9. GRAPHICS DESIGN A designer has been asked to create a border by repeating a simple pattern that has both vertical and horizontal symmetry. Create the basic pattern using a triangle and multiple transformations. Write a paragraph describing the transformations. Answers will vary. Check students' work.

In Exercises 10–11, describe two transformations that would create the image in blue. There may be more than one possible answer. For 10–11, possible answers are given.

10.

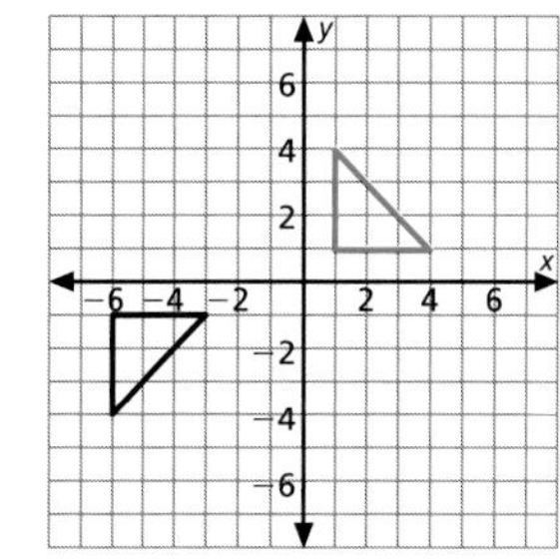

reflection over *x*-axis and translation 7 units to right.

11.

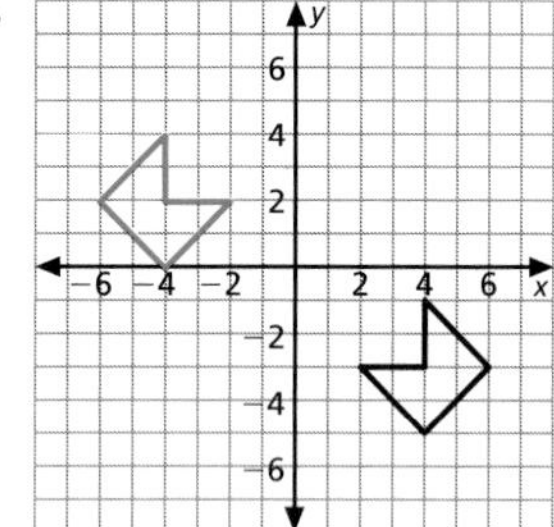

reflection over *y*-axis and translation 5 units up.

Lesson Wrap-up

QUICK ASSESSMENT

Ask the following questions to determine if students understand the content presented in this lesson.

1. Ask students to refer to Example 2. Does it matter in which order you perform the two transformations? No, the preimage could have been translated up ten units and then reflected across the *y*-axis.
2. $\triangle PQR$, with vertices $P(1, 2)$, $Q(3, 4)$ and $R(4, 2)$ is rotated 180° about the origin. How could you accomplish the same transformation without using a rotation? Reflect $\triangle PQR$ across the *x*-axis and then across the *y*-axis, or vice versa.

ASSIGNMENT GUIDE

Basic: 1–17, 21–24
Advanced: 1–20
Additional Practice: See Extra Practice Index on page 674.

ADDITIONAL ANSWERS

5.

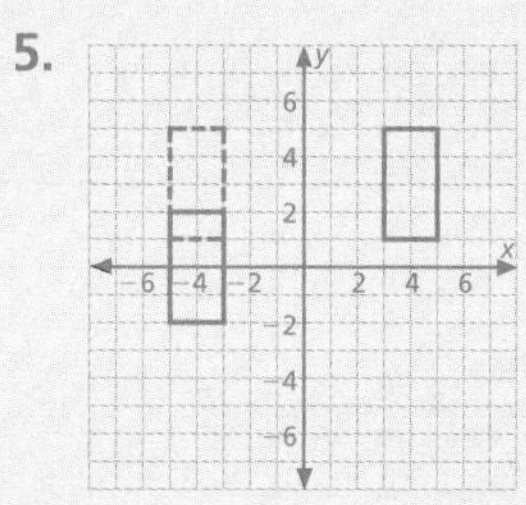

6.

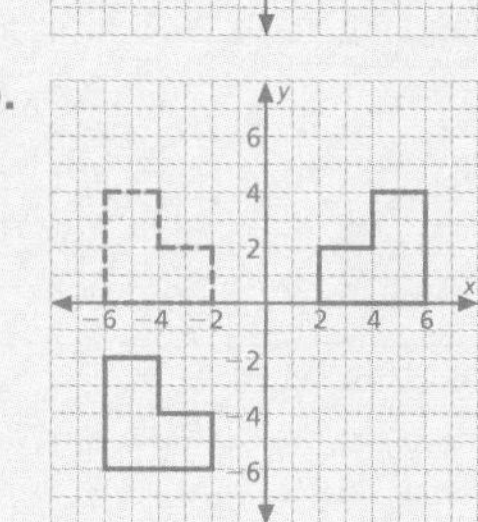

7.

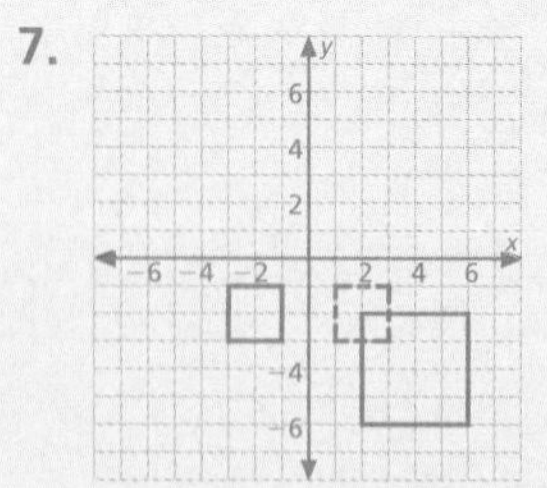

8.

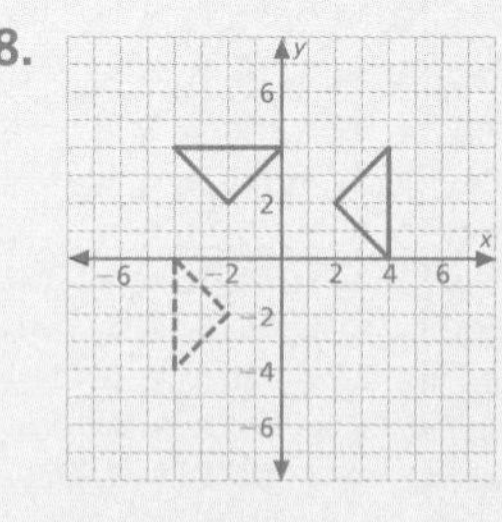

20.

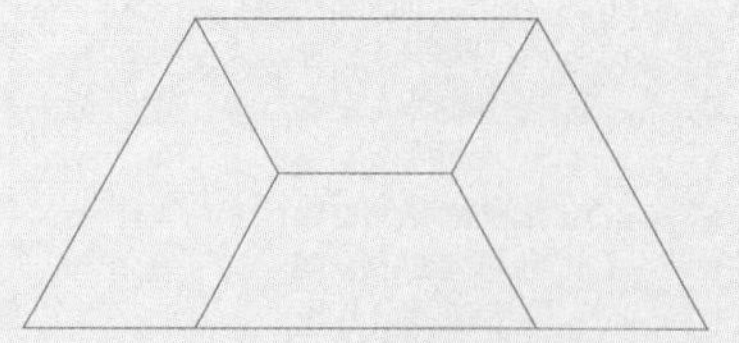

Interdisciplinary Connections

ART A tessellation is a pattern in which one or more shapes are repeated in a way that leaves no gaps and creates no overlaps. You can find many examples of tessellations in the world around you: designs using bricks and tiles, many quilt patterns, and many designs in art. Have students find examples of tessellations and name some of the composite transformations used.

Tell whether the order in which you perform each pair of transformations affects the image produced. If it does affect the image, sketch an example.

12. a translation followed by another translation no

13. a reflection followed by a rotation yes

14. a rotation followed by another rotation no

15. a translation followed by a reflection yes

16. a reflection followed by another reflection yes

17. a translation followed by a rotation yes

Extended Practice Exercises

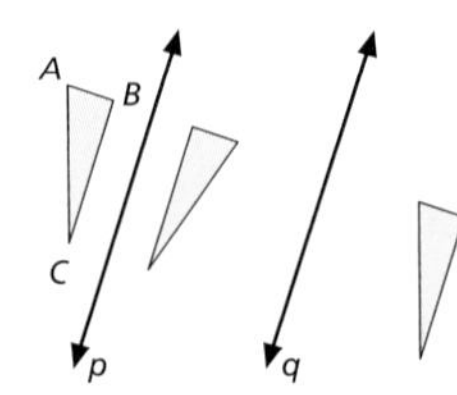

18. In the figure below, $\triangle ABC$ is reflected over line p and then reflected over line q where $p \parallel q$. What single transformation would produce the same result as a composite of reflections over two parallel lines? translation

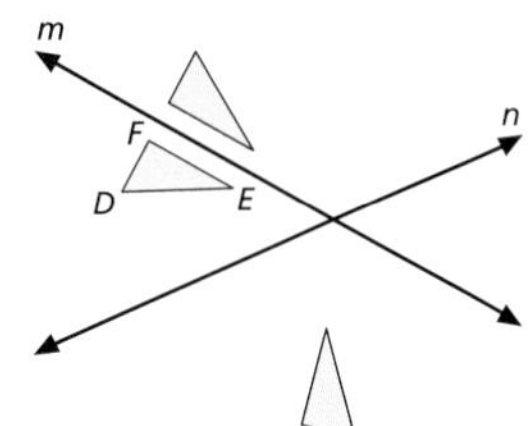

19. In the figure at the right, $\triangle DEF$ is reflected over line m and then reflected over line n, where lines m and n are intersecting lines. What kind of single transformation would produce the same result as a composite of reflections over two intersecting lines? rotation

20. Trace the trapezoid shown at the right on another sheet of paper. Make three other copies of the trapezoid. Cut out the four trapezoids and rearrange them to form a larger trapezoid that is the same shape as the smaller trapezoid. See additional answers.

Mixed Review Exercises

Determine whether each pair of triangles is similar. If the triangles are similar, give a reason: write *AA, SSS,* or *SAS.* (Lesson 7-4)

21.

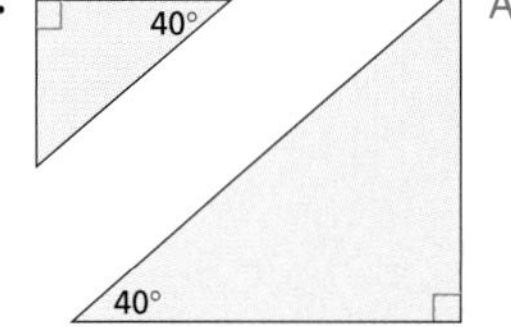

AA

22.

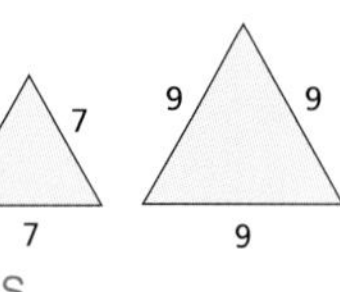

SSS

23.

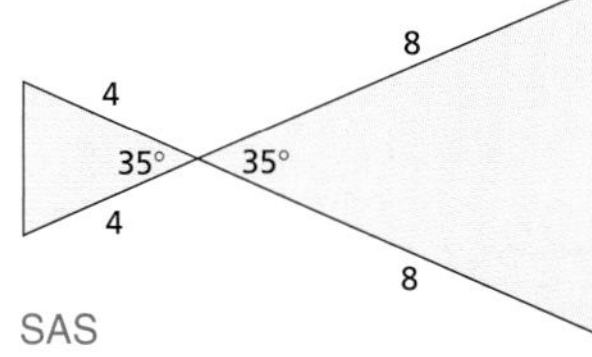

SAS

24.

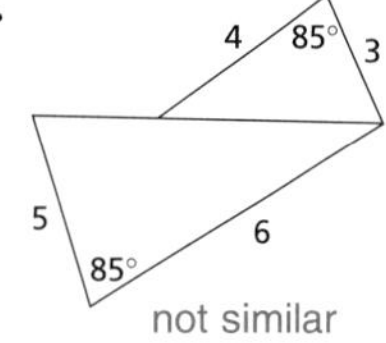

not similar

Extend the Lesson

COOPERATIVE LEARNING Have students create designs in which they use composites of transformations. They may use any shapes they like. Have them explain in their journals how they used transformations to create their designs. You may wish to have students exchange designs and identify the transformations that were used.

Extra Practice Worksheet 8-4

Name ______________ Date ______________

Extra Practice 8-4

Multiple Transformations

Exercises

For each exercise, copy the given figure onto your own grid paper. Then draw the result of the first transformation as a dashed figure and the result of the second transformation in red. Check students' graphs.

1. a translation 4 units to the left, followed by a translation 5 units down
2. a translation 6 units down, followed by a reflection over the y-axis
3. a reflection over the x-axis, followed by a dilation with center at the origin and a scale factor of 2
4. a clockwise rotation of 90° around the origin, followed by a clockwise rotation of 180° around the origin

In Exercises 5 and 6, describe two transformations that would create figure 2 as the image of figure 1. There may be more than one possible answer. Possible answers are given.

5. translation 6 units left, reflection over the x-axis
6. reflection over the y-axis, reflection over the x-axis

108 Extra Practice *Lesson 8-4*

Enrichment Worksheet 8-4

Name ______________ Date ______________

Enrichment 8-4

Nonperiodic Tessellations

In a periodic tessellation, it is possible to make any portion of the design coincide with some other portion through a translation. In a nonperiodic tessellation, this is not the case.

Example

This nonperiodic tessellation is created from diagonally-divided squares. The orientation of the diagonal (top right to bottom left, or bottom right to top left) is determined in a random fashion. So, a tracing of part of the tessellation will not necessarily match some other portion.

Exercises

1. On a separate sheet of paper, create a nonperiodic tessellation using these equilateral triangles. To make the shapes appear randomly, you use a number cube. If you roll an odd number, divide a triangle with a vertical line. If you roll an even number, divide a triangle with a slanted line.

Roll 1, 3 or 5 Roll 2, 4 or 6

2. The pentagon shown at the right will tessellate the plane. Use the pentagon to make two tessellations, one periodic and one nonperiodic.

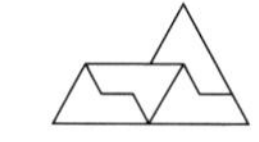

3. Create an original nonperiodic tessellation of your own.
4. A mathematical physicist named Roger Penrose has created two interesting shapes he calls *darts* and *kites*. These can be used to make a nonperiodic tessellation, but they cannot be used to make a periodic one. Make a careful copy of the two shapes on cardboard or stiff paper. Then show part of a nonperiodic tessellation using the Penrose tiles.

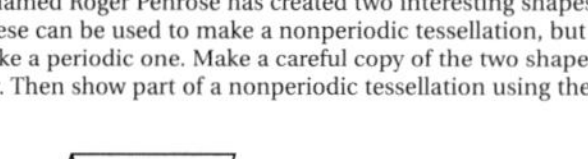

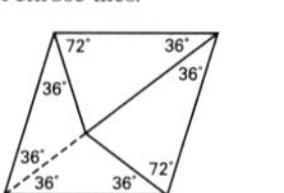

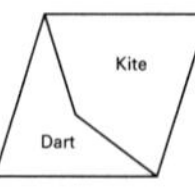

Check students' tessellations.

140 Enrichment *Lesson 8-4*

Skills Practice

Vocabulary Review

Lesson 8-3
dilation — enlargement
reduction — scale factor
center of dilation

Lesson 8-4
composite of transformations
glide reflection

Assignment Guide

All students: 1–26
Additional Practice: Refer to the Extra Practice Index on page 674 of this text.

Additional Answers

1.

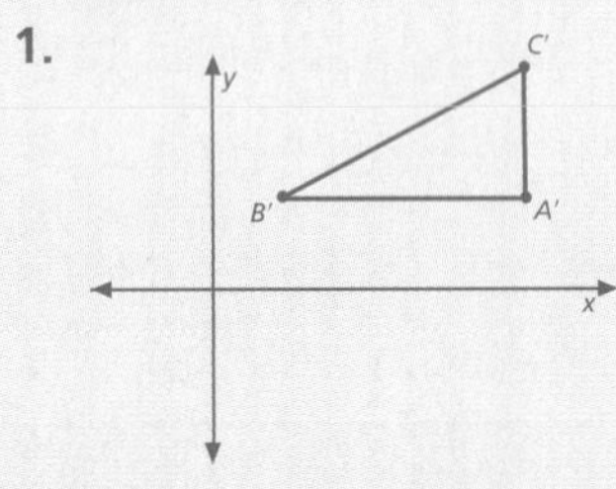

2.

3.

4.

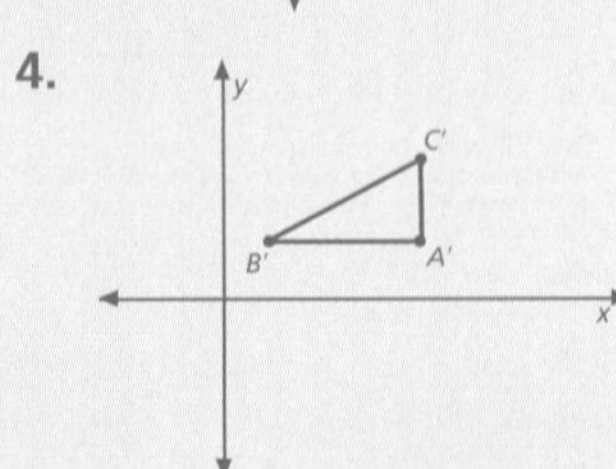

5.

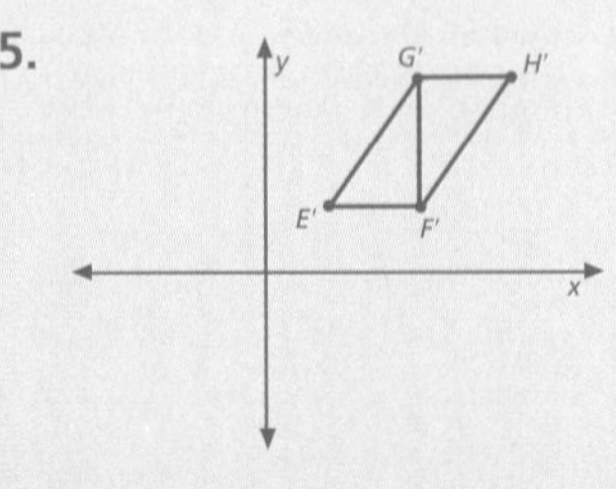

Review and Practice Your Skills

Practice ■ Lesson 8-3

Using the figure on the right, draw each dilation image on a separate coordinate grid.
For 1–4, see additional answers.

1. center of dilation is origin; scale factor is 3
2. center of dilation is origin; scale factor is $\frac{1}{2}$
3. center of dilation is A; scale factor is 2.5
4. center of dilation is B; scale factor is $\frac{3}{4}$

Using the figure on the right, draw each dilation image on a separate coordinate grid.
For 5–8, see additional answers.

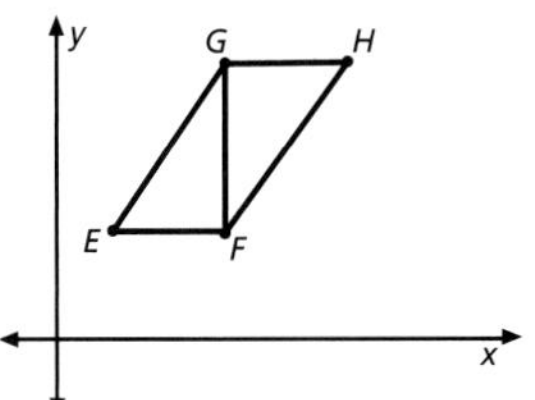

5. center of dilation is origin; scale factor is 1.5
6. center of dilation is origin; scale factor is $\frac{1}{4}$
7. center of dilation is F; scale factor is 2
8. center of dilation is G; scale factor is 3

The following two sets of points are the vertices of figures and their dilation images. Name the scale factor and the center of dilation for each.

9. $D(-5, -5)$, $E(10, 0)$, $F(-5, 0)$ Scale factor: $\frac{1}{5}$
$D'(-5, -1)$, $E'(-2, 0)$, $F'(-5, 0)$
Center of dilation: $F(-5, 0)$

10. $S(2, 2)$, $T(4, 8)$, $U(6, 4)$ Scale factor: 3
$S'(6, 6)$, $T'(12, 24)$, $U'(18, 12)$
Center of dilation: origin

Practice ■ Lesson 8-4

For each exercise, draw the result of the first transformation as a dashed figure and the result of the second transformation in red. For 11–13, see additional answers.

11. clockwise rotation of 90°; reflection across line $y = -x$

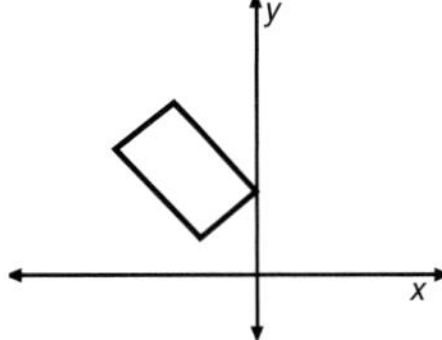

12. reflection across x-axis clockwise rotation of 270°

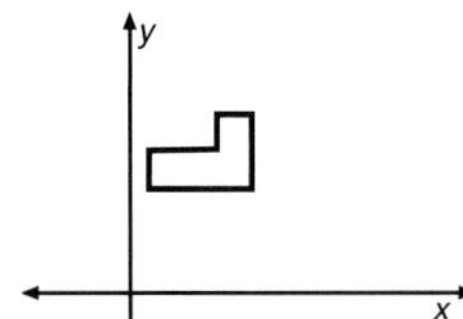

13. dilation—center at origin, scale factor of 2; translation 5 units up

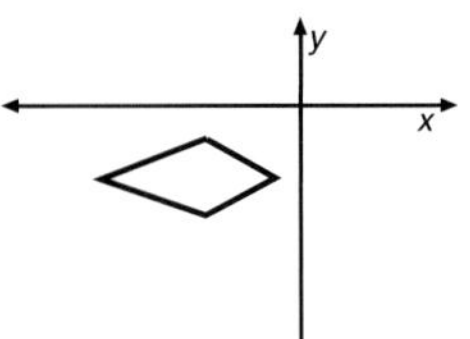

Determine the transformations necessary to create figure 2 from figure 1. There may be more than one possible answer. For 14–16, answers will vary. Sample answers are given.

14. Translation up 4 units and right 1 unit; reflection across y-axis.

15.

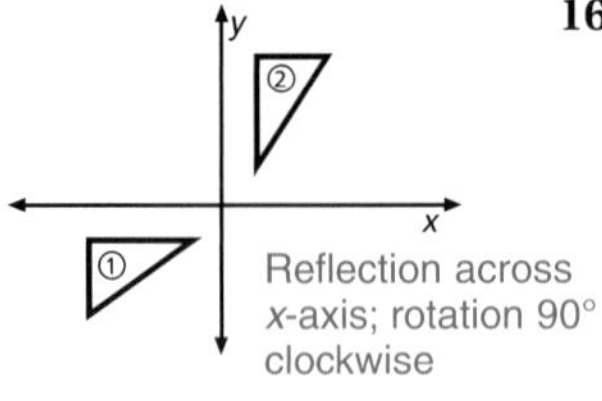

Reflection across x-axis; rotation 90° clockwise

16.

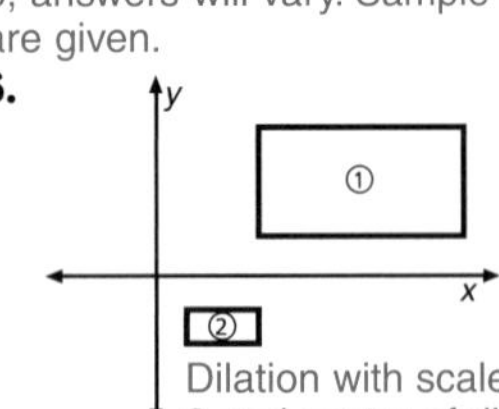

Dilation with scale factor 3 and center of dilation at origin; translation up 7 units.

6.

7.

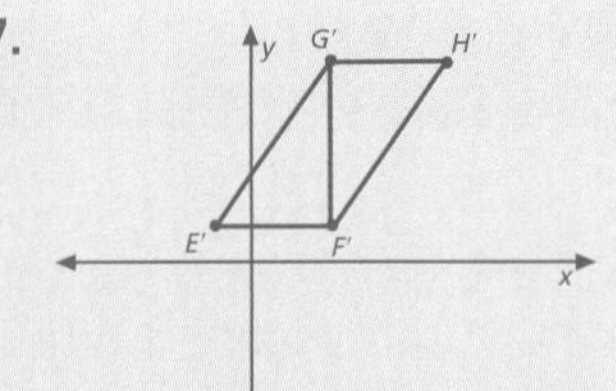

8.

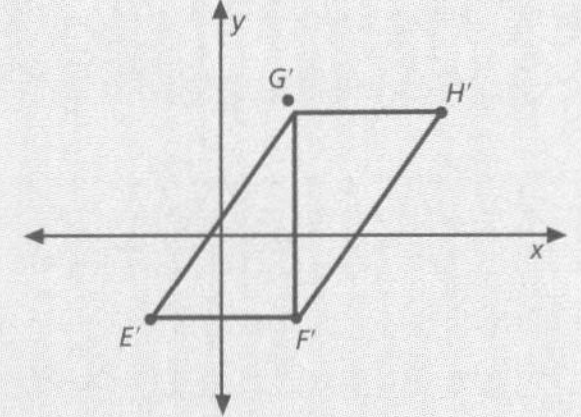

11.

12.

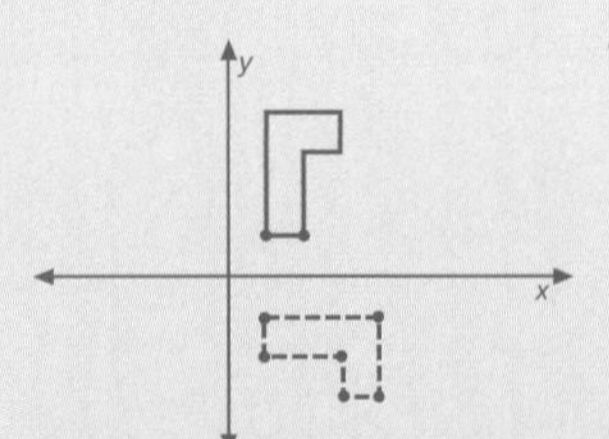

13.

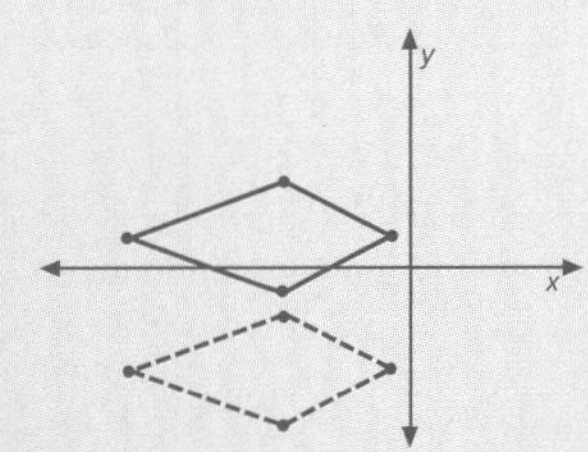

PRACTICE ■ LESSON 8-1–LESSON 8-4

Graph the image of parallelogram *PQRS* with vertices *P*(5, 0), *Q*(−1, −1), *R*(1, 3), and *S*(7, 4). Then graph its image under each transformation from the original position. (Lesson 8-1) For 17–20, see additional answers.

17. 5 units down, 6 units left 18. reflected across $y = x$ 19. reflected across $y = -2$

20. Compare the slopes of parallelogram *PQRS* in all four positions above.

Triangle *ABC* has vertices *A*(−2, 0), *B*(1, 3), and *C*(2, 7). Graph the triangle and its image after each of the following rotations about the origin. (Lesson 8-2) For 21–24, see additional answers.

21. 90° clockwise 22. 180° counterclockwise 23. 90° counterclockwise

24. Compare the slopes of triangle *ABC* in all four positions above.

The following sets of points are the vertices of figures and their dilation images. Name the scale factor and the center of dilation for each. (Lesson 8-3)

25. *D*(−5, −3), *E*(−5, 7), *F*(−1, 7)
D′(−15, −9), *E*′(−15, 21), *F*′(−3, 21)
Scale factor: 3; center of dilation: origin

26. *D*(−5, −3), *E*(−5, 7), *F*(−1, 7)
D′(−5, −18), *E*′(−5, 7), *F*′(5, 7)
Scale factor: 2.5 center of dilation: *E*(−5, 7)

Mid-Chapter Quiz

On a coordinate grid, graph parallelogram *ABCD*, with vertices *A*(1, 6), *B*(2, 9), *C*(4, 6), and *D*(3, 3) and the following transformations:

1. reflected across *x*-axis (1, −6), (2, −9), (4, −6), (3, 3)
2. translated two units right (3, 6), (4, 9), (6, 6), (5, 3)
3. △*D*′*E*′*F*′ is the image after a rotation of 270° clockwise about the origin of △*DEF* with vertices *D*(−5, −2), *E*(−2, −1), and *F*(−6, −7). Find the coordinates of *D*′, *E*′, and *F*′ and the slopes of $\overline{DF}$ and $\overline{D'F'}$. *D*′(2, −5), *E*′(1, −2), *F*′(7, −6); 5, $-\frac{1}{5}$
4. △*ABC* with vertices *A*(1, −3), *B*(6, −1), and *C*4, −5) has been rotated to △*A*′*B*′*C*′ with vertices *A*′(3, 1), *B*′(1, 6), and *C*′(5, 4). Name two rotations that could have been used. 90° counter-clockwise or 270° clockwise, both about the origin.
5. Graph parallelogram *DEFG*, with vertices *D*(−3, −1), *E*(1, 3), *F*(6, 2) and *G*(2, −2), and the dilation image of parallelogram *DEFG* if the center of dilation is *C* and the scale factor is $\frac{1}{2}$. $A'\left(4, \frac{5}{2}\right)$, $B'\left(\frac{13}{2}, 4\right)$, $C'\left(3, \frac{1}{2}\right)$
6. Find the scale factor for △*ABC*, with vertices *A*(0, 0), *B*(3, 6), and *C*(9, 3), and dilation image *A*′*B*′*C*′, with vertices *A*′(0, 0), *B*′(4, 8), and *C*′(12, 4). $\frac{4}{3}$
7. △*ABC* has vertices *A*(−2, −1), *B*(−3, −4), and *C*(−5, −3). Perform a reflection across the *y*-axis to obtain △*A*′*B*′*C*′ followed by a reflection across the *x*-axis to obtain △*A*″*B*″*C*″. *A*′(2, −1), *B*′(3, 4), *C*′(5, −3); *A*″(2, 1), *B*″(3, 4), *C*″(5, 3)
8. △*D*″*E*″*F*″ with vertices *D*″(1, 3), *E*″(6, 3) and *F*″(6, 1) is the image after two transformations of the △*DEF* with vertices *D*(−4, −3), *E*(1, −3), and *F*(1, −1). Describe the two transformations used to create △*D*″*E*″*F*″.
a translation five units right followed by reflection across the *x*-axis, or vice versa.

Chalkboard Examples

Example from Lesson 8-3

Graph △*PQR*, with vertices *P*(2, 2), *Q*(2, 6) and *R*(7, 5) and the dilation image of △*PQR* if the center of dilation is *Q* and the scale factor is $\frac{3}{4}$.
P′(2, 3), *Q*′(2, 6), $R'\left(\frac{23}{4}, \frac{21}{4}\right)$

Example from Lesson 8-4

Trapezoid *ABCD* has vertices *A*(2, 2), *B*(4, 4), *C*(6, 4) and *D*(6, 2). Perform a translation four units down to obtain trapezoid *A*′*B*′*C*′*D*′ followed by a reflection across the *y*-axis to obtain trapezoid *A*″*B*″*C*″*D*″.
A′(2, −2), *B*′(4, 0), *C*′(6, 0), *D*′(6, −2); *A*′(−2, −2), *B*′(−4, 0), *C*′(−6, 0), *D*′(−6, −2)

Mid-Chapter Quiz

Correlation Chart

Question Number	Lesson Number
1–2	8-1
3–4	8-2
5–6	8-3
7–8	8-4

ADDITIONAL ANSWERS

17–19.

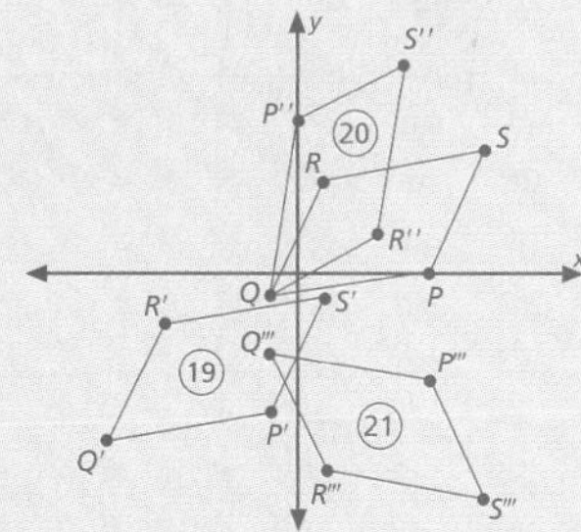

20.

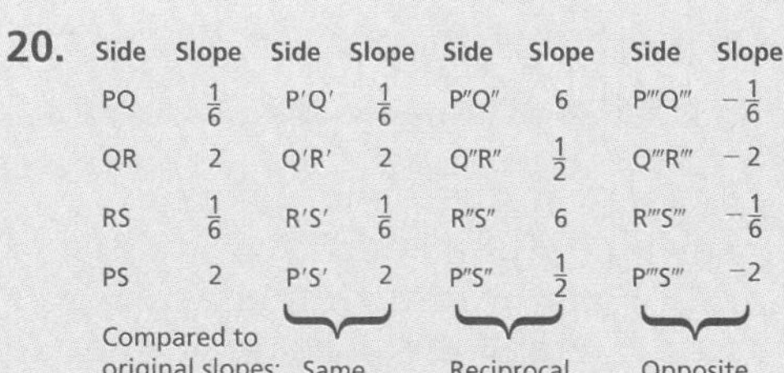

Side	Slope	Side	Slope	Side	Slope	Side	Slope
PQ	$\frac{1}{6}$	P′Q′	$\frac{1}{6}$	P″Q″	6	P‴Q‴	$-\frac{1}{6}$
QR	2	Q′R′	2	Q″R″	$\frac{1}{2}$	Q‴R‴	−2
RS	$\frac{1}{6}$	R′S′	$\frac{1}{6}$	R″S″	6	R‴S‴	$-\frac{1}{6}$
PS	2	P′S′	2	P″S″	$\frac{1}{2}$	P‴S‴	−2
Compared to original slopes:		Same		Reciprocal		Opposite	

21–23.

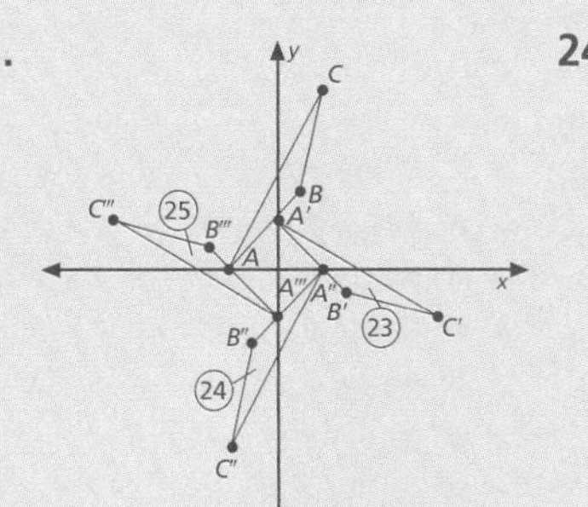

24.

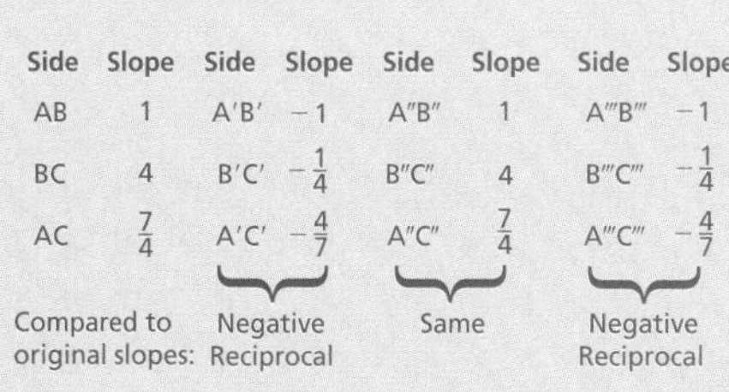

Side	Slope	Side	Slope	Side	Slope	Side	Slope
AB	1	A′B′	−1	A″B″	1	A‴B‴	−1
BC	4	B′C′	$-\frac{1}{4}$	B″C″	4	B‴C‴	$-\frac{1}{4}$
AC	$\frac{7}{4}$	A′C′	$-\frac{4}{7}$	A″C″	$\frac{7}{4}$	A‴C‴	$-\frac{4}{7}$
Compared to original slopes:		Negative Reciprocal		Same		Negative Reciprocal	

8-5 Addition and Multiplication with Matrices

Goals
- Identify matrices and elements within each matrix by rows and columns.
- Perform addition and scalar multiplication on matrices.

Applications Souvenir Sales, Cryptography, Manufacturing

Work with a partner to make a table of the following information.

Alsip and Bell work for Ski Park West. During the winter months, the company rents snowmobiles to park visitors. During December, January, and February, Alsip rented out 18, 12 and 15 snowmobiles, respectively, while Bell rented out 21, 15 and 8 during the same months.

Snowmobiles Sold

	Alsip	Bell
December		
January		
February		

1. Copy and complete the table shown at the right.
2. Make another table with the information, but reverse the rows and columns. See additional answers.

Build Understanding

A **matrix** (plural: matrices) is a rectangular array of numbers arranged into rows and columns. Usually, square brackets enclose the numbers in a matrix. An array with m rows and n columns is called an $m \times n$ (read m by n) matrix.

The **dimensions** of the matrix are m and n. The numbers that make up the matrix are the **elements** of the matrix. The information in the snowmobile problem above can be shown by two different matrices, M_1 and M_2. The titles of rows and columns are not part of the matrices.

Technology Note

You can perform operations with matrices using a graphing calculator. To define a matrix, follow these steps:

1. Access the matrix menu.
2. Select a matrix name and press EDIT.
3. Enter dimensions (*row* × *column*). Then enter the matrix elements.

Matrix M_1

	Alsip	Bell
Dec.	18	21
Jan.	12	15
Feb.	15	8

$$M_1 = \begin{bmatrix} 18 & 21 \\ 12 & 15 \\ 15 & 8 \end{bmatrix}$$

Matrix M_2

	Dec.	Jan.	Feb.
Alsip	18	12	15
Bell	21	15	8

$$M_2 = \begin{bmatrix} 18 & 12 & 15 \\ 21 & 15 & 8 \end{bmatrix}$$

Matrix M_1 has 3 rows and 2 columns. It is a 3×2 matrix. Matrix M_2 has 2 rows and 3 columns. So it is a 2×3 matrix. Although both matrices show the same information, they are not considered to be equal.

Two matrices are **equal matrices** if and only if they have the same dimensions and corresponding elements are equal.

Example 1

a. Find the dimensions of matrix C.

b. Identify the elements C_{32}, C_{21}, and C_{13}.

$$C = \begin{bmatrix} 17 & -2 & 3 \\ 0 & -1 & 5 \\ 6 & 1 & -7 \\ 5 & -4 & 2 \end{bmatrix}$$

Lesson Planning

NCTM Standards/Strands
- Patterns, Functions, & Algebra
- Geometry & Spatial Sense
- Problem Solving
- Reasoning and Proof
- Connections
- Representation

Vocabulary
matrix, dimensions, elements, equal matrices, matrix addition, scalar multiplication, scalar

Materials Needed
graphing calculators

Lesson Resources
Warm-Up Transparency 23
Reteaching 8-5
Extra Practice 8-5
Enrichment 8-5
Transparency TK-7–10

Assignment Guide
Basic: 1–27, 32–40
Enriched: 1–40

Getting Started

5-Minute Warm-Up
Solve for x and y.
1. $y = 4 + x$, $3x = 2y - 7$ $x = 1$, $y = 5$
2. $x = y + 6$, $2y - 3x = -12$ $x = 0$, $y = -6$

Introduction to Lesson 8-5
Ask students to explain how the charts are similar and how they are different. They both show the same information, but one chart has three rows and two columns, while the other chart has two rows and three columns.

Additional Answers
2.

	Dec.	Jan.	Feb.
Alsip	18	12	15
Bell	21	15	8

Teaching Strategies
Suggest that students who are having difficulty adding or multiplying matrices show their work step by step instead of doing the calculations in their heads. Suggest that they list each element to be added or multiplied and the operation. They will find they can check their work easily if they record calculations rather than relying on mental math.

Solution

a. C has 4 rows and 3 columns. So C is a 4×3 matrix.

b. C_{32} means the element in row 3, column 2.

$C_{32} = 1$, $C_{21} = 0$, $C_{13} = 3$

SOUVENIR SALES A theme park sells red, white and blue sweatshirts in small, medium and large sizes. A manager at a souvenir stand receives two shipments of sweatshirts.

Shipment 1

$$\begin{array}{l} \\ \text{Small} \\ \text{Med} \\ \text{Large} \end{array} \begin{array}{c} \begin{array}{ccc} \text{red} & \text{blue} & \text{white} \end{array} \\ \begin{bmatrix} 18 & 12 & 15 \\ 15 & 33 & 14 \\ 16 & 15 & 18 \end{bmatrix} \end{array}$$

Shipment 2

$$\begin{array}{l} \\ \text{Small} \\ \text{Med} \\ \text{Large} \end{array} \begin{array}{c} \begin{array}{ccc} \text{red} & \text{blue} & \text{white} \end{array} \\ \begin{bmatrix} 15 & 13 & 12 \\ 18 & 15 & 16 \\ 13 & 8 & 16 \end{bmatrix} \end{array}$$

To find the combined inventory, she adds the corresponding elements in each matrix. This illustrates the concept of **matrix addition**.

If two matrices M and N have the same dimensions, their sum $M + N$ is the matrix in which each element is the sum of the corresponding elements in M and N.

Math: Who, Where, When

The theory of matrices was first developed around 1858 by Arthur Cayley, a British mathematician.

Example 2

Find the sum of the matrices shown for Shipment 1 and Shipment 2 above.

Solution

Call the matrices A and B.

$$A = \begin{bmatrix} 18 & 12 & 15 \\ 15 & 33 & 14 \\ 16 & 15 & 18 \end{bmatrix} \quad B = \begin{bmatrix} 15 & 13 & 12 \\ 18 & 15 & 16 \\ 13 & 8 & 16 \end{bmatrix}$$

$$A + B = \begin{bmatrix} 18+15 & 12+13 & 15+12 \\ 15+18 & 33+15 & 14+16 \\ 16+13 & 15+8 & 18+16 \end{bmatrix} = \begin{bmatrix} 33 & 25 & 27 \\ 33 & 48 & 30 \\ 29 & 23 & 34 \end{bmatrix}$$

Check Understanding

How is the addition of matrices just like adding two numbers? How is the scalar multiplication of matrices just like multiplying two numbers?

The respective matrix elements are added. Each element of the matrix is multiplied by the scalar.

Suppose the shop owner wanted to double her total inventory for the holiday season. She could simply double each element of the matrix. This illustrates an operation on matrices called **scalar multiplication**.

A matrix can be multiplied by a constant k called a **scalar**. The product of a scalar k and matrix A is the matrix kA in which each element is k times the corresponding element in A.

Example 3

CRYPTOGRAPHY The matrix below represents numerical information that must be transmitted electronically. As the first step in encrypting the information, the matrix is multiplied by 5.

Find the product: $5 \begin{bmatrix} 8 & 12 & 10 \\ 3 & 4 & -3 \\ -2 & 0 & 6 \end{bmatrix}$

Teaching Strategies

ESL/LEP Make sure that students understand the difference between rows and columns. It may help them to think of rows as, for example, rows of flowers in a garden or rows of seats in a theater. Explain that columns are pillars that support buildings, as for example, the columns of the Parthenon in Athens, Greece. Have them find pictures of rows of columns.

Chalkboard Examples

Supplementary Example 1

a. Find the dimensions of the matrix. 4×3

$$D = \begin{bmatrix} 34 & 4 & 6 \\ 0 & 1 & 10 \\ 12 & 2 & 14 \\ 10 & 16 & 4 \end{bmatrix}$$

b. Identify the elements D_{13}, D_{22} and D_{32}. $D_{13} = 6$, $D_{22} = 1$ and $D_{32} = 2$

Supplementary Example 2

Find $A + B$.

$$A = \begin{bmatrix} 4 & -2 & 3 \\ 5 & 0 & 1 \\ 2 & 5 & -6 \end{bmatrix} \quad B = \begin{bmatrix} 0 & 1 & 3 \\ 7 & 0 & -3 \\ 4 & -2 & 5 \end{bmatrix}$$

$$A + B = \begin{bmatrix} 4 & -1 & 6 \\ 12 & 0 & -2 \\ 6 & 3 & -1 \end{bmatrix}$$

Supplementary Example 3

The matrix below must be encrypted by multiplying by 3. Find the product $3 \begin{bmatrix} 0 & 2 & 6 \\ 14 & 0 & -6 \\ 4 & -8 & 16 \end{bmatrix}$.

$$\begin{bmatrix} 0 & 6 & 18 \\ 42 & 0 & -18 \\ 12 & -24 & 48 \end{bmatrix}$$

Reteaching Worksheet 8-5

Name ______________________ Date ____________

RETEACHING 8-5

ADDITION AND MULTIPLICATION WITH MATRICES

A matrix is a rectangular array of elements, or numbers, arranged into rows and columns. The matrix at the right has four rows and three columns, so it is a 4×3 (read "four by three") matrix. A_{13} means the element in row 1, column 3. So $A_{13} = 7$.

$$A = \begin{bmatrix} 5 & -3 & 7 \\ -8 & 0 & 2 \\ 9 & -6 & 1 \\ 4 & 5 & 8 \end{bmatrix}$$

Example 1

Find the sum of matrix C and matrix D. **Solution**

$$C = \begin{bmatrix} 12 & 9 & 16 \\ 10 & 11 & 12 \\ 7 & 28 & 19 \end{bmatrix} \quad D = \begin{bmatrix} 32 & 25 & 3 \\ 16 & 34 & 26 \\ 44 & 9 & 10 \end{bmatrix} \quad C + D = \begin{bmatrix} 12+32 & 9+25 & 16+3 \\ 10+16 & 11+34 & 12+26 \\ 7+44 & 28+9 & 19+10 \end{bmatrix}$$

$$= \begin{bmatrix} 44 & 34 & 19 \\ 26 & 45 & 38 \\ 51 & 37 & 29 \end{bmatrix}$$

Example 1

Find the product.

$$7 \begin{bmatrix} 3 & 5 & 8 \\ 2 & 9 & 8 \end{bmatrix}$$

Solution

$$7 \begin{bmatrix} 3 & 5 & 8 \\ 2 & 9 & 8 \end{bmatrix} = \begin{bmatrix} 7 \cdot 3 & 7 \cdot 5 & 7 \cdot 8 \\ 7 \cdot 2 & 7 \cdot 9 & 7 \cdot 7 \end{bmatrix} = \begin{bmatrix} 21 & 35 & 56 \\ 14 & 63 & 49 \end{bmatrix}$$

EXERCISES

Identify these elements in the matrices below.

$$D = \begin{bmatrix} 7 & 3 & -1 \\ 7 & 8 & 2 \end{bmatrix} \quad E = \begin{bmatrix} 9 & -5 & 0 \\ -3 & 9 & 1 \end{bmatrix} \quad F = \begin{bmatrix} 3 & -9 & 8 \\ 6 & 5 & -1 \end{bmatrix} \quad G = \begin{bmatrix} 4 & 2 & -6 \\ -2 & 5 & 7 \end{bmatrix}$$

1. E_{23} 1
2. G_{12} 2
3. D_{21} 7
4. F_{13} 8

Refer to matrices D, E, F and G to help you find the following.

5. $E + F$ $\begin{bmatrix} 12 & -14 & 8 \\ 3 & 14 & 0 \end{bmatrix}$
6. $D + F$ $\begin{bmatrix} 10 & -6 & 7 \\ 13 & 13 & 1 \end{bmatrix}$
7. $D + G$ $\begin{bmatrix} 11 & 5 & -7 \\ 5 & 13 & 9 \end{bmatrix}$
8. $E + G$ $\begin{bmatrix} 13 & -3 & -6 \\ -5 & 14 & 8 \end{bmatrix}$
9. $10F$ $\begin{bmatrix} 30 & -90 & 80 \\ 60 & 50 & -10 \end{bmatrix}$
10. $4D$ $\begin{bmatrix} 28 & 12 & -4 \\ 28 & 32 & 8 \end{bmatrix}$
11. $5G$ $\begin{bmatrix} 20 & 10 & -30 \\ -10 & 25 & 35 \end{bmatrix}$
12. $20E$ $\begin{bmatrix} 180 & -100 & 0 \\ -60 & 180 & 20 \end{bmatrix}$
13. $F + 2G$ $\begin{bmatrix} 11 & -5 & -4 \\ 2 & 15 & 13 \end{bmatrix}$
14. $D + 10E$ $\begin{bmatrix} 97 & -47 & -1 \\ -23 & 98 & 12 \end{bmatrix}$
15. $3F + D$ $\begin{bmatrix} 16 & -24 & 23 \\ 25 & 23 & -1 \end{bmatrix}$
16. $2D + 3E$ $\begin{bmatrix} 41 & -9 & -2 \\ 5 & 43 & 7 \end{bmatrix}$

124 RETEACHING *LESSON 8-5*

Lesson Wrap-up

Quick Assessment

Ask the following questions to determine if students understand the content presented in this lesson.

1. Suppose the shop owner in Example 2 decided to double shipment 1 and triple shipment 2. Explain how you can find the total number of sweatshirts in each category she should order. Multiply matrix A by two and matrix B by three and find the sum: $2A + 3B$.
2. Using the matrices in Example 2, find the matrix that represents $2A + 3B$.

$$2A + 3B = \begin{bmatrix} 81 & 63 & 66 \\ 84 & 111 & 76 \\ 71 & 54 & 84 \end{bmatrix}$$

Assignment Guide

Basic: 1–27, 32–40
Advanced: 1–40
Additional Practice: See Extra Practice Index on page 674.

Solution

Every element in the matrix must be multiplied by 5.

$$5\begin{bmatrix} 8 & 12 & 10 \\ 3 & 4 & -3 \\ -2 & 0 & 6 \end{bmatrix} = \begin{bmatrix} 5\cdot 8 & 5\cdot 12 & 5\cdot 10 \\ 5\cdot 3 & 5\cdot 4 & 5\cdot(-3) \\ 5\cdot(-2) & 5\cdot 0 & 5\cdot 6 \end{bmatrix} = \begin{bmatrix} 40 & 60 & 50 \\ 15 & 20 & -15 \\ -10 & 0 & 30 \end{bmatrix}$$

Try These Exercises

$$M = \begin{bmatrix} 2 & -1 & 4 & 3 \\ 1 & 0 & -4 & 5 \\ 6 & -1 & 2 & 10 \end{bmatrix}$$

1. Find the dimensions of M. 3×4
2. Identify the elements M_{34} and M_{12}. 10j − 1
3. How many elements are in a 5×2 matrix? 10
4. Refer to matrix M above at the right. Find $6M$.
5. Find $C + D$ if $C = \begin{bmatrix} 3 & -5 \\ 4 & 2 \end{bmatrix}$ and $D = \begin{bmatrix} -3 & 2 \\ 1 & -2 \end{bmatrix}$. $\begin{bmatrix} 0 & -3 \\ 5 & 0 \end{bmatrix}$
6. Three schools have the following win–loss record: Appleton, 6 wins, 9 losses; Carrollton, 14 wins, 2 losses; Prestonsville, 12 wins, 5 losses. Show this information in a matrix. See additional answers.

4. $\begin{bmatrix} 12 & -6 & 24 & 18 \\ 6 & 0 & -24 & 30 \\ 36 & -6 & 12 & 60 \end{bmatrix}$

Technology Note

Exercises 10–18 can easily be solved using a graphing calculator.

First define matrices *A*, *B* and *C*.

To solve Exercise 4: Enter 2 X. Then open the MATRX menu and select 1:[A]. Press ENTER and the calculator finds the product of 2 and A.

To solve Exercise 7: Open the MATRX menu and select 1:[A]. Press +. Then access MATRX and select 2:[B]. Press ENTER and the calculator adds A and B.

Practice Exercises

Find the dimensions of each matrix.

7. $\begin{bmatrix} 2 & -5 & 3 \\ 1 & 4 & -6 \end{bmatrix}$ 2×3

8. $\begin{bmatrix} 11 & -8 & 3 & 5 \\ \frac{1}{2} & 0 & 0 & 7 \\ -8 & 0 & 0 & 0 \end{bmatrix}$ 3×4

9. $\begin{bmatrix} 0 & 3 \\ 1 & -1 \\ 5 & -5 \\ 2 & 7 \\ 8 & 1.9 \end{bmatrix}$ 5×2

Use the following matrices in Exercises 10–18.

$$A = \begin{bmatrix} 2 & -5 & 1 \\ 3 & 1 & -7 \end{bmatrix} \quad B = \begin{bmatrix} -12 & 4 & 6 \\ 8 & -2 & -4 \end{bmatrix} \quad C = \begin{bmatrix} 6 & -2 & -3 \\ -4 & 1 & 2 \end{bmatrix}$$

Find each of the following. For 10–19, see additional answers.

10. $2A$
11. $-\frac{1}{2}B$
12. $\frac{3}{2}C$
13. $A + B$
14. $A + C$
15. $A + -2B$
16. $B + 2C$
17. $A + \frac{1}{2}B$
18. **MANUFACTURING** In June, a boat manufacturer produced 12 sailboats, 18 catamarans, and 8 yachts. In July, the company produced 10 sailboats, 9 catamarans, and 9 yachts. In August, it produced 9 sailboats, 8 catamarans, and 12 yachts. Write two different 3×3 matrices to show this information. See additional answers.
19. **USING DATA** Use the Data Index on pages 632–633 to find information about the calorie count of foods. Create two different matrices showing the amount of the serving and the calorie count for white bread, whole milk, and spaghetti with meatballs and sauce. See additional answers.

20. **CHAPTER INVESTIGATION** Using the scale drawing of your roller coaster, estimate the track length of the entire coaster. Write a paragraph explaining how you arrived at your estimate. Check students' work.

Additional Answers

6.

	Wins	Losses
Appleton	6	9
Carrollton	14	2
Prestonville	12	5

10. $\begin{bmatrix} 4 & -10 & 2 \\ 6 & 2 & -14 \end{bmatrix}$

11. $\begin{bmatrix} 6 & -2 & -3 \\ -4 & 1 & 2 \end{bmatrix}$

12. $\begin{bmatrix} 9 & -3 & -\frac{9}{2} \\ -6 & \frac{3}{2} & 3 \end{bmatrix}$

13. $\begin{bmatrix} -10 & -1 & 7 \\ 11 & -1 & -11 \end{bmatrix}$

14. $\begin{bmatrix} 8 & -7 & -2 \\ -1 & 2 & -5 \end{bmatrix}$

15. $\begin{bmatrix} -13 & 5 & 1 \\ 26 & -13 & -11 \end{bmatrix}$

16. $\begin{bmatrix} 0 & 0 & 0 \\ 0 & 0 & 0 \end{bmatrix}$

17. $\begin{bmatrix} -4 & -3 & 4 \\ 7 & 0 & -9 \end{bmatrix}$

18.

	June	July	Aug.
Sailboats	12	10	9
Catamarans	18	9	8
Yachts	8	9	12

	Sailboats	Catamarans	Yachts
June	12	18	8
July	10	9	9
Aug.	9	8	12

19. $\begin{bmatrix} 1 & 70 \\ 1 & 150 \\ 1 & 330 \end{bmatrix}$ $\begin{bmatrix} 1 & 1 & 1 \\ 70 & 150 & 330 \end{bmatrix}$

21. **WRITING MATH** Describe a method for remembering the difference between rows and columns. Answers will vary.

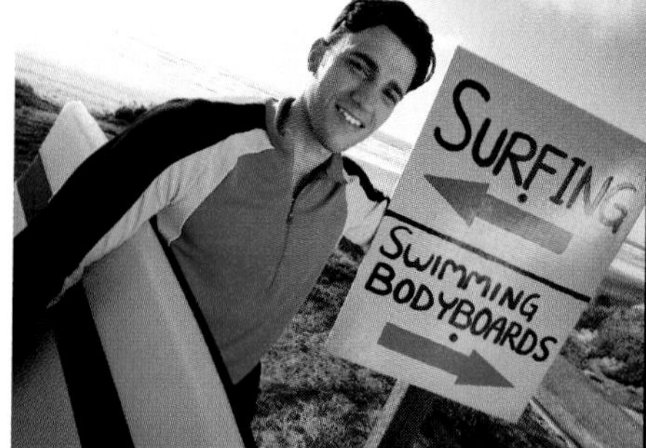

Matrix subtraction is defined by using the scalar -1. If A and B are matrices with dimensions $m \times n$, then $A - B = A + (-1)B$.

Use the above definition in Exercises 23–26.

22. $\begin{bmatrix} 2 \\ 3 \end{bmatrix} - \begin{bmatrix} -2 \\ 7 \end{bmatrix}$ $\begin{bmatrix} 4 \\ -4 \end{bmatrix}$

23. $\begin{bmatrix} 5 & -1 \\ 7 & 6 \end{bmatrix} - \begin{bmatrix} 1 & -5 \\ 4 & 3 \end{bmatrix}$ $\begin{bmatrix} 4 & 4 \\ 3 & 3 \end{bmatrix}$

24. $\begin{bmatrix} 0 & 4 & -2 \\ -5 & 1 & 3 \end{bmatrix} - \begin{bmatrix} 6 & -7 & -1 \\ 0 & -4 & 8 \end{bmatrix}$ $\begin{bmatrix} -6 & 11 & -1 \\ -5 & 5 & 5 \end{bmatrix}$

25. $[17 \quad 6 \quad 12] - [-3 \quad -5 \quad 9]$ $[20 \quad 11 \quad 3]$

Remember that two matrices are equal if and only if they have the same dimensions and all their corresponding elements are equal. Use this definition to solve for x and y.

26. $[x \quad 4y] = [y + 5 \quad 2x + 10]$
$x = 15$, $y = 10$

27. $\begin{bmatrix} x + y \\ x - y \end{bmatrix} = \begin{bmatrix} 9 \\ 4 \end{bmatrix}$ $x = \frac{13}{2}$, $y = \frac{5}{2}$

Extended Practice Exercises

28. **YOU MAKE THE CALL** Dawnae says that any two matrices can be added together. Do you agree? If not, why not?
No. Only matrices with the same dimensions can be added.

Refer to matrices R and S for Exercises 30 and 31. $R = \begin{bmatrix} 4 & 6 \\ 3 & 6 \end{bmatrix}$ $S = \begin{bmatrix} 6 & 5 \\ 1 & -3 \end{bmatrix}$

29. Find $R + S$ and $S + R$. Does addition of matrices seem to be a commutative operation? $R + S = S + R = \begin{bmatrix} 10 & 11 \\ 4 & 3 \end{bmatrix}$; yes

30. Find $R - S$ and $S - R$. Does subtraction of matrices seem to be a commutative operation? $R - S = \begin{bmatrix} -2 & 1 \\ 2 & 9 \end{bmatrix}$, $S - R = \begin{bmatrix} 2 & -1 \\ -2 & -9 \end{bmatrix}$; no

31. **POPULATION** The matrices E, W, and N, shown at the right, give the enrollments by sex and grade at East, West, and North High Schools. In each matrix, Row 1 gives the number of boys and Row 2 the number of girls. Columns 1 to 4 give the number of students in grades 9 through 12, respectively. Calculate entries in matrix T that show the total enrollment by sex and grade in the three schools. $T = \begin{bmatrix} 894 & 930 & 1011 & 949 \\ 875 & 965 & 969 & 980 \end{bmatrix}$

	9	10	11	12	
$E =$	180	220	265	250	**boys**
	205	231	255	260	**girls**
$W =$	306	300	340	310	**boys**
	290	314	270	350	**girls**
$N =$	408	410	406	389	**boys**
	380	420	444	370	**girls**

Mixed Review Exercises

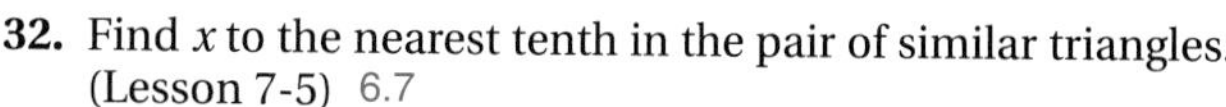

32. Find x to the nearest tenth in the pair of similar triangles. (Lesson 7-5) 6.7

2 4 10 x

Given $f(x) = -2x - 5$ and $g(x) = 3x^2$, find each value. (Lesson 2-2)

33. $f(-3)$ 1
34. $f(2)$ -9
35. $f(-8)$ 11
36. $f(5)$ -15
37. $g(-2)$ 12
38. $g(3)$ 27
39. $g(-5)$ 75
40. $g(4)$ 48

Extra Practice Worksheet 8-5

Name ______________________ Date ______________

EXTRA PRACTICE 8-5

ADDITION AND MULTIPLICATION WITH MATRICES

EXERCISES

Find the dimensions of each matrix.

1. $\begin{bmatrix} -1 & 2 & 5 & 0 \\ 3 & -3 & 4 & 1 \end{bmatrix}$ 2×4

2. $\begin{bmatrix} 2 & 10 & 6 \\ -5 & 3 & 5 \end{bmatrix}$ 2×3

3. $\begin{bmatrix} 4 & 1 \\ \frac{1}{3} & 0 \\ 5 & \frac{3}{4} \end{bmatrix}$ 3×2

4. $\begin{bmatrix} 0 & 2 & -2 \\ 6 & 6 & -7 \\ 1 & 0 & 0.5 \\ 0.8 & 7.5 & 6 \end{bmatrix}$ 4×3

Use the following matrices in Exercises 5–16.

$A = \begin{bmatrix} 2 & 5 & 1 \\ 6 & -1 & 8 \end{bmatrix}$ $B = \begin{bmatrix} -10 & 6 & 4 \\ -16 & -4 & 2 \end{bmatrix}$ $C = \begin{bmatrix} 8 & 8 & 4 \\ -4 & 20 & 12 \end{bmatrix}$

Find each of the following.

5. $-3A$ $\begin{bmatrix} -6 & -15 & -3 \\ -18 & 3 & -24 \end{bmatrix}$

6. $\frac{1}{4}C$ $\begin{bmatrix} 2 & 2 & 1 \\ -1 & 5 & 3 \end{bmatrix}$

7. $B + C$ $\begin{bmatrix} -2 & 14 & 8 \\ -20 & 16 & 14 \end{bmatrix}$

8. $C - A$ $\begin{bmatrix} 6 & 3 & 3 \\ -10 & 21 & 4 \end{bmatrix}$

9. $B + -2A$ $\begin{bmatrix} 14 & -4 & 2 \\ -28 & -2 & -14 \end{bmatrix}$

10. $\frac{1}{2}B - C$ $\begin{bmatrix} -13 & -5 & -2 \\ -4 & -22 & -11 \end{bmatrix}$

11. $2B + 3C$ $\begin{bmatrix} 4 & 36 & 20 \\ -44 & 52 & 40 \end{bmatrix}$

12. $4A - 2C$ $\begin{bmatrix} -8 & 4 & -4 \\ 32 & -44 & 8 \end{bmatrix}$

13. $B - 4C$ $\begin{bmatrix} -42 & -26 & -12 \\ 0 & -84 & -46 \end{bmatrix}$

14. $A + \frac{1}{4}C$ $\begin{bmatrix} 4 & 7 & 2 \\ 5 & 4 & 11 \end{bmatrix}$

15. $-B - C$ $\begin{bmatrix} 2 & -14 & -8 \\ 20 & -16 & -14 \end{bmatrix}$

16. $0.5A + C$ $\begin{bmatrix} 9 & 10.5 & 4.5 \\ -1 & 19.5 & 16 \end{bmatrix}$

110 EXTRA PRACTICE *LESSON 8-5*

Enrichment Worksheet 8-5

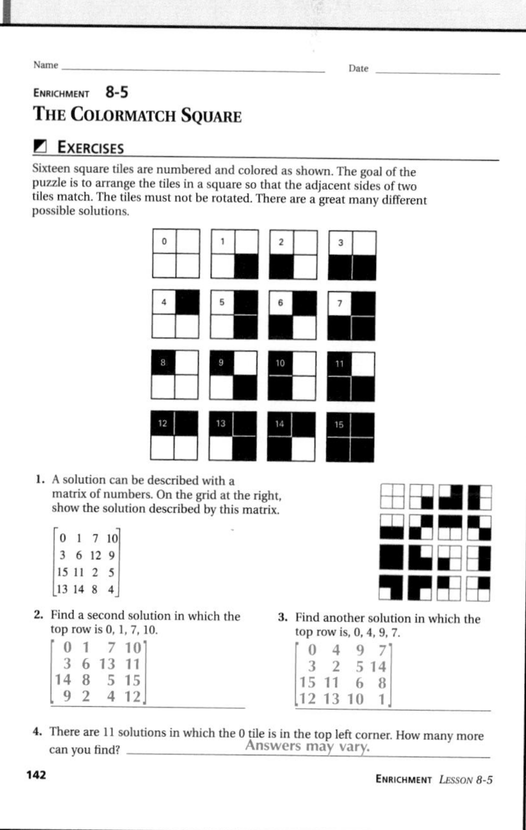

Name ______________________ Date ______________

ENRICHMENT 8-5

THE COLORMATCH SQUARE

EXERCISES

Sixteen square tiles are numbered and colored as shown. The goal of the puzzle is to arrange the tiles in a square so that the adjacent sides of two tiles match. The tiles must not be rotated. There are a great many different possible solutions.

0 1 2 3
4 5 6 7
8 9 10 11
12 13 14 15

1. A solution can be described with a matrix of numbers. On the grid at the right, show the solution described by this matrix.

$\begin{bmatrix} 0 & 1 & 7 & 10 \\ 3 & 6 & 12 & 9 \\ 15 & 11 & 2 & 5 \\ 13 & 14 & 8 & 4 \end{bmatrix}$

2. Find a second solution in which the top row is 0, 1, 7, 10.
$\begin{bmatrix} 0 & 1 & 7 & 10 \\ 3 & 6 & 13 & 11 \\ 14 & 8 & 5 & 15 \\ 9 & 2 & 4 & 12 \end{bmatrix}$

3. Find another solution in which the top row is, 0, 4, 9, 7.
$\begin{bmatrix} 0 & 4 & 9 & 7 \\ 3 & 2 & 5 & 14 \\ 15 & 11 & 6 & 8 \\ 12 & 13 & 10 & 1 \end{bmatrix}$

4. There are 11 solutions in which the 0 tile is in the top left corner. How many more can you find? Answers may vary.

142 ENRICHMENT *LESSON 8-5*

Teaching Strategies

COOPERATIVE LEARNING Have pairs of students use information from an almanac to create different matrices listing the high and low temperatures for at least six cities for at least two months. Have them show how to use their matrices to find the daily average temperature for each city for the months listed.

Lesson Planning

NCTM Standards/Strands

- Number & Operation
- Patterns, Functions, & Algebra
- Reasoning and Proof
- Representation

Vocabulary

row-by-column multiplication

Materials Needed

graphing calculators

Lesson Resources

Warm-Up Transparency 23
Reteaching 8-6
Extra Practice 8-6
Enrichment 8-6
Transparency TK-7–10, RF-31

Assignment Guide

Basic: 1–33, 40–45
Enriched: 1–45

Getting Started

5-Minute Warm-Up

Given an example of a matrix with each of the dimensions.
Answers will vary.

1. 1×4
2. 2×3
3. 4×2

Introduction to Lesson 8-6

Ask students to give the dimensions of the two matrices, and ask them what is similar about the two matrices. 1×3, 3×1; they each have the same number of rows as the other has columns.

8-6 More Operations on Matrices

Goals
- Determine dimensions of product matrices in matrix multiplication.
- Perform row-by-column multiplication of matrices.

Applications Ticket Sales, Encryption, Inventory

A theme park sells three kinds of tickets. Adults over age 18 pay \$15, students from 13 through 18 pay \$10, and children under 13 pay \$5. On one day, the park sells 280 adult tickets, 420 student tickets, and 382 children's tickets. Notice how this information can be shown by two matrices.

number of tickets

$$\begin{bmatrix} 280 & 420 & 382 \end{bmatrix}$$
adult student child

cost per ticket

$$\begin{bmatrix} \$15 \\ \$10 \\ \$5 \end{bmatrix} \begin{matrix} \text{adult} \\ \text{student} \\ \text{child} \end{matrix}$$

Write an expression to show how to compute the total receipts for the day, and then find the total receipts.
$(280 \times 15) + (420 \times 10) + (382 \times 5)$; \$10,335

Build Understanding

Matrix multiplication is done by using **row-by-column multiplication**. *You can only multiply matrices when the number of columns in the first matrix is equal to the number of rows in the second matrix.*

To multiply a row by a column, multiply the first element in the row by the first element in the column, the second element in the row by the second element in the column, and so on. (Thus, the number of elements in a row must equal the number of elements in a column.) Finally, add the products.

The product of an $m \times n$ matrix by an $n \times p$ matrix is an $m \times p$ matrix.

$$\underset{m \times n}{A} \times \underset{n \times p}{B} = \underset{m \times p}{AB}$$

Example 1

TICKET SALES Multiply the two matrices at the top of the page to find the total receipts from ticket sales for the theme park.

Solution

The first matrix is a 1×3 matrix and the second is a 3×1 matrix. So the product will be a 1×1 matrix. Use row-by-column multiplication.

$$\begin{bmatrix} 280 & 420 & 382 \end{bmatrix} \cdot \begin{bmatrix} 15 \\ 10 \\ 5 \end{bmatrix} = 280 \cdot 15 + 420 \cdot 10 + 382 \cdot 5 = 4{,}200 + 4{,}200 + 1{,}910 = 10{,}310$$

The total receipts equals \$10,310.

Teaching Strategies

Suggest that students who are having difficulty multiplying matrices use slips of paper to cover up the rows and column not being multiplied. For example, to multiply the first row of matrix A and the second column of matrix B, have students cover all but that row and that column.

Example 2

Let $M = \begin{bmatrix} 1 & 3 & 4 \\ 5 & -2 & 6 \end{bmatrix}$ and $N = \begin{bmatrix} 1 & 4 & 3 & 0 \\ 5 & -2 & 1 & 6 \\ -4 & 0 & 5 & 7 \end{bmatrix}$.

Find the dimensions of *MN*.

Solution

Because M is a 2×3 matrix and N is a 3×4 matrix, MN is a 2×4 matrix.

2×3 $\quad$ 3×4

2×4

(Notice that you could not find the product NM because the number of columns in N is not the same as the number of rows in M.)

Example 3

ENCRYPTION A business uses a coding matrix to encrypt customer account numbers. Matrix A includes the last four digits of a customer's account number. Matrix B is the coding matrix.

Let $A = \begin{bmatrix} 1 & 3 \\ 5 & -2 \end{bmatrix}$ and $B = \begin{bmatrix} 8 & 7 & 0 \\ 4 & 2 & 6 \end{bmatrix}$ Find AB.

Solution

Because A is a 2×2 matrix and B is a 2×3 matrix, the product is a 2×3 matrix. The product of row 1 and column 1 is $1(8) + 3(4) = 20$. Write 20 in row 1 and column 1 of the product matrix.

$$\begin{bmatrix} \boxed{1 \quad 3} \\ 5 \quad -2 \end{bmatrix} \begin{bmatrix} \boxed{8} & 7 & 0 \\ \boxed{4} & 2 & 6 \end{bmatrix} = \begin{bmatrix} 20 & \_\_ & \_\_ \\ \_\_ & \_\_ & \_\_ \end{bmatrix}$$

The product of row 1 of A by row 2 of B is $1(7) + 3(2) = 13$. Write 13 in row 1 and column 2 of the product.

$$\begin{bmatrix} \boxed{1 \quad 3} \\ 5 \quad -2 \end{bmatrix} \begin{bmatrix} 8 & \boxed{7} & 0 \\ 4 & \boxed{2} & 6 \end{bmatrix} = \begin{bmatrix} 20 & 13 & \_\_ \\ \_\_ & \_\_ & \_\_ \end{bmatrix}$$

The other elements in the product are formed by using this row ▭ by column ▯ pattern.

For instance, the element in the second row, third column of the product is found by multiplying the second row of A by the third column of B. This answer is shown below, along with the final result.

$$\begin{bmatrix} 1 & 3 \\ 5 & -2 \end{bmatrix} \begin{bmatrix} 8 & 7 & 0 \\ 4 & 2 & 6 \end{bmatrix} = \begin{bmatrix} 20 & 13 & 18 \\ 32 & 31 & -12 \end{bmatrix}$$

Technology Note

You will receive an error message if you attempt inappropriate multiplication of matrices using your graphing calculator.

1. Enter matrices A and B using the MATRX EDIT menu.
2. To find product AB, enter: [A]*[B] and press ENTER.
3. Try to find the product of BA by entering [B]*[A].
4. You will receive an error message reporting a dimension mismatch. The product BA cannot be found.

Teaching Strategies

COOPERATIVE LEARNING Have students work in pairs to investigate the distributive property for matrices if A is an $m \times n$ matrix, B is an $n \times p$ matrix, and C is an $n \times q$ matrix. Ask them to determine whether the following is true for matrices A, B, and C: $A \times (B + C) = (A \times B) + (A \times C)$. **Students should conclude that it is true if and only if $p = q$.**

Chalkboard Examples

Supplementary Example 1

$A = [320 \;\; 290 \;\; 452]$ $B = \begin{bmatrix} \$20 \\ \$18 \\ \$12 \end{bmatrix}$

Find the product of A and B. **\$17,044**

Supplementary Example 2

$D = \begin{bmatrix} 3 & 6 & 9 & 12 \\ 2 & 4 & 6 & 8 \end{bmatrix}$ $E = \begin{bmatrix} 6 & 5 & -3 \\ 2 & 7 & -5 \\ 0 & 2 & 0 \\ 5 & -3 & 4 \end{bmatrix}$

Find the dimensions of DE. **2×3**

Supplementary Example 3

The same business also uses a coding matrix to encrypt the last four digits of the customer's home phone number. Matrix G includes the last four digits of a customer's account number. Matrix H is the coding matrix.

$G = \begin{bmatrix} 5 & 4 \\ 8 & 0 \end{bmatrix}$ $H = \begin{bmatrix} 5 & 7 & 3 \\ 9 & 2 & 4 \end{bmatrix}$

Find GH. $GH = \begin{bmatrix} 61 & 43 & 31 \\ 40 & 56 & 24 \end{bmatrix}$

Reteaching Worksheet 8-6

Name __________ Date __________

RETEACHING 8-6

MORE OPERATIONS ON MATRICES

Matrix multiplication is done by using row-by-column multiplication. In order to multiply matrices, **the number of columns in the first matrix must be equal to the number of rows in the second matrix**.

Example

Let $A = \begin{bmatrix} 20 & 40 \\ 10 & 60 \end{bmatrix}$ and $B = \begin{bmatrix} 10 & 40 & 50 & 20 \\ 20 & 30 & 60 & 10 \end{bmatrix}$ Find AB.

Solution

The first matrix has 2 rows and 2 columns so it is a **2** $\times$ 2 matrix.
The second matrix has 2 rows and 4 columns so it is a 2 $\times$ **4** matrix.
Therefore, the product will be a **2** $\times$ **4** matrix.
To multiply a row by a column, multiply the first element in the row by the first element in the column. Then multiply the second element in the row by the second element in the column, and so on. Finally, sum the products of the multiplications.

$$AB = \begin{bmatrix} 20 \cdot 10 + 40 \cdot 20 & 20 \cdot 40 + 40 \cdot 30 & 20 \cdot 50 + 40 \cdot 60 & 20 \cdot 20 + 40 \cdot 10 \\ 10 \cdot 10 + 60 \cdot 20 & 10 \cdot 40 + 60 \cdot 30 & 10 \cdot 50 + 60 \cdot 60 & 10 \cdot 20 + 60 \cdot 10 \end{bmatrix}$$

$$\begin{bmatrix} 1000 & 2000 & 3400 & 800 \\ 1300 & 2200 & 4100 & 800 \end{bmatrix}$$

So $\begin{bmatrix} 20 & 40 \\ 10 & 60 \end{bmatrix} \cdot \begin{bmatrix} 10 & 40 & 50 & 20 \\ 20 & 30 & 60 & 10 \end{bmatrix} = \begin{bmatrix} 1000 & 2000 & 3400 & 800 \\ 1300 & 2200 & 4100 & 800 \end{bmatrix}$

EXERCISES

Refer to the matrices below. Find the product for each exercise, if possible. If it is not possible to multiply, write NP.

$A = \begin{bmatrix} 7 & -1 \\ 7 & 2 \end{bmatrix}$ $B = \begin{bmatrix} -5 & 0 \\ -3 & 9 \end{bmatrix}$ $C = \begin{bmatrix} 3 & -9 & 8 \\ 6 & 5 & -1 \end{bmatrix}$ $D = \begin{bmatrix} 4 & 2 & -6 \\ -2 & 5 & 7 \end{bmatrix}$

1. AB $\begin{bmatrix} -32 & -9 \\ -41 & 18 \end{bmatrix}$
2. CD NP
3. BC $\begin{bmatrix} -15 & 45 & -40 \\ 45 & 72 & -33 \end{bmatrix}$
4. AC $\begin{bmatrix} 15 & -68 & 57 \\ 33 & -53 & 54 \end{bmatrix}$
5. AD $\begin{bmatrix} 30 & 9 & -49 \\ 24 & 24 & -28 \end{bmatrix}$
6. BD $\begin{bmatrix} -20 & -10 & 30 \\ -30 & 39 & 81 \end{bmatrix}$
7. BA $\begin{bmatrix} -35 & 5 \\ 42 & 21 \end{bmatrix}$
8. CB NP
9. DA NP
10. DC NP

126 RETEACHING *LESSON 8-6*

Lesson Wrap-up

Quick Assessment

Ask the following questions to determine if students understand the content presented in this lesson.

1. How can you determine if row-by-column matrix multiplication is possible? Count the number of columns in the first matrix and the number of rows in the second matrix. If the numbers are the same, multiplication is possible.
2. What are the dimensions of a matrix AB when you multiply matrix A times matrix B? (the number of columns in A) × (the number of rows in B)

Assignment Guide

Basic: 1–33, 40–45
Advanced: 1–45
Additional Practice: See Extra Practice Index on page 674.

Try These Exercises

Refer to the matrices below. Find the dimensions of each product, if possible. *Do not multiply.* If not possible to multiply, write NP.

$P = \begin{bmatrix} 1 \\ 3 \\ 4 \end{bmatrix}$ $\quad Q = \begin{bmatrix} 1 & 6 & 7 & 9 \end{bmatrix}$ $\quad R = \begin{bmatrix} 2 & 3 \\ 1 & 4 \\ 6 & 2 \end{bmatrix}$ $\quad S = \begin{bmatrix} 5 \\ 3 \end{bmatrix}$

1. PQ 3 × 4
2. PR NP
3. SQ 2 × 4
4. RS 3 × 1
5. QP NP
6. SR NP
7. QS NP
8. SP NP

Find each product, if possible. If not possible, write NP.

$A = \begin{bmatrix} 10 & 18 & 5 \end{bmatrix}$ $\quad B = \begin{bmatrix} 5 \\ 6 \\ 1 \end{bmatrix}$ $\quad C = \begin{bmatrix} 6 & 1 \\ 5 & 0 \\ 3 & 3 \end{bmatrix}$

9. AB [163]
10. AC [165 25]
11. CA NP

Practice Exercises

Refer to the matrices below. Find the dimensions of each product, if possible. *Do not multiply.* If not possible to multiply, write NP.

$D = \begin{bmatrix} 1 & 3 & 5 \\ 0 & 6 & 4 \end{bmatrix}$ $\quad E = \begin{bmatrix} 0 & 4 \\ 5 & 6 \\ 6 & 1 \end{bmatrix}$ $\quad F = \begin{bmatrix} 3 \\ 4 \\ 5 \end{bmatrix}$ $\quad G = \begin{bmatrix} 2 & 3 & 6 \end{bmatrix}$

12. DE 2 × 2
13. ED 3 × 3
14. FG 3 × 3
15. GF 1 × 1
16. EG NP
17. FD NP
18. DF 2 × 1
19. GE 1 × 2

MATRICES Find each product using a graphing calculator. If not possible, write NP.

20. $\begin{bmatrix} -1 & 0 \\ 0 & -1 \end{bmatrix}\begin{bmatrix} 4 \\ 2 \end{bmatrix}$ $\begin{bmatrix} -4 \\ -2 \end{bmatrix}$
21. $\begin{bmatrix} 2 & 1 & 5 \end{bmatrix}\begin{bmatrix} -2 \\ -3 \\ 7 \end{bmatrix}$ [28]
22. $\begin{bmatrix} 1 & 0 \\ 0 & 1 \end{bmatrix}\begin{bmatrix} 3 \\ 5 \end{bmatrix}$ $\begin{bmatrix} 3 \\ 5 \end{bmatrix}$
23. $\begin{bmatrix} -2 & 4 & 0 \\ -3 & 0 & -8 \end{bmatrix}\begin{bmatrix} 1 & -1 \\ 2 & -1 \end{bmatrix}$ NP

Find each product. If not possible, write NP.

24. $\begin{bmatrix} 1 & 2 \\ 3 & 4 \end{bmatrix}\begin{bmatrix} 1 & 2 \\ -3 & 4 \end{bmatrix}$ $\begin{bmatrix} -5 & 10 \\ -9 & 22 \end{bmatrix}$
25. $\begin{bmatrix} 3 & 0 & 1 \\ 5 & 0 & 6 \end{bmatrix}\begin{bmatrix} 6 & 4 & 1 \end{bmatrix}$ NP
26. $\begin{bmatrix} -2 & 4 & 0 \\ -3 & 0 & -8 \end{bmatrix}\begin{bmatrix} -1 & -2 & -3 \\ 0 & 1 & 0 \\ 4 & 5 & 2 \end{bmatrix}$ $\begin{bmatrix} 2 & 8 & 6 \\ -29 & -34 & -7 \end{bmatrix}$

Problem Solving Tip

To prevent inappropriate application of the row-by-column multiplication method,learn this rhyme:

Row-Col is Pro
Col-Row is No

27. **INVENTORY** Find product JK to find the number of small, medium and large T-shirts in inventory at two souvenir stands.
$J = \begin{bmatrix} 2 & 3 \\ 1 & 5 \end{bmatrix}$ $\quad K = \begin{bmatrix} 3 & 5 & 1 \\ 2 & 1 & 3 \end{bmatrix}$ $\begin{bmatrix} 12 & 13 & 11 \\ 13 & 10 & 16 \end{bmatrix}$

Extend the Lesson

REAL WORLD CONNECTION Matrices are a way in which a great deal of information can be organized so that computers can perform operations on the data. By assigning each element of the matrix a name, such as C_{12}, you can quickly input any item of data exactly were you need to and then have the computer multiply or add lone elements or whole matrices in order to solve problems.

364 Chapter 8 Transformations

28. Use the rule $A(kB) = (kA)B = k(AB)$ to compute: $\begin{bmatrix} 2 & 8 \\ 4 & 2 \end{bmatrix}\left(\frac{1}{2}\begin{bmatrix} 3 & 1 \\ -2 & 1 \end{bmatrix}\right)$ $\begin{bmatrix} -5 & 5 \\ 4 & 3 \end{bmatrix}$

29. **ENCRYPTION** The data in A must be encrypted by multiplying by B. Find AB and BA.

$$A = \begin{bmatrix} 2 & 1 \\ 4 & 3 \end{bmatrix} \quad B = \begin{bmatrix} 5 & 1 \\ -3 & 2 \end{bmatrix} \quad AB = \begin{bmatrix} 7 & 4 \\ 11 & 10 \end{bmatrix} \quad BA = \begin{bmatrix} 14 & 8 \\ 2 & 3 \end{bmatrix}$$

30. **WRITING MATH** What can you conclude about the multiplication of matrices from the products in Exercise 29? The multiplication of matrices is not commutative.

31. For $A = \begin{bmatrix} 3 & 2 \\ -1 & 5 \end{bmatrix}$ and $I = \begin{bmatrix} 1 & 0 \\ 0 & 1 \end{bmatrix}$, find AI and IA. $AI = \begin{bmatrix} 3 & 2 \\ -1 & 5 \end{bmatrix}$; $IA = \begin{bmatrix} 3 & 2 \\ -1 & 5 \end{bmatrix}$

32. **WRITING MATH** What can you conclude about matrix I in Exercise 31? Matrix I is the identity matrix.

33. For $A = \begin{bmatrix} 1 & -1 \\ 0 & 3 \end{bmatrix}$, $B = \begin{bmatrix} 3 & 2 \\ 1 & 5 \end{bmatrix}$ and $C = \begin{bmatrix} 0 & 1 \\ -3 & 2 \end{bmatrix}$, show that $A(BC) = (AB)C$. $A(BC) = (AB)C = \begin{bmatrix} 9 & -4 \\ -45 & 33 \end{bmatrix}$

Extended Practice Exercises

34. If M^2 means $M \times M$, find the matrix M^2 if $M = \begin{bmatrix} 3 & 2 \\ 1 & 4 \end{bmatrix}$. $\begin{bmatrix} 11 & 14 \\ 7 & 18 \end{bmatrix}$

35. Find X^2, X^3, and X^4 if $X = \begin{bmatrix} 0 & -1 \\ -1 & 0 \end{bmatrix}$. $\begin{bmatrix} 1 & 0 \\ 0 & 1 \end{bmatrix}, \begin{bmatrix} 0 & -1 \\ -1 & 0 \end{bmatrix}, \begin{bmatrix} 1 & 0 \\ 0 & 1 \end{bmatrix}$

36. For $A = \begin{bmatrix} 1 & 0 & 0 \\ 0 & 1 & 0 \\ 0 & 0 & 1 \end{bmatrix}$, find A^3. $\begin{bmatrix} 1 & 0 & 0 \\ 0 & 1 & 0 \\ 0 & 0 & 1 \end{bmatrix}$

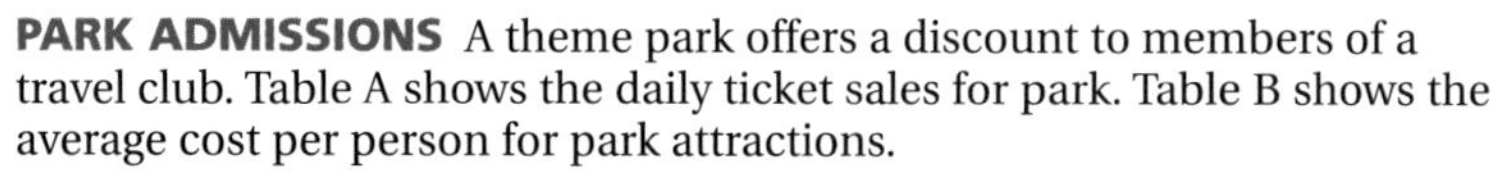

PARK ADMISSIONS A theme park offers a discount to members of a travel club. Table A shows the daily ticket sales for park. Table B shows the average cost per person for park attractions.

A.

	Club	Non-Club
Adult	2,000	1,700
Children	5,400	4,200

B.

	Admission	Food	Souvenirs
Club	\$23	\$18	\$30
Non-Club	\$26	\$20	\$35

37. Write a matrix for each table and find the product of AB to find the amount spent by park visitors by category. $\begin{bmatrix} 90{,}200 & 70{,}000 & 119{,}500 \\ 233{,}400 & 181{,}200 & 309{,}000 \end{bmatrix}$

38. What do the rows and columns of the product matrix AB represent? The product matrix shows the amounts spent by adults and children in each of the three categories.

39. **CRITICAL THINKING** Solve for x and y: $\begin{bmatrix} 6 & 2 \\ 8 & 14 \end{bmatrix}\begin{bmatrix} x \\ y \end{bmatrix} = \begin{bmatrix} 4 \\ -6 \end{bmatrix}$ $x = 1, 4 = -1$

Mixed Review Exercises

Solve each system of equations by graphing. (Lesson 6-4)

40. $\{y = -3x - 4$ $\{y = -2x - 8$ (4, −16)

41. $\{2y = 4x + 10$ $\{3y = -3x + 6$ (−1, 3)

42. $\{4x = 2y - 6$ $\{-3y = x + 5$ (−2, −1)

43. $\{5x = y + 7$ $\{-4y + x = -10$ (2, 3)

44. $\{-9(x - 3) = 6y$ $\{-4y + 28 = 8x$ (5, −3)

45. $\{3y = 2(x + 1.5)$ $\{5x = -2(-3y)$ (6, 5)

Teaching Strategies

MATH JOURNAL Suppose the theme park in the opening activity sells 315 adult tickets, 450 student tickets, and 410 children's tickets on Monday and 275 adult tickets, 320 student tickets, and 400 children's tickets on Tuesday. In their journals, have students show how to use matrices to find the total receipts for each day.

$$\begin{bmatrix} 315 & 450 & 410 \\ 275 & 320 & 400 \end{bmatrix}\begin{bmatrix} 15 \\ 10 \\ 5 \end{bmatrix} = \begin{bmatrix} 11{,}275 \\ 9{,}325 \end{bmatrix}$$

Extra Practice Worksheet 8-6

Name ______________ Date ______________

EXTRA PRACTICE **8-6**

MORE OPERATIONS ON MATRICES

EXERCISES

Refer to the matrices below. Find the dimensions of each product, if possible. *Do not multiply.* If not possible to multiply, write NP.

$R = \begin{bmatrix} 5 & 2 \\ -2 & 4 \end{bmatrix}$ $S = \begin{bmatrix} -1 \\ 2 \end{bmatrix}$ $T = \begin{bmatrix} -1 & 3 & 5 \\ 6 & 2 & 0 \end{bmatrix}$ $V = [4 \quad -5]$

1. RS 2×1
2. TV NP
3. SV 2×2
4. SR NP
5. ST NP
6. VT 1×3
7. TS NP
8. TR NP
9. RT 2×3

Find each product. If not possible, write NP.

10. $\begin{bmatrix} 1 \\ 4 \end{bmatrix}[3 \quad -1 \quad 2]$ $\begin{bmatrix} 3 & -1 & 2 \\ 12 & -4 & 8 \end{bmatrix}$

11. $\begin{bmatrix} 2 & 4 \\ 1 & 2 \end{bmatrix}\begin{bmatrix} -1 & 5 \\ 3 & 2 \end{bmatrix}$ $\begin{bmatrix} 10 & -18 \\ 5 & -9 \end{bmatrix}$

12. $\begin{bmatrix} -1 & 0 & 2 \\ 3 & 4 & 0 \end{bmatrix}\begin{bmatrix} 4 \\ 1 \\ 2 \end{bmatrix}$ $\begin{bmatrix} 0 \\ 16 \end{bmatrix}$

13. $[4 \quad -1 \quad 2][2 \quad 1 \quad 6]$ NP

14. $\begin{bmatrix} -4 & 4 \\ 2 & -2 \\ 3 & -1 \end{bmatrix}\begin{bmatrix} 6 & 1 \\ 2 & 5 \end{bmatrix}$ $\begin{bmatrix} -16 & 16 \\ 8 & -8 \\ 16 & -2 \end{bmatrix}$

15. $\begin{bmatrix} -8 & 1 & 1 & -1 \\ 0 & 0 & 5 & 2 \end{bmatrix}\begin{bmatrix} 6 & -1 \\ 2 & -1 \end{bmatrix}$ NP

16. $\begin{bmatrix} 10 & 0 \\ 9 & 1 \end{bmatrix}\begin{bmatrix} -1 \\ 2 \end{bmatrix}$ $\begin{bmatrix} -10 \\ -7 \end{bmatrix}$

17. $\begin{bmatrix} 6 & 7 \\ 1 & 2 \\ 4 & -1 \\ 3 & 0 \end{bmatrix}\begin{bmatrix} -1 \\ 0 \end{bmatrix}$ $\begin{bmatrix} -6 \\ -1 \\ -4 \\ -3 \end{bmatrix}$

18. For $X = \begin{bmatrix} 3 & 4 \\ 1 & 2 \end{bmatrix}$, $Y = \begin{bmatrix} -1 & 2 \\ 1 & 1 \end{bmatrix}$ and $Z = \begin{bmatrix} 1 & 1 \\ -2 & -2 \end{bmatrix}$, show that $(XY)Z = X(YZ)$.

$(XY)Z = X(YZ) = \begin{bmatrix} -19 & -19 \\ -7 & -7 \end{bmatrix}$

112 EXTRA PRACTICE *LESSON 8-6*

Enrichment Worksheet 8-6

Name ______________ Date ______________

ENRICHMENT **8-6**

DETERMINANTS FOR ORDER 3 MATRICES

Recall that the value of an order 2 determinant is calculated as follows.

$$\begin{vmatrix} a & b \\ c & d \end{vmatrix} = ad - cb$$

Here is how to evaluate an order 3 determinant.

$$\begin{vmatrix} a & b & c \\ d & e & f \\ g & h & i \end{vmatrix} = a\begin{vmatrix} e & f \\ h & i \end{vmatrix} - d\begin{vmatrix} b & c \\ h & i \end{vmatrix} + g\begin{vmatrix} b & c \\ e & f \end{vmatrix}$$
$$= a(ei - hf) - d(bi - hc) + g(bf - ec)$$

EXERCISES

Evaluate each determinant. Use a graphing calculator or a computer graphing program.

1. $\begin{vmatrix} 1 & 0 & 0 \\ 2 & 4 & 5 \\ -3 & 1 & 1 \end{vmatrix}$ -1

2. $\begin{vmatrix} 1 & -2 & 3 \\ 3 & 2 & 1 \\ 1 & 1 & 1 \end{vmatrix}$ 8

3. $\begin{vmatrix} 2 & -1 & 3 \\ 0 & 1 & 2 \\ 3 & -1 & 4 \end{vmatrix}$ -3

4. $\begin{vmatrix} x & y & z \\ x & y & z \\ 1 & 1 & 1 \end{vmatrix}$ 0

5. $\begin{vmatrix} x & y & z \\ 1 & 1 & 1 \\ x & y & z \end{vmatrix}$ 0

6. $\begin{vmatrix} x & 1 & x \\ y & 1 & y \\ z & 1 & z \end{vmatrix}$ 0

7. $\begin{vmatrix} x & y & 1 \\ 1 & -2 & 1 \\ 2 & 1 & -2 \end{vmatrix}$ $3x + 4y + 5$

8. $\begin{vmatrix} x & y & 1 \\ 2 & -1 & 1 \\ 1 & 2 & -2 \end{vmatrix}$ $5y + 5$

9. $\begin{vmatrix} x & y & 1 \\ -1 & 2 & 1 \\ 2 & 1 & -2 \end{vmatrix}$ $-5x - 5$

10. $\begin{vmatrix} 1 & 1 & 1 \\ a & 1 & a \\ -a & 5 & 6 \end{vmatrix}$ $-a^2 - 5a + 6$

11. $\begin{vmatrix} 1 & 1 & 1 \\ a & -1 & a \\ -a & 5 & 6 \end{vmatrix}$ $-a^2 - 7a - 6$

12. $\begin{vmatrix} 1 & 1 & 1 \\ a & -1 & a \\ -a & 5 & -6 \end{vmatrix}$ $-a^2 + 5a + 6$

144 ENRICHMENT *LESSON 8-6*

Skills Practice

Vocabulary Review

Lesson 8-5

matrix	dimensions
elements	equal matrices
matrix addition	scalar
scalar multiplication	

Lesson 8-6

row-by-column multiplication

Assignment Guide

All students: 1–68

Additional Practice: Refer to the Extra Practice Index on page 674 of this text.

Additional Answers

61–63.

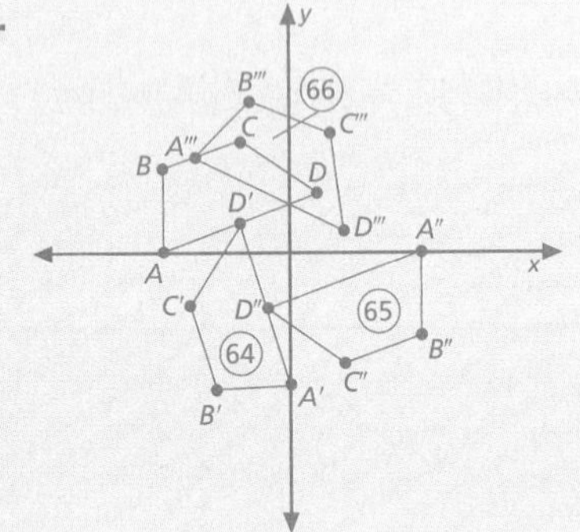

Review and Practice Your Skills

Practice ■ Lesson 8-5

Use matrices *A*, *B* and *C* to find each of the following.

$A=\begin{bmatrix}4&0&-5&7\\-3&8&-2&1\end{bmatrix}$ $B=\begin{bmatrix}-11&3&-1&6\\5&2.5&0&-6\end{bmatrix}$ $C=\begin{bmatrix}0&-8&13&-5\\1&-4&7.5&9\end{bmatrix}$

1. $3A$ $\begin{bmatrix}12&0&-15&21\\-9&24&-6&3\end{bmatrix}$

2. $A+B$ $\begin{bmatrix}-7&3&-6&13\\2&10.5&-2&-5\end{bmatrix}$

3. $C-A$ $\begin{bmatrix}-4&-8&18&-12\\4&-12&9.5&8\end{bmatrix}$

4. $2A+2C$ $\begin{bmatrix}8&-16&16&4\\-4&8&11&20\end{bmatrix}$

5. $A-C$ $\begin{bmatrix}4&8&-18&12\\-4&12&-9.5&-8\end{bmatrix}$

6. $B+A$ $\begin{bmatrix}-7&3&-6&13\\2&10.5&-2&-5\end{bmatrix}$

7. $B+\frac{1}{2}A$ $\begin{bmatrix}-9&3&-3.5&9.5\\3.5&6.5&1&-55\end{bmatrix}$

8. $A+B+C$ $\begin{bmatrix}-7&-5&7&8\\3&6.5&5.5&4\end{bmatrix}$

9. $C-B+A$

10. $5A-2B$

11. $2(A+C)$ $\begin{bmatrix}8&-16&16&4\\-4&8&11&20\end{bmatrix}$

12. $\frac{2}{3}C$ $\begin{bmatrix}0&-5\frac{1}{3}&8\frac{2}{3}&-3\frac{1}{3}\\\frac{2}{3}&-2\frac{2}{3}&5&6\end{bmatrix}$

13. $-\frac{2}{5}B$

14. $3A+2B-4C$

15. $(A+2B)-2(B+A)$

16. $-10B+20C$

17. element A_{12} 0

18. element B_{24} -6

19. dimensions of C 2×4

20. $\frac{1}{2}$(element C_{21}) $\frac{1}{2}$

9. $\begin{bmatrix}15&-11&9&-4\\-7&1.5&5.5&16\end{bmatrix}$ 10. $\begin{bmatrix}42&-6&-23&23\\-25&35&-10&17\end{bmatrix}$ 13. $\begin{bmatrix}4.4&-1.2&0.4&-2.4\\-2&-1&0&2.4\end{bmatrix}$ 14. $\begin{bmatrix}-10&38&-69&53\\-3&45&-39&-45\end{bmatrix}$ 15. $\begin{bmatrix}-4&0&5&-7\\3&-8&2&-1\end{bmatrix}$ 16. $\begin{bmatrix}110&-190&270&-160\\-30&-105&150&240\end{bmatrix}$

Solve for *x* and *y*.

21. $\begin{bmatrix}x+y\\x-y\end{bmatrix}=\begin{bmatrix}-2\\-8\end{bmatrix}$ $x=-5$; $y=3$

22. $\begin{bmatrix}3x+2y\\x-5y\end{bmatrix}=\begin{bmatrix}-4\\27\end{bmatrix}$ $x=2$; $y=-5$

23. $\begin{bmatrix}15-6x\\8x+3y\end{bmatrix}=\begin{bmatrix}3y\\13\end{bmatrix}$ $x=-1$; $y=7$

24. $[3x\quad -2.5y]=[51\quad 40]$ $x=17$; $y=-16$

25. $[x+2y\quad 3x-5y]=[-1\quad 2.5]$ $x=0$; $y=-\frac{1}{2}$

26. $\begin{bmatrix}\frac{1}{x}&\frac{1}{y+3}\end{bmatrix}=\begin{bmatrix}-\frac{3}{4}&-\frac{13}{52}\end{bmatrix}$ $x=-\frac{4}{3}$; $y=-7$

Practice ■ Lesson 8-6

Use matrices *M*–*R* to find each of the following. If not possible, write NP.

$M=\begin{bmatrix}4&-3\\2&1\end{bmatrix}$ $N=\begin{bmatrix}0&2\\-5&4\end{bmatrix}$ $P=\begin{bmatrix}6&0&10\\-1&1&7\end{bmatrix}$ $Q=\begin{bmatrix}-8&2\\1&9\\-3&5\end{bmatrix}$ $R=\begin{bmatrix}6\\-10\end{bmatrix}$

27. MN $\begin{bmatrix}15&-4\\-5&8\end{bmatrix}$

28. MP $\begin{bmatrix}27&-3&19\\11&1&27\end{bmatrix}$

29. MQ NP

30. MR $\begin{bmatrix}54\\2\end{bmatrix}$

31. NM $\begin{bmatrix}4&2\\-12&19\end{bmatrix}$

32. PM NP

33. QM $\begin{bmatrix}-28&26\\22&6\\-2&14\end{bmatrix}$

34. RM NP

35. NP $\begin{bmatrix}-2&2&14\\-34&4&-22\end{bmatrix}$

36. NQ NP

37. NR $\begin{bmatrix}-20\\-70\end{bmatrix}$

38. PN NP

39. QN $\begin{bmatrix}-10&-8\\-45&38\\-25&14\end{bmatrix}$

40. RN NP

41. PQ $\begin{bmatrix}-78&62\\-12&42\end{bmatrix}$

42. PR NP

43. QP $\begin{bmatrix}-50&2&-66\\-3&9&73\\-23&5&5\end{bmatrix}$

44. RP NP

45. QR $\begin{bmatrix}-68\\-84\\-68\end{bmatrix}$

46. RQ NP

47. $MN+PQ$ $\begin{bmatrix}-63&58\\-17&50\end{bmatrix}$

48. $MR-NR$ $\begin{bmatrix}74\\72\end{bmatrix}$

49. M^2 $\begin{bmatrix}10&-15\\10&-5\end{bmatrix}$

50. N^3 $\begin{bmatrix}-40&12\\-30&-16\end{bmatrix}$

51. NMP $\begin{bmatrix}22&2&54\\-91&19&13\end{bmatrix}$

52. PQN $\begin{bmatrix}-310&92\\-210&144\end{bmatrix}$

53. MPQ $\begin{bmatrix}-276&122\\-168&166\end{bmatrix}$

54. $-QPQ$ $\begin{bmatrix}-600&412\\186&-440\\-174&-24\end{bmatrix}$

55. QNR $\begin{bmatrix}20\\-650\\-290\end{bmatrix}$

56. $R(M+N)$ NP

57. $(M+N)R$ $\begin{bmatrix}34\\-68\end{bmatrix}$

58. $QMP+\frac{1}{2}QP$ $\begin{bmatrix}-219&27&-131\\124.5&10.5&298.5\\-37.5&16.5&80.5\end{bmatrix}$

Solve for *x* and *y*.

59. $\begin{bmatrix}6&-2\\-1&5\end{bmatrix}\begin{bmatrix}x\\y\end{bmatrix}=\begin{bmatrix}-8\\-22\end{bmatrix}$ $x=-3$; $y=-5$

60. $\begin{bmatrix}14&5\\-2&13\end{bmatrix}\begin{bmatrix}x\\y\end{bmatrix}=\begin{bmatrix}1\\-7\end{bmatrix}$ $x=\frac{1}{4}$; $y=-\frac{1}{2}$

65.

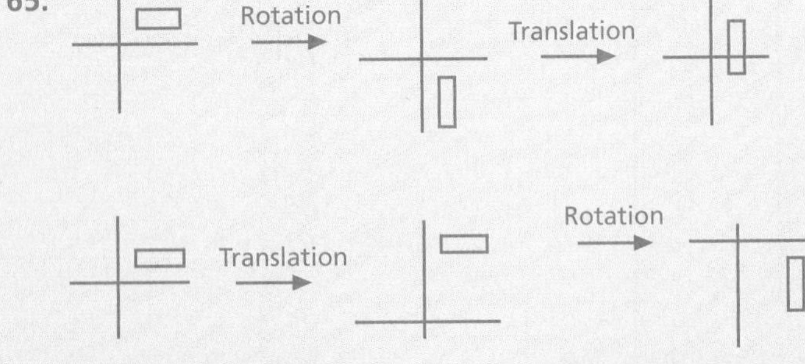

Teaching Strategies

In Lesson 8-5, students learned that matrix addition is commutative. Ask students whether matrix multiplication is commutative. Have them choose an example from the Practice Your Skills exercises to prove their answer. Scalar multiplication is commutative, but matrix multiplication is not.

Practice ■ Lesson 8-1–Lesson 8-6

Trapezoid *ABCD* has vertices $A(-5, 0)$, $B(-5, 3)$, $C(-2, 4)$, and $D(1, 2)$. Graph the trapezoid and its image after each of the following rotations about the origin from the original position. (Lesson 8-2) For 61–63, see additional answers.

61. 90° counterclockwise **62.** 180° clockwise **63.** 45° clockwise

The following sets of points are the vertices of figures and their dilation images. Name the scale factor and the center of dilation. (Lesson 8-3)

64. $P(-2, 2)$, $Q(-2, 5)$, $R(-3, 5)$, $S(-6, 2)$ scale factor: 6; center of dilation: $P(-2, 2)$
$P'(-2, 2)$, $Q'(-2, 20)$, $R'(-8, 20)$ $S'(-26, 2)$

Tell whether the order in which you perform each pair of transformations affects the image produced. If it does affect the image, sketch an example. (Lesson 8-4)

65. rotation and translation yes; see additional answers.

66. dilation (center at figure vertex) and translation no affect

67. dilation (center at the origin) and reflection no affect

68. rotation and reflection no affect

MathWorks Workplace Knowhow — Career – Aerospace Engineer

Aerospace engineers use their knowledge of structural design, aerodynamics, propulsion, thermodynamics, and acoustics to design roller coasters. They know how tight a turn can be without endangering passengers or damaging the coaster's structure. These workers use science to make a roller coaster fast, fun and safe.

You have designed a roller coaster in the shape shown at the right. This coaster is for people over 48 inches tall. The amusement park now wants you to design a children's version of the coaster on a smaller scale with a less steep first hill.

1. Find the slope between points *A* and *B*. 4
2. Find the slope between points *B* and *C*. −2
3. Replot point *B* so that the slope of $\overline{AB}$ is 2 and slope of $\overline{BC}$ is −2. What are the new coordinates for *B*? (3, 8)

Aerospace engineers are also employed to build aircraft and spacecraft and to develop military technology. These engineers apply technology learned in other industries to transportation on land, sea and air. Aerospace engineers must understand math and physics and how to use computers, calculators and other tools to test their ideas.

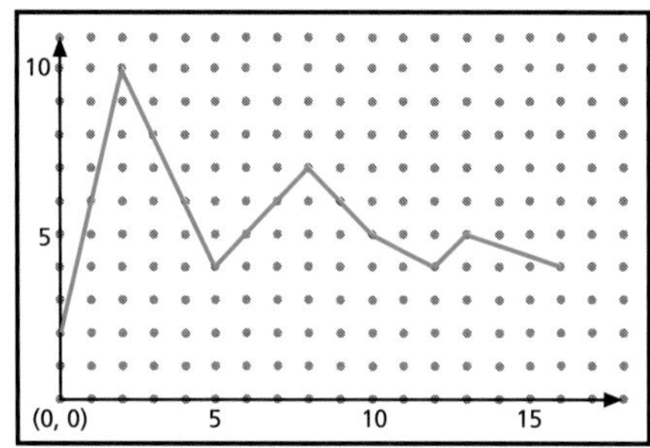

Chalkboard Examples

Example from Lesson 8-5

Find $A + B$.

$$A = \begin{bmatrix} 0 & 2 & 4 \\ 7 & 4 & -1 \\ 2 & -3 & -3 \end{bmatrix} \quad B = \begin{bmatrix} 8 & 6 & 4 \\ 3 & -8 & -2 \\ 12 & 0 & 16 \end{bmatrix}$$

$$A + B = \begin{bmatrix} 8 & 8 & 8 \\ 10 & -4 & -3 \\ 14 & -3 & 13 \end{bmatrix}$$

Example from Lesson 8-6

Find the product of the following matrices.

$$\begin{bmatrix} 2 & 13 & -1 \\ -1 & 4 & 0 \end{bmatrix} \begin{bmatrix} -3 & 6 \\ 0 & 1 \\ 5 & 14 \end{bmatrix} \quad \begin{bmatrix} -11 & 11 \\ 3 & -2 \end{bmatrix}$$

Career Opportunity

Describe the kind of work aerospace engineers do and the types of places they work. Discuss the importance of geometry and algebra in working as an engineering technician. Students should answer Questions 1–3 to better understand how aerospace engineers use mathematics in performing their job.

Students who are interested in learning more about this profession can go to learningmathmatters.com. Guidance Counselors should have information about school and training requirements.

Extend the Lesson

MATH JOURNAL Have students write a paragraph to answer the following question: How can you determine if matrix multiplication is possible? The number of columns in the first matrix must equal the number of rows in the second matrix.

Lesson Planning

NCTM Standards/Strands

- Patterns, Functions, & Algebra
- Geometry & Spatial Sense
- Reasoning and Proof
- Connections
- Representation

Vocabulary

matrices for reflections

Materials Needed

graph paper rulers

Lesson Resources

Warm-Up Transparency 24
Reteaching 8-7
Extra Practice 8-7
Enrichment 8-7
Transparency TK-7–10, RF-32

Assignment Guide

Basic: 1–24, 28–34
Enriched: 1–34

Getting Started

5-Minute Warm-up

Graph the preimage and image on a coordinate plane.
(1, 3), (6, 5), (8, 1), reflected across the x-axis (1, −3), (6, −5), (8, −1)

Introduction to Lesson 8-7

Have students create a polygon in the first quadrant with the point (2, 6) as one of its vertices. Then have them reflect the polygon as described in Steps (a) through (d). Have students share their results and discuss the similarities. Students should observe that corresponding transformations placed their polygons in the same quadrants.

8-7 Transformations and Matrices

Goals
- Represent geometric figures on the coordinate plane by matrices.
- Identify and perform transformations with matrices.

Applications Graphic Art, Ride Design

Work with a partner. Use the point (2, 6) to answer each of the following.

1. What point is (2, 6) reflected over the x-axis? (2, −6)
2. What point is (2, 6) reflected over the y-axis? (−2, 6)
3. What point is (2, 6) reflected over the line $y = x$? (6, 2)
4. What point is (2, 6) reflected over the line $y = -x$? (−6, −2)

Build Understanding

A point can be represented by a matrix, as well as an ordered pair. The ordered pair (2, 6) can be represented by the matrix $\begin{bmatrix}2\\6\end{bmatrix}$.

The element in the first row is the x-coordinate, and the element in the second row is the y-coordinate.

In general, ordered pair (x, y) is represented by the matrix $\begin{bmatrix}x\\y\end{bmatrix}$.

In a similar way, a matrix can be used to denote a polygon. Because each vertex is a point, each point can be represented by a matrix. These four matrices for the vertices can be combined into a single matrix.

Example 1

First, represent each vertex of quadrilateral *ABCD* with a 2 × 1 matrix. Then combine these matrices into a single 2 × 4 matrix.

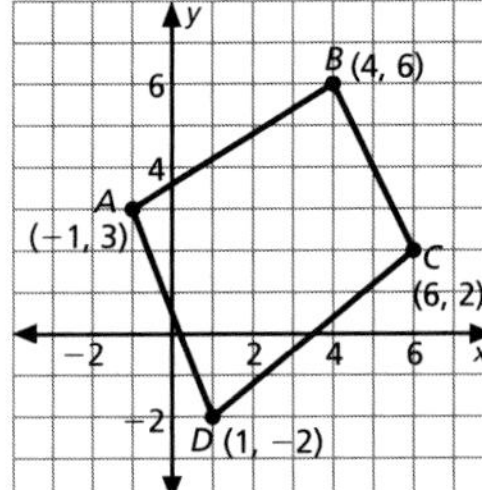

Solution

The vertices can be represented as follows:

$$\begin{bmatrix}-1\\3\end{bmatrix}, \begin{bmatrix}4\\6\end{bmatrix}, \begin{bmatrix}6\\2\end{bmatrix}, \begin{bmatrix}1\\-2\end{bmatrix}$$

Putting the column matrices into a single 2 × 4 matrix, you have the following.

$$\begin{bmatrix}-1 & 4 & 6 & 1\\3 & 6 & 2 & -2\end{bmatrix}$$

Each column refers to one vertex of the quadrilateral.

Check Understanding

What matrix represents $\triangle ABC$ if the vertices are $A(1, 4)$, $B(-2, 3)$, and $C(5, -7)$?

$\begin{bmatrix}1 & -2 & 5\\4 & 3 & -7\end{bmatrix}$

Teaching Strategies

CHALLENGE Find a matrix that produces a counterclockwise rotation of 90° followed by a reflection over the line $y = -x$ and a dilation about the origin with a scale factor of k.
$\begin{bmatrix}-k & 0\\0 & k\end{bmatrix}$

Just as points and polygons can be represented by matrices, you can represent transformations with matrices. Below is a table of matrices for reflections.

Reflection	Matrix	Reflection	Matrix
over the x-axis	$\begin{bmatrix} 1 & 0 \\ 0 & -1 \end{bmatrix}$	over the line $y = x$	$\begin{bmatrix} 0 & 1 \\ 1 & 0 \end{bmatrix}$
over the y-axis	$\begin{bmatrix} -1 & 0 \\ 0 & 1 \end{bmatrix}$	over the line $y = -x$	$\begin{bmatrix} 0 & -1 \\ -1 & 0 \end{bmatrix}$

Example 2

Find the reflection image of $\triangle ABC$ with vertices at $A(1, -2)$, $B(6, -2)$, and $C(4, -5)$ when the triangle is reflected over the line $y = x$. Use matrices.

Solution

Triangle ABC can be represented by

$$\begin{bmatrix} 1 & 6 & 4 \\ -2 & -2 & -5 \end{bmatrix}.$$

The matrix representing a reflection over the line $y = x$ is $\begin{bmatrix} 0 & 1 \\ 1 & 0 \end{bmatrix}$.

Multiply the two matrices.

$$\begin{bmatrix} 0 & 1 \\ 1 & 0 \end{bmatrix}\begin{bmatrix} 1 & 6 & 4 \\ -2 & -2 & -5 \end{bmatrix}$$

So, the image of $\triangle ABC$ is $\begin{bmatrix} -2 & -2 & -5 \\ 1 & 6 & 4 \end{bmatrix}$.

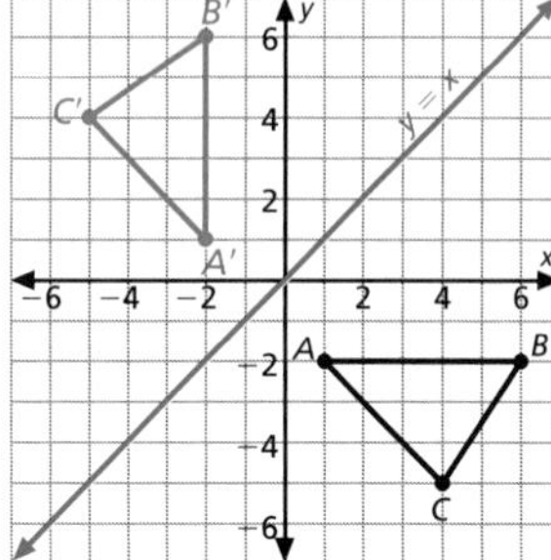

Example 3

GRAPHIC ART An artist is creating a border using design software. To create the basic pattern, she enters the coordinates for $\triangle XYZ$ with vertices $X(0, 0)$, $Y(2, -3)$, and $Z(6, -3)$. She wants to draw the triangle reflected over the y-axis. Find the coordinates of the reflection image using matrices.

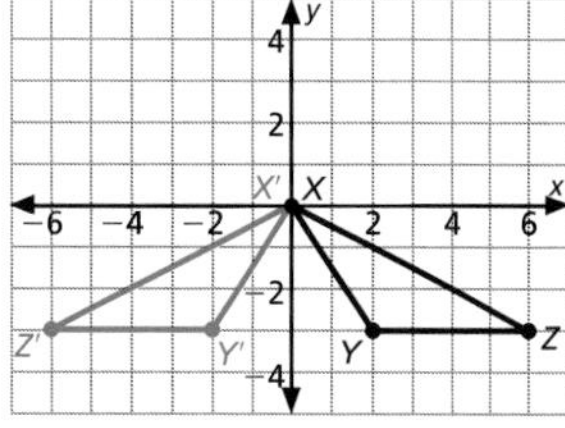

Solution

Let $\begin{bmatrix} 0 & 2 & 6 \\ 0 & -3 & -3 \end{bmatrix}$ represent the triangle.

Then multiply by $\begin{bmatrix} -1 & 0 \\ 0 & 1 \end{bmatrix}$, the matrix for a reflection over the y-axis.

$$\begin{bmatrix} -1 & 0 \\ 0 & 1 \end{bmatrix}\begin{bmatrix} 0 & 2 & 6 \\ 0 & -3 & -3 \end{bmatrix} = \begin{bmatrix} 0 & -2 & -6 \\ 0 & -3 & -3 \end{bmatrix}$$

Teaching Strategies

Suggest that students graph the preimage and image in transformations such as reflecting $\triangle ABC$ with vertices $A(-3, -2)$, $B(-1, 3)$ and $C(4, 1)$ over the x-axis, over the y-axis, over the line $y = -x$ and over the line $y = x$. This will help them appreciate how they can use matrices to perform transformations without becoming confused.

Chalkboard Examples

Supplementary Example 1

The vertices of a trapezoid are $A(-3, -2)$, $B(3, -2)$, $C(-1, 2)$ and $D(1, 1)$. What matrix represents $ABCD$?

$$\begin{bmatrix} -3 & 3 & -1 & 1 \\ -2 & -2 & 2 & 1 \end{bmatrix}$$

Supplementary Example 2

Using matrices, find the reflection image of $\triangle DEF$ with vertices at $D(-6, -2)$, $E(-5, 2)$ and $F(-3, -1)$ when the triangle is reflected over the line $y = x$.

The image of $\triangle DEF$ is $\begin{bmatrix} -2 & 2 & -1 \\ -6 & -5 & -3 \end{bmatrix}$.

Supplementary Example 3

For another pattern, the artist enters the coordinates for figure $ABCD$ with vertices $A(-7, 4)$, $B(-4, 3)$, $C(-1, 4)$ and $D(-4, 1)$. She wants to draw the figure reflected over the y-axis. Find the coordinates of the reflection image using matrices.

$$\begin{bmatrix} 7 & 4 & 1 & 4 \\ 4 & 3 & 4 & 1 \end{bmatrix}$$

Reteaching Worksheet 8-7

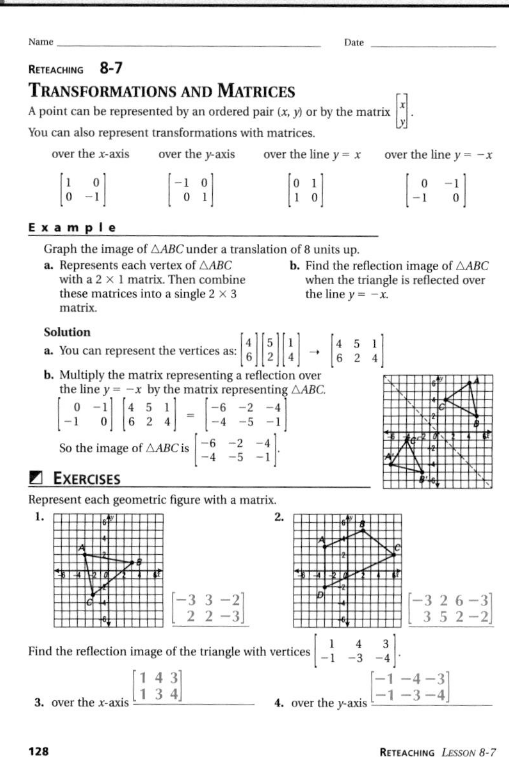

Name ______________________ Date ______________

RETEACHING 8-7

TRANSFORMATIONS AND MATRICES

A point can be represented by an ordered pair (x, y) or by the matrix $\begin{bmatrix} x \\ y \end{bmatrix}$.

You can also represent transformations with matrices.

over the x-axis	over the y-axis	over the line $y = x$	over the line $y = -x$
$\begin{bmatrix} 1 & 0 \\ 0 & -1 \end{bmatrix}$	$\begin{bmatrix} -1 & 0 \\ 0 & 1 \end{bmatrix}$	$\begin{bmatrix} 0 & 1 \\ 1 & 0 \end{bmatrix}$	$\begin{bmatrix} 0 & -1 \\ -1 & 0 \end{bmatrix}$

Example

Graph the image of $\triangle ABC$ under a translation of 8 units up.

a. Represents each vertex of $\triangle ABC$ with a 2×1 matrix. Then combine these matrices into a single 2×3 matrix.

b. Find the reflection image of $\triangle ABC$ when the triangle is reflected over the line $y = -x$.

Solution

a. You can represent the vertices as: $\begin{bmatrix} 4 \\ 6 \end{bmatrix}\begin{bmatrix} 5 \\ 2 \end{bmatrix}\begin{bmatrix} 1 \\ 4 \end{bmatrix} \rightarrow \begin{bmatrix} 4 & 5 & 1 \\ 6 & 2 & 4 \end{bmatrix}$

b. Multiply the matrix representing a reflection over the line $y = -x$ by the matrix representing $\triangle ABC$.

$$\begin{bmatrix} 0 & -1 \\ -1 & 0 \end{bmatrix}\begin{bmatrix} 4 & 5 & 1 \\ 6 & 2 & 4 \end{bmatrix} = \begin{bmatrix} -6 & -2 & -4 \\ -4 & -5 & -1 \end{bmatrix}$$

So the image of $\triangle ABC$ is $\begin{bmatrix} -6 & -2 & -4 \\ -4 & -5 & -1 \end{bmatrix}$.

EXERCISES

Represent each geometric figure with a matrix.

1. $\begin{bmatrix} -3 & 3 & -2 \\ 2 & 2 & -3 \end{bmatrix}$

2. $\begin{bmatrix} -3 & 2 & 6 & -3 \\ 3 & 5 & 2 & -2 \end{bmatrix}$

Find the reflection image of the triangle with vertices $\begin{bmatrix} 1 & 4 & 3 \\ -1 & -3 & -4 \end{bmatrix}$.

3. over the x-axis $\begin{bmatrix} 1 & 4 & 3 \\ 1 & 3 & 4 \end{bmatrix}$

4. over the y-axis $\begin{bmatrix} -1 & -4 & -3 \\ -1 & -3 & -4 \end{bmatrix}$

128 RETEACHING *LESSON 8-7*

Lesson Wrap-up

Quick Assessment

Ask the following questions to determine if students understand the content presented in this lesson.

1. Graph the transformation image that is obtained when you multiply A by M.

$$M = \begin{bmatrix} \frac{1}{2} & 0 \\ 0 & \frac{1}{2} \end{bmatrix} \quad A = \begin{bmatrix} -6 & -4 & -2 \\ 2 & 6 & 4 \end{bmatrix}$$

$$\begin{bmatrix} -3 & -2 & -1 \\ 1 & 3 & 2 \end{bmatrix}$$

2. What is the transformation? a dilation with center at (0, 0) and a scale factor of $\frac{1}{2}$

Assignment Guide

Basic: 1–24, 28–34
Advanced: 1–34
Additional Practice: See Extra Practice Index on page 674.

Additional Answers

24.

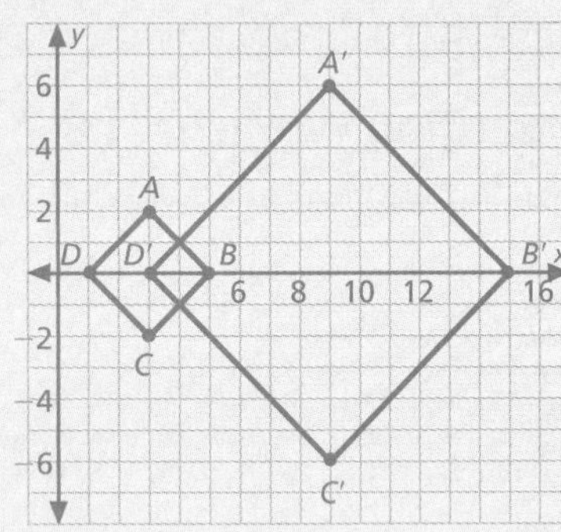

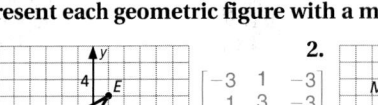

Try These Exercises

Represent each geometric figure with a matrix.

1. $\begin{bmatrix} -3 & 1 & -3 \\ 1 & 3 & -3 \end{bmatrix}$

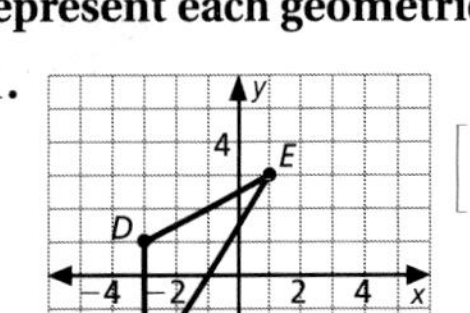

2. $\begin{bmatrix} -3 & 5 & 5 & -3 \\ 3 & 3 & -2 & -2 \end{bmatrix}$

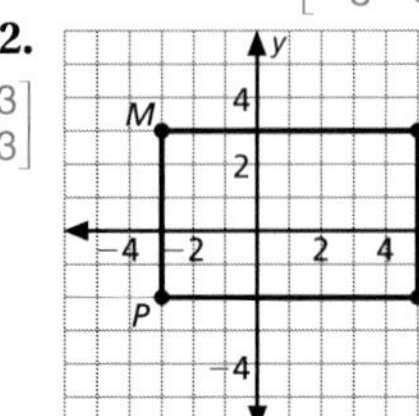

3. $\begin{bmatrix} 1 & 5 & 7 & -4 \\ 3 & 3 & -3 & -3 \end{bmatrix}$

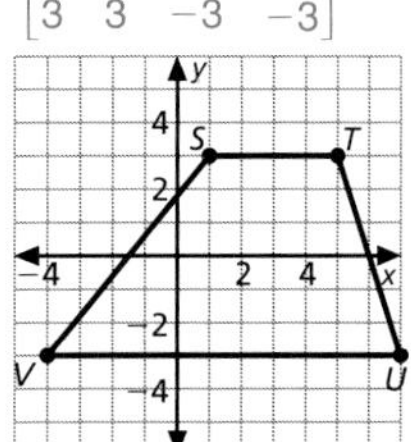

RIDE DESIGN Amusement park rides are tested using computer simulations. A triangle with vertices $\begin{bmatrix} -1 \\ 2 \end{bmatrix}$, $\begin{bmatrix} 3 \\ 7 \end{bmatrix}$, and $\begin{bmatrix} 7 \\ 3 \end{bmatrix}$ is used to represent a moving platform to which the ride's cars are attached. Find the reflection images of the triangle.

4. over the x-axis. $\begin{bmatrix} -1 & 3 & 7 \\ -2 & -7 & -3 \end{bmatrix}$
5. over the line $y = x$. $\begin{bmatrix} 2 & 7 & 3 \\ -1 & 3 & 7 \end{bmatrix}$
6. over the y-axis. $\begin{bmatrix} 1 & -3 & -7 \\ 2 & 7 & 3 \end{bmatrix}$
7. over the line $y = -x$. $\begin{bmatrix} -2 & -7 & -3 \\ 1 & -3 & -7 \end{bmatrix}$

Practice Exercises

Represent each geometric figure with a matrix.

8. $\begin{bmatrix} -2 & 4 & -4 \\ 4 & 2 & -3 \end{bmatrix}$

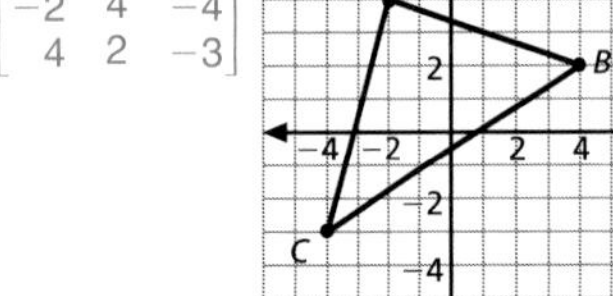

9. $\begin{bmatrix} -3 & 3 & 4 & -2 \\ 2 & 4 & 1 & -1 \end{bmatrix}$

10. $\begin{bmatrix} -2 & 2 & 2 & -2 \\ 3 & 3 & -4 & -4 \end{bmatrix}$

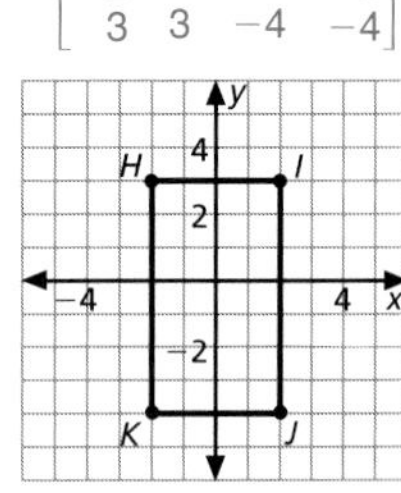

CALCULATOR Multiply matrices using a graphing calculator to find the reflection images of the quadrilateral represented by

$$\begin{bmatrix} -2 & 1 & 7 & 4 \\ 7 & 3 & 4 & 7 \end{bmatrix}$$

11. over the line $y = x$. $\begin{bmatrix} 7 & 3 & 4 & 7 \\ -2 & 1 & 7 & 4 \end{bmatrix}$
12. over the y-axis. $\begin{bmatrix} 2 & -1 & -7 & -4 \\ 7 & 3 & 4 & 7 \end{bmatrix}$
13. over the line $y = -x$. $\begin{bmatrix} -7 & -3 & -4 & -7 \\ 2 & -1 & -7 & -4 \end{bmatrix}$
14. over the x-axis. $\begin{bmatrix} -2 & 1 & 7 & 4 \\ -7 & -3 & -4 & -7 \end{bmatrix}$

Interpret each equation as indicating: ***The reflection image of point*** __?__ ***over*** __?__ ***is the point*** __?__.

15. $\begin{bmatrix} 0 & -1 \\ -1 & 0 \end{bmatrix}\begin{bmatrix} 4 \\ 4 \end{bmatrix} = \begin{bmatrix} -4 \\ -4 \end{bmatrix}$ (4, 4), line $y = x$, (−4, −4)

16. $\begin{bmatrix} -1 & 0 \\ 0 & 1 \end{bmatrix}\begin{bmatrix} 3 \\ 2 \end{bmatrix} = \begin{bmatrix} -3 \\ 2 \end{bmatrix}$ (3, 2), y-axis, (−3, 2)

17. $\begin{bmatrix} 0 & 1 \\ 1 & 0 \end{bmatrix}\begin{bmatrix} -2 \\ 4 \end{bmatrix} = \begin{bmatrix} 4 \\ -2 \end{bmatrix}$ (−2, 4), line $y = x$, (4, −2)

18. $\begin{bmatrix} 1 & 0 \\ 0 & -1 \end{bmatrix}\begin{bmatrix} 5 \\ -3 \end{bmatrix} = \begin{bmatrix} 5 \\ 3 \end{bmatrix}$ (5, −3), x-axis, (5, 3)

19. **TALK ABOUT IT** Toni says that you can produce any dilation with center at the origin and a scale factor of k using the matrix $\begin{bmatrix} k & 0 \\ 0 & k \end{bmatrix}$. Does Toni's thinking make sense? yes

370 Chapter 8 Transformations

GAME DEVELOPMENT In a hand-held game, the player must click on falling stars within a time limit. To develop the game, the programmer specifies the coordinates for a star to appear. Two seconds later, the image of the star appears under a reflection. Exercises 19–24 specify image and preimage points used in the game. For each, name the reflecting line and verify your answer by matrix multiplication.

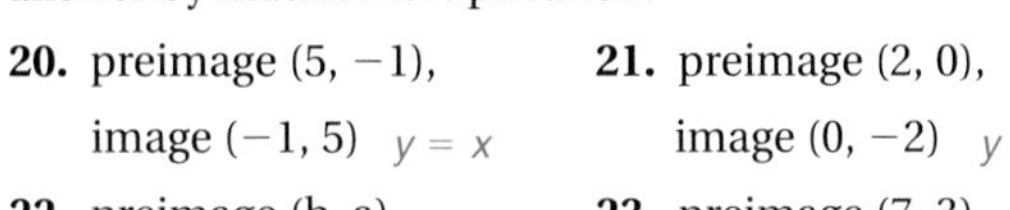

20. preimage $(5, -1)$, image $(-1, 5)$ $y = x$

21. preimage $(2, 0)$, image $(0, -2)$ $y = -x$

22. preimage (b, a), image $(b, -a)$ x-axis

23. preimage $(7, 3)$, image $(-7, 3)$ y-axis

24. **GEOMETRY SOFTWARE** Find the image of rhombus $ABCD$ under the transformation associated with matrix M. Graph both the preimage and its image using geometric-drawing software. See additional answers.

$$M = \begin{bmatrix} 3 & 0 \\ 0 & 3 \end{bmatrix} \qquad ABCD = \begin{bmatrix} 3 & 5 & 3 & 1 \\ 2 & 0 & -2 & 0 \end{bmatrix}$$

Extended Practice Exercises

25. **WRITING MATH** What type of transformation is represented by the matrix $\begin{bmatrix} -2 & 0 \\ 0 & -2 \end{bmatrix}$? Dilation with center at origin and a scale factor of 2

26. Draw any triangle in the coordinate plane. Represent it with a 2×3 matrix. Then apply each transformation below. Draw each preimage and image on a coordinate grid. Check students' drawings.

a. $\begin{bmatrix} 0 & -1 \\ 1 & 0 \end{bmatrix}$ **b.** $\begin{bmatrix} -1 & 0 \\ 0 & -1 \end{bmatrix}$ **c.** $\begin{bmatrix} 0 & 1 \\ -1 & 0 \end{bmatrix}$

27. Refer to the graphs you drew in Exercise 28.

a. Which shows a clockwise rotation of 90°? c

b. Which shows a clockwise rotation of 180°? b

c. Which shows a counterclockwise rotation of 90°? a

28. **CHAPTER INVESTIGATION** Measure the angle of descent for each hill in your roller coaster design. Estimate your coaster's top speed by comparing its features to the coasters shown in the table on page 337. Write a paragraph to justify your estimate.

Mixed Review Exercises

Find the slope and y-intercept for each line. (Lesson 6-1)

29. $y = \frac{1}{2}x - 3$ $\frac{1}{2}, -3$

30. $y = -3x + 4$ $-3, 4$

31. $2y - x = 6$ $\frac{1}{2}, 3$

32. $3(x - 4) = 5y$ $\frac{3}{5}, \frac{-12}{5}$

33. $3y - 4x - 7 = 0$ $\frac{4}{3}, \frac{7}{3}$

34. $2x = 4y + 2$ $\frac{1}{2}, -\frac{1}{2}$

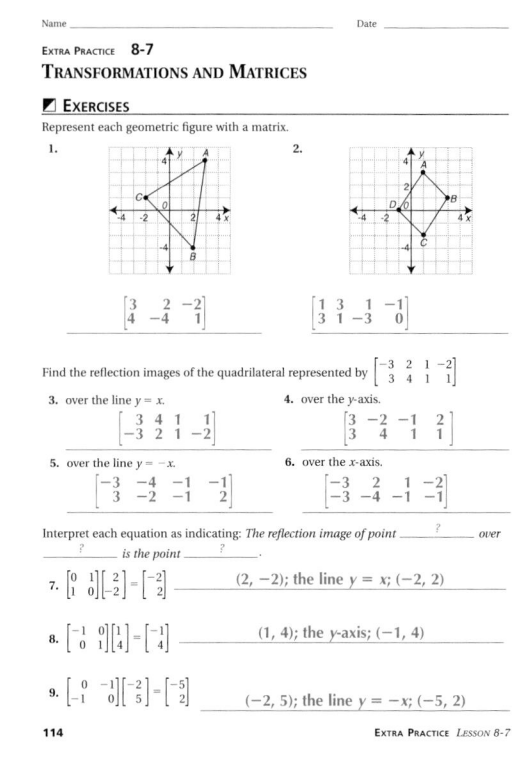

Teaching Strategies

COOPERATIVE LEARNING Have students work in pairs to find a matrix M that they can use to represent a composite of transformations. Have them use M to find transformations of a figure on the coordinate plane. possible answer: reflection over the x-axis and dilation with center at the origin and a scale factor of 4:

$$\begin{bmatrix} 4 & 0 \\ 0 & -4 \end{bmatrix} \begin{bmatrix} 1 & 3 & 3 \\ 2 & 4 & 2 \end{bmatrix} = \begin{bmatrix} 4 & 12 & 12 \\ -8 & -16 & -8 \end{bmatrix}$$

Extra Practice Worksheet 8-7

Name ______________ Date ______________

EXTRA PRACTICE 8-7

TRANSFORMATIONS AND MATRICES

EXERCISES

Represent each geometric figure with a matrix.

1. $\begin{bmatrix} 3 & 2 & -2 \\ 4 & -4 & 1 \end{bmatrix}$

2. $\begin{bmatrix} 1 & 3 & 1 & -1 \\ 3 & 1 & -3 & 0 \end{bmatrix}$

Find the reflection images of the quadrilateral represented by $\begin{bmatrix} -3 & 2 & 1 & -2 \\ 3 & 4 & 1 & 1 \end{bmatrix}$

3. over the line $y = x$. $\begin{bmatrix} 3 & 4 & 1 & 1 \\ -3 & 2 & 1 & -2 \end{bmatrix}$

4. over the y-axis. $\begin{bmatrix} 3 & -2 & -1 & 2 \\ 3 & 4 & 1 & 1 \end{bmatrix}$

5. over the line $y = -x$. $\begin{bmatrix} -3 & -4 & -1 & -1 \\ 3 & -2 & -1 & 2 \end{bmatrix}$

6. over the x-axis. $\begin{bmatrix} -3 & 2 & 1 & -2 \\ -3 & -4 & -1 & -1 \end{bmatrix}$

Interpret each equation as indicating: *The reflection image of point* ___?___ *over* ___?___ *is the point* ___?___.

7. $\begin{bmatrix} 0 & 1 \\ 1 & 0 \end{bmatrix}\begin{bmatrix} 2 \\ -2 \end{bmatrix} = \begin{bmatrix} -2 \\ 2 \end{bmatrix}$ $(2, -2)$; the line $y = x$; $(-2, 2)$

8. $\begin{bmatrix} -1 & 0 \\ 0 & 1 \end{bmatrix}\begin{bmatrix} 1 \\ 4 \end{bmatrix} = \begin{bmatrix} -1 \\ 4 \end{bmatrix}$ $(1, 4)$; the y-axis; $(-1, 4)$

9. $\begin{bmatrix} 0 & -1 \\ -1 & 0 \end{bmatrix}\begin{bmatrix} -2 \\ 5 \end{bmatrix} = \begin{bmatrix} -5 \\ 2 \end{bmatrix}$ $(-2, 5)$; the line $y = -x$; $(-5, 2)$

114 EXTRA PRACTICE *LESSON 8-7*

Enrichment Worksheet 8-7

Name ______________ Date ______________

ENRICHMENT 8-7

LINEAR SYSTEMS IN THREE VARIABLES

One application of order 3 determinants is in the solution of systems of three linear equations in three variables. Here is such a system.

$$\begin{cases} ax + by + cz = p \\ dx + ey + fz = q \\ gx + hy + iz = r \end{cases}$$

The values for x, y, and z can be found using order 3 determinants as shown below. This method is called *Cramer's Rule.*

$$x = \frac{\begin{vmatrix} p & b & c \\ q & e & f \\ r & h & i \end{vmatrix}}{\begin{vmatrix} a & b & c \\ d & e & f \\ g & h & i \end{vmatrix}} \qquad y = \frac{\begin{vmatrix} a & p & c \\ d & q & f \\ g & r & i \end{vmatrix}}{\begin{vmatrix} a & b & c \\ d & e & f \\ g & h & i \end{vmatrix}} \qquad z = \frac{\begin{vmatrix} a & b & p \\ d & e & q \\ g & h & r \end{vmatrix}}{\begin{vmatrix} a & b & c \\ d & e & f \\ g & h & i \end{vmatrix}}$$

EXERCISES

Find the solution for each system. Use a graphing calculator or a computer graphing program.

1. $\begin{cases} 3x + y + z = 14 \\ x - 2y + z = 1 \\ x + y + 2z = 15 \end{cases}$ $x = 2, y = 3, z = 5$

2. $\begin{cases} x + y - 2z = 13 \\ x - 3y = -7 \\ x + 4z = -14 \end{cases}$ $x = 2, y = 3, z = -4$

3. $\begin{cases} x + y = 1 \\ x + z = 2 \\ y + z = 3 \end{cases}$ $x = 0, y = 1, z = 2$

4. $\begin{cases} x + y + z = 1 \\ x + y - z = 2 \\ x - y + z = 3 \end{cases}$ $x = 2.5, y = -1, z = -0.5$

5. When would Cramer's Rule not be useful insolving these types of systems?
Cramer's Rule cannot be used when the determinant of the coefficients equals 0.

146 ENRICHMENT *LESSON 8-7*

Lesson Planning

NCTM Standards/Strands

- Number & Operation
- Patterns, Functions, & Algebra
- Problem Solving
- Reasoning and Proof
- Connections
- Representation

Vocabulary

profit

Materials Needed

graphing calculators (optional)

Lesson Resources

Warm-Up Transparency 24
Reteaching 8-8
Enrichment 8-8
Transparency TK-7–10

ASSIGNMENT GUIDE

Basic: 1–20
Enriched: 1–20

Getting Started

5-MINUTE WARM-UP

Find the product.

$$[0.25\ \ 0.5]\begin{bmatrix}254 & 317 & 123\\125 & 471 & 212\end{bmatrix}$$

[126 314.75 136.75]

Use the Five-step Plan and the Strategy

THE FIVE-STEP PLAN **Read**—ask questions of the students to help them understand the problem. **Plan**—guide students to related problems and previously mastered skills and strategies. **Solve**—students solve problem on their own. **Answer**—write solution in format that answers the question. **Check**—review work, check for reasonableness and review strategy used. Students will benefit from the experience of verbalizing their methods.
THE STRATEGY Use a table or list—By representing data in matrices, students can perform a series of operations in a single step.

8-8 Problem Solving Skills: Use a Matrix

Some problems can be solved by translating directly to a matrix and then performing matrix operations. Consider using matrix operations to solve problems whenever information can be easily organized into tables with corresponding elements.

Problem Solving Strategies

- Guess and check
- Look for a pattern
- Solve a simpler problem
- ✔ Make a table, chart or list
- Use a picture, diagram or model
- Act it out
- Work backwards
- Eliminate possibilities
- Use an equation or formula

Problem

BUSINESS An orchard grows Delicious, Jonathan, and Granny Smith apples. The apples are sold in boxes to two different markets. The profit is $5.85 on a box of Delicious apples, $4.25 on Jonathans, and $3.75 on Granny Smiths. The table shows the number of boxes sold.

Find the amount of profit generated by sales to each market.

Apples	Markets	
	Bill's	Jan's
Delicious	250	225
Jonathan	320	295
Granny Smith	175	190

Solve the Problem

a. Represent the market data in a 3×2 matrix as shown.

$$A = \begin{bmatrix}250 & 225\\320 & 295\\175 & 190\end{bmatrix}$$

b. Write a matrix to represent the respective profits. Think about the dimensions necessary for matrix multiplication. Since matrix A has three rows, matrix B must have three columns. The dimensions of B must be 1×3.

$$B = [5.85\ \ 4.25\ \ 3.75]$$

c. The product BA will be a 1×2 matrix that determines the profit from each market.

$$BA = [5.85\ \ 4.25\ \ 3.75]\begin{bmatrix}250 & 225\\320 & 295\\175 & 190\end{bmatrix} = [3{,}478.75\ \ 3{,}282.50]$$

The profit from Bill's market is $3,478.75. The profit from Jan's market is $3,282.50.

Extend the Lesson

REAL WORLD CONNECTION Many accountants create spreadsheets using matrices and computers. Have students do some research to discover how matrices are used on computers. If they have access to a computer, have them use a spreadsheet program and explore firsthand how useful matrices can be.

Try These Exercises

Use matrices to solve each problem.

1. **FOOD DISTRIBUTION** A farm raises two crops, which are shipped to three distributors. The table shows the number of crates shipped to distributors. A: $1,602.50; B: $1,748.25; C:$2,344.50

	Distributor		
	A	B	C
Crop 1	350	275	550
Crop 2	200	310	260

The profit on crop 1 is $2.75 per crate. The profit on crop 2 is $3.20 per crate.

Find the amount of profit from each distributor.

2. **FOOD CONCESSIONS** A large amusement park owns four bakeries, each of which produces three types of bread: white, rye and whole wheat. The bread is used to supply food vendors throughout the park. The number of loaves produced daily at each bakery is shown in the table at the right.

	Bakery			
	A	B	C	D
White	190	215	240	112
Rye	65	80	110	60
Whole wheat	205	265	290	170

By baking its own bread, the park can reduce the amount of overhead and increase its profits. The profit on each loaf of bread is 75 cents for white, 50 cents for rye, and 60 cents for whole wheat. Find the amount of profit from each bakery. A:$298; B: $360.25; C: $409; D:$216

Practice Exercises

3. **SALES** A sneaker manufacturer makes five kinds of sneakers: basketball, running, walking, cross-trainer, and tennis. The sneakers are shipped to three retail outlets. The number of pairs of sneakers shipped to each outlet is shown. A: $465; B: $571.25; C: $586.25

Profit on each pair of sneakers is as follows:

basketball	$4.50	cross-trainer	$5.25
running	$3.50	tennis	$5.00
walking	$3.75		

Find the amount of profit for each outlet.

	Outlets		
	A	B	C
Basketball	30	40	35
Running	20	25	30
Walking	15	20	25
Cross-trainer	15	15	20
Tennis	25	30	25

4. **SOUVENIRS** An amusement park sells hats, T-shirts and stuffed toys. The table on the left gives the number of each type of souvenir sold during a two-week period. The table on the right gives the price of each souvenir. Find the total amount spent on souvenirs each week. Week 1: $7504, Week 2: $7826

	Hats	T-shirts	Toys
Week 1	128	240	58
Week 2	130	215	89

Item	Price
Hats	$14
T-shirts	$18
Toys	$24

Mixed Review Exercises

Find the sum of the measures of the angles of a convex polygon with the given number of sides. (Lesson 4-7)

5. 37 6300° **6.** 52 9000° **7.** 29 4860° **8.** 45 7740°

9. 62 10,800° **10.** 40 6840° **11.** 58 10,080° **12.** 19 3060°

Chalkboard Examples

Supplementary Problem

The Art Store sells three types of art pencil sets. The profit is $8.45 on watercolor pencil sets, $6.19 on pastel pencil sets and $4.62 on charcoal pencil sets. There are two branches of the Art Store. The numbers of sales of pencil sets for each store are shown in the table below.

Set Type	Sales for Westside Branch	Sales for Downtown Branch
watercolor	132	74
pastel	56	48
charcoal	119	126

Find the amount of profit generated by each branch. **Westside, $2,011.82; Downtown, $1,504.54**

Lesson Wrap-up

Quick Assessment

When can you use matrices to solve problems. How might a computer be helpful in working with matrices? **Answers will vary.**

Reteaching Worksheet 8-8

Name ______________________ Date ______________

RETEACHING 8-8

PROBLEM SOLVING SKILLS: USE A MATRIX

You can use a matrix as a mathematical model to help you solve many different kinds of problems.

Example

A pharmacy chain sells each toothbrush for $2.50, each tube of toothpaste for $2.00 and each box of dental floss for $4.00. During the month of May, Store A sold 200 toothbrushes, 600 tubes of toothpaste and 50 boxes of dental floss; Store B sold 100 toothbrushes, 400 tubes of toothpaste and 30 boxes of dental floss; and Store C sold 50 toothbrushes, 300 tubes of toothpaste and 20 boxes of dental floss. Find the gross revenue generated by these items at each of the three stores.

Solution

Write the prices in a 1×3 matrix. Write the number of items sold in a 3×3 matrix. Then multiply. The product is a 1×3 matrix.

Brushes	Paste	Floss
[2.50	2.00	4.00]

	Store A	Store B	Store C
Brushes	200	100	50
Paste	600	400	300
Floss	50	30	20

Store A
[1900

The gross revenue generated by Store A was $1900, by Store B was $1170, and by Store C was $805.

EXERCISES

A grocery store chain shows prices of its greeting cards in a 1×3 matrix and number of cards sold daily at each of its three stores in a 3×3 matrix.

Birthday	Sympathy	Get Well
[2.00	1.50	1.00]

	Store A	Store B	Store C
Birthday	10	15	12
Sympathy	4	6	4
Get Well	13	20	16

1. Store A. $39
2. Store B. $59
3. Store C. $46
4. Stores A and B. $98
5. Stores B and C. $105
6. Stores A, B and C. $144

Find the gross revenue from the cards sold at

7. Birthday $74
8. Sympathy $21
9. Get Well $49

Find the gross revenue from each type of card sold at all three stores.

130 RETEACHING *LESSON 8-8*

Teaching Strategies

Some students may have difficulty creating matrices to solve a problem. Suggest they think about how they would solve the problem without matrices before they try to set up matrices. Also, remind them to consider the dimensions for possible matrix multiplication.

Enrichment Worksheet 8-8 not shown.

Assessment Planning

Vocabulary Review

Lesson 8-1
translations, preimage, image, transformation, reflection, line of reflection

Lesson 8-2
clockwise, counterclockwise, rotation, center of rotation, angle of rotation, origin

Lesson 8-3
dilation, enlargement, reduction, scale factor, center of dilation

Lesson 8-4
composite of transformations, glide reflection

Lesson 8-5
matrix, dimensions, elements, equal matrices, matrix addition, scalar multiplication, scalar

Lesson 8-6
row-by-column multiplication

Lesson 8-7
matrices for reflections

Lesson 8-8
profit

Assessment Options

Chapter Assessment
Alternative Assessment
Chapter Test Forms A and B
Cumulative Assessment
Achievement Test
Writing Prompts

Assignment Guide

All students: 1–18
Additional Practice: Refer to the Extra Practice Index on page 674 of this text.

Chapter 8 Review

Vocabulary ■ LESSON 8-1–LESSON 8-8

Choose the word from the list that best completes each statement.

a. matrix
b. dimensions
c. scale factor
d. preimage
e. transformation

1. A translation, reflection, rotation, or dilation is each known as a __?__ of a figure. e
2. A dilation image is obtained by multiplying the length of each side of a figure by a number called the __?__. c
3. Under a transformation, the new figure is called the image and the original figure is called the __?__. d
4. The number of rows and columns are the __?__ of the matrix. b
5. A __?__ is a rectangular array of numbers arranged into rows and columns. a

LESSON 8-1 ■ Translations and Reflections, p. 338

▶ Under a **translation**, an image is produced by sliding every point of the original figure the same distance in the same direction. Under a **reflection**, a figure is flipped over a line of reflection.

6. Graph the image of $\triangle LMN$ under a translation 3 units left. Then graph the original triangle LMN under a reflection over the x-axis. See additional answers.

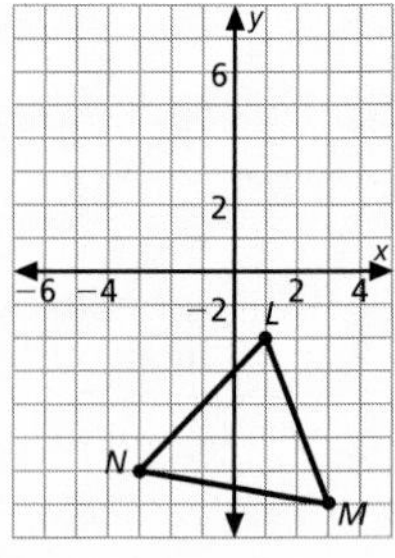

LESSON 8-2 ■ Rotations in the Coordinate Plane, p. 342

▶ Under a **rotation**, a figure is turned about a point.

7. Graph the rotation image of $\triangle QRS$ after a turn of 90° counterclockwise about the origin. See additional answers.

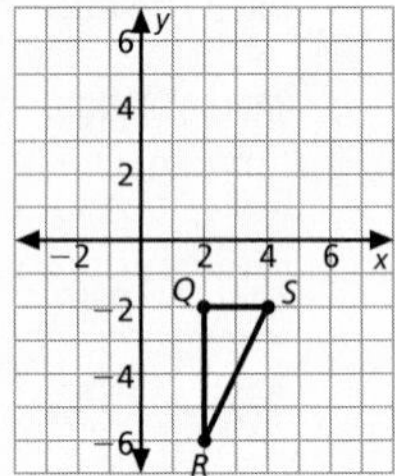

LESSON 8-3 ■ Dilations in the Coordinate Plane, p. 348

▶ A **dilation** is a transformation that produces an image of the same shape, but a different size.

8. Draw the dilation image of rectangle $ABCD$ with the center of dilation at the origin and a scale factor of $\frac{1}{3}$. See additional answers.

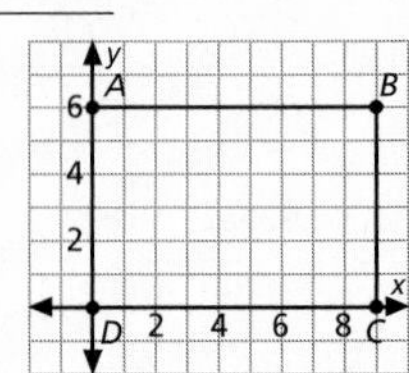

Additional Answers

6.
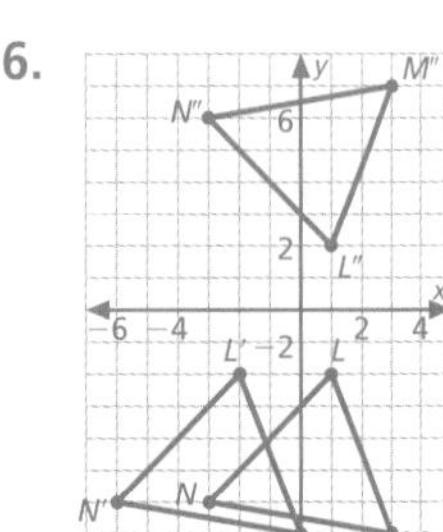

7.
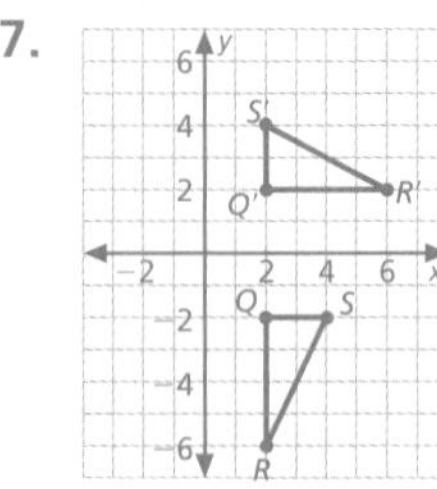

8.
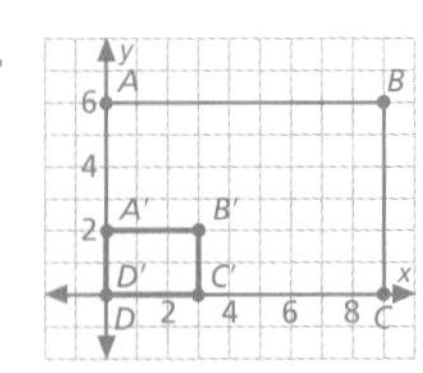

LESSON 8-4 ■ Multiple Transformations, p. 352

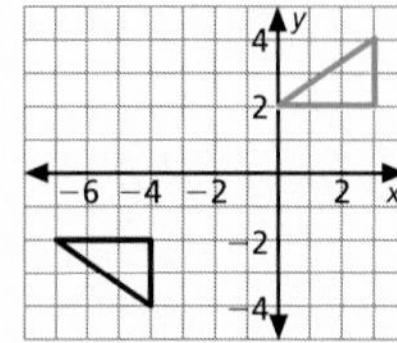

- Two or more successive transformations can be applied to a given figure. This is called a **composite of reflections**.

9. Describe the two transformations used to create the image shown in blue.
Possible answer: reflection over x-axis and translation 8 units to right.

LESSON 8-5 ■ Addition and Multiplication with Matrices, p. 358

- A **matrix** is a rectangular array of numbers arranged into rows and columns. The number of rows and columns are the **dimensions** of the matrix. The numbers that make up the matrix are the **elements** of the matrix.

10. Find the dimensions of D. $D = \begin{bmatrix} 1 & -2 & 3 & 2 \\ -4 & 5 & -6 & 4 \\ 7 & -8 & 9 & 6 \end{bmatrix}$ 3×4

11. Identify the elements D_{32}, D_{21}, and D_{13}. $-8, -4.3$

12. Find kD for $k = -2$.

12. $\begin{bmatrix} -2 & 4 & -6 & -4 \\ 8 & -10 & 12 & -8 \\ -14 & 16 & -18 & -12 \end{bmatrix}$

$A = \begin{bmatrix} -3 & 5 & 0 & 2 \\ 1 & -7 & 6 & -1 \end{bmatrix}$ $B = \begin{bmatrix} 0 & 1 & -3 & 11 \\ 13 & 2 & -4 & 1 \end{bmatrix}$ $C = \begin{bmatrix} 6 & 0 & 8 & -2 \\ 0 & 15 & -1 & 5 \end{bmatrix}$

13. $A + B - C$ $\begin{bmatrix} -9 & 6 & -11 & 15 \\ 14 & -20 & 3 & -5 \end{bmatrix}$ **14.** $C - B - A$ $\begin{bmatrix} 9 & -6 & 11 & -15 \\ -14 & 20 & -3 & 5 \end{bmatrix}$ **15.** $3(A + C)$ $\begin{bmatrix} 9 & 15 & 24 & 0 \\ 3 & 24 & 15 & 12 \end{bmatrix}$

LESSON 8-6 ■ More Operations on Matrices, p. 362

- The product of an $m \times n$ matrix and an $n \times p$ matrix is an $m \times p$ matrix.

16. Find the product AB. $A = \begin{bmatrix} 3 & -2 & -4 \\ 2 & 1 & 3 \end{bmatrix}$ $B = \begin{bmatrix} 0 & 1 \\ 2 & -2 \\ 6 & 5 \end{bmatrix}$ $\begin{bmatrix} -28 & -13 \\ 20 & 15 \end{bmatrix}$

LESSON 8-7 ■ Transformations and Matrices, p. 368

- The point represented by (x, y) can also be represented by the matrix $\begin{bmatrix} x \\ y \end{bmatrix}$.
- Polygons and transformations can also be represented by matrices.

17. Triangle DEF is represented by $\begin{bmatrix} 4 & 6 & 5 \\ 2 & 1 & 4 \end{bmatrix}$. Use the matrix $\begin{bmatrix} -1 & 0 \\ 0 & 1 \end{bmatrix}$ to find the image when $\triangle DEF$ is reflected over the y-axis. $\begin{bmatrix} -4 & -6 & -5 \\ 2 & 1 & 4 \end{bmatrix}$

LESSON 8-8 ■ Problem Solving: Use a Matrix, p. 372

- Some problems can be solved by translating to a matrix and using matrix operations.

18. The table at the right shows the number of students in a school's beginning and advanced orchestra classes. The students pay a fee for instruction books: \$5 for brass and woodwinds and \$8 for strings. Find the amount each class will spend on books.
Advanced: \$326, Beginning: \$490

	Advanced	Beginning
Brass	12	28
Strings	22	25
Woodwinds	18	30

Teaching Strategies

Have students discuss and compare the methods they have learned for finding transformations of figures on the coordinate plane. Encourage them to express which method they prefer and to explain why. You may wish to point out that knowing more than one method can be a valuable tool for checking.

Chapter 8 Test Form A, page 1

CHAPTER 8
TRANSFORMATIONS IN THE COORDINATE PLANE
ASSESSMENT FORM A, PAGE 1

Name ______________
Date ______________

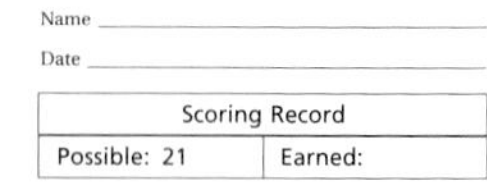

Scoring Record	
Possible: 21	Earned:

Graph each image. Then label the image points A', B' and C'.

1. Graph the image of $\triangle ABC$ under a translation 3 units up.

2. Graph the image of $\triangle ABC$ under a reflection across the y-axis.

3. Graph the image of $\triangle ABC$ under a rotation of 180° counterclockwise about the origin.

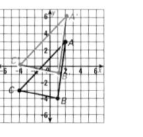

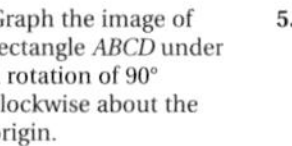

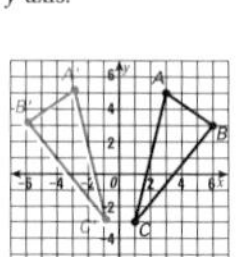

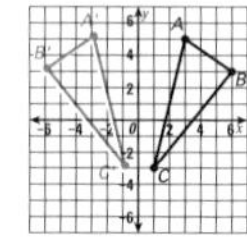

4. Graph the image of rectangle $ABCD$ under a rotation of 90° clockwise about the origin.

5. Graph the dilation image of $\triangle ABC$, using a scale factor of $\frac{1}{2}$ and the center of dilation at the origin.

6. Perform the following two transformations on $\triangle ABC$: a translation 7 units left followed by a reflection over the y-axis.

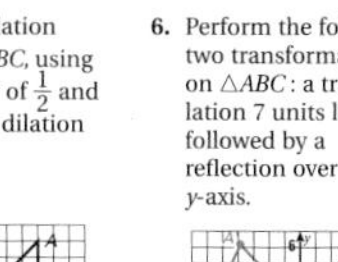

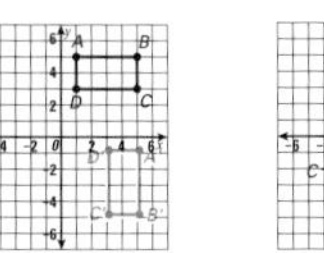

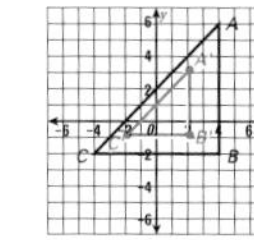

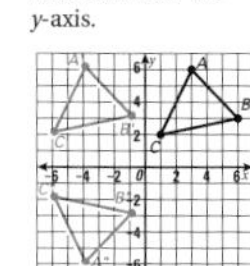

Use $x_0 = x_1 + k(x_2 - x_1)$ and $y_0 = y_1 + k(y_2 - y_1)$ to find a point that is a fractional part aof the way from $A(x_1, y_1)$ to $B(x_2, y_2)$ if k represents the fractional part. Give the coordinates.

7. $A(5, 4), B(3, 2), k = \frac{1}{2}$ (4, 3)   **8.** $A(4, 3), B(-6, -2), k = \frac{2}{5}$ (0, 1)

100   ASSESSMENT *CHAPTER 8 FORM A*

Chapter 8 Test Form A, page 2

Name ______________ Date ______________

9. Describe two transformations that could be used to create the image at the right.
Rotation of 180° followed by translation 6 units up

Use the matrices below for Exercises 10–17.

$A = \begin{bmatrix} 5 & 8 & 9 \\ 6 & 3 & 2 \end{bmatrix}$ $B = \begin{bmatrix} 9 & 3 & 6 \\ 4 & 7 & 5 \end{bmatrix}$ $C = \begin{bmatrix} 9 & 1 & 5 \\ 4 & 3 & 6 \\ 4 & 1 & 2 \end{bmatrix}$ $D = \begin{bmatrix} 9 & 2 \\ 1 & 8 \\ 4 & 5 \end{bmatrix}$

10. Find the dimension of Matrix A. 2×3 **11.** Identify the element C_{31}. 8

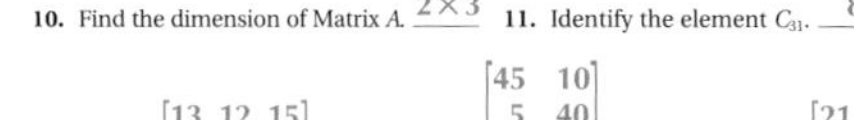

12. $A + B$ $\begin{bmatrix} 13 & 12 & 15 \\ 7 & 8 & 9 \end{bmatrix}$ **13.** $5D$ $\begin{bmatrix} 45 & 10 \\ 5 & 40 \\ 20 & 25 \end{bmatrix}$ **14.** $2B + A$ $\begin{bmatrix} 21 & 16 & 21 \\ 8 & 13 & 16 \end{bmatrix}$

15. DB $\begin{bmatrix} 74 & 46 & 68 \\ 16 & 44 & 62 \\ 37 & 41 & 59 \end{bmatrix}$ **16.** AC $\begin{bmatrix} 149 & 38 & 91 \\ 82 & 17 & 52 \end{bmatrix}$ **17.** BD $\begin{bmatrix} 100 & 78 \\ 42 & 77 \end{bmatrix}$

Use matrices to help you find the reflective image of $\triangle ABC$ represented by $\begin{bmatrix} 1 & 3 & 5 \\ 3 & 2 & 4 \end{bmatrix}$ when the triangle is reflected over the:

18. y-axis. $\begin{bmatrix} -1 & -3 & -5 \\ 3 & 2 & 4 \end{bmatrix}$ **19.** x-axis. $\begin{bmatrix} 1 & 3 & 5 \\ -3 & -2 & -4 \end{bmatrix}$

A restaurant chain shows prices of its sandwiches in a 1×3 matrix and the number sold daily at each of its three stores in a 3×3 matrix.

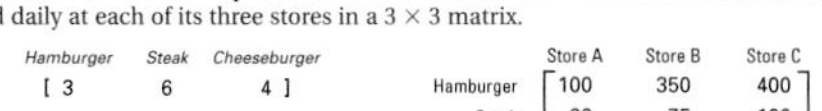

Hamburger	Steak	Cheeseburger
3	6	4

	Store A	Store B	Store C
Hamburger	100	350	400
Steak	20	75	100
Cheeseburger	50	100	150

20. Find the total gross revenue from the sales of sandwiches at all three stores. \$4920

21. Find the gross revenue from the sales of sandwiches at Store A. \$620

102   ASSESSMENT *CHAPTER 8 FORM A*

Chapter 8 Test Form B, page 1

CHAPTER 8

TRANSFORMATIONS IN THE COORDINATE PLANE

ASSESSMENT FORM B, PAGE 1

Name ____________

Date ____________

Scoring Record	
Possible: 21	Earned:

Graph each image. Then label the image points A', B' and C'.

1. Graph the image of $\triangle ABC$ under a translation 5 units left.
2. Graph the image of $\triangle ABC$ under a reflection across the x-axis.
3. Graph the image of $\triangle ABC$ under a rotation of 180° clockwise about the origin.
4. Graph the image of rectangle $ABCD$ under a rotation of 90° counterclockwise about the origin.
5. Graph the dilation image of $\triangle ABC$, using a scale factor of 2 and the center of dilation at vertex C.
6. Perform the following two transformations on $\triangle ABC$: a translation 6 units up followed by a reflection over the y-axis.

Use $x_0 = x_1 + k(x_2 - x_1)$ and $y_0 = y_1 + k(y_2 - y_1)$ to find a point that is a fractional part of the way from $A(x_1, y_1)$ to $B(x_2, y_2)$ if k represents the fractional part. Give the coordinates.

7. $A(6, 2)$, $B(4, 10)$, $k = \frac{1}{2}$ (5, 6)
8. $A(10, 5)$, $B(7, 8)$, $k = \frac{2}{3}$ (8, 7)

104 ASSESSMENT *CHAPTER 8 FORM B*

Chapter 8 Test Form B, page 2

Name ____________ Date ____________

9. Describe two transformations that could be used to create the image at the right.
Reflection across the y-axis followed by 90° clockwise rotation

Use the matrices below for Exercises 10–17.

$A = \begin{bmatrix} 1 & 2 & 8 \\ 3 & 8 & 5 \end{bmatrix}$ $B = \begin{bmatrix} 9 & 3 & 6 \\ 4 & 7 & 5 \end{bmatrix}$ $C = \begin{bmatrix} 7 & 6 \\ 2 & 3 \\ 5 & 9 \end{bmatrix}$ $D = \begin{bmatrix} 2 & 4 & 6 \\ 8 & 9 & 5 \\ 3 & 7 & 1 \end{bmatrix}$

10. Find the dimension of Matrix C. 3×2
11. Identify the element D_{32}. 7
12. $A + B$ $\begin{bmatrix} 10 & 5 & 14 \\ 7 & 15 & 10 \end{bmatrix}$
13. $3C$ $\begin{bmatrix} 21 & 18 \\ 6 & 9 \\ 15 & 27 \end{bmatrix}$
14. $2A + B$ $\begin{bmatrix} 11 & 7 & 22 \\ 10 & 23 & 15 \end{bmatrix}$
15. CB $\begin{bmatrix} 87 & 63 & 72 \\ 30 & 27 & 27 \\ 81 & 78 & 75 \end{bmatrix}$
16. AD $\begin{bmatrix} 42 & 78 & 24 \\ 85 & 119 & 63 \end{bmatrix}$
17. BD $\begin{bmatrix} 60 & 105 & 75 \\ 79 & 114 & 64 \end{bmatrix}$

Use matrices to help you find the reflective image of $\triangle ABC$ represented by $\begin{bmatrix} 2 & 1 & 3 \\ -1 & -2 & -4 \end{bmatrix}$ when the triangle is reflected over the:

18. x-axis. $\begin{bmatrix} 2 & 1 & 3 \\ 1 & 2 & 4 \end{bmatrix}$
19. y-axis. $\begin{bmatrix} -2 & -1 & -3 \\ -1 & -2 & -4 \end{bmatrix}$

An office supply store chain shows prices of its office machines in a 1×3 matrix and the number of machines sold monthly at each of its three stores in a 3×3 matrix.

Typewriter	Fax Machine	Adding Machine
200	500	70

	Store A	Store B	Store C
Typewriter	20	60	45
Fax Machine	50	100	80
Adding Machine	75	150	100

20. Find the gross revenue from the sales of these office machines sold at all three stores. $162,750
21. Find the gross revenue from the sales of these office machines sold at Store B. $72,500

106 ASSESSMENT *CHAPTER 8 FORM B*

CHAPTER 8 ASSESSMENT

Use the figure at the right for Exercises 1–3.
For 1–3, see additional answers.

1. Graph the image of $\triangle ABC$ under a translation 5 units to the right. Label the image points A', B', and C'.
2. Graph the image of $\triangle ABC$ under a reflection across the x-axis. Label the image points A'', B'', and C''.
3. Graph the image of $\triangle ABC$ under a 180° clockwise rotation about the origin. Label the image points A''', B''', and C'''.
4. Draw the dilation image of rectangle $LIMB$ with the center of dilation at the origin and a scale factor of $\frac{1}{3}$. See additional answers.

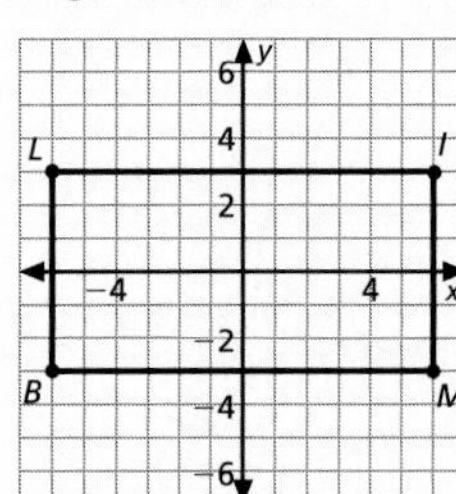

5. Describe two transformations that together could have been used to create the image shown in blue. Possible answer: reflection over x-axis and translation 8 units left.

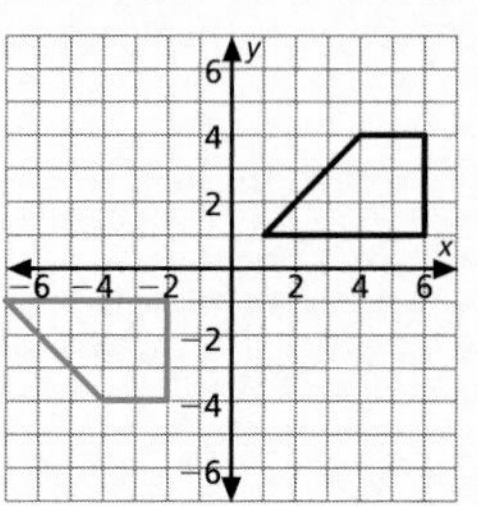

Use matrix T at the right for Exercises 6–8.

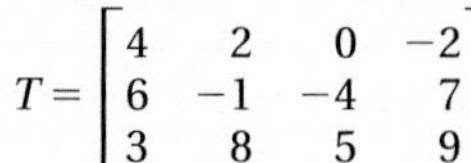

$$T = \begin{bmatrix} 4 & 2 & 0 & -2 \\ 6 & -1 & -4 & 7 \\ 3 & 8 & 5 & 9 \end{bmatrix}$$

6. Find the dimensions of T. 3×4
7. Identify the elements T_{32}, T_{23}, and T_{11}. 8, −4, 4
8. Find kT for $k = -3$. See additional answers.
9. Find the product MN: $M = \begin{bmatrix} 5 & 1 & -2 \\ 3 & 6 & 1 \end{bmatrix}$ $N = \begin{bmatrix} 4 & 3 \\ 8 & -1 \\ 0 & 7 \end{bmatrix}$ $\begin{bmatrix} 28 & 0 \\ 60 & 10 \end{bmatrix}$
10. Find the image of the rectangle with vertices $\begin{bmatrix} 3 \\ 3 \end{bmatrix}$, $\begin{bmatrix} -5 \\ 3 \end{bmatrix}$, $\begin{bmatrix} -5 \\ -2 \end{bmatrix}$, and $\begin{bmatrix} 3 \\ -2 \end{bmatrix}$ under the transformation represented by $\begin{bmatrix} 1 & 0 \\ 0 & -1 \end{bmatrix}$. $\begin{bmatrix} 3 & -5 & -5 & 3 \\ -3 & -3 & 2 & 2 \end{bmatrix}$
11. A quilt maker has three retail outlets where quilts and pillows are sold. The table shows the number of quilts and pillows sold at each outlet. Find the profit from each outlet. Use matrices.
The profit from each quilt is $90 and the profit from each pillow is $25.
A: $2,640; B: $2,275; C: $3,195

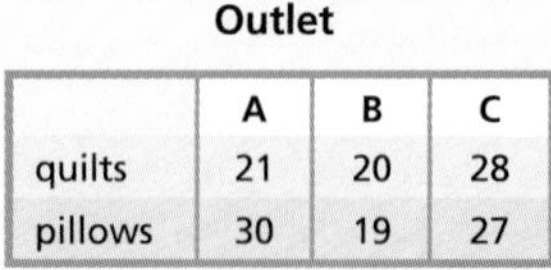

Outlet

	A	B	C
quilts	21	20	28
pillows	30	19	27

376 | Chapter 8 **Assessment**

ADDITIONAL ANSWERS

1–3.

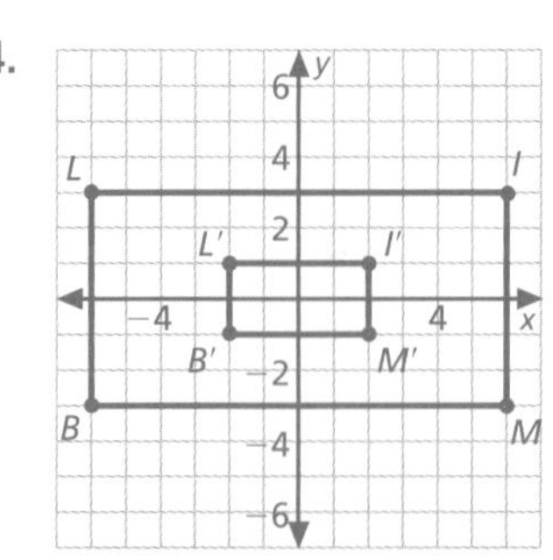

4.

8. $\begin{bmatrix} -12 & -6 & 0 & 6 \\ -18 & 3 & 12 & 21 \\ -9 & -24 & -15 & -27 \end{bmatrix}$

Chapter 8 Alternative Assessment

Critical Thinking

WALLPAPER PATTERNS Visit a store that sells wallpaper and spend some time browsing through their selection. Select at least three examples of wallpaper patterns that make use of reflections, rotations and/or dilations. If you cannot obtain a sample of the wallpaper, sketch the pattern. Then write a brief explanation of how the pattern was made, identifying the various transformations used.

Self Assessment

INVENTORY Visit a local grocery store to check the value of their inventory. Select at least three brands of five different items (brands of crackers, soups, cat food, etc.). Count the number of each on the shelves, and note the price of each. Use a matrix to find the total value of each brand. Show all your work and explain how you calculated each inventory value.

Using Other Resources

FUN RIDES Use the Internet to explore some famous roller coasters in the United States. Find maps of these rides and identify which use translations, rotations, and reflections, or combinations of these transformations. If you do not have access to the Internet, you may use the library or interview people who have visited some of these amusement parks and know the rides. From this knowledge, design your own amusement ride, incorporating at least two of the principles investigated in this chapter.

Chapter Investigation

EXTENSION Make a presentation to your class of your roller coaster. Explain why you designed the ride as you did. Display and use your marketing brochure during your presentation.

Chapter Investigation

The benchmarks and expectations for this extension are as follows.

- Students design an "out-and-back" coaster with eight hills. They make a scale drawing of their coaster on graph paper.
- Students use the scale drawing of their roller coaster to estimate the track length of the entire coaster. Write a paragraph explaining their answer.
- Students measure the angle of descent for each hill in their roller coaster design. They estimate their coaster's top speed by comparing its features to the coasters shown in the table on page 337. Write a paragraph to justify their estimate.
- Students make a presentation to their class of their roller coaster. They explain why they designed the ride as they did. They display and use their marketing brochure during their presentation.

Critical Thinking

Students may also make their own wallpaper patterns and describe them. You may wish to have students make patterns and then exchange them for analysis. **SCORING RUBRIC 5**—Student finds three examples of wallpaper that show complex transformations, correctly identifies all the transformations used, and writes a comprehensive explanation. **4**—Student finds three examples, correctly identifies all transformations used, and writes a reasonable explanation. **3**—Student finds three examples, correctly identifies most of the transformations used, and writes a fair explanation. **2**—Student finds fewer than three examples, correctly identifies some transformations, and writes a partial or incomplete explanation. **1**—Student finds one example of wallpaper that shows a simple transformation, correctly identifies the transformation used, and writes no explanation. **0**—Student makes no attempt to find or describe wallpaper transformations.

Self Assessment

You may wish to arrange a group visit to a store.
SCORING RUBRIC 5—Student inventories three or more brands of five items, correctly calculating the value of each inventory by use of a matrix, and explains the process in a well-organized manner. **4**—Student inventories items, correctly calculating the value by use of a matrix, and explains the process. **3**—Student inventories items, calculating the value by use of a matrix with a few minor errors, and explains the process. **2**—Student inventories fewer than three brands of five items, calculating the value of each inventory by use of a matrix with a few errors, and poorly explains the process. **1**—Student inventories fewer than three brands of less than five items, calculating the value of each inventory by use of a matrix with a few significant errors, and does not explain the process. **0**—Student makes no attempt to find the value of any inventory.

Using Other Resources

Students can also visit a travel agency to get information about the larger amusement parks.
SCORING RUBRIC 5—Student finds and describes in detail the transformations used in amusement park rides, designs a new, innovative ride, and points out the principles used. **4**—Student finds and describes the transformations, designs a

Scoring Rubric continued on page 378.

Cumulative Review

Cumulative review is the best way to maintain previously taught skills and concepts. This will keep students prepared for new lessons that build on previously covered skills.

Cumulative Review covers:

Lesson 1-5
Lesson 1-8
Lesson 2-2
Lesson 2-7
Lesson 3-3
Lesson 3-5
Lesson 4-1
Lesson 4-9
Lesson 5-1
Lesson 5-2
Lesson 5-7
Lesson 6-2
Lesson 6-6
Lesson 7-1
Lesson 7-4
Lesson 8-3
Lesson 8-6

Additional Answers

11. 

21. 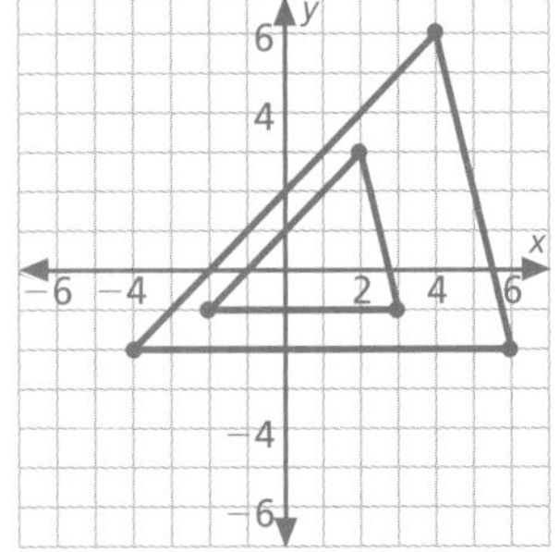

Using Other Resources Scoring Rubric continued.

new ride, and can point out the principles used. **3**—Student finds and describes some of the transformations used, designs a new ride, and points out some of the principles used. **2**—Student finds and describes a few of the transformations used in amusement park rides, designs a ride, and points out one of the principles used. **1**—Student finds and describes one of the transformations used in amusement park rides, but does not design a ride or point out principles. **0**—Student makes no attempt to describe a ride or design an original ride.

Chapters 1–8 Cumulative Review

Multiply or divide.

1. $(-6.8)(1.5)(-10)$ 102
2. $3\frac{1}{2} \div (-7)$ $-\frac{1}{2}$

Simplify.

3. $x^4 \cdot x^{-5}$ $\frac{1}{x}$
4. $(-3)^{-2}$ $\frac{1}{9}$

Given that $f(x) = 6x - 3$, find each value.

5. $f(2)$ 9
6. $f(-1)$ −9

Refer to the following list of scores.

90 78 83 86 91 95 88 90 85 84

7. Find the median. 87
8. Find the mode. 90
9. Find the mean. 87
10. In the figure, $\overleftrightarrow{FD}$ and $\overleftrightarrow{HG}$ intersect at E. Find the measure of $\angle DEG$. 70°

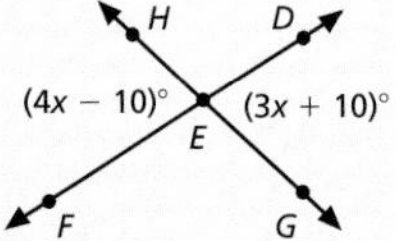

11. Draw the next figure in the pattern. Then describe the tenth figure in the pattern. See additional answers.
12. Find the value of x. 14

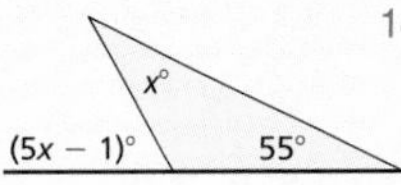

13. Find the value of x in the trapezoid. $\overline{AB}$ is a median. 40

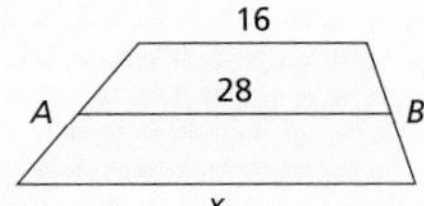

14. Change 0.7 cm to meters. 0.007 m
15. Find the area of the shaded region. 75.36 m²

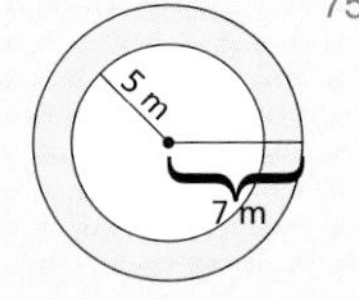

16. Find the volume. 6,912 in.³

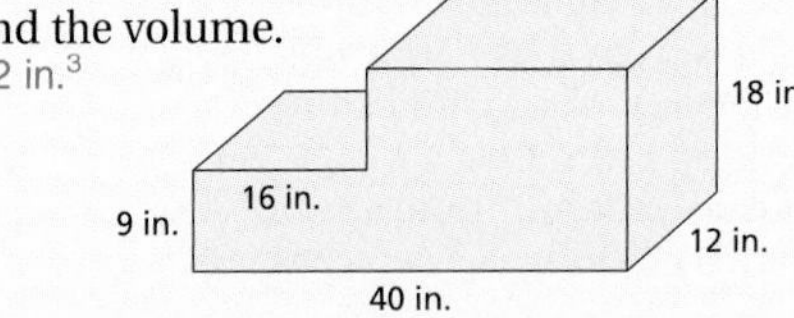

17. Find the slope of a line perpendicular to the line containing points $A(7, -2)$ and $B(-3, 4)$. $\frac{5}{3}$
18. Solve the system of equations. $\begin{cases} 7x - 2y = 22 \\ 3x + y = 15 \end{cases}$ $x = 4$, $y = 3$
19. Solve the proportion. $\frac{84}{x} = \frac{42}{50}$ 100
20. Determine if the two triangles are similar. If the triangles are similar, give a reason: Write AA, SSS, or SAS. no

65° 20° 25° 65°

21. Copy the figure, and draw a dilation with center at the origin and a scale factor of $\frac{1}{2}$. see additional answers.

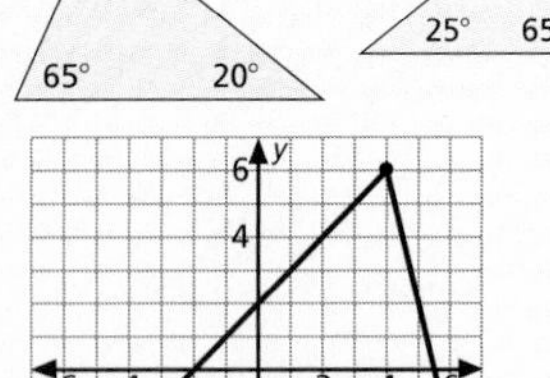

22. Find the product of matrices A and B.

$$A = \begin{bmatrix} 10 & 2 & 0 \\ 1 & 3 & 5 \end{bmatrix} \quad B = \begin{bmatrix} 6 & 4 \\ 0 & -1 \\ 3 & 2 \end{bmatrix} \quad \begin{bmatrix} 60 & 38 \\ 21 & 11 \end{bmatrix}$$

Teaching Strategies

In Exercise 16 of the Cumulative Review, students must break the figure into two rectangular prisms in order to solve for total volume. Students may benefit from making and labeling two separate sketches before they attempt to solve the problem.

CHAPTERS 1–8 CUMULATIVE ASSESSMENT

STANDARDIZED TEST PREPARATION—QUANTITATIVE COMPARISON

In each question, compare the quantity in Column A with the quantity in Column B. Select the letter of the correct answer from these choices:

A. The quantity in Column 1 is greater.

B. The quantity in Column 2 is greater.

C. The two quantities are equal.

D. The relationship cannot be determined by the information given.

Notes: In some questions, information that refers to one or both columns is centered over both columns. A symbol used in both columns has the same meaning in each column. All variables represent real numbers. Most figures are not drawn to scale.

	Column A	Column B
1. A	$(-2)(-1)(12)$	$-\|6\| \cdot \|-4\|$
2. B	$7.5 \cdot 10^{-5}$	0.00075
3. A ($x < 0$)	$(x^5)^{-2}$	$\frac{x^6}{x^3}$
4. A ($2x - 1 < 11$)	x	4
5. D (figure: transversal crossing two lines, angles 1–8)	$m\angle 7$	$m\angle 1 + m\angle 3$
6. B Lengths of the sides of a triangle are 6 m, 4 m, x m	2	x
7. B (figure: 10 cm by 10 cm square, 7 cm, 3 cm)	the probability that a point selected at random is in the shaded area	$\frac{3}{7}$
8. A (figure: box 10 mm, 5 mm, 3 mm)	number of square units in the surface area	number of cubic units in the volume
9. A	the slope of a line with equation $6y = 3x + 9$	the slope of a line with equation $x + 2y = -6$
10. A	the y intercept of a line with a slope of 2 that passes through the point $(0, 5)$	the slope of a line that passes through the points $(0, -1)$ and $(-4, -3)$
11. A The point $A(5, -6)$ is reflected across the line with equation $y = x$.	x-coordinate of image point	y-coordinate of image point

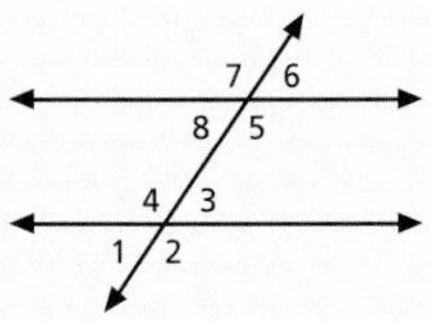

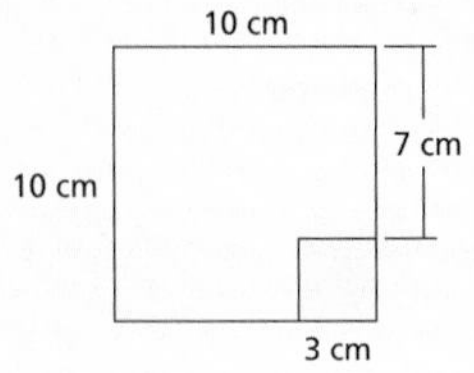

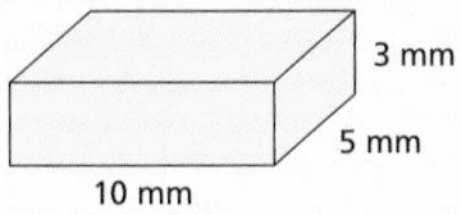

12. **CONSTRUCTED RESPONSE** Graph $\triangle RST$ and its image under a reflection across the line whose equation is $y = -x$. Compare the slopes of the corresponding sides of $\triangle RST$ and $\triangle R'S'T'$. The vertices of $\triangle RST$ are $R(-1, 3)$, $S(2, 4)$, and $T(4, 1)$.

The slopes of the corresponding sides are not related. See additional answers for graph.

STANDARDIZED TEST PREPARATION

QUANTITATIVE COMPARISON QUESTIONS Quantitative comparison questions are generally written with fewer words and require less time to compute. There are four answer choices that describe the relationships of equality and inequality. You are required to compare the quantity from Column A with the quantity from Column B and then determine their relationship. Students should carefully read the four answer choices often during the testing period.

Horizontal rules are used as separators between questions and to clarify when figures and additional information accompanies one or more problems.

Understanding the directions for quantitative comparison questions is critical for success since this type of question is not as commonplace as other types of questions.

TESTING STRATEGIES

One important key to test taking is understanding directions. On the SAT test, the quantitative comparison section can cause difficulties for students if they are unfamiliar with the question type. Go over the directions carefully with the class. Write several easy problems on the board that require little or no calculations. Have students suggest strategies for approaching this type of question.

ADDITIONAL ANSWERS

14.

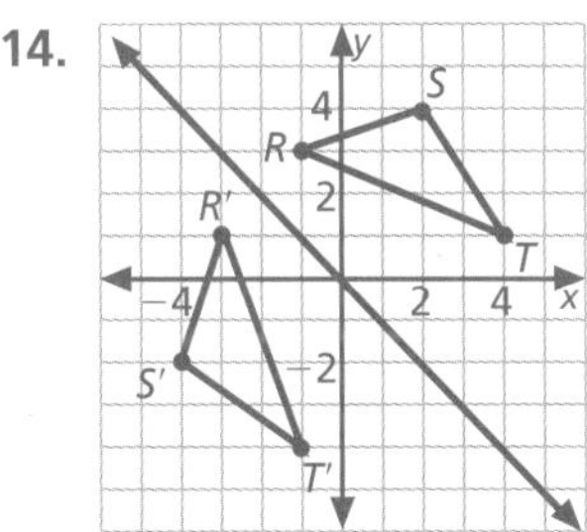

Teaching Strategies

After students have completed Exercise 5 of the Cumulative Assessment, ask them what additional information they would need in order to decide which quantity is greater. They must know the measure of any one angle in the figure. They must also know that the figure represents a transversal and two parallel lines. Remind students that when taking a test they should not assume two lines are parallel just because they appear to be parallel.

Chapter 9—Probability and Statistics

Theme: Sports
Chapter Investigation: Baseball Game Win, pages 386, 395, 399, 409, 414, 419
Careers: Team Dieticians, page 391; Physical Therapists, page 411
Data Activity: Home Run Greats

Content and Connections

Lesson, Pages	Lesson Objectives	NCTM Standards	State/Local Objectives	Interdisciplinary Connections	Real World Applications
9–1 384–387	• Find experimental and theoretical probabilities	Number Operation Patterns, Functions & Algebra Data Analysis, Stat & Prob Prob Solv; Communication Connections; Representation			Sports, Card Games, Test Taking
9–2 388–389	• Solve a problem using simulations • Act it out	Data Analysis, Stat & Prob Prob Solv; Reasoning & Proof Communication Connections; Representation			Computer Science, Marketing, Sports
9–3 392–395	• Find probabilities of compound events	Data Analysis, Stat & Prob Prob Solv; Number Operation Patterns, Functions & Algebra Reasoning & Proof Connections; Representation			Sports, Games, Business
9–4 396–399	• Find the probability of dependent and independent events	Data Analysis, Stat & Prob Number Operation Patterns, Functions & Algebra Prob Solv; Reasoning & Proof Connections; Representation		History	Sports, Surveys, Scheduling
9–5 402–405	• Find the number of permutations and combinations of a set	Data Analysis, Stat & Prob Number Operation Patterns, Functions & Algebra Prob Solv; Reasoning & Proof Communication Connections; Representation			Travel, Sports, Office Work
9–6 406–409	• Interpret and make scatter plots and boxplots	Data Analysis, Stat & Prob Number Operation Patterns, Functions & Algebra Prob Solv; Reasoning & Proof Communication Connections; Representation			Manufacturing, Sales, Sports
9–7 412–415	• Find the variance, standard deviation and 2-scores for a set of data • Use standard deviation to interpret data	Data Analysis, Stat & Prob Reasoning & Proof Number Operation Patterns, Functions & Algebra Prob Solv; Communication Connections; Representation			Sports, Test-Taking, Education

Planning and Resources

Lesson, Pages	Tools/Materials Needed	Trans-parency	Learning/Teaching Styles Options	Assignments: Basic Enriched	Additional Practice in SE	Reteaching, Extra Practice, Enrichment	Other Resources
9–1 384–387	calculator	Warm up 25	Real World Challenge	B: 1–20, 25–28 E: 1–28	Page 390–391, 401, 411, 416, 711	R: page 131 EP: page 115 E: page 151	

9–2 388–389	black spinner coins computer calculator paper bag	Warm up 25		B: 1–25 E: 1–25	Page 390–391, 401, 411, 416	R: page 133 E: page 153	SS: Teacher's Choice
9–3 392–395	six-sided dice	Warm up 25	ESL/LEP Challenge	B: 1–26, 31–42 E: 1–42	Page 400–401, 411, 416, 712	R: page 135 EP: page 117 E: page 155	SS: Teacher's Choice
9–4 396–399	coins six-sided dice	Warm up 26	ESL/LEP Visual Learners Challenge	B: 1–25, 32–34 E: 1–34	Page 400–401, 411, 416, 712	R: page 137 EP: page 119 E: page 157	
9–5 402–405	calculator	Warm up 26 Trans RF-33	Challenge	B: 1–18, 23–27 E: 1–27	Page 410–411, 417, 713	R: page 139 EP: page 121 E: page 159	
9–6 406–409	graph paper straightedge	Warm up 27 Trans RF-34, 35	Real World	B: 1–12, 16–18 E: 1–18	Page 410–411, 417, 713	R: page 141 EP: page 123 E: page 161	SS: Teacher's Choice
9–7 412–415	calculator	Warm up 27 Trans RF-37	Real World Challenge	B: 1–20, 25–30 E: 1–30	Page 417, 714	R: page 143 EP: page 125 E: page 163	

Planning and Pacing

Lesson, Pages	Lesson Title	45 min class	Assignments Basic, Enriched	90 min class	Assignments Basic, Enriched	___ min class	Assignments Basic, Enriched
9–1 384–387	AYR, Opener, Review Percents and Probability	Day 115	B: 1–20, 25–28 E: 1–28	Day 59	B: 1–20, 25–28 E: 1–28		B: 1–20, 25–28 E: 1–28
9–2 388–389	Problem Solving Skills: Simulations, R&PYS	Day 116–117	B: 1–25 E: 1–25	Day 59–60	B: 1–25 E: 1–25		B: 1–25 E: 1–25
9–3 392–395	Compound Events	Day 118	B: 1–26, 31–42 E: 1–42	Day 60	B: 1–26, 31–42 E: 1–42		B: 1–26, 31–42 E: 1–42
9–4 396–399	Independent and Dependent Events, R&PYS	Day 119–120	B: 1–25, 32–34 E: 1–34	Day 61	B: 1–25, 32–34 E: 1–34		B: 1–25, 32–34 E: 1–34
9–5 402–405	Permutations and Combinations	Day 121	B: 1–18, 23–27 E: 1–27	Day 62	B: 1–18, 23–27 E: 1–27		B: 1–18, 23–27 E: 1–27
9–6 406–409	Scatter Plots and Boxplots, R&PYS	Day 122–123	B: 1–12, 16–18 E: 1–18	Day 63	B: 1–12, 16–18 E: 1–18		B: 1–12, 16–18 E: 1–18
9–7 412–415	Standard Deviation	Day 124	B: 1–20, 25–30 E: 1–30	Day 63–64	B: 1–20, 25–30 E: 1–30		B: 1–20, 25–30 E: 1–30
Review/ Assess		Day 125–126		Day 64–66			

	Chapter 1 Days	Chapter Cumulative Days
Yearly Pacing (45 min class)	12 days	126 days
Yearly Pacing (90 min class)	8 days	66 days

Assessment Options

Assessment in Student Edition	Assessment in Teacher's Edition	Pages in Assessment Book	Software Generated Assessment
Are You Ready?, pages 380–381; Writing Math, pages,386, 387, 388, 395, 399, 405, 409, 413; Mixed Review, pages 387, 389, 395, 399, 405, 409, 415; Check Understanding pages 384, 392, 396, 402, 406, 412; Mid-Chapter Quiz, page 401, Chapter Review, page 416; Chapter Assessment, page 418; Alternative Assessment, page 419; Cumulative Review, page 420; Cumulative Assessment, page 421	5-minute Warm ups, pages 384, 388, 392, 396, 402, 406, 412; Quick Assessment, pages 386, 389, 394, 398, 404, 408, 414; Scoring Rubrics, page 419	Mid-Chapter Quiz, page 135; Test Form A, pages 137, 139; Test Form B, pages 141, 143; Cumulative Test 145, 147; Math Journal prompt, page 149, 150;	Chapter 9

Refresher Skills

The skills on these two pages are skills that have been taught in previous math courses. Continuous review of basic math skills will make stronger math students. These skills are identified as necessary to be successful in Chapter 9.

Skills Correlation Chart

Skill	Lesson Number
Percents, Decimals and Fractions	9-1, 9-2, 9-3, 9-4, 9-5, 9-6, 9-7
Measures of Central Tendency	9-6, 9-7
Histograms	9-6, 9-7

Vocabulary

mean
median
mode
range
histogram

Additional Answers

54.

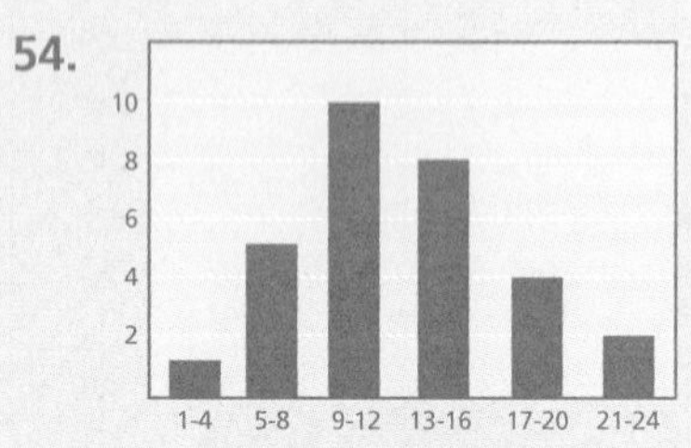

55.

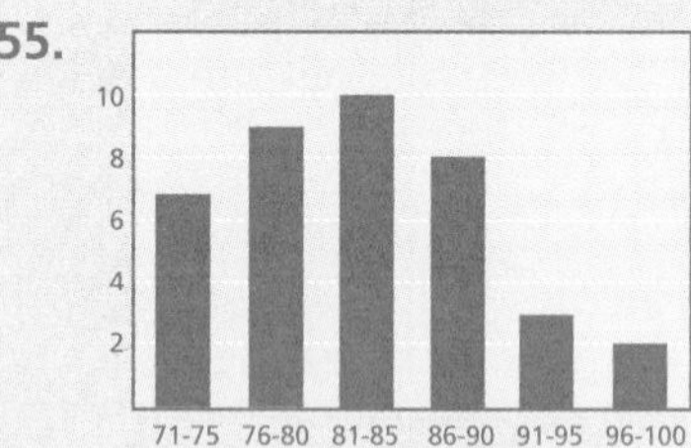

56.

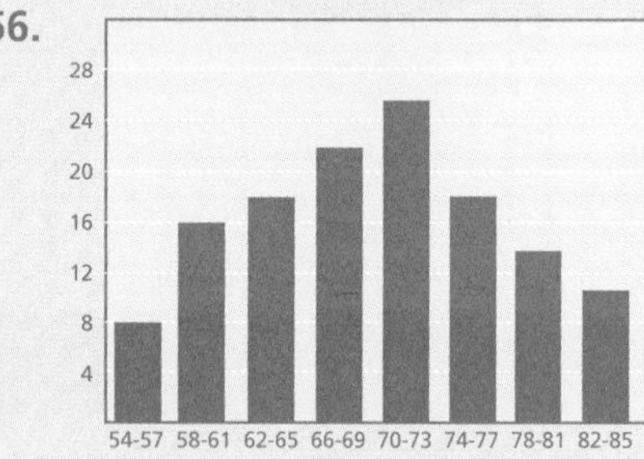

Are You Ready?

Refresh Your Math Skills for Chapter 9

The skills on these two pages are ones you have already learned. Stretch your memory and complete the exercises. For additional practice on these and more basic skills, see page 674.

Percents, Decimals and Fractions

Convert each fraction or decimal to a percent. Round to the nearest tenth if necessary.

1. $\frac{3}{4}$ 75% **2.** 0.86 86% **3.** $\frac{7}{8}$ 87.5% **4.** 0.93 93% **5.** $\frac{2}{3}$ 66.7%

6. 0.5 50% **7.** $\frac{3}{8}$ 31.5% **8.** 0.42 42% **9.** $\frac{7}{16}$ 43.8% **10.** 0.38 38%

11. $\frac{1}{6}$ 16.7% **12.** 0.64576 64.6% **13.** $\frac{6}{21}$ 28.6% **14.** 0.19823 19.8% **15.** $\frac{27}{46}$ 58.7%

Convert each percent to a decimal. Round to the nearest thousandth if necessary.

16. 46% 0.46 **17.** 83% 0.83 **18.** 29% 0.29 **19.** 15% 0.15 **20.** 12% 0.12

21. 18.76% 0.188 **22.** 9.3825% 0.094 **23.** 78.6215% 0.786 **24.** 64.93% 0.649 **25.** 21.748% 0.217

Convert each percent to a fraction in lowest terms.

26. 80% $\frac{4}{5}$ **27.** 50% $\frac{1}{2}$ **28.** 68% $\frac{17}{25}$ **29.** 92% $\frac{23}{25}$ **30.** 75% $\frac{3}{4}$

31. 64.2% $\frac{321}{500}$ **32.** 39.8% $\frac{199}{500}$ **33.** 20% $\frac{1}{5}$ **34.** 19.6% $\frac{49}{250}$ **35.** 51.9% $\frac{519}{1000}$

Measures of Central Tendency

Find the mean, median, mode and range of each set of data.

36. 74 75 79 76 77 74 78 72 71 mean = 75.1; median = 75; mode = 74; range = 8

37. 30 32 34 36 38 39 37 35 34 33 mean = 34.8; median = 34.5; mode = 34; range = 9

38. 40 48 52 47 56 49 43 55 46 48 51 mean = 48.6; median = 48; range = 16; mode = 48

39. 17 12 13 16 22 21 19 18 14 20 15 11 mean = 16.5; median = 16.5; no mode; range = 11

40. 88 87 81 92 86 87 89 90 93 91 85 92 94 mean = 88.6; median = 89; mode = 87.92; range = 13

41. 62 63 67 68 65 69 64 61 65 66 60 67 65 63 mean = 64.6; median = 65; mode = 65; range = 9

Reducing Fractions

Determine the greatest common factor of the numerator and denominator. Divide both the numerator and denominator by the factor to write the fraction in lowest terms.

42. $\frac{3}{24}$ 3; $\frac{1}{8}$ **43.** $\frac{19}{57}$ 19; $\frac{1}{3}$ **44.** $\frac{4}{52}$ 4; $\frac{1}{13}$ **45.** $\frac{2}{110}$ 2; $\frac{1}{55}$ **46.** $\frac{4}{144}$ 4; $\frac{1}{36}$ **47.** $\frac{34}{17}$ 17; 2

48. $\frac{13}{52}$ 13; $\frac{1}{4}$ **49.** $\frac{16}{48}$ 16; $\frac{1}{3}$ **50.** $\frac{25}{35}$ 5; $\frac{5}{7}$ **51.** $\frac{77}{121}$ 11; $\frac{7}{11}$ **52.** $\frac{17}{85}$ 17; $\frac{1}{5}$ **53.** $\frac{15}{75}$ 15; $\frac{1}{5}$

57.

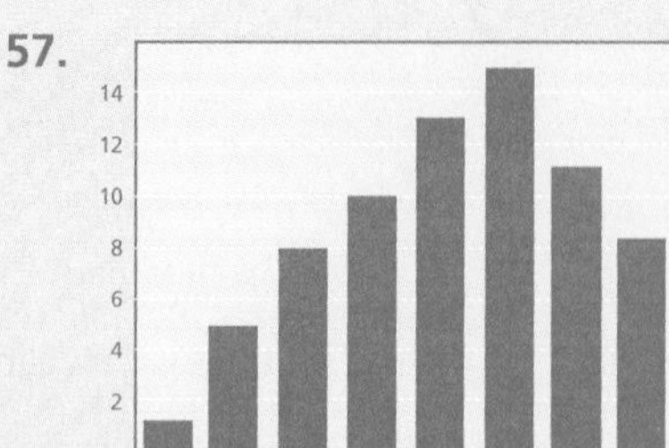

58.

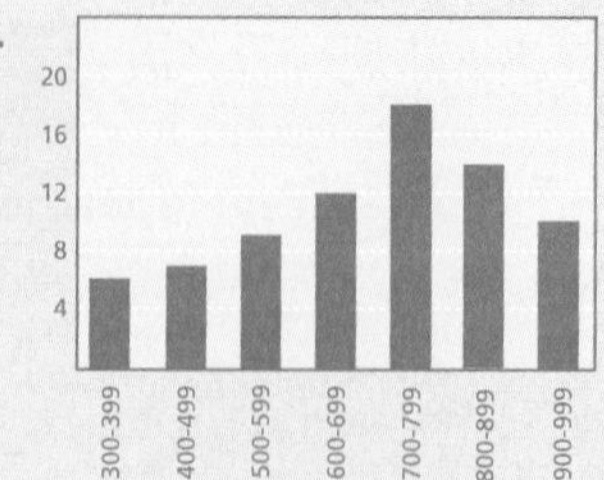

59.

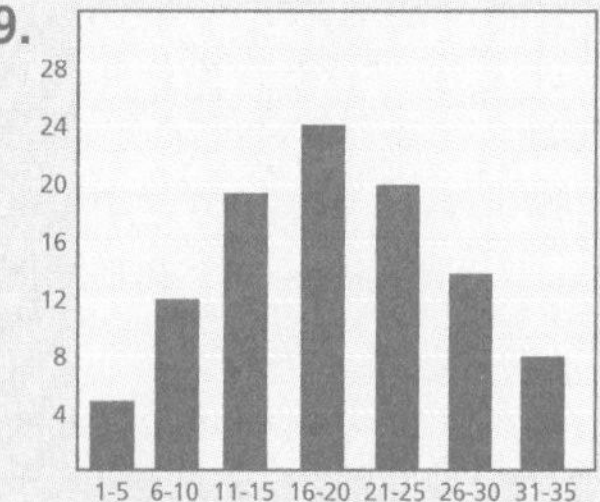

CHAPTER 9

HISTOGRAMS

Draw a histogram for each frequency chart. For 54–59, see additional answers.

54.

Interval	Tally	Frequency
21-24	||	2
17-20	||||	4
13-16	𝍸 |||	8
9-12	𝍸 𝍸	10
5-8	𝍸	5
1-4	|	1

55.

Interval	Tally	Frequency
96-100	||	2
91-95	|||	3
86-90	𝍸 |||	8
81-85	𝍸 𝍸	10
76-80	𝍸 ||||	9
71-75	𝍸 ||	7

56.

Interval	Tally	Frequency
82-85	𝍸 𝍸 |	11
78-81	𝍸 𝍸 ||||	14
74-77	𝍸 𝍸 𝍸 ||||	19
70-73	𝍸 𝍸 𝍸 𝍸 𝍸	25
66-69	𝍸 𝍸 𝍸 𝍸 ||	22
62-65	𝍸 𝍸 𝍸 |||	18
53-61	𝍸 𝍸 𝍸 |	16
54-57	𝍸 |||	8

57.

Interval	Tally	Frequency
71-80	𝍸 ||||	9
61-70	𝍸 𝍸 |	11
51-60	𝍸 𝍸 𝍸	15
41-50	𝍸 𝍸 |||	13
31-40	𝍸 𝍸	10
21-30	𝍸 |||	8
11-20	𝍸	5
1-10	|	1

58.

Interval	Tally	Frequency
900-999	𝍸 𝍸	10
800-899	𝍸 𝍸 𝍸	15
700-799	𝍸 𝍸 𝍸 ||	17
600-699	𝍸 𝍸 ||	12
500-599	𝍸 ||||	9
400-499	𝍸 ||	7
300-399	𝍸 |	6

59.

Interval	Tally	Frequency
31-35	𝍸 |||	8
26-30	𝍸 𝍸 ||||	14
21-25	𝍸 𝍸 𝍸 𝍸	20
16-20	𝍸 𝍸 𝍸 𝍸 |||	23
11-15	𝍸 𝍸 𝍸 ||||	19
6-10	𝍸 𝍸 ||	12
1-5	𝍸	5

AREA

Find the area of the shaded region of each figure. Round to the nearest hundredth if necessary.

60.

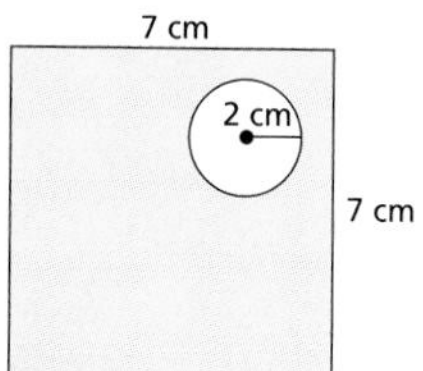

36.43 cm²

61.

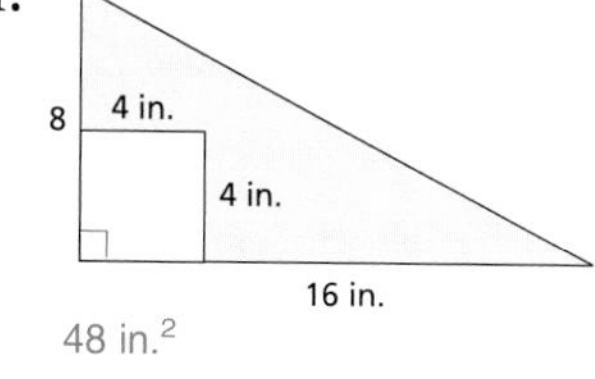

48 in.²

62.

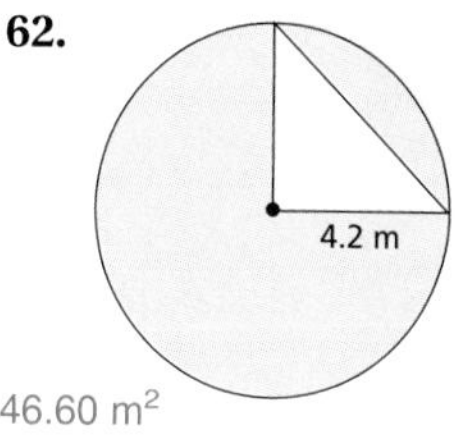

46.60 m²

Extend the Lesson

COOPERATIVE LEARNING Have students work in groups to develop a survey question, gather data from the class, and create a frequency table of survey results. Then have students create a histogram and find the mean, median, mode and range of the data.

Chalkboard Examples

Percents, Decimals and Fractions

Convert $\frac{5}{8}$ to a decimal and percent. 0.625, 62.5%

Measures of Central Tendency

Find the mean, median, mode, and range of each set of data.

16	19	15	18	21
14	12	18	20	18

mean: 17.1; median: 18; mode: 18; range: 9

Histograms

Have students work together to create a frequency chart showing shoe size. Have students create a histogram from the frequency chart.

Refresher Wrap-up

QUICK ASSESSMENT

Ask the following questions to determine if students have mastered the basic skills reviewed on these pages.

1. When is median a more useful measure of central tendency than mean? when outliers shift the mean away from the majority of data values
2. What can you learn from a data set by viewing a histogram? Answers will vary. A histogram shows how the data values are grouped. You can easily identify the range, outliers and the center of the data.

ADDITIONAL PRACTICE

All topics: Refer to the Extra Practice Index on page 674 of this text.

Chapter 9 **Are You Ready?** 381

Chapter Opener

NCTM Standards/Strands
- Number & Operation
- Patterns, Functions, & Algebra
- Data Analysis, Statistics, & Probability
- Problem Solving
- Reasoning and Proof
- Communication
- Connections
- Representation

Vocabulary
manufacturing
assembly line
technician
specification
schematics

Theme Connections
Tables and graphs are used to organize data about manufacturing and product development. Tables provide a way to organize numerical information. Many types of data are kept in regards to manufacturing. Companies are concerned with quality, performance, sales, shipping and production quotas. Graphs are often used to show trends in sales and performance. Discuss how data and graphs are used in manufacturing.

Career Opportunities
Many careers related to manufacturing make use of data and graphs. Two are highlighted in the MathWorks features. Others include product developers, packagers, designers, and sales representatives.
- Precision Assembler, page 253
- Engineering Technicians, page 273

Internet Connection

Theme Activities
Learningmathmatters.com provides web addresses to search that help students gather information about the use of math in the real world, particularly data and graphs. To search for additional addresses, begin a search using the keyword *manufacturing*. Then within that search, use such key words as *design, product safety, quality control* and *assembly line.* Students can brainstorm in small groups other key words.

CHAPTER 9

Probability and Statistics

THEME: Sports

When the Cubs send their right-handed power hitter to the plate in the ninth inning, the Mets counter with a left-handed pitcher. Why? The manager of the Mets is simply "playing the odds."

Since its inception, baseball has kept meticulous records. By studying the data managers, players, announcers and fans use the concepts of probability and chance to make predictions. Today, players and managers use computers and calculators to record and analyze data. They know that the more effectively they use statistics and probability, the better they will do their jobs.

- **Team Dietitians** (page 391) plan meals and nutritional plans for athletes. They use their knowledge of nutrition to help team members maintain overall health, muscle health and bone strength.
- **Physical Therapists** (page 411) determine exercises to strengthen muscles after injuries. Through specially designed exercise programs, they improve mobility, relieve pain, and reduce injuries.

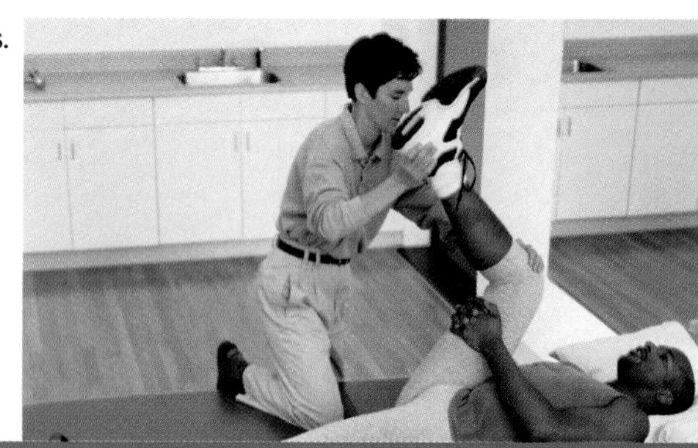

Internet Connection
www.learningmathmatters.com

Chapter Investigation
Go to learningmathmatters.com to locate additional information about manufacturing.

Extend the Lesson

ART Have students study composition in photographs by selecting one photograph published in a book or magazine. The photograph should show a variety of objects. Then have them draw a similar rectangle and make a quick sketch showing the location of objects within the photograph. Have students share their work and discuss the importance of composition in a work of art.

Home Run Greats–Then and Now

Player	Year	Home runs	Games played	At bats	Batting average	Runs batted in
Mark McGwire	1998	70	155	509	.299	147
Sammy Sosa	1998	66	159	643	.308	158
Roger Maris	1961	61	161	590	.269	142
Babe Ruth	1927	60	151	540	.356	164

Home Runs–Month by Month

Player	Mar	Apr	May	Jun	Jul	Aug	Sep	Oct
McGwire	1	10	16	10	8	10	15	0
Sosa	0	6	7	20	9	13	11	0
Maris	0	1	11	15	13	11	9	1
Ruth	0	4	12	9	9	9	17	0

Data Activity: Home Run Greats

Use the tables for Questions 1–4.

1. For each player, divide the at bats by the total of home runs and round to the nearest hundredth. Compare the unit rates. Which player had the best unit rate? McGwire = 7.27; Sosa: 9.74; Maris: 9.67; Ruth: 9. McGwire's rate is best.
2. What percent of McGwire's home runs were hit in August and September? About 36%
3. Find the average number of at bats per game for each player. Which player had the greatest average?
4. In 1998, McGwire walked 162 times. What percent of his at bats did he reach first base by drawing a walk? About 32%

Chapter Investigation

Baseball has been called "America's Pastime." In recent years, the game's appeal has become international. Using statistics and probability, fans at home can predict what will happen when the bases are loaded in the bottom of the ninth inning.

3. McGwire: 3.28 at bats per game; Sosa: 4.04; Maris: 3.66; Ruth: 3.58. Sosa's average is greatest.

Working Together

Gather baseball statistics and design your baseball simulation game using dice and percentile cards. Make a lineup and play a game. Decide whether the game's outcome matches your predictions. Use the Chapter Investigation logos to guide your group.

Data Activity

Discuss with students the importance of imports and exports to the U.S. economy. Point out that a country's natural resources often determine the types of goods that country must export and import. Discuss the importance of reading any titles or legends on a table or graph. For example, the amounts shown in this table represent millions of dollars. Have students rewrite several of the amounts in millions of dollars and determine the actual amount. Assign students Questions 1–4.

Extend the Data Activity

REAL WORLD CONNECTION Have students research what types of goods are imported and exported from their state and create a table similar to the one shown here.

Chapter Investigation

As an Overarching Problem

Display a few consumer products that students might use often. Ask students to choose one product and think about the features of the product they like and the features they would change. Have students share some of their ideas with the class. Students will continue to work on the investigation as they complete the exercises identified by the Investigation Logo that are found throughout the chapter. These exercises will guide students through the task described in *Working Together*. Encourage students to keep all of their work on the Investigation together. Have students use the suggestions in the Chapter Investigation Extension to summarize their work.

As a Chapter Project

The goal of this project is to create a drawing or prototype of an improved consumer product. Students can use the Team Project Planner on page 95 and the Project Planning Calendar on page 96 in the Enrichment text to complete the project. Benchmarks **a** and **b** should be completed after the lesson listed in parentheses has been studied. Benchmark **c** should be completed at the end of the chapter.

Project Planning Calendar

Name ______ Date ______

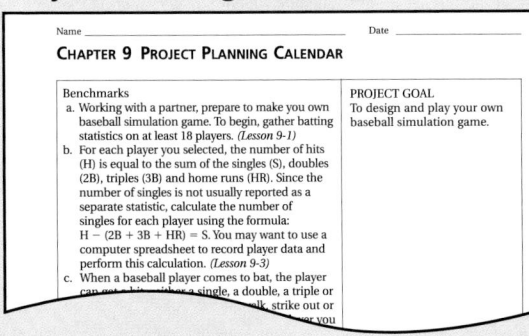

CHAPTER 9 PROJECT PLANNING CALENDAR

Benchmarks	PROJECT GOAL
a. Working with a partner, prepare to make you own baseball simulation game. To begin, gather batting statistics on at least 18 players. *(Lesson 9-1)* b. For each player you selected, the number of hits (H) is equal to the sum of the singles (S), doubles (2B), triples (3B) and home runs (HR). Since the number of singles is not usually reported as a separate statistic, calculate the number of singles for each player using the formula: H − (2B + 3B + HR) = S. You may want to use a computer spreadsheet to record player data and perform this calculation. *(Lesson 9-3)* c. When a baseball player comes to bat, the player	To design and play your own baseball simulation game.

Group Project Planner

Name ______ Date ______

CHAPTER 9 GROUP PROJECT PLANNER

Assignment ______ Objective ______

Group Members
1) ______
2) ______
3) ______
4) ______
5) ______

Assigned Roles

Deadlines Done

Lesson Planning

NCTM Standards/Strands

- Number & Operation
- Patterns, Functions, & Algebra
- Data Analysis, Statistics, & Probability
- Problem Solving
- Reasoning and Proof
- Communication
- Connections
- Representation

Vocabulary

probability   experiment
outcome
experimental probability
sample space   tree diagram
theoretical probability

Materials Needed

calculators

Lesson Resources

Warm-Up Transparency 25
Reteaching 9-1
Extra Practice 9-1
Enrichment 9-1

Assignment Guide

Basic: 1–20, 25–28
Enriched: 1–28

Getting Started

5-Minute Warm-Up

Solve for *P*.

1. $P = \frac{234}{585}$ 0.4
2. $P = \frac{(129 + 71)}{300}$ $\frac{2}{3}$ or $0.\overline{6}$
3. $P = \frac{(32 + 28 + 35)}{250}$ 0.38

Introduction to Lesson 9-1

When students have completed the experiment, you may wish to suggest that they draw a graph to show how the probability of two people sharing a birthday varies with the size of the group. Have students compare the appearance of their graphs.

9-1 Review Percents and Probability

Goals ■ Find experimental and theoretical probabilities.

Applications Sports, Card Games, Test Taking

Work with a partner. Results will vary. Statistics show that a group should have only 23–24 people for a 50% chance; for a group of 40 people, the probability is 90%.

1. Discuss: How large do you think a group must be for a 50% chance that two members will share a birthday?
2. Make a guess. Then take a survey of the students in one of your classes. Look for common birthdays. What did you learn?

Build Understanding

Recall that **probability** is the mathematics of chance. Probability is used to describe the relative frequency with which an event is likely to occur. Probabilities are reported using fractions, decimals, and percents. The greater the probability of an event, the more likely the event is to occur.

One way to find the likelihood of an event occurring in the real world is to conduct an experiment. In an **experiment**, you either take a measurement or make an observation. A probability determined by observation or measurement is called **experimental probability**. An **outcome** is the result of each trial of an experiment.

The experimental probability of an event (E) can be estimated.

$$P(E) = \frac{\text{number of favorable observations of } E}{\text{total number of observations}}$$

Example 1

RECREATION Lions fans attending a recent 3-game series were asked whether the team should have a mascot. The table shows how many fans thought it should.

Game	Attendance	In favor of mascot
Friday	681	388
Saturday	527	428
Sunday	928	786

According to these results, what is the probability that a Lions fan wants the team to have a mascot?

Solution

Use the experimental probability formula.

$$P(E) = \frac{\text{number of favorable observations of } E}{\text{total number of observations}}$$

$$P(\text{fan favoring mascot}) = \frac{1{,}602}{2{,}136} = 0.75$$

The probability that a fan interviewed at the next Lions game will favor having a team mascot is 0.75.

Extend the Lesson

ONGOING ASSESSMENT Have students solve the following item. If Sarah won 160 tennis games and lost 200, find the probability that she will win her next game. The correct solution is $\frac{4}{9}$. Students will get the answer $\frac{4}{5}$ if they compare the wins to the losses instead of comparing the wins to the total games played.

The set of all possible outcomes of an experiment is called the **sample space**.

Example 2

SPORTS A baseball team has 8 pitchers and 3 catchers. The manager is choosing a pitcher-catcher combination. How many are possible?

Solution

One way to show all possible outcomes is to use ordered pairs. For example, use the letters A–C for the catchers and the numbers 1–8 for pitchers.

(A,1)	(A,2)	(A,3)	(A,4)	(A,5)	(A,6)	(A,7)	(A,8)
(B,1)	(B,2)	(B,3)	(B,4)	(B,5)	(B,6)	(B,7)	(B,8)
(C,1)	(C,2)	(C,3)	(C,4)	(C,5)	(C,6)	(C,7)	(C,8)

There are 24 possible pitcher-catcher combinations.

Problem Solving Tip

Probability is often expressed as a percent. A *percent* is a ratio of a number compared to 100. For example, 87% means 87:100 or $\frac{87}{100}$ or 0.87.

A decimal can be converted to a percent by moving the decimal two places to the right. So, 0.4 can also be written as 40%.

Another way to show the sample space is to use a **tree diagram**.

You can use probability to predict the number of times an event will occur.

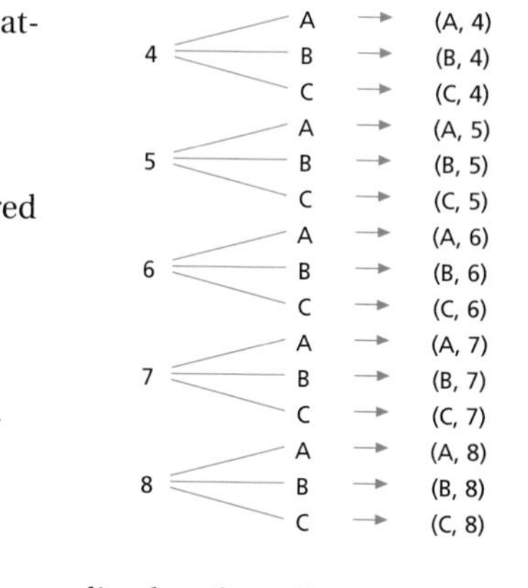

Example 3

SPORTS A softball player has had 24 hits in her first 60 times at bat. Predict her total hits in 330 at-bats.

Solution

First, use the outcomes that have already occurred to find the probability of the player getting a hit each time at bat.

$$P(\text{hit}) = \frac{24}{60} = 0.4$$

Then multiply that result by the total number of times at bat.

$$0.4(330) = 132$$

Based on the player's first 60 times at bat, you can predict that she will get 132 hits in 330 at-bats.

As you increase the number of trials in a probability experiment, the experimental probability will probably get closer to the **theoretical probability**. For example, when tossing a fair coin, $P(\text{heads}) = \frac{1}{2}$. The more often you toss the coin, the closer you will come to tossing an equal number of heads and tails.

Teaching Strategies

COOPERATIVE LEARNING Have small groups of students discuss the following. Pete and Leda toss a coin to see who bowls first. Leda has lost five times in a row. Pete says, "Don't worry. According to the law of probability, you can't possibly guess wrong six times in a row." Do you agree or disagree with Pete's statement? Why? What is the probability that Leda will win the next toss? Coins don't know who wins or loses; the probability of winning any one toss is always $\frac{1}{2}$ even after losing 100 times in a row.

Chalkboard Examples

Supplementary Example 1
Out of 1,100 people, 836 played a sport. What is the probability that a person picked at random from the group will play a sport? 0.76

Supplementary Example 2
The baseball team has 3 shortstops and 5 second basemen. How many shortstop-second baseman combinations are possible? 15

Supplementary Example 3
A softball player was at bat 93 times and made 31 hits. Predict how many hits he will make in 513 times at bat. 171

Supplementary Example 4
What is the probability of drawing a card from a standard deck that is a 5, 6, or 7? $\frac{3}{13}$

Reteaching Worksheet 9-1

Name ______________________ Date ____________

RETEACHING 9-1

REVIEW PERCENTS AND PROBABILITY

A knowledge of probability often plays a role in decision-making where an element of chance is involved, such as making weather predictions. You can find the probability of an event using the formula below. The result, a number between 0 and 1, can be written as a fraction, decimal or percent.

$$P(E) = \frac{\text{number of favorable outcomes}}{\text{number of possible outcomes}}$$

Example 1

Suppose you flip 2 pennies. What is the probability that the flip will show 1 head and 1 tail?

Solution

Make a list to find the number of possible outcomes: HH, HT, TH, TT.
There are 4 possible ourcomes.
There are 2 favorable outcomes, HT, TH.
P (one tail, one head) $= \frac{2}{4}$ or $\frac{1}{2}$, or 0.5, or 50%

Example 2

Suppose you decide to roll a die 30 times. Predict how many times you will roll a number less than 3.

Solution

Step 1: Find the probability.
There are 6 possible outcomes.
There are 2 favorable outcomes.
P(number less than 3) $= \frac{2}{6}$ or $\frac{1}{3}$
Step 2: Multiply the total rolls by the probability: $\frac{1}{3} \cdot 30 = 10$
You can predict you will roll a number less than 3 about 10 times in 30 rolls.

EXERCISES

Use the cards at the right. Then give the probability of drawing at random each of the following cards. C I N C I N N A T I

1. $P(N)$ $\frac{3}{10}$, 0.3, 30%
2. $P(C)$ $\frac{1}{5}$, 0.2, 20%
3. $P(I)$ $\frac{3}{10}$, 0.3, 30%

Suppose you flip 3 coins. What is the probability that the flip will show:

4. 3 heads? $\frac{1}{8}$, 0.125, 12.5%
5. 2 tails, 1 head? $\frac{3}{8}$, 0.375, 37.5%

Suppose you roll a die 60 times. Predict how many times you will roll each of the following.

6. a 4 10 times
7. a number greater than 4 20 times
8. a number less than 4 30 times
9. a number other than 4 50 times
10. an odd number 30 times
11. a number between 3 and 4 0 times

132 RETEACHING *LESSON 9-1*

Lesson Wrap-up

Quick Assessment

Ask the following questions to determine if students understand the content presented in this lesson.

1. Name four different ways you can express a probability. ratio, fraction, decimal, percent
2. In a probability experiment, the theoretical and experimental probabilities are far apart. Under what circumstances might this occur? when there are few trials of an experiment

Assignment Guide

Basic: 1–20, 25–28
Advanced: 1–28
Additional Practice: See Extra Practice Index on page 674.

Additional Answers

25.

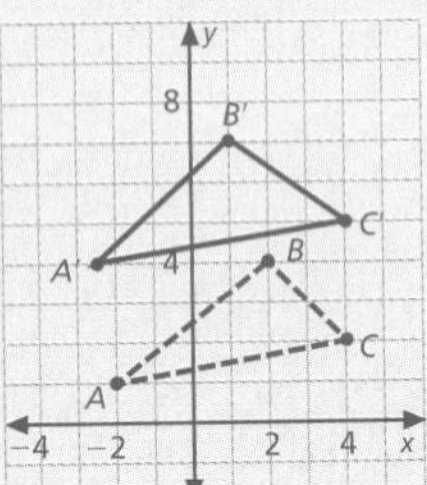

26.

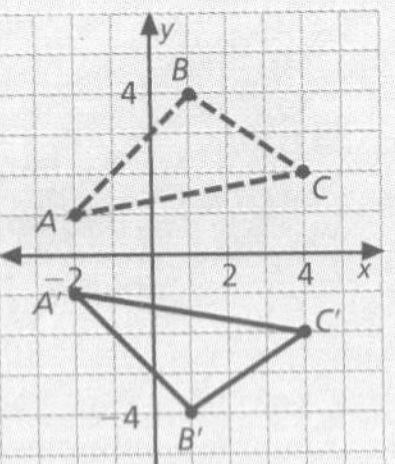

27.

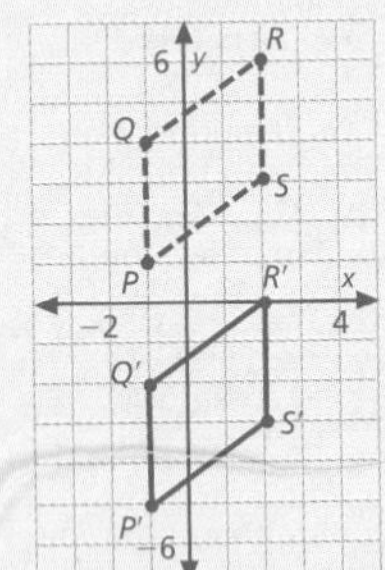

28.

The theoretical probability of an event, $P(E)$, is the ratio of the number of favorable outcomes to the number of possible outcomes in the sample space.

$$P(E) = \frac{\text{number of favorable outcomes}}{\text{number of possible outcomes}}$$

Example 4

CARD GAMES A card is picked at random from a standard deck of 52 cards. Find P(picture card).

Solution

There are 52 possible outcomes. There are 12 favorable outcomes—4 jacks, 4 queens, and 4 kings.

$P(\text{picture card}) = \frac{12}{52} = \frac{3}{13}$

> **Reading Math**
>
> **Odds** are a way of measuring success and failure. The odds in favor of an event are the ratio of the favorable outcomes to the unfavorable outcomes. For example, when you roll a die, the odds of getting a 4 are 1:5—because there is 1 way the event can occur, and 5 ways it cannot.

Try These Exercises

1. **SPORTS** Of the first 1,500 fans to pass through the turnstiles at the stadium, 1,050 had reserved seats. What is the probability that the next person through will have a reserved seat? 0.70

2. **WRITING MATH** A person flips a penny, a nickel, and a dime. Each coin can land with heads up (H) or tails up (T). Make a tree diagram to show what different outcomes are possible. (H, H, H), (H, H, T), (H, T, H), (T, H, H), (H, T, T), (T, H, T), (T, T, H), (T, T, T)
3. You roll a die 60 times. Predict the number of times you will roll an even number greater than 2. 20 times
4. **GAMES** A spinner for a game is divided into ten equal sections numbered 1 through 10. What is the probability of spinning 7 or higher? $\frac{2}{5}$

Practice Exercises

A die is rolled 100 times with the following results.

Outcome	1	2	3	4	5	6
Frequency	15	18	22	9	16	20

What is the experimental probability of rolling each of the following results?

5. a 2 0.18
6. a 6 0.2
7. a number less than 4 0.55
8. What is the theoretical probability of rolling a number less than 4? 0.5
9. **CHAPTER INVESTIGATION** Working with a partner, prepare to make your own baseball simulation game. To begin, gather batting statistics on at least 18 players. You may use statistics from the most recent baseball season or statistics from prior years. For each player, you will need to know the number At Bats (AB), Hits (H), Doubles (2B), Triples (3B), Walks (BB) and Strike Outs (SO). This information is available in the newspaper, in sports magazines or on team websites. Answers will vary.

Teaching Strategies

CHALLENGE A spinner is numbered from 1 to 8. Find the odds and the probability of a favorable outcome for each of the exercises below.

1. 3? 1 to 7; $\frac{1}{8}$ or 0.125
2. 1, 2, or 6? 3 to 5; $\frac{3}{8}$ or 0.375
3. an odd number? 4 to 4; $\frac{1}{2}$ or 0.5

386 Chapter 9 **Probability and Statistics**

List all the elements of the sample space for each of the following experiments.

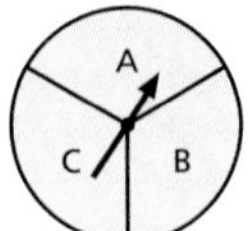

10. You flip a dime and a quarter. (H, T), (H, H), (T, H), (T, T)
11. You spin each of these spinners once. (A, 1), (A, 2), (A, 3), (B, 1), (B, 2), (B, 3), (C, 1), (C, 2), (C, 3)

Find the probability of each of the following:

12. **CARD GAMES** Drawing a black jack from a standard deck of cards. $\frac{1}{26}$
13. Rolling a die and getting a multiple of 3. $\frac{1}{3}$
14. **EDUCATION** Guessing correctly on a true-false question on a test. 0.5
15. **EDUCATION** Guessing incorrectly on a multiple-choice question with four choices. 0.75
16. Reaching into a drawer without looking and taking out a pair of black socks, when the drawer contains 3 pairs of black socks, 2 pairs of white socks, 1 pair of red socks, and 2 pairs of blue socks. 0.375
17. **SPORTS** A basketball player has made 96 free-throws in his last 128 attempts. What is the probability he will be successful in his next attempt? How many successful free-throws do you predict this player will make in 500 attempts? 0.75; 375

18. **WRITING MATH** You want to predict how many students in your school are right-handed. Describe how you would do it. Sample answer: Find the number of right-handed students in your class, express the number as a fraction or decimal, then multiply by the total number of students in your school.
19. **WEATHER** The weather forecaster predicts a 25% chance of rain in your area tomorrow. Describe what this forecast means. Of past days when weather conditions were similar to those predicted for tomorrow, it rained 25% of the time.
20. **TRANSPORTATION** A bus breaks down while traveling between two cities that are 200 miles apart. What is the probability the breakdown is within 25 miles of either city? 0.25

Extended Practice Exercises

Write whether each of the following probabilities can be determined experimentally or theoretically.

21. The probability that Player A will win when Player A plays Player B in tennis. experimentally
22. The probability that a person will win the state's lottery. theoretically
23. The probability that a family with 4 children has all boys. theoretically
24. The probability that it will snow on January 9. experimentally

Mixed Review Exercises

Graph the image of triangle *ABC* with vertices at $A(-2, 1)$ $B(4, 2)$, and $C(1, 4)$, under each transformation from the original position. (Lesson 8-11) See additional answers.

25. 3 units up
26. reflected across the x-axis

Graph the image of parallelogram *PQRS* with vertices at $P(-1, 1)$ $Q(2, 3)$, $R(2, 6)$, and $S(-1, 4)$, under each transformation from the original position. (Lesson 8-11) See additional answers.

27. 6 units down
28. reflected across the x-axis

Extend the Lesson

REAL-WORLD CONNECTION Have students write a probability problem based on statistics from the sports section of the newspaper. Then have students exchange problems and solve. You may wish to have students work in pairs or small groups and use the sports section in class to solve other students' problems. In this case, do not allow students to include all the necessary information in the problem. Instead, groups must search for at least one item of information in order to solve the problem.

Extra Practice Worksheet 9-1

Name ________ Date ________

EXTRA PRACTICE 9-1

REVIEW PERCENTS AND PROBABILITY

EXERCISES

A die is rolled 50 times with the following results.

Outcome	1	2	3	4	5	6
Frequency	8	9	10	5	10	8

What is the experimental probability of rolling each of the following?

1. a 4 $\frac{1}{10}$
2. a 1 $\frac{4}{25}$
3. a number greater than 3 $\frac{23}{50}$
4. an even number $\frac{11}{25}$
5. a number less than 5 $\frac{16}{25}$
6. an odd number $\frac{14}{25}$

List all the elements of the sample space for each of the following experiments.

7. You roll a die and toss a penny. (1, H), (2, H), (3, H), (4, H), (5, H), (6, H), (1, T), (2, T), (3, T), (4, T), (5, T), (6, T),
8. You spin a spinner with four equal sections labeled 1, 2, 3, and 4 and toss a dime. (1, H), (2, H), (3, H), (4, H), (1, T), (2, T), (3, T), (4, T),

Find the probability of each of the following.

9. Drawing a red card from a standard deck of cards. $\frac{1}{2}$
10. Tossing 3 coins and getting 2 heads and 1 tail. $\frac{3}{8}$
11. Rolling a die and getting a 1 or a 2 $\frac{1}{3}$
12. Reaching into a box without looking and taking out one red pencil, when the box contains 6 green pencils, 8 red pencils, and 5 blue pencils. $\frac{1}{19}$
13. Out of 450 people watching a parade, 200 were children. What is the probability that a person chosen at random out of this group will not be a child? $\frac{4}{9}$

116 EXTRA PRACTICE *LESSON 9-1*

Enrichment Worksheet 9-1

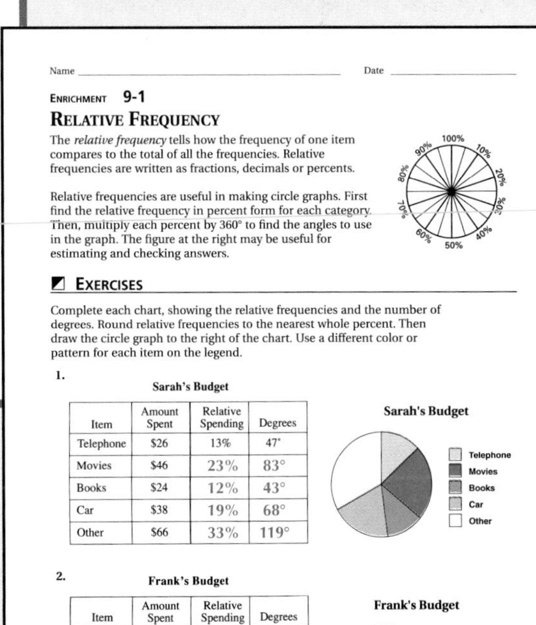

Name ________ Date ________

ENRICHMENT 9-1

RELATIVE FREQUENCY

The *relative frequency* tells how the frequency of one item compares to the total of all the frequencies. Relative frequencies are written as fractions, decimals or percents.

Relative frequencies are useful in making circle graphs. First find the relative frequency in percent form for each category. Then, multiply each percent by 360° to find the angles to use in the graph. The figure at the right may be useful for estimating and checking answers.

EXERCISES

Complete each chart, showing the relative frequencies and the number of degrees. Round relative frequencies to the nearest whole percent. Then draw the circle graph to the right of the chart. Use a different color or pattern for each item on the legend.

1. **Sarah's Budget**

Item	Amount Spent	Relative Spending	Degrees
Telephone	$26	13%	47°
Movies	$46	23%	83°
Books	$24	12%	43°
Car	$38	19%	68°
Other	$66	33%	119°

2. **Frank's Budget**

Item	Amount Spent	Relative Spending	Degrees
Insurance	$800	16%	58°
Car	$1600	32%	115°
Mortgage	$1800	36%	130°
Utilities	$600	12%	43°
Other	$200	4%	14°

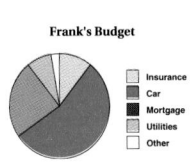

152 ENRICHMENT *LESSON 9-1*

Lesson Planning

NCTM Standards/Strands

- Patterns, Functions, & Algebra
- Data Analysis, Statistics, & Probability
- Problem Solving
- Reasoning and Proof
- Communication
- Connections
- Representation

Vocabulary

simulations   random numbers

Materials Needed

blank spinners   calculators
coins   paper bags
computers

Lesson Resources

Warm-Up Transparency 25
Reteaching 9-2

Assignment Guide

Basic: 1–25
Enriched: 1–25

Getting Started

5-Minute Warm-up

Write each as a fraction in lowest terms and as a decimal.

1. $\frac{17}{85}$ $\frac{1}{5}$; 0.2   **2.** $\frac{36}{50}$ $\frac{18}{25}$; 0.72

3. $\frac{72}{100}$ $\frac{18}{25}$; 0.72

Use the Five-step Plan and the Strategy

THE FIVE-STEP PLAN **Read**—ask questions of the students to help them understand the problem. **Plan**—guide students to related problems and previously mastered skills and strategies. **Solve**—students solve problem on their own. **Answer**—write solution in format that answers the question. **Check**—review work, check for reasonableness and review strategy used. Students will benefit from the experience of verbalizing their methods.

9-2 Problem Solving Skills: Simulations

Sometimes a probability problem is too difficult to solve theoretically or experimentally. One way to solve such a problem is to model it with a **simulation** to estimate the probability. Simulations rely on **random numbers**; these can be readily generated and recorded by a computer. You can also find random numbers by rolling dice, flipping coins, using numbered slips of paper, or spinning a spinner.

Problem Solving Strategies

- Guess and check
- Look for a pattern
- Solve a simpler problem
- Make a table, chart or list
- Use a picture, diagram or model
- ✔ Act it out
- Work backwards
- Eliminate possibilities
- Use an equation or formula

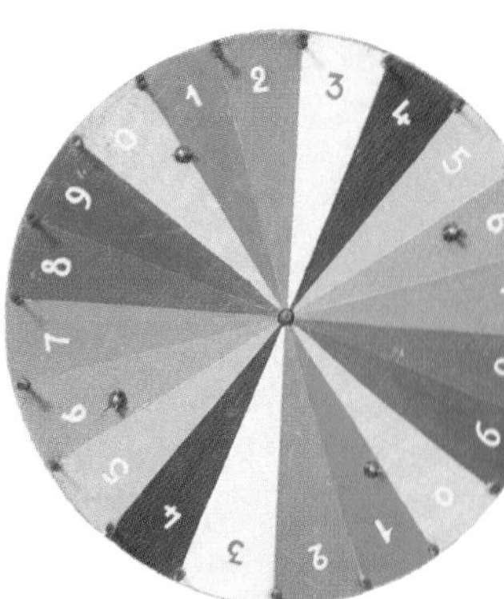

Problem

MARKETING Each box of Batter-Up Pancake Mix contains one of 5 different classic baseball cards. Assuming that the company has evenly distributed the cards among the boxes, what is the probability that you will find all 5 cards if you buy 10 boxes of Batter-Up?

Solve the Problem

Work with a partner. Use 5 slips of paper numbered 1–5; each slip represents a box of cereal. Place the slips in a paper bag. Then draw one slip of paper from the bag, record its number, and place it back in the bag. Repeat the process until you have drawn and recorded 10 slips of paper. If you have drawn each of the 5 numbers at least once, consider the outcome of your experiment to be successful. If you have not drawn every number, the outcome is unsuccessful.

Repeat the experiment 50 times, recording all results in a table. Indicate which trials are successful. Then write a ratio comparing successful outcomes to the total number of outcomes. This ratio will be an estimate of the probability of getting every card in the set when you buy 10 boxes of cereal.

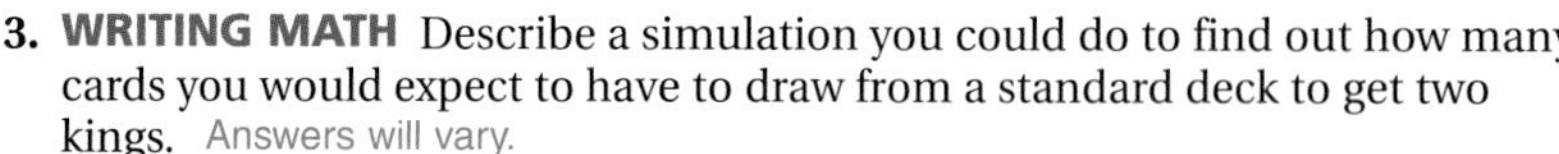

Try These Exercises

1. **COMPUTER SCIENCE** A computer generates a list of random 2-digit numbers. Zero cannot be the first digit. What probability would you expect for a number in the list to contain the digit "1"? 1:5 (18 of the 90 possible numbers)
2. **MARKETING** A candy company has placed 6 different prizes in its boxes. The prizes are uniformly distributed among the boxes of candy, only one per box. Describe a simulation you could do to estimate the probability of getting all 6 prizes in a 12-pack of candy. Answers will vary.
3. **WRITING MATH** Describe a simulation you could do to find out how many cards you would expect to have to draw from a standard deck to get two kings. Answers will vary.
4. **TEST TAKING** Suppose you are going to take a 10-question true-false test on the evolution of idiomatic phrases in Sri Lanka. You will need to guess each time, and you want to find out your chances of scoring 65% correct or better. Design a simulation to determine your chances. *Hint:* Use coin flipping. Answers will vary.

Teaching Strategies

Encourage students to find different ways to solve Exercise 4. They could record a number of trials selecting cards from a standard deck until they get two kings and then find the average. They could have a computer generate sequences of random numbers from one to 52, assign certain numbers to stand for kings, count how many numbers are used until two kings appear, repeat the experiment and average.

Practice Exercises

Five-step Plan
1 Read
2 Plan
3 Solve
4 Answer
5 Check

5. A family wants to have 3 children. Do the following simulation to determine the probability that, if they do have 3 children, all 3 will be the same sex.
   a. Use 3 coins. Let heads = girl and tails = boy. Toss the coins and record the results. Repeat the coin tosses until you have recorded 50 sets of 3 tosses each.
   b. Count the successful outcomes—those with either 3 heads or 3 tails.
   c. Write a ratio comparing successful outcomes with total outcomes. What is your experimental probability of having 3 children, all of whom are of the same sex? Answers will vary.

6. **SPORTS** One baseball player always arrives at the stadium between 5:30 PM and 6:30 PM for a night game. If batting practice always starts between 6:00 PM and 7:00 PM, what is the probability on any given night that this player will arrive before batting practice begins? Design and do a simulation to find out. Answers will vary.

7. **PROGRAMMING** A pitcher throws strikes about 60% of the time. If he throws 80 pitches, how many might be strikes? The following computer programming statements can be used to simulate 80 pitches.

```
10 S = 0                  The experiment begins with no successes.
20 FOR I = 1 TO 80        80 pitches
30 X = RND(1)             Generate a random decimal.
40 IF X < .6 THEN S = S + 1   If the decimal < 0.6, increase S by 1.
50 NEXT I                 Simulate the next pitch.
60 PRINT S                Total number of strikes.
70 END
```

   Use the program to simulate the problem. Then describe how you would adjust the program for a pitcher who throws strikes 40% of the time.
   Change line 40 to: "IF $x < .4$ THEN $S = S + 1$"

8. **TALK ABOUT IT** Petra is designing a simulation to determine the chance of guessing the correct answer on a multiple-choice test. Each item on the test has three choices. Petra plans to roll a 6-sided die to simulate random guesses. A roll of 1 or 2 will indicate choice A; a roll of 3 or 4, choice B; and a roll of 5 or 6, choice C. Will Petra's simulation work? Explain.
   Yes; the six-sided die can be used as long as each of the three choices is assigned an equal number of outcomes from the die roll.

Mixed Review Exercises

Find the slope of each line using the given information. (Lesson 6-1)

9. $A(-2, -1), B(5, 3)$ $\frac{4}{7}$
10. $C(7, 2), D(3, -2)$ 1
11. $E(1, 8), F(-3, -4)$ 3
12. $-3x + 2y = 9$ $\frac{3}{2}$
13. $4y + 2x = 8$ $-\frac{1}{2}$
14. $-12 + x = 4y$ $\frac{1}{4}$
15. $G(-3, 5), H(-3, 9)$ undefined
16. $I(-3, 5), J(3, -5)$ $-\frac{5}{3}$
17. $K(2, -1), L(-8, -1)$ 0

Solve each proportion. Find the value of x. (Lesson 7-1)

18. $\frac{5}{x} = \frac{15}{12}$ 4
19. $\frac{9}{12} = \frac{x}{20}$ 15
20. $\frac{4}{13} = \frac{16}{x}$ 52
21. $\frac{7}{22} = \frac{x}{55}$ 17.5
22. $\frac{14}{25} = \frac{x+3}{10}$ 2.6
23. $\frac{x+1}{16} = \frac{x}{9}$ $\frac{9}{7}$
24. $\frac{x+3}{12} = \frac{4}{8}$ 3
25. $\frac{16}{x+2} = \frac{8}{2-x}$ $\frac{2}{3}$

THE STRATEGY Act it out—A simulation is one way to act out what would probably happen in a given situation. Acting it out is a way to find experimental probability.

Chalkboard Examples

Supplementary Problem

Design and simulate an experiment to find the probability that a person was born on a Monday. Possible answer: Use a 1-7 spinner. Let 2 represent Monday, and spin to represent the day on which each person was born. Divide the total number of spins by the number of 2s. Results will vary.

Lesson Wrap-up

Quick Assessment

Give one example of a random event and one example of an event that is not random. random: arrow pointing to 3 on a spinner or minute hand point to 3 on a clock; not random: lining up by height or alphabetizing a list of words

Reteaching Worksheet 9-2

Name ______________________ Date ____________

RETEACHING 9-2

PROBLEM SOLVING SKILLS: SIMULATIONS

One way to solve a complex probability problem is to model it with a simulation.

Example

1. **Read** the problem carefully. — Suppose you are going to take a five-question true-false test. You want to determine the probability that you will answer at least 3 of the 5 questions correctly if you guess at each answer.

Solution

2. **Plan** what you will do. — Use a coin. Let heads stand for a correct answer and tails stand for an incorrect answer. Toss the coin 5 times for each trial and record each outcome.

3. **Solve** by carrying out the plan.

	Trial 1	Trial 2	Trial 3	Trial 4	Trial 5
Number of heads	2	4	2	3	0
Number of tails	3	1	3	2	5

4. **Answer** the question asked. — On the trials, heads showed on 3 or more coins 2 times. Based on these results, the probability of answering 3 of the 5 questions correctly is given by
   $P(3 \text{ or more correct}) = \frac{2}{5}$ ←Number of favorable outcomes / ←Number of possible outcomes

5. **Check** to see if your answer is reasonable. — Review the design of your simulation to see that you've tested what was asked for. Determine if the number of trials is sufficient.

EXERCISES Answers may vary for Exercises 1-3; check student's simulations.

Use a simulation to solve these problems. Record your results on another sheet of paper.

1. Continue the simulation in the example for 10 additional trials. Is the probability the same as was given in the example? Explain.

2. Each question of a multiple-choice test has 4 responses. If you guess at the answers to a ten-question test, what is the probability of answering 6 questions correctly?

3. A breeder has 6 dogs. Design a simulation to determine the probability of matching 3 dogs with their correct names.

134 RETEACHING *LESSON 9-2*

Teaching Strategies

Some students may have difficulty grasping the idea of simulating situations. Make sure they understand that it means to act out a similar situation. Give these students simpler problems to help them understand this concept. For example, show how tossing a coin can be used to simulate guessing answers on a true-false test or whether a boy or a girl is born.

Lesson 9-2 **Problem Solving Skills: Simulations** 389

Skills Practice

Vocabulary Review

Lesson 9-1
probability experiment
outcome
experimental probability
sample space tree diagram
theoretical probability

Lesson 9-2
simulations random numbers

Assignment Guide

All students: 1–40
Additional Practice: Refer to the Extra Practice Index on page 674 of this text.

Review and Practice Your Skills

Practice ■ Lesson 9-1

A card is picked at random from a standard deck of 52 cards. Find each theoretical probability.

1. $P(heart)$ $\frac{1}{4}$
2. $P(jack)$ $\frac{1}{13}$
3. $P(red\ card)$ $\frac{1}{2}$
4. $P(black\ two)$ $\frac{1}{26}$
5. $P(2\ or\ 3)$ $\frac{2}{13}$
6. $P(7\ of\ hearts)$ $\frac{1}{52}$
7. $P(3 \leq card \leq 8)$ $\frac{6}{13}$
8. $P(king\ of\ clubs)$ $\frac{1}{52}$

You flip a coin four times. Find each theoretical probability.

9. $P(no\ tails)$ $\frac{1}{16}$
10. $P(exactly\ one\ head)$ $\frac{1}{4}$
11. $P(2\ tails, 2\ heads)$ $\frac{3}{8}$
12. $P(4\ tails)$ $\frac{1}{16}$
13. $P(> 2\ heads)$ $\frac{5}{16}$
14. $P(0\ or\ 1\ head)$ $\frac{5}{16}$
15. $P(3\ tails)$ $\frac{1}{16}$
16. $P(1,2,3\ or\ 4\ heads)$ $\frac{15}{16}$

You roll a pair of dice and calculate the sum. Find each theoretical probability.

17. $P(7)$ $\frac{1}{6}$
18. $P(11)$ $\frac{1}{18}$
19. $P(even)$ $\frac{1}{2}$
20. $P(1)$ 0
21. $P(12)$ $\frac{1}{36}$
22. $P(4\text{ or }5)$ $\frac{2}{9}$
23. $P(6)$ $\frac{5}{36}$
24. $P(< 6)$ $\frac{5}{18}$
25. $P(10, 11, \text{ or } 12)$ $\frac{1}{6}$
26. $P(9)$ $\frac{1}{9}$
27. $P(< 11)$ $\frac{33}{36}$
28. $P(2)$ $\frac{1}{36}$

29. A basketball player has made 48 free throws in her last 72 attempts. What is the probability she will be successful on her next attempt? How many free throws do you predict she will make in 600 attempts? $\frac{2}{3}$, 400
30. A car breaks down while traveling between two cities that are 300 mi apart. What is the probability that the breakdown is within 18 mi of either city? $\frac{3}{25} = 0.12$

Practice ■ Lesson 9-2

31. A computer generates a list of random 3-digit numbers. Zero cannot be the first digit. What probability would you expect for a number in the list to contain the digit "4"? $\frac{61}{225}\left(\frac{244}{900}\right)$
32. Each box of Toasted Crunchies cereal contains a single prize. There are 4 different prizes uniformly distributed among all boxes of cereal at the production facility. Describe a simulation you could do to estimate the probability of getting all 4 prizes if you buy 10 boxes of Toasted Crunchies? Answers will vary.
33. Perform and document a simulation to determine the probability that a family has 2 children will have 1 boy and 1 girl. Answers will vary.
34. Perform and document a simulation to find the chances of scoring 50% or higher on a 5 question multiple choice test. Each question has four choices, and you guess on each question. Answers will vary.
35. Agnes walks her dog each night outside the grounds of Tellco Corporation between 7:30 and 8:00 P.M. First shift workers leave the grounds between 7:30 and 8:30 P.M. each night. Describe a simulation using two spinners that would calculate the probability of Agnes seeing first-shift workers leaving the grounds during her walk. Answers will vary.
36. Describe a simulation you could do to find out how many cards you would expect to have to draw from a standard deck to get a pair of hearts. Answers will vary.

Extend the Lesson

MATH JOURNAL Have students discuss the differences and similarities between experimental and theoretical probabilities. Students may say experimental probabilities are found using data from real-life situations while theoretical probabilities use sample spaces. Both are found by calculating the ratio of favorable outcomes to the number of total outcomes.

PRACTICE ■ LESSON 9-1–LESSON 9-2

List all the elements of the sample space for each experiment. (Lesson 9-1)

37. Tossing a quarter and a nickel.
(H, H) (T, H) (H, T) (T, T)

38. Spinning each of these two spinners:

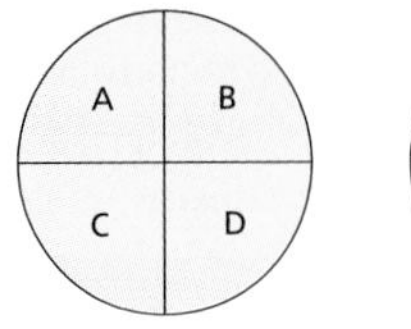

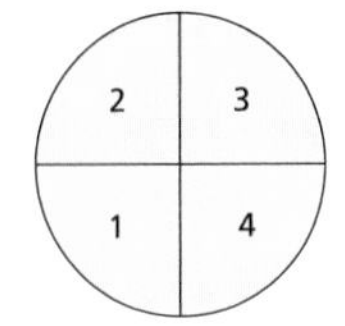

{(A, 1) (B, 1) (C, 1) (D, 1) (A, 2) (B, 2) (C, 2) (D, 2) (A, 3) (B, 3) (C, 3) (D, 3) (A, 4) (B, 4) (C, 4) (D, 4)}

39. Rolling a die and tossing a dime.
(1, H) (4, H) (1, T) (4, T) (2, H) (5, H) (2, T) (5, T) (3, H) (6, H) (3, T) (6, T)

40. A computer generates a list of random 2-digit numbers. Zero cannot be the first digit. What probability would you expect for a number in the list to be a multiple of 3? (Lesson 9-2) $\frac{1}{3}$

Career – Dietitian

MathWorks
Workplace Knowhow

Dietetics has applications in many different career fields. Clinical dietitians plan meals and nutritional plans for groups such as schools and hospitals. Community dietitians inform the public on nutritional habits to prevent disease and promote healthy lifestyles. Consultant dietitians provide advice in the areas of sanitation and safety procedures. In the sports world, nutrition is important for maintaining muscle health and bone strength.

As the dietitian for a baseball team, you need to determine whether the team members are getting enough calcium in their diets. To find out, you separate the players into three categories: infielders, outfielders, and pitchers. During a buffet, you chart the food selections of the players.

1. Find the average amount of calcium consumed by players in each group.

A. Infielders	B. Outfielders	C. Pitchers
1B–300 mg	RF–113 mg	P1–233 mg
C–220 mg	CF–197 mg	P2–212 mg
2B–186 mg	LF–262 mg	P3–184 mg
3B–216 mg	224 mg	P4–258 mg
SS–102 mg		310.75 mg
244.6 mg		

2. Your research shows that during lunch, calcium intake is $\frac{3}{4}$ of the amount consumed during the buffet. Breakfast amounts are $\frac{3}{5}$ of the buffet amount. How many milligrams of calcium is each player getting per day?
1B, 705 mg; C, 517 mg; 2B, 437.1 mg; 3B, 507.6 mg; SS, 239.7 mg; RF, 265.55 mg; CF, 462.95 mg; LF, 615.7 mg; P1, 547.55 mg; P2, 498.2 mg; P3, 432.4 mg; P4, 606.3 mg

3. If 450 mg of calcium is recommended per player per day, which players are consuming too little calcium? 2B, SS, RF, P3

Chalkboard Examples

Example from Lesson 9-1

Find the probability of rolling three standard 6-sided dice and getting three fives. $\frac{1}{216}$

Example from Lesson 9-2

A baseball card manufacturer inserts four special cards randomly in packs of baseball cards. Each pack of cards has two of the special cards. (A pack may contain two of the same special card.) Describe a simulation that could be used to determine how many packs you would expect to buy to get all four special cards. Answers will vary.

Career Opportunity

Describe the kind of work dietitians do and the types of companies that hire these workers. Have students discuss the importance of measurement and algebraic thinking in doing this type of job. Explain that dietitians must be able to follow situations and create menus based on mathematical analysis and scientific principles. Students should answer Questions 1–3 to better understand how dietitians use mathematics in performing their job.

Students who are interested in learning more about this profession can go to learningmathmatters.com. Guidance Counselors should have information about school and training requirements.

Teaching Strategies

COOPERATIVE LEARNING Have students work together to think of three situations that could be simulated. Have them choose one situation and conduct the simulation, record results and explain their conclusions.

Lesson Planning

NCTM Standards/Strands

- Number & Operation
- Patterns, Functions, & Algebra
- Data Analysis, Statistics, & Probability
- Problem Solving
- Reasoning and Proof
- Communication
- Connections
- Representation

Vocabulary

compound event
mutually exclusive events
complement

Materials Needed

six-sided dice

Lesson Resources

Warm-Up Transparency 25
Reteaching 9-3
Extra Practice 9-3
Enrichment 9-3

Assignment Guide

Basic: 1–26, 31–42
Enriched: 1–42

Getting Started

5-Minute Warm-Up

Solve for *n*.

1. $n = \frac{2}{13} + \frac{8}{52}$ $\frac{4}{13}$
2. $n = \frac{5}{52} + \frac{23}{52} - \frac{16}{52}$ $\frac{3}{13}$

Introduction to Lesson 9-3

Have pairs of students share the results. Discuss whether the game seems to be fair. Point out that in a fair game, both players have an equal chance of winning. Allow students to suggest ways in which the game could be tested for fairness.

9-3 Compound Events

Goals ■ Find probabilities of compound events.
Applications Sports, Games, Business

Play this game with a partner. Use a pair of 6-sided dice.

1. Player A rolls first. If Player A rolls a 7, Player B wins the game. If not, Player B rolls.
2. If Player B rolls a 7 or an 11, Player A wins. If not, it is Player A's turn. Continue taking turns until there is a winner.
3. Play the game several times. Do you think one player has a better chance of winning? Answers will vary.

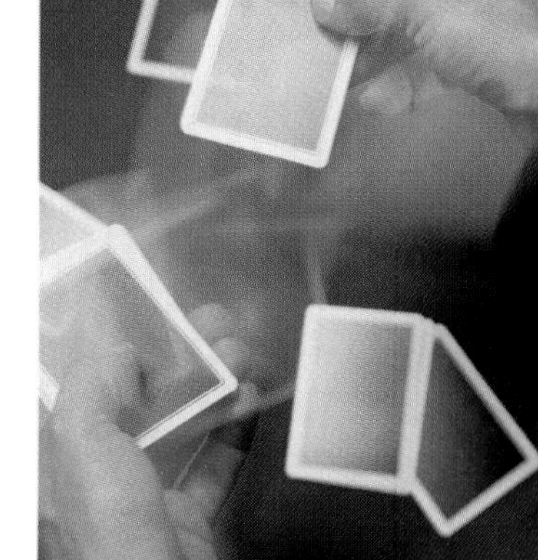

Build Understanding

A **compound event** consists of two or more simple events. Compound events may involve finding the probability of one event *and* another event occurring. Or, they may ask for the probability of one event *or* another event occurring. For example, when rolling a die, getting a number that is even (event A) *and* greater than 2 (event B) is written $P(A \text{ and } B)$. Getting a number that is even *or* greater than 2 is written $P(A \text{ or } B)$.

If two events *cannot* occur at the same time, they are called **mutually exclusive** events. For example, it is impossible to draw from a standard deck of playing cards a card that is both a heart and a club.

When two events *A* and *B* are mutually exclusive, the probability of *A* or *B* can be found using the formula $P(A \text{ or } B) = P(A) + P(B)$.

Example 1

SPORTS A standard deck of playing cards is used to simulate a baseball game. During the game, players draw a card at random. Spade number cards greater than 4 represent doubles. Home runs are represented by either a 3 or a queen.

a. Find *P*(spade and a number card greater than 4).

b. Find *P*(3 or queen).

Solution

There are 52 possible outcomes.

a. There are 6 outcomes in which spades are greater than 4: 5, 6, 7, 8, 9, and 10 of spades.

So, $P(\text{spade and number greater than 4}) = \frac{6}{52}$ or $\frac{3}{26}$.

The probability that the card will be a spade and a number greater than 4 is $\frac{3}{26}$.

Teaching Strategies

You may wish to review Venn diagrams with students to help clarify how sets of events are mutually exclusive or not mutually exclusive. Show separate circles to depict events that are mutually exclusive and intersecting circles to depict events that are not mutually exclusive. Ask students to explain what the overlap in the circles shows. the elements that both events have in common

b. A card cannot be both a 3 and a queen at the same time, so the events are mutually exclusive.

$$P(3 \text{ or queen}) = P(3) + P(\text{queen}) = \frac{4}{52} + \frac{4}{52} = \frac{8}{52} \text{ or } \frac{2}{13}$$

The probability that the card will be a 3 or a queen is $\frac{2}{13}$.

Events that can happen at the same time are *not* mutually exclusive.

Example 2

GAMES You draw a card at random from a standard deck of playing cards. Find the probability that the card is a club or a jack.

Solution

These are not mutually exclusive events, because a card can be both a club and a jack.

There are 13 clubs, so $P(\text{club}) = \frac{13}{52}$.

There are 4 jacks, so $P(\text{jack}) = \frac{4}{52}$.

However, 1 club is a jack. $P(\text{club and jack}) = \frac{1}{52}$.

$$P(\text{club or jack}) = \frac{13}{52} + \frac{4}{52} - \frac{1}{52} = \frac{16}{52} = \frac{4}{13}$$

The probability that the card is a club or a jack is $\frac{4}{13}$.

Example 2 illustrates that when two events A and B are not mutually exclusive, the probability of A or B can be found using the formula

$$P(A \text{ or } B) = P(A) + P(B) - P(A \text{ and } B)$$

Suppose you know the probability of event A. The set of outcomes in the sample space, but *not* in A, is called the **complement** of the event.

$$P(\text{not } A) = 1 - P(A)$$

Example 3

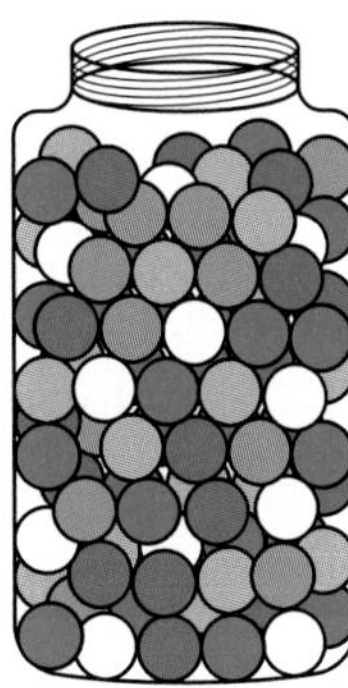

You select a marble from this jar without looking. You know $\frac{1}{5}$ of the marbles are red and $\frac{1}{5}$ are blue. What is the probability you will select *neither* the red *nor* the blue marble?

Solution

Find the probability of selecting red or blue.

$$P(\text{red or blue}) = \frac{1}{5} + \frac{1}{5} = \frac{2}{5}$$

Check Understanding

Classify each of the following pairs of events as *mutually exclusive* or *not mutually exclusive.*

1. a card drawn at random from a standard deck of cards is the 4 of clubs or the 4 of spades
2. two rolled dice show a sum of 8 or show different numbers
3. two tossed coins show two tails or two heads

1. mutually exclusive
2. not mutually exclusive
3. mutually exclusive

Chalkboard Examples

Supplementary Example 1

In another sports simulation, two six-sided dice are rolled. Find $P(\text{sum} < 5 \text{ and even})$. $\frac{1}{9}$

Supplementary Example 2

A card is drawn from a standard deck of 52 cards. Find the probability that the card is either a face card or a heart. $\frac{11}{26}$

Supplementary Example 3

The marbles in a jar are red, blue, yellow or green. You know that $\frac{1}{4}$ of the marbles are red and $\frac{1}{8}$ are green. What is the probability you will select neither a red or green marble? $\frac{5}{8}$

Teaching Strategies

ESL/LEP Students may have difficulty understanding the term mutually exclusive. Ask them questions such as the following to illustrate events that are mutually exclusive and those that are not.

"Can you be at home and climb a mountain?" no
"Can you be at home and read a book?" yes
"Can you wear glasses and have naturally black hair?" yes
"Can you have naturally black hair and naturally blond hair" no

Reteaching Worksheet 9-3

Name ______________________ Date ______________

RETEACHING 9-3

COMPOUND EVENTS

A *compound event* is a combination of two or more single events. Two such events that cannot occur at the same time, such as drawing a spade or a diamond from a standard deck of playing cards, are called *mutually exclusive events.* You can use this formula to find the probability of two mutually exclusive events, A and B.

$$P(A \text{ or } B) = P(A) + P(B)$$

To find the *complement* of the event, you can use the formula $P(\text{not } A) = 1 - P(A)$.

Example

Suppose a die is rolled. What is the probability of:

a. rolling a 3 or a 6?
b. *not* rolling a 3 or a 6?

Solution

a. Because this is a mutually exclusive event, you can use the formula to find the probability.
$P(3 \text{ or } 6) = P(3) + P(6) = \frac{1}{6} + \frac{1}{6} = \frac{2}{6} \text{ or } \frac{1}{3}$
The probability of rolling a 3 or 6 is $\frac{1}{3}$.

b. Use the formula to find the complement of $P(3 \text{ or } 6)$.
$P(\text{not } 3 \text{ or } 6) = 1 - P(3 \text{ or } 6) = 1 - \frac{1}{3} = \frac{2}{3}$
The probability of *not* rolling a 3 or 6 is $\frac{2}{3}$.

EXERCISES

Use the cards at the right. Then give the probability of drawing at random each of the following cards. P R O B A B I L I T Y

1. $P(R \text{ or } T)$ $\frac{2}{11}$
2. $P(B \text{ or } I)$ $\frac{4}{11}$
3. $P(B, I \text{ or } O)$ $\frac{5}{11}$

Suppose you roll a die. What is the probability of:

4. rolling a 2 or a 3? $\frac{2}{6}$ or $\frac{1}{3}$
5. *not* rolling a 2 or a 3? $\frac{4}{6}$ or $\frac{2}{3}$
6. rolling a 1, 3 or 5? $\frac{3}{6}$ or $\frac{1}{2}$
7. *not* rolling an odd number? $\frac{3}{6}$ or $\frac{1}{2}$

Suppose you roll two dice. Find the probability that the sum of the numbers if:

8. 5 or 12. $\frac{5}{36}$
9. *not* 5 or 12. $\frac{31}{36}$
10. 3, 6, or 9. $\frac{11}{36}$
11. *not* 3, 6, or 9. $\frac{25}{36}$

136 RETEACHING *LESSON* 9-3

Lesson Wrap-up

Quick Assessment

Ask the following questions to determine if students understand the content presented in this lesson.

1. In Example 2, why is it necessary to subtract the number of times that the events overlap? The events are not mutually exclusive.
2. Example 4 can be solved in two different ways. Describe each way. One way uses the event itself and the other way uses the complement of the event. Both are correct.

Assignment Guide

Basic: 1–26, 31–42
Advanced: 1–42
Additional Practice: See Extra Practice Index on page 674.

Additional Answers

6. (1,1), (1,2) (1,3) (1,4) (1,5) (1,6)
(2,1), (2,2) (2,3) (2,4) (2,5) (2,6)
(3,1), (3,2) (3,3) (3,4) (3,5) (3,6)
(4,1), (4,2) (4,3) (4,4) (4,5) (4,6)
(5,1), (5,2) (5,3) (5,4) (5,5) (5,6)
(6,1), (6,2) (6,3) (6,4) (6,5) (6,6)

Find the probability of *not* selecting red or blue.

$P(\text{not red or blue}) = 1 - P(\text{red or blue})$

$P(\text{not red or blue}) = 1 - \frac{2}{5}$

$P(\text{not red or blue}) = \frac{3}{5}$

The probability that you will not select the red or the blue marble is $\frac{3}{5}$.

Try These Exercises

1. A die is rolled. Find the probability of rolling a 1 or a 2. $\frac{1}{3}$
2. Two coins are tossed. Find the probability that the coins show two heads or one tail and one head. $\frac{3}{4}$
3. A card is drawn at random from a standard deck of playing cards. Find the probability that it is a 7 or a black card. $\frac{7}{13}$
4. Two dice are rolled. Find the probability that the sum of the numbers is not greater than 5. $\frac{5}{18}$
5. Each student in your class writes his or her full name on a piece of paper. The pieces are put in a box and one is chosen without looking. What is the probability that your name will not be chosen? $\frac{\text{number of pieces of paper} - 1}{\text{number of pieces of paper}}$

Practice Exercises

GAMES A player rolls two 6-sided dice.

6. List the sample space for the rolls. See additional answers.
7. Find the probability that the sum of the numbers rolled is odd and greater than 5. $\frac{1}{3}$
8. Find the probability that the sum of the numbers rolled is either 8 or 10. $\frac{2}{9}$
9. Find $P(\text{not even})$. $\frac{1}{2}$
10. Find $P(\text{not odd or sum of 6})$ $\frac{13}{36}$

SPORTS Suppose you are on a team in the midst of a horrible losing streak. Your manager decides to "shake up" the line-up. He chooses the batting order by putting nine players' names into a hat and pulling them out one by one. The player whose name is drawn first bats first, the second bats second, and so on.

11. What is the probability you will bat second or fourth? $\frac{2}{9}$
12. What is the probability you will bat fifth, or in the first third of the batting order? $\frac{4}{9}$
13. What is the probability you will bat first, or in the first third of the batting order? $\frac{1}{3}$
14. What is the probability you will bat in the last third of the batting order, or in an odd-numbered position? $\frac{2}{3}$
15. **BUSINESS** Ms. Garrett plans to select a worker at random for a special training seminar. If there are 14 workers in sales, 6 in accounting and 5 in personnel, what is the probability that the worker will be from either sales or accounting? $\frac{4}{5}$

Extend the Lesson

ONGOING ASSESSMENT Have students answer the following question. Then check their work for the common error shown below.
Two dice are rolled, and the numbers are added. Find the probability that the sum is greater than ten or even.
Students who get the answer $\frac{7}{12}$ may have followed this reasoning:
$P(>10 \text{ or even}) = P(>10) + P(\text{even}) = \frac{3}{36} + \frac{18}{36} = \frac{21}{36} = \frac{7}{12}$.
These students have neglected to subtract the overlap. The correct solution is $\frac{5}{9}$.

GAMES You spin the spinner shown. Find each probability.

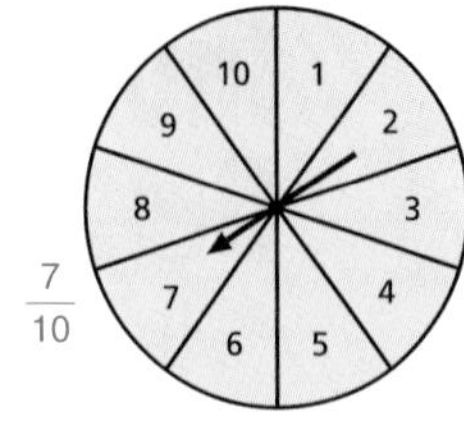

16. spinning a 2 or an odd number $\frac{3}{5}$

17. spinning an odd or an even number 1

18. spinning a multiple of 2 or a multiple of 3 $\frac{7}{10}$

CARD GAMES To begin a game, the dealer draws a card at random from a standard deck of playing cards.

19. Find the probability that the card is a 2, a 5, or a face card. $\frac{5}{13}$

20. Find the probability the card is a 7, an 8, or a red card. $\frac{15}{26}$

A card is drawn at random from a standard deck of playing cards. For each event, estimate whether the probability is closer to 1, $\frac{1}{2}$, or 0.

21. P(red or face card) $\frac{1}{2}$

22. P(2, 3, or 4) ∅

23. P(black and face card) 24

24. P(black, heart, or 7) 1

PHOTOGRAPHY A team photo album contains photos of the players by themselves, the coaching staff by themselves, and the players and the coaches together. The players are in 15 of the photos and the coaches are in 12 of the photos. In 6 photos, the players and coaches appear together.

25. How many photos are in the album? 21

26. If you open the album at random to one of the team photos, what is the probability that the photo shows only coaches? $\frac{2}{7}$

Extended Practice Exercises

27. **WRITING MATH** Why are multiples of 6 and multiples of 4 not mutually exclusive events? Sample Answer: Multiples of 12 are common to both events.

28. A pair of dice is rolled. What is the probability that the sum of the numbers is neither a 5 nor a multiple of 2? $\frac{7}{18}$

29. **SPORTS** Batting average is found by dividing hits by at-bats. In 1941, Ted Williams batted over .400 when he got 185 hits in 456 official at-bats for an average of .406. Suppose Ted Williams had 1 more at-bat in 1941. Based on his performance all season, what would you estimate the probability of his *not* getting a hit that time? 0.594

30. **CHAPTER INVESTIGATION** For each player you selected, the number of hits (H) is equal to the sum of the singles (S), doubles (2B), triples (3B), and home runs (HR). Since the number of singles is not usually reported as a separate statistic, calculate the number of singles (S) for each player using the formula: H − (2B + 3B + HR) = S. You may want to use a computer spreadsheet. Answers will vary.

Mixed Review Exercises

Evaluate each expression when $a = 5$ and $b = -4$. (Lesson 1-8)

31. a^2b^2 400

32. $a^2 + b^2$ 41

33. ab^3 −320

34. a^3b^2 2000

35. $a(a^2b^2)$ 2000

36. $ab \cdot ab^2$ −1600

37. a^3b −500

38. $(a^2 - b^2)^2$ 81

39. $(a^2)(b^2)(a^2)$ 10,000

40. $a^2 - b(ab)^2$ 1625

41. $-(b^2)$ −16

42. $(-b)^2$ 16

Extra Practice Worksheet 9-3

Name ____________ Date ____________

EXTRA PRACTICE 9-3

Compound Events

Exercises

Four coins are tossed.

1. Find the probability that two of the coins show heads and two show tails. $\frac{1}{4}$

2. Find the probability that only one coin shows a tail. $\frac{1}{4}$

3. Find the probability that all four coins show heads. $\frac{1}{16}$

4. Find P(all tails). $\frac{1}{16}$

5. Find P(1 head, 3 tails or all tails). $\frac{5}{16}$

You spin the spinner at the right.
Find each probability.

6. spinning a 3 or an even number $\frac{2}{3}$

7. spinning a prime number $\frac{1}{3}$

8. spinning a number greater than 1 $\frac{5}{6}$

9. spinning a number less than 4 $\frac{1}{2}$

A card is drawn at random from a standard deck of playing cards. For each event, estimate whether the probability is closer to 1, $\frac{1}{2}$ or 0.

10. P(a diamond or a spade) $\frac{1}{2}$

11. P(black and a 3) $\frac{1}{2}$

12. P(red and a number card) 1

13. P(face card or an ace) $\frac{1}{2}$

14. P(red, club, or a face card) 1

15. P(black and a face card) 1

118 EXTRA PRACTICE *LESSON 9-3*

Enrichment Worksheet 9-3

Name ____________ Date ____________

ENRICHMENT 9-3

The Perfect Cubes

Exercises

Two of these numbers are chosen at random by dropping a small button or other object on the grid. Find the probability that the two numbers are perfect cubes. If you need a hint, turn this page upside down. $\frac{1}{336}$

2401	49	144	3481	100	961	1156	1444
25	784	1681	81	3844	2809	121	2601
2116	1764	3600	529	729	289	225	1
1369	2916	361	3136	625	64	4	841
196	400	1936	9	2209	4096	1296	1024
3969	3025	16	2500	484	1849	676	324
1521	36	576	1225	3249	169	2704	3721
441	2025	1600	3364	1089	2304	900	256

▶ Hint

Put the numbers in the grid in order from least to greatest. What pattern is shown in the numbers?

156 ENRICHMENT *LESSON 9-3*

Teaching Strategies

CHALLENGE In one homeroom, 13 of the students are in the math club, 15 are in the drama club, and eight are in the art club. Three of the students are in all three clubs. What is the probability that a student is in the math or the drama club? $P(\text{math or drama}) = \frac{13}{30} + \frac{15}{30} - \frac{3}{30} = \frac{25}{30} = \frac{5}{6}$

Lesson Planning

NCTM Standards/Strands

- Number & Operation
- Patterns, Functions, & Algebra
- Data Analysis, Statistics, & Probability
- Problem Solving
- Reasoning and Proof
- Communication
- Connections
- Representation

Vocabulary

independent event
dependent event

Materials Needed

coins | standard dice

Lesson Resources

Warm-Up Transparency 26
Reteaching 9-4
Extra Practice 9-4
Enrichment 9-4

Assignment Guide

Basic: 1–25, 32–34
Enriched: 1–34

Getting Started

5-Minute Warm-up

Solve for *m*.

1. $m = \left(\frac{2}{3}\right)\left(\frac{3}{7}\right)$ $\frac{2}{7}$
2. $m = \left(\frac{1}{4}\right)\left(\frac{3}{13}\right)$ $\frac{3}{52}$
3. $m = \left(\frac{1}{2}\right)\left(\frac{2}{3}\right)\left(\frac{3}{10}\right)$ $\frac{1}{10}$

Introduction to Lesson 9-4

Compare results for each group, then average results for the whole class. Discuss why this average value may be a closer approximation to the theoretical probability.

9-4 Independent and Dependent Events

Goals ■ Find the probability of dependent and independent events.

Applications Sports, Surveys, Scheduling

Work with a partner. You will need a six-sided die and a coin.

1. Suppose one person rolls the die while the other tosses or flips the coin. What do you think the probability is of rolling a 3 and landing the coin heads up? Record your prediction.
2. Check your prediction by rolling the die and tossing the coin until you get both of these outcomes.
3. Share the results of your experiment with other groups.

Build Understanding

When the outcome of one event does not affect the outcome of another event, the events are said to be **independent**. To find the probability of both events occurring, multiply the probabilities of each event.

If A and B are independent events, then $P(A \text{ and } B) = P(A) \cdot P(B)$

To emphasize that A and B do not characterize a single event, sometimes $P(A$ and $B)$ is written $P(A$, then $B)$.

Example 1

A bag contains 3 white softballs, 2 yellow softballs, 3 green softballs, and 4 red softballs. You reach into the bag without looking and take out a ball. You replace it and then take out another ball at random. Find the probability that the first ball is red and the second ball is white.

Solution

Because the first ball is replaced before the second is taken, the sample space of 12 balls is the same for each event. The two events are independent. Multiply to find the probability that both will occur.

$$\begin{aligned} P(\text{red, then white}) &= P(\text{red}) \cdot P(\text{white}) \\ &= \frac{\text{number of red balls}}{\text{total number of balls}} \cdot \frac{\text{number of white balls}}{\text{total number of balls}} \\ &= \frac{1}{3} \cdot \frac{1}{4} \\ &= \frac{1}{12} \end{aligned}$$

The probability of picking red, then white, is $\frac{1}{12}$.

396 | Chapter 9 Probability and Statistics

Teaching Strategies

ESL/LEP To help students understand how independent and dependent are used in probability, ask what it means to be independent and to be dependent. to stand on your own, not to need others; to need someone, so what that person does affects you In probability, independent also means to stand on its own, and dependent means to be affected by a previous event.

When the outcome of one event *is* affected by the outcome of another, the events are said to be **dependent**.

Example 2

SPORTS Six teams—the Panthers, Tigers, Lions, Bears, Cheetahs, and Elephants—are in the lottery round for this year's draft picks. The name of each team is written on a card and placed in a box.

To determine who gets the first lottery pick, one card will be drawn at random and *not* replaced. Then a second card will be drawn at random to determine the second pick. What is the probability that the Bears get the first draft choice and the Lions get the second draft choice?

Solution

Because the first card is not replaced, the sample space for the second drawing has been changed. The second event is dependent on the first event.

Probability of first event

$$P(\text{Bears}) = \frac{\text{number of Bears cards}}{\text{total number of cards}} = \frac{1}{6}$$

Probability of second event

$$P(\text{Lions, after Bears}) = \frac{\text{number of Lions cards}}{\text{total number of cards}} = \frac{1}{5}$$

Multiply the probabilities.

$$P(\text{Lions, after Bears}) = \frac{1}{6} \cdot \frac{1}{5} = \frac{1}{30}$$

The probability of drawing the Bears first and the Lions second is $\frac{1}{30}$.

Example 3

A bag contains 3 green marbles, 2 red marbles, 4 yellow marbles, and 1 black marble. Two are taken at random from the bag. Find *P*(green, then green).

Solution

$$P(\text{first green marble}) = \frac{\text{number of green marbles}}{\text{total number of marbles}} = \frac{3}{10}$$

$$P(\text{second green marble}) = \frac{\text{number of green marbles}}{\text{total number of marbles}} = \frac{3-1}{10-1} = \frac{2}{9}$$

Multiply the probabilities.

$$P(\text{green, then green}) = \frac{3}{10} \cdot \frac{2}{9} = \frac{6}{90} = \frac{1}{15}$$

The probability of picking green, then green, is $\frac{1}{15}$.

Check Understanding

Are these events independent or dependent?

1. tossing a coin twice
2. picking two marbles from a bag without replacing the first one
3. choosing a captain and then choosing a co-captain
4. rolling three dice

1. independent
2. dependent
3. dependent
4. independent

Learning Styles

VISUAL LEARNERS To help students visualize the need of redefining the sample space after each dependent event, have them use two red and four blue centimeter cubes. To have them find P(red, then blue), ask what P(red) is. Then have them remove one red cube and find P(blue). Have them multiply the probabilities to find P(red, then blue). $\frac{1}{3}, \frac{4}{5}, \frac{4}{15}$

Chalkboard Examples

Supplementary Example 1

Using the same bag and contents, you reach into the bag without looking and take out a ball. You replace it and take out another ball at random. Find the probability that both balls are white. $\frac{1}{9}$

Supplementary Example 2

A basketball league is drafting teams for a new season. The names of ten all-star players are written on cards and placed in a box. Two of the players are Maggie and Sonya. The team called the Lasers gets the first two picks. The coach will draw one name and not replace it and then draw a second name at random. What is the probability that Maggie will be the first choice and Sonya will be the second choice? $\frac{1}{90}$

Supplementary Example 3

Using the bag of marbles in Example 3, if two marbles are taken at random from the bag, what is the probability that both will be yellow? $\frac{2}{15}$

Reteaching Worksheet 9-4

Name ______________________ Date ____________

RETEACHING 9-4

INDEPENDENT AND DEPENDENT EVENTS

A compound event can be a combination of *independent events* or *dependent events.*

Independent events: ones in which the outcome of one event has no effect on the outcome of the other. If A and B are independent events, then $P(A \text{ and } B) = P(A) \cdot P(B)$.

Dependent events: ones in which the outcome of one event depends upon the outcome of the other. If B is dependent on A, then $P(A \text{ and } B) = P(A) \cdot P(B)$, given A.

Example 1

Suppose you roll a die twice. What is the probability you will roll a 3, then 6?

Solution

The two events are independent.

$$P(3, \text{then } 6) = P(3) \cdot P(6) = \frac{1}{6} \cdot \frac{1}{6} = \frac{1}{36}$$

The probability that you will roll a 3, then 6 is $\frac{1}{36}$.

Example 2

The teacher placed Mona's name and Kay's name along with those of four other students in a bag for a random drawing. What is the probability that Mona's name will be drawn first and not replaced and Kay's name will be drawn second?

Solution

The two events are dependent.

First draw: $P(\text{Mona}) = \frac{1}{6}$ ←favorable outcomes / ←possible outcomes

Second draw: $P(\text{Kay}) = \frac{1}{5}$ ←favorable outcomes / ←possible outcomes

$$P(\text{Mona, then Kay}) = P(\text{Mona}) \cdot P(\text{Kay}) = \frac{1}{6} \cdot \frac{1}{5} = \frac{1}{30}$$

The probability that Mona's name and then Kay's name will be drawn in that order is $\frac{1}{30}$.

EXERCISES

Suppose you roll a die twice. What is the probability you will get the following results?

1. $P(4, \text{then } 5)$ $\frac{1}{36}$
2. $P(3, \text{then odd})$ $\frac{1}{12}$
3. $P(\text{even, then odd})$ $\frac{1}{4}$

A sewing basket holds 3 balls of red yarn, 2 balls of blue yarn and 5 balls of green yarn.

4. Suppose you choose each ball at random and then replace it. Find each probability.
   a. $P(\text{blue, then red})$ $\frac{3}{50}$
   b. $P(\text{green, then green})$ $\frac{1}{4}$
5. Suppose you choose each ball at random and do not replace it. Find each probability.
   a. $P(\text{green, then red})$ $\frac{3}{18}$ or $\frac{1}{6}$
   b. $P(\text{green, then blue})$ $\frac{1}{9}$

138 RETEACHING *LESSON 9-4*

Lesson Wrap-up

Quick Assessment

Ask the following questions to determine if students understand the content presented in this lesson.

1. What is the difference between independent events and dependent events? For independent events, the outcome of the second event is not affected by the outcome of the first event, while for dependent events the second outcome is affected by the first outcome.
2. Why do you need different methods to find the probability of each? The number of outcomes in dependent events changes each step of the way, while outcomes for independent events remain unchanged.

Assignment Guide

Basic: 1–25, 32–34
Advanced: 1–34
Additional Practice: See Extra Practice Index on page 674.

Additional Answers

32.
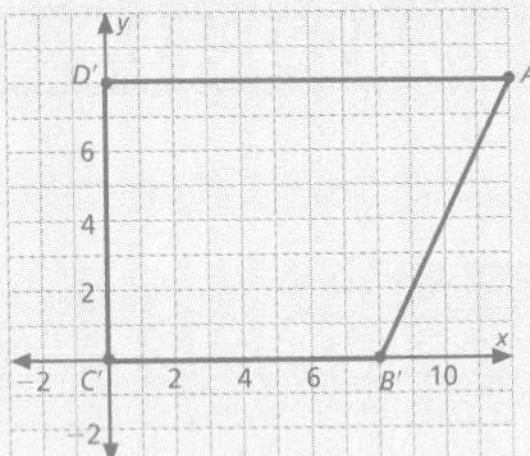

33.
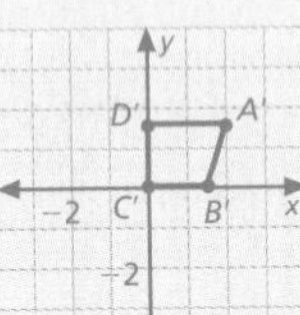

34.
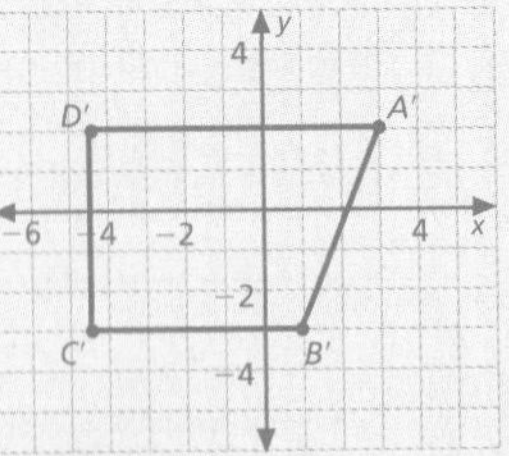

Try These Exercises

A group of numbered cards contains three 3s, four 4s, and five 5s. Cards are picked at random, one at a time, and then replaced. Find each probability.

1. P(3, then 5) $\frac{5}{48}$
2. P(4, then odd number) $\frac{2}{9}$
3. P(even, then not 4) $\frac{2}{9}$

SPORTS A baseball team has 10 pitchers, 3 catchers, 5 outfielders, and 7 infielders on its roster. Two players from this team will be chosen at random to represent the league in a tour of Japan. Find each probability.

4. P(pitcher, catcher) $\frac{1}{20}$
5. P(outfielder, infielder) $\frac{7}{120}$

SURVEYS A newspaper survey asked 100 men and 100 women whether they planned to vote for a proposed tax increase. Twenty men and 40 women said they are in support of the increase. A person from the survey is chosen at random. Find each probability.

6. What is the probability that the person chosen is in support of the tax increase? $\frac{3}{10}$
7. What is the probability that the person is a woman in support of the increase? $\frac{1}{5}$
8. What is the probability that the person is a man who is against the increase? $\frac{2}{5}$

Practice Exercises

A box contains 3 red counters, 4 yellow counters, 2 green counters, and 1 blue counter. Counters are taken at random from the box, one at a time, and then replaced. Find each probability.

9. P(red, then yellow) $\frac{3}{25}$
10. P(blue, then green) $\frac{1}{50}$
11. P(red, then not red) $\frac{21}{100}$

A drawer contains 2 pairs of black socks, 3 pairs of brown socks, a pair of beige socks, and 6 pairs of white socks. One sock at a time is taken at random from the drawer and not replaced. Find each probability.

12. P(black, then black) $\frac{1}{46}$
13. P(white, then black) $\frac{2}{23}$
14. P(beige, then white) $\frac{1}{23}$

A billboard says "EAT HERE NOW." Two letters fall off, one after the other.

15. What is the probability that both letters are vowels? $\frac{2}{9}$
16. What is the probability that the first letter is an E, and the second letter is not an E? $\frac{7}{30}$
17. You are given tickets to two soccer games at a stadium with 48,000 seats. What is the probability you will sit in Section D in the first game, and then Section A in the second game, if Section D has 4,000 seats and Section A has 3,000 seats? $\frac{1}{192}$

Maracana Stadium, Brazil

SCHEDULING Liu and Michi plan to sign up for a drawing class next term. Drawing is offered during the first 4 periods of the day, and students are assigned randomly to classes.

18. What is the probability that Liu and Michi will have drawing together? $\frac{1}{4}$
19. What is the probability that both students will have drawing during first period? $\frac{1}{16}$

Teaching Strategies

COOPERATIVE LEARNING Have pairs of students describe independent events and dependent events that involve a bag containing four \$1 bills, two \$5 bills, three \$10 bills, and one \$20 bill. For example, events might include P(\$7) with and without replacement, P(\$10 and \$5), and P(\$10, then \$5). Have pairs exchange descriptions and find probabilities.

Both spinners are spun. Find each probability.

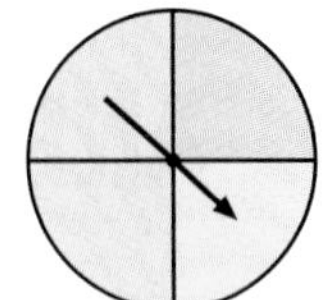

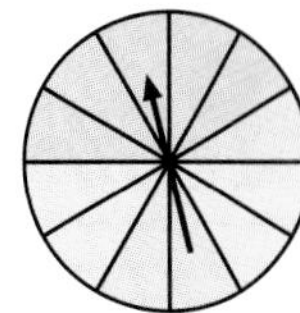

20. P(red, red) $\frac{1}{8}$

21. P(red, yellow) $\frac{1}{6}$

22. P(green, not green) $\frac{7}{48}$

A golf bag pocket contains 4 yellow golf balls, 3 white balls, 1 green ball, and 4 red balls. You pull out one ball at a time, without replacing it. Find each probability.

23. P(red, then white, then yellow) $\frac{2}{55}$

24. P(green, then red, then white) $\frac{1}{110}$

25. **WRITING MATH** Explain the difference between events that are *mutually exclusive* and those that are *independent.* Mutually exclusive events cannot occur at the same time. Independent events can occur at the same time, but they have no effect on the probability of each other.

Extended Practice Exercises

HISTORY Suppose that it is 1944, and the Homestead Grays of the Negro National League are playing the Birmingham Black Barons of the Negro American League in a "best two out of three" series. Then suppose the Grays are the favored team, and the probability they will win any individual game is $\frac{3}{4}$.

26. What is the probability that the Black Barons win a game? $\frac{1}{4}$

27. What is the probability that the Grays win in two straight games? $\frac{9}{16}$

28. What is the probability that the series goes for three games? $\frac{3}{8}$

29. What is the probability that the Homestead Grays win the series? $\frac{27}{32}$

30. **USING DATA** Use the Data Index on pages 632–633 to locate statistics on baseball. Suppose Davis had one more official at-bat in the 1943 season and Wagner had one more official at-bat in the 1948 season. What is the probability that both would have gone hitless? about 0.460

31. **CHAPTER INVESTIGATION** When a baseball player comes to bat, the player can get a hit—either a single, a double, a triple or a home run—or the player can walk, strike out, or make an out some other way. For each player you have chosen, find the probability expressed as a percent that each event will occur. For example, to find the probability that a player will walk, divide the number of walks (BB) by the number of at bats (AB) and convert the decimal to a percent. Use a spreadsheet or calculator.

Homestead Grays

Mixed Review Exercises

Copy quadrilateral *ABCD*. Then draw its dilation image. (Lesson 8-3) See additional answers.

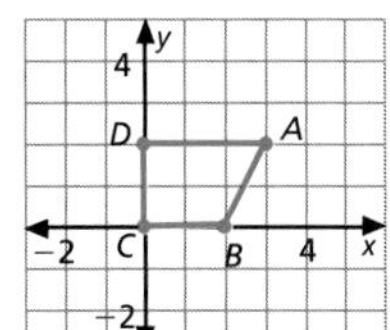

32. The center of dilation is the origin and the scale factor is 4.

33. The center of dilation is the origin and the scale factor is $\frac{3}{4}$.

34. The center of dilation is point *A* and the scale factor is 2.5.

Lesson 9-4 **Independent and Dependent Events** 399

Teaching Strategies

CHALLENGE A bag contains four red and six white tennis balls. Quang picks two balls, replaces them, and then picks two more balls. What is the probability that Quang picks two red balls and then two white balls?

$\frac{4}{10} \times \frac{3}{9} \times \frac{6}{10} \times \frac{5}{9} = \frac{2}{45}$

Extra Practice Worksheet 9-4

Name ________ Date ________

EXTRA PRACTICE 9-4

Independent and Dependent Events

Exercises

A bag contains 4 blue marbles, 6 red marbles, 8 green marbles and 2 yellow marbles. Marbles are taken at random from the bag, one at a time, and then replaced. Find each probability.

1. P(blue, then green) $\frac{2}{25}$
2. P(red, then red) $\frac{9}{100}$
3. P(yellow, then not green) $\frac{3}{50}$
4. P(not blue, then not red) $\frac{14}{25}$
5. P(blue, then not blue) $\frac{4}{25}$
6. P(green, then green) $\frac{4}{25}$
7. P(blue, then red, then green) $\frac{3}{125}$
8. P(red, then yellow, then red) $\frac{9}{1000}$

A drawer contains 3 pairs of white socks, 4 pairs of black socks, a pair of blue socks and 5 pairs of brown socks. One sock at a time is taken at random from the drawer and not replaced. Find each probability.

9. P(white, then white) $\frac{3}{65}$
10. P(blue, then black) $\frac{8}{325}$
11. P(brown, then blue) $\frac{2}{65}$
12. P(white, then black) $\frac{24}{325}$
13. P(blue, then blue) $\frac{1}{325}$
14. P(brown, then brown) $\frac{9}{65}$
15. P(blue, then white) $\frac{6}{325}$
16. P(black, the white) $\frac{24}{325}$

Both spinners at the right are spun.

Find each probability.

A B C D E F   1 2 3 4

18. P(A, 1) $\frac{1}{24}$
19. P(B, 2) $\frac{1}{24}$
20. P(a consonant, an even number) $\frac{1}{3}$
21. P(a vowel, 3) $\frac{1}{12}$

120 EXTRA PRACTICE *LESSON 9-4*

Enrichment Worksheet 9-4

Name ________ Date ________

ENRICHMENT 9-4

Tic-Tac-Toe

The game of tic-tac-toe can be analyzed by thinking of it as a game of pure chance.

The nine locations are numbered 1 to 9. Then, nine counters are also numbered 1 through 9 and put into a bag.

Player A, who plays first, draws a counter and enters an X into the corresponding location. Player B draws a counter and enters an O. The game continues until one player wins or a draw occurs.

O_1	O_2	X_1
X_2	O_3	X_3
X_4	X_5	O_4

Exercises

1. In the arrangement shown above, Player B wins. Seven other "B wins" arrangements can be obtained by reflections and rotations. Show them.

X_4 X_2 O_1 O_4 X_5 X_4 X_1 O_2 O_1 O_4 X_3 X_1
X_5 O_3 O_3 X_3 O_3 X_2 X_3 O_3 X_2 X_5 O_3 O_2
O_4 X_3 X_1 X_1 O_2 O_1 O_4 X_5 X_4 X_4 X_2 O_1

X_1 X_3 O_4 X_4 X_5 O_4 O_1 X_2 X_4
O_2 O_3 X_5 X_2 O_3 X_3 O_2 O_2 X_5
O_1 X_2 X_4 O_1 O_2 X_1 X_1 X_3 O_4

2. Show the other four arrangements in which B wins.

O_1 X_1 O_2 X_4 X_2 O_1 O_4 X_5 X_4 O_2 X_3 O_4
X_2 O_3 X_3 X_5 O_3 X_1 X_3 O_3 X_2 X_1 O_3 X_5
X_4 X_5 O_4 O_4 X_3 O_2 O_2 X_1 O_1 O_1 X_2 X_4

3. Find the number of arrangements in which A must win. 62
4. Find the number of arrangements that result in a draw. 16
5. How many equally probable arrangements are there in all? 126
6. What is the probability of a draw? 0.127

158 ENRICHMENT *LESSON 9-4*

Skills Practice

Vocabulary Review

Lesson 9-3
compound event
mutually exclusive events
complement

Lesson 9-4
independent event
dependent event

Assignment Guide

All students: 1–61
Additional Practice: Refer to the Extra Practice Index on page 674 of this text.

Review and Practice Your Skills

Practice ■ Lesson 9-3

A card is picked at random from a standard deck of 52 cards. Find each theoretical probability.

1. P(heart and face card) $\frac{3}{52}$
2. P(jack or queen) $\frac{2}{13}$
3. P(red or black card) 1
4. P(black two) $\frac{1}{26}$
5. P(2 or 3) $\frac{2}{13}$
6. P(7 of hearts) $\frac{1}{52}$
7. $P(2 \leq card \leq 5)$ $\frac{4}{13}$
8. P(king of clubs) $\frac{1}{52}$
9. P(club and (ten or king)) $\frac{1}{26}$

You flip a coin four times. Find each theoretical probability.

10. P(*exactly one head*) $\frac{5}{32}$
11. P(*2 tails, 2 heads*) $\frac{5}{16}$
12. P(*3 or 4 tails*) $\frac{5}{12}$
13. P(*> 2 heads*) $\frac{1}{2}$
14. P(*0 or 1 head*) $\frac{3}{16}$
15. P(*3 tails*) $\frac{5}{16}$

You roll a pair of dice. Find each theoretical probability.

16. P(sum = 7) $\frac{1}{6}$
17. P(sum = 11) $\frac{1}{18}$
18. P(both even) $\frac{1}{4}$
19. P(1 is rolled) $\frac{11}{36}$
20. P(4 or 5 is rolled) $\frac{5}{4}$
21. P(sum = 2) $\frac{1}{36}$
22. P(sum is odd) $\frac{1}{2}$
23. P(sum < 6) $\frac{5}{18}$
24. P(sum = 10,11, or 12) $\frac{1}{6}$
25. P(sum is even and >7) $\frac{3}{5}$
26. P(sum is odd and <11) $\frac{7}{16}$
27. P(values are equal) $\frac{1}{6}$

28. Are multiples of 4 and multiples of 9 mutually exclusive events? Explain.
No. 36 is multiple of both

29. You spin a spinner with 8 equal sectors, numbered 1–8. What is the probability of spinning a number that is neither odd nor greater than 6? $\frac{3}{8}$

Practice ■ Lesson 9-4

A drawer contains 7 red shirts, 5 blue shirts, and 4 white shirts. One shirt at a time is taken at random from the drawer and not replaced. Find each probability.

30. P(red, then blue) $\frac{7}{48}$
31. P(red, then not white) $\frac{7}{20}$
32. P(white, then blue) $\frac{1}{12}$
33. P(both white) $\frac{1}{20}$
34. P(not blue, then not red) $\frac{33}{80}$
35. P(both not blue) $\frac{121}{240}$

A box contains 4 red cards, 5 black cards, 10 green cards, and 2 blue cards. Cards are taken at random from the box, one at a time, and then replaced. Find each probability.

36. P(red, then red) $\frac{16}{441}$
37. P(red, then green, then blue) $\frac{80}{9261}$
38. P(not red, then green) $\frac{170}{441}$
39. P(black, then black, then not green) $\frac{275}{9261}$
40. P(not green in each of three draws) $\frac{1331}{9261}$
41. P(black, then not blue) $\frac{95}{441}$
42. P(red, then red, then red, then red) $\frac{256}{194,481}$
43. P(blue, then blue, then black) $\frac{20}{9261}$

44. You are given tickets to two hockey games in an arena with 18,000 seats. What is the probability that you will sit in Section B in the first game, and then Section C in the second game, if Section B has 4500 seats and Section C has 3000 seats? $\frac{1}{24}$

Extend the Lesson

COOPERATIVE LEARNING Have students work in small groups. Provide each group with index cards or slips of paper. First, each group thinks of at least 20 events related to drawing one bill at random from each of two different pockets. Each pocket contains one \$1, one \$5, one \$10, one \$20, one \$50 and one \$100 bill. For example, an event may be "sum of bills > \$50," "sum of bills a perfect square," "one bill a \$5 bill," etc. Have students write the events one per card. Then have them mix the cards and place the deck face down. Students take turns picking two cards, telling whether or not the events are mutually exclusive, and finding *P*(event A or event B).

Practice ■ Lesson 9-1–Lesson 9-4

List all the elements of the sample space for each of the following experiments. (Lesson 9-1)

45. Tossing a quarter and a nickel. {HH, HT, TH, TT}
46. Choosing a month of the year {J, F, M, A, M, J, J, A, S, O, N, D}
47. Selecting a team in the NFC Central Division {Minnesota, Green Bay, Chicago, Tampa Bay, Detroit}
48. Choosing a letter from the alphabet {A, B, C, D, E, F, G, H, I, J, K, L, M, N, O, P, Q, R, S, T, U, V, W, X, Y, Z}
49. Conduct and document a simulation using six question multiple choice test. If each question has three choices for answers, and you guess on each question, what are your chances of getting 3 or more questions correct? (Lesson 9-2) Answers will vary.

Three coins are tossed. Find each probability. (Lesson 9-3)

50. P(at least one heads) $\frac{7}{8}$
51. P(no tails) $\frac{1}{8}$
52. P(0 or 1 tails) $\frac{1}{2}$
53. P(two tails) $\frac{3}{8}$
54. P(all three coins the same) $\frac{1}{4}$
55. P(1 or 2 heads) $\frac{3}{4}$

A box contains 100 cards, numbered from 1–100. Cards are taken at random from the box, one at a time, and not replaced. Find each probability. (Lesson 9-4)

56. P(even number, then odd number) $\frac{25}{99}$
57. P(multiple of 3, then 50) $\frac{1}{297}$
58. P(45, then 45) 0
59. P(99, then 100) $\frac{1}{9900}$
60. P(number less than 40, number > 80) $\frac{39}{495}$
61. P(prime number, then prime number) $\frac{2}{33}$

Mid-Chapter Quiz

1. How many outcomes are there for an outfit chosen from three pairs of pants and five shirts? 15
2. A basketball player has made 60 out of 125 attempts. How many shots is he likely to make in 500 attempts? 240
3. A family has five children. What is P(three boys and two girls)? $\frac{5}{16}$

A card is picked at random from a standard deck of 52 cards.

4. Find P(heart and less than 5) $\frac{1}{13}$
5. Find P(heart or less than 5) $\frac{25}{52}$
6. Find P(7 or king) $\frac{2}{13}$
7. Find P(neither 5 nor diamond) $\frac{9}{13}$
8. Find P(neither 3 nor red) $\frac{6}{13}$

A bag contains seven green marbles, three red marbles and five blue marbles. A marble is picked and replaced. Then another marble is picked.

9. Find P(green, then red) $\frac{7}{75}$
10. Find P(two red marbles) $\frac{1}{25}$

From the same bag, a marble is picked and not replaced. Then another marble is picked.

11. Find P(green, then red) $\frac{1}{10}$
12. Find P(two red marbles) $\frac{1}{35}$

Chalkboard Examples

Example from Lesson 9-3

Two standard dice are rolled and the numbers are multiplied. Find P(12 or multiple of 4). $\frac{5}{12}$

Example from Lesson 9-4

There are five true-false questions in a quiz. If you guess each answer, what is the probability that you will guess four correctly? $\frac{5}{32}$

Mid-Chapter Quiz

Correlation Chart

Question Number	Lesson Number
1–3	9-1 and 9-2
4–8	9-3
9–12	9-4

Teaching Strategies

CHALLENGE A bank contains 11 dimes and 22 quarters. If two coins are shaken out at random, what is the probability that they are both quarters? $\frac{7}{16}$ What is the probability that they are both dimes? $\frac{5}{48}$

Lesson Planning

NCTM Standards/Strands

- Number & Operation
- Patterns, Functions, & Algebra
- Data Analysis, Statistics, & Probability
- Problem Solving
- Reasoning and Proof
- Communication
- Connections
- Representation

Vocabulary

fundamental counting principle
permutation
factorial notation
combination

Materials Needed

calculators

Lesson Resources

Warm-Up Transparency 26
Transparency RF-33
Reteaching 9-5
Extra Practice 9-5
Enrichment 9-5

Assignment Guide

Basic: 1–18, 23–27
Enriched: 1–27

Getting Started

5-Minute Warm-up

Solve for *y*.

1. $y = \frac{(7 \times 6 \times 5 \times 4 \times 3 \times 2 \times 1)}{(4 \times 3 \times 2 \times 1)}$ 210
2. $y = \frac{(6 \times 5 \times 4 \times 3 \times 2 \times 1)}{(4 \times 3 \times 2 \times 1)} \times (2 \times 1)$ $\frac{1}{5}$

Introduction to Lesson 9-5

If students cannot calculate exact answers, encourage them to look for patterns with smaller numbers of total letters. Point out how complicated ordered pairs or tree diagrams could become in these situations. Tell students that by the end of this lesson, they will be able to compute exact answers quickly using special formulas.

9-5 Permutations and Combinations

Goals ■ Find the number of permutations and combinations of a set.

Applications Travel, Sports, Office Work

Work with a partner to answer the questions.

You have just applied for your first set of license plates. Each license plate has 3 different letters and 3 different numbers. The letters and numbers are assigned randomly by the Department of Motor Vehicles.

1. How many arrangements of 3 different letters are possible? 15,600
2. How many arrangements of 3-digit arrangements are possible? Remember, to include 0 as a digit. 720
3. Suppose you had hoped that your plate would read "ACE 123." What is the probability that you will receive this plate? How do you know? $\frac{1}{15{,}600} \cdot \frac{1}{720} = \frac{1}{11{,}232{,}000}$

Build Understanding

Thus far, you have used either a tree diagram or a set of ordered pairs to find the number of outcomes in a sample space. But sometimes the sample space is too large to use either of these methods.

Another way to find the total number of outcomes is to use the **fundamental counting principle**. This principle can be used to count the number of ways two or more events can happen in succession. It states that, if an event A can occur in m ways and an event B can occur in n ways, then events A and B can happen in $m \cdot n$ ways.

Example 1

TRAVEL Suppose the Kenosha Comets of the All-American Girls Professional Baseball League will travel on a road trip to Grand Rapids, Peoria, and Battle Creek. They can go from Kenosha to Grand Rapids by car, train, or bus, then from Grand Rapids to Peoria by bus, train, or plane, and from there to Battle Creek by car, bus, train, or plane. Finally, from Battle Creek, they can either take the bus or the train back home. How many different routes are possible on this road trip?

Solution

Use the fundamental counting principle.

The 3 possible routes for the first leg of the trip are car, train, or bus. Then they have 3 possible routes for the second leg of the trip, 4 for the third leg of the trip, and 2 for the return trip to Kenosha.

$$3 \cdot 3 \cdot 4 \cdot 2 = 72$$

Seventy-two different routes are possible.

Interdisciplinary Connections

SCIENCE Scientists know the sequence of nitrogen bases in DNA acts as a chemical code. The four bases of DNA—adenine (A), thymine (T), cytosine (C), and guanine (G)—make up the genetic alphabet. Each unit of the genetic code is a sequence of three bases called a codon. For example, ACT and GCT are codons for different components of a protein. Thus, there are $4 \times 4 \times 4 = 64$ different possible codons.

An arrangement of items in a particular order is called a **permutation**. For the four letters M, A, T, H, there are 24 different 4-letter permutations.

MATH	MAHT	MHAT	MHTA	MTHA	MTAH
AMTH	AMHT	AHTM	AHMT	ATHM	ATMH
TAMH	TAHM	TMAH	TMHA	THAM	THMA
HAMT	HATM	HMAT	HMTA	HTAM	HTMA

You can use the fundamental counting principle to find the number of permutations of any group of items.

For each arrangement of M, A, T, H, there are 4 choices for the first letter, 3 choices for the second, 2 choices for the third, and 1 choice for the fourth.

$4 \cdot 3 \cdot 2 \cdot 1 = 24.$

$4 \cdot 3 \cdot 2 \cdot 1$ can be written in **factorial notation** as **4!**

In general, the number of permutations of n different items is written **n!** and read as **n factorial**.

Check Understanding

What is wrong with the notation ${}_4P_5$?

A group of 4 items cannot be taken 5 at a time.

Example 2

In how many different ways can you arrange your math, science, social studies, language arts, and literature anthology books in a row on a shelf?

Solution

There are five books. Find the number of permutations of five items.

number of permutations of five items = 5!

$= 5 \cdot 4 \cdot 3 \cdot 2 \cdot 1$ or 120

There are 120 different ways to line up five books on a shelf.

Sometimes you need only part of a set of items, such as selecting two of nine players. The number of permutations of n different items, taken r items at a time, with no repetitions, is written ${}_nP_r$. Use the formula below to find the number of permutations when only part of a set is used.

$${}_nP_r = \frac{n!}{(n-r)!}$$

Example 3

SPORTS Eight teams enter a tournament. How many different arrangements of first-, second-, and third-place winners are possible?

Solution

Use the formula: ${}_nP_r = \dfrac{n!}{(n-r)!}$

$${}_8P_3 = \frac{8!}{(8-3)!}$$

Cancel common factors to simplify the computation.

$$= \frac{8 \cdot 7 \cdot 6 \cdot \cancel{5} \cdot \cancel{4} \cdot \cancel{3} \cdot \cancel{2} \cdot \cancel{1}}{\cancel{5} \cdot \cancel{4} \cdot \cancel{3} \cdot \cancel{2} \cdot \cancel{1}} = 8 \cdot 7 \cdot 6 \cdot = 336$$

There are 336 ways for teams to finish first, second, and third.

Teaching Strategies

CHALLENGE Of 12 people, 3 are allergic to nuts and 9 are not. Four are randomly chosen and tested for the allergy.

1. In how many ways can the 4 be chosen? ${}_{12}C_4 = 495$
2. What is the probability that exactly one of the people in the group has the allergy? Solution: $\frac{3 \leftrightarrow {}_9C_3}{{}_{12}C_4} = \frac{3 \leftrightarrow 84}{495} = 0.51$

Chalkboard Examples

Supplementary Example 1

The coach of the Comets is planning the schedule of games for the next year. The first game can be played against either the Trojans or Reds. The second game can be played against either the Cubs, Jaguars or Tigers. The third game can be played against either the Centaurs or the Bulldogs. How many different schedules are possible for the first three games of the season? 12

Supplementary Example 2

How many different ways can you arrange six books in a row on a shelf. 720

Supplementary Example 3

Six teams enter a tournament. How many different arrangements of first-, second-, third- and fourth-place winners are possible? 360

Supplementary Example 4

Only five members from the debate club of 12 members can go to the finals. How many choices of five are there? 792

Reteaching Worksheet 9-5

Name ______________________ Date ____________

RETEACHING 9-5

PERMUTATIONS AND COMBINATIONS

A *permutation* is an ordered arrangement of a set of objects. In many situations, however, you may be interested only in the objects in a set, and not the order in which they can be arranged. A set in which order is ignored is a *combination*.

Example 1

In how many ways can 2 books be arranged on a shelf if there are 5 books to choose from?

Solution

Use this formula to find the number of permutations.

$${}_nP_r = \frac{n!}{(n-r)!}$$

$${}_5P_2 = \frac{5!}{(5-2)!} = \frac{5.4.3.2.1}{3 \cdot 2 \cdot 1} = 20$$

Given 5 books, there are 20 ways to arrange 2 books on a shelf.

Example 2

In how many different ways can a panel of 2 judges be selected from a group of 7 candidates?

Solution

Use this formula to find the number of combinations.

$${}_nC_r = \frac{n!}{(n-r)!r!}$$

$${}_7C_2 = \frac{7!}{(7-2)! \cdot 2} = \frac{7 \cdot 6 \cdot 5 \cdot 4 \cdot 3 \cdot 2 \cdot 1}{(5 \cdot 4 \cdot 3 \cdot 2 \cdot 1)(2 \cdot 1)} = \frac{42}{2} = 21$$

There are 21 ways for a panel of 2 judges to be selected from 7 candidates.

EXERCISES

Tell whether each involves a permutation or a combination. Then solve.

1. In how many ways can 3 books be arranged on a shelf if there are 6 books to choose from? permutation, 120 ways
2. In how many ways can a committee of 3 be chosen from 9 students? combination, 84 ways
3. How many kinds of pizza with 3 cheeses can be made if there are 5 cheeses available? combination, 10 ways
4. A flag locker has 8 flags. How many signals can be made by hoisting 3 flags in order, one above the other? permutation, 336 ways
5. On an examination, a student can answer any 2 of the 5 questions. In how many ways can she make a choice? combination, 10 ways

140 RETEACHING *LESSON 9-5*

Lesson Wrap-up

Quick Assessment

Ask the following questions to determine if students understand the content presented in this lesson.

1. In Example 4, how would the answer be different if each of the band members was assigned a specific instrument. Will the answer be less or greater? Why? Order matters, and so the answer involves permutation; greater since $nPr > nCr$
2. How many combinations of four items taken four at a time are there? one combination How many permutations of four items are there? 24

Assignment Guide

Basic: 1–18, 23–27
Advanced: 1–27
Additional Practice: See Extra Practice Index on page 674.

For each situation described in the preceding Examples, the order of the items in consideration is important. A set of items in *no* particular order is called a **combination**. The number of combinations of n different items, taken r times at a time, where $0 \leq r \leq n$, is written $_nC_r$. You can use the formula below to find the number of combinations of a set of items.

$$_nC_r = \frac{n!}{(n-r)!r!}$$

Math: Who, Where, When?

Although several 16th- and 17th-century mathematicians, notably Pascal and Fermat, investigated probability, Jacques Bernoulli is considered by some to be the founder of probability theory. His book, *Ars Conjectandi*, published in 1713, is the first book devoted entirely to the subject of probability. This book contains a theory of combinations, essentially the same as we understand it today, as well as the first appearance, with today's meaning, of the word *permutation*.

Example 4

How many different four-person ensembles can be chosen from a pool of 10 musicians?

Solution

There are 10 people from which to pick, 4 at a time. So, $n = 10$ and $r = 4$. Use the formula:

$$_nC_r = \frac{n!}{(n-r)!r!}$$

$$_{10}C_4 = \frac{10!}{(10-4)!4!}$$

$$= \frac{10 \cdot 9 \cdot 8 \cdot 7 \cdot \cancel{6} \cdot \cancel{5} \cdot \cancel{4} \cdot \cancel{3} \cdot \cancel{2} \cdot \cancel{1}}{(\cancel{6} \cdot \cancel{5} \cdot \cancel{4} \cdot \cancel{3} \cdot \cancel{2} \cdot \cancel{1})(4 \cdot 3 \cdot 2 \cdot 1)}$$

$$= \frac{5{,}040}{24}$$ Cancel common factors to simplify the computation.

$$= 210$$

There can be 210 different four-person ensembles.

Try These Exercises

1. **SPORTS** A manager is choosing her infield from among 4 third-base players, 3 shortstops, 2 second-base players, and 5 first-base players. How many different ways can an infield be chosen? 120

Tell whether each question involves a permutation or a combination. Then solve.

2. In how many different ways can you arrange the letters *a, c, e, g, i, k,* and *l*? permutation; 5040
3. How many different selections of three tapes can be made by a consumer choosing from among a collection of six tapes? combination; 20
4. In how many different ways can a starting lineup of 5 players be selected from a group of 12 basketball players? combination; 792
5. In how many different ways can 4 winners be chosen from 15 contestants? permutation; 32,760

Practice Exercises

6. **OFFICE WORK** A secretary has to create ten new customer files. In how many different orders can he do this? 10! or 3,628,800
7. **HIRING** Six applicants apply for two jobs. How many different outcomes are possible? 30

Teaching Strategies

Working with concrete situations may help students understand permutations and combinations. Have groups of four students make a table to find the number of ways they can have a leader and a recorder. 12 Then have them find the number of ways they can form groups of two from their group, this time using diagrams or ordered pairs. 6 Ask how the two situations are different.

404 Chapter 9 Probability and Statistics

8. Ralph is a tour guide. In how many ways can he choose 3 museums to visit from the 8 museums in a city? 56

9. In how many ways can a disk jockey play 5 of the 20 top hits? 15,504

10. **ERROR ALERT** On a test, students must choose 3 out of 5 essay questions to answer. Dale calculates that there are 60 ways to do this. What has Dale done wrong?

10. Dale has found the number of permutations instead of combinations. In this situation, order doesn't matter. There are only 10 combinations.

11. **SPORTS** Bill, Phil, and Jill are among 12 players competing for 3 spots on a table-tennis team. Every player has an equal chance of making the team. Find the probability that all three will make the team. $\frac{1}{220}$

12. What is n, if $_nP_3 = 120$? 6

13. Suppose that license plates contain three letters. What is the probability that Meg's plates will spell her name? $\frac{1}{15,600}$

14. **SPORTS** How many ways can a batting order be made for 9 players if you know that one player has already been designated to bat first and another to bat fourth? 7! or 5040

15. **WRITING MATH** Find the number of permutations of the letters in the word *shutout.* Explain how you did it. 1260; Sample answer: Divide 7! by 2! · 2! · 1! · 1! · 1!

16. Two students out of 8 will be chosen to speak at a school assembly. How many different outcomes are possible? 28

USING DATA For Exercise 17–18, use the Data Index on pages 632–633 to locate information about the All-American Girls Professional Baseball League.

17. Suppose you could interview all eight women who were batting champions of the All-American Girls Professional Baseball League to discover what playing in this league was like. How many orders for these interviews would be possible? 8!

18. If you were to interview only four of the eight women, what is the probability that you would first interview a player from either Fort Wayne or Rockford? $\frac{1}{14}$

Technology Note

Some calculators have a factorial key, marked [x!]. To find 7!, enter 7, then [x!]. On graphing calculators, you can choose the factorial function from a displayed mathematical menu. This menu may also include operations for finding permutations and combinations.

Extended Practice Exercises

19. Compare the values of $_8C_5$ and $_8C_3$. What do you notice? They are the same.

20. Find the values of $_7C_4$ and $_7C_3$. What do you notice? Both have values of 35.

21. What can you say about the sum of the number of items taken at one time for each combination shown in Exercises 18 and 19? They equal the total number of items.

22. **CRITICAL THINKING** Use what you have discovered to quickly find $_{67}C_{64}$. 47,905

Mixed Review Exercises

Add. (Lesson 8-5)

23. $\begin{bmatrix} 2 & 4 \\ 3 & 0 \end{bmatrix} + \begin{bmatrix} 1 & 3 \\ -2 & 4 \end{bmatrix}$ $\begin{bmatrix} 3 & 7 \\ 1 & 4 \end{bmatrix}$

24. $\begin{bmatrix} 6 & 0 & 3 \\ -5 & 2 & 1 \end{bmatrix} + \begin{bmatrix} 3 & -4 & -2 \\ 7 & 1 & 3 \end{bmatrix}$ $\begin{bmatrix} 9 & -4 & 1 \\ 2 & 3 & 4 \end{bmatrix}$

Find the measures of the angles. (Lesson 3-2)

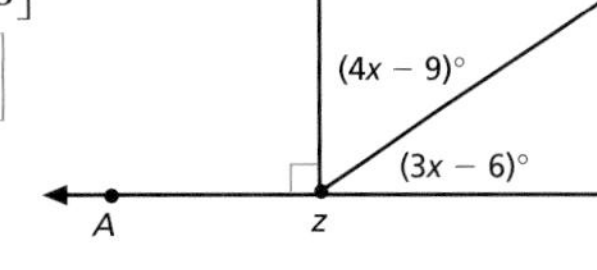

25. $\angle BZC$ 51°

26. $\angle CZD$ 39°

27. $\angle AZC$ 141°

Extend the Lesson

MATH JOURNAL Have students write the definitions of permutations and combinations and explain the difference between them. Have them write one example of each and explain how the number of outcomes can be found for each example.

Extra Practice Worksheet 9-5

Name ______ Date ______

EXTRA PRACTICE 9-5

PERMUTATIONS AND COMBINATIONS

EXERCISES

For each situation, tell whether order does or does not matter.

1. You are writing the digits of your social security number. does matter
2. Four students are selected to be on a committee. does not matter
3. You are choosing toppings for a pizza. does not matter
4. You are recording the digits of your e-mail password. does matter
5. You are listing the types of sandwiches available for lunch. does not matter

Solve.

6. A golfer has 5 balls in her bag. In how many different orders can she use these balls during the course of a game? 120
7. Three of the twenty-five semi-finalists for a contest will make the final round. How many choices of three finalists are there? 2300
8. In how many ways can three books be chosen from a box of ten books? 720
9. Mark is on vacation. He has six places he wants to visit on Tuesday, but he knows he can only make it to four in one day. In how many ways can he choose the four places to visit? 360
10. In how many different ways can you arrange the digits 0 through 9, if 0 cannot be the first or the last digit? 3,265,920
11. In how many different ways can two team captains be chosen from a team of 15 players? 210
12. Find the number of permutations of the letters in the word *computer.* 40,320
13. Ten teams enter a tournament. How many different arrangements of first-, second- and third-place winners are possible? 120
14. In how many different ways can you arrange eight videos in a row on a shelf? 40,320
15. How many different outfits can be made from five pairs of jeans, three T-shirts and three pairs of shoes? 45

122 EXTRA PRACTICE *LESSON 9-5*

Enrichment Worksheet 9-5

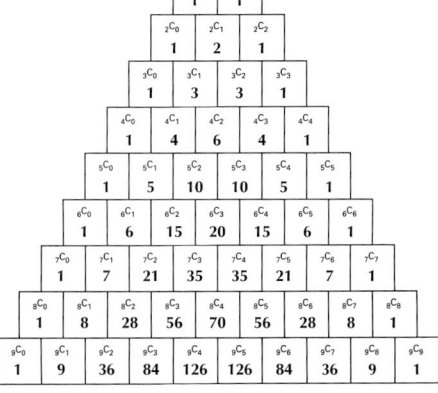

Name ______ Date ______

ENRICHMENT 9-5

COMBINATIONS: A TRIANGULAR PATTERN

EXERCISES

1. Evaluate the expression in each box of the triangle. Look for a pattern.

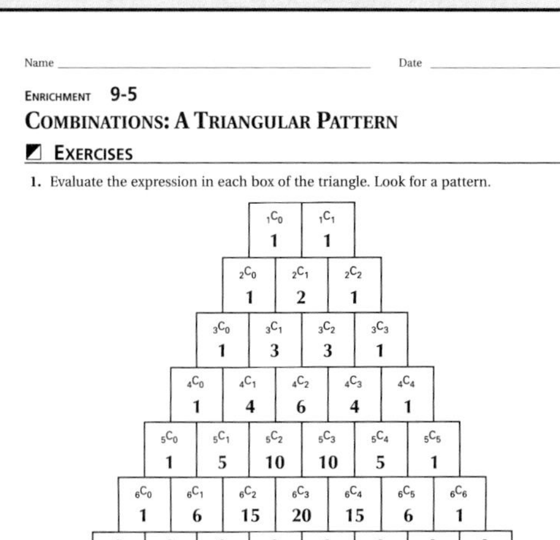

$_1C_0$ 1	$_1C_1$ 1								
$_2C_0$ 1	$_2C_1$ 2	$_2C_2$ 1							
$_3C_0$ 1	$_3C_1$ 3	$_3C_2$ 3	$_3C_3$ 1						
$_4C_0$ 1	$_4C_1$ 4	$_4C_2$ 6	$_4C_3$ 4	$_4C_4$ 1					
$_5C_0$ 1	$_5C_1$ 5	$_5C_2$ 10	$_5C_3$ 10	$_5C_4$ 5	$_5C_5$ 1				
$_6C_0$ 1	$_6C_1$ 6	$_6C_2$ 15	$_6C_3$ 20	$_6C_4$ 15	$_6C_5$ 6	$_6C_6$ 1			
$_7C_0$ 1	$_7C_1$ 7	$_7C_2$ 21	$_7C_3$ 35	$_7C_4$ 35	$_7C_5$ 21	$_7C_6$ 7	$_7C_7$ 1		
$_8C_0$ 1	$_8C_1$ 8	$_8C_2$ 28	$_8C_3$ 56	$_8C_4$ 70	$_8C_5$ 56	$_8C_6$ 28	$_8C_7$ 8	$_8C_8$ 1	
$_9C_0$ 1	$_9C_1$ 9	$_9C_2$ 36	$_9C_3$ 84	$_9C_4$ 126	$_9C_5$ 126	$_9C_6$ 84	$_9C_7$ 36	$_9C_8$ 9	$_9C_9$ 1

2. Describe the pattern formed by the numbers in the triangle.
Pascal's Triangle. Each number equals the sum of the two above it.

3. Use the triangle to write a relationship between $_nC_r$ and $_{n+1}C_{r+1}$.
$_nC_r + {_nC_r} + 1 = {_{n+1}C_{r+1}}$

160 ENRICHMENT *LESSON 9-5*

Lesson Planning

NCTM Standards/Strands

- Number & Operation
- Patterns, Functions, & Algebra
- Data Analysis, Statistics, & Probability
- Problem Solving
- Reasoning and Proof
- Communication
- Connections
- Representation

Vocabulary

scatter plot
line of best fit
trend line
positive correlation
negative correlation
box-and-whisker plot
extremes
quartile

Materials Needed

graph paper
straightedges

Lesson Resources

Warm-Up Transparency 27
Transparency RF-34, 35
Reteaching 9-6
Extra Practice 9-6
Enrichment 9-6

Assignment Guide

Basic: 1–12, 16–18
Enriched: 1–18

Getting Started

5-Minute Warm-up

Find the mean and median for each set of data.

1. 14, 3, 8, 16, 13, 7, 9 10, 9
2. 25, 8, 27, 25, 26, 9 20, 25
3. 8, 1, 9, 7, 1, 0, 9, 5 5, 6

Introduction to Lesson 9-6

Ask students to explain why the graphs they have chosen best represent their data. Have them discuss how their graphs show extremes or clusters of data.

9-6 Scatter Plots and Boxplots

Goals ■ Interpret and make scatter plots and boxplots.

Applications Manufacturing, Sales, Sports

Work in groups of 3–4 students.

Find out your classmates' favorite music performer. Make a list of ten popular music groups or artists. Using a rating scale of 1–100, survey 25 students. Make a graph to display your findings. Compare findings and graphs with classmates.
Answers will vary.

Build Understanding

Data can be presented in many ways. Graphs are useful because they can help identify characteristics of data. Recall that a histogram shows frequencies of *intervals* of data. A stem-and-leaf plot shows *all* data ordered as in a frequency table, but also visually, as in a bar graph. It shows how data are clustered.

A **scatter plot** is another type of visual display used to explore the relationship between two sets of data, represented by unconnected points on a grid.

Example 1

Factory Wages

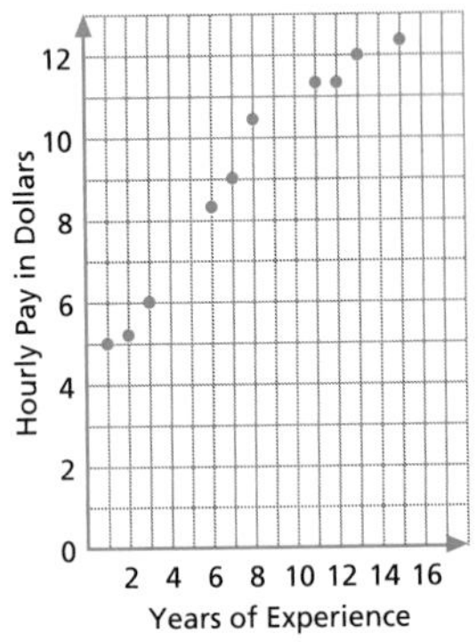

MANUFACTURING The scatter plot shows the relationship between years of experience and hourly pay at one factory.

a. Why are the scales different?

b. What does each • represent?

c. Find the average hourly pay of an employee with 8 years of experience.

d. Describe the relationship between experience and pay?

Solution

a. There are two different sets of data—average hourly pay and years of experience.

b. Each • shows the average hourly pay given the years of experience.

c. $10.50

d. Average hourly pay increases with years of work experience.

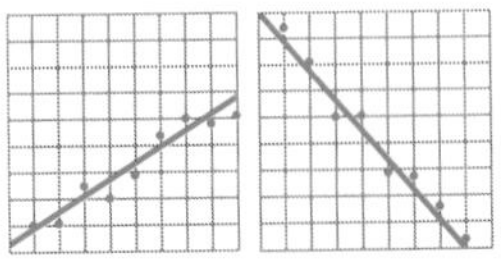

A pattern may emerge that shows a relationship between the two sets of data. If data clusters around a **line of best fit**, or **trend line**, from the bottom left upward to the top right of the graph, this shows a **positive correlation** between the sets of data. If the line slopes downward from left to right, it suggests a **negative correlation** between the data.

Interdisciplinary Connections

SCIENCE You may wish to ask students to conjecture whether there is a correlation between the weight of an object and the rate of speed at which if falls. In 1638 Galileo Galilei established that (in a vacuum, where air resistance is not a factor) all falling bodies, regardless of size, shape, or weight, descend at an equal rate of speed, and so there is no correlation.

Example 2

SALES Use the scatter plot at the right for these questions.

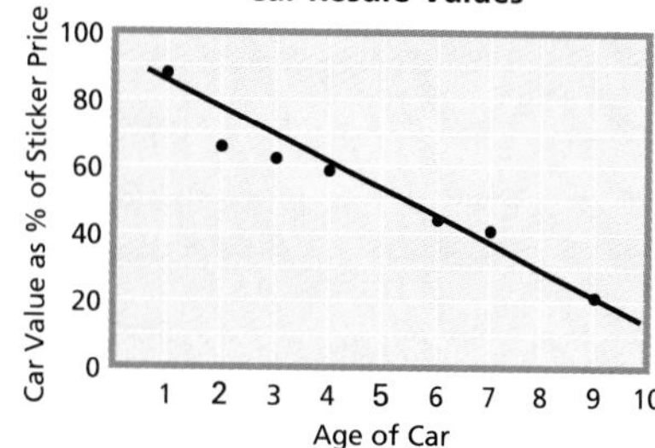

a. What can you say about the correlation between the age of a car and its resale value?

b. Predict the resale value of an 8-year-old car with an original cost of $15,000.

Solution

a. The trend line slopes downward from upper left to lower right, so there is a negative correlation between a car's age and its resale value.

b. Extend the pattern. A reasonable assumption would be for the resale value to be about 30% of the original cost for an 8-year-old car. So, a car that cost $15,000 originally might sell for about $4,500 after 8 years.

Another way to display data is with a **box-and-whisker plot**, also known as **box plot**. This plot shows how data are dispersed around a median, but does not show specific items in the data. By examining a box-and-whisker plot, you can tell if data are clustered closely together or spread far apart.

A box-and-whisker plot shows both the median and the **extremes** of a set of data. It also shows the median of the lower half of the data, called the **lower quartile**, and the median of the upper half of the data, called the **upper quartile**. Both quartiles include the median if the data contain an odd number of items.

Example 3

SPORTS Joe DiMaggio played center field for the New York Yankees for 13 years. During each year of his career, he hit the following number of home runs: 29, 46, 32, 30, 31, 30, 21, 25, 20, 39, 14, 32, and 12. Make a box-and-whisker plot for this data.

Solution

Write the data in numerical order. Find the least and greatest values, the median, the lower quartile, and the upper quartile.

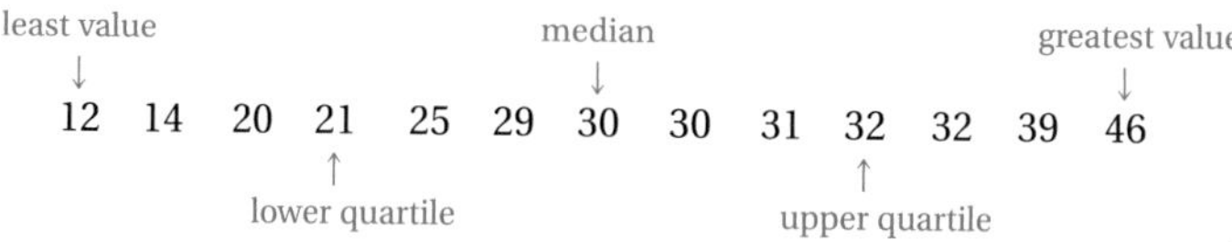

Use points to mark the values below a number line. Draw a box around the upper and lower quartiles, and a vertical line at the point for the median. Then draw *whiskers*, or line segments, from each end of the box to the least and greatest values. Finally, give your graph a title.

Dimaggio's Home Runs

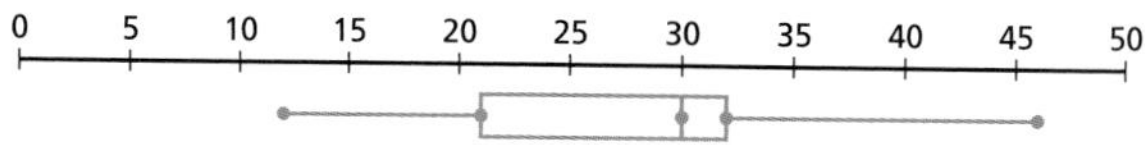

Check Understanding

In a box-and-whisker plot, what percent of a set of data is represented by the box? By the whisker to the right of the box?

50%; 25%

Chalkboard Examples

Supplementary Example 1 and 2

Using the scatter plot in Example 2, predict the resale value of a 5-year-old car with an original cost of $22,000. at 55% of original value; $12,100

Supplementary Example 3 and 4

The test scores from two classes are shown below. Make box-and-whisker plots from the data. Which class had its scores grouped more closely around its median? Class A For which class were the lowest scores clustered more closely? Class B

Class A: 14, 19, 22, 25, 24, 18, 15, 12, 16, 20, 24, 20, 18, 10, 25, 22, 24, 15, 13, 20

Class B: 10, 13, 14, 22, 24, 25, 20, 9, 14, 12, 15, 22, 24, 10, 12, 24, 24, 25, 11, 13

Extend the Lesson

ONGOING ASSESSMENT Have students find the median, the lower quartile, and the upper quartile for the following set of data: 1, 10, 2, 3, 11, 13, 4, 12. The median is 7, lower quartile is 2.5; upper quartile is 11.5. Watch for students who get a lower quartile of 3 and an upper quartile of 11. These students have found the median correctly and have then included the median in the set of data in order to find the quartiles.

Reteaching Worksheet 9-6

Name ____________________ Date __________

RETEACHING 9-6

SCATTER PLOTS AND LINES OF BEST FIT

The *scatter plot* and the *boxplot* can be used to display data.

Example 1

Use the scatter plot to predict how many cups of hot chocolate would be sold if the temperature was 45°F.

Cups Sold 40 30 20 10; 10 20 30 40 50 Temperature (%F)

Solution

Extend the pattern. About 15 cups would be sold.

Example 2

Make a boxplot for the number of cups of hot chocolate sold in September.

Number of cups: 15, 20, 30, 12, 18, 29, 17, 25.

Solution

Step 1: Write the data from least to greatest. 12 15 17 18 20 25 29 30

Step 2: Separate the data into 4 equal groups. Find the median of all the cups. Then find the medians of the lower half and the upper half. 12 15 ¦16¦ 17 18 ¦19¦ 20 25 ¦27¦ 29 30 — 1st Quartile, Median, 2nd Quartile

Step 3: Draw a box above a number line that extends from the first quartile to the third quartile. Draw a vertical line that indicates the median and extend a line from the box on either side to the highest and lowest scores. 10 15 20 25 30

EXERCISES

Use the scatter plot in Example 1. Predict how many cups you would sell at the next game if the:

1. temperature is 50°F. 10 cups
2. temperature is 0°F. 40 cups
3. On another sheet of paper, make a boxplot for these test scores: 73, 98, 47, 87, 92, 75, 93, 85. Check students' boxplots.
   a. What is the median? 86
   b. Are the data evenly distributed? No
   c. What percent of the test scores are contained in the box? 50%
   d. Write a short paragraph describing the data. Answers may vary. Possible answer: The greatest test score is 98; the least is 47. Half of the data is in the box clustered around the median 86.
4. On another sheet of paper, make boxplot for these test scores: 67, 79, 98, 83, 87, 56, 89, 75. What percent of the test scores are: Check students' boxplots.
   a. less than 71? 25%
   b. greater than 71? 75%
   c. greater than 81? 50%

142 RETEACHING *LESSON 9-6*

Lesson 9-6 **Scatter Plots and Boxplots** 407

Lesson Wrap-up

Quick Assessment

Ask the following questions to determine if students understand the content presented in this lesson.

1. How can you use a scatter plot to make predictions? Extend the line that the given points approximate to find the predicted points.
2. What does the length of the boxes and whiskers shown in a box-and-whisker plot? how the data are spread out; the longer the boxes or whiskers, the more spread out the data

Assignment Guide

Basic: 1–12, 16–18
Advanced: 1–18
Additional Practice: See Extra Practice Index on page 674.

Box-and-whisker plots can be used to compare sets of data.

Example 4

Use the box-and-whisker plots below to answer questions about the math test scores of two different classes.

Test Scores From Two Classes

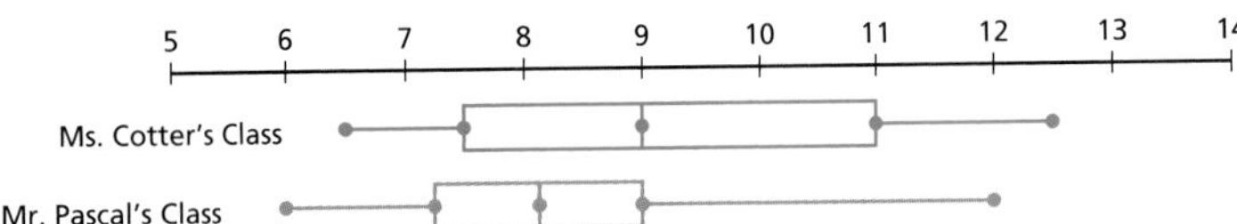

a. Which class had the higher median score?

b. What was the lower quartile in Mr. Pascal's class?

c. Which class had its scores grouped more closely around its median?

d. For which class were the lowest scores clustered more closely?

e. Which class, as a whole, scored better on the test?

Reading Math

The abbreviations Q_1, Q_2, and Q_3 are sometimes used for the lower quartile, the median, and the upper quartile. The **interquartile range** is the difference between the values of the upper and lower quartiles.

Solution

a. Cotter's, 9

b. 7.2

c. Pascal's; the range of the middle 50% of the scores is 1.8. The range for the middle 50% in Ms. Cotter's class is 3.5.

d. Cotter's; the range is 1. In Mr. Pascal's class, it is 1.2.

e. Cotter's

Try These Exercises

FITNESS Use the scatter plot at the left for Exercises 1–3.

Heights And Weights

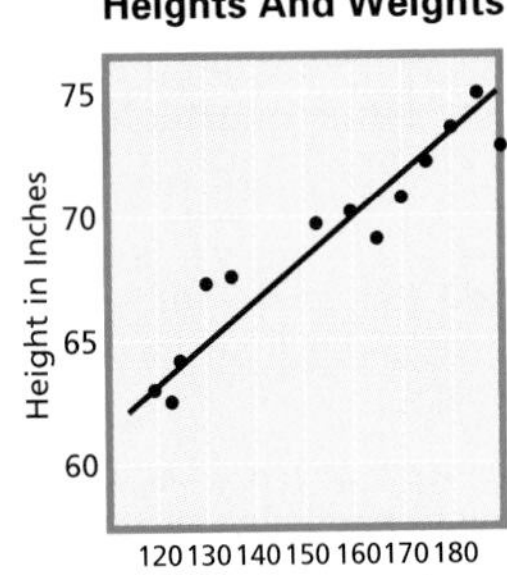

1. What is the weight of the student who is 65 in. tall? About 130 lbs
2. Does the scatter plot show a positive or negative correlation? positive
3. Give an estimate of the height of a student who weighs 145 lb. 67–68 in.
4. **SPORTS** Make a box-and-whisker plot for the following data. Check students' plots; look for median; 37.5; upper quartile: 45; lower quartile: 25; extremes: 10 and 60

TOP PRICES OF TICKETS TO SPORTING EVENTS (IN DOLLARS)

45, 55, 40, 60, 15, 25, 35, 30, 10, 40

5. Use the box-and-whisker plot you made in Exercise 4. Are the data clustered more closely above or below the median? above

Technology Note

Use a graphing calculator to make a boxplot.

1. Enter the data as a list using the STAT feature.
2. Select the STATPLOT menu and choose Plot1. Under Type, select the boxplot diagram.
3. Adjust the window dimensions if necessary and press GRAPH.
4. Use the TRACE feature to find the median and upper and lower quartiles.

Teaching Strategies

COOPERATIVE LEARNING Have pairs of students survey at least 20 people with a wide range of ages to find how much they like rock music and how much they like classical music, using a rating scale of 1–10. Have students make scatter plots to display the data and conclude if there is a correlation between age and the type of music people prefer.

Practice Exercises

SPORTS This table shows how many points a basketball player scored during his career. Use this information for Exercises 6–8.

Age	Scoring Average
23	18
24	17.5
25	22.5
26	24
27	21.5
28	26
29	23.5
30	22.5
31	27.5
32	20.5

6. Make a scatter plot. Check students' drawings.
7. What is the range of this player's scoring average? 10 points
8. Does your scatter plot show a positive correlation, a negative correlation, or no correlation? No correction

SPORTS These box-and-whisker plots show batting averages for 3 baseball teams.

Player Batting Averages

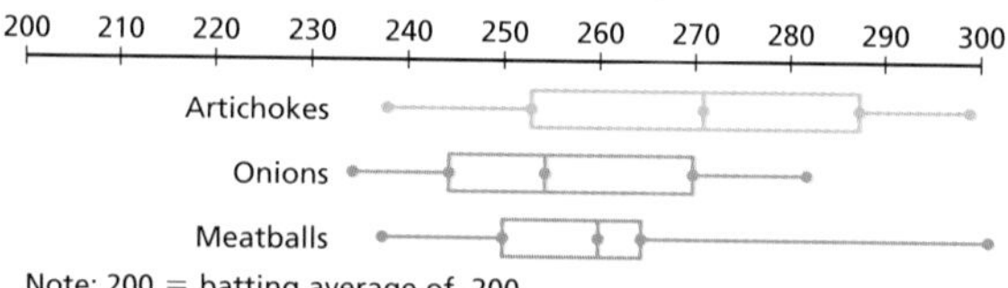

Note: 200 = batting average of .200.

9. Which team has the highest median batting average? Artichokes
10. Which team has the smallest range of batting averages? Onions

11. **WRITING MATH** Why is the right whisker for the Meatballs longer than the left whisker? There is a greater range of batting averages in the upper 25% of its players than in the lower 25%.
12. **CHAPTER INVESTIGATION** Create a 10-by-10 table for each player. Number both the columns and rows from 1 to 10. The table represents all the possible outcomes for a player at bat. Using the percents you calculated, fill in the cells of the table with appropriate abbreviations. For example, in 1998 Gary Sheffield hit a home run in 5% of his at bats. To create a table for Sheffield, you would write HR in any 5 cells. Answers will vary.

Extended Practice Exercises

Choose the graph you think works best to display the data described.
Answers will vary. Sample answers are given.

13. **MANUFACTURING** To show the relationship between the percent of polyester in an article of clothing and the price of the article of clothing scatterplot
14. To show that the test scores in your class clustered around the middle-most score box-and-whisker-plot
15. **WRITING MATH** Is it possible for the mean of a set of data to fall outside the box part of a box-and-whisker plot? Explain. Yes, when there are extreme values

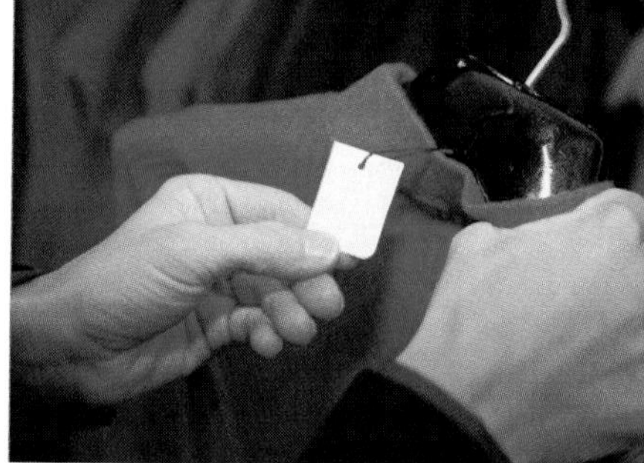

Mixed Review Exercises

Multiply. (Lesson 8-5)

16. $6 \cdot \begin{bmatrix} 6 & 3 \\ -4 & 2 \end{bmatrix}$ $\begin{bmatrix} 36 & 18 \\ -24 & 12 \end{bmatrix}$

17. $4 \cdot \begin{bmatrix} 9 & 8 \\ 7 & 1 \end{bmatrix}$ $\begin{bmatrix} 36 & 32 \\ 28 & 4 \end{bmatrix}$

18. $7 \cdot \begin{bmatrix} 7 & -3 & -5 \\ 4 & 6 & 0 \end{bmatrix}$ $\begin{bmatrix} 49 & -21 & -35 \\ 28 & 42 & 0 \end{bmatrix}$

Extend the Lesson

MATH JOURNAL Have students give examples of cases when a scatter plot or a box-and-whisker plot might help them evaluate data. Have them discuss why they might use one plot instead of another. Sample answer: A scatter plot might be used to see if there is a relationship between numbers of people at a pool and temperature, while a box-and-whisker plot might be used to see how heights of students in one class disperse about the median height.

Extra Practice Worksheet 9-6

Name ______ Date ______

Extra Practice 9-6

Scatter Plots and Lines of Best Fit

Exercises

The table at the right shows the scoring averages for the players on the Lady Tigers basketball team scored during last season.

Age of Player	Scoring Average
19	2
14	8
16	6
18	10
15	12
6	15.5
18	4.5
17	6
15	4
18	8
14	2
16	4

Use this information for Exercises 1–4.

1. Make a scatter plot on your own paper. Check students' plots.
2. What is the range of the Lady Tiger's scoring averages? 13.5
3. Does your scatter plot show a positive correlation, a negative correlation, or no correlation? no correlation
4. Can you use this data to make a prediction of a possible scoring average for an 20-year-old player? Why or why not? No, since no correlation is shown, no predictions can be made using the data.
5. Make a boxplot for the following data. Use your own paper.
   FINAL EXAM GRADES FOR MR. TRENT'S CLASS
   75, 88, 90, 94, 60, 78, 90, 99, 75, 62, 55, 80
   a. Are the data clustered more closely above or below the median? about the same
   b. What does the plot tell you about the scores? They are fairly evenly dispersed above and below the median.

124 Extra Practice Lesson 9-6

Enrichment Worksheet 9-6

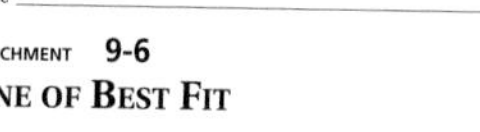

Name ______ Date ______

Enrichment 9-6

Line of Best Fit

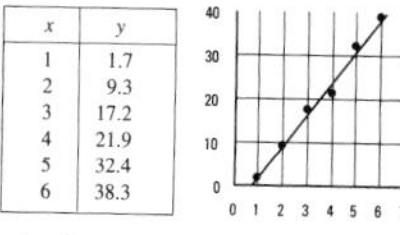

x	y
1	1.7
2	9.3
3	17.2
4	21.9
5	32.4
6	38.3

In the graph at the right, the data from the table have been plotted and the line of best fit or trend line drawn. It is hard to write the equation of this line from the graph.

However, you can use a graphing calculator to find the equation for the trend line for these data.

Use the statistics function. Begin by entering the ordered pairs (x, y). Then use the linear regression function. The calculator will display values for a, b, and r. The values of a and b are for a linear equation of the form $y = a + bx$. In this case $a = 5.5666666667$, $b = 7.342857143$ and $r = 0.9969321206$. If you round a and b to the nearest tenth, you get: $y = -5.6 - 7.3x$.

You can rewrite this in the $y = mx + b$ form that you have used in this text and get: $y = 7.3x - 5.6$.

Notice the value of r, the *correlation coefficient*, which is a measure of how good the fit is. If r is close to $+1$ or -1, the fit is very good.

Exercises

Draw the graph for each data table. Then find an approximate linear equation for the graph. (Round a and b to the nearest tenth). Then determine whether the fit is good by giving the value of r to four decimal places.

1.

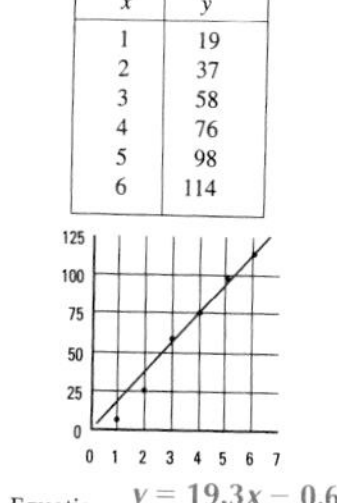

x	y
1	19
2	37
3	58
4	76
5	98
6	114

Equation: $y = 19.3x - 0.6$
$r =$ 0.9994

2.

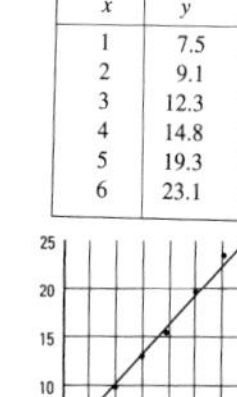

x	y
1	7.5
2	9.1
3	12.3
4	14.8
5	19.3
6	23.1

Equation: $y = 3.2x - 3.2$
$r =$ 0.9899

3.

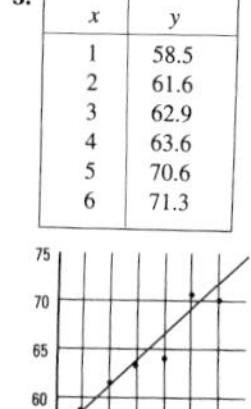

x	y
1	58.5
2	61.6
3	62.9
4	63.6
5	70.6
6	71.3

Equation: $y = 2.6x + 55.6$
$r =$ 0.9581

162 Enrichment Lesson 9-6

Skills Practice

Vocabulary Review

Lesson 9-5
fundamental counting principle
permutation
factorial notation
combination

Lesson 9-6
scatter plot
line of best fit
trend line
positive correlation
negative correlation
box-and-whisker plot
extremes
quartile

Assignment Guide

All students: 1–40
Additional Practice: Refer to the Extra Practice Index on page 674 of this text.

Additional Answers

25.

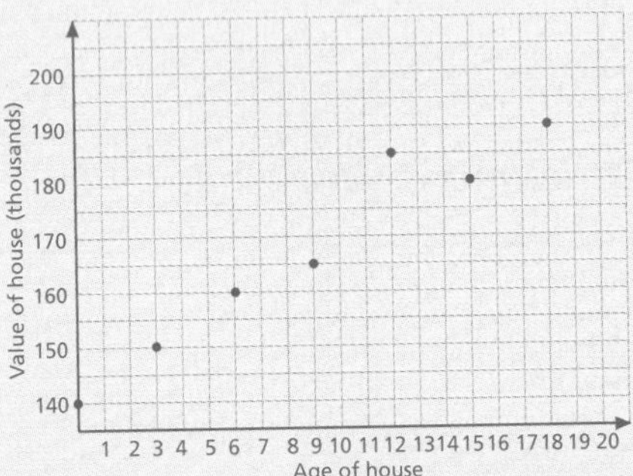

28.

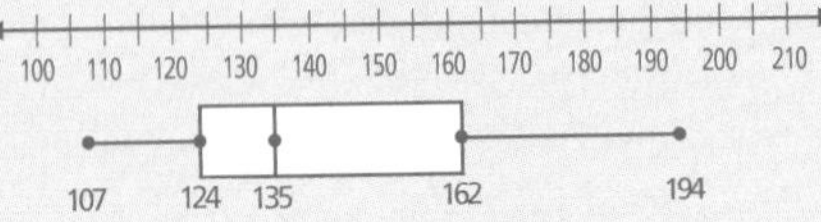

Review and Practice Your Skills

Practice ■ Lesson 9-5

Evaluate.

1. $_5P_3$ 60 **2.** $_8P_7$ 40,320 **3.** $_4P_1$ 4 **4.** $_5P_5$ 120

5. $_7P_4$ 840 **6.** $_9P_0$ 1 **7.** $_{14}P_7$ 17,297,280 **8.** $_9P_6$ 60,480

9. $_5C_3$ 10 **10.** $_9C_6$ 84 **11.** $_4C_0$ 1 **12.** $_6C_1$ 6

13. $_8C_7$ 8 **14.** $_5C_4$ 5 **15.** $_6C_2$ 15 **16.** $_{12}C_4$ 495

17. In how many ways can the positions of president, vice president, and secretary be chosen from a club containing 20 members? 6840

18. In how many ways can a committee of three people be chosen from a club containing 20 members? 1140

19. In how many ways can a volleyball coach choose 6 starters from a team of 14 players? 3003

20. In how many ways can a disc jockey play 3 of the top ten hits? 720

21. In how many ways can first-place and runner-up winners be chosen from 15 entrants in a contest? 210

22. In how many ways can the numbers 1, 2, 3, 4, and 5 be arranged in a 5 digit password? 120

23. What is n, if $_nP_2 = 72$? 9

24. Find the values of $_{10}C_4$ and $_{10}C_6$. What do you notice? 210, 210, they are equal.

Practice ■ Lesson 9-6

This table shows the appraised value of a house over time.

Age (years)	0	3	6	9	12	15	18
Value (thousands)	140	148	160	162	185	178	194

25. Make a scatter plot of the data. See additional answers.

26. What is the range of appraised values? 140,000–194,000 (54,000)

27. Does your scatter plot show a positive correlation, a negative correlation, or no correlation? positive

28. Draw a box-and-whisker plot that has the following attributes: (Lesson 9-6)
a. range of 87
b. median value of 135
c. low value of 107
d. upper quartile of 162
e. range of middle 50% of data of 38
See additional answers.

29. Draw a box-and-whisker plot that has the same value for its maximum value and its upper quartile. Define a collection of data points that would yield this type of plot. (Lesson 9-6) Answers will vary.

Teaching Strategies

COOPERATIVE LEARNING Have students work in small groups to make a box-and-whisker plot for the following set of test scores: 77, 80, 75, 73, 77, 81, 62, 87, 99, 85, 82, 81, 77, 72, 78, 83, 86, 79, 80, 78. Have students determine what the plot tells them about the scores. Have groups share their results with the class.

PRACTICE ■ LESSON 9-1–LESSON 9-6

A spinner with 8 equal sectors labeled A through H is spun. Find each probability. (Lesson 9-1)

30. *P*(spinning E) $\frac{1}{8}$ **31.** *P*(spinning vowel) $\frac{1}{4}$ **32.** *P*(spinning H) $\frac{1}{8}$

33. *P*(spinning a letter before F) $\frac{5}{8}$ **34.** *P*(spinning B or G) $\frac{1}{4}$ **35.** *P*(spinning M) 0

36. Describe a simulation you could do to find out how many cards you would expect to have to draw from a standard deck to get three clubs. (Lesson 9-2) Answers will vary.

For each situation, tell whether order does or does not matter. (Lesson 9-5)

37. You are selecting three-number combinations for school lockers. does

38. You are selecting five books to check out from the library. does not

39. You are choosing the 9 starters on a baseball team. does

40. You are choosing a 5-member committee from your leadership board. does not

Career – Physical Therapist

Physical therapists work with people who have been injured. They improve mobility, relieve pain and prevent or limit permanent physical disabilities. To relieve pain and treat injuries, physical therapists use massages, electrical stimulation, hot and cold packs and traction.

The sports world depends on physical therapists to help athletes who are injured during practices, exercise sessions or games. Some injuries require surgery, but many can be treated by rest followed by proper exercise. Physical therapists often travel with teams.

1. As a physical therapist, you have treated 236 injuries during the past year. Of the total number, 108 injuries were caused by an improper warm-up. What percent of the injuries resulted from an improper warm-up? About 46%

2. A team of 45 players suffered 15 ankle injuries during the season. If the squad is increased to 52 players, how many ankle injuries would you expect to see during a season? 17 to 18 ankle injuries

3. A physical therapist employs 4 kinds of massage, 3 kinds of baths, 1 type of electrical stimulation, and 8 exercise programs. If each injury is treated with all four types of therapies, how many combinations of therapies does this therapist offer? 96

4. You are examining the player files for 30 players who have been injured during the season. For this sport, 1 out of 3 injuries are to the knee and 1 out of 2 of these cases require surgery. What is the probability that the first file you select will belong to a player requiring knee surgery? $\frac{1}{6}$

Chalkboard Examples

Example from Lesson 9-5

What is the probability that a four-digit number formed at random from each of the digits 1, 2, 3, and 4 will be greater then 4,231? $\frac{1}{12}$

Example from Lesson 9-6

The following data tell the number of books read a month by each of six students: 3, 1, 2, 1, 8, 2. Make a box-and-whisker plot to display the data. How does the plot show distorted data? no left whisker; right whisker (25% of the data) is longer than the boxes (75% of the data).

Career Opportunity

Describe the kind of work physical therapists do and the types of places they work. Discuss the importance of geometry, probability, and algebraic thinking in working as a physical therapist. Students should answer Questions 1-4 to better understand how physical therapists use mathematics in performing their job.

Students who are interested in learning more about this profession can go to learningmathmatters.com. Guidance Counselors should have information about school and training requirements.

Extend the Lesson

MATH JOURNAL Have students answer the following question and explain their answer in a short paragraph: What kind of correlation, if any, is there between average temperature in a region and a family's heating oil bills? You may wish to have students make up data and make a scatter plot to display them. There is a negative correlation.

Lesson Planning

NCTM Standards/Strands

- Number & Operation
- Patterns, Functions, & Algebra
- Data Analysis, Statistics, & Probability
- Problem Solving
- Reasoning and Proof
- Communication
- Connections
- Representation

Vocabulary

variance standard deviation
z-score
measures of dispersion
frequency distribution
bell curve normal curve

Materials Needed

calculators

Lesson Resources

Warm-Up Transparency 27
Transparency RF-37
Extra Practice 9-7

Assignment Guide

Basic: 1–20, 25–30
Enriched: 1–30

Getting Started

5-Minute Warm-Up

Solve for *m*.

1. $m = \frac{(32 + 42 + 52 + 62)}{4}$ 21.5
2. $m = (-2 - 1)^2 + (-1 - 1)^2$ 13
3. $m = \frac{[(1 - 3)2 + (2 - 3)2 + (3 - 3)2 + (4 - 3)2]}{4}$ 1.5

Introduction to Lesson 9-7

Have students decide on the type of data to collect. Have students compare data sets for similarities and differences and discuss common characteristics for data sets representing a general population.

9-7 Standard Deviation

Goals
- Find the variance, standard deviation and z-scores for a set of data.
- Use standard deviation to interpret data.

Applications Sports, Test-Taking, Education

Work in groups of 4–5 students. Answers will vary.

1. Collect a set of data about students in class, such as heights, arm lengths, head circumference, lengths of thumbs and so on.
2. Study the data. Look for new ways to describe the data. Instead of focusing on central tendencies, study how the data are spread out, or *dispersed.*
3. Consider these questions: How much do individual values in your data differ from the greatest value? The least value? The mean, median or mode of the values?
4. Share your results with your classmates.

Build Understanding

Statistics that show how data spreads out are called *measures of dispersion.* For example, you know that the *range* of a set of data is the difference between the largest and smallest item.

Variance is another measure of dispersion. The variance of a set of numbers is the mean of the squared differences between *each* number in the set and the mean of *all* numbers in the set. For the set of numbers $x_1, x_2, \ldots x_n$, with a mean of m, use this formula.

$$m = \frac{(x_1 - m)^2 + (x_2 - m)^2 + \ldots + (x_n - m)^2}{n}$$

Example 1

SPORTS During a basketball tournament, the five starters for the Bulldogs made the following number of 3-pointers: Bowen, 3; White, 4; Fillmore, 5; Graham 6; and Bonilla, 7. Find the variance for the set of numbers.

Solution

1. Divide the sum of scores by 5 to find the mean, m. ($m = 5$)
2. Find the difference between each number and the mean. Then find the square of each difference.
3. Find the mean of all the squares in Step 2.

   $4 + 1 + 0 + 1 + 4 = 10 \qquad 10 \div 5 = 2$

number	$x - m$	$(x - m)^2$
3	$3 - 5$	$(-2)^2 = 4$
4	$4 - 5$	$(-1)^2 = 1$
5	$5 - 5$	$0^2 = 0$
6	$6 - 5$	$1^2 = 1$
7	$7 - 5$	$2^2 = 4$

The variance is 2.

The **standard deviation**, s, of a set of numbers is the square root of the variance.

Teaching Strategies

REAL WORLD CONNECTION The total number of items for which a prediction is made is called the population (all students). A sample (one class) is the set of items that represent the population. When considering a sample, statisticians have found that better estimates are obtained if you use the following formula:

$$\text{sample variance} = \frac{[(x_1 - m)^2 + (x_2 - m)^2 + \ldots + (x_n - m)^2]}{n - 1}.$$

Example 2

Find the standard deviation for the set of numbers in Example 1.

Solution

Find the square root of the variance.

$\sqrt{2} \approx 1.4$ The standard deviation is 1.4

Example 3

Molly took two tests. On which did she score better, compared with others in her class?

	Test A	Test B
Molly's score	85	80
Mean score	65	60
Standard deviation	8	10

Solution

1. Compare both of her scores with the mean. She scored 20 points higher than the mean on both tests.
2. Use the standard deviations.

In Test A, Molly's score was $\frac{20}{8}$, or 2.5 standard deviations above the mean score.

In Test B, it was $\frac{20}{10}$, or 2 standard deviations above the mean score.

Relative to her classmates, Molly scored better on Test A.

The number of standard deviations between a score and the mean score is indicated by a ***z*-score**. Molly's z-score was 2.5 on Test A. A score below the mean would have a negative z-score.

Try These Exercises

Compute the variance and standard deviation for each set of data.

1. 4, 4, 4, 4, 4 0; 0
2. 7, 3, 5, 9, 11 8; $\sqrt{8}$
3. 2.7, 4.7, 6.7, 8.7, 10.7 8; $\sqrt{8}$

Practice Exercises

Compute the variance and standard deviation for each set of data.

4. 6, 6, 6, 6 0; 0
5. 1.5, 2.5, 3.5, 4.5, 5.5 2; $\sqrt{2}$
6. 4.2, 9.2, 14.2, 19.2, 24.2 50; $\sqrt{50}$
7. 8.9, 4, 9.4, 26.5, 14.9 ≈60; $\sqrt{60}$
8. Raymond took two tests. On the first test, his score was 45, while the mean score was 55 and the standard deviation was 5. On the second test, his score was 55, while the mean score was 65 and the standard deviation was 10. On which test did Raymond score better, relative to the scores of his classmates? second test
9. On a science test taken by 28 students, the mean score was 82.5. The standard deviation for the scores was 5.3. What was the sum of all the scores? 2310
10. **WRITING MATH** What can you say about the relationship between the standard deviation of a set of scores and how spread out the scores are? Generally, the higher the standard deviation, the more spread out the scores are.

Technology Note

This computer program will compute the standard deviation for a given set of data.

```
10 DIM A(100)
20 FOR I = 1 TO 100
30 READ A(I)
40 IF A(I) = −1 THEN 70
50 LET S = S + A(I)
60 NEXT I
70 LET M = S/(I−1)
80 FOR J = 1 TO I − 1
90 LET D = A(J) − M
100 LET Q(J) = D ^ 2
110 NEXT J
120 FOR K = 1 TO I − 1
130 LET U = U + Q(K)
140 NEXT K
150 LET T = (U/(I−1)) ^ .5
160 PRINT " The Standard
    Deviation is "; T
170 DATA 86,56,98,76,−1
180 END
```

More or different data can be entered on line 160, but the −1 must be the last entry.

If you have more than 100 data, change line 10 and 20.

Teaching Strategies

CHALLENGE Use the data to find the probability that Jo will score two standard deviations above the mean or better on her next test. $\frac{1}{3}$

	Test 1	Test 2	Test 3
Jo	84	85	75
mean	78	80	80
standard deviation	3	5	2

Chalkboard Examples

Supplementary Example 1

During an 8-game basketball series, the five starters made the following number of three-pointers.

Bowen, 11; White, 7; Fillmore, 9; Graham, 6; Bonilla, 2

Find the variance of the set of numbers. 9.2

Supplementary Example 2

Find the standard deviation of the set of numbers in Supplementary Example 1.

Solution: $\sqrt{9.2} \cup 3.03$

Supplementary Example 3

Lewis scored a 60 on the first test he took, while the mean score was 65 and the standard deviation was 5. On the second test, his score was 75, while the mean score was 85 and the standard deviation was 8. On which test did Lewis score better, relative to the scores of his classmates? first test

Reteaching Worksheet 9-7

Name ______________________ Date ______________

RETEACHING 9-7

STANDARD DEVIATION

Measures of dispersion such as range, variance and standard deviation are often used to describe a set of data. To find the variance of a collection of numbers $x_1, x_2, \ldots x_n$ with a mean of m, use this formula.

$$\frac{(x_1 - m)^2 + (x_2 - m)^2 + \ldots + (x_n - m)^2}{n}$$

To find the standard variation, find the square root of the variance. The closer the standard deviation is to zero, the closer the data are grouped around the mean. As the standard deviation becomes greater, the data become more widely dispersed.

Example

Find the variance and standard deviation for this set of test scores: 95, 88, 76, 91, 85.

Solution

Step 1: Find the mean. $\frac{95 + 88 + 76 + 91 + 85}{5} = 87$

Step 2: Find the variance. $\frac{(95-87)^2 + (88-87)^2 + (76-87)^2 + (91-87)^2 + (85-87)^2}{5}$

$= \frac{(8)^2 + (1)^2 + (-11)^2 + (4)^2 + (-2)^2}{5} = \frac{206}{5} = 41.2$

Step 3: Find the standard deviation. $\sqrt{41.2} \approx 6.4$ rounded to the nearest tenth

EXERCISES

Find the variance and standard deviation for each set of data in Exercises 1–10. Round each answer to the nearest tenth if necessary.

1. 3, 5, 4, 1, 2 2; $\sqrt{2}$ or 1.4
2. 2, 8, 5, 9, 1 10; $\sqrt{10}$ or 3.2
3. 4, 6, 9, 3, 8 5.2; $\sqrt{5.2}$ or 2.3
4. 8, 2, 10, 6, 9 8; $\sqrt{8}$ or 2.8
5. 80, 77, 82, 87, 79 11.6; $\sqrt{11.6}$ or 3.4
6. 37, 46, 29, 21, 32 69.2; $\sqrt{69.2}$ or 8.3
7. 78, 68, 73, 76, 85 31.6; $\sqrt{31.6}$ or 5.6
8. 78, 78, 95, 87, 62 121.2; $\sqrt{121.2}$ or 11.0
9. 5, 7, 9, 11 5; $\sqrt{5}$ or 2.2
10. 4.5, 2.5, 4.5, 2.5 1; $\sqrt{1}$ or 1
11. In which of Exercises 1–10 were the data more closely clustered around the mean? Exercise 10
12. In which of Exercises 1–10 were the data more widely dispersed? Exercise 8
13. LaWanda's class took two tests. On the first test, her score was 76, the mean score was 85 and the standard deviation was 6. On the second, her score was 70, the mean score was 79 and the standard deviation was 8. On which test did she score better, relative to the scores of her classmates? second

144 RETEACHING *LESSON* 9-7

Lesson 9-7 **Standard Deviation** 413

A visual display that shows the relative frequency of data is called a **frequency distribution**. A histogram is often used for this purpose.

11. Find the mean, median, mode, variance, and standard deviation to the nearest whole number for the data presented below.

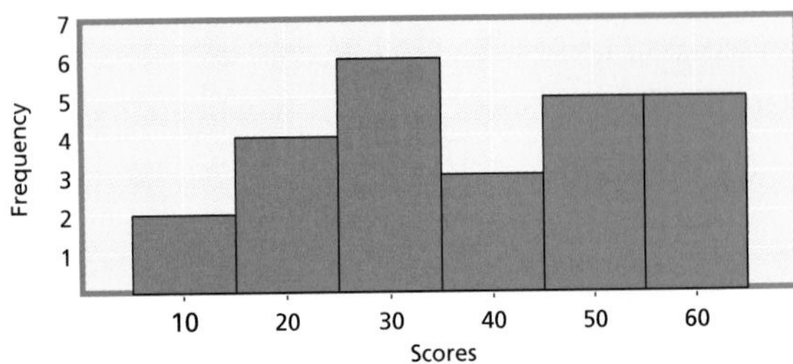

mean: 38; median: 40; mode: 30; variance: 256; standard deviation: 16

TEST TAKING Find out how well your classmates would score on a test on which they had to guess the answer to *every* question. Work in a small group of 3–4 students. Make up a 20-question multiple-choice test using the topic of obscure and unimportant sports data. Use an almanac, a sports encyclopedia, a book of records, or any other source to find facts unfamiliar to anyone in your class. Ask everybody to take the test. Then grade the test as a class. Record and analyze the results. Answers will vary.

12. Find the range, mean, median and mode for the scores.

13. Find the standard deviation.

14. Make a visual display of the scores.

Have the class retake the test. Then analyze the results.

15. Find the range, mean, median and mode for the new scores.

16. Find the standard deviation.

17. Make a visual display of the scores.

Compare both sets of results.

18. On which test did the class perform better? Explain.

19. On which test did you perform better relative to your classmates? What was your *z*-score on that test?

20. **CHAPTER INVESTIGATION** Play the baseball simulation game. Draw a baseball diamond and use coins for markers. To play, two people choose nine baseball players each. Put the players' tables in batting order.

Use 10-sided polyhedral dice or a deck of standard playing cards with the kings, queens and jacks removed. Either roll two dice or draw two cards from the deck. The first die or card indicates the row on the player's table. The second die or card indicates the column. Find the cell at the intersection of the row and column to see what happens in the game.

If the baseball player gets a hit, place a marker in the appropriate place on the baseball diamond. Keep score as you would in a real baseball game. Play nine innings. Did your team do as well as you expected.

Check students' work.

Lesson Wrap-up

Quick Assessment

Ask the following questions to determine if students understand the content presented in this lesson.

1. Stan found the variance of a data set to be -16. Why must his answer be incorrect? Variance is the sum of squared differences from the mean divided by *n*, so it must be nonnegative.
2. Suppose two sets of data have the same mean but different standard deviations s1 and s2 with $s_1 < s_2$. Which data set will show more variability? The set with standard deviation s_2 would show more variability; sets with data close to the mean have smaller standard deviations than sets with most data far from the mean.

Assignment Guide

Basic: 1–20, 25–30
Advanced: 1–30
Additional Practice: See Extra Practice Index on page 674.

Teaching Strategies

Students may be required to find the variance and standard deviation for data involving units such as centimeters or feet. Point out that the unit for the variance is the square of the unit for the original variable, but the unit for the standard deviation is the same as that of the original variable.

Extended Practice Exercises

If you were to use a smooth curved line to connect the midpoints of the histogram above, you would form a frequency distribution known as a **bell curve**.

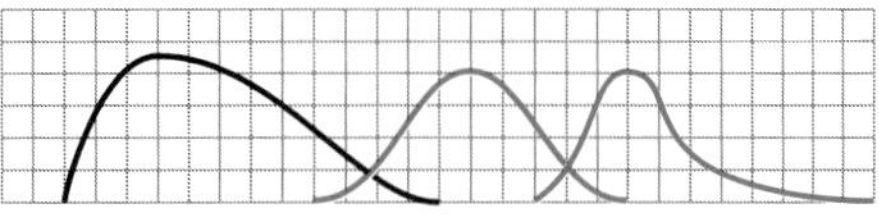

The **normal curve** is the best known frequency distribution. In a normal curve, the mean, median, and mode are the same.

Normal curves are determined by the mean and the standard deviation. In every normal curve, about 68% of the data are within one standard deviation unit of the mean. About 95% of the data are within two standard deviation units of the mean. Finally, about 99.7% of the data are within three standard deviation units.

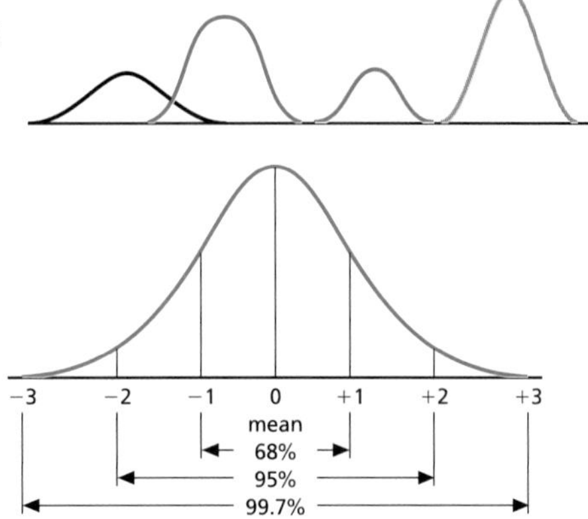

21. Suppose you drew two normal distributions on the same set of axes. Compare the appearances of these two curves if one has a greater mean than the other, but their variances are the same? The one with the higher mean is further to the right.
22. Suppose two normal curves drawn on the same set of axes have different variances but equal means. Compare the curves. The one with the greater variance is not as high as the one with the smaller variance.
23. Which of these bell curves do you think might show the distribution of scores if your class were to take a third-grade spelling test? The right one.

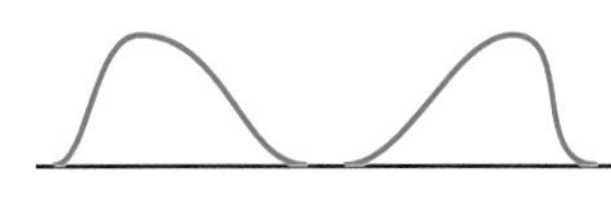

24. As items in a set of data are dispersed more and more widely from the mean, what happens to the standard deviation? It increases.

Mixed Review Exercises

Find each product. If not possible, write *NP*. (Lesson 8-6)

25. $[3 \quad 2 \quad 4] \cdot \begin{bmatrix} -1 \\ 6 \\ -5 \end{bmatrix}$ $[-11]$

26. $\begin{bmatrix} 2 & 0 \\ 1 & 5 \end{bmatrix} \cdot \begin{bmatrix} 3 & 2 \\ 8 & 4 \end{bmatrix}$ $\begin{bmatrix} 6 & 4 \\ 43 & 22 \end{bmatrix}$

27. $\begin{bmatrix} 5 & -4 \\ 3 & 1 \end{bmatrix} \cdot \begin{bmatrix} 0 & -5 & 6 \\ 2 & 8 & -4 \end{bmatrix}$ $\begin{bmatrix} -8 & -57 & 46 \\ 2 & 7 & 14 \end{bmatrix}$

Solve. (Lesson 3-1)

28. On a number line, the coordinate of point F is -4. The length of $\overline{FG}$ is 13. Give 2 possible coordinates of point G. 9, −17
29. On a number line, the coordinate of point Q is -18. The length of $\overline{QR}$ is 76. Give 2 possible coordinates of point R. 58, −94
30. Point S is between points R and T. The length of $\overline{RS}$ is twice the length of $\overline{ST}$, and $RT = 57$. Find RS and ST. $RS = 38$; $ST = 19$

Extra Practice Worksheet 9-7

Name ______________________ Date ____________

EXTRA PRACTICE 9-7

Standard Deviation

Exercises

Compute the variance and standard deviation for each set of data.

1. 8, 8, 8, 8, 8, 8, 8, 8 ___0; 0___
2. 6, 2, 4, 8, 6, 9, 5 ___5.6; 2.4___
3. 2.5, 3.5, 7.2, 4.8, 5.4, 3.3 ___2.9; 1.7___
4. 6, 7.3, 5, 7.5, 6, 2 ___4.0; 2.0___
5. 2, 1, 5, 12, 5, 15, 9 ___27; 5.2___
6. 86, 69, 90, 79, 88 ___73.3; 8.6___
7. 90, 65, 78, 92, 84 ___118.2; 10.9___
8. 15, 16, 19, 20, 24, 28 ___24.3; 4.9___
9. 12, 20, 13, 14, 18, 15 ___9.5; 3.1___
10. 10, 4, 7, 9, 15, 18, 6 ___25.1; 5.0___
11. The variance for a set of scores is 16. On the average, how much does a score deviate from the mean score? ___4___
12. The standard deviation for the scores of the students in a certain class is increasing with each test given. What is happening to the scores? ___They are dispersed more and more widely from the mean.___
13. On a math test taken by 23 students, the mean score was 86. The standard deviation for the scores was 8.2. What was the sum of all the scores? ___1978___
14. Mikal scored an 82 on a test for which the mean score was 72 and the standard deviation was 10. He scored a 78 on a second test for which the mean score was 70 and the standard deviation was 5. On which test did he score better, compared with others in his class? ___the first test___
15. Mary had a z-score of -2 on her last test. If the mean score of the last test was 78, what was Mary's score? ___76___
16. Will scored an 89 on a test for which the mean was 72. What was his z-score for the test? ___10___
17. Lina scored an 68 on a test for which the mean was 75. What was her z-score for the test? ___−7___
18. Explain the difference between a positive z-score and a negative z-score. ___A positive z-score indicates a number above the mean, while a negative z-score indicates a number below the mean.___

126 EXTRA PRACTICE *LESSON* 9-7

Enrichment Worksheet 9-7

Name ______________________ Date ____________

ENRICHMENT 9-7

Two Kinds of Means

Finding the average speed A_S for two or more trips illustrates the difference between the arithmetic mean and the harmonic mean. The arithmetic mean is used when equal times are involved; the harmonic mean is used when equal distances are involved.

Example 1

Sid drove for 3 hours at 30 mi/h and then for another 3 hours at 55 mi/h. What was his average speed?

Solution

Use the arithmetic mean.

$$A_S = \frac{30 + 55}{2}$$

$$A_S = 42.5 \text{ mi/h}$$

Example 2

Darlene drove to a friend's house at 30 mi/h. She returned on an expressway at 55 mi/h. What was her average speed?

Solution

Use the harmonic mean.

$$A_S = \frac{2}{\frac{1}{30} + \frac{1}{55}}$$

$$A_S = 38.824 \text{ mi/h}$$

Exercises

Solve each problem. Show which mean you use.

1. Jason drove for 1.5 hours at 45 mi/h and then for another 1.5 hours at 60 mi/h. What was his average speed?
   $A_S = \frac{45 + 60}{2}$; $A_S = 52.5$ mi/h
2. Mark drove to a shopping plaza at 40 mi/h. He returned during rush hour at 25 mi/h. What was his average speed?
   $A_S = \frac{2}{\frac{1}{40} + \frac{1}{25}}$; $A_S = 30.769$ mi/h
3. Ellen competed in a 200-mile bicycle race. She averaged 25 mi/h on the first half and 18 mi/h on the second half. What was her average speed for the entire race?
   $A_S = \frac{2}{\frac{1}{25} + \frac{1}{18}}$; $A_S = 20.93$ mi/h
4. Gail and Gina each drove for half of the time on a Saturday trip. Gail averaged 62 mi/h; Gina averaged 58 mi/h. What was their average speed?
   $A_S = \frac{62 + 58}{2}$; $A_S = 60$ mi/h

164 ENRICHMENT *LESSON* 9-7

Teaching Strategies

COOPERATIVE LEARNING Have students in small groups consider the following test scores: 100, 100, 78, 76, 75, 75, 73, 72, 71, 70. Have them find the mean and evaluate how well they think each student did. Then have them compute the standard deviation and evaluate the results again. Have them make a visual display of the scores. Eight of the ten students scored below average (mean = 79), and so it appears that they did not do well on the test. But when standard deviation (about 11) is used it shows that all eight of those students scored within one standard deviation of the mean, while the two top scores were almost two standard deviations from the mean. The top scores skewed the data. A box-and-whisker plot, a bell curve, or a bar graph could be used to show this.

Assessment Planning

Vocabulary Review

Lesson 9-1
probability experiment
outcome
experimental probability
sample space tree diagram
theoretical probability

Lesson 9-2
simulations random numbers

Lesson 9-3
compound event
mutually exclusive events
complement

Lesson 9-4
independent event
dependent event

Lesson 9-5
fundamental counting principle
permutation
factorial notation
combination

Lesson 9-6
scatter plot line of best fit
trend line
positive correlation
negative correlation
box-and-whisker plot
extremes quartile

Lesson 9-7
variance
standard deviation
z-score
measures of dispersion
frequency distribution
bell curve normal curve

Assessment Options

Chapter Assessment, page 418
Alternative Assessment, page 419
Chapter Test Forms A and B
Cumulative Assessment, page 420–421
Achievement Test
Writing Prompts

Assignment Guide

All students: 1–20
Additional Practice: Refer to the Extra Practice Index on page 674 of this text.

Chapter 9 Review

Vocabulary ■ LESSON 9-1–LESSON 9-7

Choose the word from the list that best completes each statement.

a. extreme
b. combination
c. positive
d. mutually exclusive
e. sample space
f. standard deviation

1. The set of all possible outcomes of an experiment is the __?__. e
2. A set of items in no particular order is called a __?__. b
3. If events cannot occur at the same time, they are __?__. d
4. An upward sloping trend line on a scatter plot suggests a __?__ correlation between the data. c
5. The whiskers of a box-and-whisker plot show __?__ of the data. a
6. The __?__ is a measure of dispersion that compares a number to the mean of a set of data. f

LESSONS 9-1 and 9-2 ■ Percents, Probability, Simulations, p. 384

▶ Divide the number of favorable observations by the number of total observations to find the **experimental probability** of an event. To find the **theoretical probability** of an event, divide the number of favorable outcomes by the number of possible outcomes.

7. Find the probability of drawing a red 7 from a standard deck of playing cards. $\frac{1}{26}$
8. A computer generates a list of random 2-digit numbers. What probability would you expect for a number in the list to have a 2? $\frac{1}{5}$

LESSON 9-3 ■ Compound Events, p. 392

▶ A **compound** event consists of two or more simple events. If A and B are **mutually exclusive** events, they cannot occur at the same time. $P(A \text{ or } B) = P(A) + P(B)$. If A and B are not mutually exclusive events, they can occur at the same time. $P(A \text{ or } B) = P(A) + P(B) - P(A \text{ and } B)$.

9. Two 1–6 spinners are spun. Find the probability that the sum of the numbers spun is 9 or less than 2. $\frac{1}{9}$
10. A card is drawn at random from a standard deck. Find the probability that it is a black card or an 8. $\frac{7}{13}$

LESSON 9-4 ■ Independent and Dependent Events, p. 396

▶ Two events are **independent** if the outcome of one does not affect the outcome of the other. If A and B are independent events, $P(A \text{ and } B) = P(A) \cdot P(B)$.

▶ Two events are **dependent** if the outcome of one affects the outcome of the other. If A and B are dependent events, $P(A \text{ and } B) = P(A) \cdot P(B)$ (given A).

11. A die is rolled two times. Find $P(3, \text{even number})$. $\frac{1}{12}$
12. A box contains 4 red marbles, 3 green marbles, 1 white marble, 2 yellow marbles, and 3 blue marbles. Two marbles are chosen at random and not replaced. Find $P(\text{red, yellow})$. $\frac{2}{39}$

Additional Answers

16.

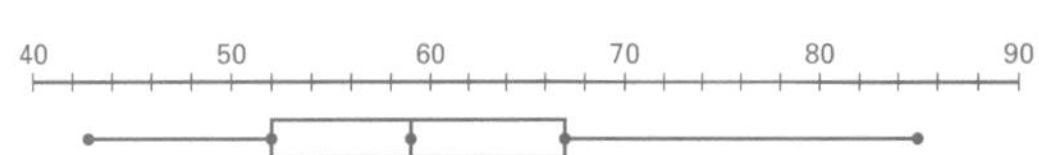

Teaching Strategies

Review with students that the formula $P(A \text{ or } B) = P(A) + P(B) - P(A \text{ and } B)$ is valid for both mutually exclusive and not mutually exclusive events. In the case that events are mutually exclusive, $P(A \text{ and } B) = 0$, so the formula becomes $P(A \text{ or } B) = P(A) + P(B)$.

LESSON 9-5 ■ Permutations and Combinations, p. 402

- The **fundamental counting principle** states that if an event A can occur in m ways and an event B can occur in n ways, then events A and B can occur in $m \cdot n$ ways.
- A **permutation** is a set of items arranged in a particular order. You can arrange a set of items in $n!$ ways. To find the number of permutations of a set of n items taken r at a time, use the formula at the right. $${}_nP_r = \frac{n!}{(n-r)!}$$
- A set of items without consideration of order is called a **combination**. To find the number of combinations of a set of n items taken r at a time, use the formula at the right. $${}_nC_r = \frac{n!}{(n-r)!\,r!}$$

13. In how many different ways can 6 pies be awarded first- through third-place prizes? 120

14. How many groups of 3 students can be chosen from a class of 20 students? 1140

LESSON 9-6 ■ Scatter Plots and Box Plots, p. 406

- A **scatter plot** displays data as unconnected points. The trend line indicates whether the items being compared have a positive correlation, a negative correlation, or no correlation.
- A **box-and-whisker plot** shows extremes of data and how data are distributed.

15. Name a situation where a scatter plot would have a negative correlation. Sketch how it might look. Check students' work.

16. Some students at Johnson High rated the performance of their basketball team from 0 to 100, with 100 as the highest. These are the ratings: 67, 71, 58, 53, 65, 73, 64, 50, 52, 74, 48, 47, 53, 82, 63, 59, 67, 85, 45, 43, and 56. Make a box-and-whisker plot of this data. See additional answers.

17. Make a box-and-whisker plot for the following set of test scores:

77, 80, 75, 73, 77, 81, 62, 87, 99, 85, 82, 81, 77, 72, 78, 83, 86, 79, 80, 78

What does the plot tell you about the scores? Median is 79.5, lower and upper quartiles are 77 and 82.5; 50% of the scores cluster around the median with the lower and upper 25% of the scores spread out due to extreme low and extreme high scores.

LESSON 9-7 ■ Standard Deviation, p. 412

- The **variance** of a set of numbers is a measure of how the data are dispersed. To find the variance of a set of numbers $x_1, x_2, \ldots x_n$, with a mean of m, use the following formula.

$$\frac{(x_1 - m)^2 + (x_2 - m)^2 + \cdots + (x_n - m)^2}{n}$$

- The **standard deviation**, s, of a set of numbers is the square root of the variance.

18. Find the variance and standard deviation for the set of numbers 4, 7, 10, 13, 16. 18; $\sqrt{18}$

19. Compute the variance and standard deviation for 3, 5, 10, 7, 5. 5.6, $\sqrt{5.6}$

20. Kendra scored 95, 90 and 95 on three tests. The class mean and standard deviation for the first test were 75 and 10; for the second, 75 and 5; and for the third, 80 and 6. On which test did Kendra do best relative to her classmates? second

Teaching Strategies

ONGOING ASSESSMENT Find the mean for 3.5, 2.5, 0, 1.5 and 4.5. Watch for students who conclude that the mean equals 3. These students have ignored the zero in the data. The correct solution is 2.4.

Chapter 9 Test Form A, page 1

CHAPTER 9

EXPANDING PROBABILITY AND STATISTICS

ASSESSMENT FORM A, PAGE 1

Name ______

Date ______

Scoring Record	
Possible: 30	Earned:

Suppose you flip two coins.

1. List the different outcomes. HH, HT, TH, TT

2. What is the probability that the flip will show two tails? $\frac{1}{4}$ or 0.25 or 25%

Suppose you roll a die 30 times. Predict the number of times you will roll each of the following.

3. a number less than 3 10 times **4.** a number other than 3 25 times

Suppose you roll a die. Find each probability.

5. P(1 or 6) $\frac{1}{3}$ **6.** P(not 2 or not 5) $\frac{2}{3}$

Suppose you draw a card at random from a standard deck of 52 playing cards. Find each probability.

7. P(jack or king) $\frac{2}{13}$ **8.** P(club or ace) $\frac{4}{13}$ **9.** P(not a spade) $\frac{3}{4}$

A marble bag holds 3 red marbles, 4 blue marbles, 2 orange marbles and 1 green marble. Suppose you choose each marble at random and then replace it. Find each probability.

10. P(red, then orange) $\frac{3}{50}$ **11.** P(blue, then green) $\frac{1}{25}$

12. P(red, then green, then red) $\frac{9}{1000}$ **13.** P(orange, then orange, then red) $\frac{3}{250}$

Suppose you choose each marble at random and do not replace it. Find each probability.

14. P(blue, then red) $\frac{2}{15}$ **15.** P(orange, then green) $\frac{1}{45}$

16. P(blue, then red, then orange) $\frac{1}{30}$ **17.** P(green, then blue, then red) $\frac{1}{60}$

Tell whether each question involves a permutation or a combination. Then solve.

18. In how many different ways can you arrange the letters *a, e, i, o, u, y*? permutation; 720 ways

19. A student may answer any 2 of 7 questions. In how many ways can she make a choice? combination; 21 ways

20. In how many ways can 3 books be arranged on a shelf if there are 5 books to choose from? permutation; 60 ways

21. In how many ways can a committee of 2 be chosen from 6 students? combination; 15 ways

114 ASSESSMENT *CHAPTER 9 FORM A*

Chapter 9 Test Form A, page 2

Name ______ Date ______

22. Use the scatter plot to predict how tall a 14-year-old girl might be.

Answers may vary.

Sample answer: about 65 in.

Height (in.) 65 60 55 50

4 6 8 10 12 14

Age

23. Make a boxplot for these scores: 37, 98, 57, 68, 88, 92, 93, 74, 76, 85, 83, 81.

30 40 50 60 70 80 90 100

Use the data from the boxplot you made to answer these questions.

24. What is the median? 82

25. What percent of the test scores are contained in the box? 50%

26. Are the data clustered more closely above or below the median? above

Find the variance and the standard deviation for each set of data.

27. Data: 3 5 4 6 2 Variance: 2 Standard deviation: $\sqrt{2}$

28. Data: 1 5 0 −3 2 Variance: 6.8 Standard deviation: $\sqrt{6.8}$

29. Data: 12 15 10 9 14 Variance: 5.2 Standard deviation: $\sqrt{5.2}$

Solve.

30. Suppose you plan to choose the answers at random on a five-question true-false test. Describe a simulation that you could do to estimate the probability of answering two of the five questions correctly.

Answers may vary. Check students' work.

116 ASSESSMENT *CHAPTER 9 FORM A*

Chapter 9 Test Form B, page 1

CHAPTER 9

EXPANDING PROBABILITY AND STATISTICS

ASSESSMENT FORM B, PAGE 1

Name ______________

Date ______________

Scoring Record	
Possible: 30	Earned:

Suppose you flip three coins.

1. List the different outcomes. HHH, HHT, HTH, THH, TTH, THT, HTT, TTT
2. What is the probability that the flip will show three heads? $\frac{1}{8}$ or 0.125 or 12.5%

Suppose you roll a die 42 times. Predict the number of times you will roll each of the following.

3. a number greater than 5 7 times
4. an even number 21 times

Suppose you roll a die. Find each probability.

5. P(2 or 3) $\frac{1}{3}$
6. P(not 5 or not 6) $\frac{2}{3}$

Suppose you draw a card at random from a standard deck of 52 playing cards. Find each probability.

7. P(spade or queen) $\frac{4}{13}$
8. P(not a jack) $\frac{12}{13}$
9. P(king or ace) $\frac{2}{13}$

A marble bag holds 4 red marbles, 3 blue marbles, 1 orange marbles and 2 green marbles. Suppose you choose each marble at random and then replace it. Find each probability.

10. P(blue, then orange) $\frac{3}{100}$
11. P(red, then red) $\frac{4}{25}$
12. P(green, then blue, then red) $\frac{3}{125}$
13. P(blue, then orange, then red) $\frac{3}{250}$

Suppose you choose each marble at random and do not replace it. Find each probability.

14. P(blue, then green) $\frac{1}{15}$
15. P(green, then red) $\frac{4}{45}$
16. P(red, then blue, then green) $\frac{1}{30}$
17. P(green, then red, then orange) $\frac{1}{90}$

Tell whether each question involves a permutation or a combination. Then solve.

18. In how many different ways can you arrange the vowels *a, e, i, o, u, y*? permutation; 720 ways
19. In how many ways can a committee of 3 be chosen from 7 students? combination; 35 ways
20. A student may answer any 3 of 6 questions. In how many ways can she make a choice? combination; 20 ways
21. In how many ways can 2 books be arranged on a shelf if there are 5 books to choose from? permutation; 20 ways

118 ASSESSMENT *CHAPTER 9 FORM B*

Chapter 9 Test Form B, page 2

Name ______________ Date ______________

22. Use the scatter plot to predict how much it would cost each month to heat a home when the temperature is 50°.

Answers may vary.

Sample answer: about $40

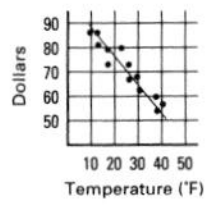

23. Make a boxplot for these scores: 77, 68, 79, 73, 78, 89, 86, 83, 75, 52, 64, 70

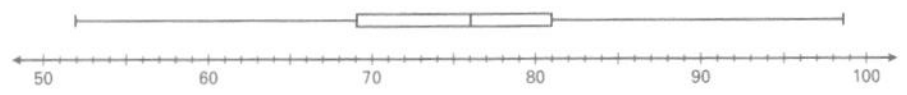

Use the data from the boxplot you made to answer these questions.

24. What is the median? 76
25. What percent of the test scores are contained in the box? 50%
26. Are the data clustered more closely above or below the median? above

Find the variance and the standard deviation for each set of data.

27. Data: 2 9 3 6 5 Variance: 6 Standard deviation: $\sqrt{6}$
28. Data: −1 3 2 1 5 Variance: 4 Standard deviation: $\sqrt{4} = 2$
29. Data: 26 37 38 29 30 Variance: 22 Standard deviation: $\sqrt{22}$

Solve.

30. A beverage company packages one of 4 different plastic cups with each box of powdered mix. Describe a simulation that you could do to estimate the probability of getting all 4 cups if you purchase 8 mixes.

Answers may vary. Check students' work.

120 ASSESSMENT *CHAPTER 9 FORM B*

CHAPTER 9 ASSESSMENT

1. A bowler got 32 strikes in her first 80 frames. Predict the probability she will not get a strike in her next frame. $\frac{3}{5}$
2. Find the probability of drawing a red king or a 5 from a standard deck of playing cards. $\frac{3}{26}$
3. A pair of dice are rolled twice. Find P(even number, odd number). $\frac{1}{4}$
4. There are 3 red T-shirts, 2 white T-shirts, 2 green T-shirts, and 1 blue T-shirt in a drawer. You reach in without looking and take out two shirts. Find P(red, green). $\frac{3}{28}$
5. How many different two-flavor ice cream cones can be chosen from a menu of 15 flavors? 105
6. Forty students are in the running for the science prize. In how many ways can a winner, a runner-up, and an alternate be chosen? 59,280

Use the scatter plot for Exercises 7 and 8.

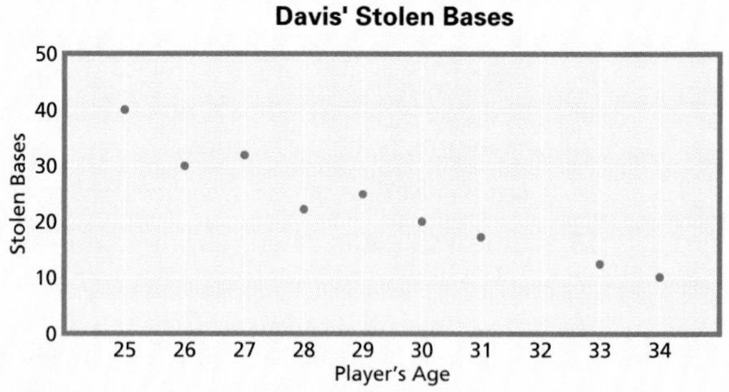

7. Does the scatter plot show a negative or a positive correlation? negative
8. Estimate the number of stolen bases. Davis will have when he is 32 years old. 14–15

Groups of students and of community leaders rated the performance of the school superintendent. The box-and-whisker plots below show the results.

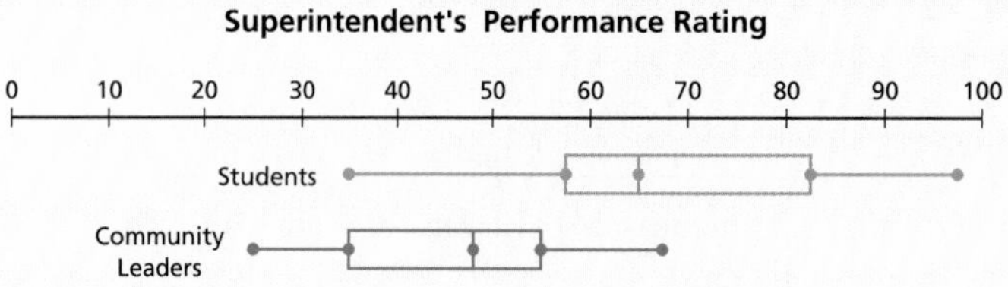

9. Which group gave the superintendent a higher median score? students
10. For which group were the scores more widely spread? students
11. Find the variance and standard deviation for the set of numbers: −10, −5, 0, 5, 10. 50; $\sqrt{50}$
12. Patty scored 75 on a test in which the mean score in her class was 60 and the standard deviation was 10. Jamaal took the same test in his class. His score was 70, the class mean score was 50, and the standard deviation was 10. Who scored better, relative to his or her classmates? Jamaal

Modeling/Using Manipulatives Scoring Rubric

SCORING RUBRIC 5—Student creates an original lesson plan, teaches a class in probability, assigns an appropriate quiz, and writes an incisive paragraph analyzing the results of the experience. **4**—Student creates an original lesson plan, teaches a class in probability, assigns an appropriate quiz, and writes a paragraph analyzing the results of the experience. **3**—Student creates a lesson plan, teaches a class in probability, assigns a reasonably appropriate quiz, and writes a paragraph analyzing the results of the experience. **2**—Student creates a partial or incomplete lesson plan, teaches a class in probability, assigns a quiz, and writes a paragraph poorly analyzing the results of the experience. **1**—Student creates a partial or incomplete lesson plan, and cannot teaches a complete class in probability, then writes a paragraph poorly analyzing the results of the experience. **0**—Student makes no attempt to create a lesson plan or teach a class.

Chapter 9 Alternative Assessment

Connections

SPORTS ANALYSIS Use resources such as the Internet, newspapers and Almanacs. Select one professional sports team. Determine the usual statistics kept for that sport, and the exact statistics for each player on the team. Select one player and find how he/she ranks on the team. Find the standard deviation for each player. Write a paragraph describing how you did your research, and how you calculated your selected player's ranking.

Cooperative Learning

GAME SHOWS Select and analyze two television game shows for the role probability plays in each. Which depends more on players' abilities? Which depends more on chance? Design your own game show that depends only on chance, or only on players' abilities. Describe how to play and the rules governing play. Write a paragraph that tells what the game's outcome relies on and defend your conclusion.

Modeling/Using Manipulatives

ROLE REVERSAL Use a standard deck of cards, number cubes and spinners to teach a class about probability. Prepare a lesson plan for your class, including questions you expect from your students, and the answers to those questions. Prepare a quiz for your students along with an answer key on a separate sheet of paper. After your class, write a paragraph describing how you thought the class went and giving yourself a grade as a teacher.

Chapter Investigation

EXTENSION Present your game to your class. After listening to everyone's presentation, compare your game to those of your classmates. List the advantages and disadvantages to your game. Make improvements to your game based on your list of disadvantages.

Chapter Investigation

The benchmarks and expectations for this extension are as follows.

- Students work with a partner to prepare to make their own baseball simulation game by gathering batting statistics on at least 18 players.
- Students compute and record the number of singles, doubles, triples, homeruns, and hits for each player.
- Students find the probability that a certain player will get a single, a double, a triple, a home run, a walk, a strike out, etc.
- Students create a 10-by-10 table for each player. They number both the columns and rows from 1 to 10 and abbreviate appropriately.
- Students play their baseball simulation game.
- Students present their game to the class. They make improvements to their game as needed.

Connections

Although any sport has a certain amount of statistics, football, baseball, hockey and basketball will have more abundant published statistics.

SCORING RUBRIC **5**—Student determines statistics for an entire team and one player, correctly calculates all standard deviations and clearly explains the process in detail. **4**—Student determines statistics and correctly calculates all standard deviations and explains the process. **3**—Student determines statistics, calculates all standard deviations with a few minor errors, and explains the process. **2**—Student determines statistics, calculates most of the standard deviations with a few minor errors, and attempts to explain the process. **1**—Student determines some statistics, calculates most of the standard deviations with significant errors, but cannot explain the process. **0**—Student makes no attempt to determine statistics or explain the process.

Cooperative Learning

You may want to assign a chance/ability game for each group to provide more diversity and give students an opportunity to compare the two kinds of shows.

SCORING RUBRIC **5**—Student compares two game shows and supports conclusions, invents an original game show and rules, then convincingly defends a position on the role of probability in the new game. **4** —tudent compares shows and supports conclusions, invents a show and rules, and defends position. **3**—Student compares two shows and supports conclusions, invents a show and rules, but poorly defends position on the role of probability. **2**—Student compares shows and unconvincingly supports conclusions, invents show and rules, but poorly defends position on probability. **1**—Student compares shows and unconvincingly supports conclusions, begins work on a new show and rules, and does not attempt to defend a position on the role of probability in the new game. **0**—Student makes no attempt to evaluate game shows or invent a new one.

Modeling/Using Manipulatives

You may want to arrange for these "teachers" to present their lesson to another class, preferably a lower grade, to better gauge their success in teaching the fundamentals of probability.

See page 418 for Scoring Rubric.

Cumulative Review

Cumulative review is the best way to maintain previously taught skills and concepts. This will keep students prepared for new lessons that build on previously covered skills.

Cumulative Review covers:

Lesson 1-7
Lesson 1-8
Lesson 2-2
Lesson 2-5
Lesson 3-2
Lesson 3-4
Lesson 4-1
Lesson 4-8
Lesson 5-1
Lesson 5-2
Lesson 6-1
Lesson 6-6
Lesson 7-1
Lesson 7-6
Lesson 8-1
Lesson 8-5
Lesson 8-6
Lesson 9-1
Lesson 9-5

Chapters 1–9 Cumulative Review

1. Evaluate x^3y^2 if $x = -5$ and $y = 3$. $-1{,}125$

2. Write 0.00078 in scientific notation. 7.8×10^{-4}

3. Given the function $f(x) = 6x^2 - 1$, find $f(-2)$. 23

4. Solve and check: $10x - 9 = 3x + 5$. 7.8×10^{-4}

5. In the figure below, find $m\angle BEC$. 23

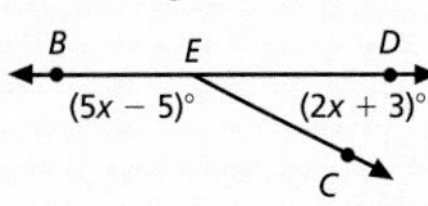

6. In the figure at the right, $\overleftrightarrow{AB} \parallel \overleftrightarrow{CD}$. Find $m\angle ECD$. 124°

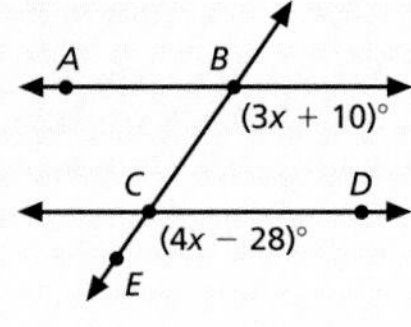

7. Find the value of x in the figure below. 48

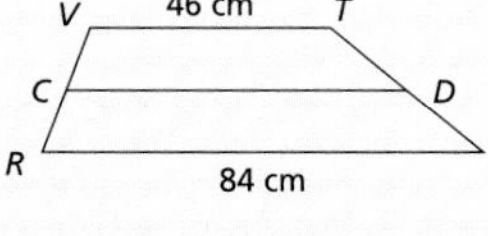

8. In the figure below, $\overline{CD}$ is a median of trapezoid $RSTV$. Find the length of $\overline{CD}$. 65 cm

9. Change 0.08 kg to grams. 80 g

10. Find the area of the shaded region in the figure below. 13.76 ft²

11. Find the slope of a line passing through points $A(0, 5)$ and $B(7, -1)$. $-\frac{6}{7}$

12. Find the slope and y-intercept of a line whose equation is $3y - x = 6$. slope, $\frac{1}{3}$; y-intercept, 2

13. Solve the following system of equations. $x = 2$, $y = -3$

$$\begin{cases} 4x + 3y = -1 \\ 5x + 2y = 4 \end{cases}$$

14. For a cleaning solution, bleach is mixed with water in the ratio 1 : 8. How much bleach should be added to 12 quarts of water to make the proper solution? 1.5 qt

15. In the figure $\overline{RS} \parallel \overline{AB}$. Find the value of x to the nearest tenth. 11.4

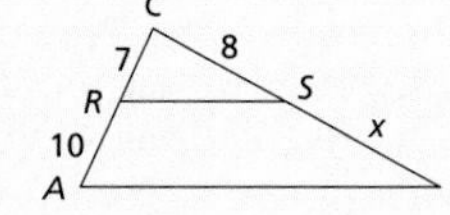

16. Graph the image of $\triangle ABC$ under a reflection over the x-axis. Then find the slopes of the sides of the image triangle.
slopes: $A'B'$, 0; $A'C'$, -1; $B'C'$; 2

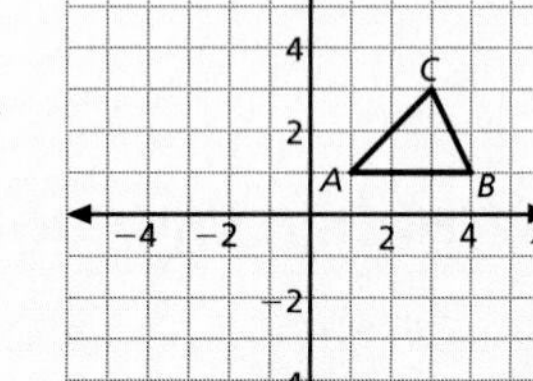

Find each sum or product

17. $\begin{bmatrix} 0 & 3 \\ 4 & 0 \end{bmatrix} + \begin{bmatrix} 0 & -5 \\ 1 & 6 \end{bmatrix}$ $\begin{bmatrix} 0 & -2 \\ 5 & 6 \end{bmatrix}$

18. $\begin{bmatrix} 1 & 3 \\ 1 & 4 \end{bmatrix}\begin{bmatrix} 3 & 5 \\ 5 & 6 \end{bmatrix}$ $\begin{bmatrix} 18 & 23 \\ 23 & 29 \end{bmatrix}$

19. What is the probability of drawing a red ace from a standard deck of cards? $\frac{1}{26}$

20. Three of 18 students will be chosen for a committee. How many different committees are possible? 816

Teaching Strategies

Refer to Exercise 8 of the Cumulative Assessment. Discuss the role of formulas in doing well on standardized tests. Point out that the SAT and ACT provide a list of formulas for student use, but students will benefit from memorizing the basic formulas to avoid wasting time searching for the formula they need.

Chapters 1–9 Cumulative Assessment

Standardized Test Preparation—Gridded Response

Solve each question. Write your answer at the top of the answer grid and fill in the ovals.

Notes: Mixed numbers such as $1\frac{1}{2}$ must be gridded as 1.5 or $\frac{3}{2}$. Grid only one answer per question. If your answer is a decimal, enter the most accurate value the grid will accommodate.

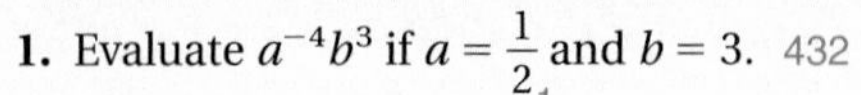

1. Evaluate $a^{-4}b^3$ if $a = \frac{1}{2}$ and $b = 3$. 432
2. Evaluate x^{-4} if $x = 2$. $\frac{1}{16}$
3. Grid the greater y-value when x is 0 or x is 2.

   $$3x + 2y = 6$$ 3
4. Solve. $5(x - 1) = 2(x + 5)$ 5
5. In the figure, find $m\angle ROZ$. 150

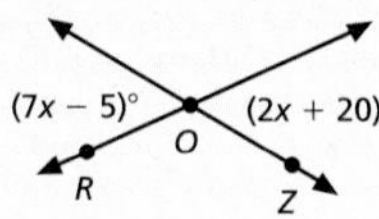

6. Find the value of x. 37

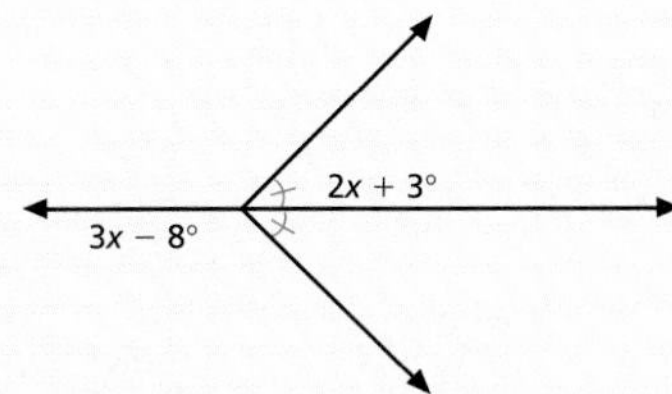

7. Find $\overline{CD}$. Round to the nearest hundredth. 4.67

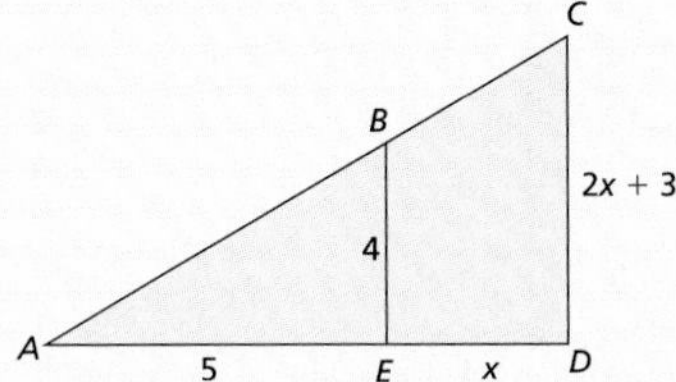

8. Find the volume of a sphere with a radius of 6 cm. 904.32
9. Two parallel lines have equations $y - 3x = 10$ and $mx + 2y = 3$. Find the value of m. 3
10. Find the area of the shaded region. 135

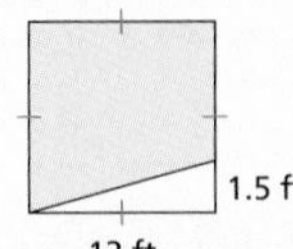

11. Solve. $\frac{12}{x} = \frac{15}{48}$ 38.4
12. Find the value of y. 70

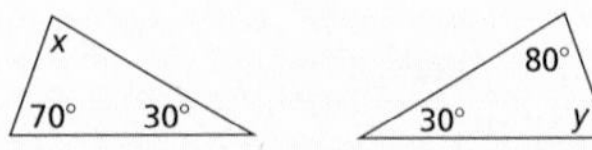

13. Identify the x-coordinate of $(3, -4)$ under a reflection over the y-axis. −3
14. Find d. $\begin{bmatrix} 3 & 0 \\ 4 & 2 \end{bmatrix} + \begin{bmatrix} 1 & 3 \\ 6 & 5 \end{bmatrix} = \begin{bmatrix} a & b \\ c & d \end{bmatrix}$ 22
15. How many four-member teams can be chosen from a group of ten? 25,200
16. A bag contains 3 blue marbles, 2 red marbles, and 7 white marbles. Find P(red, then blue). $\frac{1}{22}$
17. Describe a situation when you would want to display data with a scatter plot rather than a box-and-whisker plot. See additional answers.
18. Multiply. $\begin{bmatrix} 5 & -4 \\ 3 & 1 \end{bmatrix} \cdot \begin{bmatrix} 0 & -5 & 6 \\ 2 & 8 & -4 \end{bmatrix}$ $\begin{bmatrix} -8 & -57 & 46 \\ 2 & -7 & 14 \end{bmatrix}$

CONSTRUCTED RESPONSE Find the slope of a line parallel to the given line and a line perpendicular to the given line. Explain how you find all slopes.

19. The line containing points $A(4, 2)$ and $B(-1, -3)$ 1, −1
20. The line containing points $C(-2, -2)$ and $D(7, 7)$ 1, −1

Teaching Strategies

Discuss Exercise 17 of the Cumulative Assessment and the importance of knowing your purpose before choosing a method to display data. You may which to review standard types of graphs and discuss when each should be used. Then relate the discussion to scatter plots and box-and-whisker plots. Students may recognize that sometimes more than one type of display is necessary to meet the goals of a particular project.

Standardized Test Preparation

GRIDDED RESPONSE Questions that require you fill in grids on an answer sheet, but do not have answer choices provided are called gridded response or students-produced response. You must solve the problem and then fill in the answer on a special grid. Your answer consists of two parts-the numbers that you write in the answer boxes and the ovals that you fill in. Ovals that correspond to the digits 0–9, negative sign, fraction bars and decimal points are listed in columns under the answer boxes.

You should grid the form of the answer that you obtain from solving the problem. Enter the most accurate value the grid will accommodate. The grid will only hold numbers that range form 0–9999. The grid will not accommodate a mixed number. Change mixed numbers to either decimals or improper fractions.

Columns not needed should be left blank. It does not matter whether the extra columns are on the left or the right of the answer.

Testing Strategies

You may wish to allow students to use calculators on this assessment. Before doing so, spend some time as a class discussing the usefulness of a calculator in a test situation. Remind students that they have a limited time to answer questions. It is generally most efficient to answer questions by making calculations mentally. After completing the assessment, have students share opinions on the usefulness of the calculator. Students can share approaches to the problems and discuss which approach is most efficient.

Additional Answers

17. Possible answer: You would use a scatter plot when you want to look for a trend in data that is paired. For example, if you wanted to see if there was a trend in test scores as compared to the number of problems students solved getting ready for the test.

Chapter 10—Right Angles and Circles

Theme: Architecture
Chapter Investigation: New Design, pages 429, 439, 451, 461
Careers: Building Inspectors, page 435; Landscape Architects, page 453
Data Activity: World's Tallest Buildings

Content and Connections

Lesson, Pages	Lesson Objectives	NCTM Standards	State/Local Objectives	Interdisciplinary Connections	Real World Applications
10–1 426–429	• Find square roots • Simplify products and quotients containing radicals	Number Operation Patterns, Functions & Algebra Problem Solving Reasoning & Proof Connections; Representation			Architecture, Small Business, Construction
10–2 430–433	• Use the Pythagorean Theorem to solve problems involving right triangles	Number Operation Patterns, Functions & Algebra Reasoning & Proof Geometry & Spatial Sense Problem Solving Communication Connections; Representation			Architecture, Construction, Home Repair
10–3 436–439	• Find the lengths of the sides of 30°-60°-90° and 45°-45°-90° triangles	Patterns, Functions & Algebra Problem Solving Reasoning & Proof Number Operation Geometry & Spatial Sense Communication			Architecture, Road Planning, Plumbing
10–4 440–443	• Find measures of central and inscribed angles • Find measures of angles formed by intersecting secants and tangents	Patterns, Functions & Algebra Geometry & Spatial Sense Representation Number Operation Measurement Problem Solving Reasoning & Proof Communication			Landscape Architecture, Navigation, Surveying
10–5 446–447	• Solve a problem with circle graphs • Make a table, chart or list	Problem Solving Reasoning & Proof Communication Connections; Representation Number Operation Patterns, Functions & Algebra Geometry & Spatial Sense			
10–6 448–451	• Find lengths of chord, secant and tangent segments	Number Operation Patterns, Functions & Algebra Problem Solving Communication Representation		Art	Architecture, Art, Surveying
10–7 454–457	• Construct regular polygons using circles • Inscribe a circle in a polygon and circumscribe a circle about a polygon	Geometry & Spatial Sense Measurement Patterns, Functions & Algebra Problem Solving Reasoning & Proof Connections; Representation		Art	Architecture, Design, Art

Planning and Resources

Lesson, Pages	Tools/Materials Needed	Trans-parency	Learning/Teaching Styles Options	Assignments: Basic Enriched	Additional Practice in SE	Reteaching, Extra Practice, Enrichment	Other Resources
10–1 426–429	calculator	Warm up 28 Trans RF-38	Real World	B: 1–40, 45–50 E: 1–50	Page 434–435, 445, 453, 458, 714	R: page 145 EP: page 127 E: page 167	

10–2 430–433	graph paper scissors straightedge	Warm up 28 Trans RF-39	ESL/LEP Visual Learners Challenge	B: 1–22, 26–30 E: 1–30	Page 434–435, 445, 453, 458, 715	R: page 147 EP: page 129 E: page 169	SS: Teacher's Choice
10–3 436–439	calculator scissors compass straightedge	Warm up 28 Trans RF-40	Challenge	B: 1–15, 19–28 E: 1–28	Page 444–445, 453, 458, 715	R: page 149 EP: page 131 E: page 171	SS: Teacher's Choice
10–4 440–443	compass straightedge scissors	Warm up 29 Trans RF-41	Visual Learners Challenge	B: 1–10, 13–25 E: 1–25	Page 444–445, 453, 459, 716	R: page 151 EP: page 133 E: page 173	
10–5 446–447	calculator protractor compass straightedge	Warm up 29	Real World	B: 1–10 E: 1–10	Page 452–453, 459	R: page 153 E: page 175	
10–6 448–451	compass straightedge protractor	Warm up 30 Trans RF-42	Visual Learners	B: 1–20, 23–25 E: 1–25	Page 452–453, 459, 716	R: page 155 EP: page 135 E: page 177	SS: Teacher's Choice
10–7 454–457	calculator	Warm up 30	Prior Knowledge	B: 1–19, 21–26 E: 1–26	Page 459, 717	R: page 157 EP: page 137 E: page 179	

Planning and Pacing

Lesson, Pages	Lesson Title	45 min class	Assignments Basic, Enriched	90 min class	Assignments Basic, Enriched	___ min class	Assignments Basic, Enriched
10–1 426–429	AYR, Opener, Irrational Numbers	Day 127	B: 1–40, 45–50 E: 1–50	Day 67	B: 1–40, 45–50 E: 1–50		B: 1–40, 45–50 E: 1–50
10–2 430–433	The Pythagorean Theorem, R&PYS	Day 128–129	B: 1–22, 26–30 E: 1–30	Day 67–68	B: 1–22, 26–30 E: 1–30		B: 1–22, 26–30 E: 1–30
10–3 436–439	Special Right Triangles	Day 130	B: 1–15, 19–28 E: 1–28	Day 68	B: 1–15, 19–28 E: 1–28		B: 1–15, 19–28 E: 1–28
10–4 440–443	Circles, Angles, and Arcs, R&PYS	Day 131–132	B: 1–10, 13–25 E: 1–25	Day 69	B: 1–10, 13–25 E: 1–25		B: 1–10, 13–25 E: 1–25
10–5 446–447	Problem Solving Skills: Circle Graphs	Day 133	B: 1–10 E: 1–10	Day 70	B: 1–10 E: 1–10		B: 1–10 E: 1–10
10–6 448–451	Circles and Segments, R&PYS	Day 134–135	B: 1–20, 23–25 E: 1–25	Day 70–71	B: 1–20, 23–25 E: 1–25		B: 1–20, 23–25 E: 1–25
10–7 454–457	Construction with Circles	Day 136	B: 1–19, 21–26 E: 1–26	Day 71	B: 1–19, 21–26 E: 1–26		B: 1–19, 21–26 E: 1–26
Review/ Assess		Day 137–138		Day 72–73			

	Chapter 1 Days	Chapter Cumulative Days
Yearly Pacing (45 min class)	12 days	137 days
Yearly Pacing (90 min class)	7 days	73 days

Assessment Options

Assessment in Student Edition	Assessment in Teacher's Edition	Pages in Assessment Book	Software Generated Assessment
Are You Ready?, pages 422–423; Writing Math, pages 429, 433, 439, 443, 451, 456; Mixed Review, pages 433, 439, 443, 447, 451, 457; Check Understanding pages 426, 430, 436, 440, 448, 454; Mid-Chapter Quiz, page 445, Chapter Review, page 458; Chapter Assessment, page 460; Alternative Assessment, page 461; Cumulative Review, page 462; Cumulative Assessment, page 463	5-minute Warm ups, pages 426, 430, 436, 440, 446, 448, 454; Quick Assessment, pages 428, 432, 438, 442, 447, 450, 456; Scoring Rubrics, page 461	Mid-Chapter Quiz, page 151; Test Form A, pages 153, 155; Test Form B, pages 157, 159; Cumulative Test 161, 163; Math Journal prompt, page 165, 166;	Chapter 10

Refresher Skills

The skills on these two pages are skills that have been taught in previous math courses. Continuous review of basic math skills will make stronger math students. These skills are identified as necessary to be successful in Chapter 10.

Skills Correlation Chart

Skill	Lesson Number
Percent of a Number	10-4
Circumference and Area of Circles	10-4, 10-5, 10-6
Squares and Square Roots	10-1, 10-2, 10-3
Measures of Angles	10-2, 10-3, 10-4, 10-5, 10-6, 10-7

Vocabulary

circumference   square root

ARE YOU READY?

Refresh Your Math Skills for Chapter 10

The skills on these two pages are ones you have already learned. Review the examples and complete the exercises. For additional practice on these and more basic skills, see page 674.

PERCENT OF A NUMBER

Example Find 42% of 624.
Change the percent to a decimal: 42% → 0.42
Multiply:

$$\begin{array}{r} 624 \\ \times\ 0.42 \\ \hline 262.08 \end{array}$$

42% of 624 is 262.08.

Find the given percent of each number.

1. 15% of 500 75
2. 50% of 768 384
3. 24% of 496 119.04
4. 33% of 127 41.91
5. 64% of 481 307.84
6. 93% of 722 671.46
7. 81% of 3297 2670.57
8. 29.6% of 17.84 5.28064
9. 47% of 82.56 38.8032
10. 23.5% of 1604 376.94
11. 106% of 300 318
12. 17.8% of 296 52.688

CIRCUMFERENCE AND AREA

The formula for the circumference of a circle is $C = 2\pi r$, where r is the radius of the circle. The formula for the area of a circle is $A = \pi r^2$, where r is the radius of the circle.

Find the circumference and area of each circle. Round answers to the nearest hundredth if necessary.

13.
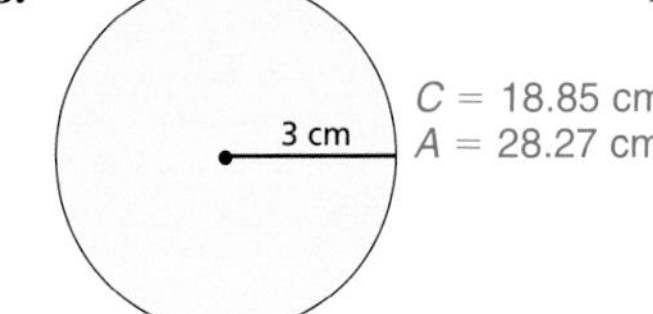

$C = 18.85$ cm; $A = 28.27$ cm^2

14.
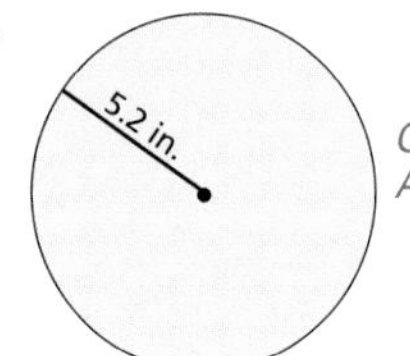

$C = 32.67$ in.; $A = 84.95$ in.^{2}

15.
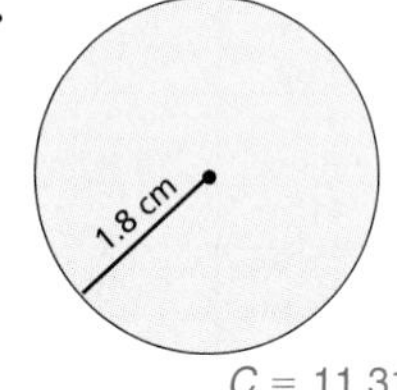

$C = 11.31$ cm; $A = 10.18$ cm^2

16.
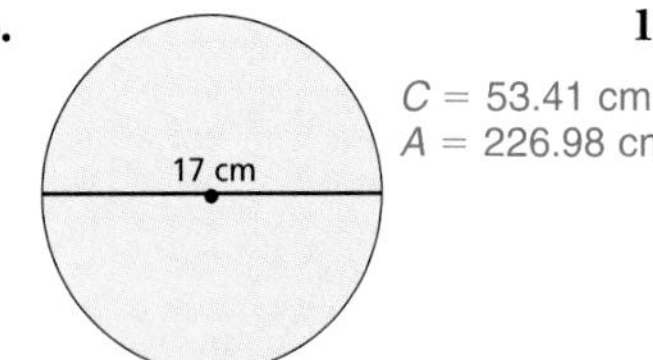

$C = 53.41$ cm; $A = 226.98$ cm^2

17.
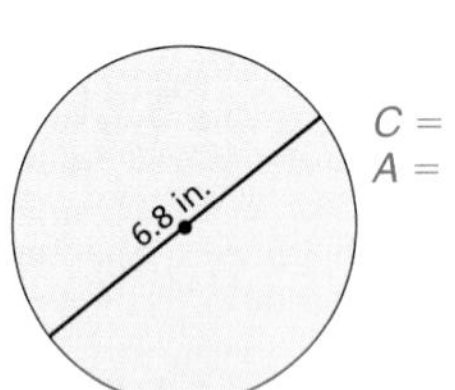

$C = 21.36$ in.; $A = 36.32$ in.^{2}

18.
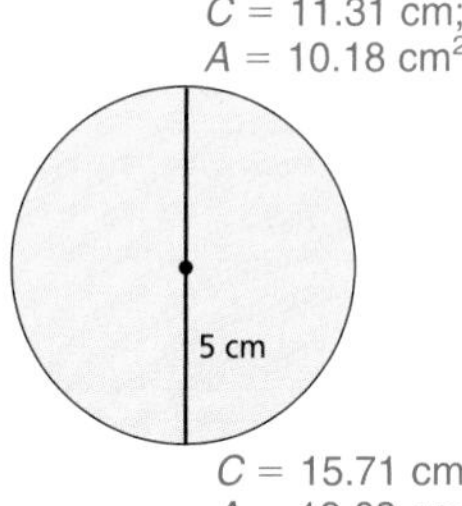

$C = 15.71$ cm; $A = 19.63$ cm^2

Teaching Strategies

Have students estimate the solutions to percent problems using the benchmarks of $25\% = \frac{1}{4}$, $50\% = \frac{1}{2}$ and $75\% = \frac{3}{4}$. For example, when finding 24% of 296, the student should expect the answer to be a little less than $\frac{1}{4}$ of 296.

CHAPTER 10

Squares and Square Roots

Simplify each expression.

19. 16^2 256	**20.** $\sqrt{144}$ 12	**21.** 24^2 576	**22.** $\sqrt{64}$ 8
23. 9^2 81	**24.** $\sqrt{529}$ 23	**25.** 15^2 225	**26.** $\sqrt{196}$ 14
27. 13^2 169	**28.** $\sqrt{289}$ 17	**29.** 30^2 900	**30.** $\sqrt{100}$ 10

Angles

Find the measure of each indicated angle.

31. $\angle AEB$ 131°
32. $\angle AED$ 49°
33. $\angle DEC$ 131°

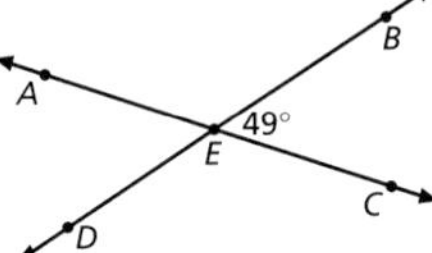

34. $\angle DCF$ 140°
35. $\angle EFB$ 130°
36. $\angle HFC$ 130°

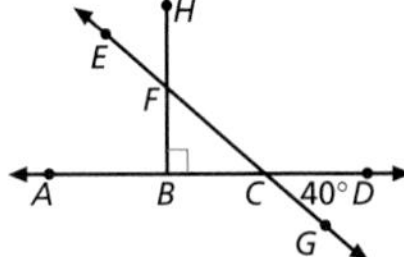

37. $\angle FEB$ 77°
38. $\angle ABG$ 103°
39. $\angle CBE$ 103°

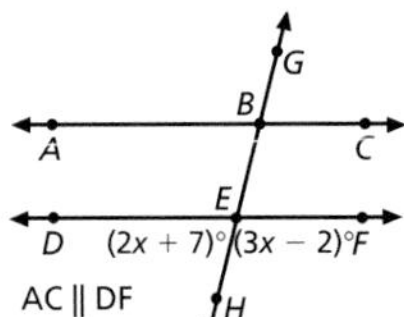

40. $\angle BAD$ 140°
41. $\angle FAE$ 65°
42. $\angle EAD$ 40°

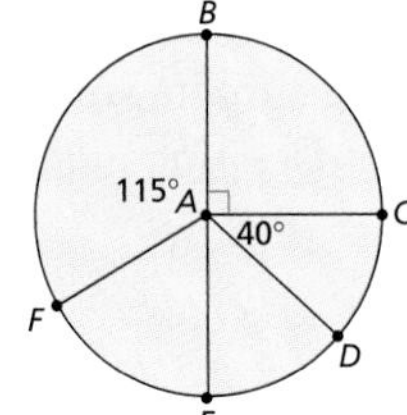

Chalkboard Examples

Percent of a Number

Find 65% of 400. 260

Circumference and Area

The radius of a circle is 4.5 cm. Find the circumference and area of the circle to the nearest hundredth.
circumference: 28.27 cm
area: 63.62 cm^2

Squares and Square Roots

Simplify each expression.
a. 16^2 256
b. $\sqrt{400}$ 20

Refresher Wrap-up

Quick Assessment

Ask the following questions to determine if students have mastered the basic skills reviewed on these pages.

1. How do you remember the formulas for area and circumference? Answers will vary. One tip is to remember that since area is measured in square units, the area formula involves squaring a number.
2. Without using a calculator, estimate the square root of 90. A good estimate is between 9 and 10 since 90 falls between 81 and 100. Actual solution to the nearest hundredth: 9.49. Encourage students to use estimation to check the reasonableness of solutions arrived at using a calculator.

Additional Practice

All topics: Refer to the Extra Practice Index on page 674 of this text.

Extend the Lesson

Have students locate and measure various angles within the classroom or on the school campus. Require students to find at least two of each type: acute, obtuse and right angles. You may wish to have students record where they found each angle. Discuss the activity. Students may notice that acute and obtuse angles are less common in construction and are often found together as supplementary angles.

Chapter 10 **Are You Ready?** 423

Chapter Opener

NCTM Standards/Strands

- Number & Operations
- Patterns, Functions, & Algebra
- Geometry & Spatial Sense
- Measurement
- Problem Solving
- Reasoning & Proof
- Communication
- Connections
- Representation

Vocabulary

topography	module
façade	design elements

Theme Connections

Tables and graphs are used to organize data about architecture and construction. Tables provide a way to organize numerical information. Many types of data are kept in regards to architecture. Architects must analyze building codes, technical specifications, cost and schedules. Discuss how data and graphs are used in architecture.

Career Opportunities

Many careers related to architecture make use of data and graphs. Two are highlighted in the Math-Works features. Others include draftspersons, construction specialists, surveyors, civil engineers, soil conservationists and urban planners.

- Construction and Building Inspectors, page 435
- Landscape Architects, page 453

Internet Connection

Theme Activities

Learningmathmatters.com provides web addresses to search that help students gather information about the use of math in the real world, particularly data and graphs. To search for additional addresses, begin a search using the keyword *architecture*. Then within that search, use such key words as *landscaping, blueprint, construction,* and *engineering*. Students can brainstorm in small groups other key words.

CHAPTER 10

Right Angles and Circles

THEME: Architecture

Imagine that you have the opportunity to design and build a new home for your family. How would you decide what design features to incorporate? Your new home must be functional to meet your family's needs. It must also be appealing to the eye to suit your family's personality and complement other buildings in the community.

Architects have been using geometric principles for centuries to design and build homes, schools and public buildings. In the search for pleasing and useful forms, they have explored and applied many important principles of geometry.

- **Building Inspectors** (page 435) ensure the safety of buildings before construction is complete. They make sure that safety guidelines and building codes are strictly followed. Inspectors may stop a project if safety standards are not met.
- **Landscape Architects** (page 453) design outdoor areas such as gardens, parks and playgrounds. Landscape architects study soil, sunlight, topography and climate when designing a landscaping plan.

Internet Connection
www.learningmathmatters.com

Chapter Investigation

Go to learningmathmatters.com to locate additional information about architecture.

The World's Tallest Buildings

Skyscraper	City	Built	Height	Stories
Central Towers	Hong Kong	1992	1227 ft	78
Petronas Towers	Kuala Lumpur	1997	1483 ft	88
Sears Tower	Chicago	1974	1450 ft	110
Jin Mao Building	Shanghai	1999	1379 ft	88
World Trade Center Towers	New York	1972	1368 ft (North) 1362 ft (South)	110
Empire State Building	New York	1931	1250 ft	102
Bank of China	Hong Kong	1989	1209 ft	70

Data Activity: The World's Tallest Buildings

Use the table for Questions 1–4.

Least—Empire State Building, about 12.3 ft per story. Greatest—Bank of China, about 17.3 ft per story.

1. Compare the number of stories to the height in feet for each building. Which building has the least amount of height per story? What has the greatest amount of height per story?
2. The framework for the Empire State Building rose at a rate of four-and-a-half stories per week. How many days were required to build the framework? about 159 days
3. The Sears Tower consists of nine 75-ft square modules which rise to staggered levels. What is the combined area of the modules in square feet? 50,626 ft^2
4. Create a bar graph of the buildings' heights with the buildings arranged according to the date they were built. Check students' work

Chapter Investigation

Often, large cities do not have available space to build new buildings or the means to tear down older buildings and replace them with new ones. As a result, many cities are hiring architects to give older buildings a facelift or makeover. A new facade is attached to the front of the building to make the building more attractive and unify design elements.

Working Together

Make a sketch of the front of your school as it now looks. Research architectural styles and select appropriate design elements. Then create a facade that could fit over the front of your school to give it a new look. Draw plans for your new design. Use the Chapter Investigation logos to guide your group.

Empire State Building

Data Activity

Discuss with students the importance of urban planning and architecture in building urban centers and promoting business. Point out that many cities take pride in their construction projects, and major skyscrapers often become landmarks as they impact the city skyline. Assign students Questions 1–4.

Extend the Data Activity

REAL WORLD CONNECTION Have students research facts about a nearby architectural landmark. Have students create a brochure advertising the site to tourists. Students may wish to create a table or graph of the data for their brochures.

Chapter Investigation

As an Overarching Problem

Display photographs of buildings that have had new facades attached to the front. If possible, show before and after photographs of the buildings, and ask students how the facade improved the building. Students will continue to work on the investigation as they complete the exercises identified by the Investigation Logo that are found throughout the chapter. These exercises will guide students through the task described in *Working Together*. Encourage students to keep all of their work on the Investigation together. Have students use the suggestions in the Chapter Investigation Extension to summarize their work.

As a Chapter Project

The goal of this project is to make and describe a proportional sketch of the subject of a photograph. Students can use the Group Project Planner on page 165 and the Project Planning Calendar on page 166 in the Enrichment text to complete the project. Benchmarks **a**, **b**, and **c** should be completed after the lesson listed in parentheses has been studied. Benchmark **d** should be completed at the end of the chapter.

Project Planning Calendar

Name ____________ Date ____________

Chapter 10 Project Planning Calendar

Benchmarks	PROJECT GOAL
a. Make a detailed sketch of the front of the main building at your school. Include any current architectural elements. Estimate measurements for the elements in the sketch. If possible, consult building plans to find the length and width of standard doors and windows. *(Lesson 10-1)* b. Research architectural styles that would complement your school grounds. Make a list of design elements that you could choose from in creating a new look for your school. *(Lesson 10-3)* c. Make a scale drawing of a new facade for a building on your school grounds. Use measurement approximations where necessary. ... and your new	To design a new facade that could fit over the front of your school to give it a new look.

Group Project Planner

Name ____________ Date ____________

Chapter 10 Group Project Planner

Assignment ____________ Objective ____________

Group Members | Assigned Roles

1) ____________
2) ____________
3) ____________
4) ____________
5) ____________

Deadlines | Done

Lesson Planning

NCTM Standards/Strands

- Number & Operations
- Patterns, Functions, & Algebra
- Problem Solving
- Reasoning & Proof
- Communication
- Connections
- Representation

Vocabulary

irrational number
radicand
rationalizing the denominator

Materials Needed

calculator

Lesson Resources

Warm-Up Transparency 28
Transparency RF-38
Reteaching 10-1
Extra Practice 10-1
Enrichment 10-1

Assignment Guide

Basic: 1–40, 45–50
Enriched: 1–50

Getting Started

5-Minute Warm-Up

Find each square root.

1. $\sqrt{9}$ 3
2. $\sqrt{25}$ 5
3. $\sqrt{100}$ 10
4. $\sqrt{121}$ 11

Introduction to Lesson 10-1

After students have completed their charts, discuss whether they think they would ever find y, such that $y^2 = 2$ exactly. Ask them why or why not.

10-1 Irrational Numbers

Goals
- Find square roots.
- Simplify products and quotients containing radicals.

Applications Architecture, Small Business, Construction

Work with a partner to explore square roots.

A calculator gives the square root of 2 as 1.4124213562. Copy and complete this chart, using a calculator.

$(1.4)^2 = $? 1.96	$(1.4142)^2 = $? 1.99996164	$(1.4142135)^2 = $? 1.999999824
$(1.41)^2 = $? 1.9881	$(1.41421)^2 = $? 1.999989924	$(1.41421356)^2 = $? 1.999999993
$(1.414)^2 = $? 1.999396	$(1.414213)^2 = $? 1.999998409	$(1.414213562)^2 = $? 1.999999999

Do you think that eventually you will find a number, y, with enough decimal places so that $y^2 = 2$ exactly? no

Technology Note

To find the square root of 2, the key sequence for most calculators is [2] [√].

For many graphing calculators, the key sequence is [√] [2] [=].

Build Understanding

The symbol $\sqrt{x}$ means the **square root** of a number, x. It is the number that multiplied by itself equals x. For example, $\sqrt{4} = 2$, because $2^2 = 4$; and $\sqrt{144} = 12$, because $12^2 = 144$. These two examples have square roots that are rational numbers.

The number $\sqrt{2}$ is an **irrational number**, which means that it is not a rational number. Therefore, it is a number that cannot be written as a fraction, a terminating decimal, or a repeating decimal. It can be represented by a nonterminating, nonrepeating decimal, and it can be approximated to any decimal place.

Example 1

ARCHITECTURE An architect is drawing a landscaping plan which includes three square cement blocks with areas of approximately 5, 31, and 98 square feet. To find the length of a side for each square, she finds the square root of each area. Find the value of each to the nearest hundredth: $\sqrt{5}$, $\sqrt{31}$ and $\sqrt{98}$.

Solution

Use a calculator. Then round to the hundredths place.

$\sqrt{5} = 2.236067977\ldots$ rounds to 2.24
$\sqrt{31} = 5.567764363\ldots$ rounds to 5.57
$\sqrt{98} = 9.899494937\ldots$ rounds to 9.90

Extend the Lesson

ONGOING ASSESSMENT Ask students to simplify $\sqrt{20}$. Watch for students who write the perfect square factor of 20 outside the radical sign instead of the square root of that perfect square factor. These students will get the answer $4\sqrt{5}$. The correct answer is $2\sqrt{5}$.

Another way to read the expression $\sqrt{2}$ is "radical 2." The number under the radical sign is called the **radicand**. There are properties of radicals that allow you to simplify them.

Theorem

The square root of a product of two numbers is the same as the product of their square roots.

$$\sqrt{a \cdot b} = \sqrt{a} \cdot \sqrt{b}$$

Example 2

Simplify each.

a. $\sqrt{32}$ **b.** $\sqrt{300}$ **c.** $\sqrt{125}$

Solution

Rewrite each radicand as a product of two numbers, so that one of them is a perfect square.

a. $\sqrt{32} = \sqrt{16 \cdot 2} = \sqrt{16} \cdot \sqrt{2} = \sqrt{4} \cdot 2$, or $4\sqrt{2}$

b. $\sqrt{300} = \sqrt{3 \cdot 100} = \sqrt{3} \cdot \sqrt{100} = \sqrt{3} \cdot 10$, or $10\sqrt{3}$

c. $\sqrt{125} = \sqrt{25 \cdot 5} = \sqrt{25} \cdot \sqrt{5} = \sqrt{5} \cdot 5$, or $5\sqrt{5}$

Numbers written in radical form can be multiplied together. First, multiply the rational factors, then the irrational factors.

Example 3

Multiply $(4\sqrt{10})(-3.1\sqrt{6})$.

Solution

Rewrite the product so that rational and irrational factors are together.

$(4\sqrt{10})(-3.1\sqrt{6}) = (4 \cdot -3.1)(\sqrt{10} \cdot \sqrt{6})$

$= -12.4\sqrt{60}$ $\quad \sqrt{10} \cdot \sqrt{6} = \sqrt{10 \cdot 6}$

The product can be simplified.

$-12.4\sqrt{60} = -12.4(\sqrt{4 \cdot 15})$

$= -12.4(2\sqrt{15})$

$= -24.8\sqrt{15}$

So, $(4\sqrt{10})(-3.1\sqrt{6}) = -24.8\sqrt{15}$.

Check Understanding

What is the product, in simplest radical form, of $\sqrt{14} \cdot \sqrt{7}$?

$7\sqrt{2}$

A similar theorem about radicals exists for quotients.

Theorem

The square root of a quotient of two numbers is equal to the quotient of their square roots.

$$\sqrt{\frac{a}{b}} = \frac{\sqrt{a}}{\sqrt{b}} \text{ where } b \neq 0$$

Teaching Strategies

Encourage students to check the reasonableness of a square root found by using a calculator. Suggest that they compare the number to the square roots of the closest perfect squares less than and greater than the radicand.

Chalkboard Examples

Supplementary Example 1
Find the value of $\sqrt{18}$ to the nearest hundredth. 4.24

Supplementary Example 2
Write in simplest radical form: $\sqrt{56}$. $2\sqrt{14}$

Supplementary Example 3
Multiply $(3\sqrt{2})(2\sqrt{10})$. $12\sqrt{5}$

Supplementary Example 4
Divide: $\frac{\sqrt{72}}{\sqrt{2}}$ 6

Reteaching Worksheet 10-1

Name ______________________ Date ____________

RETEACHING 10-1

IRRATIONAL NUMBERS

A nonterminating, nonrepeating decimal is an irrational number.
Some radicals are irrational numbers.
You can use these properties to simplify radical expressions.

$\sqrt{a \cdot b} = \sqrt{a} \cdot \sqrt{b}$ $\quad \sqrt{\frac{a}{b}} = \frac{\sqrt{a}}{\sqrt{b}}$ where $a \geq 0$ and $b > 0$.

$(a\sqrt{b})(c\sqrt{d}) = (a \cdot c)(\sqrt{b} \cdot \sqrt{d})$

Example

Simplify.

a. $\sqrt{80}$ **b.** $(3\sqrt{5})(6\sqrt{35})$ **c.** $6\sqrt{17} \div 3\sqrt{34}$

Solution

a. Rewrite the radicand as a product of two numbers, one of which is a perfect square, then simplify.

$\sqrt{80} = \sqrt{16} \cdot \sqrt{5} = 4\sqrt{5}$

So $\sqrt{80} = 4\sqrt{5}$.

b. Use the commutative and associative properties to rewrite the expression, grouping both rational factors together and both irrational factors together. Then multiply and simplify the product.

$(3\sqrt{5})(6\sqrt{35}) = (3 \cdot 6)(\sqrt{5} \cdot \sqrt{35}) = 18\sqrt{175} = 18\sqrt{25 \cdot 7} = 18(5\sqrt{7}) = 90\sqrt{7}$

So $(3\sqrt{5})(6\sqrt{35}) = 90\sqrt{7}$.

c. Write the expression in fraction form, grouping integers together and radicals together. Simplify each part of the expression and rationalize the denominator.

$\frac{6\sqrt{17}}{3\sqrt{34}} = \frac{6}{3} \cdot \frac{\sqrt{17}}{\sqrt{34}} = 2 \cdot \frac{\sqrt{17}}{\sqrt{34}} = 2\sqrt{\frac{1}{2}} = \frac{2\sqrt{1}}{\sqrt{2}} = \frac{2 \cdot 1}{\sqrt{2}} = \frac{2}{\sqrt{2}} = \frac{2\sqrt{2}}{\sqrt{2} \cdot \sqrt{2}} = \frac{2\sqrt{2}}{2} = \sqrt{2}$

So $6\sqrt{17} \div 3\sqrt{34} = \sqrt{2}$.

EXERCISES

Simplify.

1. $\sqrt{72}$ $6\sqrt{2}$
2. $\sqrt{300}$ $10\sqrt{3}$
3. $\sqrt{242}$ $11\sqrt{2}$
4. $\sqrt{56}$ $2\sqrt{14}$
5. $\sqrt{480}$ $4\sqrt{30}$
6. $\sqrt{96}$ $4\sqrt{6}$
7. $\sqrt{368}$ $4\sqrt{23}$
8. $\sqrt{768}$ $16\sqrt{3}$
9. $(4\sqrt{3})(5\sqrt{15})$ $60\sqrt{5}$
10. $(3\sqrt{2})(4\sqrt{28})$ $24\sqrt{14}$
11. $(5\sqrt{8})(5\sqrt{8})$ 200
12. $(6\sqrt{21})(4\sqrt{7})$ $168\sqrt{3}$
13. $(2\sqrt{5})(7\sqrt{20})$ 140
14. $\sqrt{24} \div \sqrt{18}$ $\frac{2\sqrt{6}}{3}$

146 RETEACHING *LESSON 10-1*

Lesson Wrap-up

Quick Assessment

Ask the following questions to determine if students understand the content presented in this lesson.

1. How can you determine whether $\sqrt{x}$ is rational or irrational? For $\sqrt{x}$ to be rational, x must be a perfect square. Otherwise, it is an irrational number.
2. Is 1.030030003 . . . an irrational number? Explain. Yes, although a pattern is formed, 1.030030003 . . . is irrational because it is neither terminating nor repeating.

Assignment Guide

Basic: 1–40, 45–50
Advanced: 1–50
Additional Practice: See Extra Practice Index on page 674.

Example 4

Find the quotient in simplest radical form of $-5\sqrt{33} \div 3\sqrt{22}$.

Solution

$$-\frac{5\sqrt{33}}{3\sqrt{22}} = \frac{-5}{3} \cdot \sqrt{\frac{33}{22}} \quad \text{Simplify } \tfrac{33}{22}.$$
$$= \frac{-5}{3} \cdot \sqrt{\frac{3}{2}}$$
$$= \frac{-5\sqrt{3}}{3\sqrt{2}}$$

In simplest radical form, denominators cannot include radicals.

$$\frac{-5\sqrt{3}}{3\sqrt{2}} = \frac{-5\sqrt{3}(\sqrt{2})}{3\sqrt{2}(\sqrt{2})} \quad \text{Multiply numerator and denominator by } \sqrt{2}.$$
$$= \frac{-5\sqrt{6}}{3\sqrt{4}}$$
$$= \frac{-5\sqrt{6}}{3 \cdot 2}, \text{ or } \frac{-5\sqrt{6}}{6}$$

The process of rewriting a quotient to delete radicals from the denominator is called **rationalizing the denominator**.

Reading Math

In everyday language, words that have "rational" as one of their parts usually relate to reasons or reasonableness. For example, rationalizing means to find a reason for something. When you say someone is being irrational, you mean that he or she is not making sense or not being reasonable.

In mathematics, words with "rational" as one of their parts always relate to ratio. Rational numbers are numbers that can be written as a ratio of two integers. Irrational numbers cannot. Rationalizing the denominator means changing it to a rational number.

Try These Exercises

CALCULATOR **Find the value to the nearest hundredth.**

1. $\sqrt{11}$ 3.32
2. $\sqrt{56}$ 7.48
3. $\sqrt{85}$ 9.22
4. $\sqrt{196}$ 14.00

Write each in simplest radical form.

5. $\sqrt{44}$ $2\sqrt{11}$
6. $\sqrt{242}$ $11\sqrt{2}$
7. $\sqrt{75}$ $5\sqrt{3}$
8. $\sqrt{48}$ $4\sqrt{3}$

Simplify.

9. $(2\sqrt{6})(7\sqrt{2})$ $28\sqrt{3}$
10. $(-3\sqrt{10})(5\sqrt{5})$ $-75\sqrt{2}$
11. $(4\sqrt{2})^2$ 32
12. $\frac{\sqrt{150}}{\sqrt{6}}$ 5
13. $\frac{2}{\sqrt{3}}$ $\frac{2\sqrt{3}}{3}$
14. $\sqrt{\frac{21}{2}}$ $\frac{\sqrt{42}}{2}$

Practice Exercises

CALCULATOR **Find the value to the nearest hundredth.**

15. $\sqrt{21}$ 4.58
16. $\sqrt{47}$ 6.86
17. $\sqrt{73}$ 8.54
18. $\sqrt{200}$ 14.14

Write each in simplest radical form.

19. $\sqrt{162}$ $9\sqrt{2}$
20. $\sqrt{500}$ $10\sqrt{5}$
21. $\sqrt{72}$ $6\sqrt{2}$
22. $\sqrt{40}$ $2\sqrt{10}$
23. $(2\sqrt{3})(4\sqrt{6})$ $24\sqrt{2}$
24. $(10\sqrt{2})(2\sqrt{6})$ $40\sqrt{3}$
25. $(3\sqrt{3})^2$ 27
26. $\frac{\sqrt{45}}{\sqrt{5}}$ 3
27. $\frac{7}{\sqrt{2}}$ $\frac{7\sqrt{2}}{2}$
28. $\sqrt{\frac{6}{5}}$ $\frac{\sqrt{30}}{5}$
29. $(2\sqrt{5})^2$ 20
30. $\frac{3\sqrt{12}}{4\sqrt{48}}$ $\frac{3}{8}$

Extend the Lesson

MATH JOURNAL Have students write the steps for simplifying the square root of any whole number, the product of two factors involving radicals, and the quotient of two numbers involving radicals. Have them simplify an example of each using the steps they have written.

31. **SMALL BUSINESS** George is starting a pet-sitting service in his backyard. He wants to enclose an area of about 6 m^2, and has decided that the best shape for the enclosure is a square. What should be the length of a side of the enclosure, to the nearest tenth of a meter? 2.4 m

32. **WRITING MATH** Is the number $\sqrt{207.36}$ rational or irrational? Explain.

32. 14.4 is a rational number because it can be written as the ratio of two integers; $\frac{144}{10}$.

33. **DATA FILE** Use the Data Index on pages 632–633 to locate information about famous rectangular archeological structures. For the Bakong Temple and the Wat Kukat Temple, express the length of a side as the square root of the area of the structure in meters. Bakong Temple, $\sqrt{4900}$ m

34. **CONSTRUCTION** A mason is building a square pedestal from brick. The entire pedestal must be about 12 ft^2 in area. To the nearest hundredth, find the length of a side of the pedestal. 3.46 ft

The expression $\sqrt[n]{x}$ means the nth root of x. It means the number which is raised to the nth power equals x. Example: $\sqrt[3]{8} = 2$, because $2^3 = 8$.

Find each of the following.

35. $\sqrt[3]{27}$ 3 36. $\sqrt[3]{125}$ 5 37. $\sqrt[4]{16}$ 2 38. $\sqrt[4]{81}$ 3

MATH HISTORY In the first century A.D., Heron of Alexandria discovered a formula for finding the area of a triangle when only the measures of its sides are known:

$$A = \sqrt{s(s-a)(s-b)(s-c)} \text{ where } s = 0.5(a + b + c).$$

Use Heron's formula to find the area of each triangle to the nearest tenth.

39. 3 cm, 5 cm, 6 cm 7.5 cm^2 40. 7 in., 12 in., 15 in. 41.2 in.^{2}

EXTENDED PRACTICE EXERCISES

Decide if each statement is true or false. If it is false, give a counterexample.

41. The product of two even numbers is always an even number. true

42. The square of a rational number is always a rational number. true

43. The product of two irrational numbers is always an irrational number. false; possible counterexample $(\sqrt{2})(\sqrt{18}) = 6$

44. **CHAPTER INVESTIGATION** Make a detailed sketch of the front of the main building at your school. Include any current architectural elements. Estimate measurements for the elements in the sketch. If possible, consult building plans to find the length and width of standard doors and windows. Check students' work.

MIXED REVIEW EXERCISES

Change each unit of measure as indicated. (Lesson 5-1)

45. 8 c = __?__ qt 2 46. 7 yd = __?__ in. 252 47. 10 km = __?__ m 10,000

48. 6 gal = __?__ pt 48 49. 14 m = __?__ dm 140 50. 7 g = __?__ kg 0.007

Extra Practice Worksheet 10-1

Name ____________ Date ____________

EXTRA PRACTICE 10-1

IRRATIONAL NUMBERS

EXERCISES

Find the value to the nearest hundredth.

1. $\sqrt{5}$ 2.24 2. $\sqrt{45}$ 6.71 3. $\sqrt{86}$ 9.27

4. $\sqrt{31}$ 5.57 5. $\sqrt{132}$ 11.49 6. $\sqrt{543}$ 23.30

7. $\sqrt{256}$ 16 8. $\sqrt{984}$ 31.37 9. $\sqrt{1235}$ 35.14

Write each in simplest radical form.

10. $\sqrt{40}$ $2\sqrt{10}$ 11. $\sqrt{600}$ $10\sqrt{6}$

12. $\sqrt{147}$ $7\sqrt{3}$ 13. $\sqrt{320}$ $8\sqrt{5}$

14. $\sqrt{216}$ $6\sqrt{6}$ 15. $\sqrt{448}$ $8\sqrt{7}$

16. $(4\sqrt{2})(3\sqrt{6})$ $24\sqrt{3}$ 17. $(12\sqrt{5})(2\sqrt{10})$ $120\sqrt{2}$

18. $(3\sqrt{8})(7\sqrt{2})$ 84 19. $(-\sqrt{12})(4\sqrt{6})$ $-24\sqrt{2}$

20. $(-3\sqrt{15})(4\sqrt{6})$ $-36\sqrt{10}$ 21. $(-5\sqrt{5})(-2\sqrt{50})$ $50\sqrt{10}$

22. $(6\sqrt{7})^2$ 252 23. $(-9\sqrt{6})^2$ 486

24. $\frac{\sqrt{3}}{\sqrt{2}}$ $\frac{\sqrt{6}}{2}$ 25. $\frac{\sqrt{72}}{\sqrt{6}}$ $2\sqrt{3}$

26. $\frac{\sqrt{14}}{\sqrt{7}}$ $\sqrt{2}$ 27. $\frac{\sqrt{5}}{\sqrt{10}}$ $\frac{\sqrt{2}}{2}$

28. $\frac{9}{\sqrt{3}}$ $3\sqrt{3}$ 29. $\frac{11}{\sqrt{7}}$ $\frac{11\sqrt{7}}{7}$

30. $\frac{4}{\sqrt{6}}$ $\frac{2\sqrt{6}}{3}$ 31. $\frac{5}{\sqrt{3}}$ $\frac{5\sqrt{3}}{3}$

32. $\sqrt{\frac{14}{3}}$ $\frac{\sqrt{42}}{3}$ 33. $\sqrt{\frac{8}{12}}$ $\frac{\sqrt{6}}{3}$

34. $\sqrt{\frac{18}{5}}$ $\frac{3\sqrt{10}}{5}$ 35. $\sqrt{\frac{3}{10}}$ $\frac{\sqrt{30}}{10}$

128 EXTRA PRACTICE *LESSON 10-1*

Extend the Lesson

REAL-WORLD CONNECTION Have students investigate the pyramids of Egypt and provide information about the areas of their bases and the lengths of the sides of those bases. For example, the Great Pyramid of Cheops has a base of 571,536 ft^2. Therefore, the length of each side of its base is $\sqrt{571,536}$ ft, or 756 ft.

Enrichment Worksheet 10-1

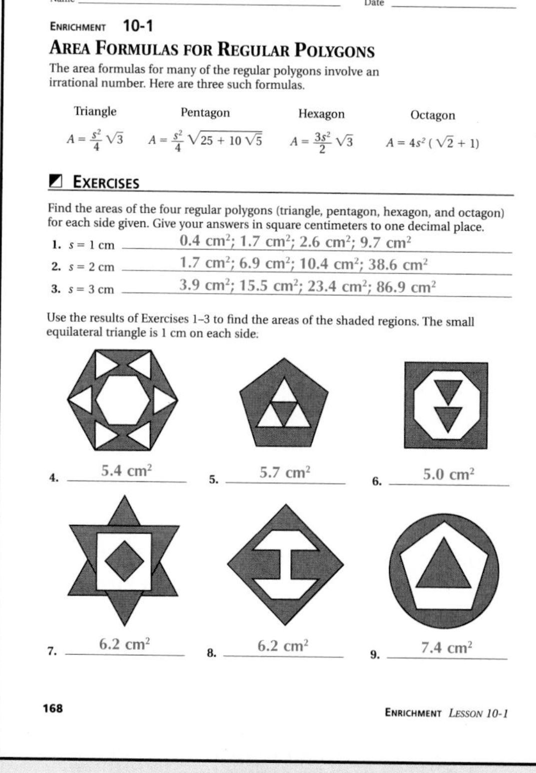

Name ____________ Date ____________

ENRICHMENT 10-1

AREA FORMULAS FOR REGULAR POLYGONS

The area formulas for many of the regular polygons involve an irrational number. Here are three such formulas.

Triangle	Pentagon	Hexagon	Octagon
$A = \frac{s^2}{4}\sqrt{3}$	$A = \frac{s^2}{4}\sqrt{25 + 10\sqrt{5}}$	$A = \frac{3s^2}{2}\sqrt{3}$	$A = 4s^2(\sqrt{2} + 1)$

EXERCISES

Find the areas of the four regular polygons (triangle, pentagon, hexagon, and octagon) for each side given. Give your answers in square centimeters to one decimal place.

1. $s = 1$ cm 0.4 cm^2; 1.7 cm^2; 2.6 cm^2; 9.7 cm^2
2. $s = 2$ cm 1.7 cm^2; 6.9 cm^2; 10.4 cm^2; 38.6 cm^2
3. $s = 3$ cm 3.9 cm^2; 15.5 cm^2; 23.4 cm^2; 86.9 cm^2

Use the results of Exercises 1–3 to find the areas of the shaded regions. The small equilateral triangle is 1 cm on each side.

4. 5.4 cm^2 5. 5.7 cm^2 6. 5.0 cm^2

7. 6.2 cm^2 8. 6.2 cm^2 9. 7.4 cm^2

168 ENRICHMENT *LESSON 10-1*

Lesson Planning

NCTM Standards/Strands

- Number & Operations
- Patterns, Functions, & Algebra
- Geometry & Spatial Sense
- Problem Solving
- Reasoning & Proof
- Communication
- Connections
- Representation

Vocabulary

Pythagorean Theorem
hypotenuse
legs
Pythagorean triples

Materials Needed

graph paper
scissors
straightedges
geometric-drawing software (optional)
computer (optional)

Lesson Resources

Warm-Up Transparency 28
Transparency RF-39
Reteaching 10-2
Extra Practice 10-2
Enrichment 10-2

Assignment Guide

Basic: 1–22, 26–30
Enriched: 1–30

Getting Started

5-Minute Warm-Up

Find the value of each.

1. 7^2 49 **2.** 12^2 144
3. $\sqrt{81}$ 9 **4.** $\sqrt{64}$ 8

Introduction to Lesson 10-2

Once students have drawn a conclusion about the relationship of the squares of the shorter sides to the square of the longest side of a right triangle, you may want them to repeat this activity with acute and obtuse triangles to see whether the same conclusions hold true.

10-2 The Pythagorean Theorem

Goals ■ Use the Pythagorean Theorem to solve problems involving right triangles.

Applications Architecture, Construction, Home Repairs

Draw any right triangle on graph paper. Make squares on each side of the triangle, as shown. Cut out the squares. Try to fit the two smaller squares on top of the larger square. You may cut up the smaller squares. What conclusion can you draw? The sum of the squares of the two shorter sides of a right triangle equals the square of the largest side.

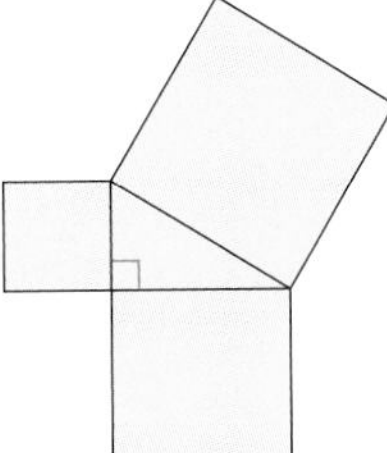

Build Understanding

In a right triangle, the shorter sides are legs, and the side opposite the right angle is the hypotenuse. The relationship between the lengths of the legs and the hypotenuse is called the **Pythagorean Theorem**.

Pythagorean Theorem

In a right triangle, the sum of the squares of the measures of the two legs is equal to the square of the measure of the hypotenuse.

$a^2 + b^2 = c^2$

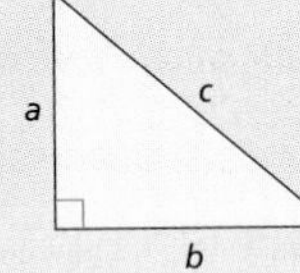

This proof of the Pythagorean Theorem uses similar triangles formed by the altitude to the hypotenuse.

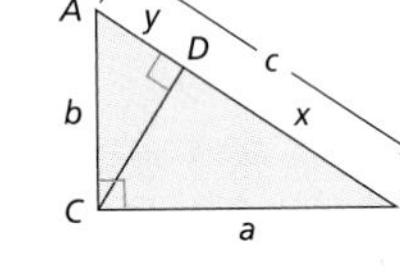

Given $\triangle ABC$, $\angle C$ is a right angle
Prove $a^2 + b^2 = c^2$

Statements	Reasons
1. $\triangle ABC$ is a right triangle.	1. Given
2. $\overline{CD}$ is perpendicular to AB.	2. There is one and only one line through a point perpendicular to a given line.
3. $\overline{CD}$ is the altitude to the hypotenuse in $\triangle ABC$.	3. Definition of altitude.
4. $\triangle ABC \sim \triangle ACD \sim \triangle CBD$	4. The altitude to the hypotenuse forms two right triangles that are similar to each other and to the original triangle.
5. $\frac{c}{a} = \frac{a}{x}, \frac{c}{b} = \frac{b}{y}$	5. Corresponding sides of similar triangles are proportional.
6. $cx = a^2$, $cy = b^2$	6. Cross products are equal.
7. $cx + cy = a^2 + b^2$	7. Addition property of equality
8. $c(x + y) = a^2 + b^2$	8. Distributive property
9. $c = x + y$	9. Segment addition postulate
10. $c(c) = a^2 + b^2$	10. Substitution
11. $c^2 = a^2 + b^2$	11. Definition of c^2.

Math: Who, Where When

The scarecrow in the classic movie The Wizard of Oz tried to recite the Pythagorean Theorem as proof of intelligence when he received his "brains." But he stated it incorrectly.

Teaching Strategies

ESL/LEP Restating the Pythagorean Theorem in words more easily understandable may be helpful. For example: If the lengths of each short side of a right triangle are squared and these squares are added, the answer is the square of the longest side.

You can use the Pythagorean Theorem to find one side of a right triangle when the other two sides are known.

Example 1

ARCHITECTURE An architect draws a right triangle, $\triangle DEF$, on a blueprint. $DE = 3$ cm and $EF = 5$ cm.

Find DF to the nearest tenth.

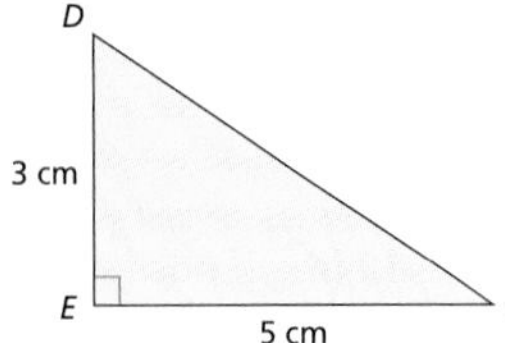

Solution

Let $x = DF$.
$$3^2 + 5^2 = x^2$$
$$9 + 25 = x^2$$
$$34 = x^2$$
$$\sqrt{34} = x$$
$$5.830951895\ldots = x$$ Use a calculator or a square root table to find $\sqrt{34}$.

So, $DF = 5.8$ cm to the nearest tenth.

GEOMETRY SOFTWARE You can use geometric-drawing software to solve problems involving right triangles. Set the distance units to either inches, centimeters, or pixels. Choose pixels when the problem involves large numbers. Use the Show Grid option to make drawing figures easier. Draw the lengths given in the problem. Use the software to check the measurement of these segments. Draw and measure any remaining segments. Use the software to find unknown lengths and angle measures as required by the problem.

Sometimes, the missing measure in the right triangle is the length of one of the legs.

Example 2

Find x to the nearest hundredth of an inch.

Solution

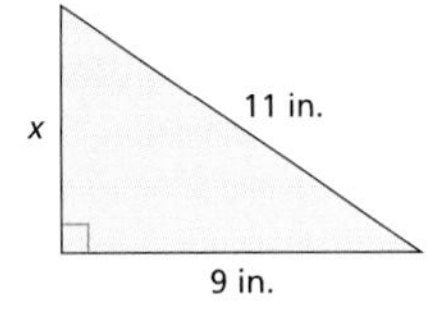

$$x^2 + 9^2 = 11^2$$
$$x^2 + 81 = 121$$
$$x^2 = 40$$
$$x = \sqrt{40}$$
$$x = 6.32455532\ldots$$ So, $x = 6.32$ in.

The converse of the Pythagorean Theorem is also true.

Converse of the Pythagorean Theorem	If the sum of the squares of the measures of two shorter sides of a triangle is equal to the square of the measure of the third side, then the triangle is a right triangle.

Example 3

Are triangles with the following lengths right triangles?

a. 2 cm, $4\sqrt{2}$ cm, 6 cm **b.** 4 in., 9 in., 10 in.

Chalkboard Examples

Supplementary Example 1

An architect draws right triangle *MNO*. Leg *MN* measures 6 cm and leg *NO* measures 5 cm. Find *MO* to the nearest tenth. 7.8 cm

Supplementary Example 2

In a right triangle, the legs measure x in. and 14 in. The hypotenuse measures 18 in. Find x to the nearest hundredth of an inch. 11.31 in

Supplementary Example 3

Is a triangle with sides 4 cm, 8 cm, and 12 cm a right triangle? No

Learning Styles

VISUAL LEARNERS Draw a right triangle, labeling its legs a and b and its hypotenuse c. Use a different color for each leg and its label. Use those same colors for the corresponding variables in the equation $a^2 + b^2 = c^2$. Then write lengths on two different sides of the triangle at a time. Have students substitute those lengths in the equation and find the missing length by referring to the colors as needed.

Reteaching Worksheet 10-2

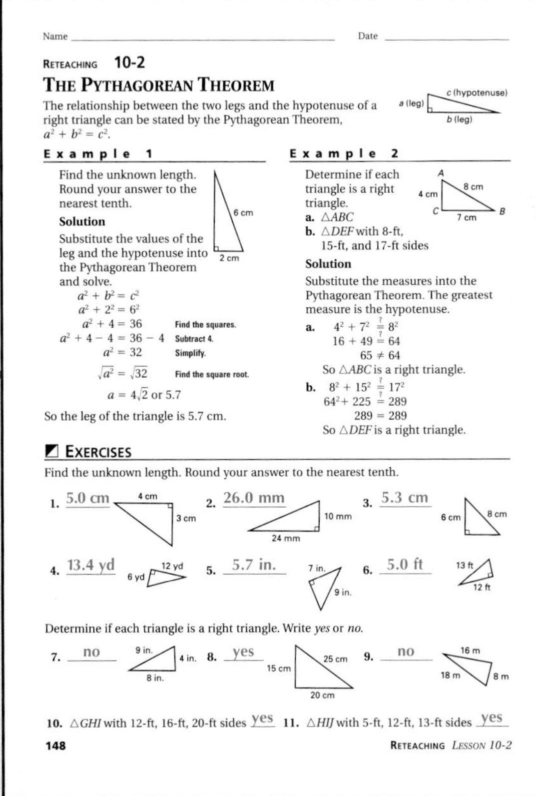

Name ______ Date ______

RETEACHING 10-2

THE PYTHAGOREAN THEOREM

The relationship between the two legs and the hypotenuse of a right triangle can be stated by the Pythagorean Theorem, $a^2 + b^2 = c^2$.

a (leg) c (hypotenuse) b (leg)

Example 1

Find the unknown length. Round your answer to the nearest tenth.

Solution

Substitute the values of the leg and the hypotenuse into the Pythagorean Theorem and solve.

$a^2 + b^2 = c^2$

$a^2 + 2^2 = 6^2$

$a^2 + 4 = 36$ Find the squares.

$a^2 + 4 - 4 = 36 - 4$ Subtract 4.

$a^2 = 32$ Simplify.

$\sqrt{a^2} = \sqrt{32}$ Find the square root.

$a = 4\sqrt{2}$ or 5.7

So the leg of the triangle is 5.7 cm.

Example 2

Determine if each triangle is a right triangle.

a. $\triangle ABC$

b. $\triangle DEF$ with 8-ft, 15-ft, and 17-ft sides

Solution

Substitute the measures into the Pythagorean Theorem. The greatest measure is the hypotenuse.

a. $4^2 + 7^2 \stackrel{?}{=} 8^2$

$16 + 49 \stackrel{?}{=} 64$

$65 \neq 64$

So $\triangle ABC$ is a right triangle.

b. $8^2 + 15^2 \stackrel{?}{=} 17^2$

$64^2 + 225 \stackrel{?}{=} 289$

$289 = 289$

So $\triangle DEF$ is a right triangle.

EXERCISES

Find the unknown length. Round your answer to the nearest tenth.

1. 5.0 cm 2. 26.0 mm 3. 5.3 cm

4. 13.4 yd 5. 5.7 in. 6. 5.0 ft

Determine if each triangle is a right triangle. Write *yes* or *no*.

7. no 8. yes 9. no

10. $\triangle GHI$ with 12-ft, 16-ft, 20-ft sides yes 11. $\triangle HIJ$ with 5-ft, 12-ft, 13-ft sides yes

148 RETEACHING *LESSON 10-2*

Lesson 10-2 **The Pythagorean Theorem** 431

Lesson Wrap-up

Quick Assessment

Ask the following question to determine if students understand the content presented in this lesson.

In your own words, what does the Pythagorean Theorem tell about a right triangle? The sum of the squares of the lengths of the legs of a right triangle is equal to the square of the length of the hypotenuse.

Assignment Guide

Basic: 1–22, 26–30
Advanced: 1–30
Additional Practice: See Extra Practice Index on page 674.

Additional Answers

23. See the figure below.
$d^2 = y^2 + h^2$ and $y^2 = l^2 + w^2$
Therefore,
$d^2 = l^2 + w^2 + h^2$
and $d = \sqrt{l^2 + w^2 + h^2}$

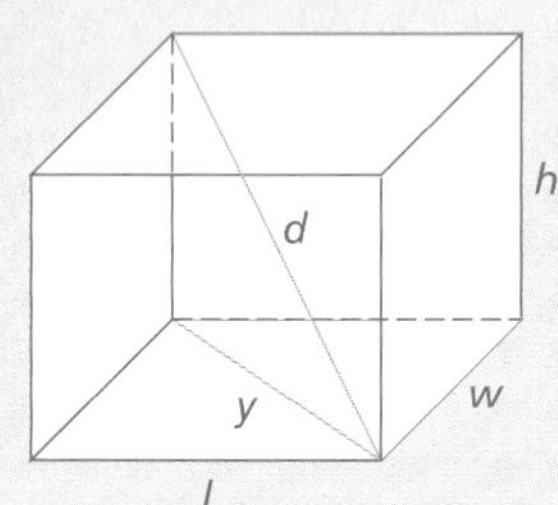

25. The system worked because the workers formed a triangle that had sides measuring 3 rope lengths, 4 rope lengths, and 5 rope lengths, [12 − (3 + 4)] = 5. Because 3, 4, and 5 form a Pythagorean triple, then they formed a right triangle.

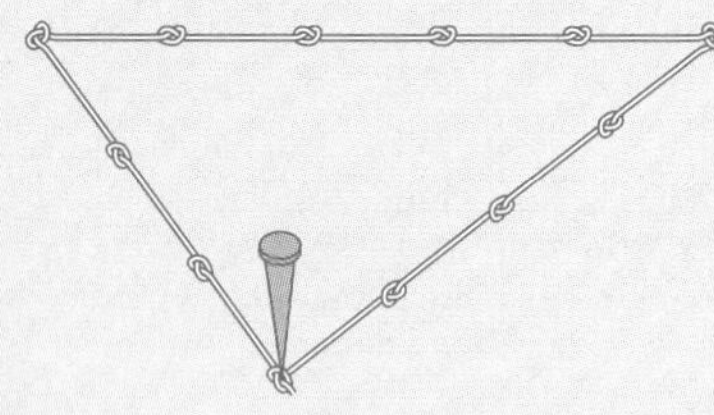

Solution

a. $2^2 + (4\sqrt{2})^2 \stackrel{?}{=} 6^2$
$4 + (16 \cdot 2) \stackrel{?}{=} 36$
$4 + 32 \stackrel{?}{=} 36$
$36 = 36$

b. $4^2 + 9^2 \stackrel{?}{=} 10^2$
$16 + 81 \stackrel{?}{=} 100$
$97 \neq 100$

Therefore, 2 cm, $4\sqrt{2}$ cm, and 6 cm form a right triangle, but 4 in., 9 in., and 10 in. do not.

Try These Exercises

Use the Pythagorean Theorem to find the unknown length. Round your answers to the nearest tenth.

1.
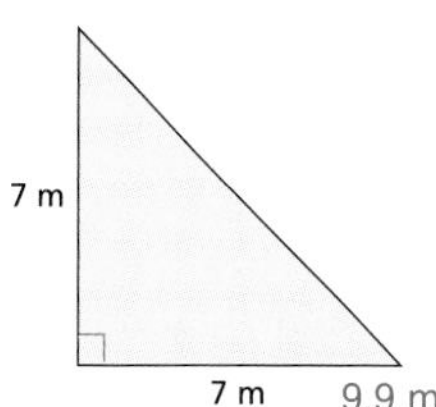

9.9 m

2.
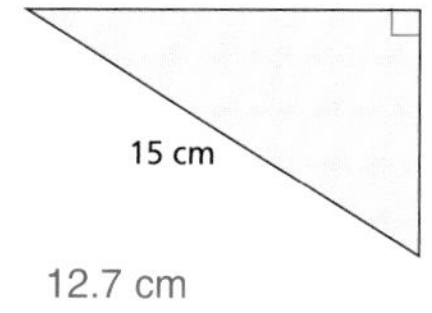

12.7 cm

3.
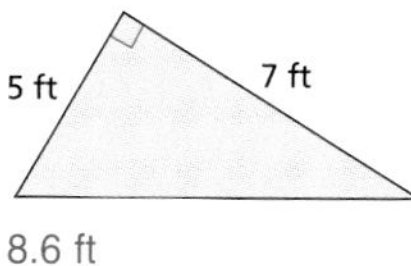

8.6 ft

4. **CONSTRUCTION** A triangular brace has sides that measure 6 cm, 8 cm, and 10 cm. Does the brace have a 90° angle? Yes

5. **HOME REPAIR** Tim is cleaning out the rain gutters on his home. He has an 18-ft ladder. If the base of the ladder is placed 5 ft from the base of the building, how far up the wall will the ladder reach? 17.3 ft

Practice Exercises

Use the Pythagorean Theorem to find the unknown length. Round your answers to the nearest tenth.

6.
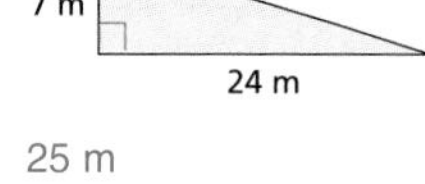

25 m

7.
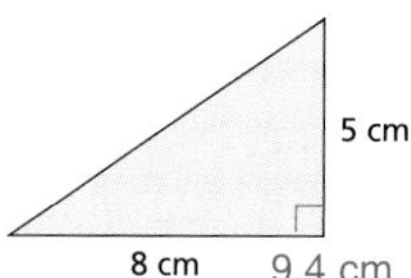

9.4 cm

8.
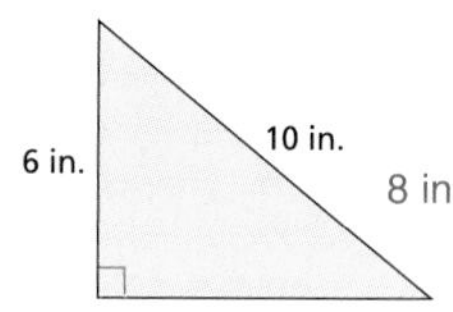

8 in.

9.
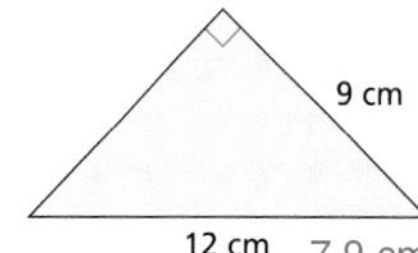

7.9 cm

10.
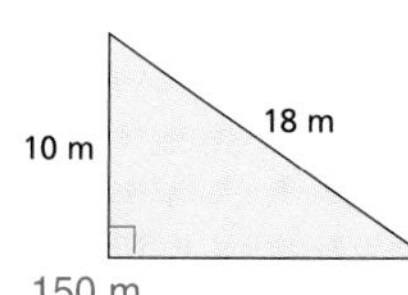

150 m

11.
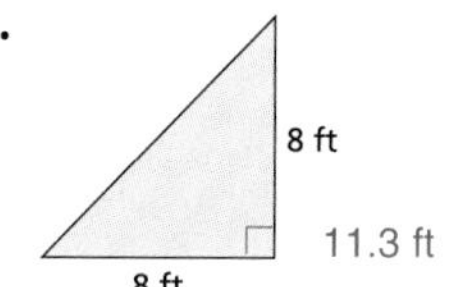

11.3 ft

Determine if the triangle is a right triangle. Write *yes* or *no*.

12. 9, 15, 12 yes

13.
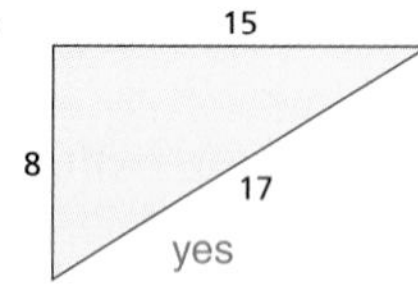

yes

14.
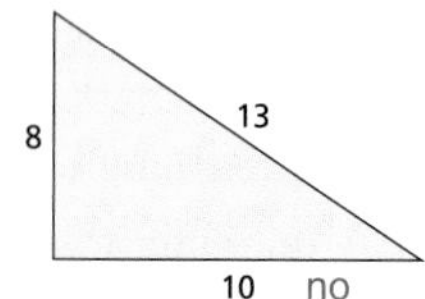

no

Extend the Lesson

COOPERATIVE LEARNING Have students work together to use the Pythagorean Theorem to find the length of an object or a distance. Have them draw diagrams of what they measured and show how they used those measures. If possible, have them measure the lengths or distances they found and compare them to their findings, using the Pythagorean Theorem.

432 Chapter 10 **Right Angles and Circles**

Solve. Round your answers to the nearest tenth.

15. **GEOMETRY SOFTWARE** Find the length of the diagonal of a rectangle with a length of 5 cm and a width of 4 cm. 6.4 cm

16. A flagpole 4 meters high is to be attached by a guy wire to a stake in the ground 1.2 meters from its base. How long must the guy wire be? 4.2 m

17. A ramp 6 yards long reaches from the loading dock to a point on the ground 4 yards from the base of the dock. How high above ground is the loading dock? 4.5 yd

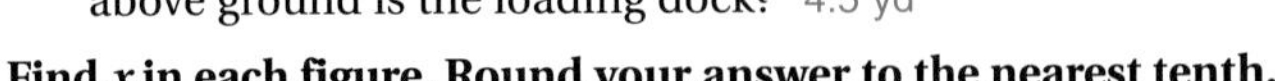

Find x in each figure. Round your answer to the nearest tenth.

18.

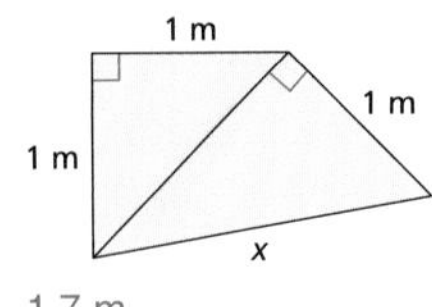

1.7 m

19.

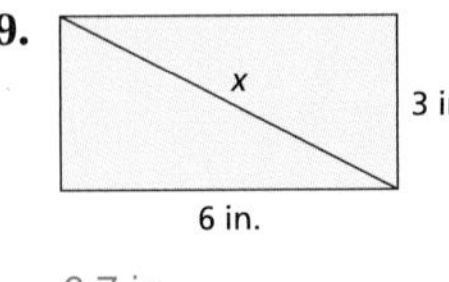

6.7 in.

20.

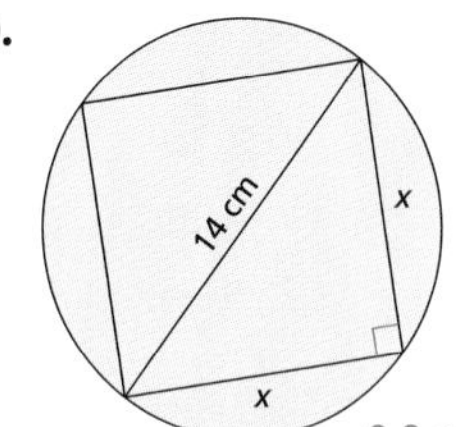

9.9 cm

21. **TALK ABOUT IT** Using s for the length of the side of a square and d for the length of the diagonal of a square, Michi has written the formula $d = s\sqrt{2}$. She says that the formula can be used to find the length of the diagonal of any square if the length of a side is known. Do you agree? Yes

22. Figure $ABCDEFGH$ is a rectangular prism. Its length, AB, is 8 cm, its width, BC, is 6 cm, and its height, BF, is 24 cm. Find the length of the diagonal, BH. (Hint: $\triangle BDH$ is a right triangle.) 26 cm

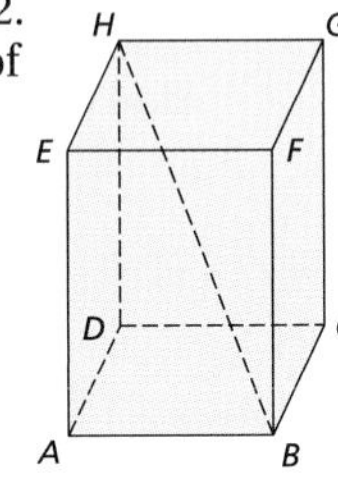

Extended Practice Exercises

23. Prove that, in any rectangular prism with length l, width w, and height h, the length of a diagonal d is given by the formula, $d = \sqrt{l^2 + w^2 + h^2}$. See additional answers.

24. Three positive integers, a, b, and c, for which $a^2 + b^2 = c^2$, are called **Pythagorean triples**. Let x and y represent two positive integers such that $x > y$. Let $a = 2xy$, $b = x^2 - y^2$, and $c = x^2 + y^2$. Try three different pairs of values for x and y. What do you notice? Any value of x and y will produce a pythagorean triple.

25. **WRITING MATH** In ancient Egypt, a pair of workers called rope stretchers would use a loop of rope divided by knots into 12 equal parts to mark off a perfect right angle in the sand. They would drive a stake through one knot. One worker would pull the rope taut at the third knot from the stake, while the other pulled the rope taut at the fourth knot on the other side of the stake. Write a paragraph and draw a diagram explaining why this system worked. See additional answers.

Mixed Review Exercises

A die is rolled 200 times with the following results: (Lesson 9-1)

Outcome	**1**	**2**	**3**	**4**	**5**	**6**
Frequency	**31**	**40**	**30**	**29**	**34**	**36**

What is the experimental probability of rolling each of the following results?

26. 3 0.15 **27.** 5 0.17 **28.** 6 0.18 **29.** 1 0.155 **30.** a number less than 5 0.65

Teaching Strategies

CHALLENGE Following completion of Exercise 25, discuss how the Pythagorean Theorem could be applied today to verify that the corner of a room or building is a square corner. Measure the sides adjacent to the corner and the diagonal opposite that corner, and find out whether they are possible lengths for the sides of a right triangle.

Extra Practice Worksheet 10-2

Name ______ Date ______

EXTRA PRACTICE 10-2

THE PYTHAGOREAN THEOREM

EXERCISES

Use the Pythagorean Theorem to find the unknown length.

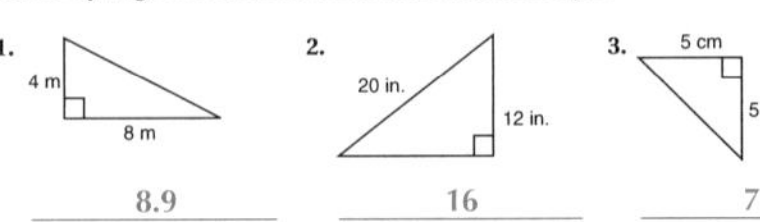

1. 8.9 2. 16 3. 7.1

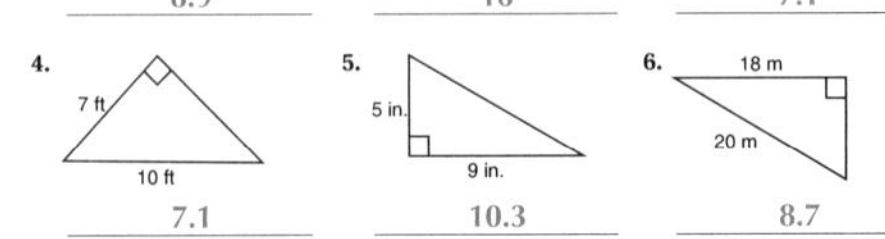

4. 7.1 5. 10.3 6. 8.7

Determine if the triangle is a right triangle. Write *yes* or *no*.

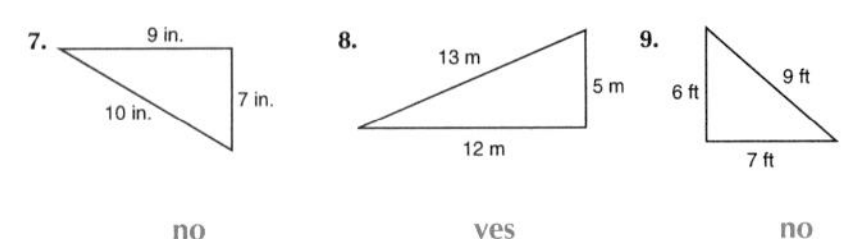

7. no 8. yes 9. no

Solve. Round your answers to the nearest tenth.

10. Find the length of the diagonal of a rectangle with a length of 6 in. and a width of 4 in. 7.2 in.

11. A ramp 8 meters long reaches from deck to a point on the ground 6 meters from the base of the deck. How high above the ground is the deck? 5.3 meters

12. A 12-foot ladder is placed against a building. The top of the ladder rests on the building 10 feet from the base of the building. How far from the base of the building is the base of the ladder? 6.6 feet

130 EXTRA PRACTICE *LESSON 10-2*

Enrichment Worksheet 10-2

Name ______ Date ______

ENRICHMENT 10-2

EXTENSIONS OF THE PYTHAGOREAN THEOREM

The exercises on this page allow you to explore two extensions of the Pythagorean Theorem.

EXERCISES

Extension 1

The square of one side of a triangle is the sum of the squares of the other two sides plus or minus twice the product of either of those sides by the projection on its line on the other side. The product is added or subtracted to yield a correct reslt depending on whether or not the angle opposite the initial side is obtuse or acute.

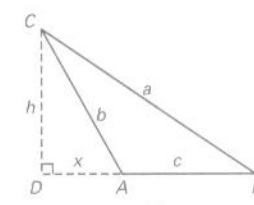

1. The case in which $\angle A$ is acute is drawn at the right. Finish this algebraic statement of Extension 1.

$a^2 = b^2 + c^2 - 2cx$

2. In the space at the right, draw the case in which $\angle A$ is obtuse. Then finish the algebraic statement of this case.

$a^2 = b^2 + c^2 + 2cx$

Extension 2

If similar figures are constructed on the sides of a right triangle (with corresponding sides used for a side of the right triangle), then the area of the polygon on the hypotenuse equals the sum of the areas of the polygons on the legs.

3. The case in which the similar figures are equilateral triangles is shown. Compute the areas to prove that this case is true.

3.9 + 6.9 = 10.8

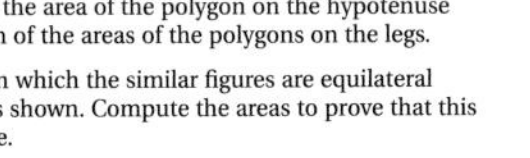

4. Draw the case in which the similar figures are right isosceles triangles with one short leg on the given triangle. Use a 3-4-5 right triangle. Then compute the areas for this case.

4.5 + 8 = 12.5

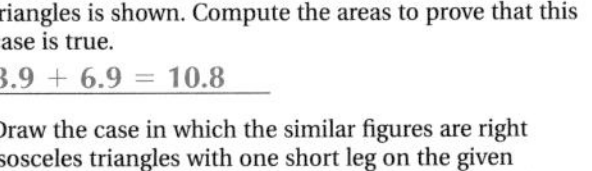

5. Show that the extension is true for semicircles. Use a 3-4-5 right triangle.

3.5 + 6.3 = 9.8

170 ENRICHMENT *LESSON 10-2*

Skills Practice

Vocabulary Review

Lesson 10-1
irrational number
radicand
rationalizing the denominator

Lesson 10-2
Pythagorean Theorem
hypotenuse
legs
Pythagorean triples

Assignment Guide

All students: 1–63
Additional Practice: See Extra Practice Index on page 674.

Review and Practice Your Skills

Practice ◼ Lesson 10-1

Find the value to the nearest hundredth.

1. $\sqrt{53}$ 7.28
2. $\sqrt{150}$ 12.25
3. $2\sqrt{7}$ 5.29
4. $\sqrt{624}$ 24.98
5. $10\sqrt{10}$ 31.62
6. $\sqrt{245}$ 15.65
7. $\sqrt{1000}$ 31.62
8. $\sqrt{37}$ 6.08

Write each in simplest radical form.

9. $\sqrt{50}$ $5\sqrt{2}$
10. $\sqrt{48}$ $4\sqrt{3}$
11. $\sqrt{72}$ $6\sqrt{2}$
12. $\sqrt{600}$ $10\sqrt{6}$
13. $\sqrt{80}$ $4\sqrt{5}$
14. $\sqrt{192}$ $8\sqrt{3}$
15. $\sqrt{88}$ $2\sqrt{22}$
16. $\sqrt{147}$ $7\sqrt{3}$
17. $\sqrt{242}$ $11\sqrt{2}$
18. $3\sqrt{20}$ $6\sqrt{5}$
19. $5\sqrt{63}$ $15\sqrt{7}$
20. $10\sqrt{75}$ $50\sqrt{3}$
21. $\left(2\sqrt{3}\right)^2$ 12
22. $\frac{\sqrt{12}}{\sqrt{3}}$ 2
23. $\sqrt{\frac{3}{4}}$ $\frac{\sqrt{3}}{2}$
24. $\left(3\sqrt{2}\right)\left(5\sqrt{10}\right)$ $30\sqrt{5}$
25. $\frac{\sqrt{10}}{\sqrt{2}}$ $\sqrt{5}$
26. $\left(8\sqrt{18}\right)\left(4\sqrt{2}\right)$ 192
27. $\sqrt{\frac{16}{7}}$ $\frac{4\sqrt{7}}{7}$
28. $\left(\sqrt{\frac{3}{8}}\right)\left(\frac{3}{\sqrt{2}}\right)\left(2\sqrt{27}\right)$ 13.5

Find each of the following.

29. $\sqrt[3]{64}$ 4
30. $\sqrt[4]{625}$ 5
31. $\sqrt[5]{32}$ 2
32. $\sqrt[5]{243}$ 3
33. $\sqrt[3]{1000}$ 10
34. $\sqrt[3]{\frac{1}{8}}$ $\frac{1}{2}$
35. $\sqrt[3]{\frac{1}{27}}$ $\frac{1}{3}$
36. $\sqrt[3]{-8}$ −2

Practice ◼ Lesson 10-2

Use Pythagorean Theorem to find the missing length. Round answers to the nearest tenth.

37.
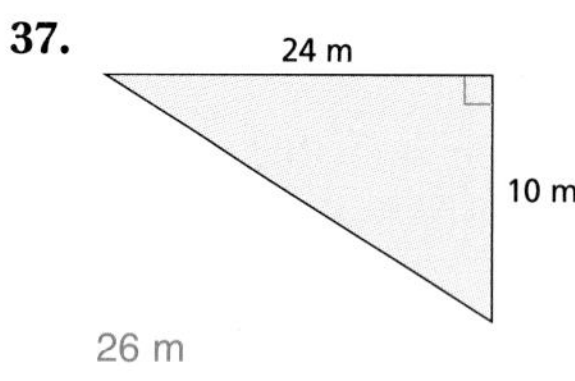

26 m

38.
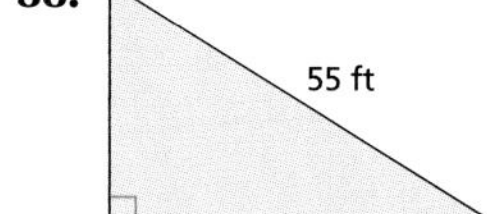

33 ft

39.
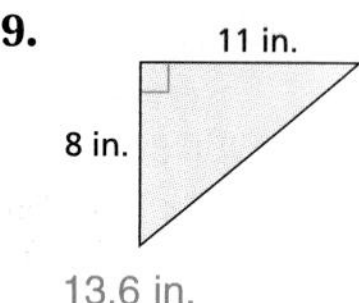

13.6 in.

40.
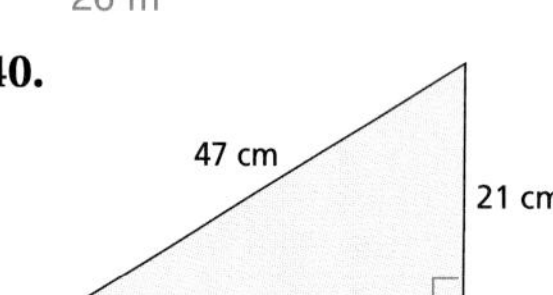

42.0 cm

41.
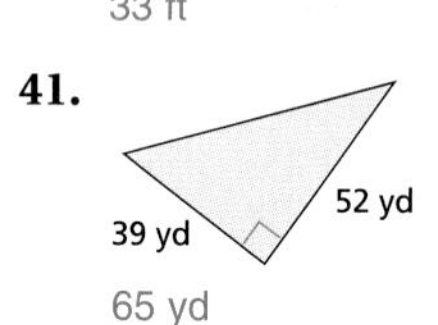

65 yd

42.
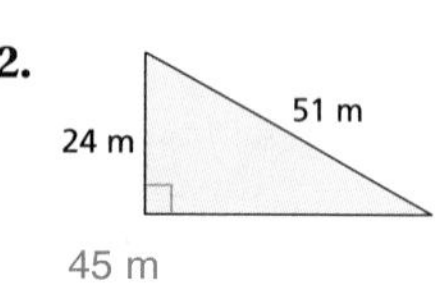

45 m

Solve. Round your answers to the nearest tenth.

43. Find the length of the diagonal of a rectangle with a length of 18 cm and a width of 24 cm. 30 cm
44. Find the width of a rectangle with a length of 30 ft and a diagonal of 34 ft. 16 ft

Extend the Lesson

CONNECT TO PRIOR KNOWLEDGE Ask students to determine whether π is a rational or irrational number and write an explanation of their decision.

PRACTICE ■ LESSON 10-1–LESSON 10-2

Write each in simplest radical form. (Lesson 10-1)

45. $\sqrt{24}$ $2\sqrt{6}$ **46.** $\sqrt{28}$ $2\sqrt{7}$ **47.** $\sqrt{40}$ $2\sqrt{10}$ **48.** $\sqrt{52}$ $2\sqrt{13}$

49. $\sqrt{\frac{1}{5}}$ $\frac{\sqrt{5}}{5}$ **50.** $\frac{4}{\sqrt{2}}$ $2\sqrt{2}$ **51.** $\frac{\sqrt{28}}{\sqrt{7}}$ 2 **52.** $\sqrt{\frac{6}{18}}$ $\frac{\sqrt{3}}{3}$

53. $8\sqrt{63}$ $24\sqrt{7}$ **54.** $5\sqrt{200}$ $50\sqrt{2}$ **55.** $11\sqrt{96}$ $44\sqrt{6}$ **56.** $-2\sqrt{72}$ $-12\sqrt{2}$

57. $(\sqrt{24})(3\sqrt{6})$ 36 **58.** $(2\sqrt{5})(5\sqrt{20})$ 100 **59.** $(7\sqrt{14})(3\sqrt{2})(8\sqrt{21})$ $235\sqrt{3}$ **60.** $\sqrt{x^2}$ x

Determine if the triangle is a right triangle. Write *yes* or *no.* (Lesson 10-2)

61.

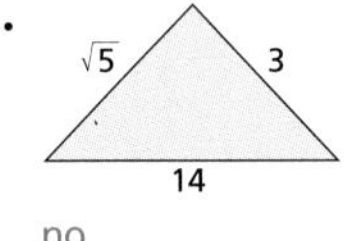

no

62.

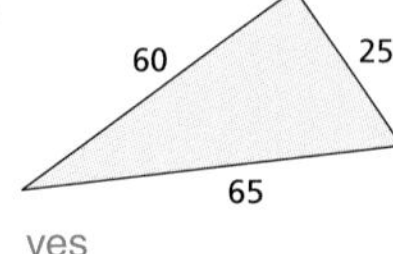

yes

63.

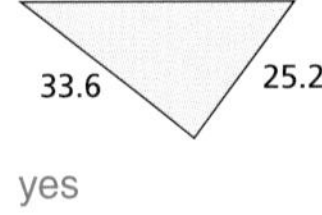

yes

Career – Construction and Building Inspectors

Construction and building inspectors make sure new structures follow building codes and ordinances, zoning regulations and contract specifications. They may also inspect alterations and repairs to existing structures.

Inspectors have the power to stop projects if safety standards and building codes are not met. To do their work, inspectors use tape measures, survey instruments, metering devices, cameras and calculators.

At a construction site, a temporary ramp has been built to assist workers in transporting materials up to the floor level. A diagram of the ramp is shown.

1. Use the Pythagorean theorem to find the length of the ramp to the nearest tenth. 19.7 ft
2. What is the measure of $\angle EFD$? 24°
3. What is the slope of the ramp? $\frac{4}{9}$
4. If DF is lengthened to 30 feet and CE remains the same, what will the new measure of EF be (to the nearest tenth)? 29.1 ft

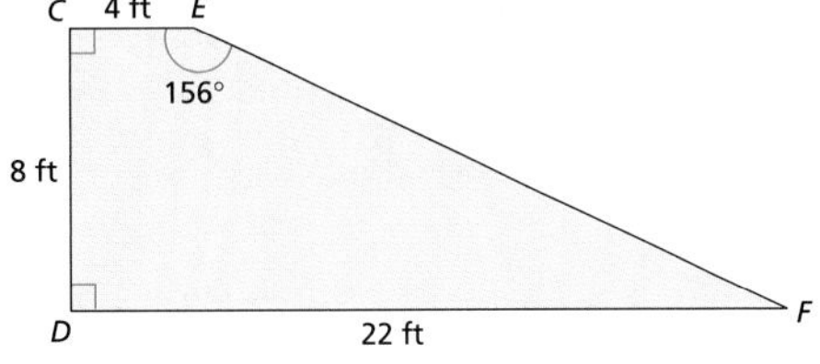

Chalkboard Examples

Example from Lesson 6-1

Write $\sqrt{\frac{18}{4}}$ in simplest radical form. $\frac{3\sqrt{2}}{2}$

Example from Lesson 6-2

A 20-foot ladder is to be used to paint a spot on a wall 18 feet above the ground. What is the farthest the base of the ladder can be placed from the wall? **8.7 ft**

Career Opportunity

Describe the kind of construction and building inspectors do and types of companies that hire these workers. Have students discuss the importance of measurement, geometry and algebraic thinking in doing this type of job. Explain that construction and building inspectors must be able to read complex plans and analyze existing structures to make sure all building codes are followed. Students should answer Questions 1–4 to better understand how construction and building inspectors use mathematics in performing their job.

Students who are interested in learning more about this profession can go to learningmathmatters.com. Guidance Counselors should have information about school and training requirements.

Extend the Lesson

MATH JOURNAL Have students list facts they can establish about a right triangle using the Pythagorean Theorem. For each, have them tell what information is needed, explain how it is used, and give an example.

Chapter 10 **Review and Practice Your Skills** 435

Lesson Planning

NCTM Standards/Strands

- Number & Operations
- Patterns, Functions, & Algebra
- Geometry & Spatial Sense
- Measurement
- Problem Solving
- Reasoning & Proof
- Communication
- Connections
- Representation

Vocabulary

45°-45°-90° triangle
30°-60°-90° triangle

Materials Needed

calculators, compasses, scissors, straightedges

Lesson Resources

Warm-Up Transparency 28
Transparency RF-40
Reteaching 10-3
Extra Practice 10-3
Enrichment 10-3

Assignment Guide

Basic: 1–15, 19–28
Enriched: 1–28

Getting Started

5-Minute Warm-Up

Simplify.

1. $\frac{6}{\sqrt{2}}$ $3\sqrt{2}$
2. $\frac{8}{\sqrt{3}}$ $\frac{8\sqrt{3}}{3}$
3. $\frac{9}{\sqrt{3}}$ $3\sqrt{3}$

Introduction to Lesson 10-3

After students have had the opportunity to make observations, ask them to share their findings, and compile a class list on the chalkboard.

10-3 Special Right Triangles

Goals ■ Find the lengths of the sides of 30°–60°–90° and 45°–45°–90° triangles.

Applications Architecture, Road Planning, Plumbing

Work with a partner. You will need a compass, straightedge and scissors.

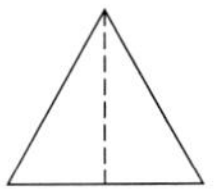

1. Construct several different equilateral triangles.
2. Cut out each triangle and fold it so that one vertex matches another.
3. Draw a line segment along the fold.
4. Examine the figures formed by the triangle and the line segment you drew. Write as many different observations about the triangle and the line segment as you can. See additional answers.

Build Understanding

All right triangles with the same angle measure are similar. This fact and the Pythagorean Theorem lead to some theorems about the relationships among the sides in two types of right triangles.

In the equilateral triangles you examined in the Introduction, you probably noticed that the segment that divided one side into two congruent segments was also perpendicular to that segment. In this way, a right triangle is formed that has a 60° angle and one leg that is half the measure of the hypotenuse.

In this 30°–60°–90° triangle, suppose the measure of the side opposite the 30° angle is s. Then the measure of the hypotenuse is $2s$. You can use the Pythagorean theorem to find the measure of the longer leg of the triangle in terms of s.

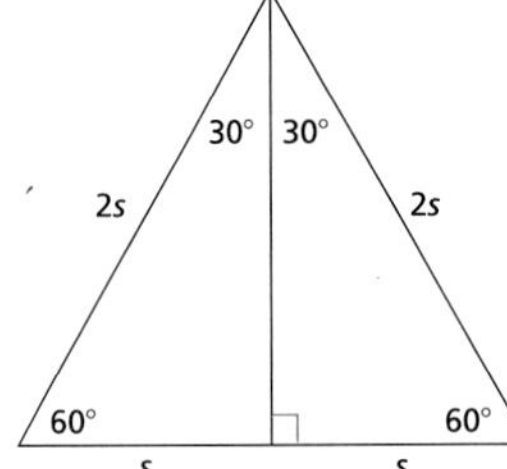

Let y represent the measure of the side opposite the 60° angle.

$$s^2 + y^2 = (2s)^2$$
$$s^2 + y^2 = 4s^2$$
$$y^2 = 3s^2$$
$$y = \sqrt{3s^2}$$
$$y = s\sqrt{3}$$

30°–60°–90° Triangle Theorem In a 30°–60°–90° triangle, the measure of the hypotenuse is two times that of the leg opposite the 30° angle. The measure of the other leg is $\sqrt{3}$ times that of the leg opposite the 30° angle.

Example 1

ARCHITECTURE A triangular support shown on a blueprint forms a 30°–60°–90° triangle. On the plans, the leg opposite the 30° angle measures 4 cm. What are the measures of the other two sides?

Extend the Lesson

ONGOING ASSESSMENT Have students solve the following problem: In a 30°-60°-90° triangle, the leg opposite the 30° angle measures 3 cm. How long is the leg opposite the 60° angle. Watch for students who arrive at an answer of 1.7 cm. These students did not recognize that multiplication by $\sqrt{3}$, rather than division by $\sqrt{3}$, is needed to find the longer side when the shorter side is given. Correct answer: 5.2 cm

Solution

In a 30°–60°–90° triangle, if the leg opposite the 30° angle is s, then the other leg is $s\sqrt{3}$, and the hypotenuse is $2s$.

So, in this triangle, the leg opposite the 30° angle is 4 cm, the other leg is $4\sqrt{3}$ cm, and the hypotenuse is 8 cm.

Problem Solving Tip

When you are given information about a geometric figure, it is best to draw a quick sketch of the figure and label any given measurements. That way, you will be able to see what relationships can help you to find the missing measure.

Example 2

In a 30°–60°–90° triangle, the hypotenuse measures 7 in. Find the measure of the other two sides to the nearest tenth.

Solution

$2s$	hypotenuse	$2s = 7$
s	side opposite 30° angle	$s = 3.5$
$s\sqrt{3}$	side opposite 60° angle	$s\sqrt{3} = (3.5)(\sqrt{3}) \approx 6.1$

So, the missing measures of this triangle are 3.5 in. and 6.1 in.

When you are given the measure of the leg opposite the 60° angle, you must remember about rationalizing denominators.

Example 3

In a 30°–60°–90° triangle, the measure of the leg opposite the 60° angle is 5 cm. Find the measures of the other two sides in simplest radical form.

5 cm
60°

Solution

You are given that $s\sqrt{3} = 5$.

$s = \frac{5}{\sqrt{3}}$ Rationalize the denominator. $s = \frac{5\sqrt{3}}{\sqrt{3} \cdot \sqrt{3}} = \frac{5\sqrt{3}}{3}$

The measure of the hypotenuse is $2s$.

Substitute the value you found for s. $2s = 2\left(\frac{5\sqrt{3}}{3}\right) = \frac{10\sqrt{3}}{3}$

So, the missing measures are $\frac{5\sqrt{3}}{3}$ cm and $\frac{10\sqrt{3}}{3}$ cm.

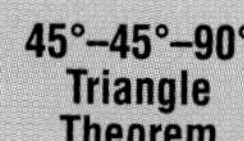

In a 45°–45°–90° triangle, the measure of the hypotenuse is $\sqrt{2}$ times the measure of a leg of the triangle.

Example 4

Find the unknown measures. **a.** **b.**

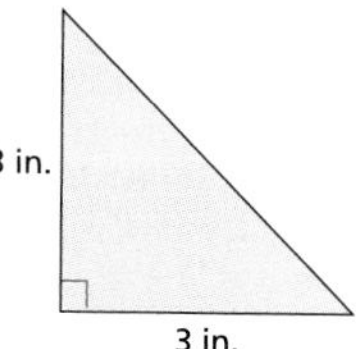

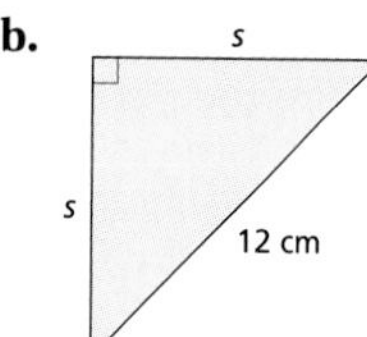

Solution

If s is the length of a side of a 45°–45°–90° triangle, then the hypotenuse is $s\sqrt{2}$.

Chalkboard Examples

Supplementary Example 1

A 30°-60°-90° triangle is shown on a set of blueprints. If the leg opposite the 30° angle measures 10 cm, what are the measures of the other two sides to the nearest tenth? 17.3 m and 20 cm

Supplementary Example 2

In a 30°-60°-90° triangle, the hypotenuse measures 9 in. Find the measure of the other two sides to the nearest tenth. 4.5 in. and 7.8 in.

Supplementary Example 3

In a 30°-60°-90° triangle the leg opposite the 60° angle is 12 in. Find the measure of the other two legs. $4\sqrt{3}$ and $8\sqrt{3}$

Supplementary Example 4

In a 45°-45°-90° triangle, the hypotenuse measures 8 m. Find the measure of the other two sides to the nearest tenth. 5.7 m

Supplementary Example 5

A surveyor marks off a 10-yard distance due north of her starting point. She then marks off 20 yards due east. How far to the nearest tenth yard is this point from the starting point? 22.4 yd

Additional Answers

4. Possible observations: the angle is bisected, the side the segment intersects is bisected by the segment, the segment is perpendicular to the side of the triangle.

Reteaching Worksheet 10-3 not shown.

Extend the Lesson

MATH JOURNAL For each type of special right triangle, have students write which one side length they would prefer to be given if their task were to find the lengths of the other sides. Have them support their answers with examples.

a. Because $s = 3$, the measure of the hypotenuse is $3\sqrt{2}$ in.

b. The measure of the hypotenuse is given as 12 cm. You can use the equation $s\sqrt{2} = 12$, and solve it for s.

$$s = \frac{12}{\sqrt{2}} = \frac{12\sqrt{2}}{\sqrt{2}\cdot\sqrt{2}} = \frac{12\sqrt{2}}{2} \quad \text{Rationalize the denominator.}$$

$s = 6\sqrt{2}$ The measure of s is $6\sqrt{2}$ cm.

Right angles are common in our everyday life. You can find examples of right angles in and around any building. Many streets intersect at right angles, and trees form right angles with the ground. Therefore, right triangles and the Pythagorean Theorem have many applications.

Example 5

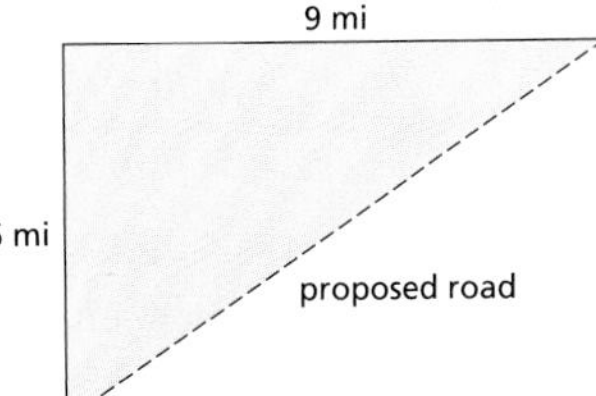

ROAD PLANNING Presently, to get from the ranger station to the park exit, you must drive 6 miles north and then 9 miles east. The park manager is considering having a new road paved to provide a more direct route. What will be the length of the new road to the nearest tenth?

Solution

The roads form a right triangle. Use the Pythagorean Theorem to find the hypotenuse, which will be the length of the new road.

Let x = the length of the new road.

$$6^2 + 9^2 = x^2$$
$$36 + 81 = x^2$$
$$117 = x^2$$
$$10.81665383 = x$$

So, the length of the new road will be 10.8 miles.

Try These Exercises

Find the unknown measures. First find each in simplest radical form and then find each to the nearest tenth.

1. (60°, 4 in.) $4\sqrt{3}$ in., 8 in.; 6.9 in., 8.0 in.
2. (60°, 10 cm) $5\sqrt{3}$ cm, 5 cm; 8.7 cm, 5.0 cm
3. (6 yd, 6 yd) $6\sqrt{2}$ yd; 8.5 yd
4. 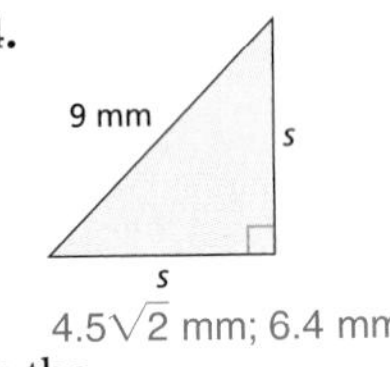

$4.5\sqrt{2}$ mm; 6.4 mm
5. A satellite dish 5 ft high is attached by a cable from its top to a point on the ground 3 ft from its base. How long is the cable to the nearest tenth? 5.8 ft
6. **INVENTIONS** Nefi has designed a small remote-controlled robot. He plans to test the robot's ability to move quickly on a rectangular piece of asphalt that measures 20 ft by 8 ft. What is the greatest distance to the nearest tenth of a foot that the robot can travel without turning? 21.5 ft

Lesson Wrap-up

Quick Assessment

Ask the following questions to determine if students understand the content presented in this lesson.

1. What relationship exists between the lengths of the sides of a triangle and its angles? The shortest side is opposite the smallest angle, and the longest side is opposite the largest angle.
2. You know that one angle in a triangle measures 90° and another measures 60°. What additional information do you need to find the measures of all sides and angles? the length of the side opposite the 60° angle

Assignment Guide

Basic: 1–15, 19–28
Advanced: 1–28
Additional Practice: See Extra Practice Index on page 674.

Additional Answers

16. If s is the length of a leg in a 45°-45°-90° triangle, then you can use the Pythagorean Theorem:
$$s^2 + s^2 = x^2$$
$$2s^2 = x^2$$
$$\sqrt{2s^2} = x$$
$$\sqrt{2}\cdot\sqrt{s^2} = x$$
$$s\sqrt{2} = x$$
17. Always; let x represent the measure of the smaller angle. Then $90 - x$ can represent the larger angle. You can solve this equation for x:
$$x = 0.5(90 - x)$$
$$x = 45 - 0.5x$$
$$1.5x = 45$$
$$x = 30$$

25. −2 −1 0

26. 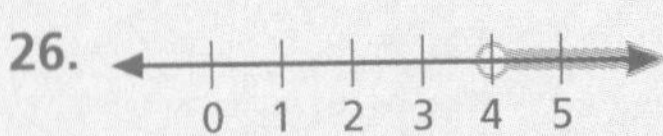

27. −6 −5 −4 −3 −2 −1 0

28. 0 1 2 3

Practice Exercises

Find the unknown measures. First find in simplest radical form and then find each to the nearest tenth.

7. 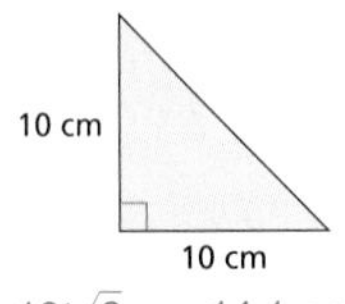

$10\sqrt{2}$ cm; 14.1 cm

8. 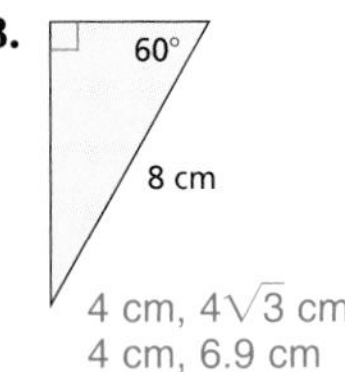

4 cm, $4\sqrt{3}$ cm; 4 cm, 6.9 cm

9. 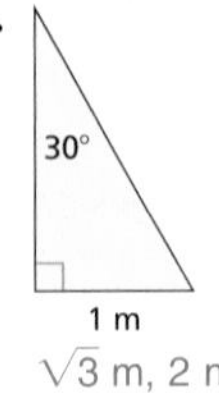

$\sqrt{3}$ m, 2 m; 1.7 m, 2 m

10. 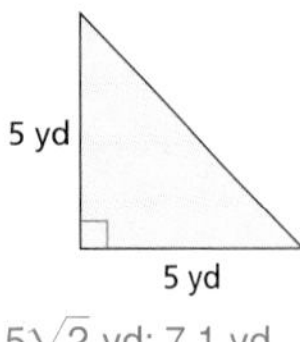

$5\sqrt{2}$ yd; 7.1 yd

Solve. Round answers to the nearest tenth.

11. The diagonal of a square measures 15 cm. Find the length of a side of the square. 10.6 cm

12. Find the measure of the altitude of an equilateral triangle with a side that measures 8 in. 43 in. or 6.9 in.

13. Find the area of an equilateral triangle with a side that measures 10 cm. $25\sqrt{3}$ cm²

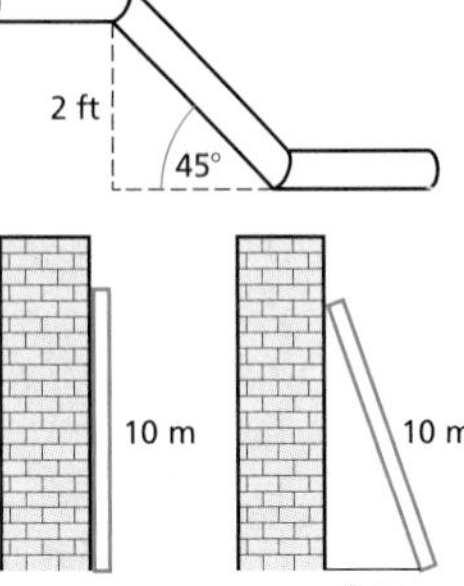

14. PLUMBING A new pipe will be connected to two parallel pipes using 45° elbows. How long must the pipe be if the two parallel pipes are 2 ft apart? 2.8 ft

15. CONSTRUCTION A beam, 10 m in length, is propped up against a building. The beam has slipped so that its base is 3 m from the wall. How far up the building does the beam reach now? 9.5 m

10 m   10 m   3 m

Extended Practice Exercises

16. Use algebra to show that, if s is the measure of a leg in a 45–45–90 triangle, then the hypotenuse measures $s\sqrt{2}$. See additional answers.

17. WRITING MATH Is this statement always, sometimes, or never true? Explain your choice. See additional answers.

If one acute angle of a right triangle is half the measure of the other acute angle, then the side opposite that angle measures half the length of the hypotenuse.

18. CHAPTER INVESTIGATION Research architectural styles that would complement your school grounds. Make a list of design elements that you could choose from in creating a new look for your school.
Check students' drawings.

Mixed Review Exercises

A card is drawn at random from a standard deck of cards. Find each probability. (Lesson 9-3)

19. The card is a 3, a 6, or a 9. $\frac{3}{13}$

20. The card is a 5 or a face card. $\frac{4}{13}$

21. The card is red and a 10. $\frac{1}{26}$

22. The card is black and a face card. $\frac{3}{26}$

23. The card is a club and a 7 or a jack. $\frac{1}{26}$

24. The card is neither a spade nor a heart. $\frac{1}{2}$

Solve each inequality and graph the solution on a number line. (Lesson 2-6)
For 25–28, see additional answers for graphs.

25. $2x + 1 > -3$ $x > -2$

26. $2(x - 1) > 6$ $x > 4$

27. $-2(x + 3) - 2 \geq 4$ $x \leq -6$

28. $\frac{1}{2}(4x - 8) \leq 2$ $x \leq 3$

Teaching Strategies

COOPERATIVE LEARNING Have students find examples of right triangles in the classroom, elsewhere in school, or at home. Ask them to measure the sides to determine whether they are 30°-60°-90° or 45°-45°-90° triangles. Then have them measure the angles to verify their determinations.

Extra Practice Worksheet 10-3

Name ______ Date ______

EXTRA PRACTICE 10-3
SPECIAL RIGHT TRIANGLES

Exercises

Find the unknown measures. First find in simplest radical form and then find each to the nearest tenth.

1. 9 m, 60° — $3\sqrt{3}$ m, $6\sqrt{3}$ m; 5.2 m, 10.4 m
2. 4 in., 30° — $4\sqrt{3}$ in., 8 in.; 6.9 in., 8 in.
3. 8 yd, 8 yd, 45° — $8\sqrt{2}$ yd; 11.3 yd
4. 60°, 10 cm — $10\sqrt{3}$ cm, 20 cm; 17.3 cm, 20 cm
5. 16 ft, 60° — 8 ft, $8\sqrt{3}$ ft; 8 ft, 13.9 ft
6. 12 m, 12 m — $12\sqrt{2}$ m; 17.0 m

Solve.

7. The diagonal of a square measures 10 cm. Find the length of a side of the square to the nearest tenth. 7.1 cm
8. The perimeter of a square is 20 ft. Find the length of the diagonal of the square to the nearest tenth. 7.1 ft
9. Find the measure of a side of an equilateral triangle with an altitude that measures 12 in. 13.9 in.

132   EXTRA PRACTICE *LESSON 10-3*

Enrichment Worksheet 10-3

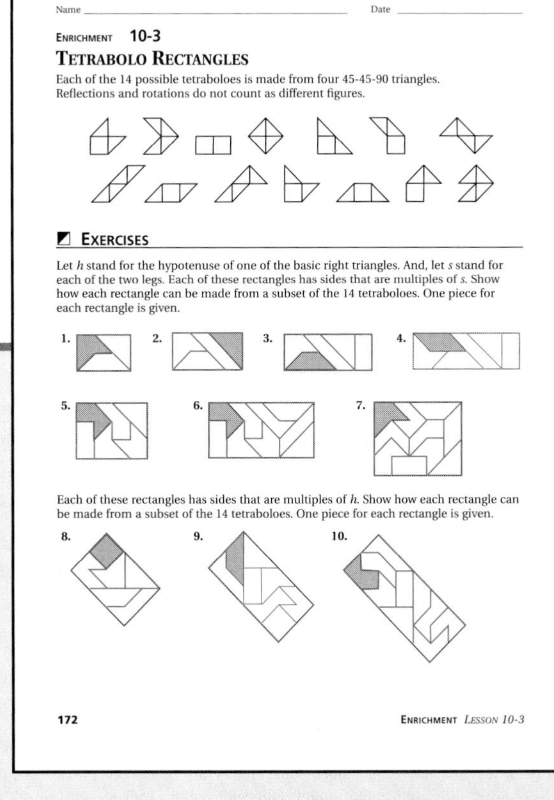

Name ______ Date ______

ENRICHMENT 10-3
TETRABOLO RECTANGLES

Each of the 14 possible tetraboloes is made from four 45-45-90 triangles. Reflections and rotations do not count as different figures.

Exercises

Let h stand for the hypotenuse of one of the basic right triangles. And, let s stand for each of the two legs. Each of these rectangles has sides that are multiples of s. Show how each rectangle can be made from a subset of the 14 tetraboloes. One piece for each rectangle is given.

1. 2. 3. 4. 5. 6. 7.

Each of these rectangles has sides that are multiples of h. Show how each rectangle can be made from a subset of the 14 tetraboloes. One piece for each rectangle is given.

8. 9. 10.

172   ENRICHMENT *LESSON 10-3*

Lesson Planning

NCTM Standards/Strands

- Number & Operations
- Patterns, Functions, & Algebra
- Geometry & Spatial Sense
- Measurement
- Problem Solving
- Reasoning & Proof
- Communication
- Connections
- Representation

Vocabulary

intercept	minor arc
major arc	inscribed angle
chord	secant
tangent	central angle

Materials Needed

compasses, scissors, straightedges, geometric-drawing software

Lesson Resources

Warm-Up Transparency 29
Transparency RF-41
Reteaching 10-4
Extra Practice 10-4
Enrichment 10-4

Assignment Guide

Basic: 1–10, 13–25
Enriched: 1–25

Getting Started

5-Minute Warm-Up

Draw a circle. Then draw and label its center, a diameter and a radius. Check students' drawings.

Introduction to Lesson 10-4

Have students share their findings. Then, after students have discussed Examples 1 through 3, have them identify which of these figures shows a central angle, which shows an inscribed angle, and what the measure of the arc intercepted by each is.

10-4 Circles, Angles, and Arcs

Goals
- Find measures of central and inscribed angles.
- Find measures of angles formed by intersecting secants and tangents.

Applications Landscape Architecture, Navigation, Surveying

Use a compass, straightedge and scissors to make these constructions:

1. Draw a circle, mark its center and cut it out. Fold the circle in half and then in quarters. Draw line segments along two creases to outline one-fourth of the circle. What kind of angle is formed at the center of the circle by the two line segments? right
2. Draw another circle and cut it out. Fold the circle in half and outline the crease to draw a diameter. Draw a line segment to connect one endpoint of the diameter with any point on the circle. Then draw another line segment to connect that point with the other endpoint of the diameter. What do you notice about the angle formed at the center of the circle? They form a right angle.

Build Understanding

In a circle, a *central angle* is an angle that has its vertex at the center of the circle. The rays of the angle are said to **intercept** an arc. $\angle ABC$ intercepts **minor arc** $\overset{\frown}{AC}$ as well as **major arc** $\overset{\frown}{AXC}$.

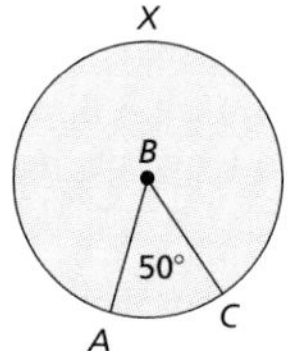

The degree measure of a minor arc is the same as the number of degrees of the corresponding central angle. Because $m\angle ABC = 50$, $m\overset{\frown}{AC} = 50$ and $m\overset{\frown}{AXC} = 360 - 50$, or 310.

Example 1

LANDSCAPE ARCHITECTURE A portion of the circumference of a circular pond will be tiled with handmade tiles donated by a local school. The portion to be tiled is represented on the plans as arc *GH*. Find $m\overset{\frown}{GH}$.

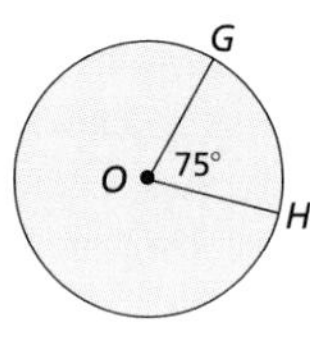

Solution

$\angle GOH$ is the central angle that intercepts $\overset{\frown}{GH}$. This angle measures 75°. Therefore, $m\overset{\frown}{GH} = 75°$.

A basic assumption about arcs is the *arc addition postulate.*

Arc Addition Postulate

If *C* is a point on an arc with endpoints *A* and *B*, then

$$\overset{\frown}{AC} + \overset{\frown}{BC} = \overset{\frown}{ACB}$$

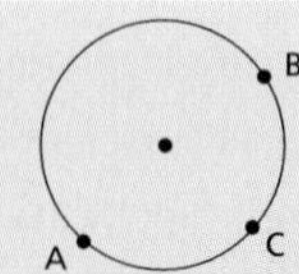

Teaching Strategies

To help students clarify the many new terms introduced in this lesson, ask them to describe how each of these pairs of terms is alike and different: central angle and inscribed angle, chord and secant, secant and tangent.

Example 2

Find the measure of a. $\widehat{PQR}$ **b.** $\widehat{SP}$

120° S O R 45° 40° P Q

Solution

a. $m\widehat{PQR} = m\widehat{PQ} + m\widehat{QR}$

$= 45 + 40 = 85$ $\quad m\widehat{PQ} = 45$, because $m\angle POQ = 45$

b. $m\widehat{SP} = 360 - (m\widehat{SR} + m\widehat{RQ} + m\widehat{QP})$

$= 360 - (120 + 40 + 45) = 155$

Therefore, $m\widehat{PQR} = 85$ and $m\widehat{SP} = 155$.

In circle P, $\angle ABC$ is an **inscribed angle**. It has its vertex, B, on the circle and intercepts minor arc AC.

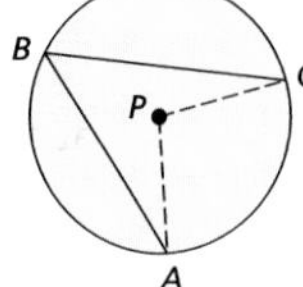

Theorem	The measure of an inscribed angle in a circle is one-half the measure of its intercepted arc.

Example 3

Find the measure of $\angle QRS$.

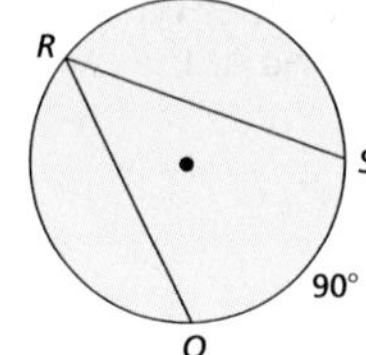

Solution

$\angle QRS$ intercepts $\widehat{QS}$. $m\widehat{QS} = 90$.

$m\angle QRS = \frac{1}{2}(90) = 45$

GEOMETRY SOFTWARE Use geometric drawing software to explore the relationship between the measures of an inscribed angle and its intercepted arc.

1. Draw a circle and inscribed angle CBD.
2. Find $m\angle\widehat{CD}$ and $m\angle CBD$.
3. Select and move point D around the circumference of the circle, observing the changes to the measures of $\widehat{CD}$ and $\angle CBD$.

$m\angle\widehat{CD}$ on $\odot AB = 50°$, $m\angle CBD = 25°$ | $m\angle\widehat{CD}$ on $\odot AB = 100°$, $m\angle CBD = 50°$

In circle O, $\overline{AB}$ is a line segment with both endpoints on the circle; it is called a **chord**. $\overleftrightarrow{CD}$ is a line that intersects the circle in two places; it is called a **secant**. $\overleftrightarrow{CF}$ is also a secant. $\overleftrightarrow{FG}$ intersects the circle in only one point; it is called a **tangent**.

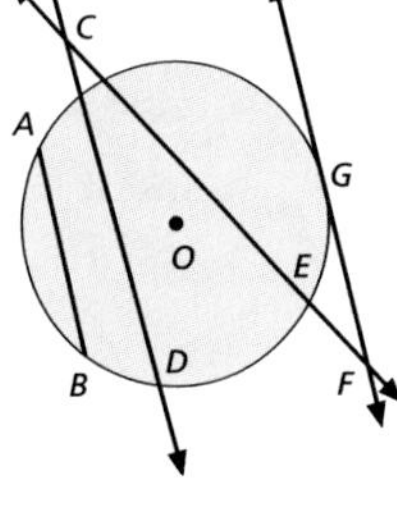

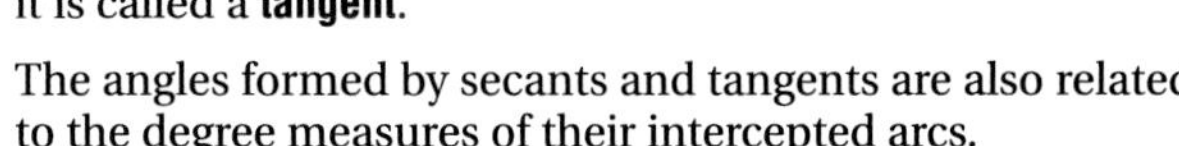

The angles formed by secants and tangents are also related to the degree measures of their intercepted arcs.

Theorem	If two secants intersect inside a circle, then the measure of each angle formed is equal to one-half the sum of the measures of the intercepted arcs.

Extend the Lesson

MATH JOURNAL Have students list the different ways in which two lines or line segments can intersect a circle to form one or more arcs. For each possibility, have students describe in words and/or diagrams the relationship of the arc(s) to the angle formed by the intersection of the lines or line segments.

Chalkboard Examples

Supplementary Example 1 and 2

Find the measure of $\widehat{ABC}$ and $\widehat{AD}$ when $\widehat{BCD} = 159°$ and $m\angle AOB = 103°$. 147°, 98°

Supplementary Example 3

Find the measure of inscribed $\angle DEF$ when $\widehat{DF} = 94°$. 47°

Supplementary Example 4 and 5

Draw a circle with secants $\overleftrightarrow{SQ}$ and $\overleftrightarrow{RT}$ intersecting at point P inside a circle. Check students' drawings.

Reteaching Worksheet 10-4

Name ______ Date ______

RETEACHING 10-4

CIRCLES, ANGLES AND ARCS

A circle contains 360 degrees.

Central angle: $\angle ECF$
Minor arc: $\widehat{EF}$
$m\angle ECF = m\widehat{EF}$
Inscribed angle: $\angle EHF$

Secants: $\overleftrightarrow{IJ}$, $\overleftrightarrow{KM}$, $\overleftrightarrow{NO}$
$m\angle KXJ = \frac{1}{2}(m\widehat{KJ}) + \frac{1}{2}(m\widehat{INM})$
$m\angle KYN = \frac{1}{2}(m\widehat{KIN} - (m\widehat{MO})$

Example

Find the value of x in each circle.

a. b. c. d.

Solution

a. Since x is the measure of a central angle, it is equal to the measure of its intercepted arc. So $x = 73°$.

b. Since x is the measure of an inscribed angle, it is equal to one-half the measure of its intercepted arc. So $x = \frac{1}{2}(156°) = 78°$.

c. Since x is the measure of an angle formed by the intersection of two secants inside a circle, it is equal to one-half the sum of the measures of its intercepted arcs. So $x = \frac{1}{2}(45° + 60°) = \frac{1}{2}(105°) = 52\frac{1}{2}°$.

d. Since x is the measure of an angle formed by the intersection of two secants outside a circle, it is equal to one-half the difference of the measures of its intercepted arcs. So $x = \frac{1}{2}(98° - 26°) = \frac{1}{2}(72°) = 36°$.

EXERCISES

Find the value of x in each circle.

1. 83° 2. 54° 3. 75°
4. 52° 5. 124° 6. 42°
7. 92° 8. 15° 9. 184°

152 RETEACHING LESSON 10-4

Lesson Wrap-up

Quick Assessment

Ask the following questions to determine if students understand the content presented in this lesson.

1. In Example 4, what are the measures of $\angle AED$, $\angle DEB$ and $\angle BEC$. 125°, 55° and 125°
2. In Example 5, how would the figure be different if $\overleftrightarrow{XZ}$ and $\overleftrightarrow{WZ}$ were both secants or both tangents? The measure of $\angle XZW$ would still be the same.

Assignment Guide

Basic: 1–10, 13–25
Advanced: 1–25
Additional Practice: See Extra Practice Index on page 674.

Additional Answers

10. The first angle intercepts an arc, say, $\overset{\frown}{XY}$. Therefore, the measure of the angle is one-half of $\overset{\frown}{XY}$. The second angle also intercepts $\overset{\frown}{XY}$. Therefore, the measure of the second angle is also one-half of $\overset{\frown}{XY}$. Because the measures of the two angles are the same quantity, the measures are equal to one another. Therefore, the angles are congruent.

Example 4

Secants $\overleftrightarrow{AB}$ and $\overleftrightarrow{CD}$ intersect at point E inside a circle. Find the measure of $\angle AEC$.

Solution

For $\angle AEC$, the two arcs intercepted by the secants are $\overset{\frown}{AC}$ and $\overset{\frown}{BD}$.

$$m\angle AEC = \frac{1}{2}(m\overset{\frown}{AC} + m\overset{\frown}{BD})$$

$$m\angle AEC = \frac{1}{2}(50 + 60)$$

$$m\angle AEC = \frac{1}{2}(110) = 55 \quad \text{So, } \angle AEC \text{ measures } 55°.$$

Theorem	If two secants, two tangents, or a tangent and a secant intersect outside a circle, then the measure of the angle formed is equal to one-half the difference of the measures of the intercepted arcs.

Example 5

Find $m\angle XZW$.

Solution

$\angle XZW$ is formed by secant $\overleftrightarrow{XZ}$ and tangent $\overleftrightarrow{WZ}$.

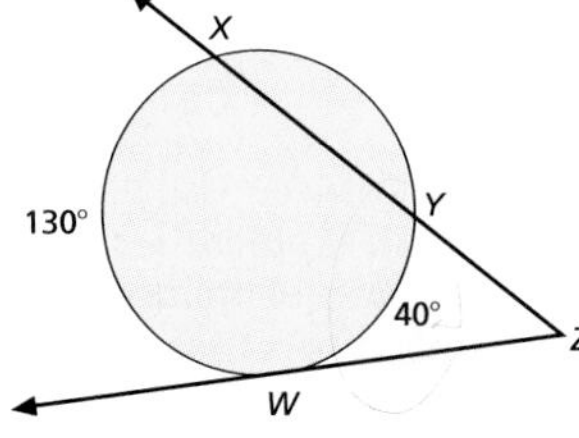

$$m\angle XZW = \frac{1}{2}(m\overset{\frown}{XW} - m\overset{\frown}{YW})$$

$$m\angle XZW = \frac{1}{2}(130 - 40)$$

$$m\angle XZW = \frac{1}{2}(90)$$

$$m\angle XZW = 45$$

So, $m\angle XZW$ is 45°.

Try These Exercises

Find x.

1.

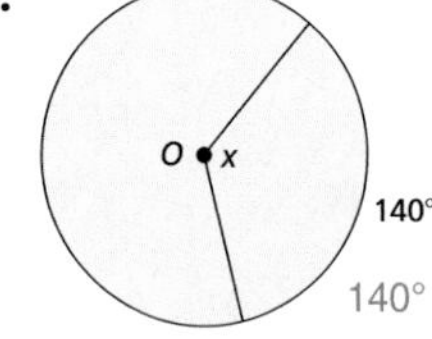

140°

2.

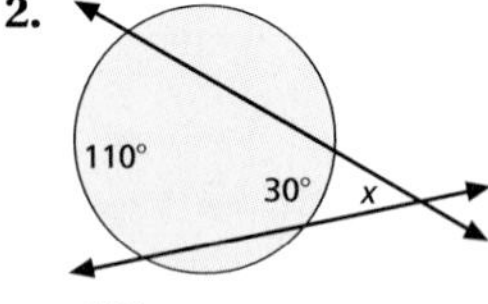

40°

3.

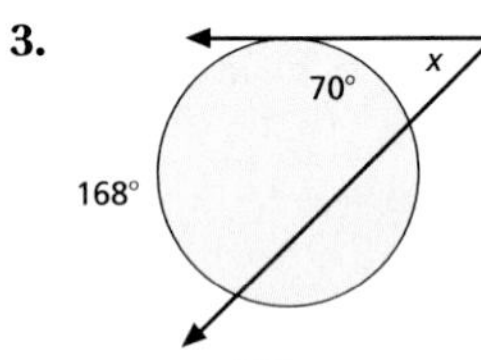

49°

Learning Styles

VISUAL LEARNERS Ask students to draw and label a central angle and an inscribed angle that intercept the same arc of a circle. Ask them to draw and label, in another circle, a chord, a secant, and a tangent. Have them draw and label each feature using a different color. If available, these learners may benefit from constructing figures using geometric-drawing software.

Practice Exercises

Find *x*.

4.

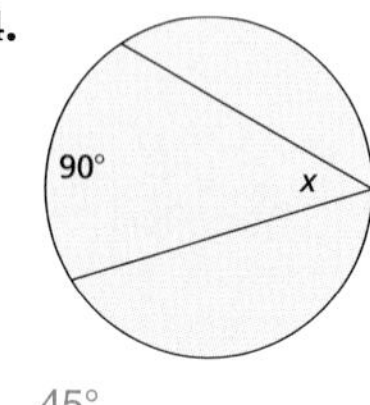

45°

5.

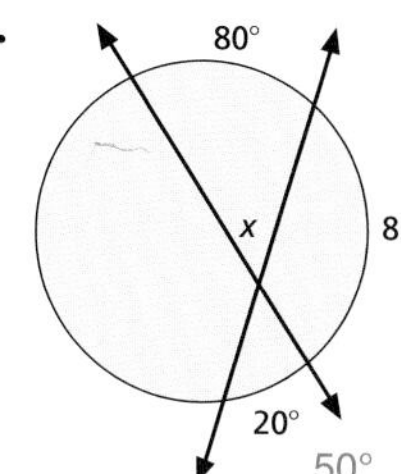

50°

6.

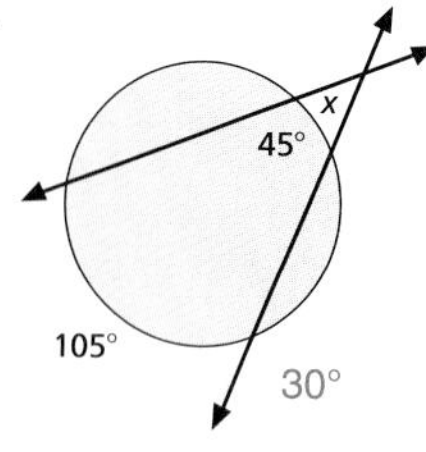

30°

7. NAVIGATION Explorers are searching for the site of an ancient shipwreck. From information found in a recently discovered diary, they draw a circle on a map with its center over a small island. They draw an inscribed angle with an intercepted arc of 80°. What is the measure of the inscribed angle? 40°

8. GEOMETRY SOFTWARE $\triangle ABC$ is an isosceles triangle inscribed in circle *O*. Draw the triangle so that $m\angle BAC = 35°$. What is the measure of $\widehat{ADC}$? 220°

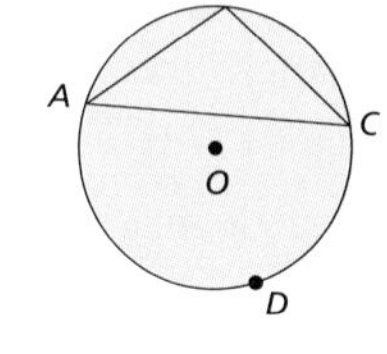

9. SURVEYING On a surveyor's map, central angle *COD* intercepts minor arc *CD* which has a measure of 100°. What is the measure of inscribed angle *CRD* if it also intercepts $\widehat{CD}$? 50°

10. WRITING MATH Write a paragraph that proves this statement: *If two inscribed angles intercept the same arc, the two angles are congruent.* See additional answers.

Extended Practice Exercises

11. State the theorem that would apply to the measure of the angles formed by two chords intersecting in a circle. When two chords intersect inside a circle, the situation is the same as when two secants intersect inside a circle.

12. Describe the locus of points of all vertices of a right triangle with the hypotenuse, $\overline{AB}$. A circle with diameter $\overline{AB}$

Mixed Review Exercises

A bag contains 5 white marbles, 8 red marbles, 7 green marbles, and 5 pink marbles. One marble at a time is taken from the bag and not replaced. Find each probability. (Lesson 9-4)

13. *P*(pink, then white) $\frac{1}{24}$

14. *P*(red, then green) $\frac{7}{75}$

15. *P*(white, then red, then pink) $\frac{1}{69}$

16. *P*(red, then pink, then green) $\frac{7}{345}$

Solve each equation. (Lesson 2-5)

17. $3(4a - 6) = -12$ $\frac{1}{2}$

18. $-7b + 6 = -b - 12$ 3

19. $-3(h + 4) = 2(3h - 3)$ $-\frac{2}{3}$

20. $5p + 3(p - 4) = -2$ $\frac{5}{4}$

21. $-2c + 6 = 2(5c - 3)$ 1

22. $\frac{1}{2}m + 3(m - 1) = -10$ -2

23. $12z + 3 - 6z + 9 = 3z$ -4

24. $0.5x + 4.2 = 0.2(x - 3)$ -16

25. $4(w - 3) + 2 = 5w - 8$ -2

Extra Practice Worksheet 10-4

Name ______________________ Date ____________

EXTRA PRACTICE **10-4**

CIRCLES, ANGLES AND ARCS

EXERCISES

Find *x*.

1. (O, x, 100°) 100°
2. (x, 90°) 45°
3. (x, 50°, 160°) 75°
4. (58°, x) 116°
5. (142°, x, 96°) 119°
6. (x, 32°, 104°) 36°

Solve.

7. An inscribed angle measures 114°. What is the measure of the minor arc it intercepts? 57°
8. A central angle intercepts a minor arc that measures 68°. What is the measure the central angle? 68°

134 EXTRA PRACTICE *LESSON 10-4*

Enrichment Worksheet 10-4

Name ______________________ Date ____________

ENRICHMENT **10-4**

THE BROKEN HEART PUZZLE

A puzzle similar to tangrams is based on dissecting a heart into nine parts as shown in the drawing. The broken heart pieces can be used for the same types of activities as tangrams: free invention, representing a specific object or copying a given shape.

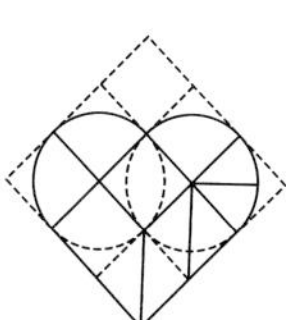

EXERCISES

Construct the broken heart puzzle. Then use all nine pieces to copy each shape shown.

1.

2.

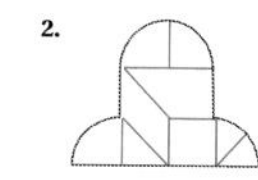

3.

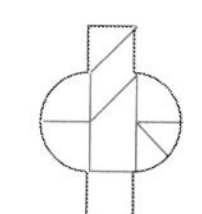

4.

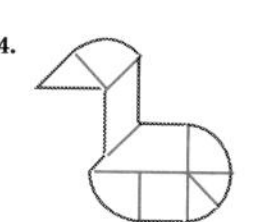

5.

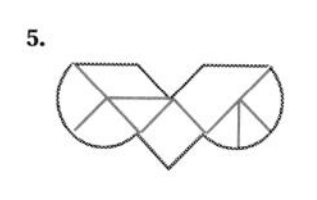

6.

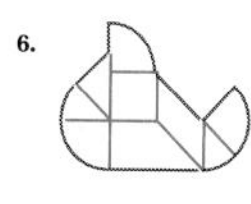

7.

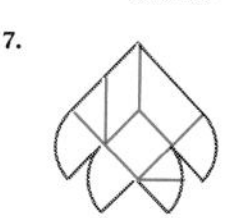

8.

9.

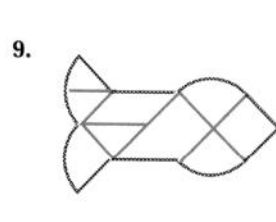

10. Use the pieces of the puzzle to create three original designs of your own.
Answers may vary. You may wish to have pairs of students exchange and solve puzzles.

174 ENRICHMENT *LESSON 10-4*

Teaching Strategies

CHALLENGE Challenge students to find a formula for the number of chords and arcs that can be formed by connecting n points on a circle. They may find it helpful to draw and count the number of chords and arcs formed by connecting 2, 3, 4 and 5 points and look for a pattern.

Skills Practice

Vocabulary Review

Lesson 10-3
45°-45°-90° triangle
30°-60°-90° triangle

Lesson 10-4
intercept
minor arc
major arc
inscribed angle
chord
secant
tangent
central angle

Assignment Guide

All students: 1–31
Additional Practice: See Extra Practice Index on page 674.

Review and Practice Your Skills

Practice ■ Lesson 10-3

Find the unknown measures. First find in simplest radical form and then find each to the nearest tenth.

1. 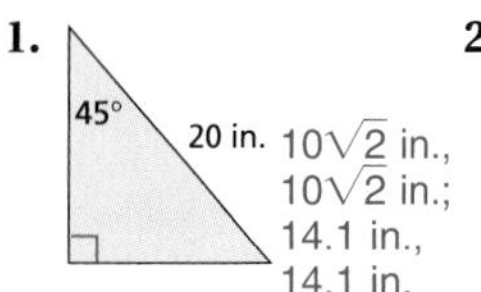

$10\sqrt{2}$ in., $10\sqrt{2}$ in.; 14.1 in., 14.1 in.

2. 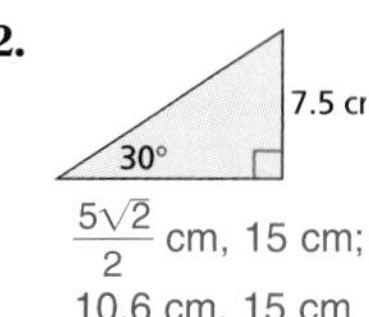

$\frac{5\sqrt{2}}{2}$ cm, 15 cm; 10.6 cm, 15 cm

3. 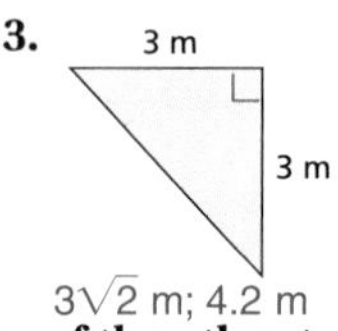

$3\sqrt{2}$ m; 4.2 m

4. 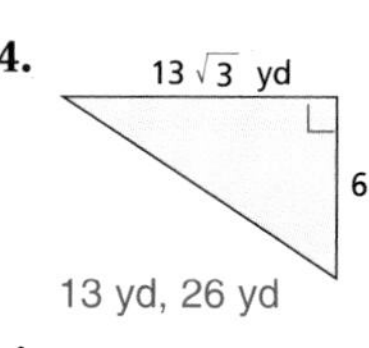

13 yd, 26 yd

For each 30°–60°–90° triangle, find the measures of the other two sides in simplest radical form.

5. side opposite 30° angle measures 11 cm $11\sqrt{3}$ cm, 22 cm

6. hypotenuse measures 48 ft 24 ft, $24\sqrt{3}$ ft

7. side opposite 60° angle measures $5\sqrt{3}$ in. 5 in., 10 in.

8. side opposite 30° angle measures $\sqrt{3}$ m 3 m, $2\sqrt{3}$ m

9. hypotenuse measures 37 ft 18.5 ft, $18.5\sqrt{3}$ ft

10. hypotenuse measures $6\sqrt{24}$ yd $6\sqrt{6}$ yd, $18\sqrt{2}$ yd

11. side opposite 60° angle measures 30 km $10\sqrt{3}$ km, $20\sqrt{3}$ km

12. hypotenuse measures $\frac{1}{3}$ cm $\frac{1}{6}$ cm, $\frac{\sqrt{3}}{6}$ cm

For each 45°–45°–90° triangle, find the measures of the other two sides in simplest radical form.

13. leg measures 44 in. 44 in., $44\sqrt{2}$ in.

14. hypotenuse measures $15\sqrt{2}$ m 15 m, 15 m

15. hypotenuse measures 8 ft $4\sqrt{2}$ ft, $4\sqrt{2}$ ft

16. leg measures $\frac{3}{\sqrt{2}}$ cm $\frac{3\sqrt{2}}{2}$ cm, 3 cm

Practice ■ Lesson 10-4

Find *x*.

17.
x
92°
46°

18. 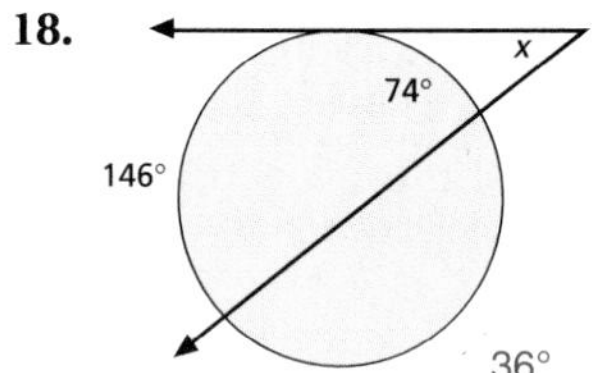

19. 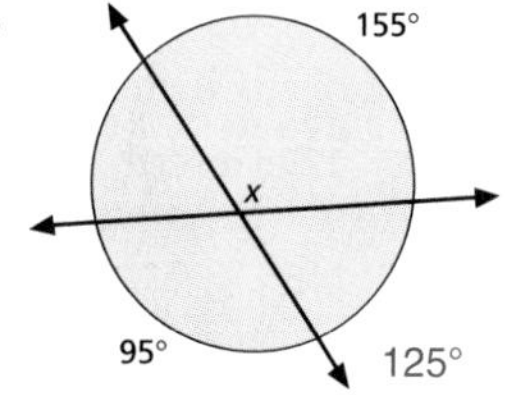

20. 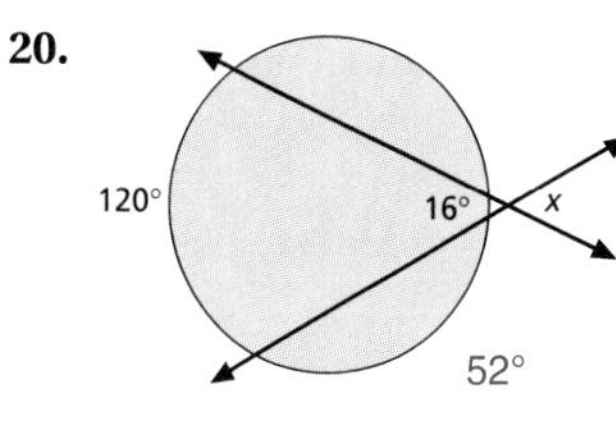

21. 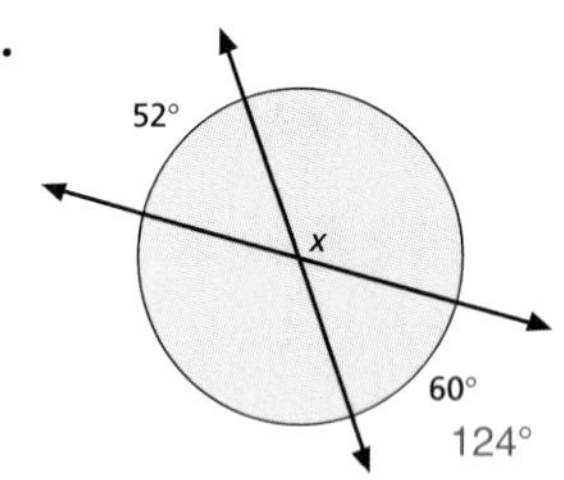

22. 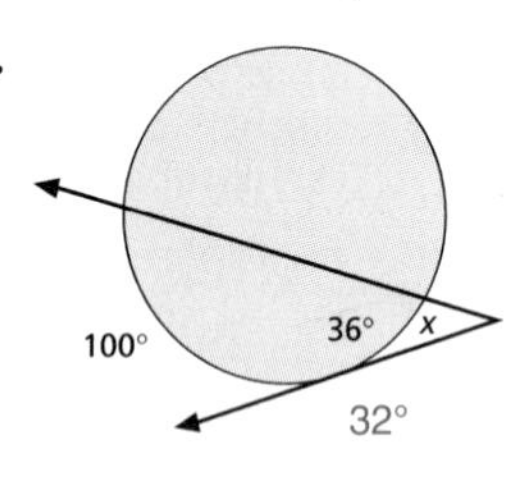

23. *True* or *false*: Two tangents can intersect inside a circle. false

24. *True* or *false*: A chord that passes through the center of a circle is called the diameter. true

Extend the Lesson

COOPERATIVE LEARNING Have students work in pairs to draw several concentric circles and draw a central angle that intersects all circles. Then ask students to compare the measure of each arc intercepted by this central angle and the length of each arc intercepted by the central angle. Have pairs share their conclusions with the class.

444 Chapter 10 **Right Angles and Circles**

Practice ■ Lesson 10-1–Lesson 10-4

Write each in simplest radical form. (Lesson 10-1)

25. $\sqrt{\frac{3}{8}}$ $\frac{\sqrt{6}}{4}$ **26.** $\frac{9}{\sqrt{3}}$ $3\sqrt{3}$ **27.** $(4\sqrt{22})(3\sqrt{33})(9\sqrt{8})$ $4752\sqrt{3}$ **28.** $\sqrt{(4)(12)(18)}$ $12\sqrt{6}$

Use the Pythagorean Theorem to find the unknown length. (Lesson 10-2)

29.

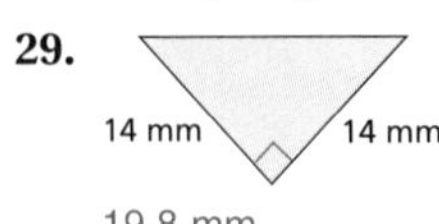

19.8 mm

30.

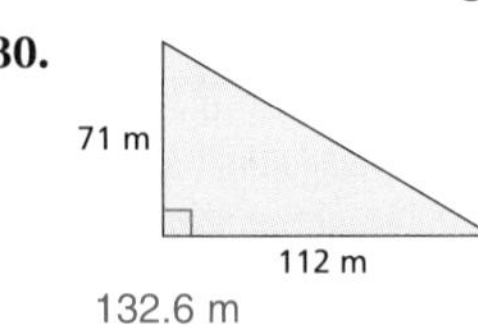

132.6 m

31.

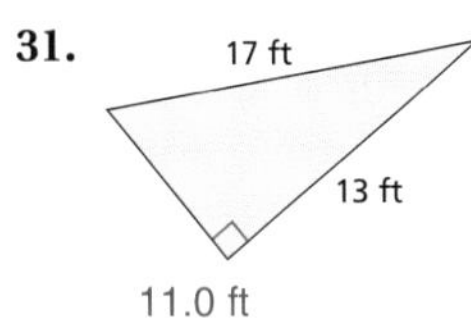

11.0 ft

Mid-Chapter Quiz

Use your calculator to find the value to the nearest hundredth.

1. $\sqrt{44}$ 6.63
2. $\sqrt{87}$ 9.33

Write each in simplest radical form.

3. $\sqrt{270}$ $3\sqrt{30}$
4. $(4\sqrt{6})(-5\sqrt{3})$ $-60\sqrt{2}$
5. $\frac{\sqrt{56}}{\sqrt{14}}$ 2

Use the Pythagorean Theorem to find the unknown length. Round your answer to the nearest tenth.

6. legs: x, 15 in.
   hypotenuse: 17 in. 8.0 in.
7. legs: 7 cm, 9 cm
   hypotenuse: x 11.4 cm

Find the missing side lengths for right triangles. Round your answers to the nearest tenth.

	Leg opposite 30° angle	Leg opposite 60° angle	Hypotenuse
8.	2 yd	3.5 yd	4 yd
9.	6 m	10.4 m	12 m

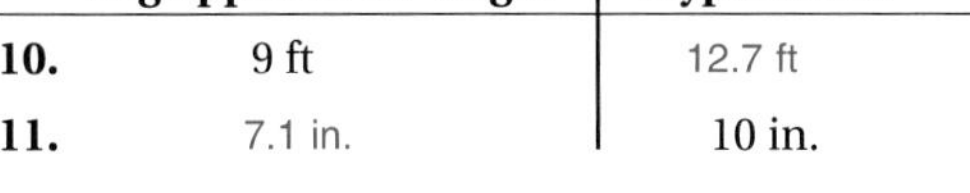

	Leg opposite 45° angle	Hypotenuse
10.	9 ft	12.7 ft
11.	7.1 in.	10 in.

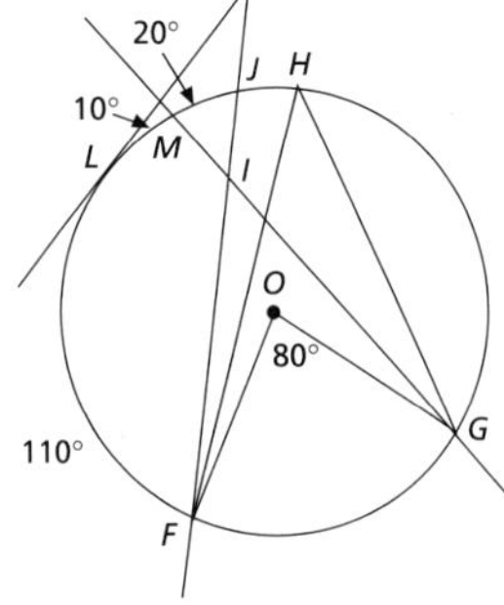

Find the measure of each.

12. arc FG 80°
13. arc FHG 280°
14. angle FHG 40°
15. angle FIG 50°
16. angle LKF 45°

Chalkboard Examples

Example from Lesson 10-3

In a 30°-60°-90° right triangle, the leg opposite the 60° angle is 5 ft. Find the missing side lengths to the nearest tenth. **2.9 ft and 5.8 ft**

Example from Lesson 10-4

Find the measure of each.

1. central angle AOC if $\overset{\frown}{AC} = 70°$ **70°**
2. inscribed angle ABC if $\overset{\frown}{AC} = 70°$ **35°**
3. arc ABC if $\overset{\frown}{AC} = 70°$ **290°**

Mid-Chapter Quiz

Correlation Chart

Question Number	Lesson Number
1–5	10-1
6–7	10-2
8–11	10-3
12–16	10-4

Extend the Lesson

MATH JOURNAL Have students describe how they could find the lengths of all three sides of 30°-60°-90° and 45°-45°-90° triangles if they are given the length of any one side.

Chapter 10 **Review and Practice Your Skills** **445**

Lesson Planning

NCTM Standards/Strands

- Number & Operations
- Patterns, Functions, & Algebra
- Geometry & Spatial Sense
- Measurement
- Problem Solving
- Reasoning & Proof
- Communication
- Connections
- Representation

Vocabulary

circle graph

Materials Needed

calculators, compasses
protractors, straightedges

Lesson Resources

Warm-Up Transparency 29
Reteaching 10-5

Assignment Guide

Basic: 1–10
Enriched: 1–10

Getting Started

5-Minute Warm-up

Solve.

1. What percent of 1,200 is 300? 25%
2. What percent of 800 is 80? 10%
3. What is 50% of 360? 180
4. What is 20% of 360? 72

Use the Five-step Plan and the Strategy

THE FIVE-STEP PLAN **Read**—ask questions of the students to help them understand the problem. **Plan**—guide students to related problems and previously mastered skills and strategies. **Solve**—students solve problem on their own. **Answer**—write solution in format that answers the question. **Check**—review work, check for reasonableness and review strategy used. Students will benefit from the experience of verbalizing their methods.

10-5 Problem Solving Skills: Circle Graphs

A circle graph is a good way to compare data that are parts of a whole. Each part of the whole can be changed to a percent of the whole. Then the percents are used to divide a circle into sectors.

Problem Solving Strategies

- Guess and check
- Look for a pattern
- Solve a simpler problem
- ✔ Make a table, chart or list
- Use a picture, diagram or model
- Act it out
- Work backwards
- Eliminate possibilities
- Use an equation or formula

Problem

URBAN PLANNING A city planner is looking at the results of a survey of housing types. She decides to make a circle graph of the data to present to the next city council meeting.

Housing type	Number
Single-family home	9,070
Two-family home	3,023
Three-to-six family home	756
Seven or more units buildings	2,267

Solve the Problem

Step 1: Add all of the data to find the total.

$9{,}070 + 3{,}023 + 756 + 2{,}267 = 15{,}116$

Step 2: Find what percent each number is of the total. Use a calculator and round percents to the nearest whole percent.

Housing type	Number	Percent of total
Single-family home	9,070	60%
Two-family home	3,023	20%
Three-to-six family home	756	5%
Seven or more units buildings	2,267	15%

Step 3: Find the central angles that correspond to each percent.

Since there are 360° in a circle, use the percents found in Step 2 to find corresponding central angle measures. For example, 60% of 360° is 216°.

Housing type	Percent of total	Central angle
Single-family home	60%	216°
Two-family home	20%	72°
Three-to-six family home	5%	18°
Seven or more units buildings	15%	54°

Step 4: Construct the graph using a compass, straightedge, and protractor.

Start with the smallest angle and work around to the largest angle. Draw a circle and one radius. Place the protractor so that the 18° mark aligns with the radius. Make a mark at 18° and draw a radius at that point.

Additional Answers

1.

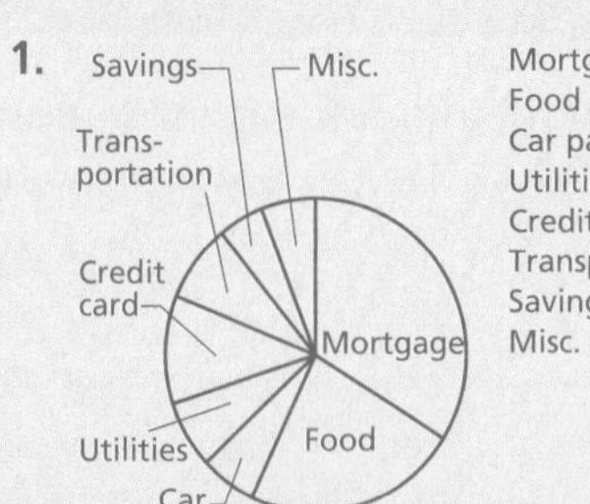

Mortgage	123°
Food	82°
Car payment	20°
Utilities	25°
Credit card	41°
Transportation	29°
Savings	16°
Misc.	25°

2.

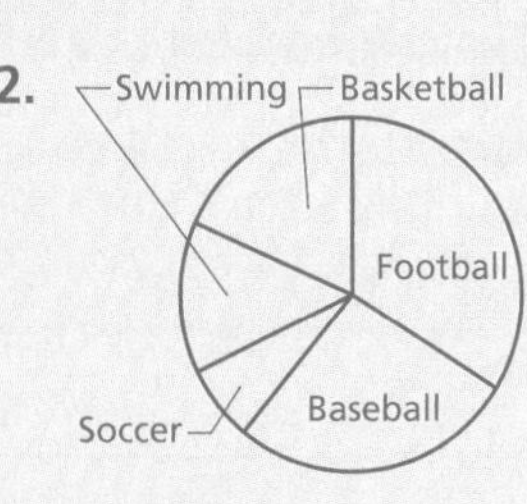

Football	123°
Baseball	98°
Soccer	26°
Swimming	51°
Basketball	62°

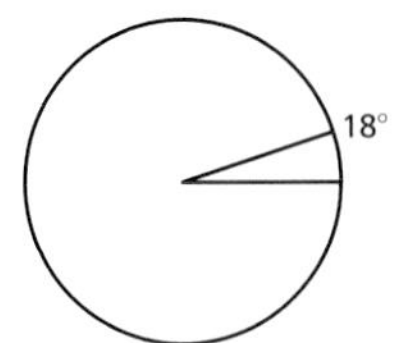

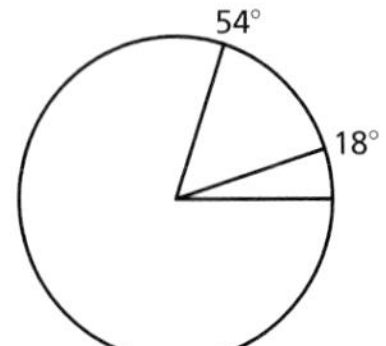

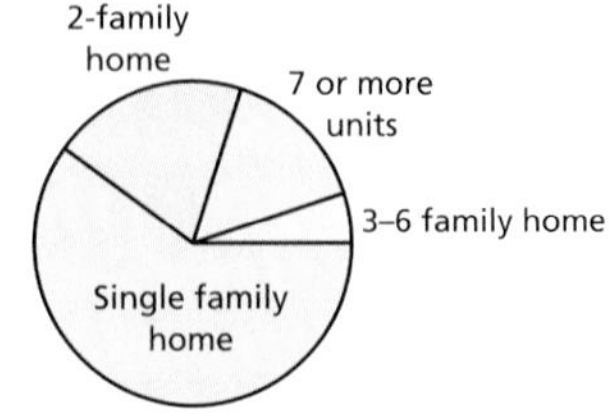

Ourtown—Housing Types

Five-step Plan

1 Read
2 Plan
3 Solve
4 Answer
5 Check

Place the protractor along the new radius. Make a mark at 54° and draw a radius at that point. Continue in this way around the circle.

Label each sector and write a title for the graph.

THE STRATEGY Use a chart—Circle graphs are helpful in understanding how parts of a whole are related to each other and to the whole.

Chalkboard Examples

Supplementary Problem

Make a circle graph for this set of data. Marie's Weekly Gross Income (rounded to the nearest dollar) Federal tax, \$41; State tax, \$7; Social Security and Medicare, \$17; Insurance, \$22; Net income, \$131 Check students' graphs. Percents in order are 19%, 3%, 8%, 10% and 60%.

Try These Exercises

Make a circle graph for each set of data. See additional answers.

1. **Rodriguez Family Monthly Budget**
   Mortgage, \$750
   Food, \$500
   Car payment, \$125
   Utilities, \$150
   Credit card payment, \$250
   Transportation, \$175
   Savings, \$100
   Miscellaneous, \$150

2. **Grace School Sports Budget**
   Football, \$12,000
   Baseball, \$9,500
   Soccer, \$2,500
   Swimming, \$5,000
   Basketball, \$6,000

Practice Exercises

For 3–5, see additional answers.

Make a circle graph for each set of data.

3. **Town Population by Age**

Under 5	3,443
5–13	4,587
14–18	2,428
19–25	4,046
26–39	7,263
40–64	1,049
Over 64	3,125

4. **Window Types Sold, Fred's Building Supply**
   Single width, 520
   Double width, 241
   Bay, 183
   Found, 27
   Semicircular, 89
   Basement, 351

5. **USING DATA** Use the Data Index on pages 632–633 to find the information needed to complete this problem. Make a circle graph for family size, according to the 1990 U.S. Census.

Mixed Review Exercises

Solve for *x* and *y*. (Lesson 8-5)

6. $[x \quad 3y] = [y + 2 \quad 2x - 1]$ (5, 3)

7. $[2x \quad y] = [y - 1 \quad 3x + 7]$ (−6, −11)

8. $\begin{bmatrix} x + 3 \\ y - 4 \end{bmatrix} = \begin{bmatrix} 3y \\ x - 5 \end{bmatrix}$ (3, 2)

9. $\begin{bmatrix} x \\ 2y \end{bmatrix} = \begin{bmatrix} y - 2 \\ 3x + 1 \end{bmatrix}$ (3, 5)

10. $\begin{bmatrix} 3x \\ 2y \end{bmatrix} = \begin{bmatrix} -2y + 1 \\ 6x - 2 \end{bmatrix}$ $\left(\frac{1}{3}, 0\right)$

Lesson Wrap-up

Quick Assessment

When is a circle graph the best way to display data? A circle graph is useful when a whole amount is subdivided into a small number of meaningful sections. It is difficult to show more than 8 sections with any degree of accuracy.

Reteaching Worksheet 10-5

Name ______________ Date ______________

RETEACHING 10-5

PROBLEM SOLVING SKILLS: CIRCLE GRAPHS

You can use a circle graph to compare data when the data are parts of a whole.

Example

Make a circle graph for this set of data.
Federal Tax Returns Processed (in thousands)

Individual	106,994	Corporate	3986
Individual estimated tax	35,489	Other(estate, gift, etc.)	47,837

Solution

- Add all the data to find the total. 106,994 + 3986 + 35,489 + 47,837 = 194,306
- Find what percent each category is of the total returns filed. Remember that the sum of the percents for each category is equal to 100%.

Individual $\frac{106{,}994}{194{,}306} \approx 55\%$ Corporate $\frac{3986}{194{,}306} \approx 2\%$

Individual estimated tax $\frac{35{,}489}{194{,}306} \approx 18\%$ Other(estate, gift, etc.) $\frac{47{,}837}{194{,}306} \approx 25\%$

- Determine the central angles by multiplying 360° by each percent calculated.

Individual $360° \cdot 0.55 \approx 198°$ Corporate $360° \cdot 0.02 \approx 7°$

Individual estimated tax $360° \cdot 0.18 \approx 65°$ Other(estate, gift, etc.) $360° \cdot 0.25 \approx 90°$

- Construct the circle graph.
  a. Use a compass to draw a circle.
  b. Draw one radius.
  c. Use a protractor to mark the proper angle degree for individual tax returns. (198°)
  d. Place the protractor along the new line and mark the proper angle degree for corporate returns. (7°)
  e. Repeat Step d for the individual estimated tax returns and other returns.

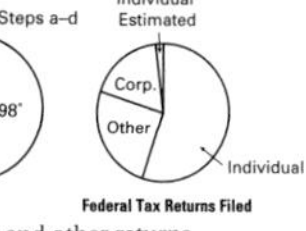

Federal Tax Returns Filed

EXERCISES

Find the percent and measure of the central angle for each set of data. Make a circle graph for each set of data on your own paper. Round each answer to the nearest whole number.

1. Media Market Value (in billions)

Publishing	\$249.2	52%	187°
Cable	88.2	18%	65°
Movie/video	29.9	6%	22°
Broadcasting	113.1	24%	86°

2. Elements in a 150-lb person (in lb)

Oxygen	97.5	65%	234°
Hydrogen	15.0	10%	36°
Nitrogen	4.5	3%	11°
Carbon	27.0	18%	65°
Other	6.0	4%	14°

154 RETEACHING *LESSON 10-5*

3.

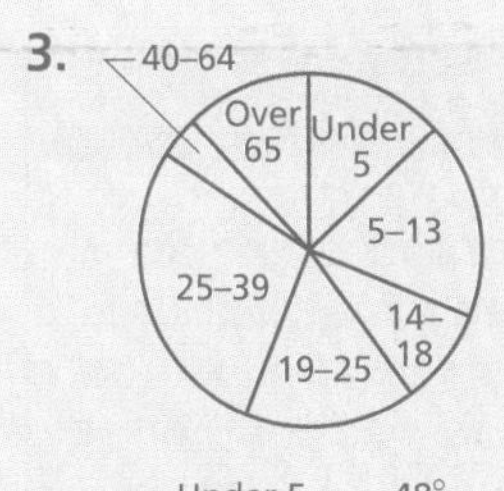

Under 5	48°
5–13	64°
14–18	34°
19–25	56°
25–39	101°
40–64	15°
Over 64	43°

4.

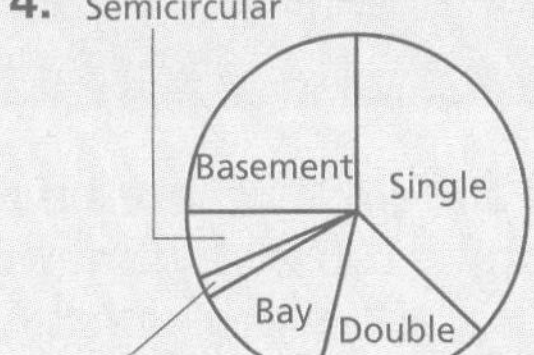

Single	133°
Double	61°
Bay	47°
Round	7°
Semicircular	23°
Basement	90°

5.

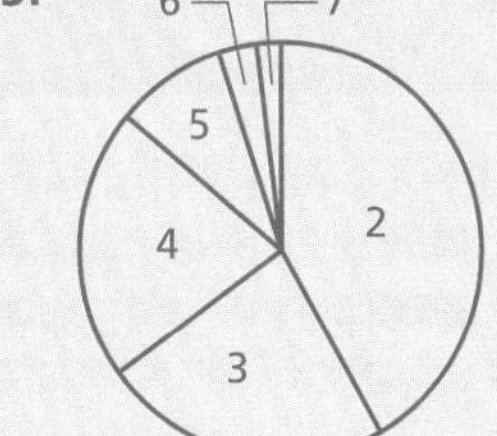

2 Persons
3 Persons
4 Persons
5 Persons
6 Persons
7 or more

Lesson 10-5 **Problem Solving Skills: Circle Graphs** 447

Lesson Planning

NCTM Standards/Strands

- Number & Operations
- Patterns, Functions, & Algebra
- Geometry & Spatial Sense
- Measurement
- Problem Solving
- Reasoning & Proof
- Communication
- Connections
- Representation

Vocabulary

secant segment
tangent segment

Materials Needed

compasses
protractors
straightedges
geometric-drawing software

Lesson Resources

Warm-Up Transparency 30
Transparency RF-42
Reteaching 10-6
Extra Practice 10-6
Enrichment 10-6

Assignment Guide

Basic: 1–20, 23–25
Enriched: 1–25

Getting Started

5-Minute Warm-up

Draw a circle. Then draw and label a chord, a secant and a tangent of the circle. Check students' drawings.

Introduction to Lesson 10-6

You may wish to review properties of similar triangles. Have students share their conclusions about the activity with the class.

10-6 Circles and Segments

Goals ■ Find lengths of chord, secant and tangent segments.
Applications Architecture, Art, Surveying

Work with a partner. You will need geometric-drawing software or a compass, ruler and protractor.

1. Draw a circle. Draw any two intersecting chords and label them as shown at the right.
2. Draw AD and EC to form two triangles. Compare the angles of the triangles. Because all three pairs of angles are congruent, the triangles are similar.
3. Using the measures of the segments, write this proportion: $\frac{AB}{BD} = \frac{BE}{BC}$.
4. Use the cross-products rule. Discuss with your partner what conclusion you can make about the products of segments of intersecting chords.
   Products of segments of intersecting chords are equal.

Build Understanding

As you can see from the above activity, any two intersecting chords can form two similar triangles. This leads to the following theorem.

Theorem	If two chords intersect in a circle, then the product of the measures of the two segments of one chord is equal to the product of the measures of the two segments of the other chord.

This theorem can be used to find a missing measure.

Example 1

Find x.

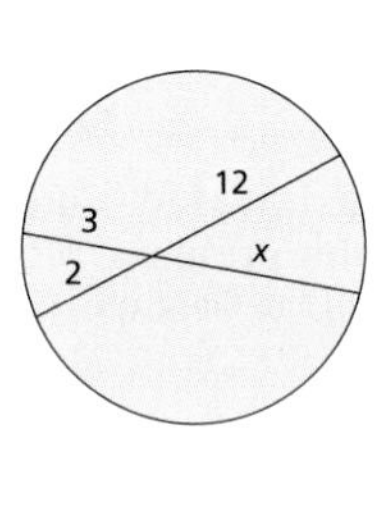

Solution

Since there are two intersecting chords, the products of the segments are equal.

$$3 \cdot x = 2 \cdot 12$$
$$3x = 24$$
$$x = 8$$

A similar type of relationship exists for intersecting secant segments. A **secant segment** intersects a circle in two points and has one endpoint on the circle and one endpoint outside the circle.

Extend the Lesson

MATH JOURNAL Have students restate each theorem presented in the lesson and explain how they would apply it to find the total length of a chord, secant segment, or tangent segment, as well as a part of each.

Theorem	If two secant segments have a common endpoint outside a circle, then the product of the measures of one secant segment and its external part is equal to the product of the other secant segment and its external part.

The following example will help you see what is meant by the length of segment and the length of the external part of the segment.

Example 2

ARCHITECTURE An architect is redesigning a museum. One of the rooms will contain a circular platform to display the skeleton of a prehistoric mammal. Steel cables represented by the two secant segments shown in the drawing to the right will be used to brace the skeleton. Find x.

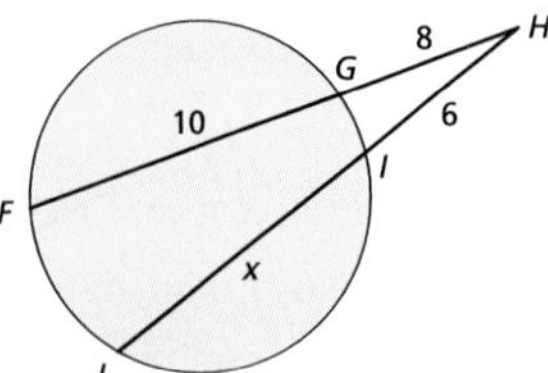

Solution

$\overline{HF}$ and $\overline{HJ}$ are secant segments. $\overline{HG}$ is the external part of $\overline{HF}$, and $\overline{HI}$ is the external part of $\overline{HJ}$. The theorem refers to the length of the entire secant segment and its external part.

$$HF \cdot HG = HJ \cdot HI$$
$$18 \cdot 8 = (6 + x)6 \quad \text{The length of } HJ \text{ is } 6 + x.$$
$$144 = 36 + 6x$$
$$108 = 6x$$
$$18 = x$$

Check Understanding

What is the length of HJ?

24

A **tangent segment** of a circle is a segment that has one endpoint on a circle and one endpoint outside the circle, and does not intersect the circle at any other point.

Theorem	If a tangent segment and a secant segment have a common endpoint outside a circle, then the square of the measure of the tangent segment is equal to the product of the measures of the secant segment and its external part.

Example 3

Find x.

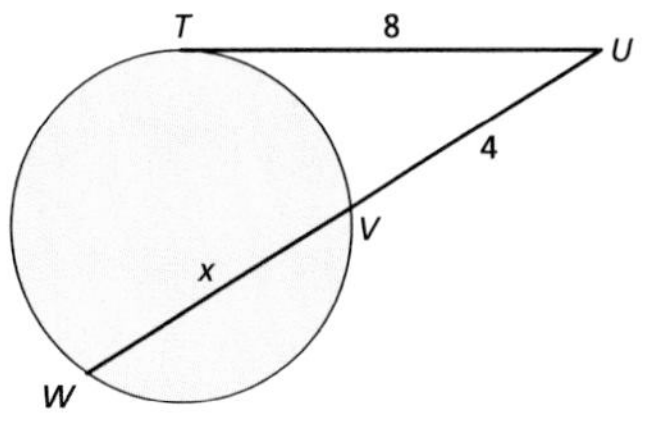

Solution

$\overline{TU}$ is a tangent segment, $\overline{UW}$ is a secant segment, and $\overline{UV}$ is its external segment.

$$(TU)^2 = UW \cdot UV$$
$$8^2 = (x + 4)4$$
$$64 = 4x + 16$$
$$48 = 4x$$
$$12 = x$$

Teaching Strategies

COOPERATIVE LEARNING Have each student draw a figure involving chords, secant segments, and/or tangent segments similar to those in this lesson. Have them label the measure of all but one of the parts. Have students exchange drawings and find the unknown measure. Then have them check their work together.

Chalkboard Examples

Supplementary Example 1 and 2

Draw a figure like the one given in Example 1. Replace the 3 with 15, the 2 with 17, and the 12 with 21. Find x. 23.8

Supplementary Example 3 and 4

Using the diagram for Example 4 on page 450, find PT if PQ = 15 cm. 7.5 cm

Reteaching Worksheet 10-6

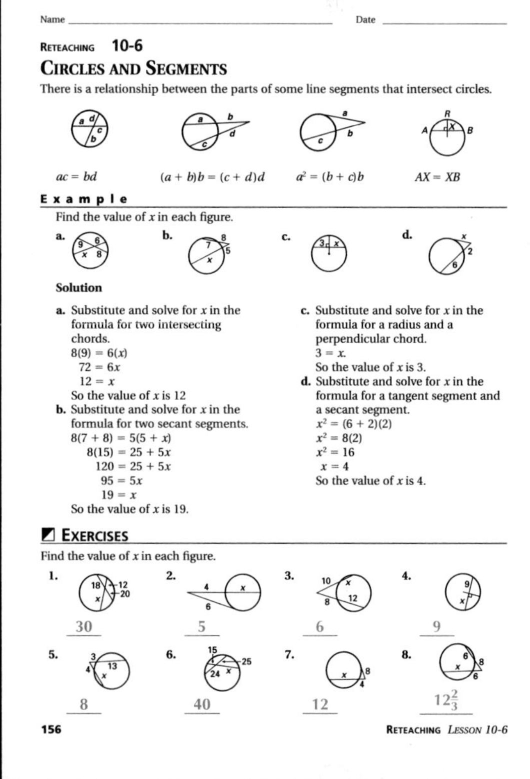

Name ______________ Date ______________

RETEACHING 10-6

CIRCLES AND SEGMENTS

There is a relationship between the parts of some line segments that intersect circles.

$ac = bd$ $\quad (a + b)b = (c + d)d \quad$ $a^2 = (b + c)b$ $\quad AX = XB$

Example

Find the value of x in each figure.

a. b. c. d.

Solution

a. Substitute and solve for x in the formula for two intersecting chords.
$8(9) = 6(x)$
$72 = 6x$
$12 = x$
So the value of x is 12

b. Substitute and solve for x in the formula for two secant segments.
$8(7 + 8) = 5(5 + x)$
$8(15) = 25 + 5x$
$120 = 25 + 5x$
$95 = 5x$
$19 = x$
So the value of x is 19.

c. Substitute and solve for x in the formula for a radius and a perpendicular chord.
$3 = x.$
So the value of x is 3.

d. Substitute and solve for x in the formula for a tangent segment and a secant segment.
$x^2 = (6 + 2)(2)$
$x^2 = 8(2)$
$x^2 = 16$
$x = 4$
So the value of x is 4.

EXERCISES

Find the value of x in each figure.

1. 30
2. 5
3. 6
4. 9
5. 8
6. 40
7. 12
8. $12\frac{2}{3}$

156 RETEACHING *LESSON 10-6*

Another interesting property of circles and chords arises from a radius perpendicular to a chord.

Theorem	If a radius of a circle is perpendicular to a chord of the circle, then that radius bisects the chord.

Example 4

In circle *O*, radius $\overline{OR}$ is perpendicular to chord $\overline{PQ}$ at *T*. Find *PT* if *PQ* = 5 cm.

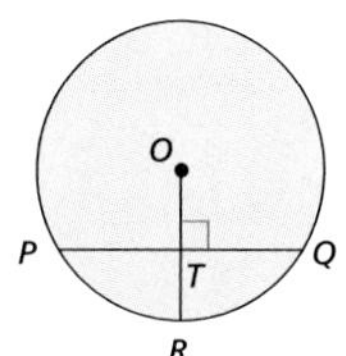

Solution

The first step is to make a diagram and label it using the given facts.

The problem states that $\overline{OR}$ is perpendicular to $\overline{PQ}$. Therefore, $\overline{OR}$ also bisects $\overline{PQ}$. If $PQ = 5$ cm, then $PT = 2.5$ cm.

Try These Exercises

Find *x*.

1. 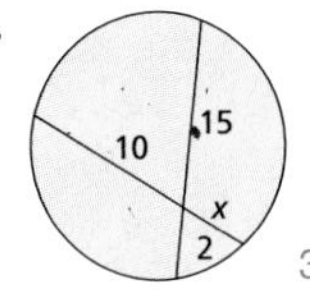

3

2. 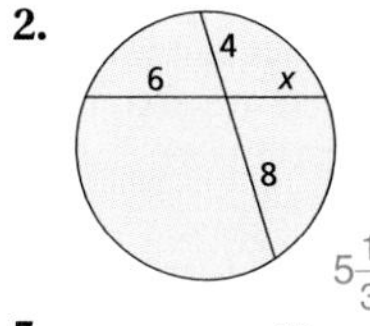

$5\frac{1}{3}$

3. 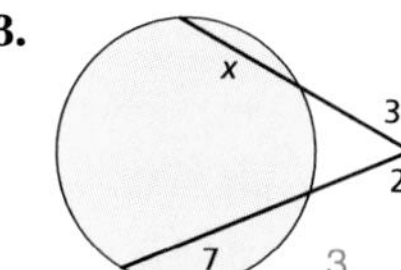

3

4. 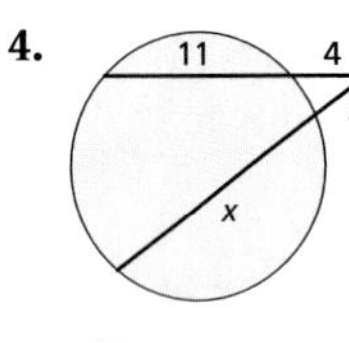

17

5. 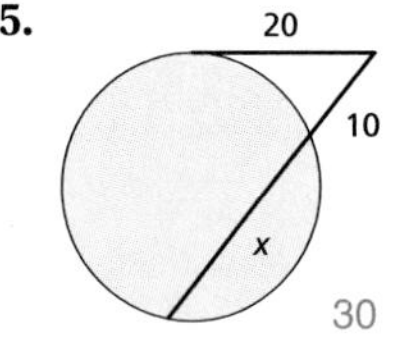

30

6. 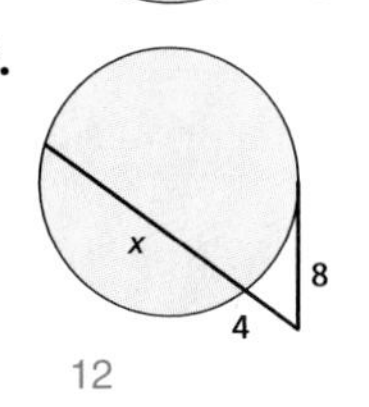

12

Practice Exercises

Find *x*.

7. 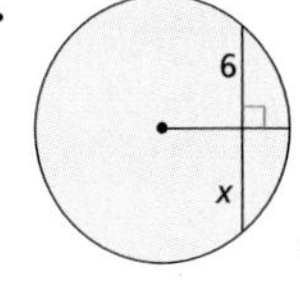

6

8. 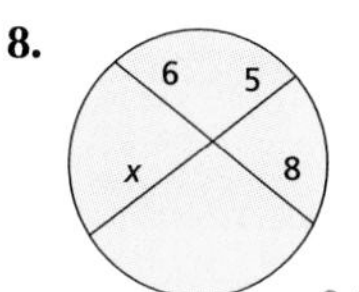

9.6

9. 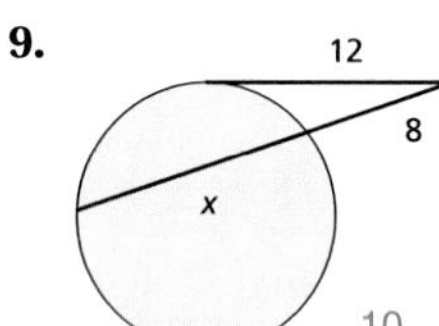

10

10. 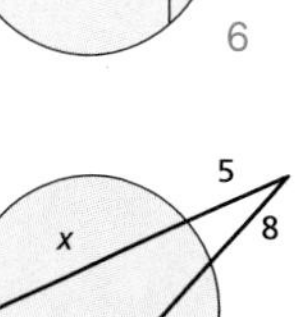

27

11. 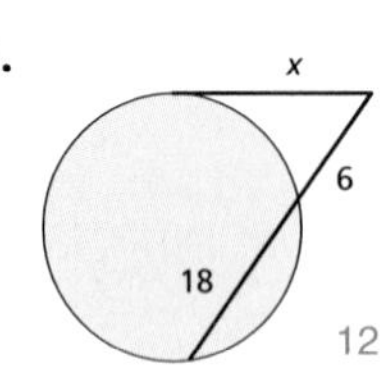

12

12. 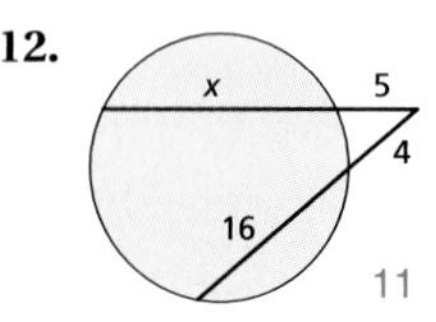

11

Lesson Wrap-up

Quick Assessment

Ask the following questions to determine if students understand the content presented in this lesson.

1. Chord *AC* consists of two segments 3 cm and 8 cm long. Chord *DE* intersects segment *AC* inside a circle and consists of two segments, 4 cm and *x* cm long. Find *x*. 6 cm
2. In Example 3, why is it possible to solve the problem using cross products? If the points were connected, they would form two similar triangles *TUV* and *UWT*.

Assignment Guide

Basic: 1–20, 23–25
Advanced: 1–25
Additional Practice: See Extra Practice Index on page 674.

Additional Answers

21. $\triangle CEB \sim \triangle CAD$ by AA similarity. $\angle C \cong \angle C$ and $\angle A \cong \angle E$, because they both intercept the same arc, *BD*.
22. Yes; the product of the secant segment and its external part is equal to the tangent segment squared. If *x* is the tangent length, then $x^2 = 3 \cdot 27$, which is 81. Because 81 is a perfect square, *x* is a rational number, 9.

Learning Styles

VISUAL LEARNERS Draw the figures in Examples 1–3 for students. Then guide them to color the segments whose measures they will multiply. For example, on the figure for Example 2, have them use a red pencil to draw a line next to *HF* and another line next to *HG*. Then have them use a green pencil to draw a line next to *HJ* and another line next to *HI*.

13. **ART** An artist is planning a circular mosaic. The design for the mosaic has two chords. Chord PR consists of two segments 3 in. and 15 in. long. Chord ST intersects $\overline{PR}$ inside the circle and consists of two segments. One segment is 5 cm in length. Find the length of the remaining segment. 9 cm

14. **SURVEYING** On a map of a city, two chords, TR and KL, intersect at point X. $TX = 4$ in., $XR = 6$ in., and $KX = 3$ in. Find the measure of $\overline{KL}$. 11 in.

Find x and y.

15.
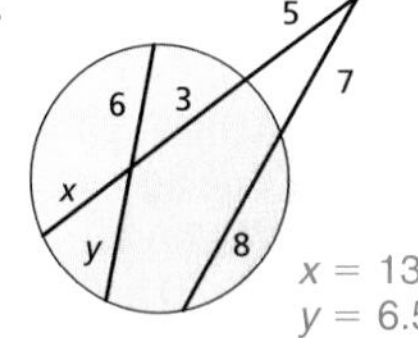

$x = 13$; $y = 6.5$

16.
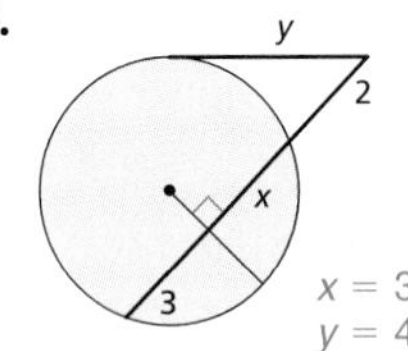

$x = 3$, $y = 4$

17.
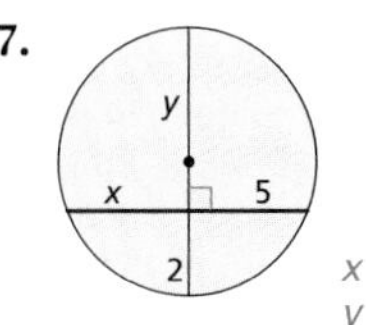

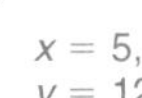
$x = 5$, $y = 12.5$

18. **GEOMETRY SOFTWARE** The distance of a chord from the center of the circle is defined as the length of a segment from the center of the circle perpendicular to the chord. Draw a large circle. Then use a centimeter ruler to draw three different chords, all the same length, at different places in the circle. Find the distance from the center of the circle to each chord. What seems to be true? They are all the same distance from the center.

Hong Kong

19. $\overline{AB}$ is a secant segment 8 cm long. Its external part is 3 cm. $\overline{AC}$ is another secant segment with an external part of 4 cm. What is its length? 6 cm

20. **CHAPTER INVESTIGATION** Make a scale drawing of a new facade for a building on your school grounds. Use measurement approximations where necessary. Display your original drawing and your new design. Check students' work.

Extended Practice Exercises

21. **WRITING MATH** Look at the figure at the right. The theorem about the products of the lengths of secant segments with a common endpoint outside the circle can be proven using similar triangles. Which triangles are similar? Explain your answer. See additional answers.

22. **WRITING MATH** A tangent segment has a common endpoint outside the circle with a secant segment. The length of the secant segment is 27 m, and the length of its external part is 3 m. Is the length of the tangent segment a rational number? Explain your answer. See additional answers.

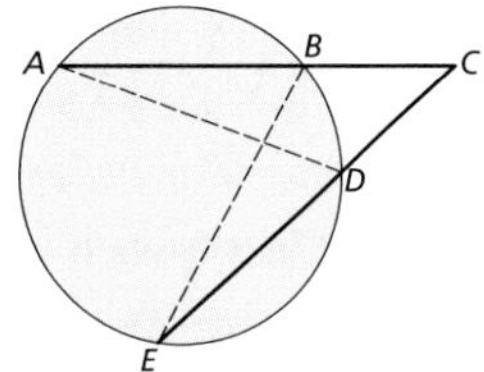

Mixed Review Exercises

Find the surface area of each figure. (Lesson 5-6)

23.
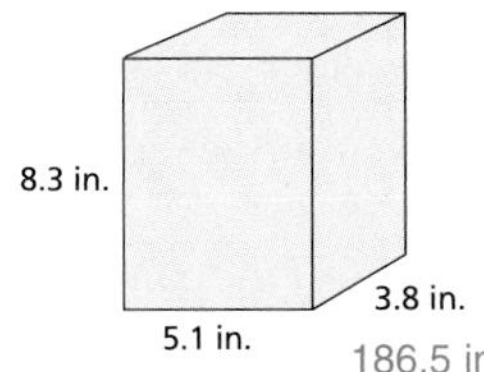

186.5 in.^{2}

24.
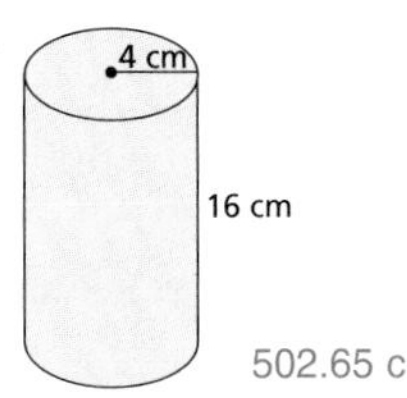

502.65 cm^2

25.
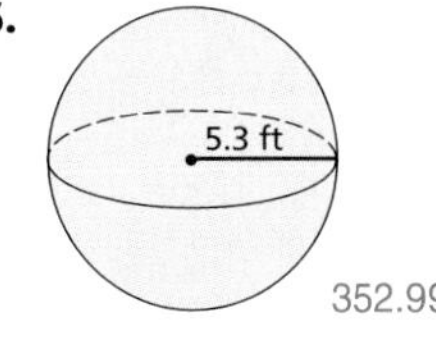

352.99 ft^2

Extra Practice Worksheet 10-6

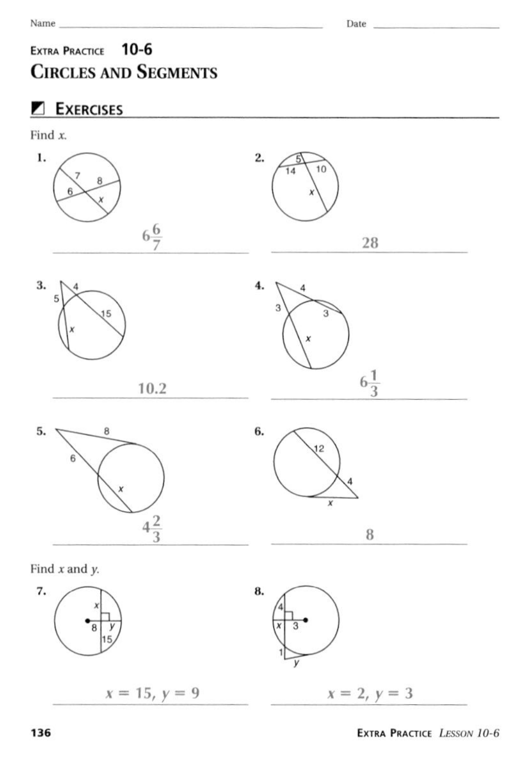

Name ______ Date ______

EXTRA PRACTICE 10-6

CIRCLES AND SEGMENTS

EXERCISES

Find x.

1. $6\frac{6}{7}$
2. 28
3. 10.2
4. $6\frac{1}{3}$
5. $4\frac{2}{3}$
6. 8

Find x and y.

7. $x = 15$, $y = 9$
8. $x = 2$, $y = 3$

136 EXTRA PRACTICE *LESSON 10-6*

Enrichment Worksheet 10-6

Name ______ Date ______

ENRICHMENT 10-6

A GEOMETRIC FALLACY: THE IMPOSSIBLE TRIANGLE

A fallacy is a demonstration or proof which leads to an impossible result. This geometric fallacy supposedly proves that a triangle can have two right angles.

EXERCISES

1. Follow along with this "proof," marking the figure as directed and completing the statements.

   Prove: A triangle can have two right angles.

   1. Draw any two intersecting circles M and N. Label the two intersection points X and Y.
   2. Draw diameters $\overline{XA}$ and $\overline{XB}$.
   3. Connect points A and B. $\overline{AB}$ intersects circle M at point D. And, $\overline{AB}$ intersects circle N at point C.
   4. Draw segments $\overline{XC}$ and $\overline{XD}$.
   5. The angles XDA and XCB must be right angles because each is inscribed in a circle.
   6. Therefore, triangle XCD has two right angles.

2. On the circles below, redraw the figure described in Steps 1–4 of the proof.

3. Use the drawing you made for Exercise 2 to explain what is wrong with the proof in Exercise 1.

   When you draw segment AB, it passes through the point Y. This point is on both circles. So, points C and D do not exist. And, $\triangle XDC$ doesn't exist either.

178 ENRICHMENT *LESSON 10-6*

Teaching Strategies

If students have difficulty understanding which measures to multiply have them redraw each figure, connecting points to form two similar triangles. Students may benefit from then drawing an additional figure showing the similar triangles separated instead of overlapping.

Skills Practice

Vocabulary Review

Lesson 10-5
circle graph

Lesson 10-6
secant segment
tangent segment

Assignment Guide

All students: 1–27
Additional Practice: See Extra Practice Index on page 674.

Additional Answers

1.
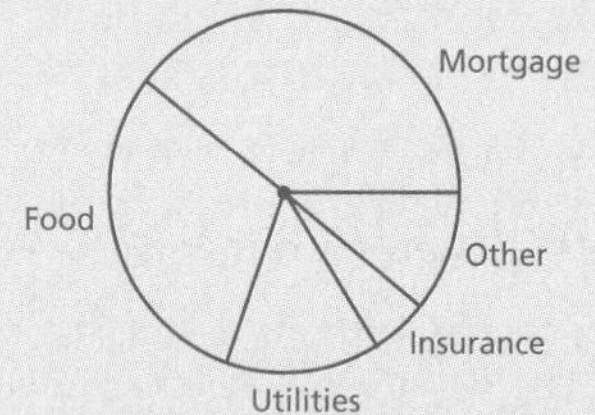

2.
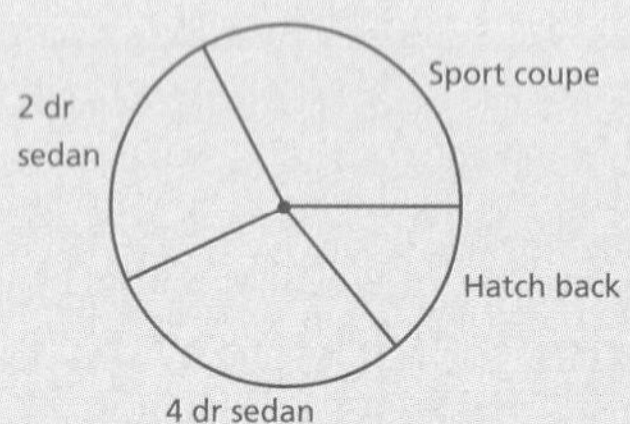

3.
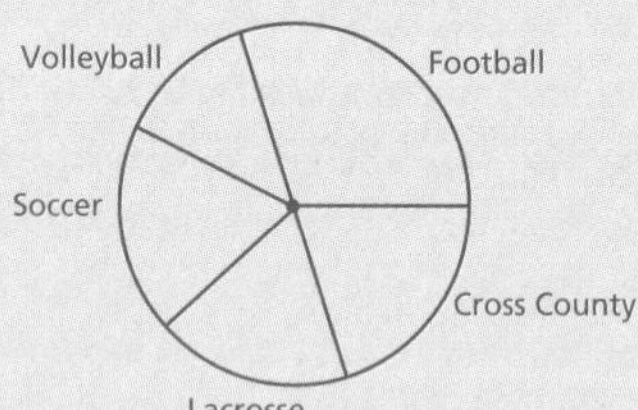

4.
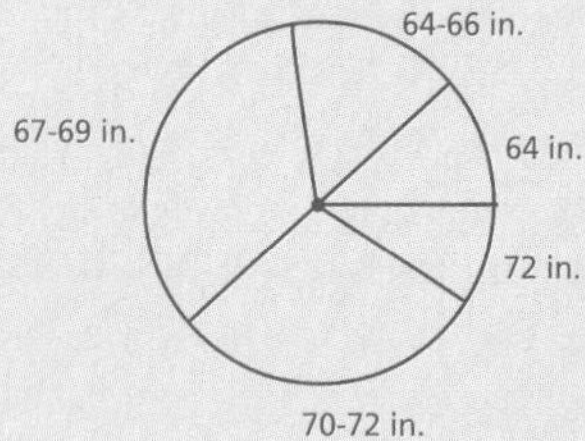

5.
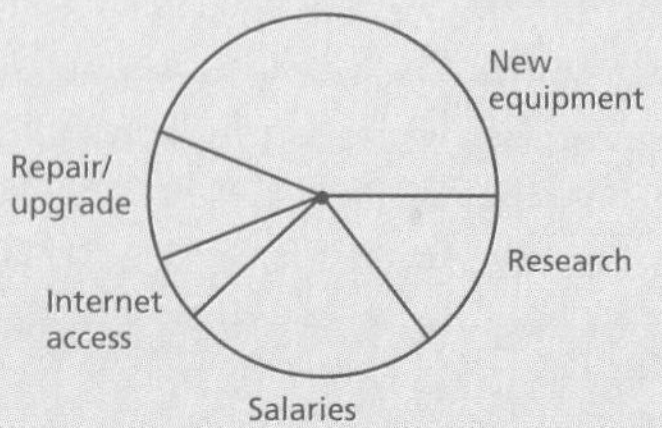

Review and Practice Your Skills

Practice ■ Lesson 10-5

Make a circle graph for each set of data. For 1–6, see additional answers.

1. Howe Family Budget

Mortgage	$860
Food	$645
Utilities	$322.50
Insurance	$107.50
Other	$215

2. Car Types Sold

Sport Coupe	50
2 Dr Sedan	35
4 Dr Sedan	45
Hatch Back	20

3. Fall Sports Athletes

Football	68
Volleyball	24
Soccer	36
Lacrosse	38
Cross Country	34

4. Heights of Freshman

<64 in.	24
64–66 in.	36
67–69 in.	72
70–72 in.	66
>72 in.	18

5. Technology Annual Budget

New equipment	$320,000
Repair/Upgrade	$80,000
Internet access	$40,000
Salaries	$180,000
Research	$100,000

6. Park Attendance by Age

Under 5	1,714
5–12	2,299
12–18	3,617
18–55	18,196
>55	2,991

Practice ■ Lesson 10-6

Find *x*.

7. 4

8.
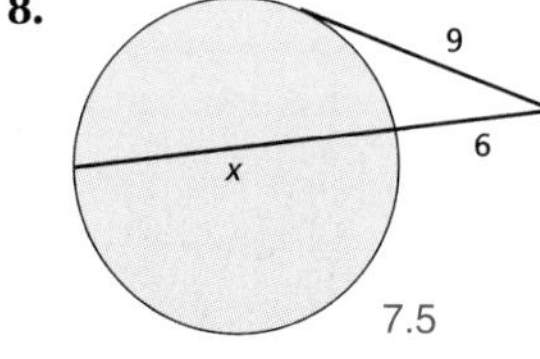
7.5

9.
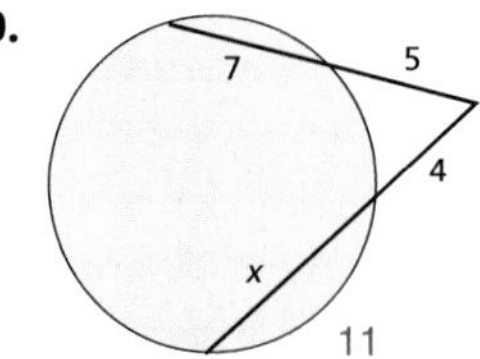
11

10.
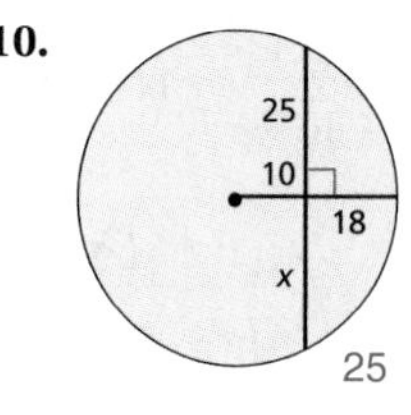
25

11.
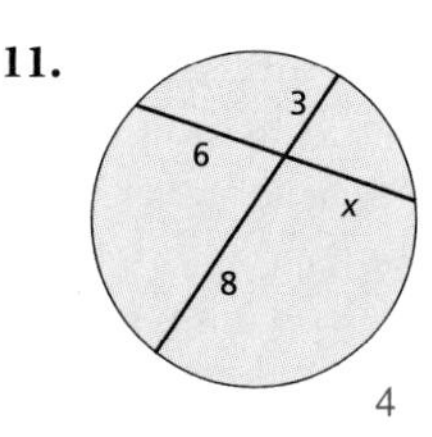
4

12.
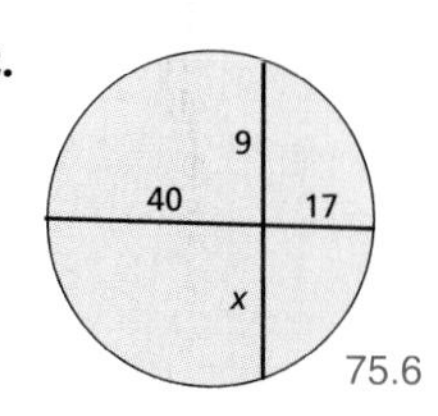
75.6

Classify each statement as *true* or *false*.

13. A radius of a circle bisects every chord of the circle. false

14. If two secant segments have a common endpoint outside a circle and their external parts are equal in length, then the chords formed by each secant inside the circle will be equal in length. true

15. Two chords of a circle will never intersect at the center of the circle. false

16. A secant and a tangent to a circle can intersect either outside or inside the circle. false

17. Chords of a circle which bisect each other are called diameters of the circle. true

6. Park Attendance by Age

Teaching Strategies

Have students complete Exercise 7. Then check to see which students got the answer 16 or 6.25. These students have failed to set up the proportion correctly. It may help these students to draw two chords forming similar triangles.

452 Chapter 10 **Right Angles and Circles**

PRACTICE ■ LESSON 10-1–LESSON 10-6

Determine if a triangle with the given sides is a right triangle. (Lesson 10-2)

18. 15 m, 36 m, 39 m yes **19.** 10 ft, 20 ft, 30 ft no **20.** 16 cm, 30 cm, 34 cm yes

21. 18 in., 30 in., 24 in. yes **22.** 4 yd, $4\sqrt{2}$ yd, $4\sqrt{3}$ yd yes **23.** 2 m, $\sqrt{17}$ m, 21 m no

Find the unknown measures. First find in simplest radical form and then find each to the nearest tenth. (Lesson 10-3)

24.

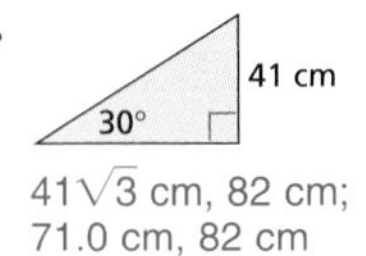

$41\sqrt{3}$ cm, 82 cm; 71.0 cm, 82 cm

25.

60°
$14\sqrt{3}$ in.

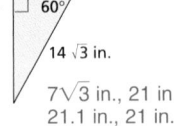

$7\sqrt{3}$ in., 21 in.; 21.1 in., 21 in.

26.

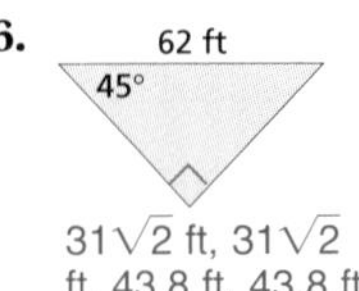

$31\sqrt{2}$ ft, $31\sqrt{2}$ ft, 43.8 ft, 43.8 ft

27.

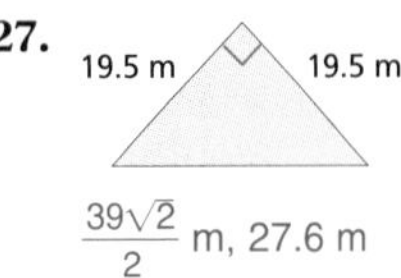

$\frac{39\sqrt{2}}{2}$ m, 27.6 m

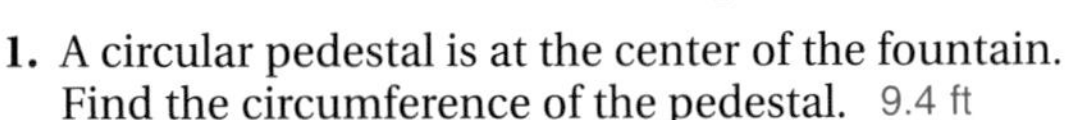

MathWorks
Workplace Knowhow

Career – Landscape Architects

Landscape architects design outdoor areas such as public parks, playgrounds, shopping centers and industrial parks. They use knowledge of the natural environment to design areas that will complement the existing surroundings.

Landscape architects study soil, sunlight, vegetation and climate. They may work with government officials and environmentalists to find ways to build new structures and roads while preserving the natural beauty and wildlife in an area.

You are designing a circular fountain for a city park. A diagram of the fountain is shown below. You have the following measurements: $DE = 4$ ft 9 in., $GE = 1$ ft 4 in. and $DF = 1$ ft 6 in. Determine all measurements to the nearest tenth of a foot. Use 3.14 for pi.

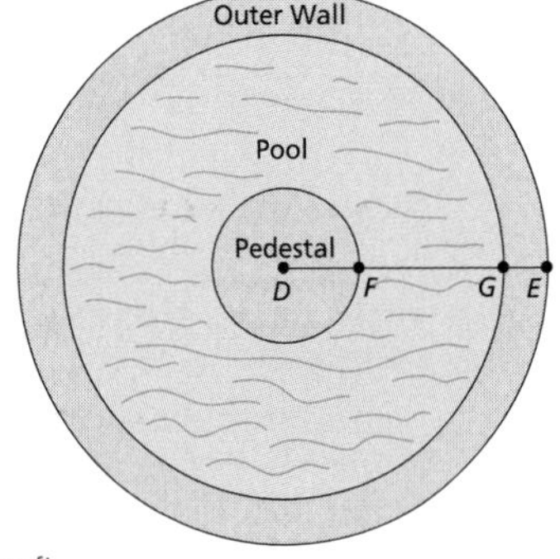

1. A circular pedestal is at the center of the fountain. Find the circumference of the pedestal. 9.4 ft
2. A wall is designed as a sitting area for park visitors. The outer edge of the wall will have a brass rim to reflect sunlight. Find the circumference of the outer edge of the wall. 29.8 ft
3. The pool, represented by the shaded area of the diagram, will be tiled. Find the area of the pool. 29.6 sq ft
4. To install the fountain, a number of square feet of shrubs and plants must be removed. Find the area of the fountain. 70.8 sq ft

Chalkboard Examples

Example from Lesson 10-5

Make a circle graph for this set of data. Round to the nearest percent.

Ramos Family Monthly Budget
Housing, $650
Food, $500
Clothing, $250
Transportation, $300
Savings, $100
Other, $500

The percents in order are 28%, 22%, 11%, 13%, 4%, 22%.

Example from Lesson 10-6

Chord *PR* consists of two segments 3 in. and 15 in. long. Chord *ST* intersects segment *PR* inside a circle and consists of two segments 5 cm and x cm long. Find *x*. 9

Career Opportunity

Describe the kind of work landscape architects do and the types of places they work. Discuss the importance of geometry, measurement and algebraic thinking in working as a landscape architect. Students should answer Questions 1–4 to better understand how landscape architects use mathematics in performing their job.

Students who are interested in learning more about this profession can go to learningmathmatters.com. Guidance Counselors should have information about school and training requirements.

Extend the Lesson

MATH JOURNAL Have students create a set of data showing how they spend their time in a 24-hour period. Then have them make a circle graph for the data. Students can share their graphs with the class and compare the times spent in common activities.

Lesson Planning

NCTM Standards/Strands

- Number & Operations
- Patterns, Functions, & Algebra
- Geometry & Spatial Sense
- Measurement
- Problem Solving
- Reasoning & Proof
- Communication
- Connections
- Representation

Vocabulary

circumscribe
inscribe

Materials Needed

compasses
straightedges
geometric-drawing software

Lesson Resources

Warm-Up Transparency 30
Reteaching 10-7
Extra Practice 10-7
Enrichment 10-7

Assignment Guide

Basic: 1–19, 21–26
Enriched: 1–26

Getting Started

5-Minute Warm-up

1. Draw an angle, and construct its bisector.
2. Draw a line segment, and construct its perpendicular bisector.

Check students' drawings.

Introduction to Lesson 10-7

After students have completed this activity, ask them how the sides of the hexagon compare to the radius of the circle.

10-7 Constructions with Circles

Goals
- Construct regular polygons using circles.
- Inscribe a circle in a polygon and circumscribe a circle about a polygon.

Applications Architecture, Design, Art

Construct a polygon using a compass and a straightedge.

1. Use a compass to draw any circle.
2. Keeping the compass open to the same radius, place the compass point anywhere on the circle and draw a small arc that intersects the circle.
3. Place the compass point on the intersection you just made and draw another arc.
4. Continue in this way around the circle. You will have drawn six points.
5. Connect each point to the one next to it with a line segment. Measure the sides and angles of the polygon you have drawn. How would you describe the figure? A regular hexagon

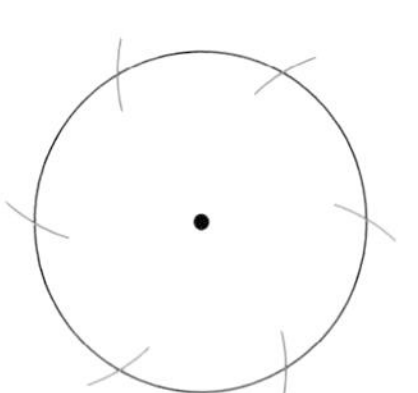

Build Understanding

Several regular polygons can be constructed using a circle. For instance, you can construct an equilateral triangle using steps 1 through 4 above to draw six evenly spaced points. Then use a straightedge to connect every other point to form the triangle. The same basic construction can be adapted to construct a regular dodecagon, a 12-sided polygon.

Example 1

Construct a regular dodecagon.

Solution

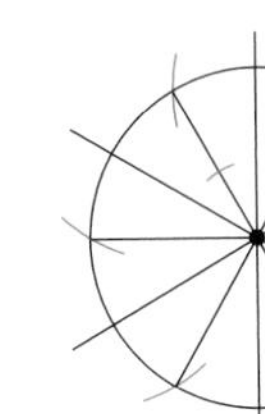

Step 1: Begin with a circle and mark off 6 equal arcs, as you did above.

Step 2: Connect each point to the center of the circle. You now have 6 central angles that are all congruent.

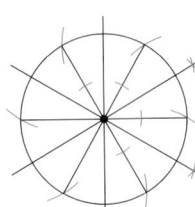

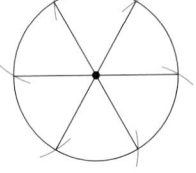

Step 3: Bisect three consecutive central angles. Extend each bisector so that it intersects the circle on two sides. You should now have 12 equally spaced points on the circle.

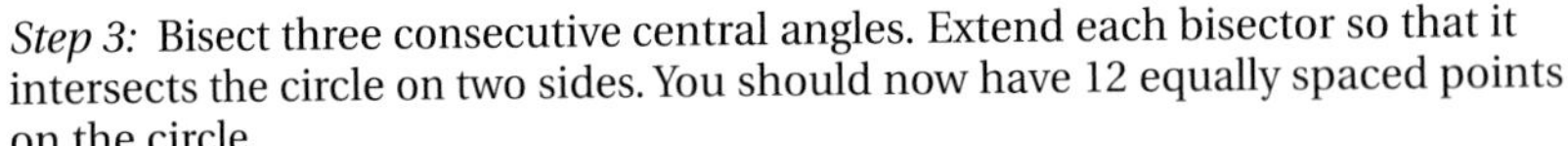

Step 4: Connect each point to the one next to it with a straight line segment. The resulting figure is a regular dodecagon.

Teaching Strategies

Point out to students that a series of very small errors in constructing regular polygons can result in starting and ending points not being the same or the sides not being the same length. Encourage them to work carefully and assure them that accurate construction takes practice.

An **inscribed polygon** has every vertex of the polygon on the same circle. Any regular polygon can be inscribed in a circle.

Example 2

Inscribe this regular pentagon in a circle.

Solution

To inscribe any regular polygon, construct the perpendicular bisector of any two of its sides. The point of intersection of the bisectors becomes the center of the circle. The radius of the circle is the distance from the center to any vertex of the polygon.

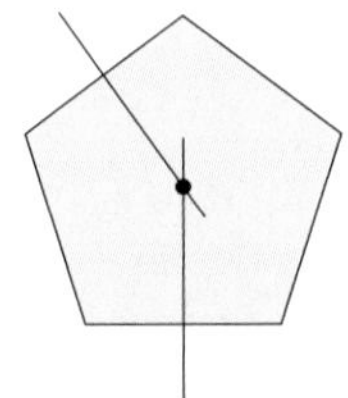

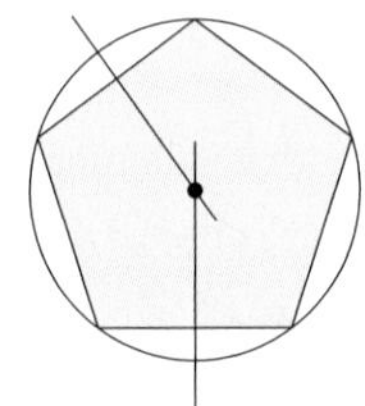

A **circumscribed polygon** of a circle has every side of the polygon tangent to the same circle. Any regular polygon can be circumscribed around a circle.

Example 3

Circumscribe a square around a circle.

Solution

Draw a square. To locate the point that will become the center of the circle, find the intersection of the perpendicular bisectors of any two sides. To find the radius of the circle, use the distance from the center of the circle along a perpendicular bisector to a side of the polygon. Draw the circle.

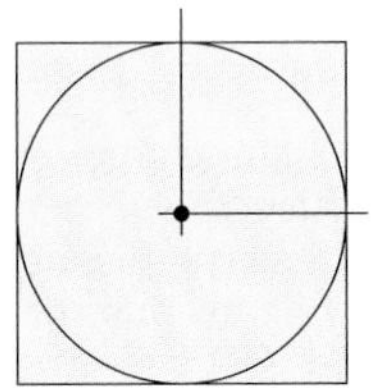

Many properties of circles and parts of circles can be used to solve real-life problems.

Example 4

ARCHITECTURE An architect is restoring an old house. She has found a part of a window that may have been used in the attic. How can she figure out the size of the original window from this fragment?

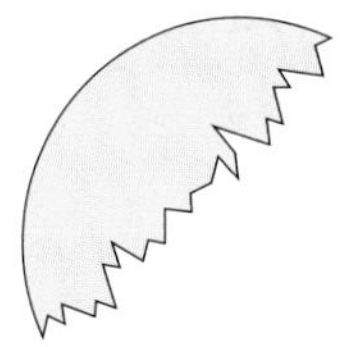

Problem Solving Tip

Use paper folding to construct a regular octagon using a circle.

1. Draw a circle and cut it out.
2. Fold the circle into halves, then fourths and finally eighths. Open the circle.

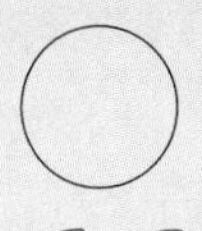

3. There are now 8 equally spaced points on the circle. Connect each point to the one next to it with a straight line segment. The resulting figure is an octagon.

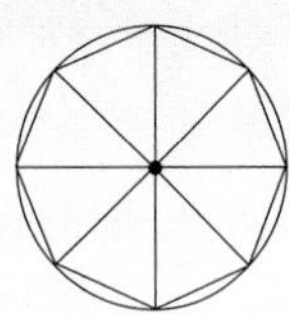

Reading Math

The word *circumscribed* comes from Latin words which mean "drawn around." The word *inscribed* means "drawn in." The Example 2 text says that the pentagon has been inscribed in the circle. It is also correct to say that the pentagon has been circumscribed around the circle.

Extend the Lesson

REAL WORLD CONNECTION Invite an architect, designer or member of your school's industrial arts department to visit your class to discuss how he or she applies the properties of circles and circle constructions to his or her work.

Chalkboard Examples

Supplementary Example 1
Construct an equilateral triangle. Check students' drawings. Students should use arcs to make 6 points and then connect every other point.

Supplementary Example 2
Inscribe a regular hexagon in a circle. Check students' drawings.

Supplementary Example 3
Circumscribe a pentagon around a circle. Check students' drawings.

Supplementary Example 4
Have students use a compass to draw an arc of at least 90°. Then have them use the method shown in Example 4 to complete the circle. Check students' drawings.

Reteaching Worksheet 10-7

Name ____________ Date ____________

RETEACHING 10-7

CONSTRUCTIONS WITH CIRCLES

A polygon is **circumscribed** about a circle if each of its **sides** is tangent to the circle.

A polygon is **inscribed** in a circle if each of its **vertices** lies on the circle.

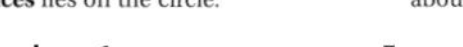

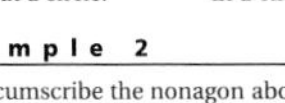

Square circumscribed about a circle. Square inscribed in a circle.

Example 1

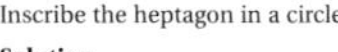

Inscribe the heptagon in a circle.

Solution

- Construct a perpendicular bisector of one of the sides of the heptagon.
- Construct a perpendicular bisector of another side.
- The point of intersection of these bisectors will be the center of the circle. Place the compass tip on this point.
- The distance from the center to any vertex will be the radius of the circle. Place the pencil tip on a vertex.
- Draw the circle.

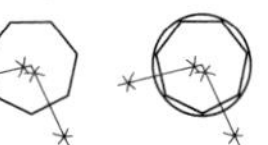

Example 2

Circumscribe the nonagon about a circle.

Solution

- Construct a perpendicular bisector of one of the sides of the nonagon. Repeat for another side.
- The point of intersection of these bisectors will be the center of the circle. Place the compass tip on this point.
- Locate the point where a bisector intersects the nonagon. The distance from the center to this point will be the radius of the circle. Place the pencil tip on this point of intersection.
- Draw the circle.

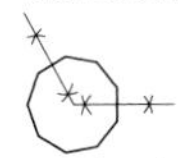

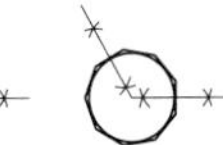

EXERCISES Check students' constructions for Exercises 1–4; final figure is shown.

Draw each circle.

1. Inscribe the triangle in a circle.

2. Inscribe the pentagon in a circle.

3. Circumscribe the hexagon about a circle.

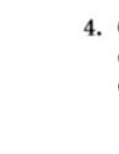

4. Circumscribe the octagon about a circle.

158 RETEACHING *LESSON 10-7*

Solution

The perpendicular bisector of a chord passes through the center of a circle. To complete the circle, begin by drawing two chords. Construct the perpendicular bisector of each. The point where the bisectors intersect must be the center of the circle. Use the center of the circle and the radius to complete the circle.

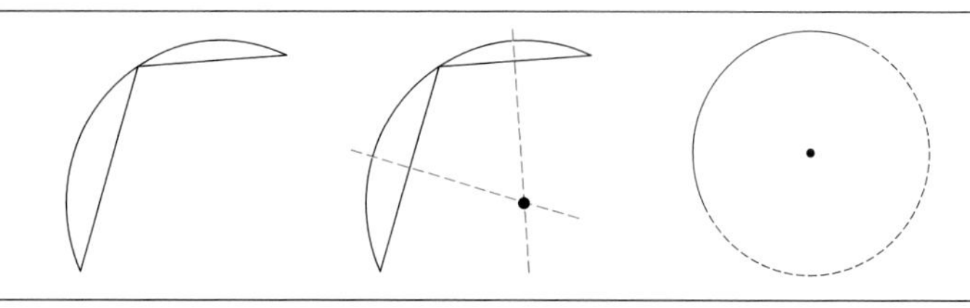

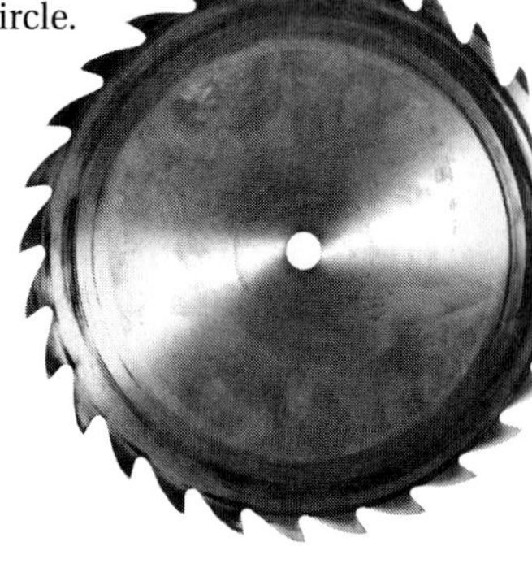

Try These Exercises

For 1–3, check students' constructions.

1. Construct a regular hexagon with a side measuring 4 cm.
2. Copy the regular octagon shown at the right. Inscribe the octagon in a circle.
3. Copy the equilateral triangle shown at the right. Circumscribe the triangle around a circle.

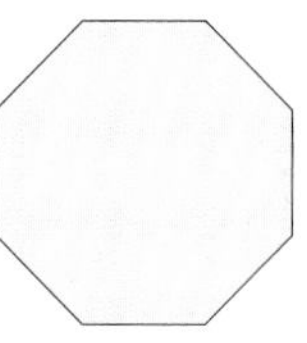

TALK ABOUT IT Discuss the following statements with a partner. Decide whether each statement is true or false. Explain your reasoning.

4. In Example 3, the circle is inscribed in the square. true
5. It is possible to inscribe a rhombus in a circle. false
6. It is impossible to inscribe a non-regular octagon in a circle. false
7. **CONSTRUCTION** Circular saws come in different sizes. A builder bought a saw at a second-hand sale. The saw blade was broken as shown in the picture at the right. Copy the drawing. Use construction to demonstrate how the builder can find the center and complete the circle in order to find out which size of replacement blade to buy. See additional answers.

8. **WRITING MATH** An architect draws a circle and its diameter. Using a compass, she constructs the perpendicular bisector to the diameter and extends it so that it intersects the circle in two places. How could the architect use the drawing to construct a square?
After completing the steps described, there are four congruent central angles and 4 equally spaced points on the circle. Connect each point to the one next to it with a straight line segment. The resulting figure is a square.

Practice Exercises

9. Construct a regular octagon. Check students' constructions.
10. Copy this regular hexagon. Inscribe the hexagon in a circle. Check students' constructions.

11. Copy this regular pentagon. Circumscribe the pentagon around a circle. Check students' constructions.

Lesson Wrap-up

Quick Assessment

Ask the following questions to determine if students understand the content presented in this lesson.

1. Have students copy the equilateral triangle in Exercise 3 and inscribe it in a circle. Check students' constructions.
2. Imagine your are a designer. Your new assignment is to design a logo that has a square inscribed in a circle and an octagon circumscribed around the same circle. Draw the logo. Check students' constructions.

Assignment Guide

Basic: 1–19, 21–26
Advanced: 1–26
Additional Practice: See Extra Practice Index on page 674.

Additional Answers

7.

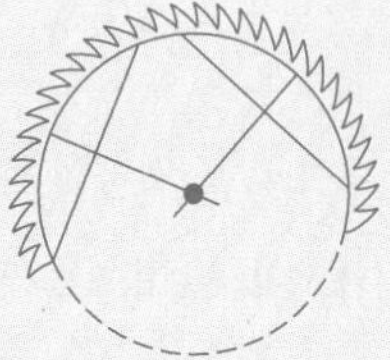

12. Draw central angles of the circle to the vertices of the pentagon. Bisect each central angle. There will be 10 points on the circle. Connect points with line segments to form a regular decagon.
13. 16-sided, 24-sided; any polygons with sides that are multiples of 6 or 4.
16.

17.

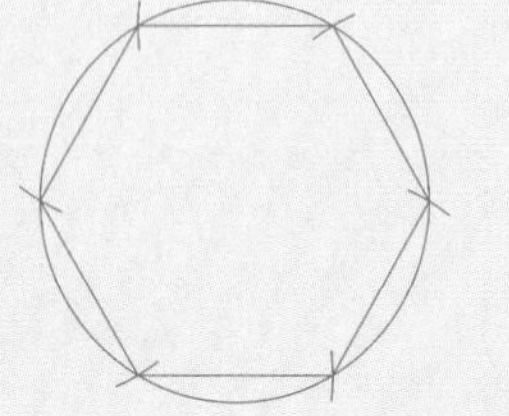

18.

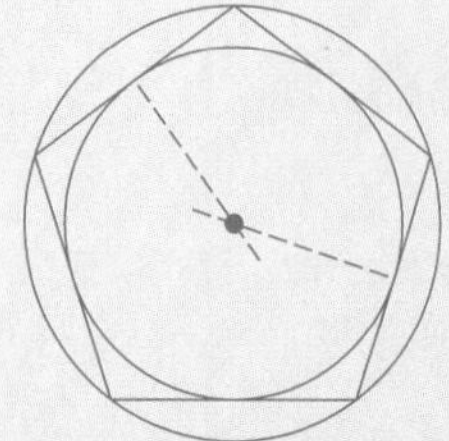

19. Mike could take the copy, find the perpendicular bisectors of two different sides of the hexagon, and use their intersections as the center of a circle. He could use a radius equal to the distance from the center of the circle to a vertex of the hexagon. Then he could construct the perpendicular bisectors of each of the other sides of the hexagon, and mark the point where these bisectors meet the circle. At this point, there should be 12 equally spaced points on the circle. He can connect these 12 points with a straightedge to draw a dodecagon.

12. Draw any regular pentagon. Describe how you can use a circle to help you construct a regular decagon from the pentagon. See additional answers.

13. Name two other regular polygons, not mentioned so far in this lesson, that can be constructed using a circle. Explain your choices. See additional answers.

14. A regular hexagon is inscribed in a circle. What is the measure of each of the six arcs of the circle? 60°

15. Copy the regular hexagon in Exercise 10. Circumscribe it around a circle. Check students' construction.

16. **ARCHITECTURE** Ming Lee is an architect. She submitted the following plan for a new house to her clients, James and Odetta Williams. The Williamses ask Ms. Lee to alter the plans so that the sun room is circular rather than square. Copy Ming Lee's plans onto graph paper, using pencil, compass, and straightedge. Inscribe a circle within the square that represents the sun room. Then erase the outline of the square. See additional answers.

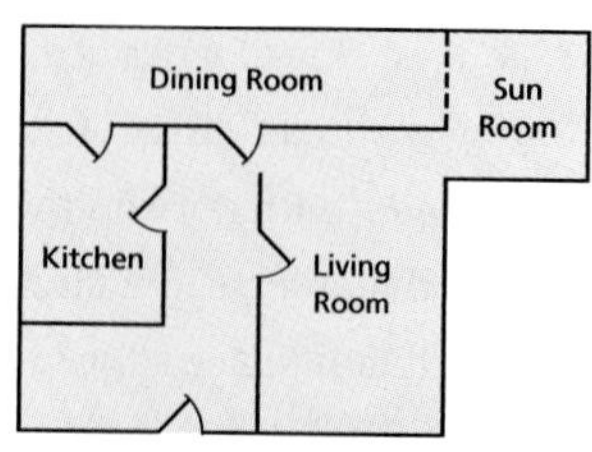

PLAN: Williams Home—First Floor

17. **DESIGN** Dave O'Brien makes and designs tiles. He has a client who wants a small tabletop covered in hexagonal tiles. Dave decides to make some sketches of different designs from which the client can choose. Construct a regular hexagon measuring 2 inches on all sides that Dave can use as a model. See additional answers.

18. **ART** Linda Soares is designing a logo for the new community center. She wants to take the pentagon at the right and inscribe it in a circle while also inscribing a circle in the pentagon. Copy the pentagon and complete Linda's design. See additional answers.

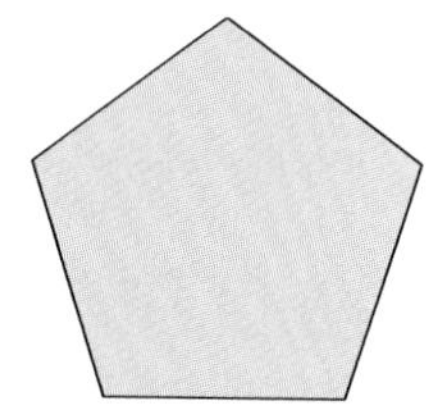

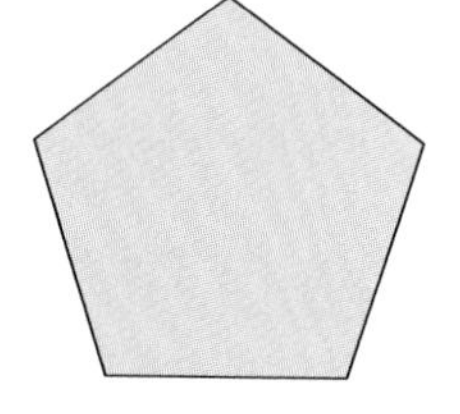

19. **ARCHITECTURE** Mike Whitehorse has a blueprint for a gazebo in the shape of a regular hexagon. He wants to change it so that it is about the same size, but has the shape of a regular dodecagon. How can he use a copy of the original blueprint to develop the other plan? See additional answers.

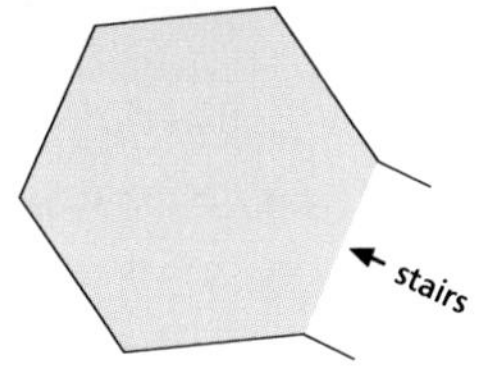

Extended Practice Exercises

20. Tell if this statement is true or false. Explain your answer. *Because any regular polygon can be inscribed in a circle, then any regular polygon can be constructed using a compass and straightedge.* See additional answers.

Mixed Review Exercises

Compute the variance and standard deviation for each set of data. Round the final answer to the nearest tenth if necessary. (Lesson 9-7)

21. 8, 9, 10, 11, 12, 13 2.9, 1.7

22. 28, 32, 31, 36, 29, 35 8.5, 2.9

23. 2, 4, 6, 3, 5, 4 1.7, 1.3

24. 12, 15, 16, 11, 15, 13 3.2, 1.8

25. 74, 73, 82, 80, 77, 79 10.25, 3.2

26. 52, 60, 65, 62, 58, 61 16.2, 4.0

20. False; the only regular polygons that can be constructed using a compass and straightedge are the equilateral triangle and polygons with sides that are multiples of 4, 5, or 6.

Extra Practice Worksheet 10-7

Name ______ Date ______

EXTRA PRACTICE 10-7

CONSTRUCTIONS WITH CIRCLES

EXERCISES

1. Copy this square onto your own paper. Inscribe the square in a circle. Check students' work.
2. Copy this regular hexagon onto your own paper. Circumscribe the hexagon around a circle. Check students' work.
3. Copy this equilateral triangle onto your own paper. Inscribe this triangle in a circle. Check students' work.
4. Copy this regular pentagon onto your own paper. Circumscribe the pentagon around a circle. Check students' work.
5. Construct a regular hexagon on your own paper. Check students' work.
6. Construct an equilateral triangle on your own paper. Check students' work.

138 EXTRA PRACTICE *LESSON 10-7*

Enrichment Worksheet 10-7

Name ______ Date ______

ENRICHMENT 10-7

MORE CONSTRUCTIONS WITH REGULAR POLYGONS

There are many different ways to construct regular polygons. The designs on this page make use of some alternate methods.

EXERCISES

Use a compass and straightedge to copy each design. The challenge will be to decide which lines or curves to draw first. Check students' constructions for Exercises 1–8.

1. 2. 3. 4. 5. 6.

7. The figure at the right includes two regular pentagons. Construct this figure.
8. The construction in Exercise 7 can be adapted to make a regular polygon with any number of sides. Adapt the construction to make a regular heptagon (7 sides).

180 ENRICHMENT *LESSON 10-7*

Assessment Planning

Vocabulary Review

Lesson 10-1
irrational number radicand
rationalizing the denominator

Lesson 10-2
Pythagorean Theorem
hypotenuse legs
Pythagorean triples

Lesson 10-3
45°-45°-90° triangle
30°-60°-90° triangle

Lesson 10-4
intercept minor arc
major arc inscribed angle
chord secant
tangent central angle

Lesson 10-5
circle graph

Lesson 10-6
secant segment tangent segment

Lesson 10-7
circumscribe inscribe

Assessment Options

Chapter Assessment, page 460
Alternative Assessment, page 461
Chapter Test Forms A and B
Cumulative Assessment, page 462–463
Achievement Test
Writing Prompts

Assignment Guide

All students: 1-22
Additional Practice: See Extra Practice Index on page 674.

Additional Answers

17. Check students' circle graphs. Percents should be Italian: 22%, Spanish: 41%, French: 33%, and Japanese: 4%.

21.

22.

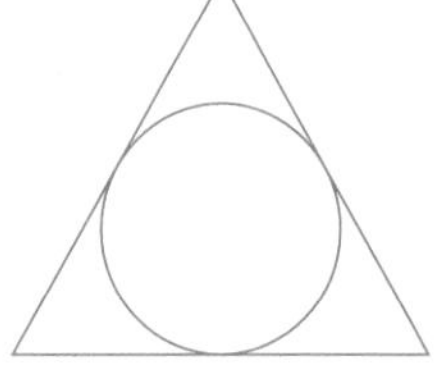

Chapter 10 Review

Vocabulary ■ LESSON 10-1–LESSON 10-7

Match the word from the list at the right with the description at the left.

1. longest side of a right triangle d
2. angle with its vertex on a circle a
3. segment with both endpoints on a circle e
4. line that intersects a circle in two points b
5. line that intersects the circle in only one point c

a. inscribed
b. secant
c. tangent
d. hypotenuse
e. chord

LESSON 10-1 ■ Irrational Numbers, p. 426

- An **irrational number** cannot be written as a fraction, terminating decimal, or repeating decimal. The square root of a number that is not a perfect square is always irrational.
- Irrational numbers can be written in simplest radical form using these theorems:
$\sqrt{a \cdot b} = \sqrt{a} \cdot \sqrt{b}$ and $\sqrt{\frac{a}{b}} = \frac{\sqrt{a}}{\sqrt{b}}$

6. Simplify $\sqrt{45}$. $3\sqrt{5}$

7. Simplify and rationalize the denominator. $\frac{\sqrt{6}}{\sqrt{8}}$ $\frac{\sqrt{3}}{2}$

LESSON 10-2 ■ The Pythagorean Theorem, p. 430

- In a right triangle, the two shorter sides are called the legs, and the longest side is called the hypotenuse. In any right triangle, the Pythagorean Theorem is true: $a^2 + b^2 = c^2$, where a and b are the measures of the legs and c is the measure of the hypotenuse.

Find the missing measure in each right triangle. Round to the nearest tenth.

8. $a = 9$ cm, $b = 12$ cm 15 cm

9. $a = 2$ in., $c = 5$ in. 4.6 in.

10. A flagpole that is 3 meters high is connected by a guy wire from its top to a stake in the ground 1.5 meters from its base. How long is the wire? 3.4 m

LESSON 10-3 ■ Special Right Triangles, p. 436

- In a 30°–60°–90° triangle, the measure of the hypotenuse is two times that of the leg opposite the 30° angle. The measure of the longer leg is $\sqrt{3}$ times the leg opposite the 30° angle.
- In a 45°–45°–90° triangle, the measure of the hypotenuse is $\sqrt{2}$ times the measure of a leg of the triangle.

Find the missing measures. Answers may be left in simplest radical form.

11.

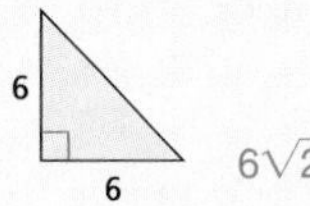

12.

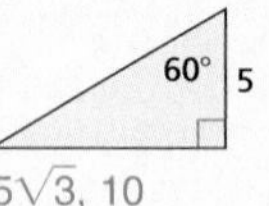

$5\sqrt{3}$, 10

13.

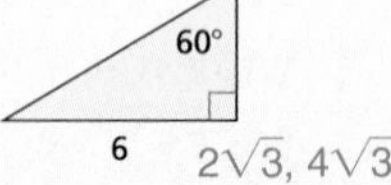

$2\sqrt{3}$, $4\sqrt{3}$

LESSON 10-4 ■ Circles, Angles, and Arcs, p. 440

- The measure of a central angle of a circle is the same as the measure of its intercepted arc.
- The measure of an inscribed angle in a circle is one-half the measure of its intercepted arc.
- If two secants intersect inside a circle, the measure of each angle formed is equal to one-half the sum of the intercepted arcs.

Find *x*.

14.

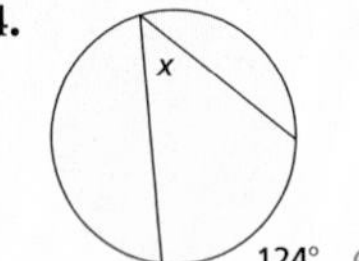

62°

15.

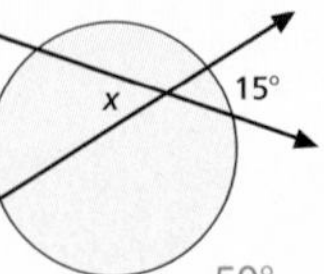

50°

16.

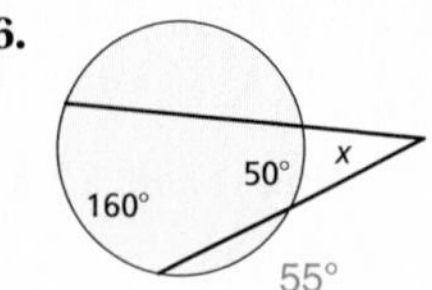

55°

LESSON 10-5 ■ Problem Solving Skills: Circle Graphs, p. 446

- Circle graphs can represent data about how a quantity is subdivided.

17. Every Elm High School student takes one foreign language. This year, 251 take Italian, 478 take Spanish, 376 take French, and 50 take Japanese. Make a circle graph for this data. See additional answers.

LESSON 10-6 ■ Circles and Segments, p. 448

- If two chords intersect in a circle, then the product of the measures of the two segments of one chord is equal to the product of the measures of the two segments of the other chord.
- If two secant segments have a common endpoint outside a circle, then the product of the measures of one secant segment and its external part is equal to the product of the other secant segment and its external part.

Find *x*.

18.

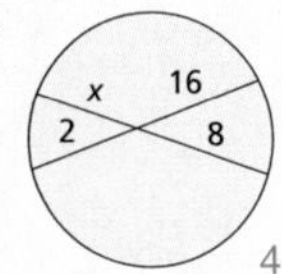

4

19.

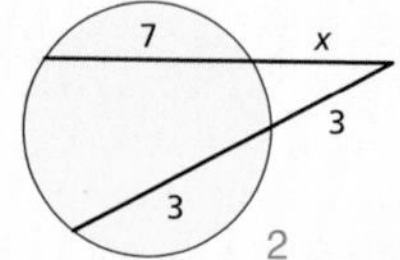

2

20.

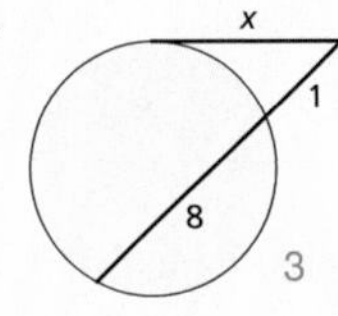

3

LESSON 10-7 ■ Constructions with Circles, p. 454

- An inscribed polygon has every vertex on the same circle.
- A circumscribed polygon has every side tangent to the same circle.

21. Draw a square. Inscribe the square in a circle. See additional answers.

22. Construct an equilateral triangle. Circumscribe the triangle around a circle. See additional answers.

Teaching Strategies

CONNECTING TO PRIOR KNOWLEDGE Have students divide an inscribed hexagon into 6 triangles by connecting the center of the circle to each vertex. The 6 triangles have a total angle measure of 1,080°. The central angles have a total measure of 360°. Each angle measure $\frac{(1{,}080° - 360°)}{6}$, or 120°. Find the measure of each angle of an inscribed pentagon, octagon, and n-gon. 108°; 135°; $\frac{180(n-2)}{n}$

Chapter 10 Test Form A, page 1

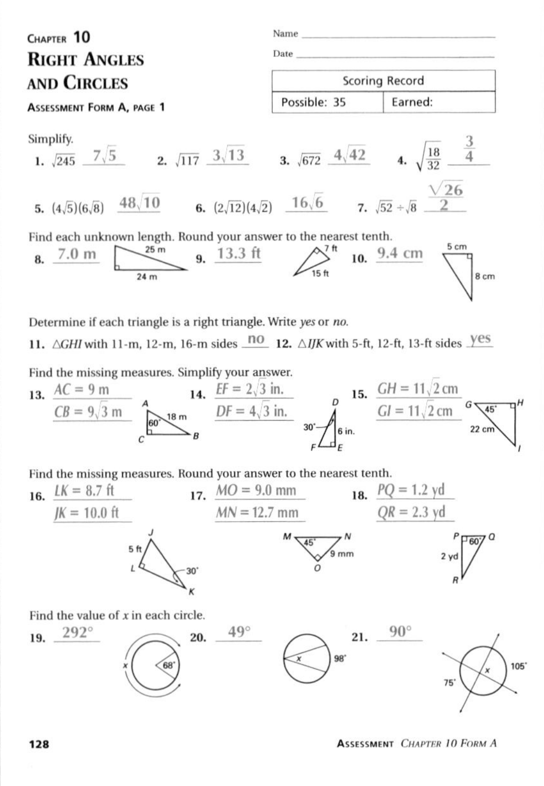

CHAPTER 10
RIGHT ANGLES AND CIRCLES
ASSESSMENT FORM A, PAGE 1

Name ______ Date ______

Scoring Record — Possible: 35 | Earned:

Simplify.

1. $\sqrt{245}$ $7\sqrt{5}$ 2. $\sqrt{117}$ $3\sqrt{13}$ 3. $\sqrt{672}$ $4\sqrt{42}$ 4. $\sqrt{\frac{18}{32}}$ $\frac{3}{4}$
5. $(4\sqrt{5})(6\sqrt{8})$ $48\sqrt{10}$ 6. $(2\sqrt{12})(4\sqrt{2})$ $16\sqrt{6}$ 7. $\sqrt{52} \div \sqrt{8}$ $\frac{\sqrt{26}}{2}$

Find each unknown length. Round your answer to the nearest tenth.

8. 7.0 m 9. 13.3 ft 10. 9.4 cm

Determine if each triangle is a right triangle. Write *yes* or *no*.

11. $\triangle GHI$ with 11-m, 12-m, 16-m sides no 12. $\triangle IJK$ with 5-ft, 12-ft, 13-ft sides yes

Find the missing measures. Simplify your answer.

13. $AC = 9$ m, $CB = 9\sqrt{3}$ m 14. $EF = 2\sqrt{3}$ in., $DF = 4\sqrt{3}$ in. 15. $GH = 11\sqrt{2}$ cm, $GI = 11\sqrt{2}$ cm

Find the missing measures. Round your answer to the nearest tenth.

16. $LK = 8.7$ ft, $JK = 10.0$ ft 17. $MO = 9.0$ mm, $MN = 12.7$ mm 18. $PQ = 1.2$ yd, $QR = 2.3$ yd

Find the value of *x* in each circle.

19. 292° 20. 49° 21. 90°

128 — ASSESSMENT CHAPTER 10 FORM A

Chapter 10 Test Form A, page 2

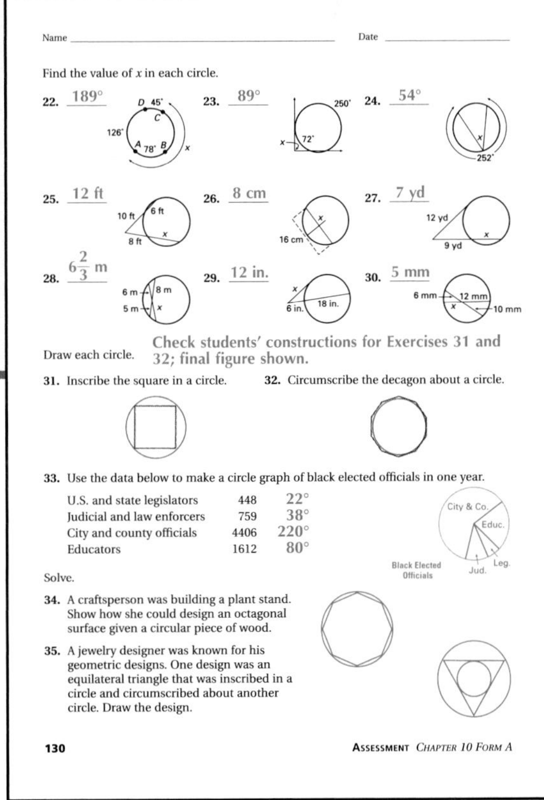

Name ______ Date ______

Find the value of *x* in each circle.

22. 189° 23. 89° 24. 54°
25. 12 ft 26. 8 cm 27. 7 yd
28. $6\frac{2}{3}$ m 29. 12 in. 30. 5 mm

Draw each circle. Check students' constructions for Exercises 31 and 32; final figure shown.

31. Inscribe the square in a circle. 32. Circumscribe the decagon about a circle.

33. Use the data below to make a circle graph of black elected officials in one year.

U.S. and state legislators	448	22°
Judicial and law enforcers	759	38°
City and county officials	4406	220°
Educators	1612	80°

Solve.

34. A craftsperson was building a plant stand. Show how she could design an octagonal surface given a circular piece of wood.

35. A jewelry designer was known for his geometric designs. One design was an equilateral triangle that was inscribed in a circle and circumscribed about another circle. Draw the design.

130 — ASSESSMENT CHAPTER 10 FORM A

Chapter 10 Test Form B, page 1

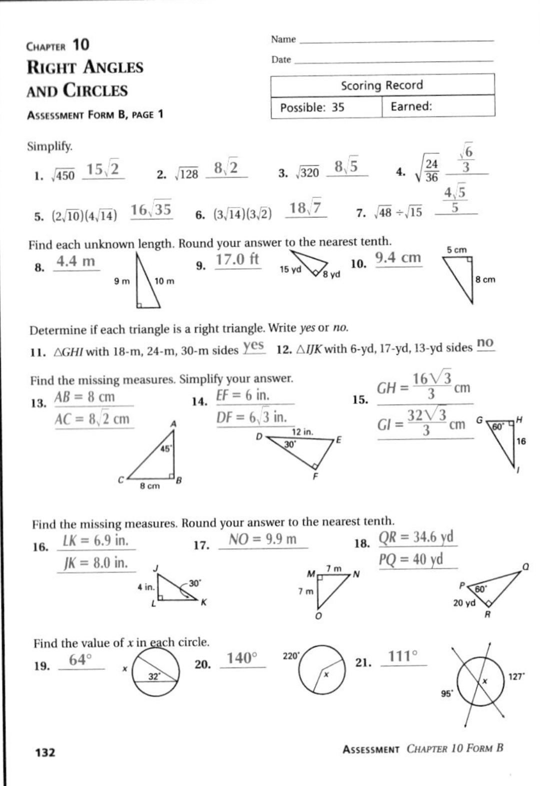
CHAPTER 10
RIGHT ANGLES AND CIRCLES
ASSESSMENT FORM B, PAGE 1

Name ______________
Date ______________

Scoring Record	
Possible: 35	Earned:

Simplify.

1. $\sqrt{450}$ $15\sqrt{2}$ 2. $\sqrt{128}$ $8\sqrt{2}$ 3. $\sqrt{320}$ $8\sqrt{5}$ 4. $\sqrt{\frac{24}{36}}$ $\frac{\sqrt{6}}{3}$

5. $(2\sqrt{10})(4\sqrt{14})$ $16\sqrt{35}$ 6. $(3\sqrt{14})(3\sqrt{2})$ $18\sqrt{7}$ 7. $\sqrt{48} \div \sqrt{15}$ $\frac{4\sqrt{5}}{5}$

Find each unknown length. Round your answer to the nearest tenth.

8. 4.4 m 9. 17.0 ft 10. 9.4 cm

Determine if each triangle is a right triangle. Write *yes* or *no*.

11. $\triangle GHI$ with 18-m, 24-m, 30-m sides yes 12. $\triangle IJK$ with 6-yd, 17-yd, 13-yd sides no

Find the missing measures. Simplify your answer.

13. $AB = 8$ cm, $AC = 8\sqrt{2}$ cm 14. $EF = 6$ in., $DF = 6\sqrt{3}$ in. 15. $GH = \frac{16\sqrt{3}}{3}$ cm, $GI = \frac{32\sqrt{3}}{3}$ cm

Find the missing measures. Round your answer to the nearest tenth.

16. $LK = 6.9$ in., $JK = 8.0$ in. 17. $NO = 9.9$ m 18. $QR = 34.6$ yd, $PQ = 40$ yd

Find the value of x in each circle.

19. 64° 20. 140° 21. 111°

132 ASSESSMENT *CHAPTER 10 FORM B*

Chapter 10 Test Form B, page 2

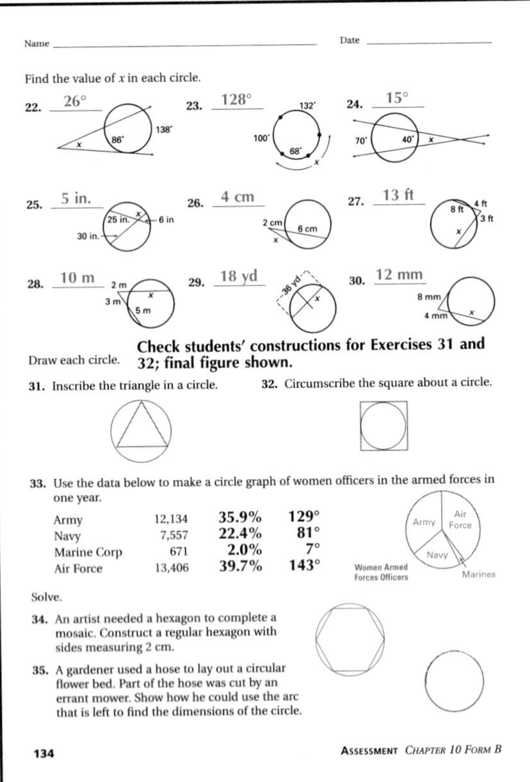
Name ______________ Date ______________

Find the value of x in each circle.

22. 26° 23. 128° 24. 15°

25. 5 in. 26. 4 cm 27. 13 ft

28. 10 m 29. 18 yd 30. 12 mm

Draw each circle. **Check students' constructions for Exercises 31 and 32; final figure shown.**

31. Inscribe the triangle in a circle. 32. Circumscribe the square about a circle.

33. Use the data below to make a circle graph of women officers in the armed forces in one year.

Army	12,134	**35.9%**	**129°**
Navy	7,557	**22.4%**	**81°**
Marine Corp	671	**2.0%**	**7°**
Air Force	13,406	**39.7%**	**143°**

Solve.

34. An artist needed a hexagon to complete a mosaic. Construct a regular hexagon with sides measuring 2 cm.

35. A gardener used a hose to lay out a circular flower bed. Part of the hose was cut by an errant mower. Show how he could use the arc that is left to find the dimensions of the circle.

134 ASSESSMENT *CHAPTER 10 FORM B*

CHAPTER 10 ASSESSMENT

Simplify. Rationalize the denominator if necessary.

1. $\sqrt{700}$ $10\sqrt{7}$ **2.** $\frac{\sqrt{75}}{\sqrt{3}}$ 5 **3.** $\frac{11}{\sqrt{3}}$ $11\frac{\sqrt{3}}{3}$

Find the unknown measures in each right triangle. Round answers to the nearest tenth. (*a* and *b* are the measures of the legs; *c* is the measure of the hypotenuse.)

4. $a = 8$ in., $b = 15$ in. 17 in. **5.** $a = 5$ m $c = 7$ m 4.9 m

Find the missing measures. Answers may be left in simplest radical form.

6.

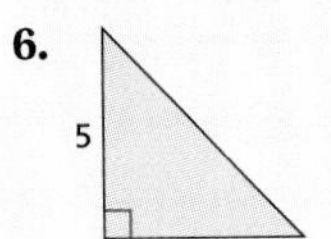

$5\sqrt{2}$

7.

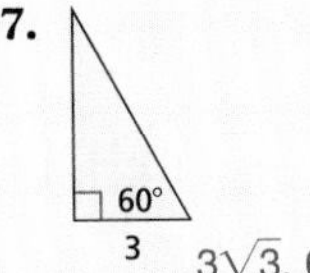

$3\sqrt{3}$, 6

8.

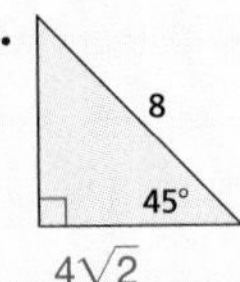

$4\sqrt{2}$

Find *x*.

9.

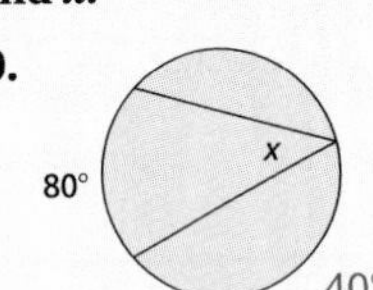

40°

10.

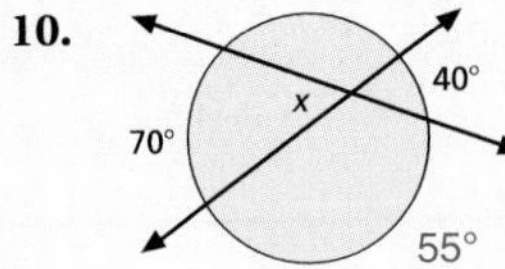

55°

11.

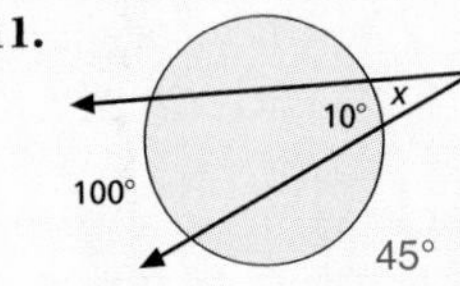

45°

12. 80° x 140°

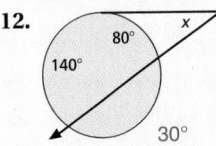

30°

13.

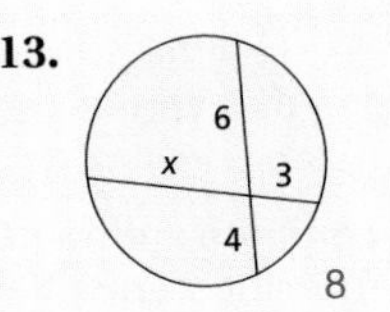

8

14.

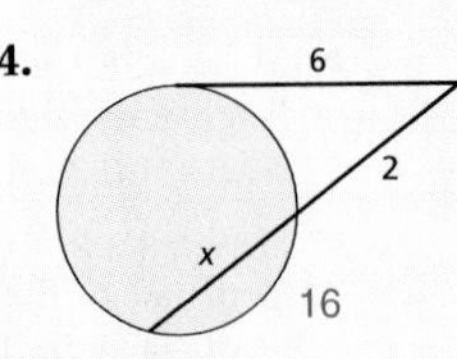

16

15.

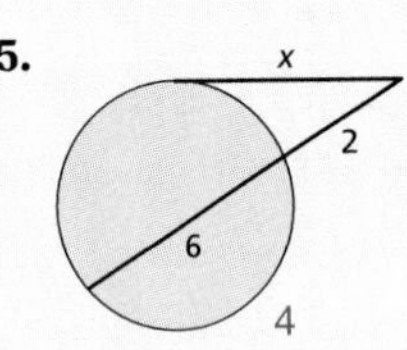

4

16.

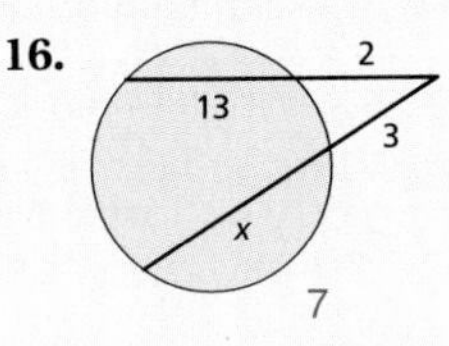

7

17.

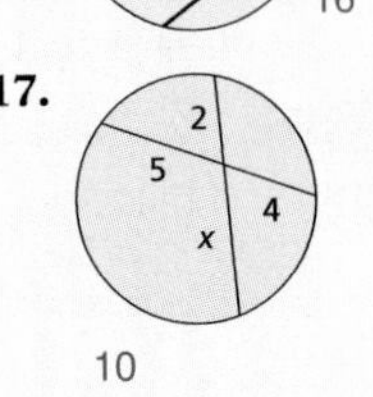

10

18. Construct a regular hexagon. Then inscribe a circle in it.
See additional answers.

Solve.

19. A 20-foot ladder is placed against a building. The base of the ladder is 4 feet from the base of the building. How high up the building does the ladder reach, to the nearest tenth of a foot? 19.6 ft

20. A survey of 200 department store customers showed that 24 had travelled more than 40 miles from home to the store, 52 travelled between 30 and 40 miles from home, 35 travelled between 20 and 30 miles, and 89 travelled less than 20 miles from home. Draw a circle graph that shows this data.
See additional answers.

460 | Chapter 10 **Assessment**

ADDITIONAL ANSWERS

18.

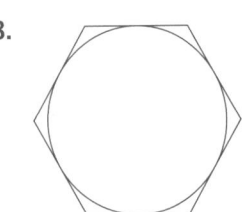

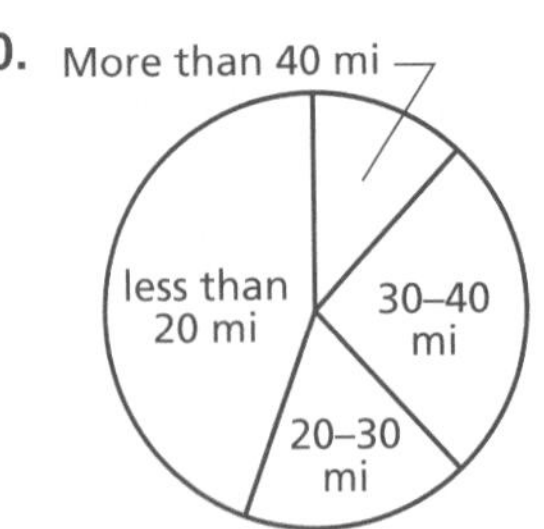

20. More than 40 mi

less than 20 mi
30–40 mi
20–30 mi

More than 40 mi—43°
30–40 mi—94°
20–30 mi—63°
less than 20 mi—160°

Chapter 10 Alternative Assessment

Technology

PRODUCT LOGO Use a computer graphics or drafting program to create a new geometric logo for a product of your choice. Write an explanation of how you chose the product, and how the logo represents the product. Name and describe the geometric shapes you used, and how the program helped you draw the shapes. If no computer program is available, draw the logo by hand and describe how you drew each portion of the logo.

Portfolio

PERSONAL POLYGONS Select 2 regular polygons. Draw each using the methods described in this chapter, and circumscribe a circle around each. Draw 2 other polygons and inscribe a circle in each. Write a step-by-step explanation of how you drew all the figures. Explain also what part of the exercise was most difficult for you, and how you took steps to overcome this difficulty.

Critical Thinking

BRAIN POWER Assume you are working as part of a construction crew. Your boss is laying out a building in preparation for digging the foundation.He asks you to help him check the diagonal of the rectangular building to insure that the 28′ by 42′ building is laid out with perfect 90° corners. Your calculator has no square root key. How can you find the length of the diagonal to the nearest $\frac{1}{16}$ of an inch? Describe the process you would use, and find the diagonal of the building.

Chapter Investigation

EXTENSION Make a presentation to your class of your design. Explain why you chose the design you did and how it will make your school more attractive.

Chapter Investigation

The benchmarks and expectations for this extension are as follows.

- Students make a detailed sketch of the front of the main building. They estimate measurements for the elements in the sketch. They consult building plans to find the length and width of standard doors and windows.
- Students research architectural styles that would complement their school grounds. They make a list of design elements that they could choose from.
- Students make a scale drawing of a new facade for a building on their school grounds. They use measurement approximations where necessary. They display their original drawing and their new design.
- Students present their design to the class. They explain why they chose the design and how it will make their school more attractive.

Technology

Encourage students to think of how the logo helps tell the public about the product.

SCORING RUBRIC 5—Student creates an innovative geometric logo for a product, draws it accurately, and can explain the steps in the process clearly and thoroughly. **4**—Student creates a geometric logo for a product, draws it accurately, and can explain the steps clearly. **3**—Student creates a geometric logo for a product, draws it fairly accurately, and can explain the steps. **2**—Student creates a geometric logo for a product, draws it fairly accurately, and can explain some of the steps. **1**—Student creates a geometric logo for a product, produces an incomplete drawing, and cannot explain the steps. **0**—Student makes no attempt to create or draw a logo.

Portfolio

You may want to assign students to choose many-sided polygons.

SCORING RUBRIC 5—Student accurately draws polygons and circles, then writes a precise and thorough explanation of the process and the difficulties. **4**—Student accurately draws polygons and circles and writes a good explanation of the process and the difficulties. **3**—Student draws polygons and circles with reasonable accuracy and writes a good explanation of the process and difficulties. **2**—Student draws polygons and circles with some accuracy and writes an adequate explanation. **1**—Student draws polygons and circles with questionable accuracy, but the explanation is incomplete. **0**—Student makes no attempt to draw polygons or write an explanation.

Critical Thinking

You may challenge students to use their methods to find the diagonal of the classroom.

SCORING RUBRIC 5—Student devises a method of finding square roots, correctly calculates the diagonal of the building, and can explain the method clearly and thoroughly. **4**—Student devises a method of finding square roots, correctly calculates the diagonal of the building, and can explain the method adequately. **3**—Student devises a method of finding square roots, makes minor mistakes in calculating the diagonal, and can explain the method. **2**—Student devises a method, makes minor mistakes calculating the diagonal, and has difficulty explaining the method. **1**—Student devises a method of finding square roots, makes significant mistakes calculating the diagonal, and cannot explain the method. **0**—Student makes no attempt to find square roots.

Cumulative Review

Cumulative review is the best way to maintain previously taught skills and concepts. This will keep students prepared for new lessons that build on previously covered skills.

Cumulative Review covers:

Lesson 1-3
Lesson 1-8
Lesson 2-3
Lesson 2-5
Lesson 3-2
Lesson 4-7
Lesson 5-2
Lesson 5-6
Lesson 6-1
Lesson 6-6
Lesson 7-2
Lesson 8-1
Lesson 8-5
Lesson 9-4
Lesson 9-7
Lesson 10-2
Lesson 10-6

Additional Answers

3.

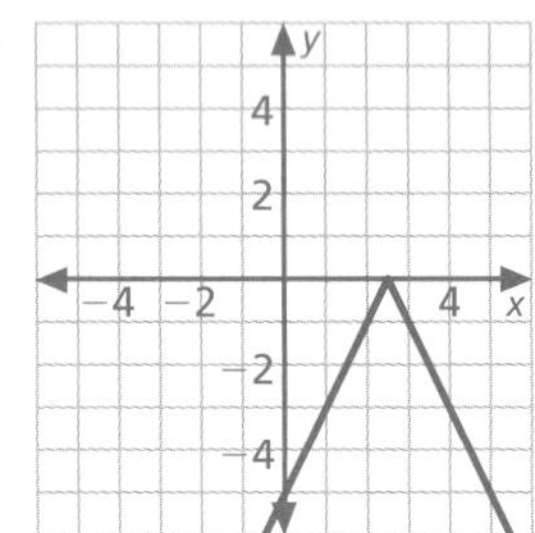

Chapters 1–10 Cumulative Review

1. Let $U = \{m, a, t, h, e, i, c, s\}$. $M = \{s, t, e, a, m\}$ and $N = \{t, h, e, m\}$. Find $M \cap N$. $\{t, e, m\}$
2. Simplify. $(a^{-2}b^{3})^{-3}$ $\frac{a^6}{b^9}$
3. Graph $f(x) = -|-2x + 5|$ See additional answers.
4. Translate the sentence into an equation using n to represent the unknown numbers. Then solve the equation for n. The quotient of a number and 3 is the same as the product of 2 and the same number increased by 4. $\frac{n}{3} = 2(n + 4); n = -4\frac{4}{5}$

Solve each equation.

5. $\frac{15}{9} = \frac{5}{6}a$ $a = 4\frac{2}{5}$
6. $-0.4x + 0.5 = 1.9 + 0.7x$ $x = 1.27$
7. In the figure at the right, $m\angle WSX$ is $(5x + 25)°$ and $m\angle TSX$ is $(2x - 15)°$. Find $m\angle RSW$. 79°
8. If you quadruple the length of the diameter of a circle, how does the area change? increased by 16 times
9. Find the unknown angle measurements. $x° = 70°$; $2x° = 140°$
10. Find the surface area of the figure. Assume all figures are regular. Round answer to nearest whole number. $48 + 8\sqrt{29}$ cm² ≈ 91.1 cm

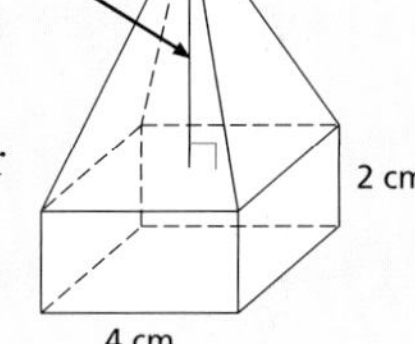

11. Find the slope of the line containing the points $R(-3, -1)$ and $T(-5, 4)$. $\frac{-5}{2}$
12. Kyle is three times as old as Andrea plus two years. In ten years, Kyle will be three years younger than twice as old as Andrea. Find the age of Kyle and Andrea in five years. Kyle: 22 yr; Andrea: 10 yr
13. The figure are similar. Find the measure of x. $x = \frac{5}{2} = 2\frac{1}{2}$
14. $\triangle ABC$ has vertices $A(3, 4)$, $B(5, 2)$ and $C(-1, -1)$. What are the vertices of $\triangle ABC$ under a reflection across the line $y = x$? $A'(4, 3)$, $B'(2, 5)$, $C'(-1, -1)$

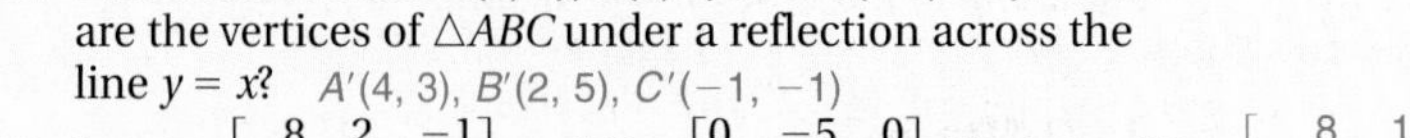

15. Let $A = \begin{bmatrix} 8 & 2 & -1 \\ -9 & 0 & 2 \end{bmatrix}$ and $B = \begin{bmatrix} 0 & -5 & 0 \\ 2 & 4 & 1 \end{bmatrix}$. Find $A - 2B$. $\begin{bmatrix} 8 & 12 & -1 \\ -13 & -8 & 0 \end{bmatrix}$
16. Find the probability of reaching in a drawer without looking and taking out a pair of white socks, when the drawer contains two pairs of black socks, eight pairs of white socks, and one pair of blue socks. $\frac{8}{11}$
17. Find the variance and standard deviation for the set of data: 3, 7, 5, 6, 6. 1.84; 1.36

Find the unknown length. Round your answer to the nearest tenth.

18.

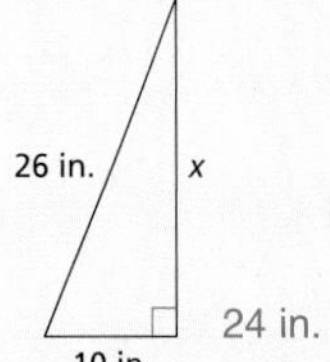

24 in.

19.

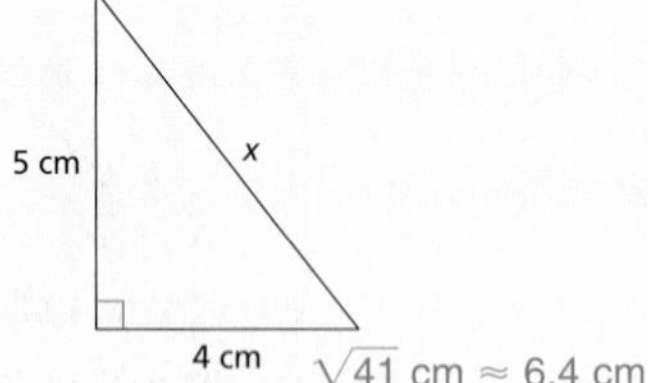

$\sqrt{41}$ cm ≈ 6.4 cm

20. Two chords of AB and CD intersect at point E. $AE = 2$ cm, $EB = 4$ cm, and $CE = 5$ cm. Find the measure of ED. $\frac{8}{5} = 1.6$ cm

Teaching Strategies

Exercise 8 in the Cumulative Review can be solved by experimenting with two circles with simple measurements. Remind students that acting out a problem, making a sketch or working with simpler numbers can be fast methods for solving a problem.

Chapters 1–10 Cumulative Assessment

Standardized Test Preparation—Standard Multiple Choice

Choose the best solution for each problem.

1. Write all the subsets of {R, A, T, E}.

D
A. {R, A}, {A, T}, {R, T}, {R, E}, {A, E}, {T, E}
B. {R, A, T, E}, {R, A, T}, {A, T, E}, {R, A, E}
C. {R}, {A}, {T}, {E}, ∅
D. all of these
E. none of these

2. Solve. $4(x - 1) = 2x + 4$

B
A. $x = 0$ **B.** $x = 4$
C. $x = \frac{4}{3}$ **D.** $x = -\frac{4}{3}$
E. none of these

3. In the figure below, $AD = 25$ and $AB = BC$. Find BC.

A

$3x + 2$ $2x - 3$
A B C D

A. 11 **B.** 9
C. 22 **D.** 3
E. none of these

4. The angles of a scalene triangle have the measures shown on the figure. What is the value of x?

C

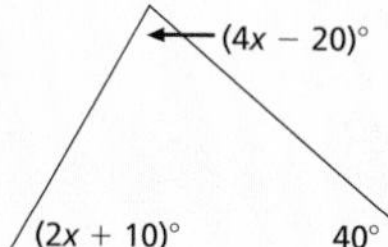

A. 60 **B.** 80
C. 25 **D.** 40
E. none of these

5. Write the ratio in lowest terms. 18 ft : 9 yd.

C
A. 2 **B.** $\frac{1}{2}$ **C.** $\frac{2}{3}$
D. $\frac{3}{2}$ **E.** $\frac{5}{3}$

6. Write an equation of a line hat is perpendicular to $2x + 3y = 1$ containing the point $P(1, -1)$.

D
A. $2y - 3x = 3$ **B.** $2y - 3x = 5$
C. $2y + 3x = 1$ **D.** $2y - 3x = -5$
E. none of these

7. Find the actual distance if the scale distance is 4.5 in. and the scale is 1\2 in. : 50 mi.

A
A. 450 mi **B.** 225 mi
C. 9 mi **D.** 2.01 mi
E. none of these

8. What is the location of A″ if you perform the following two transformations on $A(-5, -2)$: a translation of three units to the right and four units up followed by a reflection over the y-axis?

D
A. $A''(-2, 2)$ **B.** $A''(2, -2)$
C. $A''(1, 0)$ **D.** $A''(2, 2)$
E. $A''(0, -1)$

9. Let $M = \begin{bmatrix} 2 & 6 \\ 4 & -8 \end{bmatrix}$ and $N = \begin{bmatrix} -5 & 3 \\ 4 & -1 \end{bmatrix}$

B

Find $\frac{1}{2}M - N$.

A. $\begin{bmatrix} 6 & -5 \\ -3 & -1 \end{bmatrix}$ **B.** $\begin{bmatrix} 6 & 0 \\ -2 & -3 \end{bmatrix}$
C. $\begin{bmatrix} -4 & 8 \\ 7 & -7 \end{bmatrix}$ **D.** $\begin{bmatrix} 7 & 3 \\ 0 & -7 \end{bmatrix}$
E. none of these

10. The diagonal of a square measures 12 cm. Find the length of a side of the square.

B
A. $12\sqrt{2}$ cm **B.** $6\sqrt{2}$ cm
C. 12 cm **D.** 6 cm
E. $\sqrt{6}$ cm

Find x. (Lesson 10-6)

11.

18

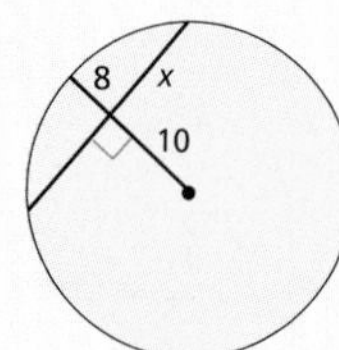

12. Possible answer: This question involves a combination since the order in which the students are chosen does not matter; there are 35 ways to select four students.

12. **CONSTRUCTED RESPONSE** Tell whether this question involves a permutation or a combination. Describe how you know which is needed. Then solve. In how many ways can four students be selected from a group of seven students?

Teaching Strategies

Review the meaning of permutation and combination. Have students suggest ways of remembering the difference. For example, when you order a combination plate at a restaurant, the order the cooks put the food on the plate does not matter. Have students brainstorm other memory tricks to remember these important definitions.

Standardized Test Preparation

STANDARD MULTIPLE CHOICE— One strategy for completing multiple-choice questions is to use mental math and eliminate unreasonable answers as choices. Often, a couple of the answer choices are values that you may get while working out the problem. For instance, if the solution requires two or three steps, one answer choice may be the value or expression that you have at the end of the first step and another choice may be the value or expression that you have at the end of the second step.

Sometimes all but the correct answer can be eliminated using only mental math and reasoning. If this is the case, verify your answer choice by working out the problem.

Testing Strategies

Caution students to read the geometric descriptions carefully. Remind them that it is never a waste of time to make a quick sketch of the figure in the description. A sketch makes it easy to see which measures are missing. Making a sketch often leads to seeing a correct method for solving the problem.

Chapter 11—Polynomials

Theme: Consumerism
Chapter Investigation: Market Profile, pages 467, 471, 481, 495, 501, 513
Careers: Brokerage Clerks, page 477; Actuaries, page 497
Data Activity: American Spending Habits

Content and Connections

Lesson, Pages	Lesson Objectives	NCTM Standards	State/Local Objectives	Interdisciplinary Connections	Real World Applications
11–1 468–471	• Write polynomials in standard form • Add and subtract polynomials	Num & Oper; Patterns, Functions & Algebra; Prob Solv; Reasoning & Proof; Connections; Representation		Art	Packaging, Transportation, Shipping
11–2 472–475	• Multiply polynomials by monomials	Num & Oper; Patterns, Functions & Algebra; Prob Solv; Reasoning & Proof; Communication Connections; Representation		Marketing	Advertising, Landscaping, Payroll
11–3 478–481	• Factor polynomials into a monomial factor and a polynomial factor	Num & Oper; Patterns, Functions & Algebra; Prob Solv; Reasoning & Proof; Communication Connections; Representation		Geometry	Manufacturing, Sculpture, Landscaping
11–4 482–485	• Multiply binomials	Num & Oper; Patterns, Functions & Algebra; Prob Solv; Reasoning & Proof; Communication Connections; Representation			Packaging, Small Business, Product Development
11–5 488–491	• Factor polynomials by grouping	Num & Oper; Patterns, Functions & Algebra; Prob Solv; Reasoning & Proof; Communication Connections; Representation			Manufacturing, Design, Sales
11–6 492–495	• Factor perfect square trinomials and difference of perfect squares • Use factoring to solve quadratic equations	Number & Operation Patterns, Functions & Algebra Prob Solv; Reasoning & Proof Connections; Representation		Art	Manufacturing, Landscaping, Art
11–7 498–501	• Factor trinomials with quadratic (x^2) coefficients of one	Number & Operation Patterns, Functions & Algebra Prob Solv; Reasoning & Proof Connections; Representation			Product Development, Construction, Chemistry
11–8 502–503	• Solve problems using the general case • Look for a pattern	Number & Operation Patterns, Functions & Algebra Prob Solv; Reasoning & Proof Connections; Representation			
11–9 506–509	• Factor trinomials of the form $ax^2 + bx + c$	Number & Operation Patterns, Functions & Algebra Prob Solv; Reasoning & Proof Connections; Representation			Small Business, Packaging, Consumerism

Planning and Resources

Lesson, Pages	Tools/Materials Needed	Trans-parency	Learning/Teaching Styles Options	Assignments: Basic Enriched	Additional Practice in SE	Reteaching, Extra Practice, Enrichment	Other Resources
11–1 468–471	pen/pencil	Warm up 31	ESL/LEP Real World	B: 1–35, 40–54 E: 1–54	Page 476–477, 487, 497, 505, 510, 717	R: page 159 EP: page 139 E: page 183	
11–2 472–475	paper/pencil	Warm up 31		B: 1–52, 55 E: 1–55	Page 476–477, 487, 497, 505, 510, 718	R: page 161 EP: page 141 E: page 185	SS: Teacher's Choice
11–3 478–481	Algeblocks	Warm up 31	Challenge	B: 1–37, 42–44 E: 1–44	Page 486–487, 497, 505, 510, 718	R: page 163 EP: page 143 E: page 187	SS: Teacher's Choice

11–4 482–485	paper/pencil	Warm up 32 Trans TK-4, RF-43	Tactile/Kinesthetic Learners Challenge	B: 1–45, 49–60 E: 1–60	Page 486–487, 497, 505, 510, 719	R: page 165 EP: page 145 E: page 189	MC: 4
11–5 488–491	paper/pencil	Warm up 32 Trans TK-4	Challenge	B: 1–38, 43–56 E: 1–56	Page 496–497, 505, 511, 719	R: page 167 EP: page 147 E: page 191	
11–6 492–495	Algeblocks	Warm up 32 Trans TK-4, RF-44	Visual Learners Challenge Prior Knowledge	B: 1–40, 43–50 E: 1–50	Page 496–497, 505, 511, 720	R: page 169 EP: page 149 E: page 193	SS: Teacher's Choice
11–7 498–501	paper/pencil	Warm up 33 Trans TK-4	Visual Learners	B: 1–53, 55–74 E: 1–74	Page 504–505, 511. 720	R: page 171 EP: page 151 E: page 195	
11–8 502–503	pencil/paper	Warm Up 33	Challenge	B: 1–16 E: 1–16	Page 504–505, 511	R: page 173 E: page 197	
11–9 506–509	pencil/paper	Warm Up 33 Trans TK-4	Prior Knowledge	B: 1–42, 45–68 E: 1–68	Page 511, 721	R: page 175 EP: page 153 E: page 199	

Planning and Pacing

Lesson, Pages	Lesson Title	45 min class	Assignments Basic, Enriched	90 min class	Assignments Basic, Enriched	___ min class	Assignments Basic, Enriched
11–1 468–471	AYR, Opener, Add and Subtract Polynomials	Day 139	B: 1–35, 40–54 E: 1–54	Day 74	B: 1–35, 40–54 E: 1–54		B: 1–35, 40–54 E: 1–54
11–2 472–475	Multiply by a Monomial	Day 139–140	B: 1–52, 55 E: 1–55	Day 74–75	B: 1–52, 55 E: 1–55		B: 1–52, 55 E: 1–55
11–3 478–481	Divide and Find Factors, R&PYS	Day 141	B: 1–37, 42–44 E: 1–44	Day 75	B: 1–37, 42–44 E: 1–44		B: 1–37, 42–44 E: 1–44
11–4 482–485	Multiply Two Binomials, R&PYS	Day 142–143	B: 1–45, 49–60 E: 1–60	Day 76	B: 1–45, 49–60 E: 1–60		B: 1–45, 49–60 E: 1–60
11–5 488–491	Find Binomial Factors in a Polynomial	Day 144	B: 1–38, 43–56 E: 1–56	Day 77	B: 1–38, 43–56 E: 1–56		B: 1–38, 43–56 E: 1–56
11–6 492–495	Special Factoring Patterns, R&PYS	Day 145–146	B: 1–40, 43–50 E: 1–50	Day 77–78	B: 1–40, 43–50 E: 1–50		B: 1–40, 43–50 E: 1–50
11–7 498–501	Factor Trinomials	Day 147	B: 1–53, 55–74 E: 1–74	Day 78	B: 1–53, 55–74 E: 1–74		B: 1–53, 55–74 E: 1–74
11–8 502–503	Problem Solving Skills: The General Case, R&PYS	Day 148	B: 1–16 E: 1–16	Day 79	B: 1–16 E: 1–16		B: 1–16 E: 1–16
11–9 506–509	More on Factoring Trinomials	Day 149	B: 1–42, 45–68 E: 1–68	Day 80	B: 1–42, 45–68 E: 1–68		B: 1–42, 45–68 E: 1–68
Review/ Assess		Day 150–151		Day 80–82			

	Chapter 1 Days	Chapter Cumulative Days
Yearly Pacing (45 min class)	13 days	151 days
Yearly Pacing (90 min class)	9 days	82 days

Assessment Options

Assessment in Student Edition	Assessment in Teacher's Edition	Pages in Assessment Book	Software Generated Assessment
Are You Ready?, pages 464–465; Writing Math, pages 470, 475, 480, 484, 490, 494, 501, 503, 509; Mixed Review, pages 471, 475, 481, 485, 491, 495, 501, 509; Check Understanding pages 468, 472, 478, 482, 488, 492, 498, 506; Mid-Chapter Quiz, page 487, Chapter Review, page 510; Chapter Assessment, page 512; Alternative Assessment, page 513; Cumulative Review, page 514; Cumulative Assessment, page 515	5-minute Warm ups, pages 468, 472, 478, 482, 488, 492, 498, 502, 516; Quick Assessment, pages 470, 474, 480, 484, 490, 494, 500, 503, 508; Scoring Rubrics, page 513	Mid-Chapter Quiz, page 167; Test Form A, pages 169, 171; Test Form B, pages 173, 175; Cumulative Test 177, 179; Math Journal prompt, page 181, 182;	Chapter 11

Refresher Skills

The skills on these two pages are skills that have been taught in previous math courses. Continuous review of basic math skills will make stronger math students. These skills are identified as necessary to be successful in Chapter 11.

Skills Correlation Chart

Skill	Lesson Number
Order of Operations	11-1, 11-2, 11-3
Simplify Exponents	11-2, 11-3, 11-4, 11-7, 11-8, 11-9
Prime Factoring	11-5, 11-6, 11-7, 11-8, 11-9

Vocabulary

order of operations
prime factor
binomial
trinomial

ARE YOU READY?

Refresh Your Math Skills for Chapter 11

The skills on these two pages are ones you have already learned. Stretch your memory and complete the exercises. For additional practice on these and more basic skills, see page 600.

ORDER OF OPERATIONS

No matter what aspect of mathematics you study, the order of operations always applies.

Example Simplify: $3(4 + 6) - 3^2 + 9 \div 3 \cdot 8$

1. First, simplify anything in parentheses and exponents.
2. Then multiply and divide from left to right.
3. Finally, add and subtract from left to right.

$3(4 + 6) - 3^2 + 9 \div 3 \cdot 8$
$3(10) - 3^2 + 9 \div 3 \cdot 8$
$3(10) - 9 + 9 \div 3 \cdot 8$
$30 - 9 + 9 \div 3 \cdot 8$
$30 - 9 + 3 \cdot 8$
$30 - 9 + 24$
$21 + 24$
45

Simplify each expression.

1. $25 \div 5 + 4 \cdot 2^2 - 15 \div 3$ 16
2. $18 \div 3 + 6 - 9 + 3 \cdot 9 \div 6$ $\frac{15}{2}$
3. $45 \cdot 3 \div 9 + 8^2 - 6^2 + 3$ 46
4. $15 \cdot 8 \div 40 - 3 + 16 - 2^2 + 18$ 30
5. $108 \div 12 \cdot 3^2 - 8 + 16 \div 2$ 81
6. $96 \div 4 + 3^2 - (5 + 3) + 11$ 36
7. $64 \div (8 \div 2) \cdot 3 - 6^2 + 4$ −20
8. $12 \cdot 9 \div 3 - 8^2 + 7^2 - (14 + 8)$ −1
9. $3 \cdot 8 \cdot 4 \cdot 2^2 \div (8 \div 4) + 17$ 209
10. $(9 \cdot 2) + (3 \cdot 4) - (4^2 \div 2) + 37$ 59

SIMPLIFY EXPONENTS

Simplify each expression. Assume that $a = 0$, $b = 0$ and $c = 0$.

11. $(a^2)(a^3)(a^4)$ a^9
12. $(a^2 \cdot a^9)^2$ a^{22}
13. $(a^2b^6c^4)^3$ $a^6b^{18}c^{12}$
14. $[(a^2)^3]^5$ a^{30}
15. $\frac{a^9}{a^4}$ a^5
16. $\frac{a^7b^6}{ab^4}$ a^6b^2
17. $\frac{a^{12}}{a^9}$ a^3
18. $(a^4b^6c^7)^8$ $a^{32}b^{48}c^{56}$
19. $(a^2)^4(a^3)^2(a)^4$ a^{18}
20. $\frac{(a^3 \cdot a^4 \cdot a^2)}{a^5}$ a^4
21. $\frac{a^9b^7c^8}{a^6b^2c^5}$ $a^3b^5c^3$
22. $\frac{(a^5b^9c^4)^3}{(a^3b^2c^3)^2}$ $a^9b^{23}c^6$

Extend the Lesson

MATH JOURNAL Have students describe the process they use to find the prime factorization of the numbers 90, 110 and 1,275.
$90 = 2 \cdot 3 \cdot 3 \cdot 5$
$110 = 2 \cdot 5 \cdot 11$
$1{,}275 = 3 \cdot 5 \cdot 5 \cdot 17$

Prime Factoring

In this chapter you will learn to factor binomial and trinomial expressions. It may be helpful to practice this "un-multiplying" skill on simpler numbers.

Examples **Find the prime factors of 36.** (Two methods are shown.)

You know that $6 \times 6 = 36$.

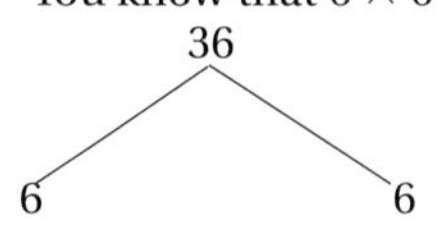

You know that $4 \times 9 = 36$.

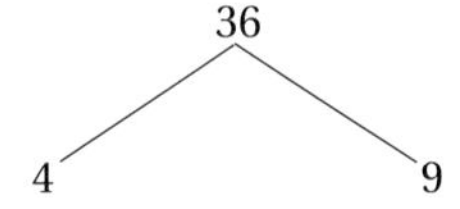

You know that $2 \times 3 = 6$.

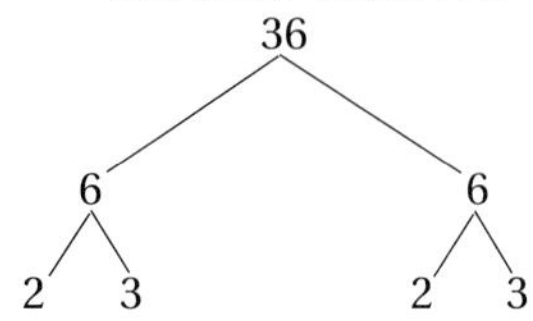

You know that $2 \times 2 = 4$ and $3 \times 3 = 9$.

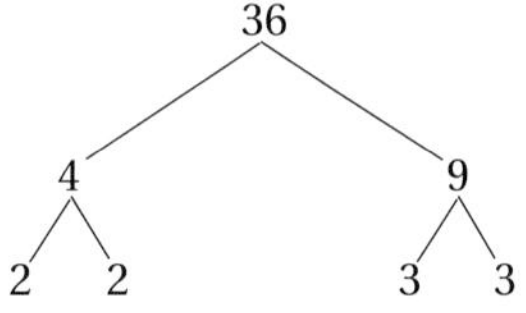

Both methods result in the same answer. Since 2 and 3 are both prime numbers, no more factoring is possible.

The prime factors of 36 are 2, 2, 3, 3. Written as a product of primes, it is $2^2 \cdot 3^2$.

Find the prime factors of 156. (Two methods are shown.)

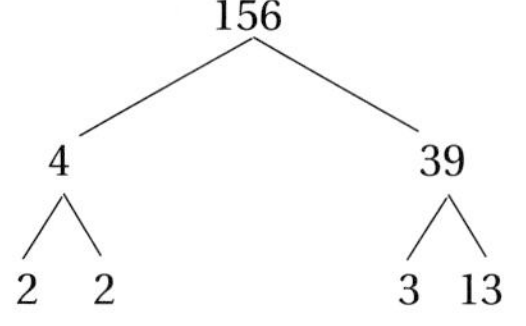

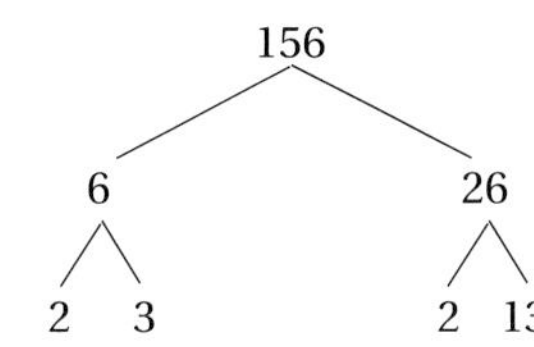

The prime factors of 156 are 2, 2, 3, 13. Written as a product of primes, it is $2^2 \cdot 3 \cdot 13$.

Find the prime factors of each number.

23. 81 3, 3, 3, 3 **24.** 74 37, 2 **25.** 100 2, 2, 5, 5 **26.** 69 3, 23

27. 58 2, 29 **28.** 44 2, 2, 11 **29.** 29 29 **30.** 68 2, 2, 17

31. 75 3, 5, 5 **32.** 32 2, 2, 2, 2, 2 **33.** 99 3, 3, 11 **34.** 84 2, 2, 3, 7

35–46. Write Exercises 23–34 as a product of primes. Use exponents when possible.

35. 3^4 **36.** $2 \cdot 37$ **37.** $2^2 \cdot 5^2$ **38.** $3 \cdot 23$

39. $2 \cdot 29$ **40.** $2^2 \cdot 11$ **41.** 29 **42.** $2^2 \cdot 17$

43. $3 \cdot 5^2$ **44.** 2^5 **45.** $3^2 \cdot 11$ **46.** $2^2 \cdot 3 \cdot 7$

Chalkboard Examples

Order of Operations

$48 \div 3 + 5^2 - (6 + 9) + 7 \cdot 2$ 40

Simplify Exponents

$(x^3)^2 (x^2)^4 (x^5)$ x^{19}

Prime Factoring

Find the prime factors of 168. 2, 2, 2, 3, 7

Refresher Wrap-up

Quick Assessment

Ask the following questions to determine if students have mastered the basic skills reviewed on these pages.

1. Why is it important to apply the order of operations each time you simplify an expression? Without the order of operations, two mathematicians could arrive at different answers although their procedures are correct.
2. Explain why $\frac{x^6}{x^2} = x^4$ instead of x^3. Writing the expression as $\frac{\cancel{x} \cdot \cancel{x} \cdot x \cdot x \cdot x \cdot x}{}$ is one way to show why the rules of exponents work.

Additional Practice

All topics: Refer to the Extra Practice Index on page 674 of this text.

Extend the Lesson

COOPERATIVE LEARNING Have students work in pairs to explore how the addition of parentheses could can affect the order of operations. For Exercises 1–10, have students change the answer by adding one pair of parentheses. Exchange lists with other teams and solve.

Chapter Opener

NCTM Standards/Strands

- Number & Operation
- Patterns, Functions, & Algebra
- Problem Solving
- Reasoning and Proof
- Communication
- Connections
- Representation

Vocabulary

consumerism marketing
apparel
demographical profile

Theme Connections

Tables and graphs are used to organize data about products, buying trends and other aspects of consumerism. Tables provide a way to organize numerical information. Graphs are often used to show trends in sales and performance of profits as well as consumers' spending habits and preferences. Discuss how data and graphs are used in marketing and consumerism.

Career Opportunities

Many careers related to consumerism make use of data and graphs. Two are highlighted in the MathWorks features. Others include advertising agents, product developers, engineers, designers, and sales managers.

- Brokerage Clerks, page 477
- Actuaries, page 497

Internet Connection

Theme Activities

Learningmathmatters.com provides web addresses to search that help students gather information about the use of math in the real world, particularly data and graphs. To search for additional addresses, begin a search using the keyword *consumerism*. Then within that search, use such key words as *marketing, advertising* and *spending habits.* Students can brainstorm in small groups other key words.

CHAPTER 11

POLYNOMIALS

THEME: Consumerism

You would probably be surprised at the number of advertisments and commercials you see daily. Nearly one-fourth of every television hour is commercial time. Some radio stations devote 1 out of every 3 minutes to advertising.

How do companies decide which products to make and sell? Across America, businesses spend millions of dollars everyday to find out what consumers want and need. Marketing executives gather data about the spending habits and patterns of consumers in every age group. Product developers design new products for specific groups of consumers, and advertisers create exciting campaigns to convince the consumer to try the new product.

- **Brokerage Clerks** (page 477) assist in the buying and selling of stocks, bonds, commodities and other types of investments. They monitor clients' accounts, make sure dividends are paid and check the accuracy of the paperwork used in making transactions.
- **Actuaries** (page 497) work for insurance companies to assemble and analyze statistical data about consumers in order to estimate the probabilities of death, sickness, injury and property loss. This information helps insurance companies predict costs and charges for insurance coverage.

Internet Connection
www.learningmathmatters.com

Chapter Investigation

Go to learningmathmatters.com to locate additional information about consumerism.

As a Chapter Project

The goal of this project is to make and describe a proportional sketch of the subject of a photograph. Students can use the Group Project Planner on page 181 and the Project Planning Calendar on page 182 in the Enrichment text to complete the project. Benchmarks **a**, **b**, **c** and **d**. should be completed after the lesson listed in parentheses has been studied. Benchmark **e** should be completed at the end of the chapter.

American Spending Habits
Average Annual Expenses Per Household

Expense item	1995	1996	1997
Food at home	$2,803	$2,876	$2,880
Food away from home	1,702	1,823	1,921
Housing	10,458	10,747	11,272
Apparel and services	1,704	1,752	1,729
Transportation	6,014	6,382	6,457
Health care	1,732	1,770	1,841
Entertainment	1,612	1,834	1,813
Insurance and pensions	2,964	3,060	3,223
Other	3,273	3,555	3,684
Total average annual expenses	**$32,262**	**$33,799**	**$34,820**

Source: Bureau of Labor Statistics, U.S. Dept. of Labor

Data Activity: American Spending Habits

Use the table for Questions 1–4.

1. In which category was there the greatest percent increase from 1995 to 1997? Food away from home
2. The government determined that there were 105,576,000 households in 1997. To the nearest million, how much was spent on apparel and services in 1997? $182,541,000,000
3. Which category demonstrated nearly a 14% increase from 1995 to 1996? Entertainment
4. To the nearest tenth, what percent of a households' total expenses were health care costs in 1997? 5.3%

Chapter Investigation

Demographics are the statistical characteristics of a particular population. Advertising decisions are often made based on the demographical profile of a market. For instance, car manufacturers generally buy commercial time during television programs that are watched by adult viewers.

Working Together

Conduct a survey to gather demographical information about your classmates. You will need to gather information about their viewing and listening preferences (television and radio), as well as their product preferences and brand loyalties. Discuss how the compiled results could be used by advertisers and manufacturers to sell products. Use the Chapter Investigation logos to guide your group.

Data Activity

Ask students how spending decisions are made in their household. How do they decide what products on which to spend their money? Discuss with students the importance to advertisers and product manufacturers of understanding the spending profiles of their customers. Explain that the table shows broad categories of expenses on which Americans spend their income. Assign students Questions 1–4.

Extend the Data Activity

REAL WORLD CONNECTION Have students choose one product and pay attention for one week to the strategies advertisers use to promote the product. In addition to commercials and print ads, students can investigate how the product is displayed in stores and how it is packaged.

Chapter Investigation

As an Overarching Problem

Display several print advertisements on a bulletin board in your room. Ask students to study the advertisements and choose the one they find most appealing. Have students share their choices with the class along with reasons why they find it most appealing. Students will continue to work on the investigation as they complete the exercises identified by the Investigation Logo that are found throughout the chapter. These exercises will guide students through the task described in *Working Together*. Encourage students to keep all of their work on the Investigation together. Have students use the suggestions in the Chapter Investigation Extension to summarize their work.

See page 466 for Chapter Investigation As a Chapter Project.

Project Planning Calendar

Name ______ Date ______

CHAPTER 11 PROJECT PLANNING CALENDAR

Benchmarks	PROJECT GOAL
a. Suppose you have developed a new product targeted for consumers your own age. Brainstorm a list of questions that can be used in a survey to find out about your classmates' shopping interests and spending habits. *(Lesson 11-1)* b. Add questions to your survey to find out how much time each day your classmates spend watching television, listening to the radio, and reading newspapers and magazines. *(Lesson 11-3)* c. Distribute the final survey to your classmates and compile the data. Use the information to create a demographic profile of your class. ...ategy for	To plan an advertising strategy for marketing a new product aimed at people your own age.

Group Project Planner

Name ______ Date ______

CHAPTER 11 GROUP PROJECT PLANNER

Assignment ______ Objective ______

Group Members Assigned Roles
1) ______
2) ______
3) ______
4) ______
5) ______

Deadlines Done

Lesson Planning

NCTM Standards/Strands

- Number & Operation
- Patterns, Functions, & Algebra
- Problem Solving
- Reasoning and Proof
- Connections
- Representation

Vocabulary

polynomial
monomial
coefficient
constant
binomial
trinomial
standard form

Materials Needed

paper/pencil

Lesson Resources

Warm-Up Transparency 31
Reteaching 11-1
Extra Practice 11-1
Enrichment 11-1

Assignment Guide

Basic: 1–35, 40–54
Enriched: 1–54

Getting Started

5-Minute Warm-up

Simplify.

1. $6a - 4a + 5a$ $7a$
2. $-8b - 2b - (-3b)$ $-7b$
3. $4c^2 + 9c^2 + (-5c^2)$ $8c^2$
4. $5a^2 + 6b^2 - 3a^2$ $2a^2 + 6b^2$

Introduction to Lesson 11-1

After students have completed the opening activity, discuss how a number written in expanded form is like a polynomial and how it is different. Have students who do not understand that $(10)^0$ is 1 complete this pattern: $(10)^3 = ?$, $(10)^2 = ?$, $(10)^1 = ?$, $(10)^0 = ?$. 1000; 100; 10; 1

11-1 Add and Subtract Polynomials

Goals
- Write polynomials in standard form.
- Add and subtract polynomials.

Applications Packaging, Transportation, Shipping

Work with a partner to answer the following questions.

Polynomials are expressions with several terms that follow patterns, such as $4x^3 + 3x^2 + 15x + 2$. Now consider the number 3,946. As you know, the digits indicate 3 thousands, 9 hundreds, 4 tens, and 6 ones. Remember that one hundred is 10^2 and one thousand is 10^3. Can you see a connection between polynomials and our place value number system?

1. The number 3,946 can be expressed as $3(10)^3 + 9(10)^2 + 4(10) + 6$. As you can see, this expression is similar to the polynomial pattern—the only difference is that a 10 is used instead of an x. Using this idea, write 62, then 832, and then 14,791 in polynomial form. $6(10) + 2$; $8(10)^2 + 3(10) + 2$; $1(10)^4 + 4(10)^3 + 7(10)^2 + 9(10) + 1$
2. Write 1,001 so that it looks like a polynomial. Omit the terms that are multiplied by zero. $1(10)^3 + 1$
3. Is it correct to say that $493 = 4(10)^2 + 9(10)^1 + 3(10)^0$? yes
4. Find the value of the polynomial $9x^3 + 7x^2 + 5x + 3$ if $x = 10$? 9753

Build Understanding

Review the words used to discuss polynomials. A simple expression with only one term is called a **monomial**. A monomial is either a number or the product of a number and one or more variables. For example, $4x^3$ is a monomial. Other monomials are 15, m, ab, and $13p^2q$. If a monomial includes any variables, the number part is called the **coefficient** of the term, and is written first. A number by itself is called a **constant**.

A polynomial is an expression that contains several monomial terms. If it has two terms, it is a **binomial**. With three terms, it is a **trinomial**. The expression $a^4 + 3b$ is a binomial; $6h^3 + 4gh + 39$ is a trinomial. Other polynomials may have more than three terms. For example, $8s^4 - 5s^3t + s^2t^2 + 6st^3 - 7t^4$ is also a polynomial.

Like terms are terms in which the variables or sets of variables are identical—though the coefficients may be different. Learn to recognize like terms, and do not be confused by unlike terms.

like terms:	$3b^2$ $15b^2$ $(-b^2)$
	$8x^3y$ $-14x^3y$ $25x^3y$
unlike terms:	$15a$ $15b$
	$15b^2$ $12b$
	$8x^3y$ $8xy^3$ $8x^3y^3$

Teaching Strategies

ESL/LEP To help students differentiate between the terms *monomial, binomial, trinomial,* and *polynomial*, have them work in small groups to make lists of common words having the prefixes *mono-*, *bi-*, *tri-*, and *poly-*. Then discuss how these words relate to *one, two, three,* and *many*, respectively.

You **simplify** a polynomial when you group and then combine all like terms.

$$4a^2 + 3bc - a^2 + 5c^2 + 9bc = (4a^2 - a^2) + (3bc + 9bc) + 5c^2$$
$$= 3a^2 + 12bc + 5c^2$$

A polynomial is in **standard form** if the terms are ordered from the greatest power of one of its variables to the least power of that variable.

$15x + 13 - 9x^2 + 2x^3 = 2x^3 - 9x^2 + 15x + 13$

To add polynomial expressions, place both expressions in parentheses with an addition sign between them, then simplify the combined expression and put it in standard form.

Problem Solving Tip

Putting polynomials in simplified and standard form will help you match the terms for adding and subtracting.

Example 1

Add: $8a^2b + 6ab^2$ to $4a^2b - 3ab^2$

Solution

$$(8a^2b + 6ab^2) + (4a^2b - 3ab^2) = 8a^2b + 6ab^2 + 4a^2b - 3ab^2$$
$$= (8a^2b + 4a^2b) + (6ab^2 - 3ab^2)$$
$$= 12a^2b + 3ab^2$$

Another way to add polynomials is to setup the problem in vertical form with like terms aligned in columns.

Example 2

PACKAGING The cost of the materials for the inner packaging of a new product is determined by the expression $10x^2 + 8xy + y^2$. The cost of the outer packaging materials is $4x^2 - 3xy + 2$. Find the total cost of the packaging.

Solution

$$\begin{array}{l} 10x^2 + 8xy + y^2 \\ 4x^2 - 3xy \qquad + 2 \\ \hline 14x^2 + 5xy + y^2 + 2 \end{array}$$

To subtract polynomials, place the expressions in parentheses with a minus sign between them, then simplify and standardize.

Think Back

Remember that to subtract an expression, you change all the signs and then add.

Example 3

Subtract:

a. $5m^2 - 2m$ from $8m^2 + m$

b. $s^2 + 3s - 4$ from $3s^2 - 5s - 3$

Solution

a. $(8m^2 + m) - (5m^2 - 2m) = 8m^2 + m + (-5m^2) + (2m)$
$= 8m^2 + (-5m^2) + m + (2m)$
$= 3m^2 + 3m$

Chalkboard Examples

Supplementary Example 1

Add $z^2 + 3z - 4$ to $2z^2 - 5z + 6$.
$3z^2 - 2z + 2$

Supplementary Example 2

The cost for materials for a project is modeled by the expression $x^2 + 3x + 5$. The cost for labor is $5x^2 - 3x - 2$. Find the total cost. Align the terms in columns before adding. $6x^2 + 3$

Supplementary Example 3

Subtract $2r - 4s$ from $4r + 5s$.
$2r + 9s$

Learning Styles

TACTILE/KINESTHETIC LEARNER Have students work in small groups using Algeblocks or algebra tiles to model adding and subtracting polynomials. To add $(3x^2 + 3x + 7)$ and $(2x^2 + x + 5)$, have them model each polynomial and count and record the total number of each type of tile. To subtract these polynomials, have them model the first polynomial, remove the tiles indicated by the second, and count and record the remaining tiles.

Reteaching Worksheet 11-1

Name __________ Date __________

RETEACHING 11-1

ADDING AND SUBTRACTING POLYNOMIALS

You can **simplify** a polynomial when you group and then combine all of its **like terms,** such as:
$5x^2 + 3xy + y^2 - 5xy + 2x^2 + 3xy + 3y^2 - 9x^2 + 7xy - 7xy = -2x^2 + xy + 4y^2$.

To add two polynomials, combine their like terms and write the polynomial in standard form.

To subtract two polynomials, add the opposite of the polynomial being subtracted to the other polynomial. Then write in standard form.

Example 1

Add $5x^2 + 3xy - y^2$ and $2x^2 + xy + 9$.

Solution

$$\begin{array}{l} 5x^2 + 3xy - y^2 + 0 \\ + 2x^2 + xy + 0y^2 + 9 \\ \hline 7x^2 + 4xy - y^2 + 9 \end{array}$$

Line up like terms. Write 0 where no like terms exist.
Combine like terms.

Example 2

Subtract $4x^2 - 3x + 2$ from $8x^2 - 5x - 8$.

Solution

First line up like terms. Then change signs and add.

$$\begin{array}{rcl} 8x^2 - 5x - 8 & = & 8x^2 - 5x - 8 \\ -(4x^2 - 3x + 2) & = & +(-4x^2 + 3x - 2) \\ \hline & & 4x^2 - 2x - 10 \end{array}$$

EXERCISES

Simplify.

1. $(3b - 6) + (4b^2 - 6b + 10)$ $4b^2 - 3b + 4$
2. $(4a + b) + (2a - 3b)$ $6a - 2b$
3. $(7m^2 + 8mn - 9) + (2m^2 - 10mn + 1)$ $9m^2 - 2mn - 8$
4. $(-3c^2 + 12cd - 7) + (5c^2 - 9cd + d)$ $2c^2 + 3cd + d - 7$
5. $(7a^2 - 3a + 5) - (-a^2 + 4a - 10)$ $8a^2 - 7a + 15$
6. $(5b^2 + 7bc - 9c^2) - (b^2 + 9bc + 2c^2)$ $4b^2 - 2bc - 11c^2$
7. $(7t^2 - 5t) - (-4t^2 + 3t - 7)$ $11t^2 - 8t + 7$
8. $(7x^2 + xy - 3y^2) - (-4x^2 + 7xy + 12)$ $11x^2 - 6xy - 3y^2 - 12$
9. $(8j^2 - 4j + 10) + (2j^2 - 8j + 2)$ $10j^2 - 12j + 12$
10. $(-4m^2 + 2mn - n^2) - (2m^2 + 3mn - 18)$ $-6m^2 - mn - n^2 + 18$
11. $(8k^3 - 6k^2 + 12) - (3k^3 + 5k + 10)$ $5k^3 - 6k^2 - 5k + 2$
12. $(5ab^2 - 2ab + 4a^2b) + (-4ab + 2a^2b - 8)$ $5ab^2 - 6ab + 6a^2b - 8$

160 RETEACHING *LESSON 11-1*

Lesson Wrap-up

Quick Assessment

Ask the following questions to determine if students understand the content presented in this lesson.

1. How can you recognize that a polynomial has been simplified and is written in standard form? All terms are unlike and ordered from greatest to least power of one variable.
2. Does each simplified polynomial have to have the same number of terms as the polynomials that are added or subtracted? No; if the sum or difference of two terms is zero, their sum or difference will not appear in the simplified polynomial.

Assignment Guide

Basic: 1–35, 40–54
Advanced: 1–54
Additional Practice: See Extra Practice Index on page 674.

b. $(3s^2 - 5s - 3) - (s^2 + 3s - 4)$
$= 3s^2 - 5s - 3 + (-s^2) + (-3s) + 4$
$= 3s^2 + (-s^2) + (-5s) + (-3s) + (-3) + 4$
$= 2s^2 - 8s + 1$

You may also setup a subtraction problem vertically.

$$\begin{array}{r} 3s^2 - 5s - 3 \\ -\ s^2 - 3s + 4 \\ \hline 2s^2 - 8s + 1 \end{array}$$

Change the signs of each term.

Try These Exercises

Write each answer as a simplified polynomial in standard form for the variable x.

1. $x + 3x^3 - 4 + x^2 + 2x^3$ $5x^3 + x^2 + x - 4$
2. $4 + x^2 + 3 - 2x^2 + 4x^2$ $3x^2 + 7$
3. Add $x^2 + 3$ to $3x^2 + 7$ $4x^2 + 10$
4. Add $7 - 2x^2$ to $5x^2 - 3$ $3x^2 + 4$
5. $(5x^2 - 7x) + (x^2 + 3x)$ $6x^2 - 4x$
6. $(4x^3 + 7) + (3x^3 - 4)$ $7x^3 + 3$
7. Subtract $3x + 4$ from $5x - 3$ $2x - 7$
8. Subtract $x - 4$ from $5 + 3x$ $2x + 9$
9. $(2x + 14) - (x - 7)$ $x + 21$
10. $(5x^2 - 5x) - (5x^2 - 5x)$ 0
11. $(15x^3 + 12x^2 - 3xy) - (8x^3 - 3x^2 + 2xy)$ $7x^3 + 15x^2 - 5xy$
12. $(x + 6x^2y - 3x - 4x^3) + (x^2y + x^2 + 5x)$ $-4x^3 + 7x^2y + x^2 + 3x$

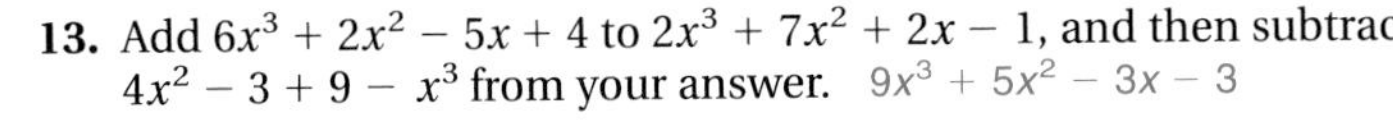

13. Add $6x^3 + 2x^2 - 5x + 4$ to $2x^3 + 7x^2 + 2x - 1$, and then subtract $4x^2 - 3 + 9 - x^3$ from your answer. $9x^3 + 5x^2 - 3x - 3$

14. **WRITING MATH** Explain how subtraction of polynomials is related to addition of polynomials. Subtraction is adding an opposite.

Practice Exercises

Simplify.

15. $(2a + 4) + (3a + 9)$ $5a + 13$
16. $(5p + q) + (2p + 2q)$ $7p + 3q$
17. $(3x^2 + 2x) + (-x^2 + 5x)$ $2x^2 + 7x$
18. $(4h - 2g + k) + (h + 3j - 2k)$ $5h - 2g + 3j - k$
19. $(5t + 7) - (3t + 2)$ $2t + 5$
20. $(3r + 2s) - (2r - s)$ $r + 3s$
21. $(4m^2 + 3n) - (-m^2 + 3n)$ $5m^2$
22. $(y^2 - 2y + 3) - (-y^2 + y - 5)$ $2y^2 - 3y + 8$
23. $(y^2 - 15x + 2x^2) - (7x - 2y^2 + x^2)$ $x^2 - 22x + 3y^2$
24. $(12r^2 - 12rs + s^2) + (3r^2 - 4s)$ $15r^2 - 12rs - 4s + s^2$
25. $(4v^2 - 9w^2) - (v^2 + 2vw + w^2)$ $3v^2 - 2vw - 10w^2$
26. $(x^4 + 3x^2 + 2x) + (8x^3 + 4x)$ $x^4 + 8x^3 + 3x^2 + 6x$
27. $(2b^2 - 15 + c) - (-c - 4b^2)$ $6b^2 + 2c - 15$
28. $(-3f^2 + 4fg + g^2) - (4f^2 - g^2)$ $-7f^2 + 4fg + 2g^2$

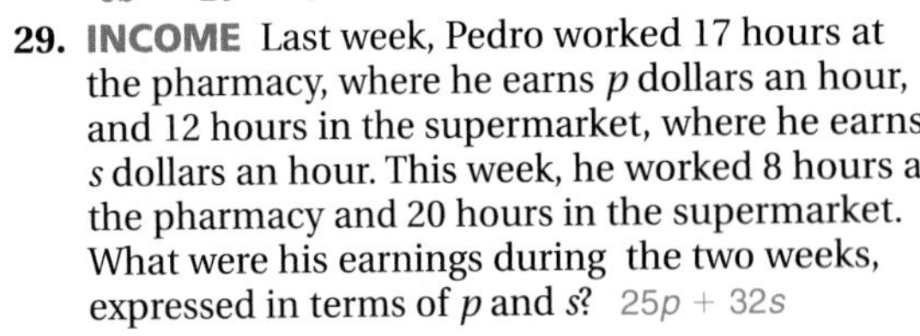

29. **INCOME** Last week, Pedro worked 17 hours at the pharmacy, where he earns p dollars an hour, and 12 hours in the supermarket, where he earns s dollars an hour. This week, he worked 8 hours at the pharmacy and 20 hours in the supermarket. What were his earnings during the two weeks, expressed in terms of p and s? $25p + 32s$

470

Math: Who, Where, When

Blaise Pascal lived in France during early colonial times (1623–1662). He was a scientist, philosopher, and mathematician who saw many connections among different disciplines. Not only did he develop ideas about arithmetic, algebra, geometry, physics, and religion, he also discovered the principle behind hydraulic brakes and invented the first calculating machine. The computer language PASCAL is named after him.

Extend the Lesson

REAL-WORLD CONNECTION Have students look through science and/or business books to find an example of a monomial, a binomial, and a trinomial. Have them copy each monomial or polynomial and tell what it represents. If it has more than one variable, have them try to identify what each variable represents.

30. **TRANSPORTATION** Airplane A uses $35d^2 + 3dr - 4r^2$ gallons of fuel to make a trip. Airplane B uses $16d^2 + 45dr - 13r^2$ gallons. How much less fuel does airplane B use than airplane A? $19d^2 - 42dr + 9r^2$

31. $(z^2 - 3z + 4) + (3z^2 + 2z - 2) - (4z^2 - z - 2)$ 4

32. $-(122 + 7x + 2x^2) + (32 - 14x + 15x^2)$ $-90 - 21x + 13x^2$

33. $(4.2a^3 - 3.6b^3 + 8.8bc^2) - (4.2a^2b - 2.1a^3 + 3bc^2 - 1.9b^3)$ $6.3a^3 - 4.2a^2b + 5 \cdot 8bc^2 + 1.76b^3$

34. $[5(10)^3 + 6(10)^2 + 3(10)^0] - [2(10)^3 + 8(10)^2 + 4(10)^1 + 10^0]$ $3(10)^3 - 2(10)^2 - 4(10)^1 + 2(10)^0$ or 2, 762

35. $2(8x^2 - 5x + 3) - (10x^2 - 16x + 2) + (13x^2 - 4)$ $19x^2 + 6x$

Extended Practice Exercises

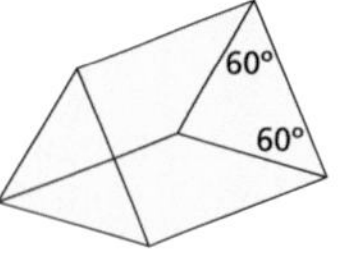

36. **ART** The prism sculpture shown at the right is being shipped to a museum exhibit. The artist plans to build a wooden frame to protect the edges of the sculpture during shipping. Each triangle side is equal to $x^2 + y$ feet and each long edge is $2x^2 - 3y$ feet. How many feet of wood will the artist need to protect the edges? $12x^2 - 3y$

37. **SHIPPING** Janine's truck starts the day with a cargo of 54 large cubic boxes measuring x feet each way. Each box contains z packages measuring 1 foot by x feet by y feet. In addition, 48 more of these packages are squeezed into the corners—the truck is totally full. At her first delivery, she drops off 12 large boxes—but she removes 3 packages from one of the boxes to keep on the truck for another customer. How much space is available on the truck after her first delivery, in terms of x and y? $12x^3 - 3xy$

38. The octal system of counting contains only eight digits. The number written 342, therefore, means only $3(8)^2 + 4(8) + 2$, not $3(10)^2 + 4(10) + 2$. Calculate $765 - 301$ in octal numbers, then convert answer to our own decimal system. $4(8)^2 + 6(8) + 4$; 464 octal; 308 decimal

39. **CHAPTER INVESTIGATION** Suppose you have developed a new product targeted for consumers your own age. What do you know about the spending habits of people in your age group? Begin development of a survey to gather demographical information about your classmates. Working with your group, brainstorm a list of questions that can be used in a survey to find out information about your classmates' shopping interests and spending habits.

Mixed Review Exercises

Find the value to the nearest hundredth. (Lesson 10-1)

40. $\sqrt{52}$ 7.21
41. $\sqrt{75}$ 8.66
42. $\sqrt{83}$ 9.11
43. $\sqrt{216}$ 14.70

Write each in simplest radical form. (Lesson 10-1)

44. $\left(4\sqrt{2}\right)^2$ 32
45. $\left(5\sqrt{7}\right)^2$ 175
46. $\dfrac{\sqrt{50}}{\sqrt{3}}$ $\dfrac{5\sqrt{6}}{3}$
47. $\dfrac{\sqrt{96}}{\sqrt{5}}$ $\dfrac{4\sqrt{30}}{5}$
48. $\left(2\sqrt{3}\right)\left(3\sqrt{2}\right)$ $6\sqrt{6}$
49. $\left(4\sqrt{6}\right)\left(3\sqrt{3}\right)$ $36\sqrt{2}$
50. $\left(5\sqrt{2}\right)\left(2\sqrt{6}\right)$ $20\sqrt{3}$
51. $\left(4\sqrt{8}\right)\left(\sqrt{7}\right)$ $8\sqrt{14}$

Write each number in scientific notation. (Lesson 1-8)

52. 0.0000000743 7.43×10^{-8}
53. 32,000,000,000 3.2×10^{10}
54. 0.000000904 9.04×10^{-7}

Extend the Lesson

ONGOING ASSESSMENT Have students simplify: $(3x^2 - 5) + (2x^2 + 1)$. Watch for students who get the answer $5x^4 - 4$. These students added both the coefficients and the exponents of the like terms with variables, rather than the coefficients only.

Extra Practice Worksheet 11-1

Name ______________________ Date ____________

EXTRA PRACTICE 11-1

Adding and Subtracting Polynomials

Exercises

Simplify.

1. $(4z + 2) + (5z - 3)$ $9z - 1$
2. $(5x + y) + (2x + 6y)$ $7x + 7y$
3. $(6 - 5m) - (8 + m)$ $-2 - 6m$
4. $(5z^2 + 4x) + (5x - 4z^2)$ $z^2 + 9x$
5. $(9m + 7m^2) - (2m - 5m^2)$ $7m + 12m^2$
6. $(x^2 + 2xy - y^2) + (4x^2 - xy + 5y^2)$ $5x^2 + xy + 4y^2$
7. $(4z^2 - 3z + 5w^3) - (2z^2 + 4z - 4w^3)$ $2z^2 - 7z + 9w^3$
8. $(6a^2 - 5b^2) - (-b^2 + 5a^2 - 12)$ $a^2 - 4b^2 + 12$
9. $(8r^2 - 6t^2) - (9r^2 - 4rt + 7t^2)$ $-r^2 + 4rt - 13t^2$
10. $(-5a^3 + 5a^2 + 6a - 1) + (3a^3 - 7a - 5)$ $-2a^3 + 5a^2 - a - 6$
11. $(z^5 - 2z^4 + 5z - 16) - (4z^2 - z + 6)$ $z^5 - 2z^4 - 4z^2 + 6z - 22$
12. $(8b^2 - 4b + 3) - (-5b + 3b^2)$ $5b^2 + b + 3$
13. $(-6r^3 + 2r^2 - 6r - 4) - (r^3 + r^2 + 6)$ $-7r^3 + r^2 - 6r - 10$
14. $(m^2 + 3m - 4) + (2m - 6) - (4m^2 - 5m)$ $-3m^2 + 10m - 10$
15. $(7w + 2w^2 - 5) - (6w - 4) + (w^3 + w^2 + 6)$ $w^3 + 3w^2 + w + 5$
16. $(3.5c^2 - 4.5c + 1.2) + (4.6c^2 + 3.2c - 2.5)$ $8.1c^2 - 1.3c - 1.3$
17. $3(4x^2 + 4x - 3) + (9x^2 - 2x + 9)$ $21x^2 + 10x$
18. $(6r^2 + 7r - 8) - 2(r^2 + r + 1)$ $4r^2 + 5r - 10$
19. $-(50 + y - y^2) + (46 - 16y + y^2)$ $2y^2 - 17y - 4$
20. $[r - (-6s)] - (r^2 + 3r + 4s)$ $-r^2 - 2r + 2s$

140 EXTRA PRACTICE *LESSON 11-1*

Enrichment Worksheet 11-1

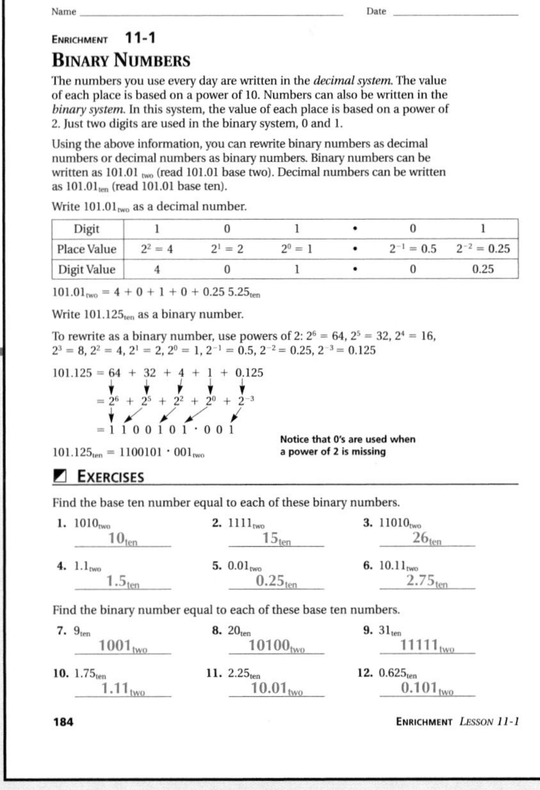

Name ______________________ Date ____________

ENRICHMENT 11-1

Binary Numbers

The numbers you use every day are written in the *decimal system*. The value of each place is based on a power of 10. Numbers can also be written in the *binary system*. In this system, the value of each place is based on a power of 2. Just two digits are used in the binary system, 0 and 1.

Using the above information, you can rewrite binary numbers as decimal numbers or decimal numbers as binary numbers. Binary numbers can be written as 101.01_{two} (read 101.01 base two). Decimal numbers can be written as 101.01_{ten} (read 101.01 base ten).

Write 101.01_{two} as a decimal number.

Digit	1	0	1	•	0	1
Place Value	$2^2 = 4$	$2^1 = 2$	$2^0 = 1$	•	$2^{-1} = 0.5$	$2^{-2} = 0.25$
Digit Value	4	0	1	•	0	0.25

$101.01_{two} = 4 + 0 + 1 + 0 + 0.25\ 5.25_{ten}$

Write 101.125_{ten} as a binary number.

To rewrite as a binary number, use powers of 2: $2^6 = 64$, $2^5 = 32$, $2^4 = 16$, $2^3 = 8$, $2^2 = 4$, $2^1 = 2$, $2^0 = 1$, $2^{-1} = 0.5$, $2^{-2} = 0.25$, $2^{-3} = 0.125$

$101.125 = 64 + 32 + 4 + 1 + 0.125$
$= 2^6 + 2^5 + 2^2 + 2^0 + 2^{-3}$
$= 1\ 1\ 0\ 0\ 1\ 0\ 1 \cdot 0\ 0\ 1$

$101.125_{ten} = 1100101 \cdot 001_{two}$

Notice that 0's are used when a power of 2 is missing

Exercises

Find the base ten number equal to each of these binary numbers.

1. 1010_{two} 10_{ten}
2. 1111_{two} 15_{ten}
3. 11010_{two} 26_{ten}
4. 1.1_{two} 1.5_{ten}
5. 0.01_{two} 0.25_{ten}
6. 10.11_{two} 2.75_{ten}

Find the binary number equal to each of these base ten numbers.

7. 9_{ten} 1001_{two}
8. 20_{ten} 10100_{two}
9. 31_{ten} 11111_{two}
10. 1.75_{ten} 1.11_{two}
11. 2.25_{ten} 10.01_{two}
12. 0.625_{ten} 0.101_{two}

184 ENRICHMENT *LESSON 11-1*

Lesson Planning

NCTM Standards/Strands

- Number & Operation
- Patterns, Functions, & Algebra
- Problem Solving
- Reasoning and Proof
- Communication
- Connections
- Representation

Vocabulary

associative property
commutative property

Materials Needed

paper/pencil

Lesson Resources

Warm-Up Transparency 31
Reteaching 11-2
Extra Practice 11-2
Enrichment 11-2

Assignment Guide

Basic: 1–52, 55
Enriched: 1–55

Getting Started

5-Minute Warm-up

Write each answer in standard form.

1. $4^2 \cdot 4^3$ 1,024
2. $(4^2)^3$ 4,096
3. $(4 \cdot 2)^3$ 512

Introduction to Lesson 11-2

Have students compare their expression in Steps (c) and (d) to their expressions in Steps (a) and (b). Ask them what property is illustrated by this. distributive property

11-2 Multiply by a Monomial

Goals ■ Multiply polynomials by monomials.

Applications Advertising, Landscaping, Payroll

Work with a partner to answer the following questions.

From your knowledge of geometry, you know that the area of a rectangle is calculated by multiplying width by length. In the diagram shown at the right:

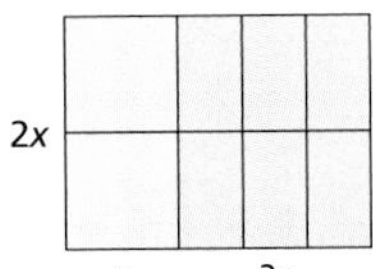

a. Express the area of the yellow section of the diagram, in terms of x. There is more than one possible answer. $(2x)(x)$ or $2x^2$

b. Express the area of the orange section of the diagram, in terms of x and y. $6xy$

c. Express the area of the whole diagram, in terms of x and y. $2x^2 + 6xy$

d. Trace the diagram and cut out the pieces. Use the pieces to form a different rectangle with the same area. Write expressions to represent the length and width of the new rectangle. How could you use the expressions to find the area? Answers will vary.

Build Understanding

When you multiply a polynomial by a monomial, the answer always has the same number of terms as the original polynomial. To understand this, begin with the idea that a monomial is a product of constants and variables. If you multiply two monomial products, you will always get another product that is a monomial.

This is clear in (**a**) above: $(2x)(x) = (2)(x)(x) = 2x^2$. (Remember the associative property of multiplication?) It may be less easy to see in (**b**): $(2x)(3y)$, because of the two coefficients in the initial expression. But by the commutative property, the expression equals $(2)(3)(x)(y)$, or $6xy$. You can also see this in the diagram above.

Example 1

Simplify.

a. $(8a)(3b)$ **b.** $(3m)(-2n)$ **c.** $(-2x)(-5x^2)$

Solution

a. $(8a)(3b) = (8)(a)(3)(b) = (8)(3)(a)(b) = 24ab$

b. $(3m)(-2n) = (3)(m)(-2)(n) = (3)(-2)(m)(n) = -6mn$

c. $(-2x)(-5x^2) = (-2)(x)(-5)(x)(x) = (-2)(-5)(x)(x)(x) = 10x^3$

When you multiply a binomial by a monomial, the answer will remain a binomial. This is because each term of the binomial must be multiplied by the monomial. There is one multiplication for each term.

Problem Solving Tip

Remember that when you multiply two negative terms together, the answer will be positive.

Teaching Strategies

Some students may find it helpful to use a column format when multiplying a polynomial by a monomial. Example 2 could be written:

$$\begin{array}{r} 3x - 4 \\ \times \quad 2x \\ \hline 6x^2 - 8x \end{array}$$

Example 2

ADVERTISING To promote a new product, a company buys $2x$ minutes of airtime. The cost of one minute of airtime is $3x - 4$. Multiply to find an expression which represents the cost of advertising the new product on television.

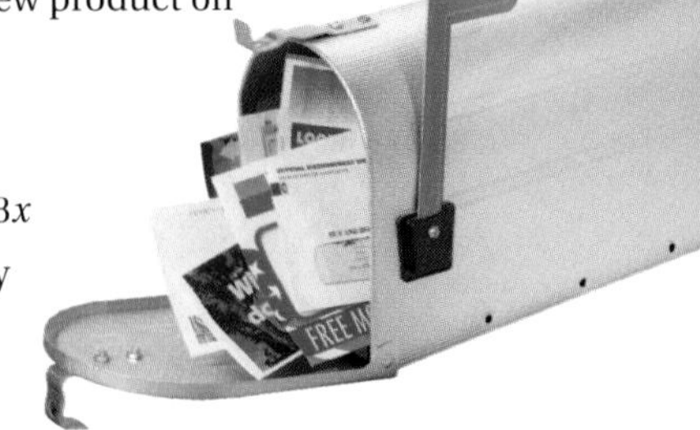

Solution

$$2x(3x - 4) = (2x)(3x) + (2x)(-4) = 6x^2 + (-8x) = 6x^2 - 8x$$

When you multiply polynomials (including trinomials) by a monomial, the answer will have the same number of terms as the polynomial. There is still one multiplication for each term.

Example 3

Simplify.

a. $3v^2(v^2 + v + 1)$

b. $12(a^2 + 3ab^2 - 3b^3 - 10)$

Solution

a. $3v^2(v^2 + v + 1) = (3v^2)(v^2) + (3v^2)(v) + (3v^2)(1)$
$= 3v^4 + 3v^3 + 3v^2$

b. $12(a^2 + 3ab^2 - 3b^3 - 10) = 12(a^2) + 12(3ab^2) + 12(-3b^3) + 12(-10)$
$= 12a^2 + 36ab^2 - 36b^3 - 120$

When you multiply $2x$ and $3y$, you first analyze each monomial into its simplest, **prime**, elements. Prime elements, including prime numbers, cannot be divided into smaller whole elements. To multiply $(2x)(3y)$, you thought $(2)(x)(3)(y)$, which was easily reorganized as $(2)(3)(x)(y)$, and then $6xy$. This type of analysis can also help you find **factors**, elements whose product is a given quantity.

Example 4

GEOMETRY List three possible sides of a rectangle with an area of $12x^2y$.

Solution

As you know, the area of a rectangle is the product of its length and width. To find a complete set of paired factors for the given area, start by analyzing its prime elements. Express the coefficient in prime numbers and separate the variables. The area $12x^2y$ is analyzed as $(2)(2)(3)(x)(x)(y)$.

Now use the analysis to find different factor pairs or sets of sides. Set up a table. The second factor contains all the elements not in the first factor.

First factor/side	Second factor/side
$(y) = y$	$(2)(2)(3)(x)(x) = 12x^2$
$(2)(2)(x) = 4x$	$(3)(x)(y) = 3xy$
$(2)(3)(x)(x) = 6x^2$	$(2)(y) = 2y$

There are many possible sets of factors.

Extend the Lesson

ONGOING ASSESSMENT Have students simplify: $3n(4n^2 - 2)$. Watch for students who get the answer $12n^3 - 2$. These students multiplied only the first term of the polynomial by the monomial instead of applying the distributive property to both terms of the polynomial.

Chalkboard Examples

Supplementary Example 1

Simplify: $(5a)(-3b)$ $-15ab$

Supplementary Example 2

The cost of ten minutes of radio airtime is determined by the expression $-x^3(-x^3 + 5x^2)$. Simplify the expression. $x^6 - 5x^5$

Supplementary Example 3

Simplify: $3g^3(2g^2 - g + 1)$
$6g^5 - 3g^4 + 3g^3$

Supplementary Example 4

A rectangle has an area of $3x^2y^2 - 2x^2$. Find one set of possible dimensions of the rectangle.
Answers may vary. One possible solution: x^2 and $3y^2 - 2$.

Reteaching Worksheet 11-2

Name ______________ Date ______________

RETEACHING 11-2

MULTIPLYING BY A MONOMIAL

When you multiply a polynomial by a monomial, the answer always has the same number of terms as the polynomial. Remember to multiply each term of the polynomial by the term of the monomial.

Example 1

Simplify $(7ab)(8c)$.

Solution

$(7ab)(8bc) =$
$(7)(a)(b)(8)(b)(c) =$
$(7)(8)(a)(b)(b)(c) =$
$56ab^2c$

Example 2

Simplify $2x^2(x^2 + 3x - 5)$.

Solution

$2x^2(x^2 + 3x - 5) =$
$2x^2(x^2) + (2x^2)(3x) - (2x^2)(5) =$
$2x^4 + 6x^3 - 10x^2$

EXERCISES

Simplify.

1. $a(abc)$ a^2bc
2 $(8xy)(9y^2z)$ $72xy^3z$
3. $(4m^2)(8mn^2)$ $32m^3n^2$
4. $3b(b - 8)$ $3b^2 - 24b$
5. $3m^2(m - 2n)$ $3m^3 - 6m^2n$
6. $x^2(a - b)$ $ax^2 - bx^2$
7. $-9d(d + 6)$ $-9d^2 - 54d$
8. $-2a^2b(3ab^2 - 7b)$ $-6a^3b^3 + 14a^2b^2$
9. $4x(x^2 + 3x - 6)$ $4x^3 + 12x^2 - 24x$
10. $7n^2(8m^2n - 7mn - 6n)$ $56m^2n^3 - 49mn^3 - 42n^3$
11. $-8a^3(3ab^2 - 2b + b^2)$ $-24a^4b^2 + 16a^3b - 8a^3b^2$
12. $-12x^2y^2(3x - 4xy + 2y)$ $-36x^3y^2 + 48x^3y^3 - 24x^2y^3$
13. $14x(4x^2 - 3x + 9)$ $56x^3 - 42x^2 + 126x$
14. $25m(-8m^2 + 6m - 4)$ $-200m^3 + 150m^2 - 100m$
15. $8abc^2(a^2bc - a^2b^3 - a)$ $8a^3b^2c^3 - 8a^3b^4c^3 - 8a^2bc^2$

162 RETEACHING *LESSON 11-2*

Lesson 11-2 **Multiply by a Monomial** 473

Lesson Wrap-up

Quick Assessment

Ask the following questions to determine if students understand the content presented in this lesson.

1. In Example 1, Part (c), what rule for exponents can be applied? The product rule for exponents: When you multiply like constants or variables, you can add their exponents.
2. In Example 3, look at the answers for Parts (a) and (b). Why is there a variable in each term of Part (a) but not in each term of Part (b)? The polynomial in Part (a) is multiplied by a monomial containing a variable; the polynomial in Part (b) is multiplied by a constant.

Assignment Guide

Basic: 1–52, 55
Advanced: 1–55
Additional Practice: See Extra Practice Index on page 674.

Additional Answers

42. In arithmetic, the single-digit number is multiplied by each place value column in the multi-digit number and the results are added. In algebra, the monomial multiplier is distributed over the terms in the polynomial and the results are added.
43. $2b^2(3a)$ $2(3ab^2)$ $3(2ab^2)$ $6(ab^2)$ $a(6b^2)$ $2a(3b^2)$ $3a(2b^2)$ $6a(b^2)$ $b(6ab)$ $2b(3ab)$ $3b(2ab)$ $6b(ab)$
44. Your classmate has added instead of multiplying the exponents. The term $(-x^2)^2$ is equal to $(-x^2)(-x^2)$ or $(-x)(-x)(-x)(-x)(-x)(-x)$ or $(-x^6)$.

Try These Exercises

Simplify.

1. $(x)(3y)$ $3xy$
2. $(a^2)(2a)$ $2a^3$
3. $(4p)(3q)$ $12pq$
4. $(3v^2)(2vw)$ $6v^3w$
5. $(-r)(-s^2)$ rs^2
6. $(-5xy)^2$ $25x^2y^2$
7. $7(x^2 + x)$ $7x^2 + 7x$
8. $2y(y + z)$ $2y^2 + 2yz$
9. $a^2(a^2 + a)$ $a^4 + a^3$
10. $4pq(p - 2r)$ $4p^2q - 8pqr$
11. $-e^2f(e + f^2)$ $-e^3f - e^2f^3$
12. $-13mn^3(2m^2 - n)$ $-26m^3n^3 + 13mn^4$
13. $a(b^2 + b - 6)$ $ab^2 + ab - 6a$
14. $3u(u^2 + uv + 2v^2)$ $3u^3 + 3u^2v + 6uv^2$
15. $-7x(x^2 - 2xy + y^2)$ $-7x^3 + 14x^2y - 7xy^2$
16. $5ef^3(h + 3j + k^2)$ $5ef^3h + 15ef^3j + 5ef^3k^2$
17. **MARKETING** A mailing list has x people from 14 to 18 years of age, y people from 19 to 25 years of age and z people from 26 to 40 years of age. A company decides to spend x dollars per person on the list to advertise its new product line. How much will the advertising cost the company? $x^2 + xy + xz$

Practice Exercises

Simplify.

18. $(2a)(3b)$ $6ab$
19. $(x^2)(3xy)$ $3x^3y$
20. $(-j)(-3jk)$ $3j^2k$
21. $(4x^3)(-3x^2y)$ $-12x^5y$
22. $(6m^2n)(5mn^2)$ $30m^3n^3$
23. $(3a^2)^2$ $9a^4$
24. $-7q(3q^2 - 5r)$ $-21q^3 + 35qr$
25. $2x^2[-(3x^2 + 2x)]$ $-6x^4 - 4x^3$
26. $5rs(3r^4 + 5s^3)$ $15r^5s + 25rs^4$
27. $-3mn^2(m^3n - m^4n^3)$ $-3m^4n^3 + 3m^5n^5$
28. $2x^2y(4x^3z - 3xz^4)$ $8x^5yz - 6x^3yz^4$
29. $8ef^2g(eg^3 - fg^3)$ $8e^2f^2g^4 - 8ef^3g^4$
30. $4abc(a^2b^3c + ab^4c^2)$ $4a^3b^4c^2 + 4a^2b^5c^3$
31. $-18lmn^4(l^2mn^3 - lm^5n)$ $-18l^3m^2n^7 + 18l^2m^6n^5$
32. $3x(x^2 + 4x - 5)$ $3x^3 + 12x^2 - 15x$
33. $ab(4e^2 - 2f + g)$ $4abe^2 - 2abf + abg$
34. $-pq^2(3p^2 - pq + 10q^2)$ $-3p^3q^2 + p^2q^3 - 10pq^4$
35. $4v^2w(3u + 2v + w^3)$ $12uv^2w + 8v^3w + 4v^2w^4$
36. $-l^4[-(3l + 5m)]$ $3l^5 + 5l^4m$
37. $7rs^3t^2(r^4st^3 - r^3s^2t^2 - r^2s^5t)$ $7r^5s^4t^5 - 7r^4s^5t^4$

Write and simplify an expression for the area of each rectangle.

38. $12a^2b^2c$

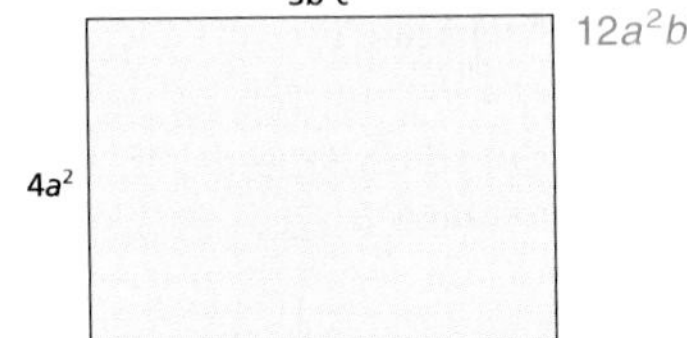

39. $4x^2 + 2xy + 2xz$

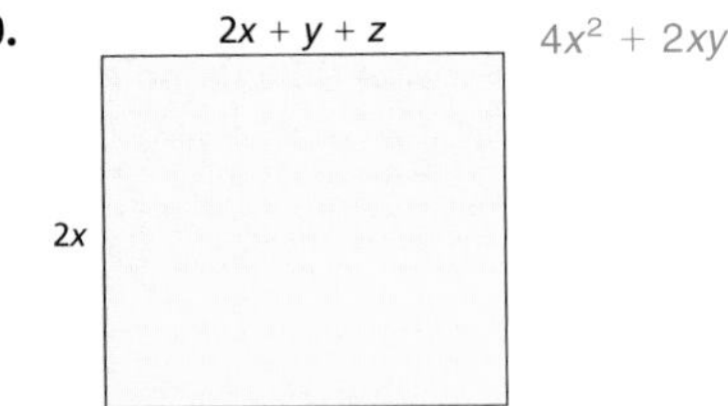

40. **PAYROLL** In 1990, a growing company employed c clerks, each of whom earned d dollars each week. The weekly pay rate increased by r dollars each year. Two years later, the number of clerks on staff had tripled. What was the total paid each week to the clerical staff in 1990? What was it in 1992? Simplify both answers if possible. cd $6cr + 3cd$

Extend the Lesson

COOPERATIVE LEARNING Have students write a problem similar to Exercises 51 and 52 based on a situation in their own lives. Then have students trade problems with partners and solve.

41. **CONSTRUCTION** A builder estimates that, for a typical office building, the height of each story is h feet from floor to floor, and the length of a building averages k feet per room. A company wants a structure that is 5 stories tall and has 12 rooms along the front; but each room is to be 3 feet longer than the standard. Estimate the area of the front wall of the building. $60hk + 180h$ sq ft

42. **WRITING MATH** How is algebraic multiplication of a monomial and a polynomial similar to arithmetic multiplication of a single-digit number and a multi-digit number? See additional answers.

43. Find the prime elements of $6ab^2$ and use them to list all positive factor pairs. (Hint: There are 12 pairs in all.) See additional answers.

44. **ERROR ALERT** A classmate says that $(-x^3)^2$ is equal to $(-x^5)$. Analyze the problem by writing the expression as the product of prime elements. What mistake has your classmate made? See additional answers.

Simplify.

45. $(x^2y)(xy^3)(xy^2)$ x^4y^5

46. $(m^2n^4)(m^4n^2) - (m^3n^3)^2$ 0

47. $(-a^3)^2 - (-a^2)^3$ $2a^6$

48. $2pq(p + q) - p^2q(2 + q)$ $2pq^2 - p^2q^2$

49. $(5x^2)(3y)(x^2 - xy + y^2)$ $15x^4y - 15x^3y^2 + 15x^2y^3$

50. $3r(2r - 5s + t) + 6s(3r - s + 2t)$ $6r^2 + 3rs + 3rt - 6s^2 + 12st$

51. **TRANSPORTATION** Alva travelled for t hours at s miles per hour, then for twice that time at $(s + 10)$ miles per hour. How many miles did she travel in all? (Remember, distance = rate × time) $3ts + 20t$

52. **LANDSCAPING** A lawn has two flower gardens with the dimensions shown below. Write an expression for the area of grass left, then simplify. $96xy - 4x$

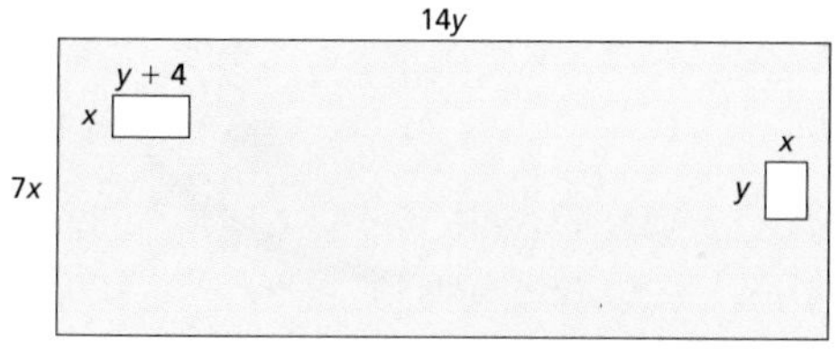

Extended Practice Exercises

53. **ARCHEOLOGY** An archaeologist finds a square-based pyramid rising in the Mexican jungle. From corner to corner, it is 60 p (paces), and from each corner to the top is 50 p. What is the total surface area of its triangular sides, expressed in terms of p? $4{,}800p^2$

54. Using the diagram on the right, find factored expressions for three areas: the shaded area, the unshaded area, and the total area. Then simplify each expression.
$2a(3x + 2y) = 6ax + 4ay$
$4b(3x + 2y) = 12bx + 8by$
$(2a + 4b)(3x + 2y) = 6ax + 12by + 4ay + 8by$

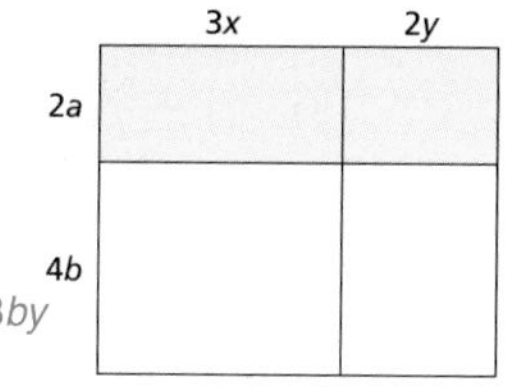

Mixed Review Exercises

55. Three brothers, named Jarius, Keshawn, and Levon play football for the Cheetahs, the Gophers, and the Goats, not necessarily in that order. Jarius scored 2 touchdowns against the Cheetahs, but none against the Goats. Keshawn hasn't played against the Cheetahs yet. For which team does each brother play? (Lesson 3-8)
Jarius plays for the Gophers, Keshawn, for the Goats, and Levon for the Cheetahs.

Extend the Lesson

MATH JOURNAL Have students write the steps for multiplying two monomials and multiplying a monomial and a polynomial. Have them include an example for each.

Extra Practice Worksheet 11-2

Name ______________________ Date __________

EXTRA PRACTICE 11-2
MULTIPLY BY A MONOMIAL

EXERCISES

Simplify.

1. $(5x)(3y)$ $15xy$
2. $(z^3)(4wxz)$ $4wxz^4$
3. $(5h^2)(6h^3k)$ $30h^5k$
4. $(8x^2y)(-2xy^2)$ $-16x^3y^3$
5. $(2a^2)^3$ $8a^6$
6. $(4mn)(3m^2n)^2$ $36m^5n^3$
7. $5w(2w^4 + 4x)$ $10w^5 + 20wx$
8. $2y^3(y^2 - 7)$ $2y^5 - 14y^3$
9. $8ab(2a^3 - b^4)$ $16a^4b - 8ab^5$
10. $-2xy^3(3x^2z - 2xy^4)$ $-6x^3y^3z + 4x^2y^7$
11. $3abc(a^2 + b^2 + 8c^2)$ $3a^3bc + 3ab^3c + 24abc^3$
12. $b^2c^3(4bc - 5b^3 + 8c^4)$ $4b^3c^4 - 5b^5c^3 + 8b^2c^7$
13. $(x^3y)(2xy)(x^3y^4)$ $2x^7y^6$
14. $(3r^2s^3)(rs)^3 - (4r^3s^4)^2$ $3r^5s^6 - 16r^6s^8$
15. $4mn(m^2 + m^3n^4) + m^3n(4m^4 + mn^3)$ $4m^3n + 4m^4n^5 + 4m^7n + m^4n^4$

Write and simplify an expression for the area of each rectangle.

16.
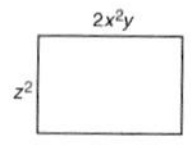

$2x^2yz^2$

17.
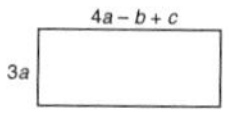

$12a^2 - 3ab + 3ac$

142 EXTRA PRACTICE *LESSON 11-2*

Enrichment Worksheet 11-2

Name ______________________ Date __________

ENRICHMENT 11-2
SQUARED RECTANGLES

A squared rectangle is one that has been divided up into squares, no two of the same size. The expression or number inside each square equals the length of its side.

EXERCISES

1. This squared rectangle is made of squares with sides that measure 36, 33, 28, 25, 16, 9, 7, 5 and 2. Let $x = 2$ and $y = 5$. Write an algebraic expression for the side of each square.

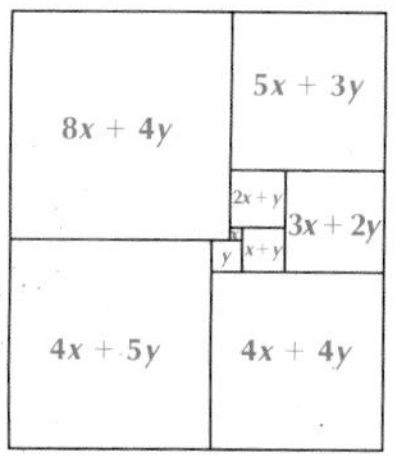

2. This squared rectangle is made of squares with sides that measure 99, 78, 77, 57, 43, 41, 34, 25, 21, 16 and 9. Let $x = 16$ and $y = 9$. Write an algebraic expression for the side of each square.

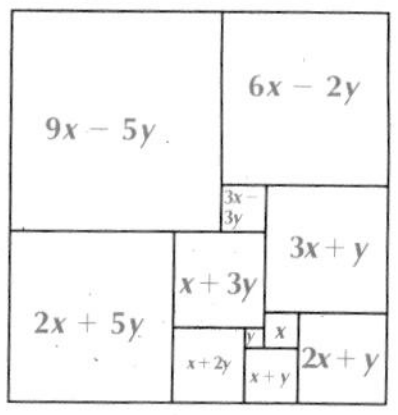

3. What must be true of x in the figure at the right? Does the figure show a squared rectangle?
The left and right sides must be equal, so $5x + 2y = x + 2y$. Solving for x gives $x = 0$. So the center square is really a point; the large rectangle is really a square divided in four quarters. Thus, it is not a squared rectangle.

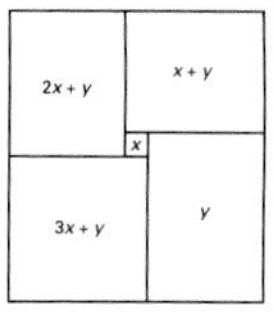

186 ENRICHMENT *LESSON 11-2*

Skills Practice

Vocabulary Review

Lesson 11-1
polynomial
monomial
coefficient
constant
binomial
trinomial
standard form

Lesson 11-2
associative property
commutative property

Assignment Guide

All students: 1–65
Additional Practice: See Extra Practice Index on page 674.

Review and Practice Your Skills

Practice ■ Lesson 11-1

Simplify.

1. $(8x + 3y) + (7y - 2x)$ $6x + 10y$
2. $(13b + 6) + (7b - 14)$ $20b - 8$
3. $(4x^2 - 9x + 6) + (12x^2 + 5x - 13)$ $16x^2 - 4x - 7$
4. $\left(\frac{1}{2}k + \frac{3}{4}g\right) + \left(\frac{3}{8}k - \frac{3}{4}h\right)$ $\frac{7}{8}k + \frac{3}{4}g - \frac{3}{4}h$
5. $(4x - 6z) - (6x - 4z)$ $-2x - 2z$
6. $(-3m + 4n - p) - (6n - 7m + p)$ $4m - 2n - 2p$
7. $(8x^3 - 5x^2 + 2x) - (6x^2 - 3x^3 + 10x)$ $11x^3 - 11x^2 + 12x$
8. $(4a^2 + ab + 7b^2) - (8ab + 5b^2)$ $4a^2 - 7ab + 2b^2$
9. $[y^2 - (-5y)] - (3y^2 + 6y + 1)$ $-2y^2 - 11y - 1$
10. $(14r^2 - 10rs + 15s^2) + (-8r^2 + 7s^2)$ $6r^2 - 10rs + 22s^2$
11. $(x^2y + xy^2) + (3x^2y - 2xy - 4xy^2)$ $4x^2y - 2xy - 3xy^2$
12. $(m^2 - 15n + 4n^2) - (8n - 3m^2 + 2n^2)$ $4m^2 - 23n + 2n^2$
13. $(3x - 2y) - (4x - 3y) + (7y - 6x)$ $-7x + 8y$
14. $-9x + (11t - 2) + (5x - 4t) - 6$ $-4x + 7t - 8$
15. $(20c^2 + 17cd) - (14d^2 + 3c^2) + 8d^2$ $17c^2 + 17cd - 6d^2$
16. $(3d^2 + 8d - 1) - (-3d^2 + 8d - 1) + (5 - 5d^2)$ $d^2 + 5$
17. Notebooks cost n cents and pens cost p cents. Julia bought 5 notebooks and 6 pens. Her brother Tim bought 7 notebooks and 3 pens. How much did their mother pay for these purchases, expressed in terms of n and p? $12n + 9p$
18. A triangle has sides of $(x - 3y)$, $(6y - 5x)$, and $(4x + 2y)$. Write and simplify an expression for the perimeter of this triangle. $-6y = (x - 3y) + (6y - 5x) + (4x + 2y)$

Practice ■ Lesson 11-2

Simplify.

19. $(3x)(-2x)$ $-6x^2$
20. $(8df)(2d^2)$ $16d^3f$
21. $(-6m)(7mn)$ $-42m^2n$
22. $(5xy^2)(x^2y)$ $5x^3y^3$
23. $(-k)(-9k^5)$ $9k^6$
24. $(8pqr)(3pr)$ $24p^2qr^2$
25. $(7s^3t^2)(4s^2t)$ $28s^5t^3$
26. $(3x^2)^2$ $9x^4$
27. $3x(4x - 10)$ $12x^2 - 30x$
28. $-2n(6n^2 - 5n)$ $-12n^3 + 10n^2$
29. $11x^2(3x^2 + 2x - 1)$ $33x^4 + 22x^3 - 11x^2$
30. $3c^2d(6d^2 - cd)$ $18c^2d^3 - 3c^3d^2$
31. $-pq(p^2q - 3pr + 7pq^3)$ $-p^3q^2 + 3p^2qr - 7p^2q^4$
32. $-2abc^2(a^2b^3c - a^2bc^2)$ $-2a^3b^4c^3 + 2a^3b^2c^4$
33. $5x(3a + 2b - 4c)$ $15ax + 10bx - 20cx$
34. $7k^2[-(5 - 4k + 6k^2)]$ $-35k^2 + 28k^3 - 42k^4$
35. $x(3x + 4) + 2(x^2 - 5x + 8)$ $5x^2 - 6x + 16$
36. $8(p^2 - 4pq + 5q^2) - 2(4p^2 + 20q^2)$ $-32pq$
37. $-4pq(p^3q + 5pr - 3pq^2)$ $-4p^4q^2 - 20p^2qr + 12p^2q^3$
38. $-6a^3bc^2(2a^3bc^2 - a^2bc)$ $-12a^6b^2c^4 + 6a^5b^2c^3$
39. $2yz(4a + 3b - 10c)$ $8ayz + 6byz - 20cyz$
40. $8k^2[-(5k^3 - 13 + 9k^2)]$ $-40k^5 + 104k^2 - 72k^4$
41. $3x^{10}y^8z(x^5yz^9 + 2xy^2z^8 - xyz^{12})$ $3x^{15}y^9z^{10} + 6x^{11}y^{10}z^9 - 3x^{11}y^9z^{13}$
42. $-3(x + 2) - 3(2 - x) + 3(x - 2) - 3[x - (-2)]$ -24

Write and simplify an expression for the area of each rectangle.

43.
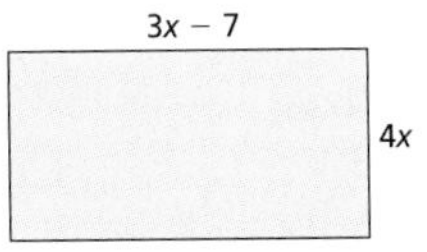

$4x(3x - 7)$; $12x^2 - 28x$

44.
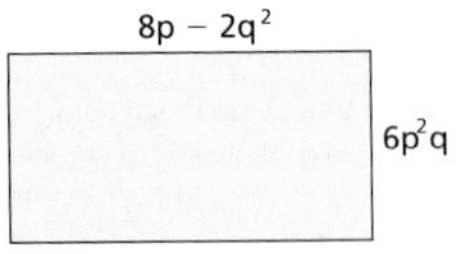

$6p^2q(8p - 2q^2)$; $48p^3q - 12p^2q^3$

45.
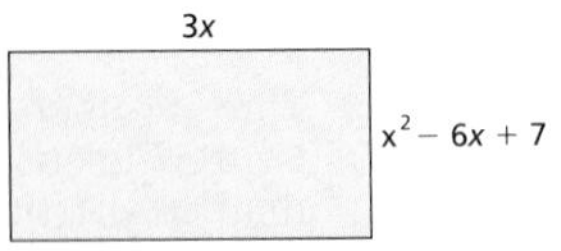

$3x(x^2 - 6x + 7)$; $3x^3 - 18x^2 + 21x$

Teaching Strategies

In items such as Exercise 5, students may find it helpful to rewrite the expression using multiplication of -1.

For example, $(4x - 6z) - (6x - 4z) = (4x - 6z) - 1(6x - 4z)$
$= 4x - 6z - 6x + 4z$
$= -2x - 2z$

476 Chapter 11 Polynomials

Practice ■ Lesson 11-1–Lesson 11-2

Simplify. (Lesson 11-1)

46. $(-5x + 2y) + (9y - 2x)$ $-7x + 11y$
47. $(15b - 6) + (-4b + 17)$ $11b + 11$
48. $(9x^2 + 4x - 6) + (13x^2 - 6x + 10)$ $22x^2 - 2x + 4$
49. $\left(\frac{1}{2}h - \frac{3}{4}g\right) + \left(\frac{3}{8}g + \frac{3}{4}h\right)$ $\frac{5}{4}h - \frac{3}{8}g$
50. $(2x - 6z) - (4x + 6z)$ $-2x - 12z$
51. $(3m - 3n + 11p) - (-5n + 8m - p)$ $-5m + 2n + 12p$
52. $(5x^3 - 8x^2 - x) - (6x + 3x^2 - 8x^3)$ $13x^3 - 11x^2 - 7x$
53. $(-4a^2 + 8ab + 12b^2) - (8ab - 12b^2)$ $-4a^2 + 24b^2$
54. $[5y^2 - (-2y)] - (5y^2 + 6y - 21)$ $-4y + 21$
55. $(6r^2 + 10rs - 13s^2) + (-8s^2 + 7r^2)$ $13r^2 + 10rs - 21s^2$
56. $(x^2y + xy^2 - 2xy) + (4x^2y - 2xy - 3xy^2)$ $5x^2y - 2xy^2 - 4xy$
57. $(-m^2 + 15n - 2n^2) - (-8n + 3m^2 - 2n^2)$ $-4m^2 + 23n$
58. $(2x + 3y) - (3x - 2y) + (x + y)$ $6y$
59. $(5x^4 - y^4) + (6x^3 + 2y^4) - (-7x^4 + 8x^3)$ $12x^4 - 2x^3 + 6x^4$

Simplify. (Lesson 11-2)

60. $(k^2)(-3k^3)$ $-3k^5$
61. $(-8p^3qr)(2pr^2)$ $-16p^4qr^3$
62. $(-2x^3)^2$ $4x^6$
63. $-2x(3x - 14)$ $-6x^2 + 28x$
64. $2n^2(5n^2 - 4n)$ $10n^4 - 8n^3$
65. $-9c^2d(-4d^2 + 3cd)$ $36c^2d^3 - 27c^3d^2$

MathWorks — Workplace Knowhow

Career – Brokerage Clerks

Brokerage clerks work for financial institutions such as brokerages, insurance companies and banks. They perform many different tasks. Purchase and sale clerks make sure that orders to buy and sell are recorded accurately and balance. Dividend clerks pay dividends to customers from their investments.

Margin clerks monitor the activity on clients' accounts, making sure clients make payments and abide by the laws covering stock purchases. Brokerage clerks often use computers to monitor all aspects of securities exchange. They use specialized software to enter transactions and check records for accuracy.

1. A client bought 60 shares of stock at x price per share and later sold 40 shares of the stock at y price. Write an expression that could be used to find the value of the client's stock after the sale. $60x - 40y$
2. A client wants to triple the number of gold certificates he owns. He has x certificates now, each worth y dollars today. Tomorrow the price of the certificates is expected to increase by z dollars. Write an expression to find the expected cost the client will pay tomorrow to triple his holdings. $2x(y + z)$
3. A client wants to buy $(x + 3)$ shares of stock for $(x + 8)$ dollars. Write an expression for the total cost of the order. $x^2 + 11x + 24$
4. A client bought $(x - 5)$ shares of stock A at a cost of $(x + 4)$ dollars. She also purchased $(x + 8)$ shares of stock B at a cost of $(x + 6)$ dollars. Write an expression to represent her total holdings of stocks A and B? $2x^2 + x - 68$

Chalkboard Examples

Example from Lesson 11-1

Simplify
$[m^3 - (-2m^2)] - (m^4 + 3m - 7)$.
$-m^4 + m^3 + 2m^2 - 3m + 7$

Example from Lesson 11-2

Simplify $x^2(x^3y^2 + y^3)$. $x^5y^2 + x^2y^3$

Career Opportunity

Describe the kind of brokerage clerks do and the types of companies that hire these workers. Have students discuss the importance of algebra and functions in doing this type of job. Explain that brokerage clerks must be able to analyze trends and determine the outcomes of complex transactions. Students should answer Questions 1–4 to better understand how brokerage clerks use mathematics in performing their job.

Students who are interested in learning more about this profession can go to learningmathmatters.com. Guidance Counselors should have information about school and training requirements.

Teaching Strategies

Refer to Exercise 34. Ask students to explain why the brackets are needed. Have students explain how to apply the order of operations in this situation.

Lesson Planning

NCTM Standards/Strands

- Number & Operation
- Patterns, Functions, & Algebra
- Problem Solving
- Reasoning and Proof
- Communication
- Connections
- Representation

Vocabulary

extracting factors
greatest common factor (GCF)

Materials Needed

Algeblocks or algebra tiles

Lesson Resources

Warm-Up Transparency 31
Reteaching 11-3
Extra Practice 11-3
Enrichment 11-3

Assignment Guide

Basic: 1–37, 42–44
Enriched: 1–44

Getting Started

5-Minute Warm-up

Write the prime elements of each.

1. $10ab$ 5, 2, a, b
2. $3x^2$ 3, x
3. $4cd^3$ 2, c, d
4. $6y^2z^2$ 2, 3, y, z

Introduction to Lesson 11-3

Have students share the strategies they used to create the other rectangles.

11-3 Divide and Find Factors

Goals ■ Factor polynomials into a monomial factor and a polynomial factor.

Applications Manufacturing, Sculpture, Landscaping

MODELING Did you realize that all monomials have factors? In fact, unless a monomial is a constant and also a prime number, it has more than one set of paired factors. What about polynomials? Can a binomial have a pair of factors? The answer is yes. The expression $4x + 2$ is equal to $1(4x + 2)$, because anything times 1 is equal to itself. Shown with Algeblocks or algebra tiles, the expression would look like this.

x	x	x	x	1	1

Are there any other paired factors of $4x + 2$? Use algebra tiles to see if you can multiply an expression by 2 and create the same area (it will be a different shape).

Now, use Algeblocks to arrange $4x^2 + 2x$ into a rectangle with one side (factor) equal to $2x$.

Build Understanding

Using Algeblocks is not the only way to find the factors of a binomial or polynomial. Another technique, called **extracting factors**, begins by determining if a polynomial has a monomial factor other than 1. Check to see if any monomial will divide evenly into every term of the polynomial. If so, you can extract the monomial factor by dividing the polynomial by that monomial factor. The quotient from that division is the second factor of the original polynomial.

Example 1

Find factors of $4x + 2$.

Solution

2 will divide $4x$ evenly, and it will also divide 2 evenly. Therefore, 2 is a factor of the polynomial. What is the other factor that pairs with 2? You can find it by dividing each term of the binomial by 2.

$$\frac{4x + 2}{2} = \frac{(2)(2)(x)}{2} + \frac{2}{2}$$
$$= 2x + 1$$

The factors are the 2 that you extracted, and $(2x + 1)$, the quotient. So, $4x + 2 = 2\,(2x + 1)$.

As you may realize, a polynomial may have more than one monomial factor.

Teaching Strategies

Refer to Example 3. Discuss when students think they will need to include the additional step shown in the example of first analyzing each monomial into its prime elements to find the GCF.

Example 2

Find the factors of $2x + 6x^2$.

Solution

You can see that 2 is a factor of both terms. You can also see that x is a factor of both terms. In addition, therefore, $(2)(x)$ or $2x$ is also a factor. In fact, $2x$ is the **greatest common factor**, or **GCF**, because it includes all the common factors. The paired factor is again found as follows.

$$\frac{2x + 6x^2}{2x} = \frac{(2)(x)}{(2)(x)} + \frac{(2)(3)(x)(x)}{(2)(x)}$$
$$= 1 + 3x$$

So, $2x + 6x^2 = 2x(1 + 3x)$.

Finding the monomial that is the GCF is very valuable for factoring a binomial.

Example 3

Find the greatest common factor of $15xy^3$ and $3x^2y^2$. Then find the paired factor that will create the expression $15xy^3 - 3x^2y^2$.

Solution

$$15xy^3 = (3)(5)(x)\quad(y)(y)(y)$$
$$3x^2y^2 = (3)\quad(x)(x)(y)(y)$$
$$(3)\quad(x)\quad(y)(y) \rightarrow 3xy^2 \text{ Greatest Common Factor}$$

$$\frac{15xy^3 - 3x^2y^2}{3xy^2} = \frac{(3)(5)(x)(y)(y)(y)}{(3)(x)(y)(y)} - \frac{(3)(x)(x)(y)(y)}{(3)(x)(y)(y)}$$
$$= 5y - x$$

This technique finds the GCF by analyzing each monomial in its simplest terms. Thus,

$$15xy^3 - 3x^2y^2 = 3xy^2(5y - x)$$

Prime elements can help with division of monomials. Analyze the dividend and the divisor into prime elements, then cancel each element they share.

Example 4

MANUFACTURING A company manufactures posters with inspirational saying. Each poster has an area of $8mn^2$ square inches. The length of each poster is $2mn$ inches. Find the width.

Solution

$$8mn^2 \div 2mn = \frac{(2)(2)\cancel{(2)}\cancel{(m)}\cancel{(n)}(n)}{\cancel{(2)}\cancel{(m)}\cancel{(n)}} = (2)(2)(n) = 4n$$

The width of the poster is $4n$.

Five-step Plan
1 Read
2 Plan
3 Solve
4 Answer
5 Check

Teaching Strategies

CHALLENGE Write the following statement on the chalkboard: The GCF of Xa^pb^q and Ya^rb^s is Zapbs. Capital letters represent coefficients. Ask students what this tells about X, Y, Z, p, q, r, and s. *Z is the GCF of X and Y; $p < r$, and $s < q$.*

Chalkboard Examples

Supplementary Example 1
Find the factors of $6y + 15$.
$3(2y + 5)$

Supplementary Example 2
Find the factors of $4y^2 + 16y$.
$4y(y + 4)$

Supplementary Example 3
Extract the greatest common factor $9e^2f^3$ and $15ef^4$ and find the paired factor that will create the expression $9e^2f^3 + 15ef^4$.
$3ef^3(3e + 5f)$

Supplementary Example 4
Another rectangular poster has an area of $6xy^2$. If the poster's width is $3xy$, what is the length of the poster? $2y$

Reteaching Worksheet 11-3

Name ______________________ Date ____________

RETEACHING 11-3

DIVISION AND FINDING FACTORS

The process of writing $6a^2 + 4a - 2ab$ as $2a(3a + 2 - b)$ is called **factoring**. The terms $2a$ and $3a + 2 - b$ are the **factors**. To factor a polynomial, first find the **greatest common factor** (GCF) of its monomial terms by following these steps.

Step 1: First find the greatest possible *number* that will divide each coefficient evenly.

Step 2: Then find each variable that is included in *all* the monomial terms.

Then find its paired factor.

Example 1

Find the greatest common factor (GCF) of $4x^2 + 8xy^2$. Then find its paired factor.

Solution

4 will divide evenly into both 4 and 8. The variable x is the only variable factor of every term. So the GCF of $4x^2 + 8xy^2$ is $4x$.

To find the paired factor, divide each term by the GCF, $4x$.

$$\frac{4x^2}{4x} + \frac{8xy^2}{4x} = x + 2y^2$$

The paired factors are $4x$ and $x + 2y^2$, so $4x^2 + 8xy^2 = 4x(x + 2y^2)$.

Example 2

Find the greatest common factor (GCF) of $10a^2 + 5ab$. Then find its paired factor.

Solution

5 will divide evenly into both $10a^2$ and $5ab$. The variable a is the only variable factor of every term. So the GCF of $10a^2 + 5ab$ is $5a$.

To find the paired factor, divide each term by the GCF, $5a$.

$$\frac{10a^2}{5a} + \frac{5ab}{5a} = 2a + b$$

The paired factors are $5a$ and $2a + b$, so $10a^2 + 5ab = 5a\,(2a + b)$.

EXERCISES

Find the GCF for each polynomial. Then find its paired factor.

1. $9x + 12$ $3(3x + 4)$
2. $a^2 + 4a$ $a\,(a + 4)$
3. $7a^3 - 14a^2$ $7a^2\,(a - 2)$
4. $15y^4 + 12y^2z$ $3y^2\,(5y^2 - 4z)$
5. $8x^5 - 5x^4 + 2x^3$ $x^3\,(8x^2 - 5x + 2)$
6. $8x^{10} - 24x^5 + 6x^4$ $2x^4\,(4x^6 - 12x + 3)$
7. $27y^5 - 9y^3$ $9y^3\,(3y^2 - 1)$
8. $a^3b^2 - 3a^4b$ $a^3b\,(b - 3a)$
9. $9y^4 - 3y^3 + y^2$ $y^2\,(9y^2 - 3y + 1)$
10. $15k^3 + 5k + 10$ $5(3k^3 + k + 2)$
11. $12x^4 + 6x^2 - 3x$ $3x\,(4x^3 + 2x - 1)$
12. $2f^3 - 18f^2 + 8f$ $-2f\,(f^2 - 9f + 4)$
13. $16m^5n + 4mn - 8mn^2$ $4\,mn\,(4m^4 + 1 - 2n)$
14. $12a^2b^2 + 8\,ab^2 + 16b^2$ $4b^2\,(3a^2 + 2a + 4)$
15. $-9y^2z - 12y^4z^3 + 15y^3z^4$ $3y^2z\,(-3 - 4y^2z^2 + 5yz^3)$

164 RETEACHING *LESSON 11-3*

Try These Exercises

Extract a monomial factor and find its paired binomial factor for the following.

1. $6x^2 + 9$ $3(2x^2 + 3)$ **2.** $2a + ab$ $a(2 + b)$ **3.** $5mn - n^2p$ $n(5m - np)$

Extract the GCF and indicate its paired binomial factor.

4. $16p - 20q$ $4(4p - 5q)$ **5.** $12x^2 + 18x$ $6x(2x + 3)$ **6.** $45a^2b - 27ab^2$ $9ab(5a - 3b)$

Extract a monomial factor and find the paired trinomial factor.

7. $7r^2 + 3rs + 2rt$ $r(7r + 3s + 2t)$ **8.** $h^2jk + jk^2l - 3klm$ $k(h^2j + jkl - 3lm)$

9. SCULPTURE A sculptor has 2 columns of marble. One is 54 inches tall, the other is 90 inches tall. He decides he can get the best value if he carves a set of identical figurines. He must use the full length of both columns, dividing them into equal pieces but making each figurine as tall as possible. How tall will the figurines be, and how many will he make? 18 inches tall, 8 figurines

10. GEOMETRY A rectangle of area $9v^2w$ has a width of $3v$. What is its length? $3vw$

Practice Exercises

Find factors for the following:

11. $6a + 8b$ $2(3a + 4b)$ **12.** $21x^2 - 35y^2$ $7(3x^2 - 5y^2)$ **13.** $15p^3 - 35q$ $5(3p^3 - 7q)$

14. $13e - 5ef$ $e(13 - 5f)$ **15.** $vw + wx$ $w(v + x)$ **16.** $8gh - 3hj$ $h(8g - 3j)$

17. $5x^2y - 2y^2$ $y(5x^2 - 2y)$ **18.** $18r^2s + 19st^2$ $s(18r^2 + 19t^2)$ **19.** $13mn^2 + 25n$ $n(13mn + 25)$

Simplify.

20. $12x3y^2 \div 6x^2y$ $2xy$ **21.** $45ef^2 \div 18ef$ $\frac{5f}{2}$

22. Find the greatest common factor of $24u^3v^2$, $6u^2v^3$ and $8uv^4$. $2uv^2$

Find the GCF and its paired factor for the following.

23. $14ab^2 + 35bc$ $7b(2ab + 5c)$ **24.** $45m^2n - 72mn$ $9mn(5m - 8)$

25. $18r^3 + 27r^2$ $9r^2(2r + 3)$ **26.** $50u^4v^2 - 100u^3v^3$ $50u^3v^2(u - 2v)$

27. $39j^7k^3l^4 - 65j^6k^5l^3 + 52j^5k^2l^6$ $13j^5k^2l^3(3j^2kl - 5jk^3 + 4l^3)$ **28.** $6a^5b - 12a^4b^2 - 9a^3b^3$ $3a^3b(2a^2 - 4ab - 3b^2)$

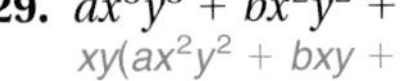

29. $ax^3y^3 + bx^2y^2 + cxy$ $xy(ax^2y^2 + bxy + c)$ **30.** $18r^5 + 45r^4s^2 - 63r^2s^4$ $9r^2(2r^3 + 5r^2s^2 - 7s^4)$

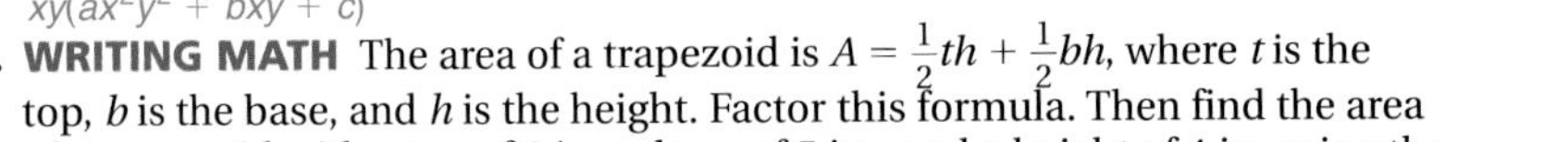

31. WRITING MATH The area of a trapezoid is $A = \frac{1}{2}th + \frac{1}{2}bh$, where t is the top, b is the base, and h is the height. Factor this formula. Then find the area of a trapezoid with a top of 6 in., a base of 5 in., and a height of 4 in. using the given formula and the factored formula. Which was easier to use? Explain. $A = \frac{1}{2}h(t + b) = 22$ sq

32. LANDSCAPING Nguyen is calculating the price of a landscaping contract using her company's formula: $P = 4r^2 + 8rs - 4rt$. For this job, $r = 2.5$, $s = 5.4$, and $t = 3.3$. Hoping to avoid a lot of multiplication, Nguyen factors the formula, and finds the math is very simple. What is her factored version of the formula, and what price does she set for the contract? $P = 4r(r + 2s - t) = 10(2.5 + 10.8 - 3.3) = 10(10) = 100$

Lesson Wrap-up

Quick Assessment

Ask the following questions to determine if students understand the content presented in this lesson.

1. How would your work be different if you were extracting a monomial factor from a trinomial instead of a binomial? The only difference would be that the monomial factor would have to be the GCF of all three terms.
2. Describe a binomial for which you could not extract a monomial factor. If the binomial has only one constant or coefficient, it is prime; if it has a constant and a coefficient or two coefficients, they are relatively prime (no common factors); in addition, if only one term has a variable, or both terms have different variables, they are prime.

Assignment Guide

Basic: 1–37, 41–58
Advanced: 1–58
Additional Practice: See Extra Practice Index on page 674.

Teaching Strategies

Refer to Exercises 23–30. Have students count the number of terms in the original polynomial and the number of terms in the paired factor to make sure they are the same.

Find a monomial and a polynomial factor. Simplify first if necessary.

33. $6x^5y^2 + 8x^4y^3 + 6x^3y^4 + 14x^2y^5 + 2xy^6$ $2xy^2(3x^4 + 4x^3y + 3x^2y^2 + 7xy^3 + y^4)$

34. $3x(y^2 + 2z) - y(3xy - 6xz^2)$ $6xz(1 + yz)$

Write, simplify, and factor an expression for each perimeter below.

35.

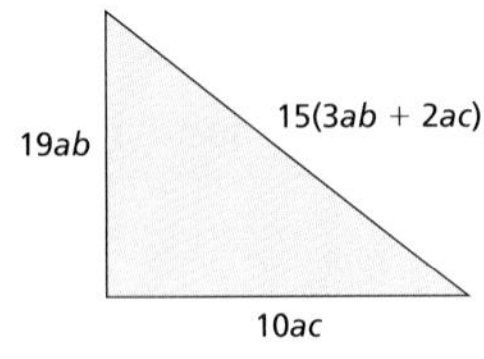

$8a(8b + 5c)$

36.

x + y
x + 2y
x
x + 3y

$2(2x + 3y)$

37.

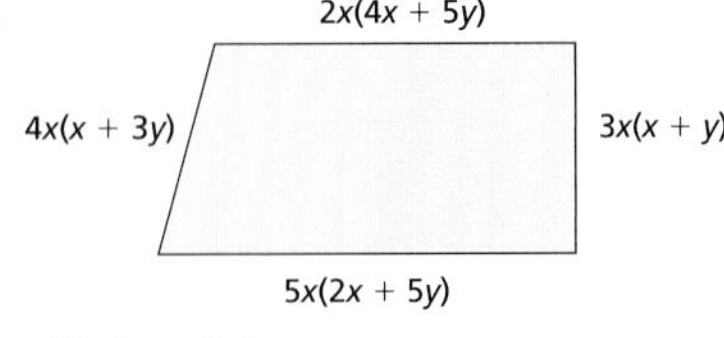

$25x(x + 2y)$

Extended Practice Exercises

38. A snail usually travels $3a$ inches every hour. However, when it is climbing out of a slippery well, it also slides back $2b$ inches each hour. The distance it has climbed after x hours is found to be $3ax - 2bx$ inches. Prove that this is exactly what you would expect by factoring this distance. $x(3a - 2b)$

39. Factor $3x^n - 2x^{(n-1)}$. $x^{(n-1)}(3x - 2)$

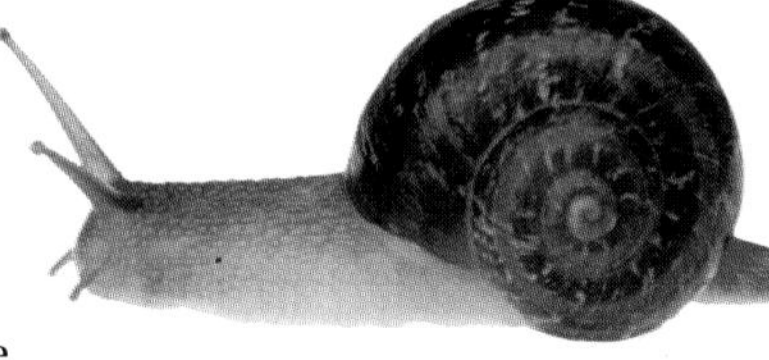

40. The sum of a series of n even numbers starting with 2 is given by the formula $S = n^2 + n$. Test the formula on $(2 + 4)$, $(2 + 4 + 6)$, and $(2 + 4 + 6 + 8)$. Next, use the formula to calculate the sum of the first 14 even numbers. Then factor the formula, and use the factored version to sum the first 17 even numbers. Note that the factored version saves a step. 14: $S = n^2 + n = (14 \times 14) + 14 = 196 + 14 = 210$
17: $S = n(n + 1) = 17 \times 18 = 306$

41. CHAPTER INVESTIGATION Continue to work on your marketing survey. What advertising methods are most effective for your age group? Add questions to your survey to find out how much time each day your classmates spend in watching television, listening to the radio, reading newspapers and magazines and traveling by car or bus. Include questions to find out which television programs, radio stations and magazines are most popular.
Check students' work.

Mixed Review Exercises

Find the unknown measures. First find each in simplest radical form, and then find each to the nearest hundredth. (Lesson 10-3)

42.

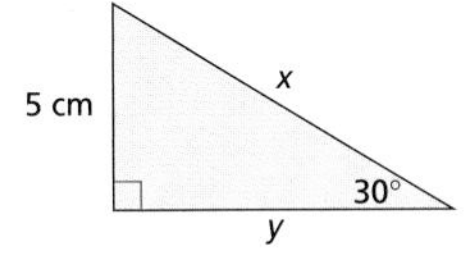

$x = 10$ cm,
$y = 5\sqrt{3} = 8.66$ cm

43.

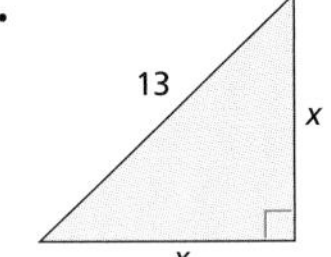

$x = \frac{13\sqrt{2}}{2} = 9.19$

44.

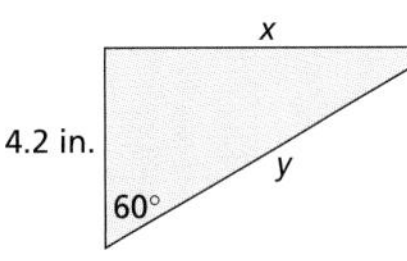

$x = 4.2\sqrt{3} = 7.27$ in.,
$y = 8.4$ in.

Extend the Lesson

ONGOING ASSESSMENT Have students find factors of $6a^2 + 3a$ by analyzing each monomial into its prime elements.
Watch for students who get the answer $3a(2a)$. These students have "cancelled out" terms, replacing their value with 0 instead of 1. The correct solution is $3a(2a + 1)$.

Extra Practice Worksheet 11-3

Name ______________________ Date ____________

EXTRA PRACTICE 11-3

DIVIDE AND FIND FACTORS

EXERCISES

Find the GCF and its paired factor for the following.

1. $8x + 4y$ $4(2x + y)$
2. $24c - 30d^2$ $6(4c - 5d^2)$
3. $5xz - z^2$ $z(5x - z)$
4. $n^2p + 6n$ $n(np + 6)$
5. $8x^2 - 12xy$ $4x(2x - 3y)$
6. $r^2s + 4r$ $r(rs + 4)$
7. $15r - 10s$ $5(3r - 2s)$
8. $12a^2 - 10a$ $2a(6a - 5)$
9. $h^3k - 3hk^3$ $hk(h^2 - 3k^2)$
10. $4n^2m^2 + 6n^4m$ $2n^2m(2m + 3n^2)$
11. $2r^2s^2 - r^2s$ $rs(2rs - r)$
12. $12a^2 + 8a^2b$ $4a^2(3a + 2b)$
13. $3x^2y - 3x^2 + 3xy$ $3x(xy - x + y)$
14. $2a^3b^2 + 3a^3b - 4a2b^3$ $a^2b(2ab + 3a - 4b^2)$
15. $20d^5 - 10cd^3 + 15c^2d^4$ $5d^3(4d^2 - 2c + 3c^2d)$

Write, simplify and factor an expression for each perimeter below.

16.

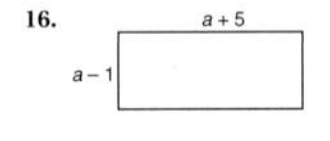

$4a + 8$; $4(a + 2)$

17.

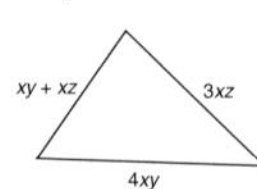

$5xy + 4xz$; $x(5y + 4z)$

18.

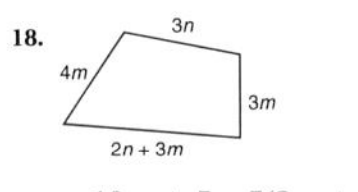

$10m + 5n$; $5(2m + n)$

19.

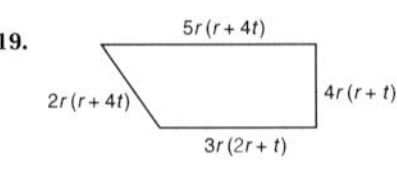

$17r^2 + 35rt$; $r(17r + 35t)$

144 EXTRA PRACTICE *LESSON 11-3*

Enrichment Worksheet 11-3

Name ______________________ Date ____________

ENRICHMENT 11-3

THE SUM OF AN ARITHMETIC SERIES

The sum of the first four even numbers starting with 2 is an example of a *series*. The terms are added together to find the sum of the series. For a series to be *arithmetic*, the difference between each pair of adjacent terms must be the same number.

If you know that a series is arithmetic, you can use the formula below to find the sum, S. In the formula, a_1 is the first term, a_n is the last term, and n is the number of terms.

$$S = \frac{n(a_1 + a_n)}{2}$$

EXERCISES

Find the sum of each arithmetic series. Show how you use the formula.

1. The natural numbers 1 through 100
$S = \frac{100(1 + 100)}{2} = 5050$
2. The counting numbers 0 through 100
$S = \frac{101(0 + 100)}{2} = 5050$
3. Multiples of 3 from 3 through 24
$S = \frac{8(3 + 24)}{2} = 108$
4. Multiples of 5 from 10 through 50
$S = \frac{9(10 + 50)}{2} = 270$
5. Even whole numbers 2 through 100
$S = \frac{50(2 + 100)}{2} = 2550$
6. Even whole numbers 2 through n
$S = \frac{n(2 + n)}{4}$
7. Odd whole numbers 1 through 99
$S = \frac{50(1 + 99)}{2} = 2500$
8. Odd whole numbers 1 through n
$S = \frac{(1 + n)^2}{4}$
9. The integers −10 through 10
$S = \frac{21(-10 + 10)}{2} = 0$
10. The integers $-x$ through x
$S = \frac{(2x + 1)(-x + x)}{2} = 0$

188 ENRICHMENT *LESSON 11-3*

Lesson Planning

NCTM Standards/Strands

- Number & Operation
- Patterns, Functions, & Algebra
- Problem Solving
- Reasoning and Proof
- Communication
- Connections
- Representation

Vocabulary

expanding
first product
outer product
inner product
last product
FOIL
cross products
difference of two squares

Materials Needed

paper/pencil

Lesson Resources

Warm-Up Transparency 32
Transparency TK-4, RF-43
Reteaching 11-4
Extra Practice 11-4
Enrichment 11-4
Multicultural Connection 4

Assignment Guide

Basic: 1–45, 49–60
Enriched: 1–60

Getting Started

5-Minute Warm-up

List the four products that could be added or subtracted to find the value of this expression: (5 − 3)(6 − 2).

(5)(6) − (5)(2) − (3)(6) + (3)(2)

Introduction to Lesson 11-4

After students have completed this activity, have them speculate about a method that could be used to find the product of two binomials.

11-4 Multiply Two Binomials

Goals ■ Multiply binomials.

Applications Packaging, Small Business, Product Development

Work with a partner to answer the following questions.

You have seen how a binomial can be multiplied and divided by a monomial. Binomials can also be multiplied (and divided) by other binomials. Look at the following diagram.

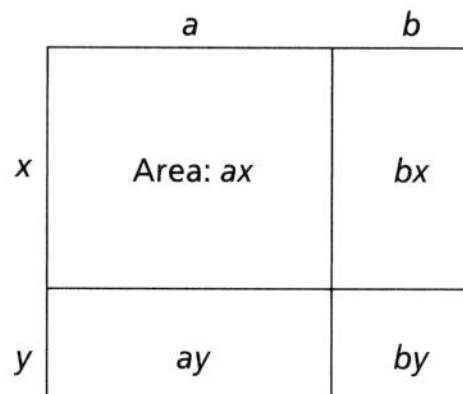

$x(a + b) = ax + bx$

$y(a + b) = ay + by$

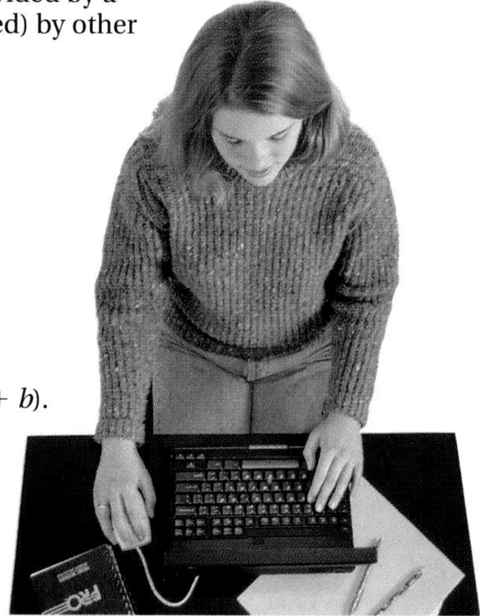

As you can see, the whole diagram represents $(x + y)(a + b)$.

1. Express the large area as a polynomial by adding the areas of all four smaller rectangles. $xa + xb + ya + yb$
2. Draw a diagram to show the expression $(2p + 4q)(l + m)$. Check students' work.
3. Express your diagram as a polynomial by adding its parts. $2pl + 2pm + 4ql + 4qm$

Build Understanding

Multiplying a binomial by another binomial starts with the idea that a binomial is the sum of two monomials. To multiply two binomials, use the distributive property twice. Multiply the second binomial separately by each term in the first binomial. Then add the answers together. This is also called **expanding** the two binomials.

Example 1

Find the product $(x + a)(2x + 3b)$.

Solution

$$(x + a)(2x + 3b) = x(2x + 3b) + a(2x + 3b)$$
$$= 2x^2 + 3bx + 2ax + 3ab$$

No further simplification is possible.

Sometimes simplification leads to a different-looking polynomial.

Teaching Strategies

Some students may find it helpful to use a column format when multiplying binomials. Example 2 could be written.

$$\begin{array}{r} x + 1 \\ \times \quad x + 5 \\ \hline 5x + 5 \\ x^2 + x \quad \\ \hline x^2 + 6x + 5 \end{array}$$

Example 2

PACKAGING The dimensions of the cover art for a new product are $x + 1$ by $x + 5$. Find the area of the cover art. Expand and simplify $(x + 1)(x + 5)$.

Solution

$$\begin{aligned}(x + 1)(x + 5) &= x(x + 5) + 1(x + 5)\\ &= x^2 + 5x + x + 5\\ &= x^2 + (5 + 1)x + 5\\ &= x^2 + 6x + 5\end{aligned}$$

The area of the cover art is $x^2 + 6x + 5$.

Now that you have seen two examples, look for a pattern. The final solutions may seem quite different, but study the second line of each answer. In each case, the first term is the product of the binomials' first terms. Describe it as the **First product**. The second term is the product of the outer pair of terms in the binomials. It can be called the **Outer product**. The third term is the product of the inner terms—the **Inner product**. And the final term is the **Last product**, the product of the last terms of the two binomials. The whole multiplication process is often called the **FOIL** process—for **F**irst, **O**uter, **I**nner, and **L**ast.

Notice that in Example 2 the inner and outer products can be simplified into a single term.

Reading About Math

The outer and innter products are also known as the **cross products**. If the binomials are placed one above the other, you can see why.

$(x + 1)$
$(x + 5)$

In each case, the first term of one binomial is multiplied by the last term of the other, making a cross.

Example 3

Expand and simplify $(y - 5)(y + 5)$.

Solution

$$\begin{aligned}(y - 5)(y + 5) &= \overset{F}{y^2} + \overset{O}{5y} - \overset{I}{5y} - \overset{L}{25}\\ &= y^2 - 25\end{aligned}$$

This multiplication produces a polynomial pattern called the **difference of two squares**. The product of two binomials that differ only in their signs is always the square of the first binomial term minus the square of the second. The cross products (the O and I terms) add to zero. In other words, $(a + b)(a - b) = a^2 - b^2$. This is true *whatever* the values of a and b.

Try These Exercises

Expand and simplify if possible.

1. $(3a + 2b)(c + 5d)$ $3ac + 15ad + 2bc + 10bd$
2. $(e - 6f)(2g - 3h)$ $2eg - 3eh - 12fg + 18fh$
3. $(l - m)(l + n)$ $l^2 + ln - lm - mn$
4. $(3r + s)(2r - 3t)$ $6r^2 - 9rt + 2rs - 3st$
5. $(2x + 5)(3x + 3)$ $6x^2 + 21x + 15$
6. $(y - 6)(y - 6)$ $y^2 - 12y + 36$
7. $(8x - y)(x + 2y)$ $8x^2 + 15xy - 2y^2$
8. $(3u - 10v)(2u + v)$ $6u^2 - 17uv - 10v^2$
9. $(p - q)(p + q)$ $p^2 - q^2$
10. $(2x + 3y)(2x - 3y)$ $4x^2 - 9y^2$

Teaching Strategies

CHALLENGE Have students use logical reasoning or guess-and-check to find each missing factor.

1. $(2a + b)(?) = 2a^2 + 5ab + 2b^2$ $a + 2b$
2. $(5 - c)(?) = 25 - c^2$ $5 + c$
3. $(d - e)(?) = 3de + df - 3e^2 - ef$ $3e + f$

Chalkboard Examples

Supplementary Example 1

Find the product $(y + 2c)(3y + d)$.
$3y^2 + dy + 6cy + 2cd$

Supplementary Example 2

The dimensions for a rectangular poster are $y + 4$ *by* $2y + 3$. Find the area of the poster.
$(y + 4)(2y + 3) = 2y^2 + 11y + 12$

Supplementary Example 3

Expand and simplify $(x - 3)(x + 3)$.
$x^2 - 9$

Reteaching Worksheet 11-4

Name ____________________ Date ____________

RETEACHING **11-4**

MULTIPLYING TWO BINOMIALS

To multiply a binomial by a binomial, multiply each term of one binomial by each term of the other and then combine like terms.

Example

Find $(a + 2b)(3a - 4b)$.

Solution

You can use the distributive property to help you multiply binomials. Be sure to combine like terms.

$(a + 2b)(3a - 4b)$
$a(3a - 4b) + 2b(3a - 4b)$
$a(3a) + a(-4b) + 2b(3a) + 2b(-4b) =$
$3a^2 + (-4ab) + 6ab + (-8b^2) =$
$3a^2 - 4ab + 6ab - 8b^2 =$
$3a^2 + 2ab - 8b^2$

The FOIL method is a shorter way to do the same multiplication. Again, be sure to combine like terms.

First terms, Outer terms, Inner terms, Last terms
$a(3a) + a(-4b) + 2b(3a) + 2b(-4b) =$
$3a^2 - 4ab + 6ab - 8b^2 =$
$3a^2 + 2ab - 8b^2$

EXERCISES

Find each product. Use the method with which you feel most comfortable.

1. $(x - 3)(x - 8)$ $x^2 - 11x + 24$
2. $(a + 3)(a + 2)$ $a^2 + 5a + 6$
3. $(m - 9)(m - 5)$ $m^2 - 14m + 45$
4. $(w + 2)(w + 9)$ $w^2 + 11w + 18$
5. $(n + 7)(n - 8)$ $n^2 - n - 56$
6. $(p - 6)(p + 9)$ $p^2 + 3p - 54$
7. $(2t - 4)(t + 3)$ $2t^2 + 2t - 12$
8. $(5s - 8)(2s - 4)$ $10s^2 - 36s + 32$
9. $(3b + 4)(2b + 7)$ $6b^2 + 29b + 28$
10. $(4 - x)(2 + x)$ $8 + 2x - x^2$
11. $(a + b)(a + b)$ $a^2 + 2ab + b^2$
12. $(4m - 2n)(3m + 6n)$ $12m^2 + 18mn - 12n^2$
13. $(3a - 2b)(3a + 2b)$ $9a^2 - 4b^2$
14. $(5c - 8d)(3c + 4d)$ $15c^2 - 4cd - 32d^2$
15. $(12x + 4y)(10x - 7y)$ $120x^2 - 44xy - 28y^2$
16. $(2x + y)(2x + y)$ $4x^2 + 4xy + y^2$
17. $(2x - y)(2x - y)$ $4x^2 - 4xy + y^2$
18. $(2x + y)(2x - y)$ $4x^2 - y^2$

166 RETEACHING *LESSON 11-4*

Lesson Wrap-up

Quick Assessment

Ask the following questions to determine if students understand the content presented in this lesson.

1. How is the distributive property applied when multiplying a binomial by another binomial? The distributive property is used twice because the second binomial is multiplied by each term in the first binomial.
2. How can you recognize when the product of two factors will be the difference of two squares? The binomial factors have the same coefficients and variables, but one factor shows a sum and the other shows a difference.

Assignment Guide

Basic: 1–45, 49–60
Advanced: 1–60
Additional Practice: See Extra Practice Index on page 674.

Additional Answers

52. $\begin{bmatrix} 11 & 6 \\ -2 & 9 \\ -3 & -6 \end{bmatrix}$ 53. $\begin{bmatrix} 4 & -3 & 3 \\ 1 & 7 & 0 \\ 2 & -3 & 7 \end{bmatrix}$

54. $\begin{bmatrix} 0 & 1 & 3 \\ 1 & -1 & 2 \\ 3 & 1 & -1 \end{bmatrix}$ 55. $\begin{bmatrix} 20 & 30 \\ 15 & 40 \\ -25 & 30 \end{bmatrix}$

56. $\begin{bmatrix} -7 & -21 \\ 28 & 14 \\ 21 & -49 \end{bmatrix}$ 57. $\begin{bmatrix} -12 & -4 \\ 32 & 24 \\ 28 & -16 \end{bmatrix}$

11. **SMALL BUSINESS** As a summer project, Andre is making hand-painted ceramic plates. The material costs \$10 for each plate, and 12 plates can be made comfortably each day. But if the work rate goes up, he uses up more materials because of mistakes. So the cost per item increases by \$1 for each plate he makes over 12. To plan his work, he needs a formula. The cost of making 12 plates each day is \$(12)(10). What is his daily cost when making $(12 + x)$ plates? Expand and simplify your answer. $x^2 + 22x + 120$

12. **WRITING MATH** Can the product of two binomials ever have more than three terms? Explain your thinking. yes; the product will contain four terms unless terms can be combined.

Practice Exercises

Simplify.

13. $(2p + 5q)(3r - 1)$ $6pr - 2p + 15qr - 5q$
14. $(7k - l)(3m - n)$ $21km - 7kn - 3lm + ln$
15. $(4a + b)(a + 3c)$ $4a^2 + 12ac + ab + 3bc$
16. $(8x - 3y)(3x - 8z)$ $24x^2 - 64xz - 9xy + 24yz$
17. $(e - 3f)(2g + 5f)$ $2eg + 5ef - 6fg - 15f^2$
18. $(6w + 7x)(y - z)$ $6wy - 6wz + 7xy - 7xz$
19. $(9p + 2q)(5p - 3r)$ $45p^2 - 27pr + 10pq - 6qr$
20. $(7a - c)(3b + c)$ $21ab + 7ac - 3bc - c^2$
21. $(5m + 6n)(m + 9n)$ $5m^2 + 51mn + 54n^2$
22. $(5 - 6n)(1 - 9n)$ $5 - 51n + 54n^2$
23. $(3x - 4)(x - 2)$ $3x^2 - 10x + 8$
24. $(3x + 4y)(x + 2y)$ $3x^2 + 10xy + 8y^2$
25. $(j - 5k)(7j - 2k)$ $7j^2 - 37jk + 10k^2$
26. $(8a + 1)(3a - 5)$ $24a^2 - 37a - 5$
27. $(8b - c)(3b + 5c)$ $24b^2 + 37bc - 5c^2$
28. $(l - 5)(7l + 2)$ $7l^2 - 33l - 10$
29. $(w - 4z)(w - 4z)$ $w^2 - 8wz + 16z^2$
30. $(x - 4)(x - 4)$ $x^2 - 8x + 16$
31. $(x + 4)(x + 4)$ $x^2 + 8x + 16$
32. $(4w + x)(4w + x)$ $16w^2 + 8wx + x^2$
33. $(a - 2)(a + 2)$ $a^2 - 4$
34. $(3b - 2)(3b + 2)$ $9b^2 - 4$
35. $(2e + 5f)(2e - 5f)$ $4e^2 - 25f^2$
36. $(10x + 3y)(10x - 3y)$ $100x^2 - 9y^2$

37. **TRANSPORTATION** Four years ago, a \$10 bill would buy x gallons of gas, and Jane's car averaged y miles per gallon. Today, the car's gas mileage has decreased by 5 miles per gallon, and a \$10 bill buys 1 gallon less. Find the difference between how far Jane could travel on \$10 in those days, compared to now. $5x + y - 5$

Expand and simplify.

38. $(4k + 1)(k + 3) - 4k^2$ $13k + 3$
39. $(3x - 4)(3x^2 + 6x - 2)$ $9x^3 + 6x^2 - 30x + 8$
40. $(7a + 3b)(6a^2 + 2ab - b^2)$ $42a^3 + 32a^2b - ab^2 - 3b^3$
41. $(p - q)(p + 2q)(2p - q)$ $2p^3 + p^2q - 5pq^2 + 2q^3$
42. $(a + b)(a + b)(a + b)$ $a^3 + 3a^2b + 3ab^2 + b^3$
43. $(a + b)^4$ $a^4 + 4a^3b + 6a^2b^2 + 4ab^3 + b^4$

Express, expand, and simplify the volumes of the two rectangular prisms shown below.

44.

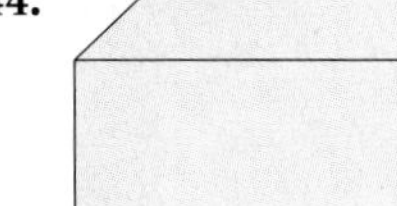

$2x^3 + 7x^2 + 6x$

45. $2y^3 + 3y^2 - 17y + 12$

y + 4
y − 1
2y − 3

Extend the Lesson

ONGOING ASSESSMENT Have students find the product $(2x + 4)(2x - 3)$. Watch for students who get the answer $2x^2 + 2x - 12$. When multiplying the first terms, these students multiplied only the variable, not the variable and the coefficient. Correct solution: $4x^2 - 2x - 12$.

484 Chapter 11 **Polynomials**

Extended Practice Exercises

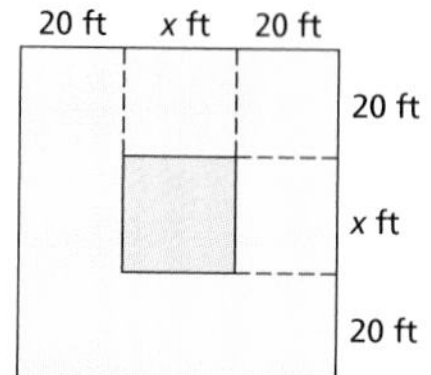

46. CONSTRUCTION A square fast-food restaurant building is surrounded by a square parking lot. The lot extends 20 ft beyond the restaurant in each direction, as shown on the map at the right. When the lot was paved, it took 4,000 sq ft of blacktop to cover it. How long is each wall of the restaurant. (Hint: Let x represent the sides of the restaurant, and make an equation.) 30 ft

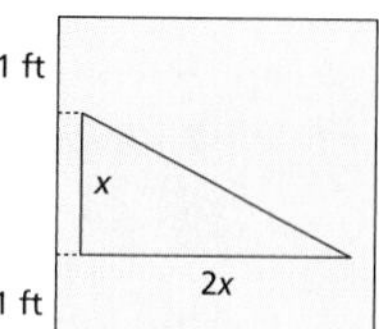

47. SEWING The skateboard club, invited to enter a local parade, decided to have a flag. Their first idea was a beige pennant to represent a street ramp. It was a right triangle, twice as wide as it was high. For visibility, they then stitched a square green background around it. As shown in the picture, the background extended one foot above and below the triangle. The green area totaled 10 square feet. About how much beige cloth did they use? (Don't worry about a seam allowance for your calculation.) $1\frac{1}{2}$ square feet

48. PRODUCT DEVELOPMENT A product engineer designs a new square handheld game. After field-testing the prototype, the engineer decides to change the shape of the game. She doubles the length and decreases the width by 4. Let s represent the length of a side on the original square. Write a polynomial to represent the area of the new rectangular game. $2s^2 - 8s$

Mixed Review Exercises

Find the volume of each. Round answers to the nearest whole number. (Lesson 5-7)

49.

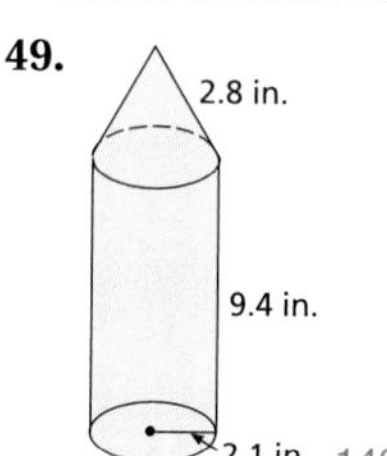

143 in.^{3}

50.

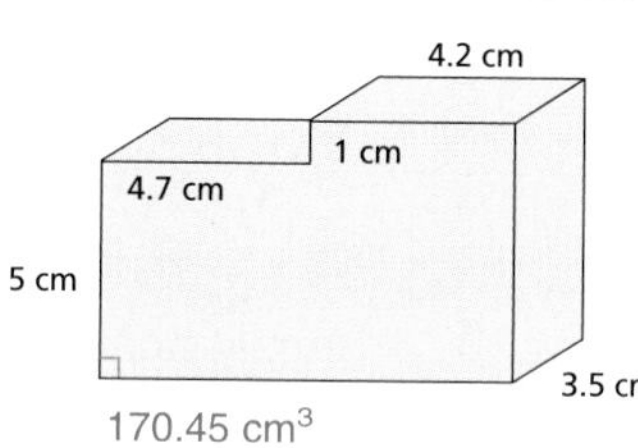

170.45 cm^3

51.

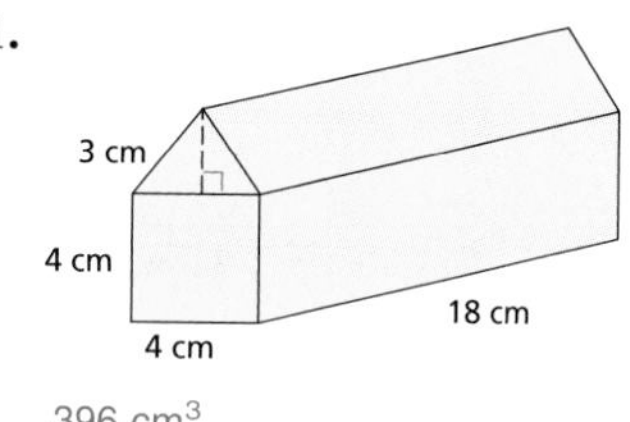

396 cm^3

Add. (Lesson 8-5) For 52–54, see additional answers.

52. $\begin{bmatrix} 6 & 2 \\ 1 & 8 \\ -3 & 2 \end{bmatrix} + \begin{bmatrix} 5 & 4 \\ -3 & 1 \\ 0 & -8 \end{bmatrix}$

53. $\begin{bmatrix} 2 & -1 & 3 \\ -2 & 3 & 1 \\ 0 & 0 & 5 \end{bmatrix} + \begin{bmatrix} 2 & -2 & 0 \\ 3 & 4 & -1 \\ 2 & -3 & 2 \end{bmatrix}$

54. $\begin{bmatrix} 2 & 1 & 2 \\ 1 & 1 & 2 \\ 2 & 1 & 1 \end{bmatrix} + \begin{bmatrix} -2 & 0 & 1 \\ 0 & -2 & 0 \\ 1 & 0 & -2 \end{bmatrix}$

Multiply. (Lesson 8-5) For 55–57, see additional answers.

55. $5 \cdot \begin{bmatrix} 4 & 6 \\ 3 & 8 \\ -5 & 6 \end{bmatrix}$

56. $7 \cdot \begin{bmatrix} -1 & -3 \\ 4 & 2 \\ 3 & -7 \end{bmatrix}$

57. $4 \cdot \begin{bmatrix} -3 & -1 \\ 8 & 6 \\ 7 & -4 \end{bmatrix}$

Find the scale length for each of the following. Round to the nearest thousandth if necessary. (Lesson 7-3)

58. actual length: 7 mi
scale is $\frac{1}{2}$ in.:3mi 1.167 in.

59. actual length: 12.4 yd
scale is 1 in.:2 yd 6.2 in.

60. actual length: 28.7 ft
scale is $\frac{1}{4}$ in.:5ft 1.435 in.

Extra Practice Worksheet 11-4

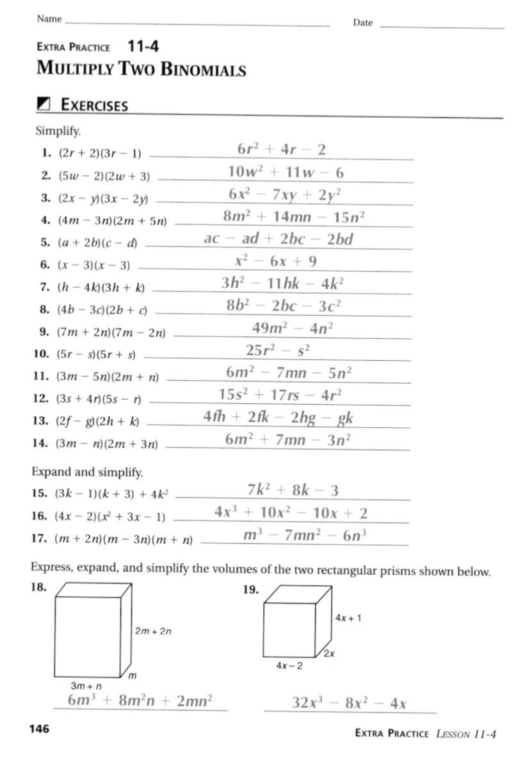

Name ______ Date ______

EXTRA PRACTICE 11-4
MULTIPLY TWO BINOMIALS

EXERCISES

Simplify.

1. $(2r + 2)(3r - 1)$ $6r^2 + 4r - 2$
2. $(5w - 2)(2w + 3)$ $10w^2 + 11w - 6$
3. $(2x - y)(3x - 2y)$ $6x^2 - 7xy + 2y^2$
4. $(4m - 3n)(2m + 5n)$ $8m^2 + 14mn - 15n^2$
5. $(a + 2b)(c - d)$ $ac - ad + 2bc - 2bd$
6. $(x - 3)(x - 3)$ $x^2 - 6x + 9$
7. $(h - 4k)(3h + k)$ $3h^2 - 11hk - 4k^2$
8. $(4b - 3c)(2b + c)$ $8b^2 - 2bc - 3c^2$
9. $(7m + 2n)(7m - 2n)$ $49m^2 - 4n^2$
10. $(5r - s)(5r + s)$ $25r^2 - s^2$
11. $(3m - 5n)(2m + n)$ $6m^2 - 7mn - 5n^2$
12. $(3s + 4r)(5s - r)$ $15s^2 + 17rs - 4r^2$
13. $(2f - g)(2h + k)$ $4fh + 2fk - 2hg - gk$
14. $(3m - n)(2m + 3n)$ $6m^2 + 7mn - 3n^2$

Expand and simplify.

15. $(3k - 1)(k + 3) + 4k^2$ $7k^2 + 8k - 3$
16. $(4x - 2)(x^2 + 3x - 1)$ $4x^3 + 10x^2 - 10x + 2$
17. $(m + 2n)(m - 3n)(m + n)$ $m^3 - 7mn^2 - 6n^3$

Express, expand, and simplify the volumes of the two rectangular prisms shown below.

18. $2m + 2n$, m, $3m + n$ — $6m^3 + 8m^2n + 2mn^2$
19. $4x + 1$, $2x$, $4x - 2$ — $32x^3 - 8x^2 - 4x$

146 EXTRA PRACTICE *Lesson 11-4*

Enrichment Worksheet 11-4

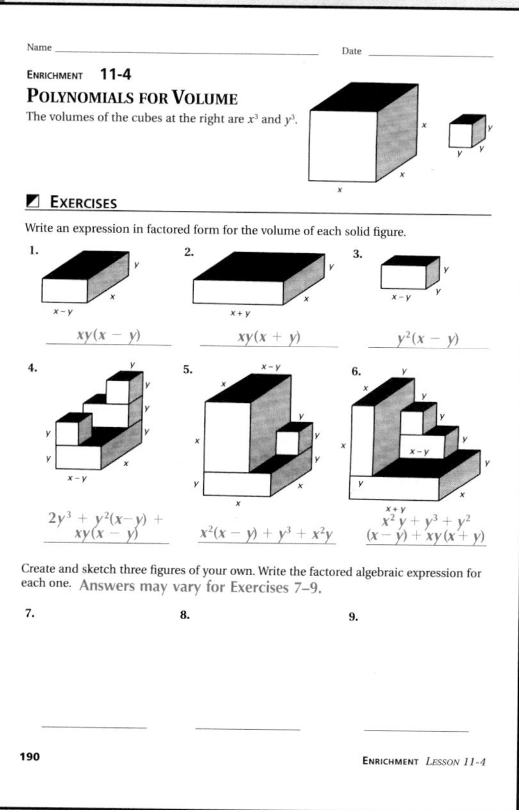

Name ______ Date ______

ENRICHMENT 11-4
POLYNOMIALS FOR VOLUME

The volumes of the cubes at the right are x^3 and y^3.

EXERCISES

Write an expression in factored form for the volume of each solid figure.

1. $xy(x - y)$
2. $xy(x + y)$
3. $y^2(x - y)$
4. $2y^3 + y^2(x - y) + xy(x - y)$
5. $x^2(x - y) + y^3 + x^2y$
6. $x^2y + y^3 + y^2(x - y) + xy(x + y)$

Create and sketch three figures of your own. Write the factored algebraic expression for each one. Answers may vary for Exercises 7–9.

7. 8. 9.

190 ENRICHMENT *Lesson 11-4*

Learning Styles

TACTILE/KINESTHETIC LEARNERS Have students model Example 2 using Algeblocks or algebra tiles. Relate FOIL to the tiles: F = large square, O = the group of rectangles, I = the single rectangle, and L = small squares. Repeat with other pairs of binomials that are the sum of a variable and a constant.

Lesson 11-4 **Multiply Two Binomials** 485

Skills Practice

Vocabulary Review

Lesson 11-3
extracting factors
greatest common factor (GCF)

Lesson 11-4
expanding
first product
outer product
inner product
last product
FOIL
cross products
difference of two squares

Assignment Guide

All students: 1–90
Additional Practice: See Extra Practice Index on page 674.

Review and Practice Your Skills

Practice ■ Lesson 11-3

Find the factors for the following.

1. $8x + 12y$ $4(2x + 3y)$
2. $6m^2 - 18n^2$ $6(m^2 - 3n^2)$
3. $7x^2 + 15x$ $x(7x + 15)$
4. $5ab + 12b$ $b(5a + 12)$
5. $2gh - ghk$ $gh(2 - k)$
6. $12pq + 24rs$ $12(pq + 2rs)$
7. $28abc - 11a^3$ $a(28bc - 11a^2)$
8. $10mn^2 - 17m^2$ $m(10n^2 - 17m)$
9. $17xy^2 + 24y^2z$ $y^2(17x + 24z)$
10. $2ab + 4bc - 8ac$ $2(ab + 2bc - 4ac)$
11. $5x^3 + 5x^2y^2$ $5x^2(x + y^2)$
12. $9r - 9r^5$ $9r(1 - r^4)$

Find the GCF and its paired factor for the following.

13. $36a + 24b$ $12(3a + 2b)$
14. $17x + 34x^2$ $17x(1 + 2x)$
15. $5ab + 10bc - 5b$ $5b(a + 2c - 1)$
16. $8mn^2 - 12m^2$ $4m(2n^2 - 3m)$
17. $18p^2q - 36pr^2$ $18p(pq - 2r^2)$
18. $14xy - 21xy^2$ $7xy(2 - 3y)$
19. $15s^2t^2 + 45s^3t$ $15s^2t(t + 3s)$
20. $24a^3b^4 + 60a^2b^3$ $12a^2b^3(2ab + 5)$
21. $4x^3 - 2x^2 + 14x$ $2x(2x^2 - x + 7)$
22. $x^2y + xy^2 + x^2y^2$ $xy(x + y + xy)$
23. $3uv - 9u^2v^2 + 3u^3v^3$ $3uv(1 - 3uv + u^2v^2)$
24. $9mn - 3m^2 + 4mn^2$ $m(9n - 3m + 4n^2)$
25. $36m^3n^5 + 72m^2n^3 + 54m^5n^2$ $18m^2n^2(2mn^3 + 4n + 3m^3)$
26. $45x^2y^2 + 65u^3v - 35s^4t^2$ $5(9x^2y^2 + 13u^3y - 7s^4t^2)$
27. $6a^2bc + 2ab^2c - 4abc$ $2abc(3a + b - 2)$
28. $15y^4z + 10y^2z^2 - 20yz$ $5yz(3y^3 + 2yz - 4)$
29. $8mnp - 20m^2np^3 + 16mn^4p^2$ $4mnp(2 - 5mp^2 + 4n^3p)$
30. $32xy^3 + 100x^2y + 2xy$ $2xy(16y^2 + 50x + 1)$

Practice ■ Lesson 11-4

Simplify.

31. $(x + 2)(x - 3)$ $x^2 - x - 6$
32. $(2x + 1)(3x - 5)$ $6x^2 - 7x - 5$
33. $(x + 2y)(2x + 3y)$ $2x^2 + 7xy + 6y^2$
34. $(3x + 2)(3x - 2)$ $9x^2 - 4$
35. $(5x + 4)(5x + 4)$ $25x^2 + 40x + 16$
36. $(7x - 4y)(8 + 3s)$ $56x + 21sx - 32y - 12s$
37. $(m - 5n)(4p + 5m)$ $4mp + 5m^2 - 20np - 25mn$
38. $(w + 3)(3 - w)$ $-w^2 + 9$
39. $(a - 6b)(3a - 5b)$ $3a^2 - 23ab + 30b^2$
40. $(2r - 7s)(5r - 3t)$ $10r^2 - 6r^2 - 35rs + 21st$
41. $(x + 6)(x - 6)$ $x^2 - 36$
42. $(8x - 3)(8x - 3)$ $64x^2 - 48x + 9$
43. $(8x - 3)(8x + 3)$ $64x^2 - 9$
44. $(a + b)(c + d)$ $ac + ad + bc + bd$
45. $(4y + 9z)(2y - 5z)$ $8y^2 - 2yz - 45z^2$
46. $(5 - 2x)(11 + 5x)$ $55 + 3x - 10x^2$
47. $(x + 1)(y + 2)$ $xy + 2y + y + 2$
48. $(10c - 13d)(2d + 3c)$ $30c^2 - 19cd - 26d^2$
49. $(9x - 1)(9x + 1)$ $81x^2 - 1$
50. $(9x + 1)(9x + 1)$ $81x^2 + 18x + 1$
51. $(8p + 8q)(8p + 8q)$ $64p^2 + 128pq + 64q^2$
52. $(x^2 + 1)(2x + 1)$ $2x^3 + x^2 + 2x + 1$
53. $(z^2 + 5)(z^2 - 5)$ $z^4 - 25$
54. $(x + 3)(3x^2 - 1)$ $3x^5 + 9x^2 - x - 3$
55. $(2r + 3s)(4r - 6s)$ $8r^2 - 18s^2$
56. $(4m + 13)(13m - 4)$ $52m^2 + 153m - 52$
57. $(-7c + 3d)(6c - 5d)$ $-42c^2 + 53cd - 15d$
58. $2(m + 17)(m - 1)$ $2m^2 + 32m - 34$
59. $x(x + 4)(x + 13)$ $x^3 + 17x^2 + 52x$
60. $(x + 5)(x - 5)(x + 5)$ $x^3 + 5x^2 - 25x - 125$
61. The dimensions of a rectangle are $(7x - 5)$ feet and $(2x + 3)$ feet. Write and simplify an expression for the area of the rectangle. $(7x - 5)(2x + 3) = 14x^2 + 11x - 15$
62. Explain the difference between $(x + 4)(x - 4)$ and $(x - 4)(x - 4)$.
$x^2 - 16$; binomial
$x^2 - 8x + 16$; trinomial

Teaching Strategies

COOPERATIVE LEARNING Have students work in pairs to write problems about area and volume using monomials and binomials as factors. Have pairs exchange problems and solve. Students should work together to check their work.

486 Chapter 11 **Polynomials**

PRACTICE ■ LESSON 11-1–LESSON 11-4

Simplify. (Lesson 11-1)

63. $(-8x + 7y) + (11y - 5x)$ $-13x + 18y$

64. $(21b - 16) - (-13b + 7)$ $34b - 23$

65. $(5x^2 + 9x - 10) + (-14x^2 + 11x + 10)$ $-9x^2 + 20x$

66. $(gh - gh^2 + 3g^2h) + (g^2h + 5gh - gh^2)$ $6gh - 2gh^2 + 4g^2h$

Simplify. (Lesson 11-2)

67. $-6x(5x - 11)$ $-30x^2 + 66x$

68. $8n^2(n^2 - 7n)$ $8n^4 - 56n^3$

69. $5x^2(4x^2 - 3x + 1)$ $20x^4 - 15x^3 + 5x^2$

70. $-9c^3d(-7d^3 + 2c)$ $63c^3d^4 - 18c^4d$

71. $-4pqr(2p^3q - 5\,pr - 3p^3q^2)$ $-8p^4q^2r + 20p^2qr^2 + 12p^4q^3r$

72. $10a^3bc^2(4a^2b^2c - 3a^2bc^3)$ $40a^5b^3c^3 - 30a^5b^2c^5$

Find the GCF and its paired factor for the following. (Lesson 11-3)

73. $-30x + 54$ $6(-5x + 9)$

74. $18m - 30n$ $6(3m - 5n)$

75. $12g + 25g^2$ $g(12 + 25g)$

76. $4a^2 - 16a$ $4a(a - 4)$

77. $45r^2st^3 + 75rs^2t^2$ $15rst^2(3rt + 5s)$

78. $-26xyz + 52x^2yz^2$ $26xyz(-1 + 2xz)$

79. $48a^3b^2 + 56a^2b^4 - 32a^4b^3$ $8a^2b^2(6a + 7b^2 - 4a^2b)$

80. $ab - abc + abcd - abcde$ $ab(1 - c + cd - cde)$

81. $8x^3 + 6x^2 + 4x + 2$ $2(4x^3 + 3x^2 + 2x + 1)$

Simplify. (Lesson 11-4)

82. $(3x + 2)(3y + 2z)$ $9xy + 6xz + 6y + 4z$

83. $(7a - b)(7 + b)$ $49a + 7ab - 7b - b^2$

84. $(5m + 9n)(-2m + 3p)$ $-10m^2 + 15mp - 18mn + 27np$

85. $(4x + 3)(4x + 3)$ $16x^2 + 24x + 9$

86. $(4x + 3)(4x - 3)$ $16x^2 - 9$

87. $(8p + 7q)(6p - 5q)$ $48p^2 + 2pq - 35q^2$

88. $(8x + y)(4y - 7x)$ $-56x^2 + 25xy + 4y^2$

89. $(2x + 1)(1 - 2x)$ $-4x^2 + 1$

90. $(a - 11b)(5a - 13b)$ $5a^2 - 68ab + 143b^2$

Mid-Chapter Quiz

1. Write $x^2y^2 + 3xy^3 + 4x^3y - 5$ in standard form for the variable x. $4x^3y + x^2y^2 + 3xy^3 - 5$

2. Write $2 - 4x^3 + 3x^2y^3 + 6$ in standard form for the variable y. $3x^2y^3 + y - 4x^3 + 2$

Simplify.

3. $(5y - 2z) - (3y - 5z)$ $2y + 3z$

4. $(3x^2 + 4x - 5) + (x^2 - 3x + 8)$ $4x^2 + x + 3$

5. $(a^2 - 5ab + 2b^2) - (ab + b^2)$ $a^2 - 6ab + b^2$

6. $(-8p)(-2q)$ $16pq$

7. $-t^4(t^2 + u)$ $-t^6 - t^4u$

8. $2v^2(3v^3 + 2v - 3)$ $6v^5 + 4v^3 - 6v^2$

9. Write and simplify an expression for the area of a rectangle that has a length of $3x$ and a width of $(x^2 - y + 4)$. $3x^3 - 3xy + 12x$

Find factors for the following.

10. $6x - 9y$ $3(2x - 3y)$

11. $6a^3b - 4a^2b^2$ $2a^2b(3a - 2b)$

12. $3km^2n + 2mn^2 - 6k^2n$ $n(3km^2 + 2mn - 6k^2)$

Simplify.

13. $(c - d)(4g + 3h)$ $4cg + 3ch - 4dg - 3dh$

14. $(12r + s)(3s - t)$ $6rs - 2rt + 3s^2 - st$

15. $(2k - 4)(2k + 4)$ $4k^2 - 16$

16. $(z + 6)(z + 6)$ $z^2 + 12z + 36$

17. $(3b - c)(2b - 3c)$ $6b^2 - 11bc + 3c^2$

18. $(x + 4)(x - 8)$ $x^2 - 4x - 32$

Chalkboard Examples

Example from Lesson 11-3

Find factors for $2p^2q - 5pq^2r + 10p^3q^2$. $pq(2p - 5qr + 10p^2q)$

Example from Lesson 11-4

Simplify $(2x + 3y)(4x - 5y)$.
$8x^2 + 2xy - 15y^2$

Mid-Chapter Quiz

Correlation Chart

Question Number	Lesson Number
1–5	11-1
6–9	11-2
11–12	11-3
6–16	11-4

Teaching Strategies

CHALLENGE How many rectangular prisms with different dimensions are possible that also have a volume of $12x^2y$? Assume that x and y represent whole numbers, and all dimensions are a whole number of units. 72

Lesson Planning

NCTM Standards/Strands

- Number & Operation
- Patterns, Functions, & Algebra
- Problem Solving
- Reasoning and Proof
- Communication
- Connections
- Representation

Vocabulary

grouping
monomial factor
polynomial factor

Materials Needed

paper/pencil

Lesson Resources

Warm-Up Transparency 32
Transparency TK-4
Reteaching 11-5
Extra Practice 11-5
Enrichment 11-5

Assignment Guide

Basic: 1-38, 43-56
Enriched: 1-56

Getting Started

5-Minute Warm-up

Find the GCF of both terms in each binomial.

1. $6a + 15b^2$ 3
2. $5x^2 - 2x^3$ x^2
3. $4y^2z + 8yz^3$ $4yz$

Introduction to Lesson 11-5

After students have completed the opening activity, ask them to share their explanations for the different types of results. Then ask groups to write examples of each type of binomial factor pair using at least one constant in each example.

11-5 Find Binomial Factors in a Polynomial

Goals ■ Factor polynomials by grouping.

Applications Manufacturing, Design, Sales

Work in groups of 2 or 3 students.

As you know, multiplying a polynomial by a monomial does not change the number of terms. The answer has exactly as many terms as the polynomial you started with. But multiplying by a binomial is not so predictable.

1. Multiply each of the following pairs, and simplify each result.

   $(a + b)(c + d)$ $ac + ad + bc + bd$

   $(a + b)(a + b)$ $a^2 + 2ab + b^2$

   $(a + b)(a - b)$ $a^2 - b^2$

2. The polynomials that result from these multiplications each have a different number of terms. Examine the three calculations and explain why there is a difference. Focus on what happens to the inner and outer products when you simplify each expression.

Build Understanding

You have seen that you can often extract a monomial factor from a polynomial. You may also be able to extract a binomial factor. Finding binomial factors is more complex, however, because of the greater variety of possible answers when you multiply by a binomial.

This lesson focuses on the $(a + b)(c + d)$ pattern you explored in the activity above. In this multiplication, the resulting polynomial has twice the terms of each polynomial that was multiplied.

When you factor a polynomial, the first step is always to look for a common monomial factor in all terms. If you find one (the GCF), extract it. The next step is to search for a binomial factor. If the number of terms in the polynomial is even, proceed as follows:

1. Group the terms in the polynomial as pairs that share a monomial factor.
2. Extract the monomial factor from each pair.
3. If the binomials that remain for each pair are identical, write this as a binomial factor of the whole expression.
4. The monomials you extracted create a second polynomial. This is the paired factor for the original expression.

Teaching Strategies

Caution students that there are some polynomials that cannot be factored; however, if they group terms in one way and cannot complete the factorization they should try grouping the terms another way instead of simply deciding the polynomial cannot be factored.

Example 1

Find factors for $4x^3 + 4x^2y^2 + xy + y^3$.

Solution

1. Check for a monomial factor for the whole expression. There is none.
2. Within the polynomial, make pairs of terms that share monomial factors.

   $(4x^3 + 4x^2y^2) + (xy + y^3)$ or $(4x^3 + xy) + (4x^2y^2 + y^3)$
3. Extract the monomial factors in each pair.

   $4x^2(x + y^2) + y(x + y^2)$ or $x(4x^2 + y) + y^2(4x^2 + y)$
4. The binomials left in each pair are identical, so they are a factor of the whole polynomial. The binomial can be extracted; the monomials create a second factor as follows.

   $(x + y^2)(4x^2 + y)$ or $(4x^2 + y)(x + y^2)$

Note that these are in fact the same expression, owing to the fact that multiplication is commutative.

Example 2

MANUFACTURING The volume of a box is $4pr - 6ps - 4qr + 6qs$. Find the dimensions of the box. (Hint: Volume is the product of three factors.)

Solution

Check for a monomial factor for the whole expression. The constant 2 can be extracted: $2(2pr - 3ps - 2qr + 3qs)$.

$2[(2pr - 3ps) - (2qr - 3qs)]$	or	$2[(2pr - 2qr) - (3ps - 3qs)]$
$= 2[p(2r - 3s) - q(2r - 3s)]$		$= 2[2r(p - q) - 3s(p - q)]$
$= 2(2r - 3s)(p - q)$		$= 2(p - q)(2r - 3s)$

Note that in the first step, the last sign had to be changed when the terms were grouped. Can you see why? There is a minus sign before the second group.

Example 3

Find factors for $2x^3 - 2x^2y - 3xy^2 + 3y^3 + xz^2 - yz^2$.

Solution

There is no shared monomial factor. Pair terms in the remaining polynomial, and factor if possible.

$(2x^3 - 2x^2y) - (3xy^2 - 3y^3) + (xz^2 - yz^2) =$

$2x^2(x - y) - 3y^2(x - y) + z^2(x - y) =$

$(x - y)(2x^2 - 3y^2 + z^2)$

Once again, note the sign changes during the grouping process.

Problem Solving Tip

There may be more than one way to pair terms. You may need to try several approaches to find the one that works best.

Extend the Lesson

MATH JOURNAL In their own words, have students write the steps for factoring a polynomial with an even number of terms. You may wish to have students include an example.

Chalkboard Examples

Supplementary Example 1

Find factors for $6ab - 9a + 4b - 6$.
$(3a + 2)(2b - 3)$

Supplementary Example 2

The volume of a rectangular prism is $12df - 3dg - 24ef + 6eg$. Find the dimensions of the prism by factoring. $3(d - 2e)(4f - g)$

Supplementary Example 3

Find factors for $2a^3 + 2a^2b - 3ab^2 + ac^2 + bc^2 - 3b^3$.
$(a + b)(2a^2 - 3b^2 + c^2)$

Reteaching Worksheet 11-5

Name ______________________ Date __________

RETEACHING 11-5

FINDING BINOMIAL FACTORS IN A POLYNOMIAL

In developing a strategy to find binomial factors, it is important to follow this strategy.

Step 1: First look for a common monomial factor.
Step 2: Within the polynomial, make pairs of terms that share monomial factors.
Step 3: Extract the monomial factors in each pair.
Step 4: When the binomials in each pair are identical, the monomials create a second binomial factor.

Example 1

Find factors for $4ac + 12bc - 2ad - 6bd$.

Solution

$4ac + 12bc - 2ad - 6bd$
$= 2(2ac + 6bc - ad + 3bd)$ Step 1
$= 2[(2ac + 6bc) - (ad + 3bd)]$ Step 2
$= 2[2c(a + 3b) - d(a + 3b)]$ Step 3
$= 2(2c - d)(a + 3b)$ Step 4

Example 2

Find factors for $ax + bx + cx + ay + by + cy$.

Solution

There is no common monomial factor. Go to Step 2.
$ax + bx + cx + ay + by + cy$
$= (ax + bx + cx) + (ay + by + cy)$ Step 2
$= x(a + b + c) + y(a + b + c)$ Step 3
$= (x + y)(a + b + c)$ Step 4

EXERCISES

Find factors for each polynomial.

1. $10ac - 4ad + 15bc - 6bd$ ____ $(2a + 3b)(5c - 2d)$
2. $ac - ad + 2bc - 2bd$ ____ $(a + 2b)(c - d)$
3. $eg + 3eh - fg - 3fh$ ____ $(e - f)(g + 3h)$
4. $12wy - 6wz + 8xy - 4xz$ ____ $2(3w + 2x)(2y - z)$
5. $24pr - 40ps - 12qr + 20qs$ ____ $4(2p - q)(3r - 5s)$
6. $3ac - 3ad + 2bc - 2bd$ ____ $(3a + 2b)(c + d)$
7. $36wy - 8wz - 9xy + 2xz$ ____ $(4w - x)(9y - 2z)$
8. $9ac + 3ad + 27bc + 9bd$ ____ $3(a + 3b)(3c + d)$
9. $8qs + 16qt - 2rs - 4rt$ ____ $2(4q - r)(s + 2t)$
10. $2ax + 2ay + 2az + 3bx + 3by + 3bz$ ____ $(2a + 3b)(x + y + z)$
11. $6ax + 4ay + 8az - 3bx - 2by - 4bz$ ____ $(2a - b)(3x + 2y + 4z)$
12. $18mr - 4ms - 6mt + 36nr - 8ns - 12nt$ ____ $2(m + 2n)(9r - 2s - 3t)$

168 RETEACHING *LESSON 11-5*

Lesson Wrap-up

Quick Assessment

Ask the following questions to determine if students understand the content presented in this lesson.

1. Can all polynomials be factored? No. Ask students to factor $a^3 - a^2 - a + 2$. Then discuss why this expression cannot be factored.
2. Which properties are used in factoring polynomials? Commutative, associative and distributive; ask students to give examples of each.

Assignment Guide

Basic: 1–38, 43–56
Advanced: 1–56
Additional Practice: See Extra Practice Index on page 674.

Try These Exercises

Find factors for the following.

1. $9wx + 6wz + 6xy + 4yz$ $(3w + 2y)(3x + 2z)$
2. $2e^2 + 14ef + 3eg + 21fg$ $(2e + 3g)(e + 7f)$
3. $18ab - 27ad - 8bc + 12cd$ $(9a - 4c)(2b - 3d)$
4. $3x^3 - 12x^2y - xy + 4y^2$ $(3x^2 - y)(x - 4y)$
5. $5rs - 40rt + 3s - 24t$ $(5r + 3)(s - 8t)$
6. $24p^3 - 18p^2q + 4pq - 3q^2$ $(6p^2 + q)(4p - 3q)$
7. $kl + mn + ml + kn$ $(k + m)(\ell + u)$
8. $8rs + 3tu + 2st + 12ru$ $(4r + t)(2s + 3u)$
9. $3mr - 8ms + 5mt - 9nr + 24ns - 15nt$ $(m - 3n)(3r - 8s + 5t)$

LANDSCAPING In the exercises below, the areas of two rectangular lawns are expressed as polynomials. Find binomial expressions for the sides (one is given).

10. Area: $6ab + 4ac + 3bd + 2cd$; side: $2a + d$ $(3b + 2c)$

11. 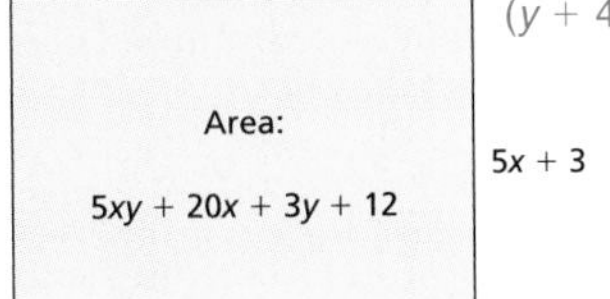

side: $5x + 3$ $(y + 4)$

12. **WRITING MATH** Suppose you are asked to factor $12pq + 8p - 3q - 2$. How would you decide the best way to group the terms? Explain your thinking. Strategies will vary. Correct factoring: $(4p - 1)(3q + 2)$

Practice Exercises

Find factors for the following.

13. $4ab + 6ad + 6bc + 9cd$ $(2a + 3c)(2b + 3d)$
14. $4a^2 + 6ab + 6ac + 9bc$ $(2a + 3b)(2a + 3c)$
15. $4qr + 12qt + sr + 3st$ $(4q + s)(r + 3t)$
16. $4q^2 + 12qs + qr + 3rs$ $(4q + r)(q + 3s)$
17. $21ef - 12eh - 7fg + 4gh$ $(3e - g)(7f - 4h)$
18. $21e^2 - 12e - 7ef + 4f$ $(3e - f)(7e - 4)$
19. $27w^2x - 18w^2z^2 - 3xy + 2yz^2$ $(9w^2 - y)(3x - 2z^2)$
20. $27 - 18z - 3y + 2yz$ $(9 - y)(3 - 2z)$
21. $2k^2l^2 + 5k^2n - 6l^2m - 15mn$ $(k^2 - 3m)(2l^2 + 5n)$
22. $2kl + 5k - 6l - 15$ $(k - 3)(2l + 5)$
23. $15tu + 20t - 6vu - 8v$ $(5t - 2v)(3u + 4)$
24. $15tu + 20tv - 6u - 8v$ $(5t - 2)(3u + 4v)$
25. $3x^2y - x^2z + 24y - 8z$ $(x^2 + 8)(3y - z)$
26. $3x^2y - x^2 + 24y - 8$ $(x^2 + 8)(3y - 1)$
27. $vy + 5vz + 3wy + 15wz + 2xy + 10xz$ $(y + 5z)(v + 3w + 2x)$
28. $6j^2m^2 - 42j^2n + 5km^2 - 35kn - 3lm^2 + 21ln$ $(m^2 - 7n)(6j^2 + 5k - 3l)$
29. $10pr - 15ps + 20pt - 2qr + 3qs - 4qt$ $(5p - q)(2r - 3s + 4t)$
30. $12a^2d + 4a^2e^2 - 6bd - 2be^2 - 15cd - 5ce^2$ $(3d + e^2)(4a^2 - 2b - 5c)$
31. $6df - 20eg - 35eh - 10ef + 12dg + 21dh$ $(3d - 5e)(2f + 4g + 7h)$

Find factors for the following.

32. $8x^2 + 4xz + 4xy + 2yz$ $2(2x + y)(2x + z)$
33. $6j^3 - 12j^2l + 3j^2k - 6jkl$ $3j(2j + k)(j - 2l)$
34. $3abd - 3abe - 3acd + 3ace$ $3a(b - c)(d - e)$
35. $3r^4 + 6r^3t - 6r^3s - 12r^2st$ $3r^2(r - 2s)(r + 2t)$

Teaching Strategies

CHALLENGE The length and width of a rectangle are both whole numbers. Find the smallest possible area of the rectangle if it is represented by $2xy - 6x + 5y - 15$. Hint: Find binomial expressions for the length and width. 1 square unit; $(2x + 5)$ and $(y - 3)$

Factoring can make calculations easier. For Exercises 36–37, calculate the value of each expression twice. First, calculate each term separately. Then factor the expressions before you calculate value.

36. DESIGN Changing the design of a computer monitor has decreased the cost of manufacturing the monitor. The change in cost is represented by the expression $8pr - 2qr - 20ps + 5qs$. Find the amount of change if $p = 2.1$, $q = 2.4$, $r = 0.5$ and $s = 1.2$. −30

37. SALES The number of units sold (in millions) of a new video game is represented by the expression $21x^2 - 14xz + 9xy - 6yz$. Find the number of sales if $x = 0.3$, $y = 0.9$ and $z = 0.2$. 2.4

38. ERROR ALERT When Monica attempts to factor $2a^2c^3 - 4a^2d - 4bc^3 + 8bd$, she gets $2a^2(c^3 - 2d) - 4b(c^3 + 2d)$. What mistake did Monica make? Monica needed to change the sign before $8bd$ when the terms were grouped.

Extended Practice Exercises

39. The area of the figure to the right is expressed as a polynomial. Find binomial expressions for the sides. (There are two possible answers.) $(2m - 6)$ and $(n - 3)$ or $(m - 3)$ and $(2n - 6)$

Area:

$2mn - 6m - 6n + 18$

SHIPPING The volumes of the boxes below are expressed as polynomials. Find expressions for the sides (one is given).

40. 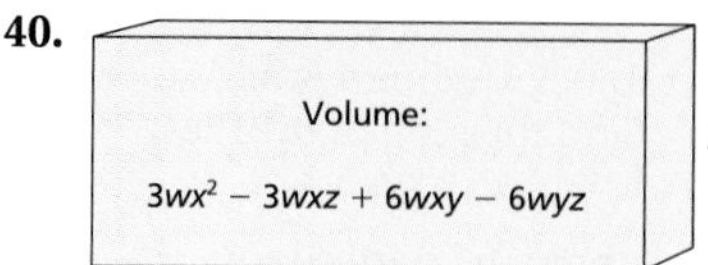

$3w(x + 2y)(x - z)$

41. 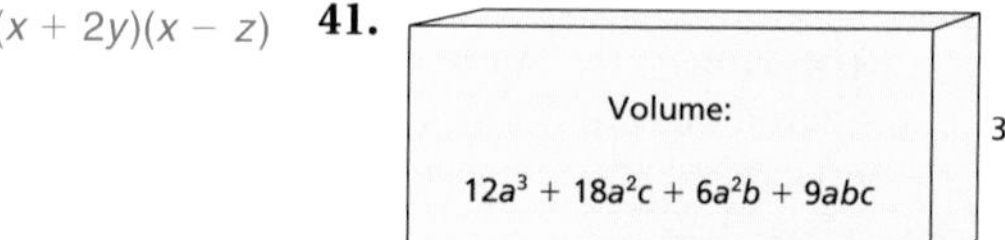

$3a(2a + b)(2a + 3c)$

42. Find the binomial expression for the base and height of this right triangle. (Hint: Remember the formula for the area of a triangle includes $\frac{1}{2}$.) The possible answers are $(4x + 2z)$ and $(2x + y)$ or $(2x + z)$ and $(4x + 2y)$

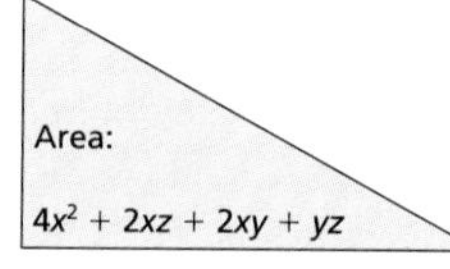

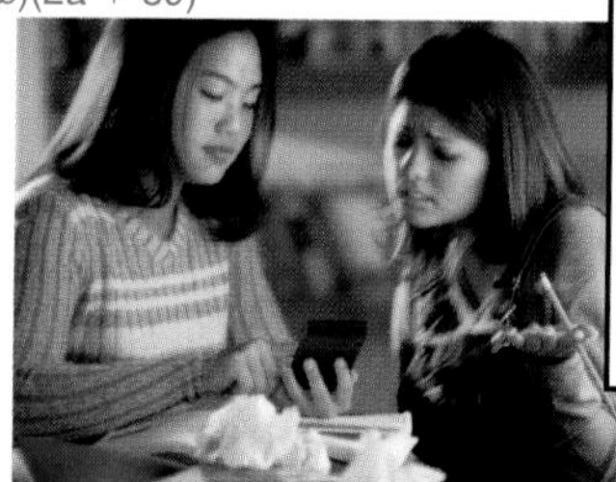

Mixed Review Exercises

Find the value of x in each. (Lesson 10-4)

43. 50°

x, 100°

44. 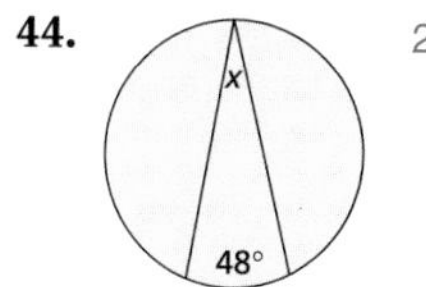

24°

45. 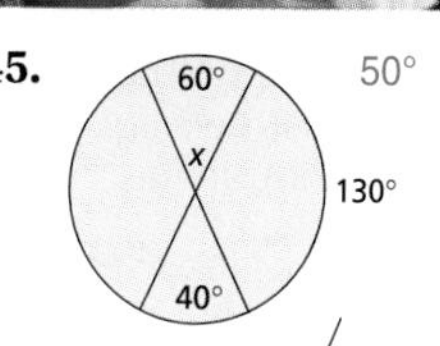

50°

46. 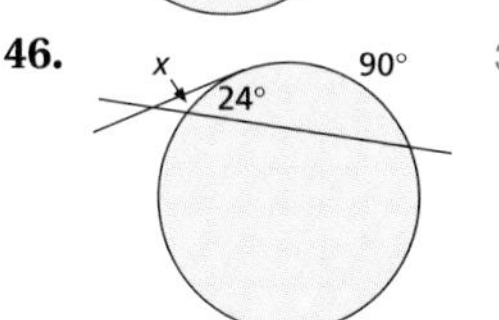

33°

47. 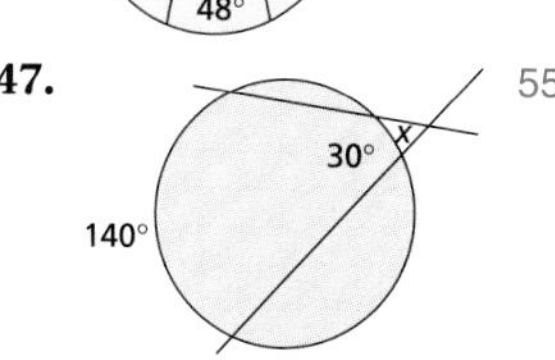

55°

48. 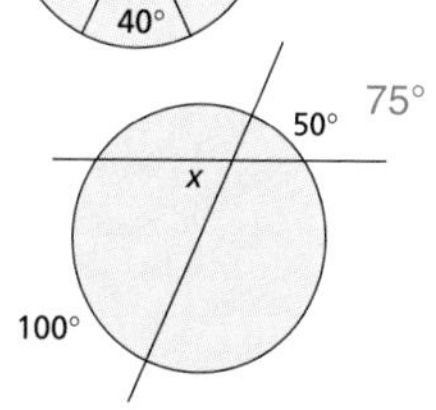

75°

Evaluate each product when $a = 4$, $b = -2$ and $c = -\frac{1}{2}$. (Lesson 1-7)

49. $7ab$ −56 **50.** $3(abc)$ 12 **51.** $2a + 3b$ 2 **52.** $-4(a)(b)(c)$ −16

53. $12bc$ 12 **54.** $-5ac$ 10 **55.** $3c - 4b$ $6\frac{1}{2}$ **56.** $2(a)(c)$ −4

Extra Practice Worksheet 11-5

Name ______________ Date ______________

EXTRA PRACTICE 11-5

FIND BINOMIAL FACTORS IN A POLYNOMIAL

EXERCISES

Find factors for the following.

1. $16ab - 20ad + 12bc - 15cd$ $(4a + 3c)(4b - 5d)$
2. $6x^2 + 2xz - 3xy - yz$ $(2x - y)(3x + z)$
3. $m^2 + mp - 4mn - 4np$ $(m - 4n)(m + p)$
4. $12r^2 + 18rt - 2rs - 3st$ $(6r - s)(2r + 3t)$
5. $35hj + 14hm - 5jk - 2km$ $(7h - k)(5j + 2m)$
6. $12w^2 - 18wz - 2xz + 3xz$ $(6w - x)(2w - 3z)$
7. $4x^2y + 3x^2z + 20y + 15z$ $(x^2 + 5)(4y + 3z)$
8. $6fh - 16fk + 3hg - 8gk$ $(2f + g)(3h - 8k)$
9. $2rs + 2rt + 2rv - s - t - 4v$ $(2r - 1)(s + t + 4v)$
10. $3x^3 + 2x^2z - x^2w + 3xy + 2yz - wy$ $(x^2 + y)(3x + 2z - w)$
11. $ab^2 - ad + 4af + b^2c - cd + 4cf$ $(a + c)(b^2 - d + 4f)$
12. $8mp^2 + 4mq - 8m - 2n^2p^2 - n^2q + 2n^2$ $(4m - n^2)(2p^2 + q - 2)$
13. $24a - 8ac - 12b + 4bc$ $4(2a - b)(3 - c)$
14. $24fdg + 18df - 8fg - 6f$ $2f(3d - 1)(4g + 3)$
15. The area of the figure at the right is expressed as a polynomial. Find the binomial expressions of the two sides. $3x - 4$ and $y + 2$

Area: $3xy + 6x - 4y - 8$

16. Find the binomial expression for the base and height of this right triangle. $4m + 10n$ and $p - 4$

Area: $4mp - 16m + 10np - 40n$

148 EXTRA PRACTICE *LESSON 11-5*

Enrichment Worksheet 11-5

Name ______________ Date ______________

ENRICHMENT 11-5

PERFECT NUMBERS

A number is *perfect* if it equals the sum of its divisors that are less than the integer itself. The divisors of 28 (excluding 28 itself) are 1, 2, 4, 7 and 14. Since 1 + 2 + 4 + 7 + 14 = 28, 28 is a perfect number.

If the sum of the divisors (excluding the integer itself) is greater than the integer, the integer is *abundant*.

If the sum of the divisors (excluding the integer itself) is less than the integer, the integer is *deficient*.

EXERCISES

Find the proper divisors of each number. Then classify it as perfect, abundant, or deficient.

	Number	Proper Divisors	Sum	Type
1.	14	1, 2, 7	10	Deficient
2.	6	1, 2, 3	6	Perfect
3.	12	1, 2, 3, 4, 6	16	Abundant
4.	20	1, 2, 4, 5, 10	22	Abundant
5.	30	1, 2, 3, 5, 6, 10, 15	42	Abundant
6.	56	1, 2, 4, 7, 8, 14, 28	64	Abundant
7.	79	1	1	Deficient
8.	148	1, 2, 4, 37, 74	118	Deficient
9.	212	1, 2, 4, 53, 106	166	Deficient
10.	496	1, 2, 4, 8, 16, 31, 62, 124, 248	496	Perfect
11.	1001	1, 7, 143	151	Deficient

12. For n = 2, 3, 5, 7, 13, 17 and 19, the expression $2^{n-1}(2^n - 1)$ gives a perfect number. Find the first seven perfect numbers.
6; 28; 496; 8, 128; 33, 550, 336; 8, 589, 869, 056; 137, 438, 691, 328

192 ENRICHMENT *LESSON 11-5*

Teaching Strategies

Ask each student to sketch a rectangle or rectangular prism and write a polynomial with four terms to describe its area or volume. Then have students trade figures and find a set of dimensions for each figure. Discuss: Could a figure have more than one set of answers? **Yes, if there is more than one way to factor the polynomial.**

Lesson Planning

NCTM Standards/Strands

- Number & Operation
- Patterns, Functions, & Algebra
- Problem Solving
- Reasoning and Proof
- Connections
- Representation

Vocabulary

perfect square trinomial
quadratic equation

Materials Needed

Algeblocks or algebra tiles

Lesson Resources

Warm-Up Transparency 32
Transparency TK-4, RF-44
Reteaching 11-6
Extra Practice 11-6
Enrichment 11-6

Assignment Guide

Basic: 1–40, 43–50
Enriched: 1–50

Getting Started

5-Minute Warm-Up

Find the positive and negative square roots of each.

1. 9 3; −3
2. 64 8; −8
3. 121 11; −11

Introduction to Lesson 11-6

Have students share the patterns they found in this activity.

11-6 Special Factoring Patterns

Goals
- Factor perfect square trinomials and differences of perfect squares.
- Use factoring to solve quadratic equations.

Applications Manufacturing, Landscaping, Art

Work with a partner to find patterns.

These two diagrams represent $(x + 2)^2$ and $(x + 5)^2$.

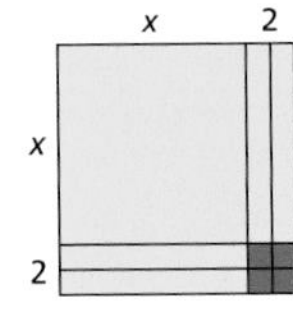

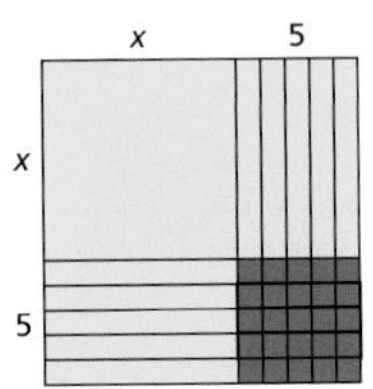

1. Express the area of each diagram as a trinomial. Do you see a pattern? $x^2 + 4x + 4$; $x^2 + 10x + 25$
2. **MODELING** Use Algebra manipulatives such as Algeblocks to build other squared binomials; for example, $(x + 1)^2$ or $(x + 3)^2$. Find the sum of the tiles and express the areas as trinomials. $(x + 1)^2 = x^2 + 2x + 1$, $(x + 3)^2 = x^2 + 6x + 9$
3. Discuss with your partner any patterns that you see. Apply the pattern to express $(x + 4)^2$ as a trinomial. The pattern is $x^2 + 2cx + c^2$, where c is the constant. $(x + 4)^2 = x^2 + 8x + 16$

Build Understanding

Finding binomial factors in polynomials with an even number of terms can be handled by pairing terms. Factoring a trinomial requires different strategies.

One strategy is to look for special patterns. You have already seen one such pattern—the difference of two squares. You can review this pattern by studying Example 3 on page 483. The activity above illustrates another pattern—the **perfect square trinomial**. Every binomial multiplied by itself fits this pattern.

Reading Math

Perfect square trinomials include squared negative binomials like $[-(a - b)]^2$, though this book does not explore all the negative options.

	Pattern of trinomial	How it relates to binomial
First term	A perfect square	The square of the binomial's first term
Last term	A perfect square	The square of the binomial's last term
Middle term	The square roots of the two perfect squares multiplied together, and then doubled	The product of the binomial's terms, multiplied by two

If you spot this pattern in a trinomial, you can always find its binomial factors.

Extend the Lesson

VISUAL LEARNERS To make a physical model of the difference of two squares, use an x-by-x square. From one corner, cut a small square y by y. Cut on the diagonal shown and arrange the pieces to form a rectangle.

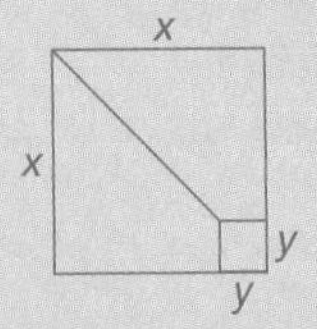

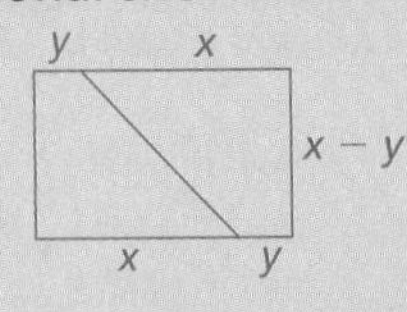

Example 1

Can you find binomial factors for the following?

a. $s^2 + 10s + 25$ **b.** $a^2 - 2ab + b^2$

Solution

a. The first term, s^2, is a perfect square. Therefore, the binomials' first terms would be s (or $-s$).

The last term, 25, is also a perfect square, so the binomials' last terms would be 5 or -5.

The middle term, $10s$, *does* equal $s \times 5 \times 2$. Therefore, the trinomial *is* a perfect square trinomial.

$s^2 + 10s + 25 = (s + 5)(s + 5)$

b. The first term, a^2, is a perfect square. Therefore, the binomials' first terms would be a (or $-a$).

The last term, b^2, is a perfect square. Therefore, the binomials' last terms would be b or $-b$.

The middle term, $(-2ab)$, is $a \times (-b) \times 2$, so the trinomial *is* a perfect square.

$a^2 - 2ab + b^2 = (a - b)(a - b)$

Check Understanding

Once again, negative options are not explored for the first quadratic terms. Is there a difference between $-(s)^2$, $(-s)^2$, and $-s^2$?

$-(s)^2 = -s^2$, but $(-s)^2 = s^2$

You may realize that the difference of two squares is also a special pattern that can be used for finding binomial factors. The difference of two squares is easy to recognize, because it is described fully by its name.

Example 2

MANUFACTURING Two rectangular metal covers have areas of $x^2 - 4$ and $25p^2 - 4q^2$. Both areas are examples of the difference of two squares. Find the dimensions of the metal covers by finding the binomial factors of each.

a. $x^2 - 4$ **b.** $25p^2 - 4q^2$

Solution

a. The first term, x^2, is a perfect square, so the first term of both binomials will be x. The second term, 4, is also a perfect square, so the binomials' second terms will be 2 and -2, respectively.

$x^2 - 4 = (x + 2)(x - 2)$

b. The first term, $25p^2$, is a perfect square, so the binomials' first terms would be $5p$ or $-5p$. The last term, $4q^2$, is a perfect square, so the binomials' second terms will be $2q$ and $-2q$.

$25p^2 - 4q^2 = (5p + 2q)(5p - 2q)$

We can use factoring to solve certain equations. Consider the equation $x^2 + 16 = 8x$. The variable x appears in an x^2-term as well as an x-term. This type of equation is called a **quadratic equation**. Solutions can be found for a quadratic equation by factoring.

Chalkboard Examples

Supplementary Example 1

Find the binomial factors for $c^2 + 6c + 9$. $(c + 3)^2$

Supplementary Example 2

Find the binomial factors for $r^2 - 49$ and $9y^2 - 4$. $(r - 7)(r + 7)$ and $(3y - 2)(3y + 2)$

Supplementary Example 3

Determine the possible solutions for $x^2 - 25$. 5 or -5

Teaching Strategies

CHALLENGE If $x^2 + y^2x + y^2$ is a perfect square trinomial, what is the value of y? 2 or -2

Reteaching Worksheet 11-6

Name ____________ Date ____________

RETEACHING **11-6**

SPECIAL FACTORING PATTERNS

You can use these patterns to help you factor a **perfect-square trinomial** or the **difference of two squares**.

Perfect square trinomial, addition:	$a^2 + 2ab + b^2 = (a + b)^2$
Perfect square trinomial, subtraction:	$a^2 - 2ab + b^2 = (a - b)^2$
Difference of two squares:	$a^2 - b^2 = (a + b)(a - b)$

Example 1

Find binomial factors for $s^2 - 10s + 25$.

Solution

The first and last terms are perfect squares: $s^2 = s \cdot s$ and $25 = 5 \cdot 5$.
The middle term is twice the square roots of the first and last terms: $2 \cdot 5 \cdot s$, or $10s$
The sign in each binomial must be negative because the middle term of the trinomial is negative and the last term is positive. So $s^2 - 10s + 25 = (s - 5)(s - 5)$, or $(s - 5)^2$.
Check by multiplying: $(s - 5)(s - 5) = s^2 - 10s + 25$.

Example 2

Find binomial factors for $x^2 - 9$.

Solution

The first and last terms are perfect squares: $x^2 = x \cdot x$ and $9 = 3 \cdot 3$.
Write the sum and difference of the square root of each term: $(x + 3)(x - 3)$.
Check by multiplying: $(x + 3)(x - 3) = x^2 - 9$.

EXERCISES

If possible, find binomial factors for each trinomial. Be sure to factor common monomial factors first. At least one does not have factors.

1. $r^2 - 16r + 64$ $(r - 8)^2$
2. $t^2 - 81$ $(t + 9)(t - 9)$
3. $w^2 + 12w + 36$ $(w + 6)^2$
4. $c^2 + 16c + 64$ $(c + 8)^2$
5. $a^2 - 16$ $(a + 4)(a - 4)$
6. $x^2 + 8x + 16$ $(x + 4)^2$
7. $9y^2 - 49$ $(3y + 7)(3y - 7)$
8. $25n^2 - 30n + 16$ no factors
9. $4b^2 - 40b + 100$ $4(b - 5)^2$
10. $9z^2 + 12z + 4$ $(3z + 2)^2$
11. $25a^2 - 60a + 36$ $(5a - 6)^2$
12. $4x^2 - 64$ $4(x + 4)(x - 4)$

170 RETEACHING *LESSON 11-6*

The logic used in solving quadratic equations is as follows. Start with the idea that if the product of two numbers or expressions is equal to zero, then at least one of the factors is equal to zero. (If $xy = 0$, then either $x = 0$ or $y = 0$.)

Example 3

Determine the possible solutions for $x^2 + 16 = 8x$.

Solution

Isolate a 0 on the right side of the equation: $x^2 - 8x + 16 = 0$.

Then factor the expression on the left side: $(x - 4)(x - 4) = 0$.

Either one or the other factor must be equal to 0. Since both factors are the same, both must be equal to zero. Solve the equation: $x - 4 = 0$, so $x = 4$.

This quadratic equation has a single answer because $x^2 - 8x + 16$ is a perfect square trinomial, and both factors are identical. When a quadratic equation has different factors, you will find more than one solution.

Try These Exercises

Find binomial factors for the following, if possible. (Two do not have such factors.)

1. $s^2 + 10s + 25$ $(s + 5)^2$
2. $4x^2 - 12xy + 9y^2$ $(2x - 3y)^2$
3. $m^2 + 8mn + 16n^2$ $(m + 4n)^2$
4. $m^2 - 8mn + 16n^2$ $(m - 4n)^2$
5. $9r^2 - 36$ $(3r + 6)(3r - 6)$
6. $25x^2 - 1$ $(5x + 1)(5x - 1)$
7. $49a^2 - 28a + 2$ none
8. $81e^2 - 8f^2$ none
9. $64u^2 - 48uv + 9v^2$ $(8u - 3v)(8u - 3v)$
10. A square is shown to have an area of $8w + 16 + w^2$. How long is each side? $(w + 4)$

11. **WRITING MATH** Describe the special pattern shown by the polynomial $p^2 - 9$. Find the binomial factors. the difference of two squares; $(p - 3)(p + 3)$

Practice Exercises

Find binomial factors for the following, if possible.

12. $p^2 - 2p + 1$ $(p - 1)^2$
13. $36a^2 + 24ab + 4b^2$ $(6a + 2b)^2$
14. $9f^2 - 49g^2$ $(3f + 7g)(3f - 7g)$
15. $4x^2 - 24xy + 27y^2$ none
16. $1 - 8x + 16x^2$ $(1 - 4x)^2$
17. $100r^2 + 220r + 121$ $(10r + 11)^2$
18. $8v^2 - 25w^2$ none
19. $9m^2 - 6mn + 9n^2$ none
20. $h^2 - 14h + 49$ $(h - 7)^2$
21. $9s^2 - 6st - t^2$ none
22. $y^2 + 2yz + z^2$ $(y + z)^2$
23. $36r^2 - s^2$ $(6r - s)(6r + s)$
24. $4a^2 - 12b^2$ none
25. $9c^2d^2 - 64e$ none
26. $4c^2 + 20cd + 25d^2$ $(2c + 5d)^2$
27. Find a monomial factor and two binomial factors for $4x^2 + 8x + 4$. $4(x + 1)^2$
28. Find a monomial factor and two binomial factors for $16v^2 - 36w^2$. $4(2v + 3w)(2v + 3w)$
29. Solve the equation $p^2 - 6p + 9 = 0$. $p = 3$
30. Solve the equation $m^2 + 25 = 10m$. $m = 5$
31. Simplify $(a - 3)(2a - 5)$, and set up an equation where your answer is equal to zero. $a = 3$ or $2\frac{1}{2}$

Lesson Wrap-up

Quick Assessment

Ask the following questions to determine if students understand the content presented in this lesson.

1. How can you tell whether a trinomial is a perfect square trinomial? The middle term must be the square root of the first term multiplied by the square root of the third term multiplied by 2.
2. Can $x^2 + y^2$ be factored? Explain. No; neither term has a common factor.

Assignment Guide

Basic: 1–40, 43–50
Advanced: 1–50
Additional Practice: See Extra Practice Index on page 674.

Additional Answers

43.

1 · 1	2 · 1	3 · 1	4 · 1	5 · 1	6 · 1	7 · 1	8 · 1
1 · 2	2 · 2	3 · 2	4 · 2	5 · 2	6 · 2	7 · 2	8 · 2
1 · 3	2 · 3	3 · 3	4 · 3	5 · 3	6 · 3	7 · 3	8 · 3
1 · 4	2 · 4	3 · 4	4 · 4	5 · 4	6 · 4	7 · 4	8 · 4
1 · 5	2 · 5	3 · 5	4 · 5	5 · 5	6 · 5	7 · 5	8 · 5
1 · 6	2 · 6	3 · 6	4 · 6	5 · 6	6 · 6	7 · 6	8 · 6
1 · 7	2 · 7	3 · 7	4 · 7	5 · 7	6 · 7	7 · 7	8 · 7
1 · 8	2 · 8	3 · 8	4 · 8	5 · 8	6 · 8	7 · 8	8 · 8

Extend the Lesson

CONNECTING TO PRIOR KNOWLEDGE The difference of two squares provides a way to multiply mentally two numbers that are the same difference from a multiple of ten. For example, to multiply 28 · 32, think (30 − 2)(30 + 2). That is $30^2 - 2^2$, which is 900 − 4, or 896. Have students try this method to multiply 46 · 54 and 67 · 73. 2,484; 4,891

Then solve the equation.

32. Solve the equation $k^2 - 16 = 0$ in two ways, one of which involves factoring. Your answers should be identical using either method. $k = 4$ or -4

33. **LANDSCAPING** A square garden with side $8x$ is planted in the center of a square lawn with side y. Write a polynomial to represent the area of the lawn. Then find two binomial factors. $y^2 - 64x^2$; $(y - 8x)(y + 8x)$

34. **ART** A mosaic in the shape of a rectangle has an area of $49x^2 - 25y^2$. Find the length and width of the rectangle if $x = 9$ in. and $y = 2$ in. 73 in. and 53 in.

35. As a part of a problem, you have to calculate: $(8.35)^2 + (1.65)^2 + (8.35)(1.65)(2)$. Can you see a fast way to do this? What is the answer? 100

Find factors for the following.

36. $3c^2x + 18cdx + 27d^2x$ $3x(c + 3d)^2$

37. $5s^3 - 20s^2t + 20st^2$ $5s(s - 2t)^2$

38. $12a^2 - 12b^2$ $12(a + b)(a - b)$

39. $3x^3y - 12xy^3$ $3xy(x + 2y)(x - 2y)$

40. **CHAPTER INVESTIGATION** Distribute the final survey to your classmates and compile the data. Use the information to create a demographic profile of your class. Discuss with your group the best way to show your findings. Work together to prepare graphs and charts.

Extended Practice Exercises

Find factors for the following.

41. $x^3 - x^2y - 2x^2 + 2xy + x - y$ $(x - y)(x - 1)^2$

42. The square floor of a shower with each side of x feet is covered with tiles measuring 1 sq ft each. Some of these tiles are removed to insert a drain. 21 tiles are left on the shower floor. Find x and y. $x = 5, y = 2$

Mixed Review Exercises

These two spinners are spun: (Lesson 9-3)

43. List the sample spaces for the spinners. See additional answers.

44. Find the probability that the sum of the numbers is odd and greater than 6. 0.41

45. Find the probability that the sum of the numbers is either 6 or 10. 0.1875

46. Find P(not an odd sum). 0.50

47. Find P(not an even sum or a sum of 9). 0.625

48. Find P(an odd sum or a sum of 8). 0.61

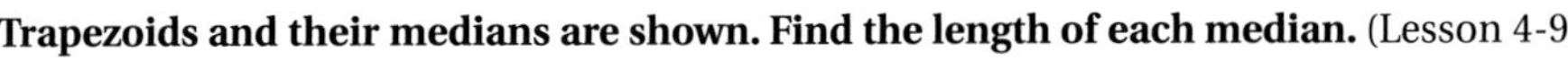

Trapezoids and their medians are shown. Find the length of each median. (Lesson 4-9)

49.

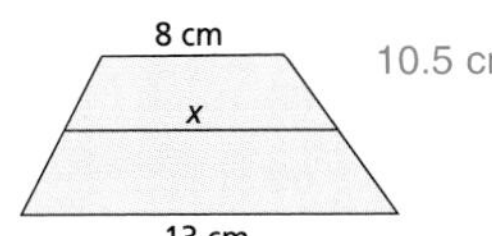

10.5 cm

50.

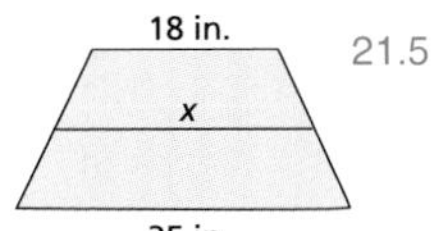

21.5 in.

Extra Practice Worksheet 11-6

Name ______________ Date ______________

EXTRA PRACTICE 11-6

SPECIAL FACTORING PATTERNS

EXERCISES

Find binomial factors for the following, if possible.

1. $p^2 + 2p + 1$ $(p + 1)^2$
2. $25x^2 - 10x + 1$ $(5x - 1)^2$
3. $w^2 - 9$ $(w + 3)(w - 3)$
4. $5s^2 - r^2$ none
5. $4m^2 - 4mn + n^2$ $(2m - n)^2$
6. $4 - 36r + 81r^2$ $(2 - 9r)^2$
7. $m^2 + 4m - 9$ none
8. $4m^2 - 25n^2$ $(2m - 5n)(2m + 5n)$
9. $16r^2 + 24rs + 9s^2$ $(4r + 3s)^2$
10. $16x^2 + 8xy + y^2$ $(4x + y)^2$
11. $25r^2 - t^2$ $(5r - t)(5r + t)$
12. $x^2 - 49y^2$ $(x + 7y)(x - 7y)$
13. $16a^2 - 12$ none
14. $9w^2 - 4$ $(3w - 2)(3w + 2)$
15. $36c^2 - 60c + 25$ $(6c - 5)^2$
16. $4t^2 - 12tr + 9r^2$ $(2t - 3r)^2$

Find a monomial factor and two binomial factors for each of the following.

17. $2x^2 - 4x - 2$ $2(x - 1)^2$
18. $12x^2 - 3$ $3(2x - 1)(2x + 1)$
19. $2x^3 + 16x^2 + 32x$ $2x(x + 4)^2$
20. $10m^2 - 40n^2$ $10(m - 2n)(m + 2n)$
21. $8r^3 - 8rt^2$ $8r(r - t)(r + t)$
22. $5w^4 - 20w^2z^2$ $5w^2(w - 2z)(w + 2z)$

Find a binomial expression for the length of each side of each square shown below.

23. Area: $x^2 + 8x + 16$ $x + 4$

24. Area: $x^2 - 10x + 25$ $x - 5$

Find two binomial expressions for the dimensions of each rectangle shown below.

25. Area: $x^2 - 4$ $x + 2$ and $x - 2$

26. Area: $x^2 - 36$ $x + 6$ and $x - 6$

150 EXTRA PRACTICE *LESSON 11-6*

Enrichment Worksheet 11-6

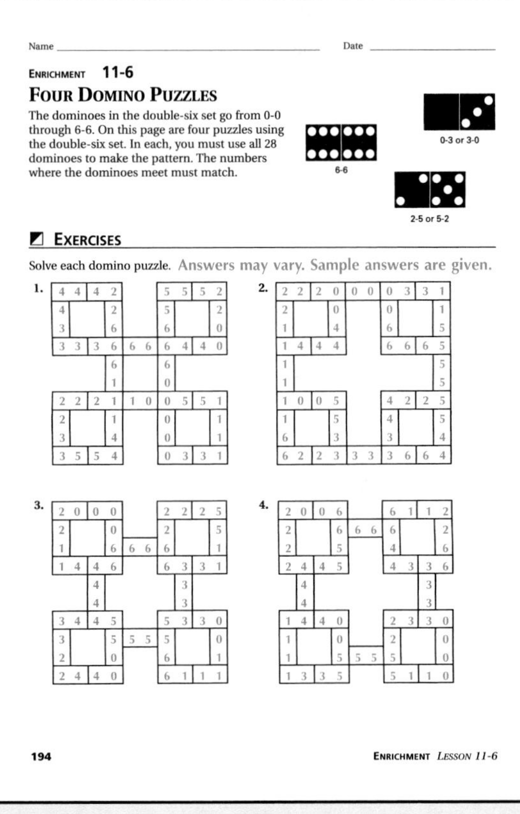

Name ______________ Date ______________

ENRICHMENT 11-6

FOUR DOMINO PUZZLES

The dominoes in the double-six set go from 0-0 through 6-6. On this page are four puzzles using the double-six set. In each, you must use all 28 dominoes to make the pattern. The numbers where the dominoes meet must match.

6-6

0-3 or 3-0

2-5 or 5-2

EXERCISES

Solve each domino puzzle. Answers may vary. Sample answers are given.

1. 2. 3. 4.

194 ENRICHMENT *LESSON 11-6*

Teaching Strategies

CHALLENGE Have students complete each to form a perfect square trinomial.

1. $a^2 - 12a + (?)$ 36
2. $(?)b^2 - 12b + 9$ 4
3. $49c^2 + (?)c + 4$ 28

Skills Practice

Vocabulary Review

Lesson 11-5
grouping
monomial factor
polynomial factor

Lesson 11-6
perfect square trinomial
quadratic equation

Assignment Guide

All students: 1–85
Additional Practice: See Extra Practice Index on page 674.

Review and Practice Your Skills

Practice ■ Lesson 11-5

1. $5(c+d)+b(c+d)$ $(5+b)(c+d)$
2. $g(f^2-8)-9(f^2-8)$ $(g-9)(f^2-8)$
3. $a(b-3)-c(b-3)$ $(a-c)(b-3)$
4. $xz+10x+yz+10y$ $(x+y)(z+10)$
5. $2h-2k+jh-jk$ $(2+j)(h-k)$
6. $x^2-x+xy-y$ $(x+y)(x-1)$
7. y^3-2y^2+3y-6 $(y^2+3)(y-2)$
8. $3a-3b+ab-a^2$ $(3-a)(a-b)$
9. $2wz-w+3-6z$ $(w-3)(2z-1)$
10. $xy+5x+2y+10$ $(x+2)(y+5)$
11. $mw-mx-nw+nx$ $(m-n)(w-x)$
12. $gh+3h^2-12h-4g$ $(h-4)(g+3h)$
13. $2x^2y-8x^2+3y-12$ $(2x^2+3)(y-4)$
14. $3wz^2+12w-z^2-4$ $(3w-1)(z^2+4)$
15. $p^2r^3-2p^2s-qr^3+2qs$ $(p^2-q)(r^3-2s)$
16. $18w^2z-3w^3+42wz^3-7w^2z^2$ $(3w^2+7wz^2)(6z-w)$
17. $w-v+wv-v^2$ $(1+v)(w-v)$
18. $8b^2-10b+4b-5$ $(2b+1)(4b-5)$
19. $x-xy-3ay^2+3ay$ $(x-3ay)(1-y)$
20. $10m^2+15mp+18mn+27np$ $(5m+9n)(2m+3p)$
21. $-9xy+6xz-6y+4z$ $(3x+2)(-3y+2z)$
22. $ax+bx+cx+2a+2b+2c$ $(x+2)(a+b+c)$
23. $xw+2yw+3zw-4x-8y-12z$ $(w-4)(x+2y+3z)$
24. $ap+aq-ar-bp-bq+br$ $(a-b)(p+q-r)$
25. $x^2-ax-bx+cx-ac-bc$ $(x+c)(x-a-b)$
26. Find the dimensions of a rectangle whose area is $mn-4m+2n-8$. $(m+2)\times(n-4)$
27. Find the dimensions of a rectangle whose area is $2g+4f-7ag-14af$. $(2-7a)\times(g+2f)$

Practice ■ Lesson 11-6

Find binomial factors of the following, if possible.

28. $x^2+10x+25$ $(x+5)(x+5)$
29. $x^2-20x+100$ $(x-10)(x-10)$
30. $m^2+16m+64$ $(m+8)(m+8)$
31. $z^2-6z+36$ not possible
32. $16d^2+40d+25$ $(4d+5)(4d+5)$
33. $36b^2-12b+1$ $(6b-1)(6b-1)$
34. $64r^2+48r+9$ $(8r+3)(8r+3)$
35. $x^2-8xy+16y^2$ $(x-4y)(x-4y)$
36. $9g^2+12gh+4h^2$ $(3g+2h)(3g+2h)$
37. w^2-144 $(w+12)(w-12)$
38. $121-p^2$ $(11+p)(11-p)$
39. c^2-9d^2 $(c+3d)(c-3d)$
40. x^2+25 not possible
41. $16u^2-81v^2$ $(4u+9v)(4u-9v)$
42. $1-4y^2$ $(1+2y)(1-2y)$
43. $25s^2-70st+49t^2$ $(5s-7t)(5s+7t)$
44. $25x^2-49y^2$ $(5x+7y)(5x-7y)$
45. $49p^2+28pq+4q^2$ $(7p+2q)(7p+2q)$
46. $49d^2-4f^2$ $(7d-2f)(7d+2f)$
47. $64m^2+176mn+121n^2$ $(8m+11n)(8m+11n)$
48. $64x^2-121z^2$ $(8x+11z)(8x-11z)$
49. $1-2a+a^2$ $(1-a)(1-a)$
50. $1-8x+64x^2$ not possible
51. $16+25v^2$ not possible
52. $25-4k^2$ $(5-2k)(5+2k)$
53. $225x^2+330xy+121y^2$ $(15x+11y)(15x+11y)$
54. $625j^2-1$ $(25j+1)(25j-1)$

Find a monomial factor and two binomial factors for each of the following.

55. $3x^2-12x+12$ $3(x-2)(x-2)$
56. $5x^2-45$ $5(x+3)(x-3)$
57. x^3-8x^2+16x $x(x-4)(x-4)$
58. $10x^2-140x+490$ $10(x-7)(x-7)$
59. $3ax^2-12a$ $3a(x+2)(x-2)$
60. $50by^2-18bx^2$ $2b(5y-3x)(5y-3x)$
61. x^4-25x^2 $x^2(x+5)(x-5)$
62. $27y^3-36xy^2+12x^2y$ $3y(3y-2x)(3y-2x)$
63. $4a^2-4b^2$ $4(a+b)(a-b)$

Find binomial factors of the following, if possible. (Lesson 11-6)

64. $121+22a+a^2$ $(11+a)(11+a)$
65. $4x^2-28xy+49y^2$ $(2x-7y)(2x-7y)$
66. $x^2+14x+196$ not possible
67. $1-256m^2$ $(1+16m)(1-16m)$
68. $9a^2-100b^2$ $(3a+10b)(3a-10b)$
69. $81c^2d^2-25b^2$ $(9cd+5b)(9cd-5b)$

Teaching Strategies

Point out the phrase "if possible" in the direction line for Exercises 28–54. Students should consider the possibility that a polynomial cannot be factored rather than force an incorrect factoring.

Practice ■ Lesson 11-1–Lesson 11-6

Simplify. (Lesson 11-1) and (Lesson 11-2)

70. $(3x + 5y - 8z) + (7y - 6x + 5z)$ $-3x + 12y - 3z$

71. $(-6n^2 + 7n - 11) + (17n^2 - 7n + 16)$ $11n^2 + 5$

72. $(xy + 2x^2 - 8y) - (4x^2 + 8y - 3xy)$ $4xy - 2x^2 - 16y$

73. $(7x + 15y) - (5x + 8y) + (2y - 4x)$ $-2x + 9y$

74. $-4a(13 - 6a^2 + 11b)$ $-52a + 24a^3 - 44ab$

75. $2xyz(3xy - 7yz + 15xz)$ $6x^2y^2z - 14xy^2z^2 + 30x^2yz^2$

76. $x(4x^2 - 9) + 2(x^3 - 7x^2 + 4x)$ $6x^3 - 14x^2 - x$

77. $3(x + 2y) - 4(2x - 5y) + 2(5x - 13y)$ $5x$

Find the GCF and its paired factor for the following. (Lesson 11-3)

78. $22x + 55y$ $11(2x + 5y)$

79. $48x^2 + 32x$ $16x(3x + 2)$

80. $13x^2y^3 - 52x^3y^2$ $13x^2y^2(y - 4x)$

81. $-4def - 8efg - 12ef$ $-4ef(d + 2g + 3)$

82. $120a^2b^3 + 24a^3b - 72a^4b^2$ $24a^2b(5b^2 + a - 3a^2b)$

83. $2sk^2 + 58sq^2 + 34sy^2$ $2s(k^2 + 29q^2 + 17y^2)$

Find factors for the following. (Lesson 11-5)

84. $xy - 2x - 4y + 8$ $(x - 4)(y - 2)$

85. $3xw + 7w - 12x - 28$ $(w - 4)(3x + 7)$

MathWorks Workplace Knowhow — Career – Actuaries

Actuaries assemble and analyze statistical data to estimate the probabilities of various types of loss. This information helps the insurance company determine how much to charge people in insurance premiums. For example, an actuary studies the effect of age on the number of driving accidents that occur. If a particular age group has more accidents than another, that group pays higher premiums.

The company must charge enough to pay all claims and still make a profit. However, if the company charges too much, customers will choose another company. Actuaries must have excellent math and statistics skills. They also need to understand economics, social trends, legislation and developments in health and medicine.

You are evaluating the risk factors involved in insuring the lives of firefighters over the course of their careers. You determine that the equation $y = -x^2 + 15x + 100$ can be used to predict risk where x equals the numbr of years a fire fighter has been on the job and y equals risk.

1. What is the base risk at the start of a firefighter's career? Use 0 for x. 100
2. Find the amount of risk a firefighter faces at 2 years, 4 years and 6 years. (Remember, to evaluate $-x^2$, square x before multiplying by -1.) 126, 144, 154
3. Make a table to show the risk for the first 10 years. At what year is the risk of insuring firefighters the highest? year 7
4. At what year does the risk come back down to 100? year 15

Chalkboard Examples

Example from Lesson 11-5

Find factors for $2j^2m^2 + 3j^2n + 4km^2 + 6kn$. $(j^2 + 2k)(2m^2 + 3n)$

Example from Lesson 11-6

Find binomial factors for $9m^2 + 12mn + 4n^2$. $(3m + 2n)^2$

Career Opportunity

Describe the kind of work actuaries do and the types of places they work. Discuss the importance of algebra, statistics and functions in working as an actuary. Students should answer Questions 1-4 to better understand how actuaries use mathematics in performing their job.

Students who are interested in learning more about this profession can go to learningmathmatters.com. Guidance Counselors should have information about school and training requirements.

Teaching Strategies

Point out that it makes factoring easier to extract the constant factor first. However, if that factor is not recognized immediately, it can still be done as a later step.

Lesson Planning

NCTM Standards/Strands

- Number & Operation
- Patterns, Functions, & Algebra
- Problem Solving
- Reasoning and Proof
- Connections
- Representation

Vocabulary

quadratic term

Materials Needed

paper/pencil
computers (optional)

Lesson Resources

Warm-Up Transparency 33
Transparency TK-4
Reteaching 11-7
Extra Practice 11-7
Enrichment 11-7

Assignment Guide

Basic: 1–53, 55–74
Enriched: 1–74

Getting Started

5-Minute Warm-up

Solve.

1. What two factors of 9 have a sum of 10? 1 and 9
2. What two factors of 24 have a sum of 11? 3 and 8

Introduction to Lesson 11-7

After students have completed this activity, have them share the patterns they found.

11-7 Factor Trinomials

Goals ■ Factor trinomials with quadratic coefficients of one.

Applications Product Development, Construction, Chemistry

Work with a partner to find factoring patterns.

A trinomial expression that does not fit a special pattern may still have binomial factors. Finding such factors requires a combination of logic and guess-and-check.

1. Start with the idea that finding factors of a trinomial is the reverse of multiplying binomials. Study these examples and look for patterns.

$(x + 3)(x + 4) = x^2 + 7x + 12$
$(x - 3)(x - 4) = x^2 - 7x + 12$
$(x - 3)(x + 4) = x^2 + x - 12$
$(x + 3)(x - 4) = x^2 - x - 12$

$(y + 5)(y + 1) = y^2 + 6y + 5$
$(y - 5)(y - 1) = y^2 - 6y + 5$
$(y + 5)(y - 1) = y^2 + 4y - 5$
$(y - 5)(y + 1) = y^2 - 4y - 5$

2. Look at the third term in each trinomial and the sign before it. How does each third term and its sign relate to the binomial factors? Answers will vary.

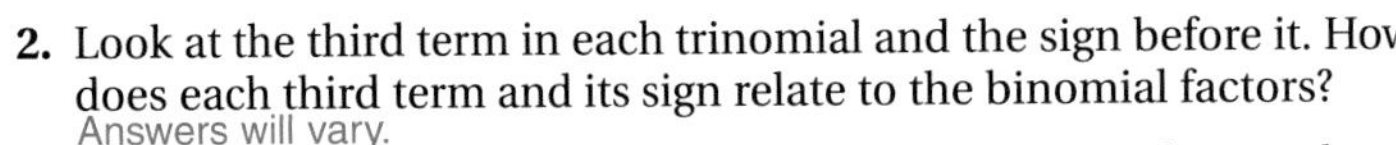

3. Look at each second term and the sign before it. How does each second term and its sign relate to the binomial factors? The second term has a coefficient equal to their sum.
4. Set up an additional example using terms and signs similar to those in the examples above. Does your example follow the patterns you have found? Answers may vary.

Reading About Math

Many trinomials have an x^2 term, an x term,and a constant. The x^2 term is called the **quadratic** term, from the Latin *quadrare*, which means "to make a square." Also, polynomials with a quadratic term as their highest power are called **quadratic polynomials**.

Build Understanding

In this lesson, you will study trinomials where the coefficient of the first term (the x^2, or **quadratic term**) is 1. This makes the pattern easier to see. From the activity above, you may have noticed the following.

a. The trinomial third term is always the product of the binomial second terms.

b. The trinomial second term is always the sum of the binomial second terms. (*Note:* When the signs in the binomials are different, this sum will *look* like a difference, because $a + (-b) = a - b$.)

c. If the sign of the trinomial third term is negative, the signs in the binomials are different. If it is positive, the signs in the binomials are the same.

d. The sign of the trinomial second term is always the same as the sign of the larger binomial second term.

With these four clues, you can find the factors of a standard form trinomial that begins with x^2.

Check Understanding

Before you count the terms, always be sure the trinomial is in standard form. Why would this be important?

There may be like terms that can be combined.

Teaching Strategies

Students may find it helpful to organize their trials of second-term constants or coefficients in a table like the following:

Paired Factors	Sum/Difference of Factors
1, 18	1 + 18 = 19
2, 9	2 + 9 = 11
3, 6	3 + 6 = 9

498 Chapter 11 Polynomials

Example 1

Find second-term constants or coefficients for the binomial factors of these polynomials.

a. $x^2 - 8x + 15$ **b.** $x^2 + 3xy - 18y^2$

Solution

a. The product of the binomial second terms is 15, and the sum (a true sum, because of the third term's positive sign) is 8. So the binomial second-term constants are 5 and 3 (because $5 \times 3 = 15$ and $5 + 3 = 8$).

The binomials will be in the form $(x \quad 5)(x \quad 3)$.

b. The product of the binomial second terms is 18, and their sum is 3. Because the third term's sign is negative, the binomial signs differ, so the sum will look like a difference.

Think of factors of 18 that have a difference of 3.

Factors	18 1	9 2	6 3
Difference	17	7	③

Stop here; 3 is the difference you want.

The coefficients will be 6 and 3. The binomials will be in the form $(x \quad 6y)(x \quad 3y)$.

The next step in finding the factors involves determining the correct signs for the binomials.

Problem Solving Tip

Making an organized list is a good strategy when the third term in the trinomial has many pairs of factors.

Example 2

In the two expressions above, complete the binomial factors by determining the signs of the second terms.

Solution

a. The second trinomial term is negative, so the larger binomial second term has a negative sign. The third trinomial term is positive, so both binomial signs are the same—both negative.

The binomial factors are $(x - 5)(x - 3)$.

b. The second trinomial term is positive, so the larger binomial second term is also positive. But the third trinomial term is negative, so the two binomial signs are different.

The binomial factors of $x^2 + 3xy - 18y^2$ are $(x + 6y)(x - 3y)$.

You can handle the numbers and signs in a single step if you wish, though this takes a little more thought.

Problem Solving Tip

As always in problem solving, you should check your solutions before you finally accept them. Whenever you identify a pair of factors, multiply them together again to be sure the product is correct.

Example 3

PRODUCT DEVELOPMENT A software company determines that the cost of producing its new financial software is a product of the number of days spent working on the project and the number of programmers assigned to the project. The total cost is represented by $x^2 - 5x - 36$. Find the binomial factors.

Chalkboard Examples

Supplementary Example 1

Find the second-term constants or coefficients for the binomial factors of these polynomials.

1. $x^2 + 6x + 8$ 4 and 2
2. $x^2 + 4xy - 21y^2$ $7y$ and $3y$

Supplementary Example 2

For the expressions above, write the binomial factors by determining the signs of the second term.

1. $x^2 + 6x + 8$ $(x + 4)(x + 2)$
2. $x^2 + 4xy - 21y^2$ $(x + 7y)(x - 3y)$

Supplementary Example 3

The total cost of a project is represented by the expression $x^2 - 3x - 40$. Find the binomial factors. $(x - 8)(x + 5)$

Extend the Lesson

MATH JOURNAL Have students describe the steps they follow to determine the second-term constants or coefficients for the binomial factors of the trinomials in this lesson. Then have them write the steps they follow to determine the signs for these second terms.

Reteaching Worksheet 11-7

Name ______________________ Date ____________

RETEACHING 11-7

FACTORING TRINOMIALS

When you factor trinomials, it is important to keep these points in mind.

- The first term in the trinomial is the product of the first terms of the binomials.
- The last term in the trinomial is the product of the second terms of the binomials.
- The coefficient of the middle term of the trinomial is the sum of the second terms of the binomials.
- If the sign of the last term of the trinomial is negative, the signs in each binomial factor are different; if the sign if positive, the signs in each binomial factor are the same—either positive or negative, depending on the sign of the trinomial's middle tern.

Example

Find binomial fctors for $a^2 - 3a - 10$.

Solution

To find the first terms, factor a^2: $(a \quad)(a \quad)$
To find the last erms, factor 10: $(a \quad 5)(a \quad 2)$
Determine the signs: $(a - 5)(a + 2)$

Since the sign of the last term in the polynomial is negative, the signs in each binomial factor must be different. Notice that = is negative and *a* is positive because $-5a + 2a = -3a$, the middle term of the trinomial.

Check your work by multiplying: $(a - 5)(a + 2) = a^2 - 3a - 10$
The answer checks. ✔

EXERCISES

Find binomial factors for each trinomial.

1. $a^2 + 5a + 6$ $(a + 3)(a + 2)$	**2.** $x^2 - 9x - 10$ $(x - 10)(x + 1)$	**3.** $m^2 + 6m - 16$ $(m + 8)(m - 2)$
4. $t^2 - 2t - 35$ $(t - 7)(t + 5)$	**5.** $s^2 + 11s + 24$ $(s + 8)(s + 3)$	**6.** $k^2 + k - 20$ $(k + 5)(k - 4)$
7. $r^2 - r - 72$ $(r - 9)(r + 8)$	**8.** $x^2 - 11x + 28$ $(x - 7)(x - 4)$	**9.** $y^2 + 8y + 12$ $(y + 6)(y + 2)$
10. $b^2 + 10b + 24$ $(b - 6)(b - 4)$	**11.** $d^2 + 9d + 20$ $(d + 4)(d + 5)$	**12.** $p^2 - 4p - 21$ $(p - 7)(p + 3)$
13. $k^2 + 2k - 120$ $(k + 12)(k - 10)$	**14.** $r^2 - r - 90$ $(r - 10)(r + 9)$	**15.** $z^2 + 5z - 150$ $(z + 15)(z - 10)$

172 RETEACHING *LESSON 11-7*

Lesson Wrap-up

QUICK ASSESSMENT

Ask the following questions to determine if students understand the content presented in this lesson.

1. How can the signs within the trinomial help you know the signs in the binomial factors? You can use logical reasoning to apply the rules of operations with signed integers to figure out which signs are necessary to produce the desired outcome.
2. When can a trinomial be factored using this method? when the product of two numbers equals the third term in the trinomial and the sum of the same two numbers equals the second term in the trinomial

ASSIGNMENT GUIDE

Basic: 1–53, 55–74
Advanced: 1–74
Additional Practice: See Extra Practice Index on page 674.

Solution

The product of the binomial second terms is (-36) and the sum is (-5). So the two binomial constants are 4 and (-9).

The binomial factors of $x^2 - 5x - 36$ are $(x + 4)(x - 9)$.

TRY THESE EXERCISES

Identify the binomial second terms when the following trinomials are factored.

1. $x^2 + 10x + 21$ 3 and 7
2. $t^2 + 9t + 20$ 4 and 5
3. $a^2 - 6ab + 8b^2$ $4b$ and $2b$
4. $m^2 - mn - 2n^2$ $2n$ and n
5. $k^2 + 5k - 6$ 6 and 1
6. $f^2 + 2fg - 15g^2$ $3g$ and $5g$

Identify second-term signs for binomial factors of the following.

7. $v^2 + 18v + 77$ + and +
8. $x^2 - 19x + 90$ – and –
9. $b^2 - 15bc - 100c^2$ – and +
10. $n^2 + n - 42$ + and –

Factor the following trinomials.

11. $c^2 + 5c + 6$ $(c + 2)(c + 3)$
12. $c^2 - 5c + 6$ $(c - 2)(c - 3)$
13. $c^2 - 5c - 6$ $(c - 6)(c + 1)$
14. $c^2 + 5c - 6$ $(c + 6)(c - 1)$

15. **MODELING** What are the sides of the rectangle you can create with one "x^2" Algeblock piece, 21 "one" tiles, and 10 "x" tiles? *Do not* experiment. Use factoring—it will save time. Then use Algeblocks to check your answer. $(x + 3)(x + 7)$

Technology Note

Computer spreadsheets allow businesses to explore decisions by using and varying data.

Coupled with a graphics program, spreadsheet formulas allow businesses to graph data as well.

Most spreadsheet applications use cell names in the data column as variables. The trinomial $x^2 + 10x + 21$ is entered as:

A2 * A2 + 10 * A2 + 21

The computer uses the value of cell A2 to calculate the expression.

PRACTICE EXERCISES

Identify binomial second-term factors for the following.

16. $p^2 + 5p + 6$ 3, 2
17. $x^2 + 12xy + 35y^2$ $7y$, $5y$
18. $h^2 - 10h + 9$ 1, 9
19. $a^2 - 7ab + 10b^2$ $5b$, $2b$
20. $c^2 + 6cd - 16d^2$ $8d$, $2d$
21. $q^2 + 2q - 63$ 9, 7
22. $r^2 - 13r + 30$ 15, 2
23. $e^2 - 7ef - 30f^2$ $10f$, $3f$

Identify binomial second-term signs for the following.

24. $x^2 + x - 12$ +, –
25. $j^2 + 12j + 27$ +, +
26. $s^2 - 18st + 17t^2$ –, –
27. $b^2 - bc - 56c^2$ +, –
28. $l^2 + 5l - 36$ +, –
29. $v^2 - 10v + 24$ –, –
30. $j^2 + 12jk + 11k^2$ +, +
31. $z^2 - 3z - 18$ –, +

Factor the following trinomials.

32. $x^2 - 25x + 24$ $(x - 1)(x - 24)$
33. $p^2 + 10pq + 24q^2$ $(p + 6q)(p + 4q)$
34. $m^2 + 5mn - 24n^2$ $(m - 3n)(m + 8n)$
35. $k^2 - 10k - 24$ $(k - 12)(k + 2)$
36. $a^2 - 2a - 24$ $(a - 6)(a + 4)$
37. $h^2 + 23h - 24$ $(h - 1)(h + 24)$
38. $r^2 + 14r + 24$ $(r + 12)(r + 2)$
39. $f^2 - 11fg + 24g^2$ $(f - 3g)(f - 8g)$
40. $p^2 + 2p - 15$ $(p + 5)(p - 3)$
41. $q^2 - 11q + 28$ $(q - 4)(q - 7)$
42. $r^2 + 21r + 20$ $(r + 1)(r + 20)$
43. $s^2 + 2st - 8t^2$ $(s - 2t)(s + 4t)$

Learning Styles

VISUAL LEARNERS Have students use Algeblocks or algebra tiles to model factoring trinomials. For example, ask them to show the tiles for $x^2 + 3x + 2$. Then have them use trial-and-error to arrange the one large square, the three rectangles, and the two small squares to form a rectangle. Ask them to write the expressions for the length and width. These are the factors of the trinomial.

44. CONSTRUCTION A rectangular trench x feet deep is being dug for the foundation of a wall. The area of the bottom is $x^2 + 34x - 35$ square feet. Compare the depth of the trench to its width and to its length.
If x narrower than depth; 35 ft longer than depth

45. WRITING MATH Can a trinomial have different sets of binomial factors? Explain your thinking. No. Only one pair of factors will satisfy all conditions.

46. CHAPTER INVESTIGATION Work with your group to develop a strategy for marketing a new product aimed at people your own age. Use the demographic profile you developed in Lesson 11-6. Suppose you can afford to run one print advertisement, one radio spot and one television commercial. Determine when and where you would run your advertisements. Give an oral presentation of your marketing strategy to your classmates. Be ready to defend your choices using the demographic data.

Factor the following.

47. $1 - 5r + 6r^2$ $(1 - 3r)(1 - 2r)$

48. $1 + 7x - 18x^2$ $(1 + 9x)(1 - 2x)$

49. $24g^2 - 10g + 1$ $(6g - 1)(4g - 1)$

50. $13a^2 - 12a - 1$ $(13a + 1)(a - 1)$

51. $5a^2x^2 - 15ax^2 + 10x^2$ $5x^2(a - 2)(a - 1)$

52. $9 + 18x - 72x^2$ $9(1 - 2x)(1 + 4x)$

53. CHEMISTRY To dilute x pounds of a chemical, you need a water tank with a volume of $3x^3 - 12x^2 - 36x$. Indicate its dimensions, in terms of x.
$3x(x + 2)(x - 6)$

Extended Practice Exercises

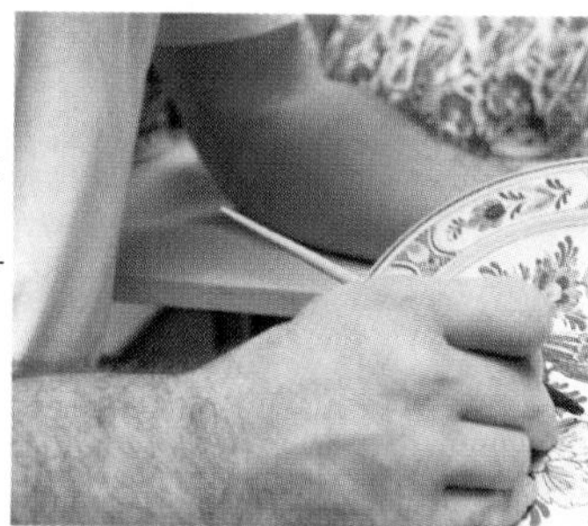

54. SMALL BUSINESS Andre receives a rush order for some hand-painted plates. But his budget for materials is limited to \$255 per day. His cost formula indicates that if he works at a rate of $(12 + x)$ plates per day, the daily cost will be $\$(x^2 + 22x + 120)$. How many plates can he make each day—maximum—to fulfill the order? (Hint: Check Lesson 11–5. Make a quadratic equation about daily cost, adjust it so that one side equals zero, then factor and reject any negative answers. Remember, the final answer will be $12 + x$.)
$(12 + 5) = 17$ plates

Mixed Review Exercises

Complete the chart in preparation for making a circle graph. *Do not make the graph.* (Lesson 10-5)

Budget Item	Percent of Total	Central Angle		
Rent—\$550	**55.**	**63.**	27.5%	99°
Food—\$415	**56.**	**64.**	20.75%	74.7°
Car Payment—\$260	**57.**	**65.**	13%	46.8°
Credit Card Payment—\$150	**58.**	**66.**	7.5%	27°
Utilities—\$210	**59.**	**67.**	10.5%	37.8°
Savings—\$115	**60.**	**68.**	5.75%	20.7°
Insurance—\$125	**61.**	**69.**	6.25%	22.5°
Misc.—\$175	**62.**	**70.**	8.75%	31.5°

Write the equation for each line. (Lesson 6-3)

71. slope $= \frac{2}{3}$, y-intercept $= -2$ $y = \frac{2}{3}x - 2$

72. passes through $(-2, -3)$ and $(4, 5)$ $y = \frac{4}{3}x - \frac{1}{3}$

73. slope $= -2$, y-intercept $= 3$ $y = -2x + 3$

74. passes through $(-3, 4)$ and $(6, -4)$ $y = -\frac{8}{9}x + \frac{4}{3}$

Extra Practice Worksheet 11-7

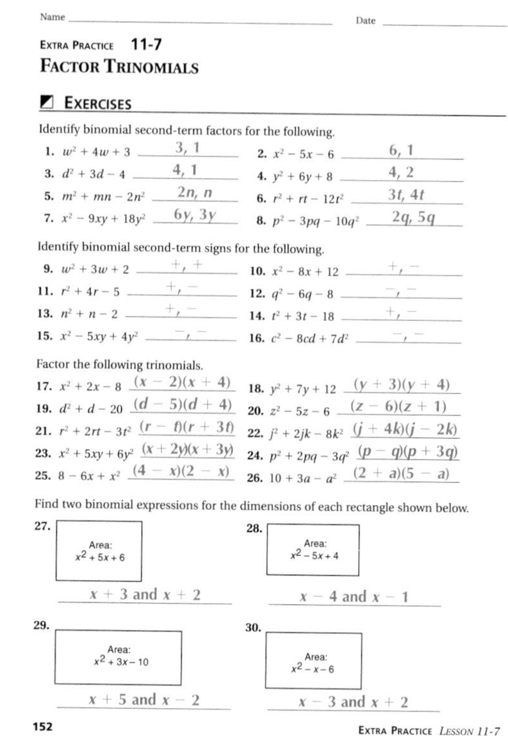

Name ______________________ Date ____________

EXTRA PRACTICE **11-7**

FACTOR TRINOMIALS

EXERCISES

Identify binomial second-term factors for the following.

1. $w^2 + 4w + 3$ 3, 1
2. $x^2 - 5x - 6$ 6, 1
3. $d^2 + 3d - 4$ 4, 1
4. $y^2 + 6y + 8$ 4, 2
5. $m^2 + mn - 2n^2$ $2n$, n
6. $r^2 + rt - 12t^2$ $3t$, $4t$
7. $x^2 - 9xy + 18y^2$ $6y$, $3y$
8. $p^2 - 3pq - 10q^2$ $2q$, $5q$

Identify binomial second-term signs for the following.

9. $w^2 + 3w + 2$ +, +
10. $x^2 - 8x + 12$ +, −
11. $r^2 + 4r - 5$ +, −
12. $q^2 - 6q - 8$ −, −
13. $n^2 + n - 2$ +, −
14. $t^2 + 3t - 18$ +, −
15. $x^2 - 5xy + 4y^2$ −, −
16. $c^2 - 8cd + 7d^2$ −, −

Factor the following trinomials.

17. $x^2 + 2x - 8$ $(x - 2)(x + 4)$
18. $y^2 + 7y + 12$ $(y + 3)(y + 4)$
19. $d^2 + d - 20$ $(d - 5)(d + 4)$
20. $z^2 - 5z - 6$ $(z - 6)(z + 1)$
21. $r^2 + 2rt - 3t^2$ $(r - t)(r + 3t)$
22. $j^2 + 2jk - 8k^2$ $(j + 4k)(j - 2k)$
23. $x^2 + 5xy + 6y^2$ $(x + 2y)(x + 3y)$
24. $p^2 + 2pq - 3q^2$ $(p - q)(p + 3q)$
25. $8 - 6x + x^2$ $(4 - x)(2 - x)$
26. $10 + 3a - a^2$ $(2 + a)(5 - a)$

Find two binomial expressions for the dimensions of each rectangle shown below.

27. Area: $x^2 + 5x + 6$ $x + 3$ and $x + 2$
28. Area: $x^2 - 5x + 4$ $x - 4$ and $x - 1$
29. Area: $x^2 + 3x - 10$ $x + 5$ and $x - 2$
30. Area: $x^2 - x - 6$ $x - 3$ and $x + 2$

152 EXTRA PRACTICE *LESSON 11-7*

Enrichment Worksheet 11-7

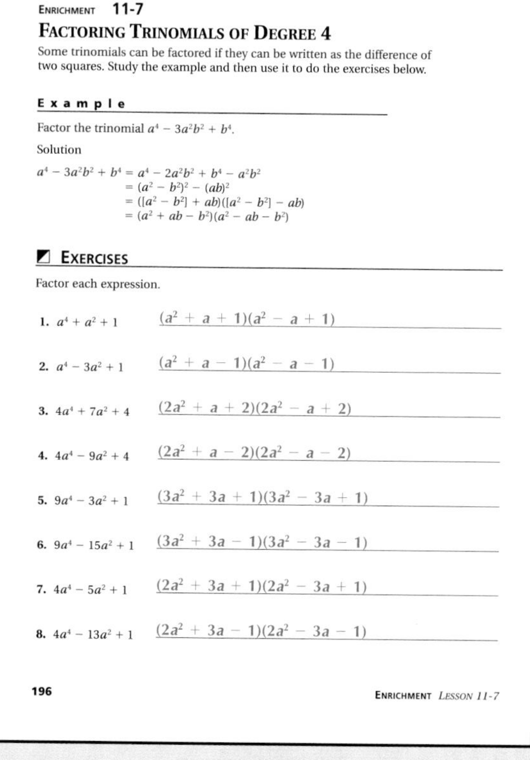

Name ______________________ Date ____________

ENRICHMENT **11-7**

FACTORING TRINOMIALS OF DEGREE 4

Some trinomials can be factored if they can be written as the difference of two squares. Study the example and then use it to do the exercises below.

Example

Factor the trinomial $a^4 - 3a^2b^2 + b^4$.

Solution

$$a^4 - 3a^2b^2 + b^4 = a^4 - 2a^2b^2 + b^4 - a^2b^2$$
$$= (a^2 - b^2)^2 - (ab)^2$$
$$= ([a^2 - b^2] + ab)([a^2 - b^2] - ab)$$
$$= (a^2 + ab - b^2)(a^2 - ab - b^2)$$

EXERCISES

Factor each expression.

1. $a^4 + a^2 + 1$ $(a^2 + a + 1)(a^2 - a + 1)$
2. $a^4 - 3a^2 + 1$ $(a^2 + a - 1)(a^2 - a - 1)$
3. $4a^4 + 7a^2 + 4$ $(2a^2 + a + 2)(2a^2 - a + 2)$
4. $4a^4 - 9a^2 + 4$ $(2a^2 + a - 2)(2a^2 - a - 2)$
5. $9a^4 - 3a^2 + 1$ $(3a^2 + 3a + 1)(3a^2 - 3a + 1)$
6. $9a^4 - 15a^2 + 1$ $(3a^2 + 3a - 1)(3a^2 - 3a - 1)$
7. $4a^4 - 5a^2 + 1$ $(2a^2 + 3a + 1)(2a^2 - 3a + 1)$
8. $4a^4 - 13a^2 + 1$ $(2a^2 + 3a - 1)(2a^2 - 3a - 1)$

196 ENRICHMENT *LESSON 11-7*

Teaching Strategies

COOPERATIVE LEARNING Have each student write, in any order, a perfect square trinomial and a trinomial like those in this lesson. Then ask students to trade trinomials and factor. Finally, have students work together to check their factoring. Discuss what methods were used to write trinomials that could be factored using the methods in the lesson.

Lesson Planning

NCTM Standards/Strands

- Number & Operation
- Patterns, Functions, & Algebra
- Problem Solving
- Reasoning and Proof
- Connections
- Representation

Vocabulary

difference of two cubes
grand product
general case

Materials Needed

pencil/paper

Lesson Resources

Warm-Up Transparency 33
Reteaching 11-8

Assignment Guide

Basic: 1–16
Enriched: 1–16

Getting Started

5-Minute Warm-up

Simplify.

1. $a(x + y)$ $ax + ay$
2. $(ax)(bx)$ abx^2
3. $(ax)^2$ a^2x^2
4. $(a + b)x$ $ax + bx$

Use the Five-step Plan and the Strategy

THE FIVE-STEP PLAN **Read**—ask questions of the students to help them understand the problem. **Plan**—guide students to related problems and previously mastered skills and strategies. **Solve**—students solve problem on their own. **Answer**—write solution in format that answers the question. **Check**—review work, check for reasonableness and review strategy used. Students will benefit from the experience of verbalizing their methods. **THE STRATEGY** Find a pattern—By using variables instead of numbers, we can more easily find the general rule to apply to mean mathematical situations.

11-8 Problem Solving Skills: The General Case

Drawing diagrams and looking at several examples are useful ways to find helpful patterns in mathematics. Another technique is to create a general case. Algebra is excellent for this. It allows you to use letters instead of numbers for an expression's coefficients. By searching for patterns formed by the letters and symbols, you can draw general conclusions that can be applied in specific situations.

Problem Solving Strategies

- Guess and check
- ✔ Look for a pattern
- Solve a simpler problem
- Make a table, chart or list
- Use a picture, diagram or model
- Act it out
- Work backwards
- Eliminate possibilities
- Use an equation or formula

Problem

Find a pattern to help discover factors of a polynomial with a quadratic (x^2) coefficient greater than 1.

Solve the Problem

Use letters instead of numbers to represent the coefficients and constants. (In this solution, a specific example is shown for comparison beside the general case.)

Step 1: Work forward from a pair of binomial factors. The letters a and b represent possible coefficients found in the first term of each monomial factor. The constants, or second term in each monomial, are represented by n_1 and n_2.

General: $(ax + n_1)(bx + n_2)$

F O I L

$= abx^2 + axn_2 + bxn_1 + n_1n_2$
$= abx^2 + (an_2 + bn_1)x + n_1n_2$

Specific: $(2x - 5)(3x + 2)$

F O I L

$= 6x^2 + 4x - 15x - 10$
$= 6x^2 + (4 - 15)x - 10$
$= 6x^2 - 11x - 10$

Step 2: Study the pattern. Carefully compare the general case to the specific example.

Think about how this pattern differs from your work with trinomials in Lesson 11-7. The sum of the second terms of the binomial factors no longer equals the coefficient of the second term of the trinomial. This is only true if the quadratic coefficient is 1.

The product of the coefficients of the F and L terms (quadratic coefficient and constant) is abn_1n_2—identical to the product of the **O** and **I** coefficients. Call this product the **grand product**.

The cross product (O and I) coefficients *multiply* to give the grand product and add to give the trinomial's second term. Apply this general rule to the specific example above.

a. Multiply 6 and 10 to find the grand product: $6 \times 10 = 60$

b. Multiply the O and I coefficients: $4 \times 15 = 60$. The product equals the grand product.

c. Add the O and I coefficients: $(+4) + (-15) = -11$. The sum equals the coefficient of the trinomial's second term.

Additional Answers

1a. $(x + n_1)(x + n_2) =$
F O I L
$x^2 + xn_2 + xn_1 + n_1n_2 =$
$x^2 + (n_2 + n_1)x + n_1n_2$

1b. $(x - 6)(x + 3) =$
F O I L
$x^2 + 3x - 6x - 18 =$
$x^2 + (3 - 6)x - 18 =$
$x^2 - 3x - 18$

2a. $(x + ay)(x + by) =$
F O I L
$x^2 + bxy + axy + aby^2 =$
$x^2 + (b + a)xy + aby^2$

2b. $(x - 9y)(x + 5y) =$
F O I L
$x^2 + 5xy - 9xy - 45 =$
$x^2 + (5 - 9)xy - 45 =$
$x^2 - 4xy - 45$

Try These Exercises

Five-step Plan
1 Read
2 Plan
3 Solve
4 Answer
5 Check

Suppose you have forgotten a useful pattern, or think you may have found a new one. Exploring a general case can be a useful strategy. As shown on the previous page, working a specific example beside the general case may help.

For 1–2, see additional answers.

1. Explore the FOIL pattern for factoring a single-variable trinomial that has a first-term coefficient of 1. Work forward from $(x + n_1)(x + n_2)$ as the general case, and $(x - 6)(x + 3)$ as a specific example.

2. Explore the FOIL pattern for factoring a double-variable trinomial with first-term coefficient of 1. Work forward from $(x + ay)(x + by)$ as the general case, $(x - 9y)(x + 5y)$ as a specific example.

Practice Exercises

For 3–5, see additional answers.

3. Using the same method, explore the pattern for perfect square trinomials. Use $(ax + by)^2$ for the general case, and select your own specific example.

4. Use the same method to explore the difference-of-two-squares pattern. (*Note*: This will *prove* that the pattern you first saw at the start of this lesson is correct for all expressions of its type.)

5. Study the following table of polynomial expansions.

Polynomial factors		Expansion
$(2x - 2)(4x^2 + 4x + 4)$	=	$8x^3 - 8$
$(2x - 1)(4x^2 + 2x + 1)$	=	$8x^3 - 1$
$(3x - 3y)(9x^2 + 9xy + 9y^2)$	=	$27x^3 - 27y^3$
$(3x - 2y)(9x^2 + 6xy + 4y^2)$	=	$27x^3 - 8y^3$
$(x - y)(x^2 + xy + y^2)$	=	$x^3 - y^3$

Notice that each expansion is a **difference of two cubes**. Now, work through the general case of $(ax - by)(a^2x^2 + abxy + b^2y^2)$, and the specific example of $(3x - 1)(9x^2 + 3x + 1)$. (*Note*: The second factor is a trinomial, so the FOIL technique will not apply. Use the original method for multiplying that you learned in Lesson 11-4.)

6. **DATA FILE** Use the Data Index on pages 632–633 to locate the section on useful mathematical data. Find the table of polynomial expansions, and study the general cases presented there. Which general case is related to the polynomial expansions listed in the table in Exercise 5 above?

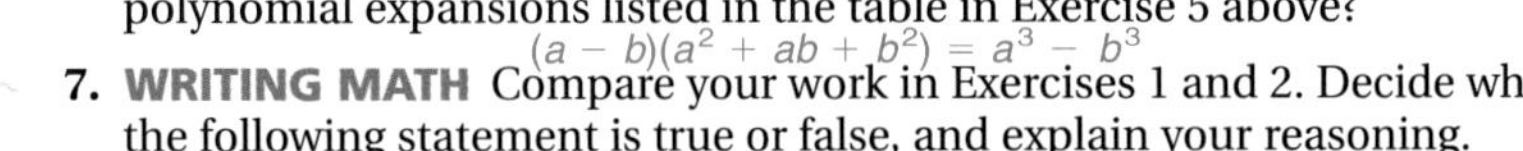

$(a - b)(a^2 + ab + b^2) = a^3 - b^3$

7. **WRITING MATH** Compare your work in Exercises 1 and 2. Decide whether the following statement is true or false, and explain your reasoning.

If you make the *y*-variable equal to 1 the single-variable pattern (Exercise 1) is a special case of the double-variable pattern (Exercise 2).

True. In a single-variable trinomial, think of the constant as a *y*-coefficient multiplied by 1.

Mixed Review Exercises

Solve each proportion. (Lesson 7-1)

8. $\frac{n}{8} = \frac{5}{12}$ $3\frac{1}{3}$
9. $\frac{3}{n} = \frac{15}{40}$ 8
10. $\frac{2x}{5} = x + \frac{1}{4}$ $\frac{5}{3}$
11. $\frac{49}{16} = x + \frac{2}{12}$ 34.75
12. $\frac{9}{3x} = \frac{6}{x} + 4$ 4
13. $\frac{8}{x} + 1 = \frac{10}{2x} - 1$ 3
14. $\frac{3n}{15} = 2n - \frac{3}{8}$ $\frac{15}{72}$
15. $\frac{4}{3n} - 1 = \frac{16}{9n} + 8$ 4
16. $\frac{16}{x} + 1 = \frac{58}{4x} + 1$ 7

3a. $(ax + by)(ax + by) =$
F O I L
↓ ↓ ↓ ↓
$a^2x^2 + abxy + abxy + b^2y^2 =$
$a^2x^2 + 2abxy + b^2y^2$

3b. Answers will vary.

4a. $(a^2x + by)(ax - by) =$
F O I L
↓ ↓ ↓ ↓
$a^2x^2 - abxy + abxy - b^2y^2 =$
$a^2x^2 - b^2y^2$

4b. Answers will vary.

5a. $(ax - by)(a^2x^2 + abxy + b^2y^2) =$
$a^3x^3 + a^2bx^2y + ab^2xy^2$
$- a^2bx^2y - ab^2xy^2 - b^3y^3 =$
$a^3x^3 - b^3y^3$

5b. $(3x - 1)(9x^2 + 3x + 1) =$
$27x^3 + 9x^2 + 3x - 9x^2 - 3x - 1 =$
$27x^3 - 1$

Chalkboard Examples

Supplementary Problem

Explain how to work forward from $(x^2 + n_1)(x^2 + n_2)$ as a general case, and $(x^2 + 4)(x^2 + 3)$ as a specific example. Check students' work.

Lesson Wrap-up

Quick Assessment

How could looking at a general case be helpful in finding patterns, and how is algebra helpful for doing this? Answers may vary. Students should point out that focusing on the general case makes the pattern stand out from the numbers. The general case makes it easier to see the pattern and distinguish it from the numbers that may be specific to only one problem.

Assignment Guide

Basic: 1–16
Advanced: 1–16
Additional Practice: See Extra Practice Index on page 674.

Reteaching Worksheet 11-8

Name ______________________ Date ____________

RETEACHING 11-8

PROBLEM SOLVING SKILLS: THE GENERAL CASE

It is often helpful to use algebra to help you create a general case when looking for patterns in mathematics.

Example

Explore the pattern when a single monomial factor can be extracted before factoring a perfect-square trinomial involving addition. Work forward from $m(2x + 3)^2$ as the specific example.

Solution

Multiply the binomial factors using the FOIL method.

	General case	Specific case
a.	$m(ax + by)^2$	$2(2x + 3)^2$
	F O I L	F O I L
b.	$m(a^2x^2 + abxy + abxy + b^2y^2)$	$2(4x^2 + 6x + 6x + 9)$
c.	$m(a^2x^2 + 2abxy + b^2y^2)$	$2(4x^2 + 12x + 9)$
d.	$ma^2x^2 + 2mabxy + mb^2y^2$	$8x^2 + 24x + 18^2$

Study the pattern beginning with the polynomials in **c** and **d**. Study the pattern in **b** and **c**.

There is a common factor—*m* in the general case and 2 in the specific case. It is important always to extract monomial factors before looking for binomial factors. The product of the **F** and **L** terms (4 and 9) in the trinomial is identical to the product of the **O** and **I** coefficients (6 and 6). The coefficients of **O** and **I** (6 and 6) add to give the second term (12) of the trinomial. Both the **F** and **L** terms are perfect squares.

EXERCISES

Explore the pattern involved when a single monomial factor can be extracted before factoring a trinomial involving the difference of two squares.

Use $m(ax + by)(ax - by)$ for the general case and $5(2x + 4)(2x - 4)$ as the specific example.

General	Specific
$m(ax + by)(ax - by)$	$5(2x + 4)(2x - 4)$
$m(a^2x^2 - axby + axby - b^2y^2)$	$5(4x^2 - 8x + 8x - 16)$
$m(a^2x^2 - b^2y^2)$	$5(4x^2 - 16)$
$ma^2x^2 + mb^2y^2$	$20x^2 - 80$

174 RETEACHING *LESSON 11-8*

Skills Practice

Vocabulary Review

Lesson 11-7
quadratic term

Lesson 11-8
difference of two cubes
grand product
general case

Assignment Guide

All students: 1–104
Additional Practice: See Extra Practice Index on page 674.

Review and Practice Your Skills

Practice ■ Lesson 11-7

Factor the following trinomials.

1. $x^2 + 7x + 6$ $(x + 6)(x + 1)$
2. $m^2 + 11m + 28$ $(m + 7)(m + 4)$
3. $d^2 + 13d + 42$ $(d + 6)(d + 7)$
4. $b^2 + 17b + 42$ $(b + 3)(b + 14)$
5. $x^2 + 16x + 28$ $(x + 14)(x + 2)$
6. $p^2 + 12p + 11$ $(p + 11)(p + 1)$
7. $x^2 - 9x + 20$ $(x - 4)(x - 5)$
8. $g^2 - 8g + 12$ $(g - 2)(g - 6)$
9. $w^2 - 10w + 21$ $(w - 3)(w - 7)$
10. $f^2 - 30f + 200$ $(f - 20)(f - 10)$
11. $x^2 - 12x + 32$ $(x - 8)(x - 4)$
12. $n^2 - 18n + 32$ $(n - 16)(n - 2)$
13. $m^2 + 3m - 54$ $(m + 9)(m - 6)$
14. $b^2 + 6b - 7$ $(b + 7)(b - 1)$
15. $c^2 + c - 20$ $(c + 5)(c - 4)$
16. $h^2 + 5h - 24$ $(h + 8)(h - 3)$
17. $t^2 + 3t - 10$ $(t + 5)(t - 2)$
18. $x^2 + 4x - 45$ $(x + 9)(x - 5)$
19. $a^2 - 2a - 48$ $(a - 8)(a + 6)$
20. $k^2 - 8k - 48$ $(k - 12)(k + 4)$
21. $p^2 - 5p - 36$ $(p - 9)(p + 4)$
22. $z^2 - 6z - 40$ $(z - 10)(z + 4)$
23. $d^2 - d - 56$ $(d - 8)(d + 7)$
24. $x^2 - 4x - 32$ $(x - 8)(x + 4)$
25. $m^2 + 11mn + 30n^2$ $(m + 6n)(m + 5n)$
26. $g^2 + 2gh + h^2$ $(g + h)(g + h)$
27. $p^2 + 17pq + 60q^2$ $(p + 12q)(p + 5q)$
28. $x^2 - 9xy + 18y^2$ $(x - 3y)(x - 6y)$
29. $r^2 - 3rs + 2s^2$ $(r - s)(r - 2s)$
30. $c^2 - 8c + 15d^2$ $(c - 3d)(c - 5d)$
31. $b^2 + 3bc - 4c^2$ $(b + 4c)(b - c)$
32. $m^2 + 8mn - 9n^2$ $(m + 9n)(m - n)$
33. $a^2 + 7ab - 18b^2$ $(a + 9b)(a - 2b)$
34. $x^2 - 11xy - 26y^2$ $(x - 13y)(x + 2y)$
35. $p^2 - 4pq - 77q^2$ $(p - 11q)(p + 7q)$
36. $g^2 - 4gh - 60h^2$ $(g - 10h)(g + 6h)$
37. $x^2 + 14x + 48$ $(x + 6)(x + 8)$
38. $z^2 + 2z - 48$ $(z + 8)(z - 6)$
39. $f^2 - 26f + 48$ $(f - 2)(f - 24)$
40. $t^2 + 22t - 48$ $(t + 24)(t - 2)$
41. $c^2 - 19cd + 48d^2$ $(c - 3d)(c - 16d)$
42. $s^2 - 13st - 48t^2$ $(s - 16t)(s + 3t)$
43. $48 - 49x + x^2$ $(48 - x)(1 - x)$
44. $p^2 + 47pq - 48q^2$ $(p + 48q)(p - q)$
45. $26x + x^2 + 48$ $(x + 2)(x + 24)$

Practice ■ Lesson 11-8

For 46–51, specific examples will vary.

46. Explore the FOIL pattern for factoring a trinomial whose factors are of the form $(n_1 - x)(n_1 - x)$. Work forward from these factors as general case, and select your own specific example. $(n_1 - x)(n_1 - x) = n_1^2 - 2n_1x + x^2$

47. Explore the FOIL pattern for factoring a trinomial whose factors are of the form $(n_1 + x)(n_1 - x)$. Work forward from these factors as general case, and select your own specific example. $(n_1 + x)(n_1 - x) = n_1^2 - x^2$

48. Explore the FOIL pattern for factoring a trinomial whose factors are of the form $(ax + y)(bx + y)$. Work forward from these factors as general case, and select your own specific example. $(ax + y)(bx + y) = abx^2 + (a + b)xy + y^2$

49. Explore the FOIL pattern for factoring a trinomial whose factors are of the form $(ax + y)(bx - y)$. Work forward from these factors as general case, and select your own specific example. $(ax + y)(bx - y) = abx^2 - y^2$

50. Explore the FOIL pattern for factoring a polynomial whose factors are of the form $(x + a)(x + a)(x + a)$. Work forward from these factors as general case, and select your own specific example from $a > 0$. $(x + a)(x + a)(x + a) = x^3 + 3ax^2 + 3a^2x + a^3$

51. Repeat Exercises #50 for $a < 0$. Answers will vary.

Teaching Strategies

Discuss: Why is it important to write polynomials in standard form before factoring? Standard form makes the FOIL pattern possible. It is difficult to see patterns when a trinomial in not in standard form.

504 Chapter 11 **Polynomials**

PRACTICE ■ LESSON 11-1–LESSON 11-8

Simplify. (Lesson 11-1)

52. $(7x - 5y - 13z) + (-4y + 6x + z)$ $13x - 9y - 12z$

53. $(-8n^2 + 9n - 13) + (13n^2 - 3n + 12)$ $5n^2 + 6n - 1$

54. $(5xy + 7x^2 - 3y) - (-4x^2 - 8y + 3xy)$ $2xy + 11x^2 + 5y$

55. $(15x + 8y) - (-5x + 8y) + (4y - 2x)$ $18x + 4y$

Simplify. (Lesson 11-2)

56. $-5a(10 - 4a^2 - 5b)$ $-50a + 20a^3 + 25ab$

57. $6xyz(xy + 8yz - 2xz)$ $6x^2y^2z + 48xy^2z^2 - 12x^2yz^2$

58. $x^2(3x^2 + 5) + 3(2x^3 - 5x^2 + x)$ $3x^4 + 6x^3 - 10x^2 + 3x$

59. $5(x + 3y) - 2(2x - 3y) + 3(5x - 7y)$ $16x$

Find the GCF and its paired factor for the following. (Lesson 11-3)

60. $78x + 39y$ $39(2x + y)$

61. $16x^2 + 60x$ $4x(4x + 15)$

62. $14x^3y - 42xy^2$ $14xy(x^2 - 3y)$

63. $-9def - 15efg - 12gde$ $-3e(3df + 5fg + 4gd)$

64. $48a^3b^2 - 24ab^5 + 72a^2b^4$ $24ab^2(2a^2 - b^3 + 3ab^2)$

65. $7sm^2 + 28sw^2 + 63sy^2$ $7s(m^2 + 4w^2 + 9y^2)$

Simplify. (Lesson 11-4)

66. $(4r + 5y)(x - 2r)$ $4rx - 8r^2 + 5xy - 10ry$

67. $(x + 9)(x + 11)$ $x^2 + 20x + 99$

68. $(8x - 5)(7x + 6)$ $56x^2 + 13x - 30$

69. $(9 - 4x)(9 + 4x)$ $81 - 16x^2$

70. $(13 - 5v)(13 - 5v)$ $169 - 130v + 25v^2$

71. $(15f + 2)(9 - 2f)$ $-20f^2 - 12f + 18$

Find factors for the following. (Lesson 11-5)

72. $xy - 5x - 4y + 20$ $(x - 4)(y - 5)$

73. $5xw - 4w + 20x - 16$ $(w + 4)(5x - 4)$

74. $24a^3 + 8a^3f + 12b + 4bf$ $(8a^3 + 4b)(3 + f)$

75. $8x^2z + 11x^2b - 40z - 55b$ $(x^2 - 5)(8z + 11b)$

76. $5x - 20y + 5z - 2ax + 8ay - 2az$ $(5 - 2a)(x - 4y + z)$

77. $6n - 21p + 42mp - 12mn$ $(3 - 6m)(2n - 7p)$

78. $5x^2 - 2xz - 15xy + 6yz$ $(x - 3y)(5x - 2z)$

79. $ax - 2bx + 7x + 5a - 10ab + 35$ $(x + 5)(a - 2b + 7)$

80. $a^2c^2 + a^3b + bc^3 + ab^2c$ $(a^2 + bc)(c^2 + ab)$

Find binomial factors of the following, if possible. (Lesson 11-6)

81. $169 + 26a + a^2$ $(13 + a)(13 + a)$

82. $9x^2 - 42xy + 49y^2$ $(3x - 7y)(3x - 7y)$

83. $x^2 + 28x + 196$ $(x + 14)(x + 14)$

84. $1 - 100m^2$ $(1 + 10m)(1 - 10m)$

85. $16a^2 - 49b^2$ $(4a - 7b)(4a + 7b)$

86. $100c^2d^2 - b^2$ $(10cd + b)(10cd - b)$

87. $x^2 - 144y^2$ $(x + 12y)(x - 12y)$

88. $x^2 - 12xy + 144y^2$ not possible

89. $25m^2 + 110mn + 121n^2$ $(5m + 11n)(5m + 11n)$

Factor the following trinomials. (Lesson 11-7)

90. $c^2 + 27c + 72$ $(c + 3)(c + 24)$

91. $b^2 - 21b + 72$ $(b - 24)(b + 3)$

92. $a^2 + ad - 72d^2$ $(a + 9d)(a - 8d)$

93. $f^2 - 17fg + 72g^2$ $(f - 8g)(f - 9g)$

94. $72 - 73x + x^2$ $(72 - x)(1 - x)$

95. $72 - 71m - m^2$ $(72 + m)(1 - m)$

96. $r^2 - 18r + 81$ $(r - 9)(r - 9)$

97. $p^2 - 24pq + 81q^2$ $(p - 27q)(p + 3q)$

98. $81x^2 + 30x + 1$ $(27x + 1)(3x + 1)$

99. $a^2b^2 - 2ab - 3$ $(ab - 3)(ab + 1)$

100. $3n^2 - 4mn + m^2$ $(3n - m)(n - m)$

101. $-20x + x^2 + 96$ $(x - 8)(x - 12)$

Use the patterns explored in Lesson 11-8 to find all values of *k* which make each polynomial factorable. (Lesson 11-8)

102. $x^2 + kx + 24$ 25, 14, 11, 10

103. $x^2 + kx - 60$ 59, 28, 17, 11, 7, 4, −4, −7, −11, −17, −28, −59

104. $x^2 + 9x + k$ $(k > 0)$ 8, 14, 18, 20

Chalkboard Examples

Example from Lesson 11-7

Find the binomial factors of $d^2 + ed - 6e^2$. $(d + 3e)(d - 2e)$

Example from Lesson 11-8

Explore the FOIL pattern for factoring a trinomial whose factors are of the form $(x + ay)(x - by)$. Work forward from these factors as general case and select your own specific example. Answers may vary. general case: $x^2 - bxy + axy - aby^2$

Teaching Strategies

Remind students who are having trouble factoring to make sure they understand completely why the particular second-term constants or coefficients were written before they think about the signs of the second terms.

Lesson Planning

NCTM Standards/Strands

- Number & Operation
- Patterns, Functions, & Algebra
- Problem Solving
- Reasoning and Proof
- Connections
- Representation

Vocabulary

complex coefficients

Materials Needed

pencil/paper

Lesson Resources

Warm-Up Transparency 33
Reteaching 11-9
Extra Practice 11-9
Enrichment 11-9
Transparency TK-4

Assignment Guide

Basic: 1–42, 45–68
Enriched: 1–68

Getting Started

5-Minute Warm-up

Analyze each product into its prime factors.

1. (4)(25) (2)(2)(5)(5)
2. (12)(8) (2)(2)(2)(2)(2)(3)
3. (6)(10) (2)(2)(3)(5)

Introduction to Lesson 11-9

After student have completed this activity, have them share their explanations and any exceptions they found.

11-9 More on Factoring Trinomials

Goals ■ Factor trinomials of the form $ax^2 + bx + c$.

Applications Small Business, Packaging, Consumerism

Work with a partner to discuss the following questions.

1. Multiply each pair of binomials. Make sure that you show the FOIL multiplication step as part of your work.
   a. $(x + 4)(x - 5)$ $x^2 - x - 20$ b. $(3x + 4)(2x - 5)$ $6x^2 - 7x - 20$ c. $(3x + 4y)(2x - 5y)$ $6x^2 - 7xy - 20y^2$
2. Compare the multiplications and their products. Describe the ways in which the examples are similar. Answers may vary. All have subtraction; All have 20 in the last term. All have x^2 in the first term.
3. Describe the ways in which the examples differ. Answers vary vary.

Build Understanding

In the previous lessons, you have factored trinomials in the form $x^2 + bx + c$ or $x^2 + bxy + cy^2$. (In each case, b and c stand for numbers or constants; only x and y are variables.)

In this lesson, you will learn to factor trinomials with a larger quadratic (x^2) term coefficient. Finding binomial factors for a trinomial that has a quadratic coefficient larger than 1 is a two-step process. First, you must identify the FOIL coefficients. Once these are found, you can use them to discover the binomial factors.

Step 1: Identify the FOIL coefficients. A standard-form trinomial already shows two possible FOIL coefficients. The coefficient of the quadratic (x^2) term will be the F-coefficient (ab in the previous lesson). The coefficient of the last trinomial term is the L-coefficient (n_1n_2 in the previous lesson).

a. Multiply these coefficients together for the grand product coefficient.

b. Find two numbers whose *product* is the grand product coefficient and whose *sum* is the middle trinomial term. These two numbers are the cross-product (O- and I-) coefficients (an_2 and bn_1).

Step 2: Analyze the FOIL coefficients to find the four binomial coefficients (a, b, n_1, and n_2). (*Note*: Four is the maximum. There may appear to be fewer if some of the binomial coefficients are the same. For example, $(2x + 3)(3x + 1)$ has two coefficients of 3.)

a. List all possible paired factors for each FOIL coefficient.

b. Inspect the pairs, and select the pair for each coefficient that gives a total set including four or fewer individual factors. These will be the binomial coefficients.

c. Figure the signs as you did in Lesson 11-7; however, instead of focusing on which is the larger of the binomial second terms, you have to decide which is the larger of the two cross products.

Teaching Strategies

Making a table like the one shown in the Teaching Strategies box in Lesson 11-7 may be helpful for some students when factoring trinomials with complex coefficients. Although students may be tempted to try to just "think" of the answer, they will have greater success if they use some method to organize the possibilities.

Example 1

Find FOIL coefficients for the trinomial $6x^2 + 29x + 35$.

Solution

The F-coefficient is 6 (the coefficient of the quadratic term). The L-coefficient is 35 (the last term coefficient or the constant). The grand product coefficient is (6)(35), or (1)(2)(3)(5)(7), or 210. The cross-product (O- and I-) coefficients add to give 29, and multiply to give 210. The numbers 14 (= 2×7) and 15 (= 3×5) are the two coefficients you need. (*Note*: At this stage, you will not be able to tell which is the inner and which is the outer coefficient.)

Example 2

Given the four FOIL coefficients above, analyze their factor pairs to find the appropriate binomial coefficients for $6x^2 + 29x + 35$.

Solution

F-coefficient (ab):	6 = (1)(6) or (2)(3)
O- and I-coefficients:	14 = (1)(14) or (2)(7)
(an_2 and n_1b)	15 = (1)(15) or (3)(5)
L-coefficient (n_1n_2):	35 = (1)(35) or (5)(7)

Among these pairs, (2)(3), (2)(7), (5)(3), and (5)(7) share only four numbers. Therefore, they are the binomial coefficients. Thus:

a. 2 and 3 (the F pair) are the x coefficients — $(2x \quad)(3x \quad)$

b. 2 and 7 are a cross-product pair — $(2x \quad)(3x \quad 7)$

c. 3 and 5 are the other cross-product pair — $(2x \quad 5)(3x \quad 7)$

d. Trinomial signs are both positive, so signs are — $(2x + 5)(3x + 7)$.

Example 3

SMALL BUSINESS Ann designs and sells bracelets. Her gross profit is represented by the expression $2x^2 - 5x - 3$. The monomial factors represent the number of bracelets sold and the selling price per bracelet. Find the monomial factors.

Solution

The F-coefficient is 2, the L-coefficient is 3. The grand product coefficient is (2)(3) = 6. The L-coefficient sign is *negative*, so you need numbers with a product of 6 and an apparent *difference* of 5. The O- and I-coefficients must be 6 and 1.

F: 2 = (1)(2) O and I: 6 = (1)(6) or (2)(3)
L: 3 = (1)(3) 1 = (1)(1)

Binomial coefficients are (2)(1), (2)(3), (1)(1), and (1)(3).

Binomial factor values are $(2x \quad 1)(x \quad 3)$.

Second trinomial sign is negative, so larger cross product (6) must be negative. Factors with signs are $(x - 3)(2x + 1)$.

Chalkboard Examples

Supplementary Example 1
Find FOIL coefficients for the trinomial $9a^2 + 6a - 8$. 9, 12, 6, 8

Supplementary Example 2
Find the binomial factors for $5x^2 + x - 18$. $(5x - 9)(x + 2)$

Supplementary Example 3
For Ann's necklaces, her profit is represented by $15y^2 - 13y + 2$ where the factors represent the number of necklaces and the cost per necklace. Find the binomial factors. $(5y - 1)(3y - 2)$

Teaching Strategies

COOPERATIVE LEARNING Have students work in pairs to find all possible whole number values for which these trinomials could be factored.
$4x^2 + ax + 2$ 6, 9
$2x^2 + ax + 12$ 10, 11, 14, 25

Reteaching Worksheet 11-9

Name ______________________ Date ____________

RETEACHING 11-9

MORE ON FACTORING TRINOMIALS

The FOIL method you used in multiplying binomials in Lesson 5 can help you factor trinomials that contain two binomial factors. When the F-coefficient is *not* 1, try different pairs of factors until you find the pair that gives the correct middle term. Always check by multiplying.

Example

Find binomial factors for $6x^2 + 11x + 3$.

Solution

F		O + I		L	
$6x^2$	+	$11x$	+	3	The F-coefficient is 6; the L-term is 3. The sum of 0 + 1 is 11.
$1 \cdot 6$		Possible		$1 \cdot 3$	The signs in the binomial factors are positive because both 0 + 1 and L are positive.
$2 \cdot 3$		factors			

Try 1 and 6 for F, 1 and 3 for L: $(x + 3)(6x^2 + 1) = 6x^2 + 19x + 3$ Doesn't check!
$(x + 1)(6x + 3) = 6x^2 + 9x + 3$ Doesn't check!
Try 2 and 3 for F, 1 and 3 for L: $(3x + 3)(2x + 1) = 6x^2 + 9x + 3$ Doesn't check!
$(2x + 3)(3x + 1) = 6x^2 + 11x + 3$ It checks! ✔

EXERCISES

Use + or − to complete each binomial factor.

1. $8x^2 - 38x + 24 = (4x - 3)(2x - 8)$
2. $15x^2 + 26x + 8 = (5x + 2)(3x + 4)$
3. $18x^2 + 16x - 2 = (9x - 1)(2x + 2)$
4. $42x^2 - 13x\ 40 = (7x - 8)(6x + 5)$
5. $12x^2 + 12x - 9 = (6x + 9)(2x - 1)$
6. $28x^2 - 5x - 3 = (7x - 3)(4x + 1)$

Find binomial factors for each trinomial. Check by multiplying.

7. $4x^2 + 8x + 3$ $(2x + 3)(2x + 1)$
8. $6x^2 - 11x + 4$ $(3x - 4)(2x - 1)$
9. $8x^2 - 8x - 6$ $(4x + 2)(2x - 3)$
10. $9x^2 - 6x - 8$ $(3x - 4)(3x + 2)$
11. $15x^2 + 14x + 3$ $(5x + 3)(3x + 1)$
12. $8x^2 - 2x - 3$ $(4x - 3)(2x + 1)$
13. $9x^2 - 18x + 8$ $(3x - 4)(3x - 2)$
14. $15x^2 + x - 2$ $(3x - 1)(5x + 2)$
15. $6x^2 - 5x - 50$ $(2x + 5)(3x - 10)$
16. $6x^2 - 17x - 14$ $(2x - 7)(3x + 2)$
17. $6x^2 + 21x - 12$ $(6x - 3)(x + 4)$
18. $10x^2 - 15x - 25$ $(5x + 5)(2x - 5)$

176 RETEACHING *LESSON 11-9*

Lesson 11-9 **More on Factoring Trinomials** 507

Lesson Wrap-up

Quick Assessment

Ask the following questions to determine if students understand the content presented in this lesson.

1. When finding the binomial coefficients for the factors of a trinomial, how do you know where to write each coefficient? Answers will vary. Students should recognize that the FOIL pattern determines the correct placement of the coefficients.
2. Have students explain their thinking in solving finding the binomial factors for $6s^2 - 5s - 4$. The solution is $(2s + 1)(3s - 4)$.

Assignment Guide

Basic: 1–42, 45–68
Advanced: 1–68
Additional Practice: See Extra Practice Index on page 674.

Try These Exercises

Find FOIL coefficients/constants for the following. (Hint: Calculate the grand product, then refactor.)

1. $3x^2 + 19x + 6$ 3, 18, 1, 6
2. $10a^2 + 7a - 12$ 10, 15, 8, 12

Given the following FOIL coefficients, identify the binomial factor coefficients.

	3.		4.	
F-coefficient	8	(2, 4)	14	(7, 2)
Cross-product coefficients	6	(2, 3)	35	(7, 5)
(O and I)	20	(5, 4)	4	(2, 2)
L-coefficient	15	(5, 3)	10	(2, 5)

Identify the correct signs for the binomial second terms.

5. $35v^2 + 11v - 6 = (7v - 2)(5v + 3)$
6. $15s^2 - 17s - 4 = (5s + 1)(3s - 4)$
7. $3a^2 - ab - 10b^2 = (3a + 5b)(a - 2b)$

Find binomial factors for the following.

8. $8m^2 - 26m + 15$ $(2m - 5)(4m - 3)$
9. $7f^2 + 4fg - 3g^2$ $(f + g)(7f - 3g)$
10. $6r^2 - r - 35$ $(3r + 7)(2r - 5)$
11. $6x^2 + 17x + 10$ $(6x + 5)(x + 2)$
12. **PACKAGING** The surface area of a rectangular package is represented by the trinomial $2x^2 - 30x + 108$. Find the possible dimensions of the package. $(2x - 12)(x - 9)$

Practice Exercises

Find FOIL coefficients for the following trinomials.

13. $3p^2 - 11p - 4$ 3, 12, 1, 4
14. $5z^2 + 17z + 6$ 5, 15, 2, 6
15. $6d^2 + 13d - 5$ 6, 2, 15, 5
16. $21a^2 - 26ab + 8b^2$ 21, 12, 14, 8
17. $10x^2 - xy - 24y^2$ 10, 15, 16, 24
18. $4n^2 + 4n - 15$ 4, 10, 6, 15

For the following FOIL coefficients, identify the appropriate binomial factor coefficients.

	19.		20.		21.		22.	
F-coefficient	3	(3, 1)	21	(3, 7)	4	(4, 1)	27	(3, 9)
Cross-product coefficients	15	(3, 5)	35	(5, 7)	24	(4, 6)	21	(3, 7)
(O and I)	2	(2, 1)	6	(3, 2)	3	(3, 1)	18	(2, 9)
L-coefficient	10	(2, 5)	10	(5, 2)	18	(3, 6)	14	(2, 7)

Place appropriate signs in these unsigned binomials.

23. $8q^2 + 22q + 15 = (2q + 3)(4q + 5)$
24. $15c^2 - 38cd + 24d^2 = (3c - 4d)(5c - 6d)$
25. $18m^2 - 9m - 20 = (3m - 4)(6m + 5)$
26. $10y^2 + 33y - 7 = (5y - 1)(2y + 7)$
27. $12j^2 - jk - k^2 = (3j - k)(4j + k)$
28. $22n^2 + 23n - 15 = (11n - 5)(2n + 3)$

Extend the Lesson

CONNECTING TO PRIOR KNOWLEDGE Have students compare the results of Steps (a) through (d) of Example 2 to the results of using grouping to factor the trinomial once the FOIL coefficients are found.

$$6x^2 + 29x + 35 = (6x^2 + 14x) + (15x + 35)$$
$$= 2x(3x + 7) + 5(3x + 7)$$
$$= (2x + 5)(3x + 7)$$

508 Chapter 11 **Polynomials**

Find binomial factors for the following trinomials.

29. $21x^2 - 22x - 8$ $(3x - 4)(7x + 2)$

30. $6p^2 + 7p - 5$ $(2p - 1)(3p + 5)$

31. $2z^2 + 11z + 12$ $(z + 4)(2z + 3)$

32. $3a^2 - 14ab + 8b^2$ $(3a - 2b)(a - 4b)$

33. $20r^2 - 20rs - 15s^2$ $(2r - 3s)(10r + 5s)$

34. $20g^2 + 13gh - 15h^2$ $(4g + 5h)(5g - 3h)$

35. $64m^2 - 16m - 15$ $(8m + 3)(8m - 5)$

36. $49x^2 + 14xy - 24y^2$ $(7x - 4y)(7x + 6y)$

Find factors for the following.

37. $18v^2x + 3vwx - 6w^2x$ $3x(2v - w)(3v + 2w)$

38. $2e^2f^2 + 60d^2f^2 + 34def$ $2f(e^2f + 30d^2f + 17de)$

39. **TRAVEL** Goods are transported by train from City A to City B. The distance between the two cities is represented by the expression $2x^2 + 7x + 3$. Factor the expression to find monomials representing the time it took to transport the goods and the train's speed. $(2x + 1)(x + 3)$

40. **WRITING MATH** What strategies do you use to determine the signs for the second terms of the monomials when factoring trinomials with quadratic coefficients larger than 1? Answers may vary.

41. Find the binomial factors for the expression $5r^2 + r - 18$. $(5r - 9)(r + 2)$

42. **CONSTRUCTION** The volume of a concrete block is $16x^2 - 20x + 6$. The height of the block is 2 ft. Find the remaining dimensions of the block. $(4x - 3)(2x - 1)$

Extended Practice Exercises

43. Solve the equation $3x^2 + 30 = 40 + x$ by writing it in standard-form equal to zero. Then factor the trinomial and state the positive and negative solutions. $3x^2 - x - 10 = 0$; $x = 2$ or $-\frac{5}{3}$

44. **BOATING** For a sailboat to fit a particular design, its right triangle sail must be 2 feet shorter than the boat along its base, and 3 times taller than the boat's length plus an extra foot. To catch enough wind, the sail area must be 124 sq ft. How long must the boat be to fit these requirements? (Hint: Make a quadratic equation that is equal to zero, then solve it by factoring.) 10 ft

Mixed Review Exercises

Write each in simplest radical form. (Lesson 10-1)

45. $\sqrt{156}$ $2\sqrt{39}$

46. $\sqrt{300}$ $10\sqrt{3}$

47. $\sqrt{16 \cdot 9}$ 12

48. $\sqrt{261}$ $3\sqrt{29}$

49. $(3\sqrt{5})(2\sqrt{7})$ $6\sqrt{35}$

50. $(4\sqrt{3})(2\sqrt{21})$ $24\sqrt{7}$

51. $(\sqrt{15})(2\sqrt{18})$ $6\sqrt{30}$

52. $(5\sqrt{5})(7\sqrt{5})$ 175

53. $\left(4\sqrt{11}\right)^2$ 176

54. $\frac{\sqrt{8}}{\sqrt{3}}$ $\frac{2\sqrt{6}}{3}$

55. $\frac{\sqrt{13}}{\sqrt{6}}$ $\frac{\sqrt{78}}{6}$

56. $\sqrt{\left(\frac{7}{3}\right)}$ $\frac{\sqrt{21}}{3}$

Given $f(x) = 3x - 2$, $g(x) = -2x + 2$, and $h(x) = 4x^2$, find each value. (Lesson 2-2)

57. $f(-2)$ -8

58. $f(3)$ 7

59. $f(-5)$ -17

60. $f(8)$ 22

61. $g(5)$ -8

62. $h(-4)$ 10

63. $g(3)$ -4

64. $g(-1)$ 4

65. $h(2)$ 16

66. $h(-3)$ 36

67. $h(4)$ 64

68. $h(-5)$ 100

Extend the Lesson

ONGOING ASSESSMENT As an alternative assessment, have students use Algeblocks or algebra tiles to show how to factor $4x - x^2$.

Extra Practice Worksheet 11-9

Name ______________ Date ______________

EXTRA PRACTICE 11-9

More on Factoring Trinomials

Exercises

Find FOIL coefficients for the following trinomials.

1. $6p^2 + p - 2$ 6, 4, 3, 2

2. $2x^2 - x - 6$ 2, 4, 3, 6

3. $9r^2 - 12r - 5$ 9, 3, 15, 5

4. $10y^2 + 9y + 2$ 10, 4, 5, 2

5. $8z^2 + 10z - 3$ 8, 12, 2, 3

6. $6s^2 - 5s + 6$ 6, 4, 9, 6

7. $14y^2 + 5y - 1$ 14, 7, 2, 1

8. $12m^2 - 2m - 10$ 12, 12, 10, 10

Place appropriate signs in these unsigned binomials.

9. $12q^2 - 17q - 5 = (3q$ __−__ $5)(4q$ __+__ $1)$

10. $10a^2 + 19a + 6 = (5a$ __+__ $2)(2a$ __+__ $3)$

11. $2x^2 - 11x + 14 = (2x$ __−__ $7)(x$ __−__ $2)$

12. $6d^2 - d - 2 = (2d$ __+__ $1)(3d$ __−__ $2)$

13. $4r^2 - 35r + 24 = (4r$ __−__ $3)(r$ __−__ $8)$

14. $15t^2 - 4t - 4 = (5t$ __−__ $2)(3t$ __+__ $2)$

Find binomial factors for the following trinomials.

15. $6x^2 - x - 2$ $(2x + 1)(3x - 2)$

16. $4y^2 - 13y - 12$ $(4y + 3)(y - 4)$

17. $10t^2 + t - 2$ $(5t - 2)(2t + 1)$

18. $2d^2 + d - 21$ $(2d + 7)(d - 3)$

19. $9h^2 - 6h - 8$ $(3h - 4)(3h + 2)$

20. $8m^2 - 34m - 9$ $(2m - 9)(4m + 1)$

21. $6r^2 + 7r - 10$ $(6r - 5)(r + 2)$

22. $12f^2 + 35f + 8$ $(3f + 8)(4f + 1)$

23. $5a^2 - 7ab - 6b^2$ $(5a + 3b)(a - 2b)$

24. $8x^2 - 10xy - 25y^2$ $(2x - 5y)(4x + 5y)$

25. $14m^2 + 19mn - 3n^2$ $(7m - n)(2m + 3n)$

26. $18w^2 - wz - 5z^2$ $(9w - 5z)(2w + z)$

27. $16h^2 + 8hk - 15k^2$ $(4h + 5k)(4h - 3k)$

28. $10x^2 - 25x - 60$ $5(x - 4)(2x + 3)$

29. $16p^2 - 4pq - 30q^2$ $2(2p - 3q)(4p + 5q)$

30. $42r^2 - 3rt - 9t^2$ $3(2r - t)(7r + 3t)$

31. $6c^3 + 11c^2 - 10c$ $c(2c + 5)(3c - 2)$

154 EXTRA PRACTICE *Lesson 11-9*

Enrichment Worksheet 11-9

Name ______________ Date ______________

ENRICHMENT 11-9

The Four Digits Problem

In the four digits problem, you use the digits 1, 2, 3 and 4 to write an expression equal to a given number. Each digit is used only once. You can use addition, subtraction, multiplication (not division), exponents, and parentheses in any way you wish. Also, you can use two digits to make one number, as in 21.

Exercises

Express each number using the digits 1, 2, 3, and 4. A few examples are given to get you started. Other answers are possible.

$1 = (3 \cdot 1) - (4 - 2)$	$18 = (2 \cdot 3) \cdot (4 - 1)$	$35 = 2^{(4+1)} + 3$
$2 = (4 - 3) + (2 - 1)$	$19 = 3(2 + 4) + 1$	$36 = 34 + (2 \cdot 1)$
$3 = (4 - 3) + (2 \cdot 1)$	$20 = 21 - (4 - 3)$	$37 = 31 + 4 + 2$
$4 = (4 - 2) + (3 - 1)$	$21 = (4 + 3) \cdot (2 + 1)$	$38 = 42 - (1 + 3)$
$5 = (4 - 2) + (3 \cdot 1)$	$22 = 21 + (4 - 3)$	$39 = 42 - (1 \cdot 3)$
$6 = 4 + 3 + 1 - 2$	$23 = 31 - (4 \times 2)$	$40 = 41 - (3 - 2)$
$7 = 3(4 - 1) - 2$	$24 = (4 + 2) \cdot (3 + 1)$	$41 = 43 - (2 \cdot 1)$
$8 = 4 + 3 + 2 - 1$	$25 = (2 + 3) \cdot (4 + 1)$	$42 = 43 - (2 - 1)$
$9 = 4 + 2 + (3 \cdot 1)$	$26 = 24 + (3 - 1)$	$43 = 42 + 1^3$
$10 = 4 + 3 + 2 + 1$	$27 = 3^2 \cdot (4 - 1)$	$44 = 43 + (2 - 1)$
$11 = (4 \cdot 3) - (2 - 1)$	$28 = 21 + 4 + 3$	$45 = 43 + (2 \cdot 1)$
$12 = (4 \cdot 3) \cdot (2 - 1)$	$29 = 2^{(4+1)} - 3$	$46 = 43 + (2 + 1)$
$13 = (4 \cdot 3) + (2 - 1)$	$30 = (2 \cdot 3) \cdot (4 + 1)$	$47 = 31 + 4^2$
$14 = (4 \cdot 3) + (2 \cdot 1)$	$31 = 34 - (2 + 1)$	$48 = 4^2 \cdot (3 \cdot 1)$
$15 = 2(3 + 4) + 1$	$32 = 4^2 \cdot (3 - 1)$	$49 = 41 + 2^3$
$16 = (4 \cdot 2) \cdot (3 - 1)$	$33 = 21 + (4 \cdot 3)$	$50 = 41 + 3^2$
$17 = 3(4 + 2) - 1$	$34 = 2 \cdot (14 + 3)$	

200 ENRICHMENT *Lesson 11-9*

Assessment Planning

Vocabulary Review

Lesson 11-1
polynomial, monomial, coefficient, constant, binomial, trinomial, standard form

Lesson 11-2
associative property
commutative property

Lesson 11-3
extracting factors
greatest common factor (GCF)

Lesson 11-4
expanding, first product, outer product, inner product, last product, FOIL, cross products, difference of two squares

Lesson 11-5
grouping, monomial factor, polynomial factor

Lesson 11-6
perfect square trinomial
quadratic equation

Lesson 11-7
quadratic term

Lesson 11-8
difference of two cubes
grand product, general case

Lesson 11-9
complex coefficients

Assessment Options

Chapter Assessment, page 512
Alternative Assessment, page 513
Chapter Tests Forms A and B
Cumulative Assessment, page 514–515
Achievement Test
Writing Prompts

Assignment Guide

All students: 1–34
Additional Practice: See Extra Practice Index on page 674.

Chapter 11 Review

Vocabulary ■ LESSON 11-1–LESSON 11-9

a. quadratic
b. like terms
c. monomial
d. prime element
e. constant

Choose the word from the list that completes each statement.

1. A __?__ cannot be divided into smaller whole elements. d
2. A __?__ equation is an equation of the format $a^2 + bx + c = 0$. a
3. A number itself is called a __?__. e
4. A simple expression with only one term is called a __?__. c
5. Terms in which the variables or sets of variables are identical even though the coefficient may be different are called __?__. b

LESSON 11-1 ■ Add and Subtract Polynomials, p. 468

▶ A *polynomial* is an expression that contains several monomial terms. It is a *binomial* if it has two terms and a *trinomial* if it has three terms.

▶ A polynomial is written in *standard form* when its terms are ordered from the greatest to the least powers of one of the variables.

▶ Simplify a polynomial by combining all like terms so that only unlike terms remain.

Simplify.

6. $(12x^2 - 5) + (3x^3 - 6x^2 + 2)$ $3x^3 + 6x^2 - 3$
7. $(4a - 3a^2 + 1) - (2a + a^2 - 5)$ $-4a^2 + 2a + 6$

LESSON 11-2 ■ Multiply By a Monomial, p. 472

▶ Use the distributive property and the rules for exponents to multiply a polynomial by a monomial.

Simplify.

8. $2x(x + 3y - z)$ $2x^2 + 6xy - 2xz$
9. $-5ab[a - (a^2b - 3b)]$ $-5a^2b + 5a^3b^2 - 15ab^2$

LESSON 11-3 ■ Divide and Find Factors, p. 478

▶ To *extract a factor*, check to see if any monomial will divide exactly into every term of the polynomial.

▶ To *factor* an expression, use the GCF and the distributive property.

Find the GCF and its paired factors for the following.

10. $81x^2y - 27x^3y^2$ $27x^2y(3 - xy)$
11. $3a^3b^2 - 6ab$ $3ab(a^2b - 2)$
12. $9a^3b + 18a^2b^2 - 6a^2b^3$ $3a^2b(3a + 6b - 2b^2)$
13. $5x^5y - 10x^4y^2 - 20x^3y^3$ $5x^3y(x^2 - 2xy - 4y^2)$

LESSON 11-4 ■ Multiply Two Binomials, p. 482

▶ To multiply two binomials, write the product of the first terms, the outer terms, the inner terms, and the last terms (FOIL), then simplify.

Simplify.

14. $(2x - 5y)(3x + 8y)$ $6x^2 + xy - 40y^2$
15. $(4a - b)(4a + b)$ $16a^2 - b^2$
16. $(m - 2n)(m - 2n)$ $m^2 - 4mn + 4n^2$

Teaching Strategies

Remind students that mastering skills such as multiplying binomials and factoring trinomials requires that they practice with many different types of problems. Students should not expect to master algebraic techniques simply by reading about them; active participation is needed.

LESSON 11-5 ■ Find Binomial Factors in a Polynomial, p. 488

► To factor a polynomial, group terms as pairs, extract the common monomial factor from each pair, and extract the identical binomial.

Factor.

17. $6a^2 + 9ab - 10ab - 15b^2$ $(3a - 5b)(2a + 3b)$

18. $14x^2 + 15y^2 - 10xy - 21xy$ $(7x - 5y)(2x - 3y)$

19. $5rt + 20ru + 2st + 8su$ $(5r + 2s)(t + 4u)$

20. $2v^2 + 3vx + 10vw + 15wx$ $(v + 5w)(2v + 3x)$

LESSON 11-6 ■ Special Factoring Pattern, p. 492

► Use these patterns to factor perfect square trinomials and polynomials that are differences of squares.

$a^2 + 2ab + b^2 = (a + b)^2$ $a - 2ab + b^2 = (a - b)^2$ $a^2 - b^2 = (a + b)(a - b)$

Factor.

21. $81m^2 - 16n^2$ $(9m - 4n)(9m + 4n)$

22. $4e^2 + 12e + 9$ $(2e + 3)^2$

23. $9x^2 - 30x + 25$ $(3x - 5)^2$

LESSON 11-7 ■ Factoring Trinomials, p. 498

► You can use four clues to factor a trinomial where the coefficient of the first term is one; the trinomial third term is always the product of the binomial second terms; the trinomial second term is always the sum of the binomial second terms; if the sign of the trinomial third term is negative (positive), the signs in the binomial are different (the same).

Factor.

24. $x^2 - xy - 6y^2$ $(x - 3y)(x + 2y)$

25. $m^2 + 3mn - 40n^2$ $(m + 8n)(m - 5n)$

26. $r^2 - 10r + 16$ $(r - 8)(r - 2)$

27. $a^2 + 8a + 15$ $(a + 5)(a + 3)$

28. $g^2 + 7g - 44$ $(g + 11)(g - 4)$

29. $m^2 - 15mn + 36n^2$ $(m - 3n)(m - 12n)$

LESSON 11-8 ■ Problem Solving Skills: The General Case, p. 502

► Studying the general case can help you identify patterns and solve problems. Use letters instead of numbers to represent the coefficients and constants.

30. Explore the pattern for finding the square of a binomial. Work from $(ax + n)^2$ as the general case and $(3x + 5)^2$ as the specific example.
$(ax + n)(ax + n) = a^2x^2 + axn + axn + n^2 = a^2x^2 + 2axn + n^2$
$(3x + 5)(3x + 5) = 9x^2 + 15x + 15x + 25 = 9x^2 + 30x + 25$

LESSON 11-9 ■ Factoring Trinomials With Complex Coefficient, p. 506

► To find binomial factors for a trinomial that has quadratic coefficients larger than one is a two-step process. First, identify FOIL coefficients. Then analyze the coefficient to find the four binomials coefficient (a, b, n_1 and m_2).

Factor.

31. $4s^2 - 4st - 15t^2$ $(2s - 5t)(2s + 3t)$

32. $15a^2 + 2ab - 8b^2$ $(5a + 4b)(3a - 2b)$

33. $28a^2 - ab - 2b^2$ $(4a + b)(7a - 2b)$

34. $30p^2 - 57pq + 18q^2$ $(5p - 2q)(6p - 9q)$

Teaching Strategies

COOPERATIVE LEARNING Have students work in pairs to write all factorable quadratic trinomials whose first term is x^2 and whose last term is 12.

$x^2 + 11x - 12$ $x^2 + x - 12$
$x^2 - 4x - 12$ $x^2 + 4x - 12$
$x^2 - 11x - 12$ $x^2 - x - 12$

Chapter 11 Test Form A, page 1

CHAPTER 11
POLYNOMIALS
ASSESSMENT FORM A, PAGE 1

Name ______
Date ______

Scoring Record	
Possible: 40	Earned:

Simplify.

1. $(x + 3) + (5x - 8)$ $6x - 5$

2. $(9n - 2) + (-6n + 1)$ $3n - 1$

3. $(3x - 4) - (2x + 7)$ $x - 11$

4. $(8y + 2) - (3y - 6)$ $5y + 8$

5. $(5x^2 + 2xy - 3y^2) + (-9x^2 + 3xy + 5y^2)$ $-4x^2 + 5xy + 2y^2$

6. $(3a^2 + 4b - c) - (-a + 5b + c)$ $3a^2 + a - b - 2c$

7. $x(xyz)$ x^2y^2

8. $(-2a^2)(5ab^2)$ $-10a^3b^2$

9. $(3w)(w^2 - 2x)$ $3w^3 - 6wx$

10. $a(b^2 - c)$ $ab^2 - ac$

11. $(5ab)(-6a^2c^2)$ $-30a^3bc^2$

12. $7x^2y^2(8x - 2xy + 3y)$ $56\,x^3y^2 - 14x^3y^3 + 21x^2y^3$

Find the greatest common factor (GCF) for each polynomial. Then find its paired factor.

13. $3x + 12y$ $3(x + 4y)$

14. $5x^2y - 4y^2$ $y(5x^2 - 4y)$

15. $3ab^3 + 4b^2c$ $b^2(3ab + 4c)$

16. $4a^2 - 6a^2b$ $2a^2(2 - 3b)$

17. $12m^4 - 8m^2n$ $4m^2(3m^2 - 2n)$

18. $8n^4 + 5n^3 - n^2$ $n^2(8n^2 + 5n - 1)$

Find each product.

19. $(x - 8)(x - 9)$ $x^2 - 17x + 72$

20. $(2b + c)(b - 2c)$ $2b^2 - 3bc - 2c^2$

21. $(3a - 2b)(4a - b)$ $12a^2 - 11ab + 2b^2$

142 ASSESSMENT *CHAPTER 11 FORM A*

Chapter 11 Test Form A, page 2

Name ______ Date ______

Find factors for each polynomial.

22. $2ac + 6ad + bc + 3bd$ $(2a + b)(c + 3d)$

23. $15wy - 5wz - 6xy + 2xz$ $(5w - 2x)(3y - z)$

24. $8qs - 20qt + 6rs - 15rt$ $(4q + 3r)(2s - 5t)$

25. $12ac + 6ad - 6bc - 3bd$ $3(2a - b)(2c + d)$

Find binomial factors for each trinomial.

26. $x^2 + 18x + 81$ $(x + 9)(x + 9)$ or $(x + 9)^2$

27. $y^2 - 18y + 81$ $(y - 9)(y - 9)$ or $(y - 9)^2$

28. $9a^2 - 16$ $(3a + 4)(3a - 4)$

29. $4y^2 + 28y + 49$ $(2y + 7)(2y + 7)$ or $(2y + 7)^2$

30. $x^2 - 9x - 10$ $(x - 10)(x + 1)$

31. $b^2 + 5b - 6$ $(b + 6)(b - 1)$

32. $y^2 - 7y - 30$ $(y - 10)(y + 3)$

33. $a^2 - 8a + 12$ $(a - 6)(a - 2)$

34. $15x^2 - 5x - 10$ $5(3x + 2)(x - 1)$

35. $2a^2 + 17a + 21$ $(2a + 3)(a + 7)$

36. $6c^2 - 21c + 15$ $3(c - 1)(2c - 5)$

37. $21x^2 - 2x - 3$ $(7x - 3)(3x + 1)$

Find each answer.

38. List all the possible whole-number sides for a rectangle whose area is $2s^2t$.
$1, 2s^2t; 2, s^2t; 2s, st; s, 2st; 2t, s^2; t, 2s^2$

39. What is the width of a rectangle with a length of $4cd$ and an area of $16c^2d$? $4c$

40. The perimeter of a square is $28r^2s$. What is the length of each side? $7r^2s$

144 ASSESSMENT *CHAPTER 11 FORM A*

Chapter 11 Test Form B, page 1

CHAPTER 11

POLYNOMIALS

ASSESSMENT FORM B, PAGE 1

Name ____________

Date ____________

Scoring Record	
Possible: 40	Earned:

Simplify.

1. $(x - 4) + (2x + 5)$ $3x + 1$
2. $(-3n - 6) + (7n - 3)$ $4n - 9$
3. $(2x + 3) - (4x - 3)$ $-2x + 6$
4. $(9y + 2) - (-2y + 6)$ $11y - 4$
5. $(-2x^2 + 5xy + 7y^2) - (-8x^2 - 9xy + y^2)$ $6x^2 + 14xy + 6y^2$
6. $(2a^2 - 9b + c) + (-4a + 10b + c)$ $2a^2 - 4a + b + 2c$
7. $a(a^2bc)$ a^3bc
8. $(7xy^2)(-4x^2y)$ $-28x^3y^3$
9. $(7x)(y^2 - 4z)$ $7xy^2 - 28xz$
10. $x(y^2 - z)$ $xy^2 - xz$
11. $(-8ab)(-6a^2c)$ $48a^3bc$
12. $3a^3b^3(5a - 3ab + 4b)$ $15a^4b^3 - 9a^4b^4 + 12a^3b^4$

Find the greatest common factor (GCF) for each polynomial. Then find its paired factor.

13. $2m + 8n$ $2(m + 4n)$
14. $7a^2b - 3ab^2$ $ab(7a - 3b)$
15. $2xy^3 + 5x^2z$ $x(2y^3 + 5xy)$
16. $3m^2 + 6m^2n$ $3m^2(1 + 2n)$
17. $10a^4 - 5a^3b^2$ $5a^3(2a - b^2)$
18. $7x^5 - 2x^4 + x^3$ $x^3(7x^2 - 2x + 1)$

Find each product.

19. $(x - 3)(x - 7)$ $x^2 - 10x + 21$
20. $(3r + s)(r - 3s)$ $3r^2 - 8rs - 3s^2$
21. $(2x - y)(5x - 2y)$ $10x^2 - 9xy + 2y^2$

146 ASSESSMENT *CHAPTER 11 FORM B*

Chapter 11 Test Form B, page 2

Name ____________ Date ____________

Find factors for each polynomial.

22. $6wy - 6wz - 2xy + 2xz$ $(6w - 2x)(y - z)$
23. $15ac + 30ad - 3bc - 6bd$ $3(5a - b)(c + 2d)$
24. $5qs + 10qt + 2rs + 4rt$ $(5q + 2r)(s + 2t)$
25. $3ac + 12ad - 6bc - 24bd$ $3(a - 2b)(c + 4d)$

Find binomial factors for each trinomial.

26. $x^2 + 4x + 4$ $(x + 2)(x + 2)$ or $(x + 2)^2$
27. $y^2 - 10y + 25$ $(y - 5)(y - 5)$ or $(y - 5)^2$
28. $4a^2 - 25$ $(2a - 5)(2a + 5)$
29. $9y^2 + 30y + 25$ $(3y + 5)(3y + 5)$ or $(3y + 5)^2$
30. $x^2 + 5x + 6$ $(x + 3)(x + 2)$
31. $b^2 - 4b + 3$ $(b - 3)(b - 1)$
32. $y^2 + 6y - 16$ $(y + 8)(y - 2)$
33. $a^2 + 8a + 7$ $(a + 7)(a + 1)$
34. $2x^2 + 24x + 54$ $2(x + 3)(x + 9)$
35. $2a^2 - 5a - 25$ $(2a + 5)(a - 5)$
36. $7c^2 + 20c - 3$ $(7c - 1)(c + 3)$
37. $16x^2 - 46x + 15$ $(8x - 3)(2x - 5)$

Find each answer.

38. List all the possible whole-number sides for a rectangle whose area is $3xy^2$.
    $1, 3xy^2; 3, xy^2; 3x, y^2; 3xy, y; 3y, xy; 3y^2, x$
39. What is the length of a rectangle with a width of $5ab$ and an area of $25a^2b^3$? $5ab^2$
40. The perimeter of a square is $32ab^2$. What is the length of each side? $8ab^2$

148 ASSESSMENT *CHAPTER 11 FORM B*

CHAPTER 11 ASSESSMENT

Simplify.

1. $(x^2 - 5x + 4) - (3x^2 + 2x - 5)$ $-2x^2 - 7x + 9$
2. $(3x^4 - 3xy + 6y^2) + (y^2 - 2x^4)$ $x^4 - 3xy + 7y^2$
3. $2(-3a^3)^2$ $18a^6$
4. $(-4a^2b)(5a - 3b + a^2b)$ $-20a^3b + 12a^2b^2 - 4a^4b^2$
5. $(5c + d)(3c - 2d)$ $15c^2 - 7cd - 2d^2$
6. $(2x + 5)(2x - 5)$ $4x^2 - 25$
7. $(6r + s)(r - 3s)$ $6r^2 - 17rs - 3s^2$
8. $(x - y)(x - y)$ $x^2 - 2xy + y^2$
9. $-5 + 4m^2 - 3m - 8m^2 + 13 + m$ $-4m^2 - 2m + 8$
10. $(5r - 6s^2 + rs) - (rs - 6s^2 - 5r)$ $10r$

Factor.

11. $25x^3yz^2 - 30xyz^3$ $5xyz^2(5x^2 - 6z)$
12. $13mn^5 - 52mn^4$ $13mn^4(n - 4)$
13. $8n^2 - 2mn - 3n^2 + 12mn$ $5n(n + 2m)$
14. $rs + 2r^2 - 10rs - 5s^2$ $(2r + s)(r - 5s)$
15. $25x^2 - 30xy + 9y^2$ $(5x - 3y)^2$
16. $4a^2 - 49b^2$ $(2a - 7b)(2a + 7b)$
17. $81x^2 + 16$ not factorable
18. $9m^2 - 24mn + 16n^2$ $(3m - 4n)^2$
19. $a^2 + 7ab - 18b^2$ $(a + 9b)(a - 2b)$
20. $x^2 - 4xy - 5y^2$ $(x - 5y)(x + y)$
21. $m^2 - 7m + 10$ $(m - 5)(m - 2)$
22. $r^2 + 8rs + 7s^2$ $(r + 7s)(r + s)$
23. $16e^2 + 2ef - 3f^2$ $(8e - 3f)(2e + f)$
24. $15x^2 - 8xy - 12y^2$ $(5x - 6y)(3x + 2y)$

Use the figure at the right for Exercises 25 and 26.

25. Write an expression for the perimeter of the rectangle. $12x + 8$
26. Write an expression for the area of the rectangle. $8x^2 - 2x - 45$

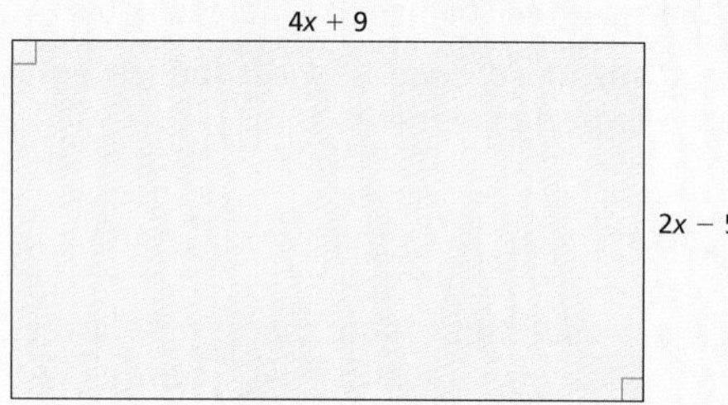

512 | Chapter 11 **Assessment**

Chapter Investigation

The benchmarks and expectations for this extension are as follows.

- Students begin to develop a survey to gather demographical information from their classmates. They write a list of questions that can be used to find out about shopping interests and spending habits.
- Students add questions to their survey to find out how much time each day their classmates spend watching television, listening to the radio, and reading newspapers and magazines.
- Students distribute the final survey to their classmates and compile the data. They use the information to create a demographic profile of their class. They work together to prepare graphs and charts.
- Students work with their group to develop a strategy for marketing a new product aimed at people their age. They determine when and where they would run advertisements
- Students decide which of the three advertisements would be most effective to sell to people their age. They present their decision to the class and explain why that type of advertisement would be most effective.

Chapter 11 Alternative Assessment

Self-Assessment

NOTEBOOK Select one typical exercise from each lesson in this chapter. Solve each exercise, including a step-by-step description of the methods you use. Then write a list of the most important things that you need to remember while solving these or other equations.

Other Resources

ALGEBRA TALK This chapter presents several new terms and combines some of the familiar algebraic symbols in new ways. Research in the library or on the Internet to learn the origins of some of these terms and symbols, such as coefficient, quadratic, algebra, trinomial, as well as exponents and factoring. Who invented these? Write a brief paragraph for these and any other terms or symbols you think are important and give examples of each.

Critical Thinking

FOILED AGAIN? Every teacher and textbook helps student learn how to multiply binomials by using the mnemonic FOIL. Does any other process, for instance IFLO, work as well? Why or why not? If IFLO or any other possible procedure works, why does everyone use FOIL? Write a brief essay to present and defend your position on this matter, giving examples.

Chapter Investigation

EXTENSION Decide which of the three advertisements, print, radio, or television, would be most effective to sell a product to people your own age. Present your decision to the class and explain why that type of advertisement would be most effective.

Self Assessment

You may wish to use new exercises to better gauge students' level of understanding.

SCORING RUBRIC **5**—Student correctly solves one exercise from each lesson, accurately describes the steps needed and writes a comprehensive list of important things to remember. **4**—Student correctly solves one exercise from each lesson, describes steps, and writes an adequate list. **3**—Student solves one exercise from each lesson with minor errors, describes steps needed and writes a partial list. **2**—Student solves one exercise from each lesson with minor errors, describes most of the steps needed and writes a partial list. **1**—Student solves one exercise from most lessons in the chapter with errors, describes some steps, and writes an incomplete list. **0**—Student makes no attempt to solve problems or describe the steps.

Using Other Resources

You may wish to go through the chapter with students to develop a master list of important terms and symbols.

SCORING RUBRIC **5**—Student researches and finds the origin of all significant algebra terms and symbols, giving a detailed history and precise examples of each. **4**—Student researches and finds the origin of many terms and symbols, giving a general history and examples of each. **3**—Student researches and finds the origin of some terms and symbols, giving a vague history and few examples. **2**—Student researches and finds the origin of a few terms and symbols, giving a vague history and few examples. **1**—Student attempts to research the origin of a few terms and symbols, but writes no history for the research. **0**—Student makes no attempt to research the algebra terms and symbols.

Critical Thinking

Encourage students to try the same multiplication and factoring of two or three different problems to be sure they draw a valid conclusion.

SCORING RUBRIC **5**—Student uses different methods to multiply binomials, reaches a logical conclusion about the use of FOIL, and convincingly supports that conclusion with several accurate examples. **4**—Student uses different methods to multiply binomials, reaches a logical conclusion about the use of FOIL, and supports that conclusion with several examples. Minor errors may exist in logic. **3**—Student uses different methods to multiply binomials with minor errors, reaches a conclusion about the use of FOIL, and supports that conclusion with a few examples. **2**—Student uses one or two methods to multiply binomials with minor errors, reaches a conclusion about the use of FOIL, but cannot support the conclusion. **1**—Student tries to multiply binomials with minor errors, reaches an illogical conclusion about the use of FOIL, and cannot support the conclusion. **0**—Student makes no attempt to investigate FOIL or other methods.

Cumulative Review

Cumulative review is the best way to maintain previously taught skills and concepts. This will keep students prepared for new lessons that build on previously covered skills.

Cumulative Review covers:

Lesson 1-4
Lesson 2-3
Lesson 2-6
Lesson 2-7
Lesson 3-6
Lesson 4-8
Lesson 5-2
Lesson 6-1
Lesson 6-8
Lesson 7-5
Lesson 8-3
Lesson 8-7
Lesson 9-4
Lesson 9-7
Lesson 10-3
Lesson 11-2
Lesson 11-4

Additional Answers

3. $-\frac{2}{3}$

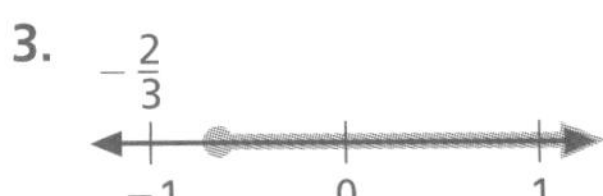

4. $-\frac{3}{8}$

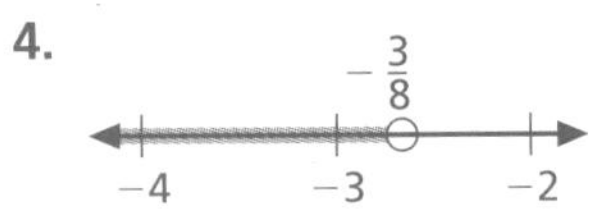

5.

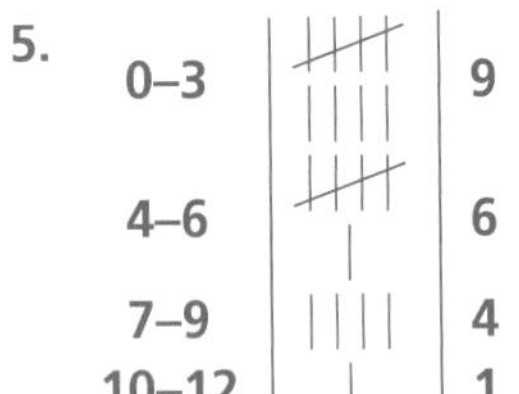

0–3	𝍸𝍸 ||||	9
4–6	𝍸 |	6
7–9	||||	4
10–12	|	1

7. If two angles are acute, the sum of the measures of the two angles is less than 90°; true; false.

10.

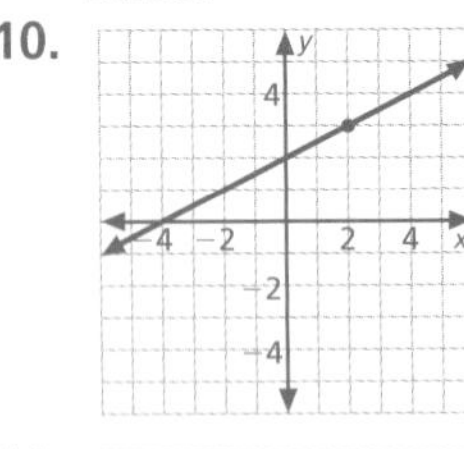

11.

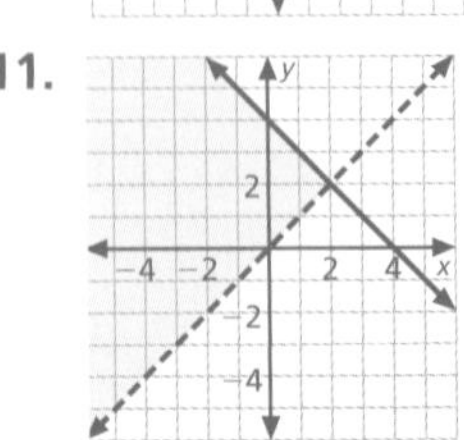

Chapters 1–11 Cumulative Review

1. Ari made the following transactions to her savings account: previous balance, \$216.95; deposit, \$29.00; deposit \$12.75; withdrawal, \$59.25; withdrawal, \$12.95. What is her new balance? \$186.50

2. Given $g(x) = -x^2$, $f(x) = 2x + 1$, and $h(x) = x^2 - 1$, find $g(a) + h(a) - |f(-1)|$. -2

Solve each inequality and graph the solution on a number line. See additional answers.

3. $\frac{2}{3}z + 1 \geq \frac{5}{9}$

4. $5(3x + 12) < 20$

A random sample of 20 student records was used to determine the average number of tardies per student during the school year. The numbers of tardies is listed below.

6	0	9	2	1	1	3	8	5	5
1	5	4	5	3	7	12	3	9	2

5. Construct a frequency table for the data. See additional answers.

6. Find the mean, median, and mode. mean = 4.55, median = 4.5, mode = 5

7. Write the converse of the statement: If the sum of the measures of two angles is less than 90°, then the angles are acute. Then tell whether each is true or false. See additional answers.

8. *ABCD* is a rectangle. Find *a* and *b*. $a = 5$ in. $b = \sqrt{91}$ in. ≈ 9.5 in.

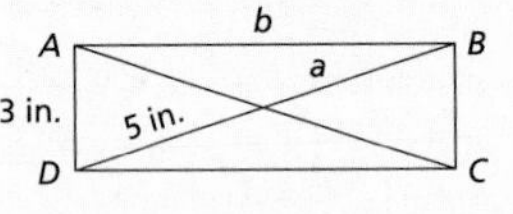

9. Find the area of the shaded region. 40π m² ≈ 125.7 m²

10. Graph the line that passes through the point *P*(2, 3) and has a slope of $\frac{1}{2}$. See additional answers.

11. Graph the solution set. $\begin{cases} y > x \\ x + y \leq 4 \end{cases}$ See additional answers.

12. Find *x* to the nearest tenth. $x = \frac{48}{5} = 9.6$

13. Draw the dilation image if the center of dilation is the origin and the scale factor is $\frac{1}{3}$. $A'(1, 1)$, $B'(3, 1)$, $C'(3, 2)$

14. Find the reflection image of the quadrilateral represented by

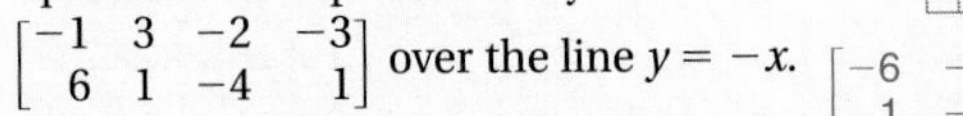

$\begin{bmatrix} -1 & 3 & -2 & -3 \\ 6 & 1 & -4 & 1 \end{bmatrix}$ over the line $y = -x$. $\begin{bmatrix} -6 & -1 & 4 & -1 \\ 1 & -3 & 2 & 3 \end{bmatrix}$

15. A group of numbered cards contains four 5s, two 3s, one 2, and three 8s. Cards are picked at random, one at a time, and then replaced. Find *P*(8, then odd). $\frac{9}{50}$

16. On a math test taken by 20 students, the mean score was 78.5. The standard deviation for the scores was 7.8. What was the sum of all the scores? 1,570

17. Find the missing measures. Leave answers in simplest radical form. $5\sqrt{3}$ m, 10 m; 60°

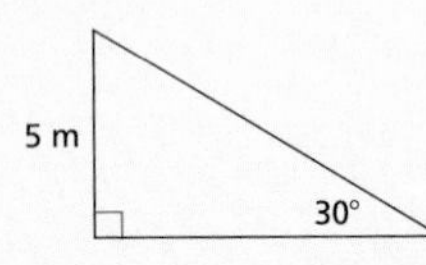

Simplify.

18. $2ab^2c(a^3bc^{-1} + a^{-1}b^2c^3)$ $2a^4b^3 + 2b^4c^4$

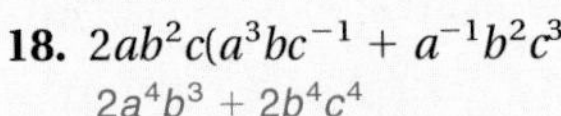

19. $(5x + 3y)(5x - 3y)$ $25x^2 - 9y^2$

Teaching Strategies

Review the vocabulary related to transformations. Have students think of memory tricks to help them remember these definitions. For example, students who have had their eyes dilated may be aware that dilation affects the size of the pupils of the eyes allowing more light to enter the eye. From this experience, they can remember that a dilation image is larger or smaller than the preimage.

Chapters 1–11 Cumulative Assessment

Standardized Test Preparation—Quantitative Comparison

In each question, compare the quantity in Column 1 with the quantity in Column 2. Select the letter of the correct answer from these choices:

A. The quantity in Column 1 is greater.

B. The quantity in Column 2 is greater.

C. The two quantities are equal.

D. The relationship cannot be determined by the information given.

Notes: In some questions, information that refers to one or both columns is centered over both columns. A symbol used in both columns has the same meaning in each column. All variables represent real numbers. Most figures are not drawn to scale.

	Column A	Column B
1. A	$g(x) = \lvert -x \rvert$ $h(x) = -2\,\lvert 4x + 1 \rvert$ $f(x) = x^2 + 5$	
	$g(f(h(-1)))$	$f(-2) + g(0) - h(0)$
2. B	8 6 0 1 12 5 8 5 7 8	
	The median of the data below	The mode of the data below
3. B	In $\triangle MNP$, $MP < NP$ and $MN > MP$.	
	$\angle M$	$\angle N$
4. A	A trapezoid and its median are shown. (3 cm, x, 5 cm)	
	x	3.5 cm
5. A	Solve. $3x + y = 5$ $2x - 5y = 9$	
	x	y

	Column A	Column B
6. A	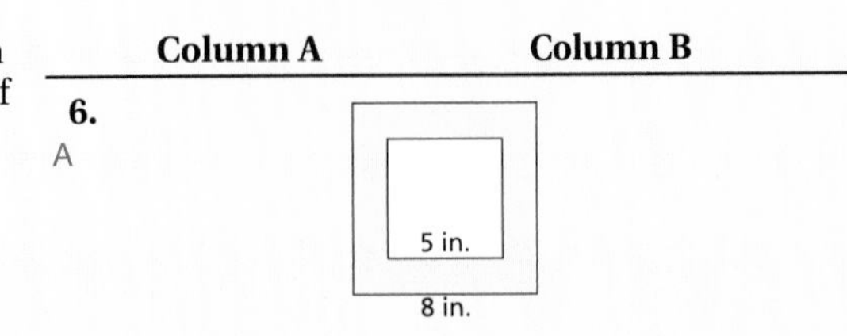	
	The probability that a point selected at random is in the shaded region.	The probability that a point selected at random is in the unshaded region.
7. B	Perform these transformations on $\triangle ABC$ with $A(0, 5)$, $B(-2, 1)$, and $C(1, -3)$. A reflection over the x-axis is followed by a translation of two units left and one unit up. Find B''.	
	x	y
8. A	$\begin{bmatrix} 2 & 3 \\ -1 & 5 \end{bmatrix} \begin{bmatrix} 0 & 1 \\ 4 & -2 \end{bmatrix} = \begin{bmatrix} a & b \\ c & d \end{bmatrix}$.	
	$1b$	$2d$
9. C		
	The probability of spinning a 5 or a multiple of 2.	The probability of spinning a 2 or an odd number.
10. A	$(4e - 3f)(4e + 3f) = xe^2 = yef + zf^2$	
	x	z

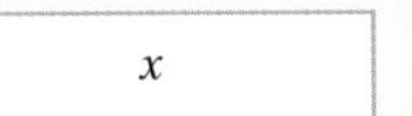

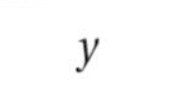

11. CONSTRUCTED RESPONSE Factor $a^2 + 2ab - 15b^2$. How is it different from factoring $a^2 + 2ab + b^2$?

$(a - 3b)(a + 5b)$; Possible Answer: $a^2 + 2ab + b^2$ is a perfect square. It factors to $(a + b)^2$. There are no coefficients on the variables.

Standardized Test Preparation

QUANTITATIVE COMPARISON QUESTIONS Quantitative comparison questions are generally written with fewer words and require less time to compute. There are four answer choices that describe the relationships of equality and inequality. You are required to compare the quantity from Column A with the quantity from Column B and then determine their relationship. Students should carefully read the four answer choices often during the testing period.

Horizontal rules are used as separators between questions and to clarify when figures and additional information accompanies one or more problems.

Understanding the directions for quantitative comparison questions is critical for success since this type of question is not as commonplace as other types of questions.

Testing Strategies

When solving quantitative comparison problems, it is important that students read the entire problem carefully before beginning. Some questions have a problem to solve before a comparison can be made. For other questions, knowing what is to be compared is important to know before solving.

For example, in Exercise 1, the comparison values tell how to handle the three equations. In Exercise 5, the equations are to be solved first before a comparison is made.

Teaching Strategies

Discuss how students may determine whether the events in a probability problem are independent or dependent. Review the basic probability situations encountered on tests: drawing a card, rolling dice, spinning a spinner, sampling from a bag or box at random and hitting a target. For each situation, have students create and solve at least one math problem.

CHAPTER 12—QUADRATIC FUNCTIONS

Theme: Gravity
Chapter Investigation: Forces of Gravity, pages 519, 523, 533, 537, 551, 555
Careers: Air Traffic Controllers, page 529; Pilots, page 549
Data Activity: How Does Gravity Affect Weight?

Content and Connections

Lesson, Pages	Lesson Objectives	NCTM Standards	State/Local Objectives	Interdisciplinary Connections	Real World Applications
12–1 520–523	• Graph parabolas or second degree equations	Patterns, Functions & Algebra Communication Problem Solving Connections Representation		Physics	Geology, Small Business, Physics
12–2 524–527	• Graph functions defined by the standard quadratic equation	Patterns, Functions & Algebra Problem Solving Reasoning & Proof Communication Representation Number Operation			Physics, Business, Astronomy
12–3 530–533	• Use factoring to solve quadratic equations	Patterns, Functions & Algebra Problem Solving Reasoning & Proof Communication Representation Connections Number Operation		Physics	Physics, Archery, Sports
12–4 534–537	• Solve quadratic equations by completing the square	Patterns, Functions & Algebra Problem Solving Reasoning & Proof Number Operation Geometry & Spatial Sense Communication Connections Representation		Art	Science, Aeronautics, Sports
12–5 540–543	• Solve equations using the quadratic formula	Patterns, Functions & Algebra Problem Solving Reasoning & Proof Number Operation Connections Representation		Physics	Aeronautics, Skydiving, Physics
12–6 544–547	• Use the Pythagorean Theorem, distance and midpoints	Patterns, Functions & Algebra Problem Solving Number Operation Communication Connections Representation			Space Exploration, Sports, Archeology
12–7 550–551	• Use graphs to write equations • Work backwards			Physics	Science, Physics

Planning and Resources

Lesson, Pages	Tools/Materials Needed	Trans-parency	Learning/Teaching Styles Options	Assignments: Basic Enriched	Additional Practice in SE	Reteaching, Extra Practice, Enrichment	Other Resources
12–1 520–523	graphing calculator graph paper	Warm up 34 Trans TK-7–10 RF-45, 46		B: 1–18, 26–34 E: 1–34	Page 528–529, 539, 549, 552, 722	R: page 177 EP: page 155 E: page 203	

12–2 524–527	graphing calculator grid paper	Warm up 34 Trans TK-7–10	ESL/LEP Real World	B: 1–20, 24–35 E: 1–35	Page 528–529, 539, 549, 552, 722	R: page 179 EP: page 157 E: page 205	SS: Teacher's Choice
12–3 530–533	graphing calculator	Warm up 34 Trans Tk-7–10	Challenge	B: 1–31, 37–50 E: 1–50	Page 538–539, 549, 553, 723	R: page 181 EP: page 159 E: page 207	SS: Teacher's Choice
12–4 534–537	Algeblocks	Warm up 35 Trans Tk-7–10		B: 1–40, 47–67 E: 1–67	Page 538–539, 549, 553, 723	R: page 183 EP: page 161 E: page 209	
12–5 540–543	paper/pencil	Warm up 35 Trans TK-7–10, RF-47	Challenge	B: 1–30, 35–48 E: 1–48	Page 548–549, 553, 724	R: page 185 EP: page 163 E: page 211	
12–6 544–547	paper/pencil	Warm up 36 Trans TK-7–10, RF-48	Prior Knowledge	B: 1–25, 29–40 E: 1–40	Page 548–549, 553, 724	R: page 187 EP: page 165 E: page 213	SS: Teacher's Choice
12–7 550–551	paper/pencil	Warm up 36 Trans TK-7–10		B: 1–14 E: 1–14	Page 553	R: page 189	

Planning and Pacing

Lesson, Pages	Lesson Title	45 min class	Assignments Basic, Enriched	90 min class	Assignments Basic, Enriched	___ min class	Assignments Basic, Enriched
12–1 520–523	AYR, Opener, Graph Parabolas	Day 152	B: 1–18, 26–34 E: 1–34	Day 83	B: 1–18, 26–34 E: 1–34		B: 1–18, 26–34 E: 1–34
12–2 524–527	The General Quadratic Function, R&PYS	Day 153–154	B: 1–20, 24–35 E: 1–35	Day 83–84	B: 1–20, 24–35 E: 1–35		B: 1–20, 24–35 E: 1–35
12–3 530–533	Factor and Graph	Day 155	B: 1–31, 37–50 E: 1–50	Day 84	B: 1–31, 37–50 E: 1–50		B: 1–31, 37–50 E: 1–50
12–4 534–537	Complete the Square, R&PYS	Day 156–157	B: 1–40, 47–67 E: 1–67	Day 85	B: 1–40, 47–67 E: 1–67		B: 1–40, 47–67 E: 1–67
12–5 540–543	The Quadratic Formula	Day 158	B: 1–30, 35–48 E: 1–48	Day 86	B: 1–30, 35–48 E: 1–48		B: 1–30, 35–48 E: 1–48
12–6 544–547	Use the Pythagorean Theorem, R&PYS	Day 159–160	B: 1–25, 29–40 E: 1–40	Day 86–87	B: 1–25, 29–40 E: 1–40		B: 1–25, 29–40 E: 1–40
12–7 550–551	Problem Solving Skills: Use Graphs to Write Equations	Day 161	B: 1–14 E: 1–14	Day 87	B: 1–14 E: 1–14		B: 1–14 E: 1–14
Review/ Assess		Day 162–163		Day 88–89			

	Chapter 1 Days	Chapter Cumulative Days
Yearly Pacing (45 min class)	12 days	163 days
Yearly Pacing (90 min class)	7 days	89 days

Assessment Options

Assessment in Student Edition	Assessment in Teacher's Edition	Pages in Assessment Book	Software Generated Assessment
Are You Ready?, pages 516–517; Writing Math, pages 522, 527, 533, 536, 543, 547, 551; Mixed Review, pages 523, 527, 533, 537, 543, 547, 551; Check Understanding pages 520, 524, 530, 534, 540, 544; Mid-Chapter Quiz, page 539, Chapter Review, page 552; Chapter Assessment, page 554; Alternative Assessment, page 555; Cumulative Review, page 556; Cumulative Assessment, page 557	5-minute Warm ups, pages 520, 524, 530, 534, 540, 544, 550; Quick Assessment, pages 522, 526, 532, 536, 542, 546, 551; Scoring Rubrics, page 555	Mid-Chapter Quiz, page 183; Test Form A, pages 185, 187; Test Form B, pages 189, 191; Cumulative Test 193, 195; Math Journal prompt, page 197, 198;	Chapter 12

ARE YOU READY?

Refresh Your Math Skills for Chapter 12

The skills on these two pages are ones you have already learned. Review the examples and complete the exercises. For additional practice on these and more basic skills, see page 674.

ADDITION AND SUBTRACTION

Making a table of ordered pairs to help graph a linear equation is a skill that will be especially useful as you learn to graph quadratic functions.

Example Graph the equation $2x + y = 8$.

Make a table of values. Generally, use -3 to 3 for the value of x.

x	y
-3	
-2	
-1	
0	
1	
2	
3	

Find the value of y for each value of x by substituting the value of x in the equation:
$2(-3) + y = 8$
$2(-2) = y = 8$
Solve each equation for y and fill in the table.

x	y
-3	2
-2	4
-1	6
0	8
1	10
2	12
3	14

With these ordered pairs you can graph the line of the equation.

Make a table of values for each equation. Do not graph the equation.

For 1–12, see additional answers.

1. $y = \frac{1}{2}x + 3$ **2.** $3y - 2x = 4$ **3.** $4x - 2y = 1$

4. $-2y + x = 3$ **5.** $y - 3x = 7$ **6.** $y + x = -6$

7. $2x - 3y = -6$ **8.** $4 + 3x = \frac{1}{2}y$ **9.** $5x + y = -8$

10. $4y + x = -7$ **11.** $8 - 3y = 2x$ **12.** $2x - 2y = -2$

GRAPHS OF FUNCTIONS

You can easily tell whether a graph is that of a function or not by the Vertical Line Test. When a vertical line is drawn through a graph of a relation, the graph is *not a function* if the vertical line intersects the graph in more than one point.

Graph of a function:

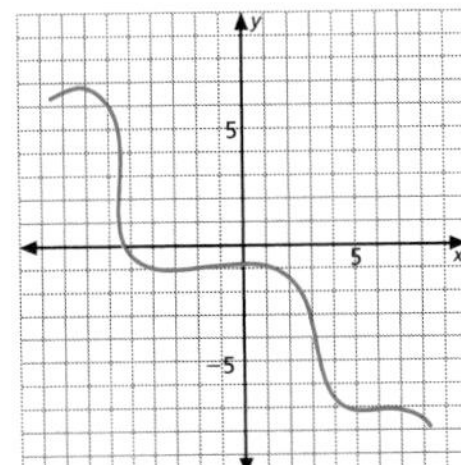

Not a graph of a function:

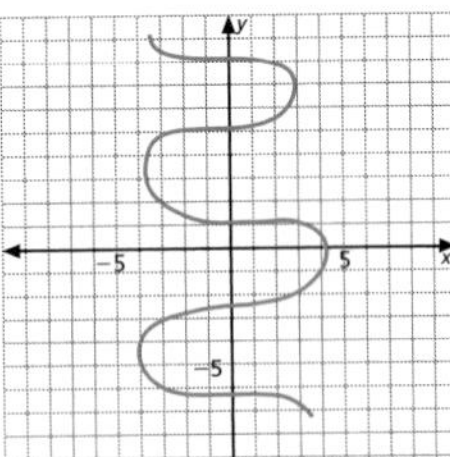

Refresher Skills

The skills on these two pages are skills that have been taught in previous math courses. Continuous review of basic math skills will make stronger math students. These skills are identified as necessary to be successful in Chapter 12.

Skills Correlation Chart

Skill	Lesson Number
Addition and Subtraction	12-1, 12-2
Graphs of Functions	12-1, 12-2, 12-3, 12-7
Pythagorean Theorem	12-6, 12-7

Vocabulary

linear equation
vertical line test
Pythagorean Theorem

ASSIGNMENT GUIDE

All students: 1–22
Additional Practice: See Extra Practice Index on page 674.

ADDITIONAL ANSWERS

1.

x	y
−3	1.5
−2	2
−1	2.5
0	3
1	3.5
2	4
3	4.5

2.

x	y
−3	−0.667
−2	0
−1	0.667
0	1.333
1	2
2	2.667
3	3.333

3.

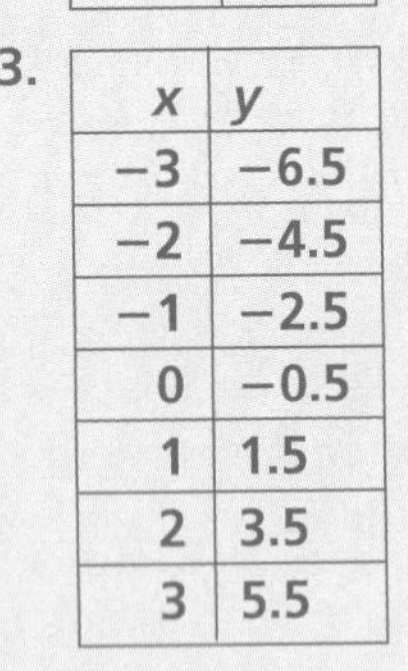

x	y
−3	−6.5
−2	−4.5
−1	−2.5
0	−0.5
1	1.5
2	3.5
3	5.5

4.

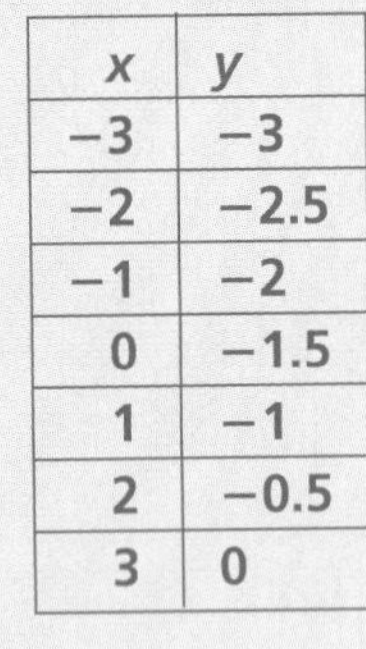

x	y
−3	−3
−2	−2.5
−1	−2
0	−1.5
1	−1
2	−0.5
3	0

5.

x	y
−3	−2
−2	1
−1	4
0	7
1	10
2	13
3	16

6.

x	y
−3	−3
−2	−4
−1	−5
0	−6
1	−7
2	−8
3	−9

7.

x	y
−3	0
−2	0.667
−1	1.333
0	2
1	2.667
2	3.333
3	4

8.

x	y
−3	−10
−2	−4
−1	2
0	8
1	14
2	20
3	26

516 Chapter 12 **Quadratic Functions**

CHAPTER 12

Use the Vertical Line Test to determine if each graph is a function.

13.

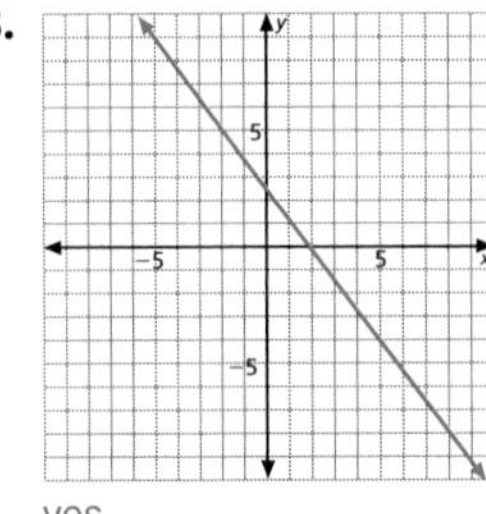

yes

14.

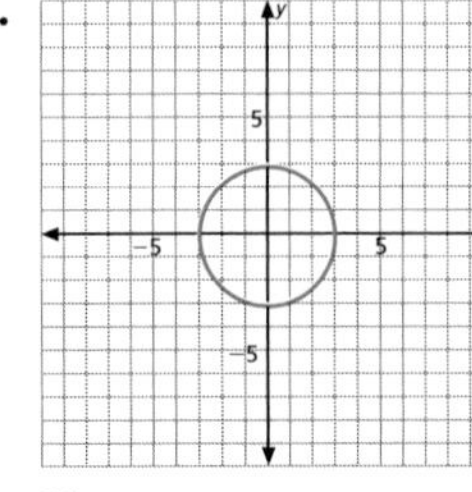

no

15.

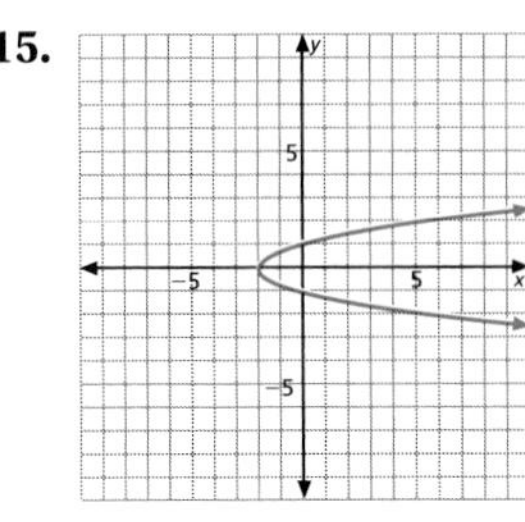

no

16.

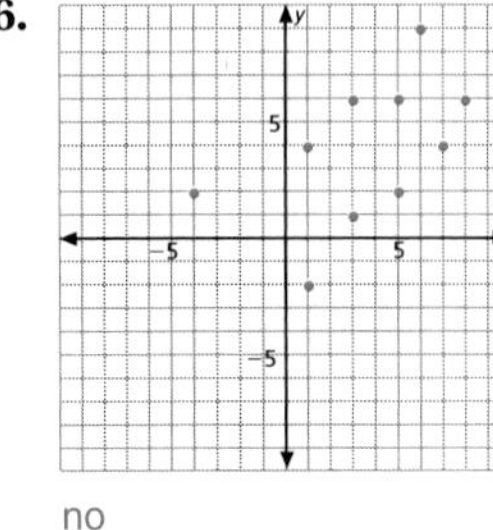

no

Pythagorean Theorem

Although you have used the Pythagorean Theorem mostly to find measures of triangles, it is a very valuable formula to know, and one you will use in real life.

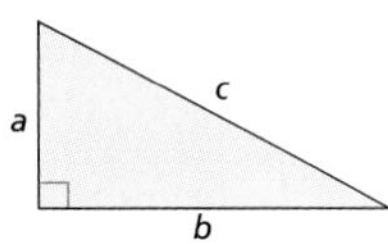

$$a^2 + b^2 = c^2$$

Find the missing measures to the nearest hundredth.

19.

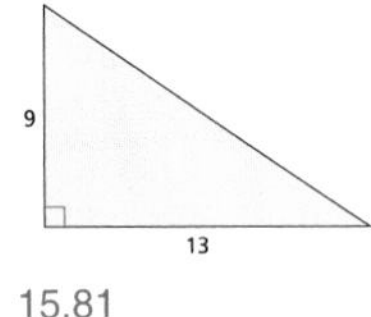

15.81

20.

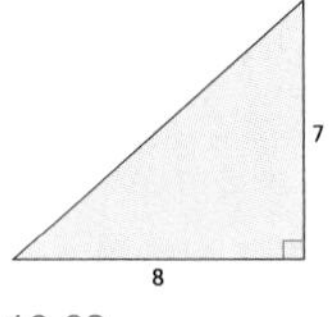

10.63

21. **22.**

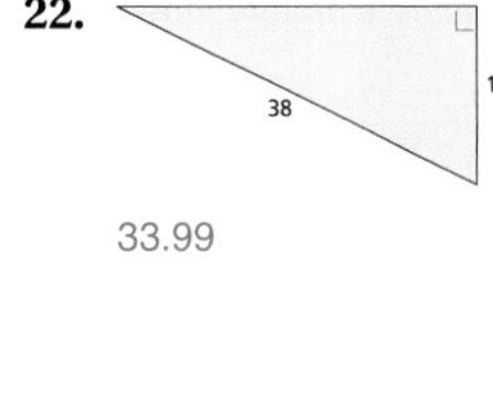

21. 12

22. 33.99

9.

x	y
−3	7
−2	2
−1	−3
0	−8
1	−13
2	−18
3	−23

10.

x	y
−3	−1
−2	−1.25
−1	−1.5
0	−1.75
1	−2
2	−2.25
3	−2.5

11.

x	y
−3	4.667
−2	4
−1	3.333
0	2.667
1	2
2	1.333
3	0.667

12.

x	y
−3	−2
−2	−1
−1	0
0	1
1	2
2	3
3	4

Chalkboard Examples

Addition and Subtraction

Make a table of values for the equation $-4y + x = 5$.

x	y
−3	−2
−2	−1.75
−1	−1.5
0	−1.25
1	−1
2	0.75
3	0.5

Graph of Functions

Draw a graph of a relation that is a function. Show the vertical line test on the graph. Draw a graph of a relation that is not a function. Show the vertical line test on the graph. Check students' graphs.

Pythagorean Theorem

The hypotenuse of a right triangle measures 7 cm. One leg of the triangle measures 4 cm. Find the length of the second leg to the nearest hundredth. 5.74 cm

Refresher Wrap-up

Quick Assessment

Ask the following questions to determine if students have mastered the basic skills reviewed on these pages.

1. How can you tell whether a graph is a function? The graph must pass the Vertical Line Test. If a vertical line intersects the graph in more than one point, the graph is not a function.
2. In your own words, explain how to use the Pythagorean Theorem to find the hypotenuse of a right triangle. Answers will vary.

Additional Practice

All topics: Refer to the Extra Practice Index on page 674 of this text.

Chapter Opener

NCTM Standards/Strands

- Number & Operation
- Patterns, Functions, & Algebra
- Geometry & Spatial Sense
- Measurement
- Problem Solving
- Reasoning and Proof
- Communication
- Connections
- Representation

Vocabulary

gravity inertia
incline

Theme Connections

Tables and graphs are often used to organize scientific data, particularly measurements. Tables provide a way to organize numerical information. In order to make sure the results of experiments are meaningful, scientists take numerous measurements and keep accurate records. Graphs are also used to show changes over time. Discuss how data and graphs are used in physics and the other sciences.

Career Opportunities

Many careers related to gravity and physics make use of data and graphs. Two are highlighted in the MathWorks features. Others include teachers, physicists, engineers, and designers.

- Air Traffic Controllers, page 529
- Pilots, page 549

Internet Connection

Theme Activities

Learningmathmatters.com provides web addresses to search that help students gather information about the use of math in the real world, particularly data and graphs. To search for additional addresses, begin a search using the keyword *gravity.* Then within that search, use such key words as *physics, astronomy, aircraft* and *rocketry.* Students can brainstorm in small groups other key words.

CHAPTER 12

QUADRATIC FUNCTIONS

THEME: Gravity

Even very small children understand an important law of physics: When you drop something, it falls. But what makes the object fall? Scientists have named the force *gravity.* Gravity is measured by hanging the object on a spring scale. This measure is called the weight of the object.

By working to understand and measure gravity, scientists have succeeded in overcoming its effects.

- **Pilots** (page 529) command airplanes, jets, helicopters and spacecraft. Pilots must understand how speed, altitude, temperature and the weight of the plane, including its contents, affect air travel.
- **Air Traffic Controllers** (page 549) ensure safe air travel by monitoring the movement of aircraft. Controllers use radar and visual observation to monitor the progress of aircraft. They work together to make sure planes stay a safe distance apart.

Internet Connection
www.learningmathmatters.com

Chapter Investigation

Go to learningmathmatters.com to locate additional information about gravity and related sciences.

As a Chapter Project

The goal of this project is to design, build and test a small gravity-powered vehicle. Students can use the Group Project Planner on page 201 and the Project Planning Calendar on page 202 in the Enrichment text to complete the project. Benchmarks **a, b,** and **c.** should be completed after the lesson listed in parentheses has been studied. Benchmark **d** should be completed at the end of the chapter.

How Does Gravity Affect Weight?

Location	Weight of 5-ton elephant
Mercury	2,842 lb
Venus	9,069 lb
Earth	10,000 lb
Moon	1,656 lb
Mars	3,803 lb
Jupiter	23,394 lb
Saturn	9,253 lb
Uranus	7,944 lb
Neptune	11,247 lb
Pluto	408 lb

Data Activity: How Does Gravity Affect Weight?

Use the table for Questions 1–4.

1. A tool weighs 2.5 lb on Earth. What is the weight of the tool on Mars? on Jupiter? 0.95075 lb; 5.8485 lb
2. An astronaut has two oxygen tanks. On Earth, Tank A weighs twice as much as Tank B. If the two tanks are transported to Saturn, Tank A's weight will be how many times the weight of Tank B? 2 times
3. At which of the locations shown in the table would you weigh less than you do on Earth? Mercury, Venus, Moon, Mars, Saturn, Uranus, Pluto
4. A bag of moon rocks weighs 225 lb on the moon. To the nearest tenth, what is the weight of the bag of rocks on Earth? 1,384.6 lb

Chapter Investigation

A child's wagon has no engine or other visible means of moving itself forward. Yet, when the wagon is positioned at the top of a steep hill and begins to move down the hill, its speed increases as it goes. The wagon is propelled by gravity.

Working Together

Design a small gravity-driven vehicle weighing no more than 10 oz. Time the vehicle's descent down an incline, recording the angle of descent and the time. Explore how changing the shape or weight of the vehicle affects its speed. Use the Chapter Investigation logos to guide your group.

Project Planning Calendar

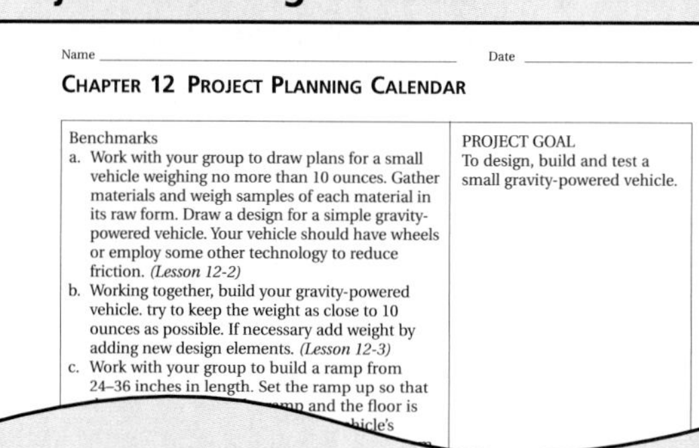

Name ______________ Date ______________

Chapter 12 Project Planning Calendar

Benchmarks	PROJECT GOAL
a. Work with your group to draw plans for a small vehicle weighing no more than 10 ounces. Gather materials and weigh samples of each material in its raw form. Draw a design for a simple gravity-powered vehicle. Your vehicle should have wheels or employ some other technology to reduce friction. *(Lesson 12-2)* b. Working together, build your gravity-powered vehicle. try to keep the weight as close to 10 ounces as possible. If necessary add weight by adding new design elements. *(Lesson 12-3)* c. Work with your group to build a ramp from 24–36 inches in length. Set the ramp up so that ... and the floor is ... hicle's	To design, build and test a small gravity-powered vehicle.

Group Project Planner

Name ______________ Date ______________

Chapter 12 Group Project Planner

Assignment ______________ Objective ______________

Group Members 1) 2) 3) 4) 5) Assigned Roles

Deadlines Done

Data Activity

Many people are confused about the meaning of weight. They think of it as the size of an object instead of the pull of gravity on a given object. The table shows how the pull of gravity differs from planet to planet within the solar system. Point out that physicists often use elephants and other amusing animals and objects to conduct thought experiments. By finding how much the elephant would weigh on each planet, we can understand more about the pull of gravity on each planet. Assign students Questions 1–4.

Extend the Data Activity

REAL WORLD CONNECTION If possible, have students go online to find news and science bulletins from NASA and other science organizations. Have students print out and share articles that discuss the pull of gravity and show its effects on objects in space.

Chapter Investigation

As an Overarching Problem

Discuss the meaning of the term gravity-powered vehicle. Ask students what will happen if such a vehicle is placed on flat ground. Ask what will happen if it is placed on slanted ground. Tell students that they will find out more about such issues as they complete this investigation. Students will continue to work on the investigation as they complete the exercises identified by the Investigation Logo that are found throughout the chapter. These exercises will guide students through the task described in *Working Together*. Encourage students to keep all of their work on the Investigation together. Have students use the suggestions in the Chapter Investigation Extension to summarize their work.

See page 518 for Chapter Investigation As a Chapter Project.

Lesson Planning

NCTM Standards/Strands

- Number & Operation
- Patterns, Functions, & Algebra
- Geometry & Spatial Sense
- Measurement
- Problem Solving
- Reasoning and Proof
- Communication
- Connections
- Representation

Vocabulary

quadratic equation
function
parabola
vertex
axis of symmetry

Materials Needed

graphing calculators
graph paper or grid paper

Lesson Resources

Warm-Up Transparency 34
Transparency TK-7–10, RF-45, 46
Reteaching 12-1
Extra Practice 12-1
Enrichment 12-1

Assignment Guide

Basic: 1–18, 26–34
Enriched: 1–34

Getting Started

5-Minute Warm-up

For $y = -3x - 4$, what is the value of y when $x =$

1. 5 -19
2. -2 2
3. 0 -4
4. 10 -34

Introduction to Lesson 12-1

As students graph each of the equations on a graphing calculator, have them sketch each on a sheet of paper. When they have drawn all six graphs, have them separate the graphs into two groups according to a rule. Ask them to write the rule they used to separate the graphs.

12-1 Graph Parabolas

Goals ■ Graph parabolas or second degree equations.
Applications Geology, Small Business, Physics

You will need a graphing calculator for this activity.

GRAPHING Use the ZOOM menu to make sure your display window is set on standard size. Then graph each of the following equations.

a. $y = -x$ **b.** $y = x^2 + 2x + 1$ **c.** $y = x + 3$
d. $y = -x^2 - 1$ **e.** $y = 2x - 2$ **f.** $y = 2x^2$

1. How can you tell by looking at an equation whether the graph will be a line or a curve? Equations that contain an x^2 term will have curved graphs.
2. Using the equations above, write two equations: one for a straight line and one for a curve?

Check Understanding

Which of the following are quadratic equations?

1. $y = x^3 + 6x^2$
2. $y = 55 - x^2$
3. $y = 4x + 16$
4. $y = 5x^2 - 9x - 1$

2 and 4

Build Understanding

In this section you will learn to graph equations that contain second degree or **quadratic** terms.

A **quadratic equation** in x contains an x^2 term, and involves no term with a higher power of x. The simplest quadratic equation is $y = x^2$.

Example 1

Graph $y = x^2$.

Solution

Find at least five ordered pairs by selecting x-values and solving the equation to find y-values.

x	−3	−1	0	1	3
y	9	1	0	1	9

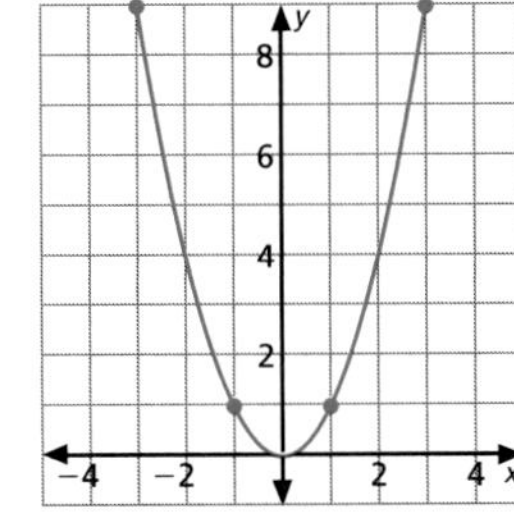

Graph the ordered pairs.

Draw a smooth curve through the points.

Because there is only one y-value for each x-value, this is the graph of a **function**. When the domain of a quadratic function is the set of real numbers, the graph is a **parabola**.

Notice that there is not only one x-value for each y-value. In fact, there are two x-values for each y-value except for point (0, 0),the lowest point on the parabola.

Teaching Strategies

Discuss the following question: Can a function in the form $y = x^2$ every have a straight line graph? Why or why not? No, since for every y-value except zero, there will be two x-values, forcing the graph to curve and have symmetric parts.

The parabola in Example 1 opens upward. For some functions, the parabolas open downward.

Example 2

Graph $y = -2x^2$.

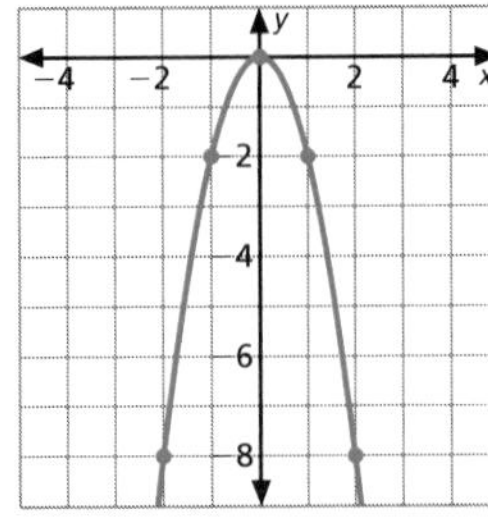

Solution

Make a table of ordered pairs.

x	−2	−1	0	1	2
y	−8	−2	0	−2	−8

Graph the points corresponding to the ordered pairs and draw a smooth curve through them.

The **vertex** is the lowest point on a parabola that opens upward, and the highest point on a parabola that opens downward. The graphs for $y = x^2$ and $y = -2x^2$ both have the point (0,0) as their vertex.

Example 3

GEOLOGY The distribution of a trace element within a geologic sample can be modeled by the equation $y = 3x^2 - 2$. Graph and locate the vertex of the parabola.

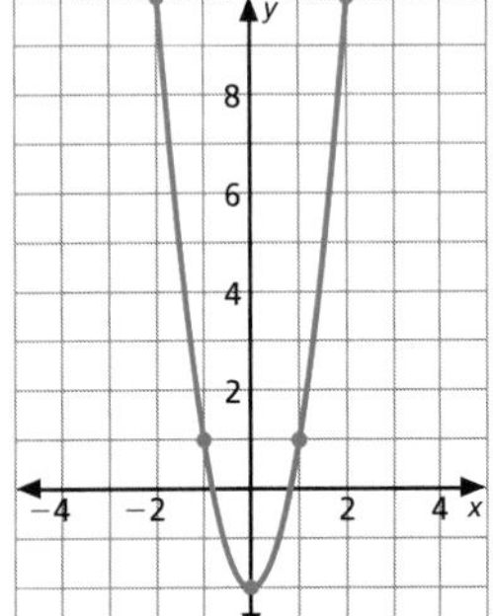

Solution

Make a table of ordered pairs.

x	−2	−1	0	1	2
y	10	1	−2	1	10

Graph the points and draw a smooth curve. Look for a y-value that has only one x-value. The vertex is $(0, -2)$.

GRAPHING You can use a graphing calculator to find the coordinates of the vertex. Key in the quadratic equation and graph. You may have to use the zoom features to adjust the size of the display. Press the TRACE key. Your calculator will place a point at the y-intercept. If the y-axis is the line of symmetry, the coordinates for the y-intercept are also the coordinates for the vertex. Use a graphing calculator to locate the vertex of the parabola in Example 3.

You can use the arrow keys to move the trace point to locate a vertex or any other point on the parabola. However, you must remember that there are limits to the display capabilities of a graphing calculator. You may have to zoom in very close in order to display the exact coordinates of a point.

The vertex lies on the line that divides the parabola in half. This line is called the **axis of symmetry** of the parabola.

In each of the three previous examples, the parabola is divided in half by the y-axis, which is the line $x = 0$. The axis of symmetry is not always $x = 0$. It is determined by the given equation.

Interdisciplinary Connections

SCIENCE The force of gravity pulls an object shot with horizontal force down in a parabolic arc. Students can try this experiment: At the same instant that one flicks a coin or checker off the end of a table, another drops an identical coin. They will both hit the floor at the same instant, since the horizontal force does not affect the pull of gravity.

Chalkboard Examples

Supplementary Example 1

Describe the graph of $y = -x^2$ and compare it with the graph of $y = x^2$. The graph of $y = -x^2$ opens downward and is a mirror image of the graph of $y = x^2$.

Supplementary Example 2

Describe the graph of $y = 3x^2$ and compare it with the graph of $y = x^2$. The graph of $x = 3x^2$ opens upward but is more narrow than the graph of $y = x^2$.

Supplementary Example 3

The distribution of a trace element in a meteorite sample is modeled by the equation $y = x^2 - 4$. Locate the vertex of the parabola. $(0, -4)$

Reteaching Worksheet 12-1

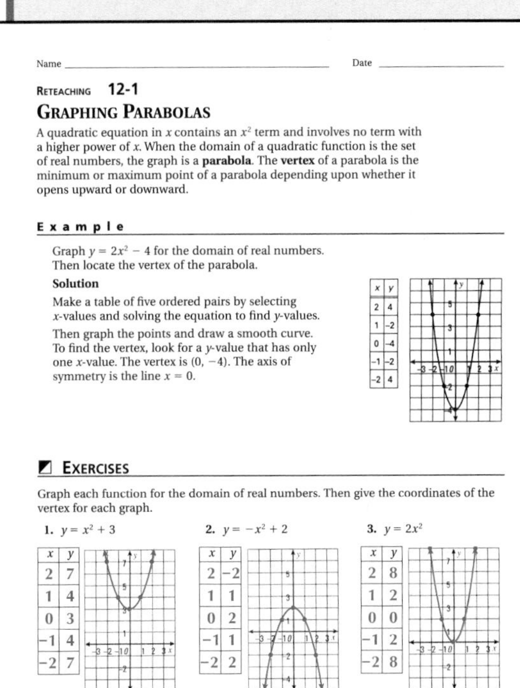
Name ____________ Date ____________

RETEACHING 12-1

GRAPHING PARABOLAS

A quadratic equation in x contains an x^2 term and involves no term with a higher power of x. When the domain of a quadratic function is the set of real numbers, the graph is a **parabola**. The **vertex** of a parabola is the minimum or maximum point of a parabola depending upon whether it opens upward or downward.

Example

Graph $y = 2x^2 - 4$ for the domain of real numbers. Then locate the vertex of the parabola.

Solution

Make a table of five ordered pairs by selecting x-values and solving the equation to find y-values.

Then graph the points and draw a smooth curve. To find the vertex, look for a y-value that has only one x-value. The vertex is $(0, -4)$. The axis of symmetry is the line $x = 0$.

x	y
2	4
1	−2
0	−4
−1	−2
−2	4

EXERCISES

Graph each function for the domain of real numbers. Then give the coordinates of the vertex for each graph.

1. $y = x^2 + 3$

x	y
2	7
1	4
0	3
−1	4
−2	7

Vertex (0, 3)

2. $y = -x^2 + 2$

x	y
2	−2
1	1
0	2
−1	1
−2	2

Vertex (0, 2)

3. $y = 2x^2$

x	y
2	8
1	2
0	0
−1	2
−2	8

Vertex (0, 0)

Ordered pairs may vary for Exercises 1–3; samples are given.

178 RETEACHING *LESSON 12-1*

Lesson Wrap-up

Quick Assessment

Ask the following questions to determine if students understand the content presented in this lesson.

1. Why does the function $y = -x^2$ open downward? For every x value, both positive and negative, the value of $-x^2$ will be negative.
2. How can you judge visually whether you have drawn the graph of a parabola correctly? Sample answer: The graph will have a line or axis of symmetry, making each half congruent with the other.

Assignment Guide

Basic: 1–18, 26–34
Advanced: 1–34
Additional Practice: See Extra Practice Index on page 674.

Additional Answers

1. 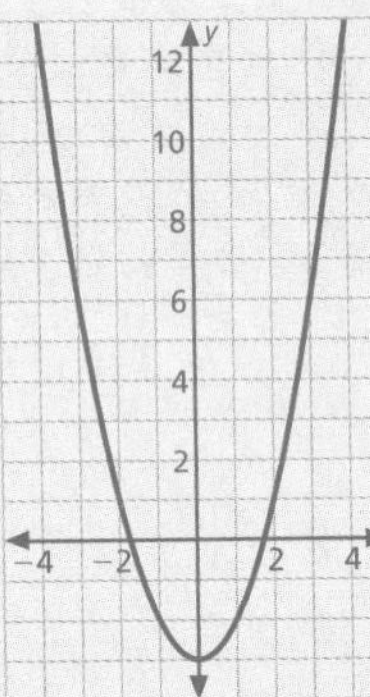

2. 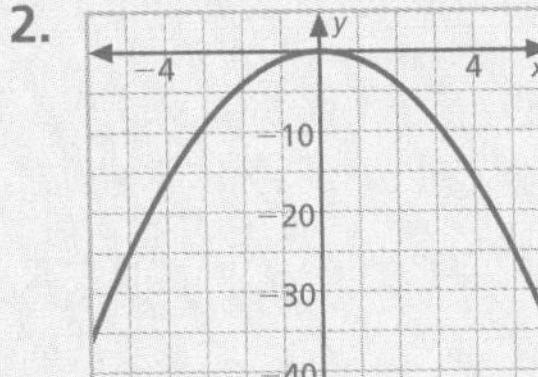

Try These Exercises

Copy and complete each table. Then draw the graph. See additional answers for graphs.

1. $y = x^2 - 3$

x	−4	−2	0	2	4
y	13	1	−3	1	13

2. $y = -x^2$

x	−6	−3	0	3	6
y	−36	−9	0	−9	−36

Graph each function for the domain of real numbers. For each graph, give the coordinates of the vertex. See additional answers for graphs.

3. $y = x^2 + 2$ (0, 2)
4. $y = -x^2 - 3$ (0, −3)
5. $y = -5x^2$ (0, 0)

6. **SMALL BUSINESS** A study shows that the daily revenue from product sales can be modeled by the equation $y = -5x^2 + 12$, where y equals the revenue in hundreds of dollars and x equals possible increases and decreases in price. Graph the equation. What is the maximum revenue? (Hint: The y-coordinate of the vertex is the maximum revenue in hundreds of dollars.) $1,200

Practice Exercises

Graph each function for the domain of real numbers. For each graph, give the coordinates of the vertex. See additional answers for graphs.

7. $y = 4x^2$ (0, 0)
8. $y = 2x^2 + 2$ (0, 2)
9. $y = -x^2 + 1$ (0, 1)

Determine if the graph of each equation below opens upward or downward.

10. $y = -3x^2 + 4$ downward
11. $y = 7x^2$ upward
12. $y = -x^2 - 10$ downward

13. **GRAPHING** The equations below are in the form $y = ax^2$. Graph each equation on a graphing calculator. How does the graph change as the value of a changes? As the absolute value of a decreases the graph becomes wider.

a. $y = 10x^2$ b. $y = 4x^2$ c. $y = 0.5x^2$
d. $y = -0.5x^2$ e. $y = -3x^2$ f. $y = -15x^2$

14. **WRITING MATH** What do you notice about the location of the vertex of a parabola formed by graphing an equation in the form $y = ax^2$? What do you notice about the location of the axis of symmetry of a parabola for the same equation? The vertex is always at the origin; The axis of symmetry is the y-axis.

15. **GRAPHING** The equations below are in the form $y = ax^2 + c$. Graph the equations on a graphing calculator. How does the graph change as the value of c decreases? The graph shifts downward.

a. $y = 2x^2 + 4$ b. $y = 2x^2 + 3$ c. $y = 2x^2 + 1$
d. $y = 2x^2 - 1$ e. $y = 2x^2 - 3$ f. $y = 2x^2 - 5$

16. **WRITING MATH** What do you notice about the location of the vertex and axis of symmetry of a parabola formed by graphing an equation in the form $y = ax^2 + c$? The vertex is always on the y-axis; the y-axis is the axis of symmetry.

17. **PHYSICS** When an object is thrown upward, the force of gravity pulls the object down in a parabolic arc. Suppose a particular arc can be represented by the equation $y = -x^2 + 18$. Find the vertex or highest point of the arc. (0, 18)

3. 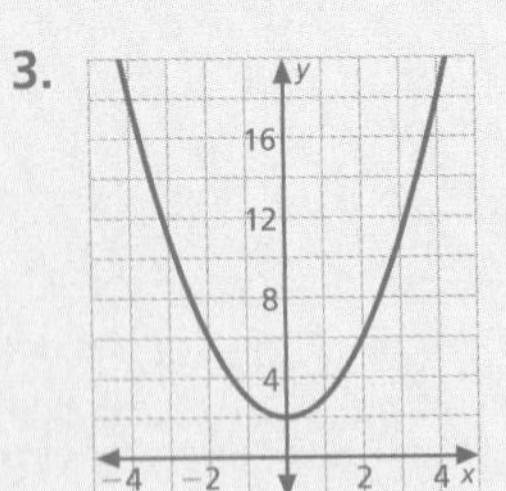

4. 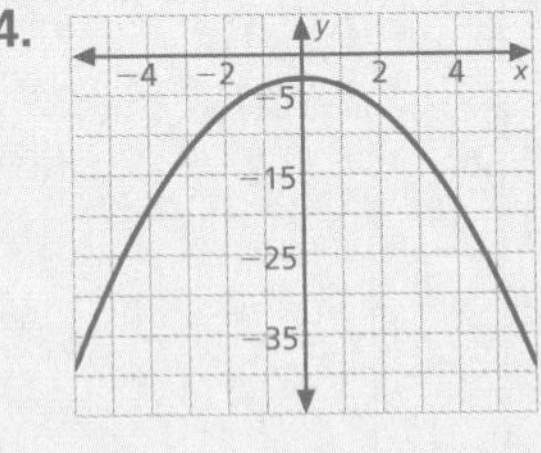

5. 7. 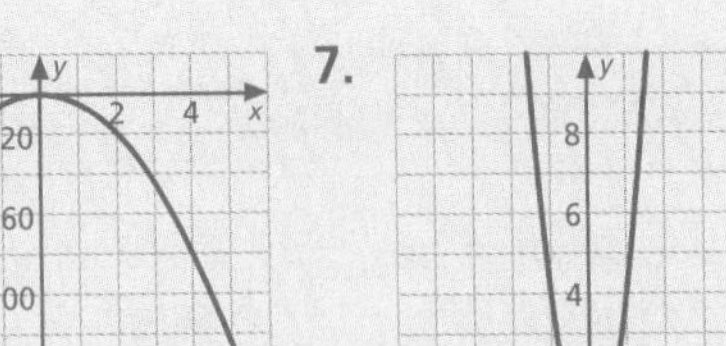

8. 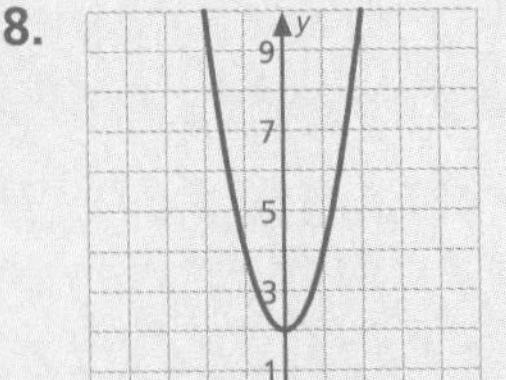

9. 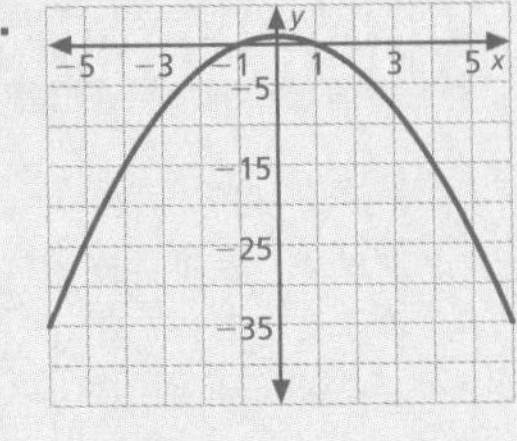

522 Chapter 12 **Quadratic Functions**

18. **PHYSICS** When thrown with a slight downwards motion, a paper airplane dips and then momentarily returns to the level at which it was released. Find the change in altitude (in inches) by finding the vertex of the equation $y = x^2 - 2$. (0, −2) The airplane dips downward 2 ft before ascending.

Extended Practice Exercises

Each graph below is in the form $y = ax^2 + c$. The value of a is +6 or −6 for one equation, and +1 or −1 for the other. Write the equation for each graph.

19. 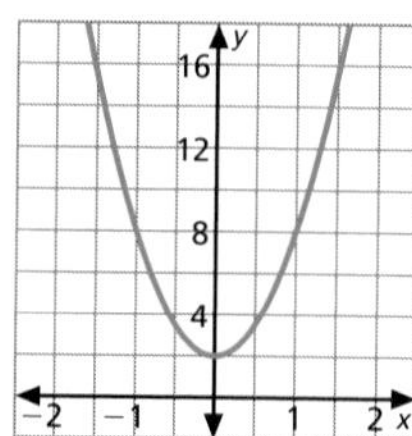

$y = 6x^2 + 2$

20. 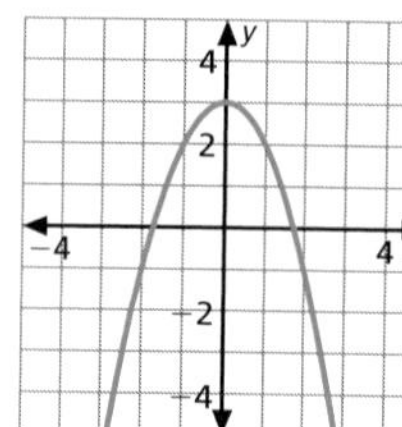

$y = -x^2 + 3$

GRAPHING Use a graphing calculator. Graph each pair of equations in the same window. Find the vertex and axis of symmetry for each pair.

21. $y = \sqrt{x}$ (0, 0)
$y = -\sqrt{x}$ x-axis

22. $y = \sqrt{-x}$ (0, 0)
$y = -\sqrt{-x}$ x-axis

The graph of a quadratic equation has either a maximum point or a minimum point. Answer the following questions.

23. What form of equation has a graph with a maximum point?
An equation in which $a < 0$; the graph opens downward and has a maximum point.

24. What form of equation has a graph with a minimum point?
An equation in which $a > 0$; the graph opens upward and has a minimum point.

25. **CHAPTER INVESTIGATION** Gravity can be used to power a vehicle on a ramp or incline. Work with your group to draw plans for a small vehicle weighing no more than 10 oz. As a first step, explore what types of materials to use to build the vehicle. Gather materials and weigh samples of each material in its raw form. Once you have made a final selection of materials, draw a design for a simple gravity-powered vehicle. Your vehicle should have wheels or employ some other technology to reduce friction.

Mixed Review Exercises

Simplify. (Lesson 11-1)

26. $(6a - 3) + (4a + 6)$
$10a + 3$

27. $(2y^2 + y) - (5y + 8)$
$2y^2 - 4y - 8$

28. $(4x^2 + 3x + 2) - (x + 4)$
$4x^2 - 2x - 2$

29. $(3b^2 + 2b) - (4b - 3)$
$3b^2 - 2b + 3$

30. $(6c + 2) + (5c - 8)$
$11c - 6$

31. $(-3d^2 + 8) - (4d - 9)$
$-3d^2 - 4d + 17$

In each triangle, $AB \parallel CD$. Find x to the nearest tenth. (Lesson 7-5)

32. 4

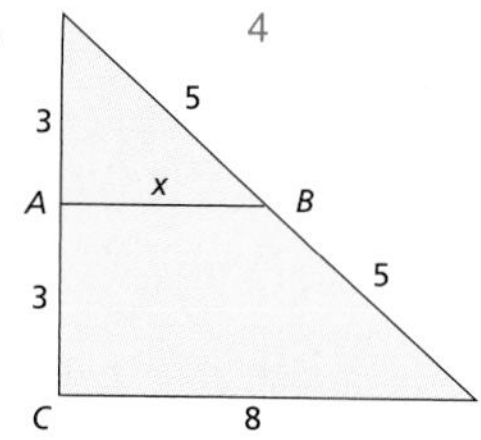

33. 11.3

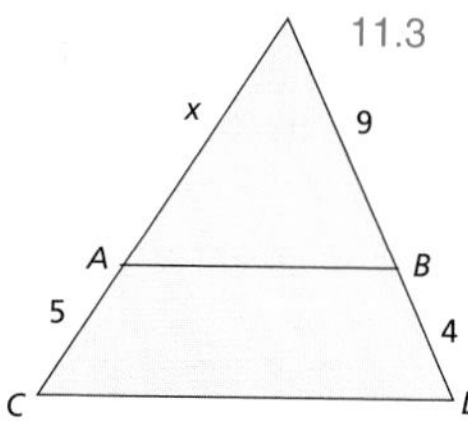

34. 6

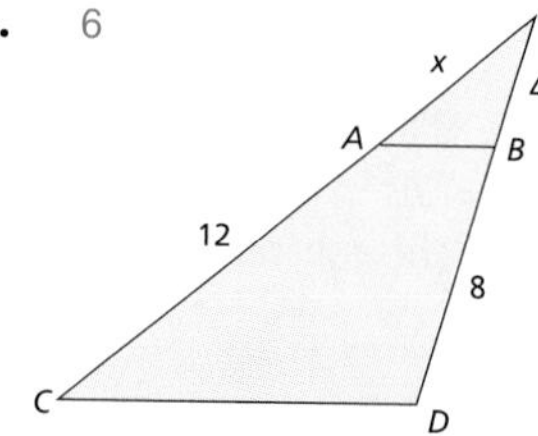

Extra Practice Worksheet 12-1

Name ______ Date ______

EXTRA PRACTICE 12-1

GRAPHING PARABOLAS

EXERCISES

Complete each table. Then draw the graph on your own paper.

1. $y = x^2 + 2$

x	−2	−1	0	1	2
y	6	3	2	3	6

2. $y = -x^2 + 3$

x	−4	−2	0	2	4
y	−13	−1	3	−1	−13

3. $y = 4x^2 - 4$

x	−2	−1	0	1	2
y	12	0	−4	0	12

4. $y = -2x^2 + 1$

x	−2	−1	0	1	2
y	−7	−1	1	−1	−7

5. $y = 2x^2 + 1$

x	−4	−2	0	2	4
y	33	9	1	9	33

6. $y = -4x^2 + 5$

x	−2	−1	0	1	2
y	−11	1	5	1	−11

Graph each function for the domain of real numbers. Use your own paper. For each graph, give the coordinates of the vertex.

7. $y = 2x^2$ (0, 0)
8. $y = 3x^2 + 2$ (0, 2)
9. $y = -x^2 + 4$ (0, 4)
10. $y = -2x^2 - 2$ (0, −2)
11. $y = 10x^2 - 6$ (0, −6)
12. $y = -6x^2 + 7$ (0, 7)

Determine if the graph of each equation below opens upward or downward.

13. $y = 4x^2 + 5$ upward
14. $y = -9x^2 - 12$ downward
15. $y = -x^2 + 1$ downward
16. $y = 3x^2 - 9$ upward
17. Suppose that when a ball is thrown upward, the arc it makes can be represented by the equation $y = -x^2 + 16$. Find the vertex or highest point on the arc.
(0, 16)

156 EXTRA PRACTICE *LESSON 12-1*

Enrichment Worksheet 12-1

Name ______ Date ______

ENRICHMENT 12-1

PREDICT THE DESIGN

EXERCISES

On the grid below, graph $y = \frac{x^2}{4a}$
for $a = 0.25, 0.5, 1, 2, 3, 4, 5, 6$ and $-a = 0.25, 0.5, 1, 2, 3, 4, 5, 6$.

Then graph $y = ax$ for $a = 0.1$ and 5 and $-a = 0.1$ and 5. Before you begin graphing, predict what the design will look like. Answers may vary.

204 ENRICHMENT *LESSON 12-1*

Extend the Lesson

MATH JOURNAL What does it mean to say that in the equation $y = x^2$, x is the independent variable and y is the dependent variable? The value of y is determined by the value of x.

Lesson Planning

NCTM Standards/Strands

- Number & Operation
- Patterns, Functions, & Algebra
- Geometry & Spatial Sense
- Problem Solving
- Reasoning and Proof
- Communication
- Representation

Vocabulary

general quadratic function
standard quadratic equation

Materials Needed

graphing calculator   grid paper

Lesson Resources

Warm-Up Transparency 34
Transparency TK-7–10
Reteaching 12-2
Extra Practice 12-2
Enrichment 12-2

Assignment Guide

Basic: 1–20, 24–35
Enriched: 1–35

Getting Started

5-Minute Warm-Up

Evaluate $\frac{-b}{2a}$ when

1. $a = 1, b = 1$ $-\frac{1}{2}$
2. $a = -1, b = 2$ 1
3. $a = 2, b = 0$ 0
4. $a = 3, b = -2$ $\frac{1}{3}$

Introduction to Lesson 12-2

Have students sketch the graph of each equation on a separate coordinate grid. After graphing all of the equations, have them form four groups according to the quadrant in which the vertex lies. Then have them complete parts (a) through (d).

12-2 The General Quadratic Function

Goals ■ Graph functions defined by the standard quadratic equation.

Applications Physics, Business, Astronomy

Work with a partner to answer the following questions.

1. **GRAPHING** The following quadratic equations are in the form $y = ax^2 + bx$. Graph the equations on a graphing calculator.

$y = x^2 + 7x$	$y = x^2 + 5x$	$y = x^2 + 2x$
$y = x^2 - 7x$	$y = x^2 - 5x$	$y = x^2 - 2x$
$y = -x^2 + 7x$	$y = -x^2 + 5x$	$y = -x^2 + 2x$
$y = -x^2 - 7x$	$y = -x^2 - 5x$	$y = -x^2 - 2x$

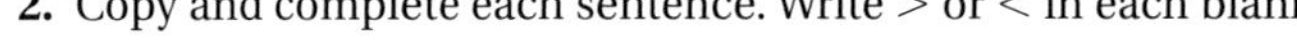

2. Copy and complete each sentence. Write $>$ or $<$ in each blank.
   a. If a is __?__ 0 and b is __?__ 0, the vertex is in quadrant III. $>$ $>$
   b. If a is __?__ 0 and b is __?__ 0, the vertex is in quadrant IV. $>$ $<$
   c. If a is __?__ 0 and b is __?__ 0, the vertex is in quadrant I. $<$ $>$
   d. If a is __?__ 0 and b is __?__ 0, the vertex is in quadrant II. $<$ $<$

Build Understanding

In Lesson 12-1, you investigated quadratic equations in the forms $y = ax^2$ and $y = ax^2 + c$. You learned that for quadratic equations, the value of y is determined by the value of x; y is a function of x. This relationship is expressed as $y = f(x)$.

Reading Math

$y = f(x)$ is read "y equals f of x"

As you have discovered, the graph of a quadratic equation is a parabola. For the parabolas you graphed in Lesson 12-1, the vertex was always located on the y-axis, and the y-axis was the axis of symmetry. Your exploration of equations in the form $y = ax^2 + bx$ showed that other locations are possible.

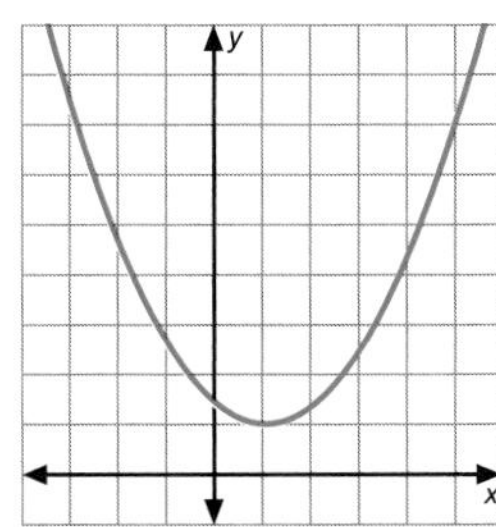

In this lesson, you will learn how to locate the vertex and axis of symmetry for any quadratic equation.

The general form of the quadratic function is defined by the **standard quadratic equation** $y = ax^2 + bx + c$, where a, b, and c are real numbers and $a \neq 0$.

The **general quadratic function** may be written $f(x) = ax^2 + bx + c$.

Check Understanding

In the standard quadratic equation, why can a not be equal to 0?

If $a = 0$, the equation is not quadratic.

Extend the Lesson

REAL WORLD CONNECTION Parabolic mirrors have a wide variety of uses. Such a mirror is formed by rotating a parabola about its line of symmetry. In an automobile headlight, light from the bulb is aimed at the mirror, and light rays are then reflected toward the road parallel to the line of symmetry.

Example 1

Graph $y = 3x^2 - 4x - 1$ on a graphing calculator. Estimate the coordinates of the vertex.

Solution

Enter the equation. Graph the function. Use the trace and zoom features to locate the coordinates of the vertex.

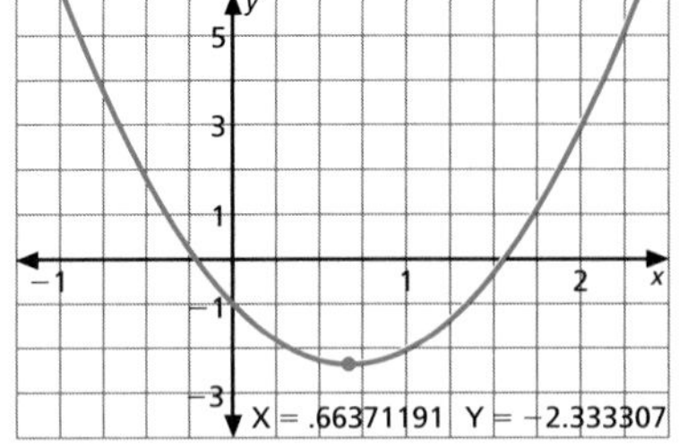

The closer you zoom in, the closer the coordinates will be to the actual values of x and y. Eventually, you may be able to see a relationship between the decimal values on the screen and a common fraction or whole number. For example, you may have arrived at the coordinates $x = .6637119$ and $y = -2.333307$.

$.6637119$ is about $\frac{2}{3}$.

2.333307 is about $2\frac{1}{3}$.

The vertex is approximately $\left(\frac{2}{3}, -2\frac{1}{3}\right)$.

For a parabola defined by the equation $y = ax^2 + bx + c$, the vertex is always at the point on the graph where the x-coordinate is $x = -\frac{b}{2a}$. The corresponding y-value can be found by substituting the x-value into the equation.

Technology Note

The *zoom* feature on a graphing calculator allows you to magnify a section of a graph. The *Zoom Box* defines the box to be enlarged.

To zoom in on the vertex:

1. Select *Box* from the *zoom* menu
2. Place the cursor on a corner of the area you want to magnify; press enter.
3. Move the cursor to the diagonal corner of the box; press enter.

Example 2

Find the coordinates of the vertex for the graph of $y = 3x^2 - 4x - 1$.

Solution

$x = -\frac{b}{2a}$

$x = -\left(\frac{-4}{2(3)}\right)$ Substitute for a and b.

$x = \frac{4}{6}$ or $\frac{2}{3}$ Simplify.

$y = 3\left(\frac{2}{3}\right)^2 - 4\left(\frac{2}{3}\right) - 1$ Substitute the x-value into the equation.

$y = 3\left(\frac{4}{9}\right) - \frac{8}{3} - 1$ Simplify.

$y = -2\frac{1}{3}$

The coordinates of the vertex for the graph of $y = 3x^2 - 4x - 1$ are $\left(\frac{2}{3}, -2\frac{1}{3}\right)$.

As you learned in Lesson 12-1, the axis of symmetry is a vertical line through the vertex of a parabola. The axis of symmetry for the graph of a quadratic function is $x = -\frac{b}{2a}$. For the graph of the function above, the axis of symmetry is $x = \frac{2}{3}$.

Check Understanding

Give the values of a, b, and c for each equation.

1. $y = 6x^2 + 4x + 5$
2. $7x^2 - 3x + 2 = 0$
3. $5 = 9x^2 - 6x$
4. $x^2 = 7$

1. 6, 4, 5
2. 7, −3, 2
3. 9, −6, −5
4. −1, 0, 7 or
5. 1, 0, −7

Teaching Strategies

COOPERATIVE LEARNING Have students work in small groups to research and report on the relationship of the parabola to the cone. They may be interested to learn that the conic sections were known as along as 2,000 years ago.

Chalkboard Examples

Supplementary Example 1

Graph $y = x^2 - x + 2$ on a graphing calculator. Estimate the coordinates of the vertex. $\left(\frac{1}{2}, 1\frac{3}{4}\right)$

Supplementary Example 2

Find the coordinates of the vertex for the graph of $y = -x^2 + 3x - 3$. $\left(\frac{3}{2}, -\frac{3}{4}\right)$

Supplementary Example 3

Graph $f(x) = 4x^2 - x + 1$. The axis of symmetry is $x = \frac{1}{8}$. The parabola opens upward.

Supplementary Example 4

The following graphs show the results of two scientific experiments. Which is the graph of a function? Graph (b) is a function.

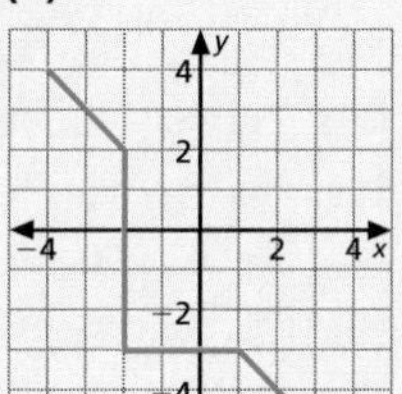

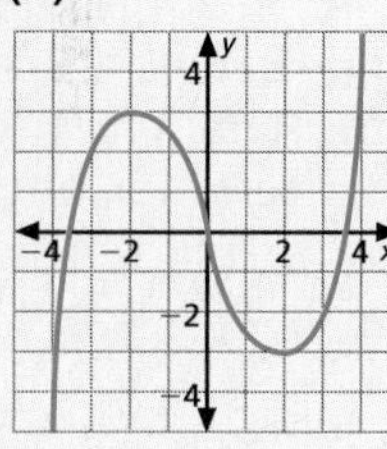

Reteaching Worksheet 12-2

Name ______ Date ______

RETEACHING 12-2

THE GENERAL QUADRATIC FUNCTION

The general quadratic function may be written as $f(x) = ax^2 + bx + c$. The general form of the quadratic function is defined by the **standard quadratic equation** $y = ax^2 + bx + c$, where a, b and c are real numbers and $a \neq 0$. You can use $x = -\frac{b}{2a}$ to find the x-value of the vertex and identify the axis of symmetry.

Example

Find the vertex and axis of symmetry for $y = x^2 - 2x + 1$. Then graph each equation.

Solution

Find the coordinates of the vertex.

$x = -\frac{b}{2a}$ $y = x^2 - 2x + 1$ Substitute 1 into the equation

$x = -\frac{-2}{2(1)}$ Substitute −2 for b and 1 for a. $y = (1)^2 - 2(1) + 1$

$y = 1 - 2 + 1$, or 0

$x = \frac{2}{2}$, or 1

x	y
−1	4
0	1
1	0
2	1
3	4

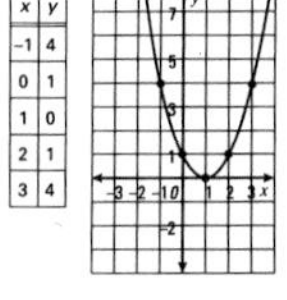

The vertex is (1, 0). The axis of symmetry is $x = 1$. Make a table of ordered pairs, graph the points and draw a smooth curve.

EXERCISES

Find the vertex and identify the axis of symmetry. Then graph each function.

1. $y = x^2 + 2x + 3$
2. $y = -2x^2 + 3$
3. $y = x^2 + 2x - 3$

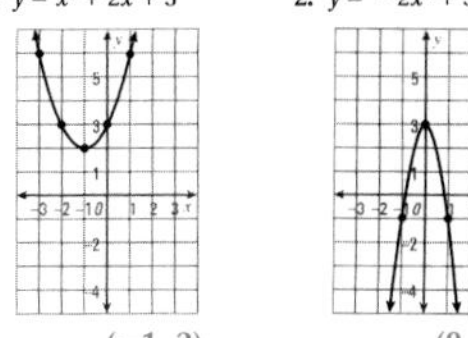

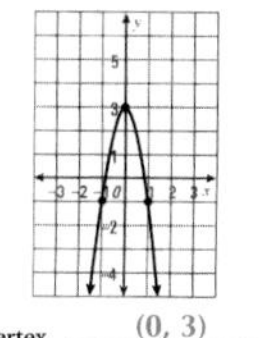

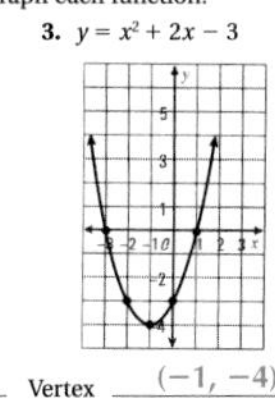

Vertex (−1, 2) Axis of symmetry $x = -1$

Vertex (0, 3) Axis of symmetry $x = 0$

Vertex (−1, −4) Axis of symmetry $x = -1$

180 RETEACHING *LESSON 12-2*

Lesson Wrap-up

Quick Assessment

Ask the following questions to determine if students understand the content presented in this lesson.

1. In the equation $y = 3x^2 - 4$, what is the equation of the line of symmetry? $x = 0$, the y-axis
2. Compare the locations of the lines of symmetry for the following two equations:
   $y = 2x^2 - 6x + 2$ and
   $y = 4x^2 - 12x - 8$
   They are the same: $x = 1$
3. How can you tell from examining a set of ordered pairs whether the relation is a function? Every x-coordinate will be matched with only one y-coordinate.

Assignment Guide

Basic: 1–20, 24, 35
Advanced: 1–35
Additional Practice: See Extra Practice Index on page 674.

Additional Answers

5.

6.

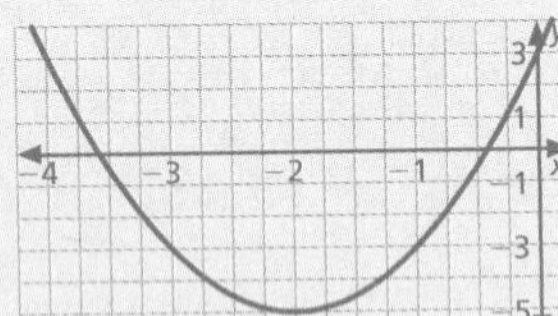

7.

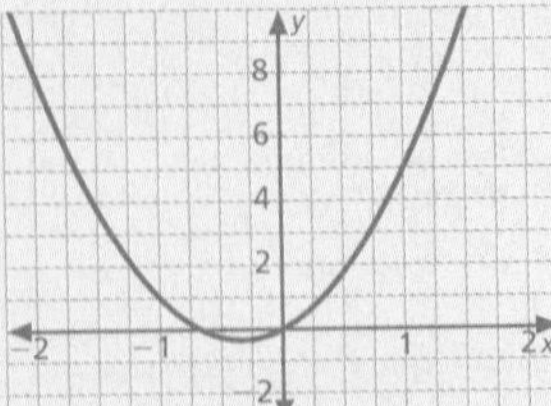

8.

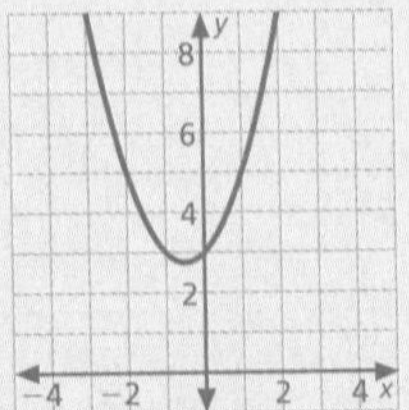

You can use the equation $x = -\frac{b}{2a}$ to graph a quadratic function.

Example 3

Graph $f(x) = -2x^2 - 3x + 1$.

Solution

Locate the vertex.

$$x = -\frac{b}{2a} = -\left(\frac{-3}{2(-2)}\right) = -\frac{3}{4}$$

$$y = -2\left(-\frac{3}{4}\right)^2 - 3\left(-\frac{3}{4}\right) + 1$$

$$= -2\left(\frac{9}{16}\right) + \left(\frac{9}{4}\right) + 1 = 2\frac{1}{8}$$

The vertex is $\left(-\frac{3}{4}, 2\frac{1}{8}\right)$. The axis of symmetry is $x = -\frac{3}{4}$.

Because a is less than 0, the parabola opens downward. Substitute values in the equation to locate a few points.

x	−1	0	1
y	2	1	−4

Use the axis of symmetry to visually locate other points. Draw a smooth curve.

As you know, a function in x assigns only one y-value for each x-value. To tell whether a graph is of a function, check to see whether each vertical line in the coordinate plane contains at most one point of the graph.

Try These Exercises

GRAPHING Estimate the coordinates of the vertex for each parabola by graphing the equation on a graphing calculator. Then use $x = -\frac{b}{2a}$ to find the coordinates.

1. $y = x^2 - 4x + 3$ (2, −1)
2. $y = 2x^2 + 3x - 1$ $\left(-\frac{3}{4}, -2\frac{1}{8}\right)$
3. $5x^2 - 2x + 5 = y$ $\left(\frac{1}{5}, 4\frac{4}{5}\right)$
4. $y + 15 = x^2 - 2x$ (1, −16)

Find the vertex and axis of symmetry. Then graph each equation. For 5–8, see additional answers for graphs.

5. $y = -3x^2 - 9x + 1$ $\left(-\frac{3}{2}, 7\frac{3}{4}\right)$; $x = -\frac{3}{2}$
6. $y = 2x^2 + 8x + 3$ (−2, −5); $x = -2$
7. $y = 3x^2 + 2x$ $\left(-\frac{1}{3}, -\frac{1}{3}\right)$; $x = -\frac{1}{3}$
8. $-x^2 + y = 3 + x$ $\left(-\frac{1}{2}, 2\frac{3}{4}\right)$; $x = -\frac{1}{2}$

Practice Exercises

GRAPHING Estimate the coordinates of the vertex for each parabola by graphing the equation on a graphing calculator. Then use $x = -\frac{b}{2a}$ to find the coordinates.

9. $y = x^2 + 6x + 4$ (−3, −5)
10. $y - 3 = 2x^2 - 12x$ (3, −15)
11. $-9 + y = -3x + 3x^2$ $\left(\frac{1}{2}, 8\frac{1}{4}\right)$
12. $y = x^2 + 4$ (0, 4)
13. $-8x + 5 = -2x^2 + y$ (2, −3)
14. $4x^2 - 16x + 1 = y$ (2, −15)

20. Graph A is not the graph of a function. The y-axis is one of several vertical lines which intersects the graph at two points. Graph B is a graph of a function. There is only one y-value for each x-value.

30.

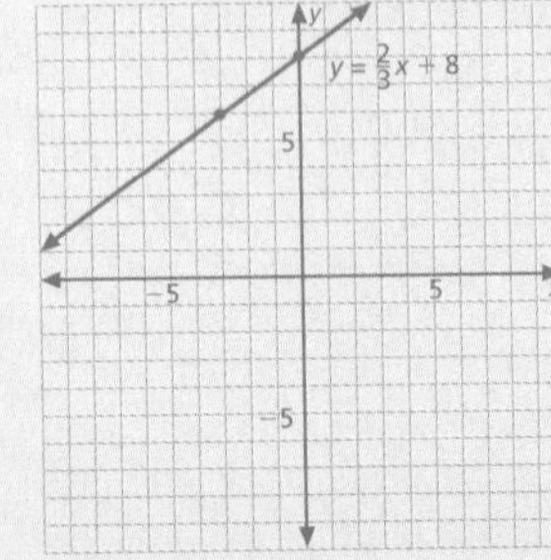

31.

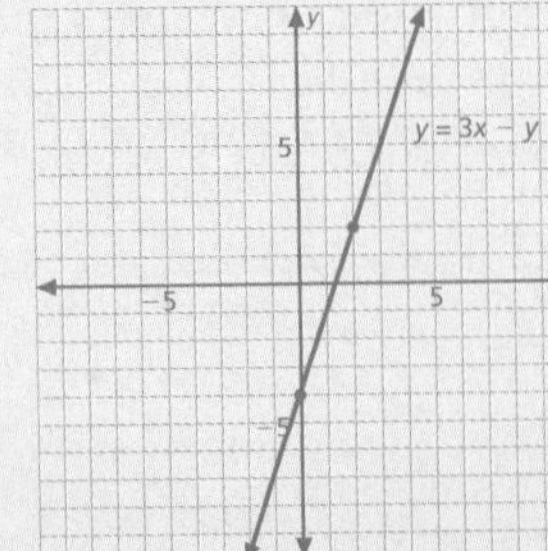

32.

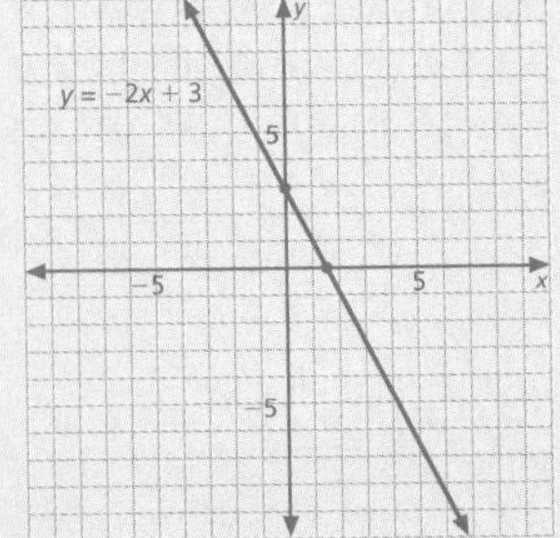

Find the vertex and axis of symmetry.

15. $y = 5x^2 - 10x + 1$ $(1, -4)$; $x = 1$

16. $x^2 + x + 3 = y$ $\left(\frac{-1}{2}, 2\frac{3}{4}\right)$; $x = -\frac{1}{2}$

17. $y = -x^2 + 1$ $(0, 1)$; y-axis

18. BUSINESS Sales projections can be represented by a quadratic equation in the form of $y = ax^2 + bx + c$ for which $c = 3$, and the axis of symmetry is $x = \frac{5}{6}$. Find the equation. $y = 3x^2 - 5x + 3$ or $y = -3x^2 + 5x + 3$

19. ASTRONOMY Observed movement in an object can be represented by a quadratic equation in the form $y = ax^2 + bx + c$ for which $c = -2$ and the vertex is $\left(-\frac{5}{8}, -\frac{7}{16}\right)$. Find the equation. $y = -4x^2 - 5x - 2$

20. WRITING MATH Study graphs A and B shown at the right. Determine if each graph is the graph of a function. Explain your reasoning.
See additional answers.

a.

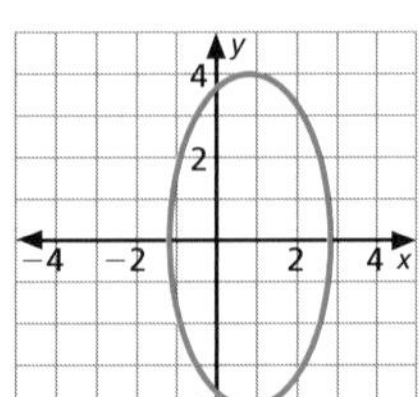

b.

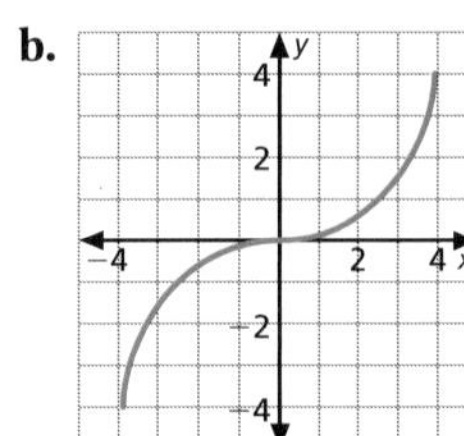

Extended Practice Exercises

PHYSICS A cannon is fired at several different angles. The paths of the cannonballs are shown on the graph.

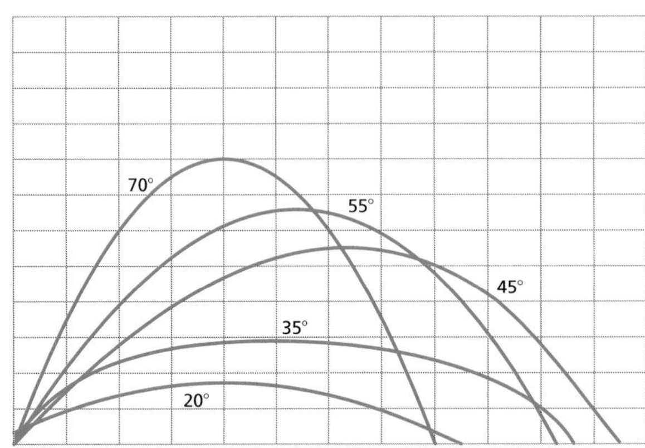

21. At what angle did the cannonball reach the highest point? 70°

22. At what angles did the cannonball travel about the same distance? 20° and 70°, 35° and 55°

23. WRITING MATH What conclusion can you draw about the maximum range of the cannon? The maximum range is reached at about 45°.

Mixed Review Exercises

Simplify. (Lesson 11-2)

24. $(7bc)(3ab)$ $21ab^2c$

25. $6s^2(s^2 + 3s + 2)$ $6s^4 + 18s^3 + 12s$

26. $(5sz^4)(3s^4z^5)$ $15s^5z^4$

27. $(2a^2)(4ac - 3bd)$ $8a^3c - 6a^2bd$

28. $(4r^2s^3)(2r^3s^2 - 8rs^4)$ $8r^5s^5 - 32r^3s^7$

29. $(25x^4y^2)(4x^3y^2 - 2x^2y^5)$ $100x^7y^4 - 50x^6y^7$

Graph each function. (Lesson 2-3) For 30–35, see additional answers for graphs.

30. $y = \frac{2}{3}x + 8$

31. $y = 3x - 4$

32. $y = -2x + 3$

33. $y = \frac{1}{2}x - 2$

34. $y = -3x - 1$

35. $y = \frac{3}{4}x + 2$

33.

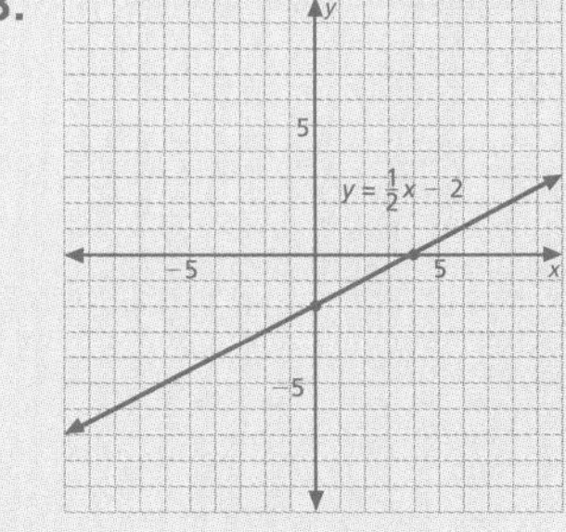

34.

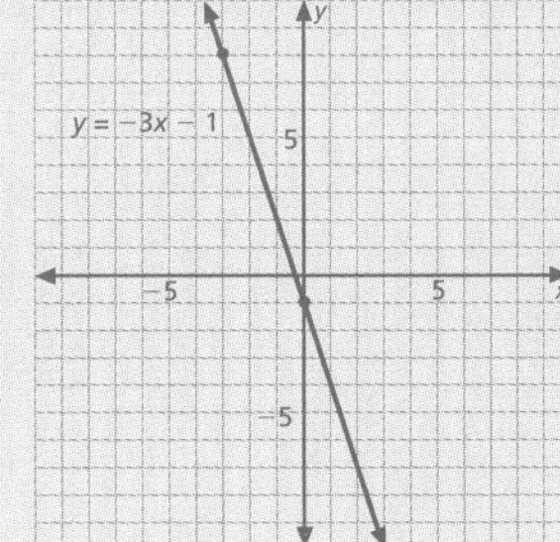

35.

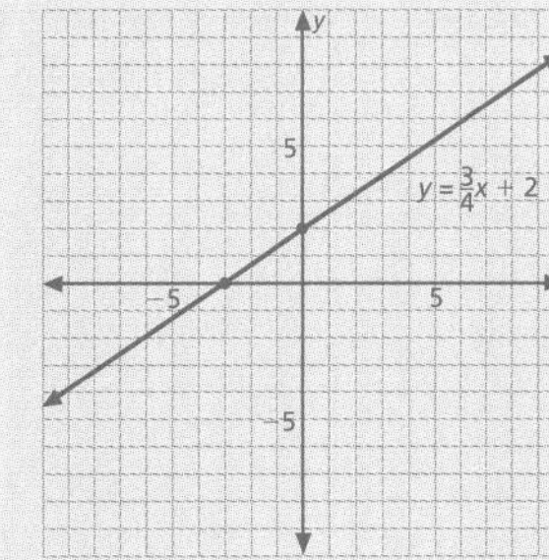

Extra Practice Worksheet 12-2

Name ____________ Date ____________

EXTRA PRACTICE 12-2

The General Quadratic Function

Exercises

Estimate the coordinates of the vertex for each parabola by graphing the equation on a graphing calculator. Then use $x = -\frac{b}{2a}$ to find the coordinates.

1. $y = x^2 + 4x + 2$ $(-2, -2)$
2. $y = 4x^2 + 8x + 4$ $(-1, 0)$
3. $y - 1 = x^2 + 10x$ $(-5, -24)$
4. $6 - y = -2x + 8x^2$ $\left(\frac{1}{8}, \frac{49}{8}\right)$
5. $y = x^2 + 6$ $(0, 6)$
6. $-10x + 4 = -4x^2 + y$ $\left(\frac{5}{4}, -\frac{9}{4}\right)$
7. $-5x + 1 = -x^2 + y$ $\left(\frac{5}{2}, -\frac{21}{4}\right)$
8. $4x^2 + 20x + 10 = y$ $\left(-\frac{5}{2}, -15\right)$

Find the vertex and axis of symmetry. Then graph each equation on your own paper.

9. $y = 2x^2 + 4x + 1$ $(-1, -1)$; $x = -1$
10. $y = x^2 + x - 2$ $\left(-\frac{1}{2}, -\frac{9}{4}\right)$; $x = -\frac{1}{2}$
11. $y = -x^2 + 4$ $(0, 4)$; $x = 0$
12. $y = 4x^2 + 5x + 1$ $\left(-\frac{5}{8}, -\frac{9}{16}\right)$; $x = -\frac{5}{8}$
13. $y = x^2 - 4$ $(0, -4)$; $x = 0$
14. $y = 6x^2 + 5x - 2$ $\left(-\frac{5}{12}, -\frac{73}{24}\right)$; $x = -\frac{5}{12}$
15. $y = -4x^2 + x - 4$ $\left(\frac{1}{8}, -\frac{63}{16}\right)$; $x = \frac{1}{8}$
16. $y = -3x^2 + 6x + 1$ $(1, 4)$; $x = 1$
17. $-x^2 + y = 4 + x$ $\left(-\frac{1}{2}, \frac{15}{4}\right)$; $x = -\frac{1}{2}$
18. $y = 6x^2 + 6x$ $\left(-\frac{1}{2}, -\frac{3}{2}\right)$; $x = -\frac{1}{2}$
19. $x^2 + 3x - 4 = y$ $\left(-\frac{3}{2}, -\frac{25}{4}\right)$; $x = -\frac{3}{2}$
20. $y = 4x^2 + x + 2$ $\left(-\frac{1}{8}, \frac{31}{16}\right)$; $x = -\frac{1}{8}$
21. Find a quadratic equation in the form $y = ax^2 + bx + c$ for which $c = 2$ and the axis of symmetry is $x = \frac{2}{3}$.
possible answer; $y = 3x^2 - 4x + 2$

158 EXTRA PRACTICE *LESSON 12-2*

Enrichment Worksheet 12-2

Name ____________ Date ____________

ENRICHMENT 12-2

Quadratic Formulas for Sequences

In some sequences, the value of each term is a function of the term number n. When this is the case, you can use a method called *finite differences* to find an expression for the nth term. Examine what happens in this table.

Term Number	Term Value	Subtract Once	Subtract Again
0	c		
		$a + b$	
1	$a + b + c$		$2a$
		$3a + b$	
2	$4a + 2b + c$		$2a$
		$5a + b$	
3	$9a + 3b + c$		$2a$
		$7a + b$	
4	$16a + 4b + c$		

Sequences in which you get a common difference after two subtractions have quadratic expressions for the value of the nth term. For example, the sequence 1, 5, 12, 22, 35 . . . gives a common difference of 3 after two subtractions. Using the table above, you can write the system $1 = c$, $3 = 2a$ and $5 = a + b + c$. In this example, solving for a, b and c gives the quadratic expression $1.5n^2 + 2.5n + 1$.

Exercises

Make drawings to find the first five figurate numbers of each type. Then, use the method above to find a quadratic expression for the nth term.

1. pentagonal numbers 1, 5, 12, 22, 35; $1.5n^2 - 0.5n$
2. hexagonal numbers 1, 6, 15, 28, 45; $2n^2 - n$
3. star numbers 1, 8, 21, 40, 65; $3n^2 - 2n$

3rd Pentagonal Number

5th Hexagonal Number

4th Star Number

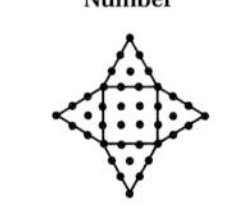

206 ENRICHMENT *LESSON 12-2*

Skills Practice

Vocabulary Review

Lesson 12-1
quadratic equation   function
parabola   vertex
axis of symmetry

Lesson 12-2
general quadratic function
standard quadratic equation

Assignment Guide

All students: 1–38
Additional Practice: See Extra Practice Index on page 674.

Additional Answers

3.

x	−3	−2	−1	0	1	2	3
y	13	8	5	4	5	8	13

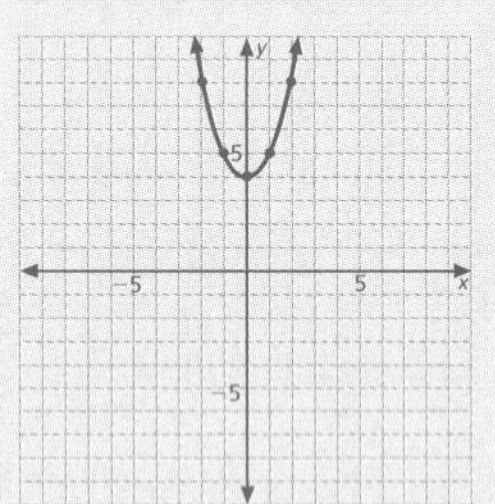

4.
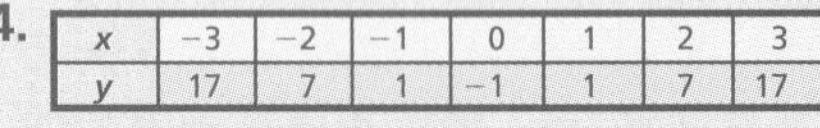

x	−3	−2	−1	0	1	2	3
y	17	7	1	−1	1	7	17

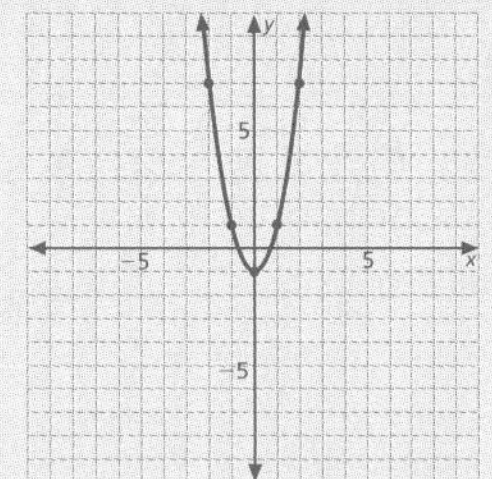

9.
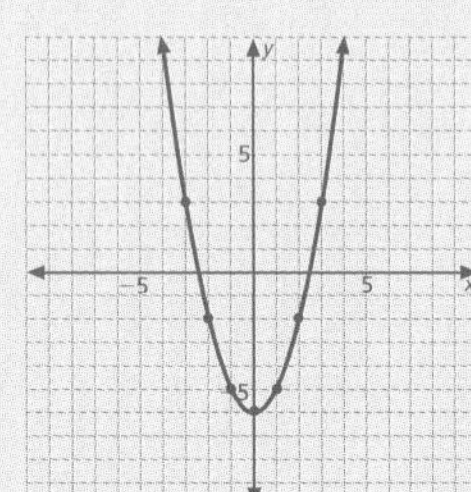

10.
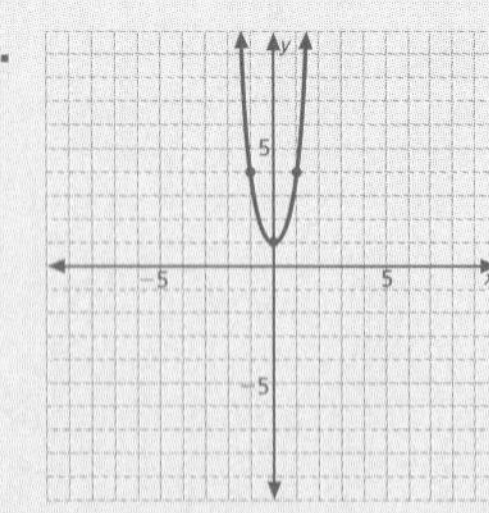

Review and Practice Your Skills

Practice ■ Lesson 12-1

1. The graph of $y = x^2$ is a function. Explain why. There is only one *y*-value for each *x*-value.
2. Describe the vertex of:
   a. a parabola that opens downward the highest point on the parabola
   b. a parabola that opens upward the lowest point on the parabola

Copy and complete each table. Then draw the graph. For 3–4, see additional answers for graphs.

3. $y = x^2 + 4$

x	−3	−2	−1	0	1	2	3
y	13						

4. $y = 2x^2 - 1$

x	−3	−2	−1	0	1	2	3
y							

Determine if the graph of each equation opens upward or downward.

5. $y = -3x^2 + 4$ downward
6. $y = 3x^2 - 8$ upward
7. $y = 2x^2 + 4$ upward
8. $y = -(x^2 + 2)$ downward

Use the graph of each function where the domain is the real numbers to name the coordinates of the vertex. For 9–12, see additional answers for graphs.

9. $y = x^2 - 6$ (0, −6)
10. $y = 3x^2 + 1$ (0, 1)
11. $y = -2x^2 + 1$ (0, 1)
12. $y = -x^2 + 9$ (0, 9)

For which of these equations is the vertex at the origin?

13. $y = x^2 + 4$ no
14. $y = x^2$ yes
15. $y = -x^2$ yes
16. $y = -x^2 + 3$ no

For which of these equations is the axis of symmetry the *y*-axis?

17. $y = \pm\sqrt{2x}$ no
18. $y = 5x^2 - 3$ yes
19. $y = \pm\sqrt{5x - 3}$ no
20. $y = -3x^2 + 1$ yes

Practice ■ Lesson 12-2

21. What is the standard quadratic equation? $y = ax^2 + bx + cx$
22. What is the general quadratic equation? $f(x) = ax^2 + bx + cx$

Estimate the coordinates of the vertex for each parabola by graphing the equation on a graphing calculator. Then use $x = -\frac{b}{2a}$ to find the coordinates.

23. $y = 2x^2 - 3x + 4$ $\left(\frac{3}{4}, \frac{23}{8}\right)$
24. $y = -3x^2 + 4x - 5$ $\left(\frac{2}{3}, -3\frac{2}{3}\right)$
25. $y = x^2 - 2x + 3$ (1, 2)
26. $y + 4x^2 = 2x - 6$ $\left(\frac{1}{4}, -5\frac{3}{4}\right)$

Find the vertex and axis of symmetry.

27. $y = -2x^2 + x - 4$ $v = \left(\frac{1}{4}, -3\frac{7}{8}\right)$; a/s $= \frac{1}{4}$
28. $y = 3x^2 + 6x - 8$ $v = (-1, -11)$; a/s $= -1$
29. $y = 2x^2 - 3x + 5$ $v = \left(\frac{3}{4}, 3\frac{7}{8}\right)$; a/s $= \frac{3}{4}$

11.
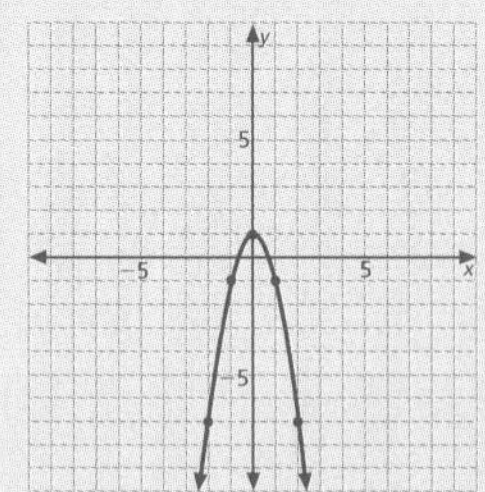

12.
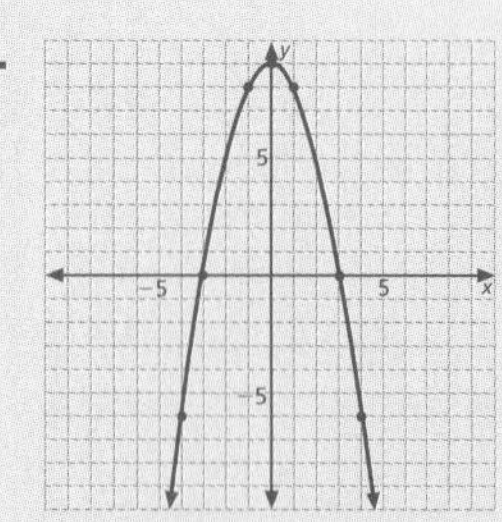

30.
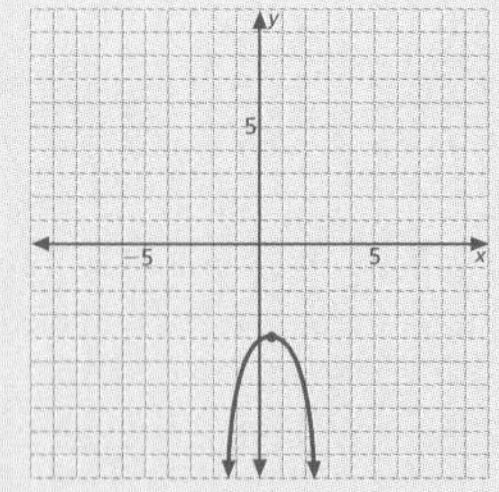

31.
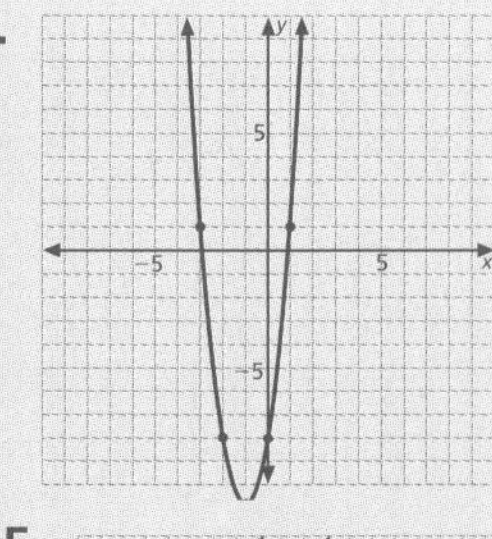

32.
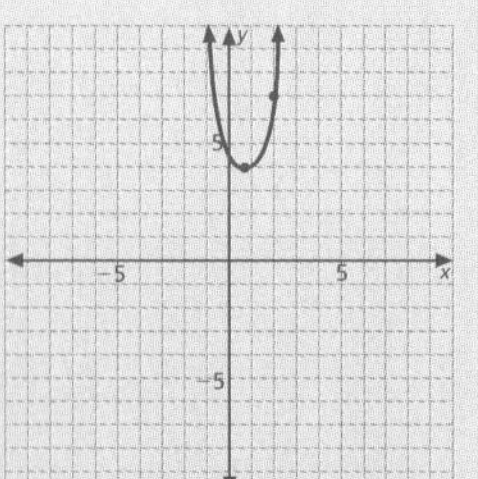

33.
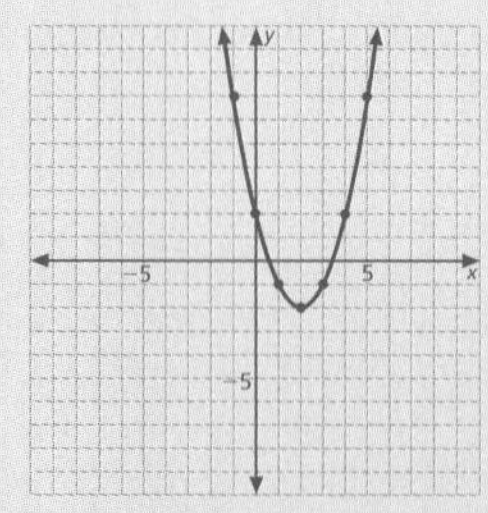

34.
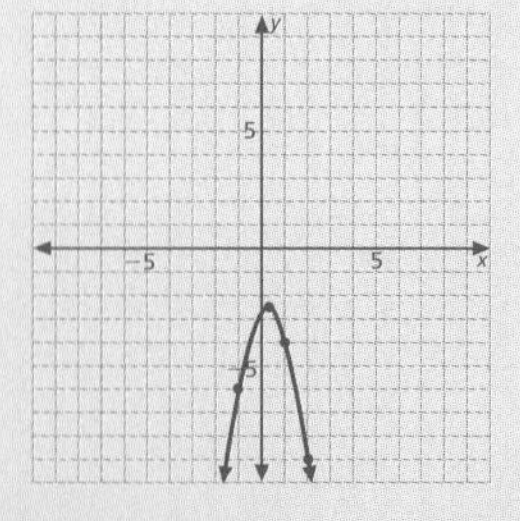

35.
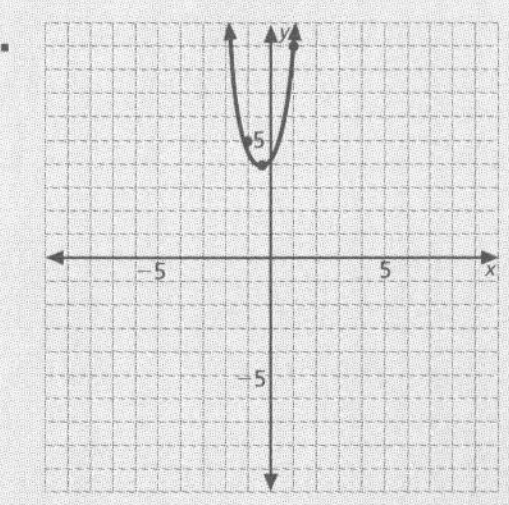

PRACTICE ■ LESSON 12-1–LESSON 12-2

Use the graph of each function to name the coordinates of the vertex. (Lesson 12-1) For 30–32, see additional answers for graphs.

30. $y = x^2 - 4x + 2$ $v = (2, -2)$

31. $y = -2x^2 + x - 3$ $v = \left(\frac{1}{4}, -2\frac{7}{8}\right)$

32. $y = 3x^2 + 2x + 4$ $v = \left(-\frac{1}{3}, 3\frac{2}{3}\right)$

Graph each equation on a graphing calculator. Give the quadrant of the vertex. (Lesson 12-2) See additional answers for graphs.

33. $y = -3x^2 + 3x - 7$ QIV

34. $y = 2x^2 + 5x + 8$ QII

35. $y = -x^2 + x + 5$ QI

Find the vertex and axis of symmetry. (Lesson 12-2)

36. $y = x^2 + 4x - 3$ $v = (-2, -7)$; a/s $= -2$

37. $y = -3x^2 - 2x + 4$ $v = \left(-\frac{1}{3}, 4\frac{1}{3}\right)$; a/s $= -\frac{1}{3}$

38. $y = 2x^2 - 3x - 5$ $v = \left(\frac{3}{4}, -6\frac{1}{8}\right)$; a/s $= \frac{3}{4}$

MathWorks — Workplace Knowhow
Career – Pilots

Pilots are the commanders of airplanes, jets, helicopters and space shuttles. Pilots transport passengers and goods, fight fires, test new aircraft, monitor traffic and crime, dust crops and conduct rescue missions. Pilots must make many calculations to fly safely. They must be able to read electronic instruments accurately. For example, they use math to calculate the speed necessary for take off. To do so, they must consider many variables such as altitude of the airport, outside temperature, weight of the plane and speed and direction of wind.

One plane left New York headed to Tokyo flying at an average speed of 375 mi/h. Another plane left New York 1 h later following the same route and flying at an average speed of 500 mi/h. If both planes followed the same course, how many hours after it left New York would the second plane catch up to the first plane?

1. Write an equation you could solve to answer the question. $375(x + 1) = 500x$
2. Use a graphing utility to graph both sides of the equation and solve the problem. Check student's graph.
3. How many miles will each plane have flown when they are equidistant from New York? 1500
4. A plane gained altitude at a rate of 1000 ft/min, descended 1300 ft, then started to climb again for $2\frac{1}{4}$ min at a rate of 800 ft/min. The total gain in altitude was 6000 ft. How long did the plane climb at a rate of 1000 ft/min? $5\frac{1}{2}$ min
5. An airplane takes off and climbs at a steady 18° angle. After flying along a path of 2 mi, how much altitude has the plane gained in feet? Round to the nearest foot. 3263 ft

Chalkboard Examples

Example from Lesson 12-1

For the equation $y = -5x^2 + 6$, name the vertex and the axis of symmetry and state whether the graph opens upward or downward. **(0, 6), *y*-axis, downward**

Example from Lesson 12-2

Find the coordinates of the vertex for the graph $y + 4 = x^2 - 4x$. **(2, −8)**

Career Opportunity

Describe the kind of work air traffic controllers do and the types of companies that hire these workers. Have students discuss the importance of measurement, geometry and algebraic thinking in doing this type of job. Explain that air traffic controllers must be able to read instruments accurately and make calculations to determine the distance between approaching aircrafts. Students should answer Questions 1–3 to better understand how air traffic controllers use mathematics in performing their job.

Students who are interested in learning more about this profession can go to learningmathmatters.com. Guidance Counselors should have information about school and training requirements.

Teaching Strategies

Errors with quadratic functions are often a result of misapplication of the order of operations. To check students' understanding, have them make a table of ordered pairs for the function: $y = 2x^2$. Watch for students who multiply before evaluating the exponent.

Chapter 12 Review and Practice Your Skills 529

Lesson Planning

NCTM Standards/Strands

- Number & Operation
- Patterns, Functions, & Algebra
- Geometry & Spatial Sense
- Problem Solving
- Reasoning and Proof
- Communication
- Connections
- Representation

Vocabulary

factoring
x-intercepts

Materials Needed

graphing calculators
grid paper (optional)

Lesson Resources

Warm-Up Transparency 34
Transparency TK-7–10
Reteaching 12-3
Extra Practice 12-3

Assignment Guide

Basic: 1–31, 37–50
Enriched: 1–50

Getting Started

5-Minute Warm-up

Factor each polynomial completely.

1. $2x^2 - 18$ $\quad 2(x + 3)(x - 3)$
2. $5x^2 - 20x$ $\quad 5x(x - 4)$
3. $12x^2 + 3x - 6$ $\quad 3(4x^2 + x - 2)$

Introduction to Lesson 12-3

As students graph the equations on their calculators, ask them to consider why the greatest number of x-intercepts possible for any parabola is two. As a general point, you may wish to remind them to check the range settings on their calculators if equations such as $y = x^2 - 25$ barely show up on their screens.

12-3 Factor and Graph

Goals ■ Use factoring to solve quadratic equations.
Applications Physics, Archery, Sports

GRAPHING Graph each equation below on a graphing calculator. How many x-intercepts, the points where the graph crosses the x-axis, does each graph have? (Hint: Use the trace feature, if necessary, to locate the x-intercepts.

a. $y = x^2 - 25$ 2
b. $y = 2x^2 + 3x - 1$ 2
c. $y = 2x^2 + 3x + 5$ 0
d. $y = x^2 - x$ 1
e. $y = x^2 + 5$ 0
f. $y = x^2 + 6x + 9$ 1

Build Understanding

The x-intercepts are the solutions to the quadratic equation. They are the x-coordinates of points for which $y = 0$. As illustrated in the above activity, there may be 0, 1, or 2 x-intercepts.

You can determine the number of x-intercepts and estimate their values by graphing the quadratic equation. You can find exact solutions by letting $y = 0$ and factoring the equation.

Example 1

Solve $y = 3x^2 - 6x$.

Solution

By graphing:

Determine the number of solutions by graphing the equation on a graphing calculator.

Locate the points where $y = 0$. Estimate the x-values for these points. Use the zoom feature to more closely estimate values. The x-values are the solutions.

The graph of the equation $y = 3x^2 - 6x$ intersects the x-axis at two points. They are located approximately at $x = 2$ and $x = 0$.

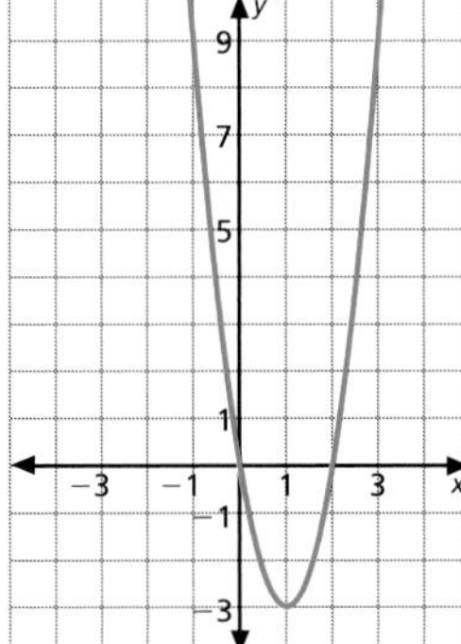

By factoring:

To solve by factoring, let $y = 0$.

$3x^2 - 6x = 0$
$3x(x - 2) = 0$ Factor.
$3x = 0, \quad x - 2 = 0$ Solve each equation.
$x = 0, \quad x = 2$

The solutions for $y = 3x^2 - 6x = 0$ are $x = 0$ and $x = 2$.

Teaching Strategies

Students may benefit from continual reinforcement of the meaning of each of the constants in the equation $y = ax^2 + bx + c$. For each equation considered, point out or elicit from students that the value of a determines the direction of opening, $\frac{-b}{2a}$ is the equation of the axis of symmetry, and c is the y-intercept.

530 Chapter 12 **Quadratic Functions**

If the vertex of the parabola is tangent to the x-axis, there is only one solution to the quadratic equation.

Example 2

Solve $x^2 + 16 = -8x + y$.

Solution

Write the equation in standard form.

$$y = x^2 + 8x + 16$$

By graphing:

Graph the equation to determine the number of solutions. Locate the point where $y = 0$. The x-value is the solution.

The graph of the equation meets the x-axis at one point, approximately $x = -4$.

By factoring:

To solve by factoring, let $y = 0$.

$$x^2 + 8x + 16 = 0$$
$$(x + 4)(x + 4) = 0$$
$$x = -4$$

The solution for $x^2 + 8x + 16 = 0$ is $x = -4$.

If the parabola for the equation does not meet the x-axis, the equation has no solutions.

Mental Math Tip

An equation in the form $ax^2 + bx + c$ in which a, b, and c are integers may be factored if ac has factors with a sum of b.

Example 3

Solve $y = -x^2 - x - 1$.

Solution

Graph the equation on a graphing calculator.

The graph of the equation does not cross the x-axis. There are no solutions to the equation.

There are no solutions for the equation $y = -x^2 - x - 1$.

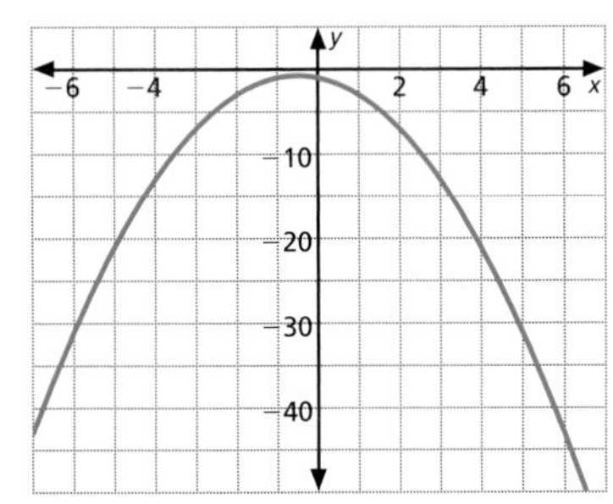

If you know the velocity with which a projectile is launched, you can find the time between launch and landing using the equation $vt - 16t^2 = 0$, where v = velocity in feet per second (ft/sec) and t = time in seconds.

Technology Note

If you have access to charting and data analysis software, you may want to investigate methods for graphing functions. For many programs, you enter data and equations on a spreadsheet. The program then draws a graph of the function.

Teaching Strategies

CHALLENGE Ask students to use the skills developed in this section to write equations that have exactly one solution. When they have written several, they can look for a pattern in the constants b and c. c must be a perfect square, b must be twice the value of the square root of c.

Chalkboard Examples

Supplementary Example 1
Solve $y = x^2 + 2x - 15$ by graphing. Check your work by factoring. $x = -5, x = 3$

Supplementary Example 2
Solve $x^2 + 4x = y + 5$ by graphing and factoring. $x = -5, x = 1$

Supplementary Example 3
Solve $-x^2 + y = x + 2$ by graphing. There are no solutions since the graph of the equation does not cross the x-axis.

Supplementary Example 4
A baseball is thrown upwards with an initial velocity of 80 feet per second. How long does the ball remain in the air? 5 seconds

Reteaching Worksheet 12-3

Name ______________________ Date ____________

RETEACHING 12-3

FACTOR AND GRAPH

You can either graph the quadratic function or find the x-intercepts (solutions) of the guadratic equation by factoring. There can be zero, one or two possible solutions.

Example 1

Use a graphing calculator to determine the number of solutions for $y = x^2 - 1$. If the equation has one or two solutions, find the exact solution(s) by factoring.

Solution

Graph the equation on a graphing calculator. The graph of the equation crosses the x-axis in two places. So there are two solutions.

To solve by factoring, let $y = 0$.
$x^2 - 1 = 0$
$(x + 1)(x - 1) = 0$
$x = -1, x = 1$

The solutions for $y = x^2 - 1$ are $x = -1$ and $x = 1$.

Example 2

Use a graphing calculator to determine the number of solutions for $x^2 + 6x = y - 9$. If the equation has one or two solutions, find the exact solution(s) by factoring.

Solution

Graph the equation on a graphing calculator. The graph of the equation crosses the x-axis in one place. So there is one solution.

Write the equation in standard form:
$y = x^2 + 6x + 9$

To solve by factoring, let $y = 0$.
$x^2 + 6x + 9 = 0$
$(x + 3)(x + 3) = 0$
$x = -3$

The solution for $y = x^2 + 6x + 9$ is $x = -3$.

EXERCISES

Use a graphing calculator to determine the number of solutions for each equation. For equations with one or two solutions, find the exact solutions by factoring.

1. $y = x^2 - 81$ $x = 9, x = -9$
2. $y = x^2 - 5x - 6$ $x = 6, x = -1$
3. $y = x^2 - 2x + 1$ $x = 1$
4. $y = x^2 + 6x - 27$ $x = 3, x = -9$
5. $y = x^2 + 2x - 15$ $x = -5, x = 3$
6. $y = x^2 - 2x - 35$ $x = 7, x = -5$
7. $x^2 = x + 56 + y$ $x = 8, x = -7$
8. $y = x^2 - x - 6$ $x = 3, x = -2$
9. $2x^2 = y - x$ $x = 0, x = -\frac{1}{2}$
10. $y + x = x^2 + 1$ no solution
11. $y + 9x = 3x^2$ $x = 0, x = 3$
12. $x^2 + 8x = 9 + y$ $x = -9, x = 1$
13. $y = \frac{1}{2}x^2 + 5x$ $x = 0, x = -10$
14. $y = \frac{1}{4}x^2 - 4$ $x = -4, x = 4$

182 RETEACHING *LESSON 12-3*

Lesson Wrap-up

Quick Assessment

Ask the following questions to determine if students understand the content presented in this lesson.

1. Why is two the maximum number of solutions for a quadratic equation? The solutions are the points were the graph crosses the x-axis. A parabola can cross the axis in only zero, one, or two places.
2. How can you tell, without drawing the graph, that a quadratic equation will have exactly one solution? The equation will have only one root or solution; the factored equation will have repeated factors.

Assignment Guide

Basic: 1–31, 37–50
Advanced: 1–50
Additional Practice: See Extra Practice Index on page 674.

Example 4

PHYSICS A football is thrown with the initial velocity of 64 ft/sec. How long does it remain in the air?

Solution

Substitute 64 for v in the formula:	$64t - 16t^2 = 0$	
Factor the equation:	$16t(4 - t) = 0$	
Set each factor equal to 0 and solve for t:	$16t = 0$	$4 - t = 0$
	$t = 0$	$t = 4$

The equation has two solutions. The first solution represents the launch time, the second represents the landing time, The football stays in the air for 4 sec.

Try These Exercises

GRAPHING Use a graphing calculator to determine the number of solutions for each equation. For equations with one or two solutions, find the exact solutions by factoring.

1. $y = x^2 + 10x + 21$ $x = -3, x = -7$
2. $y = x^2 + 5x + 6$ $x = -2, x = -3$
3. $y = x^2 + 25$ no solutions
4. $y - x^2 = 49 - 14x$ $x = 7$
5. $y - 81 = x^2 + 18x$ $x = -9$
6. $y - x^2 = -9$ $x = 3, x = -3$

Practice Exercises

GRAPHING Use a graphing calculator to determine the number of solutions for each equation. For equations with one or two solutions, find the exact solutions by factoring.

7. $y = x^2 - 100$ $x = 10, x = -10$
8. $y = x^2 + 7x$ $x = 0, x = -7$
9. $y + 9x = x^2 + 14$ $x = 2, x = 7$
10. $y - 25 = x^2 - 9x$ no solution
11. $x^2 = -y + 4$ $x = 2, x = -2$
12. $x = y - x^2$ $x = 0, x = -1$
13. $y = x^2 + x + 1$ no solution
14. $y - 25 + 10x = x^2$ $x = 5$
15. $y = x^2 + 8x - 48$ $x = 4, x = -12$

Write an equation for each problem. Then factor to solve.

16. The square of a positive integer is 20 less than 12 times the integer. Find the integer. 10 or 2
17. The square of a number exceeds the number by 30. Find the number. 6 or −5
18. The square of an integer is 5 more than 4 times the integer. Find the integer. 5 or −1
19. Seven times an integer plus 8 equals the square of the integer. Find the integer. 8 or −1

532 Chapter 12 **Quadratic Functions**

Teaching Strategies

COOPERATIVE LEARNING Have students work with a partner on this exercise. Graph these equations, each of which is in the form $y = ax^2 + bx + c$.

1. $y = x^2 + 4x + 3$
2. $y = x^2 - x - 2$
3. $y = -x^2 - x - 1$

What does the constant c represent in each equation? the y-intercept

For Exercises 20–25, use the equation $vt - 16t^2 = 0$, where v = velocity in feet per second and t = time in seconds.

20. **PHYSICS** The initial velocity of a projectile is 128 ft/sec. In how many seconds will it return to the ground? 8 sec

21. **ARCHERY** How long will an arrow with an initial velocity of 176 ft/sec remain in the air? 11 sec

22. **SPORTS** How long will a baseball thrown with an initial velocity of 96 ft/sec remain in the air? 6 sec

23. **PHYSICS** A rocket at a fireworks display was launched with the initial velocity of 208 ft/sec. It was a dud. How many seconds was it in the air before it splashed down in the lake? 13 sec

24. **PHYSICS** A projectile is launched with the initial velocity of 896 ft/sec. How many seconds will it remain in the air? 56 sec

25. **WRITING MATH** The equation $vt - 16t^2 = 0$ is of the form $y = ax^2 + bx$. From Exercises 20–23, what generalization can you make about the solutions to quadratic equations in the form $y = ax^2 + bx$?

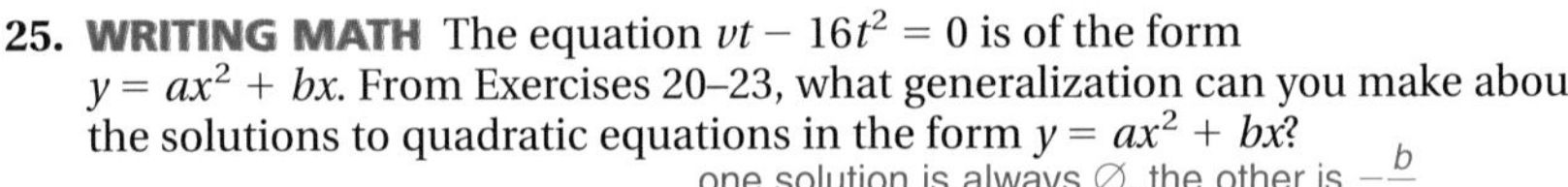

one solution is always ∅, the other is $-\frac{b}{a}$

Use your generalization from Exercise 25 to solve each of the following equations.

26. $y = x^2 - 3x$ $x = 0, x = 3$
27. $y = x^2 + 2x$ $x = 0, x = -2$
28. $y = 5x^2 + 20x$ $x = 0, x = -4$
29. $y = 2x^2 - 8x$ $x = 0, x = 4$
30. $y = 9x^2 - 72x$ $x = 0, x = 8$
31. $y = 3x^2 + 33x$ $x = 0, x = 11$

Extended Practice Exercises

The sum of the first n positive even integers is $S = n^2 + n$. How many integers must be added to give each sum?

32. 240 15
33. 2,070 45
34. 1,122 33
35. 40,200 200

36. **CHAPTER INVESTIGATION** Working together, build your gravity-powered vehicle. Try to keep the weight as close to 10 oz as possible. If necessary add weight by adding new design elements. If your vehicle has wheels, make sure the wheels turn freely. You may want to use a drop of oil or powdered graphite to lubricate the axle.

Mixed Review Exercises

Find factors for the following. (Lesson 11-3)

37. $9c - 27b$ $9(c - 36)$
38. $x^2y - x$ $x(xy - 1)$
39. $3mn^2 - 9mn$ $3mn(n - 3)$
40. $8a^2b + 32ab^2$ $8ab(a + 4b)$
41. $wx + xz$ $x(w + z)$
42. $17a^2b - 68ab$ $17ab(a - 4)$

Find the GCF and its paired factor for the following. (Lesson 11-3)

43. $36a - 63b$ $9(4a - 7b)$
44. $12ab - 8a^2b$ $4ab(3 - 2a)$
45. $10x^3 - 15x^2$ $5x^2(2x - 3)$
46. $7a^3bc^2 - 14a^2bc$ $7a^2bc(ac - 2)$
47. $27x^3y^2 - 6x^2y$ $3x^2y(9xy - 2)$
48. $72p^2q^3r - 32p^2q^4r$ $8p^2q^3r(9 - 4q)$
49. $6x^3y^3 - 9x^2y^4 + 6x^2y^3z$ $3x^2y^3(2x - 3y + 2z)$
50. $24a^3b^2c - 18ab^2c^3 + 12abc^2$ $6abc(4a^2b - 3bc^2 + 2c)$

Extra Practice Worksheet 12-3

Name ______________ Date ______________

EXTRA PRACTICE 12-3

FACTOR AND GRAPH

EXERCISES

Use a graphing calculator to determine the number of solutions for each equation. For equations with one or two solutions, find the exact solutions by factoring.

1. $y = x^2 - 25$ $x = 5, x = -5$
2. $y = x^2 + 6x$ $x = 0, x = -6$
3. $y = x^2 + 16$ no solution
4. $y = x^2 + 7x + 6$ $x = -6, x = -1$
5. $y = x^2 - 7x + 12$ $x = 3, x = 4$
6. $y - x^2 = 20 + 5x$ no solution
7. $y - 49 = x^2 + 14x$ $x = -7$
8. $y + 8x = x^2 + 16$ $x = 4$
9. $y = x^2 + 9x + 20$ $x = -5, x = -4$
10. $y = x^2 - 14x + 40$ $x = 4, x = 10$
11. $y = x^2 + 4x - 12$ $x = 2, x = -6$
12. $y + 5 = x^2 - 4x$ $x = 5, x = -1$
13. $y = x^2 + 11x + 10$ $x = -1, x = -10$
14. $y = x^2 - 9x + 8$ $x = 1, x = 8$

Write an equation for each problem. Then factor to solve.

11. The square of a positive integer is 4 more than 3 times the integer. Find the integer. 4 or −1

12. The square of a number exceeds the number by 6. Find the number. 3 or −3

13. The square of an integer is 27 more than 6 times the integer. Find the integer. 9 or −3

14. Six times an integer plus 5 equals the square of the integer. Find the integer. 1 or 5

For Exercises 15–17, use the equation $vt - 16t^2 = 0$, where v = velocity in feet per second and t = time in seconds.

15. The inititial velocity of a projectile is 112 feet per second. In how many seconds will it return to the ground? 7 s

16. How long will a football kicked with an initial velocity of 88 feet per second remain in the air? 5.5 s

17. A projectile is launched with the initial velocity of 576 feet per second. How many seconds will it remain in the air? 36 s

160 EXTRA PRACTICE *LESSON 12-3*

Teaching Strategies

Review the various ZOOM functions on graphing calculators. For some equations, students may need to reset the range and scale of the axes. Also, students may find that they can more accurately locate a point on a graph by using ZBOX instead of zooming in. Many graphing calculators also display a table of x and y values after an equation has been graphed. These tables are useful for locating particular coordinates.

Lesson Planning

NCTM Standards/Strands

- Number & Operation
- Patterns, Functions, & Algebra
- Geometry & Spatial Sense
- Problem Solving
- Reasoning and Proof
- Communication
- Connections
- Representation

Vocabulary

completing the square

Materials Needed

Algeblocks or algebra tiles

Lesson Resources

Warm-Up Transparency 35
Transparency TK-7–10
Reteaching 12-4
Extra Practice 12-4
Enrichment 12-4

Assignment Guide

Basic: 1–40, 47–67
Enriched: 1–67

Getting Started

5-Minute Warm-up

Write each expression as a perfect square.

1. $x^2 - 12x + 36$ $(x - 6)^2$
2. $x^2 + 8x + 16$ $(x + 4)^2$
3. $x^2 - 2x + 1$ $(x - 1)^2$

Introduction to Lesson 12-4

Before students begin to model the expressions using Algeblocks, review the relationship between the blocks. For the first problem, the length of a side of the large square is x, the dimensions of each rectangle are x by one, and the length of a side of each small square is one unit.

12-4 Complete the Square

Goals Solve quadratic equations by completing the square.
Applications Science, Aeronautics, Sports

MODELING Work with a partner to build equations using Algeblocks.

1. Use Algeblocks to illustrate each perfect square.

$x^2 - 6x + 9 = (x - 3)^2$ $\quad$ $x^2 + 10x + 25 = (x + 5)^2$

$x^2 + 14x + 49 = (x + 7)^2$ $\quad$ $x^2 - 2x + 1 = (x - 1)^2$

2. Each of the squares above is in the form $ax^2 + bx + c = (x + h)^2$. Discuss the following:
   a. What is the relationship between c and h? $c = h^2$
   b. What is the relationship between h and b? $h = \frac{b}{2}$
   c. What is the relationship between c and b? $c = \left(\frac{b}{a}\right)^2$

Problem Solving Tip

Use the relationships you discover for questions **a** and **b** to find **c**.

Build Understanding

Making a perfect square for an expression in the form $ax^2 + bx$ is called **completing the square**. Completing the square is another method for solving quadratic equations.

Example 1

Complete the square for $x^2 - 8x$.

Solution

MODELING Use Algeblocks to illustrate $x^2 - 8x$. Add blocks to make a perfect square. Write the expression.

$x^2 - 8x + 16$

In the activity at the top of the page, you discovered that for perfect squares in the form $ax^2 + bx + c$, the constant $c = \left(\frac{b}{2}\right)^2$. Thus, by substitution, you know $ax^2 + bx + c = ax^2 + bx + \left(\frac{b}{2}\right)^2$.

You can use this relationship to complete the square for $x^2 - 8x$.

$x^2 - 8x + \left(\frac{b}{2}\right)^2$ $\quad$ Add $\left(\frac{b}{2}\right)^2$ to complete the square.

$x^2 - 8x + (-4)^2$ $\quad$ Find $\frac{b}{2}$, $-\frac{8}{2} = -4$

$x^2 - 8x + 16$ $\quad$ Square $\frac{b}{2}$, $(-4)^2 = 16$

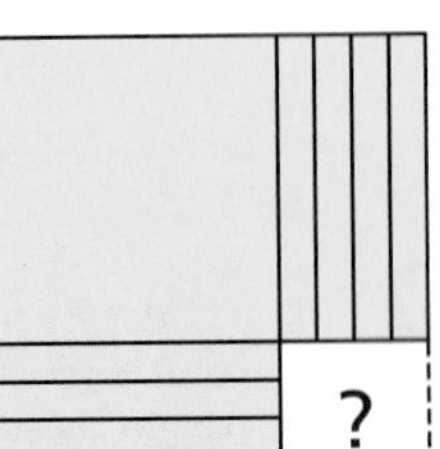

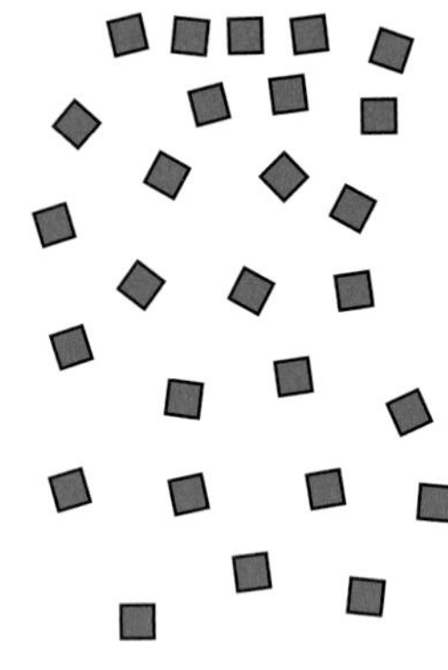

Extend the Lesson

MATH JOURNAL Have students write a paragraph explaining how an equation such as $x^2 - 4x = 5$ can have two solutions when the equation can be simplified to $(x - 2)^2 = 9$. Positive integers have two square roots, positive and negative.

You can use the process of completing the square to solve quadratic equations.

Example 2

Solve by completing the square.

$x^2 + 8x + 12 = 0$

Solution

$x^2 + 8x + 12 = 0$

$x^2 + 8x = -12$ — Add -12 to each side.

$x^2 + 8x + 16 = -12 + 16$ — Add $\left(\frac{b}{2}\right)^2$ to each side.

$(x + 4)^2 = 4$ — Factor.

$x + 4 = \sqrt{4}$ — Simplify.

$x + 4 = \pm 2$

$x = -4 + 2, x = -4 + (-2)$

$x = -2, x = -6$

The solutions to the equation $x^2 + 8x + 12 = 0$ are $x = -2, x = -6$.

Problem Solving Tip

Always check your solutions by substituting them back into the original equation.

To solve a quadratic equation by completing the square, the coefficient of the x^2 term must be 1.

Example 3

SCIENCE The number of seconds (x) that it takes a chemical reaction to begin and end can be found by solving the equation $2x^2 - 3x + 1 = 0$. Find the solutions.

Solution

$2x^2 - 3x + 1 = 0$

$x^2 - \frac{3}{2}x + \frac{1}{2} = 0$ — To make the coefficient of the x^2 term 1, multiply by $\frac{1}{2}$.

$x^2 - \frac{3}{2}x = -\frac{1}{2}$

$x^2 - \frac{3}{2}x + \frac{9}{16} = -\frac{1}{2} + \frac{9}{16}$ — Add $\left(\frac{b}{2}\right)^2$ to each side of the equation.

$\left(x - \frac{3}{4}\right)^2 = \frac{1}{16}$ — Factor.

$x - \frac{3}{4} = \sqrt{\frac{1}{16}}$ — Simplify.

$x = \frac{3}{4} \pm \frac{1}{4}$

$x = \frac{1}{2}, x = 1$

The solutions to the equation $2x^2 - 3x + 1 = 0$ are $x = \frac{1}{2}, x = 1$.

Teaching Strategies

Students may wonder when to complete the square in order to solve a quadratic equation. Remind them to try factoring first. If the equation cannot be factored, completing the square will work for every quadratic equation.

Chalkboard Examples

Supplementary Example 1

Complete the square for $x^2 - 4x$. $x^2 - 4x + 4$

Supplementary Example 2

Solve $x^2 - 12x$ by completing the square. $x^2 - 12x + 36$

Supplementary Example 3

The results of an experiment can be modeled by the equation $2x^2 - 5x + 2 = 0$. Solve by completing the square. $x = 2, x = \frac{1}{2}$

Reteaching Worksheet 12-4

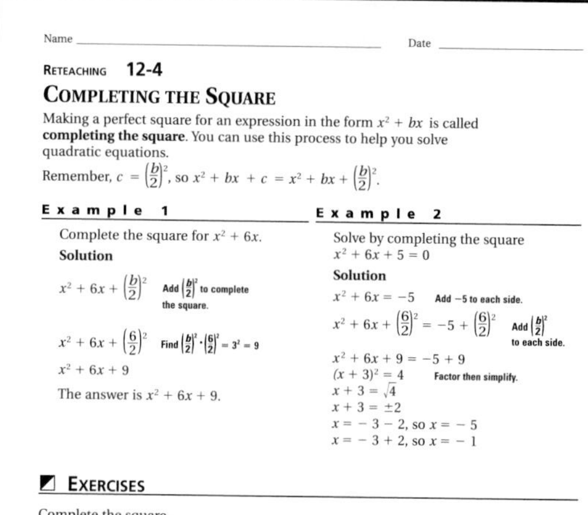

Name ______________________ Date ____________

RETEACHING 12-4

COMPLETING THE SQUARE

Making a perfect square for an expression in the form $x^2 + bx$ is called **completing the square**. You can use this process to help you solve quadratic equations.

Remember, $c = \left(\frac{b}{2}\right)^2$, so $x^2 + bx + c = x^2 + bx + \left(\frac{b}{2}\right)^2$.

Example 1

Complete the square for $x^2 + 6x$.

Solution

$x^2 + 6x + \left(\frac{b}{2}\right)^2$ — Add $\left(\frac{b}{2}\right)^2$ to complete the square.

$x^2 + 6x + \left(\frac{6}{2}\right)^2$ — Find $\left(\frac{b}{2}\right)^2 \cdot \left(\frac{6}{2}\right)^2 = 3^2 = 9$

$x^2 + 6x + 9$

The answer is $x^2 + 6x + 9$.

Example 2

Solve by completing the square $x^2 + 6x + 5 = 0$

Solution

$x^2 + 6x = -5$ — Add -5 to each side.

$x^2 + 6x + \left(\frac{6}{2}\right)^2 = -5 + \left(\frac{6}{2}\right)^2$ — Add $\left(\frac{b}{2}\right)^2$ to each side.

$x^2 + 6x + 9 = -5 + 9$

$(x + 3)^2 = 4$ — Factor then simplify.

$x + 3 = \sqrt{4}$

$x + 3 = \pm 2$

$x = -3 - 2$, so $x = -5$

$x = -3 + 2$, so $x = -1$

EXERCISES

Complete the square.

1. $x^2 + 8x$ $\underline{x^2 + 8x + 16}$
2. $x^2 - 4x$ $\underline{x^2 - 4x + 4}$
3. $x^2 + 10x$ $\underline{x^2 + 10x + 25}$
4. $x^2 + 2x$ $\underline{x^2 + 2x + 1}$
5. $x^2 - 12x$ $\underline{x^2 - 12x + 36}$
6. $x^2 - 14x$ $\underline{x^2 - 14x + 49}$

Solve by completing the square.

7. $x^2 + 6x - 7 = 0$ $\underline{x = -7, x = 1}$
8. $x^2 - 4x - 5 = 0$ $\underline{x = -1, x = 5}$
9. $x^2 - 16x - 17 = 0$ $\underline{x = -1, x = 17}$
10. $x^2 + 10x + 9 = 0$ $\underline{x = -1, x = -9}$
11. $x^2 - 12x + 11 = 0$ $\underline{x = 1, x = 11}$
12. $x^2 + 4x - 45 = 0$ $\underline{x = 5, x = -9}$
13. $x^2 + 10x + 16 = 0$ $\underline{x = -8, x = -2}$
14. $x^2 + 12x + 11 = 0$ $\underline{x = -11, x = -1}$
15. $x^2 + 24x + 44 = 0$ $\underline{x = -22, x = -2}$
16. $x^2 - 8x - 33 = 0$ $\underline{x = 11, x = -3}$

184 RETEACHING *LESSON 12-4*

Lesson Wrap-up

Quick Assessment

Ask the following questions to determine if students understand the content presented in this lesson.

1. Explain in your own words how completing the square enables you to solve a quadratic equation. **The goal is to find a value of c so that the expression $ax^2 + bx + c$ will be a perfect square.**
2. In Example 3, explain how the square root of $\frac{1}{16}$ was simplified to $\frac{1}{4}$. **The square root of a fraction is the square root of the numerator divided by the square root of the denominator.**

Assignment Guide

Basic: 1–40, 47–67
Advanced: 1–67
Additional Practice: See Extra Practice Index on page 674.

Try These Exercises

MODELING **Complete the square. Use Algeblocks if you wish.**

1. $x^2 + 4x$ $x^2 + 4x + 4$
2. $x^2 - 6x$ $x^2 - 6x + 9$
3. $x^2 - 2x$ $x^2 - 2x + 1$

Solve by completing the square. Check your solutions.

4. $x^2 - 2x - 8 = 0$ $x = 4, x = -2$
5. $x^2 + 6x = 7$ $x = -7, x = 1$
6. $2x^2 - 5x + 2 = 0$ $x = 2, x = \frac{1}{2}$
7. $x^2 + 4x - 12 = 0$ $x = -6, x = 2$

Write an equation. Then complete the square to solve the problem.

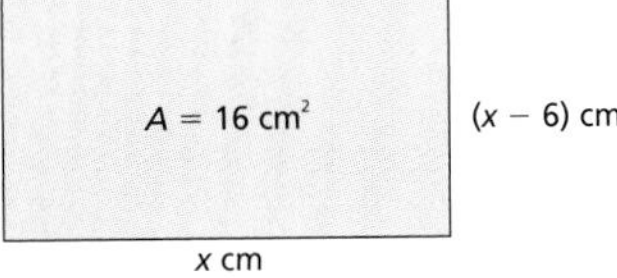

8. The width of a rectangle is 6 cm shorter than the length. The area is 16 cm^2. Find the length and width of the rectangle. 8 cm by 2 cm

9. **WRITING MATH** Explain why you cannot use the negative solution to the equation to find the answer to the problem in Exercise 8. lengths cannot be negative
10. Evan needs to solve the equation $x^2 - 2x = 15$. After completing the square, he writes $x^2 - 2x + 1 = 0$, factors the equation as $(x - 1)^2 = 0$ and solves for x. Evan is surprised when his solution, $x = 1$, doesn't check. What did he do wrong? Find the correct solutions. Evan dropped the 15 from the original equation, instead of setting the quadratic expression equal to 0. The correct solutions are $x = -3, x = 5$.

Practice Exercises

Complete the square.

11. $x^2 - 10x$ $x^2 - 10x + 25$
12. $x^2 + 20x$ $x^2 + 20x + 109$
13. $x^2 + x$ $x^2 + x + \frac{1}{4}$
14. $x^2 - 14x$ $x^2 - 14x + 49$
15. $x^2 + 18x$ $x^2 + 18x + 81$
16. $x^2 - 30x$ $x^2 - 30x + 225$
17. $x^2 + 3x$ $x^2 + 3x + \frac{9}{4}$
18. $x^2 - 16x$ $x^2 - 16x + 64$
19. $x^2 - x$ $x^2 - x + \frac{1}{4}$

Solve by completing the square. Check your solutions.

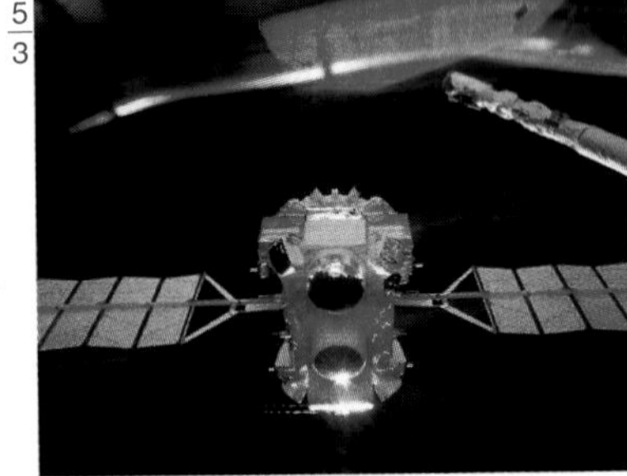

20. $x^2 - 3x - 28 = 0$ $x = -4, x = 7$
21. $3x^2 - 2x - 5 = 0$ $x = -1, x = \frac{5}{3}$
22. $9x^2 - 18x = 0$ $x = 0, x = 2$
23. $x^2 + 3x = 0$ $x = -3, x = 0$
24. $x^2 - 2x + 1 = 0$ $x = 1$
25. $2x^2 - 9x = 5$ $x = \frac{1}{2}, x = 5$
26. $x^2 - x - 12 = 0$ $x = 4, x = -3$
27. $2x^2 + 5x = 3$ $x = -3, x = \frac{1}{2}$
28. $x^2 - 6x = 7$ $x = -1, x = 7$
29. $2x^2 - 4x = 0$ $x = 0, x = 2$
30. $6x^2 - 5x = -1$ $x = \frac{1}{2}, x = \frac{1}{3}$
31. $x^2 = 6x$ $x = 0, x = 6$

Find values for c and h to complete each perfect square.

32. $x^2 + 20x + c = (x + h)^2$ $c = 100, h = 10$
33. $x^2 - 4x + c = (x + h)^2$ $c = 4, h = -2$
34. $x^2 + x - c = (x + h)^2$ $c = -\frac{1}{4}, h = \frac{1}{2}$
35. $x^2 - 3x + c = (x + h)^2$ $c = \frac{9}{4}, h = -\frac{3}{2}$
36. $x^2 - 18x + c = (x + h)^2$ $c = 81, h = -9$
37. $x^2 + 22x + c = (x + h)^2$ $c = 121, h = 11$
38. **AERONAUTICS** A length of a rectangular panel on a satellite is 4 cm greater than its width. Its area is 165 cm^2. Find its dimensions. 11 cm × 15 cm

Extend the Lesson

COOPERATIVE LEARNING Have students use the method of completing the square to explain what must be added to complete this square.
A unit square must be added:
$6^2 + 2(6) + 1 = (6 + 1)^2$.

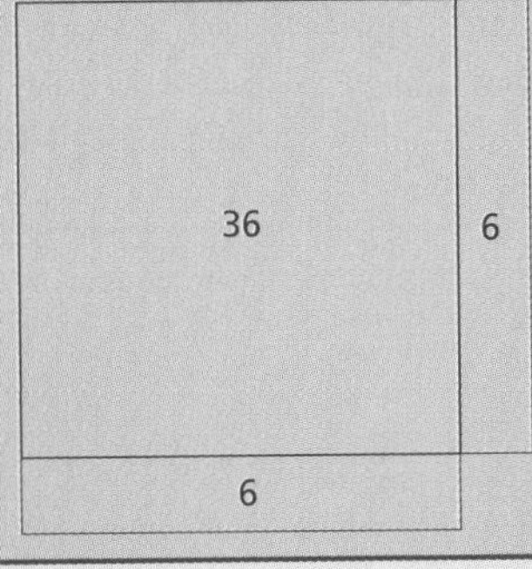

39. **PHYSICS** A ball is thrown up into the air at an initial velocity of 64 ft/sec. How long will it take the ball to reach the ground? Use the equation $64t - 16t^2 = 0$, where t equals time in seconds. 4 sec

40. **SPORTS** A length of a playing field is twice the width plus 4 ft. Its area is 2,310 ft^3. Find the length in feet of the field's boundary line. width = 33 ft, length = 70 ft, perimeter = 206 ft.

Extended Practice Exercises

Solve each problem. If necessary, write an equation and then complete the square.

41. The width of a rectangle is 2 in. less than its length. Find its dimensions if its area is 360 in.^{2}. 18 in. × 20 in.

42. If the width and length of a 4-in. by 2-in. rectangle are each increased by the same amount, the area of the rectangle will be 48 in.^{2}. Find the length and width. 8 in., 6 in.

43. A triangle with an area of 6 m^2 has a base that is 4 m longer than its height. What are the dimensions? 2 m × 6 m

44. If the height and base of a triangle with a height of 5 cm and a base of 8 cm are each decreased by the same amount, the area of the triangle will be 14 cm^2. Find the new base and height. $b = 7$ cm, $h = 4$ cm

45. **ART** A painting is 1 in. longer than it is wide. The painting and its frame have a total area of 156 in.^{2}. The frame is 1 in. wide on each side of the painting. What are the dimensions of the painting? 10 in. × 11 in.

46. **CHAPTER INVESTIGATION** Work with your group to build a ramp from 24–36 in. in length. One possibility would be to attach a length of poster board to two yard sticks, using the sticks for stability. Set the ramp up so that the angle formed by the ramp and floor is 15°. Using a stopwatch, time your vehicle's descent from the top of the ramp to the bottom. Make sure you do not push the vehicle at the starting point. Increase the ramp's incline in 5° increments. How does the increase affect your vehicle's travel time? For each angle of descent, measure the distance the vehicle travels beyond the bottom of the ramp. How does changing the incline affect this distance?

Mixed Review Exercises

Simplify. (Lesson 11-4)

47. $(3a - 2b)(2a - 4)$ $6a^2 - 12a - 4ab + 8b$
48. $(6x + 2)(4k - 3)$ $24xk - 18x + 8k - 6$
49. $(2a - 3c)(4b + 2d)$ $8ab + 4ad - 12bc - 6cd$
50. $(5x - y)(3x - 1)$ $15x^2 - 5x - 3xy + y$
51. $(3c - 2)(3c - 2)$ $9c^2 - 12c + 4$
52. $(5a + 1)(5a - 1)$ $25a^2 - 1$
53. $(4x + 2)(3x - 1)$ $12x^2 + 2x - 2$
54. $(x + 9)(2x - 4)$ $2x^2 + 14x - 36$
55. $(7a + 2b)(5a - 3b)$ $35a^2 - 11ab - 6b^2$
56. $(2c - 5b)(4c - 3b)$ $8c^2 - 26bc + 15b^2$
57. $(12x + 2y)(5x - 6y)$ $60x^2 - 62xy - 12y^2$
58. $(4b - 7y)(5b + 7y)$ $16b^2 - 49y^2$

Write each in simplest radical form. (Lesson 10-1)

59. $\sqrt{77}$ $\sqrt{77}$
60. $\sqrt{112}$ $4\sqrt{7}$
61. $\sqrt{56}$ $2\sqrt{14}$
62. $(2\sqrt{5})(3\sqrt{10})$ $30\sqrt{2}$
63. $(2\sqrt{17})(4\sqrt{22})$ $8\sqrt{374}$
64. $(5\sqrt{11})(4\sqrt{32})$ $80\sqrt{22}$
65. $\dfrac{\sqrt{18}}{\sqrt{6}}$ $\sqrt{3}$
66. $\dfrac{\sqrt{40}}{\sqrt{5}}$ $2\sqrt{2}$
67. $\sqrt{\left(\dfrac{80}{12}\right)}$ $\dfrac{2\sqrt{15}}{3}$

Learning Styles

ONGOING ASSESSMENT Have students solve the following by completing the square: $x^2 + x - 6 = 0$. Watch for students that have not added $\left(\frac{b}{2}\right)^2$ to both sides of the equation.

After completing the square, the correct step will read: $x^2 + x + \left(\frac{1}{2}\right)^2 = 6\frac{1}{4}$.

Extra Practice Worksheet 12-4

Name ____________ Date ____________

Extra Practice 12-4

Complete the Square

Exercises

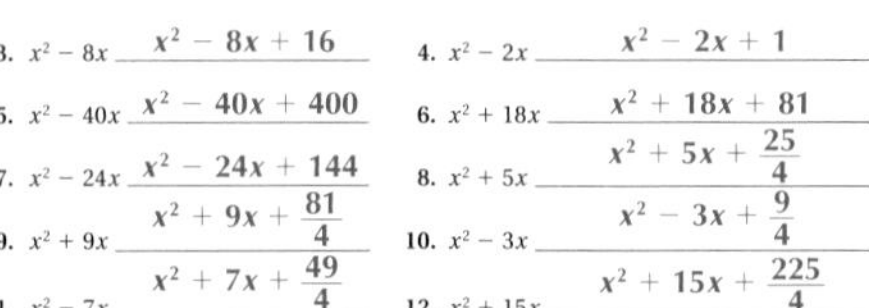

1. $x^2 + 12x$ $x^2 + 12x + 36$
2. $x^2 + 10x$ $x^2 + 10x + 25$
3. $x^2 - 8x$ $x^2 - 8x + 16$
4. $x^2 - 2x$ $x^2 - 2x + 1$
5. $x^2 - 40x$ $x^2 - 40x + 400$
6. $x^2 + 18x$ $x^2 + 18x + 81$
7. $x^2 - 24x$ $x^2 - 24x + 144$
8. $x^2 + 5x$ $x^2 + 5x + \frac{25}{4}$
9. $x^2 + 9x$ $x^2 + 9x + \frac{81}{4}$
10. $x^2 - 3x$ $x^2 - 3x + \frac{9}{4}$
11. $x^2 - 7x$ $x^2 + 7x + \frac{49}{4}$
12. $x^2 + 15x$ $x^2 + 15x + \frac{225}{4}$

Solve by completing the square. Check your solutions.

13. $x^2 - 5x + 6 = 0$ $x = 2, x = 3$
14. $2x^2 - 7x - 4 = 0$ $x = -\frac{1}{2}, x = 4$
15. $x^2 - x - 30 = 0$ $x = -5, x = 6$
16. $x^2 + 5x = 0$ $x = -5, x = 0$
17. $3x^2 + x = 2$ $x = -1, x = \frac{2}{3}$
18. $3x^2 - 6x = 0$ $x = 0, x = 2$
19. $x^2 - 9x + 20 = 0$ $x = 4, x = 5$
20. $x^2 = 5x$ $x = 0, x = 5$

Find values for c and h to complete each perfect square.

21. $x^2 + 10x + c = (x + h)^2$ $c = 25, h = 5$
22. $x^2 + 6x + c = (x + h)^2$ $c = 9, h = 3$
23. $x^2 + 8x + c = (x + h)^2$ $c = 16, h = 4$
24. $x^2 + 24x + c = (x + h)^2$ $c = 144, h = 12$
25. $x^2 + 36x + c = (x + h)^2$ $c = 324, h = -18$
26. $x^2 + 5x + c = (x + h)^2$ $c = \frac{25}{4}, h = -\frac{5}{2}$

162 Extra Practice *Lesson 12-4*

Enrichment Worksheet 12-4

Name ____________ Date ____________

Enrichment 12-4

Graphing Cubics

In a cubic equation, the first term has the exponent 3. The places where the graph crosses the x-axis give the solutions or roots of the equation. Then, the roots can be used to show a factored form.

Example

Graph this equation. Then give its roots and a factored form.
$y = -(x^3 - x^2 - 4x + 4)$

Solution

The graph is shown at the right. The three roots are -2, 1 and 2. A factored form of the equation is thus $y = -(x + 2)(x - 1)(x - 2)$.

Exercises

Graph each equation. Then give its roots and a factored form.

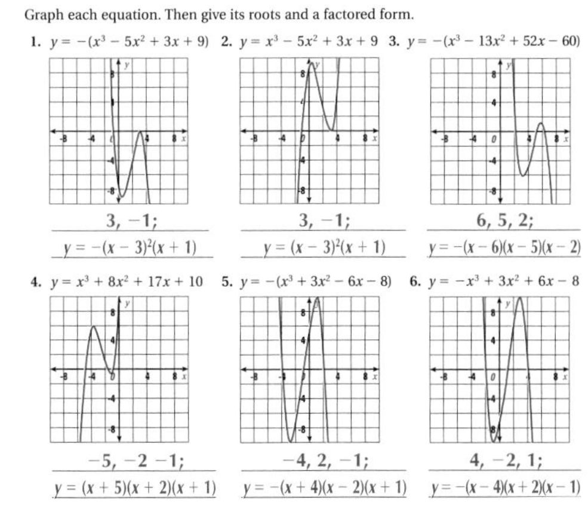

1. $y = -(x^3 - 5x^2 + 3x + 9)$ 3, −1; $y = -(x - 3)^2(x + 1)$
2. $y = x^3 - 5x^2 + 3x + 9$ 3, −1; $y = (x - 3)^2(x + 1)$
3. $y = -(x^3 - 13x^2 + 52x - 60)$ 6, 5, 2; $y = -(x - 6)(x - 5)(x - 2)$
4. $y = x^3 + 8x^2 + 17x + 10$ −5, −2 −1; $y = (x + 5)(x + 2)(x + 1)$
5. $y = -(x^3 + 3x^2 - 6x - 8)$ −4, 2, −1; $y = -(x + 4)(x - 2)(x + 1)$
6. $y = -x^3 + 3x^2 + 6x - 8$ 4, −2, 1; $y = -(x - 4)(x + 2)(x - 1)$

208 Enrichment *Lesson 12-4*

Skills Practice

Vocabulary Review

Lesson 12-3
factoring
x-intercepts

Lesson 12-4
completing the square

Assignment Guide

All students: 1–48
Additional Practice: See Extra Practice Index on page 674.

Additional Answers

31.

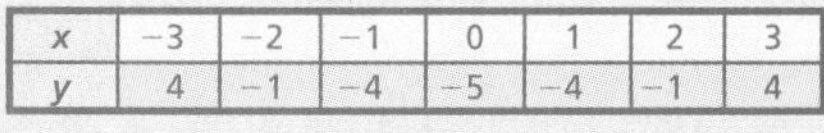

x	−3	−2	−1	0	1	2	3
y	4	−1	−4	−5	−4	−1	4

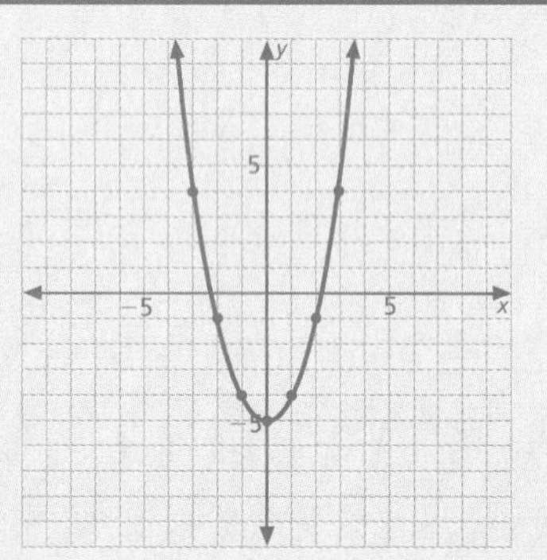

32.

x	−3	−2	−1	0	1	2	3
y	29	14	5	2	5	14	29

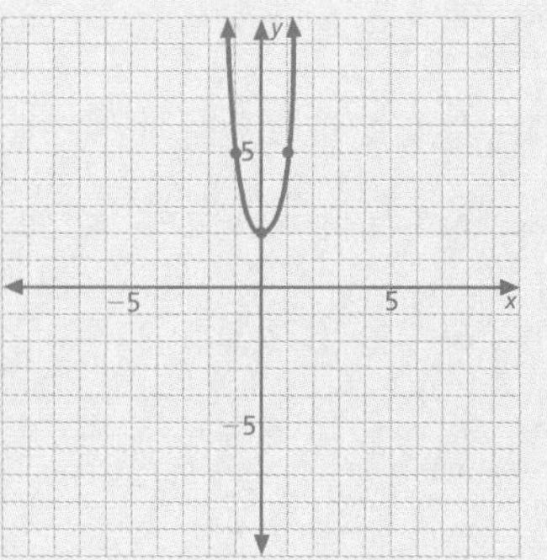

Review and Practice Your Skills

Practice ■ Lesson 12-3

Determine if each statement is *true* or *false*.

1. There are always 2 solutions to every quadratic equation. false
2. Any point at which the graph intersects the x-axis is a solution to the equation. true
3. You can find exact solutions of an equation by factoring the equation, then substituting the value of x in the equation to find y. false
4. If the parabola of the equation does not meet the x-axis, how many real solutions of the equation are there? Explain. No solutions. The only solutions of the equation are those where $y = 0$, which only occurs when the parabola meets or crosses the x-axis.

Use a graphing calculator to determine the number of solutions for each equation. For equations with one or two solutions, find the exact solutions by factoring.

5. $y = x^2 - x - 12$ $x = -3, 4$
6. $y = x^2 + 6x + 8$ $x = -2, -4$
7. $y = x^2 + 4x + 4$ $x = -2$
8. $y = x^2 - x - 6$ $x = -2, 3$
9. $y = x^2 - x - 3$ No solutions
10. $y = x^2 - 7x + 10$ $x = 2, 5$

Write an equation for each problem. Then factor to solve.

11. The square of a positive integer is 20 more than the integer. $y = x^2 - x - 20; x = 5$
12. The square of a positive integer is 9 less than 6 times the integer. $y = x^2 - 6x + 9; x = 3$
13. The square of a positive integer is 2 less than 3 times the integer. $y = x^2 - 3x + 2; x = 1, 2$
14. The square of a positive integer is 6 less than 5 times the integer. $y = x^2 - 5x + 6; x = 2, 3$

Practice ■ Lesson 12-4

15. For perfect squares in the form $ax^2 + bx + c$, what formula can be substituted for the constant c? $c = \left(\frac{b}{2}\right)^2$
16. To solve a quadratic equation by completing the square, what must be true about the x^2 term? It must have a coefficient of 1.

Complete the square.

17. $x^2 - 4x$ $x^2 - 4x + 4$
18. $x^2 + 10x$ $x^2 + 10x + 25$
19. $x^2 - 18x$ $x^2 - 18x + 81$
20. $x^2 - 14x$ $x^2 - 14x + 49$
21. $x^2 + 5x$ $x^2 + 5x + \frac{25}{4}$
22. $x^2 - 13x$ $x^2 - 13x + \frac{169}{4}$

Solve by completing the square. Check your solutions.

23. $x^2 + 6x - 3 = 0$ $x = \pm\sqrt{12} - 3$
24. $x^2 + 4x - 6 = 0$ $x = \pm\sqrt{10} - 2$
25. $x^2 - 8x + 3 = 0$ $x = \pm\sqrt{13} + 4$
26. $2x^2 + 10x = -4$ $x = \pm\sqrt{4.25} - 2.5$
27. $\frac{1}{2}x^2 - 4x + 3 = 0$ $x = \pm\sqrt{10} + 4$
28. $x^2 + 3x = 5$ $x = \pm\sqrt{7.25} - 1.5$
29. The length of a rectangle is 4 in. greater than its width. Find its dimensions if its area is 32 in.^{2}. $w = 4$ in., $\ell = 8$ in.
30. The length of a swimming pool is 3 times its width. Find its dimensions if its area is 432 ft^2. $w = 12$ ft, $\ell = 36$ ft

Learning Styles

VISUAL LEARNERS Some students may need a visual reminder that along the x-axis y will always equal zero. You may wish to have them draw a small set of axes to which to refer, calling the x-axis $y = 0$ and the y-axis $x = 0$.

PRACTICE ■ LESSON 12-1–LESSON 12-4

Copy and complete each table. Then draw the graph. (Lesson 12-1) See additional answers.

31. $y = x^2 - 5$

x	−3	−2	−1	0	1	2	3
y							

32. $y = 3x^2 + 2$

x	−3	−2	−1	0	1	2	3
y							

Find the vertex and axis of symmetry. (Lesson 12-2)

33. $y = x^2 - 3x + 4$ $v = \left(\frac{3}{2}, \frac{7}{4}\right)$; a/s $= \frac{3}{2}$

34. $y = 3x^2 + x + 3$ $v = \left(-\frac{1}{6}, \frac{35}{12}\right)$; a/s $= \frac{1}{6}$

35. $y = -2x^2 - 4x + 1$ $v = (-1, 3)$; a/s $= -1$

36. $y = -x^2 + 5x - 3$ $v = \left(\frac{5}{2}, \frac{13}{4}\right)$; a/s $= \frac{5}{2}$

37. $y = 2x^2 - 3x + 4$ $v = \left(\frac{3}{4}, \frac{23}{8}\right)$; a/s $= \frac{3}{4}$

38. $y = -x^2 - 4x - 7$ $v = (-2, -3)$; a/s $= -2$

Solve each equation by factoring or by completing the square. (Lessons 12-3–12-4)

39. $y = x^2 + 12x - 1$ $x = \pm\sqrt{37} - 6$

40. $y = x^2 - 4x + 3$ $x = 1, 3$

41. $y = x^2 + x - 12$ $x = -4, 3$

42. $y = x^2 - 5x - 14$ $x = -2, 7$

43. $y = x^2 + 10x - 4$ $x = \pm\sqrt{29} - 5$

44. $y = x^2 - 2x - 15$ $x = -3, 5$

45. $y = x^2 + 8x + 12$ $x = -6, -2$

46. $y = x^2 + 7x - 2$ $x = \pm\sqrt{14.25} - 3.5$

47. $y = x^2 - 6x - 7$ $x = -1, 7$

48. The length a rectangle is 7 inches greater than its width. Find its dimensions if its area is 144 in.^{2}. $w = 9$; $\ell = 16$

Mid-Chapter Quiz

Graph each function. Then name the vertex and the axis of symmetry, and state whether the graph opens upward or downward. See additional answers for graphs.

1. $y = 7x^2$ (0, 0); upward

2. $y = -3x^2 + 1$ (0, 1); downward

Use $x = -\frac{b}{2a}$ to find the coordinates of the vertex for each parabola.

3. $y = 2x^2 + x - 1$ $\left(-\frac{1}{4}, -1\frac{1}{8}\right)$

4. $y = -3x^2 - 2x + 2$ $\left(-\frac{1}{3}, 2\frac{1}{3}\right)$

Write the equation for the line of symmetry for each equation.

5. $y + 2x = x^2 + 4$ $x = 1$

6. $y - 3 = 3x^2 + 6x$ $x = -1$

Find the solution or solutions of each equation by factoring.

7. $y = x^2 + 2x$ −2, 0

8. $y = -x^2 - 4x - 4$ −2

9. $y = -x^2 - 3x$ −3, 0

10. $y = x^2 + 3x - 4$ −4, 1

Solve by completing the square.

11. $x^2 - 4x + 3 = 0$ $x = 1, x = 3$

12. $2x^2 + 16x - 18 = 0$ $x = -1\frac{1}{2}, x = 2\frac{1}{2}$

13. $x^2 - x = 3\frac{3}{4}$ $x = 1, x = -9$

14. $x^2 - 20x = 21$ $x = -1, x = 21$

Chalkboard Examples

Example from Lesson 12-3
Find the solution to $y = x^2 - 6x + 9$ by factoring. $x = 3$

Example from Lesson 12-4
Solve by completing the square. $x^2 - 3x = 0$ 3, 0

Mid-Chapter Quiz

Correlation Chart

Question Number	Lesson Number
1–2	12-1
3–6	12-2
7–10	12-3
11–14	12-4

ADDITIONAL ANSWERS

1.

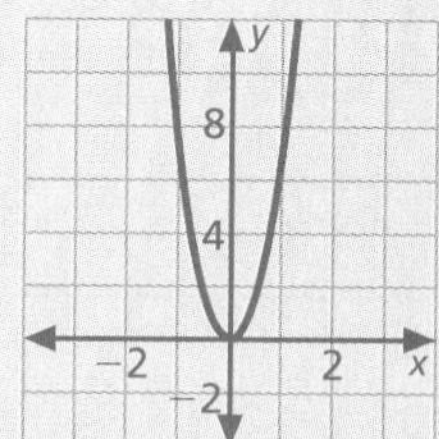

2.

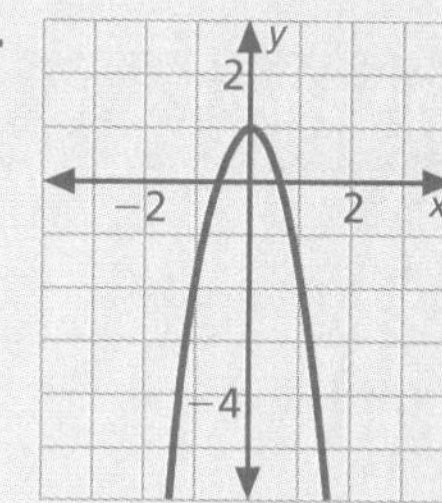

Teaching Strategies

When students are using quadratic equations to solve real-life situations, remind them to compare each factor to the problem conditions. In solving, remind them to consider which solution or solutions are appropriate. Remind them that for most real-life situations, the negative solution will have no meaning.

Chapter 12 **Review and Practice Your Skills** 539

Lesson Planning

NCTM Standards/Strands

- Number & Operation
- Patterns, Functions, & Algebra
- Problem Solving
- Reasoning and Proof
- Communication
- Connections
- Representation

Vocabulary

quadratic formula

Materials Needed

paper/pencil

Lesson Resources

Warm-Up Transparency 35
Transparency TK-7–10, RF-47
Reteaching 12-5
Extra Practice 12-5
Enrichment 12-5

Assignment Guide

Basic: 1–30, 35–48
Enriched: 1–48

Getting Started

5-Minute Warm-up

Simplify each expression.

1. $\sqrt{36}$ 6
2. $\sqrt{40}$ $2\sqrt{10}$
3. $\frac{6 - \sqrt{12}}{2}$ $3 - \sqrt{3}$

Introduction to Lesson 12-5

As students list each of the steps for completing the square, have them write or describe what each step accomplishes. Students should understand that while the method can always by used, it requires great care, since it is lengthy and provides many chances for computational error.

12-5 The Quadratic Formula

Goals ■ Solve equations using the quadratic formula.

Applications Aeronautics, Skydiving, Physics

Work in groups of 2–3 students.

By looking for relationships among the variables, coefficients and constants in equations, we can write general rules or steps for solving all equations of the same form.

1. Work together to write a list of steps for solving any quadratic equation in the form $ax^2 + bx + c = 0$ by completing the square.
2. Check your steps by using them to solve the equation $x^2 - x - 3\frac{3}{4}$.

$x = -1\frac{1}{2}$, $x = 2\frac{1}{2}$

1. Add $-c$ to each side. Multiply by $\frac{1}{a}$. Add $\frac{-b^2}{2a}$ to each side. Factor. Find square roots. Simplify.

Build Understanding

Any quadratic equation can be solved by completing the square; however, repeating the steps from the activity above for each equation you solve can be a lengthy process. Instead of repeating the steps, you can use the general quadratic equation $ax^2 + bx + c = 0$ to develop a formula for solving quadratic equations. The formula is found by solving the general quadratic equation for x.

$$ax^2 + bx + c = 0$$

$$x^2 + \frac{b}{a}x + \frac{c}{a} = 0 \qquad \text{Multiply each term by } \tfrac{1}{a}.$$

$$x^2 + \frac{b}{a}x = -\frac{c}{a} \qquad \text{Add } -\tfrac{c}{a} \text{ to each side of the equation.}$$

$$x^2 + \frac{b}{a}x + \frac{b^2}{4a^2} = \frac{b^2}{4a^2} - \frac{c}{a} \qquad \text{Half of } \tfrac{b}{a} = \tfrac{b}{2a}. \text{ Add } \left(\tfrac{b}{2a}\right)^2 \text{ or } \tfrac{b^2}{4a^2} \text{ to each side of the equation.}$$

$$\left(x + \frac{b}{2a}\right)^2 = \sqrt{\frac{b^2 - 4ac}{4a^2}} \qquad \text{Factor the left side of the equation. Combine terms on the right side of the equation.}$$

$$x + \frac{b}{2a} = \pm\sqrt{\frac{b^2 - 4ac}{4a^2}} \qquad \text{Find the square roots.}$$

$$x = -\frac{b}{2a} \pm \sqrt{\frac{b^2 - 4ac}{2a}} \qquad \text{Subtract } \tfrac{b}{2a} \text{ from each side of the equation.}$$

$$x = -\frac{b \pm \sqrt{b^2 - 4ac}}{2a} \qquad \text{Simplify.}$$

The formula for x in terms of a, b, and c is called the **quadratic formula**.

$$x = \frac{-b \pm \sqrt{b^2 - 4ac}}{2a}$$

The quadratic formula can be used to solve any quadratic equation in the form $ax^2 + bx + c = 0$, $a \neq 0$.

Teaching Strategies

Students may benefit from a discussion of the most efficient sequence to use in evaluating the quadratic formula. If students are using calculators, one option is to work from left to right by first entering the value of $-b$ and then pressing M+. They can then evaluate the radical and either add it or subtract it from the value of $-b$, using either M+ or M−. Finally, they can divide by $2a$.

Example 1

Use the quadratic formula to solve $6x^2 - 5x + 1 = 0$.

Solution

$$x = -b \pm \frac{\sqrt{b^2 - 4ac}}{2a}$$

$$x = \frac{-(-5) \pm \sqrt{(-5)^2 - 4(6)(1)}}{2(6)} \quad \text{Substitute for } a, b, \text{ and } c.$$

$$x = \frac{5 \pm \sqrt{25 - 24}}{12} \quad \text{Simplify.}$$

$$x = \frac{5 \pm \sqrt{1}}{12} = \frac{5 \pm 1}{12}$$

$$x = \frac{5 + 1}{12},\ x = \frac{5 - 1}{12}$$

$$x = \frac{6}{12} = \frac{1}{2},\ x = \frac{4}{12} = \frac{1}{3}$$

The equation $6x^2 - 5x + 1 = 0$ has two solutions, $\frac{1}{2}$ and $\frac{1}{3}$.

The radical part of the solutions is in simplified form if it contains no factors that are perfect squares.

Problem Solving Tip

Remember the product property of square roots:

For any non-negative real numbers a and b,

$\sqrt{a} \times \sqrt{b} = \sqrt{ab}$

so, $\sqrt{12} = \sqrt{4 \times 3} = \sqrt{4} \times \sqrt{3} = 2\sqrt{3}$

Example 2

Use the quadratic formula to solve $x^2 = 6x + 3$.

Solution

$$x^2 - 6x - 3 = 0 \quad \text{Write the equation in standard form.}$$

$$x = \frac{-b \pm \sqrt{b^2 - 4ac}}{2a} \quad \text{Use the quadratic formula.}$$

$$x = \frac{-(-6) \pm \sqrt{(-6)^2 - 4(1)(-3)}}{2(1)} \quad \text{Substitute for } a, b, \text{ and } c.$$

$$x = \frac{6 \pm \sqrt{36 + 12}}{2} \quad \text{Simplify.}$$

$$x = \frac{6 \pm \sqrt{48}}{2} = \frac{6 \pm \sqrt{16(3)}}{2}$$

$$x = \frac{6 \pm 4\sqrt{3}}{2} = 3 \pm 2\sqrt{3}$$

The solutions for the quadratic equation $x^2 = 6x + 3$ are $x = 3 + 2\sqrt{3}$ and $x = 3 - 2\sqrt{3}$.

You can use your knowledge of quadratic equations to solve many situations involving distance. For example, gravity acts on a freely falling object according to the formula $d = 16t^2$. Using this formula, you can find the time it takes for an object to fall a certain distance.

Now consider the path of a projectile. Velocity, the force applied to the object, and gravity act upon the object. The path of a projectile is described by the formula $d = vt - 16t^2$, where d = distance (ft), v = velocity (ft/sec) and t = time (seconds).

Chalkboard Examples

Supplementary Example 1

Use the quadratic formula to solve $x^2 - 2x - 24 = 0$. 6, −4

Supplementary Example 2

Use the quadratic formula to solve $x^2 - 8x - 2 = 0$. $4 \pm 3\sqrt{2}$

Supplementary Example 3

A model rocket is launched at a velocity of 120 feet per second.

a. How long does it take the rocket to reach its maximum height? 3.75 seconds

b. What is the maximum height reached by the rocket? 225 feet

c. From the time the rocket is launched, how many seconds does it take the rocket to return to the ground? 7.5 seconds

Extend the Lesson

COOPERATIVE LEARNING Have students work in pairs to solve this number problem. Find two consecutive numbers such that twice the square of the first is 26 more than nine times the second. 7, 8

Reteaching Worksheet 12-5

Name ______________________ Date ____________

RETEACHING 12-5

THE QUADRATIC FORMULA

The **quadratic formula** can be used to solve all types of quadratic equations in the form $ax^2 + bx + c = 0$, $a \neq 0$, and is stated in these terms:

If $ax^2 + bx + c = 0$, then $x = \frac{-b \pm \sqrt{b^2 - 4ac}}{2a}$.

Example 1

Use the quadratic formula to solve $2x^2 - 3x - 2 = 0$.

Solution

$x = \frac{-(-3) \pm \sqrt{(-3)^2 - 4(2)(-2)}}{2(2)}$

$x = \frac{3 \pm \sqrt{9 + 16}}{4}$

$x = \frac{3 \pm \sqrt{25}}{4}$

$x = \frac{3 \pm 5}{4}$

$x = \frac{3 + 5}{4}$, or 2 $\quad x = \frac{3 - 5}{4}$, or $-\frac{1}{2}$

There are two solutions: 2 and $-\frac{1}{2}$

Example 2

Use the quadratic formula to solve $x^2 - 4x - 10 = 0$.

Solution

$x = \frac{-(-4) \pm \sqrt{(-4)^2 - 4(1) - (10)}}{2(1)}$

$x = \frac{4 \pm \sqrt{16 + 40}}{2}$

$x = \frac{4 \pm \sqrt{56}}{2}$

$x = \frac{4 \pm \sqrt{4(14)}}{2} = \frac{4 \pm 2\sqrt{14}}{2} = 2 \pm \sqrt{14}$

There are two solutions: $x = 2 + \sqrt{14}$ and $x = 2 - \sqrt{14}$

EXERCISES

Use the quadrataic formula to solve each equation. Remember, some equations may have only one solution.

1. $x^2 - 7x + 10 = 0$ $x = 5, x = 2$
2. $6x^2 - 4x - 10 = 0$ $x = \frac{5}{3}, x = -1$
3. $10x^2 - 6x - 4 = 0$ $x = -\frac{2}{5}, x = 1$
4. $x^2 + 20x + 100 = 0$ $x = -10$
5. $x^2 - 4x - 12 = 0$ $x = 6, x = -2$
6. $2x^2 + 10x + 12 = 0$ $x = -2, x = -3$
7. $x^2 + 10x - 12 = 0$ $x = -5 \pm \sqrt{37}$
8. $x^2 - 6x - 2 = 0$ $x = 3 \pm \sqrt{11}$
9. $x^2 - 14x - 12 = 0$ $x = 7 \pm \sqrt{61}$
10. $x^2 + 4x - 1 = 0$ $x = -2 \pm \sqrt{5}$
11. $2x^2 + 2x - 24 = 0$ $x = 3, x = -4$
12. $x^2 - 14x + 25 = 0$ $x = 7 \pm 2\sqrt{6}$
13. $x^2 + 30x + 225 = 0$ $x = -15$
14. $2x^2 + 8x + 4 = 0$ $x = -2 \pm \sqrt{2}$

186 RETEACHING *LESSON 12-5*

Lesson 12-5 **The Quadratic Formula** 541

Lesson Wrap-up

Quick Assessment

Ask the following questions to determine if students understand the content presented in this lesson.

1. Why does an equation in which $b^2 - 4ac = 0$ have only one solution? When $b^2 - 4ac = 0$, the only term in the solution is $\frac{-b}{2a}$.
2. In the quadratic equation $x + 7 = -2x^2$, what are the values of a, b, and c. $a = -2, b = -1, c = -7$ or $a = 2, b = 1, c = 7$

Assignment Guide

Basic: 1–30, 35–48
Advanced: 1–48
Additional Practice: See Extra Practice Index on page 674.

Example 3

AERONAUTICS A model rocket is launched at a velocity of 80 ft/sec.

a. How long does it take the rocket to reach its maximum height?
b. What is the maximum height reached by the rocket?
c. How many seconds does it take the rocket to return to the ground?

Solution

The velocity is 80 feet per second, so $d = 80t - 16t^2$.

a. At its maximum height, the rocket is at the vertex of the parabola formed by the flight path. The time required to reach the maximum height is the x-coordinate of the vertex. Use $-\frac{b}{2a}$ to find the time.

$$-\frac{80}{2(-16)} = \frac{80}{32} = 2.5$$

It takes the rocket 2.5 seconds to reach its maximum height.

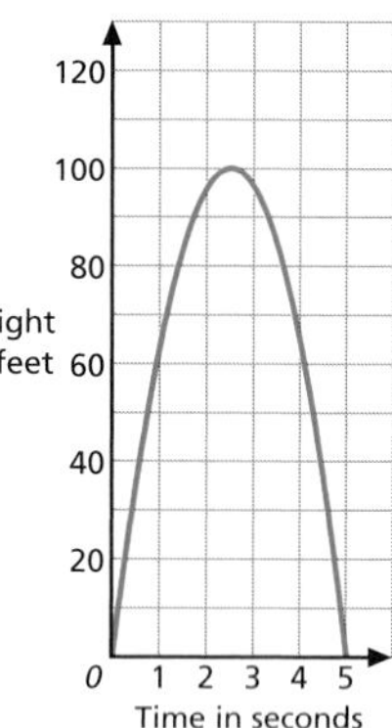

b. The maximum height is the y-coordinate of the vertex. Substitute $t = 2.5$ into the equation.

$$d = 80(2.5) - 16(2.5)^2$$
$$d = 200 - 16(6.25)$$
$$d = 200 - 100 = 100$$

The maximum height reached by the rocket is 100 feet.

c. At ground level, d is 0. To find when the rocket returns to the ground, solve the equation.

$$80t - 16t^2 = 0$$
$$5t - t^2 = 0$$
$$t(5 - t) = 0 \quad t = 0, t = 5$$

The rocket returns to ground level in 5 seconds.

Try These Exercises

Use the quadratic formula. Solve each equation.

1. $x^2 - 4x + 1 = 0$ $2 \pm \sqrt{3}$
2. $2x^2 + 15x = 8$ $\frac{1}{2}, -8$
3. $4x^2 = 8x$ $0, 2$
4. $-6x = -3x^2 - 3$ 1
5. $x^2 + 2x = 6$ $1 \pm \sqrt{7}$
6. $7x - 15 = -2x^2$ $1.5, -5$

Practice Exercises

Use the quadratic formula to solve each equation.

7. $x^2 - 9x = -14$ $2, 7$
8. $2x^2 = 6 - x$ $\frac{3}{2}, -2$
9. $4x^2 + 4x + 1 = 0$ $-\frac{1}{2}$
10. $x^2 + 2x = 11$ $-1 \pm 2\sqrt{3}$
11. $14x = 2x^2$ $0, 7$
12. $3x^2 = 84 - 9x$ $4, -7$
13. $2x^2 + 32 = 16x$ 4
14. $x^2 = 50$ $\pm 5\sqrt{2}$
15. $x^2 + 4x + 2 = 0$ $-2 \pm \sqrt{2}$
16. $x^2 + 7x = -12$ $-3, -4$
17. $45 = 2x^2 + x$ $4\frac{1}{2}, -5$
18. $x^2 - 8 = 0$ $\pm 2\sqrt{2}$

Teaching Strategies

CHALLENGE Ask students to try to solve for x in the following equation using the quadratic formula: $x^2 - 3x + 25 = 0$. They will discover that when they evaluate $b^2 - 4ac$, they will be left with a negative number beneath the radical sign and cannot find a solution among the real numbers. You may wish to mention that a solution does exist, but it involves imaginary numbers.

Choose factoring or the quadratic formula to solve each equation.

19. $x^2 + 2x = 35$ 5, −7 **20.** $2x - 4 = -x^2$ $-1 \pm \sqrt{5}$ **21.** $x^2 = -3x$ 0, −3

22. $x^2 - 2x = 7$ $1 \pm 2\sqrt{2}$ **23.** $x - 21 = -2x^2$ $-\frac{7}{2}, 3$ **24.** $x = x^2$ 0, 1

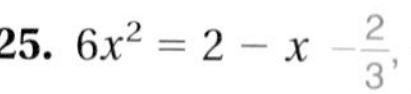

25. $6x^2 = 2 - x$ $-\frac{2}{3}, \frac{1}{2}$ **26.** $x^2 - 10x = 3$ $5 \pm 2\sqrt{7}$ **27.** $x^2 = 12x - 11$ 1, 11

28. WRITING MATH Suppose Galileo gathered the data shown. Explain how Galileo might use this data to show that free-fall distance is a function of the square of time.
The distance is equal to a constant (16) times the time squared.

Time (seconds)	Distance (feet)
0	0
1	16
2	64
3	144
4	256

$16 = 16(1^2)$
$64 = 16(2^2)$
$144 = 16(3^2)$
$256 = 16(4^2)$

29. SKYDIVING A parachutist free-falls for 20 sec. How far is the free-fall? 6,400 ft

30. PHYSICS How many times longer does a projectile launched at 200 ft/sec stay in the air than one launched at 100 ft/sec?
2 times

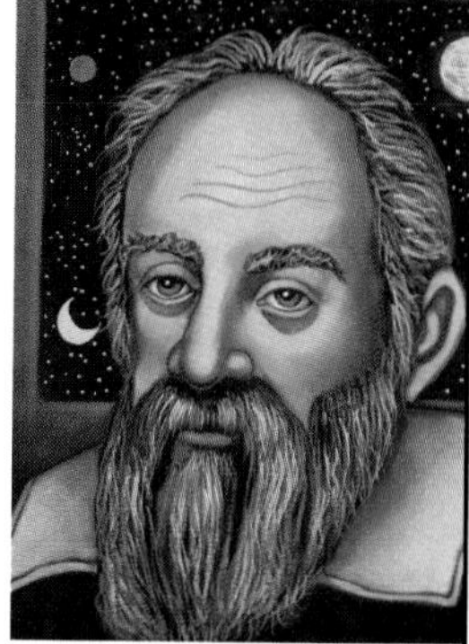

Math: Who, Where, When

Italian physicist and astronomer Galileo Galilei (1564–1642) probably did not drop cannonballs from the Leaning Tower of Pisa. However, from his research came the quadratic law of falling bodies.

Extended Practice Exercises

31. AERONAUTICS A model rocket is launched at an initial velocity of 96 ft/sec.

a. How long will it take the rocket to reach its maximum height? 3 sec

b. What is the maximum height reached by the rocket? 144 ft

c. How many seconds does it take the rocket to return to the ground? 6 sec

32. PHYSICS An object free-falls for 15 seconds; another free-falls for 30 sec. How many times farther does the second object fall than the first? 4 times

33. DATA INDEX Use the Data Index on pages 632–633 to locate information about the heights of some bridges. How long would it take a stone dropped from each bridge to reach the water below? Find your answer to the nearest tenth of a second. Firth of Forth Bridge, 3.0 sec; Verrazzano Narrows Bridge, 3.6 sec; Sydney Harbor Bridge, 3.3 sec; Tunkhannock Viaduct 3.9 sec; Garabit Viaduct, 5.5 sec; Brooklyn Bridge, 2.9 sec.

34. PHYSICS A ball is thrown up into the air from the roof of a 128-ft-tall building. The initial velocity is 64 ft/sec. How long will it take the ball to reach the ground? about 6.8 sec

Mixed Review Exercises

Find factors for the following. (Lesson 11-5)

35. $2a^2 - 5ab + 3b^2$ $(a - b)(2a - 3b)$ **36.** $18x^2 - 3xy - 6y^2$ $3(3x - 2y)(2x + y)$ **37.** $16a^2 - 16ab + 4b^2$ $4(2a - b)(2a - b)$

38. $8x^2 + 18xy + 9y^2$ $(4x + 3y)(2x + 3y)$ **39.** $6m^2 - 7mn - 20n^2$ $(3m + 4n)(2m - 5n)$ **40.** $36a^2 - 12ab + 8b^2$ $4(3a + b)(3a - 2b)$

41. $24a^3 + 16a^2b - 8ab^2$ $8a(3a + b)(a - b)$ **42.** $10z^2 - 26xz - 12x^2$ $2(z - 3x)(5z + 2x)$ **43.** $8m^2 - 10mn - 12n^2$ $2(m - 2n)(4m + 3n)$

44. $49a^3 - 28ab + 4b^2$ $(7a - 2b)(7a - 2b)$ **45.** $6x^2 + 4xy - 2y^2$ $2(3x - y)(x + 4)$ **46.** $3a^2 + 2ab - 5b^2$ $(3a + 5b)(a - b)$

Solve each system of equations. (Lesson 6-6)

47. $\begin{cases} 3x - 2y = -14 \\ 4x + 2y = 0 \end{cases}$ (−2, 4) **48.** $\begin{cases} 3y + x = 10 \\ y - 2x = 1 \end{cases}$ (1, 3)

Extra Practice Worksheet 12-5

Name ______________ Date ______________

EXTRA PRACTICE 12-5
THE QUADRATIC FORMULA

EXERCISES

Use the quadratic formula to solve each equation.

1. $x^2 - 3x = -2$ −1, −2 2. $4x^2 + 4x + 1 = 0$ $-\frac{1}{2}$
3. $2x^2 + x - 4 = 0$ $\frac{-1 \pm 33}{2}$ 4. $3x^2 + 2x - 3 = 0$ $\frac{-2 \pm \sqrt{13}}{6}$
5. $x^2 - 6x = 9$ $-3 \pm 3\sqrt{2}$ 6. $12x = 5x^2$ $0, \frac{12}{5}$
7. $x^2 = 40$ $\pm 2\sqrt{10}$ 8. $2x^2 + 6x - 5 = 0$ $\frac{-3 \pm \sqrt{19}}{2}$
9. $x^2 - 18 = 0$ $\pm 3\sqrt{2}$ 10. $16 = 4x^2 + 4x$ $\frac{-1 \pm \sqrt{17}}{2}$
11. $x^2 + 7x - 8 = 0$ 1, −8 12. $4x^2 - 6x - 1 = 0$ $\frac{3 \pm \sqrt{13}}{4}$
13. $5x^2 + 2x = 10$ $\frac{-1 \pm \sqrt{51}}{5}$ 14. $x^2 - 7x + 2 = 0$ $\frac{7 \pm \sqrt{41}}{2}$

Choose factoring or the quadratic formula to solve each equation.

15. $x^2 + 4x + 3 = 0$ −1, −3 16. $2x^2 - 7x - 15 = 0$ $5, -\frac{3}{2}$
17. $x^2 - 5x + 4 = 0$ 1,4 18. $5x^2 + 3x - 6 = 0$ $\frac{-3 \pm \sqrt{129}}{10}$
19. $x^2 = 9$ −3, 3 20. $10x + 4 = 5x^2$ $\frac{5 \pm 3\sqrt{5}}{5}$
21. $x^2 = -4x$ 0, −4 22. $7x^2 - 16x - 2 = 0$ $\frac{8 \pm \sqrt{78}}{7}$
23. $x^2 - 12 = 0$ $\pm 2\sqrt{3}$ 24. $12 = x^2 + x$ 3, −4
25. $x^2 + 6x - 5 = 0$ $-2 \pm \sqrt{14}$ 26. $3x^2 - 2x - 1 = 0$ $1, -\frac{1}{3}$

164 EXTRA PRACTICE *LESSON 12-5*

Enrichment Worksheet 12-5

Name ______________ Date ______________

ENRICHMENT 12-5
COMPLETING THE MAGIC SQUARE

EXERCISES

In a magic square, all the rows, columns and diagonals add up to the same total.

The numbers 1 through 81 can be arranged to make four nested magic squares. First, the large 9-by-9 outer square is magic. Remove its border and you get a 7-by-7 magic square. Remove its border and you get a 5-by-5 magic square. Finally, remove the border again to get a 3-by-3 magic square.

Insert the rest of the numbers 1 through 81 in the border cells to complete the nested magic squares.

5	80	59	73	61	3	63	12	13
1	2	5	3	5	2	7	2	81
4	1	3	5	2	6	3	6	78
76	5	4	3	4	4	3	2	6
7	6	3	4	4	3	4	1	75
74	6	4	4	3	4	3	1	8
67	1	4	3	5	2	5	7	15
66	5	2	5	2	5	1	6	16
69	2	23	9	21	79	19	70	77

210 ENRICHMENT *LESSON 12-5*

Teaching Strategies

CHALLENGE The formula $d = vt - 16t^2$ can be applied using units other than feet and feet per second. When using meters and meters per second, the formula is $d = vt - 5t^2$. Ask students to explain the difference and to use their ideas to rewrite the formula in other units.

Lesson Planning

NCTM Standards/Strands

- Number & Operation
- Patterns, Functions, & Algebra
- Geometry & Spatial Sense
- Measurement
- Problem Solving
- Reasoning and Proof
- Communication
- Connections
- Representation

Vocabulary

distance formula
midpoint formula

Materials Needed

paper/pencil

Lesson Resources

Warm-Up Transparency 36
Transparency TK-7–10, RF-48
Reteaching 12-6
Extra Practice 12-6
Enrichment 12-6

ASSIGNMENT GUIDE

Basic: 1–25, 29–40
Enriched: 1–40

Getting Started

5-MINUTE WARM-UP

Use the Pythagorean Theorem. Find each hypotenuse to the nearest tenth.

1. leg = 14, leg = 24 27.8
2. leg = 40, leg = 41 57.3
3. leg = 17, leg = 13 21.4

Introduction to Lesson 12-6

Have students complete Parts (1) through (4). Then have them draw another coordinate grid, labeling the x-axis and y-axis to create all four quadrants. Have them draw a right triangle with one vertical and one horizontal leg in each quadrant. Tell them to repeat the questions in Parts (2) through (4) for each triangle. Then discuss any similarities or difference among working in each of the four quadrants.

12-6 Use the Pythagorean Theorem

Goals ■ Use the Pythagorean Theorem, distance and midpoint formulas.

Applications Space Exploration, Sports, Archeology

Work with a partner.

1. Using the coordinate plane shown below, count squares to find the length and midpoint of each leg of the right triangle.

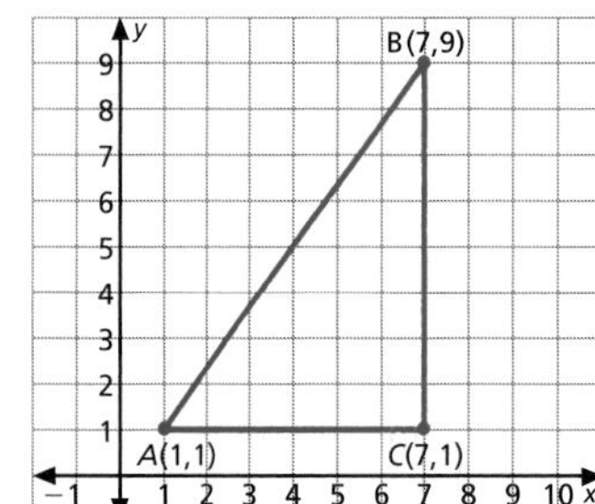

AC 6, (4,1); BC 8, (7,5)

Problem Solving Tip

From the Pythagorean Theorem, $c^2 = a^2 + b^2$. Solving for c,

$c = \sqrt{a^2 + b^2}$.

2. Note that $\overline{BC}$ is parallel to the y-axis, and $\overline{AC}$ is parallel to the x-axis. Answer these questions:
   a. How can you calculate the length of BC given the y-coordinates of B and C? See additional answers.
   b. The midpoint of $\overline{BC}$ is half the distance between endpoints B and C. How can you calculate the y-coordinate of the midpoint given the y-coordinates of B and C? See additional answers.
   c. How can you calculate the length of AC given the x-coordinates of A and C? See additional answers.
   d. How can you calculate the x-coordinate of the midpoint of $\overline{AC}$?
3. Use the Pythagorean Theorem to find the length of hypotenuse AB. 10
4. The midpoint of $\overline{AB}$ is (4, 5). What is the relationship between this point and the midpoints of the two legs? See additional answers.

BUILD UNDERSTANDING

A formula for calculating the distance, d, between any two points on the coordinate plane may be derived using the Pythagorean Theorem.

For any two points (x_1, y_1) and (x_2, y_2), a right triangle formed by drawing horizontal and vertical segments that intersect at (x_2, y_1) has legs with lengths $|x_2 - x_1|$ and $|y_2 - y_1|$.

These lengths may be substituted in the Pythagorean Theorem to find d.

$$d = \sqrt{(x_2 - x_1)^2 + (y_2 - y_1)^2}$$

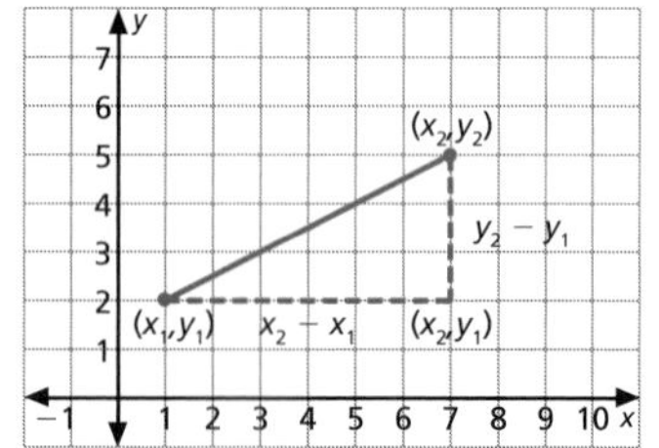

Teaching Strategies

Students may benefit from drawing line segments on centimeter graph paper. After they have calculated the length of the line segment using the distance formula, they can measure the segment using a centimeter ruler to verify their work. This will reinforce the concept of checking the reasonableness of a solution.

The distance, d, between any two points (x_1,y_1) and (x_2,y_2) may be found using the **distance formula.**

$$d = \sqrt{(x_2 - x_1)^2 + (y_2 - y_1)^2}$$

Example 1

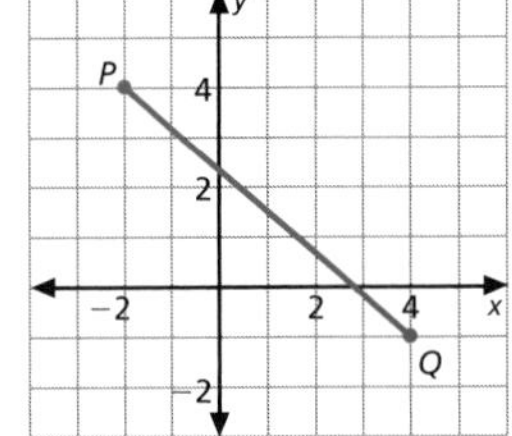

SPACE EXPLORATION Suppose the grid shown was superimposed over a photograph taken by a space probe of the Martian landscape. Calculate the distance between points $P(-2,4)$ and $Q(4,-1)$.

Solution

$$d = \sqrt{(x_2 - x_1)^2 + (y_2 - y_1)^2}$$
$$d = \sqrt{(-2 - 4)^2 + (4 - (-1))^2}$$
$$d = \sqrt{(-6)^2 + 5^2}$$
$$d = \sqrt{36 + 25} = \sqrt{61} \approx 7.8$$

The distance between points P and Q is $\sqrt{61}$ or about 7.8 units.

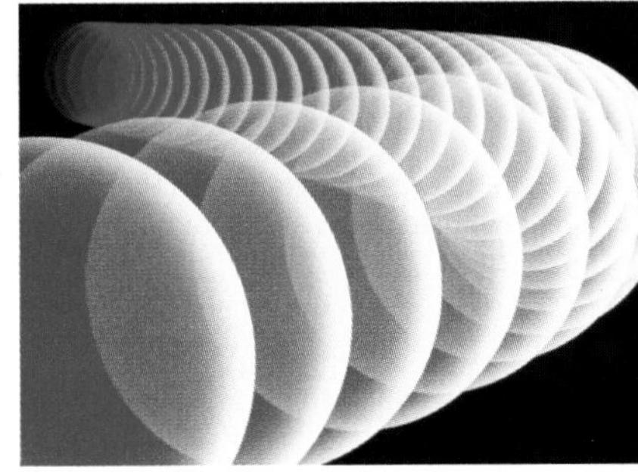

In the activity at the beginning of the lesson, you observed that for a segment parallel to the x-axis, the x-coordinate of the midpoint is $\frac{(x_1 + x_2)}{2}$. For a segment parallel to the y-axis, the y-coordinate of the midpoint is $\frac{(y_1 + y_2)}{2}$. We can use these two facts to derive the **midpoint formula.** For a line segment with endpoints (x_1, y_1) and (x_2, y_2), the coordinates of the midpoint are $\left(\frac{x_1 + x_2}{2}, \frac{y_1 + y_2}{2}\right)$.

Example 2

Find the midpoint of the line segment with endpoints $M(-2,5)$ and $N(6,-3)$.

Solution

$\left(\frac{x_1 + x_2}{2}, \frac{y_1 + y_2}{2}\right)$	Use the midpoint formula.
$\left(\frac{-2 + 6}{2}, \frac{5 + (-3)}{2}\right)$	Substitute coordinate values.
$\left(\frac{4}{2}, \frac{2}{2}\right) = (2, 1)$	Simplify.

The midpoint of line segment MN is (2, 1).

Distances on aerial and satellite photos are often estimated using a coordinate grid. Even sporting events require these calculations. Suppose you are working for a television company broadcasting a major league baseball game. A blimp is transmitting overhead shots of the field at all times. Using a computer, you could instantly overlay each visual with a grid and calculate distances for the announcers to use in the broadcast.

Additional Answers

2a. The y-coordinates are 9 and 1. The length of BC is 9 − 1 or 8 units.

2b. The y-coordinate of the midpoint is $\frac{(9 + 1)}{2} = 5$. The x-coordinate is the same as the x-coordinate of the endpoints, 7. The midpoint is (7, 5).

2c. The x-coordinates are 1 and 7. The length of AC is 7 − 1 or 6 units. The x-coordinate of the midpoint is $\frac{(1 + 7)}{2} = 4$. The y-coordinate is the same as the y-coordinate of the endpoints, 1. The midpoint is (4, 1).

4. The x-coordinate is the x-coordinate of the midpoint AC. The y-coordinate is the y-coordinate of the midpoint of BC.

Chalkboard Examples

Supplementary Example 1

A coordinate grid is drawn on a satellite photo of the Earth. Calculate the distance between points $A(-1, 3)$ and $B(-4, -6)$ to the nearest tenth unit. **about 9.5 units**

Supplementary Example 2

Find the midpoint of the line segment with endpoints $R(-7, 3)$ and $T(-5, 3)$. **(−6, 3)**

Supplementary Example 3

Using the diagram from Example 3, a fly ball is hit from home plate to point P. What is the approximate distance? **about 8.6 units**

Reteaching Worksheet 12-6

Name ______________________ Date ____________

RETEACHING 12-6

USING THE PYTHAGOREAN THEOREM

The distance, d, between any two points (x_1, y_1) and (x_2, y_2) on the coordinate plane is given by the distance formula.

$$d = \sqrt{(x_2 - x_1)^2 + (y_2 - y_1)^2}$$

The midpoint (x_m, y_m), of a line segment with endpoints (x_1, y_1) and (x_2, y_2) is given by this formula.

$$(x_m, y_m) = \left(\frac{x_1 + x_2}{2}, \frac{y_1 + y_2}{2}\right)$$

Example 1

Use the distance formula to find the distance between point $A(3, -1)$ and point $B(2, -2)$.

Solution

$$d = \sqrt{(x_2 - x_1)^2 + (y_2 - y_1)^2}$$
$$= \sqrt{(-3 - 2)^2 + (1 - -2)^2}$$
$$= \sqrt{(-5)^2 - (3)^2}$$
$$= \sqrt{25 + 9}$$
$$= \sqrt{34} \approx 5.8$$

The distance between points A and B is approximately 5.8 units.

Example 2

Use the midpoint formula to find the midpoint of the line segment with endpoints $C(-2, 1)$ and $D(6, -3)$.

Solution

$$(x_m, y_m) = \left(\frac{x_1 + x_2}{2}, \frac{y_1 + y_2}{2}\right)$$
$$= \left(\frac{-2 + 6}{2}, \frac{1 - 3}{2}\right)$$
$$= \left(\frac{4}{2}, \frac{-2}{2}\right)$$
$$= (2, -1)$$

The midpoint of $\overline{CD}$ is (2, −1).

EXERCISES

Use the distance formula to find the distance between each pair of points. If necessary, round your answer to the nearest tenth.

1. (0, 3), (−2, 0) $\sqrt{13} \approx 3.6$
2. (−3, 2), (−4, 7) $\sqrt{26} \approx 5.1$
3. (−4, 2), (−2, −3) $\sqrt{29} \approx 5.4$
4. (−2, −3), (−1, −4) $\sqrt{2} \approx 1.4$
5. (2, 7), (−3, −5) $\sqrt{169} = 13$
6. (2, 7), (3, 5) $\sqrt{5} \approx 2.2$
7. (−3, 4), (5, 2) $\sqrt{68} \approx 8.2$
8. (4, 5), (−3, −1) $\sqrt{85} \approx 9.2$
9. (4, 3), (3, −6) $\sqrt{82} \approx 9.1$

Use the midpoint formula to find the midpoint between each pair of points.

10. (4, 5), (2, 3) (3, 4)
11. (3, 5), (7, 1) (5, 3)
12. (−4, 8), (2, −4) (−1, 2)
13. (−5, 3), (1, 5) (−2, 4)
14. (6, 5), (−4, −3) (1, 1)
15. (−1, −2), (−3, −2) (−2, −2)
16. (0, −7), (8, −1) (4, −4)
17. (−9, −8), (−3, 6) (−6, −1)
16. (−10, 5), (6, −3) (−2, 1)

188 RETEACHING *LESSON 12-6*

Example 3

SPORTS On the grid shown at the right, one unit represents 30 ft. A batter hits a ball from home plate to point L. What is the distance?

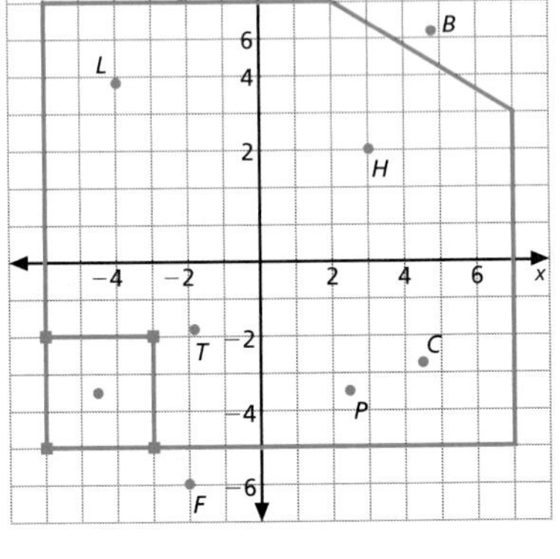

Solution

The coordinates of home plate are (−6, −5). Point L is located at approximately (−4, 5). Use the distance formula to find the distance between the points.

$$d = \sqrt{(x_2 - x_1)^2 + (y_2 - y_1)^2}$$

$$d = \sqrt{(-4 - (-6))^2 + (5 - (-5))^2}$$

$$d = \sqrt{2^2 + 10^2}$$

$$d = \sqrt{4 + 100} = \sqrt{104} \approx 10.2 \text{ units}$$

Since each unit equals 30 ft, multiply by 30: 10.2 × 30 = 306 ft. The ball traveled about 306 ft.

Try These Exercises

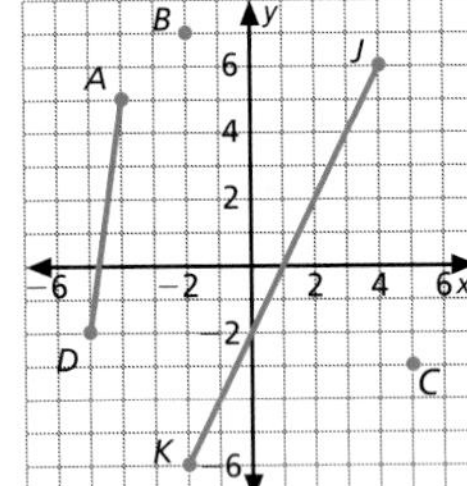

Use the graph on the right. Round distances to the nearest tenth of a unit.

1. Find the length of segment *AD*. 7.1 units
2. Find the distance between *B* and *C*. 12.2 units
3. Locate the midpoint of segment *JK*. (1, 0)

Use the distance formula to calculate the distance between each pair of points.

4. $G(-9, 4)$, $H(-5, -1)$ 6.4 units
5. $P(3, -2)$, $Q(-2, 11)$ 13.9 units
6. $M(0, 0)$, $N(6, -7)$ 9.2 units

Use the midpoint formula to calculate the midpoint between each pair of points.

7. $E(5, 5)$, $F(-3, -3)$ (1, 1)
8. $L(-10, 4)$, $M(2, 6)$ (−4, 5)
9. $T(-9, 4)$, $S(1, -1)$ (−4, 1.5)
10. **SPORTS** Use the baseball field diagram on page 545. One unit on the grid equals 30 ft. If a ball is caught at point H and thrown to second base at coordinates (−3, −2), how long is the throw? About 234 ft

Practice Exercises

Calculate the distance between each pair of points.

11. $W(20, -1)$, $X(-6, 5)$ 26.7 units
12. $C(8, 5)$, $D(-3, -6)$ 15.6 units
13. $G(-1, 5)$, $H(2, 6)$ 3.2 units
14. $P(5, 5)$, $Q(-6, 1)$ 11.7 units

Lesson Wrap-up

Quick Assessment

Ask the following questions to determine if students understand the content presented in this lesson.

1. Can the distance and midpoint formulas be used for two points, both on the same axis? Yes, for two points on the *x*-axis, y_1 and y_2 are both zero; for two points on the *y*-axis, x_1 and x_2 are both zero. Students should recognize that while the formulas can be used, it is generally not necessary.
2. How can you apply the distance formula to find the length of any two horizontal or vertical lines on the coordinate plane? For any horizontal line, the distance is $\sqrt{(x_2 - x_1)^2}$. For any vertical line, the distance is $\sqrt{(y_2 - y_1)^2}$

Assignment Guide

Basic: 1–25, 29–40
Advanced: 1–40
Additional Practice: See Extra Practice Index on page 674.

Extend the Lesson

CONNECTING TO PRIOR KNOWLEDGE Help students relate the concept of midpoint of a line to finding an average. Remind them that when finding an average, the first step is to add the values. This applies to finding a midpoint, which is really an average of the two endpoints.

Calculate the midpoint between each pair of points.

Mesa Verde National Park Cliff Dwellings, Colorado

15. $E(16, -5)$, $F(4, 1)$ (10, −2)

16. $Y(-10, -9)$, $Z(-2, -3)$ (−6, 6)

17. ARCHEOLOGY The diagram of an archeological site is drawn on a coordinate grid. An archeologist notes that standing stones are placed at $A(-3, -1)$, $B(1, 3)$, and $C(7, -3)$. The stones are connected by a low wall to form a triangle. What is the coordinate of the midpoint of the line connecting the midpoints of sides AB and AC? (0.5, −0.5)

18. What type of triangle is formed by connecting the midpoints of line segments formed by $L(-1, -2)$, $M(-1, 4)$, $N(6, -2)$? a right triangle

SPORTS Use the baseball field diagram on page 545 for Exercises 19–22. Estimate distances on the ground. One unit on the grid represents 30 ft.

19. A home run is hit from home plate to point B. What is the distance? about 467 ft

20. A foul ball is hit from home plate to point F. What is the distance? about 124 ft

21. The right fielder catches a ball at point C and throws the ball to first base at coordinates $(-3, -5)$. How long is the throw? about 234 ft

22. What is the home-run distance from home plate to point A? about 381 ft

23. DATA FILE Use the Data Index on pages 632–633 to locate a diagram of a soccer field. Using only the lengths given and a grid overlay, estimate the distance shown by each colored arrow. green: 20 yd, red: 65 yd, blue: 24 yd, purple: 44 yd

24. YOU MAKE THE CALL The point (3, 5) is the midpoint of a segment that has (7, 11) as one endpoint. Janice says there isn't enough information to find the other endpoint of the segment. Do you agree? Explain your thinking.

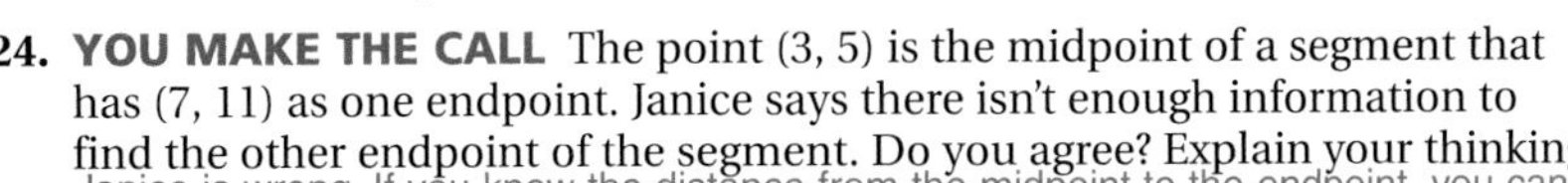

Janice is wrong. If you know the distance from the midpoint to the endpoint, you can extend that same distance from the midpoint in the opposite direction to find the other endpoint: (1, −1).

25. WRITING MATHEMATICS When finding the distance between two points (1, 4) and (8, 3), explain why it makes no difference which point you use as (x_1, y_1) and (x_2, y_2). Answers will vary. Since $(x_2 - x_1)^2 = (x_1 - x_2)^2$ and $(y_2 - y_1)^2 = (y_1 - y_2)^2$, the sum of the squared differences will be the same positive number.

Extended Practice Exercises

26. Use the distance formula to find the equation for a circle with radius 5 and center at point (0, 0). $5 = \sqrt{(x-0)^2 + (y-0)^2} \rightarrow 5 = \sqrt{x^2 + y^2} \rightarrow 25 = x^2 + y^2$

27. Use the distance formula to find the equation for a circle with radius r and center at point (0, 0). $x^2 + y^2 = r^2$

28. Find an equation for the circle with center (0, 2) and radius 8. $(x-0)^2 + (y-2)^2 = 64$, $x^2 + y^2 - 4y + 4 = 64$, $x^2 + y^2 - 4y = 60$

Mixed Review Exercises

Find binomial factors for the following, if possible. (Lesson 11-7)

29. $3x^2 + 10x - 8$ $(x + 4)(3x - 2)$

30. $2x^2 - 3x - 5$ $(2x - 5)(x + 1)$

31. $6x^2 - 26x + 24$ $2(x - 3)(3x - 4)$

32. $x^2 - 4x + 4$ $(x - 2)(x - 2)$

33. $2x^2 + 13x + 24$ not factorable

34. $15x^2 + x - 2$ $(5x + 2)(3x - 1)$

35. $12x^2 - 14xy - 6y^2$ $2(3x + 4)(2x - 3y)$

36. $28x^2 + 30xy + 8y^2$ $2(2x + y)(7x + 4y)$

37. $10x^2 - 41x + 21$ $(2x - 7)(5x - 1)$

38. $24x^2 + 4xy - 8y^2$ $4(3x + 2y)(2x - 4)$

39. $16x^2 + 24xy + 9y^2$ $(4x + 3y)(4x + 3y)$

40. $2x^2 - 16xy + 30y^2$ $2(x - 5y)(x - 3y)$

Teaching Strategies

COOPERATIVE LEARNING Have students work in pairs to solve these problems. Find the coordinates of the other endpoint of a line segment that has the following endpoint and midpoint:

1. endpoint $X(-3, 5)$, midpoint $(-6, 2)$, endpoint $Y(?, ?)$ (−9, 1)
2. endpoint $C(0, 8)$, midpoint $(3, 1)$, endpoint $D(?, ?)$ (6, −6)

Extra Practice Worksheet 12-6

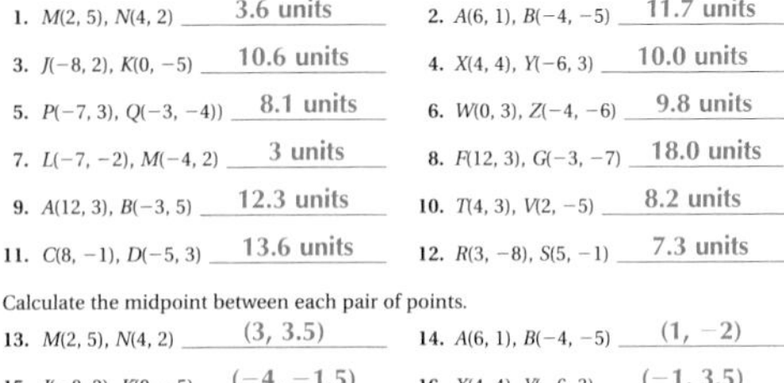

Name ______ Date ______

EXTRA PRACTICE 12-6

USE THE PYTHAGOREAN THEOREM

EXERCISES

Calculate the distance between each pair of points.

1. $M(2, 5)$, $N(4, 2)$ 3.6 units
2. $A(6, 1)$, $B(-4, -5)$ 11.7 units
3. $J(-8, 2)$, $K(0, -5)$ 10.6 units
4. $X(4, 4)$, $Y(-6, 3)$ 10.0 units
5. $P(-7, 3)$, $Q(-3, -4)$ 8.1 units
6. $W(0, 3)$, $Z(-4, -6)$ 9.8 units
7. $L(-7, -2)$, $M(-4, 2)$ 3 units
8. $F(12, 3)$, $G(-3, -7)$ 18.0 units
9. $A(12, 3)$, $B(-3, 5)$ 12.3 units
10. $T(4, 3)$, $V(2, -5)$ 8.2 units
11. $C(8, -1)$, $D(-5, 3)$ 13.6 units
12. $R(3, -8)$, $S(5, -1)$ 7.3 units

Calculate the midpoint between each pair of points.

13. $M(2, 5)$, $N(4, 2)$ (3, 3.5)
14. $A(6, 1)$, $B(-4, -5)$ (1, −2)
15. $J(-8, 2)$, $K(0, -5)$ (−4, −1.5)
16. $X(4, 4)$, $Y(-6, 3)$ (−1, 3.5)
17. $P(-7, 3)$, $Q(-3, -4)$ (−5, −0.5)
18. $W(0, 3)$, $Z(-4, 6)$ (−2, −1.5)
19. $L(-7, -2)$, $M(-4, -2)$ (−5.5, −2)
20. $F(12, 3)$, $G(-3, -7)$ (4.5, −2)
21. $A(12, 3)$, $B(-3, 6)$ (4.5, 4.5)
22. $T(4, 3)$, $V(2, -5)$ (3, −1)
23. $C(8, -1)$, $D(-5, 3)$ (1.5, 1)
24. $R(3, -8)$, $S(5, -1)$ (4, −4.5)

25. The vertices of triangle MNP are $M(5, 6)$, $N(0, 3)$ and $P(4, -2)$. What is the coordinate of the midpoint of the line connecting the midpoints of sides MN and MP? (2.25, 2.5)

26. What type of triangle is formed by connecting the midpoints of line segments formed by $R(5, 2)$, $S(-1, -3)$ and $T(2, -4)$? (2, −0.5), (0.5, −3.5), (3.5, −1)

27. Use the distance formula to find the equation for a circle with radius 4 and center at point (0, 0). $x^2 + y^2 = 16$

28. Use the distance formula to find the equation for a circle with radius 3 and center at point (2, 3). $(x - 2)^2 + (y - 3)^2 = 9$

166 EXTRA PRACTICE LESSON 12-6

Enrichment Worksheet 12-6

Name ______ Date ______

ENRICHMENT 12-6

THE PARABOLA TIPS OVER!

The graph of $y = ax^2 + bx + c$ is a parabola that points straight up or down. Equations of the form $x = ay^2 + by + c$ result in parabolas that are "sideways."

Example

Graph the equation $x = -y^2 - 8y - 8$.

Solution

With most graphing calculators and computer graphing software, you will first need to use the quadratic formula to solve the equation for y.

$y^2 + 8y + (8 + x) = 0$

$y = \frac{-8 \pm \sqrt{64 - 4(1)(8 + x)}}{2(1)}$

$y = -4 \pm \sqrt{8 - x}$

$y = -4 + \sqrt{8 - x}$

$y = -4 - \sqrt{8 - x}$

The technology will give you the graph in two pieces, one for each of these equations.

EXERCISES

Graph each equation.

1. $y = -4 \pm \sqrt{8 - x}$
2. $y = -4 \pm \sqrt{8 + x}$
3. $y = +4 \pm \sqrt{8 + 2x}$

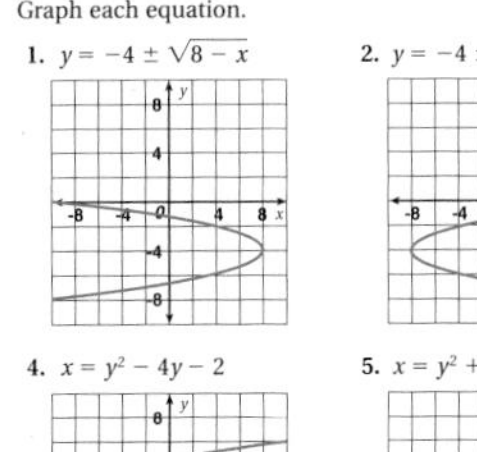

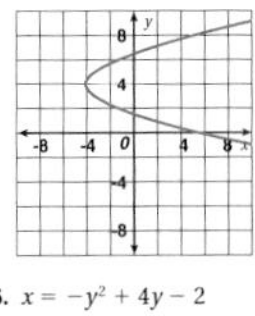

4. $x = y^2 - 4y - 2$
5. $x = y^2 + 4y - 2$
6. $x = -y^2 + 4y - 2$

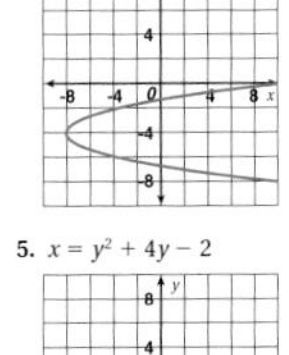

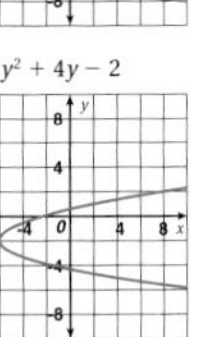

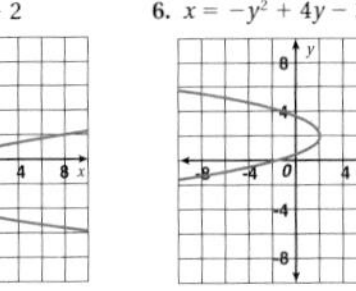

212 ENRICHMENT LESSON 12-6

Skills Practice

Vocabulary Review

Lesson 12-5
quadratic formula

Lesson 12-6
distance formula
midpoint formula

Assignment Guide

All students: 1–55
Additional Practice: See Extra Practice Index on page 674.

Review and Practice Your Skills

Practice ■ Lesson 12-5

Determine if each statement is *true* or *false*.

1. The quadratic formula was derived from the standard form of a quadratic equation. true
2. The quadratic formula cannot be used to solve equations unless they contain all terms found in the standard form of the equation. false
3. The radical part of the solution is in simplified form if it contains no factors that are perfect squares. true
4. The quadratic formula cannot be used to solve an equation when $a = 0$. Explain why. If $a = 0$, then the denominator of the fraction would be zero. Division by zero is undefined.

Use the quadratic formula to solve each equation.

5. $x^2 - 7x + 2 = 0$ $x = \frac{7 \pm \sqrt{41}}{2}$
6. $2x^2 + 5x - 3 = 0$ $x = -3, \frac{1}{2}$
7. $x^2 - 4x + 3 = 0$ $x = 3, 1$
8. $3x^2 - 4x + 1 = 0$ $x = 1, \frac{1}{3}$
9. $-2x^2 - 3x + 5 = 0$ $x = 1, -\frac{5}{2}$
10. $-x^2 + 4x + 8 = 0$ $x = 2 \pm 2\sqrt{3}$
11. $2x^2 - 7x + 2 = 0$ $x = \frac{7 \pm \sqrt{33}}{4}$
12. $4x^2 + x - 2 = 0$ $x = \frac{-1 \pm \sqrt{33}}{8}$
13. $-2x^2 + x + 5 = 0$ $x = \frac{1 \pm \sqrt{41}}{4}$

Practice ■ Lesson 12-6

14. When finding the distance on a coordinate plane, either endpoint may be designated as x_1, y_1, or x_2, y_2. Give an example that proves or disproves this theory.

Check students' examples. Because the differences are squared, either point may be x_1, y_1 or x_2, y_2.

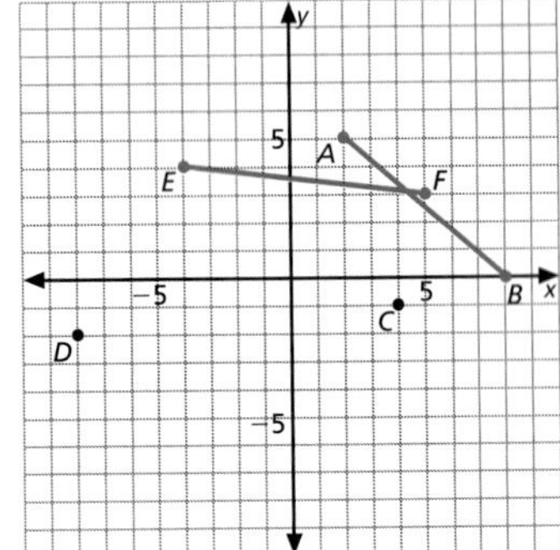

Use the graph on the right. Round distances to the nearest tenth of a unit.

15. Find the length of segment *AB*. $\sqrt{61} \approx 7.8$
16. Find the distance between *C* and *D*. 12.0
17. Find the midpoint of segment *EF*. 9.1

Calculate the distance between each pair of points.

18. $Y(1, 6), Z(-4, 3)$ 5.8
19. $W(2, -5), X(4, 2)$ 7.3
20. $U(1, -5), V(3, 7)$ 12.2
21. $S(-4, -2), T(-5, 3)$ 5.1
22. $Q(7, -2), R(-5, 3)$ 13
23. $O(3, 3), P(-3, -3)$ 8.5

Calculate the midpoint between each pair of points.

24. $A(-4, 3), B(3, -4)$ $\left(-\frac{1}{2}, -\frac{1}{2}\right)$
25. $C(1, 5), D(3, -2)$ $\left(2, \frac{3}{2}\right)$
26. $E(4, -7), F(-2, 3)$ $(1, -2)$
27. $G(5, 3), H(-5, -3)$ $(0, 0)$
28. $I(6, -1), J(-2, 5)$ $(2, 2)$
29. $K(4, 6), L(1, -3)$ $\left(\frac{5}{2}, \frac{3}{2}\right)$
30. $Z(-3, 5), Y(-2, -6)$ $m = \left(-\frac{5}{2}, -\frac{1}{2}\right)$, $d = 11.0$
31. $X(4, -5), W(-2, 1)$ $m = (1, -2)$, $d = 8.5$
32. $V(0, 3), U(5, 4)$ $m = \left(\frac{5}{2}, \frac{7}{2}\right)$, $d = 5.1$
33. $T(-1, 4), S(-2, -3)$ $m = \left(-\frac{3}{2}, \frac{1}{2}\right)$, $d = 7.1$
34. $R(2, 2), Q(-5, -5)$ $m = \left(-\frac{3}{2}, -\frac{3}{2}\right)$, $d = 9.9$
35. $P(4, -8), O(-8, 4)$ $m = (-2, -2)$, $d = 17.0$
36. Use the distance formula to find an equation for a circle with a radius of 4 and center at point (0, 0). $x^2 + y^2 = 4$

Teaching Strategies

Review students understanding of how to simplify radical expressions. Have them explain the steps they would follow to simplify $\frac{15 \pm \sqrt{50}}{5}$.

Steps: $\frac{15 \pm \sqrt{25 \times 2}}{5}$; $\frac{15 \pm \sqrt{25}\,\sqrt{2}}{5}$; $\frac{15 \pm 5\sqrt{2}}{5}$; 3 ± 2

548 Chapter 12 **Quadratic Functions**

PRACTICE ■ LESSON 12-1–LESSON 12-6

For each equation, find the vertex and axis of symmetry. (Lessons 12-1–12-2)

37. $y = x^2 + 3x - 2$ $V = \left(-\frac{3}{2}, -\frac{17}{4}\right)$; a/s $= -\frac{3}{2}$
38. $y = -2x^2 + x - 2$ $V = \left(\frac{1}{4}, -\frac{15}{8}\right)$; a/s $= \frac{1}{4}$
39. $y = 3x^2 - 4x + 2$ $V = \left(\frac{2}{3}, \frac{2}{3}\right)$; a/s $= \frac{2}{3}$
40. $y = -x^2 - 5x + 4$ $V = \left(-\frac{5}{2}, \frac{41}{4}\right)$; a/s $= -\frac{5}{2}$
41. $y = 2x^2 - 5x - 1$ $V = \left(\frac{5}{4}, -\frac{33}{8}\right)$; a/s $= -\frac{5}{4}$
42. $y = x^2 + 7x - 3$ $V = \left(-\frac{7}{2}, -\frac{61}{4}\right)$; a/s $= -\frac{7}{2}$

Solve each equation by graphing, factoring, or completing the square. (Lessons 12-3–12-4)

43. $y = x^2 - 5$ $x = \pm\sqrt{5}$
44. $x^2 - x - 6 = 0$ $x = -2, 3$
45. $x^2 + 6x + 8 = 0$ $x = 2, -4$
46. $x^2 - 6x + 2 = 0$ $x = 3 \pm \sqrt{7}$
47. $x^2 - 8x + 15 = 0$ $x = 3, 5$
48. $y = 2x^2 + 4$ no solution
49. $x^2 + 3x - 4 = 0$ $x = 1, -4$
50. $x^2 - 15x + 56 = 0$ $x = 7, 8$
51. $x^2 + 5x - 1 = 0$ $x = \frac{\pm\sqrt{21} - 5}{2}$
52. The width of a rectangular mirror is 9 inches less than its width. Find its dimensions if the area of the mirror is 360 in.^{2}. $w = 15$, $\ell = 24$

Solve each equation using the quadratic formula. (Lesson 12-5)

53. $x^2 - 4x - 3 = 0$ $x = 2 \pm \sqrt{7}$
54. $-2x^2 - 5x + 2 = 0$ $x = \frac{-5 \pm \sqrt{41}}{4}$
55. $2x^2 + 4x - 9 = 0$ $x = \frac{-2 \pm \sqrt{22}}{2}$

MathWorks Workplace Knowhow
Career – Air Traffic Controllers

Air traffic controllers manage the movement of air traffic through sections of air space. Controllers work together to monitor the movement of an aircraft from one section to another. They keep aircraft a safe distance apart and work to keep departures and arrivals on schedule. Controllers use radar and visual observation to monitor the progress of all aircraft. They also monitor the weather conditions for pilots. Air traffic controllers communicate directly with the pilot to direct the path of the flight. Together, controllers and flight crews make course corrections and respond to dangers caused by other aircraft, weather or emergency situations on the ground.

1. To determine the distance between two planes in the air, a three-coordinate system must be employed using ordered triplets (x, y, z). Each plane's position is measured in relation not only to a horizontal x-axis and a vertical y-axis, but also to a depth-measuring z-axis perpendicular to the other two axes. Place the origin (0, 0, 0) of the three axes at the O'Hare Airport control tower and measure units in miles. Using the axes and units, describe a plane with coordinates (8, 2, 5). 8 mi along x-axis, 2 mi on the y-axis and 5 mi on the z-axis
2. The space distance d between two points (x_1, y_1, z_1) and (x_2, y_2, z_2) is given by $d = \sqrt{(x_2 - x_1)^2 + (y_2 - y_1)^2 + (z_2 - z_1)^2}$. Two planes have position coordinates $(-3, 5, -4)$ and $(9, 2, 0)$. How far apart are they? 13
3. A passenger jet is midway between two private aircraft with position coordinates $(6, 2, -13)$ and $(1, -8, -5)$. Find the coordinates of the jet. Explain how you found your answer. $\left(\frac{7}{2}, -3, -9\right)$

Chalkboard Examples

Example from Lesson 12-5
Use the quadratic formula to solve $2x^2 - 12x + 3 = 0$.
Solution: $\frac{6 \pm \sqrt{30}}{2}$

Example from Lesson 12-6
Calculate the distance between the following points. Round to the nearest tenth of a unit. $A(1, -6)$ and $B(-2, 2)$ 9.4

Career Opportunity

Describe the kind of work pilots do and the types of companies that hire these workers. Have students discuss the importance of measurement, geometry and algebraic thinking in doing this type of job. Explain that pilots must be able to read instruments accurately and calculate flight paths, fuel requirements and speed for take off. Students should answer Questions 1–5 to better understand how pilots use mathematics in performing their job.

Students who are interested in learning more about this profession can go to learningmathmatters.com. Guidance Counselors should have information about school and training requirements.

Teaching Strategies

CHALLENGE Answer the questions for a triangle with the following vertices: $A(-3, 1)$, $B(2, 3)$, and $C(2, -3)$.

1. Find the length of side AB. 5.4
2. Find the length of side BC. 6
3. What are the coordinates of the midpoint of side AC? $(-0.5, -1)$
4. What kind of triangle is ABC? Isosceles

Lesson Planning

NCTM Standards/Strands

- Number & Operation
- Patterns, Functions, & Algebra
- Measurement
- Problem Solving
- Communication
- Connections

Vocabulary

parabola quadratic formula

Materials Needed

pencil/paper

Lesson Resources

Warm-Up Transparency 36
Transparency TK-7–10
Reteaching 12-7

Assignment Guide

Basic: 1–14
Enriched: 1–14

Getting Started

5-Minute Warm-up

Evaluate each formula. Round to the nearest tenth.

1. $C = \pi r^2$, when $r = 6.2$ cm
   120.7 cm
2. $V = \frac{4}{3}\pi r^3$, when $r = 4$ m
   267.9 m
3. $F = \frac{9}{5}C + 32$, when $C = 17$
   62.6

Use the Five-step Plan and the Strategy

THE FIVE-STEP PLAN Read—ask questions of the students to help them understand the problem. **Plan**—guide students to related problems and previously mastered skills and strategies. **Solve**—students solve problem on their own. **Answer**—write solution in format that answers the question. **Check**—review work, check for reasonableness and review strategy used. Students will benefit from the experience of verbalizing their methods.

12-7 Problem Solving Skills: Graphs to Equations

You have learned that the shape of the graph of a quadratic function is a parabola. You also know that quadratic equations are written in the form $y = ax^2 + bx + c$.

Suppose you were given the graph of a parabola. Can you use your knowledge to work backwards to discover the equation of the graph?

Problem Solving Strategies

- Guess and check
- Look for a pattern
- Solve a simpler problem
- Make a table, chart or list
- Use a picture, diagram or model
- Act it out
- ✔ Work backwards
- Eliminate possibilities
- Use an equation or formula

Problem

SCIENCE The results of an experiment are represented by the graph shown at the right. What is the equation of the graph?

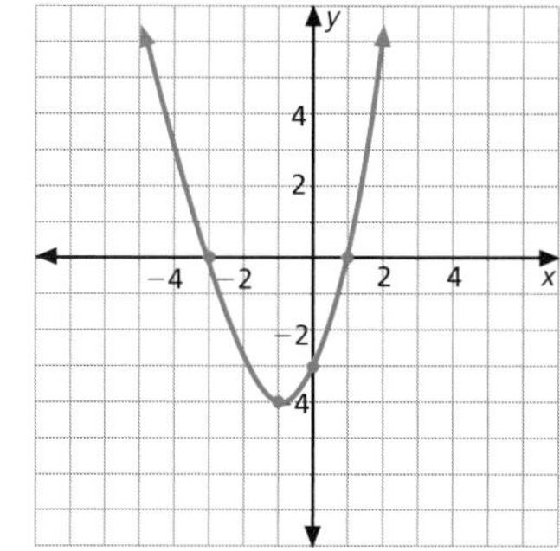

Solve the Problem

You can determine the equation of a parabola if you know three points on its graph.

Begin with the general form of a quadratic function: $y = ax^2 + bx + c$, where a, b and c represent coefficients and constants.

Locate three points on the graph. As you can see, the graph of the parabola passes through $(0, -3)$, $(-1, -4)$ and $(-3, 0)$.

Substitute the x-values and y-values for each point into the equation to create a system of three equations.

For $(0, -3)$	For $(-3, 0)$	For $(-1, -4)$
$ax^2 + bx + c = y$	$ax^2 + bx + c = y$	$ax^2 + bx + c = y$
$c = -3$	$9a - 3b + c = 0$	$a - b + c = -4$

Now use any of the methods you have learned for solving systems of equations.

$$c = -3$$
$$9a - 3b + c = 0$$
$$a - b + c = -4$$

Use $c = -3$ on the other two equations. Then solve for a.

$9a - 3b - 3 = 0 \rightarrow 9a - 3b - 3 = 0$
$a - b - c = -4 \rightarrow 3a - 3b - 9 = -12$ Multiply by 3.
$6a \quad + 6 = 12$ Subtract

Since $6a + 6 = 12$, $a = 1$. Substitute $a = 1$ and $c = -3$ into the second equation.

$$9a - 3b + c = 0$$
$$9 - 3b - 3 = 0$$
$$6 = 3b$$
$$b = 2$$

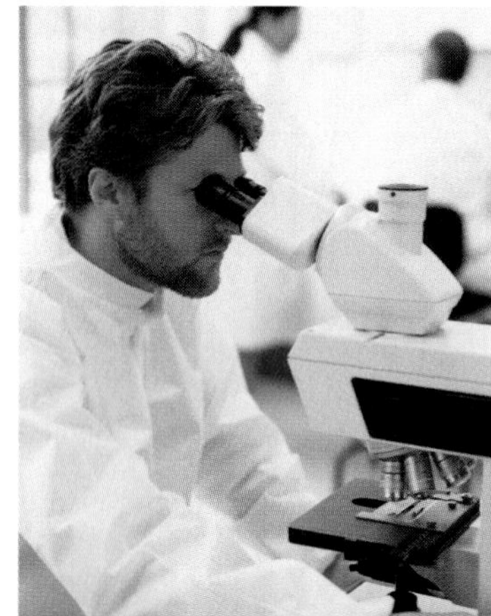

Teaching Strategies

You may wish to review how to work backward to find the equation of a line when points on the line are known. By reviewing these methods, students may be better able to see the relationship between the known points and the standard form of a quadratic equation.

Thus, $a = 1$, $b = 2$ and $c = -3$. Check these values in the third equation. Then use these values in the general quadratic form to write an equation.

The equation of the parabola is $y = x^2 + 2x - 3$.

Five-step Plan
1 Read
2 Plan
3 Solve
4 Answer
5 Check

Try These Exercises

1. The graph of a quadratic equation contains the three points $(-5, 10)$, $(0, -5)$ and $(2, 3)$. Find the equation of the parabola. $y = x^2 + 2x - 5$
2. The graph of a quadratic equation contains the three points $(2, 7)$, $(0, -1)$ and $(-1, 1)$. Find the equation of the parabola. $y = 2x^2 - 1$

Practice Exercises

3. **WRITING MATH** In the example on page 550, will you obtain the same equation if you select three different points? Explain.

3. Yes. Any three non-collinear points located on the graph will work. However, you may find it easier to solve the system of equations if you use the y-intercepts as one of the points.

4. **PHYSICS** The relationship between two variables in an experiment can be represented by the parabola shown. Using three points from the graph, find the equation of the parabola. $y = -2x^2 + 2x + 3$

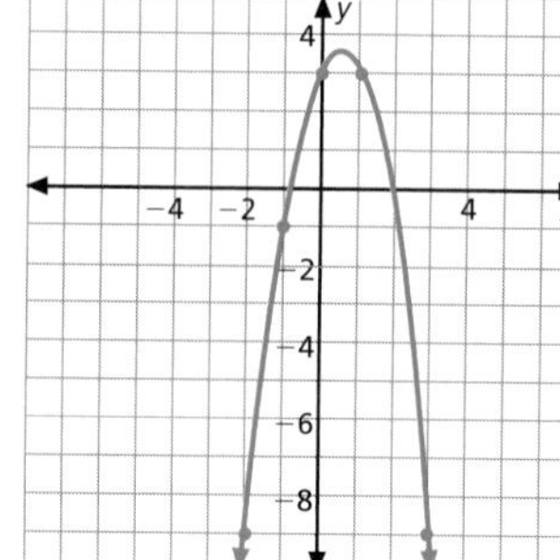

5. A graph contains the points $(1, -4)$, $(0, -3)$, $(2, -3)$, $(3, 0)$ and $(-2, 5)$. Choose three points and find the equation of the parabola. $y = x^2 - 2x - 3$

Find the equations for each parabola.

6.

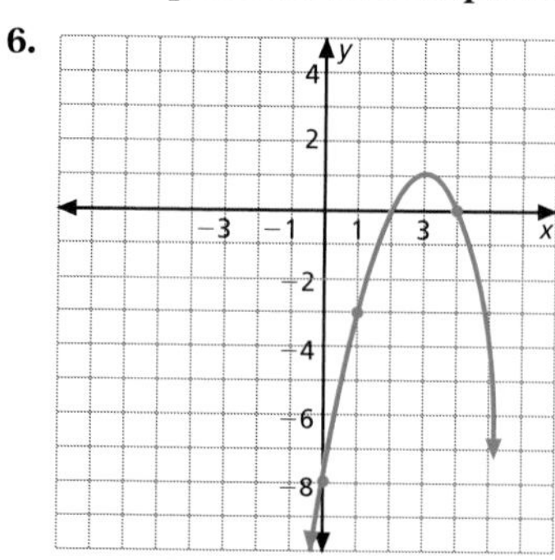

$y = -x^2 + 6x - 8$

7.

$y = x^2 - 4x + 12$

8. **CHAPTER INVESTIGATON** Calculate your vehicle's average rate of speed in feet per second for each run of the ramp. Make a poster showing a diagram of the vehicle and the vehicle's speed for various inclines of the ramp. Is there a relationship between the increase in the incline and the vehicle's speed? Answers will vary.

Mixed Review Exercises

Factor the following trinomials. (Lesson 11-9)

9. $15a^2 - 4a - 8$ $(5a - 2)(3a + 4)$
10. $a^2 + 5a - 14$ $(a - 2)(a + 7)$
11. $a^2 + 11a + 24$ $(a + 8)(a + 3)$
12. $12a^2 - 5a - 2$ $(4a + 1)(3a - 2)$
13. $3a^2 - 14a - 24$ $(a - 6)(3a + 4)$
14. $4a^2 - 16a + 16$ $4(a - 2)(a - 2)$

THE STRATEGY Work backwards—Using logical reasoning, students can retrace their steps through many math procedures to find the original starting point.

Chalkboard Examples

Supplementary Problem

The results of a scientific experiment are represented by a parabola which passes through points $(-1, 3)$, $(0, 6)$ and $(-3, 3)$. What is the equation of the parabola?
$y = x^2 + 4x + 6$

Lesson Wrap-up

Quick Assessment

Ask the following question to determine if students understand the content presented in this lesson.

When finding the equation of the graph of a parabola, why is it helpful to have one point with an x-coordinate of 0? When the x-coordinate equals 0, you know that c is equal to the y-coordinate of the same point. This simplifies solving the system of equations.

Reteaching Worksheet 12-7

Name ______________ Date ______________

RETEACHING 12-7

PROBLEM SOLVING SKILL: USE GRAPHS TO WRITE EQUATIONS

If you are given the graph of a quadratic equation or three or more points on the graph, you can find the equation.

Example

The graph of a quadratic equation contains the three points $(-2, 0)$, $(1, -3)$ and $(0, -4)$. Find the equation of the parabola.

Solution

Substitute each ordered pair into the general form of the quadratic equation, $y = ax^2 + bx + c$, to create a system of three equations.

For $(-2, 0)$	For $(1, -3)$	For $(0, -4)$
$0 = 4a - 2b + c$	$-3 = a + b + c$	$-4 = c$

Substitute $c = -4$ in the other two equations.

$0 = 4a - 2b - 4$	$-3 = a + b - 4$
$4 = 4a - 2b$	$1 = a + b$
$4 = 4a - 2b$	$2 = 2a + 2b$

$6a = 6$
$a = 1$

To find b, substitute $a = 1$ and $c = -4$ into $-3 = a + b + c$ and solve.
$-3 = 1 + b - 4$
$0 = b$

Since $a = 1$, $b = 0$ and $c = -4$, the equation is $y = x^2 - 4$.

EXERCISES

1. The graph of an equation contains the three points $(0, -3)$, $(3, 12)$ and $(-2, -3)$.
   Find the equation of the parabola. $y = x^2 + 2x - 3$
2. The graph of an equation contains the three points $(0, 4)$, $(-3, -2)$ and $(1, 10)$.
   Find the equation of the parabola. $y = x^2 + 5x + 4$
3. The graph of an equation contains the three points $(0, -6)$, $(2, 0)$ and $(-1, -6)$.
   Find the equation of the parabola. $y = x^2 + x - 6$

190 RETEACHING *LESSON 12-7*

Extend the Lesson

COOPERATIVE LEARNING If graphing calculators are available, have each pair of students write a quadratic equation and display its graph. Then using the TABLE key, list the coordinates of points on the line. Make a list of the coordinates of five points. Then exchange points with another pair and try to figure out the quadratic equation by working backwards. Students may choose which three points to use in making their calculations. Then have students use their calculators to check their results.

Assessment Planning

Vocabulary Review

Lesson 12-1
quadratic equation function
parabola vertex
axis of symmetry

Lesson 12-2
general quadratic function
standard quadratic equation

Lesson 12-3
factoring *x*-intercepts

Lesson 12-4
completing the square

Lesson 12-5
quadratic formula

Lesson 12-6
distance formula
midpoint formula

Lesson 12-7
parabola
quadratic formula

Assessment Options

Chapter Assessment, page 554
Alternative Assessment, page 555
Chapter Test Forms A and B
Cumulative Assessment, page 556-557
Achievement Test
Writing Prompts

Assignment Guide

All students: 1-25
Additional Practice: See Extra Practice Index on page 674.

Additional Answers

7. 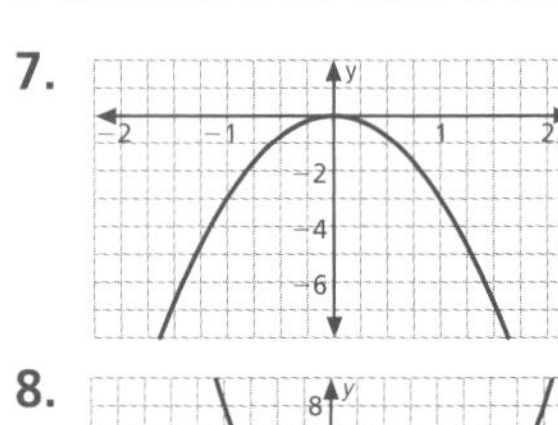

8. 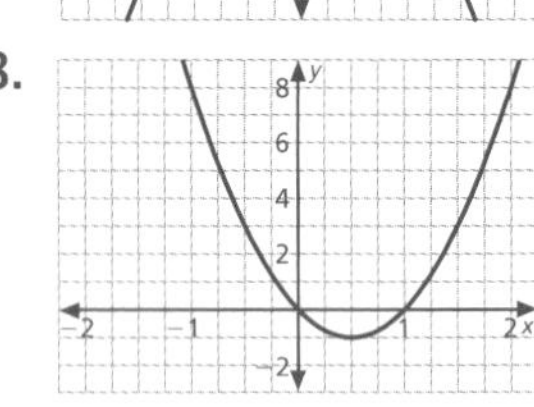

9. 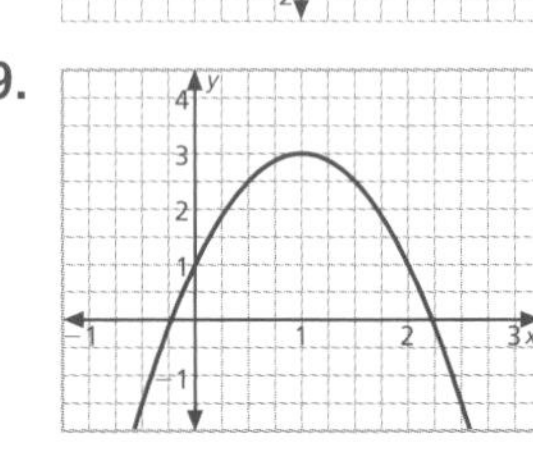

Chapter 12 Review

Vocabulary ■ LESSON 12-1–LESSON 12-7

Match the letter of the word in the list at the right with the description on the left.

- **a.** general quadratic function
- **b.** axis of symmetry
- **c.** standard quadratic equation
- **d.** vertex
- **e.** quadratic function
- **f.** function

1. an equation that contains an x^2 term and involves no term with a higher power of x e
2. the relationship in which each x-value has a unique y-value f
3. the lowest point on a parabola that opens upward or the highest point on a parabola that opens downward d
4. a function in the form $f(x) = ax^2 + bx + c$ a
5. the line that divides a parabola in half b
6. an equation in the form $y = ax^2 + bx + c$ c

LESSONS 12-1 AND 12-2 ■ Quadratic Functions, p. 520

- ▶ To graph a quadratic equation, find at least five ordered pairs by selecting x-values and solving the equation to find y-values. Locate and draw a smooth line through the points.
- ▶ To locate the vertex, use $x = -\frac{b}{2a}$ to find the x-coordinate. Then substitute the x-value into the equation to find the y-coordinate.
- ▶ The axis of symmetry for the graph passes through the vertex and is parallel to the y-axis.

Graph each function. Give the coordinates of the vertex for each graph.
See additional answers.

7. $y = -3x^2$ (0, 0)
8. $y = 4x^2 - 4x$ (0.5, −1)
9. $y = -2x^2 + 4x + 1$ (1, 3)

- ▶ A function in x assigns each x-value exactly one y-value. If each vertical line in the coordinate plane contains at most one point of the graph, the graph is of a function.

Determine if each graph is the graph of a function.

10. 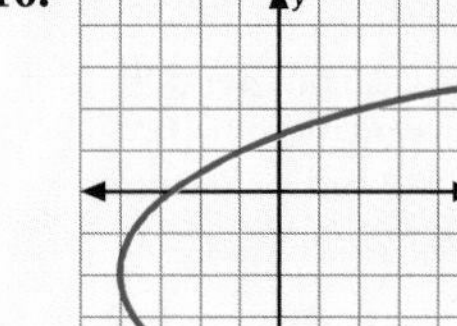
no

11. 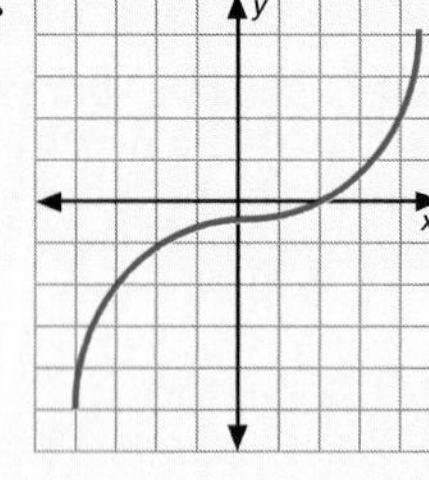
yes

12. 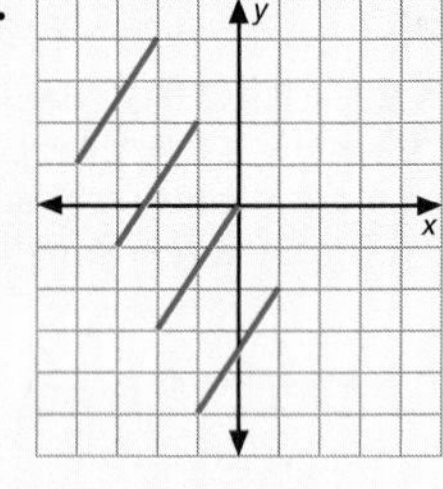
no

Extend the Lesson

COOPERATIVE LEARNING If your students have graphing calculators, they may enjoy the challenge of writing programs for the formulas in this chapter. Have them work in pairs to write the programming code for the distance formula, the midpoint formula and the quadratic formula. Each program will have three basic parts: a section where the user inputs the necessary values, a section where the formula is used and the results stored as variables, a section where the values of the variables are displayed.

LESSON 12-3 ■ Factor and Graph, p. 530

- ▶ The x-intercepts, the points for which $y = 0$, are the solutions to a quadratic equation. The number of x-intercepts and the approximate solutions may be found by graphing. To find the exact solutions, let $y = 0$, and factor the equation.

Use a graphing calculator to determine the number of solutions for each equation. Factor to solve each equation.

13. $x^2 - 14x = -49$ 7

14. $x^2 - 63 = -2x$ −9, 7

15. $3x^2 + 9x - 84 = 0$ 4, −7

LESSON 12-4 ■ Complete the Square, p. 534

- ▶ To complete the square for an expression in the form $ax^2 + bx$, add $\left(\frac{b}{2}\right)^2$.
- ▶ To solve an equation by completing the square, rewrite the equation in the form $ax^2 + bx = c$, add $\left(\frac{b}{2}\right)^2$ to each side of the equation, factor, and simplify.

Solve each equation by completing the square.

16. $x^2 - 2x - 8 = 0$ 4, −2

17. $x^2 - 8x = -15$ 3, 5

18. $x^2 + 14x = -49$ −7

LESSON 12-5 ■ The Quadratic Formula, p. 540

- ▶ The quadratic formula can be used to solve any quadratic equation in the form $ax^2 + bx + c = 0$, $a \neq 0$.

The quadratic formula is $x = \dfrac{-b \pm \sqrt{b^2 - 4ac}}{2a}$.

Use the quadratic formula to solve each equation.

19. $2x^2 - 5x = 3$ −0.5, 3

20. $6x^2 + x = 1$ $\frac{1}{3}, -\frac{1}{2}$

21. $x^2 + 2x = 6$ $-1 \pm \sqrt{7}$

LESSON 12-6 ■ Using the Pythagorean Theorem, p. 544

- ▶ The distance, d, between any two points (x_1, y_1) and (x_2, y_2) on the coordinate plane may be found using the distance formula:

$$d = \sqrt{(x_2 - x_1)^2 + (y_2 - y_1)^2}$$

The coordinates of the midpoint of a line segment with endpoints (x_1, y_1) and (x_2, y_2) are $\left(\frac{x_1 + x_2}{2}, \frac{y_1 + y_2}{2}\right)$.

Find the midpoint between each pair of points. Then, find the distance between each pair of points to the nearest tenth.

22. $L(-10, 5)$, $M(4, -9)$ (−3, −2), 19.8 units

23. $C(6, 4)$, $D(-8, -6)$ (−1, −1), 17.2 units

24. $P(0, -7)$, $Q(12, 2)$ (6, −2.5), 15 units

LESSON 12-7 ■ Problem Solving Skills, p. 550

- ▶ Work backwards to write the equation of the graph of a parabola. Use the form of the equation $y = ax^2 + bx + c$ and three points on the graph.

25. A graph of a parabola contains the points (−4, 23), (0, −1) and (2, −49). Find the quadratic equation. $y = -3x^2 - 18x - 1$

Chapter 12 Test Form A, page 1

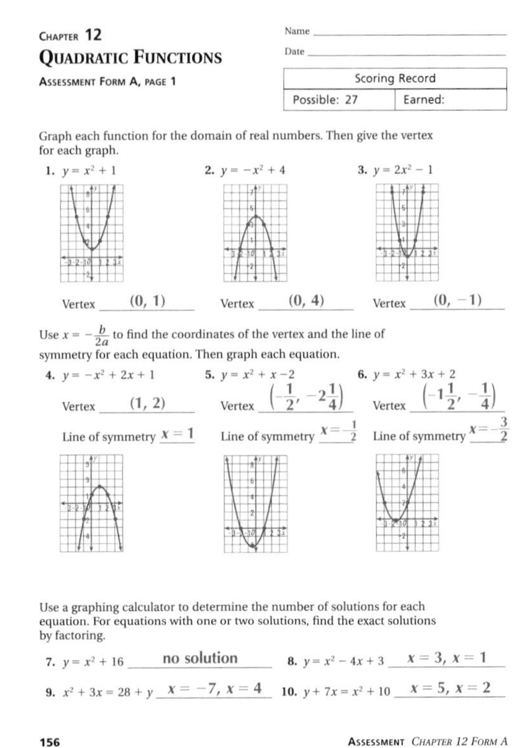
CHAPTER 12
QUADRATIC FUNCTIONS
ASSESSMENT FORM A, PAGE 1

Name ____________
Date ____________

Scoring Record	
Possible: 27	Earned:

Graph each function for the domain of real numbers. Then give the vertex for each graph.

1. $y = x^2 + 1$ Vertex (0, 1)

2. $y = -x^2 + 4$ Vertex (0, 4)

3. $y = 2x^2 - 1$ Vertex (0, −1)

Use $x = -\frac{b}{2a}$ to find the coordinates of the vertex and the line of symmetry for each equation. Then graph each equation.

4. $y = -x^2 + 2x + 1$ Vertex (1, 2) Line of symmetry $x = 1$

5. $y = x^2 + x - 2$ Vertex $\left(-\frac{1}{2}, -2\frac{1}{4}\right)$ Line of symmetry $x = -\frac{1}{2}$

6. $y = x^2 + 3x + 2$ Vertex $\left(-1\frac{1}{2}, -\frac{1}{4}\right)$ Line of symmetry $x = -\frac{3}{2}$

Use a graphing calculator to determine the number of solutions for each equation. For equations with one or two solutions, find the exact solutions by factoring.

7. $y = x^2 + 16$ no solution

8. $y = x^2 - 4x + 3$ $x = 3, x = 1$

9. $x^2 + 3x = 28 + y$ $x = -7, x = 4$

10. $y + 7x = x^2 + 10$ $x = 5, x = 2$

156 ASSESSMENT *CHAPTER 12 FORM A*

Chapter 12 Test Form A, page 2

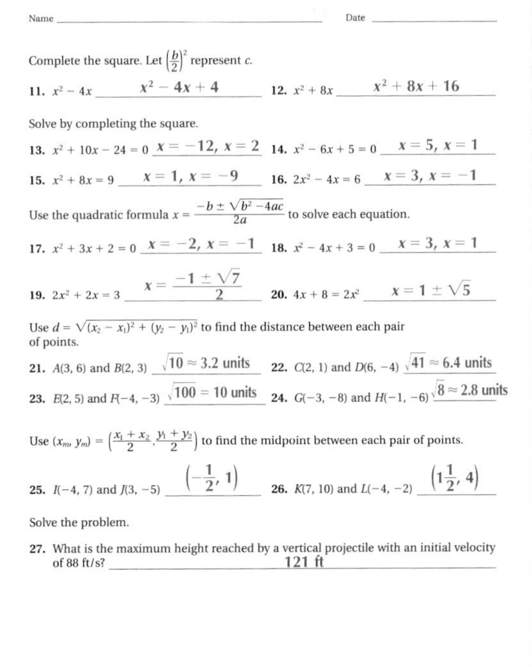
Name ____________ Date ____________

Complete the square. Let $\left(\frac{b}{2}\right)^2$ represent c.

11. $x^2 - 4x$ $x^2 - 4x + 4$

12. $x^2 + 8x$ $x^2 + 8x + 16$

Solve by completing the square.

13. $x^2 + 10x - 24 = 0$ $x = -12, x = 2$

14. $x^2 - 6x + 5 = 0$ $x = 5, x = 1$

15. $x^2 + 8x = 9$ $x = 1, x = -9$

16. $2x^2 - 4x = 6$ $x = 3, x = -1$

Use the quadratic formula $x = \frac{-b \pm \sqrt{b^2 - 4ac}}{2a}$ to solve each equation.

17. $x^2 + 3x + 2 = 0$ $x = -2, x = -1$

18. $x^2 - 4x + 3 = 0$ $x = 3, x = 1$

19. $2x^2 + 2x = 3$ $x = \frac{-1 \pm \sqrt{7}}{2}$

20. $4x + 8 = 2x^2$ $x = 1 \pm \sqrt{5}$

Use $d = \sqrt{(x_2 - x_1)^2 + (y_2 - y_1)^2}$ to find the distance between each pair of points.

21. $A(3, 6)$ and $B(2, 3)$ $\sqrt{10} \approx 3.2$ units

22. $C(2, 1)$ and $D(6, -4)$ $\sqrt{41} \approx 6.4$ units

23. $E(2, 5)$ and $F(-4, -3)$ $\sqrt{100} = 10$ units

24. $G(-3, -8)$ and $H(-1, -6)$ $\sqrt{8} \approx 2.8$ units

Use $(x_m, y_m) = \left(\frac{x_1 + x_2}{2}, \frac{y_1 + y_2}{2}\right)$ to find the midpoint between each pair of points.

25. $I(-4, 7)$ and $J(3, -5)$ $\left(-\frac{1}{2}, 1\right)$

26. $K(7, 10)$ and $L(-4, -2)$ $\left(1\frac{1}{2}, 4\right)$

Solve the problem.

27. What is the maximum height reached by a vertical projectile with an initial velocity of 88 ft/s? 121 ft

158 ASSESSMENT *CHAPTER 12 FORM A*

Teaching Strategies

After learning several methods for solving quadratic equations, students may feel confused about when to use each method. Remind students that although they have been instructed to use specific methods at times in order to assess their understanding, they are free to use any methods with future problems they encounter. Discuss which methods they find easier to use. Generally, factoring is the easiest method when the factors can be found easily.

Chapter 12 **Review** 553

Chapter 12 Test Form B, page 1

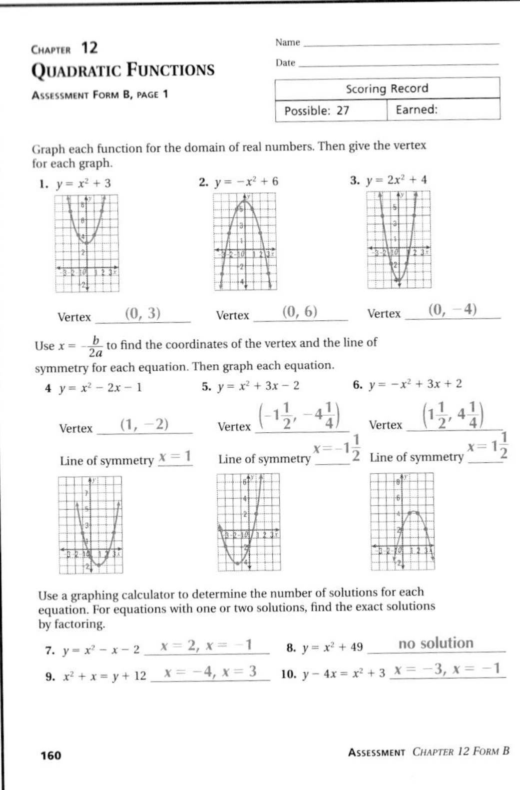

CHAPTER 12
QUADRATIC FUNCTIONS
ASSESSMENT FORM B, PAGE 1

Name ____________
Date ____________

Scoring Record	
Possible: 27	Earned:

Graph each function for the domain of real numbers. Then give the vertex for each graph.

1. $y = x^2 + 3$ Vertex (0, 3)
2. $y = -x^2 + 6$ Vertex (0, 6)
3. $y = 2x^2 + 4$ Vertex (0, −4)

Use $x = -\frac{b}{2a}$ to find the coordinates of the vertex and the line of symmetry for each equation. Then graph each equation.

4 $y = x^2 - 2x - 1$ Vertex (1, −2) Line of symmetry $x = 1$
5. $y = x^2 + 3x - 2$ Vertex $\left(-1\frac{1}{2}, -4\frac{1}{4}\right)$ Line of symmetry $x = -1\frac{1}{2}$
6. $y = -x^2 + 3x + 2$ Vertex $\left(1\frac{1}{2}, 4\frac{1}{4}\right)$ Line of symmetry $x = 1\frac{1}{2}$

Use a graphing calculator to determine the number of solutions for each equation. For equations with one or two solutions, find the exact solutions by factoring.

7. $y = x^2 - x - 2$ $x = 2, x = -1$
8. $y = x^2 + 49$ no solution
9. $x^2 + x = y + 12$ $x = -4, x = 3$
10. $y - 4x = x^2 + 3$ $x = -3, x = -1$

160 ASSESSMENT *CHAPTER 12 FORM B*

Chapter 12 Test Form B, page 2

Name ____________ Date ____________

Complete the square. Let $\left(\frac{b}{2}\right)^2$ represent c.

11. $x^2 + 2x$ $x^2 + 2x + 1$
12. $x^2 + 4x$ $x^2 + 4x + 4$

Solve by completing the square.

13. $x^2 - 8x + 7 = 0$ $x = 7, x = 1$
14. $x^2 - 12x + 11 = 0$ $x = 1, x = 11$
15. $x^2 + 13 = 14x$ $x = 13, x = 1$
16. $3x^2 + 6x = 9$ $x = -3, x = 1$

Use the quadratic formula $x = \frac{-b \pm \sqrt{b^2 - 4ac}}{2a}$ to solve each equation.

17. $x^2 + 6x - 16 = 0$ $x = -8, x = 2$
18. $x^2 - 2x + 1 = 0$ $x = 1$
19. $2x^2 + x = 6$ $x = -2, x = 1\frac{1}{2}$
20. $8x + 2 = 2x^2$ $x = 2 \pm \sqrt{5}$

Use $d = \sqrt{(x_2 - x_1)^2 + (y_2 - y_1)^2}$ to find the distance between each pair of points.

21. $A(4, 2)$ and $B(2, 5)$ $\sqrt{13} \approx 3.6$
22. $C(8, -1)$ and $D(7, 4)$ $\sqrt{26} \approx 5.1$
23. $E(-3, -1)$ and $F(1, 3)$ $\sqrt{32} \approx 5.7$
24. $G(-2, -3)$ and $H(-2, -4)$ $\sqrt{1} = 1$

Use $(x_m, y_m) = \left(\frac{x_1 + x_2}{2}, \frac{y_1 + y_2}{2}\right)$ to find the midpoint between each pair of points.

25. $I(2, -6)$ and $J(4, 2)$ (3, −2)
26. $K(3, 4)$ and $L(-2, -6)$ $\left(\frac{1}{2}, -1\right)$

Solve the problem.

27. What is the maximum height reached by a vertical projectile with an initial velocity of 120 ft/s? 225 ft

162 ASSESSMENT *CHAPTER 12 FORM B*

CHAPTER 12 ASSESSMENT

Graph each function for the domain of real numbers. Give the coordinates of the vertex for each graph. For 1–3, see additional answers for graphs.

1. $y = -2x^2$ (0, 0)
2. $y = 3x^2 - 4$ (0, −4)
3. $y + 7 = -3x^2 + 2x$ $\left(\frac{1}{3}, -6\frac{2}{3}\right)$

Determine if each graph is a function.

4. 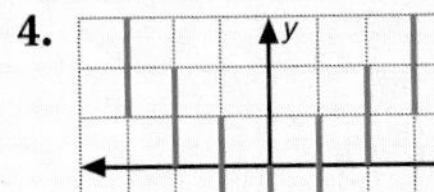
no

5. 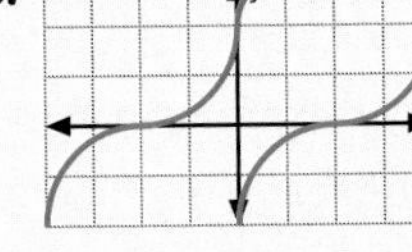
no

6. 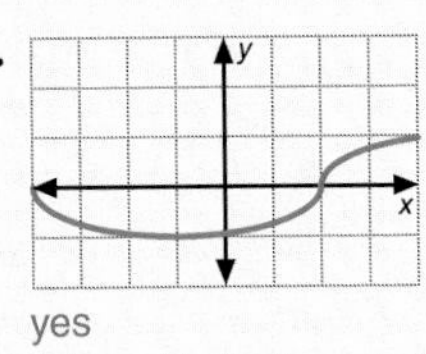
yes

Factor to solve each equation.

7. $x^2 - 6x - 16 = 0$ −2, 8
8. $x^2 + 9 = 6x$ 3
9. $4x^2 - 4x = -1$ $\frac{1}{2}$

Complete the square.

10. $x^2 + 14x$ $x^2 + 14x + 49$
11. $x^2 - 8x$ $x^2 - 8x + 16$
12. $x^2 + x$ $x^2 + x + \frac{1}{4}$

Solve by completing the square.

13. $x^2 + 3x + 2 = 0$ −2, −1
14. $x^2 - 4x = 12$ 6, −2
15. $2x^2 + 2x = 40$ 4, −5

Use the quadratic formula to solve each equation.

16. $3x^2 - 4x = 15$ $3, -1\frac{2}{3}$
17. $x^2 + 8x = -3$ $-4 \pm \sqrt{13}$
18. $24x = -x^2 - 136$ $-12 \pm 2\sqrt{2}$

Calculate the distance between each pair of points to the nearest tenth of a unit.

19. $A(1, -5), B(-3, 7)$ 12.6 units
20. $L(-6, 4), M(2, -2)$ 10 units
21. $R(0, 8), S(-3, -12)$ 20.2 units

Calculate the midpoint between each pair of points.

22. $G(-13, 9), H(7, -11)$ (−3, −1)
23. $P(10, -6), Q(-4, -7)$ (3, −6.5)
24. $C(-5, 3), D(12, 8)$ (3.5, 5.5)

Solve.

25. How long will it take a ball thrown with an initial velocity of 72 feet per second to reach the ground? 4.5 sec

ADDITIONAL ANSWERS

1. 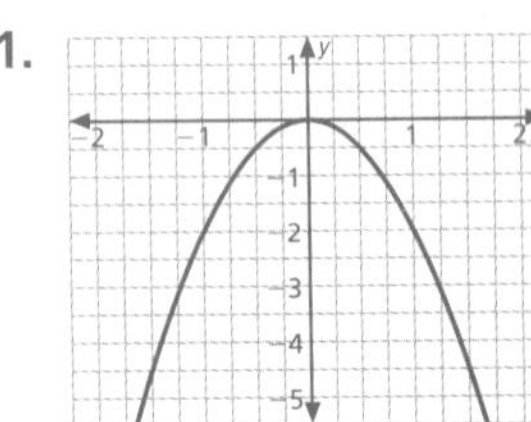

2. 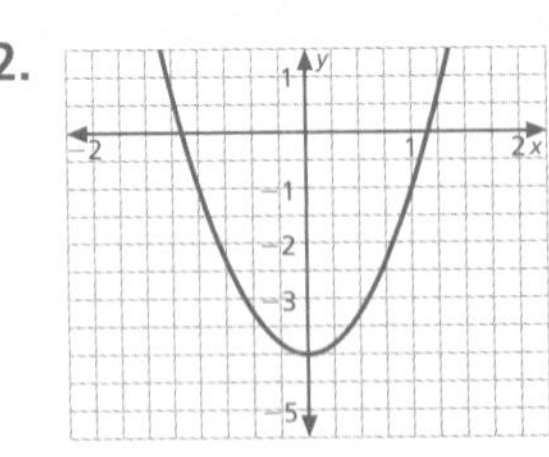

3. 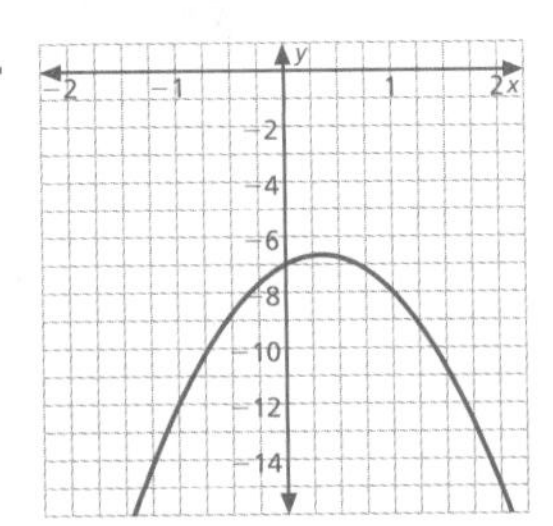

Chapter 12 Alternative Assessment

Other Resources

AS THE CROW FLIES Do crows really know a shortcut, or is this just a quaint way of using the Pythagorean Theorem? Find a place in your neighborhood where the streets form a right angle. Using this intersection as the 90° corner of a triangle, find two other places (stores, side streets, etc.) to use as the other angles of the triangle. Measure the two sides of the triangle in miles, in paces, or some approprite measure. Then use this information to find the distance the crow would fly. Find two other places to perform this experiment as well. Then write a brief paragraph describing how you found the places, the distances and the distance the crow would cover.

Portfolio

HOW-TO Make a notebook for your own reference. The notebook should show the several different methods of solving quadratic equations learned in this chapter and the instructions for how to do this. Be sure your instructions are clear and can be understood by any other math student who might want to know how to solve each kind of equation.

Critical Thinking

HANG TIME In football, *hang time* is the length of time the ball stays in the air after it has been kicked. Kickers are often rated by their hang times, longer being better. Use the formula $vt - 16t^2 = 0$ to calculate ten hang times. Assume the ball travels 50 yards on each kick. Assign different velocities (v) and find the time the ball stays in the air (t). Graph the results of your calculations. Suppose you are the coach of the team. In general terms, how would you tell your kicker to kick the ball to get the greatest hang time? Support your conclusion.

Chapter Investigation

EXTENSION Write a report about your experiment. Describe your vehicle and explain why you build it the way you did. Explain what happened when you used the ramp and changed its angle. Include all of the data you collected in your report.

Chapter Investigation

The benchmarks and expectations for this extension are as follows.

- Students work with a group to draw plans for a simple gravity-powered vehicle that weighs no more than 10 ounces.
- Students work with a group to build their gravity-powered vehicle.
- Students work with a group to build a ramp from 24-36 inches in length. They set the ramp so that the angle formed by the ramp and the floor is 15°. They time their vehicle's descent from the top to the bottom. They increase the ramp's incline in 5° increments. They measure the distance the vehicle travels beyond the bottom of the ramp.
- Students write a report that describes their vehicle. They explain what happened when they used the ramp and changed the angle.

Other Resources

This activity can also be performed using a local street map, a map of the U.S., or any map.

SCORING RUBRIC **5**—Student accurately measures three right triangles, finds the correct distance of the crow's flight, and explains clearly how this was accomplished. **4**—Student accurately measures triangles, finds the correct distance, and explains clearly. **3**—Student measures triangles, finds the distance with minor errors, and explains. **2**—Student measures fewer than three right triangles, finds the distance with minor errors, and gives an unclear explanation. **1**—Student measures fewer than three right triangles, finds the distance with significant errors, and gives a partial explanation. **0**—Student makes no attempt to solve or measure triangles.

Portfolio

You may want to provide a variety of new quadratic equations for students to use in their notebooks.

SCORING RUBRIC **5**—Student produces a complete notebook showing several different methods of solving quadratic equations, each with detailed instructions. **4**—Student produces a complete notebook showing different methods, with reasonably clear instructions. **3**—Student produces a notebook showing several different methods, with some instructions. **2**—Student produces a notebook showing a few methods, each with some instructions. **1**—Student produces a notebook showing a few methods, but without instructions. **0**—Student makes no attempt to produce a notebook.

Critical Thinking

Remind students to stay within the realm of reality on their selection of velocities to test.

SCORING RUBRIC **5**—Student accurately calculates ten or more hang times, graphs each correctly, draws a logical conclusion about the optimal kick and defends the conclusion well. **4**—Student accurately calculates ten hang times, graphs each correctly, draws a reasonable conclusion about the optimal kick and defends the conclusion. **3**—Student calculates ten hang times with minor errors, graphs each, draws a conclusion about the optimal kick and defends the conclusion. **2**—Student calculates less than ten hang times with errors, graphs most, draws a conclusion about the optimal kick, but cannot defend the conclusion. **1**—Student calculates less than ten hang times with significant errors, graphs some, but offers no conclusion. **0**—Student makes no attempt to calculate hang times.

Cumulative Review

Cumulative review is the best way to maintain previously taught skills and concepts. This will keep students prepared for new lessons that build on previously covered skills.

Cumulative Review covers:

Lesson 1-7
Lesson 1-8
Lesson 2-4
Lesson 2-5
Lesson 3-6
Lesson 4-7
Lesson 5-1
Lesson 5-7
Lesson 6-1
Lesson 6-6
Lesson 7-1
Lesson 7-4
Lesson 8-1
Lesson 8-5
Lesson 8-6
Lesson 9-3
Lesson 9-5
Lesson 10-1
Lesson 10-3
Lesson 11-4
Lesson 11-5
Lesson 11-6
Lesson 11-7
Lesson 12-5
Lesson 12-6

Additional Answers

5. If the sum of two angles is 180°, the angles are right angles. True; False.

14.

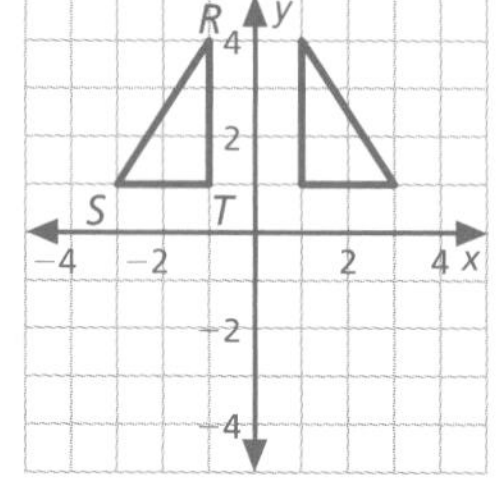

Chapters 1–12 Cumulative Review

Evaluate if $x = 5$ and $y = -6$.

1. $-2xy^2$ -360
2. y^{-3} $-\frac{1}{216}$

Solve each equation.

3. $\frac{2}{3}x = 18$ $x = 27$
4. $2(x - 5) = x + 10$ $x = 20$
5. Write the converse of the following statement. Then tell if the given statement and the converse are true or false. *If two angles are right angles, the sum of the angles is 180°.* See additional answers.
6. Find the unknown angle measure. 105°

77° 80° 98° x°

Complete.

7. 9.8 cm = __?__ m 0.098
8. 2.5 kg = __?__ g 2,500
9. Find the volume. 2,016 in.^{3}

6 in. 12 in. 18 in. 8 in. 30 in.

10. Find the slope of the line passing through points whose coordinates are (7, 4) and (−1, −3). $\frac{7}{8}$
11. Solve the system of equations. $\begin{cases} 3x + y = -11 \\ 2x + 2y = -10 \end{cases}$ $x = -3, y = -2$
12. Solve the proportion: $\frac{x}{28} = \frac{12}{16}$ 21
13. Determine if the triangles are similar. If they are, give a reason. Write *AA, SSS,* or *SAS.* yes, SSS

12 16 20 18 24 30

14. Graph a reflection of $\triangle RST$ over the y-axis. See additional answers.

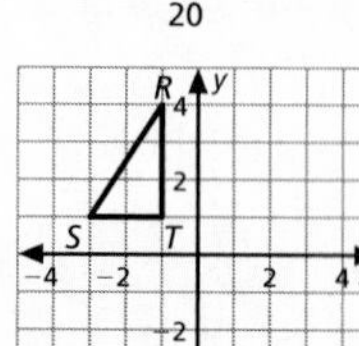

Find each matrix sum or product.

15. $\begin{bmatrix} 0 & 5 \\ 7 & 4 \end{bmatrix} + \begin{bmatrix} 5 & -3 \\ 2 & 4 \end{bmatrix}$ $\begin{bmatrix} 5 & 2 \\ 9 & 8 \end{bmatrix}$
16. $\begin{bmatrix} 1 & 0 \\ 2 & 3 \end{bmatrix} \cdot \begin{bmatrix} 3 & 4 \\ 4 & 6 \end{bmatrix}$ $\begin{bmatrix} 3 & 4 \\ 18 & 26 \end{bmatrix}$
17. A card is drawn at random from a standard deck of playing cards. Find the probability that it is a red card or a 10. $\frac{7}{13}$
18. Five people are arranged in a horizontal line for a picture. In how many ways can the people be arranged? 120

Simplify.

19. $\sqrt{12} \cdot \sqrt{3}$ 6
20. $\frac{\sqrt{200}}{\sqrt{2}}$ 10

Find the value of x.

21. 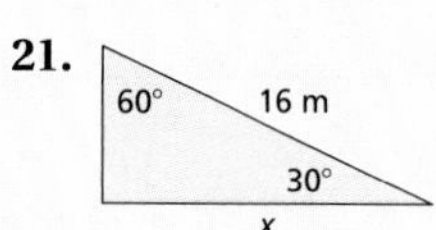

$8\sqrt{3}$

Factor completely.

22. $5x^2 + 15x$ $5x(x + 3)$
23. $4x^2 - 25$ $(2x - 5)(2x + 5)$
24. $x^2 + 2x - 15$ $(x + 5)(x - 3)$
25. $2x^2 - 3x - 2$ $(2x + 1)(x - 2)$

Teaching Strategies

Many of the items in the Cumulative Review can be solved using more than one method. Remind students that knowing more than one method is an advantage, and they should apply the method that seems easiest for that problem. For example, discuss Exercise 11. Have students explain how they solved the system of equations. Some may prefer substitution; others may prefer using multiplication and then subtraction. Discuss the importance of knowing both methods and not relying on a favorite method for all situations.

CHAPTERS 1–12 CUMULATIVE ASSESSMENT

STANDARDIZED TEST PREPARATION—GRIDDED RESPONSE

Solve each question. Write your answer at the top of the answer grid and fill in the ovals.

Notes: Mixed numbers such as $1\frac{1}{2}$ must be gridded as 1.5 or $\frac{3}{2}$. Grid only one answer per question. If your answer is a decimal, enter the most accurate value the grid will accommodate.

1. Evaluate x^{-y} if $x = -2$ and $y = 4$. $\frac{1}{16}$

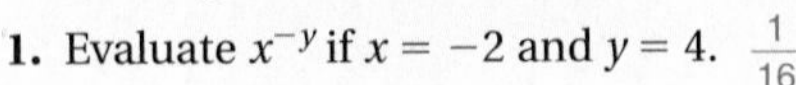

2. Find the mean of the following set of data: 86, 86, 90, 91, 92. 89

3. Which angle corresponds to $\angle 3$? 7

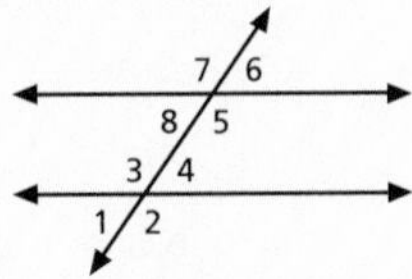

4. If *ABCD* is a trapezoid and *EF* is a median, find the length of *EF*. 22

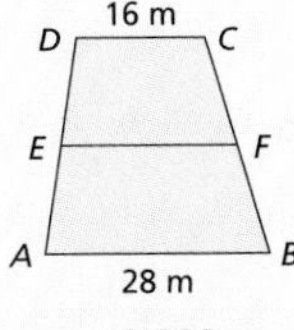

5. Complete. 8.1 mL = 0.008 L

6. Find *m*. $x - 2y = 3$ and $mx - 4y = 5$ are perpendicular. 1

7. Find *x* to the nearest tenth. 9

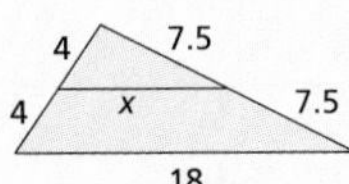

8. What is the *y* coordinate of *P*(3, −5) under a rotation of 180°? 5

9. Find *d*. $\begin{bmatrix} 1 & 2 \\ 0 & 4 \end{bmatrix} \cdot \begin{bmatrix} 1 & 0 \\ 0 & 1 \end{bmatrix} = \begin{bmatrix} a & b \\ c & d \end{bmatrix}$ 4

10. Find the probability of tossing three coins on a row and getting three heads. $\frac{1}{8}$

11. A sack contains 7 red and 3 black marbles. Two marbles are picked out at random, one at a time, then replaced. What is *P*(red, then black)? 0.21

12. Find *x*. 6

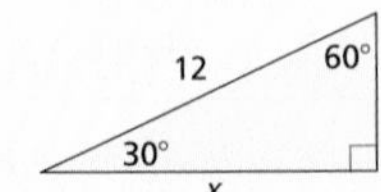

13. Factor: $9x^2 - 25 = (ax + b)(cx - d)$. Find *d*. 5

14. In order to complete the square for $x^2 - 20x$, what number would you add? 100

15. Use the quadratic formula to find the greatest value of *x* for which this equation is true. $8x^2 - 2x - 3 = 0$. $\frac{3}{4}$

Find all the unknown angles measures.)

16. $\angle B = 113°$, $\angle C = \angle D = 68°$

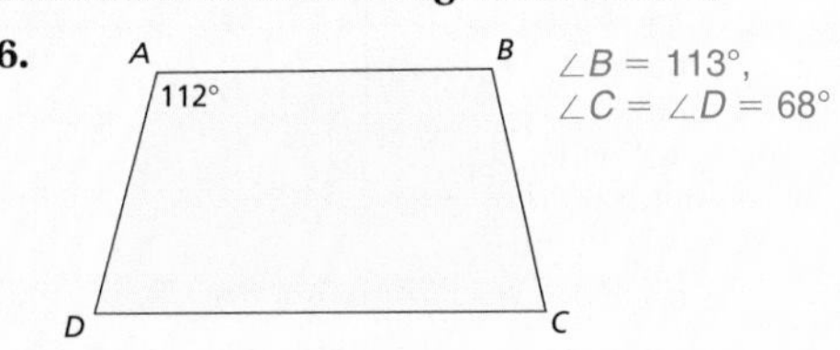

17. $\angle E = \angle F = 135°$, $\angle G = \angle H = 45°$

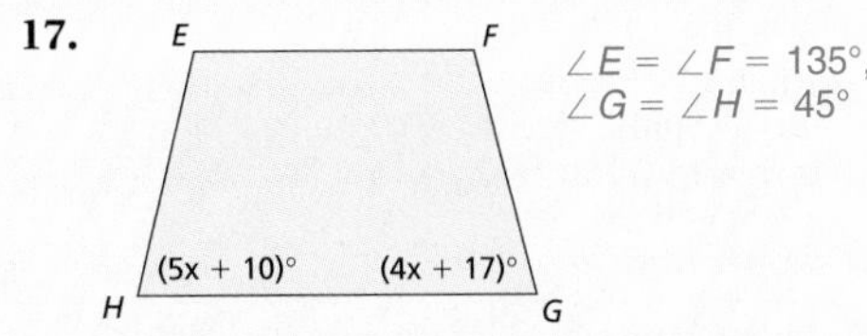

18. CONSTRUCTED RESPONSE Draw a triangle having the points *A*(8, 4), *B*(11, 8), and *C*(10, −2). Label the length of each side of the triangle. Check students' drawing. $AB = \sqrt{5}$, $BC = \sqrt{101}$, $AC = \sqrt{40}$

Teaching Strategies

Have students look in the Data File for the table entitled Polynomial Expansion of Coefficients. Have students construct polynomials for three of the patterns on the table and factor using the pattern as a guide. Discuss how recognition of patterns in mathematics makes solving problems easier.

STANDARDIZED TEST PREPARATION

GRIDDED RESPONSE Questions that require you fill in grids on an answer sheet, but do not have answer choices provided are called gridded response or students-produced response. You must solve the problem and then fill in the answer on a special grid. Your answer consists of two parts-the numbers that you write in the answer boxes and the ovals that you fill in. Ovals that correspond to the digits 0–9, negative sign, fraction bars and decimal points are listed in columns under the answer boxes.

You should grid the form of the answer that you obtain from solving the problem. Enter the most accurate value the grid will accommodate. The grid will only hold numbers that range form 0–9999. The grid will not accommodate a mixed number. Change mixed numbers to either decimals or improper fractions.

Columns not needed should be left blank. It does not matter whether the extra columns are on the left or the right of the answer.

TESTING STRATEGIES

Sometimes fractional answers cannot be entered in the grid because they require too many spaces. Point out Exercise 11. Probabilities are usually written as fractions. The reduced answer to this question requires six spaces in the grid when there are only four. Tell students to convert the fraction of a decimal. Also point out Exercise 5. Because of entering the decimal point first, there's not enough room in the grid for four digits. In this case, the decimal answer should be rounded to the nearest thousandth.

Chapter 13—Quadratic Relations

Theme: Astronomy
Chapter Investigation: Orbital Points, pages 561, 583, 597
Careers: Payload Specialist, page 571; Astronomers, page 589
Data Activity: The Solar System

Content and Connections

Lesson, Pages	Lesson Objectives	NCTM Standards	State/Local Objectives	Interdisciplinary Connections	Real World Applications
13–1 562–565	• Write equations for circles	Patterns, Functions & Algebra Reasoning & Proof Number Operation Connections Representation		Science	Astronomy, Sports, Architecture
13–2 566–569	• Relate the equation of a parabola to its focus and directix	Patterns, Functions & Algebra Problem Solving Number Operation Reasoning & Proof Connections Representation			Satellite Communications, Energy
13–3 572–573	• Solve a problem using visual thinking • Use a picture, diagram or model	Patterns, Functions & Algebra Problem Solving Geometry & Spatial Sense Reasoning & Proof Representation			
13–4 574–577	• Graph equations of ellipses and hyperbolas	Patterns, Functions & Algebra Problem Solving Reasoning & Proof Number Operation Geometry & Spatial Sense Communication Connections Representation			Astronomy, Oceanography, Communications
13–5 580–583	• Solve problems involving direct variation	Number Operation Patterns, Functions & Algebra Problem Solving Reasoning & Proof Connections Representation		Biology	Food Prices, Space Exploration, Physics
13–6 584–587	• Solve problems involving inverse variation and inverse square variation	Number Operation Patterns, Functions & Algebra Problem Solving Reasoning & Proof Connections Representation		Physics	Astronomy, Physics, Travel
13–7 590–593	• Graph quadratic inequalities	Patterns, Functions & Algebra Problem Solving Reasoning & Proof Communication Connections Representation			Astronomy, Communications, Computer Design

Planning and Resources

Lesson, Pages	Tools/Materials Needed	Trans-parency	Learning/Teaching Styles Options	Assignments: Basic Enriched	Additional Practice in SE	Reteaching, Extra Practice, Enrichment	Other Resources
13–1 562–562	graphing calculator graph paper	Warm up 37 Trans Tk-7–10, RF-49	ESL/LEP	B: 1–26, 31–45 E: 1–45	Page 570–571, 579, 589, 594, 725	R: page 191 EP: page 167 E: page 217	

13–2 566–569	graphing calculator graph paper	Warm up 37 Trans TK-7–10	Real World Challenge	B: 1–22, 27–49 E: 1–49	Page 570–571, 579, 589, 594, 725	R: page 193 EP: page 169 E: page 219	SS: Teacher's Choice
13–3 572–573	pencil/paper	Warm up 37 Trans Tk-7–10	Challenge	B: 1–16 E: 1–16	Page 578–579, 589, 594	R: page 195 E: page 221	SS: Teacher's Choice
13–4 574–577	cardboard string scissors thumbtacks graph paper graphing calculator	Warm up 38 Trans Tk-7–10, RF-50		B: 1–14, 17–22 E: 1–22	Page 578–579, 589, 595, 726	R: page 197 EP: page 171 E: page 223	
13–5 580–583	calculator	Warm up 38 Trans RF-51	Real World	B: 1–24, 31–44 E: 1–44	Page 588–589, 595, 726	R: page 199 EP: page 173 E: page 225	
13–6 584–587	calculator	Warm up 39 Trans RF-52		B: 1–16, 23–31 E: 1–31	Page 588–589, 595, 727	R: page 201 EP: page 175 E: page 227	SS: Teacher's Choice
13–7 590–593	graph paper graphing calculator	Warm up 39		B: 1–22, 27–45 E: 1–45	Page 595, 728	R: page 209 EP: page 177 E: page 229	

Planning and Pacing

Lesson, Pages	Lesson Title	45 min class	Assignments Basic, Enriched	90 min class	Assignments Basic, Enriched	___ min class	Assignments Basic, Enriched
13–1 562–565	AYR, Opener, The Standard Equation of a Circle	Day 164	B: 1–26, 31–45 E: 1–45	Day 90	B: 1–26, 31–45 E: 1–45		B: 1–26, 31–45 E: 1–45
13–2 566–569	More on Parabolas, R&PYS	Day 165–166	B: 1–22, 27–49 E: 1–49	Day 90–91	B: 1–22, 27–49 E: 1–49		B: 1–22, 27–49 E: 1–49
13–3 572–573	Problem Solving Skills: Visual Thinking	Day 167	B: 1–16 E: 1–16	Day 91	B: 1–16 E: 1–16		B: 1–16 E: 1–16
13–4 574–577	Ellipses and Hyperbolas, R&PYS	Day 168–169	B: 1–14, 17–22 E: 1–22	Day 92	B: 1–14, 17–22 E: 1–22		B: 1–14, 17–22 E: 1–22
13–5 580–583	Direct Variation	Day 170	B: 1–24, 31–44 E: 1–44	Day 93	B: 1–24, 31–44 E: 1–44		B: 1–24, 31–44 E: 1–44
13–6 584–587	Inverse Variation, R&PYS	Day 171–172	B: 1–16, 23–31 E: 1–31	Day 93–94	B: 1–16, 23–31 E: 1–31		B: 1–16, 23–31 E: 1–31
13–7 590–593	Quadratic Inequalities	Day 173	B: 1–22, 27–45 E: 1–45	Day 94	B: 1–22, 27–45 E: 1–45		B: 1–22, 27–45 E: 1–45
Review/ Assess		Day 174–175		Day 95–96			

	Chapter 1 Days	Chapter Cumulative Days
Yearly Pacing (45 min class)	12 days	175 days
Yearly Pacing (90 min class)	7 days	96 days

Assessment Options

Assessment in Student Edition	Assessment in Teacher's Edition	Pages in Assessment Book	Software Generated Assessment
Are You Ready?, pages 558–559; Writing Math, pages 564, 569, 573, 577, 582, 586, 593; Mixed Review, pages 565, 569, 573, 577, 583, 587, 593; Check Understanding pages 562, 566, 574, 580, 584, 590; Mid-Chapter Quiz, page 579, Chapter Review, page 594; Chapter Assessment, page 596; Alternative Assessment, page 597; Cumulative Review, page 598; Cumulative Assessment, page 599	5-minute Warm ups, pages 562, 566, 572, 574, 580, 584, 590; Quick Assessment, pages 564, 568, 573, 576, 582, 586, 592; Scoring Rubrics, page 597	Mid-Chapter Quiz, page 199; Test Form A, pages 201, 203; Test Form B, pages 205, 207; Cumulative Test 209, 211; Math Journal prompt, page 213, 214;	Chapter 13

Refresher Skills

The skills on these two pages are skills that have been taught in previous math courses. Continuous review of basic math skills will make stronger math students. These skills are identified as necessary to be successful in Chapter 13.

Skills Correlation Chart

Skill	Lesson Number
Circles	13-1, 13-4, 13-7
Solve Proportions	13-5, 13-6
Solve Systems of Equations by Graphing	13-7

Vocabulary

radius
diameter
circumference
cross-multiplication
system of equations

Assignment Guide

All students: 1–30
Additional Practice: See Extra Practice Index on page 674.

Are You Ready?

Refresh Your Math Skills for Chapter 13

The skills on these two pages are ones you have already learned. Review the examples and complete the exercises. For additional practice on these and more basic skills, see page 674.

Circles

Because you will be doing a lot of work with circles in this chapter, it may be helpful to review some of the basics.

Example What is the diameter and the circumference of this circle?

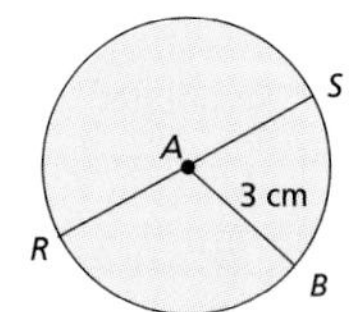

AB is a radius (r) of the circle.
RS is a diameter (d) of the circle.
A is the center point of the circle.

$d = 2r$	$C = 2\pi r$
$d = 2(3)$	$C = 2(3.14)(3)$
$d = 6$ cm	$C = 18.84$ cm

Find the diameter and circumference for each circle. Use $\pi = 3.14$. Round answers to the nearest hundredth if necessary.

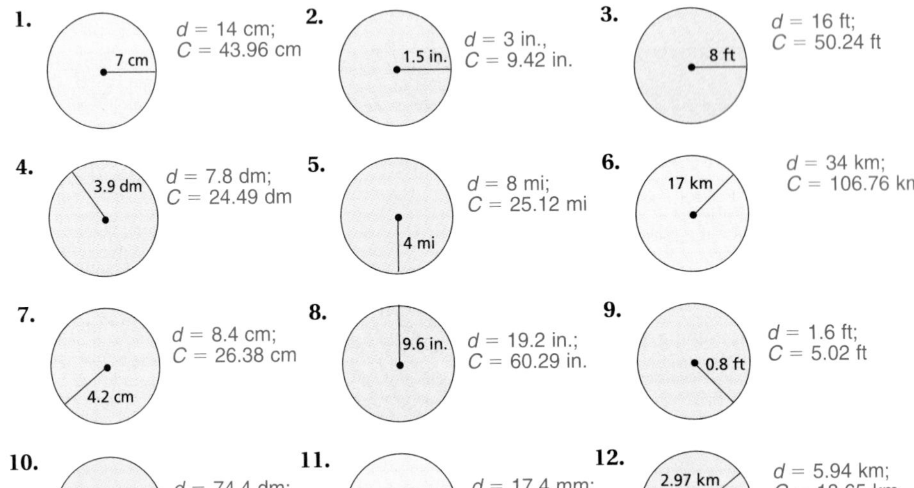

Extend the Lesson

MATH JOURNAL Have students describe a real-life situation in which proportions could be used to solve a problem. Have students write a sample problem for this situation and show the steps of the solution. If students are having difficulty thinking of a situation, remind them that proportions are often used in sports, shopping, and trip planning.

CHAPTER 13

Solve Proportions

Example Find x: $\frac{9}{12} = \frac{x}{30}$

Use cross-multiplication:

$$\frac{9}{12} = \frac{x}{30}$$
$$9 \cdot 30 = 12 \cdot x$$
$$180 = 12x$$
$$15 = x$$

Find the value of x in each proportion. Round answers to the nearest tenth if necessary.

13. $\frac{4}{12} = \frac{x}{28}$ 9.3 **14.** $\frac{9}{6} = \frac{30}{x}$ 20 **15.** $\frac{4}{5} = \frac{100}{x}$ 125

16. $\frac{8}{5} = \frac{44}{x}$ 27.5 **17.** $\frac{96}{8} = \frac{24}{x}$ 2 **18.** $\frac{108}{4} = \frac{x}{6}$ 162

19. $\frac{38}{9} = \frac{x}{14}$ 59.1 **20.** $\frac{5}{20} = \frac{7}{x}$ 28 **21.** $\frac{3}{17} = \frac{22}{x}$ 124.7

22. $\frac{x}{4} = \frac{15}{56}$ 1.1 **23.** $\frac{19}{13} = \frac{31}{x}$ 21.2 **24.** $\frac{71}{86} = \frac{x}{50}$ 41.3

Solve Systems of Equations by Graphing

Example Solve: $\begin{cases} 2x = y = 1 \\ y - x = -5 \end{cases}$

To solve, graph each equation independently on the same coordinate plane. The point at which the lines intersect is the solution.

The solution is (2, −3).

Solve each system of equations by graphing.

25. $\begin{cases} 3x - 2y = -11 \\ -2x + 3y = 9 \end{cases}$ (−3, 1) **26.** $\begin{cases} x - y = 2 \\ 3y - x = 2 \end{cases}$ (4, 2)

27. $\begin{cases} x - 2y = -2 \\ 2y - 2 = x \end{cases}$ no solution **28.** $\begin{cases} x - 6y = -1 \\ 4y + x = 1 \end{cases}$ (5, −1)

29. $\begin{cases} 2x + 2 = y \\ 3y - 5x = 3 \end{cases}$ (−3, −4) **30.** $\begin{cases} 4y - 3x = 5 \\ x + 3 = 2y \end{cases}$ (1, 2)

Chalkboard Examples

Circles

Find the diameter and circumference for a circle with a radius of 4.5 cm. Use $\pi = 3.14$. diameter = 9 cm, circumference = 28.26 cm²

Solve Proportions

Find the value of x. $\frac{92}{250} = \frac{x}{150}$
$x = 55.2$

Solve Systems of Equations by Graphing

Solve by graphing: $\begin{matrix} y = 3x + 5 \\ 2y + 6x = -2 \end{matrix}$.

The solution is (−1, 2).

Refresher Wrap-up

Quick Assessment

Ask the following questions to determine if students have mastered the basic skills reviewed on these pages.

1. How can you find the diameter of a circle when the radius is given? Multiply the radius by 2.
2. Describe the steps needed to solve the proportion: $\frac{5}{8} = \frac{x}{96}$. Multiply 5 and 96; divide by 8.

Additional Practice

All topics: Refer to the Extra Practice Index on page 674 of this text.

Extend the Lesson

COOPERATIVE LEARNING Have students work in pairs to draw four circles of different sizes using a compass. Then have students make the necessary measurements to find the circumference of each circle.

Chapter Opener

NCTM Standards/Strands

- Number & Operation
- Patterns, Functions, & Algebra
- Geometry & Spatial Sense
- Measurement
- Problem Solving
- Reasoning and Proof
- Communication
- Connections
- Representation

Vocabulary

mass
orbital speed
perihelion
aphelion

Theme Connections

Tables and graphs are used to organize astronomical data. Tables provide a way to organize specific numerical information. Many types of data are kept in regards to astronomy. When new discoveries are made, astronomers make the new data available to the public and other scientists by creating tables and charts. Discuss how data and graphs are used in astronomy and related sciences.

Career Opportunities

Many careers related to astronomy make use of data and graphs. Two are highlighted in the Math-Works features. Others include navigators, technicians, engineers, artists, and museum curators.

- Astronauts, page 571
- Astronomers, page 589

Internet Connection

Theme Activities

Learningmathmatters.com provides web addresses to search that help students gather information about the use of math in the real world, particularly data and graphs. To search for additional addresses, begin a search using the keyword *astronomy*. Then within that search, use such key words as *solar system*, *star, asteroid* and *space travel*. Students can brainstorm in small groups other key words.

CHAPTER 13

QUADRATIC RELATIONS

THEME: Astronomy

Have you ever tried counting the stars at night? On a clear night, far away from bright city lights, you can see about 3,000 stars using only your eyes. The sun, moon, stars and planets are always present.

Today, astronomers use computers, telescopes and satellites to chart the heavens. Through color-spectrum analysis, they explore the history of our galaxy and solar system.

- **Astronauts** (page 571) are pilots and scientists who travel in space. On a space shuttle, there are pilots, mission specialists and payload specialists. They are skilled in working with many different types of scientific equipment and in overseeing and conducting experiments.
- **Astronomers** (page 589) use physics and mathematics to study the universe through observation and calculation. The knowledge gained through the science of astronomy helps in related fields of navigation and space flight. Astronomers work with engineers to design and launch space probes and satellites to gather data and transmit it back to Earth.

Internet Connection
www.learningmathmatters.com

Chapter Investigation

Go to learningmathmatters.com to locate additional information about astronomy.

The Solar System

Planet	Length of year	Mass in Earth masses	Diameter	Average orbital speed
Mercury	87.97 Earth days	0.055	3,031 mi	29.8 mi/sec
Venus	224.7 Earth days	0.81	7,521 mi	21.8 mi/sec
Earth	365.26 Earth days	1	7,926 mi	18.5 mi/sec
Mars	1.88 Earth years	0.11	4,217 mi	15 mi/sec
Jupiter	11.86 Earth years	318	88,850 mi	8.1 mi/sec
Saturn	29.46 Earth years	95.18	74,901 mi	6 mi/sec
Uranus	84.01 Earth years	14.5	31,765 mi	4.2 mi/sec
Neptune	164.79 Earth years	17.14	30,777 mi	3.4 mi/sec
Pluto	248.54 Earth years	0.0022	1,429 mi	2.9 mi/sec

Data Activity: The Solar System

Use the table for Questions 1–4.

1. To the nearest million square miles, what is the surface area of the Earth? Assume the planet's shape is spherical.
   197,000,000 mi
2. Using the orbital speed and length of year in Earth days, calculate the length of Venus' orbit in miles.
   423,226,944 mi
3. An astronomer is 58 years old in Earth years. What is the astronomer's age in Martian years? 30 years old nearly 30
4. Find the circumference of Uranus to the nearest mile.
   99,742 mi

Chapter Investigation

How far is the Earth from the Sun? Because the Earth's orbit is elliptical, its distance from the Sun varies from 91.4 million miles to 94.5 million miles. The closest orbital point is called the *perihelion.* the farthest orbital point is called the *aphelion.* Astronomers have calculated each planet's aphelion and perihelion.

Working Together

Vast distances in space are difficult to imagine. Reducing these distances to a familiar scale can make them easier to visualize. Research the aphelion and perihelion for the nine planets of the Solar System. Then choose a location, such as your home or school, to represent the Sun. Using a system of maps, plot the aphelion and perihelion for each planet. Use the Chapter Investigation logos to guide your group.

Project Planning Calendar

Name ____________ Date ____________

CHAPTER 13 PROJECT PLANNING CALENDAR

Benchmarks	PROJECT GOAL
a. Research the perihelion and aphelion for each planet of the Solar System. Make a rough sketch of the planets' orbits and label the perihelion and aphelion for each planet. *(Lesson 13-2)* b. Choose a point within your school grounds or community to represent the Sun. Using a map of your school or community and an appropriate scale, plot the location of the aphelion and perihelion for each planet on the map. Make a rough sketch of the orbits of the planets. *(Lesson 13-5)* c. Select landmarks at your school or within your community to represent aphelion, or farthest ... Mark these ...	To create a map of the Solar System using familiar landmarks as reference points.

Group Project Planner

Name ____________ Date ____________

CHAPTER 13 GROUP PROJECT PLANNER

Assignment ____________ Objective ____________

Group Members | Assigned Roles
1) ____________
2) ____________
3) ____________
4) ____________
5) ____________

Deadlines | Done

Data Activity

Discuss with students the importance of comparing data to better understand the relationship among the elements in a system. Point out that we can better understand our solar system by comparing facts about the planets to the facts about the planet we know best: Earth. Discuss the terms used in the table. Assign students Questions 1–4.

Extend the Data Activity

REAL WORLD CONNECTION Have students gather additional information about one planet on the table. Then have students compile their research to make additional tables showing known facts about the solar system.

Chapter Investigation

As an Overarching Problem

Display a drawing of the Solar System for students to study. Review the names of the planets and their relationship to the Sun. Students will continue to work on the investigation as they complete the exercises identified by the Investigation Logo that are found throughout the chapter. These exercises will guide students through the task described in *Working Together*. Encourage students to keep all of their work on the Investigation together. Have students use the suggestions in the Chapter Investigation Extension to summarize their work.

As a Chapter Project

The goal of this project is create a map of the Solar System using familiar landmarks as reference points. Students can use the Group Project Planner on page 215 and the Project Planning Calendar on page 216 in the Enrichment text to complete the project. Benchmarks **a**, **b**, and **c**. should be completed after the lesson listed in parentheses has been studied. Benchmark **d** should be completed at the end of the chapter.

Lesson Planning

NCTM Standards/Strands

- Number & Operation
- Patterns, Functions, & Algebra
- Geometry & Spatial Sense
- Problem Solving
- Reasoning and Proof
- Communication
- Connections
- Representation

Vocabulary

standard equation of a circle
radius

Materials Needed

graphing calculators
graph or grid paper

Lesson Resources

Warm-Up Transparency 37
Transparency TK-7–10, RF-49
Reteaching 13-1
Extra Practice 13-1
Enrichment 13-1

Assignment Guide

Basic: 1–26, 31–45
Enriched: 1–45

Getting Started

5-Minute Warm-up

Find the distance between each pair of points.

1. (3, 5) and (9, 1) $2\sqrt{13}$
2. (−1, 0) and (6, −4) $\sqrt{65}$
3. (−2, −4) and (8, 4) $2\sqrt{41}$

Introduction to Lesson 13-1

Before students begin to study the equation for the circle with center (0, 0) and radius 4, discuss the drawing with them. Help them see that the distance of point (x, y) on the circle from (0, 0) can be thought of as the hypotenuse of the right triangle formed by moving x units along the x-axis from the origin and then y units up to the circle. The length of this segment connecting (x, y) and (0, 0) can be found using the distance formula.

13-1 The Standard Equation of a Circle

Goals ■ Write equations for circles.
Applications Astronomy, Sports, Architecture

Work with a partner. You will need a compass or geometric drawing software.

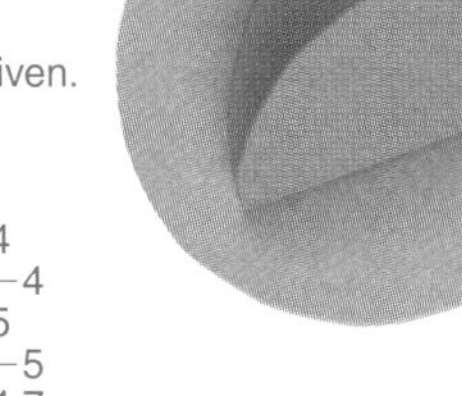

1. Draw a circle on a coordinate grid with a radius of 5 units.
Check students' work.
2. Complete the table below by estimating the missing y-coordinates for points on the circle. Then find the values of the expressions in the third and fourth columns. Note that there will be two different y-coordinates for each x-coordinate. Estimations will vary. Samples given.

x	y	x^2+y^2	$\sqrt{x^2+y^2}$	
3				4
3				−4
0				5
0				−5
1.8				4.7
1.8				−4.7

3. Explain the patterns that you see in the table. Answers will vary.

Build Understanding

You can substitute into the distance formula to find the equation for a circle with its center located at any coordinate point (h, k).

$$d = \sqrt{(x_2 - x_1)^2 + (y_2 - y_1)^2}$$
$$r = \sqrt{(x - h)^2 + (y - k)^2}$$
$$r^2 = \sqrt{(x - h)^2 + (y - k)^2}$$

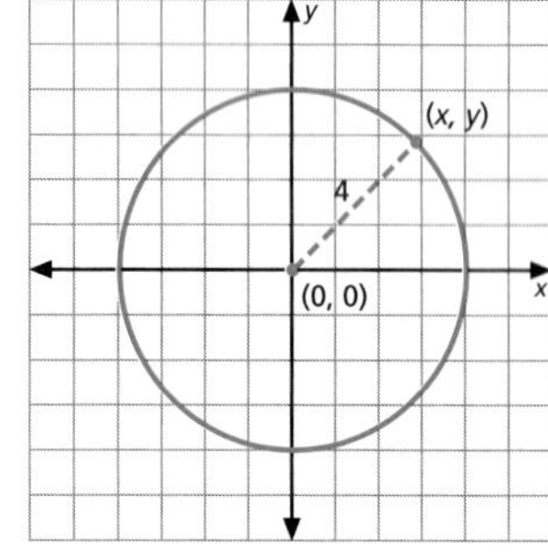

The **standard equation of a circle** is

$(x - h)^2 + (y - k)^2 = r^2, r \neq 0.$

If $h = 0$ and $k = 0$, then the standard equation reduces to $x^2 + y^2 = r^2$.

Example 1

Write an equation for a circle with radius 6 units and center (0, 0).

Solution

Because the center is at the origin, substitute in the equation $x^2 + y^2 = r^2$.

$x^2 + y^2 = 6^2$ or

$x^2 + y^2 = 36$

Extend the Lesson

ONGOING ASSESSMENT Have students find the radius and center of a circle whose equation is $(x + 3)^2 + (y + 1)^2 = 4$. Watch for students who get the incorrect answer of center (3, 1) and/or a radius of 16. These students did not realize that since the standard form of the equation calls for the subtraction of h and k, the center must be (−3, −1). Also, students may make the error of squaring the 4 instead of taking its square root.

562 Chapter 13 **Quadratic Relations**

Example 2

ASTRONOMY An astronomer is creating a computer model of a moon by entering the equation for a circle with a radius of 4 units and the center located at point (3, −2). Write the equation.

Solution

Substitute into the standard form for the equation of a circle.

$(x - h)^2 + (y - k)^2 = r^2$

$(x - 3)^2 + (y - (-2))^2 = 4^2$

$(x - 3)^2 + (y + 2)^2 = 4^2$ or

$(x - 3)^2 + (y + 2)^2 = 16$

Example 3

GRAPHING Find the radius and center of the circle $x^2 + y^2 = 9$. Then graph the circle using a graphing calculator.

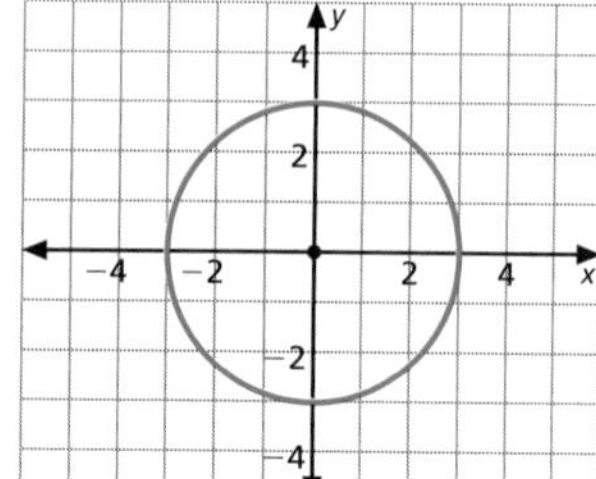

Solution

Because the equation is of the form $x^2 + y^2 = r^2$, the center is at the origin.

$x^2 + y^2 = r^2$

$x^2 + y^2 = 9$

$r^2 = 9$

$r = 3$

To graph a circle using a graphing calculator, rewrite the equation in terms of *y*.

$x^2 + y^2 = 9$

$y^2 = 9 - x^2$

$y = \pm\sqrt{9 - x^2}$

For most graphing calculators, you must enter separate formulas for the upper and lower portions of the circle. At the Y = screen, enter the functions $Y_1 = \sqrt{9 - x^2}$ and $Y_2 = -Y_1$.

Use the ZOOM menu to adjust the display so that each pixel represents a square with an equal width and height and graph. The graph is a circle with radius 3 and center at the origin.

You can also find the radius and center of a circle using the standard form for the equation of a circle.

Chalkboard Examples

Supplementary Example 1

Write an equation for a circle with radius 8 units and center (0, 0). $x^2 + y^2 = 64$

Supplementary Example 2

To create a computer animation of a sun, a programmer must find the equation of a circle with radius 6 units and center located at (−3, 4). $(x + 3)^2 + (y - 4)^2 = 36$

Supplementary Example 3

Find the radius and center of the circle $x^2 + y^2 = 4$. Then graph the circle. The center is (0, 0) and the radius is 2. Check students' graphs.

Supplementary Example 4

Find the radius and center of the circle $(x - 3)^2 + (y + 5)^2 = 28$. The center is (3, −5) and the radius is $2\sqrt{7}$.

Extend the Lesson

MATH JOURNAL Have students write an explanation of the relationships described in this lesson. They should include, in their own words, how the distance formula is used to find the length of the radius, and how the standard equation of a circle is derived from the length of the radius and the coordinates of the center point.

Reteaching Worksheet 13-1

Name ______________________ Date ____________

RETEACHING **13-1**

THE STANDARD EQUATION OF A CIRCLE

The standard equation of a circle is $(x - h)^2 + (y - k)^2 = r^2$, where $r \neq 0$. The point (x, y) lies on the circumference of the circle, and point (h, k) is the center. If $h = 0$ and $k = 0$, the standard equation reduces to $x^2 + y^2 = r^2$.

Example 1

Write an equation for each circle with the given radius and center.

a. radius 3 units and center (0, 0)

b. radius 3 units and center (−3, 2)

Solution

a. Since the center is at the origin, the equation is of the form $x^2 + y^2 = r$.

$x^2 + y^2 = r^2$

$x^2 + y^2 = 3^2$

$x^2 + y^2 = 9$

b. Substitute into the standard form for the equation of a circle.

$(x - h)^2 + (y - k)^2 = r^2$

$(x - (-3))^2 + (y - 2)^2 = 3^2$ or

$(x + 3)^2 + (y - 2)^2 = 9$

Example 2

Find the radius and the center for each circle.

a. $x^2 + y^2 = 25$

b. $(x + 2)^2 + (y - 3)^2 = 10$

Solution

a. Since the equation is of the form $x^2 + y^2 = r^2$, the center is at the origin.

$x^2 + y^2 = r^2$

$x^2 + y^2 = 25$

$r^2 = 25$, so $r = 5$

The radius is 5; the center (0, 0).

b. Substitute into the standard form for the equation of a circle.

$(x - h)^2 + (y - k)^2 = r^2$

$(x - (-2))^2 + (y - 3)^2 = 10$

$r^2 = 10$, so $r = \sqrt{10}$

The radius is $\sqrt{10}$; the center (−2, 3).

EXERCISES

Write an equation for each circle.

1. radius 5, center (3, −4)
   $(x - 3)^2 + (y + 4)^2 = 25$
2. radius 3, center (−5, 0)
   $(x + 5)^2 + y^2 = 9$

Find the radius and center for each circle.

3. $x^2 + y^2 = 144$
   $r = 12$; center (0, 0)
4. $(x - 3)^2 + (y + 6)^2 = 21$
   $r = \sqrt{21}$; center (3, − 6)
5. $(x + 4)^2 + (y + 5)^2 = 34$
   $r = \sqrt{34}$; center (−4, −5)
6. $(x - 2)^2 + (y - 4)^2 = 18$
   $r = \sqrt{18}$; center (2, 4)

192 RETEACHING *LESSON 13-1*

Lesson Wrap-up

Quick Assessment

Ask the following questions to determine if students understand the content presented in this lesson.

1. How can you tell by inspecting the equation the signs of the values h and k? Since the standard from of the equation shows the subtraction of h and k, if h or k is subtracted, h or k must be positive. If h or k is added, the variable must be negative.
2. Will the standard equation of a circle be applicable to a circle whose circle is in any of the four quadrants? Why? Yes, since each term in the standard form is squared, the radius will always be a positive integer.

Assignment Guide

Basic: 1–26, 31–45
Advanced: 1–45
Additional Practice: See Extra Practice Index on page 674.

Additional Answers

27. $x^2 + y^2 = 81$ *or* $(x - 9)^2 + y^2 = 81$
28. $(x - 2)^2 + (y + 3)^2 = 4$ or $(x - 4)^2 + (y + 3)^2 = 4$

Example 4

Find the radius and center of the circle $(x - 5)^2 + (y + 4)^2 = 18$.

Solution

Substitute into the standard form for the equation of a circle.

$(x - h)^2 + (y - k)^2 = r^2$

$(x - 5)^2 + (y - (-4))^2 = (\sqrt{18})^2$

The center (h, k) is $(5, -4)$. The radius is $\sqrt{18}$.

Try These Exercises

Write an equation for each circle.

1. radius 7 $x^2 + y^2 = 49$
   center (0, 0)
2. radius 10
   center (−3, 0)
   $(x + 3)^2 + y^2 = 100$
3. 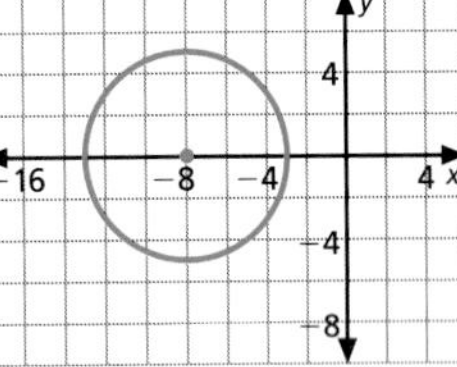

Find the radius and center for each circle.

4. $x^2 + y^2 = 100$ 10, (0, 0)
5. $x^2 + y^2 = 11$ $\sqrt{11}$, (0, 0)
6. $(x - 4)^2 + (y - 2)^2 = 49$ 7, (4, 2)
7. $(x + 8)^2 + (y - 4)^2 = 13$ $\sqrt{13}$, (−8, 4)
8. **WRITING MATH** Is the equation for a circle a function? Explain your reasoning. No, y is a quadratic term so the graph does not pass the vertical line test.

Practice Exercises

Write an equation for each circle.

9. radius 12 $x^2 + y^2 = 144$
   center (0, 0)
10. radius 9 $(x - 4)^2 + (y + 6)^2 = 81$
    center (4, −6)
11. radius 12 $(x - 9)^2 + (y - 4)^2 = 144$
    center (−9, −4)
12. radius 11 $x^2 + (y - 4)^2 = 121$
    center (0, 4)
13. **DESIGN** The position of a circular knob on a design for DVD-player is shown on the graph to the right. Find the equation of the circle. $(x - 5)^2 + (y + 1)^2 = 4$

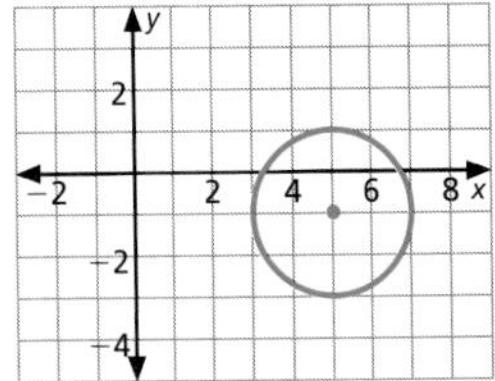

14. **SCIENCE** In a model, two metal balls are attached to the endpoints of a metal rod. The rod is attached to a machine at its center point. The machine spins the rod so that the balls move in a circle modeled by the equation $x^2 + y^2 = 144$. Find the length of the rod. 24 units

Teaching Strategies

ESL/LEP Students may benefit from drawing a circle on a coordinate grid and labeling the center (h, k), a radius, a point on the circle (x, y), and the x- and y-axes. While working the exercises, students can refer to the drawing to remind themselves which part of the circle is being referred to.

15. **SPORTS** A circular target is set up for hang glider landings. Write an equation to model a circle with a diameter of 5.2 meters and a center of (0, 0). $x^2 + y^2 = 6.76$

Find the radius and center for each circle.

16. $x^2 + y^2 = 121$ 11, (0, 0)
17. $x^2 + y^2 = 15$ $\sqrt{15}$, (0, 0)
18. $(x - 1)^2 + (y - 3)^2 = 9$ 3, (1, 3)
19. $(x + 4)^2 + (y - 2)^2 = 14$ $\sqrt{14}$, (−4, 2)
20. $(x + 5)^2 + (y + 12)^2 = 17$ $\sqrt{17}$, (−5, −12)
21. $(x - 12)^2 + (y + 6)^2 = 20$ $2\sqrt{5}$, (12, −6)
22. $x^2 + (y - 4)^2 = 17$ $\sqrt{17}$, (0, 4)
23. $(x - 9)^2 + (x + 1)^2 = 400$ 20, (9, −1)
24. $(x + 3)^2 + y^2 = 50$ $5\sqrt{2}$, (−3, 0)
25. $(x + 1)^2 + (y - 9)^2 = 81$ 9, (−1, 9)

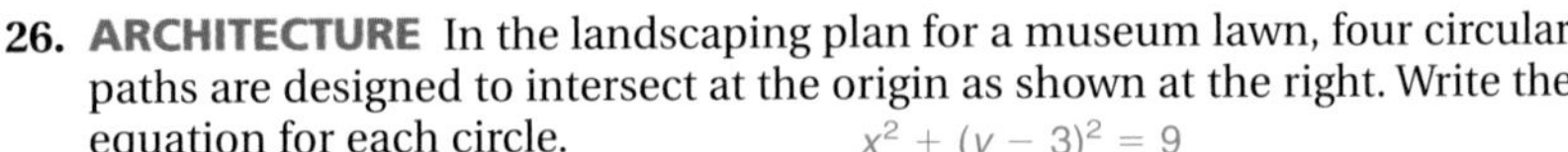

26. **ARCHITECTURE** In the landscaping plan for a museum lawn, four circular paths are designed to intersect at the origin as shown at the right. Write the equation for each circle.
$x^2 + (y - 3)^2 = 9$
$x^2 + (y + 3)^2 = 9$
$(x - 3)^2 + y^2 = 9$
$(x + 3)^2 + y^2 = 9$

Extended Practice Exercises

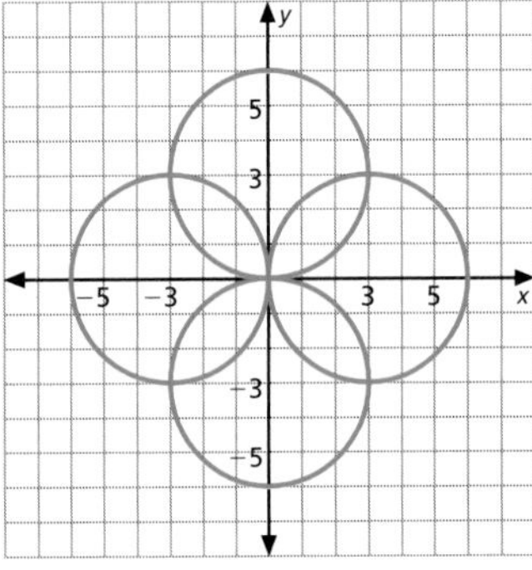

Write two equations for each circle. The endpoints of the radius are given. For 27–28, see additional answers.

27. (0, 0), (9, 0)
28. (2, −3), (4, −3)

Find the radius and center point for each equation.

29. $x^2 + y^2 = 4x + 2y + 4$ 3, (2, 1)
30. $x^2 + y^2 = 16x$ 8 (8, 0)

Mixed Review Exercises

For each function, name the coordinates of the vertex. (Lesson 12-1)

31. $y = 3x^2$ (0, 0)
32. $y = x^2 - 4$ (0, −4)
33. $y = -2x^2 + 3$ (0, 3)
34. $y = -x^2 + 2$ (0, 2)
35. $y = -4x^2 + 6$ (0, 6)
36. $y = x^2 - 5$ (0, −5)

Find the slope of the line containing the given points. (Lesson 6-1)

37. $A(-3, 4)$, $B(6, -2)$ $-\frac{2}{3}$
38. $C(3, 8)$, $D(-2, -5)$ $\frac{13}{5}$
39. $E(-4, -5)$, $F(5, 4)$ 1
40. $G(7, -3)$, $H(2, 5)$ $-\frac{5}{8}$
41. $J(-2, 6)$, $K(-2, -5)$ undefined
42. $L(4, -3)$, $M(-4, 6)$ $-\frac{9}{8}$
43. $N(5, 3)$, $P(-4, -7)$ $\frac{10}{9}$
44. $Q(-5, 3)$, $R(2, 3)$ 0
45. $S(3, 0)$, $T(-3, 3)$ $-\frac{1}{2}$

Teaching Strategies

Students can graph circles on a graphing calculator by first solving for y. Once in function form, they can graph both the positive and negative square roots of y. For example, to graph $x^2 + y^2 = 4$, solve for y: $y = \pm\sqrt{4 - x^2}$; therefore $y = +\sqrt{4 - x^2}$ and $y = -\sqrt{4 - x^2}$. Both functions combine to form the graph of the circle.

Extra Practice Worksheet 13-1

Name ______ Date ______

EXTRA PRACTICE 13-1

THE STANDARD EQUATION OF A CIRCLE

EXERCISES

Write an equation for each circle.

1. radius 3, center (0, 0) $x^2 + y^2 = 9$
2. radius 5, center (0, −5) $x^2 + (y + 5)^2 = 25$
3. radius 10, center (2, 1) $(x - 2)^2 + (y - 1)^2 = 100$
4. radius 7, center (−4, 3) $(x + 4)^2 + (y - 3)^2 = 49$
5. radius 15, center (−1, −2) $(x + 1)^2 + (y + 2)^2 = 225$
6. radius 1, center (5, −6) $(x - 5)^2 + (y + 6)^2 = 1$

7. $(x - 1)^2 + (y - 2)^2 = 25$
8. $(x + 2)^2 + (y - 2)^2 = 16$

Find the radius and center for each circle.

9. $x^2 + y^2 = 36$ 6, (0, 0)
10. $x^2 + y^2 = 12$ $2\sqrt{3}$, (0, 0)
11. $(x - 3)^2 + (y - 2)^2 = 16$ 4, (3, 2)
12. $(x + 10)^2 + (y - 4)^2 = 100$ 10, (−10, 4)
14. $(x - 15)^2 + (y + 12)^2 = 196$ 14, (15, −12)
15. $(x + 8)^2 + (y + 18)^2 = 75$ $5\sqrt{3}$, (−8, −18)
16. $(x - 6)^2 + (y + 12)^2 = 80$ $4\sqrt{5}$, (6, −12)
17. $(x + 18)^2 + (y - 14)^2 = 24$ $2\sqrt{6}$, (−18, 14)

Write two equations for each circle. The endpoints of the radius are given.

18. (0, 0), (4, 0) $x^2 + y^2 = 16$, $(x - 4)^2 + y^2 = 16$
19. (3, 5), (5, −3) $(x - 3)^2 + (y - 5)^2 = 4$, $(x - 5)^2 + (y - 3)^2 = 4$

168 EXTRA PRACTICE *LESSON 13-1*

Enrichment Worksheet 13-1

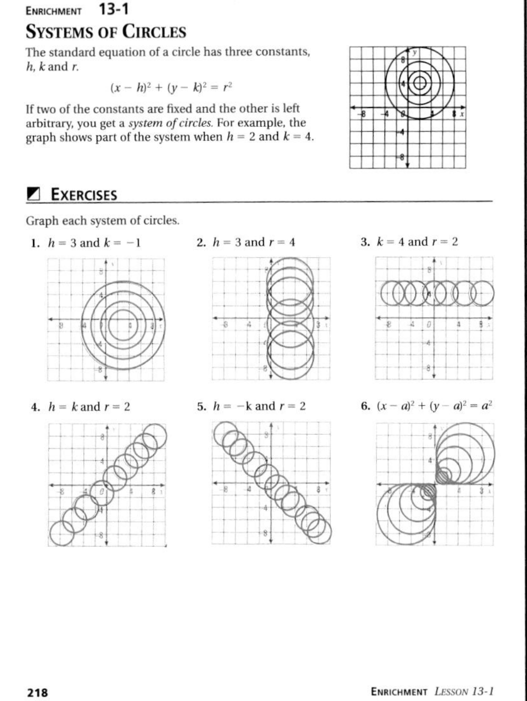

Name ______ Date ______

ENRICHMENT 13-1

SYSTEMS OF CIRCLES

The standard equation of a circle has three constants, h, k and r.

$(x - h)^2 + (y - k)^2 = r^2$

If two of the constants are fixed and the other is left arbitrary, you get a *system of circles*. For example, the graph shows part of the system when $h = 2$ and $k = 4$.

EXERCISES

Graph each system of circles.

1. $h = 3$ and $k = -1$
2. $h = 3$ and $r = 4$
3. $k = 4$ and $r = 2$
4. $h = k$ and $r = 2$
5. $h = -k$ and $r = 2$
6. $(x - a)^2 + (y - a)^2 = a^2$

218 ENRICHMENT *LESSON 13-1*

Lesson Planning

NCTM Standards/Strands

- Number & Operation
- Patterns, Functions, & Algebra
- Geometry & Spatial Sense
- Problem Solving
- Reasoning and Proof
- Communication
- Connections
- Representation

Vocabulary

focus
directrix

Materials Needed

graphing calculator   graph paper

Lesson Resources

Warm-Up Transparency 37
Transparency TK-7–10
Reteaching 13-2
Extra Practice 13-2
Enrichment 13-2

Assignment Guide

Basic: 1–22, 27–49
Enriched: 1–49

Getting Started

5-Minute Warm-up

Solve each equation for *a*.

1. $4a = 4$ 1
2. $4a = 9$ $\frac{9}{4}$
3. $4a = -8$ -2
4. $4a = -10$ $-\frac{5}{2}$
5. $4a = \frac{7}{8}$ $\frac{7}{32}$
6. $4a = -\frac{1}{2}$ $-\frac{1}{8}$

Introduction to Lesson 13-2

As students examine the drawing, stress the following points:

a. The points on the parabola from which lines have been drawn have been chosen randomly. (Students should see that those on each side of the axis of symmetry are not paired in any way.)

b. The vertical lines drawn are perpendicular to the directrix.

c. As a point on the parabola, the vertex is equidistant from the focus and the directrix.

13-2 More on Parabolas

Goals ■ Relate the equation of a parabola to its focus and directrix.

Applications Satellite Communications, Energy

Work with a partner to answer the following questions.

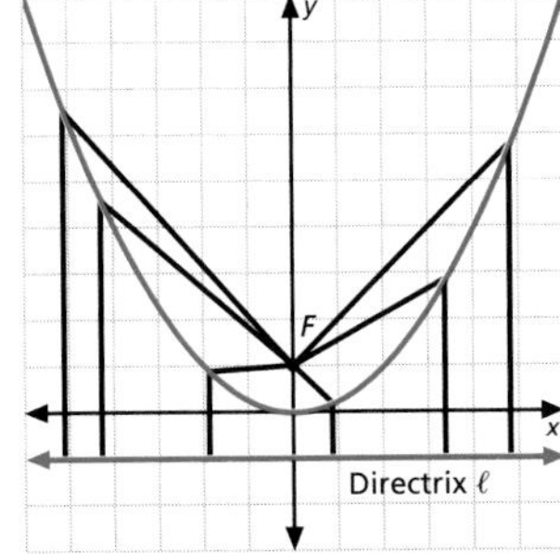

1. Study the graph shown at the right. From six randomly chosen points on the parabola, lines have been drawn to the parabola's **focus**, point F. Perpendicular lines have been drawn from these same points to line ℓ, the **directrix** of the parabola.
2. Compare the lengths of the two lines from each point.
3. Describe the relationship between the points that define the parabola and their distance from the focus and the directrix. Their distances are equal.

Build Understanding

A parabola is a set of points equidistant from a fixed point called the **focus** and a fixed line called the **directrix**.

Simple equations may be derived for parabolas that have a vertex at the origin and a directrix parallel to either the x-axis or the y-axis.

Let point $P(x, y)$ be any point on the parabola such that $FP = PD$. Then use the distance formula.

$FP = PD$

$\sqrt{(x-0)^2 + (y-a)^2} = \sqrt{(x-x)^2 + (y+a)^2}$

$(x-0)^2 + (y-a)^2 = (y+a)^2$ Square both sides. Simplify.

$x^2 + y^2 - 2ay + a^2 = y^2 + 2ay + a^2$

$x^2 - 2ay = 2ay$

$x^2 = 4ay$

When the focus $(0, a)$ is on the y-axis and the directrix is $y = -a$, the simple equation for the parabola is $x^2 = 4ay$.

In the graph above, the variable $a = 1$. When $a > 0$, the parabola opens upward. When $a < 0$, the parabola opens downward.

GRAPHING Use a graphing calculator to graph $x^2 = 4ay$ when $a = 1$ and $a = -1$. You will need to write the equation $x^2 = 4ay$ in terms of y. Notice that the y-axis remains the axis of symmetry whether the parabola opens upwards or downwards.

566 Chapter 13 **Quadratic Relations**

Extend the Lesson

REAL WORLD CONNECTION In addition to investigating the light-projecting ability of a parabola, as in Exercise 26 on page 569, students can research the effectiveness of a parabola as a receiver. Parabolic microphones are used at sporting events, parabolic antennas are becoming popular for receiving broadcast signals, and many radar antennas make use of a parabolic design.

Example 1

Find the focus and directrix of the equation $x^2 = -6y$.

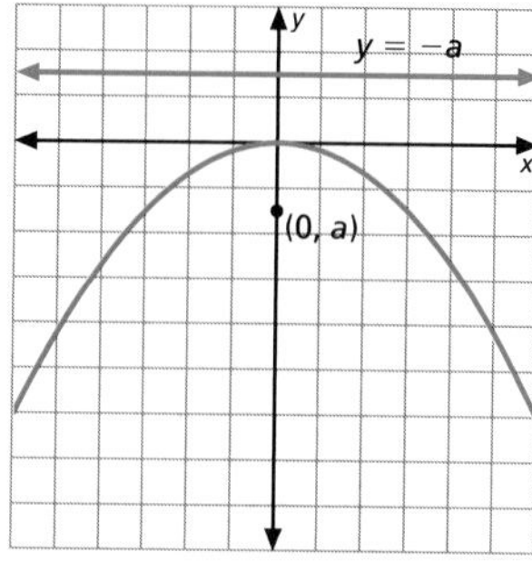

Solution

$$x^2 = -6y$$
$$x^2 = 4ay$$
$$4ay = -6y$$
$$4a = -6$$
$$a = -\frac{3}{2}$$
$$x^2 = 4\left(-\frac{3}{2}\right)y$$

Because a is negative, the parabola opens downward. The focus is located at $(0, a)$, so the focus is $\left(0, -\frac{3}{2}\right)$. The directrix is the line $y = -a$, so the directrix is $y = -\left(-\frac{3}{2}\right)$ or $y = \frac{3}{2}$.

Example 2

SATELLITE COMMUNICATIONS A parabolic satellite dish directs all incoming signals to a receiver. The receiver is located at the vertex which is at the origin and the focus is at (0, 5). Find the simple equation for the parabola.

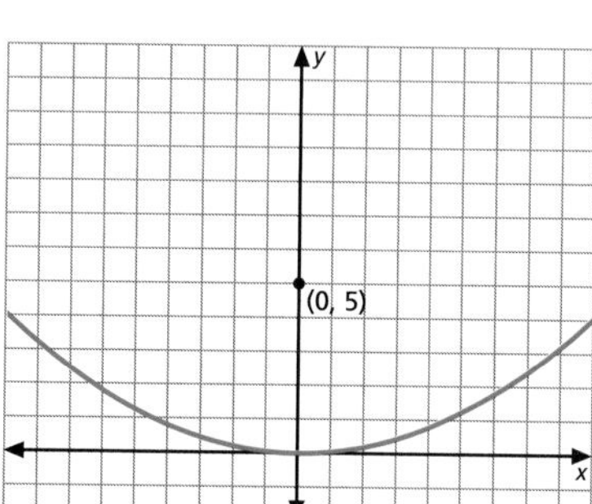

Solution

$$x^2 = 4ay$$
$$a = 5$$
$$x^2 = 4(5)y$$
$$x^2 = 20y$$

The equation is $x^2 = 20y$.

Try These Exercises

Find the focus and directrix of each equation.

1. $x^2 = 12y$ (0, 3), $y = -3$
2. $x^2 = -28y$ (0, −7), $y = 7$
3. $4x^2 - 32y = 0$ (0, 2), $y = -2$

4. **GRAPHING** Use a graphing calculator to graph the parabolas for Exercises 1–3. Using the simple equation for a parabola $x^2 = 4ay$, where the focus is $(0, a)$ and the directrix is $y = -a$, graph the directrix. How does the shape of the parabola change as the value of $4a$ increases?
The parabola becomes flatter as the value of $4a$ increases.

Teaching Strategies

Remind students that for a parabola with the vertex at the origin, if a is positive, the focus is (0, a) and the directrix must be the line $y = -a$, since that will make the vertex equidistant from the focus and directrix. Students should see that the reverse is also true: If a is negative, the focus is below the x-axis and the directrix is above the axis.

Chalkboard Examples

Supplementary Example 1

Find the focus and directrix of $x^2 = 8y$. (0, 2), $y = -2$

Supplementary Example 2

The receiver of a second satellite dish is located at the vertex which is at the origin. The focus is at (0, 3) What is the simple equation for the parabola? $x^2 = 12y$

Reteaching Worksheet 13-2

Name ______________________ Date ____________

RETEACHING 13-2

MORE ON PARABOLAS

A **parabola** consists of all those points that are equidistant from a fixed point, called the **focus**, and a fixed line called the **directrix**. The equation of the form $x^2 = 4ay$ represents a parabola with these properties:

- The vertex is (0, 0).
- The focus is $(0, a)$.
- The directrix is the line $y = -a$.
- If $a > 0$, the parabola opens upward. If $a < 0$, the parabola opens downward.

(x, y) P
(0, a)
F
y = −a
D

Example 1

Find the focus and directrix of the equation $x^2 = 2y$.

Solution

$x^2 = 4ay$ and $x^2 = 2y$ both describe a parabola with the properties described above. So $4ay = 2y$ and $a = \frac{1}{2}$.

The focus is $\left(0, \frac{1}{2}\right)$.

The directrix is $y = -\frac{1}{2}$.

Example 2

Find the simple equation for a parabola that has its vertex, at the origin and focus at (0, −4).

Solution

$x^2 = 4ay$ Substitute 24 for a.
$x^2 = 4(-4)y$ Simplify.
$x^2 = -16y$ Write the equation.

EXERCISES

Find the focus and directix of each equation.

1. $x^2 = -12y$ (0, −3); $y = 3$
2. $x^2 = 5y$ $\left(0, \frac{5}{4}\right)$; $y = -\frac{5}{4}$
3. $x^2 = -24y$ (0, −6); $y = 6$
4. $x^2 = 16y$ (0, 4); $y = -4$
5. $x^2 = -13y$ $\left(0, -\frac{13}{4}\right)$; $y = \frac{13}{4}$
6. $x^2 - 8y = 0$ (0, 1); $y = -1$

Find the simple equation for each parabola with vertex located at the origin.

7. Focus (0, −7) $x^2 = -28y$
8. Focus (0, 8) $x^2 = 32y$
9. Focus (0, 2) $x^2 = 8y$
10. Focus (0, −9) $x^2 = -36y$
11. Focus (0, −5) $x^2 = -20y$
12. Focus $\left(0, -\frac{1}{3}\right)$ $x^2 = -\frac{4}{3}y$

194 RETEACHING *LESSON 13-2*

Lesson Wrap-up

Quick Assessment

Ask the following questions to determine if students understand the content presented in this lesson.

1. Describe the axis of symmetry for a parabola with the equation $x^2 = 4ay$. Regardless of whether the parabola opens upward or downward, the y-axis is the axis of symmetry.
2. For a point on any parabola, what is the relationship between the distance to the focus and the shortest distance to the directrix? The shortest distance is a line perpendicular to the directrix; the lines are congruent.

Assignment Guide

Basic: 1–22, 27–49
Advanced: 1–49
Additional Practice: See Extra Practice Index on page 674.

Find the simple equation for each parabola with vertex located at the origin.

5. Focus $(0, 7)$ $x^2 = 28y$ **6.** Focus $(0, -3)$ $x^2 = -12y$ **7.** Focus $\left(0, \frac{3}{4}\right)$ $x^2 = 3y$

8. WRITING MATH A parabola has its vertex at the origin. Explain how to use the equation of the parabola to tell if the parabola opens up or down.
Write the equation in the form $x^2 = 4py$ and solve for p. If $p > 0$, the open end is up. If $p < 0$, the open end is down.

Practice Exercises

Find the focus and directrix of each equation.

9. $x^2 = 16y$ $(0, 4)$, $y = -4$
10. $x^2 = -10y$ $\left(0, -\frac{5}{2}\right)$, $y = \frac{5}{2}$
11. $x^2 - 20y = 0$ $(0, 5)$, $y = -5$
12. $2x^2 = -24y$ $(0, -3)$, $y = 3$
13. $16y - 4x^2 = 0$ $(0, 1)$, $y = -1$
14. $-5x^2 = -30y$ $\left(0, \frac{3}{2}\right)$, $y = \frac{-3}{2}$

ENERGY Parabolic mirrors can be used to power steam turbines to generate electricity. Three mirrors have the following focus points. Find the simple equation for each. In each case, the vertex is located at the origin.

15. Focus $(0, -2)$ $x^2 = -8y$
16. Focus $(0, 9)$ $x^2 = 36y$
17. Focus $\left(0, -\frac{1}{2}\right)$ $x^2 = -2y$

Write the simple equation for each parabola with vertex at the origin and focus and directrix shown.

18.
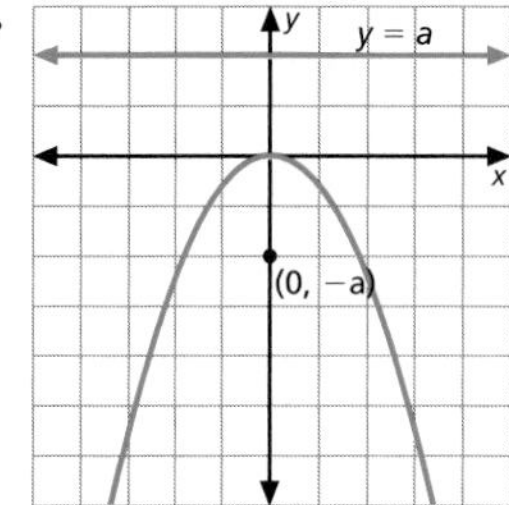

$x^2 = -4ay$

19.
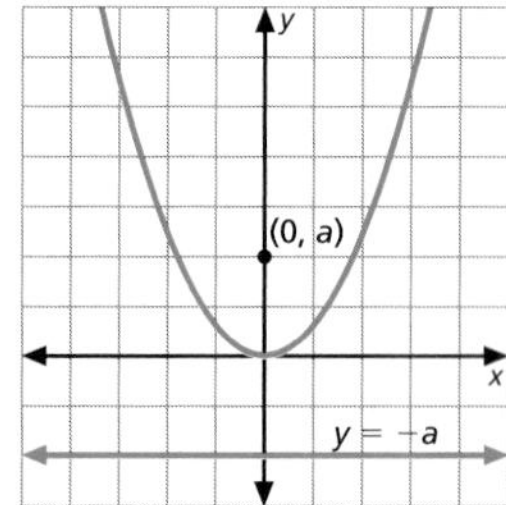

$x^2 = 4ay$

20.
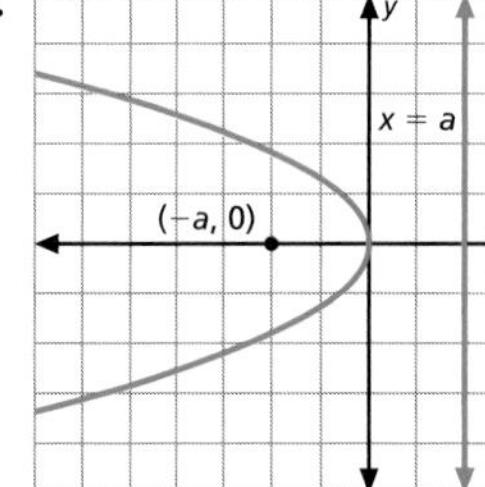

$y^2 = -4ax$

21.
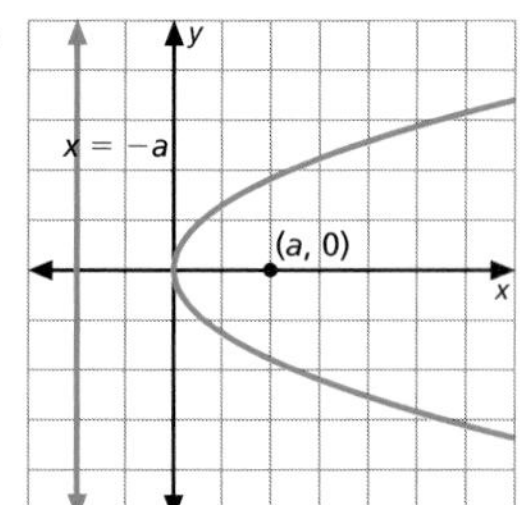

$y^2 = 4ax$

22. CHAPTER INVESTIGATION Research the perihelion (closest orbital point to the sun) and aphelion (farthest orbital point) for each planet of the Solar System. Make a rough sketch of the planets' orbits and label the perihelion and aphelion for each planet.
Check students' work.

568 Chapter 13 **Quadratic Relations**

Teaching Strategies

CHALLENGE After students have solved Exercises 20, 21 and 24, ask them to consider whether each of the equations is a function. These relationships are not functions, since each x-value has more than one y-value associated with it.

EXTENDED PRACTICE EXERCISES

The standard form for the equation of a parabola with vertex (h, k) and axis parallel to the y-axis is

$$(x - h)^2 = 4a(y - k)$$

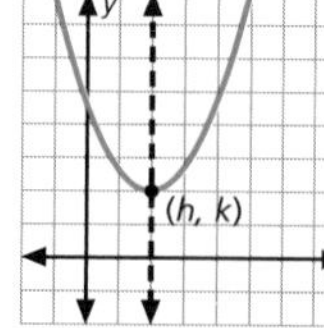

The standard form for the equation of a parabola with vertex (h, k) and axis parallel to the x-axis is

$$(y - k)^2 = 4a(x - h)$$

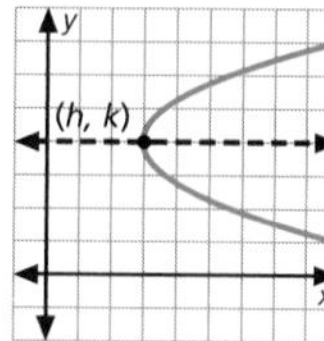

Find the equation of each parabola.

23. Focus (2, 4)
Vertex (2, 2)
$(x - 2)^2 = 8(y - 2)$

24. Focus (4, 1)
Vertex (2, 1)
$(y - 1)^2 = 8(x - 2)$

25. Focus (2, 7)
Directrix $y = -3$
$(x - 2)^2 = 12(y - 2)$

26. WRITING MATH If a light is placed at the focus of a parabola, the rays will be reflected off the parabolic surface parallel to the axis as shown. How is this concept used in a flashlight? Answers will vary. Sample: The small light bulb is surrounded by a mirrored parabolic surface that reflects light.

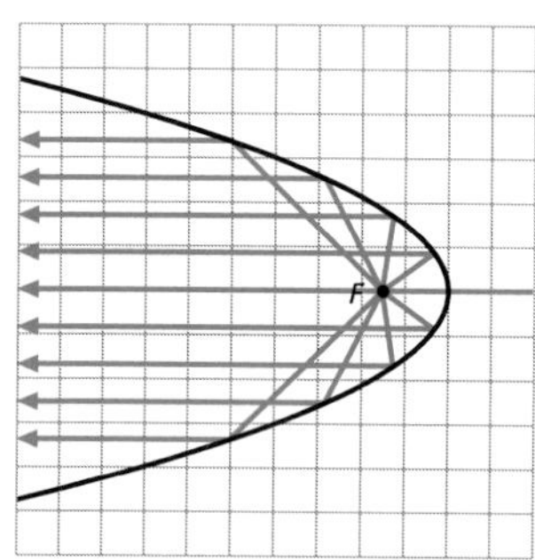

MIXED REVIEW EXERCISES

Use $x = -\frac{b}{2a}$ to find the vertex. (Lesson 12-2)

27. $y = x^2 - 3x + 6$ (1.5, 3.75)
28. $y = 2x^2 - 5x + 8$ (1.25, 4.875)
29. $y = x^2 + 3x - 4$ (−1.5, −6.25)
30. $-8 + y = 2x^2 + 2x$ (−0.5, 7.5)
31. $y = x^2 - 9$ (0, −9)
32. $-x^2 = 3x - y + 5$ (−1.5, 2.75)
33. $y - 12 = x + 2x^2$ (−0.25, 11.875)
34. $3x^2 = y - 6x + 9$ (−1, −12)
35. $y = x^2 + 4x - 3$ (−2, −7)
36. $2x - y = -x^2 + 3$ (−1, −4)
37. $4x^2 = -y - 3x + 8$ (−0.375, 8.5625)
38. $3x = -y + x^2 + 2$ (1.5, −0.25)

Let $U = \{1, 2, 3, 4, 5, 6, 7, 8, 9\}$, $P = \{1, 3, 4, 6, 7\}$ and $Q = \{2, 5, 6, 7, 8\}$. Find each union or intersection. (Lesson 1-3)

39. P' {2, 5, 8, 9}
40. Q' {1, 3, 4, 9}
41. $P \cap Q$ {6, 7}
42. $Q \cup P$ {1, 2, 3, 4, 5, 6, 7, 8}
43. $(P \cup Q)'$ {9}
44. $P' \cap Q'$ {9}
45. $Q' \cup P$ {1, 3, 4, 6, 7, 9}
46. $Q' \cap P$ {1, 3, 4}
47. $(P \cap Q)' \cup (P \cup Q)$ U
48. $(P \cup Q) \cap (P' \cup Q')$ {1, 2, 3, 4, 5, 8, 9}
49. $P' \cap (P \cup Q)$ {2, 5, 8}

Learning Styles

TACTILE/KINESTHETIC LEARNER Distribute a sheet of waxed paper approximately 12 inches long to each student. Have them fold and crease a line across the paper and mark a point about 2 inches off the line. Direct them to fold the point onto the line in at least 15 different places and, with each fold, to crease the paper. The white lines that the creases leave behind will emerge as a parabola. The curve is, in fact, a parabola according to the focus-directrix definition.

Extra Practice Worksheet 13-2

Name ______ Date ______

EXTRA PRACTICE 13-2

MORE ON PARABOLAS

EXERCISES

Find the focus and directrix of each equation.

1. $x^2 = 25y$ $\left(0, \frac{25}{4}\right), y = -\frac{25}{4}$
2. $x^2 = -4y$ $(0, -1), y = 1$
3. $x^2 = 15y$ $\left(0, \frac{15}{4}\right), y = -\frac{15}{4}$
4. $x^2 = -8y$ $(0, -2), y = 2$
5. $x^2 - 12y = 0$ $(0, 3), y = -3$
6. $x^2 + 9y = 0$ $\left(0, -\frac{9}{4}\right), y = \frac{9}{4}$
7. $50y + 2x^2 = 0$ $\left(0, \frac{25}{4}\right), y = \frac{25}{4}$
8. $-3x^2 = -27y$ $\left(0, \frac{9}{4}\right), y = -\frac{9}{4}$

Find the simple equation for each parabola with vertex located at the origin.

9. Focus (0, −2) $x^2 = -8y$
10. Focus (0, 6) $x^2 = 24y$
11. Focus $\left(0, \frac{1}{3}\right)$ $3x^2 = 4y$
12. Focus $\left(0, -\frac{1}{4}\right)$ $x^2 = -y$

Write the simple equation for each parabola with vertex at the origin and focus and directrix shown.

13.
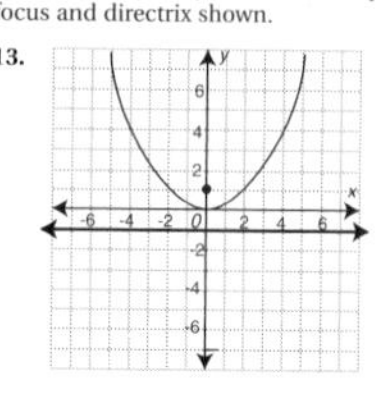

$x^2 = 4y$

14.
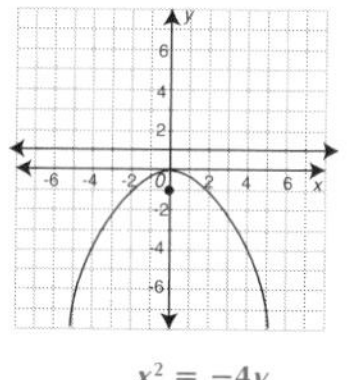

$x^2 = -4y$

Find the equation of each parabola.

15. Focus (3, 5), Vertex (3, 2) $(x - 3)^2 = 6(y - 2)$
16. Focus (2, 1), Vertex (4, 1) $(y - 1)^2 = 8(x - 4)$
17. Focus (3, 3), Vertex (2, 3) $(y - 3)^2 = 12(x - 2)$
18. Focus (1, 5), Directrix $y = -1$ $(x - 1)^2 = 4(y - 3)$

170 EXTRA PRACTICE *LESSON 13-2*

Enrichment Worksheet 13-2

Name ______ Date ______

ENRICHMENT 13-2

MECHANICAL CONSTRUCTION OF PARABOLAS

The diagram shows a way to construct a smooth parabola using a pencil, a right triangle and string. (If you don't have a plastic right triangle, a stiff cardboard right triangle may be used instead.)

The string should be the same length as segment CD in the diagram. It should not be the kind of string that stretches.

Draw the perpendicular lines shown. Then, fasten one end of the string at point F and the other end to the triangle at D.

Place a point of a pencil at point P and keep the string taught. As the triangle is moved up and down the vertical line, the point of the pencil will trace out a parabola.

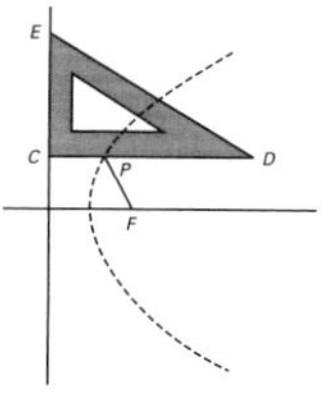

EXERCISES

1. Use quarter-inch graph paper. Experiment with different distances between the line containing segment EC and point F. Complete this chart to show your results.

Distance Between Segment EC and F (Inches)	0.50	0.75	1.00	1.25	1.50
Equation of the Resulting Parabola	$y = x^2$	$y = 0.67x^2$	$y = 0.5x^2$	$y = 0.4x^2$	$y = 0.33x^2$

2. What is the relationship between the distance of F to EC and the amount of opening of the parabola?

The greater the distance, the wider the opening.

3. Explain why the construction method works.

F is the focus and the line through $\overline{EC}$ is the directrix. As the pencil moves, $\overline{FP}$ is always equal to $\overline{CP}$. This is the definition of the parabola.

220 ENRICHMENT *LESSON 13-2*

Skills Practice

Vocabulary Review

Lesson 13-1
standard equation of a circle
radius

Lesson 13-2
focus
directrix

Assignment Guide

All students: 1–49
Additional Practice: See Extra Practice Index on page 674.

Review and Practice Your Skills

Practice ■ Lesson 13-1

1. In the standard equation for a circle, what do the variables h and k represent?
h is the x-coordinate of the center, and k is the y-coordinate of the center.
2. What is the radius of a circle?
The radius is the distance from the center to any point on the circle.
3. When is the standard equation of a circle $x^2 + y^2 = r^2$, $r \neq 0$?
when $h = 0$ and $k = 0$, when the center is at the origin.

Write an equation for each circle.

4. radius 5
center (0, 0)
$x^2 + y^2 = 25$
5. radius 3
center (2, 4)
$(x - 2)^2 + (y - 4)^2 = 9$
6.

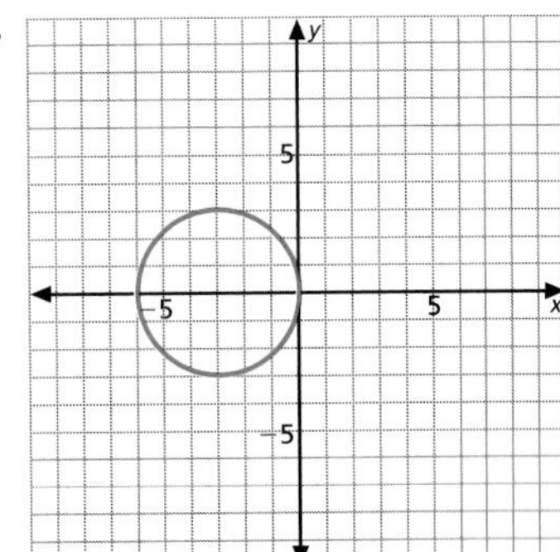

$(x + 3)^2 + y^2 = 9$

7. radius 4
center (−1, 2)
$(x + 1)^2 + (y - 2)^2 = 16$
8. radius 6
center (−2, −3)
$(x + 2)^2 + (y + 3)^2 = 36$

Find the radius and center for each circle.

9. $x^2 + y^2 = 13$ $r = \sqrt{13}$, $c = (0, 0)$
10. $(x - 2)^2 + (y + 1)^2 = 49$ $r = 7$, $c = (2, -1)$
11. $(x + 3)^2 + (y - 4)^2 = 25$ $r = 5$, $c = (-3, 4)$
12. $x^2 + (y + 2)^2 = 64$ $r = 8$, $c = (0, -2)$
13. $(x - 3)^2 + y^2 = 36$ $r = 6$, $c = (3, 0)$
14. $(x - 4)^2 + (y + 3)^2 = 74$ $r = \sqrt{74}$, $c = (4, -3)$
15. $(x + 1)^2 + y^2 = 28$ $r = \sqrt{28}$, $c = (-1, 0)$
16. $(x + 3)^2 + (y - 4)^2 = 32$ $r = \sqrt{32}$, $c = (-3, 4)$

Practice ■ Lesson 13-2

Determine if each statement is *true* or *false*.

17. When the focus is on the y-axis, and the directrix is $y = a$, the simple equation for the parabola is $x^2 = 4ay$. false
18. The focus is equidistant from most points on the parabola. false
19. When the y-coordinate of a focus is negative, the parabola opens downward. true

Find the focus and directrix of each equation.

20. $x^2 = -4y$ focus $= (0, -1)$; directrix is $y = 1$
21. $x^2 = 13y$ focus $= \left(0, \frac{13}{4}\right)$; directrix is $y = -\frac{13}{4}$
22. $x^2 = 7y$ focus $= \left(0, \frac{7}{4}\right)$; directrix is $y = -\frac{7}{4}$
23. $x^2 + 3y = 0$ focus $= \left(0, -\frac{3}{4}\right)$; directrix is $y = \frac{3}{4}$
24. $x^2 - 5y = 0$ focus $= \left(0, \frac{5}{4}\right)$; directrix is $y = -\frac{5}{4}$
25. $3x^2 + 27y = 0$ focus $= \left(0, -\frac{9}{4}\right)$; directrix is $y = \frac{9}{4}$
26. $x^2 = -8y$ focus $= (0, -2)$; directrix is $y = 2$
27. $x^2 = 5y$ 27. focus $= \left(0, \frac{5}{4}\right)$; directrix is $y = -\frac{5}{4}$
28. $x^2 = -3y$ 28. focus $= (0, -$; directrix is $y = \frac{3}{4}$
29. $x^2 - 4y = 0$ focus $= (0, 1)$; directrix is $y = -1$
30. $x^2 + 7y = 0$ focus $= \left(0, -\frac{7}{4}\right)$; directrix is $y = \frac{7}{4}$
31. $2x^2 + 12y = 0$ focus $= \left(0, -\frac{3}{2}\right)$; directrix is $y = \frac{3}{2}$

Find the simple equation for each parabola with vertex located at the origin.

32. Focus (0, 5) $x^2 = 20y$
33. Focus (0, −3) $x^2 = -12y$
34. Focus (0, 2) $x^2 = 8y$
35. Focus (0, −7) $x^2 = -28y$
36. Focus $\left(0, \frac{1}{2}\right)$ $x^2 = 2y$
37. Focus (0, 2.5) $x^2 = 10y$

Teaching Strategies

You may wish to allow students to use a compass and graph paper to draw circles described in the exercises. This may help students relate the appropriate equation to the particular circle.

PRACTICE ■ LESSON 13-1–LESSON 13-2

Without graphing, determine whether each equation is that of a circle or a parabola. (Lessons 13-1–13-2)

38. $x^2 + y^2 = r^2$ circle

39. $x^2 = 4y$ parabola

40. $x^2 = r^2 - y^2$ circle

41. $3x^2 = 9y$ parabola

42. $4x^2 = -8y$ parabola

43. $x^2 = r^2 - (y + 3)^2$ circle

Find the radius and center for each circle. (Lesson 13-1)

44. $x^2 + y^2 = 81$ $r = 9$; $c = (0, 0)$

45. $(x + 3)^2 + (y - 4)^2 = 36$ $r = 6$, $c = (-3, 4)$

46. $(x - 2)^2 + y^2 = 52$ $r = \sqrt{52}$, $c = (2, 0)$

Write an equation for each circle. (Lesson 13-1)

47. radius 7
center $(0, -3)$
$x^2 + (y + 3)^2 = 49$

48. radius 5
center $(4, 2)$
$(x - 4)^2 + (y - 2)^2 = 25$

49. radius 6
center $(-3, -1)$
$(x + 3)^2 + (y + 1)^2 = 36$

MathWorks Workplace Knowhow — Career – Payload Specialist

Today astronauts are civilian and military specialists in scientific fields as much as in engineering. One particular type of astronaut is a payload specialist. A payload specialist is a professional in the physical or life sciences and is skilled in working with equipment developed specifically for the space shuttle. The payload specialist also oversees experiments.

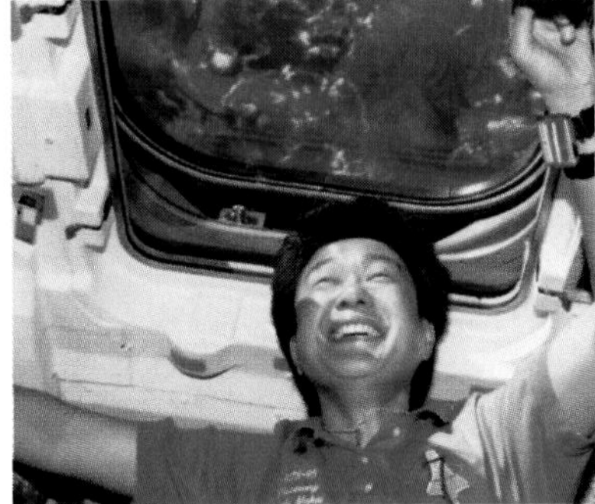

Payload specialist on space shuttle Discovery

1. In order to perform the experiment, the space shuttle must be kept ahead of the moon's orbit so that a ship, the Earth, and the moon form a right angle. The shuttle is orbiting the Earth to a distance of 325 km. The moon orbits the Earth at a distance of 384,403 as its furthest distance (the current distance). Draw a diagram of the Earth, the moon and the ship forming a right angle.
The right angle will fall at the center of the Earth. The Earth's diameter is 12,756 km. The moon's diameter is 3476 km. (Try to make your diagram look as if the moon is quite a bit farther away from the Earth than your ship. Check students' drawings.
2. How far is your ship from the center of the Earth? 6,703 km
How far is the center of the Earth from the center of the moon? 392,519 km
3. Now that you have those two distances, form the hypotenuse of the triangle you formed between the Earth, the moon, and your ship. How far are you from the moon? 384,403.1374 km

Chalkboard Examples

Example from Lesson 13-1
Write an equation for a circle with center (1, 0) and radius 4.
$(x - 1)^2 + y^2 = 16$

Example from Lesson 13-2
Find the focus and directrix of the equation $3y^2 = 12x$. $(1, 0)$, $x = -1$

Career Opportunity

Describe the kind of work astronauts do and the types of backgrounds these workers have. Have students discuss the importance of algebra, geometry and trigonometry in doing this type of job. Explain that astronauts must be able to follow complex procedures and apply their skills in scientific experiments as well as in piloting a ship in space. Students should answer Questions 1–3 to better understand how astronauts use mathematics in performing their job.

Students who are interested in learning more about this profession can go to learningmathmatters.com. Guidance Counselors should have information about school and training requirements.

Extend the Lesson

MATH JOURNAL Have students write in their own words an explanation of how each of the following is related to the equation of a parabola: its focus, its directrix, its vertex, and its axis of symmetry.

Lesson Planning

NCTM Standards/Strands

- Patterns, Functions, & Algebra
- Geometry & Spatial Sense
- Problem Solving
- Reasoning and Proof
- Connections
- Representation

Vocabulary

conics conic section

Materials Needed

pencil/paper

Lesson Resources

Warm-Up Transparency 37
Transparency TK-7–10
Reteaching 13-3

Assignment Guide

Basic: 1–16
Enriched: 1–16

Getting Started

5-Minute Warm-up

Describe the intersection when a plane makes each cut.

1. horizontally through a cube
   square
2. vertically through a cylinder
   rectangle
3. horizontally through a cone
   circle

Use the Five-step Plan and the Strategy

THE FIVE-STEP PLAN **Read**—ask questions of the students to help them understand the problem. **Plan**—guide students to related problems and previously mastered skills and strategies. **Solve**—students solve problem on their own. **Answer**—write solution in format that answers the question. **Check**—review work, check for reasonableness and review strategy used. Students will benefit from the experience of verbalizing their methods.

THE STRATEGY Use a model or picture—Being able to visualize a

13-3 Problem Solving Skills: Visual Thinking

Think of a line, *AB*, intersecting the coordinate plane at the origin. The right circular cones formed by rotating the line about the *y*-axis are used to study **conics**.

A **conic section** is formed by a plane intersecting the right circular cones.

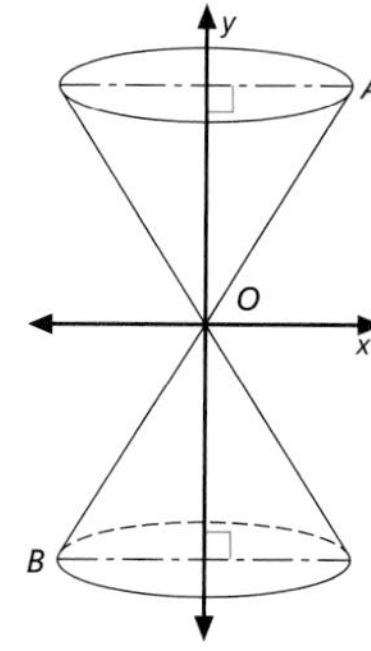

Problem Solving Strategies

Guess and check
Look for a pattern
Solve a simpler problem
Make a table, chart or list
✔ Use a picture, diagram or model
Act it out
Work backwards
Eliminate possibilities
Use an equation or formula

Problem

Draw and name the conic section formed by each plane.

a. The plane is parallel to a side of the cone and does not pass through the vertex.

b. The plane is parallel to the *y*-axis and does not pass through the vertex.

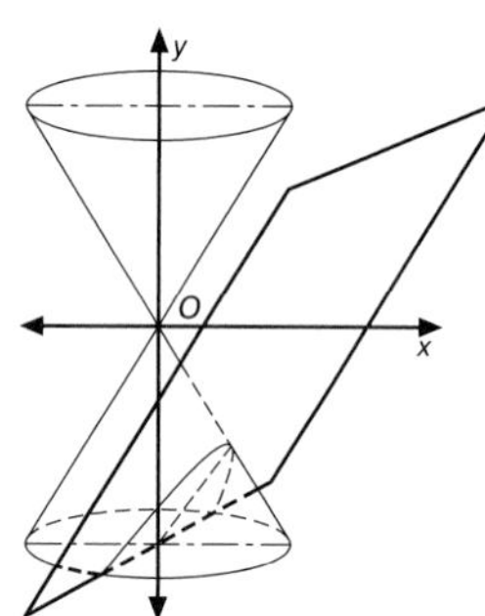

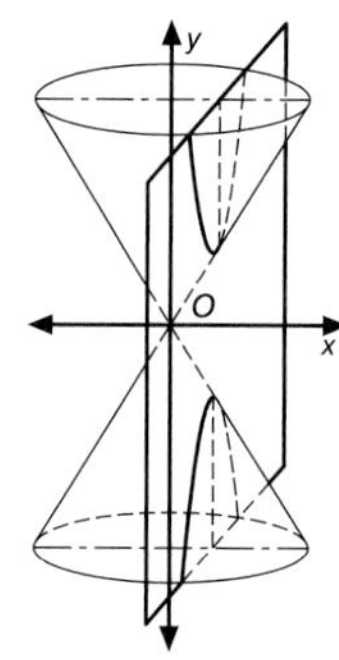

Math: Who, Where, When

Greek mathematician and geometer Apollonius of Perga studied and named the cuts made by a flat plane as it intersects a cone. About 225 B.C., he wrote *Conics*, in which he described the properties of parabolas, circles, ellipses, and hyperbolas. Conic sections were later found to represent the paths followed by planets and projectiles.

Solve the Problem

a.

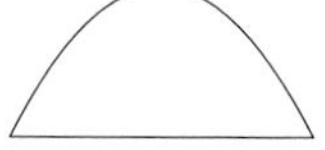

The conic section is a parabola.

b.

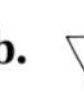

The conic section is a hyperbola.

572 Chapter 13 **Quadratic Relations**

Teaching Strategies

Students may benefit from seeing three-dimensional versions of the conic sections. Commercial models of Lucite cones showing sections are available. Alternately, a single cone can be made from clay and cut with a wire to reveal the sections. A paper cone will work if the cuts are made gently so as not to crease the paper. The outlines can then be traced.

Try These Exercises

Five-step Plan

1 Read
2 Plan
3 Solve
4 Answer
5 Check

Draw and name the conic section formed by each plane.

1. The plane is parallel to the x-axis. It does not contain the vertex. circle

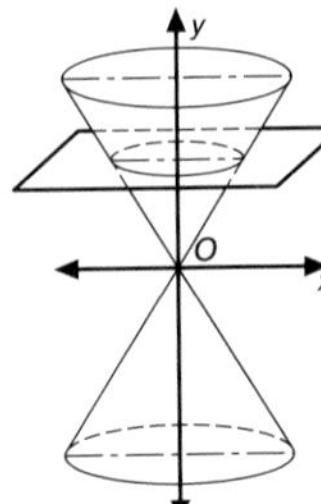

2. The plane does not contain a vertex or base. It is not parallel to the x-axis. ellipse

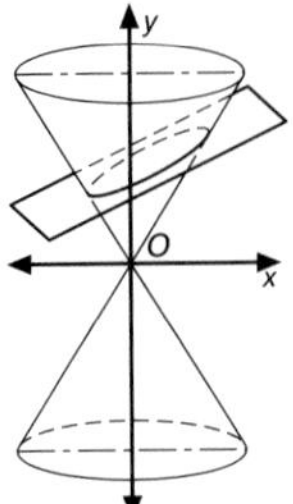

Practice Exercises

3. The plane intersects only the vertex. point

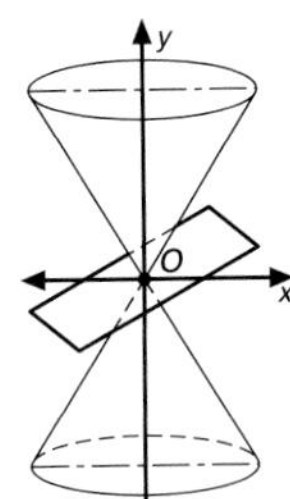

4. The plane intersects two parallel diameters. two intersecting lines.

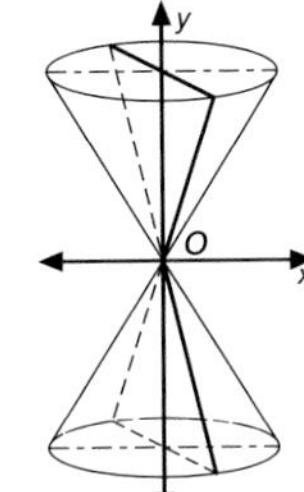

5. The plane is parallel to a side of the cone and does not pass through the vertex. parabola

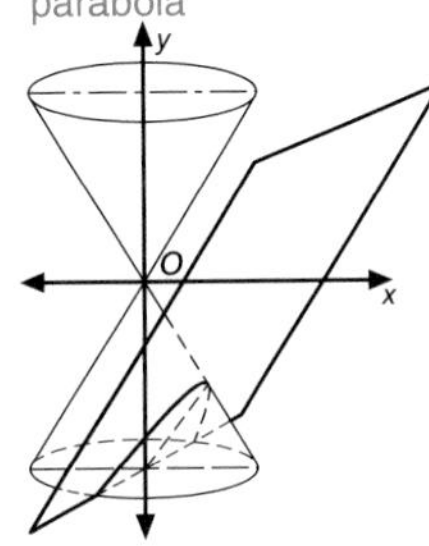

6. **DATA FILE** Use the Data Index on pages 632–633 to locate information about crystal systems. Visualize yourself standing at the intersection of the axes inside each mineral shown. Describe each face you see. See additional answers.

7. **WRITING MATH** Is it possible to intersect a double cone in such a way that two circles are formed? Explain. See additional answers.

Mixed Review Exercises

Use a graphing calculator to determine the number of solutions for each equation. For equations with one or two solutions, find the exact solutions by factoring. (Lesson 12-3)

8. $y = x^2 - 64$ 2; 8 and −8
9. $y = x^2 + x - 12$ 2; −4 and 3
10. $y = x^2 + 2x + 1$ 1; −1
11. $y = x^2 + 2x - 8$ 2; −4 and 2
12. $y = x^2 + 49$ no solution
13. $y - x^2 + 5x + 6 = 0$ 2; 6 and −1
14. $x^2 - y = 5x + 6$ 2; 6 and −1
15. $y = x^2 - 25$ 2; 5 and −5
16. $x^2 = y - 5x - 6$ 2; −2 and −3

geometric situation is an important skill. When visualization is difficult, students should draw a picture or make a model.

Chalkboard Examples

Supplementary Problem

A plane is parallel to the x-axis. Draw and name the conic section formed by the plane. circle

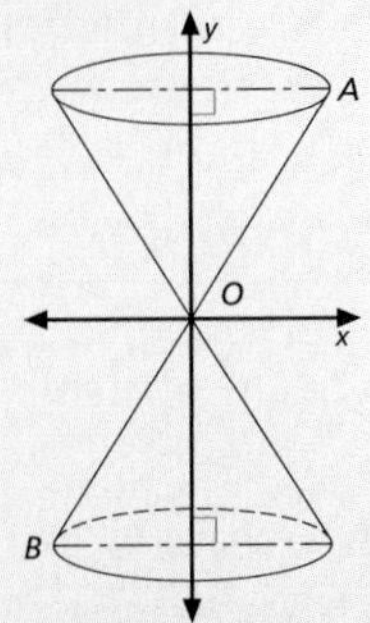

Lesson Wrap-up

Quick Assessment

Ask the following question to determine if students understand the content presented in this lesson.

Are there intersections other than the one in Exercise 4 that produce triangles? How are they obtained? An infinite number are possible, by rotating the plane about the y-axis.

Reteaching Worksheet 13-3

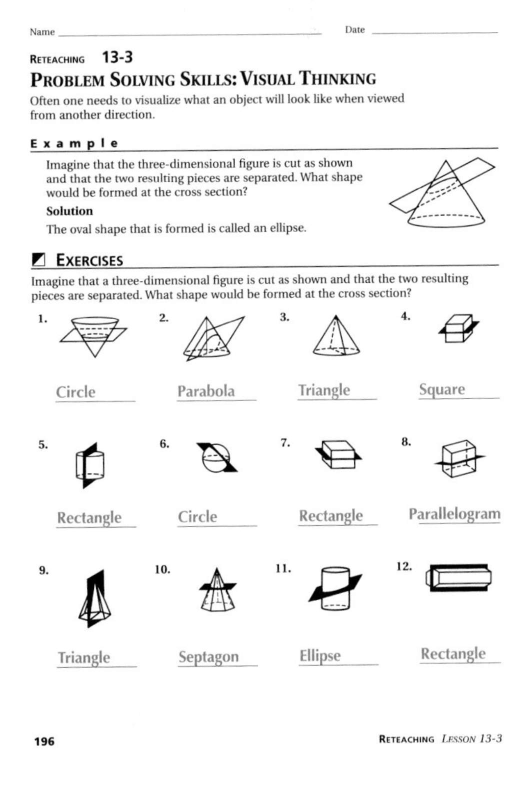

Name ______ Date ______

RETEACHING 13-3

PROBLEM SOLVING SKILLS: VISUAL THINKING

Often one needs to visualize what an object will look like when viewed from another direction.

Example

Imagine that the three-dimensional figure is cut as shown and that the two resulting pieces are separated. What shape would be formed at the cross section?

Solution

The oval shape that is formed is called an ellipse.

EXERCISES

Imagine that a three-dimensional figure is cut as shown and that the two resulting pieces are separated. What shape would be formed at the cross section?

1. Circle
2. Parabola
3. Triangle
4. Square
5. Rectangle
6. Circle
7. Rectangle
8. Parallelogram
9. Triangle
10. Septagon
11. Ellipse
12. Rectangle

196 RETEACHING *LESSON 13-3*

Additional Answers

6. Answers will vary. Samples are given:
   Halite: 6 square faces
   Zircon: 4 rectangular faces, 8 triangular faces
   Quartz: 6 rectangular faces, 12 triangular faces
   Calcite: 9 pentagonal faces
   Sulfur: 8 trapezoidal faces, 8 triangular faces
   Feldspar: 2 hexagonal faces, 4 trapezoidal faces, 4 pentagonal faces
   Rhodonite: 8 pentagonal faces, 4 rectangular faces.
7. No. A circle is formed when a plane is parallel to the circle formed by the base. Forming two circles would require a single plane to pass through both cones at the same time or to pass through a single cone twice. To do so, the plane would have to "bend," and it would no longer be considered a plane.

Lesson Planning

NCTM Standards/Strands

- Number & Operation
- Patterns, Functions, & Algebra
- Geometry & Spatial Sense
- Problem Solving
- Reasoning and Proof
- Communication
- Connections
- Representation

Vocabulary

ellipse, foci, asymptotes, hyperbola

Materials Needed

cardboard, scissors, thumbtacks, string, graph or grid paper, graphing calculators

Lesson Resources

Warm-Up Transparency 38
Transparency TK-7–10, RF-50
Reteaching 13-4
Extra Practice 13-4
Enrichment 13-4

Assignment Guide

Basic: 1–14, 17–22
Enriched: 1–22

Getting Started

5-Minute Warm-up

Find the least common denominator for each pair of fractions.

1. $\frac{1}{14}, \frac{1}{17}$ 238
2. $\frac{3}{15}, \frac{5}{18}$ 90
3. $\frac{1}{5}, \frac{1}{20}$ 20
4. $\frac{1}{9}, \frac{1}{15}$ 45
5. $\frac{1}{8}, \frac{1}{15}$ 120
6. $\frac{1}{7}, \frac{1}{12}$ 84

Introduction to Lesson 13-4

After students have drawn the two figures, distribute rulers, and have students measure several pairs of segments connecting the foci within each ellipse. This will help them verify that the sum of these two lengths is a constant for each figure.

13-4 Ellipses and Hyperbolas

Goals ■ Graph equations of ellipses and hyperbolas.

Applications Astronomy, Oceanography, Communications

Work with a partner.

You will need a piece of cardboard, two thumbtacks, string, and scissors. Place two thumbtacks in a piece of cardboard. Label their positions F_1 and F_2. Let the distance between the two points be $2c$.

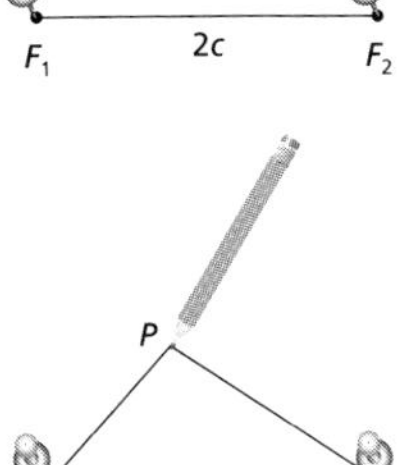

Tie the ends of a piece of string together to make a loop. Let the length of the string be $2a + 2c$ where a is any quantity greater than c. Place the loop over the tacks. With a pencil held upright, keep the string taut and draw the ellipse.

Use the same string and change the distance between F_1 and F_2. Draw another ellipse.

1. How does the shape of the ellipse change as F_1 and F_2 are farther apart? It becomes flatter.
2. For any point P on the ellipse, what is the sum of $F_1P + F_2P$? 2a
3. If F_1 and F_2 were at the same point, what figure would you draw? a circle

Check Understanding

Why must length a be greater than length c?

If $a \le c$, there would not be any slack. You could not draw the ellipse.

Build Understanding

As you discovered in the activity above, an ellipse is defined by a point moving about two fixed points. The two fixed points are called **foci**, the plural of focus. The sum of the distances from the two fixed points to any point on the ellipse is a constant, $F_1P + F_2P = 2a$.

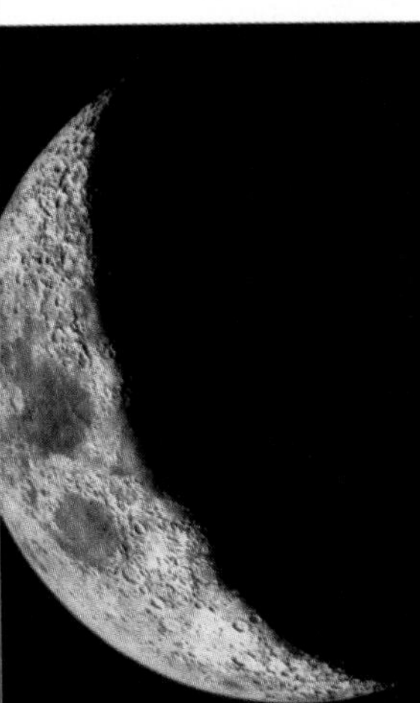

An equation for the standard form of an ellipse can be derived by placing the ellipse on a coordinate grid. Locate (0, 0) at the midpoint between F_1 and F_2. The distance from the center to each focus is c. When $x = 0$, $F_1P_1 = F_2P_1$ and $b^2 = a^2 - c^2$.

Using the distance formula, the **standard form** for the equation of an ellipse, with its center at the origin and **foci** F_1 and F_2 on the x-axis, is found to be

$$\frac{x^2}{a^2} + \frac{y^2}{b^2} = 1.$$

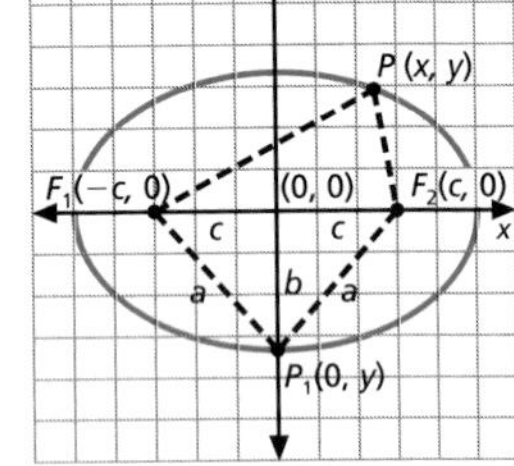

By letting $y = 0$, x-intercepts are $(a, 0)$ and $(-a, 0)$. By letting $x = 0$, y-intercepts are $(0, b)$ and $(0, -b)$.

Teaching Strategies

Students can test the equations they write for ellipses and hyperbolas by graphing the equations on a graphing calculator. It may first be necessary to solve an equation in terms of y, however, before graphing.

Example 1

Graph the equation $4x^2 + 9y^2 = 36$.

Solution

Divide both sides of the equation by 36 to change it to standard form.

$$4x^2 + 9y^2 = 36$$

$$\frac{x^2}{9} + \frac{y^2}{4} = 1$$

$$a^2 = 9,\ a = \pm 3$$

$$b^2 = 4,\ b = \pm 2$$

The x-intercepts are (3, 0) and (−3, 0). The y-intercepts are (0, 2) and (0, −2).

Locate the points and draw a smooth curve.

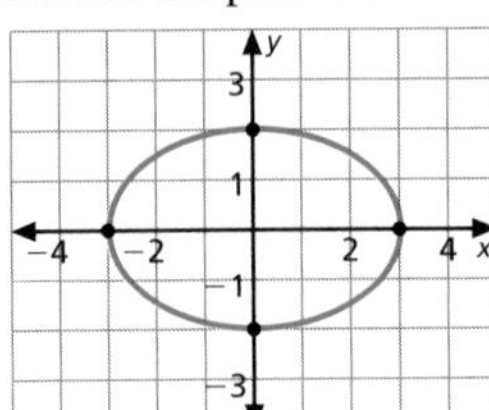

Technology Note

To graph an ellipse using a graphing calculator, rewrite the equation in terms of y.

$4x^2 + 9y^2 = 36$ becomes

$y = \pm\sqrt{-\frac{4}{9}x^2 + 4}$.

1. Enter the positive form of the equation as Y_1 to graph the upper part of the ellipse.
2. To graph the lower portion, let $Y_2 = -Y_1$.
3. Graph the ellipse. You can use the trace feature to locate the x- and y-intercepts.

Example 2

ASTRONOMY Jae is using a computer to model the orbit of a moon. After placing a grid over a drawing of the orbit, he finds that the foci of the ellipse are (4, 0) and (−4, 0) and (−5, 0) and (5, 0). He needs to enter the equation of the ellipse into the computer to finish his work. Find the equation of the ellipse.

Solution

$$\frac{x^2}{a^2} + \frac{y^2}{b^2} = 1$$

$$a = \pm 5,\ c = \pm 4$$

$b^2 = a^2 - c^2$ — Use the Pythagorean Theorem to find b^2.

$$b^2 = 25 - 16$$

$$b^2 = 9$$

$\frac{x^2}{25} + \frac{y^2}{9} = 1$ — Substitute in the standard form.

$9x^2 + 25y^2 = 225$ — Multiply by 225 (9 × 25).

The equation of the ellipse with foci (4, 0) and (−4, 0) and x-intercepts ±5 is $9x^2 + 25y^2 = 225$.

Check Understanding

Write ellipse or hyperbola for each equation.

1. $\frac{x^2}{10} + \frac{y^2}{4} = 1$
2. $\frac{x^2}{12} - \frac{y^2}{5} = 1$
3. $4x^2 - 7y^2 = 56$
4. $8x^2 + 13y^2 = 104$

1 and 4: ellipse;
2 and 3: hyperbola

Teaching Strategies

COOPERATIVE LEARNING Have students work in pairs. Ask them to imagine a pool table in the shape of an ellipse, with the only picket at one of the foci. If a ball is placed at the other focus, where against the wall must the ball be hit so that it will fall into the pocket? Regardless of where the ball hits the wall, it will rebound to the other focus and fall into the pocket.

Chalkboard Examples

Supplementary Example 1

Graph the equation $x^2 + 16y^2 = 16$. x-int: (4, 0) and (−4, 0); y-int: (0, 1) and (0, −1); Check students' work.

Supplementary Example 2

Karin is creating the visuals for a computer game. She needs to show a satellite orbiting a moon. The foci of the elliptical orbit are (4, 0) and (−4, 0) and (0, 6) and (0, −6). Find the equation of the ellipse.
$36x^2 + 16y^2 = 144$

Supplementary Example 3

Graph the hyperbola with the equation $9x^2 - 4y^2 = 36$. To locate the rectangle, use the points (2, 0), (−2, 0), (0, 3), and (0, −3). Draw lines perpendicular to the axes to form a rectangle. Draw extended diagonals of the rectangle (asymptotes), and sketch the hyperbola. Check students' graphs.

Supplementary Example 4

Find the equation for a hyperbola with center (0, 0) in which $a = 5$ and $b = 6$, and the foci are on the x-axis.
$36x^2 - 25y^2 = 900$

Reteaching Worksheet 13-4

Name ______________________ Date __________

RETEACHING 13-4

ELLIPSES AND HYPERBOLAS

The standard form for the equation of an ellipse with its center at the origin (0, 0) and foci F_1 and F_2 on the x-axis is $\frac{x^2}{a^2} + \frac{y^2}{b^2} = 1$. The standard form for the equation for a hyperbola that is symmetric around the origin (0, 0) and has foci F_1 and F_2 on the x-axis is $\frac{x^2}{a^2} - \frac{y^2}{b^2} = 1$.

Example 1

Find the equation of the ellipse with foci (3, 0) and (−3, 0) and x-intercepts (−4, 0) and (4, 0).

Solution

Substitute known values into the equation and simplify.

$\frac{x^2}{a^2} + \frac{y^2}{b^2} = 1$ — $a = \pm 4$; $c = \pm 3$; $b^2 = a^2 - c^2$, so $b^2 = 16 - 9 = 7$

$\frac{x^2}{16} + \frac{y^2}{7} = 1$ — Multiply by 112.

$7x^2 + 16y^2 = 112$ — Write the equation.

Example 2

Find the equation of a hyperbola with center (0, 0) in which $a = 5$, $b = 2$ and the foci are on the x-axis.

Solution

Substitute known values into the equation and simplify.

$\frac{x^2}{a^2} - \frac{y^2}{b^2} = 1$

$\frac{x^2}{25} - \frac{y^2}{4} = 1$ — Multiply by 200.

$4x^2 - 25y^2 = 200$ — Write the equation.

EXERCISES

Find the equation of each ellipse.

1. Foci (2, 0) and (−2, 0) and x-intercepts (−4, 0) and (4, 0) $12x^2 + 16y^2 = 192$
2. Foci (4, 0) and (−4, 0) and x-intercepts (5, 0) and (−5, 0) $9x^2 + 25y^2 = 225$
3. Foci (−5, 0) and (5, 0) and x-intercepts (−6, 0) and (6, 0) $11x^2 + 36y^2 = 396$
4. Foci (−2, 0) and (2, 0) and x-intercepts (7, 0) and (−7, 0) $45x^2 + 49y^2 = 2205$
5. Foci (3, 0) and (−3, 0) and x-intercepts (5, 0) and (−5, 0) $16x^2 + 25y^2 = 400$
6. Foci (−6, 0) and (6, 0) and x-intercepts (−7, 0) and (7, 0) $13x^2 + 49y^2 = 637$

Find the equation of each hyperbola with its center at the origin (0, 0) and foci on the x-axis.

7. $a = \pm 4$ and $b = \pm 2$ $4x^2 - 16y^2 = 64$
8. $a = \pm 3$ and $b = \pm 4$ $16x^2 - 9y^2 = 144$
9. $a = \pm 6$ and $b = \pm 4$ $16x^2 - 36y^2 = 576$
10. $a = \pm 5$ and $b = \pm 8$ $64x^2 - 25y^2 = 1600$
11. $a = \pm 9$ and $b = \pm 7$ $49x^2 - 81y^2 = 3969$
12. $a = \pm 7$ and $b = \pm 5$ $25x^2 - 49y^2 = 1225$

198 RETEACHING *LESSON 13-4*

Lesson Wrap-up

Quick Assessment

Ask the following questions to determine if students understand the content presented in this lesson.

1. How does a circle meet the definition of an ellipse? A circle is an ellipse that has both foci at the same location; any radius of the circle can be thought of as the two lines from the foci to the ellipse; the sum of any two radii will be a constant.
2. How are ellipses and hyperbolas similar? How are they different? Both have a center and two foci and are symmetrical with respect to both axes. One is the set of all points whose sum to the foci is constant; the other is the set of all points whose difference (absolute value) from the foci is constant.

Assignment Guide

Basic: 1–14, 17–22
Advanced: 1–22
Additional Practice: See Extra Practice Index on page 674.

Additional Answers

1.
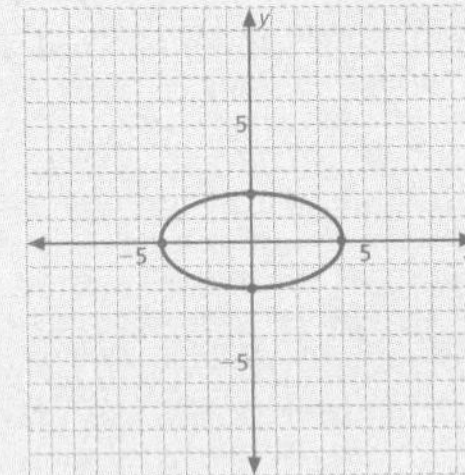

2.
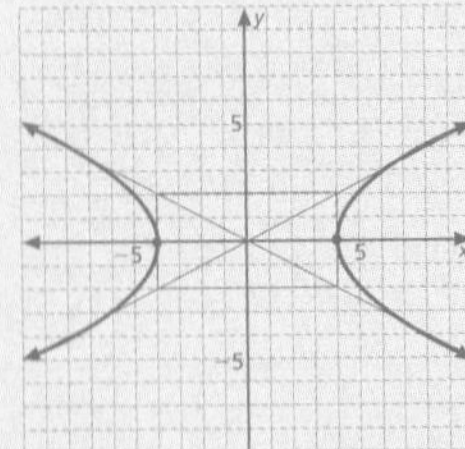

6. & 12.
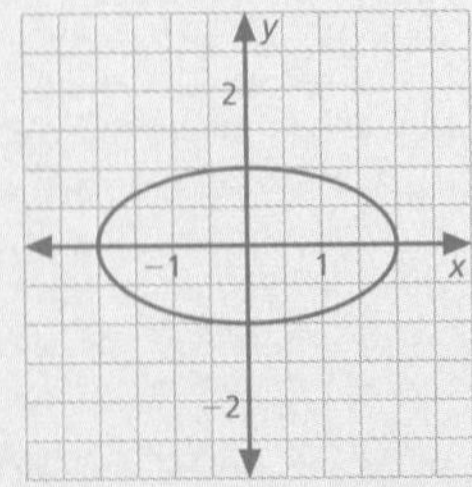

The equation for an ellipse uses the sum of the distances from a point on the ellipse to the foci. The equation for a hyperbola uses the difference of these distances. The equation of a hyperbola that is symmetric about the origin and has foci F_1 and F_2 on the x-axis is

$$\frac{x^2}{a^2} - \frac{y^2}{b^2} = 1$$

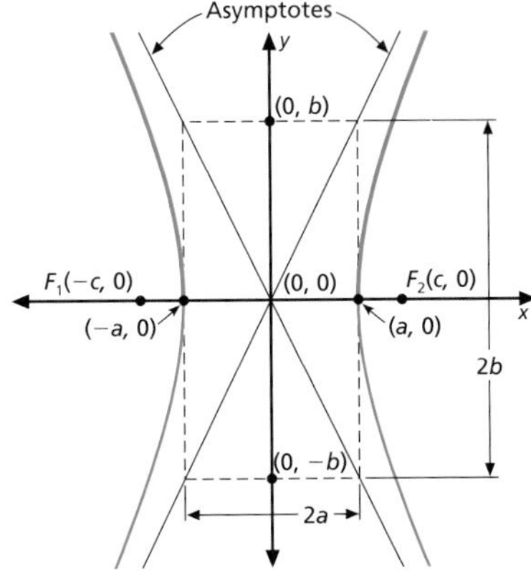

As x-values for a hyperbola become very large or very small, they approach the **asymptotes** for the hyperbola. The hyperbola never meets the asymptotes.

Asymptotes may be drawn as the diagonals of the rectangle with sides $2a$ and $2b$.

Example 3

Graph the hyperbola with equation $16x^2 - 4y^2 = 64$.

Solution

$$16x^2 - 4y^2 = 64$$

$$\frac{x^2}{4} - \frac{y^2}{16} = 1 \qquad \text{Divide by 64 to write the equation in standard form.}$$

$$a^2 = 4,\ a = \pm 2$$

$$b^2 = 16,\ b = \pm 4$$

To locate the rectangle, use the points (2, 0), (−2, 0), (0, 4), and (0, −4). Draw lines perpendicular to the axes through these points to form a rectangle. Draw extended diagonals of the rectangles; these are the asymptotes. Sketch the hyperbola.

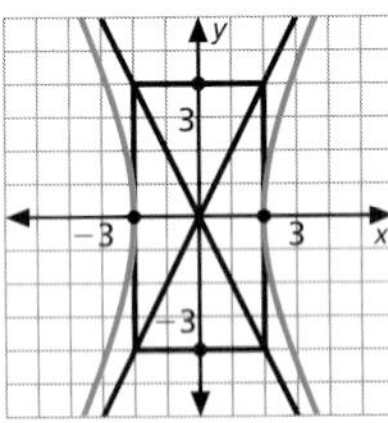

Example 4

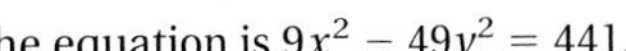

OCEANOGRAPHY An object propelled through water travels along one branch of a hyperbola with an experimental submarine at its center (0, 0) and in which $a = 7$ and $b = 3$ and the foci are on the x-axis. Find the equation for the hyperbola.

Solution

$$\frac{x^2}{a^2} - \frac{y^2}{b^2} = 1$$

$$\frac{x^2}{7^2} - \frac{y^2}{3^2} = 1$$

$$\frac{x^2}{49} - \frac{y^2}{9} = 1 \qquad \text{Substitute in the standard equation.}$$

$$9x^2 - 49y^2 = 441 \qquad \text{Multiply by 441 } (49 \times 9)$$

The equation is $9x^2 - 49y^2 = 441$.

7.

10.

11.

13.
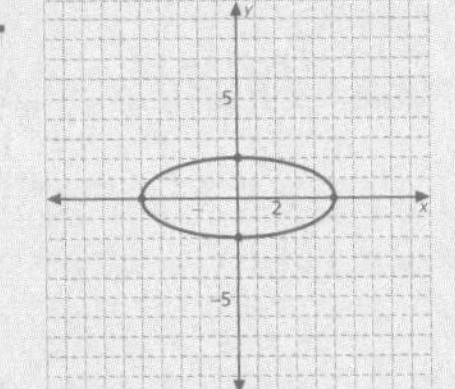

14.
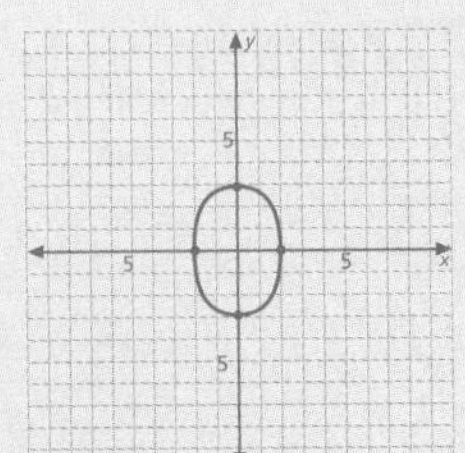

Try These Exercises

Graph each equation. For 1–2, see additional answers.

1. $4x^2 + 16y^2 = 64$
2. $4x^2 - 25y^2 = 100$

3. Find the equation of the ellipse with foci (12, 0) and (−12, 0) and x-intercepts (13, 0) and (−13, 0). $25x^2 + 169y^2 = 4{,}225$
4. Find the equation of the hyperbola with center (0, 0) and foci on the x-axis if $a = \pm 9$ and $b = \pm 6$ $36x^2 - 81y^2 = 2{,}916$

5. **WRITING MATH** Is a circle a type of ellipse? Explain your thinking. Yes. A circle is an ellipse that has both foci at the same location.

Practice Exercises

Graph each equation. For 6–7, see additional answers.

6. $9x^2 + 36y^2 = 36$
7. $9x^2 - 16y^2 = 144$

8. **COMMUNICATIONS** A communications satellite is launched into an elliptical orbit with foci (8, 0) and (−8, 0) and x-intercepts (10, 0) and (−10, 0). Find the equation of the ellipse. $9x^2 + 25y^2 = 900$
9. **ASTRONOMY** A comet follows one branch of a hyperbola with center (0, 0) and foci on the x-axis if $a = \pm 5$ and $b = \pm 11$. Find the equation of the hyperbola. $2x^2 + y^2 = 8$ $121x^2 - 25y^2 = 3{,}025$

If a hyperbola with center (0, 0) has foci on the y-axis, the equation is $\frac{y^2}{a^2} - \frac{x^2}{b^2} = 1$.

Graph each hyperbola. For 10–11, see additional answers.

10. $18y^2 - 8x^2 = 72$
11. $y^2 - x^2 = 144$

The equation for an ellipse with center (0, 0) and foci on the y-axis is $\frac{x^2}{b^2} + \frac{y^2}{a^2} = 1$.

Graph each ellipse. For 12–14, see additional answers.

12. $x^2 + 4y^2 = 4$
13. $4x^2 + 25y^2 = 100$
14. $2x^2 + y^2 = 8$

Extended Practice Exercises

The ellipse shown has center (h, k) and axes parallel to the coordinate axes.

15. Substitute in the standard form for the equation of an ellipse with its center at the origin and foci on the x-axis to find an equation for an ellipse with center (h, k). $\frac{(x-h)^2}{a^2} + \frac{(y-k)^2}{b^2} = 1$
16. Use your equation to find the center of the ellipse $16(x + 2)^2 + 9(y - 1)^2 = 144$. (−2, 1)

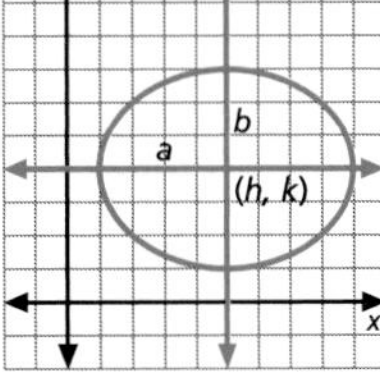

Mixed Review Exercises

Solve by completing the square. (Lesson 12-4)

17. $x^2 + 10x = 0$ $x = 0, x = 70$
18. $x^2 + 3x = 0$ $x = 0, x = -3$
19. $x^2 - 8x = 0$ $x = 0, x = 8$
20. $x^2 + 4x - 7 = 0$ $x = -2 \pm \sqrt{11}$
21. $x^2 - 6x + 2 = 0$ $x = 3 \pm \sqrt{7}$
22. $x^2 + 12x - 3 = 0$ $x = -6 \pm \sqrt{39}$

Teaching Strategies

MATH JOURNAL Have students write equations in standard form for the ellipse and the hyperbola. Then have them write why they think an ellipse is the set of all points for which the sum of the distances from two points is a constant, while a hyperbola is the set of which the difference (absolute value) of the distances from two points is a constant. In an ellipse, as one distance becomes greater, the other must become less to keep the sum a constant. In a hyperbola, as one distance becomes greater, the other must increase to keep the difference a constant.

Extra Practice Worksheet 13-4

Name ____________ Date ____________

EXTRA PRACTICE 13-4

ELLIPSES AND HYPERBOLAS

EXERCISES

Graph each equation. Use your own paper. Check students' graphs.

1. $4x^2 + 16y^2 = 25$
2. $16x^2 + 25y^2 = 81$
3. $4x^2 - 16y^2 = 64$
4. $x^2 - 9y^2 = 16$
5. Find the equation of the ellipse with foci (6, 0) and (−6, 0) and x-intercepts (8, 0) and (−8, 0). $28x^2 + 64y^2 = 1792$
6. Find the equation of the ellipse with foci (10, 0) and (−10, 0) and x-intercepts (15, 0) and (−15, 0). $125x^2 + 225y^2 = 28{,}125$
7. Find the equation of the ellipse with foci (5, 0) and (−5, 0) and x-intercepts (6, 0) and (−6, 0). $11x^2 + 36y^2 = 396$
8. Find the equation of the ellipse with foci (7, 0) and (−7, 0) and x-intercepts (9, 0) and (−9, 0). $32x^2 + 81y^2 = 2592$
9. Find the equation of the hyperbola with center (0, 0) and foci on the x-axis if $a = \pm 6$ and $b = \pm 8$. $64x^2 - 36y^2 = 2304$
10. Find the equation of the hyperbola with center (0, 0) and foci on the x-axis if $a = \pm 1$ and $b = \pm 2$. $4x^2 - y^2 = 4$
11. Find the equation of the hyperbola with center (0, 0) and foci on the x-axis if $a = \pm 5$ and $b = \pm 2$. $4x^2 - 25y^2 = 100$
12. Find the equation of the hyperbola with center (0, 0) and foci on the x-axis if $a = \pm 4$ and $b = \pm 7$. $49x^2 - 16y^2 = 784$

Graph each hyperbola. Use your own paper. Check students' graphs.

13. $16x^2 - 4y^2 = 64$
14. $x^2 - y^2 = 121$
15. $8x^2 - 2y^2 = 32$
16. $4x^2 - y^2 = 100$

Graph each ellipse. Use your own paper. Check students' graphs.

17. $x^2 + 36y^2 = 9$
18. $4x^2 + 9y^2 = 36$
19. $6x^2 + 24y^2 = 6$
20. $2x^2 + 8y^2 = 8$

172 EXTRA PRACTICE *LESSON 13-4*

Enrichment Worksheet 13-4

Name ____________ Date ____________

ENRICHMENT 13-4

SLOPE OF TANGENT LINES

A line tangent to a curve meets the curve in just one point. For the ellipse and the hyperbola, the formulas for the slope of the tangent line at a point (x_1, y_1) involve the constants a^2 and b^2.

Ellipse	Slope of Tangent
$\frac{x^2}{a^2} + \frac{y^2}{b^2} = 1$	$m = -\frac{b^2 y_1}{a^2 x_1}$
Hyperbola	**Slope of Tangent**
$\frac{x^2}{a^2} - \frac{y^2}{b^2} = 1$	$m = \frac{b^2 y_1}{a^2 x_1}$

EXERCISES

1. Choose any ellipse. Write its equation and graph it. Then pick a point on the ellipse and draw the tangent through that point.

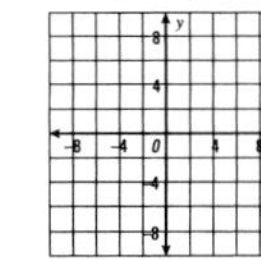

2. Choose any hyperbola. Write its equation and graph it. Then pick a point on the hyperbola and draw the tangent through that point.

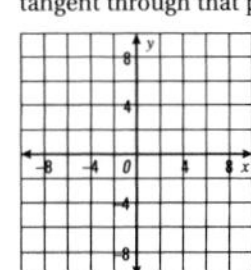

Answers may vary for Exercises 1–2.

Find the slope of the line tangent to each circle.

3. $m = -0.5$

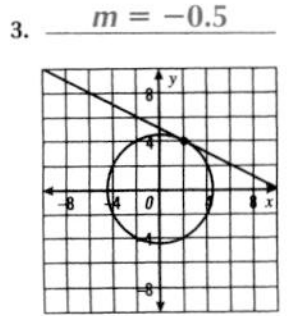

4. $m = 3$

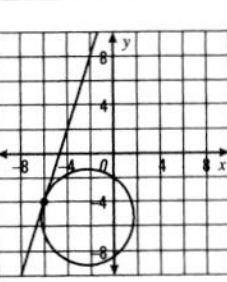

5. $m = 0.4$

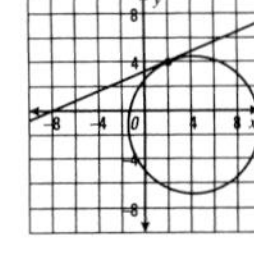

6. A circle has its center at the origin. Develop a formula for the slope of a tangent to that circle at the point (x_1, y_1). Sketch more figures if you wish. $m = -\frac{x_1}{y_1}$

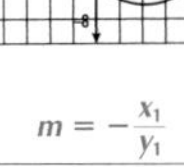

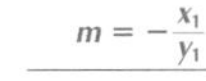

224 ENRICHMENT *LESSON 13-4*

Skills Practice

Vocabulary Review

Lesson 13-3
conics — conic section

Lesson 13-4
ellipse — foci
asymptotes — hyperbola

Assignment Guide

All students: 1–26
Additional Practice: See Extra Practice Index on page 674.

Additional Answers

5. The standard form is $\frac{x^2}{a^2} + \frac{y^2}{b^2} = 1$, where a = x-intercepts.
If $y = 0$;
$\frac{x^2}{a^2} + 0 = 1$
$\frac{x^2}{a^2} = 1$
$x^2 = a^2$
$x = \pm a$

6. The standard form is $\frac{x^2}{a^2} + \frac{y^2}{b^2} = 1$, where b = y-intercepts.
If $x = 0$;
$0 + \frac{y^2}{b^2} = 1$
$\frac{y^2}{b^2} = 1$
$y^2 = b^2$
$y = \pm b$

8.
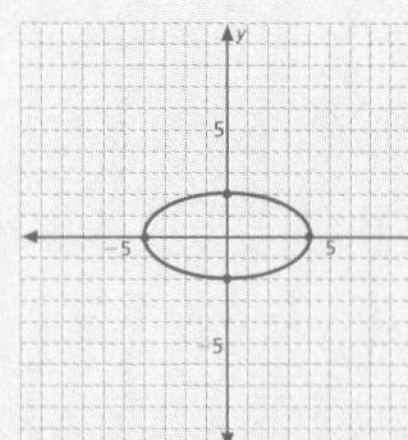

9.
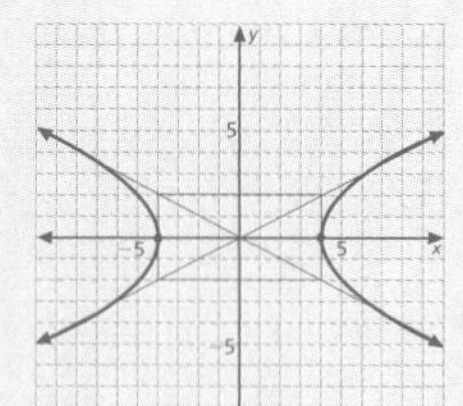

10.
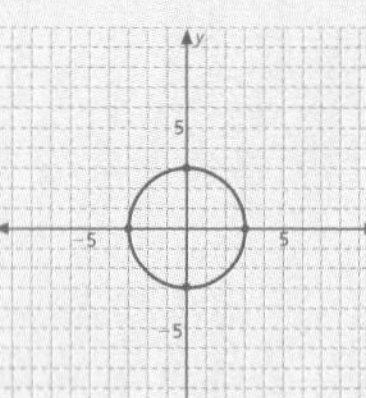

11.
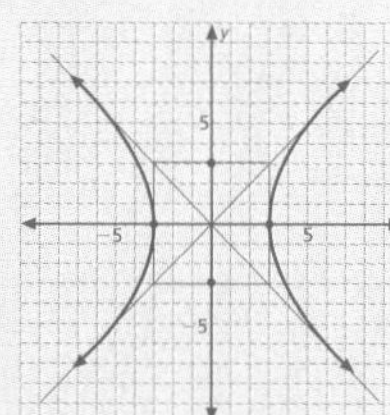

12.
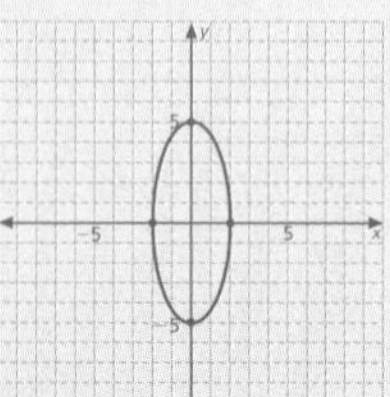

13.
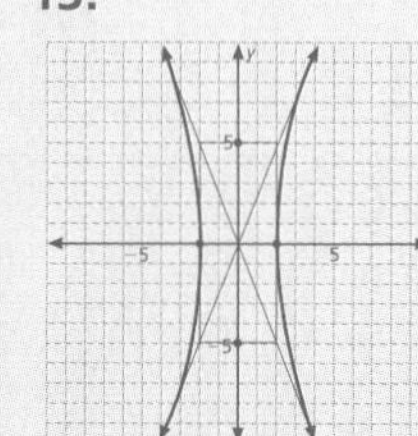

Review and Practice Your Skills

Practice ■ Lesson 13-3

Draw and name the conic section formed by each plane.

1. The plane is parallel to a side of the cone and does not pass through the vertex.

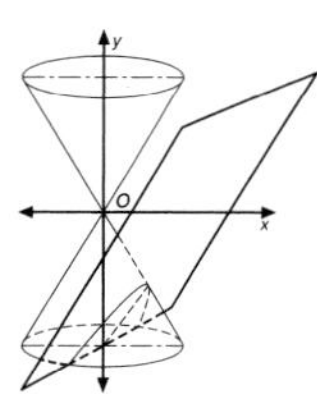

parabola

2. The plane intersects two parallel diameters.

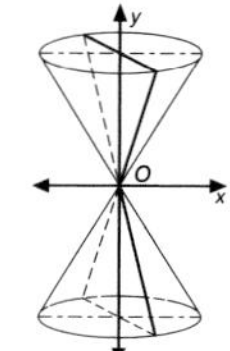

two triangles

3. The plane is parallel to the x-axis and does not contain the vertex.

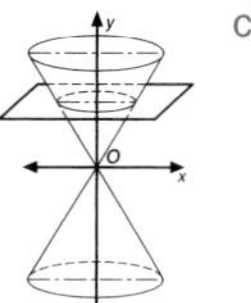

circle

4. The plane does not contain a vertex or base, and is not parallel to the x-axis.

ellipse

Practice ■ Lesson 13-4

5. Demonstrate how to determine the x-intercepts of an ellipse using the standard form of the equation for an ellipse and letting $y = 0$.
See additional answers.

6. Demonstrate how to determine the y-intercepts of an ellipse using the standard form of the equation for an ellipse and letting $x = 0$.
See additional answers.

7. The standard form of equation for and ellipse and for a hyperbola are similar. Explain how they are different. The standard equation for an ellipse uses the sum of the distances from a point on the ellipse to the foci; the equation for a hyperbola uses the differences of these distances.

Graph each equation.
For 8–13, see additional answers.

8. $4x^2 - 16y^2 = 64$
9. $4x^2 + 16y^2 = 64$
10. $9x^2 + 9y^2 = 81$
11. $9x^2 - 9y^2 = 81$
12. $25x^2 + 4y^2 = 100$
13. $25x^2 - 4y^2 = 100$

Find the equation of each ellipse.

14. foci $(5, 0)$, $(-5, 0)$
x-intercepts $(8, 0)$, $(-8, 0)$
$39x^2 + 64y^2 = 2{,}496$

15. foci $(3, 0)$ and $(-3, 0)$
x-intercepts $(10, 0)$, $(-10, 0)$
$91x^2 + 100y^2 = 9{,}100$

16. foci $(4, 0)$ and $(-4, 0)$
x-intercepts $(6, 0)$, $(-6, 0)$
$20x^2 + 36y^2 = 720$

Find the equation of each hyperbola. Assume all foci are on the x-axis.

17. center $(0, 0)$
$a = 5$, $b = 3$
$9x^2 - 25y^2 = 225$

18. center $(0, 0)$
$a = 9$, $b = 4$
$16x^2 - 81y^2 = 1{,}296$

19. center $(0, 0)$
$a = 7$, $b = 4$
$16x^2 - 49y^2 = 784$

578 | Chapter 13 **Quadratic Relations**

Teaching Strategies

COOPERATIVE LEARNING Working in small groups of 3–4 students, have students brainstorm all the ways they can think of that a plane can intersect a double cone. Remind them that if the plane is parallel to the y-axis, the cuts it makes in both cones will be identical.

PRACTICE ■ LESSON 13-1–LESSON 13-4

Draw and name the conic section formed by the plane. (Lesson 13-3)

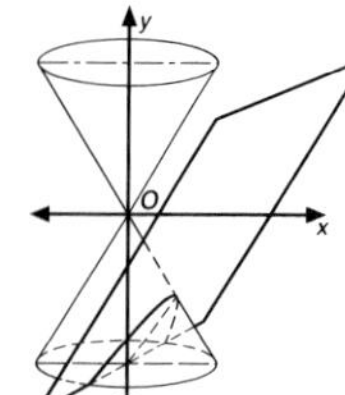

20. The plane is parallel to a side of the cone and does pass through the vertex point

Graph each equation. (Lesson 13-4) For 21–26, see additional answers.

21. $4x^2 + 4y^2 = 16$ **22.** $4x^2 - 4y^2 = 16$

23. $9x^2 + 36y^2 = 324$ **24.** $9x^2 - 36y^2 = 324$

25. $49x^2 + 16y^2 = 784$ **26.** $49x^2 - 16y^2 = 784$

Mid-Chapter Quiz

Find the radius and center for each circle.

1. $(x - 3)^2 + (y + 1)^2 = 64$ 8, (3, −1)

2. $x^2 + (y + 6)^2 = 6$ $\sqrt{6}$, (0, −6)

3. $(x + 7)^2 + (y - 9)^2 = 20$ $2\sqrt{5}$, (−7, 9)

4. $(x + 2)^2 + y^2 = 18$ $3\sqrt{2}$, (−2, 0)

Find the simple equation for each parabola with the vertex located at the origin.

5. focus (0, 2) $x^2 = 8y$ **6.** focus (0, −6) $x^2 = -24y$ **7.** focus $\left(0, \frac{2}{3}\right)$ $x^2 = \frac{8}{3}y$

8. focus $\left(0, -\frac{1}{4}\right)$ $x^2 = -y$ **9.** focus (6, 0) $y^2 = 24x$ **10.** focus (0.5, 0) $y^2 = 2x$

Write the standard equation for each ellipse.

11. foci (−2, 0) and (2, 0); x-intercepts (−5, 0) and (5, 0) $\frac{x^2}{25} + \frac{y^2}{21} = 1$

12. foci (−3, 0) and (3, 0); x-intercepts (−6, 0) and (6, 0) $\frac{x^2}{36} + \frac{y^2}{27} = 1$

Write the standard equation for each hyperbola.

13. center (0, 0); foci on x-axis; $a = +4$, $b = +7$ $\frac{x^2}{16} - \frac{y^2}{49} = 1$

14. center (0, 0); foci on x-axis; $a = +6$, $b = +8$ $\frac{x^2}{36} - \frac{y^2}{64} = 1$

15. The plane intersects the side of the cone and passes through the vertex. line

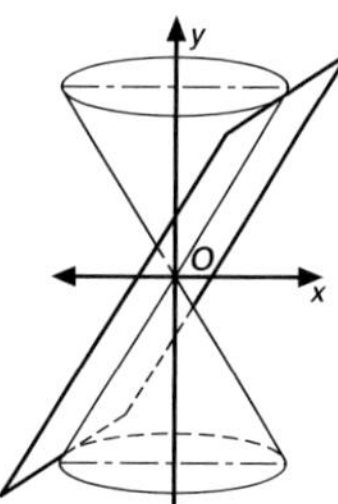

Chalkboard Examples

Example from Lesson 13-3

Describe the intersection between a plane and a cone that produces the largest circle. The plane must be parallel to and contain either base.

Example from Lesson 13-4

Find the x- and y-intercepts of the ellipse:

$\frac{x^2}{16} + \frac{y^2}{25} = 1$ x-int: ±4; y-int: ±5

Mid-Chapter Quiz

Correlation Chart

Question Number	Lesson Number
1–4	13-1
5–10	13-2
11–14	13-4

ADDITIONAL ANSWERS

21.

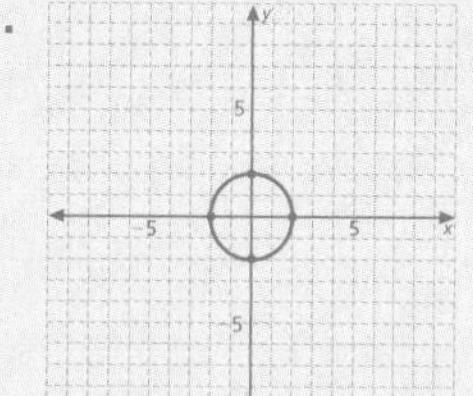

22.

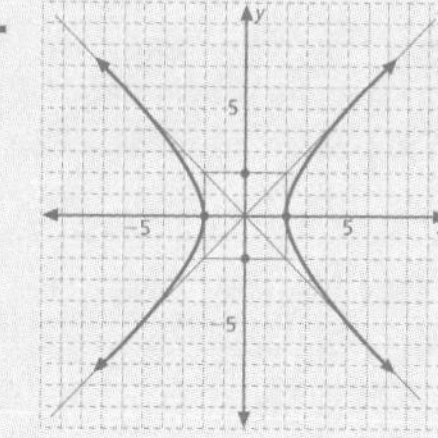

23.

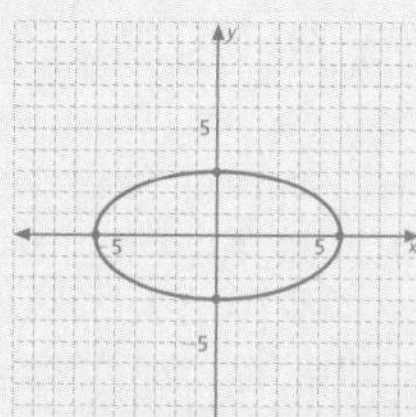

24.

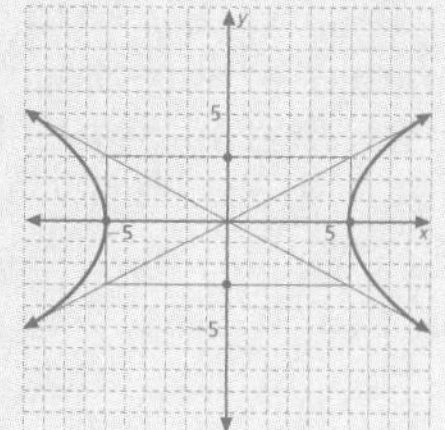

Extend the Lesson

REAL WORLD CONNECTION Point out that the rooms are sometimes built in the shape of ellipses and half-ellipsoids in order to focus sounds. A person standing at one focus can be heard whispering by someone at the other focus. Sound waves bounce off the walls and ceiling from one focal point to the other. These rooms exist in buildings such as the U.S. Capitol and the Taj Mahal in India.

25.

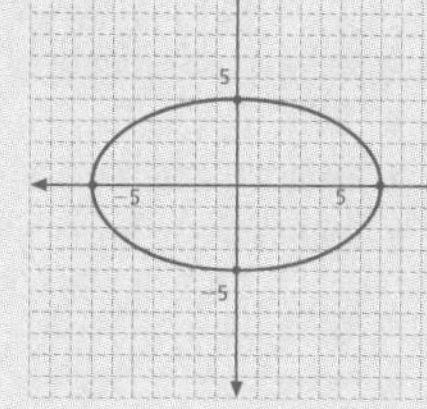

26.

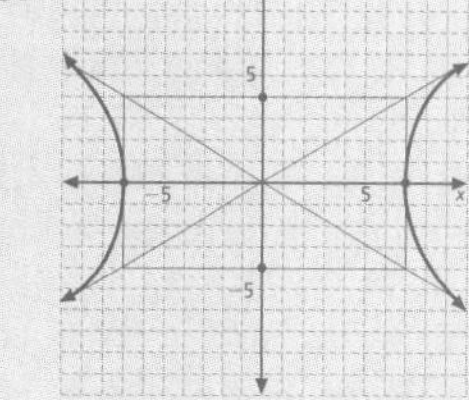

13-5 Direct Variation

Goals Solve problems involving direct variation.
Applications Food Prices, Space Exploration, Physics

Work with a partner to answer the following questions:

The table below shows the diameter and circumference of some circles.

Diameter (d)	3	6	9	12	24
Circumference (C)	9.42	18.84	28.26	37.68	75.36

a. Find the ratio $\frac{C}{d}$ for each pair of values. $\frac{C}{d} = 3.14$

b. When d doubles, what happens to C? It doubles.

c. Write an equation for C as a function of d. $C = 3.14d$

d. Graph the function. Describe the graph and find its slope. The graph is a straight line with slope 3.14.

Build Understanding

The circumference of a circle is a function of its diameter. This function can be written $C = 3.14d$. From the graph, you can see that the relationship between diameter and circumference is linear.

The relationship between diameter and circumference of a circle is an example of **direct variation**. The value of one variable increases as the other variable increases.

Direct variation can be represented by an equation in the form $y = kx$, where k is a nonzero constant and $x \neq 0$. The constant k is called the **constant of variation**. For the example above, circumference varies directly with diameter. The constant of variation is 3.14.

If y varies directly as x, the constant of variation can be found if one pair of values is known.

Reading Math

$y = kx$ is read "y is directly proportional to x" or "y varies directly as x."

Example 1

What is the equation for a direct variation when one pair of values is $x = 20$ and $y = 9$?

Solution

$y = kx$ Substitute in the equation for direct variation.

$9 = k(20)$

$\frac{9}{20} = k$

$0.45 = k$ Solve for k.

The equation is $y = 0.45x$.

Lesson Planning

NCTM Standards/Strands

- Number & Operation
- Patterns, Functions, & Algebra
- Measurement
- Problem Solving
- Reasoning and Proof
- Communication
- Connections
- Representation

Vocabulary

direct variation
constant of variation
direct square variation

Materials Needed

calculators

Lesson Resources

Warm-Up Transparency 38
Transparency RF-51
Reteaching 13-5
Extra Practice 13-5
Enrichment 13-5

Assignment Guide

Basic: 1–24, 31–44
Enriched: 1–44

Getting Started

5-Minute Warm-Up

Solve each proportion.

1. $\frac{\$1.90}{2} = \frac{x}{7}$ $x = \$6.65$
2. $\frac{51}{3} = \frac{a}{10}$ $a = 170$
3. $\frac{r}{16} = \frac{\$4.05}{3}$ $r = \$21.60$
4. $\frac{35}{3} = \frac{y}{7.5}$ $y = 87.5$

Introduction to Lesson 13-5

Reinforce the idea of direct variation. Ask students what happens to C when D is halved. For Part (d), ask students how they would interpret the fact that the graph passes through (0, 0). For a circle with a diameter of 0, the circumference is 0.

Teaching Strategies

CHALLENGE Determine whether each of the following relationships represents a direct variation. Give your reasons.

1. $2y = 5x$ yes, since $y = \frac{5}{2}x$, $k = \frac{5}{2}$
2. $3y = 2x^2$ yes, since $y = \frac{2}{3}x^2$, $k = \frac{2}{3}$

Many examples of the use of direct variation may be found in everyday situations.

Example 2

FOOD PRICES The cost of apples varies directly with weight. If 9 lb of apples cost \$4.32, how much will 17 lb of apples cost?

Solution

$y = kx$

$4.32 = k(9)$ Substitute.

$\frac{4.32}{9} = k$

$0.48 = k$ Solve for k.

$y = 0.48x$ Write the equation.

$y = 0.48(17)$ Substitute $x = 17$.

$y = 8.16$ Solve.

Seventeen pounds of apples will cost \$8.16

Check Understanding

k is sometimes called the constant of proportionality. How could this problem be solved as a proportion?

$\frac{\$4.32}{9} = \frac{x}{7}$

The equation for the area of a circle is $A = \pi r^2$. The area varies directly as the square of the radius. This is an example of **direct square variation**. The graph is a quadratic function.

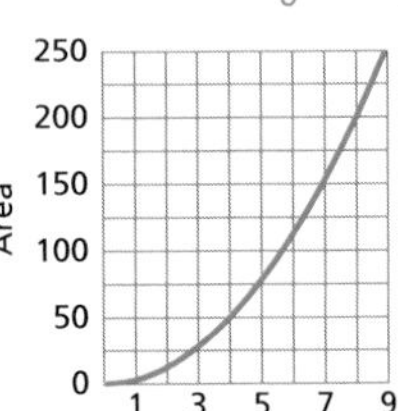

Direct square variation may be expressed in the form $y = kx^2$ where k is a nonzero constant.

Example 3

SPACE EXPLORATION An air filter used in a space vehicle is in the shape of a cube. The surface area of a cube varies directly as the square of its sides. If the surface area of the air filter is with sides 12 in. long is 864 in.^{2}, what is the surface area of an air filter in the shape of a cube with sides 11 in. long?

Solution

$y = kx^2$

$864 = k(12)^2$ Substitute in the equation.

$864 = 144k$ Solve for k.

$\frac{864}{144} = k$

$6 = k$

$y = 6x^2$ Write the equation.

$y = 6(11)^2$ Substitute 11 for x.

$y = 726$ Solve.

The surface area of a cube with sides 11 in. is 726 in.^{2}.

Extend the Lesson

COOPERATIVE LEARNING Have students work with a partner. State whether the relation is a direct variation. Then solve the problem. A taxi company charges \$1.75 plus \$0.75 per mile. If a 3 mile trip costs \$4.00, how much does an 8 mile trip cost? not a direct variation; \$7.75

Chalkboard Examples

Supplementary Example 1

What is the equation for a direct variation when one pair of values is $x = 5$ and $y = 35$? $y = 7x$

Supplementary Example 2

The cost of ground beef varies directly with weight. If 3 pounds of ground beef cost \$8.94, how much will 5 pounds cost? \$14.90

Supplementary Example 3

A packing crate is in the shape of a cube. The surface area of the crate varies directly as the square of its sides. If the surface are of a cube with sides 15 inches long is 1,350 sq in., what is the surface area of the crate with sides 28 inches? 4,704 sq in.

Reteaching Worksheet 13-5

Name ______________ Date ______________

RETEACHING 13-5

DIRECT VARIATION

Direct variation can be represented by an equation in the form $y = kx$ where k is a nonzero constant and $x \neq 0$. The constant k is called the **constant of variation**.

Example 1

Write the question for a direct variation when one pair of values is $x = 32$ and $y = 14$.

Solution

Substitute in the equation and solve for k.

$y = kx$

$14 = k(32)$

$\frac{14}{32} = k$

$0.4375 = k$

The equation is $y = 0.4375x$.

Example 2

The weight of a metal rod varies directly with its length. A rod 4 cm long weighs 12 g. How much does a rod 20 cm long weigh?

Solution

First find k.

$y = kx$

$12 = k(4)$

$3 = k$

Then solve.

$y = kx$

$y = 3x$ Substitute 3 for k.

$y = 3(20)$ Substitute 20 for x.

$y = 60$

A rod that is 20 cm long weighs 60 g.

EXERCISES

Write the equation for the direct variation for each pair of values.

1. $x = 40; y = 20$ $y = 0.5x$
2. $x = 25; y = 12$ $y = 0.48x$
3. $x = 15; y = 3$ $y = 0.2x$
4. $x = 30; y = 12$ $y = 0.4x$
5. $x = 25; y = 20$ $y = 0.8x$
6. $x = 16; y = 10$ $y = 0.625x$

Find each answer.

7. The weight of a metal rod varies directly with its length. A rod 5 cm long weighs 12 g. How much does a rod 40 cm long weight? 96 g
8. The length of a shadow at a given time is directly proportional to the height of an object. If an 8-ft tree casts a 6-ft shadow, how high is a building that casts a 90-ft shadow at the same time? 120 ft
9. A freight train can travel 260 mi in 5 hours. How far can it travel in 8 hours? 416 mi
10. The weight of earth varies directly with the weight on the moon. An astronaut weighs 210 lbs on earth and 35 lb on the moon. How much would a 180-lb astronaut weigh on the moon? 30 lb

200 RETEACHING *LESSON 13-5*

Lesson Wrap-up

Quick Assessment

Ask the following questions to determine if students understand the content presented in this lesson.

1. How does the relation $y = kx + b$ differ from direct variation? The inclusion of the constant b prevents every pair of x and y values from forming a true proportion.
2. Why must the graph of any direct variation pass through the origin? In any direct variation, when $x = 0$, $y = 0$.

Assignment Guide

Basic: 1–24, 31–44
Advanced: 1–44
Additional Practice: See Extra Practice Index on page 674.

Try These Exercises

1. What is the equation of direct variation when one pair of values is $x = 72$ and $y = 18$? $y = 0.25x$
2. If y varies directly as x and $y = 60$ when $x = 50$, find y when $x = 15$. $y = 18$
3. If y varies directly as x^2 and $y = 320$ when $x = 8$, find y when $x = 5$. $y = 125$
4. **POSTAGE** Fifteen stamps cost \$4.95. How much will 26 stamps cost? \$8.58
5. **EARNINGS** A person's income varies directly with the number of hours the person works. If the pay for 16 h is \$200, what is the pay for working 40 h? \$500
6. **CATERING** Three vegetable platters serve a party of 20 people. How many vegetable platters are needed for a party of 240 people? 36

Practice Exercises

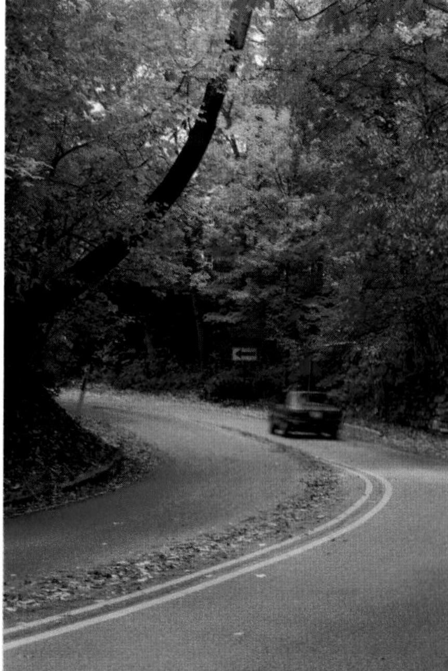

7. The distance (d) a vehicle travels at a given speed is directly proportional to the time (t) it travels. If a vehicle travels 30 miles in 45 min, how far can it travel in 2 h? 80 mi
8. **ASTRONOMY** The speed of a comet at perihelion (the closest orbital point to the sun) is 98,000 mi/h. How far will the comet travel in 30 sec? 816.67 mi
9. **BIOLOGY** The expected increase (I) of a population of organisms is directly proportional to the current population (n). If a sample of 360 organisms increases by 18, by how many will a population of 9000 increase? 450
10. **PHYSICS** The distance (d) an object falls is directly proportional to the square of time (t) it falls. If an object falls 256 ft in 4 seconds, how far will it fall in 7 sec? 784 ft
11. **SALES** The cost of ribbon is directly proportional to the length purchased. If 9.5 yd of ribbon cost \$3.42, how much will 14.75 yd cost? \$5.31
12. **SALES** A person is paid \$153 for 9 baskets of dried flowers. To earn \$450, how many baskets of dried flowers must the person produce? 27
13. **BAKED GOODS** A bakery earns \$33.75 profit on 9 cakes. How many cakes must be sold for the bakery to make a profit of \$150? 40
14. **WRITING MATH** A clothing store charges \$1.75 in sales tax on an item that costs \$25. The same store charges \$2.80 in sales tax on a \$40 item. Is sales tax an example of direct variation? Justify your answer. Yes. Sales tax varies directly with the purchase price of an item. In the example stated, the store is charging 7% sales tax on each purchase.
15. **ERROR ALERT** The distance a spring will stretch, S, varies directly with the weight, W, added to the spring. A spring stretches 1.5 in. when 12 lb are added. Paige plans to add 2 more pounds to the spring. She concludes that the spring will stretch 1.5 + 2, or 3.5 inches when the weight is added. Is Paige's thinking correct? Explain your reasoning. No, Paige's conclusion is wrong. Using the direction variation form, $y = kx$, if 14 pounds are added to the spring, it will stretch 1.75 inches.

Teaching Strategies

COOPERATIVE LEARNING Have students work together to answer the following questions. Then have the pairs share their thinking with the class.

1. Is a direct variation a function? yes, since each x-value has exactly one y-value associated with it
2. Must the graph of a direct variation always be linear? yes, because a direct variation has no variable with a power greater than 1

The area of each regular polygon varies directly as the square of its sides. One pair of values is given for each. Find the area for each regular polygon. Then find the area of each polygon if one side is 9 units.

16. Pentagon
$s = 3$
A = 15.49
A = $1.721s^2$, 139.401

17. Hexagon
$s = 5$
A = 64.95
A = $2598s^2$, 210.438

18. Octagon
$s = 7$
A = 236.57
A = $4.828s^2$, 391.068

DATA FILE The number of kilowatts of electricity used by an appliance varies directly as the time the appliance is used. Use the Data Index on pages 632–633 to find the location of information on the number of kilowatt hours of electricity used by some appliances. Then answer these questions.

19. If you watch television for 4.5 h, about how many kilowatt hours of electricity do you use? 1,035

20. If you dry your hair for 10 min, how many kilowatt hours of electricity do you use? 0.25

21. If the refrigerator runs for 2 h/day, how many kilowatt hours of electricity does it use? 10

22. If the water heater runs 2.75 h/day, how many kilowatt hours of electricity does it use? 44

23. If electricity costs 12.3¢ per kilowatt hour, find the cost for each activity in Exercises 19–22 to the nearest cent. 19. \$0.13; 20. \$0.03; 21. \$1.23; 22. \$5.41

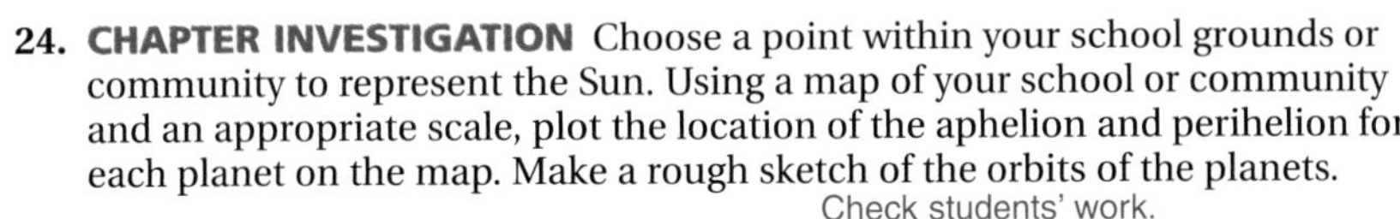

24. **CHAPTER INVESTIGATION** Choose a point within your school grounds or community to represent the Sun. Using a map of your school or community and an appropriate scale, plot the location of the aphelion and perihelion for each planet on the map. Make a rough sketch of the orbits of the planets.
Check students' work.

Extended Practice Exercises

Write *direct variation, direct square variation,* or *neither* to describe how *P* varies as *V* increases or decreases in each equation.

25. $P = IV$ direct variation

26. $P = \frac{K}{V}$ neither

27. $MP = 2V^2$ direct square variation

28. $\frac{1}{8}V^2 = P$ direct square variation

29. $\frac{P}{V} = 1$ direct variation

30. $(4V + 1) - P = -1$ direct variation

Mixed Review Exercises

Use the quadratic formula to solve each equation. (Lesson 12-5)

31. $2x^2 + 3x - 6 = 0$ $x = \frac{-3 \pm \sqrt{57}}{4}$

32. $x^2 - 4x + 3 = 0$ $x = 3, 1$

33. $x^2 - 4x - 8 = 0$ $x = 2 \pm 2\sqrt{3}$

34. $-2x^2 - 4x + 1 = 0$ $x = \frac{-2 \pm \sqrt{6}}{2}$

35. $-4x^2 - 6x + 1 = 0$ $x = \frac{-3 \pm \sqrt{13}}{4}$

36. $2x^2 - 6 = 0$ $x = \pm\sqrt{3}$

Find the value to the nearest hundredth. (Basic Math Skills)

37. $\sqrt{72}$ 8.49

38. $\sqrt{48}$ 6.93

39. $\sqrt{175}$ 13.23

40. $\sqrt{37}$ 6.08

Write each in simplest radical form. (Lesson 10-1)

41. $(2\sqrt{12})(5\sqrt{27})$ 180

42. $(2\sqrt{8})(3\sqrt{12})$ $24\sqrt{6}$

43. $\frac{\sqrt{18}}{\sqrt{8}}$ $\frac{3}{2}$

44. $\sqrt{\frac{28}{6}}$ $\frac{\sqrt{42}}{3}$

Extend the Lesson

REAL WORLD CONNECTION Have students brainstorm common life situations that involve direct variation and some that do not. Make a list on the board. Have students discuss how they can identify direct variation word problems.

Extra Practice Worksheet 13-5

Name ______________________ Date ____________

EXTRA PRACTICE 13-5
DIRECT VARIATION

EXERCISES

1. What is the equation of direct variation when one pair of values is $x = 60$ and $y = 15$? $y = 0.25x$
2. What is the equation of direct variation when one pair of values is $x = 49$ and $y = 98$? $y = 2x$
3. What is the equation of direct variation when one pair of values is $x = 72$ and $y = 54$? $y = 0.75x$
4. If y varies directly as x and $y = 24$ when $x = 20$, find y when $x = 10$. 12
5. If y varies directly as x and $y = 16$ when $x = 4$, find y when $x = 5$. 20
6. If y varies directly as x^2 and $y = 300$ when $x = 25$, find y when $x = 20$. 180
7. If y varies directly as x^2 and $y = 400$ when $x = 16$, find y when $x = 4$. 6.25
8. The distance (d) a vehicle travels at a given speed is directly proportional to the time (t) it travels. If a vehicle travels 40 miles in 60 minutes, how far can it travel in 90 minutes? 60 miles
9. The distance (d) a vehicle travels at a given speed is directly proportional to the time (t) it travels. If a vehicle travels 60 miles in 50 minutes, how far can it travel in 2 hours? 144 miles
10. The expected increase (I) of a population of organisms is directly proportional to the current population (n). If a sample of 240 organisms increases by 20, by how many will a population of 600 increase? 50
12. The expected increase (I) of a population of organisms is directly proportional to the current population (n). If a sample of 500 organisms increases by 35, by how many will a population of 6000 increase? 420
13. The distance (d) an object falls is directly proportional to the square of time (t) it falls. If an object falls 144 feet in 4 seconds, how far will it fall in 6 seconds? 324 ft
14. The distance (d) an object falls is directly proportional to the square of time (t) it falls. If an object falls 225 feet in 9 seconds, how far will it fall in 12 seconds? 400 ft

174 EXTRA PRACTICE *LESSON 13-5*

Enrichment Worksheet 12-5

Name ______________________ Date ____________

ENRICHMENT 13-5
LINE DESIGNS

Line designs are geometric patterns formed by connecting certain sequences of points. Although only straight lines are used, the resulting patterns appear curved. Often, the curves take the shape of parabolas, ellipses and hyperbolas.

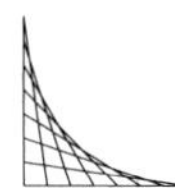

EXERCISES

Construct these line designs. The first three are based on the right angle above. The angle has sides of equal length. In Exercise 4, the right angle has one side that is twice the length of the other side.

1. 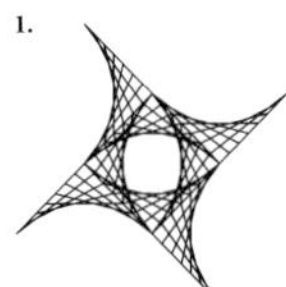
2. 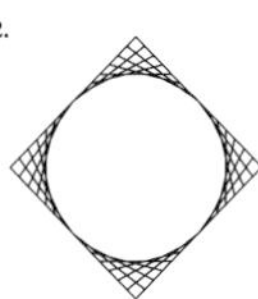
3. 
4. 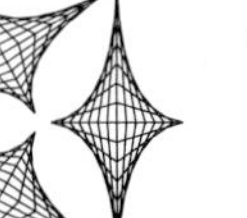

Check students' work for Exercises 1–4.

5. This final design is somewhat different. You start with a point *P* on the diameter of a circle. Draw rays from point *P*. Where each ray meets the circle, draw a perpendicular to the ray. What figure results from drawing a large number of rays and their perpendiculars?
an ellipse

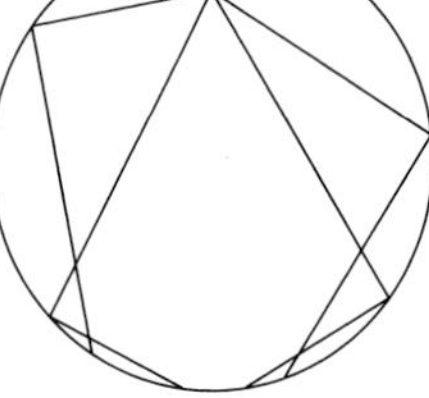

226 ENRICHMENT *LESSON 13-5*

Lesson Planning

NCTM Standards/Strands

- Number & Operation
- Patterns, Functions, & Algebra
- Data Analysis, Statistics, & Probability
- Problem Solving
- Reasoning and Proof
- Communication
- Connections
- Representation

Vocabulary

inverse variation
inverse square variation
joint variation
combined variation

Materials Needed

calculators (optional)

Lesson Resources

Warm-Up Transparency 39
Transparency RF-52
Reteaching 13-6
Extra Practice 13-6
Enrichment 13-6

Assignment Guide

Basic: 1–16, 23–31
Enriched: 1–31

Getting Started

5-Minute Warm-Up

Solve each equation in terms of *k*.

1. $y = \frac{k}{x}$ $k = xy$
2. $y = \frac{k}{x} + 1$ $k = x(y - 1)$
3. $y = \frac{k}{x} - 2$ $k = x(y + 2)$

Introduction to Lesson 13-6

Ask students to explain why the cost to each person sharing the vacation home would never reach 0, regardless of the number of renters. Students should understand that regardless of the size of n, $\frac{\$800}{n}$ can never be 0. In the same way, for $y = \frac{k}{x}$, since neither k nor x equals 0, y will never equal zero.

13-6 Inverse Variation

Goals ■ Solve problems involving inverse variation and inverse square variation.

Applications Astronomy, Physics, Travel

Work with a partner to answer the following questions.

The table shows the cost per person of renting a vacation home.

Number of people (n)	1	2	4	5	8	10
Cost per person (c)	\$800	\$400	\$200	\$160	\$100	\$80

a. Find the product of nc for each pair of values. \$800

b. When n doubles, what happens to c? It is halfed.

c. Write an equation for c as a function of n. $c = \frac{\$800}{n}$

d. Graph the function. Explain how c varies as n increases.
As n increases, c decreases.

Build Understanding

The cost per person for renting the vacation home is a function of the number of people. As the number of people increases, the cost per person decreases. The relationship between n and c is an example of **inverse variation**. The value of one variable decreases as the value of the other increases.

Reading Math

$y = \frac{k}{x}$ is read: "y is inversely proportional to x" or "y varies inversely as x."

Inverse variation may be represented by an equation in the form $y = \frac{k}{x}$, where k is a nonzero constant and $x \neq 0$. For the example above, the cost per person of renting the vacation home varies inversely as the number of people sharing the cost. The constant of variation is \$800.

The graph you drew in the activity above illustrates that the graph of an inverse variation is a hyperbola.

Most applications involve only the part of the hyperbola that lies in the first quadrant.

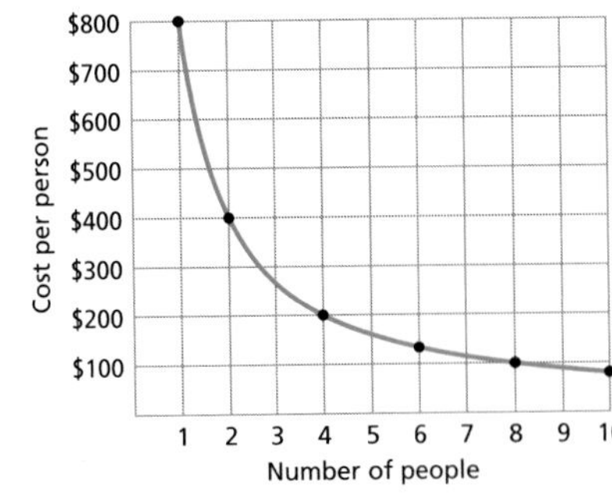

Extend the Lesson

MATH JOURNAL In their journals have students describe the similarities and differences between the equation of direct variation $y = kx$ and that of inverse variation $y = \frac{k}{x}$. In both relations, k is the constant of proportionality; in both, a change in one variable causes a change in the other; in one equation, both variables increase or decrease together; in the other, any change in one variable causes an inverse change in the other.

Example 1

Write an equation in which y varies inversely as x if one pair of values is $y = 240$ and $x = 0.4$.

Solution

$$y = \frac{k}{x}$$ Substitute in the equation for inverse variation.

$$240 = \frac{k}{0.4}$$

$$240 \times 0.4 = k$$

$$96 = k$$ Solve for k.

The equation is $y = \frac{96}{x}$.

Travel time varies inversely as travel speed. In other words, travel time decreases as speed increases.

Example 2

ASTRONOMY At its greatest distance from the sun, an asteroid travels a certain distance in 40 min while traveling at 250 mi/h. How long would it take the asteroid to travel the same distance, traveling at 400 mi/h?

Solution

$$y = \frac{k}{x}$$ Find an equation.

$$40 = \frac{k}{250}$$

$$40(250) = k$$

$$10{,}000 = k$$

$$y = \frac{10{,}000}{x}$$

$$y = \frac{10{,}000}{400}$$ Substitute 400 into the equation.

$$y = 25$$ Solve for y.

The trip will take 25 min travelling at 400 mi/h.

The table below shows how the brightness (in lumens) of a 60-watt light bulb varies with distance from the bulb.

Distance (in feet)	1	2	4	8
Brightness (in lumens)	880	220	55	13.75

This is an example of **inverse square variation**. The brightness of the light varies inversely as the square of the distance from its source.

Inverse square variation can be expressed in the form

$y = \frac{k}{x^2}$ or $x^2y = k$, where k is a nonzero constant and $x \neq 0$.

For the example above, $l = \frac{880}{d^2}$. The constant of variation is 880.

Teaching Strategies

Students may attempt to solve an inverse variation by writing a proportion. Point out that inverse variation problems can be solved by proportion; however, students must invert one side of the proportion to correspond to the inverse relationship. They can use these relationships:

$$\frac{x_1}{x_2} = \frac{y_2}{y_1} \text{ or } \frac{x_1^2}{x_2^2} = \frac{y_2}{y_1}$$

Chalkboard Examples

Supplementary Example 1

Write an equation in which y varies inversely as x if one pair of values is $y = 1.2$ and $x = 10$. $y = \frac{12}{x}$

Supplementary Example 2

An airplane travels a certain distance in 1.5 hr while traveling at 600 miles per hour. How long would it take the airplane to travel the same distance traveling at 500 miles per hour? 1.8 hr

Supplementary Example 3

The brightness of a light bulb varies inversely as the square of the distance from the source. If a light bulb has a brightness of 600 lumens at 2 feet, what will be it brightness at 16 feet? 3.75 lumens

Reteaching Worksheet 13-6

Name ______________________ Date ____________

RETEACHING 13-6

INVERSE VARIATION

Inverse variation can be represented by an equation in the form $y = \frac{k}{x}$ where k is a nonzero constant and $x \neq 0$. When the value of one variable increases, the value of the other variable decreases.

Example 1

Write the equation in which y varies inversely with x if one pair of values is $y = 64$ and $x = 0.6$.

Solution

Substitute in the equation and solve for k.

$$y = \frac{k}{x}$$

$$64 = \frac{k}{0.6}$$

$$64 \cdot 0.6 = k$$

$$38.4 = k$$

The equation is $y = \frac{38.4}{x}$.

Example 2

The resistance in an electrical circuit with constant voltage is inversely proportional to the current. If a light bulb has a resistance of 60 ohms when a current of 2 amperes flows through it, what is the resistance in another bulb when a current of 1.25 amperes flows through it?

Solution

First find k.

$$y = \frac{k}{x}$$

$$60 = \frac{k}{2}$$

$$120 = k$$

Then solve.

$$y = \frac{k}{x}$$

$$y = \frac{120}{1.25}$$

$$y = 96$$

The bulb has a resistance of 96 ohms.

EXERCISES

Write the equation in which y varies inversely with x for each pair of values.

1. $x = 0.3; y = 40$ $y = \frac{12}{x}$
2. $x = 2; y = 36$ $y = \frac{72}{x}$
3. $x = 4; y = 35$ $y = \frac{140}{x}$
4. $x = 0.9; y = 100$ $y = \frac{90}{x}$
5. $x = 0.6; y = 75$ $y = \frac{45}{x}$
6. $x = 3; y = 74$ $y = \frac{222}{x}$

Find each answer.

7. The resistance in an electrical circuit with constant voltage is inversely proportional to the current. If a light bulb has a resistance of 55 ohms when a current of 2 amperes flows through it, what is the resistance in another bulb when a current of 1.375 amperes flows through it? 80 ohms
8. The time that it takes to drive a given distance varies inversely with the rate of speed. If Rosita drives 50 mi/h for 4 hours, how long would it take her to drive the same distance at 60 mi/h? $3\frac{1}{3}$ h

202 RETEACHING *LESSON 13-6*

Lesson Wrap-up

Quick Assessment

Ask the following questions to determine if students understand the content presented in this lesson.

1. Why is the graph of an inverse variation a hyperbola? In an inverse variation, as x increases, y decreases; since y can never be 0, the graph will get closer and closer to the x-axis without ever touching it.
2. How would you use the data in the table below to determine whether the values represent an inverse variation?

x	80	120	160	200
y	60	40	30	24

Determine whether the product xy is a constant; since it is, these values represent an inverse variation.

Assignment Guide

Basic: 1–16, 23–31
Advanced: 1–31
Additional Practice: See Extra Practice Index on page 674.

Example 3

PHYSICS The brightness of a light bulb varies inversely as the square of the distance from the source. If a light bulb has a brightness of 400 lumens at 2 ft, what will be its brightness at 20 ft?

Solution

$x^2y = k$ — Substitute known values into the equation for inverse square variation.

$(2)^2(400) = k$

$4(400) = k$

$1600 = k$ — Solve.

$x^2y = 1600$ — Write the equation.

$(20)^2(y) = 1600$ — Substitute 20 for x.

$400y = 1600$

$y = 4$ — Solve.

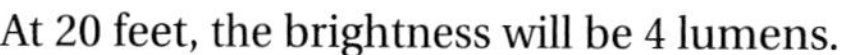

At 20 feet, the brightness will be 4 lumens.

Try These Exercises

1. Write an equation in which y varies inversely as x if one pair of values is $y = 85$ and $x = 0.8$. $y = \frac{68}{x}$
2. If y varies inversely as x and one pair of values is $y = 44$ and $x = 5$, find y when $x = 8$. $y = 27.5$
3. The amount paid by each person sharing a cab varies inversely as the number of people who share the cab. If 2 people pay $4.50 each for a ride from a restaurant to the theater, how much will the same ride cost 5 people? $1.80 each
4. If y varies inversely as the square of x and $y = 224$ when $x = 2$, find y when $x = 8$. $y = 14$
5. **PHYSICS** The brightness of a light bulb varies inversely as the square of the distance from the source. If a light bulb has a brightness of 300 lumens at 2 ft, what will be its brightness at 10 ft? 12 lumens

Practice Exercises

6. Write an equation in which y varies inversely as x if one pair of values is $y = 4550$ and $x = .05$. $y = \frac{227.5}{x}$
7. If y varies inversely as x and one pair of values is $y = 39$ and $x = 3$, find y when $x = 39$. $y = 3$
8. If y varies inversely as x and one pair of values is $y = 12$ and $x = 10$, find y when $x = 20$. $y = 6$

9. **WRITING MATH** Think of a real-life example of inverse variation. Explain how you know the type of variation the example represents. Answers may vary. Students should mention that in an inverse function, as one variable increases, the other decreases.
10. **TRAVEL** If it takes 30 min to drive from Ann's house to the museum traveling at 40 mi/h, how long will it take traveling at 50 mi/h? 24 min.

Teaching Strategies

Students who use graphing calculators to draw the graphs of inverse variations will notice that both halves of the hyperbola are visible on the screen. Remind them that since most applications of inverse variation use positive numbers, the values with which we deal can be represented by the first quadrant part of the hyperbola.

586 Chapter 13 Quadratic Relations

11. **MAGNETISM** The force of attraction between two magnets varies inversely as the square of the distance between them. When two magnets are 2 cm apart, the force is 64 newtons. What will be the force when they are 8 cm apart? 4 Newtons

12. If y varies inversely as the square of x and $y = 256$ when $x = 4$, find y when $x = 8$. $y = 64$

Write an equation of inverse variation for each.

13. Barometric pressure (p) is inversely proportional to the altitude (a). $p = \frac{k}{a}$

14. The time (t) required to fill a swimming pool is inversely proportional to the square of the diameter (d) of the hose used to fill it. $t = \frac{k}{d^2}$

15. The current (I) flowing in an electric circuit varies inversely as the resistance (R) in the circuit. $I = \frac{k}{R}$

16. The intensity of the heat from a fireplace varies inversely as the square of the distance from the fireplace. Your friend is next to the fireplace. If you only feel $\frac{1}{16}$ the amount of heat that your friend feels, how much farther are you from the fireplace than your friend? 4 times farther away

Extended Practice Exercises

When a quantity varies directly as the product of two or more other quantities, the variation is called a **joint variation**. If y varies jointly as w and x, then $y = kwx$.

When a quantity varies directly as one quantity and inversely as another, the variation is called a **combined variation**. If y varies directly as w and inversely as x, then $y = \frac{kw}{x}$.

Write an equation of joint variation for each.

17. m varies directly as s and t. $m = kst$

18. a varies jointly as c and d. $a = kcd$

19. r varies jointly as w, x, and y. $r = kwxy$

Write an equation of combined variation for each.

20. n varies directly as t and inversely as e. $n = \frac{kt}{e}$

21. v varies directly as r and inversely as the square of w. $v = \frac{kr}{w^2}$

22. d varies directly as the square of a and inversely as b. $d = \frac{ka^2}{b}$

Mixed Review Exercises

Calculate the distance between each pair of points. Find exact distances and round to the nearest hundredth if necessary. (Lesson 12-6)

23. $A(3, 2)$, $B(1, -8)$ 10.20

24. $C(-6, 2)$, $D(5, -9)$ 15.56

25. $E(3, -4)$, $F(8, -6)$ 5.39

26. $G(-4, -3)$, $H(8, -1)$ 12.17

27. $J(7, -5)$, $K(3, 8)$ 13.60

28. $L(-4, -7)$, $M(-2, -3)$ 4.47

In the figure at the right, R is the midpoint of QS. Find: (Lesson 3-3)

$4x + 1$ $6x - 4$ $7x + 3$

Q   R   S   T

29. QR 11

30. RS 11

31. ST 20.5

Extend the Lesson

COOPERATIVE LEARNING Have students write each of these geometric formulas and state whether they are examples of joint or combined variation.

1. length of a rectangle: $l = \frac{A}{w}$ combined
2. height of a cylinder: $h = \frac{V}{\pi r^2}$ combined
3. area of a parallelogram: $A = bh$ joint

Extra Practice Worksheet 13-6

Name ______________________ Date ____________

EXTRA PRACTICE 13-6

DIRECT VARIATION

EXERCISES

1. Write an equation in which y varies inversely as x if one pair of values is $x = 15$ and $y = 10$. $y = \frac{150}{x}$
2. Write an equation in which y varies inversely as x if one pair of values is $x = 24$ and $y = 0.2$. $y = \frac{4.8}{x}$
3. If y varies inversely as x and one pair of values is $y = 64$ and $x = 8$, find y when $x = 12$. $42\frac{2}{3}$
4. If y varies inversely as x and one pair of values is $y = 52$ and $x = 14$, find y when $x = 28$. 26
5. If y varies inversely as the square of x and $y = 144$ when $x = 4$, find y when $x = 12$. 16
6. If y varies inversely as the square of x and $y = 64$ when $x = 8$, find y when $x = 16$. 16
7. The brightness of a light bulb varies inversely as the square of the distance from the source. If a light bulb has a brightness of 625 lumens at 5 feet, what will be its brightness at 25 feet? 25 lumens
8. The force of attraction between two magnets varies inversely as the square of the distance between them. When two magnets are 9 centimeters apart, the force is 81 newtons. What will be the force when they are 15 centimeters apart? 29.16 newtons

Write an equation of joint variation for each.

9. r varies jointly as a and b. $r = kab$
10. q varies jointly as r, s and t. $q = krst$

Write an equation of combined variation for each.

11. x varies directly as y and inversely as z. $x = \frac{ky}{z}$
12. b varies directly as f and inversely as the square g. $x = \frac{kf}{g^2}$

176   EXTRA PRACTICE *LESSON 13-6*

Enrichment Worksheet 13-6

Name ______________________ Date ____________

ENRICHMENT 13-6

THE FIVE CUBES PUZZLE

This puzzle involves five numbered cubes in an open box. The box has room for six cubes, but there is one empty space. When working the puzzle, the cubes may be slid around in the box. But, they cannot be taken out of the box.

The position of the cubes after any move is recorded by listing the numbers clockwise, starting at the top left. The cube-position is given under each diagram.

1	2	3
	5	4

12345

EXERCISES

Make a set of five cubes, or use square pieces of paper.

1. Start with the cubes in the position at the right. Transform them back into the 12345 order above. Record the moves you use.
31524—15243—12453—24531—23451—12345

3		1
4	2	5

31524

2. Try transforming the cubes to this arrangement. Put them back in the 12345 order before you begin. Record the moves you use.
32415—34125—53412—25341—23451—12345 or 32415—34125—53412—51342—13425—12345

3	2	4
5		1

32415

3. Can you find an arrangement that *cannot* be transformed back into the 12345 order?
Many solutions are correct. There are 60 (out of 120 in all) possible such arrangements.

4. The five cubes puzzle is a simpler version of the puzzle at the right. The increased number of cubes results in many more possible arrangements. To work this puzzle, start with the cubes in the order shown. Mix them up well. Then see how long it takes you to get them back in the beginning order!
Answers may vary.

1	2	3	4
8	7	6	5
9	10	11	12
	15	14	13

228   ENRICHMENT *LESSON 13-6*

Lesson 13-6 **Inverse Variation** 587

Skills Practice

Vocabulary Review

Lesson 13-5
direct variation
constant of variation
direct square variation

Lesson 13-6
inverse variation
inverse square variation
joint variation
combined variation

Assignment Guide

All students: 1–34
Additional Practice: See Extra Practice Index on page 674.

Additional Answers

25.

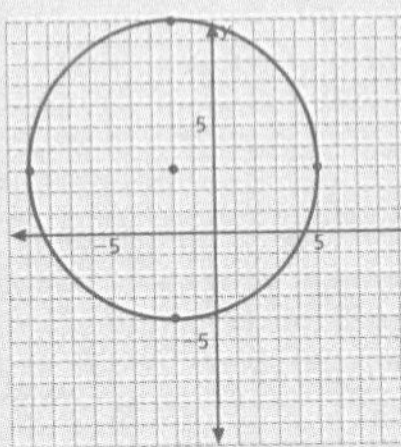

26.

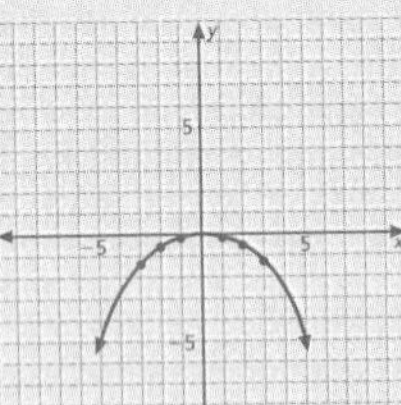

27.

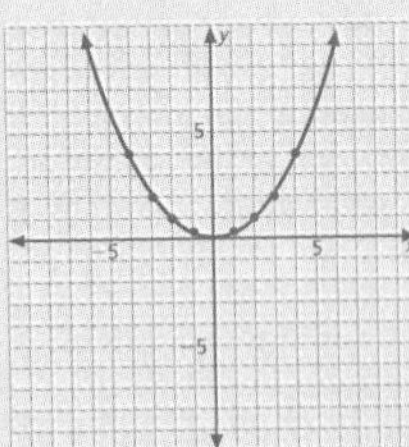

28.

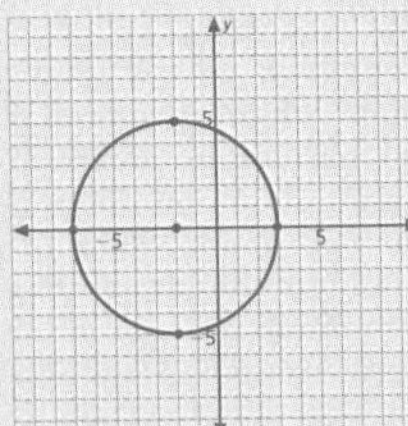

29.

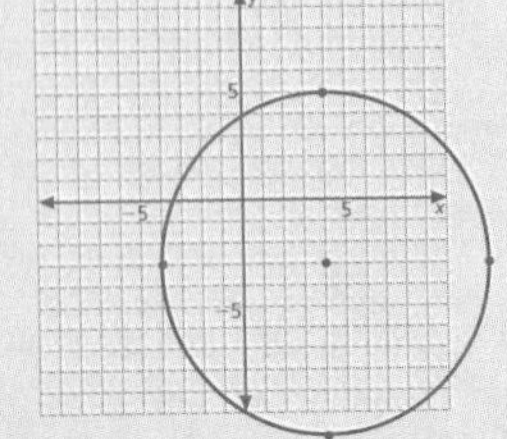

30.

Review and Practice Your Skills

Practice ■ Lesson 13-5

1. In a direct variation, if y decreases, what must be true about x? x must also decrease
2. In the equation representing direct variation, $y = kx$, what must be true about x? $x \neq 0$
3. What is the difference between a graph showing a direct variation and a graph showing a direct square variation? A direct variation graph is linear; a direct square variation graph is a quadratic function.

Write the equation of direct variation using the given values.

4. $x = 5, y = 20$ $y = 4x$
5. $x = 12, y = 8$ $y = \frac{2}{3}x$
6. $x = 9, y = 10$ $y = \frac{10}{9}x$
7. $x = 16, y = 4$ $y = \frac{1}{4}x$
8. $x = 24, y = 4$ $y = \frac{1}{6}x$
9. $x = 8, y = 18$ $y = \frac{9}{4}x$
10. If y varies directly as x and $y = 2$ when $x = 8$, find y when $x = 20$. $y = 5$
11. If y varies directly as x and $y = 5$ when $x = 35$, find y when $x = 15$. $y = 2.14$
12. If y varies directly as x and $y = 8$ when $x = 4$, find y when $x = 7$. $y = 14$
13. If y varies directly as x^2, and $y = 12$ when $x = 4$, find y when $x = 8$. $y = 48$
14. If y varies directly as x^2, and $y = 9$ when $x = 2$, find y when $x = 12$. $y = 324$
15. If y varies directly as x^2, and $y = 45$ when $x = 3$, find y when $x = 9$. $y = 405$
16. A survey showed that 7 out of 10 people liked the taste of Shimmer toothpaste. At this rate, how many people out of 2,400 would be expected to like the taste of Shimmer? 1680

Practice ■ Lesson 13-6

17. The example of an inverse variation was given as the cost of a vacation home per person. Why is the application only concerned with the part of the hyperbola that lies in the first quadrant? In real life, there are no negative people and no negative costs. Quadrant I is the only quadrant where both variables are positive.
18. Write an equation in which y varies inversely as x if one pair of values is $y = 4$ and $x = 12$. $y = \frac{48}{x}$

Find answers to the nearest hundredth.

19. Write an equation in which y varies inversely as x if one pair of values is $y = 7$ and $x = 0.9$. $y = \frac{6.3}{x}$
20. If y varies inversely as x and $y = 4$ when $x = 2$, find y when $x = 7$. $y = 56$
21. If y varies inversely as x and $y = 7$ when $x = 4$, find y when $x = 9$. $y = 3.11$
22. If y varies inversely as the square of x and $y = 9$ when $x = 2$, find y when $x = 3$. $y = 4$

Write an equation of inverse variation for each.

23. The time (t) it takes to travel from one point to another is inversely proportional to the speed (s) of the travel. $r = \frac{k}{s}$
24. The amount of oxygen (o) in the air is inversely proportional to the altitude (a). $o = \frac{k}{a}$

Teaching Strategies

COOPERATIVE LEARNING Have students work together to analyze the following data. Have them use the values in the table to determine whether y varies directly as the square of x. Have students present their findings to the class. yes

x	y
3	36
4	64
5	100
6	144

PRACTICE ■ LESSON 13-1–LESSON 13-6

Graph each equation. (Lessons 13-1–13-2) For 25–30, see additional answers.

25. $(x + 2)^2 + (y - 3)^2 = 49$ **26.** $x^2 = -5y$ **27.** $x^2 = 4y$

28. $(x + 2)^2 + y^2 = 25$ **29.** $(x - 4)^2 + (y + 3)^2 = 64$ **30.** $x^2 + (y + 5)^2 = 36$

Write the equation of direct variation using the given values. (Lesson 13-5)

31. $x = 10, y = 25$ $y = \frac{5}{2}x$ **32.** $x = 9, y = 3$ $y = \frac{1}{3}x$ **33.** $x = 5, y = 7$ $y = \frac{7}{5}x$

34. One survey showed that 7 out of 12 eligible people in Greenville planned to vote in the next election. If the population of eligible voters in Greenville is 19,824, how many could be expected to vote? 11,564

MathWorks
Workplace Knowhow

Career – Astronomer

Astronomers work in a sub-field of physics. Astronomers study things such as the birth of stars, the death of stars, natural satellites, the composition of planets, and the possibility of life on other planets. Astronomers work in observatories with very large telescopes, in universities teaching and in planetariums. Astronomers also work with engineers in the design, launch, and use of deep space probes and satellites that send astronomical data back to the Earth from planets too far away and too inhospitable for humans to visit. Precise calculations for the orbits of the satellites and probes are essential for their missions' success.

1. Suppose that you discovered a new solar system with an elliptical orbit of planets around two suns. What would be the mathematical term for the location of each sun. The locations of the suns are called foci (plural of focus)

The suns are observably 5 in. apart in scaled distance. You also observe the 4 planets' orbits for a period of six months for their measure. The point at which the planets stop moving away from the suns and start moving back toward them is the x-intercept of the ellipse when their orbits are graphed. These distances from one of the suns is as follows:

a. Planet A is 1.2 in. from a sun **b.** Planet B is 2.4 in. from a sun

c. Planet C is 3.8 in. from a sun **d.** Planet D is 5.2 in. from a sun

2. For graphing an ellipse around the origin on the coordinate place, determine x-intercepts of each planet. A: ± 3.7; B: ± 4.9; C: ± 6.3; D: ± 7.7
3. Calculate the y-intercept of each orbit. A: 2.728; B: 4.214; C: 5.783; D: 7.283
4. Write the equation for each planet's elliptical orbit. See additional answers.

Chalkboard Examples

Example from Lesson 13-5

A tank drains 8,100 gallons of water in 3 hours. How many gallons would drain in 35 minutes? 1,575 gal in 35 min

Example from Lesson 13-6

Let y vary inversely as the square of x. If $y = 16$ when $x = 2$, find y when $x = 4$. $y = 4$

Career Opportunity

Describe the kind of work astronomers do and the types of places they work. Discuss the importance of geometry, measurement, data analysis and algebraic thinking in working as an astronomer. Students should answer Questions 1–4 to better understand how astronomers use mathematics in performing their job.

Students who are interested in learning more about this profession can go to learningmathmatters.com. Guidance Counselors should have information about school and training requirements.

ADDITIONAL ANSWER

A: $7.4x^2 + 13.7y^2 = 99.8$
B: $17.76x^2 + 24.01y^2 = 423.5$
C: $33.64x^2 + 39.69y^2 = 1335.17$
D: $53.29x^2 + 59.29y^2 = 3159.56$

Teaching Strategies

For each inverse variation in the exercises, you may wish to have students find several pairs of x- and y-values. These values can be recorded in a table so that students can examine them side-by side. Seeing several pairs of values with the same product may make clearer to students the meaning of "inverse variation."

Lesson Planning

NCTM Standards/Strands

- Number & Operation
- Patterns, Functions, & Algebra
- Problem Solving
- Reasoning and Proof
- Communication
- Connections
- Representation

Vocabulary

quadratic inequality

Materials Needed

graph paper
graphing calculators (optional)

Lesson Resources

Warm-Up Transparency 39
Reteaching 13-7
Extra Practice 13-7
Enrichment 13-7

Assignment Guide

Basic: 1–22, 27–45
Enriched: 1–45

Getting Started

5-Minute Warm-up

Solve each inequality.

1. $3x + 5 > 8$ $\quad x > 1$
2. $7x - 9 < 12$ $\quad x < 3$
3. $12x - 7 < 5x$ $\quad x < 1$
4. $-2x > 6x - 4$ $\quad x < 0.5$

Introduction to Lesson 13-7

Be sure that students recognize the three sets of points into which the graph divides the plane—the curve itself, points inside the curve, and points outside the curve. For Parts (b) and (c), point out that students may begin by selecting points at random.

13-7 Quadratic Inequalities

Goals ■ Graph quadratic inequalities.

Applications Astronomy, Communications, Computer Design

Work with a partner. You may use a graphing calculator.

The graph of a quadratic equation divides the coordinate plane into three sets of points. Graph $y = x^2 + 2x + 1$ on a coordinate plane. See additional answers.

a. Find 5 ordered pairs for which $y = x^2 + 2x + 1$.

b. Find 5 ordered pairs for which $y < x^2 + 2x + 1$.

c. Find 5 ordered pairs for which $y > x^2 + 2x + 1$.

d. Use the points you found for *b* to help you locate and shade the section of the graph where $y < x^2 + 2x + 1$.

e. Use the points you found for *c* to help you locate and draw horizontal lines through the section of the graph where $y > x^2 + 2x + 1$.

Build Understanding

Just as you used linear equations to graph linear inequalities, you can use quadratic equations to graph quadratic inequalities.

Example 1

Graph $-y^2 + 4x^2 > 16$.

Solution

Graph the hyperbola $-y^2 + 4x^2 = 16$. Use the intercepts of $(2, 0)$, $(-2, 0)$, $(0, 4)$, and $(0, -4)$ to draw the rectangle. Then draw the asymptotes and sketch the hyperbola.

Because $-y^2 + 4x = 16$ is not part of the solution, the hyperbola is drawn with a dashed line.

To decide which points are part of the solution set, select points on the graph and substitute their coordinates into the equation.

Select $(0, 0)$: $-(0)^2 + 4(0)^2 \not> 16$ The point $(0, 0)$ and the region that contains it are not in the solution set.

$0 + 0 \not> 16$

Select $(-3, 1)$: $-(1)^2 + 4(-3)^2 > 16$

$-(1) + 4(9) > 16$

$35 > 16$

Select $(4, 0)$: $-(0)^2 + 4(4)^2 > 16$

$0 + 4(16) > 16$

$64 > 16$

These points and the regions they contain are in the solution set. The solution set is the shaded region shown on the graph.

Additional Answers

a–e.

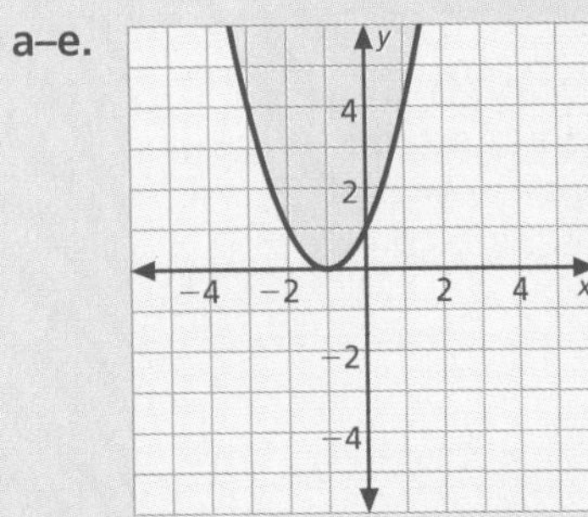

Teaching Strategies

To help students avoid confusion about the solution set of an inequality, point out that when they solve a system graphically, the intersection of the graphs includes all the solution points. However, by solving graphically, no actual solution points are named.

Systems of inequalities can be solved by finding the intersections of their graphs.

Example 2

ASTRONOMY Radio commands may be sent to a space probe during a specific portion of its flight. If commands are sent too soon or too late, they will not be received by the probe. The solution set of the following system of inequalities is used to determine when commands may be sent. Solve the system of inequalities by graphing.

$9x^2 + 4y^2 < 36$

$y \geq x - 2$

Solution

Graph the ellipse $9x^2 + 4y = 36$. The center is at the origin. The x-intercepts are $(0, -2)$ and $(0, 2)$. The y-intercepts are $(0, 3)$ and $(0, -3)$.

The points on the ellipse are not in the solution set, so the ellipse is drawn with a dashed line. Point (0, 0) is in the solution set, so points inside the ellipse are part of the solution set.

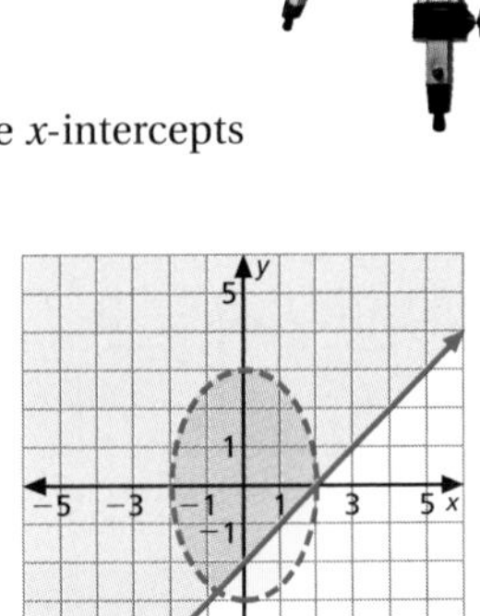

Graph $y = x - 2$. The solution set includes the line, so the line is solid. For the area above the line, $y > x - 2$, so part of this area is in the solution set.

The intersection of the two equations is shown by crosshatched lines.

Example 3

Solve this system of inequalities by graphing.

$x^2 + y^2 \leq 49$

$y > 2x^2 + 2$

Solution

Graph the circle with radius 7. The circle is in the solution set; draw the circle with a solid line. Point (0, 0) is in the solution set. The area inside the circle is in the solution set.

Graph parabola $y = 2x^2 + 2$. The parabola is not in the solution set; graph the parabola with a dashed line. Point (0, 0) is not in the solution set, so the area inside the parabola is in the solution set.

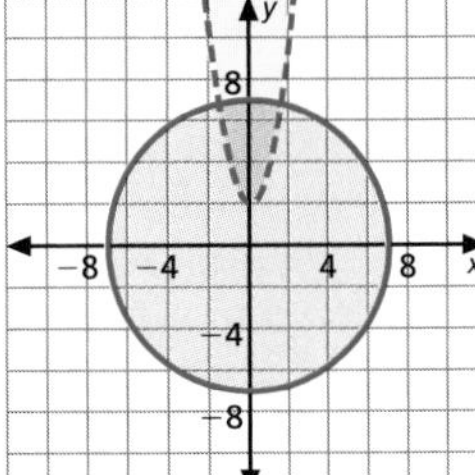

The crosshatched section shows the intersection of the two equations.

Check Understanding

How can you check to be sure that the area inside the parabola is in the solution set?

Substitute points that lie inside the parabola into the equation.

1.

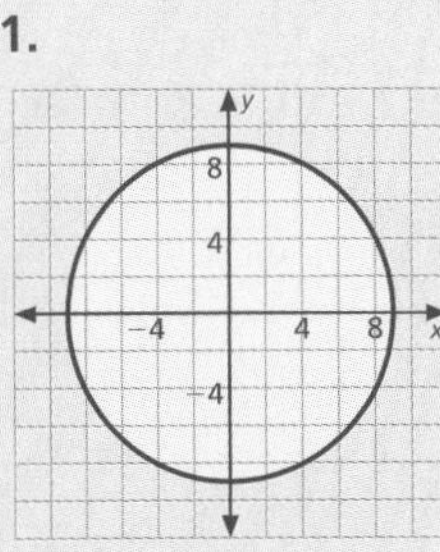

Center (0, 0), $r = 9$

2.

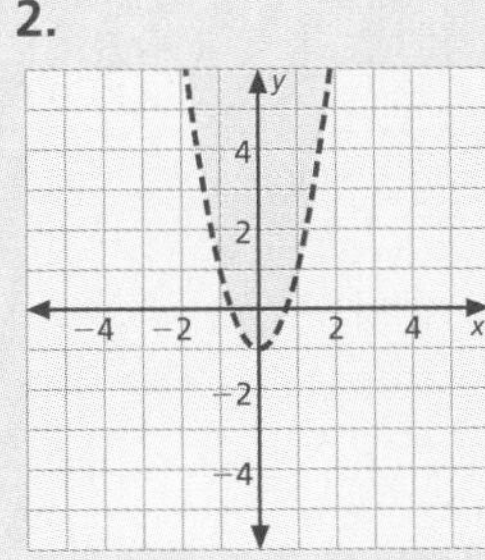

3.

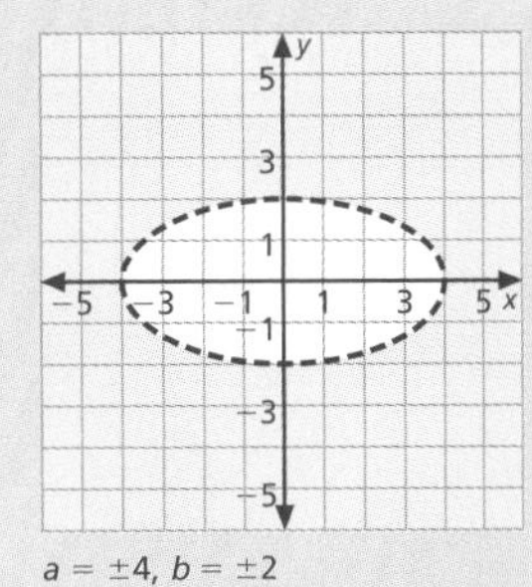

$a = \pm 4$, $b = \pm 2$

4.

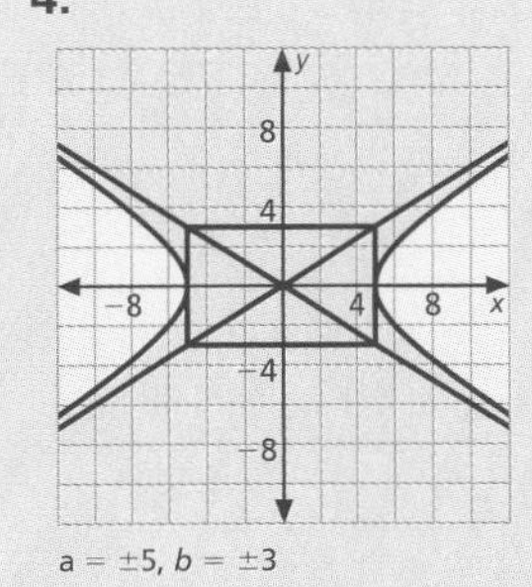

$a = \pm 5$, $b = \pm 3$

Chalkboard Examples

Supplementary Example 1

Graph $-y^2 + 9x^2 > 25$. hyperbola with x-intercepts at (3, 0) and (−3, 0) and y-intercepts at (0, 4) and (0, −4); drawn with a dashed line

Supplementary Example 2

During an experiment, measurements must be taken during a specific time frame. This time frame is indicated by the solution to this system of equations. Solve the system by graphing. $y > x^2 + 2x - 3$ $y < -x^2 + x$ Both parabolas; solution set is all points in the intersection of the two figures; both lines are drawn with dashed lines

Supplementary Example 3

Solve this system of inequalities by graphing. $x^2 + y^2 \leq 36$

$\frac{x^2}{100} + \frac{y^2}{36} < 1$

Circle with center (0, 0) and $r = 6$; circle is drawn with solid line; ellipse with center (0, 0) and x-intercepts ±10, y-intercepts ±6; ellipse is drawn with dashed line; solution set is all points inside both figures and on solid line.

Reteaching Worksheet 13-7

Name ______________________ Date ____________

RETEACHING 13-7

QUADRATIC INEQUALITIES

You can use quadratic equations like these to help you graph **quadratic inequalities**.

Circle: $x^2 + y^2 = r^2$ where $r \neq 0$ and center is at the origin (0, 0)

Parabola: $x^2 = 4ay$ with center at the origin and focus is on the y-axis

Ellipse: $\frac{x^2}{a^2} + \frac{y^2}{b^2} = 1$ with center at the origin and foci on the x-axis

Hyperbola: $\frac{x^2}{a^2} - \frac{y^2}{b^2} = 2$ with foci on the x-axis and symmetric around the origin

Example

Solve this system of inequalities by graphing: $x^2 + y^2 \geq 25$ and $y < -x^2$.

Solution

Graph the circle with a radius of 5. The circle is the solution set, so draw the circle with a solid line. Point (0, 0) and the region that contains it are not in the solution set, so the area outside the circle is in the solution set.

Graph parabola $y < -x^2$. The parabola is not in the solution set, so draw the parabola with a dashed line. Point (0, 0) is in the solution set so the area inside the parabola is in the solution set.

The cross-hatched section shows the intersection of the two equations.

EXERCISES

Graph each system of inequalities.

1. $y > x^2$ and $16x^2 + 9y^2 \leq 36$
2. $9x^2 - 4y^2 > 36$ and $x^2 + y^2 \leq 9$
3. $x^2 - y^2 < 16$ and $25x^2 + 4y^2 \leq 100$

204 RETEACHING *LESSON 13-7*

Lesson Wrap-up

Quick Assessment

Ask the following questions to determine if students understand the content presented in this lesson.

1. Why is the graph of an inequality in the form $y > ax^2 + bx + c$ drawn with a dashed line? Points on the curve are not part of the solution set.
2. Describe the solution set of a system of inequalities both algebraically and graphically. algebraically: the intersection of the solution sets of all inequalities in the system: graphically: all points on the coordinate grid that satisfy all the inequalities in the system

Assignment Guide

Basic: 1–22, 27–45
Advanced: 1–45
Additional Practice: See Extra Practice Index on page 674.

Additional Answers

5.
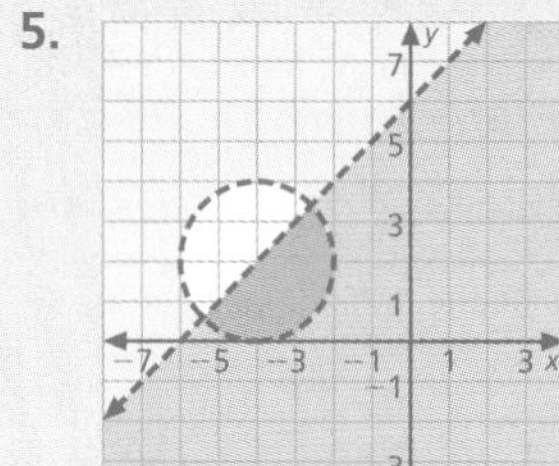
$r = 2$, Center $(-4, 2)$

6.
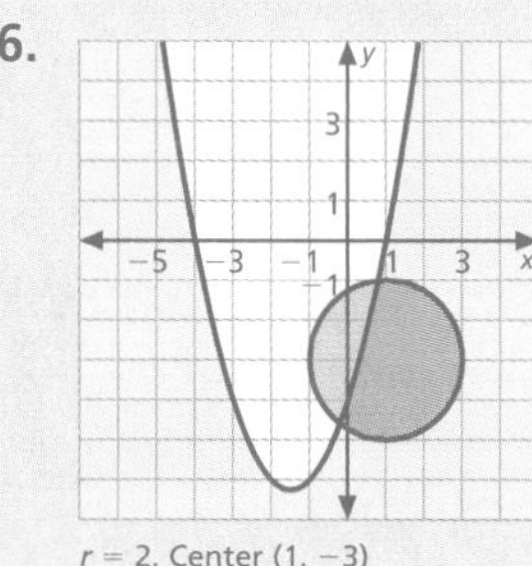
$r = 2$, Center $(1, -3)$

7.
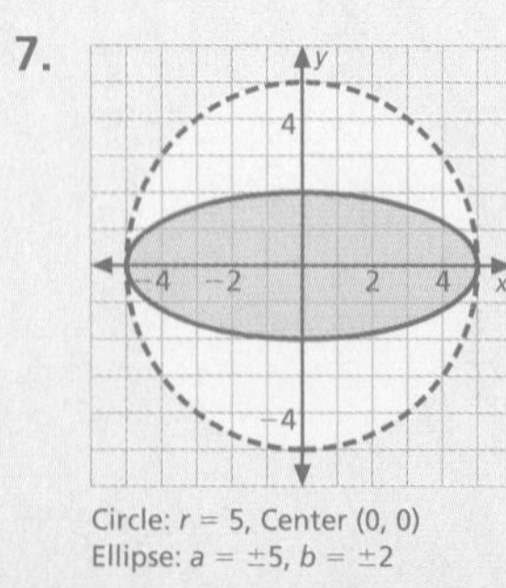
Circle: $r = 5$, Center $(0, 0)$
Ellipse: $a = \pm 5$, $b = \pm 2$

Try These Exercises

Graph each inequality. For 1–4, see additional answers.

1. $x^2 + y^2 \le 81$
2. $y < 2x^2 - 1$
3. $3x^2 + 12y^2 > 48$
4. $9x^2 - 25y^2 \ge 225$

Graph each system of inequalities. For 5–8, see additional answers.

5. $(x + 4)^2 + (y - 2)^2 > 4$
   $y < x + 6$
6. $y \le x^2 + 3x - 4$
   $(x - 1)^2 + (y + 3)^2 \le 4$
7. $4x^2 + 25y^2 \le 100$
   $x^2 + y^2 < 25$
8. $25x^2 - 4y^2 \ge 100$
   $x \ge -1$

Problem Solving Tip

All hyperbolas and ellipses in these exercises have center (0, 0).

Practice Exercises

Graph each inequality. For 9–12, see additional answers.

9. $4x^2 + 16y^2 \ge 64$
10. $(x - 5)^2 + (y + 1)^2 \ge 4$
11. $y < x^2 + x + 1$
12. $9x^2 - 16y^2 \le 144$

Graph each system of inequalities. For 13–16, see additional answers.

13. $x^2 - y^2 < 25$
    $x^2 + y^2 < 100$
14. $y > x^2 - 2x$
    $y \le x + 3$
15. $(x - 1)^2 + (y - 3)^2 > 25$
    $y < x^2 - 2x + 1$
16. $x^2 + 16y^2 \le 16$
    $x^2 + y^2 < 64$

Use the inequalities you have graphed in this lesson to help you complete each statement. Use <, >, or =.

17. If $x^2 + y^2$ is __?__ r^2, the inside of the circle is in the solution set. $<$
18. If $\frac{x^2}{a^2} + \frac{y^2}{b^2}$ is __?__ 1, the outside of the ellipse is in the solution set. $>$
19. If $ax^2 + bx + c$ is __?__ y, the section inside the parabola is in the solution set. $<$
20. **COMMUNICATIONS** The limits of a transmitter can be modeled using the system of inequalities below. Graph the system and describe the solution set.
    $x^2 + y^2 < 36$
    $\frac{x^2}{100} + \frac{y^2}{36} < 1$
    Circle with center (0, 0) and $4 = 6$; ellipse with center (0, 0) and x-intersections ± 10, y-intersections ± 6; solution set is all points inside both figures.
21. **COMPUTER DESIGN** The system of inequalities below defines the capabilities of a computer chip. Graph the system and describe the solution set.
    $y > x^2 + 2x - 3$
    $y < x^2 - x$
    Both parabolas; solution set is all points in the intersection of the two figures.

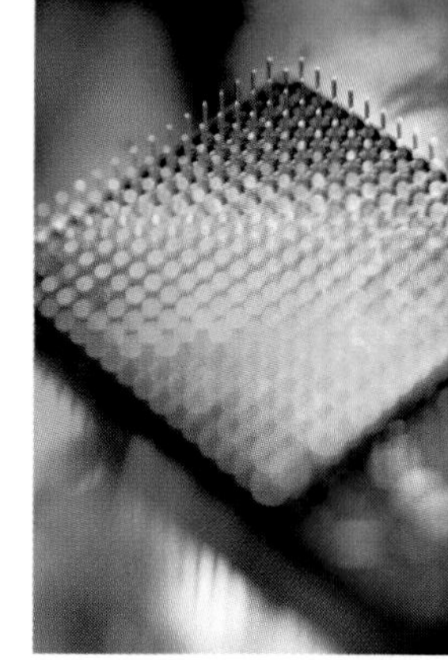

8.
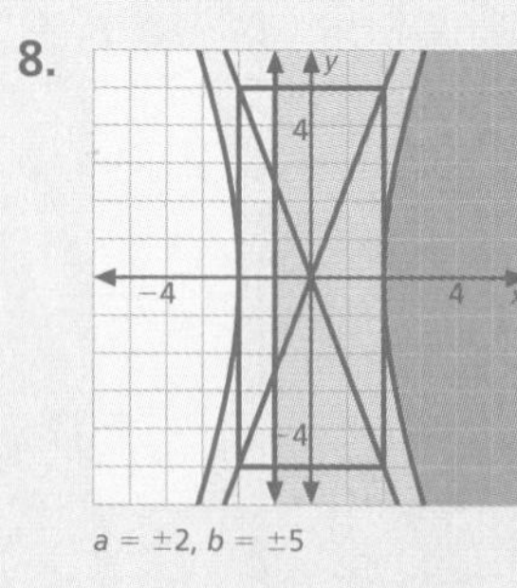
$a = \pm 2$, $b = \pm 5$

9.
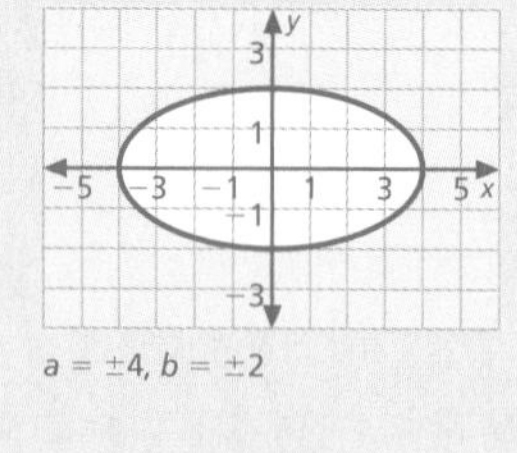
$a = \pm 4$, $b = \pm 2$

10.
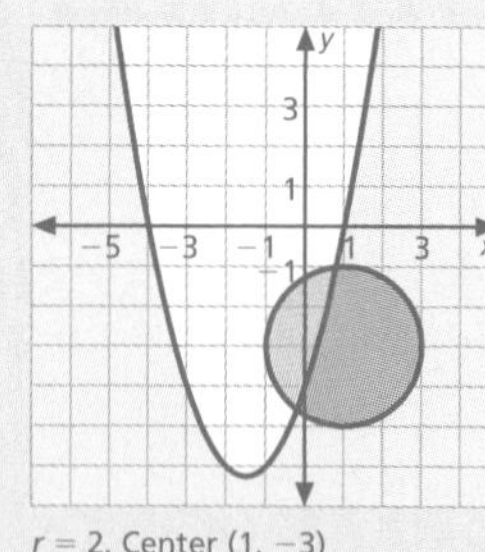
$r = 2$, Center $(1, -3)$

11.
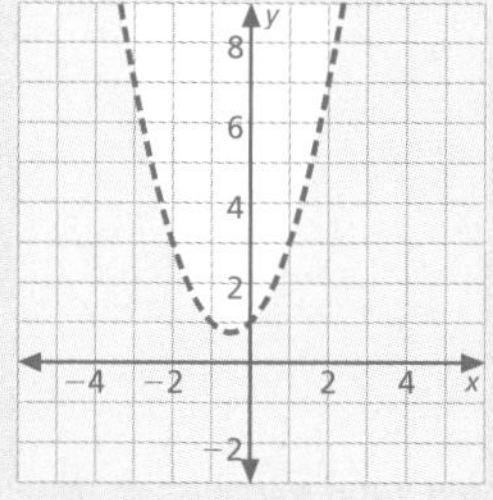

592 Chapter 13 **Quadratic Relations**

22. CHAPTER INVESTIGATION Select landmarks at your school or within your community to represent aphelion, or farthest orbital point, for each landmark. Mark these landmarks on your map. Share your findings with the class, and discuss how this activity has increased your understanding of the size of the Solar System. Check students' work.

Extended Practice Exercises

The solution to the equation $x^2 - 8x < -12$ is shown on the number lines below.

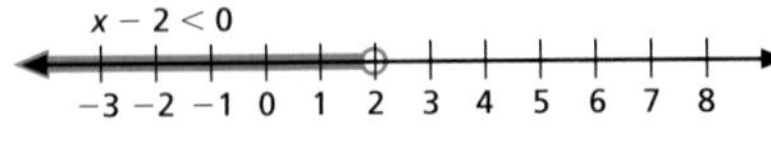

x − 6 < 0

−3 −2 −1 0 1 2 3 4 5 6 7 8

(x − 2)(x − 6) < 0

−3 −2 −1 0 1 2 3 4 5 6 7 8

23. WRITING MATH Explain this method for solving quadratic inequalities. Answers will vary.

24. Does the solution check? Try these points: 1, 3, 5, 8. yes

25. What is the solution to the set of inequalities? $2 < x < 6$

26. Use this method to solve $x^2 - 8x \leq -15$. $3 \leq x \leq 5$

Mixed Review Exercises

Calculate the midpoint between each pair of points. (Lesson 12-6)

27. $A(-6, 3)$, $B(4, -5)$ $(-1, -1)$

28. $C(3, -2)$, $D(-8, 9)$ $\left(-\frac{5}{2}, 1\right)$

29. $E(1, 0)$, $F(-3, -7)$ $\left(-1, -\frac{7}{2}\right)$

30. $G(3, 0)$, $H(-8, 2)$ $\left(-\frac{5}{2}, 1\right)$

31. $J(2, 7)$, $K(-7, 2)$ $\left(-\frac{5}{2}, \frac{7}{2}\right)$

32. $L(4, -5)$, $M(1, 3)$ $\left(\frac{5}{2}, -1\right)$

33. $N(0, 7)$, $P(3, -4)$ $\left(\frac{3}{2}, \frac{3}{2}\right)$

34. $Q(1, 6)$, $R(-6, -1)$ $\left(-\frac{5}{2}, \frac{5}{2}\right)$

35. $S(8, -3)$, $T(5, -6)$ $\left(\frac{13}{2}, -\frac{9}{2}\right)$

36. $U(-6, 4)$, $V(-3, -5)$ $\left(\frac{-9}{2}, -\frac{1}{2}\right)$

37. $W(-2, 0)$, $X(1, 6)$ $\left(-\frac{1}{2}, 3\right)$

38. $Y(3, 5)$, $Z(8, 2)$ $\left(\frac{11}{2}, \frac{7}{2}\right)$

Each figure below is a parallelogram. Find *a* and *b*. (Lesson 4-8)

39.

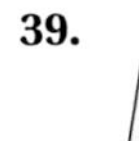

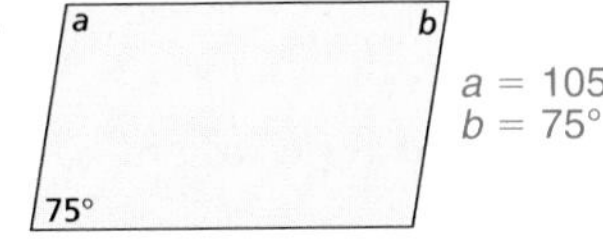

$a = 105°$, $b = 75°$

40.

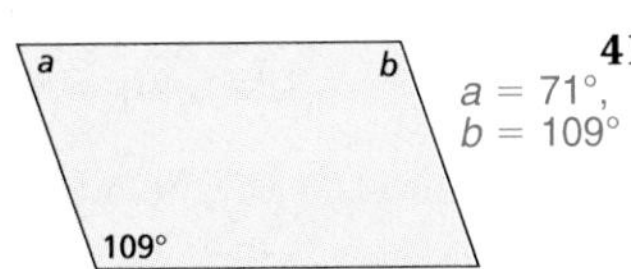

$a = 71°$, $b = 109°$

41.

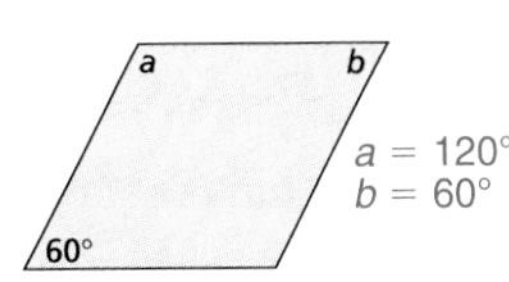

$a = 120°$, $b = 60°$

Find the volume of each figure to the nearest whole number. (Lesson 5-7)

42.

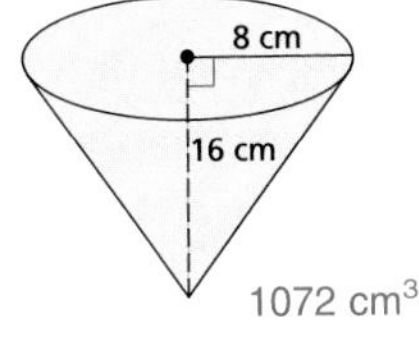

1072 cm³

43.

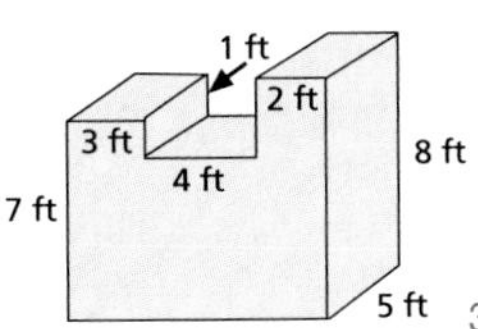

305 ft³

44.

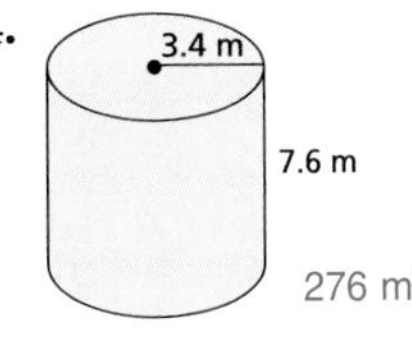

276 m³

45. Solve by completing the square: $x^2 + 20x - 1 = 0$. (Lesson 12-4) $x = 91$, $x = -111$

12.

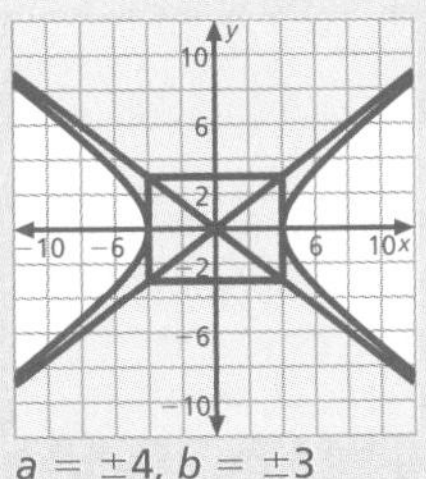

$a = \pm 4$, $b = \pm 3$

13.

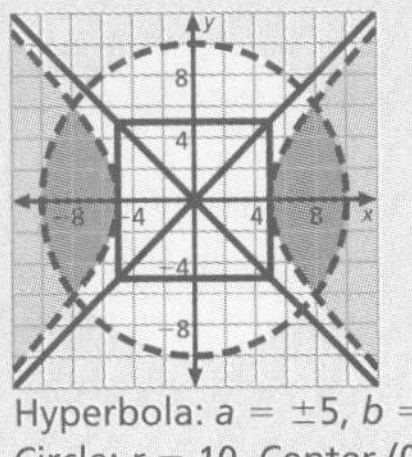

Hyperbola: $a = \pm 5$, $b = \pm 5$
Circle: $r = 10$, Center (0, 0)

14.

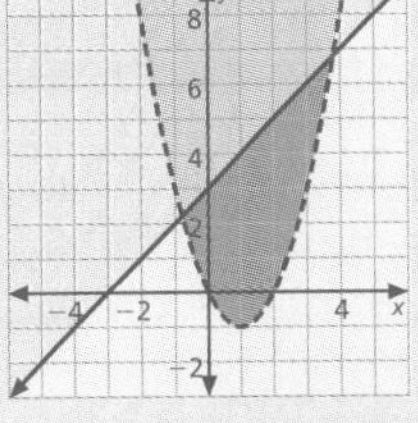

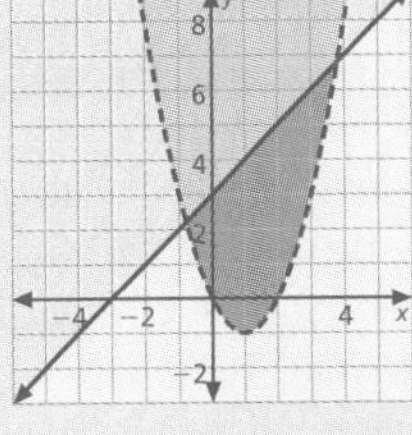

15.

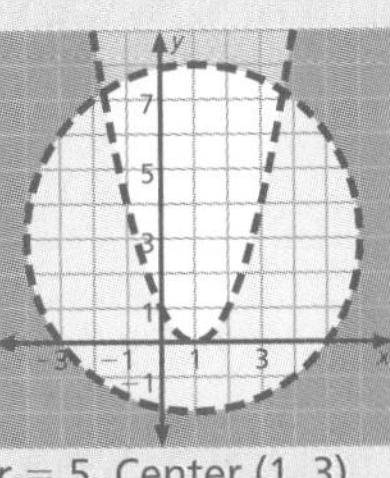

$r = 5$, Center (1, 3)

16.

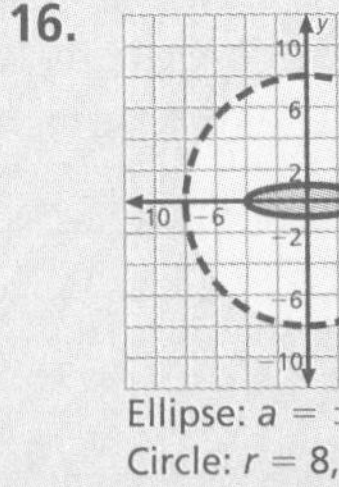

Ellipse: $a = \pm 4$, $b = \pm 1$
Circle: $r = 8$, Center (0, 0)

Extra Practice Worksheet 13-7

Name ______________________ Date ____________

EXTRA PRACTICE 13-7

Quadratic Inequalities

Exercises

Graph each inequality. Use your own paper. Check students' graphs.

1. $9x^2 + 25y^2 \leq 225$
2. $(x - 2)^2 + (y - 1)^2 > 9$
3. $y < x^2 + 4x + 4$
4. $16x^2 - 4y^2 \geq 64$
5. $x^2 + 4y^2 > 16$
6. $(x + 3)^2 + (y \_ 4)^2 - 16$
7. $y \geq x^2 - 2x - 8$
8. $8x^2 - 2y^2 \leq 32$
9. $y > x^2 + 8x + 15$
10. $(x + 2)^2 + (y + 4)^2 - 25$
11. $x^2 + y^2 > 144$
12. $y \leq x^2 - 13x - 12$

Graph each system of inequalities. Use your own paper. Check students' graphs.

13. $x^2 - y^2 < 36$
    $x^2 + y^2 > 25$
14. $y > x^2 - 3x$
    $y \leq x - 1$
15. $(x + 2)^2 + (y - 1)^2 \geq 4$
    $y > x^2 \geq 4x + 3$
16. $12x^2 + 27y^2 \leq 108$
    $x^2 + y^2 \geq 100$
17. $x^2 - y^2 > 16$
    $x^2 + y^2 < 9$
18. $y < x^2 - 2x + 1$
    $y > x + 3$
19. $(x - 2)^2 + (y +)^2 \geq 25$
    $y \leq x^2 - 9x + 20$
20. $x^2 + y^2 \leq 9$
    $x^2 - y^2 \geq 16$
21. $x^2 + y^2 \geq 16$
    $y < x - 5$
22. $4x^2 + 8y^2 \leq 64$
    $x^2 + y^2 > 64$

Complete each statement. Use <, > or =.

23. If $x^2 + y^2$ is __>__ r^2, the outside of the circle is in the solution set.
24. If $\frac{x^2}{a^2} + \frac{y^2}{b^2}$ is __<__ 1, the inside of the ellipse is in the solution set.
25. If $\frac{x^2}{a^2} - \frac{y^2}{b^2}$ is __>__ 1, the outside of the hyperbola is in the solution set.
26. If $ax^2 + bx^2 + c$ is __>__ y, the section outside the parabola is in the solution set.

178 EXTRA PRACTICE *LESSON 13-7*

Enrichment Worksheet 13-7

Name ______________________ Date ____________

ENRICHMENT 13-7

Parametric Equations

Sometimes a graph is described by two equations rather than just one. For example, the coordinates (x, y) of a moving point are often described by two equations using the variable t for time.

Equations of this type are called parametric equations. The variable t is the parameter.

Exercises

Match each graph below with its set of parametric equations. Write the letter of the correct graph above each set of equations.

1. __B__
   $x = -0.5t^2 - 3t$
   $y = -t^2 + 0.5t + 25$
   $t = -9 \ldots 8$
2. __C__
   $x = t^2 - 4t$
   $y = -0.5t^2 + 2t + 25$
   $t = -10 \ldots 6.5$
3. __D__
   $x = t^2 - 3t + 2$
   $y = t^2 + t - 1$
   $t = -6 \ldots 7$
4. __A__
   $x = -0.5t^2 - 3t - 10$
   $y = -t^2 + 0.5t - 25$
   $t = -9 \ldots 7.5$

Graph A

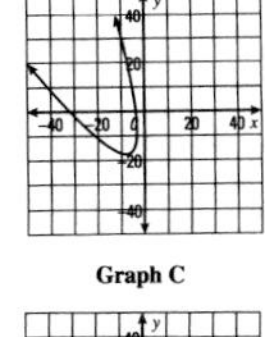

Graph B

Graph C

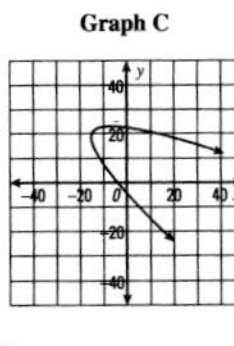

Graph D

230 ENRICHMENT *LESSON 13-7*

Assessment Planning

Vocabulary Review

Lesson 13-1
standard equation of a circle
radius

Lesson 13-2
focus
directrix

Lesson 13-3
conics
conic section

Lesson 13-4
ellipse
foci
asymptotes
hyperbola

Lesson 13-5
direct variation
constant of variation
direct square variation

Lesson 13-6
inverse variation
inverse square variation
joint variation
combined variation

Lesson 13-7
quadratic inequality

Assessment Options

Chapter Assessment, page 596
Alternative Assessment, page 597
Chapter Test Forms A and B
Cumulative Assessment, page 598–599
Achievement Test
Writing Prompts

Assignment Guide

All students: 1–30
Additional Practice: See Extra Practice Index on page 674.

Chapter 13 Review

Vocabulary ■ LESSON 13-1–LESSON 13-7

Choose the word from the list at the right that completes each statement below.

a. ellipse
b. direct variation
c. inverse variation
d. parabola

1. A set of points equidistant from a fixed point called a focus and a fixed line called a directrix is a __?__. d
2. A set of points moving about two fixed points is an __?__. a
3. A relationship in which one variable increases as the other variable increases is a __?__. b
4. A relationship in which one variable decreases as the other variable increases is an __?__. c

LESSON 13-1 ■ The Standard Equation of Circles, p. 562

▶ The equation for any circle with its center at the origin and radius r is $x^2 + y^2 = r^2$.

▶ The standard equation for a circle with its center located at any coordinate point (h, k) is $(x - h)^2 + (y - k)^2 = r^2$.

Write an equation for each circle.

5. radius 8 center (0, 0) $x^2 + y^2 = 64$
6. radius 4 center (2, 3) $(x - 2)^2 + (y - 3)^2 = 16$
7. radius 6 center (5, 0) $(x - 5)^2 + y^2 = 36$

Find the radius and center for each circle.

8. $x^2 + y^2 = 25$ 5, (0, 0)
9. $x^2 + (y - 3)^2 = 9$ 3, (0, 3)
10. $(x + 9)^2 + (y + 4)^2 = 21$ $\sqrt{21}$, (−9, −4)

LESSON 13-2 ■ More on Parabolas, p. 566

▶ When the focus $(0, a)$ is on the y-axis and the directrix is $y = -a$, the simple equation for a parabola is $x^2 = 4ay$.

Find the focus and directrix of each equation.

11. $x^2 = 20y$ (0, 5), $y = -5$
12. $-40y = 5x^2$ (0, −2), $y = 2$
13. $12x^2 - 48y = 0$ (0, 1), $y = -1$

Find the simple equation for each parabola with vertex located at the origin.

14. Focus (0, 5) $x^2 = 20y$
15. Focus (0, −4) $x^2 = -16y$
16. Focus $\left(0, \frac{1}{2}\right)$ $x^2 = 2y$

LESSON 13-3 ■ Problem Solving, p. 572

▶ You can visualize the conic section formed by a plane intersecting a cone or double cone.

17. Describe the intersection between a plane and a cone that produces the largest circle possible. The plane is parallel to and contains one of the bases.
18. Describe the intersection between a plane and a cone that produces a hyperbola. The plane is parallel to the y-axis and does not pass through the vertex.

Extend the Lesson

Have students make index cards for each of the curves studied in this chapter: circles, parabolas, ellipses and hyperbolas. Students should show a drawing and summarize the algebraic and geometric information for each. Their summaries might include a definition or description of the curve, definitions of key terms and how they relate to the equation of the curve, the standard equation of the curve, and any special forms. They might also include a specific example such as "a circle centered at the origin with radius 7 is given by $x^2 + y^2 = 49$." Encourage students to carry these cards with them and to review the names and properties often.

LESSON 13-4 ■ Ellipses and Hyperbolas, p. 574

- ▶ The standard form for the equation of an ellipse with its center at the origin and foci on the x-axis is $\frac{x^2}{a^2} + \frac{y^2}{b^2} = 1$.
- ▶ The equation for a hyperbola that is symmetric about the origin and has foci on the x-axis is $\frac{x^2}{a^2} - \frac{y^2}{b^2} = 1$.

Find an equation for each figure.

19. An ellipse with foci (8, 0) and (−8, 0) and x-intercepts (10, 0) and (−10, 0). $36x^2 + 100y^2 = 3{,}600$

20. A hyperbola with center (0, 0) and foci on the x-axis if $a = \pm 4$ and $b = \pm 7$. $49x^2 - 16y^2 = 784$

LESSON 13-5 ■ Direct Variation, p. 580

- ▶ Equations in which one variable increases as the other variable increases can be expressed as $y = kx$, where k is a nonzero constant and $x \neq 0$.
- ▶ Direct square variation is shown by the equation $y = kx^2$.

21. If y varies directly as x and $y = 75$ when $x = 7.5$, find y when $x = 5$. $y = 50$

22. If y varies directly as x^2 and $y = 51.2$ when $x = 4$, find y when $x = 9$. $y = 259.2$

23. Let y vary directly as the square of x. If $y = 45$ when $x = 3$, find y when $x = 8$. $y = 320$

LESSON 13-6 ■ Inverse Variation, p. 584

- ▶ Equations in which one variable decreases as the other variable increases can be expressed as $y = \frac{k}{x}$ where k is a nonzero constant and $x \neq 0$.
- ▶ Inverse square variation is shown by the equation $y = \frac{k}{x^2}$, or $x^2y = k$.

24. Write an equation in which y varies inversely as x if one pair of values is $y = 90$ and $x = 0.7$. $y = \frac{63}{x}$

25. If y varies inversely as the square of x and $y = 900$ when $x = 5$, find y when $x = 12$. $y = 156.25$

26. Let y vary inversely as x. If $y = 6.5$ when $x = 3$, find y when $x = 4$. $y = 4.875$

27. Let y vary inversely as the square of x. If $y = 40$ when $x = 9$, find y when $x = 6$. $y = 90$

LESSON 13-7 ■ Quadratic Inequalities, p. 590

- ▶ Substitute coordinates into the equation for a quadratic inequality to locate regions in the solution set.

Graph each inequality. For 28–30, see additional answers.

28. $x^2 + y^2 > 49$ **29.** $y \geq 2x^2 + x + 2$ **30.** $9x^2 + 36y^2 < 36$

Additional Answers

28.

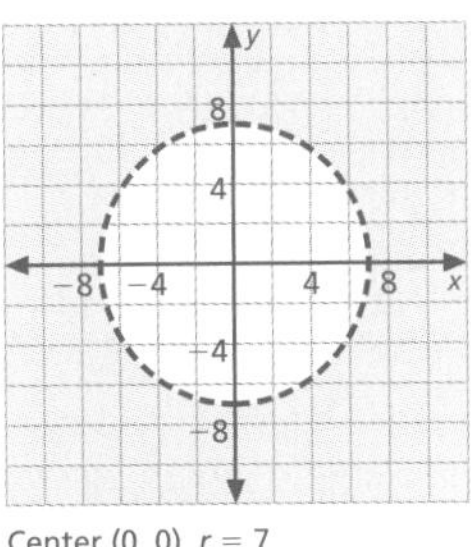

Center (0, 0), $r = 7$

29.

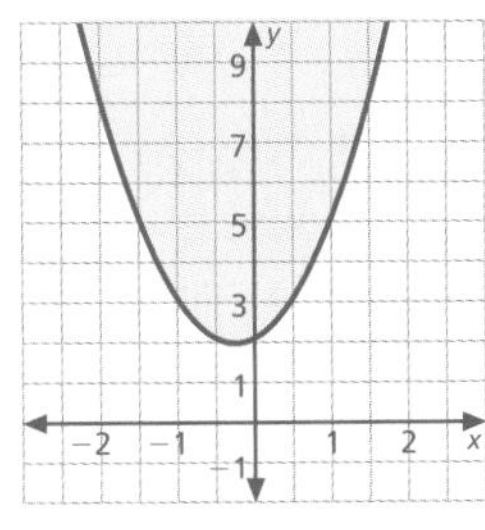

30.

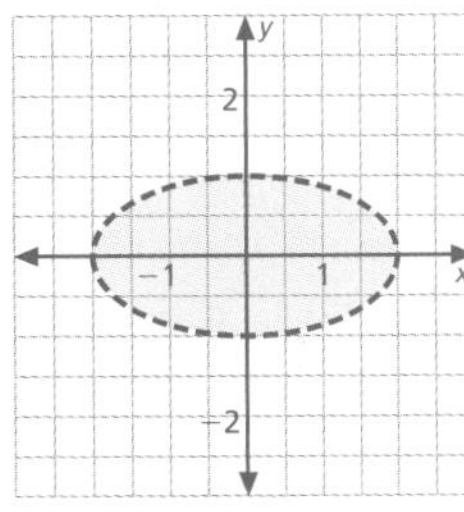

Chapter 13 Test Form A, page 1

CHAPTER 13
QUADRATIC RELATIONS
ASSESSMENT FORM A, PAGE 1

Name ____________
Date ____________

Scoring Record	
Possible: 28	Earned:

Write an equation for each circle.

1. radius 4, center (2, 5) $(x - 2)^2 + (y - 5)^2 = 16$
2. radius 7, center (−3, 4) $(x + 3)^2 + (y - 4)^2 = 49$
3. radius 5, center (0,0) $x^2 + y^2 = 25$

Find the radius and center of each circle.

4. $x^2 + y^2 = 121$ $r = 11$; center (0, 0)
5. $(x - 4) + (x + 2) = 16$ $r = 4$; center (4, −2)
6. $(x + 3) + (x + 2) = 13$ $r = \sqrt{13}$; center (−3, −2)

Find the focus and directrix of each equation.

7. $x^2 = 24y$ (0, 6); $y = -6$
8. $x^2 = -12y$ (0, −3); $y = 3$
9. $4x^2 - 32y = 0$ (0, 2); $y = -2$

Find the simple equation for each parabola with vertex located at the origin.

10. Focus (0, −6) $x^2 = -24y$
11. Focus (0, 8) $x^2 = 32y$
12. Focus $\left(0, -\frac{1}{4}\right)$ $x^2 = -y$

Tell what comic section is formed by each plane.

13. circle
14. 2 triangles
15. parabola

16. Find the equation of the ellipse with foci (2, 0) and (−2, 0) with x-intercepts (3, 0) and (−3, 0). $5x^2 + 9y^2 = 45$

17. Find the equation of each hyperbola with center at the origin (0, 0) with foci on the x-axis, in which $a = \pm 3$ and $b = \pm 5$. $25x^2 - 9y^2 = 225$

170 ASSESSMENT *CHAPTER 13 FORM A*

Chapter 13 Test Form A, page 2

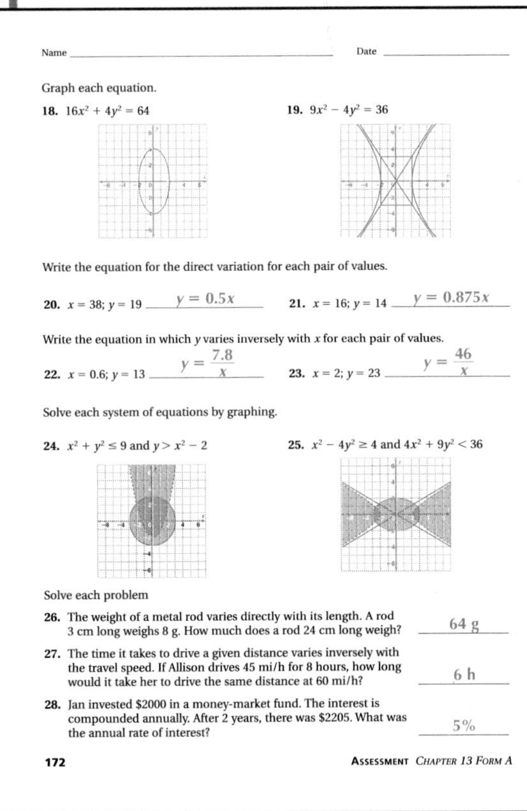

Name ____________ Date ____________

Graph each equation.

18. $16x^2 + 4y^2 = 64$
19. $9x^2 - 4y^2 = 36$

Write the equation for the direct variation for each pair of values.

20. $x = 38$; $y = 19$ $y = 0.5x$
21. $x = 16$; $y = 14$ $y = 0.875x$

Write the equation in which y varies inversely with x for each pair of values.

22. $x = 0.6$; $y = 13$ $y = \frac{7.8}{x}$
23. $x = 2$; $y = 23$ $y = \frac{46}{x}$

Solve each system of equations by graphing.

24. $x^2 + y^2 \leq 9$ and $y > x^2 - 2$
25. $x^2 - 4y^2 \geq 4$ and $4x^2 + 9y^2 < 36$

Solve each problem

26. The weight of a metal rod varies directly with its length. A rod 3 cm long weighs 8 g. How much does a rod 24 cm long weigh? 64 g

27. The time it takes to drive a given distance varies inversely with the travel speed. If Allison drives 45 mi/h for 8 hours, how long would it take her to drive the same distance at 60 mi/h? 6 h

28. Jan invested $2000 in a money-market fund. The interest is compounded annually. After 2 years, there was $2205. What was the annual rate of interest? 5%

172 ASSESSMENT *CHAPTER 13 FORM A*

Chapter 13 Test Form B, page 1

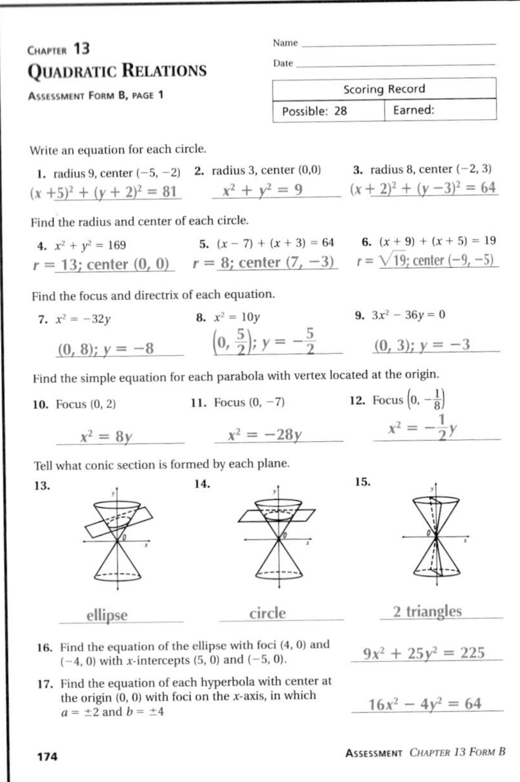
CHAPTER 13
QUADRATIC RELATIONS
ASSESSMENT FORM B, PAGE 1

Name ____________ Date ____________

Scoring Record	
Possible: 28	Earned:

Write an equation for each circle.

1. radius 9, center (−5, −2) $(x+5)^2 + (y+2)^2 = 81$
2. radius 3, center (0,0) $x^2 + y^2 = 9$
3. radius 8, center (−2, 3) $(x+2)^2 + (y-3)^2 = 64$

Find the radius and center of each circle.

4. $x^2 + y^2 = 169$ $r = 13$; center (0, 0)
5. $(x - 7) + (x + 3) = 64$ $r = 8$; center (7, −3)
6. $(x + 9) + (x + 5) = 19$ $r = \sqrt{19}$; center (−9, −5)

Find the focus and directrix of each equation.

7. $x^2 = -32y$ (0, 8); $y = -8$
8. $x^2 = 10y$ $\left(0, \frac{5}{2}\right)$; $y = -\frac{5}{2}$
9. $3x^2 - 36y = 0$ (0, 3); $y = -3$

Find the simple equation for each parabola with vertex located at the origin.

10. Focus (0, 2) $x^2 = 8y$
11. Focus (0, −7) $x^2 = -28y$
12. Focus $\left(0, -\frac{1}{8}\right)$ $x^2 = -\frac{1}{2}y$

Tell what conic section is formed by each plane.

13. ellipse
14. circle
15. 2 triangles

16. Find the equation of the ellipse with foci (4, 0) and (−4, 0) with x-intercepts (5, 0) and (−5, 0). $9x^2 + 25y^2 = 225$
17. Find the equation of each hyperbola with center at the origin (0, 0) with foci on the x-axis, in which $a = \pm 2$ and $b = \pm 4$ $16x^2 - 4y^2 = 64$

174 ASSESSMENT CHAPTER 13 FORM B

Chapter 13 Test Form B, page 2

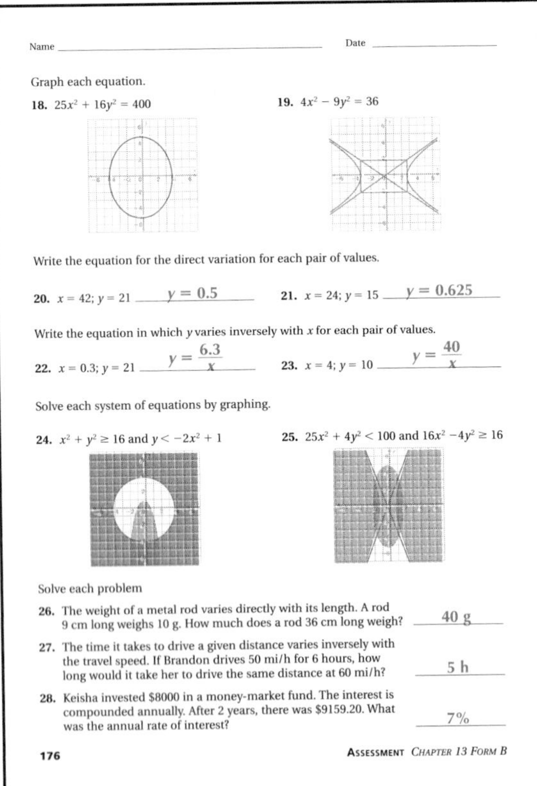
Name ____________ Date ____________

Graph each equation.

18. $25x^2 + 16y^2 = 400$
19. $4x^2 - 9y^2 = 36$

Write the equation for the direct variation for each pair of values.

20. $x = 42$; $y = 21$ $y = 0.5$
21. $x = 24$; $y = 15$ $y = 0.625$

Write the equation in which y varies inversely with x for each pair of values.

22. $x = 0.3$; $y = 21$ $y = \frac{6.3}{x}$
23. $x = 4$; $y = 10$ $y = \frac{40}{x}$

Solve each system of equations by graphing.

24. $x^2 + y^2 \geq 16$ and $y < -2x^2 + 1$
25. $25x^2 + 4y^2 < 100$ and $16x^2 - 4y^2 \geq 16$

Solve each problem

26. The weight of a metal rod varies directly with its length. A rod 9 cm long weighs 10 g. How much does a rod 36 cm long weigh? 40 g
27. The time it takes to drive a given distance varies inversely with the travel speed. If Brandon drives 50 mi/h for 6 hours, how long would it take her to drive the same distance at 60 mi/h? 5 h
28. Keisha invested $8000 in a money-market fund. The interest is compounded annually. After 2 years, there was $9159.20. What was the annual rate of interest? 7%

176 ASSESSMENT CHAPTER 13 FORM B

Chapter 13 Assessment

Write an equation for each circle.

1. radius 6, center (0, 0) $x^2 + y^2 = 36$
2. radius 10, center (1, −5) $(x - 1)^2 + (y + 5)^2 = 100$
3. radius 13, center (−1, 7) $(x + 1)^2 + (y - 7)^2 = 169$

Find the radius and center for each circle.

4. $x^2 + y^2 = 21$ $\sqrt{21}$, (0, 0)
5. $(x + 4)^2 + y^2 = 11$ $\sqrt{11}$, (−4, 0)
6. $(x - 7)^2 + (y - 6)^2 = 225$ 15, (7, 6)

Find the focus and directrix for each parabola.

7. $x^2 = 32y$ (0, 8), $y = -8$
8. $x^2 = -24y$ (0, −6), $y = 6$
9. $32y - 4x^2 = 0$ (0, 2), $y = -2$

Find the simple equation for each parabola with vertex located at the origin.

10. Focus (0, 8) $x^2 = 32y$
11. Focus (0, −2) $x^2 = -8y$
12. Focus $\left(0, -\frac{1}{4}\right)$ $x^2 = -y$

Find an equation for each figure.

13. An ellipse with foci (4, 0) and (−4, 0) and x-intercepts (5, 0) and (−5, 0). $9x^2 + 25y^2 = 225$
14. A hyperbola with center (0, 0) and foci on the x-axis if $a = \pm 5$ and $b = \pm 11$. $121x^2 - 25y^2 = 3{,}025$

Solve.

15. If y varies directly as x and $y = 36$ when $x = 15$, what is y when $x = 19$? $y = 45.6$
16. If y varies inversely as x and $y = 72$ and $x = 9$, what is y when $x = 6$? $y = 108$
17. If y varies inversely as x and $y = 144$ when $x = 6$, what is y when $x = 4$? $y = 216$
18. If y varies directly as x and $y = 360$ when $x = 12$, what is y when $x = 18$? $y = 540$
19. The total area of a picture and its frame is 456 square inches. The picture is 16 inches long and 21 inches wide. What is the width of the frame? 1.5 in.

Graph each inequality. For 20–22, see additional answers.

20. $(x + 1)^2 + (y - 6)^2 \geq 64$
21. $y > x^2 + 4x$
22. $9x^2 + 25y^2 > 225$
23. The amount of time projectiles are in the air follows the formula $vt - 16t^2 = 0$ (v = velocity, t = time). How long is a baseball in the air if thrown 64 feet per second? 4 sec
24. Write the equation of a circle having center at (2, −1) and radius of 3. $(x - 2)^2 + (y + 1)^2 = 9$ or $x^2 + y^2 - 4x + 2y - 4 = 0$
25. If y varies directly as x^2 and $y = 112$ when $x = 4$, find y when $x = 5$. 175

Additional Answers

20.

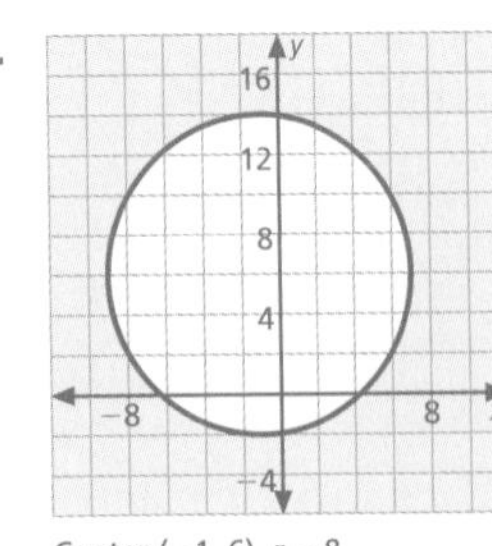

Center (−1, 6), $r = 8$

21.

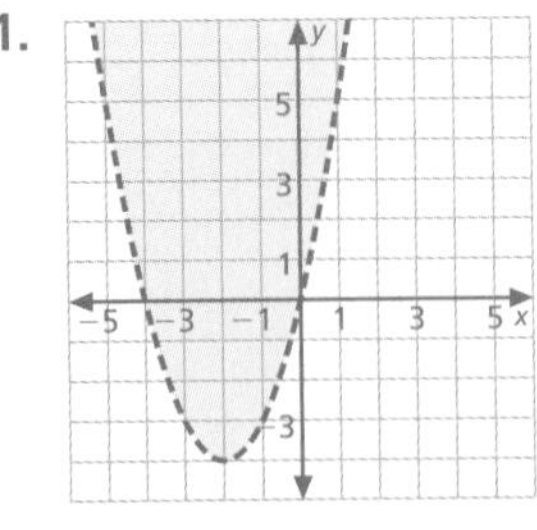

22.

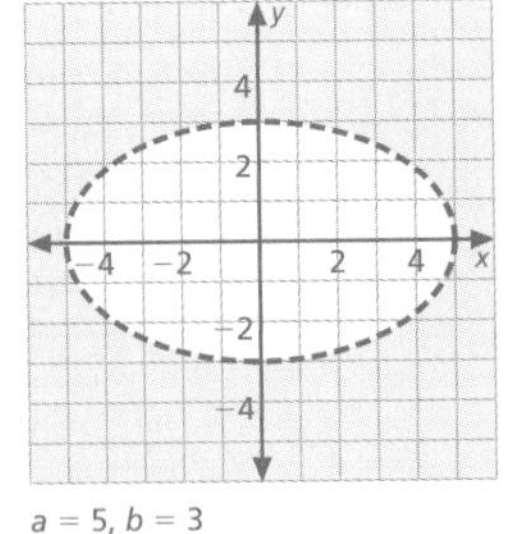

$a = 5$, $b = 3$

Chapter 13 Alternative Assessment

Cooperative Learning

GUESS THE GRAPH Work in a group. Draw a graph you would expect to be the graph of a quadratic equation. Exchange graphs to find the equation. Discuss: Did you have an idea of the equation from the shape of the graph? Did the actual equation for your graph surprise you, or were you certain of what it would be? Try again, this time trying to stump your classmates. Be sure *you* know the equation!

Technology

USES OF PARABOLAS Parabolas are a shape that is familiar to most of us, yet many people don't even know the correct name for this shape. Research to discover how modern techology uses this shape. You might look up parabolic microphones, parabolic mirrors, lasers, telescopes, satellite dishes, etc. What is the common use of the parabola in each of these instruments? How does this relate to the study of the distance of the parabola's focus on the coordinate plane? Explain and defend your conclusions.

Self-Assessment

DIFFERENT SHAPES Choose at least four different shapes of graphs you have studied in this chapter. Write four equations that result in each shape of graph chosen. What do all the equations for one shape have in common? Write a brief paragraph telling how you can identify the shape of the graph from only the equation. Defend your conclusions.

Chapter Investigation

EXTENSION Write a report to summarize your work. Be sure to include the results of your research, your map, and a description of anything you learned during your class discussion of your work.

Chapter Investigation

The benchmarks and expectations for this extension are as follows.

- Students research the perihelion and aphelion for each planet of the Solar System. They make a rough sketch of the planets' orbits and label the perihelion and aphelion for each planet.
- Students choose a point within their school to represent the Sun. They use a map of their school and plot the location of the aphelion and perihelion for each planet. They make a sketch of the orbits of the planets.
- Students select landmarks at their school to represent the aphelion for each landmark. They share their findings with the class, and discuss how this activity has increased their understanding of the size of the Solar System.
- Students write a report to summarize their work.

Cooperative Learning

Students should limit their graphs to those studied in Chapters 12 and 13.

SCORING RUBRIC **5**—Student guesses accurately the general form of the equation of a graph, calculates the exact equation precisely, and draws graphs of equations with authority. **4**—Student guesses accurately the general form of the equation, calculates the exact equation, and draws graphs of equations. **3**—Student guesses within reason the general form of the equation, calculates the equation, and draws graphs of equations. **2**—Student guesses within reason the general form of the equation, calculates the equation with minor errors, and draws graphs of one or two equations. **1**—Student guesses poorly the general form of the equation, calculates the equation with significant errors, and does not draw graphs of equations. **0**—Student makes no attempt to solve guess equations, solve, or draw graphs.

Technology

Encourage students to include drawings in their reports to solidify understanding.

SCORING RUBRIC **5**—Student reports in detail on several uses of a parabola in technology, can accurately explain how these instruments work, and draws an accurate conclusion about the similar uses of a parabola. **4**—Student reports on several uses of a parabola in technology, gives accurate explanation, and draws an accurate conclusion about the similar uses of a parabola. **3**—Student reports on several uses, can explain in generalities how these instruments work, and draws a reasonable conclusion about the uses of a parabola. **2**—Student reports on some uses of a parabola, can explain in generalities how these instruments work, but draws an illogical conclusion about the similar uses of a parabola. **1**—Student reports on one use of a parabola, can explain vaguely how the instrument works, but draws no conclusion about the similar uses of a parabola. **0**—Student makes no attempt to research parabolas in technology.

See page 598 for Self Assessment Scoring Rubric.

Cumulative Review

Cumulative review is the best way to maintain previously taught skills and concepts. This will keep students prepared for new lessons that build on previously covered skills.

Cumulative Review covers:

Lesson 1-2
Lesson 1-3
Lesson 2-2
Lesson 2-6
Lesson 3-1
Lesson 3-6
Lesson 4-6
Lesson 5-2
Lesson 5-6
Lesson 6-3
Lesson 6-4
Lesson 7-2
Lesson 8-2
Lesson 8-5
Lesson 9-4
Lesson 10-2
Lesson 10-4
Lesson 11-4
Lesson 11-9
Lesson 12-3
Lesson 13-1
Lesson 13-5

Additional Answers

4. 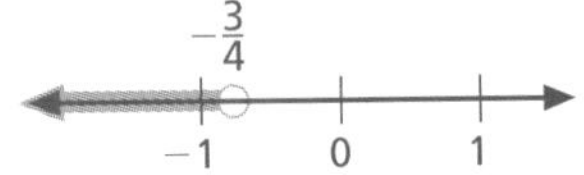

6. If points x, y, and z are collinear, they are coplanar; the original statement is false; the conversion is true.

11. 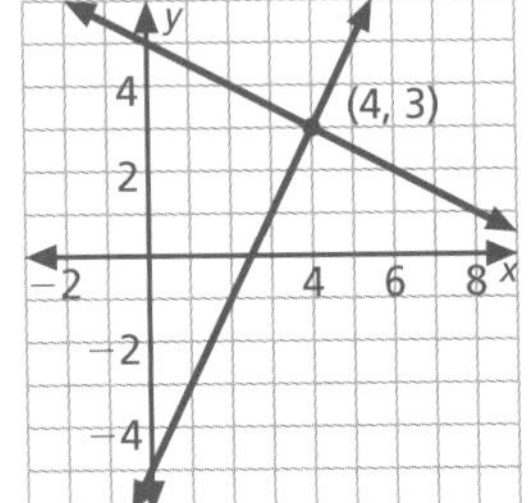

Chapters 1–13 Cumulative Review

1. Evaluate the expression $-|-(x+1)|$ when $x = -4$. -3
2. Let $A = \{M, A, J, O, R\}$, $B = \{A, D\}$ and $C = \{M, A, S, H\}$. Find $(A \cap C) \cup B$. $\{M, A, D\}$
3. Given $f(x) = 2x^2 - 3$ and $g(x) = -2x + 3$, find the value of $f(2) - g(-1)$. 0
4. Solve this inequality and graph the solution on a number line: $3(4y - 1) < -12$. See additional answers.
5. In the figure, $RT = 23$. Find ST. 11

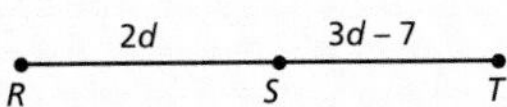

6. If points X, Y, and Z are coplanar, then they are collinear. Write the converse of this statement and tell whether the statement and the converse are true or false. See additional answers.
7. List the segments in the figure at the right in order from the shortest to the longest. $\overline{AB}$, $\overline{AD}$, $\overline{BD}$, $\overline{CD}$, $\overline{BC}$
8. If you doubled each side of a rectangle, how much would the perimeter increase? How much would the area increase? 2 times; 4 times
9. A ball has a diameter of 12 inches. What is the surface area of the ball? 144π or 452.4 in.^{2}
10. Write the equation of the line that is parallel to the line $2x - y = 3$ and passes through the point $(-2, 1)$. $2x - y = -5$
11. Graph the lines $x + 2y = 10$ and $2x - y = 5$ to find their intersection point. See additional answers.
12. $ABCD$ and $EFGH$ are similar rectangles. If $AB = 6$ in., $BC = 7$ in., and $EF = 9$ in. ($AB \simeq EF$), what is the perimeter of $EFGH$? 39 in.
13. In a counterclockwise rotation of 90°, the point (2, 5) becomes what point? (−5, 2)
14. $A = \begin{bmatrix} 2 & 3 \\ -1 & 4 \end{bmatrix}$; $B = \begin{bmatrix} 0 & -5 \\ 6 & -3 \end{bmatrix}$. Find $A - 2B$. $\begin{bmatrix} 2 & 13 \\ -13 & 10 \end{bmatrix}$
15. A box contains three red marbles, two blue, four yellow, and a green one. If you draw out a marble and return it, what is the probability of the selections being a red followed by a yellow? $\frac{3}{25}$
16. Find the length (to the nearest tenth) of a diagonal of a rectangle with sides of 4 cm and 7 cm. 8.1 cm
17. $\angle ABC$ measures 56°. Find the measure of $\angle AOC$. 112°

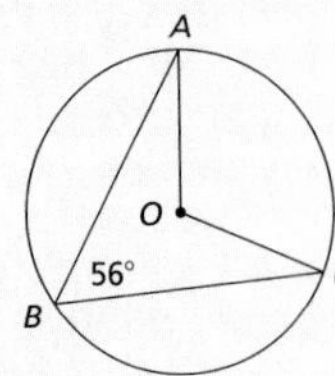

18. Multiply: $(2x - 3y)(4x + y)$. $8x^2 - 10xy - 3y^2$
19. Factor completely: $18a^2 - 9a - 20$. $(6a + 5)(3a - 4)$

Self Assessment

Students may wish to use a graphing calculator to check on the shape of the graph as they make up equations.

SCORING RUBRIC **5**—Student develops detailed and accurate rules for matching equations with the shape of their graphs, and gives precise examples for four different shapes. **4**—Student develops accurate rules for matching equations with the shape of their graphs, and gives examples for four different shapes. **3**—Student develops rules for matching equations with the shape of their graphs, and gives examples for two or three different shapes. **2**—Student develops slightly inaccurate rules for matching equations with the shape of their graphs, and gives examples for two or three different shapes. **1**—Student develops inaccurate rules for matching equations with the shape of their graphs, and gives partial or poor examples for two or three different shapes. **0**—Student makes no attempt to make rules for matching equations and shapes of graphs.

CHAPTERS 1–13 CUMULATIVE ASSESSMENT

STANDARDIZED TEST PREPARATION—STANDARD MULTIPLE CHOICE

Choose the best solution for each problem.

1. Evaluate x^3y^2 when $x = -2$ and $y = 3$.
C
A. 72 **B.** 108
C. -72 **D.** -54
E. $\frac{9}{8}$

2. Solve for x:
B $9(x + 4) - 2x = 19 - 3(x + 6)$
A. $\frac{1}{4}$ **B.** $-\frac{7}{2}$
C. -7 **D.** $\frac{37}{10}$
E. $\frac{1}{10}$

3. The sum of the measure of the complement of
C an angle and the measure of its supplement is 144°. The angle equals:
A. 72° **B.** 18°
C. 63° **D.** 27°
E. none of these

4. What type of figure do you get by joining the
D midpoints of a trapezoid?
A. rhombus **B.** rectangle
C. square **D.** parallelogram
E. none of these

5. Solve for a and b:
C $-\frac{7}{a} + \frac{5}{b} = 1; -\frac{6}{a} + \frac{4}{b} = 0$
A. $a = 29, b = 0$ **B.** $a = \frac{2}{3}, b = \frac{3}{2}$
C. $a = \frac{1}{2}, b = \frac{1}{3}$ **D.** $a = 2, b = 3$
E. none of these

6. In $\triangle XYZ$, AB and CD are parallel
C to XY. If $YB = 2$, $BD = 3$, $DZ = 4$, and $AC = 6$, how long is AZ?

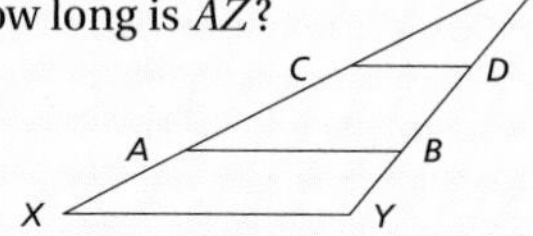

A. 18 **B.** 15
C. 14 **D.** 13
E. none of these

7. Drawing a card at random from a standard
B deck of cards, what is the probability that the card is a diamond or a face card?
A. $\frac{5}{13}$ **B.** $\frac{11}{26}$
C. $\frac{25}{52}$ **D.** $\frac{3}{52}$
E. none of these

8. What is the simplest radical form of the
B product of $(7\sqrt{3})(2\sqrt{21})$.
A. $18\sqrt{6}$ **B.** $42\sqrt{7}$
C. $14\sqrt{63}$ **D.** $9\sqrt{24}$
E. $\sqrt{231}$

9.
D

3 7 5 x

Find the value of x.
A. 4 **B.** $\frac{21}{5}$
C. $\frac{15}{7}$ **D.** $\frac{35}{3}$
E. none of these

10. Factor completely: $18x^2 + 3xy - 6y^2$.
A
A. $3(3x + 2y)(2x - y)$
B. $(9x + 6y)(2x - y)$
C. $(3x + 2y)(6x - 3y)$
D. $3(3x - 2y)(2x + y)$
E. none of these

11. Solve for z: $z^2 - 10z + 3 = 0$.
D
A. $-5 \pm \sqrt{22}$ **B.** $5 \pm 2\sqrt{22}$
C. $-5 \pm 2\sqrt{22}$ **D.** $5 \pm \sqrt{22}$
E. none of these

12. **CONSTRUCTED RESPONSE** Draw the graph of the equation $16x^2 + 25y^2 = 400$. Plot and label the foci.
See additional answers.

Teaching Strategies

A common error when factoring polynomials is to begin by looking for two binomial factors, applying the FOIL method. Remind students to look first for a common factor of all terms in the polynomial. Once they have factored out this common factor, they can proceed to find the binomial factors.

STANDARDIZED TEST PREPARATION

STANDARD MULTIPLE CHOICE— One strategy for completing multiple-choice questions is to use mental math and eliminate unreasonable answers as choices. Often, a couple of the answer choices are values that you may get while working out the problem. For instance, if the solution requires two or three steps, one answer choice may be the value or expression that you have at the end of the first step and another choice may be the value or expression that you have at the end of the second step.

Sometimes all but the correct answer can be eliminated using only mental math and reasoning. If this is the case, verify your answer choice by working out the problem.

TESTING STRATEGIES

When taking a timed test, it may not be possible to go back and check every answer after completing the test. Some questions may be more necessary to double check than others. Suggest that students mark questions with which they struggled or obtained an answer they weren't expecting. When they've completed the test, students can go back and re-work marked questions to double check their answers.

ADDITIONAL ANSWER

12.

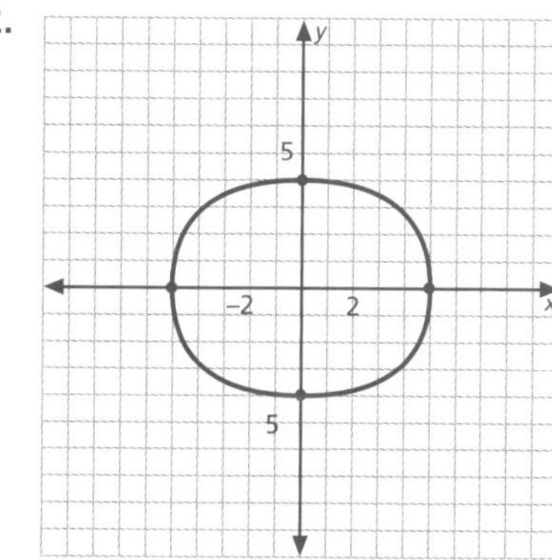

Chapter 14—Trigonometry

Theme: Navigation
Chapter Investigation: Size of Objects, pages 603, 607, 611, 629
Careers: Commercial Fishers, page 623
Data Activity: U.S. Airport Traffic

Content and Connections

Lesson, Pages	Lesson Objectives	NCTM Standards	State/Local Objectives	Interdisciplinary Connections	Real World Applications
14–1 604–607	• Identify trigonometric rations in a right triangle • Use trigonometric ratios to solve problems	Patterns, Functions & Algebra Geometry & Spatial Sense Problem Solving Reasoning & Proof Communication Connections Representation			Navigation, Construction, City Planning
14–2 608–611	• Find the length of sides and the measures of angles in right triangles	Problem Solving Reasoning & Proof Patterns, Functions & Algebra Geometry & Spatial Sense Connections Representation			Surveying, Navigation, Safety
14–3 614–617	• Solve problems using trigonometry	Patterns, Functions & Algebra Geometry & Spatial Sense Problem Solving Communication Connections Representation			Communications, Population, Flight
14–4 618–621	• Determine the period, amplitude and position of sine curves	Patterns, Functions & Algebra Number Operation Problem Solving Reasoning & Proof Communication Connections Representation			Communications, Music, Medicine
14–5 624–625	• Solve a problem by choosing a strategy	Patterns, Functions & Algebra Problem Solving Reasoning & Proof Geometry & Spatial Sense Communication, Connections, Representation		Physics	Archeology, Surveying, Geography, Science, Physics

Planning and Resources

Lesson, Pages	Tools/Materials Needed	Trans-parency	Learning/Teaching Styles Options	Assignments: Basic Enriched	Additional Practice in SE	Reteaching, Extra Practice, Enrichment	Other Resources
14–1 604–607	calculator ruler	Warm up 40 Trans RF-53, 54	Real World Visual Learners	B: 1–26, 31–43 E: 1–43	Page 612–613, 623, 626, 728	R: page 205 EP: page 179 E: page 133	
14–2 608–611	calculator	Warm up 40 Trans RF-55	Challenge	B: 1–25, 31–36 E: 1–36	Page 612, 623, 626, 729	R: page 207 EP: page 181 E: page 235	SS: Teacher's Choice
14–3 614–617	graphing calculator	Warm up 40 Trans RF-56		B: 1–42, 49–54 E: 1–54	Page 622–623, 626, 729	R: page 209 EP: page 183 E: page 237	SS: Teacher's Choice

14–4 618–621	graphing calculator	Warm up 41	Challenge	B: 1–23, 31–42 E: 1–42	Page 622–623, 627, 729	R: page 211 EP: page 185 E: page 239	
14–5 624–625	graphing calculator	Warm up 41	ESL/LEP	B: 1–27 E: 1–27	Page 627	R: page 213 E: page 241	

Planning and Pacing

Lesson, Pages	Lesson Title	45 min class	Assignments Basic, Enriched	90 min class	Assignments Basic, Enriched	____ min class	Assignments Basic, Enriched
14–1 604–607	Basic Trigonometric Ratios	Day 176	B: 1–26, 31–43 E: 1–43	Day 97	B: 1–26, 31–43 E: 1–43		B: 1–26, 31–43 E: 1–43
14–2 608–611	Solve Right Triangles	Day 177–178	B: 1–25, 31–36 E: 1–36	Day 97–98	B: 1–25, 31–36 E: 1–36		B: 1–25, 31–36 E: 1–36
14–3 614–617	Graph the Sine Function	Day 179	B: 1–42, 49–54 E: 1–54	Day 98	B: 1–42, 49–54 E: 1–54		B: 1–42, 49–54 E: 1–54
14–4 618–621	Experiment with the Sine Function	Day 180–181	B: 1–23, 31–42 E: 1–42	Day 99	B: 1–23, 31–42 E: 1–42		B: 1–23, 31–42 E: 1–42
14–5 624–625	Problem Solving Skills: Choose a Strategy	Day 182	B: 1–27 E: 1–27	Day 100	B: 1–27 E: 1–27		B: 1–27 E: 1–27
Review/ Assess		Day 183–184		Day 100–102			

	Chapter 1 Days	Chapter Cumulative Days
Yearly Pacing (45 min class)	9 days	184 days
Yearly Pacing (90 min class)	6 days	102 days

Assessment Options

Assessment in Student Edition	Assessment in Teacher's Edition	Pages in Assessment Book	Software Generated Assessment
Are You Ready?, pages 600–601; Writing Math, pages 607, 616; Mixed Review, pages 607, 611, 617, 621, 625; Check Understanding pages 604, 608, 614, 618; Mid-Chapter Quiz, page 613, Chapter Review, page 626; Chapter Assessment, page 628; Alternative Assessment, page 629; Cumulative Review, page 630; Cumulative Assessment, page 631	5-minute Warm ups, pages 604, 608, 614, 618, 624; Quick Assessment, pages 606, 610, 616, 620, 625; Scoring Rubrics, page 629	Mid-Chapter Quiz, page 215; Test Form A, pages 217, 219; Test Form B, pages 221, 223; Cumulative Test 225, 227; Math Journal prompt, page 229, 230;	Chapter 14

Refresher Skills

The skills on these two pages are skills that have been taught in previous math courses. Continuous review of basic math skills will make stronger math students. These skills are identified as necessary to be successful in Chapter 14.

Skills Correlation Chart

Skill	Lesson Number
Naming Sides of Triangles	14-1, 14-3, 14-5
Rationalizing Radicals	14-1, 14-2, 14-3, 14-4, 14-5
Special Right Triangles	14-1, 14-2, 14-5

Vocabulary

adjacent
hypotenuse
rationalize
30°-60°-90° triangle
45°-45°-90° triangle

ARE YOU READY?

Refresh Your Math Skills for Chapter 14

The skills on these two pages are ones you have already learned. Review the examples and complete the exercises. For additional practice on these and more basic skills, see page 674.

NAMING SIDES OF TRIANGLES

In the study of trigonometry, it is very important to be able to shift your focus and see the triangles form different points of view.

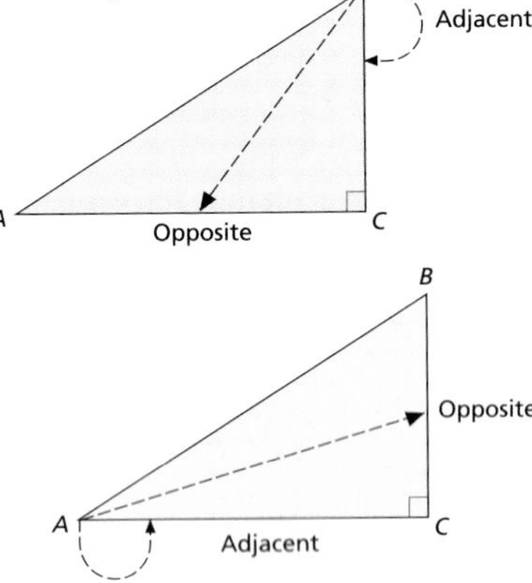

Examples Name the side of $\triangle ABC$ that is adjacent to $\angle B$.

- Adjacent means "next to." The adjacent side of an angle is never the hypotenuse (AB). Therefore, the side adjacent must be BC.

Name the side opposite $\angle B$.

- The side opposite an angle does not contain the point of the angle. Therefore, it must be AC.

Name the sides in each triangle.

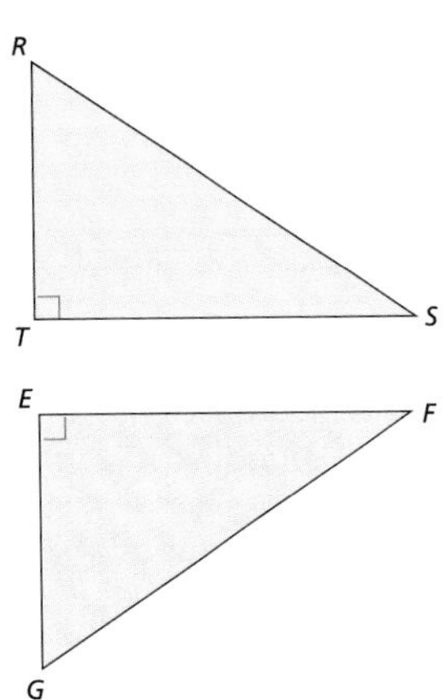

1. The side opposite $\angle S$. *RT*
2. The side adjacent to $\angle S$. *ST*
3. The side adjacent to $\angle R$. *RT*
4. The side opposite $\angle R$. *ST*

5. The side opposite $\angle F$. *EG*
6. The side adjacent to $\angle F$. *EF*
7. The side opposite $\angle G$. *EF*
8. The side adjacent to $\angle G$. *EG*

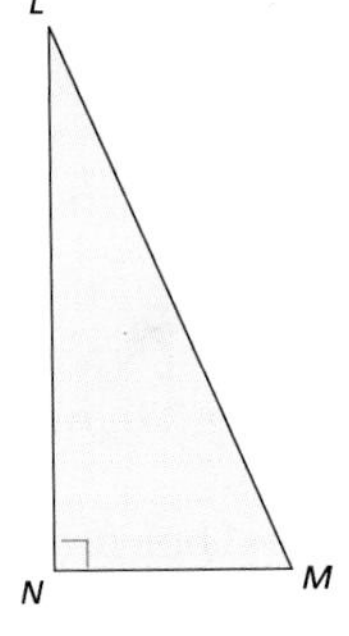

9. The side opposite $\angle M$. *LN*
10. The side opposite $\angle L$. *MN*
11. The side adjacent to $\angle L$. *LN*
12. The side adjacent to $\angle M$. *MN*

Extend the Lesson

COOPERATIVE LEARNING Have students work in pairs to draw two 45°-45°-90° triangles and two 30°-60°-90° triangles. For each triangle, have them show the measurement of one leg or the hypotenuse. Have them solve the triangles and record their answers on a separate sheet of paper. Have pairs trade triangles and solve a new set of triangles. Students can compare answers to check their work.

CHAPTER 14

Rationalizing Radicals

You will have to rationalize radicals much of the time while studying trigonometry.

Rationalize. Write in simplest radical form.

13. $\frac{\sqrt{8}}{\sqrt{3}}$ $\frac{2\sqrt{6}}{3}$ **14.** $\sqrt{\frac{32}{10}}$ $\frac{4\sqrt{5}}{5}$ **15.** $\frac{\sqrt{18}}{\sqrt{5}}$ $\frac{3\sqrt{10}}{5}$ **16.** $\sqrt{\frac{45}{3}}$ $\sqrt{15}$

17. $\frac{\sqrt{96}}{\sqrt{24}}$ 2 **18.** $\sqrt{\frac{36}{5}}$ $\frac{6\sqrt{5}}{5}$ **19.** $\frac{\sqrt{27}}{\sqrt{7}}$ $\frac{3\sqrt{21}}{7}$ **20.** $\sqrt{\frac{54}{8}}$ $\frac{3\sqrt{3}}{2}$

Special Right Triangles

- In a 30-60-90 triangle, the measure of the hypotenuse is two times that of the leg opposite the 30° angle. The measure of the other leg is $\sqrt{3}$ times that of the leg opposite the 30° angle.
- In a 45-45-90 triangle, the measure of the hypotenuse is $\sqrt{2}$ times the measure of a leg of the triangle.

Find the unknown measures.

21.

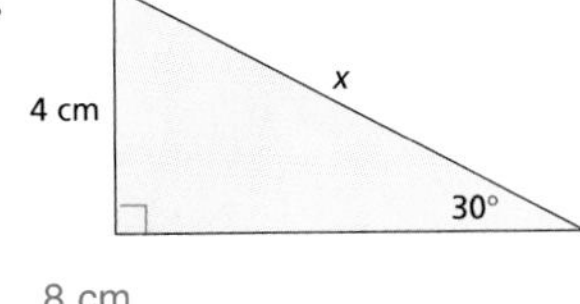

8 cm

22.

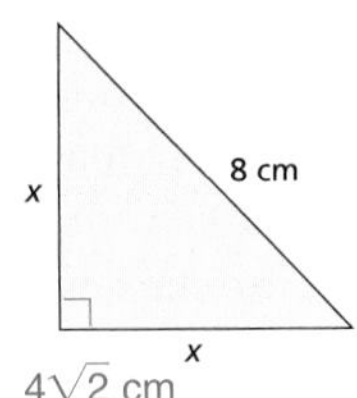

$4\sqrt{2}$ cm

23.

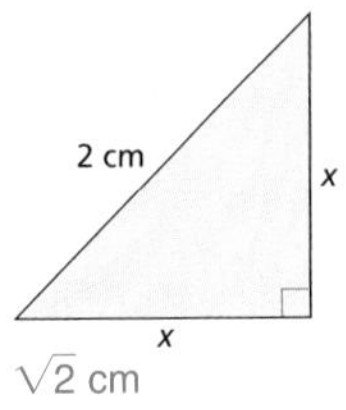

$\sqrt{2}$ cm

24.

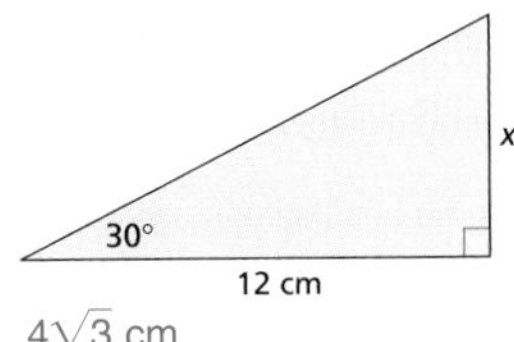

$4\sqrt{3}$ cm

25.

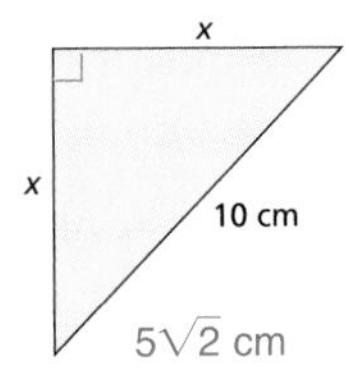

$5\sqrt{2}$ cm

26.

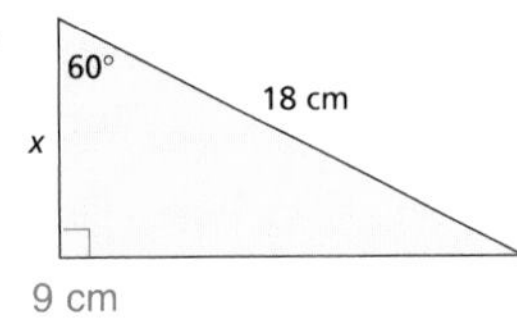

9 cm

27.

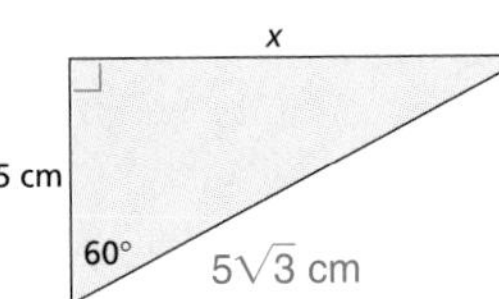

$5\sqrt{3}$ cm

28.

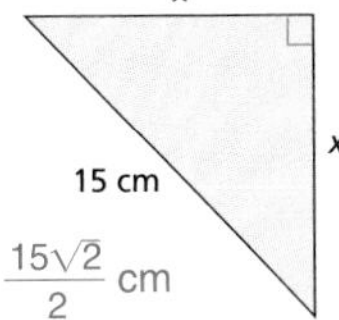

$\frac{15\sqrt{2}}{2}$ cm

29.

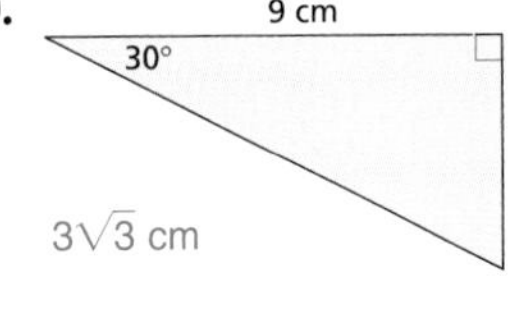

$3\sqrt{3}$ cm

Chalkboard Examples

Naming Sides of Triangles

Right triangle *NOP* where the right angle is $\angle O$.
Name:

1. the side opposite $\angle N$ *OP*
2. the side adajcent to $\angle N$ *NO*
3. the side opposite $\angle P$ *NO*
4. the side adjacent to $\angle P$ *OP*

Rationalizing Radicals

Write $\frac{\sqrt{72}}{4}$ in simplest radical form. $\frac{3}{2}$

Special Right Triangles

Right triangle *CAB* is a 45°-45°-90° triangle. The length of *AB* is 3 cm. Find the measures of *AC* and *BC*. 3 cm; $3\sqrt{2}$

Refresher Wrap-up

Quick Assessment

Ask the following questions to determine if students have mastered the basic skills reviewed on these pages.

1. Define the meaning of adjacent and opposite. Adjacent means "next to." The opposite side does not contain the point of the angle.
2. Does $\sqrt{40} = 4\sqrt{10}$? No, $\sqrt{40} = \sqrt{4} \cdot \sqrt{10} = 2\sqrt{10}$

Additional Practice

Refer to the Extra Practice Index on page 674 of this text.

Extend the Lesson

MATH JOURNAL Have students describe a real-life application in which solving special triangles could be useful. For example, a portion of a roof forms a 45°-45°-90° triangle. If a homeowner needed to replace the wood beam which forms the hypotenuse of the triangle, she could use the ratios of a 45°-45°-90° triangle to find the length of the beam.

Chapter Opener

NCTM Standards/Strands

- Number & Operation
- Patterns, Functions, & Algebra
- Geometry & Spatial Sense
- Problem Solving
- Reasoning and Proof
- Communication
- Connections
- Representation

Vocabulary

navigation
trigonometry
global positioning system
sextant
quadrant

Theme Connections

Tables and graphs are used to organize data about navigation and air travel. Tables provide a way to organize numerical information. Many types of data are kept in regards to navigation and air travel. Tables of data are used to plot courses, determine flight times, and track company performance. Discuss how data and graphs are used in navigation and air travel.

Career Opportunities

Many careers related to navigation make use of data and graphs. One area is highlighted in the MathWorks feature. Others include pilots, air traffic controllers, engineers, computer programmers, and shipboard navigators.

- Fishers, Hunters and Trappers, page 623

Internet Connection

Theme Activities

Learningmathmatters.com provides web addresses to search that help students gather information about the use of math in the real world, particularly data and graphs. To search for additional addresses, begin a search using the keyword *navigation*. Then within that search, use such key words as *ship, airplane*, and *satellite*. Students can brainstorm in small groups other key words.

CHAPTER 14

TRIGONOMETRY

THEME: Navigation

Early explorers relied on the stars and simple tools such as sextants and quadrants to find their way. How did they do it? Long ago, people realized that the stars move in predictable patterns. By keeping careful records and taking angle measurements, they discovered a way to pinpoint their location on the Earth's surface with a reasonable degree of accuracy.

In the same way, modern navigators use information from satellites and guidance computers to find their way. Even automobiles are now equipped with global positioning systems which use data from satellites to determine an automobile's exact location in case of an emergency. These advances are made possible by a branch of mathematics called trigonometry. Trigonometry, which means "triangle measurement," is the study of relationships among the sides and angles of a triangle.

Sextant

- **Commercial Fishers** (page 623) Aside from fishing duties, commercial fishers pilot small ships or boats and must be able to navigate to fishing areas. They use the stars as well as electronic equipment to pinpoint their location.

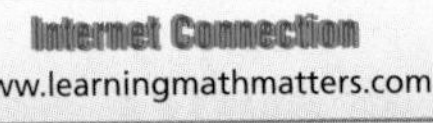

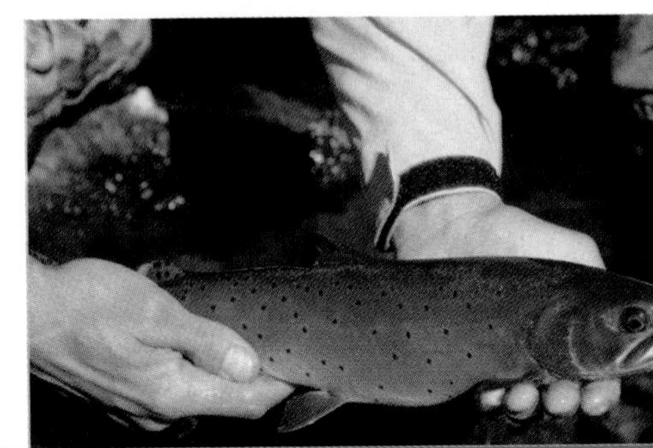

Chapter Investigation

Go to learningmathmatters.com to locate additional information about navigation.

As a Chapter Project

The goal of this project is to make a quadrant and use it to determine the height or depth of objects. Students can use the Group Project Planner on page 231 and the Project Planning Calendar on page 232 in the Enrichment text to complete the project. Benchmarks **a, b,** and **c.** should be completed after the lesson listed in parentheses has been studied. Benchmark d should be completed at the end of the chapter.

U.S. Airport Traffic, 1998

Airport	Total passengers	Change from 1997	Total freight (metric tons)	Change from 1997
Atlanta, Hartsfield (ATL)	73,474,298	7.7%	907,209	4.9%
Chicago, O'Hare (ORD)	72,485,228	3.0%	1,441,829	2.5%
Los Angeles (LAX)	61,215,712	1.8%	1,861,050	−0.7%
Dallas/Ft. Worth (DFW)	60,368,466	−0.2%	801,968	−1.1%
San Francisco (SFO)	40,060,325	−1.1%	771,931	−1.0%
Denver (DEN)	36,831,400	5.3%	447,266	2.3%
Miami (MIA)	33,935,491	−1.7%	1,793,009	1.5%
Newark (EWR)	32,512,106	5.2%	1,094,383	4.5%
Memphis (MEM)	10,063,885	−3.5%	2,368,975	6.1%

Data Activity: U.S. Airport Traffic

Use the table for Questions 1–4.

1. On average, how many passengers arrive or depart from LAX each day? 167,714
2. To the nearest ton, how many metric tons of freight were sent from Memphis in 1997? 2,232,776 T
3. If the passenger traffic for both Newark and San Francisco continue to change at the same rate, in what year would you expect Newark to have surpassed San Francisco's total passengers? 2002
4. Of the airports shown on the table, which had the greatest decrease in actual numbers of passengers from 1997 to 1998? Miami

Chapter Investigation

In the Northern Hemisphere, the stars appear to move in a circular motion around a single star named *Polaris,* commonly known as the North Star. Explorers first navigated the globe using the star and an instrument called a *quadrant.*

Working Together

Build a quadrant using a photocopy of a protractor, heavy cardboard, string, and a small weight. Use the quadrant to find the angles of elevation for several tall objects. Use trigonometric relationships to find the height of the objects. Use the Chapter Investigation logos to guide your group.

Data Activity

Discuss with students the importance of air and sea travel in carrying passengers and freight. Point out that many nations depend on imports and exports to boost their economy. In a large country such as the United States, the shipping of freight between states is crucial and is accomplished by land, sea, and air. Discuss the information in the table and the methods used to determine the percent changes from one year to the next. Assign students Questions 1–4.

Extend the Data Activity

REAL WORLD CONNECTION Have students research data regarding the shipping of goods by sea for a recent year. If students live near a harbor or major airport, you may wish to assign them to gather data about freight and passenger travel from that location.

Chapter Investigation

As an Overarching Problem

Ask students to brainstorm about how they would measure the height of a very tall object. Wrap up the discussion by telling students that in this investigation they will make a homemade quadrant that they can use to find the height or depth of different objects. Students will continue to work on the investigation as they complete the exercises identified by the Investigation Logo that are found throughout the chapter. These exercises will guide students through the task described in *Working Together*. Encourage students to keep all of their work on the Investigation together. Have students use the suggestions in the Chapter Investigation Extension to summarize their work.

See page 602 for Chapter Investigation As a Chapter Project.

Project Planning Calendar

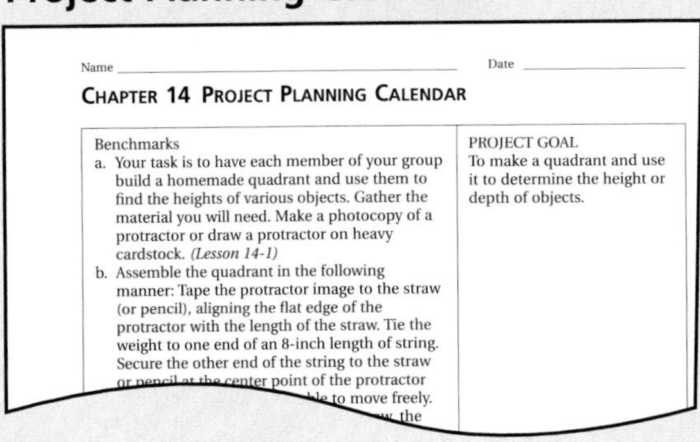

Name ______ Date ______

CHAPTER 14 PROJECT PLANNING CALENDAR

Benchmarks

a. Your task is to have each member of your group build a homemade quadrant and use them to find the heights of various objects. Gather the material you will need. Make a photocopy of a protractor or draw a protractor on heavy cardstock. *(Lesson 14-1)*

b. Assemble the quadrant in the following manner: Tape the protractor image to the straw (or pencil), aligning the flat edge of the protractor with the length of the straw. Tie the weight to one end of an 8-inch length of string. Secure the other end of the string to the straw or pencil at the center point of the protractor ... to move freely.

PROJECT GOAL
To make a quadrant and use it to determine the height or depth of objects.

Group Project Planner

Name ______ Date ______

CHAPTER 14 GROUP PROJECT PLANNER

Assignment ______ Objective ______

Group Members
1) ______
2) ______
3) ______
4) ______
5) ______

Assigned Roles

Lesson Planning

NCTM Standards/Strands
- Number & Operation
- Patterns, Functions, & Algebra
- Geometry & Spatial Sense
- Problem Solving
- Reasoning and Proof
- Communication
- Connections
- Representation

Vocabulary
trigonometric ratios
sine
cosine
tangent

Materials Needed
calculators rulers

Lesson Resources
Warm-Up Transparency 40
Transparency RF-53, 54
Reteaching 14-1
Extra Practice 14-1
Enrichment 14-1

Assignment Guide
Basic: 1–26, 31–43
Enriched: 1–43

Getting Started

5-Minute Warm-up
Solve for x. Round answers to the nearest tenth.
1. $x = \sqrt{9^2 + 12^2}$ 15
2. $x = \sqrt{14^2 - 5^2}$ 13.1
3. $x = \sqrt{32^2 + 25^2}$ 40.6

Introduction to Lesson 14-1
After finding the ratios for Parts (a), (b) and (c) and answering the questions, have students trace the drawing and draw other randomly selected vertical lines from $\overline{AB}$ to $\overline{AC}$. Have them label the new points and determine the ratios for the new triangles. Students should see that the triangles in the opening diagram were not specially chosen to the make the point.

14-1 Basic Trigonometric Ratios

Goals
- Identify trigonometric ratios in a right triangle.
- Use trigonometric ratios to solve problems.

Applications Navigation, Construction, City Planning

Work with a partner.

Measure the line segment lengths needed to calculate the following ratios. Use a calculator to evaluate the ratios.

a. $\frac{DE}{AD}$ **b.** $\frac{FG}{AF}$ **c.** $\frac{BC}{AB}$

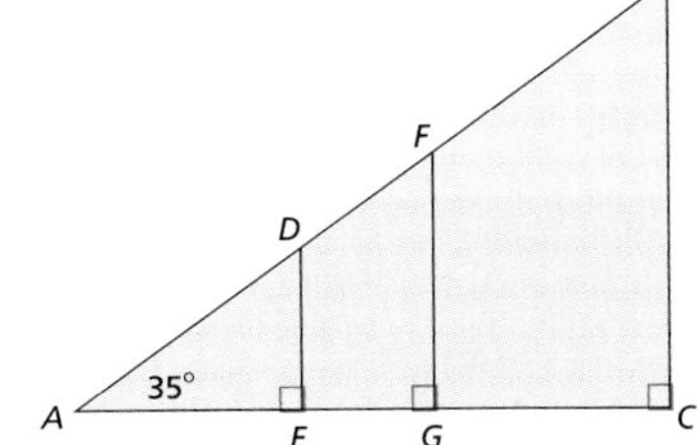

What conclusions can you draw from your results?
In a right triangle with a 35° angle, the ratio of the length of the side opposite the 35° angle to the length of the hypotenuse is approxiately 0.57.

Build Understanding

Precise measurement shows that the ratio of the lengths of two sides of a right triangle depends on the angles of the triangle—but not the lengths. For a given angle, the ratios are always the same. This fact forms the foundation for the study of trigonometry. In this lesson, you will study the most important **trigonometry ratios**: the **sine**, the **cosine**, and the **tangent**.

In a right triangle, angle A is an acute angle. Then,

$$\textbf{sine } A = \frac{\text{length of side opposite } \angle A}{\text{length of hypotenuse}}$$

$$\textbf{cosine } A = \frac{\text{length of side adjacent to } \angle A}{\text{length of hypotenuse}}$$

$$\textbf{tangent } A = \frac{\text{length of side opposite } \angle A}{\text{length of side adjacent to } \angle A}$$

Sine, cosine, and tangent are abbreviated sin, cos, and tan. Sin A means "the sine of $\angle A$."

Mental Math Tip
Use the memory device **SOH CAH TOA** (pronounced "sokatoah") to remember the trigonometric ratios.

SOH: Sine is Opposite over Hypotenuse.

CAH: Cosine is Adjacent over Hypotenuse.

TOA: Tangent is Opposite over Adjacent.

Example 1

a. Find sin M.

b. Find cos M.

c. Find tan N.

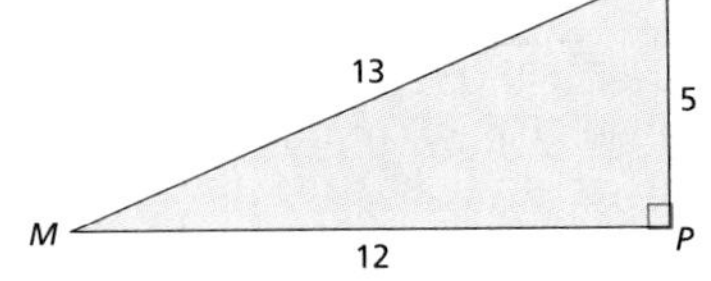

Solution

a. $\sin M = \frac{\text{opposite}}{\text{hypotenuse}} = \frac{5}{13}$

b. $\cos M = \frac{\text{adjacent}}{\text{hypotenuse}} = \frac{12}{13}$

c. $\tan N = \frac{\text{opposite}}{\text{adjacent}} = \frac{12}{5} = 2\frac{1}{5}$

Extend the Lesson

REAL WORLD CONNECTION An early use of trigonometric concepts was the measurement of shadows cast by a horizontal stick. As early as 1,000 B.C., Egyptians used the fact that the higher the sun, the longer the shadow cast by a stick attached to the side of a building. Tangent ratios could be computed from the length of the stick and the varying lengths of the shadows.

Example 2

NAVIGATION A navigator at N sights a 37° angle between a buoy at B and a landmark at L. Find sin 37°.

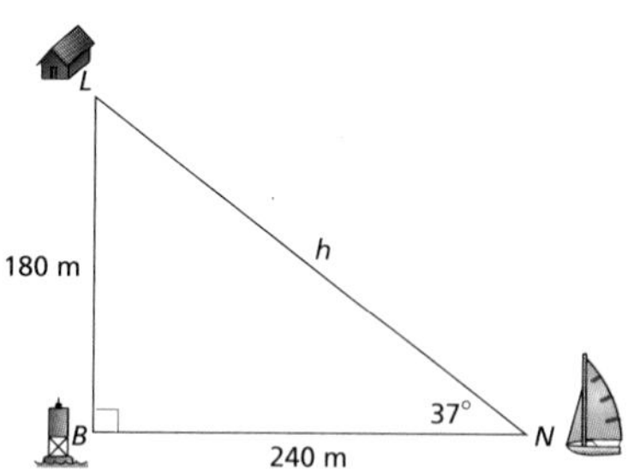

Solution

You can use the Pythagorean Theorem to find the length of the hypotenuse.

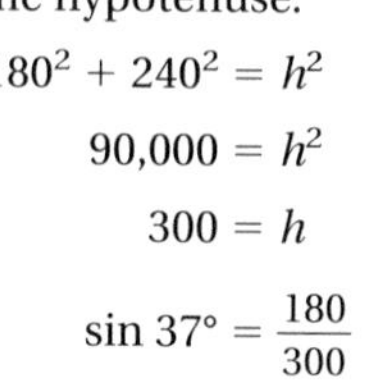

$$180^2 + 240^2 = h^2$$
$$90{,}000 = h^2$$
$$300 = h$$
$$\sin 37° = \frac{180}{300} = 0.6$$

You can use your calculator to find trigonometric ratios. Use the MODE key to set your calculator to work in the "degree" mode. Your calculator may already be set for degree mode. Scientific calculators often show the letters DEG in the display window. To check your calculator's mode, find sin 30° by pressing 30, followed by the sin key. The display should read 0.5. On a graphing calculator, press SIN and enter 30 in parentheses. Press ENTER. The display will read .5. If your calculator displays a different answer, check your manual for more information about how to change to the degree mode.

To find the angle that has a given trigonometric ratio, use the inverse function on your calculator.

Check Understanding

In the figure at the top of page 604, which side is opposite ∠*A* in △*ADE*? Which side is adjacent to ∠*AFG* in △*AFG*?

EF; *FG*

Example 3

CALCULATOR An angle has a cosine of 0.55. Find the measure of the angle to the nearest degree.

Solution

On a scientific calculator, press [.] [5] [5] [$\cos^{-1}$]. (Some calculators use [2nd] [cos] or [INV] [cos] rather than [$\cos^{-1}$].) The display should read 56.6329. . . . To the nearest degree, an angle with a cosine of 0.55 measures 57°.

Learning Styles

VISUAL LEARNERS You may wish to have students use protractors to verify solutions to problems found by use of the trigonometric ratios. This may make the relationships clearer and provide another means for checking the reasonableness of solutions.

Chalkboard Examples

Supplementary Example 1

For triangle *ABC* find (a) sin *A*, (b) cos *A* and (c) tan *B*, when *AB* = 10, *BC* = 6 and *AC* = 8. $\sin A = \frac{3}{5}$ $\cos A = \frac{4}{5}$ $\tan B = \frac{4}{3}$

Supplementary Example 2

A navigator at point *E* sights a 26° angle between a marker at *D* and a landmark at *C*. Find the hypotenuse to the nearest meter and find sin 26°. $h \approx 250$ m and $\sin 26° = \frac{110}{250} = 0.44$

Supplementary Example 3

An angle has a cosine of 0.48. Find the measure of the angle to the nearest degree. 61°

Reteaching Worksheet 14-1

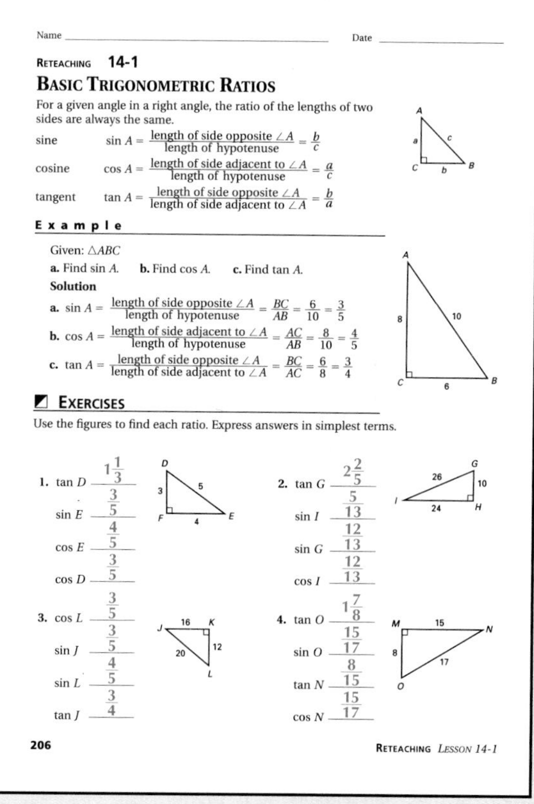

Name ______________________ Date __________

RETEACHING 14-1

BASIC TRIGONOMETRIC RATIOS

For a given angle in a right angle, the ratio of the lengths of two sides are always the same.

sine $\sin A = \frac{\text{length of side opposite } \angle A}{\text{length of hypotenuse}} = \frac{b}{c}$

cosine $\cos A = \frac{\text{length of side adjacent to } \angle A}{\text{length of hypotenuse}} = \frac{a}{c}$

tangent $\tan A = \frac{\text{length of side opposite } \angle A}{\text{length of side adjacent to } \angle A} = \frac{b}{a}$

Example

Given: △*ABC*

a. Find sin *A*. **b.** Find cos *A*. **c.** Find tan *A*.

Solution

a. $\sin A = \frac{\text{length of side opposite } \angle A}{\text{length of hypotenuse}} = \frac{BC}{AB} = \frac{6}{10} = \frac{3}{5}$

b. $\cos A = \frac{\text{length of side adjacent to } \angle A}{\text{length of hypotenuse}} = \frac{AC}{AB} = \frac{8}{10} = \frac{4}{5}$

c. $\tan A = \frac{\text{length of side opposite } \angle A}{\text{length of side adjacent to } \angle A} = \frac{BC}{AC} = \frac{6}{8} = \frac{3}{4}$

EXERCISES

Use the figures to find each ratio. Express answers in simplest terms.

1. tan *D* $1\frac{1}{3}$
sin *E* $\frac{3}{5}$
cos *E* $\frac{4}{5}$
cos *D* $\frac{3}{5}$

2. tan *G* $2\frac{2}{5}$
sin *I* $\frac{5}{13}$
sin *G* $\frac{12}{13}$
cos *I* $\frac{12}{13}$

3. cos *L* $\frac{3}{5}$
sin *J* $\frac{3}{5}$
sin *L* $\frac{4}{5}$
tan *J* $\frac{3}{4}$

4. tan *O* $1\frac{7}{8}$
sin *O* $\frac{15}{17}$
tan *N* $\frac{8}{15}$
cos *N* $\frac{15}{17}$

206 RETEACHING *LESSON 14-1*

Lesson Wrap-up

Quick Assessment

Ask the following questions to determine if students understand the content presented in this lesson.

1. In right triangle RST with hypotenuse RS, why are the values of sin R and cos S equal? $\sin R = \frac{ST}{RS}$, $\cos S = \frac{ST}{RS}$
2. Right triangle UVW with hypotenuse UV is similar to right triangle RST described above. If sin $R = 0.5$, what do you know about sin U? sin $U = 0.5$ What do you know about cos V? cos $V = 0.5$

Assignment Guide

Basic: 1–26, 31–43
Advanced: 1–43
Additional Practice: See Extra Practice Index on page 674.

Try These Exercises

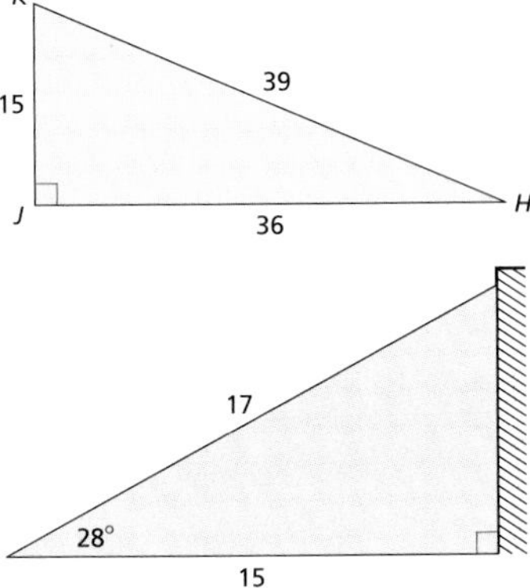

Use the figure at the right to find each ratio. Express answers in lowest terms.

1. tan K $2\frac{2}{5}$
2. cos H $\frac{12}{13}$
3. sin K $\frac{12}{13}$
4. **CONSTRUCTION** A 17-ft wire is attached near the top of a wall. The wire is then anchored to the ground 15 ft from the base of the wall. Find tan 28° to the nearest hundredth. 0.53

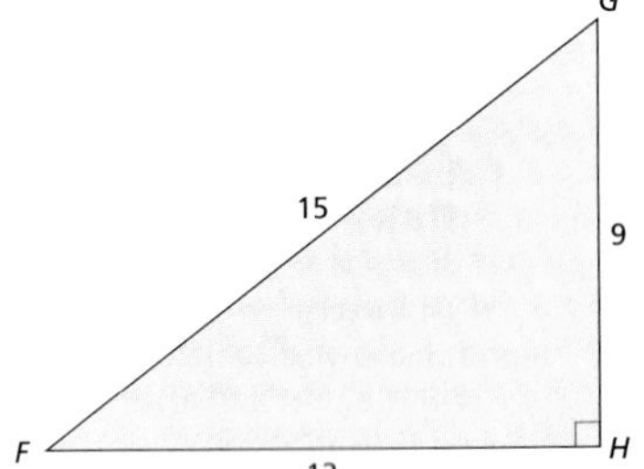

CALCULATOR Use a calculator to find each ratio. Round to the nearest ten-thousandth.

5. sin 22° 0.3746
6. cos 81° 0.1564
7. tan 52° 1.2799
8. cos 40° 0.2126
9. tan 12° 0.8988
10. sin 64° $1\frac{1}{3}$

Practice Exercises

Use the figure at the right to find each ratio.

11. tan G $\frac{3}{5}$
12. sin F $\frac{4}{5}$
13. cos F $\frac{4}{5}$
14. sin G $\frac{4}{5}$
15. tan F $\frac{3}{4}$
16. cos G $\frac{3}{5}$
17. In $\triangle ABC$, $\angle C$ is a right angle, $AB = 29$, and $AC = 21$. Find sin A and tan B to the nearest hundredth. 0.69; 1.05
18. In right triangle RST, $\angle T$ is the right angle and tan $R = \frac{9}{40}$. Write sin R and tan S as ratios. $\frac{9}{41}$; $4\frac{4}{9}$
19. **NAVIGATION** An airline navigator measures a 16° angle between the horizontal and an ocean liner. Find tan 16° to the nearest hundredth. 0.29

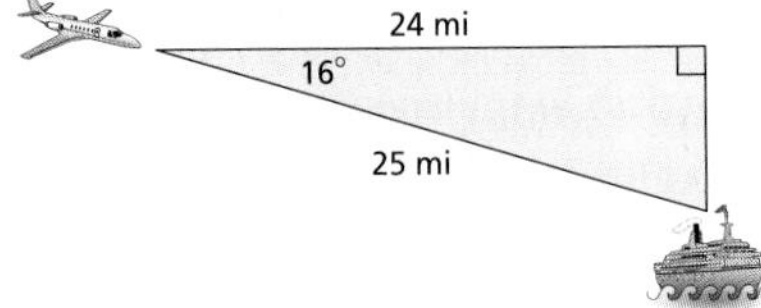

Find angles to the nearest tenth of a degree.

20. **CITY PLANNING** Filbert Street and 22nd Street in San Francisco are the nation's steepest streets. Each rises 1 ft for every 3.17 ft of horizontal distance. What angle do these streets form with the horizontal? 17.5°
21. **NAVIGATION** From a boat at sea, the distance to the top of a 2,325-ft cliff at the water's edge is 4,370 ft. What angle does a line make with the horizontal from the boat to the top of the cliff? 32.1°
22. A kite at the end of 545 ft of string is 130 ft above the ground. What angle does the kite string make with the ground? 13.8°

Teaching Strategies

For Exercises 17–18, have students draw the triangles before attempting to solve the problems. Use Exercise 18 as an opportunity to point out that, although the lengths of the sides of this triangle are 9, 40, and 41, the sine, cosine, and tangent are ratios of lengths, and there are an infinite number of possible triangles with dimensions that yield these ratios.

606 Chapter 14 **Trigonometry**

NAVIGATION The table at the right shows measurements taken at six East Coast lighthouses. Use the table for Exercises 23–24.

Location of lighthouse	Height (ft)	Distance of boat from shore (ft)
Annisquam, MA	41	233
Cape May, NJ	170	964
Fenwick Island, DE	87	493
McCellanville, SC	150	851
Millbridge, ME	123	698
Scituate, MA	25	142

23. The navigator of a ship standing 360 ft offshore by the McClellanville lighthouse measures a 23° angle to the top of the lighthouse. Write cos 23° as a ratio. $\frac{12}{13}$

24. A boat pilot standing 1120 ft offshore measures an 8° angle to the top of a lighthouse. If $\cos 8° = \frac{112}{113}$, where is the lighthouse located? McCellanville

25. WRITING MATH Suppose the top of a lighthouse measured 10° from your position offshore. If you knew the height of the lighthouse, how could you find your distance from shore? Divide the height of the lighthouse by the ratio of height to distance (or tan 10°)

26. CHAPTER INVESTIGATION Your task is to have each member of your group build a homemade quadrant and use them to find the heights of various objects. Gather the materials you will need: one plastic straw or unsharpened pencil per person, string, a weight such as a metal washer or bolt, and tape. Make a photocopy of a protractor or draw a protractor on heavy cardstock. If you choose to draw your own protractor, make and label markings for every ten degrees. Check students' work.

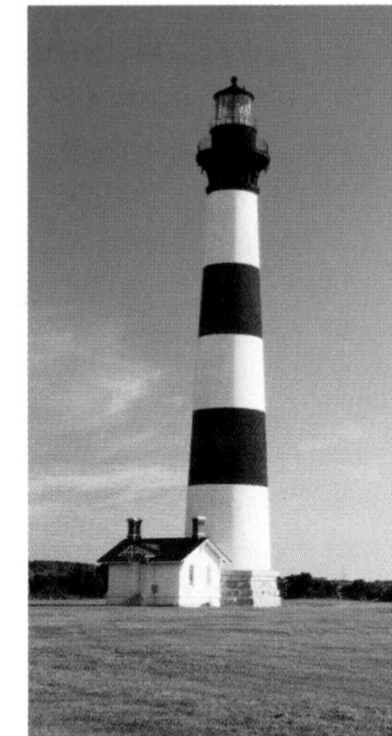

Extended Practice Exercises

Find each ratio.

27. tan 45° 1

28. cos 30° $\frac{\sqrt{3}}{2}$

29. cos 60° $\frac{1}{2}$

30. WRITING MATH True or false: In a right triangle, the sine of one acute angle equals the cosine of the other. Explain. True. In the figure, $\sin A = \frac{a}{c} = \cos B$.

Mixed Review Exercises

Write an equation for each circle. (Lesson 13-1)

31. radius 5 center (0, 0) $x^2 + y^2 = 25$

32. radius 8 center (4, 3) $(x - 4)^2 + (y - 3)^2 = 64$

33. radius 3.5 center (5, 1) $(x - 5)^2 + (y - 1)^2 = 12.25$

34. radius 10 center (−2, 0) $(x + 2)^2 + y^2 = 100$

35. radius 7 center (3, −2) $(x - 3)^2 + (y + 2)^2 = 49$

36. radius 4.7 center (−2, −2) $(x + 2)^2 + (y + 2)^2 = 22.09$

37. radius 6 center (−3, 5) $(x + 3)^2 + (y - 5)^2 = 36$

38. radius 2 center (2, −8) $(x - 2)^2 + (y + 8)^2 = y$

39. radius 7.5 center (−4, 3) $(x + 4)^2 + (y - 3)^2 = 56.25$

A bag contains 6 red marbles, 5 green marbles, 8 blue marbles and 1 white marble. Marbles are taken from the bag at random one at a time and not replaced. Find each probability. (Lesson 9-4)

40. *P*(red, then blue) $\frac{12}{95}$

41. *P*(green, then white) $\frac{1}{76}$

42. *P*(green, then blue, then red) $\frac{2}{57}$

43. *P*(white, then blue, then blue) $\frac{7}{855}$

Teaching Strategies

COOPERATIVE LEARNING Have students work in pairs to answer the following. In right triangle ABC, can each of the following be found, given the amount of information described? If the answer is yes, explain how it can be done.

1. Tan *A*, given hypotenuse *AB* = 15 cannot be found; too little information
2. m∠*A*, given hypotenuse *AB* = 15 and side *BC* = 11.8 Yes, use the Pythagorean Theorem to find side *AC* (9.3); then use the tangent ratio to find m∠*A* (52°).

Extra Practice Worksheet 14-1

Name ______ Date ______

EXTRA PRACTICE 14-1

BASIC TRIGONOMETRY RATIOS

EXERCISES

Use the figure at the right to find each ratio.

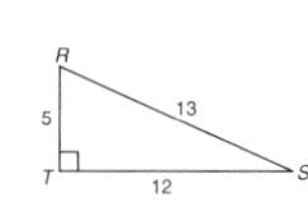

1. sin *R* $\frac{12}{13}$
2. cos *S* $\frac{12}{13}$
3. tan *R* $\frac{12}{5}$
4. sin *S* $\frac{5}{13}$
5. cos *R* $\frac{5}{13}$
6. tan *S* $\frac{5}{12}$

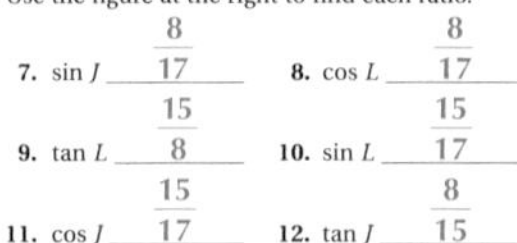

Use the figure at the right to find each ratio.

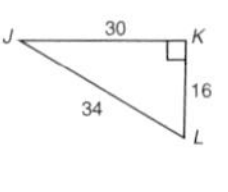

7. sin *J* $\frac{8}{17}$
8. cos *L* $\frac{8}{17}$
9. tan *L* $\frac{15}{8}$
10. sin *L* $\frac{15}{17}$
11. cos *J* $\frac{15}{17}$
12. tan *J* $\frac{8}{15}$

13. In ΔABC, ∠*C* is a right angle, *AB* = 14 and *AC* = 8. Find sin *A* and tan *B* to the nearest hundredth. 0.82; 0.70

14. In ΔABC, ∠*C* is a right angle, *BC* = 7 and *AB* = 13. Find cos *B* and tan *A* to the nearest hundredth. 0.54; 0.64

15. In right triangle *XYZ*, ∠*Z* is a right angle and $\tan X = \frac{8}{15}$. Write sin *X* and tan *Y* as ratios. $\frac{8}{17}$; $\frac{15}{8}$

16. In right triangle *XYZ*, ∠*Z* is a right angle and $\tan Y = \frac{60}{11}$. Write cos *X* and tan *X* as ratios. $\frac{60}{61}$; $\frac{11}{60}$

180 EXTRA PRACTICE *LESSON 14-1*

Enrichment Worksheet 14-1

Name ______ Date ______

ENRICHMENT 14-1

TRIG TERM PUZZLER

Use the clues below to complete the puzzle. When you have finished, the letters in the circles will spell the name of an important and useful part of mathematics.

Answer to the puzzle is "Pythagorean Theorem."

1. h y (p) o (t) e n u s e
2. (y-) a x i s
3. (t) a b l e s
4. h e i g (h) t
5. r i g (h) t
6. (a) r (e) a
7. t a n (g) e n t
8. c (o) s i n e
9. o p p (o) s i t e
10. (r) a t i o s
11. d (e) g (r) e e
12. s i n (e)
13. a d j (a) c e n t
14. t r i g o (n) o (m) e t r i c

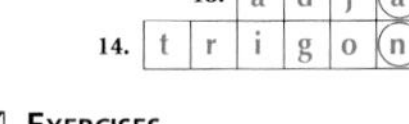

EXERCISES

1. Is never the shortest side
2. Usually is a vertical line on a coordinate grid
3. Before calculators, you needed these for trigonometry problems.
4. Needed to find the area of a triangle
5. Angle formed by perpendiculars
6. Number of units that cover a surface
7. Touching in one point
8. Adjacent side to hypotenuse
9. How the hypotenuse and the right angle are related
10. Two types are percents and cosines
11. Unit for measuring angles
12. Opposite side to hypotenuse
13. Denominator side for tangent
14. Sine, cosine and tangent

234 ENRICHMENT *LESSON 14-1*

Lesson Planning

NCTM Standards/Strands

- Number & Operation
- Patterns, Functions, & Algebra
- Geometry & Spatial Sense
- Problem Solving
- Reasoning and Proof
- Communication
- Connections
- Representation

Vocabulary

solving a right triangle
angle of elevation
angle of depression

Materials Needed

calculators

Lesson Resources

Warm-Up Transparency 40
Transparency RF-55
Reteaching 14-2
Extra Practice 14-2
Enrichment 14-2

Assignment Guide

Basic: 1–25, 31–36
Enriched: 1–36

Getting Started

5-Minute Warm-up

Right triangle *ABC* has hypotenuse *AB*, length 18; leg *AC*, length 14; and angle *A* equal to 39°. Write an equation to find the length of leg *BC*.
$BC = \sqrt{18^2 - 14^2}$; or $\sin 39° = \frac{BC}{18}$; or $\tan 39° = \frac{BC}{14}$

14-2 Solve Right Triangles

Goals ■ Find the lengths of sides and the measures of angles in right triangles.

Applications Surveying, Navigation, Safety

Work with a partner. Use the figures below to answer these questions.

a. How do you know the triangles are similar? They contain two pairs of corresponding congruent angles.

b. Write a proportion you could solve to find *x*. $\frac{1}{2} = \frac{x}{12}$

c. Solve the proportion for *x*. $x = 6$

d. Explain how you could solve the equation $\sin 30° = \frac{x}{16}$ for *x*. Write sin 30° as $\frac{1}{2}$, then multiply $\frac{1}{2}$ by 16; $x = 8$.

e. Solve for *x*: $\sin 40° = \frac{x}{16}$. Explain how you solved the equation. $\sin 40° \approx 0.6428$, $x = 16(0.6428) = 10.2848$

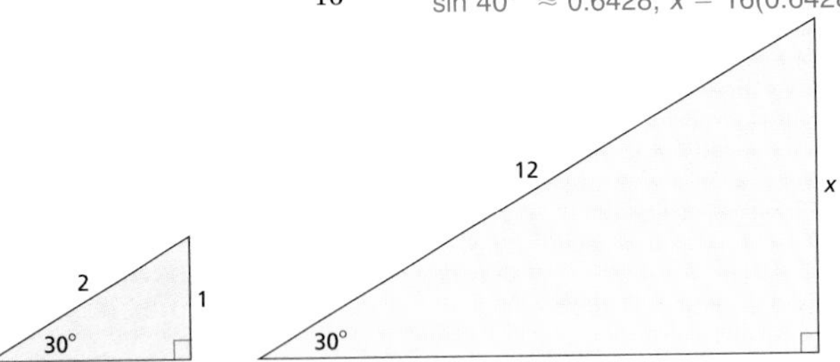

Build Understanding

You can find the measures of missing parts of a right triangle.

If you know the measure of one acute angle, you can find the measure of the other by subtracting the measure of the known angle from 90°.

If you know the lengths of two sides, you can use the Pythagorean Theorem to find the length of the third side.

If you know the measure of an angle and the length of a side, you can use trigonometric ratios and the first two rules to find the other parts of the triangle.

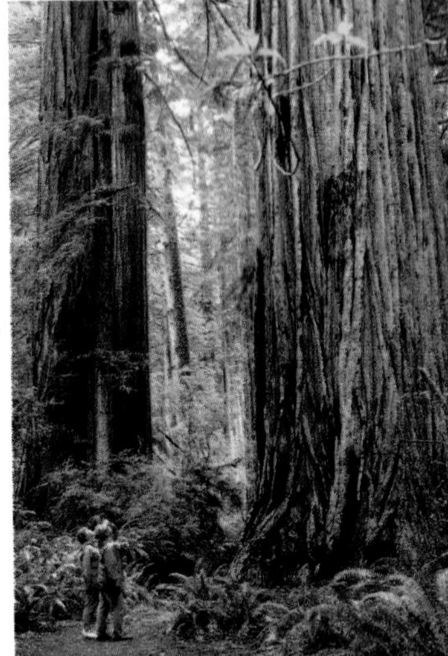

Example 1

Find the following in △*PQR*.

a. *PR* to the nearest tenth

b. m∠*P*

c. *RQ* to the nearest tenth

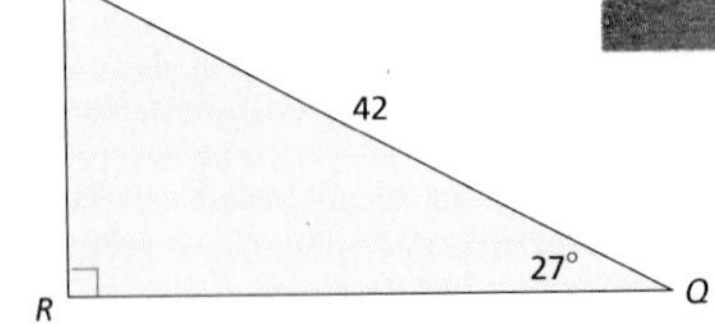

Introduction to Lesson 14-2

After students have answered Parts (a)–(e), ask these questions.

1. What other proportion could they have written for Part (b)? $\frac{1}{x} = \frac{2}{12}$
2. In Part (d), how do you know the value of sin 30° is $\frac{1}{2}$? Find the value of sin 30° in the small triangle.

After completing Part (e), ask students how they could find the missing length in either triangle. Use either sin 60° or the Pythagorean Theorem.

Teaching Strategies

CHALLENGE The angle of elevation of the treetop shown in the figure is 15° from point A. What is the length of the tree from point B to its top, point C, if the line of sight distance from A to C is 45 ft and the tree makes a 40° angle with the ground? 18 ft

C
45 ft
15°
40°
A
B

Solution

a. Decide which trigonometric ratio relates the known side PQ, the unknown side PR, and the known acute angle $\angle Q$. Think: PQ is the *hypotenuse*. PR is *opposite* $\angle Q$. The ratio that relates the hypotenuse and the side opposite an angle is the *sine*.

$$\sin Q = \frac{PR}{PQ}$$

$$\sin 27° = \frac{PR}{42}$$

$$0.4540 = \frac{PR}{42} \quad \text{calculator approximation of } \sin 27°$$

$$PR = 42(0.4540) \approx 19.1$$

b. $m\angle P = 90° - 27° = 63°$

c. Use the Pythagorean Theorem to find RQ.

$$(PR)^2 + (RQ)^2 = (PQ)^2$$
$$(19.1)^2 + (RQ)^2 = (42)^2$$
$$364.81 + (RQ)^2 = 1{,}764$$
$$(RQ)^2 = 1{,}399.19$$
$$RQ \approx 37.4$$

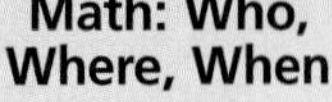

Math: Who, Where, When

Hipparchus, a Greek mathematician born about 160 B.C., is generally considered to be the creator of trigonometry. He was the first person to draw up a table of values for the sine, cosine, and tangent ratios.

Such a table of trigonometric values can be found in the Data File in the back of this book. You can use that table if a calculator is not available.

Finding the measures of all parts of a right triangle is called **solving a right triangle.**

Trigonometry problems often involve angles of depression and elevation.

An **angle of depression** is formed by a horizontal line and a line slanting downward.

An **angle of elevation** is formed by a horizontal line and a line slanting upward.

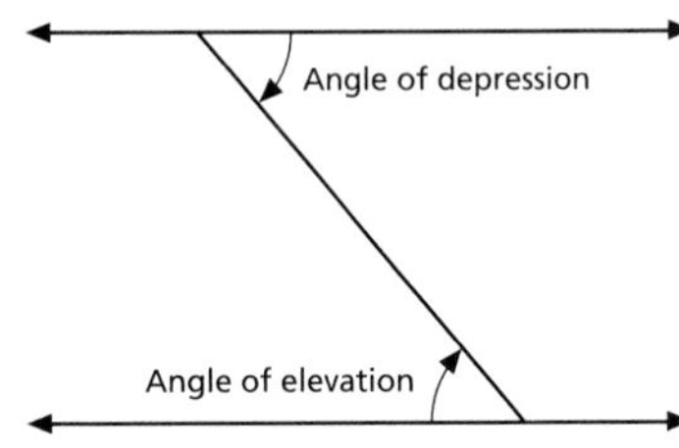

Example 2

SURVEYING A surveyor 550 ft from the base of the Howard Libbey redwood, the world's tallest tree, in Humboldt County, CA. The tree is 362 ft tall. Find the angle of elevation of the top of the tree from the spot where the surveyor is standing.

Solution

The angle of elevation is $\angle A$, formed by the horizontal line of the ground and the line slanting to the top of the tree. BC is opposite $\angle A$, and AC is adjacent to $\angle A$. The trigonometric ratio relating opposite and adjacent is the tangent.

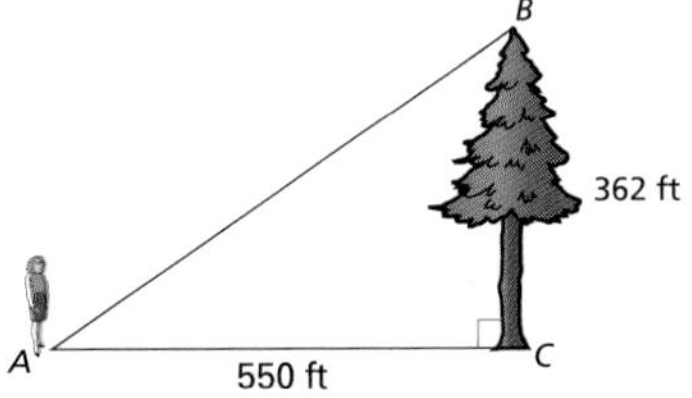

$$\tan \angle A = \frac{362}{550} \approx 0.6582$$

$$\tan \angle A \approx 33.4°$$

The angle of elevation is approximately 33.4°.

Chalkboard Examples

Supplementary Example 1

Right triangle QRS has the following measures: Hypotenuse QR has length 43; $m\angle R = 58°$. Find the following in triangle QRS. Round to the nearest tenth.

a. RS to the nearest tenth 22.8

b. QS to the nearest tenth 36.5

c. $m\angle Q$ 32°

Supplementary Example 2

The angle of elevation from a point on the ground to the top of a tree 48 feet away is 51°. How tall is the tree? 59.3 ft

Teaching Strategies

Refer to Example 2. Be sure students understand that if the problem had asked for the angle of depression of the point to the top of the tree, the answer would be the same as in the example. Ask students what geometric principle accounts for this. Parallel lines cut by a transversal create congruent alternate interior angles.

Reteaching Worksheet 14-2

Name ______________________ Date ____________

RETEACHING 14-2

SOLVING RIGHT TRIANGLES

A right triangle is solved by finding the measures of all its angles and sides. You can find the missing parts of a right triangle by using trigonometric ratios (sine, cosine, tangent) along with the Pythagorean Theorem and the sum of the angles property.

$$\text{sine} = \frac{\text{length of opposite side}}{\text{length of hypotenuse}} \qquad \text{cosine} = \frac{\text{length of adjacent side}}{\text{length of hypotenuse}}$$

$$\text{tangent} = \frac{\text{length of opposite side}}{\text{length of adjacent side}}$$

Example

Find the following in $\triangle ABC$.

a. $m\angle B$

b. $m\angle C$

c. AC to the nearest tenth

A
6 m
B
C
17 m

Solution

a. Determine which trigonmetric ratio relates the length of the known side ($\overline{AB}$ and $\overline{CB}$) and the unknown acute angle ($\angle B$). The ratio is the cosine.

$$\cos B = \frac{\text{length of side adjacent to } \angle B}{\text{length of hypotenuse}} = \frac{AB}{BC} = \frac{6}{17} \approx 0.3529$$

Use your calculator or the Table of Trigonometric Ratios to find the measure of the angle with a cosine value of about 0.3529.
The $m\angle B \approx 69°$.

b. $m\angle C = 180° - (90° + 69°) \approx 21°$. So the measure of $\angle C$ is about 21°.

c. Use the Pythagorean Theorem to find AC: $(AC)^2 + (AB)^2 = (CB)^2$.

$$(AC)^2 + 6^2 = 17^2$$
$$(AC)^2 + 36 = 289$$
$$(AC)^2 = 253$$
$$AC \approx 15.9$$

EXERCISES

For each triangle, find the measures of line segments to the nearest tenth and angles to the nearest degree.

1. DF 6.7 ft
$m\angle D$ 42°
$m\angle E$ 48°

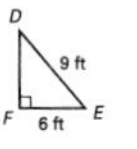

2. $m\angle H$ 35°
GI 2.9 cm
GH 4.1 cm

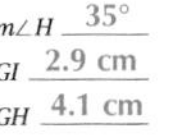

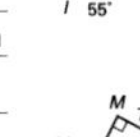

3. JK 7.3 yd
JL 2.1 yd
$m\angle K$ 17°

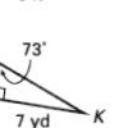

4. $m\angle O$ 27°
$m\angle N$ 63°
ON 15.7 m

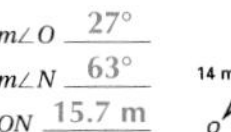

208 RETEACHING *LESSON 14-2*

Lesson Wrap-up

Quick Assessment

Ask the following questions to determine if students understand the content presented in this lesson.

Students may use a trigonometric values table to answer the following questions.

1. If the tangent of an angle of a right triangle is 1, what do you know about the triangle? It is an isosceles right triangle.
2. If the cosine of an angle of a right triangle is less then 0.5, what do you know about the angle? The angle is greater then 60°.
3. If the sine of an angle is equal to 1, what kind of angle must it be? a right angle

Assignment Guide

Basic: 1–25, 31–36
Advanced: 1–36
Additional Practice: See Extra Practice Index on page 674.

Try These Exercises

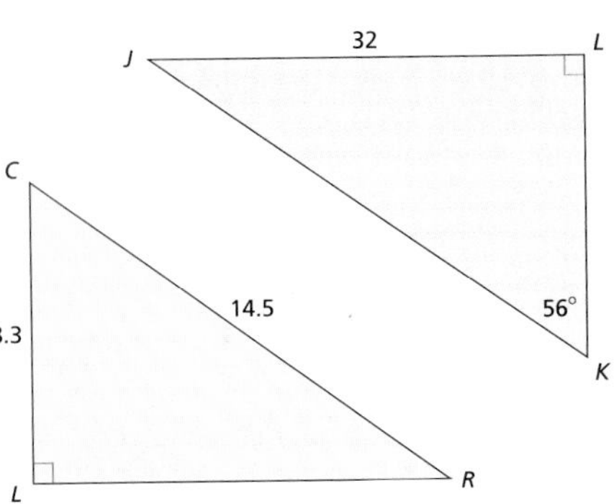

Find the following in $\triangle JKL$.

1. LK to the nearest tenth 21.6
2. JK to the nearest tenth 38.6
3. $m\angle J$ 34°

Find the following in $\triangle CRL$.

4. LR to the nearest tenth 11.9
5. $m\angle C$ to the nearest degree 55°
6. $m\angle R$ to the nearest degree 35°
7. **NAVIGATION** A ship's sonar detects a submarine 880 ft below a point on the ocean's surface 1450 ft dead ahead of the ship. To the nearest degree, find the angle of depression of the submarine. 31°

8. **SAFETY** Safety experts recommend that a ladder be placed at an angle of about 75° to the ground. Meg is using a 15-ft ladder. How far from the base of the wall should she place the foot of the ladder? Round the distance to the nearest tenth ft. 3.9 ft

Practice Exercises

Find line segments to the nearest tenth and angles to the nearest degree.

CALCULATOR Use a calculator to find the following in $\triangle ABC$.

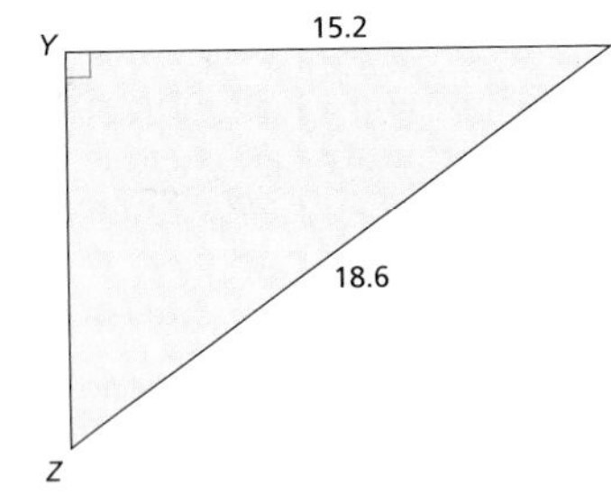

9. BC 3.0
10. $m\angle B$ 53°

Find the following in $\triangle XYZ$.

11. YZ 10.7
12. $m\angle Z$ 55°
13. $m\angle X$ 35°

Y 15.2 X
18.6
Z

14. Find the angle of elevation of the top of the 1250-ft Empire State Building from a point 850 ft from the base. 56°
15. The two legs of a right triangle measure 23.5 and 27.9. Solve the triangle. hypotenuse: 36.5; angles: 40°, 50°, 90°
16. In a right triangle, the leg adjacent to a 77° angle has a length of 87. Solve the triangle. angles: 13°, 90°; legs: 376.8; hypotenuse: 386.7
17. **AIR TRAFFIC CONTROL** From an airport runway, the angle of elevation of an approaching plane is 12.8°. If the plane's altitude is 2400 ft, how far is the plane from the runway? 10,832.8 ft

Extend the Lesson

MATH JOURNAL Have students sketch a right triangle in their journals, and depending on the shape, ask them to name the measure of one angle other than the right angle and the length of one side. Then have them describe the steps they would take to solve the triangle.

18. NAVIGATION From the top of a cliff, the angle of depression of a ship at sea is 8.8°. If the direct-line distance from the clifftop to the ship is 2.3 mi, how high is the cliff? 0.4 mi

19. BOATING The foot of a right-triangular sail is 64 in. long. The angle at the top of the sail measures 23°. Find the length of the luff of the sail. 150.8 in.

luff

foot

20. FOREST MANAGEMENT A ranger in a fire tower spots a fire at an angle of depression of 4°. The tower is 36 m tall. How far from the base of the tower is the fire? 514.8 m

21. A silo casts a shadow 42 ft long. The angle of the sun is 38°. How tall is the silo? 32.8 ft

22. DATA FILE Use the Data Index on pages 632–633 to find the location of information on highest and lowest continental altitudes. A ship's navigator at sea level in Cook's Inlet, Alaska, sights the summit of Mount McKinley at a 6.8° angle of elevation. How many miles is it from the ship to the summit of the mountain? 32.5 mi

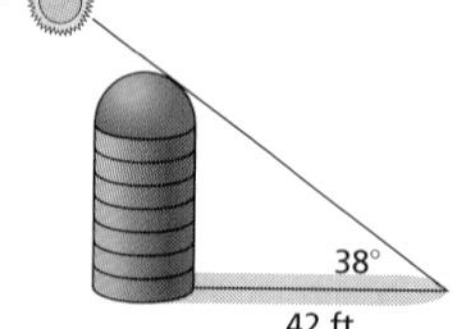

23. FLIGHT A helicopter ascends 150 ft vertically, then flies horizontally 420 ft. Find the angle of elevation of the helicopter as seen by an observer at the takeoff point. 20°

24. COMMUNICATIONS Orlando and Ryan are taking measurements related to the installation of a TV tower. Orlando measures a 62° angle of elevation to the top of the 950-ft TV tower. Find the angle of elevation for Ryan, standing 80 ft farther from the tower than Orlando. 58°

25. CHAPTER INVESTIGATION Assemble the quadrant in the following manner: Tape the protractor image to the straw (or pencil), aligning the flat edge of the protractor with the length of the straw. Tie the weight to one end of a 8-in. length of string. Secure the other end of the string to the straw or pencil at the center point of the protractor base. The string should be able to move freely. As you sight an object through the straw, the string will indicate the angle of ascent or descent on the protractor scale.
Check students' work.

Extended Practice Exercises

Can you solve a right triangle from the given information? Answer *yes, no,* or *sometimes.*

26. two sides yes

27. three angles no

28. one side no

29. one leg and one angle sometimes; the angle must be acute

30. A pike is directly beneath a trout in a lake. The direct-line distance from an angler to the trout is 35 ft. The angle of depression of the trout is 20°. The angler's direct-line distance to the pike is 42 ft. The angle of depression of the pike is 24°. How far below the trout is the pike? 5.1 ft

Mixed Review Exercises

Find the focus and directrix of each equation. (Lesson 13-2)

31. $x^2 = 4y$ $(0, 1);\ y = -1$

32. $x^2 = -5y$ $\left(0, -\frac{5}{4}\right);\ y = \frac{5}{4}$

33. $x^2 = 8y$ $(0, 2);\ y = -2$

34. $x^2 - 6y = 0$ $\left(0, \frac{3}{2}\right);\ y = -\frac{3}{2}$

35. $x^2 - 10y = 0$ $\left(0, \frac{5}{2}\right);\ y = -\frac{5}{2}$

36. $x^2 = 18y$ $\left(0, \frac{9}{2}\right);\ y = -\frac{9}{2}$

Extra Practice Worksheet 14-2

Name ______ Date ______

EXTRA PRACTICE 14-2

SOLVE RIGHT TRIANGLES

EXERCISES

For all exercises on this page, find the length of all line segments to the nearest tenth and all angles to the nearest degree.

Find the following in ΔABC.

1. BC 6.0
2. $m\angle A$ 30°
3. $m\angle B$ 60°

Find the following in ΔXYZ.

4. $m\angle X$ 48°
5. YZ 17.8
6. ZX 23.9

Find the following in ΔMNP.

7. $m\angle M$ 64°
8. MP 6.0
9. NM 13.6

10. The two legs of a right triangle measure 24.2 and 36.4. Solve the triangle. 43.7; 56°; 34°

11. In a right triangle, a leg measures 9.6 and the hypotenuse measures 15.2. Solve the triangle. 11.8; 51°; 39°

12. In a right triangle, the leg adjacent to a 56° angle has a length of 24. Solve the triangle. 34°; 35.6; 42.9

13. In a right triangle, the leg opposite a 32° angle has a length of 18. Solve the triangle. 58°; 28.8; 34.0

14. From the top of a building, the angle of depression of a car is 5.2°. If the direct-line distance from the top of the building to the car is 1500 ft, how tall is the building? 136.0 ft

182 EXTRA PRACTICE LESSON 14-2

Enrichment Worksheet 14-2

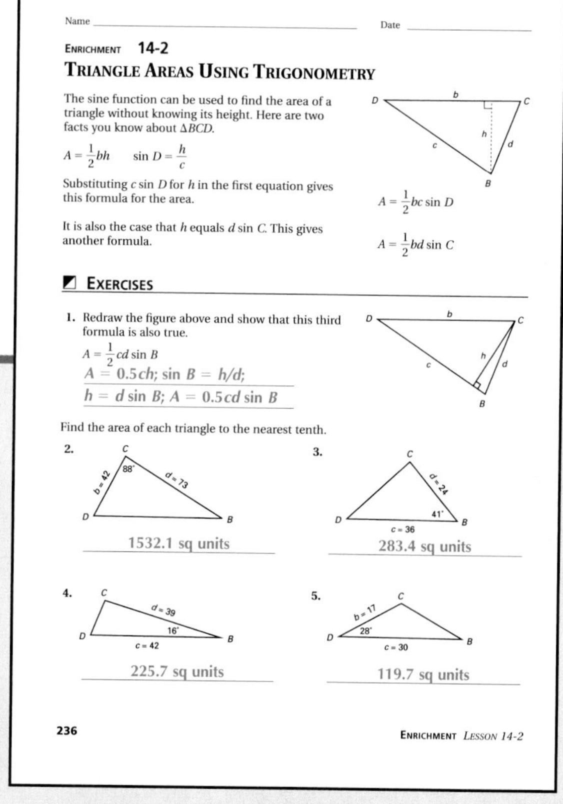

Name ______ Date ______

ENRICHMENT 14-2

TRIANGLE AREAS USING TRIGONOMETRY

The sine function can be used to find the area of a triangle without knowing its height. Here are two facts you know about ΔBCD.

$A = \frac{1}{2}bh$ $\sin D = \frac{h}{c}$

Substituting $c \sin D$ for h in the first equation gives this formula for the area. $A = \frac{1}{2}bc \sin D$

It is also the case that h equals $d \sin C$. This gives another formula. $A = \frac{1}{2}bd \sin C$

EXERCISES

1. Redraw the figure above and show that this third formula is also true.
$A = \frac{1}{2}cd \sin B$
$A = 0.5ch;\ \sin B = h/d;$
$h = d \sin B;\ A = 0.5cd \sin B$

Find the area of each triangle to the nearest tenth.

2. 1532.1 sq units
3. 283.4 sq units
4. 225.7 sq units
5. 119.7 sq units

236 ENRICHMENT LESSON 14-2

Teaching Strategies

COOPERATIVE LEARNING Have students work in pairs to solve this problem. A gift box in the shape of a cube has an edge 18 in. Can an umbrella 32 in. in length be packed in the box? No; to the nearest tenth, the greatest length in the cube is the diagonal, 31.2 in.

Skills Practice

Vocabulary Review

Lesson 14-1
trigonometric ratios sine
cosine tangent

Lesson 14-2
solving a right triangle
angle of depression
angle of elevation

Assignment Guide

All students: 1–35
Additional Practice: Refer to the Extra Practice Index on page 674 of this text.

Additional Answers

14. The Pythagorean Theorem deals with the relationship of the lengths of the sides in a right triangle. Knowing two of the three measures, you can use this Theorem to find the third length.
15. The sum of the angles of a triangle is 180°. If one angle is a right angle, you know it is 90°. Therefore, the other two angles must equal 180°–90°. If you know one of the acute angles you can use this information to find the missing measure.

Review and Practice Your Skills

Practice ■ Lesson 14-1

1. What must be true about a triangle in order to apply trigonometric ratios? It must be a right triangle.

Determine if each statement is true or false.

2. The sine is calculated by dividing the length of the opposite side by the length of the hypotenuse. true
3. The tangent is calculated by dividing the length of the hypotenuse by the length of the side adjacent. false
4. The cosine is calculated by dividing the length of the side adjacent by the length of the hypotenuse. true

Use the figure at the right to find each ratio. Express answers in lowest terms.

C, 10, 6, A, 8, B

5. tan A $\frac{3}{4}$
6. sin C $\frac{4}{5}$
7. cos C $\frac{3}{5}$
8. tan C $\frac{4}{3}$
9. sin A $\frac{3}{5}$
10. cos A $\frac{4}{5}$

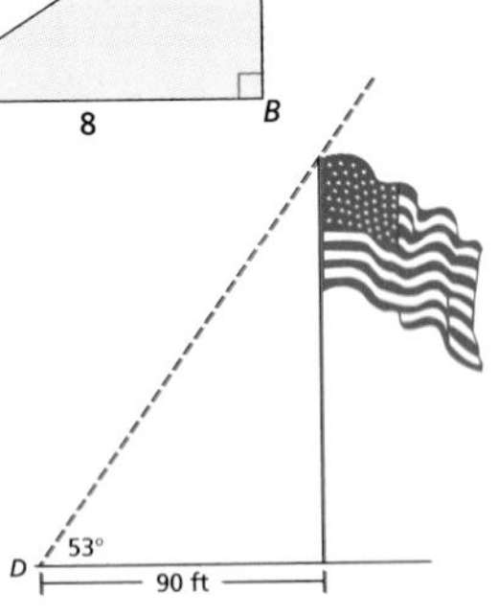

11. A 120 foot flagpole casts a shadow of 90 feet when the sun is at a 53° angle from the horizontal. Write sin D as a ratio. $\frac{4}{5}$
12. In $\triangle ABC$, $\angle C$ is a right angle, $AB = 18$ and $AC = 12$. Find sin A and tan B to the nearest hundredth. sin A = 0.75; tan B = 0.89
13. In $\triangle RST$, $\angle T$ is a right angle, $RS = 10$, and $ST = 5$. Find cos S and tan R to the nearest hundredth. cos S = 0.50; tan R = 0.58

Practice ■ Lesson 14-2

For 14–15, see additional answers.

14. Explain how you can use the Pythagorean Theorem to find the length of a side of a right triangle if you know the lengths of two sides.
15. Explain how to find the measure of an unknown acute angle in a right triangle.

Find the following in $\triangle NOP$.

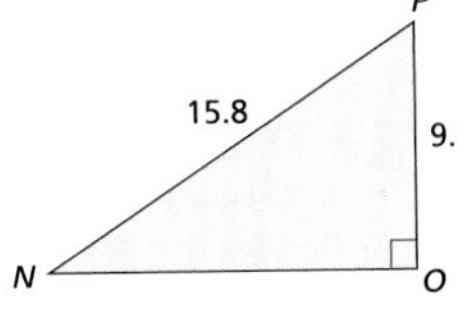

16. NO to the nearest tenth 12.8
17. m$\angle N$ to the nearest degree 36°
18. m$\angle P$ to the nearest degree 54°

Find the following in $\triangle RST$.

T, 48, 30°, R, S

19. RT to the nearest tenth 24.0
20. RS to the nearest tenth 41.6
21. m$\angle T$ to the nearest degree 60°
22. To the nearest tenth, find the angle of elevation of the top of the 1,454-ft Sears Tower from a point 750 feet from the base. 62.7°

Teaching Strategies

Discuss the table of trigonometric values. Point out that sin 90° is given as 1, but no value is shown for tan 90°. Ask students to explain these entries in terms of the definitions of sine and tangent. In a unit circle, $\sin A = \frac{\text{side opposite } A}{\text{hypotenuse}}$. At 90°, the side opposite A would equal the hypotenuse. Thus, $\frac{\text{hypotenuse}}{\text{hypotenuse}} = 1$. Tan $A = \frac{(\text{side opposite})}{(\text{side adjacent})}$. As A approaches 90°, the side opposite approaches a limit of 1. The side adjacent approaches a limit of zero. A fraction with a denominator of zero is undefined. Thus, there is no value for the tangent of 90°.

612 Chapter 14 Trigonometry

PRACTICE ■ LESSON 14-1–LESSON 14-2

In the figure at the right, use the Pythagorean Theorem to find the missing measure to the nearest hundredth. Then find each ratio. (Lesson 14-1)

23. sin L $\frac{4}{7}$ **24.** cos L $\frac{11.49}{14}$

25. cos M $\frac{4}{7}$ **26.** tan L $\frac{8}{11.49}$

27. sin M $\frac{11.49}{14}$ **28.** tan M $\frac{11.49}{8}$

29. In $\triangle DEF$, $\angle F$ is a right angle, $DE = 30$, and $EF = 24$. Find sin D and tan E to the nearest hundredth. sin $D = 0.8$; tan $E = 0.75$

30. In $\triangle LMN$, $\angle N$ is a right angle, $LM = 84$, $MN = 62$. Find cos L and tan M to the nearest hundredth. cos $L = 0.67$; tan $M = 0.91$

Find the following in $\triangle ABC$: (Lesson 14-2)

31. m$\angle C$ to the nearest degree 66°

32. BC to the nearest tenth 14.6

33. AB to the nearest tenth 32.9

34. Find the angle of elevation to the top of a 150-ft flagpole from a point 45 feet from the base of the pole. 73.3°

35. The direct-line distance from the top of the slope to the ski lodge is 2,500 ft. The top of the slope is 1,050 ft above the level of the lodge. What is the angle of depression from the top of the slope to the lodge? 24.8

Mid-Chapter Quiz

Solve.

1. A ladder on a fire truck is 75 ft long. If it makes a 45° angle with a building, what is the greatest height the ladder can reach up the side of the building? 53 ft

2. When viewed from a distance of 32 ft along the ground, the top of a flagpole can be seen at an angle of 39°. What is the height of the flagpole? 26 ft

3. Right triangle ABC has hypotenuse AC. Right triangle DEF with hypotenuse DF is similar to ABC, and angle D corresponds to angle A. If DE measures 15 and EF measures 20, what is the value of tan A? 1.33

4. An office worker on the fourteenth floor of a building sights a friend on the street. The angle of depression is 35°, and the fourteenth floor is 135 ft in the air. How far is the friend from the building? 192.8 ft

5. The pilot of a plane flying east sights another plane ahead of him at an angle of elevation of 18°. The line of sight distance between the planes is 1850 m. At how much greater altitude is the lead plane than the trailing one? 571.7 m

Chalkboard Examples

Example from Lesson 14-1

A plane flying at 1,600 ft above the ground is 14,800 ft from the beginning of a runway. At what angle above the horizontal runway would a traffic controller even with the beginning of the runway have to look to see the plane? 6.2°

Example from Lesson 14-2

Solve right triangle *CFE* having the following measures: Leg *FE* has length 19.9; leg *CE* has length 12.8. Round to the nearest tenth.
m$\angle C = 57°$; m$\angle F = 33°$; $CF = 23.7$

Mid-Chapter Quiz

Correlation Chart

Question Number	Lesson Number
1–3	14-1
4–5	14-2

Teaching Strategies

Students should use scientific calculators whenever possible. You may wish to have students use more than one trigonometric function to find the same solution. For example, if one acute angle and one leg are known, the other leg can be found by using the tangent of the known angle or the tangent of the unknown acute angle, after it is determined.

Lesson Planning

NCTM Standards/Strands

- Number & Operation
- Patterns, Functions, & Algebra
- Geometry & Spatial Sense
- Problem Solving
- Reasoning and Proof
- Communication
- Connections
- Representation

Vocabulary

initial side
reference angle
terminal side

Materials Needed

graphing calculators

Lesson Resources

Warm-Up Transparency 40
Reteaching 14-3
Extra Practice 14-3
Enrichment 14-3
Transparency RF-56

Assignment Guide

Basic: 1–42, 49–54
Enriched: 1–54

Getting Started

5-Minute Warm-up

In right triangle *ABC*, with right angle *C*, angle *A* = 40°.

1. What is the measure of angle *B*? 50°
2. What is the value of sin *A*? 0.6428
3. What is the value of tan *A*? 0.8391

Introduction to Lesson 14-3

Before using calculators, review the concept of angles as rotations for angles between 0° and 360°. Have students name the angle produced by each of the following fractions of a counterclockwise rotation about the origin, beginning on the *x*-axis: $\frac{1}{8}, \frac{1}{4}, \frac{3}{8}$, and $\frac{1}{2}$. 45°; 90°; 135°; 180°

14-3 Graph the Sine Function

Goals ■ Solve problems using trigonometry.
Applications Communications, Population, Flight

Work with a partner. You will need a calculator.

1. Choose several angles with measures in each of the given ranges below.
   a. 90°–180° **b.** 180°–270° **c.** 270°–360°
2. Draw a grid like the one shown below. Graph your results of parts **a**, **b**, and **c**.

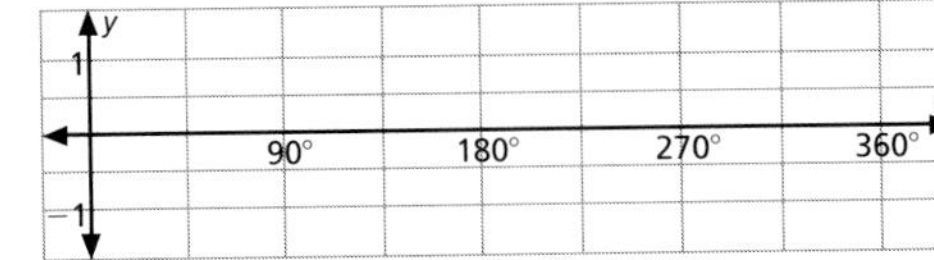

3. Study each range of angles. Describe any patterns you observe in the signs (+ and −) of the sine ratio and the cosine ratio.
   a: sine +, cosine −; b: sine −, cosine −; c: sine −, cosine +

Build Understanding

Until now, you have dealt only with acute angles in your work with trigonometric ratios. In this lesson, you will learn to find trigonometric ratios of angles with measures greater than 90°. To solve these problems, you will need to apply the relationships that hold in 30°–60°–90° and 45°–45°–90° right triangles.

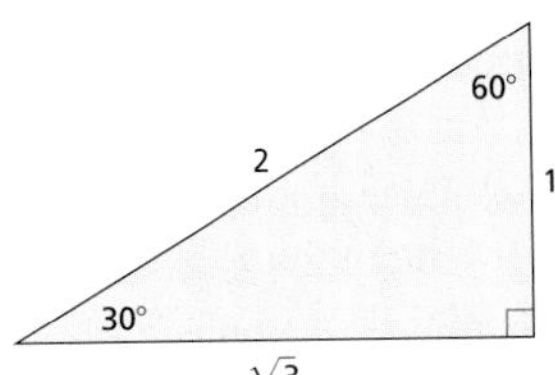

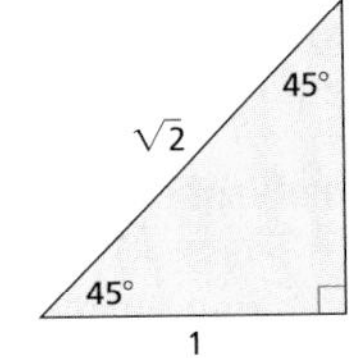

Example 1

Find sin 240°.

Solution

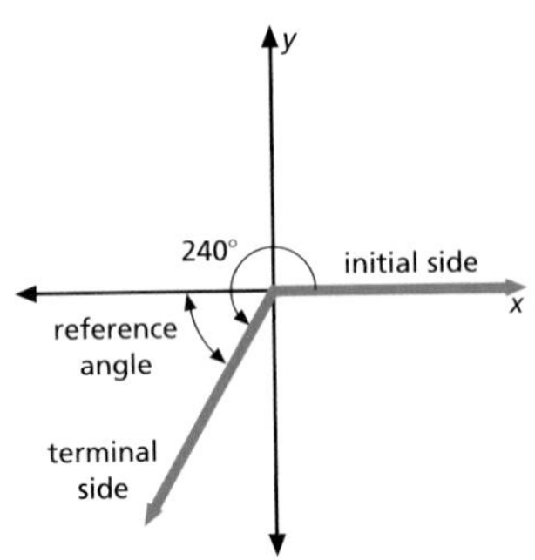

Sketch the angle on a coordinate plane. Use the positive *x*-axis as the **initial side**, and the ray resulting from a 240° counterclockwise rotation of the positive *x*-axis as the **terminal side**. The **reference angle** is the acute angle formed by the *x*-axis and the terminal side.

Teaching Strategies

COOPERATIVE LEARNING Have students work in pairs to solve this problem: *For any angle A, there is a relationship between its tangent and its sine and cosine.* Determine the relationship and explain its derivation.

$\tan A = \frac{\sin A}{\cos A}$

$$\frac{\text{opposite}}{\text{adjacent}} = \frac{\frac{\text{opposite}}{\text{hypotenuse}}}{\frac{\text{adjacent}}{\text{hypotenuse}}};\ \frac{\text{opposite}}{\text{hypotenuse}} \leftrightarrow \frac{\text{hypotenuse}}{\text{adjacent}} = \frac{\text{opposite}}{\text{adjacent}}$$

The reference angle measures $240° - 180° = 60°$. Complete a triangle with the terminal side as hypotenuse by drawing a perpendicular to the x-axis. The triangle is a 30°–60°–90° right triangle. In this example, the leg lengths are negative because the legs were drawn by moving in negative directions from the x- and y-axes. The length of the terminal side is always considered to be positive.

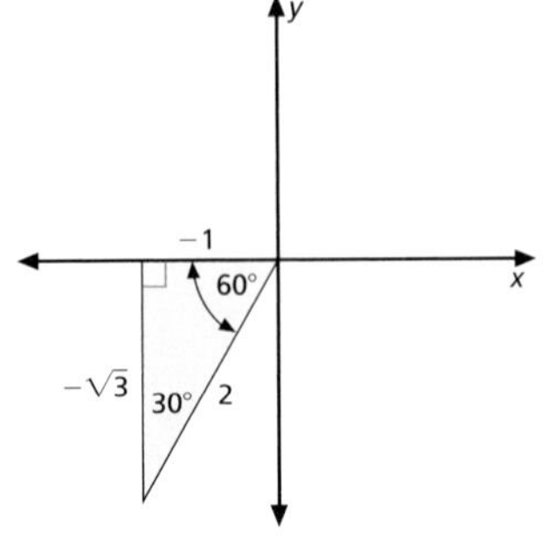

To find sin 240°, find the sine of the reference angle.

$$\sin 240° = \frac{\text{opposite}}{\text{hypotenuse}} = \frac{-\sqrt{3}}{2}$$

Technical Note

Use your calculator to check trigonometric ratios that you find using reference angles. In Example 1, $\frac{-\sqrt{3}}{2} \approx -0.8660$. Check using the **sin** key:

240 **sin** **=**

−0.8660254

Example 2

Find sin 495°.

Solution

To form an angle of 495°, the initial side must complete a 360° rotation, then continue an additional 135°. The reference angle measures $180° - 135° = 45°$. The triangle formed is a 45°–45°–90° right triangle. The leg adjacent to the 45° angle measures -1 relative to the x-axis. The leg opposite the angle measures $+1$ relative to the y-axis. The terminal side is always positive.

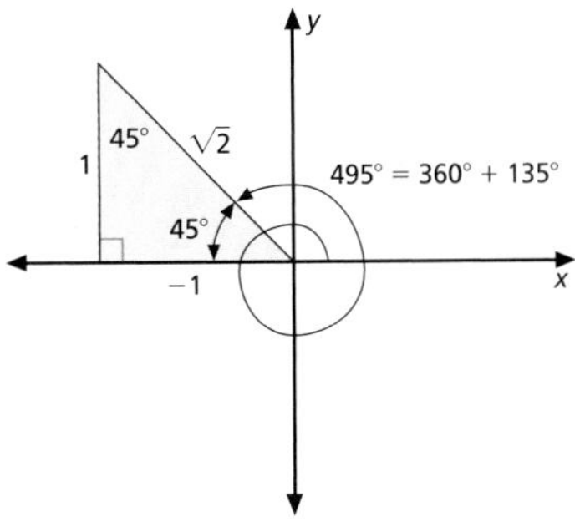

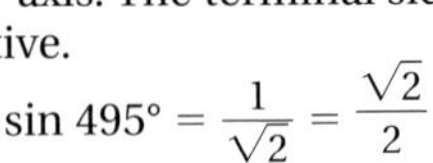

$$\sin 495° = \frac{1}{\sqrt{2}} = \frac{\sqrt{2}}{2}$$

Example 3

COMMUNICATIONS A researcher is developing new technology to assist ships at sea to send urgent communications. Ships will use various tones and patterns to send messages. The electrical impulses produced by the tones are modeled by sine curves. Graph the sine curve $y = \sin x$ for $0° \leq x \leq 360°$.

Solution

Use your calculator to make a table of ordered pairs.

x	0	30	45	60	90	120	135	150	180	210	225	240	270	300	315	330	360
$\sin x$	0	.50	.71	.87	1.0	.87	.71	.50	0	−.50	−.71	−.87	−1.0	−.87	−.71	−.50	0

Graph the points using $\sin x$ as the y-coordinate. Draw a smooth curve through the points. The graph is called a **sine curve**.

Chalkboard Examples

Supplementary Example 1

Find sin 210° by drawing a reference angle. $-\frac{1}{2}$

Supplementary Example 2

Find cos 315° by drawing a reference angle. $\frac{\sqrt{2}}{2}$

Supplementary Example 3

Another class of ships will use electrical impulses modeled by sin curves with a range $y = \sin x$ for $0° \leq x \leq 180°$. Graph the sine curve. Check students' work. If calculators are used, the Xmin should be set to 0 and the Xmax to 180.

Teaching Strategies

All graphs of the equation y = sin x are congruent. However, graphing calculators may display seemingly dissimilar graphs. Remind students to check the range and scale settings on their calculators if there appear to be discrepancies.

Reteaching Worksheet 14-3

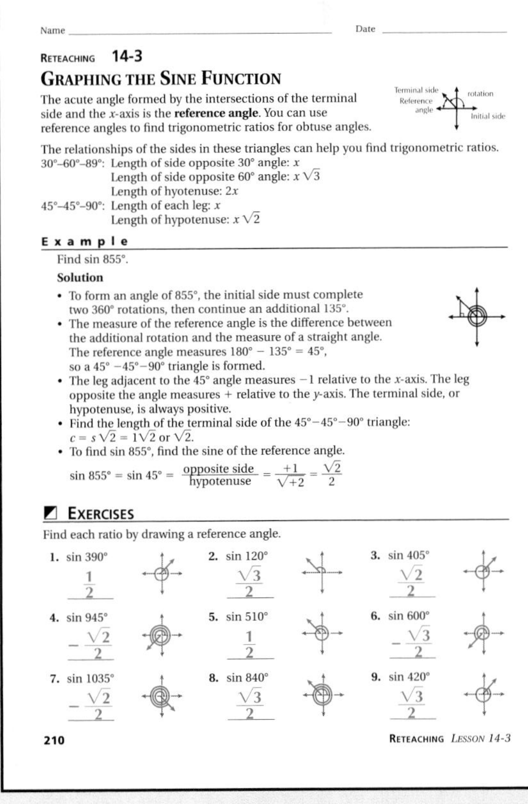

Name ______________________ Date ____________

RETEACHING **14-3**

GRAPHING THE SINE FUNCTION

The acute angle formed by the intersections of the terminal side and the x-axis is the **reference angle**. You can use reference angles to find trigonometric ratios for obtuse angles.

The relationships of the sides in these triangles can help you find trigonometric ratios.

30°–60°–89°: Length of side opposite 30° angle: x
Length of side opposite 60° angle: $x\sqrt{3}$
Length of hyotenuse: $2x$

45°–45°–90°: Length of each leg: x
Length of hypotenuse: $x\sqrt{2}$

Example

Find sin 855°.

Solution

- To form an angle of 855°, the initial side must complete two 360° rotations, then continue an additional 135°.
- The measure of the reference angle is the difference between the additional rotation and the measure of a straight angle. The reference angle measures $180° - 135° = 45°$, so a $45° - 45° - 90°$ triangle is formed.
- The leg adjacent to the 45° angle measures -1 relative to the x-axis. The leg opposite the angle measures + relative to the y-axis. The terminal side, or hypotenuse, is always positive.
- Find the length of the terminal side of the $45° - 45° - 90°$ triangle: $c = s\sqrt{2} = 1\sqrt{2}$ or $\sqrt{2}$.
- To find sin 855°, find the sine of the reference angle.

$$\sin 855° = \sin 45° = \frac{\text{opposite side}}{\text{hypotenuse}} = \frac{+1}{\sqrt{+2}} = \frac{\sqrt{2}}{2}$$

EXERCISES

Find each ratio by drawing a reference angle.

1. sin 390° $\frac{1}{2}$
2. sin 120° $\frac{\sqrt{3}}{2}$
3. sin 405° $\frac{\sqrt{2}}{2}$
4. sin 945° $-\frac{\sqrt{2}}{2}$
5. sin 510° $\frac{1}{2}$
6. sin 600° $-\frac{\sqrt{3}}{2}$
7. sin 1035° $-\frac{\sqrt{2}}{2}$
8. sin 840° $\frac{\sqrt{3}}{2}$
9. sin 420° $\frac{\sqrt{3}}{2}$

210 RETEACHING *LESSON 14-3*

Lesson Wrap-up

Quick Assessment

Ask the following questions to determine if students understand the content presented in this lesson.

1. From the graph of $y = \sin x$, what do you predict would be the next angle beyond 360° whose sine would have a value of 0? 540°
2. What is the next angle greater than 90° whose sine is 1? 450°
3. What is the next angle greater than 270° whose sine is -1? 630°
4. Explain the pattern that can be used to extend the answers to Questions 2 and 3. Add multiples of 360° to 90° and to 270°.

Assignment Guide

Basic: 1–42, 49–54
Advanced: 1–54
Additional Practice: See Extra Practice Index on page 674.

Additional Answers

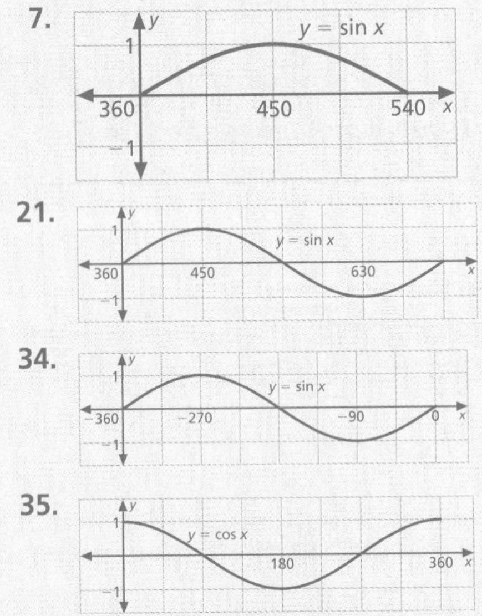

36. Both curves reach 1 as maximum values of y and –1 as minimum values. Both curves repeat themselves every 360°. The curves, in fact, appear identical, except for the fact that each is shifted 90° to the right or left of the other.

43a. Graph the equations on the same set of axes. Points of intersection of the curves represent solutions of the system of equations.

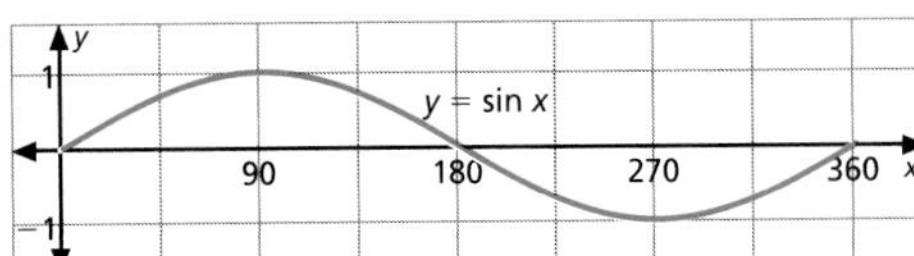

GRAPHING If you have a graphing calculator, you can easily graph sine curves. To graph $y = \sin x$, make sure your calculator is in degree mode. To show the range of x-values, set the Xmin at 0 and the Xmax at 360. Use -1.5 (Ymin) and 1.5 (Ymax) as the range of y-values. In the graph above, the x-scale is set at 45 and the y-scale is at 1. You are ready to graph the function.

Try These Exercises

Find each ratio by drawing a reference angle.

1. sin 135° $\frac{\sqrt{2}}{2}$
2. sin 300° $\frac{-\sqrt{3}}{2}$
3. sin 210° $-\frac{1}{2}$
4. sin 405° $\frac{\sqrt{2}}{2}$
5. sin 660° $\frac{-\sqrt{3}}{2}$
6. sin 855° $\frac{\sqrt{2}}{2}$

7. **POPULATION** A sine curve models the population growth for wildlife in a wooded area. Graph $y = \sin x$ for $360° \le x \le 540°$. See additional answers.

8. **WRITING MATH** Explain how you would find the reference angle and draw the right triangle for a 1050° angle. This initial side must complete two 360° rotations, then continue an additional 330°. The reference angle measures 360° − 330° = 30°. Draw a perpendicular to the x-axis to form a 30°-60°-90° right triangle. The leg adjacent to the 30° angle measures 3. The leg opposite measures −1. The hypotenuse measures 2.

Practice Exercises

Find each ratio by drawing a reference angle.

9. sin 225° $-\frac{\sqrt{2}}{2}$
10. sin 330° $-\frac{1}{2}$
11. sin 120° $\frac{\sqrt{3}}{2}$
12. sin 150° $\frac{1}{2}$
13. sin 315° $-\frac{\sqrt{2}}{2}$
14. sin 480° $\frac{\sqrt{3}}{2}$
15. sin 585° $-\frac{\sqrt{2}}{2}$
16. sin 690° $-\frac{1}{2}$
17. sin 675° $-\frac{\sqrt{2}}{2}$
18. sin 930° $-\frac{1}{2}$
19. sin 765° $\frac{\sqrt{2}}{2}$
20. sin 1,200° $\frac{\sqrt{3}}{2}$

21. **GRAPHING** Use a graphing calculator to graph $y = \sin x$ for $360° \le x \le 720°$. Use $-\frac{1}{5}$ and 1.5 as your range of y-values. See additional answers.

Find each ratio by drawing a reference angle.

22. cos 120° $-\frac{1}{2}$
23. tan 225° 1
24. cos 330° $\frac{\sqrt{3}}{2}$
25. tan 240° $\sqrt{3}$
26. tan 660° $-\sqrt{3}$
27. cos 765° $\frac{\sqrt{2}}{2}$
28. sin (−60°) $-\frac{\sqrt{3}}{2}$
29. sin (−45°) $-\frac{\sqrt{2}}{2}$
30. sin (−30°) $-\frac{1}{2}$
31. cos (−315°) $\frac{\sqrt{2}}{2}$
32. tan (−840°) $\sqrt{3}$
33. cos (−1,230°) $-\frac{\sqrt{2}}{2}$

34. Graph $y = \sin x$ for $-360° \le x \le 0°$. See additional answers.
35. Graph $y = \cos x$ for $0° \le x \le 360°$. See additional answers.

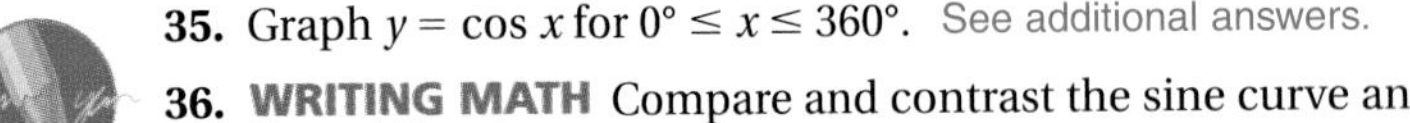

36. **WRITING MATH** Compare and contrast the sine curve and the cosine curve. See additional answers.

Solve for values of x with $0° \le x \le 360°$.

37. $\sin x = 0$ 0°, 180°, 360°
38. $\cos x = -1$ 180°
39. $\sin x = 1$ 90°, 270°
40. $\cos x = 0$ 90°, 270°

41. **FLIGHT** Use a graph of the equation $y = \sin x$ to estimate the value of sin 68°. Then use the value to find the length of a kite string pitched at a 68° angle to the ground if the kite is 450 feet above the ground. ≈485 feet

42. **CHAPTER INVESTIGATION** Work in small groups to determine the height (or depth) of five objects on your school grounds or nearby community. Check students' work. Follow these steps:

   a. Stand at a particular point.

   b. Measure the distance from the point to the base of the object.

   c. Take an angle measure from the point using the quadrant.

   d. Use your knowledge of trigonometry to determine the height of the object. Remember to take into account the distance from your eye level to the ground.

Share the results of your activity with the class. If more than one group measured the same object, compare measurements. Discuss how to account for any discrepancies.

Extended Practice Exercises

43. a. Describe a method involving the graphs of $y = \sin x$ and $y = \cos x$ that you could use to solve the equation $\sin x = \cos x$. See additional answers.

   b. Use your method to solve the equation for $0° \le x \le 360°$. $x = 45°$ and $x = 225°$

Solve for values of x with $0° \le x \le 360°$.

44. $\sin x = -\frac{1}{2}$ 210°, 330°

45. $\cos x = -\frac{\sqrt{3}}{2}$ 150°, 210°

46. $\tan x = -1$ 135°, 315°

47. $\tan x = \sqrt{3}$ 60°, 240°

48. The trademark on a 26-in. radius bicycle tire is 25.5 in. from the center of the wheel. The spoke touching the trademark is horizontal. Find the height of the trademark above the ground after the wheel turns through an angle of 495°. 44.0 in.

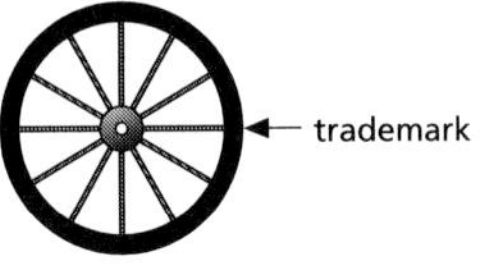

Mixed Review Exercises

Find the equation of each ellipse. (Lesson 13-4)

49. foci: (2, 0) and (−2, 0) $3x^2 + 4y^2 = 48$
x-intercepts: (4, 0) and (−4, 0)

50. foci: (2, 0) and (−2, 0) $5x^2 + 9y^2 = 45$
x-intercepts: (3, 0) and (−3, 0)

51. foci: (6, 0) and (−6, 0) $5x^2 + 9y^2 = 405$
x-intercepts: (9, 0) and (−9, 0)

52. foci: (5, 0) and (−5, 0) $11x^2 + 36y^2 = 396$
x-intercepts: (6, 0) and (−6, 0)

53. foci: (7, 0) and (−7, 0) $32x^2 + 81y^2 = 2592$
x-intercepts: (9, 0) and (−9, 0)

54. foci: (4, 0) and (−4, 0) $3x^2 + 4y^2 = 192$
x-intercepts: (8, 0) and (−8, 0)

Teaching Strategies

For Exercises 44–45, ask students to predict, without looking at a sine curve, the next angle that will have the same trigonometric value as the one given. Then have them look at a graph and determine whether their answers make sense. $\sin x = -\frac{1}{2}$: 570°; $\cos x = -\frac{\sqrt{3}}{2}$; 510°

Extra Practice Worksheet 14-3

Name ______ Date ______

EXTRA PRACTICE 14-3
GRAPH THE SINE FUNCTION

EXERCISES

Find each ratio by drawing a reference angle.

1. sin 765° $\frac{\sqrt{2}}{2}$
2. sin 600° $-\frac{\sqrt{3}}{2}$
3. sin 945° $-\frac{\sqrt{2}}{2}$
4. sin 840° $\frac{\sqrt{3}}{2}$
5. sin 1050° $-\frac{1}{2}$
6. sin 510° $\frac{1}{2}$
7. sin (−390)° $-\frac{1}{2}$
8. sin (−675)° $\frac{\sqrt{2}}{2}$
9. cos 405° $\frac{\sqrt{2}}{2}$
10. tan 300° $-\sqrt{3}$
11. cos 480° $-\frac{1}{2}$
12. cos 510° $\frac{\sqrt{3}}{2}$
13. tan 600° $\sqrt{3}$
14. cos (−675)° $\frac{\sqrt{2}}{2}$
15. tan (−480)° $\sqrt{3}$
16. tan 960° $\sqrt{3}$
17. Graph $y = \cos x$ for $360° \le x \le 720°$. Use your own paper. Check students' graphs.

Solve for values of x with $0° \le x \le 360°$.

18. $\sin x = \frac{\sqrt{3}}{2}$ 60°, 120°
19. $\cos x = -\frac{1}{2}$ 120°, 240°
20. $\tan x = 1$ 45°, 225°
21. $\tan x = -\sqrt{3}$ 120°, 300°
22. $\sin x = \frac{1}{2}$ 30°, 150°
23. $\cos x = \frac{\sqrt{3}}{2}$ 30°, 330°

184 EXTRA PRACTICE *LESSON 14-3*

Enrichment Worksheet 14-3

Name ______ Date ______

ENRICHMENT 14-3
RADIAN MEASURE

When you have used a scientific calculator you may have noticed that there are two types of measurements for angles. One is degrees; the other is radians.

To understand radians, think of a unit circle. The radius equals 1 unit, and the circumference is 2π. There are 360° in a circle, so 2π radians is defined to equal 360°.

In the figure, arc AB has a length equal to the radius r. And, $\angle AOB$ has a measure of 1 radian. To find the number of degrees in 1 radian, use ratio and proportion.

$\frac{2\pi \text{ radians}}{360°} = \frac{1 \text{ radian}}{n°}$ $n \approx 57.3°$

A r r O 57.3° r B

Example 1

Change 60° to radians.

Solution

Multiply by π over 180.

$60 \cdot \frac{\pi}{180} = \frac{\pi}{3}$

Example 2

Change $\frac{\pi}{4}$ radians to degrees.

Solution

Convert π to 180°.

$\frac{\pi}{4} = \frac{180°}{4} = 45$

EXERCISES

Change each degree measure to radians.

1. 30° $\frac{\pi}{6}$
2. −120° $\frac{-2\pi}{3}$
3. 270° $\frac{3\pi}{2}$
4. 22.5° $\frac{\pi}{8}$

Change each radian measure to degrees.

5. $\frac{\pi}{2}$ 90°
6. $\frac{-5\pi}{2}$ −450°
7. 3π 540°
8. $\frac{3\pi}{4}$ 135°

Graph each equation for $-2\pi \le \theta \le 2\pi$.

9. $y = \sin \theta$

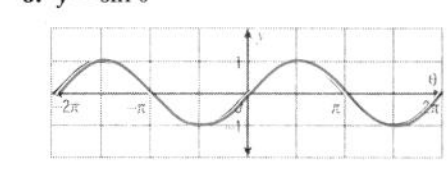

10. $y = \cos \theta$

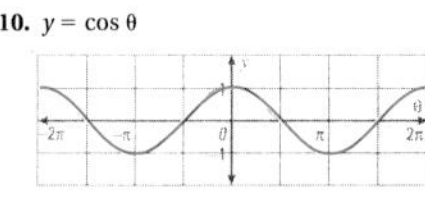

238 ENRICHMENT *LESSON 14-3*

Lesson Planning

NCTM Standards/Strands

- Number & Operation
- Patterns, Functions, & Algebra
- Geometry & Spatial Sense
- Problem Solving
- Reasoning and Proof
- Communication
- Connections
- Representation

Vocabulary

periodic function
amplitude
period

Materials Needed

graphing calculators

Lesson Resources

Warm-Up Transparency 41
Reteaching 14-4
Extra Practice 14-4
Enrichment 14-4

Assignment Guide

Basic: 1–23, 31–42
Enriched: 1–42

Getting Started

5-Minute Warm-Up

Complete each table of values.

1.

x	−2	−1	0	1	2
x^2	4	1	0	1	4

2.

x	0	30	60	90
sin x	0	0.5	0.87	1

Introduction to Lesson 14-4

After students have completed Parts 1 and 2, lead a discussion of why the graphs change as they do. For Part 1, students should be able to explain what happens to the graph of $y = nx^2$, where n is a constant, for $n > 1$. For Part 2, ask students to describe the graph of $y = (x + n)^2$.

14-4 Experiment with the Sine Function

Goals ■ Determine the period, amplitude and position of sine curves.

Applications Communications, Music, Medicine

Work with a partner to answer the following questions. You will need a graphing calculator.

1. Graph each equation on a graphing calculator. How does the coefficient of x^2 affect the shape of the graph? See additional answers.

   a. $y = x^2$ **b.** $y = 2x^2$ **c.** $y = 3x^2$

2. Graph each equation. How does the constant in parentheses affect the position of the graph? See additional answers.

   a. $y = x^2$ **b.** $y = (x + 1)^2$ **c.** $y = (x + 2)^2$

Build Understanding

Recall the sine curve you studied in the last lesson.

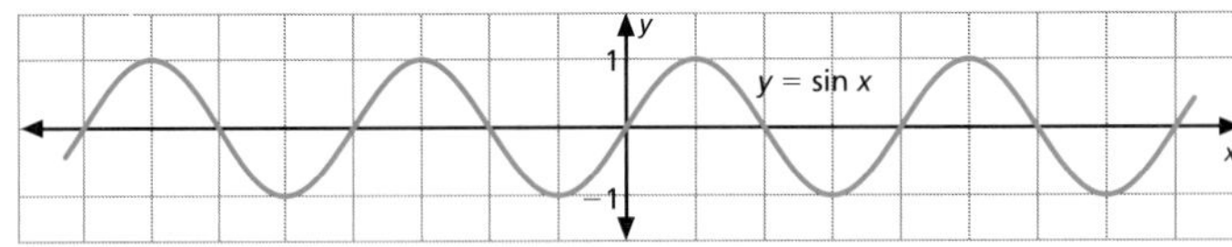

Notice that the shape of the curve repeats itself in every 360° interval along the x-axis. If you were to pick up a 360° section of the curve, you would find it congruent to the curve in each adjacent 360° section. Functions with repeating patterns like this are called **periodic functions**. The **period** of the function is the length of one complete cycle of the function. The period of the graph of $y = \sin x$ is 360°.

Example 1

COMMUNICATIONS A tone transmitted to a ship at sea produces a sound wave with the equation $y = \sin 2x$. State the period.

Solution

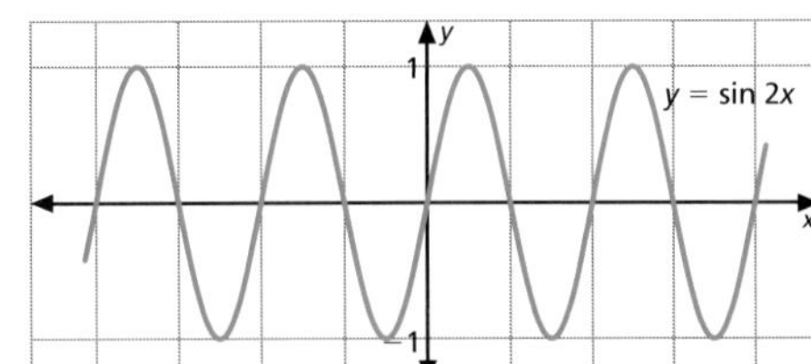

The effect of the coefficient 2 in the equation is to compress the sine curve horizontally. The period of $y = \sin 2x$ is 180°, half the period of $y = \sin x$.

Example 1 suggests the following rule.

The period of the graph of $y = \sin nx$ is $\frac{360°}{n}$.

Teaching Strategies

CHALLENGE Write T (true) or F (false) for each problem. Construct graphs, or use a graphing calculator.

1. 180° is a solution to the equation: $-\sin x = 2 \sin x$. true
2. 90° is a solution to the equation: $\sin 2x = \frac{1}{2} \sin x$. false
3. 225° is a solution to the equation: $-\sin x = -2 \sin 2x$. false

Look again at the graph of $y = \sin x$ on the opposite page. Notice that the maximum value of y is 1 and the minimum value of y is -1. The **amplitude** of a periodic function is half the difference between its maximum and minimum y-values.

The amplitude of $y = \sin x$ is $\frac{1}{2}(1 - [-1]) = 1$.

Amplitude is a measure of the height and depth of a curve.

Example 2

Graph $y = 2 \sin x$. State the amplitude.

Solution

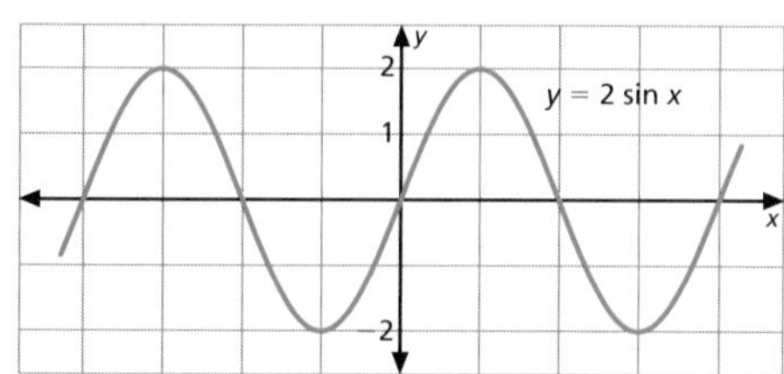

The graph of $y = 2 \sin x$ is twice as tall and twice as deep as the graph of $y = \sin x$. The amplitude is $\frac{1}{2}(2 - [-2]) = 2$.

Notice that, in Example 2, the amplitude is the same as the coefficient in the equation $y = 2 \sin x$. This suggests the following rule.

The amplitude of the graph of $y = n \sin x$ is n.

GRAPHING Graph $y = 2 \sin x$ using a graphing calculator. You may need to change the values in the display window in order to see the entire curve. Use the amplitude to set the values for the y-minimum and y-maximum. Set the y-maximum equal to or greater than the amplitude (n), and the y-minimum equal to or less than $-n$.

Example 3

Graph $y = \sin x + 1$. Describe the position of the graph.

Solution

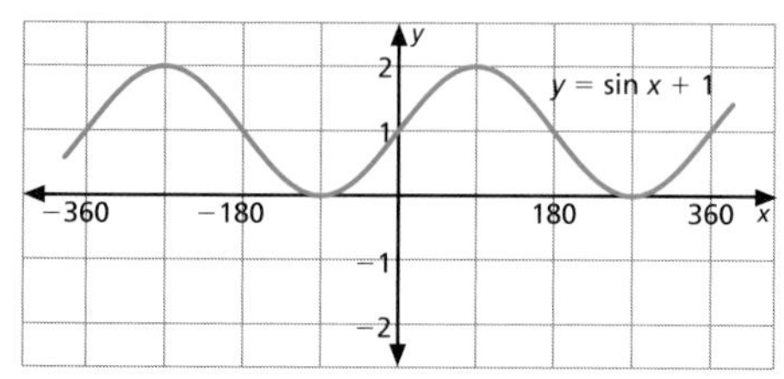

The graph of $y = \sin x + 1$ is the graph of $y = \sin x$ raised 1 unit above its normal position.

Example 3 suggests the following rule.

The graph of $y = \sin x \pm n$ is the graph of $y = \sin x$ raised or lowered n units.

Check Understanding

Without drawing a graph, find the period and amplitude of the graph of $y = 3 \sin 4x - 2$ and describe the position of the graph.

period 90°; amplitude 3; the graph is the graph of $y = 3 \sin 4x$ lowered 2 units.

Teaching Strategies

Have students use a mirror to test the sine curves they have drawn in this lesson for line symmetry and then describe the results. The curves have symmetry about a line parallel to the y-axis and passing through either a minimum or maximum point.

Chalkboard Examples

Supplementary Example 1

The tone produces a sound wave with the equation $y = \sin 3x$. Graph the equation and state the period. Check students' graphs. The period is 120°.

Supplementary Example 2

Graph $y = 3 \sin x$. State the amplitude. Check students' graphs. The amplitude is 3.

Supplementary Example 3

Graph $y = \sin x - 2$. Describe the position of the graph. Check students' graphs. The graph of $y = \sin x - 2$ is the graph of $y = \sin x$ lowered 2 units below its normal position.

Additional Answers

1. The vertex of the parabola remains the same. But as the coefficient increases, the slope of the parabola increases, squeezing it tighter around the y-axis.
2. The shape of the parabola remains the same. But the vertex shifts to the left by an amount equal to the constant.

Reteaching Worksheet 14-4

Name ______________________ Date ____________

RETEACHING 14-4

EXPERIMENTING WITH THE SINE FUNCTION

A periodic function repeats the same pattern over and over.

The length of one complete copy is the **period**.

The period of a graph $y = \sin nx$ is $\frac{360°}{n}$.

The **amplitude** of a periodic function is a measure of the height and depth of a curve.

The amplitude of the graph $y = n \sin x$ is n.

The **position** of the graph of a periodic function is the relationship between the graph and the standard graph of the function.

The position of $y = \sin x \pm n$ is the graph of $y = \sin x$ raised or lowered n units.

Example

State the period and the amplitude for the graph of the equation $y = 6 \sin 5x + 4$. Then describe its position.

Solution

Use the rules to state the period, amplitude and position of each graph.

Find the period by substituting 5 for n in $y = \sin nx$. The period is $\frac{360}{n}$. Since $\frac{360}{5} = 72$, the period is 72°.

Find the amplitude by substituting 6 for n in $y = n \sin x$. The amplitude is 6 units.

Find the position of the graph by substituting + 4 for n in $y = \sin x \pm n$. The position of the graph is raised 4 units above its normal position.

EXERCISES

State the period and the amplitude for the graph of each equation. Describe its position.

1. $y = 3 \sin 5x - 2$
   Period 72°
   Amplitude 3
   Position down 2 units
2. $y = 5 \sin 3x + 1$
   Period 120°
   Amplitude 5
   Position up 1 unit
3. $y = 3 \sin 4x - 3$
   Period 90°
   Amplitude 3
   Position down 3 units
4. $y = 4 \sin 1.5x + 5$
   Period 240°
   Amplitude 4
   Position up 5 units

212 RETEACHING *LESSON 14-4*

Lesson Wrap-up

Quick Assessment

Ask the following questions to determine if students understand the content presented in this lesson.

1. In the function $y = n \sin mx$, which coefficient determines the period of the function? m
2. Compare the amplitudes of these functions: $y = \sin x$, $y = 2 \sin x$, and $y = \frac{1}{2} \sin x$. The amplitude of $y = 2 \sin x$ is twice that of $y = \sin x$. The amplitude of $y = \frac{1}{2}x$ is $\frac{1}{2}$ that of $y = \sin x$ and $\frac{1}{4}$ that of $y = 2 \sin x$.

Assignment Guide

Basic: 1–23, 31–42
Advanced: 1–42
Additional Practice: See Extra Practice Index on page 674.

Additional Answers

1.–3.

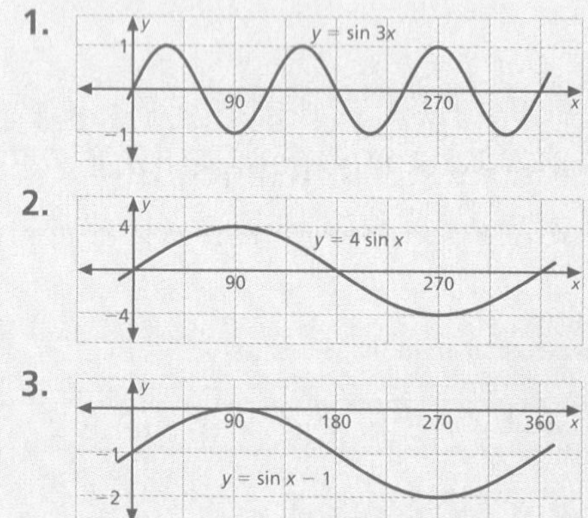

4. period 360°; amplitude 3; the graph is the graph of $y = 3\sin x$ raised 2 units.
5. period 180°; amplitude 2; the graph is the graph of $y = 2\sin 2x$ lowered 5 units.

7.–8.

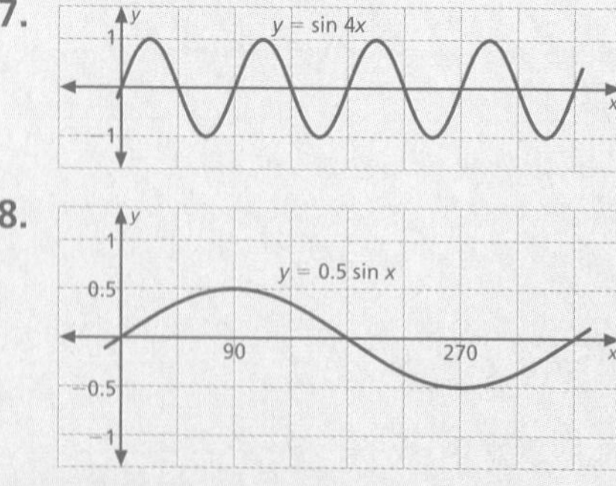

Try These Exercises

For 1–3, see additional answers for graphs.

1. Graph $y = \sin 3x$. State the period. period = 120°
2. Graph $y = 4 \sin x$. State the amplitude. amplitude = 4
3. Graph $y = \sin x - 1$. Describe the position of the graph. The graph is the graph of $y = \sin x$ lowered 1 unit from the origin.

State the period and amplitude of the graph of each equation and describe the position of the graph. See additional answers.

4. $y = 3 \sin x + 2$
5. $y = 2 \sin 2x - 5$
6. **YOU MAKE THE CALL** To graph $y = \sin x + 3$, Cynthia enters $y = \sin (x + 3)$ on her graphing calculator. Has Cynthia made a mistake? Explain your thinking. Yes. The expression $\sin x + 3$ is the sum of 3 and the sin of x. In the expression $\sin (x + 3)$, the sin is found of the sum x and 3. The graphs of the two expressions are not the same.

Practice Exercises

For 7–9, see additional answers for graphs.

7. Graph $y = \sin 4x$. State the period. The period is 90°.
8. Graph $y = 0.5 \sin x$. State the amplitude. The amplitude is 0.5.
9. Graph $y = \sin x + 3$. Describe the position of the graph. The graph is the graph of $y = \sin x$ raised 3 units.

State the period and amplitude of the graph of each equation and describe the position of the graph. See additional answers.

10. $y = 2 \sin x - 1$
11. $y = 4 \sin 3x + 2.5$

Tell if the function is periodic. If it is, state the period.

12. 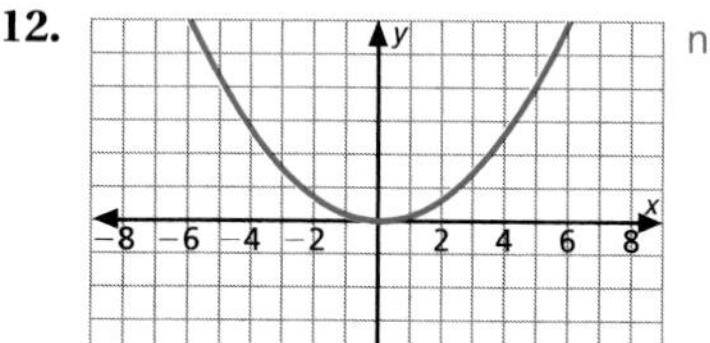 no

13. 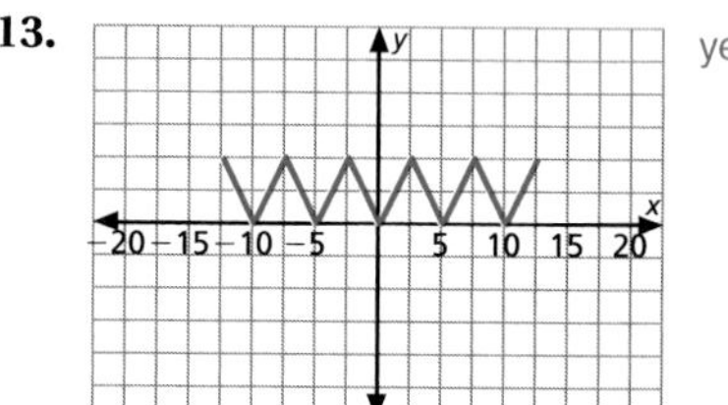 yes; 5

14. 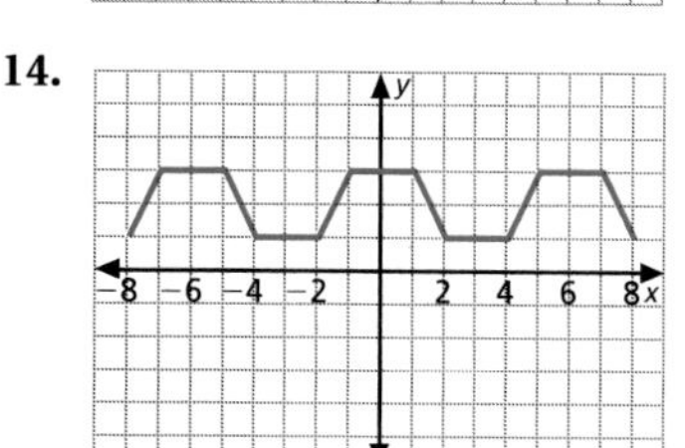 yes; 6

15. 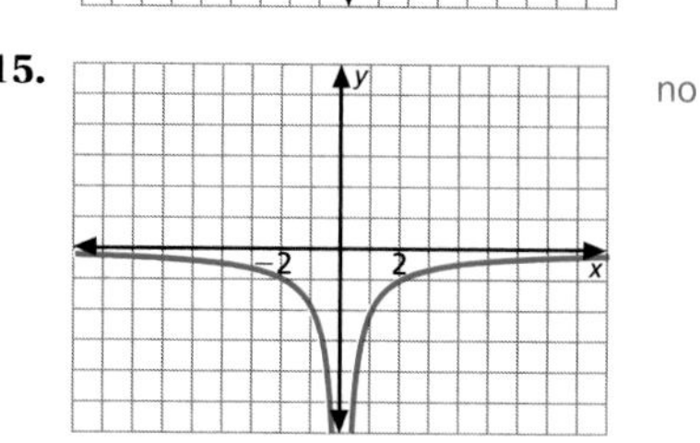 no

16. **MUSIC** A note played on a musical instrument produces a sound wave with the equation $y = 3 \sin 4x + 3$. State the period and amplitude, and describe the position of the graph. period: 90°; amplitude: 3; graph is the graph of $y = 3 \sin 4x$ raised 3 units.
17. **MEDICINE** An animal's heart rate can be modeled by a sine curve that has a period of 540° and an amplitude of $\frac{3}{2}$. Write the equation. $y = \frac{3}{2} \sin \frac{2}{3}x$

9.

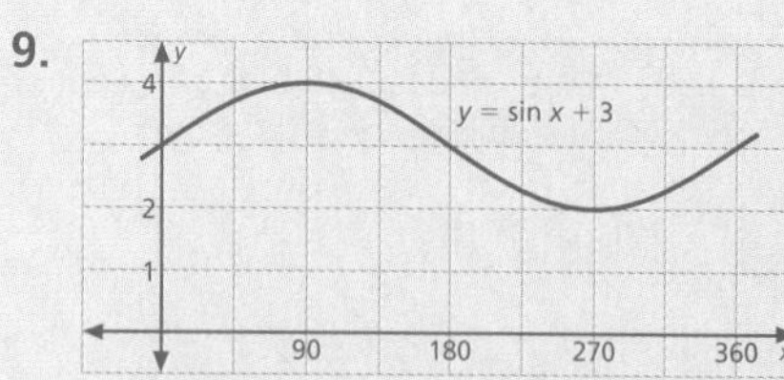

10. period 360°; amplitude 2; the graph is the graph of $y = 2\sin x$ lowered 1 unit.
11. period 120°; amplitude 4; the graph is the graph of $y = 4\sin 3x$ raised 2.5 units.

24.–25.

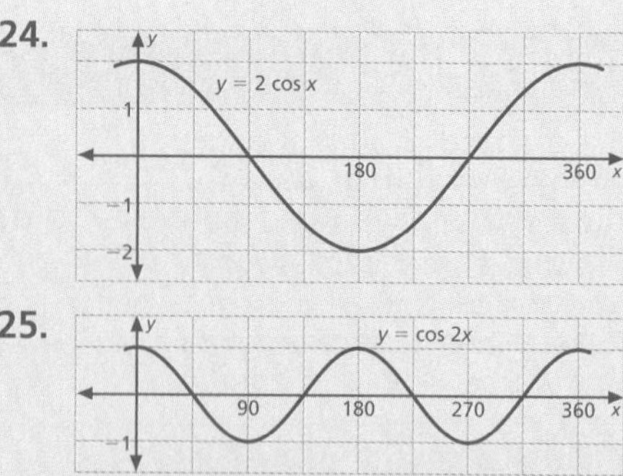

620 Chapter 14 Trigonometry

Find the period of the graph of each equation.

18. $y = \sin \frac{1}{2}x$ 720° **19.** $y = \sin \frac{3}{5}x$ 600° **20.** $y = \sin 1\frac{5}{7}x$ 210°

21. Find an equation of a graph involving the sine function that has a period of 630° and an amplitude of 8. $y = 8 \sin \frac{4}{7}x$

State the equation whose graph is shown.

22.

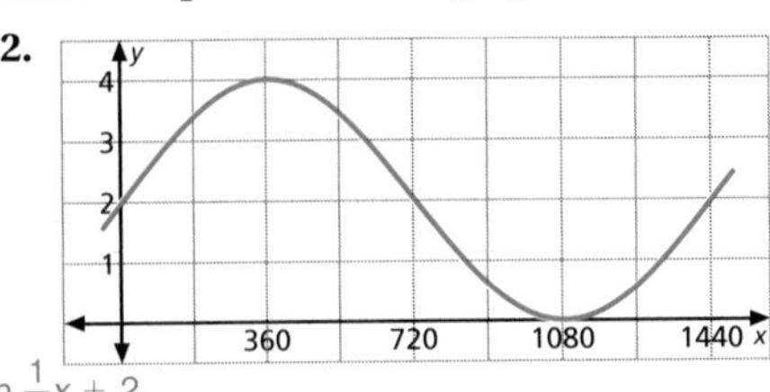

$y = 2 \sin \frac{1}{4}x + 2$

23.

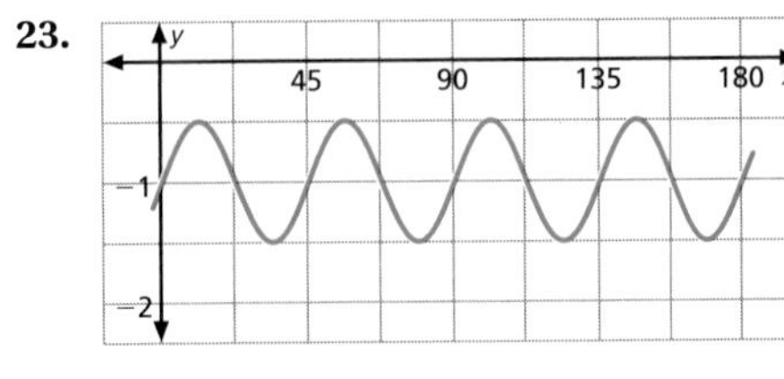

$y = \frac{1}{2} \sin 8x - 1$

Extended Practice Exercises

Graph the equation. See additional answers.

24. $y = 2 \cos x$ **25.** $y = \cos 2x$ **26.** $y = \cos x + 1$

27. WRITING MATH Write rules you can use to find the period, amplitude, and position of a graph involving the cosine function.
See additional answers.

Make a table of ordered pairs and graph the equation.
See additional answers.

28. $y = -\sin x$ **29.** $y = \sin (x + 90°)$

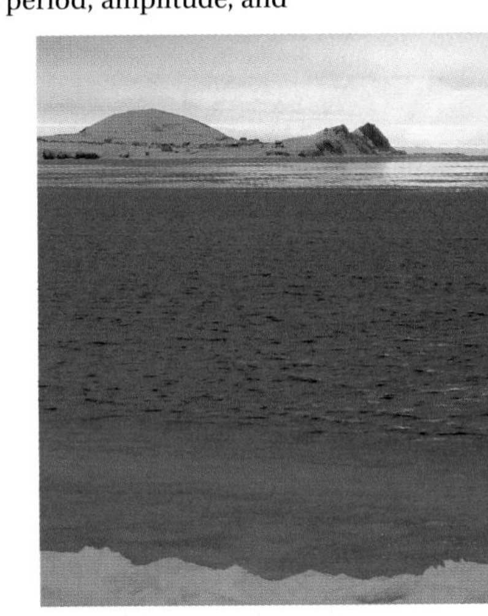

30. WILDLIFE MANAGEMENT The equation $l = 50{,}000 + 48{,}000 \sin 90t$ approximates the number of lemmings on an arctic island on January 1 of a year t years after January 1, 1980.

a. Find the maximum number of lemmings.
98,000

b. What year was the maximum first reached?
1981

c. Find the minimum number of lemmings.
2000

d. What year was the minimum first reached?
1983

Mixed Review Exercises

Solve each variation. (Lesson 13-5)

31. If y varies directly as x, and $y = 9$ when $x = 6$, find y when $x = 27$. 40.5

32. If y varies directly as x, and $y = 8$ when $x = 3$, find y when $x = 45$. 120

33. If y varies directly as x, and $y = 7$ when $x = 2$, find y when $x = 38$. 133

Factor the following trinomials. (Lesson 11-7)

34. $x^2 + 3x - 10$ $(x + 5)(x - 2)$ **35.** $2x^2 + 5x - 12$ $(2x - 3)(x + 4)$ **36.** $x^2 + 2x - 35$ $(x + 7)(x - 5)$

37. $4x^2 - 8x - 5$ $(2x + 1)(2x - 5)$ **38.** $x^2 - 6x + 9$ $(x - 3)(x - 3)$ **39.** $3x^2 + 3x - 6$ $3(x + 2)(x - 1)$

40. $3x^2 + 14x - 5$ $(3x - 1)(x + 5)$ **41.** $x^2 - 64$ $(x - 8)(x + 8)$ **42.** $2x^2 - 6x - 56$ $2(x - 7)(x + 4)$

26.

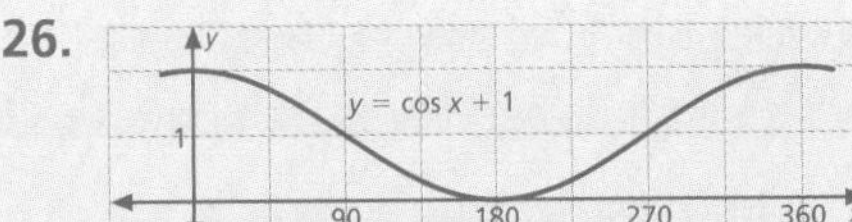

28.

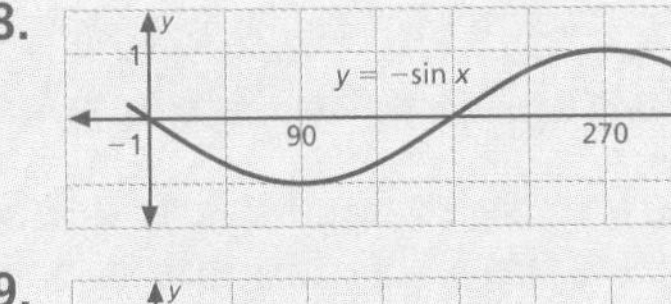

27. For the graph of the equation $y = m\cos nx \pm p$, the period is $\frac{360}{n}$, the amplitude is m, and the graph is the graph of $y = m\cos nx$ raised or lowered p units.

29.

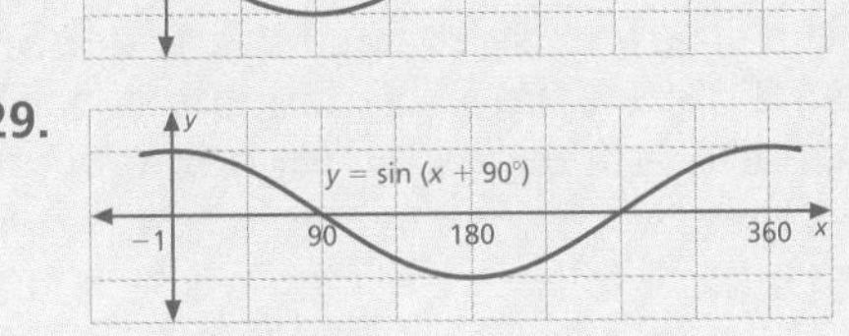

Extra Practice Worksheet 14-4

Name ______ Date ______

EXTRA PRACTICE 14-4
EXPERIMENT WITH THE SINE FUNCTION

EXERCISES

Check students' graphs.

1. Graph $y = \sin 3x$. Use your own paper. State the period. 120°
2. Graph $y = 2 \sin x$. Use your own paper. State the amplitude. 2

State the period and amplitude of the graph of each equation and describe the position of the graph.

3. $y = \sin x - 5$ period 360°; amplitude 1; the graph is the graph of $y = \sin x$ lowered 5 units
4. $y = 3 \sin x + 1$ period 360°; amplitude 3; the graph is the graph of $y = 3 \sin x$ raised 1 unit
5. $y = 0.5 \sin 2x - 2$ period 180°; amplitude 0.5; the graph is the graph of $y = 0.5 \sin x$ lowered 2 units
6. $y = 2 \sin 4x + 3.5$ period 90°; amplitude 2; the graph is the graph of $y = 2 \sin 4x$ raised 3.5 units

Tell if the function is periodic. If it is, state the period.

7. no 8. yes; 4

Find the period of the graph of each equation.

9. $y = \sin \frac{1}{4}x$ 1440° 10. $y = \sin \frac{2}{5}x$ 900°
11. $y = \sin \frac{3}{2}x$ 240° 12. $y = \sin \frac{4}{3}x$ 270°

186 EXTRA PRACTICE LESSON 14-4

Enrichment Worksheet 14-4

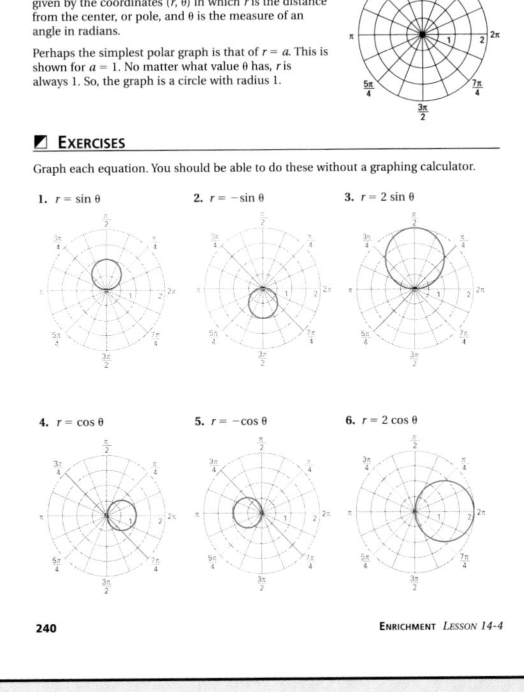

Name ______ Date ______

ENRICHMENT 14-5
POLAR COORDINATES

In a polar coordinate system, the location of a point is given by the coordinates (r, θ) in which r is the distance from the center, or pole, and θ is the measure of an angle in radians.

Perhaps the simplest polar graph is that of $r = a$. This is shown for $a = 1$. No matter what value θ has, r is always 1. So, the graph is a circle with radius 1.

EXERCISES

Graph each equation. You should be able to do these without a graphing calculator.

1. $r = \sin \theta$ 2. $r = -\sin \theta$ 3. $r = 2 \sin \theta$
4. $r = \cos \theta$ 5. $r = -\cos \theta$ 6. $r = 2 \cos \theta$

240 ENRICHMENT LESSON 14-4

Skills Practice

Vocabulary Review

Lesson 14-3
initial side
reference angle
terminal side

Lesson 14-2
periodic function
amplitude
period

Assignment Guide

All students: 1–59
Additional Practice: Refer to the Extra Practice Index on page 674 of this text.

Review and Practice Your Skills

Practice ■ Lesson 14-3

Determine if each statement is *true* or *false*.

1. The initial side of an angle always rests on the positive x-axis. true
2. The terminal side of an angle is found in the second quadrant. false
3. The reference angle is the acute angle formed by the y-axis and the terminal side of the angle. false

Find each ratio by drawing a reference angle.

4. $\sin 270°$ -1
5. $\sin 120°$ $\frac{\sqrt{3}}{2}$
6. $\sin 150°$ $\frac{1}{2}$
7. $\sin 300°$ $\frac{-\sqrt{3}}{2}$
8. $\sin 315°$ $\frac{-\sqrt{2}}{2}$
9. $\sin 585°$ $\frac{-\sqrt{2}}{2}$

Find each ratio by drawing a reference angle.

10. $\tan 210°$ $\frac{\sqrt{3}}{3}$
11. $\cos 240°$ $-\frac{1}{2}$
12. $\tan 135°$ -1
13. $\sin 330°$ $-\frac{1}{2}$
14. $\tan 300°$ $-\sqrt{3}$
15. $\cos 225°$ $\frac{-\sqrt{2}}{2}$

Solve for values of x with $0° \leq x \leq 360°$.

16. $\sin x = -1$ 270°
17. $\cos x = \frac{1}{2}$ 60°, 300°
18. $\sin x = -\frac{1}{2}$ 330°, 210°
19 $\cos x = -\frac{1}{2}$ 120°, 240°
20. $\sin x = 1$ 90°
21. $\cos x = 1$ 0°, 360°

Practice ■ Lesson 14-4

Define each term.

22. period The length of one complete cycle of the function.
23. amplitude Half the difference between the maximum and minimum y-values.
24. Use the graph of $y = \sin 4x$ to state the period. 90°
25. Use the graph of $y = 2 \sin x$ to state the amplitude. 2

Find the period of the graph of each equation.

26. $y = 2 \sin 5x$ 72°
27. $y = \sin 3x - 2$ 120°
28. $y = 3 \sin x + 4$ 360°

Find the period of the graph of each equation.

29. $y = 2 \sin 10x$ 36°
30. $y = \sin 6x - 1$ 60°
31. $y = 5 \sin x - 2$ 360°
32. $y = 3 \sin 4x - 1$ 90°
33. $y = \sin 40x + 3$ 9°
34. $y = 2 \sin 5x + 1$ 72°

State the period and amplitude of the graph of each equation and describe the position of the graph.

35. $y = 2 \sin x$ 360°, 2, centered on y-axis
36. $y = \sin 4x$ 90°, 1, centered on y-axis
37. $y = 2 \sin 3x + 1$ 120°, 2, shifted 1 unit up
38. $y = 3 \sin 6x - 3$ 60°, 3, shifted 3 units down
39. $y = 5 \sin 2x + 2$ 180°, 5, shifted 2 units up
40. $y = 2 \sin 5x - 1$ 72°, 2, shifted 1 unit down

Teaching Strategies

Remind students that since the trigonometric functions indicate division, if either the numerator or denominator is negative, the value will be negative. If both are either positive or negative, the value will be positive.

Practice ■ Lesson 14-1–Lesson 14-4

Use the figure at the right to find each ratio. (Lesson 14-1)

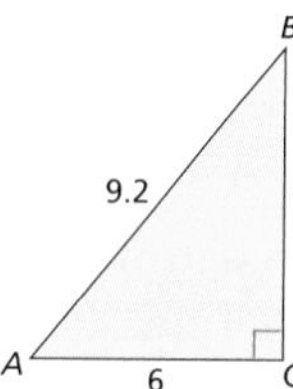

41. sin A $\frac{7}{9.2}$ **42.** cos B $\frac{7}{9.2}$
43. sin B $\frac{6}{9.2}$ **44.** tan A $\frac{7}{6}$
45. tan B $\frac{6}{7}$ **46.** cos A $\frac{6}{9.2}$

47. To the nearest tenth, find the angle of elevation to the top of a 90-ft tree from a point 40 feet from the base of the tree. (Lesson 14-2) 66.0°

Find each ratio by drawing a reference angle. (Lesson 14-3)

48. sin 30° $\frac{1}{2}$ **49.** sin 135° $\frac{\sqrt{2}}{2}$ **50.** sin 210° $-\frac{1}{2}$
51. cos 270° 0 **52.** tan 225° 1 **53.** cos 45° $\frac{\sqrt{2}}{2}$

Solve for values of x with $0° \leq x \leq 360°$. (Lesson 14-3)

54. $\sin x = \frac{-\sqrt{3}}{2}$ 240°, 300° **55.** $\sin x = \frac{\sqrt{3}}{2}$ 60°, 120° **56.** $\sin x = \frac{\sqrt{2}}{2}$ 45°, 135°
57. $\cos x = \frac{\sqrt{3}}{2}$ 30°, 330° **58.** $\tan x = -1$ 135°, 315° **59.** $\cos x = \frac{\sqrt{2}}{2}$ 45°, 315°

MathWorks
Workplace Knowhow

Career – Commercial Fisher

Commercial fishers are captains, deckhands, or boatswains (supervisor of the deckhands). Aside from fishing duties, the fishers aboard a fishing boat must be able to navigate to the fishing areas. Today this is mainly accomplished through the use of electronic equipment that pinpoints the ship's position on the surface of the Earth according to man-made satellites orbiting the planet. Before these devices were invented, sailors navigated according to the stars.

To avoid rocks near a shoreline, an experienced fisher uses the stars to know when to turn and make an arc to shore. When under the correct star, the fisher is 2.5 mi from shore. The arc begins there and ends 2.5 mi from the shore point where the arc began. His route is a quarter of a circle with a radius of 2.5 mi.

1. How many miles does the fisher travel in his arc? 3.925 mi (π = 3.14)
2. On a coordinate plane,name the x, y coordinates of the beginning and ending points of the arc. (0, ±2.5) (±2.5, 0) or (±2.5, 0), (0, ±2.5) depending on direction
3. What is the equation of the whole circle of which the fisher traveled one quarter? $x^2 + y^2 = 6.25$, center = (0, 0)

Chalkboard Examples

Example from Lesson 14-3

Solve for the values of x in which $0° < x < 360°$.

1. $\sin x = \frac{1}{2}$ **30°, 150°**

2. $\cos x = \frac{1}{2}$ **60°, 300°**

Example from Lesson 14-4

State the period and amplitude of $y = 3 \sin 4x + 3$. Describe the position of the graph. **period: 90°; amplitude: 3; graph is the graph of $y = 3 \sin 4x$ raised 3 units**

Career Opportunity

Describe the kind of work hunters, fishers, and trappers do and the types of companies that hire these workers. Have students discuss the importance of measurement, geometry, trigonometry and algebraic thinking in doing this type of job. Explain that hunters, fishers, and trappers must be able to read maps and think in terms of three-dimensional space as they plot course and locate positions from instrument readings. Students should answer Questions 1–3 to better understand how hunters, fishers and trappers use mathematics in performing their job.

Students who are interested in learning more about this profession can go to learningmathmatters.com. Guidance Counselors should have information about school and training requirements.

Teaching Strategies

Students who are having difficulty distinguishing between the graphs of different equations may benefit from using a table of values of each graph they are drawing. A graphing calculator should only be used after the values have been determined, not as a substitute for determining the ordered pairs that make up the graph.

14-5 Problem Solving Skills: Choose a Strategy

In this book, you have studied a variety of problem solving strategies. Experience in applying these strategies will help you decide which will be most appropriate for solving a particular problem. Sometimes, only one strategy will work. In other cases, any one of several strategies will offer a solution. There may be times when you will want to use two different approaches to a problem to be sure the solution you found is correct. For certain problems, you will need to use more than one strategy to find the solution.

Problem Solving Strategies

- Guess and check
- Look for a pattern
- Solve a simpler problem
- Make a table, chart or list
- Use a picture, diagram or model
- Act it out
- Work backwards
- Eliminate possibilities
- Use an equation or formula

Problems

Solve. Name the strategy you used to solve each problem. Find lengths to the nearest tenth and angles to the nearest degree.

1. **ARCHEOLOGY** The Great Pyramid at El Giza, Egypt, measures 755 ft on a side. The faces stand at a 52° angle to the ground. The top 30 ft of the pyramid has been destroyed. How tall was the pyramid when it was first built? 483.2 ft

52°
755 ft

2. A certain acute angle has the same sine as the cosine of 24°. Find the acute angle. 66°

3. **SURVEYING** Two surveyors standing 2.8 mi apart each measure the angle of elevation of the top of a mountain. The surveyor nearer the mountain, standing 3.4 mi from the base of the peak, measures an angle of 26°. Find the angle of elevation measured by the other surveyor. 15°

B
A 2.8 mi D 3.4 mi C

4. **GEOGRAPHY** The formula $l = 69.81 \cos d$ gives the length in miles, l, of one degree of longitude on the Earth's surface, where d is the latitude in degrees.

   a. Find the length of one degree of longitude at latitude 42° N. 51.9 mi

   b. At what northern hemisphere latitude is the length of one degree of longitude 19.2 mi? 74° N

Lesson Planning

NCTM Standards/Strands

- Number & Operation
- Patterns, Functions, & Algebra
- Geometry & Spatial Sense
- Problem Solving
- Reasoning and Proof
- Communication
- Connections
- Representation

Vocabulary

angle of elevation
angle of descent
latitude
longitude

Materials Needed

graphing calculators

Lesson Resources

Warm-Up Transparency 41
Reteaching 14-5

Assignment Guide

Basic: 1–27
Enriched: 1–27

Getting Started

5-Minute Warm-up

Solve for the unknown.

1. $\tan 40° = \frac{x}{103}$ $x = 118.5$
2. $\cos 71° = \frac{m}{81.3}$ $m = 26.5$
3. $\sin 40, = \frac{16}{r}$ $r = 24.9$
4. $\tan 63, = \frac{35}{e}$ $e = 17.8$

Use the Five-step Plan and the Strategy

THE FIVE-STEP PLAN Read—ask questions of the students to help them understand the problem. **Plan**—guide students to related problems and previously mastered skills and strategies. **Solve**—students solve problem on their own. **Answer**—write solution in format that answers the question. **Check**—review work, check for reasonableness and review strategy used. Students will benefit from the experience of verbalizing their methods.

Teaching Strategies

ESL/LEP Some of the problems in this section present mathematical and common vocabulary that may be difficult for some students. You may want to have these students work with others who can help them to understand the language of the problems before they begin to solve the problems.

5. **SCIENCE** This sine wave appeared on a laboratory oscilloscope screen. The technician then generated a wave congruent to this one, but 3 units lower on the screen. Find the equation of the second wave.
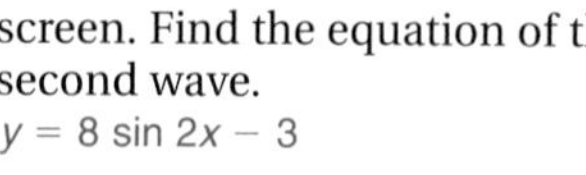
$y = 8 \sin 2x - 3$

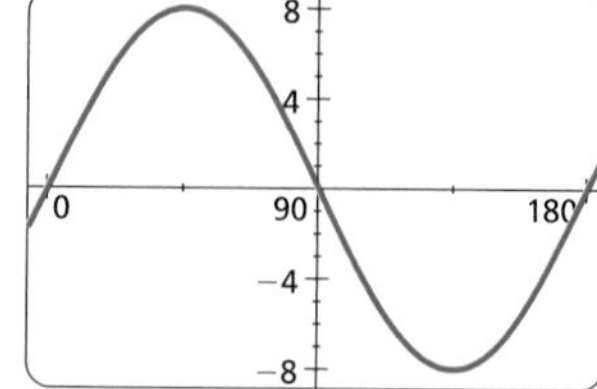

Five-step Plan

1 Read
2 Plan
3 Solve
4 Answer
5 Check

GEOGRAPHY The world's longest aerial ropeway ascends from the city of Merida, Venezuela (altitude: 5379 ft) to the summit of Pico Espejo (altitude: 15,629 ft). The ropeway is 42,240 ft in length.

6. Find the horizontal distance from the lower end of the ropeway to the point directly under the summit of Pico Espejo. 40,977.5 ft
7. Find the angle of elevation of the ropeway. 14°

8. **PHYSICS** A spring bounces up and down in such a way that the height in inches, h, of the weight at the end of the spring is given by $h = 6 \sin 360t$, where t is the number of seconds after the spring reaches the position shown for the first time.

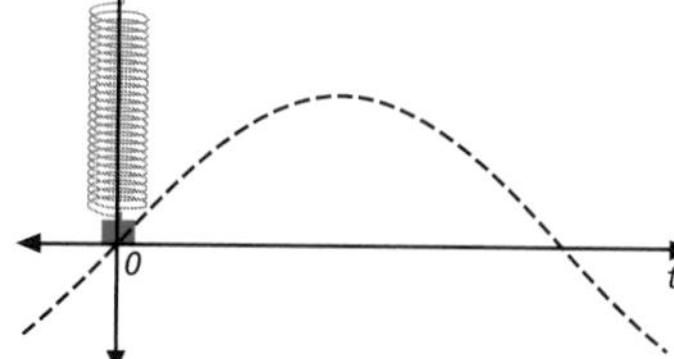

a. How high will the weight be after 5.1 sec? 3.5 in.
b. What is the maximum height reached by the weight? 6 in.
c. When is the maximum height first reached? 0.25 sec
d. What is the minimum height reached by the weight? −6 in.
e. When is the minimum height first reached? 0.75 sec

Mixed Review Exercises

Round answers to the nearest hundredth if necessary. (Lesson 13-6)

9. If y varies inversely as x, and $y = 3$ when $x = 42$, find y when $x = 27$. 4.67
10. If y varies inversely as x, and $y = 9$ when $x = 8$, find y when $x = 13$. 5.54
11. If y varies inversely as x, and $y = 15$ when $x = 2$, find y when $x = 8$. 3.75
12. If y varies inversely as x, and $y = 30$ when $x = 9$, find y when $x = 14$. 19.29
13. If y varies inversely as x, and $y = 28$ when $x = 7$, find y when $x = 52$. 3.77
14. If y varies inversely as x, and $y = 18$ when $x = 3$, find y when $x = 38$. 1.42

Simplify. (Lesson 11-2)

15. $(x - 5)(x + 8)$ $x^2 + 3x - 40$
16. $(2x - 7)(3x + 4)$ $6x^2 - 13x - 28$
17. $(-3x + 2)(-x + 8)$ $3x^2 - 26x + 16$
18. $(6x - 4)(x + 4)$ $6x^2 + 20x - 16$
19. $(2x - 9)(2x + 9)$ $4x^2 - 81$
20. $4(x - 3)(x - 4)$ $4x^2 - 28x + 48$
21. $3(2x - 1)(x + 6)$ $6x^2 + 33x - 18$
22. $(4x + 7)(3x + 2)$ $12x^2 + 29x + 14$
23. $(2x - 9)(3x + 8)$ $6x^2 - 11x - 72$

Compute the variance and standard deviation for each set of data. Round answers to the nearest hundredth if necessary. (Lesson 9-7)

24. 7, 8, 8, 5, 6, 8, 9, 7 1.44; 1.20
25. 12, 11, 13, 17, 15, 13, 12, 14 3.23; 1.80
26. 21, 23, 20, 25, 25, 29, 27, 26 8.00; 2.83
27. 15, 16, 19, 17, 18, 10, 4, 28 42.36; 6.51

THE STRATEGY Choose a strategy—There is often no best strategy for a problem. Good problem solvers keep an open mind and try various strategies until they are successful.

Chalkboard Examples

Supplementary Problem

A rectangular field is cut by a straight fence between two vertices. The fence is 85 ft long and makes a 63° angle with the shorter side. To the nearest tenth of a foot, what is the perimeter of the field?
$P = 228.7$ ft

Lesson Wrap-up

Quick Assessment

Ask the following question to determine if students understand the content presented in this lesson.

How do you determine what strategies may be useful in each problem-solving situation?
Answers will vary. Have students describe the strategies they used in each exercise.

Reteaching Worksheet 14-5

Name ______________________ Date ______________

RETEACHING 14-5

Problem Solving Skills: Choose a Strategy

Once you understand a problem, you must then make a plan to help you solve the problem. Part of your plan will involve deciding upon the strategy or strategies you will use. Some of the strategies you have used in this book include the following:

Choose a Method and an Operation	Using a Spreadsheet
Using Logical Reasoning	Find a Pattern
Solve a Simpler Problem	Using Linear Programming
Use Indirect Measurement	Use a Matrix
Using Statistics	Study the General Case
Making a Circle Graph	Choose a Formula
Solve a Quadratic Equation	

Other strategies you may choose include: Write and Solve an Equation, Reading and Interpreting Graphs, Guess and Check, Draw a Diagram, Work Backwards, Make an Organized List and Make a Model.

Exercises

Solve. Tell what strategy or strategies you used. Round each answer to the nearest tenth if necessary. **Strategies may vary for Exercises 1–7.**

1. Jason joins 2 triangular quilt pieces to form a square with sides of 4 in. What is the length of the side of each triangle that will be joined together to make the square? 5.7 in.
2. The difference in Carrie's age and Carter's age is 5. The sum of their ages is 31. Carrie is the oldest. Find the age of each person. Carter—13, Carrie—18
3. Huang, José and Sam ran a 100-m dash. José did not come in second; Huang did not come in third. José time was 2 seconds faster than that of the oldest boy in the race. In what order did the boys finish the race? 1. José, 2. Huang, 3. Sam
4. A cable car ascends from the Skier's Lodge—altitude of 2315 m—to the top of the mountain—altitude 3459 m. The length of the cable is 2168 m. What is the distance from the lodge to the point directly under the summit of the mountain? 1841.6 m
5. The sun is shining on a tree 40 ft tall. The sun's rays form a 35° angle with the ground. How long is the shadow cast by this tree? 57.1 ft
6. Suppose there are 25 students in your class. If each student shakes hands with every other student once and only once, how many handshakes will there be? 300 handshakes
7. Tiffany rode her bicycle 8 mi south from her home to visit her grandmother and then 6 mi west to go to the library. From this point, there is a direct route home. How many miles is it from her home to the library? 10 mi

214 RETEACHING *Lesson 14-5*

Teaching Strategies

Remind students of the importance of checking to make sure their answers seem reasonable. For example, in Exercise 4, students could refer to a standard globe to test the reasonableness of their solutions. Mistakes are often made when students rely on graphing calculators to solve problems. When graphing, encourage students to find several values using paper and pencil and then check the graph on their calculator to make sure theses values lie on the graph.

Assessment Planning

Vocabulary Review

Lesson 14-1
trigonometric ratios
sine
cosine
tangent

Lesson 14-2
solving a right triangle
angle of depression
angle of elevation

Lesson 14-3
initial side
reference angle
terminal side

Lesson 14-4
periodic function
amplitude
period

Lesson 14-5
angle of elevation
angle of descent
latitude
longitude

Assessment Options

Chapter Assessment, page 628
Alternative Assessment, page 629
Chapter Test Forms A and B
Cumulative Assessment, page 630–631
Achievement Test
Writing Prompts

Assignment Guide

All students: 1–18
Additional Practice: Refer to the Extra Practice Index on page 674 of this text.

Chapter 14 Review

Vocabulary ■ LESSON 14-1–LESSON 14-5

Choose the word or phrase from the list that best completes each statement.

a. angle of depression
b. angle of elevation
c. tangent
d. vertex
e. cosine
f. sine

1. __?__ is opposite over adjacent. c
2. __?__ is formed by a horizontal line and a line slanting upward. b
3. __?__ is adjacent over hypotenuse. d
4. __?__ is formed by a horizontal line and a line slanting downward. a
5. __?__ is opposite over hypotenuse. e

LESSON 14-1 ■ Basic Trigonometric Ratios, p. 604

▶ The ratio of the lengths of two sides of a right triangle depends on the angles of the triangle.

▶ For a given angle, the ratios are always the same.

In a right triangle, $\angle A$ is an acute angle. Then,

$$\sin A = \frac{\text{length of side opposite } \angle A}{\text{length of hypotenuse}}$$

$$\cos A = \frac{\text{length of side adjacent to } \angle A}{\text{length of hypotenuse}}$$

$$\tan A = \frac{\text{length of side opposite } \angle A}{\text{length of side adjacent to } \angle A}$$

6. Find sin *D*. $\frac{3}{5}$
7. Find cos *F*. $\frac{3}{5}$
8. Find tan *F*.2 $\frac{4}{3}$

LESSON 14-2 ■ Solving Right Triangles, p. 608

▶ You can find the measures of the angles and sides of a right triangle.

a. Given the measure of one acute angle, find the measure of the other by subtracting the measure of the known angle from 90°.

b. Given the lengths of two sides, use the Pythagorean Theorem to find the length of the third side.

c. Given an angle and length of a side, use trigonometric ratios to find other parts of the triangle.

▶ Trigonometry problems may involve angles of depression and elevation.

a. An angle of depression is formed by a horizontal line and a line slanting downward.

b. An angle of elevation is formed by a horizontal line and a line slanting upward.

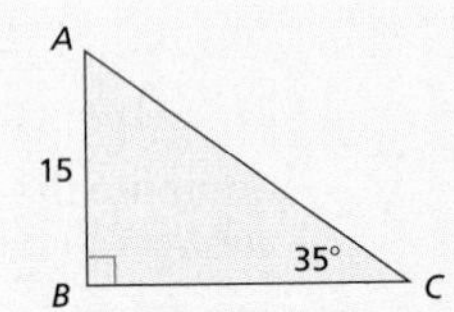

Find the following in $\triangle ABC$.

9. *BC* ≈21.4
10. m$\angle A$ 55°
11. *AC* 26.1

Additional Answers

16.

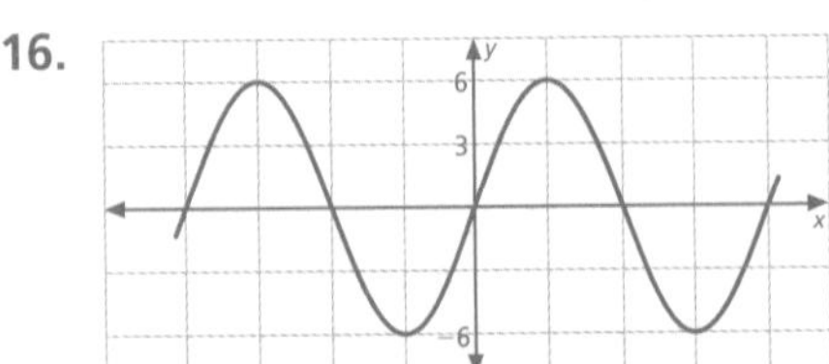

17.

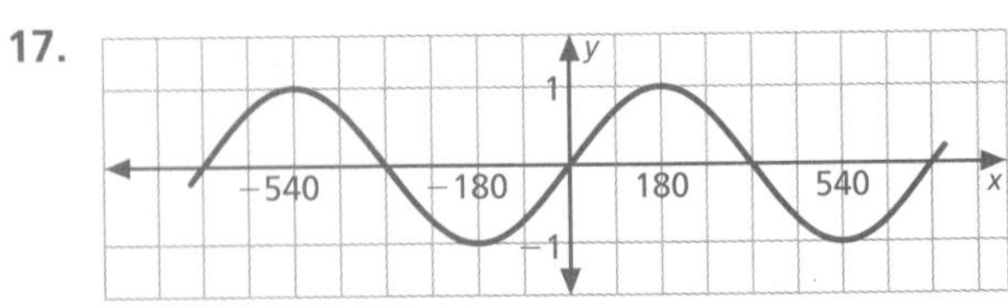

12. A tower casts a shadow 55 ft long. Measuring from the end of the shadow, Brad determines that the angle of the sun is 43°. How tall is the tower? 51.3 ft

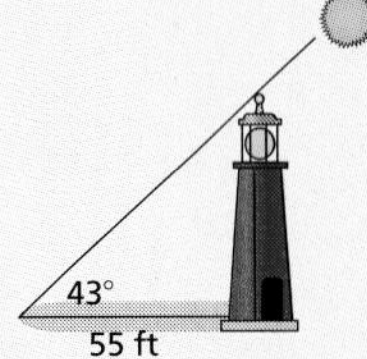

LESSON 14-3 ■ Graphing the Sine Function, p. 614

► The reference angle is the acute angle formed by the x-axis and the terminal side.

The reference angle measures 210° − 180° = 30°. Draw a triangle with the terminal side as the hypotenuse by making a line perpendicular to the x-axis. The triangle is a 30°–60°–90° right triangle. In this case, the leg lengths are negative because they were drawn by moving in negative directions from the x- and y-axis. The length of the terminal side is always considered positive.

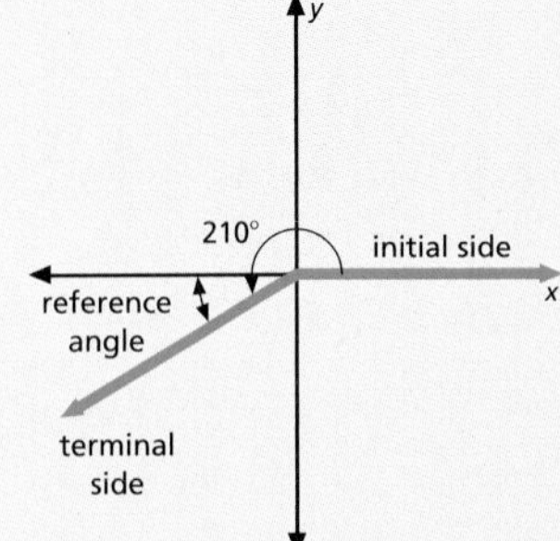

$$\sin 210° = \frac{\text{opposite}}{\text{hypotenuse}} = -\frac{1}{2}$$

Find each ratio by drawing a reference angle.

13. $\sin 120°$ $\frac{\sqrt{3}}{2}$ 14. $\tan 225°$ 1 15. $\cos 960°$ $-\frac{1}{2}$

LESSON 14-4 ■ Experimenting with the Sine Function, p. 618

► Functions with repeating patterns are called periodic functions. The period of a function is the length of one complete copy of the function.

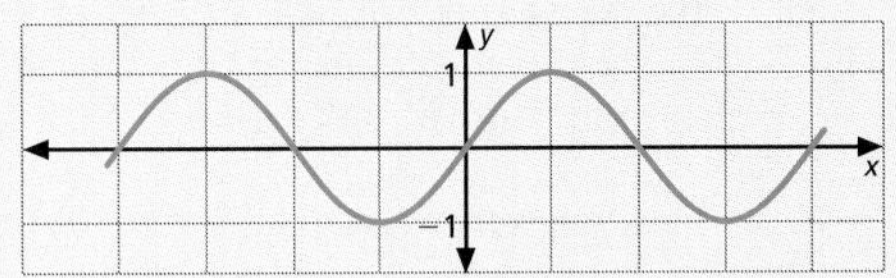

The period of $y = \sin x$ is 360°.

The period of $y = \sin nx$ is $\frac{360°}{n}$.

► The amplitude of a period function is half the difference between its maximum and minimum y-values. It is a measure of the height and depth of a curve.

For 16–17, see additional answers for graphs.

16. Graph $y = 6 \sin x$. State the amplitude. amplitude = 6

17. Graph $y = \sin \frac{1}{2}x$. State the period. period = 720°

LESSON 14-5 ■ Problem Solving, p. 624

► Experience in applying strategies will help you solve a particular problem. Sometimes one strategy will work. There may be times that you will need to use more than one strategy to find a solution.

18. A steamboat paddlewheel with a radius of 40 in. makes one complete revolution every 4 sec. Half the wheel is submerged. Point P on the rim of the wheel is at the water line. Where in relation to the surface of the water will P be 17.2 sec from now?
38 in. above surface

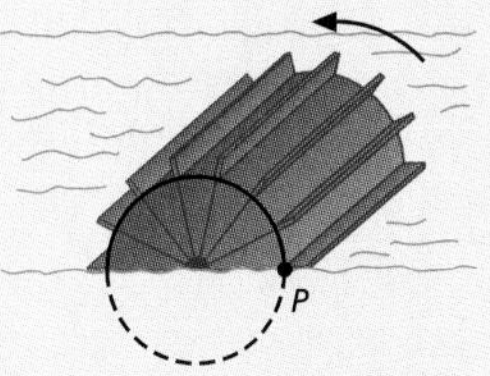

Teaching Strategies

CHALLENGE For the triangle below, suppose you are given sides z and y and the measure of the included angle X. You are not given h. Find a formula for the area A of the triangle in terms of z, y, and angle X. (*Hint:* Find an expression to substitute for h).

$A = \frac{1}{2}yz \sin X$

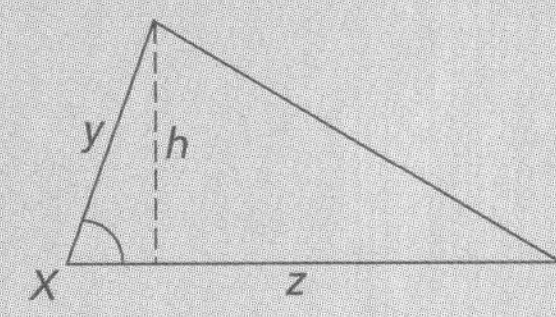

Chapter 14 Test Form A, page 1

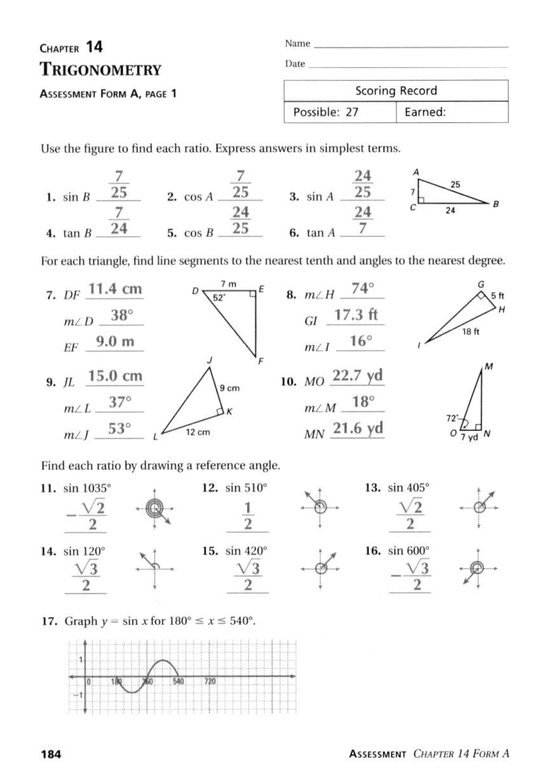

CHAPTER 14
TRIGONOMETRY
ASSESSMENT FORM A, PAGE 1

Name ______
Date ______

Scoring Record	
Possible: 27	Earned:

Use the figure to find each ratio. Express answers in simplest terms.

1. $\sin B$ $\frac{7}{25}$ 2. $\cos A$ $\frac{7}{25}$ 3. $\sin A$ $\frac{24}{25}$

4. $\tan B$ $\frac{7}{24}$ 5. $\cos B$ $\frac{24}{25}$ 6. $\tan A$ $\frac{24}{7}$

For each triangle, find line segments to the nearest tenth and angles to the nearest degree.

7. DF 11.4 cm
$m\angle D$ 38°
EF 9.0 m

8. $m\angle H$ 74°
GI 17.3 ft
$m\angle I$ 16°

9. JL 15.0 cm
$m\angle L$ 37°
$m\angle J$ 53°

10. MO 22.7 yd
$m\angle M$ 18°
MN 21.6 yd

Find each ratio by drawing a reference angle.

11. $\sin 1035°$ $-\frac{\sqrt{2}}{2}$ 12. $\sin 510°$ $\frac{1}{2}$ 13. $\sin 405°$ $\frac{\sqrt{2}}{2}$

14. $\sin 120°$ $\frac{\sqrt{3}}{2}$ 15. $\sin 420°$ $\frac{\sqrt{3}}{2}$ 16. $\sin 600°$ $-\frac{\sqrt{3}}{2}$

17. Graph $y = \sin x$ for $180° \le x \le 540°$.

184 ASSESSMENT *CHAPTER 14 FORM A*

Chapter 14 Test Form A, page 2

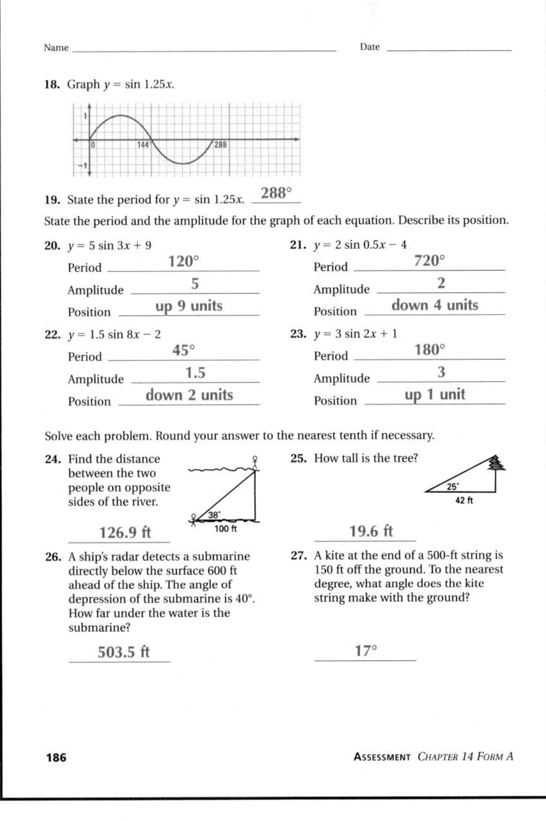

Name ______ Date ______

18. Graph $y = \sin 1.25x$.

19. State the period for $y = \sin 1.25x$. 288°

State the period and the amplitude for the graph of each equation. Describe its position.

20. $y = 5 \sin 3x + 9$
Period 120°
Amplitude 5
Position up 9 units

21. $y = 2 \sin 0.5x - 4$
Period 720°
Amplitude 2
Position down 4 units

22. $y = 1.5 \sin 8x - 2$
Period 45°
Amplitude 1.5
Position down 2 units

23. $y = 3 \sin 2x + 1$
Period 180°
Amplitude 3
Position up 1 unit

Solve each problem. Round your answer to the nearest tenth if necessary.

24. Find the distance between the two people on opposite sides of the river.
126.9 ft

25. How tall is the tree?
19.6 ft

26. A ship's radar detects a submarine directly below the surface 600 ft ahead of the ship. The angle of depression of the submarine is 40°. How far under the water is the submarine?
503.5 ft

27. A kite at the end of a 500-ft string is 150 ft off the ground. To the nearest degree, what angle does the kite string make with the ground?
17°

186 ASSESSMENT *CHAPTER 14 FORM A*

Chapter 14 **Review** 627

Chapter 14 Test Form B, page 1

CHAPTER 14
TRIGONOMETRY
ASSESSMENT FORM B, PAGE 1

Name ______
Date ______

Scoring Record	
Possible: 27	Earned:

Use the figure to find each ratio. Express answers in simplest terms.

1. sin A $\frac{60}{61}$ 2. cos A $\frac{11}{61}$ 3. sin C $\frac{11}{61}$
4. tan C $\frac{11}{60}$ 5. cos C $\frac{60}{61}$ 6. tan A $\frac{60}{11}$

For each triangle, find line segments to the nearest tenth and angles to the nearest degree.

7. DF 16.6 m; $m\angle E$ 67°; $m\angle F$ 23
8. $m\angle G$ 75°; GI 6.5 ft; HI 24.1 ft
9. $m\angle J$ 37; $m\angle L$ 53°; JK 16 mm
10. MN 8 yd; NO 3.9 yd; $m\angle N$ 61°

Find each ratio by drawing a reference angle.

11. sin 390° $\frac{1}{2}$
12. sin 1140° $\frac{\sqrt{3}}{2}$
13. sin 225° $-\frac{\sqrt{2}}{2}$
14. sin 945° $-\frac{\sqrt{2}}{2}$
15. sin 570° $-\frac{1}{2}$
16. sin 660° $-\frac{\sqrt{3}}{2}$

17. Graph $y = \sin x$ for $540° \le x \le 900°$.

188 ASSESSMENT *CHAPTER 14 FORM B*

Chapter 13 Test Form B, page 2

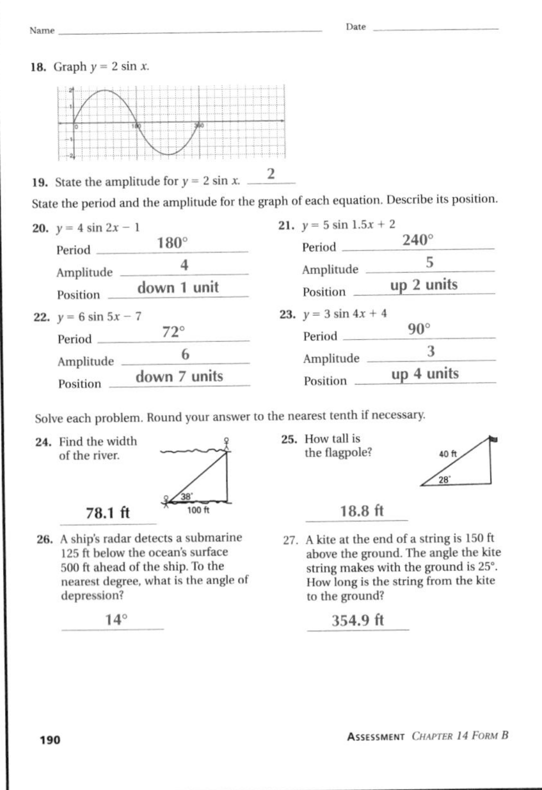
Name ______ Date ______

18. Graph $y = 2 \sin x$.

19. State the amplitude for $y = 2 \sin x$. 2

State the period and the amplitude for the graph of each equation. Describe its position.

20. $y = 4 \sin 2x - 1$
Period 180°
Amplitude 4
Position down 1 unit

21. $y = 5 \sin 1.5x + 2$
Period 240°
Amplitude 5
Position up 2 units

22. $y = 6 \sin 5x - 7$
Period 72°
Amplitude 6
Position down 7 units

23. $y = 3 \sin 4x + 4$
Period 90°
Amplitude 3
Position up 4 units

Solve each problem. Round your answer to the nearest tenth if necessary.

24. Find the width of the river. 78.1 ft

25. How tall is the flagpole? 18.8 ft

26. A ship's radar detects a submarine 125 ft below the ocean's surface 500 ft ahead of the ship. To the nearest degree, what is the angle of depression? 14°

27. A kite at the end of a string is 150 ft above the ground. The angle the kite string makes with the ground is 25°. How long is the string from the kite to the ground? 354.9 ft

190 ASSESSMENT *CHAPTER 14 FORM B*

CHAPTER 14 ASSESSMENT

Use the figure at the right to find each ratio.

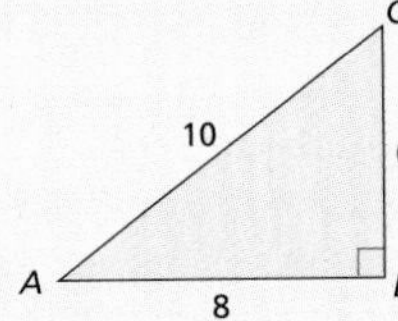

1. sin A $\frac{6}{10} = \frac{3}{5}$
2. cos C $\frac{6}{10} = \frac{3}{5}$
3. sin C $\frac{8}{10} = \frac{4}{5}$
4. tan A $\frac{6}{8} = \frac{3}{4}$
5. cos A $\frac{8}{10} = \frac{4}{5}$
6. tan C $\frac{8}{6} = 1\frac{1}{3}$

Find the following in $\triangle CAT$.

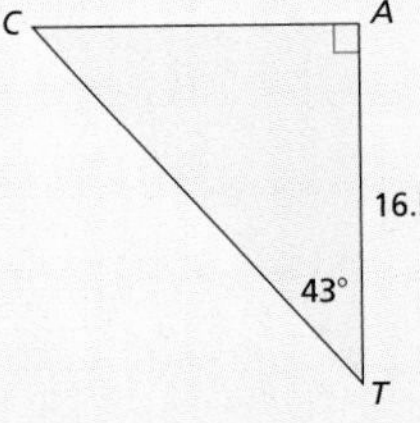

7. CT to the nearest tenth. 22.6
8. m$\angle C$ 47°
9. CA to the nearest tenth. 15.4

10. The world's largest tree is the General Sherman at Sequoia National Park. It's about 273 ft tall. If you're standing 64 ft from the base, what is the angle of elevation? ≈77°

Find each ratio by drawing a reference angle.

11. sin 120° $\frac{\sqrt{3}}{2}$
12. cos 240° $-\frac{1}{2}$
13. cos (−45°) $\frac{\sqrt{2}}{2}$
14. tan 300° $-\sqrt{3}$
15. sin (−780°) $\frac{-\sqrt{3}}{2}$
16. tan 405° 1

Graph. State the period and amplitude. For 17–19, see additional answers for graphs.

17. $y = \frac{3}{2} \sin x$ $\frac{3}{2}$; 360°
18. $y = \cos 3x$ 1; 120°
19. $y = 3 \sin x - 2$ 3; 360°

Sound is caused from continuous vibrations. You can think of a sine graph when describing sound. When the amplitude of sound vibrations is large, the sound is more intense.

20. Suppose the graph $y = 3 \sin x$ describes a sound vibration. To make the sound more intense, the amplitude grows to 8. Write and graph the equation that describes the intense sound. See additional answers.

ADDITIONAL ANSWERS

17.
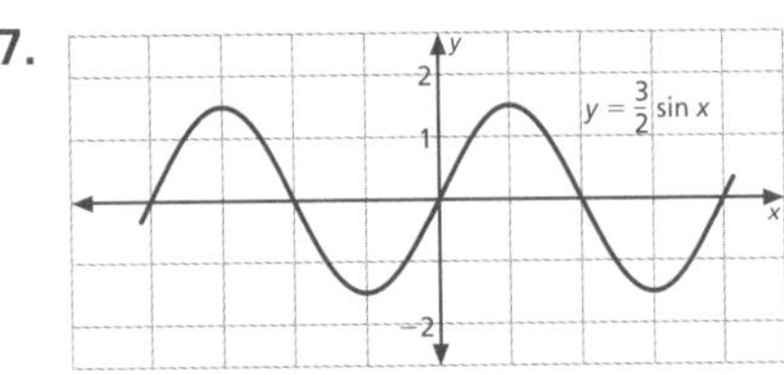

18.
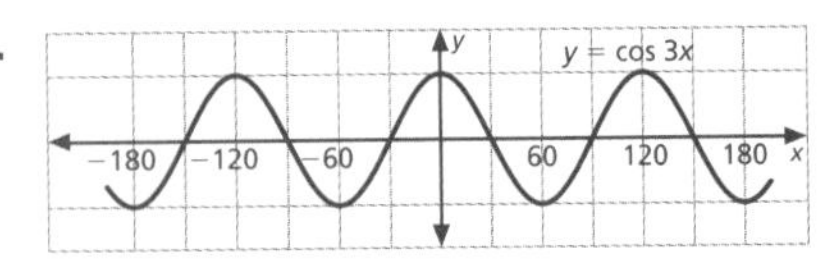

19.
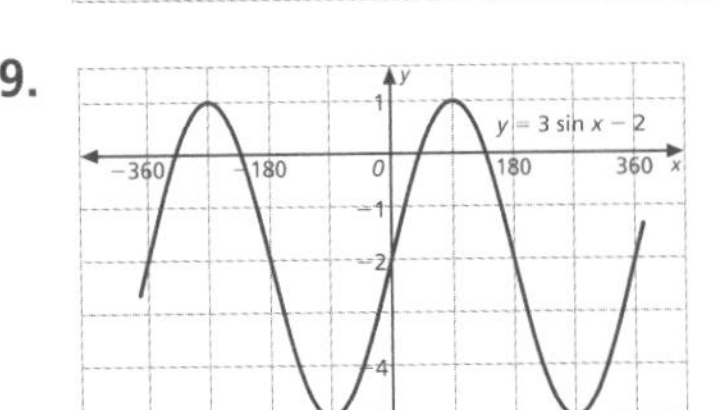

20.
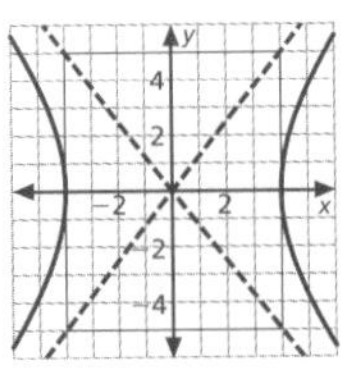

Chapter 14 Alternative Assessment

Connections

NAVIGATION Research to discover the altitude commercial jetliners must reach for flight (assuming it's a long flight) and the land miles it takes the plane to reach that altitude. Then find the angle of elevation the plane travels and the distance it travels from take off to the point where it reaches cruising altitude. Find this data for several different makes and/or sizes of planes, if possible. Explain how you found the information and how you calculated the angle and distance.

Cooperative Learning

SCHOOL FLAG Work in a group. Measure the distance from a convenient point (the front door, the edge of the parking lot) to the bottom of the school flagpole. Determine the height of the top of the flag. Then calculate the angle of elevation and the distance from the point selected on the ground to the top of the flag. What happens to these measures if the flag is flown at half-mast? If a state or school flag is flown below the US flag, find the angle of elevation and distance for the top of that flag as well.

Critical Thinking

SINE WAVES Use a graphing calculator to experiment with sine waves. Graph several different sines to discover what causes an increase or decrease in amplitude, as well as a shorter or longer period. Write a brief explanation of your conclusions, giving examples that support your theories. Include drawings of different sine graphs that you looked at in your investigation.

Chapter Investigation

EXTENSION Make a scale drawing to show how you determined the height or depth of one of the objects you measured. Label each part of your drawing and include all measurements.

Chapter Investigation

The benchmarks and expectations for this extension are as follows.

- Students gather the materials to build a homemade quadrant.
- Students assemble the quadrant.
- Students work with a partner or in a small group to determine the height (or depth) of five objects on their school grounds or nearby community.
- Students make a scale drawing to show how they determined the height or depth of one of the objects they measured. They label each part of their drawing and include all measurements.

Connections

Students will have to assume that the planes fly in a straight line from take off to cruising altitude.

SCORING RUBRIC 5—Student finds the information, correctly and accurately calculates the angle and distance, and clearly explains the process in detail. **4**—Student finds the information, correctly and accurately calculates the angle and distance, and explains the process. **3**—Student finds the information, correctly calculates the angle and distance with minor errors, and explains the process. **2**—Student finds the information, calculates the angle and distance with minor errors, and offers a partial or incomplete explanation of the process. **1**—Student finds some of the information, calculates the angle and distance with significant errors, and offers no explanation of the process. **0**—Student makes no attempt to find information.

Cooperative Learning

If a flagpole is not available, you can use a point of a window on a building or have students work independently using a flagpole or similar object located off the school grounds.

SCORING RUBRIC 5—Student accurately measures distances, correctly calculates the angle and distance, and makes correct assumptions about the flag at half mast. **4**—Student accurately measures distances, calculates the angle and distance, and makes reasonable assumptions. **3**—Student measures distances, calculates the angle and distance with minor errors, and makes reasonable assumptions. **2**—Student measures distances, calculates the angle or the distance with significant errors, and makes no assumptions. **1**—Student measures distances, offers incomplete calculations of the angle or the distance with significant errors, and makes no assumptions. **0**—Student makes no attempt to measure or calculate the angle or distance.

See page 630 for Critical Thinking Scoring Rubric.

CUMULATIVE REVIEW

Cumulative review is the best way to maintain previously taught skills and concepts. This will keep students prepared for new lessons that build on previously covered skills.

Cumulative Review covers:

Lesson 1-3
Lesson 1-7
Lesson 2-5
Lesson 3-3
Lesson 4-1
Lesson 5-2
Lesson 5-7
Lesson 6-1
Lesson 6-6
Lesson 8-5
Lesson 8-6
Lesson 9-1
Lesson 9-7
Lesson 10-1
Lesson 11-1
Lesson 11-4
Lesson 11-6
Lesson 11-7
Lesson 11-9
Lesson 12-1
Lesson 12-6
Lesson 13-1
Lesson 13-4
Lesson 13-5
Lesson 14-1
Lesson 14-2

Critical Thinking

Encourage students to organize a list of sines to graph before actually graphing. This will help them discover a pattern more easily than working with random graphs.

SCORING RUBRIC 5—Student investigates sine waves in a well-organized fashion, and draws insightful conclusions about how the amplitude and period of a sine wave are affected. 4—Student investigates sine waves in an organized fashion, and draws correct conclusions about how the amplitude and period of a sine wave are affected. 3—Student investigates sine waves, and draws some correct conclusions about how the amplitude and period of a sine wave are affected. 2—Student investigates sine waves, but draws few correct conclusions about how the amplitude and period of a sine wave are affected. 1—Student attempts to investigate sine waves, but draws no correct conclusions about how the amplitude and period of a sine wave are affected. 0—Student makes no attempt to investigate sine waves

CHAPTERS 1–14 CUMULATIVE REVIEW

Let $U = \{1, 2, 3, 4, 5, 6, 7, 8, 9\}$, $A = \{2, 4, 6, 8\}$, $B = \{1, 3, 5, 7, 9\}$, and $C = \{4, 5, 6, 7\}$

Find each intersection or union.

1. $A \cap B$ { }
2. $A' \cap C'$ {1, 3, 9}
3. $A \cup B$ {1, 2, 3, 4, 5, 6, 7, 8, 9}
4. $A \cup (C \cap B)$ {2, 4, 5, 6, 7, 8}

Simplify.

5. $(x^4)^3 \cdot x^5$ x^{17}
6. $x^7 \div x^{-11}$ x^{18}

Solve and check.

7. $6x - 8 = 12.4$ $x = 3.4$
8. $5(x + 3) = 2x$ $x = -5$
9. $14x - 8 = 5x + 7$ $x = 1\frac{2}{3}$

10. Find the midpoint of $\overline{MQ}$. O

11. Find the midpoint of $\overline{LT}$. P

L M N O P Q R S T
−4 −3 −2 −1 0 1 2 3 4

12. Find the area of the shaded region. 8 cm²

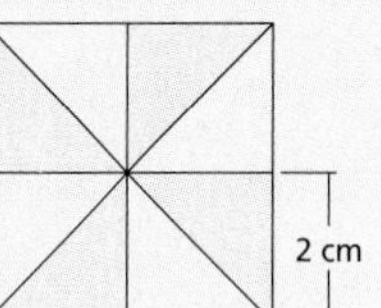

13. Find the volume of the cone below. Use 3.14 for π. Round your answer to the nearest whole number. 340 mm³

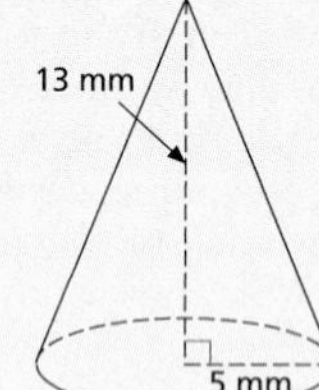

Solve the system of equations.

14. $\begin{cases} -5x + y = 12 \\ 5x + 8y = 6 \end{cases}$ (−2, 2)

15. $\begin{cases} 6x - 3y = 2 \\ x + 2y = -3 \end{cases}$ $-\frac{1}{3}, -\frac{4}{3}$

16. Find the sum. $\begin{bmatrix} 7 & 6 \\ -3 & 8 \end{bmatrix} + \begin{bmatrix} 1 & -6 \\ -3 & -4 \end{bmatrix}$ $\begin{bmatrix} 8 & 0 \\ -6 & 4 \end{bmatrix}$

17. Find the product. $3\begin{bmatrix} 6 & -3 & 4 \\ -2 & 9 & 12 \\ 6 & 0 & 5 \end{bmatrix}$ $\begin{bmatrix} 18 & -9 & 12 \\ -6 & 27 & 36 \\ 18 & 0 & 15 \end{bmatrix}$

18. Find the probability of rolling a die and getting an odd number. $\frac{1}{2}$

19. Find the standard deviation for the set of numbers 2, 4, 6, 8, 10. 2.83

20. Write (32)(27) in simplest radical form. $12\sqrt{6}$

21. Find $\sqrt[4]{625}$. 5

Simplify.

22. $(y^2 - 8x + 3x^2) - (5x - 3y^2 + x^2)$ $2x^2 - 13x + 4y^2$

23. $(4x + 3y)(x - 2y)$ $4x^2 - 5xy - 6y^2$

Factor.

24. $81x^2 - 16y^2$ $(9x - 4y)(9x + 4y)$

25. $x^2 - 14x + 45$ $(x - 9)(x - 5)$

26. $6x^2 - x - 12$ $(3x + 4)(2x - 3)$

Use the figure to find each ratio.

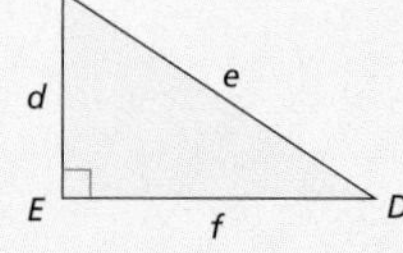

27. $\sin F$ f/e
28. $\cos D$ f/e
29. $\tan D$ d/e

30. If y varies directly as x^2 and $y = 100$ when $x = 5$, what is y when $x = 7$? 196

Teaching Strategies

Remind students that there are several methods for solving systems of equations, as shown in Exercises 18 and 19 of the Cumulative Review. Students should examine the systems and choose the easiest method for solving. For instance, some students may prefer the substitution method and automatically begin calculations; however, Exercise 18 can be easily solved using the addition/subtraction method.

Chapters 1–14 Cumulative Assessment

Standardized Test Preparation—Quantitative Comparison

In each question, compare the quantity in Column 1 with the quantity in Column 2. Select the letter of the correct answer from these choices:

A. The quantity in Column 1 is greater.

B. The quantity in Column 2 is greater.

C. The two quantities are equal.

D. The relationship cannot be determined by the information given.

Notes: In some questions, information that refers to one or both columns is centered over both columns. A symbol used in both columns has the same meaning in each column. All variables represent real numbers. Most figures are not drawn to scale.

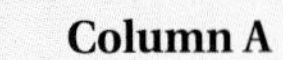

	Column A	Column B

1. $6(2x - 8) = 5x - 6$
A

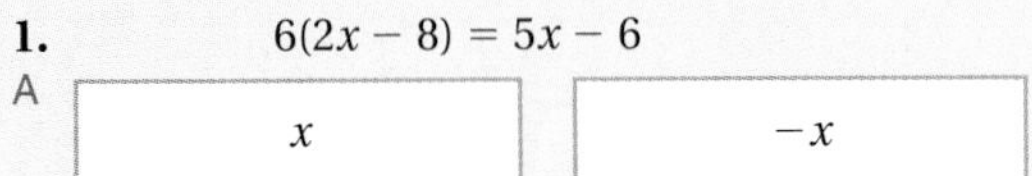

	Column A	Column B
1.	x	$-x$

2. 28, 30, 30, 32, 45
A

	Column A	Column B
2.	the mean of the set of data	the mode of the set of data

3.
B

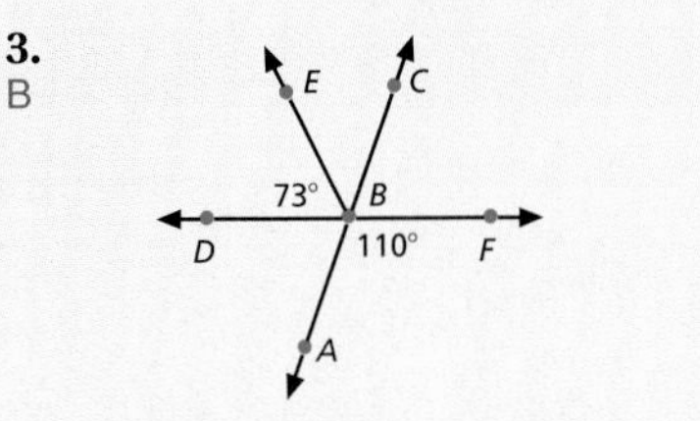

	Column A	Column B
3.	$\angle CBE$	$\angle CBF$

4. Solve the system of equations:
B
$-3x + 4y = 9$
$2x + y = 5$

	Column A	Column B
4.	x	y

5. Find $P'(x, y)$ of $P(-3, -5)$ after a 90° counterclockwise rotation.
A

	Column A	Column B
5.	x	$-y$

6.
A

(Right triangle with legs 12 cm and hypotenuse 15 cm)

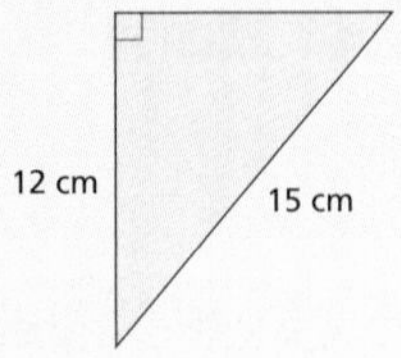

12 cm
15 cm

	Column A	Column B
6.	$15 - x$	$15 - 12$

7. Find the midpoint of the line segment with endpoints (3, 4) and (−2, 2).
B

	Column A	Column B
7.	x	y

8. The equation for a circle with radius $\sqrt{2}$ and center $(-6, 3)$ is $(x + k)^2 + (y + j)^2 = m$
A

	Column A	Column B
8.	k	j

9.
B

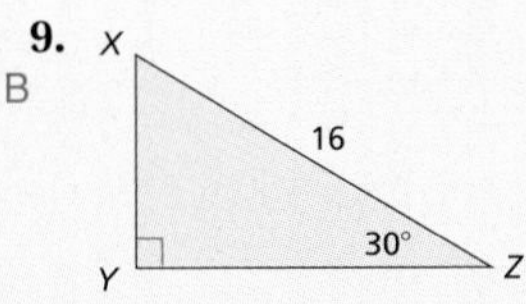

	Column A	Column B
9.	XY	YZ

10. $(4x - 2y)(3x + 6y) = ax^2 + bxy - cy^2$
D

	Column A	Column B
10.	b	y

11. $6x^2 + 2x + 15 = (ax + b)(cx + d)$
C

	Column A	Column B
11.	a	b

12. CONSTRUCTED RESPONSE Find the sine, cosine, and tangent of 120° in radical form.

$\sin 120° = \frac{\sqrt{3}}{2}$

$\cos 120° = -\frac{1}{2}$

$\tan 120° = -\sqrt{3}$

Standardized Test Preparation

QUANTITATIVE COMPARISON QUESTIONS Quantitative comparison questions are generally written with fewer words and require less time to compute. There are four answer choices that describe the relationships of equality and inequality. You are required to compare the quantity from Column A with the quantity from Column B and then determine their relationship. Students should carefully read the four answer choices often during the testing period.

Horizontal rules are used as separators between questions and to clarify when figures and additional information accompanies one or more problems.

Understanding the directions for quantitative comparison questions is critical for success since this type of question is not as commonplace as other types of questions.

Testing Strategies

One key test-taking strategy that is often overlooked has to do with sleep and nutrition. Tell students not to stay up too late the night before a test to insure a minimum of 8 hours sleep. Students may need more sleep if they are used to getting more. Tell students they also need to give themselves plenty of time in the morning so they are not rushed and can have a nutritionally balanced breakfast.

Teaching Strategies

When a problem describes a real-world situation, students should draw a diagram or make a sketch if one is not provided. This important problem-solving strategy makes it easier to see whether an answer makes sense. This is especially important when taking standardized tests. The few seconds it will take to make a sketch will save time in the long run.

Data File Index

The *Data File* contains interesting information presented in tables and graphs for your use in solving problems. The data is organized into the following categories: animals, architecture, earth science, economics, entertainment, health and fitness, science, sports, and United States. You will need to refer to this material often as you work through this book. Whenever you come upon the words "USING DATA" at the beginning of an exercise, refer to the *Data Index* to help you locate the information you will need to complete the exercise. For a quick review of commonly used symbols, formulas and other information, you may wish to refer to the last part of the *Data File*, Mathematics.

Data File

ANIMALS

Major U.S. Public Zoological Parks

Zoo	Budget (millions)	Attendance (millions)	Species	Major attractions
Arizona-Sonora Desert Museum (Tucson)	5.5	0.6	1500	Desert grasslands, Hummingbird aviary, Mountain Woodlands
Bronx (N.Y.C.)	38.0	2.0	560	Jungle World, Baboon Reserve, Asian Rain Forest
Chicago (Brookfield)	41.0	2.1	400+	7 Seas Panorama, Tropic World, Habitat Africa, The Swamp
Cincinnati	16.7	1.3	750	Blakely's Barn, Wings of the World, Jungle Trails
Cleveland	10.0	1.3	599	Rain Forest, Wildlife of the Great Lakes, Wolf Wilderness
Dallas	7.4	0.5	387	Wilds of Africa, Chimpanzee Forest, Gorilla Exhibit
Denver	9.0	1.7	600	Tropical Discovery, Primate Panorama, Northern Shores
Detroit	10.5	1.1	270	Penguinarium, Great Apes of Harambee, Interpretive Gallery
Houston Zoological Gardens	12.0	1.4	730	Pigmy Hippo Habitat, World of Primates, Asian Elephants
Los Angeles	17.2	1.3	400	Tiger Falls, Great ape families, Walk through aviary
Miami Metrozoo	7.0	0.5	250	African Plains, Children's Zoo, Koalas, Komodo Dragons
Milwaukee	14.0	1.4	300	Aquatic and Reptile Center, wolf woods, bear dens
National (Wash., D.C.)	24.0	3.0	500	Amazonia, Panda Exhibit, Komodo Dragons
Philadelphia	15.5	1.1	395	Carnivore Kingdom, white lions, red pandas
Riverbanks (Columbia, S.C.)	5.4	0.9	546	Aquarium Reptile Complex, Botanical Garden, Coral Reef
St. Louis	21.0	2.6	740	Big Cat Country, The Living World, Jungle of the Apes
San Diego	72.4	3.0	800	Tiger River, Gorilla Tropics, Polar Bear Plunge, Giant Pandas
San Diego (Wild Animal Park)	35.6	1.6	300+	Heart of Africa, Mombassa Lagoon, Hidden Jungle
San Francisco	12.0	0.9	270	Primate Discovery Center, Aye-Aye exhibit, Gorilla World
Toledo	9.0	0.9	525	Hippoquarium, aquarium, African Savanna, Children's Zoo

Top 10 American Kennel Club Registrations

Breed	Rank	Number registered
Labrador Retriever	1	149,505
Rottweiler	2	89,867
German Shepherd Dog	3	79,076
Golden Retriever	4	68,993
Beagle	5	56,946
Poodle	6	56,803
Dachshund	7	48,426
Cocker Spaniel	8	45,305
Yorkshire Terrier	9	40,216
Pomeranian	10	39,712

Some Endangered Mammals of the U.S.

Common name	Scientific name	Range
Ozark big-eared bat	*Plecotus townsendii ingens*	MO, OK, AZ
Brown or grizzly bear	*Ursus arctos horribilis*	48 conterminous* states
Columbian white-tailed deer	*Odocoileus virginianus leucurus*	WA, OR
San Joaquin kit fox	*Vulpes macrotis mutica*	CA
Southeastern beach mouse	*Peromyscus polionotus phasma*	FL
Ocelot	*Felis pardalis*	TX, AZ
Southern sea otter	*Enhydra lutris hereis*	WA, OR, CA
Florida panther	*Felis concolor coryi*	LA, AR east to SC, FL
Utah prairie dog	*Cynomys parvidens*	UT
Morro Bay kangaroo rat	*Dipodomys heermanni morroensis*	CA
Carolina northern flying squirrel	*Glaucomys sabrinus coloratus*	NC, TN
Hualapai Mexican vole	*Microtus mexicanus hualpaiensis*	AZ
Red wolf	*Canis rufus*	Southeast to central TX

* means – enclosed within one common boundary

The data on this topic are notoriously unreliable because of the many inherent difficulties of timing the movement of most animals–whether running, flying, or swimming–and because of the absense of any standardization of the method of timing, of the distance over which the performance is measured, or of allowance for wind conditions. The most that can be said is that a specimen of the species below has been timed to have attained as a maximum the speed given.

mi/h		mi/h	
219.5	Spine-tailed swift	37	Dolphin
180*	Peregrine falcon	36	Dragonfly
120*	Golden eagle	35	Flying fish
96.29	Racing pigeon	35	Rhinoceros
88	Spurwing goose	35	Wolf
70†	Cheetah	33	Hawk head moth
60	Pronghorn antelope	32	Giraffe
60	Mongolian gazelle	32	Guano bat
57	Quail	30	Blackbird
57	Swordfish	28	Grey heron
53	Partridge	25	California sea lion
45	Red kangaroo	24	African elephant
45	English hare	23	Salmon
40	Red fox	22.8	Blue whale
40	Mute swan	22	Leatherback turtle
38	Swallow	22	Wren
		20	Monarch butterfly

*Stooping
†Unable to sustain a speed of over 44 mi/h over 500 yd.

Architecture

Bridges of the World

Bridge	Height (feet)
Firth of Forth Bridge, Scotland	148
Verrazzano-Narrows Bridge, New York	213
Sydney Harbor Bridge, Australia	171
Tunkhannock Viaduct, Pennsylvania	240
Garabit Viaduct, France	480
Brooklyn Bridge, New York	135

Sydney Harbor Bridge

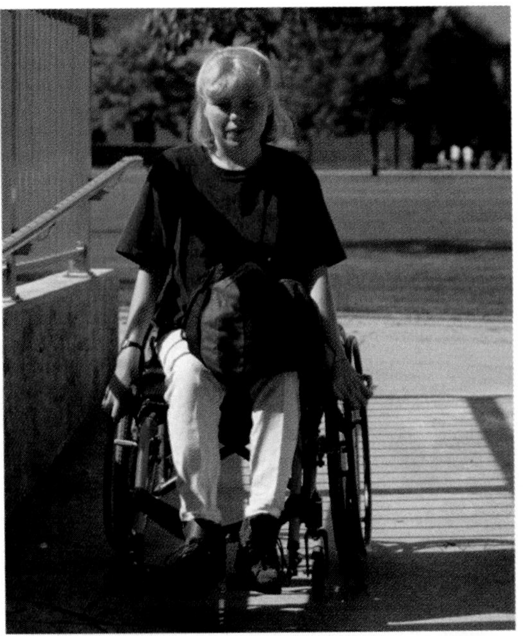

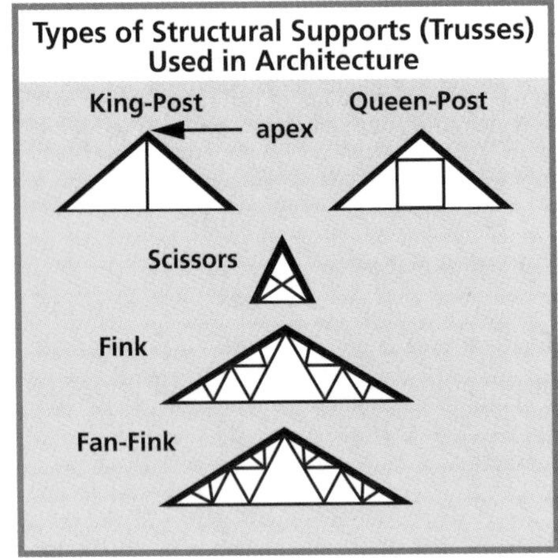

Standards for Slope and Safety

	Maximum slope
Ramps–wheelchair	0.125
Ramps–walking	0.3
Driveway or street parking	0.22
Stairs	0.83

Notable Tall Buildings of the World

Building	City	Year completed	Stories	Height (meters)	Height (feet)
Sears Tower	Chicago	1974	110	443	1454
World Trade Center, North	New York	1972	110	417	1368
World Trade Center, South	New York	1973	110	415	1362
Empire State Building	New York	1931	102	381	1250
Central Plaza	Hong Kong	1992	78	374	1227
Bank of China Tower	Hong Kong	1988	72	368	1209
Amoco Building	Chicago	1973	80	346	1136
John Hancock Center	Chicago	1968	100	344	1127

Noted Rectangular Structures

Structure	Country	Length (meters)	Width (meters)
Parthenon	Greece	69.5	30.9
Palace of the Governors	Mexico	96	11
Great Pyramid of Cheops	Egypt	230.6	230.6
Step Pyramid of Zosar	Egypt	125	109
Temple of Hathor	Egypt	290	280
Cleopatra's Needle (base)	England*	2.4	2.3
Ziggurat of Ur (base)	Middle East	62	43
Guanyin Pavilion of Dule Monastery	China	20	14
Izumo Shrine	Japan	10.9	10.9
Kibitsu Shrine (main)	Japan	14.5	17.9
Kongorinjo Hondo	Japan	21	20.7
Bakong Temple, Roluos	Cambodia	70	70
Ta Keo Temple	Cambodia	103	122
Wat Kukut Temple, Lampun	Thailand	23	23
Tsukiji Hotel	Japan	67	27

*Gift to England from Egypt

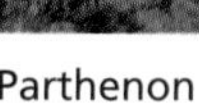
Parthenon

Housing Units-Summary of Characteristics and Equipment, by Tenure and Region: One Recent Year

(In thousands of units, except as indicated. Based on the American Housing Survey)

Item	Total housing units	Sea-sonal	Year-round units							
			Occupied							
			Total	Owner	Renter	Northeast	Midwest	South	West	Vacant
Total units	**99,931**	**3,182**	**88,425**	**56,145**	**32,280**	**18,729**	**22,142**	**30,064**	**17,490**	**8,324**
Percent distribution	100.0	3.2	88.5	56.2	32.3	21.2	25.0	34.0	19.8	8.3
Units in structure:										
Single family detached	60,607	1,834	55,076	46,703	8,373	9,368	14,958	19,984	10,766	3,697
Single family attached	4,514	64	4,102	2,211	1,890	1,431	814	1,212	645	349
2-4 units	11,655	134	10,217	1,996	8,221	3,324	2,515	2,426	1,952	1,304
5-9 units	5,134	73	4,372	344	4,029	903	967	1,408	1,094	689
10-19 units	4,558	99	3,760	261	3,500	818	766	1,360	817	699
20-49 units	3,530	146	2,913	287	2,627	904	603	651	756	470
50 or more units	3,839	135	3,230	438	2,792	1,540	661	583	446	474
Mobile home or trailer	6,094	698	4,754	3,906	848	440	860	2,440	1,014	642

EARTH SCIENCE

Some Principal Rivers of the World

River	Length (miles)
Amazon	4000
Arkansas	1459
Columbia	1243
Danube	1776
Ganges	1560
Indus	1800
Mackenzie	2635
Mississippi	2340
Missouri	2540
Nile	4160
Ohio	1310
Orinoco	1600
Paraguay	1584
Red	1290
Rhine	820
Rio Grande	1900
St. Lawrence	800
Snake	1038
Thames	236
Tiber	252
Volga	2194
Zambezi	1700

Average Daily Temperatures (°F)

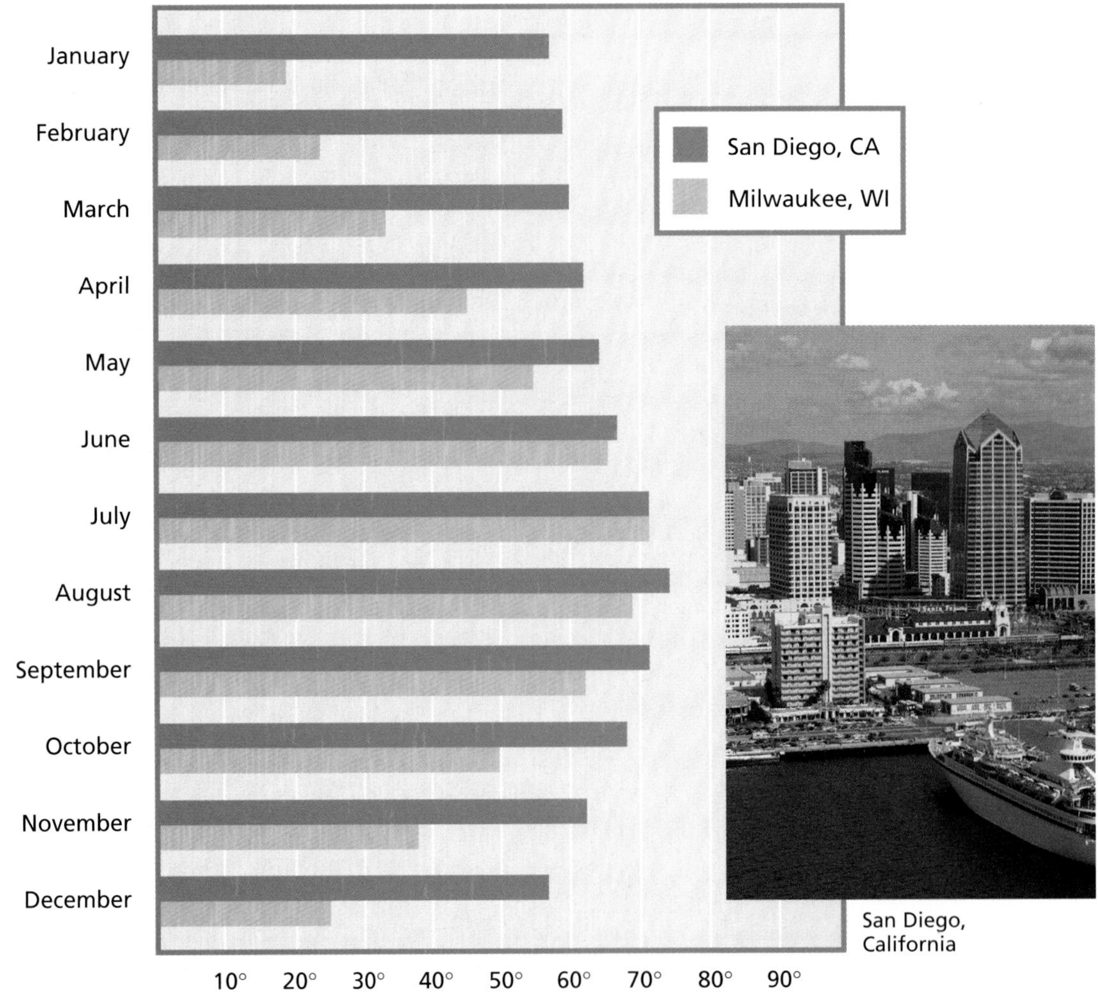

San Diego, California

Size and Depth of the Oceans

Ocean	Square miles	Greatest depth (feet)
Pacific	63,800,000	36,161
Atlantic	31,800,000	30,249
Indian	28,900,000	24,441
Arctic	5,400,000	17,881

Measuring Earthquakes

The energy of an earthquake is generally reported using the Richter scale, a system developed by American geologist Charles Richter in 1935, based on measuring the heights of wave measurements on a seismograph.

On the Richter scale, each single-integer increase represents 10 times more ground movement and 30 times more energy released. The change in magnitude between numbers on the scale can be represented by 10^x and 30^x, where x represents the change in the Richter scale measure. Therefore, a 3.0 earthquake has 100 times more ground movement and 900 times more energy released than a 1.0 earthquake.

Richter scale	
2.5	Generally not felt, but recorded on seismometers.
3.5	Felt by many people.
4.5	Some local damage may occur.
6.0	A destructive earthquake.
7.0	A major earthquake. About ten occur each year.
8.0 and above	Great earthquakes. These occur once every five to ten years.

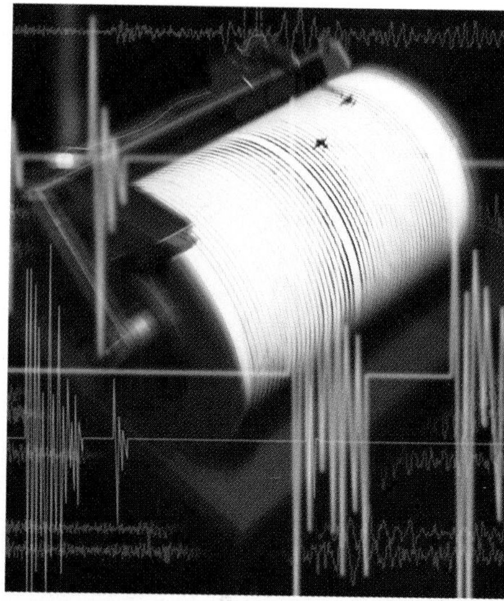
Richter Scale

Mount McKinley, Alaska

Highest and Lowest Continental Altitudes

Continent	Highest point	Feet of elevation	Lowest point	Feet below sea level
Asia	Mount Everest, Nepal-Tibet	29,028	Dead Sea, Israel-Jordan	1,312
South America	Mount Aconcagua, Argentina	22,834	Valdes Peninsula, Argentina	131
North America	Mount McKinley, Alaska	20,320	Death Valley, California	282
Africa	Kilmanjaro, Tanzania	19,340	Lake Assai, Djibouti	512
Europe	Mount El'brus, USSR	18,510	Caspian Sea, USSR	92
Antarctica	Vinson Massif	16,864	Unknown	–
Australia	Mount Kosciusko, New South Wales	7,310	Lake Eyre, South Australia	52

ECONOMICS

Money Around the World

Country	Basic monetary unit	Chief fractional unit	Coin and paper denominations
Australia	dollar	cent	100, 50, 20,10, 5,and 2 dollar notes; 2 and 1 dollar coins; 50, 20, 10, 5, 2, and 1 cent coins
Canada	dollar	cent	1000, 500, 100, 50, 20, 10, 5, 2, and 1 dollar notes; 1 dollar coin; 25 10, 5, and 1 cent coins
France	franc	centime	500, 200, 100, 50, and 20 franc notes; 10, 5, 2, and 1 franc coins; 50, 20, 10, and 5 centime coins
India	rupee	paise	1000, 500, 100, 40, 20 10, 5, 2, and 1 rupee notes; 2 and 1 rupee coins; 50 25, and 20 paise coins
Japan	yen	100 sen (not used)	10,000, 5000 and 1000 yen notes; 500, 100, 50, 10, 5, and 1 yen coins
Mexico	peso	centavo	50,000, 20,000, 10,000, 5000, 2000, 1000 and 500 peso notes; 500, 200, 100, 50, 20, 10, 5, 2, and 1 peso coins
Netherlands	guilder	cent	1000, 250, 100, 50, 25, 10, and 5 guilder notes; 5, 2.5, and 1 guilder coins; 25, 10, 5 cent coins
Sudan	pound	piasters, milliemes (1000 milliemes)	50, 20, 10, 5, and 1 pound notes; 50 and 25 piaster notes; 50, 10, 5, and 2 piaster coins; 10, 5, 2, and 1 millieme coins

In the United States the basic monetary unit is the dollar and the chief fractional unit is the penny. One dollar - 100 pennies. Unless noted otherwise, the basic monetary unit equals 100 chief fractional units for the countries listed above.

United States Foreign Trade
(millions of dollars)

Country	U.S. exports		U.S. imports	
	1992	1996	1992	1996
Canada	90,423	134,210	100,724	155,893
France	14,579	14,455	14,658	18,646
Japan	46,856	67,607	96,858	115,187
Mexico	40,469	56,791	35,588	74,297
Venezuela	5,316	4,749	8,180	13,173

State General Sales and Use Taxes, 1997

State	Percent rate	State	Percent rate	State	Percent rate
Alabama	4	Kentucky	6	Ohio	5
Arizona	5	Louisiana	4	Oklahoma	4.5
Arkansas	4.625	Maine	6	Pennsylvania	6
California	6	Maryland	5	Rhode Island	7
Colorado	3	Massachusetts	5	South Carolina	5
Connecticut	6	Michigan	6	South Dakota	4
D.C.	5.75	Minnesota	6.5	Tennessee	6
Florida	6	Mississippi	7	Texas	6.25
Georgia	4	Missouri	4.225	Utah	4.75
Hawaii	4	Nebraska	5	Vermont	5
Idaho	5	Nevada	6.5	Virginia	3.5
Illinois	6.25	New Jersey	6	Washington	6.5
Indiana	5	New Mexico	5	West Virginia	6
Iowa	5	New York	4	Wisconsin	5
Kansas	4.9	North Carolina	4	Wyoming	4
		North Dakota	5		

NOTE: Alaska, Delaware, Montana, New Hampshire, and Oregon have no statewide sales and use taxes.

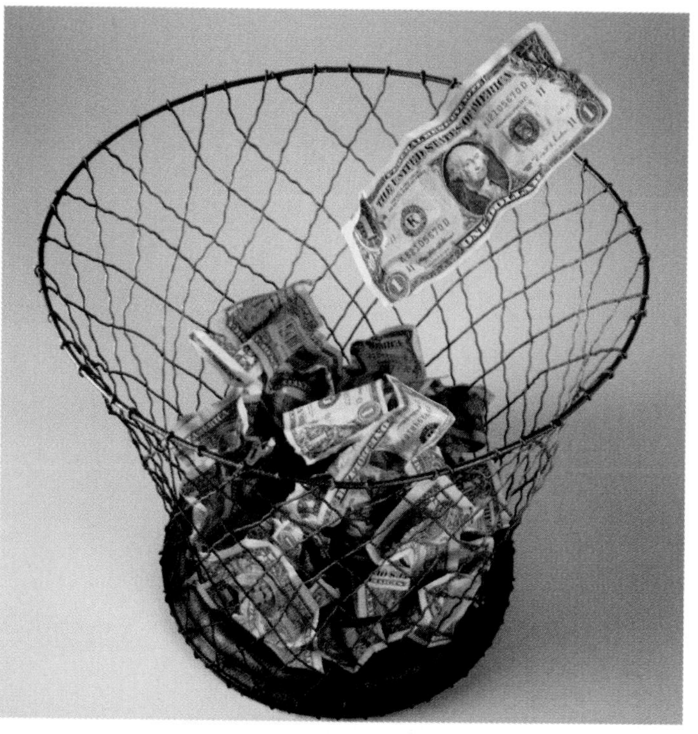

The Shrinking Value of the Dollar

Year	5-lb flour	1-lb round steak	1-qt milk	10-lb potatoes
1890	$0.15	$0.12	$0.07	$0.16
1910	0.18	0.17	0.08	0.17
1930	0.23	0.43	0.14	0.36
1950	0.49	0.94	0.21	0.46
1970	0.59	1.30	0.33	0.90
1975	0.98	1.89	0.45	0.99
1995	1.20	3.20	0.74	3.80
1999	1.70	2.93	0.83	2.82

The average retail cost of certain foods in selected years, 1890-1999

ENTERTAINMENT

All-Time Bestselling Children's Books

Hardcover	
1. *The Poky Little Puppy*, Janette Sebring Lowrey, 1942	14,000,000
2. *The Tale of Peter Rabbit*, Beatrix Potter, 1902	9,331,266
3. *Tootle*, Gertrude Crampton, 1945	8,055,500
4. *Saggy Baggy Elephant*, Kathrun and Byron Jackson, 1947	7,098,000
5. *Scuffy the Tugboat*, Gertude Crampton, 1955	7,065,000
6. *Pat the Bunny*, Dorothy Kunhardt, 1940	6,146,543
7. *Green Eggs and Ham*, Dr. Seuss, 1960	6,065,197
8. *The Cat in the Hat*, Dr. Seuss, 1957	5,643,731
9. *The Littlest Angel*, Charles Tazewell, 1946	5,424,709
10. *One Fish, Two Fish, Red Fish, Blue Fish*, Dr. Seuss, 1960	4,822,331
Paperback	
1. *Charlotte's Web*, E.B. White, 1974	7,894,103
2. *The Outsiders,* S.E. Hinton, 1968	7,798,000
3. *Tales of a Fourth Grade Nothing,* Judy Blume, 1976	6,371,000
4. *Shane,* Jack Schaeffer, 1983	6,161,000
5. *Are You There, God? It's Me, Margaret,* Judy Blume, 1972	6,015,000
6. *Where the Red Fern Grows,* Wilson Rawls, 1974	5,625,000
7. *A Wrinkle in Time,* Madeleine L'Engle, 1973	5,617,000
8. *Island of the Blue Dolphin,* Scott O'Dell, 1971	5,513,000
9. *Little House on the Prairie,* Laura Ingalls Wilder, 1971	5,291,059
10. *Little House in the Big Woods,* Laura Ingalls Wilder, 1971	5,227,120

20 Top Grossing Feature Films

Title	Distributor	Box-Office gross (millions)
*Titantic	Paramount	$584
Star Wars, released 1977	20th Century Fox	$461
E.T. The Extra-Terrestrial	Universal	$400
Jurassic Park	Universal	$357
Forrest Gump	Paramount	$330
The Lion King	Buena Vista	$313
Return of the Jedi	20th Century Fox	$307
Independence Day	20th Century Fox	$306
The Empire Strikes Back	20th Century Fox	$290
Home Alone	20th Century Fox	$285
Jaws	Universal	$260
Batman	Warner Brothers	$251
Men in Black	Columbia	$250
Raiders of the Lost Ark	Paramount	$242
Twister	Warner Brothers	$242
Beverly Hills Cop	Paramount	$235
The Lost World: Jurassic Park	Universal	$229
Ghostbusters	Columbia	$221
Mrs. Doubtfire	20th Century Fox	$219
Ghost	Paramount	$218

* indicates a movie still in release

25 Contemporary Entertainers

Name	Birth date	Birth place
Banderas, Antonio	8/10/60	Malaga, Spain
Berry, Halle	8/14/68	Cleveland, OH
Brooks, Garth	2/7/62	Tulsa, OK
Bullock, Sandra	7/26/67	Arlington, VA
Carrey, Jim	1/17/62	Jackson Point, Canada
Cosby, Bill	7/12/37	Philadelphia, PA
Dion, Celine	3/30/68	Charlemagne, Quebec
Estefan, Gloria	9/1/57	Havana, Cuba
Garcia, Andy	4/12/56	Havana, Cuba
Gibson, Mel	1/3/56	Peekskill, NY
Hanks, Tom	7/9/56	Oakland, CA
Houston, Whitney	8/9/63	East Orange, NJ
Joel, Billy	5/9/49	Bronx, NY
Letterman, David	4/12/47	Indianapolis, IN
McEntire, Reba	3/28/55	McAlester, OK
Midler, Bette	12/1/45	Paterson, NJ
Nicholson, Jack	4/28/37	Neptune, NJ
Norwood, Brandy	2/11/79	Macomb, MS
O'Donnell, Rosie	3/21/62	Commack, NY
Phillips, Lou Diamond	2/17/62	Philippines
Schwarzenegger, Arnold	7/30/47	Graz, Austria
Seinfeld, Jerry	4/29/55	New York, NY
Washington, Denzel	12/28/54	Mt. Vernon, NY
Willis, Bruce	3/19/55	W. Germany
Winfrey, Oprah	1/29/54	Kosciusko, MS

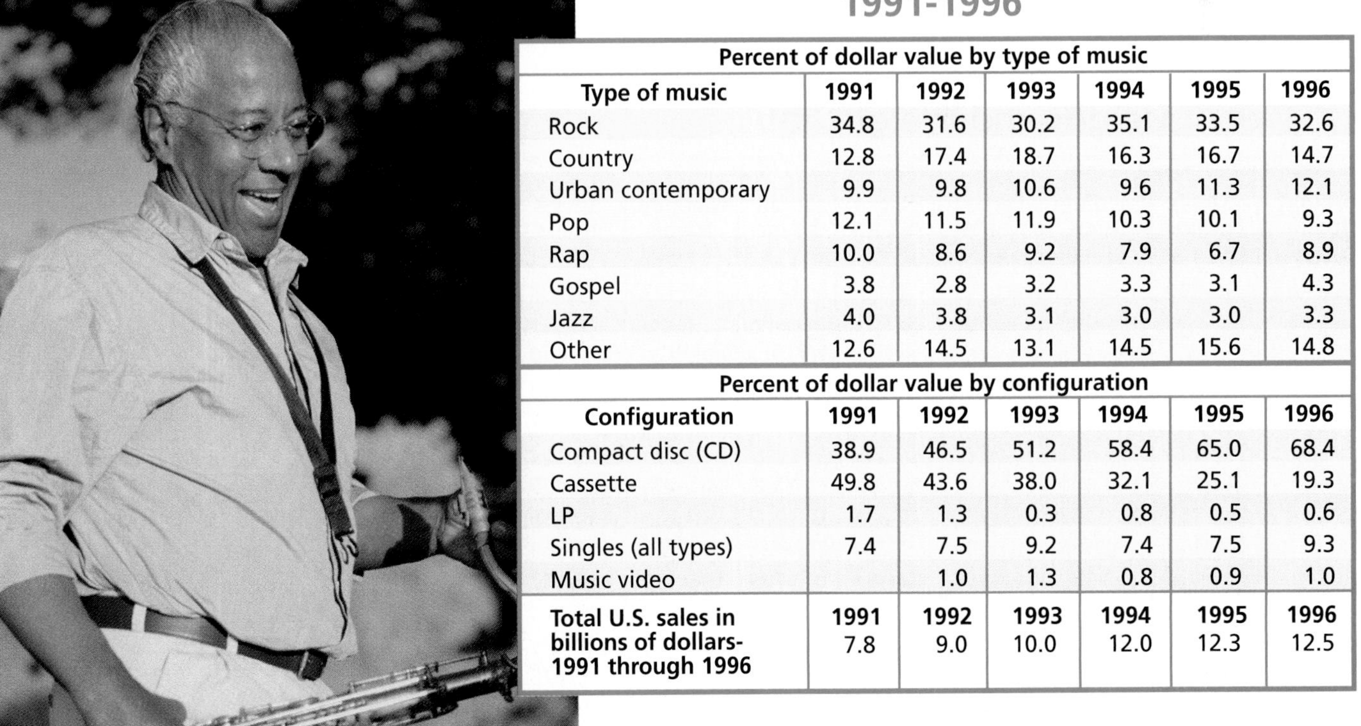

Sales of Recorded Music 1991-1996

Percent of dollar value by type of music						
Type of music	**1991**	**1992**	**1993**	**1994**	**1995**	**1996**
Rock	34.8	31.6	30.2	35.1	33.5	32.6
Country	12.8	17.4	18.7	16.3	16.7	14.7
Urban contemporary	9.9	9.8	10.6	9.6	11.3	12.1
Pop	12.1	11.5	11.9	10.3	10.1	9.3
Rap	10.0	8.6	9.2	7.9	6.7	8.9
Gospel	3.8	2.8	3.2	3.3	3.1	4.3
Jazz	4.0	3.8	3.1	3.0	3.0	3.3
Other	12.6	14.5	13.1	14.5	15.6	14.8
Percent of dollar value by configuration						
Configuration	**1991**	**1992**	**1993**	**1994**	**1995**	**1996**
Compact disc (CD)	38.9	46.5	51.2	58.4	65.0	68.4
Cassette	49.8	43.6	38.0	32.1	25.1	19.3
LP	1.7	1.3	0.3	0.8	0.5	0.6
Singles (all types)	7.4	7.5	9.2	7.4	7.5	9.3
Music video		1.0	1.3	0.8	0.9	1.0
Total U.S. sales in billions of dollars-1991 through 1996	**1991** 7.8	**1992** 9.0	**1993** 10.0	**1994** 12.0	**1995** 12.3	**1996** 12.5

HEALTH & FITNESS

Calorie Count of Selected Dairy Products, Breads, Pastas, Snacks, Fruits and Juices

Food	Approximate amount	Food energy (kcal)
Apple, raw	1	80
Apple juice	1 cup	120
Banana	1	100
Bread, white	1 slice	70
Butter or margarine	1 tbsp	100
Cheese, American	1 oz	105
Cheese, cottage	1 cup	235
Corn flakes	1 cup	95
Crackers, saltine	4	50
Lemonade	1 cup	105
Macaroni with cheese	1 cup	430
Milk, skim	1 cup	85
Milk, whole	1 cup	150
Oatmeal	1 cup	130
Orange	1	65
Orange juice	1 cup	120
Pizza, cheese	1 medium slice	145
Raisins	1/2 oz package	40
Sherbert	1 cup	270
Spaghetti with meatballs	1 cup	330

Height and Weight Tables

Men					Women				
Height ft	in.	Small frame	Medium frame	Large frame	Height ft	in.	Small frame	Medium frame	Large frame
5	2	128-134	131-141	138-150	4	10	102-111	109-121	118-131
5	3	130-136	133-143	140-153	4	11	103-113	111-123	120-134
5	4	132-138	135-145	142-156	5	0	104-115	113-126	122-137
5	5	134-140	137-148	144-160	5	1	106-118	115-129	125-140
5	6	136-142	139-151	146-164	5	2	108-121	118-132	128-143
5	7	138-145	142-154	149-168	5	3	111-124	121-135	131-147
5	8	140-148	145-157	152-172	5	4	114-127	124-138	134-152
5	9	142-151	148-160	155-176	5	5	117-130	127-141	137-155
5	10	144-154	151-163	158-180	5	6	120-133	130-144	140-159
5	11	146-157	154-166	161-184	5	7	123-136	133-147	143-163
6	0	149-160	157-170	164-188	5	8	126-139	136-150	146-167
6	1	152-164	160-174	168-192	5	9	129-142	139-153	149-170
6	2	155-168	164-178	172-197	5	10	132-145	142-156	152-173
6	3	158-172	167-182	176-202	5	11	135-148	145-159	155-176
6	4	162-176	171-187	181-207	6	0	138-151	148-162	158-179

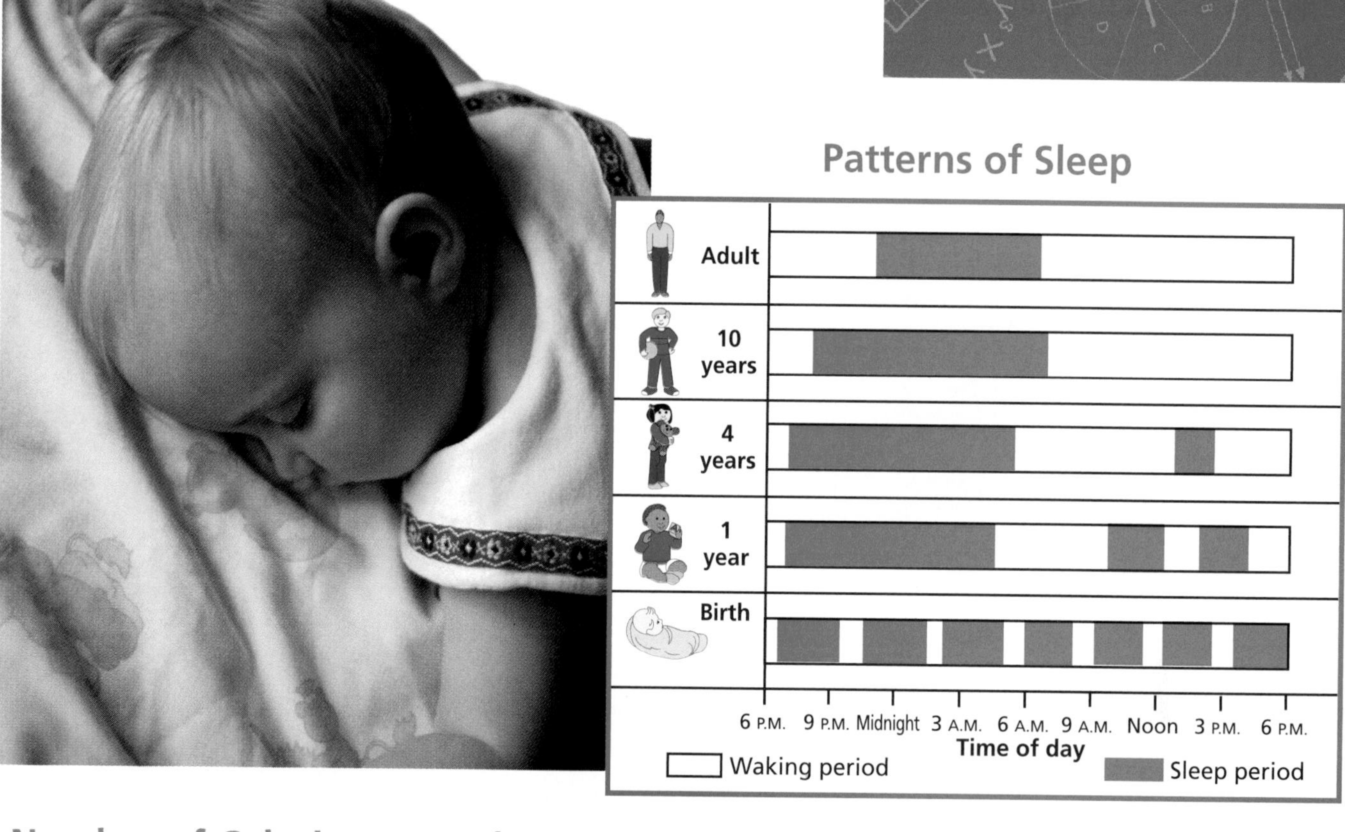

Number of Calories Burned Per Hour by People of Different Body Weights

Exercise	Calories burned per hour		
	110 lb	154 lb	198 lb
Martial arts	620	790	960
Racquetball (2 people)	610	775	945
Basketball (full-court game)	585	750	910
Skiing–cross country (5 mi/h)	550	700	850
downhill	465	595	720
Running–8-min mile	550	700	850
12-min mile	515	655	795
Swimming–crawl, 45 yd/min	540	690	835
crawl, 30 yd/min	330	420	510
Stationary bicycle–15 mi/h	515	655	795
Aerobic dancing–intense	515	655	795
moderate	350	445	540
Walking–5 mi/h	435	555	675
3 mi/h	235	300	365
2 mi/h	145	185	225
Calisthenics–intense	435	555	675
moderate	350	445	540
Scuba diving	355	450	550
Hiking–20-lb pack, 4 mi/h	355	450	550
20-lb pack, 2 mi/h	235	300	365
Tennis–singles, recreational	335	425	520
doubles, recreational	235	300	365
Ice skating	275	350	425
Roller skating	275	350	425

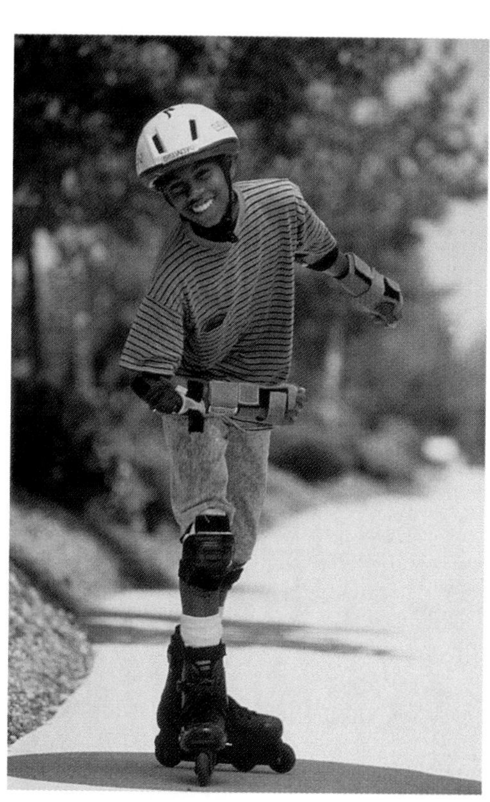

SCIENCE

Carbon Atom

Quantities of Hazardous Waste Burned in U.S.

Incinerator type	Number of facilities	Quantity of hazardous waste burned, lb/yr
Commercial incinerators	17	1.3 billion*
Captive/on-site incinerators	154	2.3 billion*
Cement kilns	25-30	1.8 billion**
Aggregate kilns	6	1.2 billion**
Boilers/other furnaces	900+	1.0 billion**
Total	1100+	7.6 billion

* Highum 1990. Data from review of states' capacity assurance plans.
** Holloway 1990. Data from U.S. EPA's Office of Solid Waste and Emergency Response.

Half-lives of Select Substances

Substance	Half-life
uranium-238	4.5 billion yr
carbon-14	5730 yr
radium-226	1620 yr
strontium-90	28 yr
hydrogen-3	12.3 yr
polonium-210	138 days
thorium-234	25 days
iodine-131	8 days
bismuth-210	5 days
radium-222	4 days
sodium-24	15 h
lead-212	10.6 h
nitrogen-13	10 min
polonium-194	0.7 sec

Blood Types

Type	Percent of population
A Positive	34%
A Negative	6%
B Positive	9%
B Negative	2%
AB Positive	4%
AB Negative	1%
O Positive	38%
O Negative	6%

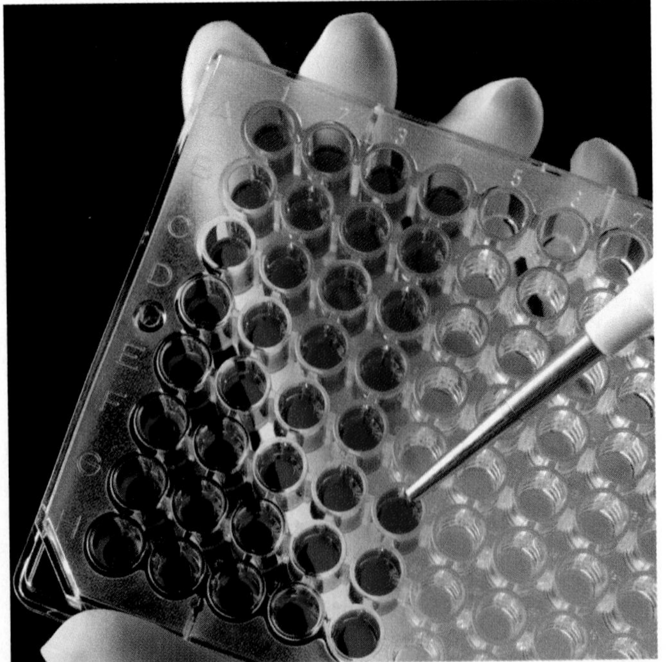

Silicon

Percent of Main Elements Making Up Earth's Crust

Oxygen	47%	Calcium	4%
Silicon	28%	Magnesium	2%
Aluminum	8%	Sodium	3%
Iron	5%	Potassium	3%

Crystal Systems

Crystals can be classified into seven systems. To classify systems, imaginary lines called crystal axes are used. The axes are drawn through the center of the crystal. Below is an example of the arrangements of the faces, the axes and of a mineral from each system.

Cubic (or isometric) system

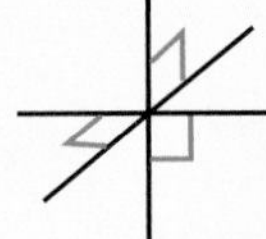

3 equal axes, all at right angles.

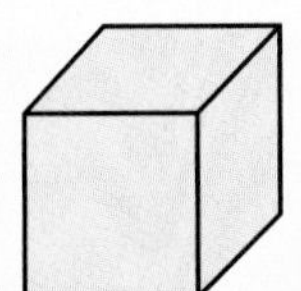

Halite (rock salt)

Tetragonal system

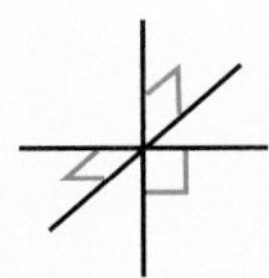

3 axes, one longer than the other two, all at right angles.

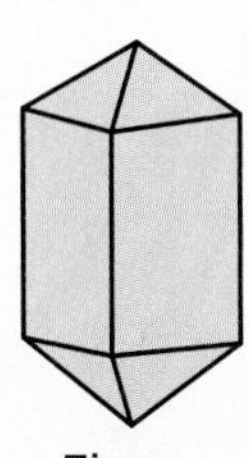

Zircon

Hexagonal system

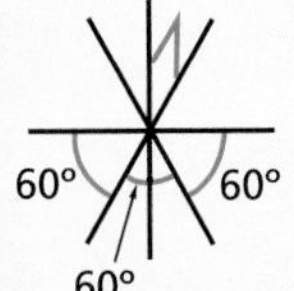

1 vertical axis, longer or shorter than 3 equal horizontal axes at 60°.

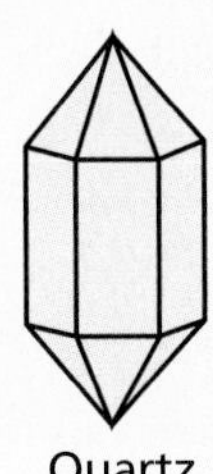

Quartz

Triagonal (or rhombohedral) system

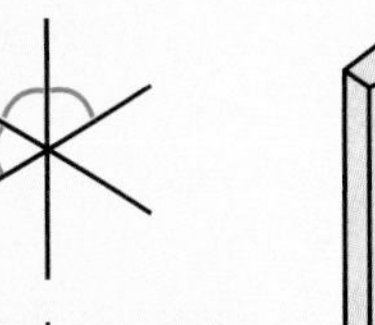

3 equal axes set obliquely at equal angles to each other.

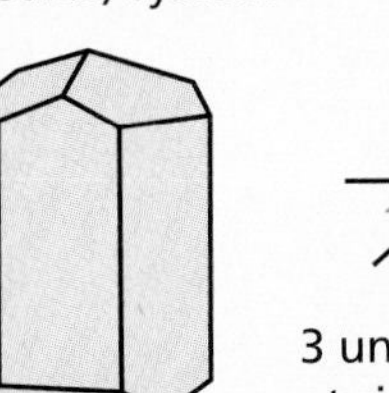

Calcite

Orthorhombic system

3 unequal axes set at right angles.

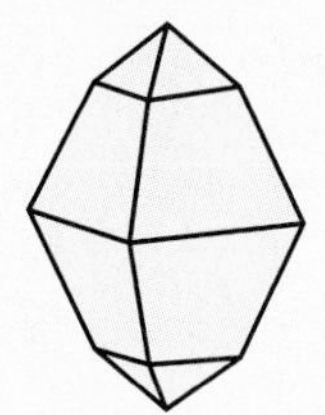

Orthorhombic sulphur

Monoclinic system

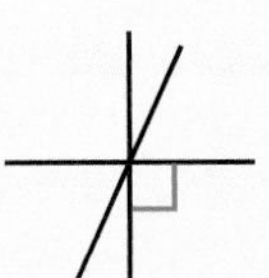

3 unequal axes; 2 at right angles, the third set obliquely.

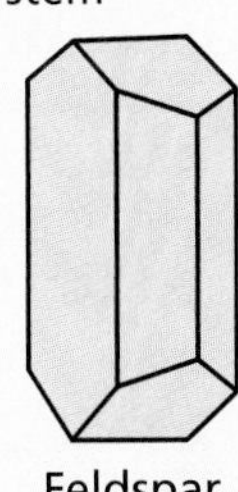

Feldspar

Triclinic system

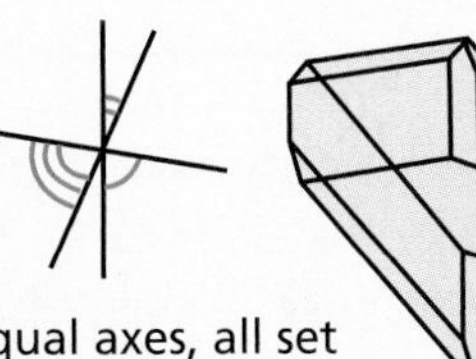

3 unequal axes, all set obliquely at unequal angles to each other.

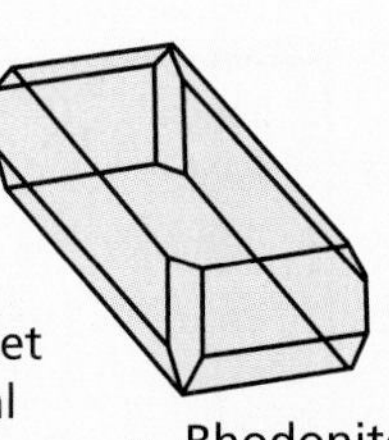

Rhodonite

SPORTS

All-American Girls Professional Baseball League Batting Champions, 1943-1954

South Bend, Indiana Blue Sox

Year	Player, Team	At-bats	Average
1943	Gladys Davis, Rockford	349	.332
1944	Betsy Jochum, South Bend	433	.296
1945	Helen Callahan, Fort Wayne	408	.299
1946	Dorothy Kamenshek, Rockford	408	.316
1947	Dorothy Kamenshek, Rockford	366	.306
1948*	Audrey Wagner, Kenosha	417	.312
1949	Doris Sams, Muskegon	408	.279
1950	Betty Weaver Foss, Fort Wayne	361	.346
1951	Betty Weaver Foss, Fort Wayne	342	.368
1952	Joanne Weaver, Fort Wayne	314	.344
1953	Joanne Weaver. Fort Wayne	410	.346
1954	Joanne Weaver, Fort Wayne	333	.429

*First year overhand pitching was allowed.
Source: *A Whole New Ballgame*, Sue Macy, Henry Holt and Company, New York, 1993.

Olympic Record Times for 400-m Freestyle Swimming, in Minutes

Year	1924	1928	1932	1936	1948	1952	1956	1960	1964
Male	5:04.2	5:01.6	4:48.4	4:44.5	4:41.0	4:30.7	4:27.3	4:18.3	4:12.2
Female	6:02.2	5:42.8	5:28.5	5:26.4	5:17.8	5:12.1	4:54.6	4:50.6	4:43.3

Year	1968	1972	1976	1980	1984	1988	1992	1996
Male	4:09.0	4:00.27	3:51.93	3:51.31	3:51.23	3:46.25	3:45.00	3:47.97
Female	4:31.8	4:19.44	4:09.89	4:08.76	4:07.10	4:03.85	4:07.18	4:07.25

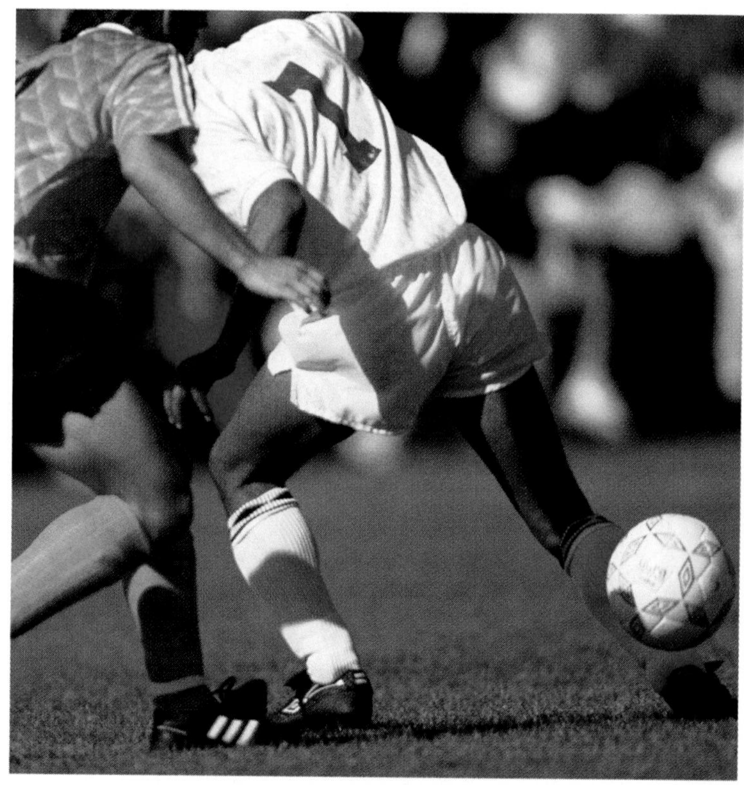

Sizes and Weights of Balls Used in Various Sports

Type	Diameter (centimeters)	Average weight (grams)
Baseball	7.6	145
Basketball	24.0	596
Croquet ball	8.6	340
Field hockey ball	7.6	160
Golf ball	4.3	46
Handball	4.8	65
Soccer ball	22.0	425
Softball, large	13.0	279
Softball, small	9.8	187
Table tennis ball	3.7	2
Tennis ball	6.5	57
Volleyball	21.9	256

Soccer Playboard

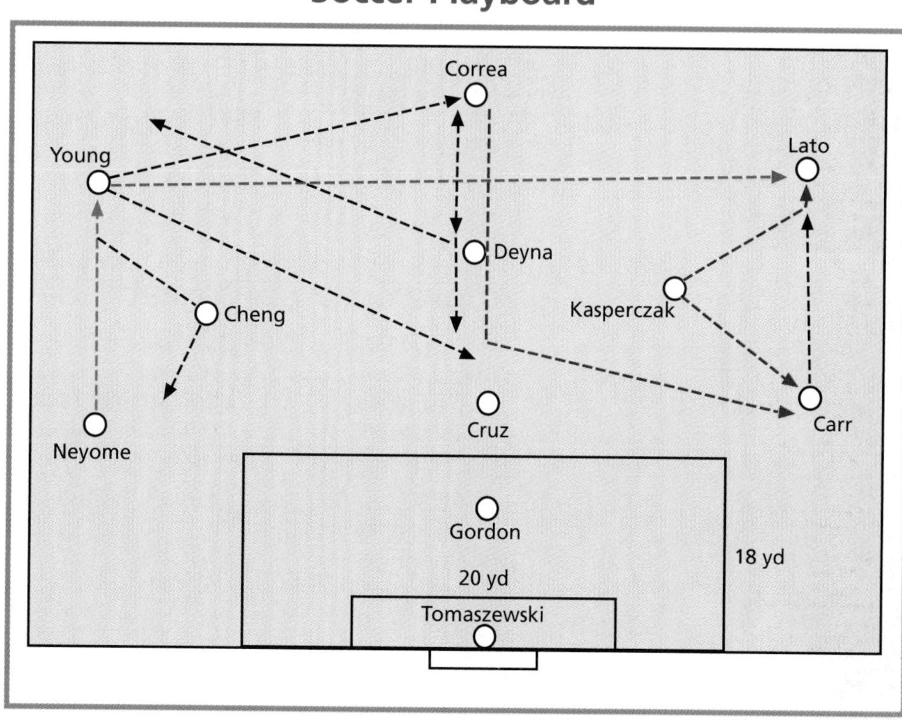

UNITED STATES

U.S. Cities

Population size	Number of cities	Total population (millions)
1 million or more	7	19.1
500,000-1 million	17	11.5
250,000-500,000	37	13.2
100,000-250,000	125	18.4
50,000-100,000	300	20.6
25,000-50,000	575	19.9
10,000-25,000	1,323	20.7
Under 10,000	16,868	28.7

Pittsburgh, Pennsylvania

Family Size*

	Total
Number of families	66,090
Two persons	27,606
Three persons	15,353
Four persons	14,026
Five persons	5,938
Six persons	1,997
Seven or more persons	1,170
Total persons	209,515
Average per family	3.17

*Numbers in thousands except for averages.

Approximate Kilowatt Usage of Some Appliances

Appliance	Kilowatts per hour
Light bulbs	0.001 per watt
Electric blanket	0.07
Stereo	0.1
Color television	0.23
Hair dryer	1.5
Refrigerator	5.0
Iron	1.0
Freezer	3.0
Water heater	16.0

1996 Population and Number of Representatives, By State
(Total Population 265,283,783)

State	Apportionment population	Number of representatives based on the 1996 census
U.S. Total*	264,740,570	436
Alabama	4,273,084	7
Alaska	607,007	1
Arizona	4,428,068	6
Arkansas	2,509,793	4
California	31,878,234	52
Colorado	3,822,676	6
Connecticut	3,274,238	6
Delaware	724,842	1
Florida	14,399,985	23
Georgia	7,353,225	11
Hawaii	1,183,723	2
Idaho	1,189,251	2
Illinois	11,846,544	20
Indiana	5,840,528	10
Iowa	2,851,792	5
Kansas	2,572,150	4
Kentucky	3,883,723	6
Louisiana	4,350,579	7
Maine	1,243,316	2
Maryland	5,071,604	8
Massachusetts	6,092,352	10
Michigan	9,594,350	16
Minnesota	4,657,758	8
Mississippi	2,716,115	5
Missouri	5,358,692	9
Montana	879,372	1
Nebraska	1,652,093	3
Nevada	1,603,163	2
New Hampshire	1,162,481	2
New Jersey	7,987,933	13
New Mexico	1,713,407	3
New York	18,184,774	31
North Carolina	7,322,870	12
North Dakota	643,539	1
Ohio	11,172,782	19
Oklahoma	3,300,902	6
Oregon	3,203,735	5
Pennsylvania	12,056,112	22
Rhode Island	990,225	2
South Carolina	3,698,746	6
South Dakota	732,405	1
Tennessee	5,319,654	9
Texas	19,128,261	30
Utah	2,000,494	3
Vermont	588,654	1
Virginia	6,675,451	11
Washington	5,532,939	9
West Virginia	1,825,754	3
Wisconsin	5,159,795	9
Wyoming	481,400	1

* Total population, not including the District of Columbia

Capitol Hill
Washington, D.C.

MATHEMATICS

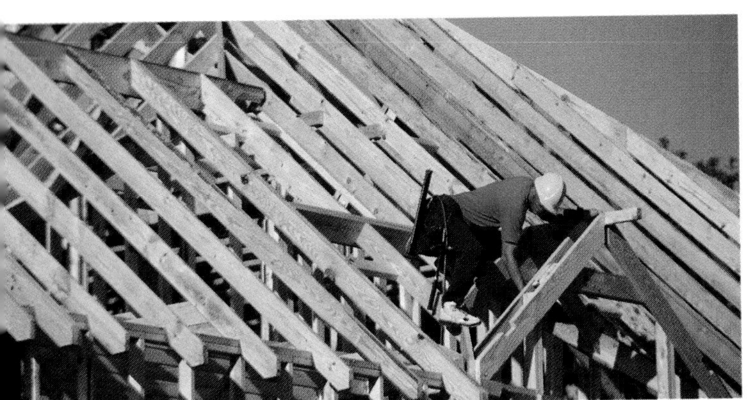

Measurement: Customary Units

Length

12 inches (in.)	=	1 foot (ft)
3 feet or 36 inches	=	1 yard (yd)
1,760 yards or 5,280 feet	=	1 mile (mi)
6,076 feet	=	1 nautical mile

Area

144 square inches (in.^{2})	=	1 square foot (ft^2)
9 square feet	=	1 square yard (yd^2)
4,840 square yards	=	1 acre (A)

Volume

1,728 cubic inches (in.^{3})	=	1 cubic foot (ft^3)
27 cubic feet	=	1 cubic yard (yd^3)

Capacity

8 fluid ounces (fl oz)	=	1 cup
2 cups	=	1 pint (pt)
2 pints	=	1 quart (qt)
4 quarts	=	1 gallon (gal)

Weight

16 ounces (oz)	=	1 pound (lb)
2,000 pounds	=	1 ton (T)

Temperature

32°F	=	freezing point of water
98.6°F	=	normal body temperature
212°F	=	boiling point of water

Metric/Customary Comparisons

5 centimeters is about the same length as 2 inches.
1 meter is slightly longer than 1 yard.
5 kilometers is about the same length as 3 miles.

Formulas/Useful Equations

Right Triangles

Pythagorean Theorem:
In a right triangle, as shown.

$$a^2 + b^2 = c^2 \text{ or } c = \sqrt{a^2 = b^2}$$

Trigonometric Ratios:

$\sin\angle A = \frac{a}{c}$ $\sin\angle B = \frac{b}{c}$

$\cos\angle A = \frac{b}{c}$ $\cos\angle B = \frac{a}{c}$

$\tan\angle A = \frac{a}{b}$ $\tan\angle B = \frac{b}{a}$

Temperature

°C: degrees Celsius
°F: degrees Fahrenheit

$°C = \frac{5}{9}(F - 32)$
$°F = \frac{9}{5}(C + 32)$

Distance

d: distance r: rate t: time

$d = rt$ $r = \frac{d}{t}$ $t = \frac{d}{r}$

On a coordinate plane, the distance between points $P(x_1, y_1)$ and $Q(x_2, y_2)$ is

$$PQ = \sqrt{(x_2 - x_1)^2 + (y_2 - y_1)^2}$$

Law of Exponents

$a^m \cdot a^n = a^{m+n}$
$(a^m)^n = a^{m \cdot n}$
$\frac{a^m}{a^n} = a^{m-n}$

Probability

Factorial: $n! = n(n-1)(n-2)\ldots(1)$
Permutations: ${}_nP_n = n!$
${}_nP_r = \frac{n!}{(n-r)!}$
Combinations: ${}_nC_r = \frac{n!}{(n-r)!\,r!}$

Probability

I: interest
p: principal
t: time (in years)
A: total amount
r: rate

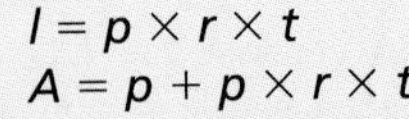

$I = p \times r \times t$
$A = p + p \times r \times t$

Table of Squares and Approximate Square Roots

n	n^2	$\sqrt{n}$	n	n^2	$\sqrt{n}$	n	n^2	$\sqrt{n}$	n	n^2	$\sqrt{n}$
1	1	1.0000	26	676	5.0990	51	2601	7.1414	76	5776	8.7178
2	4	1.4142	27	729	5.1962	52	2704	7.2111	77	5929	8.7750
3	9	1.7321	28	784	5.2915	53	2809	7.2801	78	6084	8.8318
4	16	2.0000	29	841	5.3852	54	2916	7.3485	79	6241	8.8882
5	25	2.2361	30	900	5.4772	55	3025	7.4162	80	6400	8.9443
6	36	2.4495	31	961	5.5678	56	3136	7.4833	81	6561	9.0000
7	49	2.6458	32	1024	5.6569	57	3249	7.5498	82	6724	9.0554
8	64	2.8284	33	1089	5.7446	58	3364	7.6158	83	6889	9.1104
9	81	3.0000	34	1156	5.8310	59	3481	7.6811	84	7056	9.1652
10	100	3.1623	35	1225	5.9161	60	3600	7.7460	85	7225	9.2195
11	121	3.3166	36	1296	6.0000	61	3721	7.8102	86	7396	9.2736
12	144	3.4641	37	1369	6.0828	62	3844	7.8740	87	7569	9.3274
13	169	3.6056	38	1444	6.1644	63	3969	7.9373	88	7744	9.3808
14	196	3.7417	39	1521	6.2450	64	4096	8.0000	89	7921	9.4340
15	225	3.8730	40	1600	6.3246	65	4225	8.0623	90	8100	9.4868
16	256	4.0000	41	1681	6.4031	66	4356	8.1240	91	8281	9.5394
17	289	4.1231	42	1764	6.4807	67	4489	8.1854	92	8464	9.5917
18	324	4.2426	43	1849	6.5574	68	4624	8.2462	93	8649	9.6437
19	361	4.3589	44	1936	6.6332	69	4761	8.3066	94	8836	9.6954
20	400	4.4721	45	2025	6.7082	70	4900	8.3666	95	9025	9.7468
21	441	4.5826	46	2116	6.7823	71	5041	8.4261	96	9216	9.7980
22	484	4.6904	47	2209	6.8557	72	5184	8.4853	97	9409	9.8489
23	529	4.7958	48	2304	6.9282	73	5329	8.5440	98	9604	9.8995
24	576	4.8990	49	2401	7.0000	74	5476	8.6023	99	9801	9.9499
25	625	5.0000	50	2500	7.0711	75	5625	8.6603	100	10000	10.0000

MATHEMATICS

Symbols

$=$	is equal to
$\neq$	is not equal to
$\cong$	is congruent to
$\sim$	is similar to
$\approx$	is approximately equal to
$\parallel$	is parallel to
$\perp$	is perpendicular to
$>$	is greater than
$<$	is less than
$\geq$	is greater than or equal to
$\leq$	is less than or equal to
( )	parentheses: "Do this operation first."
3^2	exponent (2); base (3)
%	percent
$0.\overline{3}$	repeating decimal
π	pi: $\approx \frac{22}{7}$ or 3.14
$\sqrt{\ }$	square root
$\lvert x \rvert$	absolute value of x
$\overleftrightarrow{AB}$	line AB

$\overline{AB}$	line segment AB
$\overrightarrow{AB}$	ray AB
$\angle ABC$	angle ABC
∟	right angle
$^\circ$	degrees
$A \to A'$	point A maps onto point A'
$(1, -2)$	coordinates of a point where $x = 1$ and $y = -2$
$\in$	is an element of
$\notin$	is not an element of
$\{\}$	set inclusion; empty set
$\subseteq$	is a subset of
$\not\subseteq$	is not a subset of
$\varnothing$	empty set
U	universal set
A'	the complement of set A
$A \cup B$	the union of set A and set B
$A \cap B$	the intersection of set A and set B
$p \to q$	If p, then q
$\sim p$	not p

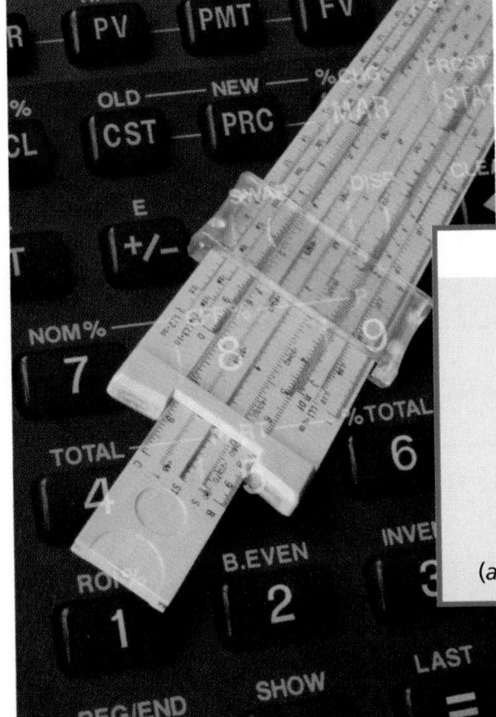

Polynomial Expansion of Coefficients

Polynomial Factors		Expansion
$(a + b)^0$	=	1
$(a + b)^1$	=	$a + b$
$(a + b)^2$	=	$a^2 + 2ab + b^2$
$(a + b)^3$	=	$a^3 + 3a^2b + 3ab^2 + b^3$
$(a + b)^4$	=	$a^4 + 4a^3b + 6a^2b^2 + 4ab^3 + b^4$
$(a + b)^5$	=	$a^5 + 5a^4b + 10a^3b^2 + 10a^2b^3 + 5ab^4 + b^5$
$(a - b)(a + b)$	=	$a^2 - b^2$
$(a - b)(a^2 + ab + b^2)$	=	$a^3 - b^3$
$(a + b)(a^2 - ab + b^2)$	=	$a^3 + b^3$
$(a - b)(a + b)(a^2 + b^2)$	=	$a^4 - b^4$

Measurement: Metric Units

Length		
1 centimeter (cm)	=	10 millimeters (mm)
10 centimeters or 100 millimeters	=	1 decimeter (dm)
100 centimeters	=	1 kilometer (km)
Area		
100 square millemeters (mm^2)	=	1 square centimeter (cm^2)
10,000 square centimeters	=	1 square meter (m^2)
100 square meters	=	1 are (a)
10,000 square meters	=	1 hectare (ha)
Volume		
1,000 cubic millimeters (mm^3)	=	1 cubic centimeter (cm^3)
1,000 cubic centimeters	=	1 cubic decimeter (dm^3)
1,000,000 cubic centimeters	=	1 cubic meter (m^3)
Capacity		
1,000 millimeters	=	1 liter (L)
1,000 liters	=	1 kiloliter (kL)
Mass		
1,000 milligrams (mg)	=	1 gram (g)
1,000 grams	=	1 kilograms (kg)
1,0000 kilograms	=	1 metric ton (T)
Temperature		
0°C	=	freezing point of water
37°C	=	normal body temperature
100°C	=	boiling point of water

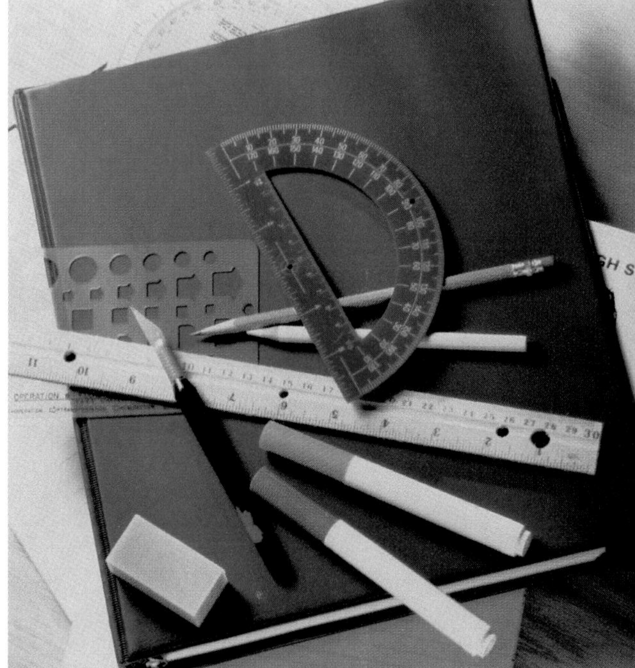

Manipulative Reference Guide

Manipulatives are models to bridge between the concrete world and the abstractions that are expressed in mathematics. Physical and pictorial models that are represented with manipulatives will clarify your thinking and develop mental images that will enhance your understanding of symbolic notation that you encounter in mathematics.

Algeblocks

The Algeblocks Class Set provides enough blocks for four groups of 2 to 4 students per cooperative learning groups. There are seven types of blocks representing unit constants (green), variables x, x^2, x^3, xy, y and y^2, (yellow, light orange, and dark orange), and three Algeblocks Mats which are Basic Mat, Quadrant Mat and Sentences Mat.

By using the three mats, you are able to model the addition and subtraction with the placement of the appropriate blocks in the positive and negative section of the Basic Mat. The basic concept is adding zero to be able to remove blocks that are the same. The remaining blocks are the correct answer.

Using the Quadrant Mat, you are able to see that multiplication and division is an area problem. By creating a rectangular area, where the sides are the factors and the area the product. The inverse would be that the quotient is the area and one side is the divisor.

Finally, the Sentence Mat gives the basis of solving equations. The equations are placed on the mat. Observe that positive and negative section of the sentence is placed correctly.

To solve, remember that you are able to place the same block on both sides with the purpose to obtain zeros until you have the variable on one side and a number on the other.

You will need to communicate mathematically, that is, using verbal, written, drawings, sketches, and physical representation in order to be successful in mathematics.

Your goal is the find models that will give you visual representations of the concepts that will easy to remember and quickly to model.

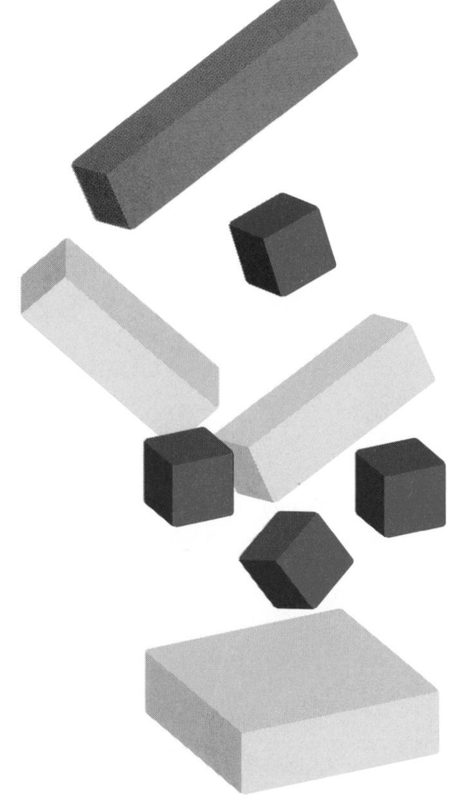

Manipulative Reference Guide

Tangram

This is a puzzle or an all-purpose manipulative that you are able to explore numbers and geometric figures. You are able to describe and relate number and figures to an object in your hand. Your teacher can supply you with a tangram pattern for you.

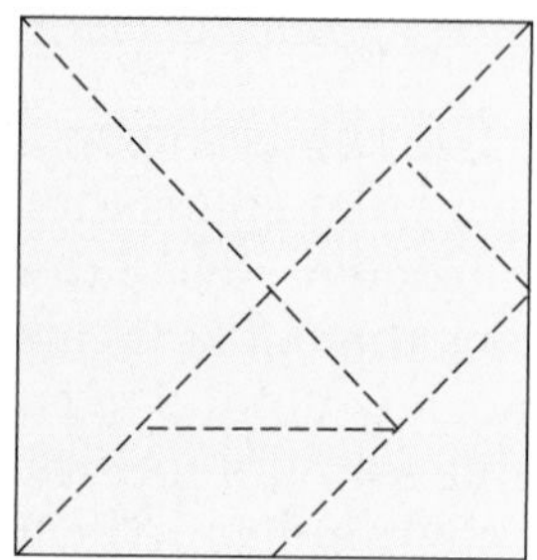

Geoboards

With Geoboards or grid paper, you are able to explore many ideas and concepts without fear of being wrong. The questions about perimeter and area are easily created and measured. You are able to change quickly and determine if the change is better or worse.

Questions you can consider include:

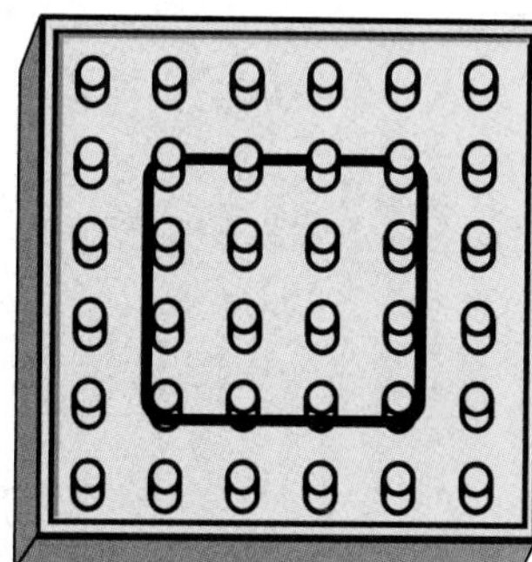

1. What is the perimeter of each shape?
2. Can you make other shapes with the same perimeter?
3. How many squares of 4 units on a side can you construct?
4. Can you construct a square with 2 square units in it?
5. Construct a triangle with an area of: 1 square unit, 2 square units, 3 square units, 4 square units, 5 square units, and 7 square units.
6. Construct a parallelogram with an area of 3 square units.
7. Construct a rectangle with a area of 4 square units that is twice as long as it is wide.
8. Construct a hexagon with an area of: 4 square units or 3 square units.

You can us an alternative technique for finding the estimate of areas by applying Pick's rule, $A = \frac{1}{2}b + i + 1$ where b is the number of lattice points on the boundary and i is the number of interior lattice points (lattice points are the pegs on the geoboard).

This method, you may use on any irregular shape. The grid that you place over the area must correspond to the scale that you are measuring the area. This works nicely on maps that do not have straight boundaries.

Manipulative Reference Guide

Geometric models (Polydrons)™

The five solids shown are the only regular solids that can be constructed. Recall that regular means that all faces are the same. It is helpful to construct a table for reference with the left column headed with name of solid, the next columns being name of the face, number of faces, number of faces per vertex, number of vertices.

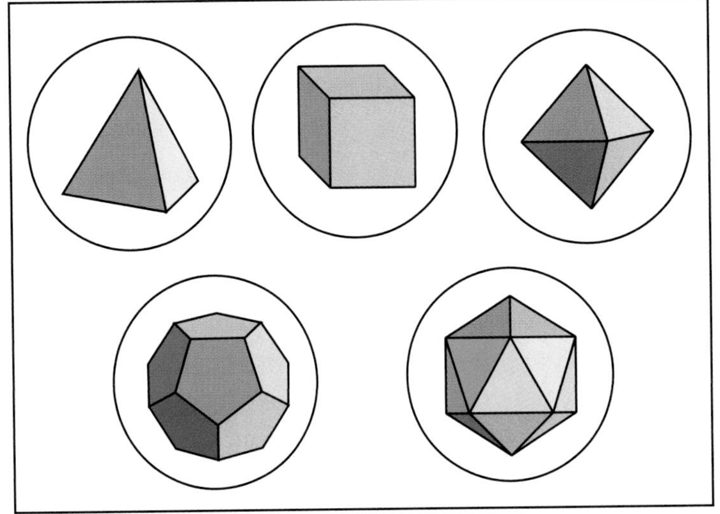

Random Number Generators

Flipping Coins

Investigation of how coins fall need to be approached in very carefully. First you need to keep a record of the outcomes of each event (heads or tails) and the order in which the outcomes occurs. The object is you are trying to see if there is a pattern that would allow you to determine the next flip.

You should know that the probability of heads or tails on one flip is 50%, but what is the probability of have two heads in a row or five heads in a row? Does the theory of multiplying the probabilities agree with flipping a coin many times.

1. Keep a record of flipping a coin one hundred times (remember, the order is important).
2. How many streaks are there?
3. What was the longest streak?
4. Do you think if you flip the coin again, you would get the same number of streak?
5. What is the smallest size streak you would expect?

Number Cubes (dice)/Spinners

Predicting the outcomes with number cubes or spinners is exciting and challenging. Studying the result of a series of experiments will lead you to a better understanding of probability and statistics.

Since a cube has six sides, the probability of any one number is $\frac{1}{6}$. Rolling two dice, the probability of these outcomes changes. Recall the probability of having a five is the number of fives over the total number of rolls. It is helpful to make a list of the probability of all the different outcomes.

Mirrors and Miras™

Investigating reflections can be a fun and exciting method of discovering symmetry of figures and shapes. A Mira is a see-through piece of Plexiglass tht enables you to draw an image behind the semi-transparent piece of Plexiglass. It is similar to tracing a figure, except the shape is reversed left to right. You are able to bisect angles, find midpoints of line segments, and even construct a circle with practice.

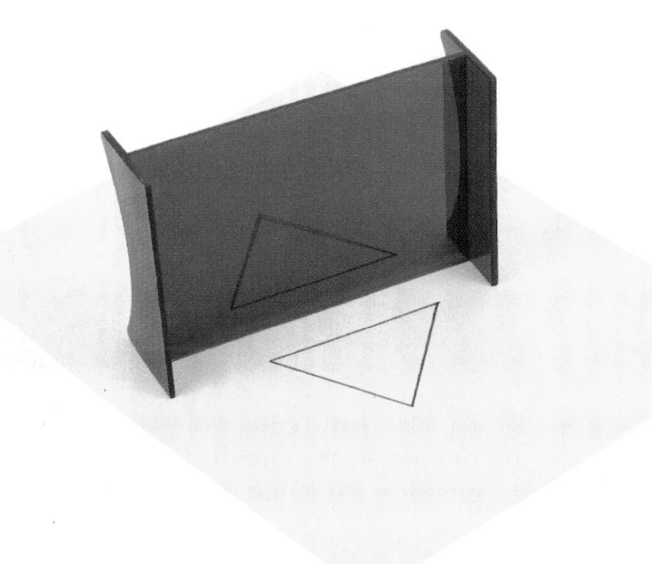

TECHNOLOGY REFERENCE GUIDE

GRAPHING CALCULATORS

Using Texas Instrument TI-73 and TI-83 Graphing Calculator

For additional features and instruction, consult the User's Manual.

THE KEYBOARD The feature accessed when a key is pressed is shown in white on the key. To access the features in blue above each key, first press the 2nd key. To access what is in white above the keys, first press ALPHA.

CALCULATIONS Calculations are performed in the Home Screen. This screen may be returned to at any time by pressing 2nd QUIT. The calculator evaluates according to the order of operations. Press ENTER to calculate. For $3 + 4 \times 5$ ENTER, the result is 23. To replay the previous line, press 2nd ENTER. Use the arrow keys to edit the expression.

Displaying Graphs

GRAPH FEATURE To enter an equation to be graphed, press Y=. Enter an equation such as $y = 2x + 3$ using the X key for x. Then press GRAPH to display the graph in the viewing window of graphs.

VIEWING WINDOW To set the range values for the viewing window, press WINDOW. That will define the left-right sides by setting the x-values and the top-bottom by setting the y-values.

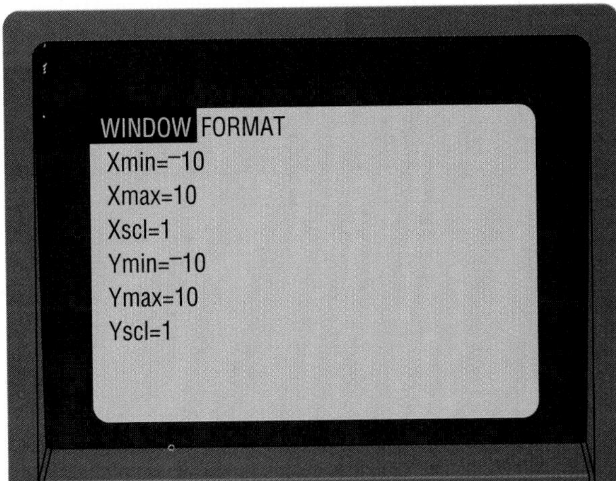

TRACE FEATURE Pressing TRACE places the cursor directly on the graph and shows the x- and y-coordinates of the point where the cursor is located. The cursor can be moved along the graph by using the right and left arrow keys.

TABLE FEATURE Press 2nd TBLSET to set up the table. Then press 2nd TABLE to see a table of values for each equation selected.

ZOOM FEATURE Press ZOOM and then press 6 for Standard viewing window. By pressing ZOOM and then 1 (ZBox), by moving the cursor to upper left hand corner of the area to be viewed closer, ENTER, moving the cursor to the lower right corner, and ENTER, the calculator will graph the selected area in a zoomed window. On the ZOOM menu, option 8 will set the window values for a "Friendly Window," that is the values will be easier to understand.

Statistics

ENTERING DATA Enter data into lists by pressing STAT 1 (Edit).

CALCULATING STATISTICS Return to the Home Screen by pressing 2nd QUIT. To calculate the mean of LIST 1, press 2nd LIST ▶ (MATH) 3 (mean) 2nd L1 ENTER. To calculate the median, choose 4 instead of 3. To calculate statistics, press STAT ▶ (CALC) 1 (1-Var Stats) ENTER {on the TI-83} or press 2nd LIST ▶ (CALC) 1 (1-Var Stats) ENTER{on the TI-73}. Choose 2 to calculate two variable statistics and 5 to calculate linear regression. To see the lower quartile of a boxplot, press VARS 5.

GRAPHING DATA To graph the data, press 2nd STATPLOT and choose one of the plots: Scatter Plot , an xyLine Plot, a Box Plot, Modified Box Plot, or a Histogram (in the TI-73, there are two additional options a pictograph or a circle graph).Then choose the data lists that would be used by the plot. Then press GRAPH to draw the graph.

Using the Casio CFX-9800-G Graphing Calculator

For additional features and instruction, consult the User's Manual.

THE KEYBOARD The feature accessed when a key is pressed is shown in white on the key. To access the features in gold above each key, first press the SHIFT key. To access what is in red above the keys, first press ALPHA

THE MAIN MENU This is the screen you see when you first turn the calculator on. Highlight by using the arrow keys the menu item to be used and press EXE, or press the number to choose the menu item. You can access the Main Menu at any time by pressing the MENU key.

PERFORMING CALCULATIONS Calculations are performed by pressing 1 (for COMPutations) in the Main Menu. The calculator evaluates according to the order of operations. Press EXE (for EXEcute) to calculate. For $3 + 4 \times 5$ EXE, the result is 23. You can replay the previous line by press ◀. Use the arrow keys to edit.

Technology Reference Guide

Display Graphs

COMP MODE Press 1 in the Main Menu. To enter an equation, press GRAPH. Enter an equation such as $y = 2x + 3$ using the X, θ, T key for x. Then press EXE to display the graph in the viewing window.

GRAPH MODE Press 6 (GRAPH) in the Main Menu. Then press AC. Use the up or down arrows to choose a location to store the equation. Enter the equation as above. Then press F6 (DRW) to display the graph in the viewing window.

VIEWING WINDOW To set range values for the viewing window, press RANGE. You can use F1 (INIT) to set standard values for a viewing window.

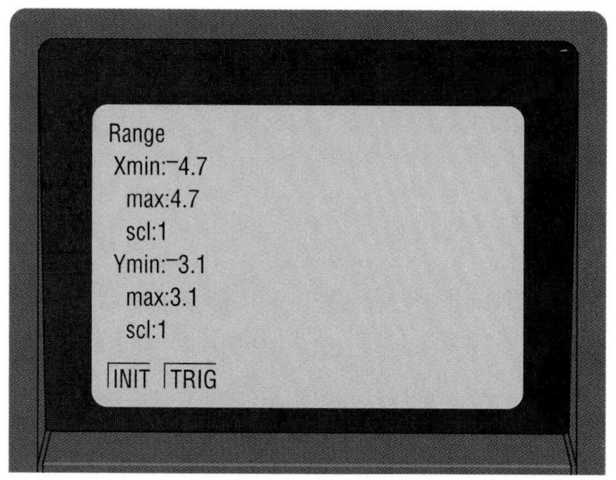

ZOOM FEATURE Press SHIFT F2 (ZOOM) to access this feature. Press F1 (BOX) to highlight a particular area by using the arrow keys starting in the upper left corner and ending in the lower right and zoom in on that part of the graph. Press F5 (AUT) to set range values for a friendly window, which gives coordinate readout in friendly numbers.

TRACE FEATURE Pressing SHIFT F1 (TRACE) places the cursor directly on the graph and shows the x- and y-coordinates of the point where the cursor is located. The cursor can be moved along the graph by using the left and right arrow keys.

TABLE FEATURE Press 8 (TABLE) in the Main Menu. Press AC to clear the screen. Then press F1 (RANGE FUNC). Select the function. Then press F5 (RNG) to set up the table. Press F6 (TBL) to see the table of values. Press SHIFT QUIT to return to a previous screen.

Statistical Features

ENTERING DATA Data may be entered into a list by pressing LIB and selecting Statistics.

CALCULATING STATISTICS To calculate mean, median, upper, and lower quartiles of the data entered in C1, press the Stats key and use the arrow keys to access all the information.

GRAPHING DATA To graph your data, press the green key and then PLOT. Then press the Choose Key. Select BoxWhisker or Histogram and ENTER. Then press the green key to VIEWS. Select Auto Scale.

Interactive Geometry

Interactive Geometry is a mathematics software program that provides a high level of interactivity with geometric ideas and objects. It allows you to explore properties and concepts in geometry. To be interactive means that the objects drawn and relationships between objects will remain true to geometry axioms and definitions. Students are able to drag points, lines, and any object about the worksheet. The procedures that are found in your investigations in Interactive Geometry software packages are not Geometric Proofs, but are strong indicators that there may be a Geometry Theorem.

The Geometer's Sketchpad™

There are many features available in Geometer's Sketchpad. The following are four features that you should be skilled at and use most of the time in any exploration.

Freehand Tools:

In Euclid Geometry, we have basic tools that are used for creating points, a compass to draw circles, and a straightedge to construct segments, rays, or lines. These are the basic Freehand tools plus a tool that will label objects that have been constructed.

Point Tool (an arrow)

Circle Tool (a circle)

Segment Tool (a line)

Hand (Pointing) Tool (labels)

In the exploring these Freehand Tools, note that the point created has a bold outline, which indicates that the point is selected. Being selected is an important concept that needs to be mastered for other features of Geometer's Sketchpad.

Objects on the worksheet are mathematically related to each other. If a circle has a point on the circumference, using the Point Tool grab the point (place the Pointing Tool on the point, a click and hold) but the point will only move on the circle.

Basic Construction

In creating a single object in Geometer's Sketchpad, the mathematical relationship of the object is maintained and followed as the object is changed. Use the Line Segment tool to construct a triangle. Label the three vertices (points) and the three segments connecting these vertices (using the Hand Tool).

Using the Point Tool (arrow icon) click and hold a single vertex. Drag the point around the worksheet, notice that the triangle changes but stays a triangle.

Click on one side of the triangle to select it. Selection indicators for line segments are two little black squares that will appear on the segment.

While the segment is selected, click on the **Construct** menu. Note that only two choices are available: **Point On Object** and **Point At Midpoint** plus the **Help**. There are several other menu options that are "grayed-out" since these options do not apply to a line segment.

Choose **Point At Midpoint**, a midpoint on the line segment will appear. Repeat the procedure with a second side of the triangle. Use the Segment tool, construct a line segment between midpoints.

Basic Measurement

Select the two points at the midpoints (the second point is selected by holding the shift key down when clicking on the second point). When the two points are selected, you can measure the distance between them by clicking on **Measure** menu. Again, most of the options are grayed-out.

Repeat this measure for the distance of the side of the triangle that is parallel to the midpoint line segment. You see that this command will be **Distance**. On the worksheet, these measurements will appear.

Using the Point Tool (an arrow), grab a vertex point and drag (click, hold and move). As the point is dragged about the worksheet, the measurements will change.

Using the Texas Instruments TI-92 Geometry

For additional features and instructions, consult your user's manual.

THE KEYBOARD The feature accessed when a key is pressed is shown in white on the key. To access the features in yellow above a key, first press the 2ND key, then the key desired. To access what is in green above the keys, first press ◆, then the desired key. To access capital letters on the keyboard, use the ↑ key. The CAPS key allows all letters to be entered uppercase.

CALCULATIONS Calculations are performed in the Home Screen. This screen may be returned to at any time by pressing APPS, then 1 or ◆ HOME. The calculator evaluates according to the order of operations. Press ENTER to calculate. For $3 + 4 \times 5$ ENTER, the result is 23. You can replay a previous calculation by highlighting it with the arrow keypad and pressing ENTER. Use the arrow keypad or ← key to edit.

CONSTRUCTIONS Constructions are performed in the Geometry Screen. A new geometry screen may be accessed at any time by pressing APPS, 8, 3. In the NEW dialog box press the down arrow. Name your construction using up to 8 characters, letters only and then press ENTER twice.

POINTS To construct a new point, press F2 then 1. Position the cursor with the arrow keypad and press ENTER. To construct a point on an object already on the screen, press F2 then 2. Select the place on the object where you want your point with the arrow keypad and press ENTER.

To find the point of intersection of two objects already on the screen, press F2, then 3. Select each object with the arrow keypad and press ENTER. To construct the midpoint of a segment, press F4 then 3. Select the segment with the arrow keypad and press ENTER.

LINES, RAYS AND SEGMENTS To construct a new line, press F2 then 4. Designate the two points through which the line will pass with the arrow keypad and press ENTER after each. To construct a new ray, press F2, then 6. Designate the endpoint and another point through which the ray will pass. Move the cursor with the arrow keypad and press ENTER after each. To construct a new line segment, press F2 then 5. Designate the endpoints of the segment and press ENTER after each.

PARALLEL LINES To construct a line parallel to another line, ray or segment, press F4 then 2. Select the line, ray, or segment with the arrow keypad and press ENTER. Then select the point through which the parallel line will pass with the arrow keypad and press ENTER.

Geometry Reference Guide

Geometry Basics

All geometric figures are made up of at least one point.

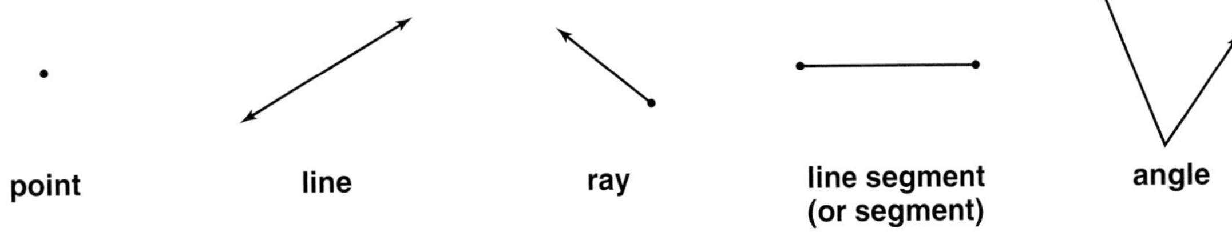

About Lines

Lines in a plane can be either parallel to each other or they can intersect each other.

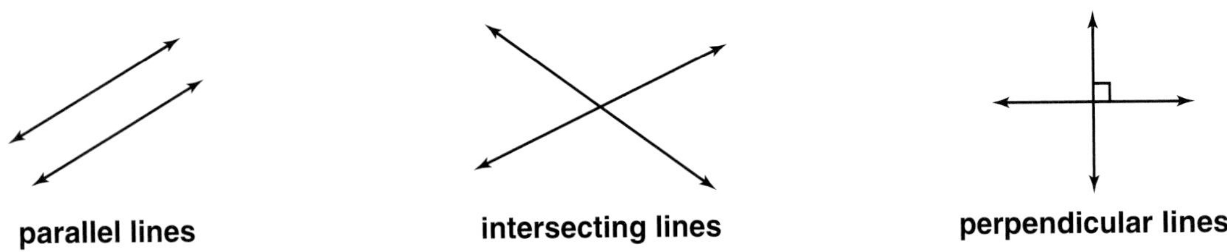

About Angles

Angles are measured in degrees.

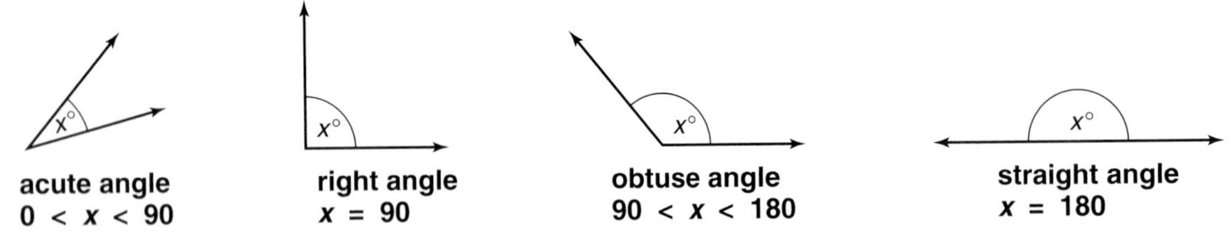

Complementary and Supplementary Angles

Two angles are complementary if the sum of their measures is exactly 90°.

Two angles are supplementary if the sum of their measures is exactly 180°.

Geometry Reference Guide

About Geometric Solid Figures

Geometric solid figures are made up of plane polygons.

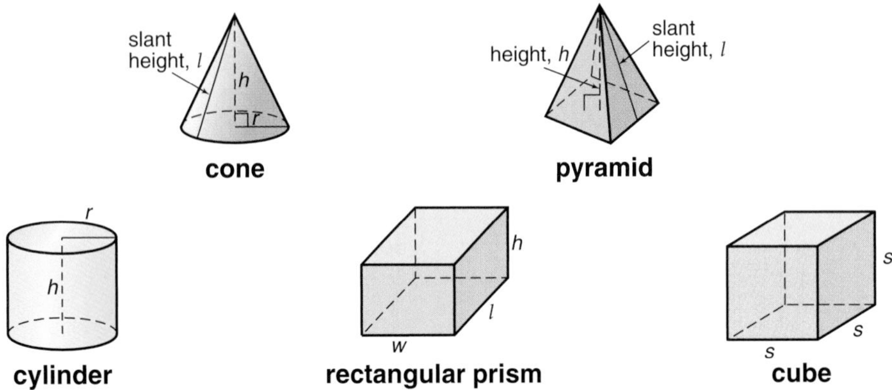

Base

The cone and the pyramid have one base. The cylinder and the prism have two bases and they are parallel. The cone and cylinder have circular bases. The base of a pyramid or prism can be any polygonal shape.

Lateral Surface

The lateral surface is the side or sides of the solid figure other than a base. The cone and cylinder have one lateral surface. The lateral surface of a pyramid is made up of triangles. The lateral surface of a right prism is made up of rectangles.

Slant Height

The slant height of a cone is measured from the vertex of the cone to the edge of its base. The slant height of a pyramid is measured from the vertex to the center of one side of the base.

Formulas

Total surface area of a right circular cone	$SA = \pi r(l + r)$
Volume of a cone	$V = \frac{1}{3}\pi r^2 h$
Total surface area of a right cylinder	$SA = 2\pi r(r + h)$
Volume of a cylinder	$V = \pi r^2 h$
Total surface area of a rectangular prism	$SA = 2(lw + lh + wh)$
Volume of a rectangular prism	$V = lwh$
Total surface area of a cube	$SA = 6s^2$
Volume of a cube	$V = s^3$

Geometry Reference Guide

About Other Polygons

Polygons are plane figures made up of segments and angles. Triangles and four-sided figures are also polygons.

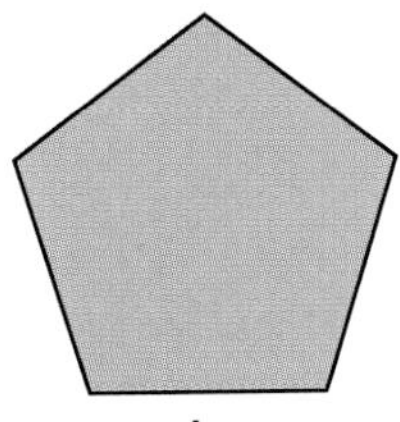
pentagon

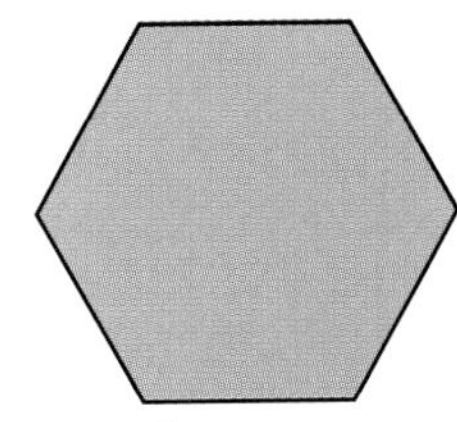
hexagon

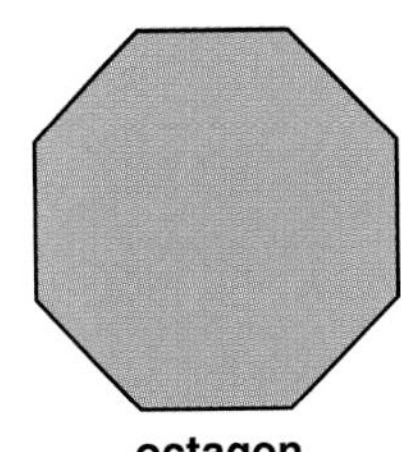
octagon

Perimeter Formulas

In the following formulas, l = length, w = width, s = side, and P = perimeter.

Perimeter of a rectangle $P = 2l + 2w$

Perimeter of a square $P = 4s$

Area Formulas

In the following formulas, b = base, B = long base, h = height, l = length, w = width, s = side, and A = area.

Area of a parallelogram $A = bh$

Area of a rectangle $A = lw$

Area of a square $A = s^2$

Area of a trapezoid $A = \frac{1}{2}(B + b)h$

Area of a triangle $A = \frac{1}{2}bh$

About Circles and Spheres

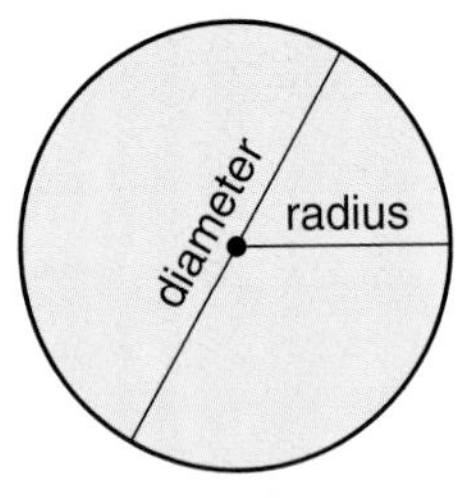

circle

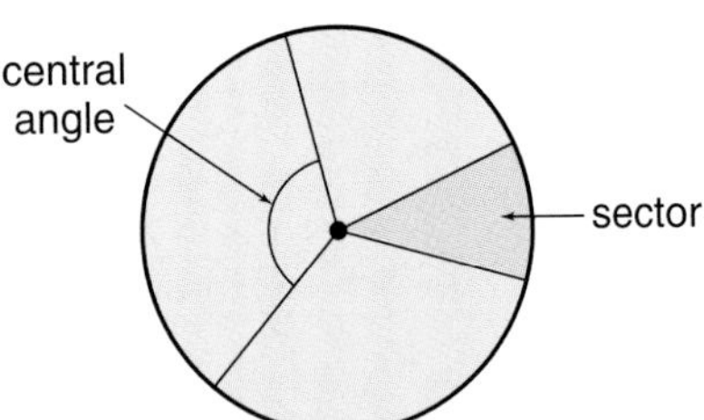

circle parts

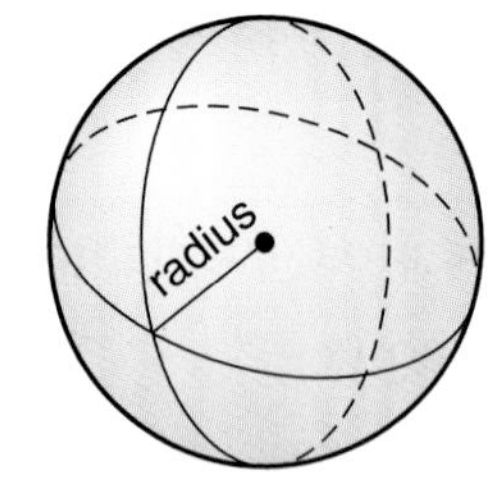

sphere

Circle Formulas

Circumference of a circle $C = 2\pi r$ or $C = \pi d$

Area of a circle $A = \pi r^2$

Sphere Formulas

Surface area of a sphere $SA = 4\pi r^2$

Volume of a sphere $V = \frac{4}{3}\pi r^3$

Geometry Reference Guide

About Triangles

Triangles are three-sided plane figures. They can be classified according to the measures of their sides or their angles.

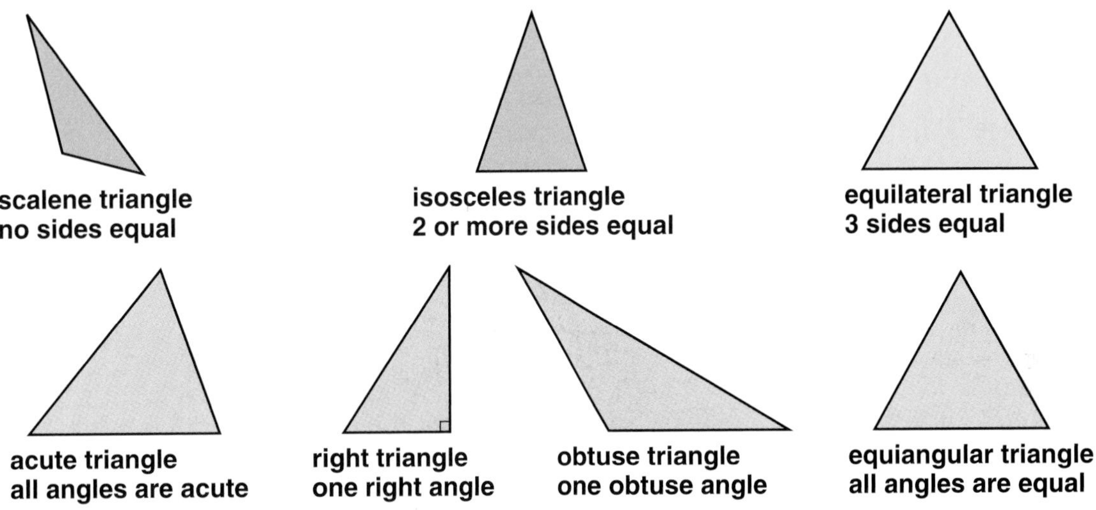

About Quadrilaterals

Quadrilaterals are four-sided plane figures. Each figure in the diagram has all the properties of the figures preceding it, including the properties listed with that figure.

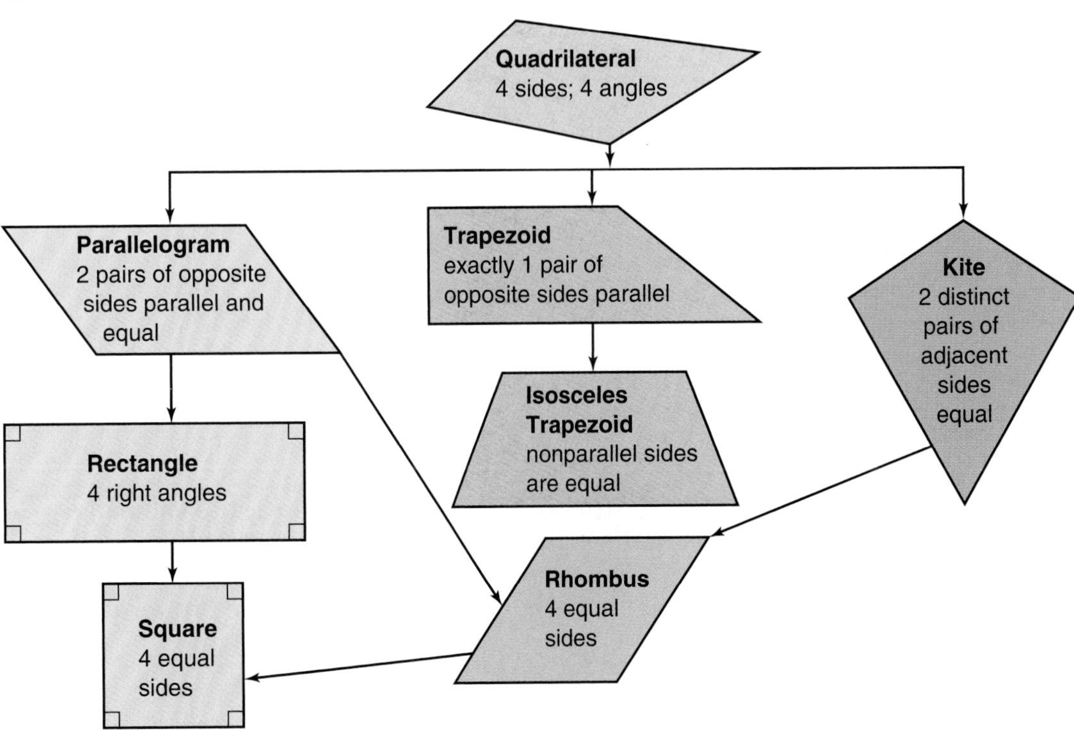

ALGEBRA REFERENCE GUIDE

Measures of Central Tendency

The *mean* is appropriate to use when all the data are approximately equal.

> The mean or *average* is the sum of the data divided by the number of items of data.

The **range** is the difference between the greatest and least values of data. When the range is large compared to the values themselves, the median may better represent the data.

> The median is the middle value when the data are arranged in numerical order. When there are two middle values, the median is the average of the two.

The *mode*, like the median, may be appropriate when the mean is not.

> The mode is the element that occurs most often in the set. A set may have no mode, one mode, or several modes. If a set of data has two modes, the set is bimodal.

Order of Operations

Use the order of operations to evaluate expressions.

Order of Operations

1. Perform operations within grouping symbols first.
2. Perform all calculations involving exponents.
3. Multiply or divide in order from left to right.
4. Add or subtract in order from left to right.

Linear Equations

When one side of an equation has a variable with a coefficient of 1, the value of the expression on the other side of the equal symbol is the solution of the equation. Use the properties that follow to solve linear equations. When linear equations contain parentheses, you may need to use the distributive property. Use more than one property when the equation contains more than one operation or variables on both sides of the equal symbol.

Algebra Reference Guide

Properties of Equality	For all real numbers a, b, and c:	
	If $a = b$, then $a + c = b + c$.	Addition
	If $a = b$, then $ca = cb$.	Multiplication
	If $a = b$, then $a - c = b - c$.	Subtraction
	If $a = b$ and $c \neq 0$, then $\frac{a}{c} = \frac{b}{c}$.	Division

Distributive Property	For all real numbers a, b, and c:
	$a(b + c) = ab + ac$ and $(b + c)a = ba + ca$

Linear Functions and Their Graphs

A linear function is a function that can be represented by a non-vertical line. The graphs show transformations of the parent linear function $y = x$.

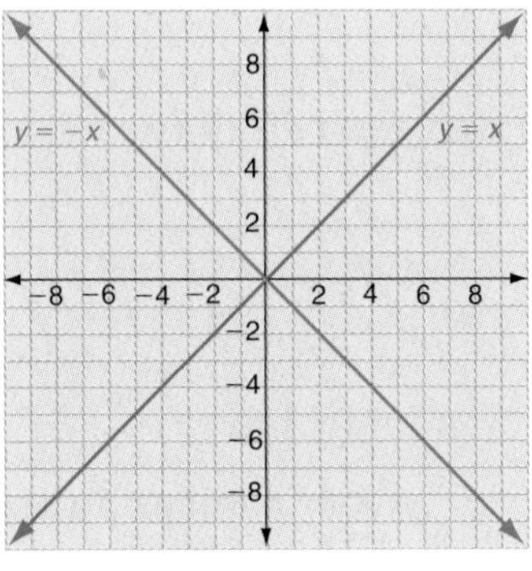

Reflection

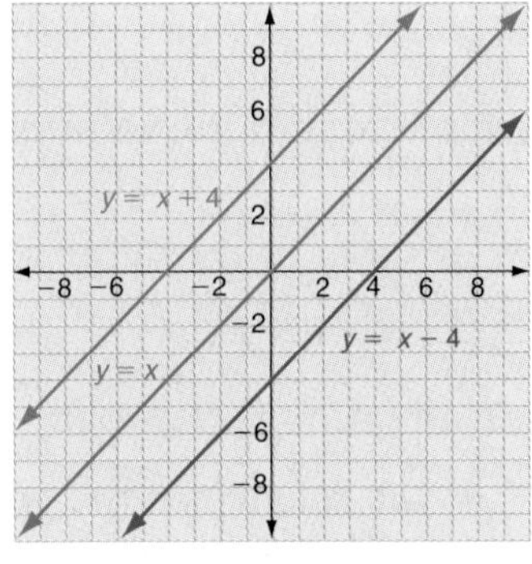

Vertical Translation

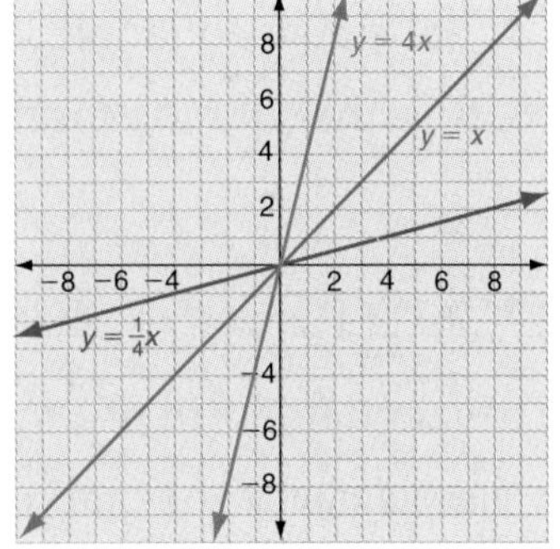

Change in Slope

Linear Inequalities

You can solve an inequality by getting the variable alone on one side of the inequality symbol, just as you do with equations. Use the following properties.

Properties of Inequality	For all real numbers a, b, and c:
	If $a > b$, then $a + c > b + c$ and $a - c > b - c$.
	If $a < b$, then $a + c < b + c$ and $a - c < b - c$.
	If $a > b$ and $c > 0$, then $ac > bc$ and $\frac{a}{c} > \frac{b}{c}$.
	If $a > b$ and $c < 0$, then $ac <$ bc and $\frac{a}{c} < \frac{b}{c}$.

Algebra Reference Guide

Systems of Linear Equations

An ordered pair that is a solution of all the equations in a system of equations in two variables is a solution of that linear system. The system can be solved by graphing both lines and locating their intersection, by using the substitution method, or by using the elimination method.

Substitution Method

1. Solve one equation for one variable in terms of the other variable.
2. Substitute for that variable in the other equation.

Elimination Method

1. Write the equations in standard form, $Ax + By = C$.
2. If no coefficients are identical or opposite coefficients, multiply one or both equations so that you have identical or opposite coefficients for one of the variables.
3. Add or subtract equations so that you have one equation in one variable.
4. Solve that equation.
5. Use the solution for that variable to find the solution for the other variable.

Quadratic Functions and Their Graphs

The standard form of a quadratic equation is $ax^2 + bx + c = 0$ and may have

- **one solution** if the graph of its corresponding function touches the x-axis at only one point.
- **two solutions** if the graph of its corresponding function crosses the x-axis at two points.
- **no solutions** if the graph of its corresponding functionneither touches or crosses the x-axis.

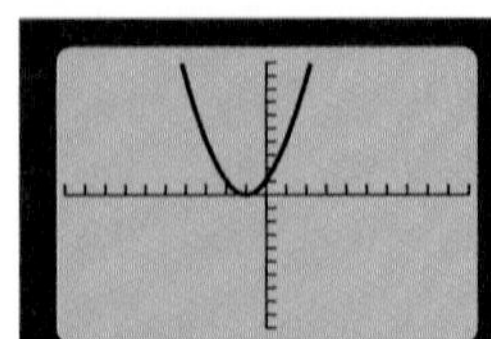

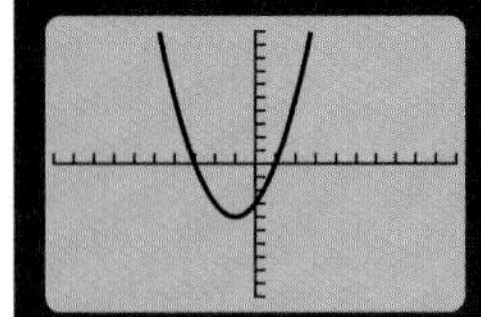

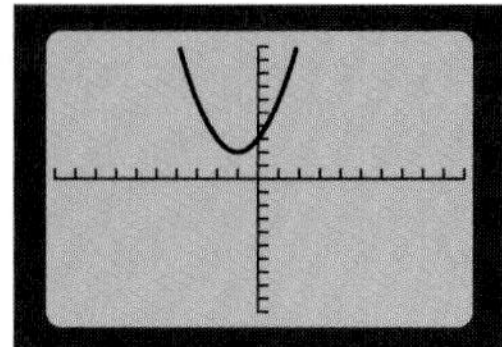

You can find the solution to a quadratic equation by using the *quadratic formula.*

Quadratic Formula	For a quadratic equation of the form $ax^2 + bx + c = 0$ where a, b, and c are real numbers and $a \neq 0$, $$x = \frac{-b \pm \sqrt{b^2 - 4ac}}{2a}$$

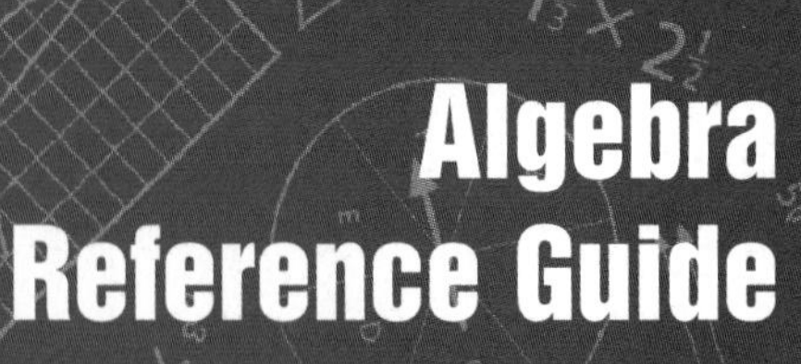

Radical Expressions

A radical expression is in simplest form when the radicand (expression under the radical symbol) contains no perfect square factors other than 1 and no fractions. A simplified expression cannot have a denominator that contains a radical. Use these properties to simplify radical expressions.

To eliminate a radical expression from a denominator, *rationalize the denominator* by multiplying the numerator and denominator by the same expression.

Like radicals are radical expressions with the same radicand and are combined in a way similar to combining like terms.

Solving Radical Equations

A **radical equation** has a radical with a variable in the radicand. To solve radical equations use the principle of squaring.

Principles of Squaring	If the equation $a = b$ is true, then the equation $a^2 = b^2$ is also true.

Rational Expressions

To add or subtract expressions with the same denominator, add or subtract the numerators. If the denominators are different, find the *least common denominator* (LCD). Use the LCD to rewrite each expression as an equivalent rational expression having the LCD as its denominator.

Properties of Square Roots	For all real numbers a, b, where $a \geq 0$:		
	If $b \geq 0$, then	$\sqrt{ab} = \sqrt{a} \cdot \sqrt{b}$	Product Property
	If $b > 0$, then	$\sqrt{\frac{a}{b}} = \frac{\sqrt{a}}{\sqrt{b}}$	Quotient Property

Rational Equations

A **rational equation** has one or more rational expressions. Use these steps to solve.

1. Multiply both sides of the equation by the LCD to eliminate denominators.
2. Solve the resulting equation.
3. Check each solution in the original equation. A value that makes a denominator of an expression in the original equation equal to zero is extraneous, and therefore, not a solution.

Extra Practice Index

Basic Math

Chapter 1

Chapter 2

Chapter 3

Chapter 4

Chapter 5

Chapter 6

Extra Practice Index

Chapter 7 — page

Chapter 8 — page

Chapter 9 — page

Chapter 10 — page

Chapter 11 — page

Chapter 12 — page

Chapter 13 — page

Chapter 14 — page

Place Value and Order

Example 1

Write 2,345,678.9123 in words.

Solution

The place value chart shows the value of each digit. The value of each place is ten times the place to the right.

millions	hundred thousands	ten thousands	thousands	hundreds	tens	ones	.	tenths	hundredths	thousandths	ten-thousandths
2	3	4	5	6	7	8		9	1	2	3

The number shown is *two million, three hundred forty-five thousand, six hundred seventy-eight and nine thousand one hundred twenty-three ten-thousandths.*

Example 2

Use < or > to make this sentence true. **6 ■ 2**

Solution

Remember, < means "less than" and > means "greater than." So, 6 > 2.

Extra Practice Exercises

Write each number in words.

1. 3647 three thousand, six hundred forty-seven
2. 6,004,300.002 six million, four thousand, three hundred and two thousandths
3. 0.9001 nine thousand one ten-thousandths

Write each of the following as a number.

4. two million, one hundred fifty thousand, four hundred seventeen 2, 150, 417
5. five thousand, one hundred twenty and five hundred two thousandths 5120.502
6. nine million, ninety thousand, nine hundred and ninety-nine ten-thousandths 9,090,900.0099

Use < or > to make each sentence true.

7. 9 ■ 8 >
8. 164 ■ 246 <
9. 63,475 ■ 6,435 >
10. 52 ■ 50 >
11. 5.39 ■ 9.02 <
12. 43.94 ■ 53.69 <

Multiply Whole Numbers and Decimals

To multiply whole numbers, find each partial product and then add.

When multiplying decimals, locate the decimal point in the product so that there are as many decimal places in the product as the total number of decimal places in the factors.

Example 1

Multiply 2.6394 by 3000.

Solution

$$\begin{array}{r} 2.6394 \\ \times\ \ 3000 \\ \hline 7918.2000 \end{array} \text{ or } 7918.2$$

Zeros after the decimal point can be dropped because they are not significant digits.

Example 2

Multiply 3.92 by 0.023.

Solution

3.92	2 decimal places
×0.023	+ 3 decimal places
1176	
+7840	
0.09016	5 decimal places

The zero is added before the nine, so the product will have five decimal places.

Extra Practice Exercises

Multiply.

1. 36 × 45 1620
2. 500 × 30 15,000
3. 17,000 × 230 3,910,000
4. 6.2 × 8 49.6
5. 950 × 1.6 1520
6. 3.652 × 20 73.04
7. 179 × 83 14,857
8. 257 × 320 82,240
9. 8560 × 275 2,354,000
10. 467 × 0.3 140.1
11. 2.63 × 183 481.29
12. 0.758 × 321.8 243.9244
13. 49.3 × 1.6 78.88
14. 6.859 × 7.9 54.1861
15. 794.4 × 321.8 255,637.92
16. 0.08 × 4 0.32
17. 0.062 × 0.5 0.031
18. 0.0135 × 0.003 0.0000405
19. 21.6 × 3.1 66.96
20. 8.76 × 0.005 0.0438
21. 5.521 × 3.642 20.107482
22. 5.749 × 3.008 17.292992
23. 8.09 × 0.18 1.4562
24. 89,946 × 2.85 256,346.1
25. 6.31 × 908 5729.48
26. 391.05 × 25 9776.25
27. 35,021 × 76.34 2,673,503.14

Divide Whole Numbers and Decimals

Dividing whole numbers and decimals involves a repetitive process of estimating a quotient, multiplying and subtracting.

```
              34     ←quotient
divisor → 7)239      ← dividend
            21↓      ←3 × 7
             29      ← Subtract. Bring down the 9.
             28      ←4 × 7
              1      ← remainder
```

Example 1

Find: 283.86 ÷ 5.7

Solution

When dividing decimals, move the decimal point in the divisor to the right until it is a whole number. Move the decimal point the same number of places in the dividend. Then place the decimal point in the answer directly above the new location of the decimal point in the dividend.

```
                         49.8
5.7.)283.8.6   →    57)2838.6
                       228
                        558
                        513
                         45 6
                         45 6
                            0
```

If answers do not have a remainder of 0, you can add 0's after the last digit and continue dividing.

Extra Practice Exercises

Divide.

1. 72 ÷ 6 12
2. 6000 ÷ 20 300
3. 26,568 ÷ 8 3321
4. 5.6 ÷ 7 0.8
5. 120 ÷ 0.4 300
6. 936 ÷ 12 78
7. 3.28 ÷ 4 0.82
8. 0.1960 ÷ 5 0.0392
9. 1968 ÷ 0.08 24,600
10. 16 ÷ 0.04 400
11. 1525 ÷ 0.05 30,500
12. 109.94 ÷ 0.23 478
13. 0.6 ÷ 24 0.025
14. 7.924 ÷ 0.28 28.3
15. 32.6417 ÷ 9.1 3.587
16. 24 ÷ 0.6 40
17. 1784.75 ÷ 29.5 60.5
18. 0.01998 ÷ 0.37 0.054
19. 7.8 ÷ 0.3 26
20. 12,000 ÷ 0.04 300,000
21. 820.94 ÷ 0.02 41,047
22. 89,946 ÷ 28.5 3156
23. 15 ÷ 0.75 20
24. 7.56 ÷ 2.25 3.36
25. 0.19176 ÷ 68 0.00282
26. 0.168 ÷ 0.48 0.35
27. 5.1 ÷ 0.006 850
28. 55,673 ÷ 0.05 1,113,460
29. 84.536 ÷ 4 21.134
30. 261.18 ÷ 10 26.118
31. 134,554 ÷ 0.14 961,100
32. 90,294 ÷ 7.85 11,502.42038
33. 59,368 ÷ 47.3 1255.137421
34. 11,633.5 ÷ 439 26.5
35. 28.098 ÷ 14 2.007
36. 16.309 ÷ 0.09 181.2111111
37. 55.26 ÷ 1.8 30.7
38. 8276 ÷ 0.627 13,199.36204
39. 10,693 ÷ 92.8 115.2262931
40. 48.8 ÷ 1.6 30.5
41. 27,268 ÷ 34 802
42. 546.702 ÷ 0.078 7009

Multiply and Divide Fractions

To multiply fractions, multiply the numerators and then multiply the denominators. Write the answer in simplest form.

Example 1

Multiply $\frac{2}{5}$ and $\frac{7}{8}$.

Solution

$$\frac{2}{5} \times \frac{7}{8} = \frac{2 \times 7}{5 \times 8} = \frac{14}{40} = \frac{7}{20}$$

To divide by a fraction, multiply by the reciprocal of that fraction. To find the reciprocal of a fraction, invert (turn upside down) the fraction. The product of a fraction and its reciprocal is 1. Since $\frac{2}{3} \times \frac{3}{2} = \frac{6}{6}$ or 1, $\frac{2}{3}$ and $\frac{3}{2}$ are reciprocals of each other.

Example 2

Divide $1\frac{1}{5}$ by $\frac{2}{3}$.

Solution

$$1\frac{1}{5} \div \frac{2}{3} = \frac{6}{5} \div \frac{2}{3} = \frac{6}{5} \times \frac{3}{2} = \frac{6 \times 3}{5 \times 2} = \frac{18}{10} \textit{ or } 1\frac{4}{5}$$

Extra Practice Exercises

Multiply or divide. Write each answer in simplest form.

1. $\frac{2}{3} \div \frac{5}{6}$ $\frac{4}{5}$
2. $\frac{3}{5} \times \frac{10}{12}$ $\frac{1}{2}$
3. $\frac{5}{8} \div \frac{1}{4}$ $2\frac{1}{2}$
4. $\frac{1}{2} \times \frac{2}{3}$ $\frac{1}{3}$
5. $\frac{2}{3} \times \frac{1}{2}$ $\frac{1}{3}$
6. $\frac{3}{4} \times \frac{5}{8}$ $\frac{15}{32}$
7. $\frac{1}{2} \div \frac{2}{3}$ $\frac{3}{4}$
8. $\frac{2}{3} \div \frac{1}{2}$ $1\frac{1}{3}$
9. $\frac{3}{4} \div \frac{5}{8}$ $1\frac{1}{5}$
10. $2\frac{2}{3} \div 1\frac{3}{5}$ $1\frac{2}{3}$
11. $1\frac{1}{5} \times 2\frac{1}{4}$ $2\frac{7}{10}$
12. $3\frac{1}{3} \times 1\frac{1}{10}$ $3\frac{2}{3}$
13. $5\frac{2}{5} \div 2\frac{4}{7}$ $2\frac{1}{10}$
14. $2\frac{4}{7} \div 5\frac{2}{5}$ $\frac{10}{21}$
15. $2\frac{4}{7} \times 5\frac{2}{5}$ $13\frac{31}{35}$
16. $1\frac{7}{8} \div 1\frac{7}{8}$ 1
17. $\frac{3}{4} \times \frac{2}{3} \times 1\frac{5}{8} \times 2\frac{2}{3}$ $2\frac{1}{6}$
18. $7\frac{1}{2} \div 2\frac{1}{4}$ $3\frac{1}{3}$
19. $6\frac{2}{3} \times 4\frac{1}{2} \times 5\frac{3}{8}$ $161\frac{1}{4}$
20. $11\frac{5}{9} \times 6\frac{1}{12}$ $70\frac{8}{27}$
21. $\frac{25}{42} \div \frac{5}{21}$ $2\frac{1}{2}$
22. $\frac{13}{18} \div \frac{8}{9}$ $\frac{13}{16}$
23. $\frac{3}{8} \times \frac{11}{12} \times \frac{16}{33}$ $\frac{1}{6}$
24. $\frac{51}{56} \div \frac{17}{24}$ $1\frac{2}{7}$

Add and Subtract Fractions

To add and subtract fractions, you need to find a common denominator and then add or subtract, renaming as necessary.

Example 1

Add $\frac{3}{4}$ and $\frac{5}{6}$.

Solution

$$
\begin{array}{rcl}
\frac{3}{4} = \frac{3}{4} \times \frac{3}{3} = & \frac{9}{12} \\
+\frac{5}{6} = \frac{5}{6} \times \frac{2}{2} = & +\frac{10}{12} \\
\hline
 & \frac{19}{12}
\end{array}
$$

Add the numerators and use the common denominator.

Then simplify. $\frac{19}{12} = 1\frac{7}{12}$

Example 2

Subtract $1\frac{3}{5}$ from $5\frac{1}{2}$.

Solution

$$
\begin{array}{rcrcr}
5\frac{1}{2} & = & 5\frac{5}{10} & = & 4\frac{15}{10} \\
-1\frac{3}{5} & = & -1\frac{6}{10} & = & -1\frac{6}{10} \\
\hline
 & & & & 3\frac{9}{10}
\end{array}
$$

You can not subtract $\frac{6}{10}$ from $\frac{5}{10}$, so rename again.

Extra Practice Exercises

Add or subtract.

1. $\frac{1}{5} + \frac{1}{10}$ $\frac{3}{10}$
2. $\frac{2}{3} + \frac{1}{3}$ 1
3. $\frac{5}{8} + \frac{3}{4}$ $1\frac{3}{8}$
4. $\frac{6}{7} - \frac{2}{7}$ $\frac{4}{7}$
5. $\frac{3}{4} - \frac{1}{3}$ $\frac{5}{12}$
6. $\frac{5}{8} - \frac{1}{4}$ $\frac{3}{8}$
7. $2\frac{1}{2} + 3\frac{1}{2}$ 6
8. $6\frac{5}{8} + 3\frac{7}{8}$ $10\frac{1}{2}$
9. $3\frac{2}{3} + 4\frac{1}{2}$ $8\frac{1}{6}$
10. $2\frac{3}{4} - 1\frac{1}{4}$ $1\frac{1}{2}$
11. $5\frac{1}{8} - 3\frac{7}{8}$ $1\frac{1}{4}$
12. $1\frac{1}{3} - \frac{2}{3}$ $\frac{2}{3}$
13. $6\frac{1}{2} + 5\frac{7}{9}$ $12\frac{5}{8}$
14. $9\frac{2}{5} - 1\frac{1}{8}$ $8\frac{11}{40}$
15. $7\frac{2}{3} + 6\frac{1}{5}$ $13\frac{13}{15}$
16. $8\frac{1}{10} - 5\frac{2}{3}$ $2\frac{13}{30}$
17. $6\frac{1}{2} - 5\frac{3}{5}$ $\frac{9}{10}$
18. $10\frac{5}{8} - 9\frac{3}{4}$ $\frac{7}{8}$
19. $1\frac{1}{5} + 2\frac{1}{3} + 5\frac{1}{4}$ $8\frac{47}{60}$
20. $9\frac{2}{3} + 4\frac{3}{5} + 6\frac{1}{2}$ $20\frac{23}{30}$
21. $10\frac{7}{8} + 3\frac{3}{4} + 6\frac{1}{2} + 2\frac{5}{8}$ $23\frac{3}{4}$

Fractions, Decimals and Percents

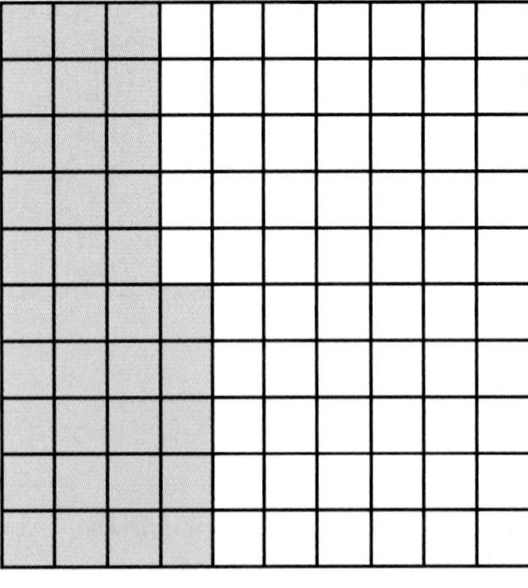

Percent means per hundred. Therefore, 35% means 35 out of 100. Percents can be written as equivalent decimals and fractions.

$35\% = 0.35$	Move the decimal point two places to the left.
$35\% = \frac{35}{100}$	Write the fraction with a denominator of 100.
$= \frac{7}{20}$	Then simplify.

Example 1

Write $\frac{3}{8}$ as a decimal and as a percent.

Solution

$\frac{3}{8} = 0.375$	Divide to change a fraction to a decimal.
$0.375 = 37.5\%$	To change a decimal to a percent move the decimal point two places to the right and insert the percent symbol.

Percents greater than 100% represent whole numbers or mixed numbers.

$200\% = 2$ or 2.00 $\qquad$ $350\% = 3.5$ or $3\frac{1}{2}$

Extra Practice Exercises

Write each fraction as a decimal and as a percent.

1. $\frac{1}{2}$ 0.5; 50%
2. $\frac{1}{4}$ 0.25; 25%
3. $\frac{3}{4}$ 0.75; 75%
4. $\frac{9}{10}$ 0.9; 90%
5. $\frac{3}{10}$ 0.3; 30%
6. $\frac{1}{25}$ 0.04; 4%
7. $3\frac{7}{8}$ 3.875; 387.5%
8. $1\frac{1}{5}$ 1.2; 120%
9. $\frac{13}{25}$ 0.52; 52%

Write each decimal as a fraction and as a percent.

10. 0.63 $\frac{63}{100}$; 63%
11. 0.15 $\frac{3}{20}$; 15%
12. 0.4 $\frac{2}{5}$; 40%
13. 2.35 $2\frac{7}{20}$; 235%
14. 10.125 $10\frac{1}{8}$; 1012.5%
15. 0.625 $\frac{5}{8}$; 62.5%
16. 0.05 $\frac{1}{20}$; 5%
17. 0.125 $\frac{1}{8}$; 12.5%
18. 0.3125 $\frac{5}{16}$; 31.25%

Write each percent as a decimal and as a fraction.

19. 10% 0.10; $\frac{1}{10}$
20. 12% 0.12; $\frac{3}{25}$
21. 100% 1; 1
22. 150% 1.5; $1\frac{1}{2}$
23. 160% 1.6; $1\frac{3}{5}$
24. 75% 0.75; $\frac{3}{4}$
25. 8% 0.08; $\frac{2}{25}$
26. 87.5% 0.875; $\frac{7}{8}$
27. 0.35% 0.0035; $\frac{7}{2000}$

Multiply and Divide by Powers of Ten

To multiply a number by a power of 10, move the decimal point to the right. To multiply by 100 means to multiply by 10 two times. Each multiplication by 10 moves the decimal point one place to the right.

To divide a number by a power of 10, move the decimal point to the left. To divide by 1000 means to divide by 10 three times. Each division by 10 moves the decimal point one place to the right.

Example 1

Multiply 21 by 10,000.

Solution

$21 \times 10{,}000 = 210{,}000$ The decimal point moves four places to the right.

Example 2

Find 145 ÷ 500.

Solution

$$\begin{aligned} 145 \div 500 &= 145 \div 5 \div 100 \\ &= 29 \div 100 \\ &= 0.29 \end{aligned}$$

The decimal point moves two places to the left.

Extra Practice Exercises

Multiply or divide.

1. 15×100 1500
2. $96 \times 10{,}000$ 960,000
3. $1296 \div 100$ 12.96
4. $9687.03 \div 1000$ 9.68703
5. $36 \times 20{,}000$ 720,000
6. $7500 \div 3000$ 2.5
7. 9×30 270
8. 94×6000 564,000
9. $561 \div 30$ 18.7
10. $1505 \div 500$ 3.01
11. $71 \times 90{,}000$ 6,390,000
12. $9 \times 120{,}000$ 1,080,000
13. $3159 \div 10{,}000$ 0.3159
14. $1{,}000{,}000 \times 0.79$ 790.000
15. $601 \times 30{,}000$ 18,030,000
16. $75 \div 300$ 0.25
17. 4000×12 48,000
18. $14 \times 7{,}000{,}000$ 98,000,000
19. $49{,}000 \div 7000$ 7
20. $980 \div 10{,}000$ 0.098
21. $216 \div 2000$ 0.108
22. $108{,}000 \div 900$ 120
23. $72 \times 10{,}000{,}000$ 720,000,000
24. $953.16 \div 10{,}000$ 0.095316
25. $1472 \div 8000$ 0.184
26. $490{,}000 \div 700$ 700
27. $80 \times 90{,}000$ 7,200.000
28. $8001 \div 90$ 88.9
29. 50×6000 300,000
30. $950{,}000 \div 50{,}000$ 19
31. $81{,}000 \times 5$ 405,000
32. $1458 \times 30{,}000$ 43,740,000
33. $452.3 \div 10$ 45.23
34. $986{,}856.008 \div 10{,}000$ 9,868,560,080
35. $316 \times 70{,}000$ 22,120,000
36. $60 \div 1200$ 0.05

Round and Order Decimals

To round a number, follow these rules:

1. Underline the digit in the specified place. This is the place digit. The digit to the immediate right of the place digit is the test digit.
2. If the test digit is 5 or larger, add 1 to the place digit and substitute zeros for all digits to its right.
3. If the test digit is 4 or smaller, substitute zeros for it and all digits to the right.

Example 1

Round 4826 to the nearest hundred.

Solution

4$\underline{8}$26 Underline the place digit.

4800 Since the test digit is 2, substitute zeros for 2 and all digits to the right.

To place decimals in ascending order, write them in order from least to greatest.

Example 2

Place in ascending order: 0.34, 0.33, 0.39.

Solution

Compare the first decimal place, then compare the second decimal place.

0.33 (least), 0.34, 0.39 (greatest)

Extra Practice Exercises

Round each number to the place indicated.

1. 367 to the nearest ten 370
2. 961 to the nearest ten 960
3. 7200 to the nearest thousand 7000
4. 3070 to the nearest hundred 3100
5. 41,440 to the nearest hundred 41,400
6. 34,254 to the nearest thousand 34,000
7. 208,395 to the nearest thousand 208,000
8. 654,837 to the nearest ten thousand 650,000

Write the decimals in ascending order.

9. 0.29, 0.82, 0.35 0.29, 0.35, 0.82
10. 1.8, 1.4, 1.5 1.4, 1.5, 1.8
11. 0.567, 0.579, 0.505, 0.542 0.505, 0.542, 0.567, 0.579
12. 0.54, 0.45, 4.5, 5.4 0.45, 0.54, 4.5, 5.4
13. 0.0802, 0.0822, 0.00222 0.00222, 0.0802, 0.0822
14. 6.204, 6.206, 6.205, 6.203 6.203, 6.204, 6.205, 6.206
15. 88.2, 88.1, 8.80, 8.82 8.80, 8.82, 88.1, 88.2
16. 0.007, 7.0, 0.7, 0.07 0.007, 0.07, 0.7, 7.0

ADDITIONAL ANSWERS

Extra Practice 1-1

2. {Jan., March, May, July, Aug., Oct., Dec.}
4. {Saturday, Sunday}
11. {*o, n, e*}, {*o, n*}, {*o, e*}, {>*n, e*}, {*o*}, {*n*}, {*e*}, ∅

Extra Practice 1–2

7.
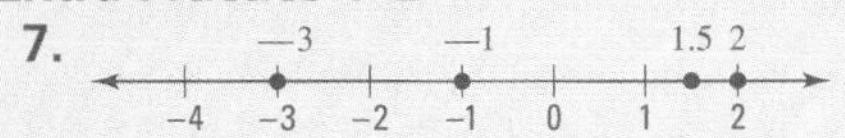

8.
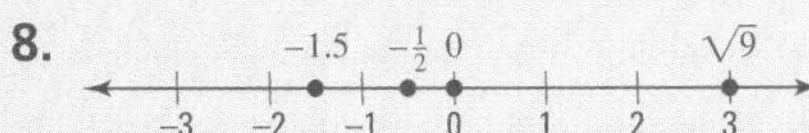

9.
$-\sqrt{3}$ -0.3 $1\frac{3}{4}$ $2\frac{1}{3}$

10.
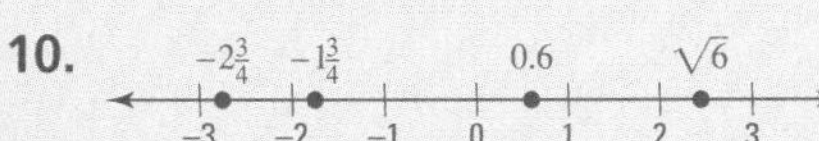

11.

12.

13.

14.

EXTRA PRACTICE

Chapter 1

Extra Practice 1–1 • The Language of Mathematics • pages 6–9

Define each set using roster notation.

1. odd natural numbers greater than 6 {7, 9, 11, . . .}
2. months having 31 days See additional answers.
3. integers between 2 and 3 ∅
4. days beginning with the letter *S* See additional answers.

Determine if each statement is *true* or *false*.

5. $7 \in \{x \mid x \text{ is a negative integer}\}$ false
6. $15 \in \{-3, 0, 3, 6, \ldots\}$ true
7. $\{a, h, t\} \subset \{m, a, t, h\}$ true
8. $\{-4\} \subset$ natural numbers false

Write all the subsets of each set.

9. $\{p\}$ $\{p\}, \varnothing$
10. $\{h, t\}$ $\{h, t\}, \{h\}, \{t\}, \varnothing$
11. $\{o, n, e\}$ See additional answers.

Which of the given values is a solution of the equation?

12. $n - 8 = -3$; 5, −5 5
13. $d + 2 = -2$; 4, −4 −4
14. $3a + 5 = 8$; −1, 0, 1 1
15. $\frac{4c}{3} = 4$; −12, 3, 12 3
16. $k + 5 = -5$; 0, −5, −10 −10
17. $c - 7 = -10$; 17, 3, −3 −3

Use mental math to solve each equation.

18. $x + 7 = 4$ −3
19. $n - 6 = 3$ 9
20. $7q = -28$ −4
21. $\frac{c}{-6} = 6$ −36
22. $n - 5 = -5$ 0
23. $\frac{1}{4} + d = \frac{3}{4}$ $\frac{1}{2}$
24. Henry saved \$36 less than Alan. Henry saved \$57. Use the equation $57 = A - 36$ and the values {91, 93, 99} for *A*. Find *A*, the amount of money Alan saved. \$93

Extra Practice 1–2 • Real Numbers • pages 10–13

Determine if each statement is *true* or *false*.

1. 3.16 is a rational number. true
2. 0.121212. . . is an irrational number. false
3. $\sqrt{8}$ is a real number. true
4. $-\sqrt{16}$ is an integer. true
5. $\sqrt{5\frac{5}{8}}$ is not a real number. false
6. $-\frac{15}{16}$ is a rational number. true

Graph each set of numbers on a number line. 7–14. See additional answers.

7. $\{-3, -1, 1.5, 2\}$
8. $\left\{-1.5, -\frac{1}{2}, 0, \sqrt{9}\right\}$
9. $\left\{-\sqrt{3}, -0.3, 1\frac{3}{4}, 2\frac{1}{3}\right\}$
10. $\left\{-2\frac{3}{4}, -1\frac{3}{4}, 0.6, \sqrt{6}\right\}$
11. whole numbers less than 1
12. real numbers less than 3
13. real numbers from −3 to 2 inclusive.
14. real numbers greater than or equal to −2

Extra Practice

Extra Practice 1–3 • Union and Intersection of Sets • pages 16–19

Refer to the diagram. Find the sets by listing the members.

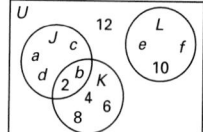

1. $J \cup K$ {a, b, c, d, 2, 4, 6, 8}
2. $J \cap L$ ∅
3. J' {e, f, 4, 6, 8, 10, 12}
4. K' {a, c, d, e, f, 10, 12}
5. $(J \cap K)'$ See additional answers.
6. $(J \cup K)'$ {e, f, 10, 12}
7. $K \cap L$ ∅
8. $K \cup L$ {b, e, f, 2, 4, 6, 8, 10}

U: J = {a, c, d, b, 2}; K = {b, 2, 4, 6, 8}; L = {e, f, 10}; 12

Let U = {g, r, a, p, h}, B = {g, r, a, p}, and C = {r, a, p}. Find each set, union, or intersection.

9. C' {g, h}
10. $B \cup C$ {g, r, a, p}
11. $B \cap C$ {r, a, p}
12. B' {h}
13. $(B \cup C)'$ {h}
14. $(B \cap C)'$ {g, h}
15. $B' \cup C$ {r, a, p, h}
16. $B' \cap C$ {h}
17. Let X = {l, i, g, h, t} and Y = {t, r, o, u, g, h}. Find $X \cap Y$. {g, h, t}
18. Let R = {−4, −2, 0, 2, 4} and S = {−2, 4, 10}. Find $R \cup S$. {−4, −2, 0, 2, 4, 10}
19. Let P = {0, 6, 12} and Q = {0, 3, 6, 9, 12}. Find $P \cap Q$. {0, 6, 12}

Use the replacement set of real numbers for the solution set for each compound inequality. 20–26. See additional answers.

20. $x \leq 0$ or $x \geq 1$
21. $x \leq 1$ and $x \geq 1$
22. $x > 4$ and $x \leq -1$
23. $x \geq 0$ and $x \geq 2$
24. $x < -4$ or $x < -1$
25. $x > 3$ or $x \leq 1$
26. Tondra's car stays in first gear until it reaches a speed of 12 mi/h. Graph the speeds at which her car is in first gear.

Extra Practice 1–4 • Addition, Subtraction and Estimation • pages 20–23

Add or subtract.

1. $-8 + (-37)$ −45
2. $-46 + 17$ −29
3. $-22 - 23$ −45
4. $18 - (-18)$ 36
5. $-16.4 + 9.3$ −7.1
6. $-68.9 + 70$ 1.1
7. $-2.1 + (-16.2)$ −18.3
8. $-4.3 - 5.7$ −10
9. $-2\frac{7}{8} - 1\frac{3}{8}$ $-4\frac{1}{4}$
10. $-6\frac{1}{4} + 5\frac{3}{4}$ $-\frac{1}{2}$
11. $-3\frac{5}{8} - 2\frac{1}{6}$ $-5\frac{19}{24}$
12. $-7\frac{2}{3} + 3\frac{1}{4}$ $-4\frac{5}{12}$
13. $-9.5 + (-11.7) + 8.6 + 0.4$ −12.2
14. $-19 + 21 + 16 + (-24)$ −6
15. $-8\frac{4}{5} + \left(-6\frac{7}{10}\right) + \left(-3\frac{1}{5}\right)$ $-18\frac{7}{10}$
16. $-42 + 29 + (-16) + 39$ 10

Evaluate each expression when $x = 24$ and $y = -18$.

17. $x + y$ 6
18. $x - y$ 42
19. $y - x$ −42
20. $-x + y$ −42

Evaluate each expression when $a = -3$ and $b = 1.8$.

21. $a - b$ −4.8
22. $-a - b$ 1.2
23. $-a + b$ 4.8
24. $a + b$ −1.2
25. Alfonse makes the following transactions to his savings account. Previous balance, $564.82; Withdrawal, $125; Deposit, $152.68; Deposit, $38.95; Withdrawal, $75. What is his new balance? $556.45

Additional Answers

Extra Practice 1-3

5. {a, c, d, e, f, 4, 6, 8, 10, 12}

20.

21.

22. ∅

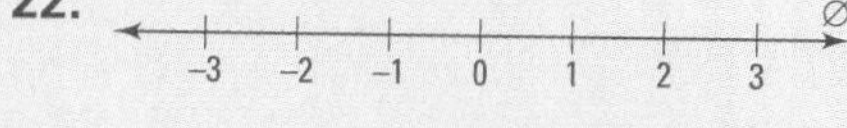

23.

24.

25.

26.

Extra Practice

Extra Practice 1–5 • Multiplication and Division • pages 26–29

Perform the indicated operations.

1. $6.7(-2.8)$ -18.76
2. $(-3.2)(-1.4)$ 4.48
3. $\left(3\frac{1}{2}\right)\left(-2\frac{1}{3}\right)$ $-8\frac{1}{6}$
4. $\left(-6\frac{1}{4}\right)\left(-1\frac{1}{2}\right)$ $9\frac{3}{8}$
5. $2\frac{2}{3} \div \left(-\frac{4}{9}\right)$ -6
6. $\left(-3\frac{1}{3}\right) \div 2$ $-1\frac{2}{3}$
7. $(1.05) \div (0.35)$ -3
8. $(2.25) \div (-15)$ -0.15
9. $3\frac{7}{8} + (5)(-6)$ $-26\frac{1}{8}$
10. $-4\frac{5}{16} + (-3)(-4)$ $7\frac{11}{16}$
11. $7.6 \div 1.9 - 4.1$ -0.1
12. $(-9.1) \div (-7) - 1.3$ 0
13. $5\frac{1}{3} \div (-4) + 6$ $4\frac{2}{3}$
14. $3\frac{1}{8} \div \left(-\frac{5}{8}\right) - 1$ -6
15. $-75 \div (-10) + \left(-2\frac{1}{2}\right)$ 5
16. $(17) \div (-2) + 2\frac{1}{2}$ -6
17. $(-64) \div \left(-\frac{2}{3}\right) + (-96)$ 0

Evaluate each expression when $r = -3$, $s = 1.5$, and $t = \frac{4}{5}$.

18. $r - s$ -4.5
19. $r + t$ $-2\frac{1}{5}$
20. $r + s$ -1.5
21. $r - t$ $-3\frac{4}{5}$
22. rs -4.5
23. $t(r + s)$ -1.2
24. $r + st$ -1.8
25. $(r + s) \div t$ -1.875
26. Nat earns \$6.40 per hour for each hour in his 32-h work week. For each hour over 32 h, he earns $1\frac{1}{2}$ times his hourly pay. How much will he earn if he works 42 h in one week? \$300.80
27. Gloria earns \$7.50 per hour and $1\frac{1}{2}$ times that amount for each hour she works over 32 h in a week. One week she earned \$273.75. How many hours of overtime did she work? 3 h

Extra Practice 1–7 • Distributive Properties and Properties of Exponents • pages 34–37

Use the distributive property to find each product.

1. $6.8 \cdot 7 + 6.8 \cdot 93$ 680
2. $2.7 \cdot 8 + 2.7 \cdot 12$ 54
3. $23 \cdot 16 - 23 \cdot 6$ 230
4. $101 \cdot 27$ 2,727
5. $35\left(2\frac{3}{7}\right)$ 85
6. $24\left(20\frac{1}{8}\right)$ 483

Evaluate each expression when $m = -2$ and $n = 5$.

7. m^2 4
8. $m^2 - n^2$ -21
9. n^3 125
10. mn^2 -50
11. m^3 -8
12. $2mn^2$ -100
13. $2m^2n$ 40
14. $(m - n)^2$ 49
15. $(n - m)^2$ 49
16. $(m^2 - 1)^3$ 27
17. $(-2mn)^2$ 400
18. $-2mn^2$ 100

Simplify.

19. $2^8 \cdot 2^6$ 2^{14}
20. $\frac{x^7}{x^4}$ x^3
21. $y^3 \cdot y^3$ y^6
22. $(y^3)^3$ y^9
23. $\left(\frac{1}{n}\right)^8$ $\frac{1}{n^8}$
24. $m^{10} \cdot m^{15}$ m^{25}
25. $(x^3)(x^4)(x^5)$ x^{12}
26. $(x)(x^2)(x^2)$ x^5

Evaluate mentally each sum or product when $j = 4.5$, $k = 2$, and $l = 0$.

27. $10jk$ 90
28. $67j^2l$ 0
29. $-5jk$ -45
30. $(5.5 - j)(11k)$ 22
31. $(2j - l)k^2$ 36
32. $(j + 0.5)(k + 2)$ 20
33. $(jk)^2 - 80$ 1
34. $(3.5 + k)(j - 5.5)$ -5.5
35. $jk^3(2j - 9)$ 0

686 Extra Practice

Extra Practice

Extra Practice 1–8 • Exponents and Scientific Notation • pages 38–41

Simplify.

1. $(-1)^{-4} + (-1)^{-5}$ 0 **2.** $(-1)^{-4} + (-1)^{-6}$ 2 **3.** $c^{-18} \div c^{-6}$ c^{-12}

4. $n^{-4} \cdot n^3$ n^{-1} **5.** $x^{-4} \cdot x^{-3}$ x^{-7} **6.** $y^{-3} \cdot y^{-3}$ y^{-6}

Evaluate each expression when $r = 3$ and $s = -3$.

7. r^{-3} $\frac{1}{27}$ **8.** s^{-3} $-\frac{1}{27}$ **9.** $(rs)^{-2}$ $\frac{1}{81}$ **10.** $r^2 \cdot r^{-2}$ 1

11. $s^{-3} \cdot s^2$ $-\frac{1}{3}$ **12.** $r^{-2}r^{-2}r^4$ 1 **13.** $s^3s^{-2}r^{-3}$ $-\frac{1}{9}$ **14.** r^3rs^{-3} -3

Write each number in scientific notation.

15. 4,700 4.7×10^3 **16.** 66,800 6.68×10^4 **17.** 1,410,000 1.41×10^6

18. 218,000 2.18×10^5 **19.** 0.0571 $5.71 + 10^{-2}$ **20.** 0.00178 1.78×10^{-3}

21. 0.00082 $8.2 + 10^{-4}$ **22.** 0.971 9.71×10^{-1} **23.** 0.0000000505 5.05×10^{-8}

Write each number in standard form.

24. 1.76×10^5 176,000 **25.** 2.6×10^4 26,000

26. 4.9×10^{-2} 0.049 **27.** 5.04×10^{-6} 0.00000504

Solve. Write your answer in scientific notation.

28. The distance from Earth to the Sun is about 93,000,000 miles. Write this distance in scientific notation. 9.3×10^7 miles

29. The speed of light is 3.00×10^{10} meters per second. How far does light travel in 1 hour? Write the answer in scientific notation. 1.08×10^{14} meters

Chapter 2

Extra Practice 2–1 • Patterns and Iterations • pages 52–55

Determine the next three terms in each sequence.

1. 1, 5, 9, 13, ____, ____, ____ 17, 21, 25 **2.** 31, 26, 21, 16, ____, ____, ____ 11, 6, 1

3. −5, −3, −1, 1, ____, ____, ____ 3, 5, 7 **4.** 25, 18, 11, 4, ____, ____, ____ −3, −10, −17

5. $9, 3, 1, \frac{1}{3}$, ____, ____, ____ $\frac{1}{9}, \frac{1}{27}, \frac{1}{81}$ **6.** $\frac{1}{10}, \frac{1}{5}, \frac{2}{5}, \frac{4}{5}$, ____, ____, ____ $\frac{8}{5}, \frac{16}{5}, \frac{32}{5}$

7. 2, 8, 18, 32, ____, ____, ____ 50, 72, 98 **8.** −8, −6, −4, −2, ____, ____, ____ 0, 2, 4

9. 1.5, 3, 4.5, 6, ____, ____, ____ 7.5, 9, 10.5 **10.** −1, −2, −4, −7, ____, ____, ____ −11, −16, −22

Draw the iteration diagram for each sequence. Calculate the output for the first 7 iterations.

11. 128, 64, 32, 16, . . . 8, 4, 2 **12.** −10, −7, −4, −1, . . . 2, 5, 8

13. 1, 6, 36, 216, . . . 1,296; 7,776; 46,656 **14.** 12, 9.5, 7, 4.5, . . . 2, −0.5, −3

Extra Practice 687

Additional Answers

Extra Practice 2-2
1–16.

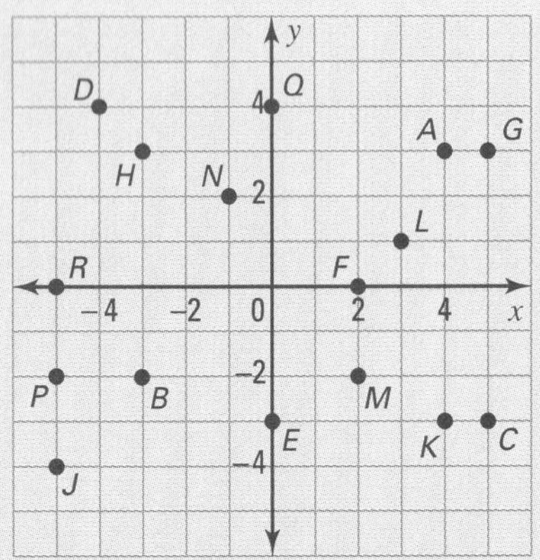

Extra Practice 2-3
1.

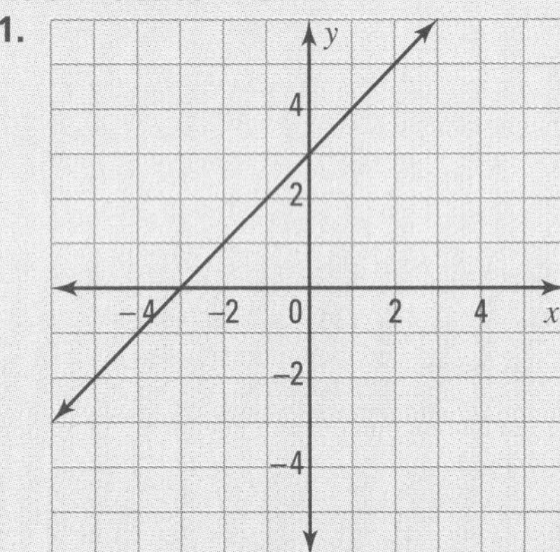

Extra Practice

Extra Practice 2-2 • The Coordinate Plane, Relations, and Functions • pages 56–59

Graph each point on a coordinate plane. 1–16. See additional answers.

1. $A(4, 3)$	**2.** $B(-3, -2)$	**3.** $C(5, -3)$	**4.** $D(-4, 4)$
5. $E(0, -3)$	**6.** $F(2, 0)$	**7.** $G(5, 3)$	**8.** $H(-3, 3)$
9. $J(-5, -4)$	**10.** $K(4, -3)$	**11.** $L(3,1)$	**12.** $M(2, -2)$
13. $N(-1, 2)$	**14.** $P(-5, -2)$	**15.** $Q(0, 4)$	**16.** $R(-5, 0)$

Given $f(x) = 3x - 2$, evaluate each function.

17. $f(1)$ 1	**18.** $f(0)$ −2	**19.** $f(-1)$ −5	**20.** $f(-3)$ −11
21. $f(2)$ 4	**22.** $f(-2)$ −8	**23.** $f(-6)$ −20	**24.** $f(6)$ 16

Write each as a set of ordered pairs. Give the domain and range.

25.

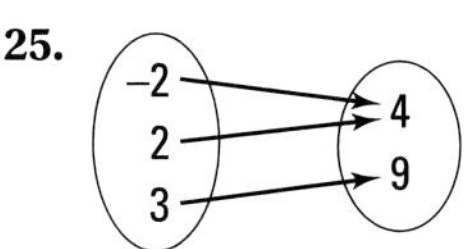

{(−2, 4), (2, 4), (3, 9)}; Domain: −2, 2, 3; Range: 4, 9

26.

x	3	5	7	9
y	2	4	6	8

{(3, 2), (5, 4), (7, 6), (9, 8)}; Domain: 3, 5, 7, 9; Range: 2, 4, 6, 8

27.

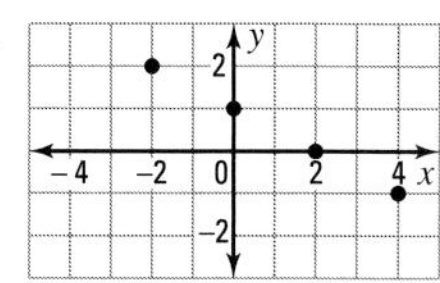

27. {(−2, 2), (0, 1), (2, 0), (4, −1)}; Domain: −2, 0, 2, 4; Range: −1, 0, 1, 2

28. José charges $3 for the first hour of baby-sitting and then $5 per hour for each additional hour. The function that describes how he is paid is $f(x) = 3 + 5(x - 1)$ where x is the number of hours he works. How much does he earn if he works 7 hours? $33

Extra Practice 2-3 • Linear Functions • pages 62–65

Graph each function. Check students' graphs. See sample answer in additional answers.

1. $y = x + 3$	**2.** $y = x - 3$	**3.** $y = x$
4. $f(x) = x + 5$	**5.** $f(x) = x - 5$	**6.** $f(x) = -x$
7. $y = 2x + 2$	**8.** $f(x) = 2x - 2$	**9.** $f(x) = -4$
10. $y = 4$	**11.** $y = 0$	**12.** $f(x) = 3x$
13. $y = -x + 5$	**14.** $y = -3x + 2$	**15.** $f(x) = -2 - 3$
16. $f(x) = 4x - 6$	**17.** $f(x) = -3x + 3$	**18.** $y = -x - 1$

Evaluate $f(x) = |3x + 4|$ for the given value of x.

19. $f(1)$ 7	**20.** $f(-1)$ 1	**21.** $f(7)$ 25
22. $f(0)$ 4	**23.** $f(-6)$ 14	**24.** $f(-2)$ 2

Evaluate $g(x) = |-5x - 3|$ for the given value of x.

25. $g(0)$ 3	**26.** $g(1)$ 8	**27.** $g(2)$ 13
28. $g(-3)$ 12	**29.** $g(-4)$ 17	**30.** $g(-2)$ 7

Extra Practice 2–4 • Solve One-Step Equations • pages 66–69

Solve each equation.

1. $m + 17 = 45$ 28
2. $9x = -54$ −6
3. $17 + d = -5$ −22
4. $-16 = j - 2$ −14
5. $-24 = c + 9$ −33
6. $16n = -12$ $-\frac{3}{4}$
7. $-8x = 96$ −12
8. $0.8a = 0.72$ 0.9
9. $b + 0.8 = 0.72$ −0.08
10. $36 = -\frac{4}{9}c$ −81
11. $\frac{5}{8}x = 10$ 16
12. $51 = \frac{3}{5}x$ 85
13. $13.24 = x - 4.2$ 17.44
14. $-8.6 = m + 2.15$ −10.75
15. $j + \frac{3}{8} = 1$ $\frac{5}{8}$

Translate each sentence into an equality using n to represent the unknown number. Then solve the equation for n.

16. When n is increased by 18, the result is −12. $n + 18 = -12; -30$

17. When a number is decreased by 7, the result is −4. $n - 7 = -4; 3$

18. The quotient of a number and 7 is 0.6. $\frac{n}{7} = 0.6; 4.2$

19. The product of 4 and a number is the same as the square of −6. $4n = (-6)^2; 9$

20. The difference between a number and 13 is 14. $n - 13 = 14; 27$

21. One fourth of −64 is the same as the product of 2 and some number. $\frac{1}{4}(-64) = 2n; -8$

22. Liya decided to save $8 per week for the next 4 weeks so that her savings would total $100. Let n represent the amount she has before she begins saving. Write an equation that illustrates the situation. Then solve the equation. $n + 8(4) = 100; n = 68$

Extra Practice 2–5 • Solve Multi-Step Equations • pages 72–75

Solve each equation and check the solution.

1. $6n + 5 = 23$ 3
2. $4n - 3 = 17$ 5
3. $-55 = 8x - 7$ −6
4. $-36 = 5x + 4$ −8
5. $-3j + 16 = -11$ 9
6. $-2n - 17 = 17$ −17
7. $2(3d - 4) = 10$ 3
8. $-4(2x + 1) = 4$ −1
9. $-3(2x - 3) = -9$ 3
10. $8x - 7 = 2x + 5$ 2
11. $-3x + 24 = 5x - 24$ 6
12. $3c + 5 = 7c - 7$ 3
13. $-3x - 1 = -2x - 1$ 0
14. $2k + 3 + 3k = 1 + 7k$ 1
15. $-4a + 7 - 2a = -11$ 3
16. $\frac{1}{2}(16k + 10) = -11$ −2
17. $4(1.5 - x) = 14$ −2
18. $4(3c - 2) = -38 + 6$ −2

Translate each sentence into an equation. Then solve.

19. Five more than 4 times a number is 33. Find the number. $4n + 5 = 33; 7$

20. Two less than 3 times a number is 13. Find the number. $3n - 2 = 13; 5$

21. When 20 is decreased by twice a number, the result is 8. Find the number. $20 - 2n = 8; 6$

22. Keisha bought 3 report binders that had the same price. The total cost came to $11.97, which included $.57 sales tax. Write and solve an equation to find out how much each binder cost. $3b + 0.57 = 11.97; \$3.80$

Extra Practice 689

Additional Answers

Extra Practice 2-6

1. 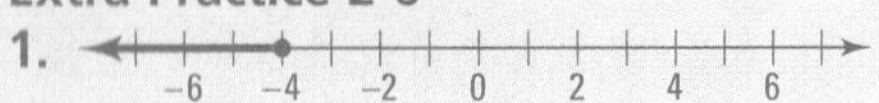

2. −6 −4 −2 0 2 4 6

17. 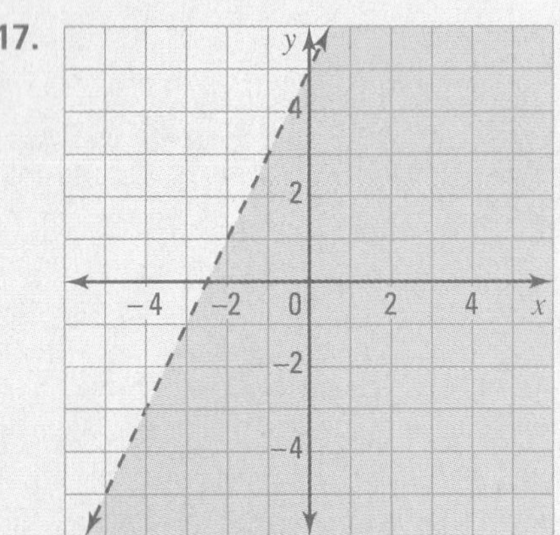

26. 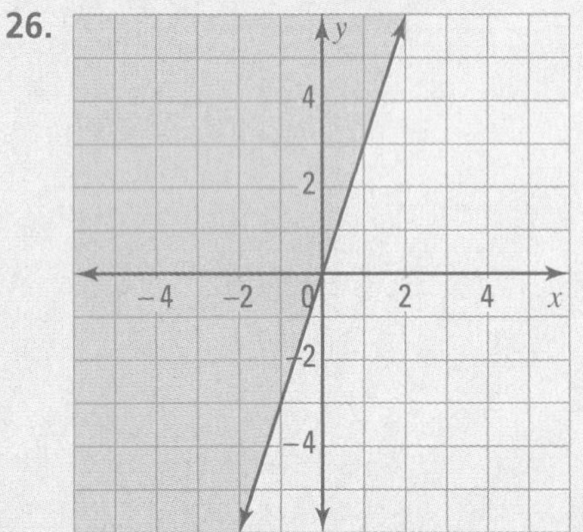

Extra Practice 2–7

1. **HOURS OF TELEVISION WATCHED ON FRIDAY**

Hours of Television	Tally	Frequency
0	卌 ||	7
1	卌 |	6
2	卌 |	6
3	卌	5
4	|||	3
5	||	2
6	|	1

3. **VIDEO RENTAL PRICES**

Price	Tally	Frequency
\$1.50 – \$1.74	||	2
\$1.75 – \$1.99	|||	3
\$2.00 – \$2.24	卌	5
\$2.25 – \$2.49	||||	4
\$2.50 – \$2.74	卌	5
\$2.75 – \$2.99	|||	3
\$3.00 – \$3.24	||||	4
\$3.25 – \$3.49	||||	4

Extra Practice

Extra Practice 2–6 • Solving Linear Inequalities • pages 76–79

Solve each inequality and graph the solution on a number line. Check students' graphs. See sample graphs in additional answers.

1. $3a + 2 \le -10$ $a \le -4$
2. $7n - 2 > 19$ $n > 3$
3. $\frac{1}{2}n - 7 < -6$ $n < 2$
4. $\frac{1}{3}c + 8 \ge 10$ $c \ge 6$
5. $10 - 3r \le 7$ $r \ge 1$
6. $7 > 2a - 5$ $a < 6$
7. $33 \le -7n - 2$ $n \le -5$
8. $-19 < 14 - 11c$ $c < 3$
9. $2 \ge -18 - 5t$ $t \ge -4$
10. $\frac{2}{3}x + 8 \le 10$ $x \le 3$
11. $2(3w + 4) < -28$ $w < -6$
12. $3(4c - 2) \le 18$ $c \le 2$
13. $2a + 5 \ge 8a - 7$ $a \le 2$
14. $2n - 13 > 11n + 14$ $n < -3$
15. $2 < \frac{2}{3}(9 - 6a)$ $a < 1$
16. $12 > \frac{4}{9}(18 - 9c)$ $c > -1$

Graph each inequality on the coordinate plane. Check students' graphs. See sample graphs in additional answers.

17. $y < 2x + 5$
18. $y \le -2x - 3$
19. $y > -x - 1$
20. $x + y \le 5$
21. $x - y < 3$
22. $y < \frac{1}{2}x - 1$
23. $2x + 4y \ge 8$
24. $x - 2y < 10$
25. $1 \ge 2x - \frac{1}{2}y$
26. A pet store charges a minimum of \$3 per hour to take care of a person's pet. The inequality that describes how the store charges is $y \ge 3x$ where x is the number of hours and y is the amount of money charged. Graph the inequality. See additional answers.

Extra Practice 2–7 • Data and Measures of Central Tendency • pages 82–85

Thirty families were randomly sampled and surveyed as to the number of hours they watched television on a typical Friday. The results are listed below.

5	0	3	1	2	0	1	4	2	2
2	1	0	4	6	1	1	3	3	3
0	3	5	2	1	0	2	0	0	4

1. Construct a frequency table for these data. See additional answers.
2. Find the mean, median, and mode of the data. 2.03, 2, 0

As part of her research for a term paper on home entertainment, Lydia surveyed video stores to find the cost of renting a movie for one day. The results are listed below.

\$2.50	\$2.86	\$1.99	\$2.00	\$3.10	\$2.15	\$1.55	\$2.83	\$3.49	\$2.69
\$1.85	\$3.14	\$2.62	\$3.35	\$3.32	\$2.45	\$2.12	\$1.99	\$2.05	\$2.90
\$2.49	\$3.07	\$1.68	\$2.33	\$3.00	\$2.60	\$2.00	\$3.25	\$2.25	\$2.50

3. Construct a frequency table for these data. Group the data into intervals. See additional answers.
4. Which interval contains the median of the data? \$2.50–\$2.74

Extra Practice

Extra Practice 2–8 • Display Data • pages 86–89

The weights in pounds of the 30 students who tried out for the Snyder High School football team were as follows:

145	160	172	129	149	202	183	176	170	169
157	146	177	200	162	164	168	165	150	161
145	171	173	162	164	164	166	175	181	179

1. Construct a stem-and-leaf plot to display the data. See additional answers.
2. Identify any outliers, clusters, and gaps in the data. See additional answers.
3. Find the mode of the data. 164
4. Find the median of the data. 165.5

On a test measuring reasoning aptitude on a scale of 0 to 100, a class of 30 students received the following scores.

59	38	48	75	78	81	52	45	55	62
91	56	39	47	80	55	72	60	58	60
63	70	65	52	42	72	70	50	47	55

5. Construct a stem-and-leaf plot to display the data. See additional answers.
6. Identify any outliers, clusters,and gaps in the data. no outliers; cluster:42-78; no gaps
7. Find the mode of the data. 55
8. Find the median of the data. 58.5
9. A newspaper took a random survey of its readers about the number of miles they travel to and from work each day. The data are recorded in this frequency table. Construct a histogram of the data. See additional answers.

DISTANCES TRAVELED TO AND FROM WORK

Miles	Frequency
0 – 9	24
10 – 19	16
20 – 29	10
30 – 39	20
40 – 49	18
50 – 59	7
60 – 69	5

Chapter 3

Extra Practice 3–1 • Points, Lines, and Planes • pages 104–107

Use the figure at the right for Exercises 1–4. Which postulate justifies your answer?

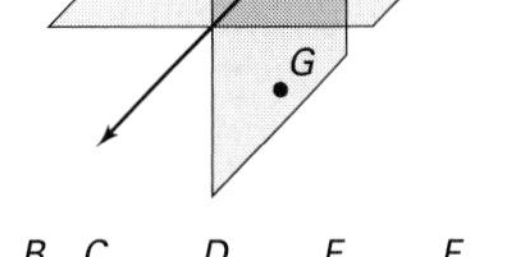

1. Name two points that determine line *l*. points *D* and *E*; Postulate 1
2. Name three points that determine plane *A*. points *D*, *E*, and *F*; Postuate 2
3. Name three lines that lie in plane *A*. $\overleftrightarrow{DE}$, $\overleftrightarrow{EF}$, $\overleftrightarrow{DF}$; Postulate 3
4. Name the intersection of planes *A* and *B*. $\overleftrightarrow{DE}$ or line *l*; Postulate 4

Use the number line at the right for Exercises 5–8. Find each length.

5. $\overline{AD}$ 5
6. $\overline{EC}$ 4
7. $\overline{FB}$ 7
8. $\overline{EF}$ 2
9. In the figure below, $RT = 85$. Find RS. 43

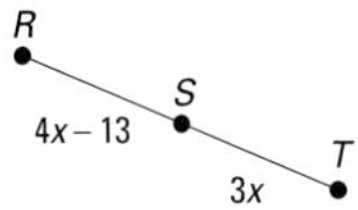

10. In the figure below, $LN = 79$. Find ML. 35

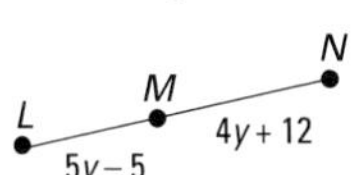

Additional Answers

Extra Practice 2-8

1.

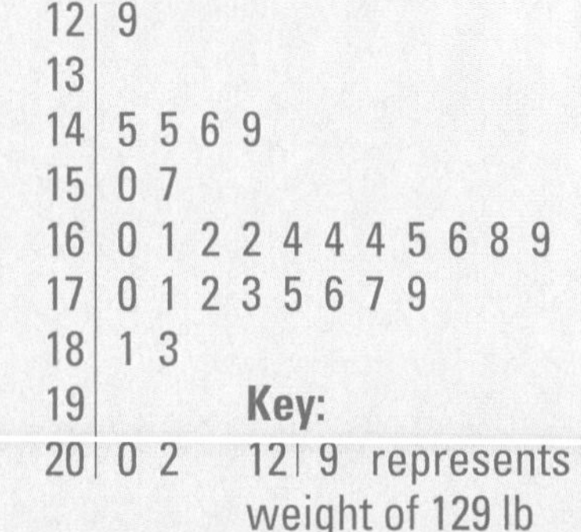

WEIGHTS OF STUDENTS TRYING OUT FOR FOOTBALL
12 | 9
13 |
14 | 5 5 6 9
15 | 0 7
16 | 0 1 2 2 4 4 4 5 6 8 9
17 | 0 1 2 3 5 6 7 9
18 | 1 3
19 |
20 | 0 2
Key: 12 | 9 represents a weight of 129 lb

2. outliers: 129, 200, 202; cluster: 145–183; gaps: between 130 and 144, between 184 and 199

5.

APTITUDE TEST SCORES
3 | 8 9
4 | 2 5 7 7 8
5 | 0 2 2 5 5 5 6 8 9
6 | 0 0 2 3 5
7 | 0 0 2 2 5 8
8 | 0 1
9 | 1
Key: 3 | 8 represents a score of 38

9.

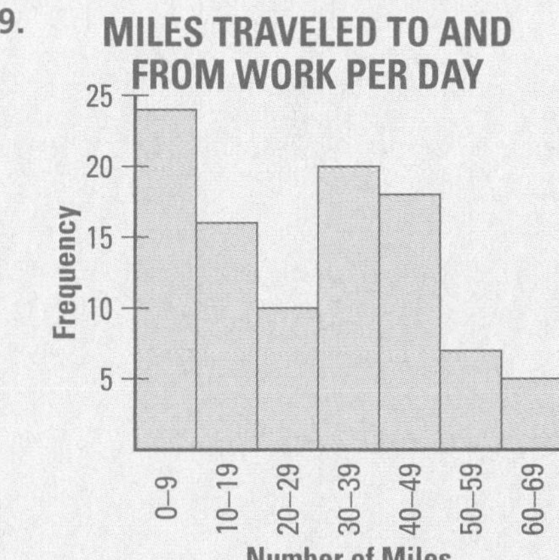

Extra Practice

Extra Practice 3–2 • Types of Angles • pages 108–111

Exercises 1–4 refer to the protractor at the right.

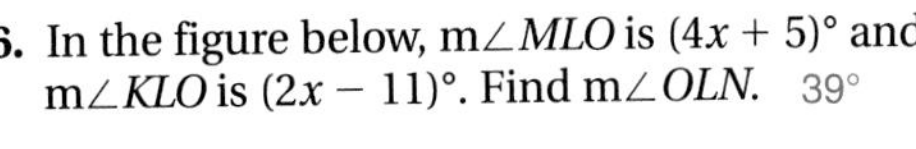

1. Name the straight angle. $\angle BAF$
2. Name the three right angles. $\angle BAD$, $\angle DAF$, $\angle CAE$
3. Name all the acute angles. Give the measure of each. $\angle BAC = 75°$, $\angle CAD = 15°$, $\angle DAE = 75°$, $\angle EAF = 15°$
4. Name all the obtuse angles. Give the measure of each. $\angle BAE = 165°$, $\angle CAF = 105°$
5. In the figure below, m$\angle QRS$ is $x°$ and m$\angle SRT$ is $5x°$. Find m$\angle SRT$. 75°

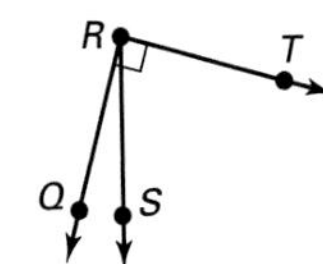

6. In the figure below, m$\angle MLO$ is $(4x + 5)°$ and m$\angle KLO$ is $(2x - 11)°$. Find m$\angle OLN$. 39°

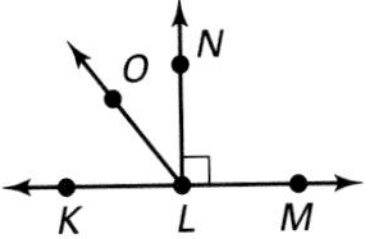

7. An angle measures 47°. What is the measure of its complement? 43°

Extra Practice 3–3 • Segments and Angles • pages 114–117

Exercises 1–4 refer to the figure at the right.

G H I J K L M N

−4 −3 −2 −1 0 1 2 3

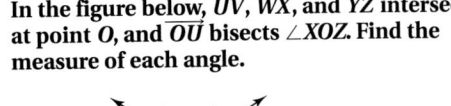

1. Name the midpoint of $\overline{GK}$. I
2. Name the segment whose midpoint is point H. $\overline{GI}$
3. Name all the segments whose midpoint is point J. $\overline{IK}$, $\overline{HL}$, $\overline{GM}$
4. Assume that point O is the midpoint of $\overline{GN}$. What is its coordinate? −0.5

In the figure below, $\overleftrightarrow{JM}$ and $\overleftrightarrow{OL}$ intersect at point K, and $\overrightarrow{KN}$ bisects $\angle OKM$. Find the measure of each angle.

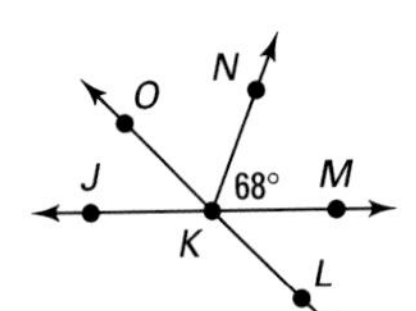

5. $\angle OKM$ 136°
6. $\angle LKJ$ 136°
7. $\angle MKL$ 44°
8. $\angle JKO$ 44°

In the figure below, $\overleftrightarrow{UV}$, $\overleftrightarrow{WX}$, and $\overleftrightarrow{YZ}$ intersect at point O, and $\overrightarrow{OU}$ bisects $\angle XOZ$. Find the measure of each angle.

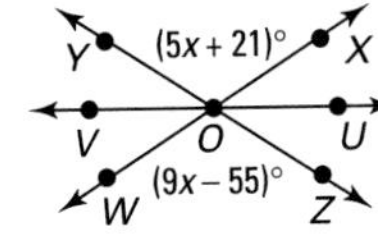

9. $\angle ZOW$ 116°
10. $\angle WOY$ 64°
11. $\angle XOU$ 32°
12. $\angle WOV$ 32°

In the figure at the right, point F is the midpoint of $\overline{EG}$. Find the length of each segment.

$4a + 6$ $7a$ $5a$

E F G H

13. $\overline{EF}$ 14
14. $\overline{EG}$ 28
15. $\overline{GH}$ 10
16. $\overline{EH}$ 38

692 Extra Practice

Extra Practice

Extra Practice 3–4 • Constructions and Lines • pages 118–121

In the figure at the right, $\overleftrightarrow{AD} \parallel \overleftrightarrow{HE}$ and $\overleftrightarrow{BF} \perp \overleftrightarrow{GC}$. Find the measure of each angle.

1. $\angle AJB$ 47°
2. $\angle JKI$ 47°
3. $\angle CJD$ 43°
4. $\angle JIK$ 43°
5. $\angle GIK$ 137°
6. $\angle GJD$ 137°
7. $\angle BKE$ 133°
8. $\angle FKI$ 133°

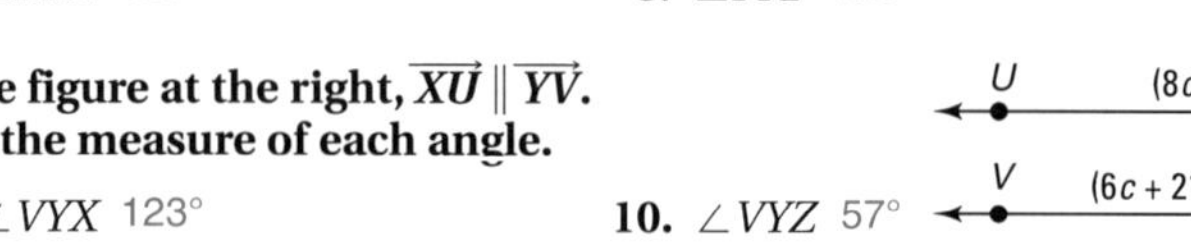

In the figure at the right, $\overleftrightarrow{XU} \parallel \overleftrightarrow{YV}$. Find the measure of each angle.

9. $\angle VYX$ 123°
10. $\angle VYZ$ 57°
11. $\angle UXW$ 123°
12. $\angle UXY$ 57°

13. Compare parallel and skew lines. Parallel lines are two coplanar lines that do not intersect. Skew lines are noncoplanar lines.

Extra Practice 3–5 • Inductive Reasoning • pages 124–127

Draw the next figure in each pattern. Then describe the twelfth figure in the pattern.

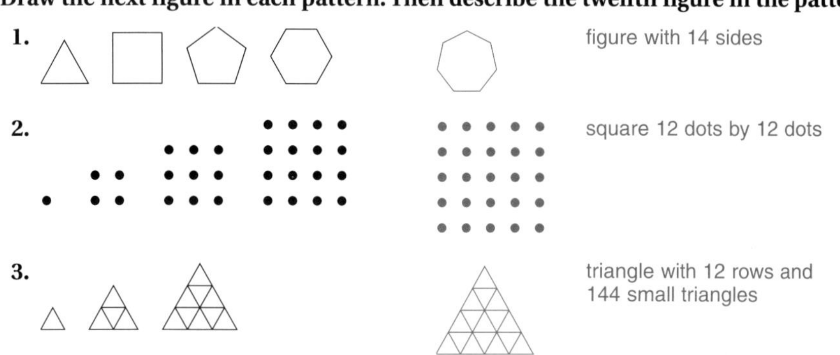

1. figure with 14 sides
2. square 12 dots by 12 dots
3. triangle with 12 rows and 144 small triangles

The figures below show one, two, three and four segments drawn inside a triangle.

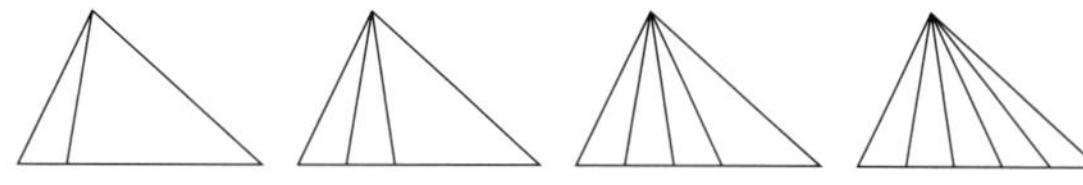

4. In each figure, the segments divide the interior of the triangle into regions. How many regions are formed in each of the figures shown? 2, 3, 4, and 5
5. Find the number of regions that would be formed when twelve segments are drawn through a triangle. 13 regions

Extra Practice 693

Additional Answers

Extra Practice 3-6

1.

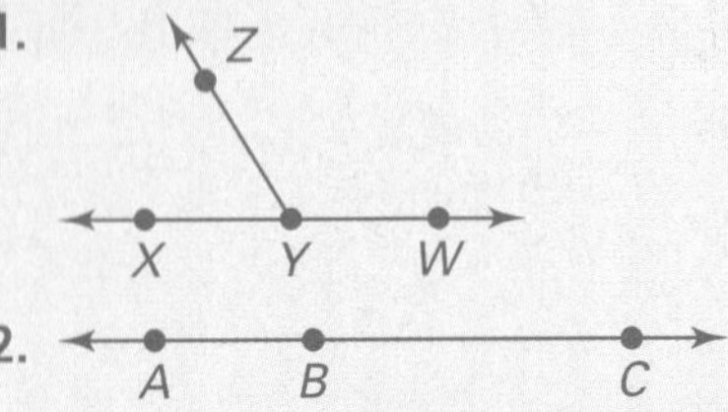

2. A B C

3.

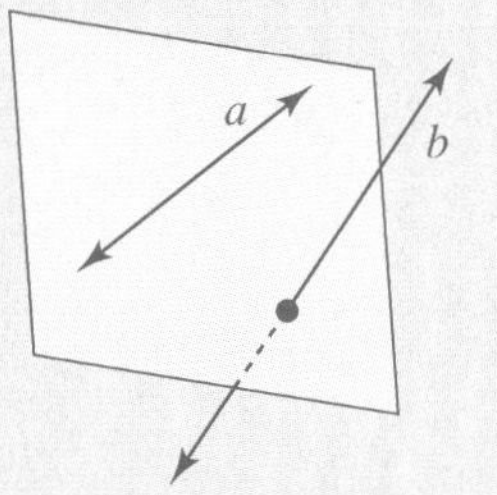

Extra Practice

Extra Practice 3–6 • Conditional Statements • pages 128–131

Sketch a counterexample that shows why each conditional is false. 1–3. See additional answers.

1. If m∠XYZ + m∠ZYW = 180°, then $ZY \perp XW$.
2. If point B is between points A and C, then B is the midpoint of $\overline{AC}$.
3. If two lines are not parallel, then they intersect.

Write the converse of each statement. Then tell whether the given statement and its converse are true or false.

4. If the sum of the measures of two angles is 180° then the angles are supplementary. true; If two angles are supplementary, then the sum of the measure is 180°; true
5. If two lines are parallel, then they intersect. false; If two lines intersect, then they are parallel; false
6. If $\overrightarrow{BC}$ and $\overrightarrow{BA}$ are opposite rays, then B is the midpoint of $\overline{AC}$. false; If B is the midpoint of $\overline{AC}$, then $\overrightarrow{BC}$ and $\overrightarrow{BA}$ are opposite rays; true

Write each definition as two conditionals and as a single biconditional.

7. Parallel lines are coplanar lines that do not intersect. Conditionals: If two coplanar lines are parallel, then they do not intersect; if two coplanar lines do not intersect, then they are parallel. Biconditional: Two coplanar lines are parallel if and only if they do not intersect.
8. Supplementary angles are two angles whose sum is 180°. Conditionals: If two angles are supplementary, then their sum is 180°; if the sum of two angles is 180°, then they are supplementary. Biconditional: Two angles are supplementary if and only if their sum is 180°.

Extra Practice 3–7 • Deductive Reasoning and Proof • pages 134–137

1. **Given:** m∠1 = m∠4
   ∠1 and ∠2 are complementary
   ∠3 and ∠4 are complementary
   Prove: ∠2 = ∠3

Statements	Reasons
1. ∠1 and ∠2 are __?__ complementary ∠3 and ∠4 are __?__ complementary	1. given
2. ∠1 + ∠2 = 90° ∠3 + ∠4 = __?__ 90°	2. definition complementary angles
3. ∠1 + ∠2 = ∠3 + ∠4	3. __?__ substitution property
4. ∠1 = ∠4	4. __?__ given
5. ∠2 = ∠3	5. __?__ subtraction postulate

2. **Given:** m∠TSW = m∠TWS
   Prove: m∠TSR = m∠TWS

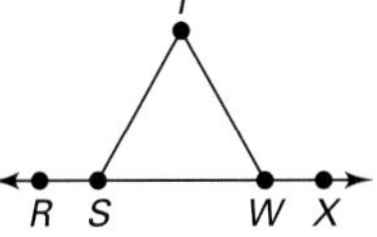

Statements	Reasons
1. ∠TSR is supplementary to __?__ ∠TSW ∠TWX is supplementary to __?__ ∠TWS	1. __?__ definition of supplementary angles
2. m∠TSW = m∠TWS	2. __?__ given
3. m∠TSR = __?__ m∠TWX	3. __?__ If two angles have equal measure, then their supplements have equal measure.

694 Extra Practice

Chapter 4

Extra Practice 4–1 • Triangles and Triangle Theorems • pages 150–153

Find the value of x in each figure.

1.  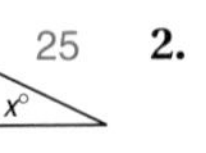 25

2. 30 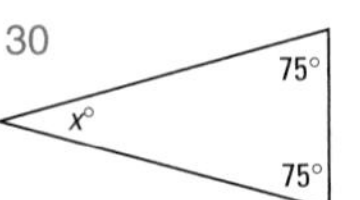

3. 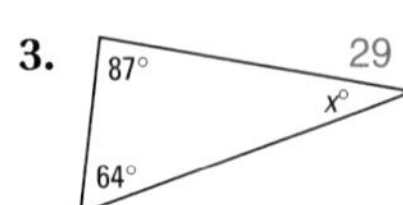 29

4. 125 

5. 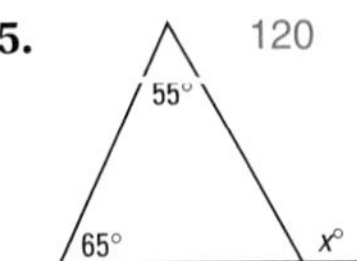 120

6. 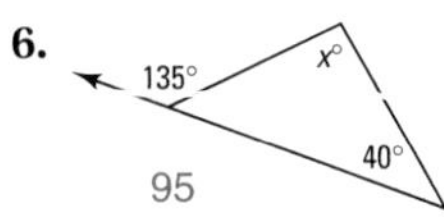 95

7. 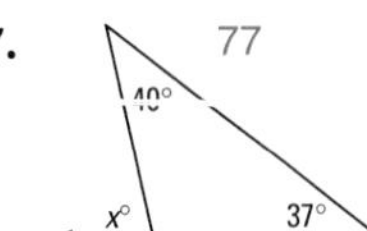 77

8. 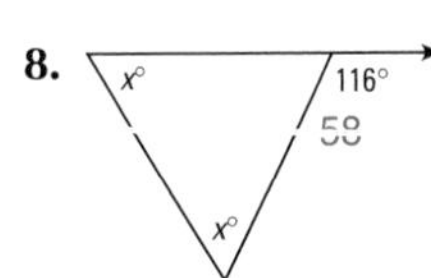 58

9. In the figure below, $ED \perp DF$. Find m$\angle DFE$. 55°

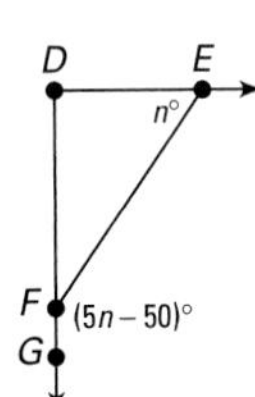

10. In the figure below, $AB \parallel JK$. Find m$\angle JCK$. 97°

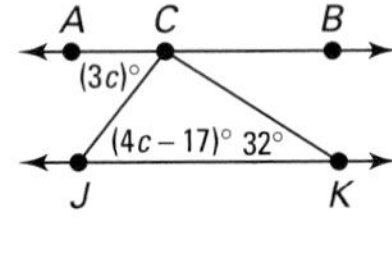

11. In the figure below, $DE \parallel RT$; m$\angle SDF$ = m$\angle SFD$. Find m$\angle SFE$. 100°

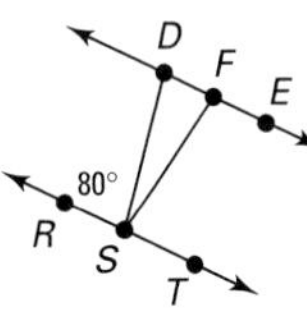

Extra Practice 4–2 • Congruent Triangles • pages 154–157

1. Copy and complete this proof.
Given: $\overline{AB} \cong \overline{CB}$; $\overline{DB}$ bisects $\angle ABC$.
Prove: $\triangle ABD \cong \triangle CBD$

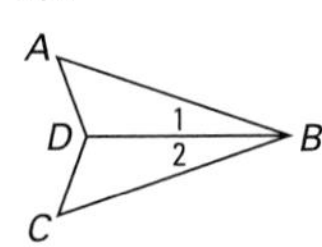

Statements	Reasons
1. ? $\overline{AB} \cong \overline{CB}$; $\overline{DB}$ bisects $\angle ABC$.	1. ? given
2. m$\angle 1$ = m$\angle 2$ or $\angle 1 \cong \angle 2$	2. ? definition of angle bisector
3. ? $\overline{DB} \cong \overline{DB}$	3. ? reflexive property
4. $\triangle ABD \cong \triangle CBD$	4. ? SAS postulate

Write a two-column proof. 2–3. See additional answers.

2. **Given:** $\overline{RS} \cong \overline{VT}$; $\overline{RV} \cong \overline{ST}$
Prove: $\triangle RSV \cong \triangle TVS$

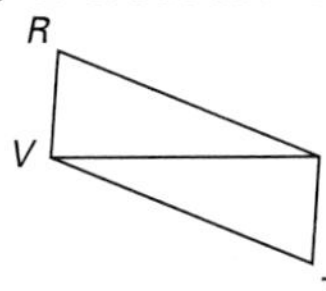

3. **Given:** $\overline{XV}$ and $\overline{WT}$ intersect at point Y; $\overline{XY} \cong \overline{VY}$; Y is the midpoint of $\overline{WT}$.
Prove: $\triangle WXY \cong \triangle VTY$

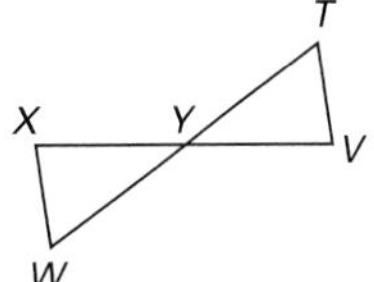

Additional Answers

Extra Practice 4-2

2.

Statements	Reasons
1. $RS \cong TV$; $VR \cong ST$	1. given
2. $VS \cong VS$	2. reflexive property
3. $\triangle RSV \cong \triangle TVS$	3. SSS postulate

3.

Statements	Reasons
1. XV and WT intersect at point Y; $XY \cong VY$; Y is the midpoint of WT.	1. given
2. $\angle WYX$ and $\angle TYV$ are vertical angles.	2. definition of vertical angles
3. $\angle WYZ \cong \angle TYV$	3. vertical angles theorem
4. $WY \cong TY$	4. definition midpoint
5. $\triangle WXY \cong \triangle VTY$	5. SAS postulate

Extra Practice

Extra Practice 4–3 • Congruent Triangles and Proofs • pages 160–163

Find the value of n in each figure.

1. 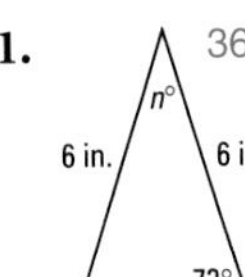

2. 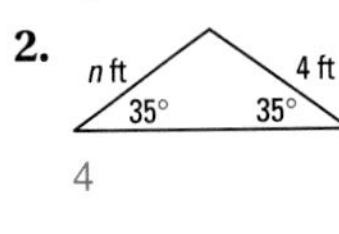

3. 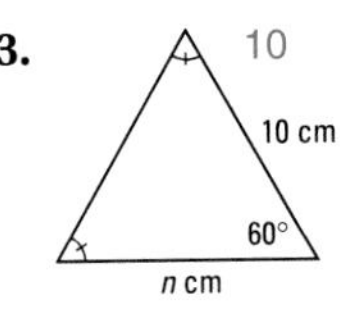

4. 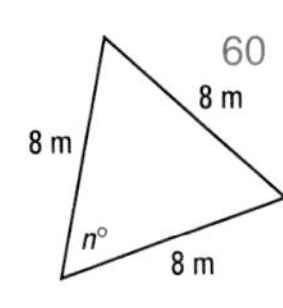

Copy and complete the proof.

5. Given: Point B is the midpoint of AC and ED.
Prove: $\angle E \cong \angle D$

Statements	Reasons
1. ? B is the midpoint of $\overline{AC}$ and $\overline{ED}$.	1. ? given
2. ? $AB = BC$; $EB = BD$	2. midpoint theorem
3. $BC = CB$; $BD = DB$	3. reflexive property
4. $AB = CB$, or $\overline{AB} \cong \overline{CB}$; $EB = DB$, or $\overline{EB} \cong \overline{DB}$	4. ? transitive property
5. $\angle 1$ and $\angle 2$ are ? vertical angles	5. ? definition of vertical angles
6. ? $\cong$? $\angle 1$, $\angle 2$	6. ? vertical angles theorem
7. $\triangle ABE \cong \triangle CBD$	7. ? SAS postulate
8. ? $\angle E \cong \angle D$	8. ? CPCTC

Extra Practice 4–4 • Altitudes, Medians, and Perpendicular Bisectors • pages 164–167

Trace each triangle onto a sheet of paper. Sketch all the altitudes and all the medians.

1. 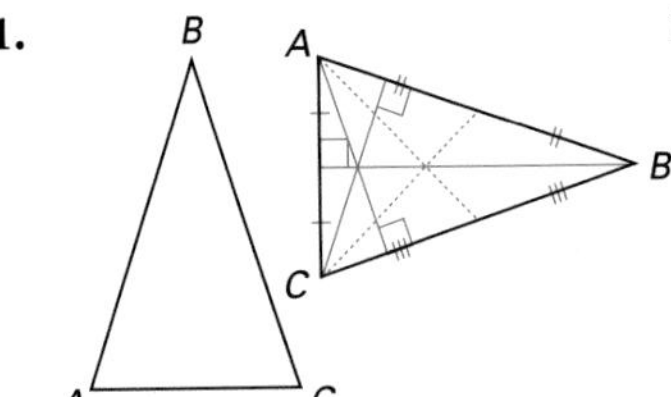

2. 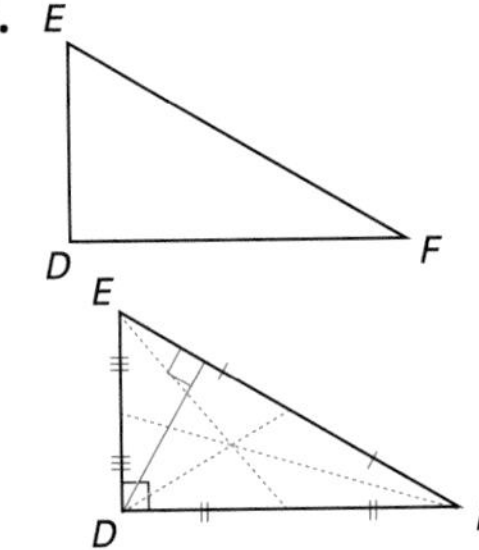

3. 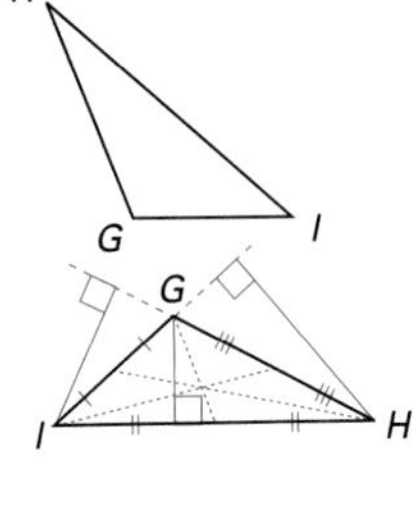

Exercises 4–9 refer to $\triangle PQR$ at the right with altitude $\overline{QT}$. Tell whether each statement is *true* or *false*.

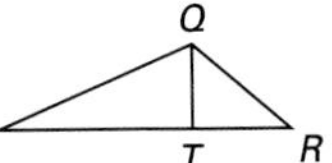

4. $\overline{QT} \perp \overline{PR}$ true

5. $\overline{TQ} \cong \overline{PT}$ false

6. $\overline{PT} \cong \overline{TR}$ false

7. $m\angle PTQ = m\angle RTQ$ true

8. $m\angle QTP = 90°$ true

9. $\angle P \cong \angle R$ false

696 Extra Practice

Extra Practice

Extra Practice 4–6 • Inequalities in Triangles • pages 172–175

Can the given measures be the lengths of the sides of a triangle?

1. 3 m, 6 m, 8 m yes
2. 9 ft, 7 ft, 2 ft no
3. 18 in., 13 in., 34 in. no
4. 15 cm, 15 cm, 15 cm yes
5. 2.4 yd, 6.7 yd, 3.9 yd no
6. $3\frac{1}{2}$ ft, $3\frac{1}{4}$ ft, $6\frac{1}{2}$ ft yes
7. 6 mm, 5 mm, 4 mm yes
8. 3 mi, 2 mi, 1 mi no
9. 2 yd, 5 ft, 72 in. yes

In each figure, give the ranges of possible values for *x*.

10. 
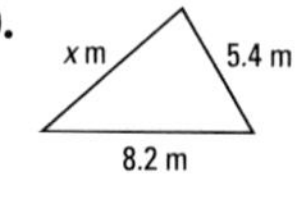

$2.8 < x < 13.6$

11. 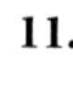
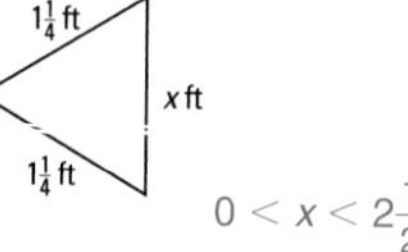

$0 < x < 2\frac{1}{2}$

12. 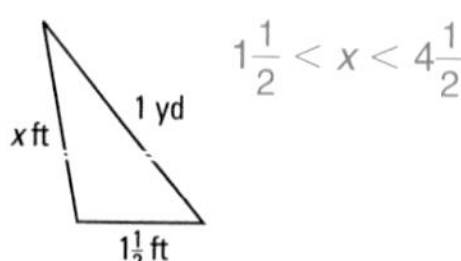

$1\frac{1}{2} < x < 4\frac{1}{2}$

13. In $\triangle FGH$, $\overline{FG} > \overline{GH}$ and $\overline{HF} > \overline{FG}$. Which is the largest angle of the triangle? $\angle G$

14. In $\triangle ABC$, $\overline{BC} = 18$, $\overline{AB} = 16.5$, and $\overline{AC} = 14$. List the angles of the triangle in order from largest to smallest. $\angle A$, $\angle C$, $\angle B$

15. In $\triangle PQR$, $m\angle P = 73°$, $m\angle Q = 57°$, and $m\angle R = 50°$. List the sides of the triangle in order from longest to shortest. $\overline{QR}$, $\overline{PR}$, $\overline{PQ}$

Extra Practice 4–7 • Polygon and Angles • pages 178–181

Find the unknown angle measure or measures in each figure.

1. 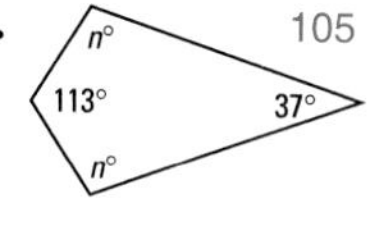

105

2. 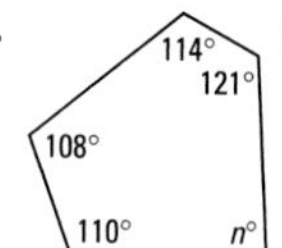

87

3. 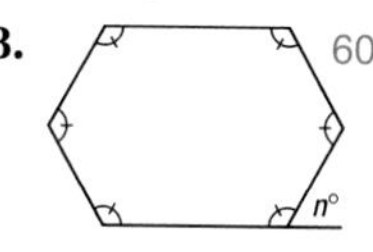

60

4. 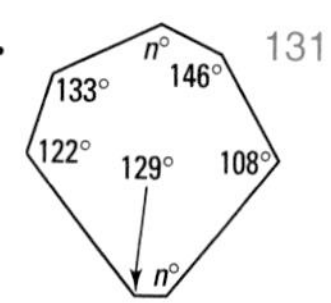

131

5. Find the measure of each interior angle of a regular heptagon. $128\frac{4}{7}$

6. Find the measure of each interior angle of a regular decagon. 144

7. Find the sum of the measures of the interior angles of a regular polygon with 16 sides. 2,520

8. Find the sum of the measures of the interior angles of a regular polygon with 20 sides. 3,240

Extra Practice 697

Extra Practice

Extra Practice 4–8 • Special Quadrilaterals: Parallelograms • pages 182–185

In Exercises 1–6, the figure is a parallelogram. Find the values of *a*, *b*, *c*, and *d*.

1. 38, 26, 105, 75

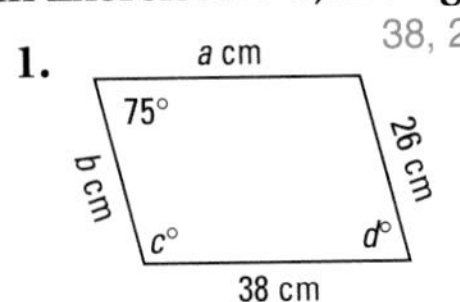

2. 6, 55, 90, 35

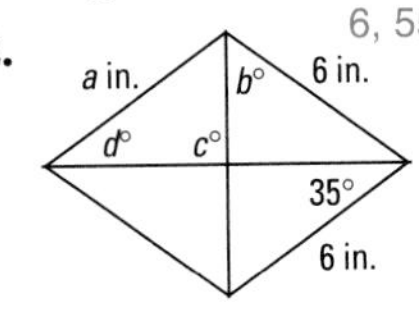

3. 90, 34, 90, 56

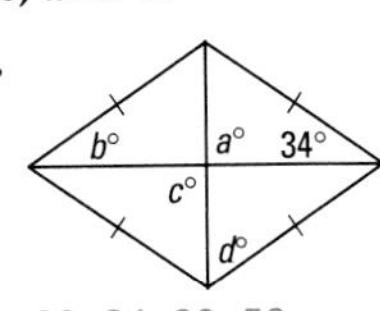

4. $\overline{AC} = 9\frac{3}{4}$ ft; $\overline{AD} = 4\frac{1}{2}$ ft

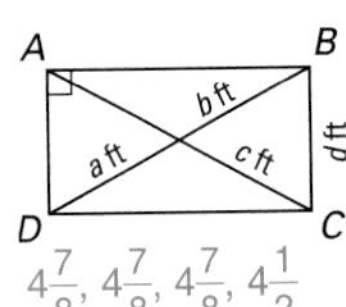

$4\frac{7}{8}, 4\frac{7}{8}, 4\frac{7}{8}, 4\frac{1}{2}$

5. $\overline{VS} = 18$ m; $\overline{RS} = 14$ m; $\overline{RT} = 12$ m

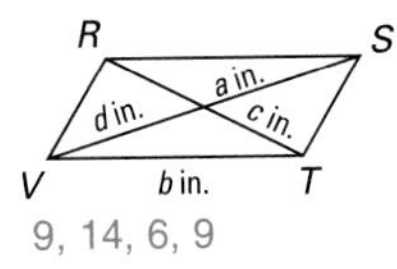

9, 14, 6, 9

6. $\overline{HG} = 5$ yd; $\overline{EG} = 8.6$ yd; $\overline{FG} = 5.4$ yd

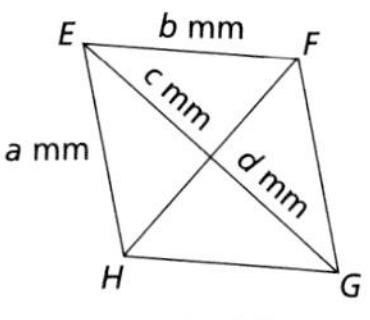

5.4, 5, 4.3, 4.3

Tell whether each statement is *true* or *false*.

7. A rhombus is a parallelogram. true

8. Every parallelogram is a quadrilateral. true

9. A square is a rectangle. true

10. Diagonals of a rectangle bisect each other. true

11. Diagonals of a square are perpendicular. true

12. Opposite sides of a square are parallel. true

Extra Practice 4–9 • Special Quadrilaterals: Trapezoids • pages 188–191

A trapezoid and its median are shown. Find the value of *n*.

1. 33

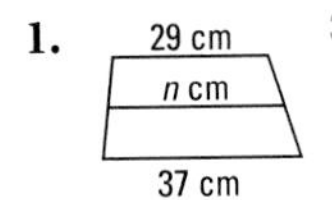

2. 2.6

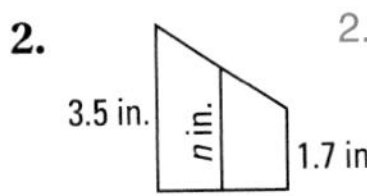

3. 128

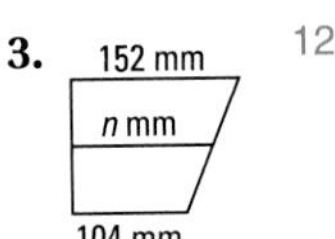

4. 8

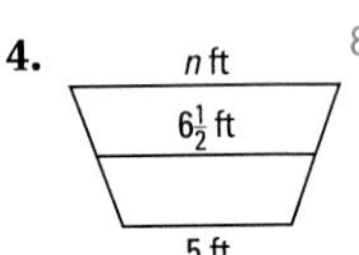

5. 12

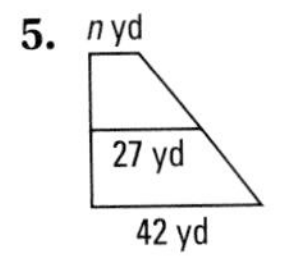

6. 23

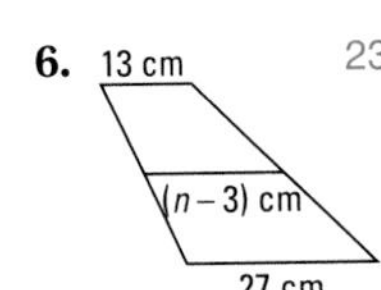

7. 0.6

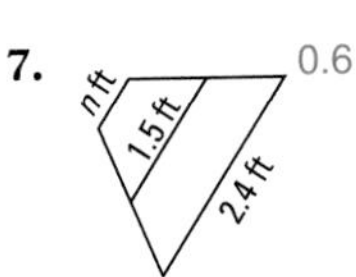

8. $3\frac{2}{3}$

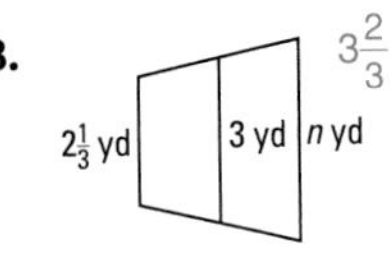

The given figure is a trapezoid. Find all the unknown angle measures.

9.

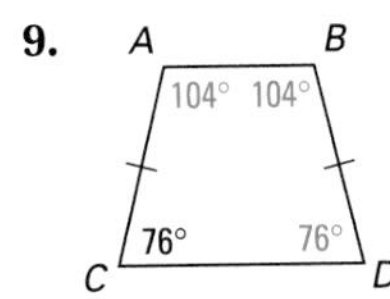

10.

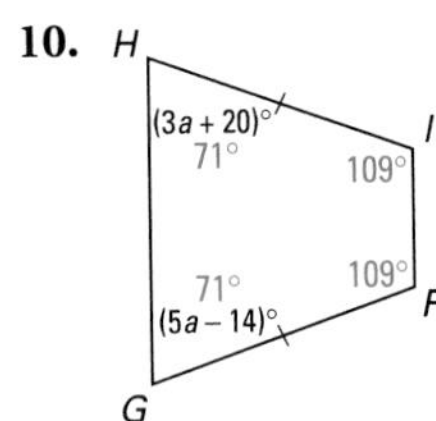

11.

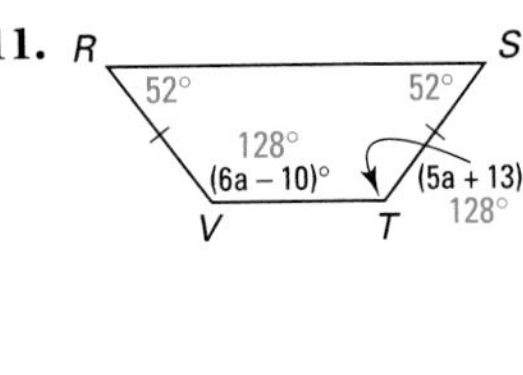

698 Extra Practice

Extra Practice

Chapter 5

Extra Practice 5–1 • Ratios and Units of Measure • pages 202–205

Complete.

1. 12 qt = __?__ c 48
2. 312 in. = __?__ yd __?__ ft 8.2
3. 3 gal = __?__ fl oz 384
4. 1.8 T = __?__ oz 57,600
5. 0.7 cm = __?__ m 0.007
6. 500 mg = __?__ g 0.5
7. 0.003 kg = __?__ g 3
8. 5.9 mL = __?__ L 0.0059
9. 3 gal = __?__ c 48
10. 6.4 L = __?__ mL 6,400
11. 31 ft = __?__ yd $10\frac{1}{3}$
12. 4.37 km = __?__ m 4,370

Name the best customary unit for expressing the measure of each.

13. weight of a computer lb
14. height of a seat in.
15. length of a room ft

Name the best metric unit for expressing the measure of each.

16. capacity of a cooler L
17. mass of a box of cereal g
18. length of a building m

Write each ratio in lowest terms.

19. 27 m:45 m 3:5
20. 60 g to 420 g 1 to 7
21. 30 min/6 h $\frac{1}{12}$

Find each unit rate.

22. 220 mi in 4 h 55 mi per h
23. $16 for 320 prints $0.05/print
24. 15 L in 3 min 5 L per min
25. Which is the better buy, 6 grapefruit for $1.80, or 8 grapefruit for $2.56? 6 for $1.80
26. In 2 h 20 min Suzanne biked 14 mi. What was her biking rate? 6 mi/h

Extra Practice 5–2 • Perimeter, Circumference, and Area • pages 206–209

1. What is the perimeter of a regular hexagon with 6-cm sides? 36 cm
2. What is the circumference of a circle with a radius of 5.4 m? 33.912 m
3. Find the base of a triangle if area = 42 cm^2 and height = 8 cm. 10.5 cm

Find the area of the shaded region of each figure.

4. 46 in.^{2}

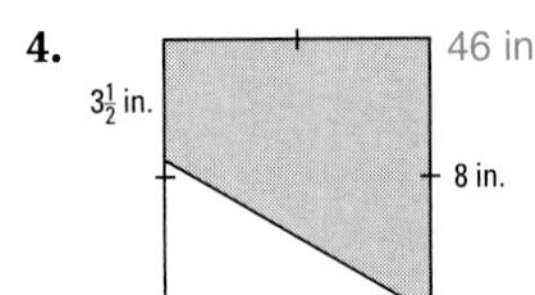

5. 50.24 cm^2

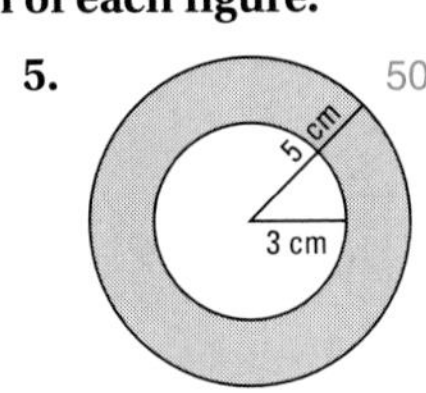

6. 16.5 m^2

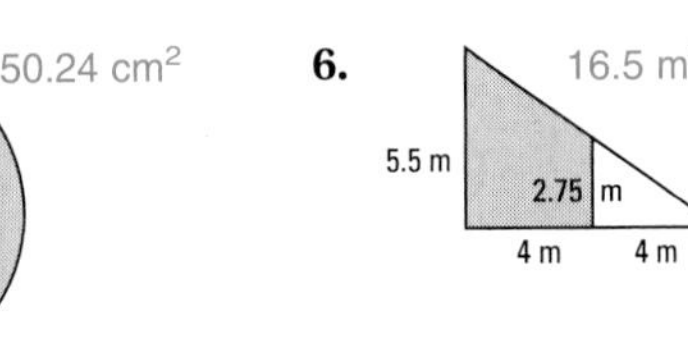

7. If you triple the length of the radius of a circle, how does the area change? It is multiplied by 9.

Extra Practice 699

Extra Practice

Extra Practice 5–3 • Probability and Area • pages 212–215

A standard deck of playing cards has 52 cards. A card is drawn at random from a shuffled deck. Find each probability.

1. $P(\text{king})$ $\frac{1}{13}$ **2.** $P(\text{black card})$ $\frac{1}{2}$ **3.** $P(\text{red face card})$ $\frac{3}{26}$

Find the probability that a point selected at random in each figure is in the shaded region.

4. $\frac{3}{5}$

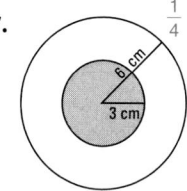

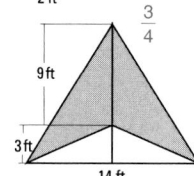

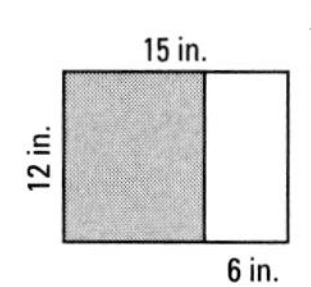

5. $\frac{4}{81}$

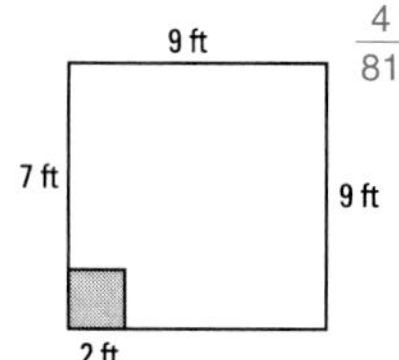

6. $\frac{5}{9}$

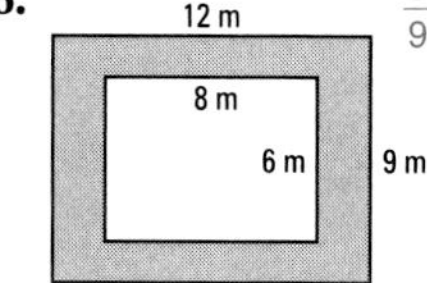

7. $\frac{1}{4}$

6 cm
3 cm

8. $\frac{3}{4}$

9 ft
3 ft
14 ft

9. $\frac{1}{32}$ or 0.03125

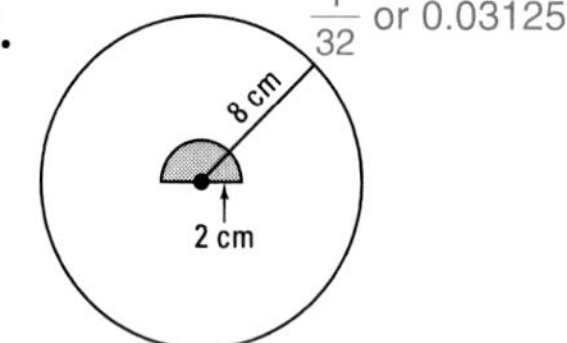

10. Suppose Mrs. O'Malley left her purse within her 1,500 ft^2 apartment. What is the probability it is in the 15-ft by 12-ft kitchen? $\frac{3}{25}$

Extra Practice 5–5 • Three-dimensional Figures and Loci • pages 220–223

Name the polyhedra shown below. Then state the number of faces, vertices, and edges each has.

1.

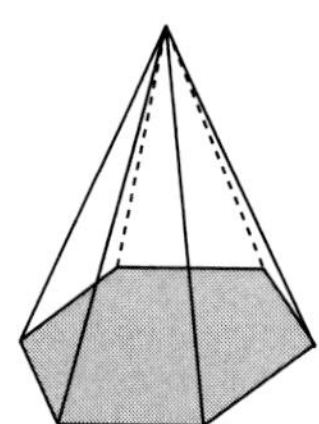

hexagonal pyramid; 7, 7, 12

2.

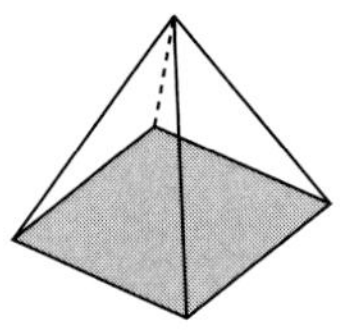

square pyramid; 5, 5, 8

3.

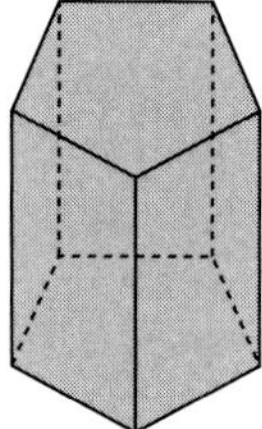

pentagonal prism; 7, 10, 15

Draw the figure. Check students' drawings

4. right rectangular prism **5.** right cylinder **6.** sphere

7. A figure has 5 triangular faces and 1 pentagonal face. What is the figure? pentagonal pyramid

700 Extra Practice

Extra Practice

Extra Practice 5–6 • Surface Area of Three-dimensional Figures • pages 224–227

Find the surface area of each figure. Assume that all pyramids are regular pyramids. Use $\pi = 3.14$. Round answer to the nearest whole number.

1.
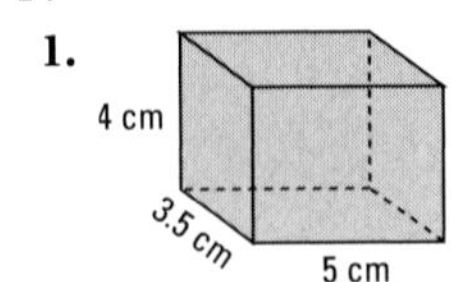

103 cm²

2.
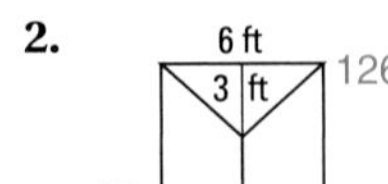

126 ft²

3.
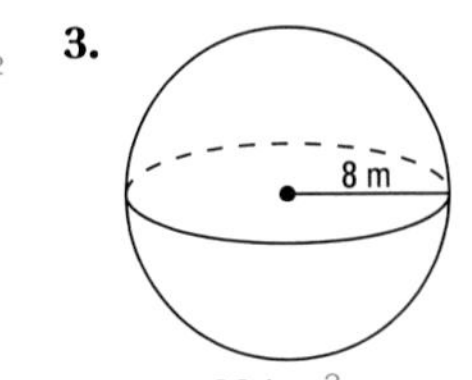

804 m²

4.
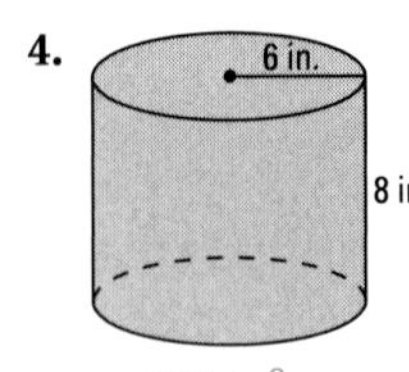

528 in.²

5. What is the surface area of a cone 8 cm across with a slant height of 5.6 cm? about 120 cm²

Extra Practice 5–7 • Volume of Three-dimensional Figures • pages 230–233

Find the volume to the nearest whole number. Use 3.14 for π.

1.
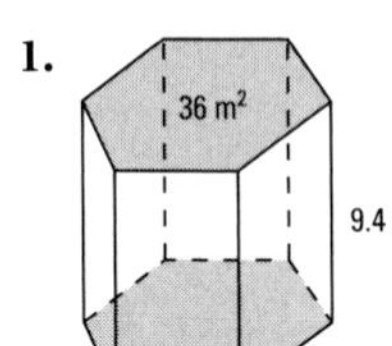

338 m³

2.
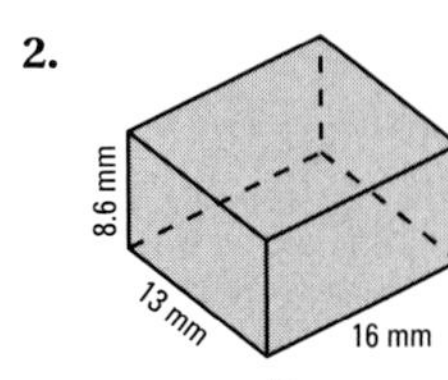

1,789 mm³

3.
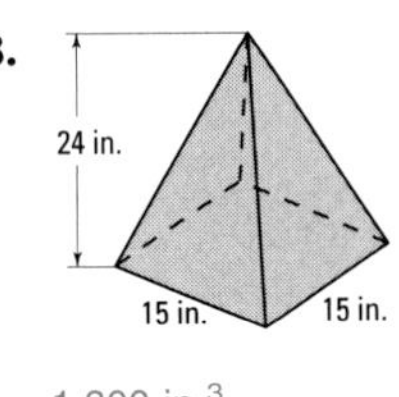

1,800 in.³

4.
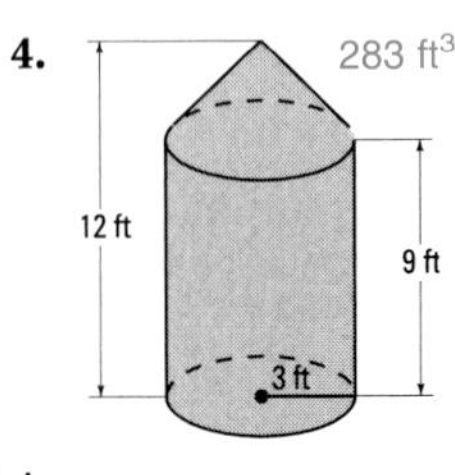

283 ft³

5. How many cubic centimeters of water cana fish tank hold, if the tank is a rectangular prism 60 cm long, 40 cm wide, and 25 cm high? 60,000 cm³

Chapter 6

Extra Practice 6–1 • Slope of a Line and Slope-intercept Form • pages 244–247

Find the slope of the line containing the given points.

1. $C(3, -1)$ and $D(0, -1)$ $m = 0$
2. $M(-2, 4)$ and $N(5, 6)$ $m = \frac{2}{7}$
3. $S(-5, 0)$ and $T(4, -3)$ $m = -\frac{1}{3}$
4. $X(-5, -3)$ and $Z(5, 5)$ $m = \frac{4}{5}$
5. $J(6, 2)$ and $K(0, 18)$ $m = -\frac{8}{3}$
6. $P(-7, -3)$ and $Q(-2, 17)$ $m = 4$
7. $Q(4, -1)$ and $R(-5, -3)$ $m = \frac{2}{9}$
8. $E(-3, -2)$ and $F(-4, 2)$ $m = -4$
9. $J(6, -4)$ and $K(-4, -4)$ $m = 0$

Graph the line that passes through the given point P and has the given slope. 10–12. See additional answers.

10. $P(-1, 4)$, $m = \frac{1}{3}$
11. $P(5, -2)$, $m = -\frac{3}{4}$
12. $P(-2, -3)$, $m = \frac{3}{2}$

Find the slope of the line.

13. $4x - 6y = 12$ $m = \frac{2}{3}$
14. $4x - 5y = -15$ $m = \frac{4}{5}$
15. $8x - y = 2$ $m = 8$
16. $-x - 2y = 8$ $m = -\frac{1}{2}$
17. $5x = -2y + 7$ $m = -\frac{5}{2}$
18. $2x = 20 - 6y$ $m = -\frac{1}{3}$
19. Find the slope of a ramp that rises 8 ft for every 120 ft of horizontal run. $m = \frac{1}{15}$

Additional Answers

Extra Practice 6-1

10.
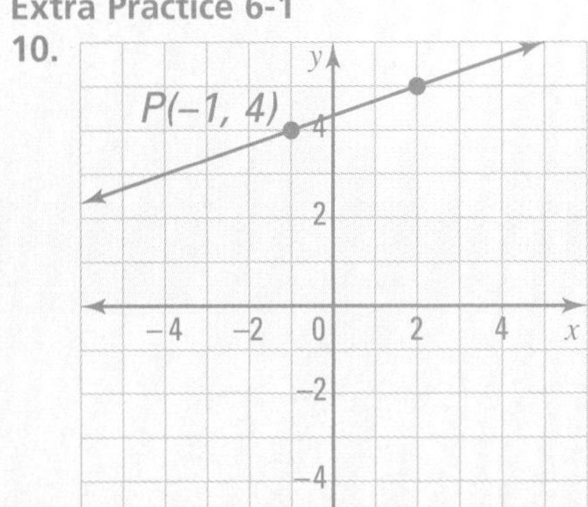

11.
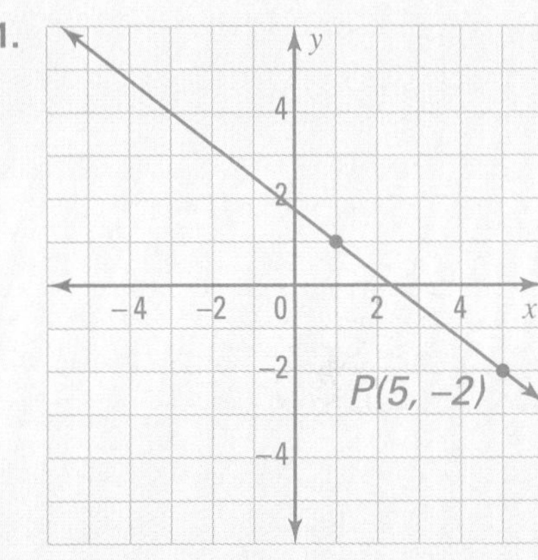

12.
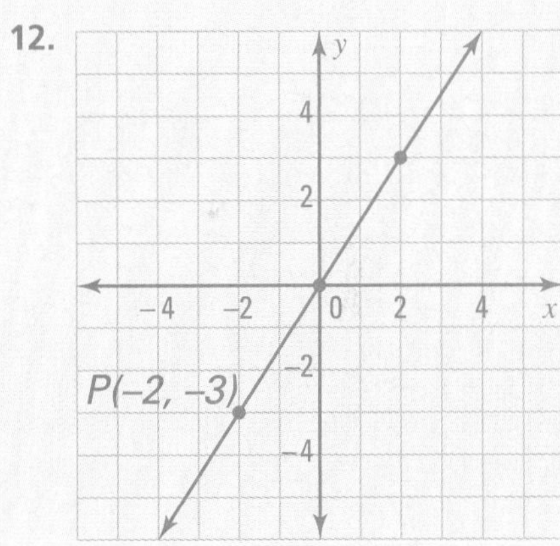

Additional Answers

Extra Practice 6-1 (continued)

32.

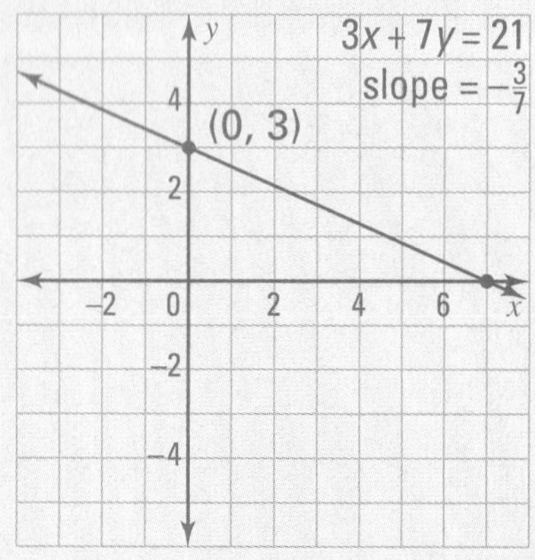

33.

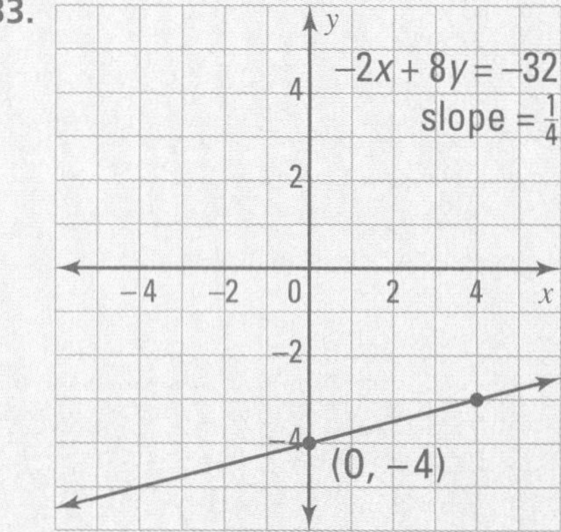

34.

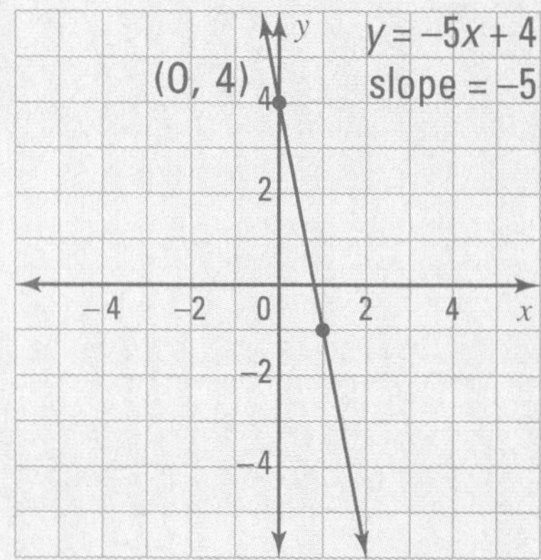

Extra Practice

Find the slope and y-intercept for each line.

20. $y = \frac{2}{3}x - 5$ $m = \frac{2}{3}, b = -5$ **21.** $y = -x + 9$ $m = -1, b = 9$ **22.** $y = 12x$ $m = 12, b = 0$

23. $2x + 8y = 16$ $m = -\frac{1}{4}, b = 2$ **24.** $-5x + 7y = 35$ $m = \frac{5}{7}, b = 5$ **25.** $\frac{1}{2}x + 4y = -24$ $m = -\frac{1}{8}, b = -6$

Write an equation of the line with the given slope and y-intercept.

26. $m = 2, b = 6$ $y = 2x + 6$ **27.** $m = \frac{3}{4}, b = 0$ $y = \frac{3}{4}x$ **28.** $m = -3, b = -9$ $y = -3x - 9$

29. $m = -\frac{1}{8}, b = \frac{1}{2}$ $y = -\frac{1}{8}x + \frac{1}{2}$ **30.** $m = 0, b = -7$ $y = -7$ **31.** $m = -1, b = \frac{5}{9}$ $y = -x + \frac{5}{9}$

Graph each equation. 32-34. See additional answers.

32. $3x + 7y = 21$ **33.** $-2x + 8y = -32$ **34.** $y = -5x + 4$

35. Each week, the Weekly News prints 400 newspapers plus 20% of the total newspapers sold the previous week. The number of papers sold last week was 420. Write an equation to show how many newspapers will be printed this week. Solve. If 450 newspapers are sold this week, how many will be printed next week? $n = 400 + \frac{1}{5}x$, 484, 490

Extra Practice 6–2 • Parallel and Perpendicular Lines • pages 248–251

Find the slope of a line parallel to the given line and a line perpendicular to the given line.

1. The line containing (2, 3) and (4, 9) $m = 3, m = -\frac{1}{3}$
2. The line containing (−1, 7) and (2, −3) $m = -\frac{10}{3}, m = \frac{3}{10}$
3. The line containing (0, 9) and (3, −6) $m = -5, m = \frac{1}{5}$
4. The line containing (−4, −3) and (0, −7) $m = -1, m = 1$
5. The line containing (−3, 0) and (−5, −3) $m = \frac{3}{2}, m = -\frac{2}{3}$
6. The line containing (4, −2) and (7, −6) $m = -\frac{4}{3}, m = \frac{3}{4}$

Determine whether each pair of lines is *parallel, perpendicular,* or *neither.*

7. The line containing points $C(-2, 5)$ and $D(5, 9)$
   The line containing points $E(2, 2)$ and $F(6, -5)$ perpendicular
8. The line containing points $M(-4, -2)$ and $N(3, -8)$
   The line containing points $O(-6, 3)$ and $P(1, -3)$ parallel

9. $7x - y = 4; -14x + 2y = 6$ parallel **10.** $\frac{1}{2}x + 5y = 20; 2x + 10y = 15$ neither

11. $x - y = 3; 3x - 4y = 9$ neither **12.** $4y = -x + 14; 8x - 2y = -10$ perpendicular

13. $4y + 10 = 6x; 3x - 2y = 10$ parallel **14.** $6x - 10y = 20; 10x + 6y = 24$ perpendicular

15. Plot and connect the points $A(4, 5)$, $B(-4, 1)$, $C(2, -3)$, and $D(-1, 8)$. Determine if $ABCD$ is a square. not a square

702 | Extra Practice

Extra Practice 6–3 • Write Equations for Lines • pages 254–257

Write an equation of the line with the given slope and *y*-intercept.

1. $m = 4, b = -1$ $y = 4x - 1$
2. $m = -2, b = 7$ $y = -2x + 7$
3. $m = -\frac{3}{4}, b = 0$ $y = -\frac{3}{4}x$
4. $m = \frac{1}{5}, b = -5$ $y = \frac{1}{5}x - 5$
5. $m = -8, b = \frac{1}{2}$ $y = -8x + \frac{1}{2}$
6. $m = -\frac{2}{3}, b = -\frac{4}{9}$ $y = -\frac{2}{3}x - \frac{4}{9}$
7. $m = \frac{3}{4}, b = -4$ $y = \frac{3}{4}x - 4$
8. $m = \frac{5}{3}, b = \frac{3}{4}$ $y = \frac{5}{3}x + \frac{3}{4}$
9. $m = -5, b = 0$ $y = -5x$

Write an equation of the line with the given information.

10. $m = 3, A(3, -7)$ $y = 3x - 16$
11. $m = -1, B(-5, 2)$ $y = -x - 3$
12. $m = -\frac{2}{3}, C(3, 3)$ $y = -\frac{2}{3}x + 5$
13. $m = \frac{1}{2}, D(-4, -6)$ $y = \frac{1}{2}x - 4$
14. $m = 5, E(8, -2)$ $y = 5x - 42$
15. $m = -6, F(-3, -9)$ $y = -6x - 27$

16. 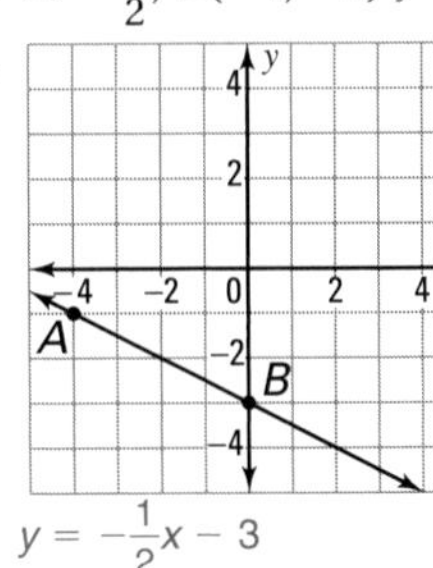
$y = -\frac{1}{2}x - 3$

17. 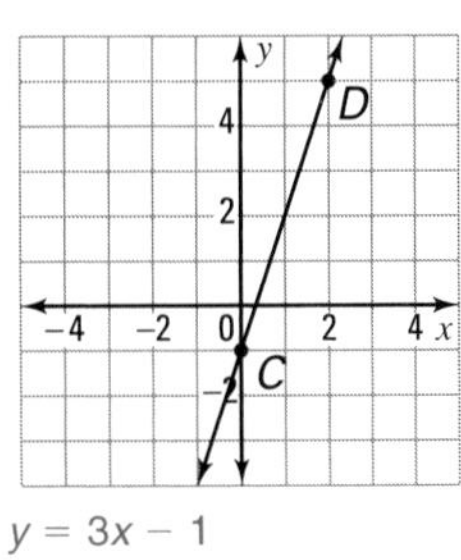
$y = 3x - 1$

18. 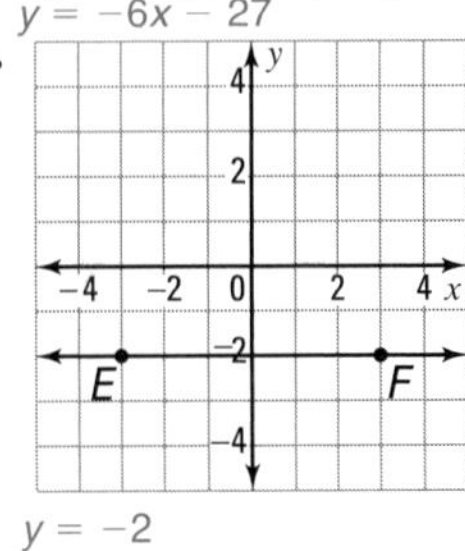
$y = -2$

Extra Practice 6–4 • Systems of Equations • pages 258–261

Determine the solution of each system of equations.

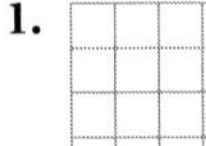

1. 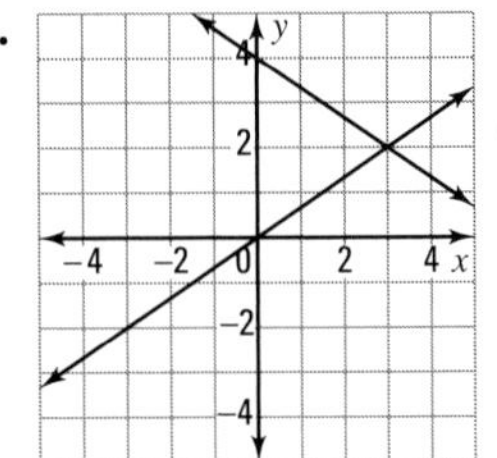
(3, 2)

2. 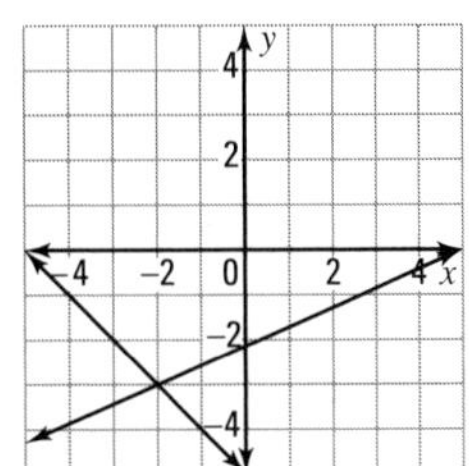
(−2, −3)

3. 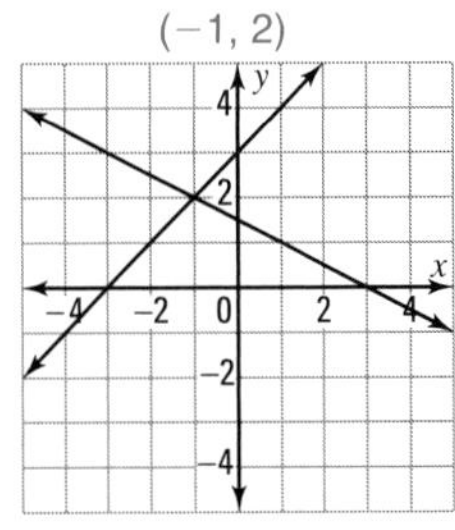
(−1, 2)

Solve each system of equations by graphing. Check students' graphs. See sample answer in additional answers.

4. $\begin{cases} y = 3x - 1 \\ y = -x + 3 \end{cases}$ (1, 2)
5. $\begin{cases} \frac{1}{2}y = 2x + 2 \\ y = 3x + 2 \end{cases}$ (−2, −4)
6. $\begin{cases} x - y = -3 \\ 2x - y = -4 \end{cases}$ (−1, 2)
7. $\begin{cases} 2x + y = 3 \\ 2x - 2y = -6 \end{cases}$ (0, 3)
8. $\begin{cases} x + y = 3 \\ 3x - y = 5 \end{cases}$ (2, 1)
9. $\begin{cases} x = 3y + 6 \\ y = -2x + 5 \end{cases}$ (3, −1)

Additional Answers

Extra Practice 6-4

4. 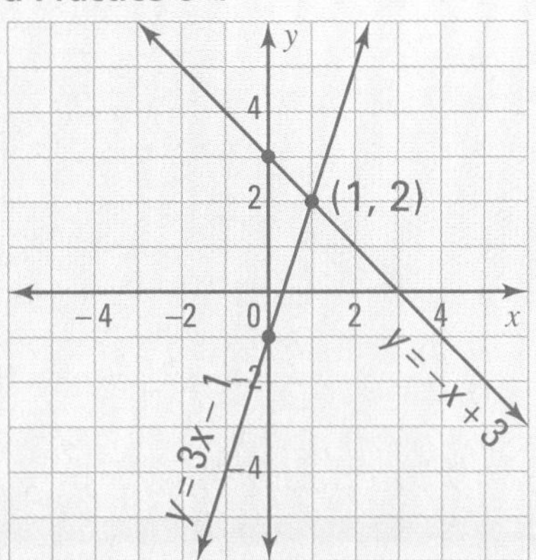

Additional Answers

Extra Practice 6-8

4. $y \le \frac{1}{3}x - 2$
   $y \le -\frac{2}{3}x + 3$

5. $y \ge 2x + 3$
   $y \ge -\frac{1}{2}x - 2$

6. $y < x + 4$

7.

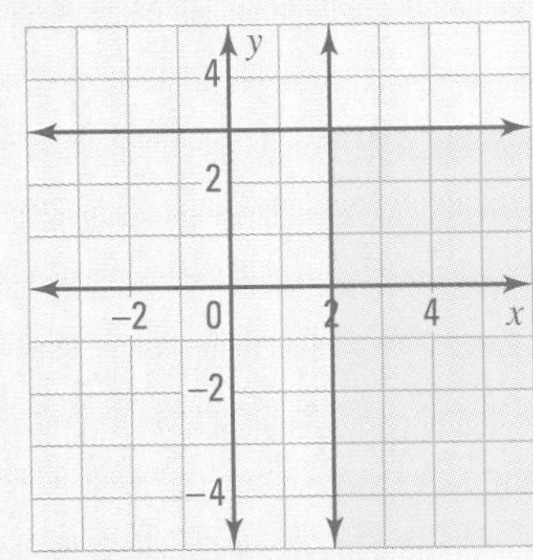

9.

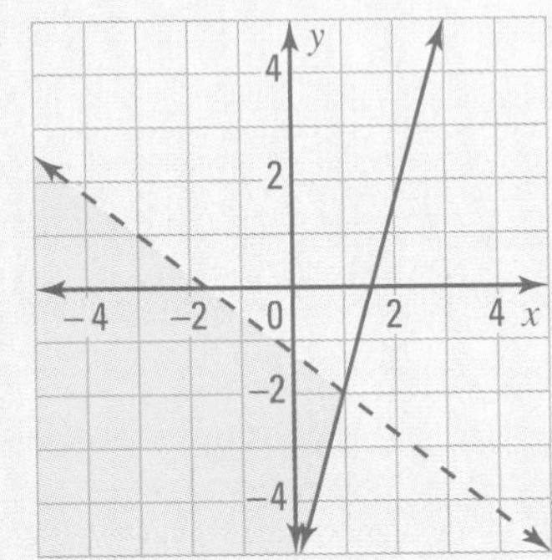

Extra Practice

Extra Practice 6–5 • Solve Systems by Substitution • pages 264–267

Solve and check each system of equations by the substitution method.

1. $\begin{cases} x + 2y = 5 \\ 4x - 4y = 8 \end{cases}$ (3, 1)
2. $\begin{cases} 6x + y = 4 \\ -x + 4y = 16 \end{cases}$ (0, 4)
3. $\begin{cases} 2x + 3y = 4 \\ 5x - 2y = -9 \end{cases}$ (−1, 2)
4. $\begin{cases} x + 3y = 5 \\ 4x + 8y = 16 \end{cases}$ (2, 1)
5. $\begin{cases} -3x + 8y = -18 \\ \frac{1}{2}x - y = 3 \end{cases}$ (6, 0)
6. $\begin{cases} 6x + 3y = 3 \\ -x + 2y = -13 \end{cases}$ (3, −5)
7. $\begin{cases} y = 3x + 8 \\ x - 3y = 8 \end{cases}$ (−4, −4)
8. $\begin{cases} \frac{1}{2}x + y = 8 \\ -3x + 6y = 0 \end{cases}$ (8, 4)
9. $\begin{cases} y = 4x + 2 \\ x + y = -3 \end{cases}$ (−1, −2)
10. $\begin{cases} y = 3x + 9 \\ x = 8y - 3 \end{cases}$ (−3, 0)
11. $\begin{cases} 4x + 3y = 3 \\ -6y = 3 - 10x \end{cases}$ $\left(\frac{1}{2}, \frac{1}{3}\right)$
12. $\begin{cases} y = -6x \\ 3x + y = 2 \end{cases}$ $\left(-\frac{2}{3}, 4\right)$
13. The perimeter of a rectangle is 96 in. If the length is three times the width, find the dimensions of the rectangle. 12 in. by 36 in.

Extra Practice 6–6 • Solve Systems by Adding and Multiplying • pages 268–271

Solve each system of equations. Check the solutions.

1. $\begin{cases} x + y = 5 \\ -x + y = 1 \end{cases}$ (2, 3)
2. $\begin{cases} 6x + y = 13 \\ 4x - y = -3 \end{cases}$ (1, 7)
3. $\begin{cases} x + 5y = 2 \\ y = x + 4 \end{cases}$ (−3, 1)
4. $\begin{cases} 6x + 3y = -15 \\ -2x + 4y = 0 \end{cases}$ (−2, −1)
5. $\begin{cases} 2x = 3y - 8 \\ y = 6x \end{cases}$ $\left(\frac{1}{2}, 3\right)$
6. $\begin{cases} y = 3x - 14 \\ 2x + 3y = 2 \end{cases}$ (4, −2)
7. $\begin{cases} y = -x + 1 \\ x = y + 15 \end{cases}$ (8, −7)
8. $\begin{cases} \frac{1}{3}x + y = 7 \\ x - 2y = 1 \end{cases}$ (9, 4)
9. $\begin{cases} -3x + 2y = -5 \\ 4y = -8 + 4x \end{cases}$ (1, −1)
10. $\begin{cases} 4y = 3x + 3 \\ \frac{1}{3}x = y - 2 \end{cases}$ (3, 3)
11. $\begin{cases} 5y = 10x - 5 \\ 3x + 2y = -9 \end{cases}$ (−1, −3)
12. $\begin{cases} x = 6y - 16 \\ 3y = x + 10 \end{cases}$ (−4, 2)
13. Andrew has 25 coins with a total value of \$3.05. The coins are all nickels and quarters. How many nickels and how many quarters does he have? 16 nickels, 9 quarters

Extra Practice 6–8 • Systems of Inequalities • pages 276–279

Determine whether the given ordered pair is a solution to the given system of inequalities.

1. (2, 1); $\begin{cases} 2x + 5y \le 4 \\ -x + 8y \le 4 \end{cases}$ no
2. (−3, 4); $\begin{cases} 3x + 3y > 2 \\ 2x - 6y \le 5 \end{cases}$ yes
3. (1, 5); $\begin{cases} 4x - y < 5 \\ -x + 3y > 0 \end{cases}$ yes

Write a system of linear inequalities for the given graph. See additional answers.

4.

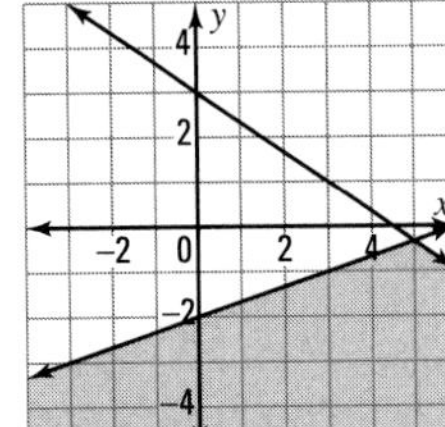

5.

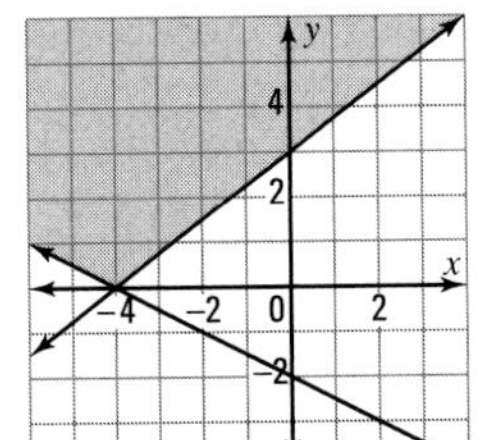

6.

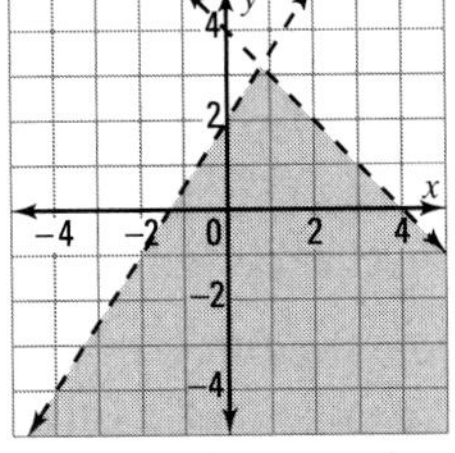

Graph the solution set of the system of linear inequalities. Check students' graphs. See sample graphs in additional answers.

7. $\begin{cases} x \ge 2 \\ y \le 3 \end{cases}$
8. $\begin{cases} y < 2x + 1 \\ y > -x + 6 \end{cases}$
9. $\begin{cases} y - 3x \ge -5 \\ y + \frac{2}{3}x < -1 \end{cases}$

Extra Practice

Chapter 7

Extra Practice 7–1 • Ratios and Proportions • pages 296–299

Is each pair of ratios equivalent? Write *yes* or *no*.

1. 4:8, 12:24 yes **2.** 14:18, 9:7 no **3.** $\frac{2.7}{3.6}, \frac{6}{8}$ yes **4.** $\frac{1.5}{2.4}, \frac{10}{25}$ no

5. 10 to 7, 30 to 14 no **6.** 12 to 8, 9 to 6 yes **7.** $\frac{18}{10}, \frac{54}{30}$ yes **8.** $\frac{4}{5}, \frac{2}{10}$ no

Solve each proportion.

9. $\frac{2}{9} = \frac{16}{x}$ 72 **10.** $\frac{2}{6} = \frac{11}{x}$ 33 **11.** $\frac{a}{16} = \frac{9}{1.8}$ 80 **12.** $\frac{7.2}{6} = \frac{n}{5}$ 6

13. $8.4{:}12 = 2.1{:}x$ 3 **14.** $6{:}1.9 = n{:}7.6$ 24 **15.** $\frac{15}{y} = \frac{4.8}{6.4}$ 20 **16.** $\frac{7}{8} = \frac{k}{12}$ $10\frac{1}{2}$

Use a calculator to solve these proportions.

17. $\frac{126}{21} = \frac{120}{x}$ 20 **18.** $\frac{154}{231} = \frac{x}{99}$ 66 **19.** $\frac{x}{325} = \frac{429}{165}$ 845 **20.** $\frac{137}{x} = \frac{118}{354}$ 411

21. A recipe for a sport drink calls for 3 parts cranberry juice to 8 parts lime juice. How much cranberry juice should be added to 20 pt of lime juice? $7\frac{1}{2}$ pt

22. Cashew nuts cost \$3 for 0.25 lb. How much will $1\frac{1}{2}$ lb cost? \$18

23. Two college roommates share the cost of an apartment in a ratio of 5:6. The total monthly rent is \$825. What is each person's share? \$450; \$375

24. Julie mixes vanilla coffee beans and hazelnut coffee beans ina 3:5 ratio. She wants to make 2 lb of the mixture. How many ounces of vanilla beans does she need? 12

Extra Practice 7–2 • Similar Polygons • pages 300–303

Determine if the polygons are similar. Write *yes* or *no*.

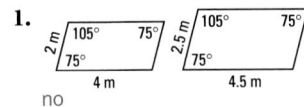

1. 105° 75° 75° 2 m 4 m; 105° 75° 75° 2.5 m 4.5 m
no

2.
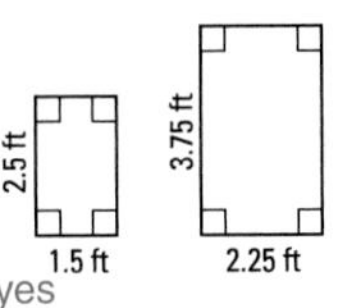

yes

3.
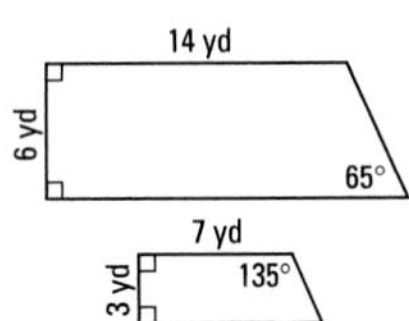

no

Find the measure of *x* in each pair of similar figures.

4.
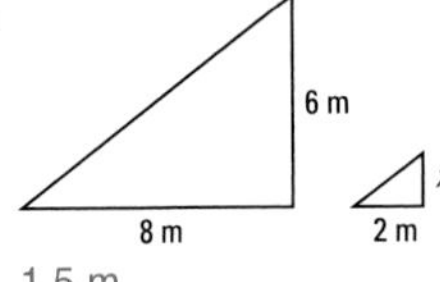

1.5 m

5.
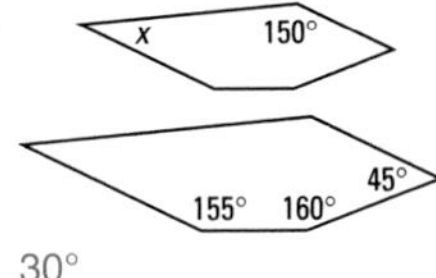

30°

6.
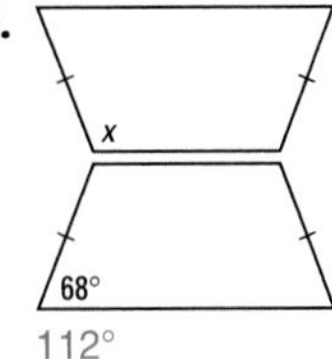

112°

7. Draw any acute angle. Copy the angle using a straightedge and compass. Check students' work.

8. A photograph that measures 5 in. by 8 in. is enlarged so that the 8 in. side measures 10 in. How long is the 5-in. side in the enlargement? $6\frac{1}{4}$ in.

Extra Practice 705

Additional Answers

Extra Practice 7-4

7. The rays of the sun form the same angle with the ground. Therefore, $\angle O \cong \angle R$. Because $\angle N$ and $\angle Q$ are both right angles, $\angle N \cong \angle Q$. Therefore, the triangles are similar by AA Similarity.
8. The line of sight and the ground form $\angle A$, and $\angle A$ is congruent to itself. Because $\angle AEB$ and $\angle ADC$ are both right angles, $\angle AEB \cong \angle ADC$. Therefore, the triangles are similar by AA Similarity.

Extra Practice

Extra Practice 7–3 • Scale Drawings • pages 306–309

Find the actual length of each of the following.

1. scale length is 5 cm
   scale is 2 cm:10 m 25 m
2. scale distance is 6.25 cm
   scale is 2.5 cm:10 m 25 m
3. scale length is $10\frac{1}{2}$ in.
   scale is $\frac{1}{4}$ in.:1 ft 42 ft

Find the scale length for each of the following.

4. actual length is 15 ft
   scale is $\frac{1}{4}$ in.:1 ft $3\frac{3}{4}$ in.
5. actual distance is 300 mi
   scale is 2 cm:50 mi 12 cm
6. actual distance is 1.5 mi
   scale is 1 in.:0.6 mi 2.5 in.

Find the actual distance using the map.

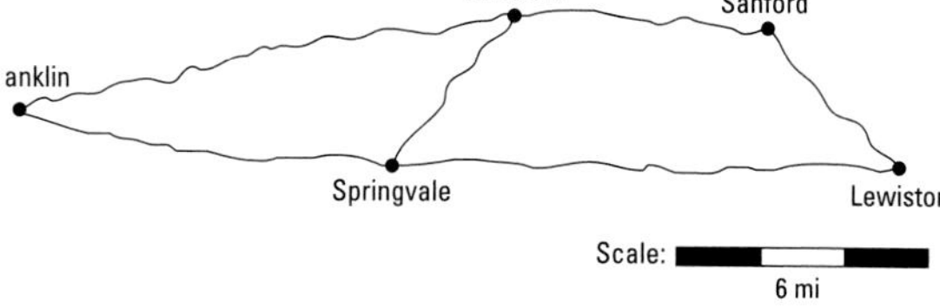

7. Franklin to Springvale 9 mi
8. Sanford to Lewiston 4 mi
9. Hillsboro to Franklin 12 mi
10. Hillsboro to Lewiston 10 mi
11. Springvale to Lewiston 12 mi
12. Sanford to Franklin 18 mi

Extra Practice 7–4 • Postulates for Similar Triangles • pages 310–313

Determine whether each pair of triangles is similar. If the triangles are similar, give a reason: write AA, SSS, or SAS.

1. 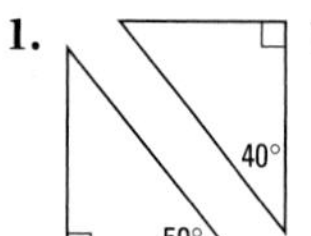

yes; AA

2. 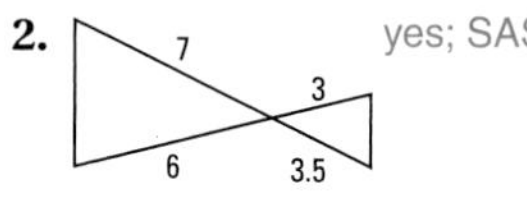

yes; SAS

3. 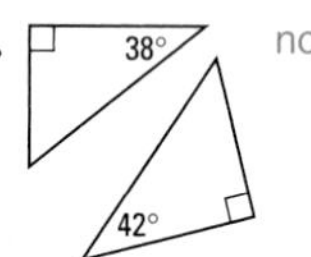

no

4.
yes; AA

5. 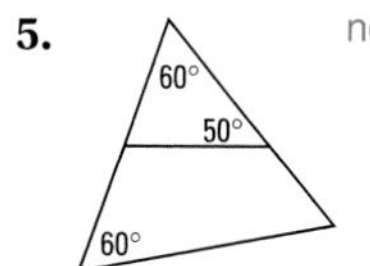

no

6. 
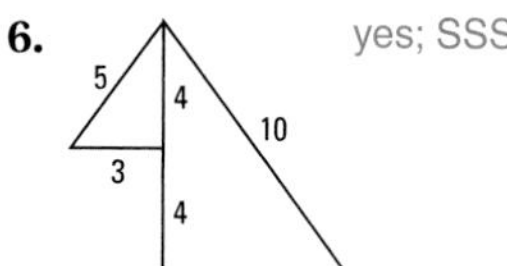

yes; SSS

7. The drawing at the right shows a smokestack and its shadow and a flagpole and its shadow. Explain why $\triangle PQR \sim \triangle MNO$. See additional answers.

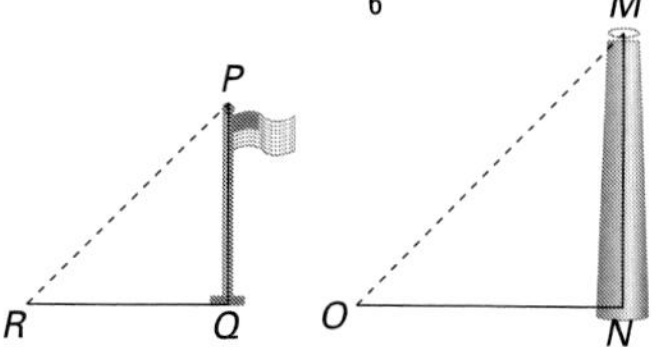

8. The drawing shows the line of sight of a person looking from the top of a building past the top of a tree down to an object (A) on the ground. Explain why $\triangle AEB \sim \triangle ADC$. See additional answers.

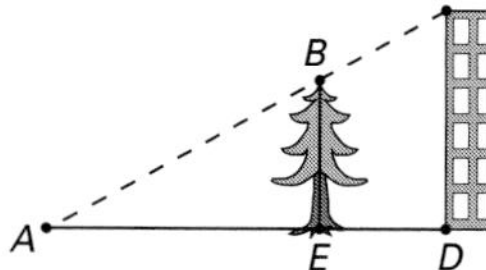

706 Extra Practice

Extra Practice

Extra Practice 7–5 • Triangles and Proportional Segments • pages 316–319

1. Copy and complete this proof.
   Given: $\triangle PQR \sim \triangle STV$; $\overline{PV} \cong \overline{VR}$; $\overline{SW} \cong \overline{WV}$
   Prove: $QV/TW = PR/SV$

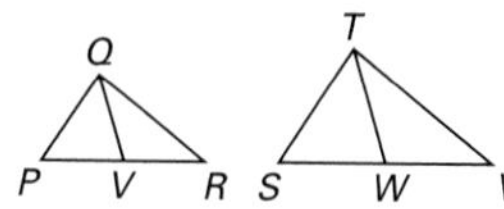

Statements	Reasons
1. $\triangle PQR \sim \triangle STV$; $\overline{PV} \cong \overline{VR}$; $\overline{SW} \cong \overline{WV}$	1. ? given
2. V is the midpoint of $\overline{PR}$ W is the midpoint of $\overline{SV}$	2. ? definition of midpoint
3. $\overline{QV}$ is a median of $\triangle PQR$ $\overline{TW}$ is a median of $\triangle STV$	3. ? definition of median
4. $QV/TW = PR/SV$	4. ? medians of $\sim$ triangles are in the same proportion as corresponding sides

Find x in each pair of similar triangles to the nearest tenth.

2. 46.7

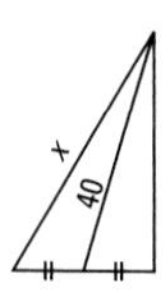

3. 4.3

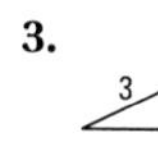

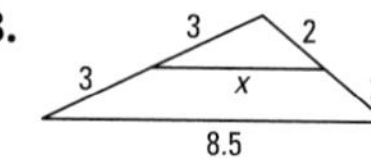

4. 8.5

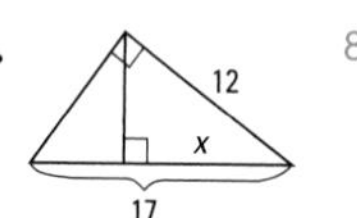

5. These cross sections of tents are similar triangles. If the support pole of the smaller tent is 4 ft, how tall is the support pole for the larger tent? 4.8 ft

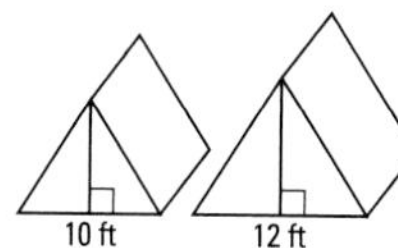

Extra Practice 7–6 • Parallel Lines and Proportional Segments • pages 320–323

In each figure, $\overline{AB} \parallel \overline{CD}$. Find the value of x to the nearest tenth.

1. 2.3

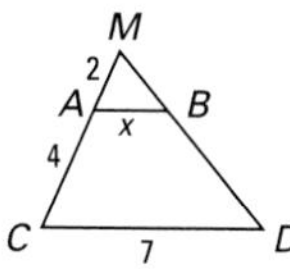

2. 7.3

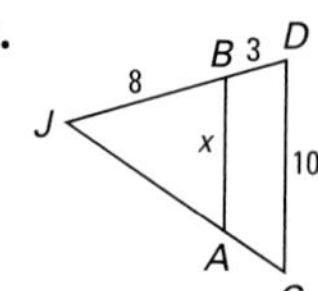

3. 6.7

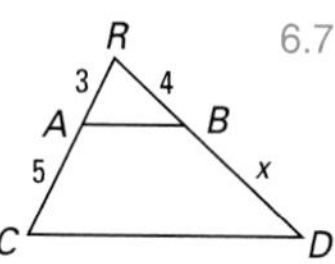

4. 2.4

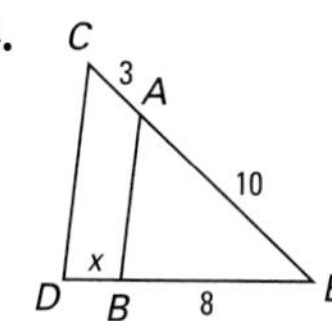

5. 14.5

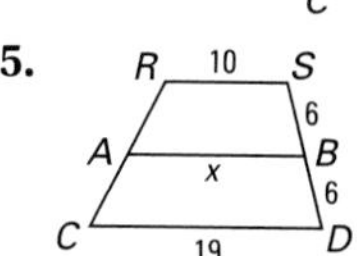

6. 5.5

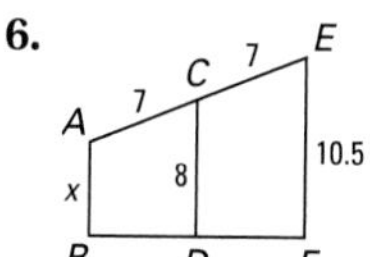

7. This map shows a vacant plot of land that is to be developed by creating four new equally-spaced north-south streets between Elm and Birch Streets. Copy the map and construct the points where the new streets would intersect Spruce Street.
See additional answers.

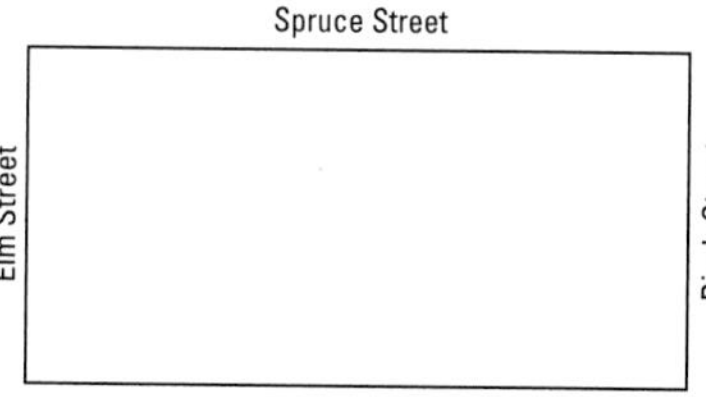

Additional Answers

Extra Practice 7-6

7.

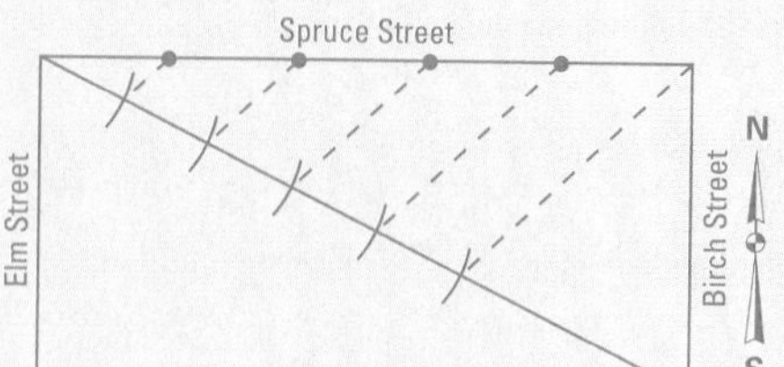

Additional Answers

Extra Practice 8-1

1. 2.

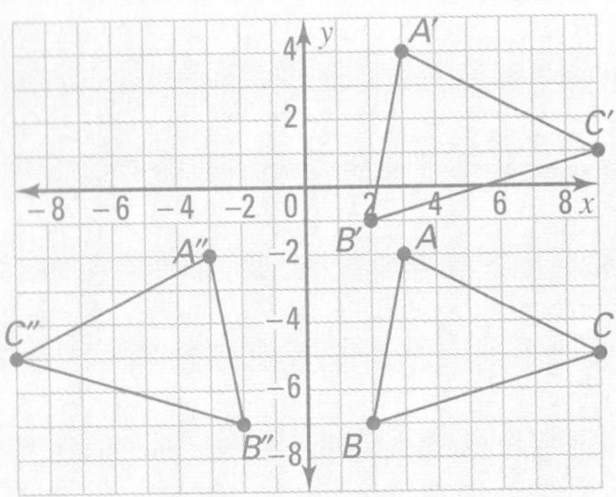

4.
5.

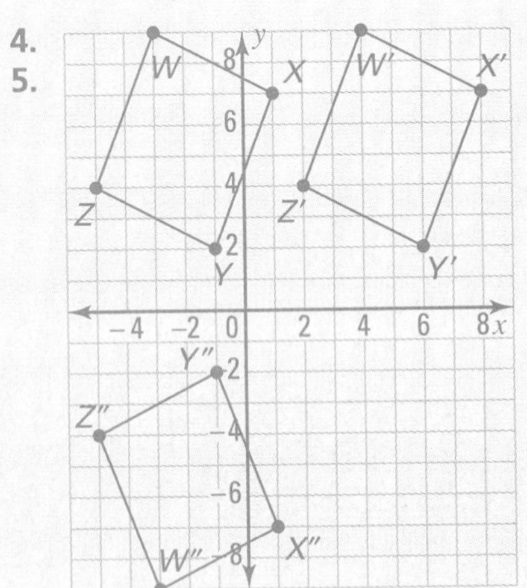

7.

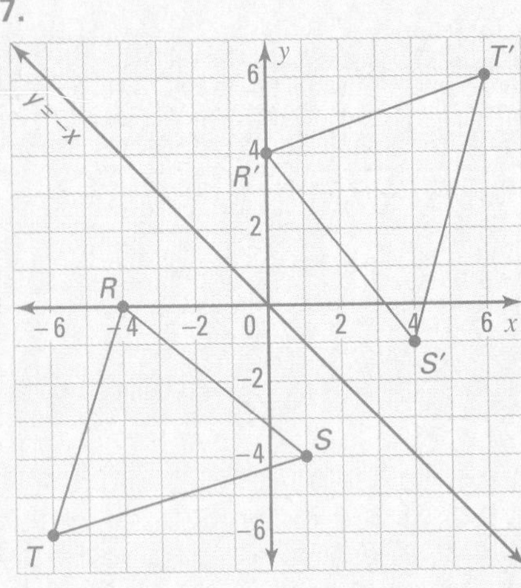

8.

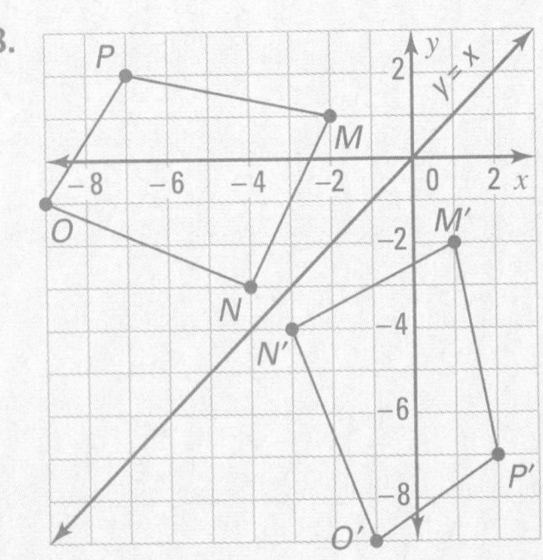

Extra Practice

Chapter 8

Extra Practice 8–1 • Translations and Reflections • pages 338–341

On a coordinate plane, graph triangle ABC with vertices $A(3, -2)$, $B(2, -7)$, and $C(9, -5)$. Then graph its image under each transformation from the original position. 1–8. See additional answers for graphs.

1. 6 units up ($A'B'C'$)
2. reflected across the y-axis ($A''B''C'$)
3. Compare the slopes of all the sides of $\triangle ABC$ in both positions above.
$\overline{AB} = 5$, $\overline{A'B'} = 5$, $\overline{A''B''} = -5$, $\overline{BC} = \frac{2}{7}$, $\overline{B'C'} = \frac{2}{7}$, $\overline{B''C''} = -\frac{2}{7}$, $\overline{AC} = -\frac{1}{2}$, $\overline{A'C'} = -\frac{1}{2}$, $\overline{A''C''} = \frac{1}{2}$

On a coordinate plane, graph figure $WXYZ$ with vertices $W(-3, 9)$, $X(1, 7)$, $Y(-1, 2)$, and $Z(-5, 4)$. Then graph its image under each transformation from the original position.

4. 7 units right ($W'X'Y'Z'$)
5. reflected across the x-axis ($W''X''Y''Z''$)
6. Compare the slopes of $\overline{WX}$, $\overline{W'X'}$, and $\overline{W''X''}$. $-\frac{1}{2}, -\frac{1}{2}, \frac{1}{2}$
7. On a coordinate plane, graph $\triangle RST$ with vertices $R(-4, 0)$, $S(1, -4)$ and $T(-6, -6)$. Graph its image under a reflection across the line with equation $y = -x$.
8. On a coordinate plane, graph figure $MNOP$ with vertices $M(-2, 1)$, $N(-4, -3)$, $O(-9, -1)$, and $P(-7, 2)$. Graph its image under a reflection across the line with equation $y = x$.

Extra Practice 8–2 • Rotations in the Coordinate Plane • pages 342–345

For each figure, draw the image after the given rotation about the origin. Then calculate the slope of each side before and after the rotation. See additional answers.

1. Use the rule $(-x, -y)$ for a 180° clockwise rotation.
2. Use the rule $(-y, x)$ for a 90° counterclockwise rotation.
3. Use the rule $(y, -x)$ for a 90° clockwise rotation.

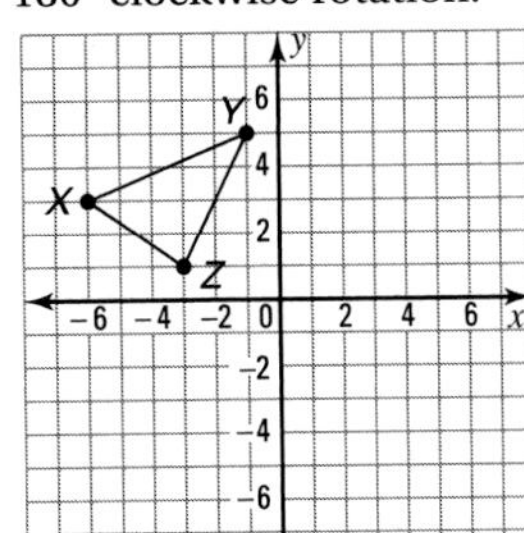

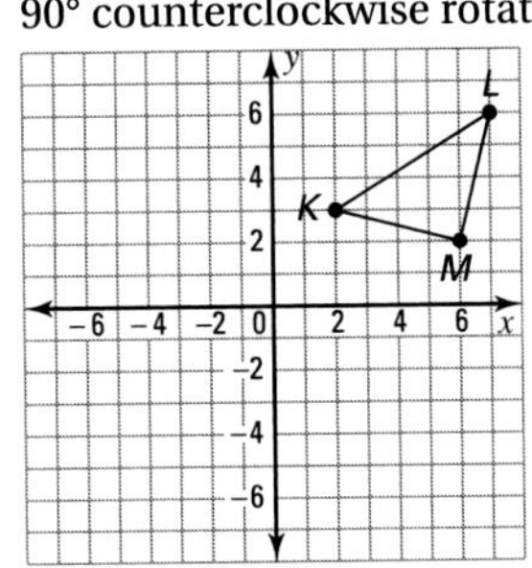

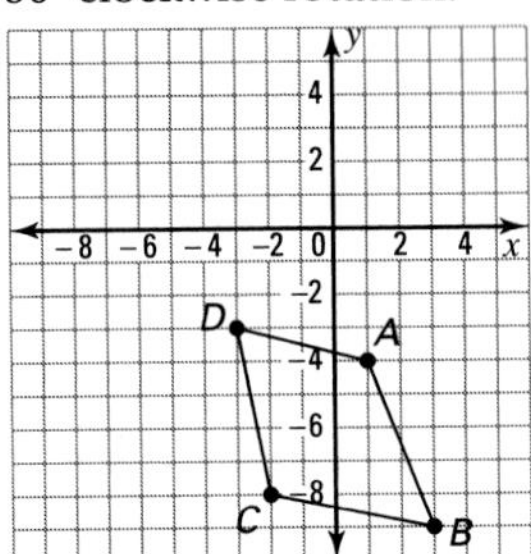

4. Triangle XYZ is rotated twice about the origin, as shown in the table below. Compare the slopes to determine how much of a rotation was completed each time. first 90°; second, 180°

Original Position		After Rotation 1		After Rotation 2	
side	slope	side	slope	side	slope
YZ	−2	$Y'Z'$	$\frac{1}{2}$	$Y''Z''$	−2
XY	$\frac{1}{3}$	$X'Y'$	−3	$X''Y''$	$\frac{1}{3}$
XZ	5	$X'Z'$	$-\frac{1}{5}$	$X''Z''$	5

Extra Practice 8–2

1.

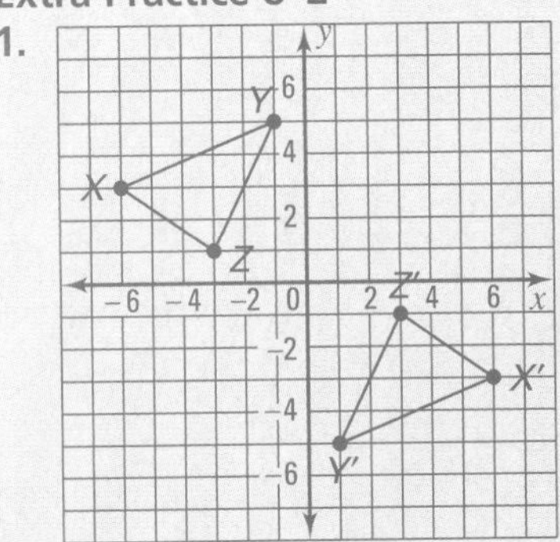

2.

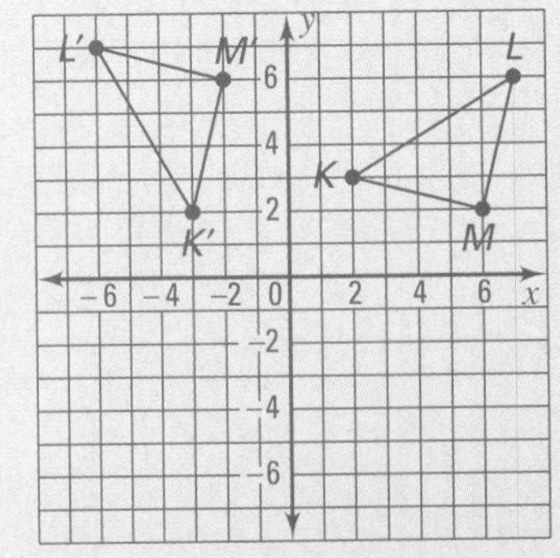

3.

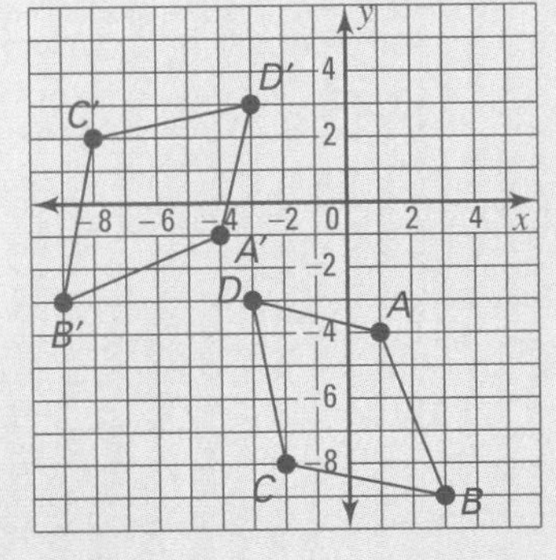

Extra Practice

Extra Practice 8-3 • Dilations in the Coordinate Plane • pages 348–351

Copy each graph on graph paper. Then draw each dilation image. See additional answers.

1. Center of dilation: origin
   Scale factor: 2

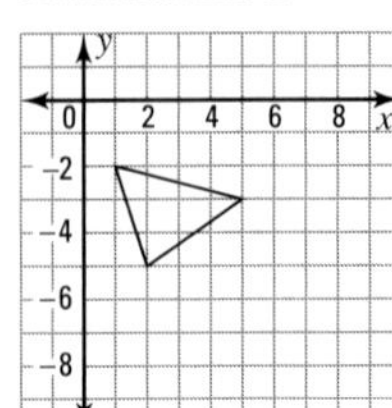

2. Center of dilation: point A
   Scale factor: 3

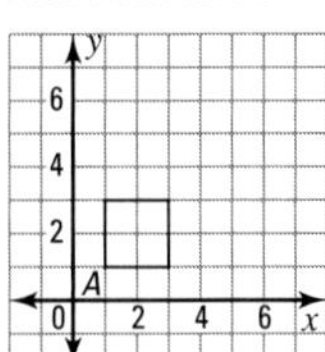

3. Center of dilation: origin
   Scale factor: $\frac{1}{2}$

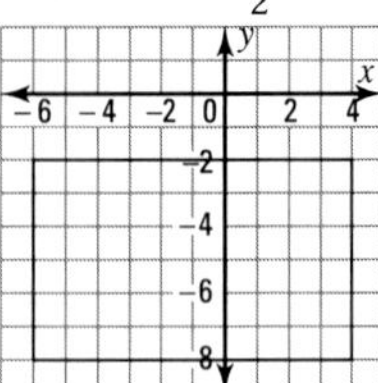

The following sets of points are the vertices of figures and their dilation images. For each two sets of points, give the scale factor.

4. $A(2, 0)$, $B(6, 0)$, $C(4, 4)$
   $A'(4, 0)$, $B'(12, 0)$, $C'(8, 8)$ scale factor = 2
5. $R(-2, 1)$, $S(-2, -7)$, $T(-10, 1)$
   $R'(-2, 1)$, $S'(-2, -2)$, $T'(-6, 1)$ scale factor = $\frac{1}{2}$
6. $J(-8, -3)$, $K(-5, -3)$, $L(-5, -7)$, $M(-8, -7)$
   $J'(-8, -3)$, $K'(1, -3)$, $L'(1, -15)$, $M'(-8, -15)$
   scale factor = 3
7. $D(2, -4)$, $E(8, -4)$, $F(8, -7)$, $G(2, -7)$
   $D'(6, -6)$, $E'(8, -6)$, $F'(8, -7)$, $G'(6, -7)$
   scale factor = $\frac{1}{3}$

Extra Practice 8-4 • Multiple Transformations • pages 352–355

For each exercise, draw the result of the first transformation as a dashed figure and the result of the second transformation in red. 1–3. See additional answers.

1. a reflection over the x-axis followed by a translation 6 units to the left.

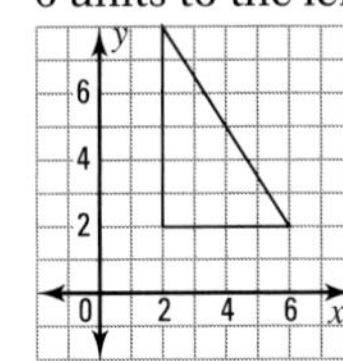

2. a clockwise rotation of 90° about the origin, followed by a reflection over the y-axis.

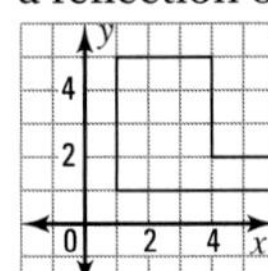

3. a counterclockwise rotation of 180° about the origin, followed by a dilation with center at the origin and a scale factor of 2.

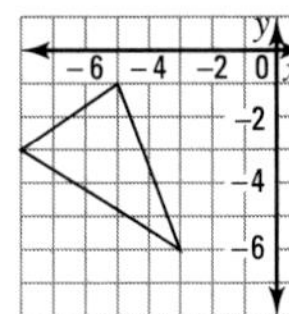

Determine the transformations necessary to create figure 2 from figure 1. There may be more than one possible answer. 4–6. See additional answers.

4.

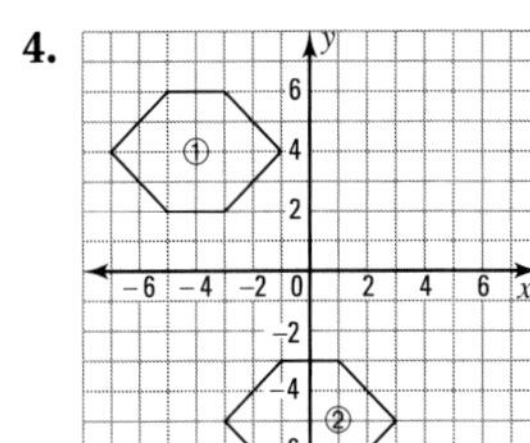

5.

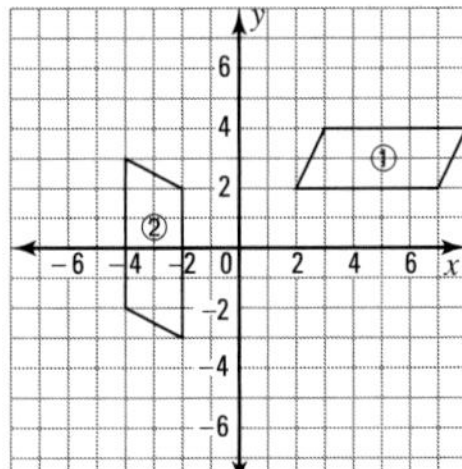

6.

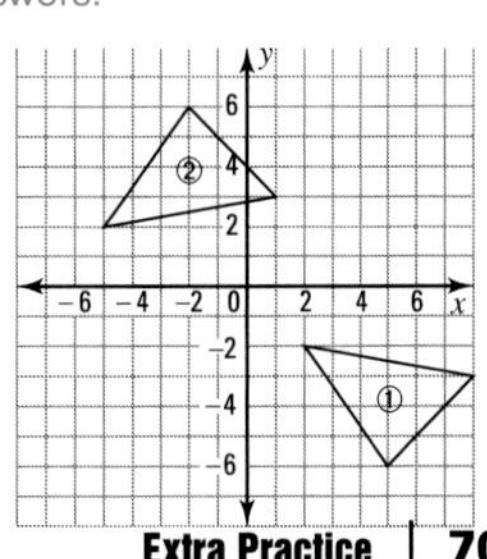

Additional Answers

Extra Practice 8-3

1.

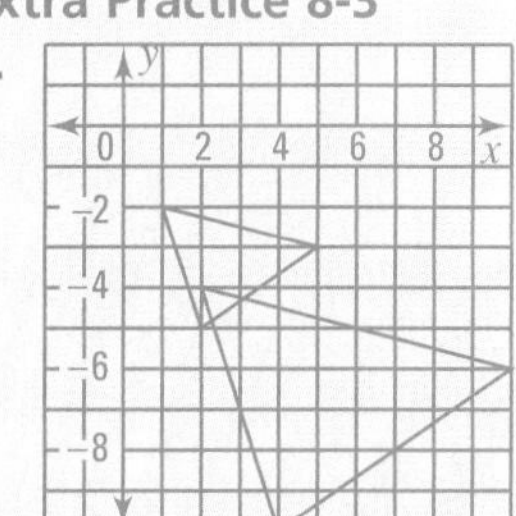

2.

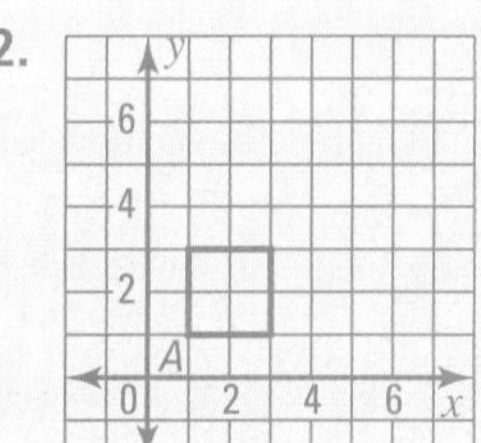

3.

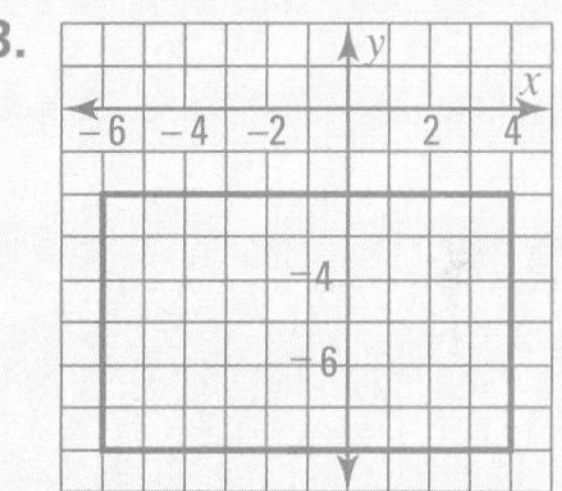

Extra Practice 8-4

1.

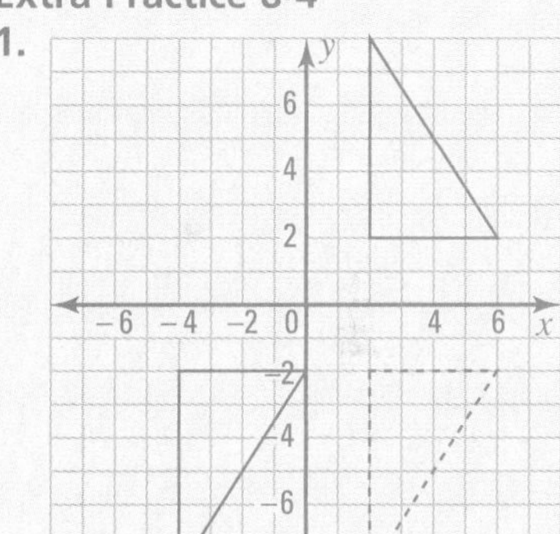

2.

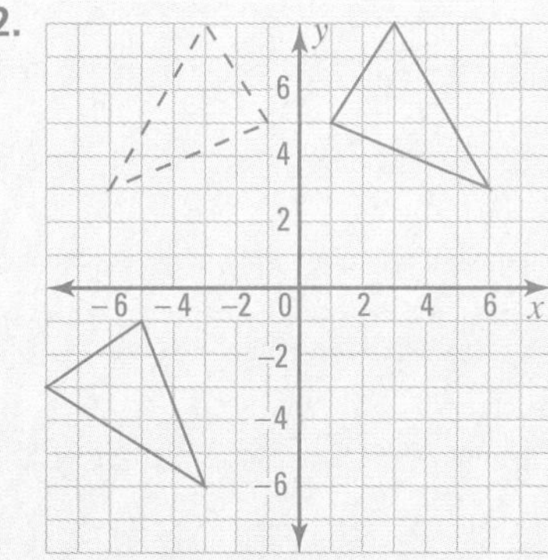

3.

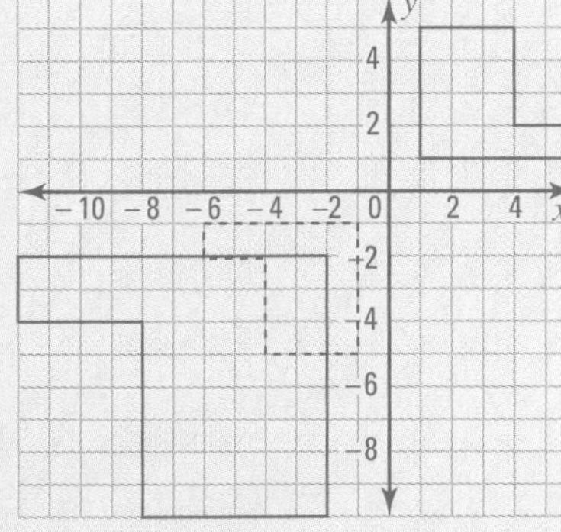

4. possible answer: translation 4 units to the right followed by a translation 9 units down
5. possible answer: a 90° counter-clockwise rotation followed by a translation 5 units down
6. possible answer: reflection over the x-axis and translation 7 units to the left

Additional Answers

Extra Practice 8-5

4. $\begin{bmatrix} 5 & 4 \\ -3 & -3 \\ -2 & 7 \end{bmatrix}$ 5. $\begin{bmatrix} -4 & 9 \\ 5 & 2 \\ -4 & 2 \end{bmatrix}$

6. $\begin{bmatrix} 13 & -5 \\ -6 & 5 \\ -6 & -7 \end{bmatrix}$ 7. $\begin{bmatrix} -33 & 15 \\ 21 & 0 \\ 6 & 3 \end{bmatrix}$

8. $\begin{bmatrix} 1 & 0 \\ \frac{1}{2} & \frac{5}{2} \\ -2 & -3 \end{bmatrix}$ 9. $\begin{bmatrix} 2 & -3 \\ -\frac{4}{3} & 1 \\ 0 & -\frac{8}{3} \end{bmatrix}$

10. $\begin{bmatrix} -2 & 9 \\ 6 & 7 \\ -8 & -4 \end{bmatrix}$ 11. $\begin{bmatrix} 5 & -5 \\ -10 & -15 \\ 10 & 17 \end{bmatrix}$

12. $\begin{bmatrix} 8 & -\frac{1}{2} \\ -5 & -\frac{3}{2} \\ -2 & 3 \end{bmatrix}$

13. $\begin{bmatrix} 28 & 30 & 14 \\ 24 & 20 & 25 \\ 16 & 15 & 27 \end{bmatrix}$

Extra Practice

Extra Practice 8–5 • Addition and multiplication with Matrices • pages 358–361

Find the dimensions of each matrix.

1. $\begin{bmatrix} 2 & 3 \\ 4 & 5 \end{bmatrix}$ 2 × 2
2. $\begin{bmatrix} 3 \\ 6 \\ 9 \\ 5 \end{bmatrix}$ 4 × 1
3. $\begin{bmatrix} 6 & 1 \\ 4 & 8 \\ 9 & 5 \end{bmatrix}$ 3 × 2

Use the following matrices in Exercises 4–12.

$J = \begin{bmatrix} 2 & 0 \\ 1 & 5 \\ -4 & -6 \end{bmatrix}$ $K = \begin{bmatrix} -6 & 9 \\ 4 & -3 \\ 0 & 8 \end{bmatrix}$ $L = \begin{bmatrix} 11 & -5 \\ -7 & 0 \\ -2 & -1 \end{bmatrix}$

Find each of the following. 4–13. See additional answers.

4. $K + L$
5. $J + K$
6. $J + L$
7. $-3L$
8. $\frac{1}{2}J$
9. $-\frac{1}{3}K$
10. $2J + K$
11. $L + -3J$
12. $\frac{1}{2}K + L$

13. Tyler Junior High School ordered school pennants. The seventh grade ordered 28 black, 24 white, and 16 green. The eighth grade ordered 30 black, 20 white, and 15 green. The ninth grade ordered 14 black, 25 white, and 27 green. Write two different 3 × 3 matrices to show this information.

Extra Practice 8–6 • More Operations and Matrices • pages 362–365

Refer to the matrices below. Find the dimensions of each product, if possible. *Do not multiply.* If not possible to multiply, write NP.

$A = \begin{bmatrix} 4 & 8 \\ 6 & 1 \\ 0 & 5 \end{bmatrix}$ $B = [5 \quad 1 \quad 3]$ $C = \begin{bmatrix} 4 & 5 & 3 \\ 8 & 9 & 6 \end{bmatrix}$ $D = \begin{bmatrix} 3 & 9 \\ 5 & 7 \end{bmatrix}$

1. AB NP
2. AC 3 × 3
3. AD 3 × 2
4. BC NP
5. CD NP
6. DC 2 × 3
7. BA 1 × 2
8. CA 2 × 2

Find each product. If not possible, write NP.

9. $\begin{bmatrix} 1 \\ 4 \\ -2 \end{bmatrix} [-5 \quad -1 \quad 3]$ $\begin{bmatrix} -5 & -1 & 3 \\ -20 & -4 & 12 \\ 10 & 2 & -6 \end{bmatrix}$
10. $[4 \quad -1 \quad 0] \begin{bmatrix} 2 & 6 \\ 0 & -3 \\ 7 & 1 \end{bmatrix}$ $[8 \quad 27]$
11. $\begin{bmatrix} 3 \\ -2 \end{bmatrix} [-2 \quad 0]$ $\begin{bmatrix} -6 & 0 \\ 4 & 0 \end{bmatrix}$
12. $[8 \quad -1] \begin{bmatrix} 3 & 2 \\ -2 & -1 \end{bmatrix}$ $[26 \quad 17]$
13. $\begin{bmatrix} -4 & 0 & 1 \\ 2 & 5 & -2 \\ 1 & -1 & 3 \end{bmatrix} \begin{bmatrix} 3 & 5 \\ -2 & 1 \\ 0 & -3 \end{bmatrix}$ $\begin{bmatrix} -12 & -23 \\ -4 & 21 \\ 5 & -5 \end{bmatrix}$
14. $\begin{bmatrix} 0 & -2 & -5 \\ 1 & 0 & -3 \end{bmatrix} \begin{bmatrix} 3 & -2 & -5 \\ 4 & 3 & 0 \end{bmatrix}$ NP

Extra Practice

Extra Practice 8–7 • Transformations and Matrices • pages 368–371

Represent each geometric figure with a matrix. See additional answers.

1. 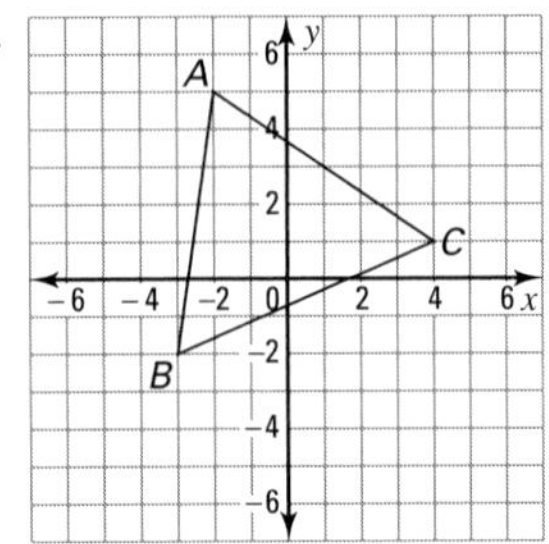

2. 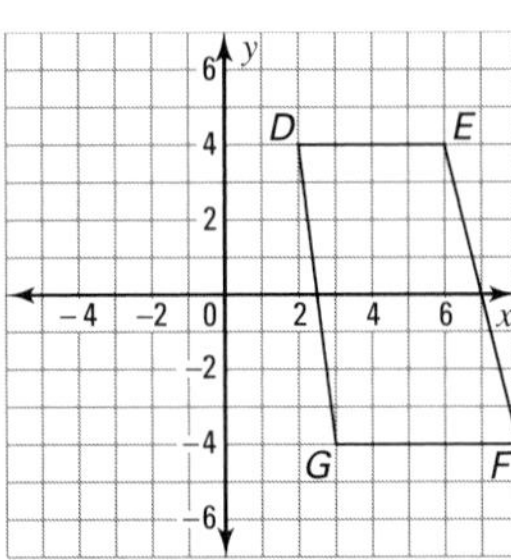

3. 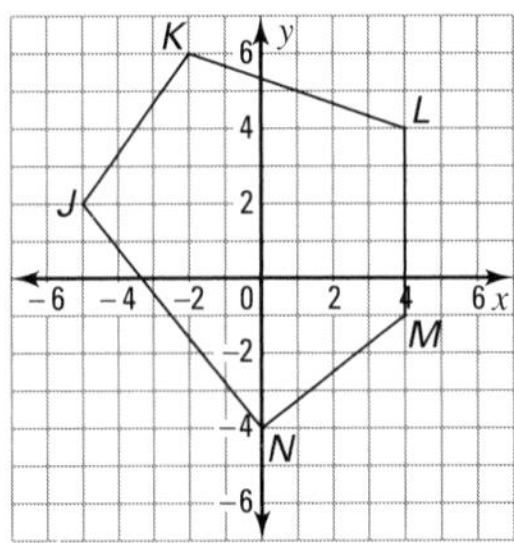

Find the reflection images of the triangle represented by $\begin{bmatrix} -2 & -5 & -7 \\ -4 & -1 & -5 \end{bmatrix}$. See additional answers.

4. over the y-axis. **5.** over the x-axis. **6.** over the line $y = -x$.

Find the reflection images of the quadrilateral represented by $\begin{bmatrix} 2 & 4 & 7 & 8 \\ -5 & -2 & -5 & -9 \end{bmatrix}$. See additional answers.

7. over the line $y = x$. **8.** over the x-axis. **9.** over the y-axis.

Interpret each equation as indicating: ***The reflection image of point __?__ over __?__ is the point __?__.***

10. $\begin{bmatrix} -1 & 0 \\ 0 & 1 \end{bmatrix}\begin{bmatrix} -3 \\ 5 \end{bmatrix} = \begin{bmatrix} 3 \\ 5 \end{bmatrix}$ (−3, 5), y-axis, (3, 5)

11. $\begin{bmatrix} 0 & -1 \\ -1 & 0 \end{bmatrix}\begin{bmatrix} -1 \\ -4 \end{bmatrix} = \begin{bmatrix} 4 \\ 1 \end{bmatrix}$ (−1, −4), $y = -x$, (4, 1)

Chapter 9

Extra Practice 9–1 • Review Percents and Probability • pages 384–387

A spinner with 8 equal sectors labeled A through H is spun 100 times with the following results.

Outcome	A	B	C	D	E	F	G	H
Frequency	8	14	12	17	9	15	14	11

What is the experimental probability of spinning each of the following results?

1. B $\frac{7}{50}$ **2.** E $\frac{9}{100}$ **3.** H $\frac{11}{100}$ **4.** C $\frac{3}{25}$

5. a letter that comes before D $\frac{17}{50}$ **6.** a letter that comes after D $\frac{49}{100}$

List all the elements of the sample space for each of the following experiments.

7. You toss a penny and a dime. (H, T), (H, H), (T, H), (T, T)

8. You spin each of these spinners once. See additional answers.

Find the probability of each of the following.

9. Drawing a ten of hearts from a standard deck of cards. $\frac{1}{52}$

10. Rolling a die and getting a prime number. $\frac{1}{2}$

Additional Answers

Extra Practice 8-7

1. $\begin{bmatrix} -2 & -3 & 4 \\ 5 & -2 & 1 \end{bmatrix}$

2. $\begin{bmatrix} 2 & 6 & 8 & 3 \\ 4 & 4 & -4 & -4 \end{bmatrix}$

3. $\begin{bmatrix} -5 & -2 & 4 & 4 & 0 \\ 2 & 6 & 4 & -1 & -4 \end{bmatrix}$

4. $\begin{bmatrix} 2 & 5 & 7 \\ -4 & -1 & -5 \end{bmatrix}$

5. $\begin{bmatrix} -2 & -5 & -7 \\ 4 & 1 & 5 \end{bmatrix}$

6. $\begin{bmatrix} 4 & 1 & 5 \\ 2 & 5 & 7 \end{bmatrix}$

7. $\begin{bmatrix} -5 & -2 & -5 & -9 \\ 2 & 4 & 7 & 8 \end{bmatrix}$

8. $\begin{bmatrix} 2 & 4 & 7 & 8 \\ 5 & 2 & 5 & 9 \end{bmatrix}$

9. $\begin{bmatrix} -2 & -4 & -7 & -8 \\ -5 & -2 & -5 & -9 \end{bmatrix}$

Extra Practice 9–1

8. (A, 2), (A, 4), (A, 6), (A, 8), (B, 2), (B, 4), (B, 6), (B, 8), (C, 2), (C, 4), (C, 6), (C, 8), (D, 2), (D, 4), (D, 6), (D, 8)

Extra Practice

Extra Practice 9–3 • Compound Events • pages 392–395

Two dice are rolled.

1. Find the probability that the sum of the numbers rolled is either 4 or 5. $\frac{7}{36}$
2. Find the probability that the sum of the numbers rolled is even and less than 7. $\frac{1}{4}$
3. Find P(not a prime). $\frac{7}{12}$
4. Find P(a sum of 8 or not prime). $\frac{7}{12}$

Ashante's Little League coach chooses the line-up by placing the 9 names into a hat and then pulling them out one by one. Find each probability.

5. batting first or third $\frac{2}{9}$
6. not batting in an even-numbered position $\frac{5}{9}$
7. batting last or in the first two thirds of the batting order $\frac{7}{9}$
8. batting second or in the first third of the order $\frac{1}{3}$

You spin this spinner. Find each probability.

9. spinning 4 or an odd number $\frac{5}{8}$
10. spinning a prime or an odd number $\frac{5}{8}$
11. spinning a prime or an even number $\frac{7}{8}$
12. spinning a prime or a number greater than 4 $\frac{3}{4}$

Extra Practice 9–4 • Independent and Dependent Events • pages 396–399

A bag contains marbles, all the same size. There are 5 red, 4 blue, 2 yellow, and 1 green. Marbles are drawn at random from the bag, one at a time, and then replaced. Find each probability.

1. P(red, then blue) $\frac{5}{36}$
2. P(blue, then yellow) $\frac{1}{18}$
3. P(green, then red) $\frac{5}{144}$
4. P(blue, then not blue) $\frac{2}{9}$
5. P(not green, then yellow) $\frac{11}{72}$
6. P(green, then not red) $\frac{7}{144}$

A box contains tennis balls. There are 4 white, 3 yellow, 1 green, and 2 pink. One ball at a time is taken at random from the box and not replaced. Find each probability.

7. P(green, then yellow) $\frac{1}{30}$
8. P(white, then pink) $\frac{4}{45}$
9. P(white, then not white) $\frac{4}{15}$
10. P(yellow, then green) $\frac{1}{30}$
11. P(green, then not white) $\frac{1}{18}$
12. P(white, then not green) $\frac{16}{45}$

A neon sign reading HOTEL CHELSEA has two of its letters go out.

13. What is the probability that both letters are vowels? $\frac{5}{33}$
14. What is the probability that the first letter is an E and the second is also an E? $\frac{1}{22}$
15. What is the probability that the first is L and the second is not L? $\frac{5}{33}$
16. You are given tickets to two concerts at a theater with 3,000 seats. What is the probability that you will sit in the orchestra section for the first concert, and then in the second balcony for the second concert, if the orchestra has 1,800 seats and the second balcony has 600 seats? $\frac{3}{25}$

712 Extra Practice

Extra Practice 9–5 • Permutations and Combinations • pages 402–405

For each situation, tell whether order does or does not matter.

1. You are recording the numbers and letters in an e-mail address. does matter
2. You are at a video store and selecting 3 movies to rent for the weekend. does not matter
3. You are selecting candidates for president, vice president, and secretary. does matter
4. You are selecting a 5-member committee from students in the class. does not matter
5. There are 5 different library books you would like to borrow, but the library allows you to borrow only 3 books at a time. How many ways can you select 3 of the books? 10
6. How many different ways can you arrange six videos in a row on a shelf? 720
7. A restaurant menu states that when you buy a dinner special you can select 3 side orders from 12 that are listed. How many ways can you do this? 220
8. Nine teams take part in an intramural volley ball tournament. How many different arrangements of first-, second-, and third-place winners are possible? 504

Extra Practice 9–6 • Scatter Plots and Box Plots • pages 406–409

This table shows the average scores of students who participated in an annual math contest.

Year	1988	1989	1990	1991	1992	1993	1994	1995	1996	1997
Average Score	8.2	7.5	7.9	8.7	5.6	6.2	7.8	6.4	7.0	7.2

1. Make a scatter plot of the data. Check students' drawings.
2. What is the range of average scores? 3.1
3. Does your scatter plot show a positive correlation, a negative correlation, or no correlation? no correlation

These box-and-whisker plots show the weights of the members of three high school football teams.

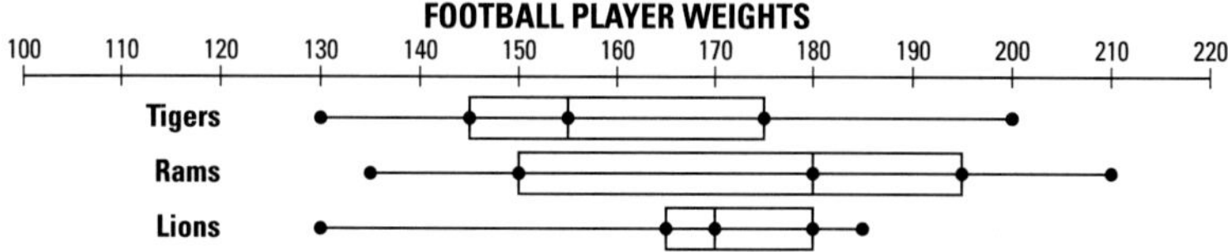

4. Which team has the highest median weight? Rams
5. What is the lower quartile of the Tigers' weights? 145
6. What is the upper quartile of the Lions' weights? 180
7. What is the median weight of the Tigers? 155
8. Which team had the smallest range of weights? the largest? Lions; Rams
9. Which team had weights most closely clustered about its median? Lions

Extra Practice 713

Extra Practice

Extra Practice 9–7 • Standard Deviation • pages 412–415

Compute the variance and standard deviation for each set of data.

1. 3, 6, 9, 12, 15 18, $\sqrt{18}$
2. 5, 5, 5, 5, 5 0, 0
3. 0.5, 2.5, 4.5, 6.5, 8.5 8, $\sqrt{8}$
4. 1, 3, 5, 7, 9 8, $\sqrt{8}$
5. 4, 7, 10, 13, 16 18, $\sqrt{18}$
6. 4, 8, 12, 16, 20 32, $\sqrt{32}$
7. 2.3, 4.3, 6.3, 8.3, 10.3 8, $\sqrt{8}$
8. 1.6, 5.6, 9.6, 13.6, 17.6 32, $\sqrt{32}$
9. 7.6, 3.4, 6, 8.3, 5.7 2.9, $\sqrt{2.9}$

Find the variance and standard deviation for each set of data.

10. The top five scores in an Olympics gymnastics trial were: 9.1, 8.5, 7.9, 8.2, and 8.5. 0.1584, $\sqrt{0.1584}$

11. The prices of lunches in five country high schools are: \$3.00, \$3.50, \$2.75, \$3.25, and \$3.75. 0.125, $\sqrt{0.125}$

12. Ping took two tests. On the first test, his score was 78, while the mean score was 72 and the standard deviation was 3. On the second test, his score was 70, while the mean score was 68 and the standard deviation was 0.5. On which test did Ping score better, relative to the scores of his classmates? the second

13. Alicia took two tests. On Test A, her score was 82, while the mean score was 90 and the standard deviation was 10. On Test B, her score was 76, while the mean score was 82 and the standard deviation was 4. On which test did Alicia score better, relative to the scores of her classmates? Test B

Chapter 10

Extra Practice 10–1 • Irrational Numbers • pages 426–429

Find the value to the nearest hundredth.

1. $\sqrt{13}$ 3.61
2. $\sqrt{30}$ 5.48
3. $\sqrt{62}$ 7.87
4. $\sqrt{150}$ 12.25
5. $\sqrt{189}$ 13.75
6. $\sqrt{270}$ 16.43
7. $\sqrt{666}$ 25.81
8. $\sqrt{106}$ 10.30

Write each in simplest radical form.

9. $\sqrt{45}$ $3\sqrt{5}$
10. $\sqrt{32}$ $4\sqrt{2}$
11. $\sqrt{147}$ $7\sqrt{3}$
12. $\sqrt{52}$ $2\sqrt{13}$
13. $\sqrt{28}$ $2\sqrt{7}$
14. $\sqrt{162}$ $9\sqrt{2}$
15. $\sqrt{125}$ $5\sqrt{5}$
16. $\sqrt{360}$ $6\sqrt{10}$
17. $(3\sqrt{5})(2\sqrt{10})$ $30\sqrt{2}$
18. $(4\sqrt{3})(2\sqrt{6})$ $24\sqrt{2}$
19. $(2\sqrt{5})^2$ 20
20. $(2\sqrt{3})(-4\sqrt{7})$ $-8\sqrt{21}$
21. $(-2\sqrt{2})(5\sqrt{8})$ -40
22. $\frac{5}{\sqrt{7}}$ $\frac{5\sqrt{7}}{7}$
23. $\frac{\sqrt{64}}{\sqrt{2}}$ $4\sqrt{2}$
24. $\frac{\sqrt{45}}{\sqrt{9}}$ $\sqrt{5}$
25. $\frac{4\sqrt{3}}{5\sqrt{6}}$ $\frac{2\sqrt{2}}{5}$
26. $-\frac{2\sqrt{6}}{5\sqrt{8}}$ $-\frac{\sqrt{3}}{5}$
27. $-\frac{3\sqrt{5}}{4\sqrt{15}}$ $-\frac{\sqrt{3}}{4}$
28. $\frac{4\sqrt{3}}{2\sqrt{8}}$ $\frac{\sqrt{6}}{2}$
29. $\frac{5\sqrt{2}}{3\sqrt{6}}$ $\frac{5\sqrt{3}}{9}$
30. $-\sqrt{\frac{5}{6}}$ $-\frac{\sqrt{30}}{6}$

31. If the area of a square is 236 square feet, find the length of each side to the nearest tenth of a foot. 15.4 ft

714 Extra Practice

Extra Practice

Additional Answers

Extra Practice 10–3
1. 6 m, 6$\sqrt{3}$ m, 10.4 m
2. 4$\sqrt{2}$ in., 5.7 in.
3. 10 yd, 5$\sqrt{3}$ yd, 8.7 yd
4. 15$\sqrt{2}$ cm, 21.2 cm
5. 6 cm, 3$\sqrt{3}$ cm, 5.2 cm
6. 12$\sqrt{2}$ in., 17.0 in.
7. 7 m, 7$\sqrt{3}$ m 12.1 m
8. 9$\sqrt{2}$ yd, 12.7 yd

Extra Practice 10–2 • The Pythagorean Theorem • pages 430–433

Use the Pythagorean Theorem to find the unknown length. Round your answers to the nearest tenth.

1. 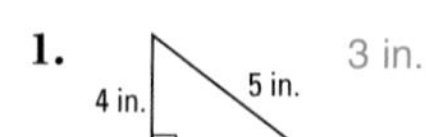

3 in.

2. 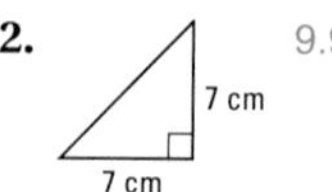

9.9 cm

3. 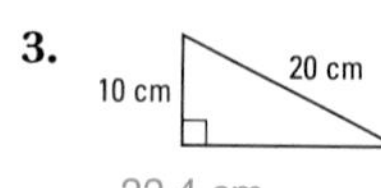

22.4 cm

4. 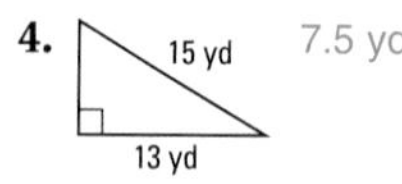

7.5 yd

5. 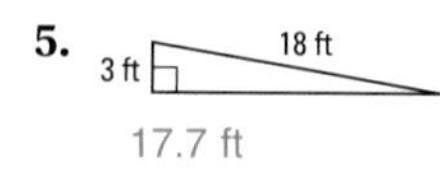

17.7 ft

6. 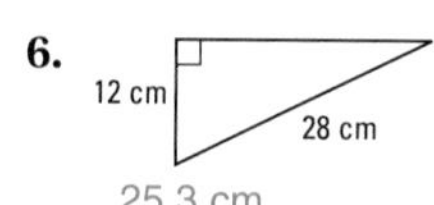

25.3 cm

7. 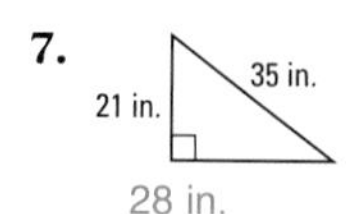

28 in.

8. 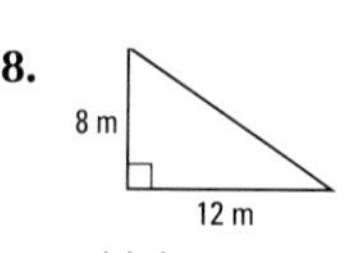

14.4 m

Determine if each figure is a right triangle. Write *yes* or *no*.

9. 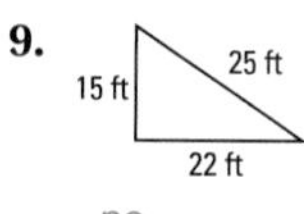

no

10. 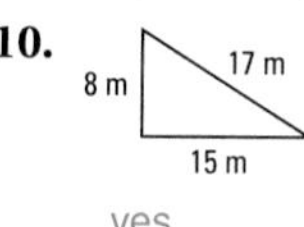

yes

11. 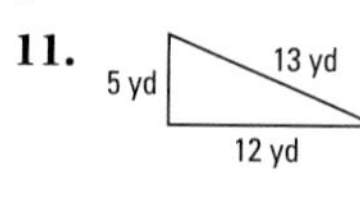

yes

12. 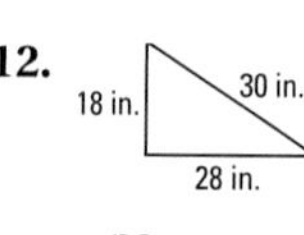

no

Solve. Round your answers to the nearest tenth.

13. Find the length of a diagonal of a rectangle with a length of 24 ft and a width of 8 ft. 25.3 ft
14. Find the width of a rectangle with a length of 9 m and a diagonal length of 11 m. 6.3 m

Extra Practice 10–3 • Special Right Triangles • pages 436–439

Find the unknown measures. First find each in simplest radical form and then find each to the nearest tenth. 1–8. See additional answers.

1. 60° 12 m

2. 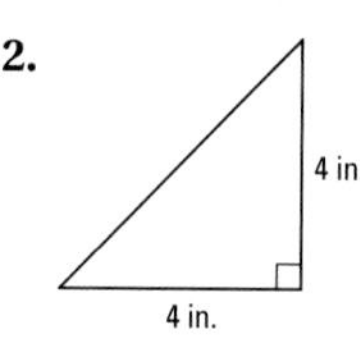

3. 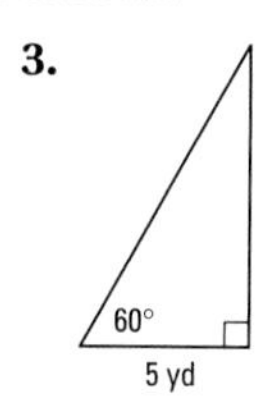

4. 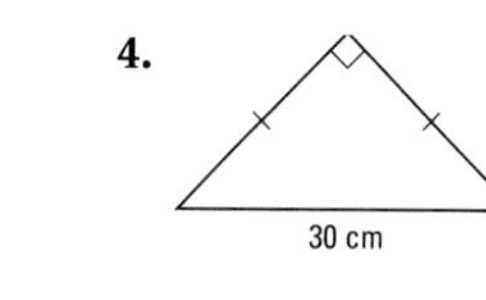

5. 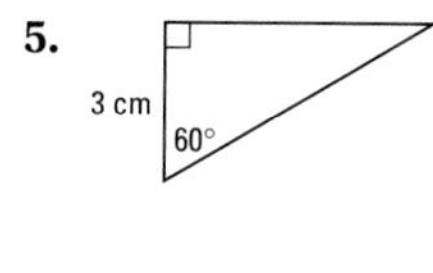

6. 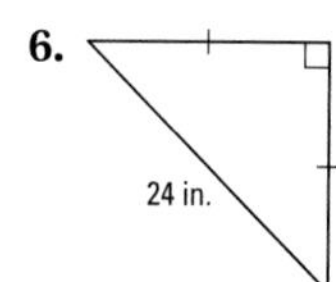

7. 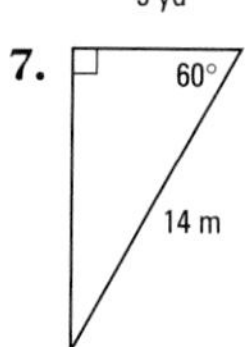

8. 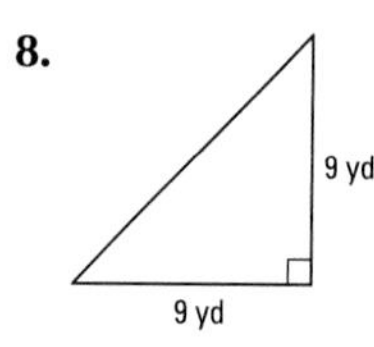

9. The diagonal of a square measures 6 cm. Find the length of a side of the square to the nearest tenth. 4.2 cm
10. The side of a square measures 10 in. Find the length of the diagonal of the square to the nearest tenth. 14.1 in.

Extra Practice

Extra Practice 10–4 • Circles, Angles, and Arcs • pages 440–443

Find *x*.

1. 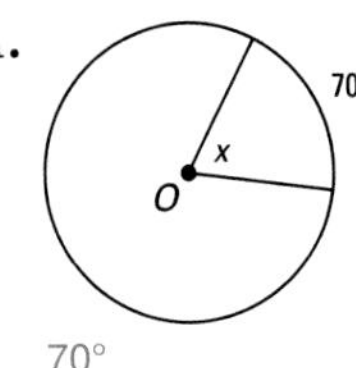

70°

2. 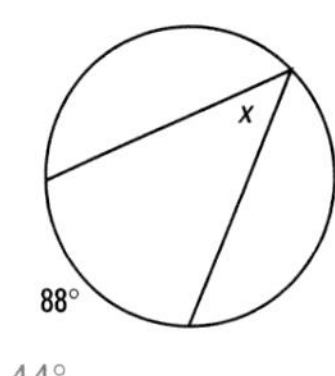

44°

3. 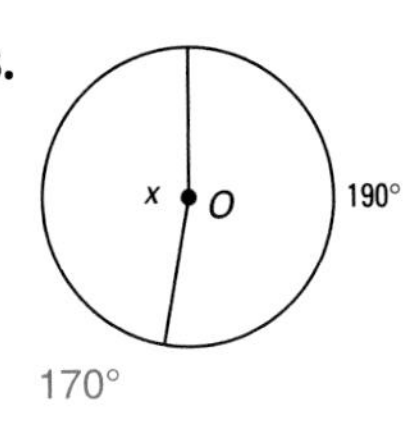

170°

4. 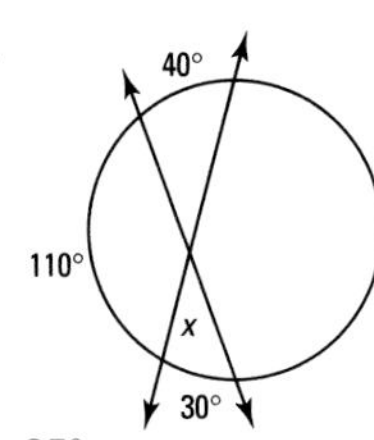

35°

5. 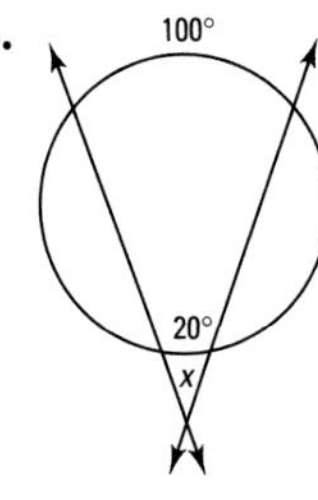

40°

6. 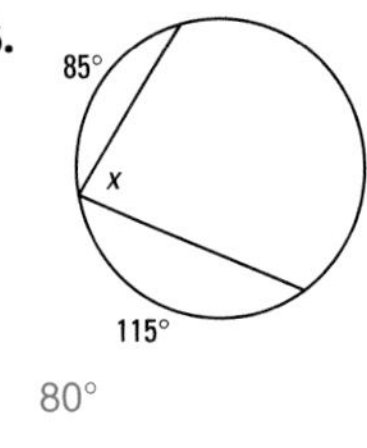

80°

7. 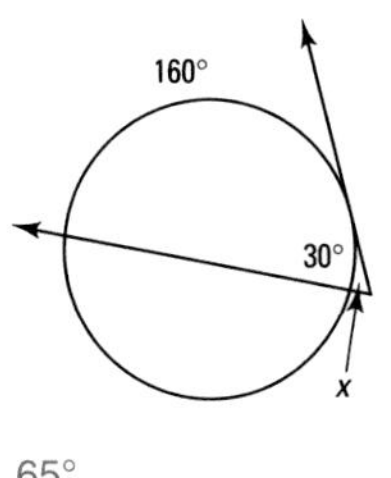

65°

8. 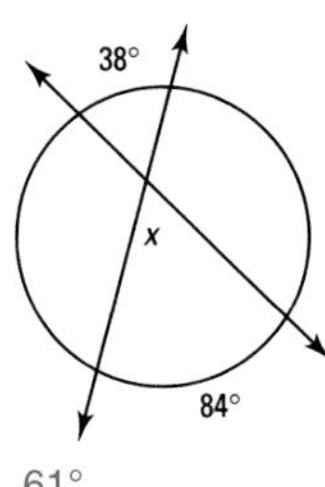

61°

9. An inscribed angle intercepts an arc of 130°. What is the measure of the inscribed angle? 65°

10. An inscribed angle measures 48°. What is the measure of the arc it intercepts? 96°

11. An inscribed angle $\angle ABC$, measures 74°. What is the measure of the major arc ABC? 212°

12. An inscribed angle $\angle JKL$, measures 50°. What is the measure of the central angle with endpoints J and L? 100°

Extra Practice 10–6 • Circles and Segments • pages 448–451

Find *x*.

1. 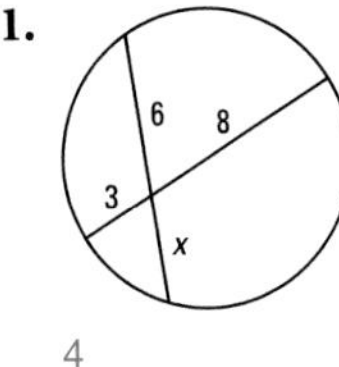

4

2. 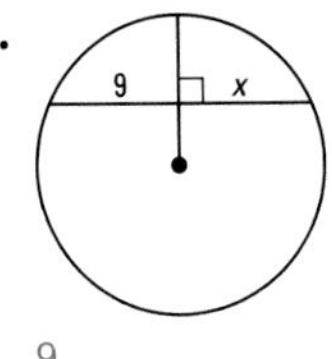

9

3. 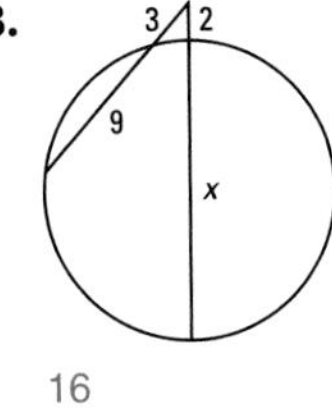

16

4. 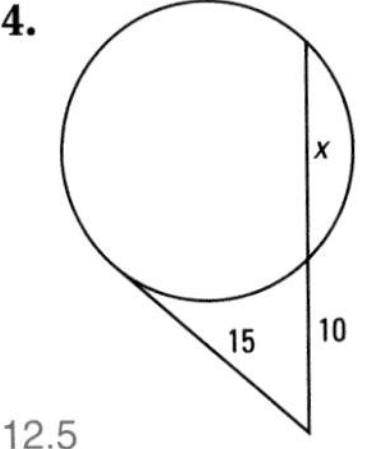

12.5

5. In circle O, two chords, $\overline{AB}$ and $\overline{CD}$ intersect at point K. $\overline{AK} = 12$ cm, $\overline{KB} = 10$ cm, and $\overline{KD} = 8$ cm. Find the measure of $\overline{CK}$. 15 cm

6. In circle O, radius $\overline{OM}$ is perpendicular to chord $\overline{JK}$ at point L. Find the measure of $\overline{JK}$ if $\overline{JL} = 18$ cm. 36 cm

716 Extra Practice

Extra Practice 10–7 • Constructions with Circles • pages 454–457

1. Construct a regular hexagon. 1–6. Check students' constructions.
2. Construct a square.
3. Copy this equilateral triangle. Inscribe the triangle in a circle.

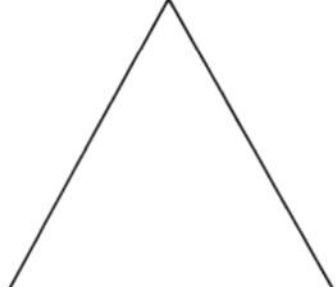

4. Copy this square. Inscribe the square in a circle.

5. Copy this regular pentagon. Inscribe the pentagon in a circle.

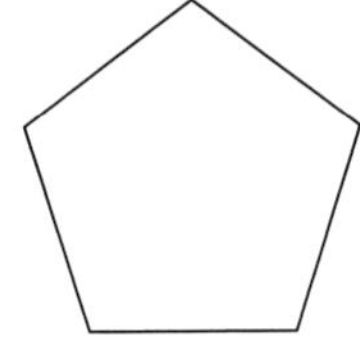

6. Copy this regular hexagon. Circumscribe the hexagon around a circle.

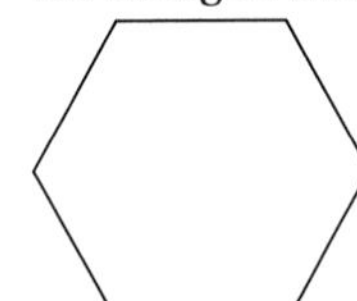

Chapter 11

Extra Practice 11–1 • Add and Subtract Polynomials • pages 468–71

Simplify.

1. $(3a + 7) + (4a + 5)$ $7a + 12$
2. $(6n + p) + (n + 7p)$ $7n + 8p$
3. $(4x^2 + 3x) + (-x^2 + x)$ $3x^2 + 4x$
4. $(-3n^2 - 4n) + (3n^2 - 4n)$ $-8n$
5. $(3j + 4k - 2) + (2j - 2k + 5)$ $5j + 2k + 3$
6. $(4x^2 + 12x + 9) + (x^2 - 2x + 1)$ $5x^2 + 10x + 10$
7. $(7a + 5) - (2a + 3)$ $5a + 2$
8. $(8x - y) - (5x + y)$ $3x - 2y$
9. $(3x^2 + 4y) - (-x^2 + y)$ $4x^2 + 3y$
10. $(a^2 - 5a - 6) - (-a^2 + a + 12)$ $2a^2 - 6a - 18$
11. $(ab - b + a) - (ab - b + a)$ 0
12. $(4d^2 - 2de + e^2) - (3d^2 - de - 3e^2)$ $d^2 - de + 4e^2$
13. $(m^2 - 16n + 3n^2) - (9n - 3m^2 + n^2)$ $4m^2 - 25n + 2n^2$
14. $(a^2 + 12b + b^2) + (4a^2 + 8b + 4b^2)$ $5a^2 + 20b + 5b^2$
15. $(5c^2 + 8cd + d^2) - (c^2 - 2c)$ $4c^2 + 8cd + 2c + d^2$
16. $(7p^2 - 5q^2) - (p^2 - 6pq + q^2)$ $6p^2 + 6pq - 6q^2$
17. Last week, Marisol worked 10 hours at her part-time job, where she earns x dollars per hour, and 35 hours at her full-time job, where she earns y dollars per hour. This week, she worked 15 hours at her part-time job and 35 hours at her full-time job. What were her earnings during the two weeks, expressed in terms of x and y? $25x + 70y$

Extra Practice 717

Extra Practice

Extra Practice 11–2 • Multiply by a Monomial • pages 472–475

Simplify.

1. $(3x)(4y)$ $12xy$
2. $(5a)(4)$ $20a$
3. $(a^2)(2ab^3)$ $2a^3b^3$
4. $(m)(-3mn)$ $-3m^2n$
5. $(-c)(-6cd)$ $6c^2d$
6. $(2p^2)(-3p^2q)$ $-6p^4q$
7. $(7x^2y)(8xy^2)$ $56x^3y^3$
8. $(4c^2)^2$ $16c^4$
9. $2x(7x^2 - 6y)$ $14x^3 - 12xy$
10. $-8n(2n^2 - 5p)$ $-16n^3 + 40np$
11. $2a^2[-(a^2 + a)]$ $-2a^4 - 2a^3$
12. $3k^2[-(3k^2 - k)]$ $-9k^4 + 3k^3$
13. $7cd(2c^2 + 3d^2)$ $14c^3d + 21cd^3$
14. $-8c^2d(c^3d - c^2d)$ $-8c^5d^2 + 8c^4d^2$
15. $3a^2b(5a^3c - 3ab^4)$ $15a^5bc - 9a^3b^5$
16. $9x^2yz(x^2y - y^3z)$ $9x^4y^2z - 9x^2y^4z^2$
17. $3jkl(j^2k^2l - jkl^3)$ $3j^3k^3l^2 - 3j^2k^2l^4$
18. $-9abc^3(a^2bc^3 - ab^4c)$ $-9a^3b^2c^6 + 9a^2b^5c^4$
19. $-15xyz(-xyz - x^2y^3z^2)$ $15x^2y^2z^2 + 15x^3y^4z^3$
20. $3d(d^3 + 2d^2 - d)$ $3d^4 + 6d^3 - 3d^2$
21. $4c(c^2 + 5c - 6)$ $4c^3 + 20c^2 - 24c$
22. $rs(5r^2 - 3rs + 4)$ $5r^3s - 3r^2s^2 + 4rs$
23. $-xy(3a^2 - 2b + c)$ $-3a^2xy + 2bxy - cxy$
24. $-rs^2(4r^2 - rs + 3s^2)$ $-4r^3s^2 + r^2+s^3 - 3rs^4$
25. $2xy^4z^3(x^3yz^2 - x^2y^2z^3 - xy^5z^2)$ $2x^4y^5z^5 - 2x^3y^6z^6 - 2x^2y^9z^5$
26. In 1995, a supermarket employed x clerks, each of whom earned y dollars per week. The weekly pay rate increased by d dollars each year. In 1998, the number of clerks quadrupled. What was the total paid each week to the staff of clerks in 1995? What was it in 1998? xy, $4xy + 12xd$

Extra Practice 11–3 • Divide and Find Factors • pages 478–481

Find the factors for the following.

1. $4x + 6y$ $2(2x + 3y)$
2. $8a^2 - 12b^2$ $4(2a^2 - 3b^2)$
3. $24n^2 + 6$ $6(4n^2 + 1)$
4. $7xy + 21x$ $7x(y + 3)$
5. $jk + jkl$ $jk(1 + l)$
6. $7pq - 21qr$ $7q(p - 3r)$
7. $7ab - 4bc$ $b(7a - 4c)$
8. $9d^2e - 5e^2$ $e(9d^2 - 5e)$
9. $13x^2y + 15yz^2$ $y(13x^2 + 15z^2)$
10. $v^2w + vw$ $vw(v + 1)$
11. $3a^5 + 3a^5 + 3a^2b^2$ $3a^2(a^3 + b^2)$
12. $4y - 4y^6$ $4y(1 - y^5)$

Find the GCF and its paired factor for the following.

13. $48a + 56b$ $8(6a + 7b)$
14. $39x + 13x^2$ $13x(3 + x)$
15. $18c^2 - 27cd^2$ $9c(2c - 3d^2)$
16. $28xy^2 - 42yz^2$ $14xy(2y - 3z^2)$
17. $30m^4n^3 - 60m^3n^4$ $30m^3n^3(m - 2n)$
18. $60x^3 + 45x^2$ $15x^2(4x + 3)$
19. $6a^3b + 12a^2b^2$ $6a^2b(a + 2b)$
20. $8x^4b^3 + 12x^3b^2$ $4x^3b^2(2xb + 3)$
21. $r^2s^2 + r^2s + rs^2$ $rs(rs + r + s)$
22. $xa^3b^3 + ya^2b^2 + 2ab$ $ab(xa^2b^2 + yab + 2)$
23. $16d^5 + 40d^4e^2 - 56d^2e^2 - 56d^2e^4$ $8d^2(d^3 + 5d^2e^2 - 7e^4)$
24. $63x^3y - 56w^3z + 28r^3t$ $7(9x^3y - 8w^3z + 4r^3t)$
25. A carpenter has two planks of wood, one 8 inches long and the other 40 inches long. She wants to use all the wood in the two planks to cut a set of small pieces, each the same size and as long as possible. How long will each cut piece be, and how many can she cut? 8 inches, 6 pieces

Extra Practice 11–4 • Multiply Two Binomials • pages 482–485

Simplify.

1. $(3a + 4b)(5c - 2)$ $15ac - 6a + 20bc - 8b$
2. $(6x - 5y)(z - 3)$ $6xz - 18x - 5yz + 15y$
3. $(2r + 7s)(r + 2t)$ $2r^2 + 4rt + 7rs + 14st$
4. $(6a - 7p)(2a + 3q)$ $12a^2 + 18aq - 14ap - 21pq$
5. $(m - 4n)(3p + 4n)$ $3mp + 4mn - 12np - 16n^2$
6. $(2x + 5y)(2z - w)$ $4xz - 2xw + 10yz - 5yw$
7. $(8r + 3s)(5r - 2t)$ $40r^2 - 16rt + 15rs - 6st$
8. $(8m - n)(2p + n)$ $16mp + 8mn - 2np - n^2$
9. $(4x + 3y)(x + 5y)$ $4x^2 + 23xy + 15y^2$
10. $(6 - 5n)(3 - 2n)$ $18 - 27n + 10n^2$
11. $(5x - 3)(x - 3)$ $5x^2 - 18x + 9$
12. $(x + 7y)(3x + y)$ $3x^2 + 22xy + 7y^2$
13. $(c - 7d)(5c - 2d)$ $5c^2 - 37cd + 14d^2$
14. $(7x + 1)(6x - 7)$ $42x^2 - 43x - 7$
15. $(6x - y)(5x + 3y)$ $30x^2 + 13xy - 3y^2$
16. $(m - 4)(3m + 5)$ $3m^2 - 7m - 20$
17. $(a + 3b)(a + 3b)$ $a^2 + 6ab + 9b^2$
18. $(c - 9)(c - 9)$ $c^2 - 18c + 81$
19. $(4x + 5y)(4x + 5y)$ $16x^2 + 40xy + 25y^2$
20. $(7c - e)(7c - e)$ $49c^2 - 14ce + e^2$
21. $(d - 4)(d + 4)$ $d^2 - 16$
22. $(5a + 2)(5a - 2)$ $25a^2 - 4$
23. $(y + 1)(y - 1)$ $y^2 - 1$
24. $(g - 6)(g + 6)$ $g^2 - 36$
25. $(4m + 1)(4m - 1)$ $16m^2 - 1$
26. $(7x + 2y)(7x - 2y)$ $49x^2 - 4y^2$
27. The dimensions of a rectangle are $(3x + 2)$ feet and $(x + 5)$ feet. Write an expression for the area of the rectangle. $3x^2 + 17x + 10$

Extra Practice 11–5 • Find Binomial Factors in a Polynomial • pages 488–491

Find factors for the following.

1. $6wy + 9wz + 4xy + 6xz$ $(3w + 2x)(2y + 3z)$
2. $10ac + 2bc + 15ad + 3bd$ $(5a + b)(2c + 3d)$
3. $rt + rv + 3st + 3sv$ $(r + 3s)(t + v)$
4. $x^2 + 7x + 6xy + 42y$ $(x + 6y)(x + 7)$
5. $8n^2 + 2np + 4nq + pq$ $(4n + p)(2n + q)$
6. $3y^2 + xy - 12yz - 4xz$ $(x + 3)(y - 4z)$
7. $n^2 - 4n + 2mn - 8m$ $(2m + n)(n - 4)$
8. $6k^2 - 8k + 3km - 4m$ $(3k - 4)(2k + m)$
9. $12a^2c - 8a^2d^2 - 3bc + 2bd^2$ $(4a^2 - b)(3c - 2d^2)$
10. $2r^3 - 4r^2t + 5rt - 10t^2$ $(2r^2 + 5t)(r - 2t)$
11. $24w^2y - 16w^2z^2 + 9xy - 6xz^2$ $(8w^2 + 3x)(3y - 2z^2)$
12. $8xz + 12x - 6yz - 9y$ $(4x - 3y)(2z + 3)$
13. $4a^2b - a^2c + 28b - 7c$ $(a^2 + 7)(4b - c)$
14. $4a^2 - 24ab + 5a - 30b$ $(4a + 5)(a - 6b)$
15. $2ac + 2ad - 3bc - 3bd$ $(2a - 3b)(c + d)$
16. $3ac^2 + 15bc^2 + 4ab + 20b^2$ $(a + 5b)(4b + 3c^2)$
17. $6ac + 9ad + 6ae + 2bc + 3bd + 2be$ $(3a + b)(2c + 3d + 2e)$
18. $16eg + 12eh + 8e^2 - 4fh - 2ef$ $(4e - f)(4g + 3h + 2e)$
19. A rectangle has an area that can be expressed as $6a^2 + 2ab + bc$. If the width can be expressed as $3a + b$, find an expression for the length. $2a + c$
20. The area of a rectangle can be expressed as $3s^2 + 2rs - 6st - 4rt$. Find expressions that might represent the dimensions of the rectangle. $2r + 3s$, $s - 2t$

Additional Answers

Extra Practice 11-7

19. $(z - 1)(z - 36)$
20. $(r + 3)(r + 12)$
21. $(x + 3)(x - 12)$
22. $(v + 2w)(v - 18w)$
23. $(f - 9)(f - 4)$
24. $(l - 2m)(l - 18m)$
25. $(g + 9)(g - 4)$
26. $(j - 3)(j + 12k)$
27. $(h - 9)(h + 4)$

Extra Practice

Extra Practice 11–6 • Special Factoring Patterns • pages 492–495

Find binomial factors of the following, if possible. (Four do not have such factors.)

1. $y^2 + 2y + 1$ $(y + 1)^2$
2. $x^2 + 18x + 81$ $(x + 9)^2$
3. $x^2 + 22x + 121$ $(x + 11)^2$
4. $a^2 - 14a + 49$ $(a - 7)^2$
5. $n^2 - 12n + 36$ $(n - 6)^2$
6. $y^2 - 7y + 49$ none
7. $d^2 - 64$ $(d + 8)(d - 8)$
8. $r^2 - 1$ $(r + 1)(r - 1)$
9. $4n^2 - 9$ $(2n + 3)(2n - 3)$
10. $9j^2 + 6j + 1$ $(3j + 1)^2$
11. $25d^2 - 20d + 4$ $(5d - 2)^2$
12. $16 - 24c + 9c^2$ $(4 - 3c)^2$
13. $4c^2 - 28cd + 49d^2$ $(2c - 7d)^2$
14. $81r^2 - s^2$ $(9r + s)(9r - s)$
15. $25p^2 - 144q^2$ $(5p + 12q)(5p - 12q)$
16. $121k^2 - 66kl + 9l^2$ $(11k - 3l)^2$
17. $64a^2 + 25b^2$ none
18. $64x^2 - 80xy + 25y^2$ $(8x - 5y)^2$
19. $25c^2 - 64d^2$ $(5c + 8d)(5c - 8d)$
20. $4x^2 + 20xy + 25y^2$ $(2x + 5y)^2$
21. $81s^2 - 50t^2$ none
22. $49p^2 - 140pq + 100q^2$ $(7p - 10q)^2$
23. $9x^2 + 36x + 64$ none
24. $64c^2 + 16c + 1$ $(8c + 1)^2$

Find a monomial factor and two binomial factors for each of the following.

25. $12x^2 - 27$ $3(2x + 3)(2x - 3)$
26. $36x^2 + 24x + 4$ $4(3x + 1)^2$
27. $x^3 - 6x^2 + 9x$ $x(x - 3)^2$
28. The expression for the area of a certain square is $14x + 49 + x^2$. Find an expression for the length of a side of the square. $x + 7$

Extra Practice 11–7 • Factor Trinomials • pages 498–501

Identify binomial second-term factors for the following.

1. $x^2 + 6x + 8$ 4, 2
2. $m^2 - 11m + 24$ 8, 3
3. $x^2 - x - 30$ 5, 6
4. $y^2 - 11y + 30$ 6, 5
5. $d^2 + 6d - 27$ 9, 3
6. $p^2 - 15p + 44$ 11, 4
7. $a^2 + 17ab + 72b^2$ $9b$, $8b$
8. $x^2 + 17xy + 42y^2$ $14y$, $3y$
9. $r^2 - 15rt + 54t^2$ $9t$, $6t$
10. $j^2 + 3jk - 54k^2$ $9k$, $6k$
11. $t^2 - tr - 30r^2$ $6r$, $5r$
12. $l^2 - 19ln + 18n^2$ $18n$, $1n$

Identify binomial second-term signs for the following.

13. $x^2 + x - 20$ +, −
14. $t^2 + 14t + 33$ +, +
15. $a^2 - a - 56$ +, −
16. $c^2 - 24c + 23$ −, −
17. $a^2 + 2ab - 15b^2$ +, −
18. $j^2 - 3jk - 10k^2$ +, −

Factor the following trinomials. 19–27. See additional answers.

19. $z^2 - 37z + 36$
20. $r^2 + 15r + 36$
21. $x^2 - 9x - 36$
22. $v^2 - 16vw - 36w^2$
23. $f^2 - 13f + 36$
24. $l^2 - 20lm + 36m^2$
25. $g^2 + 5g - 36$
26. $j^2 + 9jk - 36k^2$
27. $h^2 - 5h - 36$

28. A rectangular trench x feet deep is being dug for the foundation of a wall. The area of the bottom of the trench is $x^2 + 22x - 48$ square feet. Compare the depth of the trench to its width. Then compare the depth of the trench to its length. 2 ft narrower than depth, 24 ft longer than depth

Extra Practice 11–9 • More on Factoring Trinomials • pages 506–509

Find FOIL coefficients for the following trinomials.

1. $6x^2 + 19x + 10$ 6, 15, 4, 10
2. $8m^2 - 30m - 27$ 8, 36, 6, 27
3. $6c^2 - 11c + 3$ 6, 2, 9, 3
4. $21a^2 + 13a + 2$ 21, 7, 6, 2
5. $10x^2 - 23x + 12$ 10, 15, 8, 12
6. $18n^2 + 9n - 2$ 18, 12, 3, 2

For the following FOIL coefficients, identify the appropriate binomial factor coefficients.

	7.	8.	9.	10.
F-coefficient	6 2, 3	9 3, 3	12 4, 3	4 2, 2
Cross-product coefficients	15 5, 3	12 3, 4	16 4, 4	6 2, 3
(O and I)	8 2, 4	6 2, 3	15 5, 3	2 1, 2
L-coefficient	20 5, 4	8 2, 4	20 5, 4	3 1, 3

Place appropriate signs in these unsigned binomials.

11. $35m^2 - 6m - 9 = (7m ■ 3)(5m ■ 3)$ +, −
12. $40c^2 + 17cd - 12d^2 = (8c ■ 3d)(5c ■ 4d)$ −, +
13. $18x^2 + 55xy - 28y^2 = (2x ■ 7y)(9x ■ 4y)$ +, −
14. $30j^2 + 61j + 30 = (6j ■ 5)(5j ■ 6)$ +, +

Find binomial factors for the following trinomials. 15–26. See additional answers.

15. $25x^2 - 25x + 4$
16. $6x^2 - 5x - 4$
17. $14n^2 + 5n - 24$
18. $81y^2 - 24y - 20$
19. $15a^2 + 38ab + 24b^2$
20. $36t^2 - 19t - 6$
21. $10x^2 - 23x - 14$
22. $16x^2 - 41x + 25$
23. $8a^2 - 14ab + 3b^2$
24. $56r^2 - 6rs - 2s^2$
25. $2m^2 + 9m - 81$
26. $9k^2 + 27k + 8$
27. A rectangle has an area of $12a^2 + a - 1$. Find expressions that might be the length and width of the rectangle. $3a + 1$, $4a - 1$

Additional Answers

Extra Practice 11-9

15. $(5x - 4)(5x - 1)$
16. $(3x - 4)(2x + 1)$
17. $(7n - 8)(2n + 3)$
18. $(27y + 10)(3y - 2)$
19. $(5a + 6b)(3a + 4b)$
20. $(9t + 2)(4t - 3)$
21. $(2x + 1)(5x - 14)$
22. $(16x - 25)(x - 1)$
23. $(2a - 3b)(4a - b)$
24. $(7r + s)(8r - 2s)$
25. $(2m - 9)(m + 9)$
26. $(3k + 8)(3k + 1)$

Extra Practice

Chapter 12

Extra Practice 12–1 • Graph Parabolas • pages 520–523

Check students' graphs. Vertices and opening of the graphs are given.

Graph each function for the domain of real numbers. For each graph, give the coordinates of the vertex.

1. $y = x^2$ (0, 0) upward
2. $y = 3x^2 + 1$ (0, 1) upward
3. $y = -2x^2 - 3$ (0, −3) downward
4. $y = x^2 - 4$ (0, −4) upward
5. $y = -x^2 + 3$ (0, 3) downward
6. $y = 2x^2 - 1$ (0, −1) upward
7. $y = 4x^2 - 1$ (0, −1) upward
8. $y = -3x^2 + 1$ (0, 1) downward
9. $y = -x^2 + 2$ (0, 2) downward

Determine if the graph of each equation below opens upward or downward.

10. $y = -3x^2 - 8$ downward
11. $y = -x^2 - 4$ downward
12. $y = 4x^2 - 5$ upward
13. $y = -x^2 + 9$ downward
14. $y = 2x^2 + 5$ upward
15. $y = -9x^2 + 9$ downward
16. $y = 6x^2 - 2$ upward
17. $y = 3x^2 + 9$ upward
18. $y = -5x^2 - 9$ downward
19. $y = -5x^2 - 2$ downward
20. $y = x^2 - 16$ upward
21. $y = 5x^2 - 1$ upward

22. Given the equations (a) $y = 2x^2 - 3$ and (b) $y = 2x^2 + 3$, explain the differences and similarities in the two graphs. Possible answer. They both have the y-axis as the line of symmetry; the vertex of (a) is (0, −3), the vertex of (b) is (0, 3).

Extra Practice 12–2 • The General Quadratic Function • pages 524–527

Estimate the coordinates of the vertex for each parabola by graphing the equation on a graphing calculator. Then use $x = \frac{-b}{2a}$ to find the coordinates.

1. $y = x^2 + 4x + 9$ (−2, 5)
2. $y = -2x^2 - 8x - 8$ (−2, 0)
3. $y = 2x^2 + 12x + 13$ (−3, −5)
4. $y - 5 = 2x^2$ (0, 5)
5. $y = x^2 + 2x + 3$ (−1, 2)
6. $3x - 4 = x^2 + y$ $\left(\frac{3}{2}, -1\frac{3}{4}\right)$ (−1, −4)
7. $y = x^2 + 2x - 3$ (−1, −4)
8. $y = x^2 - \frac{1}{4}$ $\left(0, -\frac{1}{4}\right)$
9. $y = x^2 - 4x - 5$ (2, −9)

Find the vertex. Then graph each equation. Check students' graphs.

10. $y = 2x^2 + 12x + 18$ (−3, 0)
11. $y + 2 = x^2 - 4x$ (2, −6)
12. $y + 3x^2 = 1 - 6x$ (−1, 4)
13. $y = x^2 - 6x + 4$ (3, −5)
14. $y + 7 = x^2 + 5x$ $\left(-\frac{5}{2}, -13\frac{1}{4}\right)$
15. $y = x^2 - 8x + 22$ (4, 6)
16. $y = x^2 - 6x + 9$ (3, 0)
17. $y = x^2 + 6x + 7$ (−3, −2)
18. $y = x^2 - 4x - 1$ (2, −5)

19. Find a quadratic equation in the form $y = ax^2 + bx + c$ in which $c = -1$ and the vertex is (2, −5). $y = x^2 - 4x - 1$

Extra Practice 12–3 • Factor and Graph • pages 530–533

Use a graphing calculator to determine the number of solutions for each equation. For equations with one or two solutions, find the exact solutions by factoring.

1. $y = x^2 - 36$ $x = 6, x = -6$
2. $y = x^2 - x - 2$ $x = 2, x = -1$
3. $y = x^2 - x - 12$ $x = -3, x = 4$
4. $y = 2x^2 + x + 1$ no solution
5. $y + \frac{1}{4} = x^2$ $x = \frac{1}{2}, x = -\frac{1}{2}$
6. $y = x^2 - 6x + 9$ $x = 3$
7. $y + 10 = x^2 - 3x$ $x = 5, x = -2$
8. $x^2 - 4 = x + y$ no solution
9. $y = x^2 + 11x + 30$ $x = -5, x = -6$
10. $x^2 + 6x + 9 = y$ $x = -3$
11. $y = x^2 - 0.5x - 3$ $x = -1.5, x = 2$
12. $y = x^2 + 36$ no solution
13. $y = x^2 - 6x - 16$ $x = 8, x = -2$
14. $y = x^2 - x + 1$ no solution
15. $y = x^2 - 4x + 5$ no solution

Write an equation for each problem. Then factor to solve.

16. The square of a number is 6 more than the number. $y = x^2 - x - 6$; or -2
17. The square of a number is 4 more than 3 times the number. $y = x^2 - 3x - 4$; -1 or 4
18. The square of a number is 24 more than 2 times the number. $y = x^2 - 2x - 24$; 6 or -4
19. The square of a number is 27 more than 6 times the number. $y = x^2 - 6x - 27$; 9, -3

Extra Practice 12–4 • Complete the Square • pages 534–537

Complete the square.

1. $x^2 + 12x$ $x^2 + 12x + 36$
2. $x^2 + 16x$ $x^2 + 16x + 64$
3. $x^2 - 4.2x$ $x^2 - 4.2x + 4.41$
4. $x^2 + \frac{2}{3}x$ $x^2 + \frac{2}{3}x + \frac{1}{9}$
5. $x^2 - 7x$ $x^2 - 7x + 12.25$
6. $x^2 + 8x$ $x^2 + 8x + 16$
7. $x^2 + 6x$ $x^2 + 6x + 9$
8. $x^2 + x$ $x^2 + x + \frac{1}{4}$
9. $2x^2 - 5x$ $2x^2 - 5x + \frac{25}{4}$

Solve by completing the square. Check your solutions.

10. $x^2 + 12x + 11 = 0$ $x = -11, x = -1$
11. $x^2 - 2x - 3 = 0$ $x = 3, x = -1$
12. $x^2 - 8x - 9 = 0$ $x = -1, x = 9$
13. $2x^2 - 2x - 12 = 0$ $x = 3, x = -2$
14. $3x^2 + 11x - 4 = 0$ $x = -4, x = \frac{1}{3}$
15. $5x^2 = 5$ $x = 1, x = -1$
16. $x^2 + 24x + 119 = 0$ $x = -17, x = -7$
17. $x^2 - 22x + 112 = 0$ $x = 8, x = 14$
18. $x^2 - 4x - 117 = 0$ $x = -9, x = 13$
19. The width of a rectangular pool is 5 meters less than the length. The area is 24 square meters. Find the length and width. 8 m, 3 m
20. The hypotenuse of a right triangle is 25 meters. One leg is 17 meters shorter than the other. Find the lengths of the legs. 24 m, 7 m
21. The area of Harry's room is 132 square feet. The length is 1 foot more than the width. Find the length and width. 12 ft, 11 ft

Extra Practice 723

Extra Practice

Extra Practice 12–5 • The Quadratic Formula • pages 540–543

Use the quadratic formula to solve each equation.

1. $x^2 - 5x = 0$ 0, 5
2. $x^2 - 7x + 6 = 0$ 6, 1
3. $x^2 - 6x = 0$ 0, 6
4. $2x^2 + 13x + 15 = 0$ $-\frac{3}{2}, -5$
5. $x^2 + 6x + 4 = 0$ $-3 \pm \sqrt{5}$
6. $3x^2 + 8x + 5 = 0$ $-1, -\frac{5}{3}$
7. $x^2 - 2x - 15 = 0$ 5, −3
8. $x^2 - 7x - 30 = 0$ 10, −3
9. $5x^2 + 3x - 2 = 0$ $\frac{2}{5}, -1$
10. $x^2 + x = 0$ 0, −1
11. $4x^2 = 20$ $\pm\sqrt{5}$
12. $3 = 5x^2 - 8x$ $\frac{4 \pm \sqrt{31}}{5}$

Choose factoring or the quadratic formula to solve each equation.

13. $y = x^2 + 2x - 3$ $x = -3, x = 1$
14. $3x^2 - 8x = 3$ $x = 3, x = -\frac{1}{3}$
15. $x^2 = 2x = 24$ $x = 6, x = -4$
16. $y = 3x^2 + 2x - 4$ $x = \frac{-1 \pm \sqrt{13}}{3}$
17. $x^2 + 9 = 7x$ $x = \frac{7 \pm \sqrt{13}}{2}$
18. $x^2 = 2x + 15$ $x = 5, x = -3$
19. $x^2 + 6x - 16 = 0$ $x = 2, x = -8$
20. $y = 2x^2 - 7x - 4$ $x = -\frac{1}{2}, x = 4$
21. $y = x^2 + 3x - 1$ $x = \frac{-3 \pm \sqrt{13}}{2}$

Write and solve an equation for each problem.

22. Five times an integer is 4 more than the integer squared. $y = x^2 - 5x + 4$; (4, 1)

23. The square of an integer minus three times the integer equals −2. $y = x^2 - 3x + 2$; 1, 2

Extra Practice 12–6 • Use the Pythagorean Theorem • pages 544–547

Calculate the distance between each pair of points.

1. $A(9, 5), B(6, 1)$ 5 units
2. $X(0, -7), Y(3, -4)$ 4.2 units
3. $M(5, 6), N(5, -2)$ 8 units
4. $G(0, -3), H(0, 6)$ 9 units
5. $X(0, 0), Y(3, 4)$ 5 units
6. $A(1, 2), B(4, 7)$ 5.83 units
7. $K(2, 2), L(-2, -2)$ 5.7 units
8. $X(-2, 6), Y(4, 6)$ 6 units
9. $M(2, -2), N(3, 5)$ 7.1 units
10. $X(4, 3), Y(6, 7)$ 4.5 units
11. $M(9, 2), N(5, 7)$ 6.4 units
12. $A(9, 3), B(4, 1)$ 5.4 units

Calculate the midpoint between each pair of points.

13. $A(-4), 7), B(2, 3)$ (−1, 5)
14. $M(2, -5), N(-8, -9)$ (−3, −7)
15. $X(2, 2), Y(6, 6)$ (4, 4)
16. $D(4, 5), E(4, -5)$ (4, 0)
17. $A(3, 7), B(3, 11)$ (3, 9)
18. $K(2, -2), L(-2, 5)$ $\left(0, \frac{3}{2}\right)$
19. $X(4, 5), Y(6, -7)$ (5, −1)
20. $X(8, -4), Y(-3, 9)$ $\left(\frac{5}{2}, \frac{5}{2}\right)$
21. $A(3, 8), B(6, 14)$ $\left(\frac{9}{2}, 11\right)$

22. The vertices of a triangle are $A(1, 3)$, $B(8, 4)$, and $C9\ 5, 0)$. What are the lengths of the sides? side AB, 7.1 units; side AC, 5 units; side BC, 5 units

23. A triangle of ABC has vertex A at (3, 1) and vertex B at (1, 4). The length of side $\overline{BC}$ is 6 units, and the length of side $\overline{AC}$ is 5 units. Find the coordinates of vertex C. (7, 4)

24. Find the point on the x-axis that is the same distance from (1, 3) as from (8, 4). (5, 0)

25. Find the point on the y-axis that is the same distance from (2, 2) as from (2, 6). (0, 4)

724 Extra Practice

Extra Practice

Chapter 13

Extra Practice 13–1 • The Standard Equation of a Circle • pages 562–565

Write an equation for each circle. 1–6. See additional answers.

1. radius 5, center (0, 0)
2. radius 3, center (−4, 2)
3. radius 10, center (5, −1)

4. 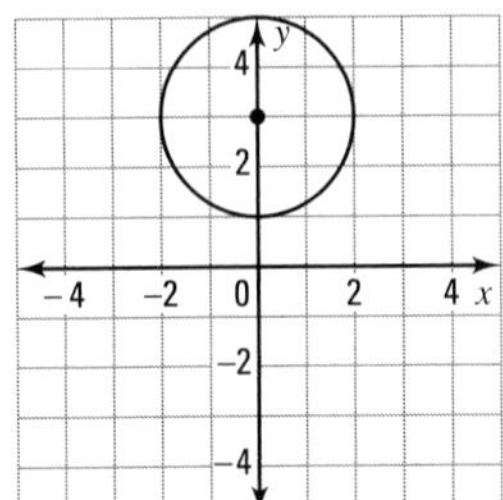

5. 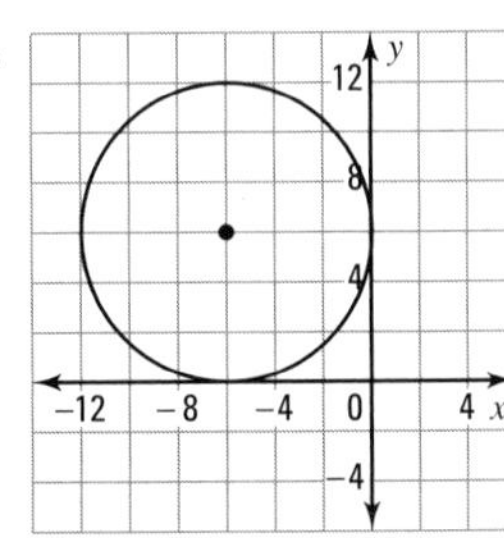

6. 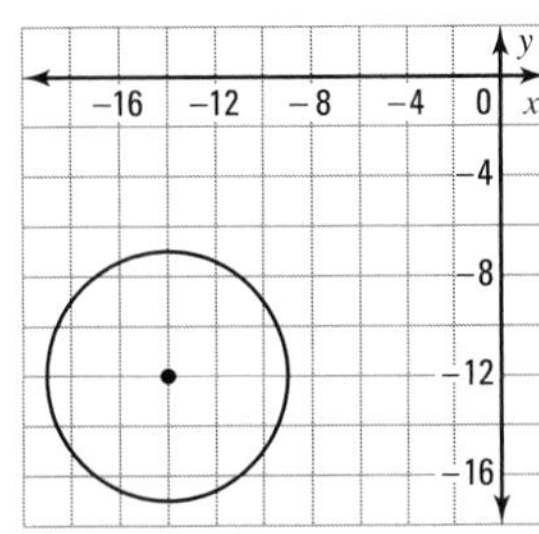

Find the radius and center of each circle.

7. $x^2 + y^2 = 64$ 8, (0, 0)
8. $x^2 + y^2 = 21$ $\sqrt{21}$, (0, 0)
9. $(x - 3)^2 + (y - 5)^2 = 49$ 7, (3, 5)
10. $(x - 4)^2 + (y + 2)^2 = 26$ $\sqrt{26}$, (4, −2)
11. $(x + 5)^2 + (y - 4)^2 = 41$ $\sqrt{41}$, (−5, 4)
12. $(x + 4)^2 + y^2 = 70$ $\sqrt{70}$, (−4, 0)
13. $x^2 + (y - 1)^2 = 6$ $\sqrt{6}$, (0, 1)
14. $(x + 5)^2 + (y + 5)^2$ 10, (−5, −5)
15. The graph of $x^2 + y^2 = 25$ is translated 4 units to the right and 3 units up. Write the equation of the circle in the new position. $(x - 4)^2 + (y - 3)^2 = 25$
16. The graph of $x^2 + y^2 = 25$ is translated 6 units to the left and 5 units up. Write the equation of the circle in the new position. $(x + 6)^2 + (y - 5)^2 = 25$

Extra Practice 13–2 • More on Parabolas • pages 566–569

Find the focus and directrix of each equation.

1. $x^2 = 8y$ (0, 2), $y = -2$
2. $x^2 = 24y$ (0, 6), $y = -6$
3. $x^2 = -8y$ (0, −2), $y = 2$
4. $x^2 = -18y$ $\left(0, -\frac{9}{2}\right)$, $y = \frac{9}{2}$
5. $x^2 - 15y = 0$ $\left(0, \frac{15}{4}\right)$, $y = -\frac{15}{4}$
6. $x^2 - 28y = 0$ (0, 7), $y = -7$
7. $2x^2 = -32y$ (0, −4), $y = 4$
8. $4x^2 = -20y$ $\left(0, -\frac{5}{4}\right)$, $y = \frac{5}{4}$
9. $8x^2 = 44y$ $\left(0, \frac{11}{8}\right)$, $y = -\frac{11}{8}$
10. $54y - 18x^2 = 0$ $\left(0, \frac{3}{4}\right)$, $y = -\frac{3}{4}$
11. $56y - 14x^2 = 0$ (0, 1), $y = -1$
12. $-7x^2 = -42y$ $\left(0, \frac{3}{2}\right)$, $y = -\frac{3}{2}$
13. $-8x^2 = 64y$ (0, −2), $y = 2$
14. $6x^2 - 48y = 0$ (0, 2), $y = -2$
15. $-9x^2 - 90y = 0$ $\left(0, -\frac{5}{2}\right)$, $y = \frac{5}{2}$

Find the simple equation for each parabola with vertex located at the origin.

16. Focus $\left(0, -\frac{9}{4}\right)$ $x^2 = -9y$
17. Focus $\left(0, \frac{15}{4}\right)$ $x^2 = 15y$
18. Focus $\left(0, \frac{2}{5}\right)$ $x^2 = \frac{8}{5}y$
19. Focus (0, 12.5) $x^2 = 50y$
20. Focus (0, 1.5) $x^2 = 6y$
21. Focus (0, −7.5) $x^2 = -30y$
22. An elevated highway is supported by a parabolic arch that can be described by the equation $x^2 = -40y$. Find the value of a. Find the value of x when $y = -10$. −10, 20

Additional Answers

Extra Practice 13-1

1. $x^2 + y^2 = 25$
2. $(x + 4)^2 + (y - 2)^2 = 9$
3. $(x - 5)^2 + (y + 1)^2 = 100$
4. $x^2 - (y - 3)^2 = 4$
5. $(x + 6)^2 + (y - 6)^2 = 36$
6. $(x + 15)^2 + (y + 12)^2 + 36$

Additional Answers

Extra Practice 13-4

1.

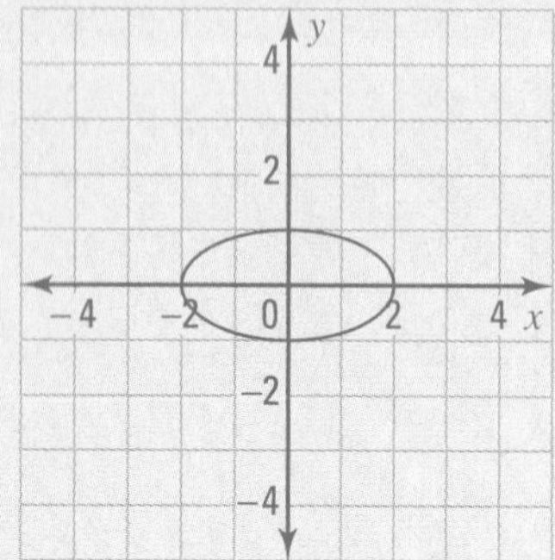

7.

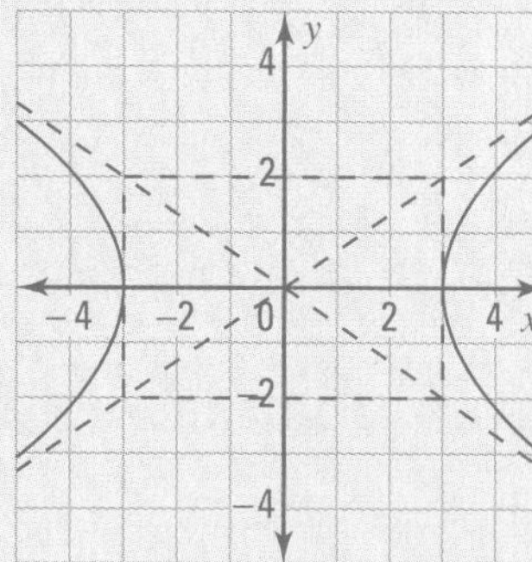

Extra Practice

Extra Practice 13–4 • Ellipses and Hyperbolas • pages 574–577

Graph each equation. Check students' graphs. See sample graphs in additional answers.

1. $x^2 + 4y^2 = 4$
2. $4x^2 + y^2 = 4$
3. $9x^2 + 4y^2 = 36$
4. $16x^2 + 9y^2 = 144$
5. $25x^2 + 4y^2 = 100$
6. $9x^2 + y^2 = 9$
7. $4x^2 - 9y^2 = 36$
8. $9x^2 - y^2 = 9$
9. $x^2 - 4y^2 = 16$
10. $25x^2 - 4y^2 = 100$
11. $x^2 - 4y^2 = 4$
12. $9x^2 + 16y^2 = 144$
13. $25x^2 + y^2 = 25$
14. $25x^2 - 9y^2 = 225$
15. $x^2 - 9y^2 = 9$
16. $x^2 + 25y^2 = 25$
17. $9x^2 - 16y^2 = 144$
18. $25x^2 + 9y^2 = 225$

19. Find the equation of the ellipse with foci (2, 0) and (−2, 0) and x-intercepts (4, 0) and (−4, 0). $3x^2 + 4y^2 = 48$
20. Find the equation of the ellipse with foci (6, 0) and (−6, 0) and x-intercepts (10, 0) and (−10, 0). $16x^2 + 25y^2 = 1{,}600$
21. Find the equation of the ellipse with foci (3, 0) and (−3, 0) and x-intercepts (5, 0) and (−5, 0). $16x^2 + 25y^2 = 400$
22. Find the equation of the hyperbola with center (0, 0) and foci on the x-axis if $a = \pm 6$ and $b = \pm 8$. $16x^2 - 9y^2 = 576$
23. Find the equation of the hyperbola with center (0, 0) and foci on the x-axis if $a = \pm 1$ and $b = \pm 4$. $16x^2 - y^2 = 16$
24. Find the equation of the hyperbola with center (0, 0) and foci on the x-axis if $a = \pm 4$ and $b = \pm 5$. $25x^2 - 16y^2 = 400$

Extra Practice 13–5 • Direct Variation • pages 580–583

Find the equation of direct variation for each pair of values.

1. $x = 54$ and $y = 18$ $y = \frac{1}{3}x$
2. $x = 4$ and $y = 12$ $y = 3x$
3. $x = 7$ and $y = 42$ $y = 6x$
4. $x = 15$ and $y = 10$ $y = \frac{2}{3}x$
5. $x = 3$ and $y = -12$ $y = -4x$
6. $x = 1.5$ and $y = 10.5$ $y = 7x$

In each of the following, y varies directly as x.

7. If $y = 7$ when $x = 3$, find y when $x = 15$. 35
8. If $y = 30$ when $x = 40$, find y when $x = 32$. 24
9. If $y = 76$ when $x = 4$, find y when $x = 5$. 95
10. If $y = 2.4$ when $x = 8$, find y when $x = 0.3$. 0.09

In each of the following, y varies directly as x^2.

11. If $y = 176$ when $x = 4$, find y when $x = 5$. 275
12. If $y = 468$ when $x = 3$, find y when $x = 12$. 7,488
13. If $y = 180$ when $x = 6$, find y when $x = 4$. 80
14. If $y = 112$ when $x = 0.4$, find y when $x = 3$. 6,300

The distance (d) a vehicle travels at a given speed is directly proportional to the time (t) it travels.

15. If a vehicle travels 24 miles in 40 minutes, how far can it travel in 2 hours? 72 miles
16. If a vehicle travels 40 miles in 50 minutes, how far can it travel in 20 minutes? 16 miles

726 | Extra Practice

17. In an electric circuit, the voltage varies directly as the current. If the voltage (v) is 90 volts when the current (c) is 15 amps, find the voltage when the current is 20 amps. 120 volts

18. The distance (d) an object falls is directly proportional to the square of the time (t) it falls. If an object falls 400 feet in 5 seconds, how far will it fall in 10 seconds? 1,600 feet

Extra Practice 13–6 • Inverse Variation • pages 584–587

For each pair of values, write an equation in which y varies inversely as x.

1. $x = 4$ and $y = 16$ $y = \frac{48}{x}$
2. $x = 6$ and $y = 60$ $y = \frac{360}{x}$
3. $x = 3.5$ and $y = 180$ $y = \frac{630}{x}$
4. $x = 0.6$ and $y = 32$ $y = \frac{19.2}{x}$
5. $x = 2.4$ and $y = 3.6$ $y = \frac{8.64}{x}$
6. $x = 3$ and $y = 1.8$ $y = \frac{5.4}{x}$

In each of the following, y varies inversely as x.

7. If $y = 13$ when $x = 2$, find y when $x = 13$. 2
8. If $y = 32$ when $x = 3$, find y when $x = 16$. 6
9. If $y = 4$ when $x = 2$, find y when $x = 6$. $\frac{4}{3}$
10. If $y = 1.4$ when $x = 5$, find y when $x = 3.5$. 2

In each of the following, y varies inversely as the square of x.

11. If $y = 9$ when $x = 2$, find y when $x = 6$. 1
12. If $y = 2$ when $x = 10$, find y when $x = 5$. 8
13. If $y = 128$ when $x = 1$, find y when $x = 4$. 8
14. If $y = 800$ when $x = 20$, find y when $x = 25$. 512

In each of the following, travel time varies inversely as travel speed.

15. If it takes 40 minutes to travel a certain distance as at a speed of 25 miles per hour, how long will it take to travel that distance at 40 miles per hour? 25 minutes
16. If it takes 20 minutes for an airplane to travel a certain distance at a speed of 120 miles per hour, how long will it take the plane to travel that distance at 100 miles per hour? 24 minutes

The brightness of a light bulb varies inversely as the square of the distance from the source.

17. If a light bulb has a brightness of 600 lumens at 2 feet, what will be its brightness at 8 feet? 37.5 lumens
18. If a light bulb has a brightness of 1,000 lumens at 3 feet, what will be its brightness at 15 feet? 40 lumens

Extra Practice 727

Additional Answers

Extra Practice 13-7

1.
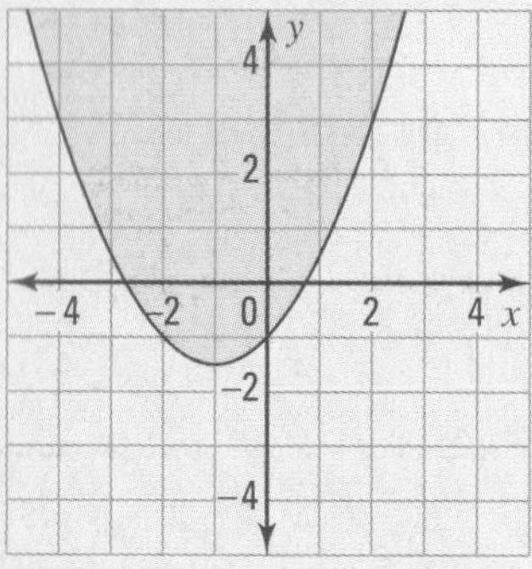

4.
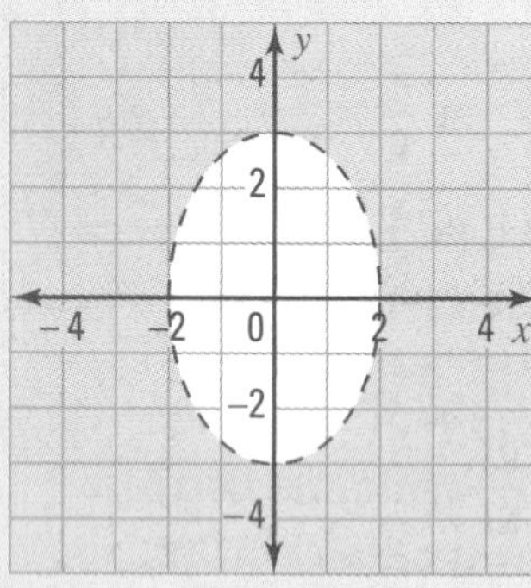

5.
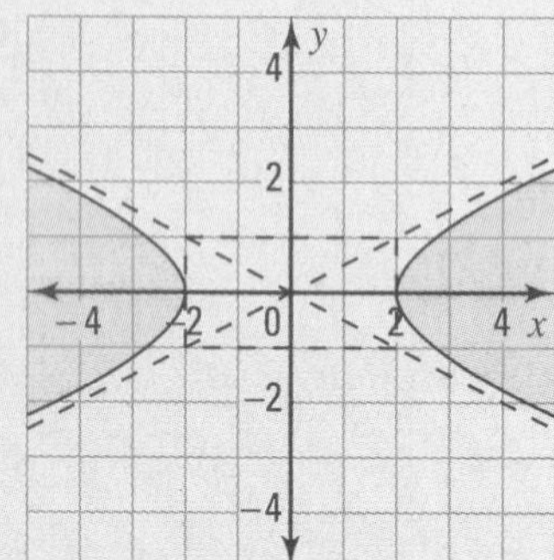

6.
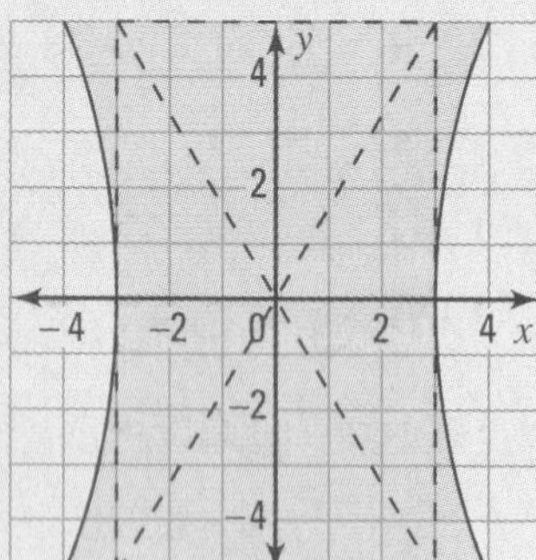

7.
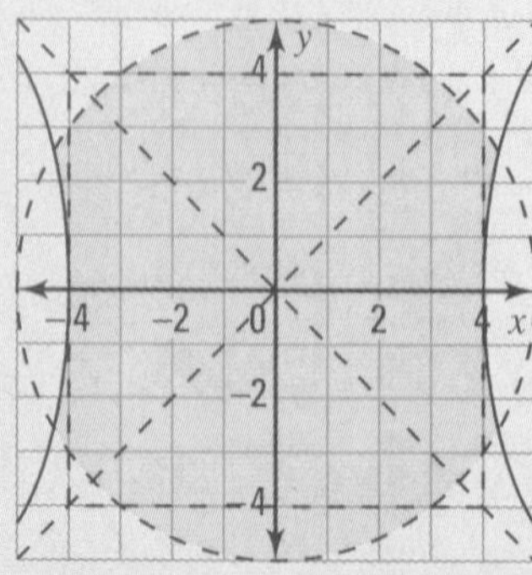

8.
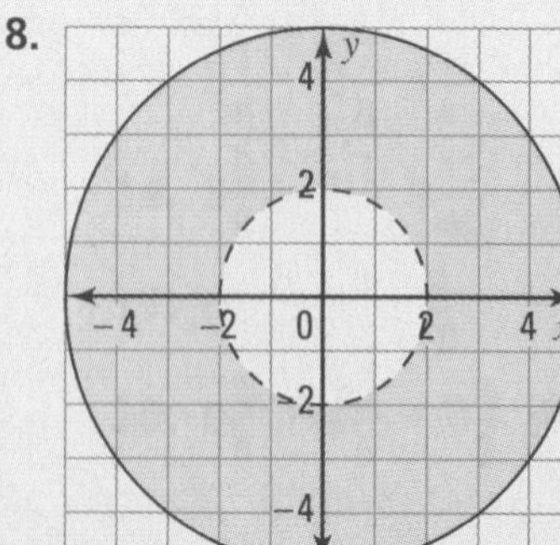

Extra Practice

Extra Practice 13–7 • Quadratic Inequalities • pages 590–593

Graph each inequality. 1–6. See sample graphs in additional answers.

1. $y \geq 0.5x^2 + x - 1$
2. $y < -x^2 + x + 3$
3. $(x - 3)^2 + (y - 2)^2 \leq 9$
4. $9x^2 + 4y^2 > 36$
5. $x^2 - 4y^2 \geq 4$
6. $25x^2 - 9y^2 \leq 225$

Graph each system of inequalities. 7–15. Check students' graphs. See sample graphs in additional answers.

7. $\begin{cases} x^2 - y^2 < 16 \\ x^2 + y^2 < 25 \end{cases}$
8. $\begin{cases} x^2 + y^2 > 4 \\ x^2 + y^2 \leq 25 \end{cases}$
9. $\begin{cases} y \geq -0.5x^2 + 2 \\ y \leq x + 3 \end{cases}$
10. $\begin{cases} y \leq -x^2 + x + 1 \\ y \geq 0.5x^2 - 3 \end{cases}$
11. $\begin{cases} x^2 + (y - 2)^2 > 16 \\ y < -x^2 - 2x + 1 \end{cases}$
12. $\begin{cases} 9x^2 + 16y^2 < 144 \\ x^2 - 9y^2 \geq 9 \end{cases}$
13. $\begin{cases} 25x^2 + 4y^2 < 100 \\ y \geq x^2 - 3 \end{cases}$
14. $\begin{cases} y \geq x^2 - 5 \\ x^2 - 4y^2 \leq 4 \end{cases}$
15. $\begin{cases} (x + 1)^2 + (y - 1)^2 \leq 16 \\ 16x^2 + 9y^2 > 144 \end{cases}$

Describe the part of the coordinate plane that is shaded.

16. $x^2 + y^2 > r^2$ the region outside the circle
17. $\frac{x^2}{a^2} + \frac{y^2}{b^2} \leq 1$ an ellipse and region inside
18. $ax^2 + bx + c < y (a > 0)$ region inside the parabola

Chapter 14

Extra Practice 14–1 • Basic Trigonometric Ratios • pages 604–607

Use the figures at the right to find each ratio.

1. $\sin A$ $\frac{1}{\sqrt{2}}$
2. $\cos A$ $\frac{1}{\sqrt{2}}$
3. $\tan A$ 1
4. $\sin B$ $\frac{1}{\sqrt{2}}$
5. $\cos B$ $\frac{1}{\sqrt{2}}$
6. $\tan B$ 1

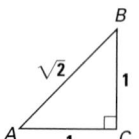

7. $\sin B$ $\frac{8}{17}$
8. $\tan B$ $\frac{8}{15}$
9. $\cos B$ $\frac{15}{17}$
10. $\sin A$ $\frac{15}{17}$
11. $\cos A$ $\frac{8}{17}$
12. $\tan A$ $\frac{15}{8}$

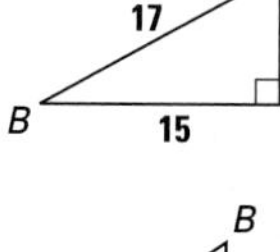

13. $\sin A$ $\frac{2}{\sqrt{13}}$
14. $\cos A$ $\frac{3}{\sqrt{13}}$
15. $\tan A$ $\frac{2}{3}$
16. $\sin B$ $\frac{3}{\sqrt{13}}$
17. $\cos B$ $\frac{2}{\sqrt{13}}$
18. $\tan B$ $\frac{3}{2}$

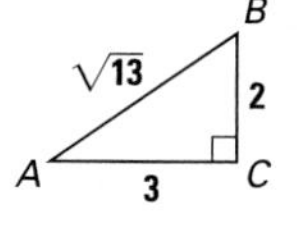

19. A tree is 75 ft tall. You stand x ft away from the tree, and the tangent of the angle to the top is about 1.5. How far are you from the tree? 50 ft

20. To find the height of a building 36 ft tall, you stand 30 ft away and measure the angle to the top. The angle is 52°. What is the tangent to the nearest tenth? 1.2

21. A ladder is leaning against a building. The foot of the ladder is 20 ft away from the building, and the top is 15 ft from the ground. How long is the ladder? 25 ft

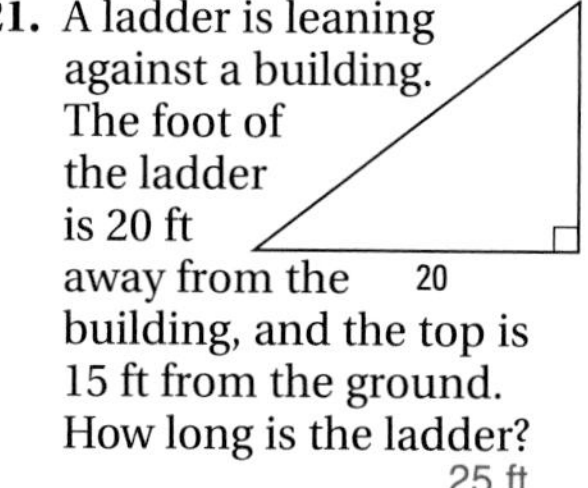

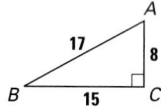

Extra Practice

Extra Practice 14–2 • Solve Right Triangles • pages 608–611

Find line segments to the nearest tenth and angles to the nearest degree.

Find the following in $\triangle LMN$.

1. $\overline{LM}$ 9.0
2. $m\angle N$ 37°

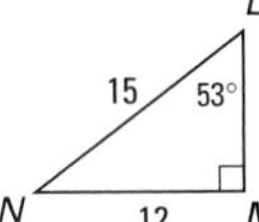

Find the following in $\triangle XYZ$.

6. $\overline{XY}$ 9.4
7. $m\angle X$ 32°
8. $m\angle Y$ 58°

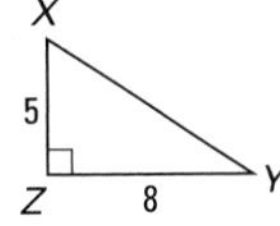

Find the following in $\triangle ABC$.

3. $m\angle B$ 42°
4. AC 9.0
5. $m\angle A$ 48°

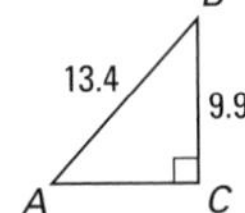

Find the following in $\triangle JKL$.

9. KL 5.3
10. JK 11.3
11. $m\angle K$ 62°

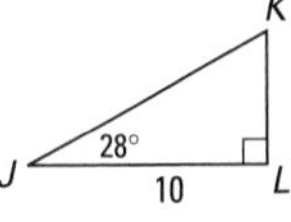

12. Find the angle of elevation of the top of the 1,046-ft John Hancock Building in Chicago from a point 700 ft from the base. 56°

Extra Practice 14–3 • Graph the Sine Function • pages 614–617

Find each ratio by drawing a reference angle.

1. sin 300° $-\frac{\sqrt{3}}{2}$
2. sin 210° $-\frac{1}{2}$
3. sin 135° $\frac{\sqrt{2}}{2}$
4. sin 450° 1
5. sin 585° $\frac{\sqrt{2}}{2}$
6. sin 660° $-\frac{\sqrt{3}}{2}$
7. sin 360° 0
8. sin 495° $\frac{\sqrt{2}}{2}$
9. sin 420° $\frac{\sqrt{3}}{2}$
10. sin 690° $-\frac{1}{2}$
11. sin 330° $-\frac{1}{2}$
12. sin 855° $\frac{\sqrt{2}}{2}$

Extra Practice 14–4 • Experiment with the Sine Function • pages 618–621 See additional answers.

1. Graph $y = \sin 6x$. State the period. The period is 60°.
2. Graph $y = 3 \sin x$. State the amplitude. The amplitude is 3
3. Graph $y = \sin x + 2$. Describe the position of the graph.
4. Graph $y = \sin x - 2$. Describe the position of the graph.

State the period and amplitude of the graph of each equation and describe the position of the graph. For 5–10, see additional answers.

5. $y = 3 \sin x + 4$
6. $y = 1.5 \sin 2x$
7. $y = 8 \sin 2x - 3$
8. $y = 4.5 \sin 2x + 2.5$
9. $y = 1.5 \sin x + 2.5$
10. $y = 9 \sin 2x$

ADDITIONAL ANSWERS

Extra Practice 14-4

1.

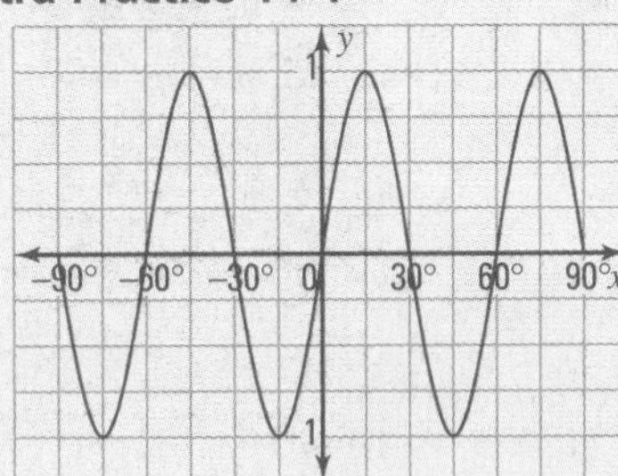

2.

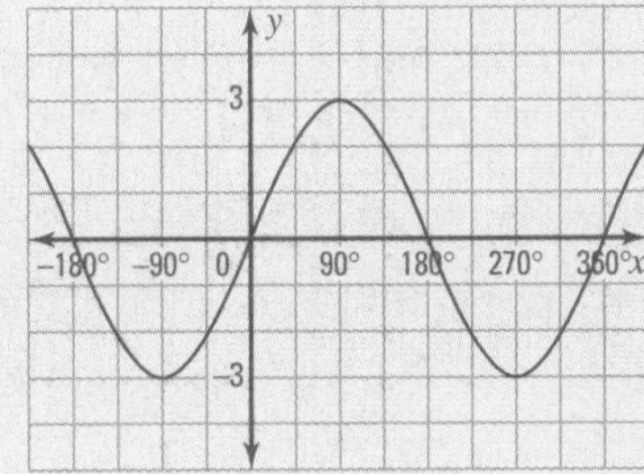

3.

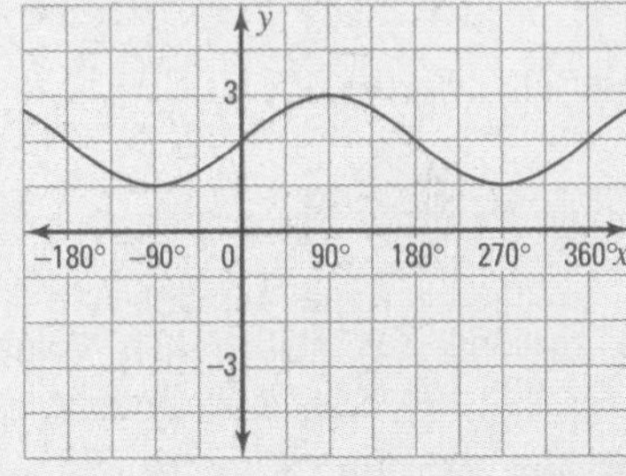

4.

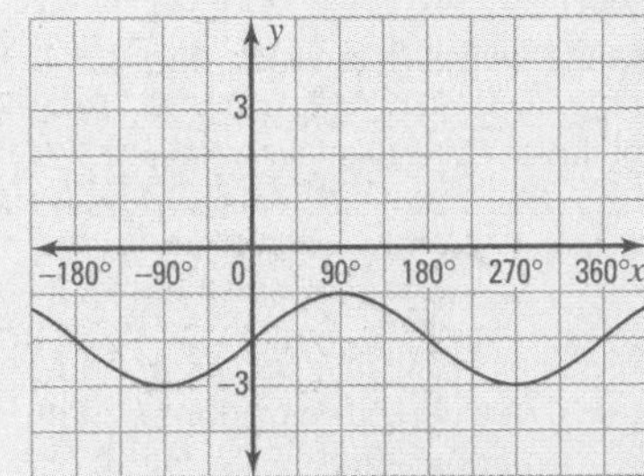

5. The period is 360°, the amplitude is 3. The graph is the graph of 3 sin *x* raised 4 units.
6. The period is 180°, the amplitude is 1.5. The graph is in normal position.
7. The period is 180°, the amplitude is 8. The graph is the same as the graph of 8 sin 2*x* lowered 3 units.
8. The period is 180°, the amplitude is 4.5. The graph is the same as the graph 4.5 sin 2*x* raised 2.5 units.
9. The period is 360°, the amplitude is 1.5. The graph is the same as the graph 1.5 sin *x* raised 2.5 units.
10. The period is 180°, the amplitude is 9. The graph is in normal position.

SELECTED ANSWERS

CHAPTER 1

Lesson 1-1, pages 6–9

1. $6 \in \{1, 3, 5, 7, 9\}$ 3. -1 5. 11 7. No. There are 8. The subsets include $\varnothing$, the members alone and any possible combinations of members. 9. $\varnothing$ 11. True 13. $\{a\}\ \varnothing$ 15. $\{t, e\}, \{t, n\}, \{e, n\}, \{t, e, n\}, \{t\}, \{e\}, \{n\}$ 17. 4 19. 1 21. -6 23. -3 25. False 27. $\{\ \} \subset \{1, 2\}$ or $\varnothing \subset \{1, 2\}$ 29. $x = \{\ \}$ 31. $\frac{1}{12}$ 33. 6 35. 8 37. 32 39. 6, -6 41. $P = \{\ \}$ or $\varnothing$ 43. $8 \in R$

Lesson 1-2, pages 10–13

1. False 3. False 5.

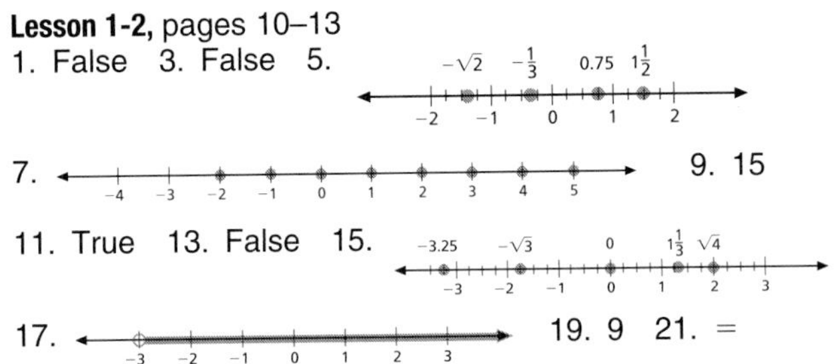

7. 9. 15 11. True 13. False 15. 17. 19. 9 21. $=$ 23. $>$ 25. $\{x|x$ is a real number, and $x < 4\}$ 27. $32 < 212$ 29. 284 million $>$ 126 million 31. Volga $<$ Mackenzie or $2{,}914 < 2{,}635$. 33. 35. Answers will vary. 37. -10 39. -2 41. 3 43. 0

Review and Practice Your Skills, pages 14–15

1. $\varnothing, \{10\}, \{15\}, \{10, 15\}$ 3. -5 5. 8, -8 7. 6, -6 9. 33 nickels 11. 8 13. -5.5 15. $\varnothing \subset \{b, u, g\}$ 17. true 19. false 21. true 23. false 25. 8 27. 8 29. 31.

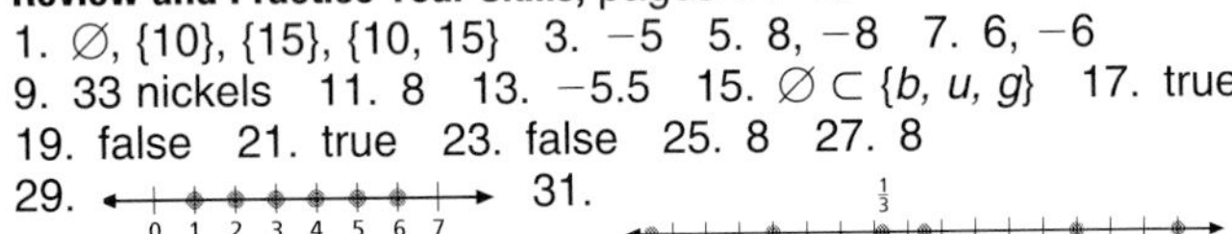

33. $<$ 35. $<$ 37. $>$ 39. 6, m and i 41. false 43. false 45. true 47. true 49. irrational, real 51. integer, rational, real 53. irrational, real

Lesson 1-3, pages 16–19

1. $\{1, 2, 3, 4, a, b, c\}$ 3. $\{1, 2\}$ 5.

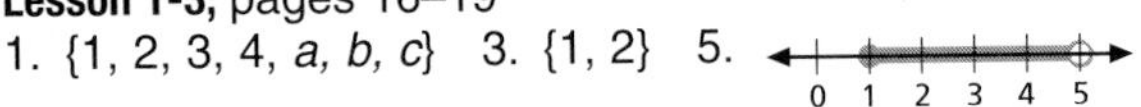

7. {ban, can, fan, man, pan, ran, tan} 9. {ban, can, man, ran, van, wan} 11. $\{c\}$ 13. $\{c\}$ 15. $\{c, h\}$ 17. $\{\ \}$ or $\varnothing$ 19. $\{x|x \le 1$ or $x \ge 2\}$ 21. $\{1, 2, 4, 5, 7, 8\}$ 23. $\{0, 1, 2, 3, 4, 5, 6, 9\}$ 25. $\{1, 9\}$ 27. $\{0, 1, 3, 5, 7, 9\}$ 29. $\{3, 5\}$ 31. $\{1\}$ 33. 3 35. True 37. False 39. True 41. 43. 45.

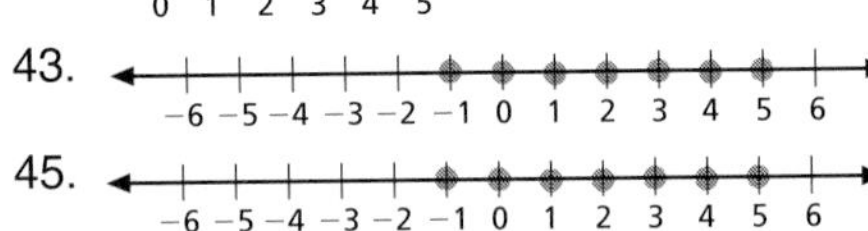

Lesson 1-4, pages 20–23

1. -27 3. $11\frac{1}{4}$ 5. -3.6 7. 3.6 9. 235 11. Yes, round the amounts and add: $5 + $8 + $2 + $9 = $24 13. -3.9 15. -24.4 17. -8 19. 43 21. -11 23. $>$ 25. $<$ 27. 5.7 ft 29. 39 31. $\{1, 2, 5, 7, 8, 9\}$ 33. $\{0, 3, 4, 5, 6, 7\}$ 35. $\{1, 8\}$ 37. 3 39. -9 41. -3 43. -3 45. 20 47. 38 49. 38 51. $\frac{13}{2}$ or 6.5 53. 12 55. 0 57. -24

Review and Practice Your Skills, pages 24–25

1. $\{p, u, t, e, r\}$ 3. $\{e, r\}$ 5. $\varnothing$ 7. $\{p, u, t, e, r\}$ 9. $\{c, o, m, p, u, t, e, r\}$ 11. 13. 15. 17. 19. 15 21. 200 23. 5 25. -67 27. 455 29. -115 31. 21.3 ft 33. No 35. false 37. false 39. false 41. true 43. -15 45. $\{c, l, e, a, n, y, r\}$ 47. $\{g, j, p, v, z\}$ 49. $\varnothing$ 51. 9.605 53. 48 55. 7.4

Lesson 1-5, pages 26–29

1. -4.93 3. -1 5. -32 7. 5.1 9. -1.2 11. 2 13. $8\frac{1}{8}$ 15. $-4°$F 17. -21.84 19. 40 21. -500 23. 0.4 25. 4.334 27. -6 29. -10.56 31. -6.1 33. -0.2 35. 8.1 37. no; 22 39. 193.75 calories 41. $+\ \div$ 43. $\div\ \div\ \div$ 45. $\frac{1}{1000} \times 16 = \frac{16}{1000} = \frac{4}{250} = \frac{2}{125}$ 47. -21 49. -4 51. -24 53. $-\frac{3}{2}$ 55. -18 57. 0 59. $\{0, 2, 3, 5, 6, 8\}$ 61. $\{1, 2, 3, 4, 5, 6, 7, 9\}$ 63. $\{1, 3, 5, 7, 9\}$ 65. $\{0, 1, 7, 9\}$

Lesson 1-6, pages 30–31

1. 2; 4; 7; 14 3. 8; 16; 19; 38 5. Store the value 4 for L. Enter these formulas using the variable L for length. Perimeter 2L + 2L * 0.5, Area: L * L * 0.5. You may have written the formulas differently. The solutions should be perimeter 12 ft and area 8 sq ft. 7. 133.65 in.^{2} 9. 1.65 yd^2 11. less than 13. 9 sq ft 15. -9.89 17. -30.91 19. 22.60 21. -2.98

Review and Practice Your Skills, pages 32–33

1. -21.84 3. $\frac{1}{10}$ 5. -1 7. -8.64 9. $\frac{9}{64}$ 11. $-37\frac{1}{2}$ 13. $-3\frac{5}{9}$ 15. $-2\frac{1}{3}$ 17. $\frac{5}{8}$ 19. Letting H = the number of hours worked, Ben's pay is equal to 8.82H + (H − 40)(1.5)(8.82). 21. $317.52 23. $273.42 25. 1, 6, -11 27. 2, 7, -12 29. $\varnothing, \left\{\frac{1}{2}\right\}, \left\{\frac{3}{5}\right\}, \left\{\frac{1}{2}, \frac{3}{5}\right\}$ 31. $\varnothing, \{0\}$ 33. $\{4, 8, 12, 16, 20, \ldots\}$ 35. $\{x|x$ is a natural number or a negative non-multiple of 4$\}$ 37. $\{0\}$ 39. Yes 41. 938.91 43. 65 45. $\frac{1}{3}$

Lesson 1-7, pages 34–37

1. 160 3. 284 5. -64 7. -576 9. c^4 11. c^{15} 13. a^8b^{12} 15. x^{11} 17. 4,346 19. 9 21. 64 23. 96 25. -8 27. 2^{25} 29. r^9 31. $\frac{1}{d^9}$ 33. a^{25} 35. 35 37. 28 39. c^{18} 41. $1{,}000^4$ and $10{,}000^3$ both equal 10^{12} 43. 27 in.^{3}, 125 cm^3, e^3 cm^3, g^3 in.^{3} 45. True 47. 2,000 49. $259.45 51. $292.60

Lesson 1-8, pages 38–41

1. $\frac{1}{16}$ 3. r^{-4} or $\frac{1}{r^4}$ 5. m^{-21} or $\frac{1}{m^{21}}$ 7. d^{17} 9. $\frac{1}{256}$ 11. $-\frac{4}{27}$ 13. 5.9×10^{-5} 15. 360,000 17. 0.000000209 19. 0 21. $\frac{1}{y^9}$ 23. $\frac{1}{8}$ 25. $\frac{1}{4}$ 27. $3.56 \cdot 10^{-3}$ 29. $2.7 \cdot 10^6$ 2,700,000 31. $7.5 \cdot 10^5$ 33. $1.5 \cdot 10^4$. If the powers are not equal, the number with the greatest power is equal. If the powers are equal, compare the mixed decimal portions of the numbers. 35. m^{40} 37. 2.45×10^0

39. 5,000,000 41. 9, a^n 43. $\frac{b^3}{a^2}$ 45. $a^{12}b^{22}$ 47. 16 49. -75 51. 135 53. 9261 55. 27 57. 16 59. m^{12} 61. $\frac{1}{m^7}$ 63. m^8 65. m^{23} 67. m^{52} 69. m^{32}

Chapter 1 Review, pages 42–43
1. c 3. a 5. 1 7.

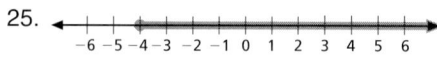

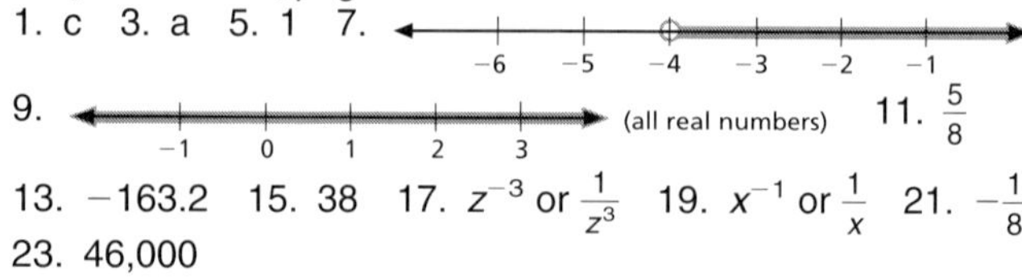

9. (all real numbers) 11. $\frac{5}{8}$

13. -163.2 15. 38 17. z^{-3} or $\frac{1}{z^3}$ 19. x^{-1} or $\frac{1}{x}$ 21. $-\frac{1}{8}$ 23. 46,000

Chapter 2

Lesson 2-1, pages 52–55
1. Add 4; 1, 5, 9 3. Subtract 3; -7, -10, -13 5. Multiply by 3; 162, 486, 1,458 7. 17, 17, 23 9. 21, 31, 43 11. 125, 216, 343 13. Add 5.2 to continue the pattern. 15. 300; 450; 1,012.50; 1,518.75
17.

19. \$1500 21. \$637.50, \$785.31, \$944.21, \$1,115.03, \$1,298.66
23.

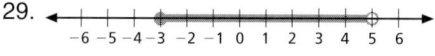

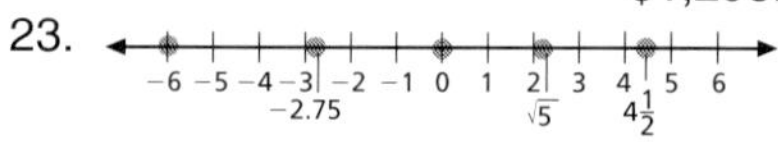

25.

27.

29.

31.

33.

Lesson 2-2, pages 56–59
1.–3.

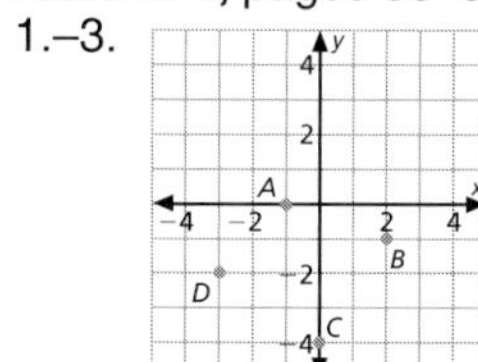

5. -2 7. -7 9. no; domain: $\{-3, -2, 0, 1\}$; range: $\{-3, -1, 0, 1\}$
11.–13.

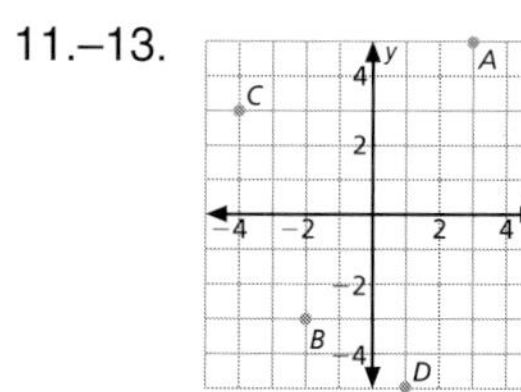

15. -17 17. 7 19. $\{(-2, -4), (0, -2), (2, 0), (4, 2)\}$; domain: $\{-2, 0, 2, 4\}$; range: $\{-4, -2, 0, 2\}$ 21. 49 23. 17 25. 48 27. $1 + b$ 29. c^3 31. No 33. Yes 35. \$230 37. {0, 8, 14, 20} 39. {0, 3, 5, 6, 14, 18, 20} 41. {0, 3, 6, 14, 20} 43. {0, 3, 6, 8, 10, 20}

Review and Practice Your Skills, pages, pages 60–61
1. 80, -160, 320 3. 23, 33, 45 5. -7, -10, -13 7. $\frac{16}{7}$, $\frac{32}{7}$, $\frac{64}{7}$ 9. 0.125, 0.0625, 0.03125
11. ; 5, 10, 15

13. ; 4, 1, $\frac{1}{4}$

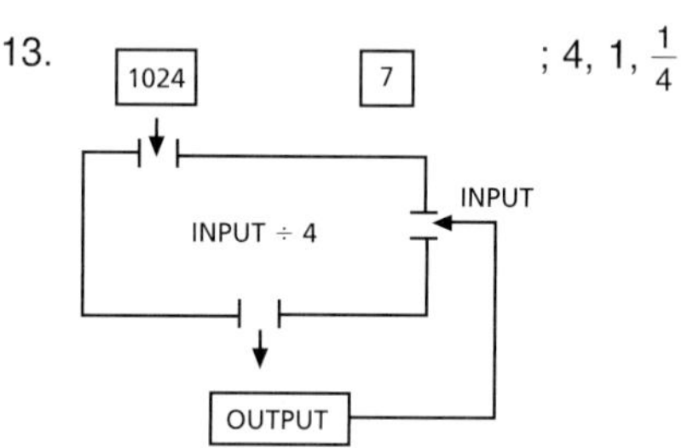

15. ; 15, -15, 15

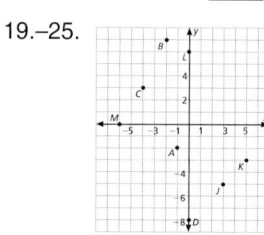

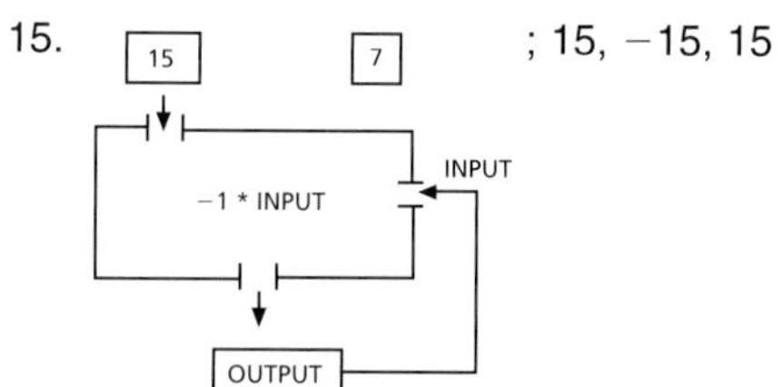

17. ; \$11,500; \$13,150; \$14,965; \$16,961.50; \$19,157.65

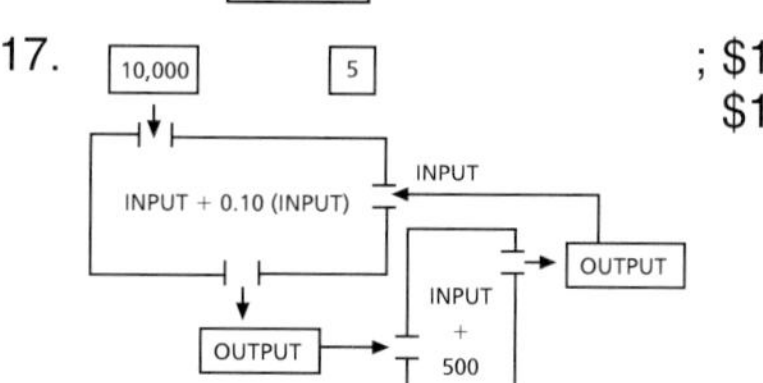

19.–25.

27. 25 29. -24 31. 0 33. 156 35. D: {0, 3, 4}, R: {0, 6, 12, 18}, Not a function 37. (0, 3), $(-1, 4)$, (1, 4), $(-2, 7)$, (2, 7); D: $\{-2, -1, 0, 1, 2\}$, R: {3, 4, 7} 39. -136, 17, 170 41. *ZYXWV*, *ZYXWVU*, *ZYXWVUT* 43. 10^5, -10^6, 10^7 45. 12 47. $1\frac{1}{2}$ 49. II

Lesson 2-3, pages 62–65
1. $(3 - x = 3$ 3. 1 5. \$60,

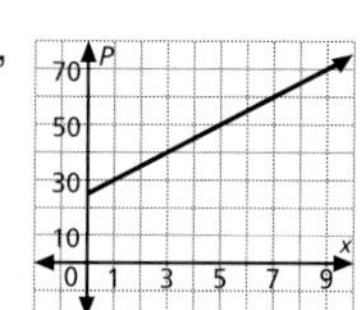

Selected Answers

7. 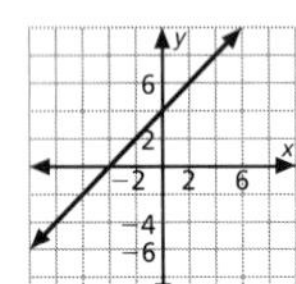
9. 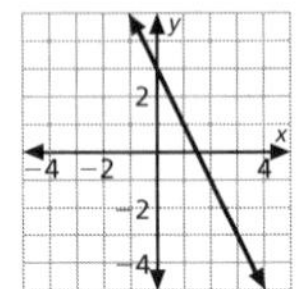
11. 1

13. 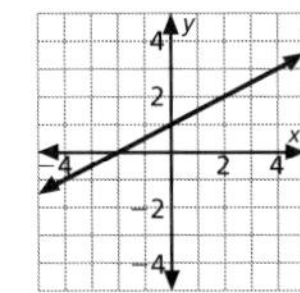
15. 7
17. 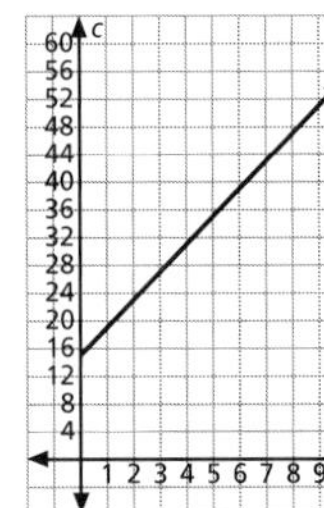

19. 8
21. 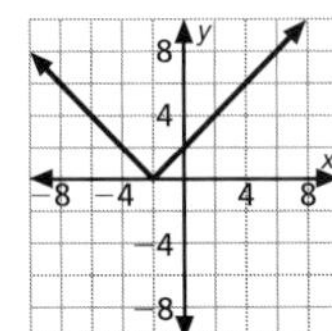
23. 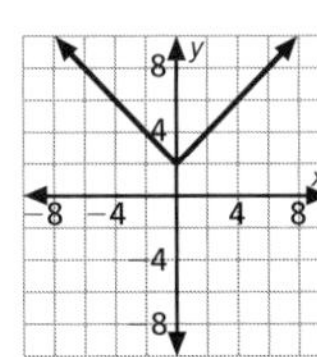

25. 216 27. 12 29. 23 31. 13,000 33. 16,000
35. \$170.00 37. \$170.00

Lesson 2-4, pages 66–69

1. 14 3. -9 5. -56 7. $\frac{1}{4}$ 9. -2.5
11. $n + 15 = \left(\frac{1}{2}\right)(72)$ 13. $n - 26 = -9$ 15. 22 17. -25
19. -0.4 21. -9 23. 9.34 25. -40 27. 5.6 29. -2
31. 25.5 33. -10 35. 6, -6 37. 0 39. 11 41. 40
43. Answers will vary; $|x| = -3$ 45. $0.6p = \$101.25$; \$168.75 47. -8 49. -23 51. 90 53. -13 55. 12

Review and Practice Your Skills, pages 70–71

1. 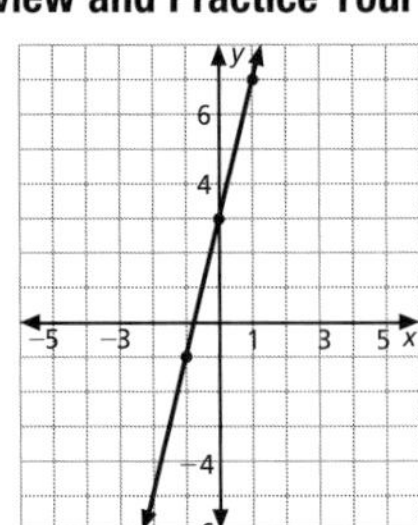
3. 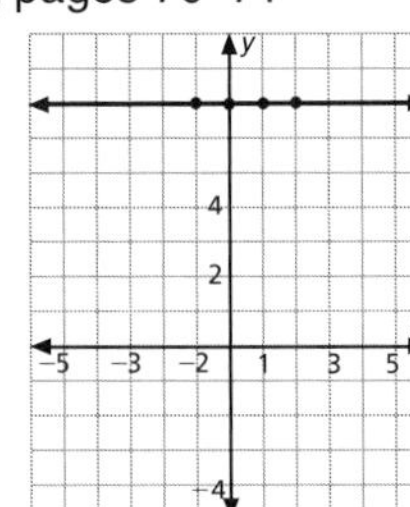
5. 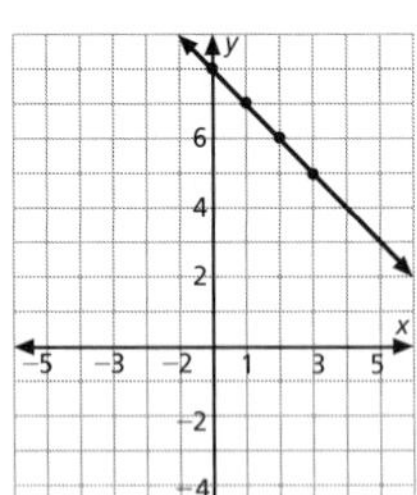
7. 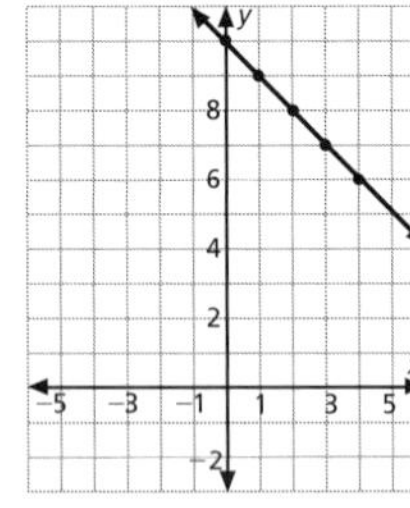

9. 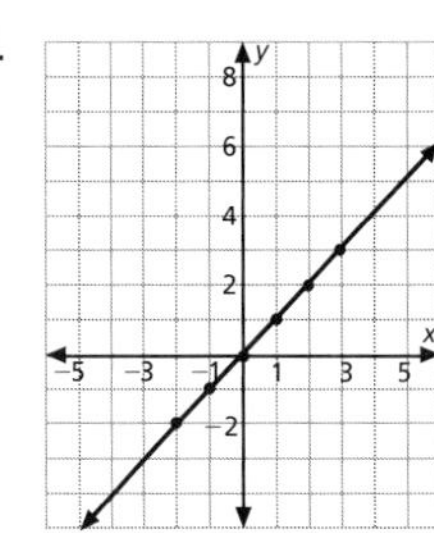

11. 13 13. -10
15. 7 17. \$177
19. 3 21. 3
23. -6 25. $1\frac{1}{4}$
27. $\frac{1}{4}$ 29. 26 31. -102.3
33. -81 35. -70 37. $112\frac{2}{3}$ 39. -36 41. 2 43. $-\frac{1}{4}$
45. -36 47. $-\frac{4}{3}$ 49. -2.2 51. -16 53. 3, -8, -19
55. 70, 44, 18
57. 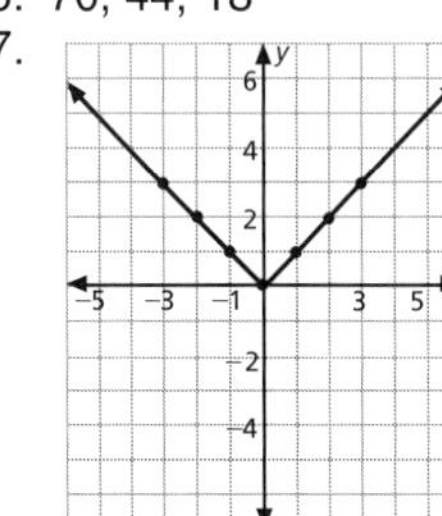

59. $n + 13 = -29$, $n = -42$ 61. $\frac{n}{-4} = 11$, $n = -44$
63. $2\frac{3}{8}$ 65. -4

Lesson 2-5, pages 72–75

1. 6 3. -6 5. -46 7. -11 9. Yes, for example, $\frac{1}{3}x$ is the same as $\frac{x}{3}$. 11. 3 13. -7 15. 4 17. -2 19. 2
21. -1 23. 9 25. 3 27. -1 29. -3.5 31. 11
33. \$29.75 35. Answers may vary. 37. Since $1 \neq 3$, no real numbers satisfy this equation. 39. 72 41. 13 43. 216
45. -27 47. 96 49. 100 51. 4.5×10^{-8}
53. 3.9×10^{-11} 55. 2.6×10^{-9}

Lesson 2-6, pages 76–79

1.  ; $m > 5$
3. ; $c > -16$
5. 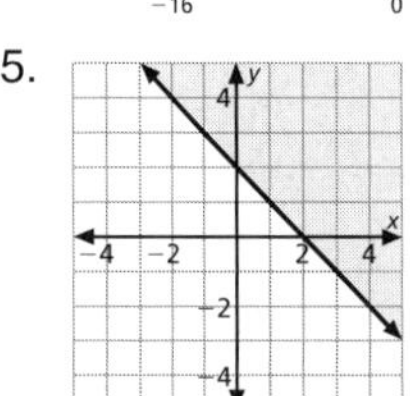
7. 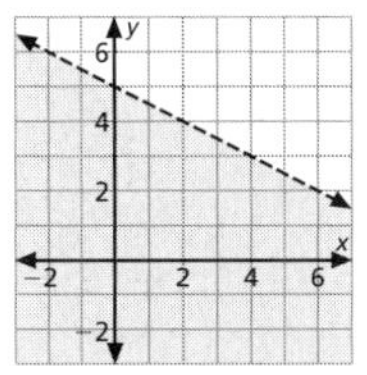
9. 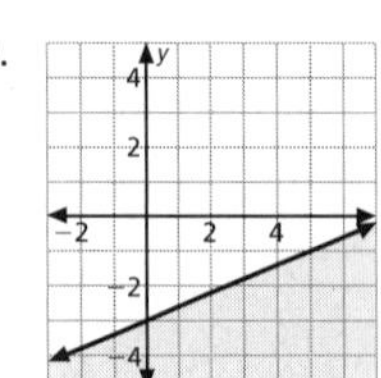

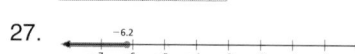

11. 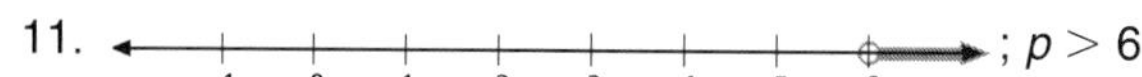 ; $p > 6$

13. ; $a > 7$

15. ; $z > 3$

17. ; $e < -\frac{3}{2}$

19. 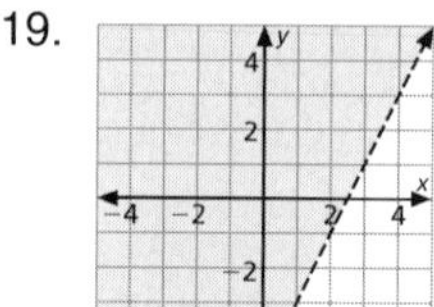

21. 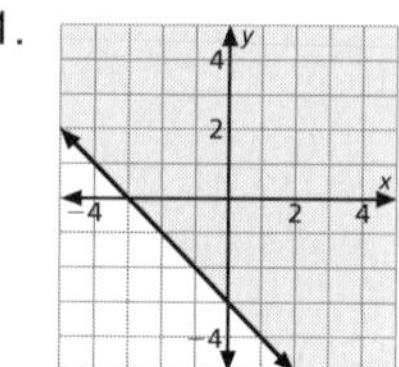

23. 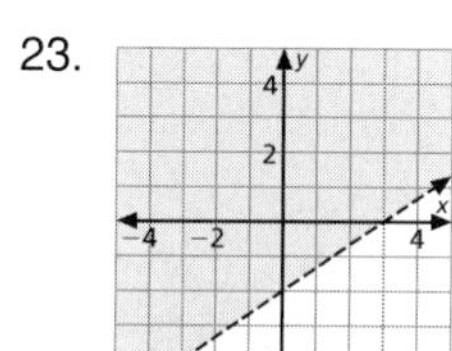

25. $h < 18$ h

27.

29.

31. $y \leq x$

33. $x + y \geq 4$ 35. Students should discuss and compare the solutions as they appear on both a number line and a coordinate plane. 37. rule: $\times$ 4; 1024; 4096; 16,384 39. rule: + 3; 16, 19, 22 41. rule: $\times$ −3; −243, 729, −2187

Review and Practice Your Skills, pages 80–81

1. 9 3. −3 5. 0 7. 20 9. 3.6 11. 84 13. −16 15. 7 17. 45 19. $-2\frac{4}{7}$ 21. 4 23. 12.5 25. $6 + 5n = -29$, $n = -7$ 27. $\frac{1}{3}n + \frac{1}{2} = 3\frac{1}{2}$, $n = 9$ 29. $3(n + 2) = -27$, $n = -11$ 31. ; $g < -12$

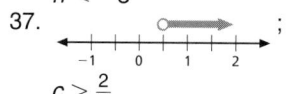

33. 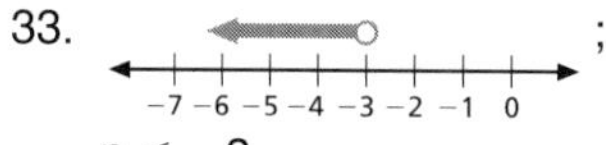 ; $n < -3$

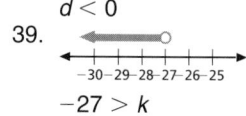

35. ; $d < 0$

37. ; $c > \frac{2}{3}$

39. ; $-27 > k$

41. ; $-8.5 > h$

43. 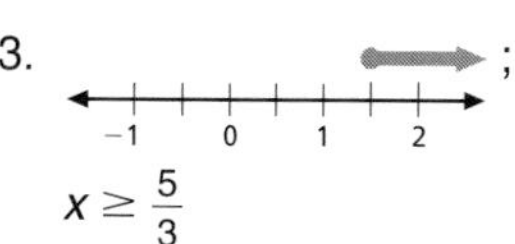 ; $x \geq \frac{5}{3}$

45. ; $m > 4$

47. 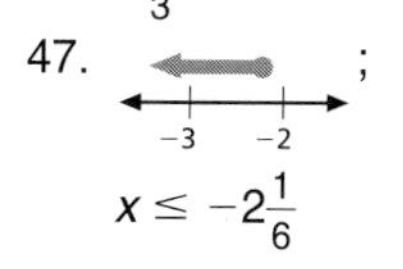 ; $x \leq -2\frac{1}{6}$

49. 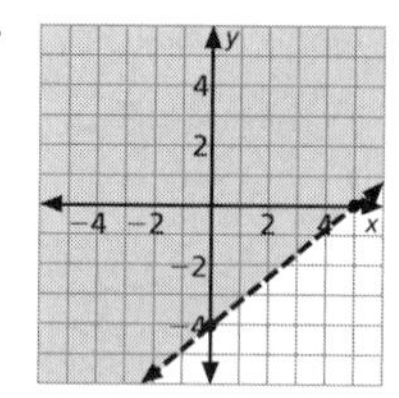

51.  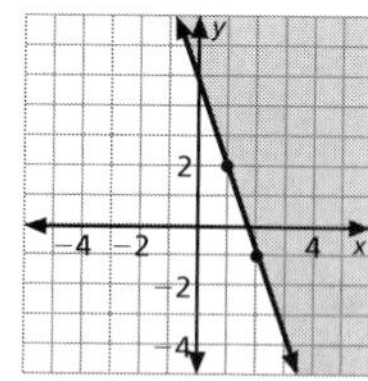

53.  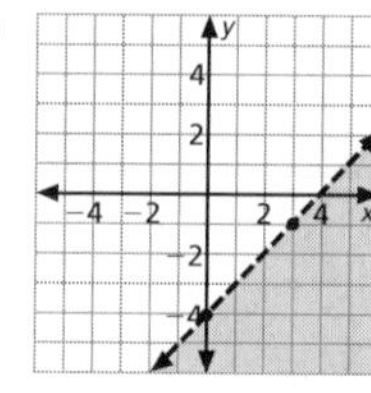

55.  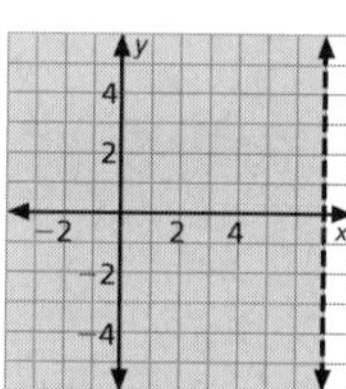

57. 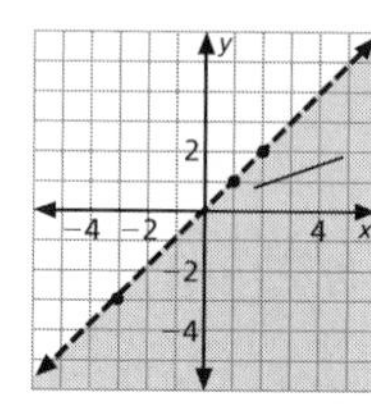

59. 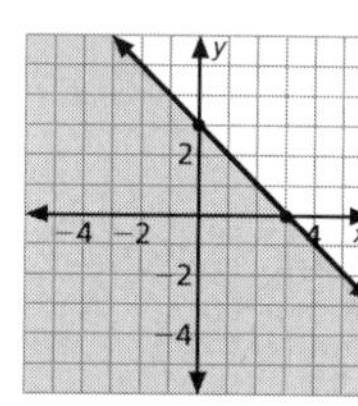

61. 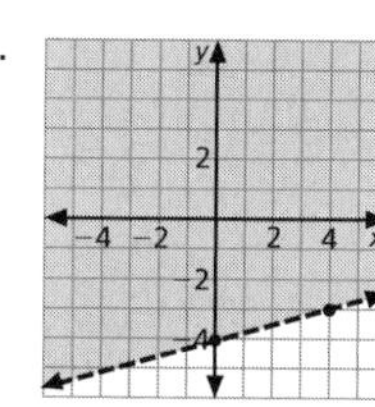

63. 2.4, 2.7, 3.0
65. −8.4, −10.8, −13.2
67. 1.21212, 1.1212121, 1.2121212

69. 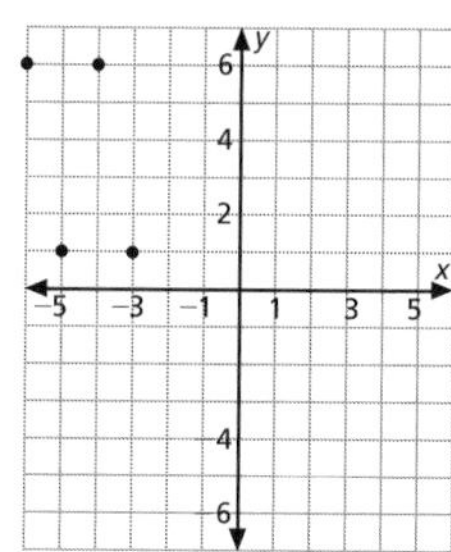 ; (−3, 1), (−4, 6), (−5, 1), (−6, 6); D: {−6, −5, −4, −3}; R: {1, 6}; Yes, it is a function.

71. 3 73. $-2\frac{2}{5}$ 75. −98 77. −22 79. −24

Lesson 2-7, pages 82–85

1. NUMBER OF ABSENCES PER STUDENT

Absences	Tally	Frequency
0	||	2
1	|||||	5
2	||||	4
3	|||	3
4	|||	3
5	|	1
6		0
7		0
8	|	1
9	|	1

3. median 5. 390–399 7. 1.9, 2, 1 9. \$2.90–\$2.99 11. 74 13. 100–109 15. greater than 17. 5 19. 26 21. 0 23. 16 25. −2

Selected Answers

Lesson 2-8, pages 86–89

1. Aptitude Test Scores

Stem	Leaves
1	6
2	7 8 8
3	2 3 4 8 8 8
4	0
5	0 2 5 6 6 7 9
6	0 0 1 2 5 6 9
7	1 1 4 5
8	6

Key
1 | 6 represents a score of 16 on a scale of 0–100

3. 38 5. 51.9

7. Time Spent on Homework

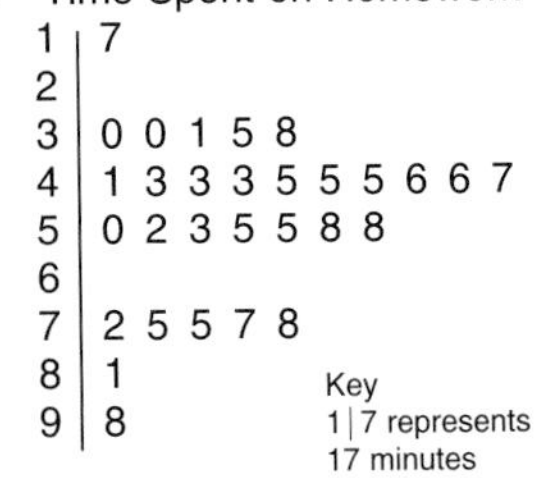

Stem	Leaves
1	7
2	
3	0 0 1 5 8
4	1 3 3 3 5 5 5 6 6 7
5	0 2 3 5 5 8 8
6	
7	2 5 5 7 8
8	1
9	8

Key
1 | 7 represents 17 minutes

9. 43 and 45 11. 52.1

13.

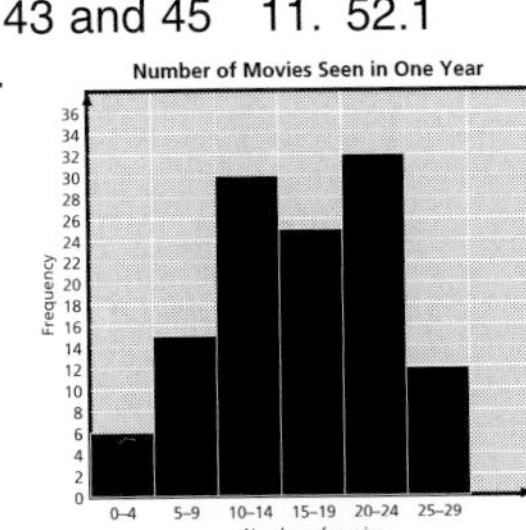

15. 21 17. 60.00–79.99 19. Answers will vary.

21.

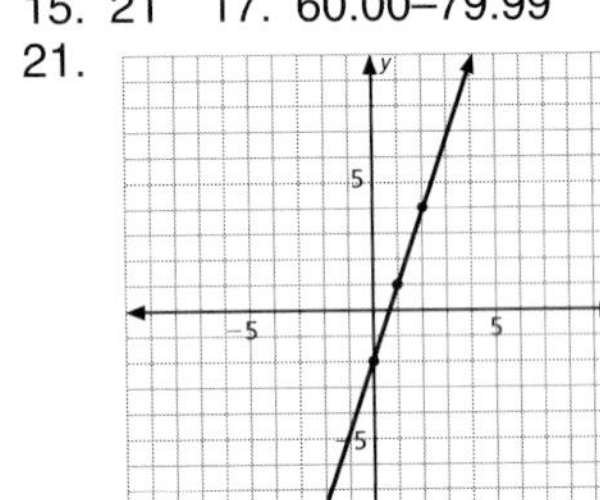

23.

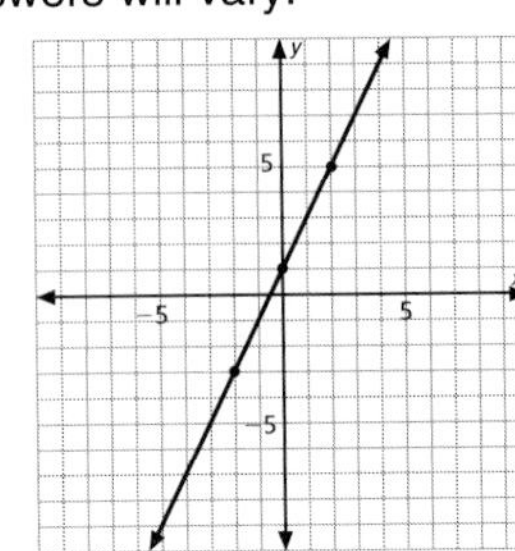

25.

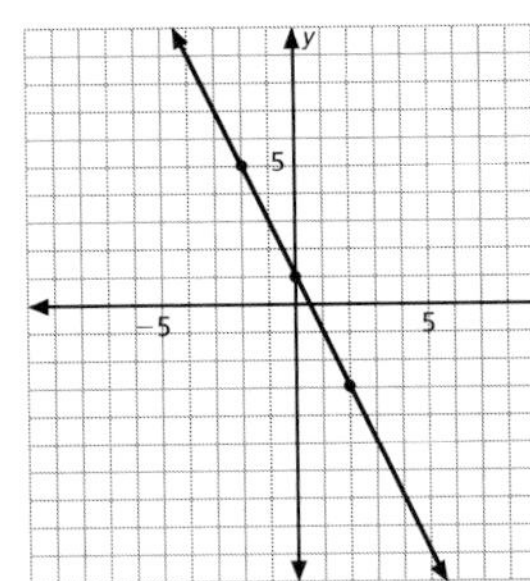

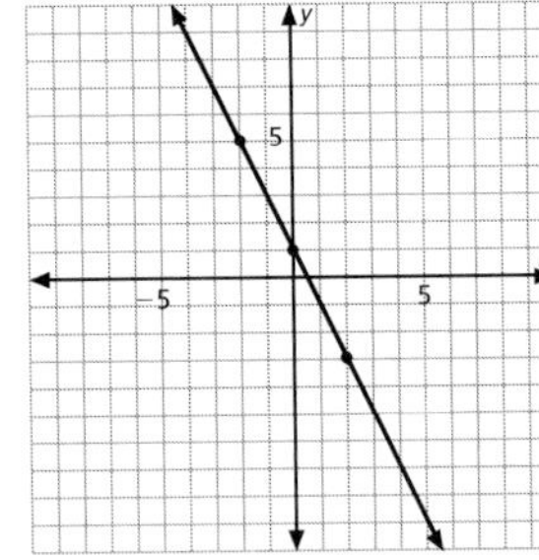

27. 41 29. −10 31. 8
33. 4.8 35. −15

Review and Practice Your Skills, pages 90–91

1. 84, 81, none 3. 240, 235, 235 and 210 5. 76.5, 75, 70
7. 201–210 : 12, 211–220 : 7, 221–230 : 5, 231–240 : 3
9. 211–220; 211–220 11.

Stem	Leaves
20	1 2 5 6 7 8 8
21	0 0 0 0 0 5 5 5 8 8 8 8
22	4 5 8
23	0 0 4 5 8

13.

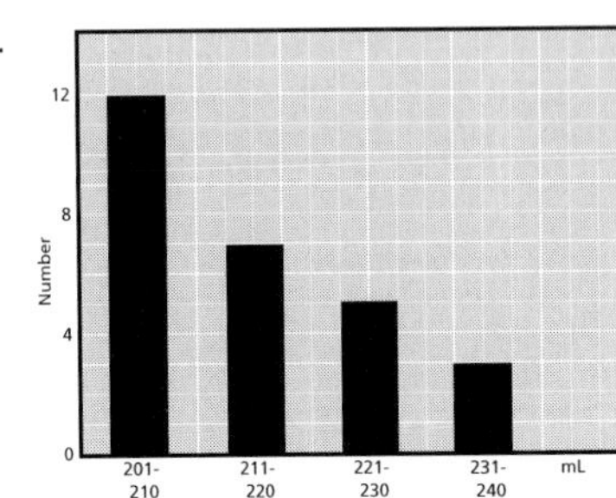

15. 23 17. 13–15 19. mean: increased by 3 hours; median: increased by 3 hours; mode: increased by 3 hours.
21. 67, 95, 123 23. Q, U, Y
25–31.

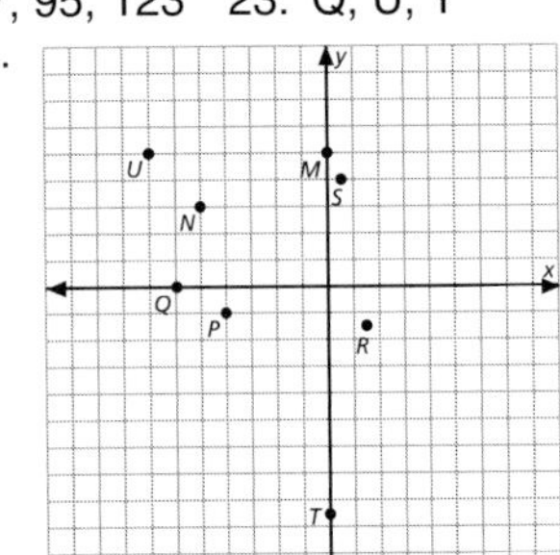

25. II
27. *x*-axis
29. I
31. II

33.

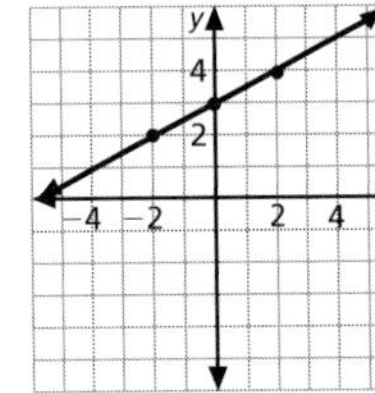

35.

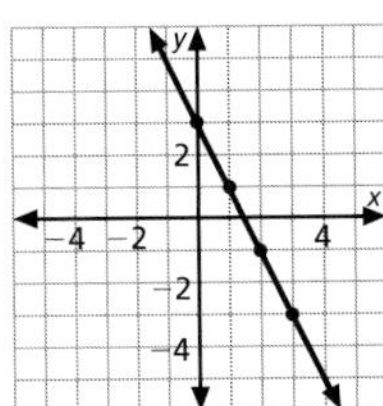

37.

39. 23 41. $-4\frac{1}{2}$
43. −150
45. $2\frac{2}{3}$ 47. −19
49. −6

51. −7 53. ; $x > 5\frac{3}{4}$

55. ; $g \geq 6$

57.

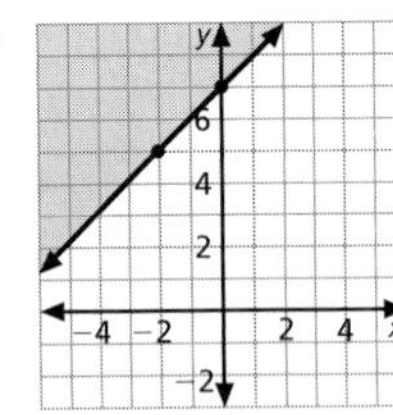

59. 29.1, 30, 22 61. mean; it is influenced by the salaries of president and group manager and will be higher than median salary

Lesson 2-9, pages 92–93

1. Yes; the horizontal scale does not have uniform increments starting with zero—although one line is twice the length of the other, it does not show twice the number of hits.
3. Different size intervals on the vertical scale; different

horizontal distances. 5. Graphs will vary. 7. 5 9. $2\frac{1}{3}$
11. $\frac{2}{3}$ 13. 7 15. −2.5 17. 16 19. 10 21. 17 23. 0.1

Chapter 2 Review, pages 94–95

1. c 3. b 5. 1, $\frac{1}{4}$, $\frac{1}{16}$, divide by 4 7. 16, −32, 64, multiply by −2 9. function; domain: {0, 1, 2}; range; {−1, 0, 1} 11. −9 13. −19

15. 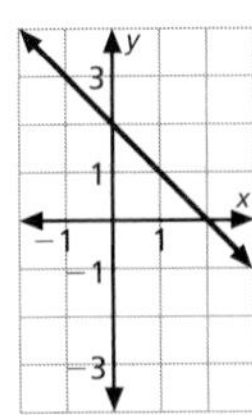

17. $a = -8$

19.

21.

23. 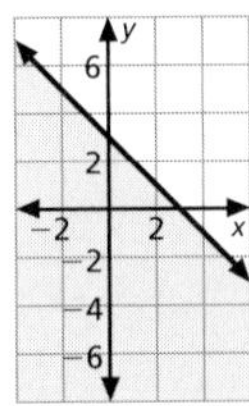

25.

Test scores	Tally	Frequency
51–60	|	1
61–70	|	1
71–80	||	2
81–90	~~||||~~ ||||	9
91–100	|||	3

27.

5	9
6	2
7	8
8	0 1 2 5 5 5 6 6
9	0 0 2 3 4

29. Answers will vary.

Chapter 3

Lesson 3-1, pages 104–107

1. There are three possible answers: $\overline{XY}$ (or $\overline{YX}$), $\overline{XZ}$ (or $\overline{ZX}$), $\overline{RV}$ (or $\overline{VR}$), and $\overline{YZ}$ (or $\overline{ZY}$). Each answer is justified by Postulate 3. 3. 34 5. points R and S; Postulate 1. 7. $\overline{RS}$ (or $\overline{SR}$); Postulate 4. 9. 5 11. 6 13. 56 15. −25 and 9 17. $PN = 34$ and $NQ = 17$ 19. Yes

21. 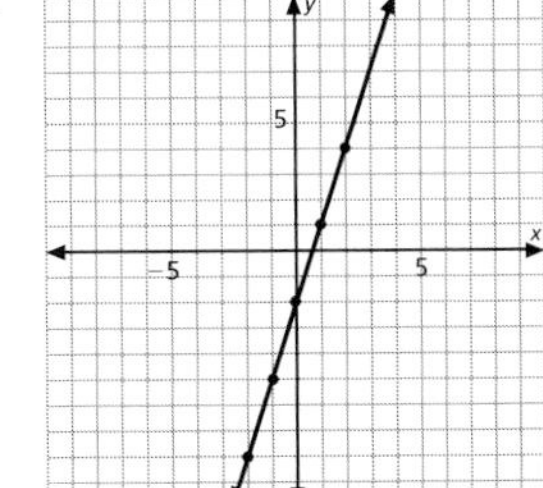

23.

25.

27. 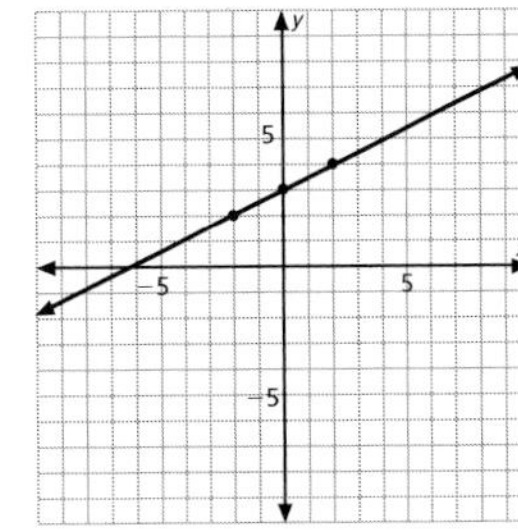

29. −2 31. −4 33. −4

Lesson 3-2, pages 108–111

1. 120°; obtuse 3. 120°; obtuse 5. 180°; straight 7. 79° 9. Answers will vary. Both postulates refer to the pairing of real numbers with geometric figures in a systematic way; both involve taking the absolute value of the difference of two real numbers. 11. $\angle MOR$ 13. $\angle MOQ$, 125° and $\angle NOR$, 145° 15. 72° 23. 115° 25. 67° 27. never 29. 41 31. 1 33. −3200 35. −508 37. −40,000 39. 1521 41. 2.4×10^{13}

Review and Practice Your Skills, pages112–113

1. 5 3. 6 5. 60 7. −18, 8 9. False 11. True 13. 12° 15. 84° 17. 1° 19. 48° 21. 97° 23. 54° 25. 149.5° 27. 162° 29. $\angle QTU = 90°$; $\angle QTR = 10°$; $\angle QTS = 90°$; $\angle RTS = 80°$ 31. $\angle XWV = 90°$; $\angle YWZ = 90°$; $\angle XWY = 90°$ 33. 104° 35. U, V, Z 37. $\overline{UY}$, $\overline{UV}$, $\overline{UW}$, $\overline{YV}$, $\overline{YW}$, $\overline{VW}$ 39. 176 41. 75°

Lesson 3-3, pages 114–117

1. point J 3. point K 5. 119°; 61° 7. $\overline{VX}$ 9. −0.5 11. 146° 13. 142° 15. 19° 17. 14 19. 49 21. 118° 23. 62° 25. 121° 27. 45° 29. E and S 31. Since $m\angle YXW + m\angle WXV + m\angle VXU = 176°$, $\angle AXB$ is an obtuse angle. Therefore, $\overrightarrow{XY}$ and $\overrightarrow{XU}$ are not opposite rays; False 33. Since $\angle TXV$ is a right angle, $\angle TXU$ and $\angle UXV$ are complementary, and $\angle TXU = 90° - 43° = 47°$. Since $\angle TXU$ and $\angle UXV$ are not equal in measure, $\overrightarrow{XU}$ does not bisect $\angle TXV$.; False 35. No information is given to indicate that $XZ = XV$, so it is not possible to identify X as the midpoint of $\overline{ZV}$; Cannot tell. 37. 2.5 39. −14 41. rule: × −3; −243, 729, −2187 43. rule: − 7.5; 62.5, 55, 47.5 45. rule: + 4, ÷ 2; 19.5, 9.75, 13.75

Lesson 3-4, pages 118–121

1. 90° 3. 53° 5. 97° 9. 51° 11. 39° 13. 141° 15.–17. 19. Answers will vary. 21. yes

Review and Practice Your Skills, pages 122–123

1. D 3. $\overline{BD}$ 5. 108° 7. 144° 9. 14 11. 28 13. false 19. 90° 21. 49° 23. 139° 25. 49° 27. true 29. 128 31. 54° 33. $\overline{CD}$, $\overline{EJ}$, $\overline{FH}$ 35. $\overline{AD}$, $\overline{BC}$, $\overline{DF}$, $\overline{CH}$

Lesson 3-5, pages 124–127

1. ; sixteenth figure: triangle with 16 dots on each side 3. 66

5. Next figure: The fourteenth figure will be a rectangular arrangement of 210 points, with 14 points along one side of the rectangle and 15 points along the other.

7. 22 9. 153 11. $f(n) = n^2 + 1$ 13. Answers will vary.

Selected Answers

15. 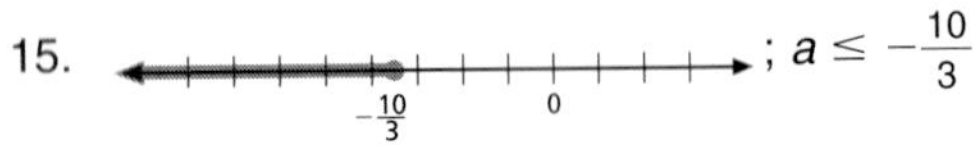 ; $a \le -\frac{10}{3}$

17. ; $c < -12$

19. ; 1. $x < \frac{1}{4}$

21.

Interval	Tally	Frequency
90-99	|||	3
80-89	卌 |	6
70-79	卌 |||	8
60-69	|||	3
50-59	|	1
40-49	|	1

23.

Interval	Tally	Frequency
85-87	卌 ||||	9
82-84	卌 |	6
79-81	卌	5
76-78	||	2
73-75	|	1
70-72	||	2

Lesson 3-6, pages 128–131

1. False; it is possible that the two lines are noncoplanar. 3. See additional answers. 5. If an error is charged to the shortstop, then the shortstop made a bad throw to first base. False. 7.

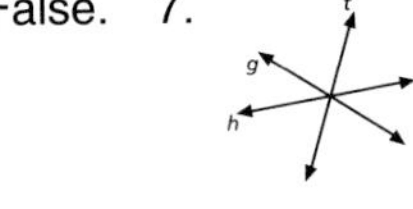

9.

11. Converse; If points J, K, and L are collinear, then they are coplanar. The given statement is false. Its converse is true. 13. Converse: If two angles are complementary, then the sum of their measures is 90º, Both the given statement and its converse are true. 15. Converse: If two lines do not intersect, then they are perpendicular. Both the given statement and its converse are false. 17. Conditionals: If a point is the midpoint of a segment, then it divides the segment into two segments of equal length; if a point divides a segment into two segments of equal length, then it is the midpoint of the segment. Biconditional: A point is the midpoint of a segment if and only if it divides the segment into two segments of equal length. 19. Conditionals: If a line is a transversal, then it intersects two or more coplanar lines in different points; if a line intersects two or more coplanar lines in different points, then it is a transversal. Biconditional: A line is a transversal if and only if it intersects two or more coplanar lines in different points. 21. *t, Q* 23. *P, X* 25. *X, Y* 27. Write given definitions as two conditionals: *If two angles are vertical angles, then their sides form opposite rays is true.* However, *If the sides of two angles form opposite rays, then the angles are vertical angles is false.* Here is a counterexample in which the sides of ∠1 and ∠2 form a pair of opposite rays, but the angles are not vertical angles.

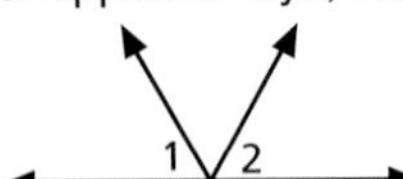

For this reason, it is necessary to define vertical angles as two angles whose sides form two pairs of opposite rays.

29. Write the given definitions as two conditionals: *If two angles are adjacent angles whose exterior sides form a right angle, then they are complementary* is true. However, *If two angles are complementary, then they are adjacent angles whose exterior sides form a right angle* is false. Complementary angles are not necessarily adjacent. 31. 72

Review and Practice Your Skills, pages 132–133

1. 19 units horizontal; 10 units vertical 3. 10 units long 5. same as second figure, but with ten lines. 7. same as second figure. 9. Answers will vary. 11. Answers will vary. 13. If $AB = 2(AC)$, then C is the midpoint of $\overline{AB}$. true, false 15. If two lines are perpendicular, then they intersect to form right angles. If two lines intersect to form right angles, then they are perpendicular. Two lines are perpendicular if and only if they intersect to form right angles. 17. −19, 7 19. 148°, 32° 21. 164° 23. 164°

Lesson 3-7, pages 134–137

1. given; parallel lines postulate; m∠2 = m∠3; transitive property of equality 3. Answers will vary. 5. m∠1 = m∠3; given; definition of complementary angles; transitive property of equality; Statement: 1: ∠1 is complementary to ∠2. ∠3 is complementary to ∠2. Statement 2: *m*∠3 + *m*∠2 = 90°. Statement 3: m∠3 + m∠2. Statement 4: *m*∠1 = *m*∠3.

7. Given: $k \parallel m$, $l \parallel m$
Prove: $k \parallel l$

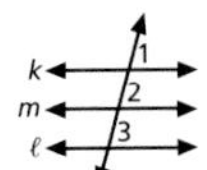

Statements	Reasons
1. $k \parallel m$, $l \parallel m$	1. given
2. $m\angle 1 = m\angle 2$; $m\angle 3 = m\angle 2$	2. parallel lines post.
3. $m\angle 1 = m\angle 3 =$	3. trans. property of eq.
4. $k \parallel l$	4. corres. Angles post.

9. Given: $l \parallel m$, $k \perp l$
Prove: $k \perp m$

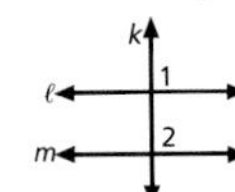

Statements	Reasons
1. $l \parallel m$, $k \perp l$	1. given
2. ∠1 is a right angle	2. def. of perpendicular lines
3. $m\angle 1 = 90º$	3. def. of right angle
4. $m\angle 1 = m\angle 2$	4. parallel lines postulate
5. $m\angle 2 = 90º$	5. substitution property
6. ∠2 is a right angle	6. def. of right angle
7. $k \perp m$	7. def. of perpendicular lines

11. 0.0000000146 13. 70,200,000 15. 0.0000000059 17. 21,000 19. 0.000397 21. 5,120,000,000 23. 33°, 123° 25. 15°, 105° 27. 7°, 97° 29. 29°, 119°

Lesson 3-8, pages 138–139

1.

	Cory	Srey	Molly	Mao
Des Moines	x	x	o	x
Pittsburg	x	o	x	x
Santa Clara	o	x	x	x
Seattle	x	x	x	o

3. Ned—Miami; Carina—Dallas; Pedro—San Francisco; and Eva—San Diego

5.–11.

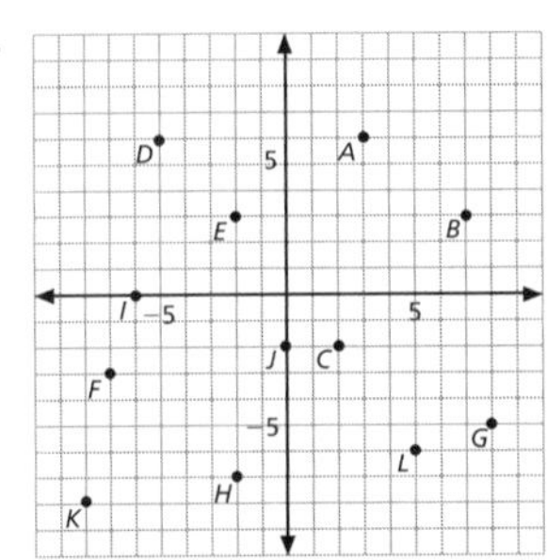

13. 1

Chapter 3 Review, pages 140–141
1. d 3. c 5. b 7. 13° 9. 110° 11. ∠2 and ∠5, ∠3 and ∠6, ∠4 and ∠7 13. ∠1 and ∠6 15. If two angles are supplementary, the sum of the measures is 180°. Both statements are true. 17. Maro: Wisconsin; Sue: California; Stephanie: Florida

CHAPTER 4

Lesson 4-1, pages150–153
1. 45° 3. 26° 5. 153° 7. 71 9. 63 11. 38°, 76°, 66°
13. m∠F = 64°; m∠G = 26°; m∠H = 90°
15.

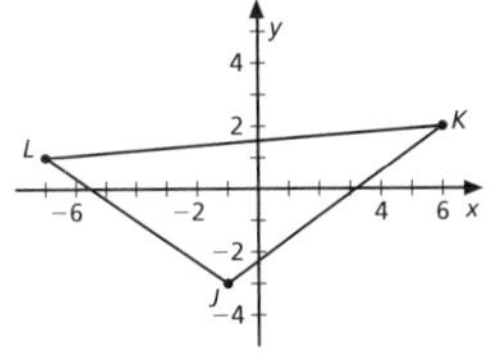

; scalene, obtuse

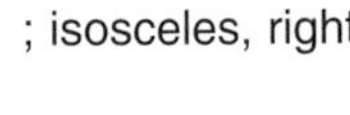

17.

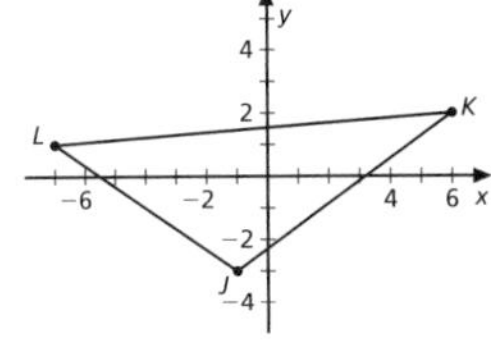

; isosceles, right

19. never 21. sometimes 23. 52 25. mean: $7\frac{1}{6}$; median: $7\frac{1}{2}$; mode: 8

Lesson 4-2, pages 154–157
1. $\overline{RQ} \cong \overline{RS}$; $\overline{RT}$ bisects ∠QRS given; angle bisector; $\overline{RT} \cong \overline{RT}$; reflexive; SAS postulate
3. Proofs may vary. A sample proof is given.

Statements	Reasons
1. $\overline{AB} \cong \overline{CB}$; $\overline{EB} \cong \overline{DB}$ $\overline{AD}$ and $\overline{CE}$ intersect at point B.	1. given
2. ∠ABE and ∠CBD are vertical angles.	2. definition of vertical angles
3. ∠ABE ≅ ∠CBD	3. vertical angle theorem
4. △ABE ≅ △CBD	4. SAS postulate

5. △ECB; ASA postulate 7. △MNK; SAS postulate
9. (−7, −11) or (−7, 15)
11. Proofs may vary. A sample proof is given.
Given: $\overline{AB} \cong \overline{XY}$; $\overline{BC} \cong \overline{YZ}$ ∠B and ∠Y are right angles.
Proof: △ABC ≅ △XYZ

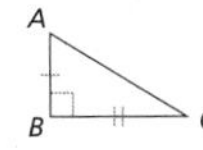

13. 104° 15. 76°

Review and Practice Your Skills, pages 158–159
1. 95° 3. 23° 5. 140° 7. false 9. false
11. ; 13. ;

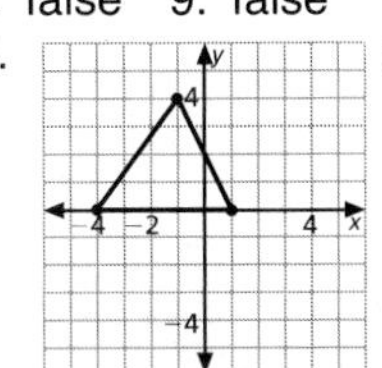

right isosceles scalene acute
15. $\overline{RS} \cong \overline{VU}$, $\overline{RT} \cong \overline{VT}$, $\overline{ST} \cong \overline{UT}$, ∠SRT ≅ ∠UVT, ∠RTS ≅ ∠VTU, ∠TSR ≅ ∠TUV, △RST ≅ △VUT
17. Always 19. Sometimes 21. △NOM; SAS postulate

Lesson 4-3, pages 160–163
1. $\overline{FG} \cong \overline{HJ}$; $\overline{FG} \perp \overline{FH}$; $\overline{JH} \perp \overline{FH}$; given; ⊥ lines; right ∠; m∠1 = m∠2, or ∠1 ≅ ∠2; $\overline{FH} \cong \overline{HF}$; SAS postulate; ∠J ≅ ∠G; CPCTC 3. 12 5. 3 7. 45 9. There are two base angles. If each measures 70°, the vertex angle must measure 40°. 11. ∠1 ≅ ∠2; ∠4 ≅ ∠5; ∠3 ≅ ∠6; ∠8 ≅ ∠7; ∠XZY ≅ ∠XYZ
13. 15.

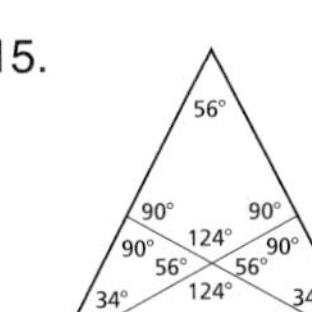

17. isosceles triangle
19. ∠ABC
21. ∠FBC = 115°, ∠ABD = 155°

Lesson 4-4, pages 164–167
1.

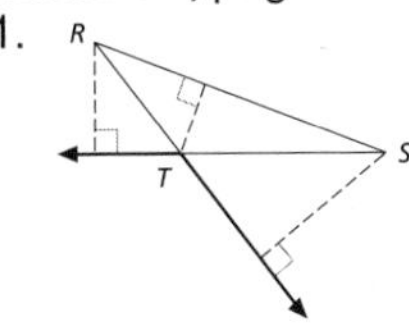

3. true 5. false 7. The altitudes are concurrent lines.
9.
11. false 13. true 15. false
17. false
19. The three distances are equal. 21. In a right triangle, the two sides that are perpendicular (the legs) are two altitudes of the triangle. It is never true that a side of a triangle is also a median. 23. Answers will vary. Possible responses: △PQR is isosceles; $\overline{QR} \cong \overline{PR}$; $\overline{QS} \cong \overline{PS}$; ∠RQP ≅ ∠RPQ; ∠QRS ≅ ∠PRS; △QRS ≅ ∠PRS; $\overline{QT}$ is an altitude of △PQR; RS is an altitude of △PQR; $\overline{RS}$ is a median of △PQR; $\overline{RS}$ is a perpendicular bisector of $\overline{QP}$; $\overline{RS}$ lies on the bisector of ∠QRP; ∠RSQ, ∠RSP, ∠QTP, and ∠**QTR** are right angles.
25. D 27. $\overline{DF}$, $\overline{CG}$, $\overline{BH}$, $\overline{AI}$ 29. −15 31. −21 33. 0
35. 10

Review and Practice Your Skills, pages 168–169
1. 8 cm 3. 46° 5. ∠3 ≅ ∠4 ∠2 ≅ ∠6 ∠1 ≅ ∠5; ∠3 ≅ ∠9 ∠2 ≅ ∠8 ∠1 ≅ ∠7; ∠3 ≅ ∠10 ∠2 ≅ ∠12 ∠1 ≅ ∠11; ∠4 ≅ ∠9 ∠6 ≅ ∠8 ∠5 ≅ ∠7; ∠4 ≅ ∠10 ∠6 ≅ ∠12 ∠5 ≅ ∠11; ∠4 ≅ ∠10 ∠8 ≅ ∠12 ∠7 ≅ ∠11

Selected Answers

7. false (must be "included angle") 9. false 15. true 17. false 19. true 21. 71° 23. 28° 25. 112°

Lesson 4-5, pages 170–171
1. Assume that the triangle *can* be an obtuse triangle 3. two; *r; s; x; y,* one; contradictory; false; true 5. *Step 1:* Assume that there are two lines through point *P* perpendicular to the given line. In particular, in the figure to the right, assume that $m \perp l$ and $n \perp l$. *Step 2:* By definition of right angle, $m\angle 1 = 90°$ and $m\angle 2 = 90°$ By the transitive property of equality, $m\angle 1 = m\angle 2$. By the corresponding angles postulate, since $m\angle 1 = m\angle 2$, it follows that *m_n*. By definition of intersecting lines, since *m* and *n* each pass through point *P*, *m* and *n* are intersecting lines. *Step 3:* The last two statements in *Step 2* are contradictory. Therefore, the assumption that there can be two lines perpendicular to a given line through a point outside the line is false. The given statement must be true. 7. 5 9. 9 11. 19

Lesson 4-6, pages 172–175
1. between 3 ft and 15 ft. 3. between 1 ft and 13 ft. 5. $\angle K, \angle M, \angle L$ 7. $\overline{XY}, \overline{YZ}, \overline{XZ}$ 9. yes 11. no 13. yes 15. $\overline{DF}$; $\overline{EF}$ 17. $\overline{VW}, \overline{UV}$ or $\overline{UW}$ 19. $0 < x < 11$ 21. $\angle C$ 23. Assume that $AB + AC \not> BC$. Then, by the property of comparison, one of these two statements must be true:
$AB + AC = BC$ or $AB + AC < BC$
If $AB + AC = BC$, then there is a path connecting points *B* and *C* that is equal in length to $\overline{BC}$; this contradicts the shortest path postulate. Similarly, if $AB + AC < BC$, then there must be a path connecting points *B* and *C* that is shorter than $\overline{BC}$; this also contradicts the shortest path postulate. Therefore, the assumption $AB + AC \not> BC$ must be false. It follows that the desired conclusion, $AB + AC < BC$, is true.
25. $\overline{BC}, \overline{AB}, \overline{AC}, \overline{CD}, \overline{AD}$ 27. $2 < z < 12$ 29. A right triangle can have only one right angle, and it cannot have an obtuse angle. Therefore, the one right angle is the largest angle. By the unequal angles theorem, the side opposite that angle, the hypotenuse, is the longest side. 31. When the top angle = 60° 33. When the top angle < 60° 35. $f(x) = x^2 - (x - 1)$ 37. 3.71×10^{11} 39. 2.56×10^{11} 41. 8.9×10^{12}

Review and Practice Your Skills, pages 176–177
1. If a triangle is not isosceles, then it *is* equilateral. 3. Assume they are *not* equal in measure. 5. Assume they *do* intersect. 7. Answers will vary. 9. Answers will vary. 11. yes 13. no 15. $0 < x < 14$ 17. $\overline{PM}, \overline{MO}, \overline{PO}, \overline{MN}, \overline{NO}$ 19. $\overline{XY}, \overline{WY}, \overline{YZ}, \overline{WZ}, \overline{XW}$ 21. false 23. true 25. false 27. 270 29. 7.5 m 31. 53°

Lesson 4-7, pages 178–181
1. 101 3. 136 5. 36° 7. 129 9. 132 11. 72 13. 2,880° 15. 15° 17. 30; 60; 60 19. 40 21. 20 23. Although the faces are all regular polygons, there are two different types of faces, pentagons and hexagons.
25. $\frac{(n-2)180}{n}$ 27. Approaches 180° 29. right scalene 31. acute isosceles

Lesson 4-8, pages 182–185
1. 68; 112 3. 64° 5. 90° 7. 90° 9. 45; 135; 42; 28 11. $1\frac{5}{8}$; $1\frac{5}{8}$, $1\frac{1}{2}$; $1\frac{5}{8}$ 13. true 15. false 17. false 19. No; the angles that are equal in measure are not opposite angles. 21. Yes, the figure is a square, and a square is, by definition, a parallelogram.
23. Given: *ABCD* is a parallelogram.
Prove: $m\angle B = m\angle D$

Statements	Reasons
1. *ABCD* is a parallelogram	1. given
2. $\overline{AB} \parallel \overline{DC}$; $\overline{AD} \parallel \overline{BC}$	2. definition of $\parallel$-ogram
3. $m\angle 5 = m\angle 7$, or $\angle 5 \cong \angle 7$; $m\angle 6 = m\angle 8$, or $\angle 6 \cong \angle 8$	3. If two $\parallel$ lines are cut by a tran., then alt. int. $\angle$s are = in measure.
4. $\overline{AC} \cong \overline{CA}$	4. reflexive property
5. $\triangle ABC \cong \triangle CDA$	5. ASA postulate
6. $\angle B \cong \angle D$, or $m\angle B \cong m\angle D$	6. CPCTC

25. Sample response: Draw the parallelogram and both diagonals. The goal is to show that *BD* bisects *AC*, and that *AC* bisects *BD*. By the definition of parallelogram, $AB \parallel DC$ and $AD \parallel BC$. When parallel lines are cut by a transversal, alternate interior angles are equal in measure, so there are four pairs of angles that are equal in measure: $\angle 1$ and $\angle 3$; $\angle 2$ and $\angle 4$; $\angle 5$ and $\angle 7$; and $\angle 6$ and $\angle 8$. Because the parallelogram-side theorem has been proved, it is known that $AB = CD$ and $AD = CB$. Therefore, by the ASA postulate, there are two pairs of congruent triangles: $\triangle BEA$ and $\triangle DEC$, and $\triangle BEC$ and $\triangle DEA$. Because corresponding parts of congruent triangles are congruent, it follows that $AE \cong CE$, or $AD = DE$, and that $DE \cong BE$, or $DE = BE$. Therefore, point *E* is the midpoint of both *AC* and *BD*. It follows that *BD* bisects *AC* and *AC* bisects *BD*. 27. $\angle ABG$ and $\angle FEH$, $\angle GBC$ and $\angle DEH$ 29. $\angle ABH$ and $\angle GEF$, $\angle CBH$ and $\angle GED$ 31. yes; domain: {2, 3, 4, 5, 6}; range: {4.5, 6.5, 8.5, 10.5, 12.5}

Review and Practice Your Skills, pages 186–187
1. 143° 3. 60° 5. 18° 7. 360° 9. 9 11. 24 13. true 15. true 17. $a = 115°$; $b = 65°$; $c = 37$ cm; $d = 43$ cm 19. $a = 7.5$ m $b = 6.8$ m $c = 133°$ $d = 47°$ 21. yes, diagonals bisect each other. 27. 45° 29. 60° 31. 100 cm $< x <$ 900 cm 33. 100° 35. 43°

Lesson 4-9, pages 188–191
1. 16 3. 3 5. $m\angle C = 75°$; $m\angle D = 75°$; $m\angle E = 105°$; $m\angle F = 105°$ 7. 3.6 9. 9 11. 7 13. $m\angle T = 83°$; $m\angle V = 97°$; $m\angle U = 97°$; $m\angle W = 83°$ 15. Answers may vary. *Possible likeness*: Each has an endpoint at the midpoint of a side of the figure. *Possible differences:* A median of a triangle has one endpoint that is also a vertex of the figure, whereas the median of a trapezoid does not; a triangle has three medians, whereas a trapezoid has only one.
17. quadrilateral 19. trapezoid, quadrilateral 21. isosceles triangle, trapezoid, quadrilateral 23. no; yes; yes; yes; yes; no; no 25. no; yes; yes; yes; yes; no; no 27. no; no; yes; no; yes; no; yes

29. Given: *ABCD* is a trapezoid.
Prove: *ADC* and *DAB* are supplementary.

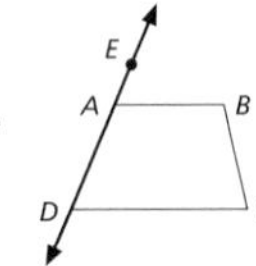

Statements	Reasons
1. *ABCD* is a trapezoid	1. given
2. $\overline{AB} \parallel \overline{DC}$	2. definition of trapezoid
3. $m\angle EAB + m\angle ADC$	3. corr. $\angle s$ postulate
4. $m\angle EAB + m\angle DAB = 180°$	4. angle addition postulate
5. $m\angle ADC + m\angle DAB = 180°$	5. substitution property
6. $\angle ADC$ and $\angle DAC$ are supplementary	6. Definition of supplementary angles

31. 9 33. 5 35. 8

Chapter 4 Review, pages 192–193
1. d 3. c 5. b 7. 70 9. $\triangle AES \cong \triangle BET$, SAS
11. $\triangle XYZ \cong \triangle QZY$, SSS 13. 45
15. Sample answer:

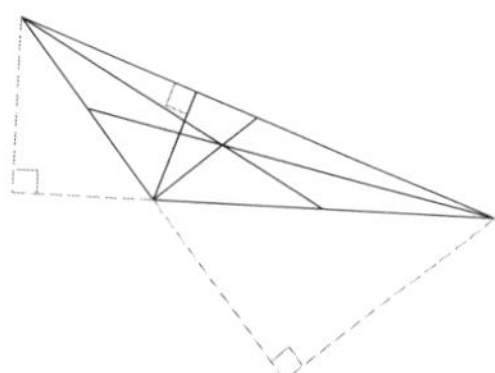

17. No; $10 + 8 < 19$ 19. 30 cm

Chapter 5

Lesson 5-1, pages 202–205
1. 32 c 3. 3,500 mL 5. $\frac{2}{5}$ 7. 9:1 9. $1\frac{1}{4}$ ft per day
11. 32 13. 256 15. 0.004 17. 6 19. lb 21. KL
23. 2:5 25. 1:3 27. \$0.06 per copy 29. 4-L 31. ±0.5 m; ± 0.05 m 33. 8,000 billion qt 35. 2 mi/h 37. 48°; 70°; 62°

Lesson 5-2, pages 206–209
1. 10 m; 7 m^2 3. 11 cm; 10 cm^2 5. 40 cm 7. 4.8 cm
9. about 62.8 m^2 11. It triples too. 13. about 144 ft
15. 3,600 ft^2 17. Luis entered 3.14 for π. Irene used the π key on her calculator. 21. Answers will vary. 23. Sample answers: Area of wall space, not including windows, doors, etc.; number and price of cans of paint needed based on the area one can of paint covers; painting speed and number of painters involved.
25. The circle ($r = 15.9$ ft and $A = 793.8$ ft^2) always has more area than the square (625 ft^2). 27. yes 29. yes 31. yes
33. yes 35. 5 37. 639. 6

Review and Practice Your Skills, pages 210–211
1. 12 3. 0.3625 5. 7030 7. 40:9 9. 7:5 11. 5:6
13. b 15. c 17. 1 gal for \$1.79 19. $P = 36$ in.
21. $C = 94.2$ in. 23. $A = 204$ ft^2 25. 96 in.^{2} 27. 27.93 mi^2 29. 832 31. 22, 2 33. 0.0004 35. 1:10 37. 30:1
39. 1:1000 41. 200 L barrel 43. $P = 124.95$ yd; $A = 961.625$ yd^2

Lesson 5-3, pages 212–215
1. $\frac{3}{16}$ 3. $\frac{1}{4}$ 5. $\frac{1}{13}$ 7. $\frac{3}{26}$ 9. $\frac{2}{3}$ 11. $\frac{3}{5}$ 13. about 0.785
15. Possible answer: about 0.009 or $\frac{9}{1,000}$ 17. No. That actual probability is $\frac{1}{9}$, which is $\left(\frac{1}{3}\right)^2$ 19. likely 21. likely
23. about 0.56 or $\frac{14}{25}$ 25. $\frac{6}{31}$ 27. The map or actual dimensions of the Palaestra and the total map or actual area of the ruins.
29.

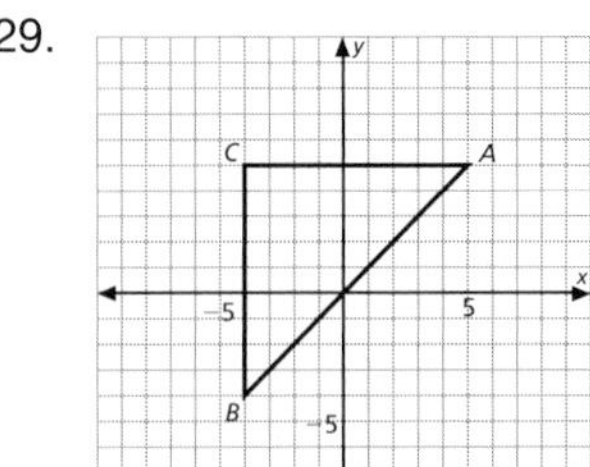

right isosceles

31.

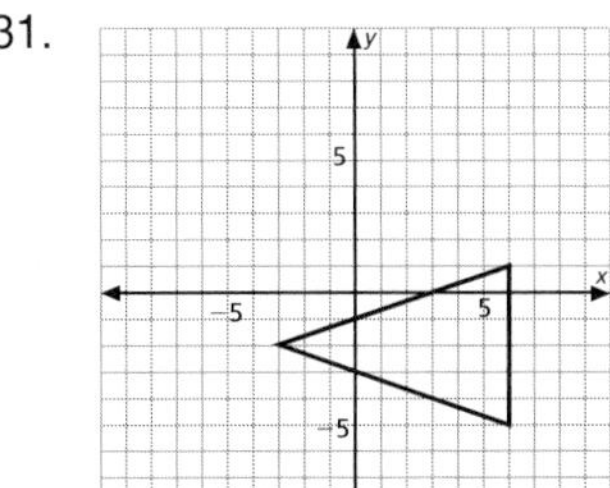

acute isosceles

33. −33 35. −48 37. 10.9 39. 39.5

Lesson 5-4, pages 216–217
1. 12,150 ft^2 3. 396 ft^2 5. 126.5 m^2 7. at least 230 ft long, at least 150 ft wide. 9. 7740°, 172° 11. 4680°, 167.14° 13. 5760°, 169.41° 15. 7200°, 171.43°

Review and Practice Your Skills, pages 218–219
1. $\frac{6}{25} = 0.24$ 3. $\frac{7.065}{27} \approx 0.262$ 5. $\frac{9}{28.26} = 0.318$
7. $\frac{1}{10} = 0.1$ 9. unlikely 11. Answers will vary. Most students will answer "likely." 13. 320 m^2 15. 46.25 ft^2
17. 1:500 19. 1:2 21. $P = 67.1$ m; $A = 235.5$ m^2
23. $\frac{2}{5} = 0.40$

Lesson 5-5, pages 220–223
1. Triangular prism; bases *ABC* and *EFD*; Remaining answers will vary. Sample answers include parallel edges $\overline{AB}$ and $\overline{EF}$; intersecting faces *BCDF* and *ABC*; intersection edges $\overline{EF}$ and $\overline{FD}$. 3. Triangular pyramid; base *LNO*; no parallel edges; Remaining answers will vary. Sample answers include intersecting faces *LMO* and *LNO*; intersecting edges $\overline{OM}$ and $\overline{OL}$. 5. a coplanar line halfway between them. 7. hexagonal prism; 8; 12; 18 11. The sum of the faces and vertices is 2 more than the number of edges. Possible rule: $e = f + v - 2$.
13. 15. a cylindrical surface 19. a square
21. 45° 23. 2.9 in.
25. $\frac{2}{5}$ 27. 6 29. 3

Selected Answers

Lesson 5-6, page 224–227
1. 158 cm^2 3. 336 m^2 5. 222 in.^{2} 7. 452 ft^2 9. 166.4 m^2 11. 476 in.^{2} 13. 735 m^2 15. the area of the base 17. About 24,727.5 mm^2 19. a) SA is 4 times as great; b) SA is $\frac{1}{9}$ times as great. 21. $1\frac{2}{3}$ qt 23. 45

Review and Practice Your Skills, pages 228–229
1. rectangular prism; 6 faces, 8 vertices, 12 edges
3. oblique cone; faces, vertices, and edges are undefined for cones. 5. Answers will vary. 7.

9.   11. 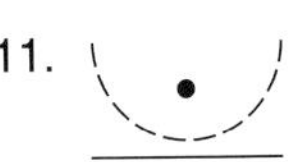 13. 240 cm^2 15. 272 ft^2 17. 1306.2 m^2

19. 9 times as large 21. 65 $\frac{\text{mi}}{\text{h}}$ 23. 34 $\frac{\text{gal}}{\text{min}}$
25. $\frac{43}{400} = 0.215$ 27. 8478 in.^{2}

Lesson 5-7, pages 230–233
1. 1,309 mm^3 3. 3,561 cm^3 5. 6 cm 7. 1,570 cm^3 9. 340 m^3 11. about 354 m^3 13. Rectangular prism has volume twice that of triangular prism 15. V = 3,840 yd^3 17. One answer: Area of base of Cheops is smaller than the area of the base of Cholula. 19. It is tripled. 21. Tetragonal system is combination of rectangular prism and two square pyramids; find volume of separate shapes and add. 23. the $0.40 can 25.

27. 36.94 ft; 80 ft^2 29. 14.54 in.; 8.22 in.^{2}

Chapter 5 Review, pages 234–235
1. b 3. e 5. c 7. 3,000 9. $P = 98$ mm; $A = 315$ mm^2 11. 1:100 13. triangular prism; 5 faces; 6 vertices; 9 edges 15. 164.9 cm^2 17. 177.1 in.^{2} 19. 135 cm^3 21. 6,188.9 mm^3

Chapter 6

Lesson 6-1, pages 244–247
1. $-\frac{1}{3}$ 3. undefined 5.
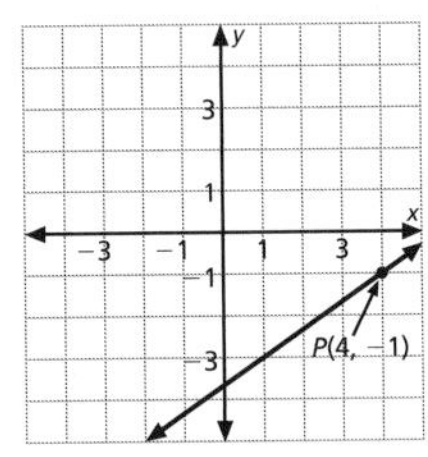

7. $m = -6, b = 4$ 9. $m = -\frac{1}{10}, b = \frac{1}{5}$

11. $y = 4x - 2$ 13.
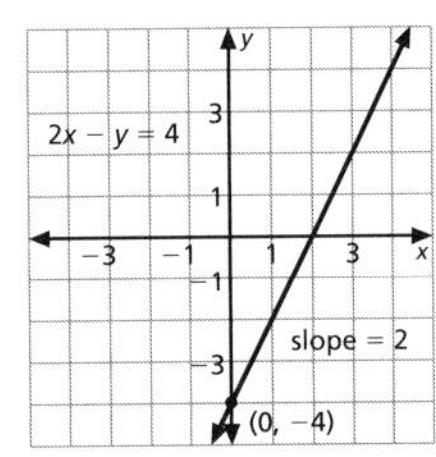

15. −2 17. $\frac{3}{7}$ 19–21.
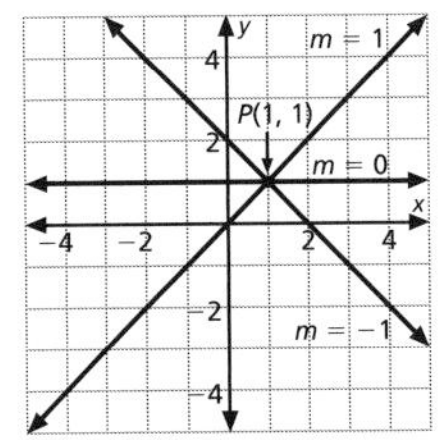

23. 2; yes 25. $m = 4, b = 0$ 27. $m =$ undefined, $b =$ none 29. $m =$ undefined, $b =$ none 31. $y = -5x + 4$ 33. $y = 6x$ 35.
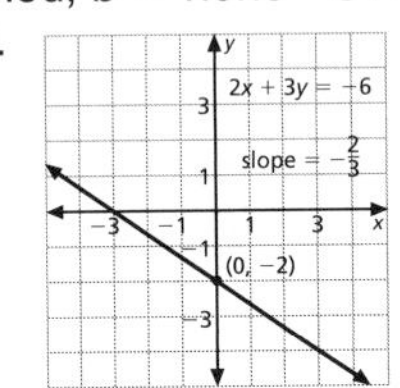

37.
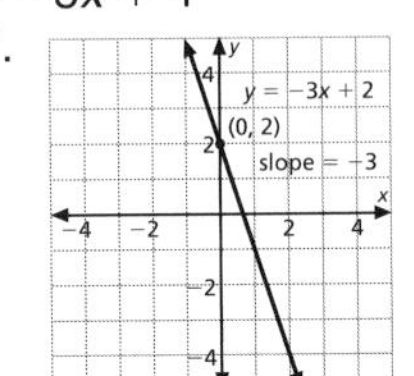

39.

41. $3,000 43. slope $= a$, y-intercept $= r - m$
45. $y = -\frac{A}{B}x - \frac{C}{B}$
47. No, since there is no change in x, the line must be vertical and the slope is undefined. 49. 62 mi/h 51. $1.19 per lb 53. 4 mi/h 55. $1.67 per jar 57. 14, −14 59. 4, −4 61. 26, −26 63. 15, −15

Lesson 6-2, pages 248–251
1. $-\frac{2}{9}, \frac{9}{2}$ 3. $4, -\frac{1}{4}$ 5. neither 7. neither 9. $-4, \frac{1}{4}$ 11. $-\frac{5}{6}, \frac{6}{5}$ 13. $-7, \frac{1}{7}$ 15. neither 17. no 19. parallel 21. 23 23. $-\frac{2}{5}$

27.
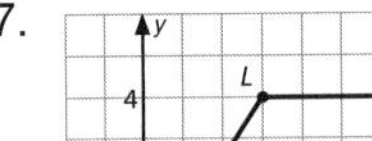

29. slope of $\overline{LN} = -\frac{3}{4}$ and slope of $\overline{EI} = \frac{3}{8}$; $\left(-\frac{3}{4}\right)\left(\frac{3}{8}\right) \neq -1$, so $\overline{LN}$ is 31. 48 cm^2 33. 5.25 cm^2

Review and Practice Your Skills, pages 252–253

1. 1 3. $-\frac{5}{4}$ 5. $-\frac{4}{9}$ 7. 0 9. -6

11.

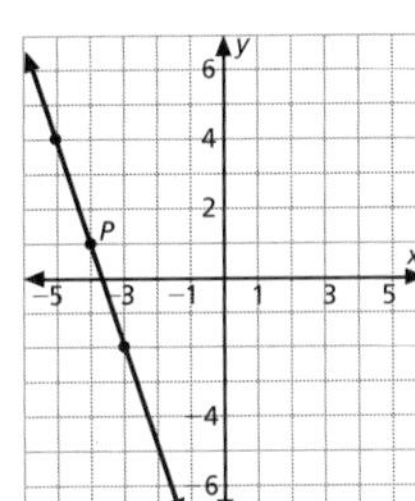

13.

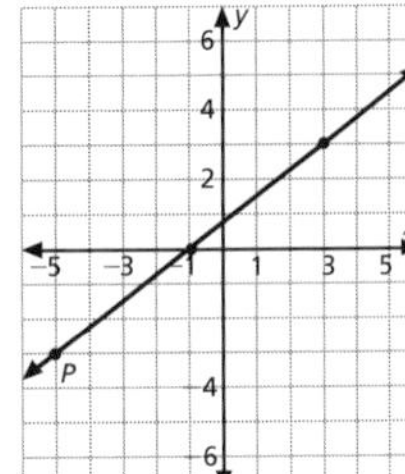

15.

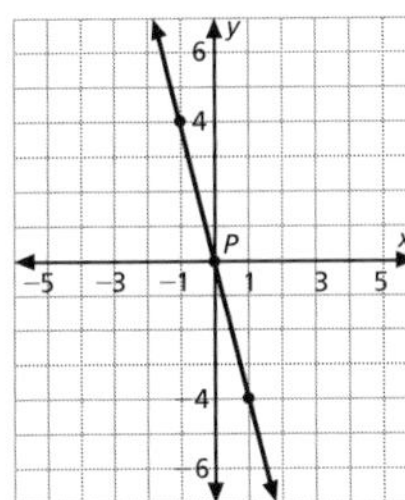

17. $m = -5, b = 9$ 19. $m = \frac{2}{3}, b = -1$ 21. $m = -\frac{2}{3}, b = 4$ 23. $m = -\frac{1}{3}, b = \frac{1}{4}$ 25. $y = 15$ 27. $y = -\frac{7}{2}x - 3$ 29. $y = -20x + 5$

31.

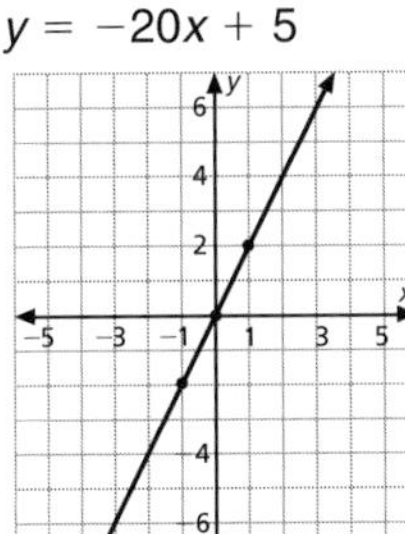

33.

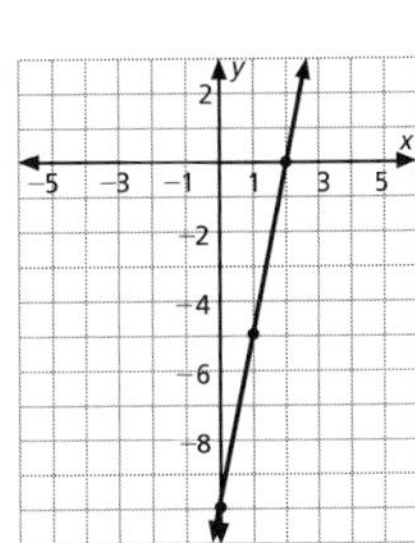

35.

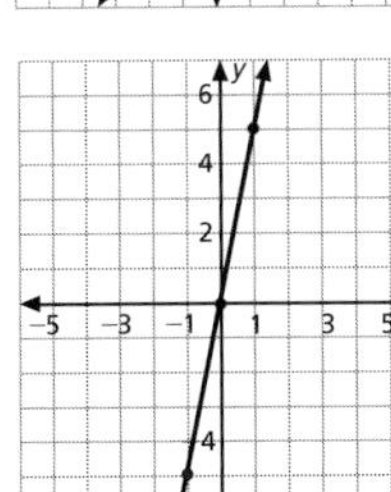

37. $-2, \frac{1}{2}$ 39. 1, -1 41. $-\frac{5}{13}, \frac{13}{5}$ 43. $-3, \frac{1}{3}$ 45. $-\frac{12}{5}, \frac{5}{12}$ 47. perpendicular 49. parallel 51. neither

53. undefined 55. $\frac{3}{4}$ 57. $m = -\frac{1}{4}, b = 7$ 59. $m = 3, b = -4$ 61. $m = \frac{1}{4}, b = 9$ 63. $-\frac{19}{9}, \frac{9}{19}$ 65. undefined, 0 67. $-\frac{1}{8}$, 8

Lesson 6-3, pages 254–257

1. $y = -3x - 2$ 3. $y = 7x + 2$ 5. $y = -5x - 1$ 7. $y = 9$ 9. $y = -\frac{3}{4}x + \frac{11}{4}$ 11. $y = \frac{2}{3}x - 3$ 13. $y = \frac{-1}{3x} - \frac{2}{3}$ 15. $y = \frac{-1}{2}x - \frac{2}{5}$ 17. $y = \frac{-3}{13}x + \frac{12}{13}$ 19. $x = 2$ 21. $d = 2t$ 23. $9y + 2x = 32$ and $y + 6x = 18$ 25. $y = 0.01px + F$ where x = sales y = pay 27. 96 cm^2 29. {2, 3, 5, 6, 7, 8, 11, 12} 31. {4, 9, 13} 33. {1, 4, 9, 10, 13, 14}

Lesson 6-4, pages 258–261

1. yes 3.

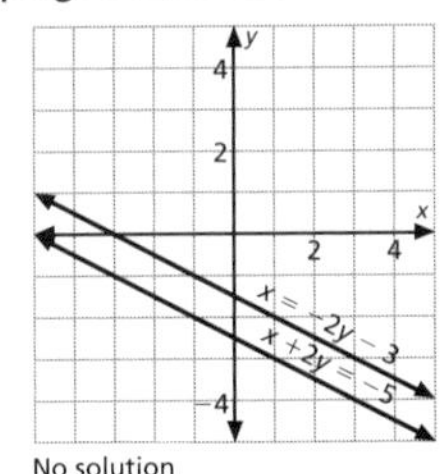

No solution

5.

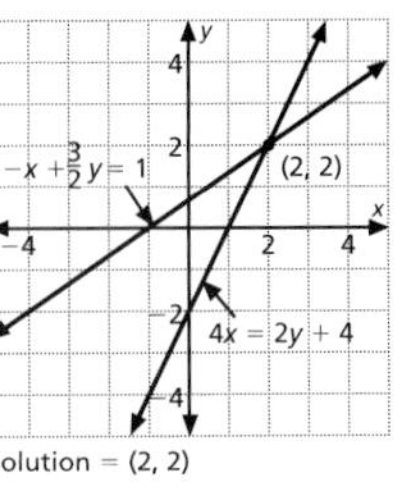

Solution = (2, 2)

7. (3, 2) 9.

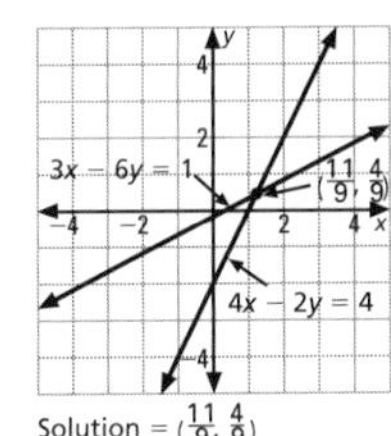

Solution = $(\frac{11}{9}, \frac{4}{9})$

11.

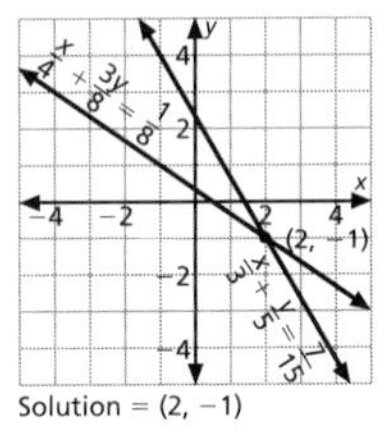

Solution = (2, −1)

13. food, 7; personal care, 3 15. \$10,092.50 19. no solution

21.

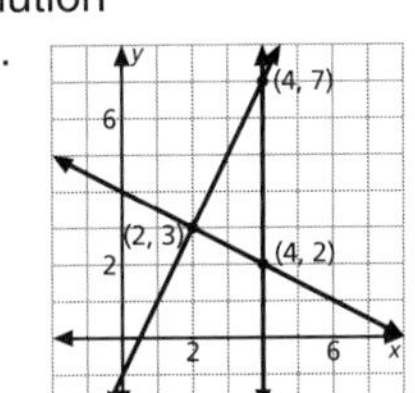

Solutions = (2, 3), (4, 2), (4, 7)

23. 15.48 in.^{2} 25. 75.98 in.^{2} 27. 190,000,000 29. 6,520,000,000,000

Review and Practice Your Skills, pages 262–263

1. $y = 4.5x - 7.5$ 3. $y = -4$ 5. $y = -\frac{5}{4}x + 13$ 7. $y = x + 0.5$ 9. $y = 300x + 530$ 11. $y = 2x + 7$ 13. $y = \frac{7}{2}x + 2$ 15. $y = -4x + 6$ 17. $x = 13$ 19. $y = \frac{1}{2}x + 7$ 21. $x = 3$ 23. $y = 3x + 2$ 25. $y = 9$ 27. $y = \frac{3}{4}x + 1$ 29. $y = \frac{3}{2}x - 12$ 31. $(-3, -4)$ 33. (8, 8) 35. parallel lines (no solution) 37. (9, 4) 39. $(4, -2)$ 41. 1 43. 5, -8 45. $m = 1, b = -3\frac{1}{2}$ 47. $m = \frac{3}{2}, b = -7$ 49. $m = 0, b = -12$ 51. parallel 53. perpendicular 55. $y = -x - 8$ 57. $y = 36$ 59. $y = 3x + 15$ 61. (45, 36)

Selected Answers

Lesson 6-5, pages 264–267
1. $(-1, 2)$ 3. $(-1, -2)$ 5. length 24 cm; width, 15 cm 7. $(4, 0)$ 9. $(1, -2)$ 11. $(5, -3)$ 13. Since it doesn't matter which variable you solve for first, choose the variable that will be easy to isolate in one of the equations.
15. \$1.19, \$0.79 17. $\left(\frac{4}{3}, \frac{6}{5}\right)$ 19. 1975 21. $\left(\frac{1}{2}, -\frac{1}{3}\right)$
23. $(-1, 1, -2)$ 25. 15 nickels, 9 dimes, 3 quarters 27. 12 29. 36 31. -72 33. -9408

Lesson 6-6, pages 268–271
1. $(4, 5)$ 3. $(-3, 1)$ 5. 13 adults, 43 children
7. $7\frac{1}{2}$ months 9. $(-1, 1)$ 11. $(4, 3)$ 13. $\left(0, \frac{1}{9}\right)$
15. $(-2, 4)$ 17. wheat, 950 acres; barley, 250 acres
19. $(2, 3)$ 21. $(3, 9)$ 23. $(-4, 5)$ 25. 15 and 18
27. $(10, 21)$ 29. $(-1, 2, -1)$ 31. 29,000,000 33. 35,500; 15,837,000,000 35. 15,300; 2,942,000,000
37. 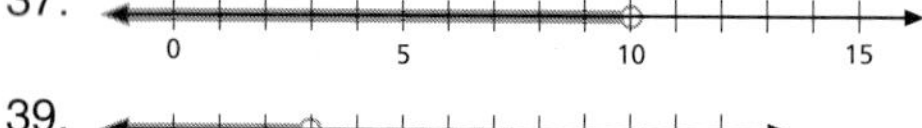

39.

41.

43.

Review and Practice Your Skills, pages 272–273
1. $(15, 3)$ 3. $(3, 2)$ 5. $\left(\frac{1}{3}, 2\right)$ 7. $(2, -7)$ 9. $(2, -3)$
11. $(0, 2)$ 13. $(7, -1)$ 15. $\left(-1, \frac{2}{7}\right)$ 17. bagel, \$0.55; peach, \$0.35 19. $(-1, 1)$ 21. $(-2, 0)$ 23. $(-2, 1)$
25. $(-3, -2)$ 27. $(-1.5, 0.25)$ 29. $\left(-\frac{1}{2}, 1\right)$ 31. $\left(-11\frac{2}{3}, 5\right)$
33. $(0, -5)$ 35. no solution
37. 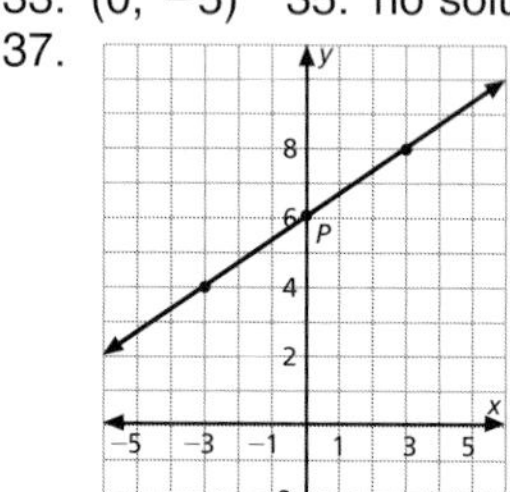
39. 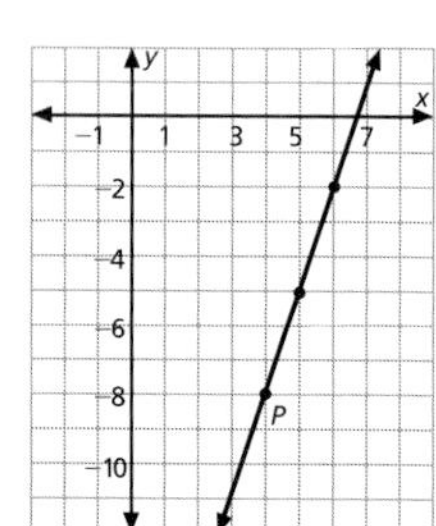
41. 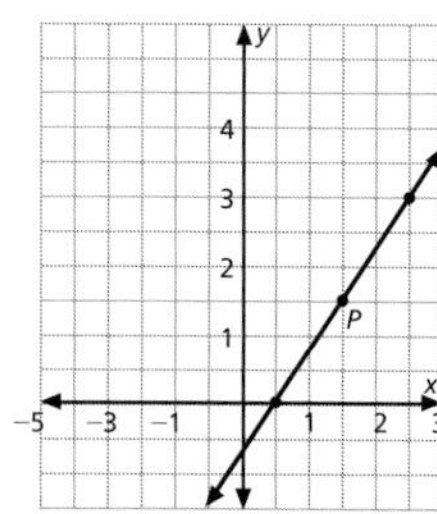
43. perpendicular 45. neither
47. $y = -3x + 29$
49. $y = 1.5x + 6$
51. $y = \frac{-1}{5}x + 1$
53. $(3, -1)$ 55. $(-2, -1)$

Lesson 6-7, pages 274–275
1. $\begin{bmatrix} 5 & 1 \\ -3 & 4 \end{bmatrix} \begin{bmatrix} x \\ y \end{bmatrix} = \begin{bmatrix} 6 \\ 2 \end{bmatrix}$; det = 23 Solution $\left(\frac{22}{23}, \frac{28}{23}\right)$
3. $\begin{bmatrix} 4 & -7 \\ -2 & 1 \end{bmatrix} \begin{bmatrix} x \\ y \end{bmatrix} = \begin{bmatrix} 2 \\ -4 \end{bmatrix}$; det $= -10$ Solution $\left(\frac{13}{5}, \frac{6}{5}\right)$
5. $\begin{matrix} 25D + 0.35M = 230 \\ 35D + 0.25M = 250 \end{matrix}$ $\begin{bmatrix} 25 & 0.35 \\ 35 & 0.25 \end{bmatrix} \begin{bmatrix} D \\ M \end{bmatrix} = \begin{bmatrix} 230 \\ 250 \end{bmatrix}$
det $= -6$; $D = 5$ days; $M = 300$ miles
7. 41 in. 9. 129 in. 11. 58.4 in.

Lesson 6-8, pages 276–279
1. yes 3. $y < 4$, $x \geq 3$
5. 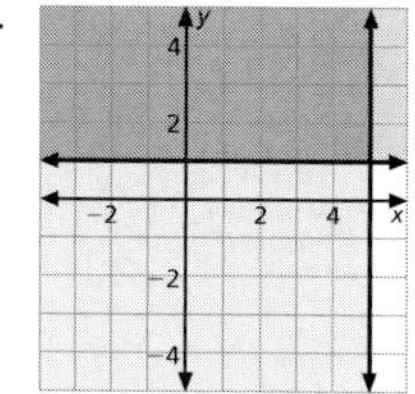
7. 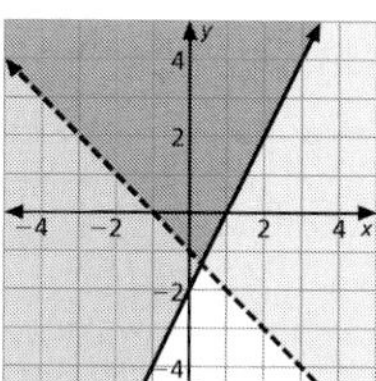
9. no 11. $y \geq x - 3$; $y \geq \frac{-3}{4}x + 4$
13. 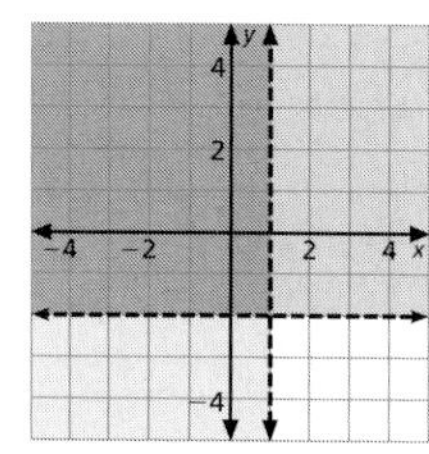
15. 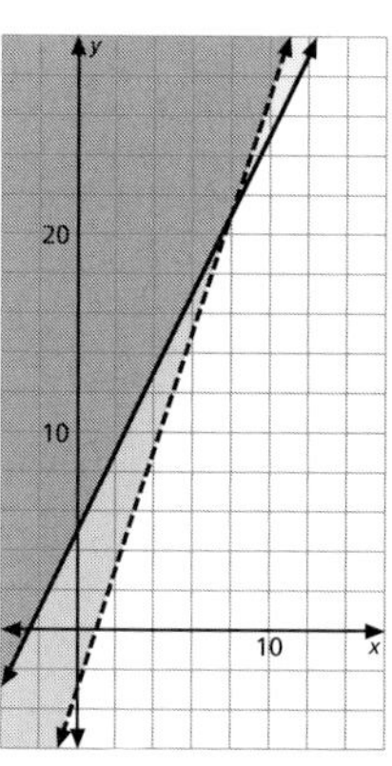
17. Solutions of $y < 2x + 3$ do not include solutions where $y = 2x + 3$. 19. $x \geq 0$; $y \leq x + 2$; $y < \frac{-3}{2}x + 5$
21. $y \geq x + 44$; $y \geq 2x - 16$
23. 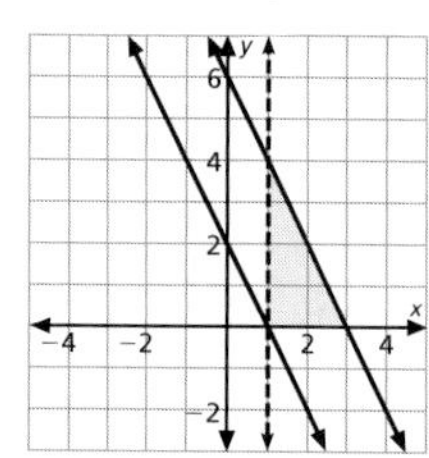
25. 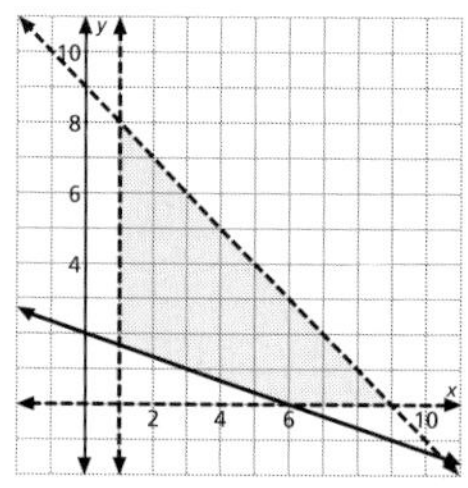
27. 78 29. 4800

Review and Practice Your Skills, pages 280–281
1. $(1, -5)$ 3. $(1, 3)$ 5. $(7, 8)$ 7. $(-1, 2)$ 9. $\left(0, -\frac{7}{3}\right)$
11. $(6, 9)$ 13. $(2.8, 0.8)$ 15. $(-2, 5)$ 17. yes 19. no

21. 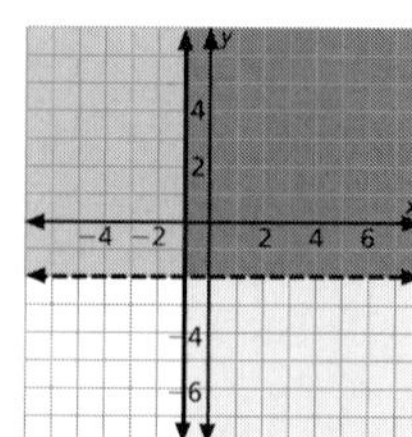
23. 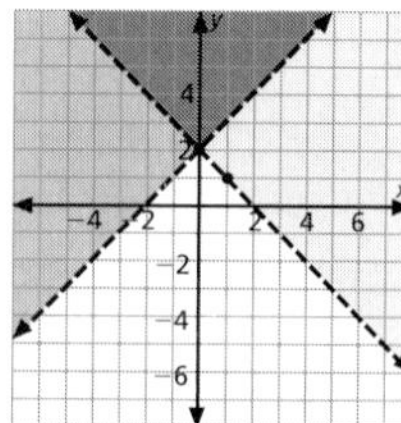

25. 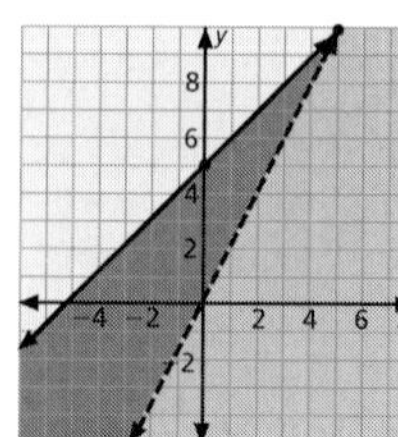
27. 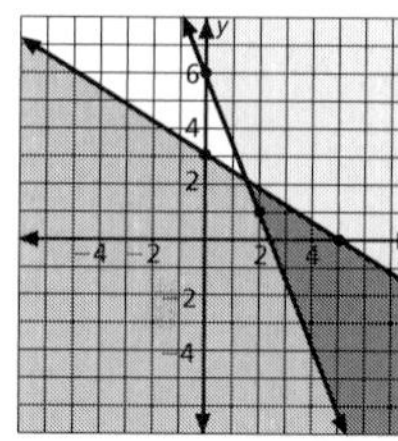

29. 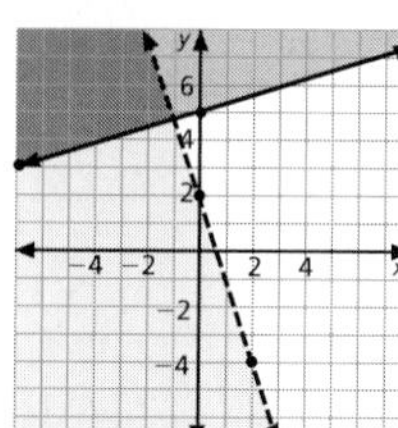
31. 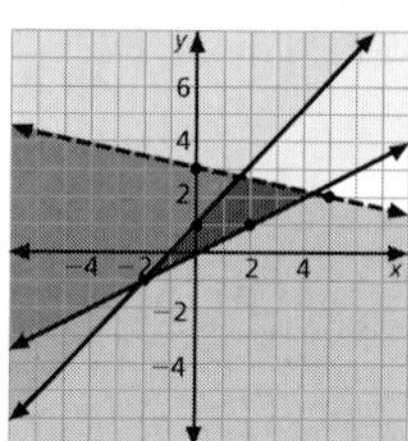

33. 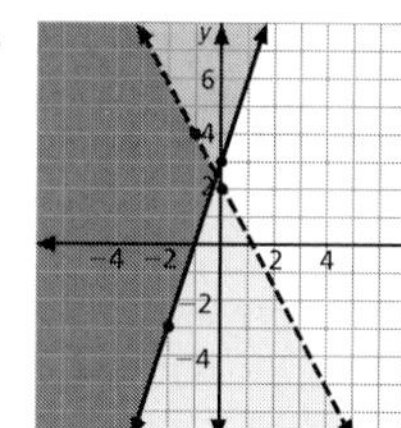

35. $m = \frac{5}{7}$, $b = -11$ 37. $m = \frac{5}{7}$, $b = 11$ 39. $m = -3$, $b = 13$ 41. parallel 43. perpendicular

45. $y = \frac{2}{3}x - 13$

47. $y = \frac{-2}{5}x + 7$

49. $y = 2.5x + 11$ 51. (5, 0) 53. infinite number of solutions 55. no solution 57. (4, 2) 59. (−5, 0) 61. (−8, 3) 63. (5, −4) 65. (−2, −4) 67. (2, −4) 69. (3, −1) 71. (−6, −43) 73. (1, 2)

75. 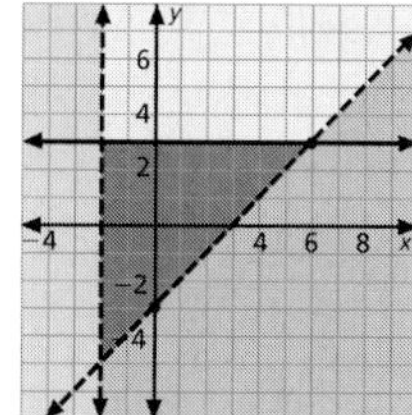

Lesson 6-9, pages 282–285

1. no 3. yes 5. maximum 225 at (15, 0)

7. 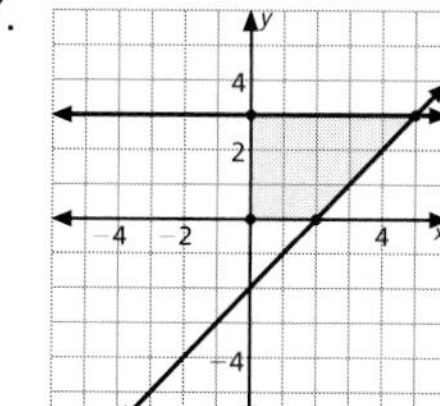

Max P = 28 at (5, 3)

9. maximum 140 at (5, 15); minimum 60 at (0, 10) 11. maximum 26.5 at (20, 2); minimum 3 at (0, 4)
13. These constraints insure that the feasible region will be confined to the first quadrant. This eliminates the possibility of a negative coordinate that would skew the results of certain computations and applications.. 15. (0, 8), (4, 0), (8, 0)
17. minimum 12 at (0, 6)

19. 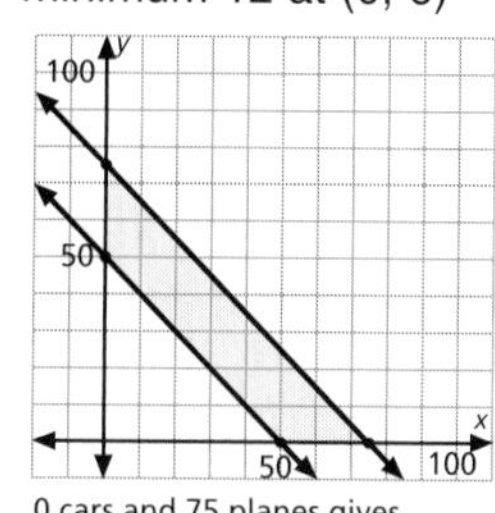

0 cars and 75 planes gives maximum profit of $2,625

21. $x \geq 15$; $y \geq 10$; $x + y \leq 50$ 23. They should sell 15 T-shirts and 10 sweatshirts. 25. Plant 1100 acres of Iceberg and 2500 acres of Romaine to yield a maximum profit of $845,000. 27. −142.272 29. 25.5 31. −12

Chapter 6 Review, pages 286–287

1. c 3. a 5. −11 7. (0, 1) 9. $y = -2x + 1$ 11. neither 13. $y = -x + 5$ 15. $y = -\frac{4}{3}x - \frac{5}{3}$

17. 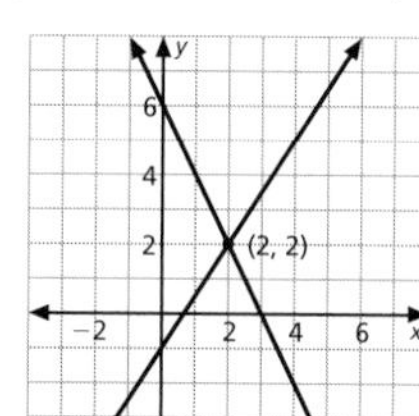

19. (−1, 0)

21. (3, 2) 23. (2, −1)

25. 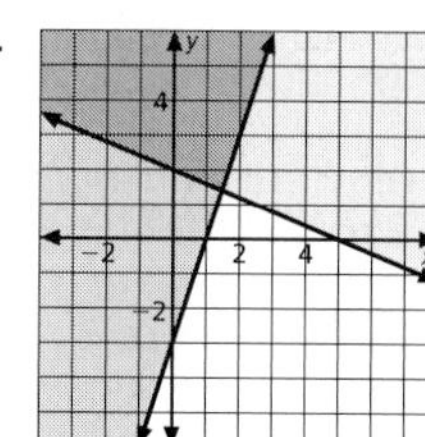

27. 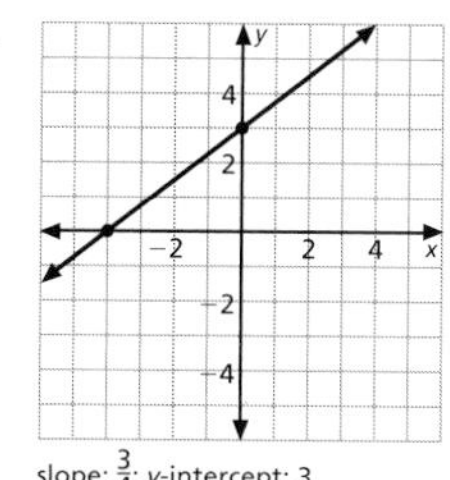

slope: $\frac{3}{4}$; y-intercept: 3

Chapter 7

Lesson 7-1, pages 296–299

1. yes 3. yes 5. 2 7. $900, $1200 9. no 11. no 13. 2.1 15. 988 17. 630 19. 1,400 CDs 21. 7.5 pounds 23. 100 : 10 = 30 : 3 25. 3 27. 5 29. 8 hours 31. 36°, 54° 33. The terms in the two ratios are not written in the same order. The correct proportion is either $\frac{3}{10} = \frac{48}{x}$ or $\frac{10}{3} = \frac{x}{48}$. 35. 6 37. $\frac{6}{7}$ 39. $-\frac{5}{3}$ 41. $\frac{3}{2}$ 43. undefined 45. $\frac{7}{4}$ 47. $-\frac{5}{6}$ 49. $0 < x < 15$ cm 51. $0 < x < 27$ cm 53. $0 < x < 12$ dm 55. $0 < x < 51$ in.

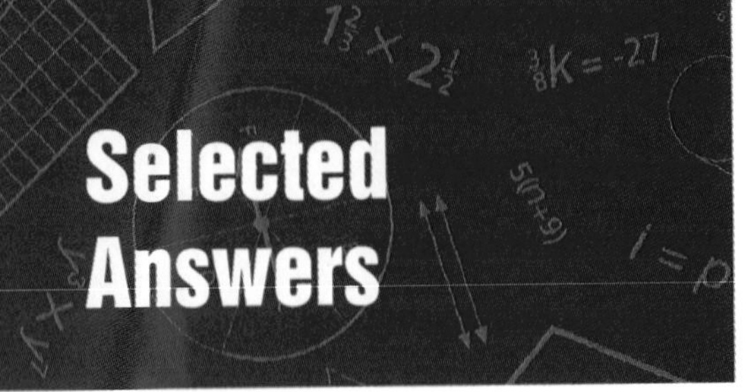

Selected Answers

Lesson 7-2, pages 300–303
1. yes 3. 100° 5. no 7. $6\frac{2}{3}$ ft 9. Corresponding angles are between corresponding sides. 11. 4.5 in. 13. 3 : 4 15. $7\frac{5}{9}$ by $9\frac{4}{9}$ 17. Answers will vary. (One possible answer involves two rhombi with different angles) 19. −3 21. $-\frac{3}{2}$ 23. $\frac{2}{3}$ 25. −2 27. 5

Review and Practice Your Skills, pages 304–305
1. no 3. yes 5. yes 7. 24 9. 5 11. 100 13. 180 15. $\frac{4}{3}$ 17. 0.036 19. $1\frac{1}{4}$ cups 21. 40.5^6 and 49.5° 23. yes 25. no 27. 5 m 29. 65 cm

Lesson 7-3, pages 306–309
1. 12 ft 3. 1.25 cm 5. 40 mi 7. 32 m 9. 7.5 m 11. 0.5 m 13. 10 mi 15. 18 mi 17. St. Lawrence, 1 in.; Columbia, about $1\frac{1}{2}$ in. 21. 15 ha 23. $\| = -\frac{5}{4}, \perp = \frac{4}{5}$

Lesson 7-4, pages 310–313
1. no 3. yes; AA 5. yes; AA 7. The rays of the sun form the same angle with the ground for both the pole and tree. Therefore, $\angle A \cong \angle D$. Because $\angle B$ and $\angle E$ are both right angles, $\angle B \cong \angle E$. Therefore, the triangles are similar by AA Similarity. 9. Yes; because $\angle 1 \cong \angle 2$, $\angle D \cong \angle D$, there are 2 pairs of congruent angles. The AA Similarity Postulate applies. 11. always 13. sometimes 15. $y = \frac{1}{2}x - 1$ 17. $y = x + \frac{1}{2}$ 19. $y = \frac{-9}{7}x + \frac{1}{7}$ 21. $y = 2x + 4$ 23. $y = -2x - 2$ 25. 11 in.

Review and Practice Your Skills, pages 314–315
1. 240 km 3. 0.02 ft 5. 172.5 mi 7. 5.55 yd 9. 1027 mi 11. 12.5 in. 13. 40 cm 15. 15 mm 17. 4 ft 19. 18 ft by 12 ft 21. yes, AA 23. no 25. no 27. −58.5 29. 6 31. 6 33. 960 cm 35. ~40 miles 37. ~56 miles

Lesson 7-5, pages 316–319
1. 4.5 3. 3.3 5. given; definition of Altitude; altitudes of similar triangles in the same proportion as corresponding sides 7. 3.2 9. The given triangles are similar, so $RW{:}AK = RY{:}AL$. Because RX is half the length of RY, and AB is half the length of AL, then RX and AB have the same ratio, $\angle R \cong \angle A$, because they are corresponding parts of similar triangles. You now have $\triangle WRX \sim \triangle KAB$ by SAS Similarity Postulate. Then, $WX{:}KB = WR{:}KA$, because they are corresponding parts of similar triangles. 11. 6.125 xy 13. 33.3 cm 15. $\left(\frac{-5}{3}, \frac{-4}{3}\right)$ 17. 6 19. $\frac{3}{7}$

Lesson 7-6, pages 320–323
1. 2.5 3. 6 7. 3 11. 8.6 13. $yw : xy = zw : xz$ 15. BE 17. congruent base angles mean triangle is isosceles 19. $\left(\frac{7}{5}, \frac{11}{5}\right)$ 21. $\left(\frac{-1}{10}, \frac{43}{10}\right)$ 23. $\left(\frac{27}{5}, \frac{3}{5}\right)$ 25. mean: 10; median: 9.5; mode: 10

Review and Practice Your Skills, pages 324–325
1. 24.5 m 3. 6 ft 5. 27 m 7. 20 in. 9. 46.5 m 11. 4 in. 13. 21 cm 15. 130° 17. true 23. 5 25. −6.5 27. 5.5 29. 11.4 m

Lesson 7-7, pages 326–327
1. Answers may vary. 3. 8.4 m 5. Answers will vary. 7. (5.5, 8.5) 9. (1, 2) 11. $\left(\frac{39}{16}, \frac{-1}{16}\right)$ 13. 20 : 36

Chapter 7 Review, pages 328–329
1. c 3. d 5. 4.375 7. 7.5 9. 10 m 11. $52\frac{1}{2}$ mi 13. 3.8 15. 24.8 17. 12 ft

CHAPTER 8

Lesson 8-1, pages 338–341

1. 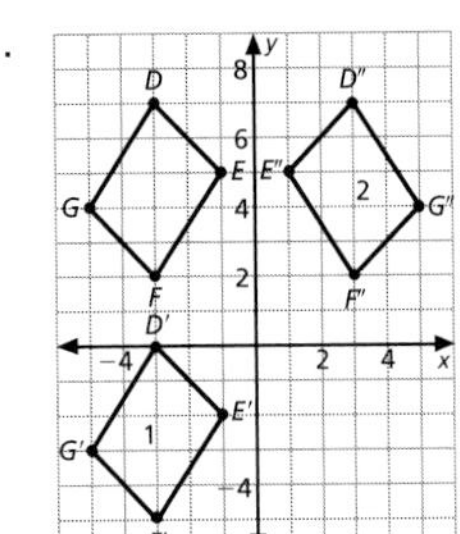

3. $-1, -1, 1, \frac{3}{2}, \frac{3}{2}, -\frac{3}{2}$ 5. If the figure on one side of a line is the reflection of the figure on the opposite side, the line is a line of symmetry.

7. 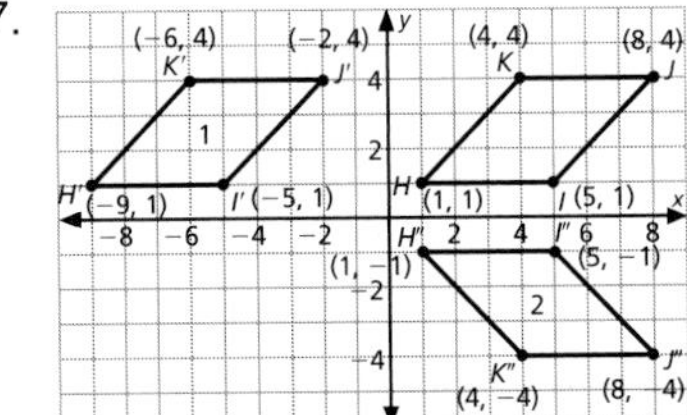

9. 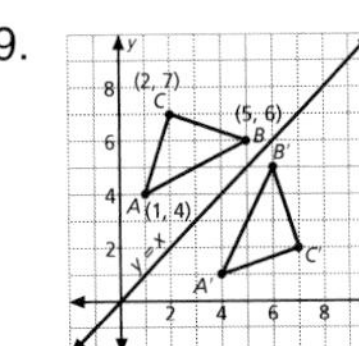

side	AB	BC	CA	$A'B'$	$B'C'$	$C'A'$
slope	$\frac{1}{2}$	$-\frac{1}{3}$	3	2	−3	$\frac{1}{3}$

11. 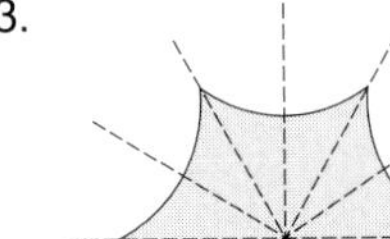

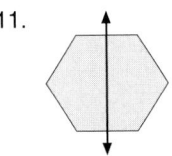

13. 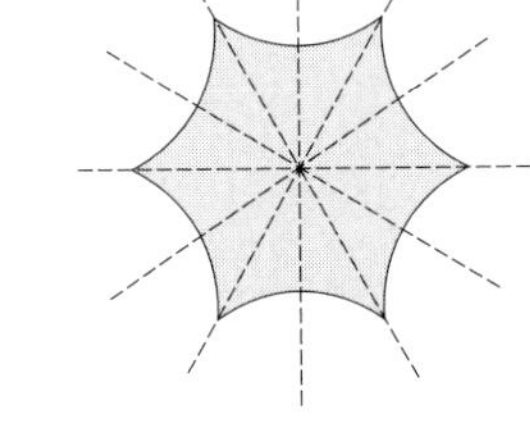

15.

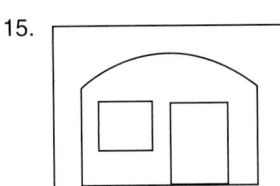

17. The slopes of translated figures are equal. 19. 15 21. 18 23. 2 25. 4 27. −2 29. 5 31. *l* 33. $\overline{DF}, \overline{CG}, \overline{BH}, \overline{AI}$

Lesson 8-2, pages 342–345

1 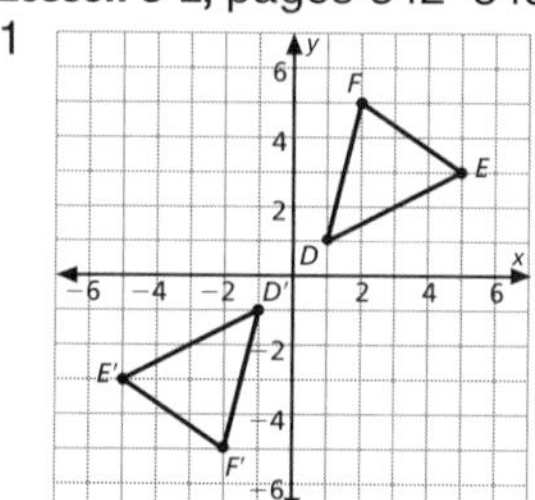

; the slopes are equal.

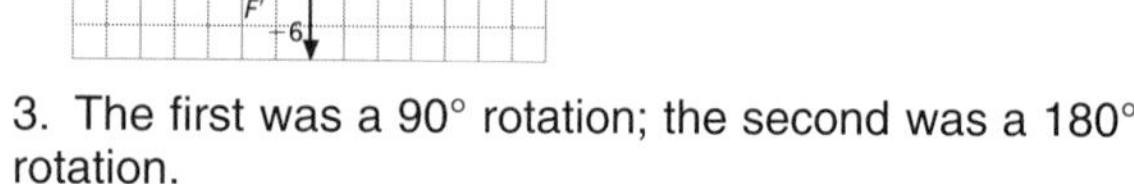
3. The first was a 90° rotation; the second was a 180° rotation.

5. 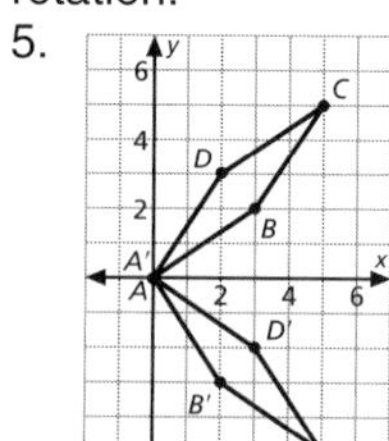

7. 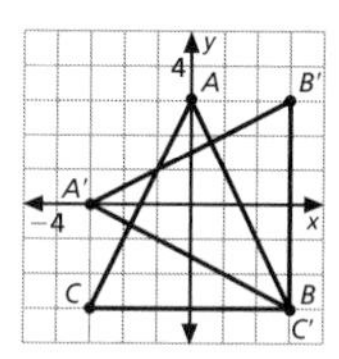

9. Triangle 6 11. Triangle 5 13. Triangle 8
15. The slopes of corresponding sides are equal; the product of corresponding slopes is −1; the product of corresponding slopes is −1. 17. 10 19. $\frac{1}{7}$ 21. $-\frac{6}{7}$ 23. $\frac{6}{5}$ 25. $-\frac{2}{3}$
27. undefined slope 29. $-\frac{3}{4}$

Review and Practice Your Skills, pages 346–347

1–3. 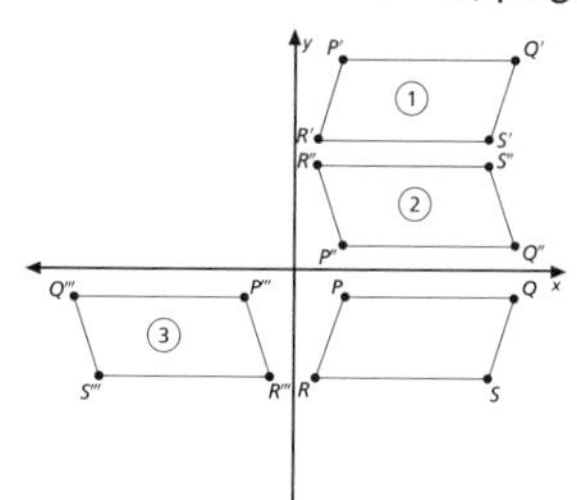

5.–7. 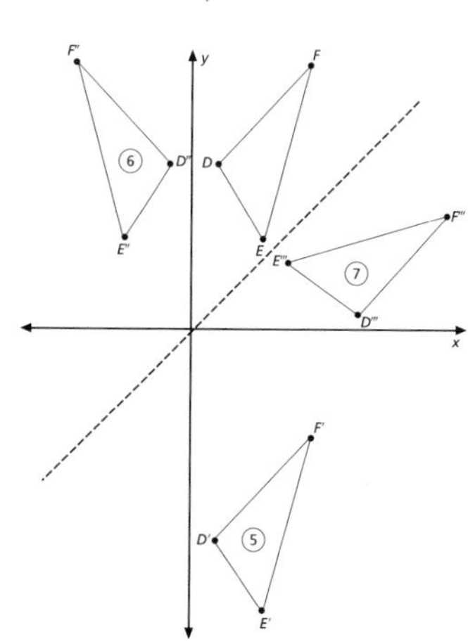

9.–11. 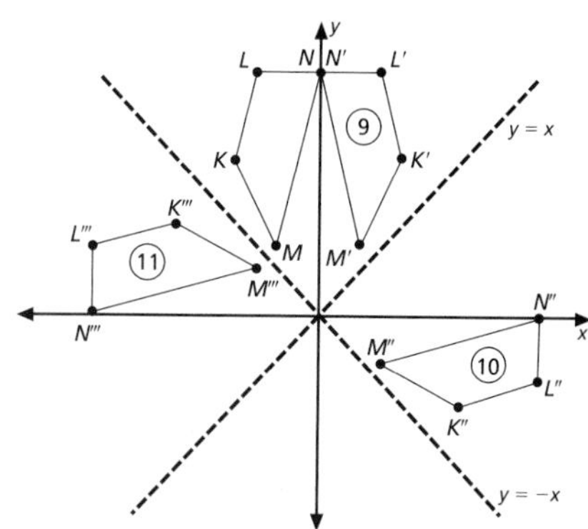

13.–15. 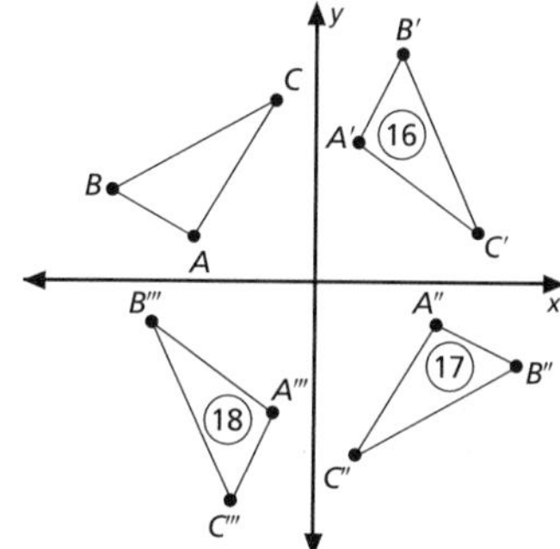

17. 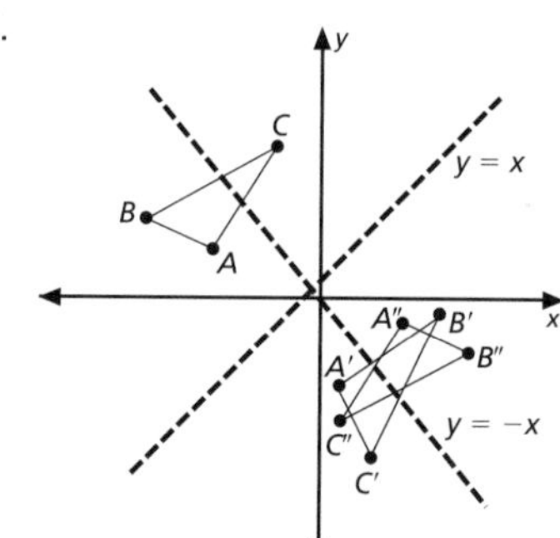

19.–21. 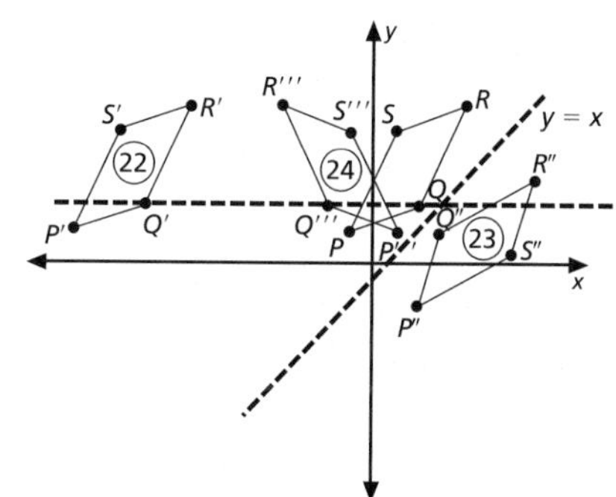

Selected Answers

23.–25.

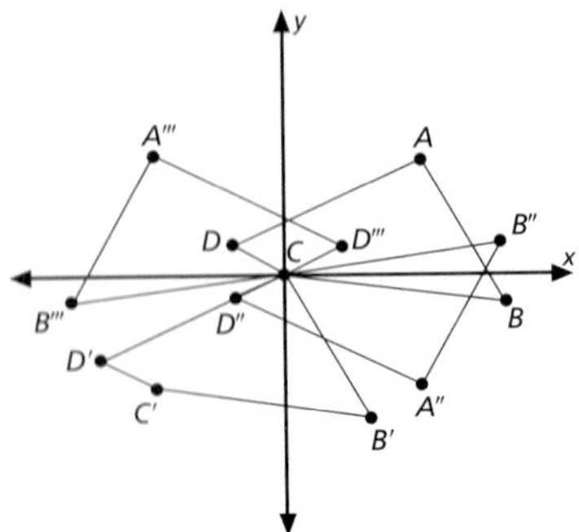

27.–29.

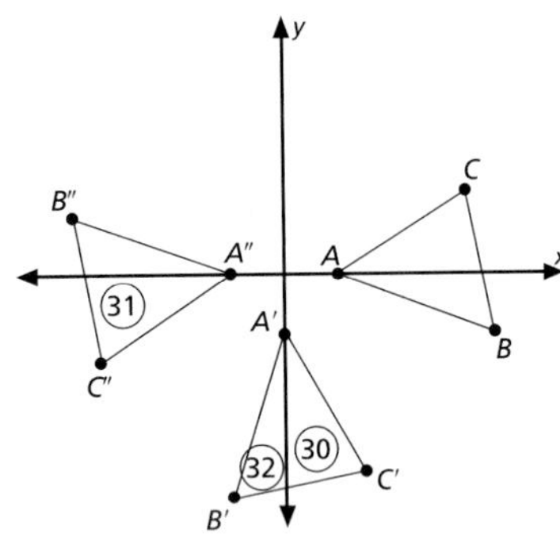

Lesson 8-3, pages 348–351

1. square $A'B'C'D'$ 3. 4 times as long.

5. 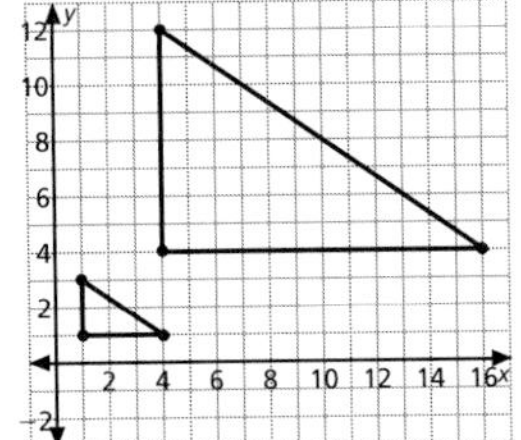

7. 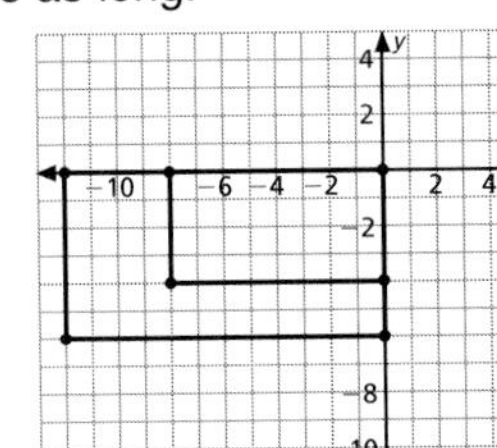

9. 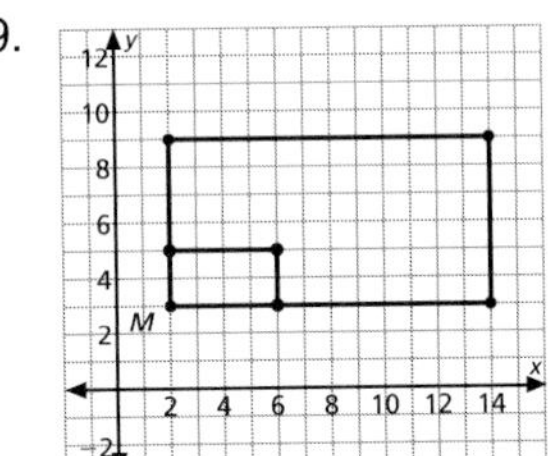

11. 1.25 13. 3 17. 24 square units; 96 square units; 6 square units; 4 times as large; $\frac{1}{4}$ as large 19. 48 ft 21. 60 km 23. 54 m 25. 81 m 27. 102 km 29. yes; domain: {−2, −1, 0, 1}; range: {−2, −1, 0, 1} 31. yes; domain: {−3, −2, 2, 3}; range: {−2, −1, 0, 1}

Lesson 8-4, pages 352–355

1. 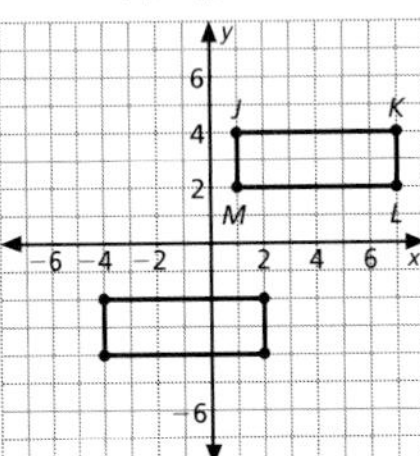

3. reflection over *x*-axis and translation 8 units right

5. 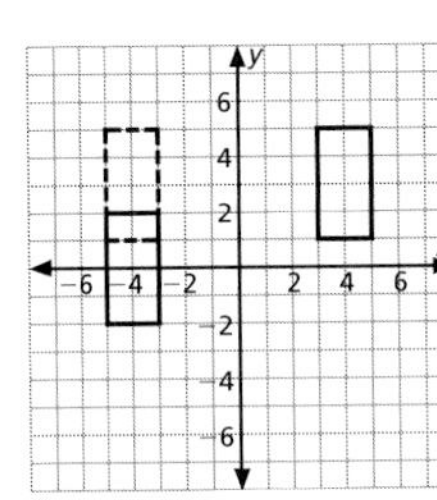

7. 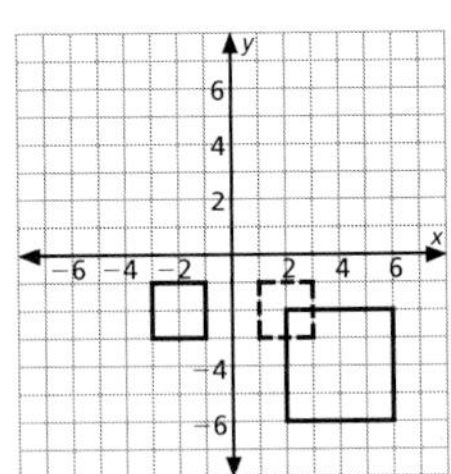

9. Answers will vary. 11. Possible answers are given; reflection over *y*-axis and translation 5 units up. 13. yes 15. yes 17. yes 19. rotation 21. AA 23. SAS

Review and Practice Your Skills, pages 356–357

1. 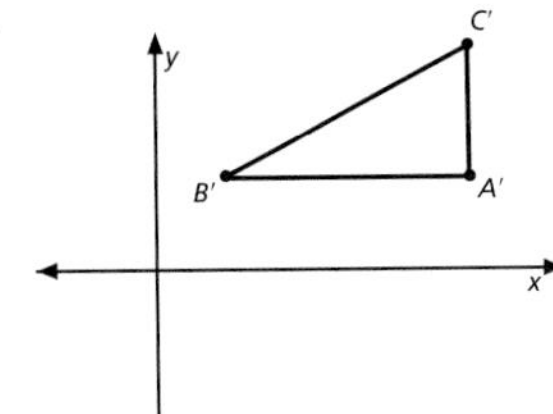

3. 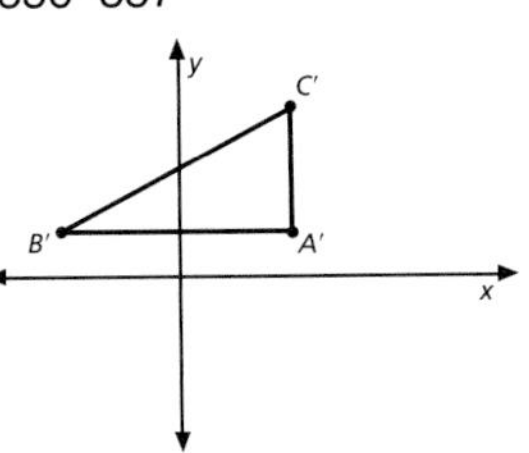

5. 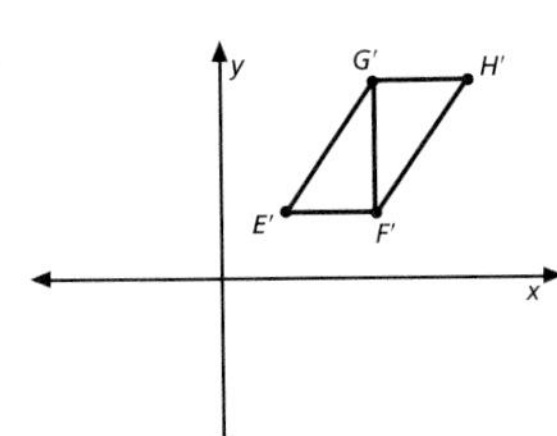

7. 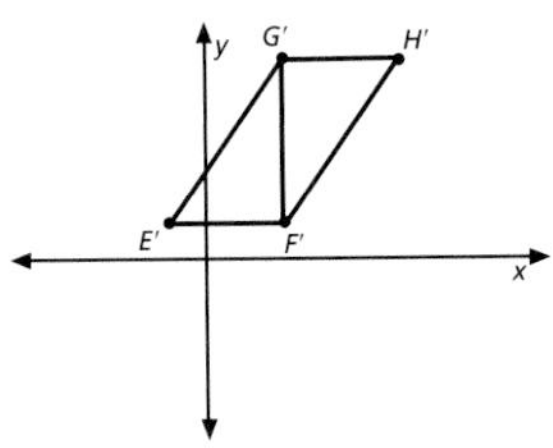

9. Scale factor: $\frac{1}{5}$; Center of dilation: $F(-5, 0)$

11. 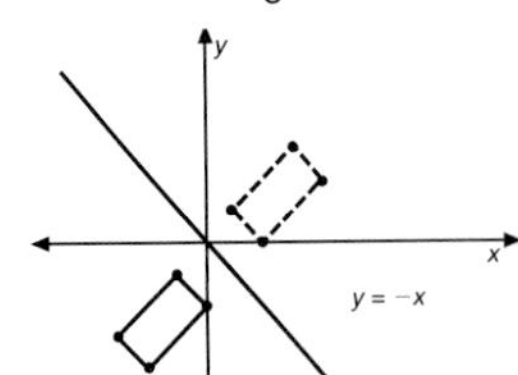

13. 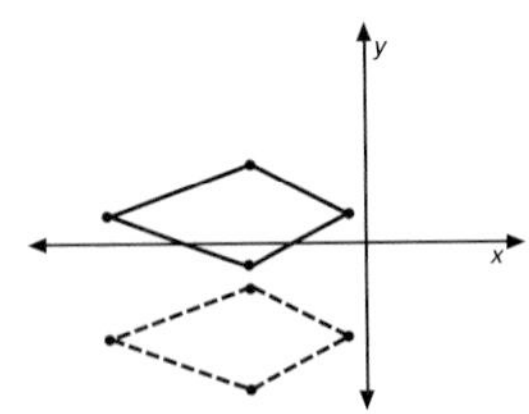

15. Answers will vary. Sample answers are given. Reflection across *x*-axis; rotation 90° clockwise

17–19.

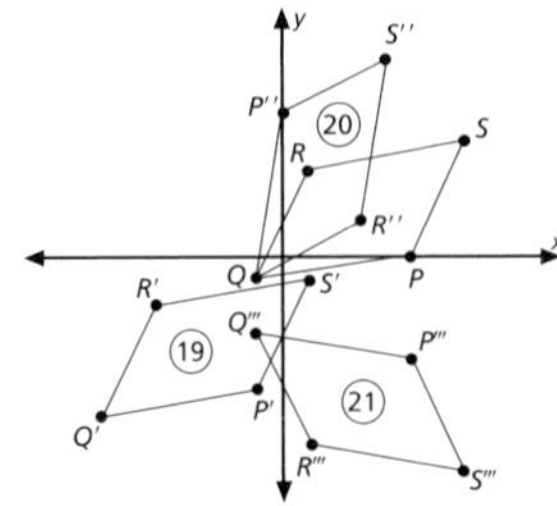

21–23.

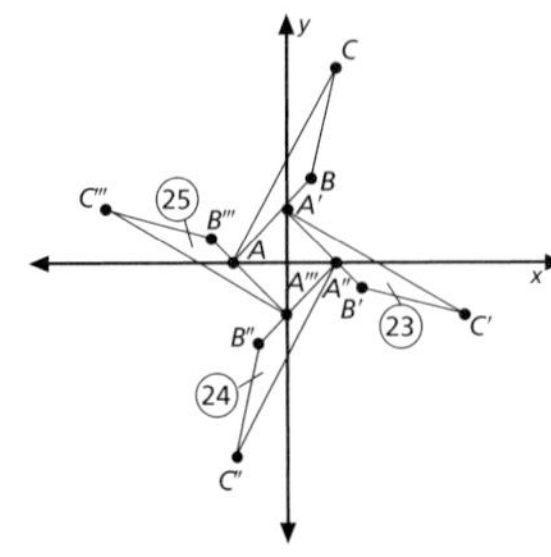

25. Scale factor: 3; center of dilation: origin

Lesson 8-5, pages 358–361

1. 3×4 3. 10 5. $\begin{bmatrix} 0 & -3 \\ 5 & 0 \end{bmatrix}$ 7. 2×3 9. 5×2

11. $\begin{bmatrix} 6 & -2 & -3 \\ -4 & 1 & 2 \end{bmatrix}$ 13. $\begin{bmatrix} -10 & -1 & 7 \\ 11 & -1 & -11 \end{bmatrix}$

15. $\begin{bmatrix} -13 & 5 & 1 \\ 26 & -13 & -11 \end{bmatrix}$ 17. $\begin{bmatrix} -4 & -3 & 4 \\ 7 & 0 & -9 \end{bmatrix}$

19. $\begin{bmatrix} 1 & 70 \\ 1 & 150 \\ 1 & 330 \end{bmatrix}$ $\begin{bmatrix} 1 & 1 & 1 \\ 70 & 150 & 330 \end{bmatrix}$ 21. Answers will vary. 23. $\begin{bmatrix} 4 & 4 \\ 3 & 3 \end{bmatrix}$

25. [20 11 3] 27. $x = \frac{13}{2}$, $y = \frac{5}{2}$

29. $R + S = S + R = \begin{bmatrix} 10 & 11 \\ 4 & 3 \end{bmatrix}$; yes

31. $T = \begin{bmatrix} 894 & 930 & 1011 & 949 \\ 875 & 965 & 969 & 980 \end{bmatrix}$ 33. 1 35. 11

37. 12 39. 75

Lesson 8-6, pages 362–365

1. 3×4 3. 2×4 5. NP 7. NP 9. [163] 11. NP 13. 3×3 15. 1×1 17. NP 19. 1×2 21. [28]

23. NP 25. NP 27. $\begin{bmatrix} 12 & 13 & 11 \\ 13 & 10 & 16 \end{bmatrix}$ 29. $AB = \begin{bmatrix} 7 & 4 \\ 11 & 10 \end{bmatrix}$

$BA = \begin{bmatrix} 14 & 8 \\ 2 & 3 \end{bmatrix}$ 31. $AI = \begin{bmatrix} 3 & 2 \\ -1 & 5 \end{bmatrix}$; $IA = \begin{bmatrix} 3 & 2 \\ -1 & 5 \end{bmatrix}$

33. $A(BC) = (AB)C = \begin{bmatrix} 9 & -4 \\ -45 & 33 \end{bmatrix}$ 35. $\begin{bmatrix} 1 & 0 \\ 0 & 1 \end{bmatrix}$, $\begin{bmatrix} 0 & -1 \\ -1 & 0 \end{bmatrix}$, $\begin{bmatrix} 1 & 0 \\ 0 & 1 \end{bmatrix}$

37. $\begin{bmatrix} 90{,}200 & 70{,}000 & 119{,}500 \\ 233{,}400 & 181{,}200 & 309{,}000 \end{bmatrix}$ 39. $x = 1, 4 = -1$

41. $(-1, 3)$ 43. $(2, 3)$ 45. $(6, 5)$

Review and Practice Your Skills, pages 366–367

1. $\begin{bmatrix} 12 & 0 & -15 & 21 \\ -9 & 24 & -6 & 3 \end{bmatrix}$ 3. $\begin{bmatrix} -4 & -8 & 18 & -12 \\ 4 & -12 & 9.5 & 8 \end{bmatrix}$

5. $\begin{bmatrix} 4 & 8 & -18 & 12 \\ -4 & 12 & -9.5 & -8 \end{bmatrix}$ 7. $\begin{bmatrix} -9 & 3 & -3.5 & 9.5 \\ 3.5 & 6.5 & 1 & -55 \end{bmatrix}$

9. $\begin{bmatrix} 15 & -11 & 9 & -4 \\ -7 & 1.5 & 5.5 & 16 \end{bmatrix}$ 11. $\begin{bmatrix} 8 & -16 & 16 & 4 \\ -4 & 8 & 11 & 20 \end{bmatrix}$

13. $\begin{bmatrix} 4.4 & -1.2 & 0.4 & -2.4 \\ -2 & -1 & 0 & 2.4 \end{bmatrix}$ 15. $\begin{bmatrix} -4 & 0 & 5 & -7 \\ 3 & -8 & 2 & -1 \end{bmatrix}$ 17. 0

19. 2×4 21. $x = -5$; $y = 3$ 23. $x = -1$; $y = 7$

25. $x = 0$; $y = -\frac{1}{2}$ 27. $\begin{bmatrix} 15 & -4 \\ -5 & 8 \end{bmatrix}$ 29. NP 31. $\begin{bmatrix} 4 & 2 \\ -12 & 19 \end{bmatrix}$

33. $\begin{bmatrix} -28 & 26 \\ 22 & 6 \\ -2 & 14 \end{bmatrix}$ 35. $\begin{bmatrix} -2 & 2 & 14 \\ -34 & 4 & -22 \end{bmatrix}$ 37. $\begin{bmatrix} -20 \\ -70 \end{bmatrix}$ 39. $\begin{bmatrix} -10 & -8 \\ -45 & 38 \\ -25 & 14 \end{bmatrix}$

41. $\begin{bmatrix} -78 & 62 \\ -12 & 42 \end{bmatrix}$ 43. $\begin{bmatrix} -50 & 2 & -66 \\ -3 & 9 & 73 \\ -23 & 5 & 5 \end{bmatrix}$ 45. $\begin{bmatrix} -68 \\ -84 \\ -68 \end{bmatrix}$ 47. $\begin{bmatrix} -63 & 58 \\ -17 & 50 \end{bmatrix}$

49. $\begin{bmatrix} 10 & -15 \\ 10 & -5 \end{bmatrix}$ 51. $\begin{bmatrix} 22 & 2 & 54 \\ -91 & 19 & 13 \end{bmatrix}$ 53. $\begin{bmatrix} -276 & 122 \\ -168 & 166 \end{bmatrix}$

55. $\begin{bmatrix} 20 \\ -650 \\ -290 \end{bmatrix}$ 57. $\begin{bmatrix} 34 \\ -68 \end{bmatrix}$ 59. $x = -3$; $y = -5$

61–63.

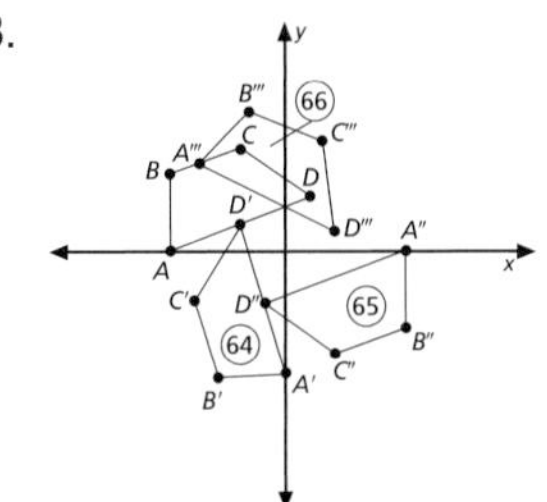

65. yes;

Rotation → Translation →

67. no affect

Lesson 8-7, pages 368–371

11. $\begin{bmatrix} -3 & 1 & -3 \\ 1 & 3 & -3 \end{bmatrix}$ 3. $\begin{bmatrix} 1 & 5 & 7 & -4 \\ 3 & 3 & -3 & -3 \end{bmatrix}$ 5. $\begin{bmatrix} 2 & 7 & 3 \\ -1 & 3 & 7 \end{bmatrix}$

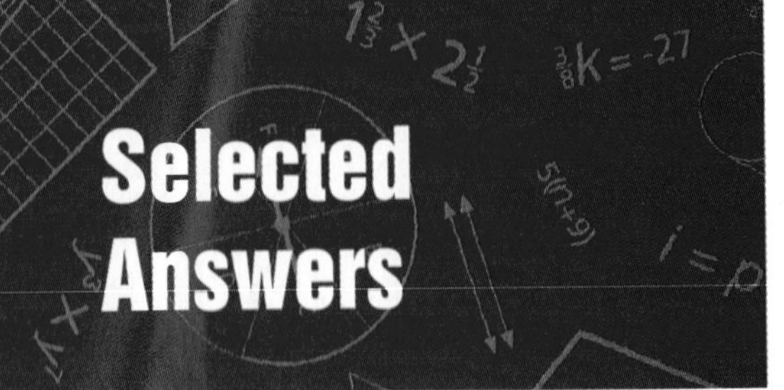

Selected Answers

7. $\begin{bmatrix} -2 & -7 & -3 \\ 1 & -3 & -7 \end{bmatrix}$ 9. $\begin{bmatrix} -3 & 3 & 4 & -2 \\ 2 & 4 & 1 & -1 \end{bmatrix}$ 11. $\begin{bmatrix} 7 & 3 & 4 & 7 \\ -2 & 1 & 7 & 4 \end{bmatrix}$

13. $\begin{bmatrix} -7 & -3 & -4 & -7 \\ 2 & -1 & -7 & -4 \end{bmatrix}$ 15. (4, 4), line $y = x$, $(-4, -4)$ 17. $(-2, 4)$, line $y = x$, $(4, -2)$ 19. yes 21. $y = -x$ 23. y-axis 25. Dilation with center at origin and a scale factor of 2 27. c; b; a 29. $\frac{1}{2}$, -3 31. $\frac{1}{2}$, 3 33. $\frac{4}{3}$, $\frac{7}{3}$

Lesson 8-8, pages 372–373
1. A: \$1,602.50; B: \$1,748.25; C:\$2,344.50 3. A: \$465; B: \$571.25; C: \$586.25 5. 6300° 7. 4860° 9. 10,800° 11. 10,080°

Chapter 8 Review, pages 374–375
1. e 3. d 5. a 7.

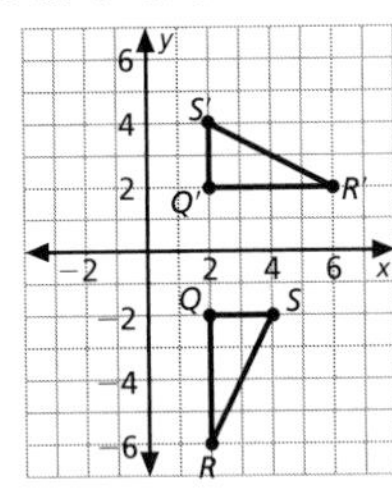

9. Possible answer: reflection over x-axis and translation 8 units to right. 11. -8, -4.3

13. $\begin{bmatrix} -9 & 6 & -11 & 15 \\ 14 & -20 & 3 & -5 \end{bmatrix}$ 15. $\begin{bmatrix} 9 & 15 & 24 & 0 \\ 3 & 24 & 15 & 12 \end{bmatrix}$ 17. $\begin{bmatrix} -4 & -6 & -5 \\ 2 & 1 & 4 \end{bmatrix}$

CHAPTER 9

Lesson 9-1, pages 384–387
1. 0.70 3. 20 times 5. 0.18 7. 0.55 9. Answers will vary. 11. (A, 1), (A, 2), (A, 3), (B, 1), (B, 2), (B, 3), (C, 1), (C, 2), (C, 3) 13. $\frac{1}{3}$ 15. 0.75 17. 0.75; 375 19. Of past days when weather conditions were similar to those predicted for tomorrow, it rained 25% of the time. 21. experimentally 23. theoretically
25.

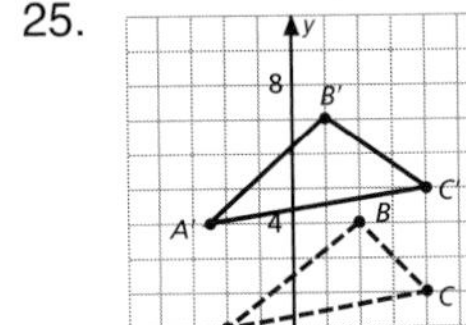

27.

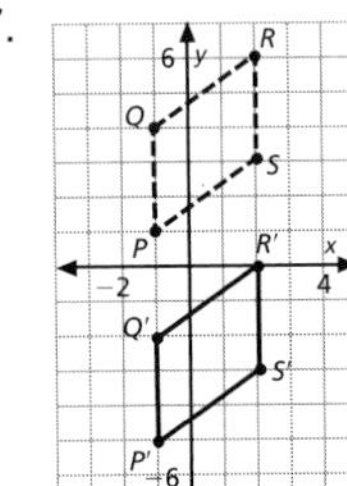

Lesson 9-2, pages 388–389
1. 1:5 (18 of the 90 possible numbers) 3. Answers will vary. 5. Answers will vary. 7. Change line 40 to: "IF $x < .4$ THEN $S = S + 1$" 9. $\frac{4}{7}$ 11. 3 13. $-\frac{1}{2}$ 15. undefined 17. 0 19. 15 21. 17.5 23. $\frac{9}{7}$ 25. $\frac{2}{3}$

Review and Practice Your Skills, pages 390–391
1. $\frac{1}{4}$ 3. $\frac{1}{2}$ 5. $\frac{2}{13}$ 7. $\frac{6}{13}$ 9. $\frac{1}{16}$ 11. $\frac{3}{8}$ 13. $\frac{5}{16}$ 15. $\frac{1}{16}$ 17. $\frac{1}{6}$ 19. $\frac{1}{2}$ 21. $\frac{1}{36}$ 23. $\frac{5}{36}$ 25. $\frac{1}{6}$ 27. $\frac{33}{36}$ 29. $\frac{2}{3}$, 400 31. $\frac{61}{225}\left(\frac{244}{900}\right)$ 33. Answers will vary. 35. Answers will vary. 37. (H, H) (T, H) (H, T) (T, T) 39. (1, H) (4, H); (1, T) (4, T); (2, H) (5, H); (2, T) (5, T); (3, H) (6, H); (3, T) (6, T)

Lesson 9-3, pages 392–395
1. $\frac{1}{3}$ 3. $\frac{7}{13}$ 5. $\frac{\text{number of pieces of paper} - 1}{\text{number of pieces of paper}}$ 7. $\frac{1}{3}$ 9. $\frac{1}{2}$ 11. $\frac{2}{9}$ 13. $\frac{1}{3}$ 15. $\frac{4}{5}$ 17. 1 19. $\frac{5}{13}$ 21. $\frac{1}{2}$ 23. 24 25. 21 27. Sample Answer: Multiples of 12 are common to both events. 29. 0.594 31. 400 33. -320 35. 2000 37. -500 39. 10,000 41. -16

Lesson 9-4, pages 396–399
1. $\frac{5}{48}$ 3. $\frac{2}{9}$ 5. $\frac{7}{120}$ 7. $\frac{1}{5}$ 9. $\frac{3}{25}$ 11. $\frac{21}{100}$ 13. $\frac{2}{23}$ 15. $\frac{2}{9}$ 17. $\frac{1}{192}$ 19. $\frac{1}{16}$ 21. $\frac{1}{6}$ 23. $\frac{2}{55}$ 25. Mutually exclusive events cannot occur at the same time. Independent events can occur at the same time, but they have no effect on the probability of each other. 27. $\frac{9}{16}$ 29. $\frac{27}{32}$
33.

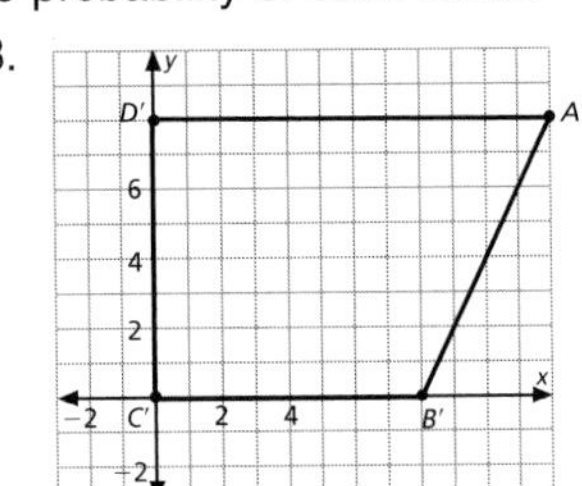

Review and Practice Your Skills, pages 400–401
1. $\frac{3}{52}$ 3. 1 5. $\frac{2}{13}$ 7. $\frac{4}{13}$ 9. $\frac{1}{26}$ 11. $\frac{5}{16}$ 13. $\frac{1}{2}$ 15. $\frac{5}{16}$ 17. $\frac{1}{18}$ 19. $\frac{11}{36}$ 21. $\frac{1}{36}$ 23. $\frac{5}{18}$ 25. $\frac{3}{5}$ 27. $\frac{1}{6}$ 29. $\frac{3}{8}$ 31. $\frac{7}{20}$ 33. $\frac{1}{20}$ 35. $\frac{121}{240}$ 37. $\frac{80}{9261}$ 39. $\frac{275}{9261}$ 41. $\frac{95}{441}$ 43. $\frac{20}{9261}$ 45. {HH, HT, TH, TT} 47. {Minnesota, Green Bay, Chicago, Tampa Bay, Detroit} 49. Answers will vary. 51. $\frac{1}{8}$ 53. $\frac{3}{8}$ 55. $\frac{3}{4}$ 57. $\frac{1}{297}$ 59. $\frac{1}{9900}$ 61. $\frac{2}{33}$

Lesson 9-5, pages 402–405
1. 120 3. combination; 20 5. permutation; 32,760 7. 30 9. 15,504 11. $\frac{1}{220}$ 13. $\frac{1}{15{,}600}$ 15. 1260; Sample answer: Divide 7! by $2! \cdot 2! \cdot 1! \cdot 1! \cdot 1!$ 17. 8! 19. They are the same. 21. They equal the total number of items.
23. $\begin{bmatrix} 3 & 7 \\ 1 & 4 \end{bmatrix}$ 25. 51° 27. 141°

Lesson 9-6, pages 406–409
1. About 130 lbs 3. 67–68 in. 5. above 7. 10 points 9. Artichokes 11. There is a greater range of batting averages in the upper 25% of its players than in the lower 25%. 13. Answers will vary. Sample answers are given; scatterplot 15. Yes, when there are extreme values
17. $\begin{bmatrix} 36 & 32 \\ 28 & 4 \end{bmatrix}$

Review and Practice Your Skills, pages 410–411
1. 60 3. 4 5. 840 7. 17,297,280 9. 10 11. 1 13. 8 15. 15 17. 6840 19. 3003 21. 210 23. 9
25.

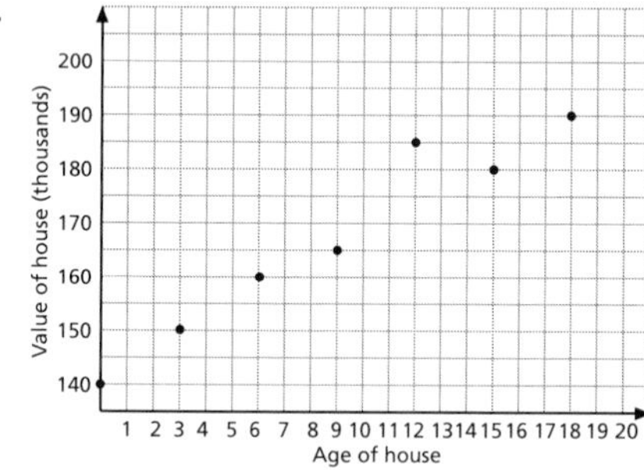

27. positive 29. Answers will vary. 31. $\frac{1}{4}$ 33. $\frac{5}{8}$ 35. 0 37. does 39. does

Lesson 9-7, pages 412–415
1. 0; 0 3. 8; $\sqrt{8}$ 5. 2; $\sqrt{2}$ 7. ≈60; $\sqrt{60}$ 9. 2310 11. mean: 38; median: 40; mode: 30; variance: 256; standard deviation: 16 13. Answers will vary. 15. Answers will vary. 17. Answers will vary. 19. Answers will vary. 21. The one with the higher mean is further to the right. 23. The right one. 25. [−11] 27. $\begin{bmatrix} -8 & -57 & 46 \\ 2 & 7 & 14 \end{bmatrix}$ 29. 58, −94

Chapter 9 Review, pages 416–417
1. e 3. d 5. a 7. $\frac{1}{26}$ 9. $\frac{1}{9}$ 11. $\frac{1}{12}$ 13. 120
17. Median is 79.5, lower and upper quartiles are 77 and 82.5; 50% of the scores cluster around the median with the lower and upper 25% of the scores spread out due to extreme low and extreme high scores. 19. 5.6, $\sqrt{5.6}$

Chapter 10

Lesson 10-1, pages 426–429
1. 3.32 3. 9.22 5. $2\sqrt{11}$ 7. $5\sqrt{3}$ 9. $28\sqrt{3}$ 11. 32 13. $\frac{2\sqrt{3}}{3}$ 15. 4.58 17. 8.54 19. $9\sqrt{2}$ 21. $6\sqrt{2}$ 23. $24\sqrt{2}$ 25. 27 27. $\frac{7\sqrt{2}}{2}$ 29. 20 31. 2.4 m 33. Bakong Temple, $\sqrt{4900}$ m 35. 3 37. 2 39. 7.5 cm^2 41. true 43. false; possible counterexample $(\sqrt{2})(\sqrt{18}) = 6$ 45. 2 47. 10,000 49. 140

Lesson 10-2, pages 430–433
1. 9.9 m 3. 8.6 ft 5. 17.3 ft 7. 9.4 cm 9. 7.9 cm 11. 11.3 ft 13. yes 15. 6.4 cm 17. 4.5 yd 19. 6.7 in. 21. Yes 23. See the figure below. $d^2 = y^2 + h^2$ and $y^2 = l^2 + w^2$; Therefore, $d^2 = l^2 + w^2 + h^2$ and $d = \sqrt{l^2 + w^2 + h^2}$.

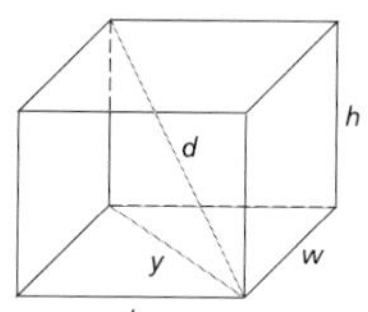

25. The system worked because the workers formed a triangle that had sides measuring 3 rope lengths, 4 rope lengths, and 5 rope lengths, [12 − (3 + 4)] = 5. Because 3, 4, and 5 form a Pythagorean triple, then they formed a right triangle. 27. 0.17 29. 0.155

Review and Practice Your Skills, pages 434–435
1. 7.28 3. 5.29 5. 31.62 7. 31.62 9. $5\sqrt{2}$ 11. $6\sqrt{2}$ 13. $4\sqrt{5}$ 15. $2\sqrt{22}$ 17. $11\sqrt{2}$ 19. $15\sqrt{7}$ 21. 12 23. $\frac{\sqrt{3}}{2}$ 25. $\sqrt{5}$ 27. $\frac{4\sqrt{7}}{7}$ 29. 4 31. 2 33. 10 35. $\frac{1}{3}$ 37. 26 m 39. 13.6 in. 41. 65 yd 43. 30 cm 45. $2\sqrt{6}$ 47. $2\sqrt{10}$ 49. $\frac{\sqrt{5}}{5}$ 51. 2 53. $24\sqrt{7}$ $44\sqrt{6}$ 57. 36 59. $235\sqrt{3}$ 61. no 63. yes

Lesson 10-3, pages 436–439
1. $4\sqrt{3}$ in., 8 in.; 6.9 in., 8.0 in. 3. $6\sqrt{2}$ yd; 8.5 yd 5. 5.8 ft 7. $10\sqrt{2}$ cm; 14.1 cm 9. $\sqrt{3}$ m, 2 m; 1.7 m, 2 m 11. 10.6 cm 13. $25\sqrt{3}$ cm^2 15. 9.5 m 17. Always; let x represent the measure of the smaller angle. Then $90 - x$ can represent the larger angle. You can solve this equation for x:

$x = 0.5(90 - x)$
$x = 45 - 0.5x$
$1.5x = 45$
$x = 30$

19. $\frac{3}{13}$ 21. $\frac{1}{26}$ 23. $\frac{1}{26}$ 25. −2 −1 0 ; $x > -2$

27. −6 −5 −4 −3 −2 −1 0 ; $x \leq -6$

Lesson 10-4, pages 440–443
1. 140° 3. 49° 5. 50° 7. 40° 9. 50° 11. When two chords intersect inside a circle, the situation is the same as when two secants intersect inside a circle. 13. $\frac{1}{24}$ 15. $\frac{1}{69}$ 17. $\frac{1}{2}$ 19. $-\frac{2}{3}$ 21. 1 23. −4 25. −2

Review and Practice Your Skills, pages 444–445
1. $10\sqrt{2}$ in., $10\sqrt{2}$ in.; 14.1 in., 14.1 in. 3. $3\sqrt{2}$ m; 4.2 m 5. $11\sqrt{3}$ cm, 22 cm 7. 5 in., 10 in. 9. 18.5 ft, $18.5\sqrt{3}$ ft 11. $10\sqrt{3}$ km, $20\sqrt{3}$ km 13. 44 in., $44\sqrt{2}$ in. 15. $4\sqrt{2}$ ft, $4\sqrt{2}$ ft 17. 46° 19. 125° 21. 124° 23. false

Selected Answers

25. $\frac{\sqrt{6}}{4}$ 27. $4752\sqrt{3}$ 29. 19.8 mm 31. 11.0 ft

Lesson 10-5, pages 446–447

1.

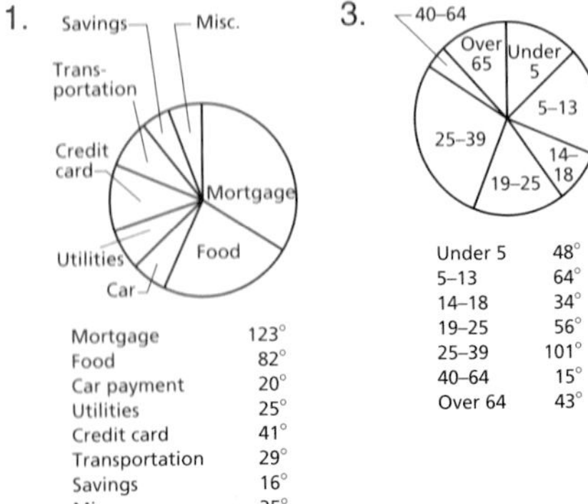

Mortgage	123°
Food	82°
Car payment	20°
Utilities	25°
Credit card	41°
Transportation	29°
Savings	16°
Misc.	25°

3.

Under 5	48°
5–13	64°
14–18	34°
19–25	56°
25–39	101°
40–64	15°
Over 64	43°

5.

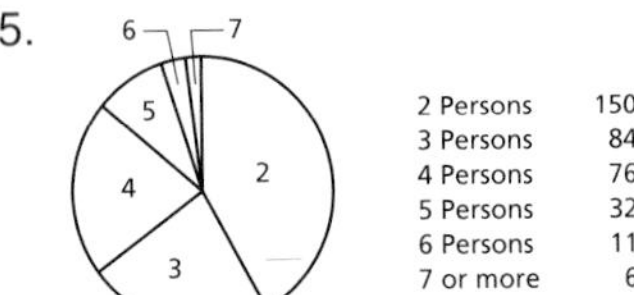

2 Persons	150°
3 Persons	84°
4 Persons	76°
5 Persons	32°
6 Persons	11°
7 or more	6°

7. $(-6, -11)$ 9. $(3, 5)$

Lesson 10-6, pages 448–451

1. 3 3. 3 5. 30 7. 6 9. 10 11. 12 13. 9 cm 15. $x = 13$; $y = 6.5$ 17. $x = 5$, $y = 12.5$ 19. 6 cm 21. $\triangle CEB \sim \triangle CAD$ by AA similarity. $\angle C \cong \angle C$ and $\angle A \cong \angle E$, because they both intercept the same arc, BD.

Review and Practice Your Skills, pages 452–453

1.

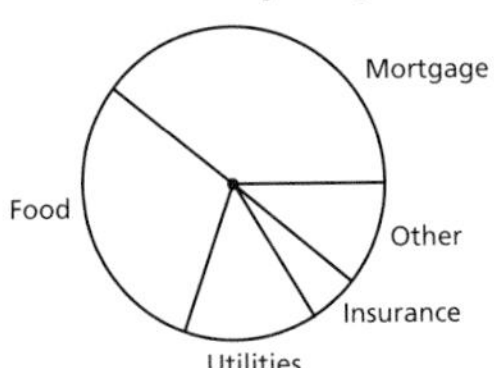

3.

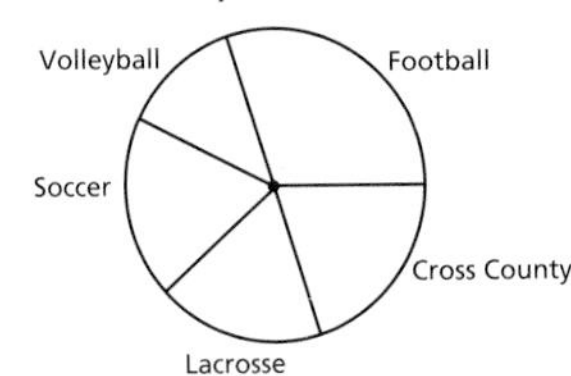

5.

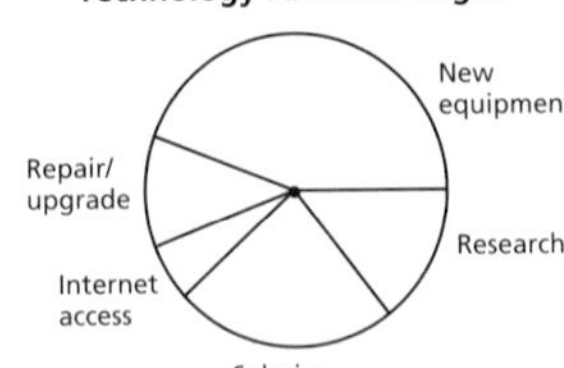

7. 4 9. 11 11. 4 13. false 15. false 17. true 19. no 21. yes 23. no 25. $7\sqrt{3}$ in., 21 in.; 21.1 in., 21 in. 27. $\frac{39\sqrt{2}}{2}$ m, 27.6 m

Lesson 10-7, pages 454–457

5. false 7.

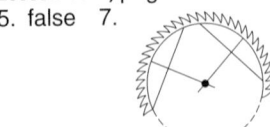

13. 16-sided, 24-sided; any polygons with sides that are multiples of 6 or 4.

17.

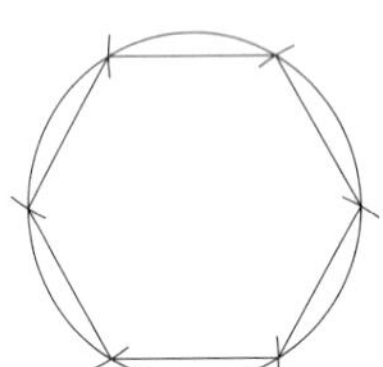

19. Mike could take the copy, find the perpendicular bisectors of two different sides of the hexagon, and use their intersections as the center of a circle. He could use a radius equal to the distance from the center of the circle to a vertex of the hexagon. Then he could construct the perpendicular bisectors of each of the other sides of the hexagon, and mark the point where these bisectors meet the circle. At this point, there should be 12 equally spaced points on the circle. He can connect these 12 points with a straightedge to draw a dodecagon. 21. 2.9, 1.7 23. 1.7, 1.3 25. 10.25, 3.2

Chapter 10 Review, pages 458–459

1. d 3. e 5. c 7. $\frac{\sqrt{3}}{2}$ 9. 4.6 in. 11. $6\sqrt{2}$ 13. $2\sqrt{3}, 4\sqrt{3}$ 15. 50° 17. Percents should be Italian: 22%, Spanish: 41%, French: 33%, and Japanese: 4%. 19. 2 21.

Chapter 11

Lesson 11-1, pages 468–471

1. $5x^3 + x^2 + x - 4$ 3. $4x^2 + 10$ 5. $6x^2 - 4x$ 7. $2x - 7$ 9. $x + 21$ 11. $7x^3 + 15x^2 - 5xy$ 13. $9x^3 + 5x^2 - 3x - 3$ 15. $5a + 13$ 17. $2x^2 + 7x$ 19. $2t + 5$ 21. $5m^2$ 23. $x^2 - 22x + 3y^2$ 25. $3v^2 - 2vw - 10w^2$ 27. $6b^2 + 2c - 1529.$ $25p + 32s$ 31. 4 33. $6.3a^3 - 4.2a^2b + 5 \cdot 8bc^2 + 1.76b^3$ 35. $19x^2 + 6x$ 37. $12x^3 - 3xy$ 41. 8.66 43. 14.70 45. 175 47. $\frac{4\sqrt{30}}{5}$ 49. $36\sqrt{2}$ 51. $8\sqrt{14}$ 53. 3.2×10^{10}

Lesson 11-2, pages 472–475

1. $3xy$ 3. $12pq$ 5. rs^2 7. $7x^2 + 7x$ 9. $a^4 + a^3$ 11. $-e^3f - e^2f^3$ 13. $ab^2 + ab - 6a$ 15. $-7x^3 + 14x^2y - 7xy^2$ 17. $x^2 + xy + xz$ 19. $3x^3y$ 21. $-12x^5y$ 23. $9a^4$ 25. $-6x^4 - 4x^3$ 27. $-3m^4n^3 + 3m^5n^5$ 29. $8e^2f^2g^4 - 8ef^3g^4$ 31. $-18l^3m^2n^7 + 18l^2m^6n^5$ 33. $4abe^2 - 2abf + abg$ 35. $12uv^2w + 8v^3w + 4v^2w^4$ 37. $7r^5s^4t^5 - 7r^4s^5t^4$ 39. $4x^2 + 2xy + 2xz$ 41. $60hk + 180h$ sq ft 43. $2b^2(3a)$ $2(3ab^2)$ $3(2ab^2)$ $6(ab^2)$
$a(6b^2)$ $2a(3b^2)$ $3a(2b^2)$ $6a(b^2)$
$b(6ab)$ $2b(3ab)$ $3b(2ab)$ $6b(ab)$
45. x^4y^5 47. $2a^6$ 49. $15x^4y - 15x^3y^2 + 15x^2y^3$ 51. $3ts + 20t$ 53. $4{,}800p^2$ 55. Jarius plays for the

Selected Answers

Gophers, Keshawn, for the Goats, and Levon for the Cheetahs.

Review and Practice Your Skills, pages 476–477

1. $6x + 10y$ 3. $16x^2 - 4x - 7$ 5. $-2x - 2z$ 7. $11x^3 - 11x^2 + 12x$ 9. $-2y^2 - 11y - 1$ 11. $4x^2y - 2xy - 3xy^2$ 13. $-7x + 8y$ 15. $17c^2 + 17cd - 6d^2$ 17. $12n + 9p$ 19. $-6x^2$ 21. $-42m^2n$ 23. $9k^6$ 25. $28s^5t^3$ 27. $12x^2 - 30x$ 29. $33x^4 + 22x^3 - 11x^2$ 31. $-p^3q^2 + 3p^2qr - 7p^2q^4$ 33. $15ax + 10bx - 20cx$ 35. $5x^2 - 6x + 16$ 37. $-4p^4q^2 - 20p^2qr + 12p^2q^3$ 39. $8ayz + 6byz - 20cyz$ 41. $3x^{15}y^9z^{10} + 6x^{11}y^{10}z^9 - 3x^{11}y^9z^{13}$ 43. $4x(3x - 7)$; $12x^2 - 28x$ 45. $3x(x^2 - 6x + 7)$; $3x^3 - 18x^2 + 21x$ 47. $11b + 11$ 49. $\frac{5}{4}h - \frac{3}{8}g$ 51. $-5m + 2n + 12p$ 53. $-4a^2 + 24b^2$ 55. $13r^2 + 10rs - 21s^2$ 57. $-4m^2 + 23n$ 59. $12x^4 - 2x^3 + 6x^4$ 61. $-16p^4qr^3$ 63. $-6x^2 + 28x$ 65. $36c^2d^3 - 27c^3d^2$

Lesson 11-3, pages 478–481

1. $3(2x^2 + 3)$ 3. $n(5m - np)$ 5. $6x(2x + 3)$ 7. $r(7r + 3s + 2t)$ 9. 18 inches tall, 8 figurines 11. $2(3a + 4b)$ 13. $5(3p^3 - 7q)$ 15. $w(v + x)$ 17. $y(5x^2 - 2y)$ 19. $n(13mn + 25)$ 21. $\frac{5f}{2}$ 23. $7b(2ab + 5c)$ 25. $9r^2(2r + 3)$ 27. $13j^5k^2l^3(3j^2kl - 5jk^3 + 4l^3)$ 29. $xy(ax^2y^2 + bxy + c)$ 31. $A = \frac{1}{2}h(t + b) = 22$ sq 33. $2xy^2(3x^4 + 4x^3y + 3x^2y^2 + 7xy^3 + y^4)$ 35. $8a(8b + 5c)$ 37. $25x(x + 2y)$ 39. $x^{(n-1)}(3x - 2)$ 43. $x = \frac{13\sqrt{2}}{2} = 9.19$

Lesson 11-4, pages 482–485

1. $3ac + 15ad + 2bc + 10bd$ 3. $l^2 + ln - lm - mn$ 5. $6x^2 + 21x + 15$ 7. $8x^2 + 15xy - 2y^2$ 9. $p^2 - q^2$ 11. $x^2 + 22x + 120$ 13. $6pr - 2p + 15qr - 5q$ 15. $4a^2 + 12ac + ab + 3bc$ 17. $2eg + 5ef - 6fg - 15f^2$ 19. $45p^2 - 27pr + 10pq - 6qr$ 21. $5m^2 + 51mn + 54n^2$ 23. $3x^2 - 10x + 8$ 25. $7j^2 - 37jk + 10k^2$ 27. $24b^2 + 37bc - 5c^2$ 29. $w^2 - 8wz + 16z^2$ 31. $x^2 + 8x + 16$ 33. $a^2 - 4$ 35. $4e^2 - 25f^2$ 37. $5x + y - 5$ 39. $9x^3 + 6x^2 - 30x + 8$ 41. $2p^3 + p^2q - 5pq^2 + 2q^3$ 43. $a^4 + 4a^3b + 6a^2b^2 + 4ab^3 + b^4$ 45. $2y^3 + 3y^2 - 17y + 12$ 47. $1\frac{1}{2}$ square feet 49. 143 in.^{3} 51. 396 cm^3 53. $\begin{bmatrix} 4 & -3 & 3 \\ 1 & 7 & 0 \\ 2 & -3 & 7 \end{bmatrix}$ 55. $\begin{bmatrix} 20 & 30 \\ 15 & 40 \\ -25 & 30 \end{bmatrix}$ 57. $\begin{bmatrix} -12 & -4 \\ 32 & 24 \\ 28 & -16 \end{bmatrix}$ 59. 6.2 in.

Review and Practice Your Skills, pages 486–487

1. 4(2x 1 3y) 3. $x(7x + 15)$ 5. $gh(2 - k)$ 7. $a(28bc - 11a^2)$ 9. $y^2(17x + 24z)$ 11. $5x^2(x + y^2)$ 13. $12(3a + 2b)$ 15. $5b(a + 2c - 1)$ 17. $18p(pq - 2r^2)$ 19. $15s^2t(t + 3s)$ 21. $2x(2x^2 - x + 7)$ 23. $3uv(1 - 3uv + u^2v^2)$ 25. $18m^2n^2(2mn^3 + 4n + 3m^3)$ 27. $2abc(3a + b - 2)$ 29. $4mnp(2 - 5mp^2 + 4n^3p)$ 31. $x^2 - x - 6$ 33. $2x^2 + 7xy + 6y^2$ 35. $25x^2 + 40x + 16$ 37. $4mp + 5m^2 - 20np - 25mn$ 39. $3a^2 - 23ab + 30b^2$ 41. $x^2 - 36$ 43. $64x^2 - 9$ 45. $8y^2 - 2yz - 45z^2$ 47. $xy + 2y + y + 2$ 49. $81x^2 - 1$ 51. $64p^2 + 128pq + 64q^2$ 53. $z^4 - 25$ 55. $8r^2 - 18s^2$ 57. $-42c^2 + 53cd - 15d$ 59. $x^3 + 17x^2 + 52x$ 61. $(7x - 5)(2x + 3) = 14x^2 + 11x - 15$ 63. $-13x + 18y$ 65. $-9x^2 + 20x$ 67. $-30x^2 + 66x$ 69. $20x^4 - 15x^3 + 5x^2$ 71. $-8p^4q^2r + 20p^2qr^2 + 12p^4q^3r$ 73. $6(-5x + 9)$ 75. $g(12 + 25g)$ 77. $15rst^2(3rt + 5s)$ 79. $8a^2b^2(6a + 7b^2 - 4a^2b)$ 81. $2(4x^3 + 3x^2 + 2x + 1)$ 83. $49a + 7ab - 7b - b^2$ 85. $16x^2 + 24x + 9$ 87. $48p^2 + 2pq - 35q^2$ 89. $-4x^2 + 1$

Lesson 11-5, pages 488–491

1. $(3w + 2y)(3x + 2z)$ 3. $(9a - 4c)(2b - 3d)$ 5. $(5r + 3)(s - 8t)$ 7. $(k + m)(\ell + u)$ 9. $(m - 3n)(3r - 8s + 5t)$ 11. $(y + 4)$ 13. $(2a + 3c)(2b + 3d)$ 15. $(4q + s)(r + 3t)$ 17. $(3e - g)(7f - 4h)$ 19. $(9w^2 - y)(3x - 2z^2)$ 21. $(k^2 - 3m)(2l^2 + 5n)$ 23. $(5t - 2v)(3u + 4)$ 25. $(x^2 + 8)(3y - z)$ 27. $(y + 5z)(v + 3w + 2x)$ 29. $(5p - q)(2r - 3s + 4t)$ 31. $(3d - 5e)(2f + 4g + 7h)$ 33. $3j(2j + k)(j - 2l)$ 35. $3r^2(r - 2s)(r + 2t)$ 37. 2.4 39. $(2m - 6)$ and $(n - 3)$ or $(m - 3)$ and $(2n - 6)$ 41. $3a(2a + b)(2a + 3c)$ 43. 50° 45. 50° 47. 55° 49. -56 51. 2 53. 12 55. $6\frac{1}{2}$

Lesson 11-6, pages 492–495

1. $(s + 5)^2$ 3. $(m + 4n)^2$ 5. $(3r + 6)(3r - 6)$ 7. none 9. $(8u - 3v)(8u - 3v)$ 11. the difference of two squares; $(p - 3)(p + 3)$ 13. $(6a + 2b)^2$ 15. none 17. $(10r + 11)^2$ 19. none 21. none 23. $(6r - s)(6r + s)$ 25. none 27. $4(x + 1)^2$ 29. $p = 3$ 31. $a = 3$ or $2\frac{1}{2}$ 33. $y^2 - 64x^2$; $(y - 8x)(y + 8x)$ 35. 100 37. $5s(s - 2t)^2$ 39. $3xy(x + 2y)(x - 2y)$ 41. $(x - y)(x - 1)^2$

43. $1 \cdot 1$ $2 \cdot 1$ $3 \cdot 1$ $4 \cdot 1$ $5 \cdot 1$ $6 \cdot 1$ $7 \cdot 1$ $8 \cdot 1$
$1 \cdot 2$ $2 \cdot 2$ $3 \cdot 2$ $4 \cdot 2$ $5 \cdot 2$ $6 \cdot 2$ $7 \cdot 2$ $8 \cdot 2$
$1 \cdot 3$ $2 \cdot 3$ $3 \cdot 3$ $4 \cdot 3$ $5 \cdot 3$ $6 \cdot 3$ $7 \cdot 3$ $8 \cdot 3$
$1 \cdot 4$ $2 \cdot 4$ $3 \cdot 4$ $4 \cdot 4$ $5 \cdot 4$ $6 \cdot 4$ $7 \cdot 4$ $8 \cdot 4$
$1 \cdot 5$ $2 \cdot 5$ $3 \cdot 5$ $4 \cdot 5$ $5 \cdot 5$ $6 \cdot 5$ $7 \cdot 5$ $8 \cdot 5$
$1 \cdot 6$ $2 \cdot 6$ $3 \cdot 6$ $4 \cdot 6$ $5 \cdot 6$ $6 \cdot 6$ $7 \cdot 6$ $8 \cdot 6$
$1 \cdot 7$ $2 \cdot 7$ $3 \cdot 7$ $4 \cdot 7$ $5 \cdot 7$ $6 \cdot 7$ $7 \cdot 7$ $8 \cdot 7$
$1 \cdot 8$ $2 \cdot 8$ $3 \cdot 8$ $4 \cdot 8$ $5 \cdot 8$ $6 \cdot 8$ $7 \cdot 8$ $8 \cdot 8$

45. 0.1875 47. 0.625 49. 10.5 cm

Review and Practice Your Skills, pages 496–497

1. $(5 + b)(c + d)$ 3. $(a - c)(b - 3)$ 5. $(2 + j)(h - k)$ 7. $(y^2 + 3)(y - 2)$ 9. $(w - 3)(2z - 1)$ 11. $(m - n)(w - x)$ 13. $(2x^2 + 3)(y - 4)$ 15. $(p^2 - q)(r^3 - 2s)$ 17. $(1 + v)(w - v)$ 19. $(x - 3ay)(1 - y)$ 21. $(3x + 2)(-3y + 2z)$ 23. $(w - 4)(x + 2y + 3z)$ 25. $(x + c)(x - a - b)$ 27. $(2 - 7a) \times (g + 2f)$ 29. $(x - 10)(x - 10)$ 31. not possible 33. $(6b - 1)(6b - 1)$ 35. $(x - 4y)(x - 4y)$ 37. $(w + 12)(w - 12)$ 39. $(c + 3d)(c - 3d)$ 41. $(4u + 9v)(4u - 9v)$ 43. $(5s - 7t)(5s + 7t)$ 45. $(7p + 2q)(7p + 2q)$ 47. $(8m + 11n)(8m + 11n)$ 49. $(1 - a)(1 - a)$ 51. not possible 53. $(15x + 11y)(15x + 11y)$ 55. $3(x - 2)(x - 2)$ 57. $x(x - 4)(x - 4)$ 59. $3a(x + 2)(x - 2)$ 61. $x^2(x + 5)(x - 5)$ 63. $4(a + b)(a - b)$ 65. $(2x - 7y)(2x - 7y)$ 67. $(1 + 16m)(1 - 16m)$

Selected Answers

69. $(9cd + 5b)(9cd - 5b)$ 71. $11n^2 + 5$ 73. $-2x + 9y$ 75. $6x^2y^2z - 14xy^2z^2 + 30x^2yz^2$ 77. $5x$ 79. $16x(3x + 2)$ 81. $-4ef(d + 2g + 3)$ 83. $2s(k^2 + 29q^2 + 17y^2)$ 85. $(w - 4)(3x + 7)$

Lesson 11-7, pages 498–501
1. 3 and 7 3. $4b$ and $2b$ 5. 6 and 1 7. + and + 9. − and + 11. $(c + 2)(c + 3)$ 13. $(c - 6)(c + 1)$ 15. $(x + 3)(x + 7)$ 17. $7y, 5y$ 19. $5b, 2b$ 21. 9, 7 23. $10f, 3f$ 25. +, + 27. +, − 29. −, − 31. −, + 33. $(p + 6q)(p + 4q)$ 35. $(k - 12)(k + 2)$ 37. $(h - 1)(h + 24)$ 39. $(f - 3g)(f - 8g)$ 41. $(q - 4)(q - 7)$ 43. $(s - 2t)(s + 4r)$ 45. No. Only one pair of factors will satisfy all conditions. 47. $(1 - 3r)(1 - 2r)$ 49. $(6g - 1)(4g - 1)$ 51. $5x^2(a - 2)(a - 1)$ 53. $3x(x + 2)(x - 6)$ 55. 27.5% 57. 13% 59. 10.5% 61. 6.25% 63. 99° 65. 46.8° 67. 37.8° 69. 22.5° 71. $y = \frac{2}{3}x - 2$ 73. $y = -2x + 3$

Lesson 11-8, pages 502–503
1.*a.* $(x + n_1)(x + n_2) =$
F O I L
↓ ↓ ↓ ↓
$x^2 + xn_2 + xn_1 + n_1n_2 =$
$x^2 + (n_2 + n_1)x + n_1n_2$

1*b.* $(x - 6)(x + 3) =$
F O I L
↓ ↓ ↓ ↓
$x^2 + 3x - 6x - 18 =$
$x^2 + (3 - 6)x - 18 =$
$x^2 - 3x - 18$

3.a. $(ax + by)(ax + by) =$
F O I L
↓ ↓ ↓ ↓
$a^2x^2 + abxy + abxy + b^2y^2 =$
$a^2x^2 + 2abxy + b^2y^2$

3b. Answers will vary.

5.*a.* $(ax - by)(a^2x^2 + abxy + b^2y^2) = a^3x^3 + a^2bx^2y + ab^2xy^2 - a^2bx^2y - ab^2xy^2 - b^3y^3 = a^3x^3 - b^3y^3$
5.*b.* $(3x - 1)(9x^2 + 3x + 1) =$
$27x^3 + 9x^2 + 3x - 9x^2 - 3x - 1 = 27x^3 - 1$
7. True. In a single-variable trinomial, think of the constant as a *y*-coefficient multiplied by 1. 9. 8 11. 34.75 13. 3 15. 4

Review and Practice Your Skills, pages 504–505
1. $(x + 6)(x + 1)$ 3. $(d + 6)(d + 7)$ 5. $(x + 14)(x + 2)$ 7. $(x - 4)(x - 5)$ 9. $(w - 3)(w - 7)$ 11. $(x - 8)(x - 4)$ 13. $(m + 9)(m - 6)$ 15. $(c + 5)(c - 4)$ 17. $(t + 5)(t - 2)$ 19. $(a - 8)(a + 6)$ 21. $(p - 9)(p + 4)$ 23. $(d - 8)(d + 7)$ 25. $(m + 6n)(m + 5n)$ 27. $(p + 12q)(p + 5q)$ 29. $(r - s)(r - 2s)$ 31. $(b + 4c)(b - c)$ 33. $(a + 9b)(a - 2b)$ 35. $(p - 11q)(p + 7q)$ 37. $(x + 6)(x + 8)$ 39. $(f - 2)(f - 24)$ 41. $(c - 3d)(c - 16d)$ 43. $(48 - x)(1 - x)$ 45. $(x + 2)(x + 24)$ 47. Specific examples will vary; $(n_1 + x)(n_1 - x) = n_1^2 - x^2$ 49. Specific examples will vary; $(ax + y)(bx - y) = abx^2 - y^2$ 51. Answers will vary. 53. $5n^2 + 6n - 1$ 55. $18x + 4y$ 57. $6x^2y^2z + 48xy^2z^2 - 12x^2yz^2$ 59. $16x$ 61. $4x(4x + 15)$ 63. $-3e(3df + 5fg + 4gd)$ 65. $7s(m^2 + 4w^2 + 9y^2)$ 67. $x^2 + 20x + 99$ 69. $81 - 16x^2$ 71. $-20f^2 - 12f + 18$ 73. $(w + 4)(5x - 4)$ 75. $(x^2 - 5)(8z + 11b)$

77. $(3 - 6m)(2n - 7p)$ 79. $(x + 5)(a - 2b + 7)$ 81. $(13 + a)(13 + a)$ 83. $(x + 14)(x + 14)$ 85. $(4a - 7b)(4a + 7b)$ 87. $(x + 12y)(x - 12y)$ 89. $(5m + 11n)(5m + 11n)$ 91. $(b - 24)(b + 3)$ 93. $(f - 8g)(f - 9g)$ 95. $(72 + m)(1 - m)$ 97. $(p - 27q)(p + 3q)$ 99. $(ab - 3)(ab + 1)$ 101. $(x - 8)(x - 12)$ 103. 59, 28, 17, 11, 7, 4, −4, −7, −11, −17, −28, −59

Lesson 11-9, pages 506–509
1. 3, 18, 1, 6 3. (2, 4); (2, 3); (5, 4); (5, 3) 5. −; + 7. +; − 9. $(f + g)(7f - 3g)$ 11. $(6x + 5)(x + 2)$ 13. 3, 12, 1, 4 15. 6, 2, 15, 5 17. 10, 15, 16, 24 19. (3, 1); (3, 5); (2,1); (2, 5) 21. (4, 1); (4, 6); (3,1); (3, 6) 23. +; + 25. −; + 27. −; + 29. $(3x - 4)(7x + 2)$ 31. $(z + 4)(2z + 3)$ 33. $(2r - 3s)(10r + 5s)$ 35. $(8m + 3)(8m - 5)$ 37. $3x(2v - w)(3v + 2w)$ 39. $(2x + 1)(x + 3)$ 41. $(5r - 9)(r + 2)$ 43. $3x^2 - x - 10 = 0$; $x = 2$ or $-\frac{5}{3}$ 45. $2\sqrt{39}$ 47. 12 49. $6\sqrt{35}$ 51. $6\sqrt{30}$ 53. 176 55. $\frac{\sqrt{78}}{6}$ 57. −8 59. −17 61. −8 63. −4 65. 16 67. 64

Chapter 11 Review, pages 510–511
1. d 3. e 5. b 7. $-4a^2 + 2a + 6$ 9. $-5a^2b + 5a^3b^2 - 15ab^2$ 11. $3ab(a^2b - 2)$ 13. $5x^3y(x^2 - 2xy - 4y^2)$ 15. $16a^2 - b^2$ 17. $(3a - 5b)(2a + 3b)$ 19. $(5r + 2s)(t + 4u)$ 21. $(9m - 4n)(9m + 4n)$ 23. $(3x - 5)^2$ 25. $(m + 8n)(m - 5n)$ 27. $(a + 5)(a + 3)$ 29. $(m - 3n)(m - 12n)$ 31. $(2s - 5t)(2s + 3t)$ 33. $(4a + b)(7a - 2b)$

Chapter 12

Lesson 12-1, pages 520–523
1.
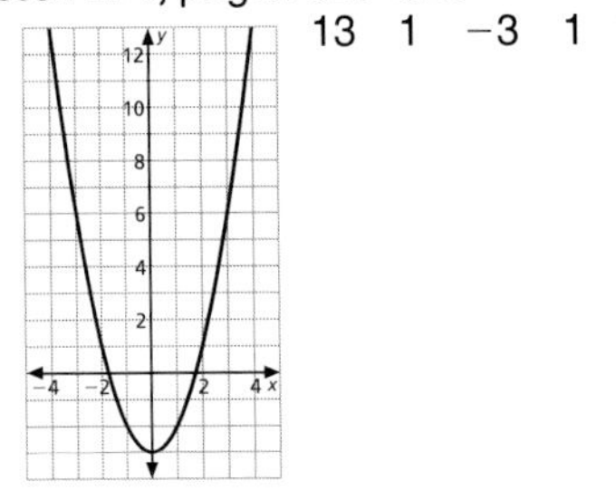
13 1 −3 1 13

3.
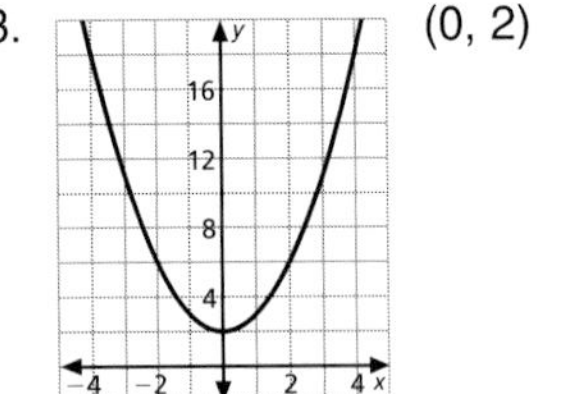
(0, 2)

5.
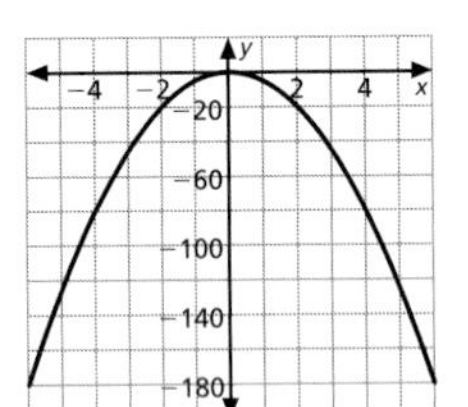
(0, 0)

7. 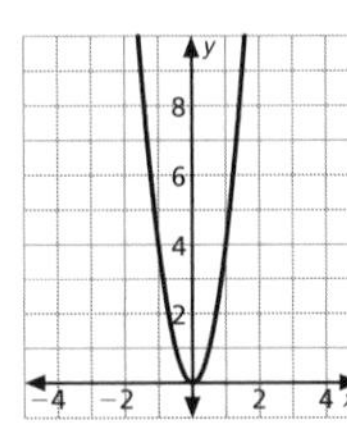 (0, 0) 9. 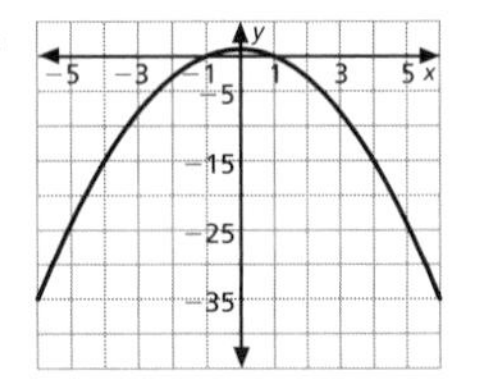 (0, 1)

11. upward 13. As the absolute value of *a* decreases the graph becomes wider. 15. The graph shifts downward. 17. (0, 18) 19. $y = 6x^2 + 2$ 21. (0, 0) *x*-axis 23. An equation in which $a < 0$; the graph opens downward and has a maximum point. 27. $2y^2 - 4y - 8$ 29. $3b^2 - 2b + 3$ 31. $-3d^2 - 4d + 17$ 33. 11.3

Lesson 12-2, pages 524–527

1. (2, 21) 3. $\left(\frac{1}{5}, 4\frac{4}{5}\right)$

5. 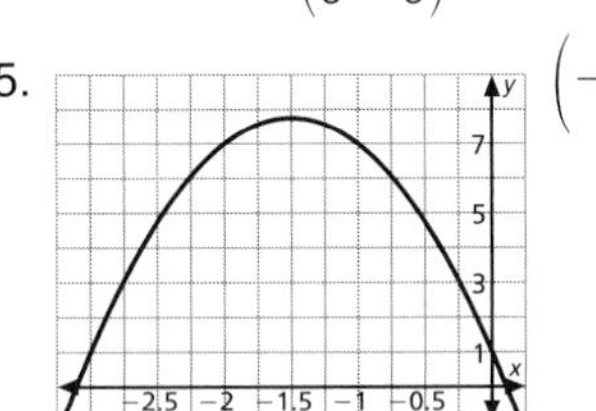 $\left(-\frac{3}{2}, 7\frac{3}{4}\right)$; $x = -\frac{3}{2}$

7. 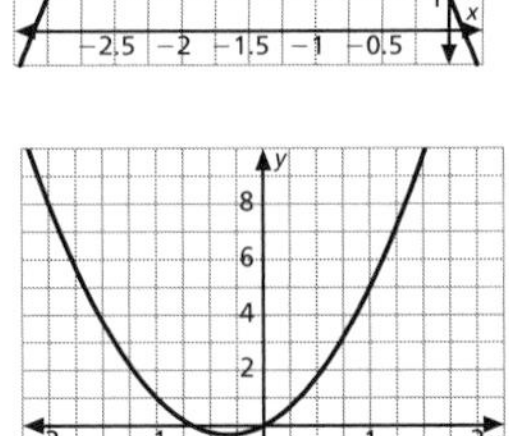 $\left(-\frac{1}{3}, -\frac{1}{3}\right)$; $x = -\frac{1}{3}$

9. (−3, −5) 11. $\left(\frac{1}{2}, 8\frac{1}{4}\right)$ 13. (2, −3) 15. (1, −4); $x = 1$ 17. (0, 1); *y*-axis 19. $y = -4x^2 - 5x - 2$ 21. 70° 23. The maximum range is reached at about 45°. 25. $6s^4 + 18s^3 + 12s$ 27. $8a^3c - 6a^2bd$ 29. $100x^7y^4 - 50x^6y^7$

31. 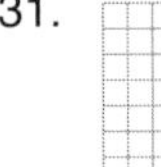 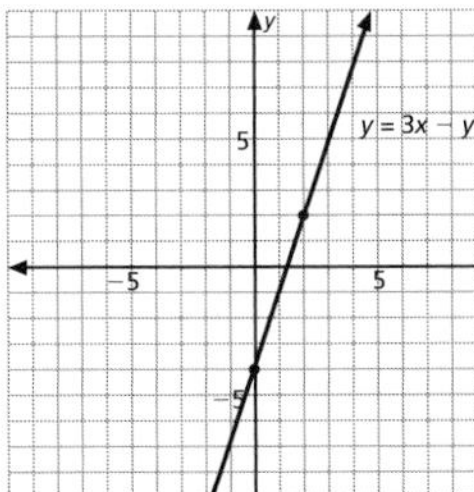 33. 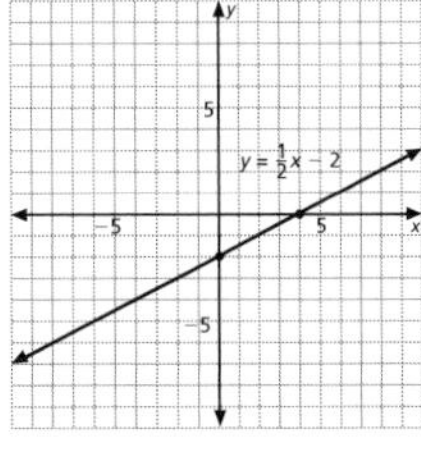

35. 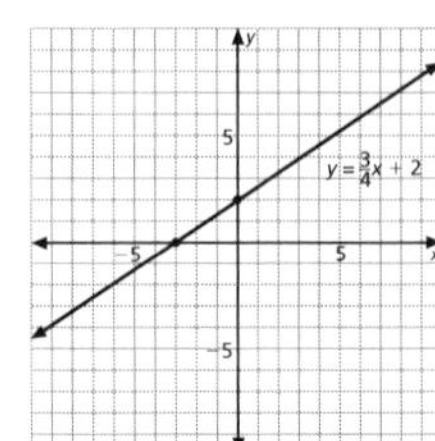

Review and Practice Your Skills, pages 528–529

1. There is only one *y*-value for each *x*-value.

3.

x	−3	−2	−1	0	1	2	3
y	13	8	5	4	5	8	13

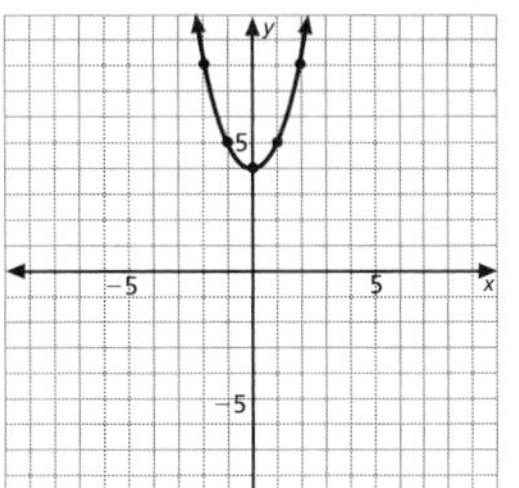

5. downward 7. upward

9. (0, 26) 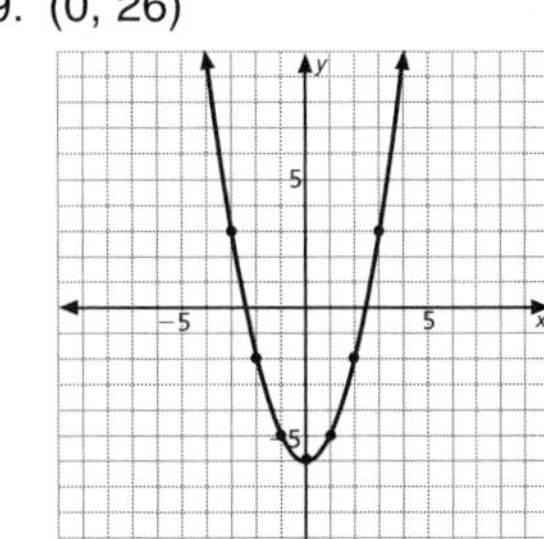

11. (0, 1) 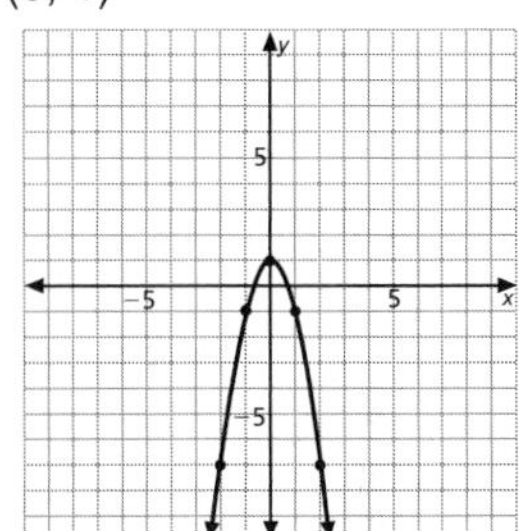

13. no 15. yes 17. no 19. no 21. $y = ax^2 + bx + cx$ 23. $\left(\frac{3}{4}, \frac{23}{8}\right)$ 25. (1, 2) 27. $v = \left(\frac{1}{4}, -3\frac{7}{8}\right)$; a/s $= \frac{1}{4}$ 29. $v = \left(\frac{3}{4}, 3\frac{7}{8}\right)$; a/s $= \frac{3}{4}$

31. $v = \left(\frac{1}{4}, -2\frac{7}{8}\right)$ 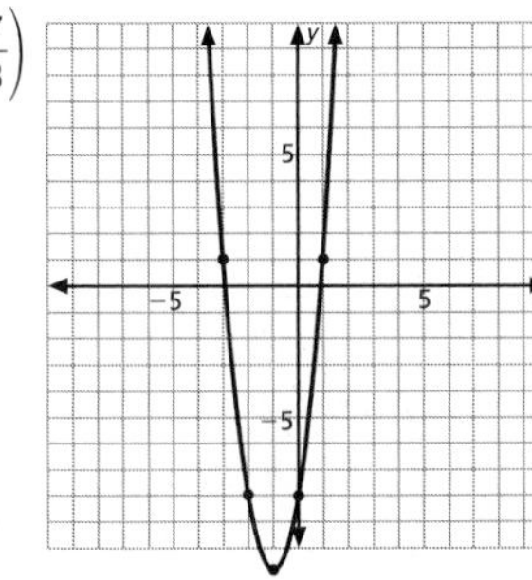

33. 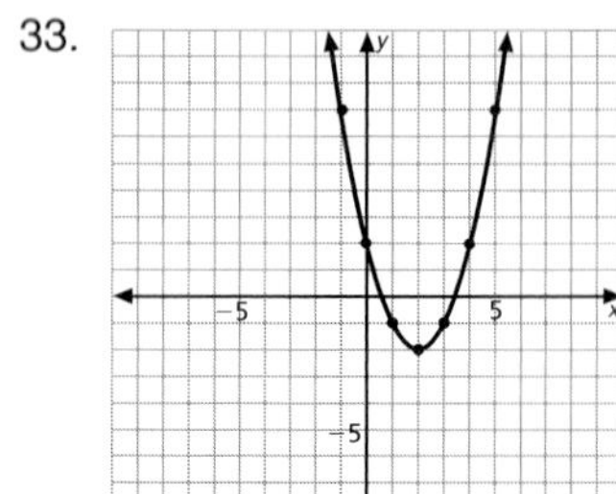 QIV

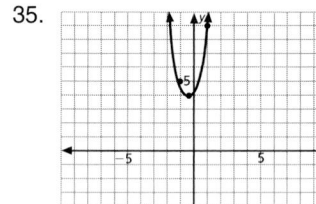

35. 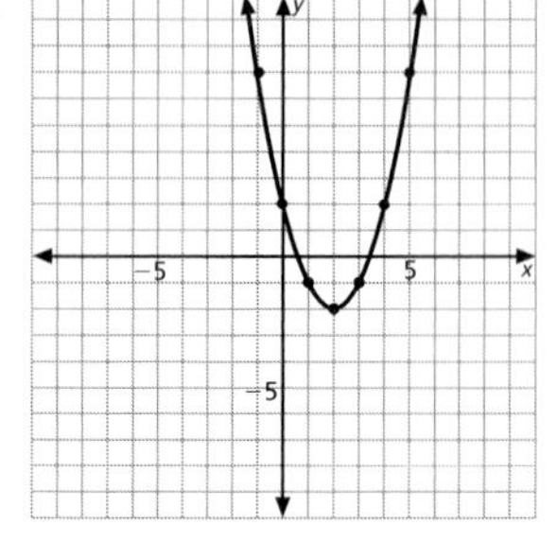 QI 37. $v = \left(-\frac{1}{3}, 4\frac{1}{3}\right)$; a/s $= -\frac{1}{3}$

Lesson 12-3, pages 530–533

1. $x = -3, x = -7$ 3. no solutions 5. $x = -9$ 7. $x = 10, x = -10$ 9. $x = 2, x = 7$ 11. $x = 2, x = -2$ 13. no solution 15. $x = 4, x = -12$ 17. 6 or -5 19. 8 or -1 21. 11 sec 23. 13 sec 25. one solution is always $\varnothing$, the other is $-\frac{b}{a}$ 27. $x = 0, x = -2$ 29. $x = 0, x = 4$ 31. $x = 0, x = 11$ 33. 45 35. 200 37. $9(c - 36)$ 39. $3mn(n - 3)$ 41. $x(w + z)$ 43. $9(4a - 7b)$ 45. $5x^2(2x - 3)$ 47. $3x^2y(9xy - 2)$ 49. $3x^2y^3(2x - 3y + 2z)$

Lesson 12-4, pages 534–537

1. $x^2 + 4x + 4$ 3. $x^2 - 2x + 1$ 5. $x = -7, x = 1$ 7. $x = -6, x = 2$ 9. lengths cannot be negative 11. $x^2 - 10x + 25$ 13. $x^2 + x + \frac{1}{4}$ 15. $x^2 + 18x + 81$ 17. $x^2 + 3x + \frac{9}{4}$ 19. $x^2 - x + \frac{1}{4}$ 21. $x = -1, x = \frac{5}{3}$ 23. $x = -3, x = 0$ 25. $x = \frac{1}{2}x = 5$ 27. $x = -3, x = \frac{1}{2}$ 29. $x = 0, x = 2$ 31. $x = 0, x = 6$ 33. $c = 4, h = -2$ 35. $c = \frac{9}{4}, h = -\frac{3}{2}$ 37. $c = 121, h = 11$ 39. 4 sec 41. 18 in. $\times$ 20 in. 43. 2 m $\times$ 6 m 45. 10 in. $\times$ 11 in. 47. $6a^2 - 12a - 4ab + 8b$ 49. $8ab + 4ad - 12bc - 6cd$ 51. $9c^2 - 12c + 4$ 53. $12x^2 + 2x - 2$ 55. $35a^2 - 11ab - 6b^2$ 57. $60x^2 - 62xy - 12y^2$ 59. $\sqrt{77}$ 61. $2\sqrt{14}$ 63. $8\sqrt{374}$ 65. $\sqrt{3}$ 67. $\frac{2\sqrt{15}}{3}$

Review and Practice Your Skills, pages 538–539

1. false 3. false 5. $x = -3, 4$ 7. $x = -2$ 9. No solutions 11. $y = x^2 - x - 20; x = 5$ 13. $y = x^2 - 3x + 2; x = 1, 2$ 15. $c = \left(\frac{b}{2}\right)^2$ 17. $x^2 - 4x + 4$ 19. $x^2 - 18x + 81$ 21. $x^2 + 5x + \frac{25}{4}$

23. $x = \pm\sqrt{12} - 3$ 25. $x = \pm\sqrt{13} + 4$ 27. $x = \pm\sqrt{10} + 4$ 29. $w = 4$ in., $\ell = 8$ in.

31.

x	-3	-2	-1	0	1	2	3
y	4	-1	-4	-5	-4	-1	4

33. $v = \left(\frac{3}{2}, \frac{7}{4}\right)$; a/s $= \frac{3}{2}$

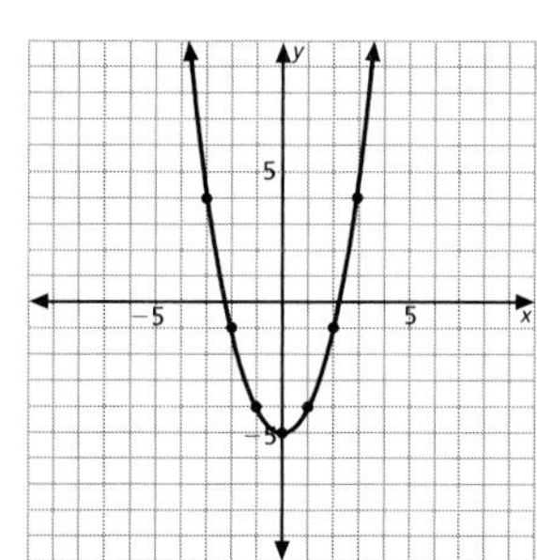

35. $v = (-1, 3)$; a/s $= -1$ 37. $v = \left(\frac{3}{4}, \frac{23}{8}\right)$; a/s $= \frac{3}{4}$ 39. $x = \pm\sqrt{37} - 6$ 41. $x = -4, 3$ 43. $x = \pm\sqrt{29} - 5$ 45. $x = -6, -2$ 47. $x = -1, 7$

Lesson 12-5, pages 540–543

1. $2 \pm \sqrt{3}$ 3. 0, 2 5. $1 \pm \sqrt{7}$ 7. 2, 7 9. $-\frac{1}{2}$ 11. 0, 7 13. 4 15. $-2 \pm \sqrt{2}$ 17. $4\frac{1}{2}, -5$ 19. 5, -7 21. 0, -3 23. $-\frac{7}{2}, 3$ 25. $-\frac{2}{3}, \frac{1}{2}$ 27. 1, 11 29. 6,400 ft 31.a. sec 31.b. 144 ft 31.c. 6 sec 33. Firth of Forth Bridge, 3.0 sec; Verrazzano Narrows Bridge, 3.6 sec; Sydney Harbor Bridge, 3.3 sec; Tunkhannock Viaduct 3.9 sec; Garabit Viaduct, 5.5 sec; Brooklyn Bridge, 2.9 sec. 35. $(a - b)(2a - 3b)$ 37. $4(2a - b)(2a - b)$ 39. $(3m + 4n)(2m - 5n)$ 41. $8a(3a + b)(a - b)$ 43. $2(m - 2n)(4m + 3n)$ 45. $2(3x - y)(x + 4)$ 47. $(-2, 4)$

Lesson 12-6, pages 544–547

1. 7.1 units 3. (1, 0) 5. 13.9 units 7. (1, 1) 9. $(-4, 1.5)$ 11. 26.7 units 13. 3.2 units 15. $(10, -2)$ 17. $(0.5, -0.5)$ 19. about 467 ft 21. about 234 ft 23. green: 20 yd, red: 65 yd, blue: 24 yd, purple: 44 yd 25. Answers will vary. Since $(x_2 - x_1)^2 = (x_1 - x_2)^2$ and $(y_2 - y_1)^2 = (y_1 - y_2)^2$, the sum of the squared differences will be the same positive number. 27. $x^2 + y^2 = r^2$ 29. $(x + 4)(3x - 2)$ 31. $2(x - 3)(3x - 4)$ 33. not factorable 35. $2(3x + 4)(2x - 3y)$ 37. $(2x - 7)(5x - 1)$ 39. $(4x + 3y)(4x + 3y)$

Review and Practice Your Skills, pages 548–549

1. true 3. true 5. $x = \frac{7 \pm \sqrt{41}}{2}$ 7. $x = 3, 1$ 9. $x = 1, -\frac{5}{2}$ 11. $x = \frac{7 \pm \sqrt{33}}{4}$ 13. $x = \frac{1 \pm \sqrt{41}}{4}$ 15. $\sqrt{61} \approx 7.8$ 16. 12.0 17. 9.1 19. 7.3 21. 5.1 23. 8.5 25. $\left(2, \frac{3}{2}\right)$ 27. (0, 0) 29. $\left(\frac{5}{2}, \frac{3}{2}\right)$ 31. $m = (1, -2), d = 8.5$ 33. $m = \left(-\frac{3}{2}, \frac{1}{2}\right), d = 7.1$ 35. $m = (-2, -2), d = 17.0$ 37. $v = \left(-\frac{3}{2}, -\frac{17}{4}\right)$; a/s $= -\frac{3}{2}$ 39. $v = \left(\frac{2}{3}, \frac{2}{3}\right)$; a/s $= \frac{2}{3}$

41. $v = \left(\frac{5}{4}, -\frac{33}{8}\right)$; a/s $= -\frac{5}{4}$ 43. $x = \pm\sqrt{5}$ 45. $x = 2, -4$ 47. $x = 3, 5$ 49. $x = 1, -4$ 51. $x = \frac{\pm\sqrt{21} - 5}{2}$ 53. $x = 2 \pm \sqrt{7}$ 55. $x = \frac{-2 \pm \sqrt{22}}{2}$

Lesson 12-7, pages 550–551
1. $y = x^2 + 2x - 5$ 3. Yes. Any three non-collinear points located on the graph will work. However, you may find it easier to solve the system of equations if you use the y-intercepts as one of the points. 5. $y = x^2 - 2x - 3$ 7. $y = x^2 - 4x + 12$ 9. $(5a - 2)(3a + 4)$ 11. $(a + 8)(a + 3)$ 13. $(a - 6)(3a + 4)$

Chapter 12 Review, pages 552–553
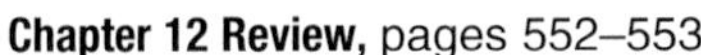
1. e 3. d 5. b 7. (0, 0)

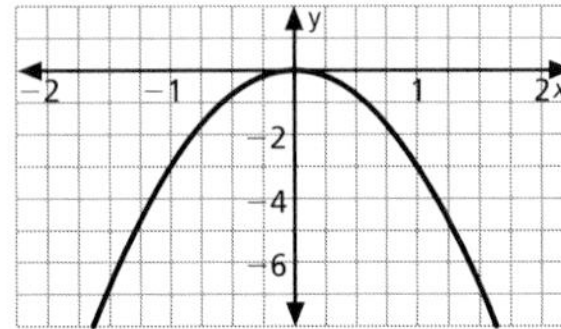

9. (1, 3)

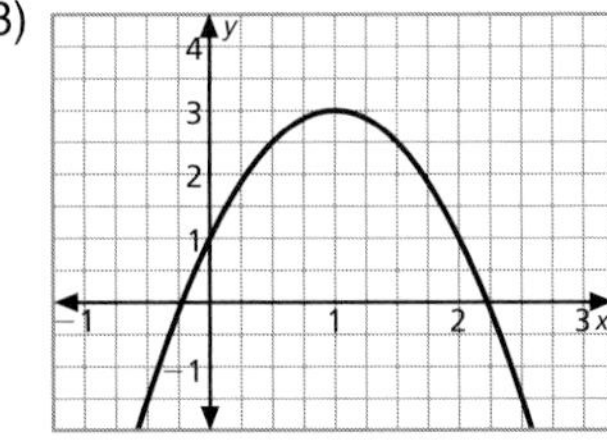

11. yes 13. 7 15. 4, −7 17. 3, 5 19. −0.5, 3 21. $-1 \pm \sqrt{7}$ 23. (−1, −1), 17.2 units 25. $y = -3x^2 - 18x - 1$

Chapter 13

Lesson 13-1, pages 562–565
1. $x^2 + y^2 = 49$ 5. $\sqrt{11}$, (0, 0) 7. $\sqrt{13}$, (−8, 4) 9. $x^2 + y^2 = 144$ 11. $(x - 9)^2 + (y - 4)^2 = 144$ 13. $(x - 5)^2 + (y + 1)^2 = 4$ 15. $x^2 + y^2 = 6.76$ 17. $\sqrt{15}$, (0, 0) 19. $\sqrt{14}$, (−4, 2) 21. $2\sqrt{5}$, (12, −6) 23. 20, (9, −1) 25. 9, (−1, 9) 27. $x^2 + y^2 = 81$ *or* $(x - 9)^2 + y^2 = 81$ 29. 3, (2, 1) 31. (0, 0) 33. (0, 3) 35. (0, 6) 37. $\frac{-2}{3}$ 39. 1 41. undefined 43. $\frac{10}{9}$ 45. $-\frac{1}{2}$

Lesson 13-2, pages 566–569
1. (0, 3), $y = -3$ 3. (0, 2), $y = -2$ 5. $x^2 = 28y$ 7. $x^2 = 3y$ 9. (0, 4), $y = -4$ 11. (0, 5), $y = -5$ 13. (0, 1), $y = -1$ 15. $x^2 = -8y$ 17. $x^2 = -2y$ 19. $x^2 = 4ay$ 21. $y^2 = 4ax$ 23. $(x - 2)^2 = 8(y - 2)$ 25. $(x - 2)^2 = 12(y - 2)$ 27. (1.5, 3.75) 29. (−1.5, −6.25) 31. (0, −9) 33. (−0.25, 11.875) 35. (−2, −7) 37. (−0.375, 8.5625) 39. {2, 5, 8, 9} 41. {6, 7} 43. {9} 45. {1, 3, 4, 6, 7, 9} 47. *U* 49. {2, 5, 8}

Review and Practice Your Skills, pages 570–571
1. h is the x-coordinate of the center, and k is the y-coordinate of the center. 3. when $h = 0$ and $k = 0$, when the center is at the origin. 5. $(x - 2)^2 + (y - 4)^2 = 9$ 7. $(x + 1)^2 + (y - 2)^2 = 16$ 9. $r = \sqrt{13}$, $c = (0, 0)$ 11. $r = 5$, $c = (-3, 4)$13. $r = 6$, $c = (3, 0)$ 15. $r = \sqrt{28}$, $c = (-1, 0)$ 17. false 19. true 21. focus $= \left(0, \frac{13}{4}\right)$; directrix is $y = -\frac{13}{4}$ 23. focus $= \left(0, -\frac{3}{4}\right)$; directrix is $y = \frac{3}{4}$ 25. focus $= \left(0, -\frac{9}{4}\right)$; directrix is $y = \frac{9}{4}$ 27. focus $= \left(0, \frac{5}{4}\right)$; directrix is $y = -\frac{5}{4}$ 29. focus $=$ (0, 1); directrix is $y = -1$ 31. focus $= \left(0, -\frac{3}{2}\right)$; directrix is $y = \frac{3}{2}$ 33. $x^2 = -12y$ 35. $x^2 = -28y$ 37. $x^2 = 10y$ 39. parabola 41. parabola 43. circle 45. $r = 6$, $c = (-3, 4)$ 47. $x^2 + (y + 3)^2 = 49$ 49. $(x + 3)^2 + (y + 1)^2 = 36$

Lesson 13-3, pages 572–573
1. circle 3. point 5. parabola 7. No. A circle is formed when a plane is parallel to the circle formed by the base. Forming two circles would require a single plane to pass through both cones at the same time or to pass through a single cone twice. To do so, the plane would have to "bend," and it would no longer be considered a plane. 9. 2; −4 and 3 11. 2; −4 and 2 13. 2; 6 and −1 15. 2; 5 and −5

Lesson 13-4, pages 574–577
1.

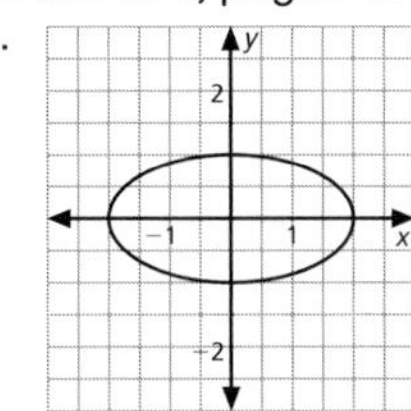

3. $25x^2 + 169y^2 = 4{,}225$ 5. Yes. A circle is an ellipse that has both foci at the same location.

7.

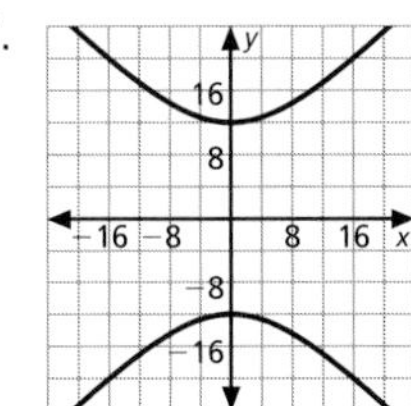

9. $121x^2 - 25y^2 = 3{,}025$

11.

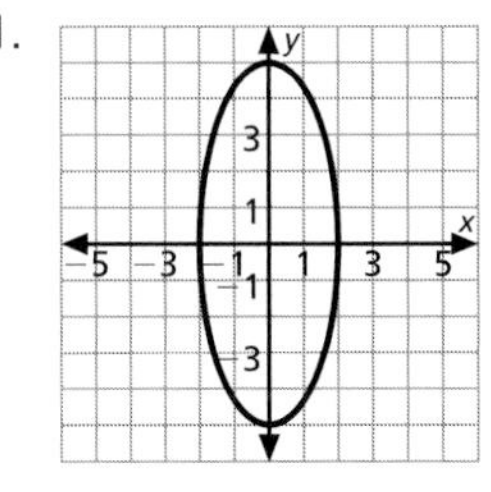

13.

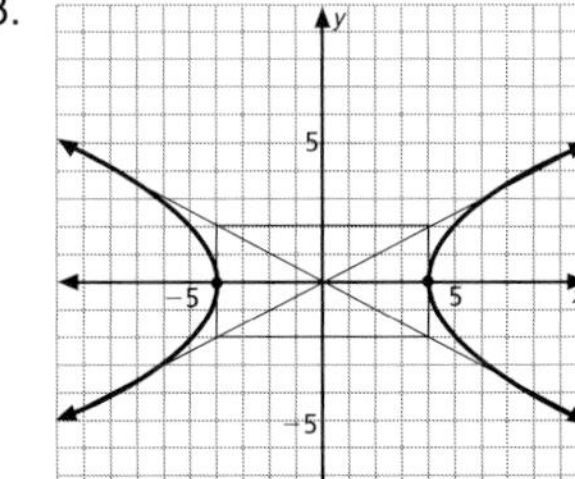

15. $\frac{(x - h)^2}{a^2} + \frac{(y - k)^2}{b^2} = 1$ 17. $x = 0, x = 70$ 19. $x = 0, x = 8$ 21. $x = 3 \pm\sqrt{7}$

Selected Answers

Review and Practice Your Skills, pages 578–579

1. parabola 3. circle 5. The standard form is $\frac{x^2}{a^2} + \frac{y^2}{b^2} = 1$, where $a = x$-intercepts. If $y = 0$; $\frac{x^2}{a^2} + 0 = 1$; $\frac{x^2}{a^2} = 1$; $x^2 = a^2$; $x = \pm a$ 7. The standard equation for an ellipse uses the sum of the distances from a point on the ellipse to the foci; the equation for a hyperbola uses the differences of these distances.

9.

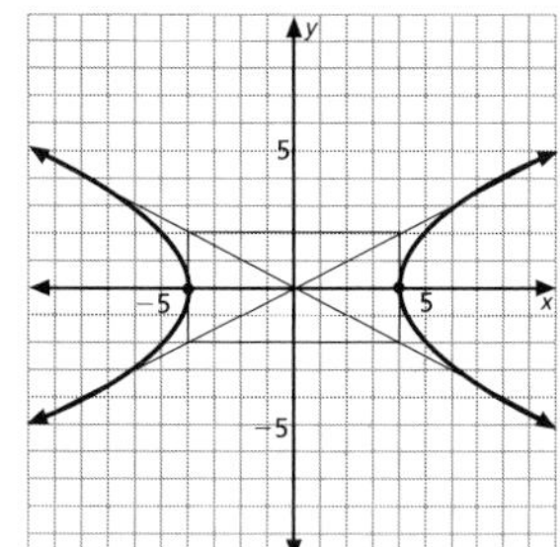

11.

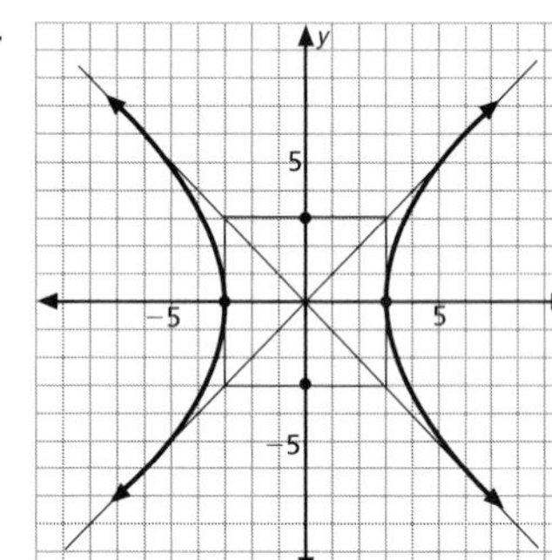

13.

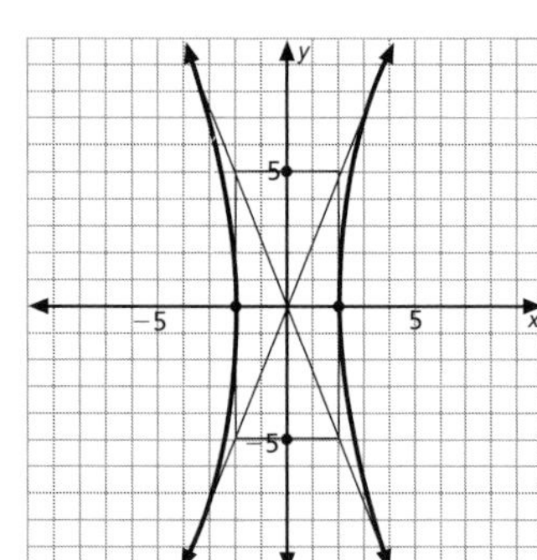

15. $91x^2 + 100y^2 = 9{,}100$ 17. $9x^2 - 25y^2 = 225$
19. $16x^2 - 49y^2 = 784$

21.

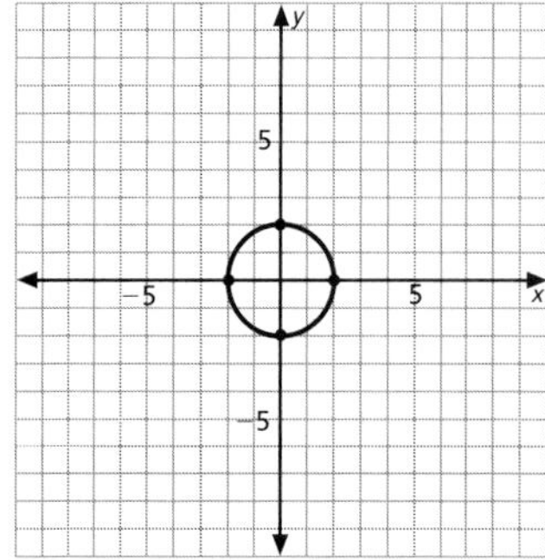

23.

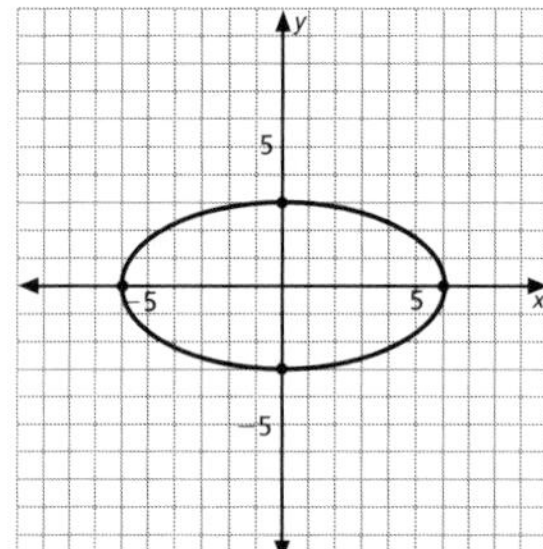

25.

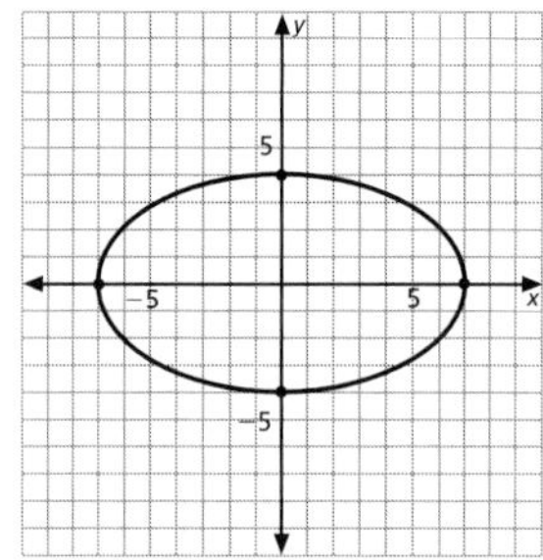

Lesson 13-5, pages 580–583

1. $y = 0.25x$ 3. $y = 125$ 5. $500 7. 80 mi 9. 450
11. $5.31 13. 40 15. No, Paige's conclusion is wrong. Using the direction variation form, $y = kx$, if 14 pounds are added to the spring, it will stretch 1.75 inches.
17. $A = 2598s^2$, 210.438 19. 1,035 21. 10
23. 19. $0.13; 20. $0.03; 21. $1.23; 22. $5.41
25. direct variation 27. direct square variation
29. direct variation 31. $x = \frac{-3 \pm \sqrt{57}}{4}$ 33. $x = 2 \pm 2\sqrt{3}$
35. $x = \frac{-3 \pm \sqrt{13}}{4}$ 37. 8.49 39. 13.23 41. 180 43. $\frac{3}{2}$

Lesson 13-6, pages 584–587

1. $y = \frac{68}{x}$ 3. $1.80 each 5. 12 lumens 7. $y = 3$
9. 24 min. 11. 4 Newtons 13. $p = \frac{k}{a}$ 15. $I = \frac{k}{R}$
17. $m = kst$ 19. $r = kwxy$ 21. $v = \frac{kr}{w^2}$ 23. 10.20
25. 5.39 27. 13.60 29. 11 31. 20.5

Selected Answers

Review and Practice Your Skills, pages 588–589
1. x must also decrease 3. A direct variation graph is linear; a direct square variation graph is a quadratic function.
5. $y = \frac{2}{3}x$ 7. $y = \frac{1}{4}x$ 9. $y = \frac{9}{4}x$ 11. $y = 2.14$ 13. $y = 48$
15. $y = 405$ 17. In real life, there are no negative people and no negative costs. Quadrant I is the only quadrant where both variables are positive. 19. $y = \frac{6.3}{x}$ 21. $y = 3.11$
23. $r = \frac{k}{s}$

25.
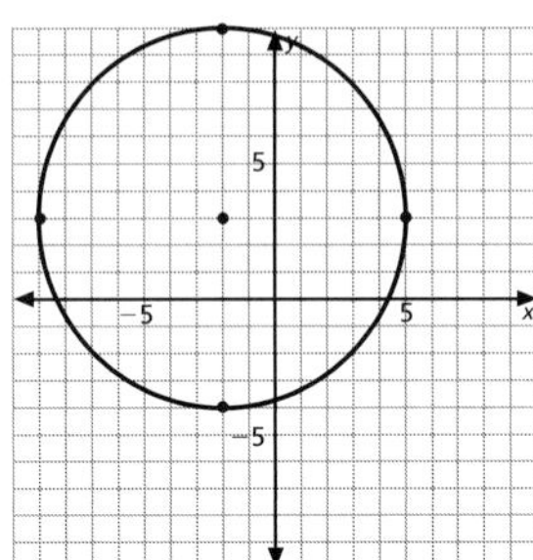

27.
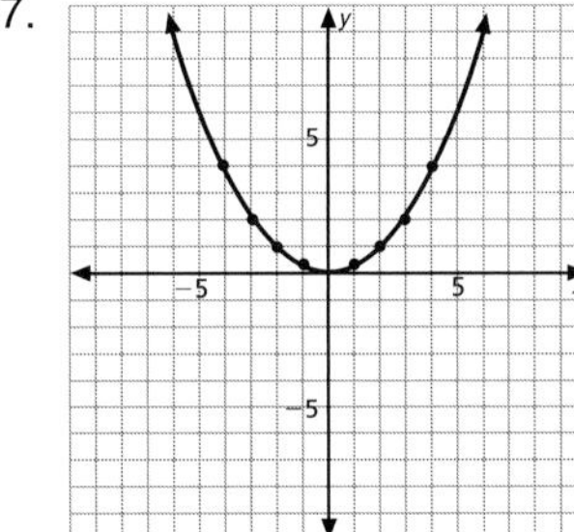

29.
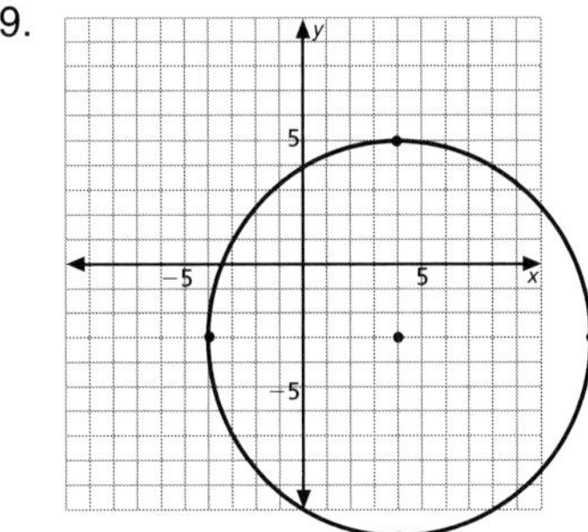

31. $y = \frac{5}{2}x$ 33. $y = \frac{7}{5}x$

Lesson 13-7, pages 590–593

1.
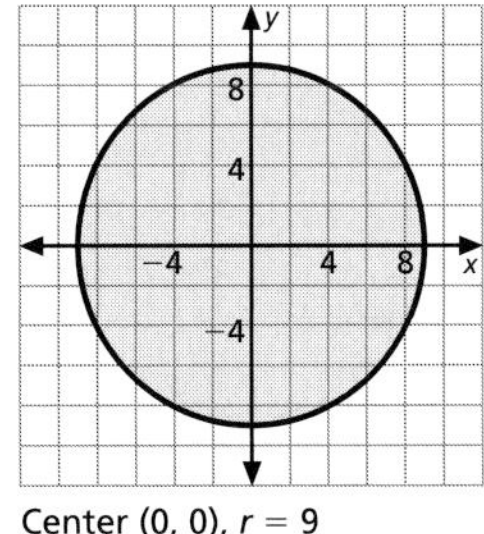
Center (0, 0), $r = 9$

3.
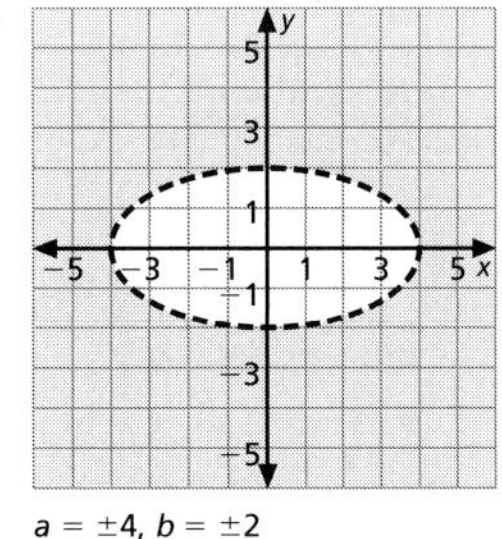
$a = \pm 4$, $b = \pm 2$

5.
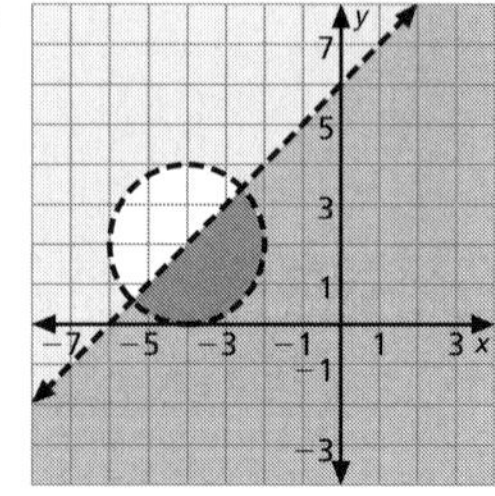
$r = 2$, Center $(-4, 2)$

7.
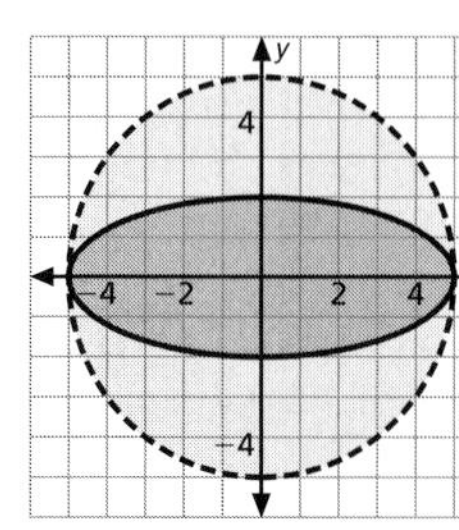
Circle: $r = 5$, Center (0, 0)
Ellipse: $a = \pm 5$, $b = \pm 2$

9.
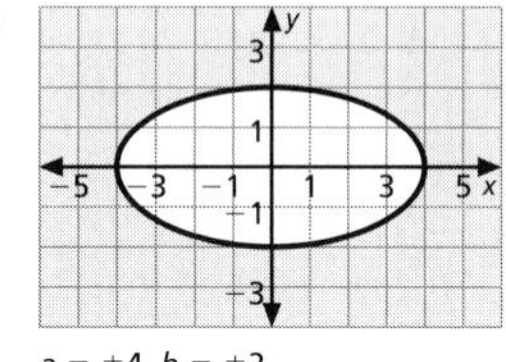
$a = \pm 4$, $b = \pm 2$

11.
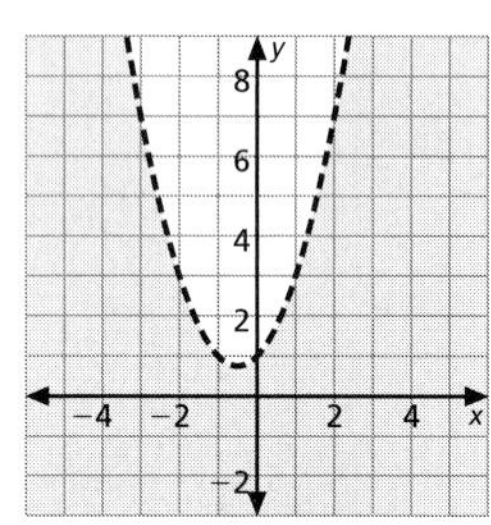

13.
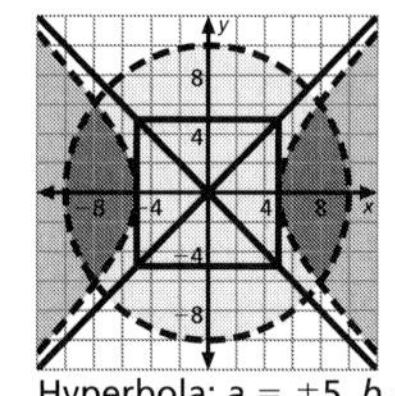
Hyperbola: $a = \pm 5$, $b = \pm 5$
Circle: $r = 10$, Center (0, 0)

15.
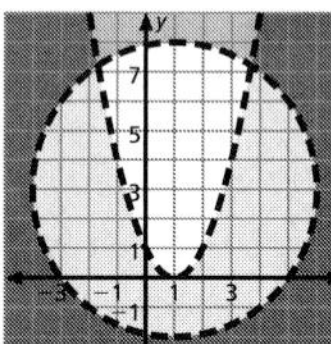
$r = 5$, Center (1, 3)

17. $<$ 19. $<$ 21. Both parabolas; solution set is all points in the intersection of the two figures.
23. Answers will vary. 25. $2 < x < 6$ 27. $(-1, -1)$
29. $\left(-1, \frac{7}{2}\right)$ 31. $\left(-\frac{5}{2}, \frac{7}{2}\right)$ 33. $\left(\frac{3}{2}, \frac{3}{2}\right)$ 35. $\left(\frac{13}{2}, -\frac{9}{2}\right)$
37. $\left(-\frac{1}{2}, 3\right)$ 39. $a = 105°$, $b = 75°$ 41. $a = 120°$; $b = 60°$
43. 305 ft^3 45. $x = 91$, $x = -111$

Chapter 13 Review, pages 594–595
1. d 3. b 5. $x^2 + y^2 = 64$ 7. $(x - 5)^2 + y^2 = 36$
9. 3, (0, 3) 11. (0, 5), $y = -5$ 13. (0, 1), $y = -1$
15. $x^2 = -16y$ 17. The plane is parallel to and contains one of the bases. 19. $36x^2 + 100y^2 = 3{,}600$ 21. $y = 50$
23. $y = 320$ 25. $y = 156.25$ 27. $y = 90$
29.
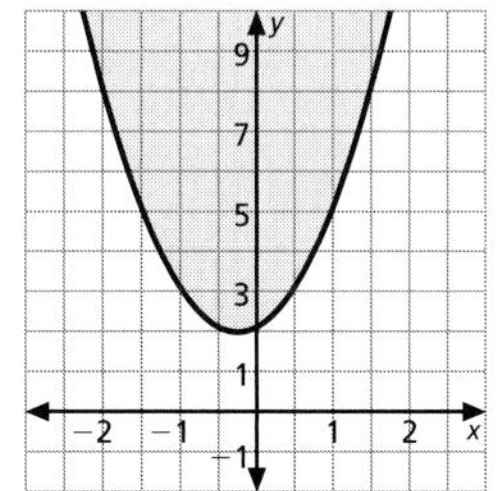

Selected Answers

Chapter 14

Lesson 14-1, pages 604–607

1. $2\frac{2}{5}$ 3. $\frac{12}{13}$ 5. 0.3746 7. 1.2799 9. 0.2126 11. $1\frac{1}{3}$ 13. $\frac{4}{5}$ 15. $\frac{3}{4}$ 17. 0.69; 1.05 19. 0.29 21. 32.1° 23. $\frac{12}{13}$ 25. Divide the height of the lighthouse by the ratio of height to distance (or tan 10°) 27. 1 29. $\frac{1}{2}$ 31. $x^2 + y^2 = 25$ 33. $(x - 5)^2 + (y - 1)^2 = 12.25$ 35. $(x - 3)^2 + (y + 2)^2 = 49$ 37. $(x + 3)^2 + (y - 5)^2 = 36$ 39. $(x + 4)^2 + (y - 3)^2 = 56.25$ 41. $\frac{1}{76}$ 43. $\frac{7}{855}$

Lesson 14-2, pages 608–611

1. 21.6 3. 34° 5. 55° 7. 31° 9. 3.0 11. 10.7 13. 35° 15. hypotenuse: 36.5; angles: 40°, 50°, 90° 17. 10,832.8 ft 19. 150.8 in. 21. 32.8 ft 23. 20° 27. no 29. sometimes; the angle must be acute 31. (0, 1); $y = -1$ 33. (0, 2); $y = -2$ 35. (0, 5\2); $y = -\frac{5}{2}$

Review and Practice Your Skills, pages 612–613

1. It must be a right triangle. 3. false 5. $\frac{3}{4}$ 7. $\frac{3}{5}$ 9. $\frac{3}{5}$ 11. $\frac{4}{5}$ 13. cos S = 0.50; tan R = 0.58 15. The sum of the angles of a triangle is 180°. If one angle is a right angle, you know it is 90°. Therefore, the other two angles must equal 180°–90°. If you know one of the acute angles you can use this information to find the missing measure. 17. 36° 19. 24.0 21. 60° 23. $\frac{4}{7}$ 25. $\frac{4}{7}$ 27. $\frac{11.49}{14}$ 29. sin D = 0.8; tan E = 0.75 31. 66° 33. 32.9° 35. 24.8

Lesson 14-3, pages 614–617

1. $\frac{\sqrt{2}}{2}$ 3. $-\frac{1}{2}$ 5. $\frac{-\sqrt{3}}{2}$

7. 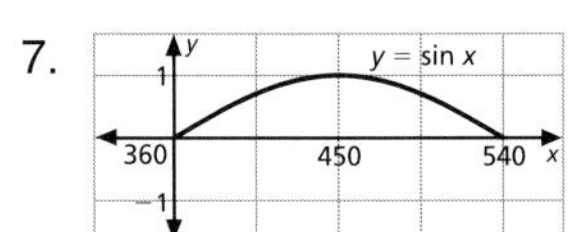

9. $-\frac{\sqrt{2}}{2}$

11. $\frac{\sqrt{3}}{2}$ 13. $-\frac{\sqrt{2}}{2}$ 15. $\frac{-\sqrt{2}}{2}$ 17. $-\frac{\sqrt{2}}{2}$ 19. $\frac{\sqrt{2}}{2}$

21. 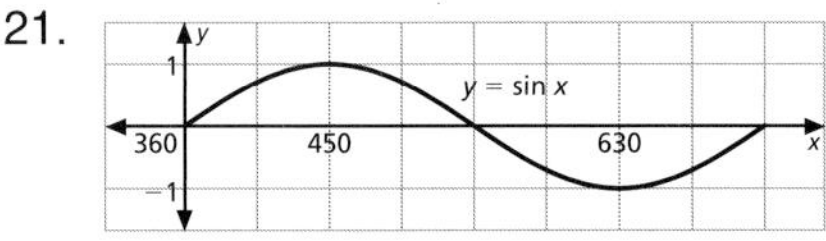

23. 1 25. $\sqrt{3}$

27. $\frac{\sqrt{2}}{2}$ 29. $-\frac{\sqrt{2}}{2}$ 31. $\frac{\sqrt{2}}{2}$ 33. $\frac{-\sqrt{2}}{2}$

35. 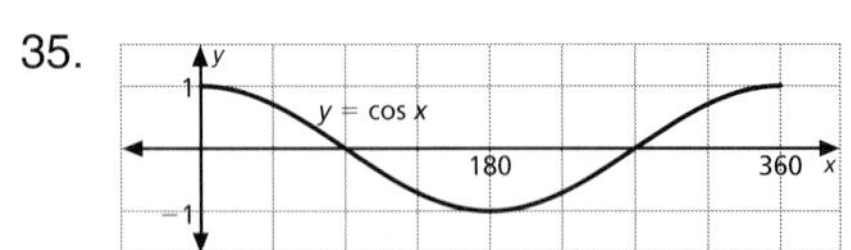

37. 0°, 180°, 360° 39. 90°, 270° 41. ≈485 feet 43.a. Graph the equations on the same set of axes. Points of intersection of the curves represent solutions of the system of equations. 43.b. x = 45° and x = 225° 45. 150°, 210° 47. 60°, 240° 49. $3x^2 + 4y^2 = 48$ 51. $5x^2 + 9y^2 = 405$ 53. $32x^2 + 81y^2 = 2592$

Lesson 14-4, pages 618–621

1. period = 120° 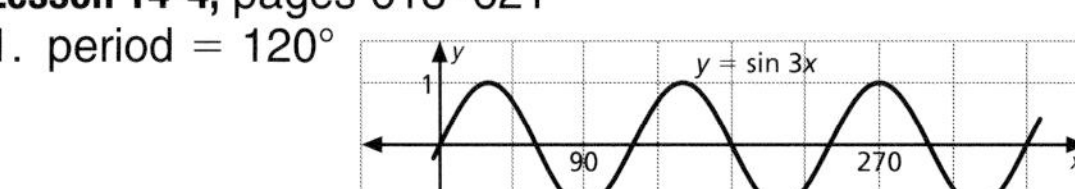

3. 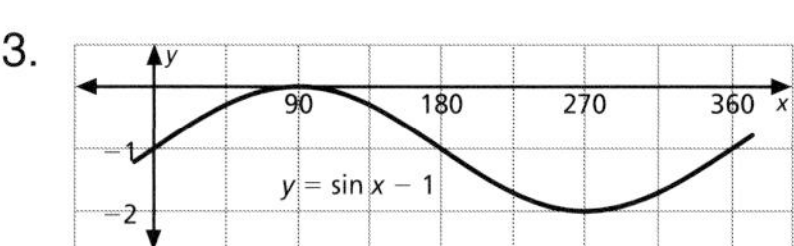

The graph is the graph of $y = \sin x$ lowered 1 unit from the origin. 5. period 180°; amplitude 2; the graph is the graph of $y = 2\sin 2x$ lowered 5 units.

7. 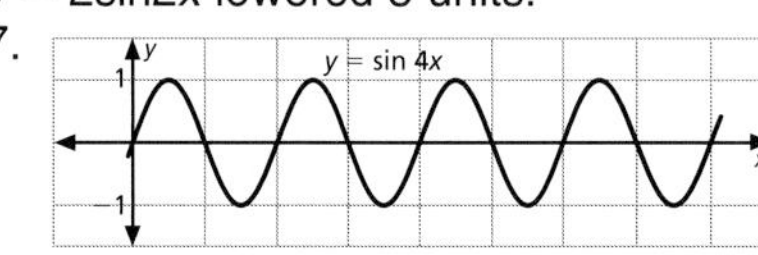

The period is 90°.

9. 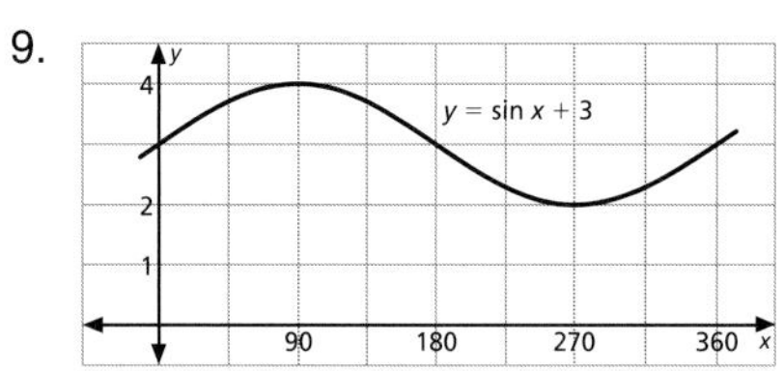

The graph is the graph of $y = \sin x$ raised 3 units. 11. period 120º; amplitude 4; the graph is the graph of $y = 4\sin 3x$ raised 2.5 units. 13. yes; 5 15. no 17. $y = \frac{3}{2}\sin\frac{2}{3}x$ 19. 600° 21. $y = 8\sin\frac{4}{7}x$ 23. $y = \frac{1}{2}\sin 8x - 1$

25. 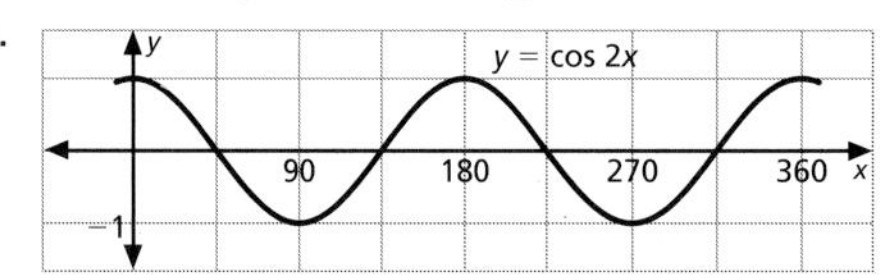

27. For the graph of the equation $y = m\cos nx \pm p$, the period is $\frac{360}{n}$, the amplitude is m, and the graph is the graph of $y = m\cos nx$ raised or lowered p units.

29. 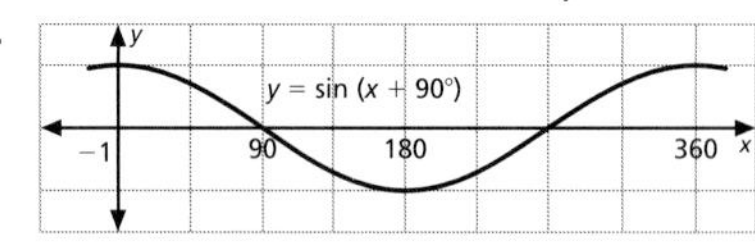

31. 40.5 33. 133 35. $(2x - 3)(x + 4)$ 37. $(2x + 1)(2x - 5)$ 39. $3(x + 2)(x - 1)$ 41. $(x - 8)(x + 8)$

Review and Practice Your Skills, pages 622–623

1. true 3. false 5. $\frac{\sqrt{3}}{2}$ 7. $\frac{-\sqrt{3}}{2}$ 9. $\frac{-\sqrt{2}}{2}$ 11. $-\frac{1}{2}$

13. $-\frac{1}{2}$ 15. $\frac{-\sqrt{2}}{2}$ 17. 60°, 300° 19. 120°, 240°

21. 0°, 360° 23. Half the difference between the maximum and minimum *y*-values. 25. 2 27. 120° 29. 36°
31. 360° 33. 9° 35. 360°, 2, centered on *y*-axis
37. 120°, 2, shifted 1 unit up 39. 180°, 5, shifted 2 units up

41. $\frac{7}{9.2}$ 43. $\frac{6}{9.2}$ 45. $\frac{6}{7}$ 47. 66.0° 49. $\frac{\sqrt{2}}{2}$ 51. 0

53. $\frac{\sqrt{2}}{2}$ 55. 60°, 120° 57. 30°, 330°135°, 315°

59. 45°, 315°

Lesson 14-5, pages 624–625

9. 4.67 11. 3.75 13. 3.77 15. $x^2 + 3x - 40$
17. $3x^2 - 26x + 16$ 19. $4x^2 - 81$ 21. $6x^2 + 33x - 18$
23. $6x^2 - 11x - 72$ 25. 3.23; 1.80 27. 42.36; 6.51

Chapter 14 Review, pages 626–627

1. c 3. d 5. e 7. $\frac{3}{5}$ 9. ≈ 21.4 11. 26.1 13. $\frac{\sqrt{3}}{2}$

15. $-\frac{1}{2}$

17.

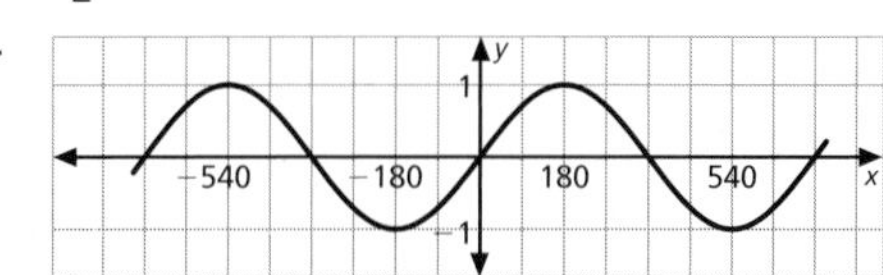

period = 720°

GLOSSARY

English / Español

A

absolute value (p. 12) The distance of any number, x, from zero on the number line. Absolute value is represented by $|x|$.

valor absoluto (p. 12) La distancia a la que se encuentra un número, x, del cero en la recta numérica. Se representa: $|x|$.

absolute value function (p. 63) The function that states: $g(x) = |x| = \{x \text{ if } x \geq 0, -x \text{ if } x < 0\}$.

función de valor absoluto (p. 63) La función que afirma que: $g(x) = |x| = x \text{ si } x < 0, -x \text{ si } x < 0$.

acute triangle (p. 150) A triangle with three acute angles that measure less than 90°.

triángulo acutángulo (p. 150) Un triángulo con tres ángulos que miden menos de 90°.

additional property of equality (p. 66) If two expressions are equal, the same number may be added to each expression and the resulting sums will be equal. For example, if $a = b$, then $a + c = b + c$ and $c + a = c + b$.

propiedad aditiva de la igualdad (p. 66) Para todos los números reales a, b, y c, si $a = b$, entonces $a + c = b + c$ y $c + a = c + b$.

addition property of inequality (p. 76) For all numbers a, b, and c, if $a < b$, then $a + c < b + c$ and $c + a < c + b$. If $a > b$, then $a + c > b + c$.

propiedad aditiva de la desigualdad (p. 76) Para todos los números reales a, b, y c, si $a < b$ entonces $a + c < b + c$ y $c + a < c + b$. Si $a > b$, entonces $a + c > b + c$.

addition property of opposites (p. 21) The sum of a number and its opposite equals 0.

propiedad aditiva de los opuestos (p. 21) La suma de un número y su opuesto es igual a 0. $a + -a = 0$

additive inverse (p. 21) The opposite of a number. The additive inverse of a is $-a$, and the additive inverse of $-a$ is a.

aditivo inverso (p. 21) El opuesto de un número. El aditivo inverso de a es $-a$ y el aditivo inverso de $-a$ es a.

adjacent angles (p. 109) Two angles in the same plane that have a common vertex and a common side but no common interior points.

ángulos adyacentes (p. 109) Dos ángulos en el mismo plano que tienen un vértice común y un lado común pero ningún punto interno común.

alternate exterior angles (p. 120) Two nonadjacent exterior angles on opposite sides of the transversal.

ángulos alternos externos (p. 120) Dos ángulos exteriores no adyacentes en lados opuestos a la transversal.

alternate interior angles (p. 120) Two nonadjacent interior angles on opposite sides of the transversal.

ángulos alternos internos (p. 120) Dos ángulos internos no adyacentes en lados opuestos a la transversal.

altitude of a triangle (p. 164) A perpendicular segment from a triangle's vertex to the opposite side.

altura de un triángulo (p. 164) Un segmento perpendicular desde el vértice de un triángulo al lado opuesto.

amplitude (p. 619) In a periodic function, half the difference between its maximum and minimum y-values.

amplitud (p. 619) En una función periódica, la mitad de la diferencia entre los valores máximo y mínimo de y.

angle (p. 108) The figure formed by two rays that have a common endpoint.

ángulo (p. 108) La figura formada por dos rayos que tienen un punto extremo en común.

angle of depression (p. 609) The acute angle formed by a horizontal line and a line slanting downward.

ángulo de depresión (p. 609) El ángulo agudo formado por una línea horizontal y una línea que se inclina hacia abajo.

angle of elevation (p. 609) The acute angle formed by a horizontal line and a line slanting upward.

ángulo de elevación (p. 609) El ángulo agudo formado por una línea horizontal y una línea que se inclina hacia arriba.

angle of rotation (p. 342) In a rotation, the amount of turn expressed as a fractional part of a whole turn or as the angle of rotation in degrees.

ángulo de rotación (p. 342) En una rotación, la cantidad de una vuelta expresada como una parte fraccional de una vuelta completa.

area (p. 206) The amount of surface a figure covers.

área (p. 206) La cantidad de superficie que una figura cubre.

associative property of addition (p. 21) Changing the grouping of terms does not change the sum. For example, $a + (b + c) = (a + b) + c$.

propiedad asociativa de la adición (p. 21) Para todos los números reales a, b, y c, $(a + b) + c = a + (b + c)$.

associative property of multiplication (p. 27) Changing the grouping of terms does not change the product. For example, $a(bc) = (ab)c$.

propiedad asociativa de la multiplicación (p. 27) Para todos los números reales a, b, y c, $a(bc) = (ab)c$.

English

asymptotes (p. 576) The values that hyperbolic functions approach but never reach.

axis of a cylinder (p. 220) The segment joining the centers of the two bases.

B

base angles of a trapezoid (p. 188) The two consecutive angles of a trapezoid that share a base. Every trapezoid has two pairs of base angles.

base of a prism (p. 220) One of the two identical parallel faces of a prism.

bell curve (p. 415) A frequency distribution that consists of a smooth curved line connecting the midpoints of a histogram. In a normal distribution of data, the curve is shaped like a bell.

biconditional statement (p. 129) A statement in the *if-and-only-if* form. In the biconditional "P if and only if Q," P is both a necessary condition and a and sufficient condition for Q.

binomial (p. 468). A polynomial with two terms.

bisector of an angle (p. 115) A ray that divides an angle into two congruent adjacent angles.

bisector of a segment (p. 114) Any line, segment, ray, or plane that intersects the segment at its midpoint.

boundary (p. 77) The line separating two half-planes in the coordinate plane.

box-and-whisker plot (p. 407) A means of visually displaying data that shows the median of a set of data, the median of each half of data, and the least and greatest value of the data.

C

cells (p. 30) Areas that are formed by the vertical columns and horizontal rows on a spreadsheet, and in which data can be stored and manipulated.

center of rotation (p. 342) The point about which the figure is rotated in a rotation.

chord (p. 441) A segment with both endpoints on the circle.

circle (p. 562) In a plane, the set of all points that are a given distance from a fixed point. That fixed point is the center of the circle.

circle graph (p. 20) A means of displaying data that represents items as parts of a whole circle; these parts are called sectors.

circumference (p. 206) The distance around a circle.

Español

asíntotas (p. 576) Los valores a que las funciones hiperbólicas se acercan pero nunca alcanzan.

eje de simetría de un cilindro (p. 220) El segmento que une los centros de las bases.

B

ángulos de base de un trapezoide (p. 188) Los dos ángulos consecutivos de un trapezoide que comparten una base.

base de un prisma (p. 220) Una de las dos caras paralelas idénticas de un prisma.

curva de campana (p. 415) Una distribución de frecuencia que consiste de una línea curva lisa que conecta los puntos medios de un histograma. En una distribución normal de datos, la curva tiene la forma de una campana.

proposición bicondicional (p. 129) Una proposición en la forma de *si y solamente si.* En la bicondicional "P si y solamente si Q", P es ambos una condición necesaria y suficiente para Q.

binomio (p. 468) Un polinomio con dos términos.

bisectriz de un ángulo (p. 115) Un rayo que divide un ángulo en dos ángulos adyacentes congruentes.

bisectriz de un segmento (p. 114) Cualquier línea, segmento, rayo o plano que interseca el segmento en su punto medio.

frontera (p. 77) La línea que separa dos semiplanos en el plano coordenado.

diagrama de bloque (p. 407) Un medio visual de representar datos que muestra la mediana de un conjunto de datos, la mediana de cada mitad de datos y el valor menor y mayor de los datos.

C

celdas (p. 30) Áreas que las forman las columnas verticales y las filas horizontales en una hoja de cálculos.

centro de rotación (p. 342) El punto alrededor del cual se gira una figura.

cuerda (p. 441) Un segmento con ambos puntos extremos en el mismo círculo.

círculo (p. 562) En un plano, el conjunto de todos los puntos que están equidistantes de un punto fijo, que se llama el centro del círculo.

diagrama de círculo (p. 20) Un medio de mostrar datos que representan artículos como partes de un círculo entero; estas partes se llaman sectores.

circunferencia (p. 206) La distancia alrededor un círculo.

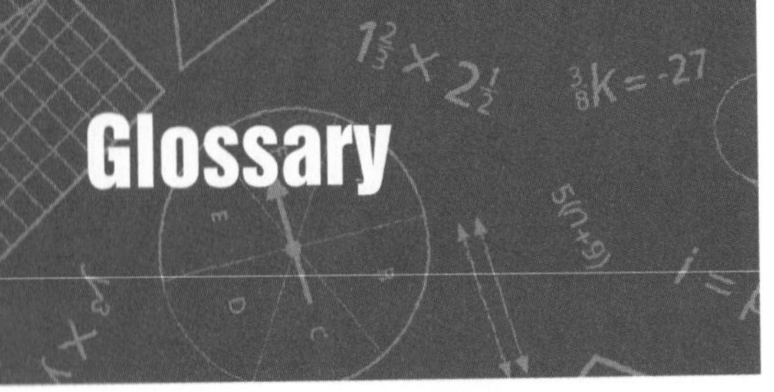

Glossary

English

circumscribed polygon (p. 455) A polygon with all sides tangent to the same circle.

closed half-plane (p. 77) The graph of either half-plane *and* the line that separates them.

cluster sampling (p. 6) Statistical sampling in which the members of the population are randomly selected from particular parts of the population and then surveyed in groups, not individually.

clusters (p. 87) Isolated groups of values on a stem-and-leaf plot.

coefficient (p. 468) The numerical, nonvariable portion of a monomial.

collinear points (p. 104) Points that lie on the same line.

column matrix (p. 275) An array of only one column.

combination (p. 417) A set of items chosen from a larger set without regard to order.

combined variation (p. 587) A variation in which a quantity varies directly as one quantity and inversely as another.

commutative property of addition (p. 21) Changing the order of two or more terms that are added does not change the sum. For example, $a + b = b + a$.

commutative property of multiplication (p. 27) Changing the order of two or more terms that are being multiplied does not change the product. For example, $ab = ba$.

complementary angles (p. 109) Two angles whose measures have a sum of 90°.

complement of an event (p. 393) The set of all outcomes in the sample space, not in A, when A is a subset of U. This is symbolized as A'.

completing the square (p. 534) Making a perfect square for an expression in the form $ax^2 + bx$.

composite of transformations (p. 352) Two or more successive transformations applied to a given figure.

compound event (p. 392) An event consisting of two or more simple events.

concave polygon (p. 178) A polygon with at least one diagonal that contains points in the exterior of the polygon.

concurrence (p. 165) The intersection of two or more lines at one point.

conditional statement (p. 128) An *if-then* statement having two parts—a hypothesis and a conclusion.

cone (p. 221) A three-dimensional figure with a curved surface and one circular base.

Español

polígono circunscrito (p. 455) Un polígono con todos los lados tangente al mismo círculo.

semiplano cerrado (p. 77) La gráfica de uno de los dos semiplanos y la línea que los separa.

muestra por conglomerado (p. 6) Una muestra de estadística en que los miembros de la población se seleccionan al azar desde sectores particulares de la población y se inspeccionan en grupos, no individualmente.

conglomerados (p. 87) Grupos aislados de valores en un diagrama de tallo y hoja.

coeficiente (p. 468) La porción numérica, no variable de un monomio.

puntos colineales (p. 104) Puntos que están en la misma línea.

matriz de columna (p. 275) Un arreglo numérico con solamente una columna.

combinación (p. 417) Un conjunto de artículos seleccionados de un conjunto más grande sin considerar el orden.

variación combinada (p. 587) Una variación en que una cantidad varía directamente como una cantidad e inversamente como otra.

propiedad conmutativa de la adición (p. 21) Para todos los números reales a, b, y c, $a + b = b + a$.

propiedad conmutativa de la multiplicación (p. 27) Para todos los números reales a, b, y c, $ab = ba$.

ángulos complementarios (p. 109) Dos ángulos cuyas medidas tienen una suma de 90°.

complemento de un suceso (p. 393) El conjunto de todos los resultados en el espacio de muestra, no en A, cuando A es un subconjunto de U.

completar el cuadrado (p. 534) Hacer un cuadrado perfecto para una expresión en la forma de $ax^2 + bx$.

compuesto de transformaciones (p. 352) Dos o más transformaciones consecutivas aplicadas a una figura determinada.

suceso compuesto (p. 392) Un suceso que consiste de dos o más sucesos simples.

polígono cóncavo (p. 178) Un polígono con por lo menos una diagonal que contiene puntos en el exterior del polígono.

incidencia (p. 165) La intersecciín de dos o más líneas en un punto.

proposición condicional (p. 128) Una proposición *si - entonces* con dos partes, una hipótesis y una conclusión.

cono (p. 221) Una figura tridimensional con una superficie curva y una base circular.

English

Español

congruent (p. 154) Term used to describe figures with the same size and shape.

congruente (p. 154) El término usado para describir figuras con el mismo tamaño y forma.

congruent angles (p. 154) Angles that have the same measure.

ángulos congruentes (p. 154) Ángulos que tienen la misma medida.

congruent line segments (p. 154) Line segments that have the same measure.

segmentos congruentes (p. 154) Segmentos que tienen la misma medida.

congruent triangles (p. 154) Triangles whose vertices can be paired in such a way that all angles and sides of one triangle are congruent to corresponding angles and corresponding sides of the other.

triángulos congruentes (p. 154) Triángulos con ángulos correspondientes congruentes y lados correspondientes congruentes.

conic section (p. 572) The section formed by a plane intersecting two circular cones whose vertices are at the origin.

sección cónica (p. 572) La sección formada por un plano que interseca dos conos circulares cuyos vértices están en el origen.

conjecture (p. 124) A conclusion reached through inductive reasoning.

conjetura (p. 124) Una conclusión que se deriva mediante el razonamiento inductivo.

consecutive sides (p. 178) Two sides of a polygon that have a common vertex.

lados consecutivos (p. 178) Dos lados de un polígono que tienen un vértice común.

consecutive vertices (p. 178) The endpoints of any side of a polygon.

vértices consecutivos (p. 178) Los puntos extremos de cualquier lado de un polígono.

constant (p. 468) A monomial that contains no variables.

constante (p. 468) Un monomio que no tiene variables.

construction (p. 118) A precise drawing of a geometric figure made with the aid of only two tools: a compass and an unmarked straightedge.

construcción (p. 118) Un dibujo de una figura geométrica hecha con la ayuda de solamente dos herramientas: un compás y una regla sin marcas.

convenience sampling (p. 6) Statistical sampling in which members of a population are selected because they are readily available, and all are surveyed.

muestra de conveniencia (p. 6) Una muestra de estadística en que los miembros de una población se seleccionan porque están fácilmente disponibles y todos son entrevistados.

converse (p. 129) A statement formed by interchanging the hypothesis and conclusion of a conditional statement.

converso (p. 129) Una declaración formada al intercambiar la hipótesis y la conclusión de una proposición condicional.

converse of the Pythagorean Theorem (p. 431) If the sum of the squares of the measures of two sides of a triangle is equal to the square of the measure of the third side, then the triangle is a right triangle.

converso del teorema pitagórico (p. 431) Si la suma de los cuadrados de dos lados de un triángulo es igual al cuadrado del tercer lado, entonces el triángulo es un triángulo rectángulo.

convex polygon (p. 178) A polygon with no diagonals that contain points in the exterior of the polygon.

polígono convexo (p. 178) Un polígono sin diagonales que contienen puntos en el exterior del polígono.

coordinate (p. 105) The real number paired with each point.

coordenada (p. 105) El número real que se aparea con un punto.

coordinate of the point (p. 105) The real number that corresponds to a point. An ordered pair of numbers associated with a point on a grid are the coordinates of the point.

coordenada del punto (p. 105) El número real asociado con un punto en la recta numérica. Un par ordenado de números asociado con un punto en el plano.

coordinate plane (p. 56) A two-dimensional mathematical grid system consisting of two perpendicular number lines, called the x-axis and the y-axis. The point where the axes intersect is the origin.

plano coordenado (p. 56) Un sistema bidimensional que consiste de dos rectas numéricas perpendiculares, llamadas *eje de x* y *eje de y*.

coplanar points (p. 104) Points that lie on the same plane.

puntos coplanares (p. 104) Puntos que están en el mismo plano.

corollary (p. 161) A statement that follows directly from a theorem.

corolario (p. 161) Una declaración que sigue directamente a un teorema.

Glossary

English	Español
corresponding angles (p. 120) Two angles in corresponding positions relative to two lines cut by a transversal. Also, angles in the same position in congruent or similar polygons.	**ángulos correspondientes** (p. 120) Dos ángulos en posiciones correspondientes relativos a dos líneas intersecadas por una transversal. También, ángulos en la misma posición en polígonos congruentes o similares.
corresponding sides (p. 154) Sides in the same position in congruent or similar polygons.	**lados correspondientes** (p. 154) Lados en la misma posición en polígonos congruentes o similares.
cosine (p. 604) In a right triangle, the cosine of acute $\angle A$ is equal to: $\frac{\text{length of side adjacent to } \angle A}{\text{length of hypotenuse.}}$	**coseno** (p. 604) En un triángulo rectángulo, el coseno del ángulo agudo A es igual a: $\frac{\text{la longitud de lado adyacente a } \angle A}{\text{la longitud de la hipotenusa}}$
counterexample (p. 128) An instance that satisfies the hypothesis but not the conclusion of the conditional statement. A single counterexample proves that the conditional statement is false.	**contraejemplo** (p. 128) Un ejemplo que satisface la hipótesis pero no la conclusión de una proposición condicional.
cross section (p. 223) The two-dimensional figure formed when a three-dimensional shape is cut with a plane.	**sección transversal** (p. 223) La figura bidimensional formada cuando una figura tridimensional es cortada por un plano.
customary units (p. 202) Units of measurement commonly used in the United States. Distance is measured in inches, feet, yards, and miles; mass is measured in ounces, pounds, and tons; and volume is measured in pints, quarts, and gallons.	**unidades inglesas** (p. 202) Las unidades de medida usualmente usadas en los Estados Unidos.
cylinder (p. 220) A three-dimensional shape made up of a curved region and two congruent circular bases that lie in parallel planes.	**cilindro** (p. 220) Una figura tridimensional que consiste de una región curva y dos bases circulares congruentes que están en planos paralelos.

D

English	Español
data (p. 82) Factual information used as a basis for reasoning, discussion, or calculation.	**datos** (p. 82) Información objetiva usada como base para razonar, discusión o calcular.
deductive reasoning (p. 134) A process of reasoning in which the truth of the conclusion necessarily follows from the truth of the premises.	**razonamiento deductivo** (p. 134) Un proceso de razonar en donde la verdad de la conclusión necesariamente sigue la verdad de las premisas.
dependent events (p. 397) Events whose outcomes affect one another.	**sucesos dependientes** (p. 397) Sucesos cuyos resultados se afectan uno a otro.
dependent system (p. 259) Two lines whose graphs coincide and thus have an infinite set of solutions.	**sistema dependiente** (p. 259) Dos líneas cuyas gráficas coinciden y tienen un conjunto infinito de soluciones.
dependent variables (p. 57) The elements of the range; also called the output values of a function.	**variables dependientes** (p. 57) Los elementos del alcance; también llamados los valores de salida de una función.
determinant (p. 274) An array of numbers arranged in rows and columns. A determinant is usually enclosed by straight lines.	**determinante** (p. 274) Un conjunto de números arreglado en filas y columnas.
diagonal (p. 178) A segment that joints two nonconsecutive vertices of a polygon.	**diagonal** (p. 178) Un segmento que une dos vértices no consecutivos de un polígono.
difference of two squares (p. 483) A binomial in which both terms are perfect squares and in which the second term is subtracted from the first. For example, $y^2 - 25$; its factors are $(y + 5)$ and $(y - 5)$.	**diferencia de dos cuadrados** (p. 483) Un binomio en que ambos términos son cuadrados perfectos y el segundo término se resta del primero. En $y^2 - 25$, los factores son $(y + 5)$ y $(y - 5)$.

English

dilation (p. 348) A transformation that produces an image that is the same shape as the original figure but a different size.

directrix (p. 566) A line whose distance from a point on a parabola is equal to the distance from the same parabolic point to the focus.

direct square variation (p. 581) A function written in the form $y = kx^2$ where k is a nonzero constant.

direct variation (p. 580) A function that can be written in the form $y = kx$, where k is a nonzero constant. In a direct variation, the value of one variable increases as the other variable increases.

distance (p. 105) The absolute value of the difference between the coordinates of any two points.

distance formula (p. 544) For any points $P_1(x_1, y_1)$ and $P_2(x_2, y_2)$, the distance between P_1 and P_2 is given by: $P_1P_2 = \sqrt{(x_2 - x_1)^2 + (y_2 - y_1)^2}$.

distributive property (p. 34) Each factor outside parentheses can be used to multiply each term within the parentheses. $a(b + c) = ab + ac$

domain of a relation (p. 94) The set of all possible values of the x-coordinates for a relation.

Español

dilatación (p. 348) Una transformación que produce una imagen que tiene la misma forma que la figura original pero un tamaño diferente.

directriz (p. 566) Una línea cuya distancia desde un punto en una parábola es igual a la distancia desde el mismo punto parab¢óico al foco.

variación cuadrada directa (p. 581) Una función escrita en la forma $y = kx^2$, donde k es una constante no igual a cero.

variación directa (p. 580) Una función que puede escribirse en la forma $y = kx$, donde k es una constante no igual a cero.

distancia (p. 105) El valor absoluto de la diferencia entre las coordenadas de dos puntos.

fórmula de distancia (p. 544) Para los puntos $P_1(x_1, y_1)$ y $P_2(x_2, y_2)$, la distancia entre P_1 y P_2 se da por: $P_1P_2 = \sqrt{(x_2 - x_1)^2 + (y_2 - y_1)^2}$.

propiedad distributiva (p. 34) Cada factor fuera del paréntesis puede usarse para multiplicar cada término dentro del paréntesis. Por ejemplo, $a(b + c) = ab + ac$.

dominio de una relación (p. 94) El conjunto de todos los valores posibles de las coordenadas de x para una relación.

■ E ■

edge (p. 220) The set of linear points at which two faces of a polyhedron intersect.

ellipse (p. 574) In a plane, the figure created by a point moving about two fixed points, called foci. The sum of the distance from the foci to any point on the ellipse is a constant, $F_1P + F_2P = 2A$.

empty set (p. 6) A set containing no elements. The symbol for the empty set is $\varnothing$. Also called the null set.

enlargement (p. 348) A dilated image that is larger than the original figure.

equation (p. 7) A mathematical statement that two numbers or expressions are equal.

equiangular triangle (p. 150) A triangle with three congruent angles.

equilateral triangle (p. 150) A triangle with three congruent sides.

equivalent ratios (p. 296) Two ratios that can both be named by the same fraction.

expanding binomials (p. 482) The multiplication and subsequent simplification of two binomials.

experiment (p. 384) An activity that is used to produce data that can be observed and recorded.

experimental probability (p. 384) The probability of an event determined by observation or measurement.

arista (p. 220) El conjunto de puntos lineales en donde dos caras de un poliedro se cruzan.

elipse (p. 574) En un plano, la figura creada por un punto que se mueve alrededor de dos puntos fijos, llamado los focos. La suma de la distancia desde los focos a cualquier punto sobre la elipse es una constante, $F_1P + F_2P = 2A$.

conjunto vacío (p. 6) Un conjunto que no contiene elementos. El símbolo para el conjunto vacío es $\varnothing$.

ampliación (p. 348) Una imagen dilatada que es más grande que la figura original.

ecuación (p. 7) Una declaración matemática en que dos números o expresiones son iguales.

triángulo equiangular (p. 150) Un triángulo con tres ángulos congruentes.

triángulo equilátero (p. 150) Un triángulo con tres lados congruentes.

razones equivalentes (p. 296) Dos razones que pueden ambas ser nombradas por la misma fracción.

expansión del binomio (p. 482) La multiplicación y simplificación subsiguiente de dos binomios.

experimento (p. 384) Una actividad que se usa para producir datos que se pueden observar y registrar.

probabilidad experimental (p. 384) La probabilidad de un suceso determinado por una observación o medida.

Glossary

English

exponent (p. 34) A superscripted number showing how many times the base is used as a factor. For example, in 2^4 4 is the exponent.

exponential form (p. 34) A number written with a base and an exponent. For example, the exponential form of (2)(2)(2)(2) is 2^4.

exterior angle of a polygon (p. 179) An angle both adjacent to and supplementary to an interior angle of a polygon.

exterior angle of a triangle (p. 151) An angle that is both adjacent to and supplementary to one of a triangle's interior angles.

exterior angles (p. 120) The angles formed by a transversal that are not between two coplanar lines.

extremes (p. 296) The first and last terms of a proportion. (p. 407) In statistics, the data gathered that varies most from the median. On a box-and-whisker plot, extremes are represented by the ends of the "whiskers" that are not adjacent to the "box."

Español

exponente (p. 34) Un número superescrito que muestra cuántas veces la base se usa como un factor. Por ejemplo, en 2^4, 4 es el exponente.

forma exponencial (p. 34) Un número escrito con una base y un exponente. Por ejemplo, la forma exponencial de (2)(2)(2)(2) es 2^4.

ángulo exterior de un polígono (p. 179) Un ángulo adyacente y suplementario al ángulo interno de un polígono.

ángulo exterior de un triángulo (p. 151) Un ángulo que es adyacente y suplementario a uno de los ángulos interiores de un triángulo.

ángulos exteriores (p. 120) Los ángulos formados por una transversal que no están entre dos líneas coplanares.

extremos (p. 296) Los primeros y últimos términos de una proporción. (p. 407) En estadísticas, los datos reunidos que más varian de la mediana.

F

face (p. 220) The surface of a polyhedron.

factorial (p. 403) The product of all whole numbers from n to 1. Written as $n!$.

factors (p. 473) Elements whose product is a given quantity.

finite set (p. 6) A set whose elements can be counted or listed.

foci (p. 574) In an ellipse, the two fixed points whose combined distances to any point on the ellipse is constant.

focus (p. 566) The fixed point whose distance from a point on a parabola is equal to the distance from the same parabolic point to the directrix.

frequency distribution (p. 414) A visual display that shows the relative frequency of data.

frequency table (p. 82) A method of recording data that shows how often an item appears in a set of data.

function (p. 56) A set of ordered pairs in which each element of the domain is paired with exactly one element in the range.

function notation (p. 57) The notation that represents the rule associating the input value (independent variable) with the output value (dependent variable). The most commonly used function notation is the "f of x" notation, written $f(x)$.

fundamental counting principle (p. 402) The principle that states: If there are two or more stages of an activity, the total number of possible outcomes is the product of the number of possible outcomes for each stage of the activity.

cara (p.220) La superficie de un poliedro.

factorial (p. 403) El producto de todos los números enteros desde n a 1. Escrito como $n!$.

factores (p. 473) Elementos cuyo producto es una cantidad determinada.

conjunto finito (p. 6) Un conjunto cuyos elementos pueden contarse o enumerarse.

focos (p. 574) En una elipse, los dos puntos fijos cuyas distancias combinadas a cualquier punto sobre la elipse es constante.

foco (p. 566) Los puntos fijos cuya distancia desde un punto en una parábola es igual a la distancia desde el mismo punto parabólico a la directriz.

distribución de frecuencia (p. 414) Una muestra visual que demuestra la frecuencia relativa de datos.

tabla de frecuencia (p. 82) Un método de registrar datos que muestra la frecuencia con que un artículo aparece en un conjunto de datos.

función (p. 56) Un conjunto de pares ordenados en donde cada elemento del dominio se aparea con exactamente un elemento del alcance.

notación de función (p. 57) La notación que representa la regla que asocia el valor de entrada (variable independiente) con el valor de salida (variable dependiente). La notación más comúnmente usada es la notación "f de x", escrita $f(x)$.

principio fundamental de conteo (p. 402) El principio que afirma: Si hay dos o más etapas de una actividad, el número total de resultados posibles es el producto del número de resultados posibles para cada etapa de la actividad.

English / Español

G

gaps (p. 87) Large spaces between values on a stem-and-leaf plot.

separacione (p. 87) Espacios grandes entre los valores en un diagrama de tallo y hoja.

general quadratic function (p. 524) A quadratic function written in the form $f(x) = ax^2 + bx + c$, where a, b, and c are real numbers, and $a \neq 0$.

función cuadrática general (p. 524) Una función cuadrática escrita en la forma $f(x) = ax^2 + bx + c$, donde a, b, y c son números reales, y $a \neq 0$.

greatest common factor (GCF) (p. 479) The greatest integer that is a factor of two or more integers. The GCF of two or more monomials is the greatest common numerical factor and the least power of the common variable factors.

máximo factor común (MFC) (p. 479) El entero más grande que es un factor de dos o más enteros. El MFC de dos o más monomios es el factor numérico común más grande y la potencia menor de los factores variables comunes.

greatest possible error (GPE) (p. 202) Half of the smallest unit used to make a measurement.

máximo error posible (p. 202) La mitad de la unidad más pequeña usada para hacer una medida.

H, I

half-plane (p. 77) The graphed region showing all solutions to a linear inequality. A half-plane that includes its boundary is a closed half-plane; a half-plane that does not include its boundary is an open half-plane.

semiplano (p. 77) La región que muestra todas las soluciones de una desigualdad lineal.

histogram (p. 87) A type of bar graph used to visually display frequencies. In a histogram the bars usually represent grouped intervals of numbers.

histograma (p. 87) Un tipo del diagrama de barra usado para visualmente mostrar las frecuencias.

hyperbola (p. 572) A curve formed by the intersection of a double-right circular cone with a plane that cuts both halves of the cone.

hipérbola (p. 572) Una curva formada por la intersección de un cono recto doble con un plano que corta ambas mitades del cono.

hypotenuse (p. 175) The side opposite the right angle in a right triangle.

hipotenusa (p. 175) El lado opuesto al ángulo recto en un triángulo rectángulo.

identity property of addition (p. 21) The sum of any number and zero is that number. For example, $a + 0 = 0 + a = a$.

propiedad aditiva de la identidad (p. 21) La suma de cualquier número y cero es ese número. Por ejemplo, $a + 0 = 0 + a = a$.

identity property of multiplication (p. 27) The product of any number and 1 is that number. for example, $a \times 1 = 1 \times a = a$.

propiedad multiplicativa de identidad (p. 27) El producto de un número y 1 es ese número. $a \times 1 = 1 \times a = a$.

image (p. 338) The new figure resulting from a translation.

imagen (p. 338) La nueva imagen que resulta de una traslación.

included angle (p. 155) In a triangle, the term used to describe an angle's relative position to the two sides that form it.

ángulo incluido (p. 155) En un triángulo, el término usado para describir la posición relativa de un ángulo a los dos lados que lo forman.

included side (p. 155) In a triangle, the term used to describe a side's relative position to the two angles common to it.

lado incluido (p. 155) En un triángulo, el término usado para describir la posición relativa de un lado a los dos ángulos comunes a él.

inconsistent system (p. 258) Two lines that are parallel and thus have no solutions.

sistema inconsistente (p. 258) Dos líneas que son paralelas y no tienen soluciones.

independent events (p. 396) Events whose outcomes are not affected by one another.

sucesos independientes (p. 396) Sucesos cuyos resultados no se afectan el uno al otro.

independent system (p. 258) Two linear equations that intersect at only one point.

sistemas independientes (p. 258) Dos ecuaciones lineales que se intersecan en un solo punto.

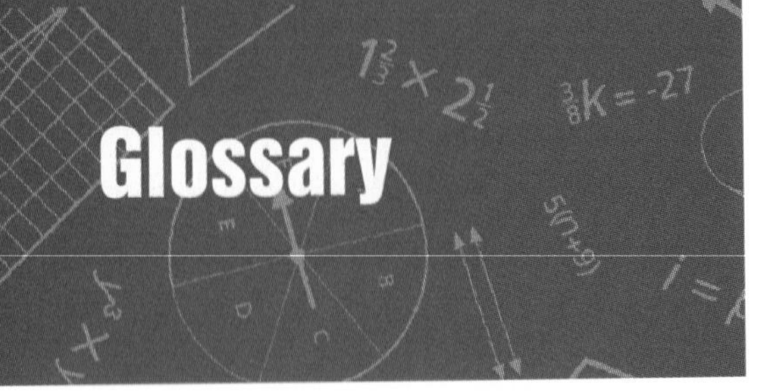

Glossary

English

independent variables (p. 57) The elements of the domain; also called the input values of a function.

indirect measurement (p. 326) The calculation of such a measurement that is difficult to measure directly. Using similar triangles is one method of indirect measurement.

indirect proof (p. 170) A proof in which one begins with the desired conclusion and assumes that it is not true. One then reasons logically until reaching a contradiction of the hypothesis or of a known fact.

inductive reasoning (p. 124) Logical reasoning where the premises of an argument provide some, but not absolute, support for the conclusion.

inequality (p. 11) A mathematical sentence that contains one of the symbols $<$, $>$, $\leq$, or $\geq$.

infinite set (p. 6) A set whose elements cannot be counted or listed.

initial side (p. 614) In the x and y coordinate plane, the side of an angle from which degree measurement begins.

inscribed polygon (p. 455) A polygon with all vertices lying on the same circle.

integers (p. 10) The set of whole numbers and their opposites.

interior angles (p. 120) The angles between coplanar lines that have been intersected by a transversal.

interior angles of a polygon (p. 178) The angles determined by the sides of a polygon.

interior angles of a triangle (p. 150) The angles determined by the sides of a triangle.

interquartile range (p. 408) The difference between the values of the first and third quartiles.

intersection of geometric figures (p. 104) The set of points common to two or more figures.

inverse operation (p. 66) An operation that undoes what the previous operation did; used when simplifying and/or solving a math sentence. For example, addition and subtraction are inverse operations.

inverse square variation (p. 585) A function that can be written in the form $y = \frac{k}{x^2}$ or $x^2y = k$, where k is a nonzero constant $x \neq 0$.

inverse variation (p. 584) A function that can be written in the form $y = \frac{k}{x}$, where k is a nonzero constant and $x \neq 0$.

irrational numbers (p. 10) A number that cannot be written as a fraction, a terminating decimal, or a repeating decimal. Examples are π and $\sqrt{2}$.

isosceles triangle (p. 150) A triangle with at least two congruent sides.

Español

variables independientes (p. 57) Los elementos del dominio; también se llaman los valores de entrada de la función.

medida indirecta (p. 326) El cálculo de una medida que es difícil de medir directamente. El usar triángulos similares es un método de medida indirecta.

prueba indirecta (p. 170) Una demostración en que se comienza con la conclusión deseada y se presume que no es cierta. Entonces se razona lógicamente hasta alcanzar una contradicción de la hipótesis.

razonamiento inductivo (p. 124) Razonamiento lógico donde las premisas de un argumento proveen algunos apoyos para la conclusión.

desigualdad (p. 11) Una frase matemática que contiene uno de los símbolos $<$, $>$, $\leq$, o $\geq$.

conjunto infinito (p. 6) Un conjunto cuyos elementos no pueden contarse o enumerarse.

lado inicial (p. 614) En el plano coordenado de x e y, el lado de un ángulo desde donde comienza la medida en grados.

polígonos inscritos (p. 455) Polígonos cuyos vértices están en un mismo círculo.

enteros (p. 10) El conjunto de números enteros y sus opuestos.

ángulos interiores (p. 120) Los ángulos entre líneas coplanares que han sido intersecadas por una transversal.

ángulos interiores de un polígono (p. 178) Los ángulos determinados por los lados de un polígono.

ángulos interiores de un triángulo (p. 150) Los ángulos determinados por los lados de un triángulo.

alcance intercuartil (p. 408) La diferencia entre el primer y tercer cuartiles.

intersección de figuras geométricas (p. 104) El conjunto de puntos comunes a dos o más figuras.

operaciones inversas (p. 66) Una operación que deshace lo qué la operación previa hizo; se usa al simplificar y/o resolver una frase matemática.

variación cuadrada inversa (p. 585) Una función que puede escribirse en la forma $y = \frac{k}{x^2}$ ó $x^2y = k$, donde k es una constante no igual a cero y $x \neq 0$.

variación inversa (p. 584) Una función que puede escribirse en la forma $y = \frac{k}{x}$, donde k es una constante no igual a cero y $x \neq 0$.

números irracionales (p. 10) Un número que no puede escribirse como una fracción, un decimal finito o un decimal periódico.

triángulo isósceles (p. 150) Un triángulo con por lo menos dos lados congruentes.

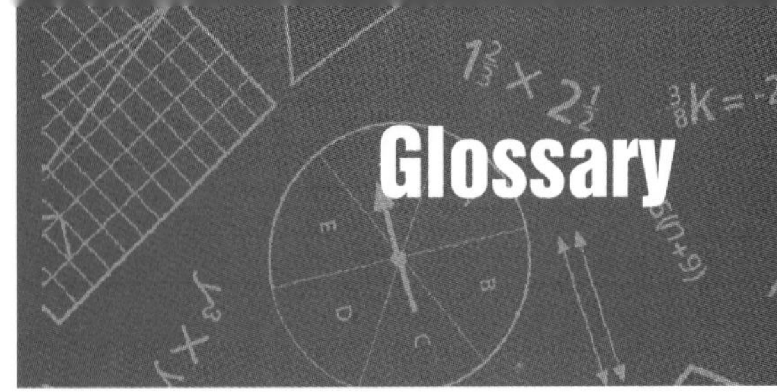

English

iteration (p. 53) A process that is continually repeated. For example, the iterative process of multiplying by 2 is used to create the numerical sequence 1, 2, 4, 8 . . . .

■ J, K, L ■

joint variation (p. 587) A variation in which a quantity varies directly as the product of two or more other quantities.

lateral edges (p. 220) The set of linear points at which lateral faces meet; these edges may be parallel, intersecting, or skew.

lateral faces (p. 220) The faces of prisms and pyramids that are not bases.

least common multiple (LCM) (p. 479) The least integer that is a multiple of two or more numbers. The LCM of two or more monomials is the product of the coefficients' LCM and the greatest power of all variable factors.

like terms (p. 468) Terms in which the variable or sets of variables are identical, though the numeric coefficients may be different.

line (p. 104) A set of points that extends infinitely in opposite directions.

linear equation (p. 62) An equation for which the graph is a line. Linear equations may be written in the form $Ax + By = C$, where A, B, and C are real numbers and A and B are not both zero.

linear function (p. 62) A function that can be represented by a linear equation. When graphed, a linear function yields a straight line.

linear programming (p. 282) A method used to solve business-related problems involving linear inequalities. Also used to find the maximum or minimum of an expression involving a solution to the system of inequalities.

line of best fit (p. 406) The line that can be drawn near most of the points on a scatter plot that shows a relationship between two sets of data; also called a trend line.

line of reflection (p. 338) The line over which a figure is reflected or flipped.

line of symmetry (p. 338) A line on which a figure can be folded, so that when one part is reflected over that line it matches the other part exactly.

line segment (p. 105) A part of a line containing two endpoints and all points in between.

lower quartile (p. 407) In statistics, the median of the lower half of the data gathered.

Español

iteración (p. 53) Un proceso que se repite continuamente.

variación conjunta (p. 587) Una variación en donde una cantidad varía directamente como el producto de dos o más otras cantidades.

aristas laterales (p. 220) El conjunto de puntos lineales en donde las caras laterales se encuentran.

caras laterales (p. 220) Las caras de prismas y pirámides que no son bases.

múltiplo común mínimo (MCM) (p. 479) El entero menor que es un múltiplo de dos o más números. El MCM de dos o más monomios es el producto de los MCM de los coeficientes y la potencia mayor de todos los factores variables.

términos semejantes (p. 468) Términos en que la variable o el conjunto de variables son idénticos, aunque los coeficientes numéricos pueden ser diferentes.

línea (p. 104) Un conjunto de puntos que se extiende infinitamente en direcciones opuestas.

ecuaciones lineales (p. 62) Una ecuación cuya gráfica es una línea. Las ecuaciones lineales pueden escribirse en la forma $Ax + By = C$, donde A, B y C son los números reales y A y B no son ambos cero.

funciones lineales (p. 62) Una función que puede ser representada por una ecuación lineal. Su gráfica es un línea recta.

programación lineal (p. 282) Un método que se usa para resolver problemas relacionados con los negocios que contienen desigualdades lineales. También se usa para encontrar el máximo o el mínimo de una expresión que contiene una solución al sistema de desigualdades.

línea de mejor ajuste (p. 406) La línea que puede dibujarse cerca de la mayoría de los puntos en un diagrama de dispersión que muestra una relación entre dos conjuntos de datos.

línea de reflejo (p. 338) La línea sobre la cual una figura se refleja.

línea de simetría (p. 338) Una línea sobre la cual una figura puede doblarse, de tal manera que al reflejar una parte sobre esta línea coincide exactamente con la otra parte.

segmento de línea (p. 105) Parte de una línea que contiene dos puntos extremos y todos los puntos entre ellos.

cuartil inferior (p. 407) En estadísticas, la mediana de la mitad inferior de los datos reunidos.

English | Español

M

major arc (p. 440) An arc that is larger than a semicircle.

arco mayor (p. 440) Un arco que es más grande que un semicírculo.

matrix (p. 275) (*plural: matrices*) A rectangular array of numbers arranged into rows and columns. Usually, square brackets enclose the numbers in a matrix.

matriz (p. 275) Un conjunto rectangular de números arreglados en filas y columnas.

maximum value (p. 283) The greatest solution to an inequality or a system of inequalities.

valor máximo (p. 283) La solución más grande de una desigualdad o de un sistema de desigualdades.

mean (p. 83) The sum of the data divided by the number of data. Also known as the arithmetic average.

media (p. 83) La suma de los datos dividido entre el número de datos. También se conoce como el promedio aritmético.

measures of central tendency (p. 83) Statistics or measurements used to describe a set of data. Examples of these are the mean, the median, and the mode.

medidas de tendencia central (p. 83) Medidas usadas para describir un conjunto de datos. Ejemplos son la media, la mediana y el modo.

median (p. 83) The middle value of the data when the data are arranged in numerical order.

mediana (p. 83) El valor medio de los datos cuando los datos se arreglan en orden numérico.

median of a trapezoid (p. 188) The segment that joins the midpoints of a trapezoid's legs.

mediana de un trapezoide (p. 188) El segmento que une los puntos medios de los lados del trapezoide.

median of a triangle (p. 164) A segment with endpoints that are a vertex of a triangle and the midpoint of the opposite side.

mediana de un triángulo (p. 164) Un segmento cuyos puntos extremos son un vértice del triángulo y el punto medio del lado opuesto.

metric units (p. 202) Units of measurement that are based on multiples of 10. In the metric system, three basic units of measurement exist; the meter (to measure length), the gram (to measure mass), and the liter (to measure volume). Relative size of these units is indicated by the prefixes "milli" $\left(\frac{1}{1{,}000}\right)$, "centi" $\left(\frac{1}{100}\right)$, "deci" $\left(\frac{1}{10}\right)$, "deka" (10), "hecto" (100), and "kilo" (1,000).

unidades métricas (p. 202) Unidades de medidas que se basan en los múltiplos de 10. Existen tres unidades básicas: el metro (mide longitud), el gramo (mide masa) y el litro (mide volumen).

midpoint (p. 114) The point that divides a segment into two congruent segments.

punto medio (p. 114) El punto que divide un segmento en dos segmentos congruentes.

midpoint formula (p. 545) For a line segment with endpoints (x_1, y_1) and (x_2, y_2), the coordinates of the midpoint are $\left(\frac{x_1 + x_2}{2}, \frac{y_1 + y_2}{2}\right)$.

fórmula del punto medio (p. 545) Para un segmento con puntos extremos (x_1, y_1) y (x_2, y_2) las coordenadas del punto medio son $\left(\frac{x_1 + x_2}{2}, \frac{y_1 + y_2}{2}\right)$.

minor arc (p. 440) An arc that is smaller than a semicircle; the degree measure of a minor arc is the same as the number of degrees in the corresponding central angle.

arco menor (p. 440) Un arco que es menor que un semicírculo.

mode (p. 83) The number that occurs most often in a set of data.

moda (p. 83) El número que ocurre más frecuentemente en un conjunto de datos.

monomial (p. 468) An expression that is either a single number, a variable, or the product of a number and one or more variables with whole-number exponents.

monomio (p. 468) Una expresión que es o un entero, una variable, o el producto de un número y una o más variables con exponentes enteros.

multiplication property of equality (p. 67) When two expressions are equal, each expression may be multiplied by the same number, and the resulting products will be equal. For example if $a = b$, then $ac = bc$.

propiedad multiplicativa de la igualdad (p. 67) Para todos los números reales a, b, y c, si $a = b$, entonces $ac = bc$.

English / Español

multiplication property of −1 (p. 27) Multiplying any real number by −1 results in a product that is the additive inverse of the number: $-1(a) = -a$, and $a(-1) = -a$.

propiedad multiplicativa de −1 (p. 27) Para todos los números reales a, b, y c, $-1(a) = -a$, y $a(-1) = -a$.

multiplication property of zero (p. 27) The product of any term and 0 is 0. For example $a \times 0 = 0 \times a = 0$.

propiedad multiplicativa de cero (p. 27) El producto de cualquier término y 0 es 0: $a \times 0 = 0 \times \text{a} = 0$.

multiplicative inverses (p. 27) Two numbers whose product is one; also called a reciprocal.

inversos multiplicativos (p. 27) Dos números cuyo producto es uno; también llamado recíproco.

multiplicative property of inequality (p. 76) For all real numbers a, b, and c, if $a < b$ and $c > 0$, then $ac < bc$; if $a < b$ and $c < 0$, then $ac > bc$; if $a > b$ and $c > 0$, then $ac > bc$; if $a > b$ and $c < 0$, then $ac < bc$.

propiedad multiplicativa de la desigualdad (p. 76) Para todos los números reales a, b, y c, si $a < b$ y $c > 0$, entonces $ac < bc$; si $a < b$ y $c < 0$, entonces $ac > bc$; si $a > b$ y $c > 0$, entonces $ac > bc$; si $a > b$ y $c < 0$, entonces $ac < bc$.

mutually exclusive (p. 392) Term used to describe events that cannot occur at the same time.

mutuamente exclusivo (p. 392) Término usado para describir sucesos que no pueden ocurrir a la misma vez.

N

***n* factorial** (p. 403) The number of permutations of n different items; n factorial is written $n!$

***n* factorial** (p. 403) El número de permutaciones de n artículos diferentes; n factorial se escribe $n!$.

negative correlation (p. 406) The inverse relationship between two sets of data. On a scatter plot, a negative correlation is evident if the trend line (line of best fit) slopes downward from the top left to the bottom right corner of the graph.

correlación negativa (p. 406) La relación inversa entre dos conjuntos de datos.

negative reciprocals (p. 248) Two fractions or ratios whose product is −1.

recíprocos negativos (p. 248) Dos fracciones o razones cuyo producto es −1.

noncollinear points (p. 104) Points that do not lie on the same line.

puntos no colineales (p. 104) Puntos que no están en la misma línea.

noncoplanar points (p. 104) Points that do not lie on the same plane.

puntos no coplanares (p. 104) Puntos que no están en el mismo plano.

normal curve (p. 415) The symmetrical bell-shaped curve resulting from a normal distribution of data. In a normal curve, the mean, median, and mode are the same.

curva normal (p. 415) La curva simétrica acampanada que resulta de una distribución normal de datos. En una curva normal, la media, mediana y la moda son iguales.

null set (p. 6) A set containing no elements. The symbol for the null is $\varnothing$.

conjunto nulo (p. 6) Un conjunto que no contiene elementos. El símbolo para el conjunto nulo es $\varnothing$.

O, P

obtuse angle (p. 109) An angle whose measure is greater than 90° but less than 180°.

ángulo obtuso (p. 109) Un ángulo cuya medida es mayor de 90° pero menor de 180°.

open half-plane (p. 77) The region on either side of a line on a coordinate plane.

semiplano abierto (p. 77) La región en cualquier lado de una línea en el plano coordenado.

opposite angles (p. 182) Two angles in a quadrilateral that do not share a common side.

ángulos opuestos (p. 182) Dos ángulos en un cuadrilátero que no comparten un lado común.

opposite of the opposite property (p. 12) The opposite of the opposite of any real number is the number. For example, $-(-n) = n$.

propiedad del opuesto de la opuesto (p. 12) Lo opuesto de lo opuesto de cualquier número real es el número. $-(-n) = n$.

opposite sides (p. 182) Two sides of a quadrilateral that do not share a common vertex.

lado opuesto (p. 182) Dos lados de un cuadrilátero que no comparten un vértice común.

ordered pair (p. 56) Two numbers named in a specific order.

par ordenado (p. 56) Dos números nombrados en un orden específico.

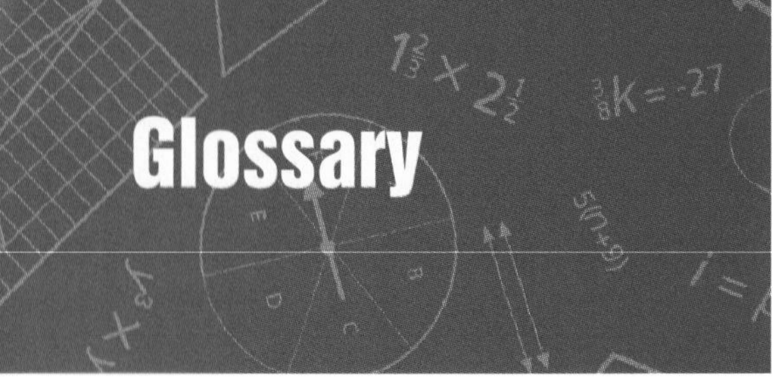

Glossary

English

order of operations (p. 62) Rules followed to evaluate expressions.

origin (p. 56) The point where the *x*-axis and *y*-axis intersect in the coordinate plane.

outcome (p. 384) The result of each trial of an experiment.

outliers (p. 16) Data values that are much greater or much less than most of the other values on a stem-and-leaf plot.

parabola (p. 520) The locus of points whose distance from the focus is equal to the distance from a fixed line (the directrix).

parallel lines (p. 119) Coplanar lines that do not intersect.

parallelogram (p. 182) A quadrilateral whose two pairs of opposite sides are parallel.

pattern (p. 52) An arrangement of numbers in a particular order; also called a sequence.

perfect square trinomial (p. 492) A trinomial that results from squaring a binomial.

perimeter (p. 206) The distance around a polygon.

period (p. 618) The length of one complete cycle of a periodic function.

periodic function (p. 618) A function that, when graphed, forms repeating patterns.

permutation (p. 403) An arrangement of items in a particular order.

perpendicular bisector (p. 165) Any perpendicular line, ray, or segment that intersects a segment at its midpoint.

perpendicular lines (p. 119) Two lines that intersect to form adjacent right angles.

plane (p. 104) An infinite set of points extending in all directions along a flat surface.

Platonic solids (p. 221) The five polyhedrons studied by the Greek scholar, Plato. Each of the polyhedrons has faces that are congruent regular polygons.

point (p. 104) A specific location in space having no dimensions. A point is represented by a dot, that is named by a letter.

point-slope form (p. 254) A common way in which the equation of a line can be expressed; used when the slope of a line (m) and the coordinates of any point (x_1, y_1) on the line are known. It is written: $y - y_1 = m(x - x_1)$.

polygon (p. 178) A closed plane figure formed by joining three or more line segments at either endpoints. Each segment or side of the polygon intersects exactly two other segments, one at each endpoint.

Español

orden de las operaciones (p. 62) Reglas que se siguen para evaluar expresiones.

origen (p. 56) El punto donde el eje de *x* y el eje de *y* se intersecan en el plano coordenado.

resultado (p. 384) El resultado de cada ensayo de un experimento.

datos extremos (p. 16) Valores de datos que son mucho mayores o menores que la mayoría de los otros valores en un digrama de tallo y hoja.

parábola (p. 520) Los lugares geométricos de puntos cuya distancia desde el foco es igual a la distancia desde una línea fija (la directriz).

líneas paralelas (p. 119) Líneas coplanares que no se cruzan.

paralelogramo (p. 182) Un cuadrilátero con dos pares de lados opuestos paralelos.

patrón (p. 52) Un arreglo de números en un orden particular; también llamado una sucesión.

trinomio cuadrado perfecto (p. 492) Un trinomio que resulta al elevar un binomio al cuadrado.

perímetro (p. 206) La distancia alrededor de un polígono.

período (p. 618) La longitud de un ciclo completo de una función periódica.

función periódica (p. 618) Una función cuya gráfica forma patrones que se repiten.

permutación (p. 403) Un arreglo de artículos en un orden particular.

mediatriz (p. 165) Cualquier línea, rayo o segmento perpendicular que cruza un segmento en su punto medio.

líneas perpendiculares (p. 119) Dos líneas que se intersecan para formar ángulos rectos adyacentes.

plano (p. 104) Un conjunto infinito de puntos que se extienden en cuatro direcciones a lo largo de una superficie plana.

sólidos platónicos (p. 221) Los cinco poliedros estudiados por el griego Platón, y cuyas caras son polígonos regulares congruentes.

punto (p. 104) Un lugar específico en el espacio que no tiene dimensiones.

forma de punto y pendiente (p. 254) Una manera común en que la ecuación lineal puede expresarse; se usa cuando la pendiente de una línea (m) y las coordenadas de cualquier punto (x_1, y_1) en la línea se conocen. Se escribe: $y - y_1 = m(x - x_1)$.

polígono (p. 178) Una figura cerrada plana formada al unir tres o m†s segmentos en sus puntos extremos. Cada segmento o lado del polígono interseca exactamente dos otros segmentos, uno en cada extremo.

Glossary

English

polyhedron (p. 220) (*plural: polyhedra*) A closed three-dimensional figure made of the only polygons.

polynomial (p. 468) An algebraic expression that is the sum of monomials. A polynomial is in standard form when its terms are ordered from the greatest to the least powers of one of the variables.

population (p. 82) The total number of people occupying a region or making up a whole.

positive correlation (p. 406) The direct relationship between two sets of data. On a scatter plot, a positive correlation is evident if the trend line (line of best fit) slopes from the bottom left upward to the top right of the graph.

postulate (p. 105) A statement accepted as truth without proof.

power of a product rule (p. 35) To find the power of a product, find the power of each factor and multiply. For example, $(ab)^m = a^m b^m$.

power of a quotient rule (p. 35) To find the power of a quotient, find the power of each number and divide. For example $\left(\frac{a}{b}\right)^m = \frac{a^m}{b^m}$.

power rule (p. 35) To raise an exponential number to a power, multiply exponents. For example, $(a^m)^n = a^{mn}$.

precision (p. 202) The accuracy of a measurement. Precision is relative to the unit of measurement used; the smaller the unit of measure, the more precise the measurement.

preimage (p. 338) The original figure of a translation.

prism (p. 220) A polyhedron that has two identical parallel bases and whose other faces are all parallelograms.

probability (p. 212) The chance of likelihood that an event will occur. The probability of an event can be expressed as a ratio:

$$P(\text{any event}) = \frac{\text{number of favorable outcomes}}{\text{number of possible outcomes}}.$$

An impossible event has a probability of zero. A certain event has a probability of one.

product rule (p. 35) to multiply numbers with the same base, add the exponents. For example, $(a^m)(a^n) = a^{m+n}$.

proportion (p. 296) An equation stating that two ratios are equivalent.

pyramid (p. 220) A polyhedron with only one base. The other faces are triangles that meet at a vertex.

Pythagorean Theorem (p. 430) In any right triangle, the square of the length of the hypotenuse c is equal to the sum of the squares of the lengths of the legs a and b. The Pythagorean Theorem is expressed as $c^2 = a^2 + b^2$.

Pythagorean triples (p. 433) Any three positive integers, a, b, and c, for which $a^2 + b^2 = c^2$.

Español

poliedro (p. 220) Una figura tridimensional cerrada formada solamente por polígonos.

polinomio (p. 468) Una expresión algebraica que es la suma de monomios.

población (p. 82) El número total de habitantes de una región que constituyen su totalidad.

correlación positiva (p. 406) La relación directa entre dos conjuntos de datos.

postulado (p. 105) Una declaración aceptada como verdadera sin prueba.

regla de la potencia de un producto (p. 35) Para todos los números reales a y b y para el entero positivo m, $(ab)^m = a^m b^m$.

regla de la potencia de un cociente (p. 35) Para todos los números reales a y b, $b \neq 0$ y para el entero positivo m, $\left(\frac{a}{b}\right)^m = \frac{a^m}{b^m}$.

regla de poder (p. 35) Para cualquier número a y los enteros positivos m y n, $(a^m)^n = a^{mn}$.

precisión (p. 202) La exactitud de una medida.

pre-imagen (p. 338) La figura original de una traslación.

prisma (p. 220) Un poliedro que tiene dos bases paralelas idénticas y cuyas otras caras son paralelogramos.

probabilidad (p. 212) La posibilidad de que un suceso ocurra. P(cualquier suceso) = número de resultados favorables/número de resultados posibles. Un suceso imposible tiene una probabilidad de cero. Un suceso seguro tiene una probabilidad de uno.

regla de producto (p. 35) Para un número a y los enteros positivos m y n, $a^m a^n = a^{m+n}$.

proporción (p. 296) Una ecuación que afirma que dos razones son equivalentes.

pirámide (p. 220) Un poliedro con una base y caras triangulares que se encuentran en un vértice.

teorema pitagórico (p. 430) En cualquier triángulo rectángulo, el cuadrado de la longitud de la hipotenusa c es igual a la suma de los cuadrados de las longitudes de los catetos a y b. El teorema pitagórico se expresa como $c^2 = a^2 + b^2$.

triples pitagóricos (p. 433) Enteros positivos a, b y c, para el cual $a^2 + b^2 = c^2$.

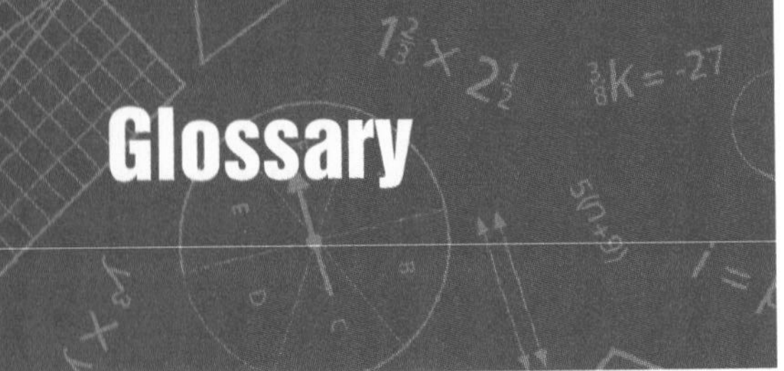

Glossary

English

Q, R

quadrant (p. 56) One of the four regions formed by the axes of the coordinate plane.

quadratic equation (p. 493) An equation of the form $Ax^2 + Bx + C = 0$, where A, B, and C are real numbers and A is not zero.

quadratic term (p. 498) In a quadratic expression, the term that contains the squared variable.

quartiles (p. 38) The three values which divide an ordered set of data into four equal parts. The *first quartile* is the median of the lower half of the data. The *third quartile* is the median of the upper half. The *second quartile* is another name for the median of the entire set of data.

quotient rule (p. 36) To divide numbers with the same base, subtract the exponents. For example, $a^m \div a^n = a^{m-n}$.

radicand (p. 427) The number under a radical sign. For example, 2 is the radicand of $\sqrt{2}$.

random sampling (p. 6) Statistical sampling in which each member of the population has an equal chance of being selected.

range (p. 56) The difference between the greatest and least values in a set of data.

range of function (p. 57) The set of all possible values of y for the function $y = f(x)$.

range of relation (p. 56) The set of all possible y-coordinates for a relation.

rate (p. 204) A ratio that compares two different kinds of quantities.

ratio (p. 202) A comparison of two numbers, a and b, represented in one of the following ways: $a{:}b$, $\frac{a}{b}$, or a to b.

rationalizing the denominator (p. 428) The process of rewriting a quotient to delete radicals from the denominator.

rational number (p. 10) A number that can be expressed in the form $\frac{a}{b}$, where a and b are any integers and $b \neq 0$.

ray (p. 105) Part of a line that starts at one endpoint and extends without end in one direction.

real numbers (p. 10) The set of rational and irrational numbers together.

reciprocals (p. 27) Two numbers that have a product of one.

rectangle (p. 183) A parallelogram that has four right angles.

reduction (p. 348) A dilated image that is smaller than the original figure.

reference angle (p. 614) In the coordinate plane, the acute angle formed by the x-axis and the terminal side.

Español

cuadrante (p. 56) Una de las cuatro regiones formadas por los ejes del plano coordenado.

ecuación quadrática (p. 493) Una ecuación de la forma $Ax^2 + Bx + C = 0$, donde A, B y C son números reales y A no es cero.

término cuadrático (p. 498) En una expresión cuadrática, el término que contiene la variable cuadrada.

cuartiles (p. 38) Los tres valores que dividen a un conjunto ordenado de datos en cuatro partes iguales.

regla del cociente (p. 36) Para un número a y los enteros positivos m y n, $\frac{a^m}{a^n} = a^{m-n}$

radicando (p. 427) El número debajo de símbolo radical.

muestra aleatoria (p. 6) Una muestra estadística en que cada miembro de la población tiene una oportunidad igual de ser seleccionado.

alcance (p. 56) La diferencia entre los valores mayores y menores en un conjunto de datos.

alcance de una función (p. 57) El conjunto de todos los valores posibles de y para la función $y = f(x)$.

alcance de una relación (p. 56) El conjunto de todas las posibles coordenadas de y para una relación.

tasa (p. 204) Una razón que compara dos tipos diferentes de cantidades.

razón (p. 202) Una comparación de dos números, a y b. Se representan en una de las maneras siguientes: $a{:}b$, $\frac{a}{b}$ o a al b.

racionalizar el denominador (p. 428) El proceso de escribir un cociente eliminando los radicales del denominador.

número racional (p. 10) Un número que puede expresarse en la forma $\frac{a}{b}$, donde a y b son enteros y $b \neq 0$.

rayo (p. 105) Parte de una línea que comienza en un punto y se extiende sin fin en una dirección.

números reales (p. 10) El conjunto de números racionales e irracionales.

recíprocos (p. 27) Dos números cuyo producto es uno.

rectángulo (p. 183) Un paralelogramo que tiene cuatro ángulos rectos.

reducción (p. 348) Una imagen dilatada que es más pequeña que la figura original.

ángulo de referencia (p. 614) En el plano coordenado, el ángulo agudo formado por el eje de x y el lado terminal.

English

reflection (p. 338) A transformation in which a figure is reflected, or flipped, over a line of reflection. A reflected image is congruent to the preimage but oriented in the opposite direction.

reflexive property (p. 34) Any number is equal to itself. For example, $a = a$.

regular polygon (p. 179) A polygon that is both equilateral and equiangular.

rhombus (p. 183) A parallelogram that has four congruent sides.

right triangle (p. 150) A triangle that has one right angle.

rotation (p. 342) A transformation in which a figure is rotated, or turned, about a point.

row-by-column multiplication (p. 362) A method by which two matrices are multiplied together. Using this method, matrices can be multiplied together only when the number of columns in the first matrix is equal to the number of rows in the second matrix.

row matrix (p. 275) An array of only one row.

Español

reflexión (p. 338) Una transformación en que una figura se voltea sobre una línea de reflejo.

propiedad reflexiva (p. 34) Para todos los números reales a, $a = a$.

polígono regular (p. 179) Un polígono que es equilateral y equiangular.

rombo (p. 183) Un paralelogramo que tiene cuatro lados congruentes.

triángulo rectángulo (p. 150) Un triángulo que tiene un ángulo recto.

rotación (p. 342) Una transformación en que se gira una figura alrededor de un punto.

multiplicación por fila y columna (p. 362) Un método en donde dos matrices se multiplican. Usando este método, las matrices se pueden multiplicar únicamente cuando el número de columnas en la primera matriz es igual al número de filas en la segunda matriz.

matriz de fila (p. 275) Un arreglo con sólo una fila.

S

sample (p. 82) A representative portion of a population, often used for statistical study.

sample space (p. 385) The set of all possible outcomes of an event.

scaler (p. 359) The constant by which a matrix is multiplied.

scaler multiplication (p. 359) The multiplication of a matrix by a constant.

scale drawing (p. 306) A drawing that represents an object. All lengths in the drawing are proportional to actual lengths in the object. The scale of the drawing is the ratio of the size of the drawing to the actual size of the object.

scale factor (p. 348) The number that is multiplied by the length of each side of a figure to create an altered image in a dilation.

scalene triangle (p. 150) A triangle with no congruent sides and no congruent angles.

scatter plot (p. 406) A method of visually displaying the relationship between two sets of data. The data are represented by unconnected points on a grid.

scientific notation (p. 39) A system for writing a very large or very small number as the product of a factor that is greater than or equal to one and less than 10 and a second factor that is a power of 10. For example, 496,000,000 written in scientific notation is 4.96×10^8.

secant (p. 441) A line that intersects a circle in two places.

muestra (p. 82) Una porción representativa de una población, frequentemente usada para el estudio estadístico.

espacio muestral (p. 385) El conjunto de todos los resultados posibles de un suceso.

escalar (p. 359) La constante por la cual una matriz se multiplica.

dibujo a escala (p. 306) Un dibujo que representa un objeto.

factor de escala (p. 348) El número que es multiplicado por la longitud de cada lado de una figura para crear una imagen alterada en una dilatación.

triángulo escaleno (p. 150) Un triángulo sin lados congruentes ni ángulos congruentes.

diagrama de dispersión (p. 406) Un método visual que muestra la relación entre dos conjuntos de datos.

notación científica (p. 39) Un sistema para escribir n£meros grandes o pequeños como el producto de un factor mayor o igual a uno y menos de 10 y un segundo factor que es una potencia de 10. Por ejemplo, 496,000,000 escrito en notación científica es 4.96×10^8.

secante (p. 441) Una línea que interseca un círculo en dos lugares.

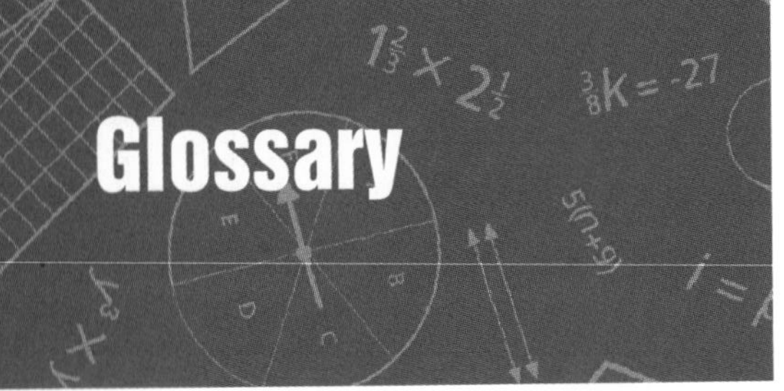

English

secant segment (p. 448) A segment intersecting a circle in two points, having one endpoint on the circle and one endpoint outside the circle.

sequence (p. 52) An arrangement of numbers according to a pattern.

set (p. 6) A well-defined collection of items. Each item is called an element, or member, of the set.

similar figures (p. 294) Figures that have the same shape but not necessarily the same size.

simplify (p. 34) For a variable expression, the performance of as many of the indicated operations as possible.

simulation (p. 388) A model used to estimate the probability of an event.

sine (p. 604) In a right triangle, the sine of acute $\angle A$ is equal to: $\frac{\text{length of side opposite } \angle A}{\text{length of hypotenuse}}$.

skew lines (p. 119) Noncoplanar lines that do not intersect and are not parallel.

slope (p. 244) The ratio of the vertical change of a line (rise) to its horizontal change (run).

slope-intercept form (p. 245) A linear equation in the form $y = mx + b$, where m represents the slope, and b represents the y-intercept.

solution (p. 7) A replacement set for a variable that makes a mathematical sentence true.

space (p. 104) The set of all possible points.

sphere (p. 221) A three-dimensional figure consisting of the set of all points that are a given distance from a given point, called the center of the sphere.

spreadsheet (p. 30) A computer application that simplifies preparation of tables.

square (p. 183) A parallelogram that has four right angles and four congruent sides.

square matrix (p. 275) An array with the same number of rows and columns.

square root (p. 426) One of two equal factors of a number. A number a is a square root of another number b if $a^2 = b$.

standard deviation (p. 417) The square rot of the variance of a set of numbers.

standard equation of a circle (p. 562) The equation of a circle with its center at any coordinate point (h, k) is $(x - h)^2 + (y - k)^2 = r^2$, where r is the circle's radius. If the circle's center is at the origin, the standard equation reduces to $x^2 + y^2 = r^2$.

standard equation of an ellipse (p. 574) The equation of an ellipse with its center at the origin is: $\frac{x^2}{a^2} + \frac{y^2}{b^2} = 1$.

Español

segmento de secante (p. 448) Un segmento que interseca un círculo en dos puntos.

sucesión (p. 52) Un arreglo de números según un patrón.

conjunto (p. 6) Una colección bien definida de artículos que se le llama un elemento, o miembro.

figuras semejantes (p. 294) Figuras que tienen la misma forma pero no el mismo tamaño.

simplificar (p. 34) Para una expresión variable, el hacer todas las operaciones indicadas que sean posibles.

simulación (p. 388) Un modelo usado para estimar la probabilidad de un suceso.

seno (p. 604) En un triángulo rectángulo, el seno del ángulo agudo A es igual a: $\frac{\text{longitud del lado opuesto al } \angle A}{\text{longitud de la hipotesusa}}$

líneas alabeadas (p. 119) Líneas no coplanares que no se intersecan ni son paralelas.

pendiente (p. 244) La razón del cambio vertical de una línea (subida) a su cambio horizontal (recorrido).

forma de pendiente e intersección (p. 245) Una ecuación lineal en la forma $y = mx + b$, donde m representa la pendiente y b representa el intercepto de y.

solución (p. 7) Un conjunto de reemplazo para una variable que hace que una frase matemática sea cierta.

espacio (p. 104) El conjunto de todos los puntos posibles.

esfera (p. 221) Una figura tridimensional que consiste del conjunto de todos los puntos equidistantes de un punto fijo, llamado el centro de la esfera.

hoja de cálculos (p. 30) Una aplicación de computadora que simplifica la preparación de tablas.

cuadrado (p. 183) Un paralelogramo que tiene cuatro ángulos rectos y cuatro lados congruentes.

matriz cuadrada (p. 275) Un arreglo numérico con el mismo número de filas y columnas.

raíz cuadrada (p. 426) Uno de dos factores iguales de un número. Un número a es una raíz cuadrada de otro número b si $a^2 = b$.

desviación estándar (p. 417) La raíz cuadrado de la varianza de un conjunto de números.

ecuación estándar de un círculo (p. 562) La ecuación de un círculo con su centro en cualquier punto coordenado (h, k) es $(x - h)^2 + (y - k)^2 = r^2$, donde r es el radio del círculo. Si el centro del círculo está en el origen, la ecuación estándar se reduce a $x^2 + y^2 = r^2$.

ecuación estándar de un elipse (p. 574) La ecuación de un elipse con su centro en el origen es: $\frac{x^2}{a^2} + \frac{y^2}{b^2} = 1$.

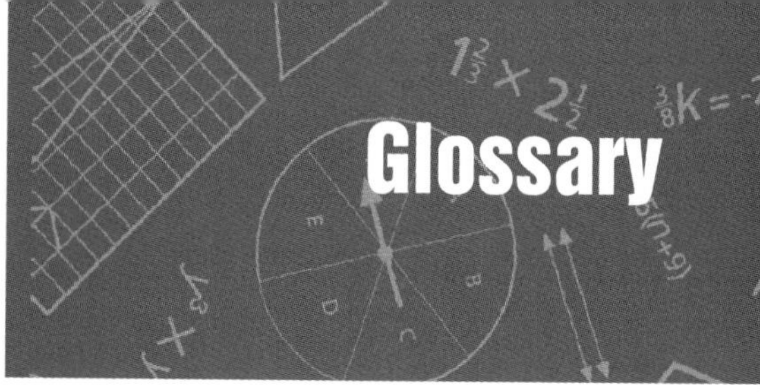

English

standard equation of a hyperbola (p. 576) The equation of a hyperbola with its center at the origin is: $\frac{x^2}{a^2} - \frac{y^2}{b^2} = 1$.

standard quadratic equation (p. 524) A quadratic equation written in the form $y = ax^2 + bx + c$, where a, b, and c are real numbers and $a \neq 0$.

statistics (p. 82) A branch of mathematics that involves the study of data, specifically the methods used to collect, organize, and interpret data.

stem-and-leaf plot (p. 86) A method of displaying data in which certain digits are used as stems and the remaining digits are used as leaves.

subset (p. 6) If every element of set A is also an element of set B, then A is called a subset of B.

substitution property (p. 34) If expressions are equivalent, they may be substituted for one another in any statement. For example, if $a = b$, then b can be substituted for a or a can be substituted for b in any statement.

supplementary angles (p. 109) Two angles whose measures have a sum of 180°.

surface area (p. 224) The sum of the areas of all the faces of a three-dimensional figure.

symmetric property (p. 34) The expressions on either side of an equals sign are equivalent and can thus be switched without affecting the equation.

systematic sampling (p. 82) Statistical sampling in which members of a population that has been ordered in some way are selected according to a pattern.

system of equations (p. 258) Two or more linear equations with the same variables.

system of linear inequalities (p. 276) Two or more linear inequalities that can be solved by graphing.

Español

ecuación estándar de una hipérbola (p. 576) La ecuación de una hipérbola con su centro en el origen es: $\frac{x^2}{a^2} - \frac{y^2}{b^2} = 1$.

ecuación cuadrática estándar (p. 524) Una ecuación cuadrática en la forma $y = ax^2 + bx + c$; a, b, y c son números reales y $a \neq 0$.

estadisticas (p. 82) Una rama de las matemáticas que comprende el estudio de datos, específicamente los métodos usados para coleccionar, organizar e interpretar datos.

diagrama de tallo y hoja (p. 86) Un método para mostrar datos en que ciertos dígitos se usan como los tallos y los dígitos restantes se usan como las hojas.

subconjunto (p. 6) Si cada elemento de un conjunto A es también un elemento del conjunto B, entonces A se llama un subconjunto de B.

propiedad de sustitiución (p. 34) Para todos los números reales, si $a = b$, entonces b se puede sustituir por a o a puede sustituirse por b en cualquiera declaración.

ángulos suplementarios (p. 109) Dos ángulos cuyas medidas tienen una suma de 180°.

área de superficie (p. 224) La suma de las áreas de todas las caras de una figura tridimensional.

propiedad simétrica (p. 34) Para todos los números reales a y b, si $a = b$, $b = a$.

muestra sistemática (p. 82) Una muestra de estadística en que miembros de una población, que se han ordenado de alguna manera, se seleccionan según un patron.

sistema de ecuaciones (p. 258) Dos o más ecuaciones lineales con las mismas variables.

sistema de desigualdades lineales (p. 276) Dos o más desigualdades lineales que se pueden resolver usando gráficas.

T

tangent (p. 604) In a right triangle, the tangent of acute $\angle A$ is equal to:

$$\frac{\text{length of side opposite } \angle A}{\text{length of side adjacent to } \angle A} \text{ or } \frac{\text{sine } \angle A}{\text{cosine } \angle A}.$$

tangent of a circle (p. 441) A line that intersects a circle in only one point.

tangent segment (p. 449) A segment with one endpoint on a circle and one endpoint outside the same circle.

terminal side (p. 614) In the coordinate plane, the side of an angle that is not the initial side; the terminal side is the side to which one measure degrees from the initial side.

tangente (p. 604) En un triángulo rectángulo, la tangente del ángulo agudo A es igual a:

$$\frac{\text{longitud del lado opuesto al } \angle A}{\text{longitud del lado adjacente al } \angle A} \text{ ó } \frac{\text{sin } \angle A}{\text{cos } \angle A}$$

tangente de un círculo (p. 441) Una línea que interseca un círculo en un solo punto.

segmento tangente (p. 449) Un segmento con un extremo sobre un círculo y el otro extremo fuera del mismo círculo.

lado terminal (p. 614) En el plano coordenado, el lado de un ángulo que no es el lado inicial; el lado terminal es el lado en el que se miden los grados desde el lado inicial.

Glossary

English

terms (p. 52) The parts of a variable expression that are separated by addition or subtraction signs. Terms that have identical variable parts are called like terms. Terms that have different variable parts are called unlike terms.

theorem (p. 114) A statement whose truth can be proven.

theoretical probability (p. 385) The probability of an event, $P(E)$, assigned by determining the number of favorable outcomes and the number of possible outcomes in the sample space:

$$P(E) = \frac{\text{number of favorable outcomes}}{\text{number of possible outcomes}}$$

tolerance (p. 202) In manufacturing, the amount that a piece may vary from its specified size.

transformation (p. 338) A way of moving or changing the size of a geometric figure in the coordinate plane. The new figure is referred to as the image of the original figure, and the original is referred to as the preimage of the new figure.

transitive property of equality (p. 34) If two expressions are equivalent, and a third expression is equivalent to the second expression, then the third expression is also equivalent to the first. For example, if $a = b$ and $b = c$, then $a = c$.

transitive property of inequality (p. 77) The property that states: For real numbers a, b, and c, if $a < b$ and $b < c$, then $a < c$. Similarly, if $a > b$ and $b > c$, then $a > c$.

translation (p. 338) A change in position of a figure such that all the points in the figure slide exactly the same distance and in the same direction at once. A translated image is congruent to the preimage.

transversal (p. 120) A line that intersects at least two coplanar lines in different points, producing interior and exterior angles.

trapezoid (p. 188) A quadrilateral that has exactly one pair of parallel sides.

tree diagram (p. 385) A diagram that shows all the possible outcomes in a sample space.

trend line (p. 406) A line that can be drawn near most of the points on a scatter plot that shows the relationship between two sets of data; also called the line of best fit. A trend line that slopes upward to the right indicates a positive correlation between the sets of data, while a trend line that slopes downward to the right indicates a negative correlation.

triangle (p. 150) A polygon formed by three line segments joining three noncollinear points.

trinomial (p. 468) A polynomial with three terms.

Español

términos (p. 52) Las partes de una expresión variable que son separadas por signos de sustracción o adición.

teorema (p. 114) Una declaración cuya verdad puede probarse.

probabilidad teórica (p. 385) La probabilidad de un suceso,

$$P(E) = \frac{\text{número de resultados favorables}}{\text{número de resultados posibles}}.$$

tolerancia (p. 202) En la manufactura, la cantidad que algo puede variar de su tamaño específico.

transformación (p. 338) Una manera de mover o cambiar el tamaño de una figura geométrica en el plano coordenado.

propiedad transitiva de la igualdad (p. 34) Para los números reales a, b y c, si $a = b$ y $b = c$, entonces $a = c$.

propiedad transitiva de desigualdad (p. 77) Para números reales a, b, y c, si $a < b$ y $b < c$, entonces $a < c$. De la misma manera, si $a > b$ y $b > c$, entonces $a > c$.

traslación (p. 338) Un cambio en la posición de una figura tal que todos los puntos en la figura se deslizan exactamente a la misma distancia y en la misma dirección.

transversal (p. 120) Una línea que interseca por lo menos dos líneas coplanares en puntos diferentes produciendo ángulos interiores y exteriores.

trapezoide (p. 188) Un cuadrilátero que tiene exactamente un par de lados paralelos.

diagrama de árbol (p. 385) Un diagrama que muestra todos los resultados posibles en un espacio de muestra.

línea de tendencia (p. 406) Una línea que puede dibujarse cerca de la mayoría de los puntos en un diagrama de dispersión que muestra la relación entre dos conjuntos de datos.

triángulo (p. 150) Un polígono formado por tres segmentos que unen tres puntos no colineales.

trinomio (p. 468) Un polinomio con tres términos.

English / Español

U, V, W

unit price (p. 204) A ratio comparing the price of an item to the unit of its measure.

precio de unidad (p. 204) Una razón que compara el precio de un artículo a la unidad de su medida.

unit rate (p. 204) A rate that has a denominator of one unit.

tasa de unidad (p. 204) Un valor que tiene un denominador de una unidad.

unlike terms (p. 468) Terms in which the variables or sets of variables are not identical.

términos diferentes (p. 468) Términos en que las variables o los conjuntos de variables no son idénticos.

upper quartile (p. 407) In statistics, the median of the upper half of the data.

cuartil superior (p. 407) En estadísticas, la mediana de la mitad superior de los datos.

variable (p. 7) A symbol, usually a letter, used to represent a number.

variable (p. 7) Un símbolo, comúnmente una letra, usado para representar un número.

variance (p. 412) For a set of numbers, the mean of the squared differences between *each* number in the set and the mean of *all* numbers in the set.

varianza (p. 412) Para un conjunto de números, la media de las diferencias cuadradas entre *cada* número en el conjunto y la media de *todos* los números en el conjunto.

vertex angle (p. 160) In an isosceles triangle, the angle opposite the base and adjacent to the two legs.

ángulo de vértice (p. 160) En un triángulo isósceles, el ángulo opuesto a la base y adyacente a los dos lados.

vertex of a polygon (p. 178) The point at which two sides of a polygon meet.

vértice de un polígono (p. 178) El punto donde dos lados de un polígono se encuentran.

vertex of a polyhedron (p. 220) The point at which three or more edges of a polyhedron intersect.

vértice de un poliedro (p. 220) El punto donde tres o más aristas de un poliedro se intersecan.

vertex of a triangle (p. 150) The point at which two sides of a triangle meet.

vértice de un triángulo (p. 150) El punto donde dos lados de un triángulo se encuentran.

vertical angles (p. 115) The angles that are not adjacent to each other when two lines intersect. Vertical angles are congruent.

ángulos verticales (p. 115) Los ángulos que no son adyacentes uno al otro cuando dos líneas se intersecan. Los ángulos verticales son congruentes.

vertical line test (p. 57) A test used to determine whether or not a graph is a function. It states: When a vertical line is drawn through the graph of a relation, the graph is not a function if the vertical line intersects the graph in more than one point.

prueba de verticalidad de línea (p. 57) Una prueba que se usa para determinar si una gráfica es una función o no. Cuando una línea vertical se dibuja a través de la gráfica de una relación, la gráfica no es una función si la línea vertical cruza la gráfica en más de un punto.

volume (p. 230) A measure of a the number of cubic units needed to fill a region of space.

volumen (p. 230) Una medida del número de unidades cúbicas necesarias para llenar una región de espacio.

X, Y, Z

***x*-coordinate** (p. 56) The first number in an ordered pair. The x-coordinate determines the horizontal location of a point in a coordinate plane. Also called the abscissa.

coordenada de x (p. 56) El primer número en un par ordenado. La coordenada de x determina la ubicación horizontal de un punto en un plano coordenado. También se le llama la abscisa.

***y*-coordinate** (p. 56) The second number in an ordered pair. The y-coordinate determines the vertical location of a point in a coordinate plane. Also called the ordinate.

coordenada de y (p. 56) El segundo número en un par ordenado. La coordenada de y determina la ubicación vertical de un punto en un plano coordenado. También se le llama la ordenada.

***y*-intercept** (p. 245) The y-intercept of a line is the y-coordinate of the point where the line intersects the y-axis.

intercepto de y (p. 245) El intercepto de y de una línea es la coordenada de y del punto donde la línea cruza el eje de y.

***z*-score** (p. 413) The number of standard deviations between a score and the mean score.

calificación z (p. 413) El número de desviaciones estándares entre una calificación y la calificación media.

INDEX

A

B

Index

Index

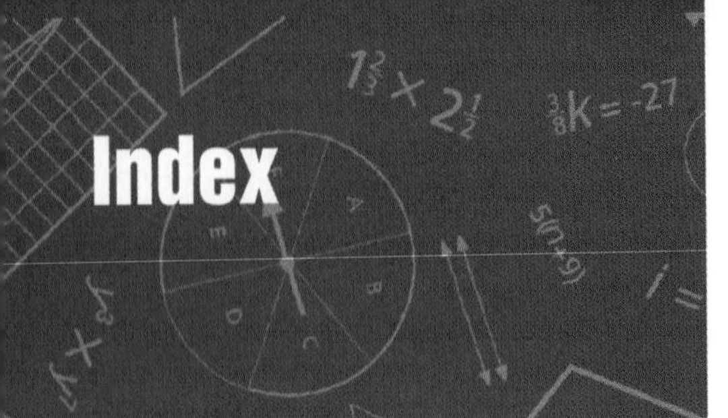

E

F

G

Index

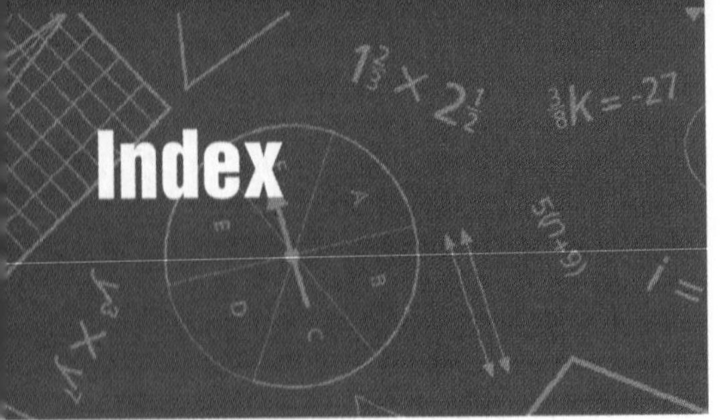

Index

Index

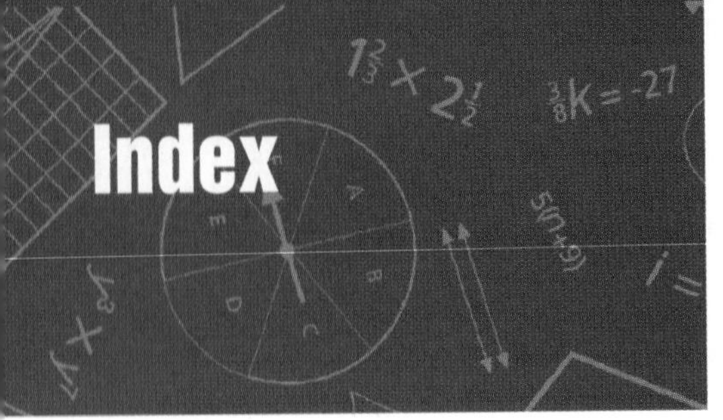

Index

Index

Index

Index

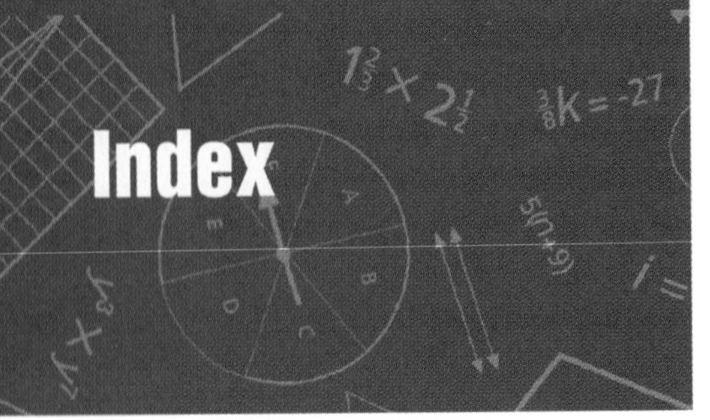
Index

Photo Credits

p. 4: © PhotoDisc, Inc.; p. 4: © PhotoDisc, Inc.; p. 5: © PhotoDisc, Inc.; p. 5: © PhotoDisc, Inc.; p. 6: © PhotoDisc, Inc.; p. 7: Duncan Smith for Artville; p. 8: © PhotoDisc, Inc.; p. 9: © PhotoDisc, Inc.; p. 11: © Corbis; p. 13: © PhotoDisc, Inc.; p. 16: EyeWire; p. 17: © PhotoDisc, Inc.; p. 18: © PhotoDisc, Inc.; p. 19: © Corbis/William James Warren; p. 20: © PhotoDisc, Inc.; p. 22: © Corbis; p. 26: © PhotoDisc, Inc.; p. 28: © PhotoDisc, Inc.; p. 29: © PhotoDisc, Inc.; p. 30: © PhotoDisc, Inc.; p. 34: © PhotoDisc, Inc.; p. 35: © PhotoDisc, Inc.; p. 36: © PhotoDisc, Inc.; p. 38: © PhotoDisc, Inc.; p. 41: © PhotoDisc, Inc.; p. 50: © PhotoDisc, Inc.; p. 51: Joe Pagliaro; p. 51: © PhotoDisc, Inc.; p. 52: © PhotoDisc, Inc.; p. 54: © PhotoDisc, Inc.; p. 54: © PhotoDisc, Inc.; p. 57: © PhotoDisc, Inc.; p. 59: © PhotoDisc, Inc.; p. 65: © PhotoDisc, Inc.; p. 69: © PhotoDisc, Inc.; p. 69: © PhotoDisc, Inc.; p. 74: © PhotoDisc, Inc.; p. 79: © PhotoDisc, Inc.; p. 83: © PhotoDisc, Inc.; p. 84: © PhotoDisc, Inc.; p. 85: © PhotoDisc, Inc.; p. 85: © PhotoDisc, Inc.; p. 87: © PhotoDisc, Inc.; p. 88: © PhotoDisc, Inc.; p. 89: © PhotoDisc, Inc.; p. 90: © PhotoDisc, Inc.; p. 102: © PhotoDisc, Inc.; p. 103: © PhotoDisc, Inc.; p. 103: Cartesia; p. 104: © PhotoDisc, Inc.; p. 104: © PhotoDisc, Inc.; p. 107: © PhotoDisc, Inc.; p. 108: © PhotoDisc, Inc.; p. 111: © PhotoDisc, Inc.; p. 114: © PhotoDisc, Inc.; p. 116: © PhotoDisc, Inc.; p. 119: © PhotoDisc, Inc.; p. 121: © PhotoDisc, Inc.; p. 124: © PhotoDisc, Inc.; p. 124: © PhotoDisc, Inc.; p. 127: © PhotoDisc, Inc.; p. 128: © PhotoDisc, Inc.; p. 130: © Corbis; p. 136: © PhotoDisc, Inc.; p. 138: © PhotoDisc, Inc.; p. 139: © PhotoDisc, Inc.; p. 149: © PhotoDisc, Inc.; p. 149: © PhotoDisc, Inc.; p. 150: © PhotoDisc, Inc.; p. 150: © PhotoDisc, Inc.; p. 150: © PhotoDisc, Inc.; p. 153: © PhotoDisc, Inc.; p. 155: © PhotoDisc, Inc.; p. 157: © PhotoDisc, Inc.; p. 160: © PhotoDisc, Inc.; p. 163: © PhotoDisc, Inc.; p. 165: © PhotoDisc, Inc.; p. 167: © PhotoDisc, Inc.; p. 171: © PhotoDisc, Inc.; p. 173: © PhotoDisc, Inc.; p. 175: © PhotoDisc, Inc.; p. 181: © PhotoDisc, Inc.; p. 184: © PhotoDisc, Inc.; p. 185: © PhotoDisc, Inc.; p. 200: © PhotoDisc, Inc.; p. 200: © PhotoDisc, Inc.; p. 201: © PhotoDisc, Inc.; p. 203: © PhotoDisc, Inc.; p. 204: © PhotoDisc, Inc.; p. 206: © PhotoDisc, Inc.; p. 209: © PhotoDisc, Inc.; p. 213: © PhotoDisc, Inc.; p. 214: © PhotoDisc, Inc.; p. 215: © PhotoDisc, Inc.; p. 217: © PhotoDisc, Inc.; p. 223: © Corbis; p. 224: © PhotoDisc, Inc.; p. 226: © PhotoDisc, Inc.; p. 227: © Corbis; p. 230: © Corbis; p. 232: © PhotoDisc, Inc.; p. 233: © PhotoDisc, Inc.; p. 242: © PhotoDisc, Inc.; p. 242: © PhotoDisc, Inc.; p. 243: © PhotoDisc, Inc.; p. 243: Joe Pagliaro; p. 247: © PhotoDisc, Inc.; p. 248: © PhotoDisc, Inc.; p. 250: Studio Fourteen for Artville; p. 255: © PhotoDisc, Inc.; p. 257: © PhotoDisc, Inc.; p. 261: © PhotoDisc, Inc.; p. 266: © PhotoDisc, Inc.; p. 267: © PhotoDisc, Inc.; p. 270: © PhotoDisc, Inc.; p. 271: © PhotoDisc, Inc.; p. 275: © PhotoDisc, Inc.; p. 276: © PhotoDisc, Inc.; p. 279: © PhotoDisc, Inc.; p. 282: © PhotoDisc, Inc.; p. 284: © PhotoDisc, Inc.; p. 285: © PhotoDisc, Inc.; p. 294: © PhotoDisc, Inc.; p. 295: © PhotoDisc, Inc.; p. 295: © PhotoDisc, Inc.; p. 297: © Corbis; p. 299: © PhotoDisc, Inc.; p. 301: © PhotoDisc, Inc.; p. 301: © PhotoDisc, Inc.; p. 301: © PhotoDisc, Inc.; p. 301: © PhotoDisc, Inc.; p. 303: © PhotoDisc, Inc.; p. 306: © PhotoDisc, Inc.; p. 307: © PhotoDisc, Inc.; p. 307: © Corbis; p. 313: © PhotoDisc, Inc.; p. 319: © PhotoDisc, Inc.; p. 319: © PhotoDisc, Inc.; p. 321: © PhotoDisc, Inc.; p. 327: © Corbis; p. 336: Joe Pagliaro; p. 336: © PhotoDisc, Inc.; p. 337: © PhotoDisc, Inc.; p. 338: © PhotoDisc, Inc.; p. 339: EyeWire; p. 340: © PhotoDisc, Inc.; p. 342: © PhotoDisc, Inc.; p. 344: © PhotoDisc, Inc.; p. 350: © Corbis; p. 352: © PhotoDisc, Inc.; p. 355: © PhotoDisc, Inc.; p. 360: © PhotoDisc, Inc.; p. 361: © Corbis; p. 362: © Corbis; p. 365: © PhotoDisc, Inc.; p. 371: © PhotoDisc, Inc.; p. 373: © PhotoDisc, Inc.; p. 382: © PhotoDisc, Inc.; p. 383: © PhotoDisc, Inc.; p. 383: © PhotoDisc, Inc.; p. 383: © PhotoDisc, Inc.; p. 384: © PhotoDisc, Inc.; p. 386: © PhotoDisc, Inc.; p. 388: © PhotoDisc, Inc.; p. 392: © PhotoDisc, Inc.; p. 394: © PhotoDisc, Inc.; p. 398: © PhotoDisc, Inc.; p. 399: © Charles Harris; Pittsburg Courier/Corbis; p. 402: © PhotoDisc, Inc.; p. 403: © PhotoDisc, Inc.; p. 405: © Corbis; p. 406: EyeWire; p. 409: © PhotoDisc, Inc.; p. 424: © PhotoDisc, Inc.; p. 425: © PhotoDisc, Inc.; p. 426: © PhotoDisc, Inc.; p. 429: © PhotoDisc, Inc.; p. 430: © Corbis; p. 433: © PhotoDisc, Inc.; p. 438: © PhotoDisc, Inc.; p. 440: © PhotoDisc, Inc.; p. 443: © PhotoDisc, Inc.; p. 446: © PhotoDisc, Inc.; p. 447: © PhotoDisc, Inc.; p. 451: © PhotoDisc, Inc.; p. 454: © PhotoDisc, Inc.; p. 456: © PhotoDisc, Inc.; p. 457: Kieth Thomas Productions for Artville; p. 466: © PhotoDisc, Inc.; p. 466: © PhotoDisc, Inc.; p. 467: © PhotoDisc, Inc.; p. 467: © PhotoDisc, Inc.; p. 469: Burke & Triolo for Artville; p. 470: © PhotoDisc, Inc.; p. 473: Burke & Triolo for Artville; p. 475: © PhotoDisc, Inc.; p. 480: © PhotoDisc, Inc.; p. 481: Burke & Triolo for Artville; p. 483: © PhotoDisc, Inc.; p. 488: Cha Cha Royale for Artville; p. 490: © PhotoDisc, Inc.; p. 491: © PhotoDisc, Inc.; p. 493: © PhotoDisc, Inc.; p. 495: © PhotoDisc, Inc.; p. 498: © Corbis; p. 501: © PhotoDisc, Inc.; p. 502: Carltons for Artville; p. 506: © PhotoDisc, Inc.; p. 509: © PhotoDisc, Inc.; p. 518: © PhotoDisc, Inc.; p. 518: © PhotoDisc, Inc.; p. 519: © PhotoDisc, Inc.; p. 519: © PhotoDisc, Inc.; p. 520: © PhotoDisc, Inc.; p. 522: © PhotoDisc, Inc.; p. 530: © PhotoDisc, Inc.; p. 532: © PhotoDisc, Inc.; p. 537: © PhotoDisc, Inc.; p. 540: © PhotoDisc, Inc.; p. 543: Christina De Musie for Artville; p. 546: © PhotoDisc, Inc.; p. 550: © PhotoDisc, Inc.; p. 560: ©

Photo Credits

PhotoDisc, Inc.; p. 561: © PhotoDisc, Inc.; p. 561: © PhotoDisc, Inc.; p. 564: © PhotoDisc, Inc.; p. 566: © PhotoDisc, Inc.; p. 568: © PhotoDisc, Inc.; p. 572: © Corbis; p. 574: © PhotoDisc, Inc.; p. 576: © PhotoDisc, Inc.; p. 581: © PhotoDisc, Inc.; p. 582: © PhotoDisc, Inc.; p. 584: © PhotoDisc, Inc.; p. 586: © PhotoDisc, Inc.; p. 590: © PhotoDisc, Inc.; p. 592: © PhotoDisc, Inc.; p. 602: © PhotoDisc, Inc.; p. 602: © PhotoDisc, Inc.; p. 603: © PhotoDisc, Inc.; p. 603: © PhotoDisc, Inc.; p. 605: © PhotoDisc, Inc.; p. 606: © Corbis; p. 608: © Corbis; p. 610: © Corbis; p. 617: © PhotoDisc, Inc.; p. 618: © PhotoDisc, Inc.; p. 620: © PhotoDisc, Inc.; p. 621: © Corbis; p. 624: © PhotoDisc, Inc.; p. 634: © PhotoDisc, Inc.; p. 635: © PhotoDisc, Inc.; p. 635: © PhotoDisc, Inc.; p. 636: © PhotoEdit; p. 636: © PhotoDisc, Inc.; p. 637: © PhotoEdit; p. 638: © PhotoDisc, Inc; p. 638: © PhotoDisc, Inc; p. 638: © PhotoDisc, Inc.; p. 639: © PhotoDisc, Inc.; p. 639: © PhotoDisc, Inc.; p. 640: © PhotoDisc, Inc.; p. 640: © PhotoDisc, Inc.; p. 641: © PhotoDisc, Inc.; p. 641: © PhotoDisc, Inc.; p. 642: © PhotoDisc, Inc.; p. 642: © PhotoDisc, Inc.; p. 643: © PhotoDisc, Inc.; p. 643: © PhotoDisc, Inc.; p. 644: © PhotoDisc, Inc.; p. 644: © PhotoDisc, Inc.; p. 645: © PhotoDisc, Inc.; p. 645: © PhotoEdit; p. 646: © PhotoDisc, Inc.; p. 646: © PhotoDisc, Inc.; p. 646: © PhotoDisc, Inc.; p. 647: © PhotoDisc, Inc.; p. 647: © PhotoDisc, Inc.; p. 647: © PhotoDisc, Inc.; p. 648: © Corbis; p. 648: © PhotoDisc, Inc.; p. 649: © PhotoDisc, Inc.; p. 649: © PhotoDisc, Inc.; p. 650: © PhotoDisc, Inc.; p. 650: © PhotoDisc, Inc.; p. 650: © PhotoDisc, Inc.; p. 651: © PhotoDisc, Inc.; p. 651: © PhotoDisc, Inc.; p. 652: © PhotoDisc, Inc.; p. 652: © PhotoDisc, Inc.; p. 653: © PhotoDisc, Inc.; p. 653: © PhotoDisc, Inc.; p. 654: © PhotoDisc, Inc.; p. 654: © PhotoDisc, Inc.; p. 655: © PhotoDisc, Inc.; p. 655: © PhotoDisc, Inc.